COMPUTATIONAL MODELLING OF CONCRETE STRUCTURES

PROCEEDINGS OF THE EURO-C CONFERENCE 2003,
ST. JOHANN IM PONGAU, AUSTRIA, 17–20 MARCH 2003

Computational Modelling of Concrete Structures

Edited by

N. Bićanić
University of Glasgow, UK

R. de Borst
Delft University of Technology, The Netherlands

H. Mang
Vienna University of Technology, Austria

G. Meschke
Ruhr-University Bochum, Germany

A.A. BALKEMA PUBLISHERS LISSE / ABINGDON / EXTON (PA) / TOKYO

Published by: A.A. Balkema, a member of Swets & Zeitlinger Publishers
www.balkema.nl and www.szp.swets.nl

ISBN 90 5809 536 3

Printed in The Netherlands

Computational Modelling of Concrete Structures, Bićanić et al. (eds)
© 2003 Swets & Zeitlinger, Lisse, ISBN 90 5809 536 3

Table of Contents

Constitutive modelling

Modelling of multiphysics phenomena and time effects

Modelling of structural performance

Modelling of seismic and cyclic behaviour

Computational Modelling of Concrete Structures, Bićanić et al. (eds)
© 2003 Swets & Zeitlinger, Lisse, ISBN 90 5809 536 3

Foreword

These Proceedings comprise both the invited and contributed papers presented at the EURO-C 2003 Conference held in St Johann im Pongau Austria, from 17th to 20th March 2003. The current EURO-C series of conferences (Innsbruck 1994, Badgastein 1998) was created as a successor to the ICE 1984 conference in Split, Croatia, SCI-C 1990 conference in Zell am See, Austria and the Concrete Mechanics Colloquia held in Delft in 1981 and 1987.

The main objectives of the EURO-C conference series remain true to the original intention to bring together both researchers and practising engineers engaged in computational modelling of concrete structures, in order to review and discuss recent research efforts and assess their applicability and robustness. Many formulations currently used in engineering practice can be traced back to earlier conferences in the EURO-C series or its predecessors.

Computational modelling of concrete and other cementitious materials continues as a very active research field. Increasing emphasis on the life-time oriented design has motivated the development of advanced multi-field models for durability analyses of concrete structures. The overall robustness of the computational algorithms and the mathematical rigour have increased significantly, ensuring more reliable numerical predictions.

We would like to thank members of the Technical Advisory Panel (Graham Baker, Zdenek Bažant, Sarah Billington, Alberto Carpinteri, Josef Eibl, Marco di Prisco, Hideyuki Hiori, Milan Jirasek, Bhushan Karihaloo, Chris Leung, Jacky Mazars, Javier Oliver, Jerzy Pamin, Gilles Pijaudier-Cabot, Jean Marie Reynouard, Jan Rots, Tada-aki Tanabe, Franz-Josef Ulm and Kaspar William) for their support and substantial efforts in the reviewing process of more than 150 submitted abstracts. Such a rigorous selection ensured the very high quality of presented papers, which we strive to maintain for the EURO-C series.

Conference papers are subdivided into five sections – (A) *Computational advances and multiscale modelling*, (B) *Constitutive modelling*, (C) *Modelling of multiphysics phenomena and time effects*, (D) *Modelling of structural performance and* (E) *Modelling of seismic and cyclic behaviour.*

We sincerely hope that the EURO-C 2003 Proceedings will continue to serve as a major reference text in identifying novel formulations and capabilities in computational modelling of concrete structures and in assessing their applicability and relevance in practice.

Nenad Bićanić
René de Borst
Herbert Mang
Günther Meschke

Glasgow, Delft, Vienna, Bochum
January 2003

Computational advances and multiscale modelling

Computational Modelling of Concrete Structures, Bićanić et al. (eds)
© 2003 Swets & Zeitlinger, Lisse, ISBN 90 5809 536 3

Keynote paper: Discrete *vs* smeared crack models for concrete fracture: bridging the gap

René de Borst & Joris J.C. Remmers
Koiter Institute Delft, Delft University of Technology, Delft, Netherlands

Alan Needleman
Brown University, Division of Engineering, Providence, Rhode Island

Marie-Angèle Abellan
LTDS-ENISE – UMR CNRS 5513, Saint-Etienne, France

ABSTRACT: Discrete and smeared crack models for concrete fracture are discussed in a historical perspective. It is argued that these two computational approaches, originally conceived as very different, can be brought together by exploiting the partition-of-unity property of finite element shape functions. The cohesive segments method, which exploits this partition-of-unity property, exhibits advantages of both the discrete and smeared crack approaches, and is capable of describing the transition from distributed micro-cracking to a dominant crack. The versatility of the cohesive methodology is shown by incorporating water diffusion and ion transport into the formulation.

1 INTRODUCTION

The numerical simulation of concrete fracture was initiated in the late 1960s by Ngo & Scordelis (1967) and by Rashid (1968), who introduced discrete crack and smeared crack models, respectively. The discrete crack model is aimed at simulating the initiation and propagation of dominant cracks. In contrast, the smeared crack model is based on the idea that in concrete, due to its heterogeneity and the presence of reinforcement, many small cracks nucleate which only in a later stage of the loading process link up to form one or more dominant cracks. Since each individual crack is not numerically resolved, the smeared crack model captures the deterioration process through a constitutive relation, thus smearing out the cracks over the continuum.

Over the years, strong opinions have been expressed regarding both approaches. Nevertheless, it seems that recent developments in computational mechanics have narrowed the gap between these apparently disparate formulations. Indeed, we will argue that the cohesive segments method, which we will describe, embodies characteristics of both approaches and can behave as either.

We begin by giving a concise overview of some historical developments of the discrete and the smeared crack approaches. Mathematical deficiencies are noted which have surfaced for smeared crack models and a now well-accepted solution – gradient enhancement – is briefly discussed. A recent development in computational mechanics, namely the use of the partition-of-unity property of finite element shape functions, turns out to be key for modeling the coalescence of distributed cracks into one or more dominant cracks. This is first shown for a gradient enhanced continuum damage model and then for the cohesive segments model. Since water and chloride ion transport play an important role in the deterioration and failure of concrete structures, we conclude with a formulation incorporating models for these phenomena into a cohesive framework.

2 DISCRETE *VS* SMEARED CRACK MODELS

2.1 *The discrete crack approach*

The discrete crack approach to concrete fracture is intuitively appealing: a crack is introduced as a geometric entity. Initially, this was implemented by letting a crack grow when the nodal force at the node ahead of the crack tip exceeded a tensile strength criterion. Then, the node is split into two nodes and the tip of the crack is assumed to propagate to the next node. When the tensile strength criterion is violated at this node, it is split and the procedure is repeated, as sketched in Figure 1 (Ngo & Scordelis 1967).

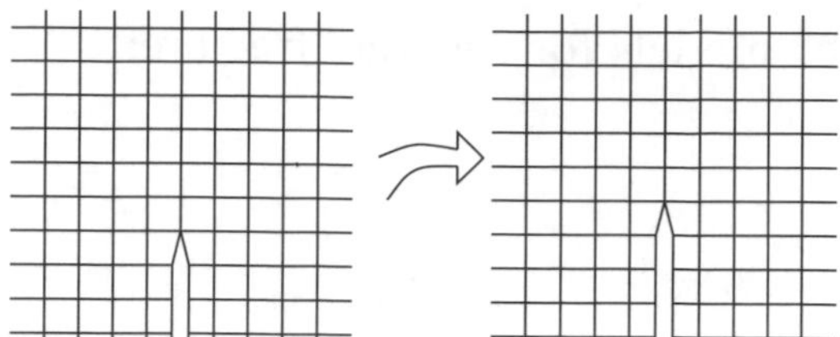

Figure 1. Early discrete crack modelling.

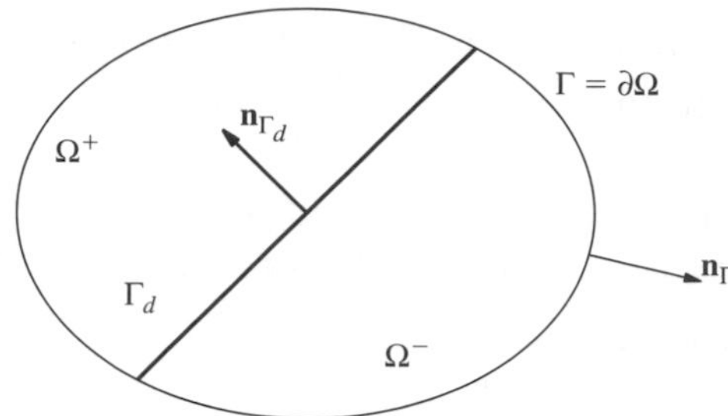

Figure 2. Body composed of continuous displacement fields at each side of the discontinuity Γ_d.

The discrete crack approach in its original form has several disadvantages. Cracks are forced to propagate along element boundaries, so that a *mesh bias* is introduced. Automatic remeshing allows the mesh bias to be reduced, if not eliminated, and sophisticated computer codes with remeshing were developed by Ingraffea and co-workers, e.g. Ingraffea & Saouma (1985). Nevertheless, a computational difficulty, namely, the continuous change in topology, is inherent in the discrete crack approach and is to a certain extent even aggravated by remeshing procedures.

The change in topology was to a large extent alleviated by the advent of meshless methods, such as the element-free Galerkin method (Belytschko *et al.* 1994). Indeed, successful analyses have been carried out using these methods, but disadvantages including difficulties with robust three-dimensional implementations, the large computational demand compared with finite element methods, the somewhat ad-hoc manner in which the support of a node is changed in the presence of a crack (Hegen 1996) and the need to still employ a background "mesh" of integration cells in order to ensure a sufficiently accurate evaluation of the load vectors and stiffness matrices, have limited the use of these methods.

A finite element method accommodating the propagation of discrete cracks through elements was proposed by Belytschko and his co-workers (Belytschko & Black 1999, Moës *et al.* 1999), exploiting the partition-of-unity property of finite element shape functions (Babuska & Melenk 1997). Since finite element shape functions φ_i form partitions of unity, $\sum_{i=1}^{n} \varphi_i = 1$ with n the number of nodal points, a field u can be interpolated as

$$u = \sum_{i=1}^{n} \varphi_i \left(\bar{a}_i + \sum_{j=1}^{m} \psi_j \tilde{a}_{ij} \right) \tag{1}$$

with $\bar{a}_i$ the "regular" nodal degrees-of-freedom, ψ_j the enhanced basis terms, and $\tilde{a}_{ij}$ the additional degrees-of-freedom at node i which represent the amplitude of the jth enhanced basis term ψ_j.

A crack can be regarded as introducing a discontinuity in the displacement field. A displacement field $\mathbf{u}$ that contains a single discontinuity at Γ_d (see Figure 2) can be written as the sum of two continuous displacement fields $\bar{\mathbf{u}}$ and $\tilde{\mathbf{u}}$ separated by a Heaviside function $\mathcal{H}_{\Gamma_d}$:

$$\mathbf{u} = \bar{\mathbf{u}} + \mathcal{H}_{\Gamma_d} \tilde{\mathbf{u}} \tag{2}$$

The displacement decomposition in eq. (2) has a structure similar to the interpolation in eq. (1). This can be seen directly by rewriting and specializing eq. (1) as:

$$\mathbf{u} = \mathbf{N}(\bar{\mathbf{a}} + \mathcal{H}_{\Gamma_d} \tilde{\mathbf{a}}) = \mathbf{N}\bar{\mathbf{a}} + \mathcal{H}_{\Gamma_d} \mathbf{N}\tilde{\mathbf{a}} = \bar{\mathbf{u}} + \mathcal{H}_{\Gamma_d} \tilde{\mathbf{u}} \tag{3}$$

where $\mathbf{N}$ contains the standard shape functions, and $\bar{\mathbf{a}}$ and $\tilde{\mathbf{a}}$ collect the conventional and the additional nodal degrees-of-freedom, respectively. Accordingly, the partition-of-unity property of finite element shape functions can be used in a straightforward fashion to incorporate discontinuities, and thus discrete crack models, in a manner that preserves the discontinuous character of cracks.

In the original contributions of Belytschko & Black (1999) and Moës *et al.* (1999) the discrete crack was introduced using linear elastic fracture mechanics, thus necessitating special functions to simulate the near-tip singularity. For this reason, close to the crack tip the enhanced basis terms were introduced as

$$\begin{aligned} \boldsymbol{\psi} = \big(&\sqrt{r}\cos(\theta/2)\,, \sqrt{r}\sin(\theta/2), \\ &\sqrt{r}\sin(\theta/2)\sin(\theta), \sqrt{r}\cos(\theta/2)\sin(\theta) \big)^{\mathrm{T}} \end{aligned} \tag{4}$$

where r is the distance from the crack tip and θ is measured from the current direction of crack propagation. Away from the crack tip Heaviside functions were employed as discussed above.

2.2 *The smeared crack approach*

In a smeared crack approach, the nucleation of one or more cracks in the area that is attributed to an integration point is translated into a deterioration of the current stiffness and strength at that integration point. Generally, when the combination of stresses satisfies a specified criterion, e.g. the major principal stress reaching the tensile strength f_t, a crack is initiated. This implies that at the integration point where the stress, strain and history variables are monitored, the isotropic stress-strain relation is replaced by an orthotropic elasticity-type relation with the n, s-axes being axes of orthotropy; n is the direction normal to the crack and s is the direction tangential to the crack. In early studies, the orthotropic relation was defined as (Rashid 1968):

$$\begin{pmatrix} \sigma_{nn} \\ \sigma_{ss} \\ \sigma_{ns} \end{pmatrix} = \begin{bmatrix} 0 & 0 & 0 \\ 0 & E & 0 \\ 0 & 0 & 0 \end{bmatrix} \begin{pmatrix} \varepsilon_{nn} \\ \varepsilon_{ss} \\ \varepsilon_{ns} \end{pmatrix} \tag{5}$$

for a plane-stress situation. Eq. (5) shows that both the normal stiffness and the shear stiffness across the crack were set equal to zero upon cracking. Consequently, all effects of lateral contraction/expansion disappear.

With $\boldsymbol{\sigma}_{ns} = (\sigma_{nn}, \sigma_{ss}, \sigma_{ns})^T$ and $\boldsymbol{\varepsilon}_{ns} = (\varepsilon_{nn}, \varepsilon_{ss}, \varepsilon_{ns})^T$ we can also write the orthotropic elastic stiffness relation in the n, s-coordinate system as

$$\boldsymbol{\sigma}_{ns} = \mathbf{D}^s \varepsilon_{ns} \tag{6}$$

where the secant stiffness matrix $\mathbf{D}^s$ is defined as:

$$\mathbf{D}^s = \begin{bmatrix} 0 & 0 & 0 \\ 0 & E & 0 \\ 0 & 0 & 0 \end{bmatrix} \tag{7}$$

If we introduce ϕ as the angle from the x-axis to the s-axis, we can relate the components of $\boldsymbol{\varepsilon}_{ns}$ and $\boldsymbol{\sigma}_{ns}$ to those in the global x, y-coordinate system via standard transformation matrices $\mathbf{T}_\varepsilon(\phi)$ and $\mathbf{T}_\sigma(\phi)$: $\boldsymbol{\varepsilon}_{ns} = \mathbf{T}_\varepsilon(\phi)\boldsymbol{\varepsilon}_{xy}$ and $\boldsymbol{\sigma}_{ns} = \mathbf{T}_\sigma(\phi)\boldsymbol{\sigma}_{xy}$. The local secant stiffness relation (6) then transforms into a secant stiffness relation in the global x, y-coordinate system:

$$\boldsymbol{\sigma}_{xy} = \mathbf{T}_\sigma^{-1}(\phi)\mathbf{D}^s\mathbf{T}_\varepsilon(\phi)\boldsymbol{\varepsilon}_{xy} \tag{8}$$

When ϕ changes continuously, e.g. to keep the direction of the crack orthogonal to the direction of the major principal stress, the *rotating smeared crack model* is obtained (Cope *et al.* 1980). The approach with ϕ fixed at crack initiation is known as the *fixed smeared crack model*. Referring to this angle as ϕ_0, we have instead of eq. (8):

$$\boldsymbol{\sigma}_{xy} = \mathbf{T}_\sigma^{-1}(\phi_0)\mathbf{D}^s\mathbf{T}_\varepsilon(\phi_0)\boldsymbol{\varepsilon}_{xy} \tag{9}$$

Because of ill-conditioning, use of eq. (5) can induce convergence difficulties. Also, physically unrealistic and distorted crack patterns may be obtained (Suidan & Schnobrich 1973). For this reason a reduced shear modulus $\beta G (0 \leq \beta \leq 1)$ was inserted into the stiffness relation:

$$\mathbf{D}^s = \begin{bmatrix} 0 & 0 & 0 \\ 0 & E & 0 \\ 0 & 0 & \beta G \end{bmatrix} \tag{10}$$

The use of the shear retention factor β not only reduces numerical difficulties, but also improves the capability of fixed smeared crack models, because it can be regarded as a representation of some effects of aggregate interlocking and friction within the crack.

Setting the stiffness normal to the crack in eq. (10) equal to zero gives a sudden drop in stress from the tensile strength f_t to zero on crack initiation. This can cause numerical problems. A gradual decrease of the tensile carrying capacity, given by

$$\mathbf{D}^s = \begin{bmatrix} \mu E & 0 & 0 \\ 0 & E & 0 \\ 0 & 0 & \beta G \end{bmatrix} \tag{11}$$

gives results that are physically more appealing and computations that are numerically more stable. In eq. (11), μ is a factor which gradually decreases from one to zero as a function of the normal strain ε_{nn}, $\mu = \mu(\varepsilon_{nn})$. The introduction of the reduced normal stiffness μE was physically motivated by the argument that, especially in reinforced concrete, the area attributed to an integration point contains a number of cracks and that due to the bond between concrete and reinforcing steel, the intact concrete between the cracks adds stiffness which would be underestimated by a sudden drop to zero of the tensile strength (the so-called *tension-stiffening effect*).

Finite element models with embedded discontinuities provide an elegant way to implement smeared-crack models (Ortiz *et al.* 1987, Belytschko *et al.* 1988, Simo *et al.* 1993). Indeed, the embedded discontinuity approaches enhance the deformational capabilities of the elements, especially when the standard Bubnov-Galerkin approach is replaced by a Petrov-Galerkin method, which properly incorporates the discontinuity kinematics (de Borst *et al.* 2001). At the expense of obtaining a nonsymmetric stiffness matrix, the high local strain gradients inside crack bands are captured more accurately. However, a true discontinuity is not obtained because the kinematics of the embedded crack band are diffused over the element when the governing equations are cast in a weak format, either via a Bubnov-Galerkin or via a Petrov-Galerkin procedure. Indeed, Wells (2001) and Mosler & Meschke (2002) have proved the equivalence of embedded discontinuity approaches and conventional smeared crack models. Consequently, the embedded discontinuity approaches inherit many of the disadvantages of conventional smeared crack models, including the sensitivity of crack propagation to the direction of mesh lines.

2.3 *Cohesive surface models*

A serious limitation to the use of linear elastic fracture mechanics for the analysis of (reinforced) concrete structures is that one of the underlying assumptions, namely that the fracture process zone is small compared to the structural dimensions, is often not met. A further limitation is that, in classical fracture mechanics, crack propagation is assumed to occur from a pre-existing crack-like flaw. Observations on cracking in concrete show that the heterogeneous character of concrete promotes crack branching, crack arrest by hard particles and/or by reinforcing bars that is followed by crack nucleation and growth at other locations. Furthermore, a multitude of microscopic initial cracks often exists in concrete due to the exothermal processes that occur during the maturing and hardening of concrete and due to the shrinkage that occurs as a consequence of moisture diffusion and the subsequent water loss. These phenomena cannot be captured using classical fracture mechanics approaches.

In contrast to conventional engineering fracture mechanics, the cohesive surface methodology permits the analysis of fracture processes in which there is no dominant flaw even though in the original contributions of Dugdale (1960) and Barenblatt (1962) a dominant flaw was assumed present. In Hillerborg *et al.* (1976) and Needleman (1987) the cohesive framework was extended to situations without an initial crack.

In the cohesive surface formulation, constitutive relations are specified independently for the bulk material and for one or more cohesive surfaces. The cohesive constitutive relation embodies the failure characteristics of the material and characterizes the separation process. The bulk and cohesive constitutive relations together with appropriate balance laws and boundary (and initial) conditions completely specify the problem. Fracture, if it takes place, emerges as a natural outcome of the deformation process without introducing any additional failure criterion. The simplest cohesive constitutive relation is one where the cohesive surface traction is a function of the displacement jump across the cohesive surface. For ductile fracture (Hutchinson & Evans 2000), the most important parameters of the cohesive surface model appear to be the tensile strength f_t and the work of separation or fracture energy $\mathcal{G}_c$

$$\mathcal{G}_c = \int \sigma \, \mathrm{d}u \tag{12}$$

with σ and u the stress and the displacement across the fracture process zone. From dimensional considerations, this introduces a characteristic length in the model. For more brittle decohesion relations, i.e. when the decohesion law stems from micro-cracking as in concrete or ceramics, the *shape* of the stress-separation relation plays a much larger role and is sometimes even more important than the value of the tensile strength f_t (Rots 1986).

When fracture takes place along well-defined interfaces as, for example, in a lamellar solid (Schellekens & de Borst 1994), the placement of cohesive surfaces is clear. Also when the crack path is known in advance from experiments, accurate predictions can be obtained, since the interface elements can be placed along the known crack path (Rots 1991). However, for a solid that is homogeneous on the scale modeled, the placement of cohesive surfaces is problematic. In Xu & Needleman (1994) the cohesive surface spacing was made identical to the mesh spacing. Although this allows for the simulation of complex crack phenomena such as crack branching and crack initiation away from the crack tip, it is not mesh-independent. Since the interface elements are aligned with the element boundaries, the orientation of cracks is restricted to a limited number of predefined angles. In addition, if, as in Xu & Needleman (1994), the cohesive surfaces are taken to have a nonzero initial compliance, the presence of the cohesive surfaces contributes to the overall compliance of the body. Consequently, the overall compliance depends on the mesh and an ill-posed problem results. To overcome difficulties associated with initially compliant cohesive surfaces, Camacho & Ortiz (1996) used initially rigid cohesive surfaces in conjunction with adaptive mesh refinement. However, other difficulties are introduced with initially rigid cohesive surfaces, at least in modeling dynamic crack growth, Falk *et al.* (2001), Papoulia & Vavasis (2003).

In another approach, Wells & Sluys (2001), Wells *et al.* (2002) and Moës & Belytschko (2002), have used the partition-of-unity property of finite element shape functions to introduce cohesive surfaces in continuum finite elements. Using the kinematics of eqs (1)–(3), which were used by Belytschko & Black (1999) and Moës *et al.* (1999) in the context of linear–elastic fracture mechanics, the jump at the discontinuity Γ_d is obtained as

$$\mathbf{v} = \tilde{\mathbf{u}}|_{\mathbf{x}\in\Gamma_d} \tag{13}$$

and the tractions at the discontinuity are given as $\mathbf{t}_d = \mathbf{t}_d(\mathbf{v})$. It is often assumed that: (i) the mode-I component in this constitutive relation has a dominant role and the mode-II component is zero; and (ii) an exponentially decaying function captures the mode-I response,

$$\mathbf{t}_d = \begin{pmatrix} f_t \exp\left(-\frac{f_t}{\mathcal{G}_c}\kappa\right) \\ 0 \end{pmatrix} \tag{14}$$

with κ a history parameter, which records the largest value of a (material-dependent) function of the relative displacements $\mathbf{v}$ over the discontinuity. A key feature of the method is the possibility of extending the cohesive crack during the calculation in an arbitrary direction, independent of the structure of the underlying finite element mesh. Since the cohesive surfaces are added during the calculation, there is no need for a high initial stiffness to minimize the effect of increasing the initial compliance of the medium due to the presence of cohesive surface elements. As a consequence, numerical anomalies, such as stress oscillations at the interface (Schellekens & de Borst 1993) or spurious stress wave reflections are avoided.

Although the cohesive surface model is essentially a discrete approach, it can be transformed into a smeared formulation by distributing the fracture energy $\mathcal{G}_c$ over the width of an element (Pietruszczak & Mróz 1981, Bažant & Oh 1983). When the fracture energy is smeared out over the width of the area in which the crack *localizes*, we obtain

$$\mathcal{G}_c = \iint \sigma \, \mathrm{d}\varepsilon(n) \, \mathrm{d}n \tag{15}$$

with n the coordinate orthogonal to the crack direction. For low-order elements the strains are constant over the width of the element and one obtains:

$$\mathcal{G}_c = w \int \sigma \, \mathrm{d}\varepsilon \tag{16}$$

The length scale introduced into the formulation is a numerical length scale, the element size w.

3 FROM CONTINUA TO DISCONTINUA

3.1 *Continuum damage modeling*

The modeling of concrete fracture using continuum damage mechanics initiated in the 1980s, e.g. Mazars (1984), Mazars & Pijaudier-Cabot (1989). In its simplest isotropic form, the governing equations of a continuum-damage based constitutive relation are given by (Lemaitre & Chaboche 1990):

$$\boldsymbol{\sigma} = (1 - \omega)\mathbf{D}^e : \boldsymbol{\varepsilon} \tag{17}$$

with $\mathbf{D}^e$ the initial isotropic elastic stiffness tensor and ω a scalar-valued damage variable, with the initial value of zero for an intact material and increasing to one at the complete loss of material load carrying capacity. To make a proper distinction between loading and unloading, damage models are, as are smeared crack models, equipped with a loading function

$$f = \varepsilon_{eq} - \kappa \tag{18}$$

with ε_{eq} the equivalent strain, which is a function of the strain tensor $\boldsymbol{\varepsilon}$, $\varepsilon_{eq} = \varepsilon_{eq}(\boldsymbol{\varepsilon})$, and κ a history parameter which starts at a threshold value κ_0 and is updated by the requirement that during damage growth $f = 0$. The set of equations is then completed by a material function $\omega = \omega(\kappa)$, which gives the dependence of the material degradation on the history.

Any directional dependence of damage evolution can be incorporated by degrading Young's modulus E in a preferential direction. A possible loading function that accounts for directional dependence is

$$f = \varepsilon_{nn} - \kappa \tag{19}$$

which is nothing but a principal strain criterion, while the total stress-strain relation (17) is generalized to eq. (6),

$$\boldsymbol{\sigma}_{ns} = \mathbf{D}^s \boldsymbol{\varepsilon}_{ns}$$

with

$$\mathbf{D}^s = \begin{bmatrix} (1-\omega_1)E & 0 & 0 \\ 0 & E & 0 \\ 0 & 0 & (1-\omega_2)G \end{bmatrix} \tag{20}$$

with $\omega_1 = \omega_1(\kappa)$ and $\omega_2 = \omega_2(\kappa)$. Identifying the factor $1 - \omega_1$ as the normal reduction factor μ and $1 - \omega_2$ as the shear retention factor β, cf. eq. (11), the smeared crack model is recovered as it is used in the numerical analysis of concrete structures.

3.2 *Higher-order continua*

Continuum damage models, including the class of smeared crack models described in Section 2, suffer from loss of ellipticity beyond a certain level of accumulated damage, see e.g. Peerlings *et al.* (2002). As a consequence, the rate boundary value problem ceases to be well-posed, which typically results in an infinite number of possible solutions. A numerical solution just "picks" a solution from this available solution space, which results in an excessive mesh dependency.

To regularize the solution, higher-order continua have been proposed. In the context of damage models, nonlocal models in an integral format (as in Pijaudier-Cabot & Bažant 1987) or in a differential format (as in Peerlings *et al.* 1996, 1998) have been suggested. An anisotropic version has been published by Kuhl *et al.* (2000). In such a case, the loading function for crack initiation and evolution (19) is replaced by

$$f = \gamma_{nn} - \kappa \tag{21}$$

with γ_{nn} a nonlocal generalization of ε_{nn}, or, identically

$$f = \gamma_{xx} \cos^2 \phi + \gamma_{yy} \sin^2 \phi + \gamma_{xy} \sin \phi \cos \phi - \kappa \tag{22}$$

The nonlocal strains γ_{xx}, γ_{yy} and γ_{xy} are derived by solving the set of Helmholtz equations (Peerlings *et al.* 1996, Kuhl *et al.* 2000):

$$\begin{aligned} \gamma_{xx} - c\nabla^2\gamma_{xx} &= \varepsilon_{xx} \\ \gamma_{yy} - c\nabla^2\gamma_{yy} &= \varepsilon_{yy} \\ \gamma_{xy} - c\nabla^2\gamma_{xy} &= \varepsilon_{xy} \end{aligned} \tag{23}$$

or

$$\boldsymbol{\gamma} - c\nabla^2\boldsymbol{\gamma} = \boldsymbol{\epsilon} \tag{24}$$

with c a material parameter with the dimension length squared, which sets the length scale in the continuum formulation. Typically, the natural boundary conditions

$$\mathbf{n}_\Gamma \cdot \nabla_\gamma = \mathbf{0} \tag{25}$$

with $\Gamma = \partial\Omega$ the boundary of the body Ω, have been adopted in numerical analyses, but the physical motivation for these boundary conditions remains an open issue.

3.3 *Incorporating discontinuities*

When $\omega_1 = 1$, a discrete crack arises and the displacement field takes the form (eq. (2)):

$$\mathbf{u} = \bar{\mathbf{u}} + \mathcal{H}_{\Gamma_d}\tilde{\mathbf{u}}$$

with $\mathcal{H}_{\Gamma_d}$ the Heaviside function separating the Ω^+-domain from the Ω^- domain ($\Omega = \Omega^+ \cup \Omega^-$), Figure 2. Using eq. (2) and assuming small strains, the strain field follows by straightforward differentiation:

$$\boldsymbol{\varepsilon} = \nabla^s\bar{\mathbf{u}} + \mathcal{H}_{\Gamma_d}\nabla^s\tilde{\mathbf{u}} + \delta_{\Gamma_d}(\tilde{\mathbf{u}} \otimes \mathbf{n}_{\Gamma_d})^s \tag{26}$$

where the superscript s denotes a symmetrized operator and δ_{Γ_d} is the Dirac delta function placed at the discontinuity Γ_d. Consistent with the decomposition (2) we partition the field that describes the nonlocal strain measure as:

$$\boldsymbol{\gamma} = \bar{\boldsymbol{\gamma}} + \mathcal{H}_{\Gamma_d}\tilde{\boldsymbol{\gamma}} \tag{27}$$

where, as emphasized by Peerlings *et al.* (2002), the boundary condition (25) must also be applied at the internal boundary Γ_d.

3.4 *Governing equations and weak forms*

The gradient-enhanced damage model as summarized in eqs (6), (20), (21), (24), (25) and the kinematic assumption (26) are complemented by the standard equation of momentum

$$\nabla \cdot \boldsymbol{\sigma} = \mathbf{0} \tag{28}$$

with $\boldsymbol{\sigma}$ the Cauchy stress tensor. The description of the initial-value problem is then completed by the addition of the boundary conditions

$$\mathbf{n}_\Gamma \cdot \boldsymbol{\sigma} = \mathbf{t}_p, \quad \mathbf{u} = \mathbf{u}_p \tag{29}$$

on complementary parts of the boundary $\partial\Omega_t$ and $\partial\Omega_u$, with $\Gamma = \partial\Omega = \partial\Omega_t \cup \partial\Omega_u, \partial\Omega_t \cap \partial\Omega_u = \emptyset$, $\mathbf{t}_p$ being the prescribed external traction and $\mathbf{u}_p$ the prescribed displacements.

For obtaining the weak form of eqs (24) and (28), we recast (28) in matrix-vector format:

$$\mathbf{L}^T\boldsymbol{\sigma} = \mathbf{0} \tag{30}$$

where $\mathbf{L}$ is a matrix that contains differential operators, e.g., for the two-dimensional case

$$\mathbf{L} = \begin{bmatrix} \frac{\partial}{\partial x} & 0 \\ 0 & \frac{\partial}{\partial y} \\ \frac{\partial}{\partial y} & \frac{\partial}{\partial x} \end{bmatrix} \tag{31}$$

In the spirit of a standard Bubnov-Galerkin approach, we assume test functions for the displacements and the nonlocal strains as follows:

$$\boldsymbol{\eta} = \bar{\boldsymbol{\eta}} + \mathcal{H}_{\Gamma_d}\tilde{\boldsymbol{\eta}} \tag{32}$$

and

$$\boldsymbol{\zeta} = \bar{\boldsymbol{\xi}} + \mathcal{H}_{\Gamma_d}\tilde{\boldsymbol{\xi}} \tag{33}$$

Substitution into eqs (24) and (30) and integrating over the domain Ω leads to the corresponding weak forms:

$$\int_\Omega (\bar{\boldsymbol{\eta}} + \mathcal{H}_{\Gamma_d}\tilde{\boldsymbol{\eta}})^T\mathbf{L}^T\boldsymbol{\sigma}\,\mathrm{d}\Omega = 0 \tag{34}$$

and

$$\int_\Omega (\bar{\boldsymbol{\xi}} + \mathcal{H}_{\Gamma_d}\tilde{\boldsymbol{\xi}})^T(\gamma - c\nabla^2\gamma - \boldsymbol{\varepsilon})\mathrm{d}\Omega = 0 \tag{35}$$

We next apply the divergence theorem, use the external boundary conditions (25) and (29), eliminate the Heaviside functions by changing the integration domain from Ω to Ω^+ and eliminate the Dirac delta functions by transforming the volume integral into a surface integral:

$$\begin{aligned} &\int_\Omega (\mathbf{L}\bar{\boldsymbol{\eta}})^T\boldsymbol{\sigma}\,\mathrm{d}\Omega + \int_{\Omega+} (\mathbf{L}\tilde{\boldsymbol{\eta}})^T\boldsymbol{\sigma}\,\mathrm{d}\Omega + \int_{\Gamma_d} \tilde{\boldsymbol{\eta}}^T\mathbf{t}_d\,\mathrm{d}\Gamma \\ &= \int_\Gamma (\bar{\boldsymbol{\eta}} + \mathcal{H}_{\Gamma_d}\tilde{\boldsymbol{\eta}})^T\mathbf{t}_p\,\mathrm{d}\Omega \end{aligned} \tag{36}$$

and

$$\int_{\Omega} \bar{\zeta}^T \gamma \, d\Omega + \int_{\Omega+} \tilde{\zeta}^T \gamma \, d\Omega + \int_{\Omega} c(\nabla \bar{\zeta})^T \nabla \gamma \, d\Omega + \int_{\Omega+} c(\nabla \tilde{\zeta})^T \nabla \gamma \, d\Omega + \int_{\Gamma_d} c\tilde{\zeta}^T \mathbf{n}_{\Gamma_d}^T \nabla \gamma \, d\Gamma = \int_{\Omega} \bar{\zeta}^T \boldsymbol{\varepsilon} \, d\Omega + \int_{\Omega+} \tilde{\zeta}^T \boldsymbol{\varepsilon} \, d\Omega \tag{37}$$

where $\mathbf{t}_d = \hat{\mathbf{L}}^T \boldsymbol{\sigma}$ is the traction at the discontinuity Γ_d with

$$\hat{\mathbf{L}} = \begin{bmatrix} n_x & 0 \\ 0 & n_y \\ n_y & n_x \end{bmatrix} \tag{38}$$

and n_x, n_y the components of the vector $\mathbf{n}_{\Gamma_d}$ which is normal to the discontinuity.

After complete decohesion of the bulk material the crack or internal discontinuity that arises is taken to be stress free. Accordingly, for the interface traction

$$\mathbf{t}_d \equiv \hat{\mathbf{L}}^T \boldsymbol{\sigma} = 0 \tag{39}$$

Thus, the third integral in eq. (36) cancels. Similarly, the fifth integral of eq. (37) cancels because of the boundary condition (25), which, as has been emphasized by Peerlings *et al.* (2002) must also hold at the internal boundary Γ_d. Therefore, eqs (36) and (37) simplify to:

$$\int_{\Omega} (\mathbf{L}\bar{\eta})^T \boldsymbol{\sigma} \, d\Omega + \int_{\Omega+} (\mathbf{L}\tilde{\eta})^T \boldsymbol{\sigma} \, d\Omega = \int_{\Gamma} (\bar{\eta} + \mathcal{H}_{\Gamma_d} \tilde{\eta})^T \mathbf{t}_p \, d\Omega \tag{40}$$

and

$$\int_{\Omega} \bar{\boldsymbol{\zeta}}^T \gamma \, d\Omega + \int_{\Omega+} \tilde{\boldsymbol{\zeta}}^T \boldsymbol{\gamma} \, d\Omega + \int_{\Omega} c(\nabla \bar{\boldsymbol{\zeta}})^T \nabla \gamma \, d\Omega + \int_{\Omega+} c(\nabla \tilde{\boldsymbol{\zeta}})^T \nabla \boldsymbol{\gamma} \, d\Omega = \int_{\Omega} \bar{\boldsymbol{\zeta}}^T \boldsymbol{\varepsilon} \, d\Omega + \int_{\Omega+} \tilde{\boldsymbol{\zeta}}^T \boldsymbol{\varepsilon} \, d\Omega \tag{41}$$

3.5 *Spatial discretization and linearization*

In a Bubnov-Galerkin sense the trial functions $\mathbf{u}$ and γ and the test functions η and ζ are discretized in the same space:

$$\mathbf{u} = \mathbf{N}(\bar{\mathbf{a}} + \mathcal{H}_{\Gamma_d} \tilde{\mathbf{a}}) \tag{42}$$

$$\boldsymbol{\gamma} = \mathbf{H}(\bar{\mathbf{g}} + \mathcal{H}_{\Gamma_d} \tilde{\mathbf{g}}) \tag{43}$$

$$\boldsymbol{\eta} = \mathbf{N}(\bar{\mathbf{w}} + \mathcal{H}_{\Gamma_d} \tilde{\mathbf{w}}) \tag{44}$$

$$\boldsymbol{\zeta} = \mathbf{H}(\bar{\mathbf{z}} + \mathcal{H}_{\Gamma_d} \tilde{\mathbf{z}}) \tag{45}$$

where the partition-of-unity property of the shape functions contained in $\mathbf{N}$ and $\mathbf{H}$ has been exploited. The arrays $\bar{\mathbf{a}}$ and $\bar{\mathbf{g}}$ contain the nodal values of the underlying continuous fields $\bar{\mathbf{u}}$ and $\bar{\gamma}$, while the arrays $\tilde{\mathbf{a}}$ and $\tilde{\mathbf{g}}$ contain the nodal values of the underlying continuous fields $\tilde{\mathbf{u}}$ and $\tilde{\gamma}$. The arrays $\bar{\mathbf{w}}, \bar{\mathbf{z}}, \tilde{\mathbf{w}}$ and $\tilde{\mathbf{z}}$ contain the discrete values related to the respective test functions. Inserting eqs (44) and (45) into eqs (40) and (41) and requiring that the result holds for all admissible $\bar{\mathbf{w}}, \bar{\mathbf{z}}, \tilde{\mathbf{w}}$ and $\tilde{\mathbf{z}}$ gives, using the notations $\mathbf{B} = \mathbf{LN}$ and $\mathbf{B}_H = \mathbf{LH}$:

$$\int_{\Omega} \mathbf{B}^T \boldsymbol{\sigma} \, d\Omega = \int_{\Gamma} \mathbf{N}^T \mathbf{t}_p \, d\Gamma \tag{46}$$

$$\int_{\Omega+} \mathbf{B}^T \boldsymbol{\sigma} \, d\Omega = \int_{\Gamma} \mathcal{H}_{\Gamma_d} \mathbf{N}^T \mathbf{t}_p \, d\Gamma \tag{47}$$

$$\mathbf{K}_{\bar{g}\bar{g}} \bar{\mathbf{g}} + \mathbf{K}_{\bar{g}\tilde{g}} \tilde{\mathbf{g}} = \int_{\Omega} \mathbf{H}^T \boldsymbol{\varepsilon} \, d\Omega \tag{48}$$

$$\mathbf{K}_{\tilde{g}\bar{g}} \bar{\mathbf{g}} + \mathbf{K}_{\tilde{g}\tilde{g}} \tilde{\mathbf{g}} = \int_{\Omega+} \mathbf{H}^T \boldsymbol{\varepsilon} \, d\Omega \tag{49}$$

where

$$\mathbf{K}_{\bar{g}\bar{g}} = \int_{\Omega} (\mathbf{H}^T \mathbf{H} + c\mathbf{B}_H^T \mathbf{B}_H) \, d\Omega \tag{50}$$

$$\mathbf{K}_{\bar{g}\tilde{g}} = \int_{\Omega+} (\mathbf{H}^T \mathbf{H} + c\mathbf{B}_H^T \mathbf{B}_H) \, d\Omega \tag{51}$$

An isotropic version of the above formulation is given in de Borst & Abellan (2002), where the tangent stiffness matrix needed for the iterative solution of the set of nonlinear equations (46)–(49) is also given.

4 A UNIFICATION: COHESIVE SEGMENTS

4.1 *Model description*

The physics of crack initiation and crack growth in a heterogeneous quasi-brittle material are illustrated in Figure 3 (van Mier 1997). The heterogeneity of the material, i.e., the presence of particles of different sizes and stiffnesses leads to a complex stress field where new cracks nucleate ("a" in Figure 3) and existing cracks branch ("b" in Figure 3). Smeared crack models do not properly capture these processes of crack initiation, growth, coalescence and branching, because essential features are lost in the smoothing process.

More detail can be preserved if we model the initiation, growth and eventual coalescence of the cracks at the mesoscopic level of observation in Figure 3 *separately*. Hitherto this could not be carried out, not only because of the high computational effort that this would require, but also because a suitable numerical framework was lacking. The exploitation of the partition-of-unity property of finite element shape

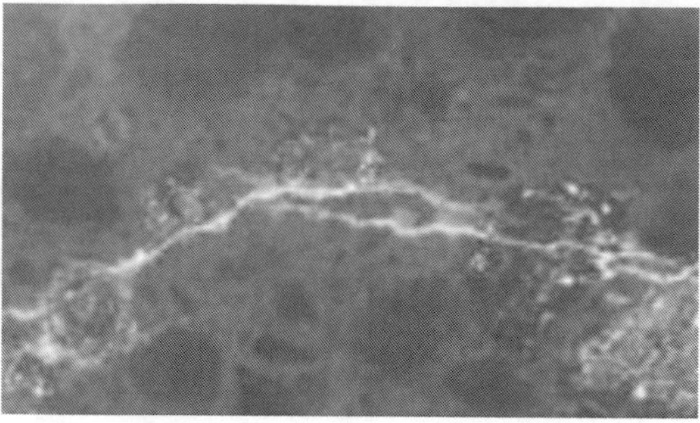

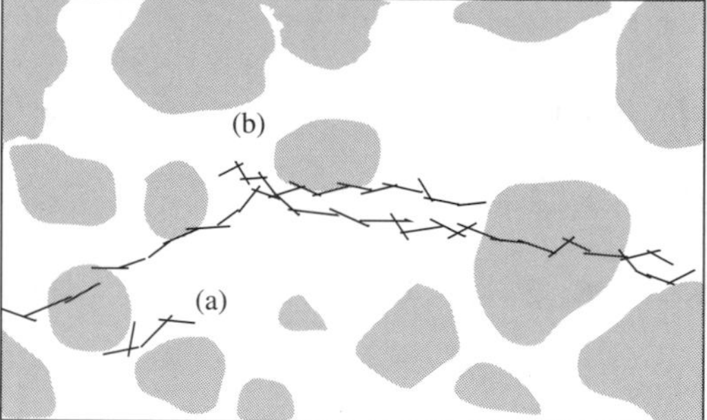

Figure 3. Experimentally observed "diffuse" crack pattern (van Mier 1997) and possible numerical representation.

functions can make such calculations feasible. One can define *cohesive segments* that can arise at arbitrary locations and in arbitrary directions and allow for the resolution of complex crack patterns including crack nucleation at multiple locations, followed by growth and coalescence (Remmers *et al.* 2002). This is quite similar to the numerical implementation of a continuous cohesive crack that runs through an existing finite element mesh without bias as in Wells & Sluys (2001) and Wells *et al.* (2002).

When the criterion for the initiation of decohesion is met, a cohesive segment is inserted through the integration point. In the applications so far, its direction has been taken to be orthogonal to the direction of the major principal stress. The segment is taken to extend throughout the element to which the integration point belongs and into the neighbouring elements, see Figure 4. Subsequently, the evolution of the crack segment is governed by a decohesion constitutive relation in the discontinuity, cf. eq. (14).

The cohesive segments approach inherits many of the advantages of the method that exploits the partition-of-unity property of finite element shape functions to describe continuous cohesive crack growth. Examples are the insensitivity of the direction of crack propagation to the structure of the underlying discretization and bypassing the need to define a high initial stiffness at the interface, which can result in numerical artefacts such as interface traction

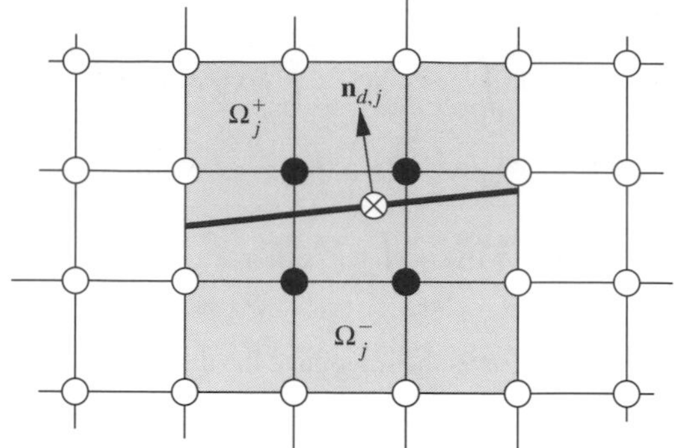

Figure 4. A single cohesive segment in a quadrilateral mesh. The segment passes through an integration point (⊗) where the fracture criterion is violated. The solid nodes contain additional degrees of freedom that determine the magnitude of the displacement jump. The gray shade denotes the elements that belong to the patch that is influenced by the cohesive segment.

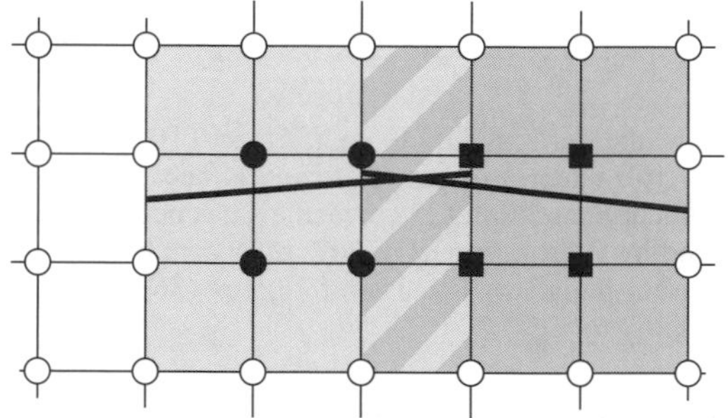

Figure 5. Interaction of two cohesive segments. The hatched elements have two sets of additional degrees of freedom. The solid circles denote the nodes that contain the additional degrees of freedom for the segment on the left and the solid squares denote those on the right.

oscillations and spurious wave reflections. Also, overlapping cohesive segments can behave as a continuous crack. This is illustrated in Figure 5, where a new cohesive segment is added in front of an existing segment. The displacement jump of the new crack segment is supported by a set of additional degrees of freedom, and is added to the nodes of the central element of the new segment (denoted by the solid squares). This set of additional degrees of freedom has no relation with the degrees of freedom associated with the other segment. The hatched elements are part of the patches that belong to both segments. In these elements two additional sets of degrees of freedom can be distinguished, although each set of degrees of freedom is added to a different set of nodes.

Because the crack is not taken as a single entity *a priori* in the cohesive segments approach, the method can also naturally simulate distributed cracking which frequently occurs in a heterogeneous solid, Thus, the cohesive segments approach embraces both extremes, distributed cracking with crack nucleation, growth and eventual coalescence at multiple locations as well as the initiation and propagation of a single, dominant crack without needing special assumptions. What is needed is a decohesion constitutive relation which then implicitly specifies the conditions for crack nucleation and the crack propagation direction. In this fashion, the cohesive segments method can be regarded as a bridge between two historically opposing methodologies for the simulation of concrete fracture.

4.2 *Governing equations*

A key feature of the cohesive segments approach is the possible emergence of multiple cohesive segments in a domain. Consider a domain Ω which contains m discontinuities $\Gamma_{d,j}$, $j = 1, \ldots, m$. Each discontinuity splits the domain in two parts, denoted as Ω_j^- and Ω_j^+, such that $\Omega_j^- \cup \Omega_j^+ = \Omega$, $j = 1, \ldots, m$. Generalizing eq. (2), the displacement field can be written as the sum of $m + 1$ continuous displacement fields $\bar{\mathbf{u}}$ and $\tilde{\mathbf{u}}_j$ (Daux *et al.* 2000, Belytschko *et al.* 2001):

$$\mathbf{u} = \bar{\mathbf{u}} + \sum_{j=1}^{m} \mathcal{H}_{\Gamma_{d,j}} \tilde{\mathbf{u}}_j \tag{52}$$

with $\mathcal{H}_{\Gamma_{d,j}}$ separating the continuous displacement fields $\bar{\mathbf{u}}$ and $\tilde{\mathbf{u}}_j$. Denoting $\delta_{\Gamma_{d,j}}$ as the Dirac delta function placed at $\Gamma_{d,j}$, the strain field follows by differentiation

$$\epsilon = \nabla^s \bar{\mathbf{u}} + \sum_{j=1}^{m} \left(\mathcal{H}_{\Gamma_{d,j}} \nabla^s \tilde{\mathbf{u}}_j + \delta_{\Gamma_{d,j}} \left(\tilde{\mathbf{u}}_j \otimes \mathbf{n}_{\Gamma_{d,j}} \right)^s \right) \tag{53}$$

Analogous to the damage mechanics introduction of a stress-free, single discrete crack at full damage, one proceeds by defining test functions for the displacements in a Bubnov-Galerkin sense

$$\eta = \bar{\eta} + \sum_{j=1}^{m} \mathcal{H}_{\Gamma_{d,j}} \tilde{\eta}_j \tag{54}$$

substitutes them into the momentum equation (30) and integrates over the domain Ω:

$$\int_{\Omega} \left(\bar{\eta} + \sum_{j=1}^{m} \mathcal{H}_{\Gamma_{d,j}} \tilde{\eta}_j \right)^T \mathbf{L}^T \sigma \, \mathrm{d}\Omega = 0 \tag{55}$$

Following a standard procedure, one applies the divergence theorem, uses the external boundary conditions (29), eliminates the Heaviside functions by changing the integration domain from Ω to Ω_j^+ and eliminates the Dirac delta functions by transforming the volume integral into a surface integral:

$$\int_{\Omega} (\mathbf{L}\bar{\eta})^T \sigma \, \mathrm{d}\Omega + \sum_{j=1}^{m} \left(\int_{\Omega_j^+} (\mathbf{L}\tilde{\eta}_j)^T \sigma \, \mathrm{d}\Omega + \int_{\Gamma_{d,j}} \tilde{\eta}_j^T \mathbf{t}_{d,j} \, \mathrm{d}\Gamma \right)$$

$$= \int_{\Gamma} \left(\bar{\eta} + \sum_{j=1}^{m} \mathcal{H}_{\Gamma_{d,j}} \tilde{\eta}_j \right)^T \mathbf{t}_p \, \mathrm{d}\Omega \tag{56}$$

with $\mathbf{t}_{d,j}$ the interface traction at Γ_j. Discretizing the trial functions $\mathbf{u}$ and the test functions η in a fashion similar to eqs (42) and (44), i.e.

$$\mathbf{u} = \mathbf{N} \left(\bar{\mathbf{a}} + \sum_{j=1}^{m} \mathcal{H}_{\Gamma_{d,j}} \tilde{\mathbf{a}}_j \right) \tag{57}$$

$$\eta = \mathbf{N} \left(\bar{\mathbf{w}} + \sum_{j=1}^{m} \mathcal{H}_{\Gamma_{d,j}} \tilde{\mathbf{w}}_j \right) \tag{58}$$

and requiring that the result holds for all admissible $\bar{\mathbf{w}}$ and $\tilde{\mathbf{w}}_j$, $j = 1, \ldots, m$ gives

$$\int_{\Omega} \mathbf{B}^T \sigma \, \mathrm{d}\Omega = \int_{\Gamma} \mathbf{N}^T \mathbf{t}_p \, \mathrm{d}\Gamma \tag{59}$$

$$\begin{aligned}
\int_{\Omega_1^+} \mathbf{B}^T \sigma \, \mathrm{d}\Omega + \int_{\Gamma_1} \mathbf{N}^T t_{d,1} \, \mathrm{d}\Omega &= \int_{\Gamma} \mathcal{H}_{\Gamma_{d,1}} \mathbf{N}^T \mathbf{t}_p \, \mathrm{d}\Gamma \\
&\cdots \\
\int_{\Omega_j^+} \mathbf{B}^T \sigma \, \mathrm{d}\Omega + \int_{\Gamma_j} \mathbf{N}^T t_{d,j} \, \mathrm{d}\Omega &= \int_{\Gamma} \mathcal{H}_{\Gamma_{d,j}} \mathbf{N}^T \mathbf{t}_p \, \mathrm{d}\Gamma \\
&\cdots \\
\int_{\Omega_m^+} \mathbf{B}^T \sigma \, \mathrm{d}\Omega + \int_{\Gamma_m} \mathbf{N}^T t_{d,m} \, \mathrm{d}\Omega &= \int_{\Gamma} \mathcal{H}_{\Gamma_{d,m}} \mathbf{N}^T \mathbf{t}_p \, \mathrm{d}\Gamma
\end{aligned} \tag{60}$$

The linearization needed for the incremental-iterative solution procedure and aspects regarding the implementation are described in Remmers *et al.* (2002).

4.3 *Examples*

Some simple examples are used to illustrate the capabilities of the cohesive segments method. The behaviour of the bulk material is assumed to be described by isotropic linear elasticity and only mode-I separation is considered with the decohesion relation given by eq. (14). We note, however, that the cohesive surface formulation readily accommodates mixed-mode crack growth.

We consider the double-cantilever beam of Figure 6 with an initial delamination with length 1 mm, which is modeled as a series of overlapping traction-free

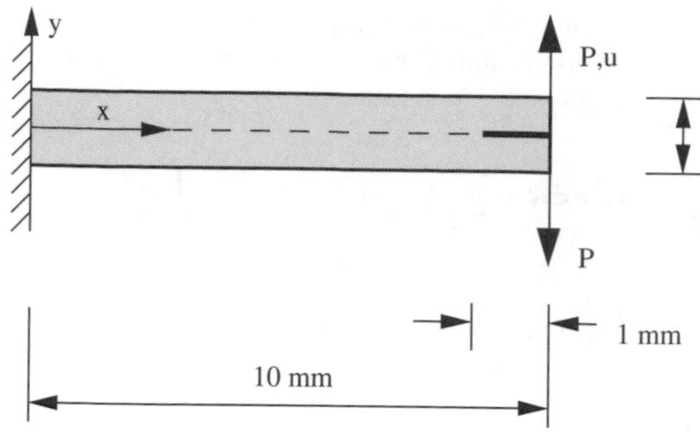

Figure 6. Double-cantilever beam with an initial notch subjected to peel forces P.

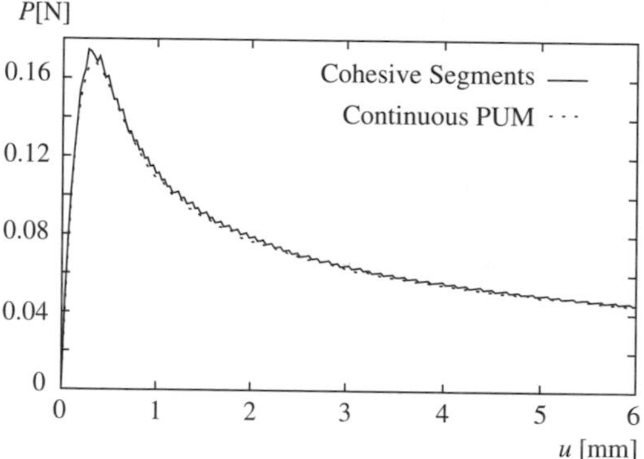

Figure 7. Load-displacement curve of double-cantilever beam.

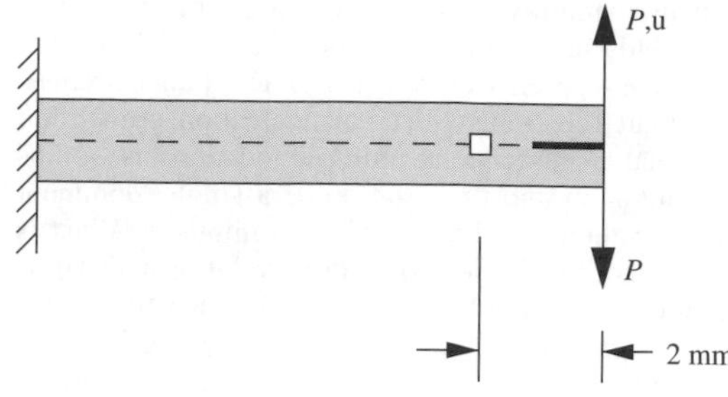

Figure 8. Geometry of double-cantilever beam with a small cavity.

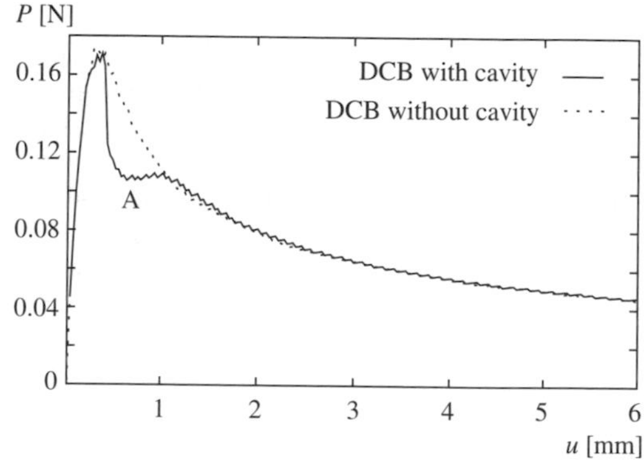

Figure 9. Load-displacement curve for double-cantilever beam with cavity.

crack segments inserted in the mesh before starting the analysis. The beam is loaded by peel forces P. The two layers of the beam have identical elastic properties: Young's modulus $E = 100$ N/mm^2 and a Poisson's ratio $\nu = 0.3$. The tensile strength of the weak adhesive is $f_t = 1.0$ N/mm^2 and the fracture energy is $\mathcal{G}_c = 0.1$ N/mm. The specimen is analyzed with a mesh of 200×7 elements.

The computed load-displacement curve is shown in Figure 7. In the present case of continuous crack growth, the cohesive segments approach gives results that are similar to those obtained with a method in which a continuous crack was modeled using the partition-of-unity property of the finite element shape functions. The computed values of the absorbed energy, the area under the load-displacement curves, are virtually the same for the two calculations. In this calculation there is a large stress concentration in a relatively small area around the crack tip. To capture these high peak stresses accurately, and therefore to add a new segment at the correct stage of loading, a rather fine mesh must be used.

A slightly modified version of the previous example is used to illustrate the ability of the cohesive segments method to simulate crack nucleation at a free edge of a structure. The double-cantilever beam now has a small cavity of length 0.25 mm at a distance $a_1 = 2$ mm, see Figure 8. During loading, the crack at the initial notch propagates to the cavity. On further loading, a new crack nucleates at the opposite side of the cavity and continues along the weak interface. This transition is indicated by the steep drop in the load-displacement curve in Figure 9. The position of cohesive segments immediately after the nucleation of the second crack, marked "A" on the load-displacement curve, is shown in Figure 10.

5 EXTENSION TO MULTI-PHASE MEDIA

Problems in the durability of concrete structures can involve the coupling of the set of equations that describe the stress evolution and sets of equations that describe diffusion-type processes, e.g., water or ion transport. For this reason, we outline an extension of the cohesive surface methodology, cast in the numerical framework of the partition-of-unity method, to include such diffusion-type problems. For simplicity,

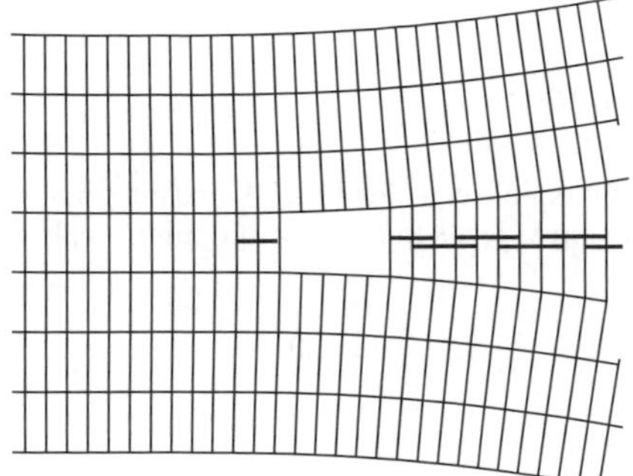

Figure 10. Position of cohesive segments, denoted by the heavy lines, in the double-cantilever beam with cavity. The individual segments are slightly shifted in a vertical sense for sake of proper visibility.

we restrict attention to a two-phase medium consisting of a porous solid and water, and to the case of a single cohesive crack. Conceptually, the extension to multi-phase media and to cohesive segments is straightforward.

5.1 *Standard two-phase medium*

We assume a standard two-phase medium (e.g. Jouanna & Abellan 1995, Lewis & Schrefler 1998) subject to the restriction of small displacement gradients. Changes in concentration are also assumed to be small. Furthermore, the assumption is made that both phases are incompressible. With these assumptions and neglecting inertia forces, we have the balance of momentum (28)

$$\nabla \cdot \boldsymbol{\sigma} = 0$$

with $\boldsymbol{\sigma}$ now the Cauchy stress tensor of the total medium, and the balance of mass:

$$\nabla \cdot \dot{\mathbf{u}} + \nabla \cdot (n_f \mathbf{w}_f) + Q^{-1}\dot{p} = 0 \tag{61}$$

with $\mathbf{w}_f$ the fluid velocity relative to the velocity of the skeleton $\dot{\mathbf{u}}$, n_f the fluid volumic fraction, p the fluid pressure and

$$Q^{-1} = \frac{1-n_f}{K_s} + \frac{n_f}{K_f} \tag{62}$$

with K_s and K_f the bulk moduli of the skeleton and the fluid, respectively. The governing equations (28) and (61) are complemented by the boundary conditions (29)

$$\mathbf{n}_\Gamma \cdot \boldsymbol{\sigma} = \mathbf{t}_p, \quad \mathbf{u} = \mathbf{u}_p$$

on complementary parts of the boundary $\partial\Omega_t$ and $\partial\Omega_u$, with $\Gamma = \partial\Omega = \partial\Omega_t \cup \partial\Omega_u$ and $\partial\Omega_t \cap \partial\Omega_u = \emptyset$, and

$$n_f \mathbf{w}_f = \mathbf{q}_p, \quad p = p_p \tag{63}$$

on complementary parts of the boundary $\partial\Omega_q$ and $\partial\Omega_p$, with $\Gamma = \partial\Omega = \partial\Omega_q \cup \partial\Omega_p$ and $\partial\Omega_q \cap \partial\Omega_p = \emptyset$.

5.2 *Constitutive assumptions*

We consider the case that a diaphragm with a permeability k_d is placed at the discontinuity in the displacement. As a consequence, the fluid pressure can be discontinuous across Γ_d and, similar to eq. (2), we have:

$$p = \bar{p} + \mathcal{H}_{\Gamma_d}\tilde{p} \tag{64}$$

It is noted that this assumption is different from that of Armero & Callari (1999), who adopt a smooth pressure field (and therefore $p = \bar{p}$) and is also different from that of Larsson & Larsson (2000), who assume that a regularized Dirac distribution is added to the continuous pressure field at the location of the discontinuity in the displacement field. For the fluid flow, gradients of the pressure need to be computed. Differentiating eq. (64), we obtain:

$$\nabla p = \nabla \bar{p} + \mathcal{H}_{\Gamma_d}\nabla\tilde{p} + \delta_{\Gamma_d}\tilde{p}\mathbf{n}_{\Gamma_d} \tag{65}$$

The above kinematics and the balance equations for momentum and mass have to be complemented by constitutive equations. The solid phase is assumed to be rate independent and to obey an incremental-linear constitutive relation. Thus, for the bulk material we have,

$$\mathrm{d}\boldsymbol{\sigma}_s = \mathbf{D} : \mathrm{d}\boldsymbol{\epsilon} \tag{66}$$

with $\mathbf{D}$ denoting the tangential material tensor in the bulk material, d( ) denoting an increment, and $\boldsymbol{\sigma}_s$ denoting the Cauchy stress tensor in the bulk:

$$\boldsymbol{\sigma}_s = \boldsymbol{\sigma} + p\mathbf{I} \tag{67}$$

Inserting the stress decomposition (67) together with the decompositions (2) and (64) into the constitutive relation (66), gives

$$\mathrm{d}\boldsymbol{\sigma} = \mathbf{D} : (\nabla^s \mathrm{d}\bar{\mathbf{u}} + \mathcal{H}_{\Gamma_d}\nabla^s \mathrm{d}\tilde{\mathbf{u}}) - (\mathrm{d}\bar{p} + \mathcal{H}_{\Gamma_d}\mathrm{d}\tilde{p})\mathbf{I} \tag{68}$$

since the Dirac delta function vanishes away from the discontinuity Γ_d. Assuming that the fluid flow in the porous medium can be described sufficiently accurately by Darcy's relation,

$$\mathbf{w}_f = -\frac{k_f}{n_f}\nabla p \tag{69}$$

k_f being the permeability of the bulk material, the behaviour of the bulk of the fluid-saturated medium is completely captured.

From the decohesion relation in the interface (14) a tangential stiffness relation can be derived:

$$\mathrm{d}\mathbf{t}_d = \mathbf{T} \cdot \mathrm{d}\mathbf{v} \tag{70}$$

where

$$\mathbf{T} = \frac{\partial \mathbf{t}_d}{\partial \mathbf{v}} \tag{71}$$

In a similar fashion, a discrete equivalent of Darcy's relation can be defined for the fluid flow $\mathbf{q}_d$ at the interface as:

$$\mathbf{n}_{\Gamma_d} \cdot \mathbf{q}_d = -k_d(p^+ - p^-) = -k_d \tilde{p}|_{\mathbf{x}\in\Gamma_d} \tag{72}$$

where it is recalled that k_d is the permeability of the diaphragm that is presumed to coincide with the displacement discontinuity Γ_d and p^+ and p^- are the pressures in the Ω^+ and Ω^- domains, respectively. For an impervious boundary, $k_d = 0$, which, according to eq. (72), implies that $\mathbf{n}_{\Gamma_d} \cdot \mathbf{q}_d = 0$. Conversely, ideal permeability requires that $k_d \to \infty$, so that $\mathbf{n}_{\Gamma_d} \cdot \mathbf{q}_d$ can only be bounded if $p^+ - p^- = 0$, which implies that no discontinuity can exist in the pressure field and the formulation of Armero & Callari (1999) is recovered.

Another possible boundary condition for the fluid flow at an internal boundary Γ_d is

$$\mathbf{n}_{\Gamma_d} \cdot \mathbf{q}_d = q_d|_{\mathbf{x}\in\Gamma_d} \tag{73}$$

Such a boundary condition would represent the existence of a drain (or a line source) with a capacity q_d per unit length.

5.3 *Weak format, discretization and time integration*

To arrive at the weak form of the balance equations, we recall the matrix-vector form of the momentum balance (30) and cast the mass balance in a similar format, to give

$$\mathbf{L}^T \boldsymbol{\sigma} = \mathbf{0}$$

and

$$\mathbf{m}^T \mathbf{L}\dot{\mathbf{u}} + \mathbf{m}^T \mathbf{L} n_f \mathbf{w}_f + Q^{-1}\dot{p} = 0 \tag{74}$$

where, for the two-dimensional case, $\mathbf{m}^T = (1, 1, 0)$. Following a standard Bubnov-Galerkin approach, we assume test functions for the displacements and the pressures as:

$$\eta = \bar{\eta} + \mathcal{H}_{\Gamma_d}\tilde{\eta}$$

and

$$\zeta = \bar{\zeta} + \mathcal{H}_{\Gamma_d}\tilde{\zeta} \tag{75}$$

Substitution into eqs (30) and (74) and integrating over the domain Ω leads to the corresponding weak forms:

$$\int_\Omega (\bar{\eta} + \mathcal{H}_{\Gamma_d}\tilde{\eta})^T \mathbf{L}^T \boldsymbol{\sigma}\, \mathrm{d}\Omega = 0$$

and

$$\int_\Omega (\bar{\zeta} + \mathcal{H}_{\Gamma_d}\tilde{\zeta})(\mathbf{m}^T \mathbf{L}\dot{\mathbf{u}} + \mathbf{m}^T \mathbf{L} n_f \mathbf{w}_f + Q^{-1}\dot{p})\, \mathrm{d}\Omega = 0 \tag{76}$$

Using the standard procedure of applying the divergence theorem, using the external boundary conditions (29) and (63), eliminating the Heaviside functions by changing the integration domain from Ω to Ω^+ and eliminating the Dirac delta functions by transforming the volume integral into a surface integral, the momentum balance takes the form:

$$\int_\Omega (\mathbf{L}\bar{\eta})^T \boldsymbol{\sigma}\, \mathrm{d}\Omega + \int_{\Omega+} (\mathbf{L}\tilde{\eta})^T \boldsymbol{\sigma}\, \mathrm{d}\Omega + \int_{\Gamma_d} \tilde{\eta}^T \mathbf{t}_d\, \mathrm{d}\Omega = \int_\Gamma (\bar{\eta} + \mathcal{H}_{\Gamma_d}\tilde{\eta})^T \mathbf{t}_p\, \mathrm{d}\Omega$$

and from eq. (76):

$$\begin{aligned} &-\int_\Omega \bar{\zeta}\mathbf{m}^T \mathbf{L}\dot{\mathbf{u}}\, \mathrm{d}\Omega - \int_{\Omega+} \tilde{\zeta}\mathbf{m}^T \mathbf{L}\dot{\mathbf{u}}\, \mathrm{d}\Omega \\ &+ \int_\Omega \nabla\bar{\zeta}^T n_f \mathbf{w}_f\, \mathrm{d}\Omega + \int_{\Omega+} \nabla\tilde{\zeta}^T n_f \mathbf{w}_f\, \mathrm{d}\Omega \\ &+ \int_{\Gamma_d} \tilde{\zeta}\mathbf{n}_{\Gamma_d}^T \mathbf{q}_d\, \mathrm{d}\Gamma - \int_\Omega \bar{\zeta} Q^{-1}\dot{p}\, \mathrm{d}\Omega - \int_{\Omega+} \tilde{\zeta} Q^{-1}\dot{p}\, \mathrm{d}\Omega \\ &= \int_\Gamma (\bar{\zeta} + \mathcal{H}_{\Gamma_d}\tilde{\zeta})\mathbf{n}_\Gamma^T \mathbf{q}_p\, \mathrm{d}\Gamma \end{aligned} \tag{77}$$

We discretize the trial functions $\mathbf{u}$ and p and the test functions η and ζ as:

$$\mathbf{u} = \mathbf{N}(\bar{\mathbf{a}} + \mathcal{H}_{\Gamma_d}\tilde{\mathbf{a}}) \tag{78}$$

$$p = \mathbf{H}(\bar{\mathbf{p}} + \mathcal{H}_{\Gamma_d}\tilde{\mathbf{p}}) \tag{79}$$

$$\eta = \mathbf{N}(\bar{\mathbf{w}} + \mathcal{H}_{\Gamma_d}\tilde{\mathbf{w}}) \tag{80}$$

$$\zeta = \mathbf{H}(\bar{\mathbf{z}} + \mathcal{H}_{\Gamma_d}\tilde{\mathbf{z}}) \tag{81}$$

Inserting eqs (80) and (81) into eqs (36) and (77) and requiring that the result holds for all admissible $\bar{\mathbf{w}}$, $\bar{\mathbf{z}}$, $\tilde{\mathbf{w}}$

and $\tilde{\mathbf{z}}$ gives:

$$\int_{\Omega} \mathbf{B}^T \boldsymbol{\sigma} \, \mathrm{d}\Omega = \int_{\Gamma} \mathbf{N}^T \mathbf{t}_p \, \mathrm{d}\Gamma \tag{82}$$

$$\int_{\Omega+} \mathbf{B}^T \boldsymbol{\sigma} \, \mathrm{d}\Omega + \int_{\Gamma_d} \mathbf{N}^T \mathbf{t}_d \, \mathrm{d}\Gamma = \int_{\Gamma} \mathcal{H}_{\Gamma_d} \mathbf{N}^T \mathbf{t}_p \, \mathrm{d}\Gamma \tag{83}$$

$$\begin{aligned} &-\int_{\Omega} \mathbf{H}^T \mathbf{m}^T \dot{\mathbf{u}} \, \mathrm{d}\Omega + \int_{\Omega} \nabla \mathbf{H}^T n_f \mathbf{w}_f \, \mathrm{d}\Omega \\ &-\int_{\Omega} \mathbf{H}^T Q^{-1} \dot{p} \, \mathrm{d}\Omega = \int_{\Gamma} \mathbf{H}^T \mathbf{n}_{\Gamma_d}^T \mathbf{q}_p \, \mathrm{d}\Gamma \end{aligned} \tag{84}$$

$$\begin{aligned} &-\int_{\Omega+} \mathbf{H}^T \mathbf{m}^T \dot{\mathbf{u}} \, \mathrm{d}\Omega + \int_{\Omega+} \nabla \mathbf{H}^T n_f \mathbf{w}_f \, \mathrm{d}\Omega \\ &-\int_{\Omega+} \mathbf{H}^T Q^{-1} \dot{p} \, \mathrm{d}\Omega + \int_{\Gamma_d} \mathbf{H}^T \mathbf{n}_{\Gamma_d}^T \mathbf{q}_d \, \mathrm{d}\Gamma \\ &= \int_{\Gamma} \mathcal{H}_{\Gamma_d} \mathbf{H}^T \mathbf{n}_{\Gamma_d}^T \mathbf{q}_p \, \mathrm{d}\Gamma \end{aligned} \tag{85}$$

For use in a Newton-Raphson solution method, (82)–(85) need to be linearized. To this end, the stress and the pressure are decomposed as

$$\boldsymbol{\sigma}_j = \boldsymbol{\sigma}_{j-1} + \mathrm{d}\boldsymbol{\sigma} \tag{86}$$

and

$$p_j = p_{j-1} + \mathrm{d}p \tag{87}$$

with the subscripts $j-1$ and j signifying the iteration numbers. Substituting these decompositions into the discrete set of equations (82)–(85), utilizing the kinematic relation (26), the stress–strain relation (68) for the bulk material and the traction–relative displacement relation (70) at the interface, using Darcy's relation for the fluid flow in the porous medium (69), its discrete equivalent (72) at the interface and the expression (65) for the pressure gradient, and using the interpolations for the displacement and the pressure according to eqs (78) and (79), leads to the following set of equations linearized at iteration $j-1$:

$$\begin{aligned} &\begin{bmatrix} \mathbf{0} & \mathbf{0} & \mathbf{0} & \mathbf{0} \\ \mathbf{0} & \mathbf{0} & \mathbf{0} & \mathbf{0} \\ \mathbf{K}_{\bar{a}\bar{p}}^T & \mathbf{K}_{\bar{a}\tilde{p}}^T & \mathbf{K}_{\bar{p}\bar{p}}^{(1)} & \mathbf{K}_{\bar{p}\tilde{p}}^{(1)} \\ \mathbf{K}_{\bar{a}\tilde{p}}^T & \mathbf{K}_{\bar{a}\tilde{p}}^T & \mathbf{K}_{\bar{p}\tilde{p}}^{(1)} & \mathbf{K}_{\bar{p}\tilde{p}}^{(1)} \end{bmatrix} \begin{pmatrix} \dot{\bar{\mathbf{a}}} \\ \dot{\tilde{\mathbf{a}}} \\ \dot{\bar{\mathbf{p}}} \\ \dot{\tilde{\mathbf{p}}} \end{pmatrix} \\ &+ \begin{bmatrix} \mathbf{K}_{\bar{a}\bar{a}} & \mathbf{K}_{\bar{a}\tilde{a}} & \mathbf{K}_{\bar{a}\bar{p}} & \mathbf{K}_{\bar{a}\tilde{p}} \\ \mathbf{K}_{\bar{a}\tilde{a}} & \mathbf{K}_{\tilde{a}\tilde{a}} & \mathbf{K}_{\bar{a}\tilde{p}} & \mathbf{K}_{\bar{a}\tilde{p}} \\ \mathbf{0} & \mathbf{0} & \mathbf{K}_{\bar{p}\bar{p}}^{(2)} & \mathbf{K}_{\bar{p}\tilde{p}}^{(2)} \\ \mathbf{0} & \mathbf{0} & \mathbf{K}_{\bar{p}\tilde{p}}^{(2)} & \mathbf{K}_{\tilde{p}\tilde{p}}^{(2)} \end{bmatrix} \begin{pmatrix} \mathrm{d}\bar{\mathbf{a}}_j \\ \mathrm{d}\tilde{\mathbf{a}}_j \\ \mathrm{d}\bar{\mathbf{p}}_j \\ \mathrm{d}\tilde{\mathbf{p}}_j \end{pmatrix} \\ &= \begin{pmatrix} \mathbf{f}_{\bar{a}}^{ext} - \mathbf{f}_{\bar{a},j-1}^{int} \\ \mathbf{f}_{\tilde{a}}^{ext} - \mathbf{f}_{\tilde{a},j-1}^{int} \\ \mathbf{f}_{\bar{p}}^{ext} - \mathbf{f}_{\bar{p},j-1}^{int} \\ \mathbf{f}_{\tilde{p}}^{ext} - \mathbf{f}_{\tilde{p},j-1}^{int} \end{pmatrix} \end{aligned} \tag{88}$$

with $\mathbf{f}_{\bar{a}}^{ext} \ldots\ldots\ldots \mathbf{f}_{\tilde{p}}^{ext}$ given by the right-hand sides of eqs (82)–(85), respectively, and with $\mathbf{f}_{\bar{a}}^{int} \ldots\ldots\ldots \mathbf{f}_{\tilde{p}}^{int}$ given by the left-hand sides of eqs (82)–(85). The stiffness matrices are defined as:

$$\mathbf{K}_{\bar{a}\bar{a}} = \int_{\Omega} \mathbf{B}^T \mathbf{D} \mathbf{B} \, \mathrm{d}\Omega \tag{89}$$

$$\mathbf{K}_{\bar{a}\tilde{a}} = \int_{\Omega+} \mathbf{B}^T \mathbf{D} \mathbf{B} \, \mathrm{d}\Omega \tag{90}$$

$$\mathbf{K}_{\tilde{a}\tilde{a}} = \int_{\Omega+} \mathbf{B}^T \mathbf{D} \mathbf{B} \, \mathrm{d}\Omega + \int_{\Gamma_d} \mathbf{N}^T \mathbf{T} \mathbf{N} \, \mathrm{d}\Gamma \tag{91}$$

$$\mathbf{K}_{\bar{a}\bar{p}} = -\int_{\Omega} \mathbf{B}^T \mathbf{m} \mathbf{H} \, \mathrm{d}\Omega \tag{92}$$

$$\mathbf{K}_{\bar{a}\tilde{p}} = -\int_{\Omega+} \mathbf{B}^T \mathbf{m} \mathbf{H} \, \mathrm{d}\Omega \tag{93}$$

$$\mathbf{K}_{\bar{p}\bar{p}}^{(1)} = -\int_{\Omega} \mathbf{H}^T Q^{-1} \mathbf{H} \, \mathrm{d}\Omega \tag{94}$$

$$\mathbf{K}_{\bar{p}\tilde{p}}^{(1)} = -\int_{\Omega+} \mathbf{H}^T Q^{-1} \mathbf{H} \, \mathrm{d}\Omega \tag{95}$$

$$\mathbf{K}_{\bar{p}\bar{p}}^{(2)} = -\int_{\Omega} \nabla \mathbf{H}^T k_f \nabla \mathbf{H} \, \mathrm{d}\Omega \tag{96}$$

$$\mathbf{K}_{\bar{p}\tilde{p}}^{(2)} = -\int_{\Omega+} \nabla \mathbf{H}^T k_f \nabla \mathbf{H} \, \mathrm{d}\Omega \tag{97}$$

$$\begin{aligned} \mathbf{K}_{\tilde{p}\tilde{p}}^{(2)} = &-\int_{\Omega+} \nabla \mathbf{H}^T k_f \nabla \mathbf{H} \, \mathrm{d}\Omega \\ &-\int_{\Gamma_d} \mathbf{H}^T k_d \mathbf{H} \, \mathrm{d}\Gamma \end{aligned} \tag{98}$$

To carry out the time integration in (88) a backward finite difference scheme is adopted:

$$\left(\frac{\mathrm{d}(\,\cdot\,)}{\mathrm{d}t} \right)^{t+\Delta t} = \frac{(\,\cdot\,)^{t+\Delta t} - (\,\cdot\,)^t}{\Delta t} \tag{99}$$

where Δt is the time increment, while $(\,\cdot\,)^t$ and $(\,\cdot\,)^{t+\Delta t}$ denote the unknowns at t and $t+\Delta t$, respectively. Furthermore, a decomposition of the arrays that contain the nodal displacements and nodal pressures is adopted similar to that of the stresses and the pressures, cf. eqs (86) and (87), $(\,\cdot\,)_j = (\,\cdot\,)_{j-1} + \mathrm{d}(\,\cdot\,)$. Combining this with eq. (99) yields

$$\left(\frac{\mathrm{d}(\,\cdot\,)}{\mathrm{d}t} \right)^{t+\Delta t} = \frac{(\,\cdot\,)_{j-1}^{t+\Delta t} + \mathrm{d}(\,\cdot\,) - (\,\cdot\,)^t}{\Delta t} \tag{100}$$

Substituting this identity into eq. (88) for all arrays finally gives

$$\begin{bmatrix} \mathbf{K}_{\bar{a}\bar{a}} & \mathbf{K}_{\bar{a}\tilde{a}} & \mathbf{K}_{\bar{a}\bar{p}} & \mathbf{K}_{\bar{a}\tilde{p}} \\ \mathbf{K}_{\bar{a}\tilde{a}}^{T} & \mathbf{K}_{\tilde{a}\tilde{a}} & \mathbf{K}_{\tilde{a}\bar{p}} & \mathbf{K}_{\tilde{a}\tilde{p}} \\ \mathbf{K}_{\bar{a}\bar{p}}^{T} & \mathbf{K}_{\tilde{a}\bar{p}}^{T} & \mathbf{K}_{\bar{p}\bar{p}}^{(1)} + \Delta t \mathbf{K}_{\bar{p}\bar{p}}^{(2)} & \mathbf{K}_{\bar{p}\tilde{p}}^{(1)} + \Delta t \mathbf{K}_{\bar{p}\tilde{p}}^{(2)} \\ \mathbf{K}_{\bar{a}\tilde{p}}^{T} & \mathbf{K}_{\tilde{a}\tilde{p}}^{T} & \mathbf{K}_{\bar{p}\tilde{p}}^{(1)} + \Delta t \mathbf{K}_{\bar{p}\tilde{p}}^{(2)} & \mathbf{K}_{\tilde{p}\tilde{p}}^{(1)} + \Delta t \mathbf{K}_{\tilde{p}\tilde{p}}^{(2)} \end{bmatrix}$$

$$\begin{pmatrix} \mathrm{d}\bar{\mathbf{a}}_j \\ \mathrm{d}\tilde{\mathbf{a}}_j \\ \mathrm{d}\bar{\mathbf{p}}_j \\ \mathrm{d}\tilde{\mathbf{p}}_j \end{pmatrix} = \begin{pmatrix} \mathbf{f}_{\bar{a}}^{ext} - \mathbf{f}_{\bar{a},j-1}^{int} \\ \mathbf{f}_{\tilde{a}}^{ext} - \mathbf{f}_{\tilde{a},j-1}^{int} \\ \Delta t \mathbf{f}_{\bar{p}}^{ext} - \mathbf{f}_{\bar{p},j-1}^{int,t+\Delta t} \\ \Delta t \mathbf{f}_{\tilde{p}}^{ext} - \mathbf{f}_{\tilde{p},j-1}^{int,t+\Delta t} \end{pmatrix} \tag{101}$$

with

$$\begin{aligned} \mathbf{f}_{p,j-1}^{int,t+\Delta t} &= \mathbf{K}_{\bar{a}\bar{p}}^{T} \left(\bar{\mathbf{a}}_{j-1}^{t+\Delta t} - \bar{\mathbf{a}}^{t} \right) + \mathbf{K}_{\tilde{a}\bar{p}}^{T} \left(\tilde{\mathbf{a}}_{j-1}^{t+\Delta t} - \tilde{\mathbf{a}}^{t} \right) \\ &+ \mathbf{K}_{\bar{p}\bar{p}}^{(1)} \left(\bar{\mathbf{p}}_{j-1}^{t+\Delta t} - \bar{\mathbf{p}}^{t} \right) + \mathbf{K}_{\bar{p}\tilde{p}}^{(1)} \left(\tilde{\mathbf{p}}_{j-1}^{t+\Delta t} - \tilde{\mathbf{p}}^{t} \right) \\ &+ \Delta t \mathbf{K}_{\bar{p}\bar{p}}^{(2)} \bar{\mathbf{p}}_{j-1}^{t+\Delta t} + \Delta t \mathbf{K}_{\bar{p}\tilde{p}}^{(2)} \tilde{\mathbf{p}}_{j-1}^{t+\Delta t} \end{aligned} \tag{102}$$

$$\begin{aligned} \mathbf{f}_{\tilde{p},j-1}^{int,t+\Delta t} &= \mathbf{K}_{\bar{\mathbf{a}}\tilde{p}}^{T} \left(\bar{\mathbf{a}}_{j-1}^{t+\Delta t} - \bar{\mathbf{a}}^{t} \right) + \mathbf{K}_{\tilde{a}\tilde{p}}^{T} \left(\tilde{\mathbf{a}}_{j-1}^{t+\Delta t} - \tilde{\mathbf{a}}^{t} \right) \\ &+ \mathbf{K}_{\bar{p}\tilde{p}}^{(1)} \left(\bar{\mathbf{p}}_{j-1}^{t+\Delta t} - \bar{\mathbf{p}}^{t} \right) + \mathbf{K}_{\tilde{p}\tilde{p}}^{(1)} \left(\tilde{\mathbf{p}}_{j-1}^{t+\Delta t} - \tilde{\mathbf{p}}^{t} \right) \\ &+ \Delta t \mathbf{K}_{\bar{p}\tilde{p}}^{(2)} \bar{\mathbf{p}}_{j-1}^{t+\Delta t} + \Delta t \mathbf{K}_{\tilde{p}\tilde{p}}^{(2)} \tilde{\mathbf{p}}_{j-1}^{t+\Delta t} \end{aligned} \tag{103}$$

6 CONCLUDING REMARKS

Discrete and smeared crack approaches have historically been considered as opposing methods for analyzing concrete fracture. Indeed, both approaches have their domain of application: discrete crack models are appropriate for modeling one or more dominant cracks, while smeared crack models can simulate the diffuse cracking patterns that arise due to the heterogeneity of concrete and the presence of reinforcement. The partition-of-unity property of finite-element shape functions enables the placement of cohesive surfaces of finite size at arbitrary locations and in arbitrary directions in finite element discretizations. This property enables diffuse crack patterns to be modeled. On the other hand, since these *cohesive segments* can overlap within a finite element, a single, continuous crack can be represented, thus enabling the simulation of the development and growth of a dominant crack without any ad hoc assumptions concerning initiation and growth. In this sense, the cohesive segments method can be viewed upon as a bridge between the traditional discrete and smeared crack models used in concrete fracture. In addition, many practically important damage or failure problems in concrete structures involve water or ion transport. We have also shown how the partition-of-unity approach provides a natural framework for analyzing processes involving the coupling of stress driven fracture with such diffusion processes.

REFERENCES

Armero, F. & Callari, C. 1999. An analysis of strong discontinuities in a saturated poro-plastic solid. *Int. J. Num. Meth. Eng.* 46: 1673–1698.

Babuska, I. & Melenk, J.M. 1997. The partition of unity method. *Int. J. Num. Meth. Eng.* 40: 727–758.

Barenblatt, G.I. 1962. The mathematical theory of equilibrium cracks in brittle fracture. *Adv. Appl. Mech.* 7: 55–129.

Bažant, Z.P. & Oh, B. 1983. Crack band theory for fracture of concrete. *RILEM Mat. Struct.* 16: 155–177.

Belytschko, T., Fish, J. & Engelman, B.E. 1988. A finite element with embedded localization zones. *Comp. Meth. Appl. Mech. Eng.* 70: 59–89.

Belytschko, T., Lu, Y.Y. & Gu, L. 1994. Element-free Galerkin methods. *Int. J. Num. Meth. Eng.* 37: 229–256.

Belytschko, T. & Black, T. 1999. Elastic crack growth in finite elements with minimal remeshing. *Int. J. Num. Meth. Eng.* 45: 601–620.

Belytschko T., Moës N., Usui S. & Parimi C. 2001. Arbitrary discontinuities in finite elements. *Int. J. Num. Meth. Eng.* 50: 993–1013.

Camacho, G.T. & Ortiz, M. 1996. Computational modeling of impact damage in brittle materials. *Int. J. Solids Struct.* 33: 2899–2938.

Cope, R.J., Rao, P.V., Clark, L.A. & Norris, P. 1980. Modelling of reinforced concrete behaviour for finite element analysis of bridge slabs. In C. Taylor, E. Hinton & D.R.J. Owen (eds), *Numerical Methods for Non-Linear Problems*: 457–470. Swansea: Pineridge Press.

de Borst, R., Wells, G.N. & Sluys, L.J. 2001. Some observations on embedded discontinuity models. *Eng. Comp.* 18: 241–254.

de Borst, R. & Abellan, M.-A. 2002. A numerical frame-work for continuum–discontinuum transition. *Arch. Mech.*, in press.

Daux, C., Moës, N., Dolbow, J., Sukumar, N. & Belytschko T. 2000. Arbitrary branched and intersecting cracks with the extended finite element method. *Int. J. Num. Meth. Eng.* 48: 1741–1760.

Dugdale, D.S. 1960. Yielding of steel sheets containing slits. *J. Mech. Phys. Sol.* 8: 100–108.

Falk, M.L., Needleman, A. & Rice, J.R. 2001. A critical evaluation of cohesive zone models of dynamic fracture. *Journal de Physique IV* Pr5: 43–50.

Hegen, D. 1996. Element-free Galerkin methods in combination with finite element approaches. *Comp. Meth. Appl. Mech. Eng.* 135: 143–166.

Hillerborg, A., Modéer, M. & Petersson, P.E. 1976. Analysis of crack formation and crack growth in concrete by means of fracture mechanics and finite elements. *Cement Concrete Res.* 6: 773–782.

Hutchinson, J.W. & Evans, A.G. 2000. Mechanics of materials: top-down approaches to fracture. *Acta Mat.* 48: 125–135.

Ingraffea, A.R. & Saouma, V. 1985. Numerical modelling of discrete crack propagation in reinforced and plain concrete. In *Fracture Mechanics of Concrete*: 171–225. Dordrecht: Martinus Nijhoff Publishers.

Jouanna, P. & Abellan, M.-A. 1995. Generalized approach to heterogeneous media. In A. Gens, P. Jouanna & B.A. Schrefler (eds), *Modern Issues in Non-Saturated Soils*: 1–128. Wien – New York: Springer–Verlag.

Kuhl, E., Ramm, E & de Borst, R. 2000. An anisotropic gradient damage model for quasi-brittle materials. *Comp. Meth. Appl. Mech. Eng.* 183: 87–103.

Larsson, J. & Larsson, R. 2000. Localization analysis of a fluid-saturated elastoplastic porous medium using regularized discontinuities. *Mech. Cohes.-frict. Mater.* 5: 565–582.

Lemaitre, J. & Chaboche, J.L. 1990. *Mechanics of Solids Materials*. Cambridge: Cambridge University Press.

Lewis, R.W. & Schrefler, B.A. 1998. *The Finite Element Method in the Static and Dynamic Deformation and Consolidation of Porous Media*, Second Edition. Chichester: John Wiley & Sons.

Mazars, J. 1984. *Application de la mécanique de l'endommagement au comportement non linéaire et à la rupture du béton de structure*. These d'Etat. Paris: Université Paris VI.

Mazars, J. & Pijaudier-Cabot, G. 1989. Continuum damage theory – application to concrete. *ASCE J. Eng. Mech.* 115: 345–365.

Moës, N., Dolbow, J. & Belytschko, T. 1999. A finite element method for crack growth without remeshing. *Int. J. Num. Meth. Eng.* 46: 131–150.

Moës, N. & Belytschko, T. 2002. Extended finite element method for cohesive crack growth. *Eng. Fract. Mech.* 69: 813–833.

Mosler, J. & Meschke, G. 2002. A comparison of embedded discontinuity approaches with fracture energy based smeared crack models. In H.A. Mang, F.G. Rammerstorfer & J. Eberhardsteiner (eds), *Proceedings of the Fifth Congress on Computational Mechanics (WCCM V)*: Paper 81054. Vienna: Vienna University of Technology.

Needleman, A. 1987. A continuum model for void nucleation by inclusion debonding. *J. Appl. Mech.* 54: 525–531.

Ngo, D. & Scordelis, A.C. 1967. Finite element analysis of reinforced concrete beams. *J. Amer. Concr. Inst.* 64: 152–163.

Ortiz, M., Leroy, Y. & Needleman, A. 1987. A finite element method for localized failure analysis. *Comp. Meth. Appl. Mech. Eng.* 61: 189–214.

Papoulia, K.D. & Vavasis, S.A. 2003. Time-continuous cohesive interface finite elements in explicit dynamics. *Int. J. Num. Meth. Eng.*, in press.

Peerlings, R.H.J., de Borst, R., Brekelmans, W.A.M. & de Vree, H.P.J. 1996. Gradient-enhanced damage for quasi-brittle materials. *Int. J. Num. Meth. Eng.* 39: 3391–3403.

Peerlings, R.H.J., de Borst, R., Brekelmans, W.A.M. & Geers, M.G.D. 1998. Gradient-enhanced modelling of concrete fracture. *Mech. Coh.-frict. Mat.* 3:323–342.

Peerlings, R.H.J., de Borst, R., Brekelmans, W.A.M. & Geers, M.G.D. 2002. Localisation issues in local and non-local continuum approaches to fracture. *Eur. J. Mech. A/Solids* 21: 175–189.

Pietruszczak, S. & Mróz, Z. 1981. Finite element analysis of deformation of strain softening materials. *Int. J. Num. Meth. Eng.* 17: 327–334.

Pijaudier-Cabot, G. & Bažant, Z.P. 1987. Nonlocal damage theory. *ASCE J. Eng. Mech.* 113: 1512–1533.

Rashid, Y.R. 1968. Analysis of reinforced concrete pressure vessels. *Nucl. Eng. Des.* 7: 334–344.

Remmers, J.J.C., de Borst, R. & Needleman, A. 2002. A cohesive-segments method for the simulation of crack growth. *Comp. Mech.*, in press.

Rots, J.G. 1986. Strain-softening analysis of concrete fracture specimens. In F.H. Wittmann (ed.), *Fracture Toughness and Fracture Energy of Concrete*: 137–148. Amsterdam: Elsevier Science Publishers.

Rots, j.G. 1991. Smeared and discrete representations of localized fracture. *Int. J. Fracture* 51: 45–59.

Schellekens, J.C.J. & de Borst, R. 1993. On the numerical integration of interface elements. *Int. J. Num. Meth. Eng.* 36: 43–66.

Schellekens, J.C.J. & de Borst, R. 1994. Free edge delamination in carbon-epoxy laminates: a novel numerical/experimental approach. *Compos. Struct.* 28: 357–373.

Simo, J.C., Oliver, J. & Armero, F. 1993. An analysis of strong discontinuities induced by softening relations in rate-independent solids. *Comp. Mech.* 12: 277–296.

Suidan, M. & Schnobrich, W.C. 1973. Finite element analysis of reinforced concrete. *ASCE J. Struct. Div.* 99: 2109–2122.

van Mier, J.G.M. 1997. *Fracture Processes of Concrete*. Boca Raton, Florida: CRC Press.

Wells, G.N. 2001. *Discontinuous Modelling of Strain Localisation and Failure.* Dissertation. Delft: Delft University of Technology.

Wells, G.N. & Sluys, L.J. 2001. A new method for modeling cohesive cracks using finite elements. *Int. J. Num. Meth. Eng.* 50: 2667–2682.

Wells, G.N., de Borst, R. & Sluys L.J. 2002. A consistent geometrically non-linear approach for delamination. *Int. J. Num. Meth. Eng.* 54: 1333–1355.

Xu, X.P. & Needleman, A. 1994. Numerical simulations of fast crack growth in brittle solids. *J. Mech. Phys. Solids* 42: 1397–1434.

Computational Modelling of Concrete Structures, Bićanić et al. (eds)
© 2003 Swets & Zeitlinger, Lisse, ISBN 90 5809 536 3

Keynote paper: Models and algorithms for localized failure

Milan Jirásek
Swiss Federal Institute of Technology at Lausanne (EPFL), Switzerland

ABSTRACT: This paper deals with various aspects of the computational analysis of strain localization and failure in quasibrittle materials. It gives a general overview of discretization techniques used for the simulation of propagating cohesive displacement discontinuities and discusses their advantages and drawbacks. Attention then shifts to regularized softening continuum models. After a general discussion, various types of integral-type nonlocal damage and plasticity formulations are described and their properties are compared. Finally, the structure of the tangent stiffness matrix for nonlocal damage models is outlined, and stress return algorithms for nonlocal plasticity are briefly commented on.

1 INTRODUCTION

The behavior of quasibrittle materials (such as concrete, rock, tough ceramics, or ice) subjected to increasing mechanical solicitations is characterized by diffuse microcracking that later localizes in relatively narrow zones, referred to as the fracture process zones. The localization of strain and damage eventually leads to a gradual development of macroscopic stress-free cracks. Despite a considerable progress in the past two decades, theoretical modeling and computational resolution of the localization process up to structural failure still remains a challenging issue of contemporary solid mechanics. The purpose of this paper is to summarize certain latest developments in this area and to outline the current trends and future prospects.

Narrow zones of highly concentrated evolving microdefects can be modeled in many different ways. A natural classification can be based on the regularity of the kinematic description. The character of the displacement field and of the corresponding strain field for three fundamental classes of models is illustrated in Fig. 1, which for simplicity refers to the one-dimensional setting. The fracture process zone is represented either by a single point at which the displacement field has a jump (Fig. 1a), or by a finite interval. In the latter case, the strain field either has a jump at the boundary of the process zone (Fig. 1b), or remains continuous (Fig. 1c).

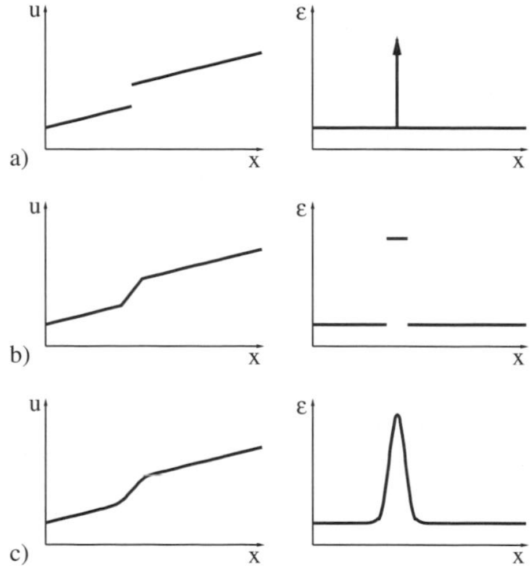

Figure 1. Representation of the process zone by a) a strong discontinuity, b) a band of localized strain separated by two weak discontinuities, c) a continuous profile of localized strain. Left column: displacement profile; right column: strain profile.

In multiple dimensions, the foregoing characteristics of the kinematic description must be appropriately generalized. Models belonging to the first class admit the presence of a *strong discontinuity*, i.e., of a curve (in two dimensions) or surface (in three dimensions) across which the displacement field has a jump. In general, the discontinuity is assumed to transmit some cohesive tractions that are related to the opening and sliding component of the displacement jump by a special traction-separation law. Physically, such a discontinuity can be considered as a cohesive crack or slip line. The tractions transmitted by a cohesive crack are usually assumed to vanish when the crack opening exceeds a certain limit, and the crack then becomes stress-free.

Models belonging to the second class use a continuous description of the displacement field but admit the presence of *weak discontinuities*, i.e., of curves or surfaces across which certain components of the strain field have jumps. Typically, such weak discontinuities form at the internal boundaries that separate a band or layer of softening material from the surrounding material that undergoes unloading. The thickness of the softening band can be considered as a material property independent of the spatial discretization (e.g., finite element mesh), or as a measure of the minimum possible localization pattern that can be resolved on a given mesh. In the former case, the band thickness is considered as an intrinsic material length and the softening stress-strain law can be uniquely defined, while in the latter case the softening modulus must be adjusted according to the spatial discretization and the model can be interpreted as a partially regularized form of a cohesive crack model, with a strong discontinuity approached in the limit as the mesh is refined.

Finally, models belonging to the third class are characterized by *continuity* of both the displacement field and the strain field, and they represent the process zone as a band or layer of softening material, in which the strain gradually increases from the minimum value on the boundary of the band to the maximum value at the center. Such fully regularized models are obtained by proper enhancements of the standard continuum theory, e.g., by the incorporation of nonlocal averages or higher-order gradients of internal variables.

2 MODELS WITH COHESIVE STRONG DISCONTINUITIES

The simplest example of a propagating displacement discontinuity is a stress-free crack with a sharp tip, at which the stress field has a singularity. If the material remains linear elastic, the process zone collapses to a single point and the model does not possess any characteristic length. Such models are suitable for the description of quasibrittle materials only on a very large scale, when the actual process zone is negligible with respect to the characteristic dimensions of the structure. Scaling of nominal strength is then given by a power law. On smaller scales, the model must be refined by introducing a cohesive traction-separation law, which removes the singularity and leads to the formation of a process zone with a finite length but still zero thickness. The length of the cohesive process zone is related to the characteristic length $l_c = EG_F/f_t^2$ that is set by the elastic modulus E, fracture energy G_F (area under the traction-separation curve), and tensile strength f_t (stress at which the discontinuity is initiated). Owing to the presence of a characteristic length, the model can reproduce a transitional size effect of a non-power type.

Computational resolution of discrete cracks is usually based on one of the following discretization techniques:

1. remeshing,
2. element edge methods, where all element interfaces are potential discontinuities,
3. elements with embedded discontinuities (EED), based on the enhanced assumed strain technique,
4. the extended finite element method (XFEM), where the discontinuities are introduced through a discontinuous partition of unity.

2.1 *Remeshing*

In remeshing methods, the finite element model is reconstructed at each step so that the element faces, or element edges in two dimensions, correspond to the crack or other discontinuity. The nodes located on these faces or edges are doubled and distributed to both sides of the discontinuity, so that the approximation of the motion, and hence displacement, can be fully discontinuous across the crack. In early versions of this method, the entire mesh was reconstructed at each step of crack growth, but more recent methods limit the reconstruction to the immediate vicinity of the crack. If stress singularities are present around crack tips, the accuracy can be increased by using special elements, such as the quarter-node elements in which the singularity of the Jacobian at one of the nodes produces a singularity of the stress approximation (Henshell and Shaw 1975).

One advantage of this method is that a growing crack software can be constructed almost entirely from parts that are already available: a mesher and a standard finite element program. However, in three dimensions, there is still a need for some type of a representation of the crack to drive the mesher.

The disadvantages are that

1. for nonlinear problems with path dependence, it is necessary to project results between different meshes many times;
2. output of time histories at selected points is more difficult;
3. the entire issue of data management becomes more difficult because the solution is associated with many meshes.

Nevertheless, this methodology has reached a high state of maturity. For example, it has been applied to growth of single cracks in three-dimensional elastostatic bodies (Martha et al. 1993; Carter et al. 1997). However, application to problems such as dynamic fracture or fragmentation appears overwhelmingly difficult; even nonlinear crack growth or the growth of several interacting cracks poses a significant challenge in meshing and projections.

2.2 *Incipient discontinuity methods*

Instead of tracing the precise path of a growing crack and inserting a new discontinuous segment (with separated nodes) each time the crack propagates, it is possible to place a potential cohesive segment at the interface between each two adjacent finite elements. This is the approach pursued by Needleman, Ortiz and coworkers (Xu and Needleman 1994; Camacho and Ortiz 1996; Pandolfi et al. 1999; Ortiz and Pandolfi 1999). It is computationally efficient especially in dynamic simulations of multiple cracks up to fragmentation. However, the theoretical foundations of such models do not seem to be completely understood.

For example, it is not clear how to link the parameters of the cohesive law in these models to measurable material properties. Traditional cohesive crack models represent the entire nonlinear process zone by a single discontinuity, which means that the opening of the "fictitious crack" represents the cumulative effect of microcracking integrated across the width of the process zone, and the area under the traction-separation curve should correspond to the macroscopically measured fracture energy of the material. Similarly, cohesive zone models for degrading interfaces are again related to a well-defined physical entity with measurable properties. Models with potential cohesive discontinuities interspersed throughout the entire body can be calibrated to match the experimental results with a given mesh size and structure, but it is expected that on finer meshes the total area of activated discontinuity segments per unit volume would be larger, and if the same traction-separation law is used, the energy dissipation would increase as a function of the mesh density. This was recognized and taken into account for instance in the so-called LCPC model for concrete, developed by Rossi and Wu (1992), who proposed a scaling law that links the model parameters to the size of the finite elements. Of course, the exact form of such a law may be material-dependent and it remains to be clarified what kind of scaling should be used, e.g., for metallic materials.

2.3 *Elements with embedded discontinuities*

Approximation techniques allowing a semiexplicit resolution of a discontinuity with an arbitrary trajectory running across the finite element mesh were inspired by the work of Ortiz et al. (1987), who enriched the standard finite element approximation by functions that made it possible to capture one weak discontinuity line crossing a finite element. The formulation proposed by Belytschko et al. (1988) could capture a softening band between two parallel weak discontinuity lines within a single element. The idea was soon extended to strong (displacement) discontinuities (Dvorkin et al. 1990; Klisinski et al. 1991; Olofsson et al. 1994; Simo and Oliver 1994).

A systematic classification and critical evaluation of embedded discontinuity models within a unified framework was presented by Jirásek (2000a). The models were classified into three main groups, called statically optimal symmetric (SOS), kinematically optimal symmetric (KOS), and statically and kinematically optimal nonsymmetric (SKON). The SOS formulation imposes a natural stress continuity condition, but it does not properly reflect the kinematics of a completely open crack. On the other hand, the KOS formulation describes the kinematic aspects satisfactorily, but it leads to an awkward relationship between the stress in the bulk of the element and the tractions across the discontinuity. Optimal performance is achieved with the nonsymmetric SKON formulation, which combines the natural stress continuity condition with a reasonable representation of the complete separation at the late stages of the fracturing process.

The optimal combination of static and kinematic equations for elements with embedded discontinuities first appeared in (Dvorkin et al. 1990), even though their exact nature is not easily understood from that paper. A very similar quadrilateral element was constructed by Klisinski et al. (1991), based on simple and instructive physical considerations. In a later paper (Olofsson et al. 1994), the same technique was applied to a constant-strain triangle. A general version of the SKON formulation for an arbitrary type of parent element was outlined in a short paper by Simo and Oliver (1994) and fully described by Oliver (1996).

Virtually all these models deal with nonconforming interpolations, i.e., compatibility is satisfied only in the weak sense. This makes it possible to treat the added degrees of freedom that correspond to discontinuous enrichments as internal ones and eliminate them on the element level. The main advantage is that only standard degrees of freedom (nodal displacements) are kept on the global, structural level, and the number of global equilibrium equations and structure of the stiffness matrix do not change when the crack propagates and enrichments are added to new elements. However, there is a price to pay for this convenience. A detailed analysis of the behavior of a single element with an embedded discontinuity reveals that, in order to guarantee uniqueness of the element response to any prescribed history of nodal displacements, severe restrictions must be placed not only on the element size (to prevent a non-unique response known as "local snapback") but also on the element shape (Jirásek 2000b). These restrictions become even more severe in the presence of multiple discontinuities in one element (needed to describe crack branching) and in three dimensions. When they are violated, the numerical algorithm evaluating the nodal forces cannot be expected to be robust and converge for all possible loading histories (even if the load is applied in very small incremental steps), and divergence on the element level occurs as soon as one of the discontinuities is introduced in an unfavorable position with respect to the basic element. Another inconvenience is that the tangent stiffness matrix of the embedded element is in general nonsymmetric, even if the material stiffnesses of the continuous material and of the cohesive discontinuity are symmetric.

2.4 *Extended finite elements*

Elements with embedded discontinuities, discussed in the previous section, provide a better kinematic description of discontinuous displacement fields than pure continuum models that smear the displacement jumps uniformly over the entire element, but they still have certain limitations. Due to the high sensitivity of elements with embedded discontinuities to the position of the discontinuity, it is practically impossible to extend the technique to three dimensions without using special numerical tricks (Wells 2001), and the beauty and simplicity of the original idea are lost.

A new class of methods with attractive properties has emerged recently, based on partitions of unity (Melenk and Babuška 1996; Babuška and Melenk 1997). A similar idea appears in the so-called manifold method, developed in the context of discontinuous deformation analysis (Shi 1992; Shi 1995; Tsay et al. 1999). The general idea is that the approximation space spanned by a partition of unity (e.g., by the standard finite element shape functions) is enriched by products of the standard basis functions with special functions that are constructed, e.g., from the analytical solution of the problem under some simplifying assumptions. This permits the incorporation of *a priori* knowledge about the character of the problem and its solutions. The enriched displacement approximation is written in the form

$$\boldsymbol{u}(\boldsymbol{x}) = \sum_{i=1}^{N_{nod}} N_i(\boldsymbol{x}) \left(\boldsymbol{d}_i + \sum_{j \in \mathcal{L}_i} G_j(\boldsymbol{x}) \boldsymbol{e}_{ij} \right) \tag{1}$$

where N_{nod} is the number of nodes of the finite element model; N_i, $i = 1, 2, \ldots, N_{nod}$, are the standard shape functions; $\boldsymbol{d}_i$, $i = 1, 2, \ldots, N_{nod}$, are the standard displacement degrees of freedom; $G_j, j = 1, 2, \ldots, m$, are the global enrichment functions; $\mathcal{L}_i \subset \{1, 2, \ldots, m\}$ is the set of integers that indicate which enrichment functions are activated at node i; and $\boldsymbol{e}_{ij}$ are the additional degrees of freedom associated with node i and enrichment function j. The key trick is that the global enrichment functions G_j are multiplied by the nodal shape functions N_i. The resulting local enrichment functions N_iG_j inherit from G_j good approximation properties and from N_i a limited support. Consequently, the enrichment has a local character and the resulting stiffness matrix is sparse (with a proper renumbering it remains banded). The standard approximation is usually enriched only locally in a certain region of interest, e.g., in a localization zone or a boundary layer, and the newly added degrees of freedom $\boldsymbol{e}_{ij}$ can be associated with nodes of the existing mesh, without the need for changing the topology. Owing to the partition-of-unity property of the standard shape functions (their sum is equal to one at any point $\boldsymbol{x}$), the enrichment functions G_j can be reproduced exactly.

In the context of localized failure simulations, this method was first applied to linear elastic fracture mechanics, with the enrichment consisting of the singular near-tip asymptotic fields and simple Heaviside functions (Moës et al. 1999). The method was later called the eXtended Finite Element Method (X-FEM). It can efficiently handle three-dimensional cracks (Sukumar et al. 2000) and even branching and intersecting cracks (Daux et al. 2000). A big advantage of this technique is that the displacement interpolation is conforming, with no incompatibilities between elements, and that the strains on both sides of a stress-free crack are fully decoupled.

The concept is also applicable to cohesive crack models. Wells and Sluys (2001) enriched the interpolation functions by products of the Heaviside function with standard finite element shape functions that correspond to the nodes of those elements that are intersected by the crack. Moës and Belytschko (2002) added special non-singular enrichments around the crack tip, motivated by asymptotic analysis of the strain field at the tip of a cohesive crack, while Mariani and Perego (2002) used enrichments by quadratic polynomials which, when multiplied by the linear shape functions, make it possible to approximate the typical cusp-like shape of the crack tip by a cubic displacement discontinuity.

Formulations based on the enhanced assumed strain technique and on the partition-of-unity concept have been compared by Jirásek and Belytschko (2002). It has been shown that PUM-XFEM formulations overcome the difficulties associated with the nonconforming interpolation of the displacement jump used by embedded discontinuity models. They also restore the symmetry of the stiffness matrix, which is lost in the SKON formulation of elements with embedded discontinuities due to the fact that the internal equilibrium conditions are not work-conjugate to the kinematic equations. The implementation of extended finite elements is, however, some-what more difficult, because it is necessary to add new global degrees of freedom during the simulation and to refine the integration scheme in the enriched area around the crack.

A big advantage of the PUM-XFEM formulations is that the displacement interpolation is conforming, with no incompatibilities between elements, and that the strain on both sides of a stress-free crack is fully decoupled, which was not the case for most of the previous elements with embedded discontinuities. The added degrees of freedom are global, but they can be assigned to the existing nodes of the basic finite element mesh, without any need for topology changes. Such degrees of freedom are easily inserted into the global set of equations and the resulting stiffness matrix preserves its banded character.

3 REGULARIZED SOFTENING CONTINUA

3.1 *General overview*

Full regularization of the localization problem can be achieved by a suitable generalization of the underlying continuum theory. Generalized continua in the broad sense can be classified according to the following criteria:

1. Generalized kinematics.
   (a) Continua with microstructure.
   (b) Continua with nonlocal strain.
2. Generalized constitutive equations.
   (a) Material models with gradients of internal variables and, in some cases, also with gradients of thermodynamic forces conjugate to those variables.
   (b) Material models with nonlocal internal variables and, in some cases, also with nonlocal thermodynamic forces conjugate to those variables.

All generalized continuum theories mentioned above abandon the assumption that the stress at a given material point is uniquely determined by the history of strain and temperature at that point. Continua with microstructure introduce additional kinematic quantities that characterize the deformation state of a material "point", e.g., the microcurvature tensor or the strain gradient. If their histories at a given point are known, the corresponding work-conjugate quantities (generalized stress measures) can be determined from the constitutive equations. So the constitutive laws are still applied locally. For the second class of generalized continuum theories, the stress at a given point depends on the distribution of an internal variable in a neighborhood of that point. For explicit models with gradients of internal variables, this influence is limited to an infinitesimal neighborhood, and therefore they are classified as *weakly nonlocal.* For models with nonlocal internal variables, the dependence involves a finite neighborhood or even the entire body, and they are called *strongly nonlocal.* This includes not only models that define the nonlocal field by weighted spatial averaging, but also for implicit gradient models that define the nonlocal field as a solution of a differential equation.

A common feature of all these generalized continuum theories is that they set a certain length scale that can be related to the material microstructure. The length scale is controled by one or more material parameters with the dimension of length, which are either explicitly present in the governing equations, or can be deduced from a combination of material parameters with different physical dimensions (e.g., of the standard stiffness and the higher-order stiffness in strain-gradient elasticity). The presence of an intrinsic length scale means that the body cannot be decomposed into arbitrarily small material "particles" that still keep all the properties of the basic material. From this point of view, all of the generalized theories mentioned here, including continua with microstructure, can be considered as nonlocal. Due to the presence of a length scale, the generalized continuum models are capable of describing size effects of the transitional type, more general than the simple power-type scaling resulting from models with no characteristic length; see (Bažant 2002) for more details.

To limit the scope of this paper, we focus on models with weighted spatial averages of internal variables, called also the integral-type nonlocal models.

3.2 *Nonlocal averaging*

Even though the idea of a nonlocal continuum has a much longer history, nonlocal material models of the integral type were first exploited as localization limiters in the 1980s. After some preliminary formulations based on the concept of an imbricate continuum (Bažant 1984), the nonlocal damage theory emerged (Pijaudier-Cabot and Bažant 1987). Nonlocal formulations were then developed for a number of constitutive theories, including softening plasticity, smeared cracking, microplane models, etc. The basic concept seems to be sound, but the details of the formulation for a given material model are still to a large extent ambiguous. Formulations applying nonlocal averaging to different variables often give similar results at the onset of localization but their behaviors at later stages of the deformation process may be dramatically different and may exhibit some pathologies (Jirásek 1998; Jirásek and Rolshoven 2003). From the practical point of view, the most serious deficiency is that no sufficiently general nonlocal model for complex materials such as concrete seems to be available. Existing formulations usually yield satisfactory results for a specific class of failure mechanisms (e.g., for tensile failure), but they are hard to extend to arbitrary loading scenarios. It is even questionable whether the general case can be covered using an isotropic nonlocal averaging scheme with a single characteristic length (Ožbolt and Bažant 1996). A rigorous derivation of the proper form of the averaging operator in the proximity of a boundary is also lacking.

Nonlocal constitutive theories admit that the local state of the material at a given point may not be sufficient to evaluate the stress at that point. This can be physically justified by the fact that no real material is an ideal continuous medium, and on a sufficiently small scale the effects of heterogeneity and discontinuous microstructure become nonnegligible. For metals, this scale is in the order of microns, but for concrete and other highly heterogeneous composite materials, it is substantially larger. If the strain distribution is sufficiently smooth, as is often the case in the elastic regime, the standard local theory provides a good approximation and no important deviations from the actual behavior can be observed. After the onset of strain localization, the characteristic wave length of the deformation field becomes much shorter and this activates the nonlocal effects. For this reason, nonlocal theories that aim at regularizing the localization problem usually neglect the nonlocal elastic effects and apply nonlocal averaging only to an internal variable (or thermodynamic force) linked to the dissipative processes.

The nonlocal counterpart of a local field $f(x)$ is given by the integral

$$\bar{f}(\boldsymbol{x}) = \int_V \alpha(\boldsymbol{x},\boldsymbol{\xi}) f(\boldsymbol{\xi})\,\mathrm{d}\boldsymbol{\xi} \tag{2}$$

in which V is the spatial domain occupied by the body of interest (finite or infinite), and α is a suitable weight function. For an infinite body made of a macroscopically homogeneous material, the weight function depends only on the distance $\|\boldsymbol{x}-\boldsymbol{\xi}\|$ and can be expressed as $\alpha(\boldsymbol{x},\boldsymbol{\xi}) = \alpha_\infty(\|\boldsymbol{x}-\boldsymbol{\xi}\|)$ where α_∞ is a nonnegative function. For a finite body, the weight function is usually adjusted such that the nonlocal field corresponding to a constant local field remains constant even in the vicinity of a boundary. This is guaranteed if the weight function satisfies the normalizing condition

$$\int_V \alpha(\boldsymbol{x},\boldsymbol{\xi})\,\mathrm{d}\boldsymbol{\xi} = 1 \quad \forall \boldsymbol{x}\in V \tag{3}$$

and the simplest way to achieve that is to set

$$\alpha(\boldsymbol{x},\boldsymbol{\xi}) = \frac{\alpha_\infty(\|\boldsymbol{x}-\boldsymbol{\xi}\|)}{\int_V \alpha_\infty(\|\boldsymbol{x}-\boldsymbol{\zeta}\|)\mathrm{d}\boldsymbol{\zeta}} \tag{4}$$

Commonly used nonlocal weight functions are the Gauss-like function

$$\alpha_\infty^{\mathrm{Gauss}}(r) = \frac{1}{l\sqrt{2\pi}}\exp\left(-\frac{r^2}{2l^2}\right) \tag{5}$$

and the truncated polynomial function

$$\alpha_\infty^{\mathrm{poly}}(r) = \begin{cases} \frac{15}{16R}\left(1-\frac{r^2}{R^2}\right)^2 & \text{if } |r|\le R \\ 0 & \text{if } |r|\ge R \end{cases} \tag{6}$$

The weight function always contains at least one parameter with the dimension of length, e.g., l in (5) or R in (6). This parameter incorporates in the simplest possible way some information about the microstructure and often controls the size of the localized plastic zone. The value of the parameter is related to the intrinsic material length, which is dictated by the size and spacing of dominant heterogeneities.

The Gauss-like weight function (5) has an unbounded support, which means that the nonlocal interaction theoretically takes place at an arbitrary long distance. Since the decay of $\alpha_\infty^{\mathrm{Gauss}}$ with increasing r/l is very fast, in practical applications it is possible to truncate the function at a finite distance. The truncated polynomial function (6) has a bounded support and vanishes at distances r exceeding the interaction radius R.

Equation (2) defines the usual, or primal, nonlocal averaging, denoted by a bar. The thermodynamically motivated nonlocal plasticity models proposed by Svedberg (1996) and Polizzotto et al. (1998) exploit the concept of dual nonlocal averaging, denoted here by a tilde and defined by

$$\tilde{f}(\boldsymbol{x}) = \int_V \alpha(\boldsymbol{\xi},\boldsymbol{x}) f(\boldsymbol{\xi})\,\mathrm{d}\boldsymbol{\xi} \tag{7}$$

This definition ensures that the duality condition

$$\int_V \bar{f}(\boldsymbol{x})\, g(\boldsymbol{x})\,\mathrm{d}\boldsymbol{x} = \int_V f(\boldsymbol{x})\,\tilde{g}(\boldsymbol{x})\,\mathrm{d}\boldsymbol{x} \tag{8}$$

holds for all fields f and g for which the integrals make sense. If $\alpha(\boldsymbol{x},\boldsymbol{\xi})$ depends only on $\|\boldsymbol{x}-\boldsymbol{\xi}\|$, there is no difference between the standard (primary) averaging and the dual averaging; in other words, the averaging operator is self-adjoint. If the weight function is scaled according to (4), the symmetry is broken. In this case, the dual averaging does not preserve a constant field.

3.3 *Nonlocal damage models*

A natural constitutive framework for the description of fracturing materials is provided by continuum damage mechanics. As a representative example, consider the simple isotropic damage model, described by the following set of equations:

$$\boldsymbol{\sigma} = (1-\Omega)\boldsymbol{D}_e : \boldsymbol{\varepsilon} \tag{9}$$

$$\Omega = \omega(Y_{max}) \tag{10}$$

$$Y_{max}(t) = \max_{\tau\le t} Y(\tau) \tag{11}$$

$$Y = \tfrac{1}{2}\boldsymbol{\varepsilon} : \boldsymbol{D}_e : \boldsymbol{\varepsilon} \tag{12}$$

Table 1. Overview of nonlocal damage formulations.

Formulation	Isotropic damage model	General model
$\omega(\overline{\varepsilon})$	$\boldsymbol{\sigma} = [1 - \omega(\overline{\varepsilon})]\boldsymbol{D}_e : \varepsilon$	$\boldsymbol{\sigma} = \boldsymbol{D}_s(\overline{\varepsilon}) : \varepsilon$
$\overline{Y}$	$\boldsymbol{\sigma} = [1 - \omega(\overline{Y(\varepsilon)})]\boldsymbol{D}_e : \varepsilon$	$\boldsymbol{\sigma} = \boldsymbol{D}_s(\boldsymbol{\Omega}(\overline{\boldsymbol{Y}(\varepsilon)})) : \varepsilon$
$\overline{\omega}$	$\boldsymbol{\sigma} = [1 - \overline{\omega(\varepsilon)}]\boldsymbol{D}_e : \varepsilon$	$\boldsymbol{\sigma} = \overline{\mathbf{D}_s(\varepsilon)} : \varepsilon$
$\overline{\gamma}$	$\boldsymbol{\sigma} = [1 + \overline{\gamma(\varepsilon)}]^{-1}\boldsymbol{D}_e : \varepsilon$	$\boldsymbol{\sigma} = [\boldsymbol{C}_e + \overline{\boldsymbol{C}_i(\varepsilon)}]^{-1} : \varepsilon$
$\overline{s}$	$\boldsymbol{\sigma} = \boldsymbol{D}_e : \varepsilon - \overline{\omega(\varepsilon)\boldsymbol{D}_e : \varepsilon}$	$\boldsymbol{\sigma} = \boldsymbol{D}_e : \varepsilon - \overline{\boldsymbol{s}(\varepsilon)}$
$\overline{\Delta s}$	$\dot{\boldsymbol{\sigma}} = (1 - \omega)\boldsymbol{D}_e : \dot{\varepsilon} - \overline{\dot{\omega}\boldsymbol{D}_e : \varepsilon}$	$\dot{\boldsymbol{\sigma}} = \boldsymbol{D}_u : \dot{\varepsilon} - \overline{\dot{\boldsymbol{s}}(\varepsilon, \dot{\varepsilon})}$
$s(\overline{\varepsilon})$	$\boldsymbol{\sigma} = \boldsymbol{D}_e : \varepsilon - \omega(\overline{\varepsilon})\boldsymbol{D})_e : \overline{\varepsilon}$	$\boldsymbol{\sigma} = \boldsymbol{D}_e : \varepsilon - \boldsymbol{s}(\overline{\varepsilon})$

In the above, $\boldsymbol{\sigma}$ is the stress tensor, ε is the strain tensor, $\boldsymbol{D}_e$ is the elastic material stiffness tensor, and Ω is the damage variable that grows from zero (virgin state) to one (fully damaged state) depending on Y_{max}, which is the maximum value of the damage energy release rate Y ever attained in the previous history of the material up to the current state.

A number of nonlocal damage formulations giving local response in the linear elastic range have been proposed in the literature. All of them act as localization limiters, and at the onset of localization they give almost the same response. The latter, though, is not true for large postpeak deformations. At late stages of the softening process, a complete fracture must be simulated, which means that the stress must be reduced to zero. Jirásek (1998) showed that this is not true for some types of nonlocal averaging, and found that averaging of different variables gives rather different responses. The nonlocal damage formulations that he considered are summarized in Table 1. For easy reference, individual formulations are denoted by symbols $\overline{Y}, \overline{\omega}$ etc.; see the first column of the table. The central column presents in a compact form the nonlocal stress-strain law for the isotropic damage model whose local version is described by (9)–(12). The right column shows a possible generalization to anisotropic damage (top part) or to a completely general inelastic model (bottom part).

Beside the symbols already defined, the following notations are used: $\gamma = \Omega/(1 - \Omega) =$ compliance variable, $\boldsymbol{D}_s =$ damaged (secant) stiffness matrix, $\boldsymbol{D}_u =$ unloading stiffness matrix, $\boldsymbol{\Omega} =$ damage tensor, $\boldsymbol{Y} =$ tensor of damage energy release rates work-conjugate to $\boldsymbol{\Omega}$, $\boldsymbol{C}_e = \boldsymbol{D}_e^{-1} =$ elastic compliance matrix, $\boldsymbol{C}_i =$ inelastic compliance matrix, and $\boldsymbol{s} = \boldsymbol{D}_e\varepsilon - \boldsymbol{\sigma} =$ in-elastic stress. An overdot denotes differentiation with respect to time.

The load-displacement diagrams generated in a uniaxial tensile test by different nonlocal damage formulations are shown in Fig. 2. The (local) damage evolution law $\Omega = \omega(Y_{max})$ is constructed such that the response remains linear up to the peak stress and the softening curve is exponential, asymptotically approaching the horizontal axis. The dashed curve in Fig. 2 corresponds to the unstable solution with uniform strain. The actual stable solutions are characterized by a nonuniform strain distribution, with strain increments localized into a finite interval, the size of which is controlled by the characteristic length. The initial response, right after the onset of

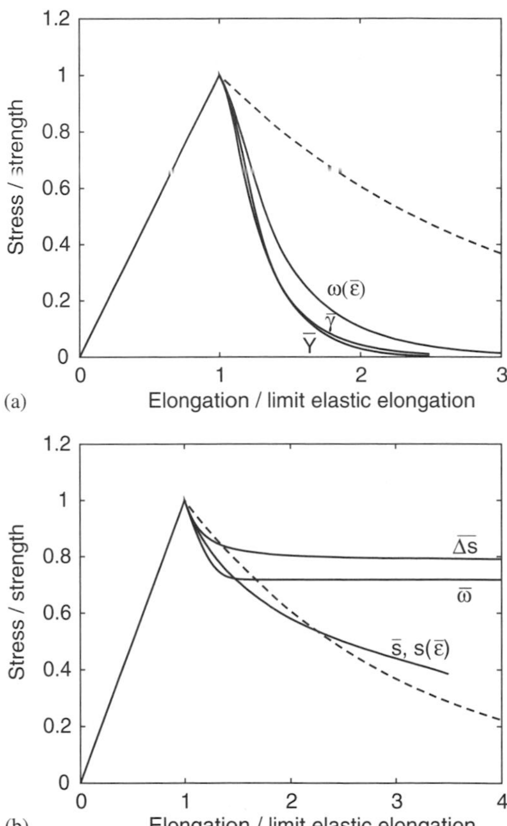

Figure 2. Comparison of postpeak load-displacement curves of a tensioned bar calculated by nonlocal damage models with averaging applied to different variables: (a) non-locking formulations $\omega(\overline{\varepsilon}), \overline{Y}$ and $\overline{\gamma}$; (b) locking formulations $\overline{\omega}, \overline{s}, \overline{\Delta s}$ and $s(\overline{\varepsilon})$.

localization, is about the same for all the models considered here. At later stages of softening, only the formulations labeled $\omega(\bar{\varepsilon})$, $\overline{Y}$ and $\overline{\gamma}$ give a reasonable behavior, with the residual strength approaching zero as the applied elongation is increased (Fig. 2a). The other formulations lead to locking effects and sometimes fail to converge (Fig. 2b). Thus it can be concluded that the complete fracture is correctly reproduced by models that average the equivalent strain, the energy release rate, or the compliance variable. The evaluation of the inelastic stress from the nonlocal strain, same as the averaging of the damage variable, inelastic stress, or inelastic stress increment, leads to spurious residual stresses and to an expansion of the softening zone across the entire bar.

The basic model with damage evolution driven by the damage energy release rate (12) is simple and appealing from the theoretical point of view, since it can be formulated within the framework of generalized standard materials (Halphen and Nguyen 1975). However, it is not suitable for quasibrittle materials, because it gives the same response in tension and in compression. To emphasize the effect of tension on the propagation of cracks, Mazars (1984) proposed to link the damage to the so-called equivalent strain, defined as the norm of the positive part of the strain tensor. He developed an isotropic damage model for concrete with two damage parameters, ω_t and ω_c, which correspond to uniaxial tension and uniaxial compression, respectively, and both depend on the maximum previously reached value of the equivalent strain. For a general stress state, the actual damage parameter is interpolated according to the current values of principal stresses. The nonlocal formulation of Mazars' model was refined by Saouridis (1988) and Saouridis and Mazars (1992), following the basic idea of Pijaudier-Cabot and Bažant (1987). The averaged quantity was the equivalent strain, which corresponds to a natural generalization of formulation $\overline{Y}$.

The equivalent strain $\tilde{\varepsilon}$ is a scalar measure of the strain level, and under uniaxial stress it could be defined as the normal strain in the direction of loading. However, since damage in many materials propagates much more easily under tension than under compression, the definition of $\tilde{\varepsilon}$ is usually modified such that the influence of tension is emphasized. As a first approximation, one may assume that compressive strains do not lead to any damage, and relate $\tilde{\varepsilon}$ to positive strains only. For instance, Mazars (1984) defined the equivalent strain as

$$\tilde{\varepsilon} = \sqrt{\sum_{I=1}^{3} \langle \varepsilon_I \rangle^2} \tag{13}$$

where $\varepsilon_I, I =$ 1, 2, 3, are the principal strains, and the brackets $\langle \ldots \rangle$ denote the positive part, i.e., $\langle \varepsilon \rangle = \varepsilon$ for $\varepsilon > 0$ and $\langle \varepsilon \rangle = 0$ for $\varepsilon \leq 0$. Another suitable definition is

$$\tilde{\varepsilon} = \frac{1}{E} \max_{I=1,2,3} \sigma_{eI} \tag{14}$$

where $\sigma_{eI}, I =$ 1, 2, 3, are the principal values of the effective stress tensor $\boldsymbol{\sigma}_e = \boldsymbol{D}_e : \varepsilon$. This definition corresponds to a Rankine-type failure surface, because in the elastic range the effective stress coincides with the actual (nominal) stress. In the damage law (10), Y_{max} is replaced by a certain internal variable κ that represents the maximum level of the equivalent strain $\tilde{\varepsilon}$ reached in the previous history of the material point. Equation (11) is thus rewritten accordingly, with Y_{max} replaced by κ and Y replaced by $\tilde{\varepsilon}$.

3.4 *Nonlocal plasticity models*

Even though the physical origins of cracking and of plastic yielding are quite different, the reduction of strength due to propagating and interacting microcracks can formally be described by plasticity models with decreasing yield stress. In the standard (local) version of the flow theory of plasticity with isotropic hardening or softening, the yield function typically has the form

$$f(\boldsymbol{\sigma}, \kappa) = F(\boldsymbol{\sigma}) - \sigma_Y(\kappa) \tag{15}$$

where κ is a scalar hardening-softening variable, $F(\boldsymbol{\sigma})$ is the equivalent stress (e.g., the von Mises equivalent stress for J_2-plasticity, or the maximum principal stress for Rankine plasticity), and σ_Y is the current yield stress. The evolution of the yield stress as a function of the hardening variable is described by the hardening law. This law is for future use conveniently written as

$$\sigma_Y(\kappa) = \sigma_0 + h(\kappa) \tag{16}$$

where σ_0 is the initial yield stress and h is the hardening function. The derivative $H = \mathrm{d}h/\mathrm{d}\kappa$ is called the plastic modulus. The fundamental equations of associated elastoplasticity include also the elastic stress-strain law,

$$\boldsymbol{\sigma} = \boldsymbol{D}_e : (\varepsilon - \varepsilon_p) \tag{17}$$

the flow rule,

$$\dot{\varepsilon}_p = \dot{\lambda} \boldsymbol{f}(\boldsymbol{\sigma}) \tag{18}$$

and the loading-unloading conditions

$$\dot{\lambda} \geq 0, \quad f(\boldsymbol{\sigma}, \kappa) \leq 0, \quad \dot{\lambda} f(\boldsymbol{\sigma}, \kappa) = 0 \tag{19}$$

Here, ε_p is the plastic strain tensor, $\dot{\lambda}$ is the rate of the plastic multiplier, and $\boldsymbol{f} \equiv \partial f / \partial \boldsymbol{\sigma}$ is the gradient of the yield function, defining the direction of plastic flow.

Table 2. Overview of nonlocal plasticity formulations.

Formulation	Stress-strain law	Softening law
Local	$\boldsymbol{\sigma} = \boldsymbol{D}_e : (\varepsilon - \varepsilon_p)$	$\sigma_Y = \sigma_0 + h(\kappa)$
Basic nonlocal		$\sigma_Y = \sigma_0 + h(\bar{\kappa})$
Vermeer–Brinkgreve		$\sigma_Y = \sigma_0 + h(m\bar{\kappa} + (1-m)\kappa)$
		$\sigma_Y = \sigma_0 + h(\kappa) + H_m[\bar{\kappa} - \kappa]$
Geers et al.		$\sigma_Y = [1 - \omega_p(\bar{\kappa})](\sigma_0 + H\kappa)$
Nonlocal Gurson		$\sigma_Y = [1 - \omega_p(\overline{p(\kappa)})](\sigma_0 + H\kappa)$
Eringen (strain space)	$\boldsymbol{\sigma} = \boldsymbol{D}_e : (\bar{\varepsilon} - \bar{\varepsilon}_p)$	
Eringen (stress space)	$\boldsymbol{\sigma} = \boldsymbol{D}_e : (\bar{\varepsilon} - \varepsilon_p)$	
Bažant–Lin	$\boldsymbol{\sigma} = \boldsymbol{D}_e : (\varepsilon - \bar{\varepsilon}_p)$	
Borino et al.		$\sigma_Y = \sigma_0 + \widetilde{h(\bar{\kappa})}$
Svedberg–Runesson		$\sigma_Y = \sigma_0 + h_L(\kappa) + \bar{h}_{NL} - h_{NL}$, where $h_{NL} = h_{NL}(\bar{\kappa} - \kappa)$
Nilsson	$\boldsymbol{\sigma} = \boldsymbol{D}_e : (\varepsilon - \bar{\varepsilon}_p)$	$\sigma_Y = \sigma_0 + h(\bar{\kappa})$

The hardening variable κ reflects the changes in the microstructure induced by plastic flow. Its rate is usually related to the plastic strain rate by the strain-hardening hypothesis,

$$\dot{\kappa} = \|\dot{\varepsilon}_p\| \tag{20}$$

A number of nonlocal formulations of softening plasticity have been scrutinized by Jirásek and Rolshoven (2003), with attention to their localization properties, boundary effects, and thermodynamic admissibility. The models are summarized in Table 2. Of course, this simple table provides only the basic orientation; it cannot cover all the specific details of different formulations. The names attached to the models are not always standard but they refer, to the best of the author's knowledge, to the first publication in which the key idea of the respective model appeared. The table displays only those stress-strain laws and softening laws that differ from the standard local equations given in the first row. The evolution equations (flow rule, evolution law for the softening variable, and loading-unloading conditions) usually keep the standard form (18)–(20), with the exception of the models due to Eringen (1981) and Bažant and Lin (1988), which work with the yield condition in strain space.

The main results of the comparative study (Jirásek and Rolshoven 2003) can be summarized as follows:

1. The nonlocal plasticity models proposed by Eringen do not act as localization limiters. The first model with averaging of stress (Eringen 1981) does not prevent localization into a set of measure zero, and the second model with averaging of strain (Eringen 1983) does not allow any localization at all (in the one-dimensional setting).
2. Models with the yield stress dependent only on the nonlocal cumulative plastic strain (basic nonlocal formulation and models of Borino et al. (1999) and Nilsson (1997)) provide only a partial regularization and are essentially equivalent to a cohesive zone model. Plastic strain is localized into a set of zero measure but, in contrast to the local formulation, the global structural response in terms of the load-displacement diagram and work spent during the failure process is captured correctly. In finite element simulations in multiple dimensions, such models are likely to exhibit mesh-induced directional bias, since the plastic yielding would localize into one layer of elements.
3. Models that combine in a suitable way the effect of the local and nonlocal cumulative plastic strain on the current yield stress act as true localization limiters and lead to a nonzero size of the localized plastic zone. This is true for the model of Vermeer and Brinkgreve (1994), models motivated by ductile damage (integral-type version of the implicit gradient plasticity model due to Geers et al. (2001) and nonlocal extension of the Gurson model proposed by Leblond et al. (1994)), and the thermodynamically motivated model of Svedberg and Runesson (1998). The model proposed by Bažant and Lin (1988) has similar properties. At the first bifurcation from a uniform strain state, all these models are essentially equivalent. The subsequent evolution of the plastic strain profile and the stress transmitted by the plastic zone depend on the specific form of the softening law.
4. If the softening process needs to be simulated until the complete loss of material integrity (zero residual yield stress), the model should be selected with great care. At late stages of the softening process, certain formulations produce pathological effects such as stress locking or spatial expansion of the plastic zone. Such formulations have a limited scope and they should be combined with appropriate tools for the description of highly localized strain and of the transition to fracture. Only the Vermeer–Brinkgreve model and the ductile

damage models seem to be suitable for a pure continuum-based description of the complete failure process.

5. If the nonlocal weight function is scaled in the proximity of boundaries, as is routinely done, the solution with a plastic region localized at the boundary usually dissipates much less energy than if the plastic region localizes inside the body. Since the solution that would actually occur is that with the steepest descent of the post-peak branch of the load-displacement diagram (Bažant and Cedolin 1991), the boundary acts as a weak layer that attracts localization. Whether this is physically realistic depends on the actual structure of the material in the boundary layer.

4 NUMERICAL ASPECTS OF NONLOCAL MODELS

4.1 *Nonlocal tangent stiffness*

In general terms, the structure of the tangent operator for nonlocal damage models was first discussed by Pijaudier-Cabot and Huerta (1991). A consistent derivation and algorithmic implementation of the tangent stiffness matrix was presented by Jirásek and Patzák (2002). The final formula can be written in the engineering notation as

$$\boldsymbol{K} = \boldsymbol{K}_u - \sum_{p,q} \omega_p' k_p w_p w_q \alpha_{pq} \boldsymbol{B}_p^T \boldsymbol{\sigma}_p^e \boldsymbol{\eta}_q^T \boldsymbol{B}_q \tag{21}$$

where

$$\boldsymbol{K}_u = \sum_p w_p (1 - \Omega_p) \boldsymbol{B}_p^T \boldsymbol{D}_e \boldsymbol{B}_p \tag{22}$$

In the above, subscripts p and q refer to the integration points of the finite element model, w_p are the corresponding integration weights (reflecting also the size of the contributing volume around each integration point), $\boldsymbol{B}$ is the usual B-matrix containing the derivatives of the shape functions, k is the loading-unloading indicator, equal to 1 if $\tilde{\varepsilon} = \kappa$ (loading) and to 0 if $\tilde{\varepsilon} < \kappa$ (unloading), ω' denotes the derivative of the damage function ω with respect to its argument κ, $\alpha_{pq} = \alpha(\boldsymbol{x}_p, \boldsymbol{x}_q)$ is the nonlocal weight function evaluated for integration points number p and q, $\boldsymbol{D}_e$ is the elastic material stiffness matrix, $\boldsymbol{\sigma}^e = \boldsymbol{D}_e \varepsilon$ is the column matrix with the effective stress components, and $\boldsymbol{\eta} = \partial\tilde{\varepsilon}/\partial\varepsilon$ is the column matrix with derivatives of the equivalent strain with respect to the engineering strain components.

The first term in (21), $\boldsymbol{K}^u$, represents the secant stiffness matrix valid if all the material points are in the unloading regime (when $k_p = 0$ for all p). The double sum in (21) provides a correction of the secant stiffness due to additional damage growth. The individual terms in the sum represent the contributions of nonlocal interaction between integration points $\boldsymbol{x}_p$ and $\boldsymbol{x}_q$ to the overall stiffness. Of course, the sum needs to be taken only over those pairs of integration points whose distance is smaller than the interaction radius R, because for all the other pairs α_{pq} vanishes. Each pair of interacting Gauss points contributes to only a small block of the global stiffness matrix, with rows corresponding to internal forces at the element that contains point $\boldsymbol{x}_p$, and with columns corresponding to nodal displacements at the element that contains point $\boldsymbol{x}_q$. The size of this block is therefore equal to the size of the usual element stiffness matrix. This means that the complete stiffness matrix can be assembled from much smaller matrices, similar to the usual assembly procedure. The difference is that the code numbers associated with the rows are in general different from the code numbers associated with the columns. This can be handled by a minor modification of the usual assembly routine. Compared to the local case, the bandwidth increases due to the nonlocal interaction, and the global stiffness matrix is in general not symmetric.

As follows from (21), the nonlocal tangent stiffness matrix has several particular properties that have to be taken into account when selecting an economical storage scheme and an efficient solver.

First of all, the stiffness matrix is nonsymmetric, but this is often the case for local models as well. A symmetric tangent stiffness of a local damage model is obtained only if the loading function is expressed in terms of the thermodynamic forces conjugate to the internal variables and the evolution laws are postulated as normality rules. Only the model relating damage to the damage energy release rate meets these criteria. Any other definition of equivalent strain gives linearly independent $\boldsymbol{b}^\sigma$ and $\boldsymbol{b}^\eta$, and the symmetry is lost already in the local case. Associated models with nice symmetry properties are appealing from the theoretical point of view, they limit the number of constitutive parameters and facilitate the numerical implementation, but they do not always capture all the essential features of the real material behavior. Therefore, the lack of symmetry cannot be considered as a substantial drawback specific to nonlocal formulations.

A more important complication caused by nonlocality is the evolutive character of the profile of the stiffness matrix. For the class of material models considered here, the elastic response remains local, and so the initial distribution of nonzero entries in the stiffness matrix is the same as in the standard local case. When the damage threshold is exceeded and the damage zone starts evolving, new nonzero entries appear due to the activated interaction between the Gauss points belonging to different elements, and the profile of the stiffness matrix must be dynamically adapted. Of course, one can also allocate the storage space for

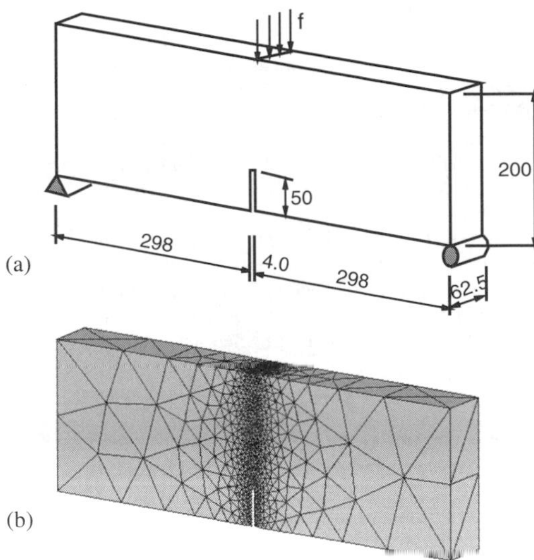

Figure 3. Three-point bending test simulated in three dimensions: (a) geometry and loading, (b) finite element mesh.

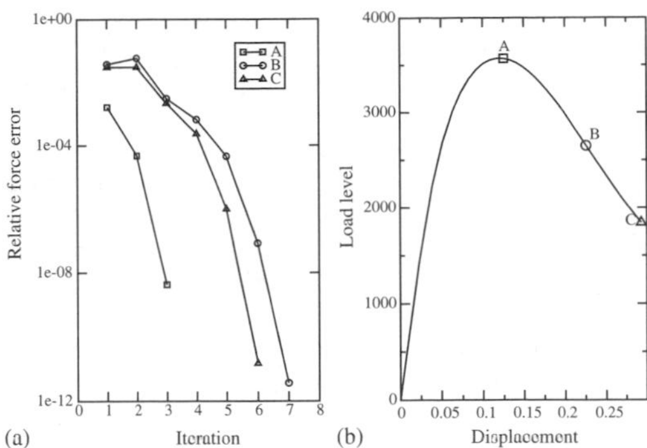

Figure 4. Three-point bending test simulated in three dimensions: (a) convergence characteristics for tangent stiffness, (b) load-displacement diagram

all possible nonzero entries from the very beginning, but this would be a big waste of resources because the process zone usually extends over only a small part of the structure and most of the allocated entries would remain zero throughout the entire simulation. If the final extent of the process zone is known in advance, one could allocate only those entries that will later become nonzero. However, such *a priori* information is available only in some academic examples but not in general applications.

An efficient and versatile solution scheme must be based on dynamic memory allocation. During the evolution of the process zone, the bandwidth of the stiffness matrix increases, but the matrix still remains sparse. This must be taken into account when selecting the most appropriate solver.

The consistent tangential stiffness matrix can be exploited in the global equilibrium iteration procedure. The resulting acceleration of the convergence rate is illustrated by the example of a three-point bending test simulated in three dimensions using the nonlocal isotropic damage model with a Rankine-type definition of equivalent strain (14) and with exponential softening; see (Jirásek and Patzák 2002) for details. The geometry and loading are shown in Fig. 3a. The finite element mesh contains 6461 nodes and 33158 constant-strain tetrahedral elements; see Fig. 3b.

The nonlinear response is analyzed in 15 increments of applied displacements at the loaded edge. Solution strategies based on the secant stiffness matrix (SSM) and tangent stiffness matrix (TSM) with either direct or iterative solvers are exploited. The direct solver is based on LU decomposition using the generalized "skyline" storage scheme and the profile optimization algorithm proposed by Sloan (1986). The iterative solver is based on the Generalized Minimum Residual Method (GMRES) using the compressed row (CompRow) storage scheme and preconditioning by an incomplete LU factorization (ILU) with fill-up (Saad 1996).

Fig. 4a shows the evolution of error (norm of the out-of-balance forces) during equilibrium iteration with the TSM. The curves correspond to three typical steps, denoted as A, B and C. The convergence rate is quadratic, which confirms that the tangent stiffness is indeed consistent, and convergence is achieved within 3 to 7 iterations.

For a high required accuracy (relative tolerance of the out-of-balance forces set to 10^{-6}), the complete analysis with a secant stiffness and a direct solver takes 27 hours and 37 minutes (of the total user time), the analysis with a tangent stiffness and a direct solver takes 136 hours and 41 minutes, but the analysis with a tangent stiffness and an iterative solver takes only 15 hours and 23 minutes. The simulations have been run on a Pentium III workstation with 1 GB of memory, running at 933 MHz under the Linux operating system.

4.2 *Stress return algorithm for nonlocal plasticity*

Incremental finite element analysis requires the implementation of a procedure for the evaluation of the increments of stress and internal variables (plastic strain and softening variable) that correspond to a given increment of strain, starting from a state at which all the state variables are known. For local plasticity models, this problem can be solved at each Gauss point independently. For nonlocal plasticity models that incorporate the weighted spatial average of the softening variable into the softening law, the yield stress at one point becomes dependent on the plastic flow at the neighboring Gauss points, and the stress evaluation must be done for all plastic Gauss points simultaneously.

A simple iterative algorithm for nonlocal stress return was proposed by Strömberg and Ristinmaa (1996) and improved by Rolshoven and Jirásek (2001). Recently, Benvenuti and Tralli (2003) presented a comparative study with an evaluation of various approaches and additional improvements. In the context of nonlocal Drucker-Prager model with nonassociated flow rule, the numerical aspects of nonlocal plasticity models are discussed and some examples presented in another paper in this volume (Rolshoven and Jirásek 2003).

ACKNOWLEDGEMENTS

The author would like to thank Professors Zdeněk P. Bažant and Ted Belytschko of Northwestern University and Bořek Patzák of the Czech Technical University, as well as his graduate student Simon Rolshoven, for stimulating discussions.

REFERENCES

Babuška, I. and J. M. Melenk (1997). The partition of unity method. *International Journal for Numerical Methods in Engineering* 40, 727–758.

Bažant, Z. P. (1984). Imbricate continuum and its variational derivation. *Journal of Engineering Mechanics, ASCE* 110, 1693–1712.

Bažant, Z. P. (2002). *Scaling of Structural Strength.* London: Hermes-Penton.

Bažant, Z. P. and L. Cedolin (1991). *Stability of Structures.* New York and Oxford: Oxford University Press.

Bažant, Z. P. and F.-B. Lin (1988). Nonlocal yield-limit degradation. *International Journal for Numerical Methods in Engineering* 26, 1805–1823.

Belytschko, T., J. Fish, and B. E. Engelmann (1988). A finite element with embedded localization zones. *Computer Methods in Applied Mechanics and Engineering* 70, 59–89.

Benvenuti, E. and A. Tralli (2003). Iterative LCP solvers for nonlocal loading-unloading conditions. *International Journal for Numerical Methods in Engineering.* Submitted.

Borino, G., P. Fuschi, and C. Polizzotto (1999). A thermodynamic approach to nonlocal plasticity and related variational approaches. *Journal of Applied Mechanics, ASME* 66, 952–963.

Camacho, G. T. and M. Ortiz (1996). Computational modelling of impact damage in brittle materials. *International Journal of Solids and Structures* 33, 2899–2938.

Carter, B. J., C.-S. Chen, A. R. Ingraffea, and P. A. Wawrzynek (1997). A topology-based system for modeling 3d crack growth in solid and shell structures. In *Proceedings of the Ninth International Congress on Fracture ICF9*, Sydney, Australia, pp. 1923–1934. Elsevier Science Publishers.

Daux, C., N. Moës, J. Dolbow, N. Sukumar, and T. Belytschko (2000). Arbitrary branched and intersecting cracks with the extended finite element method. *International Journal for Numerical Methods in Engineering* 48, 1741–1760.

Dvorkin, E. N., A. M. Cuitiño, and G. Gioia (1990). Finite elements with displacement interpolated embedded localization lines insensitive to mesh size and distortions. *Computer Methods in Applied Mechanics and Engineering* 90, 829–844.

Eringen, A. C. (1981). On nonlocal plasticity. *International Journal of Engineering Science* 19, 1461–1474.

Eringen, A. C. (1983). Theories of nonlocal plasticity. *International Journal of Engineering Science* 21, 741–751.

Geers, M. G. D., R. A. B. Engelen, and R. J. M. Ubachs (2001). On the numerical modeling of ductile damage with an implicit gradient-enhanced formulation. *Revue européenne des éléments finis* 10, 173–191.

Halphen, B. and Q. S. Nguyen (1975). Sur les matériaux standards généralisés. *Journal de Mécanique* 14, 39–63.

Henshell, R. D. and K. G. Shaw (1975). Crack tip finite elements are unnecessary. *International Journal for Numerical Methods in Engineering* 9, 495–507.

Jirásek, M. (1998). Nonlocal models for damage and fracture: Comparison of approaches. *International Journal of Solids and Structures* 35, 4133–4145.

Jirásek, M. (2000a). Comparative study on finite elements with embedded cracks. *Computer Methods in Applied Mechanics and Engineering* 188, 307–330.

Jirásek, M. (2000b). Conditions of uniqueness for finite elements with embedded cracks. In *Proceedings of the Sixth International Conference on Computational Plasticity*, Barcelona. CD-ROM.

Jirásek, M. and T. Belytschko (2002). Computational resolution of strong discontinuities. In *Proceedings of the 5th World Congress on Computational Mechanics*, Vienna, Austria.

Jirásek, M. and B. Patzák (2002). Consistent tangent stiffness for nonlocal damage models. *Computers and Structures* 80, 1279–1293.

Jirásek, M. and S. Rolshoven (2003). Comparison of integral-type nonlocal plasticity models for strain-softening materials. *International Journal of Engineering Science.* In press.

Klisinski, M., K. Runesson, and S. Sture (1991). Finite element with inner softening band. *Journal of Engineering Mechanics, ASCE* 117, 575–587.

Leblond, J. B., G. Perrin, and J. Devaux (1994). Bifurcation effects in ductile metals incorporating void nucleation, growth and interaction. *Journal of Applied Mechanics, ASME Bifurcation effects in ductile metals with nonlocal damage*, 236–242.

Mariani, S. and U. Perego (2002). Extended finite element method for quasi-brittle fracture. *International Journal for Numerical Methods in Engineering.* Submitted.

Martha, L., P. A. Wawrzynek, and A. R. Ingraffea (1993). Arbitrary crack propagation using solid modeling. *Engineering with Computers* 9, 63–82.

Mazars, J. (1984). Application de la mécanique de l'endommagement au comportement non linéaire et à la rupture du béton de structure. Thèse de Doctorat d'Etat, Université Paris VI., France.

Melenk, J. M. and I. Babuška (1996). The partition of unity finite element method: Basic theory and applications.

Computer Methods in Applied Mechanics and Engineering 39, 289–314.

Moës, N. and T. Belytschko (2002). Extended finite element method for cohesive crack growth. *Engineering Fracture Mechanics* 69, 813–833.

Moës, N., J. Dolbow, and T. Belytschko (1999). A finite element method for crack growth without remeshing. *International Journal for Numerical Methods in Engineering* 46, 131–150.

Nilsson, C. (1997). Nonlocal strain softening bar revisited. *International Journal of Solids and Structures* 34, 4399–4419.

Oliver, J. (1996). Modelling strong discontinuities in solid mechanics via strain softening constitutive equations. part 1: Fundamentals. part 2: Numerical simulation. *International Journal for Numerical Methods in Engineering* 39, 3575–3624.

Olofsson, T., M. Klisinski, and P. Nedar (1994). Inner softening bands: A new approach to localization in finite elements. In H. Mang, N. Bićanić, and R. de Borst (Eds.), *Computational Modelling of Concrete Structures*, pp. 373–382. Pineridge Press.

Ortiz, M., Y. Leroy, and A. Needleman (1987). A finite element method for localized failure analysis. *Computer Methods in Applied Mechanics and Engineering* 61, 189–214.

Ortiz, M. and A. Pandolfi (1999). Finite-deformation irreversible cohesive elements for three-dimensional crack-propagation analysis. *International Journal for Numerical Methods in Engineering* 44, 1267–1282.

Ožbolt, J. and Z. P. Bažant (1996). Numerical smeared fracture analysis: Nonlocal microcrack interaction approach. *International Journal for Numerical Methods in Engineering* 39, 635–661.

Pandolfi, A., P. Krysl, and M. Ortiz (1999). Finite element simulation of ring expansion and fragmentation: The capturing of length and time scales through cohesive models of fracture. *International Journal of Fracture* 95, 279–297.

Pijaudier-Cabot, G. and Z. P. Bažant (1987). Nonlocal damage theory. *Journal of Engineering Mechanics, ASCE* 113, 1512–1533.

Pijaudier-Cabot, G. and A. Huerta (1991). Finite element analysis of bifurcation in nonlocal strain softening solids. *Computer Methods in Applied Mechanics and Engineering* 90, 905–919.

Polizzotto, C., G. Borino, and P. Fuschi (1998). A thermodynamic consistent formulation of nonlocal and gradient plasticity. *Mechanics Research Communications* 25, 75–82.

Rolshoven, S. and M. Jirásek (2001). On regularized plasticity models for strain-softening materials. In R. de Borst, J. Mazars, G. Pijaudier-Cabot, and J. G. M. van Mier (Eds.), *Fracture Mechanics of Concrete Structures*, Lisse, pp. 617–624. Balkema.

Rolshoven, S. and M. Jirásek (2003). Numerical aspects of nonlocal plasticity with strain softening. In *Proceedings of Euro-C*, Rotterdam. Balkema.

Rossi, P. and X. Wu (1992). Probabilistic model for material behaviour analysis and appraisement of concrete structures. *Magazine of Concrete Research* 44, 271–280.

Saad, Y. (1996). *Iterative methods for sparse linear systems*. Boston: PWS Publishing Company.

Saouridis, C. (1988). *Identification et numérisation objectives des comportements adoucissants: Une approche multiéchelle de l'endommagement du béton*. Ph. D. thesis, Université Paris VI.

Saouridis, C. and J. Mazars (1992). Prediction of the failure and size effect in concrete via a biscale damage approach. *Engineering Computations* 9, 329–344.

Shi, G. (1992). Modeling rock joints and blocks by manifold method. In *Rock Mechanics, Proceedings of the 33rd U.S. Symposium*, Santa Fe, New Mexico, pp. 639–648.

Shi, G. H. (1995). Numerical manifold method. In *Working Forum on the Manifold Method of Mathematical Analysis*, pp. 1–180.

Simo, J. C. and J. Oliver (1994). A new approach to the analysis and simulation of strain softening in solids. In Z. P. Bažant, Z. Bittnar, M. Jirásek, and J. Mazars (Eds.), *Fracture and Damage in Quasibrittle Structures*, London, pp. 25–39. E & FN Spon.

Sloan, S. W. (1986). An algorithm for profile and wavefront reduction of sparse matrices. *International Journal for Numerical Methods in Engineering* 23, 239–251.

Strömberg, L. and M. Ristinmaa (1996). FE-formulation of a nonlocal plasticity theory. *Computer Methods in Applied Mechanics and Engineering* 136, 127–144.

Sukumar, N., N. Moës, B. Moran, and T. Belytschko (2000). Extended finite element method for three-dimensional crack modeling. *International Journal for Numerical Methods in Engineering* 48, 1549–1570.

Svedberg, T. (1996). A thermodynamically consistent theory of gradient-regularized plasticity coupled to damage. Licentiate thesis, Chalmers University of Technology.

Svedberg, T. and K. Runesson (1998). Thermodynamically consistent nonlocal and gradient formulations of plasticity. In A. Brillard and J. F. Ganghoffer (Eds.), *Nonlocal Aspects in Solid Mechanics, EUROMECH Colloquium* 378, Mulhouse, France, pp. 32–37.

Tsay, R.-J., Y.-J. Chiou, and W.-L. Chuang (1999). Crack growth prediction by manifold method. *Journal of Engineering Mechanics, ASCE* 125, 884–890.

Vermeer, P. A. and R. B. J. Brinkgreve (1994). A new effective non-local strain measure for softening plasticity. In R. Chambon, J. Desrues, and I. Vardoulakis (Eds.), *Localisation and Bifurcation Theory for Soils and Rocks*, Rotterdam, pp. 89–100. Balkema.

Wells, G. N. (2001). *Discontinuous modelling of strain localisation and failure*. Ph. D. thesis, Delft University of Technology, The Netherlands.

Wells, G. N. and L. J. Sluys (2001). A new method for modelling cohesive cracks using finite elements. *International Journal for Numerical Methods in Engineering* 50, 2667–2682.

Xu, X.-P. and A. Needleman (1994). Numerical simulations of fast crack growth in brittle solids. *Journal of the Mechanics and Physics of Solids* 42, 1397–1434.

Computational Modelling of Concrete Structures, Bićanić et al. (eds)
© 2003 Swets & Zeitlinger, Lisse, ISBN 90 5809 536 3

A new approach to strong embedded discontinuities

J. Alfaiate
Instituto Superior Técnico and ICIST, Dept. Eng. Civil, Lisboa, Portugal

A. Simone, L.J. Sluys
Delft University of Technology, Dept. of Civil Eng. and Geosciences, Delft, The Netherlands

ABSTRACT: In this paper, a new formulation for strong embedded discontinuities is introduced. The displacement jumps are obtained at additional nodes, located at the discontinuity surface. However, these nodes are not internal to the parent element as assumed in previous embedded formulations; instead, the new nodes and the corresponding degrees of freedom are global. In consequence: i) the jump displacement field no longer has to be homogeneous within each element and ii) the jump displacement field is continuous across element boundaries. The present formulation is compared with other approaches, namely the discrete approach with interface elements, other strong embedded discontinuity approaches and the partition of unity method. Simple examples are presented and compared with the results obtained from experiments and with some of the other formulations.

1 INTRODUCTION

In recent works, significant effort has been undertaken to model cracks as displacement discontinuities within a continuum. The embedded discontinuity formulation is an example of these contributions. Departing from the weak formulation, where discontinuities are modeled as finite width bands and the displacement field remains continuous (Sluys 1998), strong embedded discontinuities were introduced in which the kinematics of a discontinuous displacement field is approximated (Armero and Garikipati 1996; Oliver 1996a; Wells and Sluys 2001b). However, in most of the formulations shown in literature, the displacement jumps are considered to be constant within each parent finite element and, as a result, the jumps are not continuous across the element boundaries. In alternative approaches, such as the partition of unity method (Wells and Sluys 2001a), continuous jumps across element boundaries can be modeled, but the concept of embedded discontinuities is no longer addressed. Instead, the jumps are modeled by additional degrees of freedom located at the standard element nodes.

In this paper, strong discontinuities are embedded in finite elements to describe fracture in concrete. A new numerical formulation is introduced in which the displacement jumps at the discontinuities do not need to be constant a function within each parent element. Similar to previous embedded formulations, the jumps are evaluated at additional nodes, located at the discontinuity. However, these nodes are not internal to the parent element; instead, the new nodes and the corresponding degrees of freedom are global. In consequence, continuity of the crack path is enforced and the displacement jumps are continuous across element boundaries. This formulation is compared with different approaches, namely with the discrete approach, in which interface elements are inserted to model the discontinuities (called discrete-interface approach hereafter), as well as with the standard non-symmetric embedded discontinuity approach and the extended finite element and partition of unity methods. Simple examples are presented and compared with the results obtained from experiments and with some of the other formulations.

2 KINEMATICS OF A DISCONTINUITY

Consider a domain Ω, with boundary $\partial\Omega$, where a discontinuity surface Γ_d is supposed to exist. The total displacement field is the sum of a regular part $\hat{\mathbf{u}}$ on Ω and a discontinuous part corresponding to the displacement jump $[\![\mathbf{u}]\!]$, localized at the discontinuity surface Γ_d:

$$\mathbf{u}(\mathbf{x}) = \hat{\mathbf{u}}(\mathbf{x}) + \mathcal{H}_{\Gamma_d}[\![\mathbf{u}(\mathbf{x})]\!], \tag{1}$$

where $\mathcal{H}_{\Gamma_d}$ is defined as

$$\mathcal{H}_{\Gamma_d} = H_{\Gamma_d} - (1-r) \qquad 0 \leq r \leq 1 \tag{2}$$

and $H_{\Gamma d}$ is the Heaviside function at the discontinuity Γ_d,

$$H_{\Gamma_d} = \begin{cases} 1 & \text{if } \mathbf{x} \in \Omega^+ \\ 0 & \text{otherwise.} \end{cases} \tag{3}$$

Similar to other works (Klisinski et al. 1991; Ohlsson and Olofsson 1997; Lotfi and Shing 1995), the scalar parameter r defines how the jump is transmitted to the domain Ω: if $r = 1$ the jump is fully transmitted from Ω^- to Ω^+

The infinitesimal strain field is obtained from the continuous part of the displacement field:

$$\hat{\boldsymbol{\varepsilon}} = \nabla^s \hat{\mathbf{u}} \quad \text{in } \Omega \backslash \Gamma_d \tag{4}$$

where $(\,\cdot\,)^s$ refers to the symmetric part of $(\,\cdot\,)$. The total strain in the body is given by:

$$\boldsymbol{\varepsilon} = \nabla^s \mathbf{u} = \nabla^s \hat{\mathbf{u}} + \mathcal{H}_{\Gamma_d}(\nabla^s [\![\mathbf{u}]\!]) \quad \text{in } \Omega \backslash \Gamma_d \tag{5}$$

$$\boldsymbol{\varepsilon} = \nabla^s \mathbf{u} = \underbrace{\nabla^s \hat{\mathbf{u}} + \mathcal{H}_{\Gamma_d}(\nabla^s [\![\mathbf{u}]\!])}_{\text{bounded}} + \underbrace{\delta_{\Gamma_d}([\![\mathbf{u}]\!] \otimes \mathbf{n})^s}_{\text{unbounded}} \quad \text{in } \Omega \tag{6}$$

where $\otimes$ denotes a dyadic product and $\delta_{\Gamma d}$ is the Dirac-delta function along surface Γ_d. Both the displacement field and the strain field are continuous in Ω^- and Ω^+, since the unbounded term in equation (6) vanishes in $\Omega \backslash \Gamma_d = \Omega^- \cup \Omega^+$.

3 VARIATIONAL FORMULATION

In this section, the variational formulation is reviewed as a common basis for the discrete-interface approach, the strong embedded discontinuity approach and the partition of unity method. In the following, similar to earlier works (Simo et al. 1993; Lotfi and Shing 1995; Wells 2001), the principle of virtual work extended for a cracked body is obtained as a particular case of the three-field Hu-Washizu variational statements (Washizu 1982) applied to a body containing an internal discontinuity surface.

The governing field equations are imposed separately in $\Omega \backslash \Gamma_d$ and at Γ_d. According to equations (4) and (6) and assuming that the essential boundary conditions are satisfied, the following weak form can be obtained:

$$\begin{aligned} &-\int_{\Omega \backslash \Gamma_d} (\nabla^s \delta \mathbf{u}) : \boldsymbol{\sigma}\, d\Omega + \int_{\Omega \backslash \Gamma_d} \delta \mathbf{u} \cdot \mathbf{b}\, d\Omega \\ &-\int_{\Gamma_t} \delta \mathbf{u} \cdot \bar{\mathbf{t}}\, d\Gamma + \int_{\Omega \backslash \Gamma_d} \delta \boldsymbol{\varepsilon} : (\boldsymbol{\sigma} - \boldsymbol{\sigma}(\boldsymbol{\varepsilon}))\, d\Omega \\ &+\int_{\Gamma_d} (\delta \mathbf{u}^+ - \delta \mathbf{u}^-) \cdot \mathbf{t}^+ d\Gamma = 0. \end{aligned} \tag{7}$$

where $\mathbf{b}$ are the body forces, $\bar{\mathbf{u}}$ are the prescribed displacements, $\bar{\mathbf{t}}$ are the tractions at the boundary, and $\mathbf{t}$ are the tractions at Γ_d. The compatible variations of the total displacements and strains are $\delta\mathbf{u}$ and $\delta\boldsymbol{\varepsilon}$, respectively. Recalling equation (1) and taking the variations of the total displacements equal to

$$\delta \mathbf{u} = \delta \hat{\mathbf{u}} + \mathcal{H}_{\Gamma_d} \delta [\![\mathbf{u}]\!], \tag{8}$$

the variations of the strain equal to

$$\delta \boldsymbol{\varepsilon} = \nabla^s \delta \hat{\mathbf{u}} + \mathcal{H}_{\Gamma_d} \nabla^s \delta [\![\mathbf{u}]\!] \quad \text{in } \Omega \backslash \Gamma_d, \tag{9}$$

and the variations of the jumps given by

$$\delta [\![\mathbf{u}]\!] = \delta \mathbf{u}^+ - \delta \mathbf{u}^- \quad \text{at } \Gamma_d, \tag{10}$$

results in two variational statements given by (taking first $\delta[\![\mathbf{u}]\!] = \mathbf{0}$ and then $\delta\hat{\mathbf{u}} = \mathbf{0}$):

$$\begin{aligned} &-\int_{\Omega \backslash \Gamma_d} (\nabla^s \delta \hat{\mathbf{u}}) : \boldsymbol{\sigma}(\boldsymbol{\varepsilon})\, d\Omega + \int_{\Omega \backslash \Gamma_d} \delta \hat{\mathbf{u}} \cdot \mathbf{b}\, d\Omega \\ &+\int_{\Gamma_t} \delta \hat{\mathbf{u}} \cdot \bar{\mathbf{t}}\, d\Gamma = 0, \end{aligned} \tag{11}$$

$$\begin{aligned} &-\int_{\Omega} \mathcal{H}_{\Gamma_d}(\nabla^s \delta [\![\mathbf{u}]\!]) : \boldsymbol{\sigma}(\boldsymbol{\varepsilon})\, d\Omega + \int_{\Omega} \mathcal{H}_{\Gamma_d} \delta [\![\mathbf{u}]\!] \cdot \mathbf{b}\, d\Omega \\ &+\int_{\Gamma_t} \mathcal{H}_{\Gamma_d} \delta [\![\mathbf{u}]\!] \cdot \bar{\mathbf{t}}\, d\Gamma + \int_{\Gamma_d} \delta [\![\mathbf{u}]\!] \cdot \mathbf{t}^+ d\Gamma = 0. \end{aligned} \tag{12}$$

The stress field in the continuum depends upon the infinitesimal strain $\hat{\boldsymbol{\varepsilon}}$,

$$\boldsymbol{\sigma} = \boldsymbol{\sigma}(\hat{\boldsymbol{\varepsilon}}). \tag{13}$$

Substituting equation (13) into (11) and (12) reads

$$\begin{aligned} &-\int_{\Omega \backslash \Gamma_d} (\nabla^s \delta \hat{\mathbf{u}}) : \boldsymbol{\sigma}(\hat{\boldsymbol{\varepsilon}})\, d\Omega + \int_{\Omega \backslash \Gamma_d} \delta \hat{\mathbf{u}} \cdot \mathbf{b}\, d\Omega \\ &+\int_{\Gamma_t} \delta \hat{\mathbf{u}} \cdot \bar{\mathbf{t}}\, d\Gamma = 0, \end{aligned} \tag{14}$$

$$\begin{aligned} &-\int_{\Omega} \mathcal{H}_{\Gamma_d}(\nabla^s \delta [\![\mathbf{u}]\!]) : \boldsymbol{\sigma}(\hat{\boldsymbol{\varepsilon}})\, d\Omega + \int_{\Omega} \mathcal{H}_{\Gamma_d} \delta [\![\mathbf{u}]\!] \cdot \mathbf{b}\, d\Omega \\ &+\int_{\Gamma_t} \mathcal{H}_{\Gamma_d} \delta [\![\mathbf{u}]\!] \cdot \bar{\mathbf{t}}\, d\Gamma + \int_{\Gamma_d} \delta [\![\mathbf{u}]\!] \cdot \mathbf{t}^+ d\Gamma = 0. \end{aligned} \tag{15}$$

where $\Gamma_t^+ = \Gamma_t \cap \delta\Omega^+$ and $\Gamma_t^- = \cap\, \delta\Omega^-$. In equation (15), the tractions are obtained from a separate traction-jump law at the discontinuity, such that $\mathbf{t} = \mathbf{t}([\![\mathbf{u}]\!])$.

Equation (14) is the usual principle of virtual work obtained for a continuum. It is interesting to note that, taking $r = 1$ in equation (2), this formulation is the same as the one adopted in the partition of unity method (Wells and Sluys 2001a; Wells 2001;

Simone et al. 2000), in which case the second variational statement in equation (15) can be interpreted as the principle of virtual work applied to subdomain Ω^+ More generally, if r is taken between 0 and 1, separate contributions in Ω^+ and Ω^- are taken into account, similar to the work presented in Lotfi and Shing (1995) and Ohlsson and Olofsson (1997). Furthermore, summing equations (14) and (15), the principle of virtual work applied to a body crossed by a discontinuity is recovered (Malvern 1969), i.e.,

$$-\int_{\Omega\backslash\Gamma_d} (\nabla^s \delta\mathbf{u}) : \boldsymbol{\sigma}(\hat{\boldsymbol{\varepsilon}})\, d\Omega + \int_{\Gamma_d} \delta[\![\mathbf{u}]\!] \cdot \mathbf{t}^+ d\Gamma$$

$$+\int_{\Omega\backslash\Gamma_d} \delta\mathbf{u} \cdot \mathbf{b}\, d\Omega + \int_{\Gamma_t} \delta\mathbf{u} \cdot \bar{\mathbf{t}}\, d\Gamma = 0, \qquad (16)$$

which is the weak form usually adopted in the discrete-interface approach.

It is also interesting to compare the adopted variational formulation with the one introduced by Simo et al. (1993) and used in many other works, namely in Armero and Garikipati (1996), Oliver (1996a), Wells and Sluys (2001b), Jirásek and Zimmermann (2001), Alfaiate et al. (2001a). In this case, two variational statements are also introduced: i) the principle of virtual work in equation (14) and ii):

$$\frac{1}{l_d}\int_{\Gamma_d} \mathbf{t}\, d\Gamma - \frac{1}{\Omega}\int_{\Omega} \boldsymbol{\sigma} \cdot \mathbf{n}\, d\Omega = \mathbf{0}, \qquad (17)$$

where l_d is the length of the discontinuity Γ_d. In equation (17), the traction continuity condition is imposed in an averaged sense. It can be shown that, in the present formulation, the traction continuity condition is also enforced in a weak form through equation (15), although in a different way. Departing from the corresponding strong form:

$$\mathbf{t} - \boldsymbol{\sigma} \cdot \mathbf{n} = \mathbf{0}, \qquad (18)$$

pre-multiplying both sides of (18) by the variations of the jumps and integrating along Γ_d results in

$$\int_{\Gamma_d} \delta[\![\mathbf{u}]\!] \cdot \mathbf{t}\, d\Gamma - \int_{\Gamma_d} \delta[\![\mathbf{u}]\!] \cdot (\boldsymbol{\sigma} \cdot \mathbf{n})\, d\Gamma = 0. \qquad (19)$$

Taking, for example, $r = 1$ and applying the divergence theorem to subdomain Ω^+,

$$\int_{\Omega^+} (\nabla^s \delta[\![\mathbf{u}]\!]) : \boldsymbol{\sigma}\, d\Omega = \int_{\Gamma_t^+} (\delta[\![\mathbf{u}]\!]) \cdot \bar{\mathbf{t}}\, d\Gamma$$

$$\int_{\Gamma_d} (\delta[\![\mathbf{u}]\!]) \cdot (\boldsymbol{\sigma}^+ \cdot \mathbf{n}^+) d\Gamma - \int_{\Omega^+} (\delta[\![\mathbf{u}]\!]) \cdot (\nabla \cdot \boldsymbol{\sigma}) d\Omega, \qquad (20)$$

we get equation (15). In section 4, we shall see that a symmetric formulation can be obtained from equations (14) and (15) (as in the partition of unity method), instead of the non-symmetric formulation introduced by Simo et al. (1993).

4 FINITE ELEMENT APPROXIMATION

Consider a finite element discretisation of the 2D domain Ω. Assume that one element is crossed by a straight discontinuity Γ_d, which divides Ω in two subdomains Ω^+ and Ω^-. A local frame (s, n) is introduced such that $s(\mathbf{x})$ is aligned with Γ_d and $[n]$ is the normal to the discontinuity (fig. 1).

Recall equation (1). Assume that the jump $[\![\mathbf{u}]\!]$,

$$[\![\mathbf{u}]\!] = [\![\mathbf{u}(s(\mathbf{x}))]\!], \qquad (21)$$

which occurs along surface Γ_d within the element, is a linear function of s:

$$[\![\mathbf{u}(s)]\!] = q\mathbf{e}_s + (C + ks)\mathbf{e}_n, \qquad (22)$$

where q, C and k are constants and $\mathbf{e}_s$ and $\mathbf{e}_n$ are unit vectors along directions s and n, respectively.

For all $\mathbf{x} \in \Omega^+$, $r[\![\mathbf{u}(s)]\!]$ will be added to the displacement field $\hat{\mathbf{u}}$, whereas for all $\mathbf{x} \in \Omega^-$, $(1 - r)[\![\mathbf{u}(s)]\!]$ will be subtracted from the displacement field $\hat{\mathbf{u}}$. In (fig. 1-right) the total displacement field is depicted (where $\hat{\mathbf{u}}$ and q are neglected for clarity). If the jump is assumed to be constant across Γ_d, it is sufficient to adopt one internal node to represent the displacement jump. However, if higher order functions are considered for $[\![\mathbf{u}]\!]$, more nodes are required: two nodes for a linear function, three nodes for a quadratic function and so on. In the example above, the additional two nodes, i and j, are located at the intersection of Γ_d with the edges of the element (if a quadratic function were adopted, a third node would be located at the midpoint of Γ_d).

In matrix form, for each finite element e with n nodes, the following approximation of the displacement field in (1) is adopted:

$$\begin{aligned} \hat{\mathbf{u}}^e &= \mathbf{N}^e(\mathbf{x})\hat{\mathbf{a}}^e && \text{in } \Omega^e\backslash\Gamma_d^e \\ [\![\mathbf{u}]\!]^e &= \mathbf{N}_w^e[s(\mathbf{x})]\mathbf{w}^e && \text{at } \Gamma_d^e \end{aligned} \qquad (23)$$

where $\mathbf{N}^e$ contains the usual element shape functions, $\hat{\mathbf{a}}^e$ are the nodal degrees of freedom associated with $\hat{\mathbf{u}}$, $\mathbf{N}_w^e$ are the shape functions used to approximate

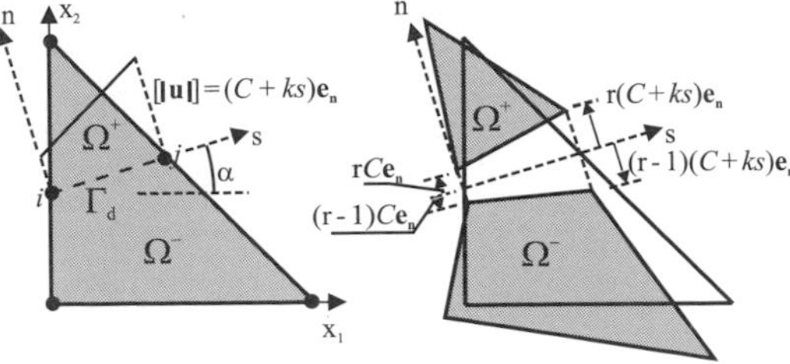

Figure 1. Displacement jump in a triangle crossed by a discontinuity.

the jumps $[\![\mathbf{u}]\!]$ and $[\![\mathbf{w}^e]\!]$ are the degrees of freedom associated with $[\![\mathbf{u}]\!]$. In equations (23), if the number of nodes used to approximate the jumps is n_w, $\mathbf{N}_w$ is a $(2 \times 2n_w)$ matrix: if $n_w = 1$, $\mathbf{N}_w$ is the identity matrix; if $n_w = 2$, $\mathbf{N}_w$ contains linear shape functions and so on.

It is assumed that the total displacement field, $\mathbf{u} = \hat{\mathbf{u}} + \mathcal{H}_{\Gamma_d}[\![\mathbf{u}]\!]$, is approximated by the usual shape functions:

$$\mathbf{u}^e = \mathbf{N}^e(\mathbf{x})\mathbf{a}^e \quad \text{in } \Omega^e \backslash \Gamma_d^e \tag{24}$$

where $\mathbf{a}^e$ are the nodal degrees of freedom associated with total displacements $\mathbf{u}$. Since the approximated displacement field is continuous, the contribution of the jump across the discontinuity Γ_d into $\mathbf{u}^e$ must also be continuous. This can be achieved by projecting the jump to the n element nodes such that:

$$\mathcal{H}_{\Gamma_d}[\![\mathbf{u}]\!]^e \approx \mathbf{N}^e\mathbf{H}\tilde{\mathbf{a}}^e. \tag{25}$$

In equation (25), $\mathbf{H}$ is as $(2n \times 2n)$ diagonal matrix, which corresponds to the implementation of the $\mathcal{H}_{\Gamma_d}$ function defined in equation (2): components of $\mathbf{H}$ related to nodes in Ω^{e+} are equal to $(0 \le r \le 1)$ and components related to nodes in Ω^{e-} are equal to $(r-1)$. The enhanced degrees of freedom $\tilde{\mathbf{a}}^e$ are given by

$$\tilde{\mathbf{a}}^e = \mathbf{M}_w^e\mathbf{w}^e. \tag{26}$$

In equation (26), $\mathbf{M}_w^e$ is a $(2n \times 2n_w)$ matrix which generates rigid body motions for subdomains Ω^{e+} and Ω^{e-}. In the following, matrix $\mathbf{M}_w^e$ is derived for a linear function $[\![\mathbf{u}(s(\mathbf{x}))]\!]$.

For each element node k ($k \in \{1,2,3\}$ in the example of fig. 1), the additional displacements $\tilde{\mathbf{a}}^{(k)}$ are given by:

$$\tilde{\mathbf{a}}^{(k)} = \left\{ \begin{array}{c} \tilde{a}_s^{(k)} \\ \tilde{a}_n^{(k)} \end{array} \right\} = \left\{ \begin{array}{c} -\theta x_n^{(k)} + w_s^i \\ \theta x_s^{(k)} + w_n^i \end{array} \right\}. \tag{27}$$

In equation (27), θ is the rotation of Ω^{e+} relative to Ω^{e-}, given by:

$$\theta = \frac{w_n^j - w_n^i}{l_d}, \tag{28}$$

$x_s^{(k)}$ and $x_n^{(k)}$ are the coordinates of node k in the local frame and (w_s^i, w_n^i), (w_s^j, w_n^j) are the jump components at nodes i and j, respectively. Expressing the components of $\tilde{\mathbf{a}}^{(k)}$ in the global frame (x_1, x_2), leads to the definition of matrix $\mathbf{M}_w^e$, whose $(2 \times 2n_w)$ components related to node k are:

$$\mathbf{M}_w^{(k)^T} = \begin{bmatrix} \frac{l_d-(x_2^{(k)}-x_2^i)\sin\alpha}{l_d} & \frac{x_1^{(k)}-x_1^i)\sin\alpha}{l_d} \\ \frac{(x_2^{(k)}-x_2^i)\cos\alpha}{l_d} & \frac{l_d-(x_1^{(k)}-x_1^i)\cos\alpha}{l_d} \\ \frac{(x_2^{(k)}-x_2^i)\sin\alpha}{l_d} & -\frac{(x_1^{(k)}-x_1^i)\sin\alpha}{l_d} \\ -\frac{(x_2^{(k)}-x_2^i)\cos\alpha}{l_d} & \frac{(x_1^{(k)}-x_1^i)\cos\alpha}{l_d} \end{bmatrix}, \tag{29}$$

where l_d is the length of the discontinuity Γ_d and α is the angle between coordinates s and x_1 (fig. 1).

Finally, the displacement field in each element is given by:

$$\begin{aligned} \mathbf{u}^e &= \mathbf{N}^e(\hat{\mathbf{a}}^e + \mathbf{H}\tilde{\mathbf{a}}^e) = \mathbf{N}^e(\hat{\mathbf{a}}^e + \mathbf{H}\mathbf{M}_w^e\mathbf{w}^e) \quad \text{in } \Omega^e \\ [\![\mathbf{u}]\!]^e &= \mathbf{N}_w^e[s(\mathbf{x})]\mathbf{w}^e \quad \text{at } \Gamma_d^e. \end{aligned} \tag{30}$$

The infinitesimal strain field is approximated by:

$$\hat{\boldsymbol{\varepsilon}}^e = \mathbf{L}\mathbf{N}^e\hat{\mathbf{a}}^e = \mathbf{B}^e(\mathbf{a}^e - \mathbf{H}\mathbf{M}_w^e\mathbf{w}^e), \tag{31}$$

where $\mathbf{L}$ is usual differential operator. The incremental stress field is

$$d\boldsymbol{\sigma}^e = \mathbf{D}^e d\hat{\boldsymbol{\varepsilon}}^e = \mathbf{D}^e\mathbf{B}^e(d\mathbf{a}^e - \mathbf{H}\mathbf{M}_w^e\, d\mathbf{w}^e). \tag{32}$$

The tractions are obtained from the traction-jump law at the discontinuity. In incremental format this reads:

$$d\mathbf{t}^e = \mathbf{T}^e d[\![\mathbf{u}]\!]^e = \mathbf{T}^e\mathbf{N}_w^e\, d\mathbf{w}^e \quad \text{at } \Gamma_d^e. \tag{33}$$

Discretising equations (14) and (15) by means of the field approximations given in equations (23 to 33), we obtain:

$$\begin{bmatrix} \mathbf{K}_{aa}^e & -\mathbf{K}_{aw}^e \\ -\mathbf{K}_{wa}^e & \mathbf{K}_{ww}^e + \mathbf{K}_d^e \end{bmatrix} \left\{ \begin{array}{c} d\mathbf{a}^e \\ d\mathbf{w}^e \end{array} \right\} = \left\{ \begin{array}{c} d\mathbf{f}_{ext}^e \\ d\mathbf{f}_{w,ext}^e \end{array} \right\}, \tag{34}$$

where

$$\mathbf{K}_{aa}^e = \int_{\Omega^e} \mathbf{B}^{eT}\mathbf{D}^e\mathbf{B}^e\, d\Omega \tag{35}$$

$$\mathbf{K}_{aw}^e = \int_{\Omega^e} \mathbf{B}^{eT}\mathbf{D}^e\mathbf{B}_w^e\, d\Omega \tag{36}$$

$$\mathbf{K}_{wa}^e = \mathbf{K}_{aw}^{eT} \tag{37}$$

$$\mathbf{K}_{ww}^e = \int_{\Omega^e} \mathbf{B}_w^{eT}\mathbf{D}^e\mathbf{B}_w^e\, d\Omega \tag{38}$$

$$\mathbf{K}_d^e = \int_{\Gamma_d} \mathbf{N}_w^{eT}\mathbf{T}^e\mathbf{N}_w^e\, d\Gamma, \tag{39}$$

$$\mathbf{B}_w^e = \mathbf{B}^e\mathbf{H}\mathbf{M}_w^e, \tag{40}$$

and

$$d\mathbf{f}_{ext}^e = \int_{\Omega} \mathbf{N}^{eT} d\mathbf{b}^e d\Omega + \int_{\Gamma_t} \mathbf{N}^{eT} d\bar{\mathbf{t}}^e d\Gamma, \tag{41}$$

$$d\mathbf{f}_{w,ext}^e = \mathbf{0} \tag{42}$$

From equations (34) to (39), it is clear that a symmetric formulation is obtained, as long as matrices $\mathbf{D}^e$ and $\mathbf{T}^e$ are symmetric.

At this stage, two different approaches can be adopted:

i) either the additional nodes are considered as internal nodes in the element, or
ii) the new nodes are global.

In the former approach, which can be considered a natural extension of the usual embedded discontinuity formulation, condensation of the corresponding additional degrees of freedom can be performed at element level and the bandwidth of the global stiffness matrix is kept fixed during the calculations. Nevertheless, no continuity of the jumps across the element boundaries is achieved in this case.

In approach ii), which is adopted here, the additional nodes are global and therefore the bandwidth of the global stiffness matrix increases during the calculations. Nevertheless, for $n_w \geq 2$ continuity of the jumps at the discontinuities is automatically enforced across the element boundaries. Moreover, since the jump is continuously distributed over the entire element Ω^e, the elements remain conform.

As previously stated in de Borst et al. (2001), this formulation is kinematically similar to a smeared approach. Nevertheless, it should be stressed that the constitutive relation adopted for the continuum is not modified; instead, a discrete relation is enforced separately at Γ_d (equation (33)), whose location and corresponding displacement jumps are known (Alfaiate and Sluys 2002).

5 MATERIAL AND NUMERICAL MODELS

In this Section, the material and numerical models are described. A linear elastic bulk behaviour is adopted, whereas a localized damage model is used for the traction-jump law at the discontinuity. An isotropic traction-jump law is adopted (Alfaiate, Wells, and Sluys 2001a):

$$\mathbf{t} = (1 - d)\mathbf{T}_{el}\mathbf{w}, \tag{43}$$

where $0 \leq d \leq 1$ is a scalar damage variable and $\mathbf{T}_{el}$ is the *elastic* constitutive tensor in which non-diagonal terms are zero and diagonal terms are penalty functions used to prevent overlapping of crack faces under crack closure. The evolution of damage is given by:

$$d = d(\kappa) = 1 - \exp\left(-\frac{f_t}{G_F}\kappa\right), \tag{44}$$

where κ is a scalar variable taken equal to the maximum positive normal jump component:

$$\kappa = \max\langle w_n\rangle^+, \quad \kappa \geq 0, \quad \kappa \geq 0, \tag{45}$$

f_t is the tensile strength and G_F is the fracture energy. A loading function is defined as

$$f = w_n - \kappa. \tag{46}$$

Only mode I opening is considered, i.e. discontinuities open perpendicularly to the direction of the maximum principal stress σ_I whenever

$$\sigma_I = f_t. \tag{47}$$

No shear tractions are allowed at the discontinuities during crack evolution.

A new embedded discontinuity always crosses the entire parent element. Only one straight discontinuity is allowed in each element; the corresponding direction is defined perpendicular to the direction of σ_I, which is obtained at the central integration point of the parent element. Crack path continuity is enforced according to an algorithm similar to the one presented in Alfaiate et al. (2001a).

The opening criterion in equation (47) should be verified at the crack tip. The jump at the crack tip, as well as the jump ahead of the crack tip (within the same element), are enforced to be practically zero through the use of the penalty functions in $\mathbf{T}_{el}$. A new discontinuity is introduced in a next element whenever the crack tip approaches the element boundary. At this point, where a new node is defined, a correct account of the stress state must be obtained, in order to properly fulfill the opening criterion. Different criteria have been proposed in literature to obtain the stresses at the crack tip, which give rise to a more or less localized crack evolution criterion, depending on the degree of mesh refinement and on the order of the parent elements. In particular, the following examples are mentioned:

i) the accurate definition of the stresses near the crack tip is approximated by using special functions taken from linear elastic fracture mechanics (Ingraffea 1989; Carpinteri et al. 1989; Möes et al. 1999);
ii) the mean values of the stress components obtained at the nearest integration points of the crack tip define the stress state adopted for crack evolution (Alfaiate et al. 1997; Alfaiate and Pires 1999), or
iii) a non-local stress state is adopted near the crack tip, in which case the averaging support is extended beyond the element size (Wells 2001).

Here, similar to ii), the stress components at each crack node are taken as the average stress components obtained at the two nearest integration points, located at each side of the common element boundary (see fig. 2).

6 NUMERICAL EXAMPLES

The first example is a constant strain triangle, submitted to tension. Due to the asymmetry, the horizontal discontinuity, introduced at the centroid, opens at one of the edges only, as depicted in fig. 3 where the deformed meshes obtained from both the embedded discontinuity example and a discrete-interface test are presented. In fig. 4, a perfect match between the load-displacement curves obtained with the two approaches is presented. Note that a homogeneous jump field across the element would be obtained with the

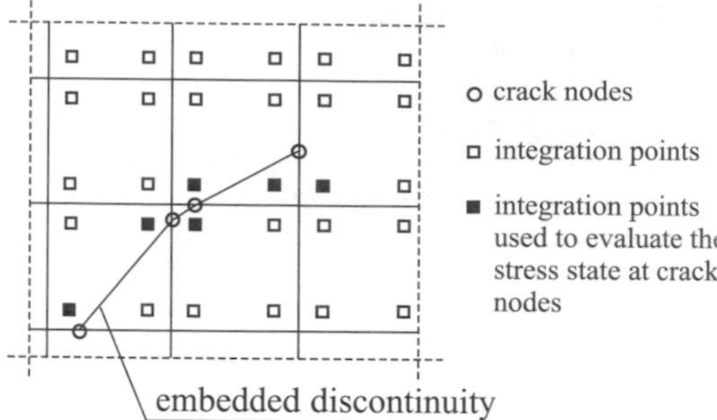

Figure 2. The stress components at each crack node are taken as the average stress components obtained at the two nearest integration points.

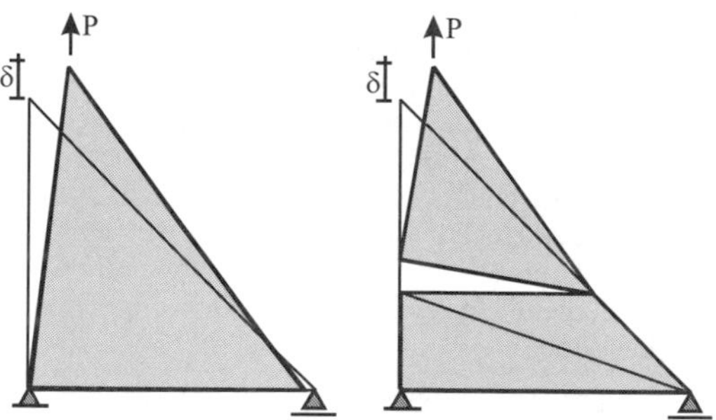

Figure 3. Asymmetric triangle submitted to tension.

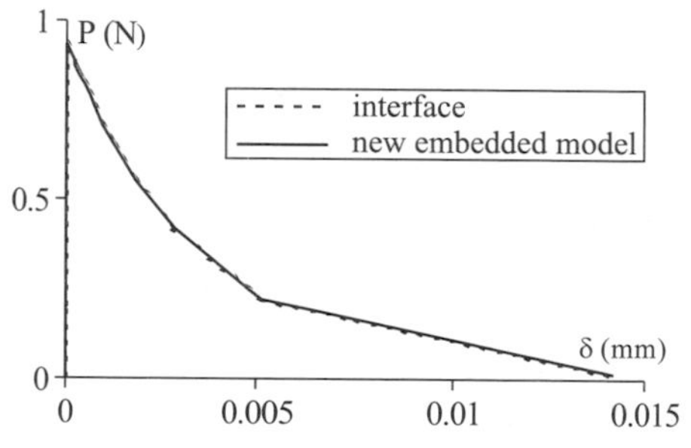

Figure 4. Load displacement curves corresponding to fig. 3.

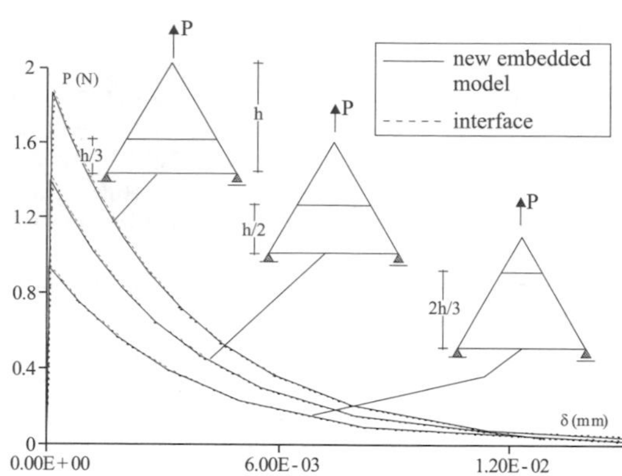

Figure 5. Symmetric triangle submitted to tension.

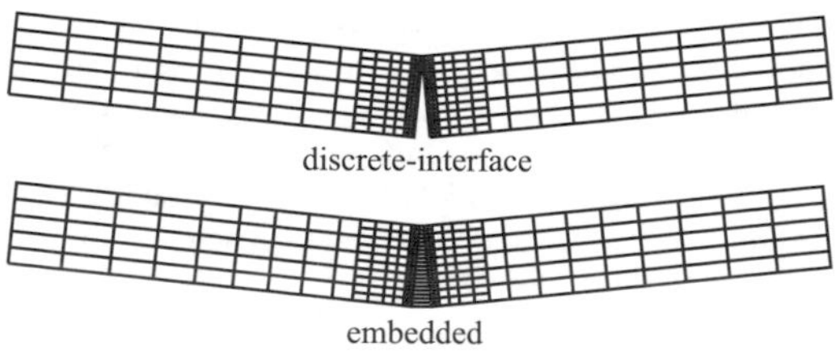

Figure 6. Three point bending beam: deformed meshes.

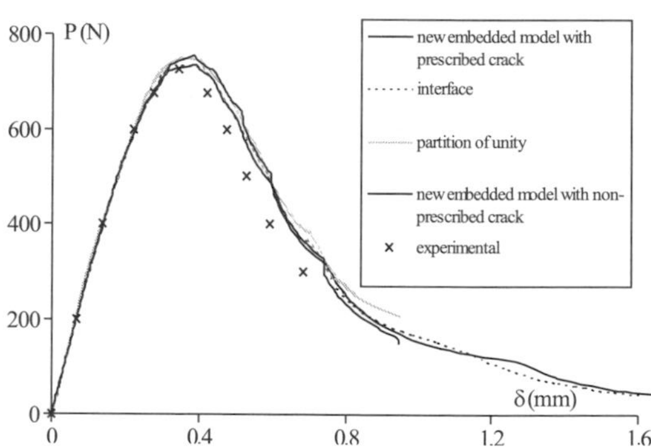

Figure 7. Three point bending beam: load-displacement curves.

non-symmetric formulation used in Armero and Garikipati (1996), Oliver (1996a), Wells and Sluys (2000), Wells and Sluys (2001b), Alfaiate et al. (2001a), yielding more dissipation of energy.

Next, a symmetric triangle is submitted to tension. Depending on the position of the prescribed discontinuity, different maximum loads are obtained, exactly in the same way as in a discrete-interface approach. In fig. 5, the load-displacement curves obtained with the new embedded model and with the interface approach are presented, which show almost complete similarity. Note that, if the non-symmetric formulation was adopted (Oliver 1996a), since it does not depend on the crack length, the same peak load (as well as load-displacement curve) would be obtained in all three examples.

The three point bending beam test is another example in which the discrete-interface and the new embedded approach have been compared. The beam dimensions are: length = 2 m, thickness = 0.05 m and depth = 0.2 m, with a 0.1 m notch depth. The adopted material parameters are: $f_t = 3.33$ MPa, Young modulus E = 30 GPa and fracture energy $G_F = 115$ N/m. In fig. 6 the deformed meshes obtained with both approaches are presented. In fig. 7, the load-displacement curves numerically obtained are presented and compared to the experimental result presented in Petersson (1981). First, a prescribed vertical crack path is introduced in both the embedded and the discrete-interface numerical examples. In this case, the location and orientation of the discontinuities are given *a priori*, such that the crack path is enforced along the midspan, both in the embedded and in the interface analyses. Then, a solution with non-prescribed embedded discontinuities is also obtained. In the latter example, as a first attempt of a simplified opening criterion, the stress state at the crack tip is obtained at the centre of the element ahead of the crack tip. As a result, compared to the prescribed solutions, the crack tip always opens later and a higher peak load than the experimental one is obtained. The result obtained with the partition of unity method is also shown, where the same opening criterion is adopted, thus leading to a solution with delayed crack propagation similar to the non-prescribed one.

The last test presented is the single edge notched beam. The beam dimensions and boundary conditions are shown in fig. 9. The adopted material parameters are: f_t = 2.8 MPa, Young modulus E=35 GPa and fracture energy GF=100 N/m. In Schlangen (1993), Schlangen and van Mier (1993), experimental results were obtained with this beam, in which the load is controlled such that a monotonic increasing of the sliding of the notch (crack mouth sliding displacement, CMSD) is enforced. In the numerical analyses, load control is performed using an arc length method where only the relative sliding displacement of the notch (CMSD) is taken into account in the constraint equation.

Four different numerical solutions are presented: in the first three analyses a structured mesh with four node isoparametric elements is used, whereas in the fourth analysis an unstructured mesh is adopted with constant strain triangles. In all cases, continuity of the crack path is enforced.

In fig. 8 the load-CMSD curves obtained with a structured mesh with quadrilateral elements are compared to the experimental load-CMSD curve. In test number one (curve 1), a numerical solution with a non-prescribed crack path was obtained adopting the simplified opening criterion mentioned for the three-point bending beam. It is clear that the solution obtained with this criterion deviates from the experimental result. In test number two the crack path was prescribed (curve 2). In this case, the information relative to the location and orientation of the cracks, which had to be known *a priori*, was taken from the first test. In test number three (curve 3), a better account of the stresses at the crack tip was adopted by means of the opening criterion defined in section 5. This criterion gives rise to a load-CMSD curve similar to the curve obtained with the prescribed crack, both curves agreeing well with the experimental one. In fig. 9 the crack path obtained from test number three is shown. In fig. 10 the corresponding deformed mesh is presented. Both the crack path and the deformed mesh agree well with experimental observation (Schlangen and van Mier 1993; Schlangen 1993), in spite of the fact that the adopted mesh is relatively coarse (about 350 elements).

In fig. 11 the load-CMSD curve obtained with the unstructured mesh (curve 4) is compared to the experimental curve. In this figure, the result taken from Alfaiate et al. (2001a) is also presented (curve 5). This result was obtained with the non-symmetric formulation (Armero and Garikipati 1996; Oliver 1996b; Oliver 1996a; Wells and Sluys 2000; Wells and Sluys

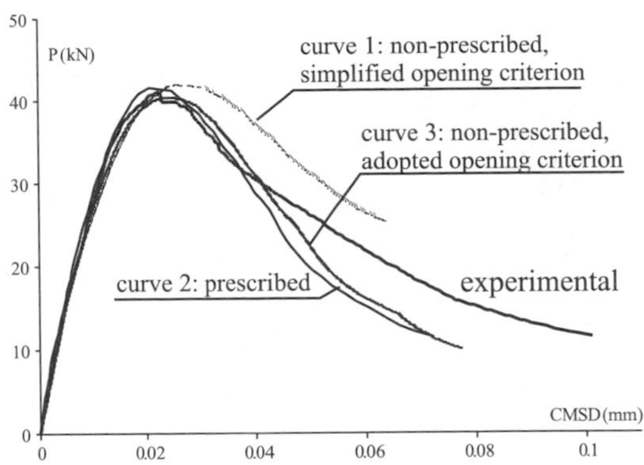

Figure 8. Structured mesh: load-CMSD curves.

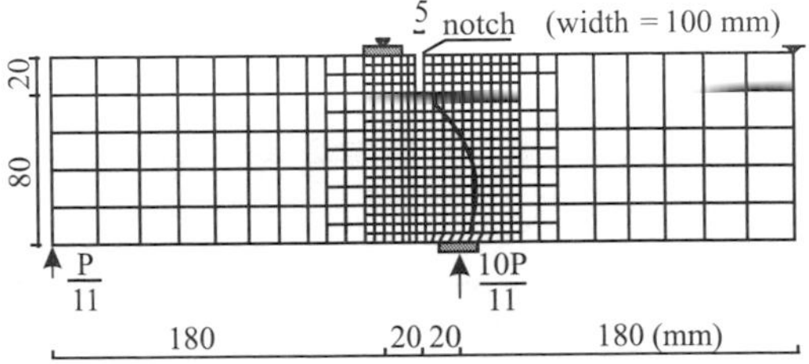

Figure 9. Single end notched beam: crack path with structured mesh from test number three.

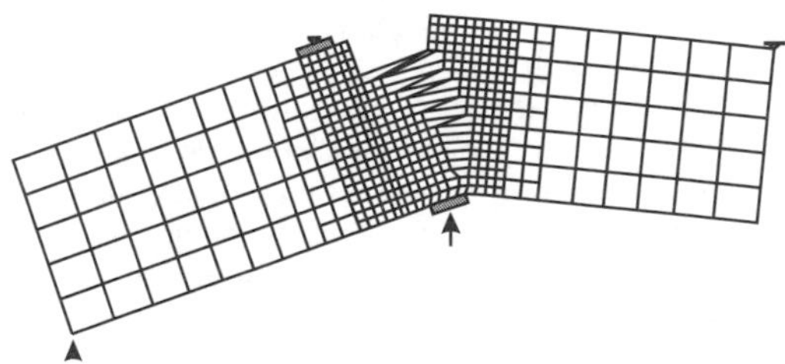

Figure 10. Single edge notched beam: deformed mesh obtained from test number three.

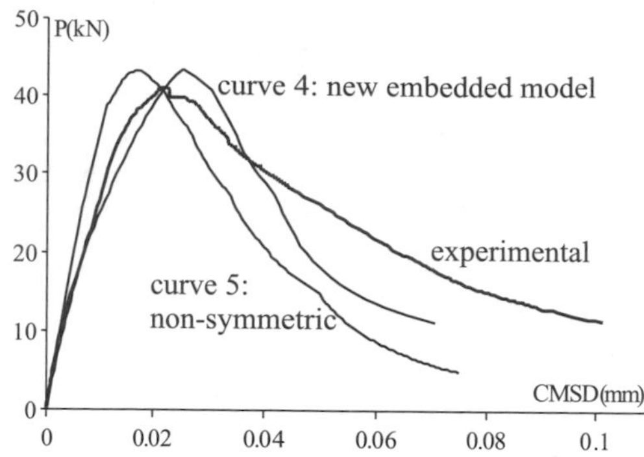

Figure 11. Unstructured mesh: load-CMSD curves.

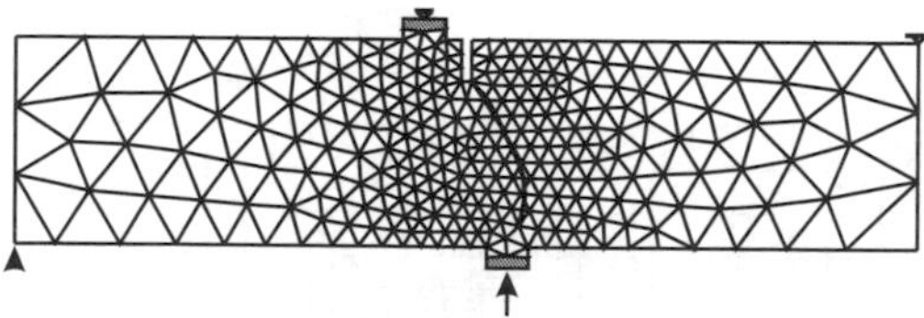

Figure 12. Single edge notched beam: crack path obtained with unstructured mesh.

2001b) with the enforcement of crack path continuity. It is interesting to see that the ascending branch is better approximated with the new formulation (curve 4) than with the non-symmetric approach (curve 5), although the peak load is still larger than the experimentally observed peak load. However, it should be mentioned that CST elements can not provide a correct definition of the stresses at crack nodes, unless a much refined mesh is used, which is not the case here. In fig. 12 the crack path obtained from test number four is shown.

Finally, it is interesting to note that curves 2 to 5 in figs. 8 and 11 show a more brittle softening branch than the experimental curve. As shown in Alfaiate et al. (2001a), Alfaiate et al. (2001b), the response becomes more ductile if a mixed mode fracture criterion is adopted, i.e., if shear tractions are allowed in the discontinuities during crack opening.

7 CONCLUSIONS

In this paper, a new numerical implementation of embedded discontinuities with non-homogeneous jumps across the elements was introduced. The comparison with other embedded approaches, the discrete-interface approach and the partition-of-unity method was carried out. The conclusions are:

1. in the adopted formulation, the jumps are approximated by additional degrees of freedom obtained at additional nodes, which are located at the discontinuity surface;
2. these degrees of freedom can be considered either as global or local:
   (a) in the former case, which is adopted in the paper, continuous jumps across element boundaries are obtained, at the cost of an increasing bandwidth of the stiffness matrix;
   (b) in the latter case (an extension of previous embedded discontinuity approaches), the jumps are considered as internal degrees of freedom, which can be condensed out at element level, thus keeping the stiffness bandwidth constant. However, in this case, continuity of the displacement jumps across element boundaries can not be enforced;
3. the variational formulation adopted consists of the principle of virtual work applied to a body crossed by a discontinuity, as previously presented by Malvern (1969);
4. it is shown that the same variational formulation is adopted in other approaches, such as the discrete-interface approach, the partition of unity method or extended finite elements, and the embedded formulation introduced by Lotfi and Shing (1995);
5. it is also shown that, similar to the work of Simo et al. (1993), in which a non-symmetric formulation is introduced, the traction continuity condition is also imposed in a weak form, which yields a symmetric formulation;
6. in the adopted formulation, the jumps are projected (*smeared*) to the element nodes using regular shape functions, as done in other embedded approaches and in the partition of unity method;
7. in the partition of unity method, the approximation of the enhanced displacement field is introduced directly at the existing element nodes; in the present approach, the enhanced degrees of freedom at the nodes are derived from the jumps measured at the discontinuity surface Γ_d, which is *embedded* in the parent element;
8. the formulation adopted depends not only on the derivatives of the enhanced degrees of freedom (symmetric and non-symmetric formulations with constant jumps), but also on the relative displacements, comparable to the partition of unity method;
9. promising results were obtained with relatively coarse structured and non-structured meshes which compare well with experimental results and with the results obtained from analyses with interface elements.

REFERENCES

Alfaiate, J. and E. B. Pires (1999). A discrete crack numerical model. In A. Carpinteri and M. Aliabadi (Eds.),

Computational fracture mechanics in concrete techonology, pp. 133–162. Southampton, United Kingdom: WIT Press/ Computational Mechanics Publications.

Alfaiate, J., E. B. Pires, and J. A. C. Martins (1997). A finite element analysis of non-prescribed crack propagation in concrete. *Computers and Structures* 63(1), 17–26.

Alfaiate, J. and L. J. Sluys (2002). Analysis of a compression test on concrete using strong embedded discontinuities. In H. Mang, F. Rammer-storfer, and J. Eberhardsteiner (Eds.), *WCCM V, Fifth World Congress on Computational Mechanics*, http://wccm.tuwien.ac.at, Vienna, Austria.

Alfaiate, J., G. N. Wells, and L. J. Sluys (2001a). On the use of embedded discontinuity elements with crack path continuity for mode I and mixed mode fracture. *Engineering Fracture Mechanics* 69(6), 661–686.

Alfaiate, J., G. N. Wells, and L. J. Sluys (2001b). Strong embedded discontinuities for simulating fracture in quasi-brittle materials. In R. de Borst, J. Mazars, G. Pijaudier-Cabot, and J. G. M. van Mier (Eds.), *Fracture Mechanics of Concrete Structures – FRAMCOS4*, Paris, France, pp. 749–756.

Armero, F. and K. Garikipati (1996). An analysis of strong discontinuities in multiplicative finite strain plasticity and their relation with the numerical simulation of strain localization. *International Journal of Solids and Structures* 33(20–22), 2863–2885.

Carpinteri, A., S. Valente, and P. Bocca (1989). Mixed mode choesive crack propagation. In K. Salama, K. Ravi-Chandar, D. Taplin, and P. R. Rao (Eds.), *7th Int. Conf. on Fracture (ICF-7)*, New York, USA, pp. 2243–2257. Pergamon Press.

de Borst, R., G. N. Wells, and L. J. Sluys (2001). Some observations on embedded discontinuity models. *Engineering Computations* 18(1–2), 241–254.

Ingraffea, A. (1989). Shear cracks. In L. Elfgren (Ed.), *Fracture mechanics of concrete structures - from theory to applications, report of the Technical Committee 90-FMA Fracture Mechanics of Concrete-Applications*, pp. 231–233. London, United Kingdom: Chapman and Hall.

Jirásek, M. and T. Zimmermann (2001). Embedded crack model: I. basic formulation. *International Journal for Numerical Methods in Engineering* 50, 1269–1290.

Klisinski, M., K. Runesson, and S. Sture (1991). Finite element with inner softening band. *ASCE Journal of Engineering Mechanics* 117(3), 575–587.

Lotfi, H. R. and P. B. Shing (1995). Embedded representation of fracture in concrete with mixed finite elements. *International Journal for Numerical Methods in Engineering* 38(8), 1307–1325.

Malvern, L. E. (1969). *Introduction to the Mechanics of a Continuous Medium.* New Jersey: Prentice-Hall International.

Möes, N., J. Dolbow, and T. Belytschko (1999). A finite element method for crack growth without remeshing. *International Journal for Numerical Methods in Engineering* 46(1), 131–150.

Ohlsson, U. and T. Olofsson (1997). Mixed-mode fracture and anchor bolts in concrete analysis with inner softening bands. *ASCE Journal of Engineering Mechanics* 123, 1027–1033.

Oliver, J. (1996a). Modelling strong discontinuities in solid mechanics via strain softening constitutive equations. Part 2: Numerical simulation. *International Journal for Numerical Methods in Engineering* 39(21), 3601–3623.

Oliver, J. (1996b). Modelling strong discontinuities in solid mechanics via strain softening constitutive equations. Part 1: Fundamentals. *International Journal for Numerical Methods in Engineering* 39(21), 3575–3600.

Petersson, P. E. (1981). *Crack growth and development of fracture zones in plain concrete and similar materials.* Ph. D. thesis, Lund Institute of Technology, Sweden.

Schlangen, E. (1993). *Experimental and numerical analysis of fracture processes in concrete.* Ph. D. thesis, Delft University of Technology.

Schlangen, E. and J. G. van Mier (1993). Mixed-mode fracture propagation: a combined numerical and experimental study. *Fracture and damage of concrete and rock*, 166–175.

Simo, J. C., J. Oliver, and F. Armero (1993). An analysis of strong discontinuities induced by strain-softening in rate-independent inelastic solids. *Computational Mechanics* 12, 277–296.

Simone, A., J. J. C. Remmers, and G. N. Wells (2000). An interface element based on the partition of unity. Technical Report CM2001.007, Technical University of Delft, Delft, The Netherlans.

Sluys, L. J. (1998). Modelling of crack propagation with embedded discontinuity elements. In H. Mihashi and K. Rokugo (Eds.), *Fracture Mechanics of Concrete Structures – FRAMCOS3*, Gifu, Japan, pp. 843–860. AEDIFICATIO.

Washizu, K. (1982). *Variational Methods in Elasticity and Plasticity* (Third ed.). Oxford: Pergamon Press Ltd.

Wells, G. (2001). *Discontinuous modelling of strain localisation and failure.* Ph. D. thesis, Delft University of Technology.

Wells, G. N. and L. J. Sluys (2000). Analysis of slip planes in three-dimensional solids. *Computer Methods in Applied Mechanics and Engineering* 190(28), 3591–3606.

Wells, G. N. and L. J. Sluys (2001a). A new method for modelling cohesive cracks using finite elements. *International Journal for Numerical Methods in Engineering* 50(2), 2667–2682.

Wells, G. N. and L. J. Sluys (2001b). Three-dimensional embedded discontinuity model for brittle fracture. *International Journal of Solids and Structures* 38(5), 897–913.

Computational Modelling of Concrete Structures, Bićanić et al. (eds)
© 2003 Swets & Zeitlinger, Lisse, ISBN 90 5809 536 3

A dynamically consistent gradient model derived from a discrete microstructure: higher-order stiffness and higher-order inertia

Harm Askes, Andrei V. Metrikine & Herwald Naaktgeboren
Faculty of Civil Engineering and Geosciences, Delft University of Technology, GA Delft, The Netherlands

ABSTRACT: Higher-order gradient models are derived from a discrete microstructure via continualisation procedures. Distinction is made between standard continualisation, which inevitably lead to anomalies in the dispersive behaviour of the resulting continuum, and a newly proposed alternative continualisation, by which these anomalies can be avoided completely. The new continualisation method results in a continuum description that is dynamically consistent, in the sense that each higher-order stiffness term is accompanied by a higher-order inertia term. Furthermore, the static version of the model is able to eliminate singularities from the strain field.

1 INTRODUCTION

It is well established that standard continuum models, in which the constitutive equations are formulated in algebraic relations between stresses and strains without the inclusion of higher-order (spatial or temporal) terms, are not capable to describe phenomena that are driven by processes on a smaller scale. An important example of such a phenomenon is wave dispersion in granular media: due to the heterogeneity of the material, the high-frequency waves travel with a lower phase velocity as compared to the low-frequency waves, by which the wave is distorted during propagation. Microstructural influences need to be considered if wave dispersion is to be modelled (Kunin 1982; Kunin 1983; Chang and Gao 1995; Mühlhaus and Oka 1996).

Several strategies exist to include these influences. Firstly, every microstructural component can be modelled separately, either by means of a discrete representation (masses, springs, dash-pots, etc.) or by means of a classical continuum formulation, the latter case leading to an *inhomogeneous* continuum. Secondly, the microstructural effects can be accounted for in an averaged sense through an *enhanced* but otherwise *homogeneous* continuum. The enhancement of the continuum can be achieved in various ways (Metrikine and Askes 2002):

Phenomenological approach: the enhancement can be postulated in the form of additional terms in the energy functional or the constitutive equation.

Statistical approach: average values of the state variables can be computed from an inhomogeneous continuum, such that enhanced field equations are obtained.

Continualisation/Homogenisation: an enriched continuum is obtained by continualising a discrete medium or by homogenising an inhomogeneous continuum.

Through the enhancement procedure, higher-order spatial or temporal terms appear in the field equations. They are accompanied by additional material parameters such as internal length scales (Triantafyllidis and Aifantis 1986; Pijaudier-Cabot and Bažant 1987; Lasry and Belytschko 1988) or internal time scales (Vardoulakis and Aifantis 1994; Chen and Fish 2001). For a straightforward identification and evaluation of these additional parameters it is desired that a close link with the underlying microstructure is maintained. This can best be achieved if the enhanced continuum is obtained via continualisation or homogenisation.

Below, homogeneous higher-order continua are derived from discrete media via continualisation procedures. The format of the resulting continua is that of a higher-order strain-gradient model, i.e. higher-order gradients of the strain enter the constitutive equation. In Section 2 strain-gradient models from the literature are discussed, and a set of requirements are formulated that should be met by gradient models. Special attention is focussed on a realistic behaviour in dynamics. Section 3 treats the derivation of gradient models from a discrete medium, via the standard continualisation as well as via an alternative continualisation.

As shown in Section 4, the standard continualisation leads to anomalies in dynamics, while these anomalies can be completely avoided if the alternative continualisation is employed. Section 5 presents examples in statics and dynamics.

2 REQUIREMENTS FOR GRADIENT MODELS

In the literature, different motivations have instigated the addition of higher-order gradients to the continuum. For instance, discrete media have been continualised whereby the higher-order terms in the continuum were used to obtain a better approximation of the dispersive properties of the corresponding discrete medium (Chang and Gao 1995; Mühlhaus and Oka 1996; Suiker et al. 2001). In that case, the relation between stresses σ and strains ε takes the following form:

$$\sigma = D(\varepsilon + c\,\nabla^2 \varepsilon) \tag{1}$$

where D contains the elastic moduli and c is a (positive) higher-order material parameter with the dimension of length squared. Through the continualisation process, c can be related univocally to the microstructural properties.

Conversely, higher-order terms have been used in order to eliminate singularities from the strain field (Altan and Aifantis 1997; Aifantis 1999) or to regularise localised deformation patterns (Triantafyllidis and Aifantis 1986; Aifantis 1987; Mühlhaus and Aifantis 1991; Pamin 1994; Peerlings et al. 1996). With the restriction to elastic models and to second-order strain gradients, the constitutive equation is generically written as

$$\sigma = D(\varepsilon - c\nabla^2 \varepsilon) \tag{2}$$

where the only difference with Equation (1) concerns with the sign of the higher-order term.

The differences between the two classes of higher-order models have been studied in (Altan and Aifantis 1997; Unger and Aifantis 2000a; Unger and Aifantis 2000b; Askes et al. 2002). In short, the models according to Equation (1) have a close link with the underlying microstructure, but uniqueness and stability are not guaranteed, whereas the models according to Equation (2) are stable and uniqueness is guaranteed, but the link with the microstructure is less obvious. Both classes of models exhibit undesirable behaviour in dynamics: use of Equation (1) results in imaginary phase velocities and instabilities for the high-frequent waves, whereas use of Equation (2) leads to infinite group velocities and an unrealistic propagation of energy (Askes et al. 2002).

Obviously, a higher-order gradient model is preferred that combines the advantages of the two above classes of models, while the dynamic behaviour should be improved. Summarising, the following requirements are imposed on the performance of a higher-order gradient model:

unconditional stability: in the energy functional, all self-adjoint terms (kinetic or potential) must be preceded by a positive sign;
unconditional well-posedness: a unique solution must exist;
direct link with microstructure: the homogeneous continuum must be derived from a discrete medium or from an inhomogeneous continuum;
realistic behaviour in statics: local perturbations must remain local and singularities in the strain field must be eliminated;
realistic behaviour in dynamics: no imaginary phase velocities should occur, and the group velocity should never exceed the wave velocity of the classical continuum.

3 CONTINUALISATION PROCEDURES

In this Section, higher-order continuum models will be derived from a discrete medium. First, the discussion will be limited to one-dimensional configurations, of which the discrete model is shown in Figure 1. At the end of this Section, a two-dimensional extension of the higher-order continuum model is discussed briefly.

3.1 *Standard continualisation*

The equation of motion for particle x_n in Figure 1 is given by

$$M\ddot{x}_n + K(2x_n - x_{n-1} - x_{n+1}) = 0 \tag{3}$$

A periodic medium is assumed: all particles have mass M, all springs between two particles have spring stiffness K, and all inter-particle distances equal l. In the standard continualisation approach, it is assumed that the continuum displacement $u(x,t)$ equals the discrete displacement $x_n(t)$. Furthermore, for the neighbouring particles it is assumed that $x_{n\pm 1}(t) = u(x \pm l, t)$. A Taylor series expansion is used to rewrite $u(x \pm l, t)$ in terms of $u(x,t)$ and higher-order derivatives thereof:

$$u(x \pm l, t) = u(x,t) \pm l\frac{\partial u(x,t)}{\partial x} + \frac{1}{2}l^2\frac{\partial^2 u(x,t)}{\partial x^2} \pm \frac{1}{6}l^3\frac{\partial^3 u(x,t)}{\partial x^3} + \frac{1}{24}l^4\frac{\partial^4 u(x,t)}{\partial x^4} + \cdots \tag{4}$$

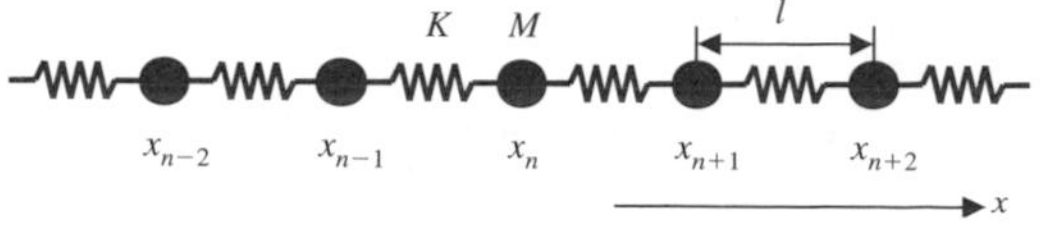

Figure 1. One dimensional discrete medium.

Substituting these relations into Equation (3) yields

$$\rho\ddot{u}(x,t) = E\left(\frac{\partial^2 u(x,t)}{\partial x^2} + \frac{1}{12}l^2\frac{\partial^4 u(x,t)}{\partial x^4} + \frac{1}{360}l^4\frac{\partial^6 u(x,t)}{\partial x^6} + \cdots\right) \tag{5}$$

where the continuum mass density ρ and Young's modulus E are related to the discrete material properties via $\rho = M/Al$ and $E = Kl/A$, with A the cross-sectional area of the configuration. Via $\rho\ddot{u}(x,t) = \partial\sigma/\partial x$ and the kinematic equation $\varepsilon = \partial u(x,t)/\partial x$, the stress-strain relation that underlies Equation (5) can be written as

$$\sigma = E\left(\varepsilon + \frac{1}{12}l^2\frac{\partial^2\varepsilon}{\partial x^2} + \frac{1}{360}l^4\frac{\partial^4\varepsilon}{\partial x^4} + \cdots\right) \tag{6}$$

Clearly, the second-order truncation of Equation (6) belongs to the category of Equation (1), and it exhibits the disadvantages outlined in Section 2. Furthermore, the dynamic behaviour of *any* truncation of series 6 is problematic, as illustrated in Section 4.

3.2 *Alternative continualisation*

Above, a one-to-one relation has been assumed between the displacement of the discrete medium and the displacement of the continuum at position n, whereas Taylor series have been applied to positions $n \pm 1$. In this manner, all truncation errors are involved with the neighbouring particles. Alternatively, it is possible to distribute the truncation error more uniformly over particles $n-1, n$ and $n+1$. Namely, the one-to-one relation between $x_n(t)$ and $u(x,t)$ is dropped and replaced by (Metrikine and Askes 2002)

$$u(x,t) = \frac{a x_{n-1}(t) + x_n(t) + a x_{n+1}(t)}{1+2a} \tag{7}$$

In other words, the continuous displacement $u(x,t)$ is taken as the spatial average of the displacement of three neighbouring particles, whereby the central particle is assigned a unit weight and the two outer particles contribute with weight a. This weighting parameter a is dimensionless; setting $a = 0$ leads to the original continualisation procedure as outlined above. It is assumed that $0 \le a \le 1$.

For backsubstitution into Equation (3) an explicit expression of $x_n(t)$ in terms of $u(x,t)$ is needed, therefore $x_n(t)$ is written as a power series, i.e.

$$x_n(t) = l^0 f_0(x,t) + l^1 f_1(x,t) + l^2 f_2(x,t) + \cdots \tag{8}$$

For the particle displacements at positions $n \pm 1$ Taylor series expansions for all f_i are used, i.e.

$$f_i(x \pm l, t) = f_i(x,t) \pm l\frac{\partial f_i(x,t)}{\partial x} + \frac{1}{2}l^2\frac{\partial^2 f_i(x,t)}{\partial x^2} \pm \frac{1}{6}l^3\frac{\partial^3 f_i(x,t)}{\partial x^3} + \frac{1}{24}l^4\frac{\partial^4 f_i(x,t)}{\partial x^4} + \cdots \tag{9}$$

Substitution of these expressions into Equation (7) gives an equality where different powers of the particle distance l appear on the left-hand-side and on the right-hand-side. Setting each multiplier of a power of l equal to zero separately leads to a set of equations as (Metrikine and Askes 2002)

$$l^0: \quad u = \frac{f_0(1+2a)}{1+2a} \tag{10a}$$

$$l^1: \quad 0 = f_1(1+2a) \tag{10b}$$

$$l^2: \quad 0 = a\frac{\partial^2 f_0}{\partial x^2} + f_2(1+2a) \tag{10c}$$

$$l^3: \quad 0 = \frac{\partial^2 f_1}{\partial x^2} + f_3(1+2a) \tag{10d}$$

$$l^4: \quad 0 = \frac{a}{12}\frac{\partial^4 f_0}{\partial x^4} + a\frac{\partial^2 f_2}{\partial x^2} + f_4(1+2a) \tag{10e}$$

etc., where (x,t)-notation has been dropped. Solving these equations sequentially, it follows that all odd f_i vanish, while for the even f_i it is found that

$$f_0 = u \tag{11a}$$

$$f_2 = -\frac{a}{1+2a}\frac{\partial^2 u}{\partial x^2} \tag{11b}$$

$$f_4 = \frac{a}{12}\frac{10a-1}{(1+2a)^2}\frac{\partial^4 u}{\partial x^4} \tag{11c}$$

etc. Next, these expressions are substituted into Equations (8) and (9), which can in turn be substituted into Equation (3). This gives

$$\rho\frac{\partial^2}{\partial t^2}\left(u - m_2 l^2\frac{\partial^2 u}{\partial x^2} + m_4 l^4\frac{\partial^4 u}{\partial x^4} - \cdots\right) = E\frac{\partial^2}{\partial x^2}\left(u - s_2 l^2\frac{\partial^2 u}{\partial x^2} + s_4 l^4\frac{\partial^4 u}{\partial x^4} - \cdots\right) \tag{12}$$

in which the (dimensionless) higher-order coefficients are expressed in terms of a as

$$m_2 = \frac{a}{1+2a} \tag{13a}$$

$$m_4 = \frac{a(10a-1)}{12(1+2a)^2} \tag{13b}$$

$$s_2 = \frac{10a-1}{12(1+2a)} \tag{13c}$$

$$s_4 = \frac{1-56a+244a^2}{360(1+2a)^2} \tag{13d}$$

Note that Equation (5) is retrieved from Equation (12) by setting $a = 0$.

As can be seen from Equation (12), not only higher-order stiffness terms (preceded by s_2, s_4, etc.) have appeared due to the continualisation process, but also higher-order inertia terms (preceded by m_2, m_4, etc.) More precisely, the two types of terms always appear in pairs: every higher-order stiffness term of a certain power of l is accompanied by a higher-order inertia term of the same power of l. The higher-order inertia terms contain mixed space-time derivatives of the displacement, cf. (Vardoulakis and Aifantis 1994; Rubin et al. 1995; Chen and Fish 2001).

According to conventional terminology, a so-called *second-order* model is obtained if a truncation after the l^2-terms is made. The underlying stress-strain relation can then be found as

$$\sigma = E\left(\varepsilon - s_2 l^2 \frac{\partial^2 \varepsilon}{\partial x^2} + m_2 \frac{l^2}{c_e^2}\frac{\partial^2 \varepsilon}{\partial t^2}\right) \tag{14}$$

in which $c_e = \sqrt{E/\rho}$ is the classical wave velocity. Whereas the higher-order stiffness term is preceded by an internal length scale squared, the higher-order inertia term is preceded by an internal time scale squared.

3.3 *Two-dimensional formulation*

For reasons of simplicity and clarity, the above derivations have been made for the one-dimensional case. However, the same alternative continualisation procedure can be applied to multi-dimensional media. A host of possibilities emerge for the discrete medium that is considered, e.g. hexagonal or rectangular configurations, the inclusion of translational springs in longitudinal/transverse directions and/or rotational springs, etc. In principle, all these discrete media are anisotropic; however, isotropy of the resulting continuum is normally preferred. Below, it will be briefly sketched how an isotropic continuum can be obtained using the newly proposed continualisation method.

Starting point is the hexagonal lattice as shown in Figure 2 (left). Only translational springs in the longitudinal direction are considered. A classical continuum derived from this lattice is known to be isotropic (Kunin 1983). However, this does not hold for higher-order continua derived from this discrete structure, irrespective of whether the standard continualisation or the alternative continualisation is used. In order to arrive at an isotropic medium, the next ring of neighbouring particles is considered, as depicted in Figure 2 (right). If the stiffness of the long-range springs is taken as 1/9 of the stiffness of the short-range springs, then both continualisation approaches lead to an isotropic continuum.

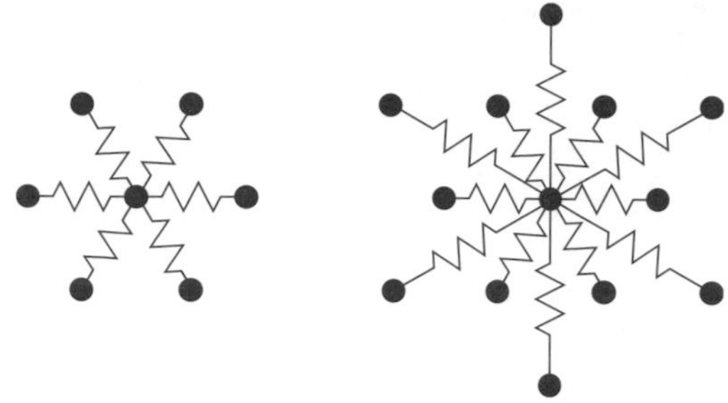

Figure 2. Nearest neighbour interaction (left) and next nearest neighbour interaction (right).

The resulting equations of motion, using tensor notation and with a truncation after the second-order terms, then read

$$\rho\ddot{u}_i - \rho l^2 \frac{3a(\sqrt{3}+1)}{4a(\sqrt{3}+3)+2}\ddot{u}_{i,kk} = \frac{E}{2}(u_{i,jj} + 2u_{j,ij}) \\ -El^2\left(\frac{2a(11\sqrt{3}+9)-1}{32(a(2\sqrt{3}+6)+1)}u_{i,jj} + \frac{4a(10\sqrt{3}+6)-4}{32(a(2\sqrt{3}+6)+1)}u_{j,ij}\right)_{kk} \tag{15}$$

where the first index denotes the corresponding displacement component, an index following a comma denotes a spatial derivative and a summation over repeated indices is implied. Note that both higher-order terms are ∇^2-versions of the classical terms. The weighting parameter a is taken inversely proportional with the radial distance from the central particle.

More details on the derivation and a thorough investigation of the two-dimensional field equations will be given in a forthcoming contribution.

4 DISPERSION ANALYSIS

An analysis of wave dispersion is performed by inserting a trial solution of the general harmonic form $u = \hat{u}\exp(i(\omega t - kx))$ into Equation (12). Here, $\hat{u}$ is the amplitude of the wave, ω is the frequency and k is the wave number. This yields

$$\frac{\omega^2}{\omega_0^2} = l^2k^2 \frac{1 + s_2 l^2 k^2 + s_4 l^4 k^4}{1 + m_2 l^2 k^2 + m_4 l^4 k^4} \tag{16}$$

in which $\omega_0^2 = K/M = c_e^2/l^2$ is the squared frequency of the discrete medium as depicted in Figure 1. For a realistic dispersive behaviour, two constraints are imposed on the model (Metrikine and Askes 2002):

1. The frequency must not be imaginary, otherwise the model will be unstable. This implies that the right-hand-side in Equation (16) must be positive.

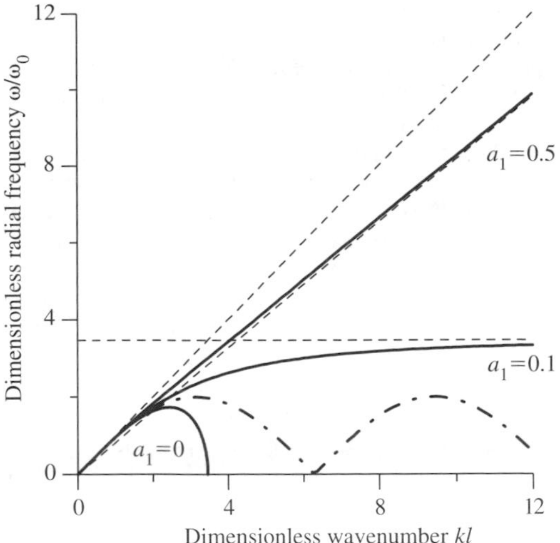

Figure 3. Dispersion curves for second-order model.

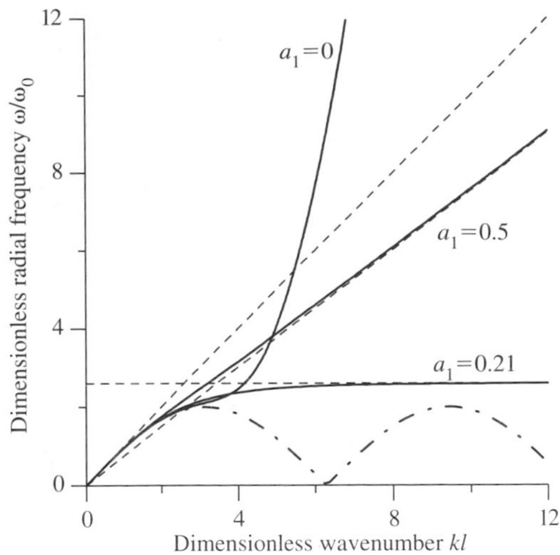

Figure 4. Dispersion curves for fourth-order model.

2. The group velocity $c_{\text{gr}} = \partial\omega/\partial k$ must not exceed the classical wave velocity c_e, in order to avoid an unrealistically fast propagation of energy.

Closer inspection of Equation (13) reveals that for the second-order model (i.e. setting $m_4 = s_4 = 0$) the first constraint is met by requiring $a > 0.1$, since it is assumed that a is non-negative. Similarly, it is found that for the range $0.1 < a < 1$ the group velocity never exceeds the classical wave velocity. In Figure 3 a number of dispersion curves for the second-order model are plotted, taking various values for a. The dash-dotted line represents the dispersion curve for the underlying discrete model. Taking $a = 0$, the results of the standard continualisation method are retrieved, and it can be seen that imaginary frequencies are obtained for $kl > \sqrt{12}$. Thus, for $a = 0$ the waves with $kl > \sqrt{12}$ are unstable. Instabilities are found for every value of a in the range $0 \leq a < 0.1$. Taking $a = 0.1$ leads to a horizontal asymptote (denoted by a dashed line in Figure 3), which implies that frequencies higher than this asymptotic value do not propagate. Taking larger values of a leads to non-horizontal asymptotes, which are located *below* the straight line $\omega = kc_e$ that corresponds to the classical continuum (also denoted by a dashed curve in Figure 3).

The same investigations are performed for the fourth-order model (i.e. taking into account the m_4-term and the s_4-term in Equation (16)). After some algebra, it is found that imaginary frequencies are avoided by taking $a \geq (14 + 3\sqrt{15})/122 \approx 0.21$. Again, this lower bound with the original upper bound $a = 1$ leads to group velocities that never exceed the classical wave velocity. Figure 4 shows the dispersion curves of the fourth-order model for a range of values of a. Again, the dash-dotted curve denotes the discrete model. Taking $a = 0$ leads to the dispersion curve of the standard continualisation method, and it can be seen that an infinite slope of the curve (i.e. an infinite group velocity) is obtained for $kl \to \infty$. Although not as calamitous as the imaginary frequencies that occur in the second-order model with $a = 0$ infinite group velocities are nonetheless undesirable. They occur for the range $0 \leq a < (29 - 12\sqrt{5})/242 \approx 0.009$, whereas imaginary frequencies and instabilities are encountered for the fourth-order model in the range $0.009 \approx (29 - 12\sqrt{5})/242 < a < (14 + 3\sqrt{15})/122 \approx 0.21$ (see (Metrikine and Askes 2002) for further details). Taking $a = 0.21$ a horizontal asymptote is obtained, and larger values of a lead to non-horizontal asymptotes.

5 EXAMPLES

Next, the response of the models is assessed in boundary-value problems and initial-value problems. The one-dimensional equations for the static case can be solved analytically in a straightforward manner, whereas Laplace transforms can be used for the dynamic case. However, it is also of interest to investigate the behaviour of the model in a numerical context. To this end, an Element-Free Galerkin (EFG) implementation has been made. EFG shape functions can be easily formulated with an arbitrary order of continuity (Belytschko et al. 1994; Belytschko et al. 1996), therefore the EFG method is pre-eminently suited to discretise higher-order gradient models (Askes et al. 2000; Askes et al. 2002; Askes and Metrikine 2002). For details on the EFG implementation as well as on the formulation of the boundary conditions the reader is referred to (Metrikine and Askes 2002; Askes and Metrikine 2002).

5.1 *Statics*

A bar with a central imperfection as shown in Figure 5 is studied. Static loading is assumed, i.e. the inertia terms have been neglected from the field equations. Furthermore, it has been taken that $L/w = 10$ and $L/l = 10$, with $2L$ the length of the bar and $2w$ the width of the imperfect zone, in which the cross-sectional area is reduced by 10%.

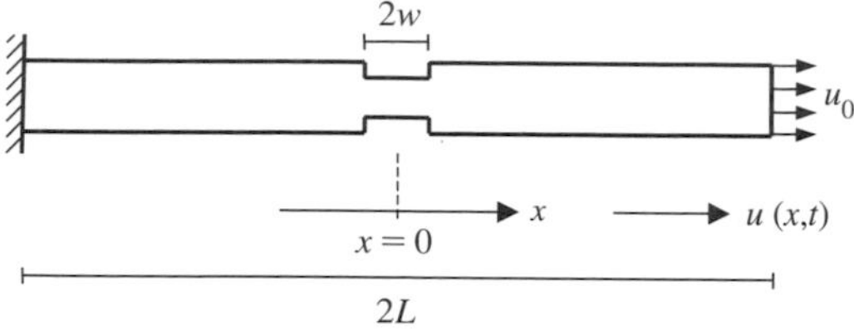

Figure 5. Static loading of finite bar with central imperfection.

In Figures 6 and 7 the numerically obtained strain profiles are plotted along the bar for the second-order model and the fourth-order model, respectively. Values for a have been taken that lie within the bounds derived through dispersion analysis, see Section 4. Taking $a = 0.1$ in the second-order model eliminates all higher-order effects, since then $s_2 = 0$, cf. Equation (13). This value of a leads to the piecewise linear response of the classical continuum and will not be considered here. All stable values of a lead to strain profiles in which the discontinuity at the imperfection is smoothened. Larger values of a lead to more smoothening of the imperfection. This holds for both the second-order model and the fourth-order model.

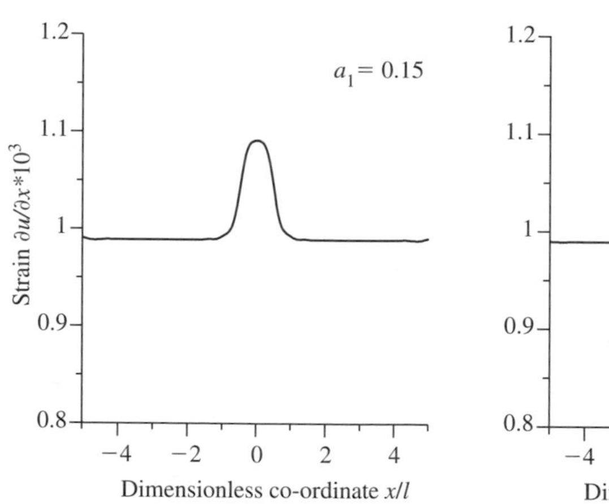

Figure 6. Static response of second-order model – $a = 0.15$ (left) and $a = 0.5$ (right).

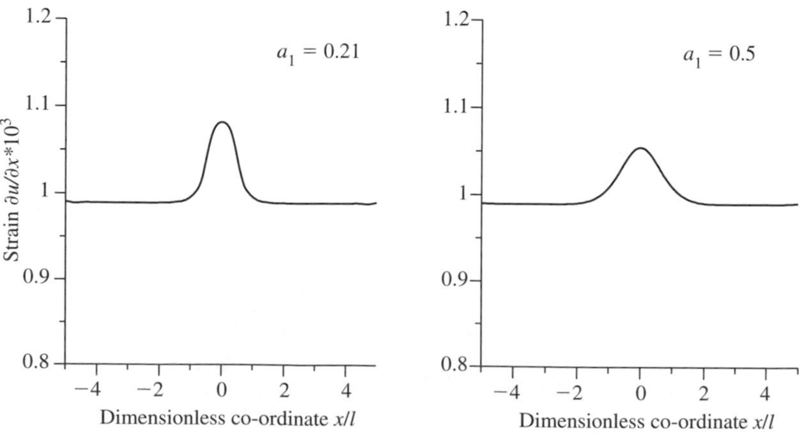

Figure 7. Static response of fourth-order model – $a = 0.21$ (left) and $a = 0.05$ (right).

Qualitatively, there are no differences between the response of the second-order model and that of the fourth-order model.

The smoothening behaviour of the models is similar to that of the gradient elasticity model proposed by Aifantis, see e.g. (Altan and Aifantis 1997; Aifantis 1999). It has been shown that models of this type can be used in strain-softening computation *without further enhancement*, i.e. the higher-order strain gradient that is present in the elastic part of the model suffices to guarantee well-posedness in the entire loading process, and no additional amendments are necessary (Chang et al. 2002). In contrast, many strain-gradient models from the literature used to simulate softening phenomena employ the higher-order terms in the history variable or in the state variable that drives the inelastic processes (Pijaudier-Cabot and Bažant 1987; Pamin 1994; Peerlings et al. 1996).

5.2 *Dynamics*

A semi-infinite bar under dynamic loading conditions is studied, see Figure 8. In a classical continuum, the depicted impact load generates a Dirac-type strain field and a Heaviside-type displacement field. However, in the higher-order models as developed above, wave dispersion is present as discussed in Section 4.

In Figures 9 and 10 the analytically obtained propagating displacement profiles are plotted for the second-order model and the fourth-order model, respectively.

Figure 8. Impact loading of semi-infinite bar.

Different values for a have been selected, each of which belongs to the stable intervals (i.e. no imaginary frequencies are triggered). Three different time instants are shown, and it can be clearly seen that in all cases the wave shape changes with ongoing time. This is the manifestation of wave dispersion. The wave dispersion is more pronounced for *smaller* values of a in the second-order model; for the fourth-order model the situation is less clear. This is somewhat in contrast with the static case, where gradient effects are more strongly present for *larger* values of a.

Considering the response of the second-order model (Figure 9), it is observed that the displacement field is much more wavy when $a = 0.1$ than when $a = 0.5$. It must be realised that $a = 0.1$ is the lower bound on a, and for $a = 0.1$ the phase velocity $c = \omega/k$ approaches zero for $kl \to \infty$. This means that the higher the frequency of the wave, the lower the propagation velocity. Indeed, as can be seen from Figure 9 (left), the frequency of the response decreases moving from the point of excitation ($x = 0$) towards the wave front.

For the fourth-order model the response of the case $a = 0$ has been plotted in Figure 10 (right). Although this value of a is outside the range as motivated in Section 4, it is still worthwhile to study the effect of the infinitely high group velocity. Here, the phase velocity $c = \omega/k$ grows unboundedly with the wave number, and indeed the classical wave velocity $c_e = \sqrt{E/\rho}$ is exceeded for the higher wave numbers. As such, the high-frequency waves that are triggered by the Dirac-type impact loading travel (much) faster than the wave front. Although it does not have a destabilising influence on the overall response, this high-frequency "messenger" is physically not realistic. On the other hand, the analysis with $a = 0.5$ yields realistic results that are in close correspondence with the results obtained for the second-order model.

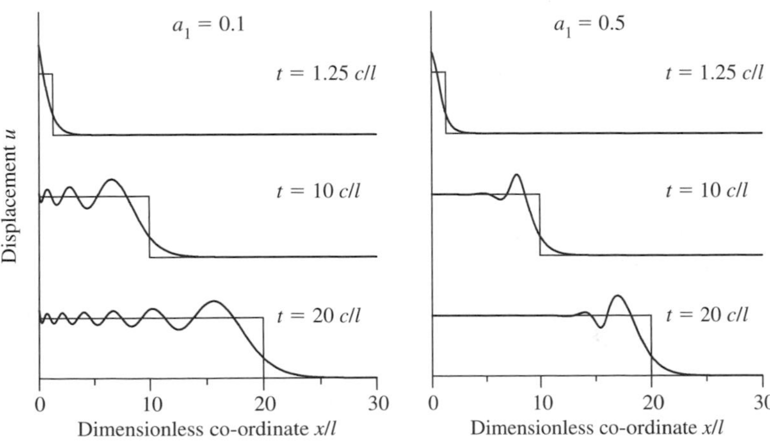

Figure 9. Dynamic response of second-order model (thick line) and of classical continuum (thin line) – $a = 0.1$ (left) and $a = 0.5$ (right).

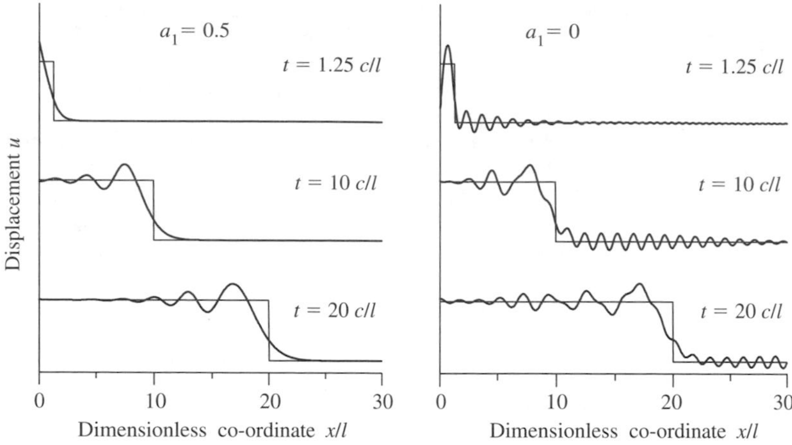

Figure 10. Dynamic response of fourth-order model (thick line) and of classical continuum (thin line) – $a = 0.5$ (left) and $a = 0$ (right).

6 CONCLUSIONS

In this contribution, a new continualisation method has been presented by which higher-order gradient models can be derived from a discrete medium. Thus, a close link with the underlying microstructure is maintained. Key difference with standard continualisation procedures is that here a nonlocal relation between the discrete degrees of freedom and the continuous displacement field is assumed. This nonlocality is reflected in a dimensionless scalar weighting parameter. Other parameters that enter the continuum equations of motion are the macroscopic stiffness, the macroscopic mass density and a length parameter related to the characteristic size of the microstructure.

A peculiarity of the resulting models is that each higher-order stiffness term is accompanied by a higher-order inertia term; as such, the models are denoted as *dynamically consistent*. Whereas the higher-order stiffness terms suffice to eliminate singularities from the strain field, the higher-order inertia terms are needed to preserve a physically realistic dispersive behaviour. Thus, the proposed models combine the advantages of earlier higher-order gradient models from the literature.

The weighting parameter that is introduced in the new continualisation procedure becomes manifest as a higher-order time scale that accompanies the higher-order inertia term. In a previous contribution, a procedure has been suggested by which this parameter can be measured (Askes and Metrikine 2002). It is noted that in statics the gradient effects are largest for large values of the weight, whereas in dynamics gradient effects are largest for small values of the weight.

The derived second-order model and fourth-order model behave qualitatively the same, in statics as well as in dynamics. On the one hand, this can be considered as an advantage of the proposed continualisation method (especially compared to the standard continualisation procedure), since the various truncations converge monotonically. On the other hand, it should be concluded that there is little added value of the fourth-order model over the second-order model. Thus, for numerical implementation the second-order model is the appropriate choice due to less strict continuity requirements.

In a future contribution, the derived second-order model will be scrutinized in a two-dimensional context. Suitable spatial and temporal discretisation schemes must be devised for use in a numerical context, whereby meshless methods or finite element methods should be considered for the spatial discretisation.

REFERENCES

Aifantis, E. (1987). The physics of plastic deformation. *International Journal of Plasticity 3*, 211–247.

Aifantis, E. (1999). Gradient deformation models at nano, micro, and macro scales. *ASME Journal of Engineering Materials and Technology 121*, 189–202.

Altan, B. and E. Aifantis (1997). On some aspects in the special theory of gradient elasticity. *Journal of the Mechanical Behavior of Materials* 8, 231–282.

Askes, H. and A. Metrikine (2002). One-dimensional dynamically consistent gradient elasticity models derived from a discrete microstructure. Part 2: Static and dynamic response. *European Journal of Mechanics A/Solids 21*, 573–588.

Askes, H., J. Pamin, and R. de Borst (2000). Dispersion analysis and Element-Free Galerkin solutions of second and fourth-order gradient-enhanced damage models. *International Journal for Numerical Methods in Engineering 49*, 811–832.

Askes, H., A. Suiker, and L. Sluys (2002). A classification of higher-order strain gradient models — linear analysis. *Archive of Applied Mechanics 72*, 171–188.

Belytschko, T., Y. Krongauz, D. Organ, M. Fleming, and P. Krysl (1996). Meshless methods: an overview and recent developments. *Computer Methods in Applied Mechanics and Engineering 139*, 3–47.

Belytschko, T., Y. Y. Lu, and L. Gu (1994). Element-free Galerkin methods. *International Journal for Numerical Methods in Engineering 37*, 229–256.

Chang, C., H. Askes, and L. Sluys (2002). Higher-order strain/higher-order stress gradient models derived from a discrete microstructure, with application to fracture. *Engineering Fracture Mechanics 69*, 1907–1924.

Chang, C. and J. Gao (1995). Second-gradient constitutive theory for granular material with random packing structure. *International Journal of Solids and Structures 32*, 2279–2293.

Chen, W. and J. Fish (2001). A dispersive model for wave propagation in periodic heterogeneous media based on homogenization with multiple spatial and temporal scales. *ASME Journal of Applied Mechanics 68*, 153–161.

Kunin, I. (1982). *Elastric media with microstructure. Volume 1: One-dimensional model*s. Berlin: Springer.

Kunin, I. (1983). *Elastric media with microstructure. Volume 2: Three-dimensional model*s. Berlin: Springer.

Lasry, D. and T. Belytschko (1988). Localization limiters in transient problems. *International Journal of Solids and Structures 24*, 581–597.

Metrikine, A. and H. Askes (2002). One-dimensional dynamically consistent gradient elasticity models derived from a discrete microstructure. Part 1: Generic formulation. *European Journal of Mechanics A/Solids* 21, 555–572.

Mühlhaus, H.-B. and E. Aifantis (1991). A variational principle for gradient plasticity. *International Journal of Solids and Structures* 28, 845–857.

Mühlhaus, H.-B. and F. Oka (1996). Dispersion and wave propagation in discrete and continuous models for granular materials. *International Journal of Solids and Structures 33*, 2841–2858.

Pamin, J. (1994). *Gradient-dependent plasticity in numerical simulation of localization phenomena*. Dissertation, Delft University of Technology.

Peerlings, R., R. de Borst, W. Brekelmans, and J. de Vree (1996). Gradient enhanced damage for quasi-brittle materials. *International Journal for Numerical Methods in Engineering 39*, 3391–3403.

Pijaudier-Cabot, G. and Z. Bažant (1987). Nonlocal damage theory. *ASCE Journal of Engineering Mechanics 113*, 1512–1533.

Rubin, M., P. Rosenau, and O. Gottlieb (1995). Continuum model of dispersion caused by an inherent material characteristic length. *Journal of Applied Physics 77*, 4054–4063.

Suiker, A., R. de Borst, and C. Chang (2001). Micro-mechanical modelling of granular material. Part 1: derivation of a second-gradient micro-polar constitutive theory. *Acta Mechanica 149*, 161–180.

Triantafyllidis, N. and E. Aifantis (1986). A gradient approach to localization of deformation. I. Hyperelastic materials. *Journal of Elasticity 16*, 225–237.

Unger, D. and E. Aifantis (2000a). Strain gradient elasticity theory for antiplane shear cracks. Part I: Oscillatory displacements. *Theoretical and Applied Fracture Mechanics 34*, 243–252.

Unger, D. and E. Aifantis (2000b). Strain gradient elasticity theory for antiplane shear cracks. Part II: Monotonic displacements. *Theoretical and Applied Fracture Mechanics 34*, 253–265.

Vardoulakis, I. and E. Aifantis (1994). On the role of microstructure in the behavior of soils: effects of higher order gradients and internal inertia. *Mechanics of Materials 18*, 151–158.

Computational Modelling of Concrete Structures, Bićanić et al. (eds)
© 2003 Swets & Zeitlinger, Lisse, ISBN 90 5809 536 3

A micromechanical approach to the strength criterion of composite materials

J.F. Barthélémy & L. Dormieux
LMSGC, ENPC, Champs-Sur-Marne, France

E. Lemarchand
LML, Villeneuve d'Ascq, France

ABSTRACT: At the microscopic scale, concrete can be considered as a frictional matrix (cement paste) surrounding rigid inclusions (aggregate or sand inclusions). A theoretical approach to the strength criterion of such a composite material has been recently proposed in (Lemarchand et al. 2002). This approach exploits a nonlinear homogenization scheme based on the definition of an equivalent nonlinear behavior for the matrix. Whereas it provides good results as regards the strength criterion, it is not completely satisfactory as regards the local flow rule of the Drucker-Prager frictional type matrix. Our contribution aims at improving this approach in order to ensure the consistency of the local strains with the flow rule of a Drucker-Prager material and, in particular, the dilatancy which characterizes such a flow rule. For this purpose, a nonlinear equivalent viscoelastic behavior is introduced and a nonlinear homogenization scheme is implemented so as to determine the macroscopic strength criterion. Assuming perfect adherence between matrix and inclusions, the method yields a macroscopic strength criterion of Drucker-Prager type with increased friction angle and cohesion.

1 INTRODUCTION

Many micromechanical models have been developed to exhibit the elastic properties of a composite material (see Zaoui 2002). Eshelby's solution of the inhomogeneity problem proves to be an efficient tool to derive estimates of the homogenized properties in the framework of linear elasticity. The determination of the macroscopic strength criterion of a composite material is fundamentally a nonlinear problem which can not be exactly solved on any kind of geometry of microstructure. The purpose of this paper is to take advantage of a nonlinear homogenization technique based on an equivalent viscoelastic behavior of the matrix to estimate the macroscopic strength criterion of a material reinforced by rigid inclusions. After presenting the main lines of the theoretical background, two methods of nonlinear homogenization are implemented on a r.e.v. with spherical rigid inclusions and macroscopic strength criteria are derived.

2 THEORETICAL BACKGROUND

2.1 *Limit analysis problem*

We consider a representative element volume (r.e.v.) Ω of a two-phase material (see Fig. 1) composed of

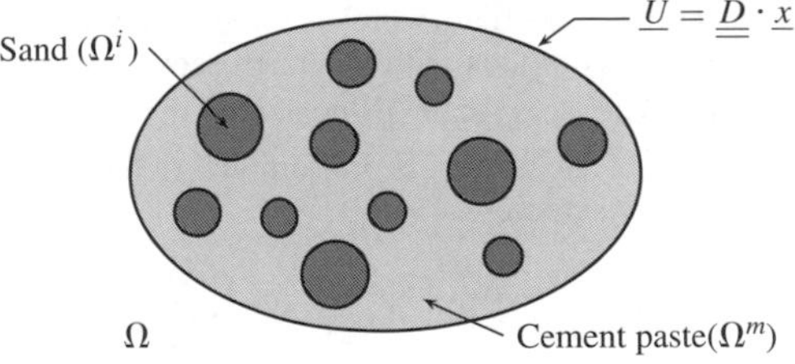

Figure 1. Representative element volume (r.e.v.) Ω.

a matrix phase Ω^m (cement paste) surrounding rigid inclusions Ω^i (sand). φ denotes the volume fraction of inclusions ($\varphi = |\Omega^i|/|\Omega|$).

The Cauchy stress tensor field at the microscopic scale is denoted by $\underline{\underline{\sigma}}$. The strength of the inclusions is infinite. No velocity jump between the inclusions and the matrix is allowed. The strength criterion of the matrix is the one of Drucker-Prager i.e. the set G of admissible stress states is defined by:

$$\underline{\underline{\sigma}} \in G \quad \Leftrightarrow \quad g(\underline{\underline{\sigma}}) = \sigma_d + T(\sigma_m - H) \leq 0 \qquad (1)$$

with $\sigma_m = \mathrm{tr}\,\underline{\underline{\sigma}}/3$, $\underline{\underline{\sigma}}_d - \underline{\underline{\sigma}} - \sigma_m \underline{\underline{1}}$ and $\sigma_d = \sqrt{\underline{\underline{\sigma}}_d : \underline{\underline{\sigma}}_d}$. The relevance of Drucker-Prager criterion implies that the friction coefficient T is subjected to the limitation $T \leq \sqrt{3/2}$ see (Desrues 2002)). As G is a convex domain, it can be characterized by its support function $\pi(\underline{\underline{d}}) = \sup\{\underline{\underline{\sigma}} : \underline{\underline{d}} | \underline{\underline{\sigma}} \in G\}$, which expression for the Drucker-Prager criterion is:

$$\begin{cases} \pi(\underline{\underline{d}}) = +\infty & \text{si } d_v < T\, d_d \\ \pi(\underline{\underline{d}}) = H d_v & \text{si } d_v \geq T\, d_d \end{cases} \tag{2}$$

where $d_v = \mathrm{tr}\,\underline{\underline{d}}$, $\underline{\underline{d}}_d = \underline{\underline{d}} - 1/3 d_v \underline{\underline{1}}$ and $d_d = \sqrt{\underline{\underline{d}}_d : \underline{\underline{d}}_d}$. The r.e.v. is submitted to homogeneous strain rate boundary conditions i.e. the set of kinematically admissible velocity fields is defined by:

$$\mathcal{K}(\underline{\underline{D}}) = \{\underline{u} : \Omega \to \mathbb{R}^3 | \, \underline{u} = \underline{\underline{D}} \cdot \underline{x} \quad \text{on } \partial\Omega \\ \text{and } \underline{u} = \underline{0} \text{ in } \Omega^i\} \tag{3}$$

The limit analysis problem aims at characterizing the set G^{hom} of macroscopic stress states which comply with the microscopic strength. Using the notation $\langle \chi \rangle_\omega$ for the average of χ over the domain ω, G^{hom} is defined as:

$$G^{hom} = \left\{ \underline{\underline{\Sigma}} \,|\, \exists \underline{\underline{\sigma}}(\underline{x}), \quad \begin{vmatrix} \mathrm{div}\, \underline{\underline{\sigma}} = 0 \\ \underline{\underline{\Sigma}} = \langle \underline{\underline{\sigma}} \rangle_\Omega \end{vmatrix}, \\ \forall \underline{x} \in \Omega^m \quad g(\underline{\underline{\sigma}}(\underline{x})) \leq 0 \right\} \tag{4}$$

In (4), $\underline{\underline{\Sigma}}$ denotes the macroscopic stress state and $\underline{\underline{\sigma}}$ is a microscopic stress field, which is statically admissible with $\underline{\underline{\Sigma}}$ and complies with the microscopic strength criterion. The stress state $\underline{\underline{\Sigma}}$ located on the boundary ∂G^{hom} where the normal is proportional to the virtual strain rate $\underline{\underline{D}}$ satisfies:

$$\underline{\underline{\Sigma}} : \underline{\underline{D}} = \inf_{\underline{u} \in \mathcal{K}(\underline{\underline{D}})} \langle \pi(\underline{\underline{d}}) \rangle_\Omega \tag{5}$$

where $\underline{\underline{d}} = (\underline{\underline{\mathrm{grad}}}\, \underline{u} + {}^t\underline{\underline{\mathrm{grad}}}\, \underline{u})/2$ and $\pi(\underline{\underline{d}}) = 0$ in Ω^i. Finding an optimal velocity field in (5) for every direction of $\underline{\underline{D}}$ boils down to the definition of the macroscopic strength criterion.

2.2 *Equivalent viscoelastic problem*

Resolving the minimization problem (5) for a r.e.v. with a random distribution of inclusions is not an easy task. However, it is possible to define a ficticious viscoelastic problem, which will formally provide a solution for the problem (5). Even if this ficticious problem can not be solved either, one can obtain an approximation of its solution by means of nonlinear homogenization techniques.

Thus, this paragraph is devoted to the definition of a ficticious well adapted viscoelastic behavior for the matrix. The relation between $\underline{\underline{\sigma}}$ and $\underline{\underline{d}}$ is derived from a convex potential $\psi_a(\underline{\underline{d}})$ where a is a parameter tending towards 0.

For each value of a, a viscoelastic problem can be defined:

$$\begin{cases} \mathrm{div}\, \underline{\underline{\sigma}} = 0 & \text{in } \Omega \\ \underline{\underline{d}} = (\underline{\underline{\mathrm{grad}}}\, \underline{u} + {}^t\underline{\underline{\mathrm{grad}}}\, \underline{u})/2 & \text{in } \Omega \\ \underline{\underline{\sigma}} = \partial\psi_a / \partial\underline{\underline{d}} & \text{in } \Omega^m \\ \underline{u} = \underline{\underline{D}} \cdot \underline{x} & \text{on } \partial\Omega \\ \underline{u} = \underline{0} & \text{in } \Omega^i \end{cases} \tag{6}$$

A solution of (6) is denoted by $(\underline{u}_a, \underline{\underline{d}}_a, \underline{\underline{\sigma}}_a)$. ψ_a chosen in order to satisfy two conditions:

1. $\underline{\underline{\sigma}}_a = \partial\psi_a / \partial\underline{\underline{d}}$ should be located on the boundary of the microscopic criterion.
2. ψ_a tends towards π defined in (2) when a tends towards 0.

Condition (1) ensures that the macroscopic stress state $\underline{\underline{\Sigma}} = \lim_{a \to 0} \langle \underline{\underline{\sigma}}_a \rangle_\Omega$ belongs to G^{hom} while condition (2) implies that it is located on the boundary of G^{hom}. Indeed, the convexity of ψ_a gives:

$$\psi_a(\underline{\underline{d}}_a) \leq \frac{\partial\psi_a}{\partial\underline{\underline{d}}}(\underline{\underline{d}}_a) : \underline{\underline{d}}_a \quad (= \underline{\underline{\sigma}}_a : \underline{\underline{d}}_a) \tag{7}$$

Using Hill's lemma, (7) implies:

$$\langle \psi_a(\underline{\underline{d}}_a) \rangle_\Omega \leq \langle \underline{\underline{\sigma}}_a \rangle_\Omega : \underline{\underline{D}} \tag{8}$$

$\underline{\underline{d}}_a$ tends towards a tensor $\underline{\underline{d}}$ deriving from a kinematically admissible velocity field $\underline{u}$ and it is assumed that $\langle \psi_a(\underline{\underline{d}}_a) \rangle_\Omega$ tends towards $\langle \pi(\underline{\underline{d}}) \rangle_\Omega$. When a tends towards 0, it follows that (8) becomes:

$$\langle \pi(\underline{\underline{d}}) \rangle_\Omega \leq \underline{\underline{\Sigma}} : \underline{\underline{D}} \tag{9}$$

Since $\underline{\underline{\Sigma}}$ belongs to G^{hom}, (9) proves that $\underline{\underline{\Sigma}}$ is located on the boundary of G^{hom} where the normal is proportional to $\underline{\underline{D}}$ and $\underline{u}$ is an optimal mechanism in the problem (5).

We shall choose a convex potential ψ_a depending on d_v and d_d. Thus, the mean and deviatoric stresses respectively read $\sigma_m = \partial\psi_a / \partial d_v$ and $\sigma_d = \partial\psi_a / \partial d_d$. Hence, $g(\partial\psi_a / \partial\underline{\underline{d}}) = 0$ means:

$$\frac{\partial\psi_a}{\partial d_d} + T\left(\frac{\partial\psi_a}{\partial d_v} - H\right) = 0 \tag{10}$$

which solution writes in the form:

$$\psi_a(d_v, d_d) = f_a(\delta) + H\, d_v \tag{11}$$

with $\delta = d_v - T\, d_d$ and f_a a function to be defined. If f_a is a convex decreasing function, it can be shown

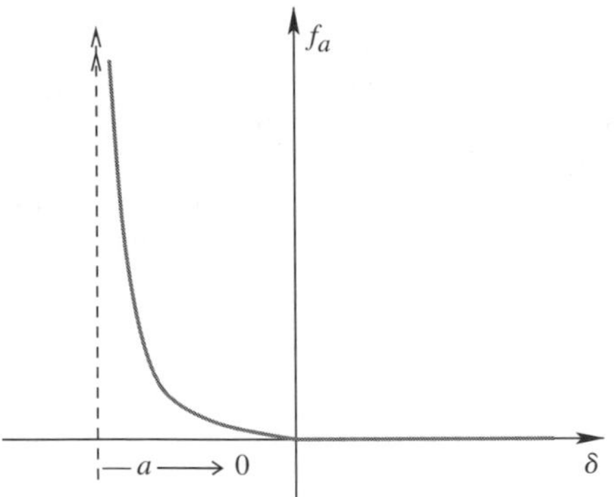

Figure 2. Function f_a.

that ψ_a is convex. Besides, as ψ_a must tend towards π given in (2) (condition (2)), one builds a family of decreasing convex functions $(f_u)_{u\in\mathbb{R}_+}$ defined on$]-a, +\infty[$, of class at least C^2 such that $f_a(\delta) = 0$ when $\delta \geq 0$ and $\lim_{\delta\to -a^+} f_a(\delta) = +\infty$ (see Fig. 2).

2.3 *Nonlinear homogenization approach*

The stress-strain relationship derived from the potential (11) can be put in the following particular form, involving shear and bulk moduli μ and k and a spherical prestress $\sigma_p \underline{\underline{1}}$:

$$\underline{\underline{\sigma}} = \underbrace{\frac{f_a'(\delta)}{\delta}}_{k} d_v \underline{\underline{1}} + \underbrace{\left(\frac{-f_a'(\delta)T}{d_d}\right)}_{2\mu} \underline{\underline{d}}_d + \underbrace{H - \frac{f_a'(\delta)\, d_d T}{\delta}}_{\sigma_p} \underline{\underline{1}} \tag{12}$$

The nonlinearity of the behavior is taken into account through the dependence of k, μ and σ_p on d_v and d_d. Except for the stress state $\underline{\underline{\sigma}} = H\underline{\underline{1}}$, δ is a negative number. This ensures that k and μ in (12) are positive. Moreover, it can also be deduced from the definition of f_a that, if $d_d \neq 0$:

$$\lim_{a\to 0} \delta = 0 \;\Leftrightarrow\; d_v \underset{a\to 0}{\approx} T\, d_d \tag{13}$$

The resolution of the problem (6) leads to a non-uniform strain rate field $\underline{\underline{d}}$. Therefore the moduli and prestress defined in (12) are non-uniform. For this reason, the classical homogenization schemes can not be directly applied because they only deal with heterogeneous materials composed of homogeneous phases. The idea is then to turn the problem into a simplified one that can take advantage of the linear homogenization schemes. This strategy is divided into three steps (see Suquet 1997):

1. The concept of effective strain rate is introduced. It aims at capturing in a simplified manner the effect of the loading on the nonlinear stiffness. The effective strain rate tensor, denoted by $\underline{\underline{d}}^{eff}$ is expected to be an appropriate average of $\underline{\underline{d}}$ over the matrix. It therefore depends on the loading $\underline{\underline{D}}$. More precisely, $k(\underline{\underline{d}}(\underline{x}))$, $\mu(\underline{\underline{d}}(\underline{x}))$ and $\sigma_p(\underline{\underline{d}}(\underline{x}))$ which depend on the location in the matrix are replaced by the uniform values $\mu(\underline{\underline{d}}^{eff})$, $k(\underline{\underline{d}}^{eff})$ and $\sigma_p(\underline{\underline{d}}^{eff})$. In other words, the nonlinear material with heterogeneous stiffness is replaced by a homogeneous "equivalent" linear material, which stiffness however depends on the loading applied to the r.e.v.: $\mu^{eq} = \mu(\underline{\underline{d}}^{eff})$, $k^{eq} = k(\underline{\underline{d}}^{eff})$ and $\sigma_p^{eq} = \sigma_p(\underline{\underline{d}}^{eff})$.
2. Linear homogenization schemes can be applied to the determination of the overall response of the r.e.v. made up on the "equivalent" material introduced in the first step together with the inclusions. The loading is defined by the macroscopic strain tensor $\underline{\underline{D}}$ and the equivalent microscopic prestress σ_p^{eq}. The microscopic state equation of the matrix reads:

$$\underline{\underline{\sigma}} = k^{eq} d_v \underline{\underline{1}} + 2\mu^{eq} \underline{\underline{d}}_d + \sigma_p^{eq} \underline{\underline{1}} \tag{14}$$

while $\underline{\underline{d}} = \underline{\underline{0}}$ in the inclusions. The solution of the equivalent linear viscoelastic problem can be seen as the superposition of those of two elementary problems respectively defined by ($\underline{\underline{D}}^{(1)} = \underline{\underline{0}}$, $\sigma_p^{eq(1)} = \sigma_p^{eq}$) and ($\underline{\underline{D}}^{(2)} = \underline{\underline{D}}$, $\sigma_p^{eq(2)} = 0$). In the case of rigid inclusions, the solution of the first one simply reads $\underline{\underline{D}}^{(1)} = \underline{\underline{d}}^{(1)} = \underline{\underline{0}}$ in the matrix as well as in the inclusions and $\underline{\underline{\Sigma}}^{(1)} = \underline{\underline{\sigma}}^{(1)} = \sigma_p^{eq} \underline{\underline{1}}$. In the second one, the local strain rate field $\underline{\underline{d}}^{(2)}$ is related to $\underline{\underline{D}}$ by means of a concentration tensor $\mathbb{A}(\underline{x})$:

$$\underline{\underline{d}}^{(2)}(\underline{x}) = \mathbb{A}(\underline{x}) : \underline{\underline{D}} \quad (\langle \mathbb{A} \rangle_\Omega = \mathbb{I}) \tag{15}$$

The local constitutive law reads:

$$\underline{\underline{\sigma}}^{(2)}(\underline{x}) = \mathbb{C}(\underline{x}) : \underline{\underline{d}}^{(2)}(\underline{x}) \quad \text{with} \begin{cases} \mathbb{C} = \mathbb{C}^{eq} & \text{in } \Omega^m \\ \mathbb{C}^i \to \infty & \text{in } \Omega^i \end{cases} \tag{16}$$

The homogenized elasticity tensor $\mathbb{C}^{hom}(k^{eq}, \mu^{eq})$ relates $\underline{\underline{\Sigma}}^{(2)} = \langle \underline{\underline{\sigma}}^{(2)} \rangle_\Omega$ to $\underline{\underline{D}}$. Indeed, combination of (15) and (16) yields:

$$\underline{\underline{\Sigma}}^{(2)} = \mathbb{C}^{hom} : \underline{\underline{D}} \quad \text{with } \mathbb{C}^{hom} = \langle \mathbb{C} : \mathbb{A} \rangle_\Omega \tag{17}$$

Special attention should be paid to the definition of the tensors $\mathbb{C}$ and $\mathbb{A}$ within the rigid inclusions. According to the assumption of rigid behavior, the first one tends towards infinity and the second one towards zero. An indetermination thus appears in (17). This problem can be solved by considering first finite values for the moduli of the inclusions, $\mathbb{A}$ being consequently a non-zero tensor in Ω^i, and then taking the limit of the equation of $\mathbb{C}^{hom}$ (17)

when the moduli of the inclusions tend towards infinity. In other words, $\mathbb{C}^{hom}$ takes the form:

$$\mathbb{C}^{hom} = (1-\varphi)\,\mathbb{C}^{eq} : \langle \mathbb{A} \rangle_{\Omega^m} + \varphi \lim_{\mathbb{C}^i \to \infty} \mathbb{C}^i : \langle \mathbb{A} \rangle_{\Omega^i} \tag{18}$$

The solution of the problem defined by $\underline{\underline{D}}$ and σ_p^{eq} is $\underline{\underline{\Sigma}}^{(1)} + \underline{\underline{\Sigma}}^{(2)}$ and reads:

$$\underline{\underline{\Sigma}} = \mathbb{C}^{hom}(k^{eq}, \mu^{eq}) : \underline{\underline{D}} + \sigma_p^{eq} \underline{\underline{1}} \tag{19}$$

3. Actually k^{eq}, μ^{eq} and σ_p^{eq} nonlinearly depend on the macroscopic strain rate $\underline{\underline{D}}$. The nonlinear relations between those equivalent properties and $\underline{\underline{D}}$ result from two coupled equations: the first one is provided by the definition of $\underline{\underline{d}}^{eff}$ as a function of $\underline{\underline{D}}$ and $\mathbb{C}^{eq}$ and the second one is the nonlinear expression of $\mathbb{C}^{eq}$ with respect to $\underline{\underline{d}}^{eff}$. Thus, the relation (19) can be read in the form:

$$\underline{\underline{\Sigma}} = \mathbb{C}^{hom}(\underline{\underline{D}}) : \underline{\underline{D}} + \sigma_p^{eq}(\underline{\underline{D}}) \underline{\underline{1}} \tag{20}$$

The application of the second step is quite straightforward. In contrast, all the difficulty lies in the choice of an appropriate effective strain rate (first step). Two different definitions of the effective strain rate will be examined in the sequel.

3 DETERMINATION OF THE HOMOGENIZED STRENGTH

3.1 *Choice of the effective strain rate (step 1)*

3.1.1 *Classical secant method*

The simplest equivalent homogeneous material is associated with the effective strain $\underline{\underline{d}}^{eff}$ defined as the strain average in the matrix:

$$\underline{\underline{d}}^{eff} = \langle \underline{\underline{d}} \rangle_{\Omega^m} = \frac{1}{|\Omega^m|} \int_{\Omega^m} \underline{\underline{d}}(\underline{x})\, \mathrm{d}\Omega \tag{21}$$

The corresponding effective strain invariants $d_v^{eff} = \overline{d}_v$ and $d_d^{eff} = \overline{d}_d$ write:

$$\overline{d}_v = \langle d_v \rangle_{\Omega^m}; \quad \overline{d}_d = \sqrt{\langle \underline{\underline{d}}_d \rangle_{\Omega^m} : \langle \underline{\underline{d}}_d \rangle_{\Omega^m}} \tag{22}$$

Since the inclusions are rigid, the macroscopic strain rate is simply related to the strain rate average within the matrix $\underline{\underline{D}} = (1-\varphi)\langle \underline{\underline{d}} \rangle_V = (1-\varphi)\underline{\underline{d}}^{eff}$ which implies:

$$\overline{d}_v = \frac{D_v}{1-\varphi}; \quad \overline{d}_d = \frac{D_d}{1-\varphi} \tag{23}$$

It can be observed on (23) that the macroscopic deviatoric and volume strain rates have uncoupled effects on the classical effective strain rates. Actually the choice of such effective strain rates can be misleading. Indeed, in the simple problem of a composite sphere made up of a rigid core and a deformable shell submitted to an isotropic loading $\underline{\underline{D}} = 1/3 D_v \underline{\underline{1}}$, the macroscopic volume strain rate D_v plays a role on the local shear strain level, which is not taken into account in (23).

3.1.2 *Modified secant method*

The modified secant method proposed by Suquet in (Suquet 1997) is based on an energy approach. The effective invariants in the matrix are defined as quadratic averages of the invariants fields over Ω^m. Those effective invariants are denoted by $\overline{\overline{d}}_v$ and $\overline{\overline{d}}_d$.

$$\overline{\overline{d}}_v = \sqrt{\langle d_v^2 \rangle_{\Omega^m}}; \quad \overline{\overline{d}}_d = \sqrt{\langle \underline{\underline{d}}_d : \underline{\underline{d}}_d \rangle_{\Omega^m}} \tag{24}$$

The expressions of $\overline{\overline{d}}_v$ and $\overline{\overline{d}}_d$ are obtained by differentiation of the macroscopic elastic volume energy density (Kreher 1990). In the case of rigid inclusions and local and global isotropy, one can show that those expressions are:

$$\begin{aligned} \frac{1}{2}(1-\varphi)\overline{\overline{d}}_v^2 &= \frac{\partial \mu^{hom}}{\partial k^{eq}} D_d^2 + \frac{1}{2}\frac{\partial k^{hom}}{\partial k^{eq}} D_v^2 \\ (1-\varphi)\overline{\overline{d}}_d^2 &= \frac{\partial \mu^{hom}}{\partial \mu^{eq}} D_d^2 + \frac{1}{2}\frac{\partial k^{hom}}{\partial \mu^{eq}} D_v^2 \end{aligned} \tag{25}$$

It is interesting to note that the effective strain rates defined in (25) do not depend on σ_p^{eq}. This remark is also valid for those given by (23). This comes from the fact that the exact solution $\underline{\underline{d}}$ of the linear problem $\underline{\underline{D}}$ and σ_p^{eq} only depends on $\underline{\underline{D}}$.

Unlike the classical secant method (23), the modified secant method (25) requires a linear homogenization scheme to define the effective invariants. Indeed, this stage needs an estimate scheme for the determination of the derivatives of $k^{hom}(k^{eq}, \mu^{eq})$ and $\mu^{hom}(k^{eq}, \mu^{eq})$.

According to (25), and in contrast to $\overline{d}_d$, the deviatoric effective strain $\overline{\overline{d}}_d$ is able to capture the influence of the macroscopic spherical part D_v on the local deviatoric strains. As regards the spherical effective strain $\overline{\overline{d}}_v$, it should be emphasized that the definition provides a positive quantity. In addition, as discussed in (Dormieux et al. 2001) and (Lemarchand et al. 2002), such a definition also hides possible sign changes within the matrix. Nevertheless, in the present problem, the matrix has a dilatant behavior, which is taken into account through the condition $\delta \approx 0$ (13). Hence, the sign of d_v is uniform and positive throughout the matrix.

3.2 *Linear homogenization scheme (step 2)*

The macroscopic stress state is given by equation (20), in which we have to choose an appropriate linear homogenization scheme providing $\mathbb{C}^{hom}$ as a function of the equivalent moduli κ^{eq} and μ^{eq} of the matrix. As the microstructure is made up of a connected matrix and spherical inclusions, Mori-Tanaka scheme should be relevant to derive an estimate of $\mathbb{C}^{hom}$. In the case of spherical rigid inclusions, Mori-Tanaka scheme leads to the following homogenized moduli:

$$\begin{aligned} k^{hom} &= \frac{3k^{eq} + 4\mu^{eq}\varphi}{3(1-\varphi)} \\ \mu^{hom} &= \mu^{eq}\frac{k^{eq}(6+9\varphi) + \mu^{eq}(12+8\varphi)}{6(k^{eq}+2\mu^{eq})(1-\varphi)} \end{aligned} \tag{26}$$

3.3 *Macroscopic strength criterion (step 3)*

In the isotropic case, the macroscopic stress tensor given by equation (20) reads:

$$\Sigma_m = k^{hom}D_v + \sigma_p^{eq}; \quad \Sigma_d = 2\mu^{hom}D_d \tag{27}$$

with k^{hom} and μ^{hom} defined in equation (26). It is recalled that k^{eq}, μ^{eq} and σ_p^{eq}, which are defined in (12), depend on the choice of the effective strain rate. Nevertheless, for any of the three methods proposed in the previous section, the limit (13) exploited in (12) gives:

$$\lim_{a\to 0}\frac{\mu^{eq}}{k^{eq}} = \lim_{a\to 0} -\frac{T\delta^{eff}}{2d_d^{eff}} = 0 \tag{28}$$

In particular, this implies that $\lim_{a\to 0}\partial\mu^{hom}/\partial k^{eq} = 0$. Therefore, it can be shown that, asymptotically, for a tending towards 0, the two effective volume strain rate are identical:

$$d_v^{eff} = \overline{\overline{d}}_v = \overline{d}_v = D_v/(1-\varphi)(\geq 0) \tag{29}$$

It must be emphasized that the macroscopic dilatancy i.e. the positive sign of D_v is the consequence of the microscopic flow rule in the matrix satisfying (13).

Replacing k^{hom} by its estimate (26) in the first relation of (27) and exploiting the relation (29) and the definitions (12) of k^{eq}, μ^{eq} and σ_p^{eq}, it follows:

$$\Sigma_m - H = f_a'(\delta^{eff})\left(1 - \frac{2}{3}\varphi T^2\right) \tag{30}$$

The notation δ^{eff} means that each of the three effective strain rates mentioned above can be used until equation (30) because of (29). The end of the determination of the macroscopic strength criterion needs information about the choice of the effective strain rates.

- *Classical secant method*
  The limit (28) exploited in (26) implies:

$$\mu^{hom} \underset{a\to 0}{\approx} \mu^{eq}\frac{1+\frac{3}{2}\varphi}{1-\varphi} \tag{31}$$

With the help of (31), (12) and (23), the second equation of (27) writes:

$$\Sigma_d = -f_a'(\delta^{eff})\left(1 + \frac{3}{2}\varphi\right)T \tag{32}$$

The combination of (30) and (32) yields:

$$\Sigma_d = -\underbrace{\frac{1+\frac{3}{2}\varphi}{1-\frac{2}{3}\varphi T^2}T}_{\overline{T}_a^{hom}}(\Sigma_m - H) \tag{33}$$

(33) appears as a condition to be satisfied by any macroscopic stress tensor $\underline{\underline{\Sigma}}$ located on the boundary of G^{hom}. Thus, (33) can be interpreted as the macroscopic strength criterion. It proves to be of the Drucker-Prager type.

Hence, for a Drucker-Prager matrix reinforced by rigid inclusions, $\overline{T}_a^{hom}$ is the estimate of the homogenized friction coefficient obtained with the help of the classical secant method. Besides, according to (33), the homogenized cohesion can be put in the form $\overline{C}_a^{hom} = H\overline{T}_a^{hom}$.

- *Modified secant method*
  The same kind of reasoning based on the modified effective strain rates leads to the following macroscopic strength criterion:

$$\Sigma_d = -\underbrace{\sqrt{\frac{1+\frac{3}{2}\varphi}{1-\frac{2}{3}\varphi T^2}}T}_{\overline{\overline{T}}_a^{hom}}(\Sigma_m - H) \tag{34}$$

By analogy with the classical secant method, $\overline{\overline{T}}_a^{hom}$ and $\overline{\overline{C}}_a^{hom} = H\overline{\overline{T}}_a^{hom}$ will respectively denote the estimates of the homogenized friction coefficient and cohesion of the r.e.v. in the framework of the modified secant method.

The two criteria found in (33) and (34) give an estimate of the reinforcement due to the presence of adherent inclusions. Indeed, the friction coefficients $\overline{T}_a^{hom}$ and $\overline{\overline{T}}_a^{hom}$ and the cohesions $\overline{C}_a^{hom}$ and $\overline{\overline{C}}_a^{hom}$ are greater than those of the matrix T and $C = HT$ (see Fig. 3).

4 CONCLUSION

The two methods of nonlinear homogenization presented in this paper allow one to estimate the macroscopic strength criterion of a r.e.v. made up of a

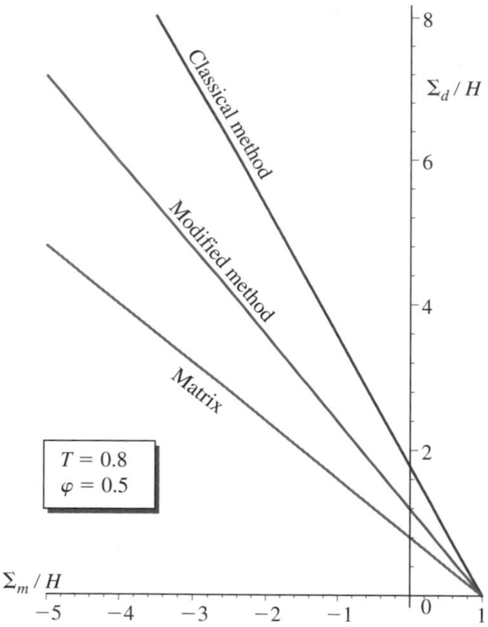

Figure 3. Strength criteria.

Drucker-Prager-type matrix and rigid inclusions. As expected, both methods lead to a reinforcement of the matrix. Indeed, in each case, the macroscopic strength criterion estimate takes the form of a Drucker-Prager criterion with a friction coefficient and a cohesion larger than those of the pure matrix. Our approach, not only takes into account the microscopic strength criterion, but also the local flow rule (here the dilatancy of the matrix). Indeed, the approach is based on the formal equivalence between a limit analysis problem and a ficticious viscoelastic problem. From a theoretical point of view, the velocity fields which are solutions of the limit analysis problem and the viscoelastic one are identical. Practically, the solutions of the latter can be estimated with the techniques of nonlinear homogenization, which are based on the effective strain concept.

While keeping the line of reasoning presented in this paper, the estimate of the macroscopic strength and of the velocity at free plastic flow can be improved as regards the effective strain concept. Indeed, instead of using a uniform effective strain in the matrix, the latter can be divided in subdomains, with specific values of the effective strain in each. In fact, our predictions may also overestimate the reality since possible limitations in the strength of the matrix-inclusion interface have been disregarded. Indeed, the rigid inclusions have been considered as adherent to the matrix whereas it is well known that the matrix-inclusion interface may present defects and flaws (allowing for example velocity jumps).

REFERENCES

Barthélémy, J.-F. and L. Dormieux (2002). Détermination du critère de rupture macroscopique d'un milieu poreux par homogénéisation non linéaire. *Cr. R. Acad. Sci. Paris Série IIb* (To be published).

Desrues, J. (2002). Limitations du choix de l'angle de frottement pour le critère de plasticité de Drucker-Prager. *RFGC*, 853–862.

Dormieux, L., A. Molinari, and D. Kondo (2001). Micromechanical approach to the behavior of poroelastic materials. *J. Mech. Phys. Solids 50*, 2203–2231.

Hashin, Z. and S. Shtrikman (1963). A variational approach to the theory of the elastic behavior of multiphase materials. *J. Mech. Phys. Solids 11*, 127–140.

Kreher, W. (1990). Residual stresses and stored elastic energy of composites and polycristals. *J. Mech. Phys. Solids 38*, 115–128.

Lemarchand, E., F.-J. Ulm, and L. Dormieux (2002, August). The effects of inclusions on the friction coefficient of highly filled composite materials. *JEM 128*(8), 876–884.

Mori, T. and K. Tanaka (1973). Average stress in the matrix and average elastic energy of materials with misfitting inclusions. *Acta Metall. 21*, 571–574.

Suquet, P. (1997). Effective behavior of nonlinear composites. *Continuum micromechanics, Springer*, 197–264.

Zaoui, A. (1997). *Matériaux hétérogènes et composites*. Cours de l'École Polytechnique.

Zaoui, A. (2002, August). Continuum micromechanics: Survey. *JEM 128*(8), 808–816.

Computational Modelling of Concrete Structures, Bićanić et al. (eds)
© 2003 Swets & Zeitlinger, Lisse, ISBN 90 5809 536 3

Computational strategies for time-dependent behaviour of ECC

W.P. Boshoff & G.P.A.G. van Zijl
Engineering Faculty, Department of Civil Engineering, University of Stellenbosch, Stellenbosch, South Africa

ABSTRACT: In search of improved performance of structural concrete, Engineered Cementitious Composites (ECC), have been introduced over the last decade. The fibres address the weakness of concrete in tension. Pseudo strain hardening is achieved by crack bridging by the short fibres, allowing more cracks to initiate in the matrix. In this manner multiple, fine cracks arise, leading to the strain hardening and enhanced toughness. The research program on which this paper reports, aims at capturing FRC time-dependent behaviour with computational strategies ranging from phenomenological, macroscale modelling to multi-scale (meso-macro) modelling. In this paper, the first results of macroscale modelling are reported. Time dependence is included by two rate processes, namely bulk creep and a crack rate dependence in the fracture process zone. Parametric studies are performed on the effect of bulk creep as well as the crack rate dependence.

1 INTRODUCTION

Invented in the early 1990's, ECC has developed at a high rate in the past decade, both in terms of applications and material development. ECC – engineered cementitious composites – consist of fibres and a cementitious matrix, usually mortar or cement paste. The fibres are mostly polymeric, also in the application discussed in this paper. Both constituents, as well as the interface between them, have time-dependent behaviour, contributing to the time dependence of the composite. This time-dependence of ECC is the main focus of this paper.

ECC is a high performance, ultra ductile fibre reinforced composite material that embodies micromechanics-based design driven concepts (Li, 2002). It encompasses fibre-bridging properties across matrix cracks, allowing tensile stress to be transmitted across a crack as it opens. Upon further deformation multiple cracks arise, accompanied with higher tensile resistance. This pseudo strain hardening behaviour is the main distinction between ECC and fibre-reinforced concrete (FRC) and leads to a higher structural resistance in bending. At the peak strength, fibre breakage and pull-out lead to localisation, marking the onset of softening, as shown schematically in Figure 1.

The micro-mechanical approach relates macroscopic properties to the microstructure of the composite (Li et al., 1991). This tool enables material optimisation, which will make it cost effective, for instance by computing the critical fibre content, i.e. the minimum amount of fibres which still leads to strain hardening.

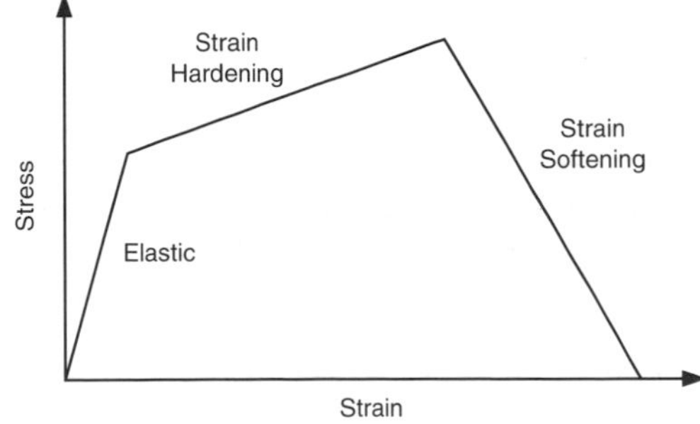

Figure 1. Typical behaviour of ECC.

However, there is evidence that ECC material has higher permeability, causing increased creep (Rouse and Billington, 2002). This source of time dependence, together with a crack mouth opening rate, which considers the viscous processes in the fracture zone, accounts for reduced life expectancy under sustained load (Zhou, 1992; van Zijl et al., 2001). This raises the question whether the strain hardening Computational strategies for time-dependent behaviour of ECC feature of ECC should be relied upon in design, as the superior behaviour may be short-lived. To begin to address this issue, a macroscopic approach to model ECC has been followed to predict the time to failure under sustained load. A finite element model, which considers the main features of ECC mechanical behaviour, including strain hardening and strain softening, has been implemented and employed to simulate experimental results of uniaxial tension and three-point-bending tests on ECC specimens. Subsequently, the time-dependent response has been predicted, including the time to failure under a sustained load.

The current research program to characterise and accurately model ECC time-dependent behaviour, aims at a multi-level approach, in which the micro/meso-scale is considered at each macroscopic material point (for example Kouznetsova et al., 2001). Thereby, the sources of time-dependence can be isolated, for instance bond-slip of the fibres, the creep of the fibres and matrix shrinkage, the source of Coulomb-friction increased slip resistance by providing confining pressure (Stang, 1997).

However, in this paper results of a macroscopic approach, the first phase towards characterising and understanding ECC time dependence, are reported. The time dependence of the matrix and the matrix-fibre interface is ascribed to moisture migration, which causes time-dependent phenomena like shrinkage, creep and crack mouth-opening rate (CMOR) dependence. These features, together with a continuum plasticity-based formulation to model cracking, have been incorporated in the model along the lines described by van Zijl et al. (2001b). It is shown that this model can capture the global response of ECC beams subjected to three-point-bending. Subsequently, the time-dependent response under a sustained load, which causes multiple cracking in the central part of the beams, has been predicted. The role of CMOR is highlighted, as mechanism to introduce the correct time scale of fracture.

2 THE MODELLING APPROACH

The focus of this paper is on the macro-level and, therefore, a phenomelogical approach has been followed. A continuum plasticity model, which incorporates creep through visco-elasticity and a CMOR-dependence (van Zijl et al., 2001b) has been employed. To account for the hardening-softening response of ECC under uniaxial tension, the Rankine limit function was hardened and softened according to the calibration data obtained from uniaxial tensile tests performed at Stellenbosch University recently, Figures 2, 3. The mix used consists of a water-cement ratio that was set at 0.35 and a ratio of aggregate: cement of 1. Silica fume was used as a substitute for 30% of cement, while another 20% was substituted by fly ash. A fibre volume of 1.5%, polyvinyl alcohol (PVA) was employed, with average length 12.5 mm and diameter 40 μm. The compression strength was found to be 54 MPa at 28 days using standard test methods.

A simple multi-linear approach has been followed to capture the tensile behaviour, as shown in Figure 3. It is assumed that the elastic behaviour is limited by the

Figure 2. The uniaxial tensile test setup, showing the clamps on opposite ends of the ECC specimen.

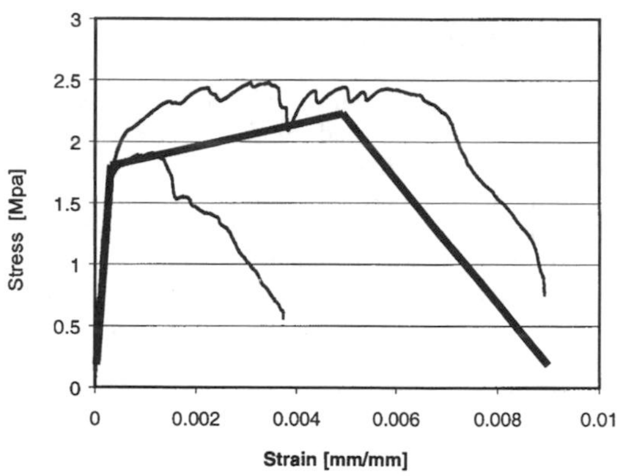

Figure 3. ECC experimental uniaxial tensile response, also showing the employed linear approximating hardening-softening response.

matrix cracking resistance. Linear strain hardening is modelled, followed by linear softening after the peak resistance. As can be seen, it is difficult to determine with precision the first matrix cracking point from the graph. The assumption is made that this point is the first position that deviates from the linear-elastic response. The values derived from the experimental results for the simplified model are shown in Table 1.

For the subsequent prediction of three-point-bending response of ECC beams, the inelastic strain limit of the strain hardening phase has been taken as 0.5% (Figure 3). In the course of the three-point-bending analyses it has been found that this value was estimated too low and, therefore, increased to 1% for

Table 1. ECC specimens uniaxial tensile test results.

	Test 1	Test 2
First Cracking Strength [MPa]	1.64	1.77
Maximum Strength [MPa]	2.43	1.89
Strain at maximum strength [mm/mm]	0.60%	0.12%
Strain softening displacement [mm]	0.9	0.7

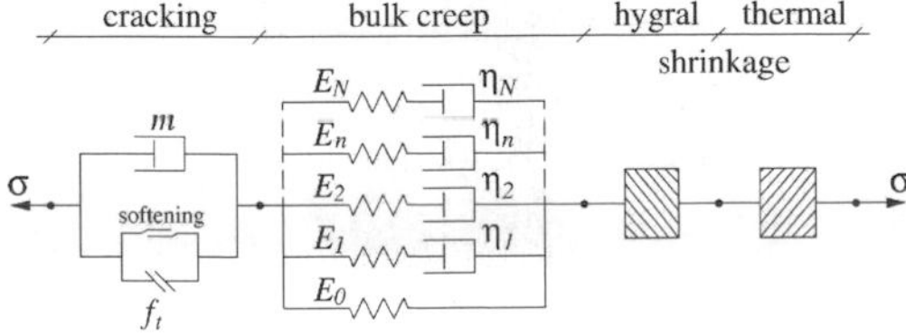

Figure 4. Schematic representation of constitutive model, including a Maxwell chain for bulk creep simulation (van Zijl et al., 2001b).

the rest of the parametric study. This value is in better agreement with results of Kabele (1999) and Li et al. (2001), who reported values ranging between 1.4% and 2%. Linear softening has been assumed to occur over a total deformation of 0.8 mm, which is the average of the two experimental results shown in Figure 3 and Table 1. Both Kabele (1999) and Li et al. (2001) report a limiting value of 6 mm for the softening branch, approximately half the average fibre length. However, this high value could not be achieved in the tensile tests in this study.

3 CREEP

Creep is inherent to any type of cementitious composite, hence its significance in the time-dependent constitutive model of ECC. It is closely linked to viscous processes in the cement-based composite. Linear visco-elastic behaviour is assumed here, represented by a Maxwell chain – Figure 4.

There is evidence of higher permeability of ECC compared with ordinary concrete, going a long way towards explaining the larger creep deformations measured on ECC specimens than on similar fibreless specimens (Rouse and Billington, 2002). Their creep results, Figure 5, have been employed to compute the Maxwell chain element stiffnesses E_i and viscosities η_i. A least squares fit of a ten-element Maxwell chain formulation to the creep data was performed. Thereby, creep model parameters have been obtained with which the creep and relaxation responses shown in Figures 5 and 6 have been computed. A creep coefficient of approximately 1.6 is evident. This value is not excessive for cement-based material, but was nevertheless found to be larger than for specimens of the same mix proportions, but excluding fibres, cured

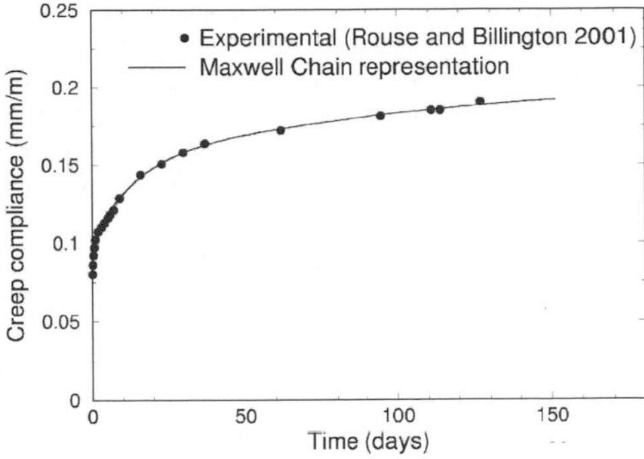

Figure 5. ECC creep behaviour employed.

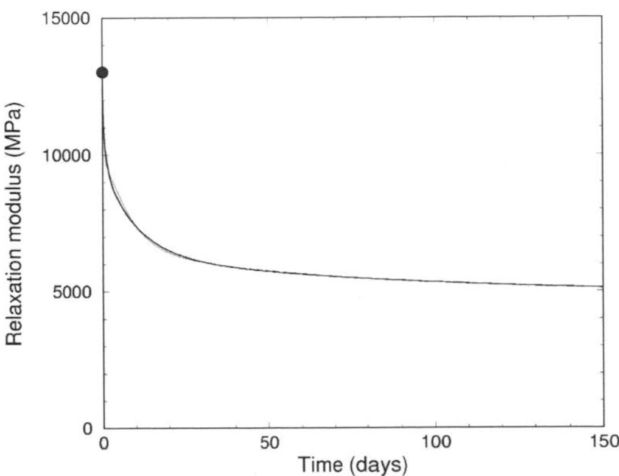

Figure 6. ECC relaxation predicted with fitted Maxwell chain.

and tested in similar conditions (Rouse and Billington, 2002).

4 CRACK MOUTH OPENING RATE

It has been shown that the inclusion of a crack mouth opening rate dependence is not merely a numerical artefact to regularize the continuum description of localisation, but is in fact an important and integral factor when analysing the time-dependent behaviour of cementitious composites (van Zijl et al., 2001). CMOR models the rate-dependence of the breakage of bonds in the fracture process zone. The role of the rate can be illustrated by performing three-point bending creep failure experiments on concrete beams, noting the time to failure under the sustained loads. Only by considering the rate process in the fracture process zone, the correct time scale is introduced, enabling simulation of measured time to failure.

As stated previously, the time-dependence of cement-based material cannot be ignored. This is governed by the ever-changing microstructure and the ever-present moisture in the pores. The creep test results of Rouse and Billington (2002) reveal an increased time-dependence, explained by the higher permeability of ECC. It is postulated that the CMOR-dependence, argued to be driven by moisture migration in the fracture process zone, plays a significant role in the time-dependent response of ECC. Therefore, it is considered imperative to activate a CMOR-dependence to obtain the correct time scale in creep fracture. In Figure 4, a CMOR model is schematised as a cracking viscosity m. Here, a CMOR model based on the rate process theory for the crack rate dependence presented by Wu and Bažant (1993) is employed, as reformulated by van Zijl et al. (2001) to read

$$\sigma_{cr} = \sigma_{cr}(\kappa)\left[1 + k \cdot \mathbf{sinh}^{-1}\left(\frac{\dot{\kappa}}{\dot{\kappa}_r}\right)\right] \tag{1}$$

where $\sigma_{cr}(\kappa)$ is the rate independent hardening-softening cracking stress, governed by the state parameter κ, which is the principal plastic strain in the Rankine formulation employed. $\dot{\kappa}_r$ is a reference, low cracking strain rate and k a model parameter chosen in combination with the reference cracking rate to produce a rate-dependent peak resistance enhancement in agreement with measured responses at various loading rates. In order to calculate the CMOR properties, inverse analysis of the tests at various loading rates must be performed. In previous studies, such parameter calculations were performed from measured three-point-bending results of plain concrete beams (Zhou, 1992). In the absence of such measurements for ECC, a 10% increase in peak tensile resistance at the tensile deformation rate of $0.15\,\text{mm} \cdot \text{minute}^{-1}$ ($2.5\,\mu\text{m} \cdot \text{s}^{-1}$) has been assumed. With this assumption, the model parameters $k = 0.025$ and $\dot{\kappa}_r = 10^{-7}\,\text{s}^{-1}$ was calculated and employed for the analyses reported in this paper. Note that the "static" tensile resistance has been reduced accordingly, to reproduce the measured tensile response at the mentioned deformation rate when the CMOR-dependence is activated. Clearly, experiments must be performed to characterise these parameters for ECC. At this stage the model and the assumed parameters serve to illustrate the time-scale of fracture processes in ECC.

5 MODEL VERIFICATION

The same mixture of ECC was used to prepare three-point bending specimens. The tests were also performed at Stellenbosch University on a Zwick machine. The experiments were performed under displacement (rate) control at 0.15 mm/min. The dimensions of the specimens were depth 40 mm and width 100 mm. The supports were 300 mm apart. The point load was applied on top using a 30 mm wide steel plate with thickness 5 mm. The test setup is shown in Figure 7.

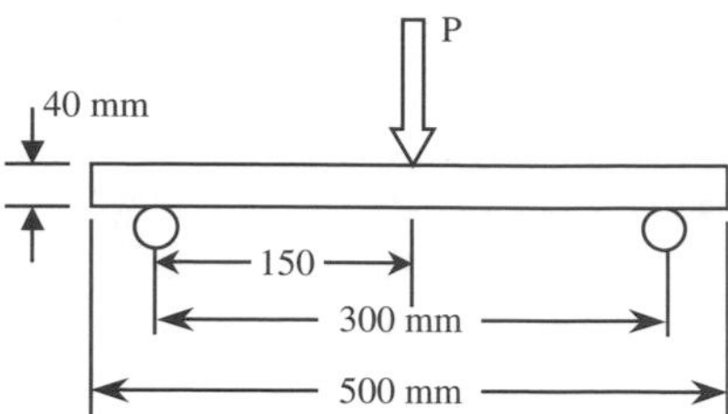

Figure 7. Three-point bending test setup.

The finite element representation was done using two-dimensional, plane stress elements. The dimensions are the same as for the experimental specimens and three-point bending setup. The steel plate was also incorporated in the model, with the point load acting on top of it, to avoid local inelasticity caused by a point load applied directly to a single node on the ECC beam model.

Note that, for the case of bending, nonlinear response in compression is expected to be activated by the high performance in tension of ECC. Therefore, a compressive limit Hill-type function has been modelled in the analyses, based on the formulation by Lourenço et al. (1998). The maximum compressive strength of 54 MPa has been prescribed, but deviation from elastic compressive response has been considered to occur from one third of the ultimate strength. To model this nonlinear response, parabolic hardening has been assumed until the ultimate strength has been reached at an inelastic compressive strain of 0.2%, with subsequent degraded compressive resistance (softening) to zero at an inelastic compressive strain of about 2%, in accordance with the values employed by Kabele (1999).

Three analyses were performed to study mesh objectivity and the influence of symmetry constraints. The first analysis was with a coarse mesh comprising elements of 2×2 mm. The second analysis was done with the same element size (2×2 mm), but only considering one symmetrical half of the model. Nearly identical results were computed for these analyses. Hereby, it was concluded that a symmetric finite element model is adequate.

A third analysis was performed with a refined mesh – 1×1 mm elements – also exploiting symmetry to study the effect of mesh refinement on the computed response. The results of the refined mesh proved to be exactly the same as that of the coarser equivalent. Therefore, it can be concluded that a symmetric mesh using 2×2 mm elements is a sufficient representation and has subsequently been employed for all the analyses. Note that in these preliminary analyses CMOR was not yet activated. This boundary

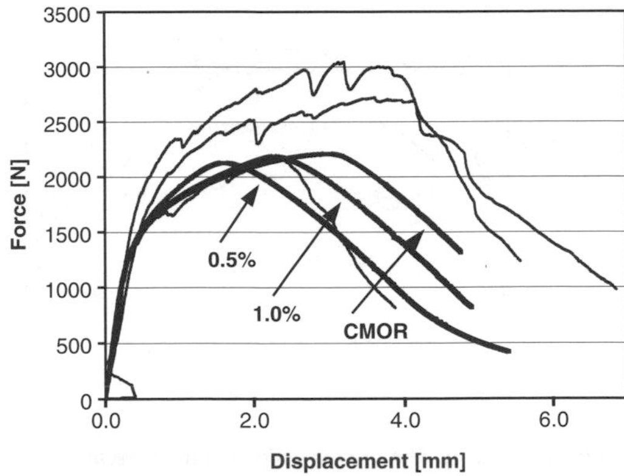

Figure 8. Measured and computed three-point bending responses.

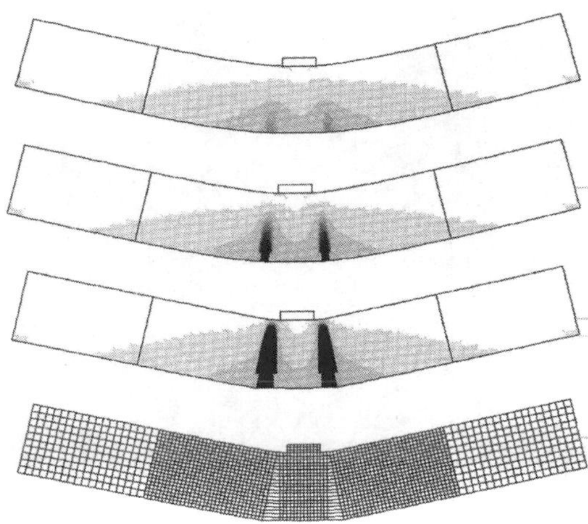

Figure 9. Computed principal plastic strain contours and deformation at: Top: 92% of the peak load in the ascending (loading) branch, Second from top: 92% of the peak load in the descending (global softening) branch, Third from top: final stage, showing localisation in two major cracks.

value problem governs the crack location, avoiding the computational in objectivity with regards to crack spacing and orientation.

The model parameters, as determined in section 2, have been employed for the three-point bending analysis. As mentioned there, the 0.5% strain hardening limiting strain did not reproduce the measured response, as shown in Figure 8. This value was increased to 1%, in line with reported values by Kabele (1999) and Li et al. (2001). The computed response with this adjusted strain hardening parameter is also shown in Figure 8, revealing the sensitivity of the global response to this model parameter.

The latter result is in better agreement with the measured responses. Thus, the 1% strain hardening was used for the remainder of the analyses. Note the lower gradient of the computed response with the strain hardening limit set at 1%. This is due to the larger strain/deflection needed before the unmodified peak stress is reached. Also this lower gradient appears in better agreement with the measured responses.

Subsequently, the CMOR has been activated. Despite compensating for the rate-increased tensile strength by lowering the cracking rate independent tensile strength with the estimated rate-induced enhancement, the higher three-point bending resistance shown in Figure 8 with CMOR activated has been computed. The increased resistance is attributed to the difficulty in estimating the cracking strain rate $\dot{\kappa}_r$ over the total load-deformation response, which clearly influences the rate-enhanced response, resulting in a different (higher in this case) total amount of inelastic energy dissipated for the case which considers CMOR.

In the model a smeared cracking approach has been used. This complicates the prediction of crack widths and spacing. Instead, contours of κ, the principal plastic strain, are shown in Figure 9 at various stages of loading.

Wide spread cracking is computed, indicated by the cloud of plastic strain in Figure 9 (top). Subsequently, localisation starts in two major cracks, Figure 9 (second from top) and propagates upwards, leaving only a small compressive zone in the final stages of the analyses, Figure 9 (third from top). Note that the level of damage/plastic strain is arrested in the damaged region outside the areas of localisation as soon as localisation sets in. The measured crack pattern is blown up in Figure 10, showing the closely spaced, multiple cracks in the pre-peak response region. The post-peak response is introduced by localisation in a single crack, Figure 11. The location of the localised crack is captured accurately by the model, although the negligence of material and geometrical imperfection has led to the computation of two cracks, the one a mirror image of the other.

The approximation by the model is found to be accurate enough to act as base for the intended purpose, namely to study the behaviour of ECC on the long term and the time-dependent factors acting within.

6 PARAMETRIC STUDY

The section that follows describes the parametric study performed to show the effect of creep and CMOR on the long-term response of ECC beams subjected to three-point bending. This is done by deriving both a creep and a fracture envelope from computed results, as well as studying creep fracture.

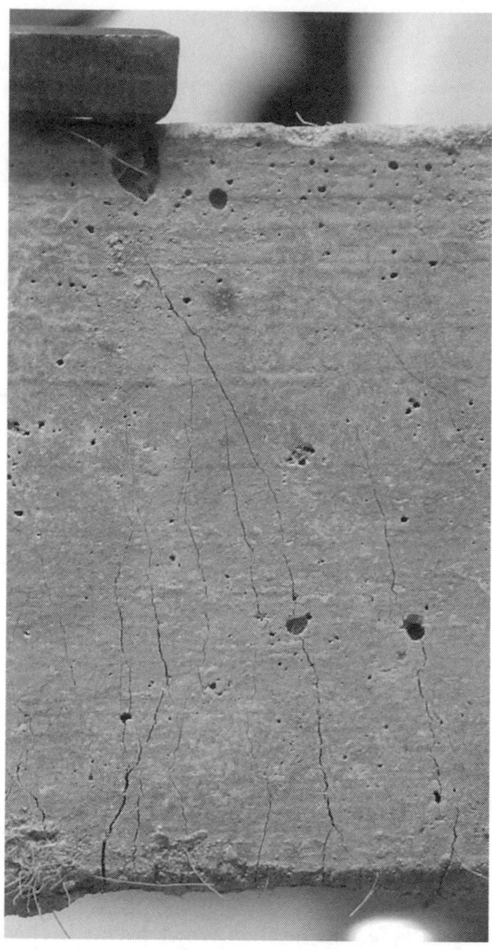

Figure 10. Crack pattern observed in the experiment before the peak resistance was reached, i.e. before localisation in a single macrocrack.

Figure 11. Observed crack localisation beyond peak resistance.

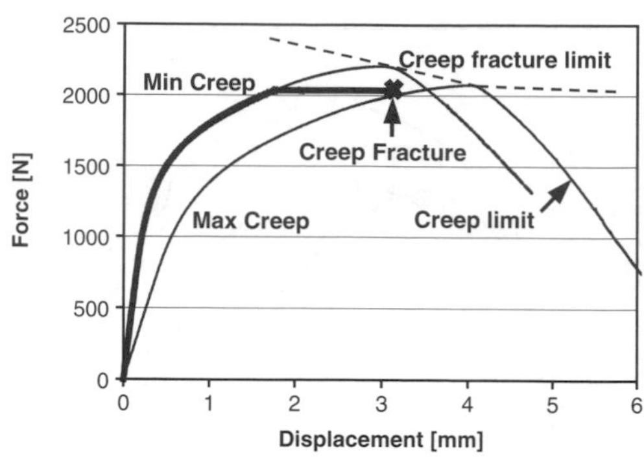

Figure 12. Creep fracture results excluding CMOR.

The first step was to determine the creep envelopes. This was done by performing two static, displacement-controlled analyses, the first with the total initial stiffness, simulating time-independent, instant response. The second analysis employed the effective stiffness modulus after all relaxation has occurred in the Maxwell chain dashpots, i.e. the stiffness of the chain element without a dashpot. The results of these two analyses form the upper and lower bound respectively of the creep and fracture envelopes (van Zijl et al., 2001). This is shown in Figure 12 for the case where CMOR is not activated, i.e. only bulk creep is considered. To test whether the use of these two analyses to form the creep fracture envelope is justified was tested by performing a creep analysis at a sustained load level of 92% of the peak resistance. This computed response is also shown in Figure 12. Failure under the sustained load is marked with an X in the figure, confirming the fracture envelope. The point of failure was reached when convergence of the internal resistance to the applied load could not be obtained any longer. In order to determine whether it was the correct failure point the analysis was continued beyond point X under displacement control. This produced subsequent global softening, confirming that the correct point of failure under the sustained load has been obtained.

The two extreme, displacement-controlled creep analysis cases, as well as the sustained load analyses have been repeated with CMOR activated – Figure 13. Failure under the sustained load, set at 92% of the peak instant resistance, has been computed to fall within the drawn envelope, as indicated by the X in Figure 13.

An important aspect, which is imperative to capture accurately if structural durability is assessed, is the time scale of fracture of ECC. An indication of this time scale is given by the predicted time it takes for the macro-crack to propagate under the sustained load until failure. For the case where CMOR is not considered, this time to failure has been computed as 1500 days. In recent research (van Zijl et al., 2001)

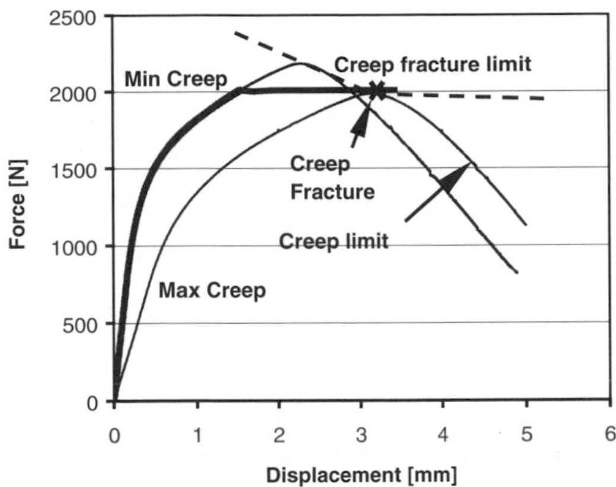

Figure 13. Creep fracture results including CMOR.

it was shown that this is a gross overestimation of the measured time to (creep) fracture of similar, but plain concrete beams subjected to three-point bending, tested by Zhou (1992). No experimental evidence to the contrary exists for ECC yet. Nevertheless it is postulated that the cracking rate dependence in the fracture process zone, as simulated by the CMOR formulation in equation (1), more realistically introduces the true fracture time scale. In the current analyses, consideration of CMOR has resulted in the computed time to failure of 1.7 days.

These results are crucial to show the importance of CMOR as a time-dependent factor in ECC and its long-term behaviour.

7 CONCLUSIONS

Even though extensive research has been done on ECC, very little is known about the time dependence of the material. This paper has explored the effect of the time dependence on the global behaviour of these materials computationally, as prediction for experiments envisaged for the near future.

All the analyses done in this paper indicate that these dependencies may indeed play a governing role, as shown for the case of creep fracture. This is particularly the case when the crack rate dependence is considered in the fracture process zone, as included in this model via CMOR.

In this study a phenomenological approach has been chosen as a first approach to simulate ECC response on a macroscopic level. This simplifies characterisation of some model parameters, for instance the time-independent constitutive law employed. However, the CMOR parameters have to be found through inverse analysis. In meso-level analysis, which is planned for a later stage in this research project as part of the multi-level modelling approach, it will hopefully not be necessary to do in verse analyses to obtain these time-dependent parameters.

It can be concluded that the doubt whether the strain hardening phase in ECC tensile response should perhaps not be relied upon entirely in design, due to the danger of delayed fracture, has not been erased by this preliminary study. At a sustained load of 92% of the peak structural resistance, creep fracture is predicted to occur within a few days. The very mechanism of enhanced durability through crack distribution, structural resistance increase through local material pseudo strain hardening and general larger toughness, has been shown to lead to eventual localisation and collapse at this load level. However, for the particular creep evolution considered in this study, a sustained load level below this value should reach the creep limit, instead of the fracture limit, theoretically ensuring infinite structural life at sustained load levels below approximately 90% of the peak instantaneous resistance.

REFERENCES

Kouznetsova, V., Brekelmans, W.A.M. and Baaijens, F.P.T. 2001. An approach to micro-macro modeling of heterogeneous materials. *Computational Mechanics*, vol 27, 2001, pp. 37–48.

Kabele, P. 1999. Performance of engineered cementitious composites in repair and retrofit: analytical estimates. H.W. Reinhardt and A. Naaman (eds), *High Performance Fiber Reinforced Composites*, Rilem: 617–627.

Li, V.C., Wang, Y. and Backer, S. 1991. A micromechanical model of tension-softening and bridging toughening of short random fiber reinforced brittle matrix composites. *J. Mech. Phys. Solids* 39(5): 607–625.

Li, V.C., Wang, Y. and Wu, C. 2001. Tensile strain hardening behavior of polyvinyl alcohol engineered cementitious composite (PVA-ECC). *ACI Materials Journal* 98(6): 483–492.

Lourenço, P.B., Rots, J.G. and Blaauwendraad, J. 1998. Continuum model for masonry: parameter estimation and validation. *ASCE J. Eng. Mechanics* 124(6): 642–652.

Rouse, J.M. and Billington, S.L. 2002. Creep, Shrinkage and Permeability of Ductile Cement-based Composites, in preparation for *ACI Materials Journal.*

Stang, H. 1997. Interface Properties in Extruded FRC-Materials, *Brittle Matrix Composits*: 3–13, Cambridge and Warsaw, Woodhead Publishers Limited, Bigraf.

Van Zijl, G.P.A.G., De Borst, R. and Rots, J.G. 2001. The role of crack rate dependence in the long-term behaviour of cementitious materials. *International Journal of Solids and Structures* 38: 5063–5079.

Van Zijl, G.P.A.G., De Borst, R. and Rots, J.G. 2001b. A numerical model for the time-dependent cracking of cementitious materials. *Int. J. Numerical Methods in Engineering* 52(7): 637–654.

Wu, Z.S., Bažant, Z.P. 1993. Finite element modelling of rate effect in concrete fracture with influence of creep. Z.P. Bažant and I. Carol (eds), *Creep and Shrinkage of Concrete*, E. & F.N.Spon, London: 427–432.

Computational Modelling of Concrete Structures, Bićanić et al. (eds)
© 2003 Swets & Zeitlinger, Lisse, ISBN 90 5809 536 3

Multiscale modelling of granular materials

I.M. Gitman, H. Askes, L.J. Sluys & M. Stroeven
Faculty of Civil Engineering and Geosciences, Delft University of Technology, Delft, Netherlands

ABSTRACT: The multiscaling analysis of the behaviour of granular material is considered in this paper. On the macro-level material is homogeneous, while on the meso-level material is considered to be heterogeneous. The problem of the size of the meso-level is of interest in the context of the paper. The influence of the size of the meso-level is analysed by means of the concept of a Representative Volume Element. A procedure, based on a statistical criterion, is proposed to find the appropriate size of the unit cell on the meso-level.

1 INTRODUCTION

To obtain a better understanding of the physical mechanisms granular materials such as of concrete fracture under static loading conditions a multiscale analysis is proposed in this paper. The analysis of the formation of cracks and the ability to predict the conditions (mechanical, environmental), which make the micro- and macro-cracks grow from their initial state to dimensions leading to failure of a structure, is of paramount importance.

According to the different interests of a particular investigation there are different approaches in modelling the same physical problem, each of them have their own advantages and disadvantages. One such technique is to consider the problem only on the highest (or so-called macroscopic) level; this way is efficient but the results are smoothed out and the influence of mesoscopic and microscopic action is sometimes neglected; as a consequence more detailed analyses at a smaller scale have been performed. The results in this case are more accurate, however, the computational costs of including such detail is prohibitive, especially when considering full scale structural applications. To overcome this limitation a combined multiscale approach of the considering problem is investigated here.

The multiscale strategy is widely used in the solid mechanic's literature. In his work, Doghri (2002) described the procedure of meso-macro modelling for the polycristals, Kouznetsova et al. (2001) and Geers et al. (2001) used this procedure for the modelling of heterogeneous materials, such as porous aluminium.

The multiscale analysis is based on the following procedure: the problem is solved on the macro level, where at each macro-level integration point a meso-level analysis is performed. Next, the results of the meso-level investigation are averaged through homogenisation techniques and the results of the macro-level analysis are corrected. In a multiscale approach, an apparent question is what the dimensions should be of the meso-level sample that is considered.

The size of the meso-level cell is defined with the help of a representative volume element (RVE), thus the term "*size of RVE*" is interchargeable with "*size of the meso-level cell*". The question of the RVE size is important because it will justify the use of a macro-meso strategy, i.e. in the case where the size of RVE is large or even similar in magnitude to the area attributable to an integration point in the macro level, there is no reason to use the macro-meso approach, a macro investigation is sufficient. However if the size of the RVE is smaller, the idea of using a macro-meso scheme for solving the problem is applicable.

In Section 2 the multiscale strategy is presented. The question "*how the RVE size is determined*" is of great importance. A multitude of methods are available in the literature. Drugan and Willis (1996) proposed the quantitative estimates of minimum RVE size for elastic composites, Borbely et al. (2001) also suggest a way to define the size of the RVE. An alternative method of RVE size measurement is described in detail in Section 4.

2 MULTISCALE STRATEGY

2.1 *Macro-level*

On the macro level the material is assumed to have a homogeneous structure, so the properties of the material are averaged. There are several methods of averaging/homogenization (Vosbeek 1994). Below the procedure of averaging is described.

The mechanical loading is applied at macro-level and it should be in equilibrium with the internal forces which are computed from the stresses at meso-level.

2.2 *Macro-meso connection*

The idea of a macro-meso connection is, that after analysis of the structure on the macro-level, the regions of critical activity can be found, and in each integration point of these regions the meso level is considered by means of an RVE.

2.3 *Meso-level*

On the meso-level a heterogeneous material is considered, which implies that in each integration point (of the macro-level) a three-phase material is investigated. The first phase are aggregates (stiff); the second phase is the cement matrix (less stiff) and the third phase is the interfacial transition zone (weakest zone). Each of these materials have their own properties (E – Young's modulus, ν – Poisson's ratio etc.). The cracking mechanism in the meso level is caused by mechanical loading and by the influence of moisture transport.

The formulation of the problem on the meso-level looks as follows:

- equilibrium equation $\nabla \cdot \sigma = f$
- constitutive equation $\sigma = D_e \varepsilon$

whereby the macroscopic strain field is translated into essential boundary conditions in terms of the vertex displacements of the meso-level cell.

2.4 *Meso-macro connection*

As a constitutive equation on the macro-level the meso-macro connection take place, thus instead of explicit formulation of the $\sigma - \varepsilon$ relation, data from the meso-level are considered.

On the macro-level, as it was mentioned above, homogeneous material is considered, on the other hand the meso-level material has a heterogeneous character. So, to keep the meso-macro relation consistent the procedure of homogenisation should be carried out.

One of the homogenisation techniques based on averaging. The average value of stresses in the RVE can be computed via

$$\langle \sigma_{meso} \rangle = \frac{1}{V_{RVE}} \int_{\Omega} \sigma_{meso} dV \tag{1}$$

The average value of the stress in the meso level is equal to the value of the stress in the macro level in the considered integration point:

$$\sigma_{macro} = \langle \sigma_{meso} \rangle \tag{2}$$

With obtained values of stresses in each integration point the further analysis of the macro-level is to be carried out.

Here, we consider the meso-scale and the macro-scale in the fracture process. Fracture at the meso-level can take place in the aggregate, in the cement paste or in the interfacial transition zone (ITZ). The constitutive behaviour of the different phases at the meso-level can be modelled. A gradient damage model is used to describe the behaviour of the material on the meso-level.

3 GRADIENT DAMAGE MODEL

Together with an advantage of simplicity in application, a local damage model has big disadvantages: the scale of the underlying micro-structure is not included as a parameter of the model. In other words, a direct link with the underlying micro-structure is missing. This causes mesh sensitivity problems. To avoid this problem it has been proposed to use a gradient damage model in the implicit gradient-enhancement formulation (Peerlings 1999):

$$\overline{\varepsilon}_{eq} - c\nabla^2 \overline{\varepsilon}_{eq} = \varepsilon_{eq} \tag{3}$$

where c has the dimension of length squared and corresponds to the internal length scale parameter. In terms of the present report, on the meso-level c represents the information from the micro-level (i.e. three different values as the three phase material on the meso-level is considered).

4 RESULTS

To carry out a proper investigation of the material behaviour with a help of the multiscale strategy it is important to know the size of the meso-level. As it was mentioned above we will understand the size of the meso-level in terms of the RVE size.

4.1 *Size of 2-D Representative Volume Element*

The Representative Volume Element (RVE) is widely used in nowadays mechanics (Aidun et al. 1999), (Ashihmin and Povyshev 1995), (Behrens et al. 1999), (Fraldi and Guarracino 2001), etc. A definition of the RVE can be written as follows:

The RVE must be chosen "sufficiently large" compared to the microstructural size for the approach to be valid and the RVE is the smallest material volume element of the composite for which the usual spatially constant "overall modulus" macroscopic constitutive representation is a sufficiently accurate model to represent mean constitutive response (Drugan and Willis 1996).

4.1.1 *Procedure description*

Several methods are available in the literature in order to determine the RVE size. Bulsara et al. (1999) in their work used a simulation scheme which generated statistically similar realizations of the actual microstructure of a ceramic-matrix composite on the basis of a radial distribution function obtained by a stereological method and image analysis. Next, they then conducted a systematic investigation of the RVE size with respect to the transverse damage initiation for one fiber volume fraction. The loading cases considered are a tensile load applied transverse to fibers and residual stresses caused by thermal cooling.

Ashihmin and Povyshev (1995) determined the statistical properties of stress with the use of an imitation model. The model is finite-element based. They obtained the statistical criterion for the representative volume determination in metals.

Here we propose a method to determine the size of the RVE. The idea of this method is as follows: for each value of aggregate density distribution on the meso level, a series of numerical experiments for different size of the unit cell are made, and for each unit cell size different aggregate locations (with the given value of aggregate density distribution) are considered. Then a statistical analysis, which is based on the Chi-square criterion, is used to determine the size of the RVE.

4.1.2 *Density distribution of aggregates*

As mentioned above, the idea of the proposed method is to find the unique size of the RVE for different types of material (specifically for different types of concrete). For this reason three different values of aggregate density distribution are assumed, Fig. 1.

4.1.3 *Chi-square criterion*

For the procedure of finding the size of RVE with a giving value of aggregates density distribution the Chi-square criterion is used. This reads

$$\chi^2 = \sum_{i=1}^{n} \frac{(\sigma_i - \langle\sigma\rangle)^2}{\langle\sigma\rangle} \tag{4}$$

where σ_i is the average value of the stress in the current unit cell, $\langle\sigma\rangle$ is the average value of σ_i in all unit cells, n is the number of unit cells under consideration, i.e. there are n unit cells with the same size but different aggregate distribution. So the idea of the method is starting from the smallest unit cell size, then take the size of a unit cell two times larger, and so on until χ^2 from (4) is less than the table value of Chi-square coefficient for a given reliability coefficient and the parameter of the degree of freedom, find the dependence of the Chi-square via n. According to the initially known reliability coefficient and the parameter of the degree of freedom find the table value of the Chi-square coefficient and, with a help of the curve, corresponding size of the unit cell, which will be considered as a size of RVE with a given value of aggregates density distribution, Fig. 2. Thus, for material with the aggregate density distribution 30% according to the graph the size of the RVE should be at least 16, for material with the aggregates density distribution 45% the size of the RVE should be more than 23 and for material with the aggregates density distribution equal to 60% the size of the RVE should be not less than 14. It should be mentioned, that for material with the aggregate density distribution equal to 0% or 100% corresponds to the purely homogeneous material, and in following considerations the size of the RVE of those materials is 0.

Note It is necessary to make clear what is the minimum value of the unit cell (starting size); van Vliet (2000) in his experiments used the RVE size of 3–5 times size of the biggest aggregate. Here, the value of V_{min} is taken as two times the size of the biggest aggregate. This hypothesis is widely used in the literature: Drugan and Willis (1996), Drugan (2000) in

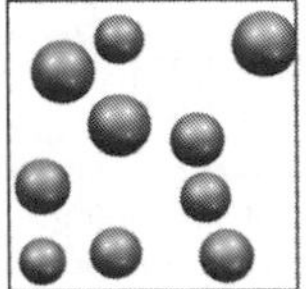
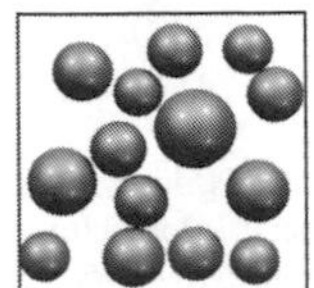

Figure 1. Aggregates density distribution a) 30% b) 45% c) 60%.

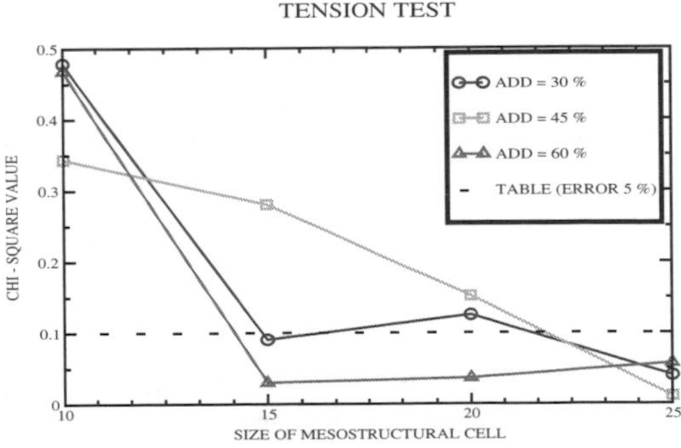

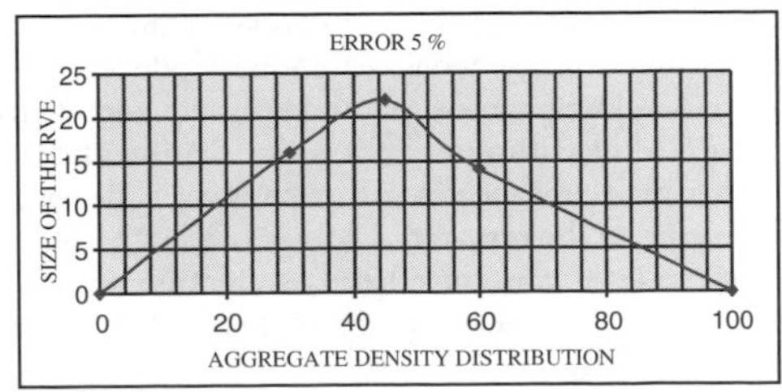

Figure 2. Tension test.

their articles, working with elastic composites, derived quantitative estimates for the minimum RVE size, so they have shown that the minimum RVE size is at most twice the reinforcement diameter.

4.2 *Multiscale results*

As a test of the multiscale procedure, in the concept of the following paper, the analysis of the material behaviour on the macro-level according to the different size of the RVE (on the meso-level) is considered. The response of the material is analysed in the terms of the reaction forces on the macro-level with a given loading. The analysis was carried out for the material with the aggregates density distribution equal to 30%. Knowing that the size of the RVE for the considering aggregates density distribution should be at least 16, the sizes 10 and 25 were taken as a tested parameters. Thus, for each size the series of the aggregates realisations were made and for each realisation the multiscaling analysis was carried out. Next, the variance analysis of the reaction forces on the macro level shows that the deviation in the case of the size of the unit cell equal to 10 is $S = 6.65$ and $S = 0.37$ in case of the size of the unit cell equal to 25. The concept of RVE implies that the size of the RVE should be big enough to lead to the same response for different realisations. Thus, it is concluded that the unit cell with size 10 is not an RVE, while on the contrary the unit cell with size 25 is an RVE. This result is correlated with the results obtained earlier and presented in Fig. 2.

5 CONCLUSIONS

In this paper the modelling of the behaviour of a three phase material is carried out by means of a multiscale approach.

In the following report the procedure of solving above problem is built. The focus of the procedure is multiscale modelling, the main idea is first to consider the structure on the macro-level, assuming that the structure is homogeneous, find the regions of high activity, and second, to consider meso-level, where the structure has heterogeneous character of three phase material, with full set of each material parameters, and third using homogeneous technique feedback to the macro-level with mesostructural accurate knowledge.

Via dealing with above procedure, the authors faced another problem: the size of meso-level is necessary parameter for the research. The size of the structure on the meso-level was considered to identify, according to definition, with a size of the representative volume element (RVE).

The procedure of finding the size of the RVE was based on a statistical/numerical study. Numerical results of the series of test were treated with the help of a Chi-square criterion. The results of these investigations allow to make a conclusion about the dependence of the RVE size from the parameters of material and also from the type of mechanical loading.

REFERENCES

Aidun, J., T. Trucano, D. Lo, and R. Fye (1999). Representative volume size: a comparison of statistical continuum mechanics and statistical physics. Technical report, Sandia National Laboratories, New Mexico.

Ashihmin, V. and I. Povyshev (1995). Statisticheskie zakonomernosti raspredeleniia napriajenii v polikristalah. *Matematicheskoe modelirovanie sistem i protsessov 3,* 11–18.

Behrens, A., H. Just, and D. Landgrebe (1999). Prediction of cracks in multistage cold forging operations by finite-element-simulations with integrated damage criteria. Technical report, Institute for Production engineering and University of the Federal Armed Forces, Hamburg.

Borbely, A., H. Biermann, and O. Hartmann (2001). Fe investigation of the effect of particle distribution on the uniaxial stress–strain behaviour of particulate reinforced metal-matrix composites. *Materials Science and Engineering A 313,* 34–45.

Bulsara, V., R. Talreja, and J. Qu (1999). Damage initiation under transverse loading of unidirectional composites with arbitrarily distributed fibers. *Composites science and technology 59,* 673–682.

Doghri, I. (2002). Multi-scale mechanics of solid materials: modeling and computation. Technical report, Universite catholique de Louvain (UCL).

Drugan, W. (2000). Micromechanics-based variational estimates for a higher-order nonlocal constitutive equation and optimal choice of moduli for elastic composites. *J. Mech. Phys. Solids 48,* 1359–1387.

Drugan, W. and J. Willis (1996). A micromechanics-based nonlocal constitutive equation and estimates of representative volume element size for elastic composites. *J. Mech. Phys. Solids 44*(4), 497–524.

Fraldi, M. and F. Guarracino (2001). A general three-dimensional approach to the analysis of porous media with locally variable volume fraction. In *European Conference on Computational Mechanics.*

Geers, M., V. Kouznetsova, and W. Brekelmans (2001). Nonlocal macroscopic constitutive behaviour for the incorporation of microstructural mechanics. In *European Conference on Computational Mechanics.*

Kouznetsova, V., W. Brekelmans, and F. Baaijens (2001). An approach to micro-macro modeling of heterogeneous materials. *Computational Mechanics 27,* 37–48.

Peerlings, R. (1999). *Enhanced damage modelling for fracture and fatigue.* Ph. D. thesis, Technical University Eindhoven.

van Vliet, M. (2000). *Size effect in tensile fracture of concrete and rock.* Ph. D. thesis, Delft University of Technology.

Vosbeek, P. (1994). *A micromechanical approach to deformation and failure of discrete media.* Ph. D. thesis, Technical University Eindhoven.

Computational Modelling of Concrete Structures, Bićanić et al. (eds)
© 2003 Swets & Zeitlinger, Lisse, ISBN 90 5809 536 3

Investigations on crack propagation in reinforced concrete structures

Ulrich Häussler-Combe
Institute of Concrete Structures and Building Materials, University of Karlsruhe, Germany

ABSTRACT: A numerical method to investigate crack propagation in reinforced concrete structures is presented. This method is based upon the element-free Galerkin method. It takes into account nonlinear concrete constitutive behaviour, nonlinear concrete fracture mechanics and nonlinear bond between concrete and reinforcement. An application with a simple reinforced beam is described, and results for propagation of discrete crack patterns together with crack width results are given.

1 INTRODUCTION

Behaviour of reinforced concrete is strongly influenced by the formation of cracks. Cracks start already under working conditions, cause an adaption of stiffness, allow for a redistribution of stresses and a relaxation of constraints, and in the end contribute to the robustness of reinforced concrete structures. Nevertheless cracks have to be controlled with respect to real crack width in order to protect reinforcing steel against corrosion, to prevent leakage if necessary, and to assure a proper structure appearance. So a realistic modelling and computation of crack formations is essential.

Within Finite-Element methods (FE) cracking of concrete has in most cases been treated by the concept of smeared cracks. So the discontinuity of displacements, which arises from cracking, is approximated by a continuous displacement field with a high gradient. Furthermore characteristic lengths have to be introduced as material parameters, in order to model energy dissipation which is implied with cracking. So conventional FE-methods give an estimation of crack formations, but the exact crack geometry cannot be evaluated except for special cases.

As a new development extensions of the FE-method have been proposed under the general framework of partitions of unity, in particular the extended FE-method (Belytschko et al. 2001), where FE-interpolations allow to describe discontinuities in a direct way. For this purpose also the element-free Galerkin method (EFG) (Belytschko et al. 1996) is given as alternative. As both methods seem to be in an early stage of development a well founded selection between these two alternatives is not possible at the moment. The EFG-method in combination with a FE-approach for the reinforcement was used for the investigations described in this paper.

These investigations based upon concrete constitutive behaviour, concrete fracture mechanics and bond between concrete and reinforcement, which are described in Section 2. Section 3 gives an overview about the EFG-method and its particular implementation, especially regarding crack propagation and coupling with FE rebar-elements. An application example is given in Section 4 and finally some conclusions are drawn in Section 5. A detailed description of all these approaches is given by (Häussler-Combe 2001).

2 REINFORCED CONCRETE BEHAVIOUR

2.1 *Concrete constitutive behaviour*

Concrete is quite a complex material. Among others overviews describing concrete behaviour are given by (Bažant and Planas 1997; Chen and Saleeb 1994; Karihaloo 1995; Eibl and Ivanyi 1976). Some major characteristics may be summarized with

1. formation of micro cracks resulting in
   - stiffness reduction (damage),
   - load induced anisotropy,
2. permanent deformations upon unloading (plasticity),
3. pressure dependent material limit states,
4. volume dilatancy under compression when approaching the limit states,
5. post critical regime with increasing deformations and decreasing stresses,
6. localization phenomena.

Following the current state of knowledge, the major items 5 and 6 can hardly be captured with the classical local continuum approach (de Borst 2001). Here these

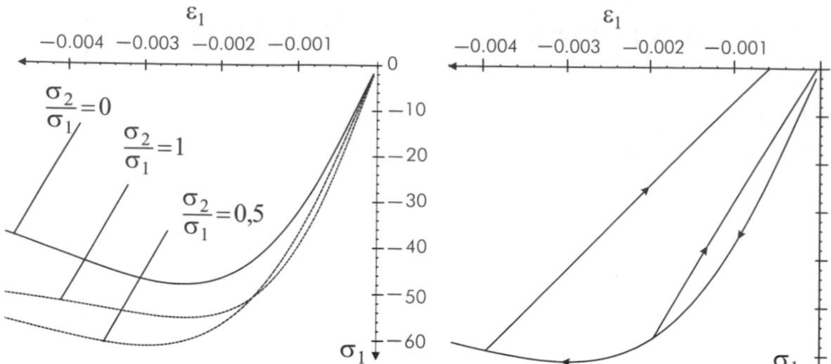

Figure 1. Concrete biaxial behaviour a) monotonic compression with different confining pressures b) unloading with confining pressure $\sigma_1/\sigma_3 = 0.5$.

particular items will be carried over to the structural level using the fictitious crack model (FCM) as proposed by (Hillerborg et al. 1976). This is described in Section 2.2.

On the other hand the major items 1–4 can be dealt with the classical continuum method, e.g. by combining elastoplastic and damage approaches. Assuming small displacements the basic approach is given by

$$\dot{\epsilon} = \dot{\epsilon}_e + \dot{\epsilon}_p + \dot{\epsilon}_d \tag{1}$$

where $\dot{\epsilon}$ means the total strain increment and $\dot{\epsilon}_e, \dot{\epsilon}_p, \dot{\epsilon}_d$ the elastic, plastic and damage contributions. This leads to

$$\dot{\sigma} = \boldsymbol{C}_e \cdot (\dot{\epsilon} - \dot{\epsilon}_p - \dot{\epsilon}_d) \tag{2}$$

with the elastic constitutive stiffness $\boldsymbol{C}_e$. Now it is assumed, that the plastic strain increments have a general form

$$\dot{\epsilon}_p = \boldsymbol{F}_p \cdot \dot{\epsilon} \tag{3}$$

and furthermore the damage strain increments have a general form

$$\dot{\epsilon}_d = \boldsymbol{F}_d \cdot (\dot{\epsilon} - \dot{\epsilon}_p) \tag{4}$$

Combining these equations one gets

$$\dot{\sigma} = \boldsymbol{C}_e \cdot (\boldsymbol{I} - \boldsymbol{F}_d) \cdot (\boldsymbol{I} - \boldsymbol{F}_p) \cdot \dot{\epsilon} \tag{5}$$

with the unit matrix $\boldsymbol{I}$. The plastic and damage strain generators $\boldsymbol{F}_p, \boldsymbol{F}_d$ are each built with the well known plasticity formalism. For the plastic part this may be summarized as follows basing upon a strain-based approach:

- a flow condition $F_p(\epsilon, q_p)$ with one (or more) interval variables q_p,
- a flow rule $\dot{\epsilon}_p = \dot{\lambda}_p \cdot \partial G_p/\partial \epsilon$ with a potential $G_p(\epsilon, q_p)$,
- a Kuhn-Tucker condition $F_p \leq 0, \dot{\lambda}_p \geq 0, F_p\dot{\lambda}_p = 0$ distinguishing loading from unloading,
- evolution laws $q_p = W_p(\epsilon_p)$ for the internal variables.

The derivation of $\boldsymbol{F}_p$ with these basic assumptions follows well known procedures and is omitted here. Formal description of damage follows the same guidelines with particular functions F_d, G_d, W_d. The multiplier λ_d is identified as the well known damage parameter D in case of scalar damage. Suitable forms for all these functions are given by (Schmidt-Hurtienne 2001) and used in the investigations described in this paper.

Simulations of biaxial material testing using this constitutive law are given in Figure 1. Material properties are chosen as for a concrete grade C 40/50 according to the German standard DIN 1045. Figure 1a shows uniaxial stress-strain curves for different confining lateral pressures under monotonic compression. It can be seen, that the strength, i.e. the highest stress reached during the whole loading process is depending on pressure. Figure 1b shows the case with a confining pressure of $\sigma_1/\sigma_3 = 0.5$, where unloading takes place at different points. So the effects of decreasing stiffness, i.e. damage and permanent displacements with a stress-free state, i.e. plasticity can be observed.

Further results for biaxial behaviour are given in Figure 2, where Figure 2a shows the longitudinal stress depending on the volumetric strain. The dilatancy effect, i.e. increasing volume when approaching the limit state is obvious. Furthermore Figure 2b shows the biaxial tension case, where maximum tension strength reaches roughly 10% of the compression strength. This value is not enforced with an explicit tension limit condition, but results implicitely from the damage formulation. Finally Figure 3 shows the computed biaxial stress limit states. These stress states, which cannot be

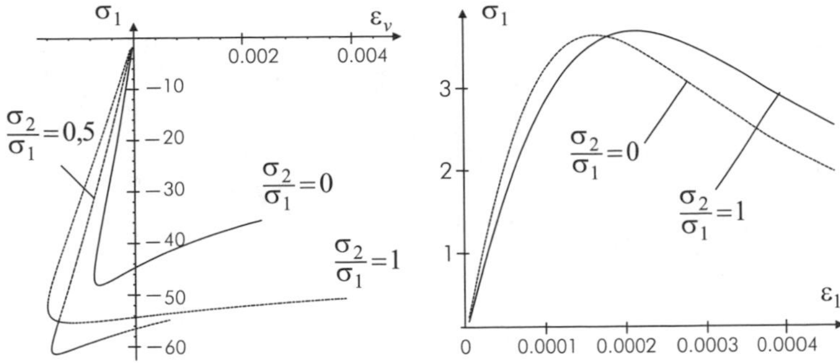

Figure 2. Concrete biaxial behaviour a) volumetric strain in case of monotonic compression b) tension.

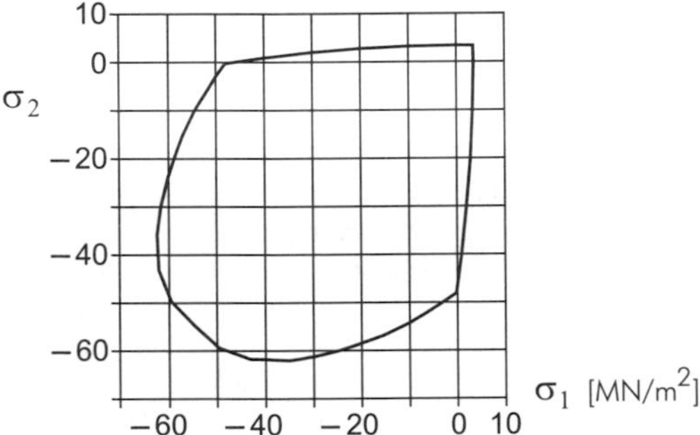

Figure 3. Concrete biaxial stress limit states.

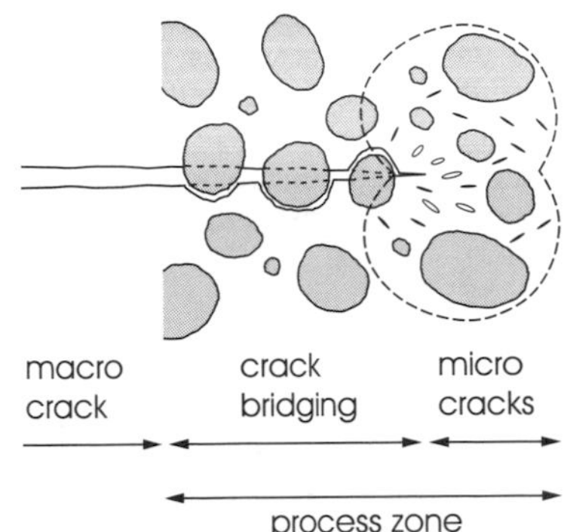

Figure 4. Concrete process zone.

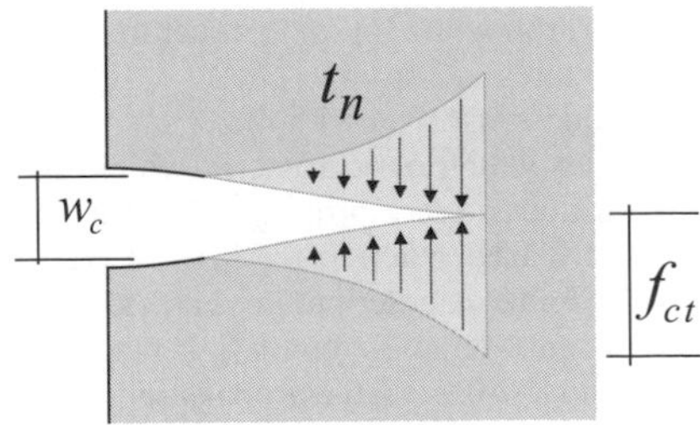

Figure 5. Fictitious crack model.

exceeded, also follow from the damage law and are not explicitely prescribed.

Regarding postcritical softening behaviour, as kann be seen in Figures 1a and 2b, this results from the continuum constitutive behaviour and shows up in the material testing simulation, but is not exploited in the structural computations: softening tension behaviour is replaced by the FCM, and compression softening states are not reached in the range of structural and loading parameters assumed here.

This approach uses scalar damage. So the constitutive law is not able to model load induced anisotropy. An anisotropic damage form as given by (Carol et al. 2001) might describe load induced anisotropy, but this has not been investigated here.

2.2 *Fracture mechanics*

In contrast to brittle materials cracking of concrete affects a larger area around a crack tip, the so called process zone. Here we have an extensive microcracking and a gradual formation of a macro crack with free boundaries, see Figure 4. Due to the heterogeneous structure of concrete a mixture of partially connected matrix areas and gradual debonding between aggregates and matrix is developed, where quite complex cohesive and frictional mechanisms take effect.

An approach to model this behaviour is given by the fictitious crack model (FCM) (Hillerborg et al. 1976). This model introduces forces which are transmitted over a fictitious crack width, see Figure 5. This starts with concrete tensile strength f_{ct} in the fictitious crack tip and reduces to zero at the critical crack width w_c. A traction-displacement relationship

$$t_n = t_n(w) \qquad (6)$$

covers the range between zero crack width and the critical crack width. A number of experimental investigations were performed to determine this relationship.

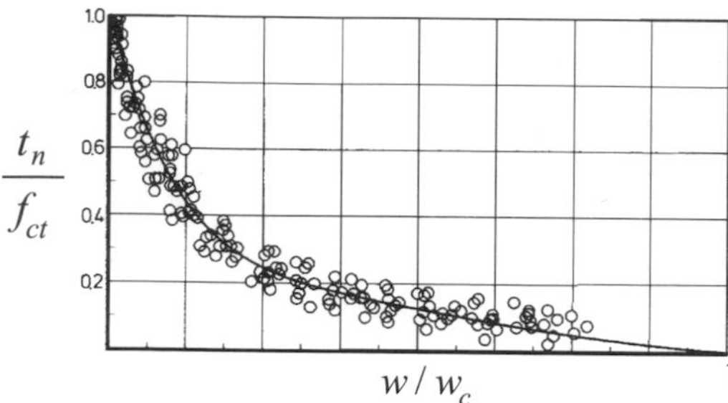

Figure 6. Crack stress depending on crack width.

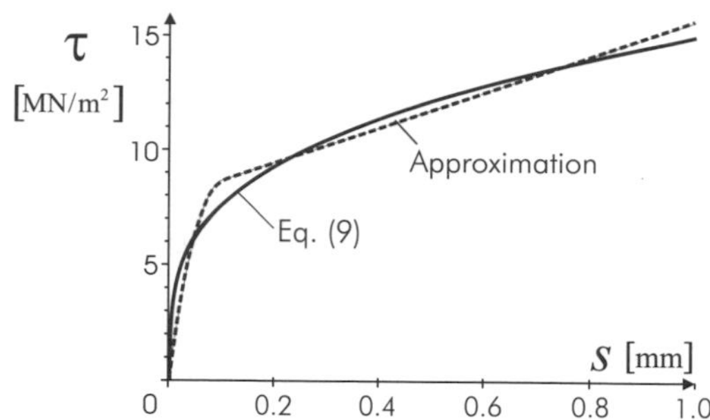

Figure 7. Bond law Eq. (9) and approximation.

A typical result (Cornelissen et al. 1986) is given in Figure 6. As crack tractions reduce with increasing crack widths a nonlinear tension softening effect occurs. For the numerical calculations the values shown in Figure 6 are approximated by a straight line.

Integration of crack tractions along crack width results in a crack energy per unit crack area:

$$G_f = \int_0^{w_c} t_n \, \mathrm{d}w. \tag{7}$$

The crack energy G_f is assumed to be another material constant. As cracking or forming of new concrete surfaces is an irreversible process, crack energy is dissipative and contributes to a ductile structural behaviour. As the size of the process zone and the amount of dissipated energy is more a material property than a structural property, this contribution to ductility is more pronounced in small structures compared to large structures.

A major question arises with the crack propagation criteria: under which conditions, in which direction and how far does a crack propagate? Regarding concrete a widely used criterion for Mode-I cracking is given by the Rankine criterion, i.e. a crack propagates when the concrete tensile strength f_{ct} is reached by the largest principal stress and the propagation direction is perpendicular to the direction of the largest principal stress. Crack propagation should occur as long as the concrete tensile strength is exceeded.

2.3 *Bond behaviour*

Bond between concrete and reinforcement plays a crucial role for structural behaviour. Bond comprises three major mechanisms, which are adhesion between concrete and steel surfaces, furthermore friction after adhesion resistance has been overcome, and finally as major contribution the bearing of reinforcement ribs against concrete (Wicke 1999). Especially the latter introduces a high complexity in a comprehensive modelling and mathematical treatment of bond. Currently this has to be simplified by bond laws, i.e. relations combining pullout forces and pullout displacements or slips respectively of reinforcement bars relative to the confining concrete. Such bond laws have a format

$$\tau = f(s) \tag{8}$$

where τ means bond stress and s means slip.

Assuming usage of standardized reinforcement with ribs the bond behaviour mainly depends upon concrete strength, and to some extent upon the reinforcements position and orientation within the formwork. Considering the case of a lower longitudinal bending reinforcement and covering the stage before debonding a relationship

$$\tau = 0.3 \cdot f_c \cdot s^{0.3} \quad \tau, f_c \text{ in } [\mathrm{MN/m^2}], \ s \text{ in } [\mathrm{mm}] \tag{9}$$

may be used (König and Tue 1996), where the influence of concrete strength is given by the compressive strength f_c. This relation is based on experimental investigations, and insofar bound to dimensions as given in Eq. (9). It is shown in Figure 7 together with an approximation, which is used in the numerical calculations.

3 COMPUTATIONAL METHOD

3.1 *Basic considerations*

Considering the effects discussed above

- nonlinear constitutive behaviour,
- nonlinear concrete fracture mechanics,
- changing geometry arising from crack propagation,
- nonlinear bond characteristics

a numerical method is mandatory to gain results. In particular regarding crack propagation the element-free Galerkin method (EFG) (Belytschko et al. 1996) is chosen as numerical method. The EFG-method allows a direct modelling of displacement discontinuities and propagating cracks, and so it is suitable to model nonlinear concrete fracture mechanics.

The EFG-method has strong similarities compared to the FE-method, and roughly offers the same possibilities to model complex geometries and physical and geometrical nonlinearites. Furthermore with a domain decomposition approach the EFG-method may be coupled with the FE-method within one model. This is used here to model the reinforcement by FE-bar elements, which are coupled through special bond elements to the EFG-domain, which models the concrete. This nonlinear coupled model may be solved by standard well-known methods, e.g. time stepping or line search algorithms combined with solvers for nonlinear algebraic equations like Newton-Raphson or BFGS.

3.2 *Basics of the EFG-method*

Numerical methods are based on approximating a field $u(\boldsymbol{x})$ with $u^h(\boldsymbol{x})$ in a domain Ω, $\boldsymbol{x} \in \Omega$, where $u^h(\boldsymbol{x})$ depends upon a finite number of nodal values u_I nodes $\boldsymbol{x}_I$:

$$u_h(\boldsymbol{x}) = \sum_{I=1}^{N} \varphi_I(\boldsymbol{x}) \cdot u_I, \quad I = 1 \ldots N. \tag{10}$$

For reasons of efficiency the shape functions $\varphi_I(\boldsymbol{x})$ should have a local support, i.e. regarding an arbitrary point $\boldsymbol{x}$ only a minor subset of all shape functions should be nonzero in this particular point. This is a very general approach which covers finite element interpolations and meshless interpolations. The basic difference comes from the fact, that a meshless node does not belong to a finite element, but to a collection of other nodes. Regarding an arbitrary point $\boldsymbol{x}$, this point has a collection of nodes in its immediate neighbourhood, and the nodal values u_I contribute to $u^h(\boldsymbol{x})$. But with meshless interpolations these nodes are not ordered in a sequence as with finite elements, and their number and position may vary to a large extent. A well-known type of meshless interpolation is given by the Moving-Least-Squares (MLS) interpolation (Lancaster and Salkauskas 1981). A MLS shape function may be given by

$$\varphi^{mls}(\boldsymbol{x}) = \boldsymbol{p}^T(\boldsymbol{x}) \cdot \boldsymbol{A}^{-1}(\boldsymbol{x}) \cdot \boldsymbol{B}(\boldsymbol{x}) \tag{11}$$

with

$$\begin{aligned} \boldsymbol{p}(\boldsymbol{x}) &= (1 \;\; x \;\; y)^T \\ \boldsymbol{P} &= [\boldsymbol{p}(x_1) \cdots \boldsymbol{p}(\boldsymbol{x}_N)] \\ \boldsymbol{A}(\boldsymbol{x}) &= \boldsymbol{P}^T \cdot \boldsymbol{W}(\boldsymbol{x}) \cdot \boldsymbol{P} \\ \boldsymbol{B}(\boldsymbol{x}) &= \boldsymbol{P}^T \cdot \boldsymbol{W}(\boldsymbol{x}) \end{aligned}$$

$$\boldsymbol{W}(\boldsymbol{x}) = \begin{bmatrix} w_1 & & & & \\ & \ddots & & & \\ & & w_I & & \\ & & & \ddots & \\ & & & & w_N \end{bmatrix} \tag{12}$$

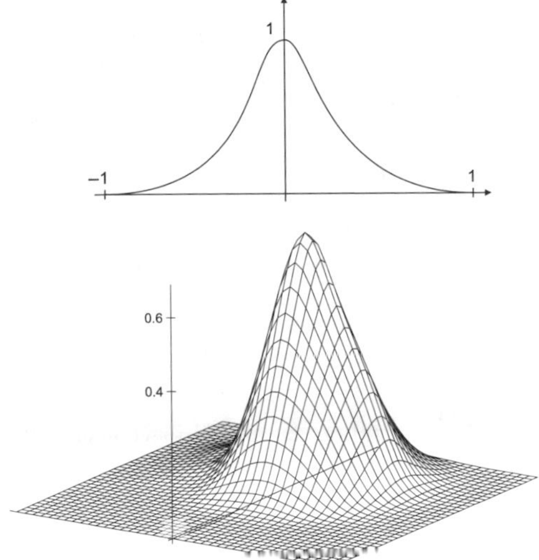

Figure 8. Typical weighting function and MLS-shape function.

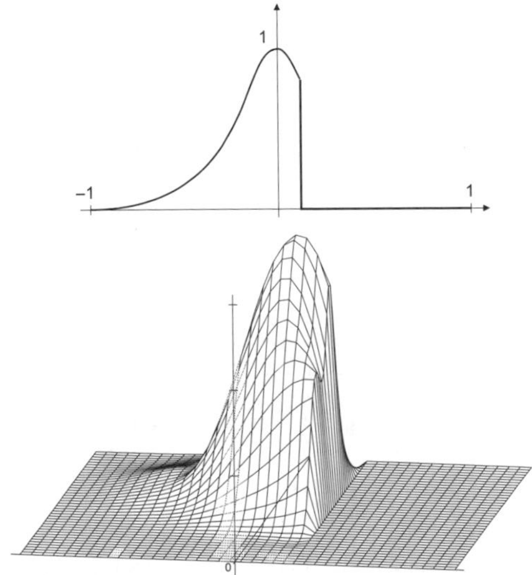

Figure 9. Weighting function and MLS-shape function for discontinuity.

for plane problems. This interpolation is to a major extent ruled by a weighting function $w_I(\boldsymbol{x} - \boldsymbol{x}_I)$, which determines the influence of a node $\boldsymbol{x}_I$ upon the point $\boldsymbol{x}$ actually under consideration. A bell shaped function is an appropriate choice and leads to MLS shape functions of similar geometry, see Figure 8.

The domains of influence need not to be complete circles as used for the example in Figure 8. Furthermore a domain of influence may be cut by a line of discontinuity, see Figure 9. So the influence of the node beyond the line of discontinuity is reduced to zero by reducing the weighting function to zero beyond the line of discontinuity, and in this way the MLS-interpolation allows for a direct interpolation of displacement discontinuities along prescribed contours.

Regarding structural problems one has to consider equilibrium equations, boundary conditions, and constitutive laws. Equilibrium equations and force boundary conditions on the boundary part Γ_t are covered by a weak form

$$\int_{\Omega} \delta\boldsymbol{\epsilon} \cdot \boldsymbol{\sigma} \, \mathrm{d}\Omega = \int_{\Omega} \delta\boldsymbol{u} \cdot \boldsymbol{b} \, \mathrm{d}\Omega + \int_{\Gamma_t} \delta\boldsymbol{u} \cdot \boldsymbol{t} \, \mathrm{d}\Gamma \tag{13}$$

under the assumption that the displacements fulfill the kinematic boundary conditions on the remaining boundary part Γ_u:

$$\boldsymbol{u} = \bar{\boldsymbol{u}} \quad \text{on } \Gamma_u. \tag{14}$$

In meshless interpolations this condition needs a special treatment, e.g. with a penalty approach or Lagrange-multipliers. Using the element-free MLS-Interpolation

$$\begin{aligned} \boldsymbol{\epsilon} &= \boldsymbol{B}^{mls} \cdot \boldsymbol{u}_I \\ \boldsymbol{u} &= \boldsymbol{N}^{mls} \cdot \boldsymbol{u}_I \end{aligned} \tag{15}$$

in Eq. (13) leads to the element-free Galerkin method:

$$\int_{\Omega} \boldsymbol{B}^{mls^T} \cdot \boldsymbol{\sigma} \, \mathrm{d}\Omega = \int_{\Omega} \boldsymbol{N}^{mls^T} \cdot \boldsymbol{b} \, \mathrm{d}\Omega + \int_{\Gamma_t} \boldsymbol{N}^{mls^T} \cdot \boldsymbol{t} \, \mathrm{d}\Gamma \tag{16}$$

This basic approach is the same as with FE-methods. So nonlinearities, e.g. from nonlinear material behaviour $\boldsymbol{\sigma} = \boldsymbol{\sigma}(\boldsymbol{\epsilon})$ may also be considered in the same way. Finally Eq. (16) results in a system of nonlinear algebraic equations

$$\boldsymbol{r}(\boldsymbol{u}_I) = \boldsymbol{p}, \tag{17}$$

where internal nodal forces $\boldsymbol{r}$ have to equal external nodal forces $\boldsymbol{p}$, and the unknown displacements $\boldsymbol{u}_I$ are determined with a nonlinear equations solver. Obviously there are strong similarities between the EFG-method and FE-methods. The differences arise from the different type of interpolation in the shape function matrices $\boldsymbol{N}$ and $\boldsymbol{B}$.

An implementation aspect concerns the numerical evaluation of the integrals in Eq. (16). A special background mesh has to be used for this purpose, introducing "finite" elements by the backdoor. This may cause a discussion, whether this method is truly meshless or not. As the primary intention here is to gain results and the EFG-method is used as tool, this discussion is not contributed.

3.3 *Coupling with finite elements*

While it is possible to use the EFG-method also for one-dimensional structural elements like rebars it does not offer any advantages in this case, but leads to some disadvantages, which are not described in details here. So the FE-method with truss elements is used to model the reinforcement. Both approaches are coupled by embedding truss-elements in the EFG-domain, whereby finite elements and EFG-domain have their own independent node sets. So a node is used either for a truss element or for the concrete continuum, but not for both at the same time. A local coordinate of a rebar shall be given by the length s in its longitudinal direction. In a first view the FE-domain and the EFG-domain provide independent displacements $\boldsymbol{u}_{fe}(s)$ and $\boldsymbol{u}_{efg}(s)$ along the rebar location. These displacements have to be made dependent. This can be reached by coupling forces

$$\begin{aligned} \boldsymbol{t}_{fe}(s) &= \boldsymbol{C} \cdot [\boldsymbol{u}_{fe}(s) - \boldsymbol{u}_{efg}(s)] \\ \boldsymbol{t}_{efg}(s) &= -\boldsymbol{C} \cdot [\boldsymbol{u}_{fe}(s) - \boldsymbol{u}_{efg}(s)], \end{aligned} \tag{18}$$

which are assumed to be linear in a first approach with coupling springs collected in $\boldsymbol{C}$. These forces together form a self equilibrium system, but each contribution can be applied to its particular domain, i.e. $\boldsymbol{t}_{efg}$ as tractions on the EFG-domain with Eq. (16)

$$\begin{aligned} \int_{\Omega_{efg}} \boldsymbol{B}^{mls^T} \cdot \boldsymbol{\sigma} \, \mathrm{d}\Omega = {} & \int_{\Gamma_{t,efg}} \boldsymbol{N}^{mls^T} \cdot \boldsymbol{t} \, \mathrm{d}\Gamma \\ & + \int_{s} \boldsymbol{N}^{mls^T} \cdot \boldsymbol{t}_{efg}(s) \, \mathrm{d}\Gamma, \end{aligned} \tag{19}$$

where body forces have been neglected, and in the same way $\boldsymbol{t}_{fe}$ as tractions on the FE-domain

$$\begin{aligned} \int_{\Omega_{fe}} \boldsymbol{B}^{fe^T} \cdot \boldsymbol{\sigma} \, \mathrm{d}\Omega = {} & \int_{\Gamma_{t,fe}} \boldsymbol{N}^{fe^T} \cdot \boldsymbol{t} \, \mathrm{d}\Gamma \\ & + \int_{s} \boldsymbol{N}^{fe^T} \cdot \boldsymbol{t}_{fe}(s) \, \mathrm{d}\Gamma, \end{aligned} \tag{20}$$

which altogether leads to a coupled system, which can be solved by standard methods.

The special properties of bond between concrete and reinforcement are modelled by choosing the components of the spring $\boldsymbol{C}$ in Eq. (18) in an appropriate way: components in the transverse direction of a rebar are made very stiff – effectively a penalty approach – and components in the longitudinal direction may obey a bond law as defined by Eq. (9). This introduces a nonlinear spring, but this nonlinearity can be solved in the same way as all other nonlinearities.

3.4 *Implementation of crack propagation*

From a more abstract point of view crack propagation using the fictitious crack model as described in Section 2.2 leads to changing boundary geometry – due to the extension of crack faces – and to nonlinear boundary conditions – due to crack tractions depending on crack width or displacements respectively. The latter leads to a displacement dependent right side $\boldsymbol{p}(\boldsymbol{u}_I)$ of Eq. (17), which has to be solved by the nonlinear algebraic equation solver.

More conceptual complexity is caused by the changing boundary geometry, as this is dependent on the crack propagation criteria, see Section 2.2, and so on the stress state of the structure. This can be treated with an explicit approach while using small load increments. An implicit approach requires another iteration loop between the load stepping loop and the nonlinear algebraic equation solution, where an appropriate strategy has to be used regarding crack length and crack direction increments (Häussler-Combe 2001).

4 EXAMPLE

This approach for a numerical method consisting of the EFG-method to model nonlinear constitutive behaviour, nonlinear concrete fracture mechanics and crack propagation, in connection with the FE-method to model reinforcing bars while considering nonlinear bond behaviour is applied to a basic problem of reinforced concrete, a simple reinforced beam with four point bending. System and discretization are shown in Figure 10. Plain strain conditions and a unit width are assumed. Loading is given with two point loads. Material properties are chosen as for a concrete grade C 40/50 according to the German standard DIN 1045. The reinforcement is chosen with a diameter of 15 mm and a spacing $a = 15$ cm. The crack energy is assumed with a value $G_f = 100\,\text{Nm/m}^2$ and the tensile strength with $f_{ct} = 3\text{N/mm}^2$. In the frame of the fictitious crack model as described in Section 2.2 this leads to a critical crack width of $w_c = 0.07$ mm.

The point loads are incrementally increased in small steps, so that the development of deformations, stresses and the propagation of cracks can be followed during the loading history. According to beam theory a constant moment is given between the two point loads, with constant stresses along the beam axis. A first distinct point of the loading history is given, when the concrete tensile strength is reached at the bottom side of the beam. It is assumed, that the weakest strength is given in beam midspan with a slight variation. So crack initiation and propagation starts in midspan according to the Rankine criterion as described in Section 3.4. The load factor is set to $LF = 1.0$ at this point. With a slightly increasing load further cracks are initiated in the constant moment area. Due to symmetry always a pair of cracks with a symmetric location initiates in the calculations. A first pair is initiated directly below the two point loads, so that a group of three cracks is given at a calculated load factor of $LF = 1.033$, see Figure 11a.

In a fully developed crack tensile stresses have been transferred from concrete into the reinforcement. So concrete tensile stresses vanish in the vicinity of crack faces. They gradually build up with increasing distance from the crack faces due to bond between concrete and reinforcement. Generally concrete tensile stresses are reduced in the sections between cracks. So the load has to be increased to initiate further cracks.

The results are given in Figure 11b–e for a sequence of load steps. The following statements can be made basing upon the computational results:

- An initiated crack does not grow instantaneously. A considerable increasing of load is necessary, to reach a value near its final length. This may be explained with the effects of the crack tractions introduced by the fictitious crack model: these tractions delay crack propagation, especially in the early stages.

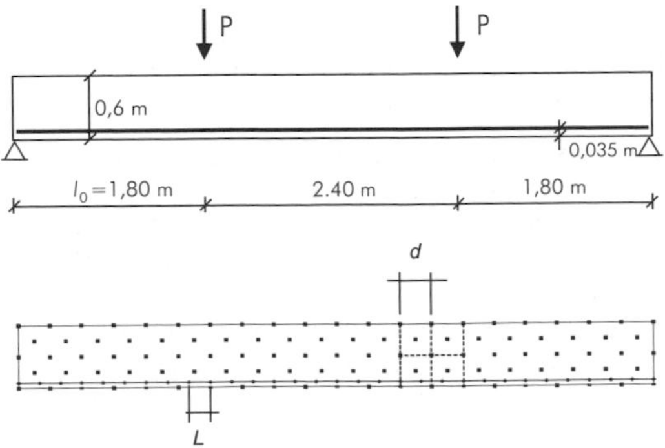

Figure 10. System and discretization.

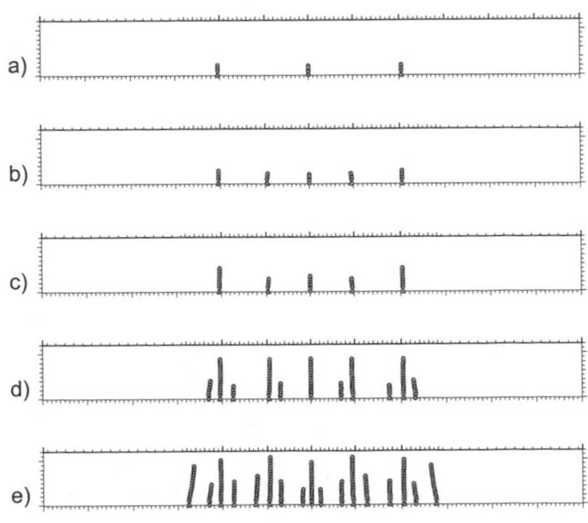

Figure 11. Crack propagation at different loading steps a) $LF = 1.033$ b) $LF = 1.100$ c) $LF = 1.167$ d) $LF = 1.300$ e) $LF = 1.433$.

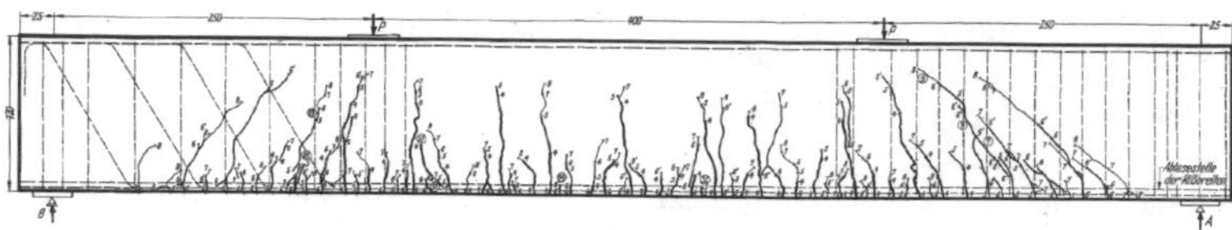

Figure 12. Crack formations of a simple beam under constant moments.

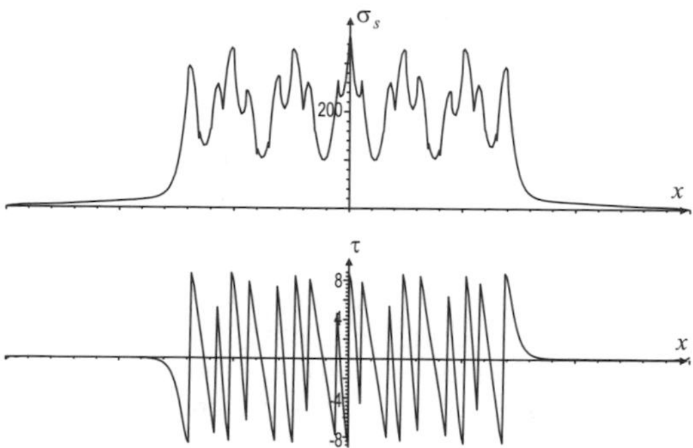

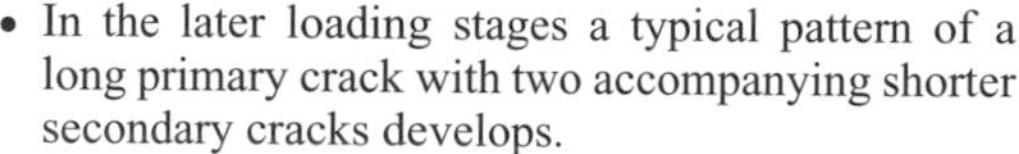

Figure 13. Reinforcement at load factor $LF = 1.433$ a) normal stresses b) bond stresses.

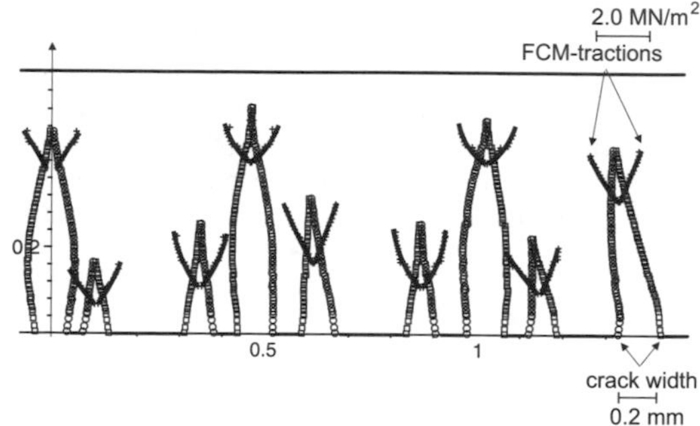

Figure 14. Crack width and crack tractions at load factor $LF = 1.433$.

- In the later loading stages a typical pattern of a long primary crack with two accompanying shorter secondary cracks develops.

The latter effect can also be observed in experiments, e.g. as performed by (Rüsch and Rehm 1963), see Figure 12, and is also explained with simple mechanical models (Eibl and Ivanyi 1976). Comparing Figure 11e and Figure 12 the overall computational and experimental crack formations agree quite well regarding the area of constant moment, as agreement can be expected with a more or less homogeneous experimental beam. In the lateral areas considerable differences are given, as shear cracks develop in the experimental beam, which is equipped with stirrups. Such stirrups have not been regarded in the computational model.

The computational crack pattern shown in Figure 11e also describes the final pattern, where only small changes will take place with a further increase of loading.

Beneath crack propagation the computational model allows to capture the behaviour of the longitudinal reinforcement and its bond with concrete. This can be seen in Figure 13, where the computed normal stresses of the reinforcement and the bond stresses are shown. As to expect high stress peaks arise directly in the cracked sections, regarding both reinforcement normal stresses and bond stresses. A difference is given with respect to signs: reinforcement normal stresses are always positive while bond stresses change sign in the cracked section.

Furthermore crack widths can directly be evaluated in the computations, as the EFG-method allows to model displacement discontinuities along the crack lines shown in Figure 12. Actually two opposite crack faces belong to every shown crack line, and each of these crack faces is computed independent from its counterpart by the EFG-method. Their common distance gives the crack width. The computed crack width results are given in Figure 14. Due to the systems and results symmetry only the right side of the cracked area is shown. The size of each crack width has been scaled by a large factor compared to the beams height, the used scaling for the crack width is also indicated in Figure 14. Following the fictitious crack model a crack width smaller than the critical crack width of $w_c = 0.07$ mm should be connected with crack tractions. These have also been regarded in the numerical model as described in Section 3.4. The computed crack tractions are also shown in Figure 14, graphically attached around the crack tips. The following statements can be made:

- Computed crack spacing has mean values of about 0.2 m, which corresponds to values, which are experimentally determined, and also to generally used models for crack width evaluation, see (König and Tue 1996).
- The computed maximum crack width reaches a value up to 0.2 mm. This also corresponds to generally evaluated values, see again (König and Tue 1996).
- One may notice, that the primary long cracks are tied together by the reinforcement and slightly

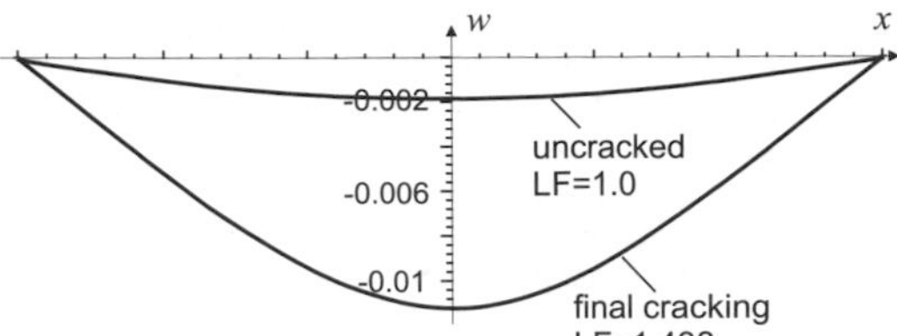

Figure 15. Computed beam deflections.

widen up between the lower reinforcement and the crack tips. This is not of major practical importance, but confirms the plausibility of the numerical calculations.
- The tip values of the crack tractions roughly start with the concrete tensile strength f_{ct} and decrease to zero at the critical crack width w_c according to the assumed FCM-law. The areas affected by crack tractions are larger in case of the shorter secondary cracks.

Regarding crack tractions, it can be seen from the numerical results, that they give a considerable contribution to the overall ductility and to a less extent to the ultimate load-carrying capacity.

Finally the beam deflections are given as a further computational result. Two deflection curves, i.e. the displacements of the lower beam side are shown in Figure 15, with the deflection of the uncracked beam just before initiation of cracking and with the deflection, when the final crack pattern is reached. Taking the increased load into account, the difference in displacements between the uncracked and the cracked beam is given by a factor 4–5, which is in accordance with general experience.

5 CONCLUSIONS

It can be shown, that the described numerical approach is basically able to analyze typical phenomena of reinforced concrete like nonlinear constitutive behaviour, crack propagation, nonlinear concrete fracture mechanics and nonlinear bond in quite a realistic way. This has been shown for the case of beam bending. Cases with shear loading may principally be treated with the same method: shear reinforcement can be modelled in the same way as longitudinal reinforcement, formally by introducing a number of separated FE-domains.

A major problem lies in the extension of the fictitious crack model from Mode-I cracking to Mode-II cracking. In the approach used here a shear component has to be added in the crack tractions. Several theoretical approaches have been proposed for a general treatment of cracking including shear (Jirasek and Zimmermann 2001), but the experimental foundation seems to be missing. Principally any type of relationship between crack tractions and crack width can be incorporated in the method described here.

A further open questions arises for the case, when concrete is under high compression beyond its ultimate strength in the softening regime. Loss of ellipticity in the static case concerns the EFG-method in the same way as the FE-method. An accepted remedy is given by an nonlocal extension of the continuum constitutive law. Regarding the EFG-method this has been investigated (Askes 2000), and is transferable to the given frame.

Finally it has to be noted, that the extended finite element method mentioned in the introduction seems to be a promising numerical approach to deal with discontinuities in solids. Its application to reinforced concrete in the way described here seems to be open at the moment and needs investigations.

REFERENCES

Askes, H. (2000). *Advanced spatial discretization strategies for localised failure*. Ph.D. thesis, Technical University Delft.

Bažant, Z. and J. Planas (1997). *Fracture and Size Effect in Concrete and Other Quasibrittle Materials.* CRC Press.

Belytschko, T., Y. Kongrauz, D. Organ, and P. Krysl (1996). Meshless Methods: an Overview and Recent Developments. *Comp. Meths. Appl. Mech. Engng. 139*, 3–47.

Belytschko, T., N. Moës, S. Usui, and C. Parimi (2001). Arbitrary discontinuities in finite elements. *Int. J. Numer. Meth. Engng. 50*, 993–1013.

Carol, I., E. Rizzi, and K. Willam (2001). On the formulation of anisotropic elastic degradation. I. Theory based on a pseudo-logarithmic damage tensor rate, II. Generalized pseudo-Rankine model. *Int. J. Solids Structures 38*, 491–546.

Chen, W. and A. Saleeb (1994). *Constitutive Equations For Engineering Materials, Volume 2: Plasticity and Modeling.* Amsterdam: Elsevier Science.

Cornelissen, H., D. Hordijk, and H. Reinhardt (1986). Experimental determination of crack softening characteristics of normal and lightweight concrete. *HERON 31*, 45–56.

de Borst, R. (2001). Some recent issues in computational failure mechanics. *Int. J. Numer. Meth. Engng. 52*, 63–95.

Eibl, J. and G. Ivanyi (1976). Studie zum Trag- und Verformungsverhalten von Stahlbeton, Heft 260. Technical report, DAfStB.

Häussler-Combe, U. (2001). Elementfreie Galerkin-Verfahren. Grundlagen und Einsatzmöglichkeiten zur Berechnung von Stahlbetontragwerken. University of Karlsruhe. Habilitation thesis.

Hillerborg, M., M. Modeer, and P. E. Petersson (1976). Analysis of crack formation and crack growth in concrete by means of fracture mechanics. *Cement and Concrete Research*, 773–782.

Jirasek, M. and T. Zimmermann (2001). Embedded crack model: I. Basic formulation. *Int. J. Numer. Meth. Engng. 50*, 1269–12903.

Karihaloo, B. (1995). *Fracture Mechanics and Structural Concrete*. Harlow, England: Longman Scientific & Technical Ltd.

König, G. and N. Tue (1996). Grundlagen und Bemessungshilfen für die Rissbreitenbeschränkung im Stahlbeton und Spannbeton, Heft 466. Technical report, DAfStB.

Lancaster, P. and K. Salkauskas (1981). Surfaces generated by moving least square methods. *Math. Comput. 37.*

Rüsch, H. and G. Rehm (1963). Versuche mit Betonformstählen, Heft 140. Technical report, DAfStB.

Schmidt-Hurtienne, B. (2001). *Ein dreiaxiales Schädigungsmodell für Beton unter Einschluss des Dehnrateneffekts unter Einschluss des Dehnrateneffekts*. Ph.D. thesis, University of Karlsruhe.

Wicke, M. (Ed.) (1999). *Structural Concrete – Textbook on Behaviour, Design and Performance, Volume 1*, Lausanne. International Federation for Structural Concrete (fib).

Computational Modelling of Concrete Structures, Bićanić et al. (eds)
© 2003 Swets & Zeitlinger, Lisse, ISBN 90 5809 536 3

Analysis of concrete fracture using a heterogeneous continuum model

V. Mechtcherine & H.S. Müller
Institute for Concrete Structures and Building Materials, University of Karlsruhe

ABSTRACT: Deformation controlled uniaxial tension tests on dog-bone shaped specimens and notched specimens as well as three-point bend tests performed in earlier studies were analysed numerically using a stochastic distribution of the properties of concrete, i.e. of the finite elements. An extensive numerical investigation was carried out in order to study the sensitivity of this approach, which was defined as a heterogeneous continuum model, with regard to the coefficient of variation of the input material properties, the mesh size and the applied FE formulation for cohesive cracks ("fixed crack" or "rotating crack"). On the basis of this comparative analysis a suitable approach for the application of the model was developed which allows a realistic simulation of fracture processes in different types of concrete and other cementitious materials. Additionally, it was shown, that the heterogeneous continuum model is an appropriate tool for the analysis of different fracture phenomena such as the localisation of failure or the effect of the rotatability of loading plates in uniaxial tension tests.

1 INTRODUCTION

Numerical modelling of concrete failure and fracture phenomena in concrete as well as other cementitious materials became a field of an intensive research in the recent years. With respect to the accuracy and the efficiency of corresponding models some few still open questions have to be focused. One of these questions is investigated in this paper: How the heterogeneity of cementitious materials can be taken into consideration in the most realistic way using commercially available finite element programs?

The chosen approach, which was defined as a heterogeneous continuum model, was to introduce a stochastic distribution of the properties of concrete, i.e. of the finite elements. This can be done, e.g. by supplementing the existing FE programs, which provide a cohesive crack formulation, by an appropriate user-subroutine. By means of this subroutine the parameter of the cohesive crack model can be defined for individual elements or groups of elements. In fact, a comparable approach to simulate the heterogeneity of concrete had been already used by Wittmann (Wittmann 1994). However, no systematic study concerning the specific properties of the model had been performed.

In order to investigate the sensitivity of the heterogeneous continuum model with regard to different parameters the deformation controlled uniaxial tension tests on dog-bone shaped concrete prisms and notched concrete prisms as well as three-point bend tests (see Figure 1) carried out in the earlier studies (Mechtcherine 2000, Mechtcherine & Müller 1997, 1998) were analysed numerically using this approach. The simulation of the experiments with different set-ups enabled to study the various aspects of the specific behaviour of the model and to verify its suitability for the simulation of some important fracture mechanical phenomena.

2 NUMERICAL MODEL

For the description of the behaviour of concrete subjected to tensile stresses the crack band model (Bazant 1983) was chosen, which is applied in numerous advanced FE programs. In this model cracks are regarded as smeared over the entire continuum finite element. In this study a linear σ-ε relation was used to describe the stress-strain behaviour of the material prior to the formation of cracks. It is defined by the tensile strength $f_{t,inp}$ (inp = input) and the modulus of elasticity $E_{0,inp}$ (Figure 2). For better understanding of the results of the calculations a simple linear softening behaviour has been chosen. The critical strain $\varepsilon_{cr,inp}$ depends on the characteristic element length h and can be calculated from the critical crack opening $w_{cr,inp}$ using the following formula: $\varepsilon_{cr,inp} = w_{cr,inp}/h$. Accordingly, the area below the entire σ-ε relation is equal to $G_{F,inp}/h$ (see Figure 2), where $G_{F,inp}$ is the part of the fracture energy corresponding to the area under the first, steeper part of the softening curves measured in the fracture mechanical experiments (Mechtcherine 2000, Mechtcherine & Müller 1998).

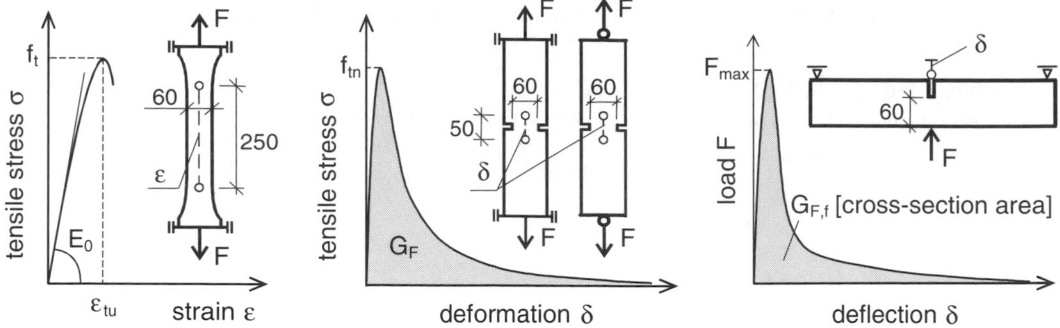

Figure 1. Schematic view of the geometry of the 3 types of specimens investigated, and typical corresponding relations obtained from the tests (geometrical data in [mm]) (Mechtcherine 2000, Mechtcherine & Müller 1998).

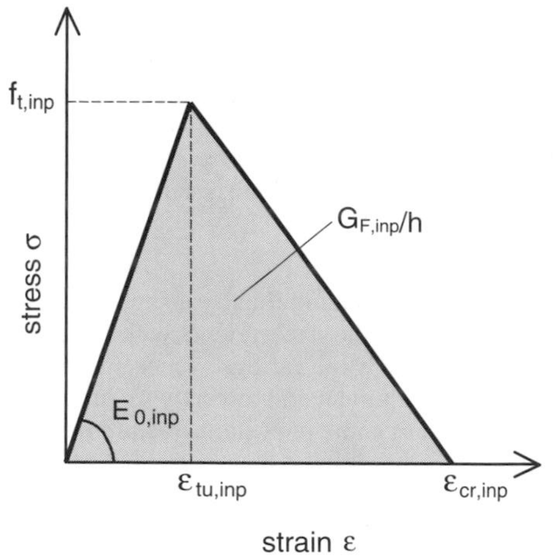

Figure 2. Applied crack band model for the description of the behaviour of concrete subjected to tension.

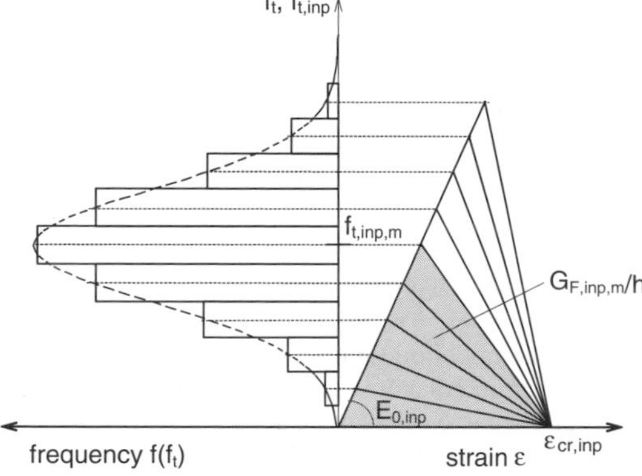

Figure 3. Assignment of the constitutive relations to the groups of finite elements.

The heterogeneity was introduced considering the tensile strength $f_{t,inp}$ to be an independent random variable following a Gaussian distribution (see Figure 3). The modulus of elasticity $E_{0,inp}$ and the critical strain $\varepsilon_{cr,inp}$ were kept constant for all finite elements. For a better handling, the elements were subdivided into nine groups according to their tensile strength. Own material laws were assigned to each group. Details may be found in (Mechtcherine 2000).

The choice of the input values has been made following the experimental results for one of the normal strength concretes investigated (Mechtcherine & Müller 1998). For the calculations presented in this paper the modulus of elasticity obtained from the uniaxial tension tests on dog-bone specimens has been chosen: $E_{0,inp} = 37.1$ GPa. The critical crack opening $w_{cr,inp} = 0.03$ mm was defined under consideration of the crack opening, which results from the extrapolation of the first steeper part of the stress-crack opening relation measured in the uniaxial tension tests on notched specimens with non-rotatable boundaries. The designation of the mean value for the tensile strength $f_{t,inp,m}$ was less straightforward. Preliminary investigations showed, that the $f_{t,inp,m}$-value has to be somewhat higher than the average value of tensile strength $f_t = 4.2$ MPa obtained from the uniaxial tension tests on dog-bone specimens for the concrete under consideration, in order to arrive approximately this experimental value. For these investigations the input value $f_{t,inp,m} = 5.0$ MPa was chosen. The input values of the fracture energy $G_{F,inp}$ were related to the corresponding $f_{t,inp}$-values by the following formula: $G_{F,inp} = f_{t,inp} \cdot w_{cr,inp}/2$.

The FE calculations were performed under the assumption of plane stress conditions. The areas of the test specimens where a crack may develop were modelled by means of four-point square elements. Since a one-point integration scheme was applied, the characteristic length of the finite elements h corresponded to the side length of the elements (further indicated as "mesh size" or "element size" h).

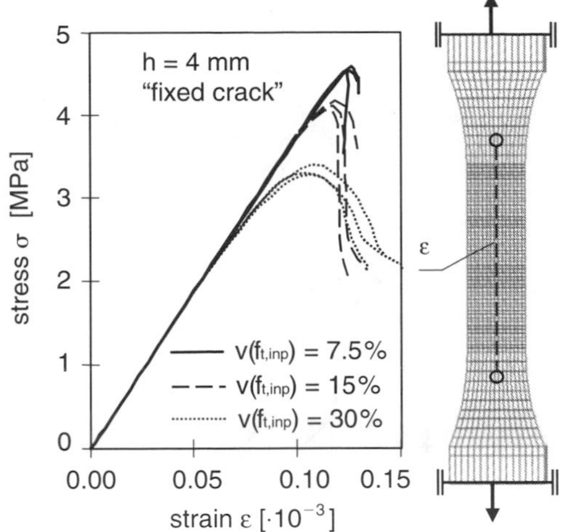

Figure 4. Effect of the coefficient of variation of the tensile strength $f_{t,inp}$ on the shape of the calculated σ–ε relations by the simulation of the tension tests on dog-bone shaped prisms.

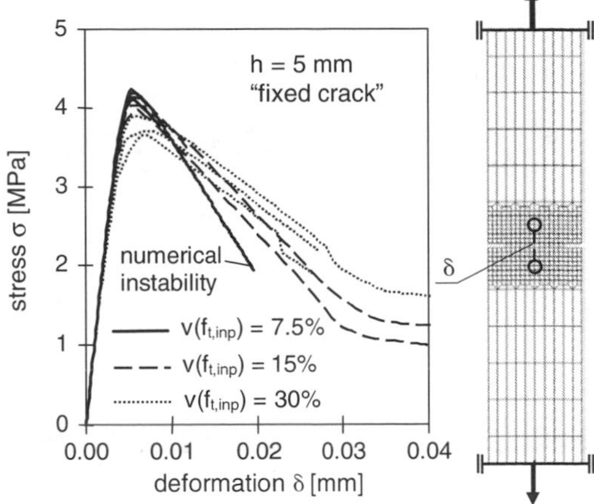

Figure 5. Effect of the coefficient of variation of the tensile strength $f_{t,inp}$ on the shape of the calculated σ–δ relations by the simulation of the tension tests on notched prisms.

3 PARAMETER ANALYSIS

3.1 *Coefficient of variation of the input tensile strength*

For the investigation of the effect of the scattering of the input tensile strength $f_{t,inp}$ on the behaviour of the heterogeneous continuum model the simulation of the uniaxial tension tests on dog-bone specimens and notched specimens with non-rotatable boundaries was performed using three different coefficients of variation $v(f_{t,inp})$ of 7.5, 15 and 30%. With each of these $v(f_{t,inp})$-values three independent simulations were completed for both types of test. The discretisation was carried out with a mesh size of h = 4 mm in the case of the dog-bone prisms and h = 5 mm in the case of the notched prisms (see Figure 4, right). In both analyses the widely used "fixed crack" formulation for smeared cracks was applied.

Figures 4 and 5 show the calculated stress-strain and stress-deformation relations, respectively. It is apparent that an increase of the coefficient of variation $v(f_{t,inp})$ leads to a significant decrease of the calculated values of the tensile strength $f_{t,cal}$ as well as of the net tensile strength $f_{tn,cal}$. The effect is more pronounced for the $f_{t,cal}$-values obtained from the simulation of tests on dog-bone shaped prisms. Concerning the ascending branch of both σ-ε and σ-δ relations higher $v(f_{t,inp})$-values result in an apparent loss of its linearity long before reaching the tensile strength. Further, the descending branch of both relations becomes shallower indicating a more ductile fracture behaviour of the simulated material. It is worthy to note, that the σ-δ relations have a bend at a deformation of approx. 0.03 mm (= $w_{cr,inp}$!) and become significantly more shallow at higher deformations (Figure 5). The reason for this phenomenon is most likely the so-called "stress-locking" effect (Rots 1988), which might be rather pronounced in the present case considering the curved crack path (compare Figure 11).

3.2 *Mesh size*

In order to study the mesh size sensitivity of the model again a simulation of the uniaxial tension tests with non-rotatable boundaries were performed using the "fixed crack" formulation. Three different mesh sizes h of 2, 4 and 8 mm were used for the discretisation of the dog-bone shaped prisms, while the notched prisms were descretised by the elements with a side length of 2.5 or 5 mm. For each combination of parameters simulations with three independent distributions of the $f_{t,inp}$-values were carried out. The coefficient of variation of the input tensile strength was $v(f_{t,inp}) = 30\%$.

The results of the calculations are presented in Figures 6 and 7. With an increasing mesh size the tensile strength $f_{t,cal}$ and the ultimate strain $\varepsilon_{tu,cal}$ obtained from the simulation of the tests on dog-bone prisms decrease, whereas the net tensile strength $f_{tn,cal}$ from the simulation of the experiments on notched specimens increases. Further calculations performed using lower values of $v(f_{t,inp})$ showed similar tendencies, however the effect of the mesh size was less pronounced (Mechtcherine 2000). Furthermore, the additional simulation of the three-point bend tests on notched beams using a coarser mesh provided higher values of the net bending strength $f_{ft,cal}$ comparing

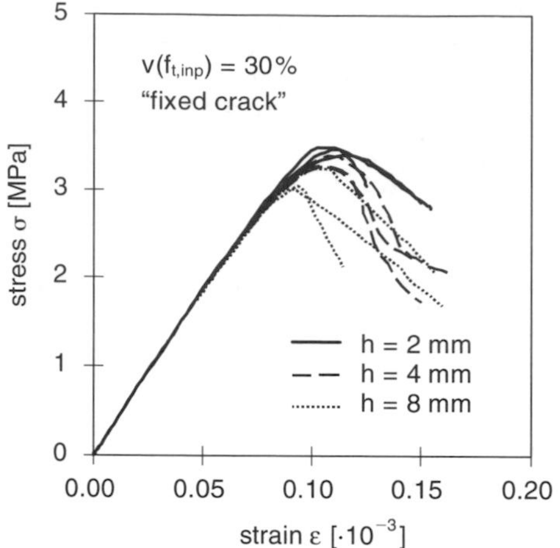

Figure 6. Effect of the mesh size on the shape of the calculated σ–ε relations by the simulation of the tension tests on dog-bone shaped prisms.

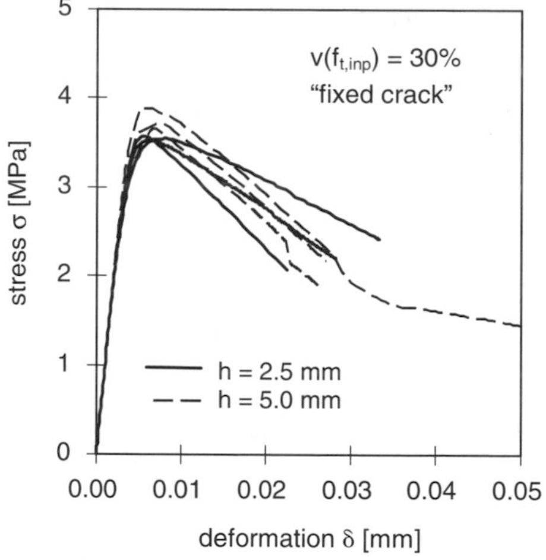

Figure 7. Effect of the mesh size on the shape of the calculated σ–δ relations by the simulation of the tension tests on notched prisms.

to the corresponding calculations with a finer mesh (Mechtcherine 2000).

3.3 *Formulation of the smeared crack*

In a series of calculations the "rotating crack" formulation for smeared cracks was used alternatively to the "fixed crack" formulation. Applying the approach "fixed crack" the coaxiality between stress and strain in a finite element is not maintained after the formation of a crack: The principal stress axes remain fixed to the orientation at the moment of cracking (at a principal stress $\sigma = f_{t,inp}$), while the principal strain axes

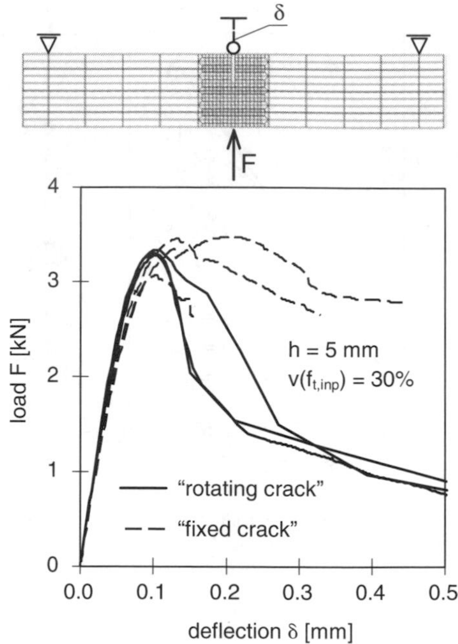

Figure 8. Effect of the finite element formulation of cracks as "fixed" or "rotating" on the calculated F–δ relations by the simulation of the three-point bend tests on notched beams.

can change their orientation (Rots 1988). The "rotating crack", in the contrary, changes his orientation in the manner that the principal stresses and strains during entire calculation always coincide (Willam 1989).

The numerical analysis of different experiments showed that effects of the formulation of smeared cracks were obtained the most pronounced in simulations of bend tests, especially for higher coefficients of variation $v(f_{t,inp})$. Figure 8 presents the load-deflection relations obtained from the calculations with a mesh size $h = 5$ mm and a coefficient of variation $v(f_{t,inp}) = 30\%$. The descending branches of the F-δ curves obtained using the "rotating crack" approach are clearly below those of the corresponding relations from the simulation by means of the "fixed crack" formulation. This can be traced back to the effect of an apparent "strengthening" due to the loss of the coaxiality between the principal stress and strain axes during the simulation using the "fixed crack" formulation.

4 PHENOMENOLOGICAL STUDIES

4.1 *Localisation of failure*

Though the phenomenon of localisation of failure belongs to the key topics of the fracture mechanics of cementitious materials, it is still insufficiently investigated, both theoretically and experimentally.

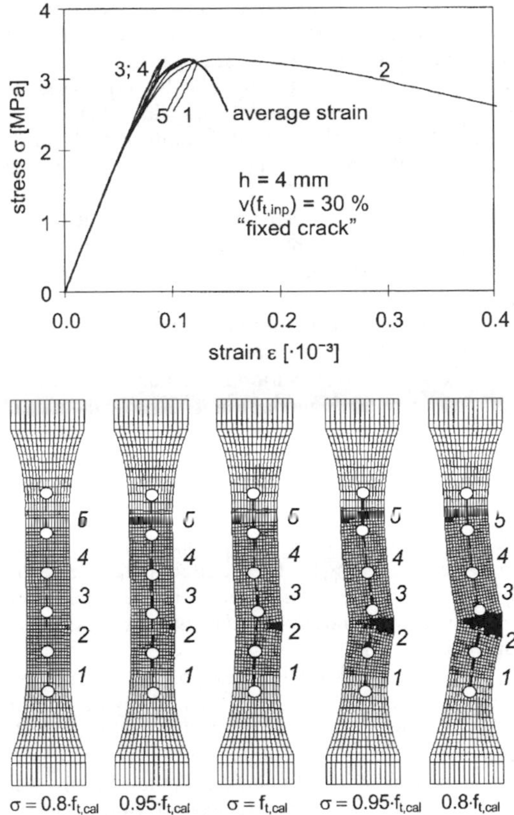

Figure 9. Calculated $\sigma-\varepsilon$ relations for different locations on the axis of the vertical symmetry of a dog-bone shaped prism subjected to tension (above), calculated deformation of dog-bone shaped prisms and the positions of the virtual gauges (below).

Therefore, it was essential to check the suitability of the heterogeneous continuum model for the investigation of this phenomenon. In order to do that, the uniaxial tension tests on dog-bone shaped prisms were simulated using a mesh size h of 4 mm and a coefficient of variation $v(f_{t,inp})$ of 30%. Besides the average strain, which was calculated over the length of a virtual gauge of 250 mm (compare with Figures 1 and 4), also local strains at different locations on the axis of vertical symmetry of the specimen were considered (Figure 9).

The results of the numerical simulation (Figure 9) correspond very well with own experimental results and the results obtained by Heilmann et al. (Heilmann et al. 1969), who used specimens of similar geometry. The $\sigma-\varepsilon$ relation for the average strain shows a deviation from the linearity beginning at a stress level of approx. 80% of the tensile strength $f_{t,cal}$. In the sector,

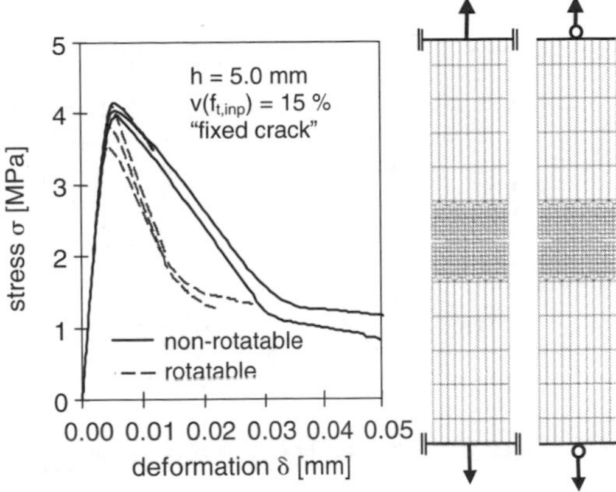

Figure 10. Effect of the rotatability of the boundaries at tension tests on the shape of the calculated $\sigma-\delta$ relations.

where the main crack develops (here: the sector of the gauge 2 in Figure 9), the deviation of the $\sigma-\varepsilon$ curve from the linearity can be observed already at stresses of approx. 60% of the $f_{t,cal}$-value. Besides this main crack further cracks start to grow (here: in the sectors 1 and 5), however they close again after the calculated tensile strength is reached. The undamaged concrete shows to a great extent a linear-elastic behaviour (see the curves for the sectors 3 and 4).

4.2 *Effect of the rotatability of the loading plates*

The test set-up has a significant influence on the obtained material parameters (Mechtcherine 2000, Mechtcherine & Müller 1998). The experiments show that the uniaxial tension tests with rotatable boundaries provide always lower values of the tensile strength and fracture energy as the corresponding tests with non-rotatable boundaries (Mechtcherine & Müller 1998, Van Mier et al. 1996). In order to analyse this phenomenon numerically the heterogeneity of concrete must be considered in some way (Mechtcherine 2000, Mechtcherine & Müller 1998). Here, the suitability of the developed heterogeneous continuum model to study such phenomena was checked.

The uniaxial tension tests on notched prisms with rotatable and non-rotatable loading plates were simulated using a mesh size h of 5 mm and a coefficient of variation $v(f_{t,inp})$ of 15%. Figure 10 gives the calculated stress-deformation relations, indicating that the $\sigma-\delta$ curves for the tests with rotatable boundaries are below the corresponding relations for the tests with non-rotatable loading plates. Further, they have a steeper descending branch, which indicates a more brittle failure of concrete in the tests with rotatable boundaries. The shape of the calculated σ-δ curves corresponds very well to the characteristic shape of

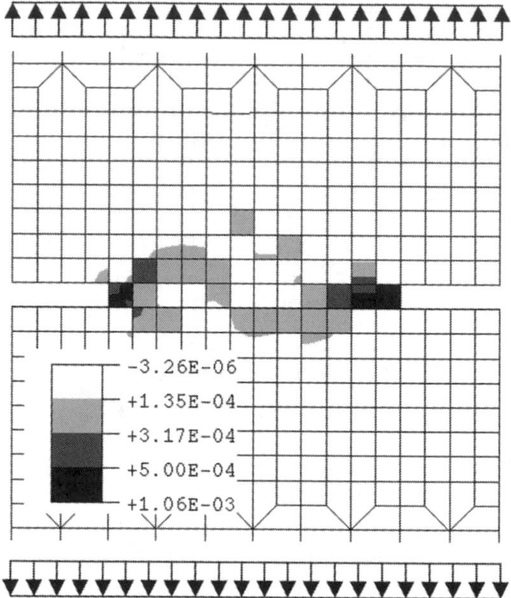

Figure 11. Distributions of the strains $\varepsilon \geq \varepsilon_{tu,inp,m}$, when the tensile strength $f_{tn,cal}$ is reached, in the simulation of the uniaxial tension tests on notched prisms with non-rotatable boundaries.

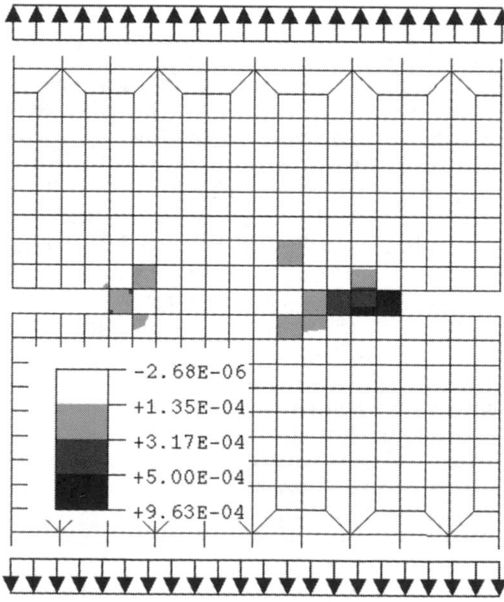

Figure 12. Distributions of the strains $\varepsilon \geq \varepsilon_{tu,inp,m}$, when the tensile strength $f_{tn,cal}$ is reached, in the simulation of the uniaxial tension tests on notched prisms with rotatable boundaries.

the stress-deformation relations observed in the experiments (Mechtcherine 2000, Mechtcherine & Müller 1998). The average tensile strength $f_{tn,cal}$ obtained from the simulated test with rotatable boundaries is 5.9% lower than the mean $f_{tn,cal}$-value obtained by the simulation of the tests with non-rotatable loading plates. This result is also in agreement with the experimental results, which provided a difference of 6.3% between the corresponding values of the tensile strength f_{tn} measured by means of the set-ups under consideration (Mechtcherine 2000, Mechtcherine & Müller 1998).

For a comprehensive interpretation of these findings, it is helpful to compare the degree of damage of concrete and the distribution of damage considering both tests set-ups. Figures 11 and 12 show the calculated distributions of the strains ε exceeding the average ultimate strain $\varepsilon_{tu,inp,m} = f_{t,inp,m}/E_{0,inp} = 0.135 \cdot 10^{-3}$ in the simulated tests both with non-rotatable and rotatable boundaries, respectively. In the experiment with non-rotatable loading plates the damage of concrete by reaching the tensile strength $f_{tn,cal}$ extends over the entire ligament, whereas in the experiment with rotatable boundaries the damage concentrates on one ligament side.

The crack formation on the right side of the ligament and the rotation of the loading plates result in an increasing eccentricity of the loading, which leads to the formation of pronounced stress gradients. This explains the lower values of the tensile strength obtained when using the test set-up with rotatable boundaries. Further, in the experiment with rotatable loading plates the process zone is significantly shorter. As a result, a more brittle failure of the specimen in the tests with this set-up is observed compared to the experiments with non-rotatable boundaries. Finally, it should be mentioned that the shape of the calculated crack paths corresponds very well to the shape of the crack profiles measured optically on the failure surfaces from the uniaxial tension tests (Mechtcherine 2000, Mechtcherine & Müller 1997).

5 DISCUSSION

The most important question to be discussed in connection with the presented heterogeneous continuum model concerns the ability of the model to reproduce the specific behaviour of heterogeneous cementitious materials.

The scattering of the tensile strength $f_{t,inp}$ is one of the main parameters of the model. The scatter range can be associated with the degree of the heterogeneity of a cementitious material. The variation of the mechanical properties in cementitious materials depends on the properties of the individual components of the material structure – the mortar, the aggregate and the bond between mortar and aggregate – and on their part of the total volume. For instance, a normal weight concrete with a lower water-cement ratio has, because of a smaller difference in the strength of its components, a more homogeneous structure as a concrete with the same aggregate type but a higher water-cement ratio. A well-cured concrete possesses also a more homogeneous structure in comparison with a concrete, which has been exposed to desiccation at early age.

An increase of the coefficient of variation $v(f_{t,inp})$ in the numerical calculations provided a decrease of the calculated tensile strengths $f_{t,cal}$ und $f_{tn,cal}$, a higher degree of the non-linearity of the stress–strain- and stress-deformation relations as well as an increase of the ductility of the simulated material. In the experiments similar relationships between the degree of the heterogeneity and the material behaviour have been observed (Mechtcherine 2000, Mechtcherine & Müller 1997). However, this correlation does not yet enable to derive precise recommendations concerning the choice of the coefficient of variation $v(f_{t,inp})$. This can be traced back mainly to the restrictions connected with a two-dimensional discretisation of the specimens. By using this kind of discretisation each finite element has to represent the material property (e.g. the tensile strength f_t) for the entire depth of the specimen or structural member. Logically, to each respective finite element a $f_{t,inp}$-value must be assigned, which corresponds to the average tensile strength of concrete over the entire depth at the individual location. Therefore, the coefficient of variation $v(f_{t,inp})$ should decrease with the increasing depth of the structural member under consideration. On the other hand, when in the simulation the same degree of the non-linearity of the σ-ε-relation should be reached as observed in the experiments, than a large scattering of the input values should be chosen. However, this can lead to an underestimation of the $f_{t,cal}$-values from the simulation of the tension tests on unnotched prisms and an overestimation of the corresponding $f_{tn,cal}$-values from the calculation using notched prisms (compare Figure 3: for $v(f_{t,inp}) = 30\%$ $f_{t,cal} < f_{tn,cal}$, which is in disagreement with the experimental results, e.g. (Mechtcherine 2000, Mechtcherine & Müller 1997, 1998). The reason for a too strong decrease of the tensile strength $f_{t,cal}$ with increasing scattering of the $f_{t,inp}$-values is the formation of apparently weak areas (compare Figure 6), which lead to a premature failure. In the simulation of the tests on notched prisms the probability that the notched cross-section will be significantly weakened by an unfavourable combination of the properties of a number of neighbouring elements is rather small in comparison with the dog-bone shaped prisms.

A further topic under investigation was the effect of the mesh size on the results of the calculation. Usually, the mesh sensitivity of a model is regarded as a disadvantage. However, it might be also a benefit, if the mesh size helps to consider a characteristic size of the material structure (for mortar or concrete it is the maximum grain size) on condition that the variation of the mesh size reproduces correctly the actual effect of the material structure on the test results. In the simulation of the uniaxial tension tests on dog-bone shaped prisms an increase of the calculated tensile strength $f_{t,cal}$ was observed as the mesh size decreases, while the simulation of the tests on notched specimens with a finer mesh provided lower $f_{tn,cal}$-values. Based on the assumption of a correlation between the mesh size and the maximum grain size these results correspond well with the experimental findings (Brameshuber 1988, Hordijk 1991). However, the number of the performed experiments on the effect of the maximum grain size is to small yet, to develop a precise relationship between the maximum grain size and the mesh size needed to its "modelling".

6 SUMMARY AND CONCLUSIONS

In this study the properties of a heterogeneous continuum model for numerical analysis of the failure behaviour of concrete were investigated. From the obtained results following conclusions can be drawn:

1. The coefficient of variation of the input tensile strength $f_{t,inp}$ can be used to simulate the degree of the heterogeneity of cementitious materials. An increase of $v(f_{t,inp})$-values leads to a decrease of the calculated tensile strength $f_{t,cal}$ or the net tensile strength $f_{tn,cal}$ and to a more ductile failure behaviour. This is in agreement with the experimental results from tests on concretes with different degrees of the heterogeneity. The specific effects connected with a two-dimensional discretisation should be considered by choosing an appropriate $v(f_{t,inp})$-value.
2. By using a finer finite element mesh the simulation of the tension tests on unnotched specimens provides higher values of the calculated tensile strength $f_{t,cal}$, while for the notched prisms lower values of the net tensile strength $f_{tn,cal}$ were derived. This mesh sensitivity can be considered as a crude approach to take into account the effect of the maximum grain size, which has been observed experimentally. For concrete and similar cementitious materials an increase of the maximum grain size results in a decrease of the tensile strength f_t and an increase of the net tensile strength f_{tn}.
3. For the finite element formulation of cracks the "rotating crack" approach is more suitable within the heterogeneous continuum model than the "fixed crack" approach. The loss of the coaxiality between the principal stress and strain axes by "fixed cracks"

during simulation provide an apparent "strengthening" of the modelled material at larger deformations (or crack openings). This effect increases due to the stochastic distribution and different orientation of the individual cracks.
4. The developed numerical model provides a powerful tool to study different fracture phenomena connected with the heterogeneity of cementitious materials.

REFERENCES

Bazant, Z. P., Oh, B. H. 1983. Crack band theory for fracture of concrete. *Materials and Structures*, Vol. 16, pp. 155–177.

Brameshuber, W. 1988. *Fracture mechanical properties of young concrete*. Doctoral thesis, University of Karlsruhe, in German.

Heilmann, G., Hilsdorf, H. K., Finsterwalder, K. 1969. Strength and deformation of concrete subjected to tensile stresses. *Publications of DAfStb*, No. 203, Ernst & Sohn Verlag, Berlin, in German.

Hordijk, D. A. 1991. *Local approach to fatigue of concrete*. Doctoral thesis, Delft University of Technology.

Mechtcherine, V. 2000. *Investigations on crack development in concrete*. Doctoral thesis, University of Karlsruhe, in German.

Mechtcherine, V., Müller, H. S. 1997. *Fracture mechanical and fractological investigations on normal and high strength concrete*. Brittle Matrix Composites – BMC 5, A. M. Brand, V. C. Li and I. H. Marshall (eds.), Woodhead Publ. Ltd./Cambridge – Bigraf/Warshaw, pp. 231–240.

Mechtcherine, V., Müller, H. S. 1998. *Effect of the test set-up on fracture mechanical parameters of concrete*. Fracture Mechanics of Concrete Structures, H. Mihashi and K. Rokugo (eds.), Aedificatio Publishers, pp. 377–386.

Rots, J. G. 1988. *Computational modeling of concrete fracture*. Doctoral thesis, Delft University of Technology.

Van Mier, J. G. M., Schlangen, E., Vervuurt, A. 1996. Tensile cracking in concrete and sandstone: Part 2 – Effect of boundary rotations. *Materials and Structures*, Vol. 29, pp. 87–96.

Willam, K., Pramono, E., Sture, S. 1989. *Fundamental issues of smeared crack models*. Fracture of Concrete and Rock, S. P. Shah and S. E. Swartz (eds.), Springer Verlag, New York, pp. 142–157.

Wittmann, F. H., Slowik, V., Alvaredo, A. M. 1994. Probabilistic aspects of fracture energy of concrete. *Materials and Structures*, Vol. 27, pp. 499–504.

Computational Modelling of Concrete Structures, Bićanić et al. (eds)
© 2003 Swets & Zeitlinger, Lisse, ISBN 90 5809 536 3

Strong discontinuity modeling of cracking in concrete

J. Oliver, E. Samaniego, E. Chaves
ETS Enginyers de Camins, Canals i Ports de Barcelona, Technical University of Catalonia (UPC), Campus Nord UPC, Barcelona, Spain

A. Huespe
CIMEC -Conicet-Unl, Instituto Intec, Guemes Santa Fe, Argentina

ABSTRACT: The ingredients to model the onset and development of cracking, in plain or reinforced concrete, in a strong discontinuity scenario are given. After describing the analytical foundations of the strong discontinuity approach (SDA), such as the strong discontinuity kinematics, softening regularization and the discontinuous bifurcation analysis, numerical issues like finite elements with embedded discontinuities and propagation algorithms are presented. A viscous perturbation based algorithm, to unfold bifurcations due to simultaneous crack arrest/activation for multiple crack cases, is described as well. Numerical examples show the performance of the presented methodology.

1 CRACK MODELING IN A STRONG DISCONTINUITY SETTING

1.1 *The strong discontinuity approach to fracture mechanics*

The strong discontinuity approach (see Oliver et al. 2001 for an overview) appears as an alternative to discrete fracture mechanics (Bazant & Planas 1998) to model the onset and development of cracks at quasi-brittle fracturing media. The main features of the approach are the following:

1. The use of a *continuum format* for the kinematics i.e.: the strains $\boldsymbol{\epsilon}(\mathbf{x})$ are computed as the symmetric gradient of the displacement field $\mathbf{u}(\mathbf{x})$, even when $\mathbf{u}$ exhibits jumps across the crack path and, therefore, $\boldsymbol{\epsilon}(\mathbf{x})$ becomes unbounded (see Figure 1). The displacement and strain fields read:

$$\begin{aligned} \mathbf{u}(\mathbf{x}) &= \bar{\mathbf{u}}(\mathbf{x}) + H_S[\![\mathbf{u}]\!](\mathbf{x}) \\ \boldsymbol{\epsilon} = \nabla^s\mathbf{u} &= \underbrace{\bar{\boldsymbol{\epsilon}}}_{\substack{regular \\ (bounded)}} + \underbrace{\delta_S([\![\mathbf{u}]\!] \otimes \mathbf{n})^S}_{singular\ (unbounded)} \end{aligned} \quad (1)$$

where H_S and δ_S stand, respectively, for line/surface Heaviside's step function and Dirac's delta function placed on the discontinuity path S in the body Ω, $\bar{\mathbf{u}}(\mathbf{x})$ is the regular (continuous) displacement field, $[\![\mathbf{u}]\!](\mathbf{x})$ is the displacement jump and $\mathbf{n}(\mathbf{x})$ is the unit normal to the discontinuity path S. This is known in the literature as *strong discontinuity kinematics* (Simo, Oliver & Armero 1993).

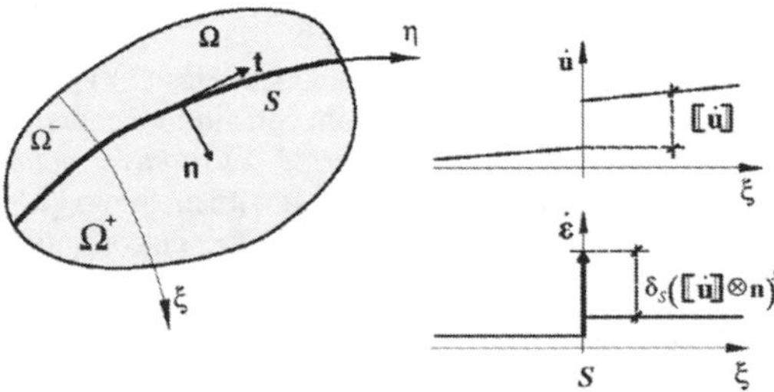

Figure 1. Strong discontinuity kinematics.

2. Consideration of standard (continuum stress-strain) rate-independent non-linear constitutive equations equipped with strain softening i.e.:

$$\sigma = \boldsymbol{\Sigma}(\boldsymbol{\epsilon}, H) \quad (2)$$

where $H \leq 0$ stands for *the continuum softening modulus* that rules the evolution of the elastic domain. The rate version of equation (2) reads:

$$\dot{\sigma} = \mathbb{C}^{tang}(\sigma, H) : \dot{\boldsymbol{\epsilon}} \quad (3)$$

where $\mathbb{C}^{tang}$ is the tangent constitutive tensor.

3. The distributional character of the inverse of the *continuum* softening modulus H in terms of the

discrete softening modulus $\overline{H}$ (which can be related to the fracture properties of the material) i.e.:

$$\frac{1}{H} = \delta_S \frac{1}{\overline{H}} \tag{4}$$

Regularized versions (more suitable for computational treatments) of equations (1) and (4) are obtained via a k-regularized sequence of Dirac's delta functions:

$$\delta_S(\mathbf{x}) \equiv \lim_{k-o} \frac{\mu_S(\mathbf{x})}{k} \tag{5}$$

where $\mu_S(\mathbf{x})$ is the collocation function on $S(\mu_S(\mathbf{x}) = 1$ for $\mathbf{x} \in S$; $\mu_S(\mathbf{x}) = 0$ otherwise) and k is a regularization parameter as small as permitted by the machine precision. In this context equation (4) turns out to be the so-called *softening regularization condition* (Oliver 2000):

$$\begin{aligned} H &= k\overline{H} \quad \forall\, \mathbf{x} \in S \\ H &= \infty \quad \text{otherwise (elastic unloading)} \end{aligned} \tag{6}$$

Under those conditions it can be shown (Oliver 2000) that:

a) Equation (4) makes compatible, for the constitutive equation (2), bounded values of the stresses $\boldsymbol{\sigma}$ with the unbounded strains emerging from equation (1) at the discontinuity interface S.
b) As the strong discontinuity kinematics (1) is activated, the original continuum constitutive equation (2) is projected into *a discrete constitutive model* (relating the traction $\mathcal{T} = \boldsymbol{\sigma} \cdot \mathbf{n}$ with the jump $[\![\mathbf{u}]\!]$) that is *automatically fulfilled* at the discontinuity interface S (Oliver 2000) i.e.:

$$\mathcal{T} = \mathcal{F}([\![\mathbf{u}]\!], \overline{H}) \tag{7}$$

Equation (7) provides a clear link with the nonlinear decohesive Fracture Mechanics (Bazant & Planas 1998) while keeping the continuum format of the analysis.

1.2 *Onset and propagation of a crack*

Failure Mechanics tools, like *the discontinuous bifurcation analysis* (Willam 2000), provide a rigorous methodology to detect the onset of a crack in a given material point $\mathbf{x}$, based on the properties of the (directional) localization tensor $\mathbb{Q}^{loc}$:

$$\mathbb{Q}^{loc}(\mathbf{x}, \mathbf{n}, t) = \mathbf{n} \cdot \mathbb{C}^{tang}(\boldsymbol{\sigma}(\mathbf{x}, t), H) \cdot \mathbf{n} \tag{8}$$

The singularity of $\mathbb{Q}^{loc}$ can be shown to be associated to the appearance of a discontinuous bifurcation of the stress/strain fields in the neighborhood of $\mathbf{x}$. Therefore the bifurcation time $t_B(\mathbf{x})$, that triggers the discontinuous kinematics (1), and the (normal to) propagation direction $\mathbf{n}$ (see Figure 1) are determined by the first time that the following condition is fulfilled for some $\mathbf{n}$:

$$\det(\mathbb{Q}^{loc}(\mathbf{x}, \mathbf{n}, t_B)) = 0 \tag{9}$$

Simplified alternatives are frequently used by associating t_B to the onset of the non-linear material behavior and $\mathbf{n}$ to the direction of the first principal stress at that time. Although less rigorous than the previous procedure, this seems to provide accurate enough results for quasibrittle materials like concrete.

1.3 *Finite element technology*

For numerical simulation purposes finite elements with embedded discontinuities, compatible with the strong discontinuity kinematics (1), are required. In the last years several families of such elements have been developed. Their common feature is that the standard underlying element is enriched with additional displacement/strain fields that make it compatible with the presence of a discontinuity embedded in the element. In this context, two different families can be distinguished:

- Elemental enrichment (see for instance Oliver 1996, Armero & Garikipati 1996).
- Nodal enrichment (see for instance Wells & Sluys 2001).

1.3.1 *Elemental enrichment*

The support of the enriching discontinuity mode is elemental (see Figure 2) and, in consequence, the additional internal discontinuous degrees of freedom (two per element in 2D cases) can be condensed at the elemental level. The displacement jump is element-wise constant.

1.3.2 *Nodal enrichment*

The support of the enriching discontinuous mode is the same than for the nodal shape functions of the underlying element (see Figure 3). The regular nodes of the enriched set of elements are increased with additional degrees of freedom (two per node in 2D cases) whose interpolation provides a (varying inside the element) discontinuous displacement field. The additional degrees of freedom can not be condensed at element level.

1.4 *Discontinuity tracking algorithms*

Constructing the additional discontinuous modes, in finite elements with embedded discontinuities discussed in section 1.3, require the definition of the set of enriched elements (see Figure 2) and the position of the discontinuity path inside every element of that set. Procedures and algorithms devoted to that purpose

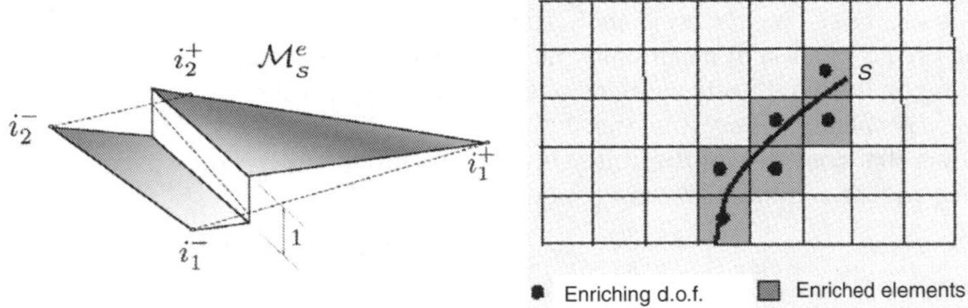

Figure 2. Elemental based enrichment. Left: discontinuous enriching displacement mode. Right: enriched domain crossed by the discontinuity path.

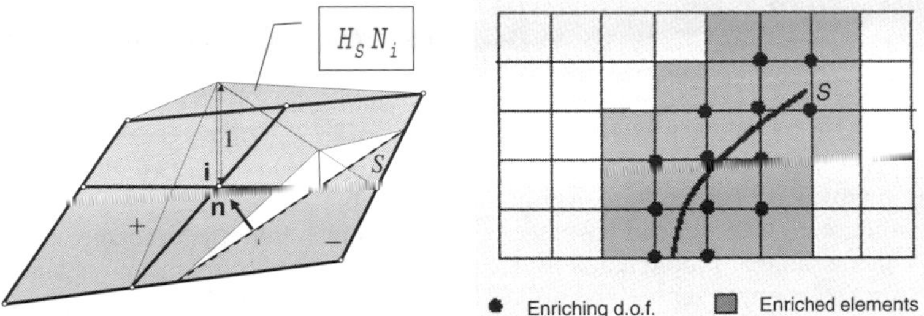

Figure 3. Nodal based discontinuous enrichment. Left: discontinuous enriching displacement mode. Right: enriched domain crossed by the discontinuity path.

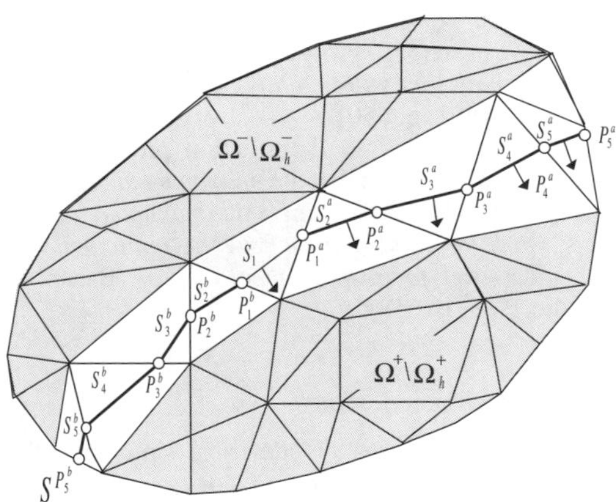

Figure 4. Discontinuity tracking in a 2D problem using a propagation algorithm. The original segment S_1 in the root element propagates in both senses by means of straight elemental segments S_i^a and S_i^b.

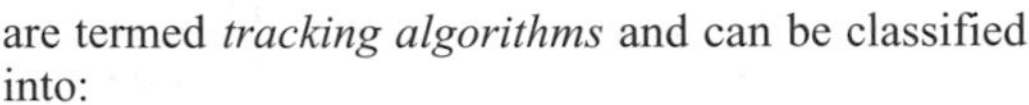

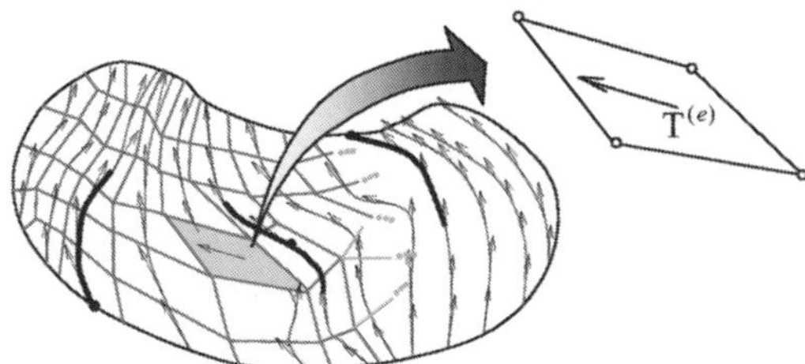

Figure 5. Discontinuity tracking in a 2D problem using a global algorithm. The algorithm traces the envelopes of the propagation vector field $\mathbf{T}(\mathbf{x}) \perp \mathbf{n}(\mathbf{x})$ and then selects some of them as propagating cracks.

are termed *tracking algorithms* and can be classified into:

- *Propagation algorithms* (see Figure 4): based on the recursive propagation from the *root element* (the first to fulfill the bifurcation condition (9)) to the neighbor elements by means of element-wise straight segments (in 2D), or planes (in 3D) orthogonal to the elemental normal vector **n**. Although very intuitive and direct to implement their application to multiple cracks or to 3D cases is cumbersome.
- *Global tracking algorithms* (see Figure 5): They aim at tracing, *at once*, all the lines (in 2D), or surfaces (in 3D), candidates to be discontinuity paths (cracks). These lines, or surfaces, are determined as that family of curves or surfaces that are orthogonal to the vector field $\mathbf{n}(\mathbf{x})$ provided by the bifurcation condition (9). Once this family is known those lines passing through the center of a root element are selected to contain a propagating crack.

In Oliver et al. (2002) a *heat conduction-like algorithm* to provide those envelopes is presented. The algorithm is based on the solution of a stationary heat conduction problem, on the same finite element mesh as for the mechanical problem, at every time step of the analysis. It is shown that considering the following, directional point dependent, conductivity tensor, $\mathbb{K}(\mathbf{x})$:

$$\mathbb{K}(\mathbf{x}) = \mathbf{T}(\mathbf{x}) \otimes \mathbf{T}(\mathbf{x}) \tag{10}$$

for 2D cases and,

$$\mathbb{K}(\mathbf{x}) = \mathbf{S}(\mathbf{x}) \otimes \mathbf{S}(\mathbf{x}) + \mathbf{T}(\mathbf{x}) \otimes \mathbf{T}(\mathbf{x}) \tag{11}$$

for 3D cases, where $\mathbf{S}(\mathbf{x})$ and $\mathbf{T}(\mathbf{x})$ are any couple of vector fields, orthogonal to the field $\mathbf{n}(\mathbf{x})$ (which can be trivially computed) the solution of the heat conduction problem in Figure 6 provides a temperature distribution $\theta(\mathbf{x})$ whose isolines are the intended envelopes. In order to avoid either a singular problem or a trivial (constant temperature) solution, arbitrary, and different, values of the temperatures should be prescribed at two points of the continuum.

Such a type of global algorithms, although involving a (on the other hand very simple) multifield problem, appears suitable for dealing with multiple cracks and 3D problems.

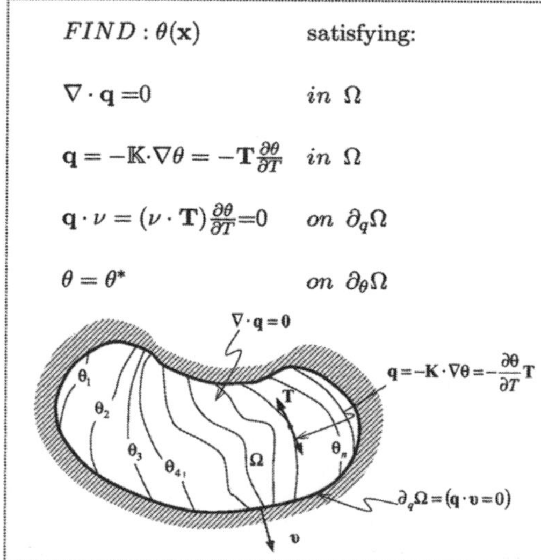

Figure 6. 2D algorithm to compute the envelopes of the propagation field as the isothermal lines for a heat conduction problem.

2 MULTIPLE CRACKING TREATMENT

Cracking in plain and reinforced concrete often exhibits multiple cracks that grow separately but interacting to each other. Those cracks could be roughly classified into:

- *Primary cracks*, that grow along a relevant part of the deformation process and substantially contribute to the whole dissipation of the structure.
- *Secondary cracks*, that are active only for little range of the deformation process and consume a small amount of energy.

Even if secondary cracks are neglected, it is often observed in concrete the presence of several primary cracks during relevant parts of the deformation process. Moreover, for the ideal model that supplies the equilibrium paths in the action-response solutions space, any of the possible combinations in terms of activation or arrest of those cracks, determines a possible equilibrium path emanating as a bifurcated branch from the fundamental one (see Figure 7, left). Upon these circumstances numerical simulation becomes highly difficult and cumbersome and makes the general theoretical problem almost intractable.

However, experimentally observed cracking in concrete suggests that there exist a physical preference for some of those combinations that determines the actual crack pattern. This preference should be translated into a physical principle *or fundamental law that determines the path that will actually be followed* (Bazant & Cedolin 1991, p. 650).

For adiabatic (slow) deformation processes this principle turns out to be the *maximum dissipation principle* which translates into the *maximum second order work for load control* and the *minimum second order work for displacement control* (Bazant & Cedolin 1991, p. 658).

2.1 *Bifurcation unfolding*

For practical purposes in numerical simulation the maximum dissipation criterion does not avoid the

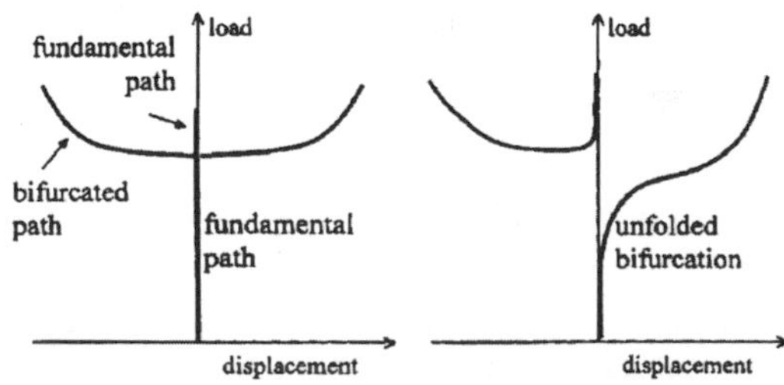

Figure 7. Bifurcation unfolding. Left: unperturbed solution. Right: perturbed solution (from Gastebled et al. 2000).

heavy task of carrying out a bifurcation analysis and search for the most dissipative evolution of the cracking at every time step of the analysis (Gastebled et al. 2000). On the other hand, one could resort to perturbation techniques to unfold the bifurcation in the solution space and, therefore, assure a unique stable path (see Figure 7, right). The problem there is that the type of perturbation determines, in turn, the type of unfolding and, thus, the obtained solution (imperfection sensitive solution).

However, one could *postulate* a type of perturbation to produce the correct unfolding according to the criterion above. The physical meaning and experimental fitting of the obtained numerical results would then assess the correctness of that assumption. This is the procedure, based on a *discrete viscous perturbation at the crack path*, which will be described in next sections.

2.2 *Discrete viscous perturbation in the crack path*

2.2.1 *Symmetric formulation*

Nodal enrichment presented in section 1.3.2 can be cast into the following variational setting:

Let us consider the boundary value problem on the solid Ω sketched in Figure 8, exhibiting a strong discontinuity $[\![\mathbf{u}]\!]$ at the discontinuity path S, and the corresponding virtual work principle:

$$\underbrace{\int_\Omega \boldsymbol{\sigma} : \underbrace{\nabla^S \boldsymbol{\eta}}_{\delta\boldsymbol{\varepsilon}} dV}_{\text{Internal Virtual Work } (\delta\mathbb{W}^{int})} = \underbrace{\int_\Omega \mathbf{b}\cdot \underbrace{\boldsymbol{\eta}}_{\delta\mathbf{u}}\, dV + \int_{\Gamma_\sigma} \mathbf{t}^* \cdot \underbrace{\boldsymbol{\eta}}_{\delta\mathbf{u}}\, d\Gamma}_{\text{External Virtual Work } (\delta\mathbb{W}^{ext})} \quad \forall \boldsymbol{\eta} \in \mathbb{V} \tag{12}$$

where $\mathbb{V}$ stands for the space of, kinematically admissible, virtual displacements:

$$\begin{aligned}\mathbb{V} := \{&\boldsymbol{\eta}(\mathbf{x}) = \overline{\boldsymbol{\eta}} + H_S\tilde{\boldsymbol{\eta}} | \overline{\boldsymbol{\eta}}(\mathbf{x})|_{\mathbf{x}\in\Gamma_u} \\ &= \tilde{\boldsymbol{\eta}}(\mathbf{x})|_{\mathbf{x}\in\Gamma_u} = \mathbf{0}\}\end{aligned} \tag{13}$$

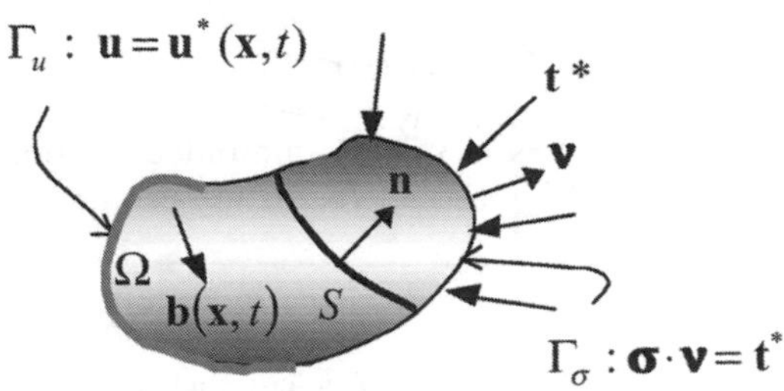

Figure 8. Boundary value problem.

Some algebraic manipulations show that the Euler-Lagrange equations fot the variational principle (12) are:

$$\nabla \cdot \boldsymbol{\sigma} + \mathbf{b} = \mathbf{0} \quad \text{in } \Omega/S \rightarrow \text{momentum balance} \tag{14}$$

$$\boldsymbol{\sigma} \cdot \boldsymbol{\nu} - \mathbf{t}^* = \mathbf{0} \quad \text{in } \Gamma_\sigma \rightarrow \text{prescribed tractions} \tag{15}$$

$$\boldsymbol{\sigma}^-_{\Omega/S}\cdot\mathbf{n} - \boldsymbol{\sigma}^+_{\Omega/S}\cdot\mathbf{n} = \mathbf{0} \quad \text{in } S \rightarrow \begin{cases}\text{outer traction} \\ \text{continuity}\end{cases} \tag{16}$$

$$\boldsymbol{\sigma}_S\cdot\mathbf{n} - \boldsymbol{\sigma}^+_{\Omega/S}\cdot\mathbf{n} = \mathbf{0} \quad \text{in } S \rightarrow \begin{cases}\text{inner traction} \\ \text{continuity}\end{cases} \tag{17}$$

Let us now consider a *perturbed problem* characterized by the addition of a set of viscous surface forces $\tilde{\mathbf{t}}(\mathbf{x}, t)$, at the discontinuity interface S, given by:

$$\tilde{\mathbf{t}}(\mathbf{x}, t) = \overline{\gamma}[\![\dot{\mathbf{u}}]\!](\mathbf{x}, t) \quad \forall\, \mathbf{x} \in S \tag{18}$$

where $\overline{\gamma} > 0$ is a perturbing (very small) viscosity and the upper dot stands for time (or pseudo time) derivative. The virtual work principle (12) reads now:

$$\underbrace{\int_\Omega \boldsymbol{\sigma} : \nabla^S \boldsymbol{\eta}\, dV}_{\text{Internal Virtual Work}} = \underbrace{\int_\Omega \mathbf{b}\cdot\boldsymbol{\eta}\, dV + \int_{\Gamma_\sigma} \mathbf{t}^*\cdot\boldsymbol{\eta}\, d\Gamma + \int_S \underbrace{\overline{\gamma}[\![\dot{\mathbf{u}}]\!]}_{\tilde{\mathbf{t}}}\cdot\boldsymbol{\eta}\, dS}_{\text{External Virtual Work}} \quad \forall \boldsymbol{\eta} \in \mathbb{V} \tag{19}$$

and the corresponding Euler-Lagrange equations are:

$$\nabla \cdot \boldsymbol{\sigma} + \mathbf{b} = \mathbf{0} \quad \text{in } \Omega/S \rightarrow \text{momentum balance} \tag{20}$$

$$\boldsymbol{\sigma} \cdot \boldsymbol{\nu} - \mathbf{t}^* = \mathbf{0} \quad \text{in } \Gamma_\sigma \rightarrow \text{prescribed tractions} \tag{21}$$

$$\boldsymbol{\sigma}^-_{\Omega/S}\cdot\mathbf{n} - \boldsymbol{\sigma}^+_{\Omega/S}\cdot\mathbf{n} = \overline{\gamma}[\![\dot{\mathbf{u}}]\!] \quad \text{in } S \rightarrow \begin{cases}\text{perturbed} \\ \text{outer traction} \\ \text{continuity}\end{cases} \tag{22}$$

$$\boldsymbol{\sigma}_S\cdot\mathbf{n} - \boldsymbol{\sigma}^+_{\Omega/S}\cdot\mathbf{n} = \overline{\gamma}[\![\dot{\mathbf{u}}]\!] \quad \text{in } S \rightarrow \begin{cases}\text{perturbed} \\ \text{inner traction} \\ \text{continuity}\end{cases} \tag{23}$$

From equations (14) to (23) it appears that, as the viscosity $\overline{\gamma}$ tends to zero, the original problem is recovered. However the introduction of the viscous perturbation $\overline{\gamma}$ has a relevant consequence on the uniqueness of the problem. In fact, let us consider two possible solutions whose differences, in terms of the displacements, strains and stresses at a given time t, are respectively: $\Delta\mathbf{u}(t)$, $\Delta[\![\mathbf{u}]\!](t)$, $\Delta\boldsymbol{\epsilon}(t)$ and $\Delta\boldsymbol{\sigma}(t)$, which satisfy (from subtraction of equation (19) for both solutions):

$$\int_\Omega \Delta\boldsymbol{\sigma} : \nabla^S(\boldsymbol{\eta})dV - \int_S \overline{\gamma}\Delta[\![\dot{\mathbf{u}}]\!]\cdot\boldsymbol{\eta}\, dS\, dV = 0 \quad \forall\, \boldsymbol{\eta} \in \mathbb{V} \tag{24}$$

Let us now consider the *bifurcation time* t_B where a bifurcation, *in terms of different combinations of arrest/activation of cracks*, takes place. This means that there are (at least) two possible solutions in terms of the *jump evolution* at S:

$$\begin{aligned} &[\![\dot{\mathbf{u}}]\!]^{(1)}(\mathbf{x}_S, t_B) \neq [\![\dot{\mathbf{u}}]\!]^{(2)}(\mathbf{x}_S, t_B) \quad \text{for some } \mathbf{x}_S \in S \\ &\Delta[\![\dot{\mathbf{u}}]\!](\mathbf{x}_S, t_B) = [\![\dot{\mathbf{u}}]\!]^{(2)} - [\![\dot{\mathbf{u}}]\!]^{(1)} \neq \mathbf{0} \\ &\quad \text{for some } \mathbf{x}_S \in S \end{aligned} \tag{25}$$

emerging from the same fundamental branch at t_B, and, therefore,

$$\Delta\mathbf{u}(t_B) = \Delta\boldsymbol{\epsilon}(t_B) = \Delta\boldsymbol{\sigma}(t_B) = \mathbf{0} \tag{26}$$

Inserting now equation (26) into equation (24) and for the particular choice $\boldsymbol{\eta}(\mathbf{x})\Delta[\![\dot{\mathbf{u}}]\!](\mathbf{x}, t_B) \in \mathbb{V}$ one can write:

$$\int_S \overline{\gamma}\Delta[\![\dot{\mathbf{u}}]\!] \cdot \Delta[\![\dot{\mathbf{u}}]\!]\, dS = \int_S \overline{\gamma}\|\Delta[\![\dot{\mathbf{u}}]\!]\|^2\, dS = 0 \tag{27}$$

and, therefore,

$$\Delta[\![\dot{\mathbf{u}}]\!](\mathbf{x}, t_B) = \mathbf{0} \quad \forall\, \mathbf{x} \in S \tag{28}$$

which is in contradiction with equation (25) and, in consequence, *all bifurcations in terms of different combinations of crack arrest/activation are precluded.* This states the beneficial effects of the discrete viscous perturbation (18) in unfolding that type of bifurcations.

2.2.2 *Unsymmetric formulation*

Some types of finite element formulations (the so-called statically and kinematically optimal formulations (Oliver 1996, Armero & Garikipati 1996), belonging to the family of elemental enrichment discussed in section 1.3.1, can not be cast into a complete Galerkin type setting. In these formulations, the inner traction continuity equation (17) is locally weighted.

In this case the viscous perturbation can be directly introduced in that equation under the form of equation (23). Then, by subtracting two possible solutions inserted in this equation one can write:

$$\begin{aligned} &\Delta\boldsymbol{\sigma}_S(\mathbf{x}, t) \cdot \mathbf{n} - \Delta\boldsymbol{\sigma}^+_{\Omega/S}(\mathbf{x}, t) \cdot \mathbf{n} \\ &\quad = \overline{\gamma}\Delta[\![\dot{\mathbf{u}}]\!](\mathbf{x}, t) \ \text{ in } S \end{aligned} \tag{29}$$

and, since at the bifurcation time t_B:

$$\Delta\boldsymbol{\sigma}^+_{\Omega/S}(\mathbf{x}, t_B) = \Delta\boldsymbol{\sigma}_S(\mathbf{x}, t_B) \cdot \mathbf{n} = \mathbf{0} \ \forall\, \mathbf{x} \in S \tag{30}$$

$$\Delta[\![\dot{\mathbf{u}}]\!](\mathbf{x}_S, t_B) \neq \mathbf{0} \text{ for some } \mathbf{x}_S \in S \tag{31}$$

substituting equation (31) into equation (29) for $t = t_B$ one gets:

$$\overline{\gamma}\, \Delta[\![\dot{\mathbf{u}}]\!](\mathbf{x}, t_B) = \mathbf{0} \quad \forall\, \mathbf{x} \in S \tag{32}$$

which, for $\overline{\gamma} \neq 0$, contradicts equation (31) this precluding any bifurcation of the considered type. Therefore the same benefits, to unfold crack opening/arrest bifurcations, than in the variationally consistent case should be expected.

2.2.3 *Time stepping. Critical time step*

In the context of a time (or pseudo-time) advancing algorithm, since progression along the equilibrium path in the action-response space is discrete, it is not possible to guarantee (for a given length of the time step $\ell_t(\Delta t)$) the uniqueness of the algorithmic response for the unfolded problem (see Figure 9). However it is possible to determine the critical length of the time step $\ell_t^{crit}(\Delta t^{crit})$ to assure algorithmic uniqueness.

Let us consider the space-time discretized version of the problem, in rate version for the elemental enrichment case of section 1.3.1, at time $t + \Delta t$:

$$\underbrace{\begin{bmatrix} \mathbf{K}_{dd} & \mathbf{K}_{d\alpha} \\ \mathbf{K}_{\alpha d} & \mathbf{K}_{\alpha\alpha} \end{bmatrix}}_{\mathbf{K}_{t+\Delta t}} \underbrace{\begin{bmatrix} \dot{\mathbf{d}}_{t+\Delta t} \\ \dot{\alpha}_{t+\Delta t} \end{bmatrix}}_{\dot{\mathbf{a}}_{t+\Delta t}} = \underbrace{\begin{bmatrix} \dot{\mathbf{f}}_{t+\Delta t} \\ \mathbf{0} \end{bmatrix}}_{\dot{\mathbf{F}}_{t+\Delta t}} \tag{33}$$

where $\mathbf{d}$ stands for the regular (nodal) degrees of freedom and $\alpha = [\alpha^{(1)}, \alpha^{(1)}, \alpha^{(1)} \ldots \alpha^{(n)}]^T$ are the internal degrees of freedom corresponding to the n enriched (crossed by the cracks) elements. The corresponding perturbed problem according to equation (23) reads:

$$\underbrace{\begin{bmatrix} \mathbf{K}_{dd} & \mathbf{K}_{d\alpha} \\ \mathbf{K}_{\alpha d} & \mathbf{K}_{\alpha\alpha} + \dfrac{\overline{\gamma}}{\Delta t}\mathbf{1} \end{bmatrix}}_{\widetilde{\mathbf{K}}_{t+\Delta t}} \underbrace{\begin{bmatrix} \dot{\mathbf{d}}_{t+\Delta t} \\ \dot{\alpha}_{t+\Delta t} \end{bmatrix}}_{\dot{\mathbf{a}}_{t+\Delta t}} = \underbrace{\begin{bmatrix} \dot{\mathbf{f}}_{t+\Delta t} \\ \dfrac{\overline{\gamma}}{\Delta t}\dot{\alpha}_t \end{bmatrix}}_{\dot{\mathbf{F}}_{t+\Delta t}} \tag{34}$$

A non-unique solution at time $t+\Delta t$ is characterized by the existence of several solutions $\dot{\mathbf{a}}^{(i)}_{t+\Delta t}$ $i = 1, 2 \ldots$ for equation (34) whose differences $\Delta\dot{\mathbf{a}}_{t+\Delta t} \neq \mathbf{0}$ fulfill:

$$\widetilde{\mathbf{K}}_{t+\Delta t} \cdot \Delta\dot{\mathbf{a}}_{t+\Delta t} = \mathbf{0} \tag{35}$$

and pre-multiplying times $\Delta\dot{\mathbf{a}}_{t+\Delta t}$:

$$\Delta\dot{\mathbf{a}}_{t+\Delta t} \cdot \widetilde{\mathbf{K}}_{t+\Delta t} \cdot \Delta\dot{\mathbf{a}}_{t+\Delta t} = 0 \tag{36}$$

which states the singularity of $\widetilde{\mathbf{K}}_{t+\Delta t}$. Therefore, that value of $\Delta t = \Delta t^{crit}$ that guarantees positive definiteness of $\widetilde{\mathbf{K}}_{t+\Delta t}$ precludes any bifurcation. Since

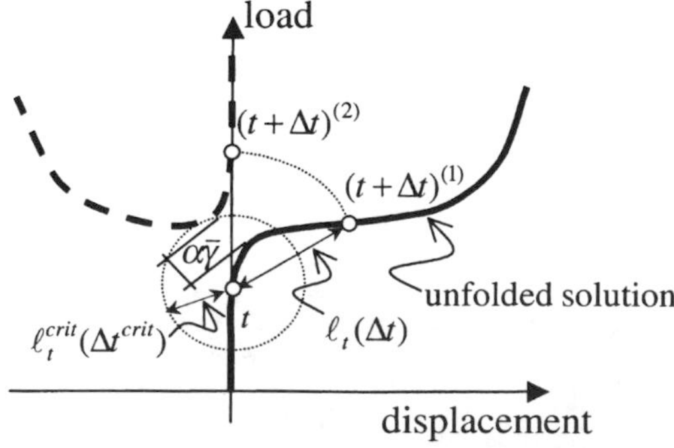

Figure 9. Algorithmic bifurcation for the unfolded problem. For $\ell_t(\Delta t) > \ell_t^{crit}(\Delta t^{crit})$ the algorithmic solution may not be unique.

sub-matrix $\mathbf{K}_{dd}$ in equation (34) can be shown to be positive, an approximate value of Δt^{crit} can be obtained by imposing positive definiteness of the symmetric part of all the diagonal sub-matrices associated to the enriching degrees of freedom α, i.e.:

$$symm(\mathbf{K}_{\alpha\alpha}^{(e)}) + \frac{\overline{\gamma}}{\Delta t}\mathbf{1} \rightarrow \text{positive def.} \quad \forall\, e \in \{1,2 \ldots n_e\} \tag{37}$$

or, equivalently,

$$\lambda_{min}^{sym(e)} + \frac{\overline{\gamma}}{\Delta t} > 0 \quad \forall\, e \in \{1,2 \ldots n_e\} \tag{38}$$

where $\lambda_{min}^{symm(e)}$ stands for the minimum eigenvalue of $symm(\mathbf{K}_{\alpha\alpha}^{(e)})$. From equation (38) the following value for Δ^{crit} emerges:

$$\Delta^{crit} = \min_e \left(\frac{\overline{\gamma}}{\langle -\lambda_{min}^{symm(e)} \rangle} \right) \quad \forall\, e \in \{1,2 \ldots n_e\} \tag{39}$$

where $\langle \bullet \rangle$ stands for the ramp function.

2.3 *Crack shielding*

In the context of multiple cracks propagating in a real concrete structure it is barely observed two *active primary cracks* crossing each other. However this is a case that can occur during the non-linear iterative procedure in the numerical model unless this possibility is precluded. Besides, if the finite element is not designed to bear multiple cracks this can result in a sort of numerical locking and a subsequent loss of convergence. To circumvent this problem, in the numerical examples presented below the elements inside a *shielding zone* (whose width is defined by the user) around an existing propagating crack can not be crossed by any external crack (see Figure 13). This prevents the formation of a number of spurious secondary cracks, around the primary ones, that give rise to numerical difficulties but do not provide any gain in the physical insight of the problem.

3 NUMERICAL SIMULATIONS

The modeling scheme outlined in previous sections has been implemented in a general multipurpose finite element code (Cervera et al. 2001). It has been used to simulate 2D fracture problems in concrete involving the onset and propagation of several primary cracks. A continuum damage model equipped with linear and exponential strain softening (Oliver, 2000), regularized according to equation (6), has been chosen to model the tensile cracking. Here, the material model is slightly modified, by precluding damage evolution whenever the principal stresses are negative, which avoid the material degradation in compressive stress states as should be expected in concrete.

Triangular and quadrilateral finite elements with elemental enrichment have been considered.

3.1 *Double notched four points bending test*

The four-point bending test shown in Figure 10, on a concrete specimen, has been analyzed by using a plane stress conditions.

The experiment corresponds to one of the tests carried out by Bocca et al. (1991), who particularly remarked the mixed mode fracture character of the crack opening mode. They observed that the structure collapsed due two the formation of two primary cracks as displayed in Figure 11. The experimental critical load was 34460 N, lower than the one obtained by numerical simulation adopting a linear softening law for the damage model (Figure 12).

In Figure 13, the crack pattern obtained at different steps of the analysis is shown. The black regions correspond to the shielding zone where crack propagation

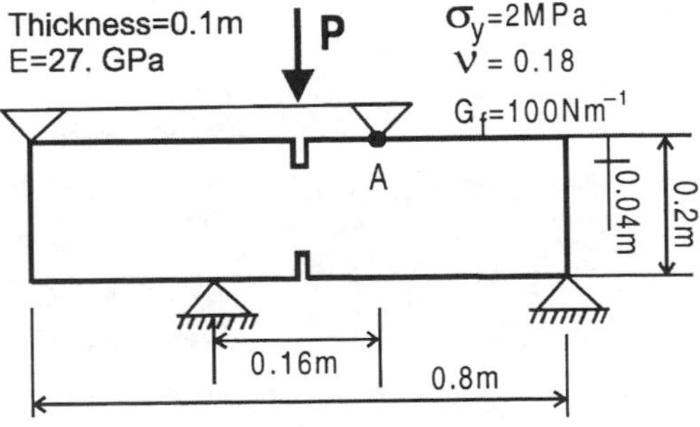

Figure 10. Four-point bending test for a double-notched beam, E: Young' modulus, ν: Poisson's ratio, σ_y: tensile strength. G_f: fracture energy.

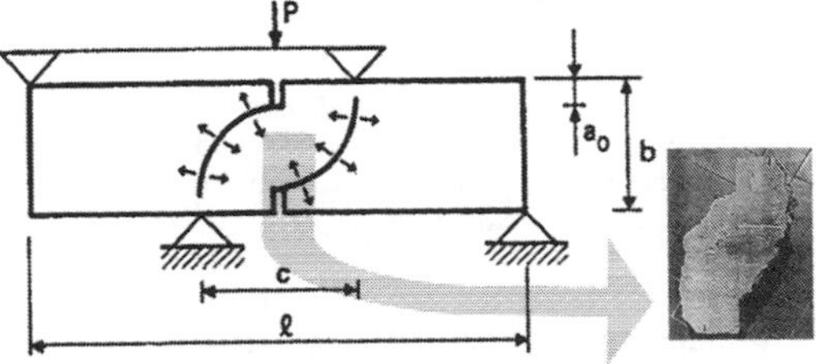

Figure 11. Four-point bending test for a double-notched beam, Left: observed cracks. Right: detail of the concrete specimen between cracks after the test (from Bocca P., Carpintieri A., Valente S., 1991).

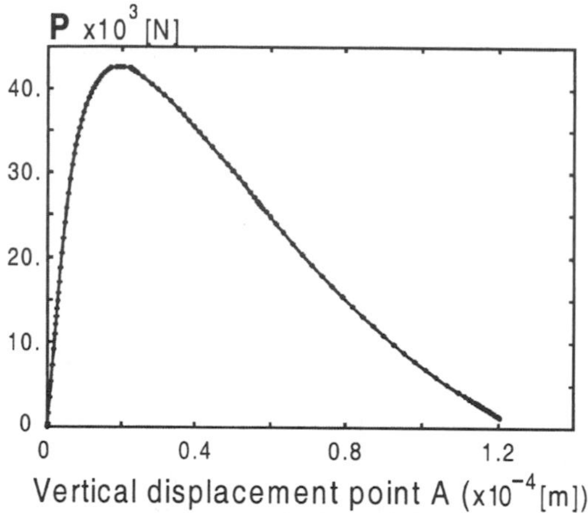

Figure 12. Load vs. vertical displacement of point A for the four point bending test.

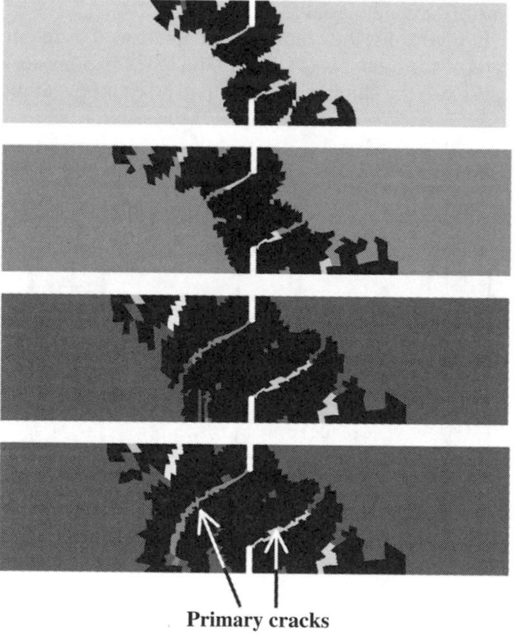

Figure 13. Double notched beam. From top to bottom: evolution of the crack pattern (elements crossed by the cracks are highlighted in gray. Black area around the crack paths corresponds to the shielding zone).

is forbidden. Every new crack sets its own shielding zone.

Two primary symmetric cracks are observed to propagate from the roots (at the notches) toward the load application points. These curves are in agreement with the experimental results.

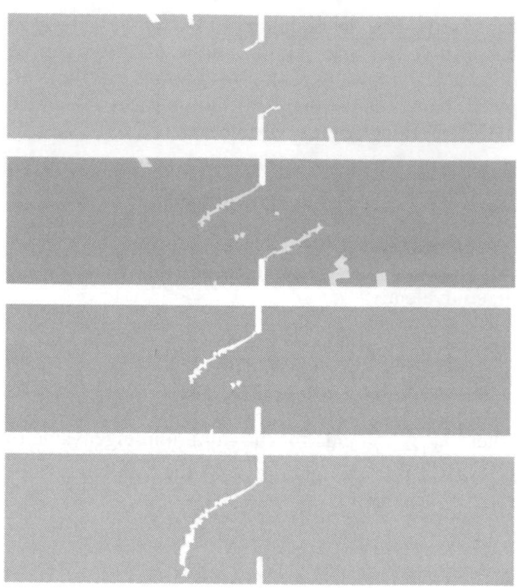

Figure 14. Double notched beam. From top to bottom: evolution of the active cracks for increasing times of the analysis (elements crossed by the active cracks are highlighted in white. The second and third pictures correspond, respectively, to load levels before and after the peak load).

Figure 14 shows the active (opening) crack pattern for different steps of the analysis. A white line means that the crack is opening, in contrast with the arrested cracks, which are not signaled. At initial stages, before the limit load is reached, it is observed that several cracks, both primary and secondary, are opening. After the peak load stage, only one of the primary cracks remains active. This fact states the presence of a bifurcation point, in the equilibrium curve of the perfect structure (at the peak of the load-displacement curve in Figure 12), and the fundamental path (corresponding to two active cracks) is unfolded by the viscous perturbation method described in section 2.2 into a one-active-crack path. Any numerical perturbation (introduced for instance by the finite element mesh) will select which one of the two primary cracks is remaining active.

3.2 *Steel slab debonding from a concrete specimen*

Next example corresponds to the simulation of the debonding of a steel slab from a concrete bulk by stretching the slab from both ends as it is shown in Figure 15.

This problem is an adaptation, to the plane strain case, of a rebar debonding axisymmetric problem presented in Rots (1988). Although the problem is not exactly the same in both cases, since the geometrical

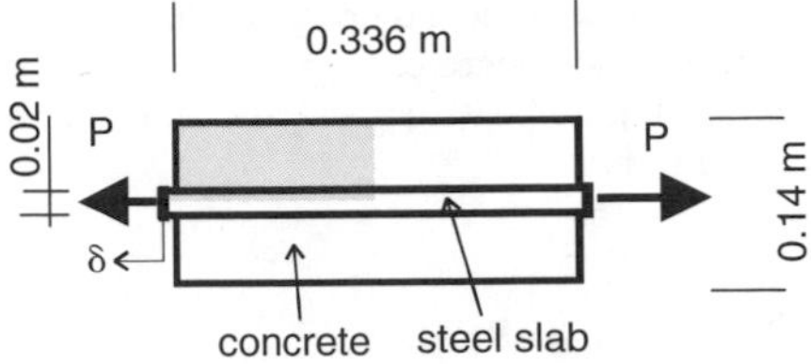

Figure 15. Debonding of a steel slab from a concrete block in plane strain. The discrete model corresponds to a quarter of the structure.

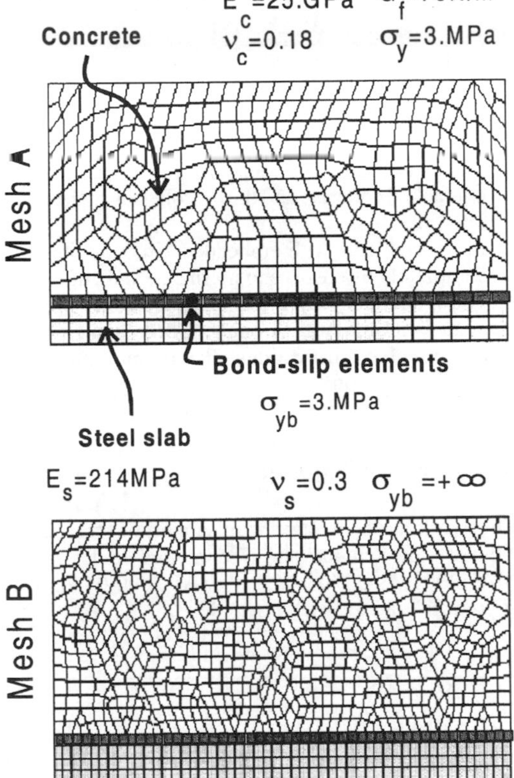

Figure 16. Steel slab in a concrete block. Finite element meshes.

dimensions and material constants are the same one should expect similar results for the structural behavior and crack pattern.

The bond-slip mechanism that rules the transmission of the tangential stresses to the concrete is modeled by a thin layer of an ideally plastic material (J2 plasticity) at the steel-concrete interface (see Figure 16). The stiffness of this bonding material is equal to that of the steel slab. The concrete is modeled in a similar way than in the previous example. The material properties (Young' modulus, E_c,

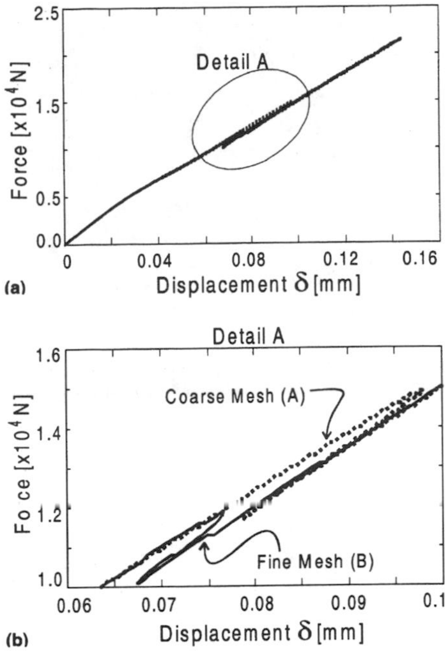

Figure 17. Steel slab in a concrete block; (a) Load vs. displacement curve at the end-face of the slab; (b) detail of the curve showing the snap-back caused by the complete softening of the primary crack.

Poisson's ratio, ν_c, tensile strength, σ_c, and fracture energy, G_f) of concrete, as well as steel parameters and the yield stress, σ_{yb}, of the bounding material are displayed in Figure 16. Two different finite element meshes A (coarse) and B (fine) have been used for the numerical simulations.

Figure 17-(a) shows the evolution of the load P, applied at the end-face of the steel slab, vs. the displacement at the same point. In Figure 17-(b) it is shown, in detail, the typical snap-back behavior of this type of structures, for both meshes.

Secondary transverse cracks can be observed in an initial diffuse failure stage, before the snap-back of the equilibrium path takes place (see Figure 18-(a)). They initially nucleate at the end-face of the specimen but, as the loading increases, they spread out through the total steel-concrete interface length. An initial primary crack onsets and propagates along the concrete section, up to the complete softening of the material in that region (see Figure 18-(b)). The subsequent loss of structural loading capacity then triggers the severe snap-back displayed in the equilibrium path in Figure 17.

At this stage, the structure is still able to hold an increasing external loading transmitted through the steel slab to other sections of the concrete bulk. In fact, a subsequent loading after the snap-back induces

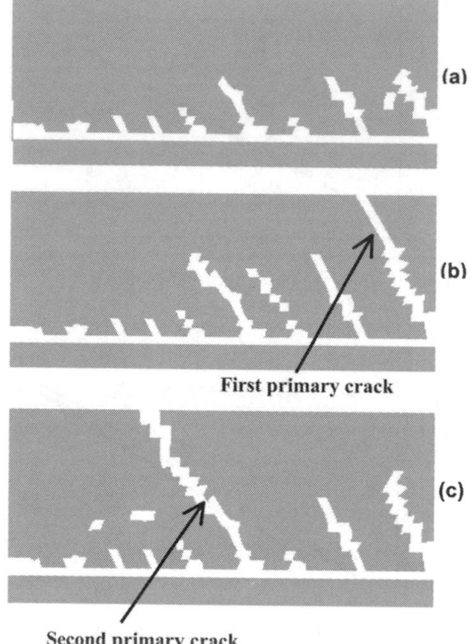

Figure 18. Steel slab in a concrete block. From (a) to (c): evolution of the active cracks for increasing times of the analysis (a) at an initial stage, (b) before the snap-back point and (c) after the snap-back point (elements crossed by the active cracks are highlighted in white).

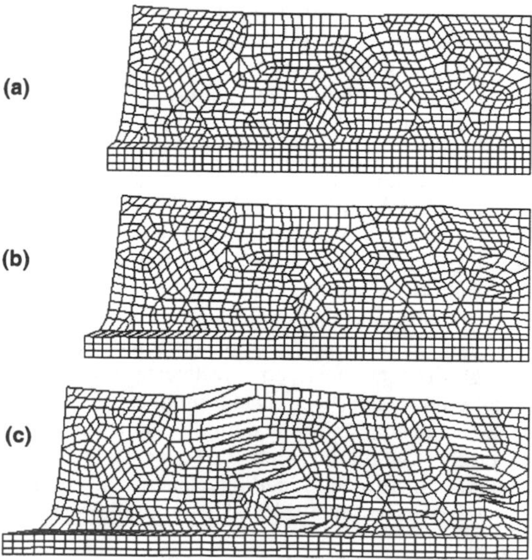

Figure 19. Steel slab in a concrete block. From (a) to (c): Deformed (amplified) meshes at identical times than in Figure 18.

the propagation of a second primary crack (see Figure 18-(c)) across the concrete.

Finally, Figure 19 shows the corresponding deformed meshes.

4 CONCLUDING REMARKS

Along this work different ingredients to model multiple cracking in concrete structures by resorting to the strong discontinuity approach have been tackled. The resulting methodology endows the numerical simulation of cracking of concrete some interesting features, i.e.:

- A clear connection of the continuum format simulation with a discrete one based on Fracture Mechanics settings.
- The same (continuum) non-linear constitutive model is used to model both the damage at the uncracked concrete and the de-cohesion at the crack interface.
- Finite elements with embedded discontinuities are completely insensitive to mesh size and alignment.
- The model can deal with multiple cracks simultaneously progressing through the body and to unfold the possible bifurcations due to multiple crack arrest/activation.

ACKNOWLEDGMENTS

This work has been done with support of the Spanish Ministry of Science and Technology under grants MAT-2001-3863-C03-03 and MAT-2000-0436. This support is gratefully acknowledged.

REFERENCES

Armero F. and Garikipati K., 1996. An analysis of strong discontinuities in multiplicative finite strain plasticity and their relation with the numerical simulation of strain localization in solids, *Int. J. Solids and Structures* (33): 2863–2885

Bazant Z.P., Cedolin L., 1991. *Stability of Structures*, Oxford University Press

Bazant Z.P., Planas J., 1998. *Fracture and size effect in concrete and other quasibrittle materials*, CRC Press

Bocca P., Carpintieri A., Valente S., 1991. Mixed Mode Fracture of Concrete, *International J. of Solids & Structures* 27, (9), 1139–1153

Cervera M., Agelet de Saracibar C., Chiumenti M.., 2001. COMET: a multipurpose finite element code for numerical analysis in solid mechanics. Technical University of Catalonia (UPC)

Gastebled O.J., May I.M., 2000. Bifurcation in the numerical solution of softening mechanisms, *Computers and Structures*, (78): 745–755

Oliver J., 1996. Modeling strong discontinuities in solid mechanics via strain softening constitutive equations. Part 2: Numerical simulation, *International Journal for Numerical Methods in Engineering*., (39):3601–3623

Oliver J., 2000. On the discrete constitutive models induced by strong discontinuity kinematics and continuum constitutive equations. *International Journal of Solids and Structures*, 37:7207–7229

Oliver J., Huespe A.E., Pulido M.D.G., Chaves E., 2001. From continuum mechanics to fracture mechanics: the strong discontinuity approach. *Engineering Fracture Mechanics*, 69(2): 113–136

Oliver, J., Huespe, A.E., Samaniego, E., Chaves, E.W.V. 2002. On Strategies for Tracking Strong Discontinuities in Computational Failure Mechanics, *Proceedings of the Fifth World Congress on Computational Mechanics (WCCM V)*, Editors: Mang, H.A.; Rammerstorfer, F.G.; Eberhardsteiner, J., Publisher: Vienna University of Technology, Austria, ISBN 3-9501554-0-6, http://wccm.tuwien.ac.at

Rots J.G. Computational modeling of Concrete Fracture, PhD Thesis, Delft University of Technology, September 1988

Simó J.C., Oliver J., Armero F., 1993. An analysis of strong discontinuities induced by strain-softening in rate-independent inelastic solids. *Computational Mechanics* 12: 277–296

Wells G.N., Sluys L.J., 2001. A new method for modeling cohesive cracks using finite elements. *International Journal for Numerical Methods in Engineering*, 50,2667–2682

Willam K. 2000. Constitutive models for engineering materials. *Encyclopedia of Physical Science & Technology* 3rd edition, Academic Press, New York, (3):603–633

Computational Modelling of Concrete Structures, Bićanić et al. (eds)
© 2003 Swets & Zeitlinger, Lisse, ISBN 90 5809 536 3

A computational approach for three-dimensional probabilistic discrete cracking in concrete

C.N.M. Paz, L.F. Martha
Department of Civil Engineering and Technology Group on Computer Graphics (Tecgraf) Pontifical Catholic University of Rio de Janeiro (PUC-Rio), Brazil Rua Marquês de São Vicente, RJ/Brazil

J.L.D. Alves, N.F.F. Ebecken, E.M.R. Fairbairn, and A.L.G.A. Coutinho
COPPE/UFRJ, Program of Civil Engineering Federal University of Rio de Janeiro, Brazil

ABSTRACT: This work presents a probabilistic crack approach based on the Monte Carlo method, implemented in a 3D fully parallelized finite element code (Paz, 2000). The cracking scheme used is the discrete crack approach. In this approach the heterogeneity of the material is taken into account by considering the properties to vary spatially following a normal distribution determined by the mean and the standard deviation of the considered material properties.

Fracturing is modeled by 3D interface elements generated in a previously defined region within the mesh (Paz 2000). The interface elements are triangular base prisms connecting adjacent faces of neighboring tetrahedra. These elements simulate crack opening through relative displacements between the triangular faces.

1 INTRODUCTION

1.1 *Probabilistic model*

Concrete cracking depends on several relevant factors such as water/cement ratio, casting and curing, loading conditions, etc. Due to the random distribution of constituents and initial defects, the heterogeneity governs the overall cracking behavior and related size effects on concrete fracture. The probabilistic crack approach based on the direct Monte Carlo method developed by Rossi and co-workers 1994a, b 1996,1997, takes this stochastic process into account by assigning in finite elements analysis, randomly distributed material properties (tensile strength, Young's modulus) to both the solid elements and the interface elements (figure 1). The stochastic process is introduced at the local scale of the material, by considering that cracks are created within the concrete with different energy dissipation depending on the spatial distribution of constituents and initial defects.

The local material behavior in concrete is assumed to obey a perfect elastic brittle behavior, so that the random distribution of local cracking energies can be replaced by a random distribution of local strengths. Therefore, solid elements are elastic, while interface elements are considered elastic-brittle.

Thus, the present probabilistic model involves a number of mechanic properties of the material to be determined, which constitutes the modeling data. From a large number of direct tensile tests it was found that a normal law describes rather well the experimental distribution (Rossi *et al.* 1994b). These characteristics are: $f_{ct,\mu}$ and E_μ, the means of the tensile strength and of the Young's modulus respectively; $f_{ct,\sigma}$ and E_σ, the standard deviations of the tensile strength and of the Young's modulus respectively. The following analytical expressions were proposed:

$$f_{ct,\mu} = 6.5(V_t/V_g)^{-a}; \qquad f_{ct,\sigma}/f_{ct,\alpha\mu} = 0.35(V_t/V_g)^{-b} \quad (1)$$

$$E_\mu = E \quad E_\sigma/E = 0.15(V_t/V_g)^{-c} \quad (2)$$

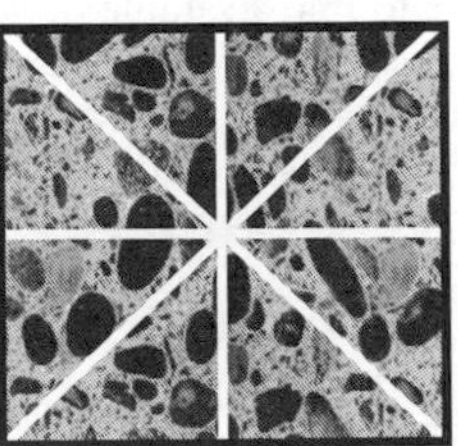

Figure 1. This stochastic process into account by assigning in finite elements analysis, randomly distributed material properties.

where V_t is the volume of the two finite elements contiguous to an individual contact element of the mesh; V_g is the volume of the coarsest aggregate; E is the average Young's modulus that does not exhibit significant volume effects. For cylinder specimens, whose dimensions are 160 mm in diameter and 320 mm high constants a, b and c are related to the compressive strength f_c given by the relations,

$$a = 0.25 - 3.6 \times 10^{-3}(f_c) + 1.3 \times 10^{-5}(f_c)^2$$

$$b = 4.5 \times 10^{-2} + 4.5 \times 10^{-3}(f_c) - 1.8 \times 10^{-5}(f_c)^2$$

$$c = 0.116 + 2.7 \times 10^{-3}(f_c) - 3.4 \times 10^{-6}(f_c)^2 \quad (3)$$

In these expressions the compressive strength f_c represents the quality of the concrete matrix, while the volume of the coarsest aggregate V_g, refers to the elementary material heterogeneity.

Equations (1) to (3) show that the smaller the scale of observation, the larger the fluctuation of the local mechanical properties, and thus the (modeled) heterogeneity of the matter. In other words, the finer the mesh, the greater the modeled heterogeneity in terms of Young's modulus and tensile strength.

Let m_v the number of tetrahedra and m_i the number of interface elements in a given mesh. Also let the set of all Young's modulus of the solid elements be denoted by $\mathbf{E}$ and $\mathbf{f}_{ct}$ that of all tensile strength of interface elements.

According to Rossi *et al.* 1994b these individual local tensile strengths and Young's modulus are represented by normal distributions having the densities:

$$g_f(f_{ct}) = \frac{1}{f_{ct,\sigma}\sqrt{2\pi}} \exp\left[-\frac{1}{2}\left(\frac{f_{ct},\mu}{f_{ct},\sigma}\right)^2\right] \quad (4)$$

$$g_E(E) = \frac{1}{E_\sigma\sqrt{2\pi}} \exp\left[-\frac{1}{2}\left(\frac{E - E_\mu}{E_\sigma}\right)^2\right] \quad (5)$$

where g_f (f_{ct}) and g_E (E) are density functions for the tensile strength f_{ct} and the Young's modulus E, respectively, and x_μ and x_σ denote the mean and standard deviation of the distribution of quantity x. For the problem at hand, it is possible to find a sample of m_i values $f_{ct,ii}$, each value corresponding to an interface element, and m_v values E_{iv}, each value corresponding to a volume element, by using a standard routine for generation of random numbers for a given normal distribution (Press *et al.*, 1992).

The problem with this approach is that these statistical moments are not known a priori for the characteristic volume of the finite elements used in the analysis. However, some methods have been proposed to determine these parameters by means of inverse analysis using neural networks Paz 2000, Farbairn 1999, 2000a, b.

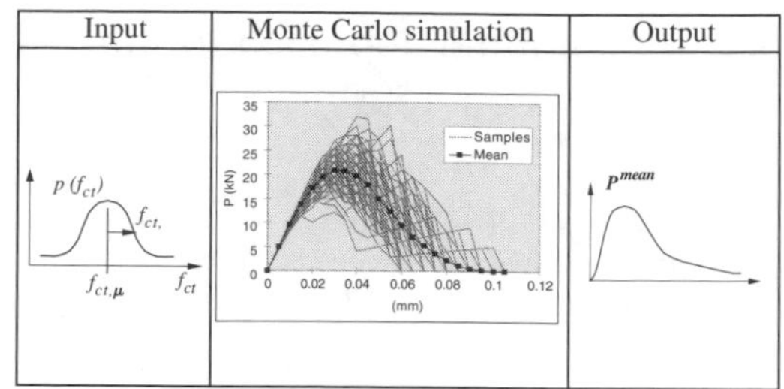

Figure 2. Monte Carlo simulation.

The solution for this probabilistic approach is obtained by means of a Monte Carlo simulation. As depicted in figure 2, in a Monte Carlo simulation a number of n samples are generated for a given normal distribution and some characteristic responses of the structure. A characteristic response may be for example a stress crack-width σ–w curve. Let the jth samples correspond to the jth σ–w curve. This jth σ–w curve is composed of discrete values, σ_k^j and w_k^j, where the superscript j indicates the sample and the subscript k the discrete value of the σ–w curve. The discrete w values are assumed known and the response is defined exclusively by the values of σ_k^j. The mean curve composed by pairs σ_k^{mean}, w_k then simply reads

$$\sigma_k^{mean,j} = \frac{1}{j}\sum_{l=1}^{j}\sigma_k^l \quad (6)$$

The Monte Carlo simulation is stopped when

$$\left|\sigma_k^{mean,j} - \sigma_k^{mean,j-1}\right| \leq tol \quad (7)$$

where tol is the prescribed tolerance to check the convergence of the procedure. Where convergence is reached the number of samples is set to $n = j$. This total number of samples n, corresponding to a Monte Carlo converged simulation, clearly depends on tol, which is a measure of the accuracy required by the analysis. It also depends on the heterogeneity of the material represented by the standard deviation. The more heterogeneous is the material the greater is the number of samples necessary to obtain a converged solution. Our experience in this field indicates that 15 to 30 samples are sufficient to obtain a converged σ–w curve.

2 DISCRETE CRACKING: 3D INTERFACE ELEMENTS

The finite element cracking model is a discrete model for which volume elements are always elastic and cracking occurs in elastic-brittle (almost rigid brittle) contact elements placed between two neighbor

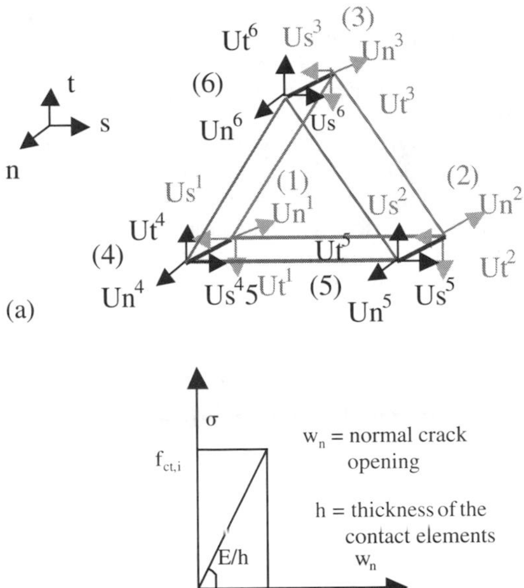

Figure 3. (a) An interface element and its degrees of freedom in a local system; (b) Elastic-brittle contact law.

surfaces of the volume elements. The 3D interface elements (Paz 2000) depicted in Figure 3(a) can be thought as triangular base prisms connecting adjacent faces of neighboring tetrahedra.

These elements are formulated to represent relative displacements between the triangular faces to simulate crack opening.

The constitutive law of the 3D interface element is defined by equation (8) for a non cracked elastic state, characterized by $\sigma_n < f_{ct,i}$. When the tensile strength is exceeded, the elements reach a cracked stage and module E_c and G_c are set to zero as shown in figure 3b.

$$\Delta\sigma = \mathbf{D}_{cr}\,\Delta\mathbf{w} = \begin{Bmatrix} \Delta\sigma_n \\ \Delta\sigma_s \\ \Delta\sigma_t \end{Bmatrix} = \begin{bmatrix} E_c/h & 0 & 0 \\ 0 & G_c/h & 0 \\ 0 & 0 & G_c/h \end{bmatrix} \begin{Bmatrix} \Delta w_n \\ \Delta w_s \\ \Delta w_t \end{Bmatrix} \tag{8}$$

In equation (8) the subscript n normal stands for, while stand for s and t tangential indicating the direction respective to the crack plane, w are the relative displacements between the two faces of the interface element, h is the width of interface element, E_c and G_c are respectively the normal (Young's) and the shear modulus along crack plane.

Equations (8) and figure 3 define the elastic-britle constitutive behavior. The thickness h plays the role of a penalization parameters and should be conveniently chosen not to affect the solution.

The kinematic relation for the interface element is given by:

$$\Delta\mathbf{w} = \mathbf{B}\Delta\mathbf{a}_l^e \tag{9}$$

$\Delta\mathbf{w}$ is the crack opening incremental and $\Delta\mathbf{a}_l^e$ is the vector incremental nodal displacements for the interface element.

Applying a standard displacement based F. E. formulation, the resulting tangent stiffness matrix for the interface element is given by:

$$\mathbf{K}_{\text{Intf}}^{\text{e}} = \int_{\Omega} \mathbf{B}^T \mathbf{D}_{cr} \mathbf{B}\, d\Omega \tag{10}$$

The interface elements are generated contiguous to the faces of selected tetrahedra elements. This selection is performed by the user, defining a 3D box inside the mesh that contains the target elements.

Remark Our experience indicates that to increase robustness of nonlinear solution process we have to limit only one interface element to "crack" at each nonlinear iteration.

3 3D INTERFACE ELEMENTS MESH GENERATION

Fracturing is modeled by 3D interface elements generated in a previously defined cracked region. This selection is performed by the user defining a 3D box inside the mesh that contains the target elements. The interface elements are generated (Paz 2000) contiguous to the faces of the selected tetrahedra.

The procedure initially establishes the neighborhoods of the faces, then it maps how many elements share each node. Later it creates the nodes necessary to the interface elements, all nodes having the same coordinates of the neighbor node. After this all nodes with the same coordinates are visited, and the first element of the loop takes the existing node of the initial mesh of tetrahedra (see figure 4a) and, for the other elements that share this node, it is introduced a new node numbering (see figure 4b) Later connectivities are created, introducing the interface elements according to an ordering previously established.

Interface elements with collapsed nodes provide continuity to the elements outside the cracked region (see figure 4c). These elements are implemented using an artifice that allows the use of elements with six nodes. This artifice consists of multiple references for the same collapsed nodes duplicating the node numbering for the elements outside the 3D box.

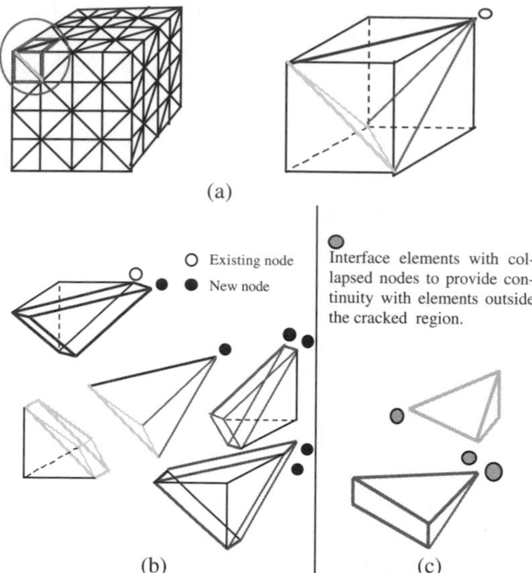

Figure 4. (b, c) 3D Interface elements mesh generation.

4 COMPUTATIONAL STRATEGIES; PARALLEL VECTOR PROCESSOR (PVP)

4.1 *Solution of equilibrium equations and the inexact Newton Method*

Traditional finite element technology for non linear problems involves the repeated solution of systems of sparse linear equations by a direct solution method, that is, some variant of Gauss elimination. The updating and factorization of the sparse global stiffness matrix can result in extremely large storage requirements and a very large number of floating point operations.

In this paper we employ an Inexact Newton method (Kelley 1995), to solve large-scale three-dimensional incremental elastic-brittle problems. In the Inexact Newton Method, at each non linear iteration, a linear system of finite element equations is approximately solved by the preconditioned conjugate gradient method (PCG).

The implementation of global matrix-vector products are easily parallelized in different computer architectures, performing element level products followed by global assembly. This type of implementation is often referred to element-by-element (EBE) schemes. Matrix-vector products computed by EBE schemes are memory intensive, requiring more operations than the product with the assembled matrix, because element matrices have many overlapping non-zero entries. However, particularly for large scale non linear problems EBE methods have been very successful, because they handle large sparse matrices in a simple and straightforward manner. Besides, efficient preconditioners may be derived keeping the same data structure. For a recent review of such topics see Hughes 1987.

When solving iteratively the finite element system of linear equations, it is straightforward to employ inexact versions of the standard Newton-like methods. In this case, tolerances for the inner iterative driver may be adaptively selected to minimize computational effort towards the solution, giving rise to the following algorithm:

Given u_{tol}, r_{tol}, η_i relative and residual tolerance.
Compute stiffness tetrahedra matrix $\mathbf{K}_{Tetra}$
do k = 1, 2 . . . , number of load increments do
  Compute external forces vector
  $\mathbf{F}^k_{ext} = \mathbf{F}^K_{nodal} + \mathbf{F}_{volume} + \mathbf{F}_{\bar{\sigma}} - \left(\mathbf{K}_{tetra}\,\bar{\mathbf{U}}_k + \mathbf{K}_{Intef}\,\bar{\mathbf{U}}_k\right)$
  do i = 1, 2 . . ., while convergence
    Compute internal forces vector,
    $\mathbf{F}^i_{\text{int}} = \left(\mathbf{F}^i_{\text{int}}\right)_{Tetra} + \left(\mathbf{F}^i_{\text{int}}\right)_{Intf}$
    Compute residual vector,
    $\psi^i = \mathbf{F}^i_{\text{int}} - \mathbf{F}^i_{ext}$
    Update stiffness interface matrix $\mathbf{K}^i_{Intf}$
    $\mathbf{A}^i = \mathbf{K}_{Tetra} + \mathbf{K}^i_{\text{Intf}}$
    Compute tolerance for iterative driver, $\boldsymbol{\eta}_i$
    Solver: $\mathbf{A}^i \;\; \Delta\mathbf{u} = \psi^i$ for tolerance $\boldsymbol{\eta}_i$
    Update solution,
    $\mathbf{U} = \mathbf{U} + \Delta\mathbf{u}$
    if $\dfrac{\| \Delta\mathbf{u} \|}{\| \mathbf{U} \|} \leq utol$ and $\dfrac{\| \Delta\psi^i \|}{\| \mathbf{F}^k_{ext} \|} \leq_{rtol}$ then
    convergence
  end while i.
end do k.

Note that in $\mathbf{F}^k_{ext}$ we account for nodal forces, body forces and prescribed displacements and stresses $\mathbf{U}$, $\boldsymbol{\sigma}$. The total internal forces vector $\mathbf{F}^i_{\text{int}}$ the sum of the solid elements internal forces $\left(\mathbf{F}^i_{\text{int}}\right)_{Tetra}$ plus the interface element internal forces vector $\left(\mathbf{F}^i_{\text{int}}\right)_{Intf}$. The total stiffness matrix is the sum of the continuum matrix $\mathbf{K}_{Tetra}$ plus the interface matrix $\mathbf{K}^i_{Intf}$ updated at each non linear iteration.

We adopted a simple nodal block-diagonal preconditioner. Therefore, the most expensive computational kernel in the linear solver is the matrix-vector product. According to the above algorithm, an approximate solution is obtained when the Inexact Newton termination criterion is satisfied, that is, when,

$$\left\| \mathbf{A}^i \Delta \mathbf{u} - \Psi^i \right\| \leq \eta_i \left\| \Psi^i \right\| \tag{11}$$

We selected η_i as suggested by Kelley [11], based on a measure of how far the nonlinear iteration is from the solution, that is,

$$\eta_i^A = \min\left(\eta_{\max}, \gamma \frac{\|\psi^i\|^2}{\|\psi^o\|^2}\right), \quad 0 < \gamma < 1 \tag{12}$$

If η_i^A is uniformly limited away from 1, and taking $\eta_i = \max\left(\eta_{\min}, \eta_i^A\right)$ Kelley [11] has shown general convergence properties when Equation (12) is used. To avoid that η_i^A be too small when the nonlinear iteration is away from the solution, Kelley also suggests the following modification,

$$\eta_i^B = \left\{ \begin{array}{ll} \min\left(\eta_{\max}, \eta_i^A\right), & if \quad \gamma\, \eta_{i-1}^2 < 0.1 \\ \min\left(\eta_{\max},\right. & if \quad \gamma\, \eta_{i-1}^2 \geq 0.1 \\ \quad \left.\max\left(\eta_i^A, \gamma\, \eta_{i-1}^2\right)\right), & \end{array} \right\} \tag{13}$$

In some cases $\|\psi_i\|$ can be very small, well beyond the required accuracy, resulting in undesired work. To remedy this over solving Kelley 1995 proposes to compute η_i^c using;

$$\eta_i^C \left(\eta_{\max}, \max\left(\eta_i^B, 05\, rtol \frac{\left\|\mathbf{F}^k_{ext}\right\|}{\left\|\psi^i\right\|}\right)\right) \tag{14}$$

and finally taking $\eta_i = \max\left(\eta_{\min}, \eta_i^C\right)$.

Our experience indicates that selecting $\eta_{\max} = 0.1$ *and* $10^{-3} \leq \eta_{\min} \leq 10^{-6}$ for *utol* and *rtol* in the usual range, that is, 10^{-3} to 10^{-2}, is enough for practical engineering computations. Typical values for γ and g are 0.5 and 0.1, respectively.

4.2 *Matrix-vector products element-by-element, EBE*

In the present implementation matrix vector products in EBE, PCG are computed as:

$$\mathbf{Ap} = \sum_{i=1}^{N_{tetra}} \left(\mathbf{K}_{Tetra}\, \mathbf{p}_i\right) + \sum_{j=1}^{N_{Intf}} \left(\mathbf{K}_{Intf}\, \mathbf{p}_j\right) \tag{15}$$

where, N_{tetra} is the number of tetrahedra, N_{intf} is the number of interface elements, $\mathbf{K}_{Tetra}$ and $\mathbf{K}_{Intf}$ are respectively element matrices for the tetrahedra and interface; $\mathbf{p}_i$ and $\mathbf{p}_j$ are the components of $\mathbf{p}$ restricted to the degrees of freedom of two element type.

Stiffness matrices for tetrahedra are computed and stored at the beginning of the analysis since they are elastic.

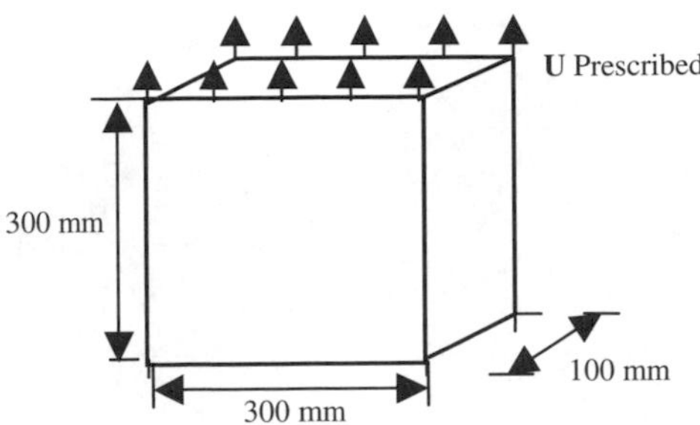

Figure 5. Uniaxial tension specimen geometry, dimensions, load, boundary conditions.

Stiffness matrix for interface elements are updated at every nonlinear iteration.

The arrays of element stiffness matrices are stored taking into account their symmetry; in the case of the element tetrahedra 78 coefficients are stored and for the interface element only 18 coefficients are stored, exploring the particular structure of the discrete gradient operator.

The mesh coloring algorithm of Hughes 1987 was extended to block both solid and interface elements into disjoint groups thus enabling full vectorization an parallelization of the operations involved in equation (15).

5 NUMERICAL EXAMPLE

The analized example is direct tension tests (figure 5). This numerical example was analyzed by Rossi et al 1997 in a bidimensional program.

The numerical experiment were controlled by a field of uniform displacements applied at the top of the test specimen in 30 incremental steps. $\Delta u = 4.0 \times 10^{-3}$. The boundary conditions restrain the degrees of freedom in the vertical at the bottom. Concrete with maximum aggregate diameter of 10.00 mm was used. Its average tensile strength and Young's modulus at the age of 28 days were: $f_{ct} = 3.0$ Mpa and $E_c = 30{,}000$ Mpa.

The final mesh figure 6, has 4,756 elements, where 2,000 are tetrahedra and 2,726 are interface elements, and node numbers 4,181. The final mesh figure7, has 3,292 elements, where 2,000 are tetrahedra and 1,292 are interface elements and node numbers 2,131.

6 COMPUTATIONAL PERFORMANCE

A detailed vector performance analysis is obtained by the summary of the PERFVIEW's Report presented in figures 13 and 14. The CPU time of the vectorized single processor run for CRAY T90 are 1.48 hours and 1.02 hours (mesh (a) and (b) respectively).

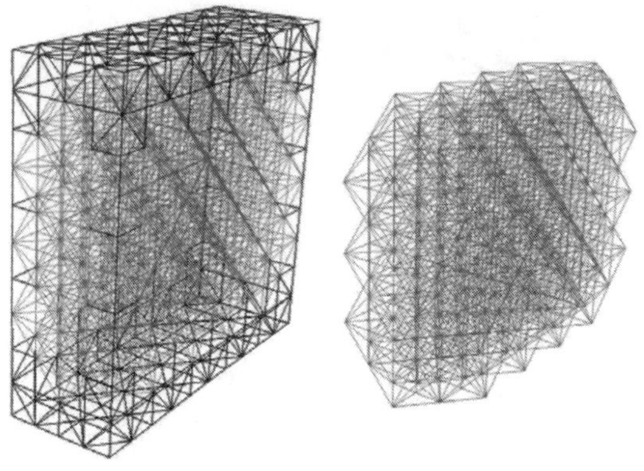

Figure 6. Representations of the computational mesh for the simulation of a direct tension test. It is also shown aside the resulting mesh for the interface elements considering (mesh a).

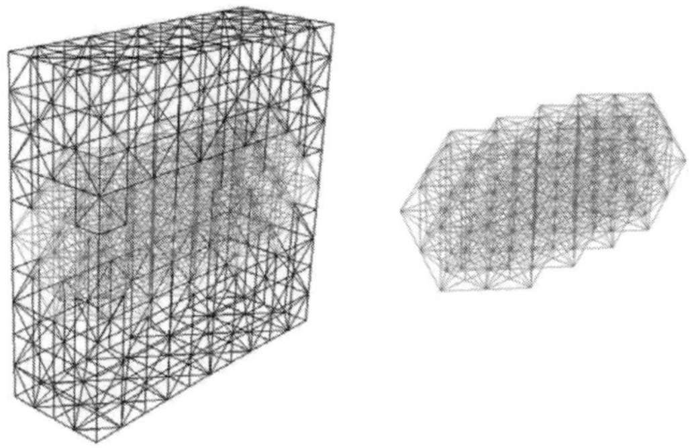

Figure 7. Representations of the computational mesh for the simulation of a direct tension test. It is also shown aside the resulting mesh for the interface elements considering (mesh b).

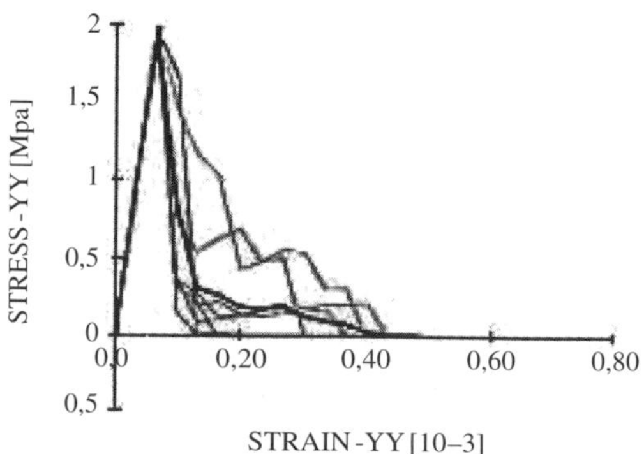

Figure 8. Curve stress–strain Rossi et al. 1997.

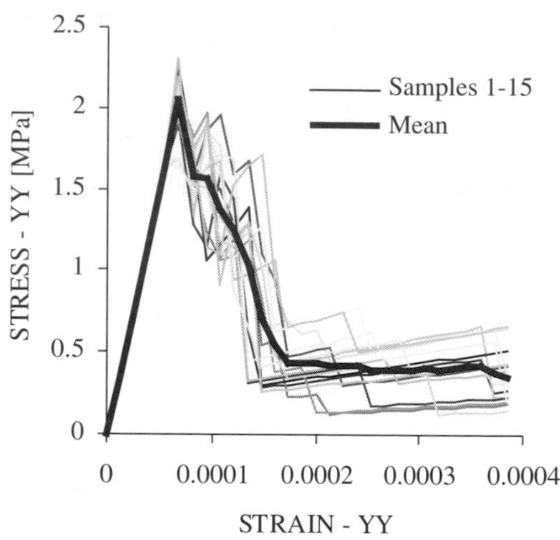

Figure 9. Results for the complete Monte Carlo simulation (mesh a).

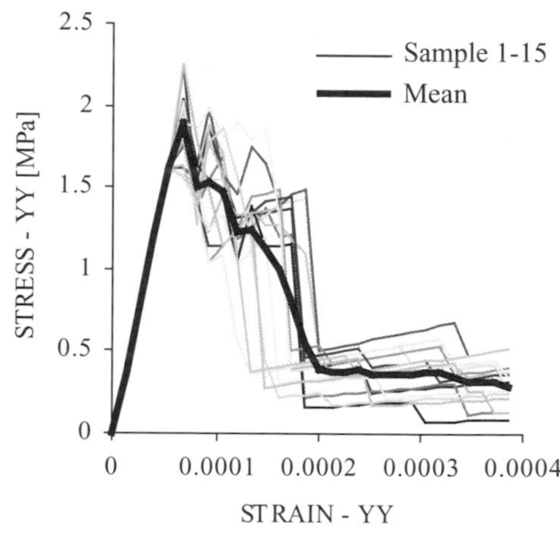

Figure 10. Results for the complete Monte Carlo simulation (mesh b).

Figure 11. Crack evolution for numerical simulation (mesh a).

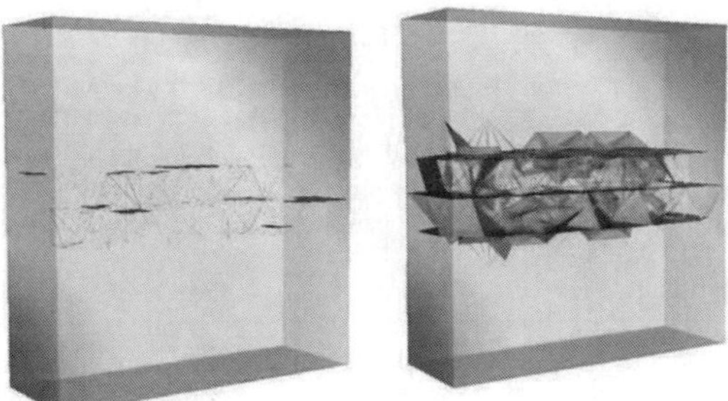

Figure 12. Crack evolution for numerical simulation (mesh b).

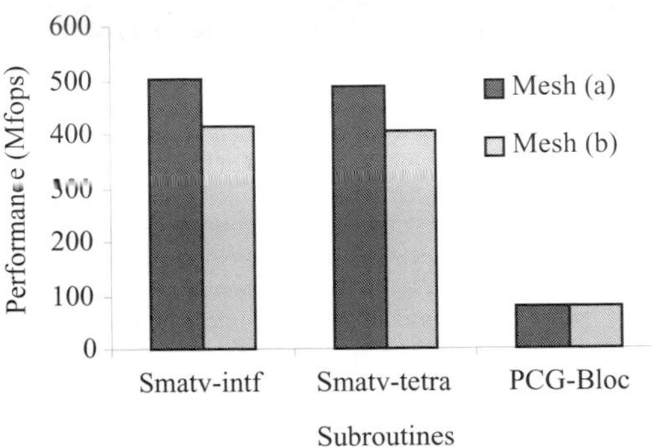

Figure 13. Performance analysis – the top three subroutines – the 2 meshes.

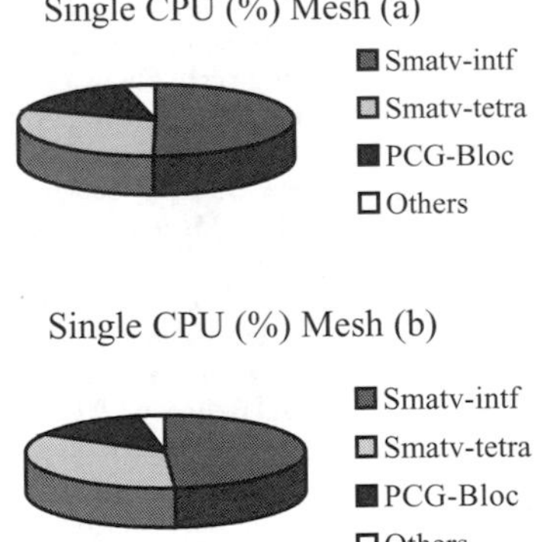

Figure 14. Performance analysis – single CPU (%) – the top three subroutines.

This table list for a single CPU run, the Mflop/s rates for the three top routines. The routines **Smatv-fint** and **Smatv-tetra** are respectively responsible for the matrix-vector products for the interface and tetrahedra elements, routines are the computations kernels the routine **PCG-block,** implements the iterative solver the nodal block diagonal Preconditioned Conjugate Gradient (PCG).

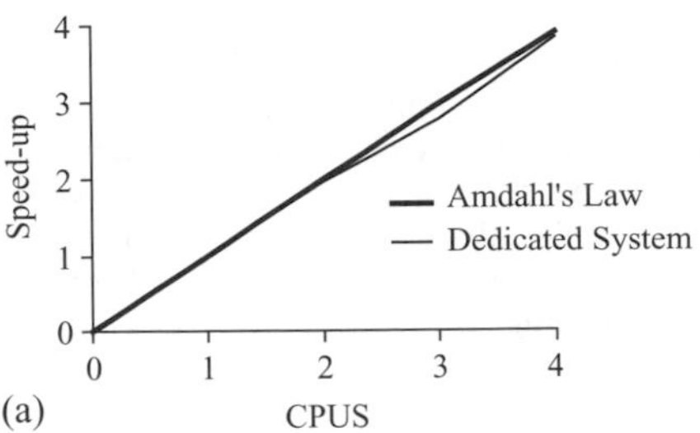

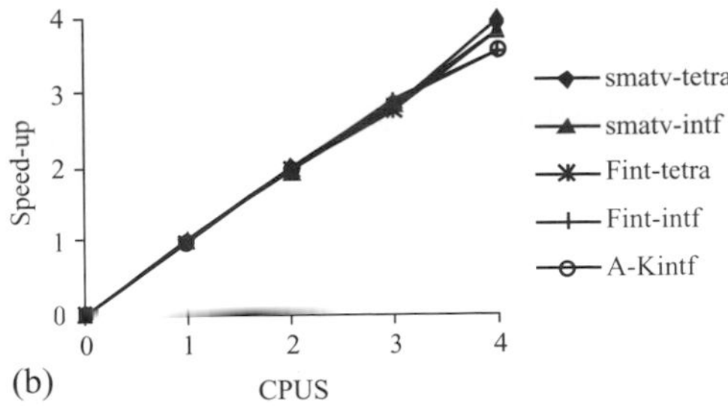

Figure 15. (a) Program Summary of the ATEXPERT's Report (b) ATEXPERT's Report-The top 5 subroutines on a Dedicated Speedup.

The code achieved good vectorization on the CRAY T90 for a mesh (a). The top three subroutines are consume the major CPU utilization in the whole analysis.

The parallel performance is shown figures 15 (a, b) as obtained from a summary ATEXPERT, report autotasking performance tool. The top five subroutines assigned for parallel analysis are presented for the direct tension test. The routines **Fint-tetra** and **Fint-intf** evaluate respectively the internal force vector of the interface and tetrahedra elements. The routine **Kintf** computer update interface elements stiffness.

According to ATEXPERT tool this program appears to be 99.2 percent parallel and 0.8 percent serial. Amdahl's Law predicts the program could expect to achieve a 3.9 times speedup on 4 CPU's. A 3.8 speedup is predicted with 4 CPU's on dedicated system.

7 CONCLUDING REMARKS

This paper presented the optimized implementation of Rossi's a probabilistic model for the simulation of cracking in concrete structures. This model is based on the assumption that some particularities of the cracking behavior of concrete, such as strain softening, cracking evolution and size-effects are derived from the heterogeneous characteristics of the material.

The probabilistic methodology presented in this paper corresponds to the 3D analysis of a strongly nonlinear material that develops cracks. In adition, the finite elements analysis must be called several times within a Monte Carlo simulation. Therefore, the code needs to be optimized in such a way that the simulation time does not exceed a practical limit.

The code achieved a very good level for both parallel performance and vetorization. The most demanding routines, which implement the matrix-vector-multiply computational kernel for the interface and tetrahedral elements, are "fully" parallelized (~99%) and responsible for over 80% of CPU time. The results emphasizes the suitability of the implemented code on the parallel-vector machine, CRAY T90 for 2 CPU's, which presented a flop rate of 614 Mflop/s and a parallel speed-up of 3.8 for 4 CPU's.

Extensive use of element-by-element techniques within the computational kernels comprised in the iterative solution drivers provided a natural way for achieving high Flop rates and good parallel speed-up's. Furthermore, element-by-element techniques, avoid completely the formation and handling of large sparse matrices. Therefore, the computational strategies presented herein provide a natural way to deal with more complex scenarios, particularly those involving three-dimensional problems.

ACKNOWLEDGMENTS

The authors are indebted to the Computer Graphics Technology Group TECGRAF/PUC-Rio, High Performance Computing Center NACAD/COPPE/UFRJ, and the Laboratory of Computational Methods in Engineering of the Program of Civil Engineering LAMCE/COPPE/UFRJ. CESUP/UFRGS is gratefully acknowledged for the computer time provided in the CRAY T90. This work was partially supported by CNPq grants N° 150039/01-8(NV).

REFERENCES

Coutinho A.L.G.A., Martins M.A.D., Alves J.L.D., Landau L., and Moraes A. 2001. Edge-based finite element tecniques for nonlinear solid mechanics problems, *Int. J. for Numerical Methods in Engineering*, 50 (9), 2050–2068.

Fairbairn E.M.R., Ebecken N.F.F., Goulart E. and Paz C.N.M. 1998. Probabilistic modelling of Concrete cracking using Monte Carlo and Neural networks to solve the inverse problem, *Computational Structures Technology 1998 & Engineering Computational Technology 1998 (Edinburgh), Editor: B.H.V. Topping, Advances in Engineering Computational Technology – Neural computing for engineering computations,* ISBN 0-948749-55-5, 215–220.

Fairbairn E.M.R., Paz C.N.M., Ebecken N.F.F., and Ulm F.-J., 1999. Use of neural network for fitting of probabilistic scaling model parameter, *Int. J. Fracture*, 95, 315–324.

Fairbairn E.M.R., Ebecken N.F.F., Paz C.N.M. and Ulm F.-J., 2000a. Determination of probabilistic parameters of concrete: solving the inverse problem by using artificial neural networks, Computers and Structures, 78, 497–503.

Fairbairn E.M.R., Debeux V.J.C., Paz C.N.M. and Ebecken N.F.F. 2000b. Applications of probabilistic approach to the analisysis of gravite dam centrifuge, *Test, 8th ASCE Specialty Conference on probabilistic Mechanics and Structural Reability. 24–26 July, USA, CD ROMM Proceedings* PMC 2000–216.

Hughes T.J.R. 1987. Algorithms for parabolic problems, element-by-element (EBE) implicit methods. In: The finite element method analysis, Chapter 8, p 483–489, *New Jersey*: Prentice – Hall.

Kelley C.T. 1995. Iterative methods for linear and nonlinear equations, frontiers in applied mathematics, SIAM ? Society for industrial and applied mathematics, Philadelphia.

Paz C.N.M. 2000. Development and implementation probabilistic model for 2D and 3D discrete cracking concrete in parallel computing, D.Sc. Thesis, Dept, of Civil Engineering, COPPE/UFRJ, Federal University of Rio de Janeiro, Brazil [in Portuguese].

Paz C.N.M., Martha L.F., Fairbairn E.M.R., Alves J.L.D., Ebecken N.F.F. and Coutinho A.L.G.A. 2002a. "Parallel implementation and development of a probabilistic model for 3D discrete concrete cracking", *WCCM Fifth Congress on Computational Mechanics July 7–12, Vienna, Austria Eds: H.A. Mang, F.G. rammerstorfer, J. Eberhardsteiner.* ISBN 3-9501554-0-6. http://wccm.tuwiwn.ac.at.

Paz C.N.M., Martha L.F., Fairbairn E.M.R., Alves J.L.D., Ebecken N.F.F. and Coutinho A.L.G.A. 2002b. Probabilistic model 3D discrete cracking concrete in parallel computing, *VECPAR 2002, 5t International meeting, high performance computing for computational science, University of Porto FEUP, Porto, Portugal June 26–28.*

Press W.H., Teukolski S., Vetterling W.T. and Flannery B. 1992. Numerical recipes, *Cambridge University Press.*

Rossi P., Wu X., Maou F. le, and Belloc A. 1994a. Scale effect on concrete in tension, *Materials and Structures*, 27 (172), 437–444.

Rossi P., Wu X., Maou F. le, and Belloc 1994b. Scale effect on concrete in tension, *Materials and Structures*, 27 (172), 437–444.

Rossi P., Ulm F.-J., and Hachi F. 1996. Compressive behavior of concrete: physical mechanisms and modeling, *Journal of Engineering Mechanics ASCE*, 122 (11), 1038–1043.

Rossi P. and Ulm F.-J. 1997. Size effects in the biaxial behavior of concrete: physical mechanisms and modeling, *Materials and Structures*, 30 (198), 210–216.

Computational Modelling of Concrete Structures, Bićanić et al. (eds)
© 2003 Swets & Zeitlinger, Lisse, ISBN 90 5809 536 3

Adaptive simulation of quasibrittle failure

Bořek Patzák
Czech Technical University (ČVUT), Prague, Czech Republic

Milan Jirásek
Swiss Federal Institute of Technology (EPFL), Lausanne, Switzerland

ABSTRACT: This paper presents an adaptive mesh refinement technique suitable for the resolution of highly localized damage in concrete and other quasibrittle materials. Objectivity of the description of softening is ensured by using regularized material models based on the concept of nonlocal averaging, which is applied to isotropic and anisotropic damage formulations. The distributions of strain and internal variables produced by such regularized models are continuous, which facilitates the projection of information from one finite element mesh onto another. The paper comments on important components of a complete adaptive strategy: the error indicator, refinement rules, mesh generator, transfer operators for the mapping of displacements and internal variables, and consistency recovery. The efficiency of the proposed strategy is illustrated by two examples that treat straight as well as curved crack trajectories.

1 INTRODUCTION

In quasibrittle materials such as concrete, rocks, ice, tough ceramics or certain composites, failure is preceded by a gradual development of a nonlinear fracture process zone and by localization of strain. Realistic failure analysis of quasibrittle structures requires the consideration of progressive damage due to microcracking, which is usually modeled by a constitutive law with strain softening. In the context of a standard Boltzmann continuum, the boundary value problem becomes ill-posed due to the loss of ellipticity of the governing differential equations. Numerically obtained results are not objective with respect to the discretization, and the total energy consumed by the fracture process tends to zero as the computational grid is refined.

Advanced regularization methods introduce an additional material parameter – the characteristic length, which is related to the size and spacing of major inhomogeneities and controls the width of the numerically resolved fracture process zone. Such methods can be based on nonlocal averaging, higher-order gradient theories, continua with microstructure, or viscous regularization. They serve as localization limiters, in the sense that they impose a nonzero limit on the minimal width of the process zone and prevent strain localization into a set of measure zero.

In this paper, attention is focused on nonlocal continuum damage models of the integral type. Such models replace a suitably chosen local quantity by its nonlocal counterpart, obtained as a weighted average over a certain neighborhood (domain of influence) of each material point. The size of this domain and the shape of the nonlocal weight function determine the characteristic length imposed by the model, which should be related to the material microstructure. The nonlocal quantity is then inserted into the original constitutive equation. Besides acting as a localization limiter, the nonlocal approach can substantially reduce the mesh-induced directional bias and improve convergence of the incremental-iterative solution strategies.

An accurate resolution of bands of highly localized strain typically requires very fine computational grids. The efficiency of the analysis can be greatly increased by using adaptive techniques that adjust the mesh during the simulation, depending on the intermediate localization pattern and its evolution. The proposed approach is based on *h*-adaptivity, i.e., on the adjustment of the element size, keeping the order of the elements constant. An alternative technique, based on enrichment of the standard finite element interpolation by special functions in the spirit of the partition-of-unity method, is presented elsewhere (Patzák and Jirásek 2003).

2 ISOTROPIC DAMAGE MODEL

2.1 *Local formulation*

As a prototype of a continuum damage model, consider the simple one-parameter isotropic damage model described by the stress-strain law

$$\boldsymbol{\sigma} = (1-\omega)\boldsymbol{D}_e\varepsilon \tag{1}$$

damage law

$$\omega = f(\kappa) \tag{2}$$

and loading-unloading conditions

$$g(\varepsilon,\kappa) \equiv \tilde{\varepsilon}(\varepsilon) - \kappa \leq 0, \quad \dot{\kappa} \geq 0, \quad \dot{\kappa}g(\varepsilon,\kappa) = 0 \tag{3}$$

in which $\boldsymbol{\sigma}$ is the column matrix of stress components, ε is the column matrix of engineering strain components, $\boldsymbol{D}_e$ is the elastic material stiffness matrix, ω is a scalar damage variable, $\tilde{\varepsilon}$ is the equivalent strain (to be specified later), and κ is a history variable that corresponds to the maximum level of equivalent strain ever reached in the previous history of the material. Function g controls the shape of the elastic domain in the strain space, and function f controls the shape of the stress-strain curve. It is often assumed that the damage growth begins only when a certain elastic limit is exceeded, i.e., that $f(\kappa) = 0$ for $\kappa \leq \varepsilon_0$ where ε_0 is the strain at the elastic limit. For $\kappa > \varepsilon_0$, the function f is monotonically increasing and it reaches the limit value $f = 1$ either at a finite value $\kappa = \kappa_f$, or asymptotically as $\kappa \to \infty$.

The equivalent strain $\tilde{\varepsilon}$ is a scalar measure of the strain level, and under uniaxial stress it could be defined as the normal strain in the direction of loading. However, since damage in many materials propagates much more easily under tension than under compression, the definition of $\tilde{\varepsilon}$ is usually modified such that the influence of tension is emphasized. As a first approximation, one may assume that compressive strains do not lead to any damage, and relate $\tilde{\varepsilon}$ to positive strains only. For instance, Mazars (1984) defined the equivalent strain as

$$\tilde{\varepsilon} = \sqrt{\sum_{I=1}^{3} \langle \varepsilon_I \rangle^2} \tag{4}$$

where $\varepsilon_I, I = 1,2,3$, are the principal strains, and the brackets $\langle \ldots \rangle$ denote the positive part, i.e., $\langle \varepsilon \rangle = \varepsilon$ for $\varepsilon > 0$ and $\langle \varepsilon \rangle = 0$ for $\varepsilon \leq 0$.

In the present study, we define the equivalent strain as

$$\tilde{\varepsilon} = \frac{1}{E} \max_{I=1,2,3} \sigma_{eI} \tag{5}$$

where $\sigma_{eI}, I = 1,2,3$, are the principal values of the effective stress $\boldsymbol{\sigma}_e = \boldsymbol{D}_e\varepsilon$. This definition corresponds to a Rankine-type failure surface, because in the elastic range the effective stress coincides with the actual (nominal) stress. Damage is initiated if the maximum principal stress reaches a critical level, given by the uniaxial tensile strength $f_t = E\varepsilon_0$. The damage law

$$f(\kappa) = \begin{cases} 0 & \text{if} \kappa \leq \varepsilon_0 \\ 1 - \dfrac{\varepsilon_0}{\kappa} \exp\left(-\dfrac{\kappa - \varepsilon_0}{\varepsilon_f - \varepsilon_0}\right) & \text{if} \kappa > \varepsilon_0 \end{cases} \tag{6}$$

is constructed such that the softening part of the uniaxial stress-strain curve is exponential. Here, ε_f is a model parameter that controls the post-peak slope. The tangent modulus just after the peak is $E_t = -f_t/(\varepsilon_f - \varepsilon_0)$.

2.2 *Nonlocal formulation*

In a broad sense, nonlocal formulations are those that contain a material parameter with the dimension of length, i.e., those that are not invariant with respect to spatial scaling. The internal length scale can be imposed by weighted spatial averaging, by incorporating gradients of strain or of internal variables, or by introducing additional kinematic fields independent of the displacement field. Here we focus on nonlocal formulations of the integral type, sometimes called strongly nonlocal, which replace selected internal variables by their weighted averages over a spatial neighborhood of each material point.

If $f(\boldsymbol{x})$ is a certain "local" field, its nonlocal counterpart is defined by the spatial integral

$$\bar{f}(\boldsymbol{x}) = \int_V \alpha(\boldsymbol{x},\boldsymbol{\xi}) f(\boldsymbol{\xi}) \mathrm{d}\boldsymbol{\xi} \tag{7}$$

where $\alpha(\boldsymbol{x},\boldsymbol{\xi})$ is a given weight function, also called the attenuation function. In an infinite, isotropic and macroscopically homogeneous body, the strength of the interaction depends only on the distance $r = ||\boldsymbol{x} - \boldsymbol{\xi}||$ between the "source" point, $\boldsymbol{\xi}$, and the "receiver" point, x. In the vicinity of a boundary, the weight function is usually scaled such that the nonlocal operator does not alter a uniform field, i.e., such that

$$\int_V \alpha(\boldsymbol{x},\boldsymbol{\xi}) \mathrm{d}\boldsymbol{\xi} = 1 \tag{8}$$

This can be achieved by setting

$$\alpha(\mathbf{x},\boldsymbol{\xi}) = \frac{\alpha_0(||\boldsymbol{x} - \boldsymbol{\xi}||)}{\int_V \alpha_0(||\boldsymbol{x} - \boldsymbol{\zeta}||) \mathrm{d}\boldsymbol{\zeta}} \tag{9}$$

where $\alpha_0(r)$ is a nonnegative function, monotonically decreasing for $r \geq 0$. It is often taken as the truncated polynomial bell-shaped function

$$\alpha_0(r) = \left\langle 1 - \frac{r^2}{R^2} \right\rangle^2 \tag{10}$$

where R is a parameter related to the intrinsic material length. Note that $\alpha_0(r)=0$ for $r > R$. Since R corresponds to the maximum distance of point $\boldsymbol{\xi}$ that affects the nonlocal average at point $\boldsymbol{x}$, it is called the interaction radius.

A number of nonlocal damage formulations have been published, starting from the seminal paper by Pijaudier-Cabot and Bažant (1987). In the present study, we use the isotropic damage model with the nonlocal equivalent strain, i.e., in the loading-unloading conditions (3) we replace the equivalent strain $\tilde{\varepsilon}$ by its weighted average

$$\bar{\varepsilon}(\boldsymbol{x}) = \int_V \alpha(\boldsymbol{x},\boldsymbol{\xi})\tilde{\varepsilon}(\boldsymbol{\xi})\,\mathrm{d}\boldsymbol{\xi} \tag{11}$$

3 ANISOTROPIC DAMAGE MODEL

3.1 *Local formulation*

The concept of isotropic damage is appropriate for materials weakened by voids, but if the physical source of damage is the initiation and propagation of microcracks, isotropic stiffness degradation can be considered only as a first rough approximation. More refined damage models take into account the highly oriented nature of cracking, which is reflected by the anisotropic character of the damaged stiffness or compliance matrices.

A number of anisotropic damage formulations have been proposed in the literature. Here we use a model outlined by Jirásek (1999), which is based on the principle of energy equivalence (Cordebois and Sidoroff 1979) and on the construction of the inverse integrity tensor by integration of a scalar over all spatial directions. The general framework for this class of models was established by Carol and Bažant (1997). Since the model uses certain concepts from the microplane theory, it is called the microplane-based damage model (MDM).

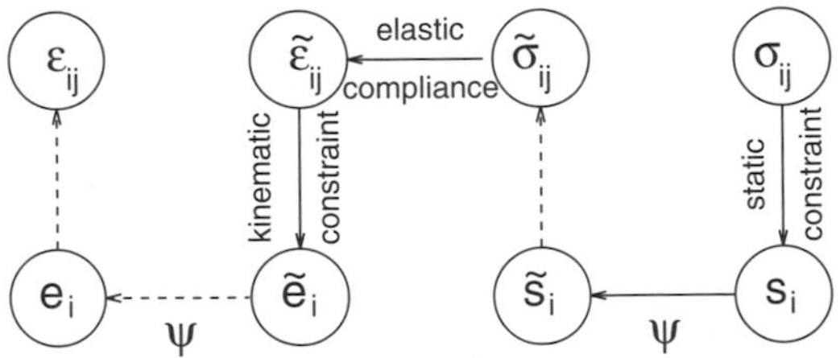

Figure 1. Structure of microplane-based anisotropic damage model.

The general structure of the MDM model is schematically shown in Fig. 1 and the basic equations are summarized in Table 1. Here, ε and σ are the (nominal) second-order strain and stress tensors with components ε_{ij} and σ_{ij}; $\boldsymbol{e}$ and $\boldsymbol{s}$ are first-order strain and stress tensors with components e_i and s_i, which characterize the strain and stress on "microplanes" of different orientations given by a unit vector $\boldsymbol{n}$ with components n_i; ψ is a dimensionless compliance parameter that is a scalar but can have different values for different directions $\boldsymbol{n}$; the symbol δ denotes a virtual quantity; and a superimposed tilde denotes an effective quantity, which is supposed to characterize the state of the intact material between defects such as microcracks or voids.

Combining the basic equations, it is possible to show that the components of the damaged material compliance tensor are given by

$$C_{ijkl} = M_{pqij}M_{rskl}C^e_{pqrs} \tag{12}$$

where C^e_{pqrs} are the components of the elastic material compliance tensor,

$$M_{ijkl} = \tfrac{1}{4}(\psi_{ik}\delta_{jl} + \psi_{il}\delta_{jk} + \psi_{jk}\delta_{il} + \psi_{jl}\delta_{ik}) \tag{13}$$

are the components of the so-called damage effect tensor, and

$$\psi_{ij} = \tfrac{3}{2\pi}\int_\Omega \psi n_i n_j \mathrm{d}\Omega \tag{14}$$

are the components of the second-order inverse integrity tensor. The integration domain Ω is the unit hemisphere. In practice, the integral over the unit hemisphere is evaluated by summing the contribution from a finite number of directions, according to one of the numerical integration schemes that are used by microplane models (Bažant and Oh 1986).

The scalar variable ψ characterizes the relative compliance in the direction given by the vector $\boldsymbol{n}$. If ψ is the same in all directions, the inverse integrity tensor evaluated from (14) is equal to the unit second-order tensor (Kronecker delta) multiplied by ψ, the damage effect tensor evaluated from (13) is equal to the symmetric fourth-order unit tensor multiplied by ψ, and the damaged material compliance tensor evaluated from (12) is the elastic compliance tensor multiplied by ψ^2. The factor multiplying the elastic compliance tensor in the isotropic damage model is $1/(1-\omega)$, and

Table 1. Basic equations of microplane-based anisotropic damage model.

$\tilde{\boldsymbol{e}} = \tilde{\boldsymbol{\varepsilon}}\cdot\boldsymbol{n}$	$\boldsymbol{s}^T = \psi\boldsymbol{s}$	$\boldsymbol{s} = \boldsymbol{\sigma}\cdot\boldsymbol{n}$
$\tilde{\boldsymbol{\sigma}} : \delta\tilde{\boldsymbol{\varepsilon}} = \frac{3}{2\pi}\int_\Omega \tilde{\boldsymbol{s}}\cdot\delta\tilde{\boldsymbol{e}}\mathrm{d}\Omega$	$\delta\boldsymbol{s}\cdot\boldsymbol{e} = \delta\tilde{\boldsymbol{s}}\cdot\tilde{\boldsymbol{e}}$	$\delta\boldsymbol{\sigma} : \varepsilon = \frac{3}{2\pi}\int_\Omega \delta_s\cdot\boldsymbol{e}\mathrm{d}\Omega$
$\tilde{\boldsymbol{\sigma}} = \frac{3}{2\pi}\int_\Omega(\tilde{\boldsymbol{s}}\otimes\boldsymbol{n})_{\mathrm{sym}}\mathrm{d}\Omega$	$\boldsymbol{e} = \psi\tilde{\boldsymbol{e}}$	$\varepsilon = \frac{3}{2\pi}\int_\Omega(\boldsymbol{e}\otimes\boldsymbol{n})_{\mathrm{sym}}\mathrm{d}\Omega$

so ψ corresponds to $1/\sqrt{1-\omega}$. In the initial undamaged state, $\psi = 1$ in all directions. The evolution of ψ is governed by the history of the projected strain components. In the simplest case, ψ is driven by the normal strain $e_N = \varepsilon_{ij} n_i n_j$. Analogy with the isotropic damage model leads to the damage law

$$\psi = f(\kappa) \tag{15}$$

and loading-unloading conditions

$$g(e_N, \kappa) \equiv e_N - \kappa \leq 0, \quad \kappa \geq 0, \quad \dot{\kappa} g(e_N, \kappa) = 0 \tag{16}$$

in which κ is a history variable that represents the maximum level of normal strain in the given direction ever reached in the previous history of the material. An appropriate modification of the exponential softening law (6) leads to the damage law

$$f(\kappa) = \begin{cases} 1 & \text{if} \kappa \leq e_0 \\ \sqrt{\dfrac{\kappa}{e_0} \exp\left(\dfrac{\kappa - e_0}{e_f - e_0}\right)} & \text{if} \kappa > e_0 \end{cases} \tag{17}$$

where e_0 is a parameter controlling the elastic limit, and $e_f > e_0$ is another parameter controlling ductility. Note that softening in a limited number of directions does not necessarily lead to softening on the macroscopic level, because the response in the other directions remains elastic. Therefore, e_0 corresponds to the elastic limit but not to the state at peak stress.

If the MDM model is used in its basic form described above, the compressive strength turns out to depend on the Poisson ratio and, in applications to concrete, its value is too low compared to the tensile strength. The model is designed primarily for tensile-dominated failure, so the low compressive strength is not considered as a major drawback. Still, it is desirable to introduce a modification that would prevent spurious compressive failure in problems where moderate compressive stresses appear. The desired effect is achieved by redefining the projected strain e_N as

$$e_N = \frac{\varepsilon_{ij} n_i n_j}{1 - \frac{m}{Ee_0} \sigma_{kk}} \tag{18}$$

where m is a nonnegative parameter that controls the sensitivity to the mean stress, σ_{kk} is the trace of the stress tensor, and the normalizing factor Ee_0 is introduced in order to render the parameter m dimensionless. Under compressive stress states (characterized by $\sigma_{kk} < 0$), the denominator in (18) is larger than 1, and the projected strain is reduced, which also leads to a reduction of damage. A typical recommended value of parameter m is 0.05.

3.2 *Nonlocal formulation*

Nonlocal formulation of the MDM model is based on averaging of the inverse integrity tensor. This roughly corresponds to the nonlocal isotropic damage model with averaging of the compliance variable $\gamma = \omega/(1-\omega)$, which does not cause any spurious locking effects. In equation (13) for the evaluation of the damage effect tensor, the inverse integrity tensor is replaced by its weighted average with components

$$\overline{\psi}_{ij}(\boldsymbol{x}) = \int_V \alpha(\boldsymbol{x}, \boldsymbol{\xi}) \psi_{ij}(\boldsymbol{\xi}) \mathrm{d}\boldsymbol{\xi} \tag{19}$$

By fitting a wide range of numerical results, it has been found that the parameters of the nonlocal MDM model can be estimated from the measurable material properties using the formulas

$$\lambda_f = \frac{EG_f}{Rf_t^2} \tag{20}$$

$$\lambda = \frac{\lambda_f}{1.47 - 0.0014\lambda_f} \tag{21}$$

$$e_0 = \frac{f_t}{(1-m)E(1.56 + 0.006\lambda)} \tag{22}$$

$$e_f = e_0[1 + (1-m)\lambda] \tag{23}$$

where E is Young's modulus, G_f is the fracture energy, f_t is the uniaxial tensile strength, m is the compressive correction factor, typically chosen as $m = 0.05$, and R is the radius of nonlocal interaction reflecting the internal length of the material.

4 MESH-ADAPTIVE TECHNIQUES

4.1 *General structure of the adaptive procedure*

Nonlocal models lead to smooth solutions with a continuous variation of strain. However, to resolve narrow bands of highly localized strains, it is necessary to use sufficiently fine computational grids. Fortunately, the mesh must be fine only in the process zone, while the remaining part of the structure can be reasonably represented by a coarser mesh. The localization pattern is in general not known in advance, and it is extremely tedious to construct suitably refined meshes "by hand". Efficiency of the analysis can be greatly increased by an adaptive technique, which automates the whole process.

The basic components of the adaptive procedure include

- an error estimator or indicator,
- a remeshing criterion,
- transfer algorithms for primary unknowns and internal variables, and
- a mesh generator interface.

The general algorithm of nonlinear adaptive analysis can be described as follows: After reaching the equilibrium state (corresponding to a given load increment) and updating the solution state, an *a posteriori* error estimation is performed, in order to evaluate the error distribution. Then, a remeshing criterion uses the information about the error distribution and determines the subsequent activity. If the obtained error level is still acceptable, the analysis continues with the next load increment on the current mesh. If the evaluated error exceeds a limit defined by the user, the required mesh density is determined from the error distribution and a new spatial discretization is generated.

In a truly adaptive approach, after generating a new discretization, the data structures corresponding to the newly generated mesh are created, and the transfer of displacements and internal variables from the old mesh to the new one is performed. After the mapping, the internal variables are used together with the strain computed from the mapped displacements to update the internal state of each integration point on the new mesh (to achieve local consistency). Once the transfer has been finished, the old discretization is deleted and the mapped configuration is brought into global equilibrium by iteration at constant value of the loading parameter (e.g., of the applied displacement). Afterwards, the solution continues with the next load increment.

Another possibility is to restart the analysis from the initial state after the new discretization has been generated (Jirásek and Zimmermann 1997; Rodríguez-Ferran and Huerta 2000). This approach does not require the transfer of the current state from the old discretization to the new one, but from the computational point of view is less efficient compared to the truly adaptive approach, especially if the remeshing is done frequently.

4.2 *Error estimators and indicators*

For linear problems, error analysis of the finite element solution can be developed in a mathematically rigorous way (Babuška and Rheinboldt 1978; Zienkiewicz and Zhu 1987). In the nonlinear range, however, rigorous error estimates can be constructed only for a restricted class of problems. A general theory is not available since there are various sources and forms of nonlinearities. In the linear elastic case the error arises essentially from the discretization of the domain (so-called spatial error). In the nonlinear case, the error depends on the time discretization for history-dependent solids, and a part of the error is always induced by the incremental-iterative technique. The path dependency renders the problem more complex and, consequently, a reliable error estimation becomes more difficult, especially for nonconventional theories of enriched continua. Nevertheless, considerable progress has been made in recent years. For instance, Rodríguez-Ferran and Huerta (2000) proposed a sophisticated error estimator for nonlocal damage models. Ladevèze and coworkers (Ladevèze and Leguillon 1983; Gallimard et al. 1996) developed *a posteriori* estimators based on the error in the constitutive relation, and Comi and Perego (2002) adapted this technique to their nonlocal damage theory (Comi and Perego 2001). However, the implementation of these complicated estimators requires a considerable effort.

A simple and convenient alternative to rigorous error estimators is provided by heuristic error indicators. They are often based on physical intuition and insight into the problem at hand. Error indicators are typically based on variables that characterize the nonlinear dissipative process, e.g., on the inelastic strain. In the present work we use directly the damage variable ω for isotropic damage models, and the maximum principal value of the damage tensor for anisotropic damage models. Therefore, we will use the expression "damage indicator" instead of "error indicator".

4.3 *Remeshing criterion*

A high value of the damage indicator allows to identify the process zone and to determine the required mesh size. The maximum allowable mesh size inside the process zone is related to the nonlocal interaction radius R. It should be a fraction of R, to make sure that the nonlocal interactions between individual integration points are properly activated. The minimum requirement needed for nonlocal averaging to work reasonably well is to have at least four remote integration points influencing the given point in an equivalent one-dimensional situation. Extending this simple one-dimensional rule, the corresponding maximum element size can be estimated. For example, for quadratic elements with 3×3 integration points the maximum size allowed in the process zone is $h_1 = 3 \times 2R/4 = 1.5R$. The optimal recommended value that allows to accurately represent the localized character of the solution and is still feasible from the computational point of view is about 10 integration points per width of the process zone. The proposed remeshing criterion, based on the damage indicator ω, uses element size h_0 for $\omega \leq \omega_0$ and element size h_1 for $\omega \leq \omega_1$, with linear interpolation in between. Here, ω_0 is typically selected as damage at peak stress in a uniaxial test (i.e., zero for models that remain linear elastic up to the peak), ω_1 is taken, e.g., as the damage level corresponding to softening to 50% of the peak stress, h_0 is a typical element size in a coarse mesh, and h_1 is the element size that enables good nonlocal interaction.

As an additional improvement, the remeshing criterion based on the damage indicator is combined with an error estimator for elastic problems, e.g., with the Zienkiewicz-Zhu estimator (Zienkiewicz and Zhu 1987). This allows to optimize the mesh in the elastic or slightly damage part of the domain. The transition of the mesh size between the linear elastic part and the nonlinear damaged part can be handled by the mesh generator or by a suitable smoothing technique.

4.4 *Transfer operators*

The truly adaptive approach requires the mapping of displacements and internal history variables (in the present case of the damage variable or damage tensor), allowing to continue the analysis from the current state, instead of restarting the analysis from scratch after each mesh refinement. Mapping of the primary unknowns (nodal displacements) is usually done using the shape function projections. Mapping of the damage variables is more intricate. It can be done using one of the following techniques:

- CPT – closest point transfer (Fig. 2a)
- SFT – shape function projection transfer (Fig. 2b)
- LST – least-square projection transfer (Fig. 2c)

A comparative study undertaken by the present authors (Jirásek and Patzák 2003) revealed that SFT leads to an artificial damage diffusion, i.e., to an expansion of the damaged zone after the mapping, while CPT and LST preserve the size of the damaged zone. CPT results into lower stress oscillations after mapping, but when these oscillations are removed by equilibrium iterations (at constant loading parameter), the original shape of the damage profile is best reproduced by LST. It can be concluded that SFT is not a suitable mapping technique for the damage variable, and that LST is probably the most accurate technique but computationally more expensive than CPT, which provides acceptable results at low cost.

Spurious damage diffusion produced by SFT is confirmed in a two-dimensional example of a four-point shear test on a single-edge-notched specimen; see the damage patterns in Fig. 3.

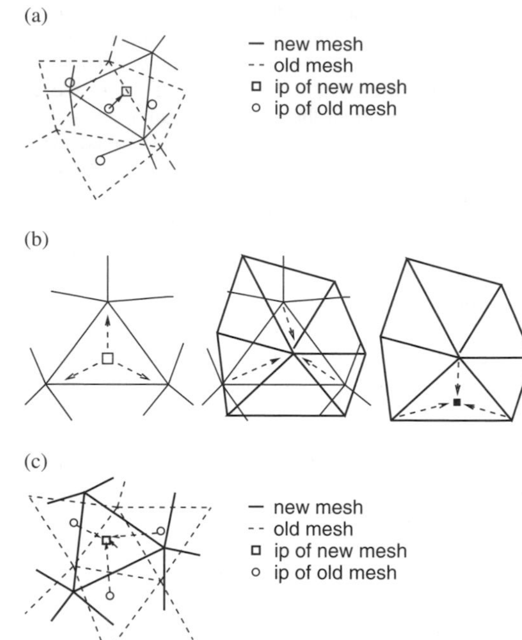

Figure 2. Mapping techniques: (a) closest point transfer, (b) shape function projection, (c) least square projection.

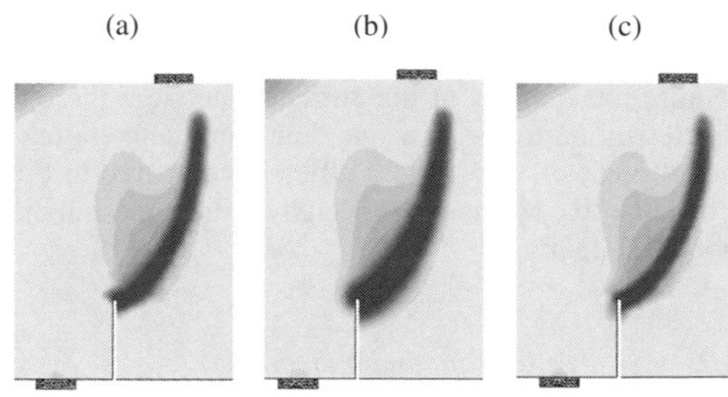

Figure 3. Four-point shear test – comparison of damage patterns for different transfer operators: (a) CPT, (b) SFT, (c) LST.

5 EXAMPLES

The proposed adaptive methodology has been successfully implemented by the first author into his object-oriented finite element system OOFEM (Patzák and Bittnar 2001). To illustrate its performance, we present two examples of failure simulations.

5.1 *Three-point bending beam*

The first example illustrates the analysis of a three-point bending test. The nonlocal formulation of the isotropic damage model with the Rankine-like equivalent strain measure (5) and the exponential damage law (6) has been used. The constitutive properties have been set to: Young's modulus of elasticity $E = 30$ GPa, Poisson's ratio $\nu = 0.3$, limit elastic strain $\varepsilon_0 = 0.0001$, equivalent strain corresponding to the fully damaged state $\varepsilon_f = 0.012$. The bell-shaped nonlocal average function (10) has been used with nonlocal averaging radius R = 10 mm. The damage-based error indicator has been used together with the remeshing criterion based on a linear interpolation scheme. The mesh size has been set to $h_0 = 40$ mm in the elastic regions and to $h_1 = 10$ mm in the damaged regions.

The initial coarse mesh contained only 24 nodes and 28 constant strain elements. Only four remeshing steps were necessary to obtain the final mesh, containing

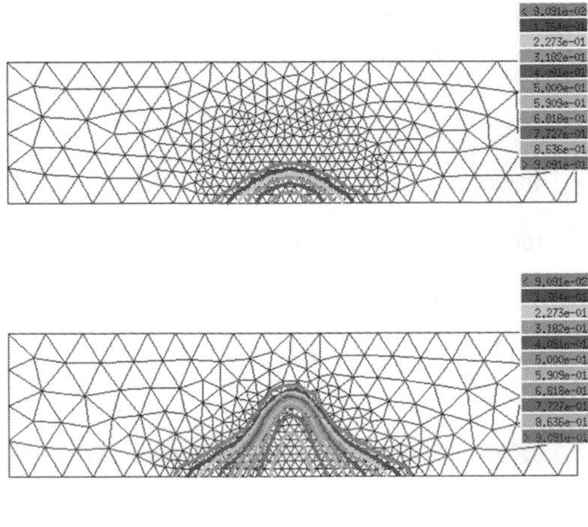

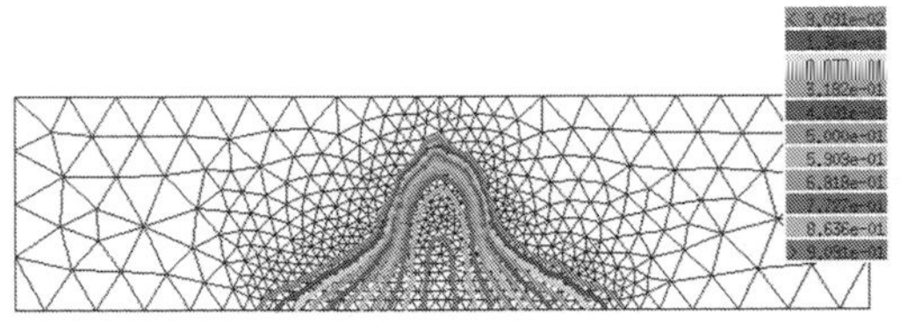

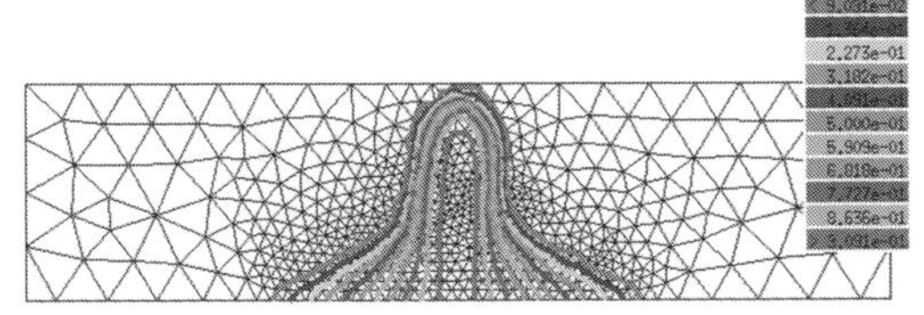

Figure 4. Adaptive analysis of damage evolution in a three-point bending test.

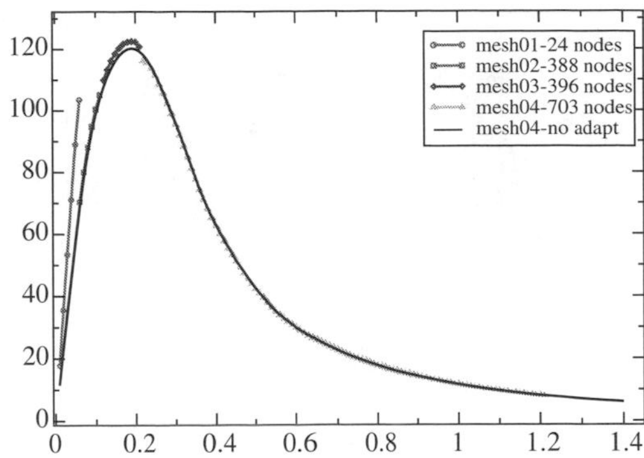

Figure 5. Load-displacement curve for three-point bending test.

703 nodes and 1328 elements. The initial damage pattern is rather diffuse because there is no initial notch, but later it gets localized, see Fig. 4. The loading has been controlled by a prescribed displacement under the applied force. The resulting load-displacement diagram is shown in Fig. 5. In this plot, the parts of diagram are distinguished as corresponding to different adaptive steps. The load-displacement diagram obtained by a standard (non-adaptive) analysis on the final mesh is also shown. A very good agreement has been obtained, demonstrating the quality of the transfer operators. The differences in the initial part of the load-displacement diagram are mainly due to the fact that the initial meshes are too coarse and thus naturally lead to a much stiffer response than the final fine mesh.

5.2 *DEN specimen with curved cracks*

As another example of damage propagation along a curved trajectory, we present the adaptive simulation of the double-edge-notched (DEN) specimen that was tested by Nooru-Mohamed (1992) using the experimental setup in Fig. 6a. The specimen can be subjected to a combination of shear and tension (or compression). Nooru-Mohamed performed his experiments for a number of loading paths, some of them even nonproportional. Among the most interesting loading scenarios were paths 4a, 4b and 4c. During the first stage, the specimen was loaded by an increasing "shear" force, P_s, while keeping the "normal" force, P, at zero. After reaching a certain load level, the type of loading was changed. During the second stage, the force P_s was kept constant and the test was controled by increasing the "normal" displacement δ. For path 4a the change of loading occured at $P_s = 5\,\text{kN}$, for path 4b at $P_s = 10\,\text{kN}$, and for path 4c at the maximum shear force that the specimen could sustain, $P_s = P_s^{max} = 27.5\,\text{kN}$. In all the cases, the failure pattern consisted of two macroscopic cracks propagating from the notches in an inclined direction. For path 4a, these cracks were almost horizontal and close to each other (Fig. 6b top), while for path 4c they were highly curved and farther apart (Fig. 6b bottom).

Failure of the DEN specimen under loading paths 4a and 4c has been simulated using the nonlocal version of the anisotropic damage model described in Section 3. The material parameters have been deduced from the data provided by Nooru-Mohamed (1992): compressive strength measured on cubes $f_c = 46.24\,\text{MPa}$ for path 4a and $f_c = 46.19\,\text{MPa}$ for path 4c, and splitting tensile strength $f_s = 3.67\,\text{MPa}$ for path 4a and $f_s = 3.78\,\text{MPa}$ for path 4c. The compressive strength is slightly above the value that corresponds to concrete C-30 according to the CEB-FIP Model Code (1991). Interpolation between the values of tensile strength and Young's modulus corresponding to concretes C-30 and C-40 gives $f_t = 3\,\text{MPa}$ and $E = 29\,\text{GPa}$. The value of tensile strength is in agreement with the empirical formula $f_t = 0.8 f_s$. The fracture energy is considered by the same value as by (di Prisco et al. 2000), i.e.,

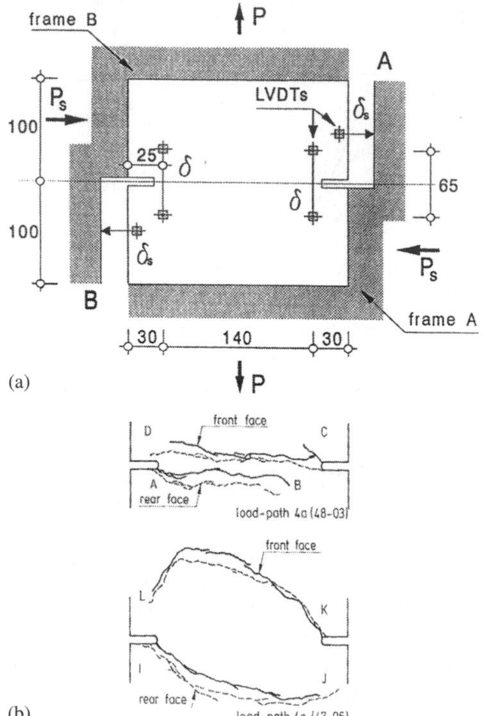

Figure 6. Curved crack propagation in a DEN specimen: (a) experimental setup and (b) observed crack patterns (reproduced from Nooru-Mohamed 1992).

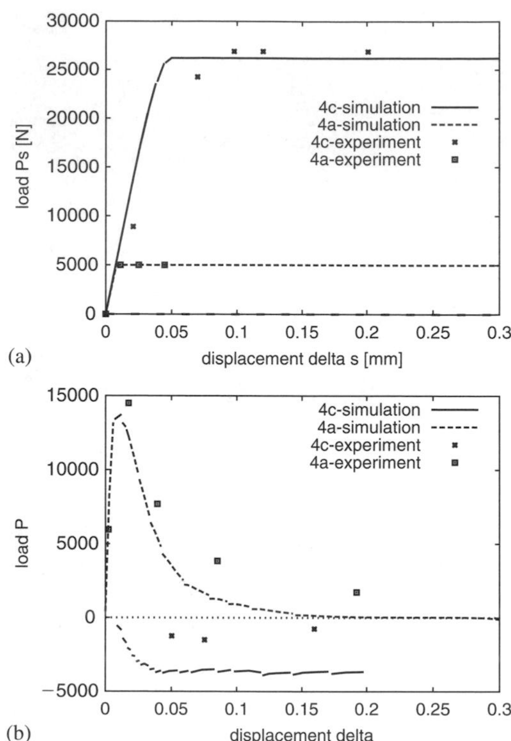

Figure 7. Measured and simulated load-displacement curves in terms of (a) shear force P_s and displacement δ_s, (b) normal force P and displacement δ.

$G_f = 110\,\text{J/m}^2$, and the interaction radius is set to R = 5 mm.

The adaptive analysis with parameters $\omega_0 = 0.6$, $h_0 = 8\,\text{mm}$, $\omega_1 = 0.8$ and $h_1 = 2.8\,\text{mm}$ required 13 remeshings for path 4a and 23 remeshings for path 4c. The evolving meshes with the isolines of damage and strain are shown in Figs. 8 and 9. It is clear that for both loading paths the numerical prediction is in an excellent agreement with the experimental results. Even the highly curved cracks generated by path 4c are reproduced very accurately.

The load-displacement curves, plotted in Fig. 7, reveal certain discrepancies. For low and moderate levels of the shear force P_s, the response is linear, but the measured displacements δ_s exceed the computed displacements by as much as 50%. To get a perfect agreement in this range, the elastic modulus used in the simulation would need to be reduced to about 20 GPa. It is not very likely that the actual elastic modulus was really that low. The measured displacement was probably not only due to the elastic deformation of the specimen. The fact that the displacements are underpredicted is a general problem observed in other simulations of this test published in the literature; see e.g. di Prisco et al. (2000).

Leaving aside the difference between the experimental and numerical compliances, one can say that the agreement between the test and the simulation is very satisfactory. For loading path 4a, the response to the shear loading is linear and the normal loading generates a tensile reaction force increasing up to 15 kN in the test and 14 kN in the simulation. For loading path 4c, the first clearly visible deviation from linearity during the shear loading appears at $P_s = 20$ kN in both test and simulation, and the peak shear force is 27.5 kN in the test and 26.2 kN in the simulation. The simulation correctly predicts that the reaction force generated by increasing the normal displacement at constant shear force is compressive. The maximum magnitude of the normal force is 1.5 kN in the test and below 4 kN in the simulation, which should not be considered as a huge relative error because these forces are very small compared to the previous loading case. The compressive force generated by an extension of the specimen can be explained by the fact that the resistance of the specimen was already exhausted by the shear force, and adding a tensile force under load control would lead to an unstable crack propagation.

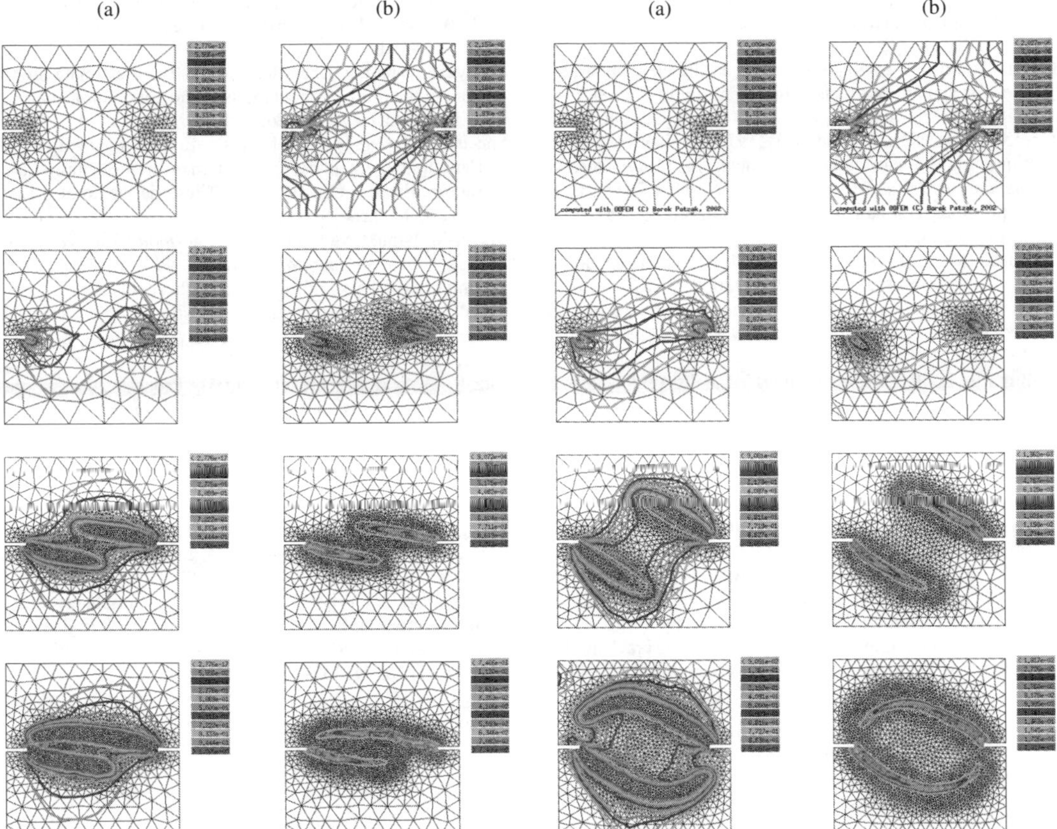

Figure 8. Loading path 4a: evolving meshes with isolines of (a) damage and (b) strain.

Figure 9. Loading path 4c: evolving meshes with isolines of (a) damage and (b) strain.

6 CONCLUDING REMARKS

In this paper we have presented and discussed various aspects of mesh-adaptive techniques applied to nonlocal continuum models with softening. Intuitively it is clear that adaptive mesh refinement can provide a good balance between accuracy and speed of the simulation, but the components of an adaptive strategy must be selected with care. For instance, we have shown that certain transfer operators for the mapping of internal variables can lead to an artificial diffusion of the process zone. Another important factor is the element size in the damage process zone, which must be small enough to permit nonlocal interaction of each Gauss point with a sufficient number of its neighbors. But even on a fine mesh, the actual curved crack trajectory would not be reproduced correctly if the material model does not properly take into account the anisotropic character of damage; see Jirásek and Patzák (2003) for further details.

ACKNOWLEDGEMENTS

Financial support of the Swiss Commission for Technology and Innovation under project CTI 4424.1 and of the Ministry of Education of the Czech Republic under project MSM 210000003 is gratefully acknowledged.

REFERENCES

Babuška, I. and W. C. Rheinboldt (1978). Aposteriori error estimates for finite element method. *International Journal for Numerical Methods in Engineering 12*, 1597–1615.

Bažant, Z. P. and B.-H. Oh (1986). Efficient numerical integration on the surface of a sphere. *Zeitschrift für angewandte Mathematik und Mechanik 66*, 37–49.

Carol, I. and Z. P. Bažant (1997). Damage and plasticity in microplane theory. *International Journal of Solids and Structures 34*, 3807–3835.

CEB (1991). *CEB-FIP Model Code 1990, Design Code*. London: Thomas Telford.

Comi, C. and U. Perego (2001). Numerical aspects of nonlocal damage analyses. *Revue européenne des éléments finis 10*, 227–242.

Comi, C. and U. Perego (2002, May). Finite element strategies for damage assessment up to failure. In *Proceedings of the 6th National Congress SIMAI*, Chia Laguna, Italy.

Cordebois, J. P. and F. Sidoroff (1979). Anisotropie élastique induite par endommagement. In *Comportement mécanique des solides anisotropes*, Number 295 in Colloques internationaux du CNRS, Grenoble, pp. 761–774. Editions du CNRS.

di Prisco, M., L. Ferrara, F. Meftah, J. Pamin, R. de Borst, J. Mazars, and J. Reynouard (2000). Mixed mode fracture in plain and reinforced concrete: some results on benchmark tests. *International Journal of Fracture 103*, 127–148.

Gallimard, L., P. Ladevèze, and J. P. Pelle (1996). Error estimation and adaptivity in elastoplasticity. *International Journal for Numerical Methods in Engineering 39*, 189–217.

Jirásek, M. (1999). Comments on microplane theory. In G. Pijaudier-Cabot, Z. Bittnar, and B. Gérard (Eds.), *Mechanics of Quasi-Brittle Materials and Structures*, Paris, pp. 55–77. Hermès Science Publications.

Jirásek, M. and B. Patzák (2003). Adaptive resolution of localized damage in quasibrittle materials. *Journal of Engineering Mechanics, ASCE*. In press.

Jirásek, M. and T. Zimmermann (1997). Rotating crack model with transition to scalar damage: I. Local formulation, II. Nonlocal formulation and adaptivity. LSC Internal Report 97/01, Swiss Federal Institute of Technology, Lausanne, Switzerland.

Ladevèze, P. and D. Leguillon (1983). Error estimate procedure in the finite element method and application. *Journal of Numerical Analysis 20*, 485–509.

Mazars, J. (1984). Application de la mécanique de l'endommagement au comportement non linéaire et à la rupture du béton de structure. Thèse de Doctorat d'Etat, Université Paris VI., France.

Nooru-Mohamed, M. B. (1992). *Mixed-mode fracture of concrete: An experimental approach*. Ph. D. thesis, Delft University of Technology, The Netherlands.

Patzák, B. and Z. Bittnar (2001). Design of object oriented finite element code. *Advances in Engineering Software 32*, 759–767.

Patzák, B. and M. Jirásek (2003). Process zone resolution by extended finite elements. *Engineering Fracture Mechanics*. In press.

Pijaudier-Cabot, G. and Z. P. Bažant (1987). Nonlocal damage theory. *Journal of Engineering Mechanics, ASCE 113*, 1512–1533.

Rodrǵuez-Ferran, A. and A. Huerta (2000). Error estimation and adaptivity for nonlocal damage models. *International Journal of Solids and Structures 37*, 7501–7528.

Zienkiewicz, O. C. and J. Z. Zhu (1987). A simple error estimator and adaptive procedure for practical engineering analysis. *International Journal for Numerical Methods in Engineering 24*, 337–357.

Computational Modelling of Concrete Structures, Bićanić et al. (eds)
© 2003 Swets & Zeitlinger, Lisse, ISBN 90 5809 536 3

Description of thermo-mechanical behaviour of reinforced concrete

S. Pietruszczak
McMaster University, Hamilton, Ontario, Canada

A. Winnicki
Cracow University of Technology, Cracow, Poland

ABSTRACT: In this paper a mathematical framework is outlined for modelling the thermo-mechanical behaviour of reinforced concrete. The formulation is an extension of recent work of the present authors (2002). The constitutive model is developed by employing simple volume averaging schemes for the macroscopic stress/strain rates. Both, a homogeneous deformation mode (prior to cracking) as well as a localized mode, associated with formation of macrocracks, is considered. In the latter case, the representative volume comprises the reinforced fractured zone and the "intact material". The stiffness of the reinforcing network is evaluated by considering the individual steel bars to be rigidly embedded in the adjacent concrete matrix. Numerical examples are provided illustrating the performance of the proposed framework under the mechanical loads as well as the thermal gradients.

1 INTRODUCTION

For large concrete structures, such as dams, nuclear containment buildings, pressure vessels, etc., the reinforcement cannot be modelled in a discrete way. Hence, in this case, the presence of reinforcement needs to be accounted for in an average sense by considering the material as a composite medium. In the last few decades several different constitutive models have been proposed for the description of reinforced concrete. Majority of them fall into the category of the so-called "smeared crack" models. They have proven to be efficient in the analysis of practical engineering problems (Bangash, 1989), however, their most serious deficiency is the lack of the distinction between the mechanical properties within the fractured zone and those of the intact material in-between the cracks. This leads to problems in specification of material functions/parameters which, in fact, cannot be uniquely defined. In addition, the finite element analysis may display a sensitivity to the discretization of the domain of interest.

The approach pursued here is based on the concept of "embedded discontinuity" applied at the level of the constitutive relation. In this way the internal length parameter is explicitly invoked and the objectivity of the numerical solution is assured. Such an approach has been successfully used in the past for the analysis of plain concrete (Pietruszczak, Xu, 1995) and has recently been extended for the description of the reinforced concrete (Pietruszczak, Winnicki, 2002). In the present paper the latter formulation is further enhanced by incorporating the thermo-mechanical coupling, which takes into account different thermal properties of the constituents involved.

The paper is written in the following sequence. First a general mathematical formulation is provided. Both, a homogeneous deformation mode (prior to cracking) as well as a localized mode, associated with formation of macrocracks, is considered. Later, the framework is applied to examine the characteristics of reinforced concrete panels under both mechanical and thermal loading conditions.

2 MATHEMATICAL FORMULATION

Referring to the previous work of the present authors (2002), the reinforced concrete is treated as a composite material comprising the concrete matrix (m) and two orthogonal families of reinforcement (set *1* and *2*, respectively). The problem is formulated in two stages. *Phase I* deals with the homogeneous deformation mode prior to cracking of the concrete matrix, whereas *Phase II* involves a localized deformation associated with formation of macrocracks.

(i) Phase I – prior to cracking
The problem is formulated in the frame of reference $\mathbf{x}^*$, such that x_2^* and x_3^* are along the axes of

reinforcement. The average macroscopic stress/strain rates for the composite body are defined through integration over the volume of the representative element (Hill, 1963), i.e.

$$\dot{\sigma}^* = \eta_1\dot{\sigma}_1^* + \eta_2\dot{\sigma}_2^* + (1-\eta_1-\eta_2)\dot{\sigma}_m^*$$
$$\dot{\varepsilon}^* = \eta_1\dot{\varepsilon}_1^* + \eta_2\dot{\varepsilon}_2^* + (1-\eta_1-\eta_2)\dot{\varepsilon}_m^* \quad (1)$$

In the equations above, η_1 and η_2 represent the volumetric fractions of the respective sets of reinforcement, whereas $\dot{\sigma}_k^*, \dot{\varepsilon}_k^*$ ($k = 1, 2, m$) are the volume averages of stress/strain rates in the constituents involved. Assume now that both these local fields are homogeneous within themselves, so that unique constitutive relations can be assigned to them. Furthermore, in order to simplify the formulation, assume that the coefficients of thermal expansion for both the concrete (α_m) and the reinforcing steel (α_1,α_2) are approximately the same, i.e. $\alpha_m \approx \alpha_1 = \alpha_2 = \alpha$. The latter constraint is justified, as both these coefficients are typically in the range of 8–14 ($\times 10^{-6}$/°C). The stress-strain relations for the individual constituents can now be written in a general form

$$\dot{\varepsilon}_1^* - \dot{\varepsilon}_T = [C_1^*]\dot{\sigma}_1^*;\;\; \dot{\varepsilon}_2^* - \dot{\varepsilon}_T = [C_2^*]\dot{\sigma}_2^*$$
$$\dot{\varepsilon}_m^* - \dot{\varepsilon}_T = [C_m^*]\dot{\sigma}_m^*;\;\; \dot{\varepsilon}_T = \alpha\dot{T}\delta \quad (2)$$

where δ is the Kronecker's delta, $\dot{T}$ is the temperature gradient and $[C]$'s are the compliance operators. It is further assumed that the reinforcing steel is a linearly elastic material, while the behaviour of concrete matrix is described by an elastoplastic formulation developed in the article by Pietruszczak & Xu (1995).

The local stress rate averages can be related to the overall macroscopic measure $\dot{\sigma}^*$ (Hill, 1963), i.e.

$$\dot{\sigma}_1^* = [B_1]\dot{\sigma}^*;\;\; \dot{\sigma}_2^* = [B_2]\dot{\sigma}^*;\;\; \dot{\sigma}_m^* = [B_m]\dot{\sigma}^* \quad (3)$$

The operators $[B]$ have been estimated here based on a simple approach similar to that pursued by Wung & Dvorak (1985). The main assumption employed is that the diameter of the reinforcement is negligible as compared to the spacing. The problem is formulated by imposing some explicit kinematic constraints on the axial deformation of both phases (i.e. reinforcement and concrete), while the interaction in the transverse direction is largely neglected. As a result, the local stress/strain fields remain uniform. The details on the specification of the operators $[B]$ are provided in Pietruszczak & Winnicki (2002). It needs to be pointed out that the form of the matrices $[B]$ is not affected by the change in temperature, as the thermal coefficients of both constituents are assumed here to be the same.

Substitution of eqs.(2) and (3) into the second equation in (1) gives the average constitutive relation for the considered composite medium in the form

$$\dot{\varepsilon}^* = [C^*]\dot{\sigma}^* + \alpha\dot{T}\delta;$$
$$[C^*] = \{\eta_1[C_1^*][B_1] + \eta_2[C_2^*][B_2] + \eta_m[C_m^*][B_m]\} \quad (4)$$

In eq.(4) all the operators are referred to the local system $\mathbf{x}^*$ attached to the reinforcement. Apparently, the macroscopic stress/strain rates can be transformed to an arbitrarily chosen global Cartesian system $\mathbf{x}$, so that

$$\dot{\varepsilon} = [C]\dot{\sigma} + \alpha\dot{T}\delta;\;\; [C] = [T]^T[C^*][T] \quad (5)$$

where $[T]$ is the transformation matrix, such that $\dot{\sigma}^* = [T]\dot{\sigma}$.

Eq.(5) defines the constitutive relation, which governs the response of the composite prior to the onset of localized deformation, i.e. formation of macrocracks in the concrete matrix. The general conditions under which localized modes arise have been investigated by a number of researchers (e.g., Rudnicki & Rice, 1975; Ortiz, 1987). In reinforced concrete structures, the formation of cracks is typically associated with the *tensile* stress regime. In this case, the inception of localization is commonly defined in terms of a tensile fracture criterion, while the localization plane is assumed to be orthogonal to the direction of the major tensile stress. Once a macrocrack forms, the formulation of the problem follows the procedure outlined in part (*ii*) below.

(ii) *Phase II – after formation of a macrocrack*

A representative volume of the material, comprises now the "intact" reinforced concrete intercepted by a macrocrack of a given orientation $\boldsymbol{n}$. The latter represents a composite medium within itself as it consists of a zone of fractured concrete reinforced with steel bars. It should be emphasized that the thickness of the fractured zone, h, cannot be directly identified with the visible crack width. In fact, the presence of reinforcement in the neighbourhood of the crack triggers a further local damage in the adjacent concrete matrix, thereby resulting in widening of the crack on the macroscale. Thus, h is interpreted here as the thickness of the damaged concrete zone, which in case of reinforced concrete is of the same order of magnitude as the diameter of the reinforcement (e.g., CEB/FIP code, 1993).

The formulation of the problem invokes again a volume averaging procedure for specifying the macroscopic stress/strain rates, i.e.

$$\dot{\sigma} = \mu_i\dot{\sigma}_i + \mu_f\dot{\sigma}_f;\;\; \dot{\varepsilon} = \mu_i\dot{\varepsilon}_i + \mu_f\dot{\varepsilon}_f \quad (6)$$

Here, i refers to the intact material outside the localization zone, f denotes the material in the fractured zone and μ's represent the corresponding volume fractions. All tensor quantities are referred to the global coordinate system and the local fields are again considered as homogeneous within themselves.

For both constituents involved, appropriate constitutive relations are now assigned. For the "intact" material, outside the fractured zone, the relation established in *Phase I*, viz. eq.(5), is used. In the fractured zone, the constitutive law is expressed in terms of traction $\dot{\boldsymbol{t}}$ and velocity discontinuities $\dot{\boldsymbol{g}}$. It is assumed here that the thermal gradients affect the behaviour of the network of reinforcement, while the thermal expansion of the degraded concrete is neglected. The general form of the governing constitutive relations is

$$\dot{\boldsymbol{\varepsilon}}_i = [C]\dot{\boldsymbol{\sigma}}_i + \alpha\dot{T}\boldsymbol{\delta};\ \ \dot{\boldsymbol{g}} = [K]^{-1}\dot{\boldsymbol{t}} + \alpha\dot{T}\boldsymbol{\Psi} \tag{7}$$

where $[K]$ is the stiffness of the fractured zone and $\boldsymbol{\Psi}$ is an unspecified yet operator, which depends on the characteristics of the reinforcing network. The strain rate in the fractured zone may be conveniently defined in terms of velocity discontinuities $\dot{\boldsymbol{g}}$, as a symmetric part of a dyadic product

$$[\dot{\varepsilon}_f] = \frac{1}{2h}(\boldsymbol{n}\dot{\boldsymbol{g}}^T + \dot{\boldsymbol{g}}\boldsymbol{n}^T);\ \Rightarrow \dot{\boldsymbol{\varepsilon}}_f = \frac{1}{h}[N]\dot{\boldsymbol{g}} \tag{8}$$

where

$$[N]^T = \begin{bmatrix} n_1 & 0 & 0 & n_2 & 0 & n_3 \\ 0 & n_2 & 0 & n_1 & n_3 & 0 \\ 0 & 0 & n_3 & 0 & n_2 & n_1 \end{bmatrix} \tag{9}$$

and n_1 through n_3 are the components of a unit vector defining the orientation of the fractured zone.

The equilibrium requires that the traction along the discontinuity plane remains continuous. Thus,

$$\dot{\boldsymbol{t}} = [N]^T\dot{\boldsymbol{\sigma}}_i \tag{10}$$

Substituting eq.(10), together with the second equation in (7), into eq.(8) leads to

$$\dot{\boldsymbol{\varepsilon}}_f = \frac{1}{h}[N]\dot{\boldsymbol{g}} = \frac{1}{h}[N][K]^{-1}[N]^T\dot{\boldsymbol{\sigma}}_i + \frac{1}{h}\alpha\dot{T}[N]\boldsymbol{\Psi} \tag{11}$$

Introducing now the expressions for $\dot{\boldsymbol{\varepsilon}}_f$, eq.(11), and $\dot{\boldsymbol{\varepsilon}}_i$, eq.(7), in the strain decomposition (6), and noting that $\mu_f \ll \mu_i$ while the stress rates in both constituents are of the same order of magnitude (so that $\dot{\boldsymbol{\sigma}} \simeq \mu_i\dot{\boldsymbol{\sigma}}_i \simeq \dot{\boldsymbol{\sigma}}_i$), the following macroscopic constitutive relation is obtained

$$\dot{\boldsymbol{\varepsilon}} = ([C] + \mu[N][K]^{-1}[N]^T)\dot{\boldsymbol{\sigma}} + \alpha\dot{T}(\boldsymbol{\delta} + \mu[N]\boldsymbol{\Psi}) \tag{12}$$

In the above expression $\mu = \mu_f/h$ represents the ratio of the area of the fractured zone to the volume of the sample. Thus μ is, in fact, independent of h.

In order to define the properties of the fractured zone, viz. $[K]$ and $\boldsymbol{\Psi}$, it is convenient to choose a local base $\boldsymbol{x}'$, such that x_1' and x_2' axes are confined to the localization plane and x_3' is normal to it. With reference to this base, eq.(11) takes the form

$$\dot{\boldsymbol{\varepsilon}}_f' = \frac{1}{h}[N'][K']^{-1}[N']^T\dot{\boldsymbol{\sigma}}_i' + \frac{1}{h}\alpha\dot{T}[N']\boldsymbol{\Psi}' \tag{13}$$

where, in view of $\boldsymbol{n}' = \{0,0,1\}^T$, the operator $[N']$ is now composed of 0's and 1's. Denoting by $[Q]$ the transformation matrix form $\boldsymbol{x}$ to $\boldsymbol{x}'$ system, one can write

$$\dot{\boldsymbol{\sigma}}_i' = [Q]\dot{\boldsymbol{\sigma}}_i;\ \ \dot{\boldsymbol{\varepsilon}}_f = [Q]^T\dot{\boldsymbol{\varepsilon}}_f' \tag{14}$$

so that eq.(13) becomes

$$\dot{\boldsymbol{\varepsilon}}_f = \frac{1}{h}[M][K']^{-1}[M]^T\dot{\boldsymbol{\sigma}}_i' + \frac{1}{h}\alpha\dot{T}[M]\boldsymbol{\Psi}' \tag{15}$$

where $[M] = [Q]^T[N']$. Comparing now eqs (11) and (15), it is evident that

$$[N][K]^{-1}[N]^T = [M][K']^{-1}[M]^T;\ \ [N]\boldsymbol{\Psi} = [M]\boldsymbol{\Psi}' \tag{16}$$

Thus, the constitutive relation (12) may be expressed in an equivalent form

$$\dot{\boldsymbol{\varepsilon}} = ([C] + \mu[M][K']^{-1}[M]^T)\dot{\boldsymbol{\sigma}} + \alpha\dot{T}(\boldsymbol{\delta} + \mu[M]\boldsymbol{\Psi}') \tag{17}$$

Let us focus now on the specification of the operators $[K']$ and $\boldsymbol{\Psi}'$ appearing in the above representation.

(iii) Specification of mechanical characteristics of the fractured zone

As mentioned earlier, the fractured zone is a composite medium within itself. The primary constituents here are the damaged concrete, *c*, and the two families of reinforcement *1* and *2*. The stiffness operator is derived by assuming that the velocity discontinuity $\dot{\boldsymbol{g}}'$ is the same in the primary constituents, while the traction vector is obtained through the volume averaging

$$\dot{\boldsymbol{t}}' = \eta_1\dot{\boldsymbol{t}}_1' + \eta_2\dot{\boldsymbol{t}}_2' + (1 - \eta_1 - \eta_2)\dot{\boldsymbol{t}}_c' \tag{18}$$

The properties of the reinforcement network are assessed by considering the individual bars as beams of a characteristic length l, rigidly embedded in the adjacent intact material.

Consider the first family of reinforcement and refer the problem to the local coordinate system $\mathbf{x}^*$ introduced earlier in the part (*i*). Denote by $\boldsymbol{e}_i^*(i = 1,2,3)$ the corresponding base vectors, and assume that $\boldsymbol{e}_3^*$ is along the axis of the reinforcement. With reference to this coordinate system, the constitutive relation can be written in a general form

$$\dot{\boldsymbol{t}}_1^* = [K_1^*](\dot{\boldsymbol{g}}^* - \alpha\dot{T}l\boldsymbol{e}_3^*);\ \ l = h(\boldsymbol{n}^T\boldsymbol{e}_3^*)^{-1} \tag{19}$$

where $\dot{\boldsymbol{t}}_1^*$ is the traction on the plane perpendicular to the reinforcement axis. Following now the Timoshenko's beam theory, the elastic stiffness operator $[K_1^*]$ can be defined as

$$[K_1^*] = \frac{E}{l}\begin{bmatrix} \gamma & 0 & 0 \\ 0 & \gamma & 0 \\ 0 & 0 & 1 \end{bmatrix}; \quad \gamma = \frac{\phi}{2(1+\phi)(1+v)} \tag{20}$$

where $\phi = 3/2(1+v)\left(\frac{d}{l}\right)^2$.

In the above equations, E and v represent the Young's modulus and Poissons' ratio of steel, d is the diameter of an individual member and l is the characteristic length as defined in eq.(19).

In order to define the material characteristics in the local coordinate system $\boldsymbol{x}'$ associated with the fractured zone, the relation (19) needs to be transformed from $\boldsymbol{x}^*$ to $\boldsymbol{x}'$ bases. The transformation rules for both the velocity discontinuity $\dot{\boldsymbol{g}}^*$ and the traction vector on the localization plane, take the form

$$\dot{\boldsymbol{g}}' = [\hat{T}]\dot{\boldsymbol{g}}^*; \quad \dot{\boldsymbol{t}}_1' = (\mathbf{n}^T\boldsymbol{e}_3^*)[\hat{T}]\dot{\boldsymbol{t}}_1^* \tag{21}$$

Substituting now eqs.(21) into eq.(19) one obtains after some algebraic manipulations

$$\begin{aligned} \dot{\boldsymbol{t}}_1' &= [K_1']\{\dot{\boldsymbol{g}}' - \alpha\dot{T}h(\boldsymbol{n}^T\boldsymbol{e}_3^*)^{-1}\boldsymbol{e}_3'\}; \\ [K_1'] &= (\boldsymbol{n}^T\boldsymbol{e}_3^*)[\hat{T}][K_1^*][\hat{T}]^T \end{aligned} \tag{22}$$

Apparently, following the same procedure, a similar expression can be obtained for the second family of reinforcement. In this case

$$\dot{\boldsymbol{t}}_2' = [K_2']\{\dot{\boldsymbol{g}}' - \alpha\dot{T}h(\boldsymbol{n}^T\boldsymbol{e}_2^*)^{-1}\boldsymbol{e}_2'\} \tag{23}$$

It should be noted that in the formulation presented above the reinforcement is considered as elastic. The notion of the yield criterion for the reinforcing network is addressed in details in Pietruszczak & Winnicki (2002). This criterion is again formulated in a discrete sense, by considering the bending characteristics of the structural system.

Finally, the properties of the degraded concrete have been described by adopting a phenomenological framework of strain-softening plasticity. In addition, the elastic stiffness degradation has been accounted for by introducing a scalar damage parameter. As mentioned earlier, the thermal effects have been neglected here as insignificant compared to those of the reinforcement network, so that the general form of the constitutive relation is

$$\dot{\boldsymbol{t}}_c' = [K_c']\dot{\boldsymbol{g}}' \tag{24}$$

The details on the specification of the stiffness operator $[K_c']$ are provided in the original reference.

Given the expressions (22)–(24), the average constitutive relation for the fractured zone can be established by employing the decomposition (18). Substituting eqs.(22)–(24) into eq.(18) results in

$$\begin{aligned} \dot{\boldsymbol{t}}' = [K']\dot{\boldsymbol{g}}' - \alpha\dot{T}h\{&\eta_1[K_1'](\boldsymbol{n}^T\boldsymbol{e}_3^*)^{-1}\boldsymbol{e}_3' \\ &+\eta_2[K_2'](\boldsymbol{n}^T\boldsymbol{e}_2^*)^{-1}\boldsymbol{e}_2'\} \end{aligned} \tag{25}$$

where

$$[K'] = \eta_1[K_1'] + \eta_2[K_2'] + (1-\eta_1-\eta_2)[K_c'] \tag{26}$$

It should be noted now that in view of the definition of $[K_1']$, eq.(22), and that of $[K_1^*]$, eq.(20), the following identity can be established

$$\begin{aligned} [K_1']\boldsymbol{e}_3' &= (\boldsymbol{n}^T\boldsymbol{e}_3^*)[\hat{T}][K_1^*][\hat{T}]^T\boldsymbol{e}_3' \\ &= \cdots = Eh^{-1}(\boldsymbol{n}^T\boldsymbol{e}_3^*)^2\boldsymbol{e}_3' \end{aligned} \tag{27}$$

Thus, eq.(25) can be recast in the form

$$\dot{\boldsymbol{t}}' = [K'](\dot{\boldsymbol{g}}' - \alpha\dot{T}\boldsymbol{\Psi}') \tag{28}$$

where

$$\boldsymbol{\Psi}' = [K']^{-1}\{\eta_1E_1(\boldsymbol{n}^T\boldsymbol{e}_3^*)\boldsymbol{e}_3' + \eta_2E_2(\boldsymbol{n}^T\boldsymbol{e}_2^*)\boldsymbol{e}_2'\} \tag{29}$$

Equations (26) and (29) define the operators $[K']$ and $\boldsymbol{\Psi}'$ appearing in the constitutive relation (17), and thereby complete the formulation of the problem.

3 NUMERICAL EXAMPLES

In this part, the results of numerical simulations are provided, illustrating the performance of the framework outlined in Section 2. A comprehensive discussion on specification of material parameters as well as modelling of mechanical response in the context of a series of pure shear tests, is provided in Pietruszczak & Winnicki (2002). The focus here is primarily on the aspects of the behaviour of RC panels subjected to thermal gradients.

The first set of simulations reported in this section pertains to mechanical tests carried out by Vecchio & Collins (1982). The tests involved RC panels subjected to pure shear at different reinforcement ratios and different concrete strengths. Here, two particular tests, i.e. those for panels PV10 and PV16, have been chosen at random, in order to identify the basic trends in the mechanical response. Figures 1 and 2 present the stress–strain characteristics for both these tests. The panels have the volume fractions of reinforcement of $\{\eta_1 = 0.01785, \eta_2 = 0.00999\}$ and $\eta_1 = \eta_2 = 0.0074\}$, respectively. The remaining material parameters are

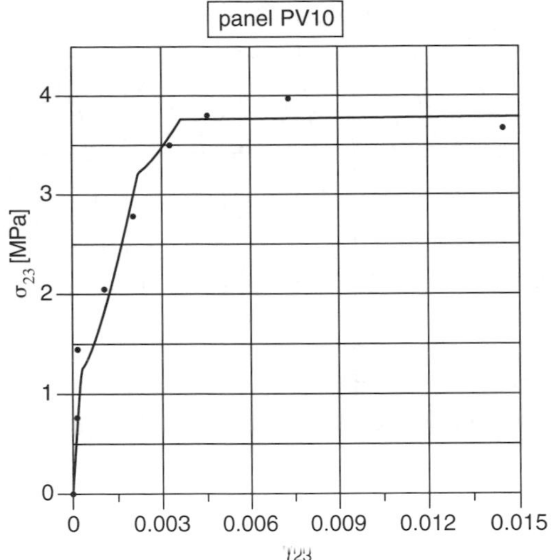

Figure 1. Panel PV10 tested in the pure shear (after Pietruszczak & Winnicki, 2002).

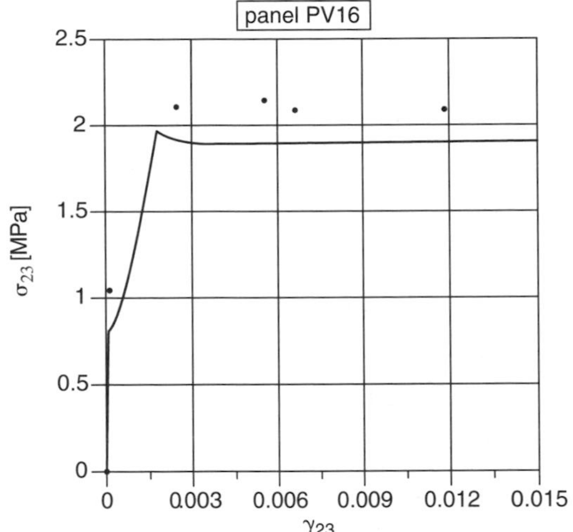

Figure 2. Panel PV16 tested in the pure shear (after Pietruszczak & Winnicki, 2002).

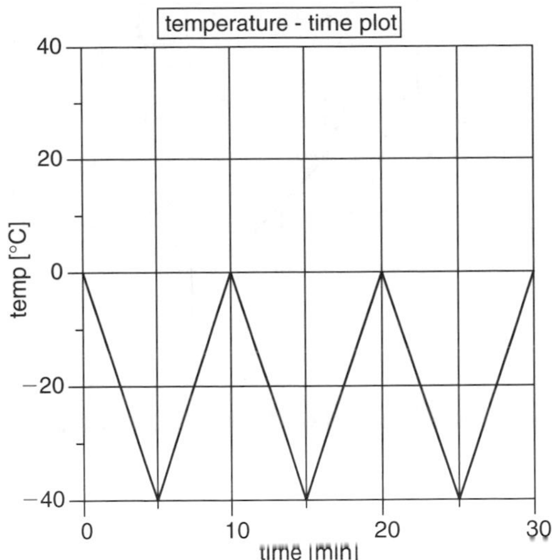

Figure 3. Time history of temperature.

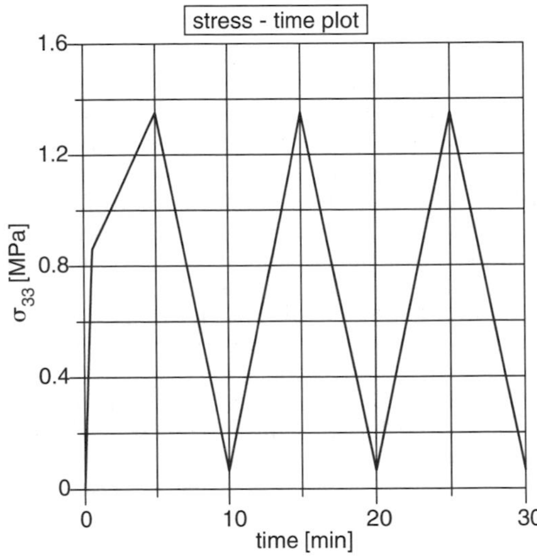

Figure 4. Stress evolution in time domain corresponding to loading history in Figure 3.

the same as those specified in the original reference. The numerical predictions are shown here with solid lines, whereas the dots represent the experimental data. The characteristics clearly depict the different stages of the deformation process. The linear branch is associated with homogeneous deformation prior to cracking. The subsequent non-linear branch corresponds to localized deformation (i.e. *Phase II*), whereas a *plateau* represents the yielding of reinforcement. The numerical results appear to be fairly consistent with the experimental data.

The next set of simulations involves the panel PV16 (i.e., $\eta_1 = \eta_2 = 0.0074, \alpha = 8 \times 10^{-6}/°\mathrm{C}$) subjected to thermal gradients under a kinematic constraint of zero displacement along x_3 -axis, with x_1 and x_2 directions unrestraint. No explicit experimental data is available here, therefore the performance of the model can only be assessed on purely qualitative basis. The results shown in Figures 4 and 5 correspond to the time history of temperature as indicated in Figure 3. The imposed temperature fluctuations are confined to the interval $\{0, -40°\mathrm{C}\}$.

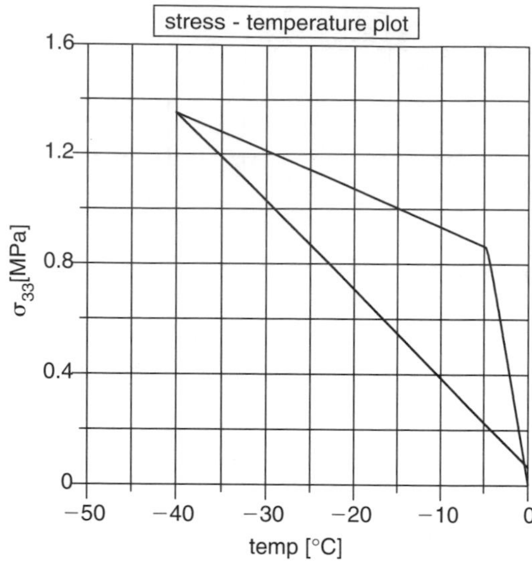

Figure 5. Stress–temperature relation corresponding to loading history in Figure 3.

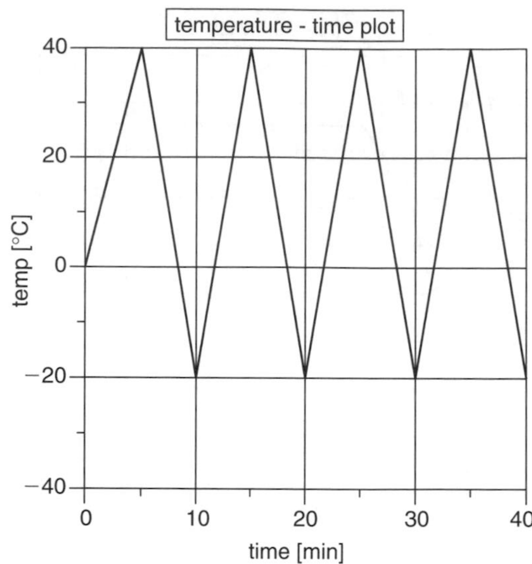

Figure 6. Time history of temperature.

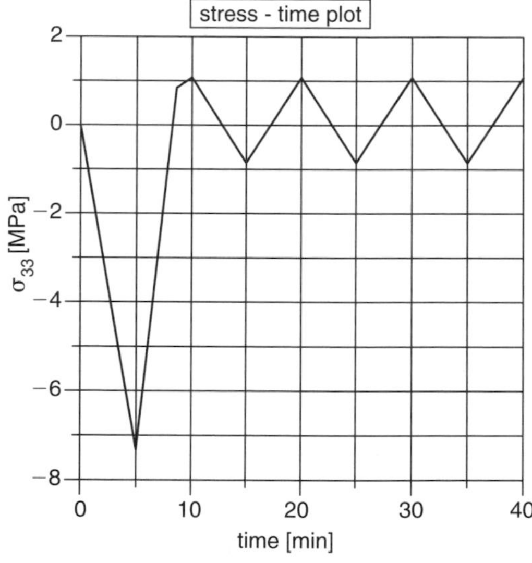

Figure 7. Stress evolution in time domain corresponding to loading history in Figure 6.

The reduction in temperature triggers uniaxial tension in both constituents. In the first loading cycle, the cracking in concrete occurs leading to a decrease in stiffness, Figure 4. During the unloading stage (i.e. increase in temperature back to zero) the stiffness is partly recovered (crack closing). Afterwards, the stress cycles stabilize and the consecutive loading/unloading branches have a similar stiffness, which is an intermediate of that corresponding to *Phase I* and *Phase II*, respectively. The model does not render the hysteretic loops as no irreversible deformation on unloading is accounted for. The evolution of the global stiffness of the panel is governed here by the plasticity formulation for the fractured zone, augmented by a scalar damage parameter (Pietruszczak & Winnicki, 2002). A complete stress-temperature history, which corresponds to the considered loading process, is shown in Figure 5.

The results presented in Figures 6–8 pertain to a similar loading process, which now involves the temperature cycles in the range {+40°C, −20°C}. In this case, a compressive stress is generated in both constituents in the first loading cycle, Figure 7. During the unloading phase, the axial stress in concrete becomes tensile triggering the initiation of cracking. This is evidenced by the reduction in stiffness, as shown in Figure 7. During subsequent cycles the stress trajectories stabilize again and the composite remains in *Phase II* (consecutive crack closing/opening).

Finally, the results shown in Figures 9–11 correspond to a coupled mechanical and thermal loading program. In the first stage, the panel is loaded in axial tension up to a prescribed value of longitudinal strain, Figure 9. In the next stage, the axial strain is kept constant while the sample is subjected to a temperature gradient, as indicated in Figure 10. The resulting evolution of axial stress is reported in Figure 11.

During the stage of mechanical loading, the cracking in the concrete matrix occurs, Figure 11. In the first thermal cycle, as the temperature increases, the unloading takes place and the axial stress drops. However, in the course of the subsequent cycle the axial stress in the primary constituents increases again and

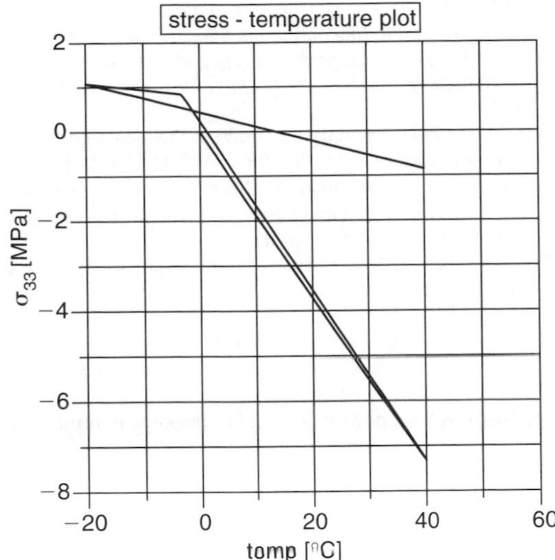

Figure 8. Stress–temperature relation corresponding to loading history in Figure 6.

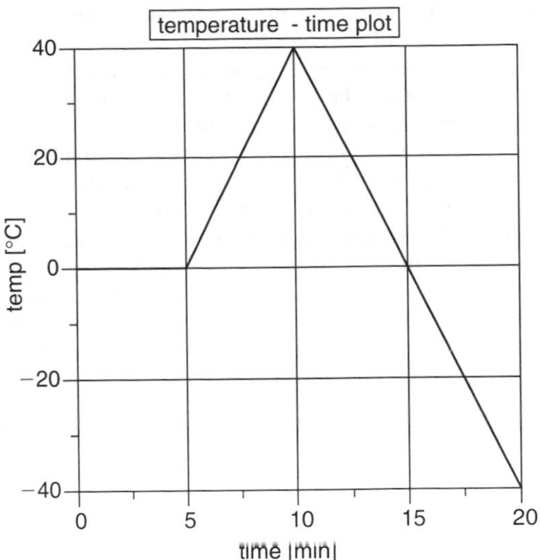

Figure 10. Time history of temperature.

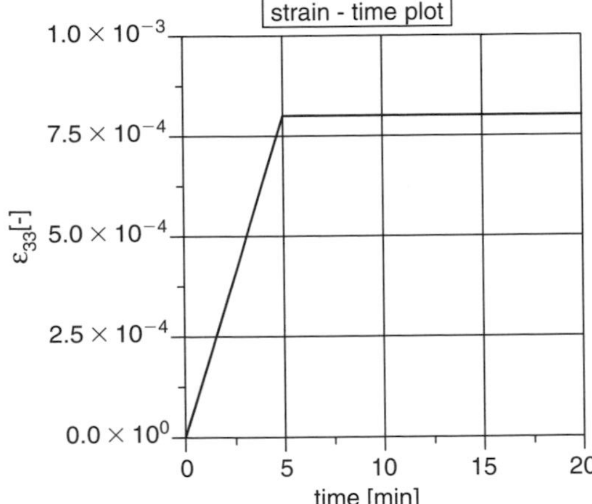

Figure 9. Imposed strain history.

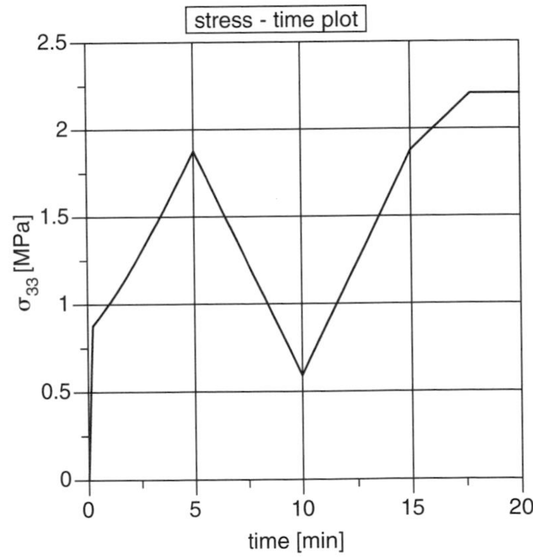

Figure 11. Stress evolution in time domain corresponding to loading history in Figures 9 and 10.

eventually the yielding of steel commences. It is interesting to note that the ultimate stress attained by the panel during the considered loading history is the same as the strength under axial tension imposed by purely mechanical means.

4 FINAL REMARKS

In this paper a constitutive model has been presented for modelling of coupled thermo–mechanical behaviour of reinforced concrete. The formulation of the problem is an extension of recent work of the present authors (2002). The material has been treated here as a composite medium comprising the concrete matrix intercepted by orthogonal sets of reinforcement. A two-stage volume averaging procedure has been incorporated for dealing with a homogeneous and localized deformation modes, respectively. In the latter case, the stiffness characteristics of the reinforcement network have been assessed by considering the individual members as beams of a characteristic length. The transient effects of the thermal loading have been incorporated by taking into account different thermal properties of all the constituents.

The numerical results presented in Section 3 confirm, in general, the ability of the model to simulate the thermo–mechanical response of reinforced concrete. In view of the lack of appropriate experimental data, it is not possible to assess the model predictions in a quantitative manner. However, the obtained trends appear to be reasonable, at least in a qualitative sense, rendering the formation of macrocracks in concrete and even yielding of steel under the thermal gradients. The formulation of the problem, although quite general, can be further enhanced by incorporating the creep phenomenon in concrete.

REFERENCES

CEB/FIP (1993). CEB – FIP Model Code 90 (Final version). *Comité Euro-International du Béton*, Bulletin d'Information No. 213–214, Lausanne, Switzerland.

Bangash, M.Y.H. (1989). *Concrete and concrete structures; numerical modelling and applications*, Elsevier Publ.

Hill, R. (1963). Elastic properties of reinforced solids: some theoretical properties. *J. Mech. Phys. Solids*, Vol. 11: 357–372.

Ortiz, M. (1987). An analytical study of the localized failure modes of concrete. *Mech. Mater.*, Vol. 6: 159–174.

Pietruszczak, S. & Winnicki, A. (2002). A constitutive model for concrete with embedded sets of reinforcement, *J. Engng Mech.*, ASCE, in print.

Pietruszczak, S. & Xu, G. (1995). Brittle response of concrete as a localization problem. *Int. J. Solids Structures*, Vol. 32: 1517–1533.

Rudnicki, J.W. & Rice, J.R. (1975). Conditions for the localization of deformation in pressure sensitive dilatant materials. *J. Mech. Phys. Solids*, Vol. 23: 371–394.

Vecchio, F. & Collins M.P. (1982). The response of reinforced concrete to in – plane shear and normal stresses. *Univ. of Toronto, Dept. of Civ. Engng*, Publ. No. 82–03.

Wung, C.J. & Dvorak, G.J. (1985). Strain-space plasticity analysis of fibrous composites. *Int. J. Plasticity*, Vol. 1: 125–139.

Computational Modelling of Concrete Structures, Bićanić et al. (eds)
© 2003 Swets & Zeitlinger, Lisse, ISBN 90 5809 536 3

Combined damage-plasticity models for discontinuous fracture

K. De Proft & W.P. De Wilde
Vrije Universiteit Brussel, vakgroep Mechanica van Materialen en Constructies, Brussel, Belgium

G.N. Wells & L.J. Sluys
Delft University of Technology, Department of Civil Engineering and Geosciences, Delft, The Netherlands

ABSTRACT: In this paper, two combined damage/plasticity continuum models are rewritten in a degenerated fashion. The continuum models must be consistent with the incorporation of a displacement discontinuity. Consequently, the stress/strain relationships degenerate into traction-separation laws. The obtained degenerated models are used in a discontinuous fracture model. Within this model, displacement discontinuities are incorporated using the partition of unity property of finite element shape functions. Finally, the model is used to simulate tensile tests performed on a double-edge notched stone specimen. The computed load-deformation responses are compared with experimentally obtained curves.

1 INTRODUCTION

Many models have been presented in the literature to overcome mesh sensitivity problems when classical damage/plasticity models are used in fracture analysis. In contrast with continuum models, where kinematic fields remain continuous, discontinuous models introduce discontinuous functions into the strain field (weak discontinuity) or into the displacement field (strong discontinuity). With these discontinuous models, the continuum behaviour and the behaviour of the discontinuity can be split. Consequently, all inelastic behaviour can be concentrated at the discontinuity, while the continuum remains elastic. At the discontinuity, the constitutive law is no longer expressed in terms of stresses and strains, but in terms of tractions and separations. A constitutive relation in terms of tractions and separations can be redefined from a continuum model. The stress field must remain bounded after the incorporation of a discontinuity.

In this paper, two different formulations for combined damage-plasticity models are studied. Firstly, combined damage-plasticity models as formulated by Simo & Ju (1987) are studied and cast in a discretized form. Then, the anisotropic elastoplastic-damage model developed by Meschke *et al.* (1998) is investigated and discretized. Finally, both models are compared and their capabilities are tested.

2 KINEMATICS OF A DISCONTINUITY

The displacement field of a body Ω (see figure 1) crossed by a discontinuity can be decomposed into a continuous and a discontinuous part:

$$\mathbf{u} = \hat{\mathbf{u}} + H_{\Gamma}\tilde{\mathbf{u}} \tag{1}$$

where $\hat{\mathbf{u}}$ and $\tilde{\mathbf{u}}$ are continuous functions and H_{Γ} is the Heaviside step function ($H_{\Gamma} = 1$ if $\mathbf{x} \in \Omega^+$ and $H_{\Gamma} = 0$ if $\mathbf{x} \in \Omega^-$).

The infinitesimal strain field can be found by taking the symmetric gradient of the displacement field:

$$\boldsymbol{\varepsilon} = \nabla^s\hat{\mathbf{u}} + H_{\Gamma}\nabla^s\tilde{\mathbf{u}} + \delta_{\Gamma}(\tilde{\mathbf{u}}\otimes\mathbf{n})^s \tag{2}$$

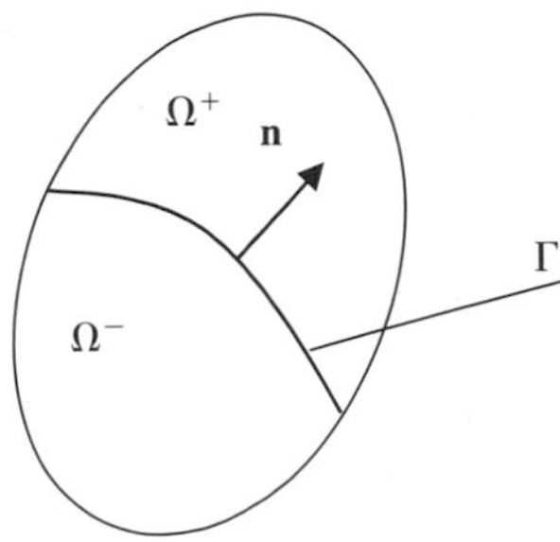

Figure 1. Body Ω crossed by a displacement discontinuity Γ.

where $\mathbf{n}$ is the normal to the discontinuity and δ_Γ is the Dirac delta distribution.

3 CONTINUUM DAMAGE/PLASTICITY MODELS

In the literature, many combined models can be found (Simo & Ju, 1987, Ju, 1988, Meschke *et al.*, 1998, De Borst *et al.*, 1999, Armero *et al.*, 2000). In this part of the paper, two formulations for combined damage/plasticity models are further elaborated: a model formulated by Simo & Ju (1987) and an anisotropic elastoplastic-damage model as formulated by Meschke *et al.* (1998).

3.1 *Model 1*

The combined damage/plasticity model formulated by Simo & Ju (1987) uses the following constitutive equation

$$\boldsymbol{\sigma} = (1-d)\,\mathbf{C}^e(\boldsymbol{\varepsilon}-\boldsymbol{\varepsilon}^p) \tag{3}$$

where $\boldsymbol{\sigma}$ is the stress tensor, d is the damage variable, $\mathbf{C}^e$ is the elastic constitutive tensor, $\boldsymbol{\varepsilon}$ is the total strain tensor and $\boldsymbol{\varepsilon}^p$ is the plastic strain tensor.

The damage is governed by a damage loading function, given by:

$$f^d = \varepsilon^{eq} - \kappa^d \tag{4}$$

where ε^{eq} is the equivalent strain measure and κ^d is a history parameter. In the literature, several equivalent strain measures are defined:

- energy release rate : $\varepsilon^{eq} = \sqrt{\frac{1}{2}\boldsymbol{\varepsilon}:\mathbf{C}:\boldsymbol{\varepsilon}}$ (5a)
- positive principal strains : $\varepsilon^{eq} = \sqrt{\sum_{i=1}^{3}\langle\varepsilon_i\rangle^2}$ (5b)

where $\langle\varepsilon_i\rangle = (\varepsilon_i + |\varepsilon_i|)/2$ and ε_i is the i-th principal strain. The history parameter κ^d represents the most severe value of the equivalent strain measure ever reached.

When the damage grows, the damage variable is updated via:

$$d = 1 - \frac{q(\kappa^d)}{\kappa^d} = 1 - \frac{\kappa_i \exp(-\mu\kappa^d)}{\kappa^d} \tag{6}$$

where κ_i is the damage threshold. Plastic deformations are assumed to occur in the undamaged material bonds, so that the plastic yield function can be expressed in the effective stress space,

$$f^p = \phi(\hat{\boldsymbol{\sigma}}) - \bar{\sigma}(\kappa^p) \tag{7}$$

where $\hat{\sigma}$ is the effective stress tensor, defined as:

$$\hat{\boldsymbol{\sigma}} = \frac{\boldsymbol{\sigma}}{1-d} \tag{8}$$

and κ^p is the internal plastic variable. Introducing the effective stress tensor in equation (3) reduces the classical elasto-plastic problem to:

$$\hat{\boldsymbol{\sigma}} = \mathbf{C}^e(\boldsymbol{\varepsilon}-\boldsymbol{\varepsilon}^p) \tag{9}$$

The plastic strain rate is defined according to the classical flow theory of plasticity,

$$\boldsymbol{\varepsilon}^p = \dot{\lambda}\frac{\partial f}{\partial \hat{\boldsymbol{\sigma}}} \tag{10}$$

where λ is the plastic multiplier.

The integration of the constitutive law is performed in two steps. Firstly, the plasticity problem is solved in the effective stress space. Secondly, the damage variable is updated and the computed effective stresses are mapped back into the homogenized stress space.

3.2 *Model 2*

The second model was introduced by Meschke *et al.* (1998). Meschke *et al.* developed an anisotropic combined damage/plasticity model based on the following constitutive relation:

$$\boldsymbol{\sigma} = \mathbf{C}(\boldsymbol{\varepsilon}-\boldsymbol{\varepsilon}^p) \tag{11}$$

where $\boldsymbol{\sigma}$ is the stress tensor, $\boldsymbol{\varepsilon}$ is the total strain tensor and $\boldsymbol{\varepsilon}^p$ is the plastic strain tensor. Microstructural deterioration is captured by the degradation of the elastic moduli $\mathbf{C}$.

The plastic strain rate and the damage strain rate are found through:

$$\dot{\boldsymbol{\varepsilon}}^{pd} = \dot{\boldsymbol{\varepsilon}}^d + \dot{\boldsymbol{\varepsilon}}^p = \dot{\lambda}\frac{\partial f(\boldsymbol{\sigma},q)}{\partial \boldsymbol{\sigma}} \tag{12}$$

A parameter β is introduced to split the effects associated with inelastic slip processes (resulting in an increase of plastic strain) and deterioration of the microstructure (resulting in an increase of compliance moduli $\mathbf{D}$):

$$\dot{\boldsymbol{\varepsilon}}^d = \dot{\mathbf{D}}\boldsymbol{\sigma} = \beta\dot{\lambda}\frac{\partial f(\boldsymbol{\sigma},q)}{\partial \boldsymbol{\sigma}} \tag{13a}$$

$$\dot{\boldsymbol{\varepsilon}}^p = (1-\beta)\dot{\lambda}\frac{\partial f(\boldsymbol{\sigma},q)}{\partial \boldsymbol{\sigma}} \tag{13b}$$

and the evolution of the compliance moduli $\mathbf{D}$ is given by:

$$\dot{\mathbf{D}} = \beta\dot{\lambda}\frac{\partial_{\boldsymbol{\sigma}} f(\boldsymbol{\sigma},q)\,\partial_{\boldsymbol{\sigma}} f^T(\boldsymbol{\sigma},q)}{\partial_{\boldsymbol{\sigma}} f^T(\boldsymbol{\sigma},q)\,\boldsymbol{\sigma}} \tag{14}$$

The integration of the constitutive equation is analogous to the classical plasticity theory. The stresses at time t_{n+1} are given by:

$$\boldsymbol{\sigma}_{n+1} = \mathbf{C}_{n+1}\left(\boldsymbol{\varepsilon}_{n+1} - \boldsymbol{\varepsilon}^p_{n+1}\right) \tag{15}$$

In the sense of an incremental approach, equation (15) can be rewritten as:

$$\boldsymbol{\sigma}_{n+1} = \mathbf{C}_n\left(\boldsymbol{\varepsilon}_{n+1} - \boldsymbol{\varepsilon}^p_n\right) - \mathbf{C}_n \Delta\boldsymbol{\varepsilon}^p + \Delta\mathbf{C}\left(\boldsymbol{\varepsilon}_{n+1} - \boldsymbol{\varepsilon}^p_{n+1}\right) \tag{16}$$

The first term in equation (16) computes the stresses as if no inelastic behaviour (damage or plasticity) is occurring. The second term in equation (16) corrects the stresses due to plastic deformations. Finally, the obtained stress is corrected with a damage corrector term. The latter term can be replaced resulting in:

$$\boldsymbol{\sigma}_{n+1} = \mathbf{C}_n\left(\boldsymbol{\varepsilon}_{n+1} - \boldsymbol{\varepsilon}^p_n\right) - \mathbf{C}_n \Delta\boldsymbol{\varepsilon}^p - \mathbf{C}_n \Delta\boldsymbol{\varepsilon}^d \tag{17}$$

so that, recalling the definition of plastic and damage strain rate in equation (13),

$$\boldsymbol{\sigma}_{n+1} = \mathbf{C}_n\left(\boldsymbol{\varepsilon}_{n+1} - \boldsymbol{\varepsilon}^p_n\right) - \mathbf{C}_n \Delta\lambda \frac{\partial f(\boldsymbol{\sigma}, q)}{\partial \boldsymbol{\sigma}} \tag{18}$$

Clearly, the updated stresses are written in a form analogous to the standard plasticity theory. The updated compliance moduli are updated as:

$$\mathbf{D}_{n+1} = \mathbf{D}_n + \beta\Delta\lambda \left.\frac{\partial_{\boldsymbol{\sigma}} f(\boldsymbol{\sigma}, q)\partial_{\boldsymbol{\sigma}} f^T(\boldsymbol{\sigma}, q)}{\partial_{\boldsymbol{\sigma}} f^T(\boldsymbol{\sigma}, q)\boldsymbol{\sigma}}\right|_{n+1} \tag{19}$$

4 DEGENERATED MODELS

In order to use the combined models in the strong discontinuity context, the constitutive relations must be rewritten in terms of displacement jumps and tractions. To obtain these degenerated equations, the constitutive equations must be consistent with an incorporated discontinuity.

4.1 *Model 1*

The constitutive relation for the combined damage/plasticity model as formulated by Simo & Ju (1987) is given in equation (3). Introducing the kinematic expression for the strain field, given in equation (2), yields

$$\boldsymbol{\sigma} = (1-d)\mathbf{C}^e\left(\nabla^s \hat{\mathbf{u}} + H_\Gamma \nabla^s \tilde{\mathbf{u}} + \delta_\Gamma (\tilde{\mathbf{u}} \otimes \mathbf{n})^s - \boldsymbol{\varepsilon}^p\right) \tag{20}$$

Assuming that the plastic strain field can be decomposed in a similar form as the total strain field and allowing only plasticity at the discontinuity, equation (20) can be rewritten as:

$$\boldsymbol{\sigma} = (1-d)\mathbf{C}^e\left(\nabla^s \hat{\mathbf{u}} + H_\Gamma \nabla^s \tilde{\mathbf{u}} + \delta_\Gamma (\tilde{\mathbf{u}} \otimes \mathbf{n})^s - \delta_\Gamma (\tilde{\mathbf{u}}^p \otimes \mathbf{n})^s\right) \tag{21}$$

The stress field must remain bounded and consequently, the Dirac delta must cancel. Oliver (2000) proposed to give the damage variable d a distributed character by making the internal damage variable κ^d unbounded which results in $\overline{\kappa}^d$. Again, inelastic behaviour is restricted to the discontinuity. Introducing the distributional character of the internal damage variable from equation (6), and substituting the damage variable in equation (20), the Dirac delta function can be eliminated,

$$\boldsymbol{\sigma}_P = \frac{q(\overline{\kappa}^d)}{\overline{\kappa}^d}\mathbf{C}^e\left((\tilde{\mathbf{u}} \otimes \mathbf{n})^s - (\tilde{\mathbf{u}}^p \otimes \mathbf{n})^s\right) \tag{22}$$

where "P" stands for a point on the discontinuity. Finally, the tractions are obtained from equation (22)

$$\mathbf{t} = \frac{q(\overline{\kappa}^d)}{\overline{\kappa}^d}\mathbf{n}\mathbf{C}^e\mathbf{n}\left(\tilde{\mathbf{u}} - \tilde{\mathbf{u}}^p\right) = (1-\omega)\mathbf{Q}^e\left(\tilde{\mathbf{u}} - \tilde{\mathbf{u}}^p\right) \tag{23}$$

where ω is the degenerated damage variable and $\mathbf{Q}^e$ is the elastic acoustic tensor. Examining equation (23), a significant similarity with the continuum model is observed. A major difference is that the elastic part has disappeared in the degenerated model, so that the separation of the discontinuity is completely inelastic. The total separation at the discontinuity is split into a recoverable damage part and an irrecoverable plastic part. Furthermore, the elastic constitutive tensor is replaced by the elastic acoustic tensor. Finally, the continuum damage variable d is replaced by the degenerated damage variable ω, which varies between – infinity and 1. The evolution of the degenerated damage variable is governed by the degenerated internal variable $\overline{\kappa}^d$, which represents the most severe value of an equivalent strain measure ever reached. The equivalent strain measure is given in terms of strains and needs to be rewritten in terms of separations.

The energy release rate strain measure, equation (5a), can be rewritten by introducing the kinematic relation for the strain field:

$$\varepsilon^{eq} = \sqrt{\frac{1}{2}\frac{1}{h}(\tilde{\mathbf{u}} \otimes \mathbf{n})^s : \mathbf{C} : \frac{1}{h}(\tilde{\mathbf{u}} \otimes \mathbf{n})^s} \tag{24}$$

where only the part relevant to the discontinuity is considered and the Dirac delta distribution has been replaced by *1/h* for $h \rightarrow 0$. Equation (24) can be replaced by:

$$\varepsilon^{eq} = \frac{1}{h}\sqrt{\frac{1}{2}\tilde{\mathbf{u}} \cdot [\mathbf{n} \cdot \mathbf{C} \cdot \mathbf{n}] \cdot \tilde{\mathbf{u}}} \tag{25}$$

and finally,

$$\Delta^{eq} = \sqrt{\frac{1}{2}\tilde{\mathbf{u}} \cdot \mathbf{Q} \cdot \tilde{\mathbf{u}}} \tag{26}$$

For mode-I cracking, the positive principal strains (equation (5b)) are often used as the equivalent strain measure. In this case, the material cracks perpendicular to the maximal principal stress direction. The equivalent strain measure (5b) can be easily degenerated as:

$$\Delta^{eq} = \Delta_n \tag{27}$$

where Δ_n is the normal separation of the discontinuity.

The integration of the degenerated constitutive law is analogous to the integration of the continuum model. In a first stage, the plasticity problem is solved in the effective stress space. Then, the damage problem is solved and the effective stresses are mapped back to the homogenized stress space.

4.2 *Model 2*

The constitutive equation for the combined damage plasticity model proposed by Meschke *et al.* (1998) is given in equation (11). Introducing the kinematic description of the strain field, given in equation (2), the constitutive relation is written as:

$$\boldsymbol{\sigma} = \mathbf{C}\left(\nabla^s \hat{\mathbf{u}} + H_\Gamma \nabla^s \tilde{\mathbf{u}} + \delta_\Gamma (\tilde{\mathbf{u}} \otimes \mathbf{n})^s - \boldsymbol{\varepsilon}^p\right) \tag{28}$$

Assuming plastic behaviour only at the discontinuity and assuming that the plastic strain field is expressed in an analogous fashion as the total strain field, equation (28) becomes:

$$\boldsymbol{\sigma} = \mathbf{C}\left(\nabla^s \hat{\mathbf{u}} + H_\Gamma \nabla^s \tilde{\mathbf{u}} + \delta_\Gamma (\tilde{\mathbf{u}} \otimes \mathbf{n})^s - \delta_\Gamma (\tilde{\mathbf{u}}^p \otimes \mathbf{n})^s\right) \tag{29}$$

In equation (29) the Dirac delta distribution is not cancelled. The only term that is possibly unbounded is the constitutive tensor. The evolution of the compliance tensor **D** is given by equation (14). Equation (14) can be integrated for a given time t,

$$\mathbf{D}(t) = \mathbf{D}_0 + \int_0^t \dot{\mathbf{D}} dt \tag{30}$$

The form of the plastic strain field requires that the plastic/damage multiplier is of the form,

$$\lambda = \delta_\Gamma \bar{\lambda} \tag{31}$$

and consequently,

$$\begin{aligned}\mathbf{D}(t) &= \mathbf{D}_0 + \delta_\Gamma \int_0^t \beta \dot{\bar{\lambda}} \frac{\partial_{\boldsymbol{\sigma}} f(\boldsymbol{\sigma}, q) \partial_{\boldsymbol{\sigma}} f^T(\boldsymbol{\sigma}, q)}{\partial_{\boldsymbol{\sigma}} f^T(\boldsymbol{\sigma}, q) \boldsymbol{\sigma}} dt \\ &= \mathbf{D}_0 + \delta_\Gamma \overline{\mathbf{D}}\end{aligned} \tag{32}$$

Introducing equation (32) into equation (29) and eliminating the Dirac delta distribution, results in,

$$\overline{\mathbf{D}}\boldsymbol{\sigma} = (\tilde{\mathbf{u}} \otimes \mathbf{n})^s - (\tilde{\mathbf{u}}^p \otimes \mathbf{n})^s \tag{33}$$

The traction vector is given by:

$$\mathbf{t} = \mathbf{n}\overline{\mathbf{C}}\mathbf{n}\left(\tilde{\mathbf{u}} - \tilde{\mathbf{u}}^p\right) = \overline{\mathbf{Q}}\left(\tilde{\mathbf{u}} - \tilde{\mathbf{u}}^p\right) \tag{34}$$

The degenerated compliance $\overline{\mathbf{D}}$ is initially zero, resulting in an infinite initial stiffness. This means that the continuum model is replaced by a rigid plastic/damage model. In this model, the elastic part has disappeared. Consequently, the total increment of the displacement jump can be written as:

$$\Delta\tilde{\mathbf{u}} = \Delta\tilde{\mathbf{u}}^d + \Delta\tilde{\mathbf{u}}^p \tag{35}$$

The increments of damage and plastic strains are given by equation (13), and can be degenerated by inserting the degenerated damage/plasticity multiplier, resulting in:

$$\Delta\tilde{\mathbf{u}}^d = \beta \Delta\bar{\lambda} \frac{\partial f(\boldsymbol{\sigma}, q)}{\partial \boldsymbol{\sigma}} \tag{36a}$$

$$\Delta\tilde{\mathbf{u}}^p = (1 - \beta) \Delta\bar{\lambda} \frac{\partial f(\boldsymbol{\sigma}, q)}{\partial \boldsymbol{\sigma}} \tag{36b}$$

and consequently, the total displacement jump is given by:

$$\Delta\tilde{\mathbf{u}} = \Delta\bar{\lambda} \frac{\partial f(\boldsymbol{\sigma}, q)}{\partial \boldsymbol{\sigma}} \tag{37}$$

Equation (37) is very similar to the strong discontinuity equations Oliver (2000) obtained when degenerating a classical elasto-plastic model. In that case, the jump was completely plastic and no return mapping algorithm was necessary.

5 NUMERICAL EXAMPLE

Both degenerated models are now used to model tensile tests of a double-edge notched stone specimen (see figure 2). The discontinuities are incorporated into finite elements using the partition of unity property of finite element shape functions (Wells & Sluys, 2001). This property allows enhancing nodes with additional degrees of freedom.

5.1 *Model 1*

To model the DEN-tensile tests, the following yield surface is used in the effective stress space (cf. equation 7):

$$f^p = \hat{T}_n - \left(H\kappa^p\right) \tag{38}$$

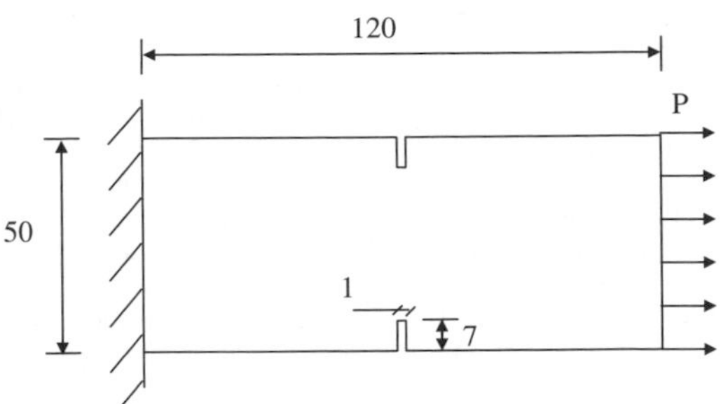

Figure 2. Double-edge notched specimen used during experiments and simulations (all dimensions in mm, thickness is 11 mm).

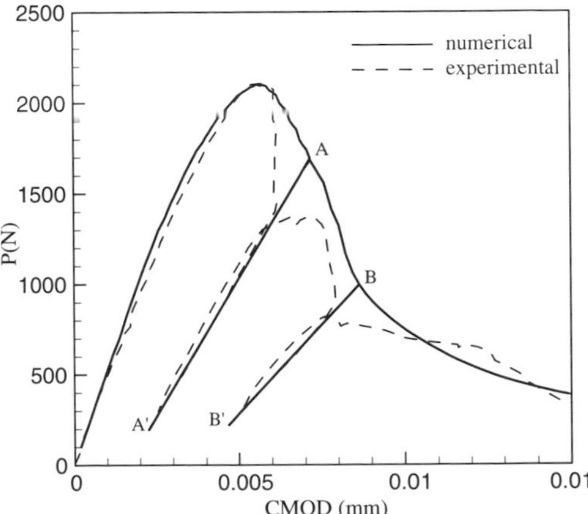

Figure 3. Computed load-deformation curve compared with experimental result for cyclic loading.

where $\hat{T}_n$ is the effective normal traction, H is the softening modulus, and κ^p is an internal variable. For the damage model, the normal separation is used as equivalent strain measure resulting in:

$$f^d = \Delta^{eq} - \kappa^d \tag{39}$$

The damage variable is defined as in equation 6:

$$\omega = 1 - \frac{q(\kappa^d)}{\kappa^d} = 1 - \frac{\kappa_i}{\kappa^d} \exp\left[-\mu\kappa^d\right] \tag{40}$$

where κ_i is the threshold for damage initiation and μ is a model parameter.

Figure 3 shows the computed load-deformation curve compared to the experimentally obtained curve. For this computation, $H = 27000\,\text{N/mm}^3$, $\mu = 400$ and $f_t = 6.5\,\text{MPa}$. Clearly, the model can capture the loading–unloading behaviour in an acceptable way. Moreover, the complete computed curve is in good agreement with the measured curve.

5.2 *Model 2*

For model 2, only a yield surface, bounding the elastic domain needs to be defined. For the example studied here, a Rankine type of surface is chosen. For the Rankine yield surface, Oliver *et al.* (1999) obtained a degenerated solution:

$$\Delta\tilde{u}_n = \frac{1}{\bar{h}} \Delta T_n \tag{41a}$$

$$\Delta\tilde{u}_t = 0 \tag{41b}$$

where T_n is the normal traction, $\tilde{u}_n$ and $\tilde{u}_t$ are the normal and tangential separation and h is the degenerated softening modulus. For an exponential softening law, equation (41) can be easily integrated.

$$t_n = f_t \exp[-\mu\tilde{u}_n] \tag{42}$$

The increment of damage and plastic deformations can be obtained with equation (36). With the updated tractions, found with equation (42), the compliance moduli can be updated using equation (14). An appropriate choice of the model parameters results in an identical behaviour as obtained with the previous described model, based on the Simo and Ju formulation. The adopted model parameters are:

- $\beta = 0.3$
- $\mu = 130$

Consequently, the same load-deformation curve is obtained (see figure 3).

6 CONCLUSIONS

In this paper, two continuum models combining plasticity and damage were developed in a degenerated form. The constitutive law was made consistent with the appearance of a displacement discontinuity. For both models studied, the elastic part of the solution is zero on the discontinuity surface. Consequently, the jump is completely inelastic and contains a recoverable damage and an irrecoverable plastic part. The model formulated by Simo & Ju (1987) introduces damage in an isotropic manner, by means of a scalar damage variable. The model proposed by Meschke *et al.* (1998) allows a more general damage evolution. The discrete form was analogous to the classical elasto-plastic degenerated model. Consequently, no return mapping algorithm is necessary. Both models were used to model cyclic loading of double-edge notched stone specimens. In the future, the models can be easily extended to more complicated plasticity and damage models.

REFERENCES

De Borst, R., Pamin, J. & Geers, M.G.D. 1999. On coupled gradient-dependent plasticity and damage theories with a view to localization analysis. *Eur. J. Mech. A/Solids* 18: 939–962.

De Proft, K. 2003. Combined experimental-computational study to discrete fracture of brittle Materials, Phd thesis, Vrije Universiteit Brussel.

Meschke, G., Lackner, R. & Mang, H.A. 1998. An anisotropic elasto-plastic-damage model for plain concrete. *International Journal for Numerical Methods in Engineering* 42: 703–727.

Oliver, J. 2000. On the discrete constitutive models induced by strong discontinuity kinematics and continuum constitutive equations. *International Journal of Solids and Structures* 37: 7207–7229.

Oliver, J., Cervera, M. & Manzoli, O. 1999. Strong discontinuities and continuum plasticity models: the strong discontinuity approach. *International Journal of Plasticity* 15: 319–351.

Simo, J.C. & Ju, J.W. 1987. Strain- and stress-based continuum damage models – I. Formulation. *International Journal of Solids and Structures* 23(7): 821–840.

Simo, J.C. & Ju, J.W. 1987. Strain- and stress-based continuum damage models – II. Computational aspects. *International Journal of Solids and Structures* 23(7): 841–869.

Wells, G.N. & Sluys, L.J. 2001. A new method for modelling cohesive cracks using finite elements. *International Journal for Numerical Methods in Engineering* 50(12): 2667–2682.

Computational Modelling of Concrete Structures, Bićanić et al. (eds)
© 2003 Swets & Zeitlinger, Lisse, ISBN 90 5809 536 3

Combined continuous/discontinuous failure of cementitious composites

A. Simone & L.J. Sluys
Faculty of Civil Engineering and Geosciences, Delft University of Technology, The Netherlands

P. Kabele
Faculty of Civil Engineering, Czech Technical University in Prague, Czech Republic

ABSTRACT: A novel numerical technique for the introduction of kinematic discontinuities is exploited for the analysis of failure in cementitious composites. Application to conventional fibre-reinforced and strain-hardening fibre-reinforced composites illustrates the procedure.

1 INTRODUCTION

Failure in composite materials ranges from quasi-brittle failure of plain concrete or conventional fibre-reinforced cement or concrete (FRC) to ductile fracture observed in high performance fibre-reinforced cement composite (HPFRCC). Although very sophisticated models are available in the literature for a micromechanical characterisation of the behaviour of composites, the difficulties related to a unique de-termination of model parameters from experimental data contrast with the basic requirements of any sound computational approach. To date, a unique and accurate characterisation is only possible for a narrow portion of engineered man-made materials. In this study, a phenomenological approach has been considered and attention will be restricted to conventional FRC and ECC (Li and Wu 1992), a high performance fibre-reinforced cement composite.

Tensile stress–strain relations of engineering materials can be retrieved through an appropriate analysis of the behaviour of specimens loaded up to tensile failure. Tensile failure of FRC specimens is characterised by localised deformations with a decreasing load-carrying capacity after the formation of the first crack. At this last stage, due to the contribution of the fibres, a certain amount of load can be still transmitted across the macrocrack. Unlike failure in FRC, the failure process of ECC tensile specimens is characterised by a uniform deformation stage due to multiple cracking process, followed by localised deformations. Multiple cracking, which is characterised by formation of numerous and almost uniformly distributed fine cracks, is accompanied, at the global level response, by an extended strain-hardening response.

Aim of this paper is to illustrate a simple, although effective, computational framework for the characterisation of the above failure processes in which the description of the continuum as well of the discrete failure phenomena is dealt with in a unified fashion. A framework is considered in which diffuse cracking develops in a discrete crack. Discontinuities in the problem fields are introduced by means of a discontinuous interpolation of the problem field exploiting the partition of unity paradigm of finite-element shape functions (Moës et al. 1999), which allows discontinuities in the problem fields to cross arbitrarily through solid elements. With the introduction of a discontinuity at the onset of localisation, softening constitutive laws across the discontinuity can be applied in a meaningful fashion. It is worth noting that the kinematic characterisation of this approach is equivalent to the one obtained through interface elements, the only difference being the advantage of avoiding costly remeshing procedures and the possibility of an activation of a discontinuity on demand (Moës et al. 1999; Wells and Sluys 2001). This combined continuous/discontinuous analysis framework allows the numerical analysis of a wide range of cementitious fibre composite materials in a unified framework.

2 NUMERICAL STRATEGIES FOR CEMENTITIOUS COMPOSITES

Different failure characterisations can be observed in cementitious composite specimens loaded up to tensile failure. In plain concrete and conventional fibre-reinforced concrete, an initial linear response is followed by a softening branch which is characterised by the widening of a single crack. In high performance fibre reinforced cement composites, an initial linear response is followed by a hardening branch characterised by multiple cracking; the softening branch observed at the global level stems from the widening of a single crack. This characterisation is typical of various composites such as SIFCON (randomly distributed short steel fibres infiltrated with cement slurry with fibre volume fraction $V_f = 5$–20%) or ECC (cementitious matrix with $V_f = 1$–2% of short random synthetic fibres).

The numerical strategy described here is based on the direct translation of the above failure characterisations into a computational framework. The determination of the (model) parameters is based on the application of a phenomenological procedure.

In composites with quasi-brittle behaviour, most of the deformation relates to crack splitting which is described by including cohesive surfaces where all the non-linearity is concentrated. The appearance of a (localised) macrocrack can be associated to softening in the global response. In case of ductile composites, the behaviour prior to softening can be described by a stress–strain curve due to the relatively homogeneous distribution of cracks in the hardening regime; the softening part of the global response can be described with cohesive surfaces.

Some strategies for the failure analysis of cementitious composites are reported in Figure 1. In this work attention is restricted to the case depicted in Figure 1a and Figure 1c. The continuous modelling refers to C, while a transition to a discontinuous range D is shown by the dashed line. The other two cases can also be considered, but regularisation techniques must be employed to resolve mesh dependence due to use of continuum softening constitutive relationships. Note, however, that the transition from a continuous to a continuous/discontinuous framework can be problematic due to the difficulties hidden in the determination of the material/model parameters (Simone et al. 2002).

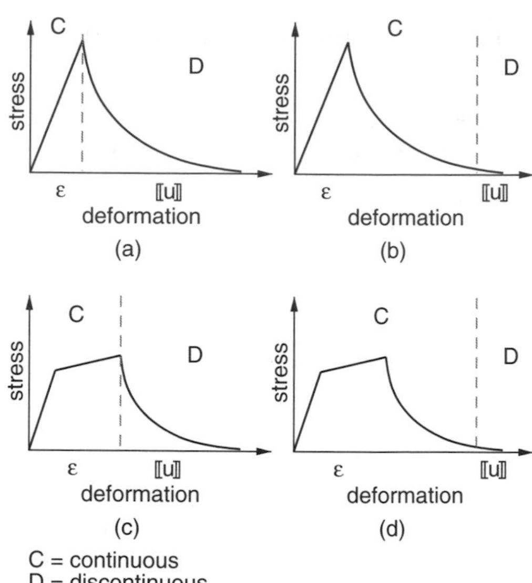

Figure 1. Strategies for failure analysis of cementitious composites: idealised tensile behaviour for quasi-brittle composites (a,b) and ductile composites (c,d).

2.1 *Constitutive modelling*

When a quasi-brittle composite is considered (FRC, cf Figure 1a), all the non-linear behaviour experienced by the tensile specimen can be described by a cohesive law expressed in terms of tractions t_n and displacement jumps (crack opening displacement-COD) w normal to the discontinuity surface. A simple cohesive law (Jenq and Shah 1986) is considered:

$$t_n(w) = \begin{cases} t_{\max}\left(1 - \dfrac{w}{w_{\max}}\right)^m & \text{if } 0 \leq w \leq w_{\max} \\ 0 & \text{if } w > w_{\max} \end{cases} \quad (1)$$

where $t_{\max}$ is defined as the maximum load divided by the cracked area (nominal tensile strength of the composite), $w_{\max}$ is the COD at zero load (theoretically equal to half of the fibre length) and the shape parameter $m > 1$ depends on the type of the fibre.

In case of ductile composites (ECC, cf Figure 1c), inelasticity develops as a consequence of multiple cracking which is initiated on planes normal to the maximum principal stress. Cracking is diffuse over a large part of the tensile specimen and the portion of total displacement to it associated is not recoverable when the specimen is unloaded to zero stress state. The composite can then be idealised as homogeneous and continuous in the multiple cracking stage and the Rankine criterion can be used to characterise inelastic behaviour before localisation (Kabele and Horii 1997). It is noted that the use of anisotropic plasticity is better suited to describe diffuse cracking with a preferential directional pattern. However, since in this study the failure mode is controlled by uniaxial extension, the use of anisotropic plasticity is not required (Fichant et al. 1999). Localisation of deformation is accounted for by means of a cohesive law as in the case of FRC.

2.2 *Discontinuities*

The partition of unity requirement of finite-element shape functions in exploited to incorporate discontinuities in the kinematic fields. When a body $\bar{\Omega}$ (cf Figure 2) is crossed by a discontinuity, the displacement field can be expressed as

$$\mathbf{u}(\mathbf{x}, t) = \hat{\mathbf{u}}(\mathbf{x}, t) + \mathscr{H}_{\Gamma_d}(\mathbf{x})\tilde{\mathbf{u}}(\mathbf{x}, t), \quad (2)$$

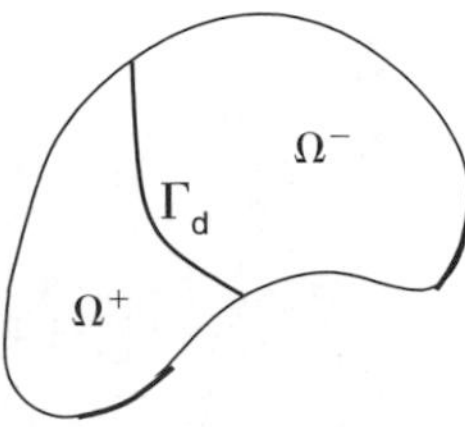

Figure 2. Body $\bar{\Omega}$ crossed by a discontinuity Γ_d.

where $\mathscr{H}_{\Gamma_d}(\mathbf{x})$ is the Heaviside function centred at the discontinuity surface Γ_d($\mathscr{H}_{\Gamma_d} = 1$ if $\mathbf{x} \in \bar{\Omega}^+$, $\mathscr{H}_{\Gamma_d} = 0$ if $\mathbf{x} \in \bar{\Omega}^-$) and $\hat{\mathbf{u}}$ and $\tilde{\mathbf{u}}$ are continuous functions on Ω. In the geometrically linear case, a similar decomposition holds for the strain field, computed as the symmetric part of the gradient of the displacement field:

$$\boldsymbol{\varepsilon} = \nabla^s \hat{\mathbf{u}} + \mathscr{H}_{\Gamma_d} \nabla^s \tilde{\mathbf{u}} \quad \text{if } \mathbf{x} \notin \Gamma_d, \tag{3}$$

where $(\cdot)^s$ refers to the symmetric part of $(\cdot)$. Following standard procedures (Wells and Sluys 2001), the discretised format of equation (2) reads, for nodes whose support is crossed by a discontinuity:

$$\mathbf{u}_h = \mathbf{N}\mathbf{a} + \mathscr{H}_{\Gamma_d}\mathbf{N}\mathbf{b}, \tag{4}$$

where $\mathbf{N}$ is an array containing standard finite-element shape functions and the global nodal degrees of freedom $\mathbf{a}$ and $\mathbf{b}$ represent, in the arrangement of equation (4), the total displacement field; the displacement jump across the discontinuity Γ_d is given by

$$[\![\mathbf{u}_h]\!] = \mathbf{N}\mathbf{b}|_{\Gamma_d}. \tag{5}$$

The strain field can be discretised in a similar fashion.

3 APPLICATIONS

The computational framework described in the previous section is used to describe the failure processes in an FRC bending test and in an ECC compact tension test. It is recalled that in this approach a discontinuity can propagate through finite elements and that the insertion of a discontinuity in an element is conditional upon the strength capacity of at least one of its integration points. The direction of the discontinuity is determined from the maximum principal stress direction. More details regarding the finite-element implementation can be found in Wells and Sluys (2001).

3.1 *Quasi-brittle failure*

The flexural behaviour of the beam depicted in Figure 3 has been analysed by Ward and Li (1990).

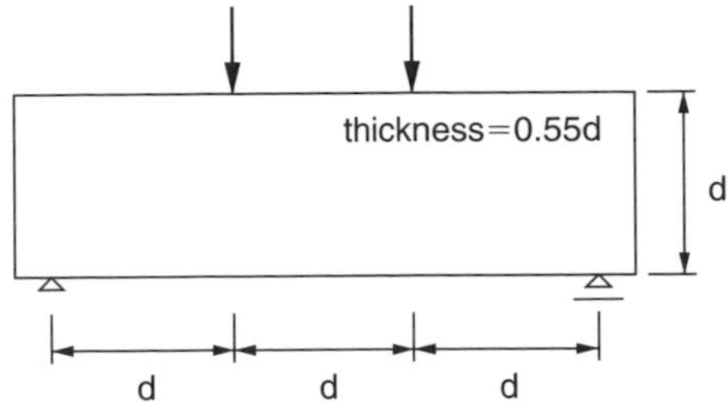

Figure 3. Geometry and loading configuration for bending test (d = 114 mm).

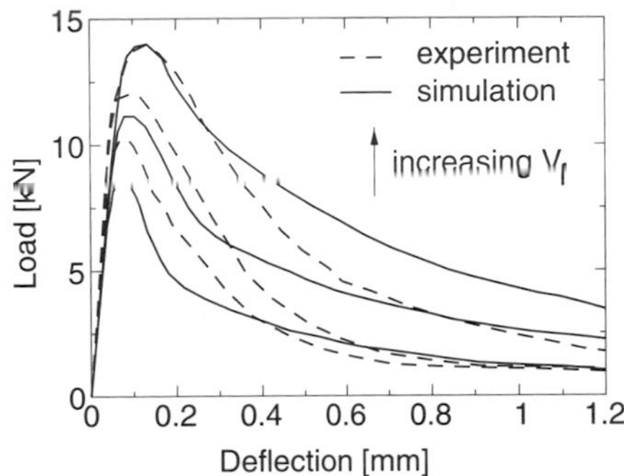

Figure 4. Load-deflection curves for different fibre volume fractions V_f = 0.5, 1.0, 1.5%.

The beam was tested with different fibre types and volume fractions. Aramid fibres with l_f = 6.4 *mm* and fibre volume fraction V_f = 0.5, 1.0, 1.5% were used in the beams considered for the numerical analyses. In the experiment, the load was applied via displacement control and the midspan deflection was measured by a linear variable differential transformer (LVDT) placed at the centre of the beam. The following material parameters were used in the simulations: Young's modulus 30 GPa, Poisson's ratio 0.2, composite tensile strength $t_{\max}$ = 1.96, 2.60, 3.25 MPa for fibre content equal to 0.5, 1.0, 1.5%, respectively. Inelasticity is introduced through the cohesive law in equation (1) with m = 2. The beam has been analysed using a structured mesh with 8–node quadrilateral elements in plane stress with an element size in the central part of the beam equal to 3 mm.

The numerical and experimental load-deflection curves are reported in Figure 4 were the applied load is plotted against the midspan deflection. A good agreement was found only for the beam with the highest fibre volume fraction (V_f = 1.5%); for the other fibre volume fraction beams the peak load was predicted lower and the post-peak response was more ductile. These results are very similar to those reported by Kullaa (1997) who used interface elements. As shown

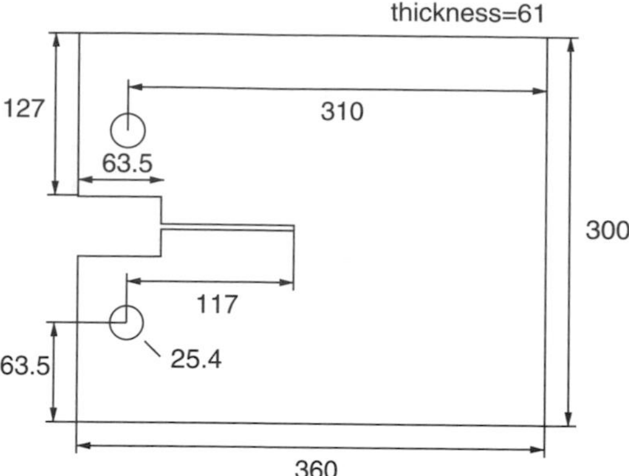

Figure 5. Geometry for compact-tension specimen (dimensions in mm).

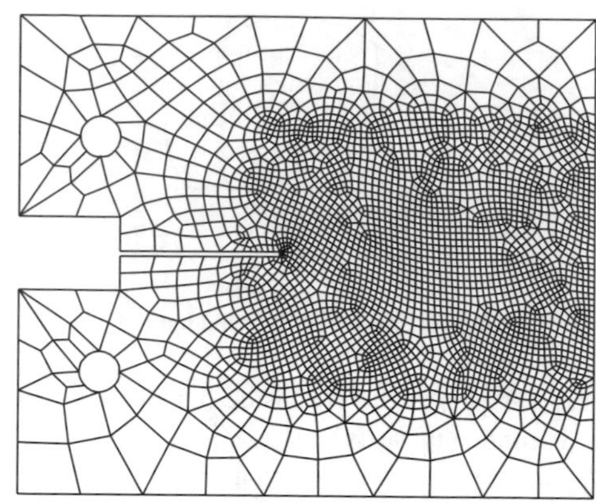

Figure 6. Mesh for compact-tension specimen.

by Wells and Sluys (2001), a possible remedy to modify the post peak regime is the introduction of a more complicated dissipation mechanism through the introduction of displacement contributions parallel to the crack. Such an approach may lead to a *better fit* between numerical and experimental curves.

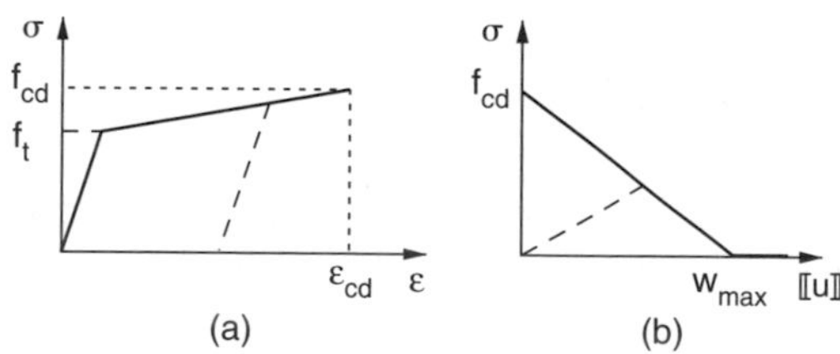

Figure 7. Stress-deformation relations for ECC: continuum (a) and discrete (b) constitutive laws. Dashed lines indicates unloading behaviour.

3.2 *Ductile failure*

A Compact-tension specimen, tested by Li and Hashida (1992), was analysed using the procedure illustrated in Section 2 (cf Figure 1c) for the case of ductile failure. The material was an ECC with 2% by volume of 12 mm long polyethylene fibres. The geometry of the specimen is depicted in Figure 5. In the experiment, the load was applied at the loading pins and was controlled using a LVDT placed near the notch tip. The following model parameters have been used for the analysis (Kabele and Horii 1997): Young's modulus 22 GPa, Poisson's ratio 0.2, matrix cracking strength $f_t = 2.2$ MPa, continuous/discontinuous transition strain value $\varepsilon_{cd} = 5.78\%$ with corresponding stress level $f_{cd} = 4.32$ MPa as depicted in Figure 7a. Inelasticity in the continuum bulk was introduced by Rankine plasticity (cf Figure 7a), while the cohesive law in equation (1) with $t_{\max} = f_{cd}$, $W_{\max} = 0.5 l_f$ and $m = 1$ was used to describe the in-elastic behaviour across the macrocrak (cf Figure 7b). In the finite-element simulation, an unstructured mesh of 2964 8–node quadrilateral elements in plane stress with medium element size in the central part of the specimen equal to 4 mm was used; the mesh is shown in Figure 6.

The simulated behaviour compares reasonably well with the experimental data as depicted in Figure 5 and the softening effect due to the 'activation' of the discrete crack is fully captured by the model.

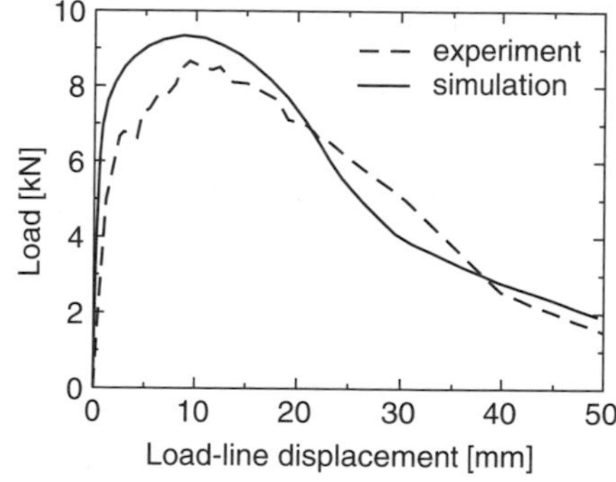

Figure 8. Load-displacement curves.

4 CONCLUSIONS

The analyses reported in this study have shown the applicability of a novel approach to describe displacement discontinuities in the analysis of cementitious composites. Two kinds of dissipative phenomena have been considered whose characterisation has been derived upon the analysis of one-dimensional tensile test loaded up to failure. In quasi-brittle composites the classical cohesive zone approach has been considered, with the use of a discrete traction law at the onset of strain localisation. For composites whose behaviour presents an extensive hardening response, inelasticity due to diffuse cracking is introduced by a continuum plasticity model; like in quasi-brittle materials, at the onset of strain localisation a cohesive law has been considered. Despite the conceptual simplicity of the approach, this framework can be useful in the qualitative description of the experimental trend. However, a precise quantitative comparison between experimental and the numerical curves was not possible and the overall agreement in the prediction was considered not satisfactory due to the simplistic constitutive description of the composite.

ACKNOWLEDGEMENTS

Financial support to AS through the BEO programme (special fund from TU Delft for excellent research) and to PK through grant GACR 106/02/0678 is gratefully acknowledged. Stimulating discussions with Pietro Lura (TU Delft) and Alberto Meda (Università di Brescia, Italy) are gratefully acknowledged.

REFERENCES

Fichant, S., C. La Borderie, and G. Pijaudier-Cabot (1999). Isotropic and anisotropic descriptions of damage in concrete structures. *Mechanics of Cohesive-Frictional Materials* 4, 339–359.

Jenq, Y. S. and S. P. Shah (1986). Crack propagation in fiber-reinforced concrete. *Journal of Structural Engineering* 112(1), 19–34.

Kabele, P. and H. Horii (1997). Analytical model for fracture behavior of pseudo strain-hardening cementitious composites. *Concrete Library International (proc. of JSCE)* 29, 105–120.

Kullaa, J. (1997). Finite element modelling of fibre-reinforced brittle materials. *HERON* 42(2), 75–95.

Li, V. C. and T. Hashida (1992). Ductile fracture in cementitious materials? In Z. P. Bažant (Ed.), *Fracture Mechanics of Concrete Structures, FramCoS* 1, London, pp. 526–535. Elsevier.

Li, V. C. and H. C. Wu (1992). Conditions for pseudo strain-hardening in fiber reinforced brittle matrix composites. *Applied Mechanics Reviews* 45(8), 390–398.

Möes, N., J. Dolbow, and T. Belytschko (1999). A finite element method for crack growth without remeshing. *International Journal for Numerical Methods in Engineering* 46(1), 131–150.

Simone, A., G. N. Wells, and L. J. Sluys (2002). Discontinuous modelling of crack propagation in a gradient-enhanced continuum. In H. A. Mang, F. G. Rammerstorfer, and J. Eberhardsteiner (eds.), *Proceedings of the Fifth World Congress on Computational Mechanics (WCCM V)*. Vienna University of Technology, Austria.

Ward, R. J. and V. C. Li (1990). Dependence of flexural behavior of fiber reinforced mortar on material fracture resistance and beam size. *ACI Materials Journal* 87(6), 627–637.

Wells, G. N. and L. J. Sluys (2001). A new method for modelling cohesive cracks using finite elements. *International Journal for Numerical Methods in Engineering* 50(12), 2667–2682.

Computational Modelling of Concrete Structures, Bićanić et al. (eds)
© 2003 Swets & Zeitlinger, Lisse, ISBN 90 5809 536 3

Ductility assessment of HP2C structures: from material modeling to model-based structural simulation

E. Chuang, H. Park, & F.-J. Ulm
Massachusetts Institute of Technology, Cambridge, Massachusetts

ABSTRACT: This paper demonstrates the suitability of a 3D two-phase model for structural simulation of high performance cementitious composites (HP2C). This is shown with finite element simulations of two HP2C applications: a flexural girder and a shear girder which have been recently tested by the US Federal Highway Administration (FHWA).

1 INTRODUCTION

High performance cementitious composites (HP2C), such as DUCTAL™, are a new generation of fiber reinforced cementitious composites (FRCC) with substantial improvements in mechanical behavior. The most important development in these HP2C materials may be the nearly elasto-plastic ductile behavior, which allows safe utilization of the tensile and shear capacity in structural elements. To exploit the ductile capacity of these materials in real world structures, one must be able to quantify the relationship between HP2C ductility at a material level and the in-situ material performance in large-scale structural applications. A short review of recent developments [Chuang & Ulm, 2002a] in the model-based assessment of the ductility of HP2C materials and structures is the focus of this paper. It is organized as follows: In the first part of this paper, we present the 3D version of a two-phase macroscopic material model for HP2C materials, which is the extension of the 1D HP2C model [Chuang & Ulm, 2002b] displayed in Figure 1. The structural validation of the model, by means of finite element simulations of two HP2C applications, a flexural girder and a shear girder recently tested by the US Federal Highway Administration (FHWA), highlights the suitability and the accuracy of the proposed HP2C model and its finite element implementation.

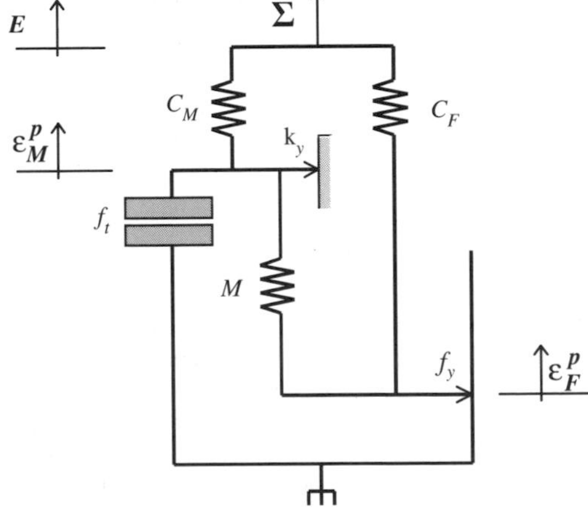

Figure 1. 1D HP2C model [Chuang & Ulm, 2002b].

2 MATERIAL MODEL AND INPUT PARAMETERS

The main challenge in modeling the behavior of fiber reinforced cementitious composites (FRCC) is to distinguish in the overall composite response the brittle behavior of the cementitious matrix from the ductile contribution of the reinforcing fibers and their interface. Previously developed FRCC models have addressed this challenge by micromechanical fracture and/or yield approaches [e.g. Leung & Li, 1989; Karihaloo et al., 1996; for a comprehensive review see Chuang & Ulm, 2002a], which allow one to show that HP2C materials, in contrast to standard FRCC materials, are an optimized mechanical match of two materials and their interactions [Chuang et al., 2002]. These micromechanical models form the backbone of the HP2C model presented here, which is macroscopic in nature. In this way, HP2C is modeled in a continuous fashion, i.e. through the various stages of damage evolution. This macroscopic approach allows for a straightforward finite element implementation of the

HP2C model which can be applied to simulate HP2C structures.

A 1D representation of the HP2C model is illustrated in Figure 1. In this model, developed in detail in [Chuang & Ulm, 2002b], a brittle-plastic matrix phase (stiffness $\mathbb{C}_M$, brittle strength f_t, plastic strength k_y) is coupled to an elasto-plastic fiber phase (stiffness $\mathbb{C}_F$ strength f_y) by means of the stiffness $\mathbb{M}$. In 3-D, the state equations of the model read:

$$\begin{pmatrix} \underline{\underline{\Sigma}} \\ \underline{\underline{\sigma}}_M \\ \underline{\underline{\sigma}}_F \end{pmatrix} = \begin{bmatrix} \mathbb{C}_M + \mathbb{C}_F & -\mathbb{C}_M & -\mathbb{C}_F \\ \mathbb{C}_M & -(\mathbb{C}_M + \mathbb{M}) & \mathbb{M} \\ \mathbb{C}_F & \mathbb{M} & -(\mathbb{C}_F + \mathbb{M}) \end{bmatrix} : \begin{pmatrix} \underline{\underline{E}} \\ \underline{\underline{\varepsilon}}^p_M \\ \underline{\underline{\varepsilon}}^p_F \end{pmatrix} \quad (1)$$

where $\underline{\underline{E}}$ is the total strain, $\underline{\underline{\varepsilon}}^p_M$ is the plastic strain in the composite matrix representing cracking, and $\underline{\underline{\varepsilon}}^p_F$ is the permanent strain associated with the irreversible deformation of the composite fiber phase. The total stress $\underline{\underline{\Sigma}}$ that is related to the external forces is the sum of the composite matrix stress $\underline{\underline{\sigma}}_M$ and the composite fiber stress $\underline{\underline{\sigma}}_F$:

$$\underline{\underline{\Sigma}} = \underline{\underline{\sigma}}_M + \underline{\underline{\sigma}}_F \quad (2)$$

2.1 *Isotropic state equations and material constants*

The general 3D constitutive model with matrix-fiber interaction involves 3×21 stiffness parameters associated with the tensors $\mathbb{C}_F$, $\mathbb{C}_M$ and $\mathbb{M}$. This number decreases significantly when considering some specific matrix behavior and specific fiber orientation systems. For example, fiber reinforced materials with random fiber orientation can be estimated to act isotropically. Accordingly, the separate behaviors of the cementitious matrix and the randomly oriented reinforcing fibers can be approximated to be isotropic. In this case, $\mathbb{C}_M$ and $\mathbb{C}_F$ can both be described with two unique scalar values:

$$\mathbb{C}_i = 3K_i\mathbb{K} + 2G_i\mathbb{J}; \quad \left\{ \begin{array}{l} K_i = \frac{C_i}{3(1-2\nu_i)} \\ G_i = \frac{C_i}{2(1+\nu_i)} \end{array} \right\} i = M, F \quad (3)$$

where $K_{ijkl} = \frac{1}{3}\delta_{ij}\delta_{kl}$ is the volumetric part of the 4th order unit tensor $\mathbb{I}$, and $\mathbb{J} = \mathbb{I} - \mathbb{K}$ is the deviatoric part; G_F and G_M are the shear moduli of the composite fiber and matrix; K_F and K_M are the bulk moduli of the composite fiber and matrix. Using the assumption of randomly oriented cracks after matrix cracking, the post-cracking stiffness behavior of the modeled material can be shown to be isotropic. For that reason, the coupling stiffness also takes an isotropic behavior.

Thus, the following form of the matrix-fiber coupling tensor $\mathbb{M}$ is chosen:

$$\mathbb{M} = 3M^v\mathbb{K} + 2M^d\mathbb{J} \quad (4)$$

where M^v and M^d are the bulk and shear matrix-fiber coupling moduli, respectively. M^v and M^d are described by:

$$M^v = \frac{M^{3D}}{3(1-2\nu_I)}; \quad M^d = \frac{M^{3D}}{2(1+\nu_I)}; \quad (5)$$

where M^{3D} is the 3D counterpart of M, the 1D coupling modulus; and ν_I is an effective Poisson's ratio for this composite coupling mechanism. Since the fiber volume content of HP2C is relatively low (1–6%), the Poisson's ratio of the composite coupling mechanism and the composite fibers can be estimated to be equal to that of the composite matrix:

$$\nu_I = \nu_F = \nu_M \quad (6)$$

By enforcing a 1D consistency condition, i.e. the uniaxial response of the 3D model must be identical to that of the 1D model, M^{3D} can be solved:

$$M^{3D} = \beta_I M + (\beta_I - 1)\frac{C_M C_F}{(C_M + C_F)} \quad (7)$$

$$\text{where } \beta_I = \frac{\left(\alpha_M^{UN} + \sqrt{\frac{2}{3}}\right)^2 (1-2\nu_M)(1+\nu_M)}{\left[3\left(\alpha_M^{UN}\right)^2(1+\nu_M) + (1-2\nu_M)\right]}$$

Thus, the state equation (1) can be described with four elasticity parameters, C_M, C_F, M, and ν_M, that can be extracted from a 1D tensile curve on HP2C-materials and one plastic strength model parameter, α_M^{UN}, which is described below.

2.2 *3D strength domain*

The strength domain $\mathcal{D}_E$ of the composite material is defined for the partial stresses $\underline{\underline{\sigma}}_M$ and $\underline{\underline{\sigma}}_F$, that is:

$$\underline{\underline{\Sigma}} \in \mathcal{D}_E$$
$$\Updownarrow$$
$$\begin{pmatrix} \underline{\underline{\sigma}}_M \in \mathcal{D}_M \Leftrightarrow F_M = \max\left[f_M\left(\underline{\underline{\sigma}}_M\right) \le 0\right] \\ \underline{\underline{\sigma}}_F \in \mathcal{D}_F \Leftrightarrow F_F = \max\left[f_F\left(\underline{\underline{\sigma}}_F\right) \le 0\right] \end{pmatrix} \quad (8)$$

where $\mathcal{D}_M$ and $\mathcal{D}_F$ denote the strength domains of the composite matrix and the composite fiber, respectively, which are expressed in 3D by a set of loading functions, F_M and F_F, that mimic the characteristic strength behavior of fiber-reinforced composite materials. The composite strength domains in this study, F_M and F_F, are represented with combinations of two types of loading functions: the tension cut-off criterion (TC) and the Drucker-Prager criterion (DP). These loading functions represent isotropic strength domains. The TC loading surface is given by:

$$f^{TC} = I_1 - \sigma^{TC} \tag{9}$$

and the DP reads:

$$f^{DP} = \alpha I_1 + \left|\underline{\underline{s}}\right| - c^{DP} \tag{10}$$

where $|\underline{\underline{s}}| = \sqrt{s_{ij}s_{ij}}$ refers to the magnitude of the stress deviator; α, σ^{TC}, and c^{DP} are material parameters.

2.2.1 Composite matrix strength domain

Following the form of the 1D HP2C model, the composite matrix is assigned a brittle-plastic 3D strength domain, the composite fiber an elasto-plastic strength domain. The composite matrix captures an elastic-brittle behavior with a higher initial (uncracked) limit, and a lower yield limit after cracking. This strength domain is described by 6 characteristic values, listed as absolute values:

1. The initial tensile strength, σ_{Mt}. By definition, this must be the same as the cracking strength $(f_t + k_y)$ of the 1D HP2C model:

$$\sigma_{Mt} = f_t + k_y \tag{11}$$

2. The initial compressive strength, σ_{Mc}.
3. The initial biaxial compressive strength, σ_{Mb}.
4. The tensile post-cracking yield strength, σ_{Mt}^{cr}. This is equivalent to the 1D composite matrix post-cracking strength k_y:

$$\sigma_{Mt}^{cr} = k_y \tag{12}$$

5. The compressive yield strength, σ_{Mc}^{cr}.
6. The biaxial compressive yield strength, σ_{Mb}^{cr}.

Before matrix cracking, the initial strength limits (σ_{Mt}, σ_{Mc}, and σ_{Mb}) govern the composite matrix loading functions. To enforce these initial strength limits, a tension cut-off criterion governs the tension-tension stress states f_M^{TC}, one Drucker-Prager criterion governs the compression-tension stress states f_M^{UN}, and another DP governs the compression-compression stress states f_M^{BI}. These criteria are expressed by:

$$f_M^{TC} = I_{1,M} - \sigma_{Mt} \leq 0 \tag{13}$$

$$f_M^{UN} = \alpha_M^{UN} I_{1,M} + \left|\underline{\underline{s}}_M\right| - c_M^{UN} \leq 0 \tag{14}$$

$$f_M^{BI} = \alpha_M^{BI} I_{1,M} + \left|\underline{\underline{s}}_M\right| - c_M^{BI} \leq 0 \tag{15}$$

where

$$\alpha_M^{UN} = \frac{\sqrt{2/3}(\sigma_{Mc}-\sigma_{Mt})}{\sigma_{Mc}+\sigma_{Mt}}; c_M^{UN} = \left(\sqrt{2/3} - \alpha_M^{UN}\right)\sigma_{Mc}$$
$$\alpha_M^{BI} = \frac{\sqrt{2/3}(\sigma_{Mb}-\sigma_{Mc})}{2\sigma_{Mb}-\sigma_{Mc}}; c_M^{BI} = \left(\sqrt{2/3} - \alpha_M^{BI}\right)\sigma_{Mc} \tag{16}$$

After cracking, the post-cracking strength parameters ($\sigma_{Mt}^{cr}, \sigma_{Mc}^{cr}$, and σ_{Mb}^{cr}) govern the strength domain. As a simplification, it is assumed that the post-cracking composite strengths are reduced by the same factor, $\gamma^{cr} = \sigma_{Mt}^{cr}/\sigma_{Mt}$:

$$\sigma_{Mc}^{cr} = \gamma^{cr}\sigma_{Mc}; \quad \sigma_{Mb}^{cr} = \gamma^{cr}\sigma_{Mb} \tag{17}$$

With this simplification, the six composite matrix strength parameters ($\sigma_{Mt}, \sigma_{Mc}, \sigma_{Mb}, \sigma_{Mt}^{cr}, \sigma_{Mc}^{cr}$, and σ_{Mb}^{cr}) are effectively reduced to four ($\sigma_{Mt}, \sigma_{Mc}, \sigma_{Mb}$, and σ_{Mt}^{cr}). The post-cracking loading functions read:

$$f_M^{TC,cr} = I_{1,M} - \sigma_{Mt}^{cr} \leq 0 \tag{18}$$

$$f_M^{UN,cr} = \alpha_M^{UN} I_{1,m} + \left|\underline{\underline{s}}_M\right| - c_M^{UN,cr} \leq 0 \tag{19}$$

$$f_M^{BI,cr} = \alpha_M^{BI} I_{1,M} + \left|\underline{\underline{s}}_M\right| - c_M^{BI,cr} \leq 0 \tag{20}$$

where $c_M^{UN,cr} = \gamma^{cr} c_M^{UN}$ and $c_M^{BI,cr} = \gamma^{cr} c_M^{BI}$. The composite matrix loading functions provided in Eqs. (13) to (20) are illustrated in the biaxial stress plane (Fig. 2) and in the $I_1 - |\underline{\underline{s}}|$ plane (Fig. 3). In its uncracked state, Eqs. (13) to (16) dictate the composite matrix strength domain, as depicted by the dashed lines in Fig. 3 in the $I_1 - |\underline{\underline{s}}|$ plane. After cracking, Eqs. (17) to (20) represent the post-cracking plastic limit, illustrated with the dotted lines in Fig. 3. In this figure, the composite matrix is loaded in uniaxial tension, uniaxial compression, and biaxial (equibiaxial) compression.

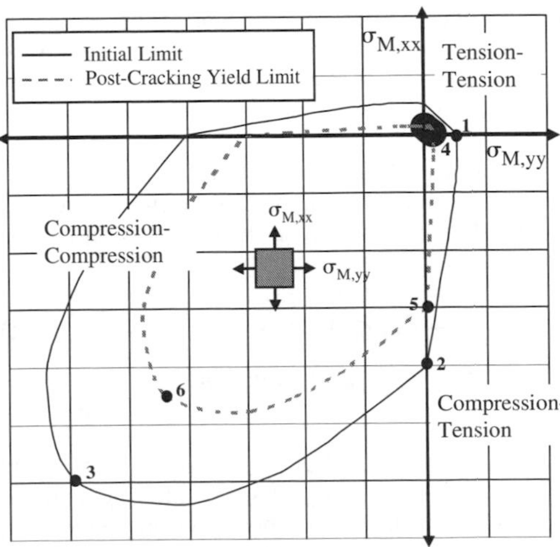

Figure 2. Plane stress composite matrix strength domain.

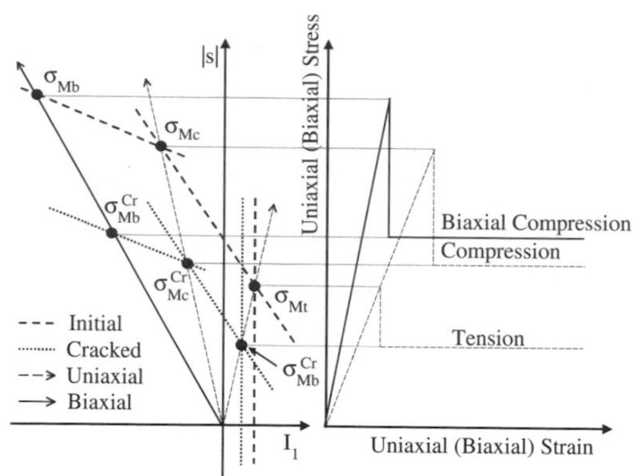

Figure 3. Composite matrix strength domain in the $I_1 - |\underline{s}|$ halfplane (left), and stress-strain response (right).

Upon achieving its initial strength limit, the composite matrix exhibits a brittle stress drop to its corresponding post-cracking strength. It is this stress drop in the composite matrix which enforces a macroscopic stress drop at first cracking.

2.2.2 Composite fiber strength domain

The elasto-plastic composite fiber is characterized by three strength values in the 3D space:

1. The tensile strength, σ_{Ft}. This, by definition, is the same as the 1D fiber strength f_y:

$$\sigma_{Ft} = f_y \tag{21}$$

2. The compressive strength, σ_{Fc}.
3. The biaxial compressive strength, σ_{Fb}.

Note that the characteristic compressive strengths of the composite fiber σ_{Fc} and σ_{Fb} are not the compressive strengths of the reinforcing fiber. Rather, one may consider the compressive strength of the composite fiber as the compressive capacity added to the overall HP2C composite strength as a result of reinforcing fibers (see, for example, Karihaloo & Lange-Kornbak, 2001).

To enforce these elasto-plastic limits, a TC is employed to limit the tension-tension stress states f_F^{TC} and a DP is utilized to limit the compression-tension stress states f_F^{DP}:

$$f_F^{TC} = I_{1,F} - \sigma_{Ft} \leq 0 \tag{22}$$

$$f_F^{DP} = \alpha_F^{DP} I_{1,F} + \left|\underline{\underline{s}}_F\right| - c_F^{DP} \leq 0 \tag{23}$$

where $\alpha_F^{DP} = \sqrt{2/3}(\sigma_{Fc} - \sigma_{Ft})/(\sigma_{Fc} + \sigma_{Ft})$ and $c_F^{DP} = (\sqrt{2/3} - \alpha_F^{DP})\sigma_{Fc}$. As a simplifying assumption, a criterion is not specifically designated to limit the composite fiber's biaxial compressive strength σ_{Fb}. Furthermore, unlike the composite matrix, the strength domain of the composite fiber is constant, regardless of the cracking state in the composite matrix.

Table 1. Model parameters of the 3D HP2C model.

	Description
C_M	Composite matrix stiffness
C_F	Composite fiber stiffness
M	Composite interface stiffness
ν_M	Poisson's ratio
f_t	Brittle tensile strength of composite matrix
k_y	Post-cracking tensile strength of composite matrix
σ_{Mc}	Initial compressive strength of composite matrix
σ_{Mb}	Initial biaxial compressive strength of composite matrix
f_y	Tensile strength of composite fiber
σ_{Fc}	Compressive strength of composite fiber

2.3 *Summary of model parameters*

The composite loading function F^{3D} is described by six model strength parameters – four composite matrix parameters ($\sigma_{Mt}, \sigma_{Mc}, \sigma_{Mb}$, and σ_{Mt}^{cr}) and two composite fiber parameters (σ_{Ft} and σ_{Fc}) – in order to capture six physically observed strength values ($\Sigma_t, \Sigma_c, \Sigma_b, \Sigma_t^{cr}, \Sigma_c^{cr}$, and Σ_b^{cr}). In addition, assuming an isotropic elastic behavior of both composite matrix and composite fiber, four additional parameters are required by the state equations. This leads to a total of $4 + 6 = 10$ unique model parameters summarized in Table 1, which can be determined through a single tensile test, a single compression test, and commonly accepted cementitious data (see Chuang & Ulm, 2002a, for details).

3 VALIDATION OF THE 3D MODEL

To validate the HP2C model and its finite element implementation, two case studies are presented: a flexure test and a shear test. These tests were carried out by the Federal Highway Administration (FHWA) at the Turner-Fairbank Highway Research Center Structures Laboratory in McLean, Virginia under the direction of Joey Hartmann and Benjamin Graybeal [FHWA, 2002]. Both of these tests were numerically simulated to gauge the accuracy and reliability of the HP2C model. The finite element simulation was validated with the experimental data with respect to three different criteria:

- Load-deflection curves. The load-deflection curves of the FHWA specimen and the FE simulation demonstrate very good correlation.
- Strain gauge measurements. The FE program provides results for the deflection of the nodes in a given mesh during loading. Strain results are calculated as the change in distance between two nodes

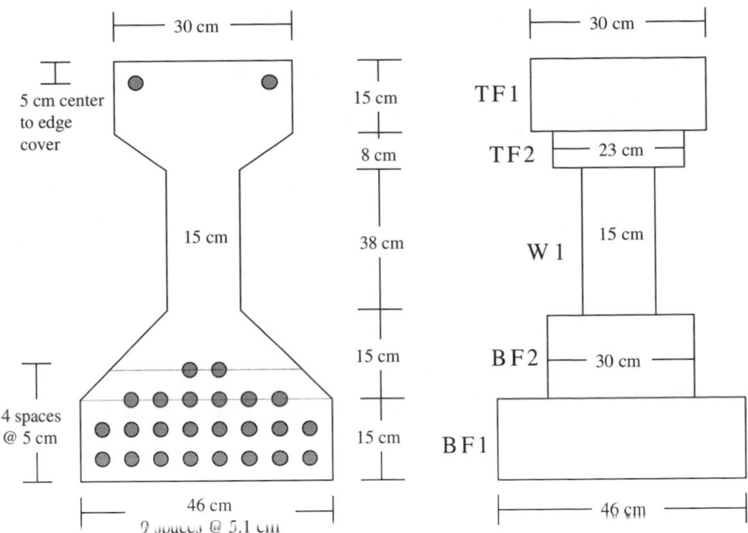

Figure 4. Cross section of the AASHTO Type II girder: (left) actual beam and (right) idealized 2D-modeling in FE simulation.

divided by the original distance between the nodes. Strain predictions obtained from the FE simulation exhibit excellent agreement with strain measurements from strain gauges placed at various locations on the FHWA specimens.

- Cracking patterns. Plastic strains in the composite matrix can be related to cracking, which occurs in the cementitious matrix of HP2C. The composite matrix plastic strains as given by the FE simulation accurately model cracking observed in the FHWA specimens.

Both tests involved AASHTO Type II girders, depicted in Fig. 4, comprised of the HP2C. The AASHTO Type II prestressed concrete girder is a 91 cm high beam with a 30 cm wide top flange, 15 cm wide web, and a 46 cm wide bottom flange. The AASHTO Type II girder is prestressed with 26 steel tendons, each 1.27 cm in diameter, composed of low relaxation steel with 1,860 MPa strength and 200 GPa stiffness [AASHTO, 2000]. Each prestressing tendon was initially loaded to 55% of its ultimate strength. Half of the tendons in the bottom flange were debonded for 91 cm from each end [FHWA, 2002]. No shear reinforcement was used in either girder.

3.1 *HP2C model parameters*

The 3-D material parameters for the HP2C are given in Table 2, labeled "HP2C Only". The tensile material parameters (C_M, C_F, M, f_t, k_y, and f_y) were derived from the tensile notched plate test provided by the HP2C manufacturer (here DUCTAL, see Chuang & Ulm, 2002b). The rationale behind using notched tensile data is that it is presumed that the notched configuration best reflects – in an average sense – HP2C structural behavior, particularly after cracking. Other parameters were solved from commonly accepted HP2C material values [Chuang & Ulm, 2002a].

Table 2. Values of HP2C model parameters used in the FHWA flexure and shear finite element simulations.

Model parameter	HP2C only	HP2C with tendons, BF1	HP2C with tendons, BF2
C_M [GPa]	53.9	53.9	53.9
$C_F^{(B)}$ [GPa]	0	6.0	4.4
M [GPa]	1.65	1.65	1.65
ν_M [1]	0.17	0.17	0.17
f_t [MPa]	0.7	0.7	0.7
k_y [MPa]	6.9	6.9	6.9
σ_{Mc} [MPa]	190	190	190
σ_{Mb} [MPa]	220	220	220
$f_y^{(B)}$ [MPa]	4.6	29.6	22.9
σ_{Fc} [MPa]	10	30	30

3.2 *Simplified modeling of prestressing*

As depicted in Fig. 4, the prestressing tendons are not explicitly simulated in the finite element model. Instead, the equivalent effect of the tendons is modeled, that is (1) the prestressing forces and (2) the contribution of the tendons to the stiffness and strength.

To capture the effect of prestressing forces, an equivalent external pressure is applied at the ends of the girder, particularly on the bottom flange, BF1

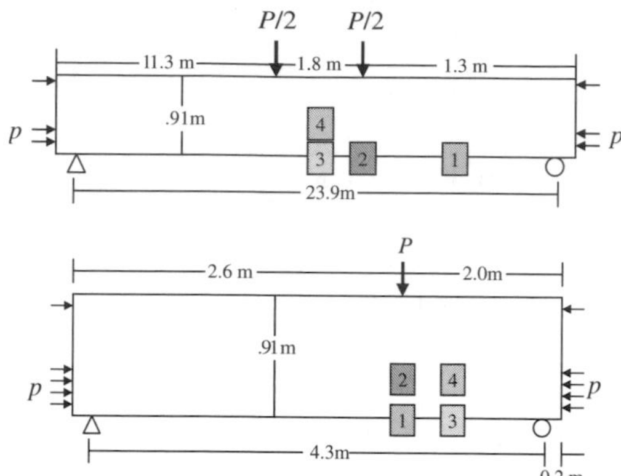

Figure 5. Loading configuration and strain gauge locations for: (top) the FHWA flexure test, (bottom) the FHWA shear test.

and BF2, and the upper part of the top flange, TF1 (see Fig. 5). The prestressing pressure p is calculated according to:

$$p = 0.55\,(c_T \sigma_T^u) \tag{24}$$

where factor 0.55 refers to the level of prestressing in the tendons. Subscript "T" refers to the prestressing tendons; c_T is the volume fraction of the tendons in a particular cross sectional subdivision i (i.e. $c_T = A_T/A_i$ where A_T is the cross sectional area of the tendons in the subdivision and A_i is the total area of the subdivision) and $\sigma_T^u = 1{,}860$ MPa is the strength of the tendons. In BF1, $c_T = 3.0\%$, in BF2, $c_T = 2.2\%$, and in TF1, $c_T = 0.6\%$; $c_T = 0\%$ in the other cross sectional areas. Thus, the magnitude of p varies along the height of the girder according to the tendon concentration c_T.

3.3 *Effective stiffness and strength of the bottom flange*

At the structural level, the bottom flange is analogous to a composite material comprised of two homogeneous phases, the HP2C and the prestressing tendons. In this case, the stiffness of the bottom flange E^B can be assigned an upper bound using the general rule of mixtures:

$$E^B = E_{HP2C} + c_T\,(E_T - E_{HP2C}) \tag{25}$$

where E_{HP2C} is the stiffness of the HP2C and $E_T =$ 200 GPa is the Young's modulus of the tendons. The rule of mixtures represents an assumption of a constant strain field in the flange. Thus, for the tendon-HP2C composite, (25) implies a perfect bond between the HP2C and the prestressing tendons. In the simulations, the stiffness change due to the tendons is enforced through the stiffness of the composite fiber phase in the bottom flange C_F^B, while the composite matrix stiffness is kept constant:

$$C_F^B = C_F + c_T\,(E_T - C_F) \tag{26}$$

The presence of prestressing fibers also affects the composite yield strength of the bottom flange. Based on yield design theory, it can be shown that a lower strength bound of a uniaxially reinforced composite is simply the lower strength of the two composite components [Ulm & Coussy, 2003], in this case:

$$\min\,(\sigma_{HP2C}^u; \sigma_T^u) \leq \Sigma^B \tag{27}$$

where Σ^B is the actual strength of the bottom flange and $\sigma_{HP2C}^u = \Sigma_2 = k_y + f_y$ is the composite yield strength of HP2C. On the other hand, an upper strength bound as given by yield design theory is the weighted average of the composite components [Ulm & Coussy, 2003]:

$$\Sigma^B \leq \sigma_{HP2C}^u + c_T\,(\sigma_T^u - \sigma_{HP2C}^u) \tag{28}$$

The lower strength bound (27) is derived from the assumption of a constant stress field, while the upper strength bound (28) is based on the assumption of an identical strain rate field in the tendons and the HP2C at plastic failure. Finally, to account for the initial prestressing level of 55% of the tendon strength (24), σ_T^u in both (27) and (28) is adjusted:

$$\min\,(\sigma_{HP2C}^u; 0.45\sigma_T^u) \leq \Sigma^B \tag{29}$$

$$\Sigma^B \leq \sigma_{HP2C}^u + c_T\,(0.45\sigma_T^u - \sigma_{HP2C}^u) \tag{30}$$

In the simulations, the upper strength bound (30) is applied to the yield strength of the composite fiber phase in the bottom flange f_y^B, while the composite matrix strength k_y is unchanged:

$$f_y^B \leq f_y + c_T\,(0.45\sigma_T^u - f_y) \tag{31}$$

The parameter values (and associated ductility parameter values) for the HP2C with a tendon effect (26) and (31) are provided in Table 2 ("HP2C with Tendons, BF1" and "HP2C with Tendons, BF2"). As shown in Table 2, the values for σ_{Fc} are also increased to ensure $\sigma_{Fc} > f_y$ and, as a consequence, $\alpha_F^{DP} \geq 0$ (see Eq. (23)), which is a requirement of the Drucker-Prager criterion (10). The new values of σ_{Fc} were chosen to be arbitrarily larger than f_y, but since no compressive yielding occurred in the bottom flange of the FHWA tests, the exact value of this parameter is irrelevant.

For each case study, two different set of material parameters were used, a lower bound set and an upper bound set. The lower bound represents an "HP2C Only" structure. That is, the effect of the prestressing tendons was neglected except for the prestressing load p. In other words, this lower bound neglects any additional dissipative mechanisms related to tendon yielding, friction at the tendon-HP2C interface, etc. In turn, the upper bound represents the maximum possible effect of the prestressing tendons on the stiffness and strength of the bottom flange (Eqs. (26) and (31)). This upper bound overestimates the yielding capacity at failure, as it neglects the possibility of bond failure as a kinematic failure mechanism.

3.4 *Application of loading*

For both case studies, the prestressing pressures and gravity loads were simultaneously applied first. Then external loads were applied in 20 to 25 load steps, graduated such that smaller load steps were applied after the onset of plasticity. For this nonlinear material, iterations were required at each load step to ensure convergence for the magnitude of the vector of residual forces to a tolerance of 0.01. A modified Newton-Raphson method in which the tangent modulus is always fixed to the initial tangent modulus was employed at each iteration step.

4 CASE STUDY: FHWA FLEXURE TEST

For the FHWA flexure test, an AASHTO Type II girder with a 23.9 m long test span was loaded in four point bending with two equal load points (total load P) located 0.9 m from the midspan. The loading configuration and beam geometry for the FHWA flexure test is illustrated in Fig. 5 (Top). Figure 6(a) displays the boundary conditions and load application locations; Figure 6(b) shows the deformed shapes obtained by finite element simulation, and Figure 6(c) the principal plastic strain in the composite matrix.

4.1 *Global results: load-deflections curves*

The load-deflection results from the FHWA experiment and the finite element simulation are graphed in Fig. 7(a). Four different FE results are plotted in the figure: the upper bound with a coarse and fine mesh (composed of respectively 1005 and 2391 elements) and the lower bound with a coarse and fine mesh. The mesh stability of the FE simulations is verified as the coarse and fine mesh for each bound converge upon the same solution. As demonstrated, the simulated upper bound correlates very closely with the actual beam behavior. This is to be expected as the loading configuration induces high tensile stresses and flexure cracks in the bottom flange which activate significant stresses in the tendons.

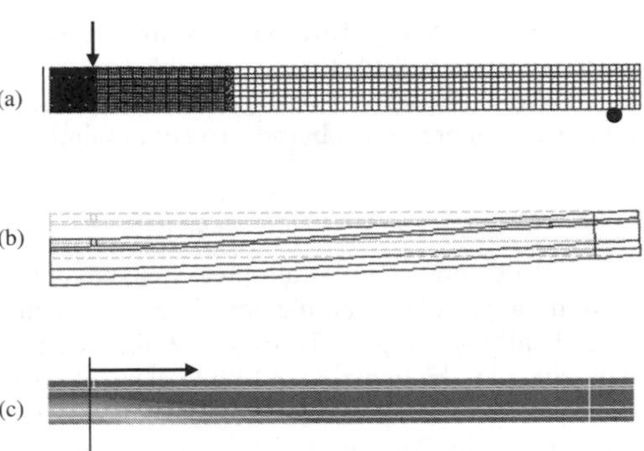

Figure 6. Model-based simulation results of FHWA flexure test: (a) fine mesh with boundary conditions; (b) deformed shape, (c) plastic composite matrix strains.

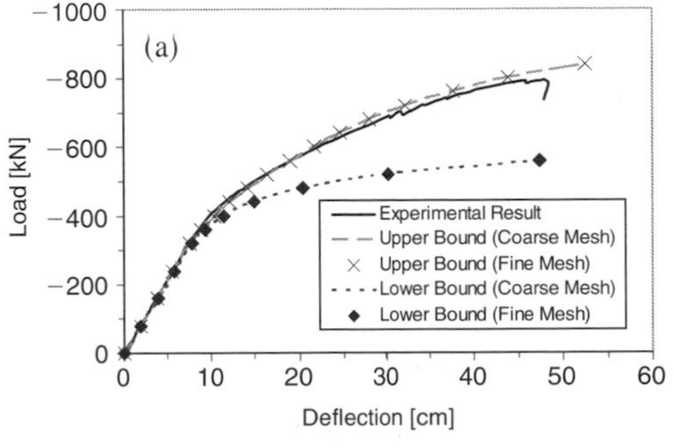

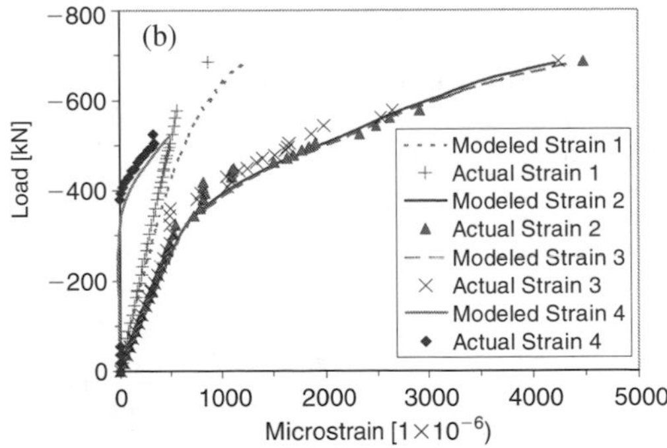

Figure 7. FHWA flexure test: (a) global force-midspan deflection curve; (b) local strain gauge measurements.

The upper bound solution and the experimental result diverge at a load of $P = -800\,\text{kN}$, when the girder showed an abrupt load loss associated with the rupture of the tendons. This confirms the assumption of a tendon stiffness and strength effect as the stresses in the tendons are high enough to induce rupture. Since the FE model does not account for brittle rupture of the prestressing tendons, this abrupt load loss does not appear in the FE simulation, which achieves a much higher ultimate strength. However, despite its inability to directly predict the ultimate strength and subsequent collapse, the upper bound provides a very good load-deflection prediction before structural failure.

4.2 *Local results: strain gauge measurements*

Figure 7(b) compares the strain measurements given by strain gauges placed on the actual beam (oriented longitudinally, see Fig. 5 (Top)) and strain measurements given by the upper bound finite element result (which was shown to give a very good load-deflection prediction in Fig. 7(a)). Both the FE and actual strain measurements are plotted to the load at which the actual strain measurement exhibits discontinuity due to local cracking. As shown, the finite element simulation is also able to reproduce local strain results thus confirming the relevance of the upper bound data set (see Table 2) for predicting the structural behavior both locally and globally.

5 CASE STUDY: FHWA SHEAR TEST

The FHWA shear test consisted of a 4.3 m long AASHTO Type II girder tested in three point bending, as illustrated in Fig. 5 (Bottom). The load P was applied off-center, 1.8 m from one of the supports, in order to induce high shear stresses in the short load span. The model-based simulations were carried out with a coarse mesh (720 elements) and a fine mesh (2878 elements); the latter is displayed in Figure 8(a) along with the applied boundary conditions and loads. The deformed shape for the fine mesh is drawn in Figures 8(b), and Figure 8(c) displays the plastic composite matrix strains.

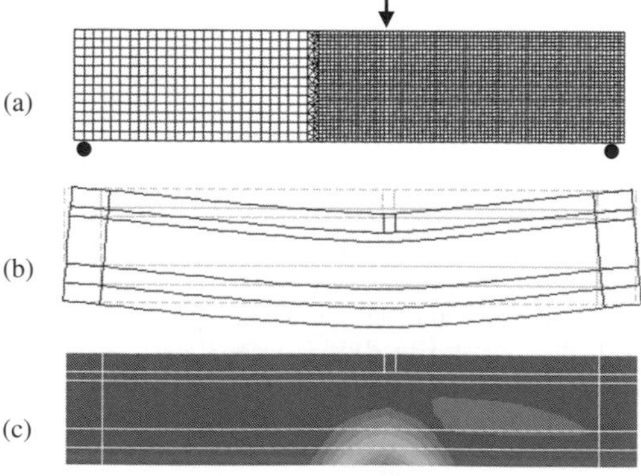

Figure 8. Model-based simulation results of FHWA shear test: (a) fine mesh with boundary conditions; (b) deformed shape, (c) plastic composite matrix strains.

5.1 *Global results: load-deflections curves*

The load-deflection results from the shear experiment and the FE results are plotted in Fig. 9(a). The experimental curve was adjusted to account for the experimetal compliance in the supports determined by Benjamin Graybeal (2002). Again, four different FE results are presented in Figure 9(a) which verify the FE mesh stability: a coarse and fine mesh for the lower bound ("HP2C Only") and a coarse and fine mesh for the upper bound (tendon strength and stiffness added to the bottom flange). The lower bound closely matches with the actual load-deflection behavior. This suggests the insignificant activation of stresses in the prestressing tendons. This supposition was substantiated by the large shear cracks which appeared in the web during pre-peak loading. Thus, it can be interpreted that the nonlinear structural behavior is mainly governed by shear cracking in the (unreinforced) web.

The FE lower bound prediction and the experimental result diverge at a load $P = -3400$ kN, when the FHWA shear beam exhibited structural failure. Nonetheless, the FE lower bound solution appears to approach this same maximum load. It is suggested that this structural failure was a result of bond failure between the tendons and the HP2C in the bottom flange. In support of this bond failure premise is the severe tendon slip that was recorded during the test just

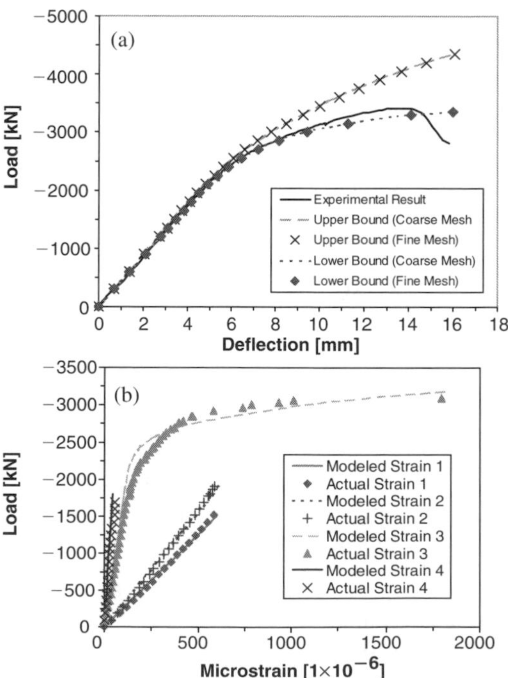

Figure 9. FHWA shear test: (a) global force-midspan deflection curve; (b) local strain gauge measurements.

prior to a deflection of 15 mm, when the girder manifests an abrupt drop in load capacity. Accordingly, the lower bound result manages to predict the shear behavior accurately since a tendon effect never fully develops due to tendon slip. Furthermore, since the beam acted as an "HP2C Only" structure, the lower bound also seems to ably predict the ultimate load capacity of the beam. The bond failure is most likely a consequence of short anchorage lengths (only 2 m on the short load span side) which do not permit proper tendon-HP2C bonding.

5.2 *Local results: strain gauge measurements*

Strain measurements given by strain gauges on the FHWA beam and the lower bound FE simulations are plotted in Figure 9(b), for the strain gauges, depicted in Figure 5 (Bottom). The predicted local strain measurements exhibit very good correlation with the actual local strain readings, thus confirming the relevance of the lower bound solution for simulation of both the local and global response of the FHWA shear girder.

5.3 *Local results: cracking patterns*

The principal composite matrix plastic strains $\varepsilon^p_{M,I}$ for the FHWA shear beam as predicted by the lower bound FE simulations are illustrated in Fig. 8(c). As shown in the figure, there is substantial flexural cracking in the bottom flange under the load point, which was detected in the FHWA shear beam. Focusing on the short load span, significant shear cracking is also evident. Fig. 10(a) and (b) represent magnitude and orientation of the composite matrix plastic strains (or shear crack

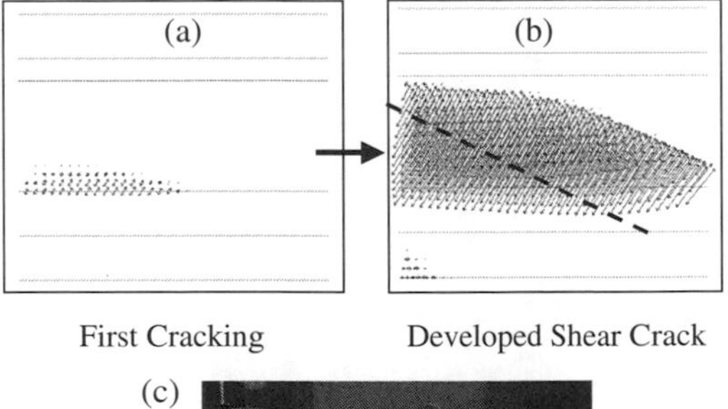

(c)

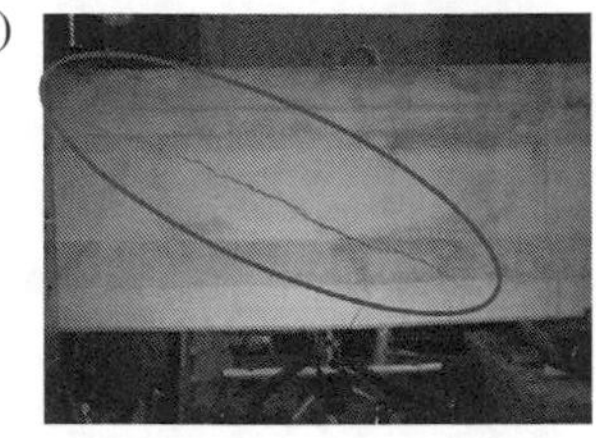

Figure 10. Model-based simulation results of composite matrix cracking: (a) crack initiation at the flange-web interface; (b) fully devloped shear cracks in the web; (c) experimental crack pattern at failure of the shear beam.

opening displacement) in the web. As physically observed, first shear cracking is predicted at the web-flange interface, alongside a large shear crack (see Fig. 10(c)). Furthermore, the fully developed plastic strain vectors suggest the orientation of crack propagation, which is perpendicular to the orientation of the crack openings (crack propagation denoted in Fig. 10(b) with dashed line). The shear cracking angle $\alpha_{cr} \approx 35°$ predicted by the FE simulation also corresponds well with to that of the large shear crack pictured in Fig. 10(c).

6 CONCLUSIONS

This paper outlines a 3-D two-phase material model for HP2C, which exhibits a clear capacity to simulate the behavior of large scale HP2C structures:

1. The model is characterized by 10 parameters of clear physical significance. In particular, the tensile results from a single notched tension test on the HP2C-material are capable of providing all the necessary tensile material parameters to execute an accurate large scale simulation.
2. The finite element simulation is able to reproduce not only global effects, such as the load-deflection behavior of girders, but also local effects, such as strains at various locations on the girder. In addition, by using a two-phase model, which distinguishes matrix cracking from fiber yielding and interface yielding or debonding mechanisms, the cracking patterns can be predicted.
3. Using an upper bound parameter set offers a simple, yet effective way to model the tendon effect in the FHWA flexure test. Conversely, utilizing a lower bound material set to simulate the FHWA shear test proves to be appropriate as tendon slip in the girder compromises the tendon effect. For both FHWA specimens, structural failure was tendon related: tendon rupture in the flexure test and tendon-bond failure in the shear test. While more sophisticated models can be employed for encompassing bond failure into concrete structure simulations [Ulm, 1994; Désir et al., 1999, Ben Romdhane & Ulm, 2002], FE simulations with a lumped tendon-HP2C behavior in the bottom flange seemingly offer accurate estimates for the ultimate loads for both of these tests.

ACKNOWLEDGMENT

This research was performed as part of a research collaboration of MIT with Lafarge-Bouygues-Rhodia Industry with Dr. Paul Acker as Program Director. The authors also gratefully acknowledge the assistance and experimental data provided by the Federal Highway Administration through Joey Hartmann.

REFERENCES

American Association of State Highway and Transportation Officials (AASHTO). (2000). Interim standard specifications for highway bridges, *American Association of State Highway and Transportation Officials,* Washington, D.C.

Ben Romdhane, M.R., and Ulm, F.-J. (2002). "Computational mechanics of the steel-concrete interface." *International Journal for Numerical and Analytical Methods in Geomechanics,* 26, 99–120.

Chuang, E., and Ulm, F.-J. (2002a). "Ductility enhancement of high performance cementitious composites and structures." *MIT-CEE Report R02-02 to the Lafarge Corporation.*

Chuang, E., and Ulm, F.-J. (2002b). "A two-phase composite model for high performance cementitious composites and structures." *Journal of Engineering Mechanics,* ASCE, Vol. 128(12), 1314–1323.

Chuang, E., Overland, M., Ulm, F.-J. (2002). "On length scales of fiber reinforced cementitious composites." *Concrete Science and Engineering,* RILEM, Vol. 4(14), 98–105.

Désir, J.M., Ben Romdhane, M.R., Ulm, F.-J., Fairbairn, E.M.R. (1999). "Steel-concrete interface: revisiting constitutive and numerical modeling." *Computers & Structures,* Vol. 71(5), 489–503.

Federal Highway Administration (FHWA). (2002). "UHPC testing: shear." Handout made available at FHWA shear test.

Graybeal, B. (2002). Personal communication.

Karihaloo, B., Wang, J., and Grzybowski, M. (1996). "Doubly periodic arrays of bridged cracks and short fibre-reinforced cementitious composites." *Journal of the Mechanics and Physics of Solids,* 44(10), 1565–1586.

Karihaloo, B., and Lange-Kornbak, D. (2001). "Optimization techniques for the design of high-performance fibre-reinforced concrete." *Structural and Multidisciplinary Optimization,* 21, 32–39.

Leung, C., and Li, V. (1989). "First cracking strength of short fiber-reinforced ceramics." *Ceram. Eng. Sci. Proc.,* 9/10, 1164–1178.

Ulm, F.-J. (1994). "Modelling of the prestressing of concrete structures: a boundary condition problem of slip at interface." *Computational Modelling of Concrete Structure,* Mang, H. et al. (eds.), Euro-C, Vol. II, 785–794.

Ulm, F.-J., and Coussy, O. (2003). *Mechanics and Durability of Solids, Volume I: Solid Mechanics,* Prentice-Hall, Upper Saddle River, New Jersey.

Computational Modelling of Concrete Structures, Bićanić et al. (eds)
© 2003 Swets & Zeitlinger, Lisse, ISBN 90 5809 536 3

Modeling of static multi-cracking fracture processes in concrete

Rena C. Yu & Gonzalo Ruiz
ETSI de Caminos, C.y P., Universidad de Castilla-La Mancha, Spain

ABSTRACT: This paper analyzes static multi-cracking fracture processes in plain concrete. We use a methodology consisting of combining cohesive theories of fracture together with a novel fragmentation algorithm to model the fracture process. The formation of cohesive surface within the solid is simulated by inserting cohesive elements at the interface between existing elements when the opening condition is attained. Such process generates high geometrical and material nonlinearities which are difficult to handle by direct solvers. As an alternative, we choose an explicit dynamic relaxation method to get the static solution. The code is validated against experimental data from concrete beams loaded at three points. The model closely follows the experimental trends. Particularly, it accurately predicts the load peak for several sizes, thus reproducing the experimentally observed size effect.

1 INTRODUCTION

The dynamic behavior of brittle materials, including glass, ceramics, rocks and concrete, which often involves complex fracture and fragmentation processes, has been thoroughly studied by Ortiz and coworkers (Camacho and Ortiz 1996; Ortiz and Pandolfi 1999; Ruiz et al. 2000; Ruiz et al. 2001). The feasibility of cohesive theories of fracture applied to the dynamic regime has also been demonstrated by (Pandolfi et al. 1999; Yu et al. 2002). However, the modeling of crack propagation within static regime has been hindered by the difficulties to find efficient and stable numerical algorithms which are able to deal with high geometric and material nonlinearities. Implicit solvers such as Newton-Raphson or quasi-Newton schemes may not provide either stable or converged solutions. Explicit dynamic relaxation method, as an alternative, avoids the use of direct solvers and makes it specially attractive in computational mechanics, since all quantities are treated as vectors and the large storage requirements associated with large matrices are avoided.

The idea behind dynamic relaxation is based on viewing the solution of a static problem as the steady-state solution of a damped wave equation. Since the transient part of the solution is not of interest, the corresponding dynamic problem can be chosen in such a way that the rate of convergence to the steady-state is maximized. This is often realized by adding fictitious inertia and damping terms to the real static problem.

First applications of dynamic relaxation method to solve linear structural problems were developed by Otter (Otter 1965) and Day (Day 1965), along with Welsh (Welsh 1967) and Wood (Wood 1967) in the 1960s. More papers (Cassell 1970; Cassell and Hobbs 1976) in the 1970s showed the efficiency and capability to solve complex engineering problems. In early 1980s, Underwood (Underwood and Park 1980) gave an thorough review on the methodology, selection of integration parameters (fictitious mass and damping matrices), as well as an adaptive algorithm to evaluate those parameters for high nonlinear problems. Following the ideas given by Underwood, Sauve and Metzger (Sauve and Metzger 1995) as well as many others showed the reliability of dynamic relaxation method in solving problems involving geometric and material nonlinearities, incompressibility conditions and creep. In the meanwhile, the performance of dynamic relaxation method has also been enhanced by parallel computing (Topping and Khan 1994; Gabbert et al. 1996) as well as convergence accelerators (Dolovich and Brodland 1995). Recently dynamic relaxation method also found its use in molecular dynamics (Pan et al. 2002) and proved itself a powerful solver as well.

We implement the ideas by Underwood for fictitious mass and critical damping coefficients. A combined convergence criteria of kinetic energy and out-of-balance forces is used. Also the computations are enhanced by guessing a critical damping coefficient of the system. Next, a brief introduction of the cohesive model is given; the formulation of the explicit dynamic relaxation method is presented afterwards and the simulations results as well as the comparisons with the experiments are discussed at the end.

2 THE COHESIVE MODEL

For completeness, in this section we summarize the main features of the cohesive law used in the calculations. An extensive account of the theory and its finite-element implementation may be found elsewhere (Ortiz and Pandolfi 1999; Camacho and Ortiz 1996). A simple class of mixed-mode cohesive laws accounting for tension-shear coupling (Camacho and Ortiz 1996; Ortiz and Pandolfi 1999; De Andrés et al. 1999), is obtained by the introduction of an effective opening displacement δ, which assigns different weights to the normal δ_n and sliding δ_S opening displacements.

$$\delta = \sqrt{\beta^2\delta_S^2 + \delta_n^2} \tag{1}$$

Assuming that the cohesive free-energy density depend on the opening displacements only through the effective opening displacement δ, a reduced cohesive law, which relates δ to an effective cohesive traction

$$t = \sqrt{\beta^{-2}|t_s|^2 + t_n^2} \tag{2}$$

where t_s and t_n being the shear and the normal tractions, respectively, can be obtained (Camacho and Ortiz 1996; Ortiz and Pandolfi 1999). It is observed that the weighting coefficient β defines the ratio between the shear and the normal critical tractions (and thus it also roughly defines the ratio of K_{IIc} to K_{Ic}). Assuming the existence of a loading envelope defining a relation between t and δ under the conditions of monotonic loading, and that unloading is irreversible, a simple and convenient type of irreversible cohesive law is furnished by the linearly decreasing envelope shown in Figure 1, where σ_c is the tensile strength and δ_c the critical opening displacement.

Cohesive theories introduce a well-defined length scale into the material description and, in consequence, are sensitive to the size of the specimen (see, for example (Bažant and Planas 1998)). The characteristic length of the material may be expressed as

$$\ell_c = \frac{EG_c}{f_{ts}^2} \tag{3}$$

where G_c is the fracture energy and f_{ts} the static tensile strength.

In calculations, only decohesion along element boundaries is allowed to occur. Initially, all element

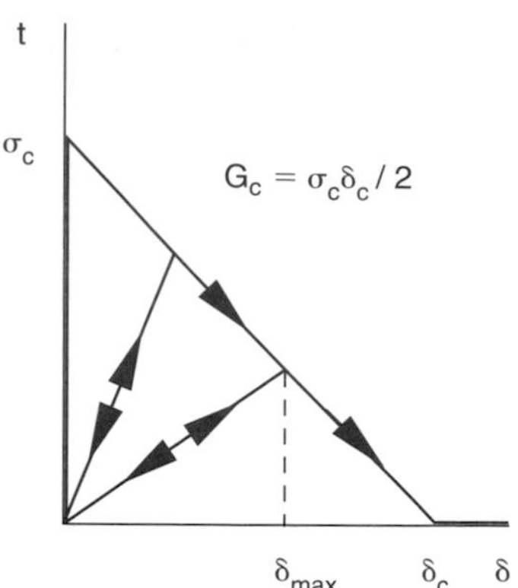

Figure 1. Linear irreversible cohesive law.

boundaries are perfectly coherent and the elements conform in the usual sense of the displacement finite element method. When the critical cohesive traction is attained at the interface between two volume elements, a cohesive element is inserted at that location using a fast algorithm (Pandolfi and Ortiz 2002). The cohesive element subsequently governs the opening of the cohesive surface.

3 FORMULATION OF THE EXPLICIT DYNAMIC RELAXATION METHOD

As we mentioned before, in calculations, the fracture surface is confined to inter-element boundaries and, consequently, the structural cracks predicted by the analysis are necessarily rough. Even though this numerical roughness in concrete can be made to correspond to the physical roughness by the simple device of choosing the element size to resolve the cohesive zone size (Ruiz et al. 2001), the nonlinearity of the solution thus induced plus the material nonlinearity is difficult to handle in static regime for traditional solvers. We choose explicit dynamic relaxation method as an alternative to tackle this situation, the formulation of this methodology is given below.

Consider the semi-discrete governing equation for a static problem at a certain step n

$$F_n^{int}(x_n) = F_n^{ext} \tag{4}$$

where x_n is the solution array (displacements), F_n^{int} and F_n^{ext} are internal and external force vectors. Following the ideas of dynamic relaxation, Eqn. 4 is transformed to a dynamic system by adding both inertia and damping terms.

$$M\ddot{x}_n + D\dot{x}_n = R_n \tag{5}$$

where M and D are the fictitious mass and damping matrices, $\ddot{x}_n$ and $\dot{x}_n$ are acceleration and velocity vectors respectively, $R_n = F_n^{ext} - F_n^{int}(x_n)$ is the out-of-balance residual forces at load step n.

The solution of Eqn. (5) is obtained by the explicit time integration method using the central difference integration scheme in two steps.

First the displacements and predictor velocities are obtained

$$x_n^{t+1} = x_n^t + \Delta t \dot{x}_n^t + 1/2 \Delta t^2 \ddot{x}_n^t \quad (6)$$

$$\dot{x}_n^{t+\Delta t/2} = \dot{x}_n^t + 1/2 \Delta t \ddot{x}_n^t \quad (7)$$

Second we update the internal force vector and obtain the accelerations and corrected velocities.

$$\ddot{x}_n^{t+1} = (M + 1/2 \Delta t D)^{-1} \quad (8)$$

$$[R(x_n^{t+1}) - D\dot{x}_n^{t+\Delta t/2}] \quad (9)$$

$$\dot{x}_n^{t+1} = \dot{x}_n^{t+\Delta t/2} + 1/2 \Delta t \ddot{x}_n^{t+1} \quad (10)$$

In order to preserve the explicit form of the time-stepping integrator, both fictitious mass M and damping D matrices are set to be diagonal. At the same time, the damping matrix is chosen to be proportional to $M, D = \xi M, \xi$ is the damping coefficient to be determined.

3.1 *Evaluation of the fictitious mass matrix*

As aforementioned, the dynamic relaxation parameters can be selected to produce fastest and most stable convergence to the static solution of the real physical system. These parameters includes the mass matrix M, the damping coefficient ξ and time step Δt.

Owing to the explicit formulation, the time step must satisfy the Courant-Fridricks-Lewy stability condition, $\Delta t \leq 2/\omega_{max}$, where ω_{max} is the highest undamped frequency of the discretized system, which can be estimated as $\omega_{max}^e = 2c_e/h_e$, where c_e is the dilationational wave speed of the material ($c_e = \sqrt{(\lambda + 2G)/\rho_e}$), h_e represents the size of the element. This condition provides a relation between the maximum admissible $\Delta t_{cr} = 2/\omega_{max}$ and the fictitious mass matrix.

$$\rho_e \geq (\lambda + 2G) \left(\frac{\Delta t_{cr}}{h_e} \right)^2 \quad (11)$$

3.2 *Estimation of the fictitious damping matrix*

In Underwood (Underwood 1983), the convergence rate of dynamic relaxation is given in terms of the spectral radius of the iterative error equations

$$R \approx \left| 1 - 2\frac{\omega_{min}}{\omega_{max}} \right| \quad (12)$$

where ω_{min} and ω_{max} are the lowest and highest frequencies of the discretized equations of motion. By maximizing the ratio $\omega_{min}/\omega_{max}$, and therefore minimizing R, a faster convergence rate can be obtained. The highest frequency ω_{max} is mesh and material dependent, whereas ω_{min} is based on the lowest participating mode of the structure corresponding to the load distribution. For models where the mode of loading corresponding to a very low structural frequency and the mesh discretization is such that the local element frequencies are high, a slower convergence rate can be expected (Sauve and Metzger 1995). By adjusting the mass matrix and choosing a suitable time step, this effect can be minimized. In the present work, the scheme to estimate the critical damping coefficient suggested by Underwood is implemented. The current value of ω_{min} is estimated at each iteration t using Rayleigh quotient

$$\omega_{min}^t = \sqrt{\frac{(\omega^t)^T K^t \omega^t}{(\omega^t)^T M^k \omega^t}} \quad (13)$$

where ω^t stands for the eigenvector associated to ω_{min}^t and M is the mass matrix at tth iteration. For nonlinear problems, K represents a diagonal estimate of the tangent stiffness matrix at the tth iteration and is given by

$$K_{ii}^t = \frac{F_i^{int}(x_n^t) - F_i^{int}(x_n^{t-1})}{x_n^t - x_n^{t-1}} = \frac{R^{t-1} - R^t}{x_n^t - x_n^{t-1}} \quad (14)$$

where R^t and R^{t-1} are the residual forces at iteration t and $t-1$.

3.3 *Convergence criteria*

In dealing with convergence problems, it is highly desirable to choose a criteria that will neither lead to unproductive cycles nor inaccurate results. In this work, taking into consideration the specific problem we are trying to solve, a combined convergence criteria is implemented. One is the ratio between the residual forces and the estimated maximum external forces, the measure that says how far is the solution from equilibrium, the other is the global kinetic energy, which measures whether the loading is static or not. This combined criteria is characterized by the following inequalities

$$\frac{\| R \|_2}{\| F_{ext} \|_2} < f_{tol} \quad (15)$$

$$\sum M \dot{x}^{t2} < ke_{tol} \quad (16)$$

where $\| \cdot \|_2$ denotes the Euclidean norm. The tolerances f_{tol} and ke_{tol} are taken as 0.001 and $10^{-8}\, N \cdot mm$

respectively. For the problems we have considered, these tolerances provide a good balance of accuracy and efficiency. The algorithm as implemented is summarized in Table 1, where ξ^0 is the damping coefficient computed in the program after the first insertion of the cohesive element takes place, or when the non-linearity of the material started to emerge. By setting the damping coefficient to this value when the solution is far away from equilibrium, while using the critical damping coefficient computed when the solution is near static, the global convergence rate is enhanced by a factor of more than 10.

Table 1. Explicit dynamic relaxation algorithm.

1. Get F_{int} from initial condition and initialize M for $\Delta t = 1.01$ for each element.
2. At iteration t (i) compute displacements at $t+1$ and velocities at $t + \Delta t/2$

$$x_n^{t+1} = x_n^t + \Delta t \dot{x}_n^t + 0.5 \Delta t^2 \ddot{x}_n^t$$

$$\dot{x}_n^{t+\Delta t/2} = \dot{x}_n^t + 0.5 \Delta t \ddot{x}_n^t$$

(ii) Update internal forces $F_n^{int}(x^t)$ and calculate residuals

$$R^t = F_n^{ext} - F_n^{int}(x^t) \tag{17}$$

(iii.1) If error norm 1 (out-of-balance residual forces) is larger than $1.1 f_{tol}$, set $\xi^t = \xi^0$, go to (iv); otherwise go to (iii.2).
(iii.2) Evaluate current damping coefficient ξ^t

$$\Delta x_n = x_n^t - x_n^{t-1}$$

$$\omega_{min}^t = \frac{(\Delta x_n^t)^T (R^t - R^{t-1})}{(\Delta x_n^t)^T M^t (\Delta x_n^t)}$$

$$\xi^t = 2\omega_{min}^t$$

(iv) Compute accelerations and velocities at $t+1$

$$\ddot{x}_n^{t+1} = (M + 1/2 \Delta t D)^{-1} [R(x_n^{t+1}) - D \dot{x}_n^{t+\Delta t/2}]$$

$$\dot{x}_n^{t+1} = \dot{x}_n^{t+\Delta t/2} + 1/2 \Delta t \ddot{x}_n^{t+1}$$

(v) Check error norm

$$\frac{||R||_2}{||F_{ext}||_2} < f_{tol}$$

$$\sum M(\dot{x}_n^t)^2 < ke_{tol}$$

If satisfied move to the next load step $n+1$
(vi) Otherwise, go to (i) and set $t = t+1$.

4 NUMERICAL ANALYSIS

We apply the dynamic relaxation method to solve the static propagation of a crack through a notched concrete beam subjected to three point bending, see Figure 2. Two specimens, one with depth $D = 75$ mm (specimen a), the other one with depth $D = 150$ mm (specimen b), both with the same thickness $B = 50$ mm are modeled. Both specimens are discretized into ten-node quadratic tetrahedral elements and have element size of 6 mm near the middle surface. The material paramerters are given in Table 2.

4.1 *Mesh dependency*

In previous studies, Camacho and Ortiz (Camacho and Ortiz 1996) have noted that the accurate description of fracture processes by means of cohesive elements requires the resolution of the characteristic cohesive length of the material. Ruiz et al. pointed out that in concrete, the element size can be made to be comparable to the maximum aggregate size, which is 5 mm in this case. Here, we compare the load curve results from a coarse mesh 1 (element size 6 mm, 3556 nodes, 2103 tetrahedron) and a finer mesh 2 (element size 3 mm, 6283 nodes, 3950 tetrahedron) in Figure 4. The loading path from mesh 1 and 2 as well the peak load are in good agreement, while the computations for mesh 1 would be much more efficient, so the following simulations are based on this mesh for specimen a and mesh of similar size for specimen b.

4.2 *Peak load comparison*

According to the work of Ruiz (Ruiz 1998), the peak load is directly related to the initial slope of the cohesive law of the material. Since our goal is to test

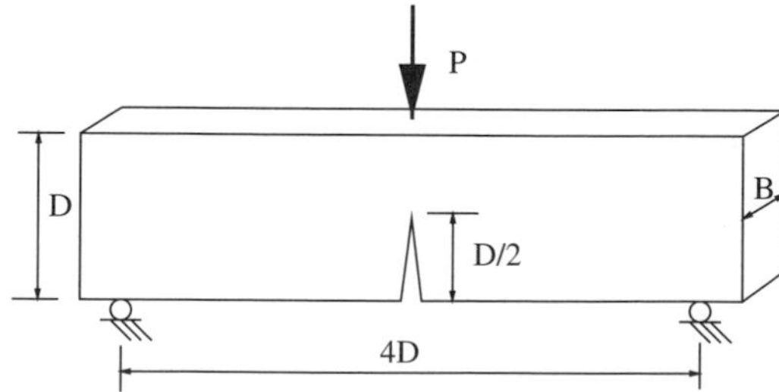

Figure 2. A concrete beam subjected to three point bending.

Table 2. Concrete mechanical properties.

f_t (MPa)	E (GPa)	G_c (N/m)	l_{ch} (mm)
3.8	30.5	62.5	130

the newly implemented dynamic relaxation method, a simple cohesive law depicted in Figure 1 is applied. The fracture paramaters are chosen to meet the initial slope (the ratio of f_{ts}/δ_c in Figure 1) suggested by Ruiz (Ruiz 1998) and the static tensile strength is according to the experiments, i.e., $f_{ts} = 3.8$ MPa, $\delta_c = 14.5\,\mu$m. This gives a fracture energy (27.55 N/m) smaller than in the experiments (62.5 N/m), which we plan to correct with a more complicated cohesive law in future work. The load versus displacement curve for specimen a and b, compared with the experimental results are shown in Figure 5. An agreement of 0.4% for specimen a and 3.0% for specimen b is achieved.

Since specimen a and b are scaled in size, we also set the load and displacement curve for Figure 5a and b proportional, the experimentally observed size effect is also seen in the simulations.

(a)

(b)

Figure 3. The meshes used in the simulations for $D = 75$ mm; element size at middle surface (a) 6 mm; (b) 3 mm.

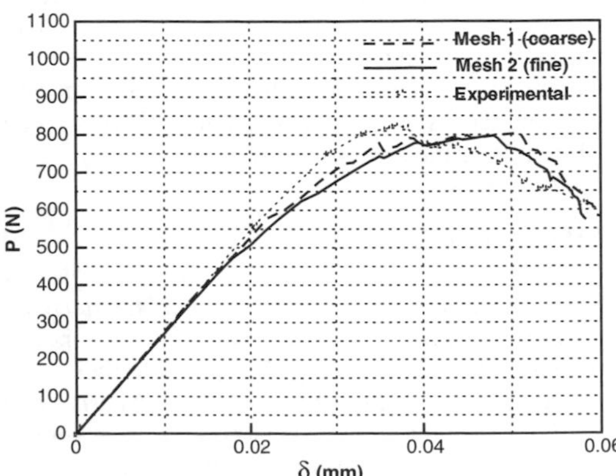

Figure 4. The load curve comparison for mesh 1 and 2.

4.3 *Fracture patterns*

Three snapshots of the fracture patterns on mid-plane for specimen a is shown in Figure 6, where the displacements have been magnified 100 times to aid visualizaton. Figure 6a is the point of peak load, Figure 6c is the moment where the beam is almost completely crushed, while Figure 6b is a point in between which shows how the fracture zone has developed. Also shown in the figures are the level contours of damage, defined as the fraction of expended fracture energy to total fracture energy per unit surface, or critical energy

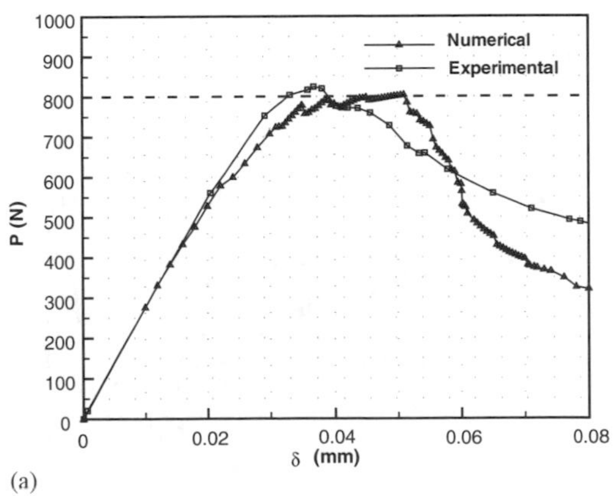

(a)

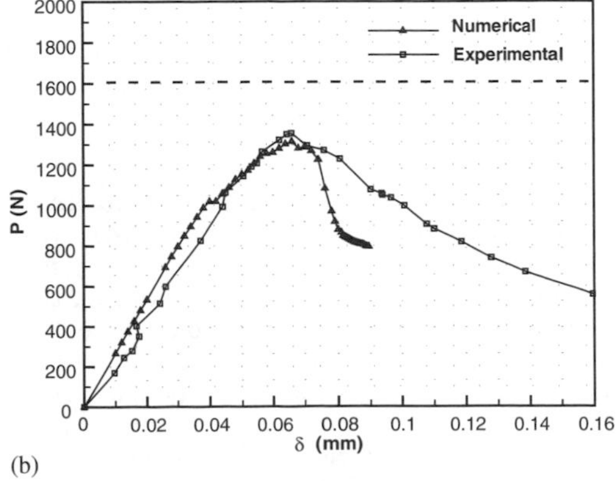

(b)

Figure 5. The peak load comparison for specimen a and b.

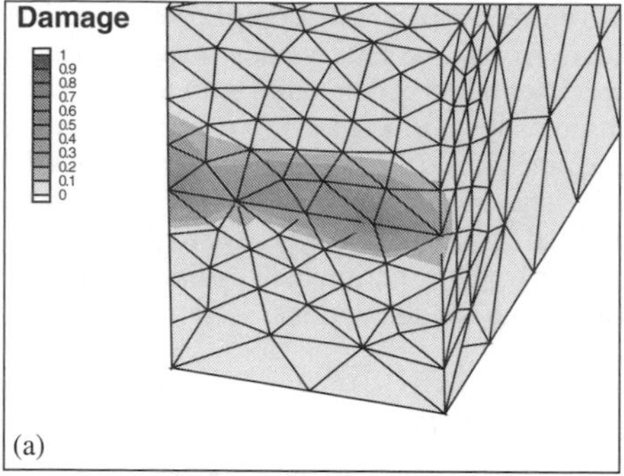

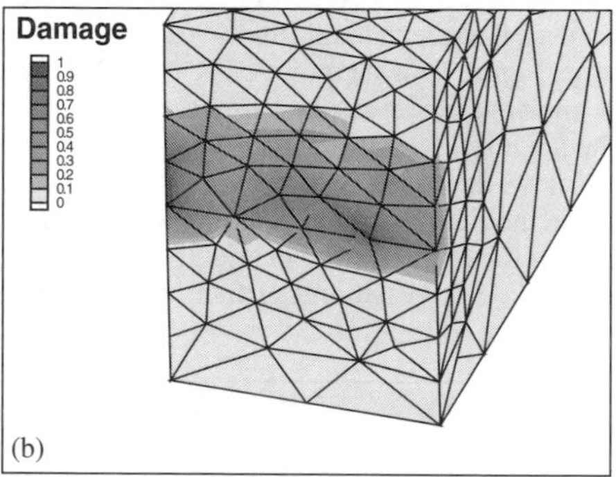

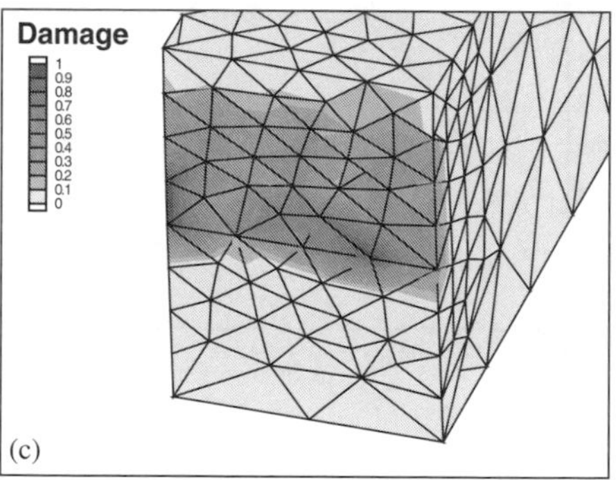

Figure 6. Snapshots of the fracture patterns developed at loading displacement (a) 0.05 mm(near peak load); (b) 0.07 mm; and (c) 0.11 mm respectively.

release rate. Thus, a damage density of zero denotes an uncracked surface, whereas a damage density of one is indicative of a fully cracked or free surface. The transition zone wherein the damage variable takes intermediate values may be regarded as the cohesive zone, and the crack front may conventionally be identified with the level contour of 1/2. It can be noticed that in the peak load, Figure 6a, the fracture zone has developed to some degree, only that the crack surfaces are not fully open yet; later on in Figure 6b, the same zone is more developed while new surfaces are open and the crack front propagates in a nonuniform way, which can only be observed in a full three-dimensional modeling. It is interesting to note that the crack front is convex in the direction of propagation, a feature which is characteristic of mode-I crack growth, the exterior of the crack front ostensibly lags behind the interior points. In Figure 6c, the crack continues to grow till the specimen is almost completely crushed and loses its strength.

5 SUMMARY AND CONCLUSIONS

We have put together the explicit dynamic relaxation method in conjunction with the cohesive theory to solve the static multi-cracking fracture process in a three-point-bend concrete beam. In calculations, the fracture surface is confined to inter-element boundary elements and, consequently, the structural cracks predicted by the analysis are necessarily rough. Even though, this numerical roughness in concrete can be made to correspond to the physical roughness by choosing the element size comparable to the aggregate size, thus-induced geometrical nonlinearity and the material nonlinearity inherent to concrete are hard to handle for traditional static solvers. Explicit dynamic relaxation method, as an alternative, avoids the use of direct solvers as well as the large storage requirements associated with large matrices, makes it specially attractive in computational mechanics (Sauve and Metzger 1995). We follow the ideas of Underwood (Underwood 1983) for fictitious mass and damping matrices and implemented the explicit dynamic relaxation method. An initial damping coefficient estimated from the system is adopted to enhance the convergence when the solution is far away from equilibrium. A combined convergence criterion of relative residual forces and global kinetic energy is chosen to provide a good balance of accuracy and efficiency. Two sizes of concrete beams were modeled to validate against the experimental results of Ruiz (Ruiz 1998). The preliminary results show the model captures the peak load accurately. Explicit dynamic relaxation method seems promising as an viable alternative in providing an efficient and accurate solutions to static problems when high nonlinearities are involved.

ACKNOWLEDGMENTS

Rena C. Yu and Gonzalo Ruiz thank the *Ministerio de Educacin, Cultura y Deporte*, Spain, for the scholarship SB2000-0191, which makes possible their joint work at the *ETSI de Caminos, C. y. P., Universidad de Castilla-La Mancha* (UCLM). Gonzalo Ruiz acknowledges financial support from the *Ministerio de Ciencia y Tecnologa*, Spain, under grant MAT2000-0705, and the UCLM, Spain, under grant 011,1075.

REFERENCES

Bažant, Z. P. and J. Planas (1998). *Fracture and Size Effect in Concrete and Other Quasibrittle Materials*. Boca Raton, Florida: CRC Press.

Camacho, G. T. and M. Ortiz (1996). Computational modelling of impact damage in brittle materials. *International Journal of Solids and Structures 33* (20–22), 2899–2938.

Cassell, A. (1970). Shell of revolution under arbitrary loading and the use of ficticious densities in dynamic relaxation. *Proc. Inst. Civ. Engrs. 45*, 65–78.

Cassell, A. and R. Hobbs (1976). Numerical stability of dynamic relaxation structural analysis. *International Journal for Numerical Methods in Engineering 10*, 1407–1410.

Day, A. (1965). An introduction to dynamic relaxation. *The Engineer 219*, 218–221.

De Andrés, A., J. L. Pérez, and M. Ortiz (1999). Elastoplastic finite-element analysis of three-dimensional fatigue crack growth in aluminum shafts subjected to axial loading. *International Journal of Solids and Structures 36*(15), 2231–2258.

Dolovich, A. and G. Brodland (1995). Convergence accleration for iterative finite-element method. *Journal of Engineering Mechanics, ASCE 121*(1).

Gabbert, U., J. Grochla, and H. Koppe (1996). Dynamic-explicit finite element simulation of complex problems in civil engineering by parallel computing. *Internationales Kolloquium über Anwendungen de Informatik und Mathematik in Architektur und Bauwesen 9*(1–4), in press.

Ortiz, M. and A. Pandolfi (1999). Finite-deformation irreversible cohesive elements for three-dimensional crack-propagation analysis. *International Journal for Numerical Methods in Engineering 44*, 1267–1282.

Otter, J. (1965). Computations for prestressed concrete reactor pressure vessels using dynamic relaxation. *Nucl. Struct. Engng. 1*, 61–75.

Pan, L., D. Metzger, and M. Niewczas (2002). The meshless dynamic relaxation techniques for simulating atomic structures of materials. *To appear*.

Pandolfi, A., P. Krysl, and M. Ortiz (1999). Finite element simulation of ring expansion and fragmentation. *International Journal of Fracture 95*, 279–297.

Pandolfi, A. and M. Ortiz (2002). An Efficient Adaptive Procedure for Three-Dimensional Fragmentation Simulations. *Engineering with Computers 18*(2), 148–159.

Ruiz, G. (1998). *Influencia del Tamao y de la Adherencia en la Armadura minima de Vigas en Flexion*. Madrid, Spain: Grupo Espanol del Hormigon.

Ruiz, G., M. Ortiz, and A. Pandolfi (2000). Three-dimensional finite-element simulation of the dynamic Brazilian tests on concrete cylinders. *International Journal for Numerical Methods in Engineering 48*, 963–994.

Ruiz, G., A. Pandolfi, and M. Ortiz (2001). Three-dimensional cohesive modeling of dynamic mixed-mode fracture. *International Journal for Numerical Methods in Engineering 52*, 97–120.

Sauve, R. G. and D. Metzger (1995). Advances in dynamic relaxation techniques for nonlinear finite element analysis. *Transactions of the ASME 117*, 170–176.

Topping, B. and A. Khan (1994). Parallel computation schemes for dynamic relaxation. *Engineering Computations 11*, 513–548.

Underwood, P. (1983). Dynamic relaxation. *Computational Methods for Transient Analysis 1*, 145–265.

Underwood, P. and K. C. Park (1980). A variable-step central difference method for structural dynamics analysis – part 2. implementation and performance evaluation. *Transactions of the ASME 23*, 259–279.

Welsh, A. (1967). Discussion on dynamic relaxation. *Proc. Inst. Civ. Engrs. 37*, 723–750.

Wood, W. (1967). Comparison of dynamic relaxation with three other iterative methods. *Engineer 224*, 683–687.

Yu, C., A. Pandolfi, D. Coker, M. Ortiz, and A. Rosakis (2002). Three-dimensional Modeling of Intersonic Shear-CRack Growth in Asymmetrically-Loaded Unidirectional Composite Plates. *International Journal of Solids and Structures 39*(25), 6135–6157.

Constitutive modelling

Computational Modelling of Concrete Structures, Bićanić et al. (eds)
© 2003 Swets & Zeitlinger, Lisse, ISBN 90 5809 536 3

Keynote paper: Non local damage model with evolving internal length based on micromechanics

K. Haidar & G. Pijaudier-Cabot
R&DO, Laboratoire de Génie Civil de Nantes Saint Nazaire, Ecole Centrale de Nantes, Nantex cedex, France

J.F. Dubé
LMGC/UMR5508 – Université de Montpellier II, Place Eugène Batailoon, Montpellier cedex, France

ABSTRACT: A modified non local damage model with evolving internal length based on micromechanics is developed in this contribution. It is shown in particular that the non local influence between two points depends on the value of damage at each point. The resulting weight function is non symmetric and truncated, which prevents a widening of the FPZ in the course of failure. The finite element implementation on a one dimensional problem is presented, focusing on the variation of the weight function. Comparisons with experimental data on model materials with controlled porosity are recalled. Acoustic emission analyses provide results with which the theoretical model is consistent qualitatively. This model opens also the path for durability mechanics analyses where it has been demonstrated that the internal length in the non local model should change.

1 INTRODUCTION

Non local continuum damage, in an integral or gradient format, is a consistent general concept for macroscopic modelling of fracture in quasi-brittle materials such as concrete, rocks, ceramics and fibre composites. The non local aspect is a requisite for a realistic description of fracture, including crack inception, crack propagation, and also size effect which is a consequence of the existence of a finite size fracture process zone.

Among the governing parameters in the non local damage model, the internal length plays a pivotal role as it controls the size of the fracture process zone. This internal length can be obtained from size effect experiments with the help of inverse analysis (Le Bellégo et al. 2002a). In most existing proposals, the internal length is a constant parameter. In fact, one could say that it is already so difficult to measure it with a relatively good accuracy that considering a variation of this parameter can hardly be envisioned without solid experimental evidences.

There are some theoretical indications, however, which suggest that the internal length should change in the course of the fracture process. With the help of micromechanics, Bazant (1990, 1992) has proposed a description of the interactions between cracks and voids in the course of failure. The resulting weight function, with which the non local damage variable is derived, should evolve during failure. Geers has observed that the internal length should change in order to describe fracture with a displacement discontinuity across the crack faces (Geers et al. 1998). If the introduction of an internal length in the constitutive relations results from non local interactions, there is also a simple reasoning which tells that without cracks there is no interaction (the internal length should be zero). In the course of the degradation process, interactions grow and then the internal length should evolve at the same time.

Recent experiments on model cementitious materials with controlled initial porosity, i.e. initial damage, have attempted to capture the evolution of the size of the fracture process zone with growing damage and growing initial porosity (Haidar 2002). Acoustic emission analyses exhibit clearly an increase of the fracture process zone with initial porosity/damage (Haidar and Pijaudier-Cabot, 2002).

This paper is an attempt to integrate these observations in a modified non local damage model. In the first part, the variation of the internal length is derived from simplified micromechanics of interacting voids. It is shown that interaction is proportional to the state of damage at each point. A modified weight function is introduced into the volume integration of the non local damage model. The non local influence of one point on another depends on the state of damage at the first one, it does not exist without damage. The constitutive model results into a formulation with an evolving internal length. The consequences on the inception of strain localisation are considered numerically with the help of a one dimensional example. Experimental results on model materials with initial porosity are recalled. They are compared, qualitatively, with the proposed new non local model in which the influence of initial damage can be also introduced. The differential approximation of the non local model is derived. It provides a slightly different formulation compared to the model proposed by Geers.

2 ORIGINAL NON LOCAL DAMAGE MODEL

In the following, we recall the general concept of the non local damage model (Mazars 1984, Pijaudier-Cabot et al. 1987), from which we will derive the modified non local damage model with evolving internal length. The scalar isotropic damage model proposed by Mazars (1984) is chosen. The stress–strain relation is:

$$\sigma_{ij} = (1-d)\Lambda_{ijkl}\varepsilon_{kl} \quad (1)$$

where σ_{ij} and ε_{ij} are the components of the stress and strain tensors respectively (i, j, k, l $\in$ [1, 3]), Λ_{ijkl} are the initial stiffness moduli, and d is the damage variable which varies between 0 and 1. The material is initially isotropic, with E and ν the initial Young's modulus and Poisson's ratio respectively. For the purpose of defining damage growth, the equivalent strain is introduced first:

$$\tilde{\varepsilon} = \sqrt{\sum_{i=1}^{3}\langle\varepsilon_i\rangle_+^2} \quad (2)$$

where $\langle\rangle_+$ is the Macauley bracket and ε_i are the principal strains. The non-local variable $\bar{\varepsilon}$ is defined:

$$\bar{\varepsilon}(x) = \frac{1}{V_r(x)}\int_\Omega \psi(x-s)\tilde{\varepsilon}(s)ds$$
$$\text{with } V_r(x) = \int_\Omega \psi(x-s)ds \quad (3)$$

where Ω is the volume of the structure, $V_r(x)$ is the representative volume at point x, and $\Psi(x-s)$ weight function:

$$\psi(x-s) = \exp\left(-\frac{4\|x-s\|^2}{l_c^2}\right) \quad (4)$$

l_c is the internal length of the non local continuum. It appears in the analytical formula of the weight function Ψ, and defines the size of the non-local strain averaging support. Previous studies have shown that the internal length l_c can be correlated to the heterogeneity the material. It is generally assumed that $l_c = 3*da$, where da is the maximum size of aggregate. l_c is considered as a constant material parameter.

$\bar{\varepsilon}$ is the variable that controls the growth of damage according to the following conditions:

$$F(\bar{\varepsilon}) = \bar{\varepsilon} - \kappa \quad (5)$$

$$\text{if } F(\bar{\varepsilon}) = 0 \text{ and } \dot{\bar{\varepsilon}} > 0$$

$$\text{then} \begin{cases} \dot{d} = h(\kappa) \\ \dot{\kappa} = \dot{\bar{\varepsilon}} \end{cases} \text{with the condition } \dot{d} \geq 0 \quad (6)$$

$$\text{else} \begin{cases} \dot{d} = 0 \\ \dot{\kappa} = 0 \end{cases}$$

κ is the softening parameter and takes the largest value of $\bar{\varepsilon}$ ever reached during the previous loading history at a given time and at the considered point in the medium. Initially $\kappa = \kappa_0$, where κ_0 is the threshold of damage. The damage evolution law h(κ) in (6), is defined in an integrated form. In particular, the damage growth in tension is:

$$d = 1 - \frac{\kappa_0(1-A_t)}{\kappa} - \frac{A_t}{\exp(B_t(\kappa - \kappa_0))} \quad (7)$$

where constants A_t, B_t are model parameters which must be obtained from a fit of experimental data.

3 SIMPLIFIED MICROMECHANICS OF VOID INTERACTION

Consider an infinite isotropic two dimensional solid (E_0, ν_0) subjected to remote uniform boundary traction producing a uniform stress field σ_∞. Our purpose is essentially qualitative, then we can consider, for more simplicity, σ_∞ as isotropic. This solid contains many microcracks caused by the loading history that are arbitrary distributed in the matrix of material. Each crack induces a modification of the local stress and strain fields. In order to evaluate this correction, we assume that the microcracks are far enough from each other. The induced stress and strain fields are similar to those generated by circular voids. Therefore, cracks are going to be replaced by circular voids in the following derivations.

3.1 *Interaction of two circular voids in an elastic medium*

Consider the medium described in figure 1. The two inclusions denoted as S_1 and S_2 are of diameter a_1

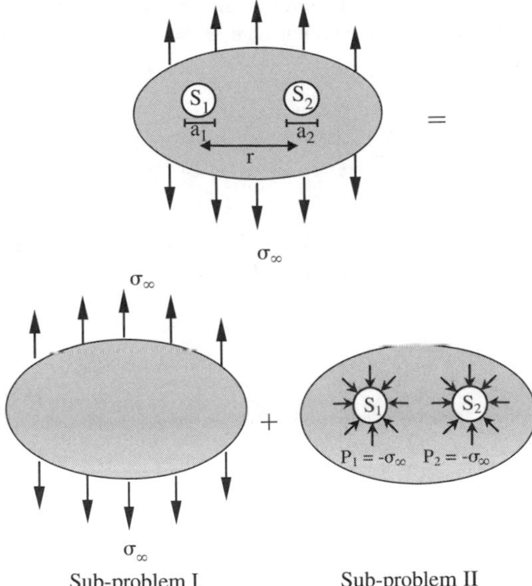

Figure 1. Superposition scheme for an isotropic solid with a two circular voids.

and a_2 respectively. Exploiting the principle of superposition (Kachanov 1987), we may decompose this problem into two sub-problems:

- Sub-problem I: the solid is considered without any inclusions and loaded by the remote traction corresponding to σ_∞.
- Sub-problem II: the remote traction σ_∞ is transformed by distributed normal and tangential forces $P_1 = P_2 = -\sigma_\infty$ acting inside each inclusion.

Consider now sub-problem II. Again, we can apply the principle of superposition (Figure 2), in order to compute the interaction stress field due to the presence of the inclusion, and subsequently the distribution of internal pressure on the crack faces.

In sub-problem II-1, the inclusion S_1 is considered alone. It is loaded by an unknown pressure P_1 which cause on the imaginary location of S_2 a normal pressure p_{21}. We have the same sub-problem II-2. To evaluate P_1 and P_2, we can write (by super-position):

$$P_1 = \overline{P_1} + p_{12} \qquad P_2 = p_{21} + \overline{P_2} \tag{8}$$

The interaction pressure have the following forms (Bazant, 1992):

$$p_{12} \propto -\overline{P_2} \times \frac{a_2^2}{r^2} \text{ and } p_{21} \propto -\overline{P_1} \times \frac{a_1^2}{r^2} \tag{9}$$

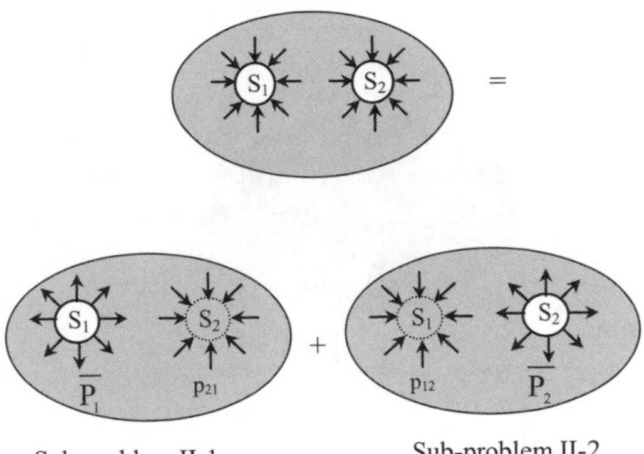

Sub-problem II-1 Sub-problem II-2

Figure 2. Superposition scheme in Subproblem II.

with the notations $l_{12} = \frac{a_2}{r}$ and $l_{21} = \frac{a_1}{r}$. We have:

$$\begin{bmatrix} P_1 \\ P_2 \end{bmatrix} = \begin{bmatrix} 1 & -l_{12}^2 \\ -l_{21}^2 & 1 \end{bmatrix} \begin{bmatrix} \overline{P_1} \\ \overline{P_2} \end{bmatrix} \tag{10}$$

Assuming $\sigma_\infty = 1$, the solution of this system of two equations is:

$$\overline{P_1} = -\left(\frac{1 + l_{12}^2}{1 - l_{21}^2 * l_{12}^2}\right) \text{ and } \overline{P_2} = -\left(\frac{1 + l_{21}^2}{1 - l_{21}^2 * l_{12}^2}\right) \tag{11}$$

with $r \gg a_1, a_2$ we can write :

$$\overline{P_1} \approx -\left(1 + \frac{a_2^2}{r^2}\right) \text{ and } \overline{P_2} \approx -\left(1 + \frac{a_1^2}{r^2}\right) \tag{12}$$

Hence, a part of the interaction between the two inclusions is neglected because they are far away from each other.

3.2 *Interaction of n circular voids in an elastic medium*

We consider the same medium as in the previous paragraph with n circular voids S_i $(i = 1 - n)$, of diameter a_i, arbitrary distributed in the matrix. In order to calculate the stress and displacement field for this problem, we use the same superposition method as the precedent paragraph (Figure 3).

The remote traction σ_∞ is transformed into distributed normal forces $P_i = -(\sigma_\infty \cdot n_i)$ acting inside each defect S_i. Where n_i is the outward normal vector to the contour Γ_i of each surface S_i. The effect of the pressure inside the defect S_j is written as the sum of

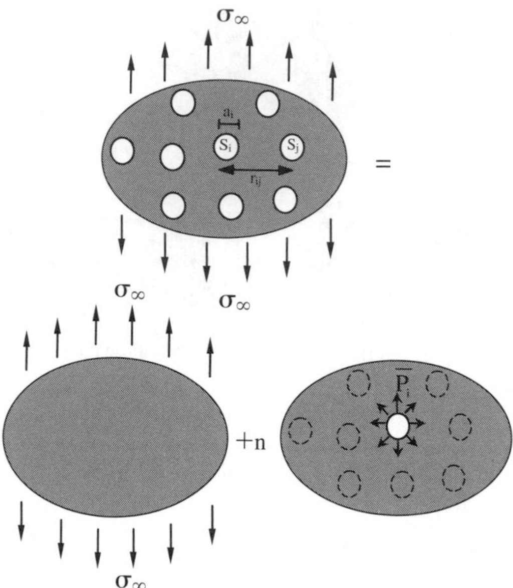

Figure 3. Superposition scheme for an isotropic solid with n circular voids.

the $p_{ij}(i \neq j)$, where p_{ij} represent the normal pressure on the imaginary location of S_i produced by S_j:

$$p_{ij} = -\overline{P_j} \times \frac{a_j^2}{r_{ij}^2} \tag{13}$$

We can replace the problem of n interacting defects by the superposition of n problems in which each defect is considered alone in the infinite solid and subjected to an unknown pressure

$$P_i = \overline{P_i} + \sum_{j=1(j\neq i)}^{n} p_{ij} = \overline{P_i} - \sum_{j=1(j\neq i)}^{n} \left(l_{ij}^2 \times \overline{P_j}\right) \tag{14}$$

where $l_{ij} = a_j/r_{ij}$ are the influence coefficient representing the average pressure at the boundary of the imaginary void S_i caused by a unit uniform pressure on defect S_j and all other defects do not exist. Taking into account the same simplifications as for the example with two voids, we obtain for the pressure the approximate expression:

$$\overline{P_i} \approx P_i \left(1 + \sum_{j=1(j\neq i)}^{n} \frac{a_j^2}{r_{ij}^2}\right) \tag{15}$$

The equilibrium condition on the contour Γ_i of each voids surface may be written as:

$$\overline{\sigma_i} n_i + \overline{P_i} = 0 \tag{16}$$

where σ_i denotes the stress field solution of the subproblem II. With the notation $\sigma_i \cdot n_i = \sigma_i$, equation (15) becomes:

$$\overline{\sigma_i} \approx \sigma_\infty \left(1 + \sum_{j=1(j\neq i)}^{n} \frac{a_j^2}{r_{ij}^2}\right) \tag{17}$$

For simplicity, we define $l_{jj} = 0$, which permits to take the sum in equation (17) over all $j = 1 - n$. We may then write:

$$\overline{\sigma_i} \approx \sigma_\infty \left(1 + \sum_{j=1}^{n} \frac{a_j^2}{r_{ij}^2}\right) \tag{18}$$

The resulting local stress field at each point x of the studied isotropic medium is:

$$\sigma(x) = \sigma_\infty + \sum_i \overline{\sigma_i} * \frac{a_i^2}{r_{ix}^2} \tag{19}$$

Substituting equation (18) into the previous equation we obtain:

$$\begin{aligned} \sigma(x) &= \sigma_\infty + \sum_i \sigma_\infty \left(1 + \sum_{j=1}^{n} \frac{a_j^2}{r_{ij}^2}\right) * \frac{a_i^2}{r_{ix}^2} \\ &= \sigma_\infty + \sum_i \sigma_\infty . \frac{a_i^2}{r_{ix}^2} + \sum_i \sum_{j=1}^{n} \frac{a_j^2}{r_{ij}^2} \frac{a_i^2}{r_{ix}^2} \sigma_\infty \end{aligned} \tag{20}$$

The third term of the previous equation is neglected because it is a product of two small quantities. Therefore, the simplified approximate local stress field reads:

$$\sigma(x) \approx \sigma_\infty + \sum_i \frac{a_i^2}{r_{ix}^2} \sigma_\infty \tag{21}$$

In the case of a non homogeneous loading (but with slow variations) σ_∞ is variable at each point. The local equivalent stress field becomes:

$$\sigma(x_k) \approx \sigma_\infty(x_k) + \sum_i \frac{a_i^2}{r_{ix}^2} \sigma_\infty(x_i) \tag{22}$$

In this equation $\sigma_\infty(x_k)$ is the macroscopic stress field and $\sigma(x_k)$ is the true stress field which take into account the amplification due to the presence of voids. Note that this equation is typically non local, with a_i^2 proportional to the damage at point x_i noted $d(x_i)$.

4 MODIFIED NON LOCAL DAMAGE MODEL

We are going now to derive the local strain field with a view to the expression which controls the growth of damage locally. For this purpose, we will try to relate the local expressions devised in the previous section with the macroscopic stress–strain relation. From equation (22) we can deduce the expression of the local deformation field as follows:

$$\varepsilon(x_k) = \frac{\sigma(x_k)}{K} = \frac{\sigma_\infty(x_k)}{K} + \sum_i \frac{a_i^2}{r_{ik}^2}\frac{\sigma_\infty(x_i)}{K} \tag{23}$$

K is the stiffness of the isotropic elastic medium. We have now to identify the terms σ_∞/K. For that, we can use the macroscopic law:

$$\sigma_\infty(x_k) = K(1 - d(x_k))\ \varepsilon_\infty(x_k) \tag{24}$$

Equation (23) becomes:

$$\varepsilon(x_k) = (1 - d(x_k))\ \varepsilon_\infty(x_k) + \sum_i \frac{a_i^2}{r_{ik}^2}(1 - d(x_i))\ \varepsilon_\infty(x_i) \tag{25}$$

Hence, the mean strain of the homogenised medium, which controls the evolution of damage, is written as:

$$\bar{\varepsilon}(x_k) = \frac{\varepsilon(x_k)}{(1 - d(x_k))} = \varepsilon_\infty(x_k) + \sum_i \frac{d(x_i)}{r_{ik}^2}\frac{(1 - d(x_i))}{(1 - d(x_k))}\varepsilon_\infty(x_i) \tag{26}$$

As we have small interactions between inclusions, then $d(x_k) \approx d(x_i)$ and the previous equation can be further simplified:

$$\bar{\varepsilon}(x_k) = \varepsilon_\infty(x_k) + \sum_i \frac{d(x_i)}{r_{ik}^2}\varepsilon_\infty(x_i) \tag{27}$$

On the macroscale, the continuum counterparts of this equation may be written according to several possibilities. For instance:

$$\bar{\varepsilon}(x) = \frac{1}{V_r(x)}\int_\Omega \psi^*(z)\frac{(1 - d(s))}{(1 - d(x))} d(s)\tilde{\varepsilon}(s)ds \tag{28}$$

or

$$\bar{\varepsilon}(x) = \frac{1}{V_r(x)}\int_\Omega \psi^*(z, d(s))\frac{(1 - d(s))}{(1 - d(x))}\tilde{\varepsilon}(s)ds \tag{29}$$

where $z = ||x - s||$, $\tilde{\varepsilon}$ is the equivalent strain, Ψ^* the weight function and $V_r(x)$ is the representative volume at point x. In the second expression, a modified weight function is introduced in the integral non local damage model:

$$\psi^*(z;(d(s)) = \exp\left(-\frac{4\|x - s\|^2}{(l_c(d(s)))^2}\right) \tag{30}$$

In this equation, the internal length is not constant. It is a function of the damage state at each point. The internal length l_c varies with damage according to a law such as:

$$l_c(d(s)) = \alpha * d(s) \tag{31}$$

α is a constant which can be taken equal to l_{c0}, the final value of the internal length and d(s) is the damage state at point s. This formulation can be looked at as the integral equivalent of the model devised by Geers et al. (1998) and can be implemented similarly.

Our formulation has several characteristics: first it is not symmetric because the weight function is expressed in terms of the value of damage at each considered point. A material point where damage is equal to zero does not influence its neighbour were damage is non zero. Conversely a material point at which damage is non zero has an influence on a point where damage is zero. Second, the weight function evolves during the failure process.

5 NUMERICAL EXAMPLE

The aim of this section is to investigate failure with the proposed formulation and to compare with the original formulation of the non local damage model. For that, a one dimensional example is considered. It is a bar with an initial imperfection in the centre loaded in simple tension as presented in figure 4. Due to this defect, localisation is triggered and it is possible to

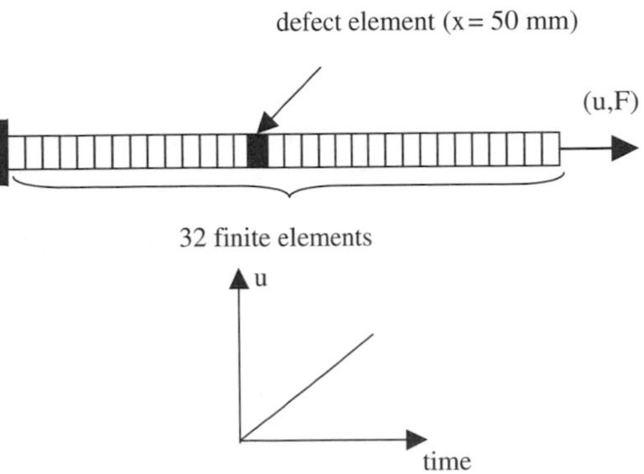

Figure 4. One dimensional bar loaded in tension.

Table 1. Characteristics of the bar and the material parameters.

Test configuration and model parameters	Quantity	Value
Geometry	Length of the bar	10 cm
	Cross-section	1 m2
Elastic parameters	Young's modulus	33000 MPa
	Young's modulus of the defect element	24750 MPa
Non local parameters	l_{c0}	0.02 m
	ε_{D0}	$1.5\ 10^{-4}$
	A_t	1
	B_t	6000

visualise the development of the damage zone. The finite element implementation of the modified constitutive model follows the same procedure as the one devised by Geers et al. (1998).

The geometry of the bar and the material parameters are provided in table 1. The bar length is 10 cm and the initial value of the internal length l_{c0} is 2 cm. The finite element mesh is made of 32 constant size elements (smaller that the internal length).

We compare in figure 5 the evolution of damage in the bar obtained with the original non local damage model with fixed internal length (a) and with the present one with an evolving internal length (b). This figure shows that the modified non local model yields a damage profile with a sharp boundary. The kinetics of damage growth is also different, with an expansion of the damage zone in the modified model. In order to understand the mechanism behind the localisation, it is necessary to analyse the non local interactions between the bar elements. We have plotted in figure 6 the evolution of the non local interactions, i.e. the contribution of each element in the integral (Eq. 29) in the modified model for two cases. The first one (a) correspond of the equivalent strain in the element where damage localises (defect element). The second one (b) corresponds to a point which is at the limit of the damage zone (element 20 at coordinate x = 62.5 mm). The non local contribution increases as damage increases and that there is, as expected, a cut off of the non local contribution in the region where damage is equal to zero. It is this cut off which yields a sharp variation of damage at the boundary of the damage zone.

6 COMPARISON WITH EXPERIMENTS

The experimental program, which is briefly recalled here, aimed at determining the mechanical characteristics (compressive strength, elastic modulus and

(a)

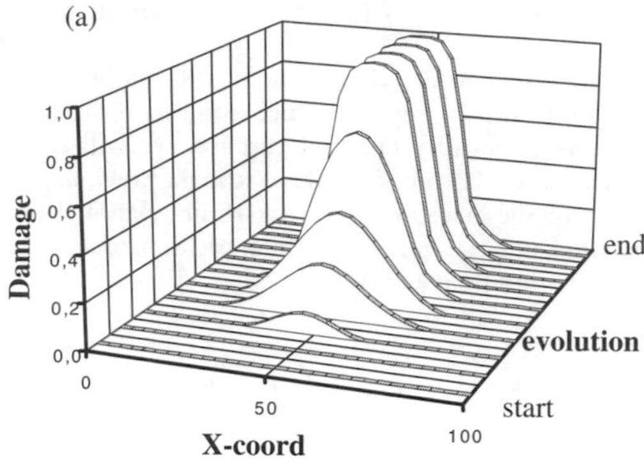

(b)

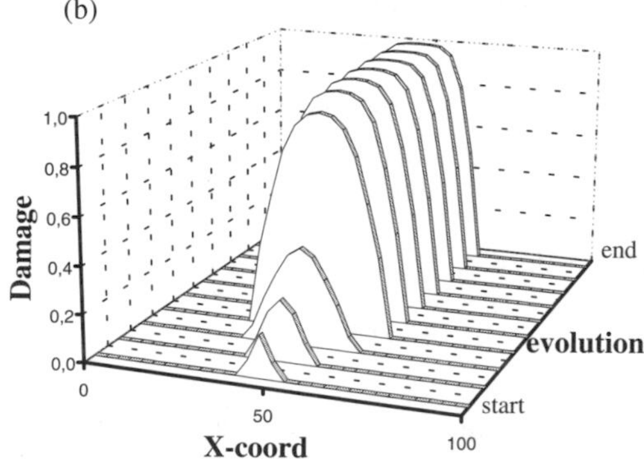

Figure 5. Damage evolution in the one dimensional tension test in the original (a) and modified (b) non local damage model.

fracture energy) of mortar specimens with a controlled microstructure. This control was achieved by adding inclusions of weak mechanical characteristics (polystyrene beads) in a matrix having a reduced porosity.

All specimens were made with a mix which consisted of ordinary Portland cement CPA-CEMI 52.5, polystyrene beads, normal-density fine sand with a maximum size of 2 mm, a superplasticizing agent (Glenium 51) and water. Expanded polystyrene spheres of 3–7 mm mean diameter were used as aggregate in the mix design. The inclusions have a mass density of 20 Kg/m^3, a Young's modulus of 7 Mpa and a compressive strength of 80 Kpa. Expanded polystyrene consists essentially of air 98%. It is hydrophobic (Sussman 1975).

Figure 7 shows the spatial distribution of polystyrene beads for the mix with the lowest density (1.4).

Four different mixes of densities 2.0, 1.8, 1.6 and 1.4, having polystyrene content *g* of 13, 22, 31 and 39% respectively, were considered for the present test programme in addition to the reference material (mortar without inclusions). All mixes have a cement/sand ratio of 0.46 and a water/cement ratio of 0.4.

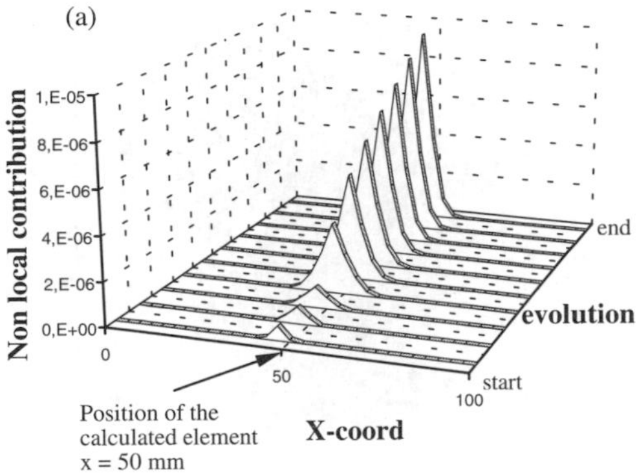

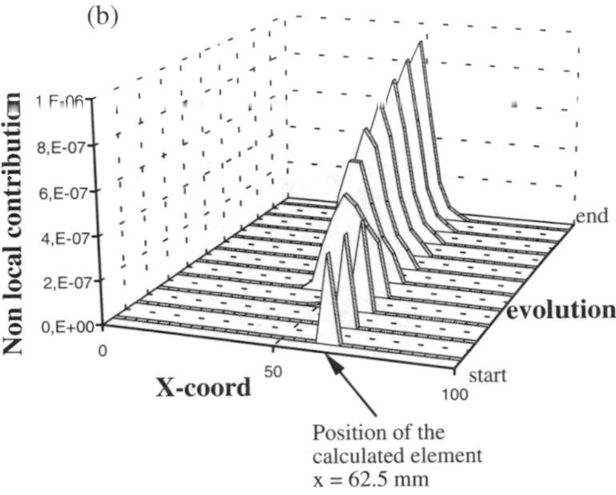

Figure 6. Evolution of the non local contribution in the modified model for the calculation of the weighted average in the middle of the damage zone (a) and at the boundary of the damage zone (b).

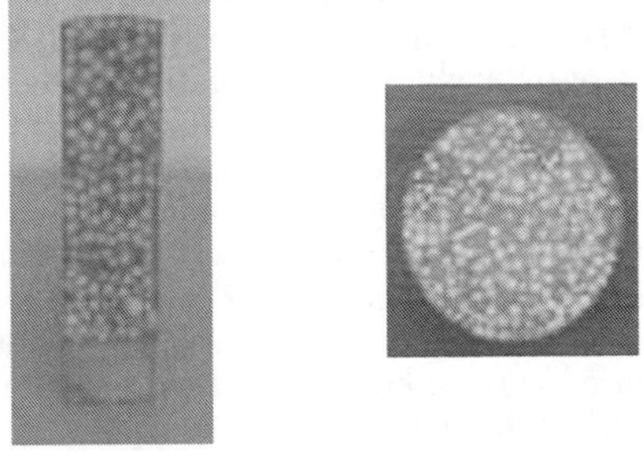

Figure 7. Distribution of polystyrene beads in 1.4 density mixture.

Size effect bending tests were performed on four different sizes of geometrically notched concrete specimens. The depths were D = 40, 80, 160 and 320 mm while the thickness was kept constant for all the specimens b = 40 mm. The length to depth ratio was L/D = 8:3 and the span-to-depth ratio was l/D = 2.5 for all specimens. One notch of depth D/6 and thickness 1.5 mm (same for all dimensions) was performed in each specimen by placing a steel plate at the midpoint perpendicular to the long direction of the mould before casting. For the acoustic emission measurement, only specimens of depth D = 160 mm were tested.

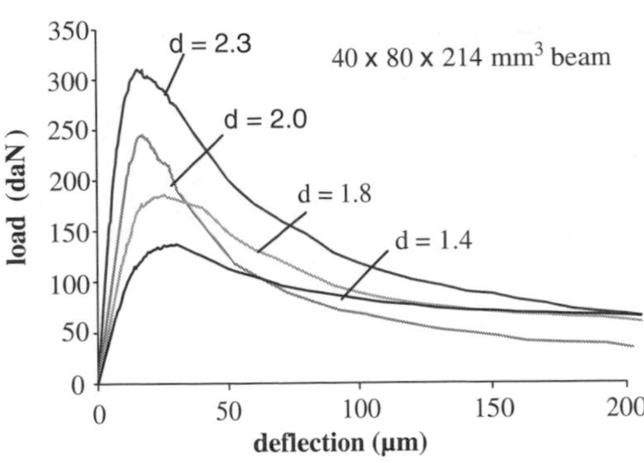

Figure 8. Influence of density on structural behavior, average load-deflection curves for different material density on 40*80*214 mm^3 beams.

The size effect tests followed the guidelines established by RILEM using a closed-loop testing machine, (a 160 KN capacity INSTRON machine), under crack mouth opening displacement (CMOD) control. Figure 8 shows the average responses for all densities and for the medium size (40 * 80 * 214 mm^3).

Note that the deflection at peak is almost independent from the density of material. This parameter (density of material) influences the mechanical behavior of beams; the lower the density, the lower the stiffness and the lower the maximal force.

The size effect tests have been interpreted with the help of Bazant's size effect law (Bazant and Planas, 1998). Results are represented in a $\log(\sigma/Bf_t) - \log(D/d_0)$ diagram (Figure 9), only for the two extreme densities 1.4 and 2.3 for more clarity. The larger the beam, the lower the relative strength. It can be noticed that the mechanical failure of the beams with polystyrene (density = 1.4) tends to adhere more to strength of material than to fracture mechanics, since the corresponding data shift right on the size effect plot. It means that the material becomes more ductile as the density decreases (increasing porosity).

Acoustic emission (AE) is an experimental tool well suited for monitoring fracture processes. It is a technique where the elastic wave generated by cracking events can be measured and processed using seismic analysis techniques. The strength of AE measurement techniques is the ability to monitor microscopic damage occurring inside the material. Several works have focused on relating acoustic emission characteristics to properties of the fracture process zone (Maji and Shah 1988), and using AE source location analysis to

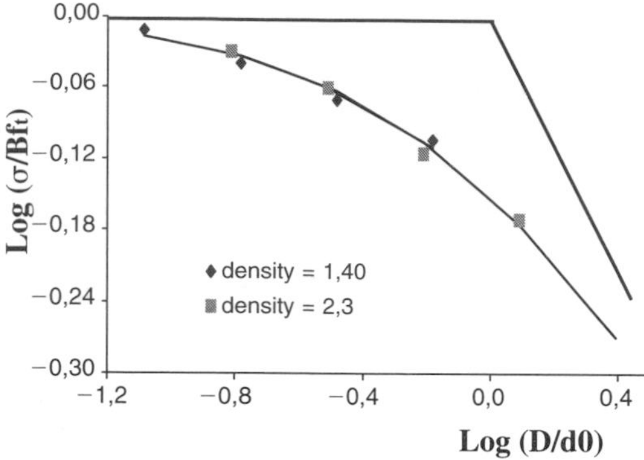

Figure 9. Bazant's size effect curves: calibration for material density 2.3 and 1.4.

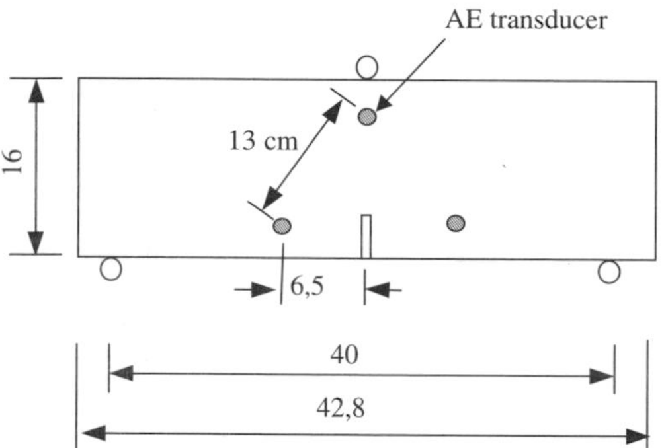

Figure 10. Specimen geometry and AE transducers placement.

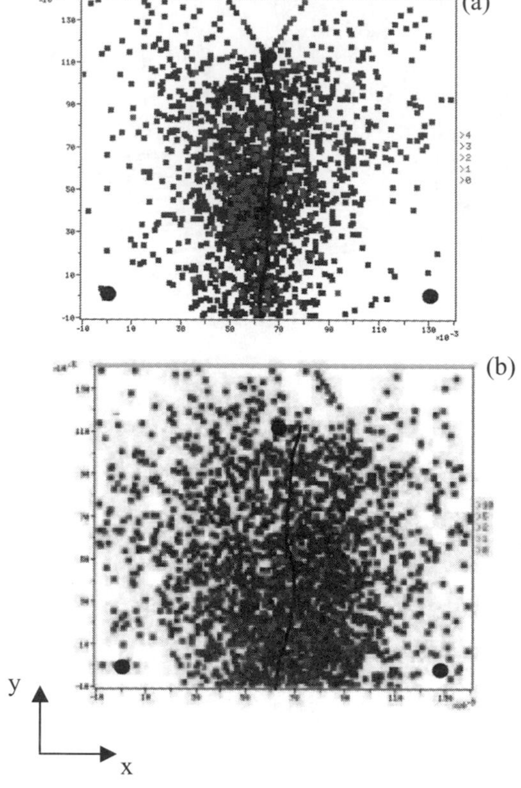

Figure 11. Cumulative location of AE events : a) material density 2.3; b) material density 1.8.

evaluate damage localization (Berthelot et al. 1987). Otsuka et al. (1998) show that the location of cracks, identified by the AE method, has a close relation to the extent of the fracture process zones.

In the present study, the primary aim is to evaluate the size of the fracture process zone, i.e. the crack band width, which is proportional to the internal length l_c of the material in the non-local damage model (Mazars and Pijaudier-Cabot 1996). Such measurements can be performed at the end of the tests and during failure. The evolution of the width with the material density is also monitored.

The AE system used was comprised of, (a) an eight-channel MISTRAS system manufactured by Physical Acoustic Corporation, (b) a general-purpose interface bus (PCI-DISP4) and (c) a PC celeron 600 MHz machine. Three piezoelectric transducers (R15/C, resonant frequency of 150 KHz) were used. Transducers were placed around the expected location of the process zone to minimize errors in the source-location program. It was placed on one side of the specimen, in a triangular array (Figure 10). A silicon adhesive was used as couplant. The detected signals were then amplified with a 40 dB gain differential amplifier (1220 A) in a frequency band from 20 to 1200 KHz. In order to overcome the back-ground noise, the signal detection threshold was set at a value of about 30 dB (value adjusted before every test).

The AE measurement was performed only on specimen (40*160*428 mm^3) and for two different material densities (2.3 and 1.8) in order to compare the effect of the density on the crack band width. The cumulative location of acoustic emission events throughout the test is shown in figure 11: the plotted points indicate the detected AE sources. The accuracy of the localisation of AE events is ±0.5 cm. As shown in the figure, the specimen with a density of 2.3 has a smaller area with AE sources than that of the specimen with a density of 1.8. In the same figure, we have plotted the observed crack path that appeared after each test at failure on the specimen surface.

In processing events, a useful approach, well suited to measure the crack band width, has been to divide the specimen into an array of rectangular elements

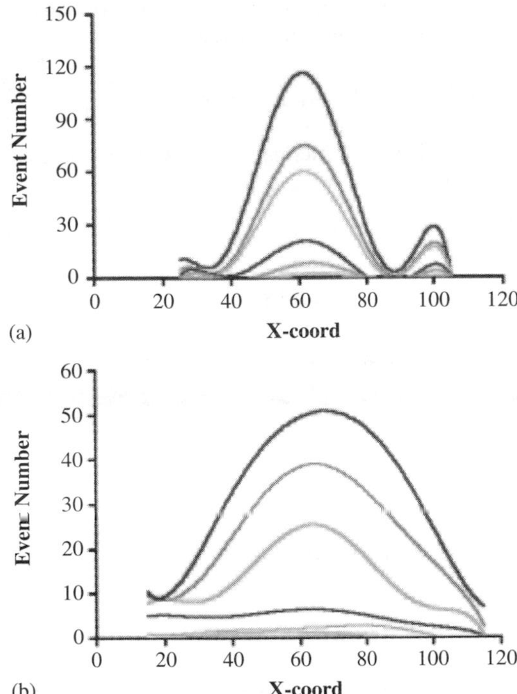

Figure 12. Evolution of the cumulative events in a cross section perpendicular to the crack propagation at different loading state: a) material density 2.3 ; b) material density 1.8.

and counting the AE events located within each element. The fineness of the network should give a sufficient number of events per grid, a (1*1) cm^2 grid area was used. This technique allows the determination of the shape of fracture process zone, in it final stage, but incremental counts (in between two loading stages) provide also the evolution of the FPZ as the degradation process develops. Figure 12 shows the development of the fracture process zone in the course of loading. It is clear on this figure that the width of the FPZ enlarges in the course of damage. Furthermore, the width is larger for the porous material compared to the reference one (without inclusions).

It is possible from these two figures to obtain the evolution of the width of the FPZ. The result is shown in figure 13. The width of the FPZ is arbitrarily defined as the intersection between the profiles given in figure 12 and an horizontal line placed at 20% of the maximum number of event (for each curve).

We can observe two different characteristics: first, the width is increasing almost linearly with the applied load; second the slope of the two curves is approximately the same, with a different value at the origin. This value can be related to the initial damage in the material with inclusions compared to reference material.

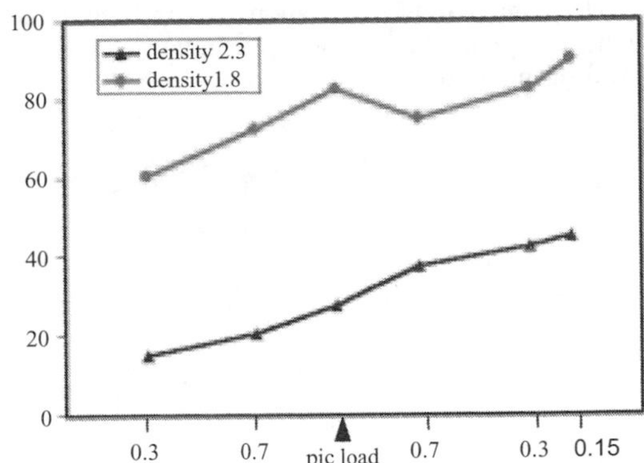

Figure 13. Evolution of the width of the fracture process zone vs the loading state for the two material density.

One could carry out the same derivation from micromechanics as in section 3, taking into account initial damage due to the voids. It can be easily demonstrated that the evolution of the internal length is similar to that in equation (31), with an initial value (when damage due to microcracks is zero) which is non zero and controlled by the amount of spherical inclusions placed in the cementitious matrix.

7 GRADIENT APPROXIMATION

As a final remark, we are going to compare the modified non local damage model derived in the previous section, to the differential formulation proposed by Geers (1998). For that, we consider the expression of the mean strain given in equation (28) and for simplicity, we assume that:

$$\frac{(1-d(s))}{(1-d(x))} \approx 1 \tag{32}$$

The non local strain is:

$$\bar{\varepsilon}(x) = \frac{1}{V_r(x)} \int_\Omega \psi^*(z).d(s)\tilde{\varepsilon}(s)ds \tag{33}$$

We replace now the damage and the local strain tensor in equation (33) by its second order Taylor expansion about $s = 0$:

$$d(s) = d(x) + d,_x\, s + d,_{xx} \frac{s^2}{2} + \ldots. \tag{34}$$

$$\varepsilon(s) = \varepsilon(x) + \varepsilon,_x\, s + \varepsilon,_{xx} \frac{s^2}{2} + \ldots.. \tag{35}$$

The combination of equations (34,35) and equation (33), yields the following definition of the non local strain tensor:

$$\bar{\varepsilon}(x) = \frac{1}{V_r(x)} \int \psi'(x-s) \begin{bmatrix} d(x).\varepsilon(x) + d(x)\varepsilon,_x s + \\ d(x)\varepsilon,_{xx} \frac{s^2}{2} + d,_x.s.\varepsilon(x) + \\ d,_x.s^2.\varepsilon,_x + d,_x.\frac{s^3}{2}.\varepsilon,_{xx} + \\ d,_{xx}.\frac{s^2}{2}.\varepsilon(x) + d,_{xx}.\frac{s^3}{2}.\varepsilon,_x + \\ d,_{xx}.\frac{s^4}{4}.\varepsilon,_{xx} \end{bmatrix} .ds \quad (36)$$

Upon the assumption that the weight function is an even function, the expression of the non local strain becomes:

$$\bar{\varepsilon}(x) = \frac{1}{V_r(x)} \int \psi'(x-s) \begin{bmatrix} d(x).\varepsilon(x) + d(x)\varepsilon,_{xx} \frac{s^2}{2} + \\ d,_x.s^2.\varepsilon,_x + d,_{xx}.\frac{s^2}{2}.\varepsilon(x) + \\ d,_{xx}.\frac{s^4}{4}.\varepsilon,_{xx} \end{bmatrix} .ds \quad (37)$$

with:

$$\frac{1}{V_r(x)} \int \psi'(x-s).d(x).\varepsilon(x).ds = \varepsilon(x) \quad (38)$$

Neglecting the terms of order 4, equation (37) becomes:

$$\bar{\varepsilon}(x) = \varepsilon(x) + \frac{1}{V_r(x)} \left[d(x)\varepsilon,_{xx} + 2.d,_x.\varepsilon,_x + d,_{xx}.\varepsilon(x) \right] \int \psi'(x-s)\frac{s^2}{2}.ds \quad (39)$$

or

$$\bar{\varepsilon} = \varepsilon(x) + c^2 \left[d.\varepsilon,_{xx} + 2.d,_x.\varepsilon,_x + d,_{xx}.\varepsilon(x) \right] \quad (40)$$

We take now the second order derivative of this non local strain:

$$\bar{\varepsilon},_x = \varepsilon,_x + c^2 \begin{bmatrix} d,_x.\varepsilon,_{xx} + d.\varepsilon,_{xxx} + 2.d,_{xx}.\varepsilon,_x + \\ 2.d,_x.\varepsilon,_{xx} + d,_{xxx}.\varepsilon(x) + d,_{xx}.\varepsilon,_x \end{bmatrix} \quad (41)$$

$$\bar{\varepsilon},_{xx} = \varepsilon,_{xx} + c^2 \begin{bmatrix} d,_{xx}.\varepsilon,_{xx} + d,_x.\varepsilon,_{xxx} + d,_x.\varepsilon,_{xxx} + d.\varepsilon,_{xxxx} \\ + 2.d,_{xxx}.\varepsilon,_x + 2.d,_{xx}.\varepsilon,_{xx} + 2.d,_{xx}.\varepsilon,_{xx} \\ + 2.d,_x.\varepsilon,_{xxx} + d,_{xxxx}.\varepsilon(x) + d,_{xxx}.\varepsilon,_x \\ + d,_{xxx}.\varepsilon,_x + d,_{xx}.\varepsilon,_{xx} \end{bmatrix} \quad (42)$$

After neglecting the spatial derivatives of damage of order higher than 2, we obtain:

$$\bar{\varepsilon},_{xx} = \varepsilon,_{xx} + c^2 \left[6.d,_{xx}.\varepsilon,_{xx} \right] \quad (43)$$

Assuming that the terms $d,_{xx} \cdot \varepsilon,_{xx}$ can be neglected, we have:

$$\bar{\varepsilon},_{xx} = \varepsilon,_{xx} \quad (44)$$

and equation (40) becomes:

$$\bar{\varepsilon} - c^2 d.\bar{\varepsilon},_{xx} = \varepsilon(x) + 2c^2 d,_x.\varepsilon,_x + c^2 d,_{xx}.\varepsilon(x) \quad (45)$$

In the above equation, the parameter c is constant, whereas it is not in the model devised by Geers. There are, however, additional terms in equation (45) which do not appear in Geers' model, and which can also provide a variation of the width of the damage localization profile as damage develops.

8 CLOSING REMARKS

Although the simplest proposal of a non local model relies on a constant value of the internal length, it can be derived from micromechanics that this length scale should not be taken as a constant. The internal length is the result of interaction among defects and it is growing with the density of defects. Qualitatively, the same variation is observed from AE analysis. Furthermore, AE analysis exhibits the influence of initial damage in model materials with controlled porosity.

This variation of the internal length scale is often neglected in mecahnical analysis. However, previous studies by Geers et al. (1998) showed that it was necessary to consider a variable length scale if one wants to capture the entire failure process and to model a discontinuous displacement field across a macrocrack, when progressive damage is complete. The model proposed in this contribution follows the same spirit, with a slightly different formulation in the gradient approximation.

The experimental data are related to some academic investigations on a model material and might not be adequate for real concrete. Nevertheless results are pointing out that when damage is not due to mechanical loads (i.e. damage due to physical, thermal and chemical effect yielding a variation of the porosity), it induces a variation of the internal length. Such variations have been observed by Le Bellégo et al. (2000, 2002b) experimentally. It is also necessary to represent them theoretically and numerically in order to capture size effect (Pijaudier-Cabot et al. 2002). The modified non local model devised in this contribution can incorporate such variations consistently.

REFERENCES

Bazant Z.P., 1990. "Why continuum Damage is Nonlocal: Michromechanics Arguments". J. Engrg. Mech. ASCE, 117, pp. 1070–1087.

Bazant Z.P., 1992. "New Nonlocal Damage Concept Based on Micromechanics of Crack Interactions." AFOSR Project on micromechanics of nonlocal damage, ASCE Journal of Engineering Mechanics, Report No. 92-7/C457n.

Bazant, Z.P., and Planas, J., 1998. Fracture and Size Effect in Concrete and other Quasibrittle Materials. CRC Press, Boca Raton, Fla.

Berthelot, J.M., and Robert, J.L., 1987. "Modeling Concrete Damage by Acoustic Emission." Journal of Acoustic Emission. Vol. 7.1 : 43–60.

Geers M.G.D., de Borst R., Brekelmans W.A.M et Peerlings R.H.J., 1998. "Strain-based Transient Gradient Damage Model for Failure Analysis". Comp. Meth. In applied Mech. And Engrg., 160, pp. 133–154.

Haidar K., 2002. "Modélisation de l'endommagement des structures en béton – approches numériques et effet de la microstructure sur les propriétés de rupture." PhD Thesis, Ecole Centrale de Nantes, France, (in French).

Haidar K. and Pijaudier-Cabot G. 2002. "The Relation Between Porosity, Chemical Degradation and Fracture in Mortar Structures." Proc. Int. Conf. on Structural Integrity and Fracture, A. Dyskin et al. Eds, Balkema, pp. 315–324, 2002.

Le Bellégo C., Gérard B., and Pijaudier-Cabot G. 2000, "Chemomechanical Effects in Mortar Beams Subjected to Water Hydrolysis", J. Engrg. Mech. ASCE, 126, 266–272.

Le Bellégo C., Dubé J.F., Pijaudier-Cabot G. et Gérard B., 2002a. "Calibration of Non Local Damage Model from Size Effect Tests". European Journal of Mechanics A/Solids, in press.

Le Bellégo C., Pijaudier-Cabot G., Gérard B., Dubé J.F., and Molez L. 2002b, "Coupled Mechanical and Chemical Damage in Calcium Leached Cementitious Structures", J. Engrg. Mech. ASCE, in press.

Mazars J., 1984. "Application de la mécanique de l'endommagement au comportement non linéaire et à la rupture du béton de structure". PhD thesis, Université Paris VI, Paris, France, (in french).

Mazars, J. and Pijaudier-Cabot, G. 1996. "From Damage to Fracture Mechanics and Conversely: a Combined Approach", Int. J. Solids & Structures, 1996, 33, 3327–3342.

Kachanov M., 1987. "Elastic Solids with Many Cracks – a Simple Method of Analysis." Int. J. solids structures, 23, 1, pp. 23–43.

Maji, A.K. and Shah, S.P., 1988. "Process Zone and Acoustic Emission in Concrete". Experimental mechanics. 28: 27–33.

Otsuka, K., Date, H., and Kurita, T., 1998. "Fracture Process Zone in Concrete Tension Specimens by X-ray and AE Techniques." Fracture Mechanics of Concrete Structures; Proceedings of FRAMCOS-3: 3-16. AEDIFICATIO Publishers.

Pijaudier-Cabot, G., and Bazant, Z.P., Non Local Damage Theory, J. Engrg. Mech. ASCE, (1987), 113, 1512–1533.

Pijaudier-Cabot G., Haidar K., Dubé J.F., and Omar M., 2002 "Non Local Damage Approaches in Durability Mechanics: Influence of the Length Scale in Coupled Problems." 5th World Congress on Computational Mechanics, H.A. Mang, F.G. Rammerstorfer, J. Eberhardsteiner Eds., Wien.

Sussman, V., 1975. "Lightweight Plastic-Aggregate Concrete." Journal. Am. Concr. Inst. Vol 72 : 321–323.

Computational Modelling of Concrete Structures, Bićanić et al. (eds)
© 2003 Swets & Zeitlinger, Lisse, ISBN 90 5809 536 3

New results in meso-mechanical modeling of concrete using fracture-based zero-thickness interface elements

D. Ciancio
Dipartimento di Ingegneria Civile ed Ambientale, University of Catania, Italy

C.M. Lopez & I. Carol
ETSECCPB, Universitat Politecnica de Catalunya, Barcelona, Spain

M. Cuomo
Dipartimento di Ingegneria Civile ed Ambientale, University of Catania, Italy

ABSTRACT: The work concerns the mesomechanical simulation of concrete specimens of different lengths subjected to compression and tension tests. The goal of the analysis is to study the influence of the heterogeneous phases of concrete on the overall behaviour of concrete. The material has been discretized in two phases: mortar and aggregate continuum elements, assumed linear elastic, and zero-thickness interface elements capable of representing cracking inserted along most of the mesh lines. In previous studies, the model was used successfully to represent simple loading scenarios on concrete specimens. In the present study the effect of the length of specimen in tension and compression is investigated. The results obtained follow the trends observed in experimental data. In the final part of the paper, some considerations on the influence of the dilatancy in the constitutive law at this level of analysis will be given.

1 INTRODUCTION

Macroscopic models of the continuum type, which have been used traditionally in structural analysis, may lead to realistic results for simple loading scenarios. However, some aspects of concrete behavior under more complex loading can only be reproduced if the first-order heterogeneities of the material are considered explicitly in the context of a mesomechanical analysis approach (Roelfstra et al. 1985, Stankowski 1990, Vonk 1992). One such a model was proposed in recent years and applied successfully to represent a variety of loading scenarios on concrete specimens in 2D such as uniaxial tension, uniaxial compression, biaxial tension-compression, Brazilian tests, basic creep, etc. (Lopez 1999, Lopez et al. 2000, Lopez et al. 2001). In the work presented, this model has been used to study the effects of specimen length in tension and compression, reproducing the well-known experimental observations reported by van Mier (van Mier 1997), and of incorporating dilatancy in the constitutive law of the interface elements employed. The numerical model is briefly described in section 2. In section 3, the results obtained for three specimens of different length subjected to uniaxial tension and compression are described and compared with experimental results. In section 4, the effects of dilatancy in the interface constitutive law are presented and discussed. Finally, in section 5, the main conclusions and future possibilities of this approach, are summarized.

2 MESOMECHANICAL MODEL

Concrete specimens are represented using 3 components: 1) a particle phase consisting of the larger aggregates with randomly generated shapes and locations; 2) an homogeneous embedding matrix phase which represents the mortar and smaller aggregates, and 3) a number of zero-thickness interface elements placed in between particles and matrix, and between matrix elements themselves, in order to represent all major potential cracks through the specimen. Using the FE analysis, the matrix and the aggregates have been discretized into triangular continuum finite elements, using a linear elastic constitutive law, with different values of mechanical parameters according to the phase of the heterogeneity they represent (aggregates are more rigid than mortar), while the weakest zone in the heterogeneity of the material is simulated by the zero thickness elements, which are equipped with a fracture-based constitutive law with elasto-plastic structure, capable to simulate the phenomenon of discontinuity in the strain field.

The interface elements represent a convenient way to implement the Hillerborg's "fictitious crack model" (Hillerborg et al. 1979). The constitutive model is based on the surface stresses (traction components σ_N, σ_T), and on the dual kinematic variables (relative displacements u_n, u_t). A fracture surface is defined in this stress space as a hyperbola, function.of the mechanical parameters (such as the tensile strength or the apparent cohesion), which evolves during the loading process according to the value of the dissipative work spent. These parameters affect the evolution of the limit surface, as shown in Figure 1. The fracture energies in mode I (G_f^I) and mode IIa (asymptotic shear-compression mode, G_f^{IIa}) are parameters of the model.

The main parameters of this interface model are: the elastic stiffnesses K_N, K_T, the initial cracking surface parameters $\chi_0, c_0, \tan\varphi_0$, the dilatancy limit σ_{dil}, the fracture energies G_f^I, G_f^{IIa}, plus some shape parameters not relevant to this case. A detailed description of the model and constitutive results compared to experimental tests on developing cracks in concrete, can be found in (Carol et al. 1997, Lopez 1999, Carol et al. 2001).

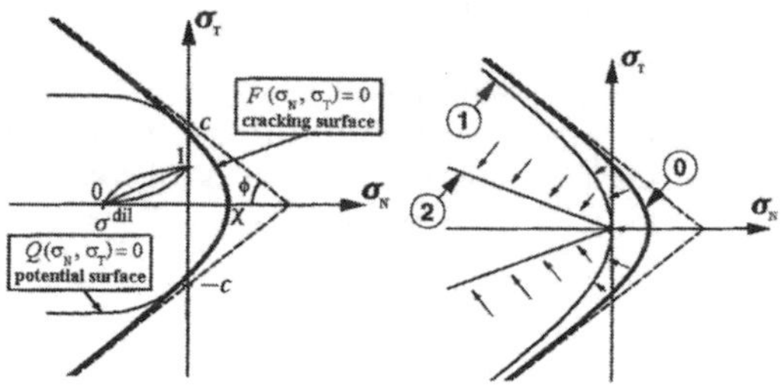

Figure 1. Fracture surface in $\sigma_N-\sigma_T$ plane, and surface evolution according to Mode I (1) and Mode IIa (2) process.

The constitutive model is implemented in the context of the general-purpose FE code DRAC developed in-house at the mechanics of material group of Dept. of Geotechnical Eng. and Geosciences, ETSECCPB-UPC.

3 EFFECT OF SPECIMEN LENGTH

3.1 *Uniaxial tension test*

As already mentioned, three concrete specimens have been considered. They all have the same base (10 cm), but different heights (5, 10 and 20 cm), as shown in Figure 2. Uniaxial loading is applied by prescribing the displacements all along the upper and bottom sides, and leaving the lateral boundaries free (i.e. only one node with lateral restraint, and the rest free to move horizontally), similar to using very stiff and frictionless loading platens. Average strain has been computed by dividing the prescribed displacement by the specimen length. Average stress has been obtained by summing all reactions along the upper or lower sides, and dividing by specimen width.

The relevant material parameters have been assumed the following values:

- Elastic parameters for the aggregates $E_a = 70000$ MPa, $\nu = 0.2$

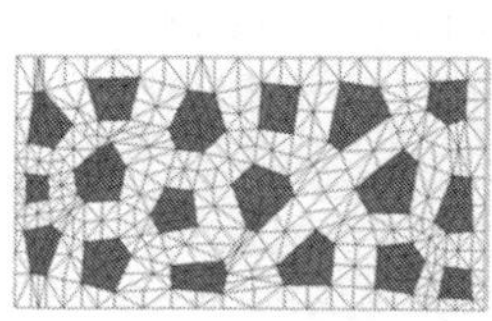
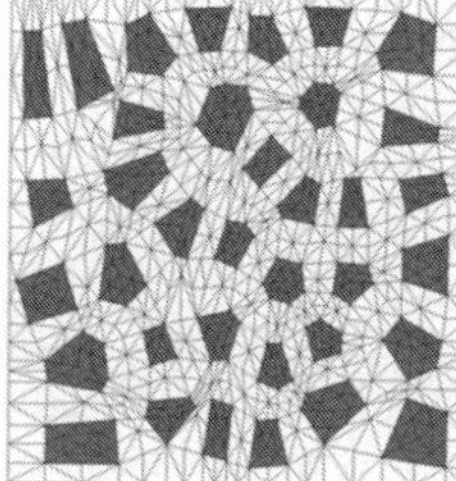
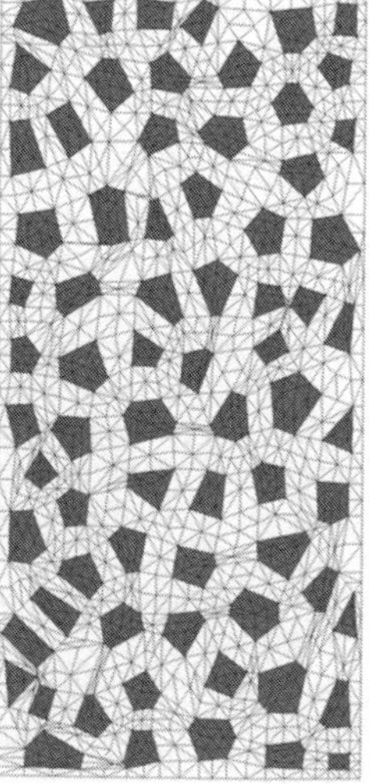

Figure 2. Concrete specimens discretization.

- Elastic parameters for the matrix $E_m = 25000$ MPa, $\nu = 0.2$
- Aggregate-mortar interface $K_N = K_T = 10^9$ MPa, $\chi_0 = 2$ MPa, $c_0 = 7$ MPa, $\tan\varphi_0 = 0.6$, $G_f^I = 0.03$ N/mm, $G_f^{IIa} = 10\ G_f^I$, $\sigma_{dil} = 40$ MPa.

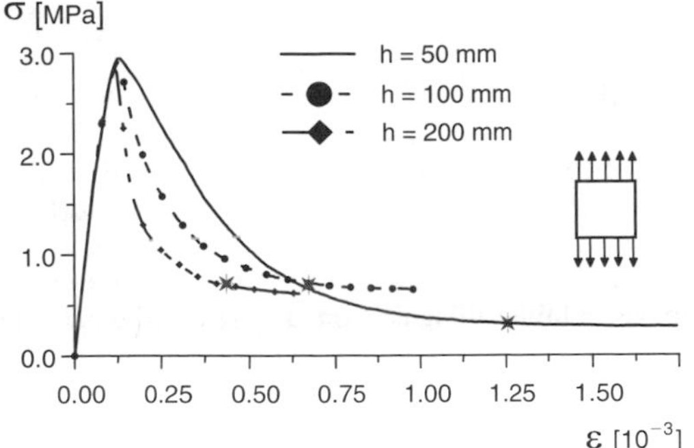

Figure 3. Stress-strain curves for 10 × 5, 10 × 10 and 10 × 20 specimens under tension loading.

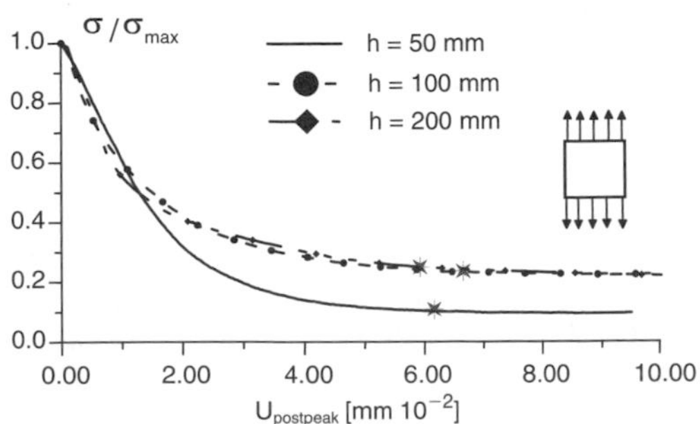

Figure 4. Stress-post peak displacement curve for 10 × 5, 10 × 10 and 10 × 20 specimens under tension loading.

- Mortar-mortar interface: same except $\chi_0 = 4$ MPa, $c_0 = 14$ MPa, $G_f^I = 0.06$ N/mm.

In Figure 3, the results of the uniaxial tension test in terms of stress-strain are shown. The increasing part of the σ–ε curve is similar for all the specimens. Just before the peak, a loss in linearity is shown, corresponding to the beginning of microcracking along the interfaces between the aggregates and the matrix. When the microcracks start to coalesce, the localization phenomenon becomes evident and macrocracks develop. This is the softening part of the σ–ε curve. Three different points of the loading process are indicated with a star symbol on the σ–ε diagram.

In Figure 4, the same results are shown in terms of σ versus u–u_{peak}. In this diagram, the same points in the loading process are indicated also with stars. Note that, while different in σ–ε, the three points represent similar states in the development of fracture, in the sense that the same energy has been spent in the three specimens even if they have different height. To better understand the meaning of the curves in Figure 4, the loading process after the peak point must be considered. From this point, the strain is no longer evenly distributed along the specimen. The deformation increases within the localization zone but it decreases in the rest of the specimen, which undergoes elastic unloading. The contribution of the elastic deformation of the "intact" continuum can be neglected compared to the value of the opening of the main crack, and so this is the meaning represented by u–u_{peak}.

In Figure 5, the cracking state within the specimens are shown, in terms of energy dissipated in fracture processes at each interface which has exceeded the initial cracking condition.

We note that fracture shape is different for each specimen. In Figure 5a, there is only one macrocrack, which is more or less horizontal; Figures 5b,c exhibit

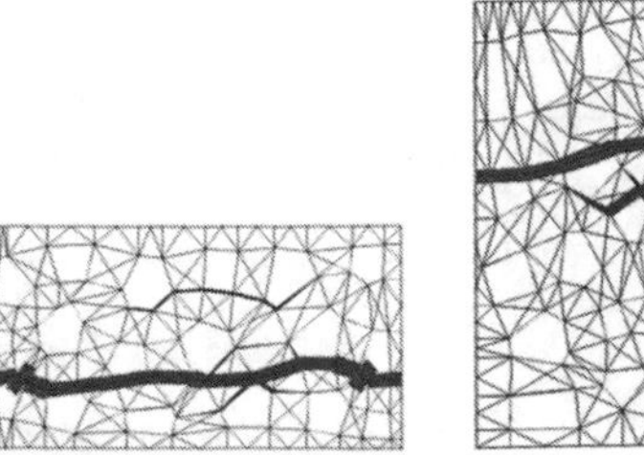
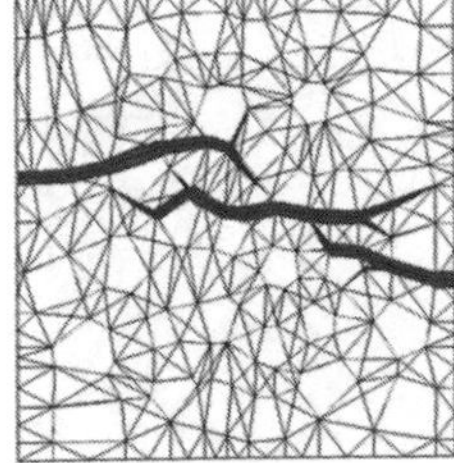
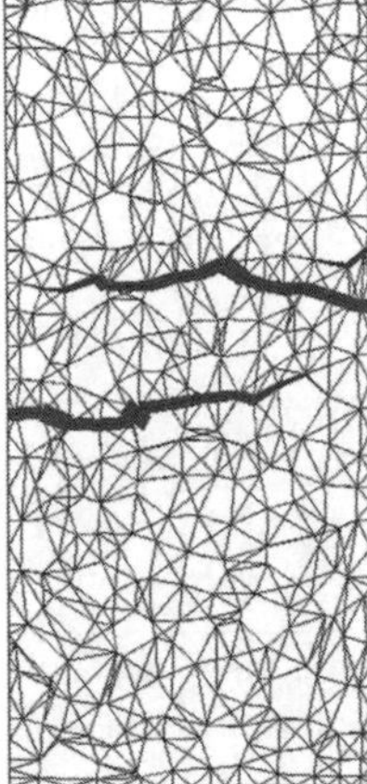

Figure 5. Spent energy along the fracture process zone.

more macrocracks (three for the 10×10 and two for the 10×20 specimen), with an equivalent length equal to approximately 1.5 times the length of the base. Since the area below the stress–post peak displacement represents the work spent during the fracture process, the 10×10 and the 10×20 specimens dissipate more energy than the 10×5, as shown in Figure 4.

3.2 *Uniaxial compression test*

The same meshes and parameters have been used to perform compression tests. Also in this case, the results are presented in term of σ–ε (Fig. 6) and σ–$u_{postpeak}$ (Fig. 7).

In Figure 6, the curve for the 10×20 specimen is not completely developed, since numerical simulations are still going on for this case. In the figure, the three curves look very similar in the pre-peak zone, but it seems that the different height of each specimens affect the response in the post-peak zone. In contrast, in Figure 7 in spite of some differences there seems to be a more clear common pattern and the three curves become more similar. This agrees well with experimental observations by van Mier (van Mier 1984, van Mier 1997) shown in Figures 8–9, and to a certain extent could be expected since the fracture process should be basically the same for the three specimens. This is clearly shown by the crack patterns shown in Figure 10, where the three specimens undergo macrocracks with similar angle and crack lengths overall, even if in each case is divided in more sections zig-zagging through the specimen.

4 DILATANCY

One of the features of concrete fracture is dilatancy, i.e. opening displacements along fracture surfaces due to the irregular arrangement of aggregates in concrete. The interface constitutive law described above exhibits dilatancy on its own defined by its "flow rule", which in tension corresponds to the normal to the surface, with normal and shear components, but in compression has its normal component progressively reduced so that it vanishes for $\sigma = \sigma_{dil}$ (see top of Fig. 11). This non-associated flow rule including dilatancy was originally proposed having in mind concrete cracks at the macroscopic level, i.e. considering concrete as a homogeneous material. In that context, the flow rule of the constitutive model was the only way to represent crack dilatancy.

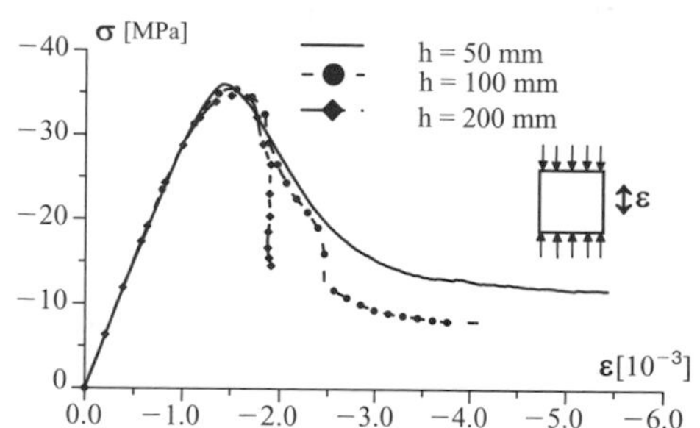

Figure 6. Stress–strain curves under compression loading.

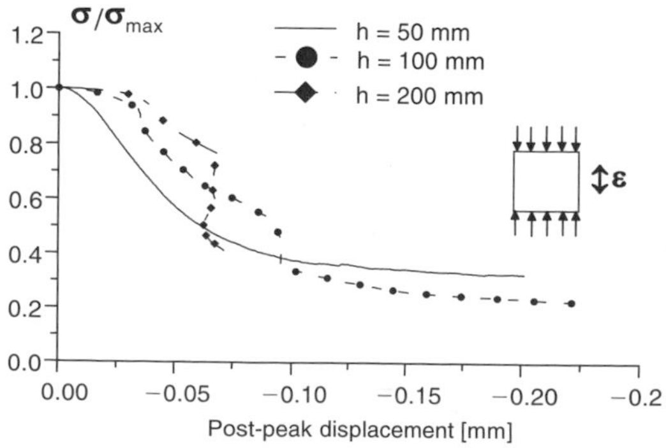

Figure 7. Dimensionless stress–post peak displacement curves under compression loading.

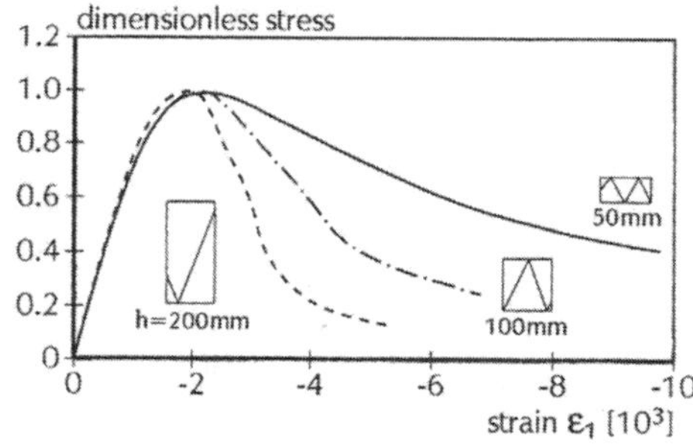

Figure 8. Experimental results by van Mier (van Mier 1997) of stress–strain curves under compression loading.

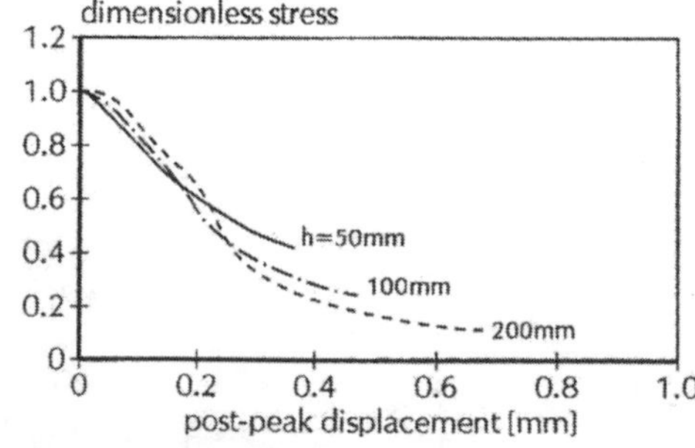

Figure 9. Experimental results by van Mier (van Mier 1997) of dimensionless stress–post peak displacement curves under compression loading.

However, in the meso-mechanical analysis described, in which aggregate particles are represented explicitly, dilatancy of cracks will also be generated by the tortuosity of the crack path itself, which has to go around particles to cross the specimen. In fact in Lopez (1999) it was shown that the overall dilatancy of a specimen in uniaxial compression is obtained almost exclusively from this effect, and thus dilatancy could be simply eliminated altogether from the interface constitutive law. This, on the other hand, could help eliminate some residual stresses in tension tests when crack is tortuous, which can be explained in terms of really long tails in the constitutive response, when dilatancy is present and the prescribed relative displacements has fixed rate of significant tangential sliding simultaneous to normal opening.

In this section, this possibility is investigated by developing a version of the interface constitutive model with no dilatancy, as represented on the bottom side of Figure 11.

This model has been implemented in the FE code DRAC, and the simplest case of tension test on the 10 × 5 cm specimen of Figure 2 has been reanalyzed. The results are shown in Figures 12–13, which can be compared to their dilatant interface counterparts in Figures 3 and 5a.

Figure 12 represents the resulting average stress-strain curves using the interface constitutive law without dilatancy, together the same curve previously obtained with dilatancy. It can be seen, how the dilatant interface produces a curve which goes down faster after peak, but then exhibits a long tail which seems not to go down to zero. In contrast, the curve obtained with the non-dilatant interface, first decreases slowly but then at some point decreases more clearly with vanishing stresses as one should expect.

In Figure 13, the crack path obtained with the non-dilatant interface (a) is compared with the one obtained in the specimen with dilatant interface (b).

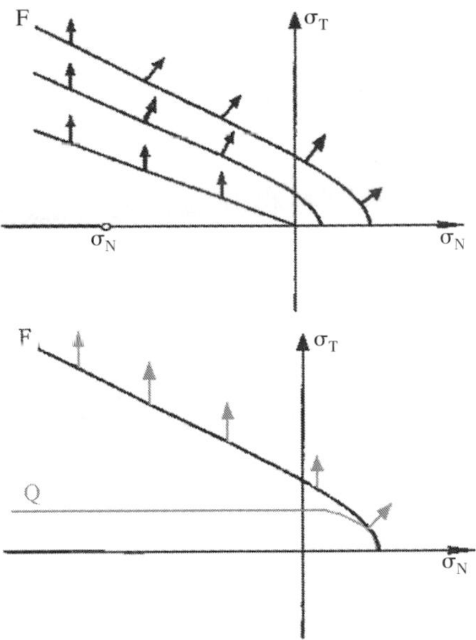

Figure 11. Fracture surface including dilatancy (on the top), and without dilatancy (on the bottom).

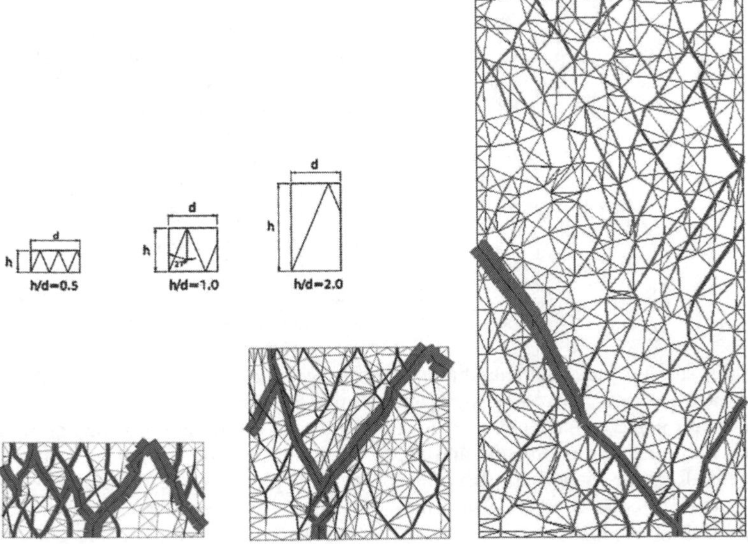

Figure 10. Fracture process under compression loading.

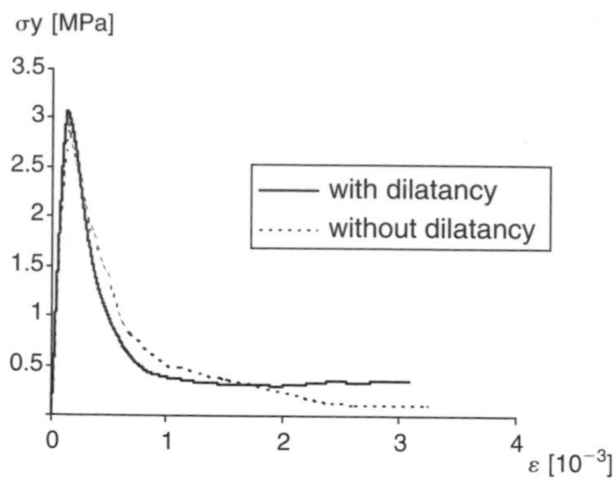

Figure 12. Mechanical behavior of a 10 × 5 cm specimen under uniaxial tension, using "dilatant" and "non dilatant" interface model.

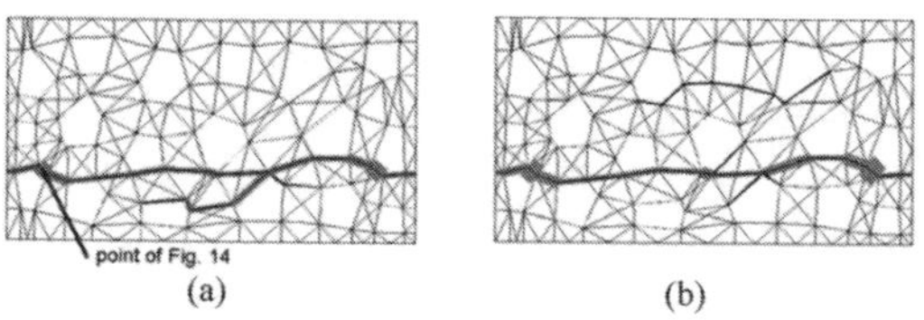

Figure 13. Comparison between non-dilatant (a) and dilatant interfaces (b).

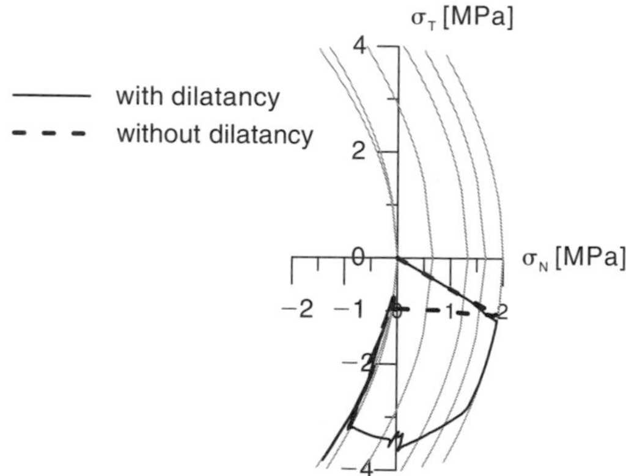

Figure 14. Fracture surface evolution and stress state in a Gauss point.

One interesting observation is that, changing this specific aspect of the interface constitutive law, the crack path has changed, with some initial branching to a lower path, which at some point is arrested and the main crack proceeds along the same final path as the dilatant interface model. However, even if the final path is the same, there are two differences to remark. One, the additional lower branch accounts for the higher crack path and additional energy spent overall.

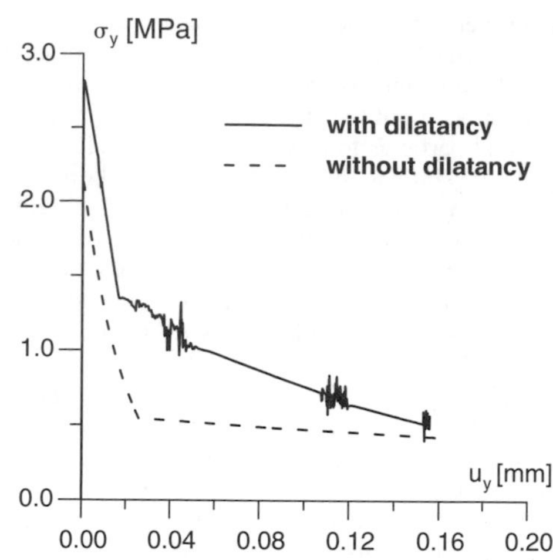

Figure 15. Stress–displacement behavior in a Gauss point.

Second, once the final crack path is formed (and the lower arrested), the macroscopic stress-strain curve has a clear drop, reducing significantly the residual values produced by the dilatant model.

This result can be explained by observing the mechanical behaviour of those interfaces that have certain slope with respect to the load direction. In Figure 14, a comparison between the non dilatant and the dilatant law is shown, in terms of the stress state evolution at a point of the fracture surface marked in Figure 13, which exibits an orientation of 51° to the x-axis.

It can be observed that, although the elastic range is similar for both processes, when the initial fracture surface is reached the behaviour of the curves is quite different. According to the dilatant law, the normal stress which is initially tensile (positive) decreases value and reverses sign until a compressive peak is reached, while the tangential stresses reach relative high values; then, both of them decrease again to zero. The presence of negative normal and high tangential stresses is due to the dilatancy effect greater than the effective opening displacement of the crack. On the other side, the stress evolution obtained using a non dilatant law, shows a monotonous decrease of the normal stresses down to zero without the development of high tangential stresses.

In Figure 15, the stress-displacement behavior along the loading direction is calculated at the same interface point.

It is worth noting that the non dilatant curve tends to zero more rapidly than the dilatant one. This means that the global result of the concrete specimen is may be highly affected by the microstructural behaviour of each interface, according to its geometrical orientation and its mechanical behaviour, and in particular by the constitutive dilatancy.

5 CONCLUDING REMARKS

In previous studies the mesomechanical model described for concrete has been shown capable of simulating a number of loading situations on 2D specimens including tension, compression, biaxial, Brazilian test, creep, etc. In this study, the model has been shown to also represent adequately the effect of length of specimen in uniaxial tension and compression, according to leading experimental results, which is a significant achievement in this field. Additionally, the possibility of using a non-dilatant interface model in this type of analysis has been explored. Still pending of further verification, the preliminary results presented seem to indicate that this type of model could solve some undesirable effects observed with the dilatant models used so far, such as too long residual tails.

ACKNOWLEDGEMENTS

The second and third authors wish to acknowledge grant MAT2000-1007 and BTE 2001-2227 from MCYT (Madrid, Spain). The first author thanks the Ministero dell'Universita e della Ricerca Scientifica e Tecnologica (MURST) for the support of her extended visit to ETSECCPB-UPC.

REFERENCES

Carol, I., Prat, P.C. & López, C.M. 1997. A normal/shear cracking model. Application to discrete crack analysis. *J. of Engineering Mechanics No. 8*: 123.

Carol, I., López, C.M. & Roa, O. 2001. Micromechanical analysis of quasi-brittle materials using fracture-based interface elements. *Int. J. Numer. Meth. in Engrg.*, Vol. 52: 193–215.

Hillerborg, A., Modéer, M. & Petersson, P.E. 1979. Analysis of crack formation and crack growth in concrete by means of Fracture Mechanics and Finite Elements. *Cement and Concrete Research*, Vol. 13: pp.195–205.

Lopez, C.M. 1999. Microstructural analysis of concrete fracture using interface elements. Application to various concretes (In Spanish). Doctoral Thesis. Universitat Politecnica de Catalunya. ETSECCCP-UPC, E-08034 Barcelona, Spain.

Lopez, C.M., Carol, I. & Aguado, A. 2000. Microstructural analysis of concrete fracture using interface elements. *European Congress of Computational Methods in Applied Sciences and Engineering – ECCOMAS 2000*, Oñate E. Owen DRJ (eds), CIMNE, Barcelona, Spain. ISBN 84-89925-70-4 (in CD-ROM).

Lopez, C.M., Carol, I. & Murcia, J. 2001. Mesostructural modeling of basic creep at various stress levels. *Creep, Shrinkage and Durability Mechanics of Concrete and other Quasi-Brittle Materials.* Elsevier Publishers. Oxford, pp.101–106

Roelfstra P.E., Sadouki H. & Wittmann F.H. 1985. Le béton numerique. *Materials and Structures*, Vol. 18: pp. 309–317.

Stankowski, T. 1990. Numerical simulation of progressive failure in particle composites. *PhD thesis*, Dept. CEAE, University of Colorado, Boulder, CO 80309-0428, USA.

Van Mier, J.G.M. 1984. Strain-softening of Concrete under Multiaxial Loading Conditions. *PhD thesis.* Eindhoven University of Technology, Eindhoven, Holanda.

Van Mier, J.G.M. 1997. Fracture Processes of Concrete. CRC Press.

Vonk, R. 1992. Softening of concrete loaded in compression. *PhD thesis*, Technische Universiteit Eindhoven, Postbus 513, 5600 MB Eindhoven, Netherlands.

Computational Modelling of Concrete Structures, Bićanić et al. (eds)
© 2003 Swets & Zeitlinger, Lisse, ISBN 90 5809 536 3

Benchmarking of concrete cracking constitutive laws: MECA project

S. Ghavamian
Electricité de France, R&D, Clamart Cedex, France

I. Carol
ETSECCPB-UPC, Jordi Girona, Barcelona, Spain

ABSTRACT: In 1999 EDF, seeking to have available a 3D constitutive model to describe the cracking of concrete to study the performance of its prestressed pressure containment vessels, initiated a series of calculations based on several cases. For this purpose a benchmark activity was launched with a pool of 13 academic and industrial teams, with the goal of performing nonlinear 3D analysis using different type of material laws for concrete. This paper is the first publication of a set of results obtained by the members of the project, and the aim of this presentation is to describe the project and provide results accompanied by some comments and facts for two of the tested structures. The ongoing project will end in 2003 with further analyses on the last specimen not presented here.

1 INTRODUCTION

The evaluation of mechanical margin safety of reinforced concrete structures is the meeting point between industrial needs and researches specialised in material behaviour. Taking advantage of recent progresses made in research to reach new solutions for industrial issues will enhance the safety of their installations and raise their competitiveness. However, this scheme suffers serious issues to guarantee reliable technological transfer of tools from research laboratories to industrial platforms. A gap exists between these two partners and the aim of the project is gather around a table partners from different institutions in order to take a step forward in enriching the capacity of industries to perform nonlinear behaviour numerical simulation.

The purpose of this benchmark is to evaluate the pertinence and robustness of several models of nonlinear behaviour (including cracking) of concrete. With this aim in mind several cases are presented in Section 2 along with the description of the geometric model as well as the parameters of the materials used, which were to be taken into consideration during calculations.

The project, started in year 2000, has been running since that time with 12 to 15 partners from both academic research teams and commercial companies. It should terminate by the end of year 2003 by completing the test cases with a final one where both the geometry and loading will be fully 3D.

2 BENCHMARK DESCRIPTION

The benchmark comprises four tests. These tests were selected in order to verify different aspects of concrete models and are the following:

A – Uniaxial tension – compression with load reversals
B – Willam's test of tension/shear with rotation
C – Three-point bending RC beam
D – Nooru-Mohammed's shear box test.

Test A verifies the essential aspects of uniaxial loading with strain-prescribed history including load reversals involving tension and compression, leading to the classical response similar to that of Figure 1.

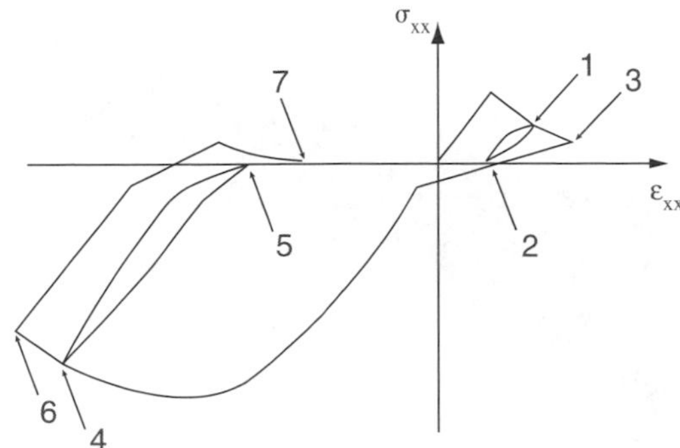

Figure 1. Expected results of Test A.

Test B is a tension-shear test with rotation of principal directions proposed by Willam, Pramono and Sture (1987), which has become a classic for checking anisotropic models. In this test, all in-plane components of strain ($\varepsilon_{xx}, \varepsilon_{yy}, \varepsilon_{xy}$) are prescribed in two stages. The first one, proportional to (1, -v, 0), represents a uniaxial tension loading of elastic material along x-axis is applied exactly until the tensile strength is reached. The second one, with increments proportional to (1, 0.5, 0.5), represents increments of biaxial tensile strain accompanied by a rotation of the principal strain axes. The rotation is faster at the beginning and then progressively slower, with a final asymptotic value of 52.02°. This is a purely numerical test (no experiments have been carried out or seem to be likely to be carried out in the near future due to practical difficulties). What can be considered typical results and an extended discussion of the results obtained by the participants are included in Section 4.

Test C is a three-point bending beam, represented in Figure 2, and details of test and results are compared in some detail in Section 5.

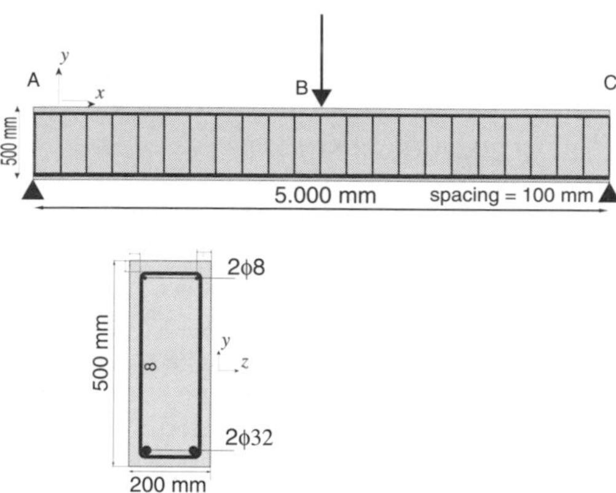

Figure 2. Beam of Test C (1973).

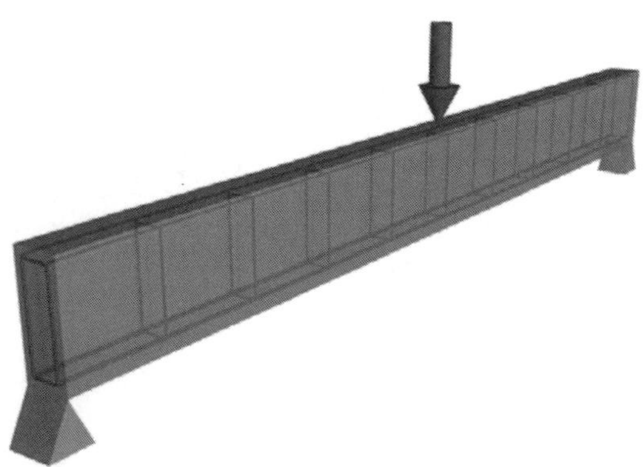

Figure 3. A 3D view of the beam of Test C (1973).

The test is expected to be run in 3D (Figure 3), and the mesh is also provided with 50 × 12 elements in the length × height dimensions, and 6 though depth.

Case C includes calculations of two subcases:

- test C1 without transversal stirrups
- test C2 with transversal stirrups.

Finally, test D consists of a series of test reported by Nooru-Mohammed (1992). A concrete unreinforced specimen is placed and glued into the "shear box" type loading platens shown in Figure 4, and subject to a variety of biaxial loading scenarios, Figure 5.

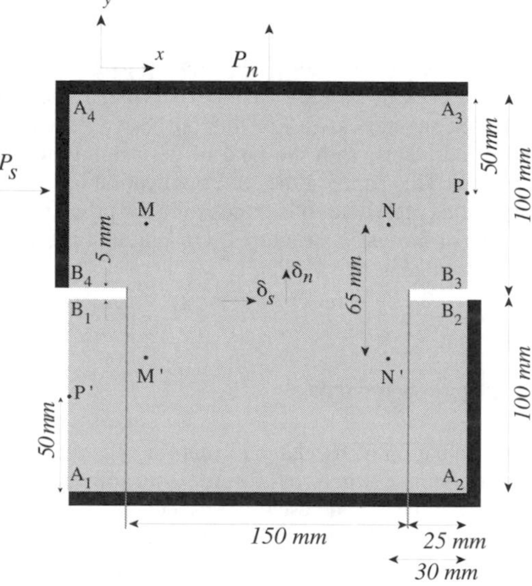

Figure 4. Nooru-Mohammed Test D (1992).

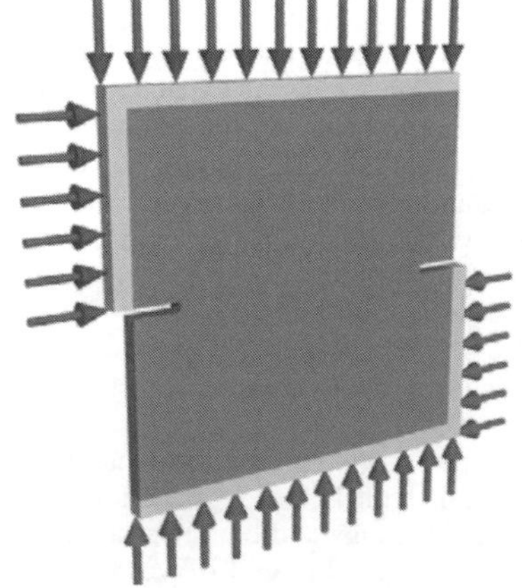

Figure 5. A 3D view of Test D loading (1992).

Four loading sequences are considered for this test:

- Test D-1a: tension is applied vertically until displacement is 0.2 mm, then reduced to zero, then shear is applied by horizontal tensile forces for constant zero vertical force.
- Test D-1b: tension is applied vertically until displacement is 0.2 mm, then reduced to 1 kN of compression and kept constant during the application of tensile shear.
- Test D-2a: horizontal compression in applied to shear the specimen at 5 kN, at zero vertical force. Then shear force is kept constant and tensile vertical force is applied.
- Test D-2b: same as 2a but with a horizontal compressive force of 10 kN.

While test case A allow to roughly estimate the uniaxial behaviour of constitutive laws under classical tension compression cycles, test B aims to reach out for the capacity of models to follow the rotation of the principal stress axis under tension and shear strain. Test C was a structural case were the robustness of the models were estimated. And finally test D was selected because of the structural aspect of the specimen, and complexity of the loading paths [2].

3 PARTICIPANTS

A total of 13 groups have participated in the benchmark. The list is given in the subsequent table:

	Acronym	Institution/Company
1	EUDIL	EUDIL, Laboratoire de Mécanique de Lille
2	INSA	URGC, INSA de Lyon
3	LGCU	LGCU, Université de Marne La Vallée
4	LGCNSN	LGNSN, R&DO - École Centrale de Nantes
5	ISA	ISA BTP, Laboratoire de Sci. App. au Génie Civil
6	LMT	LMT, ENS de Cachan
7	CUB	Dept. Civil, Env. and Arch. Eng. U. of Colorado
8	ABAQUS	ABAQUS France
9	ANSYS	Cr11 Technology Simulation (ANSYS)
10	IWB	Inst. für Werk. im Bauwessen Univ. of Stuttgart
11	LCPC	LCPC, Laboratoire Central des Ponts et Chaussés
12	CSTB	CSTB, Centre Scientifique et Tech. du Bâtiment
13	SOCOTEC	SOCOTEC Industrie

Not all participants have provided results to all the tests proposed.

4 DETAILED DISCUSSION OF TEST B (WILLAM'S TEST)

A very interesting case from the viewpoint of concrete constitutive modelling, is Test B (Willam et al, 1987). In test B, already described briefly in Section 2, the material is basically subject to two load steps. The first one is standard uniaxial tension until the peak stress is reached. The second consists of an increasing tensile major principal strain, accompanied by lower but also tensile strain in the perpendicular direction, and both of them subject to a significant rotation of the principal axes. In Figure 6a, the prescribed strain components ε_{yy}, ε_{xy} and the resulting principal in-plane strains ε_1 *and* ε_2, are plotted against the prescribed strain along the x-axis, ε_{xx}, which is always increasing in the test, and is taken in all subsequent

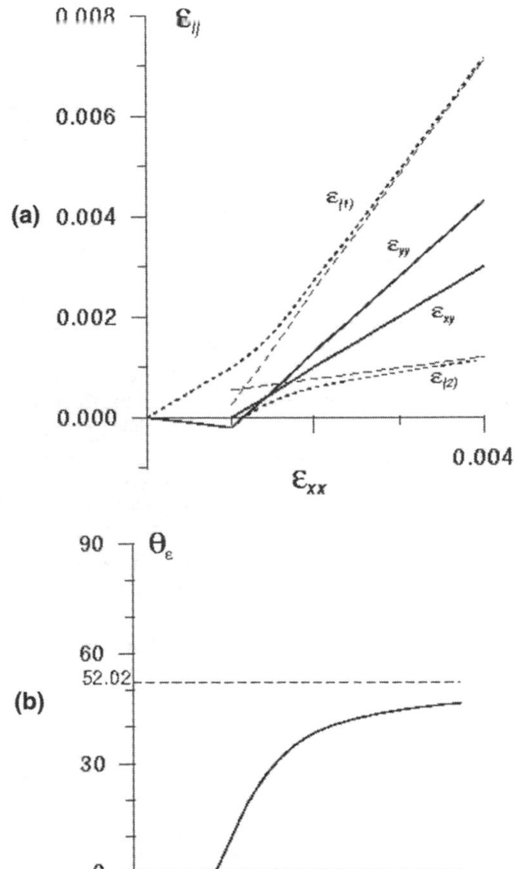

Figure 6. Components of prescribed strain in Willam's test (a) and prescribed rotation of principal strains (b).

plots as a "time variable". The angle of rotation of the major principal strain which is implied by previous prescribed values, is plotted in Figure 6b.

This test is very relevant to practical use of concrete constitutive models because this type of strain history appears very commonly in many Gauss points of FE calculations involving cracking, even if the external loads on the structure are increasing monotonously. Change in direction of applied strain at the Gauss points may be indeed caused by the change of relative stiffness between parts of the structure after the first points start cracking (for instance, this happens at the Gauss points closer to the notch of a 3-point bending beam).

Therefore, a model not behaving well under this test may cause serious problems when used systematically in a Finite Element context, even in very simple structural calculations. At the same time, this test is also important because it serves very efficiently to establish differences between isotropic and anisotropic models, and to discriminate among the latter with respect to various very specific features.

To run the calculations, the following parameters were given to the participants: $E = 32000$ MPa, $\nu = 0.2$, tensile strength 3.0 MPa and fracture energy $G_f^I = 110\,\text{J/m}^2$. The results obtained with all the models who have contributed to this test, in terms of stresses, are shown in Figure 7. In the diagrams at the top, middle and bottom, stress components σ_{xx}, σ_{yy} and τ_{xy} are plotted respectively against the same strain component ε_{xx}. Some of the participants in the benchmark have obtained results with two models. These are labelled 4a, 4b, etc. A good set of reference curves for comparison, plotted in thicker line in the three diagrams, is given by a basic isotropic damage model labelled "4a-LGCNSN-Iso".

In general a relatively large dispersion can be observed between the various models, including important differences in some essential features. Regarding the σ_{xx} component (top diagram), some models continue increasing beyond the established tensile strength. The model labelled "10" exhibits a nonlinear response before the peak, and the curves received indicate that for this model, application of part 2 of the test was delayed until the peak stress had been reached, at about twice the x-strain than the rest of the models. The pace of decay after the peak, also differs considerably from one model to another. In some of the models, a second peak or plateau is present in the σ_{xx} curve. Most of them achieve a zero residual stress between $\varepsilon_{xx} = 0.0002$ and 0.0006, although in one extreme case an spurious residual stress of about half of the peak stress is reached and maintained, and in another case the tensile stress reverses sign and even becomes increasingly compressive with values beyond the order of the tensile strength and no peak in sight.

Focusing now on the σ_{yy} diagram, differences are even larger than before regarding the peak stress, peak strain, decay pace after the peak, and again the same two models as before exhibit peculiar behaviour, one of them with a constant plateau right after the peak, and the other with a change of sign going into always-increasing compression.

Looking finally at τ_{xy}, the discrepancies between models become the largest. In most of them (but not all!) shear stress starts with the same sign as the prescribed shear strain, taken as positive. In most of them, the sign is maintained throughout the test, but in four cases, at some point in the test shear stress is reversed to negative (shear stress sign opposite to shear strain, phenomenon which we call "sear inversion") and remains negative although progressively vanishing for the rest of the test (interestingly enough, the only model which started slightly negative, soon reverses sign and then stays positive for the rest of the test!). Peak location and intensity on the positive side (and also on the negative side for those models with shear inversion) vary considerably from model to model, and the same two models as before give peculiar behaviour with a constant plateau and an always-increasing shear stress (both on the positive side).

As in any benchmark test, the ideal procedure to discriminate between the various models would be to compare with experimental results, but unfortunately these results are not available in this case. What is available at present in the literature are the results obtained by certain number of authors who have used this test for comparison of dedicated isotropic and anisotropic formulations for cracking, damage, etc. This includes, among others, Rots (1988), Oliver et al. (1990), Kroeplin and Weihe (1997), Meschke et al. (1998), Carol and Prat (1995) and Carol et al. (2001). From the study of these results, some common trends seem to indicate the following:

- First a general observation: (i) stresses in all the tests vanish for sufficiently advanced stage of the loading.
- Isotropic models exhibit smooth decays after peak of σ_{xx}, as is shown by the curve obtained by "4a-LGC-NSN Iso". In terms of σ_{yy}, and τ_{xy}, these models lead to a small peak on positive side of axis, without any sign inversion or second peaks. These models, however, represent a relatively simplistic description of the anisotropic nature of the damage and cracking in concrete.
- Calculation with more sophisticated anisotropic models of various different types (e.g. microplane, multicrack, tensorial pseudo-Rankine) seem to exhibit common characteristic features such as: (ii) small plateau or second peak at approximately the middle of the descending part of the σ_{xx} curve; (iii) larger peak σ_{yy} than given by the isotropic

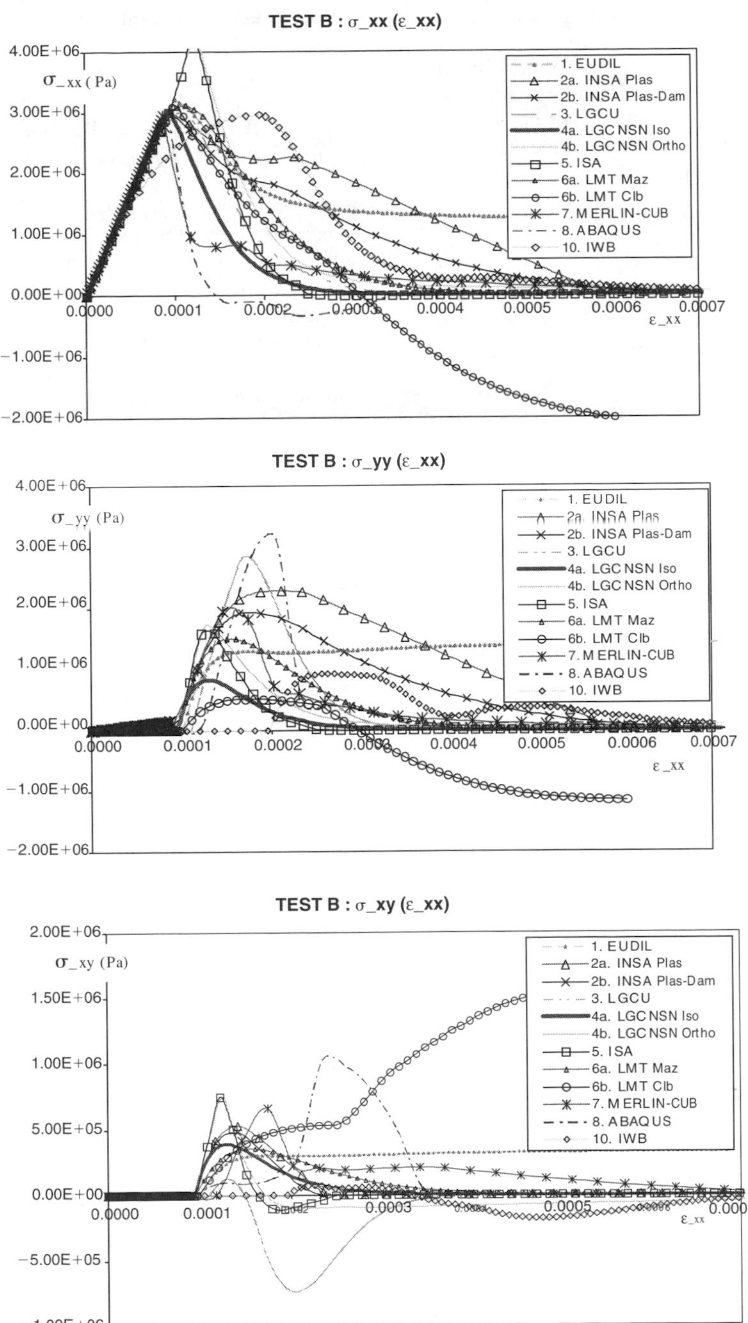

Figure 7. Results from Test B (Willam's test).

formulation, which, at some point of the test may even exceed the current σ_{xx} value; and (iv) an initial positive shear stress with a peak and then a more or less pronounced shear inversion with a negative peak and finally vanishing to zero on the negative side.

– If interpreted in terms of principal stresses (not done in these figures), (v) a well marked second peak

may appear in the evolution of the major principal stress σ_1.
- If interpreted in terms of rotation of principal directions of stress θ_σ or damage, as compared to those of prescribed strain θ_ε, (vi) anisotropic damage models tend to lead to stresses that overrotate the prescribed strains (i.e. $\theta_\sigma > \theta_\varepsilon$).

Assuming that the previous conclusions correctly represent anisotropic features of damage and cracking of concrete subject to Willam's test, the evaluation of the participating models (in terms of conditions (i)–(iv), (the only ones which can be assessed from the curves in Figure 7) leads to the following conclusions:

- Two models, "1" and "6b", can be considered totally inadequate for this behaviour in their present form since they do not satisfy general requirement (i) of vanishing stresses.
- Model "4a" is only isotropic, not satisfying conditions (ii), (iii), (iv).
- Model "8" only satisfies one of these three conditions.
- Model "2a", 2b, 3, 4b, 5 and 7, satisfy two of these three conditions.
- Only the model labelled "10" satisfies the three conditions above.

Of course these conclusions must be taken with caution for various reasons. First of all, other factors not easily measurable from the viewpoint of a benchmark, may be very important for the practical use of the models, such as number and nature of parameters, robustness of the implementation, etc. Second, and very important, the fact that no experimental information is available and probably will not be in the near future due to practical difficulties of this test, puts a serious question mark.

However, in this sense a new powerful tool is emerging that in the near future may help fill this experimental gap: the meso-mechanical modelling of concrete, in which larger aggregates are discretized explicitly on a matrix representing the overall bed of mortar and smaller aggregates. In this approach, pioneered in concrete by Roelfstra et al. (1985) and then followed with some differences by a few other authors (e.g. Stankowski 1990; Lopez et al. 2001; Carol et al. 2001), the behaviour is completed with the aggregate-mortar interfaces which can be represented with joint elements. So far, 2D specimens with a reduced number of aggregates (less than 10×10) have been analysed successfully under a variety of loading conditions including tension, compression, Brazilian, biaxial tension/compression, creep, shrinkage, etc. (Lopez et al. 2000, Lopez et al., 2001), but work is progressing rapidly and the first 3D simulations of specimens, and 2-D simulations of small structural elements or special tests such as this one or Nooru-Mohamed (test D in this benchmark) may be available very soon, throwing additional light into the model benchmarking process.

5 DETAILED DISCUSSION OF TEST C (3 POINT BENDING BEAM)

Before discussing the results on this test case it is important to give the reasons why this specimen was selected. First it represents a structural case which may emphasize more practical issues in terms of complexity and size when it comes to industrial structures. At the same time, it remains a "simple" case, since 3 point bending beams have been studied for decades, and nonlinear response of the structure is easier to apprehend.

Some of the practical aspects that may raise numerical difficulties are the loading conditions that require specific treatments. These are more specifically located on the support and mid-span loading point. To avoid any non-linearity on the support end we advised all participants to adopt linear elastic behaviour for concrete on the entire cross section of the beam composed over a length covering one element along the beam axis (see Figure 8). The support conditions being prescribed displacements u_{xyz} along a mesh line.

The second difficulty was the mid-span applied load. In reality the test is carried out by applying a vertical displacement using a hydraulic jack and by placing a steel plate between the jack piston and concrete top surface. Numerically we could adopt several solutions: represent the exact experimental conditions by modelling the steel plate; directly apply load on a line of the mesh. In both cases we noticed that regarding the mesh size that we had also imposed in this exercise, under stress concentration beneath the load loading area, concrete could exhibit early stage nonlinear behavior and raise unrealistic numerical difficulties. That is why we adopted a third solution by applying a vertical uniform displacement over the entire cross section of the beam.

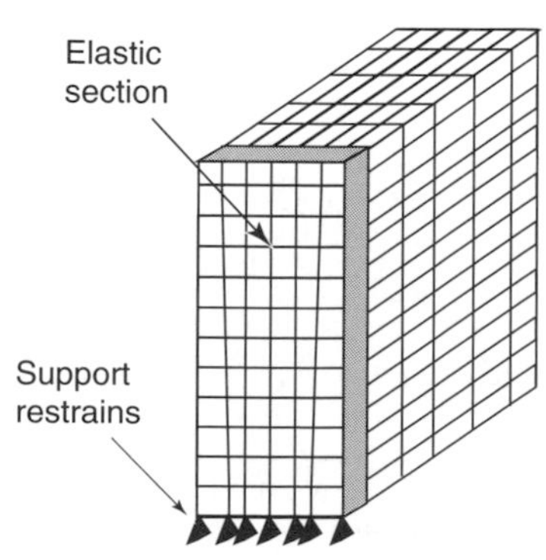

Figure 8. End support conditions.

Studies were proposed over two subcases: without stirrups (C1) and with stirrups (C2). The reason to this was to evaluate the decrease in load capacity and the additional numerical difficulties when stirrups are not considered. Thus comparison with experimental results has no sense with subcase C1.

The calculations were asked to be performed in 3D even though the structure does not exhibit any significant out of plane mechanisms. Most participants did their calculations in 3D, and a few in 2D.

The results were requested for the global (vertical support reaction and mid-span deflection) and local measures (stress and strain in steel and concrete along the beam). These are provided in Figure 9 for both subcases, along with the experimental results. The first result is the global F_y/u_y response where we can observe quite well estimates from most of the models for subcase C2. For C1 the results are widely spread around the experimental response (with stirrups). An interesting comment that can be made is that one could expect that a beam with no stirrups should fail earlier and exhibit lower load capacity compared to the same beam with stirrups. However the results found by different models are equally dispersed above and under the experimental curve.

For subcase C2 we can observe that all models exhibit the major characteristics of the beam behaviour which are: elastic response, first cracking with change in stiffness, tensile reinforcement yielding with the apparition of the ductile plateau, and finally some numerical behaviour which may in some cases be qualified as the "collapse" of the structure.

In reality this beam collapsed due to compressive crushing of concrete and the formation of a shear failure.

Local results concerned stress and strain along the tensile and compressive surfaces of the beam, as well as the tensile reinforcement bar. The most interesting measure for this structure is the tensile reinforcement strain over the mid-span cracking zone. If the collapse was due to the rupture of these bars, the analysis result on this measure would be of great importance. This was also recorded during the test, and Figure 10 indicates the strain as a function of the mid-span deflection. From subcase C1 it is clear that ignoring the presence of stirrups will provide very poor results for both global and local results. For C2 the agreement with experimental data is much more satisfying, for most constitutive laws.

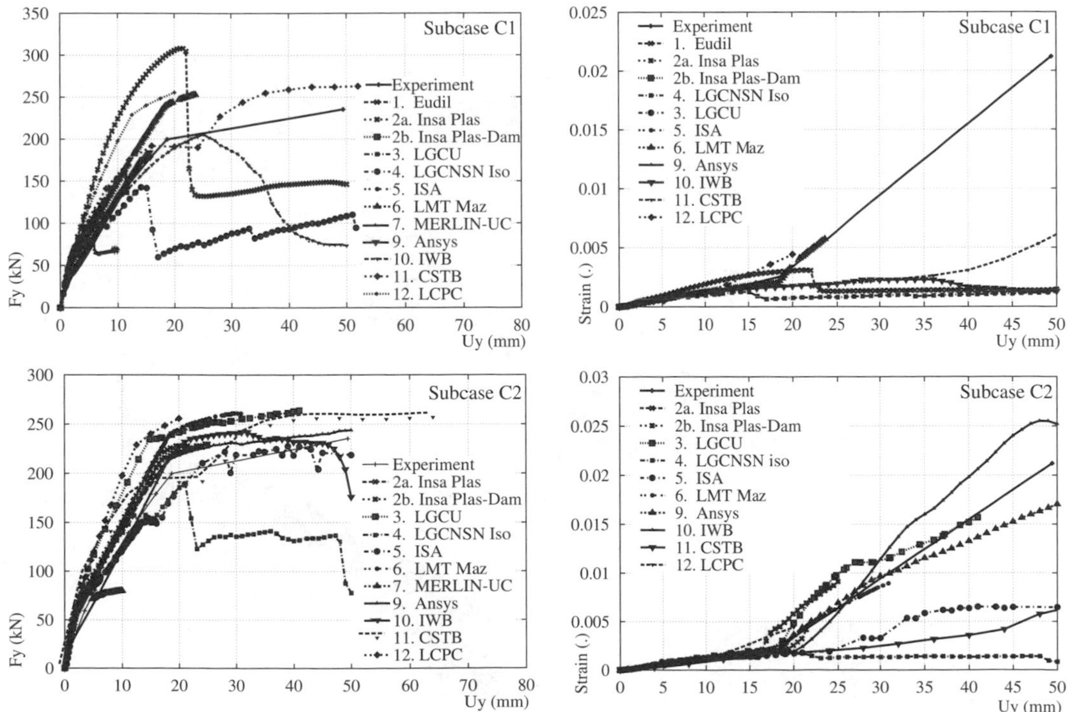

Figure 9. Global and local results from subcases C1(top) and C2 (bottom). Force/displacement (left) and reinforcement strain (at mid-span)/displacement (right). Comparisons with experimental records.

Detailed studies made on this case show that while the behaviour of a 3PB beam might seam simple, serious issues may render the task very complicated. First it concerns the isostatic equilibrium of the structure which can be lost when the entire cross section of the beam is cracked, for constitutive models that may not exhibit shear strength across cracks, and reinforcement bars are modelled as bars instead of beams. The specifications of the benchmark requested the use of bars,

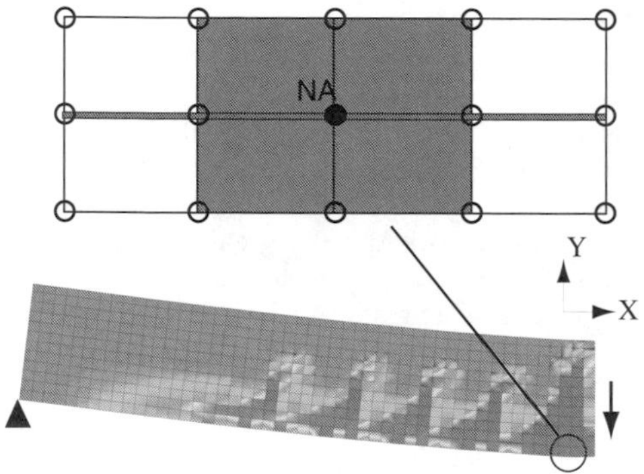

Figure 10. An example of crack pattern obtained by LGCU. Weak conditioning of stiffness matrix.

but it would probably ease the calculations if beam elements are used. Other reason that could emphasize this point is the fact that when adjacent mesh elements which may undergo severe cracking may lose stiffness in all directions, and this could raise convergence issues caused by a poor conditioning of the tangent stiffness matrix. This is shown on Figure 10 where node NA may hardly see any stiffness in U_y displacement.

This issue may also occur for all nodes supported by totally cracked elements, and slow the calculations.

The final results presented here are the crack pattern obtained from different modelling approaches, as shown on Figure 11. The comparison between graphics must be qualitative, since different constitutive models provide different type of results. Some correspond to damage (isotropic or non-isotropic) while others provide an elastic–plastic strain or principal strain measures.

It can be seen that most of them reveal the presence of stirrups by following a pattern in cracks, except for LMT model which is an isotropic damage model.

It can also be seen that in some cases the crack progression over mid-span is not the maximum along the beam. And from LGCNSN one can see that excessive damage has occurred on the compressive side of the beam which could partially explain the poor

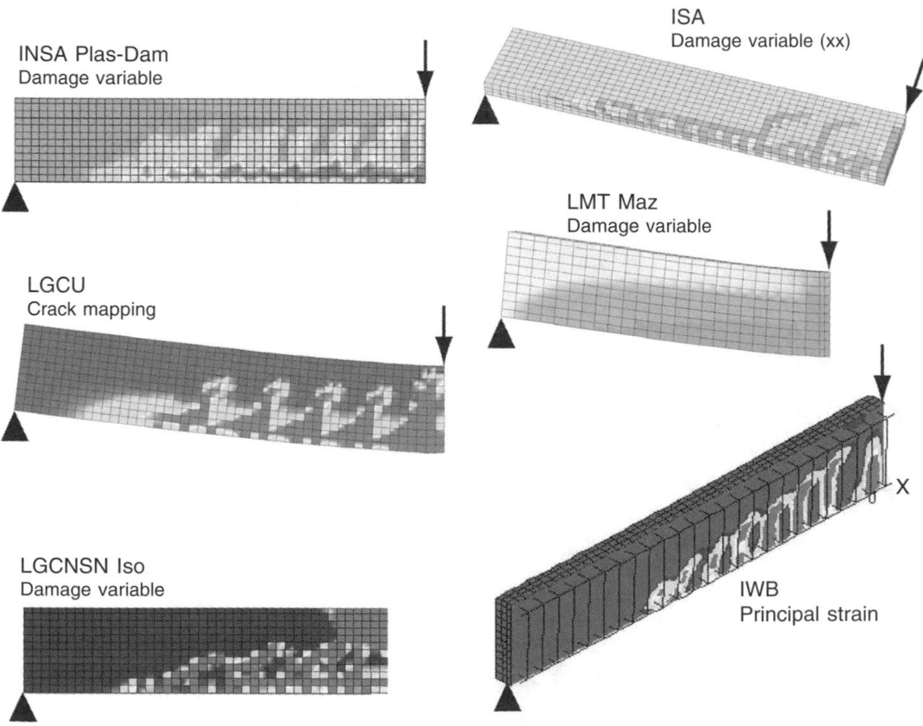

Figure 11. Subcase C2 "Crack patterns" obtained from 6 of the tested models.

quality in results compared to others. Probably this is due to early damage caused by crushing of concrete, and reveal boundary conditions not totally equivalent to other modelling.

6 CONCLUSIONS

A benchmark test has been proposed for concrete constitutive models and related implementation aspects in FE codes. It includes four tests, two purely-constitutive and two more of structural nature. A total of 13 groups from academia and industry have responded to this call, by providing most of or all the results to the test. Out of the four tests, Willam's test on the constitutive side and the 3PB RC beam analyzed as 3D continuum have been described in this paper.

Nine of the participants, some of them with two different models, have contributed to the constitutive test B, involving shear/tension with significant rotation of principal stresses. This is not a simple test which actually turns out very selective for the models, discriminating very clearly between isotropic and anisotropic behavior. By selecting four objective target features, two of the models turn out clearly inadequate since they fail to yield vanishing stresses at the end of the test. One more model is behaving well but limited by its isotropic character; one more only gets one of the remaining desired features, six get two of them, and finally only one satisfies all three of them.

For 3 point bending beam (test C) nearly all constitutive models provide good quality in results on the global nonlinear behaviour of the structure. This is mostly favoured by the presence of reinforcement bars and the isostatic character of the specimen.

However, local measures such as mid-span steel strain are more sensitive to the concrete law, but in general the results are of good quality.

ACKNOWLEDGEMENTS

We would like to express our gratitude for all the participants who accepted to undertake all these studies and provide their results with a very high quality of presentation, explanations and participate in meetings where very interesting exchange could take place.

We also thank the sponsors of this project, "Electricité de France" (EDF group) who initiated this activity providing scientific and financial support, and the French Ministry of Equipment within the framework of the RGCU national network for hosting the project and providing financial support.

The second author wishes to thank MCYT (Madrid, Spain) for funding obtained under grant MAT2000-1007. The help of Ricardo Lorefice in dealing with commercial software for the preparation of Figure 7 is also greatly appreciated.

REFERENCES

Carol, I., Lopez, C.M., Roa, O. (2001) Micromechanical analysis of quasi-brittle materials using fracture-based interface elements, *Int. J. of Numerical Methods in Engineering*, Vol. 52, No. 1–2, pp. 193–215.

Carol, I, Prat, P. (1995) A multicrack model based on the theory of multisurface plasticity and two fracture energies. In *Computational Plasticity–COMPLAS 4*, D.R.J. Owen et al., editors, Vol. 2, pp.1583–1594, Pineridge Press (UK).

Carol, I., Rizzi, E., Willam, K. (2001) On the formulation of anisotropic elastic degradation. I: Theory based on a pseudo-logarithmic damage tensor rate, and II: Generalized pseudo-Rankine model for tensile damage. *Int. J. of Solids and Structures*, Vol. 38, No. 4, pp. 491–546.

Delaplace, A., Ghavamian, S. (2003) Etat de l'art de modèles de fissuration de béton – Projet MECA. Revue Française de Génie Civil, no. Vol. 7 (to be published).

Kroeplin, B., Weihe, S. (1997) Constitutive and geometrical aspects of fracture-induced anisotropy. In *Computational Plasticity, COMPLAS 5*, E. Owen et al. editors, CIMNE, Vol. 1, pp. 255–279.

Lopez, C.M., Carol, I., Aguado, A. (2000) Microstructural analysis of concrete fracture using interface elements. In *ECCOMAS 2000*, E. Oñate et al. editors, CIMNE (in CD ROM).

López, C.M., Carol, I., Murcia, J. (2001) Mesostructural modeling of basic creep at various stress levels. In *Creep and Shrinkage of Concrete Structures –CONCREEP 6*, F.J. Ulm et al. editors, John Wiley. pp. 101–106.

Meschke, G., Macht, J., Lackner, R. (1998) A damage-plasticity model for concrete accounting for fracture-induced anisotropy. In *Computational Modelling of Concrete Structures – EUROC 98*. Mang, H. et al. editors, Balkema, pp. 3–12.

Nooru-Mohamed, M.B. (1992) *Mixed mode fracture of concrete: an experimental approach*. Doctoral thesis, Delft University of Technology.

Oliver J., Cervera, M., Oller, S., Lubliner, J. (1990) Isotropic damage models and smeared crack analysis of concrete. In *Computer-Aided Analysis and Design of Concrete Structures* – SCI-C 90, N. Bicanic and H. Mang, editors, Pineridge Press, Vol. 2, pp. 945–957.

Pera, J. (1973) Reinforced concrete redundant beams. Theoretical and experimental analysis. Thesis by a doctor-engineer, INSA Lyon.

Roelfstra, P.E., Sadouki, H., Wittmann, F.H. (1985) Le béton numerique, *Materials and Structures*, Vol. 18, pp. 309–317.

Rots J.G. (1988) *Computational Modelling of Concrete Fracture*, Doctoral thesis, Delft University of Technology.

Stankowski, T. (1990) *Numerical simulation of progressive failure in particle composites*, PhD thesis, Dept. CEAE, Univ. of Colorado, Boulder (USA).

Willam, K., Pramono, E., Sture, S. (1987). Fundamental issues of smeared crack models, *Proc. of the SEM-RILEM Int. conf. on fracture of concrete and rock*, Shah, S.P. and Swartz, S.E. editors, Society of Engineering Mechanics, pp. 192–207.

Computational Modelling of Concrete Structures, Bićanić et al. (eds)
© 2003 Swets & Zeitlinger, Lisse, ISBN 90 5809 536 3

Bond in finite element modelling of reinforced concrete

V. Cervenka, J. Cervenka & L. Jendele
Cervenka Consulting, Prague, Czech Republic

ABSTRACT: Numerical model for bond between reinforcing bars and concrete is formulated and implemented in the nonlinear finite element code for simulation of real behavior of concrete structures. It is based on one-dimensional geometry of reinforcing bars, but it considers interface surface properties, and the same physical models can be applied as in the case of the full surface interface. Any bond-slip relationship can be implemented within the model. Example analysis of a shear failure of beam is presented and the effect of bond on the beam response is demonstrated.

1 INTRODUCTION

Bond between steel reinforcing bars and concrete is an important part of the reinforced concrete mechanical system. Due to its significance for practical design bond was investigated by many researchers and technical committees (Report FIB 2000). This effort resulted in numerous empirical formulas and complex code provisions for bar anchoring length and reinforcement detailing. This knowledge is mainly utilized in practical design, but it can also serve as a basis for constitutive models in numerical analysis.

The current worldwide trend towards more extensive use of computer simulation in engineering design requires robust and sound numerical tools for modelling of concrete structures. Major features of structural behaviour can be captured by advanced nonlinear finite element techniques combined with sound material models for concrete and steel. In this context, adequate bond models for finite element analysis are missing. In fact, most computer codes assume a rigid bond. It is argued, that the friction bond mechanism is prevented by ribbed surface of reinforcing bars and thus the bond failure actually occurs within the surrounding concrete. Of course, this assumes that an appropriate constitutive model is applied to concrete itself. This argument is in contradiction with experimental observation in cases of short anchorage regions, lapped splices, smooth reinforcing bars, post-installed reinforcement, etc (FIB 2000).

The current computer codes offer possibilities of very detailed analysis of interaction between concrete and bars on a small-scale level. Such an approach was chosen by Lundgren (1999) who developed an interface bond model based on plasticity theory with fully three-dimensional features. This approach can well support the research projects. However, for practical reinforced concrete structures it is not suitable for its extremely large capacity requirements.

Authors have recently attempted to overcome the large capacity requirements of the surface interface models. The idea was to reduce the dimension of the interface problem, so that it can be applied to truss bar finite elements of discrete reinforcement. It is based on one-dimensional geometry of reinforcing bars, but it considers interface surface properties, and the same physical models can be applied as in the case of the full surface interface. The only difference is that the model assumes the bond stresses along the perimeter in the average sense. Obviously, the bond behaviour can vary along the bar length. The advantage is that the model can be used for finite element models with embedded bars and can be applied in 2D as well as 3D finite element models of concrete. The bond model is defined in form of a bond stress-slip relation, such as proposed by Bigaj (1999) or FIB (2000). Other relations can be easily adopted.

The effect of bond on structural behavior is demonstrated on simulation of shear beam experiments due to Leonhardt and Walther (1961).

2 BOND MODEL

The bond behaviour was extensively studied experimentally and is quite well understood. Two basic features can be recognized: friction behaviour between two materials, steel and concrete and crack propagation in the concrete volume adjacent to reinforcement.

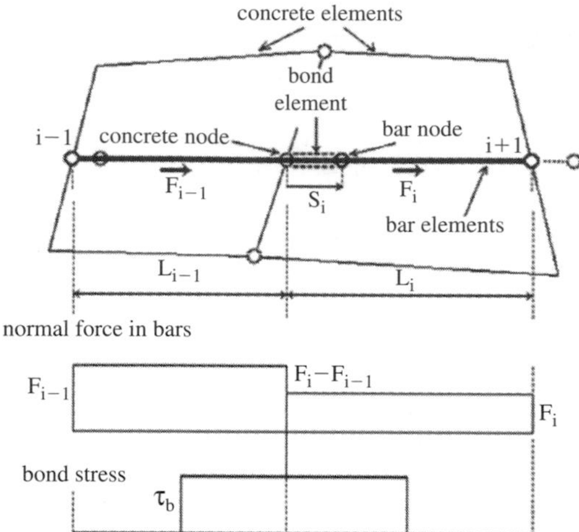

Figure 1. Bond element.

The extent of cracking depends on the level of friction resistance of the contact surface. In the most typical case of ribbed bars the friction resistance is greatly improved and the crack propagation is prevailing. This behaviour is greatly affected by confining pressure of surrounding concrete. Another important feature is relatively small size of the cracked concrete comparing to structural size.

The numerical model can be structured in two levels. First level covers the bond finite element, which supports the contact between bars and concrete. The second level is the bond-slip law on the contact surface. This law is independent of the bond finite element.

2.1 *Bond finite element*

The reinforcing bar elements are subjected to axial forces with constant distribution along the bar, Fig. 1. Bars are connected with concrete elements only on their borders. In case of perfect bond the bar nodes are exactly following the deformations of concrete. The displacements of the bar ends, denoted as "concrete nodes", are kinematically dependent on the deformation of concrete element. Therefore, there are no additional degrees of freedom in bars. The solution of bar internal forces F_i is performed within the finite element model in terms of nodal displacements u_i of concrete elements.

The bond behavior is introduced by bond elements defined around connecting points between bars and concrete elements. Bond elements connecting the bar nodes to concrete nodes represent contact links and are formulated kinematically. (They are not based on stiffness approach.) Bond slip deformations s_i are introduced as additional degrees of freedom.

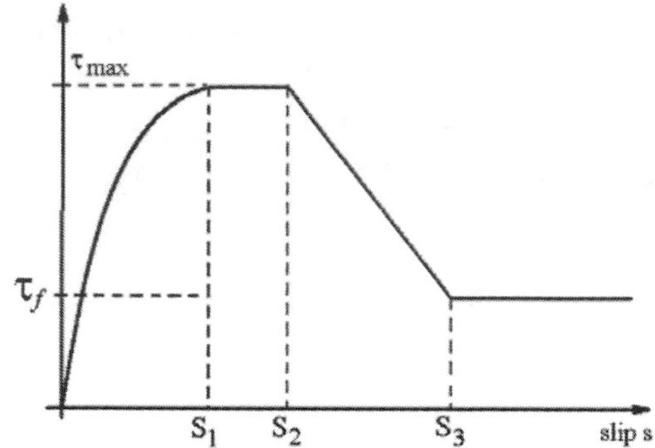

Figure 2. CEB-FIP Model Code 90 Bond model.

Bond constitutive law is defined as a bond stress-slip relation.

$$\tau_b = C_{(s_i)}^{bond} \tag{1}$$

Constant distribution of bond stress τ_b on the bar surface is assumed and the lengths of the bond element reaches to the middle of the bars on both sides. A bond slip occurs if the forces on the bond element are exceeding the resisting bond forces.

$$\|F_i - F_{i-1}\| \geq \frac{1}{2}\tau_b p(L_{i-1} + L_i) \tag{2}$$

In the above p is bar perimeter and L_i is the bar length in the bond element i. The slip displacements s_i are obtained if the equilibrium in the system consisting of bond and bar elements (K_i^{bar} is bar axial stiffness) is established.

$$F_{i-1} + k_{i-1}^{bar}(s_i - s_{i-1}) - F_i + k_i^{bar}(s_{i+1} - s_i) = \frac{1}{2}\tau_b p(L_{i-1} + L_i) \tag{3}$$

The equation system (3) for unknown slips s_i must be solved simultaneously. This solution is performed as an internal loop within the iteration of the global nonlinear solution of nodal displacements u_i.

2.2 *Bond-slip law*

The function $C_{(s_i)}^{bond}$ represents the constitutive relation of the bond slip. It is a function of slip and some parameters such as confinement and surface properties. In presented models the confinement effect is simplified and is covered only roughly by classes of confinement instead of directly including the confining stress.

Model according to CEB-FIP Model Code 90 offers the bond-slip relation as illustrated in Fig. 2. The ascending branch describes the propagation of radial cracks around reinforcement and descending branch the widening of cracks after they stabilize or reach the

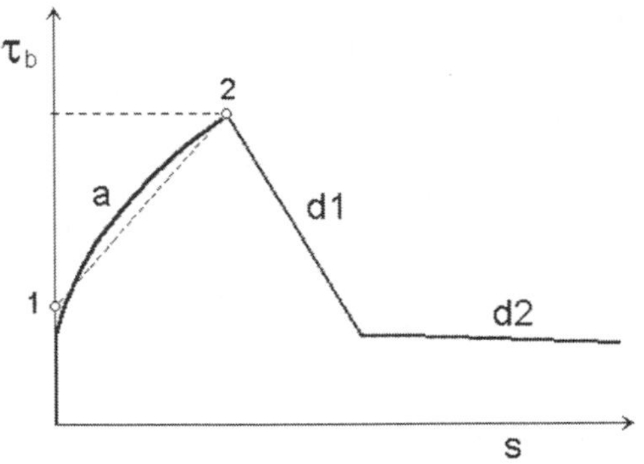

Figure 3. Bond-slip law according to Bigaj (1999).

surface. The details of the function and numerical data for specific conditions can be found in the Model Code 90. We shall mention only some important values. The ascending part has exponential form.

$$\tau_b = \tau_{\max}(s/s_1)^{\alpha}$$
$$\tau_{\max} = c\sqrt{f_c} \tag{4}$$

The slip at the peak of the ascending branch $s_1 = 1$ mm (0.6 mm) for hot rolled bars with ribs and cold drawn wires, respectively. The peak bond stress is in the range of approximately 6 to 15 MPa.

Recently developed model by Bigaj (1999) is shown in Fig. 3. The ascending part has also the exponential form:

$$\tau_b = \sqrt{f_c}A(s/D)^B \tag{5}$$

where D is bar diameter, A, B are constants and can be found in the original report, Bigaj (1999).

The authors implemented the above models in the finite element code ATENA. In this implementation the ascending branches were approximated by a piecewise linear function as shown in Fig. 3. Some experience with this application is shown in this paper.

3 EFFECT OF BOND ON SHEAR FAILURE

Authors have shown in the previous publication Cervenka (1998), that the shear failure can be well simulated by models based on fracture mechanics. Although the main features of brittle failure are covered, it was suspected, that the bond behavior may also contribute to the response. The present study offers answers to these questions. The analysis is performed by the program ATENA.

The four-point bending test by Leonhardt and Walther (1962) is analyzed. The beam properties: shear span $L_s = 1{,}62$ m, cross section $h = 320$ mm, $b = 190$ mm, reinforcement by 2 bars Ø26 mm, cover 37 mm, concrete $f_c = 28{,}5$ MPa, $f_t = 1{,}64$ MPa, $E_c = 32000$ MPa, $G_f = 62$ N/m.

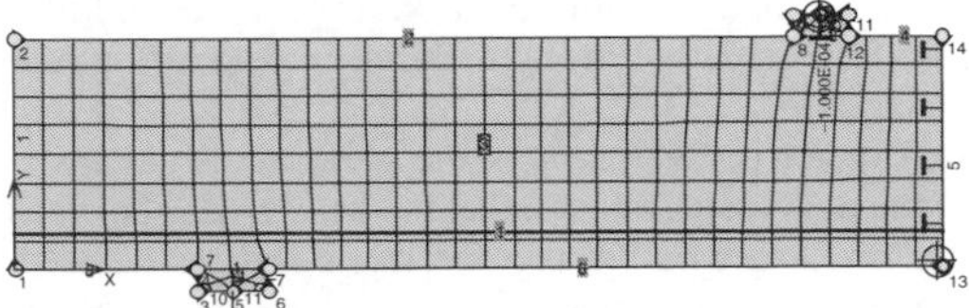

Figure 4. Numerical model of Leonhardt's beam.

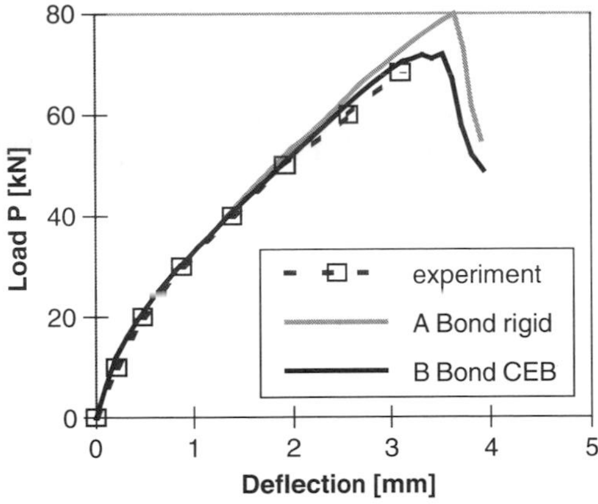

Figure 5. Load-displacement response of shear test.

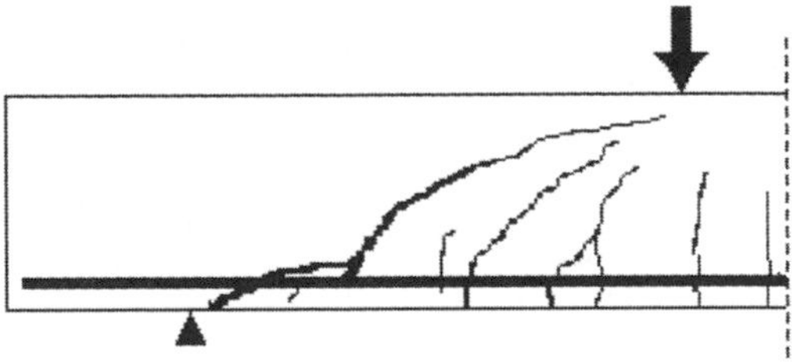

Figure 6. Crack pattern in Beam no. 5 from experiments.

The numerical model is shown in Fig. 4. The concrete is modeled by 4-node isoparametric element with four integration points. The reinforcement by embedded truss bars in uniaxial stress state. The major nonlinear behavior is due to crack propagation and is modelled by energy-based crack band method as proposed by Bazant and Oh (1983). Details of ATENA constitutive models can be found in another paper in this proceedings, Pukl et al. (2003). The bond model used is according to CEB-FIP MC90, Fig. 2.

Only symmetrical half of the beam is analyzed. The load is applied as imposed displacement on the top loading plate. Case A is modeled with perfectly rigid bond and Case B with the bond model. They are compared with experiment in Fig. 5.

The crack pattern after failure found in experiment is shown in Fig. 6. The experimental crack pattern can be compared with simulations illustrated in Fig. 7 and 8, where the effect of bond model can be also observed.

The above figures show the stress and crack states in the next load step just after the peak load. The crack patterns show only those cracks, which are (according to the crack band theory) wider than 0.05 mm. The introduction of bond model causes only minor change in the crack pattern. Nevertheless, it produces a pattern, which is closer to the experimental one. The experiment indicates some splitting cracks near reinforcement, which are better reflected by bond model.

The stiffness of the beam in the cracked state under medium loading was not significantly effected by the bond model. It can be concluded that the bond model has only minor effect on the tension stiffening, which is well covered by fracture of concrete. This conclusion can serve as an important message for practical applications.

The high bond stresses found in Case A with rigid bond are reduced in Case B, where bond slip is generated. The bond slip has also a small effect on reduction of ultimate load.

It should be also realized, that the bond laws used in this approach are relatively rough and introduce only qualitative change comparing to rigid bond.

The numerical implementation of the model is very robust and does not cause any numerical problems in normal situation with well anchored bars. Of course, in case of pull-out failure the response can be unstable and should be analyzed with due care.

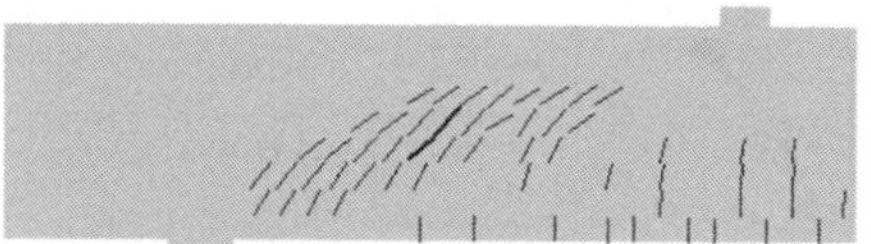

Rigid bond. Crack pattern.

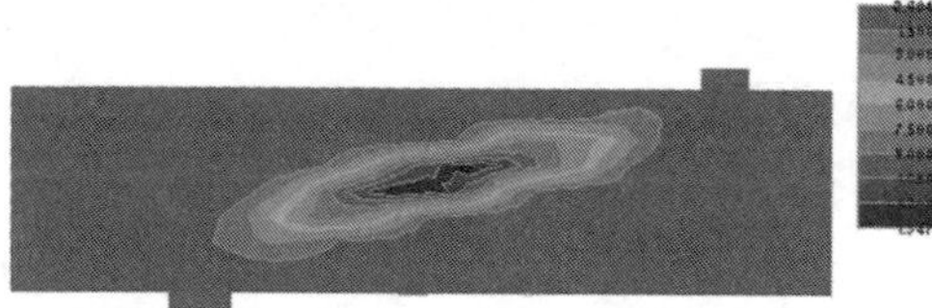

Rigid bond. Principal tensile strain iso-areas.

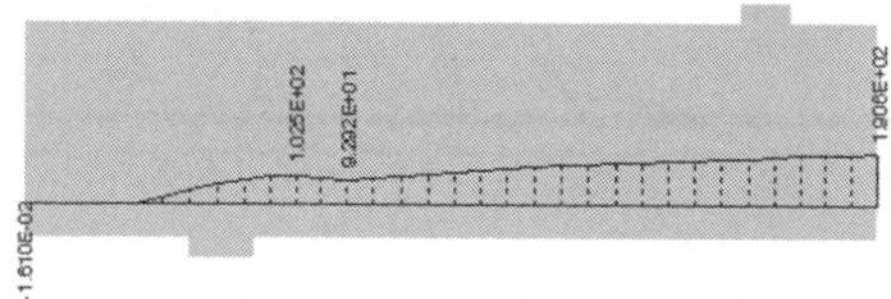

Rigid bond. Stress in reinforcing bar.

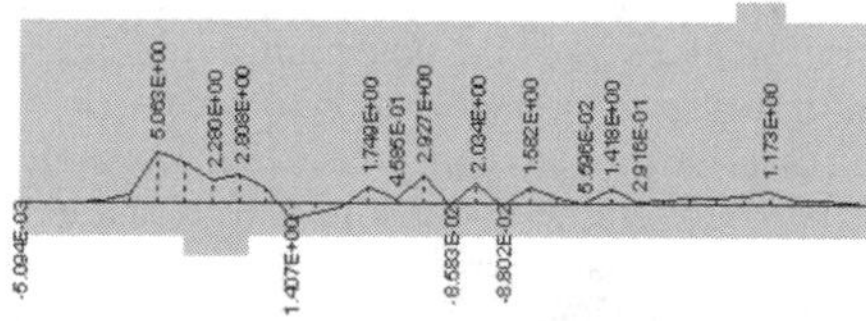

Rigid bond. Bond stress.

Rigid bond. Bond slip.

Figure 7. Case A with rigid bond. Stress state at peak.

Bond model. Crack pattern.

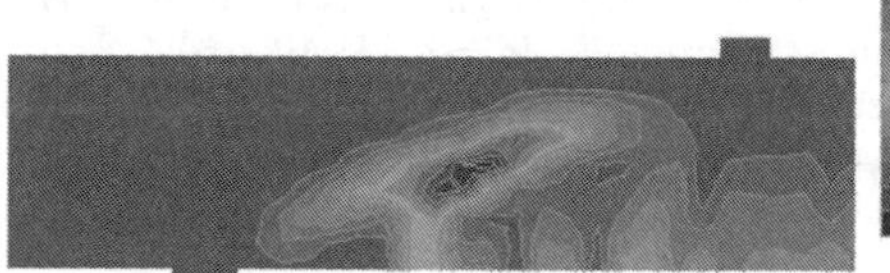

Bond model Principal tensile strain iso-areas.

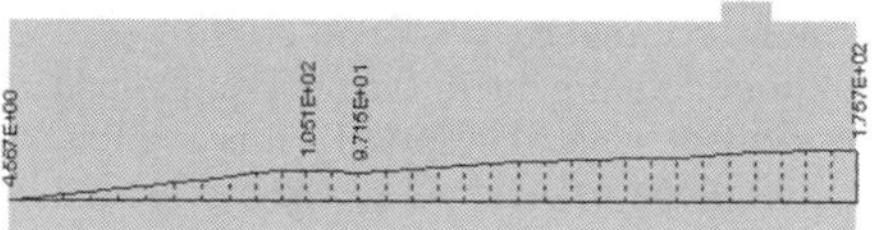

Bond model. Stress in reinforcing bar.

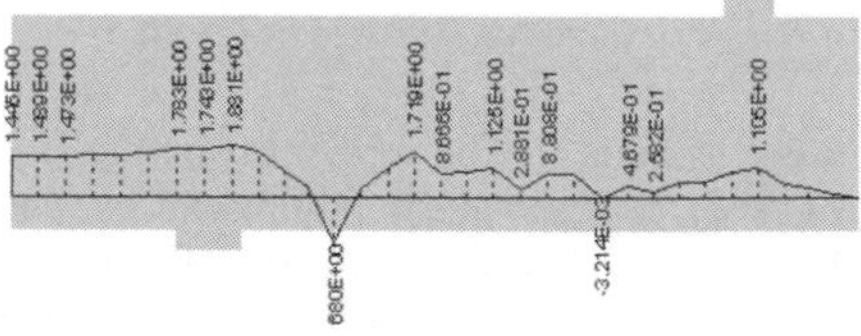

Bond model. Bond stress.

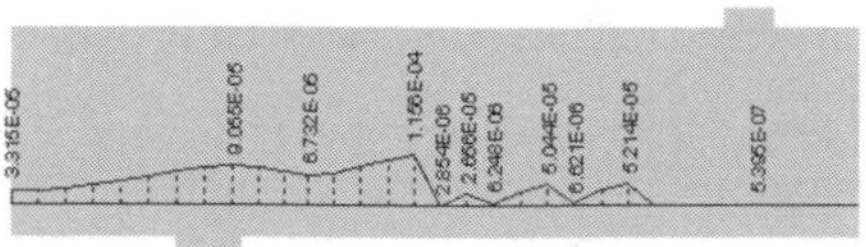

Bond model. Bond slip.

Figure 8. Case B with bond model. Stress state at peak.

4 MESH SIZE EFFECT

The mesh sensitivity is an important issue deciding about objectivity of results. In the previous study, Cervenka (1998), it was confirmed, that the brittle failure in concrete due to formation of diagonal crack was not very mesh sensitive. The reason for this was the energy-based constitutive law for crack propagation. Now we shall investigate the mesh sensitivity of bond.

The above example had medium mesh with 8 divisions through the beam height, the size of element was 40 mm. Such mesh should be sufficient for acceptable simulation of shear failure. For more detail simulation we choose the fine mesh, with 16 divisions and the element size 20 mm. The results of this analyses for both meshes are compared on load displacement diagram in Fig. 9. The crack pattern, stress localization and bond stress distribution are shown in Figs. 10, 11 and 12.

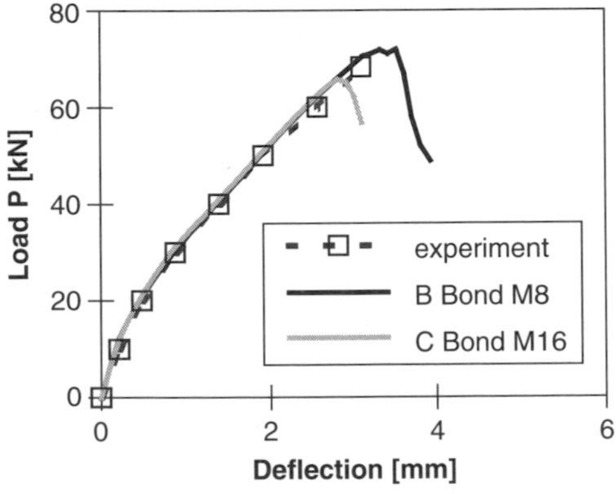

Figure 9. Effect of mesh size on load-displacement response.

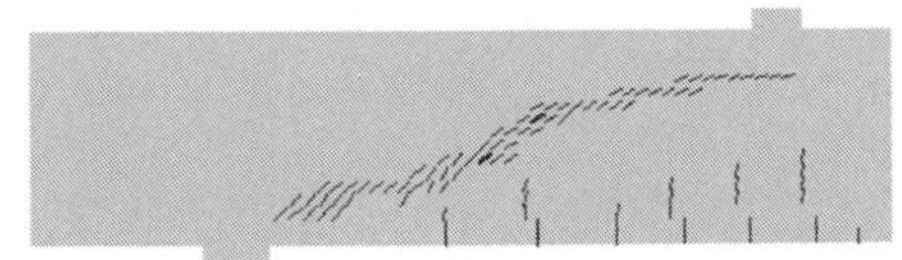

Figure 10. Fine mesh. Crack pattern after peak.

Figure 11. Fine mesh. Principal tensile strains after peak.

It can be observed, that the load displacement response follows almost identical path as the one for medium mesh. However, the peak load is slightly lower at Pmax = 65.7 kN compared to 71.9 kN for the medium mesh. It can be observed, that the bond model is more sensitive to the mesh size.

This can be expected because the formulation of the bond model is based on local values of bond stress and slip. Another factor increasing the mesh sensitivity could be in the nature of bond stresses. It is defined as a first derivative of normal stress distribution in reinforcement and thus it reflects closely all gradients created by finer meshes.

It can be also concluded that the fine mesh reproduce much better the experimental crack pattern.

By various numerical studies, which exceed the frame of this paper, it was found that the bond model must be seen in the context of introduced approximations. 3D analysis poses less constrains on the deformation fields compared to the plane stress idealization used here and allows more precisely to model cracking around reinforcement. In such analysis the element size plays an important role.

5 CONCLUSIONS

The proposed bond model is based on the bond-slip constitutive law. The complexity of bond behavior is simplified and various influences on bond behavior are reflected in the form of bond-slip relationship.

The model was applied to investigate the significance of bond on the shear failure of reinforced concrete beam. It was found that it improves the simulation of real behavior. However, in the investigated case the effect is relatively minor. This conclusion can be explained by the fact that the major effect in bond failure is due to the crack propagation, which is already included in the concrete constitutive model and only minor part of bond slip is contributed by the bond-slip model.

In most practical situations the model need not to be included. For example the tension stiffening effect, which is considered as an important feature in reinforced concrete modeling can be sufficiently covered by concrete fracture.

Further research and development is needed to improve the bond models, especially the confinement effect, and to validate its application.

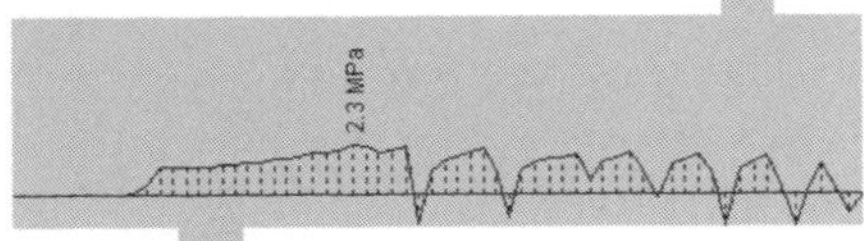

Figure 12. Fine mesh. Bond stress distribution after peak.

REFERENCES

Bažant, Z.P. & Oh, B.H. 1983. Crack band theory for fracture of concrete. *Materials and Structures* 16: 155–177.

Bigaj, A.J. 1999. *Structural Dependence of Rotation Capacity of Plastic Hinges in RC Beams and Slabs*, PhD Thesis, Delft University of Technology, ISBN 90-407-1926-8.

Cervenka, V. 1998. Simulation of shear failure modes of r.c. structures. Proceedings of the EURO-C 1998, Conference on Computational Mechanics of Concrete Structures, Badgastein, Austria, 31 March – 3 April, 1998, pp. 833–838.

FIB 2000. Bond of reinforcement in concrete, State-of-art report by Task Group Bond Models, *fib Bulletin 10*, August 2000.

Leonhardt, F. & Walther, R. 1962. Schubversuche an einfeldrigen Stahbetonbalken mit und ohne Schubbewehrung. *Deutsche Ausschuss für Stahbeton, Heft 151*, Ernst & Sohn, Berlin.

Lundgren, K. 1999. *Three-Dimensional Modelling of Bond in Reinforced Concrete.* PhD Thesis, Department of Civil Engineering, Chalmers University of Technology.

Pukl, R., Novak, D., Bergmeister, K. 2003. Reliability assessment of concrete structures, Euro-C 2003 (to be published).

Computational Modelling of Concrete Structures, Bićanić et al. (eds)
© 2003 Swets & Zeitlinger, Lisse, ISBN 90 5809 536 3

Multiaxial stress-induced thermal strains: constitutive modelling and laboratory findings

Roger S Crouch, Mihail Petkovski & Peter Waldron
Computational Mechanics Unit, Department of Civil & Structural Engineering, University of Sheffield, Sheffield, UK

ABSTRACT: This paper reports on recent findings on the deformation response of concrete subjected to elevated temperatures (up to 250°*C*) while under multiaxial compressive loads (up to 400 *MPa*). Firstly, a motivation for understanding such behaviour and constructing a unified constitutive framework is given through consideration of the structural integrity of pre-stressed concrete nuclear reactor vessels. Some remarks on de-coupling the strain rates then follows, together with an overview of some existing constitutive approaches that include stress-dependent thermal strains. The next section describes the novel apparatus (mac^{2t}) used to capture the deformation behaviour of concrete under multiaxial conditions. Using this rig, a quite different short-term behaviour is revealed when comparing the response under uniaxial compression to that under hydrostatic compression. The results suggest that it is the deviatoric proximity to the multiaxial peak stress envelop, rather than the absolute stress level, that dictates the degree of microcrack activity, which in turn influences the vapour transfer rate. The tests also confirm the appropriateness of using a tensorial (multiaxial) relationship when deciding upon the dependency of the asymptotic drying creep strains on the stress state. Finally, some remarks are made on the influence of the heat-load sequence on the pre-peak stress-strain nonlinearity and the less pronounced influence of the sequence on the peak strength.

1 PREDICTING STRUCTURAL DEFORMATIONS IN PCPVs

Prestressed concrete reactor pressure vessels (PCPV) continue to function successfully within the nuclear industry worldwide. These thick-walled structures (with numerous apertures) are designed to resit moderately high internal pressure and elevated temperature conditions throughout their working life. Because of their importance, significant safety margins and fail-safe features are built-in at the design stage to effectively eliminate any risk of a severe accident. A system of regular monitoring, assessment and maintenance ensures that these structures (many nearing the end of 40 years of service) still operate as originally intended.

It is widely understood that the strength and stiffness of concrete depends on the maximum temperature it is subjected to. Between 20–100°*C* (spanning the operational range for many reactors) concrete remains essentially undamaged, provided very rapid local heating does not occur, which could lead to high pore pressures and possible surface bursting. Above about 150°*C*, the material characteristics begin to alter, with significant damage occurring beyond 600°*C*.

As part of the concrete vessel management strategy, advanced Finite Element (FE) codes have proved valuable to engineers wishing to predict the behaviour under differing load cases. While most operational-state FE simulations may be performed assuming linearity in the response, certain cases require a nonlinear analysis. The degree of realism available to the latter depends almost exclusively upon the accuracy and robustness of the constitutive model describing the coupled thermo-hygral-elasto-viscoplastic behaviour of this complex heterogeneous material. Comprehensive 3D FE codes able to simulate pore vapour diffusion, thermal conduction and material deformation are only just emerging. In order that constitutive models may be constructed and verified, new experimental data are needed to clarify the relationship between temperature, nonlinear deformation and moisture state.

The three-dimensional short-term creep and shrinkage response of concrete is of particular interest to structural engineers considering PCPV integrity under local temperature excursions. While some carefully conducted studies have been performed under uniaxial and biaxial compressive stress conditions (6), few experimental data exist at elevated temperature under *multiaxial* stress states. In the case of uniaxial loading, the results show that the sequence in which heat-up and compressive loading occurs has a pronounced effect on irrecoverable straining in concrete. Pre-loading a specimen in uniaxial compression to $\approx 0.25 f_c$, then holding this load while heating at a rate slow enough to limit spatial variations in the temperature (say, $<0.03°C \cdot s^{-1}$ for a 100 mm cube heated uniformly on all faces) can totally suppress the thermal expansive strains, yet significant net expansion remains in a nominally identical specimen that has been heated at the same rate *and then* subjected to the same axial compressive stress. During cool down, the additional strains recorded in the first specimen seem to be not recovered. Upon second and subsequent reheating these strains are not experienced again until the previous maximum temperature is exceeded.

It is apparent that a meaningful interpretation of such behaviour in a laboratory test on a suitably representative volume can only be made through consideration of (i) the three-dimensional stress state, (ii) the moisture conditions in, and on the boundary of, the specimen (in terms of vapour transfer) and (iii) the history of loading, humidity and temperature variations.

At present there is no agreement over the name (and, more importantly, the chemo-physical explanation) for these additional creep strains observed on first heating under load. The lack of consensus has been due in part to the difficulty of unambiguously isolating the various components in laboratory experiments. As discussed below, the additional strains have been referred to as *transitional thermal creep, transient thermal creep, load-induced thermal strains, drying creep* (or the Pickett effect), *sorption creep* and *stress-induced shrinkage*. Whatever the name, this phenomenon is of interest to the operational behaviour of pre-stressed concrete nuclear reactor vessels.

It is in this light that a programme of experimental work to capture the stress-dependent thermal strains was initiated at Sheffield University in 1997 through support from the UK nuclear industry and the UK Engineering and Physical Sciences Research Council.

2 CONSTITUTIVE MODELLING: A REVIEW OF STRAIN DECOMPOSITION AND SOME FORMULATIONS

This short review does not claim to be comprehensive in its scope, however the selected findings serve to illustrate some of the approaches adopted by others. Furthermore, readers should not attach too much significance to the date associated with each work as it does not refer always to the time that the researchers first offered their ideas.

2.1 *Anderberg and Thelandersson (1976)*

Over twenty-five years ago Anderberg and Thelandersson presented a one-dimensional phenomenologically-based constitutive model for the behaviour of concrete under elevated temperature (1). A relationship was established between total strain, ε, stress, σ, time, t, temperature, T, and stress history, $\tilde{\sigma}$. The total strain was considered to be the sum of four components: (i) instantaneous stress-related strain, $\varepsilon^{\sigma}(\sigma, \tilde{\sigma}, T)$ (based on the stress strain curves obtained at constant temperature), (ii) thermal strain, $\varepsilon^{T}(T)$ (including shrinkage, measured on unstressed specimens under variable temperature), (iii) creep strain, $\varepsilon^{c}(\sigma, T, t)$ (time-dependent strain recorded at constant temperature under constant stress) and (iv) transient creep strain, $\varepsilon^{tc}(\sigma, T)$ (accounting for the effect of strains induced by temperature changes under constant stress).

$$\varepsilon = \overbrace{\varepsilon^{\sigma} + \underbrace{\varepsilon^{T}}_{\substack{free\\thermal}} + \underbrace{\varepsilon^{tc}}_{\substack{transient\\thermal\ creep}}}^{instantaneous} + \overbrace{\underbrace{\varepsilon^{c}}_{\substack{basic\\creep}}}^{time\text{-}dependent} \tag{1}$$

The *instantaneous* stress-related strain was assumed to follow a parabolic relationship with stress in the pre-peak range. The post-peak behaviour was deemed to follow the parabolic form until a *transition strain* (ε_1) is attained, whereafter a linear softening model (with stiffness $E^* = 0.8\ GPa$) was proposed. The latter is inobjective, but the authors were probably aware that the post-peak behaviour is controlled largely by specimen restraint (boundary) conditions. Anderberg and Thelandersson acknowledged that more work was needed in this area.

$$\begin{aligned} \sigma &= \sigma_u \frac{\varepsilon^{\sigma}}{\varepsilon_u}\left(2 - \frac{\varepsilon^{\sigma}}{\varepsilon_u}\right) & 0 \geq \varepsilon^{\sigma} \geq \varepsilon_1 \\ \sigma &= E^{\star}.\varepsilon^{\sigma} + \sigma_u\left(1 - \frac{\varepsilon_u.E^{\star}}{2.\sigma_u}\right)^2 & \varepsilon^{\sigma} < \varepsilon_1 \end{aligned} \tag{2}$$

where

$$\varepsilon_1 = \varepsilon_u\left(1 - \frac{E^{\star}}{E_c}\right) \tag{3}$$

The peak uniaxial stress, σ_u, was known to depend on temperature and mix details (water:cement ratio, aggregate type and relative aggregate volume). Although Anderberg and Thelandersson gave some evidence of the stress history affecting the peak strength for a given concrete mix, these were neglected in their model. Thus, for a particular mix σ_u was made to depend only on temperature. The history was shown to exhibit a more significant influence on the strain corresponding to the peak stress (ε_u)

than the peak stress itself. In the case of unstressed specimens heated without pre-stress, ε_u increases with temperature level. With a pre-stress greater than $0.3f_c$, Anderberg and Thelandersson found that ε_u essentially did not change over the temperature range 20–700°C. Their expression for ε_u read as follows

$$\varepsilon_u = \min(\varepsilon_{u0}, \bar{\varepsilon}_u - \varepsilon^{tc}) \tag{4}$$

where ε_{u0} is the strain corresponding to peak stress under ambient conditions, $\bar{\varepsilon}_u(T)$ is the temperature-dependent strain at peak stress with zero pre-stress and ε^{tc} is the transient creep strain. This implies that the strain at peak stress is reduced by $|\varepsilon^{tc}|$ due to a history of pre-stress, but is always greater than, or equal to, the peak strain under ambient conditions.

In Anderberg and Thelandersson's model, the basic creep strain was given by

$$\varepsilon^c = \beta.\frac{\sigma}{\sigma_u(T)}\sqrt{\left(\frac{t}{10.8\times 10^3}\right)}.\exp^{3.04\times 10^{-3}(T-20)} \tag{5}$$

where the coefficient $\beta = -0.53\times 10^{-3}$ and t, the time, is expressed in seconds.

The transient creep strain (occurring only on first heating) was given as a function of the free thermal strain and the relative stress magnitude, as follows

$$\varepsilon^{tc} = -2.35\frac{\sigma}{\sigma_{u0}}.\varepsilon^T \tag{6}$$

It is interesting to note that Anderberg and Thelandersson suggested that the difference between the expansive strains recorded in specimens heated at $0.017°C\cdot s^{-1}$ and $0.083°C\cdot s^{-1}$ was rather small at all load levels. These numbers are important since they influenced the maximum rate of heating used in the 2001–2002 University of Sheffield tests.

2.2 *Thelandersson (1983)*

Seven years later, Thelandersson was the first to offer a fully three-dimensional model for concrete behaviour under elevated temperature (9). Although this formulation was based on his earlier work with Anderberg, it was far more ambitions in its scope. The total strain rate was split into (i) elasto-viscoplastic mechanical and (ii) temperature and moisture-induced components.

$$\dot{\varepsilon}_{ij} = \overbrace{\dot{\varepsilon}_{ij}^{evp}}^{\textit{mechanical}} + \overbrace{\dot{\varepsilon}_{ij}^{Th}}^{\substack{\textit{temperature +}\\ \textit{moisture-induced}}} \tag{7}$$

The hardening/softening plasticity model was in the form of a Modified Cam Clay formulation, embracing the concept of a critical state of deformation. Thelandersson suggested that the temperature and moisture-induced strain components ($\dot{\varepsilon}_{ij}^{Th} = \dot{\varepsilon}_{ij}^{T} + \dot{\varepsilon}_{ij}^{h}$) could be described by

$$\begin{aligned} \dot{\varepsilon}_{ij}^{T} &= \left(\alpha(T,h)\delta_{ij} + g_{ij}(T,h,\sigma_{ij},\mathrm{sign}(\dot{T})\right)\\ \dot{\varepsilon}_{ij}^{h} &= \left(\kappa(T,h)\delta_{ij} + f_{ij}(T,h,\sigma_{ij},\mathrm{sign}(\dot{h})\right) \end{aligned} \tag{8}$$

The functions g_{ij} and f_{ij} were acknowledged as being difficult to determine explicitly, as measurements of shrinkage and expansion are invariably affected by specimen size and rate of moisture change due to differential shrinkage cracks forming. Thelandersson restricted his model to treat the response of unsealed specimens subject to temperatures above 100°C. In this way the moisture content was treated as being independent of time but uniquely dependent upon temperature. The problem then reduced to one of specifying the function g_{ij}. The simple form

$$g_{ij} = \beta(T).\sigma_{ij} \tag{9}$$

was chosen, with two alternative expressions for the dependence of β on the temperature

$$\beta(T) = \frac{\beta_1}{f_c}\alpha \tag{10}$$

where α is the coefficient of thermal expansion and $\beta_1 = 2.35$. The second form was given in terms of a dependence on the difference between the free thermal volumetric strain in the aggregate ε_a^T and the free thermal volumetric strains in the hardened cement paste ε_p^T

$$\beta(T) = \frac{\beta_2}{f_c}\frac{d(\varepsilon_a^T - \varepsilon_p^T)}{dT} \tag{11}$$

Note that $d(\varepsilon_p^T)/dT$ is contractive over the range $100 \rightarrow 600°C$.

2.3 *Bazant and Chern (1987)*

Bazant and Chern (3) presented a convincing argument (largely on the strength of the quality of their coupled thermo-mechanical diffusion simulations) that the increase in creep due to temperature *changes* (referred to as transitional thermal creep by others) is in fact the same phenomenon as the increase of creep due to changes in pore humidity; known for more than 40 years as the *Pickett effect* (or the drying creep effect). Two causes were identified for the stress-dependent creep increase under humidity and temperature changes: (i) rapid micro-diffusion of water between capillary and gel pores leading to a change in the bond rupture rate and (ii) the effect of spatially variable tensile micro-fracturing causing a stress-free body to exhibit smaller strains than the uncracked material expansion or shrinkage might suggest. Bazant and Chern considered the following decomposition of the strain rate

$$\dot{\varepsilon}_{ij} = \overbrace{\dot{\varepsilon}_{ij}^{ec}}^{\substack{\text{elastic +}\\ \text{basic creep}}} + \overbrace{\dot{\varepsilon}_{ij}^{cr}}^{\text{cracking}} + \overbrace{\dot{\varepsilon}_{ij}^{s} + \dot{\varepsilon}_{ij}^{T}}^{\substack{\text{transient drying}\\ \text{creep/shrinkage \&}\\ \text{stress-induced thermal}}} \tag{12}$$

Basic creep is not discussed further here, but the reader is directed to the B3 model for details of a fairly comprehensive formulation (2). The last two terms in the above equation were expressed as follows

$$\dot{\varepsilon}_{ij}^{s} + \dot{\varepsilon}_{ij}^{T} = \kappa_{ij}.\dot{h} + \alpha_{ij}.\dot{T} \tag{13}$$

where $\dot{h}$ is the relative humidity rate and k_{ij} and α_{ij} represent the tensorial shrinkage and expansion coefficients respectively (see Thelandersson (9)).

$$\begin{aligned} \kappa_{ij} &= -\varepsilon_0^s 3h^2.g_s\left(\delta_{ij} - \mathsf{sign}(\dot{H})r\sigma_{ij}\right) \\ \alpha_{ij} &= \bar{\alpha}\left(\delta_{ij} - \mathsf{sign}(\dot{H})\rho\sigma_{ij}\right) \end{aligned} \tag{14}$$

given

$$\dot{H} = \dot{h} + a_T.\dot{T} \tag{15}$$

a_T is a coefficient independent of $\dot{h}$ and $\dot{T}$. ε_0^s is a shrinkage constant and g_s is a given function of age which characterises the reduction in shrinkage due to continued hydration. The fundamental coefficient of thermal expansion ($\bar{\alpha}$) is considered to be independent of stress, h and $\dot{T}$ but Bazant and Chern felt that it probably depends on h.

2.4 *Khennane and Baker (1992)*

Khennane and Baker proposed a biaxial thermoplastic constitutive model using a decomposition of the strain rate that did not include basic creep effects or an explicit dependency on pore humidity (the laboratory tests simulated were of small, unsealed, specimens heated relatively slowly).

$$\dot{\varepsilon}_{ij} = \dot{\varepsilon}_{ij}^{e} + \dot{\varepsilon}_{ij}^{p} + \dot{\varepsilon}_{ij}^{T} + \dot{\varepsilon}_{ij}^{tc} \tag{16}$$

where superscripts e, p, T and tc refer to the elastic, plastic, free thermal and transient thermal creep rates respectively. In this model, the free thermal strain rate was given by

$$\dot{\varepsilon}_{ij}^{T} = \alpha.\dot{T}.\delta_{ij} \tag{17}$$

where α is the coefficient of thermal expansion and δ_{ij} is the Kronecker delta. The transient thermal creep strains followed the approach used by de Borst and Peters (5). Written in matrix form for the generalised three-dimensional case, this strain was given by (errata corrected)

$$\{\dot{\varepsilon}^{tc}\} = \frac{k.\alpha.\dot{T}}{f_c}[Q]\{\sigma\} \tag{18}$$

where

$$[Q] = \begin{bmatrix} 1 & -\nu & -\nu & 0 & 0 & 0 \\ -\nu & 1 & -\nu & 0 & 0 & 0 \\ -\nu & -\nu & 1 & 0 & 0 & 0 \\ 0 & 0 & 0 & 2(1+\nu) & 0 & 0 \\ 0 & 0 & 0 & 0 & 2(1+\nu) & 0 \\ 0 & 0 & 0 & 0 & 0 & 2(1+\nu) \end{bmatrix} \tag{19}$$

and k and v were given values of 2.33 (similar to Anderberg and Thelandersson's coefficient of 2.35) and 0.285 respectively for the concretes under investigation.

2.5 *Khoury (1996)*

Khoury summarised the body of experimental work on concrete at elevated temperatures conducted at the University of London over the period 1980–1991 in his 1996 report (8). Although the work dealt only with uniaxial compression, it provided useful new data and insight. In these studies, the total uniaxial strain observed during the first heating of loaded unsealed specimens was split into 5 components (additional terms, not discussed to here, were added for the case of cooling).

$$\varepsilon = \varepsilon^{\sigma} + \varepsilon^{ttc} + \varepsilon^{T} + \varepsilon^{s} + \varepsilon^{ck} \tag{20}$$

Khoury referred to ε^{σ} as *elastic strain* (dependent on the stress level and temperature; all nonlinear, irrecoverable and damage strains apparently being considered as crack-induced strains). The *transient creep* strains ε^{ttc} were thought to consist of three further components: (i) *transitional thermal creep*, the largest component, appearing only during first heating, and developing rapidly above 100°C, (ii) time-dependent thermal creep (considered to not be a part of basic creep, since the latter refers to the response under a constant temperature at thermal equilibrium) that develops during slow heating and (iii) drying creep; a function of moisture change, related to drying shrinkage. No test for concrete has been devised yet to isolate the three contributions in unsealed specimens. However, Khoury noted that for sealed specimens, drying creep is absent (although sufficiently large specimens may exhibit spatial variations of pore humidity.

The thermal strains were given as a function of the aggregate content as follows

$$\varepsilon^T = \varepsilon^T|_{T,V_a^{0.65}}(A.V_a - B) \tag{21}$$

where A and B are material constants appropriate to a particular aggregate type. Khoury noticed that the low relative stiffness and volume of the cement paste and relatively high expansion of the sands he tested caused the effective coefficient of thermal expansion of concrete to be close to that of the coarse aggregate.

With no underlying physical justification, the transient thermal creep (for $\sigma = 0.3f_c$) was deemed to fit a fourth-order polynomial relationship with respect to temperature

$$\begin{aligned}\varepsilon^{ttc}|^{0.3} = \; & (-43.9 + 2.7T + 6.25 \times 10^{-2}T^2 - \\ & 2.19 \times 10^{-4}T^3 + 2.77 \times 10^{-7}T^4) \\ & (0.032 + \tfrac{3.23\sigma}{f_c})(3.05 + 3.15V_a)\end{aligned} \tag{22}$$

For other pre-stress levels, the following relationship was given

$$\varepsilon^{ttc}|^{\sigma} = \varepsilon^{ttc}|^{0.3}\left(0.032 + 3.23\frac{\sigma}{f_c}\right) \tag{23}$$

where f_c is the uniaxial compressive strength under ambient conditions.

2.6 *Six component decomposition*

In the work undertaken at Sheffield University, the total strain rate tensor at a material point (ignoring non-local influences) has been interpreted as comprising 6 components, each associated with different chemo-mechanical processes. The last two components coincide with those introduced by Bazant and Chern (3).

$$\dot{\varepsilon}_{ij} = \overbrace{\dot{\varepsilon}^e_{ij} + \dot{\varepsilon}^d_{ij} + \dot{\varepsilon}^p_{ij}}^{\textit{instantaneous}} + \overbrace{\dot{\varepsilon}^c_{ij}}^{\substack{\textit{basic}\\ \textit{creep}}} + \overbrace{\dot{\varepsilon}^s_{ij} + \dot{\varepsilon}^T_{ij}}^{\substack{\textit{transient drying}\\ \textit{creep/shrinkage \&}\\ \textit{stress-induced thermal}}} \tag{24}$$

where $\dot{\varepsilon}^e_{ij}(\dot{\sigma}_{kl}, T)$ represents the elastic strain rate, $\dot{\varepsilon}^d_{ij}(\dot{\sigma}_{kl}, \omega_{mn}, T)$ the damage strain rate, $\dot{\varepsilon}^p_{ij}(\dot{\sigma}_{kl}, \varepsilon^p_{kl}, T)$ the plastic strain rate, $\dot{\varepsilon}^c_{ij}(\sigma^c_{kl}, t-\tau, T)$ the basic creep strain rate, $\dot{\varepsilon}^s_{ij}(\sigma_{kl}, \dot{T}, \dot{h})$ drying creep (considered equivalent to the *stress-induced shrinkage*) strain rate and $\dot{\varepsilon}^T_{ij}(\sigma_{kl}, \dot{T}, \dot{h})$ the (stress-dependent) thermal strain rate. Terms in parenthesis denote key independent variables controlling the rates (stress, σ; temperature, T; damage, ω; time, τ, t; moisture content, h (or alternatively, pore vapour pressure). Recall that Khoury (8) referred to the last two terms in (24) as stress-temperature-time dependent *transient creep* comprising the three components discussed earlier.

The 6-component decomposition is currently being used to construct a generalised (phenomenological) constitutive model for concrete under elevated temperature as part of a European project to develop an analytical approach to simulating the effects of ageing on concrete nuclear reactor and containment vessels. In order to construct and verify this constitutive model, a series of short-term multiaxial creep experiments at temperatures up to 250°C are to be performed in the mac^{2t} apparatus. The following section describes the rig and presents some results from recently completed studies.

3 MULTIAXIAL TEST RIG

3.1 *Load delivery and recording system*

The stiff servo-controlled compression apparatus comprises three independent nested orthogonal load frames. The two horizontal frames (X and Y, Figure 1) float on precision bearings, allowing free movement of the steel cross-heads in a horizontal plane. The third load frame (Z) is supported on an array of 4 hydraulic

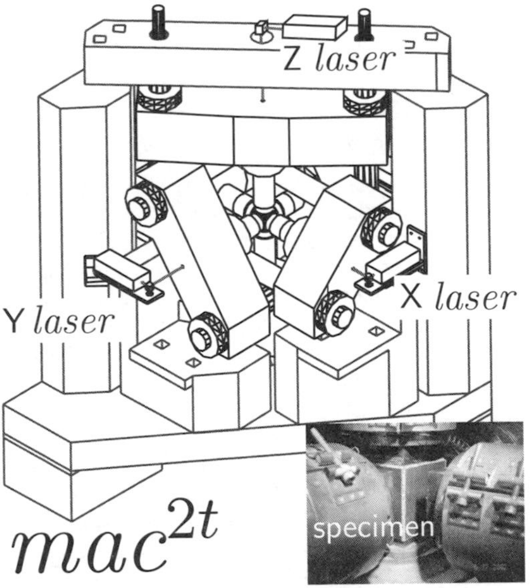

Figure 1. Multiaxial compression device with inset photograph showing cubic specimen and PFTE platens. Lasers are externally mounted with beams passing through vacuum tubes. The fourth hydraulic actuator (maintaining the centroid of the specimen at a fixed level) is not visible in this view.

actuators which constantly, automatically check and modify the position of the frame to ensure that the centre of the specimen remains at the same location during deformation. Each of the load frames houses a 4 MN hydraulic actuator within the 550 mm deep crossheads to deliver stresses up to 400 MPa on a ground 100 mm cubic specimen. Double full-bridge concentric *shear-type* load cells record the forces passing through the specimen in each of the three directions. The reaction is provided by 2×200 mm diameter tensile bars in each frame. A 200 mm ram (with strokes of 180 mm, 60 mm and 60 mm in the X, Y and Z directions respectively) delivers load to the concrete cube through a system of heat resistant, hardened steel platens and low friction 95 mm square PTFE pads.

3.2 *Displacement measurement*

Axial straining of the specimen is recorded by a system of 6 laser interferometers with a precision down to $10^{-7} m$. The laser heads are mounted outside the load frames with the light beams travelling through the centre of each frame within 18 mm diameter vacuum tubes ($10^{-5} bar$, to remove optical disturbance during hot tests) to a high grade silica corner cube positioned near the specimen face at the end of the tubes. The 0.5 mm diameter reflected light beam runs parallel to the incoming beam in the vacuum tube to return to the laser head. The tips of the tubes remain in contact with the concrete via a metal probe $\approx$ 47 mm away from the silica cube. This high precision displacement measuring system is both used to position the rig and measure strains in the specimen.

3.3 *Heating system*

mac^{2t} is designed to heat 100 mm specimens up to 350°C at a maximum rate of $0.083°C \cdot s^{-1} \cdot 1.4\,kW$ electric ceramic band heaters are wrapped around each of the 6 loading rams, close to the specimens. The heat is delivered by conduction, with continuous water cooling of the rams on the crosshead side to keep temperatures within the range 15–17°C in the load cells. The pumped water (with a maximum flow rate of $0.48\,m^3 \cdot s^{-1}$ is kept cool by means of a fan-assisted cooling tower mounted on the roof of the laboratory.

3.4 *Test control software*

The data acquisition and control system is integrated in a dual processor PC using NI DA hardware and purpose-written labVIEW software. Tests are monitored using a total of 46 sensors comprising 24 thermocouples, 3 load cells, 6 laser interferometers 6 LVDTs to record the position of the crossheads, 6 backup displacement LVDTs and a gauge to record fluid pressure in the 4th actuator. Signals from all the LVDTs, load cells and pressure transducer are conditioned by RDP units. The data acquisition rate is up to 500 samples per second. Data are averaged in packets of 100 samples and simultaneously stored on the hard disc. The servo-controlled system enables predefined multiaxial paths to be followed in mixed displacement and load control in principal strain space (to over 50000 $\mu\varepsilon$) and compressive principal stress space (up to 400 MPa).

3.5 *Specimen preparation*

All tests reported in this paper were conducted on a single concrete with 300 $kg \cdot m^{-3}$ OP cement, 99 $kg \cdot m^{-3}$ PFA, 735 $kg \cdot m^{-3}$ sand, 735 $kg \cdot m^{-3}$ 10 mm aggregate, 861 $kg \cdot m^{-3}$ 20 mm aggregate, 1.8 $kg \cdot m^{-3}$ Grace CP Darcem SP4 plasticiser and 168 $kg \cdot m^{-3}$ water. Specimens were at least 9 months old, having followed a temperature-matched curing regime during the first 10 days after casting. The cubes were cut from larger slabs and precision ground to within 0.05 mm to form right-regular specimens. The mean 28-day uniaxial compressive strength was 57.9 MPa (24+ month-old specimens showed strengths of up to 80 MPa).

4 EXPERIMENTAL RESULTS

Some preliminary results from tests conducted at elevated temperature (all heated at a rate of $0.03°C \cdot s^{-1}$ are now presented. Full details are to appear in a forthcoming publication.

4.1 *Specimen temperature profile during heating: Pre-dried and non-dried cases*

In order to investigate the differences between the average of the 24 thermocouples mounted near the surface of the steel platens and the temperature profile within the concrete cubes, three thermocouples were positioned inside the specimen along the central vertical axis (i) in the middle, (ii) 30 mm above and (iii) 30 mm below the centre of the cube. These sensors would not normally be used in a multiaxial experiment as they would disturb the stress field (and probably the vapour transfer path). Figure 2 shows the phase lag between the platen temperature (positioned midway between the heaters and the specimen faces) and the concrete, for a specimen that has not been previously subjected to drying (as is the case for all cubes in the preliminary test series). Of particular interest is the change in the Z response after a temperature of 160°C has been reached. All specimens were in a partially (largely) *sealed* state in these preliminary tests by virtue of the continuous platen-PTFE contact on the 6 faces, although the $2.5\,mm \times 2.5 \times 100\,mm$ edge strips and central 4 mm holes (for the laser probe)

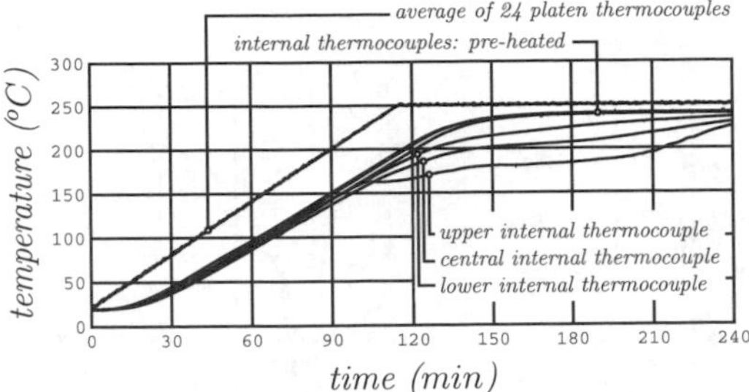

Figure 2. Temperature lag and spatial variation inside specimen compared with average platen readings. Pre-heated specimen exhibits much tighter grouping of internal thermocouple readings.

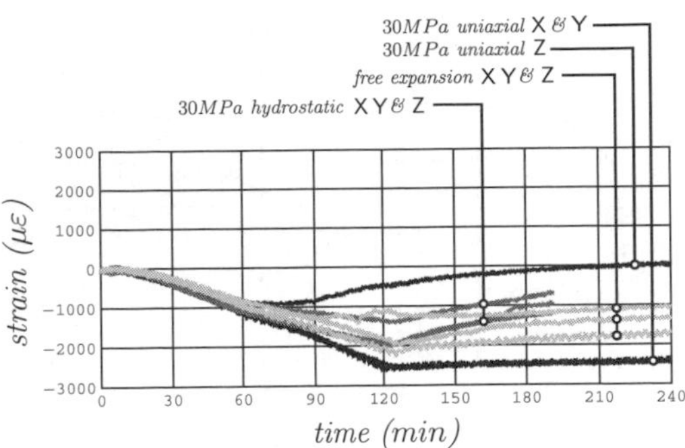

Figure 3. Expansive (and compactive) strains following pre-stress and then heating: free expansion, uniaxial pre-stress 30 MPa, hydrostatic pre-stress 30 MPa. Note that expansive strains are negative in this plot.

provide zones where water vapour may be driven-off during heating. The internal thermocouple nearest the top of the specimen (on the Z axis) records the lowest temperature. This is thought to be due to water evaporating through the central probe-hole, convecting up, and circulating around the upper Z axis laser tube, then condensing to fall back onto the specimen face. Future tests will eliminate this.

The temperature gradient is a result of the heating rate and the specimen size (100 mm being considered as the smallest acceptable dimension to represent this highly heterogeneous material) and shape (a cube is the only choice for orthogonal stressing). Additional testing is to be conducted at the slower heating rate of of $0.0083°C \cdot s^{-1}$ in 2003.

Figure 2 also shows the temperature profile obtained from the same cube in a repeat test 24 hours after the first. This now pre-dried specimen shows a very close agreement between the three internal thermocouples throughout the heating range, thus internal free-water advection and diffusion are shown to have a noticeable influence on the spatial temperature variation.

4.2 *Thermal expansion (plus shrinkage) during heating under different pre-stress conditions*

Figure 3 shows the expansive strain readings recorded during 3 tests under different pre-stress conditions: (a) with essentially no pre-stress (a nominal platen contact stress of 0.7MPa is applied throughout all multiaxial experiments in order to record the displacements and provide a conduction path for the heat), (b) with a uniaxial pre-stress of 30 MPa and (c) with a hydrostatic pre-stress of 30 MPa.

There are some differences between the strains recorded in each of the three axis (partly due to the effect discussed in the previous section; the direction showing least expansion in the unstressed test is the Z axis) however some general trends emerge. In the case of the nominally un-stressed specimen, expansive strains of 2000 $\mu\varepsilon$ are recorded in the X and Y directions by the time the platen thermocouples have reached 250°C (120 minutes). Maintaining this temperature for a further 2 hours allows some shrinkage to occur.

For the hydrostatically confined specimen, the expansive strains are of a similar but slightly lower magnitude during the heating phase. After holding the platen temperature at 250°C for a further 75 minutes, significant contractive strains appear. Previous researchers have offered somewhat contradictory statements with regards the effect of the heating rate (a number consider that $0.033°C \cdot s^{-1}$ is slow enough). However, it is evident that in these partially-sealed experiments that conclusions drawn about the presence of any transient thermal creep at the end of the heating stage (that is, at 120 minutes in these tests) will fail to identify the delayed creeping (and shrinkage).

For the uniaxially compressed specimen the effects are even more dramatic, with significant transient creep strains first mobilised in the major principal stress direction after a temperature of just over 100°C is recorded. The increased expansion in the other two directions during the heating phase is consistent with the Poisson's ratio effect. During the temperature hold period, shrinkage strains are detected in these directions.

The uniaxially stressed cube experiences a significantly higher transient thermal creep strain rate (starting at a lower temperature) when compared with the hydrostatically loaded specimen. This could be attributed to the greater (micro-crack) damage caused by the uniaxial loading (the confined stress point lies closer to the peak stress envelop than in the hydrostatic case) leading to an increase in vapour mobility.

For these two different pre-stress states, Khennane and Baker's model (7) gives a good prediction of the *final* transient thermal creep strain magnitude, but the

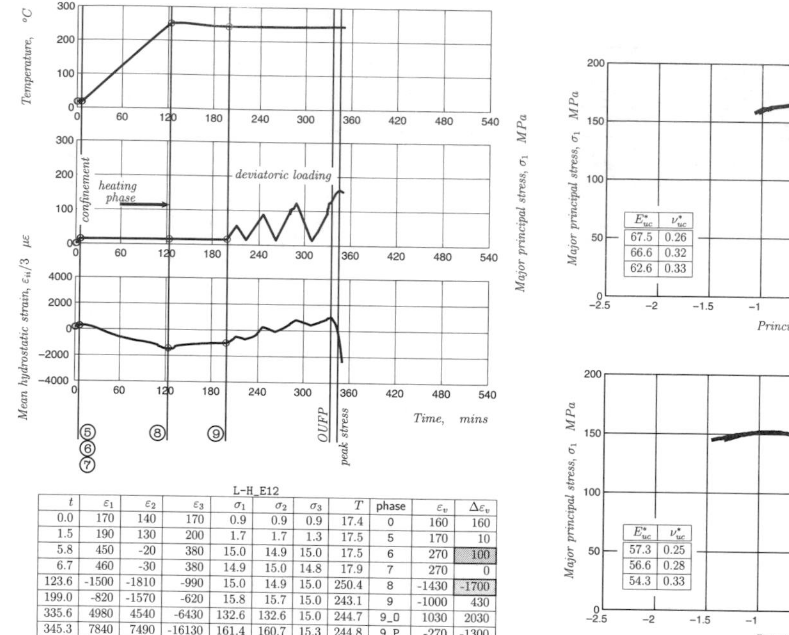

L-H_E12

t	ε_1	ε_2	ε_3	σ_1	σ_2	σ_3	T	phase	ε_v	$\Delta\varepsilon_v$
0.0	170	140	170	0.9	0.9	0.9	17.4	0	160	160
1.5	190	130	200	1.7	1.7	1.3	17.5	5	170	10
5.8	450	-20	380	15.0	14.9	15.0	17.5	6	270	100
6.7	460	-30	380	14.9	15.0	14.8	17.9	7	270	0
123.6	-1500	-1810	-990	15.0	14.9	15.0	250.4	8	-1430	-1700
199.0	-820	-1570	-620	15.8	15.7	15.0	243.1	9	-1000	430
335.6	4980	4540	-6430	132.6	132.6	15.0	244.7	9_O	1030	2030
345.3	7840	7490	-16130	161.4	160.7	15.3	244.8	9_P	-270	-1300

Figure 4. Typical result from a *load-then-heat* (to 250°C and 15 MPa hydrostatic stress) path followed by cyclic loading on the extension meridian. The table gives selected values at key intervals during test (including onset-of-unstable-fracture-propagation; attainment of minimum volume during deviatoric loading).

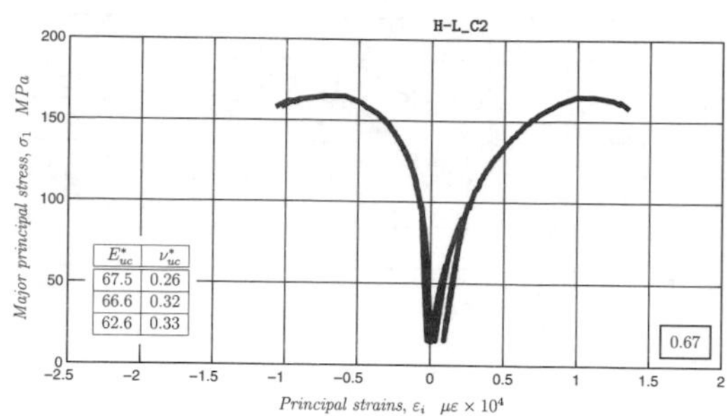

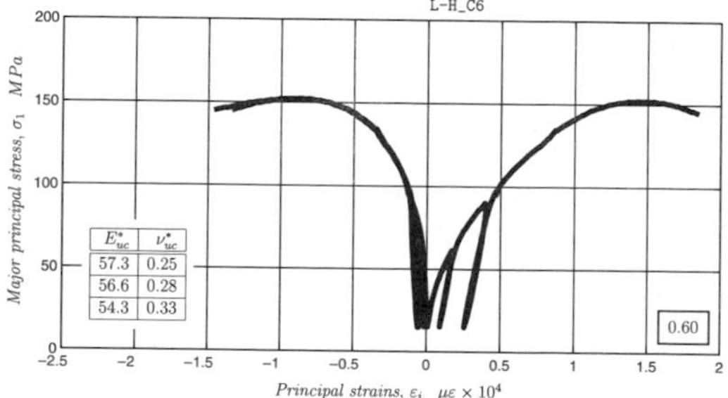

Figure 5. Cyclic compression meridian stress-strain response following *heat-then-load* (above) and *load-then-heat* (below) to 250°C and 15 MPa hydrostatic stress. The values in the inset boxes on the left are the effective elastic unloading stiffnesses and Poisson's ratios. The values in the boxes on the right are the relative stress levels at which OUFP was reached.

particular boundary conditions in these tests and the different deviatoric stress states influence the rate at which these strains are realised.

Figure 4 shows a typical set of temperature, stress, volumetric strain (all versus time) results for a specimen first subjected to a hydrostatic confinement of 15 MPa, then heated to 250°C and finally deviatorically cycled to its peak stress in the extension meridian while holding the temperature and $\sigma_2 = \sigma_3 = 15\,MPa$ constant.

The lower plot in Figure 4 illustrates the progression from volumetric expansion during heating, to contraction due to creep and shrinkage, with further contraction during deviatoric loading before final expansion as the peak stress is approached and significant internal fracturing occurs. The point at which the volume stops compacting and starts dilating has been linked to the onset-of-unstable-fracture-propagation (OUFP). It has been suggested that stresses above OUFP are attainable only of the loading rate is fast enough. The delay between achieving OUFP and the peak stress is also thought to be controlled by the relative strength of the aggregate and the hardened cement paste (a measure of material heterogeneity); low strength concretes, being more heterogeneous, show a larger stress interval between the two states leading to greater pre-peak stress-strain nonlinearity.

4.3 *Compression and extension meridian cycling*

Figure 5 shows two test results from specimens subjected to *heat-then-load* (250°C, then 15 MPa hydrostatic, upper plot) and *load-then-heat* (15 MPa hydrostatic then 250°C lower plot). The graphs show the stress-strain responses for the deviatoric loading phase in the compression meridian ($\sigma_1 > \sigma_2 = \sigma_3 \approx 15\,MPa$). Three unload-reload cycles were followed in each test. It is evident that a confining stress of 15 MPa produces a significant strength gain over uniaxial compression. The *heat-then-load* test shows a higher peak strength and reduced pre-peak nonlinearity compared to the *load-then-heat* experiment. It appears that the latter causes greater damage to the material. This is confirmed by the lower unloading stiffnesses in the second type of test (notice the progressive reduction in the stiffness for subsequent loading cycles within each test). The minor/intermediate principal strains show close agreement in both tests.

Figure 6 shows the comparable results from two tests where the initial paths were nominally identical

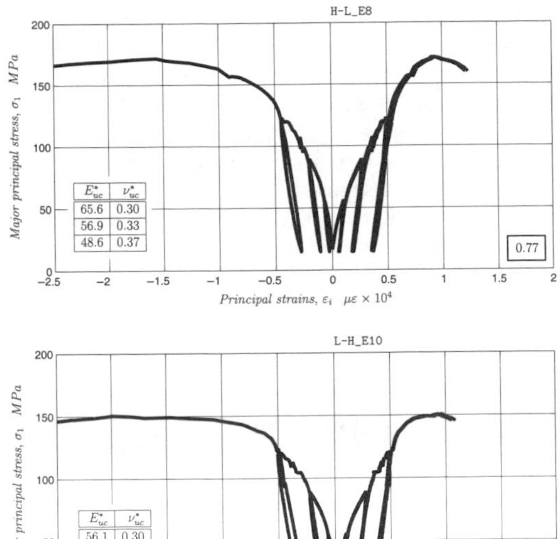

Figure 6. Cyclic extension meridian stress-strain response following *heat-the-load* (above) and *load-then-heat* (below) to 250°C and 15 *MPa* hydrostatic stress. The values in the inset boxes on the left are the effective elastic unloading stiffnesses and Poisson's ratios. The values in the boxes on the right are the relative stress levels at which OUFP was reached.

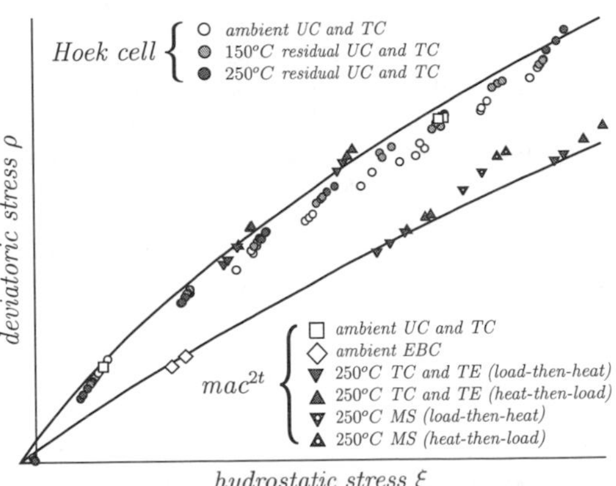

Figure 7. Peak stress data from Hoek cell, *sπce rack* and mac^{2t} devices on a single mix. Hoek cell results are all on the compression meridian whereas mac^{2t} data include results on the extension and shear meridians. Note that marginally higher strengths are consistently obtained from the *heat-then-load* experiments compared with the companion *load-then-heat* tests.

to the previous pair, however the deviatoric loading takes place in the extension meridian ($\sigma_1 = \sigma_2 > \sigma_3 \approx$ 15 *MPa*). A greater strength following the *heat-then-load* path is once again observed, but although the unloading stiffnesses are higher in this test, there is no clear trend in the degree of pre-peak nonlinearity. It is worth pointing out that the level at which OUFP appears in the extension tests is relatively higher than that seen in the compression tests. Extensional loading forces the final rupture surfaces to propagate in the plane normal to the minimum principal stress direction, whereas in the case of compression the rupture can occur in any number of planes parallel to the major principal stress direction. This leads to a more unstable response near the peak in extensional tests.

4.4 *Peak stress results*

Figure 7 collects together all the recent strength data from uniaxial, biaxial, triaxial and multiaxial tests performed at Sheffield University using the mac^{2t} rig and a 60 *mm* diameter Hoek cell. The latter was used to determine the residual triaxial strength (ambient tests at confinements up to 70 *MPa*) of the chosen mix following exposure for 48 hours to 150°*C* or 250°*C* (compared against the ambient results). Tests were performed using 4 × 60 *mm* strain gauges (2 circumferential and 2 longitudinal). The results (circular symbols in Figure 7) revealed that although the pre-heated specimens showed lower uniaxial (tension and) compression strengths, at confinements above 10 *MPa* the residual pre-heated triaxial strengths were greater than those not subjected to elevated temperatures (the hot specimens were allowed to cool slowly over a period of 48 hours). This suggests that the nature of the (micro-crack) damage suffered through pre-heating reduces the effective cohesion but appears to increase the frictional strength. More importantly, pre-heating will remove most of the free water, thus leading to lower capillary pore water pressures, which are known (through the effective stress principle) to reduce the strength.

The open squares in Figure 7 indicate the uniaxial compression and triaxial compression (50 *MPa*) results performed at ambient temperature. These agree well with the data obtained from the Hoek cell. The open diamond symbols record the ambient equal biaxial compression test results (on the extension meridian). The low biaxial strength ($\hat{\sigma}_1 = \hat{\sigma}_1 \approx 1.1 f_c$) confirms the effectiveness of carefully grinding the specimens, isolating the loading frames and using PFTE coated platen pads to prevent restraint at the platen-specimen interface.

The flat-base triangular symbols indicate the results from tests carried out at 250°*C* under triaxial compression (upper meridian) and extension (lower meridian) following *heat-then-load* (where hydrostatic pre-stress levels are either 15 *MPa* or 30 *MPa*). The inverted triangles show the corresponding results from the *load-then-heat* sequences. The triangles with enclosed circles show recent data (for a hydrostatic pre-stress of $30\,MPa$) gathered on the shear meridian. These preliminary results suggest that the peak envelop is quite closely approximated by a Mohr-Coulomb-like deviatoric section (although this criterion is not so applicable to the meridians as curvature is evident).

For each regime, three tests were performed under nominally identical conditions (stress paths, sequence, temperature, and loading rates) to identify the material variability. All data are plotted in terms of the familiar cylindrical stress invariants ξ and ρ. As noted previously, the *heat-then-load* tests consistently show higher strengths than the *load-then-heat* results. To-date no constitutive model appears to be able to reproduce this path dependency. It remains a challenge for modellers to identify an appropriate (tensorial) damage measure that can properly reflect this change in the internal structure.

5 CONCLUSIONS

1. There exists several different explanations for, and models of, the additional creep strains observed in *load-then-heat* tests above those seen in *heat-then-load* tests. It is not just the models that differ, but also the terminology. The work by Bazant and Chern (3) appears the most comprehensive.
2. A tensorial measure of stress must be used in preference to the maximum stress level when identifying the dependence of transient thermal creep strains on confinement.
3. The behaviour of concrete under elevated temperature can only be properly understood through consideration of the temporal and spatial variations of (i) the pore water content, (ii) the temperature and (iii) the stresses. Since uniform states of these measures cannot be delivered to a representative volume of concrete under short-term loading, back-analyses are required in order to interpret laboratory tests.
4. A six component strain decomposition (elastic, damage, plastic, basic creep, drying creep/shrinkage and stress-dependent thermal) offers a useful conceptual decoupling of the different mechanisms apparently operating in concrete under elevated temperature and multiaxial load.
5. A high precision multiaxial test apparatus has been constructed to generate new data for development and verification of generalised constitutive models for concrete. Preliminary results from the rig show that the (supposedly instantaneous) transient thermal creep strains require some time to become activated in partly-sealed specimens following heating at a rate greater than $0.03°C \cdot s^{-1}$. The time required to mobilise these additional strains appears to be dependent upon the deviatoric proximity of the pre-stress state to the peak stress envelop.
6. Applying a hydrostatic stress (in the range 15–30 MPa) prior to heating to $250°C$ can lead to lower multiaxial strengths, greater damage and greater pre-peak stress-strain nonlinearity compared to the situation where the concrete is heated first and then the same level of hydrostatic stress applied before deviatorically loading to failure.

REFERENCES

Anderberg, Y and Thelandersson, S, 1976, Stress and deformation characteristics of concrete at high temperature 2: Experimental investigation and material behaviour model, *Bulletin 54*, Division of Structural Mechanics and Concrete Construction, Lund Institute of Technology, 1–84.

Bazant, Z P and Baweja, S, 1995, Creep and shrinkage prediction model for analysis and design of concrete structures: Model B3, in collaboration with RILEM committee TC 107-GCS, *Materials & Structures*, 28, 357–365.

Bazant, Z P and Chern, J-C, 1987, Stress-induced thermal and shrinkage strains in concrete, *J Eng Mech*, ASCE, 113(10), 1493–1511.

Bazant, Z P and Kaplan, M F, 1996, *Concrete at high temperatures: Material properties and mathematical models*, Longman, 300–317.

de Borst, R and Peters, P P J M, 1989, Analysis of concrete structures under thermal loading, *Comp Meth Appl Mech & Eng*, 77, 293–310.

Ehm, C and Schneider, U, 1985, The high temperature behaviour of concrete under biaxial conditions, *Cement & Concrete Res*, 15, 27–34.

Khennane, A and Baker, G, 1992, Thermoplasticity model for concrete under transient temperature and biaxial stress, *Proc R Soc Lond A*, 439, 59–80.

Khoury, G A, 1996, *Performance of heated concrete: Material properties*, Imperial College report submitted to the NII and HSE, UK, 1233.

Thelandersson, S, 1983, On the multiaxial behaviour of concrete exposed to high temperature, *Nuclear Eng & Design*, 75(2), 271–282.

Computational Modelling of Concrete Structures, Bićanić et al. (eds)
© 2003 Swets & Zeitlinger, Lisse, ISBN 90 5809 536 3

Bond-slip in reinforced concrete structures, constitutive modelling and finite element implementation

N. Dominguez, F. Ragueneau & A. Ibrahimbegovic
LMT, Ecole Normale Supérieure de Cachan, France

S. Michel-Ponnelle, Sh. Ghavamian
EDF R&D/AMA, Clamart, France

ABSTRACT: In this paper, we present two principal items: a new formulation for a non-dimensional joint finite element (zero-thickness element) as well as an inelastic constitutive model for bonding phenomena between steel bars and concrete. For the first item, based on a typical four-node quadrilateral element, we developed a new 2D degenerated interface element which allows small normal strain and which is capable to treat the nonlinear tangential-normal behavior of bonding. About second item, the bonding behavior is expressed according to the small deformation and plane strain assumptions within the framework of thermodynamics and continuum damage mechanics. One of the most important contribution of this model consists in using a specific kinematics for contact allowing the implementation of standard constitutive relations.

1 INTRODUCTION

1.1 *General background*

Reinforced concrete structures are usually analyzed conservatively considering a perfect adherence between concrete and reinforcing steel bars, and that means that deformations are similar in both materials when solicited by external mechanical loads, even in dynamic loading. In reality, these structures develop different deformations according to nature of each material, and that initiates a relative slip between both surfaces in contact; as a matter of fact, a lot of failures as large and radial cracks are due to this phenomena. Therefore, a lot of experiments have been performed over last thirty years with a great variety of specimens, trying to measure displacements, elongations, slips and/or strains in concrete or embedded bar, maximal resistance before fracture, crack formation, etc.

The predictive models of the response of reinforced concrete structures subjected to mechanical loadings need physical representations of different mechanisms involved both in material behavior and element kinematics. Many efforts have been paid in describing multidimensional non-linear response of concrete and steel materials in order to have a better understanding of global behavior of reinforced concrete structures subjected to complex loading path, monotonic or cyclic, or even dynamic loads. The prediction of ultimate bearing capacity requires a more detailed modeling while dealing with multiple cracks' opening and spacing. For development of fine material constitutive equations, it is also necessary to revise the specific kinematics between steel and concrete in order to include them in a complete finite element analysis.

Different formulations have been developed in order to solve bonding problems, either in small or finite deformations. For example, contact algorithms, which use penalty methods or augmented Lagrangian regularizations, have been employed by Maker & Laursen (1994); some approaches may be able to represent the displacement discontinuity at the concrete and steel interface based on a force formulation within the beam kinematic assumption's framework (see Salari & Spacone 2001), or using some interface modeling with a zero or even non-zero element's thickness (Cox & Hermann 1998); another kind of formulation was developed by Romdhane & Ulm (2002), who have presented a model of interface that tries to link two mechanisms which govern concrete cracking in reinforced concrete structures: the activation of bond forces at the steel-concrete interface and the bridge effects of the reinforcement crossing a macro-crack. With respect to the debonding, they developed a truss element with slip degrees of freedom that represents the relative displacement between surfaces in contact.

This work focuses on development and numerical implementation of a new 2D degenerated interface element as well as the introduction of constitutive equations coupling the nonlinear tangential behavior and radial confinement effects on bonding phenomena. The model developed hereafter allows the implementation of standard constitutive relations, thanks to a particular kinematics for contact element. Avoiding the classical instability phenomena when dealing with contact forces, the nodal contribution of the bond elements results from a local integration of stresses at the Gauss point level. The stability and robustness of standard constitutive relations implementation algorithm are recovered within a joint element framework.

1.2 *Element Stiffness formulation*

According to Galerkin method, (e.g. see Hughes 1987) and applying the standard finite element concepts to equate the mechanical equilibrium leads to solving

$$k^e \cdot \Delta u = \Delta F^{\text{ext}} - \Delta F^{\text{int}} \tag{1}$$

where Δu = displacements at the interface; ΔF^{ext} = nodal applied incremental external forces; and ΔF^{int} = increment of internal nodal forces.

The stiffness of the element can be expressed as:

$$k^e = \int_\Omega B^T DB\, d\Omega = \int_V B^T DBj\, dV \tag{2}$$

where B = strain-displacement matrix; D = material constitutive relations matrix; and j = determinant of the Jacobian matrix.

To compute integrals, we can apply numerical quadrature and then

$$k^e \cong \sum_{l=1}^{n\,\text{int}} (B^T DBj)|_{\xi_l} W_l \tag{3}$$

where n int = number of integration points; ξ_l = coordinate of the lth integration point; and W_l = weight of the lth integration point.

Last equation could be simplified combining j and W_l with D, so we can write

$$\tilde{D} = j(\xi_l) W_l D \tag{4}$$

and thus

$$k^e \cong \sum_{l=1}^{n\,\text{int}} (B^T \tilde{D} B)_l \tag{5}$$

From last equation, we can identify the two principal items that we can separate in order to reformulate for the specific problem of bonding:

- The strain-displacement matrix B, that will be affected for the construction of a new joint element

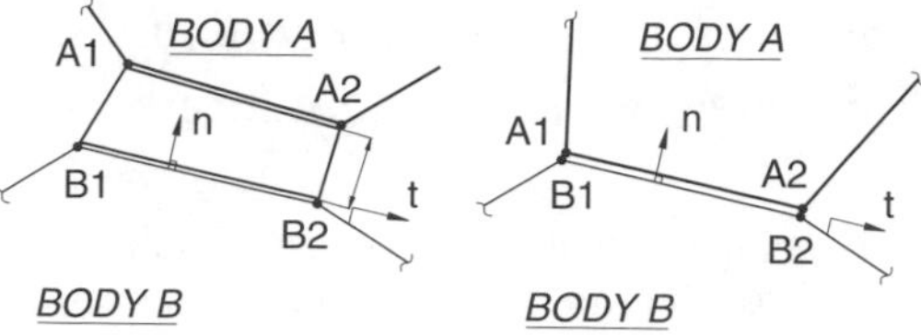

Figure 1. Degeneration of a four-node quadrilateral element in a one-dimensional element.

- The material constitutive matrix D, where the new constitutive model for bonding will be implemented

Both items will be developed and explained in the next paragraphs.

2 FORMULATION OF A NEW JOINT FINITE ELEMENT

The study of a problem of steel-concrete bonding (for example, modeling of a pull-out test) in a finite element code implies the construction of a mesh where elements may be truss (uniaxial bond) or quadrilateral solid in a 2D space. However, for our case we need to join two different materials with a contact element which should not have any dimension or thickness, and despite of all, it must be able to carry out tangential and normal stresses that would be calculated from normal and tangential strains. The domain of a straight-edged quadrilateral element is defined by the location of its four nodal points x_a^e, $a = 1, \ldots, 4$ in the 2D-space. However, this element should have only two coordinates and four nodes, that is at least two different nodes in the same coordinate; in fact, this is a case of a quadrilateral element degenerated in a one-dimensional element (see Fig. 1).

2.1 *The "h_{pen}" parameter*

Based on the work of Ibrahimbegovic & Wilson (1992), we decide to use a principle of penetration between surfaces that will allow to define a normal stress even if this one does not have any normal dimension, avoiding the classical stability drawbacks when dealing with contact forces. If we consider that concrete in contact with steel surface has a zone of asperities which could be compressed or crushed, it is possible to assume a small penetration between surfaces (see Fig. 2).

The "h_{pen}" parameter is a physical value that corresponds to the maximal penetration due to thickness of compressed-pulverized concrete. From Figure 2 it is possible to deduce the next relationships:

$$t_0^A = t^A + U_n^A \Rightarrow U_n^A = t_0^A - t^A \Rightarrow t^A = t_0^A - U_n^A \tag{6}$$

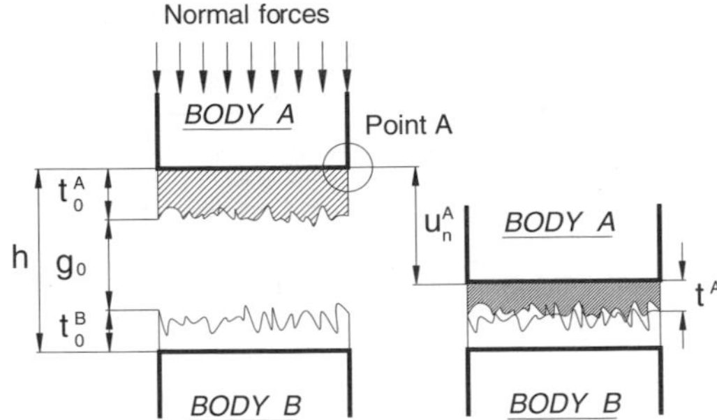

Figure 2. Initial configuration, normal displacement and penetration for crushing of asperities.

$$t_0^B = t^B + U_n^B \Rightarrow U_n^B = t_0^B - t^B \Rightarrow t^B = t_0^B - U_n^B \quad (7)$$

$$h_{pen} = t_0^A + g_0 + t_0^B \quad (8)$$

If at the moment of contact, $g_0 = 0$, and also we consider B as a rigid body, $t_0^B = 0$, the deformable thickness becomes

$$h_{pen} = t_0^A \quad (9)$$

Now, the normal strain could be expressed as

$$\varepsilon_n = \frac{U_n^A}{t_0^A} = \frac{U_n^A}{h_{pen}} \quad (10)$$

U_n being the normal displacement of a point in the concrete surface.

Once we have defined the "h_{pen}" parameter, it can be introduced into the geometric properties of the degenerated element for initialization.

2.2 *Shape functions and their derivatives*

For a typical four-node quadrilateral element, the bilinear shape functions are expressed as

$$N_a(\xi, \eta) = \frac{1}{4}(1 + \xi_a \xi)(1 + \eta_a \eta) \quad (11)$$

where $a = 1, 2, 3, 4$; ξ, η = natural coordinates; and ξ_a, η_a = nodal coordinates in ξ-space.

Until here, there is not any difference between joint element and the isoparametric four-node element. Nevertheless, the element stiffness matrix is constructed with derivatives of shape functions, which will be calculated explicitly and that implies utilization of nodal coordinates. Because coordinates in two nodes are the same, it is necessary to avoid indetermined Jacobian for the interface normal dimension.

Thus, applying the chain rule it is possible to evaluate derivatives of N_a with respect to x and y:

$$N_{a,x} = N_{a,\xi}\xi_{,x} + N_{a,\eta}\eta_{,x} \quad (12)$$

$$N_{a,y} = N_{a,\xi}\xi_{,y} + N_{a,\eta}\eta_{,y} \quad (13)$$

These relations may be presented in the following matrix form:

$$\langle N_{a,x} N_{a,y} \rangle = \langle N_{a,\xi} N_{a,\eta} \rangle \begin{bmatrix} \xi_{,x} & \xi_{,y} \\ \eta_{,x} & \eta_{,y} \end{bmatrix} \quad (14)$$

Due to difficulties in computing explicitly the terms into the matrix, it is more convenient to use the inverse relations

$$x(\xi, \eta) = \sum_{a=1}^{nen} N_a(\xi, \eta) x_a^e \quad (15)$$

$$y(\xi, \eta) = \sum_{a=1}^{nen} N_a(\xi, \eta) y_a^e \quad (16)$$

now we can get the next matrix

$$X_{,\xi} = \begin{bmatrix} x_{,\xi} & x_{,\eta} \\ y_{,\xi} & y_{,\eta} \end{bmatrix} \quad (17)$$

which is the inverse of the matrix in Equation 14; that may be rewritten in the form

$$\begin{bmatrix} \xi_{,x} & \xi_{,y} \\ \eta_{,x} & \eta_{,y} \end{bmatrix} = (X_{,\xi})^{-1} = \frac{1}{j} \begin{bmatrix} y_{,\eta} & -x_{,\eta} \\ -y_{,\xi} & x_{,\xi} \end{bmatrix} \quad (18)$$

where

$$j = \det(X_{,\xi}) = x_{,\xi} y_{,\eta} - x_{,\eta} y_{,\xi} \quad (19)$$

The components of this matrix, known as natural coordinate functions, are expressed in the following way:

$$x_{,\xi} = \sum_{a=1}^{nen} N_{a,\xi} x_a^e \quad (20)$$

$$x_{,\eta} = \sum_{a=1}^{nen} N_{a,\eta} x_a^e \quad (21)$$

$$y_{,\xi} = \sum_{a=1}^{nen} N_{a,\xi} y_a^e \quad (22)$$

$$y_{,\eta} = \sum_{a=1}^{nen} N_{a,\eta} y_a^e \quad (23)$$

The calculation of these functions depends on element's nodal coordinates, and we now add the geometrical parameter h_{pen}. Thus, for a four-node degenerated element we have on x-coordinate (or tangential direction):

$$x_0^e = x_1^e - x_2^e + x_3^e - x_4^e \tag{24}$$

$$x_\xi^e = -x_1^e + x_2^e + x_3^e - x_4^e + x_0^e \cdot \eta \tag{25}$$

$$x_\eta^e = -x_1^e - x_2^e + x_3^e + x_4^e + x_0^e \cdot \xi \tag{26}$$

On y-coordinate (or normal direction), expressions will be modified as follows :

$$y_0^e = y_1^e - y_2^e + \left(y_3^e + h_{pen}\right) - \left(y_4^e + h_{pen}\right) \tag{27}$$

$$y_\xi^e = -y_1^e + y_2^e + \left(y_3^e + h_{pen}\right) - \left(y_4^e + h_{pen}\right) + y_0 \cdot \eta \tag{28}$$

$$y_\eta^e = -y_1^e - y_2^e + \left(y_3^e + h_{pen}\right) + \left(y_4^e + h_{pen}\right) + y_0 \cdot \xi \tag{29}$$

The last expressions can be used in the calculation of shape function's derivatives, which are determined for each integration point

$$N_{a,x} = \frac{N_{a,\xi} y_{,\eta} - N_{a,\eta} y_{,\xi}}{j} \tag{30}$$

$$N_{a,y} = -\frac{N_{a,\xi} x_{,\eta} - N_{a,\eta} x_{,\xi}}{j} \tag{31}$$

Once we have determined both derivatives, it is possible to build the strain-displacement matrix B:

$$B = \begin{bmatrix} N_{a,x} & 0 \\ 0 & N_{a,y} \\ N_{a,y} & N_{a,x} \end{bmatrix} \tag{32}$$

2.3 *Numerical integration with two Gauss points*

For the numerical implementation of the model in the joint quadrilateral element, it is worth noting that two different constitutive laws have been attributed to each direction (normal and tangential), and in the case of numerical integration, we assumed a similar consideration about the number of Gauss points to introduce in order to reduce calculations; that is the reason why there were adopted two points of integration for tangential direction and only one for normal direction.

Gaussian rules for integrals in several dimensions can be developed by employing one-dimensional rules on each coordinate. So, the correspondent formulation

Table 1. Two-Point Quadrature Rule in two dimensions.

	Integration Point	
Associated values	1	2
l^1	1	2
l^2	1	1
$g(\xi_i)$	−0.57735	+0.57735
$g(\eta_i)$	0.00000	0.00000
W_i	2	2

for two dimensions is

$$\int_{-1}^{1}\int_{-1}^{1} g(\xi,\eta)\, d\xi\, d\eta \cong \sum_{l^1=1}^{n\,\text{int}^1} \sum_{l^2=1}^{n\,\text{int}^2} g\left(\tilde{\xi}_{l^1}^1, \tilde{\eta}_{l^2}^2\right) \cdot W_{l^1}^1 \cdot W_{l^2}^2 \tag{33}$$

In our case, substituting the values shown in Table 1 in Equation 32, we obtain

$$\int_{-1}^{1}\int_{-1}^{1} g(\xi,\eta)\, d\xi\, d\eta = 2g\left(-\frac{1}{\sqrt{3}}, 0\right) + 2g\left(+\frac{1}{\sqrt{3}}, 0\right) \tag{34}$$

It is now possible to have an element which deals with tri-dimensional state of stress, capable to treat the nonlinear tangential behavior coupled to the degree of radial confinement; that means as well, that surrounding radial forces can be transferred from an element to its adjacent one through this particular layer. This simple expression of such a joint element may be improved by using quadratic shape functions in order to describe in a satisfactory manner the strong gradients between concrete and steel materials.

3 FORMULATION OF AN INELASTIC CONSTITUTIVE MODEL FOR BONDING

The main purpose of this paragraph is to express a constitutive relation for bond slip accounting for some physical phenomena such as cracking and frictional sliding and their couplings. In this part we first present the thermodynamic framework and then the numerical implementation in adequacy with the finite element geometrical support previously expressed. Some results for simple state of loadings and model features are presented at the end.

3.1 *Thermodynamic formulation*

The constitutive relations which rely the stress tensor to the strain tensor should include

– cracking for an excessive tangential stress

– inelastic strain due to sliding
– hysteretic behavior due to friction
– coupling between tangential response and normal state of stress

The tensors σ and ε reduce to the following one for a plane formulation:

$$\sigma = \begin{pmatrix} \sigma_N & \sigma_\tau & 0 \\ \sigma_\tau & 0 & 0 \\ 0 & 0 & 0 \end{pmatrix} \text{ and } \varepsilon = \begin{pmatrix} \varepsilon_N & \varepsilon_\tau & 0 \\ \varepsilon_\tau & 0 & 0 \\ 0 & 0 & 0 \end{pmatrix} \quad (35)$$

where σ_N is the stress normal to the bond element, σ_τ is the stress tangent to the bond element. ε_N is the normal strain and ε_τ the tangential strain.

The thermodynamic potential, using the Helmholtz free energy is expressed as:

$$\rho\psi = \frac{1}{2}\left[\varepsilon_N E \varepsilon_N + (1-d)\varepsilon_\tau G \varepsilon_\tau + \left(\varepsilon_\tau - \varepsilon_\tau^s\right) \times Gd\left(\varepsilon_\tau - \varepsilon_\tau^s\right) + \gamma\alpha^2\right] + H(z) \quad (36)$$

where ρ is the density, E the Young's modulus, d the internal damage variable(ranging from 0 to 1) linked to cracking, G is the Poisson's modulus, ε_τ^s is the sliding strain, α is the internal variable for kinematic hardening, γ a material parameter and $H(z)$ the hardening function for damage.

For the tangential part of the behavior, one can easily recognize a classical elasto-damage coupling and a new term allowing the energy to shift from the elasto-damageable part to the frictional sliding part. The coupling between sliding and cracking is made possible thanks to the presence of the damage variable as a multiplier in the second element of the right-hand side of (35). This kind of approach has proved its efficiency in describing the 3D behavior of plain concrete (see Ragueneau *et al.* 2000).

The state laws may be derived as follows, defining as well the thermodynamic associated variables:

The normal stress, assuming an elastic behavior

$$\sigma_N = \rho\frac{\partial\psi}{\partial\varepsilon_N} = E\varepsilon_N \quad (37)$$

The total tangential stress:

$$\sigma_\tau = \rho\frac{\partial\psi}{\partial\varepsilon_\tau} = G(1-d)\varepsilon_\tau + Gd\left(\varepsilon_\tau - \varepsilon_\tau^s\right) \quad (38)$$

σ_τ^s is the tangential sliding stress, associated to the sliding strain ε_τ^s:

$$\sigma_\tau^s = -\rho\frac{\partial\psi}{\partial\varepsilon_\tau^s} = Gd\left(\varepsilon_\tau - \varepsilon_\tau^s\right) \quad (39)$$

The sliding and plastic strains being different, the thermodynamic forces associated with the total strain and the sliding strain are different. Such a formulation greatly differs from the classical plasticity-damage coupling. This choice of introducing damage into the sliding stress is guided by the idea that all inelastic phenomena in a brittle material result from the cracks' growth.

The damage energy release rate, associated with damage can be written as

$$-Y = \rho\frac{\partial\psi}{\partial d} = \frac{1}{2}\varepsilon_\tau G\varepsilon_\tau + \frac{1}{2}\left(\varepsilon_\tau - \varepsilon_\tau^s\right)G\left(\varepsilon_\tau - \varepsilon_\tau^s\right) = -(Y_d + Y_s) \quad (40)$$

The back stress for kinematic hardening is defined as

$$X = \rho\frac{\partial\psi}{\partial\alpha} = \gamma\alpha \quad (41)$$

The internal variable associated to isotropic damage hardening is defined as (La Borderie 1991):

$$Z = \rho\frac{\partial\psi}{\partial z} = H'(z) = \left[Y_0 + \frac{1}{A_d}\left(\frac{-z_i}{1+z_i}\right)\right] \quad (42)$$

where A_d is a material parameter, driving the slope of the softening branch. Damage initiating the nonlinear behavior, a damage yield function is expressed in terms of the variable Y:

$$f_d = Y_d - (Y_0 + Z) \leq 0 \quad (43)$$

where Y_0 is the initial damage threshold defined in function of the initial yield strain ε_τ^0

$$Y_0 = \frac{1}{2}\varepsilon_\tau^0 G\varepsilon_\tau^0 \quad (44)$$

The evolution laws of internal variables, in the framework of standard associate laws allow the derivation of the damage multiplier λ_d

$$\dot{d} = \dot{\lambda}_d\frac{\partial f_d}{\partial Y_d} = \dot{\lambda}_d \quad \text{and} \quad \dot{z} = \dot{\lambda}_d\frac{\partial f_d}{\partial Z} = -\dot{\lambda}_d \quad (45)$$

The damage evolution laws is obtained from the consistency condition as:

$$d = 1 - \frac{1}{1 + A_d(Y_d - Y_0)} \quad (46)$$

The sliding part of the constitutive relation is assumed to represent a plasticity-like behavior. Nonlinear kinematic hardening is considered. Initially introduced by Armstrong & Frederick (1966), it allows the formulation to overcome the major drawback of Prager's kinematic hardening law, i.e. the linearity of the state law defining the forces associated with kinematic hardening. The nonlinear terms are added in the dissipative potential. The sliding criterion takes the classical form for a Drücker-Prager yield function accounting for the effect of pressure on sliding

$$f_s = \left|\sigma_\tau^s - X\right| + cI_1 \leq 0 \tag{47}$$

where I_1 is the first invariant of the stress tensor. In our case:

$$I_1 = \frac{1}{3}Tr(\sigma) = \frac{1}{3}\sigma_N \tag{48}$$

with c a material parameter.

The initial threshold for sliding is 0. Through the maximum of plastic dissipation principal, the evolution laws are derived from the expression of a plastic potential expressed as:

$$\phi_s = \left|\sigma_\tau^s - X\right| + cI_1 + \frac{3}{4}aX^2 \tag{49}$$

where a is a material parameter. The evolution laws for the sliding strain as well as the kinematic hardening follows:

$$\dot{\varepsilon}_\tau^s = \dot{\lambda}_s \frac{\partial \phi_s}{\partial \sigma_\tau^s} \quad \text{and} \quad \dot{\alpha} = -\dot{\lambda}_s \frac{\partial \phi_s}{\partial X} \tag{50}$$

The sliding multiplier $\dot{\lambda}_s$ is numerically computed by imposing the consistency condition. The details concerning the numerical implementation are given in the next paragraph.

3.2 *Numerical implementation*

By defining two different surfaces, one can explicitly integrate the damage variable. Concerning the sliding stress, a classical implicit analysis has to be carried out. Among the different methods available for this purpose (Euler's backward or mid-point rules algorithm solved by an iterative Newton method), we chose the classical form of the so-called "return mapping" algorithm (Ortiz & Simo 1986) ensuring convergence in the most efficient way.

The yield function is linearized around the current values of the associated variables. At convergence, we impose the respect of the return onto the yield surface:

$$f_s^{i+1} = f^i + \frac{\partial f_s^i}{\partial \sigma_\tau^s}\left(\sigma_\tau^{s^{i+1}} - \sigma_\tau^{s^i}\right) + \frac{\partial f_s^i}{\partial X}\left(X^{i+1} - X^i\right) \approx 0 \tag{51}$$

Moreover we know that:

$$\dot{X} = \gamma\dot{\alpha} = -\gamma\Delta\lambda_s \frac{\partial \phi_s}{\partial X} \tag{52}$$

$$\dot{\sigma} = -Gd\dot{\varepsilon}_\tau^s = -\Delta\lambda_s Gd \frac{\partial \phi_s}{\partial \sigma_\tau^s} \tag{53}$$

One can deduce from (51), (52) and (53), the expression of the incremental sliding multiplier $\Delta\lambda_s$ at the local iteration i:

$$\Delta\lambda = \frac{f_s^i}{\frac{\partial f_s^i}{\partial \sigma_\tau^s} \cdot Gd \cdot \frac{\partial \phi_s}{\partial \sigma_\tau^s} + \frac{\partial f_s^i}{\partial X} \cdot \gamma \cdot \frac{\partial \phi_s}{\partial X}} \tag{54}$$

4 MODEL FEATURES

The aim of this chapter is to sum up the previous developments. Such a model for bond slip needs the computation of internal variables as well as the identification of material and geometrical parameters.

4.1 *Summary*

The next tables give an overall summary of any quantities involved in the modeling.

Concerning the internal variables, Table 2 gives the relation between any internal variable and its corresponding thermodynamic associated forces.

Table 3 gives the geometrical and material parameters which have to be identified on the representative elementary volume in order to perform predictive analysis at the structural scale.

Table 2. Internal variables for constitutive relations.

Internal variables	Denomination	Associate forces
ε_N	Normal strain	σ_N
ε_τ	Tangential strain	σ_τ
ε_τ^s	Tangential sliding	σ_τ^s
d	Damage	Y
z	Damage hardening	Z
α	*Kinematic hardening*	X

Table 3. Geometrical and material parameters.

Mechanism	Denomination	Parameter
Contact kinematics	Penetration	h_{pen}
Elasticity	Young' mod.	E
	Poisson ratio	ν
Damage	Brittleness	A_d
	Threshold	ε_τ^0
Sliding	Kinematic law	γ
	Non-linear hardening	a
	Lateral pressure	c

4.2 *Model response*

This paragraph aims to show some numerical results obtained at the Gauss point level to point out the main features of such a constitutive modeling for reinforced concrete bond slip under monotonic and cyclic loading.

First of all, the material parameters used for the following analysis are given in Table 4. At this stage, we do not need the identification of the h_{pen} parameter.

In Figure 3, one can appreciate the effect of an imposed lateral pressure on the nonlinear tangential behavior. The dissipated energy is much larger when the confinement increases. Figure 4 points out the effect of the coupling introduced between damage and sliding. The hysteretic dissipation during unloading increases with the state of cracking; that feature is a crucial point when dealing with predictive analysis of physical dissipation in a structure. One can remark the presence of residual slip when unloading. The cyclic response for a load reversal is plotted in Figure 5.

Table 4. Material parameters value.

Denomination	Parameter	Value
Young' mod.	E	30 000 MPa
Poisson ratio	ν	0.2
Brittleness	A_d	$3.5\ 10^{-03}$ MPa^{-1}
Threshold	ε_τ^0	$1.2\ 10^{-0.4}$
Kinematic law	γ	$1.0\ 10^{10}$ MPa
Non linear hardening	a	$4.0\ 10^{-06}$ MPa^{-1}
Late-pressure	c	1.5

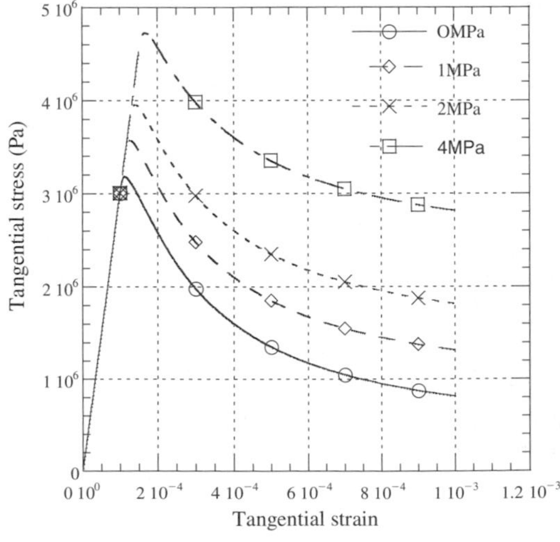

Figure 3. Monotonic response for different values of the normal stress.

5 NUMERICAL EXAMPLE

In order to point out this approach of debonding in reinforced concrete structure, a first numerical example has been treated. The experimental results for pull-out test of Bamonte *et al.* (2002), developed in order to show the importance of the size-effects on bonding, have been analysed. The specimen called ND1, has the next geometric and material properties (see Table 5).

The test has been carried out under displacement-control, measuring the relative displacement at the load-end of the bar, with a displacement rate of 5 μm/s, reaching a maximum slip of $0.4d_b$.

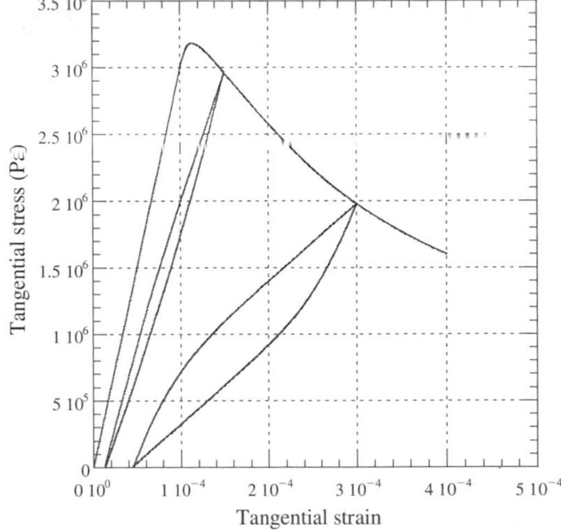

Figure 4. Non-linear response with unloading. Hysteretic behavior and inelastic slip.

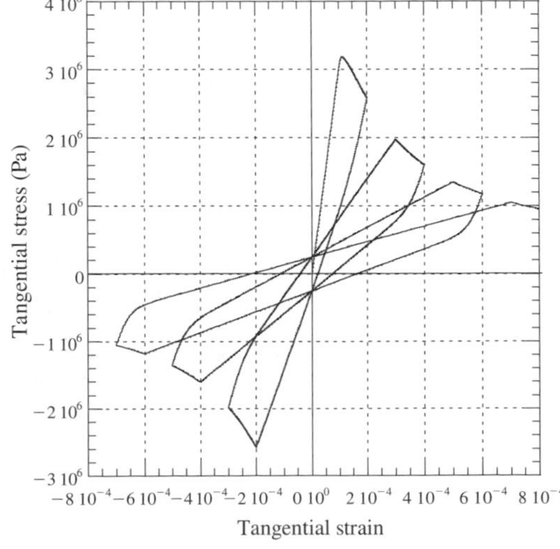

Figure 5. Cyclic response.

Table 5. Geometric and material properties for specimen ND1.

Properties	Value	Units
NSC (Normal-Strength Concrete):		
f'c	39	MPa
cement content	300	kg/m^3
water–cement ratio	0.66	–
Steel:		
fy	800–1000	MPa
Dimension		
d_b (steel bar)	5	mm
L (Length)	26.75	mm
c (concrete cover)	20	mm

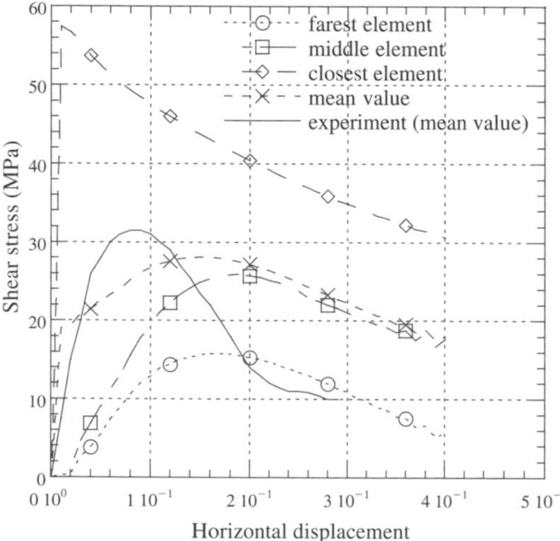

Figure 6. Displacement at the end of the bar. Comparison experiment/modeling.

Only a half of the specimen was modeled in the FEAP code, as it is shown in the displacement and stress distribution graphics. Some of our parameters were computed according to the experiments, as the elastic deformation yield point and G, the shear modulus. For the h_{pen} parameter, we have used values between 1 and 10 mm, in order to check the influence on calculations. From these values the shear stress behavior was calculated on the interface. So, the final curve correspond to the envelope of behavior curves for each interface element. Obviously, the resistance of the first element decreases almost immediately and that means some effects on the behavior of the other elements (as relaxation). However, it is very important to mention that concrete has not been yet modeled for degradation or cracking. The comparison between numerical modeling and experiment are shown in Figure 6. The shear stresses are given for 3 elements along the bar and the comparison with the experiment can be carried out by plotting the mean value of these.

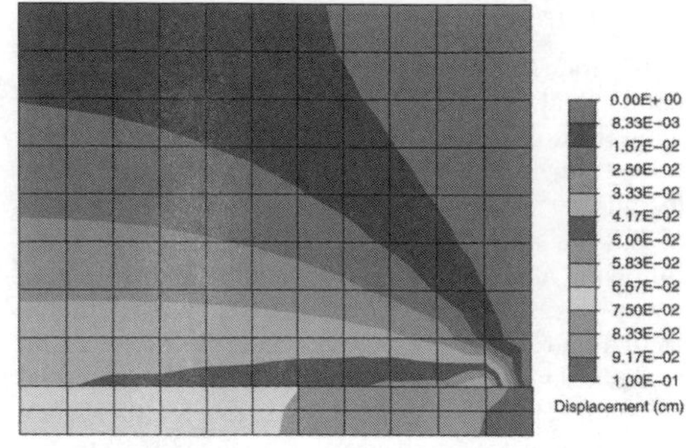

Figure 7. Horizontal displacement (cm) at the end of the steel bar.

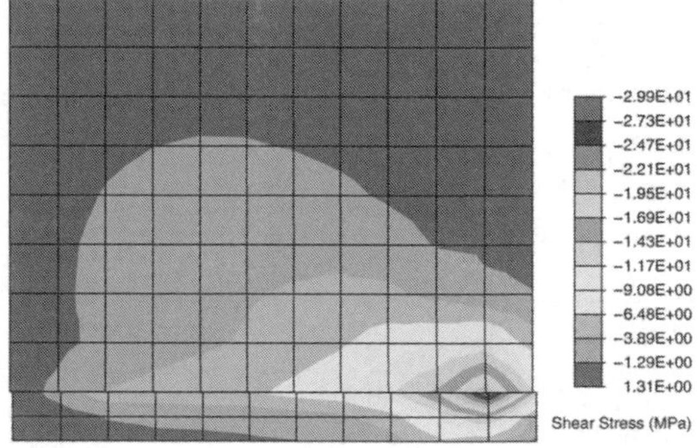

Figure 8. Shear stress (MPa) at the interface.

In Figures 7 and 8, some isovalues of horizontal displacement and shear stresses in the structure are plotted.

6 CONCLUSIONS

The response modelling of reinforced concrete structures subjected to mechanical loadings needs physical representations of the different mechanisms involved both in materials behaviour and elements kinematics. Many efforts have been paid in describing the multidimensional nonlinear response of concrete and steel in order to better understand the global behaviour of concrete structures subjected to complex loading path, cyclic or even in dynamics. The prediction of ultimate bearing capacity requires more detailed and local investigation while dealing with crack opening and spacing. Not only refine material constitutive equations have to be introduced in the analysis but also specific kinematics between steel and concrete. The main objectives of this presentation concern the expression of a particular modelling of bond phenomena in reinforced concrete structural elements. Such an element, dealing with three-dimensional state of stress, account for the coupling between the non-linear tangential behaviour and the level of radial confinement. The modelling handles not only a displacement interpolation field allowing discontinuities but also a robust and physical material constitutive behavior able to treat the interface layer degradation and frictional sliding contact.

The aim of this work focuses on the development and the numerical implementation of a new 2D degenerated interface element based on 4 double nodes discretization. The surrounding radial forces can be transmitted from an element to its adjacent one through this particular layer. The constitutive behavior, coupling damage and plasticity within the standard thermodynamics framework allows a realistic description of the complex phenomena involved in a cracked medium without neglecting the pragmatic requirement due to the related numerical implementation in a finite element analysis program. The main phenomena accounted for are:

- Symmetric cracking in the tangential direction
- Inelastic strains upon unloading including frictional sliding
- Influence of the radial stress on the tangential flow rule

The capability of such a formulation to predict the behavior up to failure of reinforced concrete structures has been exemplified by comparing numerical computations and experimental results on a simple case study where bond-slip plays a major role in the global behavior.

REFERENCES

Armstrong, P.J. & Frederick, C.O. 1966. A Mathematical Representation of the Multiaxial Bauschinger Effect', G.E.G.B., Report RD/B/N, 731.

Bamonte, P., Coronelli, D. & Gambarova, P.G. 2002. Size Effect in high-bond bars. *"Bond in Concrete: from research to standards" Proceedings of the 3rd International Symposium at the Budapest University of Technology and Economics*: 43–52.

Cox, J.V. & Hermann, L.R. 1998. Development of a plasticity bond model for steel reinforcement. *Mechanics of cohesive-frictional materials* 3: 155–180.

Hughes, T.J.R. 1987. *The Finite Element Method.* New Jersey: Prentice-Hall International Editions.

Ibrahimbegovic, A. & Wilson, E.L. 1992. Unified computational model for static and dynamic frictional contact analysis. *International journal for numerical methods in engineering* 34: 233–247.

La Borderie, Ch. 1991. Phénomènes unilatéraux dans un matériau endommageable: modélisation et application à l'analyse de structures en béton. *Ph.D. thesis:* Univ. Paris VI.

Maker, B.N. & Laursen, T.A. 1994. A finite element formulation for rod/continuum interactions: the one-dimensional slideline. *International journal for numerical methods in engineering* 37: 1–18.

Ortiz, M. & Simo, J.C. 1986. An analysis of a new class of integration algorithms for elastoplastic constitutive relations, *Int. J. Numer. Meth., Eng.* Vol. 23: 353–366.

Ragueneau, F., La Borderie, Ch. & Mazars J. 2000. Damage Model for Concrete Like Materials Coupling Cracking and Friction, Contribution towards Structural Damping: First Uniaxial Application. Mechanics of Cohesive. Frictional. Materials. Vol. 5: 607–625.

Romdhane, B. & Ulm, F.J. 2002. Computational mechanics of the steel-concrete interface. *International journal for numerical and analytical methods in geomechanics* 26: 99–120.

Salari, M.R. & Spacone, E. 2001. Finite element formulations of one-dimensional elements with bond-slip. *Engineering structures* 23: 815–826.

Computational Modelling of Concrete Structures, Bićanić et al. (eds)
© 2003 Swets & Zeitlinger, Lisse, ISBN 90 5809 536 3

Bond-slip relationship for smooth steel reinforcement

A.P. Fantilli & P. Vallini
Politecnico di Torino, Torino, Italy

ABSTRACT: In this paper a theoretical approach for the pullout mechanism is proposed, in order to improve the bond-slip relationship proposed by the CEB-FIP Model Code of smooth steel reinforcement. In particular, a reduced number of parameters and a softening branch in the τ-s model are introduced. With a view to use this relationship both for RC and SFRC elements, the dependence of the model on the size of reinforcement is also considered. Despite the limited number of parameters used in the calibration of bond-slip, there is a good agreement between the results obtained with the proposed model and the experimental ones.

1 INTRODUCTION

The post cracking response of fibre reinforced composite elements in tension is very important in modelling the mechanical behaviour of steel fibre reinforced concrete structures (SFRC). By means of the relationship between the stress on crack surfaces and the corresponding crack width (i.e. the cohesive model), it is possible to reproduce theoretically the structural response of beams subjected to bending and shear actions (Casanova et al. 1997). Moreover, as is well evident in many experimental and theoretical analyses (Hillerborg 1980), this relationship allows to model different kind of fractures in fibre reinforced members. Even if the importance of fictitious crack model for cementitious elements is known (Hillerborg et al. 1976), its definition for fibre reinforced concrete appears uneasy. In comparison with the cohesive models for plain concrete, in SFRC the fibres produce a more complicate bridging action.

In the past, several cohesive models were introduced in order to model fibre reinforced concrete (see Hu & Day 2000 for a review). In many cases, due to the simplifications adopted, they only give a qualitative definition of the stress-crack opening displacement relationship. Hillerborg (1980), for example, assumed a stiff-plastic relationship to model the bond-slip on the interface between fibres and concrete. In this way, the bond stresses are assumed constant for all slip values different from zero. On the contrary, a nonlinear relationship between the mean values of bond stresses and slips has been detected in a huge number of experimental pullout tests (Bartos 1981). Generally, in a test conducted on smooth and straight steel fibres, the pullout force reaches the peak for small slips before decreasing and approaching to an asymptotic value (Balaguru & Shah 1992). The average value of bond stresses, corresponding to a pullout force, can be very different from the local stresses measured along the anchored length. Moreover, the pullout response also depends on strength and composition of the concrete and on the form of the fibres.

The bond stress on the interface between the cementitious matrix and the reinforcement is generally estimated with a bond-slip (or τ-s) function. This re-lationship is a fundamental tool to model the structural behaviour of reinforced concrete structures (RC). The CEB-FIP Model Code (CEB 1993) furnishes a bond-slip model, which has been successfully adopted to reproduce theoretically the serviceability stage of concrete beams reinforced with ribbed steel bars (Fantilli et al. 1998). On the contrary, the τ-s relationship proposed by CEB (1993) for smooth steel rein-forcement is not able to reproduce the physical evidence. There are two main reasons for this failure. First of all the shape of the τ-s law seems inappropriate: for the cold drawn wire and hot rolled bar, the bond-slip function does not take into account the softening branch after the peak stress. When the maximum value is reached, the bond stress remains constant, regardless of the values of slip between reinforcement and surrounding concrete. This formulation cannot be used to reproduce pullout tests, which are generally considered to measure the bond stresses. As clearly shown in the experimental tests of Bazant et al. (1995), where the failure of pullout specimens is only due to the interface slip, after the peak load a softening branch appears. Before complete debonding, the pullout load seems to tend to a non-zero asymptotic value.

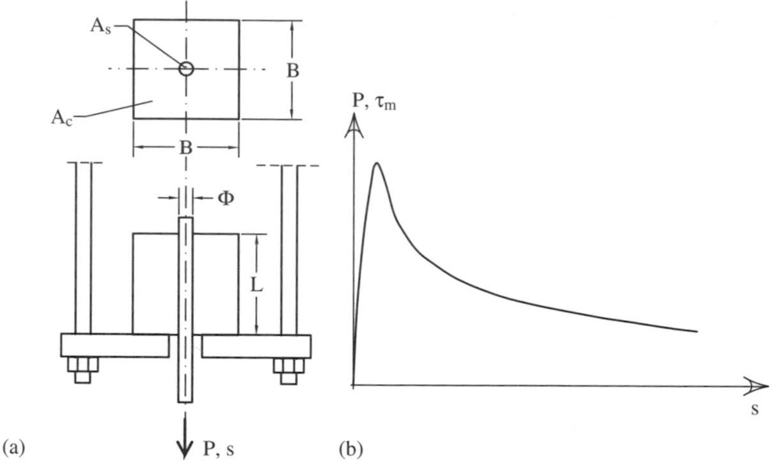

Figure 1. The pullout test: a) the specimen; b) experimental P-s or τ_m-s pullout curves.

The second reason, for the unreliability of the τ-s model proposed by CEB (1993) for smooth steel reinforcement, consists in an incorrect evaluation of the bond parameters, which are independent of the bar diameter. The tests of Bazant et al. (1995), performed on specimens having the same geometry but different dimensions, clearly show the dependence of maximum bond stress on the size of the tested elements, even when no crushing of concrete occurs. This size effect has been also confirmed in recent experimental tests (Coronelli et al. 2001), where the influence of concrete strength and bar diameter on bond stresses have been measured. Analytical models (Bazant & Desmorat 1994, Shah & Ouyang 1991) also show the correspondence between the strain energy release rate (and consequently the peak and the shape of the softening branch of pullout curves) and the scale of the pullout specimens.

Starting from these experimental and numerical results, the bond-slip relationship for smooth reinforcement proposed in CEB (1993) can be improved, by introducing a post peak softening branch and a link between the bond parameters and the dimension of bar diameter. Only a τ-s relationship accounting for on size effect can be used more effectively in modelling the bond between steel bar and concrete (in RC elements) and between steel fibres and surrounding matrix (in SFRC).

2 A ONE-DIMENSIONAL PULLOUT MODEL

The bond stress between steel bars and concrete, or fibres and cementitious matrix, is experimentally measured from pullout tests (Fig. 1a). In these tests, according to the RILEM/CEB/FIP (1970) guidelines, the anchored length cannot be too long, in order to avoid yielding of the reinforcement before the complete debonding. Moreover, in a shorter anchorage zone, bond stresses do not exhibit a considerable variation, therefore their values can be estimated with the average bond stress τ_m formula:

$$\tau_m = \frac{P}{\pi \cdot \Phi \cdot L} \tag{1}$$

where P = pullout force; Φ = diameter of reinforcement; and L = anchorage length.

Each value of P, or each average value of bond stress obtained with (1), corresponds to a value of slip s measured between reinforcement and concrete at the head of the pullout specimen of Figure 1a. In other words, by means of pullout tests it is possible to evaluate the whole τ_m-s curve of Figure 1b, and the parameters that affect the bond slip relationship (Eligehausen & Bigaj-van Vliet 1999).

On the other hand, however little the anchorage length is, the distribution of bond stresses along L cannot be considered constant, so τ_m obtained from Equation 1 is a rough estimate of the local bond stress. As a consequence, it is necessary to introduce refined approaches in order to obtain the bond slip relationship from pullout tests. Referring to the classification made by Shah & Ouyang (1991), in addition to the strength models based on Equation 1 and the energetic approaches of Bazant & Desmorat (1994) and Shah & Ouyang (1991), the so-called Cohesive Interface Models seem to reproduce adequately a pullout test. In these models, the bond stress at a point of reinforcement-matrix interface depends on the slip between the two materials. So, a suitable local bond-slip function (τ-s), which is better defined in the next paragraph, must be added to the classical equilibrium and compatibility equations (Fantilli et al. 1998).

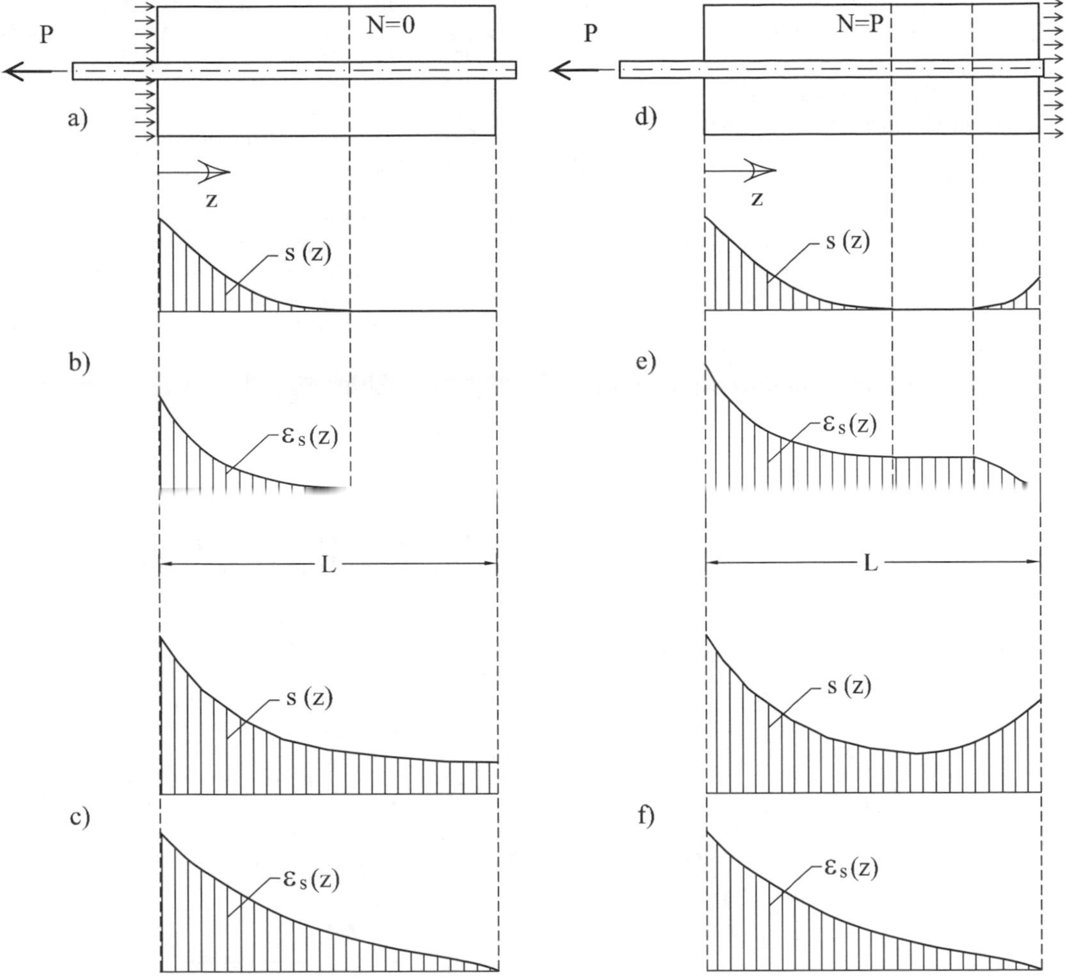

Figure 2. Slip and steel strain distributions along the bar: a)-b)-c) classical pull-out test; d)-e)-f) pullout test with concrete in tension.

In particular, with the hypotheses of linear behaviour of steel and concrete and linear strain profile in each cross-section of the pullout element, an Interface Cohesive Model consists in solving a system composed by an equilibrium equation of a generic cross-section (Eq. 2), an equilibrium equation of a portion of the naked bar (Eq. 3) and a compatibility equation between steel and concrete at the interface (Eq. 4):

$$N = A_s \cdot E_s \cdot \varepsilon_s + A_c \cdot E_c \cdot \varepsilon_c \quad (2)$$

$$\frac{d\varepsilon_s}{dz} = \frac{4 \cdot \tau}{E_s \cdot \Phi} = f_1(z) \quad (3)$$

$$\frac{ds}{dz} = \varepsilon_s - \varepsilon_c = f_2(z) \quad (4)$$

where N = normal resultant in a cross-section; E_s, E_c = Young moduli for steel and concrete; $\varepsilon_s, \varepsilon_c$ = steel and concrete strains; A_s, A_c = areas of concrete and steel in a cross-section; and z = distance from the point where the pullout force P is applied (Fig. 2).

If the bond-slip function is known, the system (Eqs. 2–4) can be solved with a predictor-corrector numerical method, since the function $f_1(z)$ and $f_2(z)$ are unknown. In the present paper, Equations 2–4 regard both the classical pullout tests, where the concrete is in compression (Fig. 2a), and pullouts with concrete in tension, which is more frequent in RC structures (Fig. 2d). In the first case $N = 0$ in each cross-section of the specimen, while $N = P$ in the last case. In order to reproduce the tests in both cases, the adopted numerical procedure consists in the evaluation of the pullout force P that corresponds to an imposed value of slip $s(z=0)$ at the head of the element. In particular:

1) Select a slip $s(z = 0)$.
2) Assume a value for the pullout force P.
3) Compute the boundary condition in $z = 0$.

$$\varepsilon_s(z=0) = \frac{P}{A_s \cdot E_s} \tag{5}$$

4) Integrate Equations 2–4 with a predictor corrector method in the domain [0, L].
5) If in any internal section of the domain ($0 < z < L$), the condition $\varepsilon_s = 0$ (Fig. 2b) in the classical pullout (or $\varepsilon_s = \varepsilon_c$ in the pullout with concrete in tension of Fig. 2e) is satisfied, the slip in this section must be equal to zero. If this situation is verified, it is possible to obtain the distributions of slips $s(z)$ and steel strains $\varepsilon_s(z)$ along the bar (Figs. 2b, 2e). Otherwise, the procedure goes back to step 3 with a new trial value of P.
6) If in the end of the element ($z = L$), the condition $\varepsilon_s = 0$ is satisfied, it is possible to obtain the distributions of slips $s(z)$ and steel strains $\varepsilon_s(Z)$ along the bar (Figs. 2c, 2f). Otherwise, the procedure goes back to step 3) with a new trial value of P.

With this procedure, by increasing monotonically the slip $s(z = 0)$, it is possible to obtain numerically the pullout curves P-s or τ_m-s as shown in Figure 1b.

It must be remarked that in the proposed one-dimensional interface cohesive model, the transversal strains in steel and concrete are neglected. On the contrary, the Poisson effect between the reinforcement and the surrounding matrix is taken into account in more complex three-dimensional interface models (Shah & Ouyang 1991).

3 THEORETICAL MODELLING OF BOND-SLIP

With a view to define both the τ-s curve and the bond-slip parameters, the previous model can be used in conjunction with the results of pullout tests. Previous cohesive interface approaches were used to that end. Wang et al. (1988) for example, neglecting the concrete strain, considered a quadratic function for the bond-slip. The parameters of the τ-s relationship were set by comparing theoretical and experimental pullout results obtained for different kind of fibres and matrix. Similarly, Naaman et al. 1991a, b, with a model able to take into account the Poisson effect, defined a τ-s relationship with a linear ascending and nonlinear descending branches. In both these cases, the bond-slip modelled only the behaviour of a fibre in a matrix, neglecting the extension to bars. Consequently, the size effect of bar diameter on bond stresses, measured in recent tests (Bazant et al. 1995, Coronelli et al. 2001), cannot be included in these bond-slip relationships. This is also true for the τ-s proposed by CEB (1993) for steel rebars and not for fibres. For this reason, in the present paper the theoretical approach for the pullout mechanism previously proposed, is used to improve the bond-slip relationship proposed by Model Code (CEB 1993) for smooth steel reinforcement. The purpose is to introduce a more reliable τ-s

a)

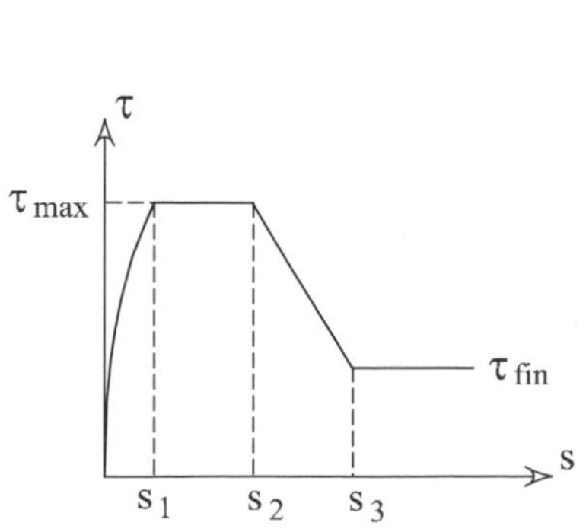

b)

	Cold drawn wire		Hot rolled bars	
	Bond conditions		Bond conditions	
	Good	All other cases	Good	All other cases
s_1	0.01 mm		0.1 mm	
s_2	0.01 mm		0.1 mm	
s_3	0.01 mm		0.1 mm	
α	0.5		0.5	
τ_{max}	$0.1\sqrt{f_c}$	$0.05\sqrt{f_c}$	$0.3\sqrt{f_c}$	$0.15\sqrt{f_c}$
τ_{fin}	τ_{max}		τ_{max}	

Figure 3. Bond-slip model propoded in CEB (1993): a) the branches of τ-s; b) the bond parameters.

relationship for smooth steel reinforcement including bars and fibres.

Actually, the τ-s relationship for smooth steel reinforcement is composed by the following branches (Fig. 3a):

$$\tau = \tau_{max} \cdot \left(\frac{s}{s_1}\right)^{\alpha} \qquad s \le s_1 = s_2 = s_3 \qquad (6)$$

$$\tau = \tau_{max} = \tau_{fin} \qquad s > s_1 = s_2 = s_3 \qquad (7)$$

where the three parameters s_1, $\tau_{\max}$ and α are defined in Figure 3b as a function of bond conditions, the type of smooth bar (hot rolled or cold drawn) and the concrete compressive strength f_c.

Experimental tests, both on bars (Bazant et al. 1995, Coronelli et al. 2001) and on fibres (Naaman et al. 1991b, Naaman & Najm 1991), measured different values of the maximum bond stress $\tau_{\max}$ and the ultimate one τ_{fin} (residual bond). In particular, after the peak, bond stresses decrease indicatively according

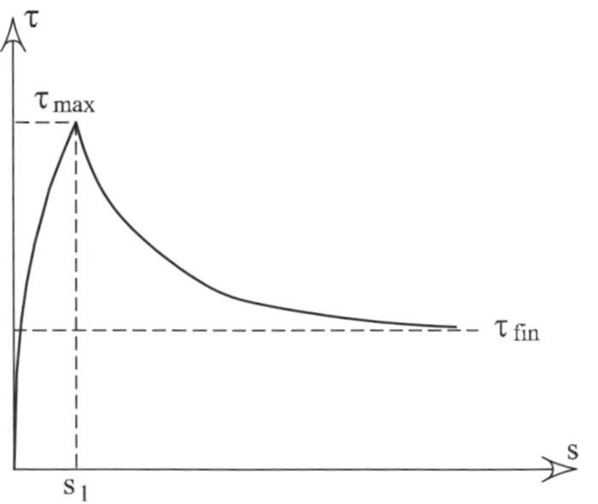

Figure 4. Proposed bond-slip relationship.

to an exponential function (Naaman et al. 1991a, b), before reaching an asymptotic value.

Therefore, Equation 6 should be replaced with the following post-peak softening branch (Fig. 4):

$$\tau = \tau_{fin} + (\tau_{max} - \tau_{fin})e^{k(s_1 - s)} \qquad s > s_1 = s_2 = s_3 \quad (8)$$

where $\tau_{\max} > \tau_{fin}$; and k = positive coefficient whose dimension is the inverse of length.

With the proposed interface cohesive model, the size effect can be taken into account by setting the bond parameters ($\tau_{\max}$, τ_{fin}, s_1, α and k in the Equations 6, 8) as a functions of the bar diameter. In the bond-slip relationship proposed by CEB (1993), only the concrete strength and the bond conditions affect these parameters (Fig. 3b).

Unfortunately, there is not a wide experimental campaign that regards pullout tests of smooth steel rebars or smooth steel fibres of SFRC. For this reason, in the present paper two different experimental researches are considered in defining the bond parameters of the relationship depicted in Figure 4. The first one, by Bazant et al. (1995), consists of pullout tests of smooth steel reinforcing bars of diameter Φ, embedded in concrete cubes having the side equal to the anchored length L (Fig. 5a). As is shown in Figure 5a, the specimens are geometrically similar but of different dimensions. These dimensions and the mechanical properties of materials are summarized in Figure 5b.

For smooth steel fibres, the experimental pullout tests of Naaman et al. (1991b) and Naaman & Najm (1991) are taken into consideration. In particular, the fibres are one by one pulled out from the mortar block, as depicted in Figure 6a. Even if the experimental campaign also regarded the deformed fibres embedded in high strength mortar, in the present work only the tests performed on the straight and smooth steel fibre are considered (Fig. 6b).

a)

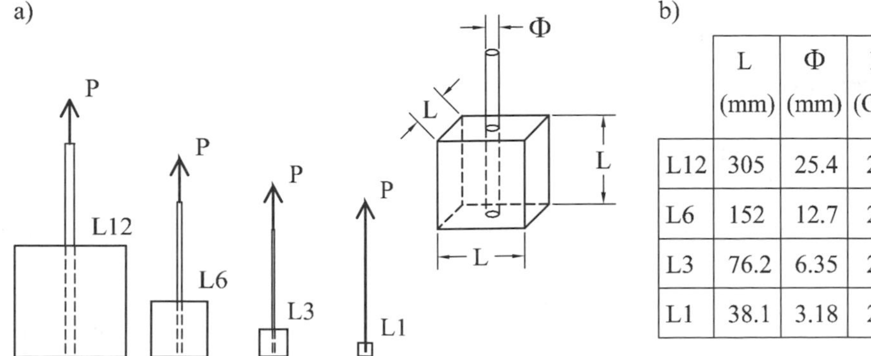

b)

	L (mm)	Φ (mm)	E_s (GPa)	E_c (GPa)	f_c (MPa)
L12	305	25.4	210	30	36
L6	152	12.7	210	30	36
L3	76.2	6.35	210	30	36
L1	38.1	3.18	210	30	36

Figure 5. Experimental pullout tests of Bazant et al. (1995): a) the specimen sizes; b) geometrical and mechanical properties of the elements.

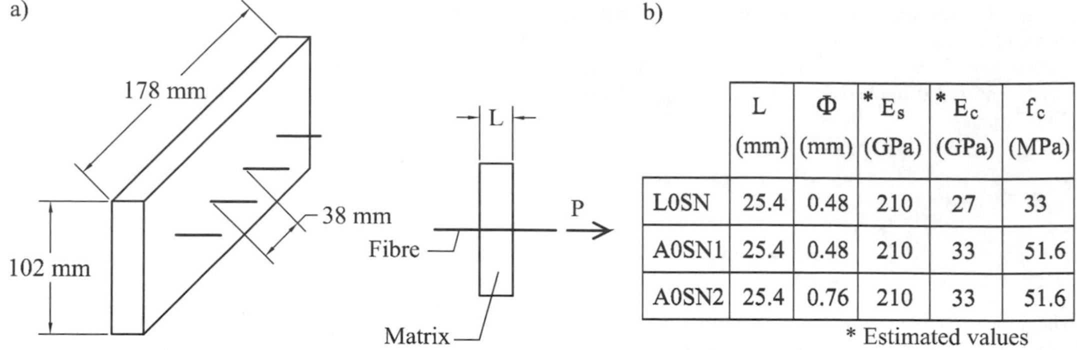

	L (mm)	Φ (mm)	* E_s (GPa)	* E_c (GPa)	f_c (MPa)
L0SN	25.4	0.48	210	27	33
A0SN1	25.4	0.48	210	33	51.6
A0SN2	25.4	0.76	210	33	51.6

* Estimated values

Figure 6. Experimental pullout tests of Naaman et al. (1991b) and Naaman & Najm (1991): a) the specimen sizes; b) geometrical and mechanical properties of the elements.

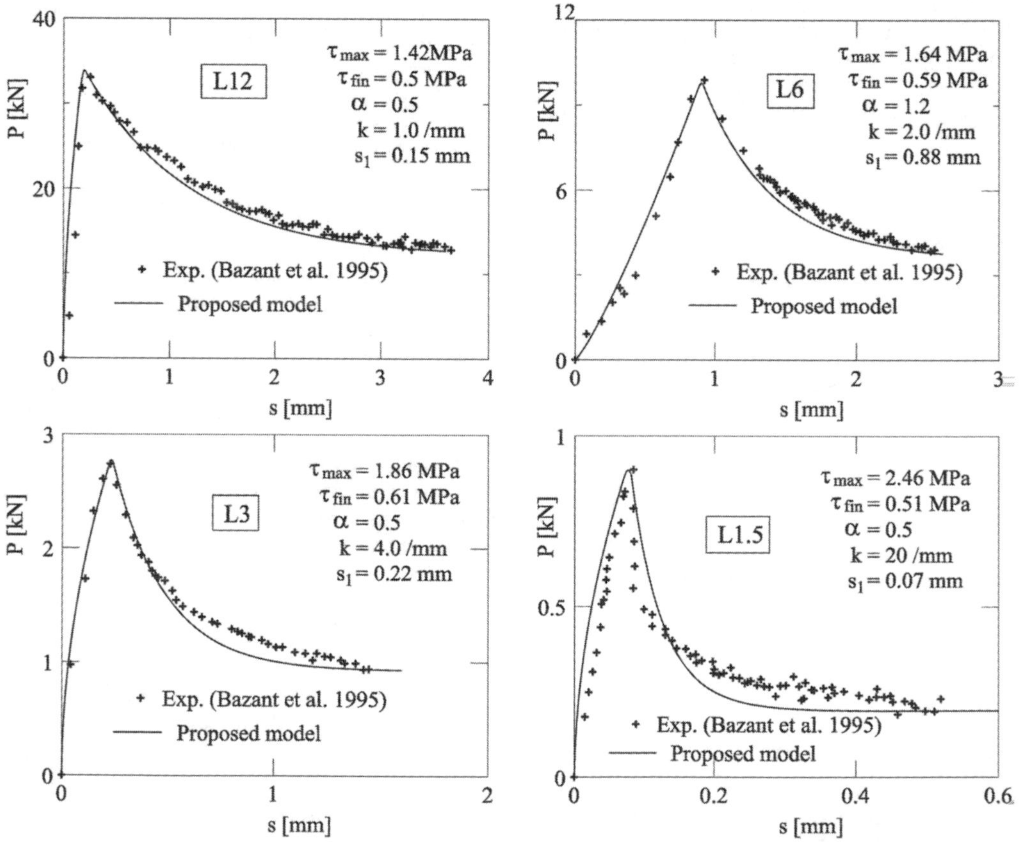

Figure 7. Comparison between numerical results and experimental pullout tests of Bazant et al. (1995).

4 THE PARAMETERS OF THE PROPOSED τ-s RELATIONSHIP

In Figures 7–8 the pullout force-displacement diagrams obtained in seven tests are considered, four of them extracted from Bazant et al. (1995) and the remaining part from Naaman et al. (1991b) and Naaman & Najm (1991). In these pictures, both the experimental results and the numerical best fitted curves are compared. The numerical results are

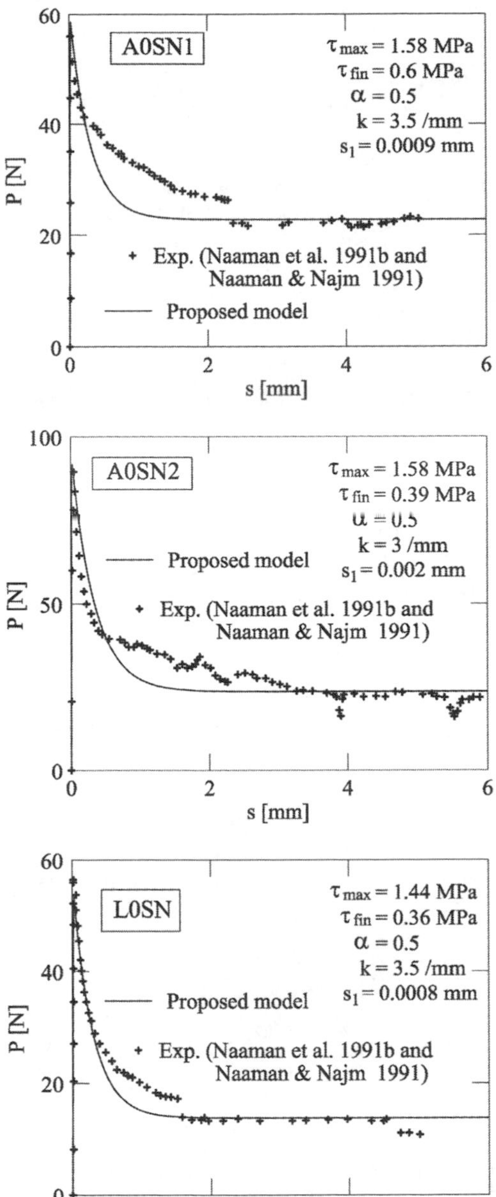

Figure 8. Comparison between numerical results and experimental pullout tests of Naaman et al. (1991b) and Naaman & Najm (1991).

obtained with the proposed model by adopting suitable values for the parameters τ_{max}, τ_{fin}, s_1, α and k of the proposed τ-s relationship (Fig. 4). Regarding the experimental results of Bazant et al. (1995), the maximum value of bond stress must be considered as a function of the bar diameter through the Bazant's size effect law:

$$\tau_{max} = \frac{A}{\sqrt{B+\Phi}}\sqrt{f_c} \tag{9}$$

where $A = 1.572$ and $B = 12.50$ if τ_{max} and f_c are measured in MPa. For hot rolled smooth steel bars in good bond conditions, the Model Code (CEB 1993) suggests the following expression of τ_{max} (Fig. 3b):

$$\tau_{max} = 0.3\sqrt{f_c} \tag{10}$$

This result can be obtained with the Equation 9 by using $\Phi = 15$ mm for the bar diameter. As is shown in Figure 9a, this value of Φ is ranked in the middle of the diameter range of currently used steel smooth bars (5 mm–25 mm). Thus, the smooth bar used in Bazant et al. (1995) must be considered hot rolled, even if the Authors do not declare the type of bar. For cold drawn wire in good bond conditions, the Model Code (CEB 1993) suggests a reduced value for τ_{max} (Fig. 3b):

$$\tau_{max} = 0.1\sqrt{f_c} \tag{11}$$

More precisely, the maximum bond stress obtained from Equation 10 must be reduced by 2/3. If the whole Equation 9 is shifted of the same quantity, the expression of τ_{max} for cold drawn wire can be obtained (Fig. 9b):

$$\tau_{max} = \left(\frac{A}{\sqrt{B+\Phi}} - 0.2\right)\sqrt{f_c} \tag{12}$$

In Figure 9b the maximum values of bond stress obtained with Equation 12 are in agreement with the ones measured in the pullout tests of smooth steel fibres (Naaman et al. 1991b and Naaman & Najm 1991). For this reason, the smooth steel fibres used by Naaman et al. (1991b) and Naaman & Najm (1991) must be considered as cold drawn wire, even if the Authors do not declare the type of fibres.

In the experimental results of Bazant et al. (1995), the final bond stress τ_{fin} remains nearly constant, independently of the variation of the bar diameter (Fig. 10a):

$$\tau_{fin} = 0.1\sqrt{f_c} \tag{13}$$

In other words, according to the experimental results of Coronelli et al. (2001), the residual bond stress must be considered a material property, since it is size-independent. Moreover, Equation 13 does not seem to be in agreement with the value suggested by the CEB (1993) for the hot rolled bars in good bond conditions (Fig. 3b).

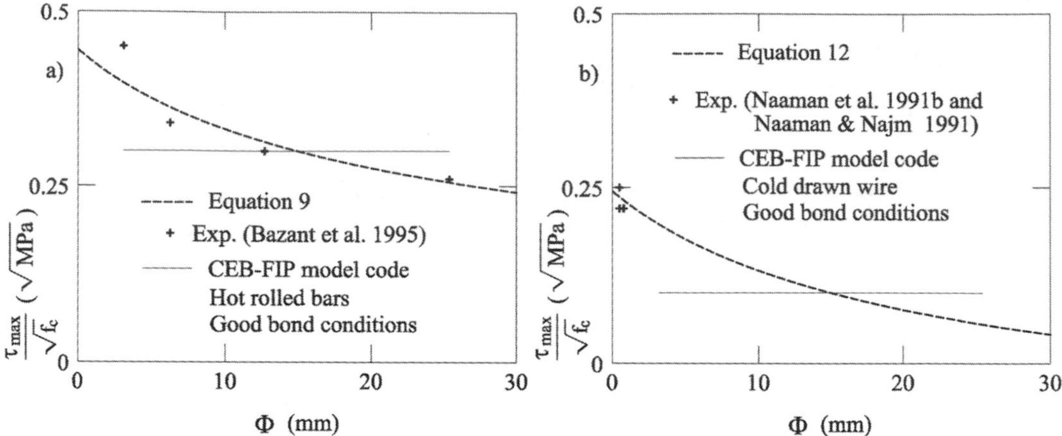

Figure 9. Evaluation of τ_{max}: a) Comparison between Equation 9, CEB-FIP model code formula for hot rolled bar and experimental pullout tests of Bazant et al. (1995); b) Comparison between Equation 12, CEB-FIP model code formula for cold drawn wire and experimental pullout tests of Naaman et al. (1991b) and Naaman & Najm (1991).

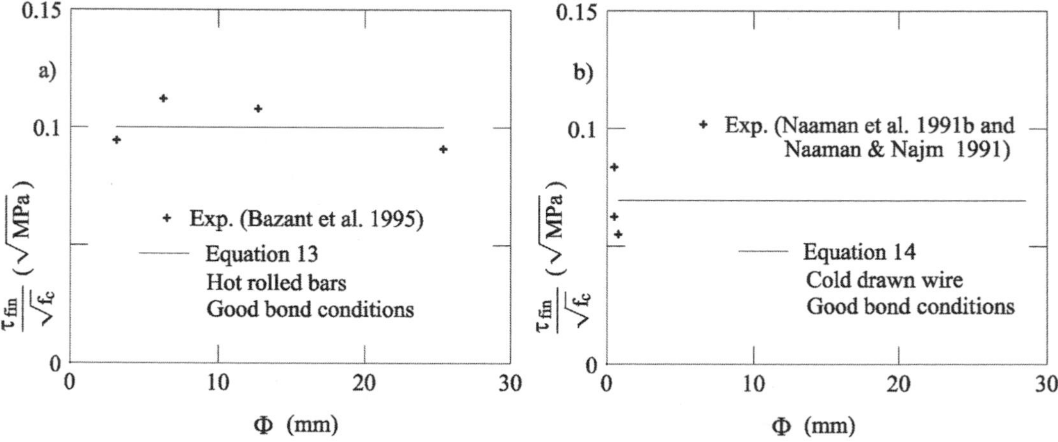

Figure 10. Evaluation of τ_{fin}: a) Comparison between Equation 13 for hot rolled bar and experimental pullout tests of Bazant et al. (1995); b) Comparison between Equation 14 for cold drawn wire and experimental pullout tests of Naaman et al. (1991b) and Naaman & Najm (1991).

Regarding the final value of bond stress measured in the pullout of fibres, from the experimental tests of Naaman et al. (1991b) and Naaman & Najm (1991) there is a wider scatter of τ_{fin}, whose average value can be evaluated with the following equation:

$$\tau_{fin} = 0.067\sqrt{f_c} \tag{14}$$

Compared to the expression proposed in the CEB-Model Code for cold drawn wire in good bond conditions (Fig. 3b), Equation 14 furnishes a lower value of τ_{fin} (about 70% lower).

In the theoretical evaluation of pullout curves of Figures 7–8, except for the test L6 of Bazant et al. (1995), the exponent α in the ascending branch of τ-s (Eq. 6) is equal to the value suggested by the Model Code ($\alpha = 0.5$). This value seems to be constant, independently of the bar diameter and of the strength of surrounding concrete.

The definition of s_1 and k is quite difficult in comparison with the estimate of τ_{max}, τ_{fin}, and α. Both these parameters are remarkably affected by the experimental evaluation of slips, whose measure is not simple in small scale specimens (Wang et al. 1988). In the case of pullout of fibres, the slip s_1 at the peak bond stress does not affect appreciably the position of the softening branch of τ-s (Fig. 4), where most of the debonding energy is released (Naaman et al. 1991b).

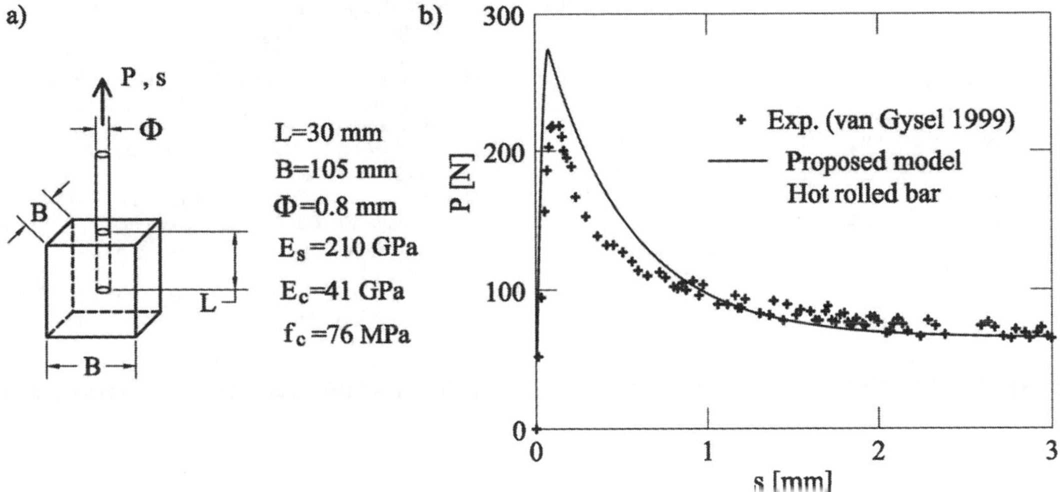

Figure 11. Pullout test on a smooth steel fibre (van Gysel 1999): a) geometrical and mechanical properties of the specimen; b) comparison between the numerical results obtained with the proposed model and the experimental ones.

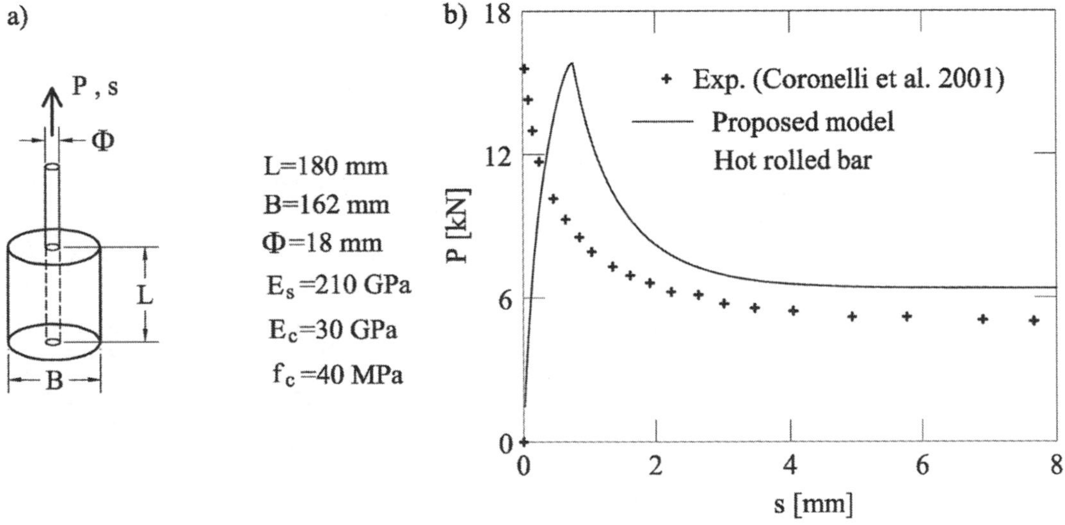

Figure 12. Pullout test on a smooth steel bar (Coronelli et al. 2001): a) geometrical and mechanical properties of the specimen; b) comparison between the numerical results obtained with the proposed model and the experimental ones.

So, in absence of an experimental campaign aimed to the correct evaluation of s_1, it is reasonable to assume the values proposed in Figure 3b by Model Code (CEB 1993). The parameter k cannot be correctly defined because it is affected by the incorrect evaluation of s_1. From the comparison between the numerical and experimental results of Figures 7–8, except for the test L1.5 of Bazant et al. (1995), k is comprised in a range of 1–4/mm, independently of the bar diameter and of the strength of surrounding concrete.

5 COMPARISON WITH OTHER EXPERIMENTAL RESULTS

The bond–slip relationship of Figure 4 and its parameters τ_{max}, τ_{fin}, s_1, α and k, evaluated on the basis of the previous indications, are now used to reproduce theoretically the pullout responses of other experimental tests. In particular, by using the proposed one-dimensional interface cohesive approach, the P-s curves of smooth hot rolled fibres of Figure 11a (van Gysel 1999) and rebars of Figure 12a (Coronelli et al. 2001) are evaluated. These curves are compared, in Figures 11b–12b, with the corresponding experimental results.

Despite the different sizes of the two specimens, there is a satisfying agreement between numerical and experimental *P-s* diagrams, especially in the case of fibres. In fact, in the pullout of fibres, there is a high scatter of experimental measures, in particular in the first part of the curve in Figure 1b, where the position of the peak is difficult to evaluate. This is due to the disuniformity of fibres, whose diameter is often larger at the ends than in the middle portion (Stroeven et al. 1978), and to the heterogeneity of the matrix, particularly relevant for smaller diameters reinforcement.

6 CONCLUSIONS

In this paper, a bond-slip relationship between smooth steel reinforcement and concrete (or mortar) has been defined. By means of a suitable evaluation of the fundamental parameters, the τ-s model suggested by CEB-FIP Model Code for rebars has been extended to fibres. To compare theoretical and experimental results, it has been necessary to introduce an exponential function for the softening branch in this relationship, whose parameters have been set as a function of the bar diameter. A good agreement between the experimental results and the numerical ones (Figs. 11, 12) has been obtained at different scale. The procedure adopted in this paper for defining the bond of smooth steel reinforcement should be also extended to other situations (e.g. plastic or deformed fibres), with the condition that further experimental campaign will be performed.

REFERENCES

Balaguru, P.N., and Shah, S.P. 1992. *Fiber-Reinforced Cement Composites*, New York, USA: McGraw-Hill, Inc.

Bartos, P. 1981. Review paper: Bond in Fibre Reinforced Cements and Concretes, *The International Journal of Cement Composites*, 3(3): 159–177.

Bazant, Z.P., & Desmorat, R. 1994. Size Effect in Fiber or Bar Pullout with Interface Softening Slip, *ASCE Journal of Engineering Mechanics*, 120(9): 1945–1962.

Bazant, Z.P., Li, Z., and Thoma, M. 1995. Identification of Stress-Slip Law For Bar or Fiber Pullout by Size Effect Tests, *ASCE Journal of Engineering Mechanics*, 121(5): 620–625.

Casanova, P., Rossi, P., and Shaller, I. 1997. Can Steel Fibers Replace Transverse Reinforcement in Reinforced Concrete Beams?, *ACI Materials Journal*, 94(5): 341–354.

CEB, 1993. CEB-FIP *Model Code 1990. Bulletin d'information no.* 203–205, London, England: Thomas Telford.

Coronelli, D., Gambarova, P.G., and Ravazzani, P. 2001. Size Effect in Steel-concrete Bond: Test Results and Modelling for Smooth Bars. In de Borst et al. (eds), *Proceedings of the fourth International Conference on Fracture Mechanics of Concrete and Concrete Structures, Cachan, France, 28 May-1 June*: 669–676.

Eligehausen, R., and Bigaj-van Vliet, A. 1999. Bond Behaviour and Models, *Structural Concrete-Text-book on Behaviour, Design and Performance*, Vol. 1, Fib bulletin no. 1: 161–187.

Fantilli, A.P., Ferretti, D., Iori, I., and Vallini, P. 1998. Flexural Deformability of Reinforced Concrete Beams, *ASCE Journal of Structural Engineering*, 124(9): 1041–1049.

van Gysel, A. 1999. A Pullout Model for Hooked End Steel Fibres. In H.W. Reinhardt and A.E. Naaman (eds), *Third International Workshop on Higher Performance Fiber Reinforced Cement Composites (HPFRCC3), Mainz, Germany, May 16-19*: 351–359.

Hillerborg, A. 1980. Analysis of Fracture by Means of the Fictitious Crack Model, Particularly for Fibre Reinforced Concrete, *The International Journal of Cement Composites*, 2(4):177–184.

Hillerborg, A., Modéer, M., and Petersson, P.E. 1976. Analysis of Crack Formation and Crack Growth in Concrete by Means of Fracture Mechanics and Finite Elements, *Cement and Concrete Research*, 6: 773–782.

Naaman, A.E., Namur, G.G., Alwan, J.M., and Najm, H.S. 1991a. Fiber Pullout and Bond Slip. I: Analytical study, *ASCE Journal of Structural Engineering*, 117(9): 2769–2790.

Naaman, A.E., Namur, G.G., Alwan, J.M., and Najm, H.S. 1991b. Fiber Pullout and Bond Slip. II: Experimental Validation, *ASCE Journal of Structural Engineering*, 117(9): 2791–2800.

Naaman, A.E., and Najm, H. 1991. Bond-Slip Mechanisms of Steel Fibers in Concrete, *ACI Materials Journal*, 88(2): 135–145.

RILEM/CEB/FIP 1970. Bond test for reinforcing steel: I beam test; II Pull-Out test, *Materials and Structures*, 3(15): 169–178.

Shah, S.P., and Ouyang, C. 1991. Mechanical Behavior of Fiber-Reinforced Cement-Based Composites, *Journal of American Ceramic Society*, 74(11): 2727–38, 2947–53.

Stroeven, P., de Haan, Y.M., and Bouter, C. 1978. Pull-out Tests of Steel Fibres. In R.N. Swamy (ed.), *Rilem Symposium on Testing and Test Methods of Fibre Cement Composites*: 345–353.

Wang, Y., Li, V.C., and Backer, S. 1988. Modelling of Fibre Pull-out from a Cement Matrix, *The International Journal of Cement Composites and Light-weight Concrete*, 10(3):143–149.

Xu, X., and Day, R. 2000. Uniaxial Stress-strain Relationship of Cementitious Composites – A Review. In P. Rossi and G. Chanvillard (eds), *Fifth RILEM Symposium on Fibre-Reinforced Concretes (FRC), Lyon, France, 13–15 September 2000*: 431–440.

Computational Modelling of Concrete Structures, Bićanić et al. (eds)
© 2003 Swets & Zeitlinger, Lisse, ISBN 90 5809 536 3

A coupled damage-viscoplasticity model for the analysis of localisation and size effects

J. F. Georgin
URGC Structures/INSA de LYON, Villeurbanne France

L. J. Sluys
Faculty of Civil Engineering and Geosciences, Delft, The Netherlands

J. M. Reynouard
URGC Structures/INSA de LYON, Villeurbanne France

ABSTRACT: A coupled damage-viscoplasticity model is studied for the analysis of localisation and size effects. On one hand, viscosity helps to avoid mesh sensitivity because of the introduction of a length scale in the model and, on the other hand, enables to represent size effects. Size effects were analysed by means of three-point bending tests (RILEM).

1 INTRODUCTION

A new coupled damage-viscoplasticity model has been developed with excellent characteristics for the modelling of localisation under static and dynamic loading conditions. Viscoplasticity is known to be a good concept for the computational modelling of failure. The introduction of viscous terms in the constitutive model introduces a length scale effect and solves mesh dependence in localisation problems (Georgin et al. 2002; Needleman 1988; Sluys 1992).

The length scale effect in the Duvaut-Lions viscoplastic model is constant (Georgin et al. 2002). For this reason, the width of the localisation band is constant and does not narrow and finally collapse into a macro-crack of zero width when the strain reaches the ultimate strain. Furthermore, since the strain rate at ultimate strain is unequal to zero we have a viscous stress component. This viscous contribution of the stress causes that we cannot obtain a stress-free crack at ultimate strain. If the strain rate is increasing, which normally takes place at crack opening, even some rehardening effects can be observed in the crack. Both the narrowing localisation zone and the stress-free crack are features which can be modelled with the coupled damage-viscoplasticity model.

Fracture energy is a significant parameter in localisation analyses representing the energy required for fracture. A RILEM recommendation specifies an experimental method for the determination of the fracture energy (Gf) of mortar and concrete by means of stable three-point bending tests on notched beams. In this work, the RILEM test is simulated with the damage-viscoplasticity model in order to validate its ability to describe the fracture process properly. If the model takes into account the amount of energy necessary to create one unit area of crack in a proper way, the area under the load mid-span displacement curve divided by area of crack must restore a consistent value of the consumed energy. In other words, the size effect must be reproduced. Size effects which are experimentally observed can be used to determine the viscosity in the model.

2 CONCRETE MODEL

The damage variable, associated to concrete failure processes, can be interpreted as the surface density of materials defects (Ju 1989; Kachanov 1986), and will be defined as the ratio between the area occupied by created micro-cracks and the overall material area. This definition states that the damage variable is a non decreasing parameter, since the reduction of the effective resisting section area will continuously increase until failure occurs.

$$\tilde{S} = (1-D)\, S \qquad (1)$$

Where D is the damage variable.

The stress-strain relationship in a coupled damage-viscoplastic medium is written as:

$$\sigma=(1-D)\mathbf{E}_0:\varepsilon^e=\mathbf{E}:(\varepsilon-\varepsilon^{vp})$$
$$\sigma=(1-D)\tilde{\sigma} \tag{2}$$

where σ is the nominal stress tensor and $\tilde{\sigma}$ is the effective stress tensor.

Viscoplasticity is formulated by means of a Duvaut-Lions approach (Sluys 1992) according to

$$\dot{\varepsilon}^{vp} = \frac{1}{\eta}\mathbf{E}_0^{-1}(\sigma-\overline{\sigma}) \tag{3}$$

in which $\overline{\sigma}$ is damage-plastic stress which result from:

$$\overline{\sigma}=(1-D)\mathbf{E}_0:\varepsilon^e=\mathbf{E}:(\varepsilon-\varepsilon^{p})$$
$$\overline{\sigma}=(1-D)\tilde{\overline{\sigma}} \tag{4}$$

The plastic response is characterized in the stress space and the yield surface is given by

$$F(\tilde{\overline{\sigma}},\boldsymbol{\kappa})\leq 0 \tag{5}$$

A non-smooth multisurface criterion is used to describe the dissymmetrical material behaviour in tension and compression of concrete (Feenstra 1993). The employed yield surfaces F_i, are function of invariants of the stress tensor $\tilde{\overline{\sigma}}$ and the hardening parameter κ. For tension, a Rankine yield function is used

$$F_t(\tilde{\overline{\sigma}},\kappa_t)=\tilde{\overline{\sigma}}_I-\tilde{\overline{\tau}}_t(\kappa_t) \tag{6}$$

and for compression a Drucker-Prager yield function is used

$$F_c(\tilde{\overline{\sigma}},\kappa_c)=J_2(\tilde{\overline{s}})+\beta_1 I_1(\tilde{\overline{\sigma}})-\beta_2\tilde{\overline{\tau}}_c(\kappa_c) \tag{7}$$

where $\tilde{\overline{\sigma}}_I$ is the major principal stress, $I_1(\tilde{\overline{\sigma}})$ is the first invariant of the stress tensor, $J_2(\tilde{\overline{s}})$ is the second invariant of the deviatoric stress tensor $\tilde{\overline{s}}$, $\beta_i(i=1,\,2)$ are two multiplying factors and $\tilde{\overline{\tau}}_x$ is the equivalent stress in tension or in compression. The cohesion capacity of the material given by $\tilde{\overline{\tau}}_x$ is expressed by an analytically convenient function that is valid for tension and compression. It is consistent with the fact that experimentally observed stress-strain curves tend to attain zero-stress level asymptotically and is chosen according to

$$\tilde{\overline{\tau}}_x = f_{x0}\left[(1+a_x)\exp(-b_x\kappa_x)-a_x\exp(-2b_x\kappa_x)\right] \tag{8}$$

in which a_x and b_x are material parameters and f_{x0} is the initial tensile ($x=t$) or compressive ($x=c$) strength.

At the multisurface corners in the stress space, the ambiguity of the plastic flow direction is removed using Koiter's rule (Koiter 1953; Maier 1969) by considering the contribution of each individual loading surface separately:

$$\dot{\varepsilon}^p=\sum_{i=1}^{i=2}\dot{\lambda}_i\frac{\partial F_i}{\partial\overline{\sigma}} \tag{9}$$

where $\dot{\lambda}_i$ is the plastic multiplier associated to the plastic potential function F_i in tension or in compression.

The hardening parameter κ is updated with the following equation:

$$\dot{\kappa}=\frac{1}{\eta}(\kappa-\overline{\kappa}) \tag{10}$$

We assume an isotropic scalar damage model. The degree of brittleness of the mechanical effect of progressive micro-cracking due to external loads is described by the single internal scalar variable which degrades the current Young's modulus of the material such as the stiffness tensor reads

$$\mathbf{E}=(1-D)\mathbf{E}_0 \tag{11}$$

The damage evolution has an exponential form

$$1-D_x=\exp(-c_x\kappa_x) \tag{12}$$

dependent on the cumulated viscoplastic strain (see Eq. (10)), where c_x is a material parameter (Lee 1998; Meftah et al. 2000; Nechnech 2000). In order to describe different behaviour under tensile (where subscript $x = t$) and compressive loading (where subscript $x = c$) as observed in test data, the mechanical damage variable is subdivided into two parts, one for tensile loading and one for compressive loading.

$$D(\kappa)=1-(1-D_c)(1-D_t) \tag{13}$$

The rate derivative of Eq (2) leads to

$$\dot{\sigma} = (1-D)\mathbf{E}_0:(\dot{\varepsilon}-\dot{\varepsilon}^{vp}) - \dot{D}\mathbf{E}_0:(\varepsilon-\varepsilon^{vp}) \tag{14}$$

Substituting Eq. (3) into Eq. (14) yields the following differential equation:

$$\dot{\sigma} + \left(\frac{1}{\eta}+\frac{\dot{D}}{1-D}\right)\sigma = (1-D)\mathbf{E}_0:\dot{\varepsilon} + \frac{1}{\eta}\overline{\sigma} \tag{15}$$

The stress update for the damage-viscoplasticity model is obtained by an Euler numerical approach where the stress rate is determined with an approximate value from Eq. (15):

$$\sigma^{t+\Delta t} = C\left(\sigma^{t} + \left(1-D^{t+\Delta t}\right)\mathbf{E}_0 : \Delta\varepsilon + \frac{\Delta t}{\eta}\bar{\sigma}^{t+\Delta t}\right) \tag{16}$$

With

$$C = \frac{1}{1+\dfrac{\Delta t}{\eta}+\dfrac{\Delta D}{1-D^{t+\Delta t}}}$$

3 REGULARISATION ASPECTS

The introduction of rate dependence in the coupled damage-plasticity model prevents the model from becoming ill-posed when strain softening takes place. It introduces a length scale parameter in the problem which is dependent on η. Both the narrowing localisation zone and the stress-free crack are features that are modelled with the coupled damage-viscoplasticity model.

If we differentiate Eq. (15) for a one-dimensional coupled damage-viscoplasticity element with respect to x and use the kinematic expression:

$$\varepsilon = \frac{\partial u}{\partial x} \tag{17}$$

and the one-dimensional equation of motion:

$$\frac{\partial \sigma}{\partial x} = \rho \frac{\partial^2 u}{\partial t^2} \tag{18}$$

with ρ the density, we obtain:

$$\eta\left(\rho\frac{\partial^3 u}{\partial t^3} - (1-D)E_0\frac{\partial^3 u}{\partial x^2 \partial t}\right) = \frac{\partial \bar{\sigma}}{\partial x} + \rho\left(1+\frac{\eta \dot{D}}{1-D}\right)\frac{\partial^2 u}{\partial t^2} \tag{19}$$

which is the wave equation for a coupled damage-viscoplastic element. We can distinguish three cases:

1. Rate independence: For this case $\eta = 0$ and the first and second term in Eq. (19) cancel. The problem is ill-posed in case of statics and dynamics.
2. Rate dependence-statics: The inertia terms one and four cancel from Eq. (19) in the static case. The behaviour is set by the remaining third-order term. The problem is well-posed but approaches the ill-posed limit when the viscosity becomes zero ($\eta \rightarrow 0$) or the material is fully damaged ($D \rightarrow 1$).

Figure 1. Geometry of the uniaxial bar.

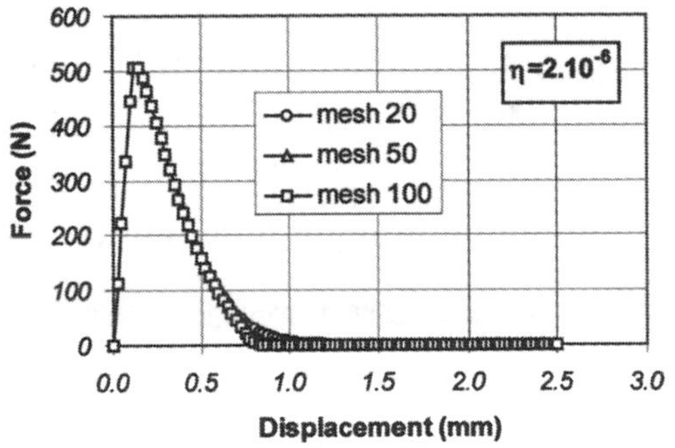

Figure 2. Uniaxial force-displacement response.

3. Rate dependence-dynamics: All terms appear in the equation (19), but the behaviour is set by the two third-order terms. The problem remains well-posed if $\eta > 0$.

From the second case it can be concluded that for the static case the regularising effect, which is constant for the viscoplastic model, decreases upon increasing damage. This results in a narrowing localisation zone and a stress drop to zero at full crack opening.

To show the regularisation effect provided by this approach, we present a numerical result of a bar (Fig. 1) which is subjected to an imposed displacement for different meshes (20, 50 and 100 elements). The damage viscoplasticity model gives the same force displacement curve for the two meshes (see Fig. 2). Furthermore, we can observe that the stress drops to zero after failure. A slight mesh dependence appears when the stress is almost zero. In this case the length scale (=regularising) effect is almost zero and the corresponding width of the localisation zone is close to the finite element size. From the stroboscopic evolution of the axial strain in Figs 3 and 4 we can see that the crack band narrows when the strain increases. The band width even becomes smaller than the finite element size, which causes that the strain is more localised for the analysis with 100 elements (Fig. 4) than for the analysis with 20 elements (Fig. 3). With the 100 elements mesh, the narrowing band width can be captured in a better way.

4 FRACTURE ENERGY – SIZE EFFECT

4.1 *Size dependence*

The ability of the damage-viscoplasticity model to describe fracture is illustrated by comparing the

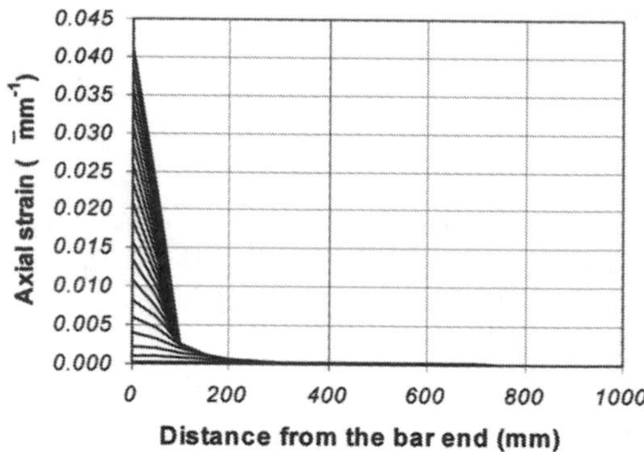

Figure 3. Axial strain gradient (mesh 20).

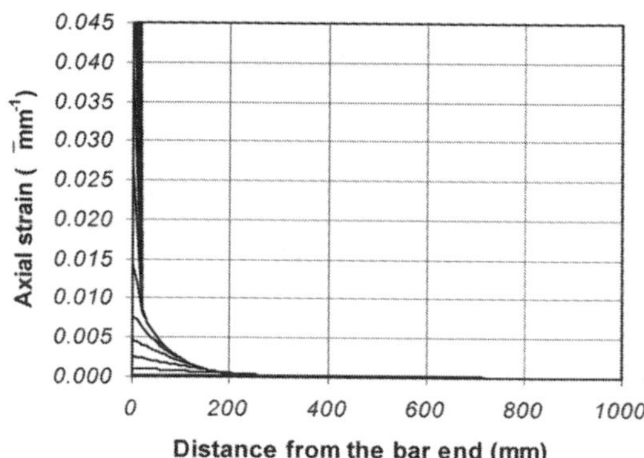

Figure 4. Axial strain gradient (mesh 100).

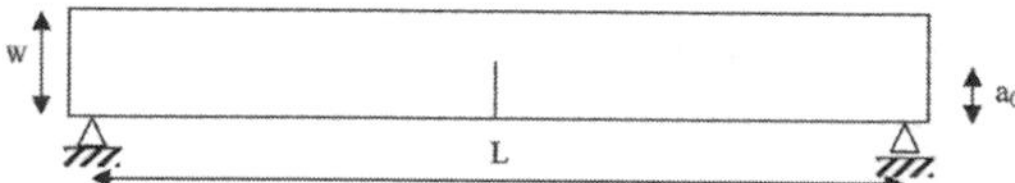

Figure 5. Three-point bending test.

Table 1. Geometrical characteristics.

Type	L (mm)	W (mm)	a_0 (mm)
A	80	10	5
B	400	50	25
C	800	100	50
D	1600	200	100
E	2400	300	150

numerical results of the three point bending test on five geometrically similar specimens (see Fig. 5 and Table 1). The material parameters used in these calculations are the Young's modulus E = 30 GPa, compression strength fc = 42 MPa, tension strength ft = 4.2 Mpa, the viscosity parameter $\eta = 2.10^{-6}$s and the volumic fracture energy gf = 0.07 N/mm^2 which is the

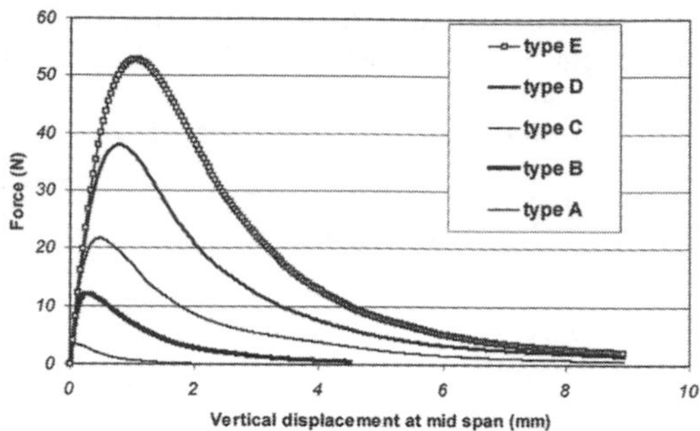

Figure 6. Force versus vertical displacement at mid span.

Table 2. Fracture energy and internal length.

Beam type	Area under F-u curve	Crack area	Gf (N/mm)	h = Gf/ gf (mm)
A	2.22	5	0.44	6.28
B	18.49	25	0.7396	10.42
C	49.308	50	0.98	14.00
D	99.75	100	0.99	14.14
E	155.73	150	1.03	14.71

area under the equivalent stress-hardening parameter curve (Eq. (8)). Results of the finite element calculation are shown in Fig. 6 in terms of forces versus displacements.

The fracture energy can be estimated by using the work of fracture method (Hillerborg 1985) from the curves shown in Fig. 6 and the following expression:

$$Gf = \frac{\text{Area under force displacement curve}}{\text{Crack area}} \tag{20}$$

Table 2 gives the obtained numerical values. We can clearly see that Gf depends on the size W for small sizes of the beam. This point was experimentally demonstrated by (Van Vliet and Van Mier 1998; Wittmann et al. 1990). Isovalues of the damage variable shown in Table 3 highlight the boundary effect which causes the lower values of Gf for the A and B specimen. Assuming that the volumic fracture energy is uniformly distributed over the localisation band, we can estimate an equivalent width h of the crack band, with the following expression (Feenstra 1993):

$$h = \frac{Gf}{gf} \tag{21}$$

All these simulations were also carried out with a different value of the volumic fracture energy gf = 0.007 N/mm^2 which is ten times less. We observe in Fig. 7 that the restored energy is also ten times

Table 3. Damage variable isovalues.

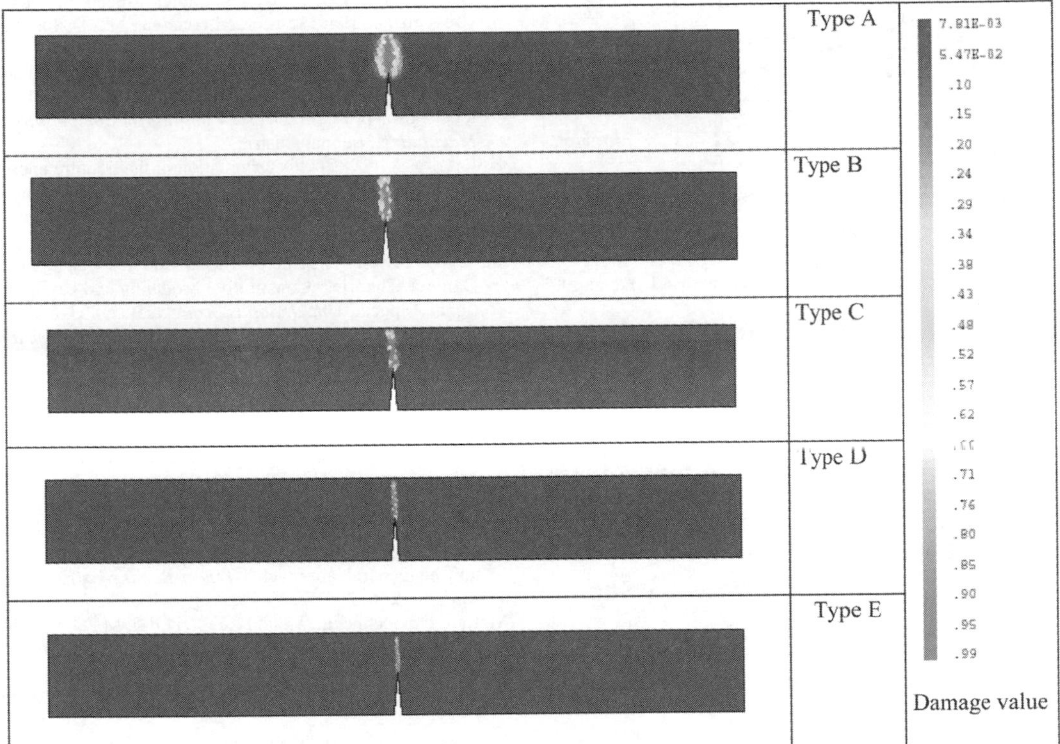

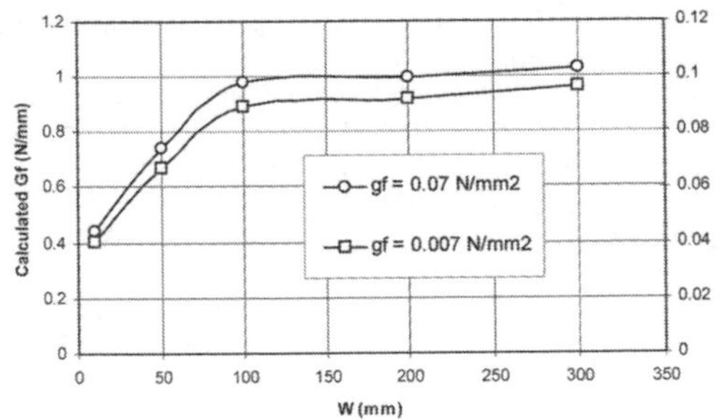

Figure 7.

less. It seems than this modelling approach taking into account a material viscosity is able to describe correctly the energy dissipation into the failure zone when Gf = f(gf) and η are determined.

4.2 *Estimation of model parameter gf and η*

The estimation of the fracture energy and the viscosity parameter must be carried out in two phases. The first one concerns the identification of the fracture energy by comparison of the measured value with the calculated one for the case of a very slow test. In a second phase, to identify the viscosity parameter, we compared results given by the damage-viscoplasticity model and experimental results in terms of the size effect law σ_{Nu} versus W (Bazant and Planas 1998). Here we assume gf = 0.007 N/mm^2.

Table 4.

Type	L (mm)	W (mm)	a_0 (mm)
A1	95	38	5
B1	190	76	25
C1	380	152	50

In order to estimate the viscosity parameter, experimental results shown in (Bazant and Gettu 1992) are used to fit numerical results. The same geometrical characteristics of the specimen are presented in Table 4. This study concerns only the fast series where the time to reach the peak load is around 1.5 seconds. In figure 8, we can see the adequate value for η is near 0.025 s^{-1} for gf = 0.007 N/mm^2. Figure 9 shows the model response in term of force versus time, which is in good agreement with experimental measurements in spite of the lack of knowledge on gf.

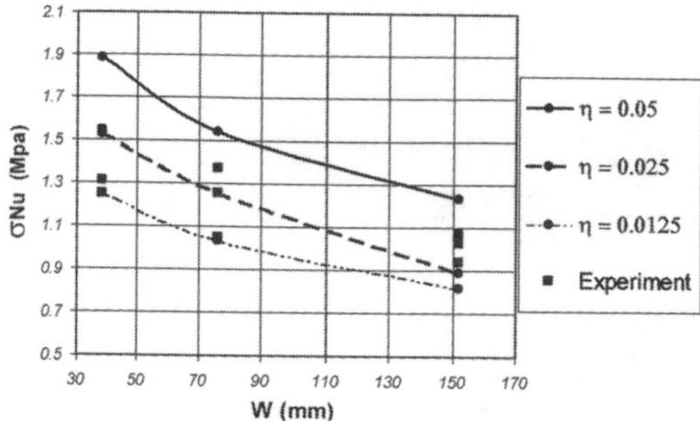

Figure 8. Size effect law estimated with A1, B1 and C1 specimens.

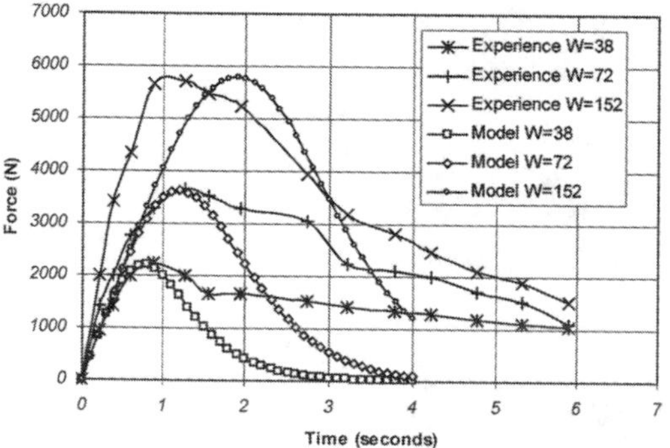

Figure 9.

5 CONCLUSION

This work shows that a coupled damage-viscoplasticity model is a good concept for the computational modeling of failure. Both the narrowing localisation zone and the stress-free crack are features which can be modelled with this approach. The numerical results on the three point bending show for a given material set (gf, η), that the dissipated energy in the localization zone of the beam firstly, depends on size W, and secondly tends to a constant value with W increasing which is consistent with experiments. The size effect under dynamic/impact loading can be analysed with this model.

REFERENCES

Bazant, Z. P., and Gettu, R. (1992). "Rate effects and load relaxation in static fracture of concrete." *ACI Materials Journal*, 89(5), 456–468.

Bazant, Z. P., and Planas, J. (1998). *Fracture and size effect in concrete and other quasibrittle materials*, CRC Press.

Feenstra, P. H. (1993). "Computational aspects of biaxial stress in plain and reinforced concrete," Ph.D, Institute of technology, Delft.

Georgin, J. F., Nechnech, W., Sluys, L. J., and Reynouard, J. M. "A coupled damage-viscoplasticity model for localisation problem." *World Congress on Computational Mechanics*, Vienna, Austria.

Hillerborg, A. (1985). "Determination of the fracture energy of mortar and concrete by means of three-point bend tests on notched beams." *50-FMC Committe fracture mechanics of concrete*, 18(106), 285–290.

Ju, J. W. (1989). "On Energy-Based Coupled Elastoplastic Damage Theories: Constitutive Modeling and Computational Aspects." *Int. J. Solids Struct.*, 25(7), 803–833.

Kachanov, L. M. (1986). *Introduction to continuum damage mechanics*, Martinus Nijhoff, Dordrecht.

Koiter, W. T. (1953). "Stress-strain relations, uniqueness and variational theorems for elastic-plastic materials with a singular yield surface." *Q. Appl. Math*, 3.

Lee, J. (1998). "Theory and implementation of plastic-damage model for concrete structures under cyclic and dynamic loading," Ph.D, University of California, Berkeley.

Maier, G. (1969). "Linear flows-laws of elastoplasticity: a unified general approach." *Lincei-Rend. Sci. Fis. Mat. Nat.*, 47, 266–276.

Meftah, F., Nechnech, W., and Reynouard, J. M. "An elasto-plastic damage model for plain concrete subjected to combined mechanical and high temperatures loads." *14th Engineering Mechanical Conference (A.S.C.E)*, Austin U.S.A, 21–24.

Nechnech, W. (2000). "Contribution à l'étude numérique du comportement du béton et des structures en béton armé soumises à des sollicitations thermiques et mécaniques couplées – Une approche thermo-élasto-plastique endommageable," Ph.D, INSA, LYON.

Needleman, A. (1988). "Material rate dependence and mesh sensitivity on localisation problems." *Comp. Meth. Appl. Mech. Eng.*, 67, 69–86.

Sluys, L. J. (1992). "Wave propagation, localisation and dispersion in softening solids," Ph.D, University of Technology, Delft.

Van Vliet, M. R. A., and Van Mier, J. G. M. "Experimental investigation of size effect in concrete under uniaxial tension." *FRAMCOS-III*, Gifu, 1923–1936.

Wittmann, F. H., Mirashi, H., and Nomura, N. (1990). "Size effect on fracture energy of concrete." *Engineering Fracture Mechanics*, 35(1/2/3), 107–115.

Computational Modelling of Concrete Structures, Bićanić et al. (eds)
© 2003 Swets & Zeitlinger, Lisse, ISBN 90 5809 536 3

Localisation of deformations in concrete subjected to compression

P. Grassl & K. Lundgren
Department of Structural Engineering, Chalmers University of Technology, Gothenburg, Sweden

ABSTRACT: In the present study a three-dimensional plasticity formulation for concrete was used to analyse the occurrence of localisations in compression both on the material point level and on the structural level. On the material point level, the influence of the position of the stress state in the deviatoric plane and of the hydrostatic pressure is studied. On the structural level two different structural parts, namely a wall subjected to uniform multiaxial compressive stress fields and a column in eccentric compression, were analysed by means of three-dimensional finite element models. The analyses show that the onset of localisation depends on the type of compressive stress fields. Hereby, the discontinuous failure criterion detects the onset of localisation. However, the size of the zone with concentrated compressive strains varies significantly during loading. Furthermore, the analyses of the eccentric compression tests show the sensitivity of how the structure is modelled.

1 INTRODUCTION

The present study deals with the modelling of concrete in triaxial compression by means of the theory ofplasticity. Concrete is a heterogeneous cohesive-frictional material, in which the stress transfer in compression is accomplished to a great extent by frictional forces. Consequently, the three-dimensional stress-strain behaviour of concrete in compression is highly pressure-sensitive, i.e. with increasing lateral confinement the load resistance and the deformation capacity increases. Moreover, the failure process of concrete in compression is accompanied by localisation of deformations in the form of shear and splitting cracks, which under certain boundary conditions evolve into so-called shear planes. The shear planes enable parts of the specimen to separate and slide with respect to each other. Hence, frictional forces contribute significantly to the force transfer. In constitutive models the macroscopic fracture process in the post-peak regime is included in the constitutive relations, in the form of the stress-strain or stress-deformation relations and the residual strength in triaxial compression.

However, also in numerical analysis using classical local continuum models it is observed that deformations may localise in regions. In finite element analyses, the localisation region depends on the size of the elements used, a phenomenon that is called mesh dependency.

In the present study it is examined how, and under which conditions, localisation occurs in continuum models. The localisation study is carried out on the material level and on the structural level. On the material point level the influence of the position of the stress state in the deviatoric plane and of the hydrostatic pressure on the occurrence of localisation is studied by means of several biaxial andtriaxial compressive stress states. On the structural level two different structural parts were analysed by means of three-dimensional finite element models. Firstly, a concrete wall is loaded by a compressive stress in axial direction. Different out of plane compressive stress states were applied leading to different homogeneous biaxial stress states. The second example is a concrete column in eccentric compression. Here the stress state is not homogenous, but a considerable strain gradient exists, which alters the type of the compression failure. This example is also related to over-reinforced beams, where the capacity of the compressive zone limits both the maximum capacity and the deformations, van Mier & Ulfkjaer (2000).

2 CONSTITUTIVE MODEL

The plasticity model used in this study consists of an alternative hardening law, which is based on the volumetric plastic strain combined with a non-linear plastic potential. This hardening law was combined with a yield surface proposed by Menétrey & Willam (1995). Both the plastic potential within the hardening law, and the yield surface, are constituted by using the unified co-ordinates in the Haigh-Westergaard stress space which are based on the stress invariants. The three unified co-ordinates ξ, ρ and θ are defined as

$$\xi = \frac{I_1}{\sqrt{3} \cdot f_c}, \quad I_1 = \delta_{ij}\sigma_{ij} \tag{1}$$

$$\rho = \frac{\sqrt{2 \cdot J_2}}{f_c}, \quad J_2 = \frac{1}{2} \cdot s_{ij}s_{ij}, s_{ij} = \sigma_{ij} - \frac{1}{3} \cdot \delta_{ij}\sigma_{kk} \tag{2}$$

$$\cos 3\theta = \frac{3\sqrt{3}}{2} \cdot \frac{J_3}{J_2^{3/2}}, \quad J_3 = \frac{1}{3} \cdot s_{ij}s_{jk}s_{ki} \tag{3}$$

2.1 *Yield surface*

A three-parameter yield surface, which depends on three stress invariants and all principal stress components, is used. It is an extension of the Hoek and Brown failure criterion and was proposed by Menétrey & Willam (1995). It has the form

$$f = 1.5 \cdot \rho^2 + q_h(\kappa) \cdot m \cdot \left[\frac{\rho}{\sqrt{6}} \cdot r(\theta, e) + \frac{\xi}{\sqrt{3}}\right] - q_h(\kappa) \cdot q_s(\kappa) \leq 0, \tag{4}$$

where m is defined as

$$m = 3 \cdot \frac{f_c^2 - f_t^2}{f_c \cdot f_t} \cdot \frac{e}{e+1} \tag{5}$$

and the elliptic function as

$$r(\theta, e) = \frac{4 \cdot (1-e^2) \cdot \cos^2\theta + (2 \cdot e - 1)^2}{2 \cdot (1-e^2) \cdot \cos\theta + (2e-1) \cdot [4 \cdot (1-e^2) \cdot \cos^2\theta + 5 \cdot e^2 - 4 \cdot e]^{1/2}} \tag{6}$$

The elliptic function $r(\theta, e)$ describes the out-of-roundness of the deviatoric section and is controlled by the eccentricity e. The three calibration parameters are the strength in uniaxial compression f_c, the strength in uniaxial tension f_t and the eccentricity e. More details concerning the calibration of the yield surface can be found in Grassl *et al.* (2002).

The surface possesses parabolic meridians and the deviatoric sections change from triangular shapes at low confinement to almost circular shapes at high confinement. In this way the change from quasi-brittle to ductile behaviour with increasing hydrostatic pressure can be described. The surface is smooth and convex, except the point of equitriaxial tension, where the parabolic meridians intersect the hydrostatic axis.

2.2 *Hardening law and non-associated flow rule*

The hardening law describes the influence of multiaxial stress states on the deformation capacity. This is achieved by combining the volumetric plastic strain as hardening parameter with a non-linear plastic potential.

The evolution of the plastic strain rate is determined by the flow rule

$$\varepsilon_{ij}^p = \lambda \cdot \frac{\partial g}{\partial \sigma_{ij}}. \tag{7}$$

The flow rule is non-associated, meaning that the form of the plastic potential differs from the form of the yield surface. It has a quadratic form and is constituted, using the co-ordinates in the Haigh-Westergaard stress space, as

$$g = -A \cdot \left(\frac{\rho}{\sqrt{q(\kappa)}}\right)^2 - B \cdot \frac{\rho}{\sqrt{q(\kappa)}} + \frac{\xi}{\sqrt{q(\kappa)}} = 0, \quad \text{if} \quad \frac{\rho}{\sqrt{q(\kappa)}} > \sqrt{\frac{2}{3}} \tag{8}$$

$$g = -C \cdot \left(\frac{\rho}{\sqrt{q(\kappa)}}\right)^2 - D \cdot \frac{\rho}{\sqrt{q(\kappa)}} + \frac{\xi}{\sqrt{q(\kappa)}} = 0, \quad \text{if} \quad \frac{\rho}{\sqrt{q(\kappa)}} \leq \sqrt{\frac{2}{3}}. \tag{9}$$

The parameters A and B, C and D are determined by means of the axial strain at maximum stress in uniaxial compression and in a triaxial compressive state. Furthermore, it is assumed that the lateral plastic strains in uniaxial tension are equal to zero. The determination of the parameters is described in Grassl *et al.* (2002).

The hardening parameter is the volumetric component of the plastic strain increment:

$$\dot{\kappa}(\varepsilon^p) = \varepsilon_v^p = \delta_{ij}\varepsilon_{ij}^p = \lambda \cdot \delta_{ij}\frac{\partial g}{\partial \sigma_{ij}} = \lambda \cdot \frac{\sqrt{3}}{\sqrt{q(\kappa)} \cdot f_c}, \tag{10}$$

where δ_{ij} is the Kronecker delta. Thus, the hardening parameter depends entirely on the plastic strain. This is one of the main differences from other plasticity theories in which the hardening variable depends on the plastic strain and the stress state. The hardening/softening law has the form

$$q(\kappa) = \left(\frac{\sigma_c(\kappa)}{f_c}\right)^2, \tag{11}$$

where σ_c is the stress in uniaxial compression and κ the hardening parameter, which is the volumetric component of the corresponding plastic strain vector. The hardening law, see Figure 1, was defined by

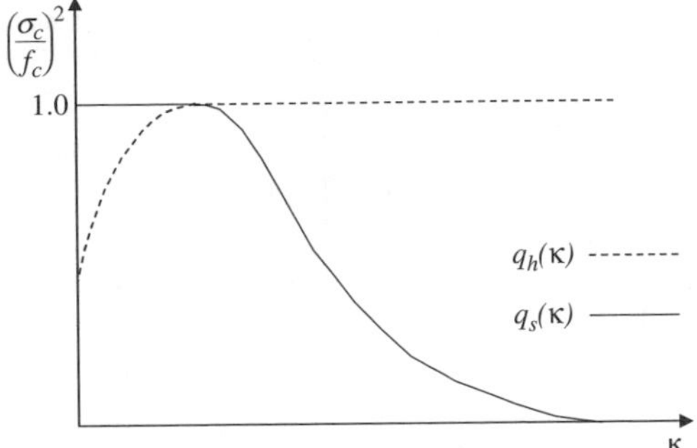

Figure 1. Split of the hardening function into a hardening and softening part.

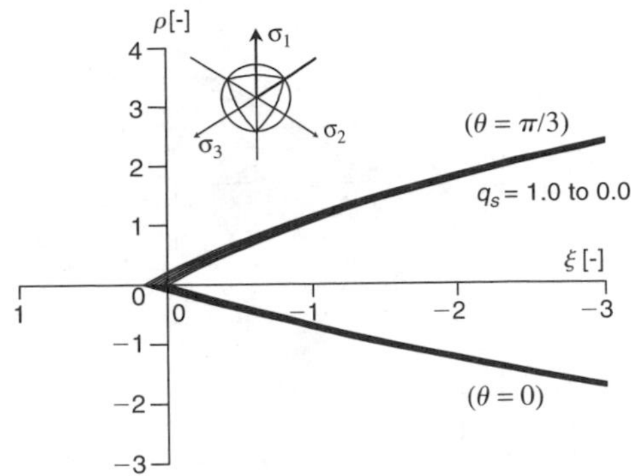

Figure 3. Evolution of the compressive meridian ($\theta = \pi/3$) and the tensile meridian ($\theta = 0$) of the yield surface in softening.

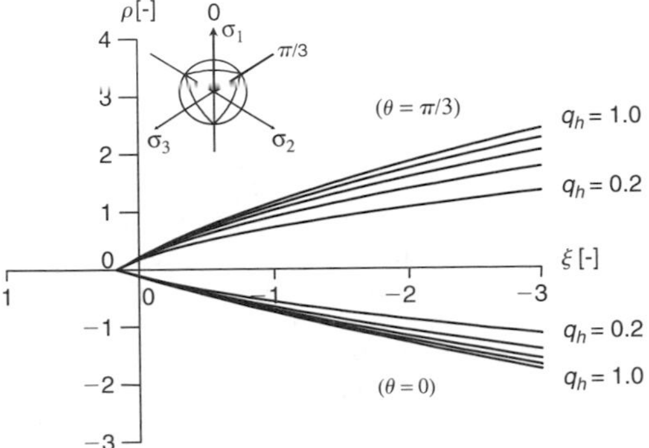

Figure 2. Evolution of the compressive meridian $\theta = \pi/3$) and the tensile meridian ($\theta = 0$) of the yield surface in hardening.

means of the uniaxial stress strain relation according to CEB (1993), which was proposed for a specimen height of 300 mm. The cohesive strength and the friction parameter m, within the formulation of the yield surface in equation (4), are uncoupled so that they can be adjusted separately to control hardening and softening, see Figures 2 and 3. Therefore, the hardening/softening function $q(\kappa)$ is split into one function concerning the hardening, $q_h(\kappa)$, and one concerning the softening, $q_s(\kappa)$:

$$q(\kappa) = q_h(\kappa) \cdot q_s(\kappa). \tag{12}$$

By means of the split, residual strength in multiaxial compression is provided, as shown in Figure 3.

The calibration of the hardening law is based on the following hypotheses:

1. The maximum stress in uniaxial compression, f_c, is reached when the volumetric strain, ε_v, is equal to zero.
2. The volumetric plastic strain at maximum stress in uniaxial compression is the value of the hardening parameter when the maximum stress is reached, and is also used for all other stress states.
3. The inclination of the total plastic strain is equal to the gradient of the plastic potential within the same state of loading.

These hypotheses seem to be valid for compressive loading. For dominant tensile loading ($\sigma_1 + \sigma_2 + \sigma_3 > 0$), however, the deformation capacity cannot be described accurately with the chosen hardening law.

3 FAILURE ANALYSIS

A failure analysis on the material point level was carried out based on the elasto-plastic stiffness tensor. The elasto-plastic stiffness tensor gives information on both the type of failure and the orientation and evolution of a possible localisation. In this study, the type of failure is distinguished into diffuse and localised; see Kang & Willam (1999). Diffuse failure is defined as a failure mode, in which, despite of strain softening, no formation of discontinuities in the deformation or strain field occurs. Localised failure is defined as a mode in which the reduction of strength on the structural level is accompanied by the formation of spatial discontinuities.

3.1 *The indicators for diffuse or localised failure*

The indicators for diffuse or localised failure are:

- The general bifurcation criterion (Hill (1958)):

$$\dot{\varepsilon}_{ij}\dot{\sigma}_{ij} = 0 \quad \text{or} \quad \dot{\varepsilon}_{ij}D^s_{ijkl}\dot{\varepsilon}_{kl} = 0 \quad \text{or} \quad det(D^s_{ijkl}) \tag{13}$$

- The limit point (stationary stress):

$$\dot{\sigma}_{ij} = D_{ijkl}\dot{\varepsilon}_{kl} = 0 \quad \text{or} \quad det(D_{ijkl}) = 0 \tag{14}$$

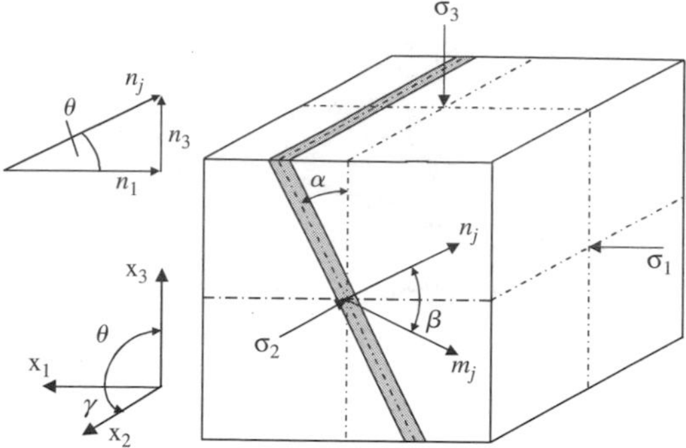

Figure 4. The localisation plane in a specimen with the normal vector n_j and the evolution vector m_j.

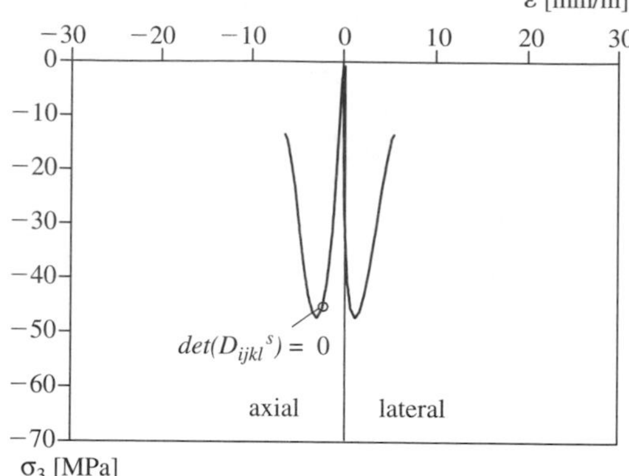

Figure 5. The axial stress versus the axial and lateral strain in uniaxial compression.

The indicators for localised failure are:

- Loss of strong ellipticity (Neilsen & Schreyer (1993)):

$$m_i Q^s_{il} m_l = 0 \quad \text{or} \quad det(Q^s_{il}) = 0. \tag{15}$$

- Classical discontinous bifurcation (Hill (1962)):

$$Q_{il} m_k = 0 \quad \text{or} \quad det(Q_{il}) = 0. \tag{16}$$

The acoustic tensor Q_{il} is defined as

$$Q_{il} = n_j D_{ijkl} n_k, \tag{17}$$

Monotonic and proportional loading, and uniform stress and strain distribution before the onset of bifurcation were assumed.

The elasto-plastic stiffness tensor was studied for uniaxial, biaxial and triaxial compression. It was, hereby, distinguished between axisymmetric and asymmetric stress states. For every load increment, the general stability criterion from equation (13) is checked. If this criterion is violated, the localisation criterion from equation (15) and (16) are checked. This is done for all the directions n_j. In Grassl *et al.* (2002) the stress-strain responses of the constitutive model are compared to experimental results in uniaxial and biaxial compression reported by Kupfer *et al.* (1969) and triaxial compression reported by Imran (1994). The input parameters obtained there were used also here for the localisation analyses.

The dominant compressive loading was applied within the x_2–x_3-plane, see Figure 4. The maximum compressive stress component was always applied in the x_3 direction. The normal vector n_j was defined as

$$n_j = \begin{pmatrix} \cos\theta\cos\gamma \\ \cos\theta\sin\gamma \\ \sin\theta \end{pmatrix}. \tag{18}$$

It was expected to detect the localisation in the x_3–x_1-plane ($\gamma = 0$), as a possible separation evolves in the direction of the smallest resistance.

3.2 *Axisymmetric loading*

As representative for axisymmetric compressive stress states, the case of uniaxial compression was analysed. The stress–strain relation for uniaxial compression is shown in Figure 5. The general bifurcation criterion was violated within the hardening regime. However neither the condition for the loss of strong ellipticity or the classical discontinous bifurcation was satisfied before the stress was equal to zero. Thus, it can be concluded that no localised failure is expected in this case.

3.3 *Deviation from axisymmetric loading*

Two other compressive stress states with a deviation from the compressive meridian were analysed. These stress states were a biaxial stress with a stress ratio of $\sigma_3/\sigma_2 = 8/1$ and a biaxial compressive stress state in plane strain condition.

The stress–strain relation for the biaxial compressive state with a stress ratio of $\sigma_3/\sigma_2 = 8/1$ is shown in Figure 6. All failure criteria were detected. Both the general stability criterion and the condition for the loss of strong ellipticity were violated and satisfied, respectively, in the hardening regime. The condition for classical discontinuous bifurcation was satisfied in the softening regime. The evolution of the unified determinant of the acoustic tensor for an angle of $\gamma = 0$ is shown in Figure 7. The angle of the localisation zone, α, was computed to 34 degrees.

The stress–strain relation for the axial compression in plane strain condition is shown in Figure 8. Both criteria for localised failure are satisfied within the hardening regime. The evolution of the unified determinant of the acoustic tensor for an angle of $\gamma = 0$ is shown in Figure 9. Also here, the angle of the localisation zone, α, was computed to 34 degrees.

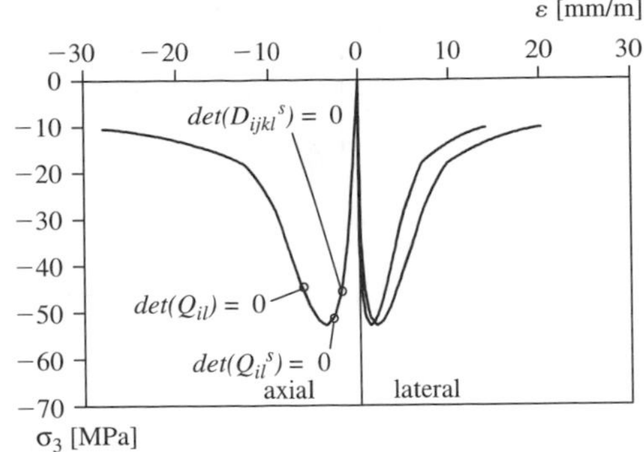

Figure 6. The axial stress versus the axial and lateral strain in biaxial compression ($\sigma_3/\sigma_2 = 8/1$).

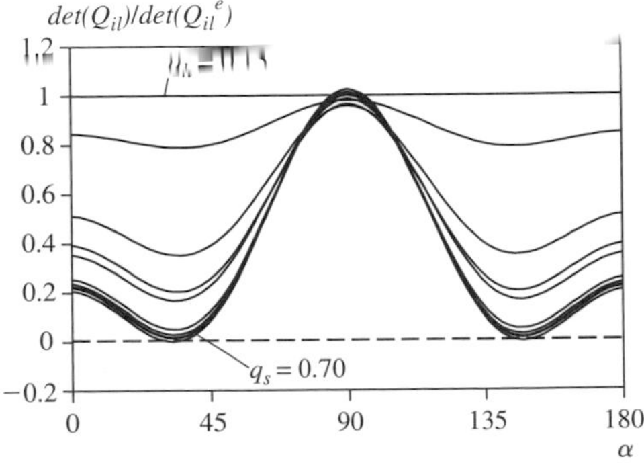

Figure 7. The evolution of the unified determinant of the acoustic tensor for biaxial compression ($\sigma_3/\sigma_2 = 8/1$) for an angle $\gamma = 0$ for $q_h = 0.15$ until $q_s = 0.7$.

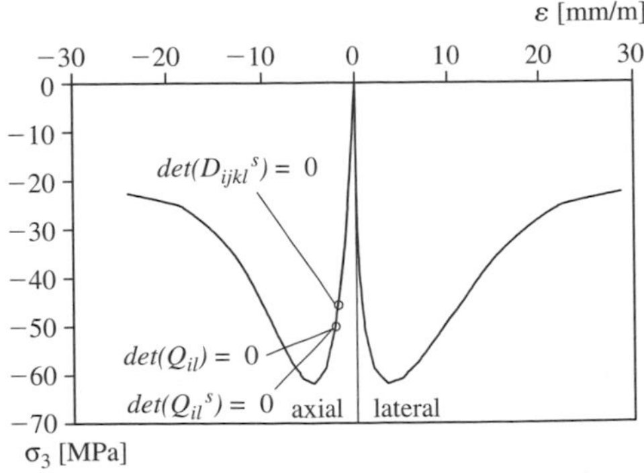

Figure 8. The axial stress versus the axial and lateral strain in biaxial compression (plane strain).

From the results of the three compressive load cases analysed, it can be seen that the location of the stress state in the biaxial stress plane is of importance. The influence of the location of the stress state in the biaxial

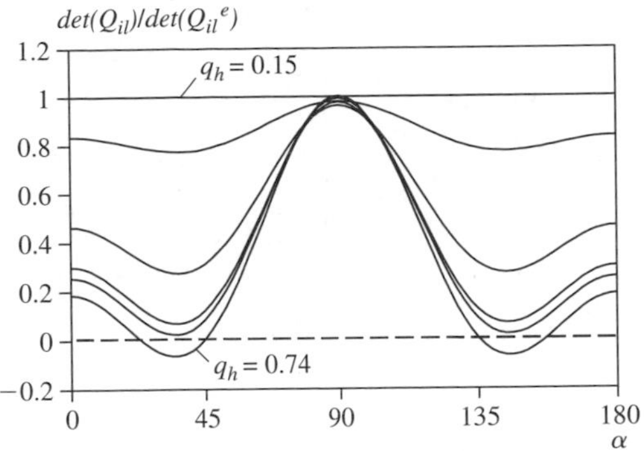

Figure 9. The evolution of the unified determinant of the acoustic tensor for biaxial compression (plane strain) for an angle $\gamma = 0$ for the hardening regime ($q_h = 0.15 - 0.7$).

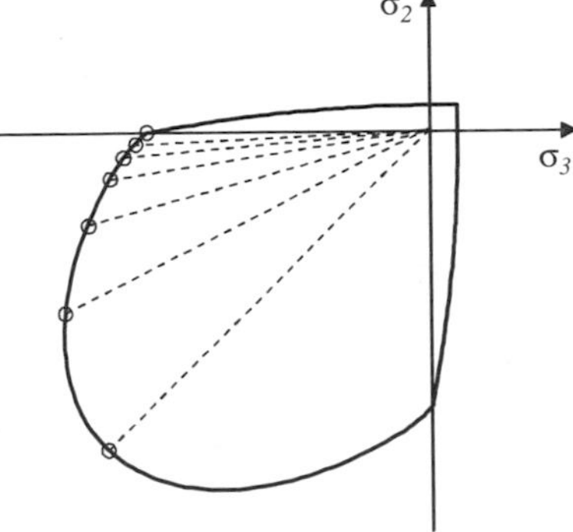

Figure 10. The different stress paths analysed in the biaxial stress plane.

plane was therefore further investigated. Seven analyses with different biaxial compressive stress ratios were carried out. The paths of the stress states in the biaxial stress plane are shown in Figure 10. The results of the analyses are presented in Figure 11 in form of the localisation angle α and the unified hardening parameter, at the points when the conditions for the classical discontinuous bifurcation and the loss of strong ellipticity were satisfied. The localisation angle α, for the condition of the classical discontinous bifurcation, varies from 32° for almost uniaxial compression to 29° for equibiaxial compression. The results of the condition for the loss of strong ellipticity are similiar, whereby the computed localisation angle is greater. Furthermore, it can be seen that the condition for classical discontinous bifurcation is satisfied later than for the loss of strong ellipticity.

Furthermore, the influence of the hydrostatic stress state on the failure type was studied. Therefore, compressive stress states with a constant angle in the

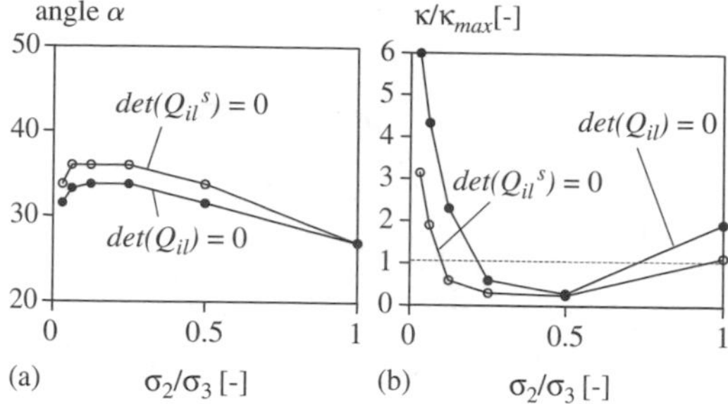

Figure 11. The angle α for the biaxial stress ratios. The unified hardening parameter at localisation.

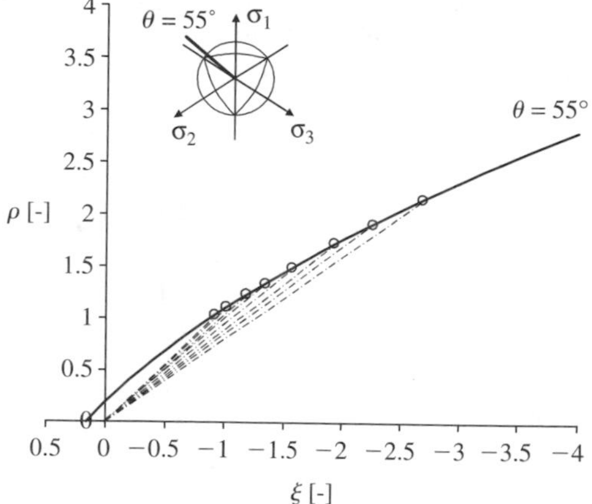

Figure 12. The triaxial stress paths in the volumetric plane with a constant angle θ.

deviatoric plane of 55° were chosen. The path of the compressive stress states in the hydrostatic plane is shown in Figure 12. The results of the analyses are shown in Figure 13. In Figure 13a the localisation angle for the different analyses is shown. In case of the condition for the classical discontinous bifurcation, localised failure was not detected for ξ less than -1.4. The hydrostatic stress has no influence of the localisation angle. In Figure 13b the unified hardening parameter at localisation is shown. Here, the difference between the two conditions of localised failure is large. The loss of strong ellipticity is detected for all analyses within the hardening regime, whereby the hydrostatic pressure has only a minor influence. However, when the condition for the classical discontinous bifurcation criterion is used, the influence of the hydrostatic stress state is significant. With increasing hydrostatic pressure, the localisation is delayed. Localisation was detected for all analyses within the softening regime. The analyses show that there is a significant difference between the two conditions for localised failure

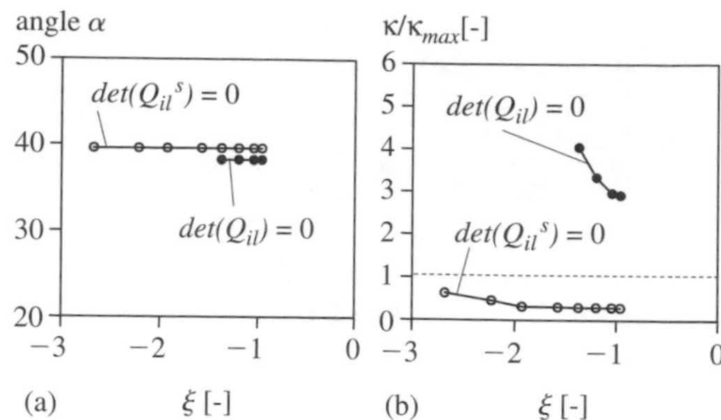

Figure 13. (a) The localisation angle versus the the length of the hydrostatic stress. (b) The unified hardening parameter versus the length of the hydrostatic stress.

in the case of both biaxial and triaxial compression. In the following section the validity of the two falure conditions is analysed by means of the finite element method.

4 FINITE ELEMENT ANALYSES

Three-dimensional finite element analyses of specimens subjected to uniform compressive stress fields and eccentric compression were carried out. For all the analyses 8-node solid brick elements were used. The element is based on linear interpolation and Gauss integration. A $2 \times 2 \times 2$ Gauss integration scheme was used.

4.1 *Uniform compressive stress fields*

The results obtained on the integration point level give information on the type of failure expected for different compressive load cases. If these failure types are obtained in the finite element analyses one of the failure indicators may be used to apply a computational procedure based on the crack band approach, see Bažant & Oh (1983), to obtain an objective solution. Therefore, it is necessary to study which failure indicator predicts the change from diffuse to localised failure on the structural level. The three biaxial compressive stress states presented in Section 3 were analysed using the finite element method; namely uniaxial compression, biaxial compression with a stress ratio of 8/1, and axial compression in plane strain condition. The finite element model is shown in Figure 14. Both homogeneous specimens, and specimens with a weakened element were analysed. The load-displacement relations of the three compressive stress states are shown in Figures 15–17. The contour plots of the principal compressive strains is shown for the three for a stress state in the post-peak regime are shown in Figure 18.

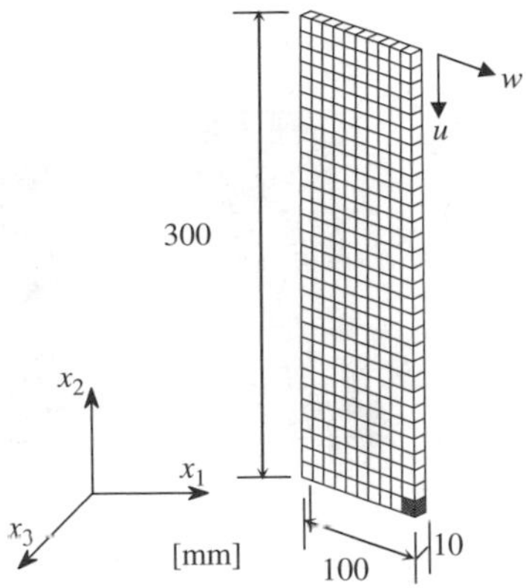

Figure 14. The finite element model. The grey marked element was weakened in some analyses.

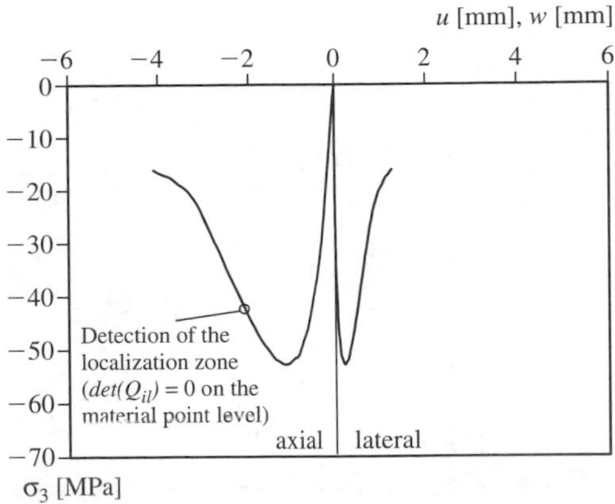

Figure 16. The load–displacement relation for biaxial compression ($\sigma_3/\sigma_2 = 8/1$). The response of the weakened specimen matches the one of the homogenous specimen.

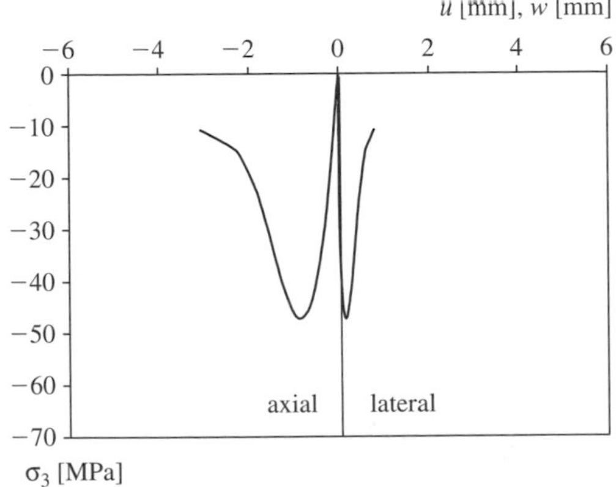

Figure 15. The load–displacement relation for uniaxial compression. The response of the weakened specimen matches the one of the homogenous specimen.

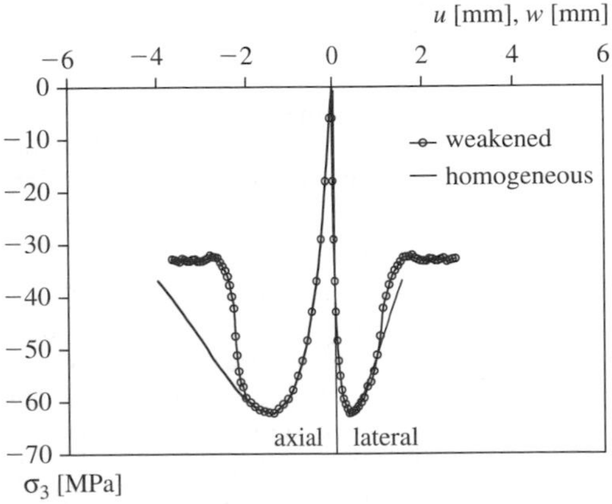

Figure 17. The load–displacement relation of the weakened and homogenous specimen for axial compression in plane strain condition.

In uniaxial compression the response of the weakened specimen matches the response of the homogenous specimen. A localisation zone was not detected. This corresponds to the findings on the material point level. Also in the case of biaxial compression ($\sigma_3/\sigma_2 = 8/1$) the response of the weakened specimen matches the one of the homogenous specimen. However, a localisation zone was detected. This zone started to form at the point when the classical discontinuous bifurcation condition on the material point was satisfied. However, this localisation zone did not lead to an elastic response in the adjacent areas of the specimen. In the third case, i.e. axial compression in plane strain condition, the localisation zone starts to evolve in the pre-peak regime. However, first in the post-peak regime the localisation zone evolves sufficiently to form a shear band, which is accompanied by a drop in the load-displacement relation, see Figure 17.

From these analyses it can be concluded that the classical discontinuous bifurcation criterion detects the onset of localisation. However, in three-dimensional finite element analyses with local material models the onset of localisation does not correspond to a classical localisation, in the meaning that adjacent parts respond elastically. On the contrary, it was observed that the size of the zone with concentrated compressive strains varies significantly during loading. It seems that the elements try to adapt to the deformation field. Thus, the localised zone is smeared out over several elements. To be able to use a computational procedure similar to the crack band approach for this type of three-dimensional analyses, it

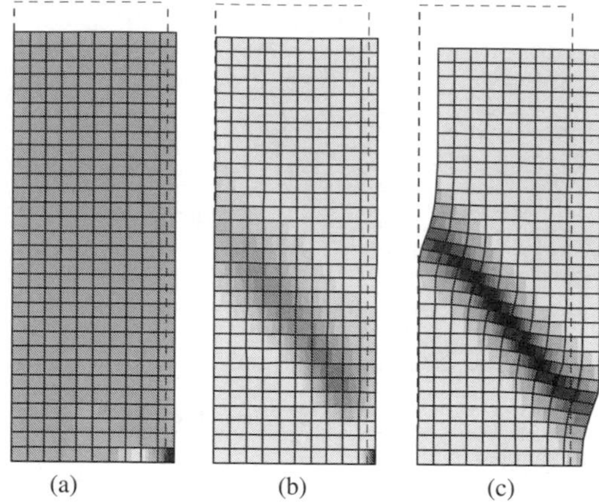

Figure 18. Contour plots of the principal compressive strains of the analyses in (a) uniaxial compression at a stress in the post-peak regime of 30 MPa, (b) biaxial compression ($\sigma_3/\sigma_2 = 8/1$) at a stress in the post-peak regime of 40 MPa and (c) axial compression in plane strain condition at a stress state in the post-peak regime of 40 MPa.

would be required to determine the condition for classic localisation on the structural level. Furthermore, the crack band width cannot be assumed to be constant.

4.2 *Eccentric compression*

One important load situation is the compressed zone in a beam. Earlier analyses carried out using other material models have shown that this can be a problem, since it is difficult to know in advance in how many elements the failure will localise, see for example Lundgren (2002). However, the developed model is mainly suited for compressive loading; therefore it was chosen to model tests with an eccentric, but mainly compressive, loading. Debernardi & Taliano (2001) carried out such tests, measuring the deformations over three parts of the loaded specimen, in order to study localisation in eccentric compression. Their tests with the three smallest eccentricities were modelled. The test specimens were 500 mm high, 100 mm deep and 150 mm wide in the mid part where the compression failure took place. Two different finite element meshes were used, one coarse and one less coarse mesh, see Figure 19, to investigate if there was any mesh dependency in the solutions. To enable a localisation of the failure, one element row was weakened in the analyses, see Figure 19. The loading plates used in the tests were modelled by use of shell elements at the upper and lower surfaces in the finite element models, assuming a thickness of 8 cm.

The load versus the average deformation over both sides is shown in Figure 20. As can be seen, the both meshes gave similar results. The response was more ductile in the analyses than in the tests, especially in the

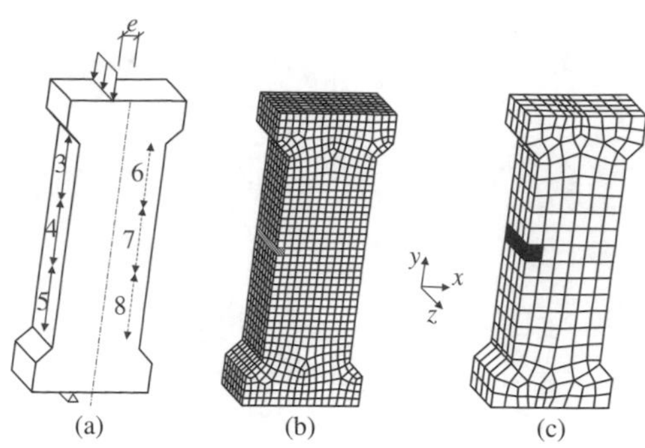

Figure 19. (a) The modelled test specimen and the arrangement of measuring instruments by Debernardi & Taliano (2001). (b) and (c) Finite element meshes used in the analyses; grey marked elements were weakened.

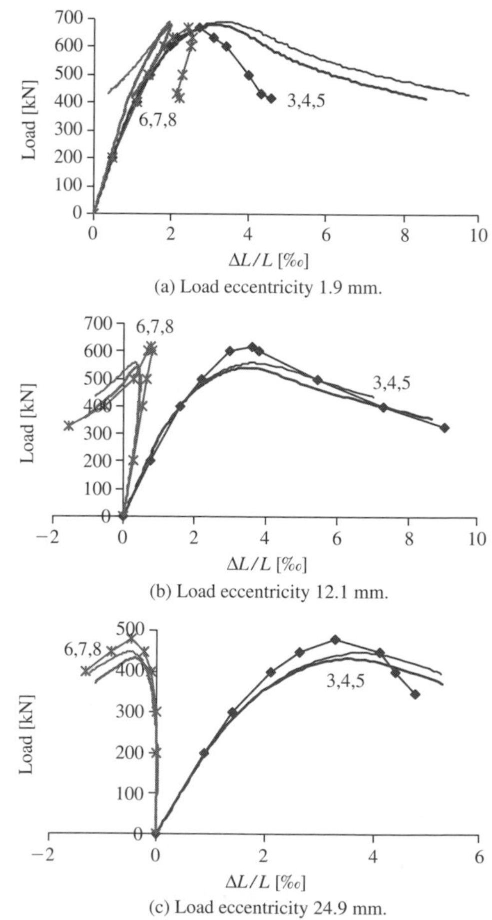

Figure 20. The load versus the average deformation; thin lines for analysis with fine mesh, thick lines for analysis with coarse mesh, lines with markers experimental results from Debernardi & Taliano (2001).

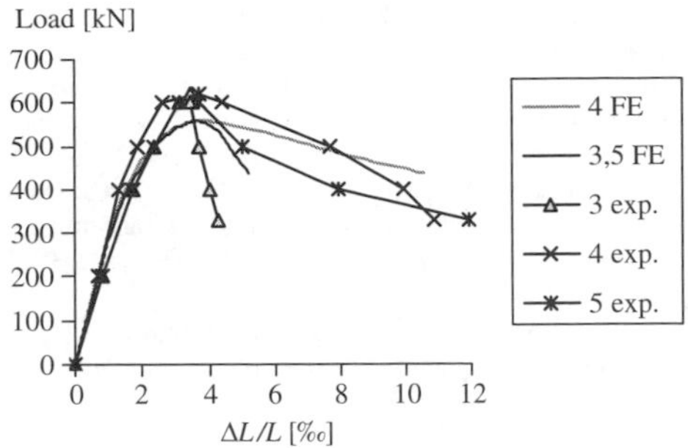

Figure 21. The load versus the deformation in the three measuring bases on the compressed side. Load eccentricity 12.1 mm, results from analysis with a fine mesh.

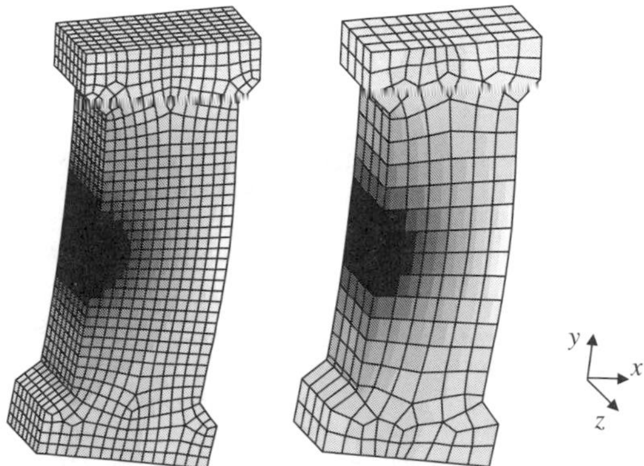

Figure 22. Contour plots of the strain in the y-direction in the analyses of test with $e = 12.1$ mm when the deformation in measuring base 3, 4, 5 is 7‰.

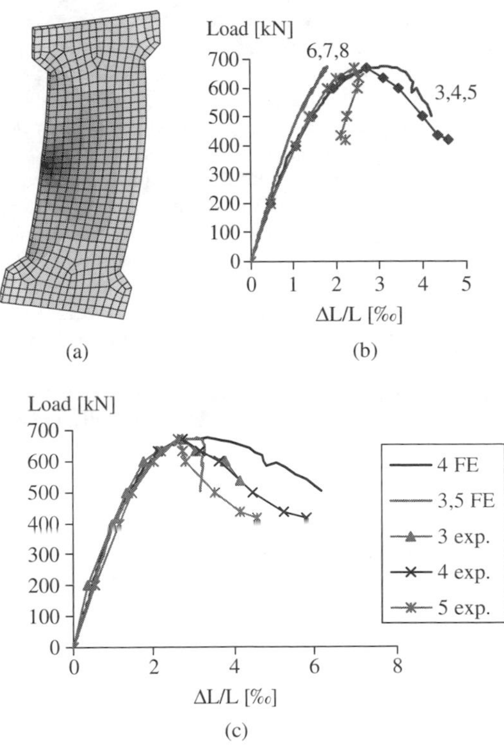

Figure 23. Results from analysis of test with $e = 1.9$ mm with one element row. (a) Contour plot of the strain in the y-direction when the deformation in measuring base 4 is 6‰. (b) The load versus the average deformation; thick lines for analysis, lines with markers experimental results from Debernardi & Taliano (2001). (c) The load versus the deformation in the three measuring bases on the compressed side.

analyses with an eccentricity close to zero. The results from the analyses showed that no classic localisation occurred. One zone in the middle part of the specimen obtained larger strains than the other parts of the specimen, but no part was unloaded. This can be seen in Figure 21, where the load versus the deformation in the three measuring bases on the compressed side from test and analysis with a load eccentricity of 12.1 mm is shown. Similar results were obtained for the other two analysed eccentricities. This corresponds with the findings from the tests, where also no unloading could be seen. In Figure 22, a contour plot of the strain in the y-direction is shown. The size of the zone with higher strains is about the same size in both analyses; i.e. it appears not to be mesh dependant.

However, when beams are analysed, it is rather common that plane stress elements are used, instead of modelling the whole width of the beam with three-dimensional solid elements. Therefore, an analysis was carried out where only one element row was used, leaving the boundaries in the z-direction free to correspond with a plane stress situation. The mesh used is shown in Figure 23, together with results from the analysis with a load eccentricity of 1.9 mm. As can be seen, the ductility in this analysis better corresponds with what was measured in the test. However, in Figure 23c it can be seen that in this analysis a classic localisation with unloading in parts of the specimen was obtained, which does not correspond with the experimental findings. Therefore, the solution becomes mesh dependant.

5 CONCLUSIONS

The combination of the failure analysis on the material point level and the three-dimensional finite element analyses indicates that the classic discontinuous bifurcation criterion detects the onset of localisation in case of uniform compressive stress fields.

The failure analysis on the material point level shows that the occurrence of localisation in compression depends very much on the position of the stress state in the deviatoric plane. For stress states on the compressive meridian localisation does not occur. However, as soon as the stress state deviates from the compressive meridian localisation occurs. For great deviations the localisation is detected within the hardening regime. Furthermore, the occurrence of localisation depends significantly on the hydrostatic pressure. With increasing hydrostatic pressure the behaviour becomes more ductile and, therefore, the occurrence of localisation delayed. The finite element analyses for uniform compressive stress fields show that at the onset of localisation the adjacent parts of the localisation zone do not respond elastically. On the contrary, it was observed that the size of the zone with concentrated compressive strains varies significantly during loading. To be able to use a computational procedure similar to the crack band approach for this type of three-dimensional analyses, it would be required to determine the condition for classic localisation on the structural level. Furthermore, the crack band width cannot be assumed to be constant.

The analyses of the eccentric compression tests show the sensitivity of how the structure is modelled; when changing from several elements across the cross-section to one element row, there was a clear difference in the localisation behaviour.

REFERENCES

Bažant Z. & Oh B. H. 1983. Crack band theory for fracture of concrete. *Materials and Structures* 16(93): 155–177.

CEB 1993. *CEB-FIP Model Code 1990.* Lausanne, Switzerland: Bulletin d'Information 213/214.

Debernardi P. G. & Taliano M. 2001. Softening behaviour of concrete prisms under eccentric compressive forces. *Magazine of Concrete Research* 53(4): 239–249.

Grassl P., Lundgren K. & Gylltoft K. 2002. Concrete in compression: a plasticity theory with a novel hardening law. *International Journal of Solids and Structures* 39(20): 5205–5223.

Hill R. 1958. A general theory of uniqueness and stability in elastic-plastic solids. *Journal of the Mechanics and Physics of Solids* 6: 236–249.

Hill R. 1962. Acceleration waves in solids. *Journal of the Mechanics and Physics of Solids* 10: 1–6.

Imran I. 1994. *Applications of non-associated plasticity in modelling the mechanical response of concrete.* Ph.D. thesis. Toronto, Canada: Graduate Department of Civil Engineering, University of Toronto.

Kang H. D. & Willam K. J. 1999. Localization characteristics of triaxial concrete model. *Journal of Engineering Mechanics-ASCE* 125(8): 941–950.

Kupfer H., Hilsdorf H. K. & Rusch H. 1969. Behaviour of concrete under biaxial stresses. *ACI Journal* 66(52): 656–666.

Lundgren K. 2002. FE analyses and tests of lap splices in frame corners. *Structural Concrete* 3(2): 47–57.

Menétrey P. & Willam K. J. 1995. Triaxial failure criterion for concrete and its generalization. *Aci Structural Journal* 92(3): 311–318.

Neilsen M. K. & Schreyer H. L. 1993. Bifurcations in Elastic Plastic Materials. *International Journal of Solids and Structures* 30(4): 521–544.

van Mier J. G. M. & Ulfkjaer J. P. 2000. Round-Robin analysis of over-reinforced concrete beams – Comparison of results. *Materials and Structures* 33(July): 381–390.

Computational Modelling of Concrete Structures, Bićanić et al. (eds)
© 2003 Swets & Zeitlinger, Lisse, ISBN 90 5809 536 3

Calibration of non local damage model with size effect law

K. Haidar & G. Pijaudier-Cabot
R&DO, Laboratoire de Génie Civil de Nantes Saint Nazaire, Ecole Centrale de Nantes, Nantes cedex, France

J.F. Dubé
LMGC/UMR5508 – Université de Montpellier II, Montpellier cedex, France

ABSTRACT: The calibration of non-local models, which contain an internal length, has been among the major issues conditioning the implementation of this kind of failure models. Direct calibration from uniaxial testing, where the state of strain remains homogeneous throughout the specimen, is impossible. The softening law is not directly accessible, because the state of strain cannot remain homogeneous during the entire test. In the absence of local information on the displacement field and on microcracking in the fracture process zone, the calibration has to rely on inverse analysis. This paper presents such a procedure based on three point bend size effect tests on notched specimens. The complete load deflection curves are used for the identification of the constitutive relations. Manual calibration is discussed first with the help of Bazant's size effect law, which is related to peak loads only. Then, automatic calibration is described. It is shown that an optimal set of model parameters can be obtained within a reasonable number of iterations. The identification concerns generally three sizes of specimens. We show the contribution of a fourth size on the quality of the identification. The results have a dispersion, which results from the heterogeneity of the material, and from the dimensions of specimens. Finally the simulations have also their defect and we try to integrate the hazard of conception of the mesh. We look the influence of the mesh onto the parameters identified and on the size effect law.

1 INTRODUCTION

Non-local models are necessary in order to achieve a consistent modelling of failure due to strain softening in concrete (Pijaudier-Cabot and Bazant 1987, Peerlings et al. 1996. Hence, their calibration has to rely on sufficiently robust methods for an efficient implementation and an effective use of such models. Whether they are in a gradient or in an integral form, these constitutive relations incorporate an internal length, which must be calibrated from experiments in addition to other, more classical, model parameters. The fact that the models contain this internal length raised many questions in the past on the most appropriate technique for measuring it. As of today, direct experimental techniques, which would provide this internal length, have not been devised according to our knowledge. An inverse technique relies on the fact that the size of the fracture process zone in the material is independent from the size of the structure, provided it does not interfere with its boundaries. Hence, the response of geometrically similar specimens is not geometrically similar and there is a size effect. A complete explanation of this size effect can be found in the textbook by Bazant and Planas (1998). It is due to energy redistribution from the rest of the structure (whose size changes from one specimen to another) where elastic energy is stored, to the fracture process zone whose size is constant. It is only with failure models, which contain an internal length that this size effect can be described numerically. It can be the cohesive crack model in which the characteristic length is related to the length of the fracture process zone (see Bazant and Planas 1998), plasticity or damage based non local models (see e.g. Mazars et al. 1991, de Borst and Gutierrez 1999) where the characteristic length is proportional to the width of the fracture process zone approximately (Mazars and Pijaudier-Cabot 1996). We already showed the relevance of the use of the size effect law to identify the parameters materials (Le Bellego et al. 2002). Therefore, fits of size effect tests may provide the model parameters in non-local models. In this paper, we will focus on this calibration technique based on size effect tests.

In the present contribution, we will show that the model parameters entering into the description of the nonlinear response of the material in tension, including the internal length, can be obtained from a fit of one set of size effect tests only. The load deflexion curves obtained from three or four point bend notched specimens of three or four geometrically similar sizes will serve as experimental input data. The non-local damage model will be used as an illustrative example. It is recalled in section 2. Section 3 presents the experimental results used in this study and the sensitivity analysis of the constitutive relations, carried out with the help of Bazant's size effect law. An automatic calibration process discussed in section 4.

2 CONSTITUTIVE MODEL

We are going to use the non local version of the scalar damage model for which all the model parameters are recalled in this section (Mazars et al. 1991). The influence of microcracking is introduced via a single scalar damage variable d ranging from 0 to 1. The stress strain relation reads:

$$\varepsilon_{ij} = \frac{1+\nu}{E(1-d)}\sigma_{ij} - \frac{\nu}{E(1-d)}[\sigma_{kk}\delta_{ij}] \text{ or}$$

$$\sigma_{ij} = C_{ijkl}\varepsilon_{kl} = (1-d)C^0_{ijkl}\varepsilon_{kl} \tag{1}$$

E and ν are the Young's modulus and the Poisson's ratio of the undamaged material. ε_{ij} and σ_{ij} are the strain and stress components, C_{ijkl} and C^0_{ijkl} are the damaged and initial (elastic) secant stiffness of the material, and δ_{ij} is the Kronecker symbol. The evolution of damage is based on the amount of extension that the material is experiencing during the mechanical loading. In the non-local damage model, it is controlled by the average $\bar{\varepsilon}$ of an equivalent strain $\tilde{\varepsilon}$ defined as:

$$\tilde{\varepsilon} = \sqrt{\sum_{i=1}^{3} (\langle \varepsilon_i \rangle_+)^2} \tag{2}$$

where $\langle\,\rangle_+$ is the Macauley bracket and ε_i are the principal strains. In order to avoid ill-posedness due to strain softening, the mechanical model is enriched with an internal length (Mazars et al. 1984, Pijaudier-Cabot et al. 1987). The non-local variable $\bar{\varepsilon}$ is defined:

$$\bar{\varepsilon}(x) = \frac{1}{V_r(x)} \int_{\Omega} \psi(x-s)\tilde{\varepsilon}(s)ds$$

$$\text{with } V_r(x) = \int_{\Omega} \psi(x-s)ds \tag{3}$$

where Ω is the volume of the structure, $V_r(x)$ is the representative volume at point x, and $\psi(x-s)$ is the weight function:

$$\psi(x-s) = \exp\left(-\frac{4\|x-s\|^2}{l_c^2}\right) \tag{4}$$

l_c is the internal length of the non local continuum. It appears in the analytical formula of the weight function Ψ, and defines the size of the non-local strain averaging support. Previous studies have shown that the internal length l_c can be correlated to the material characteristic length, which defines the scale of the heterogeneity of the material. It is generally assumed that $l_c = 3 \times da$ (Bazant et al. 1983) where da is the maximum size of aggregate. lc is considered as a constant material parameter.

$\bar{\varepsilon}$ is the variable that controls the growth of damage according to the following conditions:

$$F(\bar{\varepsilon}) = \bar{\varepsilon} - \kappa \tag{5}$$

if $F(\bar{\varepsilon}) = 0$ and $\dot{\bar{\varepsilon}} > 0$

$$\text{if } F(\bar{\varepsilon}) = 0 \text{ and } \dot{\bar{\varepsilon}} > 0$$

$$\text{then} \begin{cases} \dot{d} = h(\kappa) \\ \dot{\kappa} = \dot{\bar{\varepsilon}} \end{cases} \text{with the condition } \dot{d} \geq 0 \tag{6}$$

$$\text{else} \begin{cases} \dot{d} = 0 \\ \dot{\kappa} = 0 \end{cases}$$

κ is the softening parameter and takes the largest value of $\bar{\varepsilon}$ ever reached during the previous loading history at a given time and at the considered point in the medium. Initially $\kappa = \kappa_0$, where κ_0, is the threshold of damage. In order to capture the difference of mechanical responses of the material in tension and in compression, the damage variable is split into two parts:

$$d = \alpha_t d_t + \alpha_c d_c \tag{7}$$

d_t and d_c are the damage variables in tension and compression respectively. They are combined with the weight coefficients α_t and α_c defined as functions of the principal values of the strains ε^t_{ij} and ε^c_{ij} due to positive (σ^t_{kl}) and negative (σ^c_{kl}) stresses:

$$\varepsilon^t_{ij} = C^{-1}_{ijkl}\sigma^t_{kl},\ \varepsilon^c_{ij} = C^{-1}_{ijkl}\sigma^c_{kl} \tag{8}$$

The stress tensor is separated into positive and negative parts according to a polar decomposition. The weight coefficients are:

$$\alpha_t = \sum_{i=1}^{3}\left(\frac{\langle \varepsilon^t_i \rangle \langle \varepsilon_i \rangle_+}{\tilde{\varepsilon}^2}\right)^{\beta},\quad \alpha_c = \sum_{i=1}^{3}\left(\frac{\langle \varepsilon^c_i \rangle \langle \varepsilon_i \rangle_+}{\tilde{\varepsilon}^2}\right)^{\beta} \tag{9}$$

Note that in these expressions strains labeled with single indicia are principal strains. In uniaxial tension

$\alpha_t = 1$ and $\alpha_t = 0$. In uniaxial compression $\alpha_t = 0$ and $\alpha_c = 1$. The evolution of damage is provided in an integrated form, as a function of the variable κ:

$$
\begin{aligned}
d_t &= 1 - \frac{\kappa_0(1-A_t)}{\kappa} - \frac{A_t}{\exp(B_t(\kappa-\kappa_0))} \\
d_c &= 1 - \frac{\kappa_0(1-A_c)}{\kappa} - \frac{A_c}{\exp(B_c(\kappa-\kappa_0))}
\end{aligned}
\tag{10}
$$

Standard values of the model parameters in the damage have been given by Mazars (1984). Same as in the paper by Carmeliet (1999), we will focus attention on the response of the model in tension only. It means that in the course of all the calculations $d_c = 0$ and that the model parameters considered for the identification are the Young's modulus E, the Poisson's ratio ν, the initial damage threshold κ_0, constants A_t and B_t, and the internal length l_c. The parameter β acts on the shear response of the material and it is usually considered to be constant (here $\beta = 1.06$). The elastic constants are easily obtained from uniaxial compression tests and will be considered to be known in the present study, although the Young's modulus could be obtained from the fit of the experimental data.

Hence, the calibration should aim at obtaining four parameters: κ_0, A_t, B_t and the internal length l_c. In the case of uniaxial tension it is possible to provide a meaning to some of them, as it helps for manual calibration or initial guesses in an automatic calibration process. In a first approximation, κ_0 is the peak strain. $(1-A_t)$ is the ratio between the residual stress for large strains and the peak stress. The residual stress is zero when $(1-A_t)=0$ and when $(1-A_t)=1$ the uniaxial response looks like an elastic perfectly plastic one. B_t controls the softening slope. The larger B_t, the faster the stress decreases in the post peak regime.

Note that the general calibration technique discussed in this paper is not dependent on the type of non-local model considered. The uniaxial tension responses of most models rely on an average number of four model parameters. Obviously, the sensitivity of each model parameter on the computational results remains strongly dependent on the exact mathematical form of the constitutive relation. The scalar damage model, which is widely used for quasi-brittle materials, is considered as an illustrative example here, and a procedure similar as the one depicted in the next sections, could be equally applied with another type of constitutive relation.

3 CALIBRATION

The calibration of the model parameters in the non local damage model is facing the difficulty of getting at the same time the parameters involved in the evolution of damage κ_0, A_t, B_t and the internal length l_c. The first set is a point-wise piece of information while the influence of the internal length appears in a boundary value problem with non homogeneous strains. Theoretically, it could be possible to calibrate the evolution of damage on tests in which the strain distribution remains homogeneous over the specimen. This is in fact impossible because whenever strain softening occurs, the strain and damage distributions happen to become strongly non homogeneous due to bifurcation and strain localisation during the failure process.

We are going first to recall the experimental data from which the calibration process will be carried out. Manual calibration from these data and sensitivity analyses will be discussed afterwards.

3.1 *Experimental results*

Three point bend experiments have been carried out by Haidar (2002) on notched geometrically similar specimens of various height $D = 40$, 80, 160, and 320 mm, of length $L = 2.675D$, and of thickness $b = 40$ mm kept constant. The span-to-height ratio is $l/D = 2.5$, and the notch length is $a_0 = D*5/6$ (Figure 1). These are mortar specimens with a water-cement ratio of 0.4 and a cement-sand ratio 0.46 (all by weight). The maximal sand grain size is $d_a = 3$ mm. Figure 2 shows the load deflection responses for the four different sizes.

Experimental and computational data will be also interpreted with the help of Bazant's size effect law, in one of its simplest form. The nominal strength σ is obtained with the formula:

$$
\sigma = \frac{3}{2}\frac{Fl}{b(5D/6)^2} \tag{11}
$$

where F is the peak load. The size effect law reads:

$$
\sigma = \frac{Bf_t'}{\sqrt{1+D/D_0}} \tag{12}
$$

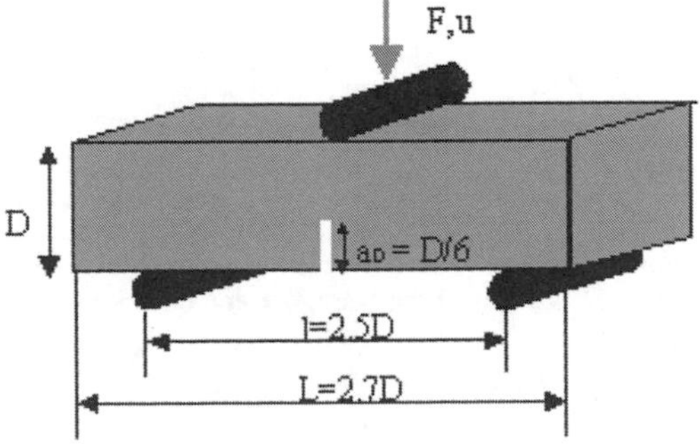

Figure 1. Geometry of the three points bend size effects tests.

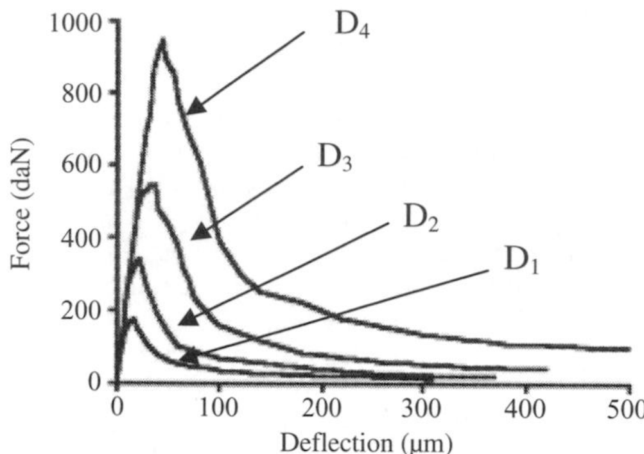

Figure 2. Three point bend testing of notched beams – experimental results for $D_1 = 40$ mm, $D_2 = 80$ mm, $D_3 = 160$ mm, $D_4 = 320$ mm sizes.

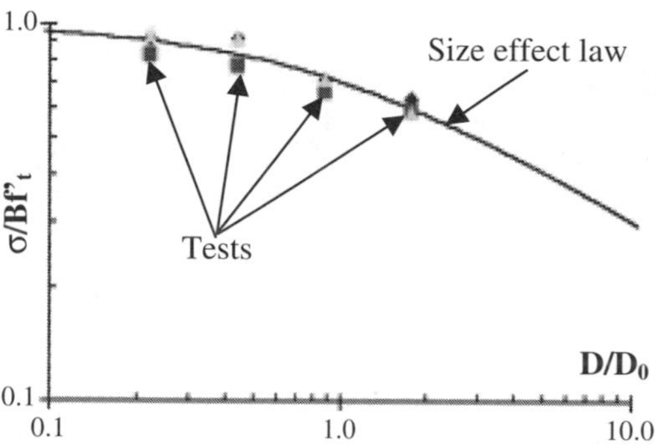

Figure 3. Experimental results of size effect in three points bend tests.

D_0 is a characteristic size, f_t' is the tensile strength of the material, and B is a geometry-related parameter. In a log–log plot, a strength of material criterion is represented by a horizontal curve and a linear elastic fracture mechanics (LEFM) criterion is depicted by a line with the slope –1/2. The two lines cut each other at the abscissa $D/D_0 = 1$. D_0 and Bf_t' are obtained from a linear regression.

$$\frac{1}{\sigma^2} = aD + c \quad \text{with } Bf_t' = \frac{1}{c} \text{ and } D_0 = \frac{c}{a} \tag{13}$$

Figure 3 shows the fit of the size effect law (Eq. 12) obtained from the average of the experimental data in Figure 2. The larger the beam, the lower the nominal strength.

3.2 *Geometrical variability*

Despite the care that one can take to realize the specimens, we have a certain dispersion of dimensions. Each size of specimens is the subject of 3 tests. This does not make it possible to draw up precise statistics but gives an idea on the generated dispersion. We measured all the specimens tested in order to establish the variability of dimensions. On the whole we thus have 12 series of values of L, H, D, b. We could note that each dimension had a dispersion of ±0.5 mm around its average. We assume a standard deviation of 0.5 mm for each dimension. We then calculate the average value and the standard deviation of $1/\sigma^2$ by integrating these standard deviations. We then compare the computed values with those obtained during the tests. The average of $1/\sigma^2$ used in the size effect laws is:

$$E\left(\frac{1}{\sigma^2}\right) \propto E\left(\frac{b^2D^4}{F^2l^2}\right) > \frac{E(b^2)E(D^4) + \text{cov}(b^2, D^4)}{E(F^2)E(L^2) + \text{cov}(F^2, l^2)} \tag{14}$$

where $E(X)$ is the average of X, $\text{cov}(X, Y)$ is the covariance between X and Y. We assume that dimensions are not correlated even if this assumption is strong for the correlation between F and l. We can then write the average and the variance of $1/\sigma^2$:

$$E\left(\frac{1}{\sigma^2}\right) \propto E\left(\frac{b^2D^4}{F^2l^2}\right) > \frac{E(b^2)E(D^4)}{E(F^2)E(l^2)} \text{ and } E(X^2) = E(X)^2 + \text{var}(X) \tag{15}$$

$$\text{var}\left(\frac{b^2D^4}{F^2l^2}\right) = \frac{\left[E(b^2D^4)^2\,\text{var}(F^2l^2) + E(F^2l^2)^2\,\text{var}(b^2D^4)\right]}{\left[E(F^2l^2)^2\left(E(F^2l^2)^2 + \text{var}(F^2l^2)\right)\right]} \tag{16}$$

with:

$$\text{var}(XY) = E(X)^2\,\text{var}(Y) + E(Y)^2\,\text{var}(X) + \text{var}(X)\,\text{var}(Y)$$

if X and Y are not correlated and

$$\text{var}(X^2) = \text{var}(X)\left[3E(X)^2 + 2\,\text{var}(X)\right]$$

Table 1 presents the averages and the standard deviations for the tests and the results of these equations. D is replaced here by the useful height between the top of the beam and the notch D-a_0. By comparing the results of these equations with those of the tests we obtain an average of the relative deviation of 1.23% on the averages and 13.22% on the standard deviations.

We see well that the errors resulting from the variations of dimensions are of the same order as the variability of the answer F nominal strength while being lower. Thus they do not generate dispersions on the results, which would be necessary to take into account. Figure 4 show that the function $1/\sigma^2$ is well described by a linear regression. We framed the values of the tests by the averages more or less two once standard deviation.

Table 1. Averages and standard deviations for the tests and the analysis of variability.

Size (mm)	Tests		Computations	
	Average	Standard deviation	Average	Standard deviation
40	$4.48\ 10^{-2}$	$5.82\ 10^{-3}$	$4.44\ 10^{-2}$	$5.22\ 10^{-3}$
80	$4.78\ 10^{-2}$	$9.68\ 10^{-3}$	$4.65\ 10^{-2}$	$7.55\ 10^{-3}$
160	$7.34\ 10^{-2}$	$6.78\ 10^{-3}$	$7.30\ 10^{-2}$	$5.89\ 10^{-3}$
320	$9.77\ 10^{-2}$	$8.45\ 10^{-3}$	$9.71\ 10^{-2}$	$7.82\ 10^{-3}$

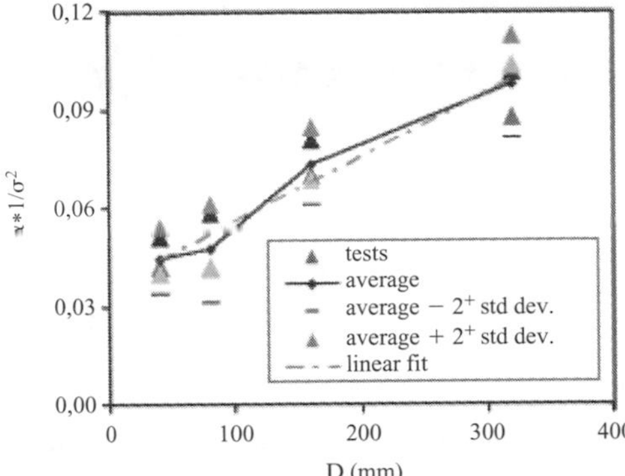

Figure 4. Evolution of the function proportional to $1/\sigma^2$ according to the size D.

3.3 *Automatic calibration*

The calibration of the model parameters from inverse analysis relies first on an accurate finite element representation of each size of specimen. Experience shows that the accuracy of the finite element discretisation in the fracture process zone depends on the ratio of the size of the finite element to the internal length. Therefore, geometrically similar finite element meshes should not be used for the calibration. Figure 5 shows an example of set of bidimensional finite element meshes.

In the present application, the model parameters are optimised so that the three computational results are as close as possible to the experimental results.

The optimisation process uses the responses of the three specimen sizes simultaneously. We do not integrate the effects into the identification not to force the results to follow the size effect law. A Levenberg-Marquardt algorithm has been implemented. The algorithm performs the minimisation of the following functional $\mathcal{J}(\vec{P})$:

$$\Im(\vec{P}) = \frac{1}{2}\sum_{i=1}^{3}\left\{\left(\frac{R^i(\vec{P}) - R^i_{\text{exp}}}{Max(R^i_{\text{exp}})}\right)^T \cdot \left(\frac{R^i(\vec{P}) - R^i_{\text{exp}}}{Max(R^i_{\text{exp}})}\right)\right\} \quad (17)$$

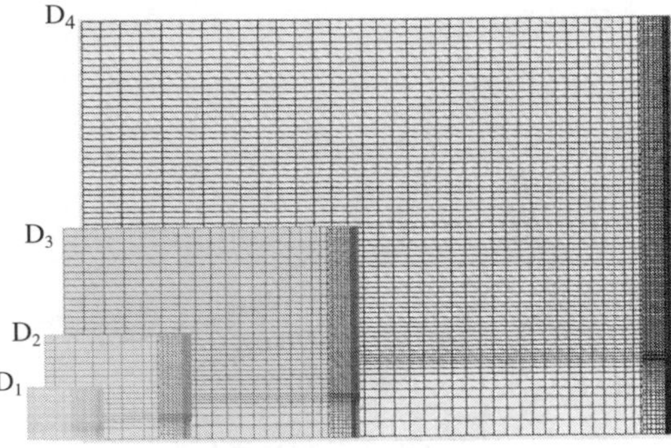

Figure 5. 2D finite element meshes for sizes D_1, D_2, D_3 and D_4.

where R^i_{exp} is the experimental response for the size $i (i \in [1,4])$, and $R^i(\vec{P}$ is the numerical response for size i corresponding to the vector of model parameters $\vec{P}$. Several definitions can be used for the responses. In the present optimisation scheme we use a simple one: it is the stress computed according to equation (11) from the applied load measured for 50 values of the deflection equally spaced and covering the experimental data range. Because the experimental responses and numerical ones have rather different values from one specimen size to another, it is necessary to express them in a non-dimensional way (or to use a weighted functional). This is the reason why the error between the numerical and experimental responses is divided by the maximal experimental response for each size. The functional is minimised within 3 iterations approximately. We observed that the error reaches a minimum and remains constant afterwards. At this stage, the calculation was stopped. The optimisation process is rather fast because it starts from a relatively correct initial set of model parameters.

We use two strategies. The first consists in determining the parameters starting from the response of the four tests. The second uses coarser meshes and identifies the parameters starting from the response of the beams D_2 and D_3. One can assume that the classical parameters can be identified from a test data and that the second test data makes it possible to determine the characteristic length. The meshes are such that at the tip of the notch the density of mesh remains unchanged.

Figure 6 shows the comparison between the minimum and maximum test data and the computations for the optimal fit with used the four sizes.

This result is transposed into the size effect plot in Figure 7. We shows the comparison between the higher and lower delimiters of the tests data and the computations for the optimal fit. The tests data are represented by the average more or less the standard deviation. The

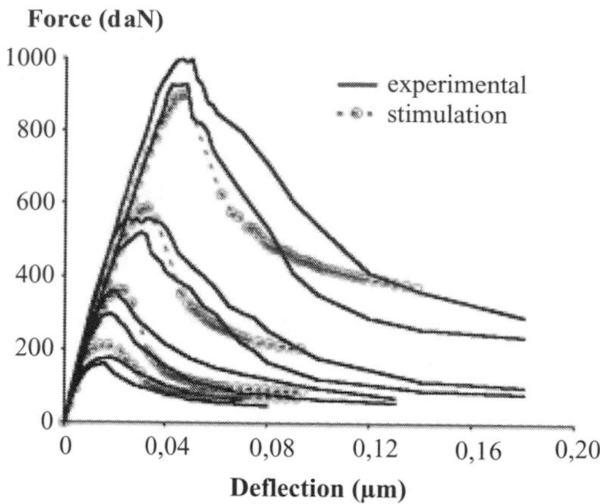

Figure 6. Automatic calibration of model parameters. The dotted lines are the model responses and the plain lines are, for each size, the minimum and maximum curves obtained from the experimental data.

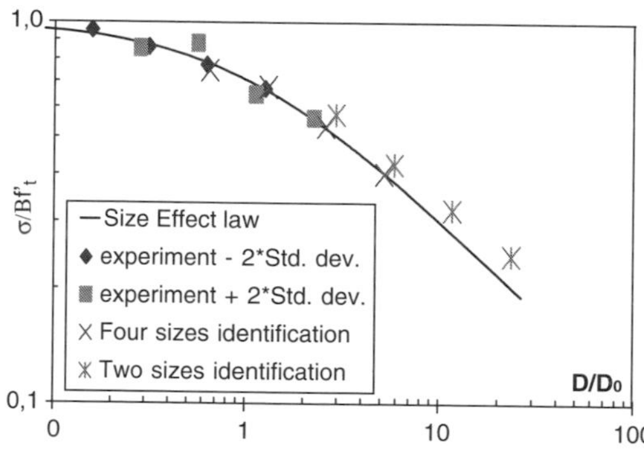

Figure 7. Size effect predictions for optimal fits. Experiment is the average of the tests data. Identification are the numerical result with four fine or two coarse meshes.

numerical predictions are identified from the experimental results by using two or four sizes. Each series of point is traced with its own parameters Bf_t' and D_0. The two sizes identification used also a coarser mesh (Figure 8). We can see that the identification using coarser meshes and with two sizes is much worse than that obtained with the four sizes and the fine meshes.

The obtained set of material data is optimal with respect to the load deflection curves. But with respect to the size effect law, it is not since. It was not possible to place the computed points in between the offer and lower bounds in Figure 7.

4 CONCLUSION

The development of non-local continuum models with local strains, which has been observed in the literature over the past decade, corresponds to the aim of providing sound and computationally robust failure continuum models. Failure is predicted with a finite – non-zero – energy dissipated. From the analysis of size effect tests, it seen that dispersion obtained does not come from the quality of realization of the specimens but from dispersion inherent to material. But the structural size effect, observed for most quasi-brittle materials, is well captured. The dispersion of the tests data affects little the variation of the size effect law. Their calibration, however, is facing the usual difficulties involved in inverse analysis. Automatic calibration by the minimisation of a quadratic functional and a Levenberg–Marquardt algorithm provides an optimal set of model parameters. In spite of a weak dispersion of the tests, the identification is not very good in comparison with the law of scale effect. However we can say that the number of curves used for the identification is of primary importance on this one.

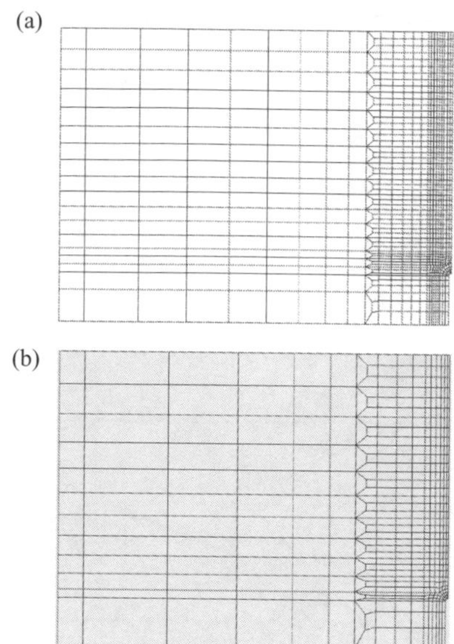

Figure 8. Finite element meshes for the beam D_2. (a) fine mesh, (b) coarse mesh.

REFERENCES

Bazant Z.P. and Planas J. (1998), Fracture and size effect in concrete and other quasi-brittle materials, CRC Press.

Carmeliet J. (1999), Optimal estimation of gradient damage parameters from localisation phenomena in quasi-brittle materials, Int. J. Mech. Cohesive Frict. Mats., 4, 1–16.

de Borst, R. and Gutierrez, M. (1999), A unified framework for concrete damage and fracture models including size effects, Int. J. Fracture, 95, 261–277.

Haidar K.H. (2002), Modélisation de l'endommagement des structures en béton – approches numériques et effet de la microstructure sur les propriétés de rupture, Thése de doctorat de l'Ecole Centrale de Nantes et de l'Univ. de Nantes, 2002.

Le Bellego C., Dubé J.F., Pijaudier-Cabot G., Gérard B. (2002), Calibration of non local damage from size effect tests, European J. of Mech., in press.

Mazars J. (1984), Application de la mécanique de l'endommagement au comportement non linéaire et á la rupture du béton de structure, Thèse de Doctorat ès Sciences, Université Paris 6, France.

Mazars J. and Pijaudier-Cabot G. (1996), From damage to fracture mechanics and conversely: a combined approach, Int. J. Solids. Struct., 33, 3327–3342.

Mazars J., Pijaudier-Cabot G. and Saouridis C. (1991), Size Effect and Continuous Damage in Cementitious Materials, International Journal of Fracture, 51, 159–173.

Peerlings R.H., de Borst R., Brekelmans W.A.M., de Vree J.H.P. (1996), Gradient enhanced damage for quasi-brittle materials, Int. J. Num. Meth. Engrg., 39, 3391–3403.

Pijaudier-Cabot G., and Bazant Z.P. (1987), Non local damage theory, J. of Engrg. Mech. ASCE, 113, 1512–1533.

Computational Modelling of Concrete Structures, Bićanić et al. (eds)
© 2003 Swets & Zeitlinger, Lisse, ISBN 90 5809 536 3

Constitutive model for highly ductile fiber-reinforced cement-based composites under cyclic load

Tong-Seok Han, Peter H. Feenstra* & Sarah L. Billington
Cornell University, Ithaca, NY, USA

ABSTRACT: A class of ductile fiber-reinforced cement-based composite referred to as Engineered Cementitious Composites (ECC) are being investigated for new design as well as retrofitting of structures in seismic regions. ECC is highly ductile and is characterized by strain-hardening in tension to strains of over 3% and by unique cyclic loading behavior. To predict accurately the structural performance of ECC components under cyclic and seismic loading, a robust constitutive model is needed for structural scale simulations. In this paper, a constitutive model based on total-strain is proposed and applied to simulate structural component tests. The model in particular captures ECC's unique reversed cyclic loading behavior. The simulation results show that the implemented model is robust and reasonably accurate in simulating ECC structural components reinforced with steel and fiber-reinforced polymer bars.

1 INTRODUCTION

Engineered Cementitious Composites (ECC) are a type of high performance material that exhibits multiple, fine cracks upon loading in tension as a result of steady-state cracking (Marshall et al. 1985; Marshall and Cox 1988; Li and Leung 1992). ECC materials are composed of Portland cement, water, silica fume or fly ash, fine sand, and roughly 2% by volume of high modulus, high aspect ratio polymeric fibers. ECC displays a much higher tensile ductility, tensile (strain) hardening behavior, and energy dissipation than traditional concrete and many fiber-reinforced concrete materials (summary in Li 1998).

Since the ductility of ECC materials is significantly larger than that of conventional concrete, applications of ECC materials to structures under severe loading conditions are being investigated. Most of the research to date on ECC has focused on experimental investigations of the material and structural components. A simulation framework which can verify, validate, or predict the performance of the structural members using ECC material has been explored by very few researchers for cyclic analysis of structural members.

The primary objective of the research presented here is to develop a constitutive model which can be used to simulate structural components with ECC under cyclic and seismic loading. In particular, the constitutive model must be efficient, and robust for large-scale simulations. This paper focuses on models for cyclic loading.

The total strain-based rotating crack model based on the model described in Feenstra et al. (1998), is implemented for 2D plane stress elements. The monotonic failure envelope of the ECC constitutive model behavior for the current formulation is similar to the ECC model proposed by Kabele (2001). However, the model presented here differs from latter in that the proposed model uses the co-axial concept, and the unloading and reloading behavior adopted is based on reversed cyclic load experiments on ECC. A strain-based criterion for failure initiation is used in the implementation. The Poisson effect is included in the model through an equivalent strain concept.

2 COMPUTATIONAL METHODOLOGY

The constitutive model for the ECC material is implemented in 2D using a co-axial rotating crack model with two orthogonal cracks described in Feenstra et al. (1998). The reduction of stiffness due to cracking/crushing is assumed to be only dependent on the loading direction, and the tensile and compressive damage is not coupled in the current formulation.

2.1 *Modeling the material behavior*

The local stresses are evaluated as function of the local strain vector using a stress function, $\mathcal{F}$,

$$\sigma = \mathcal{F}(\varepsilon). \tag{1}$$

* Stanford University, Stanford, CA, USA

Note that the function $\mathcal{F}(\varepsilon)$ describes the tensile and compressive regime of the stress–strain space.

Loading–unloading is modeled using internal state variables, α, with which Equation 1 becomes:

$$\boldsymbol{\sigma} = \mathcal{F}(\boldsymbol{\varepsilon}, \boldsymbol{\alpha}). \tag{2}$$

The stress state is determined by the stress function, $\mathcal{F}$, as a function of the current strain and the history parameter in the associated principal direction.

Poisson's effect is accounted for using an equivalent strain concept (Selby and Vecchio 1993), and the implementation is described in Han et al. (2002).

The stress function is based on the general response of uni-axial reversed cyclic load experimental data (Kesner and Billington 2001), see Figure 1. The experimental results are idealized so that the model is more suitable for numerical calculations. The failure envelopes in tension and compression used in this study are similar to the envelopes of the reversed cyclic load experiments shown in Figure 2.

The failure envelopes of the ECC shown in Figure 2 can be expressed as Equations 3 and 4. The tensile failure envelope shown in Figure 2(a) (a multi-linear curve) can be written as:

$$\mathcal{F}_{tensile} = \begin{cases} E\,\varepsilon & 0 \le \varepsilon < \varepsilon_{to} \\ \sigma_{to} + (\sigma_{tp} - \sigma_{to})\left(\dfrac{\varepsilon - \varepsilon_{to}}{\varepsilon_{tp} - \varepsilon_{to}}\right) & \varepsilon_{to} \le \varepsilon < \varepsilon_{tp} \\ \sigma_{tp}\left(1 - \dfrac{\varepsilon - \varepsilon_{tp}}{\varepsilon_{tu} - \varepsilon_{tp}}\right) & \varepsilon_{tp} \le \varepsilon < \varepsilon_{tu} \\ 0 & \varepsilon_{tu} \le \varepsilon \end{cases} \tag{3}$$

where E is Young's modulus.

The first cracking occurs when ε is larger than the cracking strain (ε_{t0}), followed by the strain-hardening region up to the peak tensile strain (ε_{tp}). The material softens after ε_{tp}, and the stress becomes zero at an ultimate tensile strain (ε_{tu}).

The compressive failure envelope shown in Figure 2(b) can be expressed as:

$$\mathcal{F}_{compressive} = \begin{cases} E\,\varepsilon & \varepsilon_{cp} \le \varepsilon < 0 \\ \sigma_{cp}\left(1 - \dfrac{\varepsilon - \varepsilon_{cp}}{\varepsilon_{cu} - \varepsilon_{cp}}\right) & \varepsilon_{cu} \le \varepsilon < \varepsilon_{cp} \\ 0 & \varepsilon \le \varepsilon_{cu}. \end{cases} \tag{4}$$

The compressive failure envelope is assumed to be linear up to a peak stress/strain ($\sigma_{cp}/\varepsilon_{cp}$) followed by linear softening until the strain reaches the ultimate compressive strain (ε_{cu}).

The unloading and reloading schemes shown in Figure 3 are based on observations from the reversed cyclic uni-axial test results (Figure 1) (Kesner and Billington 2001). To accurately model the material behavior, a partial unloading and reloading scheme is implemented, which is essential for seismic analysis. The unloading and reloading schemes for the strain-hardening range in tension and the softening range in compression are assumed to be governed by power laws, while the unloading in the tensile softening range is assumed to be linear.

The tensile unloading and reloading scheme ($\varepsilon \ge 0$ and $\varepsilon < \varepsilon_{tmax}$ where ε_{tmax} is the experienced maximum tensile strain) can be written as Equation 5. In the equation, a_t (≥ 1) is a constant parameter which needs to be determined by calibrating the unloading behavior of reversed cyclic experimental results. The value of ε^*_{tmax} is determined as:

$$\varepsilon^*_{tmax} = \begin{cases} \varepsilon_{tmax} & \text{for initial unloading} \\ \varepsilon_{tprl} & \text{for unloading followed by partial reloading} \end{cases}$$

$$\varepsilon^*_{tmax} \le \varepsilon_{tmax}$$

where ε_{tprl} is the maximum tensile strain during partial reloading and σ^*_{tmax} is the stress associated with ε^*_{tmax};

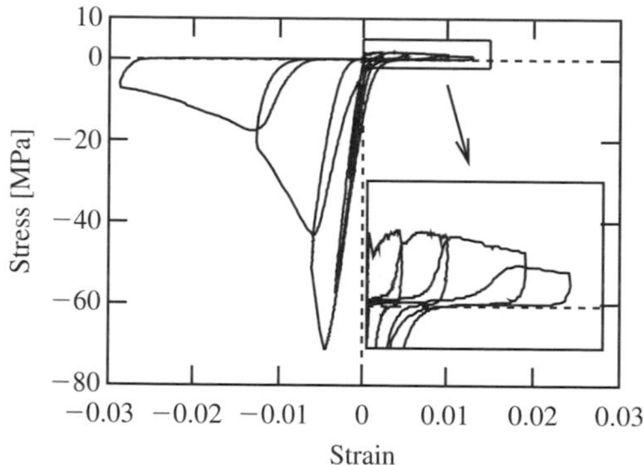

Figure 1. ECC cyclic behavior (experiment).

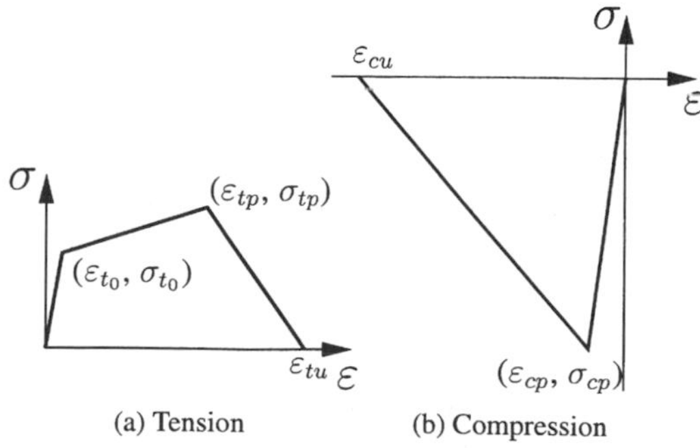

Figure 2. Schematic of ECC failure envelope.

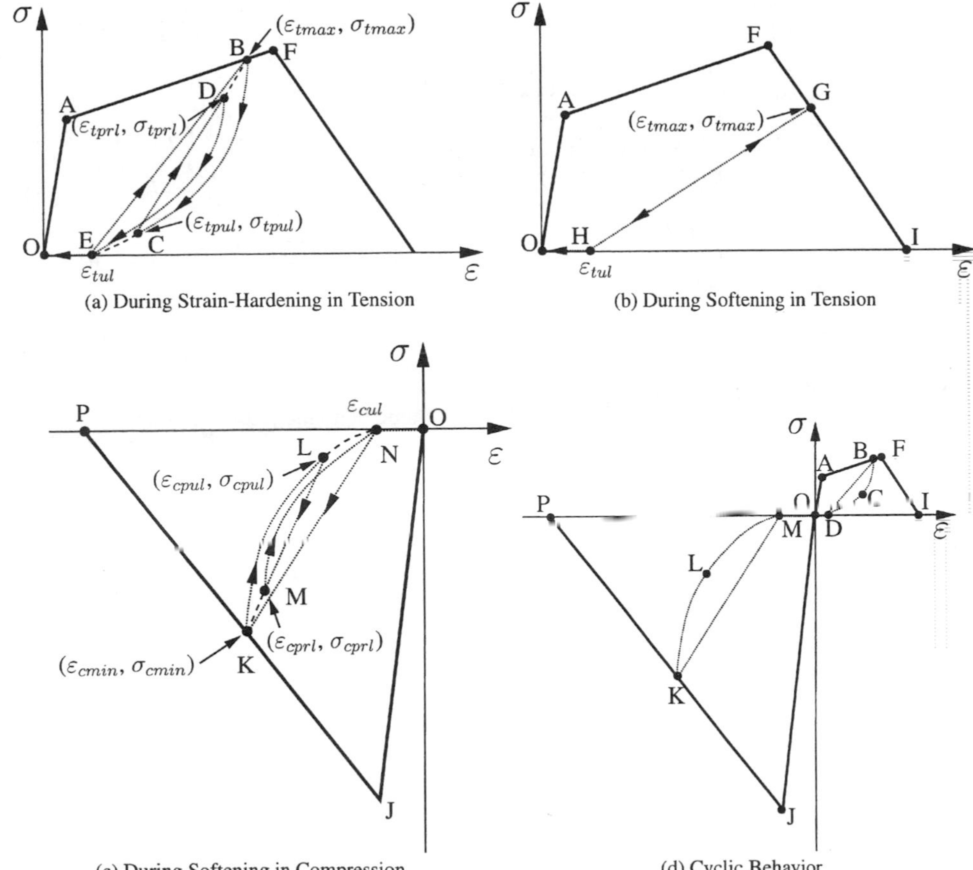

Figure 3. Schematic of ECC material behavior.

the value of ε^*_{tul} is determined by:

$$\varepsilon^*_{tul} = \begin{cases} \varepsilon_{tul} & \text{for initial unloading} \\ \varepsilon_{tpul} & \text{for unloading followed by partial reloading} \end{cases}$$

where $\varepsilon_{tul} = b_t \times \varepsilon_{tmax}$ (b_t: constant), ε_{tpul} is the minimum strain during partial unloading. σ^*_{tul} is the stress associated with the strain ε^*_{tul}. Note that $\varepsilon_{tmax}, \varepsilon_{tprl}, \varepsilon_{tpul}$ are the internal variables (α) in tension, which need to be traced to determine stresses during the unloading/reloading process.

Using the above equations, loading/unloading behavior and partial unloading/reloading behavior in tension for the strain-hardening range can be represented. For example, as in Figure 3(a):

- O→ A → B: Loading
- B→ C: Partial unloading
- C → D: Partial reloading
- D → E : Full unloading
- E → O: Assumed behavior for further unloading to origin

$$\mathcal{F}_{tensile} = \begin{cases} E\,\varepsilon & 0 \le \varepsilon_{tmax} < \varepsilon_{t0} \\ \max\left\{0, \sigma^*_{tmax}\left(\dfrac{\varepsilon - \varepsilon_{tul}}{\varepsilon^*_{tmax} - \varepsilon_{tul}}\right)^{a_t}\right\} & \varepsilon_{t0} \le \varepsilon_{tmax} < \varepsilon_{tp}, \quad \dot{\varepsilon} < 0 \\ \max\left\{0, \sigma^*_{tul} + (\sigma_{tmax} - \sigma^*_{tul})\left(\dfrac{\varepsilon - \varepsilon^*_{tul}}{\varepsilon^*_{tmax} - \varepsilon^*_{tul}}\right)\right\} & \varepsilon_{t0} \le \varepsilon_{tmax} < \varepsilon_{tp}, \quad \dot{\varepsilon} \ge 0 \\ \max\left\{0, \sigma_{tmax}\left(\dfrac{\varepsilon - \varepsilon_{tul}}{\varepsilon_{tmax} - \varepsilon_{tul}}\right)\right\} & \varepsilon_{tp} \le \varepsilon_{tmax} < \varepsilon_{tu} \\ 0 & \varepsilon_{tu} \le \varepsilon_{tmax} \end{cases} \quad (5)$$

- O → E → B: Full reloading
- B → F: Further loading.

Beyond the tensile peak strain (ε_{tp}), linear unloading/reloading is assumed as shown in Figure 3(b). While unloading, the stress–strain state follows the path G→H→O shown in Figure 3(b). The reverse process is assumed for the reloading and further loading (O→ H→G→I).

Similarly, the compressive unloading/reloading scheme ($\varepsilon < 0$ and $\varepsilon > \varepsilon_{cmin}$ where ε_{cmin} is the experienced minimum compressive strain) can be expressed as Equation 6. In the equation, a_c (≥ 1) is a constant parameter, which is determined by calibration of the unloading portion after the peak compressive strength as observed in reversed cyclic experiments. The value of $\varepsilon^{\star}_{cmin}$ is determined as:

$$\varepsilon^{*}_{cmin} = \begin{cases} \varepsilon_{cmin} & \text{for initial unloading} \\ \varepsilon_{cprl} & \text{for unloading followed} \\ & \text{by partial reloading} \end{cases}$$

$$\varepsilon^{*}_{cmin} \geq \varepsilon_{cmin}$$

where ε_{cprl} is the minimum strain during partial reloading in compression and $\alpha^{\star}_{cmin}$ is the stress that corresponds to $\varepsilon^{\star}_{cmin}$; the value of $\varepsilon^{\star}_{cul}$ in Equation 6 is determined as:

$$\varepsilon^{*}_{cul} = \begin{cases} \varepsilon_{cul} & \text{for initial unloading} \\ \varepsilon_{cpul} & \text{for unloading followed} \\ & \text{by partial reloading} \end{cases}$$

where $\varepsilon_{cul} = bc \times \varepsilon_{cmin}$ (b_c: constant) and ε_{cpul} is the minimum strain during partial unloading in compression. For example, the loading/unloading scheme in compression with a partial reloading in between can be represented as O →J→K→L→M→N→O as shown in Figure 3(c). The reloading process in compression is then O→N→K→P.

The loading, unloading, and reloading process in tension and compression under cyclic loading can be fully described using Equations 3–6. The stress–strain path for two cycles of loading in tension and compression are expressed graphically in Figure 3(d) as:

- O→A→B→C→D→O: First loading cycle in tension up to the strain-hardening region
- O→J→K→L→M→O: First loading cycle in compression to the softening region
- O→D→B→F→I→O: Second loading cycle in tension up to the complete failure

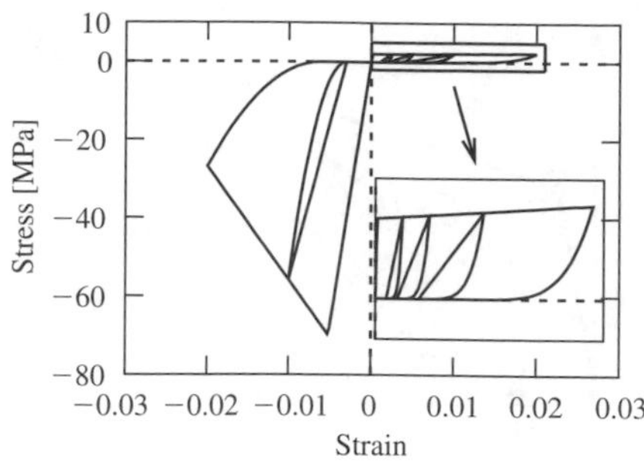

Figure 4. ECC cyclic behavior (simulation).

- O→M→K→P→O: Second loading cycle in compression up to the complete failure.

2.2 *Summary of the model and the parameters*

The constitutive model proposed in this section is relatively simple and is capable of describing the complex cyclic behavior of ECC with reasonable accuracy as will be shown in the following sections. For practical usage, the number of parameters of the constitutive model should be (a) limited, and (b) easy to determine with simple experiments. We introduced ten model parameters that can be easily determined with a uniaxial cyclic loading experiment. We have three properties to define the failure envelope of the ECC in tension ($\sigma_{t0}/\varepsilon_{t0}, \sigma_{tp}/\varepsilon_{tp}, \varepsilon_{tu}$), two properties in compression ($\sigma_{cp}/\varepsilon_{cp}, \varepsilon_{cu}$), and Poisson's ratio ($\nu$). Also, two constants for unloading behavior in tension and compression (a_t, a_c), and two constants for residual strain (b_t, b_c) should be provided.

The cyclic behavior of the developed ECC constitutive model is shown in Figure 4, and can be compared to the experimental uni-axial cyclic tension-compression response shown in Figure 1. The necessary material properties used for the simulation in Figure 4 were taken from uni-axial experiments (Table 1, Kesner and Billington 2001). The constants a_t and a_c, the parameters in unloading Equations 5 and 6 are chosen as 5 and 2, respectively. The constant parameters b_t and b_c in determining the permanent strains ε_{tul} and ε_{cul} are selected as 0.4 and 0.3. These parameters were selected through calibration with the reversed cyclic loading response as observed in the experiments.

The simulation model captures the significant features observed in the tensile region of the tests. In

$$\mathcal{F}_{compressive} = \begin{cases} E\,\varepsilon & \varepsilon_{cp} \leq \varepsilon_{cmin} < 0 \\ \min\left\{0, \sigma^{*}_{cmin}\left(\dfrac{\varepsilon - \varepsilon_{cul}}{\varepsilon^{*}_{cmin} - \varepsilon_{cul}}\right)^{a_c}\right\} & \varepsilon_{cu} \leq \varepsilon_{cmin} < \varepsilon_{cp},\ \dot{\varepsilon} > 0 \\ \min\left\{0, \sigma^{*}_{cu} + (\sigma_{cmin} - \sigma^{*}_{cmin})\left(\dfrac{\varepsilon - \varepsilon^{*}_{cul}}{\varepsilon^{*}_{cmin} - \varepsilon^{*}_{cul}}\right)\right\} & \varepsilon_{cu} \leq \varepsilon_{cmin} < \varepsilon_{cp},\ \dot{\varepsilon} \leq 0 \\ 0 & \varepsilon_{cmin} < \varepsilon_{cu} \end{cases} \quad (6)$$

Table 1. ECC material properties.

Properties	Kesner and Billington (2001)† Tension/Compression	Fischer and Li (2002)‡ Tension/Compression
Young's Modulus (E)	13.3 GPa	16.0 GPa
Cracking strain ($\varepsilon_{t0}/-$)	0.00015/–	0.00025/–
Cracking stress ($\sigma_{t0}/-$)	2 MPa/–	4 MPa/–
Peak strain ($\varepsilon_{tp}/\varepsilon_{cp}$)	0.03/ –0.00525	0.037/–0.005
Peak stress (σ_{tp}/σ_{cp})	2.5 MPa/–70 Mpa	6 MPa/–80 Mpa
Ultimate strain ($\varepsilon_{tu}/\varepsilon_{cu}$)	0.06/–0.03	0.06/–0.012
Poisson's ratio (ν)	0.15	0.15

† From reversed cyclic tension–compression tests.
‡ From monotonic tension and monotonic compression tests.

compression, the simplified unloading is expected to dissipate more energy than in the actual material. Furthermore, there are differences between the model and experimental results in the region where the response changes from tension to compression. To asses the effect of these similarities and differences, two structural component tests were simulated.

3 EXPERIMENTS

Two cantilever beam experiments reported in Fischer and Li (2002) are simulated here using the described constitutive model for ECC. Both experiments are beams fabricated with ECC with different types of reinforcement having the same layout. The first specimen (Specimen 1) had a ECC matrix with steel reinforcement, and the second specimen (Specimen 2) had a ECC matrix with fiber-reinforced polymer (FRP) reinforcement; in this case Aramid. Both specimens had cross sections of 100 mm × 100 mm and were 500 mm in height. The beams were subjected to fully reversed, quasi-static cyclic loading (Figure 5(a)).

Specimen 1 had four No. 3 bars (diameter 9.5 mm) for longitudinal reinforcement and 3 mm-diameter steel wire stirrups for transverse reinforcement. The transverse reinforcement was spaced at 25 mm from the bottom of the beam to a height of 150 mm, at 75 mm from a height of 150 mm to 450 mm, and at 25 mm from a height of 450 mm to the top of the beam. The yield strength of the steel was 411 MPa at 0.24% strain. Specimen 2 was reinforced with four FRP bars of 5 mm diameter, and each FRP bar had a failure strength of 1800 MPa at 3.4% strain. The transverse reinforcement was steel and had the same layout as in Specimen 1. The ECC material properties obtained from monotonic tension and compression tests for the cantilever beam specimens are shown in Table 1 (Fischer and Li 2002).

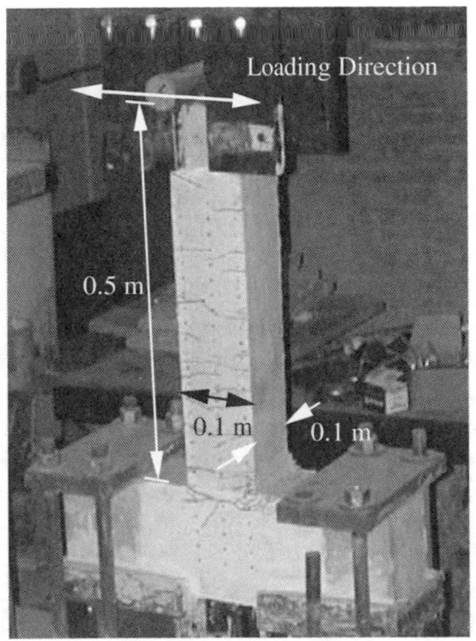

(a) Experimental Model

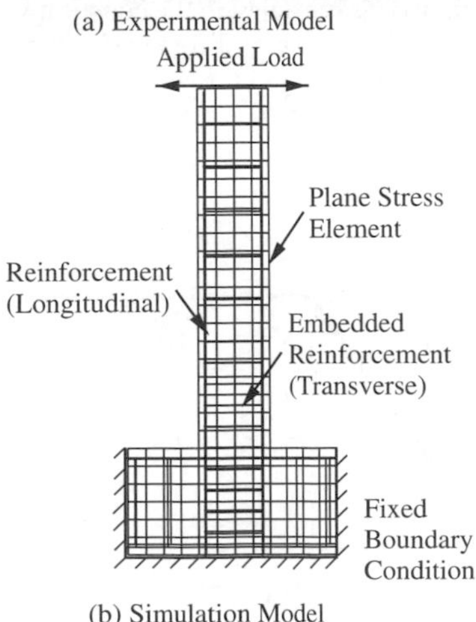

(b) Simulation Model

Figure 5. Experimental and simulation models.

4 SIMULATIONS

4.1 *Finite element models*

The experiments are simulated using 2D finite element models. Two finite element models, Models 1 and 2 for both specimens, are used to simulate the experiments. Model 2 incorporates bond-slip modeling between the ECC and the longitudinal reinforcement whereas Model 1 does not. The bond-slip model is analyzed to investigate the effect of bond-slip on hysteretic pinching behavior during cyclic loading. The basic finite element model is shown in Figure 5(b). 8-noded plane stress elements with a 3 × 3 Gauss numerical integration scheme are used for the ECC.

Model 1, which assumed perfect bond between the ECC and the reinforcement, uses embedded reinforcement elements. The constitutive model for the steel includes the Bauschinger effect using a model developed by Kwan and Billington (2001). In Model 2 the longitudinal reinforcement is modeled as truss elements, with interface elements between the truss elements and the edges of the plane stress elements to model bond-slip. The bond-slip model for the interface elements is adapted from Ayoub and Filippou (1999), and has a simplified linear relationship to the slip strength (σ_c) with a constant stress after the critical slip displacement (Δ_c) (Figure 6). Experimental results of the bond-slip relationship between the ECC material and the steel/Aramid reinforcement were not available at the time of this modeling. Therefore, representative values for σ_c and Δ_c are selected as 10 MPa and 0.1 mm, respectively. Secant unloading and reloading is assumed for cyclic bond-slip behavior. The FRP reinforcement is modeled as a linear-elastic material.

Two variations in the simulations are presented here: (1) accounting for bond-slip or not, and (2) variations in unloading/reloading in the constitutive model for the ECC. A total of three different simulation cases are outlined in Table 2 and further described next.

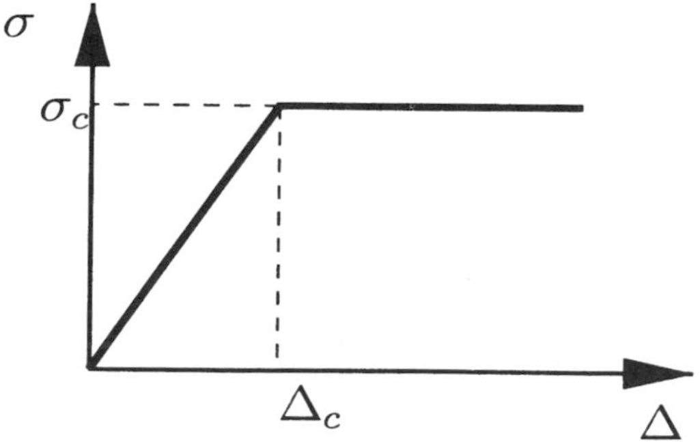

Figure 6. Bond–slip behavior.

Table 2. Analysis cases.†

Case	Bond-slip	Un(re)loading
1	Perfect bond	Realistic
2	Bond-slip	Realistic
3	Bond-slip	Secant

† Same for Specimen 1 and 2.
† Implemented un(re)loading model for the current research.

The ECC constitutive model uses the material properties from Fischer and Li (2002) (Table 1). The parameters a_t, a_c, b_t and b_c are equal to those used for the simulation of Figure 4 described in Section 2.2 (calibrated to the reversed cyclic load tests by Kesner and Billington 2001). A reduction parameter for the ECC initial modulus is introduced to consider shrinkage cracking in the structural-scale components. All three cases in Table 2 use a modulus reduction factor of 80%. This reduction factor is in the same range (i.e. between 50 and 90%) as that for concrete structures when simulated under cyclic and seismic load (D'Ambrisi and Filippou 1997; El-Attar et al. 1991; Bracci et al. 1992; Shahrooz and Moehle 1987; Aycardi et al. 1992). When the initial ECC modulus is reduced, the cracking stress and strain in tension is determined from the intersection of the original strain-hardening failure envelope and the reduced modulus. The compressive peak stress and strain after the modulus reduction is also determined from the interaction of the compressive softening envelope and the reduced modulus.

4.2 *Simulation of Specimen 1 – ECC with steel reinforcement*

Load versus displacement results of the three analysis cases as well as the experimental results for Specimen 1 are shown in Figure 7.

From all thee simulation cases, the initial beam stiffnesses at 0.6% drift are about 2 times larger than the initial beam stiffness from the experiment. Beam yield strengths from all simulation cases are at most 10% larger than the experimental value. Peak strengths at 5% drift are about 7 to 8% larger than the experimental results. The yield and peak strength of Specimen 1 is dominated not by ECC properties, but rather by the steel yield strength, which can be easily captured in the simulations with reasonable accuracy. In all cases, the hysteretic energy dissipation of the simulations is larger than that of the experiments.

Figure 7(b) shows the effect of accounting for bond-slip. In Figure 7(b), the initial stiffness of Case 2 is reduced by only 2%, and the yield strength is increased by about 3% relative to Case 1. The peak strength at 5% drift is not affected by the bond-slip modeling. However, there are more significant differences between

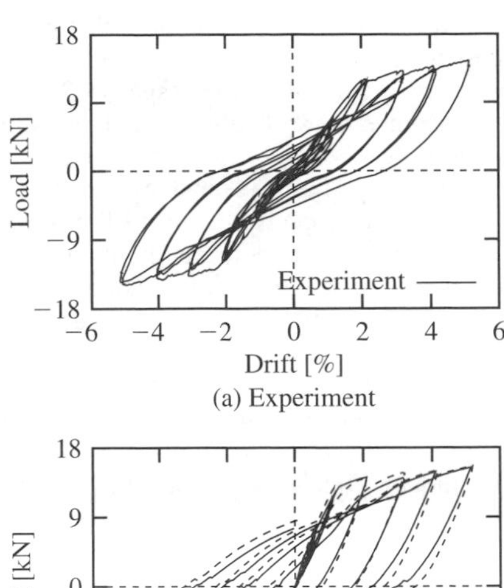

(a) Experiment

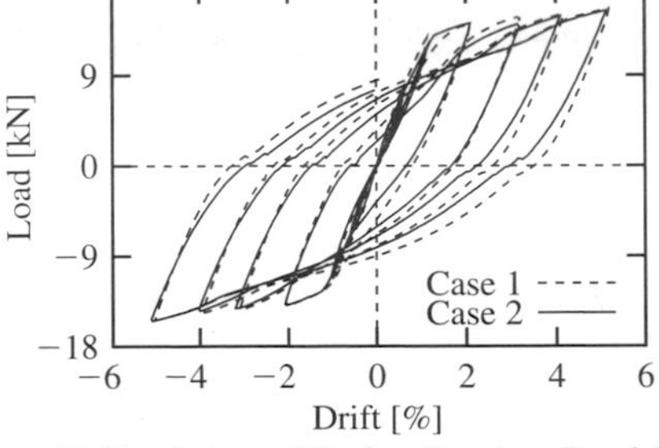

(b) Simulations of Perfect-Bond vs. Bond-Slip

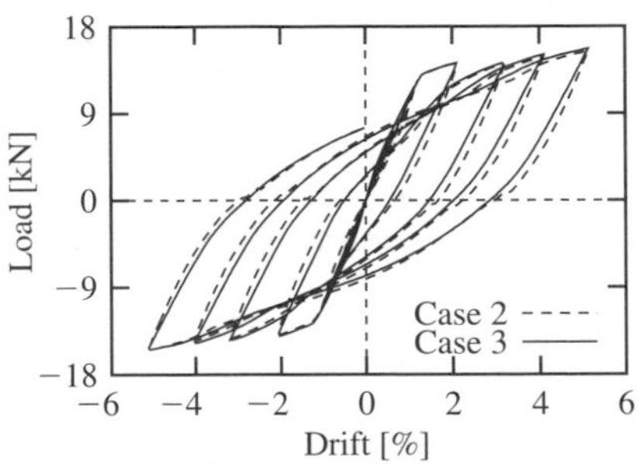

(c) Simulations for Unloading/Reloading Schemes

Figure 7. Results comparisons (load versus drift, Specimen 1).

(a) Cracking Pattern from Experiment

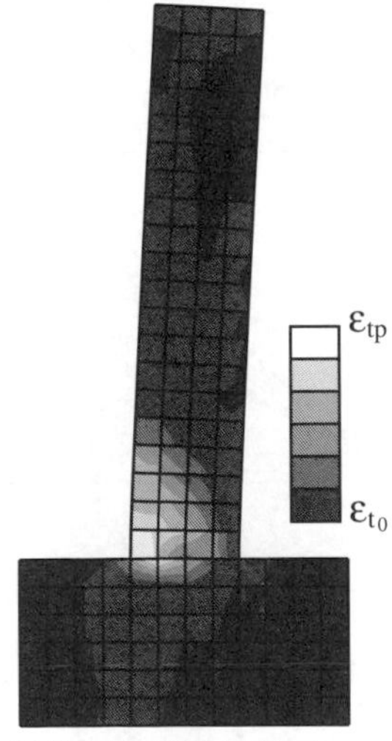

(b) Principal Strain Contour

Figure 8. Damage pattern comparison at 5% drift (Specimen 1).

the two cases in the unloading/reloading portions. As expected, Case 2 with bond-slip shows more flexible results (i.e., more pinching behavior as seen in the experiment) than Case 1.

In Figure 7(c), the effect of the implemented unloading/reloading scheme is compared with a simple secant unloading/reloading scheme. In this case where the structural behavior is dominated by the steel behavior, the effect of the ECC unloading/reloading model is not significant. Although as expected, the secant scheme results in less permanent damage at zero stress.

The cracking pattern at 5% drift from the experiment is compared with principal strains at 5% drift from the simulation (Case 1) in Figure 8. In Figure 8(b), the principal strain contour between the cracking strain (ε_{t0}) and the peak strain (ε_{tp}) is plotted. Figure 8(b) shows that strain is concentrated at the base of the beam but most areas of the beam did exceed the cracking strain (ε_{t0}). The simulation predicts that cracking is well distributed throughout the beam as was the case in the experiment.

4.3 *Simulation of Specimen 2 – ECC with FRP reinforcement*

The same modeling approach used for Specimen 1 is also used for Specimen 2. However, linear elastic

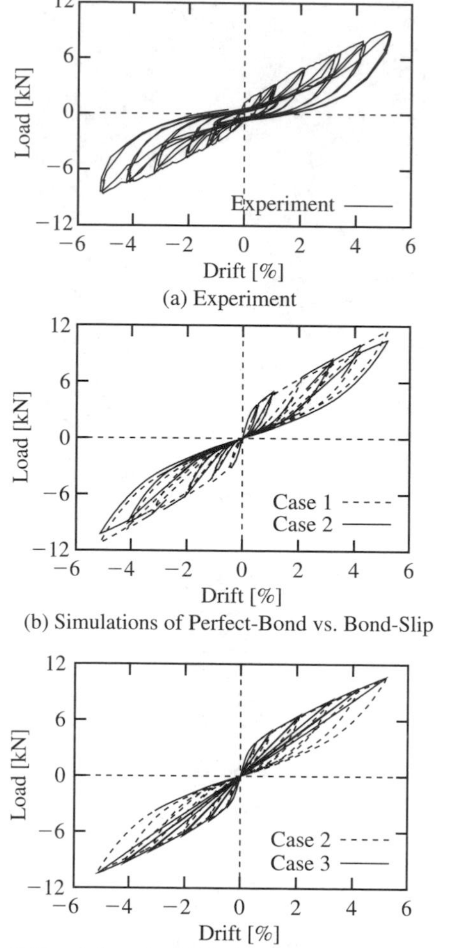

Figure 9. Results comparisons (load versus drift, Specimen 2).

material properties for the FRP reinforcement are assigned to the longitudinal reinforcement instead of steel properties. The use of elastic material properties with no failure cut-off for the FRP is reasonable since the strength of the FRP is not reached during the experiments or simulations. Three simulation cases with the same input parameter variations as in Specimen 1 (Table 2) are presented in Figure 9.

Initial stiffness from the simulations are determined at 0.2% drift. The stiffness of the all simulation models are about 30% higher than the experimental results. The peak strengths at 5% drift for Specimen 2 are 16 to 26% higher than those from the experiment. The higher peak strength from the simulations may be a result of not coupling the tensile and compressive damage in the ECC model. Coupling this damage would allow for further strength degradation of the material subjected to cyclic behavior.

As shown in Figure 9(b), the response from the simulation becomes more flexible when bond-slip is included. However, bond-slip in this case does not cause more pinching. Here, the pinching is dominated by the linear elastic response of the FRP reinforcement. In all cases, the hysteretic energy dissipation is somewhat underpredicted by the simulations.

In Figure 9(c), the effect of the modified unloading/reloading scheme compared with the secant unloading/reloading scheme can be clearly noticed. Since the FRP reinforcement does not yield, all of the nonlinearity in the beam response is due to the ECC constitutive model. This is consistent with the simulation results shown in Figure 9(c). The energy dissipation from the secant unloading/reloading model is only half of that from the modified unloading/reloading model. Also, in the secant unloading/reloading model, the unloading/reloading paths are identical, which is very different from the experimental results.

The proposed modified unloading/reloading scheme is essential in representing the realistic behavior of ECC structural members when the FRP reinforcement is used. In general, a model with more realistic unloading/reloading for the ECC is necessary where the reinforcement remains linear-elastic, such as with FRP and/or unbonded post-tensioned steel. It has been shown for example, that the proposed model can be applied to accurately simulate more complicated structural components such as unbonded post-tensioned segmental concrete bridge columns (Yoon and Billington 2002).

5 CONCLUSIONS

A co-rotational total strain-based constitutive model for ductile fiber-reinforced cementitious composites is developed. The accuracy of the model is verified by simulating structural component experiments under cyclic load. Based on the simulations, it is concluded that the model can be used for the simulation of ECC cantilever beams with reasonable accuracy. Further calibration of the simulation model will improve its accuracy. In particular, an unloading and reloading scheme that does not have to pass through the origin and the use of a more accurate compressive envelope curve will improve accuracy.

Accurate characterization of the hysteretic energy dissipation from ECC is necessary to predict the cyclic and seismic behavior of structural components. The developed model becomes most important for simulations where the ECC response heavily influences the structural response, for example when ECC structural components are combined with reinforcement that will behave predominantly in a linear elastic manner.

To improve the accuracy of the model, coupling of the tensile and compressive damage in each loading direction is suggested for further work. Such a model should be calibrated to more in-depth multi-axial ECC material tests to validate the accuracy.

ACKNOWLEDGEMENTS

The constitutive model is implemented using a user-supplied material subroutine of DIANA Release 7.2. This research was partially funded by the National Science Foundation under grant CMS-9984127 and by the Multi-Disciplinary Center for Earthquake Engineering Research (MCEER) in Buffalo, NY. MCEER is an Earthquake Engineering Research Center under the Engineering Research Centers Program of the National Science Foundation. Prof. Victor Li of the University of Michigan at Ann-Arbor and Prof. Gregor Fischer of the University of Hawaii are gratefully acknowledged for sharing experimental data and useful discussions.

REFERENCES

Aycardi, L., J. Mander, and A. Reinhorn (1992). Seismic resistance of reinforced concrete frame structures designed only for gravity loads: Part II – Experimental performance of subassemblages. Technical Report NCEER-92-0028, NCEER, Buffalo, NY.

Ayoub, A. and F. C. Filippou (1999). Mixed formulation of bond-slip problems under cyclic loads. *Journal of Structural Engineering, ASCE 125*(6), 661–671.

Bracci, J., A. Reinhorn, and J. Mander (1992). Seismic resistance of reinforced concrete frame structures designed only for gravity loads: Part III – Experimental performance and analytical study of a structural model. Technical Report NCEER-92-0029, NCEER, Buffalo, NY.

D'Ambrisi, A. and F. Filippou (1997). Correlation studies on an RC frame shaking-table specimen. *Earthquake Engineering and Structural Dynamics 26*, 1021–1040.

El-Attar, A., R. White, and P. Gergely (1991). Shake table test of a 1/8 scale three-story lightly reinforced concrete building. Technical Report NCEER-91-0018, NCEER, Buffalo, NY.

Feenstra, P. H., J. G. Rots, A. Arnesen, J. G. Teigen, and K. V. Hoiseth (1998). A 3D constitutive model for concrete based on corotational concept. In R. de Borst, N. Bićanić, H. Mang, and G. Meschke (Eds.), *Computational Modelling of Concrete Structures*, pp. 13–22. Proceedings of the EURO-C 1998 Conference on Computational Modelling of Concrete Structures, Austria.

Fischer, G. and V. C. Li (2002). Deformattion behavior of FRP reinforced ECC flexural members under reversed cyclic loading conditioned. *ACI Structural Journal*. Accepted, Feb.

Han, T.-S., P. H. Feenstra, and S. L. Billington (2002). Simulation of highly ductile fiber-reinforced cement-based composites. *ACI Structural Journal*. Accepted. Nov.

Kabele, P. (2001). Assessment of structural performance of engineered cementitious composites by computer simulation. Technical report, CTU Report 4, Vol. 5, Prague.

Kesner, K. E. and S. L. Billington (2001). Investigation of ductile cement based composites for seismic strengthening and retrofit. In R. de Borst, J. Mazars, G. Pijaudier-Cabot, and J. van Mier (Eds.), Fracture Mechanics of Concrete Structures, pp. 65–72. Proceedings of the Fourth International Conference on Fracture Mechanics of Concrete and Concrete Structures, Cachan, France, May – June.

Kwan, W.-P. and S. L. Billington (2001). Simulation of stuctural concrete under cyclic load. *ASCE Journal of Structural Engineering 127*(12), 1391–1401.

Li, V. C. (1998). Engineered cementitious composites – tailored composites through micromechanical modeling. In N. Banthia, A. Bentur, and A. Mufti (Eds.), *Fiber Reinforced Concrete: Present and the future*, pp. 64–97. Canadian Society for Civil Engineering, Montreal.

Li, V. C. and C. Leung (1992). Steady-state and multiple cracking of short random fiber composites. *Journal of Engineering Mechanics 118*(11), 2246–2264.

Marshall, D. B. and B. Cox (1988). A J-integral method for calculating steady-state matrix cracking stresses in composites. *Mechanics of Materials 7*, 127–133.

Marshall, D. B., B. Cox, and A. Evans (1985). The mechanics of matrix cracking in brittle-matrix composites. *Acta Metallographa 33*(11), 2013–2021.

Selby, R. G. and F. J. Vecchio (1993, November). Three-dimensional constitutive relations for reinforced concrete. Technical Report 93-02, Department of Civil Engineering, University of Toronto.

Shahrooz, B. M. and J. P. Moehle (1987). Experimental study of seismic response of R.C. setback buildings. Technical Report UCB/EERC-87/16, EERC, Berkeley, CA.

Yoon, J. K. and S. L. Billington (2002). Experimental and Numerical Studies of Precast Unbonded Post-tensioned Bridge Columns with Engineered Cementitious Composites. Technical Report 02-03, Cornell University, Ithaca, NY.

Computational Modelling of Concrete Structures, Bićanić et al. (eds)
© 2003 Swets & Zeitlinger, Lisse, ISBN 90 5809 536 3

Development of an inverse procedure for parameters estimates of numerical models

C. Iacono & L.J. Sluys
Faculty of Civil Engineering and Geosciences, Delft University of Technology, Delft, The Netherlands

J.G.M. van Mier
ETH Hönggerberg, Institute for Building materials, Zürich, Switzerland

ABSTRACT: Discrete and continuum models, used to study fracture of concrete and rock, rely on an accurate identification of the model parameters. The determination of the parameters, that cannot be measured directly in laboratory experiments, requires the development of a procedure in the framework of Inverse Problems. The Kalman filter methodology is adopted as inverse technique to identify the parameters in the gradient-enhanced continuum damage model. The identification process exploits the experimental data generated in the cable loaded uniaxial tensile tests performed on single-edge-notched sandstone specimens in the Microlab of Delft University of Technology. Various aspects of the Kalman filter technique have been discussed.

1 INTRODUCTION

In numerous scientific fields researchers are interested in developing models that can describe a physical system. In the course of time these models can be updated and improved as consequence of new experimental or computational tools or by means of a deeper understanding of the phenomena. Updating may be quantitative or qualitative. The Inverse problem theory concerns with the development of rules for quantitative updating.

Let S represent a physical system described by a model that contains a set of n_x model parameters assembled in a vector $\mathbf{x}$. The response of S to an external perturbation can be represented by n_y observable or measurable quantities collected in a vector $\mathbf{y}$, that depend on the set of model parameters $\mathbf{x}$ and on the perturbation applied on the system. The solution of the *forward problem* means the prediction of the response represented by $\mathbf{y}$, given $\mathbf{x}$ and the perturbation. The solution of the *inverse problem* means the determination of the model parameters $\mathbf{x}$, given the perturbation and the response (Tarantola 1987; Bui 1994).

The inverse problem treated in the present paper represents the parameter identification of an elasticity based gradient-enhanced continuum damage model used to describe fracture phenomena in quasi-brittle materials (concrete and rock). The Kalman filter technique is adopted as inverse procedure applied to the case of the cable loaded uniaxial tensile test.

A brief presentation of the gradient-enhanced continuum damage model is presented in Section 2. Section 3 is devoted to the Kalman filter technique. In Section 4 the numerical results of the studied parameters identification problem are discussed. The concluding remarks are given in Section 5.

2 CONSTITUTIVE MODEL

2.1 *Local damage model*

The basis of the constitutive model is formed by an isotropic, quasi-brittle elasticity based damage model.

The damage is represented by a scalar variable ω whose range is the interval [0,1]: if $\omega = 0$ no damage has developed in the material that remains in the initial virgin state, while $\omega = 1$ corresponds to the completely damaged material without any residual load-carrying capacity. The damage scalar variable ω is responsible for the degradation of the elastic stiffness according the following classical stress-strain relation (Lemaitre and Chaboche 1990):

$$\boldsymbol{\sigma} = (1-\omega)\boldsymbol{D}^{el}\boldsymbol{\varepsilon} \quad (1)$$

in which $\boldsymbol{D}^{el}$ represents the matrix of the virgin elastic stiffness moduli.

The beginning of the damage process is determined by the model parameter κ_i that represents a strain threshold for damage evolution. If an invariant measure of strain, the equivalent strain ε_{eq}, is defined, then for $\varepsilon_{eq} < \kappa_i$ the response of the material is linear elastic while for $\varepsilon_{eq} = \kappa_i$ the damage process starts.

A history variable κ can be defined, representing the most severe deformation undergone by the material:

$$\kappa(\varepsilon_{eq}) = max(\varepsilon_{eq}, \kappa_{max}) \tag{2}$$

where κ_{max} is the maximum value of equivalent strain occurred in the material.

The damage variable ω is a function of the history variable κ according a damage evolution law that governs the growth of damage

$$\omega = \omega(\kappa) \tag{3}$$

Whether damage growth is possible is determined on the basis of a loading function expressed in terms of the equivalent strain:

$$f(\varepsilon_{eq}) = \varepsilon_{eq} - \kappa(\varepsilon_{eq}) \tag{4}$$

The magnitude of the loading function is governed by the Kuhn-Tucker relations

$$f\dot{\kappa} = 0 \quad f \leq 0 \quad \dot{\kappa} \geq 0 \tag{5}$$

For $f < 0$ no growth of damage can take place ($\dot{\kappa} = 0$) and the response remains linear elastic. Damage increases ($\dot{\kappa} > 0$) only if the strain state satisfies the equation $f = 0$ and during this increment the consistency condition $\dot{f} = 0$ must be satisfied as well.

The complete definition of the model requires the explicit definition of the equivalent strain and of the damage evolution law (Eq. 3). Different models exist (Peerlings 1999) and here only those used in the numerical simulations discussed in the Section 4 are given. For the equivalent strain the *modified von Mises* definition (Vree et al. 1995) reads:

$$\varepsilon_{eq} = \frac{(\eta - 1)I_1'}{2\eta(1 - 2\nu)} + \frac{1}{2\eta}\sqrt{\frac{(\eta - 1)^2 I_1'^2}{(1 - 2\nu)^2} + \frac{12\eta J_2'}{(1 + \nu)^2}} \tag{6}$$

where I_1' and J_2' are the first invariant of the strain tensor and the second invariant of the deviatoric strain tensor respectively, given by

$$I_1' = \varepsilon_{xx} + \varepsilon_{yy} + \varepsilon_{zz} \tag{7}$$

$$J_2' = \frac{({\varepsilon_{xx}}^2 + {\varepsilon_{yy}}^2 + {\varepsilon_{zz}}^2 - \varepsilon_{xx}\varepsilon_{yy} - \varepsilon_{yy}\varepsilon_{zz} - \varepsilon_{zz}\varepsilon_{xx})}{3} + {\varepsilon_{xy}}^2 + {\varepsilon_{yz}}^2 + {\varepsilon_{zx}}^2 \tag{8}$$

and η is a model parameter given by the ratio of the compressive and the tensile strength of the material: $\eta = f_{cc}/f_{ct}$. The following *exponential softening* damage evolution law has been chosen

$$\omega = 1 - \frac{\kappa_i}{\kappa}[1 - \alpha + \alpha e^{-\beta(\kappa - \kappa_i)}] \tag{9}$$

where α and β are two additional model parameters that govern the softening curve. As schematicaly represented in Fig. 1, α determines the residual stress of the damaged material, and β the negative stiffness of the softening branch.

The standard local damage model presented above can reproduce the global softening behaviour of the material, but it does not contain information about the micro-structure of the material. This lack of data is reflected in what is commonly known as *mesh dependence* of the numerical simulations obtained with such a model. The width of the zone where damage localises depends on the finite element size that is used: with decreasing element size the strain localisation band decreases as well. In the limit case the growth of damage tends to localise in a portion of material of zero volume, the crack propagation becomes instantaneous and no work is necessary to produce the fracture process (zero fracture energy). This non physical behaviour, predicted by the local damage theory, is mathematically explained by a local loss of ellipticity of the set of partial differential equations that governs the rate of deformation. Mechanically the phenomenon can be seen as lack of a localization limiter related to the micro-structural properties of the material. The introduction of this localization limiter can be realized using the nonlocal approach.

2.2 *Nonlocal model: gradient-enhanced formulation*

The basic idea of the nonlocal model is that a state variable in a material point depends on state variables (and/or history) in the considered point (local action principle) and in the neighboring points. The size of this neighborhood or interaction radius depends on the characteristic micro-structure of the material and it is related to an additional model parameter known as the

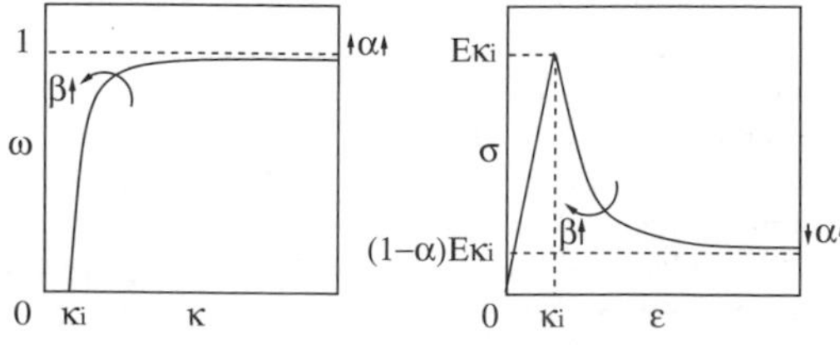

Figure 1. (a) Exponential softening damage evolution law (b) uniaxial stress-strain curve.

length scale that meets the necessity to introduce information on the microlevel in the constitutive model. The introduction of this principle of non-locality in the constitutive equations was first considered by Kröner (1967) and Eringen and Edelen (1972) and then applied to continuum damage mechanics by Bazant and Pijaudier-Cabot (1988) (1987).

In the nonlocal damage theory a nonlocal equivalent Strain $\bar{\varepsilon}_{eq}$ can be defined as a spatially averaged quantity:

$$\bar{\varepsilon}_{eq}(\boldsymbol{x}) = \frac{1}{\Psi(\boldsymbol{x})}\int_{\Omega}\Psi(\boldsymbol{y};\boldsymbol{x})\varepsilon_{eq}(\boldsymbol{y})d\Omega \tag{10}$$

where $\boldsymbol{y}$ points to the positions of the infinitesimal volume $d\Omega$ and $\Psi(\boldsymbol{x})$ is defined as

$$\Psi(\boldsymbol{x}) = \int_{\Omega}\Psi(\boldsymbol{y};\boldsymbol{x})d\Omega \tag{11}$$

As weight function $\Psi(\boldsymbol{y};\boldsymbol{x})$ the homogeneous and isotropic Gauss distribution is usually adopted:

$$\psi(\rho) = \frac{1}{(2\pi)^{3/2}l^3}exp\Big[-\frac{\rho^2}{2l^2}\Big] \tag{12}$$

where $\rho = |\boldsymbol{y} - \boldsymbol{x}|$ and the factor $(2\pi)^{-3/2}l^{-3}$ normalizes the weight function such that on R^3

$$\int_{R^3}\psi(\rho)d\Omega = 1 \tag{13}$$

The length parameter l in the Eq. 12 gives the size of the neighborhood that determines the nonlocality of the model. The constitutive equations are similar to those presented in Section 2.1 replacing properly the local equivalent strain $\varepsilon_{eq}(\mathbf{x})$ with the nonlocal counterpart $\bar{\varepsilon}_{eq}(\mathbf{x})$.

For sufficiently smooth ε_{eq}-fields, it is possible to expand the local equivalent strain into a Taylor series. Substituting the Taylor series into the integral Eq. 10, carrying out some calculus and neglecting terms of order four and higher, the following differential equation can be obtained (Bazant et al. 1984), (Lasry and Belytschko 1988), (Peerlings et al. 1995), (Peerlings et al. 1996), (Peerlings 1999):

$$\bar{\varepsilon}_{eq} - c\nabla^2\bar{\varepsilon}_{eq} = \varepsilon_{eq} \tag{14}$$

where it has been set $c = (1/2)l^2$ and $\nabla^2 = \Sigma_i \partial^2/\partial x_i^2$.

Equation 14 is the differential approximation of the integral Eq. 10 and, since the nonlocal variable $\bar{\varepsilon}_{eq}$ depends implicitly on the corresponding local variable ε_{eq}, it is referred to as the *implicit gradient-enhancement.*

The solution of Eq. 14 requires the boundary conditions. In most cases the natural boundary condition is used:

$$\boldsymbol{n}^T\nabla\bar{\varepsilon}_{eq} = 0 \tag{15}$$

The diffusion problem (Eq. 14) is solved, with the natural boundary condition (Eq. 15), in addition to the classical equilibrium problem. This set represents a fully coupled problem, in the sense that the solution of the diffusion equation is not possible without the solution of the equilibrium equations and vice versa.

The numerical implementation of the gradient-enhanced damage model in the finite element framework requires the spatial discretisation, by means of different shape functions, of the displacement field $\boldsymbol{u}$ and of the non local equivalent strain $\bar{\varepsilon}_{eq}$ (Peerlings 1999).

The model parameters can be assembled in the following vector:

$$\mathbf{x}^T = [\ E \quad \nu \quad \kappa_i \quad \alpha \quad \beta \quad c \quad \eta\ ] \tag{16}$$

where

E = Young's modulus
ν = Poisson's ratio
κ_i = strain threshold for damage initiation
α = softening curve parameter (related to the residual stress)
β = softening curve parameter (related to the stiffness of softening branch)
c = gradient parameter
η = ratio of compressive and tensile strength f_{cc}/f_{ct}

The inverse problem reduces to the identification of the $\boldsymbol{x}$ vector.

3 KALMAN FILTER TECHNIQUE

The basic notions of the mathematical formulation of the Kalman filter procedure are presented in Bolzon et al. (2002), while for detailed treatments see e.g. Kailath et al. (2000), Tarantola (1987), Bittanti et al. (1984), Catlin (1989), Bui (1994).

Let $\mathbf{y}^t_{\text{exp}}$ collect the experimental measurements of the n_{y} observable quantities at a certain "instant" t, representing the real response of the system S to an external perturbation applied on it (see Fig. 2). The variable t represents a time ordering variable and it can also be the amplitude of the imposed external action on the system. If the n_{y} quantities are experimentally measured at different "instants" t, then the laboratory test gives a flow of experimental data $\mathbf{y}^t_{\text{exp}}$ for t = 1, 2, 3 … n.

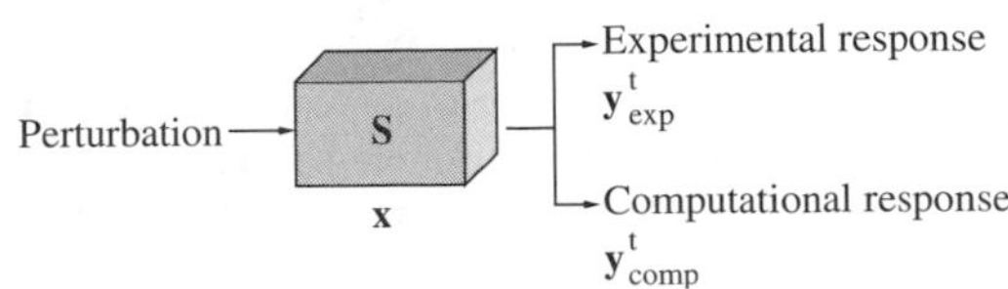

Figure 2. Schematic representation of the system response on a perturbation.

On the other hand the numerical model of S, given the model parameters vector $\mathbf{x}$, is able to compute at every "instant" t, as a solution of the forward problem, numerical values for the n_y quantities that can be collected in the vector $\mathbf{y}^t_{comp}$. This vector is the corresponding computational counterpart of $\mathbf{y}^t_{exp}$ and it depends on the models parameters $\mathbf{x}$ according to the following general relation:

$$\mathbf{y}^t_{comp} = \mathbf{h}_t(\mathbf{x}) \tag{17}$$

$\mathbf{h}_t(\mathbf{x})$ will be referred to as the *forward operator* and for the gradient-enhanced damage model the forward operator will be nonlinear. Basically the inverse problem of identifying the model parameter vector $\mathbf{x}$ is based on the idea of minimizing, at every t, the difference between the vectors $\mathbf{y}^t_{exp}$ and $\mathbf{y}^t_{comp}$.

The Kalman filter technique (referred to as KF) solves the parameters identification problem in a statistical context. The following assumptions are considered herein:

- all the random variable vectors involved follow a Gaussian distribution
- the mathematical model, i.e. the forward operator $\mathbf{h}_t(\mathbf{x})$, is considered as deterministic
- measurements uncertainties, represented by the vector $\mathbf{v}_t$, are considered as Gaussian white noises.

In other words, the measurements of the n_y quantities, collected in the $\mathbf{y}^t_{exp}$ vector, at every instant t, are characterized by an uncertainty vector $\mathbf{v}_t$ for which we assume:

$$f(\mathbf{v}_t) = ((2\pi)^{n_y}\det(\mathbf{C}^t_{exp}))^{-\frac{1}{2}} \exp\left\{-\frac{1}{2}\mathbf{v}_t^T(\mathbf{C}^t_{exp})^{-1}\mathbf{v}_t\right\} \tag{18}$$

$$E(\mathbf{v}_t) = \mathbf{0} \quad E(\mathbf{v}_t\mathbf{v}_t^T) = \mathbf{C}^t_{exp} \quad E(\mathbf{v}_t\mathbf{v}_s^T) = \mathbf{0} \quad s \neq t \tag{19}$$

where $E(\cdot)$ represents the averaging operator and $E((\cdot E(\cdot))(\cdot - E(\cdot))^T)$ the covariance operator.

Since the forward operator has been assumed as deterministic, the measurement noise $\mathbf{v}_t$ determines the difference between the experimental and computed observable quantities. Namely, considering Eq. 17:

$$\mathbf{y}^t_{exp} = \mathbf{y}^t_{comp} + \mathbf{v}_t = \mathbf{h}_t(\mathbf{x}) + \mathbf{v}_t \tag{20}$$

Besides the experimental and computed data, the KF procedure relies on a third information source: an initial "a priori" estimate of the model parameter vector $\mathbf{x}$. This initial guess (Bayesian approach) can be chosen by a hypothetical expert or can be based on engineering experience of the material under consideration. Also the a priori information on the n_x parameters is assumed to be statistically characterized by a Gaussian distribution with $\mathbf{x}_0$ mean and $\mathbf{C}_0$ covariance matrix:

$$f_{\mathbf{X}}(\mathbf{x}) = ((2\pi)^{n_x}\det(\mathbf{C}_0))^{-\frac{1}{2}}$$
$$\exp\left\{-\frac{1}{2}(\mathbf{x}-\mathbf{x}_0)^T\mathbf{C}_0^{-1}(\mathbf{x}-\mathbf{x}_0)\right\} \tag{21}$$

Substitution of the measurement noise $\mathbf{v}_t$ in Eq. 18, as difference between the experimental vector and the forward operator according to Eq. 20, gives the following conditional probability density of experimental data with respect to the parameters:

$$f_{\mathbf{y}|\mathbf{x}}(\mathbf{x}, \mathbf{y}^t_{exp}) = ((2\pi)^{n_y}\det(\mathbf{C}^t_{exp}))^{-\frac{1}{2}}$$

$$\exp\left\{-\frac{1}{2}(\mathbf{y}^t_{exp} - \mathbf{h}_t(\mathbf{x}))^T(\mathbf{C}^t_{exp})^{-1}(\mathbf{y}^t_{exp} - \mathbf{h}_t(\mathbf{x}))\right\} \tag{22}$$

Applying Bayes' theorem results in the following conditional probability density (Tarantola 1987):

$$f_{\mathbf{x}|\mathbf{y}}(\mathbf{x}, \mathbf{y}^t_{exp}) = \frac{f_{\mathbf{y}|\mathbf{x}}(\mathbf{x}, \mathbf{y}^t_{exp})f_{\mathbf{X}}(\mathbf{x})}{\mu} \tag{23}$$

where $\mu \equiv f_y(\mathbf{y}^t_{exp})$ does not depend on the model parameter vector $\mathbf{x}$, and can be seen as a normalizing factor. If Eq. 22 and Eq. 21 are substituted in Eq. 23, the following relation is obtained:

$$f_{\mathbf{x}|\mathbf{y}}(\mathbf{x}, \mathbf{y}^t_{exp}) = \frac{1}{\mu}\exp\{-S_t(\mathbf{x})\} \tag{24}$$

in which

$$2S_t(\mathbf{x}) = (\mathbf{y}^t_{exp} - \mathbf{h}_t(\mathbf{x}))^T(\mathbf{C}^t_{exp})^{-1}(\mathbf{y}^t_{exp} - \mathbf{h}_t(\mathbf{x})) +$$
$$(\mathbf{x}-\mathbf{x}_0)^T\mathbf{C}_0^{-1}(\mathbf{x}-\mathbf{x}_0) \tag{25}$$

The unknown parameter vector $\mathbf{x}$ corresponds to the maximum conditional probability density. In other words, the solution of the identification problem is the vector $\mathbf{x}$ of thc following optimization problem:

$$\max_{\mathbf{X}}\left\{f_{\mathbf{x}|\mathbf{y}}(\mathbf{x}, \mathbf{y}^t_{exp})\right\} = \min_{\mathbf{X}}\{S_t(\mathbf{x})\} \tag{26}$$

If the forward operator $\mathbf{h}_t(\mathbf{x})$ is linear, the a posteriori conditional probability density $f_{\mathbf{x}|\mathbf{y}}(\mathbf{x}, \mathbf{y}^t_{exp})$ given by Eq. 24, is a Gaussian (normal) distribution. However, this is not the case for the gradient-enhanced damage model considered here. In this case an iterative inverse procedure can still be formulated, introducing a step-by-step linearization of the forward operator and assuming a normal distribution of $f_{\mathbf{x}|\mathbf{y}}(\mathbf{x}, \mathbf{y}^t_{\text{exp}})$ within each step. For the linearization of the forward operator, the 1st-order Taylor expansion can be used around the initial guess of the model parameters $\mathbf{x}_0$:

$$\mathbf{h}_t(\mathbf{x}) \simeq \mathbf{h}_t(\mathbf{x}_0) + \mathbf{L}_0(\mathbf{x} - \mathbf{x}_0) \tag{27}$$

having set:

$$\mathbf{L}_0 \equiv \frac{\partial \mathbf{h}_t}{\partial \mathbf{x}}(\mathbf{x}_0, t) \tag{28}$$

Substituting Eq. 27 into Eq. 25 and subsequently into Eq. 24 and using matrix algebra rules and assuming that

$$\begin{aligned} \hat{\mathbf{x}} &= \hat{\mathbf{x}}_t & \hat{\mathbf{C}} &= \hat{\mathbf{C}}_t \\ \mathbf{x}_0 &= \hat{\mathbf{x}}_{t-1} & \mathbf{C}_0 &= \hat{\mathbf{C}}_{t-1} \end{aligned} \tag{29}$$

results in the following set of equations.

$$\mathbf{L}_t = \frac{\partial \mathbf{h}_t}{\partial \mathbf{x}}(\hat{\mathbf{x}}_{t-1}, t) \tag{30}$$

$$\mathbf{K}_t = \hat{\mathbf{C}}_{t-1}\mathbf{L}_t^T[\mathbf{L}_t\hat{\mathbf{C}}_{t-1}\mathbf{L}_t^T + \mathbf{C}^t_{\text{exp}}]^{-1} \tag{31}$$

$$\hat{\mathbf{x}}_t = \hat{\mathbf{x}}_{t-1} + \mathbf{K}_t(\mathbf{y}^t_{\text{exp}} - \mathbf{h}_t(\hat{\mathbf{x}}_{t-1})) \tag{32}$$

$$\hat{\mathbf{C}}_t = \hat{\mathbf{C}}_{t-1} - \mathbf{K}_t\mathbf{L}_t\hat{\mathbf{C}}_{t-1} \tag{33}$$

The tangent operator $\mathbf{L}_t$ is denoted as *sensitivity matrix* and $\mathbf{K}_t$ as *gain matrix*.

Vector $\hat{\mathbf{x}}$ is the mean value vector of the normal distribution $f_{\mathbf{x}|\mathbf{y}}$ given by Eq. 24 and it is the solution of the optimization problem represented by Eq. 26, while $\hat{\mathbf{C}}$ is the corresponding covariance matrix.

Eqs. 32–33 define a recursive procedure (schematically represented by the "KF" box in Fig. 3) that, filtering along the sequence of experimental data, gives at each step t a better estimate of the mean value of the model parameters and the corresponding covariance matrix. The initialization of the iterative scheme (i.e. for $t = 1$) requires the initial guess $\hat{\mathbf{x}}_0$ and $\hat{\mathbf{C}}_0$. The a priori information on the model parameters is the input also for the computational model (represented by the "Comp. Mod." box in Fig. 3) in order to calculate the $\mathbf{y}^1_{\text{comp}}$ vector as solution of the forward analysis. The three information sources $\mathbf{y}^1_{\text{exp}}$, $\mathbf{y}^1_{\text{comp}}$ and $\hat{\mathbf{x}}_0$ are processed by the Kalman filter scheme and, as a result, a new improved estimation $\hat{\mathbf{x}}_1$ is obtained and the next step (i.e. $t = 2$) can be analyzed.

The Kalman filter process applied to the flow of experimental data $\mathbf{y}^t_{\text{exp}}$ with their uncertainty $\mathbf{C}^t_{\text{exp}}$, starting from the a priori information $\hat{\mathbf{x}}_0$ with its uncertainty $\hat{\mathbf{C}}_0$, can symbolically be represented by:

$$[\hat{\mathbf{x}}, \hat{\mathbf{C}}] = \mathcal{F}(\hat{\mathbf{x}}_0, \hat{\mathbf{C}}_0; \mathbf{y}^t_{\text{exp}}, \mathbf{C}^t_{\text{exp}}, t = 1, 2...n) \tag{34}$$

If the forward operator $\mathbf{h}_t(\mathbf{x})$ is linear, it can be proven (Catlin 1989) (Kalman 1960) that the final estimates of the model parameters $[\hat{\mathbf{x}}, \hat{\mathbf{C}}]$ do not depend on the initial guess $[\hat{\mathbf{x}}_0, \hat{\mathbf{C}}_0]$. If, on the contrary, the forward operator is non-linear, even if not rigorously proven, this independence can be valid for the asymptotic result $[\bar{\mathbf{x}}, \bar{\mathbf{C}}]$ of an iterative application of the KF procedure on the same experimental data at the same KF steps (see Fig. 3):

$$[\bar{\mathbf{x}}, \bar{\mathbf{C}}] = \mathcal{F}^{N\to\infty}(\hat{\mathbf{x}}_{N-1}, \hat{\mathbf{C}}_{N-1}; \mathbf{y}^t_{\text{exp}}, \mathbf{C}^t_{\text{exp}}, t = 1, 2...n) \tag{35}$$

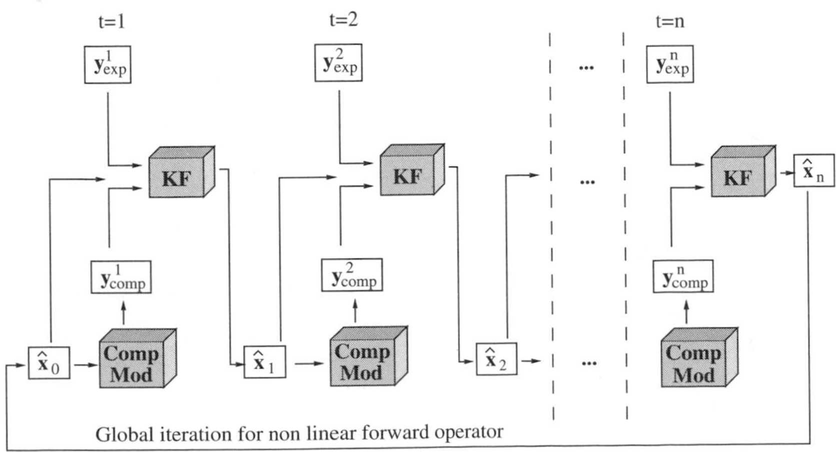

Figure 3. Parameters identification procedure.

where $\hat{\mathbf{x}}_{N-1}$ and $\hat{\mathbf{C}}_{N-1}$, final estimates of the $(N-1)$th KF process, are used as initial guess of the Nth KF process. For a finite number N of global iterations of the KF process the resulting estimate of the model parameters can depend on the initial guess since multiple local minima and corresponding attraction basins may exist. The KF procedure, in fact, implicitly minimizes a norm of the difference between the experimental and computed data which is a non-convex function of the parameter vector $\boldsymbol{x}$, and therefore possible local minima may exist. The initial guess should be selected such that a point within the right attraction basin, and also as close as possible to the absolute minimum is chosen in order to speed up the convergence of the method.

4 NUMERICAL APPLICATIONS

The parameter identification problem of the gradient-enhanced continuum damage model through the KF technique is solved exploiting the experimental data of the cable loaded uniaxial tensile test. These tests have been performed on single-edge-notched sandstone specimens in the Microlab of Delft University of Technology (van Mier and Shi 2002). The specimen geometry and dimensions are shown in Fig. 4.

The monotonically increasing external load, represented by t in the previous Section, can be identified with the relative vertical displacement of the opposite faces of the notch ("crack mouth opening displacement", CMOD). As an observable quantity, the force F applied at the cables is considered for each time-step t, i.e:

$$y^t_{\text{exp}} = [F^t_{\text{exp}}] \tag{36}$$

$$y^t_{\text{comp}} = h_t(\mathbf{x}) = [F^t_{\text{comp}}] \tag{37}$$

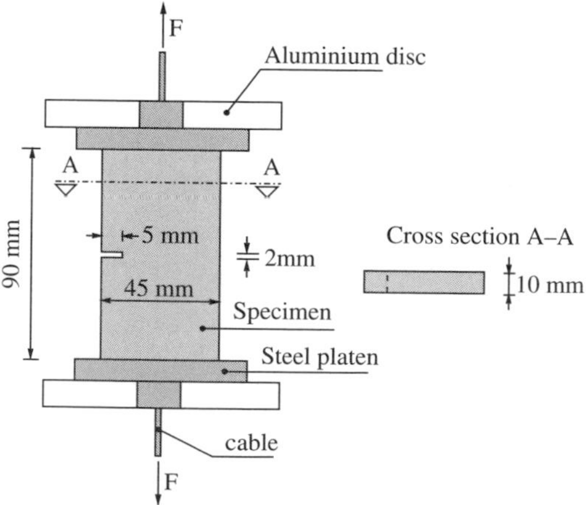

Figure 4. Specimen geometry and dimensions.

The measurement noises are taken proportional to the measured force F^t_{exp} considering measurement accuracies of ±5%:

$$C^t_{\text{exp}} = (\frac{0.05}{3} F^t_{\text{exp}})^2 \tag{38}$$

In Table 1 the experimental values for F and the relative covariance (computed according the Eq. 38) for each KF step t are listed.

The sensitivity matrix (see Eq. 30) is computed numerically according the following three points formula:

$$\frac{\partial h_{i,t}}{\partial x_j}((x_{1,t-1}), ..., (x_{n,t-1}), t) \simeq$$

$$\{h_i((x_{1,t-1}), ..., (x_{j,t-1} + \Delta x_j), ..., (x_{n,t-1}), t) -$$

$$h_i((x_{1,t-1}), ..., (x_{j,t-1} - \Delta x_j), ..., (x_{n,t-1}), t)\}/2\Delta x_j \tag{39}$$

Several forward analyses are needed for each KF time step t. In order to limit the computing time, only a subset of the model parameter vector (see Eq. 16) is identified by the KF procedure. The model parameters chosen for the KF identification process are the gradient parameter c and the softening curve parameter β, while the other model parameters are considered as a priori known and their values, collected in the Table 2, are measured by standard tests.

4.1 *Scalar KF procedure for c identification (KFI and KFII)*

The term "*scalar*" is used for a KF procedure for the identification of only one model parameter. Since

Table 1. Experimental values.

F_{exp}[N]	$C^t_{\text{exp}}[N^2]$	t	CMOD[mm]
50.107	0.70	1	0.007
46.526	0.60	2	0.037
29.715	0.25	3	0.067
20.694	0.12	4	0.097
14.021	0.05	5	0.127
10.580	0.03	6	0.157

Table 2. Values of the a priori known parameters.

	Specimen	Platens	Cables
E[MPa]	7000	210000	195000
ν	0.2	0.3	0.3
κ_i	0.3286×10^{-3}	–	–
α	0.96	–	–
η	12.0	–	–

also one observable quantity is taken into account, the entire KF procedure involves, in this case, only scalar quantities.

The first KF process, marked as KFI, concerns the identification of the gradient parameter, i.e.:

$$\hat{x} = [\hat{c}] \quad (40)$$

The β parameter is considered in this case as "a priori" known and its value is set to 150. The following values are chosen for the initial guess of the gradient parameter and its covariance (the last corresponding to an uncertainty of about 25%):

$$\hat{x}_0 = [\hat{c}_0] = 0.8 \text{ mm}^2 \quad (41)$$

$$\hat{C}_0 = 0.005 \text{ mm}^4 \quad (42)$$

In order to limit the computing time, six steps are considered for the KFI procedure (i.e. $t = 1, 2, 3, 4, 5, 6$). Three forward analysis for each KF step are necessary in order to numerically compute the derivatives of the forward operator $\mathbf{h}_t(\mathbf{x})$, considering a fixed variation of $\Delta x = 0.05 \text{ mm}^2$ for the gradient parameter.

The gradient parameter mean value estimates and the corresponding covariances during KFI are illustrated in Fig. 5 (square points). The c parameter estimate converges towards a constant value, while the covariance, related to the uncertainty on the parameter estimate, reduces progressively to zero.

In Fig. 6 (a) the global load-CMOD curves are plotted for each step of KFI (three curves per step). Since all the forward analyses are carried out sequentially without any interruption, the straight lines towards the origin have no meaning and they simply correspond to the analysis reinitialization. The experimental curve is fitted well by the numerical one computed using the KFI converged value of the gradient parameter.

An analogous KF identification process, KFII, is carried out starting from a different value of the a priori guess. The gradient parameter initial estimate is taken equal to 1.0 mm^2 and the covariance equal to 0.01 / mm^4.

As shown if Fig. 5, after the third step, the two KF process result in similar estimations of the parameter, with similar fitting of the global load-CMOD curve (see Fig. 6 (b)).

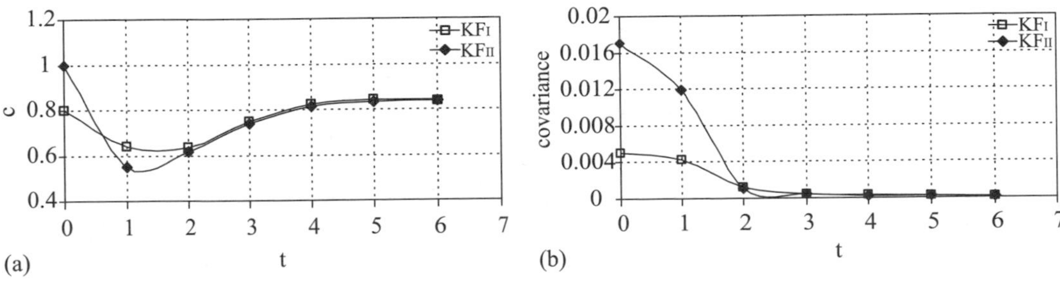

Figure 5. KFI and KFII estimations of the (a) c parameter and (b) covariance of c.

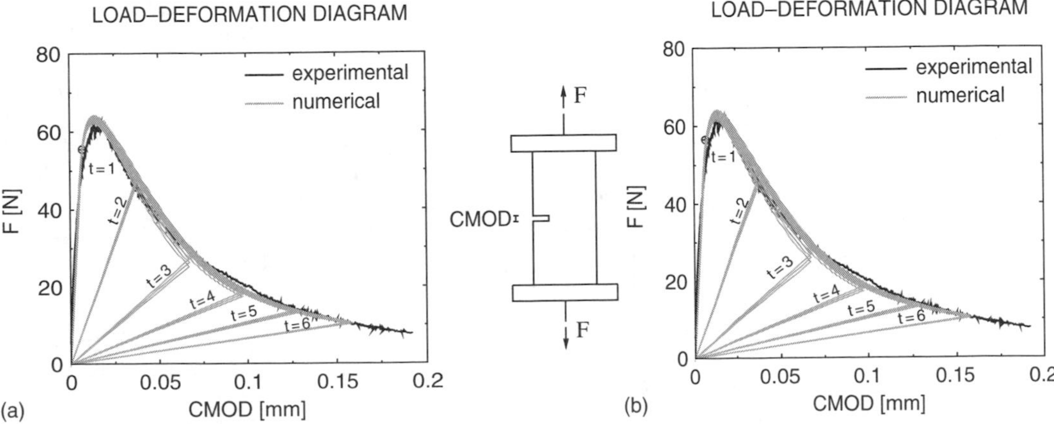

Figure 6. (a) KFI and (b) KFII fitting of the global Load-CMOD response.

4.2 *Scalar KF procedure for β identification (KF*III*)*

A "scalar" KF procedure, referred to as KFIII, is also applied for the identification of the β parameter. The gradient parameter c is considered in this case as a priori known and equal to 0.83 mm^2 (close to the previous KFs converged value). The β parameter initial estimate is taken equal to 100.0 and the covariance equal to 70.0, while a fixed variation $\Delta x = 10.0$ is applied to β at each of the three forward analyses.

The β parameter and covariance estimates during the KFIII procedure are plotted in Fig. 7. The a priori guess of β is underestimated resulting in a higher load value around the peak of the global load-CMOD curve (see Fig. 8 (a)). After the third KFIII step the final value of β is almost found. The procedure converges towards a constant value that is very close to the a priori fixed one of the two previous KF procedures.

4.3 *Bidimensional KF procedure for c and β identification (KF*IV*)*

The term "*bidimensional*" is used for a KF procedure for the identification of two model parameters. A bidimensional KF procedure, referred to as KFIV, is carried out for the identification of the c and β parameters simultaneously. In this case the vector $\mathbf{x}$ is defined as

$$\hat{\mathbf{x}}^T = \begin{bmatrix} \hat{c} & \hat{\beta} \end{bmatrix} \tag{43}$$

The initial guess for $\mathbf{x}$ and $\mathbf{C}$ are:

$$\hat{\mathbf{x}}_0^T = \begin{bmatrix} 1.0 & 200.0 \end{bmatrix} \tag{44}$$

$$\hat{\mathbf{C}}_0 = \begin{bmatrix} 0.017 & 0.0 \\ 0.0 & 270.0 \end{bmatrix} \tag{45}$$

Considering the following model parameter variation vector:

$$\Delta\mathbf{x}^T = \begin{bmatrix} 0.05 & 10.0 \end{bmatrix} \tag{46}$$

in order to numerically evaluate the sensitivity matrix, five forward analyses have been carried out for each KFIV step.

The c and β mean value estimates and the corresponding covariances during the KFIV process are plotted in Fig. 9. The simultaneous identification of the β parameter results in different c mean value estimates during KFIV in comparison with the ones obtained

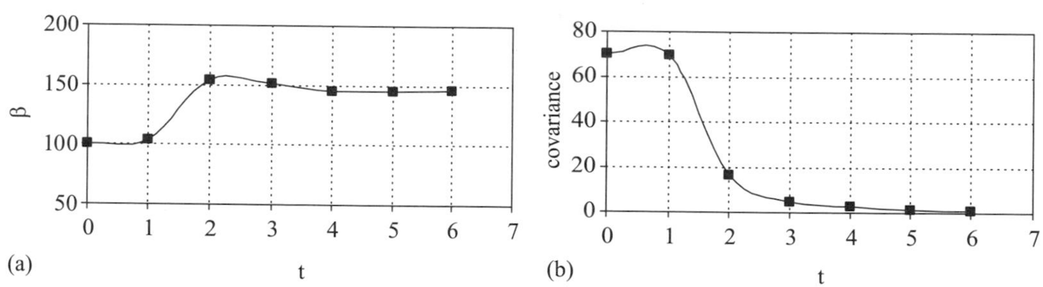

Figure 7. KFIII estimations of the (a) β parameter and (b) covariance of β.

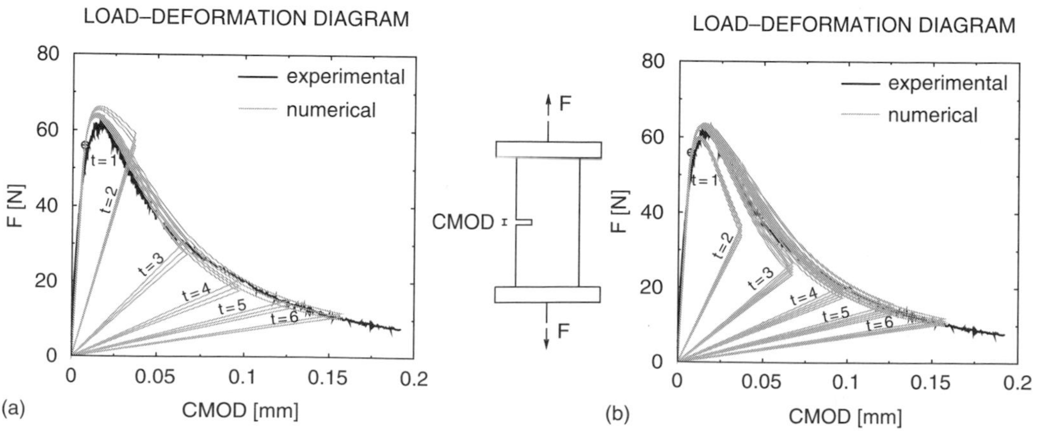

Figure 8. (a) KFIII and (b) KFIV fitting of the global Load-CMOD response.

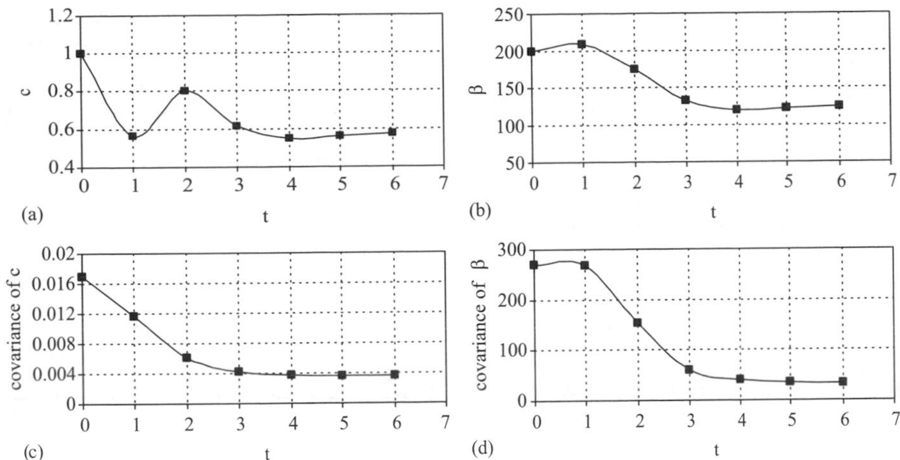

Figure 9. KFIV estimations of the (a) c parameter (b) β parameter (c) covariance of c and (d) covariance of β.

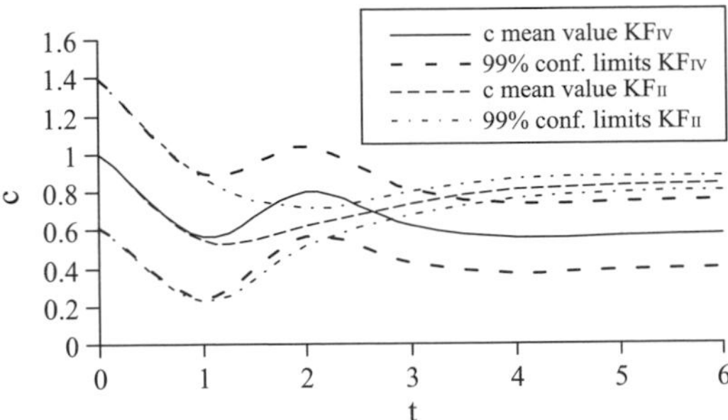

Figure 10. c mean values and corresponding 99% confidence limits for KFII and KFIV.

during KFII. It is worth noting that, for computing time reasons, a fixed number of KF steps has been considered. However in the case of the KFIV procedure the convergence of the identification process towards a constant value for c is slower than in KFI or KFII. The final value for the covariance of c, in fact, is one order of magnitude larger in the case of KFIV. Consequently a bigger uncertainty domain is present for KFIV, as illustrated in Fig. 10.

This suggests the necessity to increase the number of steps in the bidimensional identification process or to perform an iterative application of the KF procedure with the same experimental data according to Eq. 35.

In Fig. 8 (b) the global numerical and experimental load-CMOD curves are plotted.

5 CONCLUSIONS

The investigated KF inverse procedure is shown to be a powerful tool, which not only identifies the required model parameters, but also provides a quantitative assessment of the uncertainties. The experimental global response of the specimen, represented by the load-CMOD curve, is well fitted by the numerical curves obtained using the model parameters dentified by the Kalman filter procedures.

However, both the non linearity of the problem and the fact that no closed form expression can be given for the sensitivity matrix, necessitate the numerical evaluation of the derivatives of the forward operator. Hence, several forward analyses are necessary for each Kalman filter step, resulting in a time consuming procedure. This limits the number of KF steps, the number of parameters to identify simultaneously and the number of experimental quantities whose difference with the numerical corresponding quantities has to be minimized.

Different types of measurable quantities can be assembled in the $\mathbf{y}_{exp}$ vector. For structural mechanics problems they can be forces, the crack mouth opening displacements CMOD, relative displacements measured by interferometry technique, strains measured by strain gages or embedded optical fibers, width of the localisation zone etc. Hence the generality of the Kalman filter technique makes possible applications in more complex situations such as uniaxial compression, as well as size effects, etc.

REFERENCES

Bazant, Z. P., T. B. Belytschko, and T. Chang (1984). Continuum theory for strain-softening. *ASCE Journal of Engineering Mechanics 110*(12), 1666–1693.

Bazant, Z. P., and G. Pijaudier-Cabot (1988). Nonlocal continuum damage, localization instability and convergence. *Journal of Applied Mechanics 55*, 287–293.

Bittanti, S., G. Maier, and A. Nappi (1984). Inverse problem in structural elastoplasticity: a kalman filter approach. In A. Sawczuk and G. Bianchi (Eds.), *Plasticity Today*, London, pp. 311–329. Elsevier Applied Science Publ.

Bolzon, G., R. Fedele, and G. Maier (2002). Parameter identification of a cohesive crack model by kalman *Computer Methods in Applied Mechanics and Engineering 191*, 2847–2871.

Bui, H. D. (1994). *Inverse Problem in the Mechanics of Materials: an Introduction.* London: CRC Press.

Catlin, D. E. (1989). *Estimation, Control, and the Discrete Kalman Filter.* Berlin: Springer Verlag, Applied Mathematical Sciences.

Eringen, A. C. and D. G. B. Edelen (1972). On nonlocal elasticity. *International Journal of Engineering Science 10*, 233–248.

Kailath, T., A. H. Sayed, and B. Hassibi (2000). *Linear Estimation.* London: Prentice Hall.

Kalman, R. (1960). A new approach to linear filtering and prediction problems. *(ASME) J. Basic Eng. 82D*, 34–45.

Kröner (1967). Elasticity theory of materials with long range cohesive forces. *International Journal of Solids and Structures 3*, 731–742.

Lasry, D. and T. Belytschko (1988). Localisation limiters in transient problems. *International Journal of Solids and Structures 24*, 581–597.

Lemaitre, J. and J.-L. Chaboche (1990). *Mechanics of Solid Materials.* Cambridge: Cambridge University Press.

Peerlings, R. H. J. (1999). *Enhanced Damage Modelling for Fracture and Fatigue.* Ph. D. thesis, Eindhoven University of Technology, The Netherlands.

Peerlings, R. H. J., R. de Borst, W. A. M. Brekelmans, and J. H. P. de Vree (1995). Computational modelling of gradient enhanced damage for fracture and fatigue problems. In D. R. J. Owen and E. Oñate (Eds.), *Computational Plasticity, Fundamentals and Applications*, Swansea, UK, pp. 975–986. Pineridge Press.

Peerlings, R. H. J., R. de Borst, W. A. M. Brekelmans, and J. H. P. de Vree (1996). Gradient-enhanced damage for quasi-brittle materials. *International Journal for Numerical Methods in Engineering 39*, 3391–3403.

Pijaudier-Cabot, G. and Z. P. Bazant (1987). Nonlocal damage theory. *ASCE Journal of Engineering Mechanics 113*(10), 1512–1533.

Tarantola, A. (1987). *Inverse Problem Theory. Methods for Data Fitting and Model Parameter Estimation.* Southampton: Elsevier Applied Science.

van Mier, J. G. M. and C. Shi (2002). Stability issues in uniaxial tensile tests on brittle disordered materials. *International Journal of Solids and Structures 39(13/14)*, 3359–3372.

Vree, J. H. P., W. A. M. Brekelmans, and M. A. J. Gils (1995). Comparison of nonlocal approaches in continuum damage mechanics. *Computers and Structures 55*, 581–588.

Computational Modelling of Concrete Structures, Bićanić et al. (eds)
© 2003 Swets & Zeitlinger, Lisse, ISBN 90 5809 536 3

Total-local consistency condition in a plastic-damage model for concrete

A.D. Jefferson
Cardiff University, UK

ABSTRACT: An outline description of a new plastic damage model for concrete is given. This model employs a damage (or contact) tensor generated from crack planes, each of which are formed when a damage criterion is satisfied. A key element of the model is a so-called Total-Local consistency condition which is used to ensure that both local and total constitutive relationships are simultaneously satisfied. The validity of the coupled equations, that arise from this condition and which relate local crack plane strains to the total recoverable strain tensor, is studied by examining the positive definiteness of the coupling matrix. The convergence properties of the Total-Local consistency equations are examined using a finite element analysis example, in which the constitutive model, implemented in the finite element program LUSAS, is applied to the analysis of a double edge notched shear-tension test.

1 INTRODUCTION

In the 1970s and 1980s a number of plastic-cracking models were developed for the finite element analysis of concrete structures (Buyukozturk 1977, ASCE 1982, Owen et al. 1983, de Borst 1986). These models used plasticity theory to represent the compressive behaviour of concrete and various total and incremental fracture theories to simulate directional cracking on defined planes. Although there were significant differences in detail between the individual models the underlying principles were similar. Differences lay both in the details of the plasticity components and in the way in which cracks were simulated, which included the fixed crack, rotating crack and multi non-orthogonal fixed crack methods. These have been been reviewed in more recent publications (Weihe et al. 1998, Petrangelo and Ozbolt 1996). The reasons why this type of plastic-cracking model received wide interest are understandable. The idea of representing cracks with defined planes of degradation is intuitively appealing because a crack in the model represents a macroscopically observed physical crack. Plasticity naturally simulates well the compressive behaviour of concrete, which is characterised by, initially elastic and then, at higher stress levels, non-linear behaviour before a peak is reached after which softening occurs. In compressive regions, the strains causing non-linearity are largely permanent and thus can be equated to plastic strains and the peak stress states are able to be well represented by surfaces in stress space (Kupfer et al 1969, Kotsovos and Newman 1978).

However, whilst considerable success was gained with these models in that, when implemented in finite element programs they were able to simulate many aspects of the behaviour of concrete structures, they did have a number of important flaws. Amongst others, Crisfield and Wills (1989) examined the fixed crack approach and showed that it tended to produce over stiff results and to rely on the questionable device of a shear retention factor. The rotating crack model proved to be able to simulate overall softening behaviour in a smooth manner but by its very nature could not simulate post crack shear response on a crack plane. It relied on the computing Poisson's ratio from current stress and strain components such that the coaxiality of the principal stresses and strains was maintained, a device that has a questionable physical basis (Rots 1988). The multiple non-orthogonal crack theory is, in the author's opinion, the most appealing of the approaches, but this method, as previously implemented, also has a number of disadvantages which have been summarised by Feenstra and de Borst (1995). These include computational and algorithmic difficulties associated with combining plasticity and fracture and the problems of simulating state changes, e.g. crack closure. Furthermore, these models were not generally developed, at least explicitly, within thermodynamically consistent frameworks.

In recent years this type of model has received little attention, which is no doubt due in part to the above difficulties, and developments have concentrated more on other types of model and theories that have more theoretical rigour. These include damage models (Krajcinovic 1996), advanced plasticity based models (Este and Willam 1994, Feenstra and De Borst 1995), plastic-damage models Lubliner et al. 1989, Lee and Fenves 1998) and the micro-plane model (Bazant et al. 2000).

The older type of plastic-cracking models in some respects were better at simulating certain of the above characteristics than many of the new models, yet the theoretical and practical flaws are too great to ignore. However, the author believes that a model that employs modern plastic-damage theory and yet retains certain of the features of the early plastic-cracking models is an attractive proposition; not least because such a model works with defined crack planes and therefore has the potential to simulate crack opening and closing behaviour and post-crack shear behaviour on fully formed crack planes accurately.

For these reasons the author has attempted to develop such a model and address the problems of thermodynamic consistency, crack opening-closing, shear behaviour on formed crack planes and computational efficiency by use of a fully consistent D-matrix- stress recovery algorithm for finite element implementation. The resulting model and details of its implementation are described elsewhere (Jefferson 2003), however a key step in the work was the development of an approach which ensured that the total and local constitutive equations were simultaneously satisfied. The device for achieving this has been termed the Total-Local (T-L) consistency condition, and it is this condition and the associated relationships that is the subject of the present paper.

The model, named Craft, has been implemented in the finite element program LUSAS and it is this program that is used in the example presented in Section 5 of this paper.

2 OUTLINE DESCRIPTION OF MODEL

The model uses a combination of plasticity, damage and contact theories. The plasticity component uses a triaxial yield surface and plastic potential based on the compressive meridians of Lubliner et al. (1989), but are rounded in the pi-plane using the function of Willam and Warnke (1975). A deviatoric work hardening/softening model is employed that includes an additional stress parameter that performs essentially the same role as the ductility parameter of Willam and Este (1994). The model is developed in such a way that when stress states are in predominantly tensile regions, the behaviour is either elastic or damaging, and a stress state at the theoretical apex of the yield surface is not admissible, since the damage surfaces would bring the stress inside this limit point.

The model uses a damage-contact model for the crack planes that is a simplified version of the

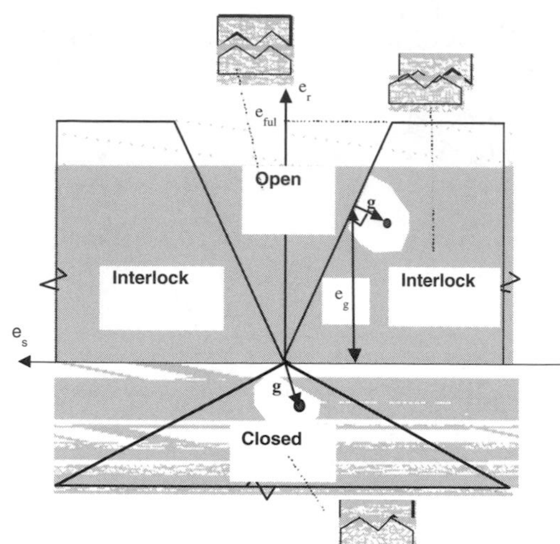

Figure 1. Contact states and embedment.

crack-plane model described in Jefferson (2002). Here a relationship is derived in terms of effective local strains $\mathbf{e}$ and local stresses $\mathbf{s}$ in which the local stress comprises two components, one the undamaged component, governed by h_c which may be thought of as $(1 - \omega)$ if ω were a scalar damage parameter, and the other a contact component that depends on the degree of embedment $\mathbf{g}$, relative to a contact surface, and the fully debonded parameter $h_f = (1 - h_c)H_f$, where H_f is a function which decreases with increased crack opening and obeys the condition that $0 \leq H_f \leq 1$. The local stress is given by

$$\mathbf{s} = \mathbf{D}_L(h_c\mathbf{e} + h_f\mathbf{g}) = \mathbf{D}_L(h_u\mathbf{I} + h_f\mathbf{\Phi}_d)\mathbf{e} \tag{1}$$

in which $\mathbf{D}_L$ is the local elastic constitutive matrix, Φ_d is a transformation matrix and $\mathbf{g}$ the compressive contact strain (termed the embedment).

h_c depends upon a local damage parameter.

The essential features of the contact part of the model are shown is Figure 1.

In the Open state the stress in the debonded component is assumed zero. In the Interlock state $\mathbf{g}$ is the normal to the Interlock contact surface, and in the Closed state $\mathbf{g}$ is equal to the local strain vector since the contact point coincides with the origin of the local strain space.

The interlock and closed functions used to identify which state is active are as follows, in which m_g is the slope of the interlock function.

$$\phi_{int}(\mathbf{e}) = m_g\, e_r - \sqrt{e_s^{\,2} + e_t^{\,2}} \tag{2a}$$

$$\phi_{cl}(\mathbf{e}) = e_r + m_g\sqrt{e_s^{\,2} + e_t^{\,2}} \tag{2b}$$

If $\phi_{cl}(\mathbf{e}) \leq 0$ State = Closed
If $\phi_{cl}(\mathbf{e}) > 0$ and $\phi_{int}(\mathbf{e}) < 0$ and $e_r < e_{ful}$ State = Interlock
If $\phi_{int}(\mathbf{e}) \geq 0$ or $e_r \geq e_{ful}$ State = Open

Returning to equation (1), this may be written in a more compact form as follows

$$\mathbf{s} = \mathbf{D}_L \mathbf{M}_x \mathbf{e} = \mathbf{D}_{ls} \mathbf{e} \tag{3}$$

The added compliance from each crack plane is given by the transformed local "fracture" strain $\mathbf{e}_f$, which is the inelastic component of $\mathbf{e}$, such that

$$\mathbf{e}_f = \left[\mathbf{M}_x^{-1}\mathbf{C}_L - \mathbf{C}_L\right]\mathbf{s} = \mathbf{C}_{lsf}\,\mathbf{s} \tag{4}$$

Applying (4) to each crack plane and introducing the subscript i on local terms to denote the crack plane number, the total stress-strain relationship is then given by

$$\boldsymbol{\sigma} = \mathbf{D}_e\left(\boldsymbol{\varepsilon} - \boldsymbol{\varepsilon}_p - \sum_{i=1}^{n} \mathbf{N}_i^T \mathbf{e}_{fi}\right) \tag{5}$$

in which $\boldsymbol{\sigma}$ = stress tensor, $\boldsymbol{\varepsilon}$ = total strain tensor, $\boldsymbol{\varepsilon}_p$ = plastic strain tensor, $\mathbf{D}_e$ = elastic constitutive tensor, $\mathbf{C}_L = \mathbf{D}_L^{-1}$, $\mathbf{N}_i$ is stress transformation matrix such that $\mathbf{s}_i = \mathbf{N}_i\boldsymbol{\sigma}$ and n = number of crack planes.

The static constraint, i.e. the local stress vector is the transformed total stress tensor, is central to the model.

3 TOTAL-LOCAL CONSISTENCY CONDITION

The T-L consistency condition ensures that the overall stress-strain relationship (5), the local stress-strain relationship for all crack planes (3) and the static constraint are all satisfied in the stress recovery computations. The condition may be written

$$\mathbf{f}_{TL_i} = \mathbf{N}_i\left[\mathbf{D}_e(\boldsymbol{\varepsilon} - \boldsymbol{\varepsilon}_p) - \sum_{j=1}^{n} \mathbf{N}_j^T\left(\mathbf{I} - \mathbf{M}_{xj}\right)\mathbf{e}_j\right] - \mathbf{D}_{ls_i}\mathbf{e}_i = \mathbf{0} \tag{6}$$

When more than one crack plane exists, equation (6) forms a set of equations that provide coupling between the crack planes. It may be noted that the equations are developed in terms of $\mathbf{e}$ rather than $\mathbf{e}_f$, since when cracks close the local compliance w.r.t $\mathbf{e}_f$ tends to infinity. The set of non-linear coupled equations (6) is solved by a Newton iterative procedure.

Expanding (6) for a case with two crack planes gives the following

$$\begin{bmatrix} \mathbf{N}_1\mathbf{D}_e\mathbf{N}_1^T(\mathbf{I}-\mathbf{M}_{x1})+\mathbf{D}_{ls1} & \mathbf{N}_1\mathbf{D}_e\mathbf{N}_2^T(\mathbf{I}-\mathbf{M}_{x2}) \\ \mathbf{N}_2\mathbf{D}_e\mathbf{N}_1^T(\mathbf{I}-\mathbf{M}_{x1}) & \mathbf{N}_2\mathbf{D}_e\mathbf{N}_2^T(\mathbf{I}-\mathbf{M}_{x2})+\mathbf{D}_{ls2} \end{bmatrix}\begin{bmatrix}\mathbf{e}_1\\ \mathbf{e}_2\end{bmatrix} = \begin{bmatrix}\mathbf{N}_1\\ \mathbf{N}_2\end{bmatrix}\mathbf{D}_e(\boldsymbol{\varepsilon}-\boldsymbol{\varepsilon}_p) \tag{7}$$

The condition is important in the derivation of a consistent algorithmic approach for the implementation of the model and also in derivations associated with demonstrating that the model is thermodynamically consistent.

4 PROPERTIES OF COUPLING MATRIX

In order to ensure that the model, when implemented, would be robust it was decided that there should always be a unique relationship between the local and Cartesian elastic strain $\mathbf{e}_i$ and the total "recoverable" strain tensor $(\varepsilon - \varepsilon_p)$ respectively. To investigate this, the determinate of the coupling matrix has been evaluated for a range of conditions. This has carried out here for the total $\mathbf{e}_i$ vs $(\varepsilon - \varepsilon_p)$ relationship, but it is noted that the incremental form of the relationship is used in the iterative solution process. However, the aim here is to establish that valid relationships exist for the basic forms of the contact and damage relationship given above.

The relationship is examined for a plane stress condition with two cracks. In the closed state the contact matrix $\mathbf{M}_x$ becomes equal to the Identity matrix and for this condition the added fracture strains are zero, and therefore the case is of no interest here, hence, only the Open and Interlock states will be examined.

In the present plane-stress condition $\mathbf{M}_x$ takes the form

$$\mathbf{M}_x = \begin{pmatrix} h_c & 0 \\ 0 & h_c \end{pmatrix} + H_f \frac{1-h_c}{1+m_g^2}\begin{pmatrix} m_g & -m_g \\ -m_g & 1 \end{pmatrix} \tag{8}$$

and the transformation matrix takes the form

$$\mathbf{N}_i = \begin{bmatrix} (rd_{i1})^2 & (rd_{i2})^2 & 2rd_{i1}rd_{i2} \\ rd_{i1}sd_{i1} & rd_{i2}sd_{i2} & rd_{i1}sd_{i2} + rd_{i2}sd_{i1} \end{bmatrix} \tag{9}$$

where $\mathbf{rd}_i$ is the unit normal vector to crack plane i, and $\mathbf{sd}_i$ is the unit vector in the shear direction, norsmal to $\mathbf{rd}_i$. In the Open state $H_f = 0$ and $\mathbf{M}_x$ only contains the first component matrix of equation (8).

Using (3), (8) and (9) the coupling matrix (10) may be written as follows

$$\mathbf{E}_m = \begin{bmatrix} \mathbf{N}_1\mathbf{D}_e\mathbf{N}_1^T(\mathbf{I}-\mathbf{M}_{x1})+\mathbf{M}_{x1}\mathbf{D}_L & \mathbf{N}_1\mathbf{D}_e\mathbf{N}_2^T(\mathbf{I}-\mathbf{M}_{x2}) \\ \mathbf{N}_2\mathbf{D}_e\mathbf{N}_1^T(\mathbf{I}-\mathbf{M}_{x1}) & \mathbf{N}_2\mathbf{D}_e\mathbf{N}_2^T(\mathbf{I}-\mathbf{M}_{x2})+\mathbf{M}_{x2}\mathbf{D}_L \end{bmatrix} \tag{10}$$

A simplification to (10) arises if the terms of $\mathbf{D}_L$ are chosen such that

$$\mathbf{D}_L = \mathbf{N}_i\mathbf{D}_e\mathbf{N}_i^T \tag{11}$$

Table 1. Coupling matrix determinates.

h_{c1}	h_{c2}	θ	$\lvert \mathbf{E}_m \rvert / E$
Interlock cases			
0.995	0.995	60°	0.198
0.0	0.0	60°	0.192
0.0	0.995	60°	0.198
0.0	0.0	30°	0.093
0.995	0.995	30°	0.198
0.0	0.995	30°	0.197
0.995	0.995	90°	0.198
0.0	0.0	90°	0.198
0.0	0.995	90°	0.198
Open cases			
0.995	0.995	60°	0.198
0.0	0.0	60°	0.0
0.0	0.995	60°	0.197
0.0	0.0	90°	0.0

In the plane-stress case equation (11) is true if

$$\mathbf{D}_L = \begin{bmatrix} E_{ef} & 0 \\ 0 & G \end{bmatrix} \quad (12)$$

in which $E_{ef} = E/(1 - \nu^2)$

With this form of $\mathbf{D}_L$ equation (10) becomes;

$$\mathbf{E}_m = \begin{bmatrix} \mathbf{D}_L & \mathbf{N}_1\mathbf{D}_e\mathbf{N}_2^T(\mathbf{I} - \mathbf{M}_{x2}) \\ \mathbf{N}_2\mathbf{D}_e\mathbf{N}_1^T(\mathbf{I} - \mathbf{M}_{x1}) & \mathbf{D}_L \end{bmatrix} \quad (13)$$

The determinates of $\mathbf{E}_m$, normalised with respect to E, are given for a number of cases with $\nu = 0.15$, $m_g = 0.4$ in Table 1.

In all cases the determinate is real positive except in the Open state case with $h_{c1} = h_{c2} = 0$, i.e. two fully formed cracks. This last case is not surprising since there is effectively no strength in the material. In this case $\mathbf{E}_m$ has one null eigen-value. Also, although not shown in the Table, it is interesting to note that in the non-orthogonal Interlock cases shown, $\mathbf{E}_m$ has two imaginary eigen-values.

Because of the null determinate, h_c in the model is never allowed to fully decay to zero, rather an exponential function is used which is asymptotic to a small residual value.

5 NUMERICAL EXAMPLE

A single example is presented in which the convergence of the T-L consistency equations is monitored at all Gaussian Integration points (GPs) in a finite element mesh. In addition, other information relating to the analysis is provided, in order to show that the example is meaningful.

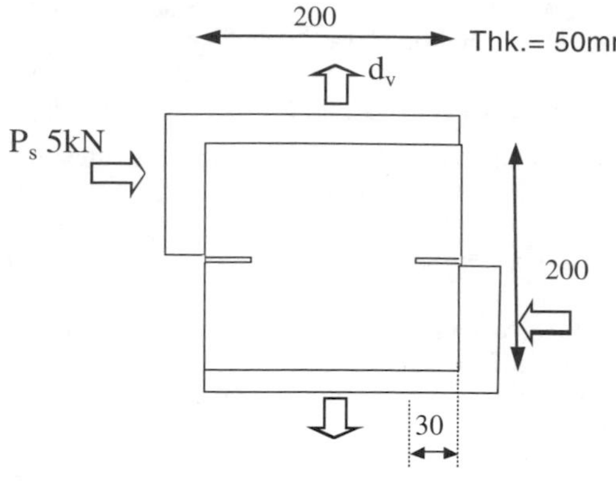

Figure 2. Testing arrangement.

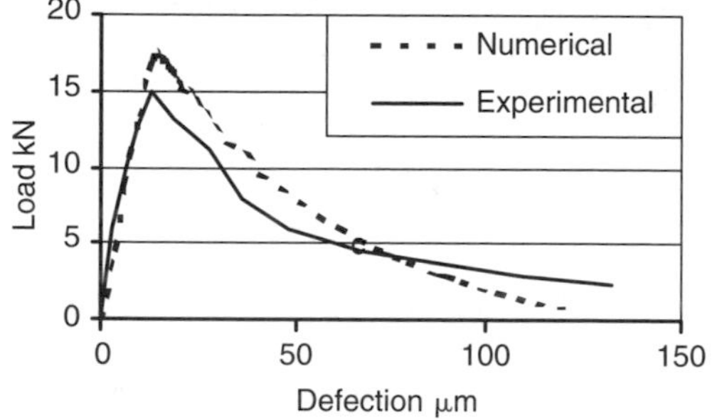

Figure 3. Vertical load-displacement response.

The example used is a double edge notched specimen tested by Nooru-Mohamed (1992), which is illustrated in Figure 2. The test was from the series 4a and was denoted 48-03 by Nooru-Mohamed. In the test the shear force P_s was applied first and then, with this kept constant at 5 kN, the vertical displacement d_v was increased using a feedback control loop.

The basic material properties used for this analysis were as follows

$$E = 33\,\text{kN/mm}^2 \quad \nu = 0.2 \quad f_c = 37\,\text{N/mm}^2$$
$$f_l = 2.7\,\text{N/mm}^2 \quad G_f = 0.1\,\text{N/mm} \quad m_g = 0.5$$

where f_c = compressive strength, f_t = tensile strength and G_f = fracture energy.

Relatively onerous convergence tolerances of 10^{-5} for both residual force and iterative displacement norms were used for the analysis. The analysis used an automatic incrementation procedure available in LUSAS. Fifty steps were used for the complete analysis with an average of four iterations per step.

A comparison between experimental and numerical responses is given in Figure 3, and an undeformed and deformed mesh plot are given in Figure 4. The exaggerated deformed mesh plot, at a normal

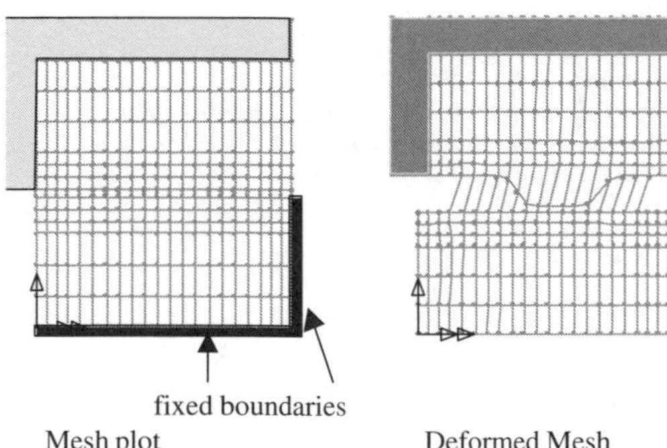

Figure 4. Mesh plots.

Table 2. T-L convergence data.

No. crks at GP	Average T-L iterations	Max. no. T-L iterations
1	2	4
2	3	8

displacement of 120 μm, shows the localised crack band staying close to the original notch plane, which is consistent with the experimental observations from this particular test. Information gathered from the analysis for the number of T-L iterations for different numbers of cracks is given in Table 2. The tolerance used for the T-L condition is $f_t * 10^{-5}$ on a norm of $\mathbf{f}_{TL}$ (see equation (6)).

Note: If no plasticity occurs convergence, w.r.t. the $\mathbf{f}_{TL}$ norm, is always achieved in one iteration for one crack, however, since a check is also carried out using an iterative norm of $\delta\mathbf{e}_i$ then a 2nd iteration always is needed to satisfy the convergence criterion w.r.t. this local strain norm.

6 CONCLUDING REMARKS

The paper has discussed a consistency condition, termed the Total-Local consistency condition, which satisfies local and total constitutive equations, together with the static constraint. This condition leads to a set of coupled equations between the total recoverable strains and the local strains on all crack planes. The coupling matrix was found to simplify if a particular form of local elastic constitutive matrix was used. In all cases examined, under plane-stress conditions, the relationship between the local and total recoverable strains was found to be unique, except in the case where there was full loss of strength and no contact in more than one direction. This led to the conclusion that damage should not be allowed to become zero. An example analysis showed that relatively few iterations are required to achieve convergence of the T-L equations, and in no case did the solution break down. Furthermore, the use of this consistency condition allowed the development of a consistent algorithm for the finite element implementation model, and the performance of this was illustrated in the example analysis in which convergence to relatively tight tolerances was achieved with 4 iterations per step, on average.

ACKNOWLEDGEMENT

I would like to thank FEA Ltd. for their support and for their help in the implementation and application of the model in their F.E. program LUSAS.

REFERENCES

ASCE Task Committee on Finite element analysis of reinforced concrete structures, 1982. *Finite element analysis of reinforced concrete structures,* ASCE.

Bazant, Z.P., Caner, F.C., Carol, I., Adley, M.D., Akers, S.A., 2000. Micro-plane model M4 for concrete. I Formulation with work conjugate deviatoric stress. *J. Eng. Mech.* ASCE 126(9), 944–953.

Buyukozturk, O., 1977. Non-linear analysis of reinforced concrete. *Computers and Structures* 7, 149–156.

Crisfield, M.A., Wills, J., 1989. The analysis of reinforced-concrete panels using different concrete models . *J. Eng. Mech.* ASCE 15(3), 578–597.

de Borst R., 1986. Computational aspects of smeared crack analysis. *Computational modelling of reinforced concrete,* edited, Hinton, E. , Owen, D.R.J., (eds) Pineridge Press, 44–83.

Este, G., Willam, K., 1994. Fracture energy formulation for inelastic behavior of plain concrete. *J. Eng. Mech.* ASCE 120(9), 1983–2011.

Feenstra, P.H., de Borst, R., 1995. A plasticity model and algorithm for mode-I cracking in concrete. *Int. J. Num. Meth. Eng.* 38, 2509–2529.

Jefferson, A.D., 2002. Tripartite cohesive crack model. *J. Eng. Mech.* ASCE, 128(6), 644–653.

Jefferson, A.D., 2003. Craft, a plastic-damage model for concrete, submitted for publication.

Kotsovos, M.D., Newman, J.B., 1979. A mathematical description of the deformable behaviour of concrete under complex loading. *Magazine of concrete research* 31, 77–90.

Krajcinovic, D., 1996. *Damage Mechanics,* Elsevier.

Kupfer, H.B., Hilsdorf, H.K., Rusch, H., 1969. Behaviour of concrete under biaxial stresses, *Journ. ACI* 66(8), 656–666.

Lee, J., Fenves, G.L., 1998. Plastic-damage model for cyclic loading of concrete structures, *J. Eng. Mech.* ASCE 124, 892–900.

Lubiner, J., Oliver, J., Oller, S., Onate, E., 1989. A plastic-damage model for concrete. *Int. J. Solids Structures* 25(3), 299–326.

Nooru-Mohamed, M.B. 1992. *Mixed-mode fracture of concrete: and experimental approach*, PhD thesis, Delft.

Owen, D.R.J., Figueriras, J.A., Damjanic, F. 1983. Finite element analysis of reinforced and prestressed concrete

structures including thermal loading. *Computer methods in Appl. Mech. and Eng.* 41, 323–366
Petrangeli, M., Ozbolt, J., 1996. Smeared crack approaches-Material modelling. *J. Eng. Mech.* ASCE 122(6), 545–554.
Rots, J.G., 1988. *Computational modelling of concrete fracture*, PhD Thesis, Delft.
Weihe, S., Kroplin, B., de Borst, R., 1998. Classification of smeared crack models based on material and structural properties. *Int. J. Solids Structures* 35(12), 1289–1308.
Willam, K., Warnke, E., 1975. Constitutive models for triaxial behaviour of concrete. *Proc. Int. Assoc. Bridge Struct. Engrg,* Report 19, Zurich, Switzerland, 1–30.

Computational Modelling of Concrete Structures, Bićanić et al. (eds)
© 2003 Swets & Zeitlinger, Lisse, ISBN 90 5809 536 3

Regularisation behaviour of the viscoplastic consistency model

J. Karolak & A. Winnicki
Cracow University of Technology, Cracow, Poland

ABSTRACT: In the paper numerical investigations of the viscoplastic Hoffman consistency model for elements made of plain concrete are presented. Two examples, namely a compressive test for the cube specimen and a cylinder splitting test are analysed using different FE meshes and for different rates of applied loading. It is shown that for the fairly fast loading processes the model displays the regularisation behaviour, whereas for the static loading strong mesh dependency is observed. The regularisation behaviour is observed for both dynamic analyses (using the Newmark integration scheme) and static computations (inertia effects are neglected and the Newton–Raphson procedure is used). Hence, it can be deducted that the regularisation behaviour of the model is connected only with its viscous properties and the wave speed in material is not important.

1 INTRODUCTION

Recently a new viscoplastic model, termed as "consistency model", has been introduced. The general idea of the model for the simplest case of the J2 viscoplasticity together with appropriate numerical algorithm has been presented by Wang (1997). According to this framework the yield function is assumed to be rate dependent and undergoes evolution with viscoplastic strain rate. The stress state is constrained to remain on the yield surface and the consistency condition is enforced. It can be shown that the consistency model with constant viscoplastic modulus coincides with the Perzyna's model in case of the simplest form of the "over-stress" function (Heeres et al. 2002, Winnicki 2001). Using the backward Euler scheme for a finite load step and enforcing the yield equation at the end of the load step an algorithm has been developed which is analogous to the Closest Point Projection Method. It has been shown that this algorithm together with the consistent stiffness matrix leads to a faster convergence of the iterative solution process in comparison with the Perzyna's model (Heeres et al. 2002)

The consistency model is adapted for the description of the concrete behaviour for dynamic loading using the Hoffman yield surface and associated flow rule. Previously the Hoffman yield surface in the rate independent format has been successfully employed for analysis of concrete structures (Bičanič et al. 1994). The isotropic form of the Hoffman yield surface is assumed and two independent internal variables are used for separate description of behaviour in compression and tension (both of them being functions of the equivalent viscoplastic strain). The model is described in detail by Winnicki et al. (2001). The Hoffman consistency viscoplastic model has been successfully used for prediction of behaviour of the well-known Koyna dam under both seismic and impact (caused by airplane crash) loading (Karolak & Winnicki 2002).

However, the above quoted papers on the Hoffman consistency model have not analysed the regularisation properties of this model in the case of the structures made of plain concrete. Hence, there is no guarantee that the presented results have been mesh independent. In the paper regularisation properties of the model is numerically analysed for two simple examples: a compression test of the cube specimen loaded by the stiff platens with friction and a cylinder splitting test (the so-scalled Brazilian test) in case of the impact loading. For both cases three different FE meshes are adopted (coarse, medium, fine). The analysis is focused on the obtained force–displacement curves and the observed patterns of shear bands. For the time being no attempt is made for a theoretical analysis of the regularisation behaviour of the Hoffman model. In the future more strict approach will be pursued analysing the properties of the acoustic tensor at the material point level. Especially, a change in the value of the determinant of the acoustic tensor depending on the assumed strain rate will be traced (similarly to the approach applied by Kang & Willam (2000) for a Duvaut–Lions concrete model). For low strain rates where the mesh dependency is observed the Hoffman model will be augmented using the "fracture energy trick".

2 NUMERICAL EXAMPLES

In this section two numerical examples are presented. The first one is a compression test for a cube specimen in the plane stress state. It is assumed that both bottom and top of the specimen are rigidly fixed (displacements are blocked in both vertical and horizontal directions) and the uniform vertical displacement is imposed at the top. This renders in the approximate way the conditions of the experimental test with load transmitted trough the stiff steel platens when friction between the specimen and the platens is high. Dimensions of the specimen are $100 \times 100 \times 100$ mm. Material properties are the same as assumed for the cylinder split test by Winnicki et al. (2001).

Three different FE meshes ("coarse" with 10×10 elements, "medium" with 20×20 elements and "fine" with 40×40 elements) are employed in computations. It is assumed that top displacement of the specimen changes linearly with the time up to the value 0.5 mm. Three velocities of the process are analysed. For the slow process the value of 0.5 mm is reached in 300 sec, which makes this case of loading a static one. For the medium speed process the maximum value of the displacement is attained in time 300e − 4 sec, whereas for the fast process the total time is 300e − 5 sec. Hence, those two cases can be qualified as dynamic ones. For every loading case the computations are made for all three meshes using the full Newton–Raphson scheme with the displacement control. In those computations inertia effects are neglected. Additionally, for the medium speed and fast processes the dynamic computations are made using the Newmark scheme for integration in time domain (also with the displacement control). Below for each loading case the reaction–time plots and displacement patterns at the end of loading process are presented. The reaction is defined as the resultant vertical force acting on the top surface of the specimen. As the time–top displacement dependency is linear the reaction–time plots can be also interpreted as the reaction–displacement plots. Because from the very beginning the stress state in the specimen is non-uniform there is no need to use any localisation trigger in the computations. For all loading processes and used meshes the pattern of localisation is the same – two shear bands are formed crossing each other at the centre of the specimen where the highly compressed zone is created. This mechanism of the specimen failure is, in authors' opinion, in agreement with the actual failure pattern appearing in the experimental test for cube specimens. However, for the static case the strong mesh dependency is observed. It is clearly seen in Figure 1 where the reaction–time plots for the "coarse", "medium" and "fine" FE meshes are described as "slow10", "slow20" and "slow40" curves, respectively.

This mesh dependency is confirmed by the displacement patterns at the end of the loading process

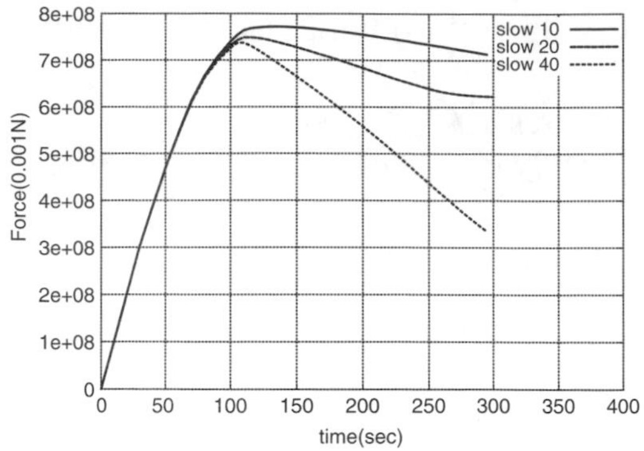

Figure 1. Reaction–time plot for slow process.

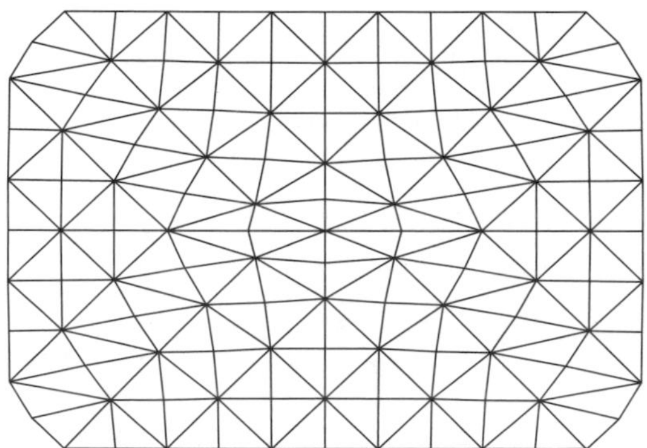

Figure 2. Displacement pattern (×50) for slow process for coarse mesh.

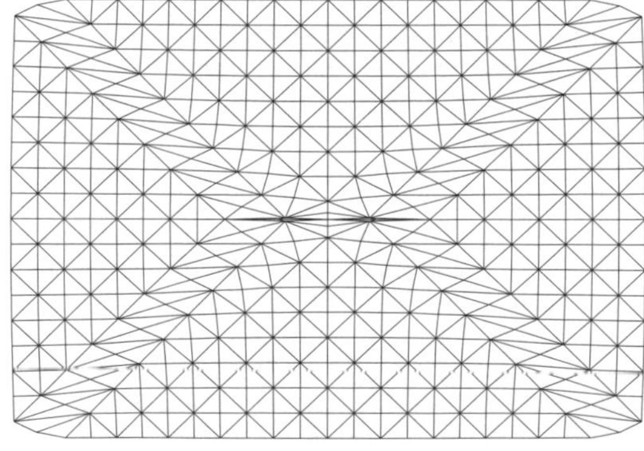

Figure 3. Displacement pattern (×50) for slow process for medium mesh.

(i.e. for the top displacement equal to 0.5 mm) which are presented in Figures 2–4. The shear bands localise in one row of the elements and the use of a finer mesh leads to decrease of the shear band width.

Figure 5 shows the reaction–time plots for the medium speed process. The curves "stat10", "stat20"

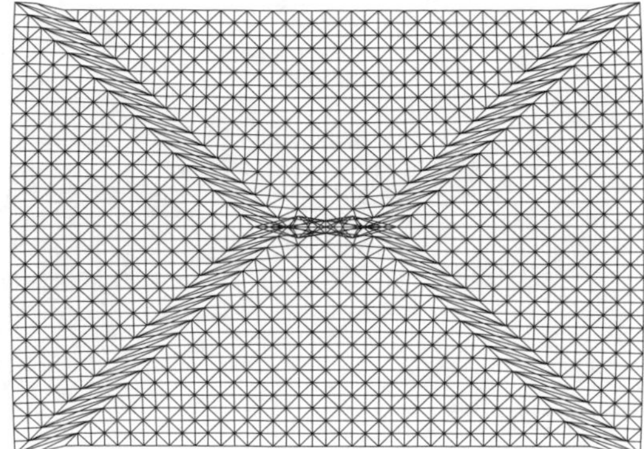

Figure 4. Displacement pattern (×50) for slow process for fine mesh.

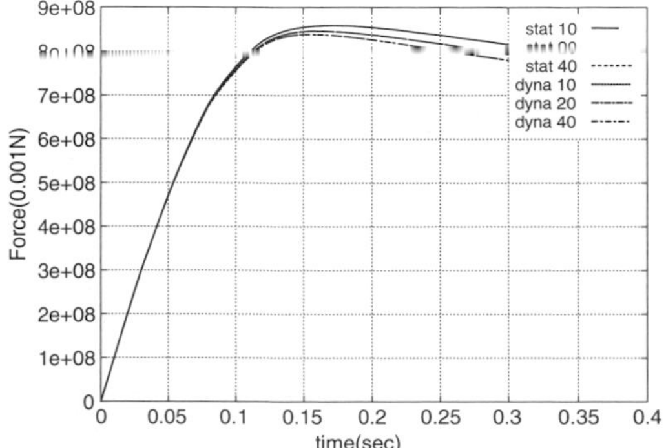

Figure 5. Reaction–time plot for medium speed process.

and "stat40" present the plots for the "coarse", "medium" and "fine" FE meshes in the case when the inertia effects are neglected (the Newton–Raphson procedure), respectively. Similarly, the curves "dyna10", "dyna20" and "dyna40" depict the reaction–time plots for the full dynamic process (the Newmark procedure).

In this case the ultimate load attained is higher than in the static case, which is due to the higher compressive strength displayed by the model for the higher strain rate (Winnicki et al. 2001). It worth noticing that there is almost no difference between the curves obtained by the Newton–Raphson and Newmark techniques. This points out clearly that inertia effects are negligible for this rate of loading. In this case there is almost no the mesh dependency and the width of the shear bands remains stable engaging the couple of element rows in the case of the "fine" mesh, which is presented in Figures 6–8 for the Newton–Raphson procedure (displacement patterns at the end of the loading process for the top displacement equal to 0.5 mm).

In Figure 9 the reaction–time plots for the fast loading process are presented (curves "fast10", "fast20"

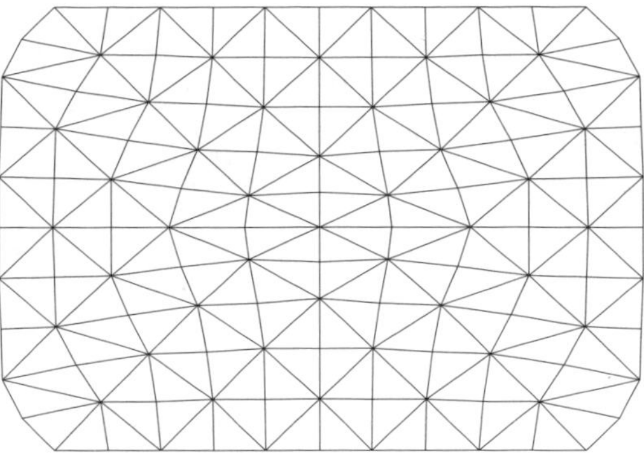

Figure 6. Displacement pattern (×50) for intermediate speed process for coarse mesh.

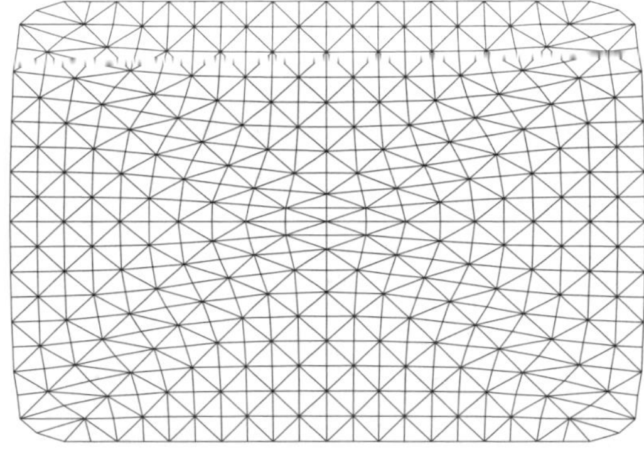

Figure 7. Displacement pattern (×50) for intermediate speed process for medium mesh.

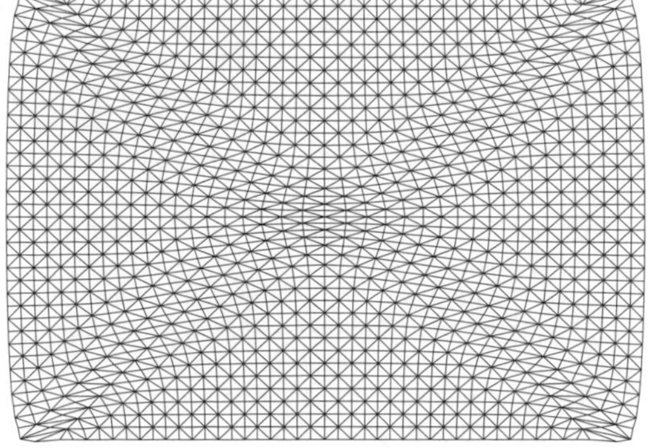

Figure 8. Displacement pattern (×50) for intermediate speed process for fine mesh.

and "fast40" for the "coarse", "medium" and "fine" FE meshes, respectively). This time, only the "static" (i.e. neglecting inertia effects) plots are shown. The "dynamic" curves (i.e. with inertia effects included using the Newmark procedure) are virtually the same.

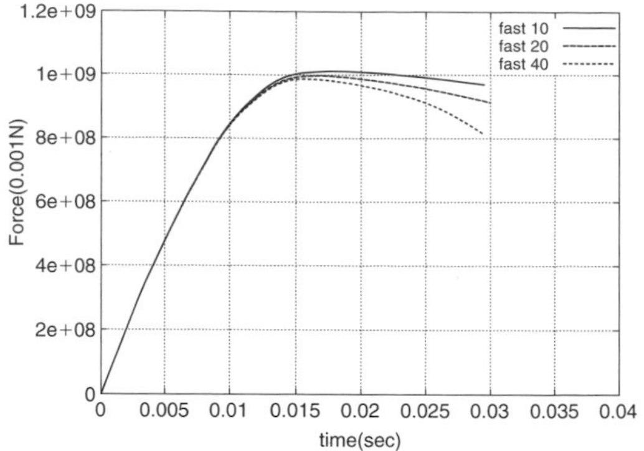

Figure 9. Reaction–time plot for fast process.

Figure 10. Displacement pattern (×50) for fast process for coarse mesh.

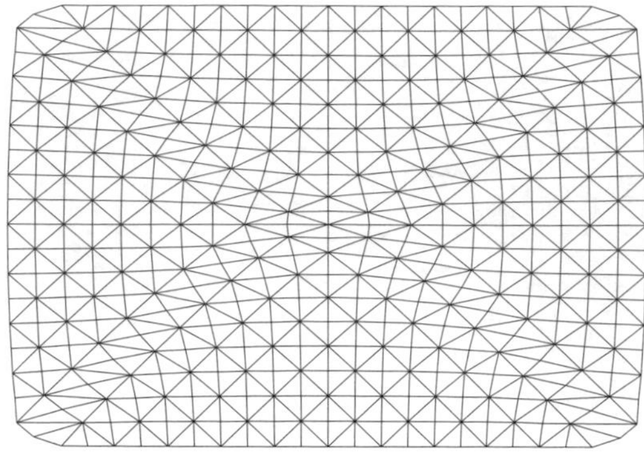

Figure 11. Displacement pattern (×50) for fast process for medium mesh.

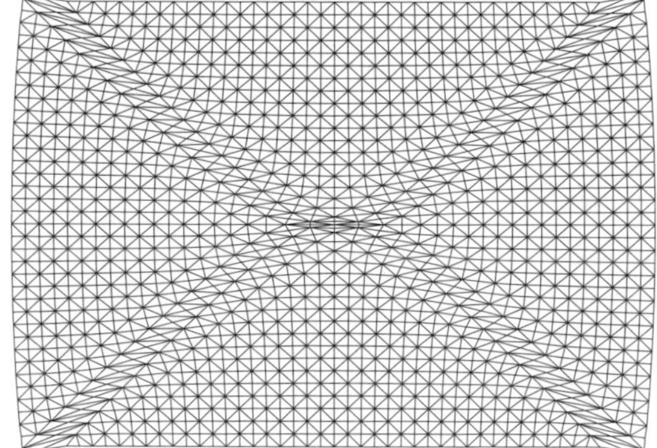

Figure 12. Displacement pattern (×50) for fast process for fine mesh.

Hence, for this loading rate the inertia effects are still negligible. The ultimate load in this case is obviously the highest one. However, the obtained results display the mesh dependency again. It is not so strong as for the slow process (static loading case) and shear bands are capable of engaging more than one row of the elements. Nevertheless, the width of the shear bands is narrower than that obtained for the medium speed process (Figures 10–12).

In authors' opinion, for the time being, it is not clear whether for the really high rates of loading the regularisation properties of the presented model are weaker or simply the width of the shear bands in that case is much smaller and, in turn, the much finer meshes have to be investigated.

As the second example the cylinder split test (or the so-called Brazilian test) already presented by Winnicki et al. (2001) is chosen. This time an analysis is made for the three different meshes in order to investigate the mesh dependency. Due to the symmetry conditions only a quarter of the specimen is analysed in plane stress state. The diameter of the whole specimen is 80 mm and thickness is 160 mm. The load is applied to the specimen via a steel platen, which is modelled as the stiff elastic material. The load is of the impact character–it increases linearly in the time domain up to the value of the 80 kN for the time equal to 30e-6 sec, then is kept constant. The assumed material parameters are the same as used by Winnicki et al. (2001). The analysis is carried out as the dynamic one using the Newmark integration scheme. Two different material models are employed–the original viscoplastic Hoffman consistency model and its rate independent variant, where the concrete compressive and tensile strengths are rate independent.

Figures 13–15 show the assumed FE meshes for the cylinder split test. In Figures 16 and 17 change of the displacement at the top of the specimen in time domain is presented. The curves "visco10", "visco20" and "visco40" are computed for the "coarse", "medium" and "fine" meshes, respectively, using the viscoplastic model. Similarly, the curves "plastic10", "plastic20" and plastic40" "are obtained for the same meshes using rate independent model.

In the analysed example the stress gradients are high and even for the purely elastic analysis finer

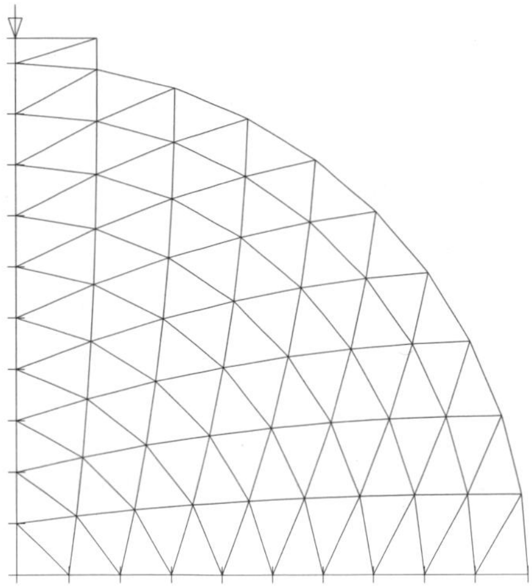

Figure 13. Coarse FE mesh for Brazilian test.

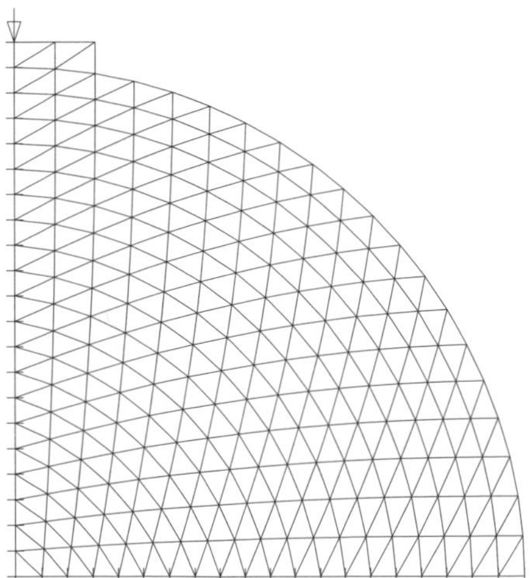

Figure 14. Medium FE mesh for Brazilian test.

meshes produce the larger values of the displacement. However, the adopted rate independent model has no regularisation properties and clearly exhibits the mesh dependency of both first and second order (using terminology proposed by Łodygowski 1996). On the other hand, the change in the values of the displacement for the different meshes in case of the viscoplastic model is definitely less pronounced and can be mainly attributed to the mesh dependency of the first order.

3 FINAL REMARKS

In this paper behaviour of the viscoplastic Hoffman consistency model for specimens made of plain

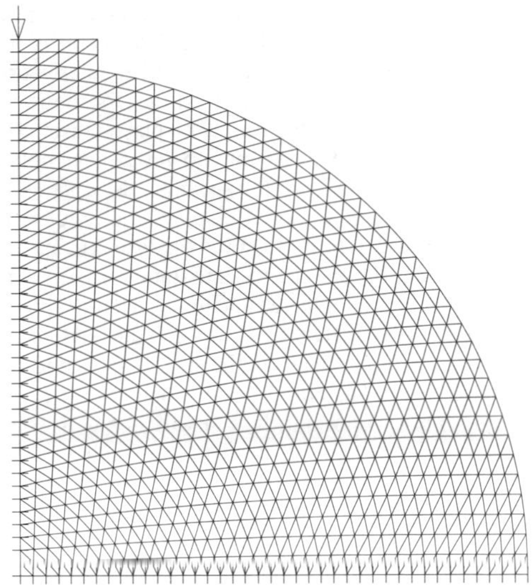

Figure 15. Fine FE mesh for Brazilian test.

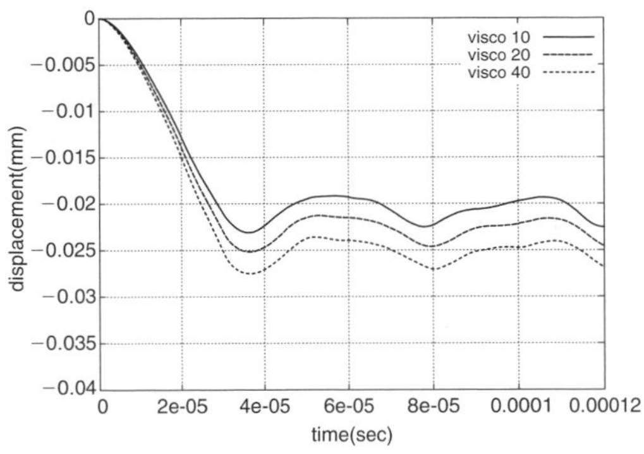

Figure 16. Displacement–time plot for viscoplastic model.

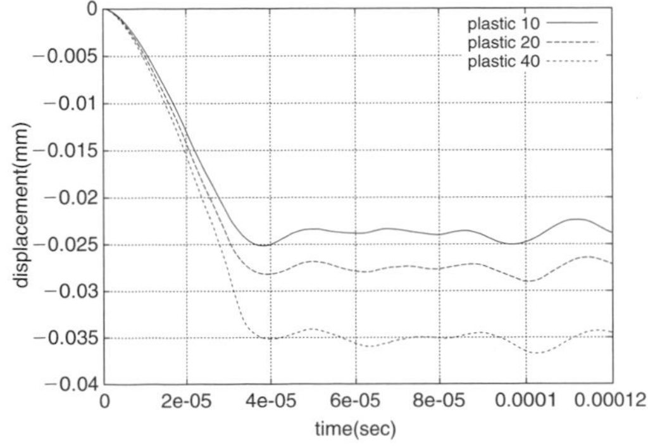

Figure 17. Displacement–time plot for rate independent model.

concrete is presented. The detailed formulation of the model is given by Winnicki et al. (2001) and is not repeated here. The paper focuses on the regularisation behaviour of the model. It is found that for the fairly fast loading processes the model displays reasonably strong regularisation behaviour, however it can not be excluded that the width of the shear bands depends on the loading rate. The model displays the regularisation behaviour for both dynamic analyses (using the Newmark integration scheme) and static computations (inertia effects are neglected and the Newton–Raphson procedure is used). Hence, it can be deducted that the regularisation behaviour of the model is connected only with its viscous properties and the wave speed in material is not important. It is at variance with some existing opinions (Łodygowski 1996) which attribute the regularisation capabilities to both viscous properties of material (relaxation time) and the wave speed. In the future more rigorous approach to the localisation problem will be pursued analysing the properties of the acoustic tensor at the material point level (Rudnicki & Rice 1975).

REFERENCES

Bičanič, N., Pearce, C.J. & Owen, D.R.J. (1994). Failure predictions of concrete like materials using softening Hoffman plasticity model. In H. Mang, N. Bičanič, R. de Borst, editors. *Computational modelling of concrete structures:* 185–98. Euro-C 1994, Vol. 1, Swansea: Prineridge.

Heeres, O.M., Suiker, A.S.J. & de Borst, R. (2002). A comparison between the Perzyna viscoplastic model and the Consistency viscoplastic model. *European J. of Mechanics, A/Solids*, Vol. 21: 1–12.

Kang, H.D. & Willam, K.J. (2000). Performance Evaluation of Elastoviscoplastic Concrete Model, *J. Engng Mech., ASCE*, Vol. 126, No 9: 995–1000.

Karolak, J. & Winnicki, A. (2002). Simulation of Dynamics Response of Concrete Structures Using Viscoplastic Hoffman Consistency Model, *5th World Congress on Comp. Mech.*, Vienna, Austria, July 7–12.

Łodygowski, T. (1996). Theoretical and numerical aspects of plastic strain localization. *Politechnika Poznańska, Rozprawy*, No. 312, Poznań, Poland.

Rudnicki, J.W. & Rice, J.R. (1975). Conditions for the localization of deformation in pressure sensitive dilatant materials. *J. Mech. Phys. Solids*, Vol. 23: 371–394.

Wang, W. (1997). Stationary and propagative instabilities in metals – a computational point of view, *Ph.D thesis, TU Delft*, Netherlands.

Winnicki, A. (2001). Viscoplastic consistency model – basic features. *ECCM-2001 European Conference on Computational Mechanics*, Cracow, Poland, June 26–29.

Winnicki, A., Pearce, C.J. & Bičanič, N. (2001). Viscoplastic Hoffman consistency model for concrete, *Computers & Structures*, Vol. 79: 7–19.

Computational Modelling of Concrete Structures, Bićanić et al. (eds)
© 2003 Swets & Zeitlinger, Lisse, ISBN 90 5809 536 3

A new constitutive model for concrete-steel bond behavior

Giovanni Di Luzio & Gianluca Cusatis
Dept. of Structural Engrg., Technical University (Politecnico) of Milan, Milan, Italy

ABSTRACT: In the present work, the load transfer mechanisms between concrete and steel reinforcement are modeled through a four node axisymmetric finite element which represents the behavior of a layer of material of vanishing thickness. The constitutive model directly relates the interface stresses (normal and tangential) to the displacements (opening and slip) and takes into account the most important physical mechanisms, including softening, confinement, etc. Numerical simulations of pull-out tests for confined and unconfined specimens are then presented, in which the concrete is modeled by through the microplane M4 model. The results of these simulations show a good agreement with the available experimental data.

1 INTRODUCTION

The modelling of the mechanical behavior of the interface between concrete and steel is a very important topic in the analysis of reinforced concrete structures. The load transfer mechanisms between concrete and reinforcement (also referred to as bond mechanisms) largely influence the performance of concrete structures under both service and failure load.

In the literature two types of models can be found for the modelling of the bond behavior: continuous models and lumped models (Hofstetter and Mang 1995). In this work the continuous approach is used. Thus, the interface is considered as in a zero-thickness layer of material which is modeled by a so-called interface element. The constitutive law of this element directly relates the interface stresses (normal and tangential) to the displacements (normal opening and tangential slip). The mathematical formulation of the constitutive behavior is based on the analysis of the physical phenomena occurring in the layer of concrete surrounding the steel bar.

The bond behavior is a combination of adhesion, friction and support of the ribs (Goto 1971; Tassios 1979; Gambarova et al. 2000). A nonlinear constitutive law able to reproduce the previous phenomena is introduced in the interface element. In order to validate the proposed model, numerical simulations of pull-out tests for confined and unconfined specimens are carried out. An axisymmetric finite element discretization is used in which the steel is assumed elastic and concrete is modeled through the microplane M4 model.

The results of numerical simulations are compared with the available experimental data and they confirm the ability of the proposed formulation to reproduce the complex phenomena occurring at the interface between steel and concrete.

2 INTERFACE ELEMENT

A classical four-node interface element for axisymmetric analysis is used and a new interface constitutive law has been formulated. Fig. 1a shows the geometry of the element. In the undeformed configuration the nodes 1 and 4 coincide with point A and the nodes 2 and 3 coincide with point B. The displacement field along the side 1–2 is related to the degrees of freedom of node 1 and node 2 and the displacements field along the side 3–4 depends on the degrees of freedom of nodes 3 and 4. We can write:

$$u_{12} = N_1 u_1 - N_2 u_2 \quad (1)$$

$$v_{12} = N_1 v_1 - N_2 v_2 \quad (2)$$

$$u_{34} = N_3 u_3 - N_4 u_4 \quad (3)$$

$$v_{34} = N_3 v_3 - N_4 v_4 \quad (4)$$

where N_1, N_2, N_3 and N_4 are linear shape functions with the following expressions:

$$N_1(s) = N_4(s) = \frac{1}{2}(1 - s) \quad (5)$$

$$N_2(s) = N_3(s) = \frac{1}{2}(1 + s) \quad (6)$$

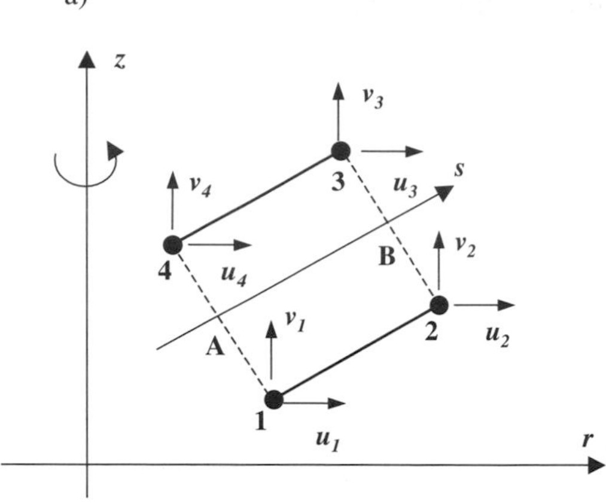

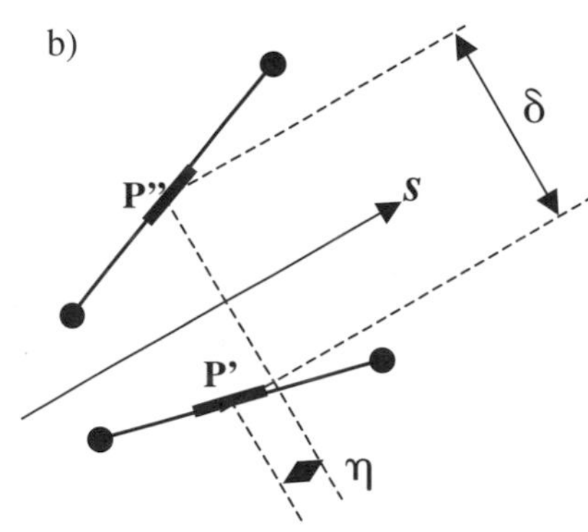

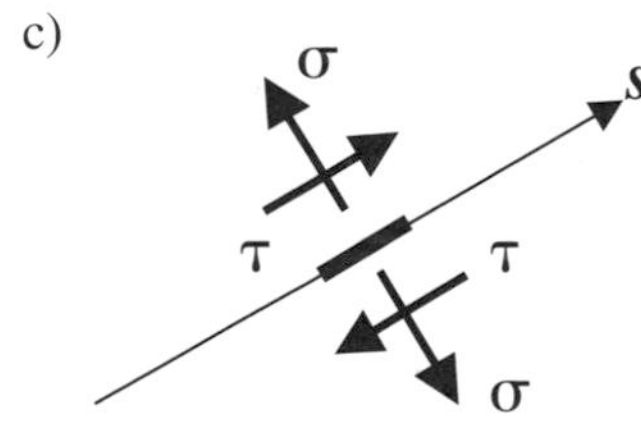

Figure 1. a) Geometry of interface element; b) definition of strains; c) definition of stresses.

In the interface element the "strains" are defined as relative displacements which can be calculated as

$$u = u_{34} - u_{12} \tag{7}$$

$$v = v_{34} - v_{12} \tag{8}$$

In a matrix form these relations take the form

$$\mathbf{u} = \mathbf{BU} \tag{9}$$

in which $\mathbf{u}^T = [u\ v]$, $\mathbf{U}^T = [u_1\ v_1\ u_2\ v_2\ u_3\ v_3\ u_4\ v_4]$ and

$$\mathbf{B} = \begin{pmatrix} -N_A & 0 & -N_B & 0 & N_B & 0 & N_A & 0 \\ 0 & -N_A & 0 & -N_B & 0 & N_B & 0 & N_A \end{pmatrix} \tag{10}$$

It is also convenient to express the "strain" field in the local coordinate system through the equation

$$\epsilon = \mathbf{Ru} = \mathbf{RBU} \tag{11}$$

where $\epsilon^T = [\gamma\delta]$ according with the notation of Fig. 1b, and

$$\mathbf{R} = \frac{1}{2L}\begin{pmatrix} r_B - r_A & z_B - z_A \\ -(z_B - z_A) & r_B - r_A \end{pmatrix} \tag{12}$$

Fig. 1c shows the work conjugate stresses $\boldsymbol{\sigma}^T = [\sigma, \tau]$.

3 INTERFACE CONSTITUTIVE BEHAVIOR

Physical mechanisms – The behavior of the interface is very complex and it depends on a combination of many phenomena as chemical adhesion, friction and the rib effect. Usually, in the analysis of bond, the exact distribution of stresses and strains at the interface is not computed and the interface behavior is taken into account by means of a relationship between the bond stresses at a certain point of a bar, averaged over the distance between two ribs, and the associated relative displacement (slip) of the bar in relation to the adjacent concrete. This slip should be regarded as an integral of the deformations in the concrete in contact with the steel bar, so that the bond-slip law cannot be regarded as a material law in the classical sense but it describes the behavior of the first layer of concrete surrounding the bar. Since this layer is quite thin compared to the dimension of usual laboratory specimens, it is possible to model it by a zero thickness finite element (interface element).

Bond of reinforcing steel in concrete is the effect of a combination of adhesion, friction and support of the ribs. In the elastic range the load is transferred from the bar to the concrete mainly by shear stresses parallel to the bar, which can be modeled by an elastic relation between the tangential stress and the tangential slip. As the load in the steel bar increases non linear phenomena occur and inclined radial cracks start to develop at the top of the ribs. At this stage the load is transferred by friction and by compressive stresses parallel to the already formed radial cracks. The bond-slip relation is not linear any longer and it is assumed to be a softening function of the tangential slip whose peak stress depends on the compressive stress in the normal direction. At the same time the inelastic part of the tangential slip is assumed to lead to a dilatancy in the normal direction due to the interlock between ribs and surrounding cracked concrete. The normal stress at the interface may be then computed through an elastic relation from the difference between the total normal opening and the dilatancy.

In this situation radial compressive stresses occur in the direction orthogonal to the bar inducing tensile stresses in the surrounding concrete. As soon as these tensile stresses reach the tensile strength, longitudinal crack initiate. After a first stable propagation, further increase of the load leads to the complete failure (splitting failure) of the connection with an unstable propagation of the splitting crack. Nevertheless, if transversal confinement is present, the longitudinal cracks or at least their unstable propagation may be prevented. In this case the final failure (pull-out failure) occurs when concrete just in front of the ribs is completely crushed. This different failure modes are a structural effect which can be simulated if the bond model is coupled with a reliable model for the concrete surrounding the steel bar.

Analytical description The interface constitutive behavior is described by the relationships between the relative displacement $\boldsymbol{\epsilon}^T = [\gamma\, \delta]$ and the bond stresses $\boldsymbol{\sigma}^T = [\tau\;\; \sigma]$. The shear law is based on the Tassios model (Tassios 1979) shown in Fig. 2a, which is characterized by the following properties:

- if the shear stress, τ, is smaller than τ_0 the behavior is elastic;
- when the shear stress reaches τ_0 the behavior becomes nonlinear because inclined cracks start to propagate. However the shear stress can still increase and the load can be transmitted by means of inclined compression bands;
- the shear stress increases until τ_2 if the splitting phenomenon is prevented. This stress value is related to the crashing of the concrete and then, after this point, the behavior becomes of softening type;
- after the crashing of concrete the strength does not go to zero because of the friction which still exists between concrete and steel.

The numerical value of τ_0 is usually very small, τ_1 is often very close to τ_2 and between τ_2 and τ_3. As far as τ is little than τ_1, the behavior may be considered as reversible. It is then reasonable to assume a simplified law of the type shown in Fig. 2b. Using the concept of stress-strain boundary (Bažant et al. 1996), the development of shear stress can be described as

$$\dot{\tau} = K_\tau \dot{\gamma} \qquad -\bar{\tau} \leq \tau \leq \bar{\tau} \tag{13}$$

where the symbols, defined in Fig. 2b, are given by

$$\bar{\tau} = \begin{cases} \tau_0 & \text{if } |\gamma_{\max}| \leq \gamma_{\mathrm{L}} \\ (\tau_0 - \tau_{\mathrm{R}}) f(\xi) + \tau_{\mathrm{R}} & \text{if } \gamma_{\mathrm{L}} < |\gamma_{\max}| \leq \gamma_{\mathrm{R}} \\ \tau_{\mathrm{R}} & \text{if } |\gamma_{\max}| > \gamma_{\mathrm{R}} \end{cases} \tag{14}$$

where $\tau_0 = \tau_L^0(1 + \beta\langle\sigma_N - \sigma_N^0\rangle), f(\xi) = 1 - e^\alpha \xi/[1 - (1 - e^\alpha)\xi]$ and $\xi = \min[(\gamma_{\max} - \gamma_L)/(\gamma_R - \gamma_L); 1]$.

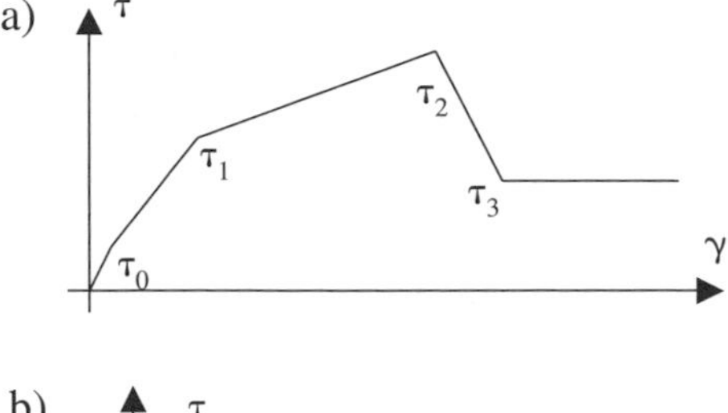

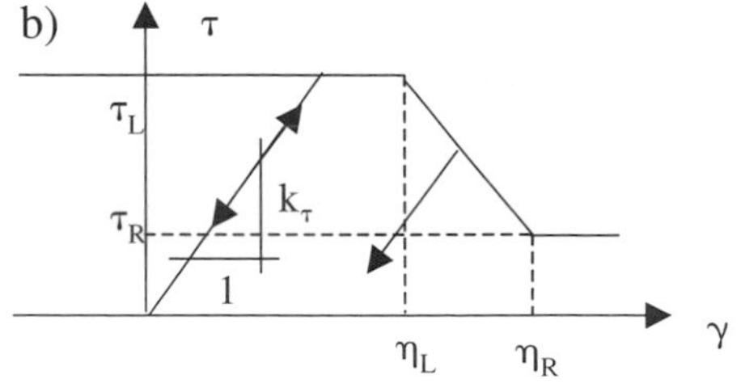

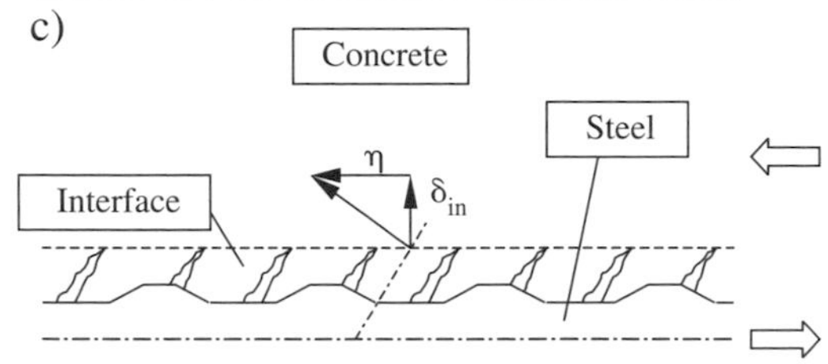

Figure 2. a) Bond-slip relationship for the Tassioss model; b) implemented bond-slip relationship; c) crack pattern in the concrete-steel interface.

In these expressions $\gamma_{\max} = \max[\gamma(t)]$ and τ_0 is assumed to be an increasing function of the compressive normal stress, in order to take into account the presence of confining pressure. The concrete layer close to the steel bar tends to expand during the pull-out process because of the inclined crack development. This effect can be taken into account in the material law relating the normal components by assuming that the normal "strain" is a sum of an elastic part and an inelastic part as

$$\dot{\delta} = \dot{\delta}_{\mathrm{el}} + \dot{\delta}_{\mathrm{in}} = \dot{\sigma}/K_\sigma + \dot{\delta}_{\mathrm{in}} \tag{15}$$

From the analysis of the cracking processes it can be assumed that

$$\dot{\delta}_{\mathrm{in}} = m|\dot{\gamma}| \tag{16}$$

The physical interpretation of the Eq. (16) is shown in Fig. 2c. By introducing Eq. (16) into Eq. (15) we get the normal stress:

$$\dot{\sigma} = K_\sigma(\dot{\delta} - m|\dot{\gamma}|) \tag{17}$$

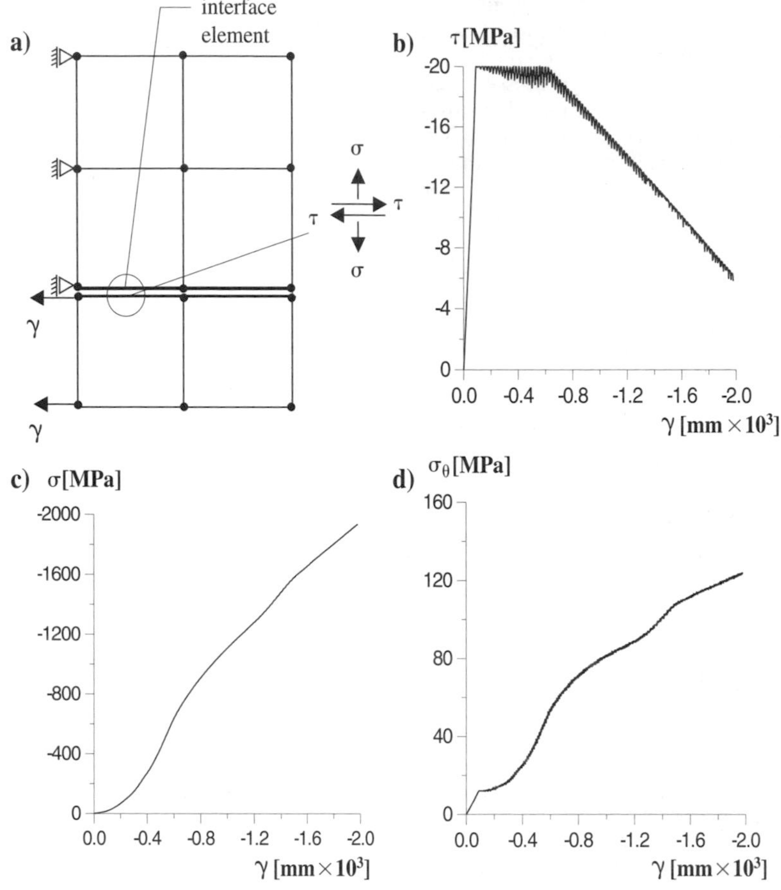

Figure 3. Numerical test for the validation of the interface element formulation: a) used mesh, b) shear stress-slip relation, c) normal stress, d) circumferential stress.

The finite element formulation and the bond-slip law described above have been implemented into the Plexis-3C (Casadei and Halleux 1999) explicit finite element code. In order to validate the model, a numerical simulation has been carried out in which the interface element is introduced between elastic four node axisymmetric elements 3a. The bond-slip curve is reproduced well (Fig. 3b) and as expected the interface is loaded in compression (Fig. 3c) while the elements adjacent to the interface have positive circumferential stress (Fig. 3d).

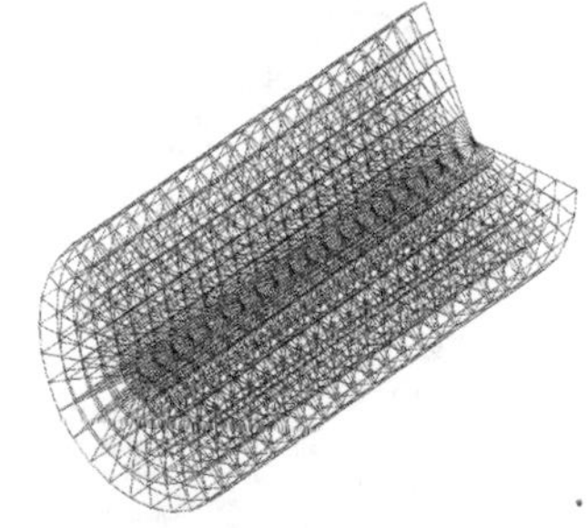

Figure 4. Axisymmetric mesh for pull-out tests.

4 NUMERICAL ANALYSIS

The simulations of pull-out tests are illustrated in this section. A cylindrical specimen (diameter = 10 cm, length = 20 cm) with a 10 cm bond length has been analyzed. The diameter of the bar is 2 cm. For numerical simulations an axisymmetric mesh has been created (Fig. 4), in which the concrete and the steel are described by 4 node elements and the interface is modeled by 4 node interface elements. The Microplane Model "M4" (Bažant et al. 2000; Caner

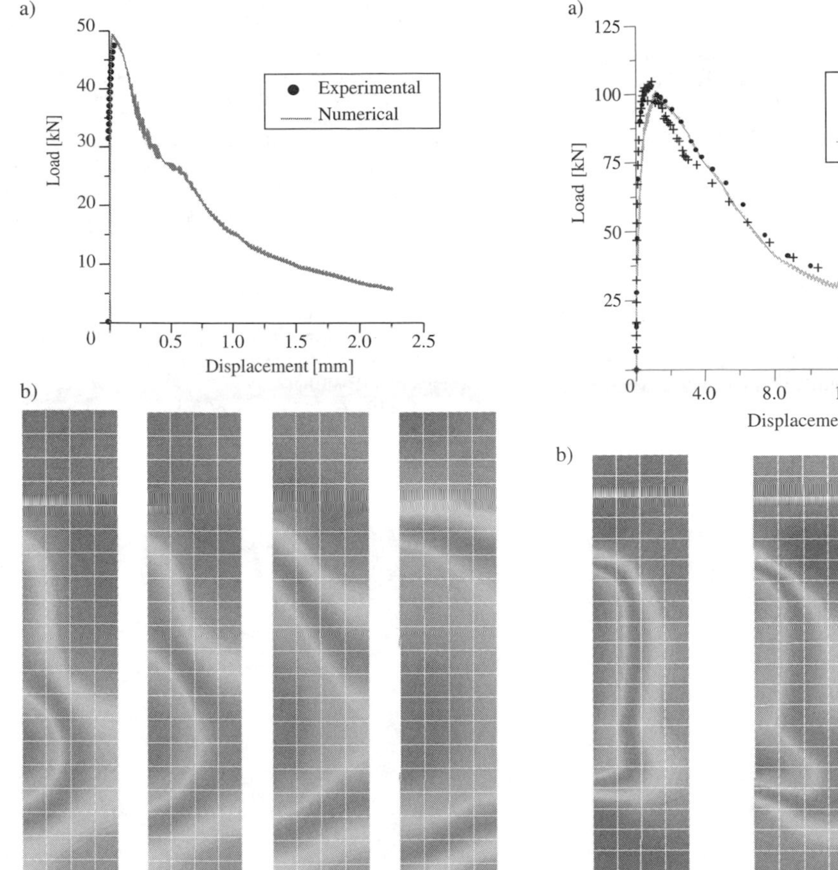

Figure 5. Simulation of pull-out test with splitting failure; a) load-displacement curve; b) evolution of circumferential stress.

Figure 6. Simulation of pull-out test with bond failure; a) load-displacement curve; b) evolution of shear stress.

and Bažant 2000; Di Luzio 2002), regularized in the spirit of the crack band model, has been adopted for the constitutive law of concrete.

Splitting failure – In Fig. 5a the comparison between the experimental and numerical load-displacement curves is shown. The agreement is quite satisfactory. In Fig. 5b we can see the evolution of the circumferential stress during the simulation. Since this component of the stress causes the longitudinal cracks which lead the specimen to failure, we can say that also the physical phenomena occurring during the experimental test are correctly represented by the numerical simulation.

Bond failure – The specimens used for this kind of simulations have the same geometry of the previous ones, but in this case the concrete is confined by a 1 cm thick steel jacket. In Fig. 6a the comparison between numerical and experimental results is shown. Once again the agreement is quite good. The numerical model is able to reproduce the increase of the peak load due to the confinement of the steel jacket. In this case the longitudinal cracks do not develop and so the failure is due to the failure of the concrete-steel interface. The evolution of the shear stress is shown in Fig. 6b.

5 CONCLUSIONS

As shown by the numerical results shown, the proposed concrete-steel interface element is capable of reproducing the different observed failure mode when there is a confinement (bond failure) or no confinement (splitting failure). This means that the interface constitutive law is able to capture the influence of the type of confinement of the surrounding concrete until failure.

ACKNOWLEDGMENT

The work of the authors were supported under European Commission grant (ANCHOR Project). The authors kindly thank the Joint Research Centre of Ispra and Ing. F. Casadei for providing the finite element code PLEXIS-3C.

REFERENCES

Bažant, Z. P., F. C. Caner, I. Carol, M. D. Adley, and S. A. Akers (2000). Microplane model M4 for concrete: I. Formulation with work–conjugate deviatoric stress. *ASCE Journal of Engineering Mechanics 126*(9), 944–953.

Bažant, Z. P., Y. Xiang, and P. C. Prat (1996). Microplane model for concrete: I. Stress–strain boundaries and finite strain. *ASCE Journal of Engineering Mechanics 122*(3), 245–254.

Caner, F. C. and Z. P. Bažant (2000). Microplane model M4 for concrete: II. Algorithm and calibration. *ASCE Journal of Engineering Mechanics 126*(9), 954–961.

Casadei, F. and J. E. Halleux (1999). *PLEXIS-3C User's Manual*. 21020 Ispra, Italy: European Commision Institute for Systems, Information and Safety Joint Research Centre.

Di Luzio, G. (2002, March). *A new nonlocal microplane model for fracture and damage of concrete*. Phd thesis, Politecnico di MIlano.

Gambarova, P. G., G. Plizzari, R. G. P., and G. Russo (2000). *Bond Modelling Including Pull-out and Splitting Failures* (FIB Task Group "Bond Models" ed.), Volume FIB Bulletin No. 10 of *Bond of Reinfocement in Concrete*, Chapter 1, pp. 1–98. Göteborb, Sweden: Convener R. Tepfers.

Goto, Y. (1971). Cracks formed in concrete around deformed tension bar. *Journal of the ACI*.

Hofstetter, G. and H. Mang (1995). *Computational Mechanincs of Reinforced Concrete Structures*. Braunschweig. Wiesbaden: VIEWER.

Tassios, T. P. (1979). *Properties of bond between concrete and stell under load cycles idealizing seismic actions*, Vol. 1. Rome: AICAP CEB Symposium.

Computational Modelling of Concrete Structures, Bićanić et al. (eds)
© 2003 Swets & Zeitlinger, Lisse, ISBN 90 5809 536 3

Scalar damage model for concrete without explicit evolution law

J. Ožbolt
Institute of Construction Materials, University of Stuttgart, Germany

S. Ananiev
Institute of Lightweight Structures and Conceptual Design, University of Stuttgart, Germany

ABSTRACT: Based on the fact that for an isotropic material model the elastic predictor and the projected stress tensors have the same eigenvectors, it is shown that the scalar damage can be obtained directly from the projection algorithm. This eliminates the difficulty of a proper definition of equivalent strain which serves as a driving force for evolution of damage in concrete. Moreover, if eigenvectors are known it is not more necessary to use invariants of the stress tensor for the formulation of loading surface. In the present model the loading surface is represented in the polynomial form. This has two advantages: (i) it automatically fulfils C_1 continuity and (ii) plane stress formulation is achieved by simply setting the third stress to zero. The performance of the model is illustrated on example of a mixed-mode fracture of concrete. It is shown that for the present example the model prediction strongly depends on the choice of the stress degradation law.

1 INTRODUCTION

Damage and plasticity theories are well established theoretical frameworks for macroscopic modelling of materials. One difficulty in the modelling of concrete like materials is due to the fact that such materials exhibit highly non-symmetrical behaviour in tension and compression. Consequently, damage formulations based on the equivalent strain concept are difficult to extend to concrete.

The problem can be partially solved by taking only the positive (tensile) part of the strain tensor as a driving force of damage (Mazars, 1986). This seems reasonable, however, the absence of the loading surface leads to mismatching of elastic and inelastic ranges with a consequence that, for instance, under biaxial compression damage takes place too early.

In plasticity-like formulation (see for instance Meschke et al., 1998) the loading surface is a starting point. Degradation of the elastic stiffness tensor is related to the derivates of the loading surface following the postulate of maximum energy dissipation functional or, for softening materials, using the postulate of its stationarity. The problem here is a validity of this postulate for concrete like materials. Even under biaxial compression the inelastic deformations do not have the same nature as plastic deformations in metals. There is no crystal slip simply because the material does not have a crystalline structure. Therefore, it is not surprising that the non-associated flow rule, which is prohibited by the above mentioned postulate, often produces better results.

Here, a scalar damage model for concrete is formulated based on a procedure that is formally the same as the plasticity-like formulation. As well known, the main purpose of the flow rule is a determination of six equations for six unknown components of plastic stain tensor. Additional unknown, the rate of plastic strains, is obtained from the consistency condition, i.e. during inelastic loading the stress state can not leave the loading surface.

Formally, one can say that if there are no plastic strains (elastic damage), there is no need for the flow rule. If damage is defined with only one unknown (scalar damage), than it can be obtained from the consistency condition. The assumption of scalar damage has a consequence that the material remains isotropic for the entire load history and that the current damage can be determined by scaling of elastic stress state back to the loading surface. The loading surface itself is parameterized using the actual values (strengths) of the uniaxial tensile and uniaxial compressive stresses. Since the loading surface is formulated in the principal stress space, their dependences on damage variable can be taken directly from experiments.

2 FORMULATION OF THE MODEL

2.1 *Loading surface*

The idea that inelastic (plastic) deformations initiate after the equivalent stress reaches a certain level (yield stress or strength of the material) can be traced back to the work of von Mises. It is reasonable to assume that for isotropic material the equivalent stress is independent of the choice of the coordinate system. This is the reason why the expression for equivalent stress was formulated using invariants of the stress tensor. Unfortunately, for concrete like materials it is not a simple task to formulate loading surface in a closed form. One possible way is a formulation in Haigh-Westergaard coordinates as it has been done, for example, by Etse (1992). The loading surface is continuously differentiable, however, it has a relatively large number of parameters, which complicates the formulation. A more serious disadvantage of this model, which hold for every plasticity model, is the fact that the tension softening is understood as accumulation of plastic strains, which is physically not correct.

Another possibility is a multi-surface formulation where more loading surfaces with a relatively low number of parameters per surface are used. The disadvantage here is that the resulting surface is non-continuously differentiable. This is from a numerical point of view not a problem since the corresponding value of plastic multiplier can be obtained using Koiter's rule. However, from the physical point of view it is not acceptable that a infinitely small rotation of stress tensor causes a finite rotation of plastic strain tensor.

When the loading surfaces are expressed in terms of stress invariants than it is difficult to connect them smoothly. Therefore, to avoid the problem of the discontinuity when skipping from one loading surface to another, in the present model a polynomial representation of the entire loading surface is employed. The components of the polynomial expression are principal stresses. The justification of not using invariants of stress tensor can be found in Simo (1992). He has shown that for isotropic material models the stress projection algorithm does not change the eigenvectors and if they are known it is not longer necessary to use stress invariants.

An additional advantage of such a formulation is that the reduction to the plane stress state is achieved by simple setting of one principal stress to zero. This removes the need for some special techniques like subiterations (Feenstra, 1993) or a projection matrix (Simo and Taylor, 1986).

For reasons of simplicity, in the following the formulation of the model is restricted to the two-dimensional state of stresses. The proposed loading surface F, which is a fourth order polyinom with seven unknown coefficients, reads:

$$F(\sigma_I,\sigma_{II}) \equiv 0$$

$$a_1\left(\sigma_I^2\cdot\sigma_{II}^2\right)+a_2\left(\sigma_I^3+\sigma_{II}^3\right)+a_3\left(\sigma_I^2\cdot\sigma_{II}+\sigma_{II}^2\cdot\sigma_I\right)+ +a_4\left(\sigma_I^2+\sigma_{II}^2\right)+a_5\sigma_I\cdot\sigma_{II}+a_6\left(\sigma_I+\sigma_{II}\right)+a_7=0 \quad (1)$$

where σ_I and σ_{II} are principal stresses and a_1 to a_7 are unknown coefficients. Seven coefficients are sufficient to realistically describe the loading function in two-dimensional stress space. They are obtained from the following seven conditions:

- The biaxial compressive strength $k_1 f_C$ is larger than the uniaxial compressive strength f_C:

$$F\left(-k_1\cdot|f_C|,-k_1\cdot|f_C|\right)=0 \quad \text{(i)}$$

- The maximal compressive strength $k_2 f_C$ is achieved when the stress in one direction is two times larger than in the other direction. The maximal compressive strength is larger than the biaxial compressive strength ($k_2 > k_1$):

$$F\left(-0.5k_2\cdot|f_C|,-k_2\cdot|f_C|\right)=0 \quad \text{(ii)}$$

$$\frac{\partial F}{\partial\sigma_I}\left(-0.5k_2\cdot|f_C|,-k_2\cdot|f_C|\right)=0 \quad \text{(iii)}$$

- The uniaxial compressive strength is f_C:

$$F\left(0,-|f_C|\right)=0 \quad \text{(iv)}$$

- The uniaxial tensile strength f_T:

$$F(f_T,0)=0 \quad \text{(v)}$$

- The Rankine tension cut-off criteria reads:

$$\frac{\partial F}{\partial\sigma_I}(f_T,0)=1 \quad \text{(vi)}$$

$$\frac{\partial F}{\partial\sigma_{II}}(f_T,0)=0 \quad \text{(vii)}$$

The above specified seven boundary conditions allow us to find seven constants from (1). The resulting loading surface is shown in Figure 1. It is similar to the curve obtained from experiments. At the points A and B the loading surface is automatically C_1 continuous because of the symmetry of the polynom with respect to the principal stresses.

2.2 *Softening rules*

Similar to the idea of Feenstra (1993), the softening rule is taken directly from the uniaxial experiment. The softening rules are understood here as dependence between tensile/compressive stress and scalar damage. The objectivity with respect to the finite element size is accounted for by scaling of fracture energy G_F to the average element size, i.e. the crack band approach is employed (Bažant and Oh, 1983). In the following the

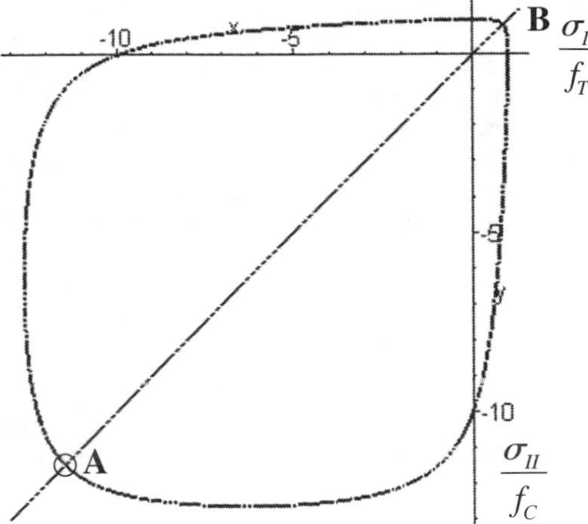

Figure 1. Loading surface defined as a fourth order polynom.

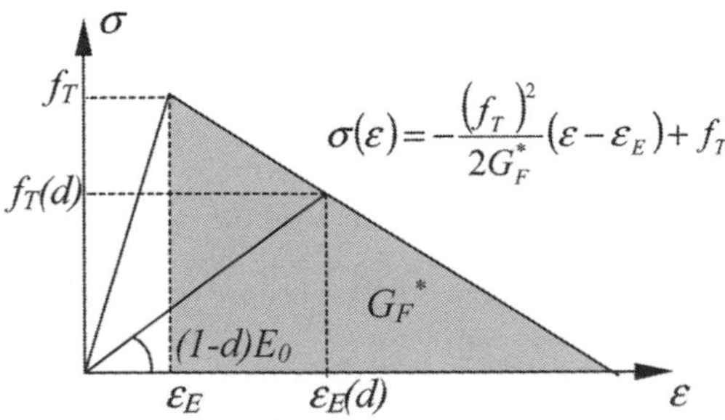

Figure 2. Linear degradation of tensile stress.

description of several different softening curves that are implemented into the model is discussed.

2.2.1 *Tension: linear degradation*

As a first approximation let's assume a linear degradation of tensile stress. The dependence between tensile stress f_T and scalar damage d can be obtained by solving an equation which states that for a given stress one has to get the same elastic strain from the damage formulation and from the degradation equation:

$$\frac{f(d)}{(1-d)E_0}=\left(f_T+\frac{f_T^2\varepsilon_E}{2G_F^*}-f(d)\right)\frac{2G_F^*}{f_T^2} \tag{2}$$

where G_F^* is the area under the softening stress-strain curve that is related to concrete fracture energy G_F and element size h (crack band approach) and E_0 is the

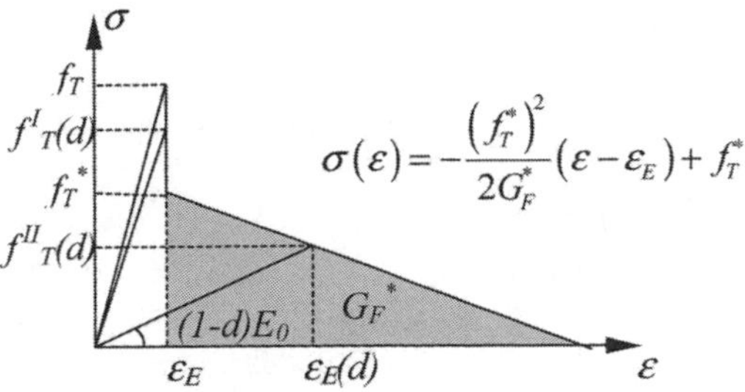

Figure 3. Discontinuous linear degradation of tensile stress.

Young's modulus. From (2) the desired dependence reads:

$$f(d)=(1-d)f_T\frac{2G_F^*E_0+f_T^2}{2G_F^*(1-d)E_0+f_T^2} \tag{3}$$

It is obvious that the softening law is not a linear function of damage.

2.2.2 *Tension: discontinuous linear degradation*

To study the influence of a sharp discontinuity in the stress-strain curve after onset of cracking, the stress-strain softening curve shown in Figure 3 is implemented into the model.

The discontinuity of the degradation law for tensile stress does not mean that the softening law as a function of damage is discontinuous as well. It reads:

$$\begin{aligned}&f(d)=(1-d)f_T \qquad (\varepsilon=\varepsilon_E)\\&f(d)=(1-d)f_T^*\frac{2G_F^*E_0+(f_T^*)^2}{2G_F^*(1-d)E_0+(f_T^*)^2} \qquad (\varepsilon>\varepsilon_E)\end{aligned} \tag{4}$$

2.2.3 *Tension: exponential degradation*

The exponential degradation law is frequently used in the literature, (Feenstra, 1993; Mazar, 1986; Pivonka et al., 2002). Unfortunately, it does not allow a closed form formulation of the softening law for $f(d)$ which somewhat complicates the formulation.

The desired softening law is now expressed in implicit form:

$$\begin{aligned}&\Theta(f,d)=0\\&f(d)+(1-d)E_0\frac{G_F^*}{f_T}\ln(f(d))-\\&-(1-d)\left(E_0\frac{G_F^*}{f_T}\ln(f_T)-f_T\right)=0\\&\frac{\partial\Theta}{\partial f}(f,d)=1+\frac{1}{f}(1-d)E_0\frac{G_F^*}{f_T}\end{aligned} \tag{5}$$

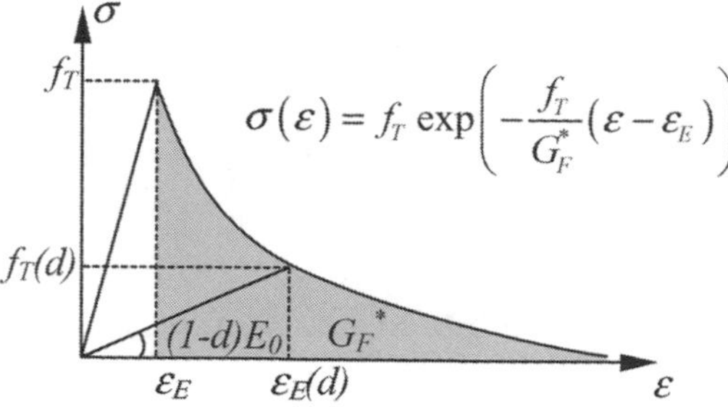

Figure 4. Exponential degradation of tensile stress.

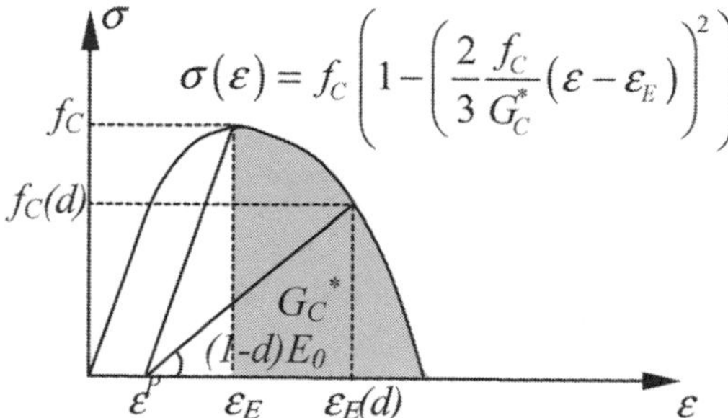

Figure 5. Parabolic degradation of compressive stress.

An analytical solution of (5) is possible using Lambert's function, but here the solution is obtained numerically using standard Newton-Rathson method:

$$f(d^*) = f_0 - \sum_i \frac{\Theta(f_{i-1}, d^*)}{\frac{\partial\Theta}{\partial f}(f_{i-1}, d^*)} \tag{6}$$

where d^* is a given value of damage for which the corresponding value of tensile stress has to be found. Equation (6) is a general solution and it makes possible to use any form of degradation law.

2.2.4 *Compression: parabolic degradation*

Damage approach describes the behaviour of concrete in tension reasonably well. However, from experimental evidence it is obvious that for dominant compressive load a plasticity plays an important role as well. Principally, a realistic model for concrete has to take into account both phenomena, i.e. damage and plasticity. When a single loading surface is used, then it is necessary to separate somehow these inelastic processes. Meschke et al. (1998) proposed to use a simple scalar multiplicator, which controls the proportion between dissipated energy during plastic and damage deformations.

In the present model, two separate and non-overlapping loading surfaces are proposed. Non-overlapping in this context means that the plastic strains are accumulated only during hardening and that the softening is modelled exclusively by scalar damage. Moreover, it is assumed that the hardening affects only compressive stress (see also Feenstra, 1993). This allows independent consideration of plasticity and damage. The proposed stress-strain relationship for uniaxial compression is illustrated in Figure 5. In the following, no hardening is accounted for ($\varepsilon^p = 0$) since the plasticity formulation is out of the scope of the present paper.

The desired softening law is obtained as a first (positive) root of the quadratic equation:

$$f(d) = -\frac{1}{2}A(d) + \frac{1}{2}\sqrt{(A(d))^2 - 4B(d)}, \tag{7}$$

where,

$$\begin{aligned} A(d) &= -2(1-d) f_C + \frac{((1-d)E_0)^2}{f_C}\left(\frac{3}{2}\frac{G_C^*}{f_C}\right)^2, \\ B(d) &= (1-d)^2\left(f_C^2 - \left(\frac{3}{2}\frac{G_C^*}{f_C}E_0\right)^2\right). \end{aligned} \tag{8}$$

3 RESOLVING EQUATION

For simplicity the formulation of the model is illustrated in Figure 6 assuming a Tresca loading surface. The actual value of damage is obtained from scaling of the elastic stress state back to the loading surface, but not to the initial one. The loading surface is parameterized using uniaxial tensile and compressive stresses, which depend on the actual level of damage. That means, the failure surface "shrinks" with increase of damage (see Figure 6).

Such scaling can be understood as a simplified form of a backward Euler integration scheme, which can be reduced to one scalar equation regarding the plastic multiplier. For example, the consistency condition for the case shown in Figure 6 reads:

$$(1-d)\left(\frac{\sigma_I^{Elastic}}{f_T(d)} + \frac{\sigma_{II}^{Elastic}}{f_C(d)}\right) = 1 \tag{9}$$

In special case, where both softening laws have linear form (3), it is possible to solve (9) analytically. The solution reads:

$$d(\sigma_I, \sigma_{II}) = \frac{CD(\sigma_I f_C - f_T f_C + \sigma_{II} f_T)}{\sigma_I (C - f_T^2) f_C D + \sigma_{II}(D - f_C^2) f_T C} \tag{10}$$

where,

$$\begin{aligned} C &= 2G_F^* E_0 + f_T^2, \\ D &= 2G_C^* E_0 + f_C^2. \end{aligned} \tag{11}$$

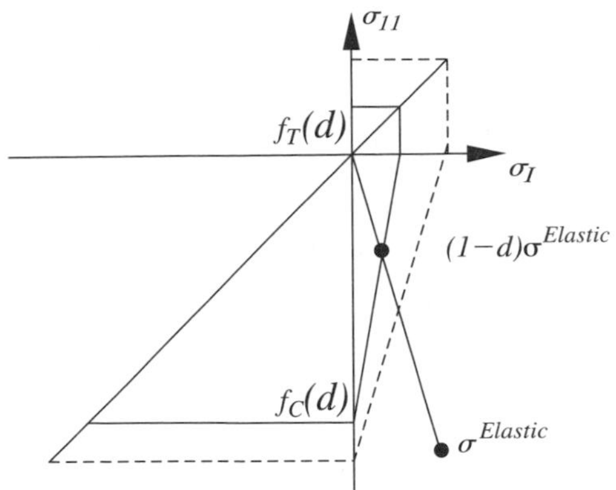

Figure 6. Projection of elastic stress state back to the loading surface.

Actually, (10) represents a damage evolution law, but it is formulated in completely different way as one usually does, i.e. without the equivalent strain concept.

4 ALGORITHMIC TANGENT OPERATOR

In the plasticity formulation the continuous tangent operator does not provide a quadratic convergence of the Newton-Rathson method during equilibrium iterations (Simo and Taylor, 1986). One has to linearize the stress state equations after projection to the loading surface.

In the presented formulation the scalar damage does not result from the integration of a certain flow rule. Consequently, there is no continuous tangent operator. The only possible tangent operator is a algorithmic one derived from the projected stress state. A general expression for scalar damage can be written as follows:

$$\boldsymbol{\sigma} = (1-d)\,\boldsymbol{D}^{el} : \boldsymbol{\varepsilon} \tag{12}$$

Differentiation of (12) gives the tangent operator:

$$\frac{\partial \boldsymbol{\sigma}}{\partial \boldsymbol{\varepsilon}} = (1-d)\,\boldsymbol{D}^{el} - \frac{\partial d}{\partial \boldsymbol{\varepsilon}} \otimes \boldsymbol{\sigma}^{Elastic} \tag{13}$$

The partial derivates of damage required in (13) can be obtained from (10). During unloading the tangent operator automatically reduces to the secant one because damage does not change, i.e. it is independent of strains. It is interesting to observe that although a relatively simple isotropic scalar damage model is considered, the tangent stiffness operator (13) is in general non-symmetric.

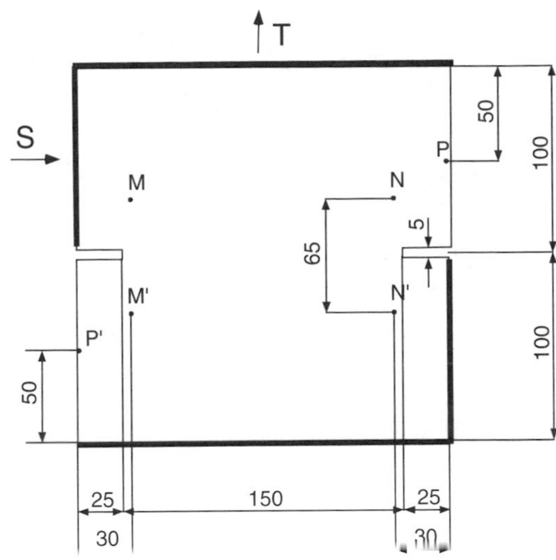

Figure 7. Geometry of the Nooru-Mohamed test specimen.

5 NUMERICAL EXAMPLE

Recent numerical studies showed that relatively complex mixed-mode failure of concrete can be objectively simulated using smeared fracture finite element analysis only if a sophisticated material model is used (Ožbolt and Reinhardt, 2003). The studies were carried out on the Double-Edge-Notched specimen tested by Nooru-Mohamed (1992) (see also Pivonka et al., 2002). It has been observed that the material model plays the most important role. To investigate the performance of the presented scalar damage model the numerical analysis is performed on the same example.

The geometry and the test set-up are shown in Figure 7.

The specimen was first loaded by shear load S. Subsequently, at a constant shear load of $S = 10\,\text{kN}$, the vertical tensile load T was applied up to failure. The load control procedure was used by moving the upper loading platens in horizontal and vertical direction, respectively. The rotation of the loading platens was restricted. During the application of the horizontal load S, the vertical load was kept at zero ($T = 0$). Upon subsequent tensile loading the shear force was kept constant. The bottom (support) platens were fixed and, the same as the upper (loading) platens, glued to the surface of the specimen. The finite element discretization is performed by plane stress quadrilateral finite elements with four integration points. The material properties are taken as: Young's modulus $E_C = 32800\,\text{MPa}$, Poisson's ratio $v = 0.2$, tensile strength $f_T = 3.0\,\text{MPa}$, uniaxial compressive strength $f_C = 38.4\,\text{MPa}$ and concrete fracture energy $G_F = 0.11\,\text{N/mm}$.

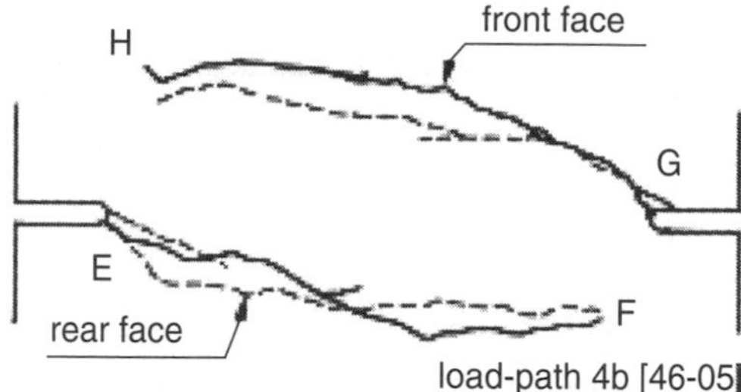

Figure 8. Crack pattern observed in the experiment.

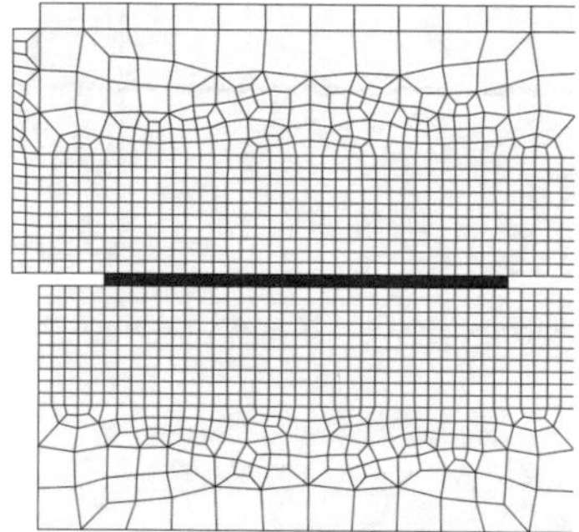

Figure 9. Crack pattern obtained by employing a linear softening law.

For the present example the compressive properties of concrete are of minor importance. For this reason only the influence of the tensile softening laws was investigated. The experimentally observed crack pattern is shown in Figure 8. As can be seen the crack pattern consists of two separate cracks.

The first softening law that was linear. The fracture energy G_F was scaled to the width of the crack-band $h = 5$ mm (element size). The observed crack pattern is shown in Figure 9.

As it can be seen, the use of the linear softening law results in an unrealistic crack pattern, i.e. the crack plotted in terms of maximal principal strains propagates along straight line and it is not curved as in the experiment. The corresponding load-displacement curve (Figure 12) indicates an overestimation of the peak resistance obtained in the experiment.

By introducing discontinuity in the degradation law for tensile stress ($f_T^*/f_T = 0.8$, see Figure 3) the crack path becomes more realistic and the calculated load-displacement curve agrees better with the experiment.

The third simulation was performed by employing a linear degradation law with extremely low fracture energy G_F. It was reduced by a factor of 150. The concrete in this case is thus assumed to be nearly a brittle material. This explains a steep decrease of the softening branch of the load-displacement curve and consequently significant underestimation of the ultimate load (see Figure 12). However, as can be seen from Figure 11 the calculated crack pattern agrees well with the experimental one.

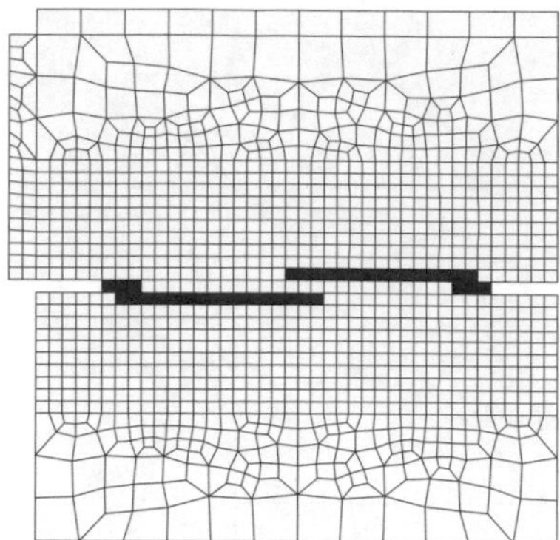

Figure 10. Crack pattern obtained by the use of discontinuous linear degradation.

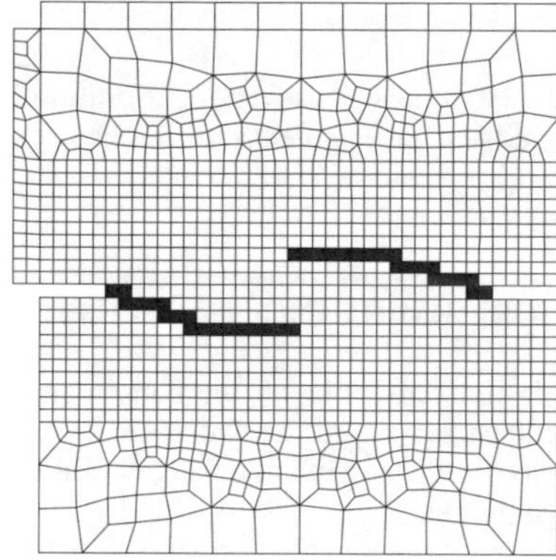

Figure 11. Crack pattern obtained assuming a brittle material.

To get an insight in the behaviour of the model it is useful to draw the softening laws based on the linear degradation of tensile stress (see Figure 2) for different values of the crack band width.

In the present example the size of the finite elements (crack band width) was $h = 5$ mm which leads to a rather ductile stress-strain relationship that is almost elasto-plastic. Consequently, with such stress-strain law the model is not able to predict the crack path correctly. With a more realistic stress-strain relationship, i.e. by adopting the shape such that it is similar to an realistic stress-crack opening law, the predicted crack path becomes more objective. This shows that the present model is sensitive to the choice of the stress-strain curve, which is actually not a problem of the model but a limitation of the crack band approach. Namely, in a limit case ($h \rightarrow 0$) the solution yields to the plasticity solution.

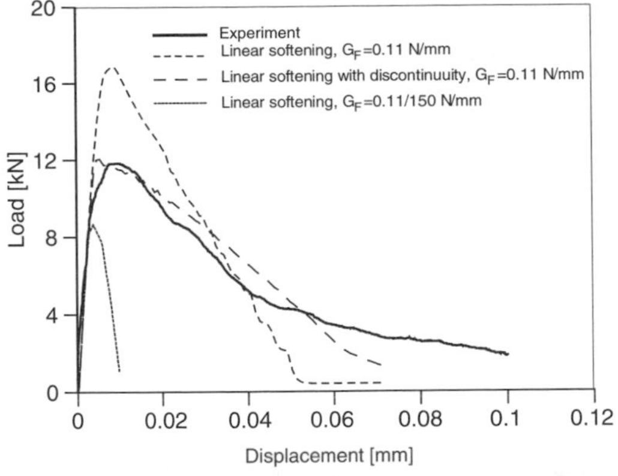

Figure 12. Load-displacement curves obtained in all three numerical simulations as well as the experimentally obtained curve.

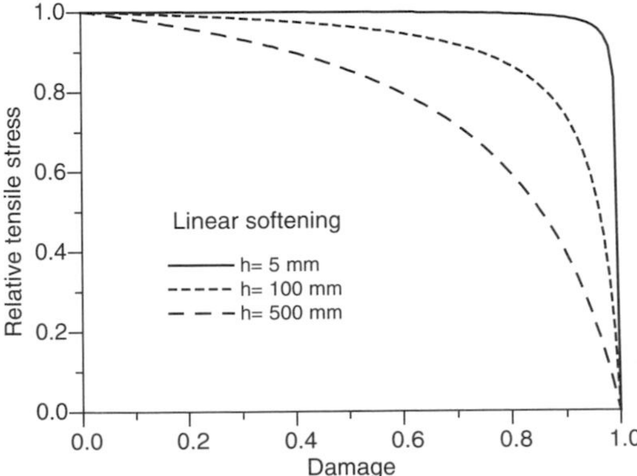

Figure 13. Dependency of the stress degradation on the damage for three different crack band widths.

The three curves shown in Figure 13 correspond to three different widths of the crack band (element size): $h = 5$, 100 and 500 mm. It can be seen that for small elements the stress does not reduce significantly until a very high level of damage. From the damage point of view the function is unrealistic, however, for the given element size the function is required to assure correct consumption of fracture energy (crack band method). There is an obvious need for the optimisation of the shape of the softening curve such that the stress is a realistic function of damage and at the same time that the consumption of energy is objective with respect to the element size.

6 CONCLUSIONS

A scalar damage model with loading surface in closed form is proposed without any explicit evolution law based on the equivalent strain concept. For the two-dimensional problem a forth order polynom is shown to be a realistic choice for the loading surface. It is shown that the formulation of the loading surface in the principal stress space has some advantages over the formulation in terms of stress invariants. The main advantage is transparency of the model and, what is of a great importance for the calibration of the model, it is easy to identify model parameters from experiments. The influence of different softening laws on the performance of the model is discussed. For a linear softening law and for a linear form of the loading surface, implicit damage evolution law is formulated in a closed form. It is demonstrated that in spite of its simplicity, the model can realistically predict the cracking of concrete for rather complex mixed mode fracture. If the crack band method is used as a localization limiter, then for small crack band width (small elements) the constitutive law tends to be almost elasto-plastic. This leads to unrealistic predictions and is in contradiction with quasi-brittle nature of fracture of concrete. More work is needed to further improve the proposed model.

REFERENCES

Bažant, Z. P., Oh, B. H., 1983. Crack band theory for fracture of concrete. *Materials and Structures*, RILEM, 93(16): 155–177.

Etse, G., 1992. *Theoretical and numerical investigation of diffusive and localized damage in concrete*. Ph.D. Thesis, University of Karlsruhe. (in German).

Feenstra, P., 1993. *Computational aspects of biaxial stress in plain and reinforced concrete*. Ph.D. Thesis, Technical University of Delft.

Mazars, J., 1986. A description of micro- and macro-scale damage of concrete structures. *Journal of Engineering Fracture Mechanics*, 25: 729–737.

Meschke, G., Lackner, R., Mang, H., 1998. An anisotropic elastoplastic-damage model for plain concrete. *International Journal of Numerical Methods in Engineering*, 42: 703–727.

Ožbolt, J., Reinhardt, H. W., 2003. Numerical study of mixed mode fracture in concrete. *International Journal of Fracture*. Z.P., in press.

Nooru-Mohamed, M. B., 1992. *Mixed-mode fracture of concrete: an experimental approach*. Ph.D. Thesis, Technical University of Delft.

Pivonka, P., Ožbolt, J., Lackner, R., Mang, H., 2002. Comparative studies of 3D-constitutive models for concrete: application to mixed-mode fracture. *International Journal of Numerical Methods in Engineering*, (submitted).

Simo, J., 1992. Algorithms for static and dynamic multiplicative plasticity that preserve the classical return mapping schemes of the infinitesimal theory. *Computer Methods in Applied Mechanics and Engineering*, 99: 61–112.

Simo, J., Taylor, R., 1986. A return mapping algorithm for plane stress elastoplasticity. *International Journal of Numerical Methods in Engineering*, 29: 649–670.

Computational Modelling of Concrete Structures, Bićanić et al. (eds)
© 2003 Swets & Zeitlinger, Lisse, ISBN 90 5809 536 3

Modeling of concrete response under fatigue

T. Pfister*, D. Pfanner, F. Stangenberg & Y.S. Petryna
Institute for Reinforced and Prestressed Concrete Structures, Ruhr-Universität Bochum, Germany

ABSTRACT: A numerical approach to modeling the response of structural concrete under fatigue load is proposed. It aims at realistic predictions of concrete degradation in context of structural lifetime assessment. The approach is essentially based on the material model for concrete developed within the uniform framework of the elasto-plastic continuum damage theory, which is presented in the companion paper (Bockhold et al. 2003). The present contribution improves the classical fatigue curve (*S–N* curve) approaches by nonlinear evolution laws for fatigue strains and damage. Besides estimates of the fatigue failure itself, it is able to predict the intermediate material states of concrete during fatigue life. The model is statistically founded and accounts for uniaxial loading conditions. Its theoretical extension to multiaxial fatigue as well as a supporting experimental program are currently in progress. The approach is illustrated by a numerical example of reinforced concrete column under cyclic loading.

1 INTRODUCTION

After rapid growth of civil engineering infrastructure built of concrete during the last five decades a certain degree of saturation can be observed nowadays. The main focus of engineering practice is continuously moving from new design to lifetime assessment and lifetime design (Sarja and Vesikari 1996; Clifton 2000). The time-invariant response of concrete may be considered as relatively well studied, there are already several standard modeling approaches, based on fracture mechanics (van Mier 1997; Bažant and Planas 1998) or continuum damage mechanics (Ortiz 1985; Mazars 1986; de Borst 1997). The time-variant response seems to be much less investigated. In order to match these new challenges mechanical sciences dealing with modeling of concrete must be able to better simulate the long-term behavior of concrete structures and related degradation mechanisms.

Fatigue of concrete is a degradation mechanism, which is rarely reported to be a sole cause of structural failure. Its influence may be rather observable in interaction with other deterioration processes. This complications make a realistic modeling of concrete response to fatigue a really challenging task.

The most research on concrete fatigue has been focused on determination of fatigue life itself, similarly to steel structures. Such efforts finally resulted in approaches able to predict a number N_f of stationary load excursions, which can be resisted until fatigue failure, see (Hsu 1981) for example. However, these approaches, attended in the literature in form of *S–N* or Wöhler curves, leave some substantial problems unsolved. It concerns first of all the intermediate material states during fatigue life and the evolution laws for strains, stresses and damage. Quite few systematic investigations can be mentioned in this context, among others (Jinawath 1974; Holmen 1979; Zhang and Wu 1997).

The present contribution proposes an empirical approach taking the evolution of fatigue damage into account. The approach, presently formulated for uniaxial loading conditions, is further shown to be able to predict concrete state for arbitrary number of load excursions. It is developed to be implemented into the finite element analysis of concrete structures.

2 TIME-INVARIANT MATERIAL MODEL FOR CONCRETE

The present approach is essentially based on the time-invariant 3D material model for concrete recently developed in (Pölling 2000; Krätzig and Pölling 2002), which is presented in the companion paper (Bockhold et al. 2003). In this contribution, the fatigue state under uniaxial compressive loading are mainly addressed to, therefore, the basic material model is summarized below with respect to the uniaxial conditions only.

The plastic strain

$$\varepsilon^{pl} = \varepsilon - \varepsilon^{e} = \varepsilon - \varepsilon^{0} - \varepsilon^{d}, \quad \text{with} \quad \varepsilon^{0} = \frac{\sigma}{E_c} \tag{1}$$

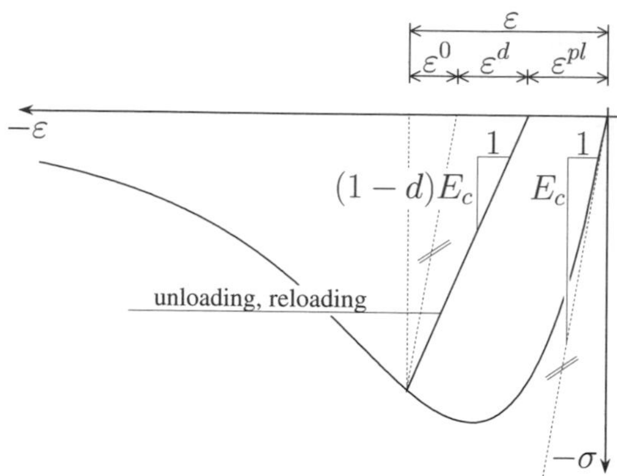

Figure 1. Strain partition and damage variable.

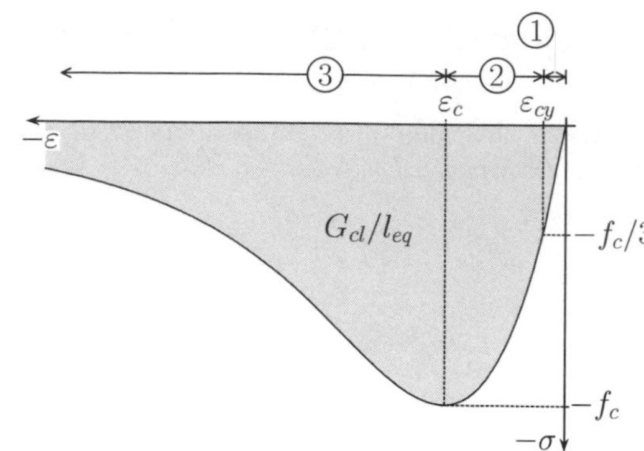

Figure 2. Stress-strain relationship for concrete under compression.

and the scalar damage variable

$$d = 1 - \frac{E}{E_c} \tag{2}$$

are selected as internal thermodynamic state variables (Fig. 1). Herein, $\varepsilon, \varepsilon^e$ denote the total and elastic strain, E, E_c stand for the instantaneous and initial elasticity modulus respectively.

The evolution of plastic and damage strains are assumed to be coupled by means of a scalar parameter b, so that

$$\varepsilon^{pl} = b\,\varepsilon^{in}, \tag{3}$$

$$\varepsilon^{d} = (1-b)\,\varepsilon^{in}. \tag{4}$$

With these variables, the stress-strain relationship then reads:

$$\sigma = \left(\varepsilon - \varepsilon^{pl}\right)(1-d)\,E_c. \tag{5}$$

The nonlinear behavior of concrete is described by the joint plastic/damage potential of the form:

$$\Phi(\sigma,\bar{\sigma}) = |\sigma| - \sigma_{ci}(\varepsilon) = 0, \tag{6}$$

where $\sigma_{ci}(\varepsilon)$ denotes the uniaxial back-stress given in an analytical form below.

The concrete response to monotonic loading generally experiences three different stages:

- linear elastic behavior under stresses below $\sigma = -1/3f_c$;
- hardening behavior for stresses $-1/3f_c \geq \sigma \geq -f_c$ characterized by extensive micro-cracking;
- softening behavior caused by the formation of a macro-crack, which can be observed in strain-controlled tests for $\varepsilon < -\varepsilon_c$.

In order to correctly reflect these types of behavior, the uniaxial stress-strain relationship $\sigma(\varepsilon)$ is subdivided into three domains, which are described below (Fig. 2).

Domain 1: linear elastic response $-\varepsilon_{cy} \leq \varepsilon$

$$\sigma_{c1} = E_c\,\varepsilon \tag{7}$$

Domain 2: hardening response $-\varepsilon_c \leq \varepsilon < -\varepsilon_{cy}$

$$\sigma_{c2} = \frac{E_{ci}\frac{\varepsilon}{f_c} + \left(\frac{\varepsilon}{\varepsilon_c}\right)^2}{1 - \left(E_{ci}\frac{\varepsilon_c}{f_c} - 2\right)\left(\frac{\varepsilon}{\varepsilon_c}\right)} f_c \tag{8}$$

with

$$E_{ci} = \frac{1}{2E_c}\left(\frac{f_c}{\varepsilon_c}\right)^2 - \frac{f_c}{\varepsilon_c} + \frac{3}{2}E_c\,. \tag{9}$$

The stress-strain relation is adopted from (CEB/FIP 1991). The elasticity modulus E_c has been replaced by E_{ci} in order to guarantee that the relation contains the point $-\varepsilon_{cy}, -f_{cy}$.

Domain 3: softening response $\varepsilon < -\varepsilon_c$

$$\sigma_{c3} = \left(\frac{2 + \gamma_c f_c \varepsilon_c}{2 f_c} + \gamma_c \varepsilon_c + \frac{\gamma_c}{2\varepsilon_c}\varepsilon^2\right)^{-1}. \tag{10}$$

In order to avoid ill-posed continuum boundary value problems due to formation of discrete cracks, the crack band and fracture energy approach (Bažant and Oh 1983; Bažant and Planas 1998) are applied. The softening branch is governed using the crushing energy G_{cl} considered as material parameter and the characteristic length l_{eq} of the finite element. Parameter r_c depending on G_{cl}, l_{eq} controls the area under the stress-strain curve.

3 FATIGUE MODEL FOR CONCRETE

3.1 *Phenomenology*

Similar to fatigue problems in steel, the interest of research on concrete fatigue has been mainly dedicated to the estimation of the fatigue life, i.e. a number N_f of load excursions with constant stress amplitudes until fatigue failure. Based on such results, the well known *S–N* curves have been determined for concrete, see (Hsu 1981) for example. The curves typically exhibit a bend at $N_f \approx 10^3$ marking a boundary between the low-cycle and high-cycle fatigue domains.

Some observations, especially those dealing with earthquake loading (Sinha et al. 1964; Karsan and Jirsa 1969), show the hysteretic loops of cyclic loading to lie "inside" the monotonic stress-strain relationships, which have been called envelope curves due to this reason. However, the contrary results of other authors, e.g. (Awad and Hilsdorf 1971; Subramaniam et al. 2000), make the envelope concept at least questionable.

From the mechanical viewpoint, fatigue tests on concrete reveal irreversible strains on the one hand and stiffness degradation on the other hand (Holmen 1979). Therefore, the same internal variables as under monotonic loading – plastic strains ε^{pl} and damage d – may be used to describe fatigue states of material. The evolution of both, the plastic strain and damage observed in experiments, follows characteristic S-shaped curves. Accordingly, the entire fatigue life can be subdivided into three periods with different dominant micro-mechanical deformation mechanisms: initiation of micro-cracks, their stable growth and finally localization to a macro-crack (Holmen 1979; Klausen 1978; Shah 1970). Independently of the stage of fatigue life, various experimental techniques conclude about the proportionality between plastic strains and damage, and thus indirectly confirm the assumption of equations (3), (4) made for monotonic loading.

3.2 *Fatigue failure and energy assumption*

Fatigue failure is phenomenologically associated with the beginning of damage localization and formation of a macro-crack in the representative volume element. After that moment, the stress σ_{max} originating from the fatigue loading can no longer be resisted by the concrete. The material state may be considered qualitatively analogous to the state at the failure point $\sigma = -f_c$ under monotonic loading.

This mechanical observation is used to define the fatigue failure criterion by means of the following energy assumption. It is postulated that the energy required to cause a certain degree of material damage is independent of the way the energy was exhausted. This means that the damage caused by monotonic loading equals the damage due to fatigue loading if the energy dissipated for damage growth is equal in both cases:

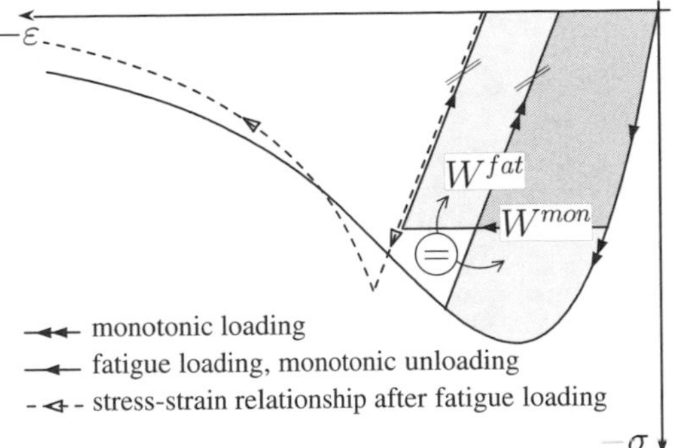

Figure 3. Energy equivalence for damage under monotonic and fatigue loading.

$$W^{mon}(d,\sigma) \stackrel{!}{=} W^{fat}\left(d,\sigma^{fat},N\right). \tag{11}$$

This hypothesis seems to be questionable at first sight due to the fact that much more external energy is usually dissipated in fatigue tests until failure compared to the energy in monotonic tests. The reason of such a difference has been explained by ultrasonic measurements of (Spooner and Dougill 1975). They show that only a part of the applied external energy is dissipated in micro-cracks, the rest is exhausted for plastic deformations and heat.

The energy assumption is illustrated in Fig. 3 for the same damage state reached under monotonic and cyclic loading.

3.3 *Fatigue life of concrete*

Various approaches to define the fatigue life of concrete have been critically analyzed in (Pfanner 2002). According to this analysis, the approach by (Hsu 1981) is selected as the most advantageous one. It matches a lot of experimental data, defines different fatigue curves for low-cycle and high-cycle fatigue and additionally accounts for deformation rate:

low-cycle:

$$\frac{\sigma_{max}}{f_c} = 1.2 - 0.2\,R - 0.133\,(1-0.779\,R)\cdot$$

$$\cdot\log N_f - 0.053\,(1-0.445\,R)\log T, \tag{12}$$

high-cycle:

$$\frac{\sigma_{max}}{f_c} = 1.0 - 0.0662\,(1-0.556\,R)\log N_f -$$

$$-0.0294\log T. \tag{13}$$

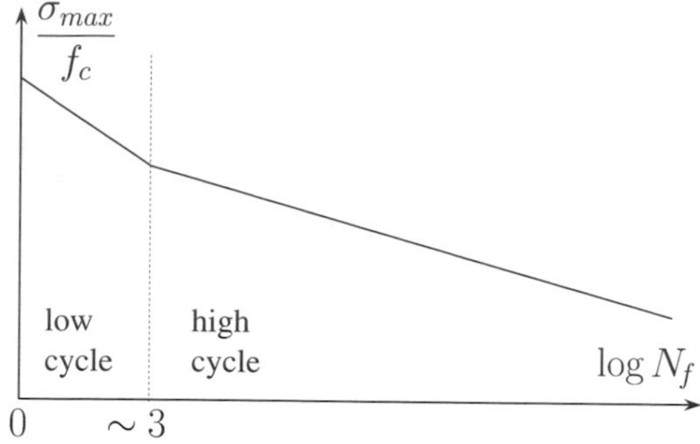

Figure 4. S–N curve.

Herein, R denotes the ratio between the maximum and minimum stress magnitude, while T stands for the duration of one load cycle. The approach results in the well known S–N curves as shown exemplarily in Fig. 4.

3.4 *Fatigue strain evolution*

In classical approaches taking nonlinear or time dependent behavior into account, strain is usually calculated by the time integration of a strain rate:

$$\varepsilon(t) = \int_{t=t_0}^{t_1} \dot{\varepsilon}\, dt.$$

This presumes material laws to be formulated in the rate form, that is a complex problem in case of arbitrary load histories. Such a problem becomes, however, much easier for constant stress amplitudes of cyclic loading, the case well documented in experiments, see (Holmen 1979) for example. Then, fatigue strains can be determined for each lifetime instant empirically.

On the basis of statistical analysis of numerous experimental data from the literature, (Pfanner 2002) has developed such an empirical model to predict fatigue strains, which is adopted here.

As mentioned above, the strain evolution curves under constant stress amplitudes exhibit a typical S-shaped character. They can be generally subdivided into three domains according to the normalized number of load excursions N/Nf (Fig. 6). Each curve begins at the strain ε_0 determined from the stress-strain relationship under monotonic loading:

$$\varepsilon_0 = \begin{cases} \dfrac{\sigma_{max}}{E_c} & : \ |\sigma_{max}| \leq \dfrac{f_c}{3} \\ \varepsilon_c^2 \left[\sqrt{Y_0^2 + \frac{4\sigma_{max}}{\varepsilon_c^2 f_c}} - Y_0\right] & : \ |\sigma_{max}| > \dfrac{f_c}{3} \end{cases} \quad (14)$$

with

$$Y_0 = \frac{E_{ci}}{f_c} \left[\left(\frac{1}{f_c} - \frac{2}{E_{ci}\,\varepsilon_c} \right) \sigma^{fat} + 1 \right] \quad (15)$$

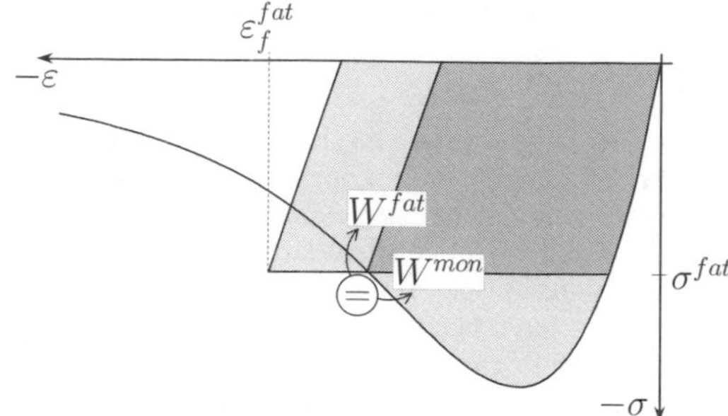

Figure 5. Determination of the failure strain ε_f^{fat}.

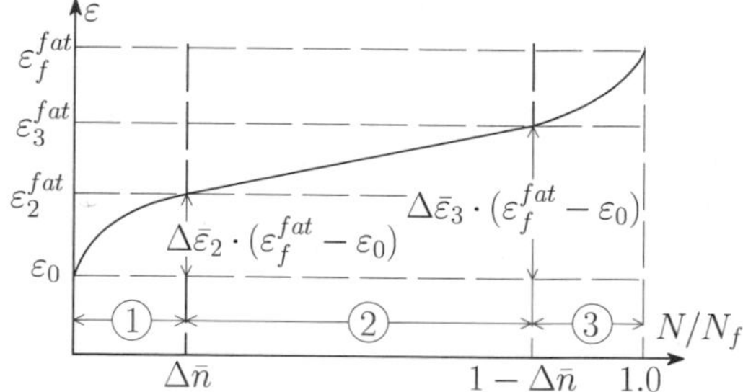

Figure 6. Evolution of fatigue strain over related number of load cycles.

The failure strain can be expressed using the energy hypothesis as follows:

$$\varepsilon_f^{fat} = \frac{1}{\sigma^{fat}} \left[g_{tot}^{da}\left(d_c^{da}\left(\sigma^{fat}\right)\right) - g_{c,asc}\left(\varepsilon_{c0}\right)\right] + \varepsilon_{c0}. \quad (16)$$

In this equation g_{tot}^{da} stands for the whole area under the stress-strain curve, $g_{c,asc}$ for the area under the ascending branch. It follows from the energy assumption adopted above that the area under the curve for monotonic load until $\sigma = \sigma^{fat}$ and that under the curve for fatigue load are equal (Fig. 5).

The boundary points between the domains are determined as portions $\Delta\bar{n}$ and $(1 - \Delta\bar{n})$ of the fatigue life N/Nf (Fig. 6). For concrete under compression they have been identified as:

$$\Delta\bar{n} = 0.2; \qquad 1 - \Delta\bar{n} = 0.8. \quad (17)$$

The corresponding strains ε_2^{fat}, ε_3^{fat} are determined with respect to the initial strain ε_0 and failure strain ε_f^{fat}:

$$\varepsilon_2^{fat} = \varepsilon_0 + \Delta\bar{\varepsilon}_2 \left(\varepsilon_f^{fat} - \varepsilon_0\right), \quad (18)$$

$$\varepsilon_3^{fat} = \varepsilon_0 + \Delta\bar{\varepsilon}_3 \left(\varepsilon_f^{fat} - \varepsilon_0\right). \quad (19)$$

Herein, $\Delta_{\bar{\varepsilon}2,3}$ denote the corresponding relative strain increments

$$\Delta\bar{\varepsilon}_2 = \kappa_R \left[0.3\left(\frac{\sigma_{max}}{f_c}\right)^2 + 0.46\right] \tag{20}$$

and

$$\Delta\bar{\varepsilon}_3 = \kappa_R \left[-0.3\left(\frac{\sigma_{max}}{f_c}\right)^2 + 0.8\right], \tag{21}$$

with

$$\kappa_R = 1 + R\left(\frac{\sigma_{max}}{f_c}\right)^{10}. \tag{22}$$

which are estimated on the basis of experimental data obtained by (Gaede 1962; Awad and Hilsdorf 1971; Holmen 1979).

Under such conditions, we are able to define a monotonic evolution function

$$J(N) = 1.0 + \Delta J_1(N) + \Delta J_2(N) + \Delta J_3(N) \tag{23}$$

with the following summands:

$$\Delta J_1(N) = \left(\frac{\varepsilon_f^{fat}}{\varepsilon_0} - 1\right) \cdot \frac{\Delta\bar{\varepsilon}_2\left[\left(\Delta\bar{n}^2 - 1\right)\Delta\bar{\varepsilon}_2 + \left(1 - \Delta\bar{n}\right)\Delta\bar{\varepsilon}_3\,\Delta\bar{n}\right]}{\Delta\bar{\varepsilon}_2 - \Delta\bar{n}\Delta\bar{\varepsilon}_3} \cdot \left[\left(\frac{\Delta\bar{n}\Delta\bar{\varepsilon}_3 - \Delta\bar{\varepsilon}_2}{\Delta\bar{n}^2\left(\Delta\bar{\varepsilon}_2 - \Delta\bar{\varepsilon}_3\right)}\frac{N}{N_f} + 1\right)^{-1} - 1\right], \tag{24}$$

$$\Delta J_2(N) = \left(\frac{\varepsilon_f^{fat}}{\varepsilon_0} - 1\right)\left(\Delta\bar{\varepsilon}_3 - \Delta\bar{\varepsilon}_2\right)\frac{N}{N_f}, \tag{25}$$

$$\Delta J_3(N) = \left(\frac{\varepsilon_f^{fat}}{\varepsilon_0} - 1\right)\left(1 - \Delta\bar{\varepsilon}_3\right) \cdot \left(\frac{N}{N_f}\right)^{\log_{1-\Delta\bar{n}}\left[\frac{\Delta\bar{n}\left(\Delta\bar{\varepsilon}_2 - \Delta\bar{\varepsilon}_3\right)}{\Delta\bar{\varepsilon}_3 - 1}\right]}. \tag{26}$$

This function defines fatigue strains for arbitrary given numbers N of load cycles:

$$\varepsilon^{fat} = J(N) \cdot \varepsilon_0. \tag{27}$$

The approach proposed above is appropriate for stationary stress amplitudes $\sigma_{max} = const$, $\sigma_{min} = \text{const}$. In reality, the maximum/minimum stress permanently changes during the fatigue life as a result of damage accumulation and stress redistribution. In this case, corrected values of stresses must be used at each new lifetime increment. Such a correction as well as the choice of time increments cause numerical difficulties, which are discussed in (Pfanner 2002). Therefore, the evolution model for fatigue strains will be reformulated in a consistently rate form. A supporting experimental program and corresponding theoretical developments are already in progress.

3.5 *Damage evolution*

In contrast to classical evolution laws of continuum damage mechanics, the present model for fatigue damage is not formulated in the rate form like

$$\dot{d} = \Phi_d(\sigma, \varepsilon^p, d, N),$$

but is based on the energy assumption adopted above: the energy dissipated in damage is independent of the loading type. Thus, the area in between the stress-strain loop under monotonic loading traced for a damage state d is assumed to be equal to the area under the envelope stress-strain path under fatigue loading for the same damage state d (Fig. 3).

Thus, the damage state d for a given fatigue life instant N/Nf can be estimated from an equivalent reference state under monotonic loading defined by

$$d = 1 - \frac{\sigma(\varepsilon)}{\varepsilon_c E_c\,(1 - b) + b\,\sigma(\varepsilon)} \tag{28}$$

leading to the following nonlinear equations:

$$\frac{1}{2}\frac{\left(\sigma^{fat}\right)^2 - \left(\sigma_{c3}(\varepsilon)\right)^2}{(1 - d)\,E_c} = \left[1 - (1 - d)^{\kappa_c}\right] g_{cl2}^{da}(d), \tag{29}$$

with

$$g_{cl2}^{da}(d) = g_{cl2} - g_{c,desc}^{da}(d), \tag{30}$$

$$g_{c,desc}^{da} = \int_{\bar{\varepsilon}=\varepsilon}^{-\varepsilon_c} \sigma_{c3}(\bar{\varepsilon})\,d\bar{\varepsilon}. \tag{31}$$

Such a nonlinear problem cannot be solved for d in closed form, even for the known stress-strain relation. It is, therefore, solved iteratively according to (Pfanner 2002).

3.6 *Material degradation*

The uniaxial stress-strain relation for concrete under compression is generally defined by four material parameters: the elasticity modulus E_c, the compression strength f_c, the failure strain ε_c and the crushing energy $g_{cl} = G_{cl}/l_{eq}$ (Fig. 2). The material degradation during fatigue life will obviously influence this stress-strain relation. In order to define the new one for each lifetime instant, the following approach is developed.

The stress-strain curve is assumed to keep its analytical form during the entire lifetime, i.e. it is defined by four parameters, which are considered as degenerated pendants to the material parameters mentioned above. Each stress-strain curve starts from the current plastic strain ε^{pl}.

The current elasticity modulus is defined by the damage extent as follows:

$$E = (1 - d)E_c. \tag{32}$$

The reduction of compressive strength occurs in case of reaching the energy that is dissipated in the diffuse area under monotonic loading. As the damage and the dissipated energy correspond directly to each other, the transition may be expressed in terms of damage variable d_c. The boundary between diffuse and localized dissipated energy under monotonic loading corresponds to the damage at compressive strength ($\sigma = -f_c$):

$$d_c = 1 - \frac{f_c}{\varepsilon_c E_c\,(1-b) + b\,f_c}\,. \tag{33}$$

The change of compressive strength can then be determined as follows:

$$f_c^{fat} = \begin{cases} f_c & : \; d^{fat} \le d_c \\ \dfrac{f_c}{\sqrt{\sigma_{c3}^2 + 2\,E_c\,(1-d^{fat})\,\Delta g_{cl2}^{fat}}} & : \\ d^{fat} > d_c \end{cases}\,. \tag{34}$$

Herein, Δg_{cl2}^{fat} stands for the first part of the residual volume-specific crushing energy g_{cl2}^{fat}, that corresponds to the ascending branch of the stress-strain curve (Fig. 7). The second, complementary part is determined by

$$g_{cl1}^{fat} = \frac{1}{2} f_c^{fat}\left[\varepsilon_c^{fat}(1-b) + b\frac{f_c^{fat}}{E_c\,(1-d^{fat})}\right], \tag{35}$$

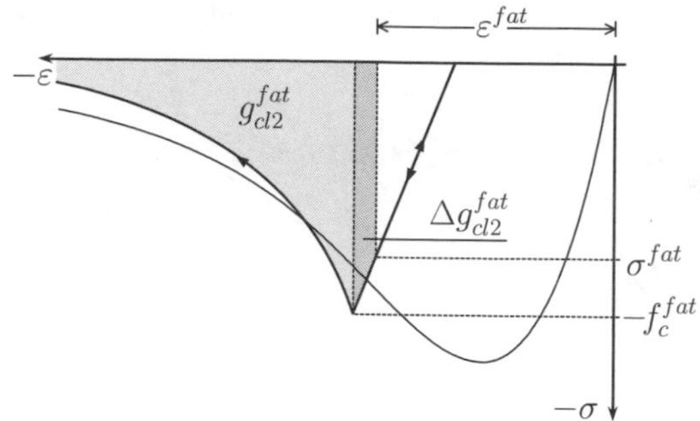

Figure 7. Estimation of residual energy under fatigue.

so that the entire crushing energy in the damaged state results in:

$$G_{cl}^{fat} = \begin{cases} G_{cl} & : \; d^{fat} \le d_c \\ l_{eq}\left(g_{cl2}^{fat} + g_{cl1}^{fat}\right) & : \; d^{fat} > d_c \end{cases}\,. \tag{36}$$

The strain, at which the concrete would fail under monotonic loading from the current fatigue state, is estimated to:

$$\varepsilon_c^{fat} = \begin{cases} \varepsilon_c + b\left(\varepsilon_{cd} - \dfrac{\sigma_{c3}}{E_c}\right) & : \quad d^{fat} \le d_c \\ -\varepsilon_{cd} + b\left(\varepsilon_{cd} - \dfrac{\sigma_{c3}}{E_c}\right) + \dfrac{f_c^{fat} - \sigma_{c3}}{E_c^{fat}} & : \\ d^{fat} > d_c \end{cases}\,. \tag{37}$$

4 FINITE ELEMENT SIMULATION

The simulation of structural response under fatigue loading by means of the finite element method is performed in two generalized steps. First, the vector of external loading is step-by-step increased up to a given level $\lambda_{max}\mathbf{P}$. The solution of the tangent stiffness relation on system level

$$\mathbf{K}_T\Delta\mathbf{u} \;=\; \lambda\mathbf{P} - \mathbf{F}_I. \tag{38}$$

is obtained by one of the standard incremental-iterative approaches (Bathe 1996; Zienkiewicz and Taylor 2000a; Zienkiewicz and Taylor 2000b). Herein, $\mathbf{P}$, $\Delta\mathbf{u}$ denote the vector of external nodal forces and the vector-increment of nodal displacements respectively, λ stands for load factor, and $\mathbf{K}_T$, $\mathbf{F}_I$ indicate the tangent stiffness matrix and the vector of internal forces respectively.

In this state at time instant t_0 with displacements $\mathbf{u}_n$, the equilibrium of external and internal forces reads:

$$\lambda_{max}\mathbf{P} \;=\; \mathbf{F}_{I(n)}. \tag{39}$$

This initial state is characterized on material level by the stress and strain, the internal variables and current values of the material parameters listed below:

$$\sigma_{max(n)}, \varepsilon_{(n)}, \varepsilon^{p}_{(n)}, d_{(n)}, E_{c(n)}, f_{c(n)}, \varepsilon_{c(n)}, G_{cl(n)}.$$

In the second generalized step, the long-term structural response is simulated for selected time increments Δt_{n+1} and corresponding number of load excursions ΔN_{n+1}.

The lifetime simulation begins by prediction of the damaged state of material at time instant $t_{n+1} = t_n + \Delta t_{n+1}$. That is made for $N_{n+1} = N_n + \Delta N_{n+1}$ load cycles with constant stress amplitudes $\sigma_{max(n)}$ by means of the fatigue model described above:

$$\sigma_{max(n+1)} = \sigma_{max(n)}, \varepsilon_{(n+1)}, \varepsilon^{p}_{(n+1)}, d_{(n+1)}. \tag{40}$$

This trial state along with the material parameters

$$E_{c(n+1)}, f_{c(n+1)}, \varepsilon_{c(n+1)}, G_{cl(n+1)}, \tag{41}$$

degraded as a result of fatigue damage, obviously does not satisfy the global equilibrium on system level at time instant t_{n+1}. The corresponding residual reads:

$$\mathbf{R}_{(n+1)} = \lambda_{(n+1)}\mathbf{P} - \mathbf{F}_{I(n+1)} \neq \mathbf{0}. \tag{42}$$

The equilibrium is restored by the time-invariant Newton-Raphson iterations indicated by $(i+1)$

$$\mathbf{K}^{(i+1)}_{T(n+1)}\Delta\mathbf{u}^{(i+1)}_{(n+1)} = \lambda_{(max)}\mathbf{P} - \mathbf{F}^{(i+1)}_{I(n+1)}, \tag{43}$$

which are performed until a given tolerance limit with respect to the residual R^{i+1}_{n+1} is reached:

$$\|\mathbf{R}^{(i+1)}_{(n+1)}\| \leq \mathrm{TOL}_R. \tag{44}$$

At the end of the time-invariant iteration process, a new value of the maximum stress $\sigma^{(k)}_{\max(n+1)}$ is obtained. It is used to correct the fatigue predictions (40), (41) of the material state corresponding to the assumption of stationary stress amplitudes. Then, the entire solution procedure recurs anew until the prediction of the maximum stress converges:

$$r_\sigma = \sigma^{(k+1)}_{max(n+1)} - \sigma^{(k)}_{max(n+1)} \leq \mathrm{TOL}_\sigma. \tag{45}$$

After that, a new lifetime step $t_{n+2} = t_{n+1} + \Delta t_{n+2}$ can be performed. This incremental-iterative procedure with recursive corrections is illustrated in Fig. 8. It has been recently applied to lifetime assessment of concrete structures, for example in (Pfanner 2002; Petryna et al. 2002).

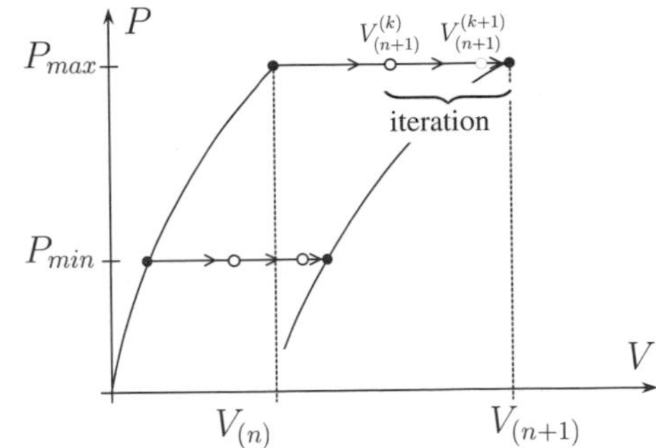

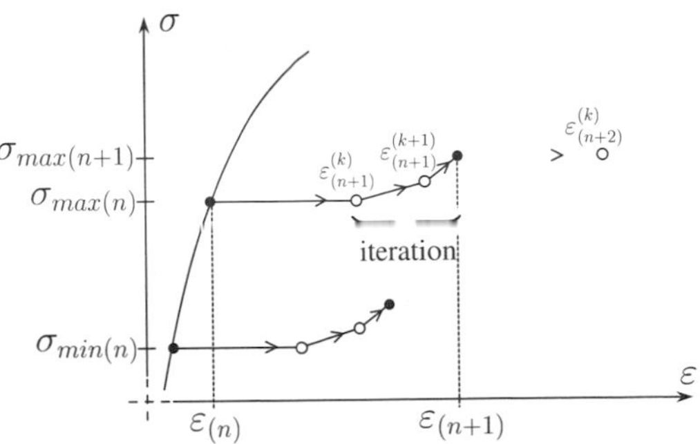

Figure 8. Fatigue simulation on system and material level.

In spite of several advantages, the approach is found to be very sensitive to the choice of lifetime steps Δt, which can considerably affect fatigue predictions. Strictly speaking, it is not completely objective and independent on the time discretization. The related problems are discussed in detail by (Pfanner 2002). An improvement of the approach with respect to the consistent model formulation and time integration procedure is currently in progress.

5 NUMERICAL EXAMPLE

As a numerical example, a reinforced concrete column under eccentric normal force cyclically varying between P_{max} and P_{min} is considered. The geometry, load, cross-section data as well as the finite element model are given in Fig. 9. The complete set of material and model parameters are taken according to (Pfanner 2002).

All cross-sections along the column remain under conditions of non-symmetrical compression. The column fails under the monotonically increased load at $Pu = 3.79\,MN$ due to the compression failure of concrete. The corresponding load-displacement diagram is depicted in Fig. 10 for $N = 0$.

The fatigue loading is applied with the frequency of 1 HZ and the amplitude $P_{max} = 0.6\,P_u$. In order to demonstrate the influence of fatigue damage on

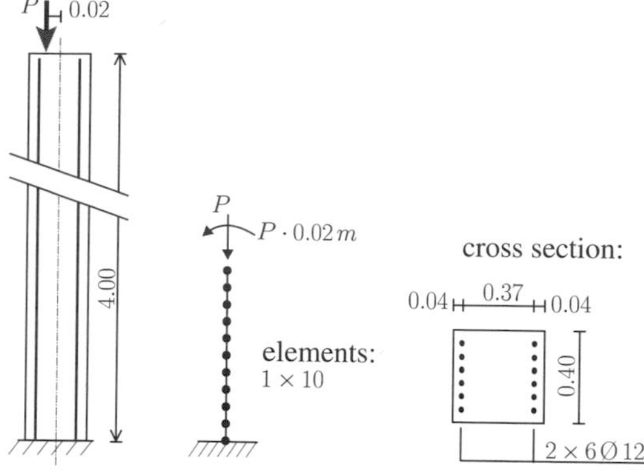

Figure 9. Geometry, load and FE discretization of the column.

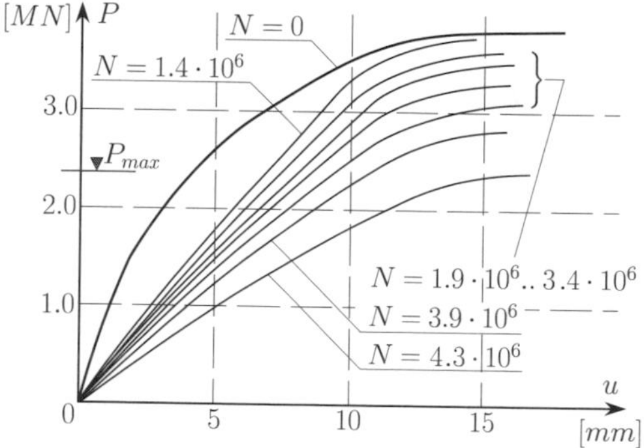

Figure 10. Load-displacement diagrams for different lifetime instants.

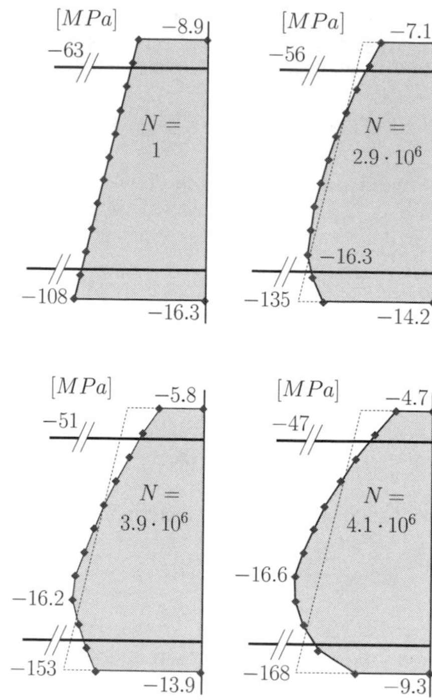

Figure 11. Stress distribution over the cross-section at different fatigue life instants.

both material and structural degradation, the fatigue loading has been a few times interrupted after a certain number of load excursions. Then, the column has been subjected to monotonic loading until collapse. The corresponding load-displacement diagrams for $N = 1.4 \cdot 10^6$ to $N = 4.3 \cdot 10^6$ are presented in Fig. 10. It can be seen that fatigue damage first of all affects structural stiffness. The response curves lose their curvature, become almost linear and simultaneously reduce their slope, the larger fatigue damage is. This is a typical influence of concrete fatigue on structural behavior.

The reason of such mechanical phenomena can be explained by redistribution of stresses within the cross-section due to fatigue damage and by a certain "saturation" effect, when no additional stress can be withstood. This effect is clearly visible in Fig. 11, which shows the stress distribution over the cross-section close to support for different instants of fatigue life.

During the first load cycle ($N = 1$), a linear stress distribution is observed with the maximum stress in the lower concrete fiber. Naturally, this fiber experience the largest damage during fatigue loading, so that after a certain lifetime period it can withstand less stresses, than the neighboring, less damaged fibers. As a result, the maximum stress drifts inside the cross-section. Such stress and damage redistributions obviously change the stiffness of each cross-section and weaken global resistance.

Fig. 11 also exhibits the effect of nonlinear damage evolution during fatigue life. The lowest concrete layer fails due to fatigue after $3.9 \cdot 10^6$ load cycles. The ensuing damage localization causes a rapid drop of stress in this fiber during a short time period between $N = 3.9 \cdot 10^6$ and $N = 4.1 \cdot 10^6$. The change of stress during this period is significantly greater than that during the entire fatigue loading before. This dramatic reduction of the cross-section resistance naturally results in the reduction of the global carrying capacity observed in Fig. 10 for $N \geq 3.9 \cdot 10^6$.

6 CONCLUSION, OUTLOOK

A new fatigue damage model for concrete has been developed on empirical basis of numerous experimental measurements and fundamental energy assumptions. It is able to predict not only the fatigue life of concrete itself, but even more important, the intermediate material states for arbitrary time instants. Such predictions enable to trace structural response during fatigue life and to study the influence of fatigue on diverse failure mechanisms. This covers an important aspect of lifetime assessment of concrete structures little discussed in the literature.

The model is currently developed for uniaxial loading conditions. Its extension onto the multiaxial states and improvement with respect to consistent thermodynamical formulation is in progress. These theoretical developments are supported by a selective experimental program.

ACKNOWLEDGEMENT

Financial support of the German National Science Foundation (DFG) within the Special Research Center SFB398 at the Ruhr-Universität Bochum is gratefully acknowledged.

REFERENCES

Awad, M. E. and H. K. Hilsdorf (1971). Strength and deformation characteristics of plain concrete subjected to high repeated and sustained loads. Technical report, Urbana, University of Illinois.

Bathe, K. J. (1996). *Finite element procedures*. London, New Jersey: Prentice-Hall International, Inc.

Bažant, Z. P. and B. H. Oh (1983). Crack band theory for fracture of concrete. *Materiaux et Constructions 16*(93), 155–177.

Bažant, Z. P. and J. Planas (1998). *Fracture and Size Effect in Concrete and Other Quasibrittle Materials*. New Directions in Civil Engineering. Boca Raton, Boston, London, New York: CRC Press.

Bockhold, J., W. B. Krätzig, R. Pölling, and Y. S. Petryna (2003). Material model for concrete under monotonic load and creep. In *Proceedings of the Euro-C 2003 Conference, St. Johann, Austria*.

CEB/FIP (1991). CEB-FIP Model Code 1990 (Design Code).

Clifton, J. R. (2000). Service-life prediction – State-of-the-Art Report, ACI 365.1R-00. Technical report, American Concrete Institute, Farmington Hills.

de Borst, R. (1997). Some recent developments in computational modelling of concrete fracture. *International Journal of Fracture* 86, 5–36.

Gaede, K. (1962). Versuche über die Festigkeit und die Verformung von Beton bei Druck-Schwellbeanspruchung. *DAfStb* 144.

Holmen, J. O. (1979). *Fatigue of concrete by constant and variable amplitude loading*. Ph. D. thesis, The Norwegian Institute of Technology, Trondheim.

Hsu, T. T. C. (1981). Fatigue of plain concrete. *ACI Journal* 78, 292–305.

Jinawath, P. (1974). *Cumulative fatigue damage of plain concrete in compression*. Ph. D. thesis, University of Leeds.

Karsan, I. D. and J. O. Jirsa (1969). Behavior of concrete under compressive loadings. *J. Struct. Div. (ASCE)* 95(ST12), 2543–2563.

Klausen, D. (1978). *Festigkeit und Schädigung von Beton bei häufig wiederholter Beanspruchung*. Ph. D. thesis, Technische Hochschule Darmstadt.

Krätzig, W. B. and R. Pölling (2002). An elastoplastic damage theory for reinforced concrete with minimum number of material parameters. *Computers and Structures (submitted)*.

Mazars, J. (1986). A description of micro- and macroscale damage of concrete structures. *Engng. Fract. Mech.* 25(5/6), 729–737.

Ortiz, M. (1985). A constitutive theory for the inelastic behavior of concrete. Mech. Mater. 4, 67–93.

Petryna, Y. S., D. Pfanner, F. Stangenberg, and W. B. Krätzig (2002). Reliability of reinforced concrete structures under fatigue. *Reliability Engineering and System Safety* 77(3), 253–261.

Pfanner, D. (2002). *Zur Degradation von Stahlbetonbauteilen unter Ermüdungsbeanspruchung*. Ph. D. thesis, Ruhr-Universität Bochum.

Pölling, R. (2000). *Eine praxisnahe, schädigungs-orientierte Materialbeschreibung von Stahlbeton für Strukturanalysen*. Ph. D. thesis, Ruhr-Universität Bochum.

Sarja, A. and E. Vesikari (1996). Durability design of concrete structures. Report of RILEM Technical Committee 130-CSL. Technical report, RILEM, E & FN Spon, London.

Shah, S. P. (1970). Fracture of concrete subjected to cyclic and sustained loading. *ACI J.*

Sinha, B. P., K. H. Gerstle, and L. G. Tulin (1964). Stress-strain relations for concrete under cyclic loading. *J. ACI* 61(2), 195–211.

Spooner, D. C. and J. W. Dougill (1975). A quantitive assessment of damage sustained in concrete during compresion. *Mag. Conc. Res.* 27(92), 151–160.

Subramaniam, V. K., J. S. O'Neil, E. F. Popovics, and S. P. Sha (2000). Crack propagation in flexural fatigue of concrete. *J. Engng. Mech. (ASCE)* 126(9), 891–898.

van Mier, J. G. M. (1997). *Fracture Processes of Concrete: Assessment of Material Parameters for Fracture Models*. Boca Raton: CRC Press.

Zhang, B. and K. Wu (1997). Residual fatigue strength and stiffness of ordinary concrete under bending. *Cem. Conc. Res.* 27(1), 115–126.

Zienkiewicz, O. C. and R. L. Taylor (2000a). *The finite element method* (5 ed.), Volume 1: The Basis. Oxford: Butterworth-Heinemann.

Zienkiewicz, O. C. and R. L. Taylor (2000b). *The finite element method* (5 ed.), Volume 2: Solid Mechanics. Oxford: Butterworth-Heinemann.

Computational Modelling of Concrete Structures, Bićanić et al. (eds)
© 2003 Swets & Zeitlinger, Lisse, ISBN 90 5809 536 3

Numerical aspects of nonlocal plasticity with strain softening

Simon Rolshoven & Milan Jirásek
LSC, FENAC, Swiss Federal Institute of Technology at Lausanne (EPFL), Switzerland

ABSTRACT: A regularized plasticity model for the strain-softening behavior of concrete under compression is proposed, the corresponding numerical algorithms are developed, and numerical examples are presented. The plasticity model of the integral nonlocal type is an extension of the classical Drucker-Prager model, which is based on a pressure-dependent yield criterion and a non-associated flow rule. Regularization is achieved by replacing the local softening variable in the softening law by a linear combination of the local softening variable and its nonlocal average. The stress-return algorithm of this model leads to a nonlinear complementarity problem with coupling among the Gauss points, which can be solved in an iterative fashion. Numerical examples of a compression test under plane strain show no spurious dependence on the computational grid. The angle under which a shear band develops agrees well with analytical predictions. The loading process up to complete failure can be simulated without spurious locking effects.

1 INTRODUCTION

Classical plasticity theories based on material models that are *simple* in the sense of Noll (1972) fail to provide an objective description of localized material failure. For sufficiently strong softening or non-associated flow rules, ellipticity of the governing differential equations is lost, and the boundary value problem becomes ill-posed. The problem admits solutions with plastic strain rate localized into a set of zero measure, i.e., a curve in two dimensions or a surface in three dimensions. The actual width of the zone of localized plastic strain is related to the heterogeneous material microstructure and can be correctly predicted only by models that have a parameter with the dimension of length. The intrinsic length scale is absent from standard theories of plasticity, and should be introduced by an appropriate enhancement. A properly formulated enhancement has a regularizing effect, i.e., it acts as a localization limiter that restores the well-posedness of the boundary value problem.

In the previous work of the authors (Jirásek and Rolshoven 2003), a number of nonlocal plasticity formulations have been compared, with attention to their localization properties and their capabilities to model the softening path up to complete failure without any spurious effects. A nonlocal model with the softening law dependent on a linear combination of local and nonlocal cumulative plastic strain, initially proposed by Vermeer and Brinkgreve (1994), has been found to be particularly suitable for this purpose and is selected for further development in the two- and three-dimensional settings.

The present article demonstrates the applicability of this nonlocal concept and presents the general format of the stress-return algorithm. The non-associated Drucker-Prager plasticity model is chosen as a simple example. This model reflects some of the fundamental properties of concrete under compression, e.g. the pressure dependence of the yield limit and the dilatant plastic flow. Of course, it would require additional refinements if the dependence of the yield limit on the third invariant or the nonlinear behavior under highly confined compression should be captured. Nevertheless, the present simple model permits an assessment of the nonlocal concept and of its applicability as a localization limiter under compression.

2 NONLOCAL DRUCKER-PRAGER PLASTICITY MODEL

2.1 *Local model*

A simple plasticity model that takes into account the cohesive-frictional behavior of concrete can be based on a yield function

$$f(\boldsymbol{\sigma}, \tau_Y) = F(\boldsymbol{\sigma}) - \tau_Y \tag{1}$$

with the pressure-dependent equivalent stress

$$F(\boldsymbol{\sigma}) = c_\phi I_1 + \sqrt{J_2} \tag{2}$$

This special case of the Burzyński (1929) yield function was later made popular by Drucker and Prager (1952). As usual, σ is the stress tensor, τ_Y is the yield stress under pure shear, and I_1 and J_2 are the first invariant and second deviatoric invariant of the stress tensor. The friction coefficient c_ϕ is a positive parameter that controls the influence of the pressure on the yield limit, important for cohesive-frictional materials such as concrete, soils or other geomaterials. The flow rule is derived from the plastic potential

$$g(\boldsymbol{\sigma}) = c_\psi I_1 + \sqrt{J_2} \tag{3}$$

where c_ψ is the dilatancy coefficient. An associated model with $c_\phi = c_\psi$ would overestimate the dilatancy of concrete, so the dilatancy coefficient is usually chosen smaller than the friction coefficient. The local model is described by the equations

$$\boldsymbol{\sigma} = \boldsymbol{D} : (\boldsymbol{\varepsilon} - \boldsymbol{\varepsilon}_p) \tag{4}$$

$$\tau_Y = h(\kappa) \tag{5}$$

$$\dot{\boldsymbol{\varepsilon}}_p = \dot{\lambda}\frac{\partial g}{\partial \boldsymbol{\sigma}} = \dot{\lambda}\left(c_\psi \boldsymbol{\delta} + \frac{\boldsymbol{s}}{2\sqrt{J_2}}\right) \tag{6}$$

$$\dot{\kappa} = \sqrt{\frac{2}{3}}\,\|\dot{\boldsymbol{\varepsilon}}_p\| \tag{7}$$

$$\dot{\lambda} \geq 0, \quad f(\boldsymbol{\sigma}, \tau_Y) \leq 0, \quad \dot{\lambda} f(\boldsymbol{\sigma}, \tau_Y) = 0 \tag{8}$$

which represent the linear elastic law, hardening law, evolution laws for plastic strain and hardening variable, and the loading-unloading conditions. In the above, $\boldsymbol{D}$ is the elastic stiffness tensor, ε is the strain tensor, ε_p is the plastic strain tensor, λ is the plastic multiplier, $\boldsymbol{\delta}$ is the unit second-order tensor, $\boldsymbol{s}$ is the deviatoric stress tensor, κ is the hardening variable, and a superior dot marks the derivative with respect to time. The flow rule has the form given in Eq. (6) at all points of the conical yield surface with the exception of its vertex, located on the hydrostatic axis. Since this state corresponds to hydrostatic tension, it is hardly reached in a compressive test and we exclude it from further considerations.

We will pay special attention to the case of softening, with a decreasing function $h(\kappa)$, and so we will call κ the softening variable and (5) the softening law. For the present model, the evolution of the softening variable can be explicitly linked to the plastic multiplier. Substituting the flow rule (6) into Eq. (7) and computing the norm leads to

$$\dot{\kappa} = k\dot{\lambda} \tag{9}$$

with a constant parameter $k = \sqrt{1/3 + 2c_\psi^2}$, so the softening variable is proportional to the plastic multiplier. For $c_\phi = c_\psi = 0$, the associated J_2-plasticity model is recovered as a special case.

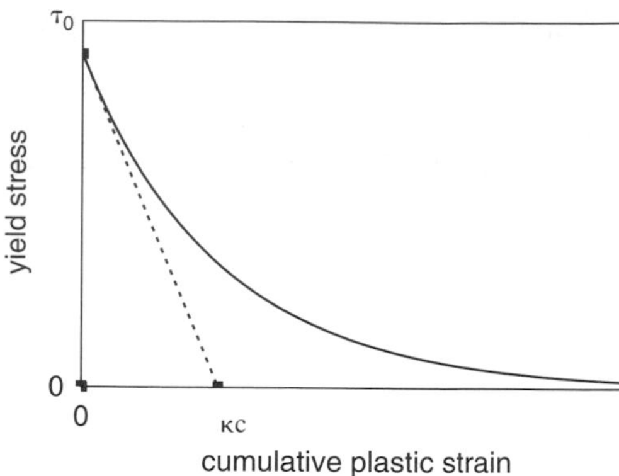

Figure 1. Exponential softening function.

In the simplest case of linear softening, the softening function is a bilinear function of κ, given by

$$h(\kappa) = \langle \tau_0 + H\kappa \rangle \tag{10}$$

where τ_0 is the initial yield stress, $H < 0$ is the softening modulus, and $\langle \cdots \rangle$ is the positive-part operator, defined as $\langle x \rangle = \max(0, x)$. An exponential softening function

$$h(\kappa) = \tau_0\, \mathrm{e}^{-\kappa/\kappa_c} \tag{11}$$

is used here for a more realistic description of softening; the meaning of parameter κ_c is clear from Fig. 1.

2.2 *Nonlocal extension*

The nonlocal plasticity model first proposed by Vermeer and Brinkgreve (1994) belongs to the class of regularized models that use the nonlocal average of an internal variable related to the inelastic process. The nonlocal average of a local field $\kappa(\boldsymbol{x})$ is defined by the integral

$$\bar{\kappa}(\boldsymbol{x}) = \int_{\mathcal{V}} \alpha(\boldsymbol{x}, \boldsymbol{\xi})\kappa(\boldsymbol{\xi})\mathrm{d}\boldsymbol{\xi} \tag{12}$$

where α is a suitable weight function and $\mathcal{V}$ is the spatial domain representing the body (finite or infinite). For an infinite body made of a macroscopically homogeneous material, the weight function depends only on the distance $r = \|\boldsymbol{x} - \boldsymbol{\xi}\|$ and can be expressed as $\alpha(\boldsymbol{x}, \boldsymbol{\xi}) = \alpha_\infty(\|\boldsymbol{x} - \boldsymbol{\xi}\|)$ where α_∞ is a function of r. The weight function is scaled such that the average operator does not modify a constant local field. For a finite body, the weight function is usually adjusted such that the nonlocal field corresponding to a constant local field remains constant even in the vicinity of a boundary. This is guaranteed if the weight function satisfies the normalizing condition

$$\int_{\mathcal{V}} \alpha(\boldsymbol{x}, \boldsymbol{\xi})\mathrm{d}\boldsymbol{\xi} = 1 \quad \forall \boldsymbol{x} \in \mathcal{V} \tag{13}$$

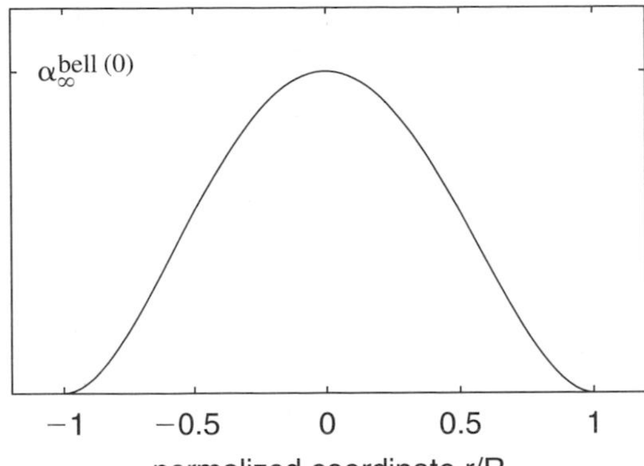

Figure 2. Nonlocal weight function $\alpha_\infty^{\text{bell}}(r)$.

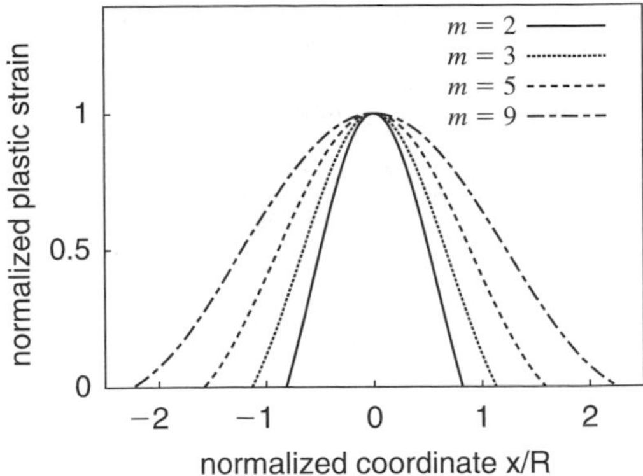

Figure 3. Influence of parameter m and interaction radius R on the shape of the plastic strain profile and the size of localized zone for the one-dimensional case.

and the simplest way to achieve that is to set

$$\alpha(\boldsymbol{x},\boldsymbol{\xi}) = \frac{\alpha_\infty(\|\boldsymbol{x}-\boldsymbol{\xi}\|)}{\int_V \alpha_\infty(\|\boldsymbol{x}-\boldsymbol{\zeta}\|)\mathrm{d}\boldsymbol{\zeta}} \tag{14}$$

In this paper, we use the bell-shaped truncated polynomial function

$$\alpha_\infty^{\text{bell}}(r) = \begin{cases} \dfrac{1}{c}\left(1-\dfrac{r^2}{R^2}\right)^2 & \text{if } |r| \le R \\ 0 & \text{if } |r| \ge R \end{cases} \tag{15}$$

plotted in Fig. 2. The length scale is set by parameter R called the interaction radius. The scaling factor c depends on the problem dimension, and is given by $c = 16R/15$ for one dimension, $\pi R^2/3$ for two dimensions, and $32\pi R^3/105$ for three dimensions.

For plasticity, the natural choice of the internal variable to be averaged is the softening variable κ, but simply replacing κ in the softening function by its nonlocal average $\bar{\kappa}$ does not completely regularize the model. The nonlocal formulation of Vermeer and Brinkgreve (1994) evaluates the yield stress from a linear combination of the local and nonlocal softening variable

$$\hat{\kappa} = m\bar{\kappa} + (1-m)\kappa \tag{16}$$

substituted into the softening law, which now reads

$$\tau_{\text{Y}} = h(\hat{\kappa}) \tag{17}$$

Parameter m controls the weight of the two variables in the linear combination. All other equations of the local model, including the loading-unloading conditions in the local format, remain the same. A bifurcation analysis of the present nonlocal model in the one-dimensional setting has been performed by Planas et al. (1996), cf. also pp. 496-498 in Bažant and Planas (1998). The localized plastic region is of finite size if and only if $m > 1$. For a fixed value of m, its width is controlled by the interaction radius R, see Fig. 3. Any value of $m > 1$ could in principle be used, but for practical reasons it is advantageous to take neither very large values nor values close to 1. Here, $m = 2$ will be used. In this case, the width of the plastic zone in the one-dimensional setting is approximately 1.64 R.

3 STRESS-RETURN ALGORITHM

3.1 *Local stress-return algorithm*

Incremental finite element analysis requires the implementation of a procedure for the evaluation of the stress and plastic strain increments that correspond to a given increment of strain. In the local case, the algorithmic form of the elastic law and the softening law

$$\boldsymbol{\sigma}_{(n+1)} = \boldsymbol{\sigma}_{(n)} + \boldsymbol{D} : (\Delta\boldsymbol{\varepsilon} - \Delta\boldsymbol{\varepsilon}_{\text{p}}) \tag{18}$$

$$\tau_{\text{Y}\,(n+1)} = h\left(\kappa_{(n)} + \Delta\kappa\right) \tag{19}$$

are simply obtained from Eqs. (4) and (5) if the values at the end of the step, denoted by subscript $n+1$, are expressed as the sum of the value at the end of the previous step, denoted by subscript n, and the increment, denoted by Δ. Using a backward Euler difference scheme for the time discretization of the flow rule, the increment of plastic strain is obtained from

$$\Delta\boldsymbol{\varepsilon}_{\text{p}} = \Delta\lambda\,\boldsymbol{g}\left(\boldsymbol{\sigma}_{(n+1)}\right) \tag{20}$$

which is the algorithmic form of Eq. (6).

Since the flow rule is associated in the deviatoric plane, the deviatoric stress at the end of the step

$s_{(n+1)}$ is a scalar multiple of the deviatoric "trial" stress s_{trial} computed from the trial stress state $\sigma_{\text{trial}} = \sigma_{(n)} + D : \Delta\varepsilon$. This simplifies the expression for the flow direction at the end of the step

$$\begin{aligned} g\left(\sigma_{(n+1)}\right) &= c_\psi \delta + \frac{s_{(n+1)}}{2\sqrt{J_2\left(s_{(n+1)}\right)}} \\ &= c_\psi \delta + \frac{s_{\text{trial}}}{2\sqrt{J_2\left(s_{\text{trial}}\right)}} \\ &= g\left(\sigma_{\text{trial}}\right) = g_{\text{trial}} \end{aligned} \tag{21}$$

which is directly evaluated from the trial stress state. The value of the yield function at the end of the step,

$$\begin{aligned} f\left(\sigma_{(n+1)}, \tau_{Y(n+1)}\right) &= F\left(\sigma_{(n+1)}\right) - \tau_{Y(n+1)} \\ &= F\left(\sigma_{\text{trial}} - \Delta\lambda D : g_{\text{trial}}\right) - h\left(\kappa_{(n)} + k\Delta\lambda\right) \\ &= f_{(n+1)}(\Delta\lambda) \end{aligned} \tag{22}$$

can now be written as a function of the unknown increment of the plastic multiplier $\Delta\lambda$ only, because all other variables are known at the beginning of the step. The subscript at $f_{(n+1)}$ indicates the dependency of the function on the internal variables and the strain increment.

The numerical solution must satisfy the algorithmic form of the loading-unloading conditions

$$\Delta\lambda \geq 0, \quad f_{(n+1)} \leq 0, \quad \Delta\lambda f_{(n+1)} = 0 \tag{23}$$

This is in general a nonlinear complementarity problem (NLCP) for the unknown plastic multiplier $\Delta\lambda$. In the elastic region, characterized by $\Delta\lambda = 0$, the stress state must lie inside the elastic domain at the end of the step, i.e. $f_{(n+1)} \leq 0$. In the plastic region, characterized by $\Delta\lambda > 0$, the yield condition $f_{(n+1)} = 0$ must be fulfilled at the end of the step. For local plasticity models, the NLCP can be solved at each Gauss point independently using the conventional stress-return algorithms, i.e. the NLCP is uncoupled.

Evaluating the equivalent stress that appears in the definition (22) of the function $f_{(n+1)}$ using the formulae (2) and (21), we obtain for sufficiently small increments $\Delta\lambda$ a linear expression

$$\begin{aligned} &F\left(\sigma_{\text{trial}} - \Delta\lambda D : g_{\text{trial}}\right) \\ &\quad = F_{\text{trial}} - (9c_\phi c_\psi K + G)\Delta\lambda \\ &\quad = F_{(n+1)}(\Delta\lambda) \end{aligned} \tag{24}$$

where $F_{\text{trial}} = F(\sigma_{\text{trial}})$, K is the bulk modulus of elasticity and G is the shear modulus.

If the softening function $h(\kappa)$ is linear as well, then the yield function can be expressed as

$$\begin{aligned} f_{(n+1)}(\Delta\lambda) &= F_{\text{trial}} - (9c_\phi c_\psi K + G)\Delta\lambda \\ &\quad - \tau_0 - H\left(\kappa_{(n)} + k\Delta\lambda\right) \\ &= f_{\text{trial}} - (9c_\phi c_\psi K + G + Hk)\Delta\lambda \end{aligned} \tag{25}$$

where $f_{\text{trial}} = f(\sigma_{\text{trial}}, \tau_Y(\kappa_{(n)}))$ is the trial value of the yield function. For the Drucker-Prager model with linear softening, the value of the yield function at the end of the step, $f_{(n+1)}$, depends on the increment of the plastic multiplier $\Delta\lambda$ in a linear way, and (23) becomes a linear complementarity problem (LCP).

3.2 *Nonlocal stress-return algorithm*

For the nonlocal model, the difference in the governing equations is that the local softening variable in the softening function is replaced by $\hat{\kappa} = m\bar{\kappa} + (1-m)\kappa$, with $m > 1$. The algorithmic form of the softening law is then

$$\tau_{Y(n+1)} = h\left(\hat{\kappa}_{(n)} + \Delta\hat{\kappa}\right) \tag{26}$$

and the nonlocal softening variable $\Delta\bar{\kappa}$ in the expression for $\Delta\hat{\kappa} = m\Delta\bar{\kappa} + (1-m)\Delta\kappa$ leads to coupling of the CP, because the yield stress at one Gauss point now depends on the softening variable at the Gauss points in the neighborhood.

To develop the stress-return algorithm for the nonlocal model, column matrices that collect the values of a quantity at all Gauss points of the structure are introduced. They are represented by bold face symbols, e.g. $\Delta\boldsymbol{\kappa}$ is a column matrix carrying the values of $\Delta\kappa$ of all Gauss points, and $\Delta\bar{\boldsymbol{\kappa}}, \Delta\hat{\boldsymbol{\kappa}}, \boldsymbol{h}, \boldsymbol{f}_{(n+1)}, \boldsymbol{f}_{\text{trial}}, \boldsymbol{F}_{\text{trial}}$ are defined in a similar fashion. The discretized nonlocal averaging operator is represented by the square matrix $\boldsymbol{A}$, so the increment of the nonlocal softening variable is given by $\Delta\bar{\boldsymbol{\kappa}} = \boldsymbol{A}\Delta\boldsymbol{\kappa}$. Since the increment of the softening variable, $\Delta\kappa$, is proportional to the increment of the plastic multiplier $\Delta\lambda$, Eq. (9), $\Delta\lambda$ can be replaced by $\Delta\kappa$ in the algorithmic form of the loading-unloading conditions

$$\Delta\boldsymbol{\kappa} \geq \boldsymbol{0}, \quad \boldsymbol{f}_{(n+1)} \leq \boldsymbol{0}, \quad \Delta\boldsymbol{\kappa}^{\mathrm{T}} \cdot \boldsymbol{f}_{(n+1)} = 0 \tag{27}$$

now written in a compact form for the whole body. Here, the operators $\leq$ and $\geq$ have to be interpreted for each component of the corresponding column matrices separately. Eq. (27) is a coupled NLCP for the plastic multipliers of all Gauss points $\Delta\boldsymbol{\kappa}$, and it must be solved simultaneously for all Gauss points. Same as for the local model, the elastic region is characterized by $\Delta\kappa = 0$ and the stress state must lie inside the elastic domain at the end of the step, i.e. $f_{(n+1)} \leq 0$, while in the plastic region, characterized by $\Delta\kappa > 0$, the yield condition $f_{(n+1)} = 0$ must be fulfilled at the end of the step.

Consider first the case of linear softening. Similar to Eq. (25), but with a different softening law (17) and the constant $C = (9c_\phi c_\psi K + G)/k$, the yield condition can be written as

$$f_{(n+1)}(\Delta\kappa, \Delta\hat{\kappa}) = f_{\text{trial}} - C\Delta\kappa - H\Delta\hat{\kappa} = 0 \tag{28}$$

for one Gauss point in the plastic zone, and the yield function now depends also on $\Delta\hat{\kappa}$. In the compact format written for all Gauss points in the plastic zone, this can be expressed as

$$\begin{aligned} \boldsymbol{f}_{(n+1)}(\Delta\boldsymbol{\kappa}) &= \boldsymbol{f}_{\text{trial}} - C\Delta\boldsymbol{\kappa} - H\Delta\hat{\boldsymbol{\kappa}} \\ &= \boldsymbol{f}_{\text{trial}} - C\Delta\boldsymbol{\kappa} - H\left[m\boldsymbol{A} + (1-m)\boldsymbol{I}\right]\Delta\boldsymbol{\kappa} \\ &= \boldsymbol{f}_{\text{trial}} - \boldsymbol{B}\Delta\boldsymbol{\kappa} = \boldsymbol{0} \end{aligned} \tag{29}$$

where

$$\boldsymbol{B} = [C + (1-m)H]\boldsymbol{I} + mH\boldsymbol{A} \tag{30}$$

is a constant matrix, and $\boldsymbol{f}_{(n+1)}(\Delta\boldsymbol{\kappa})$ is a function of the increment of the softening variable at all Gauss points. If the plastic zone is known, the increments of the softening variable can be solved from the linear system of equations $\boldsymbol{B}\Delta\boldsymbol{\kappa} = \boldsymbol{f}_{\text{trial}}$, where all the matrices are reduced such that they contain only entries that correspond to the yielding Gauss points. However, for the nonlocal model, the plastic zone is not known at the beginning of the step. Due to the coupled structure of the CP, the size of the elastic domain depends on the softening variable at all Gauss points of the structure, and an iterative technique must be used to solve the coupled LCP. With a proper renumbering of the Gauss points, the matrix $\boldsymbol{B}$ is banded but typically has a large bandwidth, so a direct solution procedure would be quite expensive anyhow.

In a typical iteration number $k+1$, the yield condition $\boldsymbol{f}^{(k+1)}_{(n+1)} = \boldsymbol{0}$ is solved for the set of "plastic" Gauss points for which the trial stress lies outside the elastic domain evaluated at iteration k. The superscript indicates the iteration of the stress-return algorithm. The subscript $n+1$ corresponding to the global incremented step will be omitted in the following for brevity. Of course, the set of plastic Gauss points may change from one iteration to the next. A Gauss point is considered as plastic for iteration $k+1$, if

$$f^{(k)} = F_{\text{trial}} - \tau_Y\left(\hat{\kappa}_{(n)} + \Delta\hat{\kappa}^{(k)}\right) > 0 \tag{31}$$

Instead of solving the system of equations $\boldsymbol{B}\Delta\boldsymbol{\kappa} = \boldsymbol{f}_{\text{trial}}$ directly, we use an iterative technique based on an additive split of the linear operator $\boldsymbol{B} = \boldsymbol{B}_1 + \boldsymbol{B}_2$, with an easy-to-invert matrix $\boldsymbol{B}_1$. In each iteration of the nonlocal stress return algorithm, the equation

$$\boldsymbol{B}_1\Delta\boldsymbol{\kappa}^{(k+1)} = \boldsymbol{f}_{\text{trial}} - \boldsymbol{B}_2\Delta\boldsymbol{\kappa}^{(k)} \tag{32}$$

is solved for $\Delta\boldsymbol{\kappa}^{(k+1)}$. For each Gauss point, the computed increment $\Delta\kappa^{(k+1)}$ is checked for admissibility, and is set to zero if $\Delta\kappa^{(k+1)} < 0$, because negative increments of the softening parameter violate the loading-unloading conditions. Convergence is achieved if each Gauss point fulfills the loading-unloading conditions at the end of the step. For elastic Gauss points with $\Delta\kappa^{(k+1)} = 0$, it is checked that the stress state lies inside the elastic domain, i.e. $f^{(k+1)} \leq 0$. For plastic Gauss points with $\Delta\kappa^{(k+1)} > 0$, the yield condition $f^{(k+1)} = 0$ does not need to be satisfied exactly, but only approximately with a prescribed tolerance.

Several splits with a diagonal matrix $\boldsymbol{B}_1$ can be found in the literature. In the context of J_2-plasticity, Strömberg and Ristinmaa (1996) proposed a split into $\boldsymbol{B}_1 = (C + H)\boldsymbol{I}$ and $\boldsymbol{B}_2 = mH(\boldsymbol{A} - \boldsymbol{I})$. Even in the one-dimensional case, this split does not converge for all physically admissible values of the model parameters. If softening is too strong, convergence is lost. As an improvement, the split into the local part $\boldsymbol{B}_1 = [C + (1-m)H]\boldsymbol{I}$ and nonlocal part $\boldsymbol{B}_2 = mH\boldsymbol{A}$ has been proposed by Rolshoven and Jirásek (2001). Though convergent, this split is not optimal in terms of performance.

Benvenuti and Tralli (2003) compared different splitting techniques and their convergence properties. As the optimal technique, they identified the split of $\boldsymbol{B}$ into its diagonal part

$$\boldsymbol{B}_1 = [C + (1-m)H]\boldsymbol{I} + mH\boldsymbol{A}^{\text{d}} \tag{33}$$

and the remainder

$$\boldsymbol{B}_2 = mH\boldsymbol{A}^{\text{r}} \tag{34}$$

where $\boldsymbol{A}^{\text{d}}$ and $\boldsymbol{A}^{\text{r}} = \boldsymbol{A} - \boldsymbol{A}^{\text{d}}$ are the diagonal part and the remainder of $\boldsymbol{A}$, respectively.

The foregoing analysis allows for an iterative solution of the LCP, where at each Gauss point and in each iteration, a linear equation for $\Delta\kappa^{(n+1)}$ needs to be solved. To understand the physical meaning of this iterative technique and for an extension to the case of nonlinear softening, it is useful to rewrite the equation that has to be solved for one Gauss point. Using the identity

$$\begin{aligned} \boldsymbol{A}^{\text{r}}\Delta\boldsymbol{\kappa}^{(k)} &= \left(\boldsymbol{A} - \boldsymbol{A}^{\text{d}}\right)\Delta\boldsymbol{\kappa}^{(k)} \\ &= \Delta\bar{\boldsymbol{\kappa}}^{(k)} - \boldsymbol{A}^{\text{d}}\Delta\boldsymbol{\kappa}^{(k)} \end{aligned} \tag{35}$$

Eq. (32) can be written for the split (33)–(34) and one specific Gauss point as

$$f_{\text{trial}} - C\Delta\kappa^{(k+1)} - H\left\{m\left(\Delta\bar{\kappa}^{(k)} - A^{\text{d}}\Delta\kappa^{(k)}\right)\right.$$

$$\left.+\left[1 - m(1 - A^{\text{d}})\right]\Delta\kappa^{(k+1)}\right\} = 0 \quad (36)$$

where A^{d} is the diagonal element of the discretized nonlocal averaging operator $\boldsymbol{A}$ for this Gauss point. Comparing Eq. (36) to the yield condition at the end of the step, Eq. (28), we arrive at the following interpretation of the iterative scheme: For each Gauss point that is assumed to be plastic, the updated approximation $\Delta\kappa^{(k+1)}$ of the increment of the softening variable is computed from the yield condition at that point, assuming that the increments of the softening variable at all other Gauss points are temporarily frozen (i.e., set equal to their values $\Delta\kappa^{(k)}$ determined in the previous iteration).

For a nonlinear softening function, the CP is nonlinear. A standard way to proceed would be to solve an iterative series of LCP's. Since each LCP again involves an iterative solution, this technique would be quite expensive.

In the present implementation, an alternative technique has been used. It is basically the same as for the case of linear softening, except for the iterative strategy that solves the yield condition for the plastic Gauss points. For a Gauss point in the plastic region, the yield condition at the end of the step

$$f_{(n+1)}\left(\Delta\kappa, \Delta\hat{\kappa}\right)$$

$$= F_{\text{trial}} - C\Delta\kappa - h\left(\hat{\kappa}^{(n)} + \Delta\hat{\kappa}\right) = 0 \quad (37)$$

is similar to Eq. (28). The yield condition can be written in the compact form for all plastic Gauss points

$$\boldsymbol{f}_{(n+1)}\left(\Delta\boldsymbol{\kappa}\right)$$

$$= \boldsymbol{F}_{\text{trial}} - C\Delta\boldsymbol{\kappa} - \boldsymbol{h}\left(\hat{\boldsymbol{\kappa}}^{(n)} + \Delta\hat{\boldsymbol{\kappa}}\right) = \boldsymbol{0} \quad (38)$$

where $\boldsymbol{h}$ is now a nonlinear operator.

In the spirit of the iterative technique developed for the case of linear softening, we propose an iterative technique for the solution of this nonlinear system of equations. Each equation of the system of equations is solved independently, i.e., the updated increment of the softening variable at each Gauss point is computed from the yield condition at that point, with increments of the softening variable at all other Gauss points temporarily frozen.

After some substitutions, the equation to be solved for one Gauss point reads

$$F_{\text{trial}} - C\Delta\kappa^{(k+1)} - h(\kappa^{\star(k)} + a\Delta\kappa^{(k+1)}) = 0 \quad (39)$$

where $a = 1 - m(1 - A^{\text{d}})$ and

$$\kappa^{\star(k)} = \hat{\kappa}_{(n)} + m\left[\Delta\bar{\kappa}^{(k)} - A^{\text{d}}\Delta\kappa^{(k)}\right] \quad (40)$$

This is a nonlinear equation for the unknown $\Delta\kappa^{(k+1)}$, and it can be solved by a Newton iteration.

This concept for the solution of the NLCP has been successfully implemented and typically converges in a few iterations, even though a formal convergence proof has not been obtained yet. The same procedure could be used for nonlocal plasticity models for which the equivalent stress $F_{(n+1)}$ cannot be expressed as a linear function of $\Delta\kappa$.

Regarding the implementation of the nonlocal stress-return algorithm for nonlinear softening, we can take advantage of the fact that Eq. (39) resembles the yield condition in the standard stress-return algorithm. The only difference lies in the meaning of the variables and the modified softening function. The implementation of this model has been carried out in the finite element code OOFEM (Patzák and Bittnar 2001). Due to the object-oriented character of this code, the stress-return algorithm of the local model can easily be reused in its original format.

Another aspect of the implementation concerns the nonlocal average $\Delta\bar{\kappa}^{(k)}$, which has to be evaluated before each iteration $k+1$. The nonlocal average of the softening variable in a Gauss point that has at least one plastic "neighbor" changes from one iteration to the next. Neighbors are those Gauss points which influence the nonlocal average at a given Gauss point, i.e. their distance to this point is less than the interaction radius. On the contrary, at elastic Gauss points without plastic neighbors, the nonlocal average of the softening variable does not change. In the present implementation, this has been used to speed up the algorithm: Elastic Gauss points without plastic neighbors are excluded from nonlocal averaging. In a typical simulation, the actual process zone is small compared to the structure, and the nonlocal average needs to be computed only in the process zone and at the neighboring Gauss points.

4 SHEAR BANDING EXAMPLES

4.1 *Test geometry and material parameters*

The numerical model is used to simulate a compression test under plane strain. A rectangular specimen with dimensions $B = 600$ mm and $L = 1200$ mm between smooth rigid plates is subjected to uniform displacement of the top, while the lateral displacement is unconstrained but plane strain conditions are enforced in the third direction. Initially, a homogeneous biaxial stress state develops in the specimen. At the onset of yielding, formation of a shear band is possible if ellipticity is lost. For the local model, this is the case if the hardening modulus is smaller than or equal to a certain critical value that can be determined by analysis of the acoustic tensor, cf. Rudnicky and Rice (1975).

For plane strain, the critical hardening modulus and the direction of a shear band have been derived analytically by Runesson et al. (1991). The principal components of the deviatoric stress s_1, s_2 and s_3 are ordered such that s_3 corresponds to the out-of-plane direction and $s_1 \geq s_2$. The orientation of the shear band is specified by the angle θ between the second principal axis and the normal to the band. Under certain mild conditions that are satisfied in the current examples, the critical hardening modulus normalized by the elastic modulus is

$$\frac{H^{\rm crit}}{E} = \frac{3}{4k(1-\nu)} \times \left[2\left(c_\phi - c_\psi\right)^2 - (1-\nu)\left(\frac{s_3}{\sqrt{J_2}} + c_\phi + c_\psi\right)^2 \right] \tag{41}$$

The orientation of the shear band can be determined from the relation

$$\tan^2\theta = -\frac{s_1 + \nu s_3 + (1+\nu)\sqrt{J_2}\left(c_\phi + c_\psi\right)}{s_2 + \nu s_3 + (1+\nu)\sqrt{J_2}\left(c_\phi + c_\psi\right)} \tag{42}$$

and it depends only on the sum of the friction and dilatancy coefficients and on the stress state.

Under plane strain, plasticity models generally exhibit locking when standard low-order finite elements are used, cf. de Borst and Groen (1995). For isochoric plasticity models such as J_2-plasticity, this phenomenon is known as volumetric locking. In the future, a proper technique to deal with locking will be implemented on the element level. A simple way to avoid locking for a straight shear band is to use an oriented mesh. If the mesh lines in a triangular mesh are aligned with the direction of the shear band, there is no locking effect. For comparison, we consider also meshes with a misalignment between the mesh lines and the band orientation.

The finite element meshes used here consist of a region with an automatically generated mesh and a region where the finite elements are arranged into layers, either oriented in the direction of the expected shear band, or deviating from that direction. The oriented region is chosen larger than the expected width of the localized plastic zone for the nonlocal model. Four different meshes are used: a coarse mesh with one layer of finite elements across the oriented region, a medium mesh with five layers, a fine mesh with ten layers, and a very fine mesh with twenty layers. The thickness of one layer of finite elements is respectively 170 mm, 34 mm, 17 mm and 8.5 mm. Along the oriented band, the number of elements is kept constant.

The material parameters adopted for the examples are: $E = 30$ GPa, $\nu = 0.2$, $\tau_0 = 20.2$ MPa, $c_\phi = 0.1$, $c_\psi = 0.05$, $\kappa_c = 5.7 \times 10^{-3}$, $R = 50$ mm, and $m = 2$. These parameters are close to those that characterize concrete, but slightly adapted so as to obtain bifurcation right at the onset of yielding. The adopted value of κ_c corresponds to the critical softening modulus of $H/E = -0.12$, evaluated from Eq. (41) for the state at the onset of yielding. To trigger the formation of a shear band, the yield stress of the central elements at the lower end of the oriented region is reduced by 1.5% (in two elements for the coarse and medium meshes and in four for the fine and very fine meshes).

At the onset of yielding, the stress state in the specimen is homogeneous. Denoting by σ the nominal stress in the vertical direction (negative for compression), the principal values of the deviatoric stress can be evaluated as $s_1 = -(1+\nu)\sigma/3, s_2 = (2-\nu)\sigma/3$, and $s_3 = (2\nu - 1)\sigma/3$, and $s_3 = (2\nu - 1)\sigma/3$, and the second deviatoric invariant is $J_2 = (1 - \nu + \nu^2)\sigma^2/3$. For the given set of material parameters, the direction of the shear band obtained from Eq. (42) is $\theta = 0.82$, which is approximately 47°.

For all load-displacement diagrams, the load is normalized by the product of the width of the specimen and the initial yield strength, $B\tau_0$, and the top displacement by the height of the specimen, L.

4.2 *Results on meshes aligned with the bifurcation direction*

In a first series of simulations, the orientation of the refined zone is chosen so as to coincide with the expected bifurcation angle. As amply documented in the literature, a local plasticity model leads to a strongly mesh-dependent load-displacement response, see Fig. 4. Upon mesh refinement, the numerical response becomes excessively brittle because the boundary value problem is ill-posed in the local case. Fig. 5 compares the distribution of the local softening variable κ for the local and the nonlocal model obtained on the medium mesh. It demonstrates the well-known fact that, for the local model, the plastic zone tends to localize into one layer of finite elements.

The load-displacement curves obtained with the nonlocal model show good convergence upon mesh refinement, see Fig. 6. For the coarse mesh, the distance of the Gauss points is larger than the interaction radius, and the result is the same as for the local model. This shows that a minimum mesh density is necessary to activate the nonlocal interaction.

For the other meshes, medium, fine and very fine, the load-displacement curves are practically identical

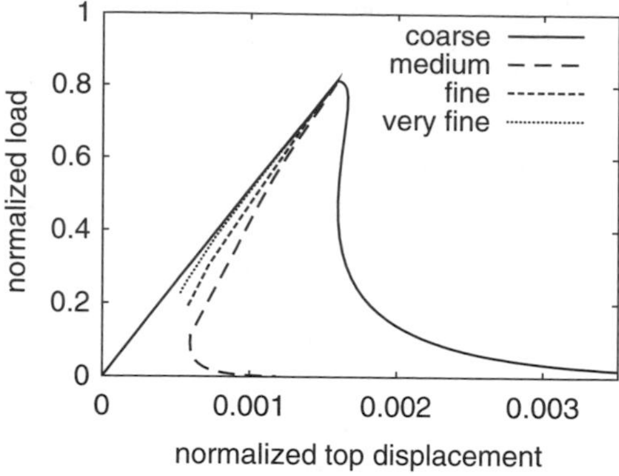

Figure 4. Load-displacement curve for the local model and four different meshes.

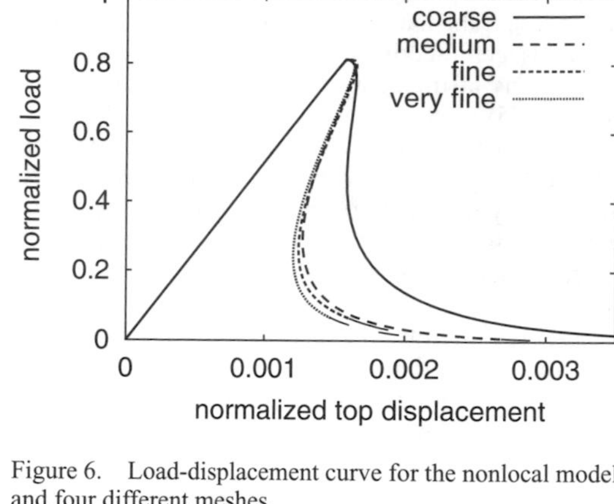

Figure 6. Load-displacement curve for the nonlocal model and four different meshes.

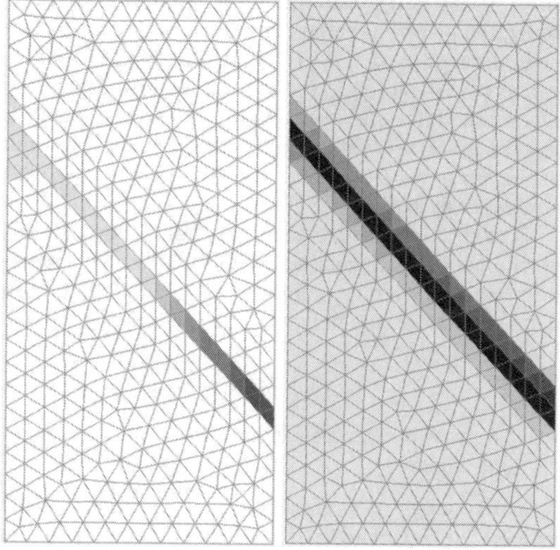

Figure 5. Distribution of the local softening variable κ for the local (left) and nonlocal (right) model on the medium mesh.

at early stages of the softening process, but some deviation is observed at later stages. During the loading process, the stress state of the specimen changes, and the bifurcation angle can deviate from the angle valid for the elastic limit state. The shear band is then no longer perfectly aligned with the computational grid, and mesh locking effects, though small, are present. These lead to a slightly more rigid response of the coarser meshes.

The distribution of the local softening variable κ for the four different meshes at two thirds of the peak load (in the post-peak range) is plotted in Fig. 7. Except for the coarse mesh, for which plastic strain localizes in one layer of finite elements, the size and shape of the plastic zone exhibit no spurious dependence on the

Figure 7. Distribution of the local softening κ variable for the nonlocal model, obtained on the coarse mesh (top left), medium mesh (top right), fine mesh (bottom left), and very fine mesh (bottom right).

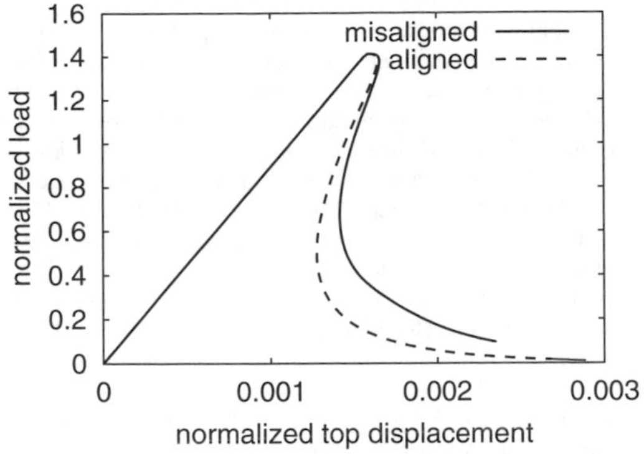

Figure 8. Load-displacement curve for the nonlocal model on the misaligned and the aligned medium mesh.

mesh density. The plastic zone extends over approximately 85 mm, which corresponds well to the size of the plastic zone of $1.64R = 82$ mm in the uniaxial case. In conclusion, the three finer meshes give practically the same results. It is thus sufficient to design the finite element mesh so as to obtain a plastic zone that extends over three to four layers.

4.3 *Results on a mesh not aligned with the bifurcation direction*

In the second example, both the local and the nonlocal model are tested on the medium mesh with the oriented region deviating by 8° from the expected shear band. The imperfection is now situated in the topmost layer of the refined region, so that the shear band can evolve within the oriented region. The load-displacement curves obtained with the nonlocal model on meshes aligned and misaligned with the expected bifurcation direction are compared in Fig. 8. Initially, the response is quite similar, but at later stages of the loading process the misaligned mesh gives a stiffer response, caused by the locking effect of the finite elements.

Fig. 9 shows the distribution of the local softening variable κ at one third of the peak load (in the post-peak range). For the local model, the plastic zone is still confined to one layer of finite elements, and follows the prescribed direction of the grid. In contrast to that, the influence of the mesh on the orientation of the shear band is negligible for the nonlocal model. The direction of the shear band is still captured correctly.

Two effects are at the origin of mesh-induced directional bias. First, in a mesh of standard finite elements without any remedy for locking, the mesh lines are preferential directions for weak discontinuities. Second, even for elements without locking, a local model tends to localize in one layer of finite elements because the dissipated energy is smaller in this case.

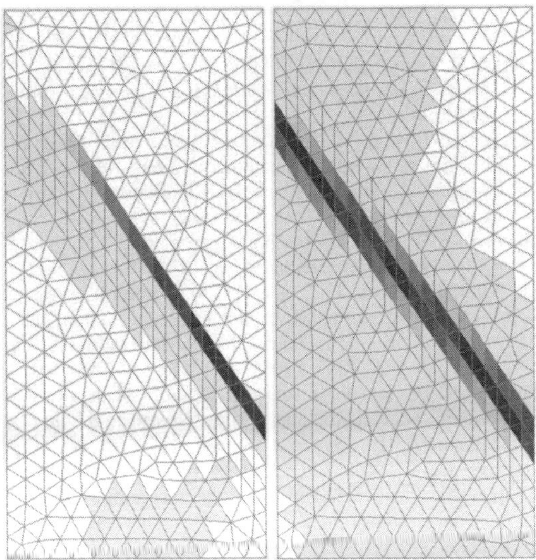

Figure 9. Distribution of the local softening variable κ for the local (left) and nonlocal (right) model on the medium mesh not aligned with the expected shear band.

5 CONCLUSIONS

The adopted nonlocal formulation properly regularizes the Drucker-Prager model with strain softening and non-associated flow. The results match the expectations that could be drawn from the one-dimensional study of the nonlocal model.

The iterative technique for the solution of the coupled NLCP describing the loading-unloading conditions performs well. Even though convergence is only linear, it typically reduces the error in the yield criterion by half an order of magnitude per iteration. Sufficient accuracy can thus be achieved within only a few iterations. This is true for the linear as well as the exponential softening function.

The numerical results demonstrate the capability of the model to correctly predict the size and shape of a shear band. Locking effects due to the element formulation are milder for the nonlocal model, which apparently relaxes the constraints dictated by the plastic flow direction. The bifurcation direction is correctly captured even on a mesh that is not aligned with the orientation of the shear band.

ACKNOWLEDGMENT

The results presented in this paper have been obtained in research project 2100-057062.99/1 supported by the Swiss National Science Foundation.

REFERENCES

Bažant, Z. P. and J. Planas (1998). *Fracture and Size Effect in Concrete and Other Quasibrittle Materials.* Boca Raton: CRC Press.

Benvenuti, E. and A. Tralli (2003). Iterative LCP solvers for nonlocal loading-unloading conditions. *International Journal for Numerical Methods in Engineering.* Submitted for publication.

Burzyński, W. (1929). Über die Anstrengungs-hypothesen. *Schweizerische Bauzeitung, Zürich 94*, 259.

de Borst, R. and A. E. Groen (1995). Some observations on element performance in isochoric and dilatant plastic flow. *International Journal for Numerical Methods in Engineering 38*, 2887–2906.

Drucker, D. C. and W. Prager (1952). Soil mechanics and plasticity analysis of limit design. *Quarterly of Applied Mathematics 10*, 157–162.

Jirásek, M. and S. Rolshoven (2003). Comparison of integral-type nonlocal plasticity models for strain-softening materials. *International Journal of Engineering Science.* In press.

Noll, W. (1972). A new mathematical theory of simple materials. *Archive for Rational Mechanics and Analysis 48*, 1–50.

Patzáak, B. and Z. Bittnar (2001). Design of object oriented finite element code. *Advances in Engineering Software 32*, 759–767.

Planas, J., G. V. Guinea, and M. Elices (1996). Basic issues on nonlocal models: uniaxial modeling. Technical Report 96-jp03, Departamento de Ciencia de Materiales, ETS de Ingenieros de Caminos, Universidad Politécnica de Madrid, Ciudad Universitaria sn., 28040 Madrid, Spain.

Rolshoven, S. and M. Jirásek (2001). On regularized plasticity models for strain-softening materials. In R. de Borst, J. Mazars, G. Pijaudier-Cabot, and J. G. M. van Mier (Eds.), *Fracture Mechanics of Concrete Structures,* Lisse, pp. 617–624. Balkema.

Rudnicky, J. W. and J. R. Rice (1975). Conditions for the localization of deformation in pressure-sensitive dilatant materials. *Journal of the Mechanics and Physics of Solids 23*, 371–394.

Runesson, K., N. S. Ottosen, and D. Perić (1991). Discontinuous bifurcations of elastic-plastic solutions at plane stress and plane strain. *International Journal of Plasticity 7*, 99–121.

Strömberg, L. and M. Ristinmaa (1996). FE-formulation of a nonlocal plasticity theory. *Computer Methods in Applied Mechanics and Engineering 136*, 127–144.

Vermeer, P. A. and R. B. J. Brinkgreve (1994). A new effective non-local strain measure for softening plasticity. In R. Chambon, J. Desrues, and I. Vardoulakis (Eds.), *Localisation and Bifurcation Theory for Soils and Rocks,* Rotterdam, pp. 89–100. Balkema.

Computational Modelling of Concrete Structures, Bićanić et al. (eds)
© 2003 Swets & Zeitlinger, Lisse, ISBN 90 5809 536 3

Study on shape properties of concrete crack surface in frequency domain

R. Sato, & M. Ueda
Hokkaido UNIV., Hokkaido, Japan

T. Wada
Hokkaido Polytechnic College, Hokkaido, Japan

ABSTRACT: Although the finite element method is now widely used as a practical method for structural analysis of concrete structures, the mechanism by which shear stress is transferred by surfaces of cracks formed in concrete must be elucidated in order to improve the accuracy of analysis of concrete structures, which show various patterns of failure. In this paper, the shape property of a concrete crack surface is represented as "amplitude constituent" and "phase constituent" in the frequency domain by utilizing Fourier transform. And we propose a experimental model for standard shape of crack surfaces based on general tendencies of these constituents.

1 INTRODUCTION

The possibility of applying "the shape property of a concrete crack surface" to constitutive models for stress transfer on crack surfaces has already been discussed in many research papers (8, for example). However, very few studies applying Fourier analysis theory have reported analyzing the actual shape of the crack. Moreover, presently, there have been only a few papers, which describe only the evaluation stage of a Fourier one-dimensional spectrum in the joint shape or the crack surface shape of concrete members. Originally, Fourier analysis theory is a dual theory, providing for the "analysis and reduction of objects intended for analysis", which included a inverse Fourier transform concept as well as a Fourier transform concept. Identification of the crack shape using inverse Fourier transform, although conceptually possible, has not been performed yet. Therefore, this suggests that research concerning the shape analysis of concrete materials based on Fourier analysis is still at a preliminary stage.

In light of the current state of this research, in the present paper a shape analysis method using "the phase constituents of a shape property from Fourier analysis" will be proposed. This method has never been dealt with when analyzing a shape using Fourier analysis. Therefore, based on these analytical results, a standard shape model of crack surfaces derived from a experimental inverse Fourier transform will be proposed and its potential discussed.

2 THE ANALYTICAL METHOD OF CONCRETE CRACK SURFACE SHAPE BY FOURIER ONE-DIMENSIONAL TRANSFORM

A crack surface introduced in concrete is usually characterized by an extremely complicated shape change. The elements constituting the crack surface do not have a clear relationship to each other in the same way that the properties of angle and depth are lacking in the so-called space domain. In this paper, Fourier analysis theory is applied to analyze this complicated shape. The basic concept of Fourier analysis is to decompose the objects intended for analysis into composite trigonometric curves and to evaluate their properties in a frequency domain by dealing with the information acquired from trigonometric function in each frequency. The analysis of concrete crack surfaces when applying Fourier analysis theory has the same meaning as seeing the shape of the crack surface as a set of composite trigonometric curves from a new point of view for a frequency domain, and therefore, is expected to produce a new interpretation, that has never been obtained previously, of this complicated shape.

2.1 *Measurement coordinate system of concrete crack surfaces*

The authors have already developed and constructed a system to measure a concrete crack surface using the structured light projection method that obtains

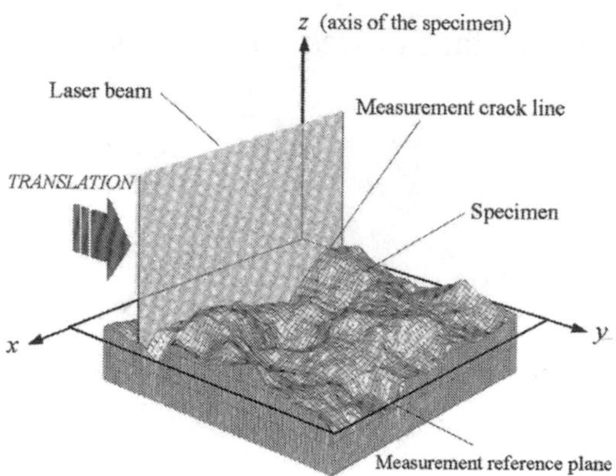

Figure 1. Coordinate system of concrete crack surfaces.

extremely accurate three-dimensional coordinate data from concrete crack surfaces. This measurement system and the details of the measurement have already been reported (6), and all of the shape analyses of concrete crack surfaces dealt with in this paper are based on these measurement data. Therefore, in this paper, these three-dimensional coordinate data will be explained.

Measuring concrete crack surfaces using the structured light projection method is shown in Figure 1. At first, a laser beam parallel to the x–z plane seen in the figure is radiated to concrete crack surfaces and the crack line shape is measured using a CCD (charge-coupled device) camera. Next, this laser is moved in a direction parallel to the y-axis and the measurement of several crack lines can be obtained each time by repeating the process described above. This measurement system can construct the three-dimensional co-ordinate data of concrete crack surfaces by connecting these lines. In this bout, these coordinate data are supposed to be given to indicate the height of a crack surface which develops against the x–y plane (hereafter called "measurement reference plane") in the direction of the z-axis.

Three-dimensional coordinate data can be acquired by A/D transformation of the crack line images measured by a CCD camera and then two kinds of contributing factors, which influence the accuracy of the coordinate data, at this stage, can be indicated. One is the transformation ratio of the data (hereafter called "measurement sensitivity") in the process of A/D transformation and the other is the "sampling interval" derived when interpreting the shape of the crack surface, which varies continuously in space as discrete digital data. The details have already been reported previously (3). The influences over these two parameters on the analytical results in analyzing the concrete crack shape have been verified and it has been concluded that a measurement sensitivity,

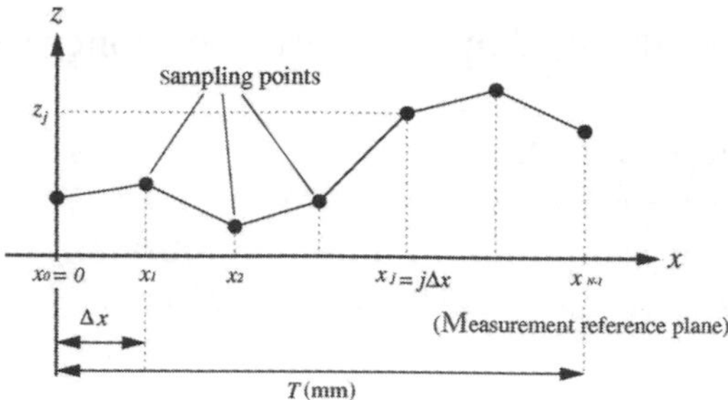

Figure 2. Cross-sectional profile for Fourier one-dimensional transform.

0.125 (mm/dot) and sampling interval, of about 1.0 (mm) are reasonable.

2.2 *Representation of the shape property of concrete crack surfaces by Fourier transform*

There are two methods used to perform the Fourier transform for three-dimensional coordinate data of the crack shape. One is a Fourier two-dimensional transform that interprets the crack surfaces with three-dimensional spatiality as the plane domain function and the other is a Fourier one-dimensional transform, which transforms the cross-sectional shape of an arbitrary plane cut out vertically to the measurement reference plane. The former transforms the relationship of all sampling points into a frequency domain and the latter represents the positioning relationship of the sampling points in the cut out cross-sectional shape as a frequency domain. In this paper, we apply to Fourier one-dimensional transform as a first step of our study.

2.2.1 *Crack cross-sectional profile for Fourier one-dimensional transform*

In Figure 2, "a crack cross-sectional profile model" is given to perform the shape analysis of this study. This is the cross-sectional shape of the concrete crack cut out from a three-dimensional coordinate of the measured crack planes described above. Naturally the cross-sectional direction is arbitrary, but this paper will discuss the cross-sectional profile cut in the x-axis direction to make subsequent descriptions concise.

The cross-sectional profile is of the discrete data of the crack line shape, which are sampled in the sampling interval Δx, with the projection length T (mm) up to the measurement reference plane. Therefore, when the number of sampling points are represented as N, T is represented as follows:

$$T = (N-1)\Delta x \tag{1}$$

In this study, this cross-sectional profile is considered to be a discrete function and the discrete Fourier

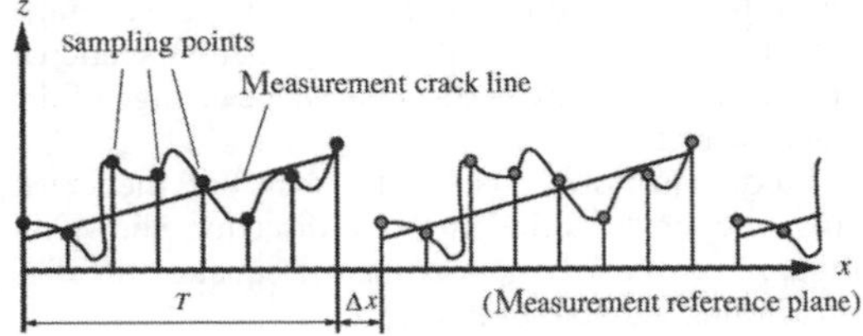

Figure 3. Features of discrete Fourier transform.

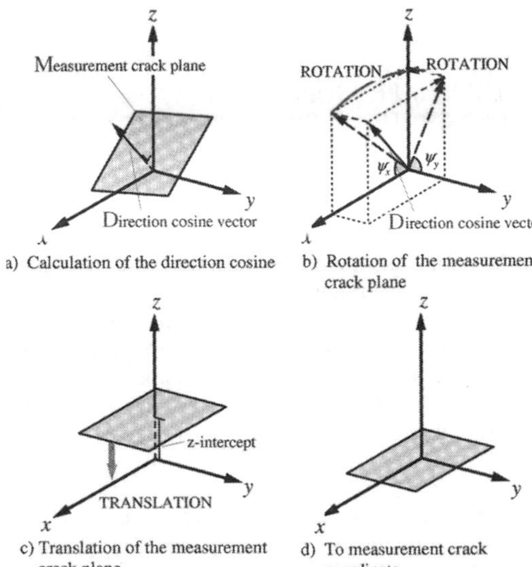

Figure 4. Transformation of coordinates.

transform is applied to the shape analysis. In this bout, the inclination of the whole cross-sectional profile against the measurement reference plane remarkably lowers the accuracy of the analytical results.

The problem described above can be explained in Figure 3. This figure indicates the features of the discrete Fourier transform and here the input cross-sectional profile is assumed to be a frequency function with the period $(T+\Delta x)$. Therefore, the big inclination of the cross-sectional profile is considered to represent points of complete discontinuity; invisible from the real crack surfaces, and this in turn influences the analytical results.

A method to solve this problem was proposed in our previous report (4) and will be applied in this paper. The overview is shown in Figure 4. In brief, with this method the regression plane, which is calculated by performing the method of least squares on all of the sampling points, is considered a "measurement crack plane", and the shape analysis of the crack surface is performed on the coordinate system "measurement crack plane coordinate" acquired by rotating and translating the measurement crack plane, as shown in the figure. However, it has been confirmed that this method has gained results in which the influence from the slope of the cross-sectional profile was minimized.

2.2.2 *The one-dimentional Fourier transform of concrete crack surfaces*

The shape of the cross-sectional profile shown in the previous section is decomposed into a trigonometric curve with different frequency (wave length), amplitude, phase by a basic equation of the discrete Fourier transform, as shown in equation (2) and is newly described in the frequency domain.

$$c_k = \frac{1}{N}\sum_{j=0}^{N-1} z_j e^{-ijk2\pi/N} \tag{2}$$

where $k = 0, 1, \dots, N-1$.

As is easily understood from equation (2), Fourier coefficient c_k is usually calculated as a complex numbers. Therefore, we consider that c_k is described as equation (3)

$$c_k = a_k + ib_k \tag{3}$$

c_k whose totals number is N, is called a Fourier coefficient derived from equations (2) and (3), and each Fourier coefficient has character information determining the property of the component wave (trigonometric function) of the crack surface shape. The index k of c_k is zero or a natural number governing space frequency f_k and its wavelength λ_k of the component wave which c_k represents. In this study, it is called "frequency index k". The relationship among frequency index k, frequency f_k, and amplitude λ_k is described in equation (4).

$$\begin{cases} f_k = \dfrac{k}{T+\Delta x} = \dfrac{k}{N\Delta x} \\ \lambda_k = \dfrac{1}{f_k} \end{cases} \tag{4}$$

As is clear from equation (4), as the value k becomes a larger Fourier coefficient, the frequency of trigonometric function (component wave) becomes higher (wavelength becomes shorter). More specifically, as the Fourier coefficient becomes a higher-frequency, it represents a more minute shape change. The most minute shape change that the discrete Fourier transform can derive, is known to depend on the sampling interval Δx. This is because the discrete Fourier transform approximates the shape of the concrete crack surface that changes continuously in a space domain with finite component waves. The largest space frequency f_u can be acquired from the equation (5) based on *Nyquist's theorem*.

$$f_u \leq \frac{1}{2\Delta x} \tag{5}$$

Moreover, since frequency f_k and frequency index k are related through the equation above (4), the value of the frequency index k can be determined from the following conditioned equation (6).

$$\frac{2k}{N} \leq 1 \quad (k, N \text{ is zero or natural number}) \tag{6}$$

Therefore, the total number of Fourier coefficients calculated from the above equation (2) is N, but an practical Fourier coefficient is determined from zero or a natural number k which satisfies equation (6). Therefore, in this study, the largest frequency index is defined as u.

As is seen from equation (3), because the Fourier coefficient c_k, given by the Fourier transform of the cross-sectional profile, is a complex number, the Fourier coefficient c_k is interpreted as a vector having intensity F_k and the direction angle ϕ_k on the complex plane with the real part a_k on the horizontal axis and imaginary part b_k on the vertical axis, as seen in the Figure 5. The F_k and ϕ_k indicate the amplitude and phase of the component waves for each frequency respectively. In short, the shape property of a concrete crack in the frequency domain can be represented, as the two constitutes above. The former is called "the amplitude constituent", the latter is called "the phase constituent".

The amplitude constituent F_k and phase constituent ϕ_k, the shape properties in the frequency domain, are easily determined from the geometric conditions in the figure. This is shown in equation (7).

$$\begin{cases} F_k = \sqrt{a_k^2 + b_k^2} \\ \phi_k = \tan^{-1}\left(\dfrac{b_k}{a_k}\right) \end{cases} \tag{7}$$

In addition, the Fourier coefficient c_0 is generally known as a "zero-frequency component" when the frequency index is $k = 0$ and is always calculated as a real value. This coefficient indicates only the average height of all the sampling points on the cross-sectional profile and is not involved in the shape change of the crack surfaces.

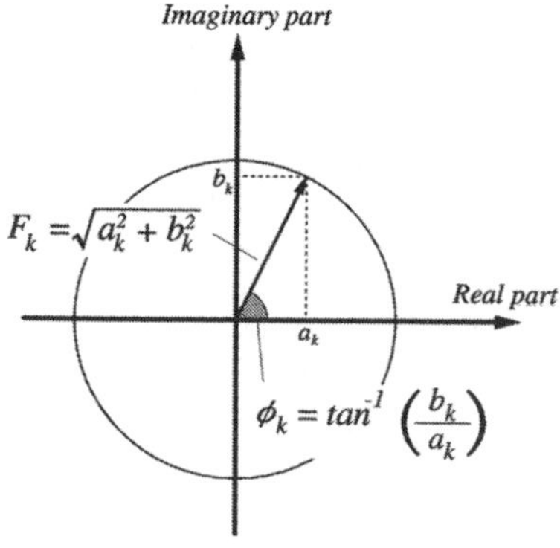

Figure 5. Fourier coefficient on complex plane.

From equations (1)–(7), the shape of the crack cross-sectional profile in the x direction shown in Figure 2 proved to be approximate to equation (8) after performing the Fourier transform.

$$z(x) = c_0 + \sum_{k=1}^{u} 2F_k \cos\left(\frac{2\pi k}{N\Delta x}x + \phi_k\right) \tag{8}$$

3 THE SHAPE PROPERTY OF CONCRETE TENSILE SURFACE IN THE FREQUENCY DOMAIN

As described in the previous section, the shape of concrete crack surfaces can be represented as two constitutes, "the amplitude constitute" and "the phase constitute" in the frequency domain.

Here, we will perform the Fourier transform on the shape of concrete crack surfaces introduced by the tensile stress mode and seek a possibility to quantify the shape property of concrete crack surface by investigating the analytical results.

3.1 *An overview of the experiment of the crack formation and its concrete properties*

We have carried out the experiment in which some concrete, of the same mix proportions, is acted on by four basic stress modes, such as tensile, split, bending, and shear, and destructed. The three-dimensional coordinate data of the crack surface's shape have been obtained through the experimental results. In this paper, among the tensile crack surfaces which are known to have the most stable stress conditions and an almost uniformly distributed load through the cracing process, *Tn*-3 that has the most stable shape, is the subject being analyzed. Table 1 indicates the mix proportion of concrete used. From the table, it is clear that the mix design of the concrete analyzed is quite similar to that which is used to reinforce concrete buildings. The concrete properties are shown in Table 2.

3.2 *Measurement and analytical conditions*

It has already been stated that cutting out the cross-sectional profile in an arbitrary direction is possible from the three-dimensional coordinates of concrete crack surfaces gained by this measurement system. However, regarding analyzing the actual shape, the cross-sectional profile parallel to the measurement x and y axis is cut out in the same intervals and the behavior of the whole crack surfaces are evaluated from two-way directions. Therefore, each analytical result is described as an "x direction" and a "y direction" separately.

On the tensile crack surface, *Tn*-3, the three- dimensional coordinate was measured under the measurement conditions shown in Table 3. Based on this, the Fourier analysis conditions are indicated in Table 4.

Moreover, x direction constituent ψ_x and y direction constituent ψ_y of the direction cosine vector and z-intercept of the measurement crack plane are shown in Table 5.

Table 1. Mix proportion of concrete.

Cement	Nominal strength	Slump	Maximum aggregate size
Normal	30 N/mm²	10 cm	20 mm
Aggregate	**Cement**	**Water–Cement ratio**	**Sand–Aggregate ratio**
Crashed stone	N	48.0%	20%

Table 2. Concrete strength.

Name	Measured slump	Compressive strength	Splitting strength
S18-II	21.7 cm	36.2 N/mm^2	2.64 N/mm^2

3.3 *Verification and speculation of an amplitude constitute in concrete crack surfaces*

Here, regarding the shape property of concrete crack surfaces in the frequency domain, the amplitude constituent, the histogram enumerating its intensity value F_k in the order from the smallest frequency index k, is called an "amplitude spectrum". The shape evaluation is carried out in the frequency domain through mutual comparison.

From the measurement conditions shown in Table 4, 61 cross-sectional profiles were cut out in the x direction and 57 cross-sectional profiles were cut out in the y direction. From the results of the Fourier analysis for the cross-sectional profiles, as examples, the 29th and the 44th amplitude spectrums in the x direction are shown in Figure 6 and in the same way, the 29th and the 44th amplitude spectrums in the y direction are shown in Figure 7. The reason why the intensity value F_0, when the frequency index is $k = 0$ among each spectrum, is represented as an outline typeface is that the zero-frequency component described above is not involved in the shape change and it has to be separated.

Each amplitude spectrum in the x direction shown in Figure 6 (a) and (b) has a peak value in the vicinity of the frequency index, $k = 1$ and the distribution is such that the intensity value becomes smaller as the frequency becomes larger. The same tendency

Table 5. Property of measurement crack plane.

Nonation	ψ_x (rad)	ψ_y (rad)	z-intercept (mm)
Tn-3	−1.56	−1.54	−1.09

Table 3. Measuring conditions.

Direction	Measurement sensitivity (mm/dot)	Sampling interval (mm)	Length T (mm)	Number of sampling points N	Number of cross-sectional profile
x	0.125	1.0	56	57	61
y			60	61	57

Table 4. Analytical conditions.

Direction	Maximum value of k u	Lowest frequency f_ℓ (1/mm)	Highest frequency f_u (1/mm)	Max wavelength λ_l (mm)	Min wavelength λ_u (mm)
x	29	0.018	0.49	57	2.04
y	30	0.016	0.49	61	2.03

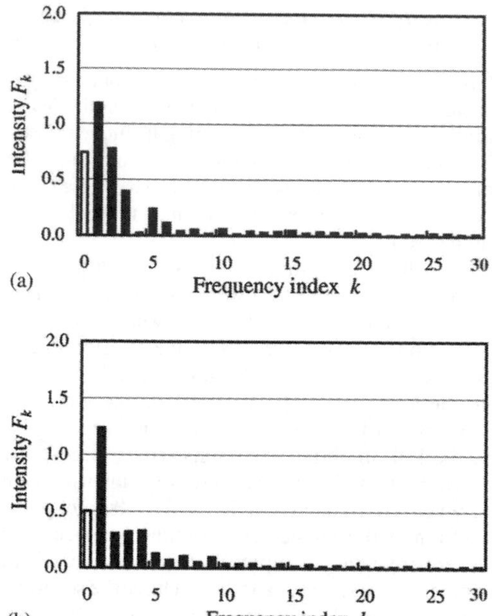

Figure 6. (a) Amplitude spectrum of the 29th cross-sectional profile (x direction). (b) Amplitude spectrum of the 44th cross-sectional profile (y direction).

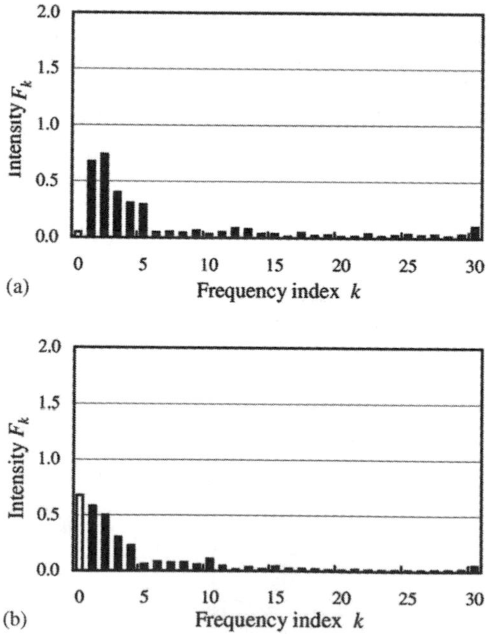

Figure 7. (a) Amplitude sspectrum of the 29th cross-sectional profile (x direction). (b) Amplitude spectrum of the 44th cross-sectional profile (y direction).

has been ascertained, not just in the y direction in Figure 7, but also in the amplitude spectra of the other cross-sectional profiles, not reported in this paper.

The following Figure 8 (a) indicates the arithmetic average of spectrums calculated from the analytical results of all of the cross-sectional profiles in the x direction. After taking the average, the amplitude spectrum is apt to decrease more gradually and have its peak at $k = 1$. This tendency is also true of the analytical results in the y direction shown in Figure 8(b). Regarding the distribution of the amplitude spectrum, from the above results, no distinction can be seen in the different analytical directions. And it seems that amplitude spectra of concrete crack surfaces indicate a "downward sloping distribution" in general.

3.4 *Verification and estimate of the phase constitutes of concrete crack surfaces*

Although, the phase constitutes given as a result of the Fourier analysis are applied to verifying the property of the objects in the field of vibration analysis, there is no example for the concrete materials. When verifying the phase constitutes in this study, not only a histogram enumerating the phase angle ϕ_k in the order from the smallest frequency – "phase spectrum", but also, the probability density of ϕ_k are taken into consideration, and then a more detailed distribution can be carried out.

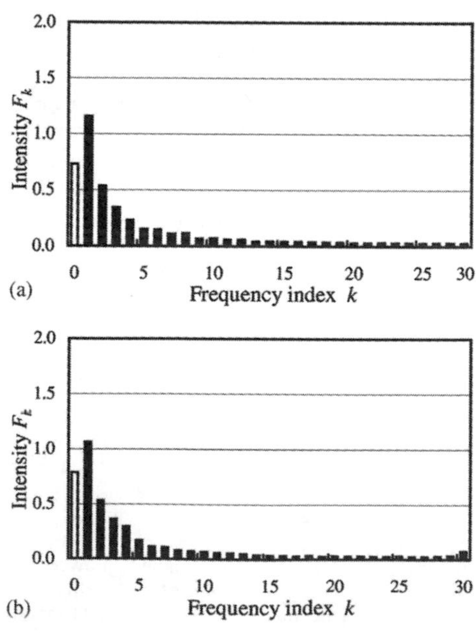

Figure 8. (a) Arithmetic average of spectra of all of the profiles (x direction). (b) Arithmetic average of spectra of all of the profiles (y direction).

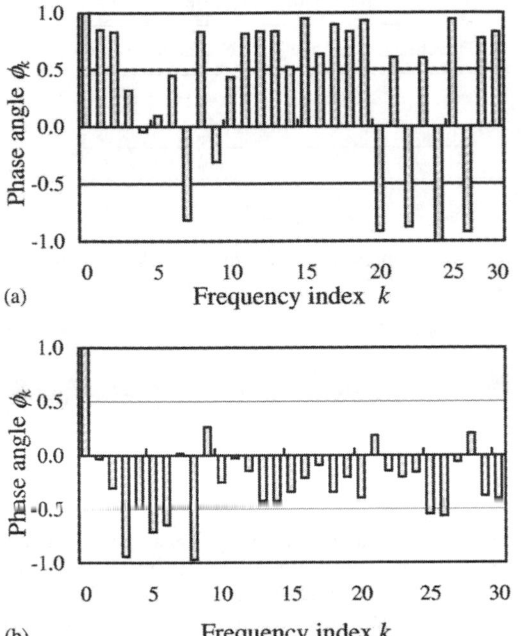

Figure 9. (a) Phase spectrum of the 29th cross-sectional profile (x direction). (b) Phase spectrum of the 44th cross-sectional profile (x direction).

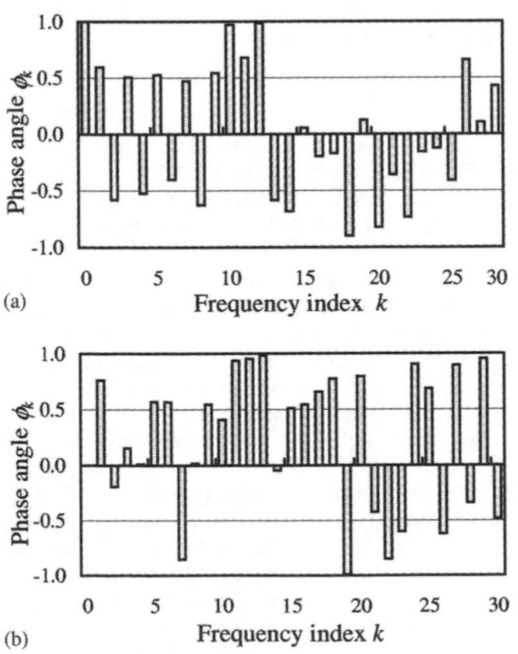

Figure 10. (a) Phase spectrum of the 29th cross-sectional profile (x direction). (b) Phase spectrum of the 44th cross-sectional profile (y direction).

3.4.1 *The transition of the phase spectrum and the probability density distribution of the phase spectrum of concrete crack surfaces*

As in the case of the analysis of the amplitude spectrum, the 29th phase spectrum from the cross-sectional profile of the tensile crack surface *Tn*-3 in the x direction is shown in Figure 9(a) and the 44th phase spectrum is shown in Figure 9(b) respectively. In Figure 9(a) and (b), the distribution of the phase spectrums, given from the object intended for analysis, the cross-sectional profile shows a change bearing poor similarity to each other. Therefore, there is no clear distinction as is seen in the amplitude spectrum. The same tendency has been observed not just in the analytical results of the y direction in Figure 10(a) and (b), but also in the analytical results of all of the other cross-sectional profiles, not reported in this paper.

The probability density distribution of the phase angle ϕ_k calculated to read the entire tendency related to the phase angle of all of the cross-sectional profiles is indicated in Figure 11(a) and (b).

From the Figure, in each analytical direction, it is evident that the phase angles ϕ_k exist all across the domain from $-\pi$ to π, but it seems hard to recognize distinct characteristic such as symmetric distribution.

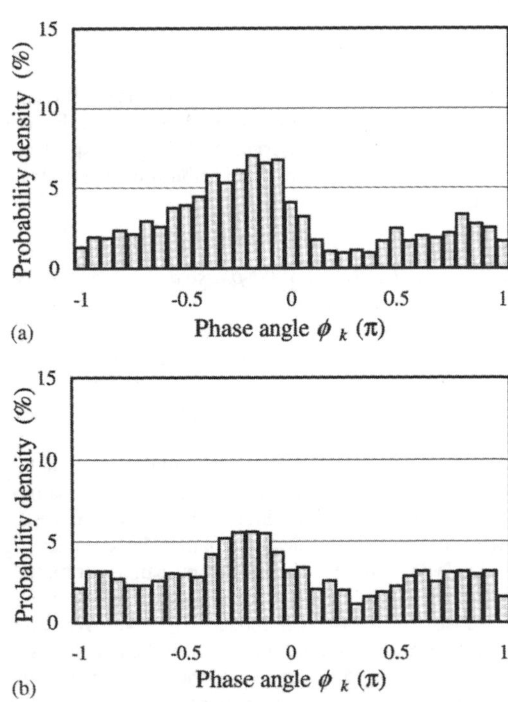

Figure 11. (a) Probability density distribution of the phase angle (x direction). (b) Probability density distribution of the phase angle (y direction).

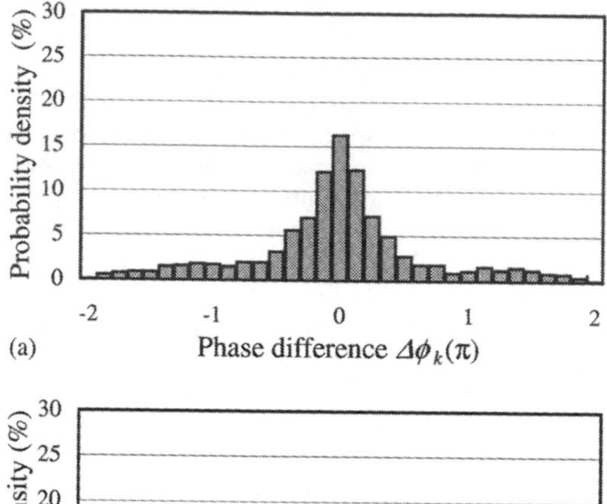

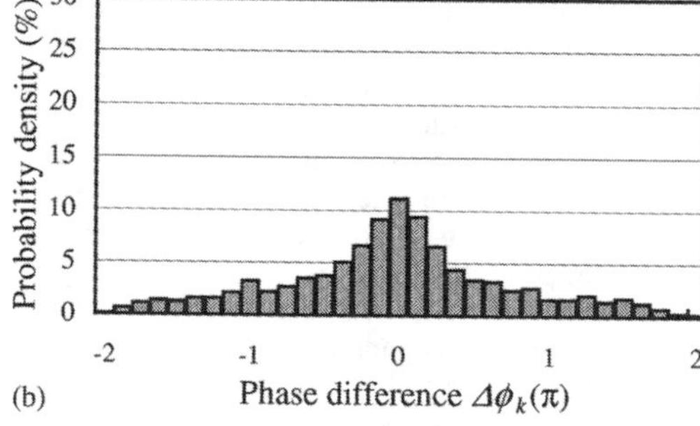

Figure 12. (a) Probability density distribution of the phase difference (x direction). (b) Probability density distribution of the phase difference (y direction).

When summarizing therein before, regarding the phase constituents of concrete crack surfaces, a distinct tendency couldn't be observed with the phasespectrum and the probability density distribution of the phase angle ϕ_k.

3.4.2 *The transition of the phase difference and its probability density*

In the previous section, the phase constituents of the shape of concrete crack surfaces changed randomly in the frequency domain and consequently further estimate will be needed to quantify the character of this constituent.

Therefore, in this section, the relationship of the neighboring phase angles in the frequency domain is focused on and by introducing the phase difference $\Delta\phi_k$, the probability density of the phase angle ϕ_k, which is dependent on the frequency, can be estimated.

Figure 12 indicates the probability density distribution of the phase difference $\Delta\phi_k$ given from the analytical results in the x direction on the tensile crack surface Tn-3. From Figure 12, the phase difference $\Delta\phi_k$ indicates the symmetric distribution with the peak at 0 and an almost normal distribution. This tendency is seen in the phase difference in the y direction in Figure 12(b) and it is concluded that the character of

Table 6. Property of phase difference $\Delta\phi_k$.

Direction	Average	Standard deviation
x	$0.0045\,\pi$	$0.44\,\pi$
y	$-0.0214\,\pi$	$0.42\,\pi$

the phase constituent of concrete crack surfaces,

$$\Delta\phi_k = \phi_{k+1} - \phi_k \tag{9}$$

where $k = 1, 2, \ldots, u-1$ which cannot be obtained when only the phase angle is ϕ_k, can now be analyzed.

Moreover, the average value and the standard deviation of the phase difference $\Delta\phi_k$, gained from the probability density distribution in each analytical direction, can be described in Table 6. Since these values are close, the symmetric distribution and the probability density distribution seem to resemble one another regardless of the analytical direction.

Regarding the description above, although the phase constituent of the shape of the concrete crack surfaces indicates a random behavior in a frequency domain, it is suggested that the introduction of the phase difference $\Delta\phi_k$ provides the possibility of representing a charastic.

4 PRESENTATION OF THE STANDARD SHAPE OF CONCRETE CRACK SURFACES BY INVERSE FOURIER ONE-DIMENSIONAL TRANSFORM

As has already been stated, the shape of concrete crack surfaces has two properties such as "download sloping amplitude spectrum" and "random phase spectrum" in a frequency domain. the phase constituents are recognized as a means to represent the property because of "phase difference based on probability density with rather normal distribution". In other words, however, the shape of concrete crack surfaces is difficult to grasp the property in a space domain, but the complicated changes are possible to be identified numerically in a frequency domain. Here, the adequacy of the shape property of concrete crack surfaces will be verified. Moreover, regarding the potential of developing a standard shape model of crack surfaces with this analytical method, the standard shape models of crack surfaces constructed experimentally by the inverse Fourier transform will be explained a little more.

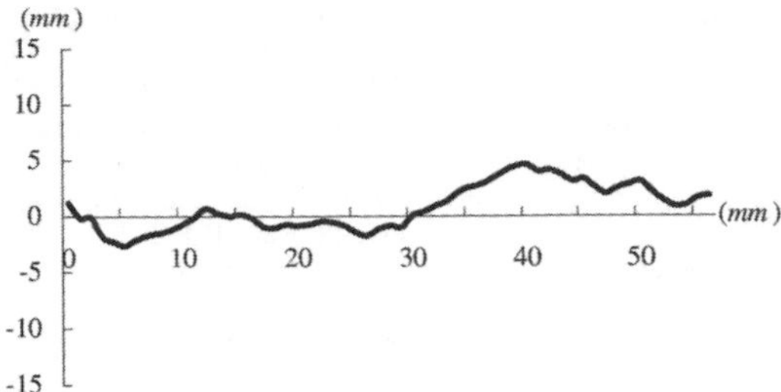

Figure 13. Example of standard shape model of concrete crack surface constructed by inverse Fourier transform.

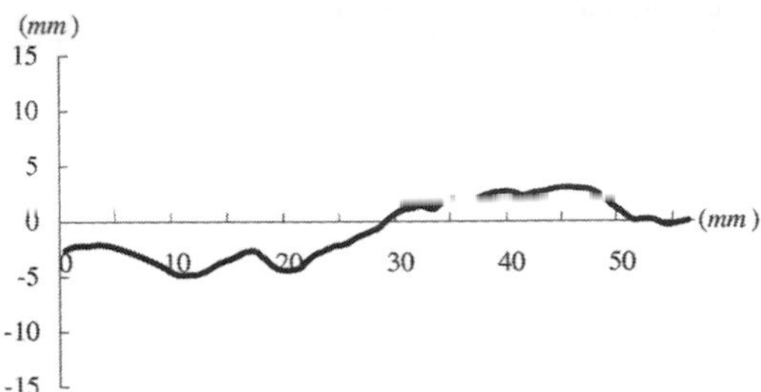

Figure 14. Typical shape of real cross-sectional profile.

4.1 *Identification of the Fourier coefficient having the frequent property of concrete crack surfaces*

The Fourier transform is a mathematically inverse calculation. In short, if the Fourier coefficient $_I c_k$, including the shape property in a frequency domain is identified, the property can be resolved to a space domain using discrete inverse Fourier one-dimensional transform, as shown in equation (10). As can be seen, the shape of the cross sectional profile in a space domain was transformed into a frequency domain.

$$_I z_j = \sum_{k=0}^{N-1} {_I c_k} e^{ijk2\pi/N} \tag{10}$$

Where $j = 0, 1, \ldots, N-1$.

The shape property of concrete crack surfaces, described in this paper, is the shape of the actual cross sectional profile accurately measured by the structured light projection method, which is newly described in a frequency domain using the Fourier transform. In other words, the amplitude spectrum shown in Figure 8 and the probability density distribution of phase difference $\Delta\phi_k$ shown in Figure 12, represents the shape property of 61 cross sectional profiles in the x direction and 57 cross sectional profiles in the y direction in a frequency domain.

If this is assumed to be the shape property of a general cross sectional profile, it seems natural that the result given by inverse Fourier transforming the Fourier coefficient $_I c_k$ indicates behavior close to the actural cross sectional profile, in a space domain.

Therefore, we resolve the Fourier coefficients $_I c_k$ by substituting the F_k given from the amplitude spectrum indicating downward spoling distribution in Figure 8 and the ϕ_k based on the probability density distribution of phase difference $\Delta\phi_k$ shown in Figure 12 in to equation (12). Figure 13 shows an example that obtained by inversely Fourier transforming the coefficients resolved by the process described above. Figure 14 shown next to it is an example of the actual cross-sectional profile given from the three-dimensional coordinate data on the tensile crack surface *Tn*-3. Both of them look as if they have many similarities such as large swells and detailed shape changes seen locally. This fact seems to argue that the standard crack shape has sufficient value to represent a qualitative tendency of the actural concrete crack surface.

5 CONCLUSIONS

In this paper, the shape property of crack surfaces is represented by applying the one-dimensional Fourier transform to the shape analysis of the actual three-dimensional coordinate data of the tensile concrete crack surface. And the standard shape model of concrete crack surfaces is proposed experimentally through verifying the physical quantity. A summary of these results is as follows.

1. The amplitude spectra of the concrete tensile crack surface has a downward sloping distribution, and it peaks when the frequency index is $k = 1$, after that, it has a tendency to gradually decrease.
2. The phase angle ϕ_k indicates a random character, which has no regularity, but the random character is quantified as a useful analytical index by introducing the phase difference $\Delta\phi_k$.
3. The probability density of the phase difference $\Delta\phi_k$, described above, bears the character rather normal distribution. Moreover, the tendencies are similar regardless of the analytical direction, and this fact agrees with the analytical result of only the amplitude constituent already reported.
4. Based on the results discussed so far, it has been concluded that the standard shape model of concrete crack surfaces constructed experimentally by the inverse Fourier transform does not smooth down the local roughness, but instead, provides a relatively reliable model of an actual recreation of the crack surface.

REFERENCES

Katori, K., Hayashi, S., Makitani, T. & Ushigaki, K. 1998. Estimate of the shear strength of construction joint by using its surface roughness and characteristic of slip displacement of joint under shear forces-studies on shear behavior of precast joint. Journal of Structural and Construction Engineering, AIJ. No. 507. pp. 107–116.

Meyberg, K. & Vachenauer, P. translated by Oikawa, M. 1998. Ḧohere Mathematik ed., 2.

Sato, R., Wada, T., Sato, R. & Ueda, M. 2001. Fast Fourier One-dimensional analysis of concrete crack surface. Fracture Mechanics of Concrete Structures, Proceeding, FRAMCOS-4, pp. 423–430.

Sato, R., Wada, T., Sato, R. & Ueda, M. 2002. Analysis of the shapes of concrete crack surface by the structured light projection method using fast fourier transform. Journal of Structural Engineering, Vol. 48B, pp. 357–366.

Sato, R., Wada, T., Sato, R. & Ueda, M. 2002. Analysis of concrete crack surfaces introduces by various kinds of stress using fast Fourier Transform. Proceedings of the Japan Concrete Institute, JCI No. 24, No. 2, 2 pp. 133–138.

Wada, T., Sato, R., Chiharu, C. & Ueda, M. 1998. Development of measurement of the concrete crack surface by laser beam and proposal of 2-diemnsional analytical method of the measured image – a basic study on shape properties of the concrete crack surface Part 1. Journal of Structural and Construction Engineering, AIJ No. 504, pp. 179–188.

Wada, T., Sato, R., Chiharu, C. & Ueda, M. 1998. 2-Dimensional shape analyses of the concrete crack surfaces introduced by various kind of stress – a basic study on shape properties of the concrete crack surface Part2. Journal of Structural and Construction Engineering, AIJ No. 504, pp. 81–86.

Walraven, J.C., Reinhardt, H.W. 1981. Theory and experiments on the mechanical behavior of cracks in plain and reinforced concrete subjected to shear loading. HERON Vol. 26, No. 1A.

Computational Modelling of Concrete Structures, Bićanić et al. (eds)
© 2003 Swets & Zeitlinger, Lisse, ISBN 90 5809 536 3

Two- and three-dimensional gradient damage-plasticity simulations of cracking in concrete

Jerzy Pamin, Adam Wosatko & Andrzej Winnicki
Faculty of Civil Engineering, Cracow University of Technology, Cracow, Poland

ABSTRACT: The paper deals with a numerical simulation of fracture processes in plain and fiber-reinforced concrete. The applied continuum model combines the gradient damage theory formulated in the strain space with the hardening plasticity theory formulated in the space of effective stresses. The formulation incorporates a projection operator which accounts for the crack closing phenomenon. The model is implemented in the FEAP finite element package. Two- and three-dimensional simulations of localized deformation are performed.

1 INTRODUCTION

While for structures one- or two-dimensional idealization usually suffices, for certain structural elements and experimental specimens three-dimensional analysis of stress and failure is necessary. The paper deals with a numerical simulation of fracture processes in plain and fiber-reinforced by means of a gradient-enhanced continuum model. Monotonically increasing or reversed static loading is considered.

The applied model combines the gradient damage theory formulated in the strain space (Peerlings et al. 1996) with the hardening plasticity theory formulated in the space of effective stresses (de Borst et al. 1999; Pamin and de Borst 1999). In particular, the Burzyński-Drucker-Prager plastic flow theory with linear isotropic hardening is adopted as the plastic part of the model. The classical equilibrium problem is augmented by an additional averaging equation for a selected equivalent strain measure, so that a coupled problem is obtained. The formulation presented in this paper incorporates a projection operator which accounts for the crack closing phenomenon.

Owing to the gradient enhancement, the model produces results without spurious discretization sensitivity, while it remains algorithmically robust. The finite element formulation involves an independent discretization of the averaged strain measure and leads to two-field finite elements. The elements and the constitutive model are implemented in the FEAP finite element package (Taylor 2001).

The paper is focused on the modelling problems related to the crack closing phenomenon and on the extension of the analysis to three-dimensions. So far, quadratic interpolation functions have been used for the displacements and linear for the averaged strain measure. Now, other combinations of polynomials are considered. In Section 2 the constitutive theory is given and in Section 3 the gradient regularization and finite element algorithm are briefly discussed.

Section 4 contains the numerical simulations. A two-dimensional analysis of localized deformation in a cantilever beam illustrate the performance of the model, and in particular of the damage-plasticity coupling and the projection operator. The possible interpolation orders for the two discretized fields are discussed for the classical tensile bar benchmark test performed in 2D and 3D.

2 CONSTITUTIVE MODEL

We start the discussion of constitutive relations by selecting a form of coupling between plasticity and damage theories. We combine a plasticity theory formulated in the stress space with a damage theory formulated in the strain space. The theories have a simple, isotropic format (intrinsic anisotropy is neglected). As will be shown in Section 3, the damage component of the theory is made gradient-dependent in order to ensure that numerical simulations of strain localization give meaningful results.

Considering the damage evolution we distinguish the actual body with strains $\boldsymbol{\epsilon}$ and stresses $\boldsymbol{\sigma}$ and its fictitious undamaged counterpart with stresses $\hat{\boldsymbol{\sigma}}$ and strains $\hat{\boldsymbol{\epsilon}}$. The fictitious counterpart represents the undamaged "skeleton" of the body, and the stresses $\hat{\boldsymbol{\sigma}}$ acting on it are called effective. We adopt the postulate that the strains observed in the actual body and in its undamaged representation are equal (Simo and Ju 1987; Ju 1989):

$$\epsilon = \hat{\epsilon}, \tag{1}$$

and that the stresses are related as follows:

$$\boldsymbol{\sigma} = \hat{\boldsymbol{\sigma}} - \boldsymbol{\sigma}^{\mathrm{d}}, \tag{2}$$

with the effective stress

$$\hat{\boldsymbol{\sigma}} = \boldsymbol{E} : \boldsymbol{\epsilon}^{\mathrm{e}}, \tag{3}$$

and the so-called damage stress $\boldsymbol{\sigma}^{\mathrm{d}}$ which describes the change in the stress state due to damage represented by one scalar parameter ω:

$$\boldsymbol{\sigma}^{\mathrm{d}} = \omega \boldsymbol{E} : \boldsymbol{\epsilon}^{\mathrm{e}}. \tag{4}$$

Here $\boldsymbol{E}$ is the elastic stiffness operator and $\boldsymbol{\epsilon}^{\mathrm{e}}$ is the elastic strain tensor. The measure ω accounts for the degradation of the elastic stiffness. It is a function of a damage history parameter κ^{d} and grows from zero to one as κ^{d} grows from the damage threshold κ_{o} to an ultimate value κ_{u}.

The motivation for coupling the model with plasticity is to incorporate the physically observed irreversible strains. Assuming the standard additive decomposition of strain rate into an elastic and a plastic part, the elastic strain rate can be written as:

$$\dot{\boldsymbol{\epsilon}}^{\mathrm{e}} = \dot{\boldsymbol{\epsilon}} - \dot{\lambda} \boldsymbol{m}(\hat{\boldsymbol{\sigma}}), \tag{5}$$

where λ is the plastic multiplier and $\boldsymbol{m}$ is the plastic flow direction tensor in the effective stress space (non-associated flow rule is admitted). The two theories are coupled by relation (4), in which the elastic strain ϵ^{e} follows from time integration of (5).

Further, we define a damage activation function which limits the elasto-plastic behaviour of the material in the strain space:

$$f^{\mathrm{d}} = \tilde{\epsilon} - \kappa^{\mathrm{d}} = 0. \tag{6}$$

Here $\tilde{\epsilon}$ is an equivalent strain measure. During the damage evolution the history parameter κ^{d} is equal to the largest value of $\tilde{\epsilon}$ reached in the loading history.

Moreover, we define the yield condition in the effective stress space as follows

$$f^{\mathrm{p}} = \tilde{\sigma}(\hat{\boldsymbol{\sigma}}) - \sigma_{\mathrm{y}}(\kappa^{\mathrm{p}}) = 0, \tag{7}$$

where $\tilde{\sigma}$ is an equivalent stress function, σ_{y} is the yield strength and isotropic hardening is assumed. Proportionality between the plastic multiplier λ and the plastic strain measure κ^{p} p is also assumed for simplicity. Standard loading/unloading conditions are imposed with each of these two limit state criteria.

If we invoke the local plastic consistency condition $\dot{f}^{\mathrm{p}} = 0$ in order to compute $\dot{\lambda}$ and substitute it into Equation (5), then after standard rearrangements we obtain the tangential relation between the strain and effective stress:

$$\dot{\hat{\boldsymbol{\sigma}}} = \boldsymbol{E}^{\mathrm{ep}} : \dot{\boldsymbol{\epsilon}}, \tag{8}$$

with the classical elasto-plastic tensor $\boldsymbol{E}^{\mathrm{ep}}$. The combination of the rate form of Equation (3) and Equation (8) gives the elastic strain tensor in terms of the total strain:

$$\dot{\boldsymbol{\epsilon}}^{\mathrm{e}} = \boldsymbol{E}^{-1} : \boldsymbol{E}^{\mathrm{ep}} : \dot{\boldsymbol{\epsilon}}. \tag{9}$$

This simple formulation has been successfully used not only for the analysis of isotropic materials, but also, with appropriate equivalent strain measures, for materials exhibiting strain or stress induced anisotropy like concrete or soils (Peerlings et al. 1998). However, this formulation does not cover the crack closure phenomenon which occurs for compression following tension (cf. Fig. 1). Damage does not grow during unloading, hence for a pure damage model the constant (secant) stiffness

$$\boldsymbol{E}_{sec} = (1-\omega)\boldsymbol{E} \tag{10}$$

results in unloading to the origin, i.e. no residual strains remain after unloading, which means that

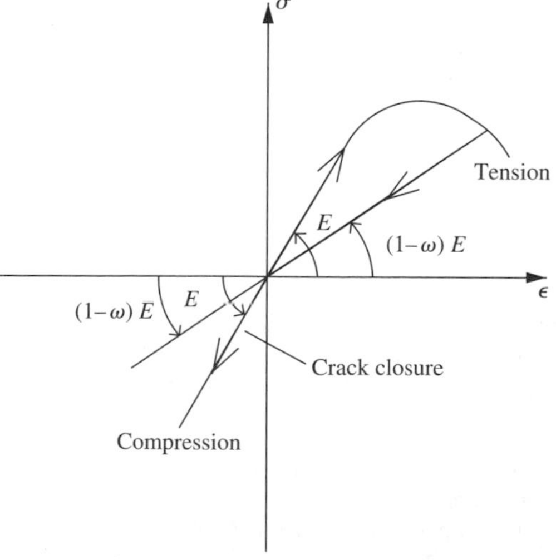

Figure 1. Idealization of crack closing phenomenon in damage (without irreversible strains).

micro-cracks and micro-voids are assumed to close completely. Now, the damage history should not influence the elastic behaviour under compression. In order to capture this crack closure effect (or, more generally, a different stiffness in compression and tension) a very simple mechanism has been devised (Mazars 1984; Simo and Ju 1987; LaBorderie 1991; Hansen 1993). First the tensile part of the elastic strain tensor is separated from the total elastic strain tensor using a projection operator $\boldsymbol{P}^{+}$:

$$\boldsymbol{\epsilon}^{e+} = \boldsymbol{P}^{+} : \boldsymbol{\epsilon}^{e}, \tag{11}$$

then it is assumed that only this part is responsible for damage:

$$\boldsymbol{\sigma}^{d} = \omega \boldsymbol{E} : \boldsymbol{\epsilon}^{e+}. \tag{12}$$

In turn, only the tensile part of the damage stress is allowed

$$\boldsymbol{\sigma}^{d+} = \boldsymbol{P}^{+} : \boldsymbol{\sigma}^{d}. \tag{13}$$

Hence, the stress state becomes

$$\boldsymbol{\sigma} = \hat{\boldsymbol{\sigma}} - \boldsymbol{\sigma}^{d+} = \boldsymbol{E} : \boldsymbol{\epsilon}^{e} - \omega \boldsymbol{P}^{+} : \boldsymbol{E} : \boldsymbol{P}^{+} : \boldsymbol{\epsilon}^{e} \tag{14}$$

or

$$\boldsymbol{\sigma} = \boldsymbol{E}_{sec} : \boldsymbol{\epsilon}^{e}, \tag{15}$$

where the secant stiffness is now

$$\boldsymbol{E}_{sec} = \boldsymbol{E} - \omega \boldsymbol{P}^{+} : \boldsymbol{E} : \boldsymbol{P}^{+}. \tag{16}$$

For triaxial tension (i.e. for the state when all principal values of the strain tensor are larger than zero) we have

$$\boldsymbol{P}^{+} : \boldsymbol{\epsilon}^{e} = \boldsymbol{\epsilon}^{e}; \quad \boldsymbol{P}^{+} : \boldsymbol{\sigma}^{D} = \boldsymbol{\sigma}^{D}, \tag{17}$$

and the relationship between stress and elastic strain simplifies to Equation (2), i.e. the secant stiffness matrix takes the form (10). On the contrary, for triaxial compression we obtain no damage

$$\boldsymbol{P}^{+} : \boldsymbol{\epsilon}^{e} = \boldsymbol{0}; \quad \boldsymbol{E}_{sec} = \boldsymbol{E}, \tag{18}$$

i.e. the material behaves elastically even if in the past it has undergone a damage process in tension.

For the other cases (between triaxial tension and triaxial compression) the general form of the stiffness matrix given by Equation (16) is valid. Due to its nature, the stress–strain relationship becomes anisotropic (strain induced anisotropy via the projection operator $\boldsymbol{P}^{+}$). The key point in the above approach is a proper formulation of the projection operator which transforms the strain tensor into its tensile part.

The idea of the tensile strain projection can be summarized as follows (Mazars 1984; LaBorderie 1991): transform the strain tensor to its principal directions, choose the principal values which are larger than zero and make a new strain tensor (called positive or tensile) composed of these values, and finally transform this strain tensor to the original coordinate system. It remains to derive the projection operator which is necessary to build the stiffness matrix. For this purpose we first represent the strain tensor as the explicit function of its principal values using the spectral decomposition formula:

$$\boldsymbol{\epsilon}^{e} = \sum_{i=1}^{3} \epsilon_i^{e} \boldsymbol{k}_i \otimes \boldsymbol{k}_i \tag{19}$$

in which $\boldsymbol{k}_i$ are unit vectors along principal directions. The principal strains ϵ_i^{e} can be represented as

$$\epsilon_i^{e} = \boldsymbol{\epsilon}^{e} : (\boldsymbol{k}_i \otimes \boldsymbol{k}_i). \tag{20}$$

Next we preserve only the tensile part of the strain tensor using the spectral decomposition and the Heaviside function $h(\cdot)$

$$\boldsymbol{\epsilon}^{e+} = \sum_{i=1}^{3} h(\epsilon_i^{e}) \epsilon_i^{e} \boldsymbol{k}_i \otimes \boldsymbol{k}_i. \tag{21}$$

Inserting the formulae for the principal values we eventually obtain

$$\boldsymbol{\epsilon}^{e+} = \sum_{i=1}^{3} h(\epsilon_i^{e}) (\boldsymbol{k}_i \otimes \boldsymbol{k}_i \otimes \boldsymbol{k}_i \otimes \boldsymbol{k}_i) : \boldsymbol{\epsilon}^{e} \tag{22}$$

or in a compact form

$$\boldsymbol{\epsilon}^{e+} = \boldsymbol{P}^{+} : \boldsymbol{\epsilon}^{e}, \tag{23}$$

where projection operator $\boldsymbol{P}^{+}$ appears as

$$\boldsymbol{P}^{+} = \sum_{i=1}^{3} h(\epsilon_i^{e}) (\boldsymbol{k}_i \otimes \boldsymbol{k}_i \otimes \boldsymbol{k}_i \otimes \boldsymbol{k}_i). \tag{24}$$

It is easy to show that for the triaxial tension case (i.e. when $h(\epsilon_1) = h(\epsilon_2) = h(\epsilon_3) = 1$) the tensile strain tensor is equal to the full strain tensor, i.e. Equations (17) hold, although the projection operator $\boldsymbol{P}^{+}$ is not equal to the 4th-order identity tensor. Moreover, as required, for triaxial compression (i.e. when $h(\epsilon_2) = h(\epsilon_2) = h(\epsilon_3) = 0$) the projection operator becomes zero, so that Equations 18 are satisfied.

The projection operator as given by Equation (24) was probably introduced for the first time in (Ortiz 1985) and hence will be called $\boldsymbol{P}^{+}_{Ortiz}$, although it was there defined for the stress tensor.

The above formula for the projection operator is not unique. Another form was proposed in (Simo and Ju 1987) and afterwards simplified in (Hansen 1993), hence it will be called $\boldsymbol{P}^+_{SJH}$. First, second order tensor $\boldsymbol{Q}^+$ is introduced as

$$\boldsymbol{Q}^+ = \sum_{i=1}^{3} h(\epsilon_i)\boldsymbol{k}_i \otimes \boldsymbol{k}_i \tag{25}$$

and the projection operator is defined in indicial notation as follows

$$P^+_{ijkl} = Q^+_{ik} Q^+_{jl}. \tag{26}$$

After some tedious but simple tensor algebra the projection tensor can be written as:

$$\boldsymbol{P}^+_{SJH} = \boldsymbol{P}^+_1 + \boldsymbol{P}^+_2 + \boldsymbol{P}^+_3 + \boldsymbol{P}^+_4, \tag{27}$$

where the first term $\boldsymbol{P}^+_1$ is exactly equal to the whole operator $\boldsymbol{P}^+_{Ortiz}$ in Equation (24) and the other terms depend on mixed products of $h(\epsilon^e_i)$ and $h(\epsilon^e_j)$, $j \neq i$ and the respective principal direction vectors. However, for triaxial tension and compression this operator leads to the same results as $\boldsymbol{P}^+_{Ortiz}$ (Winnicki 2000). In fact, for triaxial tension tensor $\boldsymbol{Q}^+$ degenerates to the 2nd order identity tensor, and hence the $\boldsymbol{P}^+_{SJH}$ operator degenerates to the 4th order identity tensor. Therefore, the SJH operator could be regarded as a superior one.

Next, we define equivalent strain measures $\tilde{\epsilon}$ used in this paper. An important issue is whether $\tilde{\epsilon}$ should be a measure of total strains, their elastic part or some other function of elastic and plastic strains (Ju 1989). If a full coupling exists, for instance for plastically induced damage in metals, plastic strains also contribute to $\tilde{\epsilon}$. Usually, and in this paper, it is assumed that the equivalent strain measure is a function of elastic strains only $\tilde{\epsilon}(\boldsymbol{\epsilon}^e)$. This means that the damage growth is driven by the effective stresses, see Equation (3), and the coupling of plasticity and damage is weaker.

If the projection operator is not included in the damage formulation, the modified von Mises definition of the equivalent strain measure (de Vree, Brekelmans, and van Gils 1995) is employed. It takes into account the ratio k between the compressive and tensile strength and prevents damage under mild compression. If the projection operator is incorporated, we use the classical elastic energy release measure with $\boldsymbol{\epsilon}$ substituted by $\boldsymbol{\epsilon}^{e+}$:

$$\tilde{\epsilon} = \sqrt{\frac{1}{E}\boldsymbol{\epsilon}^e : \boldsymbol{P}^+ : \boldsymbol{E} : \boldsymbol{P}^+ : \boldsymbol{\epsilon}^e}. \tag{28}$$

In Figure 2 the strain measures are compared for plane-stress conditions, $\tilde{\epsilon} = 0.001$, $k = 10$ and $\nu = 0.2$.

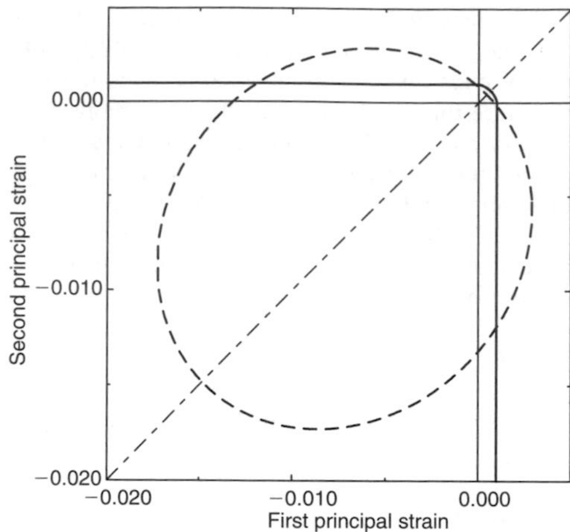

Figure 2. Comparison of equivalent strain measure definitions: Equation (28) – solid, modified von Mises – dashed.

Further, the derivative of the strain measure with respect to the strain tensor is computed as:

$$\boldsymbol{s} = \frac{\partial \tilde{\epsilon}}{\partial \boldsymbol{\epsilon}} = \frac{\partial \tilde{\epsilon}}{\partial \boldsymbol{\epsilon}^{e+}} : \frac{\partial \boldsymbol{\epsilon}^{e+}}{\partial \boldsymbol{\epsilon}^e} : \frac{\partial \boldsymbol{\epsilon}^e}{\partial \boldsymbol{\epsilon}}, \tag{29}$$

with

$$\frac{\partial \boldsymbol{\epsilon}^{e+}}{\partial \boldsymbol{\epsilon}^e} = \boldsymbol{P}^+ + \frac{\partial \boldsymbol{P}^+}{\partial \boldsymbol{\epsilon}^e} : \boldsymbol{\epsilon}^e; \qquad \frac{\mathrm{d}\boldsymbol{\epsilon}^e}{\mathrm{d}\boldsymbol{\epsilon}} = (\boldsymbol{E})^{-1} : \boldsymbol{E}^{ep}. \tag{30}$$

Upon differentiation of Equation (15) we now obtain the rate stress–strain relationship in the form

$$\dot{\boldsymbol{\sigma}} = \boldsymbol{E}_{tan} : \dot{\boldsymbol{\epsilon}} - \hat{\boldsymbol{\sigma}}^+ \dot{\omega}, \tag{31}$$

where

$$\boldsymbol{E}_{tan} = (\boldsymbol{E}_{sec} - \omega \boldsymbol{E}^*) : \boldsymbol{E}^{-1} : \boldsymbol{E}^{ep}, \tag{32}$$

$\hat{\boldsymbol{\sigma}}^+$ and $\boldsymbol{E}^*$ are defined as:

$$\hat{\boldsymbol{\sigma}}^+ = \boldsymbol{P}^+ : \boldsymbol{E} : \boldsymbol{P}^+ : \boldsymbol{\epsilon}^e, \tag{33}$$

$$\boldsymbol{E}^* = \boldsymbol{\epsilon}^e : \frac{\partial \boldsymbol{P}^+}{\partial \boldsymbol{\epsilon}^e} : \boldsymbol{E} : \boldsymbol{P}^+ + \boldsymbol{P}^+ : \boldsymbol{E} : \frac{\partial \boldsymbol{P}^+}{\partial \boldsymbol{\epsilon}^e} : \boldsymbol{\epsilon}^e, \tag{34}$$

and the relation of elastic and total strain rates (9) has been used. The rate of damage during its evolution ($\kappa^d = \tilde{\epsilon}$) is computed as

$$\dot{\omega} = \frac{\partial \omega}{\partial \kappa^d} \frac{\partial \tilde{\epsilon}}{\partial \boldsymbol{\epsilon}} : \dot{\boldsymbol{\epsilon}}, \tag{35}$$

and during unloading we have $\dot{\omega} = 0$. Substituting the following definition:

$$H = \frac{\mathrm{d}\omega}{\mathrm{d}\kappa^d} \tag{36}$$

and Equation (29) into Equation (31) we eventually obtain the tangent stress–strain relation

$$\dot{\boldsymbol{\sigma}} = \left(\boldsymbol{E}_{tan} - H\hat{\boldsymbol{\sigma}}^{+} \otimes \boldsymbol{s}\right) : \dot{\boldsymbol{\epsilon}}. \tag{37}$$

If crack closing is neglected then in Equation (37) we have $\boldsymbol{E}_{tan} = (1-\omega)\boldsymbol{E}^{\text{ep}}$ and $\hat{\boldsymbol{\sigma}}^{+} = \hat{\boldsymbol{\sigma}}$.

3 GRADIENT ENHANCEMENT AND FINITE ELEMENTS

In the gradient-enhanced formulation the plasticity theory remains standard and the damage theory is made nonlocal. It is emphasized that in this combination the unstable material behaviour is caused by damage and the plasticity model is hardening.

Following (Peerlings et al. 1996), the damage evolution in the gradient-enhanced model is governed by the following damage loading function:

$$f^{\text{d}} = \bar{\epsilon} - \kappa^{\text{d}} = 0, \tag{38}$$

where the averaged (nonlocal) strain measure $\bar{\epsilon}$ satisfies the Helmholtz equation:

$$\bar{\epsilon} - c\nabla^2\bar{\epsilon} = \tilde{\epsilon}. \tag{39}$$

The damage measure ω is computed from the damage evolution law as a function of the history parameter κ^{d} which is now the largest nonlocal strain measure $\bar{\epsilon}$ attained at a material point in the deformation history. The parameter $c > 0$ has a unit of length squared and is related to an internal length scale. It is assumed here to be constant, although, with some modifications in the formulation, it can be made a function of $\tilde{\epsilon}$ or $\bar{\epsilon}$ (Geers 1997).

The finite element implementation of the model is based on the weak-form equilibrium equation

$$\int_V (\boldsymbol{L}\boldsymbol{v})^{\text{T}}\boldsymbol{\sigma}\,\text{d}V = \int_V \boldsymbol{v}^{\text{T}}\boldsymbol{b}\,\text{d}V + \int_S \boldsymbol{v}^{\text{T}}\boldsymbol{t}\,\text{d}S, \tag{40}$$

where we now use the Voigt's notation and the superscript T is the transpose symbol. A weak-form of Equation (39) is obtained using Green's formula and the non-standard boundary condition $(\nabla\bar{\epsilon})^{\text{T}}\boldsymbol{\nu} = 0$:

$$\int_V [w\bar{\epsilon} + c(\nabla w)^{\text{T}}\nabla\bar{\epsilon}]\text{d}V = \int_V w\tilde{\epsilon}\text{d}V. \tag{41}$$

In Equations (40–41) $\boldsymbol{L}$ is a differential operator matrix, $\boldsymbol{v}$ and w are suitable weighting functions, $\boldsymbol{b}$ is the body force vector, $\boldsymbol{t}$ is the traction vector. In the ensuing two-field formulation the average strain measure must be discretized in addition to the displacements. C^0-continuity suffices for all shape functions.

The coupling to plasticity influences only the equilibrium equation (40), while Equation (41) is exactly the same as for pure gradient damage.

We write Equations (40) and (41) for iteration $i+1$ of the incremental-iterative algorithm and decompose the stress vector, the strain measure and its averaged version as

$$\boldsymbol{\sigma}^{(i+1)} = \boldsymbol{\sigma}^{(i)} + \text{d}\boldsymbol{\sigma},\ \tilde{\epsilon}^{(i+1)} = \tilde{\epsilon}^{(i)} + \text{d}\tilde{\epsilon},\ \bar{\epsilon}^{(i+1)} = \bar{\epsilon}^{(i)} + \text{d}\bar{\epsilon} \tag{42}$$

to obtain:

$$\int_V (\boldsymbol{L}\boldsymbol{v})^{\text{T}}\text{d}\boldsymbol{\sigma}\,\text{d}V = f_{ext} - f_{int} \tag{43}$$

$$-\int_V w\,\text{d}\tilde{\epsilon}\,\text{d}V + \int_V [w\,\text{d}\bar{\epsilon} + c(\nabla w)^{\text{T}}\nabla(\text{d}\bar{\epsilon})]\,\text{d}V = \int_V w\,\tilde{\epsilon}^{(i)}\,\text{d}V - \int_V [w\bar{\epsilon}^{(i)} + c(\nabla w)^{\text{T}}\nabla\bar{\epsilon}^{(i)}]\,\text{d}V. \tag{44}$$

with

$$f_{ext} = \int_S \boldsymbol{v}^{\text{T}}\boldsymbol{t}\,\text{d}S + \int_V \boldsymbol{v}^{\text{T}}\boldsymbol{b}\text{d}V, \tag{45}$$

$$f_{int} = \int_V (\boldsymbol{L}\boldsymbol{v})^{\text{T}}\boldsymbol{\sigma}^{(i)}\,\text{d}V. \tag{46}$$

Equation (31) is now invoked in an incremental form:

$$\text{d}\boldsymbol{\sigma} = \boldsymbol{E}_{tan}\text{d}\boldsymbol{\epsilon} - \hat{\boldsymbol{\sigma}}^{+}\text{d}\omega \tag{47}$$

and, when Equation (38) holds, the damage increment $\text{d}\omega$ can be computed as

$$\text{d}\omega = H\text{d}\bar{\epsilon}, \tag{48}$$

with H defined in Equation (36) The increment of strain measure $\text{d}\tilde{\epsilon}$ is computed as:

$$\text{d}\tilde{\epsilon} = \boldsymbol{s}^{\text{T}}\text{d}\boldsymbol{\epsilon}, \tag{49}$$

with $\boldsymbol{s}^{\text{T}}$ defined in Equation (29). Equation (48) is substituted into Equation (47) and next Equations (47) and (49) are substituted into Equations (43) and (44), + to give:

$$\int_V (\boldsymbol{L}\boldsymbol{v})^{\text{T}}\boldsymbol{E}_{tan}^{(i)}\text{d}\boldsymbol{\epsilon}\,\text{d}V - \int_V (\boldsymbol{L}\boldsymbol{v})^{\text{T}}H^{(i)}\hat{\boldsymbol{\sigma}}^{+(i)}\text{d}\bar{\epsilon}\,\text{d}V = f_{ext} - f_{int}, \tag{50}$$

$$-\int_V w\,\boldsymbol{s}^{\text{T}(i)}\text{d}\boldsymbol{\epsilon}\,\text{d}V + \int_V [w\,\text{d}\bar{\epsilon} + c(\nabla w)^{\text{T}}\nabla(\text{d}\bar{\epsilon})]\,\text{d}V = \int_V w\,\tilde{\epsilon}^{(i)}\,\text{d}V - \int_V [w\,\bar{\epsilon}^{(i)} + c(\nabla w)^{\text{T}}\nabla\bar{\epsilon}^{(i)}]\,\text{d}V. \tag{51}$$

The discretization of the incremental equations leads to two-field finite elements. The displacements and the averaged strain measure are interpolated as follows:

$$\boldsymbol{u} = \boldsymbol{N}\boldsymbol{a}, \quad \bar{\epsilon} = \boldsymbol{h}^{\mathrm{T}}\boldsymbol{e}, \tag{52}$$

so that

$$\boldsymbol{\epsilon} = \boldsymbol{B}\boldsymbol{a}, \quad \nabla\bar{\epsilon} = \boldsymbol{q}^{\mathrm{T}}\boldsymbol{e}, \tag{53}$$

with $\boldsymbol{q}^{\mathrm{T}} = \nabla \boldsymbol{h}^{\mathrm{T}}$. The respective weighting functions are discretized similarly, so that Equations (50–51) are represented by the following matrix equation:

$$\begin{bmatrix} \boldsymbol{K}_{aa} & \boldsymbol{K}_{a\bar{\epsilon}} \\ \boldsymbol{K}_{\bar{\epsilon}a} & \boldsymbol{K}_{\bar{\epsilon}\bar{\epsilon}} \end{bmatrix} \begin{bmatrix} \mathrm{d}\boldsymbol{a} \\ \mathrm{d}\boldsymbol{e} \end{bmatrix} = \begin{bmatrix} \boldsymbol{f}_{ext} - \boldsymbol{f}_{int} \\ \boldsymbol{f}_{\epsilon} - \boldsymbol{f}_{\bar{\epsilon}} \end{bmatrix}, \tag{54}$$

with the plasticity dependent sub-matrices now defined as follows:

$$\boldsymbol{K}_{aa} = \int_V \boldsymbol{B}^{\mathrm{T}}\boldsymbol{E}_{tan}^{(i)}\boldsymbol{B}\,\mathrm{d}V, \tag{55}$$

$$\boldsymbol{K}_{a\bar{\epsilon}} = -\int_V H^{(i)}\boldsymbol{B}^{\mathrm{T}}\hat{\boldsymbol{\sigma}}^{+(i)}\boldsymbol{h}^{\mathrm{T}}\,\mathrm{d}V, \tag{56}$$

and the other matrices similar to the gradient damage formulation of (Peerlings et al. 1996):

$$\boldsymbol{K}_{\bar{\epsilon}a} = -\int_V \boldsymbol{h}\boldsymbol{s}^{\mathrm{T}(i)}\boldsymbol{B}\,\mathrm{d}V, \tag{57}$$

$$\boldsymbol{K}_{\bar{\epsilon}\bar{\epsilon}} = \int_V (\boldsymbol{h}\boldsymbol{h}^{\mathrm{T}} + c\,\boldsymbol{q}\boldsymbol{q}^{\mathrm{T}})\,\mathrm{d}V, \tag{58}$$

$$\boldsymbol{f}_{\epsilon} = \int_V \boldsymbol{h}\tilde{\epsilon}^{(i)}\,\mathrm{d}V, \quad \boldsymbol{f}_{\bar{\epsilon}} = \boldsymbol{K}_{\bar{\epsilon}\bar{\epsilon}}\boldsymbol{e}^{(i)}. \tag{59}$$

In the absence of damage growth $\boldsymbol{K}_{a\bar{\epsilon}} = \boldsymbol{0}$ and the equilibrium equations in identity (54) are uncoupled from the averaging equations. The stress update algorithm is presented in Box 1, and does not differ much from the algorithm for the formulation without crack closure effects presented in (de Borst et al. 1999). Further details of the numerical implementation aspects can also be found in (de Borst et al. 1999).

Box 1. Stress update algorithm for gradient damage coupled to plasticity.

1. Solve Eq. (54) for increments of nodal displacements and averaged strain
2. At integration point compute strain increments $\Delta\boldsymbol{\epsilon}$, $\Delta\bar{\epsilon}$ and update total strains $\boldsymbol{\epsilon}$ and averaged strain measure $\bar{\epsilon}$
3. Resolve plasticity in effective stress space:
   trial stress $\hat{\boldsymbol{\sigma}}_{\mathrm{t}} = \hat{\boldsymbol{\sigma}}_{\mathrm{o}} + \boldsymbol{E}\Delta\boldsymbol{\epsilon}$, $\hat{\boldsymbol{\sigma}}_{\mathrm{o}} = \boldsymbol{E}\boldsymbol{\epsilon}^{\mathrm{e}}_{\mathrm{o}}$
   if $f^{\mathrm{p}} > 0$ - Eq. (7) then $\hat{\boldsymbol{\sigma}} = \hat{\boldsymbol{\sigma}}_{\mathrm{t}} - \Delta\lambda\boldsymbol{E}\boldsymbol{m}(\hat{\boldsymbol{\sigma}})$
   else $\hat{\boldsymbol{\sigma}} = \hat{\boldsymbol{\sigma}}_{\mathrm{t}}$
   elastic strain $\boldsymbol{\epsilon}^{\mathrm{e}} = \boldsymbol{E}^{-1}\hat{\boldsymbol{\sigma}}$
4. Resolve damage:
   projection operator $\boldsymbol{P}^{+}(\boldsymbol{\epsilon}^{\mathrm{e}})$
   strain measure $\tilde{\epsilon}(\boldsymbol{\epsilon}^{\mathrm{e}+})$
   if $f^{\mathrm{d}} > 0$ - Eq. (38) then $\kappa^{\mathrm{d}} = \bar{\epsilon}$, $\omega = \omega(\kappa^{\mathrm{d}})$
   else $\omega = \omega_{\mathrm{o}}$
5. Compute stress $\boldsymbol{\sigma} = \boldsymbol{E}_{sec}\boldsymbol{\epsilon}^{\mathrm{e}}$,
   $\boldsymbol{E}_{sec} = \boldsymbol{E} - \omega\boldsymbol{P}^{+\mathrm{T}}\boldsymbol{E}\boldsymbol{P}^{+}$
6. Compute out-of-balance forces and tangent matrices according to Equations (55-59), check convergence, if necessary, return to 1

4 NUMERICAL SIMULATIONS

4.1 *Tensile bar*

We analyze a bar with a unit cross-section subjected to tension. In order to initiate strain localization a 10% smaller cross-section (for 2D simulations) or 10% smaller value of the damage threshold κ_{o} (for 3D simulations) is adopted in the central zone $d = 10$ mm. The length of the bar is $L = 100$ mm, Young's modulus is $E = 20000$ N/mm^2. A gradient constant $c = 4$ mm^2 is adopted. In the one-dimensional problem the equivalent strain measure $\tilde{\epsilon}$ is equal to the absolute value of the axial strain ϵ. The damage growth function represents a linear softening stress–strain diagram:

$$\omega(\kappa^{\mathrm{d}}) = \frac{\kappa_{\mathrm{u}}}{\kappa^{\mathrm{d}}}\frac{\kappa^{\mathrm{d}} - \kappa_{\mathrm{o}}}{\kappa_{\mathrm{u}} - \kappa_{\mathrm{o}}}. \tag{60}$$

We adopt $\kappa_{\mathrm{o}} = 0.0001$ and $\kappa_{\mathrm{u}} = 0.0005$.

For C^0 – continuous gradient damage elements three variants of interpolation can be examined:

1. Similar quadratic interpolation of the displacements and the averaged strain: element with 6 degrees of freedom (dofs) and 2 integration points (ips) in 1D, 24 dofs and 2 × 2 ips in 2D (serendipity), 80 dofs and 2 × 2 × 2 ips in 3D;
2. Quadratic interpolation of the displacements and linear of the averaged strain: element with 5 dofs and 2 ips in 1D, 20 dofs and 2 × 2 ips in 2D, 68 dofs and 2 × 2 × 2 ips in 3D;
3. Similar linear interpolation of the displacements and the averaged strain: element with 4 dofs in 1D,

12 dofs in 2D, 24 dofs in 3D, all integrated with just 1 sampling point.

The second option requires that a different number of dofs at corner and midside nodes is admitted by the employed finite element package. The third option is applicable in 2D and 3D only with hourglass control, see e.g. (Belytschko and Bindeman 1993). In any case care must be taken with respect to volumetric locking and excessive shear stiffness, but well-established solutions such as the enhanced assumed strain approach or $\overline{B}$ formulation can be adopted just as for ordinary displacement-based elements. Figure 3 presents the load-displacement diagrams obtained for the three options with 80 elements along the bar. It turns out that, although choice 2 might be optimal, the other options are also possible and give stable results, since the analyzed problem is coupled rather than mixed. Option 3 seems especially tempting due to the low computational cost, although the quadratic interpolation is usually more reliable. Figure 4 shows the deformation and strain distribution in the three-dimensional bar. It is noted that localization occurs in a zone broader than one row of elements (or two due to symmetry), which would be the case in the absence of regularization.

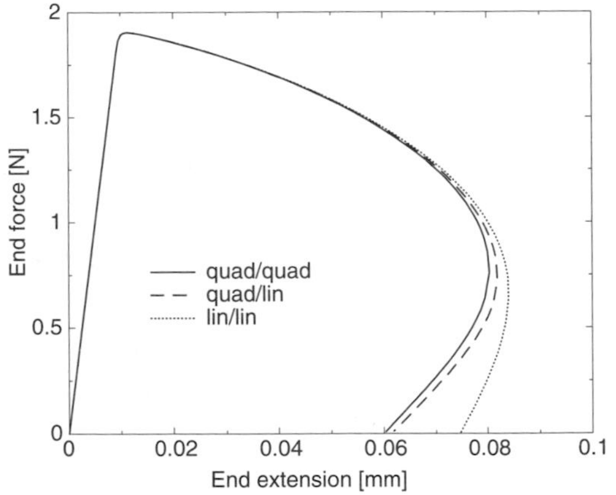

Figure 3. Tensile bar – three types of discretization.

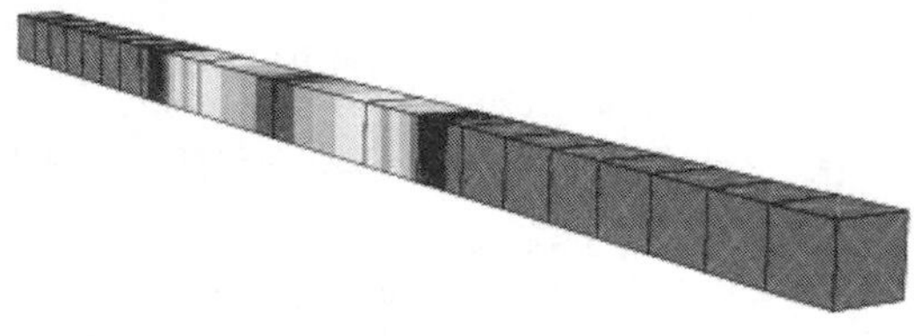

Figure 4. Tensile bar in 3D – discretization with 20 eight-noded elements (cross section 5 × 5 mm for clear post-processing).

4.2 *Cantilever subjected to load reversals*

We analyze a cantilever beam (coarsely equivalent to a half of a three-point bending test) subjected to reversed loading in plane-stress conditions. The beam is shown in Figure 5 together with the finite element mesh employed. The gradient damage elements with quadratic/linear interpolation and a 2 × 2 Gauss integration are used. The thickness of the beam is 50 mm. The vertical force is exerted along the left edge under deformation control of all vertical degrees of freedom. The nodes at the clamped edge are fully constrained.

An exponential damage function which represents exponential softening is used:

$$\omega(\kappa^{\mathrm{d}}) = 1 - \frac{\kappa_{\mathrm{o}}}{\kappa^{\mathrm{d}}}\left(1 - \alpha + \alpha e^{-\eta(\kappa^{\mathrm{d}} - \kappa_{\mathrm{o}})}\right), \tag{61}$$

where α and η are additional parameters. This function is suitable for reproducing the experimental tensile fracture in concrete, since the experimental uniaxial softening relation is exponential (Hordijk 1991). Since it approaches the complete loss of coherence ($\omega = 1$) asymptotically, singularities caused by a complete loss of strength are also avoided. If no crack closure projection is adopted, the modified von Mises definition of the equivalent strain measure is used. If crack closing is included, the equivalent strain is the elastic energy release measure. The material data are as follows: Young's modulus $E = 40000\,\mathrm{N/mm}^2$, Poisson's ratio $\nu = 0.20$, damage threshold $\kappa_{\mathrm{o}} = 0.000075$, damage function parameters $\alpha = 0.92$, $\eta = 300$, gradient influence factor $c = 4\,\mathrm{mm}^2$.

Figure 6 shows the relationships between the total force exerted at the left edge and the vertical displacement of the middle node of the edge. The load is first applied upwards which leads to damage starting at the bottom near the clamped edge. Then after an unloading stage the load changes the sign and the upper fibers are under increasing tension, followed by unloading again.

We start the analysis from the case without $\boldsymbol{P}^{+}$. During the first stage of loading damage grows at the clamped edge from the bottom in a band with an approximate width of 10 mm (see Fig. 7). Unloading

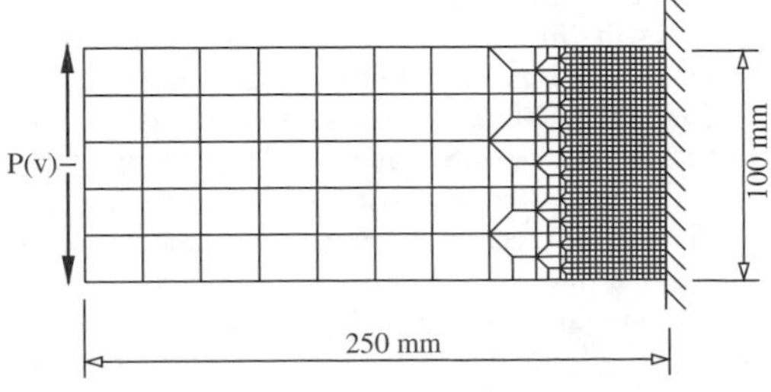

Figure 5. Cantilever beam under load reversals – discretized configuration (dimensions in *mm*).

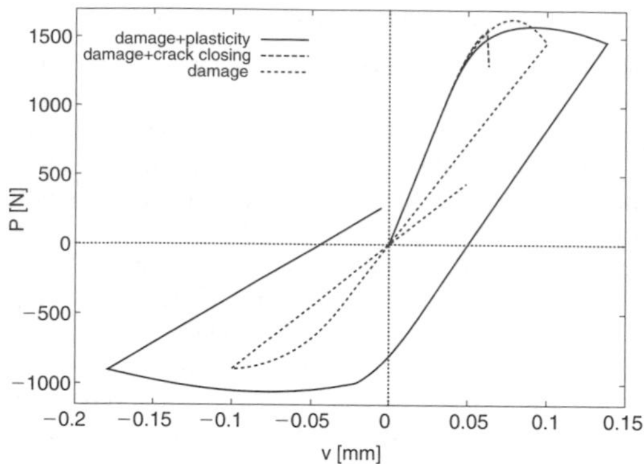

Figure 6. Load–displacement diagrams for $\nu = 0.2$: damage with crack closing projection results in divergence.

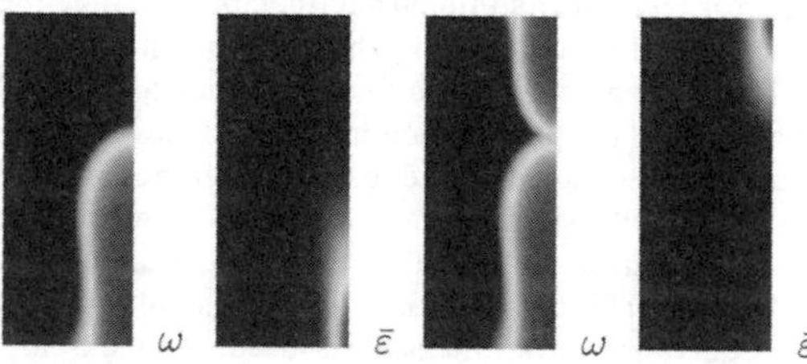

Figure 7. Pure damage without crack closing ($\nu = 0.2$): fracture in the cantilever beam at extreme loading states. Damage measure and averaged strain distributions for the maximum load upwards (left) and similar diagrams for the maximum load downwards (right).

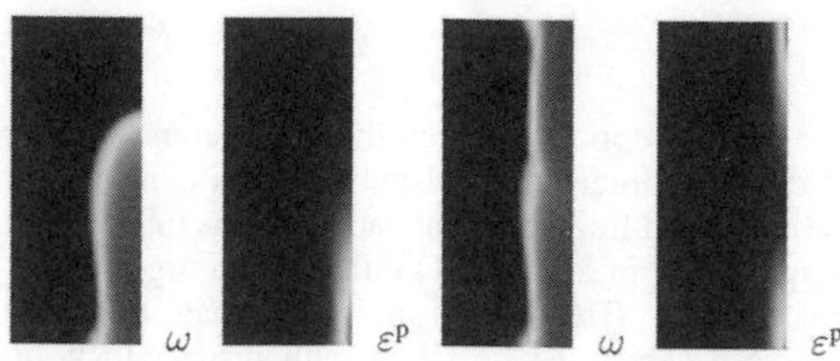

Figure 8. Damage-plasticity without crack closing ($\nu = 0{,}2$): damage and equivalent plastic strain distribution in the beam for maximum load upwards (left) and maximum load downwards (right).

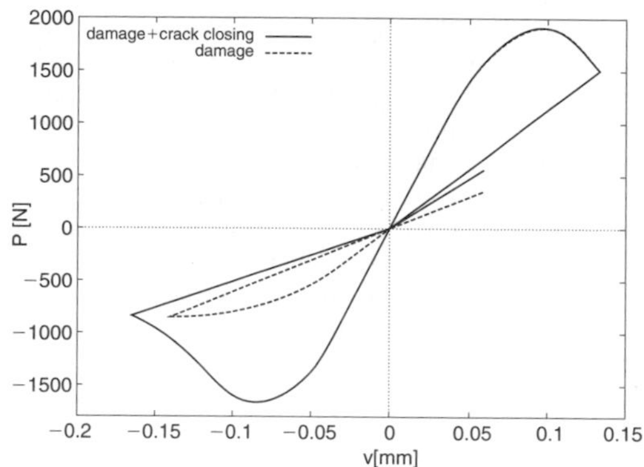

Figure 9. Load-displacement diagram for $\nu = 0.0$: proper convergence without and with the crack closing projection.

shows the stiffness degradation due to the accumulated damage, and when the origin is passed the original stiffness is not retrieved, since no crack closure takes place. The secondary damage zone develops at the top near the clamped edge at smaller stress, since the primary damage weakened the critical cross-section. In the employed model the attained damage level is remembered at a point irrespective of any change of the sign of strains, so in the secondary phase two damage zones are visible although the averaged strain measure plot shows the active zone.

Then, the plastic component is added to the model which results in irreversible strains (see Fig. 6). The Huber-Mises-Hencky criterion as a special case of the Burzyński-Drucker-Prager is used in this test with yield strength $\sigma_y = 4.5\,\mathrm{N/mm^2}$ and hardening modulus The value of σ_y is chosen such that the damage and plastic processes start almost simultaneously. It is noticed that the excessive stiffness degradation of the pure damage model (unloading to the origin) is now controlled. Figure 8 presents the fracture zones at the extreme deformation values.

The crack closure enhancement should not only make it possible to retrieve the original elastic stiffness when compression follows tensile damage at a point, but also makes it unnecessary to use the modified von Mises definitions of the strain measure. This is because with the projection operator $\boldsymbol{P}^{+}$ the simplest definition of elastic energy release rate in Equation (28) is now sensitive to the sign of loading (cf. Fig. 2), and in fact in triaxial compression no damage occurs at all. Therefore, the cantilever test is repeated with this equivalent strain measure and the crack closing effect incorporated by means of $\boldsymbol{P}^{+}_{Ortiz}$, cf. Equation (24). As noticed in (Chaboche 1992), due to the unilateral character of the crack closure phenomenon a discontinuity is introduced in the formulation and this is the cause of the encountered numerical difficulties. In the loading process after to the peak the second principal strain in many elements changes sign (the first one is larger and positive) which results in an abrupt change in the equivalent strain measure, stresses and equilibrium resulting in divergence.

The stress jump does not occur for the case with zero Poisson effect ($\nu = 0$) shown in Figure 9. Figure 10 presents the fracture zones at the moments when the largest deformation is applied in the two directions. Figure 11 shows similar diagrams for the case with the crack closing projection.

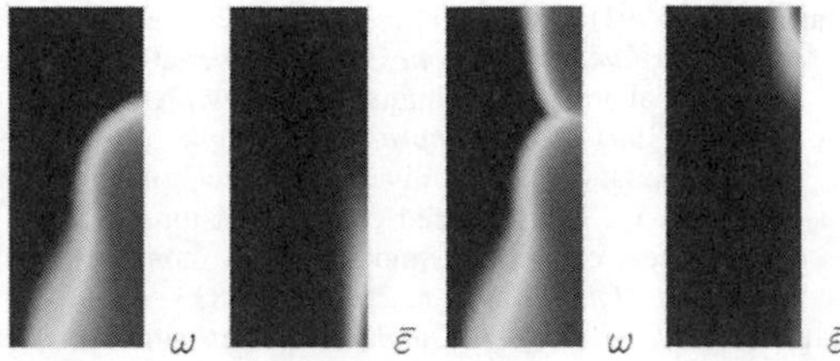

Figure 10. Pure damage without crack closing for $\nu = 0.0$: damage and averaged strain distribution in the cantilever beam for maximum load upwards (left) and maximum load downwards (right).

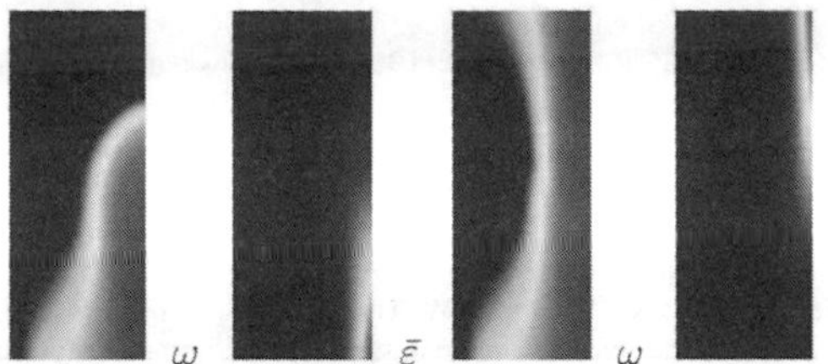

Figure 11. Damage with crack closing and $\nu = 0.0$: damage measure and averaged strain distributions for the maximum load upwards (left) and for the maximum load downwards (right).

As has been noticed in (Geers 1997), the damage predictions of the model with a constant gradient influence factor c and the exponential damage evolution law may be unrealistic for two-dimensional configurations, since the damage zone tends to broaden instead of evolving into a crack when failure is approached. The coupling to plasticity does not solve this problem. If a localized final damage pattern is to be simulated, the gradient-damage model must be improved, for instance by making the gradient influence factor variable and adopting a different damage law, see (Geers 1997).

Finally, Figure 12 shows that the results of the computations are quite similar for different elements belonging to the three groups listed in Section 4.1. In fact, for the element with linear interpolation of all fields (12 dofs) selective reduced ($2 \times 2/1$) and not uniform 1 point integration has been used.

5 CONCLUSIONS

To reproduce a realistic stiffness degradation and to obtain more versatility in modelling of damage or fracture processes a combination of gradient damage with hardening plasticity is pursued. The algorithm is augmented with a crack closing projection operator which ensures that damage occurs due to tensile and not compressive strains. The model is formulated in such

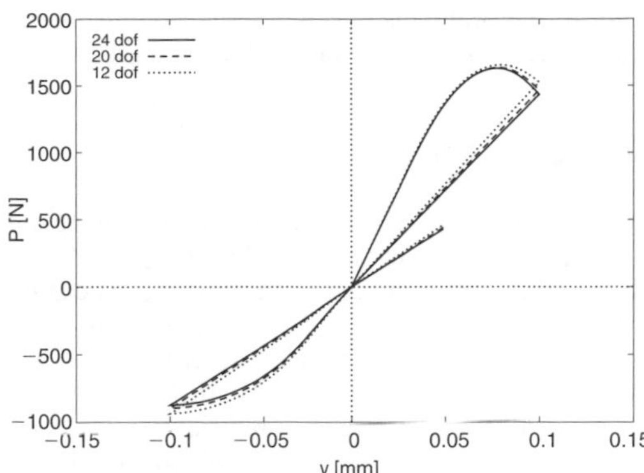

Figure 12. Load-displacement diagrams for three combinations of interpolation polynomials: quadratic/quadratic, quadratic/linear and linear/linear ($\nu = 0.2$, no $\boldsymbol{P}^{+}$).

a way that the coupling influences only the equilibrium equations, while the second differential equation, which serves the purpose of regularization, preserves the original format of gradient damage (Peerlings et al. 1996). A consistently linearized two-field finite element formulation is used and therefore a quadratic rate of convergence of the incremental-iterative algorithm is observed. It has been shown that the crack closing projection requires only a minor modification of the formerly used algorithm (de Borst et al. 1999).

The model has the following advantages: the physical interpretation of the model components is simple and convincing, there are few model parameters to determine, the elastic stiffness degradation is properly reproduced, non-associative plastic flow can be incorporated. Due to the presence of an intrinsic length scale in the formulation not only the spurious mesh sensitivity of results is avoided (de Borst et al. 1999), but also size effects can be studied.

Various two-field finite elements have been proposed for the theory and will further be developed towards an hourglass control for uniform linear interpolations and one point integration and towards locking control for full integration. Difficulties caused by the discontinuous character of the Heaviside function used to define the crack closing projection have been encountered and will be solved by a regularization of the function according to (Ekh and Runesson 2000) or a modification of the crack closing algorithm along the lines proposed in (Chaboche et al. 1995).

Further testing of the model will include the cylinder split test analyzed in 2D and 3D (cf. (Ruiz et al. 2000)). The results will also be compared with the predictions of the viscoplastic Hoffman model (Winnicki et al. 2001). Moreover, the experimental results of the dowel bar test (di Prisco et al. 1994) will be reproduced. The potential of the formulation for simulating engineering problems (e.g. reinforced concrete joints under earthquake excitations) will be assessed.

ACKNOWLEDGMENTS

The computational results were obtained using the development version of the FEAP program of R.L. Taylor. Valuable discussions with Prof. René de Borst from Koiter Institute Delft, Delft University of Technology, as well as with Dr. Andreas Menzel from Chair of Applied Mechanics, University of Kaiser-slautern are gratefully acknowledged. Financial support of the Polish Committee for Scientific Research within grant no. 7 T07E 020 18 is also acknowledged.

REFERENCES

Belytschko, T. and L. Bindeman (1993). Assumed strain stabilization of the eightnode hexahedral element. *Comput. Meth. Appl. Mech. Eng. 105*, 225–260.

Chaboche, J.-L. (1992). Damage induced anisotropy: on the difficulties associated with the active/passive unilateral condition. *Int. J. Damage Mechanics 1*(2), 148–171.

Chaboche, J.-L., P.-M. Lesne, and J.-F. Maire (1995). Continuum damage mechanics, anisotropy and damage deactivation for brittle materials like concrete and ceramic composites. *Int. J. Damage Mechanics 4*, 5–22.

de Borst, R., J. Pamin, and M. Geers (1999). On coupled gradient-dependent plasticity and damage theories with a view to localization analysis. *Eur. J. Mech. A-Solids 18*(6), 939–962.

de Vree, J., W. Brekelmans, and M. van Gils (1995). Comparison of nonlocal approaches in continuum damage mechanics. *Comput. Struct. 55*(4), 581–588.

di Prisco, M., M. Caruso, and S. Piatti (1994). On fiber role in dowel action. *Studi e Ricerche 15*, 151–194.

Ekh, M. and K. Runesson (2000). Bifurcation results for plasticity coupled to damage with mcr-effect. *Int. J. Solids Struct. 37*, 1975–1996.

Geers, M. (1997). *Experimental analysis and computational modelling of damage and fracture.* Ph.D. dissertation, Eindhoven University of Technology, Eindhoven.

Hansen, N. (1993). *Theories of elastoplasticity coupled with continuum damage mechanics.* Ph.D. dissertation, Sandia National Laboratories, Albuquerque, New Mexico.

Hordijk, D. (1991). *Local approach to fatigue of concrete.* Ph.D. dissertation, Delft University of Technology, Delft.

Ju, J. (1989). On energy-based coupled elastoplastic damage theories: constitutive modeling and computational aspects. *Int. J. Solids Struct. 25*(7), 803–833.

LaBorderie, C. (1991). *Phénomènes unilatéraux dans un matérial endommageable.* Ph.D. dissertation, Université Paris 6, Paris.

Mazars, J. (1984). *Application de la mécanique de l'edommagement au comportement non linéaire et à la rupture du béton de structure.* Ph.D. dissertation, Université Paris 6, Paris.

Ortiz, M. (1985). A constitutive theory for the inelastic behaviour of concrete. *Mech. Mater. 4*, 67–93.

Pamin, J. and R. de Borst (1999). Stiffness degradation in gradient-dependent coupled damage-plasticity. *Arch. Mech. 51*(3–4), 407–433.

Peerlings, R., R. de Borst, W. Brekelmans, and J. de Vree (1996). Gradient-enhanced damage for quasi-brittle materials. *Int. J. Numer. Methods Engng 39*, 3391–3403.

Peerlings, R., R. de Borst, W. Brekelmans, and M. Geers (1998). Gradient-enhanced damage modelling of concrete fracture. *Mech. Cohes.-frict. Mater. 3*, 323–342.

Ruiz, G., M. Ortiz, and A. Pandolfi (2000). Three-dimensional finite-element simulation of the dynamic Brazilian tests on concrete cylinders. *Int. J. Numer. Methods Engng 48*, 963–994.

Simo, J. and J. Ju (1987). Strain- and stress-based continuum damage models – I. Formulation, II. Computational aspects. *Int. J. Solids Struct. 23*(7), 821–869.

Taylor, R. (2001). FEAP – A Finite Element Analysis Program, Version 7.4, User manual. Technical report, University of California at Berkeley, Berkeley.

Winnicki, A. (2000). Crack closing effect in damage mechanics. In W. Biliński (Ed.), *Proc. XII Conf. Computer Methods in Design and Analysis of Hydrostructures*, Cracow, pp. 155–164. Cracow University of Technology.

Winnicki, A., C. Pearce, and N. Bićanić (2001). Viscoplastic Hoffman consistency model for concrete. *Comput. Struct. 79*, 7–19.

Computational Modelling of Concrete Structures, Bićanić et al. (eds)
© 2003 Swets & Zeitlinger, Lisse, ISBN 90 5809 536 3

Shear-compression modeling of cement-based material

G.P.A.G. van Zijl
Faculty of Architecture, Delft University of Technology, Delft, The Netherlands &
Civil Engineering Dept., University of Stellenbosch, Stellenbosch, South Africa

ABSTRACT: Shear failure is an important, often governing mode of failure in structures, which are laterally loaded by wind and earthquake action. Compression acts in combination with the shear, caused by the vertical action of self-weight and top weight, as well as confinement by, for instance, structural frames to infill walls. In frictional, cement-based materials dilatational behaviour upon shearing is a well-known phenomenon, as opposed to isochoric plasticity of metals and plastics. Volume increase accompanies shearing. If such volume increase is prevented or resisted by confining structural elements, wedging occurs, causing pressure build-up. In case of pressure-dependent strength, another well-known characteristic of cement-based materials, a significant strength increase may accompany such confined shearing. This paper presents a dilatancy formulation for interface mechanics, specifically suitable for masonry modeling. For continuum modeling strategies, employed when potential locations of cracks are not obvious, a non-associative Rankine-type limit function is elaborated. Both strategies are applied to simulate shear wall experiments.

1 INTRODUCTION

Shear is the governing mode of failure in concrete and masonry buildings subjected to lateral loads like wind and earthquakes. Shear wall action is evoked by these loads. In multi-storey masonry buildings a degree of confinement accompanies this shearing. Thereby, any amount of vertical upward translation upon shearing is resisted, leading to wedging and associated compression. The pure shear behaviour is altered, defining an important class of problems: shear-compression.

Shearing dilatancy of cement-based materials is a well-known phenomenon, as discussed for instance in soils Roscoe et al. (1963) and Rowe (1971), in concrete (Green and Swanson 1973), rock (Michelis 1981), as well as in masonry (Van der Pluijm 1992, 1998). In these materials volume increase upon non-linear shearing occurs, as opposed to isochoric plasticity of metals and plastics. If such volume increase is prevented or resisted by confining structural elements, wedging occurs, causing pressure build-up. In case of pressure-dependent strength, which is a well-known characteristic of the mentioned cement-based, dilatational materials, a significant strength increase may accompany such confined shearing.

Shearing dilatancy has been discussed as long ago as the 19th century by Reynolds (1885). A comprehensive discussion of the dilatancy modelling in soils, but also touching on concrete and rock has been presented by Vermeer and De Borst (1984). They proposed a non-associative Mohr-Coulomb plasticity model, with a dilatancy coefficient dependent on the evolving plastic strain. However, the corners in the Mohr-Coulomb failure surface makes it unattractive for robust implementation in finite element routines. The Drucker-Prager surface overcomes this problem, but poorly approximates the Coulomb surface for cases of high friction angles, which occur in concrete and masonry. Arguing that local failure in concrete is mainly in mode I tension, whatever the structural boundary value problem, De Borst and Nauta (1985) and Rots (1988) employed multiple-fixed, or rotating smeared cracking with a principal stress criterion for crack initiation.

It has since become widely accepted that a maximum principal stress criterion is an appropriate limit function for cement-based materials like concrete and masonry. Rankine-type principal stress models were employed by Feenstra and De Borst (1996) and Lourenço et al. (1998) as limit surfaces in the tension-shear regime for continuum plasticity computational models for concrete and masonry respectively. However, such mode I type limit functions do not appropriately capture the predominantly mode II type failure phenomenon of shearing dilatancy. This is because they simulate shearing failure also by diagonal mode I cracking, but leave the inelastic deformation parallel to the cracks unchanged. Thereby, an inelastic dilatant volume change equal to the crack width times the section area is implied. This defines an effective dilatancy of unity, expressed as the ratio of the inelastic volumetric strain rate to the rate of inelastic distortion. An overestimation of inelastic volume increase upon shearing is the result. For the boundary value problem of confined shearing, the constraint of this overestimated volume increase causes diagonal tension stresses to be offset. Upon increased shearing deformation, pressure builds up unlimited, unless a compression limit surface is activated. However, there is experimental evidence that shearing dilatancy is reduced and eventually arrested with increased shearing deformation, as well as with increased confining pressure. The latter dilatancy limiting pressure is up to an order lower than the crushing strength. This renders the use of a compressive cap to limit confined shearing resistance as being artificial, confirmed by the need to employ an artificially low crushing strength to find agreement with measured confined shearing resistance.

In this paper the implied dilatancy is elaborated for a Rankine-type limit function in a continuum plasticity framework. It is shown that this constitutive model predicts unlimited strength of confined shear walls. If a compression cap is activated, employing a realistic compressive strength, a peak resistance is computed, which overestimates the experimental resistance by up to 70%. Subsequently the Rankine formulation is modified to produce isochoric inelastic behaviour. This approach is shown to give a conservative prediction of confined shear wall structural response, when realistic compression model parameters are employed.

An alternative approach, specially suited for modeling masonry, is elaborated. Use is made of interface elements to model mortar joint behaviour, while the bricks are modeled as continua. Inelastic behaviour is confined to the joints, known to be the weak points in masonry. A novel dilatancy formulation for interface shear-slipping is presented, which captures the arrest of dilatant volume increase accurately. Both the Rankine continuum plasticity formulation and the interface modeling strategy are applied to simulate masonry shear wall experiments.

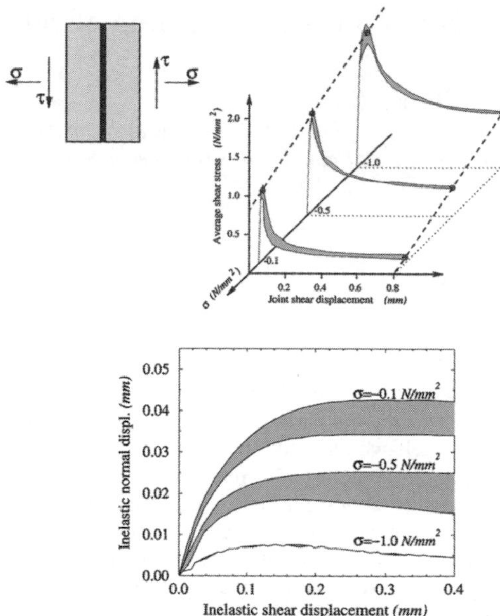

Figure 1. Masonry shear test (Van der Pluijm 1992, 1998).

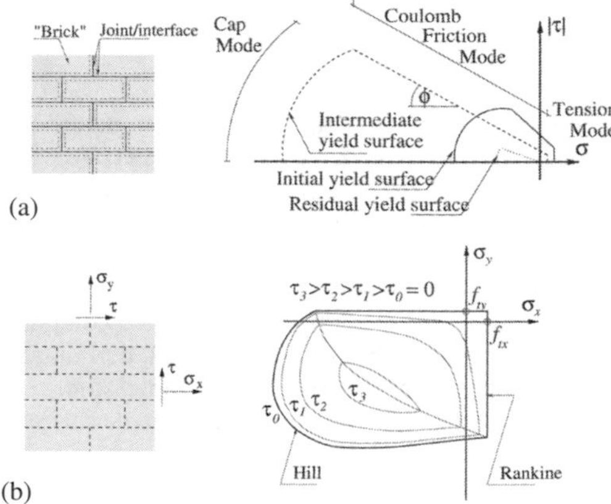

Figure 2. (a) Discrete modelling strategy and interface material law. (b) Composite modelling strategy and anisotropic Rankine-Hill material law (Lourenço and Rots 1997).

2 SHEARING DILATANCY MODELLING

In this paper the focus will be on masonry, although the principles are applicable to other cement-based materials. Figure 1 illustrates shear experiments on small masonry specimens, which were performed to characterize the Coulomb-friction character of masonry shearing resistance along bed joints.

To simulate such behaviour, two modelling strategies are employed in this paper, following Lourenço and Rots (1997); and van Zijl et al. (2001). In a *discrete cracking* strategy the continuum parts (bricks in masonry) are kept (visco-)elastic, while all cracking, slipping and crushing occur in the interfaces (mortar joints in masonry). The interface material law is shown in Figure 2. In the *continuum/composite* strategy masonry is considered to be a homogeneous continuum. In both models shear-compression behaviour is captured consistently by the shown stress limit functions.

Both the discrete and composite crack models are formulated in a plasticity framework, by which the stress rate vector is expressed as

$$\dot{\boldsymbol{\sigma}} = \boldsymbol{D}(\dot{\varepsilon} - \dot{\varepsilon}^p) \tag{1}$$

where $\boldsymbol{D}$ is the elastic stiffness matrix and $\dot{\varepsilon}$ and $\dot{\varepsilon}^p$ are the vectors of total strain rate and the plastic strain rate respectively. For cases where a tensile stress limit is exceeded, the plastic strain may be interpreted as the crack strain. Shear-slipping strain and crushing evolve respectively if a shear stress limit and a compression limit are exceeded. The first term $\boldsymbol{D}\dot{\varepsilon}$ is the trial stress, which, if it exceeds the stress limit, must be corrected by subtracting the second term $\boldsymbol{D}\dot{\varepsilon}^p$. The plastic/crack strain rate is expressed by

$$\dot{\boldsymbol{\varepsilon}}^p = \dot{\lambda}\frac{\partial g}{\partial \boldsymbol{\sigma}} \tag{2}$$

where λ is the amount of plastic flow/crack strain and g the *plastic potential.* The stress is limited by the function

$$f(\boldsymbol{\sigma}, \kappa) \leq 0 \tag{3}$$

with κ a state parameter, which describes the current level of degradation in the material. Suitable choices of limit functions f and potential functions g are made for the discrete and composite models. For the proper simulation of masonry shear-compression behaviour, it is crucial that an appropriate potential function is chosen.

2.1 *Discrete model dilatancy*

The discrete model is based on the interface material model of Lourenço and Rots (1997). It combines a tension cut-off limit and a compression cap with a Coulomb-friction limit function, see Figure 2. This model has been extended recently to capture the dilatancy upon shearing accurately (van Zijl et al. 2001). Here, a brief description of the model is given to set the scene for the dilatancy model elaboration.

In the interface model the stress and strain vectors are actually tractions and relative displacements, expressed in a plane stress formulation as

$$\boldsymbol{\sigma} = \{\sigma \tau\}^T, \quad \boldsymbol{\varepsilon} = \{uv\}^T \tag{4}$$

with σ (the traction) and u (the relative displacement) in the interface normal direction, τ (the shear traction) and v (the relative displacement) in the tangential direction. The dilatancy manifests in the shear formulation. There, the standard limit and potential functions are given by

$$\begin{aligned} f &= |\tau| + \sigma\Phi(\boldsymbol{\sigma}, \kappa) - c, \\ g &= |\tau| + \sigma\Psi(\boldsymbol{\sigma}, \kappa) - c \end{aligned} \tag{5}$$

with Φ, Ψ the friction and dilatancy coefficients respectively and c the cohesion. From the flow rule (2) and the stress derivative of the potential function (5) it follows that

$$\Psi = \frac{\dot{u}^p}{\dot{v}^p}\operatorname{sign}(\tau) \tag{6}$$

which provides a way of formulating a dilatancy coefficient from measurement data of displacement components in shear specimens. This has been done for clay brick and calcium silicate masonry specimens (Van Zijl 2000), producing the expression

$$\Psi = \begin{cases} 0, & \sigma < \sigma_u \\ \Psi_0\left[1 - \dfrac{\sigma}{\sigma_u}\right]e^{-\delta v^p}, & \sigma_u \leq \sigma < 0 \\ \Psi_0 e^{-\delta v^p}, & \sigma \geq 0. \end{cases} \tag{7}$$

Equation (7) has three parameters, which are to be obtained by a least squares fit to experimental test data. Ψ_0 is the dilatancy at zero normal confining stress and zero shear slip, σ_u is the compressive stress at which the dilatancy becomes zero and δ is the dilatancy shear slip degradation coefficient.

2.2 *Continuum/composite model dilatancy*

For the composite model the stress and strain vectors are (in plane stress).

$$\boldsymbol{\sigma} = \{\sigma_x \; \sigma_y \; \tau\}^T, \quad \boldsymbol{\varepsilon} = \{\varepsilon_x \; \varepsilon_y \; \gamma\}^T \tag{8}$$

The anisotropic Rankine limit function is given by (Lourenço et al. 1998)

$$f = \frac{(\sigma_x - \sigma_{tx}) + (\sigma_y - \sigma_{ty})}{2} + \sqrt{\left(\frac{(\sigma_x - \sigma_{tx}) - (\sigma_y - \sigma_{ty})}{2}\right)^2 + \alpha\tau^2} \tag{9}$$

The shear stress contribution to failure is controlled by α. The current admissible stresses in the material x, y directions are denoted by σ_{tx}, σ_{ty} respectively. The potential function has been chosen as

$$g = f|_{\alpha=1} \tag{10}$$

with derivative

$$\frac{\partial g}{\partial \boldsymbol{\sigma}} = \begin{Bmatrix} \frac{1}{2} + \frac{1}{4}[(\sigma_x - \sigma_{tx}) - (\sigma_y - \sigma_{ty})] \Big/ \sqrt{\left[\frac{(\sigma_x - \sigma_{tx}) - (\sigma_y - \sigma_{ty})}{2}\right]^2 + \tau^2} \\ \frac{1}{2} - \frac{1}{4}[(\sigma_x - \sigma_{tx}) - (\sigma_y - \sigma_{ty})] \Big/ \sqrt{\left[\frac{(\sigma_x - \sigma_{tx}) - (\sigma_y - \sigma_{ty})}{2}\right]^2 + \tau^2} \\ \tau \Big/ \sqrt{\left[\frac{(\sigma_x - \sigma_{tx}) - (\sigma_y - \sigma_{ty})}{2}\right]^2 + \tau^2} \end{Bmatrix} \tag{11}$$

The dilatancy implied by this choice of g can be expressed by the relation between the rate of plastic volume increase and the deviatoric/distorting strain rate as follows:

$$\Psi = \frac{\dot{\varepsilon}_1^p + \dot{\varepsilon}_2^p}{\dot{\varepsilon}_1^p - \dot{\varepsilon}_2^p} \tag{12}$$

where the plastic principal strain rates are given by

$$\dot{\varepsilon}_{1,2}^p = \frac{1}{2}\left[\dot{\varepsilon}_x^p + \dot{\varepsilon}_y^p \pm \sqrt{(\dot{\varepsilon}_x^p - \dot{\varepsilon}_y^p)^2 + \dot{\gamma}^p}\right] \tag{13}$$

From equations (2, 11–13) it follows that $\Psi = 1$. Such a large, constant dilatancy coefficient leads to a large overestimation of the volume increase. If this volume increase is constrained by the boundaries, an incorrect stress build-up is predicted in the structure. It is no simple matter to formulate a softening dilatancy for the composite model. A simple approach is to set it to zero, by choosing the potential function stress gradient as

$$\frac{\partial g}{\partial \sigma} = \left\{ \begin{array}{l} \frac{1}{2}[(\sigma_x - \sigma_{tx}) - (\sigma_y - \sigma_{ty})] \Big/ \sqrt{\left[\frac{(\sigma_x - \sigma_{tx}) - (\sigma_y - \sigma_{ty})}{2}\right]^2 + \tau^2} \\ -\frac{1}{2}[(\sigma_x - \sigma_{tx}) - (\sigma_y - \sigma_{ty})] \Big/ \sqrt{\left[\frac{(\sigma_x - \sigma_{tx}) - (\sigma_y - \sigma_{ty})}{2}\right]^2 + \tau^2} \\ 2\tau \Big/ \sqrt{\left[\frac{(\sigma_x - \sigma_{tx}) - (\sigma_y - \sigma_{ty})}{2}\right]^2 + \tau^2} \end{array} \right\} \tag{14}$$

3 INAPPROPRIATE DILATANCY MODELLING

A simple pure shear case is considered to illustrate the volume increase and strength build-up consequences of the dilatancy model employed, see Figure 3. Unit values are employed for the illustration. A pure shear load is applied to an element, which may be an interface element of the discrete modelling strategy, or a continuum element obeying the composite Rankine model. For better illustration, the latter element deformations are shown in Figure 3. Note further that the compression limit functions are not activated in order to illuminate the dilatancy behaviour.

When the shear strength is exceeded, volume increase starts if the boundary conditions allow it. This is shown in Figure 3 for the case of *no confinement*. In the post-peak phase the material degrades, accompanied by a decrease in the shearing resistance.

However, if the boundary conditions prevent the volume increase, the *confinement* case in Figure 3, a continued increase in shearing resistance is calculated even after the (virgin) shear strength has been exceeded. This is due to the build-up of compressive stresses, which, in both the interface Coulomb-friction model and the Rankine model, leads to an increased shearing resistance. If a dilatancy coefficient $\Psi = 1$ is employed for the discrete model, or the standard Rankine formulation is employed, the resistance is unlimited. This is an unrealistic result. As suggested in the previous section, a remedy is to set the dilatancy to zero. Then, the same post-peak softening response is found as for the case of *no confinement*. The unrealistic build-up of compressive stresses is avoided. However, this ignores any dilatant wedging completely. There is evidence (Van der Pluijm 1999) that wedging does occur to some extent. The interface dilatancy model has been calibrated to this effect (Van Zijl 2000) according to section 2.1. The response when this softening dilatancy formulation is employed, is also shown in Figure 3 ($\Psi = \Psi(\sigma, \kappa)$). An increased peak resistance is found, but a softening response follows as the volume increase is arrested and the material degrades upon further shearing.

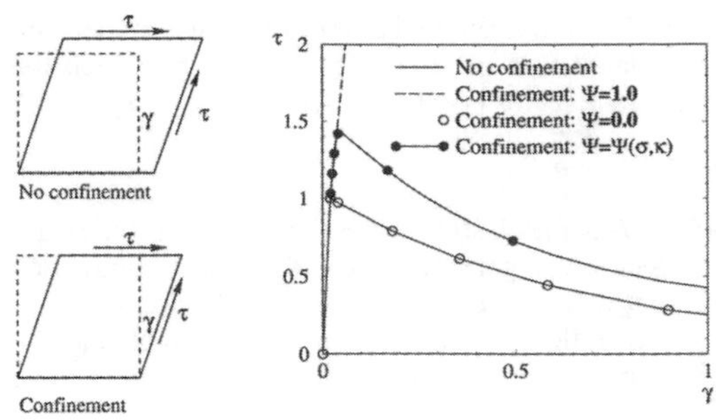

Figure 3. Dilatancy responses under pure shear.

4 CASE STUDY: MASONRY SHEAR WALL

As a case study, the results of three masonry shear wall experiments are compared with the results of numerical simulations. The three walls discussed in this paragraph were part of a larger program in which 16 walls were tested (Vermeltfoort et al. 93).

4.1 *The experiments*

Three walls, one metre square with an opening one brick wide were loaded in shear. Each wall was rigidly connected with two horizontal steel beams, which were mounted in a rigid steel frame – Figure 4. The horizontal top beam was moved horizontally by jack 1. Three vertical jacks (2–4) were used to control its position parallel to the bottom beam. The distance between the mortar joints above the bottom layer of bricks and below the top layer of bricks was 1000 *mm*. The wall was 980 *mm* long (four and a half brick lengths, including four mortar head joints). The opening was one brick wide and five layers high and situated six layers from the bottom and two brick lengths from the loaded side.

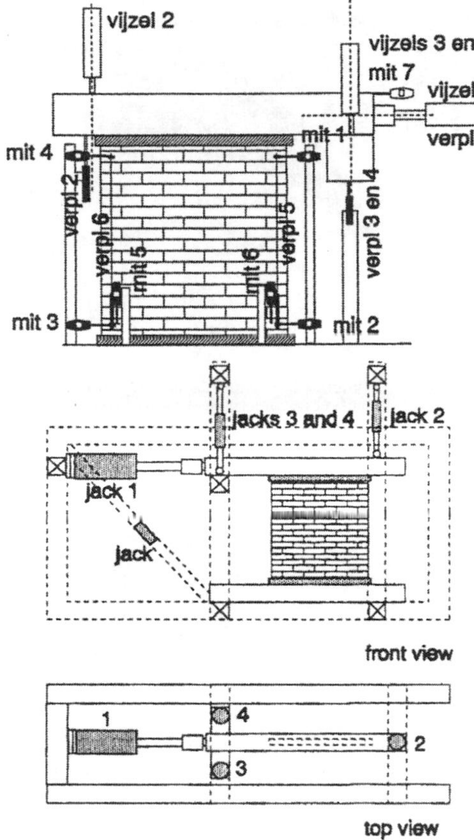

Figure 4. Shear wall experimental set-up and measurement scheme.

The walls pre-loaded with a vertical load of 30 *kN*. Then, shear was applied by deforming the wall by means of jack no 1 at a speed of $10\,mm.h^{-1}$. This deformation was measured by the LVDT marked "verpl 1" in Figure 4. After the horizontal deforming of the wall had started, the vertical jacks were used to sustain the horizontal upper beam vertical level, controlled by the LVDT's marked "verpl 2" and "verpl 3 en 4". These LVDT's measured the distance between the horizontal bottom and top beams. Unfortunately, the bending of the beams between the vetical jacks was not measured.

Figure 5 shows the crack patterns observed in the three wall tests. In Figure 6 the results of the "mit 7" measurements on the upper beam are plotted versus the horizontal load. From the figures it is apparent that the different failure mechanisms, i.e. either diagonal cracking domination (wall J1G) or a shear-slip/sliding domination (wall J3G) or some combination (wall J2G), led to significantly different global force-displacement responses.

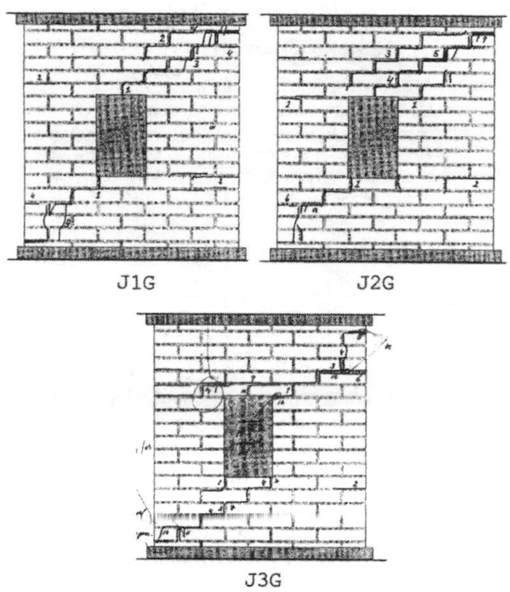

Figure 5. Failure patterns for the shear walls J1G, J2G and J3G.

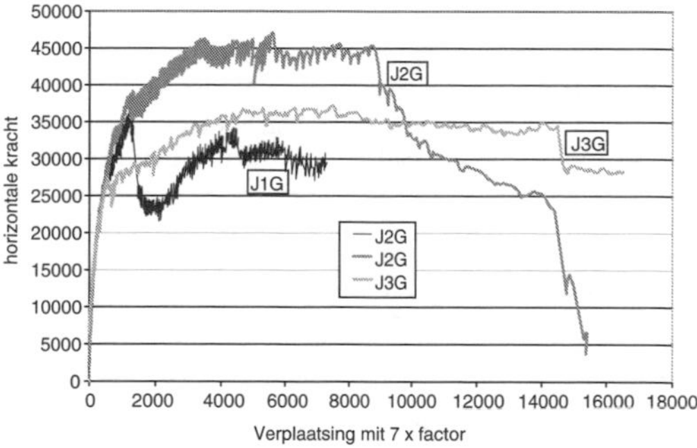

Figure 6. Displacements of wall top right corners ("mit 1").

The different measurements of horizontal sway are compared in Figure 7 for wall J2G. Displacement "verpl 1", which includes frame and jack deformation is the largest, followed by "mit 7". However, measurements by "mit 1" are the most direct indication of the wall response. However, the displacement measured by mit 1 was arrested after some damage in the wall in the vicinity of this gauge point. From the development of the cracking pattern it can be concluded that parts of the wall continue to deform, despite the mit 1 measurement data beyond that point. For comparison with computational results both the mit 1 and mit 7 measurements should be observed.

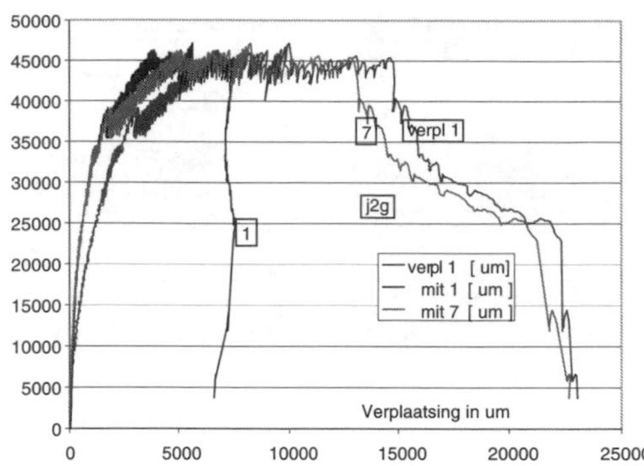

Figure 7. Load-displacement of J2G wall at top right corner (mit 1), the loading beam (mit 7) and as recorded by the horizontal actuator (verpl 1).

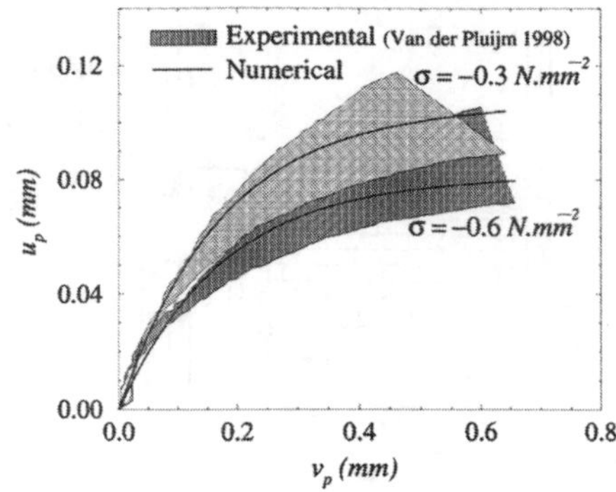

Figure 8. Computed vs. measured masonry dilatancy of a specimen with a single mortar joint (Figure 1).

Table 1. Interface model parameters.

$k_n = 83\,N.mm^{-3}$	Normal stiffness
$k_s = 36\,N.mm^{-3}$	Shear stiffness
$f_t = 0.25\,N.mm^{-2}$	Tensile strength
$c_o = 0.35\,N.mm^{-2}$	Cohesion
$G_f^I = 0.018\,N.mm^{-1}$	Tensile fracture energy
$G_f^{II} = 0.05\,N.mm^{-1}$	Shear fracture energy
$f_c = 8.5\,N.mm^{-2}$	Compressive strength
$\Phi = 0.75$	Friction coefficient
$G_c = 5.0\,N.mm^{-1}$	Compressive fracture energy
$\Psi_o = 0.6$	Initial dilatancy coefficient
$\kappa_p = 0.093$	Compressive plastic strain at f_c
$\sigma_u = -1.3\,N.mm^{-2}$	Stress at which dilatancy is zero
$\delta = 5$	Dilatancy softening gradient

4.2 *Shear wall analyses*

The behaviour of the shear walls with an opening is analysed numerically in this section. The discrete modelling approach is followed. Parameters employed are for the steel frame: Young's modulus 200 GPa, Poisson's ratio 0.3; for the bricks: Young's modulus 16.7 Gpa, Poisson's ratio 0.15; for the joints/interfaces: see Table 1. These values are taken as average values from characterizing tests on small masonry specimens of similar type as used in the shear walls (Van der Pluijm 1992, 1998; Vermeltfoort et al. 1993). These parameters produce the computed shearing response shown in Figure 8 for the masonry shear experiment shown in Figure 1.

Firstly, the discrete modelling strategy and the dilatancy equation (7) are employed. Eight-noded elements model the bricks and steel frame, while six-noded interface elements model the mortar joints. A discretisation of 2 × 4 elements per brick is employed. The initial vertical load is applied by a prescribed downward displacement of the points on the top steel beam where the vertical actuators acted

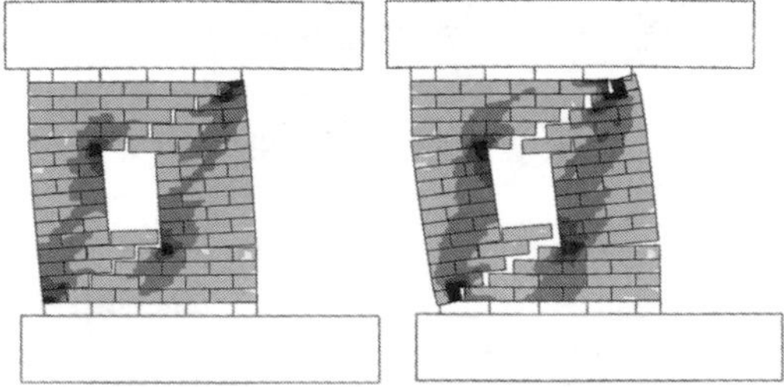

Figure 9. Computed deformations at $3\,mm$(left) and $5\,mm$(right) horizontal displacement showing compressive principal stress contours (black $-10\,N.mm^{-2}$, white zero).

in the experiments. Subsequently, the horizontal load is applied under displacement control of the frame at the right top, at a fixed rate of $10\,\text{mm.h}^{-1}$. A good agreement is obtained with the experimentally observed failure pattern in the J2G specimen, Figure 9. The horizontal force evolution with horizontal displacements at the gauge positions "mit 1" and "mit 7" (section 4.1) of the J2G specimen are compared with the experimental data in Figure 10, showing reasonable agreement.

The stiffness, peak strength and toughness are captured well. However, the unloading (negative slope) post-peak experimental response at point "mit 1" is not found numerically, but is ascribed to the damage to the gauge area of mit 1, as discussed in the previous section.

Note that the responses shown in Figure 10 are for a simulation of the true frame stiffness, as well as consideration of relaxation through a visco-elastic (Maxwell chain) model (Van Zijl et al. 2001) with an effective creep coefficient of 0.85 after 1 day of relaxation. As quantified in Figure 11, both aspects play a role in explaining the greater post-peak ductility observed experimentally than in previous analyses (Lourenço and Rots 1997, Beranek & Hobbelman 1994, De Almeida & Alfaiate 2000).

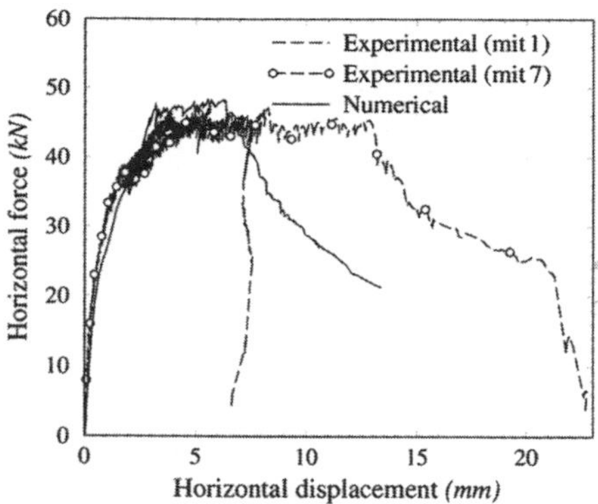

Figure 10. Comparison of numerical and experimental load-displacement responses.

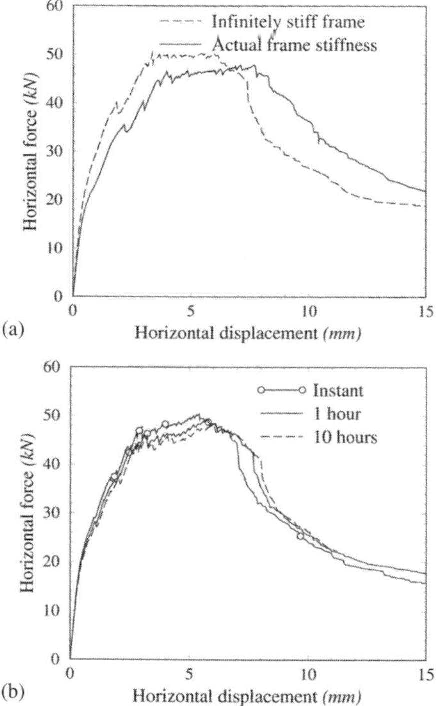

Figure 11. (a) Influence of frame stiffness on numerical shear wall response. (b) Relaxation/creep effect for various test durations.

4.3 *The role of dilatancy in the shear wall response*

To demonstrate the importance of simulating the dilatancy appropriately, the analyses have been repeated for different dilatancy formulations. In the case of discrete modelling a constant dilatancy coefficient has been assumed, with values of $\Psi = 0$ (no uplift upon shearing) and $\Psi = 0.4$. In Figure 12 the responses are compared with the response obtained before with the softening dilatancy formulation.

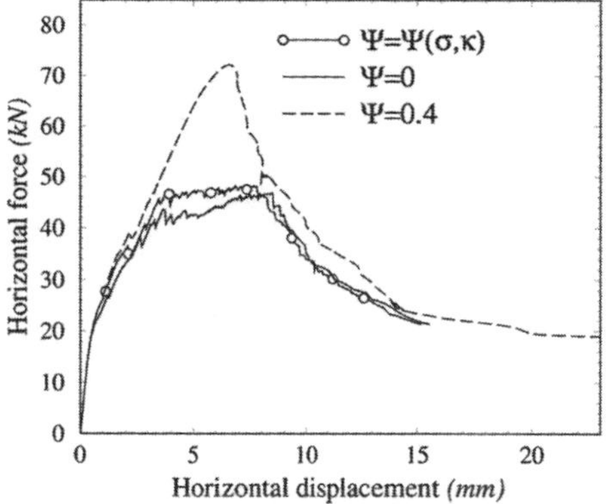

Figure 12. Influence of dilatancy model on global response of discrete model.

The case $\Psi = 0$ ignores the wedging due to interface dilatancy, but approximates the experimental response and the numerical response with a softening dilatancy reasonably. This indicates that the brick-mortar interfaces are indeed smoothened by the combination of shearing and high compressive stresses. If a constant, non-zero dilatancy is employed the peak strength is overestimated. In the case of $\Psi = 0.4$ an overestimation of 50% is computed.

The dilatancy generates high compressive stresses at the corners of the wall. Here, the compression cap of the failure criterion of Figure 2a is reached. The local compression softening in the corners leads to overall structural softening. If compression cap softening is not included, the force would be unlimited (Rots 1994).

The analyses are also performed with the composite model, to illustrate the composite dilatancy formulation, section 2.2. For the masonry composite a Young's modulus of 8 GPa and Poisson's ratio 0.15 are employed, along with the parameters in Table 2.

In Figure 13 the wall load-displacement responses are shown for the standard Rankine model with effective dilatancy $\Psi = 1$ and the modified Rankine with zero dilatancy. The peak load is overestimated by factor 2 for the former case, while the latter is in reasonable agreement with the discrete model results for $\Psi = 0$. At the time of publication the full load-displacement response was not yet analysed.

5 CONCLUSIONS

The phenomenon shearing dilatancy of cement-based materials and computational models for analysing it

Table 2. Composite masonry model parameters for shear wall analysis.

Parallel to bed joints	
$f_{tx} = 0.35\, N.mm^{-2}$	Tensile strength
$G_{fx} = 0.05\, N.mm^{-1}$	Tensile fracture energy
$f_{cx} = 8.5\, N.mm^{-2}$	Compressive strength
$G_{cx} = 20.0$	Compressive fracture energy
Perpendicular to bed joints	
$f_{ty} = 0.25\, N.mm^{-2}$	Tensile strength
$G_{fy} = 0.018\, N.mm^{-1}$	Tensile fracture energy
$f_{cy} = 8.5\, N.mm^{-2}$	Compressive strength
$G_{cy} = 15.0\, N.mm^{-1}$	Compressive fracture energy
$\kappa_p = 0.0012$	Compressive plastic strain at f_c
$h = \sqrt{2(\text{finite element area})}$	Crack band width (Lourenço 1996)

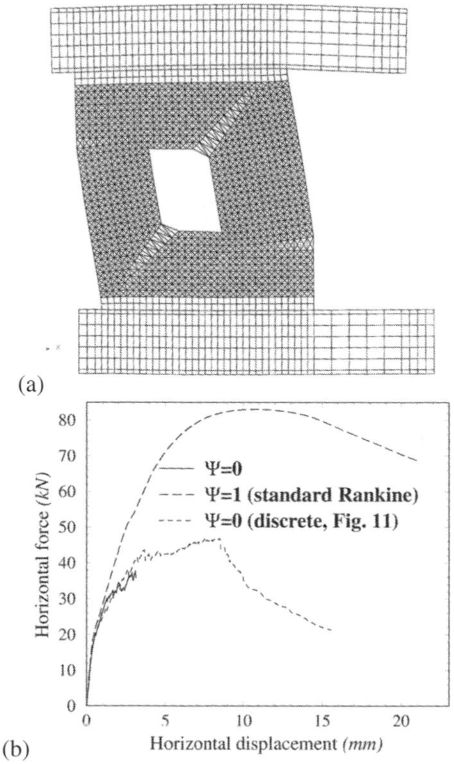

Figure 13. Shear wall response calculated with composite model. (a) Deformation at 1 *mm* horizontal displacement. (b) Influence of inappropriate dilatancy modelling.

have been discussed. The danger of unconservative predictions with these models has been highlighted. Dilatancy has been identified as the underlying source of shear resistance enhancement. If it is modelled inappropriately, the strength build-up may be largely overestimated. A safe/conservative approach is to set the dilatancy coefficient to zero, whereby the enhancement is neglected. However, the nonlinear dilatancy model developed for the discrete modelling strategy has been shown to capture the strength build-up accurately.

A continuum approach to modeling cementitious materials, which employs mode I type limit criteria, has been shown not to capture confined shearing in masonry or concrete correctly. The inherent dilatancy of such models overestimate the volume increase upon shearing. By this mechanism, pressure builds up due to the confinement, offsetting the diagonal tensile stresses, whereby the global shearing resistance is greatly overestimated.

An alternative dilatancy formulation has been derived for a Rankine continuum plasticity formulation to produce isochoric inelastic behavior, which leads to a conservative shearing resistance computation. Yet, such models cannot distinguish between mode I and II types of failure, prohibiting the generalization of such a dilatancy formulation for mode I type models. In the case elaborated, an artificial inelastic deformation evolves in the direction orthogonal to the principal strain, or parallel to the crack, which is not realistic in pure mode I behavior.

To enable distinction between the different failure modes, a discrete approach has been followed. An interface model, suitable for simulating masonry joint behavior, or cracks through bricks, has been discussed. The model combines a Coulomb-shear limit function with a tension cut-off, as well as a compression cap. Thereby, it is possible to modify the dilatancy in the shear mode, without altering the pure mode I modes. A nonlinear dilatancy model, which captures the pressure dependence, as well as the "smoothening" of the joints by large shear-slipping deformation, has been elaborated. This model has been shown to accurately simulate experimentally observed behavior of masonry confined shear walls.

REFERENCES

Beranek, W.J. and Hobbelman, G.J. 1994. Modelling of masonry as an assemblage of spheres on various scales. *Proc. 10th Int. Brick/Block Masonry Conf*: 1–20.

De Almeida, J.R. and Alfaiate, J.V. 2000. Discrete cracking of masonry walls. (CD Rom) *Proc. European Congress on Comp. Meth. in Appl. Sc. and Eng.*, Barcelona.

De Borst, R. and Nauta, P. 1985. Smeared cracking, plasticity, creep and thermal loading – a unified approach. *Engrg. Comp.* 2: 35–46.

Feenstra, P.H. and De Borst, R. 1996. A composite plasticity model for concrete. *Int. J. of Solids and Structures* 33(5): 707–730.

Green, S.J. and Swanson, S.R. 1973. Static constitutive relations for concrete. *AFWL-TR-72-244 Air Force Weapons Laboratory*, Kirtland Air Force Base, New Mexico.

Lourenço, P.B. and Rots, J.G. 1997. Multisurface interface model for analysis of masonry structures. *ASCE J. Eng. Mechanics*. 123(7): 660–668.

Lourenço, P.B., Rots, J.G. and Blaauwendraad, J. 1998. Continuum model for masonry: parameter estimation and validation. *ASCE J. Eng. Mechanics* 124(6): 642–652.

Michelis, P.N. 1998. Work-softening and hardening behaviour of granular rocks. *Rock Mechanics* 14: 187–200.

Reynolds, O. 1885. On the dilatancy of media composed of rigid particles in contact. *Phil. Mag*. 5th Ser. 20: 469.

Roscoe, K.N., Schofield, A.N. and Thurairajah, A. 1963. Yielding of clays in states wetter than critical. *Geotechnique* 13(3): 211–240.

Rots, J.G. 1988. *Computational modeling of concrete fracture*. Dissertation, Delft University of Technology, The Netherlands.

Rots, J.G. 1993. The smeared crack model for localized mode-I tensile failure. In F.H. Wittmann (ed.), *Numerical models in fracture mechanics of concrete*: 101–113.

Rots, J.G., (ed.) 1994 (in English 1997). *Stuctural masonry; an experimental/numerical basis for practical design rules*. Rotterdam: Balkema.

Rowe, P.W. 1971. Theoretical meaning and observed values of deformation parameters for soil. In *Proc. Roscoe Memorial Symp. on stress-strain behavior of soils*, Cambridge: 143–194.

Van der Pluijm, R. 1992. *Deformation controlled shear tests on masonry* (in Dutch). Report BI-92-104, TNO Building and Construction, Delft, The Netherlands.

Van der Pluijm, R. 1998. *Overview of deformation controlled combined tensile and shear tests*. Report TUE/CCO/98.20, Eindhoven Univ. of Techn., Eindhoven, The Netherlands.

Van der Pluijm, R. 1999. *Out-of-plane bending of masonry behavior and strength*. Dissertation, Eindhoven Univ. of Techn., Eindhoven, The Netherlands.

Van Zijl, G.P.A.G. 2000. *Computational modelling of masonry creep and shrinkage*. Dissertation, Delft Univ. of Techn., Delft, The Netherlands.

Van Zijl, G.P.A.G., De Borst, R. and Rots, J.G. 2001. A numerical model for the time-dependent cracking of cementitious materials. *Int. J. Num. Meth. in Engrg*. 52(7): 637–654.

Vermeer, P.A. and De Borst, R. 1984. Non-associated plasticity for soils, concrete and rock. *Heron* 29(3): 1–64.

Vermeltfoort, A.Th., Raijmakers, Th.M.J. and Janssen, H.J.M. 1993. Shear tests on masonry walls. *Proc. 6th North American Masonry Conf.*: 1183–1193.

Modelling of multiphysics phenomena and time effects

Computational Modelling of Concrete Structures, Bićanić et al. (eds)
© 2003 Swets & Zeitlinger, Lisse, ISBN 90 5809 536 3

Keynote paper: [Bio-]Chemomechanics at finer scales – A review

F.-J. Ulm, G. Constantinides, F.H. Heukamp, E.C. Silva, J.T. Germaine E. Lemarchand,*
Ch. Hellmich,† O. Bernard‡
Massachusetts Institute of Technology, Cambridge, Massachusetts, USA

ABSTRACT: Concrete, like many other materials (whether man-made, geological or biological), is a highly heterogeneous material with heterogeneities that manifest themselves at multiple scales. As new experimental techniques such as nanoindentation have provided unprecedented access to micro-mechanical properties of materials, it becomes possible to identify the mechanical effects of chemical reactions at the micro-scale, where the reactions occur, and trace these micro-chemo-mechanical effects through upscaling techniques to the macro scale. The focus of this paper is to review recent developments of a micro-chemomechanics theory which ultimately shall make it possible to capture chemomechanical deterioration processes at *the* scale where physical chemistry meets mechanics. This is illustrated through application of the theory to early-age concrete, calcium leaching, and finally to biologically mediated deterioriation processes in solid materials.

1 INTRODUCTION

One of the keys to the modeling of chemomechanical couplings in materials is the choice of an appropriate thermodynamic system for the description of the governing conservation laws, including mass, energy and entropy conservation. We owe this insight to the works of Coussy (1995), who provided the macroscopic framework for modeling chemomechanical couplings in porous materials as an extension and refinement of Biot's saturated porous media theory. The thermodynamic system in the Biot-Coussy theory of poromechanics, is the *open* macroscopic porous material, considered as a superposition in time and space of a solid phase and *n* fluid phases that exchange matter through its boundary with the outside. Chemomechanics in this theory is associated with phase change phenomena in between the fluid phases or in between the fluid phases and the solid phase (Coussy & Ulm, 1996), inducing time dependent deformation, both reversible (chemoelasticity) and irreversible (chemoplasticity). This leads, in this macroscopic theory, to an advanced internal variable theory, in which a reaction extent or a reaction degree is added to the set of standard solid mechanics state variables (strain, temperature, plastic strain, damage, etc.; see e.g. Sercombe et al. 2000a). The theory has been employed in many engineering applications to capture coupled chemomechanical phenomena in concrete mechanics: thermo-chemo-mechanical couplings in early-age concrete (Ulm & Coussy 1995, 1998; Cervera et al. 1999a,b) and shotcrete (Sercombe et al. 2000b); early-age concrete strength growth as chemoplastic hardening (Ulm & Coussy 1996; Hellmich et al. 1999a); hydration heat as thermochemical coupling (Hellmich et al. 1999b; Ulm & Coussy 2001); chemoplastic softening in rapidly heated concrete (Ulm et al. 1999a); chemoelasticity of ASR-expansion (Larive 1998; Ulm et al. 2000; Li & Coussy 2002; Li 2002); chemoplasticity of ASR-expansion (Ulm et al. 2002b); chemoporoplasticity of calcium leaching (Ulm et al. 1999b), and so on. The strength of the macroscopic theory is that it makes it possible to lump a large number of complex chemical reactions into one or several macroscopic state variables, the evolution of which can be determined with almost standard macroscopic material tests. On the other hand, this macroscopic approach has one fundamental drawback which is intrinsically related to the choice of the macroscopic system that is, the choice of chemomechanical state variables is not unique, in contrast to the chemophysical phenomena at stake which are certainly unique for any specific (i.e. identified) chemomechanical deterioration process. The dilemma is due to the fact that the chemical processes at stake typically occur at much smaller scales than the macroscopic scale, where chemical reactions express themselves through loss of stiffness, strength, frictional capacity, cohesion, deleterious expansion, and so on. This recognized limit of the macroscopic theory has prompted research into the

* Now at: LML UMR 8107, Villeneuve d'Ascq, France.
† Now at: Technical University of Vienna, Austria.
‡ Now at: Oxand SA, Montigny-sur-Loing, France.

development of a micro-chemomechanics theory. The premise of such a theory is to capture chemomechanical deterioration processes at *the* scale where physical chemistry meets mechanics. The same holds true for coupled biomechanical processes in solid materials, in which biological organisms deteriorate solid materials leading to premature failure of materials and structures (biocorrosion in sewer pipes; increased risk of bone fracture due to osteoporosis, etc.).

The aim of this review paper is two fold: (1) to review recent developments in micro-chemomechanics theory, and (2) to identify relevant macroscopic state variables from the scale where physical chemistry (or biochemistry) meets mechanics. This will be illustrated through several application of this theory to early-age concrete, concrete deterioration by calcium leaching, and finally to biologically mediated deterioration processes.

2 MICRO-CHEMOMECHANICS

Physical chemistry focuses on the chemical agents in an undeformable system. By contrast, chemomechanics at the microscale focuses on the deformable solid in the porous material subjected to a chemical attack on the solid-fluid interface. This comes to choose the solid matter, at the microscale, as the thermodynamic system, which as we will see below provides the missing link between physical chemistry (or biochemistry) and mechanics. This solid matter is subjected at a part of its boundary, say $\Gamma \subseteq \partial V_s$, to a chemical reaction which adds (precipitation) or removes (dissolution) matter.

2.1 *Chemical porosity*

The total mass of the solid enclosed in an r.e.v. $|V|$ is:

$$|V| m_s = \int_{V_s} \rho_s dV \tag{1}$$

and its variation reads:

$$\frac{dm_s}{dt} = \frac{1}{|V|}\int_{V_s}\left(\frac{d\rho_s}{dt} + \rho_s \operatorname{div}\left(\underline{u}^s\right)\right) dV + \\ + \frac{1}{|V|}\int_{\partial V_s} [[\rho_s \underline{u}^s]] \cdot \underline{n}_s da \tag{2}$$

The first term expresses the mass conservation in the solid bulk, that is at the microlevel of the solid phase; while the second term refers to the mass rate jump that is induced by precipitation ($[[\rho_s \underline{u}^s]] \cdot \underline{n}_s \geq 0$) or dissolution ($[[\rho_s \underline{u}^s]] \cdot \underline{n}_s \leq 0$) that occurs at the boundary Γ of the solid phase:

$$[[\rho_s \underline{u}^s]] \cdot \underline{n}_s = \eta \rho_c \underline{u}^c \cdot \underline{n}_s \tag{3}$$

where $\eta = 1$ on Γ, and $\eta = 0$ on $\partial V_s - \Gamma$; ρ_c is the mass density of the chemical species, and $\underline{u}^c \cdot \underline{n}_s d\Gamma \times dt$ the associated infinitesimal volume of the surface layer that is chemically added to or chemically removed from the solid phase. It is readily understood that this surface layer gives rise to a change in macroscopic porosity. More precisely, if we introduce the Lagrangian porosity:

$$\phi = \frac{V_f}{|V|} = \frac{V_t - V_s}{|V|} \simeq (1 + E_v) - \frac{V_s}{|V|} \tag{4}$$

where V_t, V_f and V_s are the volumes of the porous material, the fluid phase and the solid phase in the current (deformed) configuration, the total change of the Lagrangian porosity is:

$$\frac{d\phi}{dt} = \frac{dE_v}{dt} - \frac{V_s}{|V|}\left\langle \frac{d\varepsilon_v^s}{dt} \right\rangle_{V_s} - \frac{1}{|V|}\int_{\partial V_s} [[\underline{u}^s]] \cdot \underline{n}_s da \tag{5}$$

The first two terms in this relation represent the change in porosity due to mechanical loading, where $\langle \frac{d\varepsilon_v^s}{dt} \rangle_{V_s} = \frac{1}{V_s}\int_{V_s} \operatorname{div}(\underline{u}^s) dV$ is the volume average of the rate of relative volume change in the solid phase. In turn, the second term represents the chemically induced porosity change:

$$\frac{d\phi^c}{dt} = -\frac{1}{|V|}\int_{\partial V_s} \eta \underline{u}^c \cdot \underline{n}_s da = -\frac{1}{|V|}\int_{\Gamma} \underline{u}^c \cdot \underline{n}_s d\Gamma \tag{6}$$

The chemical porosity ϕ^c appears as an appropriate state variable to describe volume changes related to the chemical reaction that occur at the boundary Γ of the solid phase.

2.2 *Driving forces of chemomechanics*

We want to determine the driving forces of the chemomechanical processes in the porous material. Since we are interested in the solid's response, the thermodynamic system we consider is the solid phase (volume V_s). To simplify the presentation, we will assume isothermal and quasi-static evolutions. The energy transformations are expressed by the Clausius-Duhem inequality, which states that the external energy supply to the solid, dW^s_{ext}, which is not stored as free energy dW^s in the solid system, is dissipated into heat form:

$$\frac{d\mathcal{D}}{dt} = \frac{d\mathcal{W}^s_{ext}}{dt} - \frac{d\mathcal{W}^s}{dt} \geq 0 \tag{7}$$

For the chemomechanical solid system under consideration, the external energy supply has two origins:

The first is of mechanical origin due to volume forces in the solid (term of the form $\rho_s \underline{f} dV$), and surface tension $\underline{t}_s = \underline{\underline{\sigma}} \cdot \underline{n}_s$, where $\underline{\underline{\sigma}}$ is the stress tensor in the solid phase. These forces supply a work rate along the velocity field $\underline{u}^s$. Thus, by application of the theorem of virtual work rate,

$$\frac{dW^m_{ext}}{dt} = \int_{V_s} \underline{u}^s \cdot \rho_s \underline{f} dV + \int_{\partial V_s} \underline{u}^s \cdot \underline{t}_s da$$
$$= \int_{V_s} \underline{\underline{\sigma}} : \frac{d\underline{\underline{\varepsilon}}^s}{dt} dV - p \int_{\partial V_s} [[\underline{u}^s]] \cdot \underline{n}_s da \quad (8)$$

where we made use of $2d\underline{\underline{\varepsilon}}^s/dt = \text{grad}\,\underline{u}^s + {}^t\text{grad}\,\underline{u}^s$, the symmetry of the stress tensor $\underline{\underline{\sigma}}$, the local equilibrium in the solid phase, div $\underline{\underline{\sigma}} + \rho_s \underline{f} = 0$, and the stress continuity at the solid-fluid interface ∂V_s, i.e. $\underline{\underline{\sigma}} \cdot \underline{n}_s = -p\underline{n}_s$, where p is the fluid pressure assumed constant in the pore space. Applied to the r.e.v. with regular stress or strain boundary conditions, it can be shown with the help of the Hill-Mandel lemma that the external mechanical work rate (8) can be developed in the form:

$$\frac{dW^m_{ext}}{dt} = |V| \left[\underline{\underline{\Sigma}} : \frac{d\underline{\underline{E}}}{dt} + p\frac{d\phi}{dt} \right] \quad (9)$$

where $\underline{\underline{\Sigma}} = \langle \underline{\underline{\sigma}} \rangle_V$ = macroscopic stress tensor; and $d\underline{\underline{E}}/dt = \langle d\underline{\underline{\varepsilon}}/dt \rangle_V$ = macroscopic strain rate.

The second external energy supply to the solid results from the action of the chemical potential of the solute, μ^{sol}, at the boundary Γ, which provides energy to the solid along a molar flux $J_N = (\frac{\rho_c}{\mathcal{M}})\underline{u}^c \cdot \underline{n}_s = -(\frac{\rho_c}{\mathcal{M}})\underline{u}^c \cdot \underline{n}_f((\rho_c/\mathcal{M})$ = number of moles per unit solid volume of the chemical species precipitating or dissolving from the solid):

$$\frac{dW^c_{ext}}{dt} = \int_\Gamma \mu^{sol} \left(\frac{\rho_c}{\mathcal{M}}\right) \underline{u}^c \cdot \underline{n}_s d\Gamma$$
$$= -\mu^{sol} \left(\frac{\rho_c}{\mathcal{M}}\right) \frac{d\phi^c}{dt} |V| \quad (10)$$

where we assumed μ^{sol} constant along Γ.

Finally, we need to express the change in free energy of the solid, which involves at least two components: the elasticity potential ψ_s^{el} (of dimension $[\psi_s^{el}] = L^{-1}MT^{-2}$), and the chemical potential μ_c (of dimension $[\mu_c] = L^2MT^{-2} \times MOL^{-1}$) of the chemical species in the solid phase:

$$W_s = \int_{V_s} \psi_s dV = \int_{V_s} \left(\psi_s^{el} + \left(\frac{\rho_c}{\mathcal{M}}\right)\mu_c\right) dV \quad (11)$$

The elasticity potential ψ_s^{el} represents the recoverable (elastic) energy volume density stored by externally applied work into the solid phase; the chemical potential μ_c expresses the energy stored as chemical bonds in the solid phase. The total time derivative of (11) reads:

$$\frac{dW^s}{dt} = \int_{V_s} \left[\frac{d\psi_s}{dt} + \psi_s \frac{d\varepsilon_v^s}{dt} \right] dV \quad (12)$$
$$+ \int_\Gamma \left(\psi_s^{el} + \left(\frac{\rho_c}{\mathcal{M}}\right)\mu_c\right) \underline{u}^c \cdot \underline{n}_s d\Gamma$$

Last, collecting the different components from (9), (10) and (12), the dissipation of chemomechanical energy supplied to the solid is shown to be the sum of two terms (Ulm et al. 2002c):

$$\frac{d\mathcal{D}}{dt} = |V| (\varphi_1^s + \varphi^c) \geq 0 \quad (13)$$

- The first term φ_1^s represents the dissipation rate associated with mechanical deformation of the solid phase. It is similar to the one of the intrinsic solid dissipation emerging from the Biot-Coussy theory of macroscopic poromechanics (Coussy 1995), if we replace the total porosity ϕ in the theory by the mechanical porosity $\phi^m = \phi - \phi^c$:

$$\varphi_1^s = \underline{\underline{\Sigma}} : \frac{d\underline{\underline{E}}}{dt} + p\left(\frac{d\phi}{dt} - \frac{d\phi^c}{dt}\right) - \frac{d\Psi_s}{dt} \geq 0 \quad (14)$$

with:

$$\frac{d\Psi_s}{dt} = \frac{1}{|V|} \int_{V_s} \left(\frac{d\psi_s}{dt} + \psi_s \frac{d\varepsilon_v^s}{dt} \right) dV \quad (15)$$

The modeling of the deformation behavior of the porous material that occurs simultaneously with chemical reactions, can make use of the rich body of macroscopic models available in the porous media literature.

- The second term φ^c represents the dissipation related to the chemical reaction at the boundary of the solid phase, and the associated increase of the chemical porosity (defined by (6)):

$$\varphi^c = -\mathcal{A} \circ \frac{d\phi^c}{dt} = \frac{1}{|V|} \int_\Gamma A \times J_N d\Gamma \geq 0 \quad (16)$$

with

$$A = \mu^{sol} - \mu_c - \left(\frac{\mathcal{M}}{\rho_c}\right) (\psi_s^{el} + p) \quad (17)$$

where A is the chemical affinity, i.e. the driving force of the molar flux $J_N = (\frac{\rho_c}{\mathcal{M}})\underline{u}^c \cdot \underline{n}_s$. The quantity $\Delta G = \mu^{sol} - \mu_c$ is the Gibbs energy; that is the pure chemical driving force of the precipitation/dissolution process. In addition, due to the multiphase nature of the porous media and the deformability of the solid phase, two additional quantities affect the chemomechanical driving force: the local elastic energy ψ_s^{el}, and the fluid pressure p. Following standard thermodynamics, the identification of affinity $A = \Delta G - (\mathcal{M}/\rho_c)(\psi_s^{el} + p)$ as driving force implies that the local kinetic law of the precipitation or dissolution process is one which relates the molar flux J_N to the chemical affinity:

$$\text{on } \Gamma: J_N = J_N(A);\ A \times J_N(A) \geq 0 \tag{18}$$

Remarkably, since $J_N \propto \underline{u}^c \cdot \underline{n}_s < 0$ in a dissolution process, $A < 0$, so that the chemical dissipation remains locally greater or equal than zero. Hence, since $\psi_s^{el} \geq 0$ and $p \geq 0$, it follows that the elastic energy stored in the microstructure and a non-zero fluid pressure will actually increase the intensity of the dissolution process; and for a precipitation process it is the inverse. This effect of strain and pressure on the affinity must be checked, from case to case, relative to the pure chemical potential difference $\Delta G = \mu^{sol} - \mu_c$.

In summary, the brief study of the micro-to-macroscopic energy transformations for the solid as thermodynamic system, allows one to break down the chemomechanical behavior (at the macro-scale) of materials into two components: a purely poromechanics behavior (which is well established for engineering materials by the Biot-Coussy theory of poromechanics), and a chemomechanical precipitation or dissolution process, which is witnessed by the solid as a moving boundary at its boundary Γ. This split of the macroscopic response allows one to capture many coupled chemomechanical phenomena in geomaterials and biomaterials some of which are developed below.

3 APPLICATION TO CONCRETE

The application of the micro-chemomechanics theory requires a breakdown of the material into the scale where the material can be represented as a porous material. This is a particular difficult task for cementitious materials due to the high level of heterogeneity of the material that manifests itself at multiple scales. A rough breakdown of characteristic mechanical length scale of cementitious materials is displayed in Figure 1 (Constantinides & Ulm 2002a,b):

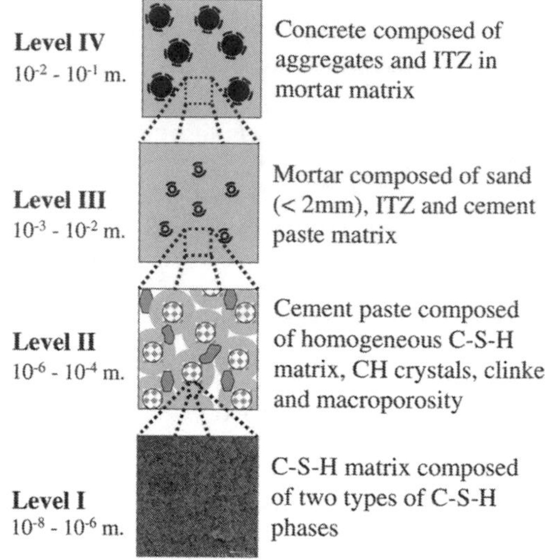

Figure 1. The four-level microstructure of concrete (Constantinides & Ulm 2002a,b).

1. Level I: The lowest level we consider is the one of the C–S–H matrix that forms at early ages by the hydration of C_3S and C_2S.[1] This level of a characteristic length scale of $10^{-8} - 10^{-6}$ m, is the smallest material length scale that is, at present, accessible by mechanical testing, i.e. nanoindentation. At this scale, it is now well established that the C–S–H exist in, at least, two different forms with different volume fractions (Tennis & Jennings 2000; Jennings 2000) and elastic properties (Acker 2001; Velez et al. 2001; Constantinides & Ulm 2002a,b). The morphology of the two types of C–S–H is correlated with two different processes of hydration of clinker compounds. During the early stages of hydration, nucleation and growth of C–S–H occurs at the surface of the cement grains, leading to the softer outer products. With the hydration progressing, the cement grains are covered by a growing layer of C–S–H and the hydration is controlled by the diffusion process through this layer. While outer products are still formed, new C–S–H is primarily formed in a space confined by the existing C–S–H layer; and these new C–S–H have a higher density, leading to an on-average higher stiffness of the inner product C–S–H. This is displayed in Figure 2 (Top) which displays the stiffness histogram of nanoindentation results of a hardened cement paste ($w/c = 0.5$). The reason for the difference in stiffness between inner and outer product is attributed to the difference in C–S–H gel porosity at a scale still below. However, this gel porosity (28%) is of a characteristic size of some water molecules (4–5 water molecules, i.e. a size on the order of 10^{-9} m), so that the water present at this scale cannot be considered as a bulk water phase (Wittmann 1982). In turn, at the considered level I, the C–S–H matrix can be considered as a two-phase solid material, composed of a stiffer inclusion phase (C–S–H_b-phase), embedded into a softer matrix phase (C–S–H_a–phase). The governing variables

[1] The cement's chemistry abbreviation will be used in this paper ($C_3S = 3 \cdot CaO \cdot SiO_2$, $C_2S = 2 \cdot CaO \cdot SiO_2$, $C_3A = 3 \cdot CaO \cdot Al_2O_3$, $C_4AF = 4 \cdot CaO \cdot Al_2O_3 \cdot Fe_2O_3$).

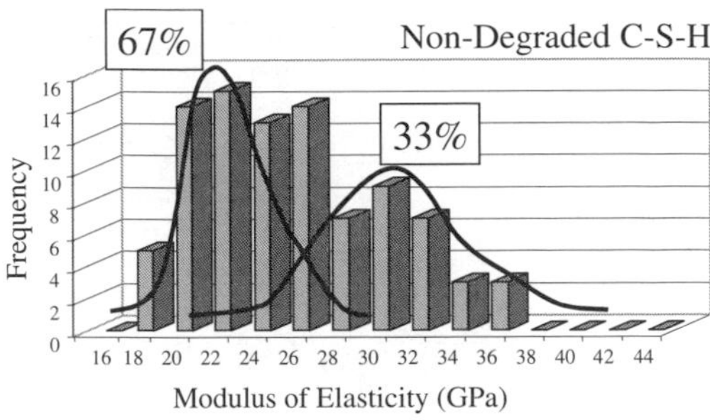

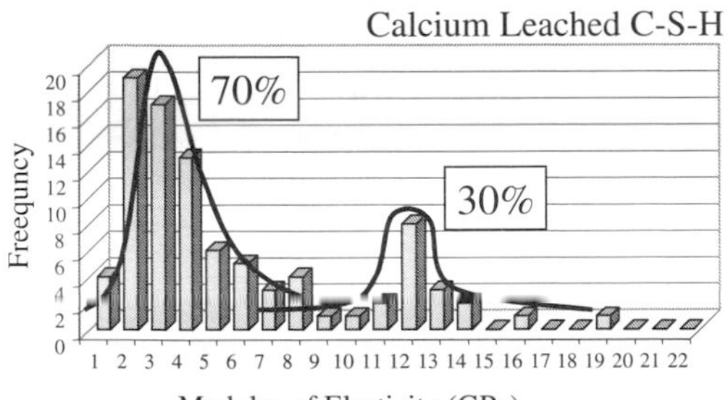

Figure 2. Histogram of elasticity moduli obtained by nanoindentation on Calcium-Silica-Hydrates (C–S–H); $w/c = 0.5$: Top: non-degraded C–S–H, Bottom: calcium leached C–S–H. The histogram allows the identification of the two types of C–S–H in cementitious materials that affect the macroscopic elasticity of concrete materials (Costantinides & Ulm 2002a,b).

at this scale are the volume fractions of the two phases, and the intrinsic material properties of the two phases.

2. Level II: The C–S–H matrix together with unhydrated cement products (i.e. the four clinker phases $X = C_3S, C_2S, C_3A, C_4AF$), large Portlandite crystals ($CH = Ca(OH)_2$), aluminates and macroporosity in the case of high water-to-cement ratio materials (roughly $w/c > 0.4$) forms the cement paste, and is referred to as Level II ($10^{-6} - 10^{-4}\,m$). It is the scale at which cementitious materials can be considered as porous materials.
3. Level III ($10^{-3} - 10^{-2}$ m) refers to mortar; that is a three phase composite material composed of a cement paste matrix, sand particle inclusions, and an Interfacial Transition Zone (ITZ). The volume fractions of these three phases are fixed in time, that is they are not affected by chemical reactions.
4. Concrete as a composite material is considered on Level IV (10^{-2}–10^{-1} m). Similar to Level III, concrete at this scale is a three-phase material composed of aggregates (>2 mm) embedded in a continuous homogeneous mortar matrix and an ITZ; and the required volume fractions are also fixed in time.

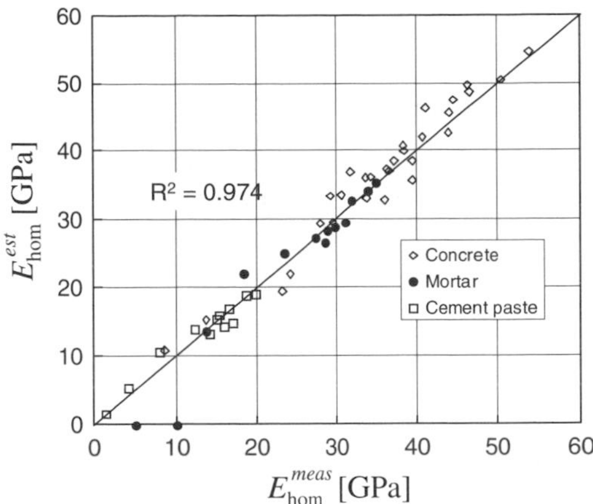

Figure 3. Predictive capability of a four-level homogenization model: predicted versus measured E–modulus (Bernard et al. 2002).

It is worth noting that the four levels described above respect the separability of scale condition; that is each scale is separated by the next one by at least one order of length magnitude. This makes it possible to apply linear and nonlinear continuum micromechanics to upscale material properties (Zaoui 2002).

3.1 *Early-age concrete*

Starting point of modeling early-age concrete is level I, that is the scale of the C–S–H matrix. At this scale, the two types of C–S–H, which are the reaction products of the hydration of the two clinker phases C_3S and C_2S, result from two different hydration processes: The low-density C–S–H_a-phase corresponding to the outer products, is formed during the nucleation and growth process; the high-density C–S–H_b-phase corresponding to the inner products, is formed during the diffusion controlled hydration reaction. The kinetics of these processes has been focus of intensive research in cement chemistry, delivering reaction degrees for C_3S and C_2S hydration. The link between physical chemistry (hydration degrees) and mechanical properties is provided by the volume fractions of the two C–S–H-phases. Use of the volume fractions in appropriate homogenization schemes that capture the strong matrix-inclusion morphology of cementitious materials (e.g. Mori & Tanaka 1973; see also Zaoui 2002), delivers the composite stiffness of the C–S–H matrix (level I), cement paste (level II), mortar (level III) and concrete (level IV); see Figure 3. While the development of this four-level homogenization scheme, presented in Bernard et al. (2002), goes beyond the scope of this paper, it is worth noting that the intrinsic stiffness of all involved phases do not change, neither in time, nor from one cement based material to another.

The only parameters that change, at early ages, are the volume fractions, which depend primarily on the w/c–ratio.

In turn, we will address here the application of the micro-chemomechanics theory to level II, which is the level where the material can be considered as a porous material. The r.e.v. at this scale is defined by the initial volume of the cement, V_c^0, and water, V_w^0, in the mixture:

$$|V_{II}| = V_c^0 + V_w^0 = V_c^0 \times \left(1 + \frac{\rho_c^0}{\rho_w^0} \times \frac{w}{c}\right) \tag{19}$$

where $\rho_c^0/\rho_w^0 = 3.15$ is the cement-to-water mass density ratio, and w/c is the water-to-cement ratio. In turn, the total Lagrangian porosity (4) at early ages reads:

$$\phi \simeq (1 + E_v) - \frac{(\sum_X V_X(t) + V_{C-S-H}(t) + V_{CH}(t) + V_A(t))}{|V_{II}|} \tag{20}$$

where $V_X(t)$ is the volume of the four clinker phases, $V_{C-S-H}(t)$ the volume of the C–S–H matrix, $V_{CH}(t)$ the volume of the Portlandite crystals, and $V_A(t)$ the volume of the aluminates. At this scale, the chemical porosity change is related to the dissolution of the reactants, i.e. $\underline{u}_X^c \cdot \underline{n}_X d\Gamma \leq 0$, and the growth of the solid hydration products (Y = C–S–H, CH, A) into the macroporosity, i.e. $\underline{u}_Y^c \cdot \underline{n}_Y d\Gamma \geq 0$. Expressed in terms of the volume fractions of the solid reactants and products, $f_i = V_i/|V_{II}|$, the change of chemical porosity reads:

$$\frac{d\phi^c}{dt} = -\left(\sum_X \frac{df_X}{dt} + \frac{df_{C-S-H}}{dt} + \frac{df_{CH}}{dt} + \frac{df_A}{dt}\right) \tag{21}$$

Figure 4 displays the evolution of these volume fractions as a function of the overall hydration degree,[2] which can be determined with hydration kinetics models of cement chemistry (Bernard et al. 2002). The key ingredient therefore for the application of the micro-chemomechanics theory turns out to be advanced

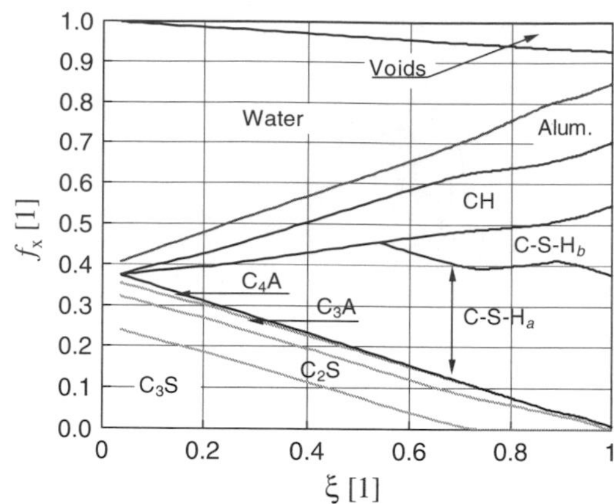

Figure 4. Evolution of volume fractions of the phases of a $w/c = 0.5$ cement paste (Level II) as a function of the hydration degree; (Bernard et al. 2002).

hydration kinetics models. It is also readily understood from this figure that the chemical porosity is almost linearly related to the hydration degree, and the rate (21) linearly to the hydration rate. In other words, the chemical porosity and the overall hydration degree can be considered as equivalent macroscopic state variables. This correlation provides an interface with macroscopic thermo-chemo-mechanical models of early-age concrete that have been developed in the last decade.

3.2 *Calcium leaching*

Calcium, which is the dominant chemical element of cement-based materials, is leached from the material by a coupled diffusion-dissolution problem (Adenot 1992; Mainguy & Coussy 2000) when concrete is in contact with water having a lower calcium concentration than some chemical equilibrium concentrations at which the probability that calcium ions in solution precipitate onto the skeleton is equal to the probability that calcium bound in the solid phase goes into solution. This process which involves tens of chemical species at very fine scales, manifests itself, at a macroscale, by a loss of mechanical performance of concrete. Calcium leaching, which is the reference scenario for the design of nuclear waste storage systems, is a very slow process. It takes some 100 years to degrade 1 m of concrete. However, this is the time-scale during which the safety of storage structures for medium- and high level nuclear waste must be ensured. The question is how?

It is instructive to trace the effects of calcium leaching through the four-level microstructure of cementitious materials displayed in Figure 1.

[2] The overall hydration degree is defined as the weighted average of the clinker hydration degrees ξ_X (Powers & Brownyard 1946; Taylor 1997):

$$\xi(t) = \frac{\sum_X m_X \xi_X(t)}{\sum_X m_X}$$

where $m_X = m_{C_3S}, m_{C_2S}, m_{C_3A}, m_{C_4AF}$ are the mass fractions of the clinker phases in the cement, which are generally provided by the cement producer, based on a chemical analysis of the cement.

Non-degraded C–S–H Matrix 10^{-6} m

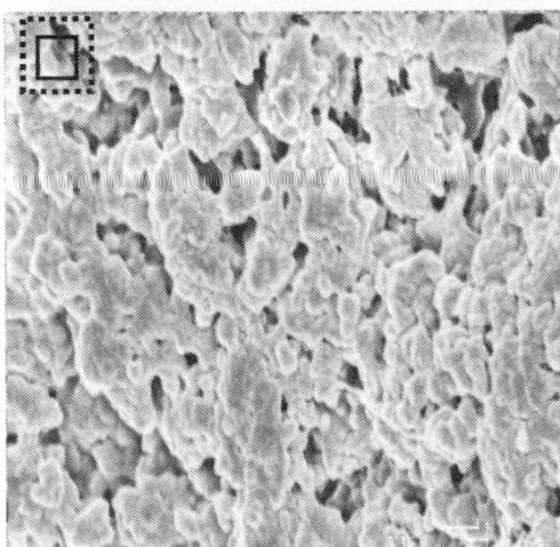

Calciumleached C–S–H Matrix 10^{-6}m

Figure 5. SEM images of C–S–H matrix ($w/c = 0.5$): Non-degraded (Top), calcium leached (Bottom). The boxes in these figures represents the effective influence zone affected by nanoindentation of a size of roughly $9 \times h$, where h is the penetration depth (Constantinides & Ulm 2002a).

3.2.1 *Level I – chemical damage of the C–S–H matrix*

Figure 2 displays the results of nanoindentation results on a non-degraded and a calciumleached C–S–H matrix, representing asymptotic (equilibrium) states of the cement-based material. The appearance of the C–S–H matrix can be grasped from the SEM images displayed in Figure 5, which also displays the effective material surface area of nanoindentation. The figures show that the highly disordered non-degraded C–S–H matrix becomes, after degradation, much more continuous, which is related to the dissolution and re-precipitation of C–S–H at a lower C/S-ratio (Adenot 1992).

Nanoindentation at this scale provides a means of assessing the effect of calcium leaching on the material properties and volume fractions of the involved chemomechanical phases. In fact, at level I the volume fraction of the two C–S–H phases is equivalent to the probability of intending on one or the other phase. This is displayed in the histograms of Figure 2. Remarkably, during calcium leaching (and in contrast to early-age concrete), the volume fraction of the two types of C–S–H phases does not change, despite the partial dissolution and re-precipitation process. However, that what changes during calcium leaching is the intrinsic elasticity of the C–S–H phases: The low-density C–S–H_α-phase has a residual stiffness of roughly 14% of its initial value, while the high-density C–S–H_b-phase has 41% residual stiffness (Table 1). This difference in chemical damage of the two types of C–S–H is readily understood from the morphology of the phases at level I. The high density C–S–H_b-phase (inner products) degrade much less than the low-density C–S–H_a-phase (outer products) coating it. This finding, which sets out a new basis for the development of sustainable cement-based materials,[3] provides strong evidence that the scale accessible by nanoindentation is the scale where physical chemistry meets mechanics.

Table 1. Elastic moduli of the C–S–H obtained by nanoindentation $w/c = 0.5$ (Mean/St.Dev.). (Constantinides & Ulm 2002a,b).

Phase	C–S–H_a	C–S–H_b
E_0 [GPa] (intact)	21.7 ± 2.2	29.4 ± 2.4
E_∞ [GPa] (leached)	3.0 ± 0.8	12.0 ± 2.4
Residual value [%]	14	41
Volume Fractions	0.7	0.3

3.2.2 *Level II – Micro-chemomechanics of calcium leaching*

The first phase to be dissolved, in course of calcium leaching, is the Portlandite phase that manifests itself as large crystals in the material system. The space previously occupied, at level II, by Portlandite crystals is added to the macroporosity; that is:

$$\frac{d\phi^c}{dt} = -\frac{1}{|V_{II}|}\int_\Gamma \underline{u}^{\mathrm{CH}} \cdot \underline{n}_s d\Gamma \geq 0 \tag{22}$$

where $\underline{u}^{\mathrm{CH}} \cdot \underline{n}_s \leq 0$ is the CH-dissolution rate at the boundary of the solid phase. Application of the micro-chemomechanis theory is straightforward (Ulm et al. 2002c): The mechanical dissipation is given

[3] It suffices, indeed, to fine tune the mix design (particularly the w/c ratio), to increase the volume fractions of the C–S–H_b-phase, to obtain cement-based materials with low chemomechanical leaching tendendancy [Constantinides & Ulm 2002a,b].

by (14), and the chemical dissipation rate by (16), reading here:

$$\varphi^c = \frac{1}{|V_{II}|}\int_\Gamma \left(\left(\frac{\rho}{\mathcal{M}}\right)_{\mathrm{CH}} \Delta G_{\mathrm{CH}} - \left(\psi_s^{el} + p\right)\right) \underline{u}^{\mathrm{CH}} \cdot \underline{n}_s d\Gamma \geq 0 \quad (23)$$

where $\Delta G_{\mathrm{CH}} = \mu_{\mathrm{Ca}^{2+}} - \mu_{\mathrm{CH}}$ is the pure chemical driving force of the dissolution, while $\psi_s^{el} + p \geq 0$ are driving forces as well, which actually accelerate the dissolution rate. It is useful to evaluate the order of magnitude of the different quantities in (23), and more precisely how the strain energy compares to the pure chemical affinity ΔG_{CH}. A rough estimate of ΔG_{CH} of Portlandite dissolution, that is:

$$\mathrm{Ca(OH)_2} \rightarrow \mathrm{Ca}^{2+} + 2\mathrm{OH}^- \quad (24)$$

is provided by considering the change in the ionic activity product $IAP = [\mathrm{Ca}^{2+}]\ [\mathrm{OH}^-]^2$ of the solution with regard to the solubility product K_{so} of the Portlandite, according to:

$$\max \Delta G_{\mathrm{CH}} = \mathcal{R}\theta \ln\left(\frac{IAP}{K_{so}}\right) \quad (25)$$

where $\mathcal{R} = 8.314510\,\mathrm{J\,mol^{-1}\,K^{-1}}$ is the universal gas constant and $\theta = 293.15\,\mathrm{K}$ is a reference temperature. In unleached cementitious systems, $\Delta G_{\mathrm{CH}} = 0$, such that $K_{so} = IAP_{eq} = [\mathrm{Ca}^{2+}]_{eq}\ [\mathrm{OH}^-]^2_{eq}$. A rough estimate of ΔG_{CH} can be obtained from (25) by considering the change in $[\mathrm{OH}^-]$ and $[\mathrm{Ca}^{2+}]$ between the equilibrium states before and after Portlandite dissolutions. Calculations by Adenot (1992) provide numerical values for the changes in concentrations of the different species in the fluid phase. The $[\mathrm{OH}^-]$ concentration changes from 5×10^{-2} mol/l to 6×10^{-3} mol/l, which corresponds to pH values of 12.7 and 11.8, respectively; while the $[\mathrm{CA}^{2+}]$ concentration changes from 2.2×10^{-2} mol/l to 3×10^{-3} mol/l. Evaluating (25) we find:

$$\max \Delta G_{\mathrm{CH}} \simeq -15\,\mathrm{kJ\,mol^{-1}} \quad (26)$$

or in mechanical units

$$\left(\frac{\rho}{\mathcal{M}}\right)_{\mathrm{CH}} \max \Delta G_{\mathrm{CH}} \simeq -450\ \mathrm{MPa} \quad (27)$$

where the Portlandite density was taken as $\rho_{\mathrm{CH}} = 2,240\,\mathrm{kg/m^3}$ and the molar mass as $\mathcal{M}_{\mathrm{CH}} = 74.1\,\mathrm{g\,mol^{-1}}$. The magnitude of the pure chemical driving force ΔG_{CH} needs to be compared to the elastic energy ψ_s^{el} and pressure p. Given small deformations, the elastic free energy ψ_s^{el} is smaller than 1 MPa, and the maximum fluid pressure level in the interstitial pore solution never exceeds 10 MPa. It therefore appears that both strain energy and pressure are at least one order of magnitude smaller than the pure chemical driving force, and that their effect on the chemical affinity can be neglected. This allows us to develop the chemical dissipation (23) in the form:

$$\varphi^c = -\mathcal{A} \times \frac{d\phi^c}{dt} \geq 0; \mathcal{A} = \left(\frac{\rho}{\mathcal{M}}\right)_{\mathrm{CH}} \Delta G_{\mathrm{CH}} \quad (28)$$

Following standard thermodynamics, the dissolution kinetics is described by a law that links the "mechanical" macro-affinity $\mathcal{A}$ to the the change of porosity. For calcium leaching, it has been recognized that the dissolution occurs much faster compared to the diffusion process of calcium through the porosity, so that the dissolution can be considered as instantaneous. Such an instantaneous dissolution process is captured by the following dissolution law:

$$\mathcal{A} \leq 0; \frac{d\phi^c}{dt} \geq 0; \mathcal{A} \times \frac{d\phi^c}{dt} = 0 \quad (29)$$

Implemented in numerical simulation tools (finite elements or finite volumes), such an instantaneous dissolution law leads to sharp dissolution fronts (Mainguy & Coussy 2000).

3.2.3 *Level III – chemical softening of friction and interface properties of mortar*

The effect of calcium leaching on the mechanical behavior of mortar (level III) is quite different. The material, at this scale, is not a porous material in the sense of the poromechanics theory, but rather composed of a continuous cement paste matrix and (almost) rigid inclusions that have a common interface: the Interfacial Transition Zone (ITZ). Figure 6 displays SEM-images of the ITZ in the vicinity of a sand grain in a leached mortar; prior to load application (Top), and after high confinement loading (Bottom). It is readily understood from these figures that the main effect of calcium leaching on the mechanical behavior relates to the role of inclusions and of the interface properties. While many researchers have addressed by means of linear homogenization models the effect of inclusions and the ITZ properties on the elastic stiffness of mortars and concrete (Garboczi 1993; Li et al. 1999), the study of the strength properties by means of nonlinear continuum micromechanics is more recent (Lemarchand et al. 2002; Heukamp et al. 2002). The underlying idea of nonlinear micro-mechanics is to represent the real nonlinear behavior of the composite phases and the composite by a secant representation, to homogenize the secant moduli, and analyze the homogenization result for infinite strains that define the strength limit case. We owe this secant method

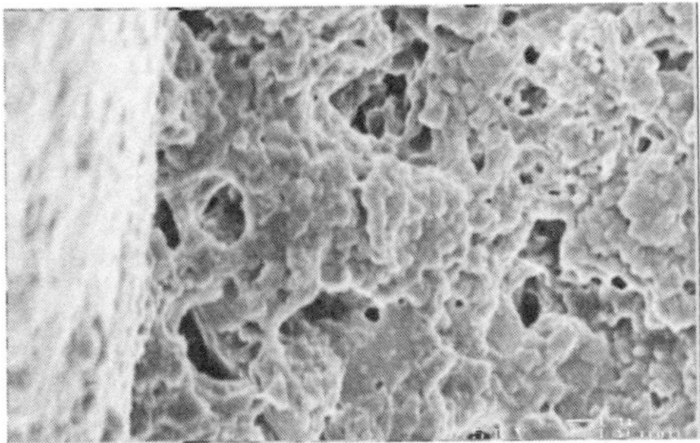

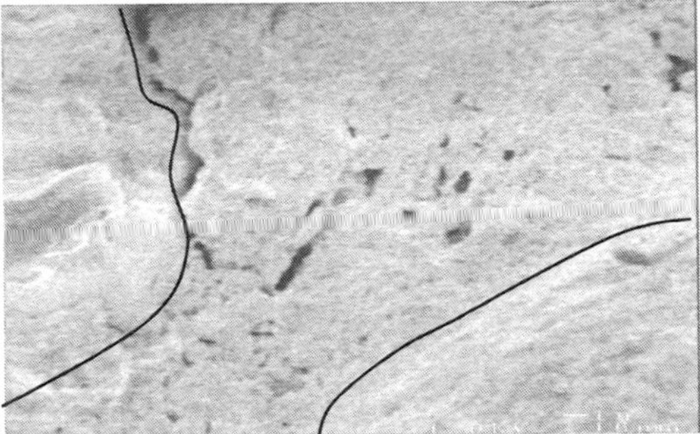

Figure 6. The Interfacial Transition Zone in leached mortars: Top – before load application; Bottom – after high pressure confinement (Heukamp & Ulm 2002).

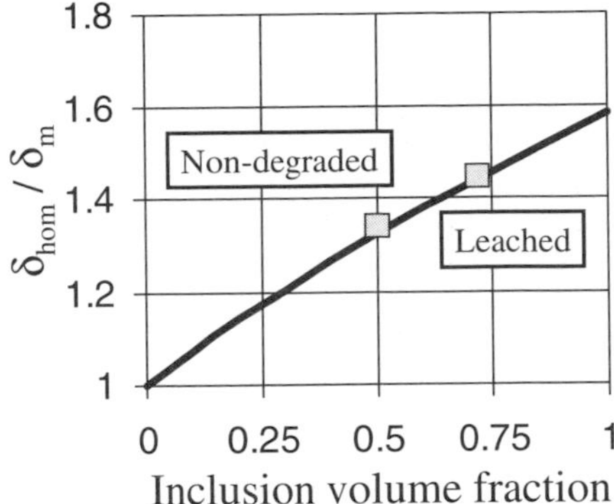

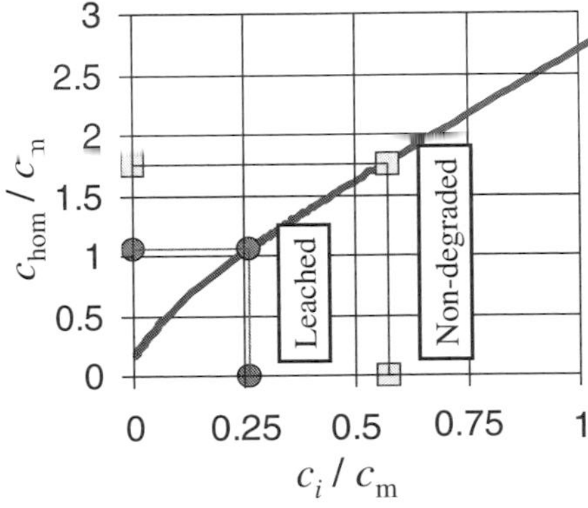

Figure 7. Results from nonlinear continuum micromechanics: Top – The effect of inclusions on the friction coefficient of highly filled composite materials (Lemarchand et al. 2002); Bottom – The effect of interfacial properties on the cohesion of highly filled composite materials (Heukamp et al. 2002b).

to Suquet (1998), and further developments of the method by Dormieux et al. (2002). Applied to mortar (level III) and concrete (level IV), or more generally to pressure sensitive materials, we distinguish:

- At high confinement, Figure 6 (Bottom) shows that the ITZ vanishes during load application. The composite material, in its limit state, is represented as a two phase material system of a deformable matrix and rigid inclusions. For this two-phase system, application of the secant method delivers estimates of the frictional enhancement of highly filled cementitious materials (Lemarchand et al. 2002):

$$\frac{\delta_{\text{hom}}}{\delta_m} = \sqrt{1 + \frac{3}{2} f_s} \tag{30}$$

where $f_s = V_s/|V_{III}|$ is the inclusion (sand) volume fraction, δ_{hom} is the composite friction coefficient, and δ_m the friction coefficient of the (cement paste) matrix. Relation (30) predicts an increase of the friction coefficient of the matrix due to the presence of rigid inclusions. This is readily understood from Figure 6 (Bottom): the presence of rigid inclusions lead to a local confinement of the deformable matrix that is much higher than the macroscopic confinement applied to the composite. Figure 7 (Top) displays the predictive capability of (30) for unleached and leached mortar. The experimental values of the friction coefficients of mortar and cement paste were determined from triaxial compressive tests (Heukamp et al. 2001; Heukamp et al. 2002a; Ulm et al. 2002; Heukamp & Ulm 2002).
- At low confinement, the cohesion properties of mortars and concrete are governed by the interfacial properties. Application of the secant method to a three phase representation of the composite materials, delivers the cohesion of the composite as a function of the volume fractions of the inclusions, f_s, and the ITZ, f_i, and the interface-to-matrix cohesion ratio $\chi = c_i/c_m$; that is (Heukamp et al. 2002b; Heukamp & Ulm 2002):

$$\frac{c_{\text{hom}}}{c_m} = \mathcal{F}(f_s, f_i, c_i/c_m) \tag{31}$$

While relation (31) does not permit an analytical representation, the inverse application of the model allows one to assess the chemical softening of the interfacial properties of mortar and concrete. This is displayed in Figure 7 (Bottom). The input parameters are the volume fractions of the inclusions and of the interface (the first is known from the mix design, here $f_s = 0.5$, the second from the SEM-investigation, here $f_i = 0.3$), and the mortar-to-cement paste cohesion ratio that can be determined from uniaxial compression and uniaxial tension test results (here $(c_{\mathrm{hom}}/c_m)^0 = 1.76$ for the non-degraded materials, and $(c_{\mathrm{hom}}/c_m)^\infty = 1.05$ for the asymptotically leached materials). Use of these values in relation (31) provides an assessment of the interface-to-matrix cohesion ratio $\chi = c_i/c_m$ for the two-asymptotic states:

$$\chi^0 = 0.57; \quad \chi^\infty = 0.26 \tag{32}$$

The values highlight the strong chemical softening of the ITZ-cohesion, which is much higher than the chemical softening of the matrix. This is readily understood from the chemical composition of the ITZ, which is known to be a zone of higher Portlandite concentration. This leads, after leaching, to a material that is much more sensitive, at failure, to the ITZ properties.

In summary, nonlinear continuum micromechanics provides a rational means of determining the asymptotic states of the chemical softening of the strength parameters undergone by mortar (level III) and concrete (level IV) in the course of calcium leaching.

3.2.4 *Summary: multiscale chemomechanics of calcium leaching of concrete*

The underlying idea of the chemomechanics approach for calcium leaching presented here is to (1) experimentally and theoretically assess at multiple scales the two asymptotic states of aging of cement-based materials, i.e. the initial intact non-degraded material and final homogeneously leached material states, and (2) to bridge these asymptotic states by means of a multiscale constitutive modeling of the chemomechanical deterioration kinetics. We borrow this approach from physical chemistry, in which thermodynamic equilibrium states are assessed, but employ it here for chemomechanical properties and deformation behavior of the solid.

The multiscale representation of cementitious materials suggests the following three chemomechanical equilibrium states: the intact material, an intermediate state defined by the dissolution of Portlandite at level II, and finally the asymptotically leached material defined by the decalcification of the C–S–H matrix at level I. Table 2 summarizes the poroelastic properties of these three material states for a $w/c = 0.5$ cementitious material. The input parameters are only

Table 2. Summary of the values for the poroelastic properties for a $w/c - 0.5$ cementitious material (Heukamp & Ulm 2002): k = bulk modulus, μ = shear modulus, b = Biot coefficient, M = Biot modulus, B = Skempton coefficient.

Properties	Intact	Intermediate	Leached
Level I			
k^I_{hom} [GPa]	15.2	15.2	2.7
μ^I_{hom} [GPa]	9.6	9.6	1.7
Level II			
k^{II}_{hom} [GPa]	15.4	11.2	1.7
μ^{II}_{hom} [GPa]	9.4	7.3	1.1
b^{II} [1]	0.07	0.26	0.37
M^{II} [GPa]	1170	124	20
B^{II} [GPa]	0.32	0.32	0.98
Level III			
K^{III}_{hom} [GPa]	15.1	12.3	2.6
μ^{III}_{hom} [GPa]	10.2	8.6	1.9
b^{III} [1]	0.04	0.19	0.35
M^{III} [GPa]	2340	247	40.4
B^{III} [GPa]	0.15	0.17	0.61

Table 3. Summary of the values for the strength properties. The asymptotic values for level II and III were determined experimentally, while the intermediate states and the level I values were determined with nonlinear continuum micromechanics (Heukamp & Ulm 2002): δ = friction coefficient; c = cohesion.

Properties	Intact	Intermediate	Leached
High Confinement			
Level I			
δ [1]	0.85	0.85	0.71
c [GPa]	18.01	18.01	1.26
Level II			
δ [1]	0.82	0.71	0.56
c [GPa]	17.11	14.5	1.15
Level III			
δ [1]	1.02	0.94	0.81
c [GPa]	9.82	9.13	0.96
Low Confinement			
Level I			
δ [1]	1.67	1.67	1.26
c [GPa]	2.2	2.2	1.05
Level II			
δ [1]	1.62	1.39	0.99
c [GPa]	2.09	1.8	0.79
Level III			
δ [1]	1.43	1.34	0.91
c [GPa]	3.67	3.41	0.83

nanoindentation results and volume fractions of the involved phases. Similarly, Table 3 summarizes the strength parameters of the material. Combined with chemoporoplasticity models that allow modeling the

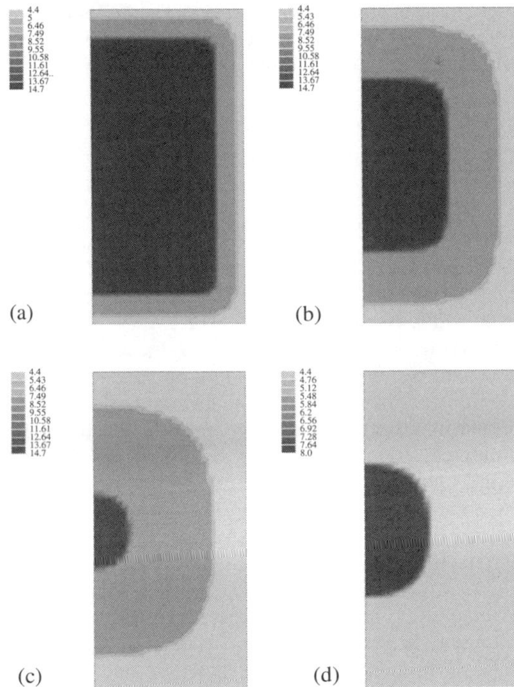

Figure 8. Results from the leaching calculation with the 6M ammonium nitrate solution. (a) 7days, (b) 35 days, (c) 84 days and (d) 182 days of leaching. The result values are solid calcium concentrations in mol/l in the mid-span beam section [Results with CESAR-LCPC@ MIT]; (Heukamp & Ulm 2002).

simultaneous chemical damage, chemical softening and plastic softening (Coussy & Ulm 1996; Ulm et al. 1998; Ulm et al. 1999b; Heukamp & Ulm 2002), the prediction and anticipation of the deterioration of the load bearing capacity of materials and structures subjected to calcium leaching becomes possible.

By way of application, Figures 8 and 9 display results from model-based simulations of residual 4-point bending strength tests presented by Schneider & Chen (1998). The tests were performed on mortar beams produced at a water-cement ratio of $w/c = 0.5$ from a Type I Portland cement and a fine sand. The beams have a size of 40 mm× 40 mm × 160 mm, and were subjected to accelerated calcium leaching. At different times between 7 days and 821 days, specimens were taken from the bath and tested in four-point bending. The results are reported by Schneider & Chen as a plot of the ratio β/β_{28} versus time. β is the maximum equivalent tensile stress in the beam calculated under the assumption of a linear stress distribution over the beam section:

$$\beta = \frac{6M}{bh^2} \tag{33}$$

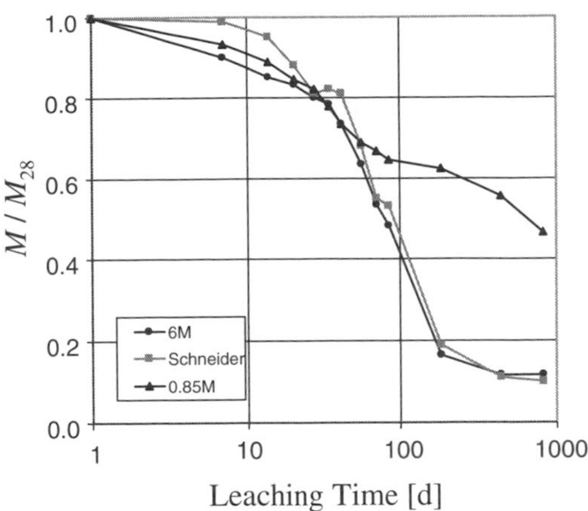

Figure 9. Model based simulations of the change in bending strength with two different leaching boundary conditions in comparison with the experimental data by Schneider & Chen (1998); [simulation results with CESAR-LCPC@MIT].

where M is the bending moment, b and h are the width and height of the beam section, respectively. The index 28 refers to the strength measured on intact specimens after 28 days of curing. The β_{28} value is 8.3 MPa, corresponding to a bending moment of 88.53 Nm. Figure 8 displays the results of the two-front dissolution-diffusion simulation for different leaching times; and 9 displays the normalized bending moment capacity as a function of the leaching time, for two molar concentrations of the ammonium nitrate solution: 0.85 M and 6 M. Schneider & Chen (1998) reported to have used a 10% ammonium nitrate solution, corresponding to a concentration of 68 g NH_4NO_3 per kilogram of solution, that is a 0.85 M ammonium nitrate solution. However, the model-based chemomechanics simulation results indicate that the one which matches best the overall load-bearing capacity in time is leaching in a 6M ammonium nitrate solution. Given the almost perfect correlation of experimental versus model-based simulation results of the bending strength decay, it is likely that the actual leaching conditions of Schneider & Chen's test may well correspond to a 6 M solution.

4 CONCLUDING REMARKS: FROM CHEMOMECHANICS TO BIO-CHEMOMECHANICS

Concrete, like many other materials (whether man-made, geological or biological) are highly heterogeneous materials, with heterogeneities that manifest themselves at multiple scales. As new experimental techniques such as nanoindentation have provided unprecedented access to micro-mechanical properties of materials, it becomes possible to identify the mechanical effects of chemical reactions at the micro-scale, where the reactions occur, and trace these micro-chemomechanical effects through upscaling techniques to the macro-scale. The identification of the length scales where physical chemistry *meets* (solid) mechanics appears to us as a key to prediction and anticipation of material deterioration. This approach is not restricted to geomaterials, but equally applies to biological materials, and biodegradation processes in which cell-mediated biological processes interact with solid matter affecting the mechanical performance of the solid.

While the link between biological processes and chemical effects has been a focus of biochemistry, the integration of biological processes into a consistent framework of chemomechanics is challenging. The main difficulty arises from the very nature of biological processes: Biological processes are dynamic in nature, and not defined with respect to an equilibrium state, in contrast to both mechanical processes and chemical processes. In addition, the absence of biological conservation laws complicates the direct integration of biological processes into the constitutive modeling of the material behavior.

On the other hand, the consideration of the solid material system as thermodynamic system provides a means of addressing this problem. In this case, the solid witnesses the biological activity through the biochemical conditions generated at the solid surface where cells or biological microorganisms attach. This is, in a nutshell, the way by which micro-chemomechanics can be extended to bio-chemo-mechanics (Silva & Ulm 2002), and leads to a similar split of the overall behavior of the porous materials as in chemomechanics: the purely mechanical response is captured by the poromechanics theory (i.e. Eq. (14)), and the bio-chemomechanical response by the moving boundary problem; that is analogously to relation (16):

$$\varphi^c = -\mathcal{A} \circ \frac{d\phi^c}{dt} = \frac{1}{|V|}\int_\Gamma A \times J_N d\Gamma \geq 0 \tag{34}$$

with

$$A = \mu^{\mathrm{BGP}} - \mu_c - \left(\frac{\mathcal{M}}{\rho_c}\right)(\psi_s^{el} + p) \tag{35}$$

where μ^{BGP} is the biological generated potential on surface $\Gamma \subseteq \partial V_s$ where biological organisms attach. The pure biochemical driving force $\Delta G = \mu^{\mathrm{BGP}} - \mu_c$ is given by biochemistry, while the additional terms $\psi_s^{el} + p$ relate to the solid deformation and the liquid pressure prevailing in the porosity. The porosity generated by the biochemical activity follows from (6), and can be developed in the form:

$$\frac{d\phi^{bc}}{dt} = \frac{\mathcal{N}}{\mathcal{L}_p} \times \frac{1}{\Gamma}\int_\Gamma -\underline{u}^{bc} \cdot \underline{n}_s d\Gamma \tag{36}$$

where $\mathcal{L}_p = |V|/|\partial V_s|$ is the total volume to solid surface ratio of the porous material; $\mathcal{N} = \Gamma/|\partial V_s|$ is the solid surface fraction occupied by cells or microorganisms, while the quantity $\frac{1}{\Gamma}\int_\Gamma -\underline{u}^{bc} \cdot \underline{n}_s d\Gamma$ represents the average biologically induced solid resorption or deposition activity of cells or microorganisms. While the first parameter $\mathcal{L}_p$ relates to the morphology of the porous material, the second parameter $\mathcal{N}$ relates to the biological population density, and provides the link with biological population models. Finally, the third quantity relates to the average biochemical activity, eventually amplified by the strain energy in the solid and the interstitial pore pressure (see Eq. (35)). While the average strain energy is generally some orders of magnitude smaller than the biological generated potential, In return, the strain energy in the immediate surrounding of cracks increases due to stress concentrations to a value that often compares to the one of the biologically generated potential. This phenomenon has been suggested as one possible origin of the self-healing adaptive capacity of living tissues as bones (Silva & Ulm 2002), which are remodeled continuously during adulthood through the resorption of old bone by Osteoclasts and the subsequent formation of new bone by Osteoblasts (see Figure 10): In the course of random remodeling events by cells, the chemomechanical coupling is a nonrandom remodeling stimulus, initiating the repair of damage in bone, which at the same time reduces by resorption the risk of crack propagation.

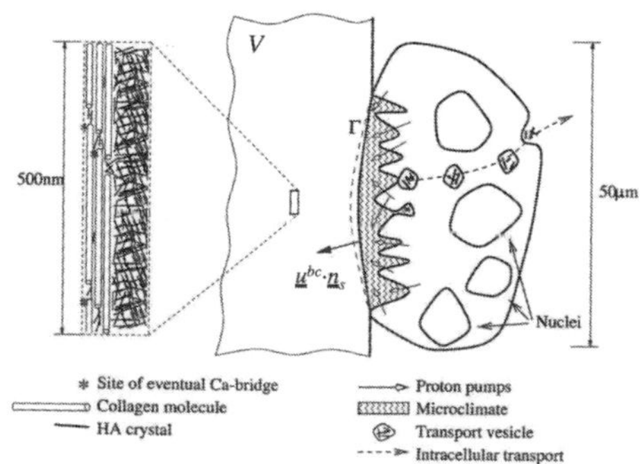

Figure 10. Biologically mediated chemomechanical dissolution process: osteoclast resorption on a bone surface (Silva & Ulm 2002).

ACKNOWLEDGMENT

Financial support of this study by the Nuclear Energy Research Initiative of the US-Department of Energy and the Esther & Harold E. Edgerton Chair at M.I.T. is gratefully acknowledged. Emilio Silva acknowledges the support through the Professor Stephen D. Senturia Fellowship for interdisciplinary studies at M.I.T. We are also grateful to the support of Dr. Christian Hellmich and Dr. Olivier Bernard through postdoctoral fellowships by the Kade Foundation, New York, administered through the Austrian Academy of Sciences, and by the Swiss Science Foundation. Finally, the results presented in this paper would not have been achieved without the fruitful collaboration of the MIT-team with Prof. Luc Dormieux (ENPC-LMSGC, France) and Prof. Olivier Coussy (Institute de Navier, France). Last, a special thanks to the CESAR-LCPC development team in Paris for maintenance of CESAR-LCPC@MIT.

REFERENCES

Acker, P. (1988). "Mechanical behavior of concrete: a physico-chemical approach", *Research report*, **LPC 152**, Laboratoire Central des Ponts et Chaussées, Paris [in French].

Acker, P. (2001). "Micromechanical analysis of creep and shrinkage mechanisms", In: F.-J. Ulm, Z.P. Bažant and F.H. Wittmann (Eds.), *Creep, shrinkage and durability mechanics of concrete and other quasi-brittle materials*, Proc. of the Sixth International Conference CONCREEP-6@MIT, Elsevier, 15–25.

Adenot, F. (1992). "Durabilité du béton : caractérisation et modélisation des processus physique et chimiques de dégradation du ciment", *PhD Dissertation*, Université d'Orléans, France.

Bernard, O., Ulm, F.-J., Lemarchand, E. (2002). "A multi-scale micromechanics–hydration model for the early-age elastic properties of cement-based materials", *Cement & Concrete Research*, In press.

Constantinides, G., Ulm F.-J. (2002a). "The elastic properties of calcium-leached cement pastes and mortars: a multi-scale investigation", *MIT-CEE Report*, **R02-01**, (MSc Dissertation, M.I.T.).

Constantinides, G., Ulm, F.-J. (2002b). "The effect of two types of C-S-H on the elasticity of cement-based materials: results from nanoindentation and micromechanical modeling", *Cement & Concrete Research*, submitted February 2002.

Cervera, M., Oliver, J., Prato, T. (1999a). "Thermo-chemomechanical model for concrete. I: hydration and aging", *Journal of Engineering Mechanics*, ASCE, **125**(9), 1018–1027.

Cervera, M., Oliver, J., Prato, T. (1999b). "Thermo-chemomechanical model for concrete. II: damage and creep, *Journal of Engineering Mechanics*, ASCE **125**(9), 1028–1039.

Coussy, O. (1995). *Mechanics of porous media*, J. Wiley & Sons, Chichester, UK.

Coussy, O., Ulm, F.-J. (1996). "Creep and plasticity due to chemo-mechanical couplings", *Archive of Applied Mechanics*, **66**, 523–535.

Dormieux, L., Molinari, A., and Kondo, D. (2002). "Micromechanical approach to the behavior of poroelastic materials." *J. Mech. Phys. Solids*, **29**(8), 2203–2231.

Garboczi, E.J. (1993). "Computational materials science of cement-based materials", *Materials and Structures*, RILEM, **26**(2), 191–195.

Hellmich, C., Ulm, F.-J., Mang, H.A. (1999a). "Multi-surface chemoplasticity. I: Material Model for shotcrete", *Journal of Engineering Mechanics*, ASCE, **125**(6), 692–701.

Hellmich, C., Ulm, F.-J., Mang, H.A. (1999b). "Consistent linearization in finite element analysis of coupled chemo-thermal problems with exo- or endothermal reactions", *Computational Mechanics*, **24**(4), 238–244.

Heukamp, F.H., Ulm, F.-J., Germaine, J.T. (2001). "Mechanical Properties of calcium leached cement pastes: Triaxial stress states and the influence of the pore pressure", *Cement & Concrete Research*, **31**(5), 767–774.

Heukamp, F.H., Ulm, F.-J. (2002). "Chemomechanics of calcium leaching of cement-based materials at different scales: The role of CH-dissolution and C-S-H degradation on strength and durability performance of materials and structures", *MIT-CEE Report*, **R02-03** (PhD Dissertation, M.I.T.).

Heukamp, F.H., Ulm, F,-J., Germaine, J.T. (2002a). "Poro-plastic properties of calcium leached cement-based materials", *Cement & Concrete Research*, In press.

Heukamp, F.H., Lemarchand, E., Ulm, F.-J. (2002b). "The effect of interfacial properties on the cohesion of highly filled composites", *Int. Journal of Solids and Structures*, submitted September 2002.

Jennings, H.M. (2000). "A model for the microstructure of calcium silicate hydrate in cement paste", *Cement & Concrete Research*, **30**(1), 101–116.

Larive, C. (1998). "A combined experimental-theoretical approach for the understanding of the alkali reaction and of its mechanical effects", *Monograph LPC*, **OA 28**, (Laboratoires des Ponts et Chaussées, Paris) [in French, partially translated into English].

Lemarchand, E., Ulm, F.-J., Dormieux, L. (2002). "Effect of inclusions on friction coefficient of highly filled composite materials", *Journal of Engineering Mechanics*, ASCE, **128**(8), 876–884.

Li, G.Q., Zhao, Y., Pang, S.S., Li, Y.Q. (1999). "Effective Young's modulus estimation of concrete", *Cement & Concrete Research*, **29**(9), 1455–1462.

Li, K. (2002). "Chemomechanics modeling of the behavior of concrete affected by the alkali-silica reaction and model-based assessment of degraded structures", *PhD Dissertation*, Ecole Nationale des Ponts et Chaussees, Paris, France [in French].

Li, K., and Coussy, O. (2002). "Concrete ASR degradation: From material modeling to structure assessment", *Concrete Science and Engineering*, RILEM, **4**(3), 35–46.

Mainguy, M., Coussy, O. (2000). "Propagation fronts during calcium leaching and chloride penetration", *Journal of Engineering Mechanics*, ASCE, **126**(3), 250–257.

Mori, T., Tanaka, K. (1973). "Average stress in matrix and average elastic energy of materials with misfitting inclusions", *Acta Metallurgica*, **21**(5), 1605–1609.

Powers, T.C., Brownyard, T.L. (1946). "Studies of the physical properties of hardened Portland cement paste", *Journal of the American Concrete Institute*, **18**, 101–132.

Schneider, U., Chen, S.-W. (1998). "The chemomechanical effect and the mechanochemical effect on high-performance concrete subjected to stress corrosion", *Cement & Concrete Research*, **28**(4), 509–522.

Sercombe, J., Ulm, F.-J., Mang, H.A. (2000a). "Consistent return mapping algorithm for chemoplastic constitutive laws with internal couplings", *Int. J. Numerical Methods in Engineering*, **47**, 75–100.

Sercombe, J., Hellmich, C., Ulm, F.-J., Mang, H.A. (2000b). "Modeling of early-age creep of shotcrete. I: Model and model parameters", *Journal of Engineering Mechanics*, ASCE, **126**(3), 284–291.

Silva, E.C., Ulm, F.-J. (2002). "A bio-chemomechanics approach to bone remodeling and fracture", In: B.L. Karihaloo (Ed.), Proc. IUTAM Symposium, *Analytical and Computational Fracture Mechanics of Non-Homogeneous Materials*, Cardiff June 2001, Kluwer Academic Publishers, Dordrecht–Boston–London, 355–366.

Suquet, P. (1997). "Effectives properties of nonlinear composites", In: P. Suquet (Ed.), *Continuum micro-mechanics*, **CISM 377**, Springer-Verlag, Wien.

Taylor, H.F.W. (1990). *Cement Chemistry*, Academic Press, New York.

Tennis, P.D., Jennings, H.M. (2000). "A model for two types of calcium silicate hydrate in the microstructure of portland cement paste", *Cement & Concrete Research*, **30**(6), 855–863.

Ulm, F.-J., Coussy, O. (1995). "Modeling of thermochemomechanical couplings of concrete at early ages". *Journal of Engineering Mechanics*, ASCE, **121**(7), 785–794.

Ulm, F.-J., Coussy, O. (1996). "Strength growth as chemoplastic hardening in early age concrete", *Journal of Engineering Mechanics*, ASCE, **122**(12), 1123–1132.

Ulm, F.-J., Coussy, O. (1998). "Couplings in early-age concrete: From material modeling to structural design", *Int. Journal of Solids and Structures*, **35**(31–32), 4295–4311.

Ulm, F.-J., Coussy, O., Hellmich, Ch. (1998). "Chemoplasticity: A review of evidence", In: R. de Borst, N. Bicanic, H. Mang and G. Meschke, Proc. Euro-C '98, *Computational Modelling of Concrete Structures*, Bad Gastein, Austria, March/April 1998, A.A. Balkema, Rotterdam, 421–439.

Ulm, F.-J., Coussy, O., Bažant, Z.P. (1999a). "The 'Chunnel' fire. I. Chemoplastic softening in rapidly heated concrete", *Journal of Engineering Mechanics*, ASCE, **125**(3), 272–282.

Ulm, F.-J., Torrenti, J.M., Adenot, F. (1999b) "Chemoporoplasticity of calcium leaching in concrete", *Journal of Engineering Mechanics*, ASCE, **125**(10), 1200–1211.

Ulm, F.-J., Coussy, O., Li, K. and Larive, C. (2000). "Thermo-chemo-mechanics of ASR-expansion in concrete structures", *Journal of Engineering Mechanics*, ASCE, **126**(3), 233–242.

Ulm, F.-J., Coussy, O. (2001). "What is a 'massive' concrete structure at early ages? – Some dimensional arguments", *Journal of Engineering Mechanics*, ASCE, **127**(5), 512–522.

Ulm, F.-J., Heukamp, F.H., Germaine, J.T. (2002a). "Residual design strength of cement-based materials for nuclear waste storage systems", *Nuclear Engineering and Design*, **211**(1), 51–60.

Ulm, F.-J., Peterson, M., Lemarchand, E. (2002b) "Is ASR-Expansion caused by chemoporoplastic dilatation?", *Concrete Science and Engineering*, RILEM, **4**(3), 47–55.

Ulm, F.-J., Lemarchand, E., Heukamp, F.H. (2002c). "Elements of chemomechanics of Calcium leaching of cement-based materials at different scales", *Journal of Engineering Fracture Mechanics*, In Press.

Velez, K., Maximilien, S., Damidot, D., Fantozzi, G., Sorrentino, F. (2001). "Determination by nanoindentation of elastic modulus and hardness of pure constituents of Portland cement clinker", *Cement & Concrete Research*, **31**(4), 555–561.

Wittmann, F.H. (1982). "Creep and shrinkage mechanisms", In: Z.P. Bažant & F.H. Wittmann (Eds.), *Creep and shrinkage of concrete structures*, J. Wiley and Sons, 129–161.

Zaoui, A. (2002). "Continuum micromechanics – a survey", *Journal of Engineering Mechanics*, ASCE, **128**(8), 808–816.

Computational Modelling of Concrete Structures, Bićanić et al. (eds)
© 2003 Swets & Zeitlinger, Lisse, ISBN 90 5809 536 3

Modeling of concrete deterioration by alkali-silica reaction

Falko Bangert, Detlef Kuhl & Günther Meschke
Institute for Structural Mechanics, Ruhr University Bochum, Germany

ABSTRACT: The paper is concerned with a macroscopic model for the description of concrete deterioration by alkali-silica reaction. Based on concepts of the Theory of Porous Media, concrete is regarded as a mixture of three superimposed and interacting constituents: the skeleton, the pore liquid and the pore gas. The skeleton in turn represents a mixture of the unreacted, un-swollen and the already reacted, swollen material. When ASR takes place, mass of the unreacted material is non-instantaneously converted into mass of the reacted material. Since the unreacted material is characterized by a smaller density, the material swells. In the proposed model the dependence of the kinetics and the magnitude of the swelling process on the moisture content is taken into account. The suitability of the model is investigated, at least partially, by means of one-dimensional parametric studies.

1 INTRODUCTION

The durability of concrete structures is significantly impaired by various types of chemical degradation processes. Recent progress in computational durability mechanics has opened the perspective for fundamental, numerical approaches to obtain not only estimates for the life-time of structures, but also to obtain a better insight into the influencing factors that are relevant quantities with regards to durability and those that are not. The present paper in concerned with the modeling of the alkali-silica reaction (ASR), which is one of the most hazardous chemical deterioration mechanisms in concrete. The alkali-silica reaction is a chemical reaction between alkali hydroxides, primarily generated from the alkalis in the cement, and reactive forms of silica within the aggregate particles, see e. g. (Glasser 1992). The reaction leads to alkali-silica gel as the reaction product. In the presence of water the gel swells, creating an increasing internal pressure. The swelling pressure causes excessive expansion, the opening and propagation of cracks as well as a drastic reduction of the mechanical properties.

Since the impact of the alkali-silica reaction is influenced by many factors (e.g. humidity, reactivity of silica), it is nearly impossible to make realistic predictions on a structural level solely based on the qualitative knowledge of the ASR chemistry. That is why in recent years several numerical models have been developed in order to characterize the observed behavior of ASR affected concrete on a material level or even a structural level. Depending on the level of observation these models follow either a mesoscopic or a macroscopic approach. A mesoscopic approach involves the analysis of a single representative aggregate particle and its vicinity, whereby the kinetics of the chemical and diffusional processes involved are described on the scale of the aggregates, see e.g. (Bažant and Steffens 2000). On the other hand, in a macroscopic approach concrete is described at the scale of laboratory specimens, see e.g. (Larive and Coussy 1996). In these models, the main characteristics of ASR are incorporated phenomenologically on the macroscopic level. Hence, they can directly be used for numerical analysis of concrete structures (Ulm et al. 2000).

In the present paper a macroscopic model for ASR affected concrete is developed. The paper is organized as follows: Section 2 gives an overview on experimental results concerning the deterioration mechanisms of concrete caused by the alkali-silica reaction. These results show, that the moisture content within the concrete has a dominant influence on the initiation and progress of ASR. Based on this observation, a coupled chemo-hygro-mechanical model is developed in Section 3 within the framework of the Theory of Porous Media. The respective material parameters of the model are fitted to experimental results in Section 4. The paper is closed with one-dimensional parametric studies in Section 5, which illustrate some important properties of the proposed model.

2 EXPERIMENTAL FINDINGS AND DETERIORATION MECHANISMS

Two main mechanisms constitute the alkali-silica reaction. Firstly, silica is dissolved from the aggregates, whereby a gel is formed and secondly, the swelling of the gel by imbibition of water, which results in the expansion and deterioration of the affected concrete.

2.1 *Dissolution of silica*

In all forms of silica the basic structural unit is a silicon ion (Si^{4+}) surrounded by four oxygen ions (O^{2-}) with the arrangement of a tetrahedron (Iler 1979). In crystalline, less reactive forms of silica, these tetrahedra are linked to produce a dense three-dimensional network. In amorphous silica, these tetrahedra are linked to produce a random, more open network with a large specific surface that exhibits substantially enhanced reactivity towards alkaline solutions like the pore liquid in concrete (Powers and Steinour 1955), which contains relatively high concentrations of hydroxyl ions (OH^-). The charge of these hydroxyl ions is mainly balanced by alkali ions (Na^+, K^+), which are usually provided by sodium oxide (Na_2O) and potassium oxide (K_2O) in the cement by dissolution during the process of hydration.

Considering the alkali-silica reaction as a multistage process, it starts with the dissolution of silica on the surface of the aggregate particles as a topochemical reaction. Dent Glasser and Kataoka identified two mechanisms that constitute the dissolution of silica (Dent Glasser and Kataoka 1981): In a first reaction the hydroxyl ions attack the siloxane groups

$$\equiv \mathrm{Si-O-Si} \equiv + \mathrm{OH^-} \rightarrow \equiv \mathrm{Si-O^-} + \mathrm{OH-Si} \equiv, \tag{1}$$

whereby silanol groups are formed. In a second step, these silanol groups react with further hydroxyl ions:

$$\equiv \mathrm{Si-OH} + \mathrm{OH^-} \rightarrow \equiv \mathrm{Si-O^-} + \mathrm{H_2O}. \tag{2}$$

The negative charges resulting from (1) and (2) attract positively charged sodium, potassium and calcium ions (Na^+, K^+, Ca^{2+}), which are present in the pore liquid. These ions diffuse towards the reaction site in sufficient numbers to balance the negative charges.

As more siloxane bonds are attacked by the dissolution process, a gel-like layer forms at the surface of the aggregates. Some silica may even pass into solution as monomeric species (H_4SiO_4, $H_3SiO_4^-$, $H_2SiO_4^{2-}$) depending on the pH value of the pore liquid (Dent Glasser and Kataoka 1981).

Dron and Brivot reproduced the dissolution mechanism of silica in experimental studies using various kinds of silica, which were stored in solutions of sodium hydroxide (Dron and Brivot 1993). Their results show, that the dissolution of silica can be interpreted as a chemical reaction, which follows a first-order kinetic law (Atkins 1998). This means that the dissolution rate decreases linearly from an initial value, where the reaction starts, to zero, at which the silica is completely dissolved. If a non-dimensional reaction extent $\xi \in [0, 1]$ is introduced as a chemical state variable (Coussy 1995), where $\xi = 0$ represents the beginning and $\xi = 1$ the end of the reaction, the kinetics of the dissolution process can be expressed by the following evolution equation:

$$\frac{\partial \xi}{\partial t} = k[1 - \xi]. \tag{3}$$

Herein, the parameter k denotes the velocity parameter, which depends on the reactivity of silica, the alkalinity of the pore liquid, the temperature and the available water. Integration of the evolution equation (3) yields:

$$\xi = 1 - \mathrm{e}^{-kt}. \tag{4}$$

2.2 *Swelling mechanism*

The gel which is formed by the aforementioned dissolution process is hydrophilic (Poole 1992). This means, that in a sufficiently humid environment the gel imbibes water, which results in the swelling of the gel. The gel may expand into pores and cracks in the cementitious skeleton. Once this free expansion is filled, the swelling is restrained and the gel exerts a pressure on the concrete skeleton.

The swelling pressure exerted by the gel is variable within a wide range depending on the humidity and the type and proportions of the reacting materials. But it is often sufficiently large to induce expansions in localized regions which, in turn, lead to the opening and propagation of cracks and to the disruption of the affected concrete. This results in a drastic reduction of the mechanical properties and consequently to structural degradation (Hobbs 1988).

By comparing the timescales of the swelling of synthetic gels on the one hand side (Struble and Diamond 1981) and of concrete specimens on the other hand side (Larive 1998), it can be concluded, that the imbibition of water by a gel is much faster than the progress of deterioration in ASR affected concrete. This leads to the conclusion, that the imbibition of water by the gel can be regarded as an instantaneous process in comparison to the formation of the gel by the dissolution of silica. Since in a multistage process the slowest process controls the overall kinetics, it is reasonable to assume, that the alkali-silica reaction (formation of gel + swelling of gel) is governed by the non-instantaneous kinetic law of the gel formation, which is represented by equation (3).

3 MACROSCOPIC MODEL FOR CONCRETE DETERIORATION BY ASR

In this section a macroscopic model for the numerical simulation of ASR affected concrete structures is developed, which is based on the Theory of Porous Media, see e.g. (Ehlers 2002; Lewis and Schrefler 1998).

3.1 *Constituents, volume fractions and densities*

It turns out from Section 2, that a sufficient amount of water is a prerequisite for the initiation of damage resulting from the alkali-silica reaction. Acting as the transport medium for ions and as a necessary compound for the formation of the swollen gel, water affects both kinetics and magnitude of ASR impact (Larive 1998). These dependencies must be taken into account in a model for ASR in order to provide realistic predictions of the concrete deterioration. Therefore, concrete is modeled as a partially saturated porous material, which is interpreted as a mixture of three superimposed and interacting constituents φ^α, namely the non-porous skeleton (index $\alpha = s$), the pore liquid (index $\alpha = l$) and the pore gas (index $\alpha = g$). The skeleton itself is a mixture of the aggregate particles, the gel resulting from the dissolution of silica and the cement paste.

Following the standard concepts of the Theory of Porous Media, it is assumed, that the constituents φ^α are homogenized over a representative volume element, which is occupied by the mixture $\varphi = \varphi^s + \varphi^l + \varphi^g$. Therefore, material points of each constituent φ^α exist at each geometrical point $\boldsymbol{x}$. Hence, the local composition of the mixture φ is described by the volume fraction ϕ^α, which is defined as the ratio of the volume element dv^α occupied by the individual constituent φ^α and the volume element dv occupied by the mixture φ:

$$\phi^\alpha = \frac{dv^\alpha}{dv}. \tag{5}$$

It follows from definition (5), that the saturation condition must hold:

$$\phi^s + \phi^l + \phi^g = 1. \tag{6}$$

Finally, the material density ϱ^α and the partial density ρ^α of the constituent φ^α are introduced:

$$\varrho^\alpha = \frac{dm^\alpha}{dv^\alpha}, \quad \rho^\alpha = \frac{dm^\alpha}{dv} = \frac{dv^\alpha}{dv}\frac{dm^\alpha}{dv^\alpha} = \phi^\alpha \varrho^\alpha. \tag{7}$$

Herein, dm^α denotes the local mass of the volume element dv^α. It should be noted, that a material incompressibility of any constituent φ^α ($\varrho^\alpha = \text{const.}$) does not result in incompressibility of the respective constituent on the macroscopic level ($\rho^\alpha = \text{const.}$), since the volume fraction ϕ^α may change.

3.2 *Kinematics*

Starting from different reference positions $\boldsymbol{X}_\alpha$ at time t_0, each constituent φ^α follows its individual motion

$$\boldsymbol{x} = \chi_\alpha(\boldsymbol{X}_\alpha, t) \tag{8}$$

with the velocity:

$$\boldsymbol{x}'_\alpha = \frac{\partial \chi_\alpha(\boldsymbol{X}_\alpha, t)}{\partial t}. \tag{9}$$

The motion of the skeleton can be described in a Langrangean setting by introduction of the displacement vector of the skeleton:

$$\boldsymbol{u}_s = \boldsymbol{x} - \boldsymbol{X}_s. \tag{10}$$

By extension of the usual Eulerian description, the motion of the pore fluids is described relatively to the deforming skeleton by the so called seepage velocities:

$$\boldsymbol{w}_l = \boldsymbol{x}'_l - \boldsymbol{x}'_s, \qquad \boldsymbol{w}_g = \boldsymbol{x}'_g - \boldsymbol{x}'_s. \tag{11}$$

Restricting the presentation to the geometrically linear case, the spatial derivative with respect to the actual position $\boldsymbol{x}$ and the reference position $\boldsymbol{X}_\alpha$ are approximately equivalent:

$$\begin{aligned} \text{grad}(\bullet) &= \frac{\partial \bullet}{\partial \boldsymbol{x}} \approx \text{Grad}_\alpha(\bullet) = \frac{\partial \bullet}{\partial \boldsymbol{X}_\alpha}, \\ \text{div}(\bullet) &= \text{tr}\left(\frac{\partial \bullet}{\partial \boldsymbol{x}}\right) \approx \text{Div}_\alpha(\bullet) = \text{tr}\left(\frac{\partial \bullet}{\partial \boldsymbol{X}_\alpha}\right). \end{aligned} \tag{12}$$

Therefore, the linear strain tensor

$$\boldsymbol{\varepsilon}_s = \tfrac{1}{2}[\text{grad}(\boldsymbol{u}_s) + \text{grad}(\boldsymbol{u}_s)^T] \tag{13}$$

is used.

3.3 *Balance equations*

In the following, the balance of mass and the balance of linear momentum are specified for each constituent φ^α.

3.3.1 *Mass balance equations*

Investigations on the role of water in the alkali-silica reaction have shown, that reactive concrete specimens do not absorb significantly more water than non-reactive ones, when they are stored under the same environmental conditions (Larive 1998; Larive et al. 2000). Thus, no specific model needs to be developed to predict water movement in ASR affected concrete and it is reasonable to neglect any mass exchange between the skeleton and the pore fluids. In doing so, the mass balance equation of the skeleton reads (Ehlers 2002)

$$(\phi^s \varrho^s)'_s + \phi^s \varrho^s \operatorname{div}(\boldsymbol{x}'_s) = \frac{\partial[\phi^s \varrho^s]}{\partial t} + \operatorname{div}(\phi^s \varrho^s \boldsymbol{x}'_s) = 0, \tag{14}$$

where $(\bullet)'_\alpha = \partial(\bullet)/\partial t + \operatorname{grad}(\bullet) \cdot \mathbf{x}'_\alpha$ denotes the material time derivative of the quantity $(\bullet)$ following the individual motion of the respective constituent φ^α. It can be verified, that in the geometrically linear case equation (14) can be re-written in the form

$$\frac{\partial[\phi^s \varrho^s]}{\partial t} + \phi_0^s \varrho_0^s \operatorname{div}(\boldsymbol{x}'_s) = 0, \tag{15}$$

where ϕ_0^s and ϱ_0^s are the initial volume fraction and initial density of the skeleton, respectively. Since in most structural engineering applications it is appropriate to emanate from small strains and small strain rates, the term

$$\operatorname{div}(\boldsymbol{x}'_S) = \operatorname{tr}\left(\frac{\partial \boldsymbol{\varepsilon}_s}{\partial t}\right) \approx 0 \tag{16}$$

can be neglected, yielding:

$$\frac{\partial[\phi^s \varrho^s]}{\partial t} = 0. \tag{17}$$

By integration of equation (17) one obtains for the volume fraction of the skeleton:

$$\phi^s = \frac{\varrho_0^s}{\varrho_s} \phi_0^s. \tag{18}$$

As already mentioned, the skeleton itself represents a mixture of the aggregates, the gel and the cement paste. When the alkali-silica reaction takes place, mass from the aggregate phase is converted into mass of the gel phase by the non-instantaneous dissolution of silica (see Section 2.1). This process is followed by the instantaneous swelling of the gel (see Section 2.2), which results in the swelling of the skeleton φ^s and finally of the whole mixture φ. The swelling of the skeleton can be described mathematically by a variation of the material density ϱ^s. Hence, an evolution equation for the material density ϱ^s is formulated next.

In order to keep the number of model parameters small and the model formulation simple, the skeleton is regarded as a binary mixture $\varphi^s = \varphi^u + \varphi^r$ of an unreacted phase φ^u and a reacted phase φ^r without making any distinction between aggregates, gel and cement paste. The unreacted phase φ^u represents the unreacted, unswollen skeleton material *before* it was affected by the alkali-silica reaction. On the other hand the reacted phase φ^r represents the reacted, swollen skeleton material *after completion* of ASR. Both constituents φ^u and φ^r have one common kinematic variable, the skeleton velocity $\boldsymbol{x}'_s$. Accordingly, the partial density of the skeleton ρ^s is assumed to by composed by:

$$\rho^s = \phi^s \varrho^s = \phi^u \varrho^u + \phi^r \varrho^r, \qquad \phi^s = \phi^u + \phi^r. \tag{19}$$

Thereby, the material densities $\varrho^u > \varrho^r$ are constant under iso-hydro-thermal conditions. In other words, the constituents φ^u and φ^r are assumed as materially incompressible. Since $\varrho^u = \text{cont.}$ and $\varrho^r = \text{const.}$, the associated mass balance equations of the constituents φ^u and φ^r result into the following volume balance equations:

$$\begin{aligned} \frac{\partial \phi^u}{\partial t} + \phi_0^u \operatorname{div}(\boldsymbol{x}'_s) &= \frac{\partial \phi^{u \to r}}{\partial t}, \\ \frac{\partial \phi^r}{\partial t} + \phi_0^r \operatorname{div}(\boldsymbol{x}'_s) &= \frac{\partial \phi^{r \leftarrow u}}{\partial t}. \end{aligned} \tag{20}$$

The terms $\varrho^u \partial\phi^{u\to r}/\partial t$ and $\varrho^r \partial\phi^{r\leftarrow u}/\partial t$ represent the mass exchange between the phases φ^u and φ^r due to the dissolution process. The conservation of mass enforces:

$$\varrho^u \frac{\partial \phi^{u\to r}}{\partial t} + \varrho^r \frac{\partial \phi^{r\leftarrow u}}{\partial t} = 0. \tag{21}$$

Since, according to equation (3) the kinetics of the dissolution process and of the mass exchange follows a first order kinetic law one may write (Atkins 1998)

$$\frac{\partial \phi^{u\to r}}{\partial t} = -k\,\phi^u, \qquad \frac{\partial \phi^{r\leftarrow u}}{\partial t} = -\frac{\varrho^u}{\varrho^r}\frac{\partial \phi^{u\to r}}{\partial t} = \frac{\varrho^u}{\varrho^r} k\,\phi^u, \tag{22}$$

whereby k denotes the velocity parameter, which strongly depends on the moisture content, see Sections 2.1 and 4.

Inserting $(22)_1$ and (16) into the volume balance equation of the unreacted material φ^u $(20)_1$ yields:

$$\frac{\partial \phi^u}{\partial t} = -k\,\phi^u. \tag{23}$$

After integration, one obtains

$$\phi^u = \phi_0^u\, \mathrm{e}^{-kt}, \tag{24}$$

with the initial volume fraction of the skeleton $\phi_0^u = \phi_0^s$. Furthermore, inserting $(22)_2$ and (16) into the volume balance equation of the reacted material φ^r $(20)_2$

$$\frac{\partial \phi^r}{\partial t} = \frac{\varrho^u}{\varrho^r} k\,\phi^u = \frac{\varrho^u}{\varrho^r} k\,\phi_0^u\, \mathrm{e}^{-kt} \tag{25}$$

yields after integration and using the initial condition $\phi_0^r = 0$:

$$\phi^r = \frac{\varrho^u}{\varrho^r}\,\phi_0^u\,[1 - \mathrm{e}^{-k\,t}]\,. \tag{26}$$

Now, the non-dimensional reaction extent ξ defined by equations (3) and (4) can be inserted in the volume fractions of the constituents φ^u and φ^r given by (24) and (26):

$$\phi^u = \phi_0^u\,[1-\xi]\,, \qquad \phi^r = \frac{\varrho^u}{\varrho^r}\,\phi_0^u\,\xi\,. \tag{27}$$

Inserting (27) into (19) yields an evolution equation for the material density of the skeleton ϱ^s as a function of the reaction extent ξ:

$$\varrho^s = \frac{\varrho^u}{1+\xi\left[\frac{\varrho^u}{\varrho^r}-1\right]}\,. \tag{28}$$

Analogous to equation (15), the mass balance equations of the pore fluids φ^β (index $\beta = l \rightarrow$ liquid phase, index $\beta = g \rightarrow$ gas phase) are given in a geometrically linear setting by

$$\frac{\partial[\phi^\beta\varrho^\beta]}{\partial t} + \phi_0^\beta\varrho_0^\beta\,\mathrm{div}(\boldsymbol{x}'_\beta) = 0\,, \tag{29}$$

with the initial values of the volume fraction ϕ_0^β and the material density ϱ_0^β. It makes sense to neglect the material compressibility of the pore liquid in comparison to the material compressibility of the pore gas ($\rightarrow \varrho^l = \varrho_0^l =$ const.). With this assumption and using the seepage velocities (11) and the assumption (16) one obtains from (29):

$$\frac{\partial\phi^l}{\partial t} + \phi_0^l\,\mathrm{div}(\boldsymbol{w}_l) = 0\,, \qquad \frac{\partial[\phi^g\varrho^g]}{\partial t} + \phi_0^g\varrho_0^g\,\mathrm{div}(\boldsymbol{w}_g) = 0\,. \tag{30}$$

3.3.2 *Momentum balance equations*

The partial momentum balances for the quasi-static case with the body forces neglected are given by:

$$\mathrm{div}(\boldsymbol{\sigma}^\alpha) + \hat{\boldsymbol{p}}^\alpha = 0\,. \tag{31}$$

Herein, $\boldsymbol{\sigma}^\alpha$ is the partial stress tensor and $\hat{\boldsymbol{p}}^\alpha$ the momentum production, which can be interpreted as the local interaction force per unit volume between φ^α and the other constituents. Thereby the following constraint

$$\hat{\boldsymbol{p}}^s + \hat{\boldsymbol{p}}^l + \hat{\boldsymbol{p}}^g = \mathbf{0} \tag{32}$$

must hold due to the overall conservation of momentum

$$\mathrm{div}(\boldsymbol{\sigma}) = 0\,, \tag{33}$$

with the overall stress tensor $\boldsymbol{\sigma} = \boldsymbol{\sigma}^s + \boldsymbol{\sigma}^l + \boldsymbol{\sigma}^g$.

3.4 *Constitutive laws*

After defining the saturation functions s^β for the pore fluids and the capillary pressure p^c, constitutive laws for the capillary pressure p^c, the stresses $\boldsymbol{\sigma}^\alpha$, the gas pressure p^g and the momentum productions $\hat{\boldsymbol{p}}^\beta$ are derived in the following.

3.4.1 *Saturation functions and capillary pressure*

The pore space is partially saturated with liquid and partially with gas. The degree of liquid and gas saturation s^β, respectively, is given by:

$$s^\beta = \frac{\phi^\beta}{\phi^l + \phi^g} = \frac{\phi^\beta}{1-\phi^s}\,. \tag{34}$$

In the capillary pores, the pore liquid and the pore gas are separated by a curved interface (meniscus). The radius of curvature of this interface depends on the pressure jump across the interface expressed by the so called capillary pressure p^c:

$$p^c = p^g - p^l\,. \tag{35}$$

In what follows, however, the capillary pressure p^c will be interpreted as macroscopic pressure representing all hygrally induced stresses acting on various scales of the nano-porous cementitious material, see e.g. (Bear and Bachmat 1991). The macroscopic capillary pressure p^c depends on the liquid saturation s^l. Here, the following expression for the capillary pressure p^c as a function of the liquid saturation s^l is used

$$p^c = p^r\left[\left[s^l\right]^{-\frac{1}{m}} - 1\right]^{\frac{1}{n}}, \tag{36}$$

which has been originally proposed for soils by (van Genuchten 1980). In equation (36), p^r, n, m, denote material parameters, which have to be determined experimentally. For concrete these parameters have been determined by (Mainguy et al. 2001).

3.4.2 *Stress tensors*

From general thermodynamical considerations, it turns out, that the stress state of the skeleton and the fluid constituents is usually separated into two parts, where the first part is governed by the skeleton deformation, while the second part is governed by the pore pressures (Ehlers 2002):

$$\begin{aligned} \boldsymbol{\sigma}^s &= \boldsymbol{\sigma}^{s'} - \phi^s\,p\,\mathbf{1}\,, \\ \boldsymbol{\sigma}^\beta &= \boldsymbol{\sigma}^{\beta'} - \phi^\beta\,p^\beta\,\mathbf{1}\,. \end{aligned} \tag{37}$$

Therein, the pore pressure p is given by Dalton's law

$$p = s^l p^l + s^g p^g\,, \tag{38}$$

where p^l denotes the unspecified effective liquid pressure, whereas the effective gas pressure p^g is related to the effective density ϱ^g by the following constitutive law for an ideal gas:

$$\varrho^g = \frac{\mathcal{M}^g}{RT} p^g . \tag{39}$$

In equation (39), $\mathcal{M}^g$ denotes the molar mass of the pore gas, R the universal gas constant and T the absolute temperature.

The overall stress tensor $\boldsymbol{\sigma}$ of the porous material is given by the sum of the partial stress tensors $\boldsymbol{\sigma}^\alpha$ according to (37):

$$\boldsymbol{\sigma} = \boldsymbol{\sigma}^{s'} + \boldsymbol{\sigma}^{l'} + \boldsymbol{\sigma}^{g'} - p\,\mathbf{1} . \tag{40}$$

However, in the Theory of Porous Media the fluid frictional stresses $\boldsymbol{\sigma}^{\beta'}$ are usually neglected ($\boldsymbol{\sigma}^{\beta'} \approx 0$), yielding the well known concept of effective stress, e.g. (Bishop and Blight 1963):

$$\boldsymbol{\sigma} = \boldsymbol{\sigma}^{s'} - p\,\mathbf{1} . \tag{41}$$

3.4.3 *Momentum productions and Dary's law*

Although the fluid frictional stresses $\boldsymbol{\sigma}^{\beta'}$ are neglected, the fluid viscosity is included via the momentum production terms $\hat{\boldsymbol{p}}^\beta$ in the partial momentum balance equations (31). These are chosen as

$$\hat{\boldsymbol{p}}^\beta = p^\beta \operatorname{grad}(\phi^\beta) - \left[\phi^\beta\right]^2 \frac{\mu^\beta}{k^\beta}\, \boldsymbol{w}_\beta , \tag{42}$$

with the dynamic viscosity μ^β and the relative permeability k^β (Ehlers 2002). In turn, the relative permeability k^β depends on the intrinsic permeability k_0 and on the non-dimensional scaling factors k^β_s and k_ϕ, which take the dependence of the relative permeability on the saturation and on the pore volume into account:

$$k^\beta = k^\beta_s \, k_\phi \, k_0 . \tag{43}$$

The intrinsic permeability k_0 represents the permeability of the fully saturated, undeformed porous material, which is independent of the saturating fluid phase. The influence of the saturation is considered according to (van Genuchten 1980)

$$\begin{aligned} k^l_s &= \sqrt{s^l}\left[1 - \left[1 - \left[s^l\right]^{\frac{1}{m}}\right]^m\right]^2 , \\ k^g_s &= \sqrt{1 - s^l}\left[1 - \left[s^l\right]^{\frac{1}{m}}\right]^{2m} . \end{aligned} \tag{44}$$

m is the same material parameter as used in the capillary pressure relation (36). The dependence of the relative permeability on the pore volume is described by (Ehlers 2002):

$$k_\phi = \frac{1 - \phi^s}{1 - \phi^s_0} . \tag{45}$$

Finally, inserting the momentum productions $\hat{\boldsymbol{p}}^\beta$ (42) into the the related momentum balance equations (31) yields Darcy's law:

$$\phi^\beta \boldsymbol{w}_\beta = -\frac{k^\beta}{\mu^\beta} \operatorname{grad}(p^\beta) . \tag{46}$$

Within a geometrically linear theory the term $\phi^\beta \boldsymbol{\omega}_\beta$ on the left hand side of equation (46) has to be replaced by $\phi^\beta_0 \boldsymbol{\omega}_\beta$. By inserting the result into the partial mass balance equations of the pore fluids (30), the seepage velocities $\boldsymbol{\omega}_\beta$ can be eliminated as primary variables.

3.4.4 *Effective stress*

To close the description of the proposed model, the effective stress $\boldsymbol{\sigma}^{s'}$ has to be specified. For the modeling of the material degradation of the skeleton (reduction of stiffness and strength) the continuum damage theory proposed by (Kachanov 1958) is employed. In the context of the present paper an isotropic damage model with a single damage parameter d and a strain based description of the damage evolution is used (Simo and Ju 1987).

According to the effective area concept (Kachanov 1958), the scalar damage parameter can be interpreted as the ratio of the damaged cross section and the initial cross section. Since the stresses in the skeleton are transferred by the intact, undamaged cross section, the effective stress reads

$$\boldsymbol{\sigma}^{s'} = [1 - d]\, \phi^s\, \mathcal{C}^s : \left[\boldsymbol{\varepsilon}_s - \left[\frac{\varrho^s_0}{\varrho^s} - 1\right]\mathbf{1}\right] , \tag{47}$$

with the elasticity tensor of the skeleton $\mathcal{C}^s = 2\,\mu^s \mathcal{I} + \lambda^s \mathbf{1} \otimes \mathbf{1}$, defined in terms of the Lamé constants μ^s and λ^s. In equation (47), the volumetric expansion resulting from ASR is considered by the term $[\varrho^s_0/\varrho^s - 1]\mathbf{1}$. Inserting the initial material density $\varrho^s_0 = \varrho^u$ and the current density ϱ^s (28) into equation (47) yields:

$$\boldsymbol{\sigma}^{s'} = [1 - d]\, \phi^s\, \mathcal{C}^s : \left[\boldsymbol{\varepsilon}_s - \left[\frac{\varrho^u}{\varrho^r} - 1\right]\xi\,\mathbf{1}\right] . \tag{48}$$

According to (Simo and Ju 1987) the evolution of the damage parameter d is described by the damage criterion

$$\Phi = \eta - \kappa \leq 0 , \tag{49}$$

where η and κ are the equivalent strain and the internal variable defining the current damage threshold.

From the Kuhn-Tucker loading/unloading conditions and the consistency condition

$$\Phi \leq 0\,, \quad \frac{\partial \kappa}{\partial t} \geq 0\,, \quad \Phi \kappa = 0\,, \quad \frac{\partial \Phi}{\partial t}\frac{\partial \kappa}{\partial t} = 0 \tag{50}$$

follows, that κ is unchanged for $\Phi < 0$ and calculated by $\kappa = \eta$ otherwise.

Finally, the equivalent strain η and the damage function d must be specified. Here, the equivalent strain measure corresponding to the Rankine criterion of maximal principal stress is used:

$$\eta = \frac{\lambda^s + \mu^s}{\mu^s[3\lambda^s + 2\mu^s]} \max(\tilde{\sigma}_i^{s'})\,, \qquad i = 1,2,3\,. \tag{51}$$

Therein, $\tilde{\sigma}_i^{s'}$ denote the eigenvalues of the undamaged effective stress tensor

$$\tilde{\boldsymbol{\sigma}}^{s'} = \frac{1}{1-d}\boldsymbol{\sigma}^{s'} = \sum_{i=1}^{3} \tilde{\sigma}_i^{s'}\, \boldsymbol{n}_i \otimes \boldsymbol{n}_i\,, \tag{52}$$

with $\boldsymbol{n}_i$ as the corresponding eigenvectors. The damage function is given by

$$d = 1 - \frac{\kappa_0}{\kappa}[1 - \alpha_1 + \alpha_1 \exp[\alpha_2[\kappa_0 - \kappa]]]\,, \tag{53}$$

where α_1, α_2 are material parameters, which control the post-peak slope of the tensile stress-strain curve, when the initial damage threshold κ_0 is exceeded.

The definition of the effective stress $\boldsymbol{\sigma}^{s'}$ and its evolution underly the important assumption, that the deterioration by ASR only takes place if the expansion $[\varrho_0^s/\varrho^s - 1]\xi\mathbf{1}$ in equation (48) is hindered (see also Section 5). This assumption, which is used for most model formulations in the literature (Pietruszczak 1996; Ulm et al. 2000; Steffens et al. 2003), implies, that the concrete degradation by ASR is mainly induced by structural effects as gradients in the ASR swelling related to a non-uniform moisture distribution.

4 MODEL CALIBRATION

The macroscopic model for ASR affected concrete proposed in Section 3 is based on material constants, which control the material behavior of the skeleton and the pore gas (ϕ_0^s, λ^s, μ^s, κ_0, α_1, α_2, $\mathcal{M}^g/[RT]$), on parameters governing the transport of the pore fluids (p^r, n, m) and on the chemical parameters ϱ^u, ϱ^r, and k. In what follows, the calibration of the densities ϱ^u, ϱ^r, and of the reaction velocity k according to experimental results is described.

Consider a stress-free expansion test ($\boldsymbol{\sigma} = \mathbf{0}$) of a reactive concrete specimen carried out at a certain temperature and humidity. Inserting equation (48) into equation (41) yields after re-arrangement:

$$\boldsymbol{\varepsilon}_s = \left[\frac{\varrho^u}{\varrho^r} - 1\right] \xi\mathbf{1} + \frac{1}{[1-d]\,\phi^s}[\boldsymbol{C}^s]^{-1} p\,\mathbf{1}\,. \tag{54}$$

For iso-hydro-thermal conditions the second part of equation (54) is constant since the pore pressure p does not change in time. Hence, in laboratory tests on ASR affected concrete, only the part of the strain tensor ε_s is measured, which is governed by the reaction extent ξ:

$$\boldsymbol{\varepsilon}_s^{\xi} = \left[\frac{\varrho^u}{\varrho^r} - 1\right] \xi\mathbf{1}\,. \tag{55}$$

Differentiation of equation (55) with respect to time results in:

$$\frac{\partial \boldsymbol{\varepsilon}_s^{\xi}}{\partial t} = \frac{\partial \boldsymbol{\varepsilon}_s^{\xi}}{\partial \xi}\frac{\partial \xi}{\partial t} = k\,[1-\xi]\left[\frac{\varrho^u}{\varrho^r} - 1\right]\mathbf{1}\,. \tag{56}$$

From equations (55) and (56) the following values are obtained for the onset ($t = 0 \Rightarrow \xi = 0$) and for the completion of the alkali-silica reaction ($t \rightarrow \infty \Rightarrow \xi = 1$):

$$\begin{aligned} \xi = 0 \;&\Rightarrow\; \boldsymbol{\varepsilon}_s^{\xi} = \mathbf{0}, & \frac{\partial \boldsymbol{\varepsilon}_s^{\xi}}{\partial t} &= k\left[\frac{\varrho^u}{\varrho^r} - 1\right]\mathbf{1}, \\ \xi = 1 \;&\Rightarrow\; \boldsymbol{\varepsilon}_s^{\xi} = \left[\frac{\varrho^u}{\varrho^r} - 1\right]\mathbf{1}, & \frac{\partial \boldsymbol{\varepsilon}_s^{\xi}}{\partial t} &= \mathbf{0}. \end{aligned} \tag{57}$$

From (57) the meaning of the chemical material parameters becomes clear. The parameter $\varrho^u/\varrho^r - 1$ represents the asymptotic volumetric strain in a stress-free expansion test. Furthermore, the parameter k controls the slope of the respective expansion-time-relation with the onset of ASR. Thus, the chemical material parameters are well-defined and can be determined by means of strain measurements on reactive concrete specimens.

Both chemical material parameters ($\varrho^u/\varrho^r - 1$ and k) depend on the concrete mix design, the type of aggregates, the temperature and the moisture content (Section 2). In particular, the moisture dependence is of interest for the proposed model. The role of moisture has been studied in detail in an extensive test campaign at the Laboratoire Central des Ponts et Chaussées (LCPC), Paris (Larive 1998). In these tests, cylindrical concrete specimens of a certain mix design were stored under different hygral conditions (immersed in water, exposed to different relative humidities, wrapped in aluminum foil), whereby the temperature was kept constant at 38°C (Larive et al. 2000). For each specimen, the macroscopic expansion and the weight

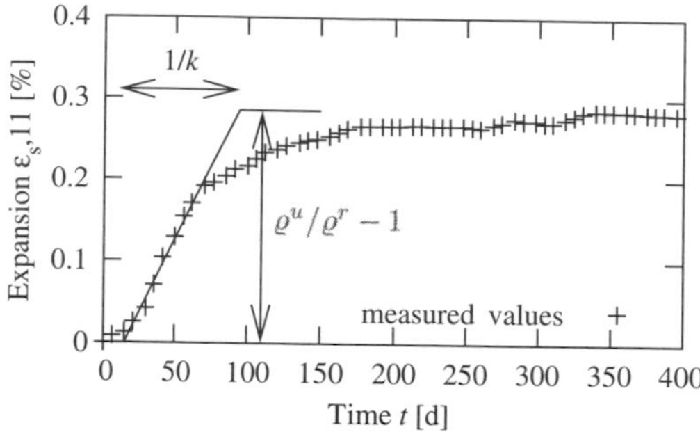

Figure 1. Typical evolution of the volumetric specimen expansion $\varepsilon_s^\xi = \varepsilon_s^\xi \mathbf{1}$ in a stress-free, iso-hydro-thermal laboratory test, see (Larive 1998).

change have been measured. Figure 1 shows a typical result of the expansion measurements, together with an illustration of the aforementioned material parameters $\varrho^u/\varrho^r - 1$ and k. It turns out from the test results, that with an increasing moisture content also the asymptotic expansion $\varrho^u/\varrho^r - 1$ and the reaction velocity k increase. In a recent study based on Larive's results, the weight change of the specimens has been converted to averaged liquid saturation within the specimens (Steffens et al. 2003). An almost linear relation between the liquid saturation s^l and the asymptotic expansion $\varrho^u/\varrho^r - 1$ as well as the inverse of the reaction velocity $1/k$ has been observed, which can be expressed by:

$$\frac{\varrho^u}{\varrho^r} - 1 = 1.27\, s^l - 0.754\ [\%], \quad \frac{1}{k} = -300\, s^l + 326\ [\text{d}]. \tag{58}$$

Hence, the chemical material parameters are known.

5 MODEL STUDIES

In this section the constitutive model for ASR affected concrete is evaluated by means of one-dimensional parametric studies, whereby in a first step pore pressure effects are neglected ($p = 0$). The material parameters are summarized in Table 1. The overall stress of the porous material according to equation (41) is re-written for the one-dimensional case. Neglecting pore pressure effects and using equation (48) results in

$$\sigma = [1 - d]\, \phi^s E^s \left[\varepsilon_s - \left[\frac{\varrho^u}{\varrho^r} - 1 \right] \xi \right], \tag{59}$$

with the Young's modulus of the skeleton $E^s = \mu^s [3\lambda^s + 2\mu^s]/[\lambda^s + \mu^s]$ and the reaction extent ξ defined in equation (4). The equivalent strain is given in analogy to equation (51) by:

$$\eta = \varepsilon_s - \left[\frac{\varrho^u}{\varrho^r} - 1 \right] \xi\,. \tag{60}$$

Table 1. Material parameters.

$\lambda^s = 1.215 \cdot 10^{10}\ \left[\frac{\text{N}}{\text{m}^2}\right]$	$\mu^s = 1.823 \cdot 10^{10}\ \left[\frac{\text{N}}{\text{m}^2}\right]$
$\alpha_1 = 0.9$	$\alpha_2 = 500$
$\kappa_0 = 1.0 \cdot 10^{-4}$	$\phi_0^s = 0.8$
$\frac{\varrho^u}{\varrho^r} - 1 = 1.27\, s^l - 0.754\ [\%]$	$\frac{1}{k} = -300\, s^l + 326\ [\text{d}]$

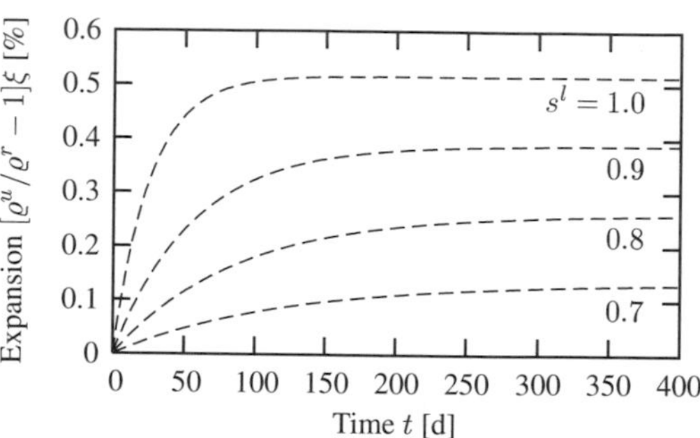

Figure 2. Expansion according to equation (61) in a stress- free, iso-hydro-thermal ASR test for different liquid saturations s^l.

The damage criterion (49), the Kuhn-Tucker loading/unloading conditions and the consistency condition (50) and the damage function (53) remain unchanged.

At first, the stress-free expansion test is revisited ($\sigma = 0$). The environmental conditions (temperature and humidity) are assumed to be constant. From equation (59) follows the stress-free material expansion due to ASR as:

$$\varepsilon_s = \left[\frac{\varrho^u}{\varrho^r} - 1 \right] \xi\,. \tag{61}$$

Inserting equation (61) into equation (60) yields $\eta = 0$. Hence, according to the model, no damage occurs under stress-free conditions. Figure 2 shows the evolution of the strain ε_s according to equation (61) for different liquid saturations s^l. Obviously, the model equations represent the effect of the moisture content on the kinetics and the magnitude of the ASR impact correctly. The reaction velocity and the asymptotic expansion increase with increasing moisture content.

The stress-free expansion test illustrates, that within the present model concrete deterioration only takes place if free expansion is hindered. This is the case either if the deformations are hindered by kinematic

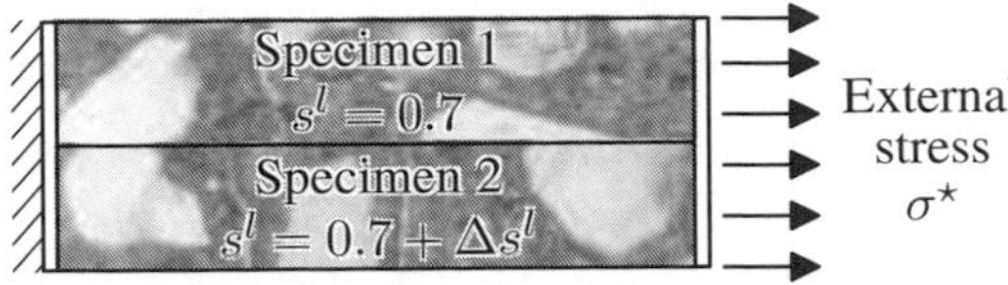

Figure 3. Arrangement of the concrete specimens in order to study the dependence of the concrete deterioration on gradients in the ASR swelling.

restraints or if there are gradients in the ASR swelling. To obtain a deeper insight into the dependence of the concrete deterioration on gradients in the ASR swelling, another one-dimensional analysis is performed. Two concrete specimens arranged in parallel with the same cross sectional area and a common rigid support at both ends are considered in this numerical study, see Figure 3. The material behavior of each specimen is described by the equations (49, 50, 53, 59, 60) together with the material parameters given in Table 1. In order to simulate structural gradients in the ASR swelling, the two concrete specimens are assumed to have different moisture contents expressed by different liquid saturations s^l. For specimen 1, the liquid saturation is assumed as $s^l = 0.7$. The liquid saturation within the second specimen is given by $s^l = 0.7 + \Delta s^l$, whereby a variation of Δs^l enforces swelling gradients of variable magnitude within the structural element shown in Figure 3. However, for both specimens the respective liquid saturation is assumed to be uniformly distributed and constant in time.

Equilibrium between the external stress σ^* (see Figure 3) and the stresses within the two concrete specimens yields:

$$\begin{aligned} \sigma^\star &= \tfrac{1}{2}[1-d_1]\,\phi_1^s\,E^s\left[\varepsilon_s - \left[\frac{\varrho_1^u}{\varrho_1^\star} - 1\right]\xi_1\right] \\ &+ \tfrac{1}{2}[1-d_2]\,\phi_2^s\,E^s\left[\varepsilon_s - \left[\frac{\varrho_2^u}{\varrho_2^\star} - 1\right]\xi_2\right]. \end{aligned} \quad (62)$$

The strain ε_s at time t is calculated on the basis of equation (62) by use of a Newton-Raphson procedure. Afterwards, the macroscopic stiffness E of the structural element shown in Figure 3

$$E = \tfrac{1}{2}[[1-d_1]\,\phi_1^s + [1-d_2]\,\phi_2^s]\,E^s. \quad (63)$$

is calculated.

Figure 4 shows the ratio of the current macroscopic stiffness E and the macroscopic stiffness E_0 of the sound material as a function of time t for different stress levels σ^* and values of Δs^l. The topmost diagram in Figure 4 illustrates, that in a macroscopically stress free experiment ($\sigma^* = 0$) the macroscopic stiffness E is drastically reduced within a few days even if the liquid saturation in the two specimens differs only by $\Delta s^l = 0.1$. The value $E/E_0 \approx 0.5$ for $t \to \infty$ can be explained as follows. For $\sigma^* = 0$ compression is induced in specimen 1 by the (restraint) ASR swelling, which is balanced by a tensile stress within specimen 2. After a while, the tensile stress exceeds the tensile strength. Finally, for $t \to \infty$, specimen 2 is fully damaged and as a result the macroscopic stiffness E is equal to the stiffness of specimen 1, which is just the half of the initial stiffness E_0. However, if both specimens show the same moisture content ($\Delta s^l = 0$), the macroscopic stiffness E remains unchanged since this scenario corresponds to the stress-free expansion test discussed above. The middle and the bottom diagram in Figure 4 illustrate the evolution of the macroscopic stiffness E if an external (compressive) stress σ^* is applied. At time $t = 0$ compressive stresses within both specimens balance the external stress σ^*. Nevertheless, as the ASR swelling evolves the compressive

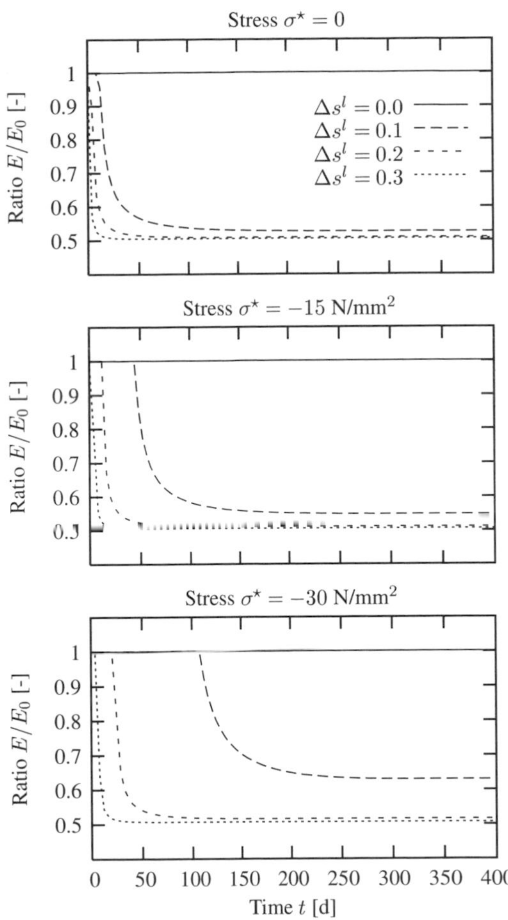

Figure 4. Reduction of the macroscopic stiffness E as function of time t for different stress levels σ^* and different values of Δs^l.

stress in specimen 2 is converted into a tensile stress. Finally, the tensile strength of specimen 2 is exceeded resulting in a degradation of the macroscopic stiffness E. The final value of the ratio E/E_0 is larger if a compressive stress σ^* is applied compared to situations, where no external stress or even a tensile stress is applied.

6 CONCLUSIONS

In this paper, a macroscopic model for ASR affected concrete was developed within the framework of the Theory of Porous Materials. The alkali-silica reaction is described by a first order reaction. The model is based on two chemical material parameters, the volumetric expansion $\varrho^u/\varrho^r - 1$ and the reaction velocity k, which can be easily determined by strain measurements on reactive concrete specimens. In particular, the dependence of these material parameters on the moisture content was addressed. It was shown, that the model allows for a realistic description of the kinetics and the magnitude of the ASR impact for varying hygral conditions.

One of the most important assumptions in the model formulation was, that damage induced by structural effects, e.g stresses induced by gradients in the ASR expansion, dominate the overall deterioration of ASR affected concrete. Hence, material damage resulting from ASR only occurs, if the ASR swelling is (at least partially) hindered. It was illustrated by means of one-dimensional model studies, that even a small inhomogeneity in the moisture distribution within a structural element results in a significant reduction of the mechanical properties due to gradients in the ASR swelling. Hence, structural effects indeed seem to play the predominant role in the deterioration by ASR and the aforementioned assumption seems to be reasonable.

ACKNOWLEDGMENT

Financial support was provided by the German National Science Foundation (DFG) in the framework of project A9 of the collaborative research center SFB 398. This support is gratefully acknowledged.

REFERENCES

Atkins, P. (1998). *Physical chemistry.* Oxford: Oxford University Press.

Bažant, Z. and A. Steffens (2000). Mathematical model for kinetics of alkali-silica reaction in concrete. *Cement and Concrete Research 30*, 419–428.

Bear, J. and Y. Bachmat (1991). *Introduction to modeling of transport phenomena in porous media.* Dordrecht, Boston, London: Kluwer Academic Publishers.

Bishop, A. and G. Blight (1963). Some aspects of effective stress in saturated and partly saturated soils. *Géotechnique 13*, 177–197.

Coussy, O. (1995). *Mechanics of porous continua.* Chichester: John Wiley & Sons.

Dent Glasser, L. and N. Kataoka (1981). The chemistry of alkali-aggregate reaction. *Cement and Concrete Research 11*, 1–9.

Dron, R. and F. Brivot (1993). Thermodynamic and kinetic approach to the alkali-silica reaction. Part 2: Experiment. *Cement and Concrete Research 23*, 93–103.

Ehlers, W. (2002). Foundations of multiphasic and porous materials. In W. Ehlers and J. Bluhm (Eds.), *Porous media – Theory, experiments and numerical applications*, pp. 3–86. Berlin: Springer.

Glasser, F. (1992). Chemistry of the alkali-aggregate reaction. In R. Swamy (Ed.), *The alkali-silica reaction in concrete*, Chapter 2, pp. 30–53. Glasgow, London: Blackie and Son Ltd.

Hobbs, D. (1988). *Alkali-silica reaction in concrete.* London: Thomas Telford.

Iler, R. (1979). *The chemistry of silica.* New York: John Wiley & Sons.

Kachanov, L. (1958). Time rupture process under creep conditions. *IVZ Acad Nauk SSR Otd Tech Nauk 8*, 26–31.

Larive, C. (1998). *Apports combinés de l'expérimentation et de la modélisation à la compréhension de l'alcali-réaction et de ses effects mécaniques*. Ph. D. thesis, Laboratoire Central des Ponts et Chaussées, Paris.

Larive, C. and O. Coussy (1996). Behaviour of AAR-affected concrete: Modelling. In *10th International Conference on Alkali-Aggregate Reaction in Concrete,* Melbourne, Austrialia, pp. 662–669.

Larive, C., A. Laplaud, and O. Coussy (2000). The role of water in alkali-silica reaction. In M. Bérubé, B. Fournier, and B. Durand (Eds.), *Alkali-aggregate reaction in concrete,* Québec, Canada, pp. 61–69. Centre de Recherche Interuniversitaire sur le Béton, Université Laval.

Lewis, R. and B. Schrefler (1998). *The finite element method in the static and dynamic deformation and consolidation of porous media* (2. ed.). Chichester: John Wiley & Sons.

Mainguy, M., O. Coussy, and V. Baroghel-Bouny (2001). Role of air pressure in drying of weakly permeable materials. *Journal of Engineering Mechanics (ASCE) 127*, 582–592.

Pietruszczak, S. (1996). On the mechanical behavior of concrete subjected to alkali-aggregate reaction. *Computers & Structures 58*, 1093–1097.

Poole, A. (1992). Introduction to alkali-silica reaction in concrete. In R. Swamy (Ed.), *The alkali-silica reaction in concrete,* Chapter 1, pp. 1–29. Glasgow, London: Blackie and Son Ltd.

Powers, T. and H. Steinour (1955). An interpretation of some published researches on the alkali-aggregate-reaction. Part I- The chemical reaction and mechanisms of expansion. *Journal of the American Concrete Institute 51*, 497–516.

Simo, J. and J. Ju (1987). Strain and stress-based continuum damage models I - Formulation. *International Journal for Solids and Structures 23*, 821–840.

Steffens, A., K. Li, and O. Coussy (2003). Aging approach to water effect on alkali-silica reaction degradation of structures. *Journal of Engineering Mechanics (ASCE) 129*(1).

Struble, L. and S. Diamond (1981). Unstable swelling behaviour of alkali silica gels. *Cement and Concrete Research 64*, 611–617.

Ulm, F.-J., O. Coussy, L. Kefei, and C. Larive (2000). Thermo-chemo-mechanics of ASR expansion in concrete structures. *Journal of Engineering Mechanics (ASCE) 126*, 233–242.

van Genuchten, M. (1980). A closed-form equation for predicting the hydraulic conductivity of unsaturated soils. *Soil Science Society of America 44*, 892–898.

Computational Modelling of Concrete Structures, Bićanić et al. (eds)
© 2003 Swets & Zeitlinger, Lisse, ISBN 90 5809 536 3

Long time behavior of concrete: creep, fracture and size effects

F. Barpi & S. Valente
Department of Structural and Geotechnical Engineering, Politecnico di Torino, Torino, Italy

ABSTRACT: This paper analyses the interaction between strain-softening and time-dependent behaviour of quasi-static fracture of concrete. In the framework of the cohesive model, a viscous element based on a fractional order rate law is coupled with a micromechanical model for the fracture process zone. This approach makes it possible to include a whole range of dissipative mechanisms in a single rheological element. Creep fracture in mode I conditions is analysed through the finite element method, the cohesive (or fictitious) crack model and a new space and time integration scheme. The comparison with creep tests executed in three point bending conditions shows a good agreement both in terms of load vs. displacement and failure lifetime. Finally, the effect of specimen size on failure lifetime, in the case of constant load tests, is examined.

1 INTRODUCTION

The long term performance of concrete structures is fundamentally affected by the behaviour of the material after cracking. It is well known that concrete presents a diffused damage zone within which microcracking increases and stresses decrease as the overall deformation increases. This results in the softening of the material in the so called *fracture process zone* (FPZ). The size of this zone can be compared with a characteristic dimension of the structure and can vary during the evolutionary process. In this context, a numerical method (based on finite or boundary elements) has to be used together with the *cohesive or fictitious* crack model as shown by Barenblatt (Barenblatt 1959), Dugdale (Dugdale 1960) and Hillerborg (Hillerborg, Modeer, and Petersson 1976).

The interaction between strain-softening and time dependent behaviour is analysed, with the emphasis on very slow or quasi-static fracture. This is the case of cracking in massive concrete structures like dams, where inertial forces can be neglected. This approach is based on a *micromechanical* model which combines time-dependent and time-independent information. One of these models was proposed by Santhikumar and Karihaloo (Santhikumar and Karihaloo 1996; Santhikumar, Karihaloo, and Reid 1998). The time-independent part of this model is based on the concept of *effective spring*, which derives from a micromechanical hypothesis for the static softening behaviour of the concrete in the fictitious process zone proposed in (Huang and Li 1989). In the present paper this approach is enhanced using a *fractional order rate* law and is applied to the numerical simulation of the three point bending tests described by Zhou (Zhou 1992).

The proposed approach can be compared with the one proposed in (Bažant and Gettu 1992) (rate dependent softening) and the one proposed in (Carpinteri, Valente, Zhou, Ferrara, and Melchiorri 1995) (stress relaxation law in the FPZ obtained by fitting experimental results).

In the last part of the paper the effect of specimen size on failure lifetime is examined. In the study of the behaviour of large structures like dams, it is important to understand how specimen lifetime is affected by its size. The cohesive model, that has been proven able to keep into account the transition from ductile to brittle behaviour of structures with increasing size, all other ratios remaining the same (size-effect on *peak load*, see (Carpinteri 1989)), has also been shown useful to study this phenomenon (size-effect on *failure life-time*). In the literature there are few papers about experiments focused on this effect: we consider here the one by Bažant and Xiang (Bažant and Xiang 1997).

2 RHEOLOGICAL MODEL

In the simplest rheological model of the linear standard viscoelastic solid (Fig. 1), the springs are characterized by linear stress-displacement relationships:

$$\sigma_1 = E_1 (\varepsilon - \varepsilon_1), \tag{1a}$$

$$\sigma_2 = E_2 \varepsilon, \tag{1b}$$

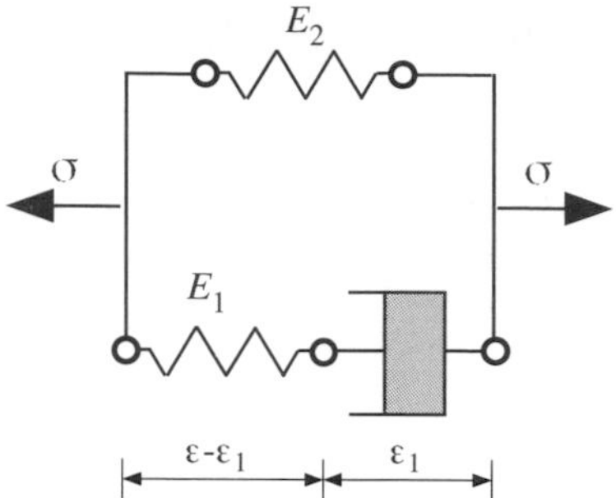

Figure 1. Rheological model.

while the dashpot is based on the following *fractional order rate* law for the internal variable ε_1 (Barpi and Valente 2002):

$$D^{\alpha}\varepsilon_1 = \frac{\mathrm{d}^{\alpha}\varepsilon_1}{\mathrm{d}t^{\alpha}} = \frac{\sigma_1}{E_1\,\tau_1^{\alpha}} = \frac{\varepsilon-\varepsilon_1}{\tau_1^{\alpha}} \quad \text{with} \quad \alpha \in (0,1). \tag{2}$$

The fractional differentiation of a function $y(t)$ is defined according to (Oldham and Spanier 1974; Carpinteri and Mainardi 1997). Equation 2 represents a generalization of the well-known Newton's constitutive law for the dashpot:

$$\sigma = \eta\frac{\mathrm{d}\varepsilon}{\mathrm{d}t} = E\,\tau\,\frac{\mathrm{d}\varepsilon}{\mathrm{d}t}. \tag{3}$$

that can be rewritten as

$$\frac{\mathrm{d}\varepsilon}{\mathrm{d}t} = D^1\varepsilon = \frac{\sigma}{E\,\tau}. \tag{4}$$

In this form, Eq. 2 and Eq. 4 can be easily compared. In Eq. 2 the classical derivative of integer order ($\alpha = 1$) of the deformation ε is replaced by a derivative of order α and, to mantain the correct dimensions, the relaxation time τ_1 is changed by τ_1^{α}.

In particular, the non integer (of order m) differential operator $D^m y(t)$ is defined as:

$$D^{-(1-\alpha)}\,y(t) = \int_0^t \Phi_{1-\alpha}(t-\bar{t})\;y(\bar{t})\mathrm{d}\bar{t}, \tag{5}$$

where

$$\Phi_{1-\alpha}(t) = \frac{t_+^{-\alpha}}{\Gamma(1-\alpha)} \quad \text{with} \quad t_+ = \begin{cases} t & \text{if} \quad t>0 \\ 0 & \text{if} \quad t<0 \end{cases}. \tag{6}$$

In the previous expression $\Gamma(x)$ represents the Euler's *Gamma* function, defined as follows:

$$\Gamma(x) = \int_0^{\infty} e^t\,t^{(x-1)}\mathrm{d}t = \lim_{n\to\infty}\frac{n^x\,n!}{x\,(x+1)\ldots(x+n)}. \tag{7}$$

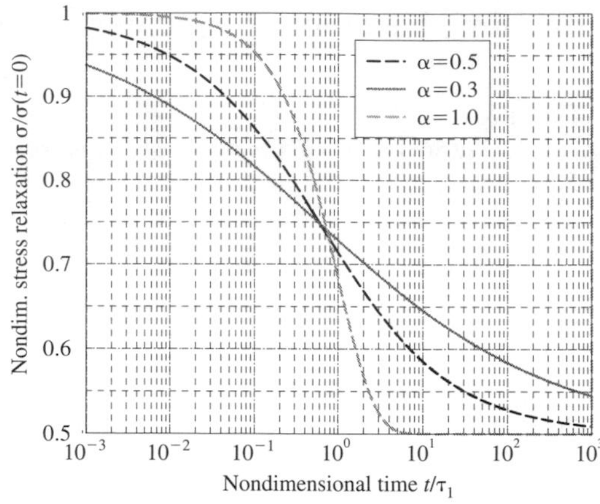

Figure 2. Stress relaxation functions ($E_1 = E_2$).

It must be noticed that Eq. 5 can also be obtained by using an hereditary model based on a *Rabotnov* fractional exponential kernel (Karihaloo 1995).

A convergent expression for the α-order fractional derivative operator D^{α} is given by:

$$\begin{aligned} D^{\alpha}y(t) &= D^1 D^{-(1-\alpha)}y(t) \\ &= \frac{\mathrm{d}}{\mathrm{d}t}\int_0^t \Phi_{1-\alpha}(t-\bar{t})\;y(\bar{t})\mathrm{d}\bar{t} \\ &= \frac{1}{\Gamma(1-\alpha)}\,\frac{\mathrm{d}}{\mathrm{d}t}\int_0^t \frac{y(\bar{t})}{(t-\bar{t})^{\alpha}}\mathrm{d}\bar{t}. \end{aligned} \tag{8}$$

In the case of $\alpha = 1$ the classical dashpot with an integer order rate law is obtained from Eq. 2. In particular, the solutions for the relaxation problem (under constant w) and for the creep problem (under constant σ) become of exponential type, with τ_1 as the *relaxation time*, and $\tau_1\frac{E_1+E_2}{E_2}$ as the *retardation time*. A comparison between the solutions corresponding to different values of α (integer and non integer) is plotted in Figs 2 and 3.

2.1 *Numerical integration of constitutive response*

A possible approximation for the fractional differentiation of a function $y(t)$ is (Oldham and Spanier 1974):

$${}^{n+1}(D^{\alpha}y) = \frac{1}{(\Delta t)^{\alpha}}\sum_{j=0}^{n} b_j(\alpha)\;{}^{n+1-j}y. \tag{9}$$

It is assumed that the spacing in time is uniform, i.e., ${}^n y = y(n\Delta t)$. The coefficients $b_j(\alpha)$ depend on the Gamma function as follows:

$$b_j(\alpha) = \frac{\Gamma(j-\alpha)}{\Gamma(-\alpha)\,\Gamma(j+1)}. \tag{10}$$

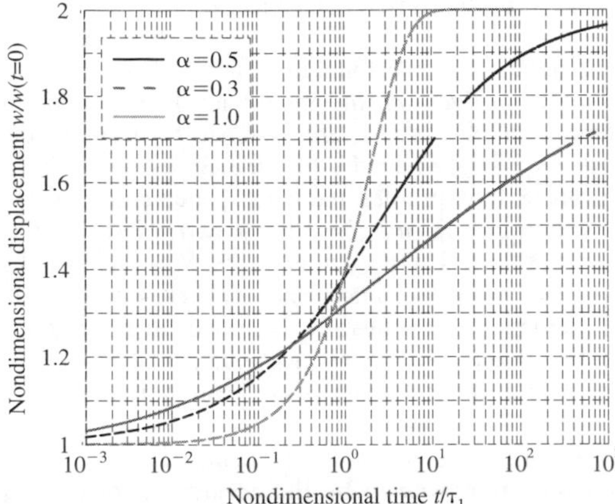

Figure 3. Creep displacement functions ($E_1 = E_2$).

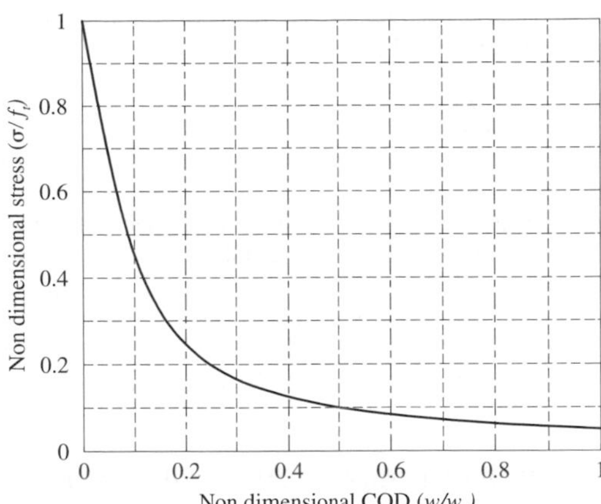

Figure 4. Cohesive stress-*COD* law ($\beta = 0.05$).

By using the recursion formula:

$$\frac{\Gamma(j-\alpha)}{\Gamma(j+1)} = \frac{(j-1-\alpha)}{j} \frac{\Gamma(j-1-\alpha)}{\Gamma(j)}, \tag{11}$$

no evaluation of Gamma function is needed and the coefficients $b_j(\alpha)$ are given by:

$$b_0(\alpha) = 1, \quad \ldots \quad b_k(\alpha) = \frac{(k-1-\alpha)}{k} b_{k-1}(\alpha), \quad \ldots \quad k = 1, \ldots, n. \tag{12}$$

For convenience, the expression in Eq. 9 can be rewritten as:

$$^{n+1}(D^\alpha y) = \frac{1}{(\Delta t)^\alpha}(^{n+1}y - {^n\overline{y}}), \tag{13}$$

where

$$^n\overline{y} = -\sum_{j=1}^{n} b_j(\alpha)\, ^{n+1-j}y, \tag{14}$$

is a known quantity at time t_{n+1}.

At this point the updated stress quantities $^{n+1}\sigma$ can be obtained by using Eq. 13 with reference to Eq. 2. The integration over time is executed with the classical *General Midpoint Rule* (Enelund, Mähler, Runesson, and Lennart Josefson 1999).

3 MICROMECHANICAL MODEL FOR THE PROCESS ZONE

In each point of the fictitious process zone a micromechanical approach to tension softening is combined with the rheological model described in Fig. 1 and Eq. 2, according to a method proposed in (Santhikumar and Karihaloo 1996; Santhikumar, Karihaloo, and Reid 1998). Tension softening behaviour appears when the damage in the material has localized along eventual fracture planes: this behaviour has been successfully modelled using two- and three-dimensional micromechanical models (Huang and Li 1989; Karihaloo 1995).

3.1 *Loading phase*

All models provide a relationship between the residual tensile stress carrying capacity σ[1] and crack opening displacement w (also called *COD*) as a function of known concrete microstructural parameters (included in the factor β), e.g. aggregate volume fraction V_f, Young's modulus E_c, ultimate tensile strength f_t and fracture toughness of the homogenized material K_{Ic}^{hom} (see Fig. 4).

According to these models, the function is assumed to be:

$$\frac{w}{w_c} = \underbrace{\frac{(K_{Ic}^{hom})^2}{E_c(1-V_f)f_t}}_{\beta} \frac{f_t}{\sigma} \left[1 - \left(\frac{\sigma}{f_t}\right)^3\right]. \tag{15}$$

3.2 *Creep phase*

During the creep phase (at constant load), the $(\sigma - w)$ law is generalized in the same way of Eq. 3; the deformations ε and ε_1 are now replaced by the crack opening

[1] From here to the end of the paper, σ indicates the *stress* in the *cohesive zone*.

displacements w and w_1. By using a fractional order rate law it is possible to write:

$$D^\alpha w_1 = \frac{d^\alpha w_1}{dt^\alpha} = \frac{\sigma_1}{K_1\,\tau_1^\alpha} = \frac{w - w_1}{\tau_1^\alpha} \quad \text{with} \quad \alpha \in (0,1), \tag{16}$$

where the constants K_1 and K_2 are defined in the following Sections: They represent the counterpart of the elastic moduli E_1 and E_2 of the classical viscoplastic model, though it should be noted that the physical dimensions of K_1 and K_2 differ from those of an elastic modulus. In the present work $K_1 = K_2$ is assumed.

4 RHEOLOGICAL AND MICROMECHANICAL MODEL INTERACTION

During the loading phase each point of the FPZ moves on the same $(\sigma - w)$ curve. Later on this condition does not hold any longer, due to the combined effect of *viscosity* and *damage*.

In order to understand how the rheological and micromechanical models interact, three *single degree of freedom* systems were presented in (Santhikumar and Karihaloo 1996; Santhikumar, Karihaloo, and Reid 1998), in the case of integer rate law.

1. In the case of the first system, the displacement discontinuity w is kept constant along time step Δt; the stress relaxation $\Delta\sigma$ occurs according to the standard viscoelastic model described. Each stress relaxation $\Delta\sigma$, through the unloading stiffness hypothesis induces an instantaneous stiffness reduction (*effective spring* concept) related to the time increment Δt.
2. In the case of the second system, the stress σ is kept constant along time step Δt; the creep displacement Δw occurs according to the standard viscoelastic model described. Each creep displacement increment Δw, through the tangential softening hypothesis, induces an instantaneous softening reduction related to the time increment Δt.
3. In the case of the third system, both stress σ and displacement discontinuity w are forced to stay on the static curve (Eq. 15). One of the two increments ($\Delta\sigma$ or Δw) occurs as predicted by the rheological model, while the other is smaller or equal to the value predicted by the rheological model.

At the end of each time step, the microcrack pattern changes and, in either case, stiffness is reduced.

In the present work, a *multiple degree of freedom* system is analyzed. Therefore, the above mentioned approach were generalized through the Finite Element method as described in the following Sections. It is worth noting that each point in the FPZ follows a different path and, hence, exhibits a different stiffness, while $K_1 = K_2$.

5 FINITE ELEMENT ANALYSIS

The continuum surrounding the process zone is assumed as *linear elastic* and all non-linear and time-dependent phenomena are assumed to occur in the *process zone*. When the fictitious crack tip (F.C.T.) advances by a pre-defined length, each point located on the crack trajectory, is split into two points. The virtual mechanical entity, acting on these two points only, is called *cohesive element*. The local behaviour of such an element follows the rules mentioned in the previous section. Each cohesive element interacts with the others only through the undamaged continuum, external to the process zone.

According to the finite element method, by taking the unknowns to be the n nodal displacement increments, $\Delta\mathbf{u}$, and assuming that compatibility and equilibrium conditions are satisfied at all points in the solid, it is possible to obtain the following system of n equations with $(n+1)$ unknowns ($\Delta\mathbf{u}$, $\Delta\lambda$ or Δt). The creep effect is incorporated by adding the pseudo-load induced by relaxation to the load vector in the equilibrium equations (Bocca, Carpinteri, and Valente 1991; Barpi, Chillè, Imperato, and Valente 1999):

$$(\mathbf{K}_T + \mathbf{C}_T)\,\Delta\mathbf{u} = \Delta\lambda\mathbf{P} + \Delta t\,\mathbf{Q}, \tag{17}$$

where:

- $\mathbf{K}_T$: positive definite tangential stiffness matrix, containing contributions from linear elastic (undamaged) elements and possible contributions from cohesive elements having (σ, w) below the curve of Fig 4. The conditions in which this possibility applies will be described later on;
- $\mathbf{C}_T$: negative definite tangential stiffness matrix, containing contributions from cohesive elements with (σ, w) on the curve of Fig 4;
- $\mathbf{P}$: the vector of external load;
- $\Delta\lambda$: maximum load multiplier which is compatible with Eq. 15 and the fictitious crack tip growth condition ($\sigma_{F.C.T.} = f_t$);
- $\mathbf{Q}$: vector of unbalanced load (or pseudo-load) due to relaxation in the process zone, related to a unitary time increment.

During the loading phase, the behaviour of the material is assumed to be time-independent ($\mathbf{Q} = \mathbf{0}$), the external load changes, $\Delta\lambda \neq 0$ and $\Delta t = 0$. On the contrary, during the sustained loading phase, the behaviour of the process zone is assumed to be time-dependent ($\mathbf{Q} \neq \mathbf{0}$), the external load is kept constant, $\Delta\lambda = 0$, and $\Delta t = 0$.

5.1 *Interaction between cohesive elements*

During the loading phase, all the stress paths in the FPZ are forced to follow the $(\sigma - w)$ law (defined Eq. 15). For the boundary condition analysed dw is *always* and *everywhere* positive. A more complex situation occurs during the next loading phase (*sustained*): the unloading stiffness approaches ∞ when w tends to 0^+. In order to avoid this difficulty, a threshold value has to be assumed for w. A cohesive element is classified as *active*, and submitted to the rheological model, when and only when its w is bigger than the threshold, assumed equal to $0.001w_c$. Otherwise the stress path is forced to follow the $(\sigma - w)$ law as it occurs during the loading phase.

According to the rheological model, for each active cohesive element, it is possible to compute the stress relaxation under constant w ($\mathrm{d}\sigma_t$) as well as the creep displacement under constant σ ($\mathrm{d}w_t$). It is important to notice that $\mathrm{d}\sigma_t$ and $\mathrm{d}w_t$ are *threshold values* computed according to the micromechanical model, while $\mathrm{d}\sigma$ and $\mathrm{d}w$ are *real values* obtained from equilibrium, compatibility and minumum free energy conditions.

The local compatibility conditions can be grouped in the following cases:

1. full relaxation only: $\mathrm{d}\sigma = \mathrm{d}\sigma_t < 0$ and $\mathrm{d}w < \mathrm{d}w_t$,
2. full creep only: $\mathrm{d}\sigma < \mathrm{d}\sigma_t < 0$ and $\mathrm{d}w = \mathrm{d}w_t$,
3. full creep with elastic increment: $\mathrm{d}\sigma = (\mathrm{d}w - \mathrm{d}w_t)K_1 > 0$ and $K_1 > 0$ and $\mathrm{d}w > \mathrm{d}w_t$,
4. full creep with softening increment: $\mathrm{d}\sigma = (\mathrm{d}w - \mathrm{d}w_t)K_1 < 0$ and $K_1 < 0$ and $\mathrm{d}w > \mathrm{d}w_t$.

In this context, the *Helmholtz free energy* is assumed as the objective function to be minimised at each time step under the constraints shown in the list above. The use of optimisation techniques in structural analysis to solve the variational inequality that occurs in elastoplasticity is well known (see (Maier 1971)). Since the loading conditions are assumed as piecewise linear, each physical time step is divided into numerous *logical* substeps that can be solved through the *Simplex method*, a classical linear programming tool. When case (3) or case (4) are applied, stiffness matrix coefficients are changed from one substep to the next (Bocca, Carpinteri, and Valente 1991; Barpi, Chillè, Imperato, and Valente 1999). Otherwise they are kept constant during all the substep iterations.

Creep rupture time is reached when the smallest eigenvalue of the tangential stiffness matrix becomes negative: this means that the external load can no longer be kept constant.

6 EXPERIMENTAL AND NUMERICAL RESULTS

The experimental tests, executed on prenotched beams, described in (Zhou 1992), were simulated numerically. The experimental procedure is based on two phases:

- the external load grows from zero to the nominal level (a fraction of the maximum load P_{max} taken equal to 0.76, 0.80, 0.85 and 0.92, see Fig. 5 and Fig. 6) under deflection control (5×10^{-6} m/s),
- the load is kept constant until the creep rupture occurs (*pre-peak* sustained bending).

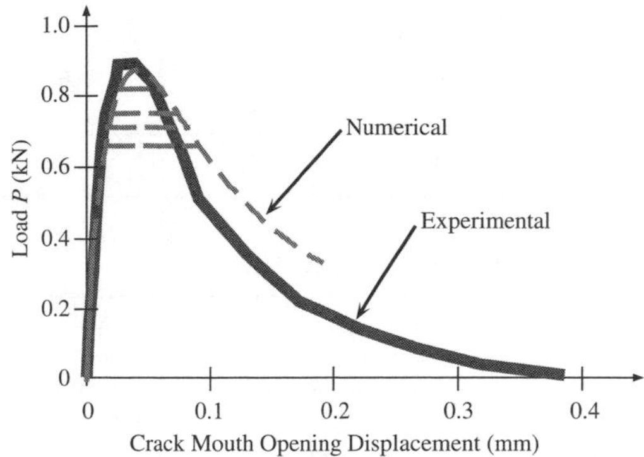

Figure 5. Three point bending test: experimental and numerical load vs. crack mouth opening displacement ($h = 10$ cm).

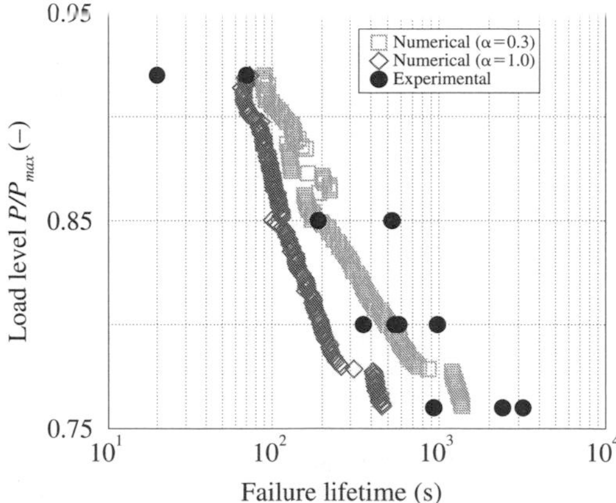

Figure 6. Three point bending test: comparison between experimental and numerical results in terms of failure lifetime ($h = 10$ cm).

These tests are usually associated with the name of *pre-peak sustained bending* tests. Of course, in order to know the maximum load $P_{max} \approx 900$ N, a number of static tests have to be previously executed. To overcome this difficulty, different authors prefer to use the so-called *post-peak* tests where the creep phase starts beyond the peak point (Carpinteri, Valente, Zhou, Ferrara, and Melchiorri 1997; Barpi, Ferrara, Imperato, and Valente 1999). The specimen dimensions are $10 \times 10 \times 80$ cm, the notch depth is 5 cm, while the material properties, as described in (Zhou 1992), are presented in Table 1.

The numerical simulations were executed using the values listed in Table 2, and neglecting the time

Table 1. Material properties.

E (GPa)	ν (–)	$\mathcal{G}_F$ (N/m)	f_t (MPa)
36	0.10	82	2.8

Table 2. Numerical parameters.

w_c (mm)	τ_1 (s)	$\Delta t/\tau_1$ (–)	β (–)	Element size (mm)
2.2×10^{-4}	150	1/50	0.05	0.0625

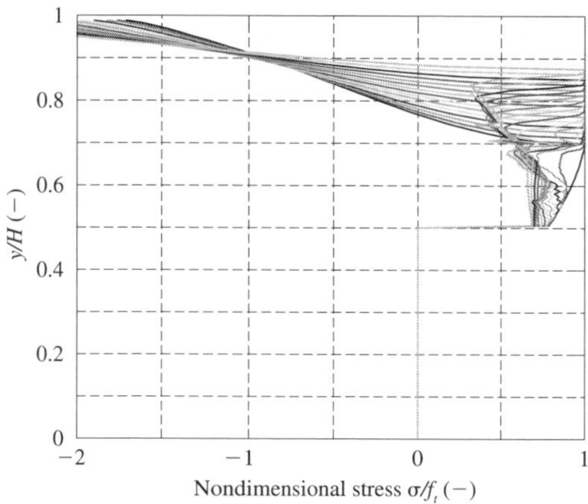

Figure 7. Stress distribution in the midspan section ($h = 10$cm, notch depth ratio equal 0.50, load level equal 0.92, dead weight effects included).

dependent behaviour of the undamaged material. As suggested in (Barpi, Chillè, Imperato, and Valente 1999) the following limit is applied to each step: $|\frac{d\sigma}{f_t}| \leq 0.01$.

Figure 5 shows the experimental and numerical load vs. crack mouth opening displacement curves for static tests as well as for sustained load tests, while Fig. 6 shows the load level vs. the logarithm of the *failure lifetime* (creep rupture time), for different values of the fractional derivative order α (0.30 and 1.00). The best fitting of the experimental results is achieved assuming $\alpha = 0.30$. Experimental and numerical results appear to be in good agreement. Finally, Fig. 7 show the stress distribution inside the process zone (the maximum value of tensile stress is f_t, according to the cohesive model) for $\frac{P_{cost}}{P_{max}} = 0.92$.

7 SIZE-EFFECTS ON FAILURE LIFETIME

In this Section the effect of specimen size on failure lifetime is analysed using the previously described

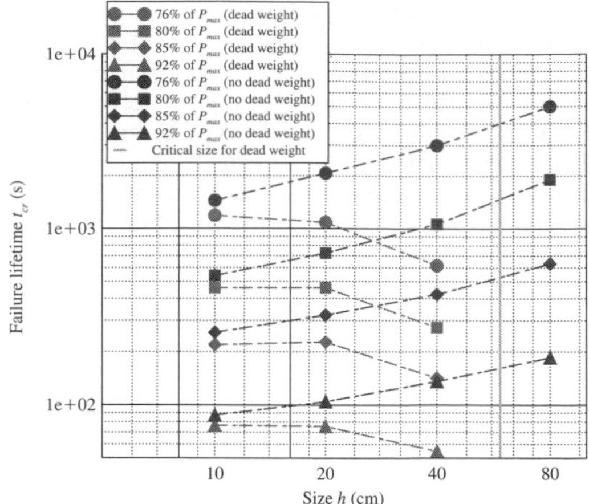

Figure 8. Failure lifetime vs. size h (log-log scale). Solid line: dead weight effects included; dashed line: without dead weight effects ($\alpha = 0.3$).

model. For each load level ($\frac{P_{cost}}{P_{max}} = 0.92, 0.85, 0.80, 0.76$) the numerical simulations were repeated varying the size (h = 10, 20, 40 and 80 cm). The failure lifetime is plotted in Fig. 8 as a function of size h in log-log scale. For the geometrical ratios analysed (notch depth ratio 0.5, span height ratio 8) and the material properties of Table 1, a critical height exists ($h \approx 59$ cm) above which no external load can be applied, since the dead weight alone is sufficient to cause crack propagation.

When size drops below a critical value, size-effect on failure lifetime appears to be negligible. After a series of experimental tests executed on single edge notched specimens, loaded in eccentric compression, Bažant and Xiang (Bažant and Xiang 1997) came to the same conclusion. When size approaches the critical value, a reduction in failure lifetime appears at all the load level analysed.

As a second hypothesis, it can be assumed that all loads are proportional to the same factor, and therefore the dead weight can be neglected. Fig. 8 shows that the effects of the new hypothesis on failure life-time are an increasing function of sizc and a dccrcasing function of load level. In the case of small size they are negligible. Under this hypothesis, failure life-time appears as an increasing function of size.

8 CONCLUSIONS

- A *linear* viscoelastic rheological element, combined with a suitable *micromechanical* model is successfully used for studying subcritical crack propagation problems.

- A *single* rheological element has been used. *Four* material properties only, namely, the elastic constants K_1 and K_2, the relaxation time τ_1 and the order α of the fractional derivative are enough to describe the phenomenon. It is possible to avoid the use of long chains of rheological elements, whose properties are difficult to determine. Moreover, a *fractional order rate* makes it possible to include a *whole spectrum* of dissipative mechanisms in a *single* viscous element.
- A new time integration scheme is used. The incremental problem is formulated as *piecewise linear* and controlled through a linear programming algorithm (*Simplex method*).
- A critical size exists, above which the dead weight alone is sufficient to cause crack propagation. At a sufficient distance from the above mentioned critical value, the size-effect on the failure lifetime appears to be negligible. Elsewhere it appears to be a decreasing function of size.
- In the case of the three point bending test, neglecting the influence of dead weight, the life-time appears to be an increasing function of time.
- The dead weight effects on failure lifetime increase with size and decrease with load level.

REFERENCES

Barenblatt, G. I. (1959). The formation of equilibrium cracks during brittle fracture: general ideas and hypotheses. *Journal of Applied Mathematics and Mechanics*, 622–636.

Barpi, F., F. Chillè, L. Imperato, and S. Valente (1999). Creep induced cohesive crack propagation in mixed mode. In D. Durban and J. R. A. Pearson (Eds.), *Non-Linear Singularities in Deformation and Flow*, The Netherlands, pp. 155–168. Kluwer Academic Publishers.

Barpi, F., G. Ferrara, L. Imperato, and S. Valente (1999). Lifetime of concrete dam models under constant loads. *Materials and Structures* 32, 103–111.

Barpi, F. and S. Valente (2002). Creep and fracture in concrete: A fractional order rate approach. *Engineering Fracture Mechanics*. In press.

Bažant, Z. P. and R. Gettu (1992). Rate effects and load relaxation in static fracture of concrete. *American Concrete Institute Journal* 89(5), 456–468.

Bažant, Z. P. and Y. Xiang (1997). Crack growth and lifetime of concrete under long time loading. *Journal of Engineering Mechanics (ASCE)* 4, 350–358.

Bocca, P., A. Carpinteri, and S. Valente (1991). Mixed-mode fracture of concrete. *International Journal of Solids and Structures* 27, 1139–1153.

Carpinteri, A. (1989). Size effects on strength, toughness and ductility. *Journal of Engineering Mechanics (ASCE)* 115, 1375–1392.

Carpinteri, A. and F. Mainardi (1997). *Fractals and Fractional Calculus in Continuum Mechanics*. Wien: Springer.

Carpinteri, A., S. Valente, F. P. Zhou, G. Ferrara, and G. Melchiorri (1995). Crack propagation in concrete specimens subjected to sustained loads. In F. H. Wittmann (Ed.), *Fracture Mechanics of Concrete Structures*, Germany, pp. 1315–1328. Aedificatio.

Carpinteri, A., S. Valente, F. P. Zhou, G. Ferrara, and G. Melchiorri (1997). Tensile and flexural creep rupture tests on partially-damaged concrete specimens. *Materials and Structures* 30, 269–276.

Dugdale, D. S. (1960). Yielding of steel sheets containing slits. *Journal of Mechanics and Physics of Solids* 8, 100–114.

Enelund, M., L. Mähler, K. Runesson, and B. Lennart Josefson (1999). Formulation and integration of the standard linear viscoelastic solid with fractional order rate laws. *International Journal of Solids and Structures* 36, 2417–2442.

Hillerborg, A., M. Modeer, and P. E. Petersson (1976). Analysis of crack formation and crack growth in concrete by means of fracture mechanics and finite elements. *Cement and Concrete Research* 6, 773–782.

Huang, J. and V. Li (1989). A meso-mechanical model of the tensile behaviour of concrete. *Composites* 20, 370–378.

Karihaloo, B. L. (1995). *Fracture Mechanics and Structural Concrete*. England: Longman Scientific and Technical.

Maier, G. (1971). Incremental plastic analysis in the presence of large displacements and physical instabilizing effects. *International Journal of Solids and Structures* 7, 345–372.

Oldham, K. B. and J. Spanier (1974). *The Fractional Calculus*. New York: Academic Press.

Santhikumar, S. and B. L. Karihaloo (1996). Time-dependent tension softening. *Mechanics of Cohesive-Frictional Materials* 1, 295–304.

Santhikumar, S., B. L. Karihaloo and G. Reid (1998). A model for ageing visco-elastic tension softening material. *Mechanics of Cohesive-Frictional Materials* 3, 27–39.

Zhou, F. P. (1992). *Time-dependent Crack Growth and Fracture in Concrete*. Ph. D. thesis, Report TVBM-1011, Lund University of Technology (Sweden).

Computational Modelling of Concrete Structures, Bićanić et al. (eds)
© 2003 Swets & Zeitlinger, Lisse, ISBN 90 5809 536 3

Coupled moisture-carbon dioxide-calcium transfer model for carbonation of concrete

B. Bary
CEA, DEN/DPC/SCCME/LECBA, France

A. Sellier
LMDC, Université P. Sabatier, France

ABSTRACT: Carbonation mechanisms of concrete are analyzed in this paper accounting for evolutions of relative humidity within the material. The model is based on macroscopic mass balance equations governing diffusion and permeation processes for the water, the carbon dioxide contained in the gaseous phase and the calcium contained in the pore solution. By using an idealized description of the main hydrated products of the cement paste, the dissolution phenomenon can be regarded as depending only on the calcium concentration in the aqueous phase. The calcite formation and the hydrates dissolution are introduced in the mass balance as source terms, and conduct to significant variations of porosity. The mass balance equations are discretized in time and space in the one-dimensional case; the simulation of a concrete wall subjected to combined drying and accelerated carbon dioxide attack is then performed, and the results are compared to experimental data in term of carbonation depth.

1 INTRODUCTION

Durability of cement-based materials is a topic of great concern in many fields of civil engineering. The prediction of the long-term behavior of concrete requires the identification and the analysis of the various deterioration mechanisms that will affect the structure. Among the numerous degradation causes, corrosion of the reinforcing steel induced by deleterious substances reaching the embedded bars is one of the most important (Saetta et al. 1995, Johannesson 1997).

Carbonation of concrete consists in the chemical reaction between carbon dioxide present in the air and the calcium contained in the pore solution, in equilibrium with the hydration products of the cement. One of the main consequences of this phenomenon is the modification of the pH in the pore solution from a standard value between 12.5 and 13.5, to a value below 9 in the carbonated zones. This drop of the pH value leads to the destruction of the protective film (passive layer) which covers the reinforcing steel, so that the metallic iron becomes available and corrosion processes can carry on. As reported by several works, the quantity of water present in the pores are of great importance with regards to the carbonation phenomenon (Saetta et al. 1995, Saetta et al. 1993); indeed, although the carbonation reactions need the presence of water for the cathode process, in high relative humidity conditions the diffusion of both carbon dioxide and oxygen is reduced and even inhibited by the water filling the pores. The most suitable range of internal relative humidity for the carbonation of concrete seems to be 40% to 80% (Saetta et al. 1995).

The object of this paper is to model the atmospheric carbonation processes of concrete when subjected to natural drying (i.e. the material is considered as an unsaturated porous medium) in isothermal conditions; for more details and further analyses of this approach, see (Bary & Sellier, submitted). The theoretical model is based on the mass balance equations for the water, the carbon dioxide flowing through the gaseous phase and the calcium concentration in the pore solution, respectively; these three species are assumed to completely define the state of the material with regard to the carbonation phenomenon. The dissolution of the main hydrates of the cement paste as well as the formation of calcite are introduced in the mass balance equations as source or sink terms; these two mechanisms lead to substantial variations of porosity, and consequently to modifications of the transfer parameters for the three variables. The mass conservation equation for the water rests on the hypothesis that the transfer mechanism is mainly provoked by gradients of capillary pressure arising from the decrease of pore water content due to drying (Baroghel-Bouny et al. 1999, Mainguy et al. 2001); moreover, a water source term takes into account the released water resulting from the dissolution of the hydrated products. The capillary pressure is directly related to the liquid phase (due to the low value of gaseous phase pressure with respect to the liquid phase one), and the moisture transport is assumed to be controlled by this sole liquid phase, the contribution of the vapor transfer being neglected, as recently proposed by Mainguy (Mainguy et al. 2001). The dissolution of the main hydrates of the cement paste is supposed to be governed by the calcium concentration in the aqueous phase; moreover, the simplification consisting in assuming that the cement paste is composed of four main hydrates with regard to the progressive decalcification processes is adopted (Faucon et al. 1998, Adenot 1992). The three resulting partial differential equations are solved numerically in the one-dimensional case by application of a classical linearization procedure consisting in a time and space discretization. The parameters and functions introduced in the formulation are identified or adjusted on existing data and experiments, including accelerated carbonation tests. In the first part of this paper the theoretical model is presented, then an accelerated carbonation test is simulated and compared to experimental results. This comparison shows an acceptable agreement in term of carbonation depth, and the numerical results highlight several phenomena of major importance regarding the carbonation process: the progressive filling of the porous space in the carbonated zone, which has a direct impact on the moisture, carbon dioxide and calcium transfers, and the quantity of solid calcium available for dissolution and calcite precipitation.

2 GOVERNING EQUATIONS

2.1 *Mass conservation of water*

Based on the assumption that the main phenomenon conducting to moisture transport is due only to the liquid flowing through the connected porosity ϕ, the expression of the mass conservation for the water takes the following form:

$$\frac{\partial \rho_l \phi S_r}{\partial t} = -div[w_l] + k_{sl} M_{H_2O} \frac{d\omega_{lS}}{dC_a} \frac{\partial C_a}{\partial t} \tag{1}$$

in which w_l denotes the mass water flux of the liquid phase:

$$w_l = -\frac{K\rho_l}{\eta} k_{rl} grad(p_l) \tag{2}$$

In these two expressions, p_l, M_{H_2O}, S_r and K are the pressure of the liquid phase, the molar mass of water, the saturation degree and the intrinsic permeability coefficient, respectively; ρ_l and η are the density and the viscosity of the water, supposed to be constant. The function k_{rl} affects K by accounting for the variations of saturation degree due to moisture transfer and porosity variation; the following form based on a formulation proposed in (van Genuchten 1980) and adopted by (Baroghel-Bouny et al. 1999) among others is retained:

$$k_{rl} = \sqrt{S_r}\left(1-\left(1-S_r^{\,b}\right)^{1/b}\right)^2 \tag{3}$$

where b is a real scalar, set to 1.65 for the numerical applications. We assume that the intrinsic permeability coefficient depends on the total porosity by the following expression, inspired from (van Genuchten 1980):

$$K = K_0\left(\frac{\phi}{\phi_0}\right)^3\left(\frac{1-\phi_0}{1-\phi}\right)^2 \tag{4}$$

where ϕ_0 and ϕ stand for the reference and current porosity, respectively; K_0 denotes the intrinsic permeability corresponding to the reference porosity ϕ_0. Assuming the pressure of the gaseous phase is negligible with respect to the liquid one, we have $p_{cap} = p_m - p_l \approx -p_l$, with p_{cap} and p_m the capillary and the gas mixture pressure, respectively. Invoking the Kelvin's law which relates the pressure p_{cap} to the relative humidity h_r by:

$$p_{cap}(h_r) = -\rho_l \frac{RT}{M_v} \ln(h_r) \tag{5}$$

and under the hypothesis of the existence of the function $h_r(S_r)$, Equation 2 can be recast as:

$$w_l = -\frac{K\rho_l}{\eta} k_{rl} p'_{cap} h'_r grad(S_r) \tag{6}$$

In Equations 5–6, M_v is the molar mass of the vapor, R and T stand for the ideal gas constant and temperature, respectively; $(f)'$ denotes the derivative of f with respect to its unique variable. The last term of Equation 1 expresses the water mass supply resulting from the progressive dissolution of the different hydrated phases retained for the description of the cement paste; ω_{ls} denotes the water concentration per unit volume of material released by dissolution of hydrated products, which is supposed to be controlled by the sole calcium concentration C_a in the pore solution (Faucon et al. 1998, Mainguy et al. 2000). The function $k_{sl}(S_r)$ describes the reduction of contact surfaces between the water and the hydrated products when the saturation degree decreases; it affects the dissolution processes by acting on the water and calcium mass supplies in the corresponding mass balance equations. This function is chosen as the one proposed by (Bazant et al. 1972):

$$k_{sl}(S_r) = \frac{1}{(1+625(1-S_r)^4)} \quad (7)$$

The current porosity ϕ, strongly dependent on both the dissolution of the hydrated products of the cement paste and the calcite formation is given by:

$$\phi = \phi_0 + V_d(t) - V_{cal}(t) \quad (8)$$

where $V_d(t)$ and $V_{cal}(t)$ are the current volume fractions corresponding to the dissolved volume of idealized hydrated products and the volume of formed calcite, respectively, per unit volume of material. These terms are expressed in the following form:

$$V_{cal}(t) = V_{CaCO_3} \int_0^t \phi(\tau) S_r(\tau) \dot{N}_{cal}(\tau) d\tau \quad (9)$$

$$V_d(t) = \int_{C_a = C_{a0}}^{C_a(t)} k_{sl}(S_r) \left(\frac{dV_{dt}(C_a)}{dC_a} \right) dC_a \quad (10)$$

in which V_{CaCO_3} is the molar volume of the calcite, V_{dt} is the volume fraction corresponding to the total dissolution of the hydrated products (supposed to be reached at total saturation, i.e. with $k_{sl} = S_r = 1$), and $\dot{N}_{cal}$ denotes the rate of calcite formation per volume of pore solution. This rate is chosen to be linearly dependent on carbon dioxide pressure in gaseous phase $p_{\hat{c}}$ and on solute calcium concentration C_a:

$$\dot{N}_{cal} = C k_{gl} p_{\hat{c}} C_a \quad (11)$$

In Equation 11, C is a kinetic temperature-dependant constant fitted on experimental data, and k_{gl} is a normalized transfer function affecting the kinetic constant in function of the saturation degree. This function is proposed in the following form:

$$k_{gl} = \frac{\bar{k}_{gl}}{\bar{k}_{gl\,\max}}, \text{ with } \bar{k}_{gl} = \frac{S_r^{\alpha}(1-S_r^{\beta})}{\left(\frac{\alpha}{\alpha+\beta}\right)^{\alpha}\left(\frac{\beta}{\alpha+\beta}\right)^{\beta}} \quad (12)$$

The choice of $\alpha = 10$ and $\beta = 6$ in Equation 12 results from the assumption of a maximum carbon dioxide transfer between gaseous phase and liquid one for a saturation degree ranging from 55 to 70%. The kinetic constant C is assessed by considering the expression of the reaction rate for the calcite formation proposed in (Papadakis et al. 1991); the estimated value of $C = 0.123\ \mathrm{Pa^{-1}s^{-1}}$ is then obtained (see for details Bary & Sellier, submitted). Finally, when introducing Equations 6, 9 and 10 into Equation 1, the mass conservation of the water is expressed in the following form:

$$\rho_l \phi \frac{\partial S_r}{\partial t} + \left[\rho_l S r k_{sl} \frac{dV_{dt}}{dC_a} - k_{sl} M_{H_2O} \frac{d\omega_{lS}}{dC_a} \right] \frac{\partial C_a}{\partial t}$$
$$- \rho_l S_r V_{CaCO3} \phi S_r \dot{N}_{cal} \quad (13)$$
$$- div\left[\frac{K\rho_l}{\eta} k_{rl} p'_{cap} h'_r grad(S_r) \right] = 0$$

2.2 *Mass conservation of the carbon dioxide*

The mass conservation of the carbon dioxide in the gaseous phase is expressed by the following equation:

$$\frac{\partial(\rho_{\hat{c}} \phi (1-S_r))}{\partial t} = -div[w_{\hat{c}}] - \mu_{gl} \quad (14)$$

where μ_{gl} is the source term describing the dissolution rate of CO_2 into the liquid phase, $\rho_{\hat{c}}$ is the density of CO_2 in the gaseous phase, $w_{\hat{c}}$ is the mass flux of the CO_2. Under the assumption of an immediate reaction between the dissolved carbon dioxide and the calcium, μ_{gl} can be directly related to the molar formation rate of calcite $\dot{N}_{cal}$ by:

$$\mu_{gl} = M_{\hat{c}} \phi S_r \dot{N}_{cal} \quad (15)$$

with $M_{\hat{c}}$ the molar mass of carbon dioxide. The hypothesis formulated in the previous section concerning the homogeneity of the gas mixture pressure p_m within the material still holds; consequently the transport of CO_2 reduces to the only diffusion in the gaseous phase. The choice of a molar-averaged velocity for expressing the energy dissipation due to diffusion transport phenomenon, adopted by (Mainguy et al. 2001, Adenakan et al. 1993) among others, leads to the following formulation for the mass flux of CO_2 in the gaseous phase:

$$w_{\hat{c}} = -f(\phi, S_r) D_{\hat{c}} \frac{M_{\hat{c}}}{RT} p_m grad\left(\frac{p_{\hat{c}}}{p_m} \right) \quad (16)$$

where $D_{\hat{c}}$ is the diffusion coefficient of CO_2 in the atmosphere and $f(\phi, S_r)$ is the resistance factor describing the tortuosity effects and the variation of space offered to the gaseous constituents. Equation 16 does not account for Knudsen effect and dispersion effects due to the heterogeneity of the velocity distribution at microscopic scale, and holds under the condition that the molar-averaged velocity of the gas mixture is negligible (which can be equivalently expressed by $p_m = constant$ within the material). It can be then reduced to:

$$w_{\hat{c}} = -f(\phi, S_r) D_{\hat{c}} \frac{M_{\hat{c}}}{RT} grad(p_{\hat{c}}) \quad (17)$$

The coefficient $fD_{\hat{c}}$ can be regarded as the macroscopic diffusion coefficient of the gaseous constituents through the porous material. Numerous expressions have been proposed by researchers for the reduction factor f involved in Equation 16; we retain the one derived by (Millington 1959) for variably saturated porous media and used by several authors, among others (Mainguy et al. 2001, Adenakan et al. 1993):

$$f(\phi, S_r) = \phi_g^{4/3}(1-S_r)^2 = \phi^{4/3}(1-S_r)^{10/3} \quad (18)$$

where ϕ_g denotes the volume fraction occupied by the gaseous phase. By assuming that the gaseous mixture is composed of ideal gases, the state equation of the CO_2 in the gaseous phase is formulated in the standard form:

$$\rho_{\hat{c}} = \frac{M_{\hat{c}}}{RT} p_{\hat{c}} \quad (19)$$

The introduction of Equations 17, 15 and 19 into Equation 14 gives the following formulation for the mass conservation of the carbon dioxide in the gaseous phase, in which appears the time derivative of the only variables defining the state of the system:

$$\begin{aligned} &(1-S_r) p_{\hat{c}} k_{sl} \frac{dV_{dt}}{dC_a} \frac{\partial C_a}{\partial t} + \phi(1-S_r) \frac{\partial p_{\hat{c}}}{\partial t} - \phi p_{\hat{c}} \frac{\partial S_r}{\partial t} \\ &+ \phi S_r \dot{N}_{cal} (RT - (1-S_r) p_{\hat{c}} V_{CaCO3}) \\ &- div(fD_{\hat{c}} grad p_{\hat{c}}) = 0 \end{aligned} \quad (20)$$

2.3 *Mass conservation of the calcium in the pore solution*

The mass balance equation for the calcium present in the pore solution is formulated by the following expression:

$$\frac{\partial \phi S_r C_a}{\partial t} + div\left[C_a \frac{w_l}{\rho_l} + w_c^{fl}\right] = \dot{C}_S \quad (21)$$

In Equation 21, the first term in the bracket is the convective part of the flux, and characterizes the calcium transport due to the movement of the water containing the solute C_a (Johannesson 1997). The second term in the bracket describes the flux of molecular diffusion of C_a in the water, and is given by:

$$w_c^{fl} = -D_a grad C_a \quad (22)$$

in which D_a is the effective (or global) diffusion coefficient of C_a through the porous material. The right-hand side of Equation 21 stands for the source (or sink) term governed by the dissolution and the calcite formation processes, and which leads to increase (or decrease) the calcium concentration in the liquid phase. By denoting C_{aS} the calcium concentration of the solid phase per unit volume of porous material, which variations are supposed to be only dependent on C_a, and making use of the definition of the calcite formation rate $\dot{N}_{cal}$ and the function k_{sl}, the source term $\dot{C}_S$ introduced in Equation 21 is given by Equation 23:

$$\dot{C}_S = -k_{sl} \frac{\partial C_{aS}}{\partial t} - \phi S_r \dot{N}_{cal} \quad (23)$$

Finally, the mass balance equation for the calcium in the liquid phase is expressed, in function of the time derivative of the state variables, in the following form:

$$\begin{aligned} &\left(\phi S_r + S_r C_a k_{sl} \frac{dV_{dt}}{dC_a} - k_{sl} \frac{dC_{aS}}{dC_a}\right) \frac{\partial C_a}{\partial t} + \\ &\phi C_a \frac{\partial S_r}{\partial t} + \phi S_r \dot{N}_{cal} \left[1 - S_r C_a V_{C_aCO_3}\right] - \\ &div\left(\frac{C_a K}{\eta} k_{rl} p'_{cap} h'_r grad(S_r) + D_a grad(C_a)\right) = 0 \end{aligned} \quad (24)$$

The global diffusion coefficient of C_a depends both on the porosity and on the saturation degree and is assumed to take the following expression:

$$D_a(S_r, \phi) = k_{sl}(S_r) D_e(\phi) \quad (25)$$

where k_{sl} is defined by Equation 7 and D_e is taken from (Tognazzi 1998):

$$D_e = 2{,}3.10^{-13} \exp(9.95\phi(t)) \text{ (in m}^2\text{/s).} \quad (26)$$

3 CHEMICAL DESCRIPTION OF THE HYDRATED PRODUCTS

3.1 *Principles*

Carbonation leads to a progressive decalcification of the hydrated cement paste. This decalcification occurs when the calcium concentration decreases in the pore water because of the calcium consumption induced by the carbonation reaction. The portlandite is first dissolved, followed by AFm, AFt and CSH which are progressively decalcified when their chemical equilibrium with the pore solution is no longer satisfied. The decalcification of cement paste has three main consequences:

- calcium released in pore water,
- hydration water released in porosity,
- porosity volume modification.

These phenomena are taken into account in the mass balance equations presented previously by several coupling terms, which are assessed by adopting a simplified chemical description of the cement paste. This description, proposed by (Adenot 1992) for leaching modeling of cement paste, consists, for an Ordinary Portland Cement (OPC), in assuming that the cement is hydrated in four main components:

- Portlandite (CH): $Ca(OH)_2$,
- CSH (Ca/Si = 1,65): 1,65 $CaO\ SiO_2\ nH_2)O$,
- Monosulphate (AFm): $3CaO\ Al_2O_3\ CaSO_3\ 12H_2O$,
- Ettringite (AFt): $3CaO\ Al_2O_3\ 3CaSO_3\ 32H_2O$.

The component quantities depend linearly on the four main oxyde component quantities present in anhydrous cement; adopting the following notation for mol number of oxyde component contents per unit volume of cement paste: $S = S_iO_2$, $C = C_aO$, $\hat{S} = SO_3$, $A + F = 2Al_2O_3 + 2Fe_2O_3$, we can assess a theoretical composition of the completely hydrated cement paste by solving the following set of equations 27:

$$\begin{cases} C = CH + 1.65CSH + 4AFm + 6AFt\ (or\ 3\ \ Hexa) \\ S = CSH \\ A + F = 2AFt(or\ 2\ Hexa) + 2AFm \\ \overline{S} = 3AFt\ (or\ 0\ Hexa) + AFm \end{cases} \quad (27)$$

3.2 *Simplified chemical equilibrium*

The calcium C_a^{2+} and the carbonates CO_3^{2-} present in pore water are combined according to Equation 28 if the equilibrium condition 29 is fulfilled:

$$C_a^{\ 2+} + CO_3^{2-} \rightarrow C_aCO_3 \quad (28)$$

$$[C_a^{\ 2+}][CO_3^{2-}] = 10^{-8.35} \quad (29)$$

The ionic specie C_a^{2+} is provided in the pore water by dissolution of the hydrated products, whereas CO_3^{2-} results from the dissolution of carbon dioxide present in the gaseous phase. The equilibrium equations governing the carbon dioxide dissolution used in our approach are taken from (Taylor 1997). When carbonation occurs, the consumption of dissolved C_a^{2+} leads to a decreasing of C_a^{2+} concentration in pore water. Firstly, this decreasing is compensated by portlandite dissolution, but after achievement of this process, C_a^{2+} are taken from Aft, AFm and CSH hydrated phases. In the case of leaching, (Faucon et al. 1998) have shown that the decalcification mechanism can be described, in the case of idealized hydrated cement paste, by a chemical zoning characterized by the C_a^{2+} ionic concentration in pore water.

In our approach, the same basic description is adopted and completed by equations 28 and 29 for reproducing carbonation phenomenon. The idealized chemical zoning used for evaluating the ionic concentration of the main species involved in carbonation description is as follows (see Bary & Sellier, submitted for details). The zone 0 represents the sound material and does not contain calcite; it is characterized by the C_a^{2+} concentration equal to 22 mol/m^3. The portlandite $Ca(OH)_2$ is linearly dissolved in zone 1 (where calcite appears at first) when C_a^{2+} varies from 22 to 21 mol/m^3. AFm is progressively dissolved in zone 2, and CSH starts to decalcify (its Ca/Si rate passes from 1.65 to 1.45). In zone 3 Aft is totally decalcified whereas the Ca/Si rate of CSH continues to decrease. In zone 4 and zone 5, CSH progressively decalcify and vanish, and in zone 6, where all the C_a^{2+} are combined in calcite, the material is considered as completely carbonated.

The calculation of the calcium concentrations in the pore water is performed in each of the zone previously defined, by using the equilibrium equations (Adenot 1992, Lea 1970, Damidot et al. 1992), the water dissociation equation and the electro-neutrality condition. Alkali appear to modify chemical concentrations only through this electro-neutrality condition. For a given alkali content and temperature, the set of nonlinear equations give the concentrations of the ionic species in pore water for each of the chemical zones (Bary & Sellier, submitted). Two remarks can be formulated regarding the obtained results: first, as long as the level of C_a^{2+} remains high in pore water, the CO_2 pressure is close to zero, which means that the CO_2 cannot reach a sound zone with a high pressure value, and consequently the consumption of C_a^{2+} and CO_2 occurs on a very localized zone. This remark is in good agreement with usual experimental observations and with assumptions formulated in several simplified carbonation modeling proposed in the literature (see e.g. Saetta et al. 1993, Papadakis et al. 1991). The second remark concerns the C_a^{2+} progressive decreasing between the sound material (zone 0, $C_a^{2+} = 22$ mol/m^3) and completely carbonated material (zone 6, $C_a^{2+} \leq 0.4$ mol/m^3); this fact confirms that carbonation effects on solid hydrated phases can actually be described by using the C_a^{2+} concentration in pore solution, as in the case of the leaching phenomenon.

3.3 *Source terms*

The assumptions concerning the solid phases present in each zone allow to assess the calcium content in residual hydrated phases during carbonation process. Thus, the source term of the calcium mass conservation equation, denoted dC_{aS}/dC_a in Equation 24, depends on C_a^{2+} and is expressed in (Bary & Sellier, submitted). The hydrate dissolution leads to a progressive decreasing of solid phase volume according to the chemical zoning description. Starting from the molar volume of each solid phase present in a zone, we can assess the total dissolved volume in function of the calcium concentration, and consequently the term dV_{dt}/dC_a appearing in the conservation equations 13, 20, 24 can be estimated explicitly (Bary & Sellier, submitted). Further, during carbonation process, the hydrated product dissolution leads to release hydration water in porosity, which modifies the degree of saturation, and consequently the connectivity of the gaseous phase and the liquid phase network. This released water is taking into account in the mass conservation equation of water 13 by the term $d\omega_{lS}/dC_a$. To assess this term we use the mole number of H_2O given by the usual chemical composition of the hydrated products. (i.e. 12 moles per mole of *AFm*, 32 moles for *AFt*, 1 for *CH*); for *CSH*, the H_2O mole number "n" depends on the Ca/Si rate. According to (Fuji et al. 1981) "n" is given by the following relationship:

$$n = Ca/Si + 0.8 \quad (30)$$

4 NUMERICAL APPLICATIONS

The three partial differential equations 13, 20 and 24 are solved numerically in the one-dimensional case; the simulation of the behaviour of a concrete wall subjected to accelerated carbonation is then carried out and analyzed.

4.1 *Numerical procedure*

The implementation of equations 13, 20 and 24 in 1-D is obtained by applying a classical linearization procedure, which consists in discretizing these equations in both time and space domains, conducting to a system of three algebraic non-linear equations for each time step. The non-linearity results from the solution-dependency of the coefficients involved in the governing equations. The space and time domains are discretized by using a finite difference scheme and the theta-method, respectively. The latter is characterized by a linear approximation of the evolution of the unknowns within each time step, i.e. when denoting $\theta = (t - t_n)/(t_{n+1} - t_n)$ with $0 \leq \theta \leq 1$, evolutions of the state variable $X(t)$ within $[t_n, t_{n+1}]$ are defined by:

$$X(t) = (1-\theta)X(t_n) + \theta X(t_{n+1}) \quad (31)$$

The coefficients of the algebraic equations which depend on the unknown variables are assessed, in the time domain $[t_n, t_{n+1}]$, by making use of the converged solution of the two previous steps at t_n and t_{n-1}; then the value of the coefficient K is given by:

$$K_{\beta_n} = (1+\beta)K_{t_n} - \beta K_{t_{n-1}} \text{ with } 0 \leq \beta \leq 1 \quad (32)$$

4.2 *Simulation of an accelerated carbonation test*

Boundary conditions: The boundary conditions of the simulated structure are defined by:

- Flux set to zero for the three species at one extremity, corresponding to symmetry conditions, and describing the conditions which prevail in the core of the structure,
- Prescribed values of both carbon dioxide and saturation degree, and flux of calcium ions set to zero at the other extremity, representing the conditions at the surface exposed to external aggression.

4.2.1 *Chemical properties*

As an example, we have simulated an accelerated carbonation test on high performance concrete specimen carried out by (Miragliotta 2000). The cement used is a *CPA* CEM I 52.5 R; its average chemical composition is given in Table 1.

Aggregates used are: sand 0/4 mm of real density 2550 Kg/m^3, and fine crushed gravel 6/14 mm of real density 2600 Kg/m^3; these aggregates are inert with regard to the carbonation process. The concrete composition is described in Table 2.

This concrete composition leads to a compressive strength at 28 days of 65.9 MPa. The water/cement ratio calculated from Table 2 is equal to 0.43 and leads to the cement paste chemical composition given on Table 3. Using Equation 27, we obtain the idealized hydrated cement paste composition shown in Table 4.

According to the chemical zoning description detailed in (Bary & Sellier, submitted) and with the

Table 1. Cement chemical composition CPA CEM I 52.5 R.

(% mass)	
Fe_2O_3	2.38
Al_2O_3	5.19
CaO	64.02
SO_3	3.5
H_2O	2.23
SiO_2	19.81

values calculated in Table 4, the source term dC_{aS}/dC_a used in the calcium mass conservation equation 24 and defined per m^3 of cement paste, is shown on Figure 1. Further, the evolutions of released water due to hydrate dissolution and volume of dissolved hydrates, corresponding to the term $d\omega_{lS}/dC_a$ and dV_{dt}/dC_a in Equation 13, are depicted on Figure 2 and Figure 3 for one m^3 of cement paste, respectively.

Table 2. Mix proportions in concrete.

(kg/m^3)	
cement	323
water	142
sand 0/4	721
Fine gravel 6/14	1157

Table 3. Composition of one liter of cement paste.

mole/m^3 of cement paste				
C	S	A+F	S	H_2O
14860	4290	1710	570	1610

Table 4. Idealized composition of one m^3 of hydrated cement paste.

Hydrates	mole/m^3 of cement paste
AFm	570
Hexa	290
CH	4350
CSH	4290

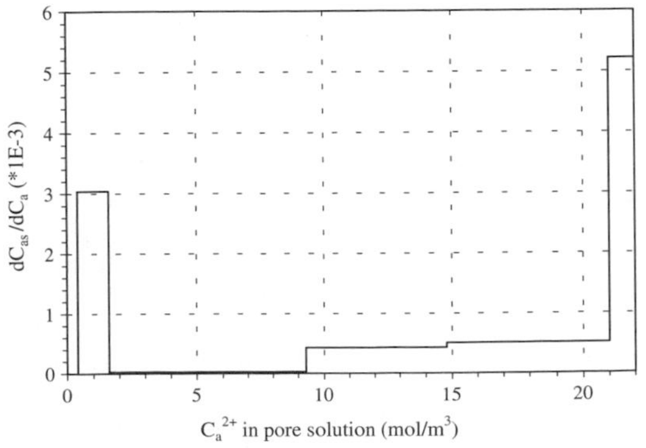

Figure 1. Dissolved calcium source used in calcium mass conservation equation.

The source terms shown on Figure 1, Figure 2 and Figure 3 are calculated for one m^3 of cement paste, but since in our case the concrete contains 25% of hydrated cement paste, these source terms must be reduced by this rate before being transposed in the mass conservation equations for concrete.

4.2.2 *Physical properties*

The relation between the relative humidity h_r and the saturation degree S_r is given by the inversion of the desorption isotherm expression adopted to reproduce the experimental results in the range $h_r = 60\%$ to $h_r = 100\%$ taken from (Baroghel-Bouny et al. 1999). This expression is given in the following simplified form:

$$S_r = Ah_r^3 + Bh_r^2 + Ch_r + D \qquad (33)$$

The coefficients appearing in Equation 33 are fitted to the values: $A = 6.43$, $B = -15.46$, $C = 12.52$ and $D = -2.50$ (see Figure 4).

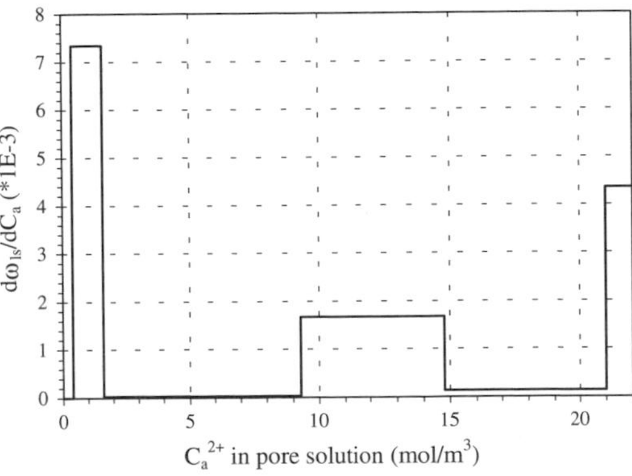

Figure 2. Released water source term for dissolved hydrates.

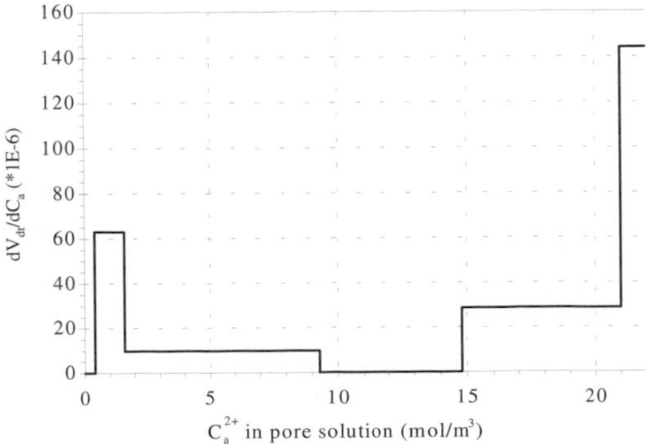

Figure 3. Source term corresponding to the volume of dissolved hydrates.

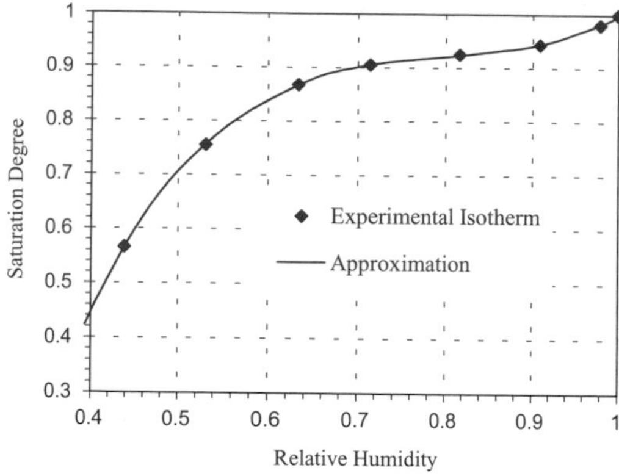

Figure 4. Experimental and approximated desorption isotherm.

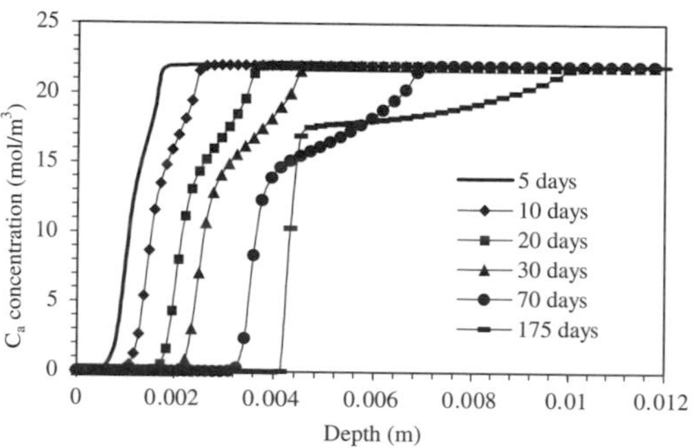

Figure 5. Calcium concentration in porosity water versus depth for different times between 5 and 175 days.

4.2.3 *Numerical Results*

The following numerical values are used for performing the simulations, some of them drawn from (Miragliotta 2000) when available: $\phi = 0.094$, $K\rho_l/\eta = 10^{-15}$ m/s, $V_{CaCO_3} = 36.910^{-6}\text{m}^3/\text{mol}, D_{\hat{c}} = 8.0\ 10^{-7}\text{m}^2/\text{s}$; the boundary conditions are $p_{\hat{c}max} = 50\,\text{kPa}$, $h_r = 65\%$ and $C_a^{2+} = 0$, whereas the initial conditions are set to $p_{\hat{c}} = 0$, $h_r = 94\%$ and $C_a^{2+} = 22\,\text{mol/m}^3$, with the length of the concrete specimen equal to 5 cm.

The Figure 5 presents the calcium concentration profiles numerically obtained at different times between 5 and 175 days. The carbonated zones correspond to the zones where the calcium concentration is closed to zero. It can be observed that in the first stage (5 days to 70 days), the carbonation front propagation is mainly governed by CO_2 diffusion in the carbonated zone. After this stage, the concentration profile is modified due to a progressive filling of the porosity by calcite, as illustrated by Figure 7. The porosity filling occurs progressively because it is conditioned by the quantity of dissolved calcium arriving from the sound

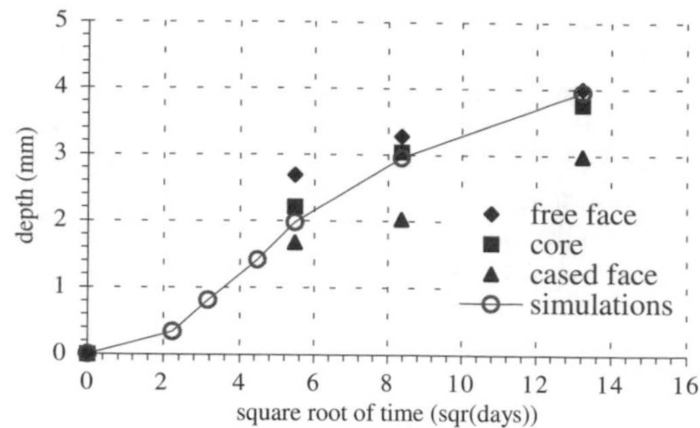

Figure 6. Carbonation depth versus square root of time.

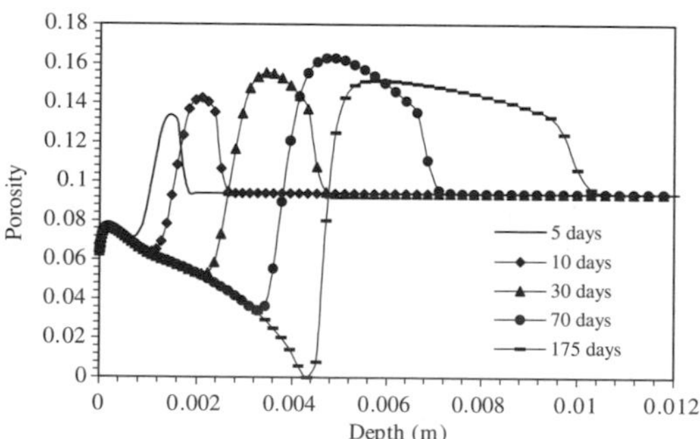

Figure 7. Simulated porosity profiles for different time between 5 and 175 days.

zone, this quantity being essentially governed by diffusion processes. The decreasing rate of the carbonation front propagation induced by the progressive porosity filling can also be observed on Figure 6, where numerical prediction as well as experimental data of the carbonation depth versus time show a propagation rate lower than the one obtained with a constant diffusion coefficient (i.e. proportional to the square root of time). It can also be remarked on Figure 5 the slight decrease of calcium concentration behind the carbonation front, which means that dissolution process (essentially reduced to the portlandite) occurs in this zone.

In the (Miragliotta 2000) study, the specimens subjected to carbonation were extracted from different parts of a same concrete block (cased face, free face without any cure, or core of the block), aiming at analyzing the effects of different porosity on the carbonation kinetic. However, numerical simulations presented here have been carried out only on the concrete corresponding to the core of the concrete block, where the initial porosity is assumed to be homogeneous. In this configuration, the model appears to correctly reproduce experimental results.

The Figure 7 shows the porosity profiles obtained numerically for different time ranging from 5 to 175 days; the face exposed to atmospheric carbonation is located at depth = 0. The progressive filling of the porosity in the carbonated zone appears clearly. We can notice that this filling increases with the carbonation front progression. Simultaneously, an increasing of the porosity occurs just ahead of the carbonation front, which results from the calcium leaching (diffusion) from the sound zone to the carbonated zone due to the calcium concentration gradient. These observations permit to conclude that the possibility of a total porosity filling depends not only on the quantity of calcium locally available in the zone where carbonation occurs, but also on the calcium dissolved in the pore solution and transported by convection and diffusion from the sound zone. The *CH* quantity, alkali contents, and transfer properties are of major importance regarding the porosity filling phenomenon.

5 CONCLUSION

The approach developed in this paper for describing the carbonation of concrete takes into account simultaneously carbon-dioxide, dissolved calcium, and liquid water transfers. The coupled partial differential equations set is obtained by using the classical mass balance principle where both diffusive and convective transfers are considered. Chemical equilibrium are introduced using a simplified chemical zoning on idealized hydrated cement paste. This method allows to estimate the source or sink terms used in mass balance equations according to the chemical composition of cement. Diffusion coefficients, gaz permeability and water permeability are affected by the dissolution or the precipitation phenomena and by the saturation degree variation. The functions introduced for linking the porosity and the saturation degree variations to the transfer coefficients are fitted on experimental data or taken from literature.

The simulation of an accelerated carbonation test and its comparison to experimental data shows an acceptable agreement in term of carbonation depth. Moreover, the numerical results point out some aspects which turn out to be of importance with regard to the carbonation mechanisms; the main of these aspects are the progressive filling of the porous space in the carbonated zone (which has in turn a direct impact on the moisture, carbon dioxide and calcium transfers properties), and the quantity of solid calcium available for dissolution and calcite precipitation. However these points have to be justified and validated by experimental investigations; some of them, in which a particular attention will be paid to the microstructure analysis of the carbonated zone, are planned to be carried out within one year. A subsequent paper will be focused on these experimental results and their confrontation to numerical simulations.

ACKNOWLEDGMENTS

Fruitful discussions and financial support from the French national agency for nuclear waste management (ANDRA) is gratefully acknowledged.

REFERENCES

Adenakan, A. E., Patzek, T., & Pruess, K. 1993. Modeling of multiphase transport of multicomponent organic contaminants and heat in the subsurface: numerical model formulation. *Wat. Res. Res.* 29(11): 3722–3740.

Adenot, F. 1992. *Durabilité des bétons: caractérisation et modélisation des processus physiques et chimiques de dégradation du ciment.* PHD thesis of the University of Orléans (in french).

Baroghel-Bouny, V., Mainguy, M., Lassabatère, T., & Coussy, O. 1999. Characterization and identification of equilibrium and transfer moisture properties for ordinary and high-performance cementitious materials. *Cement and Concrete Research* 29: 1225–1238.

Bary, B., & Sellier, A. Coupled moisture-carbon dioxide-calcium transfer model for carbonation of concrete. Submitted to: *Cement and Concrete Research.*

Bazant, Z. P., & Najjar, L. J. 1972. Nonlinear water diffusion in nonsaturated concrete. *Mat. Struc.* 5(25): pp 3–20.

Damidot, D. & Glasser, F.P. 1992. Thermodynamic investigation of CaO-Al_2O_3-$CaSO_4$-H_2O system at 25°C and the influence of Na_2O, *Cement and Concrete Research*: 22–28.

Faucon, P., Adenot, F., Jacquinot, J. F., Petit, J. C., Cabrillac, R., & Jorda, M. 1998. Long-term behaviour of cement pastes used for nuclear waste disposal: review of physico-chemical mechanisms of water degradation. *Cement and Concrete Research* 28(6): 847–857.

Fuji, K. & Kondo, W. 1981. Heterogeneous equilibrium of calcium silicate hydrate in water at 30°C, *J. Chem. Soc., Dalton Trans:* 645–651.

Johannesson, B. F. 1997. Nonlinear transient phenomena in porous media with special regard to concrete and durability. *Advanced Cement Based Materials* 6: 71–75.

Lea R. 1970. *The chemistry of cement and concrete.* 3rd edition, Arnold.

Mainguy, M., Coussy, O., & Baroghel-Bouny, V. 2001. Role of air pressure in drying of weakly permeable materials. *Journal of Engineering Mechanics* 127(6): 582–592.

Mainguy, M., Tognazzi, C., Torrenti, J. M., & Adenot, F. 2000. Modelling of leaching in pure cement paste and mortar. *Cement and Concrete Research* 30: 83–90.

Millington, R. J. 1959. Gas diffusion in porous media. *Science* 130: 100–102.

Miragliotta, R. 2000. *Modélisation des processus physico-chimiques de la carbonatation des bétons préfabriqués – Prise en compte des effets de parois.* PHD thesis of the University of la Rochelle (in french).

Papadakis, V. G., Vayenas, C. G., & Fardis, M. N. 1991. Fundamental modeling and experimental investigation of concrete carbonation. *ACI Materials Journal* 88(4): 363–373.

Saetta, A. V., Schrefler, B. A., & Vitaliani, R. V. 1993. The carbonation of concrete and the mechanism of moisture, heat and carbon dioxide flow through porous materials. *Cement and Concrete Research* 23(4): 761–772.

Saetta, A. V., Schrefler, B. A., & Vitaliani, R. V. 1995. 2-D model for carbonation and moisture/heat flow in porous materials. *Cement and Concrete Research* 25(8): 1703–1712.

Taylor, H.F.W. 1997. *Cement Chemistry,* 2nd edition, Thomas Telford Ed, ISBN 0-7277-2592-0.

Tognazzi, C. 1998. Couplage fissuration – dégradation chimique dansles matériaux cimentaires: caractérisation et modélisation, thèse de doctorat de l'Institut National des sciences Appliquées de Toulouse, soutenue le 11 décembre (in french).

van Genuchten, M.T. 1980. A closed-form equation for predicting the hydraulic conductivity of unsaturated soils. *Soil Sci. Soc. Am. Proc* 44: 892–898.

Computational Modelling of Concrete Structures, Bićanić et al. (eds)
© 2003 Swets & Zeitlinger, Lisse, ISBN 90 5809 536 3

An unified approach for the modeling of drying shrinkage and basic creep of concrete: analysis of intrinsic behaviour and structural effects

F. Benboudjema
Laboratoire de Mécanique et Technologie – École Normale Supérieure de Cachan, Cachan, France

F. Meftah
Laboratoire de Mécanique – Université de Marne La Vallée, Champs sur Marne, France

J.M. Torrenti
Institut de Radioprotection et de Sûreté Nucléaire, Fontenay-Aux-Roses Cedex, France

ABSTRACT: In this paper, a numerical modeling of the delayed behavior of concrete is presented. The model takes into account drying, shrinkage, creep and cracking phenomena. The coupling between all these phenomena is performed by using the theory of non-saturated porous media and the effective stress concept. The analysis of the long-term behavior is performed on a concrete specimen subjected to drying. Numerical simulations are compared to experimental data from the literature, in order to check the abilities of the model to describe the hydro mechanical behavior of concrete.

1 INTRODUCTION

Non-homogeneous drying induces the occurrence of cracking which interacts with concomitant creep and shrinkage mechanisms. This means that structural effect affects, simultaneously, measurements of shrinkage and creep deformations in any experiment (Granger 1996). Unfortunately, no experimental procedure allows for separating, easily, intrinsic behaviors from structural effects when shrinkage and creep are concerned. Identifying constitutive laws from experimental measurements cannot be performed in a straightforward way. It needs the use of a robust cracking model in order to quantify accurately, by means of computations, the structural part. The intrinsic behavior of the drying concrete can therefore be deduced in a sort of inverse analysis on focusing on the conventional components: drying shrinkage and basic creep.

A hydro-mechanical model is developed, including the description of drying, shrinkage, creep and cracking phenomena for concrete as a non-saturated porous medium. The modeling of drying shrinkage is based on a unified approach of creep and shrinkage. Basic creep model is based on chemo-physical mechanisms, which occur at different scales of the cement paste. Indeed, basic creep is explicitly related to the micro-diffusion of the adsorbed water between interhydrates and intrahydrates and the capillary pores, and the sliding of the C-S-H gel at the nano-porosity level. Drying shrinkage is, therefore, assumed to result from the elastic and creep response of the solid skeleton, submitted to both capillary and disjoining pressures. Furthermore, the cracking behavior of concrete is described by an orthotropic elastoplastic damage model. The coupling between all these phenomena is performed by using effective stresses which account for both external applied stresses and pore pressures.

This model has been incorporated in a finite element code. The analysis of the long-term behavior is performed on concrete specimens subjected to drying.

2 HYDRO-MECHANICAL MODELING

In this section, we present the modeling framework for concrete in partially saturated conditions. The formulation of the model is based on the mechanics of porous media, using the effective stress concept.

The vector of the total strain $\boldsymbol{\varepsilon}$ is split into 3 components:

$$\boldsymbol{\varepsilon} = \boldsymbol{\varepsilon}^{e} + \boldsymbol{\varepsilon}^{p} + \boldsymbol{\varepsilon}^{bc} \quad (1)$$

where $\boldsymbol{\varepsilon}^{e}$ = elastic strain; $\boldsymbol{\varepsilon}^{p}$ = plastic strain; and $\boldsymbol{\varepsilon}^{bc}$ = basic creep strain.

The drying shrinkage strain does not appear in the decomposition (equation 1). The modeling of drying shrinkage is based on an unified approach of creep and shrinkage. Indeed, we suggest that that drying shrinkage is driven by pore pressures. Therefore, this strain results implicitly from the elastic and creep straining of the material due to capillarity and disjoining pressure.

The modeling of each studied phenomena (drying, cracking, creep and shrinkage) is successively presented.

2.1 *Drying model*

The drying of concrete is modeled here by a diffusion-type equation, i.e. second Fick's law:

$$\dot{C} = \nabla\left(D(C)\nabla C\right) \tag{2}$$

in which C = water content; and D = the diffusivity, which varies in a strongly non-linear manner as a function of the water content. The dot represents the derivative with respect to time.

The diffusivity is calculated with the relationship derived by Xi et al. (1994):

$$D(h) = D_0\left[1 + a\left(1 - 2^{-10^{b(h-1)}}\right)\right] \tag{3}$$

where D_0, a and b are material parameters, depending upon the concrete formulation.

Eq. (2) and (3) taken into account the migration of both of liquid and vapor phases in concrete.

The relative humidity is relied on the water content by the desorption isotherm curve. The BSB model (Xi et al. 1994), called also the three-parameter BET model, is used here to calculate the desorption isotherm curve:

$$C = \frac{AkV_m h}{(1-kh)\left[1+(A-1)kh\right]} \tag{4}$$

where A, k and V_m are material parameters of the BSB model.

The drying boundary conditions are of a convective type. The exchanged flux of moisture per unit surface between the exposed faces of concrete and the ambient air $\mathbf{J}$ is as follow (Torrenti et al. 1999):

$$\mathbf{J} = \left[\beta_{cf}\left(\left(2C_0 - C_{eq}\right) - C_s\right)\right]\left(C_s - C_{eq}\right)\mathbf{n} \tag{5}$$

where β_{cf} = a constant parameter equal to 5.10^{-10} $m^4/s/l$; C_0 = initial water content; C_s = the water content on the drying face; C_{eq} = the water content corresponding to the environmental relative humidity; and $\mathbf{n}$ is the normal vector to the drying surface (oriented toward the exterior).

It is considered here that drying-induced cracking does not influence significantly the drying process. As a matter of fact, experimental results show that a non-loaded specimen and a loaded specimen in compression dry in the same manner (Lassabatère et al. 1997), even if the compressive loading prevents from pronounced microcracking. Moreover, drying-induced cracking is not important. Their opening is less than 50 μm (Sicard et al. 1992, Bisschop & van Mier 2002). Therefore, it can be expected that drying-induced cracking has little influence on the drying process.

2.2 *Cracking model*

The behavior of cracked concrete is modeled by a damage model coupled with softening plasticity, developed by the authors (Benboudjema et al. 2001). The plastic strain describes irreversible deformation observed experimentally at unloading. The accompanying stiffness degradation due to microcracks is given by the second order damage tensor $\mathbf{D}$.

The cracked material is considered to be a material, the effective surface (resistant) of which is reduced due to the cracking process (see Figure 1). The vector of the nominal stress $\boldsymbol{\sigma}$ is related to the vector of the effective stress $\tilde{\boldsymbol{\sigma}}$, which acts on the uncracked material only, by the following relationship:

$$\boldsymbol{\sigma} = (\mathbf{I} - \mathbf{D})\cdot\tilde{\boldsymbol{\sigma}} \tag{6}$$

where $\mathbf{I}$ is the second order unit tensor.

Effectives stresses are related to the elastic strain by:

$$\tilde{\boldsymbol{\sigma}} = \mathbf{E}^0\cdot\boldsymbol{\varepsilon}^e \tag{7}$$

where $\mathbf{E}^0$ is the second order elastic stiffness tensor.

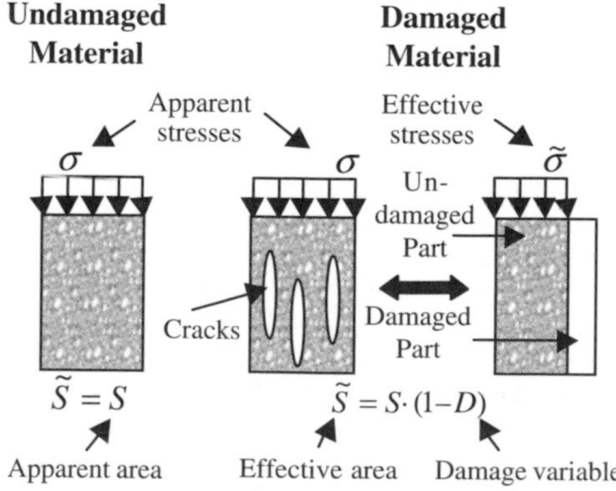

Figure 1. Definition of the damage variable.

2.2.1 *Damage evolution*

In order to describe properly difference of damage process in compression and in tension, the damage variable is separated into a compressive and a tensile one. The damage process is assumed here to be isotropic in compression and orthotropic in tension, where orthotropy is induced by cracking (see Fig. 1). Hence, a scalar damage variable D_c is used in compression, while a tensorial one $\mathbf{D}_t$ is considered in tension. The damage tensor $\mathbf{D}$ is then given by (Benboudjema 2002):

$$\mathbf{I}-\mathbf{D}=(1-D_c)(\mathbf{I}-\mathbf{D}_t) \tag{8}$$

Damage evolution is related to the cumulative plastic strains. As a matter of fact, experimental evidences show that this choice is relevant for concrete (Ju 1989). The evolution function is of exponential type (Lee & Fenves 1998, Nechnech 2000):

$$\begin{cases} 1-D_c(\kappa_c)=\exp(-c_c\kappa_c) \\ 1-D_t^{ii}(\kappa_t^{ii})=\exp(-c_t\kappa_t^{ii}) \end{cases} \tag{9}$$

where κ_c and κ_t^{ii} are the compressive and ith principal tensile cumulative plastic strains, respectively, given by Equation 15.

2.2.2 *Plastic evolution*

The coupling between damage and plasticity is based on the effective stress concept and on the hypothesis that the undamaged material behavior is elasto-plastic (Ju 1989). In order to reproduce a suitable behavior both in compression and in tension, a Drucker-Prager criterion in compression and three Rankine criteria in tension are used (see Figure 2). The use of 3 independent criteria allows for retrieving an orthotropic behavior.

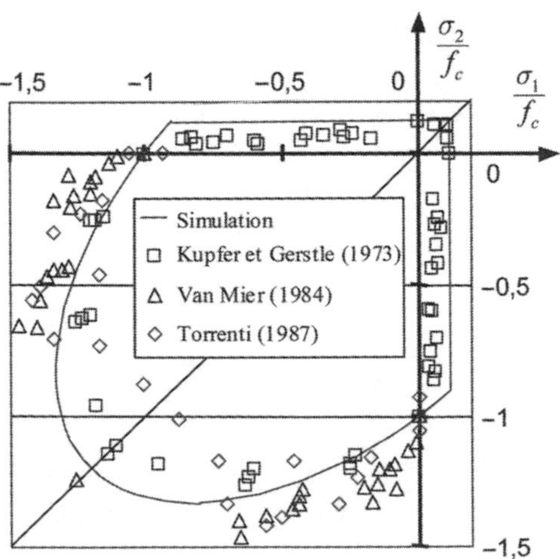

Figure 2. Drucker-Prager and Rankine criteria in the principal stress space (2D).

This choice has been previously made by many authors (Feenstra 1993, Heinfling 1998, Nechnech 2000) in the isotropic case.

The Drucker-Prager criterion is written as:

$$F_c(\tilde{\boldsymbol{\sigma}},\kappa_c)=\sqrt{3J_2(\tilde{\boldsymbol{\sigma}})}+\alpha_f I_1(\tilde{\boldsymbol{\sigma}})-\beta\tilde{\tau}_c(\kappa_c) \tag{10}$$

where $J_2(\tilde{\sigma})=$ the second invariant of the effective stress $\tilde{\sigma}$; $I_1(\tilde{\sigma})=$ the first invariant; $\tau_c=$ the nominal strength in compression; α_f and β are two material parameters.

The Rankine criteria are written as:

$$F_t^i(\tilde{\boldsymbol{\sigma}},\boldsymbol{\kappa}_t)=\tilde{\sigma}_{ii}-\tilde{\tau}_t^i(\kappa_t^{ii}) \tag{11}$$

where $\tau_t=$ the nominal strength in tension.

The nominal strengths are defined by:

$$\tilde{\tau}_x^{ii}=f_{x0}\left[(1+a_x)\exp(-b_x\kappa_x^{ii})-a_x\exp(-2b_x\kappa_x^{ii})\right] \tag{12}$$

where a_x and b_x are material parameters identified from an uniaxial test. The subscript x refers to tension (t) or compression (c).

The non-associative plastic flow theory is adopted in compression:

$$G_c=\sqrt{3J_2(\tilde{\boldsymbol{\sigma}})}+\alpha_g\, I_1(\tilde{\boldsymbol{\sigma}})-\beta\,\tilde{\tau}_c(\kappa_c) \tag{13}$$

where α_g is a material parameter which controls dilatancy of concrete.

The plastic strain rate is then obtained by Koiter assumption:

$$\dot{\boldsymbol{\varepsilon}}^p=\dot{\lambda}_t^{ii}\sum_i\frac{\partial F_t^i}{\partial\tilde{\boldsymbol{\sigma}}}+\dot{\lambda}_c\frac{\partial G_c}{\partial\tilde{\boldsymbol{\sigma}}} \tag{14}$$

where λ_x^i is the plastic multiplier associated to the plastic potential functions in compression and in tension.

The cumulated plastic strains, used as hardening/softening parameters, are defined by (in the system of the principal effective stresses for the tensile component):

$$\begin{cases} \dot{\kappa}_c=(1+2\alpha_g{}^2)^{1/2}\,\dot{\lambda}_c \\ \dot{\kappa}_t^{ii}=\dot{\lambda}_t^{ii} \end{cases} \tag{15}$$

Strain softening induces inherent mesh dependency and produces failure without energy dissipation (Bažant 1976). In order to avoid these features, the fracture energy approach, proposed by Hillerborg et al.

(1976) is used. The fracture energy density g_{fx} is related to fracture energy G_{fx} by a characteristic length l_e:

$$g_{fx} = \frac{G_{fx}}{l_e} \tag{16}$$

The characteristic length l_e is related to the size of the finite element (Feenstra 1993):

$$l_e = k_e \sqrt{A} \tag{17}$$

where k_e is a coefficient which depends upon the type of finite element.

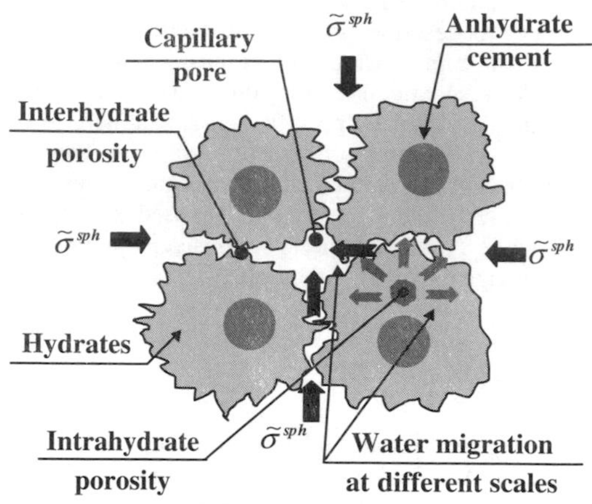

Figure 3. Proposed mechanism for the spherical creep. (Benboudjema et al. 2001)

2.3 *Basic creep model*

Basic creep of concrete is still a controversial issue. Many explanations for the mechanisms have been proposed in the literature to retrieve the collected experimental evidences (Jennings & Xi 1992). However, no theory has been universally accepted yet, although it is well admitted that water plays a fundamental role.

In this paper, a multiaxial model developed by the authors is used (Benboudjema et al. 2001), where the role of water is integrated in a original manner. In this model, the basic creep is considered to be the result of two major mechanisms. They are driven by the spherical and deviatoric components of the stress tensor, respectively. Several experimental findings prove that the splitting of the creep strain process to a spherical part and a deviatoric part is relevant (Benboudjema et al. 2001). Indeed, they showed that the spherical creep strains and the deviatoric creep strains are proportional to the spherical part and the deviatoric part of the stress tensor, respectively.

Each part of the creep strain process is therefore associated with a different chemo-physical mechanism. The decomposition of the basic creep strains vector $\boldsymbol{\varepsilon}^{bc}$ reads therefore:

$$\boldsymbol{\varepsilon}^{bc} = \boldsymbol{\varepsilon}_{bc}^{dev} + \varepsilon_{bc}^{sph}\,\mathbf{1} \tag{18}$$

where $\boldsymbol{\varepsilon}_{bc}^{sph}$ and $\boldsymbol{\varepsilon}_{bc}^{dev}$ are the spherical and the deviatoric creep strains respectively. The vector $\mathbf{1}$ reads:

$$\mathbf{1} = \begin{bmatrix} 1 & 1 & 1 & 0 & 0 & 0 \end{bmatrix}^T \tag{19}$$

It should be emphasized that such a decomposition of the creep strain (in a spherical and a deviatoric part) has been previously proposed (Bažant 1988).

2.3.1 *Spherical creep*

The spherical part is assumed to occur in the microporosity (0,01 – 50 μm range). It is associated to the migration of adsorbed water, located at the interface between hydrates and the hydrates intrinsic porosity, towards the capillary pores (Fig. 3).

This mechanism has been previously suggested by many authors (see Benboudjema 2002). As a matter of fact, several experimental findings confirms this theory.

By assuming that the behavior of the hydrated and the unhydrated cement particles are elastic and that the migration of water follows the Poiseuille equation, the adopted mechanisms lead to the following system of equations:

$$\begin{cases} \dot{\varepsilon}^{sph} = \dfrac{1}{\eta_r^{sph}}\left[h\tilde{\sigma}^{sph} - k_r^{sph}\varepsilon_r^{sph}\right] - 2\dot{\varepsilon}_i^{sph} \\ \dot{\varepsilon}_i^{sph} = \dfrac{1}{\eta_i^{sph}}\Big\langle\left[k_r^{sph}\varepsilon^{sph} - \left(k_r^{sph} + k_i^{sph}\right)\varepsilon_i^{sph}\right] \\ \qquad\qquad - \left[h\tilde{\sigma}^{sph} - k_r^{sph}\varepsilon_r^{sph}\right]\Big\rangle^+ \end{cases} \tag{20}$$

with:

$$\varepsilon^{sph} = \varepsilon_r^{sph} + \varepsilon_i^{sph} \text{ with } \langle x\rangle^+ = \frac{x + |x|}{2} \tag{21}$$

where ε_r^{sph} and ε_i^{sph} are the reversible and the irreversible spherical creep strain respectively; η_r^{sph} and η_i^{sph} are the apparent viscosities of the water at two different scales of the material (macroscopic and microscopic level, respectively). These apparent quantities depend upon the water viscosity and the connected

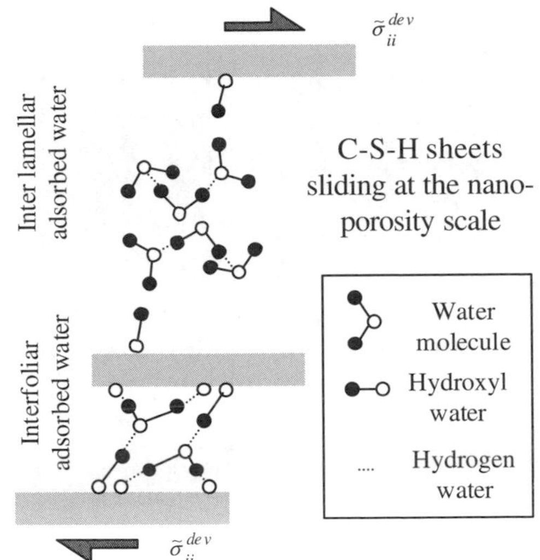

Figure 4. Mechanisms of the deviatoric creep in the C-S-H nanopores. (Benboudjema et al. 2001)

porosity geometry. Further, k_r^{sph} and k_i^{sph} are the apparent stiffness associated to the precedent viscosities and related to the stiffness of the porous material and the skeleton. $\tilde{\sigma}^{sph}$ is the spherical effective stress.

2.3.2 *Deviatoric creep*

The deviatoric part is supposed to be caused by the sliding of the C-S-H layers (see Benboudjema 2002). This phenomenon occurs in the nanoporosity (dimension of about 1 nm).

The deviatoric creep mechanism is presented in Figure 4.

As the case of the spherical basic creep, the deviatoric creep strain vector $\boldsymbol{\varepsilon}_{bc}^{dev}$ is split in a reversible part $\boldsymbol{\varepsilon}_r^{dev}$ and an irreversible part $\boldsymbol{\varepsilon}_i^{dev}$:

$$\boldsymbol{\varepsilon}_{bc}^{dev} = \boldsymbol{\varepsilon}_r^{dev} + \boldsymbol{\varepsilon}_i^{dev} \tag{22}$$

The reversible part is associated to the interfoliar adsorbed water (great adsorption energy). The irreversible part is due to the rupture of the hydrogen bridge in the interlamellar adsorbed water.

The physical mechanism of the deviatoric creep leads to constitutive relations:

$$\begin{cases} \eta_r^{dev}\dot{\boldsymbol{\varepsilon}}_r^{dev} + k_r^{dev}\boldsymbol{\varepsilon}_r^{dev}(rev) = h\tilde{\boldsymbol{\sigma}}_{ii}^{dev} \\ \eta_i^{dev}\dot{\boldsymbol{\varepsilon}}_i^{dev} = h\tilde{\boldsymbol{\sigma}}^{dev} \end{cases} \tag{23}$$

The Eq. (20) and (23) can be solved analytically for constant stresses and a constant relative humidity. The basic creep strains vector $\boldsymbol{\varepsilon}_{bc}$ can be expressed as:

$$\boldsymbol{\varepsilon}_{bc}(t) = h\,\mathbf{J}_{bc}(t)\cdot\tilde{\boldsymbol{\sigma}} \tag{24}$$

where $\mathbf{J}_{bc}$ is the basic creep compliance tensor (second order), depending upon the materials parameters (Benboudjema 2002).

2.4 *Drying shrinkage model*

Concrete is a material which is strongly hydrophilic and has an important specific surface. Indeed, it exhibits a behavior very sensitive to the hygrometric conditions.

The modeling of drying shrinkage is based on the mechanisms of disjoining pressure and capillary pressure, which seem to be predominant in the range 50–100% of relative humidity (Soroka 1979). We suppose that drying shrinkage results from the elastic and the delayed response of the solid skeleton under capillary pressure and disjoining pressure. This idea, that creep and shrinkage are similar, has been previously reported by many authors (see Benboudjema 2002). But, to the author's knowledge, only Bažant & Wu (1974) traduced this idea in the form of constitutive relations.

The capillary pore pressure p_c is derived from the Kelvin law, which states that the gaseous phase (air and water vapor, pressure p_g) and the liquid phase (water, pressure p_l) are in equilibrium. This leads to the equation:

$$p_l - p_g = p_c = \frac{\rho_l RT}{M_v}\ln(h) \tag{25}$$

where R is the gas constant; T is the temperature; ρ_l is the water density; and M_v is the water molar mass.

The expression of the disjoining pressure variation Δp_d is also obtained from equilibrium consideration. Indeed, the Gibbs free energy of the water in the capillary pore and the adsorbed water located in the hindered adsorption zone are equal. This leads to the equation (Bažant & Wittmann 1982):

$$\Delta p_d = \frac{\rho_a RT}{M_a}\ln(h) \tag{26}$$

where ρ_a is the adsorbed water density; and M_a is the adsorbed water molar mass.

It should be emphasized that the disjoining pressure is maximum at a relative humidity equal to 100%.

The pressure applied to the solid skeleton results from an average of the capillary pressure and the disjoining pressure. These effects are taken into account by the saturation degree S_l and a homogenized coefficient α_{rd}:

$$p_{sol} = \alpha_{rd} S_l p_c \tag{27}$$

The coefficient α_{rd} can be identified from a drying shrinkage test.

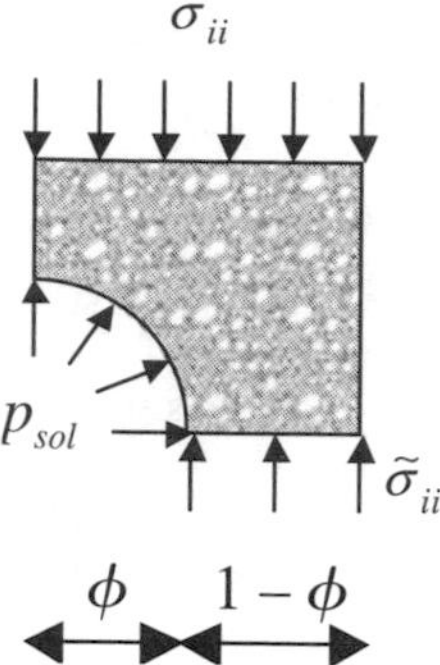

Figure 5. Effective stress concept (without cracking).

The evolution of free drying shrinkage is derived directly from the framework of the mechanics of unsaturated porous media, using the concept of effective stress.

Indeed, effective stresses $\tilde{\boldsymbol{\sigma}}$ are related to the apparent stresses $\boldsymbol{\sigma}$ and the pore pressure p_{sol}, by the following relationship (see Fig. 5):

$$\boldsymbol{\sigma} = (1-\phi)\tilde{\boldsymbol{\sigma}} - \phi p_{sol}\mathbf{1} \tag{28}$$

In the case of free drying shrinkage conditions (no restrain), apparent stresses are equal to zero and cracking does not occur. Hence, the behavior of the solid skeleton reads:

$$\tilde{\boldsymbol{\sigma}} = \mathbf{E}^{sol}\cdot(\boldsymbol{\varepsilon} - \boldsymbol{\varepsilon}^{bc}) \tag{29}$$

where $\mathbf{E}^{sol}$ is the elastic stiffness of the solid skeleton.

In the case of non constant effective stresses, basic creep strains may be evaluated by the superposition principle of Boltzmann:

$$\boldsymbol{\varepsilon}^{bc}(t) = \int_{\tau=0}^{\tau=t} \mathbf{J}^{bc}_{sol}(t-\tau)\, d\big(h(\tau)\tilde{\boldsymbol{\sigma}}(\tau)\big) \tag{30}$$

The use of Eq. 28, Eq. 29 and Eq. 30 allows for rewriting the expression of the strain, which corresponds in this case to the free drying shrinkage strain $\boldsymbol{\varepsilon}^{ds}$:

$$\boldsymbol{\varepsilon} = \boldsymbol{\varepsilon}^{ds} = \phi\left[\left(\mathbf{E}^0\right)^{-1} p_{sol} + \int_0^t \mathbf{J}^{bc}(t-\tau)\, d(hp_{sol})\right]\mathbf{1} \tag{31}$$

with:

$$\mathbf{E}^0 = (1-\phi)\mathbf{E}^{sol} \text{ and } \mathbf{J}^{bc} = (1-\phi)\mathbf{J}^{bc}_{sol} \tag{32}$$

Therefore, it is not necessary to introduce in the strain decomposition (Eq. 1) the drying shrinkage strain. Drying shrinkage results directly from the elastic and delayed response of the solid skeleton, which is included in the adopted framework (Eq. 27), as it has been suggested.

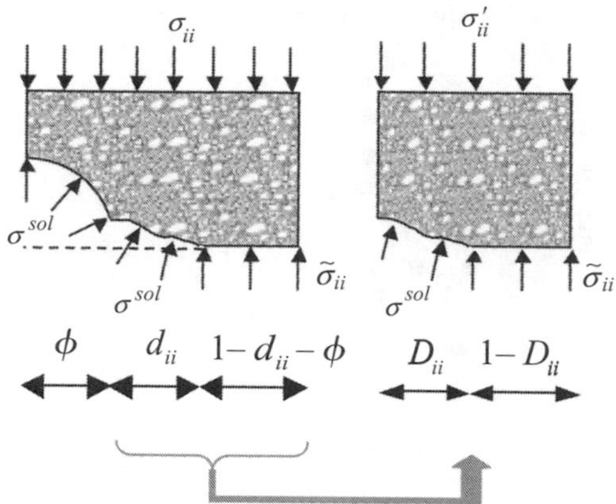

Figure 6. Effective stress concept (with cracking).

2.5 *Coupling between creep, shrinkage and cracking*

In the case where drying and cracking occur simultaneously, the surface of voids increase and the pore pressure applies on the cracks lip. Meanwhile, the cracking induces a decrease of the pore pressure effect, due to the rearrangement of water molecules. The relationship between pore pressure, the apparent stress and the effective stresses reads (see Fig. 6):

$$\boldsymbol{\sigma} = [(1-\phi)\mathbf{I} - \mathbf{d}]\cdot\tilde{\boldsymbol{\sigma}} - (\phi\mathbf{I} + \mathbf{d})\cdot\boldsymbol{\sigma}^{sol} \tag{33}$$

where $\mathbf{d}$ is a tensor which is associated to the cracking (which the terms values are between 0 and $1-\phi$). $\mathbf{d}$ is related to $\mathbf{D}$ by the relationship:

$$\mathbf{d} = (1-\phi)\mathbf{D} \tag{34}$$

The stress $\boldsymbol{\sigma}^{sol}$ applied to the solid skeleton reads:

$$\begin{cases} \sigma^{sol}_{ii} = (1-D_c)^{\alpha_c}\left(1_{ii} - (D_t)_{ii}\right)^{\alpha_t} p_{sol} \\ \sigma^{sol}_{ij} = 0 \text{ si i} \neq \text{j} \end{cases} \tag{35}$$

2.6 *Incremental-iterative solution procedure*

During a time-step, the stress and the relative humidity histories are approximated by linear functions:

$$\begin{cases} h(t) = h_n + \Delta h_n \dfrac{(t-t_n)}{\Delta t_n} \\ \tilde{\boldsymbol{\sigma}}(t) = \tilde{\boldsymbol{\sigma}}_n + \Delta\tilde{\boldsymbol{\sigma}}_n \dfrac{(t-t_n)}{\Delta t_n} \end{cases} \text{with} \begin{cases} t \in [t_n, t_{n+1}] \\ \Delta t_n = t_{n+1} - t_n \\ \Delta h_n = h_{n+1} - h_n \\ \Delta\tilde{\boldsymbol{\sigma}}_n = \tilde{\boldsymbol{\sigma}}_{n+1} - \tilde{\boldsymbol{\sigma}}_n \end{cases} \tag{36}$$

By solving the differential Equations 20 and 23 with the approximation of stresses and relative humidity (Eq. 36), the total creep strains can be expressed as (Benboudjema 2002):

$$\boldsymbol{\varepsilon}_{bc}^{n+1} = \mathbf{A}_{bc}\ \boldsymbol{\varepsilon}_{bc}^{n} + \mathbf{B}_{bc}\ \tilde{\boldsymbol{\sigma}}_{n} + \mathbf{C}_{bc}\ \tilde{\boldsymbol{\sigma}}_{n+1} \tag{37}$$

where $\boldsymbol{\varepsilon}_{bc}^{n}$ is the basic creep strains vector at time-step number n; $\mathbf{A}_{bc}$, $\mathbf{B}_{bc}$ and $\mathbf{C}_{bc}$ are tensors (of second orders) which depend only upon material parameters, relative humidity (h_n and h_{n+1}), t_n and Δt.

The effective stresses at the end of the time step number n are updated by the relationship:

$$\tilde{\boldsymbol{\sigma}}_{n+1} = \mathbf{E}_0\ \boldsymbol{\varepsilon}_{e}^{n+1} = \mathbf{E}_0\ \left(\boldsymbol{\varepsilon}^{n+1} - \boldsymbol{\varepsilon}_{p}^{n+1} - \boldsymbol{\varepsilon}_{ds}^{n+1} - \boldsymbol{\varepsilon}_{bc}^{n+1}\right) \tag{38}$$

where $\boldsymbol{\varepsilon}_{bc}^{e}$, $\boldsymbol{\varepsilon}_{bc}^{p}$ and $\boldsymbol{\varepsilon}^{n+1}$ are the elastic, plastic and total strains vectors at time step number $n+1$, respectively.

Finally, if one makes use of the Equation 38, the stresses vector at the end of the time step reads:

$$\begin{cases} \tilde{\boldsymbol{\sigma}}_{n+1} = \tilde{\boldsymbol{\sigma}}_{n+1}^{tr} - \mathbf{E}_{bc} \cdot \Delta\boldsymbol{\varepsilon}_{p}^{n+1} \\ \mathbf{E}_{bc} = \left(\mathbf{1} + \mathbf{E}_0\ \mathbf{C}_{bc}\right)^{-1}\ \mathbf{E}_0 \\ \tilde{\boldsymbol{\sigma}}_{n+1}^{tr} = \mathbf{E}_{bc} \cdot \left(\boldsymbol{\varepsilon}^{n+1} - \boldsymbol{\varepsilon}_{p}^{n} - \mathbf{A}_{bc} \cdot \boldsymbol{\varepsilon}_{bc}^{n} - \mathbf{B}_{bc} \cdot \tilde{\boldsymbol{\sigma}}_{n}\right) \end{cases} \tag{39}$$

where $\mathbf{E}_{bc}$ is the stiffness tensor, corrected by creep effect, and $\tilde{\sigma}_{n+1}^{tr}$ is the trial stress vector, corrected by creep effect. They can be calculated at the beginning of the time step, since all the involved quantities are known at this time.

Eq. (38) shows that the creep effect can be taken into account without any noticeable changes in existing return mapping algorithms for softening/hardening plasticity (Simo & Taylor 1986, Feenstra 1993). The computed stress state is therefore simultaneously affected by creep and cracking.

The governing equations of the softening plastic model are non linear. Hence, a local iterative procedure is used. During a time step, an Euler backward integration scheme is adopted. The non-linear equations are solved by the Newton-Raphson method (Simo & Taylor 1986). Therefore, the numerical solution is unconditionally stable (Chen & Schreyer 1995).

3 VALIDATION OF THE MODELING

The validation of the modeling is performed. The objective is to see if the model is able to retrieve the main characteristics of drying shrinkage. Reanalysis of the experimental results obtained by Granger (1996) is undertaken in term of drying shrinkage evolution (with respect to weight loss).

Drying shrinkage is performed on a concrete specimen (mix 1:3, 9:5, 6:0, 5) 16 cm in diameter and 100 cm in height. The strain is measured on a 50-cm base located in the central part of the specimen, which allows avoiding all boundary effects (Fig. 7). An identically-sized specimen protected from desiccation serves to measure the level of autogeneous shrinkage, which is then subtracted from total shrinkage in order to derive the actual drying shrinkage.

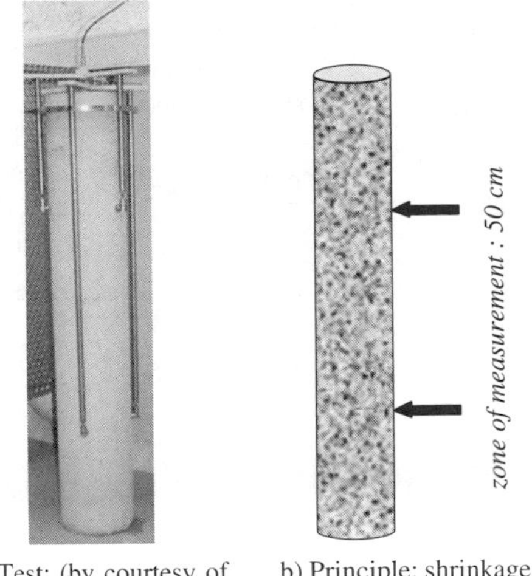

a) Test: (by courtesy of LCPC, Paris, France)

b) Principle: shrinkage is measured in the central part of the sample

Figure 7. Test description.

Basic creep tests are performed on a identical specimen. First, drying parameters are identified from weight loss measurements. Basic creep parameters are also identified from experimental results. Then drying shrinkage is simulated with the present model. As presented in previous investigations, linear drying shrinkage model:

$$\boldsymbol{\varepsilon}^{ds} = k^{ds}\dot{C}\mathbf{1} \tag{40}$$

coupled with damage or elastoplastic damage models, fails to describe accurately the whole evolution of drying shrinkage strains (Benboudjema 2002).

Numerical simulations are performed. We display in Fig. 8 the numerical and experimental evolutions of drying shrinkage strains with respect to weight loss. The evolution obtained with the linear drying shrinkage model is also reported on the same figure.

The computations show that a best agreement with the experimental results is reached with the present model, especially at the end of the evolution.

The decomposition of the drying shrinkage strain is plotted in Fig. 9. We can see that the amplitude of the obtained structural strain (inelastic strain) reaches a maximum value of about 180 μm.m^{-1} (25% of the total drying shrinkage strain), which is not negligible.

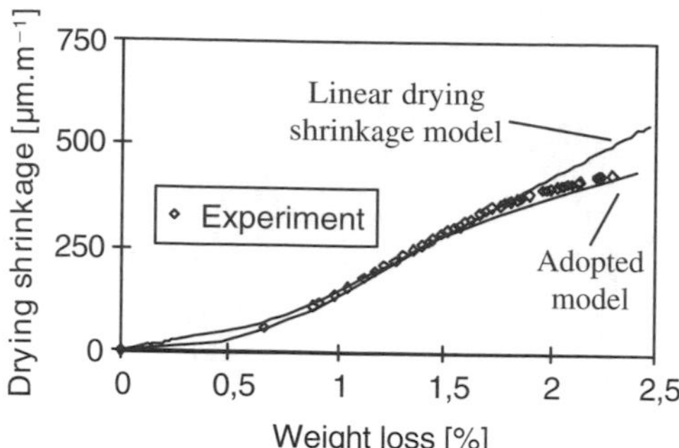

Figure 8. Drying shrinkage evolutions.

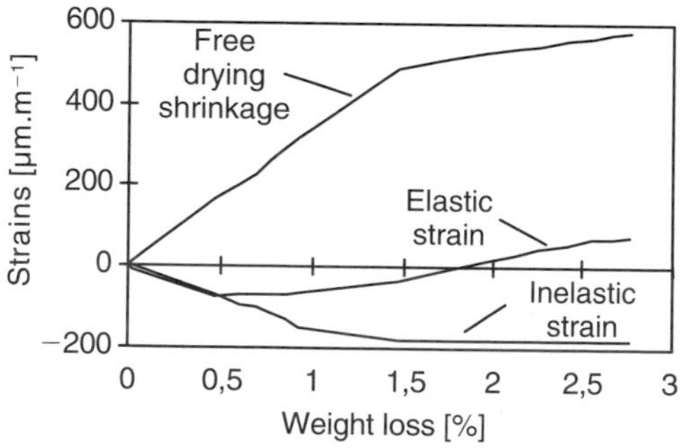

Figure 9. Decomposition of the drying shrinkage strain.

Moreover, we can confirm a number of hypotheses forwarded by various authors, namely (Fig. 9):

- the drying of concrete begins by a phase of skin cracking (see the inelastic strain) which does not give rise to drying shrinkage;
- afterwards, drying shrinkage becomes proportional to weight loss (see the free drying shrinkage).

Granger (1996) suggests that the asymptotic evolution of drying shrinkage is due to partial crack closure. We found here that this not the case, since a linear drying shrinkage model (see Fig. 8) do not succeed to reproduce this behaviour. We show through this simulation that this behaviour is a purely intrinsic effect, since the Fig. 9 shows a decrease of the free drying shrinkage.

4 CONCLUSION

An unified approach for creep and drying shrinkage has been proposed. It is suggested that drying shrinkage is a consequence of the elastic and the delayed response (basic creep) of the solid skeleton under pore pressure. The adopted modeling framework, the mechanics of unsaturated porous media, allows for modeling in a implicit fashion drying shrinkage. This model has been coupled with a cracking model (orthotropic elastoplastic damage model). A simple drying model has been also used.

Experimental results of Granger (1996) have been reanalyzed through numerical simulations. It shows that the experimental evolution of drying shrinkage can be reproduced more accurately, especially the asymptotic evolution (in the drying shrinkage – weight loss diagram) with respect to a linear drying shrinkage model.

REFERENCES

Bažant, Z.P. & Wu, S.T. 1974. Creep and shrinkage law of concrete at variable humidity. *Journal of Engineering Mechanics* 6(100): 1183–1209.

Bažant, Z.P. 1976. Instability, ductility and size effect in strain softening concrete. *Journal of Engineering Mechanics* 102: 331–344.

Bažant, Z.P. & Wittman, F.H. 1982. *Mathematical modeling of creep and shrinkage of concrete*. John Wiley & Sons Limited: New York.

Bažant, Z.P. 1988. Material models for structural creep analysis. *Mathematical Modeling of Creep and Shrinkage of Concrete*: New York: J. Wiley.

Bažant, Z.P. & Hauggaard, A.B. & Baweja, S. & Ulm, F.J. 1997. Microprestress-Solidification Theory for Concrete Creep 1: Aging and Drying Effects. *Journal of Engineering Mechanics* 123(11): 1188–1194.

Benboudjema, F. & Meftah, F. & Torrenti, J.-M., Sellier, A. & Heinfling, G. 2001. A Basic Creep Model for Concrete Subjected to Multiaxial Loads. In *4th International Conference on Fracture Mechanics of Concrete and Concrete Structures, Cachan, 28-31 Mai 2001*: 161–168. Balkema.

Benboudjema, F. 2002. Modélisation des déformations différées du béton sous sollicitions biaxiales. Application aux bâtiment réacteurs de centrales nucléaires. *PhD Thesis, University of Marne-La-Vallée*: 257 (in french).

Bisschop J. & van Mier, J.G.M. 2002. How to study drying shrinkage microcracking in cement-based materials using optical and scanning electron microscopy? *Cement and Concrete Research* 32: 279–287.

Chen, Z. & Schreyer, H.L. 1995. Formulation and computational aspects of plasticity and damage models with application to quasi-Brittle materials. *Report SAND95-0328, Sandia National Laboratories*.

Feenstra, P.H. 1993. Computational aspects of biaxial stress in plain and reinforced concrete, *PhD thesis, Delft institute of technology*. Netherlands: 149.

Granger, L. 1996. Comportement Différé du Béton dans les Enceintes de Centrales Nucléaires: Analyse et Modélisation. *Thèse de doctorat de l'ENPC*: 398. Paris: LCPC. (*in french*)

Heinfling, G. 1998. Contribution à la modélisation numérique du comportement du béton et des structures en béton armé sous sollicitations thermomécaniques à hautes températures. *Thèse de doctorat de l'INSA de Lyon*: 227. (*in french*)

Hillerborg, A. & Modeer, M. & Petersson, P.E. 1976. Analysis of crack formation and crack growth in concrete by means of fracture mechanics and finite elements. *Cement and Concrete Research* 6(6): 773–782.

Jennings, H.M. & Xi, Y. 1992. Relationships between microstructure and creep and shrinkage of cement paste. *Material Science of Concrete* III: 37–69.

Ju, J.W. 1989. On energy-based coupled elastoplastic damage theories: constitutive modeling and computational aspects. *International Journal of Solids and Structures* 25(7): 803–833.

Kupfer, H.B. & Gerstle, K.H. 1973. Behavior of concrete under biaxial stresses. *Journal of Engineering Mechanics* 99: 853-866.

Lassabatère, T. & Torrenti, J.-M. & Granger, L. 1997. Sur le couplage entre séchage du béton et contrainte appliquée. In *Actes du Colloque Saint-Venant, Paris, France*, 331–338.

Lee, J. & Fenves, G.L. 1998. Plastic-damage model for cyclic loading of concrete structures. *Journal of Engineering Mechanics* 124 (8): 892–900.

Nechnech, W. 2000. Contribution à l'étude numérique du comportement du béton et des structures en béton armé soumises à des sollicitations thermiques et mécaniques couplées: Une approche thermo-élasto-plastique endommageable. *Thèse de doctorat de l'INSA de Lyon*: 207 (*in french*).

Sicard, V. & François, R. & Ringot, E. & Pons, G. 1992. Influence of creep and shrinkage on cracking in high strength concrete. *Cement and Concrete Research* 22: 159–168.

Simo, J.C. & Taylor, R.L. 1986. A return mapping algorithm for plane stress elastoplasticity. *International Journal for Numerical Methods in Engineering* 22: 649–670.

Soroka, I. 1979. Portland cement paste and concrete. *Macmillan, Londres*.

Torrenti, J.-M. 1987. Comportement multiaxial du béton: Aspects expérimentaux et modélisation. *Thèse de doctorat de l'ENPC*: 310. (*in french*)

Torrenti, J.M. & Granger, L. & Diruy, M. & Genin P. 1999. Modeling concrete shrinkage under variable ambient conditions. *ACI Materials Journal* 96(1): 35–39.

Ulm, F.J. & Acker, P. 1998. Le Point sur le Fluage et la Recouvrance des Bétons. *Bulletin des Laboratoires des Ponts et Chaussées spécial XX*: 73–82. (*in french*)

Van Mier, J.G.M. 1984. Strain-softening of concrete under multiaxial loading conditions. PhD Dissertation, Eindhoven University of Technology, The Netherlands.

Xi, Y. & Bažant, Z.P. & Jennings, H.M. 1994. Moisture diffusion in cementitious materials – adsorption isotherms. *Advanced Cement Based Materials* 1: 258–266.

Computational Modelling of Concrete Structures, Bićanić et al. (eds)
© 2003 Swets & Zeitlinger, Lisse, ISBN 90 5809 536 3

Material model for concrete under monotonic loading and creep

J. Bockhold & Y. Petryna
Institute for Reinforced and Prestressed Concrete Structures, Ruhr-University Bochum, Germany

W.B. Krätzig
Institute for Statics and Dynamics, Ruhr-University Bochum, Germany

R. Pölling
Building contractor Mesenbrock, Dülmen, Germany

ABSTRACT: In this contribution a completely three-dimensional time-invariant model for concrete is formulated within the framework of the multi-surface elasto-plastic continuum damage theory and is shown to simulate concrete response with a good efficiency and accuracy. This model has been extended to account for long-term creep in context of the lifetime assessment of concrete structures. The model supplements the microprestress-solidification theory by nonlinear damage evolution laws. An extended experimental program is currently in progress to support this numerical model.

1 INTRODUCTION

Lifetime assessment of reinforced concrete structures has become an urgent topic of civil engineering practice. On the one hand, time-invariant material models for concrete are needed to estimate structural response to instantaneous loads. On the other hand, predictions of structural degradation during lifetime demand long-term material models, among others for concrete creep as an important factor affecting structural durability.

Numerous material models have been developed in the last decades for concrete. It is self-evident nowadays that such models should be able to reflect inelastic deformations, micro- and macro-cracking of concrete, damage localisation and some other features observed in experiments. A variety of approaches – stress- and strain-based, with discrete or smeared cracks, local and nonlocal – can be found in the literature (Bažant and Kim 1979; Han and Chen 1986; Simo and Ju 1987; Mazars and Pijaudier-Cabot 1989; Hofstetter and Mang 1995; Cervenka 1994; Ortiz 1985; de Borst et al. 1993). However, this process of theoretical understanding and modelling is still in progress. The present contribution proposes a new model based on elasto-plastic continuum damage mechanics, which satisfies these theoretical requirements and preserves simultaneously a level of simplicity essential for practical applications.

This model is then extended to account for long-term concrete behaviour on the basis of the microprestress-solidification theory (Bažant et al. 1997) widely used in academic research, e.g. (Cervera et al. 1999). This approach is supplemented by nonlinear damage evolution laws and supported by a special experimental program which is currently in progress.

2 TIME-INDEPENDENT MATERIAL MODEL FOR CONCRETE

The concrete response in experiments exhibits irreversible strains or plastic component on the one hand and stiffness degradation due to microcracking or damage component on the other hand. Therefore, the present material model for time-independent response of concrete is uniformly formulated within the multi-surface stress-based elasto-plastic continuum damage theory. The model is developed for large-scale analysis of concrete structures by means of the finite element method. For efficiency purposes, the smeared crack and fracture energy approach (Bažant and Planas 1998) are applied.

Two major aspects of concrete behaviour, namely its response to tension and compression, are considered below according to (Pölling 2002; Krätzig and Pölling 2002). In order to deal with reinforced concrete, the model is complete by the description of the reinforcement and bond behaviour, which is briefly sketched and the end of this section.

2.1 *Concrete under compression*

The plastic strain tensor ε^p, the compliance tensor $\mathbb{D}^d$ and the scalar hardening variable q are selected as thermodynamic internal variables. The corresponding stress-strain relation reads:

$$\boldsymbol{\sigma} = (\mathbb{D}^0 - \mathbb{D}^d)^{-1} : (\boldsymbol{\varepsilon} - \boldsymbol{\varepsilon}^p). \tag{1}$$

The elastic domain $\mathbb{E}_\sigma$ is defined in the stress-space:

$$\mathbb{E}_\sigma := \{(\boldsymbol{\sigma}, \alpha) | \Phi(\boldsymbol{\sigma} \otimes \boldsymbol{\sigma}, \alpha) \leq 0\}. \tag{2}$$

The plastic/damage loading is described by the following joint potential of Drucker-Prager especially suitable for plains stress:

$$\Phi_c(\boldsymbol{\sigma}, \alpha) = \frac{1}{\frac{1}{\sqrt{3}} - \mu} \left[\mu I_1 + \sqrt{J_2}\right] - \alpha_c(q_c). \tag{3}$$

In this potential I_1 and J_2 denote the first invariant of the stress tensor $\boldsymbol{\sigma}$ and the second invariant of its deviator $\mathbf{s}$, respectively, μ controls the influence of the hydrostatic stress, $\alpha_c(q_c)$ describes the hardening/softening behaviour of the material.

The differentiation of the potential with respect to $\boldsymbol{\sigma}$ delivers:

$$\frac{\partial \Phi_c}{\partial \boldsymbol{\sigma}} = \frac{1}{\frac{1}{\sqrt{3}} - \mu} \left(\mu \mathbf{I} + \frac{\mathbf{s} + \mathbf{s}^T}{4\sqrt{J_2}}\right), \tag{4}$$

where $\mathbf{I}$ denotes the unity tensor of second order.

The joint potential reflects the basic observation that the evolution of damage and plastic strains are not independent from each other (Lemaitre and Chaboche 1990). This fact is used to separate the rate of damage and plastic strains in the evolution equations by means of a simple scalar variable β:

$$\dot{\boldsymbol{\varepsilon}}^p = (1 - \beta) \frac{\partial \Phi_c}{\partial \boldsymbol{\sigma}} \dot{\lambda}, \tag{5}$$

and

$$\dot{\boldsymbol{\varepsilon}}^d = \beta \frac{\partial \Phi_c}{\partial \boldsymbol{\sigma}} \dot{\lambda}, \tag{6}$$

as originally proposed in (Meschke et al. 1998). Although this parameter has been mentioned to be of a speculative character, it may be founded by the proportionality of damage and plastic deformations observed in experiments.

The hardening/softening behaviour of concrete is usually proposed to control by an equivalent plastic strain defined in different ways (Lemaitre and Chaboche 1990; Cervera et al. 1999; Mazars and Pijaudier-Cabot 1989). In the present model, the hardening response is governed by the function $\alpha_c(q_c)$ defined to fit directly the uniaxial stress-strain relationship best documented in the literature (CEB/FIP 1991):

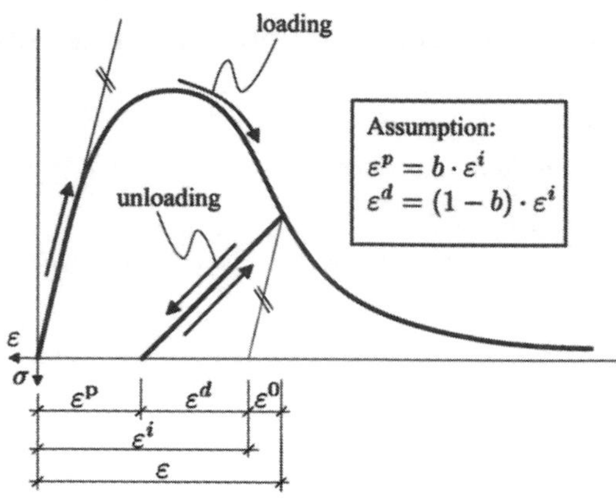

Figure 1. Definition of material parameter b.

$$\alpha_c(q_c^*) = \bar{\sigma}(\varepsilon(q_c^*)). \tag{7}$$

For this purpose, the internal variable q_c is replaced by a new one so that:

$$\dot{q}_c^* = -\frac{2\beta}{\psi(\boldsymbol{\sigma})} \dot{q}_c = \frac{2\beta}{\psi(\boldsymbol{\sigma})} \dot{\lambda}, \tag{8}$$

with

$$\psi(\boldsymbol{\sigma}) = \frac{2\left(\frac{1}{\sqrt{3}} - \mu\right)}{\frac{\mu}{I_1} + \frac{1}{3\sqrt{J_2}}}. \tag{9}$$

The parameter β is related to the characteristic values of the uniaxial stress-strain relation (Figure 1) such as the strain split factor b, the hardening/softening function $\alpha(q_c^*)$ as well as its derivative $d\alpha_c/dq_c^*$ as follows:

$$\beta(q_c^*) = \frac{1}{1 + \frac{b}{1-b}\left[1 + \frac{q_c^*}{\alpha_c(q_c^*)} \frac{d\alpha_c}{dq_c^*}\right]}. \tag{10}$$

Under assumption of isotropic damage, the rate of damage (compliance) tensor may be replaced by

$$\dot{\mathbb{D}}^{d,c} = \dot{D}_{s1}^{d,c} \mathbf{I} \otimes \mathbf{I} + \dot{D}_{s2}^{d,c} (\mathbb{I} + \bar{\mathbb{I}}), \tag{11}$$

where $\mathbb{I}$ is the unity tensor of fourth order.

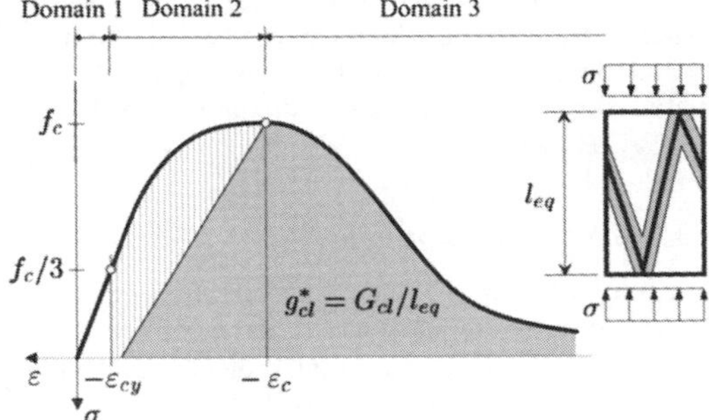

Figure 2. Uniaxial stress-strain relation of concrete under compression.

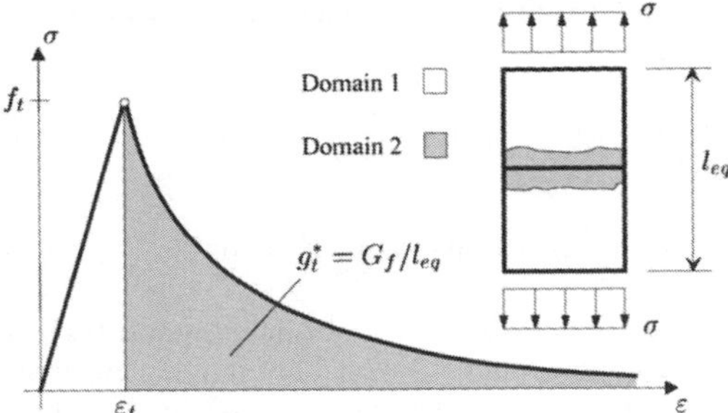

Figure 3. Uniaxial stress-strain relation of concrete under tension.

Two scalar damage variables $D^{d,c}_{s1}$ and $D^{d,c}_{s2}$ describing the evolution of isotropic damage are derived as:

$$\dot{D}^{d,c}_{s1} = \frac{\beta(q^*_c)}{\frac{1}{\sqrt{3}} - \mu}\left(\frac{\mu}{I_1} - \frac{1}{6\sqrt{J_2}}\right)\dot{\lambda}, \tag{12}$$

$$\dot{D}^{d,c}_{s2} = \frac{\beta(q^*_c)}{\frac{1}{\sqrt{3}} - \mu}\left(\frac{1}{4\sqrt{J_2}}\right)\dot{\lambda}. \tag{13}$$

Thus, the evolution equations of internal variables are regarded to the uniaxial stress-strain relation. According to numerous observations, see (Mindess 1983) in (Nechvatal 1996) for example, concrete experiences different damage mechanisms under monotonic loading: bond micro-cracks, matrix micro-cracks, their growth and coalescence leading finally to macro-cracks. Such a damage evolution is reflected in the uniaxial stress-strain relationship by distinguishing between three domains of different behaviour.

The first domain corresponds to linear-elastic response and is limited by stresses below $1/3\,f_c$:

$$\bar{\sigma}_1(\varepsilon) = E_c \cdot \varepsilon, \quad \bar{\sigma}_1 < \frac{1}{3}f_c, \tag{14}$$

where E_c denotes the Young's modulus.

The second domain describes the hardening/ softening behaviour of concrete up to the failure strain ε_c. The analytical hardening function based on experimental data is taken from (CEB/FIP 1991);

$$\bar{\sigma}_2(\varepsilon) = \frac{E_{ci}\frac{\varepsilon}{f_c} + \left(\frac{\varepsilon}{\varepsilon_c}\right)^2}{1 - \left(E_{ci}\frac{\varepsilon_c}{f_c} - 2\right)\left(\frac{\varepsilon}{\varepsilon_c}\right)} f_c. \tag{15}$$

The initial modulus of elasticity E_c in this equation has been replaced by E_{ci} to guarantee that the curve crosses the point $(-\varepsilon_{cy}, -f_{cy})$:

$$E_{ci} = \frac{1}{2E_c}\left(\frac{f_c}{\varepsilon_c}\right)^2 - \frac{f_c}{\varepsilon_c} + \frac{3}{2}E_c. \tag{16}$$

The last region is characterised by softening behaviour of concrete due to damage localisation. If no special techniques are applied, the boundary value problem becomes in this case ill-posed (de Borst et al. 1993). To avoid this difficulty, the crack band and fracture energy approach according to (Bažant and Oh 1983; Bažant and Planas 1998) are applied.

The softening branch of the stress-strain relation is governed by the volume specific localised crushing energy $g^*_{cl} = G_{cl}/l_{eq}$ depending on the crushing energy G_{cl} and characteristic length l_{eq} of the finite element:

$$\bar{\sigma}_3(\varepsilon) = -\frac{1}{\frac{2+\gamma_c f_c \varepsilon_c}{2f_c} + \gamma_c\varepsilon + \frac{\gamma_c}{2\varepsilon_c}\cdot\varepsilon^2}, \tag{17}$$

where γ_c is the parameter controlling the area under the stress-strain curve. This area is finite and can be deduced as:

$$\gamma_c = \frac{\pi^2 f_c \varepsilon_c}{2\left[\frac{G_{cl}}{l_{eq}} - \frac{1}{2}f_c\left(\varepsilon_c(1+b) + b\frac{f_c}{E_c}\right)\right]^2} > 0. \tag{18}$$

2.2 *Concrete under tension*

For concrete under tension the multi-surface stress-based anisotropic continuum damage theory is applied. Since concrete predominantly exhibits an elastic response under tensile stresses up to failure, the plastic strains are ignored.

The damage surface in each of the main stress directions is defined by the Rankine criterion:

$$\Phi_{t(i)}(\boldsymbol{\sigma}, \boldsymbol{\alpha}_t) = \xi^{(i)} - f_t \le 0, \quad i = 1,2,3. \tag{19}$$

Herein $\boldsymbol{\alpha}_t$ denotes the back-stress tensor, $\xi^{(i)}$ the i-th eigenvalue of $\boldsymbol{\xi} = \boldsymbol{\sigma} - \boldsymbol{\alpha}_t$ and f_t the tension strength of concrete.

The derivatives of this potential with respect to $\boldsymbol{\sigma}$ and $\boldsymbol{\alpha}_t$ are:

$$\frac{\partial \Phi_{t(i)}}{\partial \boldsymbol{\sigma}} = \mathbf{M}_\xi^{(i)} \quad \text{and} \quad \frac{\partial \Phi_{t(i)}}{\partial \boldsymbol{\alpha}_t} = -\mathbf{M}_\xi^{(i)}, \tag{20}$$

where $\mathbf{M}_\xi^{(i)}$ is the eigenvalue basis of the i-th eigenvalue of $\boldsymbol{\xi}$.

The softening behaviour of concrete governed by a tensorial function $\boldsymbol{\alpha}_t(\mathbf{q})$ would result in the evolution law with a fourth-order tensor $\partial\boldsymbol{\alpha}_t(\mathbf{q})/\partial\mathbf{q}$. This difficulty is eliminated by introducing the softening rule

$$\dot{\boldsymbol{\alpha}}_t = -\sum_{i=1}^{3} \dot{\lambda}_{t(i)} Z(\alpha_{t\xi}^{(i)}) \mathbf{M}_\xi^{(i)} \tag{21}$$

with a scalar function $Z(\alpha_{t\xi}^{(i)})$ in each of main stress directions and

$$\alpha_{t\xi}^{(i)} = \boldsymbol{\alpha}_t : \mathbf{M}_\xi^{(i)}. \tag{22}$$

The anisotropic damage rule according to (Govindjee et al. 1995)

$$\dot{\mathbb{D}}^{d,t} = \sum_{i=1}^{3} \lambda_{t(i)} \frac{\mathbf{M}_\xi^{(i)} \otimes \mathbf{M}_\xi^{(i)}}{\mathbf{M}_\xi^{(i)} : \boldsymbol{\sigma}} \tag{23}$$

provides the compliance tensor $\mathbb{D}$ to preserve its symmetry properties.

The Kuhn-Tucker damage loading conditions

$$\lambda_{t(i)} \geq 0; \quad \Phi_{t(i)} \leq 0; \quad \lambda_{t(i)} \cdot \Phi_{t(i)} = 0 \tag{24}$$

are used for each crack direction separately.

The softening function $Z(\alpha_{t\xi}^{(i)})$ is defined according to the exponential softening rule (Bažant and Planas 1998)

$$\sigma(\varepsilon) = f_t \cdot e^{\frac{1}{\gamma_t}(\varepsilon_t - \varepsilon)}. \tag{25}$$

The parameter

$$\gamma_t = \frac{G_f}{l_{eq} f_t} - \frac{1}{2}\frac{f_t}{E_c}, \tag{26}$$

controls the area under the stress-strain curve and depends on the volume-specific fracture energy $g_t^* = G_f / l_{eq}$.

With these assumptions, the desired softening function $Z(\alpha_{t\xi}^{(i)})$ takes the form

$$Z(\alpha_{t\xi}^{(i)}) = \frac{1 + \alpha_{t\xi}^{(i)}}{\varepsilon_t + \gamma_t \left[1 - \ln\left(1 + \alpha_{t\xi}^{(i)}/f_t\right)\right]}. \tag{27}$$

This material model for concrete requires only 9 material parameters, which can easily be determined by standard monotonic tests or approximately estimated for a given concrete strength according to (CEB/FIP 1991).

2.3 *Algorithmic formulation*

The integration of the material law, both for concrete under compression and tension, is performed by the classical General Closest Point Projection algorithm according to (Simo and Hughes 1998).

With a given material state at time t_n, the trial state at time t_{n+1} is computed under the assumption of freezing plastic flow:

$$\left.\begin{aligned}
\boldsymbol{\sigma}_{n+1}^{\text{trial}} &= \mathbb{C} : \boldsymbol{\varepsilon}_{n+1}^{e,\text{trial}}, \\
\boldsymbol{\varepsilon}_{n+1}^{e,\text{trial}} &= \boldsymbol{\varepsilon}_{n+1} - \boldsymbol{\varepsilon}_{n+1}^{p,\text{trial}}, \\
\boldsymbol{\varepsilon}_{n+1}^{p,\text{trial}} &= \boldsymbol{\varepsilon}_n^p, \quad q_{n+1}^{\text{trial}} = q_n, \\
\alpha_{n+1}^{\text{trial}} &= \alpha(q_{n+1}^{\text{trial}}), \\
\Phi_{n+1}^{\text{trial}} &= \Phi(\boldsymbol{\sigma}_{n+1}^{\text{trial}}, \alpha_{n+1}^{\text{trial}})
\end{aligned}\right\} \tag{28}$$

If the trial state belongs to the admissible (elastic) domain and $\Phi_{n+1}^{\text{trial}} \leq 0$, no plastic/damage flow occurs. Otherwise, the return-mapping corrector step is performed for internal variables. Only two variables ε^p and α are considered below for brevity.

First, the residuals of the trial state are determined as follows:

$$\mathbf{R}_{n+1} = \begin{Bmatrix} -\boldsymbol{\varepsilon}_{n+1}^p + \boldsymbol{\varepsilon}_n^p \\ -\alpha_{n+1} + \alpha_n \end{Bmatrix} + \Delta\lambda_{n+1} \begin{Bmatrix} \dfrac{\partial \Phi}{\partial \boldsymbol{\sigma}} \\ \dfrac{\partial \Phi}{\partial q} \end{Bmatrix}, \tag{29}$$

which must vanish for the solution searched. This nonlinear algebraic problem is solved by the Newton iteration algorithm. At each iteration step, the increment of the consistency parameter is calculated as follows:

$$\Delta\Delta\lambda_{n+1} = \frac{\Phi_{n+1}^{\text{trial}} - \boldsymbol{\nabla}^{\text{T}} \mathbf{A}\, \mathbf{R}_{n+1}}{\boldsymbol{\nabla}^{\text{T}} \mathbf{A}\, \boldsymbol{\nabla}}, \tag{30}$$

with

$$\boldsymbol{\nabla}^T = [\ \partial\Phi/\partial\boldsymbol{\sigma} \quad \partial\Phi/\partial q\ ], \tag{31}$$

and

$$\mathbf{A}^{-1} = \begin{bmatrix} \mathbb{C}^{-1} + \Delta\lambda \dfrac{\partial^2 \Phi}{\partial \boldsymbol{\sigma}^2} & \Delta\lambda \dfrac{\partial^2 \Phi}{\partial \boldsymbol{\sigma} \partial q} \\ \Delta\lambda \dfrac{\partial^2 \Phi}{\partial q \partial \boldsymbol{\sigma}} & -\dfrac{\mathrm{d}q}{\mathrm{d}\alpha} + \Delta\lambda \dfrac{\partial^2 \Phi}{\partial q^2} \end{bmatrix}. \tag{32}$$

The increments of the internal variables are then obtained as:

$$\begin{bmatrix}\Delta\varepsilon^p_{n+1}\\ \Delta\alpha_{n+1}\end{bmatrix} = \mathbf{R}_{n+1} + \Delta\Delta\lambda_{n+1}\begin{Bmatrix}\partial\Phi/\partial\sigma\\ \partial\Phi/\partial q\end{Bmatrix}, \tag{33}$$

which finally leads a new state with

$$\begin{aligned}\varepsilon^p_{n+1} &= \varepsilon^p_n + \Delta\varepsilon^p_{n+1},\\ \alpha_{n+1} &= \alpha_n + \Delta\alpha_{n+1},\\ \Delta\lambda_{n+1} &= \Delta\lambda_n + \Delta\Delta\lambda_{n+1}.\end{aligned} \tag{34}$$

Then the residual (29) is computed anew and the iteration step is repeated, until this residual becomes less than a tolerance limit:

$$\mathbf{R}_{n+1} \leq \mathrm{TOL}_R. \tag{35}$$

The algorithmic consistent tangent modulus is obtained by claiming the differential residual tensor to become zero: $\mathrm{d}\mathbf{R} = \mathbf{0}$. With help of the differentiated constitutive law

$$\mathrm{d}\sigma_{n+1} = \mathbb{C} : (\mathrm{d}\varepsilon_{n+1} - \mathrm{d}\varepsilon^p_{n+1}) \tag{36}$$

we finally gain after some transformations:

$$\mathbb{C}_T = \left.\frac{\mathrm{d}\sigma}{\mathrm{d}\varepsilon}\right|_{n+1} = \Xi_{n+1} - \mathbf{N}_{n+1}\otimes\mathbf{N}_{n+1}, \tag{37}$$

where

$$\Xi_{n+1} := \left[\mathbb{C}^{-1} + \Delta\lambda_{n+1}\frac{\partial^2\Phi}{\partial\sigma^2}\right]^{-1} \tag{38}$$

and

$$\mathbf{N}_{n+1} := \frac{\Xi_{n+1} : \partial\Phi/\partial\sigma}{\sqrt{\partial\Phi/\partial\sigma : \Xi_{n+1} : \partial\Phi/\partial\sigma}}. \tag{39}$$

2.4 *Extension to reinforced concrete*

For reinforcement bars, an elasto-plastic uniaxial material model with isotropic and kinematic hardening is applied. In (Pölling 2000) it has been modified by the boundary surface theory of (Dafalias and Popov 1976) in order to realistically account for the Bauschinger effect.

A stiff bond between concrete and steel reinforcement is generally assumed. The concept of bond modelling proposed by (Feenstra 1993) is adopted, but only the tension-softening part is accounted for. The complete fracture energy G_f associated to the finite element with a characteristic length l_{eq} is considered as the sum of the fracture energies of all cracks inside the element:

$$G_f = \max\left\{G_{fc};\frac{l_{eq}}{l_s}G_{fc}\right\}. \tag{40}$$

The distance between cracks in one element l_s can be calculated according to (CEB/FIP 1991) as:

$$l_s = \frac{2}{3}l_{s,\max} \text{ with } l_{s,\max} = \frac{d_s}{3.6\,\rho_{s,\mathrm{eff}}}. \tag{41}$$

In this equation, d_s denotes the cross-section of the steel bars and $\rho_{s,\mathrm{eff}}$ the effective degree of reinforcement.

3 LONG-TERM MATERIAL MODEL

The time-independent behaviour of concrete is a quite useful idealisation, however, its response generally depends on the rate of deformation, on loading and enviornmental conditions varying with time. The main interest of the present work concerns the long-term concrete response to sustained loads in context of the lifetime analysis of concrete structures. The time-independent model of concrete presented above is currently under development in order to account for basic creep with an emphasis on damage evolution.

The microprestress-solidification theory by (Bažant et al. 1997) presents two main mechanisms affecting concrete creep and is able to reflect well the evolution of creep strains both for short-time and long-time processes. However, the solidification theory provides acceptable results mainly for low stress ranges and primary creep. Experimental observations show that the secondary and tertiary creep characteristic for high stress ranges require an explicit damage variable associated with concrete micro-cracking (Chen and Wang 1993; Meyers et al. 1969). The creep failure at stresses higher than $0.75f_c$, as illustrated in Figure 4, can unfortunately not be explained within the solidification theory. Many authors mention in this

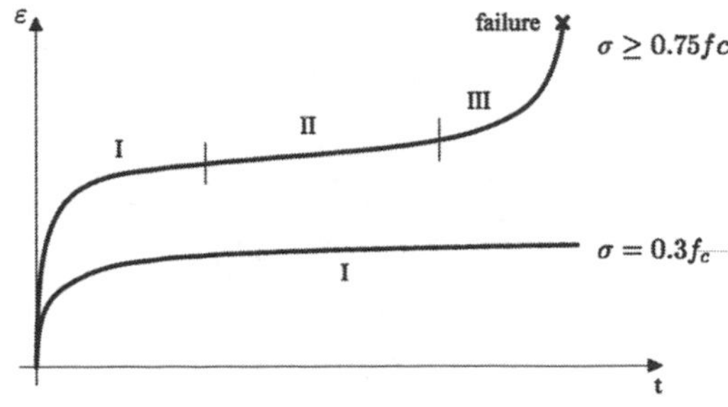

Figure 4. Creep strain evolution at different stresses.

context, that the interaction of creep with fracture and damage in concrete still remains a challenging task, see (Bažant 2001) for example. The present contribution constitutes basic elements of the long-term creep model for concrete currently under development by the authors. This model combines the microprestress-solidification theory with the approaches of continuum damage mechanics.

3.1 *Strain partitioning, strain rates and time scales*

The strain partitioning for long-term response of concrete remains preserved compared to the time-independent case. The total strain is separated into an elastic part ε^e and plastic part ε^p according to Figure 1:

$$\boldsymbol{\varepsilon} = \boldsymbol{\varepsilon}^e + \boldsymbol{\varepsilon}^p = \boldsymbol{\varepsilon}^0 + \boldsymbol{\varepsilon}^d + \boldsymbol{\varepsilon}^p. \tag{42}$$

However, both the plastic and damage strains consists now of an instantaneous component (subscript "i") and a time-dependent one (subscript "t"):

$$\begin{aligned} \boldsymbol{\varepsilon} &= \boldsymbol{\varepsilon}^0 + (\boldsymbol{\varepsilon}_i^p + \boldsymbol{\varepsilon}_i^d) + (\boldsymbol{\varepsilon}_t^p + \boldsymbol{\varepsilon}_t^d) \\ &= \boldsymbol{\varepsilon}^0 + \boldsymbol{\varepsilon}_i + \boldsymbol{\varepsilon}_t. \end{aligned} \tag{43}$$

Experiments show that creep deformations take place even for low stresses, where the time-independent behaviour is assumed to be linear-elastic, and then remain to a large extent irreversible (Bažant and Prasannan 1989; Nechvatal 1996). This fact is taken into account by introducing the visco-elastic elements into the corresponding rheological model (Figure 5) and assuming their strains completely irreversible. The creep recovery is thus neglected. This rheological model generally contains elastic, damage, plastic, visco-elastic and viscous (flow) elements, which are able to simulate both the time-invariant and long-term behaviour of concrete.

Since the strain rates for time-invariant and long-term deformation processes under sustained loads typically differ in the order of magnitude, they may be simulated independently in two time scales:

$$\dot{\boldsymbol{\varepsilon}} = \begin{cases} \dot{\boldsymbol{\varepsilon}}_0 + \dot{\boldsymbol{\varepsilon}}_i^d + \dot{\boldsymbol{\varepsilon}}_i^p & : \ \dot{\boldsymbol{\varepsilon}}_i \gg 0 \\ \dot{\boldsymbol{\varepsilon}}_0 + \dot{\boldsymbol{\varepsilon}}_t^d + \dot{\boldsymbol{\varepsilon}}_t^p & : \ \dot{\boldsymbol{\varepsilon}}_t \to 0 \end{cases}. \tag{44}$$

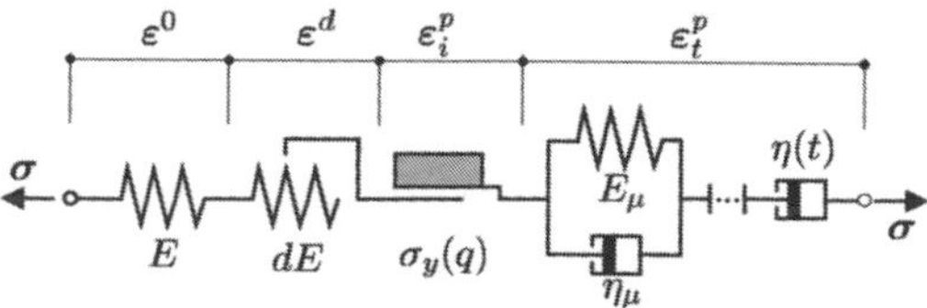

Figure 5. Rheological model for concrete creep.

The interaction of both components is assumed by the fact that they contribute to the same plastic $\varepsilon^p = \varepsilon_i^p + \varepsilon_t^p$ or damage term $d = d_i + d_t$ actually used during each simulation.

3.2 *Thermodynamic formulation*

The same internal variables as before are used for the creep model. The stress and evolution equations are incorporated within the thermodynamic framework of irreversible processes. The energy dissipation is considered in different time scales separately. The dissipation potential is splitted into the instantaneous part $\prod_{dis}^i$ and long-term one $\prod_{dis}^t$, e.g. for isotropic damage as follows:

$$\Pi_{dis} = \Pi_{dis}^t + \Pi_{dis}^i, \tag{45}$$

with

$$\Pi_{dis}^i = \boldsymbol{\sigma} : \dot{\boldsymbol{\varepsilon}}_i^p + \boldsymbol{\alpha}_i \cdot \dot{\mathbf{q}}_i + Y \cdot \dot{d}_i \geq 0, \tag{46}$$

$$\Pi_{dis}^t = \boldsymbol{\sigma} : \dot{\boldsymbol{\varepsilon}}_t^p + \boldsymbol{\alpha}_t \cdot \dot{\mathbf{q}}_t + Y \cdot \dot{d}_t \geq 0. \tag{47}$$

Herein, Y denotes the elastic strain energy density release rate, the conjugate thermodynamic variable to damage (Lemaitre and Chaboche 1990).

The stress associated with the rheological model of Figure 5 then reads:

$$\boldsymbol{\sigma} = (1 - d_i - d_t)\,\mathbb{C}^0 : (\boldsymbol{\varepsilon} - \boldsymbol{\varepsilon}_i^p - \boldsymbol{\varepsilon}_t^p). \tag{48}$$

The evolution equations for irreversible strains ε_t^p are formulated within the visco-elastic framework of the microprestress-solidification theory. The hardening and damage variables, q_t and d_t respectively, are handled analogously leading to the following generalised evolution equations:

$$\dot{\boldsymbol{\varepsilon}}_t^p = \Phi_\sigma(\boldsymbol{\sigma}, t),\ \dot{\mathbf{q}}_t = \Phi_q(\boldsymbol{\sigma}, t),\ \dot{d}_t = \Phi_d(\boldsymbol{\sigma}, t). \tag{49}$$

The potentials $\Phi_\sigma(\boldsymbol{\sigma}, t), \Phi_q(\boldsymbol{\sigma}, t), \Phi_d(\boldsymbol{\sigma}, t)$ can in part be derived directly from the solidification theory. Additionally, a special experimental program supporting their formulation within thc present creep model is currently in progress. Some principal aspects of the approach are explained below.

3.3 *Microprestress-solidification theory*

The solidification theory by (Bažant and Prasannan 1989) and especially its modification as microprestress-solidification theory by (Bažant et al. 1997) are recognised as the state-of-the-art approach to explain the physical processes influencing concrete creep. They identify two principal micro-mechanical mechanisms and describe their influence on creep strains.

According to the solidification theory, the creep strain tensor of concrete is decomposed into viscoelastic ε^v and viscous (flow) part ε^f leading to the following rate split:

$$\dot{\varepsilon}^p_t = \dot{\varepsilon}^c = \dot{\varepsilon}^v + \dot{\varepsilon}^f. \tag{50}$$

The viscoelastic component ε^v is dominant at early ages and affected by the hydration of cement. The influence of hydration on the short-term creep is taken into account by the dependence of the viscoelastic strains on the so-called viscoelastic micro-strain γ and the volume of the solidified matter $v(t)$:

$$\dot{\varepsilon}^v(t) = \frac{\phi_h(h)}{v(t)}\,\dot{\gamma}. \tag{51}$$

Herein, $\phi_h(h)$ denotes a function of the internal humidity h. The viscoelastic microstrain γ is determined as the total strain of the corresponding Kelvin or Maxwell chain with N elements. In case of the Kelvin chain depicted in Figure 5 one obtains with the nondimensional fourth-order constitutive tensor $\mathbf{G}$ from (Bažant et al. 1997):

$$\bar{\gamma} = \sum_{\mu=1}^{N} \gamma_\mu; \quad E_\mu \gamma_\mu + \eta_\mu \dot{\gamma}_\mu = \mathbf{G}\,\boldsymbol{\sigma}. \tag{52}$$

The stiffness tensors E_μ and viscosities η_μ vary proportional with time, while the retardation times $\tau_\mu = \eta_\mu / E_\mu$ remain constant.

The visco-elastic microstrain γ can be integrated by means of the exponential algorithms (Bažant and Prasannan 1989) as follows:

$$\begin{aligned} \gamma_{\mu,n+1} &= \gamma_{\mu,n} \exp(-\Delta y_\mu) \\ &+ \frac{\mathbf{G}\,\boldsymbol{\sigma}_n}{E-\mu}[1-\exp(-\Delta y_\mu)] \\ &+ \frac{1-\lambda\mu}{E_\mu}\mathbf{G}\,\Delta\boldsymbol{\sigma}_{n+1}, \end{aligned} \tag{53}$$

with

$$\Delta y_\mu = \frac{\Delta t_{n+1}}{\tau_\mu}, \quad \lambda_\mu = \frac{1-\exp(-\Delta y_\mu)}{\Delta y_\mu}. \tag{54}$$

The long-term creep is characterised by the viscous strain ε^f described by the flow element with a time-dependent viscosity connected in series to the aging Kelvin chain (Figure 5). The second principal micro-mechanical mechanism postulates the viscosity η of the flow dashpot as a function of the normalised tensile microprestress S_μ carried by the bonds across the pores in the hardened cement gel:

$$\eta(t) = \frac{\eta_0}{S_\mu}. \tag{55}$$

The long-term creep is assumed to originate from viscous shear slip between the opposite walls of micropores, at which the bonds and bridges transmitting the microprestress break and reform anew. Under these assumptions, the rate of the flow strain can be expressed as follows:

$$\dot{\varepsilon}^f = \frac{1}{\eta(S_\mu)}\,\mathbf{G}\,\boldsymbol{\sigma}. \tag{56}$$

The rate formulation of the microprestress-solidification theory is adopted in the present model and further extended to account for damage.

3.4 *Evolution of creep damage*

The kinematic evolution laws of classical damage mechanics (Lemaitre 1992) presume that damage evolves proportionally to plastic strains. Many experimental observations on concrete confirm this fact, for example, the acoustic emission tests of (Diderichs et al. 1983) and (Rossi et al. 1993) as well as systematic creep tests of (Bierwirth et al. 1994).

The basic creep process undergoes in general three different stages denominated as primary, secondary and tertiary creep (Bažant 1988) and associated with different micro-mechanical mechanisms such as bond micro-cracks, matrix micro-cracks and damage localisation. These three stages are usually distinguished from each other only for constant stress histories in a subjective manner. However, it can be done in a systematic way and independently of load histories, if an objective characteristic function able to reflect the damage evolution in material can be found.

A similar approach is long adopted in the time-independent analysis of concrete deformation: the damage evolution is governed by an equivalent plastic strain, see (Mazars and Pijaudier-Cabot 1989) for example. In this context, the elastic strain energy density release rate

$$Y = \frac{1}{2}\,\boldsymbol{\varepsilon}^e : \mathbb{C} : \boldsymbol{\varepsilon}^e \tag{57}$$

seems to be the most natural function satisfying this requirement, since it is a thermodynamically conjugate to damage itself. Its critical value Y_c at the material instability point corresponding to damage localisation is already considered as material parameter in classical damage mechanics (Lemaitre 1992).

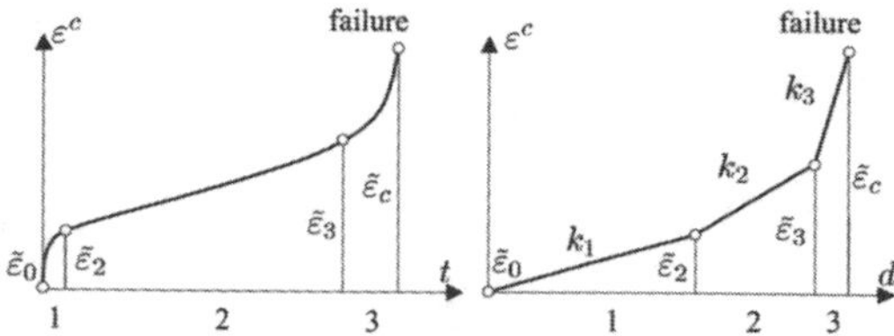

Figure 6. Evolution of creep strain and damage.

Therefore, damage evolution laws for concrete creep have been proposed to govern by means of this function (Petryna 2002). From the practical point of view, it is moreover advantageous to relate three-dimensional creep states to the uniaxial ones, which can be easily supported experimentally. Thus an appropriate equivalent strain is defined as:

$$\tilde{\varepsilon} = \sqrt{\frac{2\,Y}{E_c}} = \sqrt{\frac{1}{E_c}\,\boldsymbol{\varepsilon}^e : \mathbb{C}_0 : \boldsymbol{\varepsilon}^e}. \tag{58}$$

We additionally postulate that the outset of different creep stages can be identified by corresponding critical values $\tilde{\varepsilon}_2, \tilde{\varepsilon}_3, \tilde{\varepsilon}_c$ of this equivalent strain (Figure 6) independently of stress magnitude and load history. In order to preserve generality, the degree of proportionality between concrete damage and plastic strains is assumed to be different for each stage. The evolution law for creep damage can then be expressed in the rate form as:

$$\dot{d}_t = \frac{Y}{k_Y}\dot{\tilde{\varepsilon}} = \frac{Y}{k_Y}\Phi_d(\boldsymbol{\sigma}, t), \tag{59}$$

with

$$k_Y = \begin{cases} k_1 & : \ \tilde{\varepsilon}_0 < \tilde{\varepsilon} \le \tilde{\varepsilon}_2 \\ k_2 & : \ \tilde{\varepsilon}_2 < \tilde{\varepsilon} \le \tilde{\varepsilon}_3 \\ k_3 & : \ \tilde{\varepsilon}_3 < \tilde{\varepsilon} \le \tilde{\varepsilon}_c \end{cases} \tag{60}$$

The constants k_1, k_2, k_3 can be determined from the uniaxial creep tests.

3.5 *Time integration algorithm*

The time integration of the material law is performed within the visco-elastic incremental framework. With a known material state $\boldsymbol{\sigma}_n, \boldsymbol{\varepsilon}_n, \boldsymbol{\varepsilon}^p_{i,n}, \boldsymbol{\varepsilon}^p_{t,n}, d_n$ at time instant t_n and the increment of the total strain $\Delta\boldsymbol{\varepsilon}_{n+1}$ obtained from the finite element analysis, one can estimate a new material state at time $t_{n+1} = t_n + \Delta t_{n+1}$ using the Backward-Euler scheme as follows:

$$\begin{aligned} \boldsymbol{\varepsilon}_{n+1} &= \boldsymbol{\varepsilon}_n + \Delta\boldsymbol{\varepsilon}_{n+1}, \\ \boldsymbol{\varepsilon}^p_{i,n+1} &= \boldsymbol{\varepsilon}^p_{i,n}, \\ \boldsymbol{\varepsilon}^p_{t,n+1} &= \boldsymbol{\varepsilon}^p_{t,n} + \Delta t_{n+1}\cdot\Phi_\sigma(\boldsymbol{\sigma}_{n+1}, t_{n+1}), \\ d_{n+1} &= d_n + \Delta t_{n+1}\cdot\Phi_d(\boldsymbol{\sigma}_{n+1}, t_{n+1}), \\ \boldsymbol{\sigma}_{n+1} &= \boldsymbol{\sigma}_n + \Delta\boldsymbol{\sigma}_n. \end{aligned} \tag{61}$$

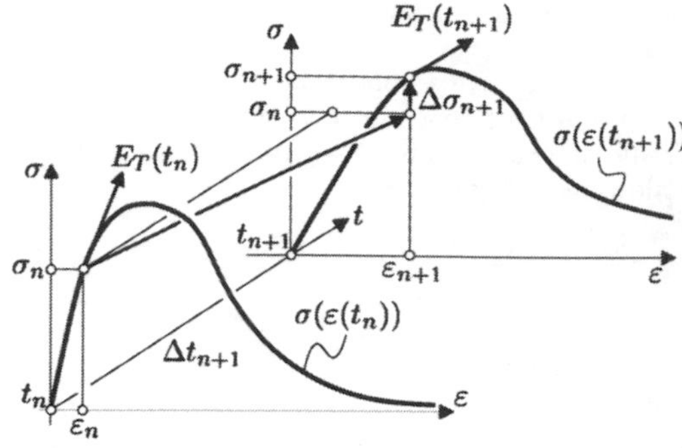

Figure 7. Time integration scheme for uniaxial material law.

The stress increment $\Delta\boldsymbol{\sigma}_n$ is taken (as a first approximation) from the last time step. This trial state does not simultaneously satisfy both the stress relation (48) and the evolution equations (49). The solution of the problem is searched for using the Newton iteration scheme with respect to the strain residual $\mathbf{R}$ as follows:

$$\begin{aligned} \mathbf{R}^{(k)}_{n+1} &= \boldsymbol{\varepsilon}_{n+1} - \boldsymbol{\varepsilon}^{p(k)}_{i,n+1} - \boldsymbol{\varepsilon}^{p(k)}_{t,n+1}, \\ &\quad - [(1 - d^{(k)}_{n+1})\,\mathbb{C}^0]^{-1} : \boldsymbol{\sigma}^{(k)}_{n+1} \neq \mathbf{0}, \\ \mathbf{R}^{(k+1)}_{n+1} &= \mathbf{R}^{(k)}_{n+1} + \frac{\partial \mathbf{R}^{(k)}_{n+1}}{\partial \boldsymbol{\sigma}} : \Delta\boldsymbol{\sigma}^{(k+1)}_{n+1} = \mathbf{0}, \\ \Delta\boldsymbol{\sigma}^{(k+1)}_{n+1} &= -\left[\frac{\partial \mathbf{R}^{(k)}_{n+1}}{\partial \boldsymbol{\sigma}}\right]^{-1} : \mathbf{R}^{(k)}_{n+1}, \\ \boldsymbol{\sigma}^{(k+1)}_{n+1} &= \boldsymbol{\sigma}^{(k)}_{n+1} + \Delta\boldsymbol{\sigma}^{(k+1)}_{n+1}. \end{aligned} \tag{62}$$

The iteration (62) is carried out until the norm of this residual reduces below a given tolerance limit:

$$||\mathbf{R}^{(k+1)}_{n+1}|| \le \mathrm{TOL}_R. \tag{63}$$

The integration algorithms is illustrated in Figure 7 for the uniaxial case.

4 NUMERICAL EXAMPLE

The efficiency and accuracy of the time-independent material model presented above is illustrated by comparison of the present numerical solution for a reinforced concrete plate with the experimental results obtained in (Jofriet and McNeice 1971). The plate is pointwise supported at four corners and subjected to a single load F at the midpoint. The material parameters listed in Figure 8 are taken from (Jofriet and McNeice 1971) and in part estimated according to (CEB/FIP 1991).

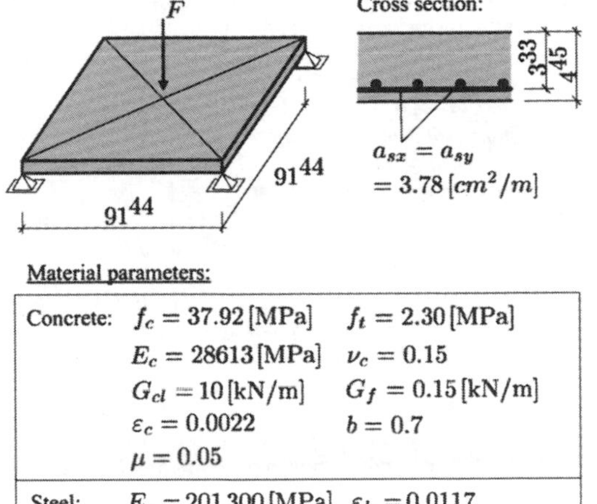

Material parameters:

Concrete:	$f_c = 37.92\,[\text{MPa}]$	$f_t = 2.30\,[\text{MPa}]$
	$E_c = 28613\,[\text{MPa}]$	$\nu_c = 0.15$
	$G_{cl} = 10\,[\text{kN/m}]$	$G_f = 0.15\,[\text{kN/m}]$
	$\varepsilon_c = 0.0022$	$b = 0.7$
	$\mu = 0.05$	
Steel:	$E_s = 201\,300\,[\text{MPa}]$	$\varepsilon_h = 0.0117$
	$f_y = 345.4\,[\text{MPa}]$	

Figure 8. Reinforced concrete plate under consideration.

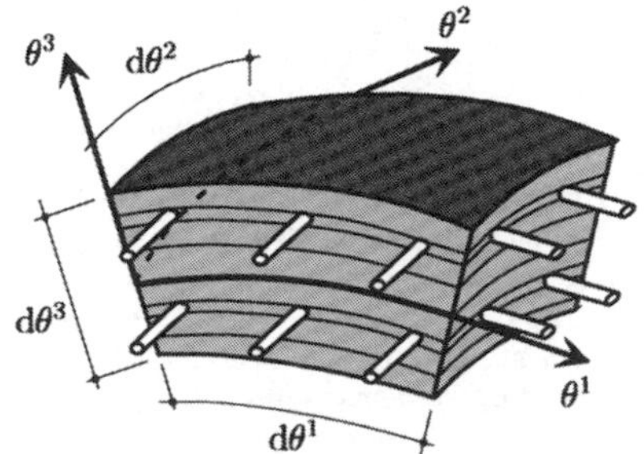

Figure 9. Multi-layered 3-D shell element.

The material model has been implemented into a 3-D finite shell element developed in (Jun 2002) and based on the enhanced assumed strain and assumed natural strain concepts. The multi-layer kinematics (Figure 9) is applied for proper descripton of transverse strains.

The numerical study is carried out within the FEMAS software (Beem et al. 1996). By accounting for the symmetry conditions, only a quarter of the plate is discretised by [5 × 5] elements with 20 concrete layers. The quasi-static loading is simulated by means of the Riks-Wempner-Wessels incremental-iterative scheme. The load-displacement diagram given in Figure 10 agrees well with the curve obtained experimentally.

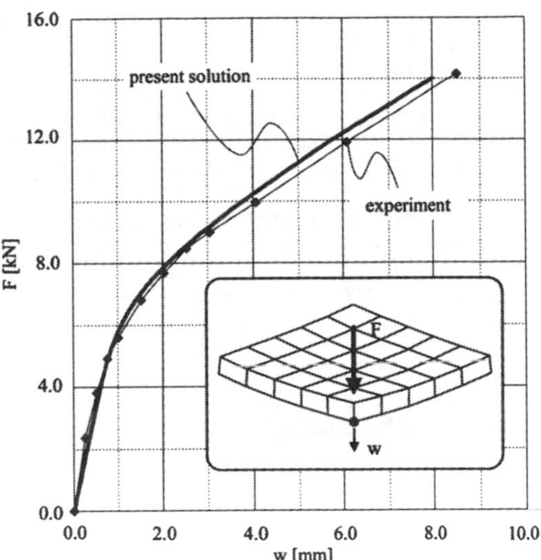

Figure 10. Comparison of numerical study with experiment.

5 CONCLUSION

Both time-invariant and long-term material models for concrete are presented for realistic simulations of structural lifetimes. A completely three-dimensional time-invariant model is formulated to account for principal features associated with concrete response such as plastic strains, damage, localisation and different behaviour under tension and compression. In spite of complexity, the model is uniformly formulated within the framework of multi-surface elasto-plastic continuum damage theory and is shown to simulate concrete response with a good efficiency and accuracy.

A characteristic feature of the model is its adjustment to the uniaxial response best documented in the literature. Such a response is often the only information at disposal and is directly referred to in design codes.

Some principal elements of a new long-term creep model are presented. It improves the microprestress-solidification theory by nonlinear damage evolution laws governed by means of the elastic strain energy density release rate. An extended experimental program aiming to support this numerical model is currently in progress.

ACKNOWLEDGEMENTS

Financial support of the German National Science Foundation (DFG) within the Special Research Center SFB398 at the Ruhr-University Bochum is gratefully acknowledged.

REFERENCES

Bažant Z. P. (1988). *Mathematical modeling of creep and shrinkage of concrete.* Chichester: John Wiley and Sons.

Bažant, Z. P. (2001). Prediction of concrete creep and shrinkage: past, present and future. *Nuclear Engineering and Design 203*, 27–38.

Bažant, Z. P., A. B. Hauggard, S. Baweja, and F.-J. Ulm (1997). Microprestress-solidification theory for concrete, I: Aging and drying effects. II: Algorithm and verification. *Journal of Engineering Mechanics (ASCE) 123*(11), 1188–1201.

Bažant, Z. P. and S.-S. Kim (1979). Nonlocal damage theory based on micromechanics and crack interactions. *Journal of Engineering Mechanics (ASCE) 105*(3), 407–428.

Bažant, Z. P. and B. Oh (1983). Crack band theory for fracture concrete. *Journal of Engineering Mechanics (ASCE) 16*(93), 155–177.

Bažant, Z. P. and J. Planas (1998). *Fracture and Size Effect in Concrete and Other Quasibrittle Materials*. New Directions in Civil Engineering. Boca Raton – Boston – London – New York: CRC Press.

Bažant, Z. P. and S. Prasannan (1989). Solidificaton theory for concrete creep. I: Formulation. II: Verification and application. *Journal of Engineering Mechanics (ASCE) 105*(3), 1691–1725.

Beem, H., C. Könke, U. Montag, and W. Zahlten (1996). FEMAS-2000 Users Manual, Release 3.0. Technical report, Institute for Statics and Dynamics, Ruhr-University Bochum.

Bierwirth, H., S. Stöckl, and H. Kupfer (1994). Versuche zum Kriechen von Beton bei mehrachsiger Beanspruchung – Auswertung auf der Basis von errechneten elastischen Anfangsverformungen. Technical report, Lehrstuhl für Massivbau, Technische Universität München.

CEB/FIP (1991). CEB-FIP Model Code 1990. *(Design Code)*.

Cervenka, J. (1994). *Discrete crack modeling in concrete structures*. Ph. D. thesis, University of Colorado, Boulder.

Cervera, M., J. Oliver, and T. Prato (1999). Thermo-chemo-mechanical model for concrete. I: Hydration and aging II: Damage and creep. *Journal of Engineering Mechanics (ASCE) 125*(9), 1018-1039.

Chen, Z. and M. L. Wang (1993). A partitioned-solution method with moving boundaries for nonlocal creep damage of concrete. In Z. P. Bažant and I. Carol (Eds.), *Creep and Shrinkage of Concrete*, pp. 393–398. RILEM. E & FN Spon.

Dafalias, Y. and E. Popov (1976). Plastic internal variables formalism of cyclic plasticity. *Journal of Applied Mechanics*, 645–651.

de Borst, R., L. J. Sluys, H.-B. Mühlhaus, and J. Pamin (1993). Fundamental issues in finite element analyses of localization of deformation. *Engineering Computations 10*, 99–121.

Diderichs, U., U. Schneider, and M. Terrien (1983). Formation and propagation of cracks and acoustic emission. In F. Wittmann (Ed.), *Fracture mechanics of concrete, Vol. 7*, pp. 157–205. Amsterdam: Elsevier Applied Science.

Feenstra, P. (1993). *Computational aspects of biaxial stress in plain and reinforced concrete*. Ph. D. thesis, University of Delft.

Govindjee, S., G. Kay, and J. C. Simo (1995). Anisotropic modelling and numerical simulation of brittle damage in concrete. *Int. J. Num. Meth. Engng. 38*, 3611–3633.

Han, D. J. and W. F. Chen (1986). Strain-space plasticity formulation for hardening-softening materials with elastoplastic coupling. *International Journal for Solids and Structures 22*(8), 935–950.

Hofstetter, G. and H. A. Mang (1995). *Computational mechanics of reinforced concrete structures*. Grundlagen und Fortschritte der Ingenieurwissenschaften. Braunschweig: Vieweg & Sohn.

Jofriet, J. C. and M. McNeice (1971). Finite element analysis of reinforced concrete slabs. *Journal of the structural division (ASCE97(ST3))*.

Jun, D. (2002). *Adaptive Strategien für nichtlineare Finite-Element-Simulationen von Schalentragwerken*. Ph. D. thesis, Institute for Structural Mechanics, Ruhr-University Bochum.

Krätzig, W. B. and R. Pölling (2002). An elasto-plastic damage theory for reinforced concrete with minimum number of material parameters. *Computers & Structures (submitted)*.

Lemaitre, J. (1992). *A course on damage mechanics*. Berlin: Springer-Verlag.

Lemaitre, J. and J.-L. Chaboche (1990). *Mechanics of solid materials*. Cambridge University Press.

Mazars, J. and G. Pijaudier-Cabot (1989). Continuum damage theory – application to concrete. *Journal of the Engineering Mechanics 115*(2), 345–365.

Meschke, G., R. Lackner, and H. A. Mang (1998). An anisotropic elastoplastic-damage model for plain concrete. *Int. J. Num. Meth. Engng. 42*, 703–727.

Meyers, B. L., F. O. Slate, and G. Winter (1969). Relationship between time-dependent deformation and microcracking of plane concrete. *ACI Journal 66*, 60–69.

Mindess, S. (1983). The application of fracture mechanics to cement and concrete: A historical review. In F. Wittman (Ed.), *Developments in Civil Engineering: Fracture Mechanics of Concrete*, pp. 1–30.

Nechvatal, D. (1996). *Normalbeton unter hohen Dauerlasten bei verhindertem Feuchteaustausch*. Ph. D. thesis, Lehrstuhl für Massivbau, Technische Universität München.

Ortiz, M. (1985). A constitutive theory for the inelastic behavior of concrete. *Mechanics of Materials 4*, 64–93.

Petryna, Y. (2002). Schädigung, Versagen und Sicherheit von Tragwerken des Konstruktiven Ingenieurbaus. Habilitationsschrift, Ruhr-Universität Bochum.

Pölling, R. (2000). *Eine praxisnahe, schädigungs-orientierte Materialbeschreibung von Stahlbeton für Strukturanalysen*. Ph. D. thesis, Institute for Structural Mechanics, Ruhr-University Bochum.

Rossi, P., N. Godart, J. Robert, and J, Gervais (1993). Investigation of the basic creep of concrete by acoustic emission. In Z. P. Bažant and I. Carol (Eds.), *Creep and shrinkage of concrete*, pp. 33-38. RILEM: E & FN Spon.

Simo, J. C. and T. J. R. Hughes (1998). *Computational Inelasticity*. Berlin: Springer-Verlag.

Simo, J. C. and J. W. Ju (1987). Strain- and stress-based continuum damage models. I: Formulation. *Int. J. Solids Structures 23* (7), 821–840.

Computational Modelling of Concrete Structures, Bićanić et al. (eds)
© 2003 Swets & Zeitlinger, Lisse, ISBN 90 5809 536 3

Numerical simulation of hydrous damage in concrete: application to a structural element

F.X. Hubert, N. Burlion & J.F. Shao
Laboratory of Mechanics of Lille, Polytech'Lille-USTL, Cité Scientifique, Bd P. Langevin, Villeneuve d'Ascq, France

ABSTRACT: Drying shrinkage of concrete induces the formation of microcracks that damage concrete structures. This damage can be characterized by deteriorating material elastic properties over the years. In this investigation, we present a modeling of this phenomenon, simple and easy to be implemented, and that should allow engineers to predict cracking as a durability indicator and to model ageing. We first present the basic hypotheses of our model, and formulate hydrous damage by using an isotropic scalar variable proportional to the decrease in water content. Uncoupling between drying-induced damage and classical mechanical damage is assumed. Three simple applications are proposed. For the 1-D drying case (cylindrical specimens), we compare the simulation of the variation in time of the dimensionless stiffness with experimental results in uniaxial compression. A good adequacy is observed. For the 2-D drying case (prismatic specimens), the knowledge of measurements of mass loss and average longitudinal shrinkage in time allows us to predict the value of drying induced stresses in a cross section, taking hydrous damage into account. A better estimate of stresses induced by drying is proposed. For the 3-D drying example (wall case), a comparison between a purely elastic calculation and a calculation which includes hydrous damage is proposed for the case of an embedded wall subjected to drying.

1 INTRODUCTION

It is well known that drying of ordinary concrete induces shrinkage that provokes tensile stresses as the main cause of the formation of cracks both near the skin (Bažant & Wittmann 1982, Acker 1988, Kkhelidj et al. 1998) (structural effect due to moisture gradients) and inside the core (Kkhelidj et al. 1998, Hearn 1999, Sadouki & Wittmann 2001, Bisschop & van Mier 2001) (difference of stiffness between aggregates and hardened cement paste retracting) of structures. Variations in surface tension of capillary pores (Wittmann 1997), variations in disjoining pressure in micropores (Bažant & Kim 1991) and capillary depression can explain this macroscopical shrinkage. A first consequence deals with the durability of concrete that cracks lower allowing faster diffusion of aggressive agents as chloride and sulphates. A second consequence showed experimentally is that drying cracks are one major cause of the deterioration of the material elastic properties over the years. For drying concrete or mortar, this deterioration can reach 20% of the Young modulus and 25% of the Poisson ratio (Yurtdas et al. 2003). So for engineers that have to design power plants, bridges or storage containers of radioactive waste, knowledge of chemical and mechanical features of concrete must be considered for long time. Our purpose is to implement a simple modeling (Hubert et al. 2003) of this phenomenon that would allow engineers to predict cracking as a durability indicator and to model ageing. We first formulate our isotropic hydrous damage model as an ageing variable (Gérard 1996, Carde 1997, Pijaudier-Cabot et al. 1998) in the frame of thermodynamics of non reversible processes. Hydrous damage variations are defined proportional to local decrease in water content. Uncoupling between drying-induced damage and classical mechanical damage is assumed. A discussion is proposed about the necessity of taking into account or not plasticity for this kind of study. Mechanical damage is described with the classical model of Mazars (Mazars 1984). The model is calibrated with experimental results where measurements of mass loss kinetics and loading-unloading uniaxial compression tests were carried out. For numerical 3-D simulation, we used the finite-element code named *Code_Aster®* developed by *Electricité De France* (EDF) for structural calculations. Classical case of a concrete wall with a non shrinking foundation submitted to drying conditions on three faces is studied. The moisture diffusion is described by a non linear equation with a diffusivity function of water content (Bažant 1972,

Granger 1995). For boundary conditions, a non-linear flux of water content is applied on the faces of the wall. Drying shrinkage is assumed locally proportional to water content. Drying induced stresses are estimated with the elasto-hydrous damageable model. Taking into account hydrous damage for numerical simulation gives a good prediction of those tensile and compressive stresses inside concrete.

2 PRESENTATION OF THE MODEL

2.1 *Drying model*

In concrete, the interactions between the cement matrix and interstitial liquid are strong. Several phenomena will occur: surface tension at the liquid-solid interface that creates tensile micro-stresses between molecules (Wittmann 1997), adsorption of water molecules on solid surface which, by repulsion between themselves, generate disjoining pressures (Bažant 1972), capillary depressions with menisci. These phenomena depend on the value of the relative humidity (RH) inside the pores. When RH is lower than 40%, surface tension at the liquid-solid interface is predominant (Wittmann 1997, Beltzung et al. 2001), and when $50\% < RH < 100\%$, there is coexistence between disjoining pressures and capillary attraction (Bažant 1972, Beltzung et al. 2001). At macroscopic level these phenomena are quite important and induce shrinkage of concrete. For instance, Beddoe (Beddoe & Lippok 1999) measured uniaxial tensile stresses on a hardened cement paste cylinder submitted to capillary suction in the region of 8 MPa, a value higher than tensile strength of classical concrete.

Drying of concrete is also a complex phenomenon. There are three types of hydrous transfers (Mainguy 2001): Darcy flow of liquid water, Darcy flow of vapor, and diffusive transfer of vapor. Considering the experimental results, Mainguy analyzed the influence of the three kinds of transfers on mass loss kinetics of one concrete. He showed that for vapor, the Darcy movements are negligible compared to diffusive movements. Then, he attested the existence of gas overpressures inside the pores, and concluded to a fast uniformisation of the vapor molar fraction (then to a very short duration of the influence of the vapor diffusion). For our purpose, a description of water transfers in liquid form with evaporation to the sides of the structure and the use of a non-linear diffusion equation expressed as a function of saturation degree or water content are sufficient to correctly describe kinetics of mass loss during time. To facilitate the implementation of our model easy to be implemented, we express the whole hydrous transfers with a coefficient of hydrous diffusion C in m^2/s, as used in the classical pattern proposed by Bažant and Kim (Bažant & Kim 1991). The expressions translating the conservation of the water mass and of the flux of relative humidity are:

$$\frac{\partial w}{\partial t} = div(C.grad\, h) \tag{1}$$

where t is time, h relative humidity inside the pores, w the water content, C the coefficient of hydrous diffusion which depends on h. The identification of the evolution of parameter C requires drying tests on concrete. The results show that C decreases with relative humidity inside the material.

The sorption-desorption isotherm is written under the $w = g(h)$ shape, and thus:

$$\frac{\partial h}{\partial t} = \frac{1}{g'(h)}\frac{\partial w}{\partial t} \tag{2}$$

in which

$$\frac{\partial w}{\partial h} = g'(h) \tag{3}$$

with $g'(h)$ hydrous capacity (slope of the desorption isotherm). For the sake of simplicity, we will only consider the case where hydration is achieved, as assumed in our experimental conditions. From relation (1), we obtain the non-linear diffusion equation which is a function of h.

$$\frac{\partial h}{\partial t} = \frac{1}{g'(h)} div(C(h) grad\, h) \tag{4}$$

What is required for this equation to be resolved is first the sorption-desorption isotherm of the studied material, second the values of the coefficient of hydrous diffusion, function of h, and finally boundary conditions. To characterize the sorption-desorption isotherm, a concrete specimen has to be submitted to various hydrous balances, depending on the external relative humidity. The result is a curve linking internal relative humidity h and water content w. It is a very long and delicate work because of possible errors in measurement, mainly for high values of relative humidity for which specimens are in an unstable hydrous state (Baroghel-Bouny 1994). The results obtained by the desorption isotherm influence a lot the numerical results of the modeling of hydrous transfers from the specimen to the outer environment.

In this first study, the value of C is assumed to be constant for 1-D and 2-D numerical simulations, which is the same as to assume that relative permeability of water remains constant. No doubt this assumption is inadequate since everybody knows that this parameter depends heavily on the saturation degree and on the damage of concrete. We stay in the phase of drying where liquid transfers are still occurring, knowing that under our weather conditions no more than 50% of the vaporisable water inside a structure can evaporate

(Acker 1988). This range of variation of water contents is a limit to our pattern because only when there is liquid water left can we assume a proportional relation between hydrous damage and mass loss. The diffusion equation becomes therefore linear. We should normally take into account the decreasing value of the coefficient of hydrous diffusion due to carbonation (Pihlajavaara 1982). In our experimental campaign, the curing lasted 28 days, and drying tests 70 days. This duration is sufficiently short to allow us to disregard carbonation effects.

2.2 *Why modeling hydrous damage?*

The drying of concrete is not uniform within structures. Whereas the outer layers are at the same relative humidity as the outside environment, the core of the specimen keeps water-saturated for a long time. There are therefore hydrous gradients. At the microstructural level, some capillary depressions, variations in disjunction pressure and superficial tensions are generated that induce imposed strains and tensile stresses. Where the hydrous gradient is high, notably near the skin, there is structural cracking as soon as hydrous tensile stresses reach the limit tensile strength of concrete (Bažant 1972, Acker 1988, Sadouki & Wittmann 2001). Cracks first appear near the skin, and then their depth and their width increase. When the front of drying progresses toward the core of the specimen, a network of parallel cracks is created, whose width and spacing are proportional to the distance of penetration of the drying? There is therefore cracking in the core of the material (Bažant & Raftshol 1982). Besides, microcracks are induced by the variation of instantaneous Young moduli (lower for the cement matrix) between the aggregates (inclusions) and the matrix composed by hardened cement paste (hcp). Their behaviors also are different: viscoelastic with damage for the hcp retracting, stiff for aggregates. This causes differential strains and a locally auto-stress state at the aggregate-cement interface. This leads to microcracking, more particularly around the aggregates (Acker 1988, Hearn 1999, Sadouki & Wittmann 2001, Bisshop & van Mier 2001). Furthermore, Bažant and Raftschol (Bažant & Raftshol 1982) showed that in order to avoid any drying crack, it was necessary to reduce the relative surrounding humidity sufficiently slowly and gradually so that there is no difference of relative humidity inside the specimen superior to 2% and to use test specimen unreasonably thin (in the region of one millimeter thick). In structures there are geometrical effects that induce a non uniform distribution of relative humidity, even if the drying process is extremely slow. In case they are suddenly exposed to outside relative humidity, only structures 0.1 mm thick would be exempt of microcracks. Which means that generally occur heterogeneous drying microcracking even if no mechanical load is applied. Some authors (Bisshop & van Mier 2001) are currently trying to quantify experimentally on the one hand the area of induced microcracks (around aggregates) and on the other hand the area of structural cracks. In this first approach, no difference will be made between the two kinds of cracking.

We assume that, for a given cross-section, cracking leads to a loss of resistant surface. We can draw a parallel with mechanical damage, of which one definition is the diminution of load resistant surface. We then formulate the existence of hydrous damage. But the comparison between those two kinds of damage cannot be extended as the size and distribution of cracks is quite different. Besides, we assume that drying-induced cracking is diffuse in the material and locally proportional to the decrease of water content. The hydrous damage is supposed to be isotropic, and uncoupled with mechanical damage. For this first pattern, we take into account hydrous damage in a simplified manner i.e. an ageing function similar to those already proposed in order to describe damage induced by leaching of concrete (Gérard 1996, Carde 1997, Pijaudier-Cabot et al. 1998), or in order to describe damage caused by thermal effects (Nechnech 2000).

The general framework is the thermodynamics of non-reversible processes, more precisely the one of isotropic damageable ageing elasticity. We will suppose that the Poisson ratio υ keeps constant. The variable of hydrous damage is assumed to be isotropic. The stiffness tensor E is defined as follows, with E_0 tensor of initial stiffness (Hubert et al. 2003):

$$E = (1 - d_m)(1 - d_h)E_0 \tag{5}$$

The damage variables d_h and d_m are internal variables, non directly observable. The state laws are:

$$\sigma = (1 - d_m)(1 - d_h)E_0 : \varepsilon^e \tag{6}$$

The uncoupling hypothesis between the evolutions of the two types of damage allows showing their individual contribution. This can be questionable since, on the one hand, the hydrous cracks modify the mechanical damage of concrete structures (Bourgeois et al. 2002) and, on the other hand, we can assume that the increase of porosity (caused by microcracking induced by mechanical loading) modifies interactions between liquid and solid and therefore the distribution of cracks linked to drying (Bažant et al. 1986, Granger et al. 1994). We must point out that concrete maturation/hydration can on the contrary increase the stiffness of the material. The coupling of both phenomena, maturation and hydrous damage, remains to be studied, as well as physico-chemical phenomena during maturation. In our model, we suppose that hydration is achieved.

Notice that Bodé showed that numerical calculation could be done on already-damaged structures (Bodé 1994). This is precisely our case, since structures will be hydrousally damaged before mechanical loading. Therefore, hydrous damage can be taken into account for numerical simulations.

2.3 *Mechanical damage evolution*

In order to describe mechanical damage, we have chosen the yield function proposed by Mazars (Mazars 1984). In this model, the intensity of local extensions is expressed by the equivalent strain $\tilde{\varepsilon}$, defined as:

$$\tilde{\varepsilon} = \sqrt{\sum_i (\varepsilon_{i+})^2} \quad \text{with} \quad \varepsilon_{i+} = \frac{\varepsilon_i + |\varepsilon_i|}{2} \tag{7}$$

and ε_i the principal deformation in the direction i. The evolution of mechanical damage is derived from the yield function:

$$f_d = \tilde{\varepsilon} - K^m(d_m) \leq 0 \tag{8}$$

with $K^m(d_m)$ the current mechanical damage threshold. The total damage is equal to a linear relation between damage caused by tensile and compressive stresses:

$$d_m = \alpha_t d_t + (1 - \alpha_t) d_c \tag{9}$$

with α_t the parameter defined as a function of stress state. In case of pure tension, $\alpha_t = 1$, and for uniaxial compression, $\alpha_t = 0$. In general its value is:

$$\alpha_t = \sum_i \frac{H(\varepsilon_i)\varepsilon_{ti}(\varepsilon_{ti} + \varepsilon_{ci})}{(\tilde{\varepsilon})^2} \tag{10}$$

where $H(x)$ is the Heaviside function of principal strain. ε_{ti} and ε_{ci} are respectively elastic strains due to tensile (positive) and compressive (negative) principal stress σ_i under a given damage state. Damage evolutions due to tensile and compressive loading are described by the following relations proposed by Mazars:

$$d_t = 1 - \frac{\varepsilon_{d0}(1 - A_t)}{\varepsilon_M} - \frac{A_t}{exp[B_t(\varepsilon_M - \varepsilon_{d0})]} \tag{11}$$

$$d_c = 1 - \frac{\varepsilon_{d0}(1 - A_c)}{\varepsilon_M} - \frac{A_c}{\exp[B_c(\varepsilon_M - \varepsilon_{d0})]} \tag{12}$$

where ε_{d0} is the initial damage limit and ε_M the maximal total equivalent tensile strain. Four parameters A_c, B_c, A_t and B_t control damage evolution kinetics respectively under tensile and compressive loading.

2.4 *Hydrous damage evolution*

In order to describe the evolution of hydrous damage, we use the same framework of formulation as that proposed by Mazars (Mazars 1984). We introduce a yield function:

$$F_h(w, d_h) = f(w) - K^h(d_h) \tag{13}$$

with $K^h(d_h)$ the current hydrous damage threshold, and $f(w)$ a function of water content. We obtain the expression of hydrous damage function of water content (Hubert et al. 2003):

$$d_h = -a.[w_{\min} - w_0] \tag{14}$$

with $w_{\min}$ the lowest water content reached in the drying-wetting history and w_0 initial water content, at saturation equal to *1*. The difficulty is that the value of the constant a has to be identified. A method of determination of this parameter is proposed in the following paragraph. We can observe from the expression of hydrous damage that both hydrous gradient and uniform decrease of water content can be modeled. Then, the case of very thin pieces with very slow decreasing of surrounding relative humidity (Bažant & Raftshol 1982) cannot be described by our model, but this case is quite rare in civil engineering. The case of increase of relative humidity wasn't studied here as it is more complex: intricate phenomena occur, like restarting of hydration with higher kinetics.

We will now examine the influence of hydric damage on uniaxial compressive behavior of concrete. In Figure 1 we plotted stress versus strain from the

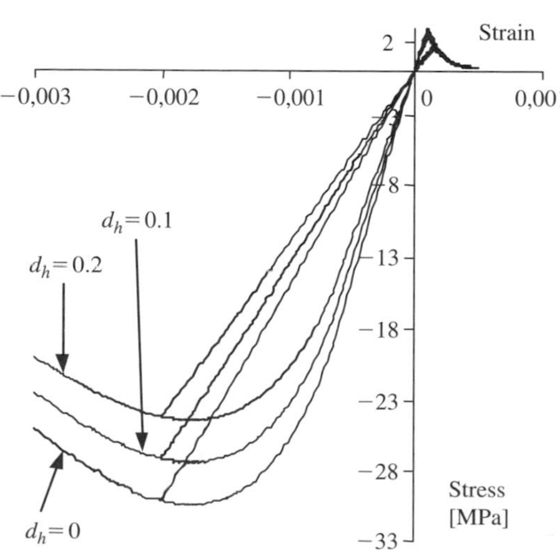

Figure 1. Effects of hydric damage on behavior of concrete.

model, using the parameters of Table 1, simulating one cycle of loading-unloading. Three curves are plotted. The first one takes into account only mechanical damage. In the other two, we can see the influence of an additional hydrous damage equal to 0.1, and 0.2. This shows that maximal tensile stress and compressive stress are locally lowered by the effects of desiccation, even if for the structure studied (specimen) the maximal strength of cylinders was increasing (Bourgeois et al. 2002). We can also see that Young modulus of concrete decreases when hydric damage increases. If we zoom in on the tensile part of the curve, in Figure 1, we can observe the same phenomenon. Tensile strength and Young modulus decrease when hydrous damage increases.

2.5 *Plasticity strain evolution*

By adopting the assumption of small strains, the total strain tensor is decomposed into an elastic part and plastic part:

$$\varepsilon_{ij} = \varepsilon_{ij}^e + \varepsilon_{ij}^p, \quad d\varepsilon_{ij} = d\varepsilon_{ij}^e + d\varepsilon_{ij}^p \tag{15}$$

The characterization of plastic deformation should take into account relevant experimental evidences. Furthermore, in cohesive geomaterials like concrete, in addition to plastic deformation, microcrack induced damage is a common phenomenon resulting in deterioration of material elastic properties. We present here a simplified model for isotropic behavior of unsaturated concrete. Material damage is represented by a scalar variable D. Accordingly, the plastic functions will be formulated with the following effective stress tensor, which couples in a simplified manner damage and plasticity:

$$\hat{\sigma} = \frac{\sigma}{1 - D} \tag{16}$$

In our case, damage is both due to hydrous and mechanical effects. So we can write:

$$\hat{\sigma} = \frac{\sigma}{(1 - d_m)(1 - d_h)} \tag{17}$$

Table 1. Parameters used for numerical simulations.

Compression	Tension
$E_0 = 24\,\text{GPa}$	$E_0 = 24\,\text{GPa}$
$\alpha_0 = 0.95$	$\alpha_0 = 0.29$
$\alpha_m = 2.5$	$\alpha_m = 6$
$b = 800$	$b = 100$
$\varepsilon_{d0} = 10^{-4}$	$\varepsilon_{d0} = 10^{-4}$
$A_c = 1.5$	$A_t = 0.7$
$B_c = 1100$	$B_t = 30000$

The yield function of plasticity is expressed by using the classic Drücker-Prager criterion ($f_{\hat{\sigma}}$):

$$(f_{\hat{\sigma}}) = \hat{q} - \alpha(\hat{p} + C_0) = 0 \tag{18}$$

where $\hat{p}$ is the effective mean stress, $\hat{q}$ the effective deviatoric stress. The parameter C_0 in (18) appears as the material cohesion in saturated conditions. The plastic hardening of concrete is described by variation of friction coefficient α in equation (18). As a coupled elastoplastic damage modelling is concerned, we assume that induced damage is responsible of material softening behaviour. Therefore, we propose that the hardening function α is continuously increasing with the total equivalent plastic strain $\bar{\varepsilon}_p$. Based on experimental data, the following particular form is determined:

$$\alpha = \alpha_r - (\alpha_r - \alpha_0)e^{-b\bar{\varepsilon}_p} \tag{19}$$

with

$$\bar{\varepsilon}_p = \int (d\varepsilon_{ij}^p d\varepsilon_{ij}^p)^{\frac{1}{2}} \tag{20}$$

The parameters α_0 and α_r are the initial and ultimate value of the hardening function and b controls the rate of plastic hardening.

On Figure 2 is presented the response of the model in tension. Evolution of plastic strain was estimated from test data obtained in direct tension (Terrien 1980). One can notice that no unilateral effect is taken into account. Comparison between the model and experimental data is proposed on Figure 3 for a uniaxial compression test. A good accordance between the two curves is observed. In this simulation, hydrous damage

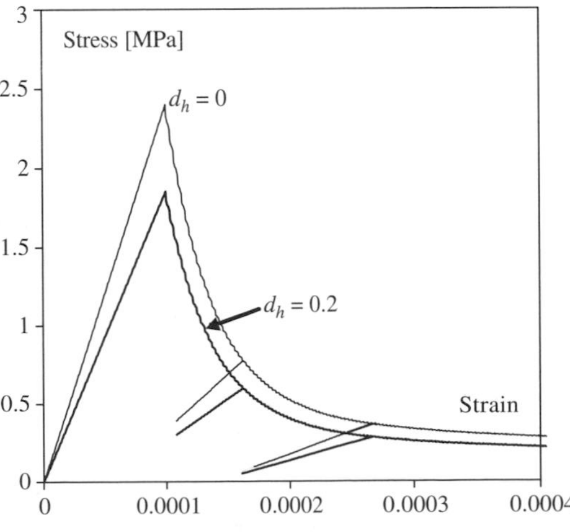

Figure 2. Elastoplastic damageable behavior of concrete in tension.

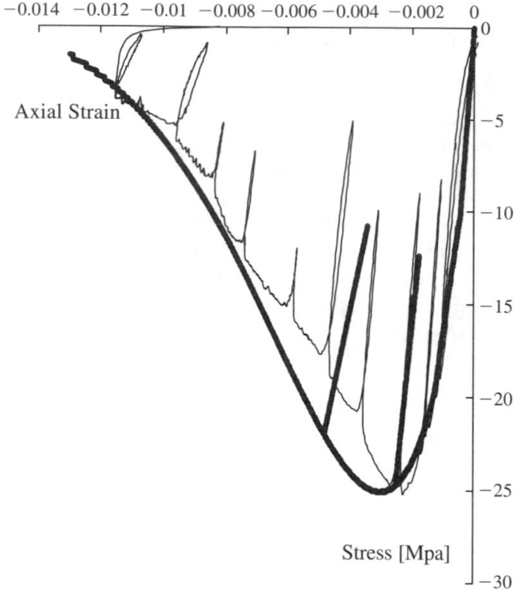

Figure 3. Elastoplastic damageable behavior of concrete in compression: comparison between experimental data (thin line) and the model (saturated case, $d_h = 0$).

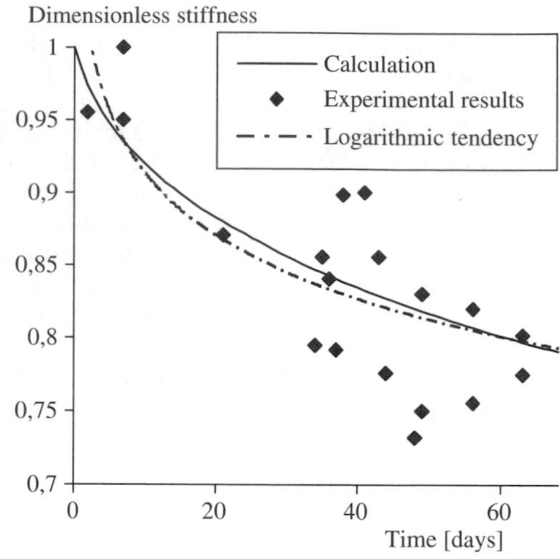

Figure 4. Elastoplastic damageable behavior of concrete in compression: comparison between experimental data and the model response.

is equal to 0, because the experimental data correspond to a saturated sample. We can notice that the mechanical damage is slightly overvalued, because it is the pilot of the curve softening, whereas plastic hardening remains positive. The used parameters for all simulations are given in Table 1. Plasticity parameters used in tension and compression are different to reproduce the fact that tension plastic strains are very small.

3 NUMERICAL SIMULATIONS

3.1 *1-D numerical test*

We developed a simplified case in order to identify the coefficient *a*, to check the validity of our hypothesis and to compare its responses to experimental results. We studied the classical one-dimensional case corresponding to the drying of a cylindrical specimen supposedly of infinite length. The calibration of the model was done from experimental measurements of mass loss in time and initial stiffness in time of concrete cylinders with 110 mm in diameter and 220 mm in height (for more detail see (Hubert et al. 2003)). For parameter *a*, we used the compression tests results showed in Figure 4. Each experimental point of the curve is issued from an average of at least two tests. We took the value reached by the dimensionless stiffness at 30 days, value issued from tendency curve of experimental points. We used the dimensionless values (ratio between initial stiffness at *t* time and initial stiffness at $t = 0$ when drying starts) of initial stiffness in order to allow comparisons between specimens with different initial stiffness. This allowed to calculate parameter $a = 5.93$. We chose the early value reached at 30 days, in order to allow long time prediction. Since we know the value of proportionality coefficient *a* between hydric damage and decrease of water content, we can now predict the evolution in time of the stiffness of the cylindrical structure. We can see in Figure 4 the dimensionless stiffness versus time. Experimental points, tendency curve and the response of the model are plotted. We can observe a good agreement with logarithmic tendency curve of experimental results. But we must note the high dispersion of experimental points. So, for engineering problems, it would be more judicious to propose a predictive envelop curve rather than one curve.

3.2 *2-D numerical test*

In order to get an evaluation of drying shrinkage in prismatic specimen, the model was also developed in a two dimensions case. We first compare the water loss kinetics issued from calculation, with the same hypothesis and parameters as assumed in the precedent section, and experimental measures on prisms with 40*40 mm^2 cross section and 160 mm height. For the specimens, the making was done at the same time that for cylindrical ones. They were maintained in the same conditions of curing, relative humidity and temperature after curing. Six prismatic specimens were made, and only submitted to drying. Losses in mass and longitudinal shrinkages were measured in time.

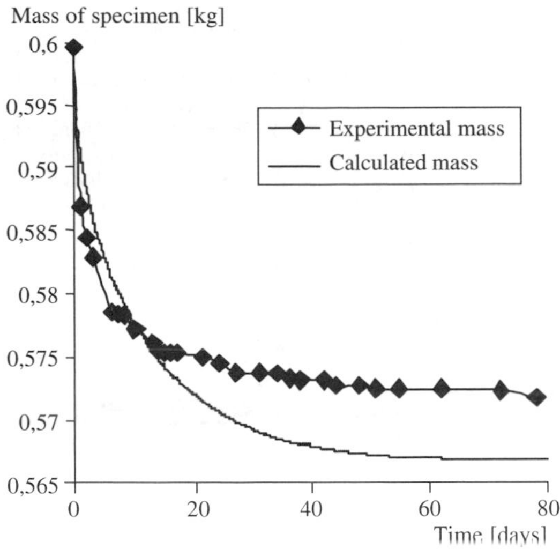

Figure 5. Loss in mass kinetics for prismatic specimens.

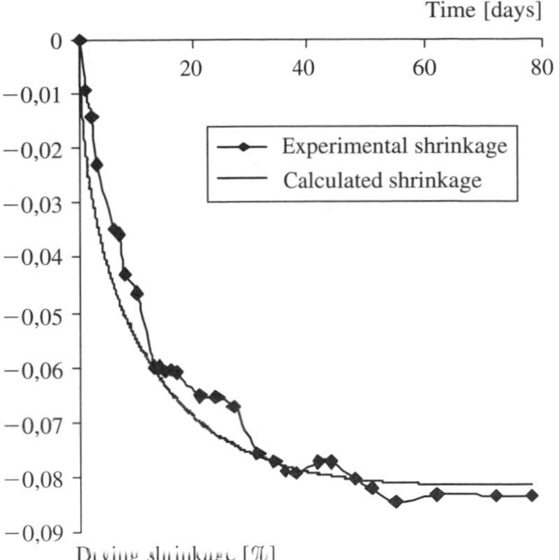

Figure 6. Average longitudinal shrinkage of prismatic specimens.

We plotted the loss in mass kinetics issued from calculation. We can see in Figure 5 that keeping the same hypothesis than for cylindrical pieces, the mass loss is over estimated. It could be explained first by the value of coefficient of hydric diffusion, assumed to be constant. It is known that when saturation decreases, relative permeability of water decreases. Second, considering the dimensions of our prisms, drying has occurred faster than for cylinders. So, liquid transfers stop more quickly, that is as we said before a limit of our modeling. Third, we took the same sorption-desorption isotherm as for cylinders. The characterization of this isotherm should have been completed experimentally, as (Baroghel-Bouny 1994) did. We would have obtained more precision in the results.

The drying shrinkage is assumed to be locally proportional to decrease in water content. So the average longitudinal shrinkage ε_{sh} due to drying can be assumed as to be proportional to loss in mass (Granger et al. 1994, Granger 1995) and can be written:

$$\varepsilon_{sh} = \frac{1}{V}\int_V Ks(w - w_0)dV \tag{21}$$

with V volume of the structure, Ks proportionality coefficient considered constant, and w_0 initial water content. In our case, considering the specimens with infinite length, we can write:

$$\varepsilon_{sh}(t) = \frac{1}{S}\int_S Ks(w(x,y,t) - w_0)dS \tag{22}$$

The value of Ks is characterised from experimental measurements of average longitudinal shrinkage. This value depends on concrete formulation, making, and external conditions of temperature and relative humidity. We chose to calibrate Ks with the value of average shrinkage measured at early age, 30 days, to give a long term predictive response of the model. From calibration, $Ks = 0.01337$. From expression (22), we plotted in Figure 6 drying shrinkage versus time. We can see in this figure that drying shrinkage kinetics are correctly described.

Drying leads to an autostress state. With this structural effect, cracks occur when tensile strength is exceeded. The apparitions of cracks release a part of those hydric stresses. We propose an estimation of drying induced stresses, taking into account structural effect of drying and microcracking due to heterogeneity. The average longitudinal drying shrinkage is known from experimental measurements. Since drying shrinkage is locally proportional to decrease in water content, we can calculate the value of free shrinkage of virtual independent fibbers of material. It gives the value of elastic strain in those virtual fibbers. So we can calculate the value of drying induced stresses in a cross section, with the same analysis as (Benboudjema et al. 2001). We plotted in the Figure 7 the stress versus transversal x-coordinate at two times: 5 days and 30 days. The origin of x-axis is located at the centre of the cross section. We can see at 30 days a great influence of hydric damage on compressive stress (almost 1 MPa), and at 5 days on tensile stress (almost 0.5 MPa). It shows that stresses are better predicted if the hydric damage is taken in counts. If we take only into count structural effect (shrinkage), we over estimate stresses in the structure because microcracking

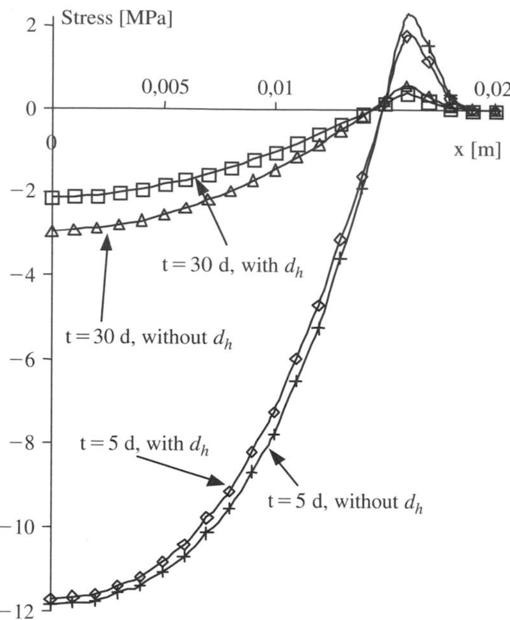

Figure 7. Estimation of stresses in a half part of a prismatic specimen, showing the influence of hydric damage d_h.

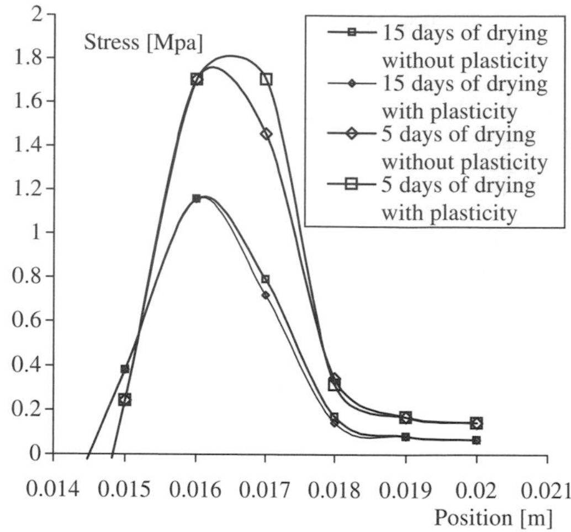

Figure 8. Estimation of stresses in a half part of a prismatic specimen, showing the influence of plasticity.

effects are not taken into account. Nevertheless, we must point that we can't quantify currently the contribution of structural effect or interaction between cement matrix and aggregates for the existence of cracking. So values of the stresses are slightly under estimated if hydric damage is used. On Figure 8, the same type of simulation is proposed for 2 times of drying (5 and 15 days): on one hand with the elastic-damageable model with plasticity and on the other hand, without consideration of plasticity. This figure presents only the prismatic sample zone which is submitted to tension, i.e. for a position close to the surface (x-position between 14 and 20 cm). We can notice that differences between both simulations are very close, what confirms the major role of damage in the case of tension loading. In compression, stresses remain small and do not activate plasticity.

3.3 *3-D numerical test*

The simulation is presented here in its simplified version: the objective is to show the interest of using hydrous damage in a 3-D calculation. We used the finite-element code "Code_Aster®" developed by *Electricité De France* (EDF) for classical case of a 3D wall with a non shrinking foundation submitted to drying conditions on three faces is studied. This example is similar to those proposed by (van Zijl et al. 2001). The moisture diffusion is described by a non linear equation with a diffusivity function of water content (Bažant 1972, Granger 1995), presented in the second section of this paper. The coefficient of hydrous diffusion, which depends on w, is calculated from expression proposed by (Granger 1995):

$$C(w) = A.\exp(B.w) \tag{23}$$

with $A = 3.10^{-13}\ m^2/s$ and $B = 0.05$.

For boundary conditions, a non-linear flux of water content is applied on the wall faces (Figure 9). This condition is expressed as (Granger 1995):

$$\varphi = 0.5^*\beta^*(w - (2w_0 - w_{eq}))^*\frac{w - w_{eq}}{(w_0 - w_{eq})^2} \tag{24}$$

where φ is the moisture flux, β is a coefficient (depends on surface convection conditions) taken equal to 5 *mm per day*, w_0 is the initial moisture content and w_{eq} the equilibrium moisture content.

Drying shrinkage is assumed locally proportional to water content. Drying induced stresses are estimated with the elasto-hydrous damageable model (no mechanical damage and no plasticity).

In Figure 9, we present the dimensions of the considered concrete wall, which is submitted to both its own weight and drying on 3 faces. To reduce calculations costs, only a quarter of the wall is used for this simulation. The used mesh is composed with 12500 quadratic elements (cubic elements with 8 nodes) with 15606 nodes (Figure 10). The bottom of the mesh is embedded. Symmetrical conditions are taken into account. Desiccation shrinkage is going to induce important stresses. After desiccation and drying shrinkage, the wall is deformed (Figure 10).

The hydrous damage remains small and provokes a diffuse microcracking. In this 3D-calculation, taking

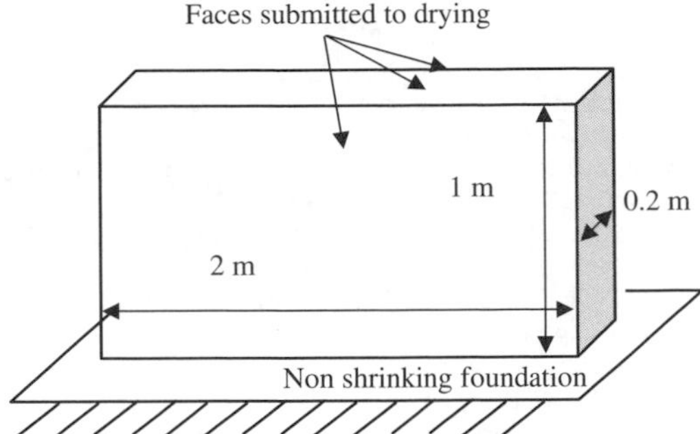

Figure 9. Geometry of the wall with a non-shrinking foundation.

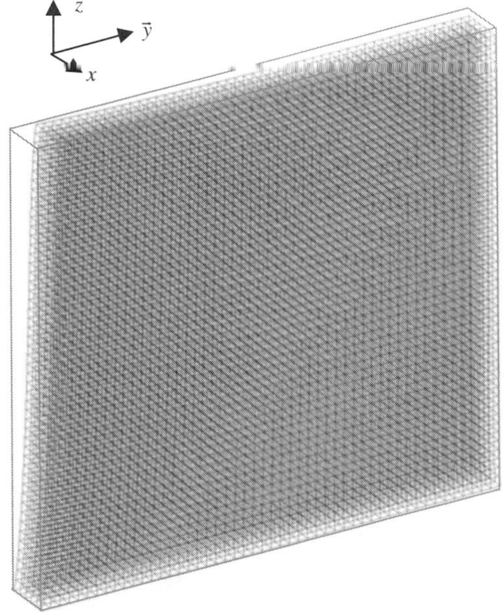

Figure 10. Deformed mesh after 12 months of drying (amplification ratio = 100).

into account hydrous damage is made by affecting locally the elastic modulus of the material with the following equations:

$$E(x,y,z,t) = (1 - d_h(x,y,z,t))E_0 \tag{25}$$

with

$$d_h(x,y,z,t) = -a.[w_{\min}(x,y,z,t) - w_0] \tag{26}$$

As drying provokes evolution of moisture content inside the concrete wall, induced variation of Young Modulus is depicted Figure 11, where E is the actual modulus, while its initial value is E_0.

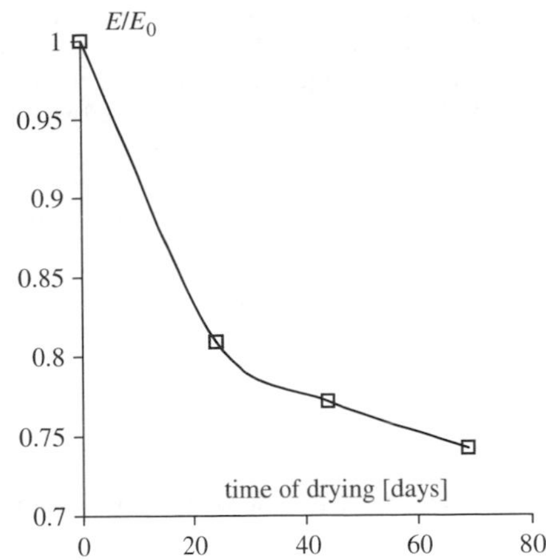

Figure 11. Variation of Young modulus with time of drying.

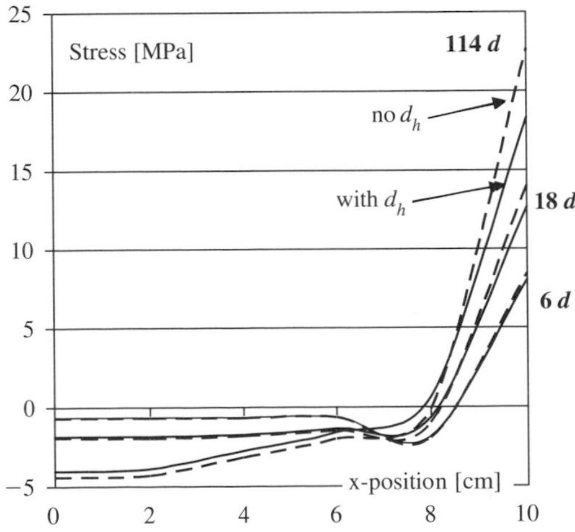

Figure 12. Stress profiles in a cross section: effects of hydrous damage.

Figure 12 shows the evolution of stress along the thickness of the wall, calculated at 0.5 m of height (z-position): the first curve corresponds to an elastic simulation ("no d_h" – dotted line) and the second ("with d_h") is drawn from a simulation where the elastic hydrous damageable model is used. Three drying times are considered: 6, 18 and 114 days. The introduction of the microcracking effect induced by wall desiccation, by means of a simple and scalar model, leads to decreases of tension stresses near the surface and of compressive stresses inside the wall. Notice that in this type of calculation, tensile stresses are overestimated because of no mechanical damage modelling. These both stress decreases can lead to minimize the effects of, on one hand, the differential contraction between the inside and the surface of the wall. This differential contraction leads to introduce some damage of mechanical origin. On the other hand, effects of creep will be minimized: in surface, tension creep will be weaker, what will increase mechanical damage. On the contrary, in the compress part of the wall, compressive stresses are also weaker, what will lead to fewer deferred strains.

4 CONCLUSIONS

We have developed a simple (and easy to be implemented for engineers) model to study the influence of desiccation and drying shrinkage on mechanical behavior of concrete. We chose from the literature a model of drying. We defined a hydric damage isotropic and proportional to local decrease in water content and modeled by a new scalar damage variable. From the elasto-damagable model of Mazars for concrete, prediction of hydrous stresses in prismatic specimen was done (2D problem), showing the influence of hydric damage and allowing for a better estimate of stress values in microcracked concrete. Plasticity has been taken into account in the model to describe more precisely the mechanical behavior of concrete. Finally, 3-D simulations have been done and show that hydrous damage reduces stresses in concrete.

In perspective in this work, it is henceforth necessary to couple mechanical damage and hydrous damage for a better simulation. Furthermore, our hypotheses are undeniably questionable, in particular for 1-D and 2-D proposed simulations. First, a non-constant coefficient of hydric diffusion will be used in a later work, taken from (Bažant et al. 1986) for instance. Second, the hydric damage was assumed to be isotropic. Experiments could show an anisotropic distribution of micro-cracks (Bourgeois et al. 2001). Third, mechanical damage and hydric damage should be coupled in order to show up their interactions and their influences on transport properties.

ACKNOWLEDGEMENT

Electricité de France (EDF) is gratefully acknowledged for the use of Code_Aster® and their help in problem resolutions (www.code-aster.org).

REFERENCES

Acker, P. 1988, Comportement mécanique du béton: apports de l'approche physico-chimique, Phd Thesis of Ecole Nationale des Ponts et Chaussées, Paris, Rapport de Recherche LPC, n° 152.

Baroghel-Bouny, V. 1994, Caractérisation microstructurale et hydrique des pâtes de ciment et des bétons ordinaires et à très hautes performances, Phd Thesis of Ecole Nationale des Ponts et Chaussées, Paris.

Bažant, Z.P. & Raftshol, W.J. 1982, Effect of cracking in drying and shrinkage specimens, *Cement and Concrete Research* 12: 209–226.

Bažant, Z.P. 1972, Nonlinear water diffusion in nonsaturated concrete, *Materials and Structures* 5: 3–20.

Bažant, Z.P. & Wittmann, F.H. 1982, *Creep and Shrinkage in Concrete Structures*, J. Wiley and Sons.

Bažant, Z.P., Sener, S. & Kim, J.K., 1986, Effect of cracking on drying permeability and diffusivity of concrete, *ACI Materials journal* 86: 351–357.

Bažant, Z.P. & Kim, J.K. 1991, Consequences of diffusion theory for shrinkage of concrete, *Materials and Structures* 24: 323–326.

Beddoe, R.E. & Lippok, R. 1999, Hygral stress in hardened cement paste, *Materials and Structures* 32: 627–634.

Beltzung, F., Wittmann, F.H. & Holzer, L. 2001, Influence of composition of pore solution on drying shrinkage, in: F.-J. Ulm, Z.P. Bažant & F.H. Wittmann (eds), *Creep, Shrinkage and Durability mechanics of Concrete and other Quasi-Brittle Materials*, Proc. of CONCREEP6@MIT, Elsevier, 39–48.

Benboudjema, F., Heinfling, G., Meftah, F., Selleir, A. & Torrenti, J.M. 2001, On the prediction of delayed strains for concrete subjected to drying and loading simultaneously, in: F.-J. Ulm, Z.P. Bažant & F.H. Wittmann (eds), *Creep, Shrinkage and Durability mechanics of Concrete and other Quasi-Brittle Materials*, Proc. of CONCREEP6@MIT, Elsevier, 245–250.

Bisschop, J. & van Mier, J.G.M. 2001, Meso-level mechanisms of drying shrinkage cracking in concrete, *Proc. of Int. RILEM Workshop on Fracture and durability, Post-Conference Workshop of Framcos IV*, Cachan, France, June.

Bodé, L. 1994, Stratégies numériques pour la prévision de la ruine des structures du génie civil, Phd Thesis of Ecole Normale Supérieure de Cachan, France.

Bourgeois, F., Burlion, N. & Shao, J.-F 2001, Elastoplasticity and anisotropic damage due to drying shrinkage in concrete, in: F.-J. Ulm, Z.P. Bažant & F.H. Wittmann (eds), *Creep, Shrinkage and Durability mechanics of Concrete and other Quasi-Brittle Materials*, Proc. of CONCREEP6 @MIT, Elsevier, 171–177.

Bourgeois, F., Burlion, N. & Shao, J.-F 2002, Modelling of elastoplastic damage in concrete due to desiccation shrinkage, *Int. J. for Numerical and Analytical Methods in Geomechanics* 26: 759–774.

Carde, C. & François, R. 1997, Aging damage model of concrete behavior during the leaching process, *Materials and Structures* 30: 465–472.

Gérard, B. 1996, Vieillissement des structures de confinement en béton: modélisation des couplages chimico-mécaniques, Phd Thesis of Ecole Normale Supérieure de Cachan, France.

Granger, L., Torrenti, J.M. & Diruy, M. 1994, Simulation numérique du retrait du béton sous hygrométrie variable, *Bulletin de Liaison des Laboratoires des Ponts-et-Chaussées* 190, rèf. 3811, 57–64.

Granger, L. 1995, Comportement différé du béton dans les enceintes de centrale nucléaire, analyse et modélisation, Phd Thesis of Laboratoire Central des Ponts et Chaussées, Paris.

Hearn, N. 1999, Effect of shrinkage and load-induced cracking on water permeability of concrete, *ACI Materials Journal* 96: 234–241.

Hubert, F.X., Burlion, N. & Shao, J.F. 2003, Drying of concrete: modeling of an hydrous damage, *Materials and structures* 36 (255), in press.

Kkhelidj, A., Loukili, A. & Bastian, G. 1998, Etude expérimentale du couplage hydro-chimique dans les bétons en cours de maturation: influence sur les retraits, *Materials and Structures* 31: 588–594.

Mazars, J. 1984, Application de la mécanique de l'endommagement au comportement non linéaire et à la rupture du béton de structure, Phd Thesis of Université Pierre et Marie Curie, Paris.

Mainguy, M., Coussy, O. & Baroghel-Bouny, V. 2001, Role of air pressure in drying of weakly permeable materials, *ASCE J. Engrg. Mech.* 127 (6): 582–592.

Nechnech, W. 2000, Un modèle thermoplastique couplé à l'endommagement pour le béton sous chargements cycliques à hautes températures, *Proc. of Forum des associations du Génie civil et urbain*, Lyon, France, 423–430.

Pihlajavaara, S.E. 1982, Estimation of drying of concrete at different relative humidities and temperatures of ambient air with special discussion about fundamental features of drying and shrinkage, in: Z.P. Bažant & F.H. Wittmann (eds), *Creep and Shrinkage in Concrete Structures*, J. Wiley and Sons, 87–108.

Pijaudier-Cabot, G., Gérard, B., Burlion, N. & Molez, L. 1998, Localisation of damage in quasi-brittle materials and influence of chemically activated damage, in : R. de Borst and E. van der Giessen (eds), *Material Instabilities in Solids*, John Wiley & Sons Ltd., 441–456.

Sadouki, H. & Wittmann, F.H. 2001, Numerical investigations on damage in cementitious composites under combined drying shrinkage and mechanical load', in: R. de Borst, J. Mazars, G. Pijaudier-Cabot and J.G.M. van Mier (eds), *Fracture Mechanics of Concrete Structures, Proc. of Int. Conf. Framcos IV*, Rotterdam: Balkema, 95–98.

Terrien, M. 1980, M., Emission acoustique et comportement mécanique post-critique d'un béton sollicité en traction, *Bulletin de Liaison des Laboratoires des Ponts-et-Chaussées* 105 (rèf. 2398): 65–72.

van Zijl, G.P.A.G., de Borst, R. & Rots, J.G. 2001, The role of crack rate dependence in the long-term behaviour of cementious materials, *Int. J. of Solids and Structures* 38: 5063–5079.

Wittmann, F.H. 1997, Le séchage et le retrait de dessiccation du béton', in *Expérimentation et Calcul en Génie Civil, Proc. of EC97* (in french), Strasbourg, France, 15–26.

Yurtdas, I., Burlion, N. & Skoczylas, F. 2003, Experimental characterisation of the drying effect on uniaxial mechanical behaviour of mortar, *Materials and Structures*, accepted for publication.

Computational Modelling of Concrete Structures, Bićanić et al. (eds)
© 2003 Swets & Zeitlinger, Lisse, ISBN 90 5809 536 3

Measuring and computational modeling of liquid transfer in fractured concrete

J. Carmeliet*, J.-F. Delerue, K. Vandersteen & S. Roels
K.U. Leuven, Department of Civil Engineering, Leuven, Belgium

ABSTRACT: Cracks in concrete are measured in 3D by microfocus X-ray computer tomography. A methodology to convert 3D measured crack data into a network of tubes, which offers the possibility to calculate liquid filling and the permeability, is presented. A numerical model for unsaturated moisture transport in fractured porous media is developed. The model combines a discrete fracture model for moisture flow in a variable aperture fracture with a finite element model for unsaturated flow in the porous matrix. To achieve a stable numerical solution, movable nodes, following the moisture front in the fracture, are introduced at the boundary of the finite element mesh. The model is validated with simulations of water infiltration in fractured brick samples as monitored with X-ray radiography.

1 INTRODUCTION

Durability of porous building materials like concrete is closely related to heterogeneity of the material and the presence of cracks. If cracks are present, preferential flow may occur, resulting in preferential wetting. This feature often leads to an increased deterioration progress, resulting in more cracks and a progressively increase of the deterioration risk. Insight in liquid flow in fractured porous building materials is a necessary step towards the understanding of deterioration mechanisms associated with preferential flow. To gain insight, numerical prediction models are important tools to be combined with artificial ageing tests and laboratory experiments. However, dealing with unsaturated flow in fractured porous media, simulation models often struggle with two difficulties: (1) detailed crack geometry information is difficult to predict, rough-walled fracture with variable aperture; (2) numerical difficulties due to the highly non-linear nature of the constitutive relationships and the strong contrast between physical properties of fracture and matrix (Gerke & van Genuchten 1993, Therrien & Sudicky 1996, McLaren et al. 2000).

In this paper, we study in a first part fluid transfer in a real crack in concrete determined by microfocus X-ray computer tomography. A methodology to convert 3D measured crack data into a network of tubes is presented. The influence of a tendency in crack aperture is highlighted.

In a second part we present a numerically stable discrete approach for modelling unsaturated flow in a fractured porous medium is presented. A 1D discrete model for dynamically simulating liquid flow in a rough-walled fracture with variable aperture is coupled with a finite element model, which solves the Richards' equation for unsaturated liquid flow in porous media. To illustrate the validity of the developed model, simulations of water infiltration in fractured brick samples are compared with experiments monitored with X-ray radiography. This technique is able to determine the moisture profiles in the matrix as well as the filling of the fracture during the experiment.

2 MEASUREMENT OF CRACKS BY MEANS OF μCT

To simulate experiments in cracked materials it is necessary to have an estimation of the dimensions of the cracks. By means of X-ray computer tomography a 3D visualization of cracked specimen is possible in a non-destructive way (Fig. 1).

X-rays sent through the sample are attenuated. For every angle the attenuation intensity is given by the law of Beer-Lambert:

$$\frac{I}{I_0} = \exp\left(- \int_0^D \mu(x) dx \right) \tag{1}$$

* Also at: TU/e, Eindhoven, Netherlands

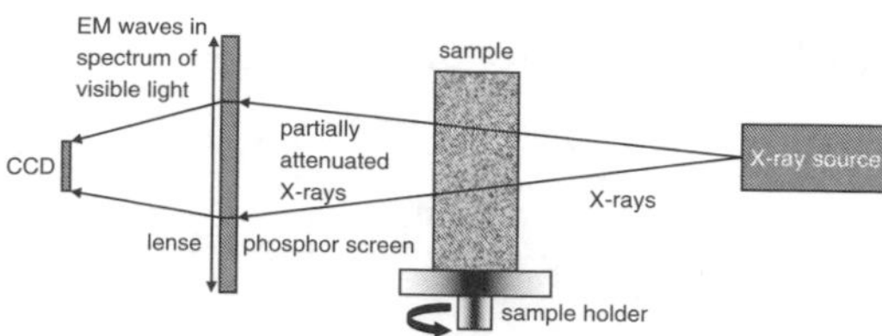

Figure 1. Principle of μCT scan.

Figure 2. CT image of concrete with macro-crack. The crack follows the boundary of the granulates and has a varying width.

where I = the intensity of X-rays after passing the object; I_0 = the initial intensity of the X-rays; μ = the attenuation coefficient; and D = the object thickness.

In CT, the intensity of the X-rays is measured over various angles. Using a reconstruction algorithm, it is possible to determine the attenuation coefficient at various places in the material, thus enabling a 3D visualization of the object. The image of a fractured concrete specimen is represented in Figure 2.

3 DETERMINATION OF A 3D HYDRAULIC NETWORK FROM A μCT IMAGE OF A CRACK

In this method we consider the crack as a pore space, which is represented by a 3D network of tubes. The conductivity of every tube is described by Poiseuille's law

$$q = \frac{\pi r^4}{8\mu}\frac{\Delta P}{L} = k(r, L)\Delta P \tag{2}$$

where ΔP is the pressure drop, r the radius, L the length, μ the viscosity, and k is the conductivity. In order to apply Poiseuille's law, each region corresponding to a pore should look as much as possible like a "tube". They should have a homogeneous diameter or aperture, and should be surrounded by the solid phase. In order to determine the tube network from the CT image we use the segmentation algorithm,

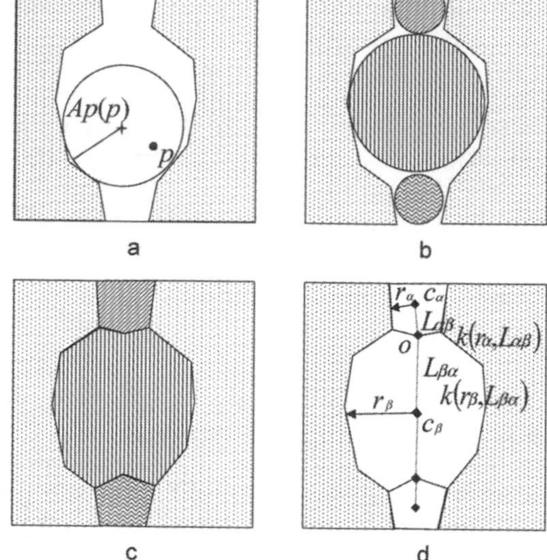

Figure 3. Illustration of the segmentation method used to convert the binary pore space into a 3D network of tubes. (a) determination of skeleton; (b–c) segmentation; (d) vectorisation.

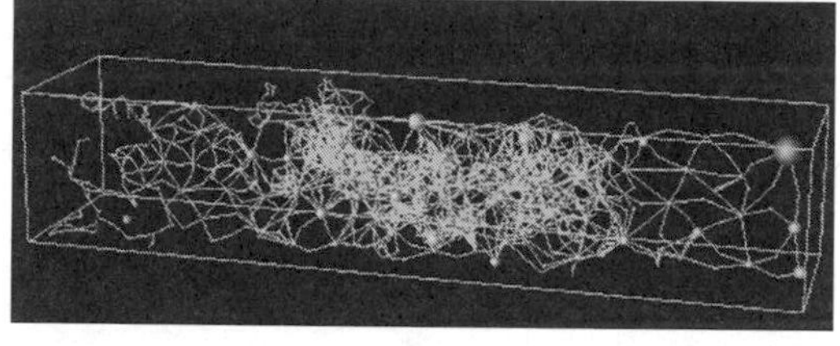

Figure 4. Network representation of the crack of figure 2. The size of the balls represents the size of the pore apertures. At the right of the specimen the crack shows a simple geometry with high apertures; at the left the crack has very low values of apertures (closed crack). In the middle of the crack the complex network represents the complex crack geometry with varying apertures.

proposed by Delerue et al. 1999[a], 1999[b]. Using a thresholding algorithm the grey-scale CT-image is converted into a binary representation where the value of each voxel (0 or 1) indicates if the voxel belongs to the void space or to the matrix. Then the skeleton and local aperture Ap for any point p of the void space is computed (Fig. 3a). The skeleton is defined as the locus of all maximal spheres (balls) inscribed in the pore space, when these spheres touch the pore space boundary at more than one point. The effective segmentation of the pore space is made using a seed-growing algorithm. The algorithm is divided in two steps: (1) non-overlapping maximum balls are first placed into the pore space using the determined skeleton (Fig. 4b) (2) subsequently, the balls are expanded

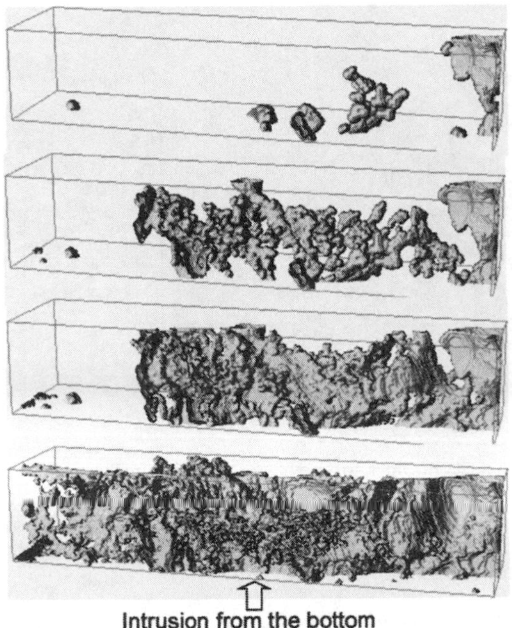

Figure 5. Filling patterns during mercury intrusion from the bottom side for increasing intrusion pressures. We observe first a preferential filling at the right side where the crack is open. With increasing pressure a horizontal fluid flow to the left of the specimen occurs.

into a pore map using a growing region algorithm (Fig. 4c). We now use the pore map to construct a graph representation of the pore space (vectorisation of the pore space. Fig. 4d). Nodes are placed at the centre of pores and links represent the connectivity between the pores. Pore apertures of half of r_α, r_β and the distances $L_{\alpha\beta}, L_{\beta\alpha}$ between centres of different pores are computed. It is assumed that each "half" link can be represented as a cylinder of length $L_{\alpha\beta}$ and radius r_α and that the flux through this cylinder can be computed using Poiseuille's law (Eq. 2). The conductivity $k_{\alpha\beta}$ of the link between pores α and β is computed using the harmonic mean between the conductivities of the two sub tubes connected by this link

$$\frac{2}{k_{\alpha\beta}} = \frac{1}{k(r_\alpha, L_{\alpha\beta})} + \frac{1}{k(r_\beta, L_{\beta\alpha})} \tag{3}$$

Figure 4 gives the computed network determined from the cracked concrete sample of Figure 2 using the segmentation method. The crack shows a trend going from high values of aperture at the left of the specimen (open crack) to very low values at the right (closed crack). In the middle a complex crack geometry is observed. Now fluid can be intruded (imbibided) in the tube network. Figure 4 gives as an example the intrusion of non-wetting mercury at different intrusion pressures from the bottom side.

The overall conductivity k can computed for different degrees of fluid filling of the crack space. We apply a pressure difference ΔP over the network and determine the global flux Q of fluid passing through the entrance (or exit) pores

$$Q_{entrance} = \sum_{\alpha \in inlet} k_{\alpha\beta}(P_\alpha - P_{in}) \tag{4}$$

The selection of the entrance and exit pores defines the direction $\vec{t}$ in which the fluid is flowing during the simulation, and thus the direction in which the permeability is computed. The overall hydraulic conductivity is deduced from the flux using Darcy's equation

$$k = \frac{Q_{exit}}{\Delta P} \tag{5}$$

4 MEASURING AND SIMULATING MOISTURE UPTAKE IN A FRACTURE – POROUS MATRIX SYSTEM

To simulate moisture uptake in a fractured porous building material, we combine a 1D discrete model for liquid flow in a fracture with a finite element model that solves the unsaturated liquid flow in the porous matrix (Fig. 6) (Vandersteen et al. 2002[a], 2002[b], Roels et al. 2003). In the discrete fracture flow model the fracture is represented as a row of nodes, connected to each other by flow channels. These flow channels are represented by parallel plates with an aperture equal to the aperture of the fracture segment they belong to. We use a moving front technique to solve the water flow problem in this system. Following Hvistendahl Karlsen et al. (1998) the progress of the moisture front in the fracture is calculated from a combination of a quasi static pressure (Eq. 6) and a Darcian flux (Eq. 7):

$$\nabla(K(\nabla P_l + \rho_l g \cos\phi)) - S = 0 \tag{6}$$

$$u = \frac{\partial z}{\partial t} = -\frac{K}{\rho_l}(\nabla P_l + \rho_l g \cos\phi) \tag{7}$$

in which K represents the permeability, P_l the liquid pressure in the fracture segment, g the gravity constant, ρ_l the liquid density (kg/m^3), the angle of the fracture segment with the vertical ($0° \leq \phi \leq 90°$) and u (m/s) the velocity of the moisture front in the fracture. The velocity u can be approximated by dz/dt, with dz the progress of the waterfront during the timestep dt (see Fig. 6). In the pressure equation a sink term S is introduced, corresponding to the capillary imbibition flow from the crack to the matrix.

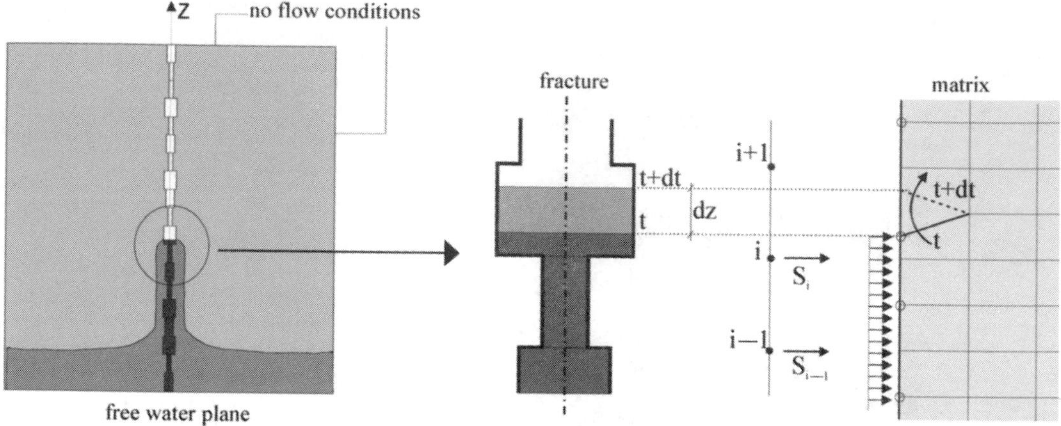

Figure 6. Coupling the discrete fracture model with the FE-model. In order to eliminate instability in the numerical calculations movable nodes (hollow dots on the FE-mesh) are introduced which follow the water front in the crack.

If a fracture segment with aperture b is filled with water, its permeability K is given by (Bear et al. 1993):

$$K = \frac{\rho_l b^2}{12\mu_l} \tag{8}$$

with μ_l the dynamic viscosity of water.

Each timestep Equation 6 and Equation 7 are solved alternatively. First, the pressure equation at time t^n is solved using finite difference schemes. To solve Equation 6 boundary conditions have to be imposed. At the entrance of the fracture ($z = 0$) a prescribed liquid pressure corresponding to a water column of 1 mm, is assumed. At the top of the waterfilled part, the liquid pressure corresponding to the capillary pressure in the active fracture segment is imposed. The capillary pressure is related to the aperture of the active fracture segment by:

$$P_c = 2\sigma_w \cos(\theta)/b \quad \text{and} \quad P_l = P_g - P_c \tag{9}$$

Here σ_w is the surface tension and θ is the contact angle. We assume that every gas pressure build-up rapidly vanishes, so that the gas pressure P_g in the non-filled part of the fracture remains at constant atmospheric pressure. Once the pressures $\mathbf{P}^n$ are known in the nodes at the interfaces of the water filled fracture segments, the waterfront movement dz in the active fracture segment is computed using Equation 7. Since the sink term S in the pressure equation is a function of the waterfront, a solution can only be achieved by iteration. Each iteration step, this sink term is determined separately by calculating the unsaturated moisture transport in the matrix. To do so, a finite element model based on the continuum approach (Bear & Bachmat 1975, Whitaker 1977) is coupled to the discrete fracture model. Since we limit to unsaturated moisture transfer under isothermal conditions the moisture flow in the matrix can be written by the following differential equation:

$$\frac{\partial w}{\partial t} = -c(P_c)\frac{\partial P_c}{\partial t} = -\nabla^T(\mathbf{k}\nabla P_c) \tag{10}$$

with w (kg/m^3) the moisture content, P_c the capillary pressure, $c(P_c)$ the moisture capacity and $\mathbf{k}(P_c)$ the unsaturated permeability of the matrix. The matrix is initially assumed to be oven dry ($t = 0$: $w \approx 0\,\text{kg/m}^3, P_c = 10^9\,\text{Pa}$). For boundary conditions, at the bottom of the matrix a similar condition as for the entrance of the fracture is applied: $z = 0, t > 0$: $P_c = 10\,\text{Pa}$. At the lateral and top matrix boundaries no flow conditions are assumed. Along the matrix-fracture interface, the capillary pressure corresponding to the calculated liquid pressure field in the crack is imposed as boundary condition. This however is one of the major problems in dual permeability models. Due to the highly non-linear state and transport properties of most porous materials, sudden changes in boundary conditions, as caused by filling of the crack, often result in unstable numerical solutions of the unsaturated moisture transfer in the matrix. In the presented model this problem is solved by introducing movable nodes in the FE-mesh. To do so, beginning point and ending point of each fracture segment correspond to a boundary node of the FE-mesh. In addition one movable node is placed inside each fracture segment, which will follow the waterfront in the active fracture segment (see Fig. 6). The movable node has two advantages. On the one hand the sink term S can more accurately be calculated each step in time. On the other hand, sudden changes in boundary conditions (from dry state to saturated state) in the FE-node corresponding to the end of the crack segment, are avoided when the waterfront reaches the end of the crack segment. This results in a stable numerical solution.

The time step $dt^{n+1} = t^{n+1} - t^n$ used in the simulations is chosen as follows:

$$dt^{n+1} = \frac{z_{active}.dt^n}{p_1.\max(dz, z_{active} / p_2)} \tag{11}$$

with z_{active} the length of the active fracture segment, dz the progress of the waterfront in the fracture during the previous timestep dt^n, and p_1 and p_2 parameters controlling the time stepping process. The parameter p_2 controls the number of timesteps per fracture segment, while parameter p_1 scales the new timestep based on the previous timestep. Since cumulative inflow during an imbibition process proportionates to square root of time, the timestep can increase during the simulation. When the waterfront reaches a following crack segment, the movable node is replaced to the middle of the previous segment and the next movable node is activated. At the same moment the water height is set to a very small value in the next fracture segment before starting the next pressure calculation. This is done in order to avoid singularities in the pressure equations.

5 EXPERIMENTAL VALIDATION OF THE TRANSPORT MODEL FOR A FRACTURE – POROUS MATRIX SYSTEM

To validate the developed model, experiments of water infiltration in fractured brick samples are simulated. During the experiments the moisture profiles in the matrix and the filling of the fracture were monitored with X-ray radiography. This allows a validation of the predicted height of rise in the fracture as well as of the moisture distribution in the matrix.

5.1 *Experimental setup*

Free water uptake is measured in fractured brick samples using X-ray radiography. Two measurements have been performed: one with an artificial fracture with a constant fracture aperture of 0.01 mm and one with a natural fracture with an aperture of approximately 0.1 mm. For the artificial fracture two identical halves of a rectangular brick sample are polished and placed together such that a parallel plate fracture is formed. The total brick sample with the artificial fracture has a height of 145 mm, a width of 70 mm and a thickness of 20 mm. To create the natural fracture a brick sample of 80 by 80 mm^2 and a thickness of 20 mm was broken. Afterwards the two halves were placed together again. The fracture aperture is determined with a crack meter at the fracture surface. The aperture was rather constant, with a mean value of 0.1 mm.

Prior to the uptake experiment, the samples were dried in an oven at 105°C. After drying, the lateral sides of the specimen are wrapped with cling film in order to avoid evaporation during the experiment. In the X-ray apparatus, the oven dry brick sample is placed on a recipient, which can be filled with water. First, an X-ray image of the oven-dry sample is taken. Then, water is poured into the recipient, resulting in a very small overpressure of 10 Pa (1 mm water column) at the base of the specimen. Images are now taken at regular timesteps while the water is being taken up by the sample. Figure 2 shows the experimental set-up, the brick sample and an image as directly obtained with the X-ray apparatus. Quantitative moisture profiles are obtained by writing Beer's law to account for the water in the sample (van Besien et al. 2002):

$$I_{wet} = I_{dry} \exp(-\frac{\mu_w w d}{\rho_w}) \tag{12}$$

with I_{dry} and I_{wet} the intensity of the attenuated X-rays through the dry and wet sample, μ_w the attenuation coefficient of water, ρ_w the density of water, d the thickness of the sample and w the moisture content. I_{dry} and I_{wet} are both measured as greyscales and depend on sample thickness, energy and moisture content. In the image-processing step, the image of the dry sample is logarithmically subtracted from the images of the wet sample, making the moisture front in the sample visible as greyscales. Multiplying these greyscales by

$$-\frac{\rho_w}{\mu_w d} \tag{13}$$

quantitative moisture profiles are obtained. It was found that for most porous materials a high energy results in a strong contrast. However, the X-ray source proved to be unstable at high tube voltages. Also, the space in the apparatus warms up too much during radiation with high energies, encouraging evaporation. For the present study, all images were taken at a tube voltage of 115 kV and a current of 0.04 mA. This was found to be a good compromise between criteria of accuracy and stability of the source (van Besien et al. 2002).

5.2 *Numerical simulations*

To solve Equation 10 the unsaturated moisture permeability $k(p_c)$ and the moisture capacity or moisture retention curve $w(p_c)$ of the matrix has to be known. Due to air entrapment the moisture content of ceramic brick during a capillary uptake experiment will only raise to the so-called capillary moisture content w_{cap}. The moisture retention curve from dry state to the capillary moisture content ($w(p_c = 0) = w_{cap}$) is experimentally determined using the pressure plate technique starting from initially capillary saturated specimen. In the hygroscopic range extra data points are obtained from sorption isotherm measurements. To properly describe the moisture retention curve in the

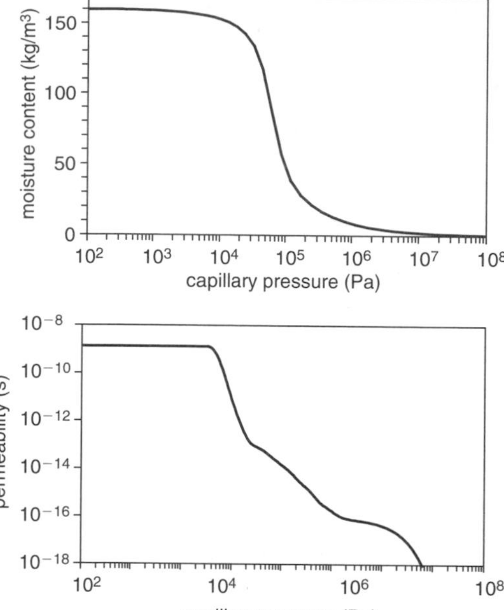

Figure 7. Material properties of ceramic brick. Top: moisture content versus capillary pressure $w(p_c)$ and bottom: moisture permeability versus capillary pressure $k(p_c)$.

hygroscopic as well as overhygroscopic range, a multimodal function of the 'van Genuchten'-type is used (Durner, 1994):

$$w(p_c) = w_{sat} \sum_{i=1}^{k} l_i \left(1 + (c_i p_c)^{n_i}\right)^{(1-\frac{1}{n_i})} \tag{14}$$

with k the number of modal functions, l_i the weighting factors ($0 < l_i \leq 1, \Sigma l_i = 1$), c_i and n_i model parameters. The relationship between unsaturated moisture permeability and capillary pressure is obtained with a multiscale network model (Carmeliet et al. 1999). The moisture retention curve and the permeability curve are shown in Figure 7.

Both the behaviour of the artificial fracture as well as of the natural fracture is simulated. For the artificial fracture an equidistant fracture aperture was assumed. For the natural fracture a variable fracture aperture, in accordance with the measurements, was applied. However, we simplified the problem by assuming a perfectly vertical fracture. Doing so, only halve the matrix has to be calculated. The predicted waterfront in the fracture and corresponding profiles in the matrix are compared with the X-ray results in Figure 8 for the sample with the artificial fracture (aperture of 0.01 mm) and in Figure 9 for the natural fracture (aperture of approximately 0.1 mm).

For the natural fracture, with an aperture of 0.1 mm (Fig. 9), the waterfront in the fracture quickly reaches

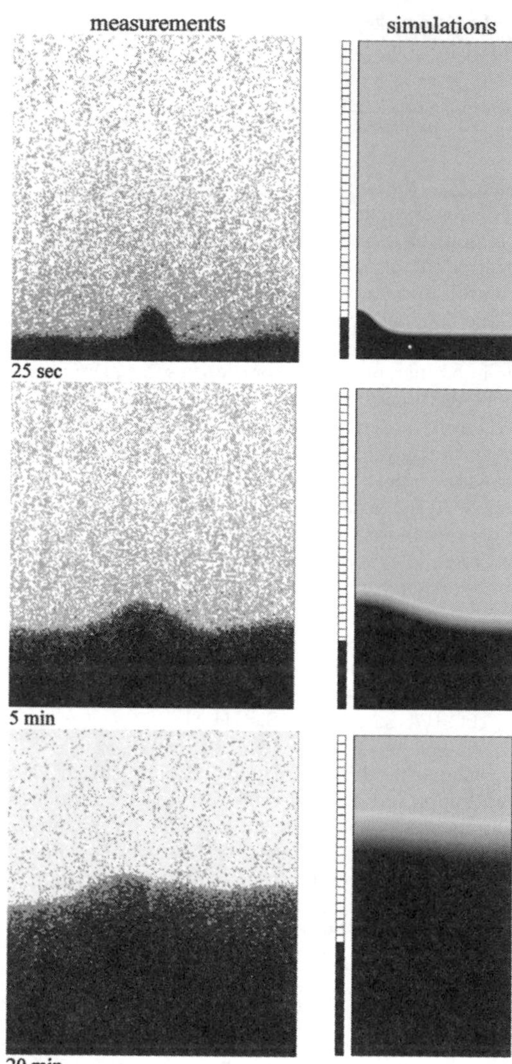

Figure 8. Measured and simulated moisture front in the artificial fractured brick sample (aperture of 0.01 mm). At the left side of the simulated profiles the predicted height of the waterfront in the fracture is presented.

the opposite side of the specimen. Afterwards, the fracture acts as an extra water source for the surrounding matrix over the total height of the specimen. As can be seen in the bottom figure, the moisture distribution is asymmetric. Moisture flow from fracture to matrix at the right side of the crack seems to be obstructed by a hair crack, running parallel to the main fracture.

The waterfront in the small fracture (Fig. 8), however, does not reach the opposite side of the specimen but stagnates at a height of approximately 1.2 cm,

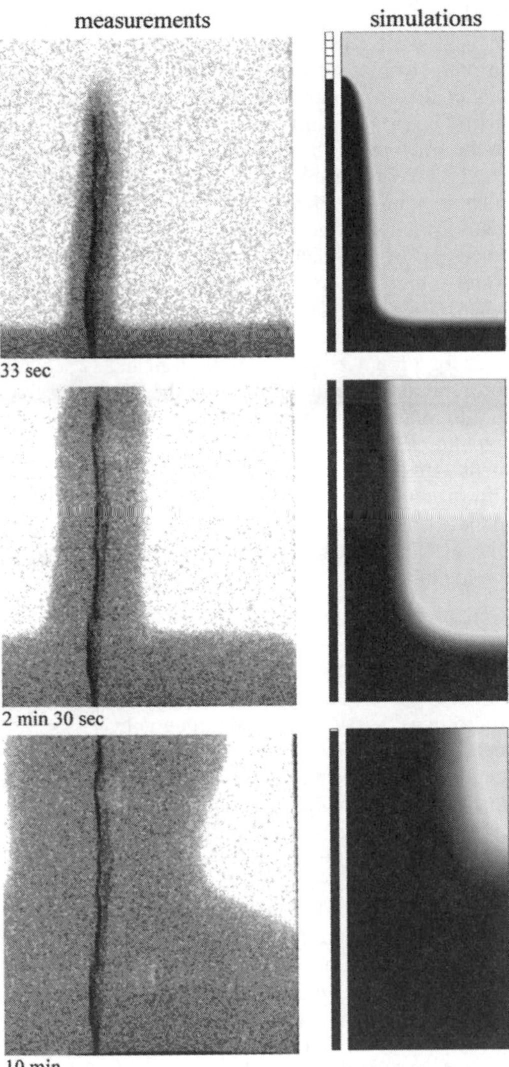

Figure 9. Measured and simulated moisture front in the naturally fractured brick sample (aperture of 0.1 mm). At the left side of the simulated profiles the predicted height of the waterfront in the fracture is presented.

which is far below 145 cm, the maximum height of rise for a fracture of 0.01 mm when no matrix flow is incorporated. This means that the inflow in the fracture is totally sucked up by the matrix and the uptake capacity of the fracture is too low to fill both matrix and fracture. As a result, the moisture profiles in the matrix run ahead of the waterfront in the fracture. One can expect that once the matrix is saturated and the flow to the matrix will drop to zero, the waterfront in the fracture will rise again.

Table 1. Material properties of blue stone (BS), concrete (C), ceramic brick (CB) and Lavoux (L).

Property	BS	C	CB	L
Mass density (kg/m^3)	2660.0	2250.0	2004.8	2048.9
Open porosity (%)	<0.50	15.80	24.00	24.25
Capil. moist. cont. (kg/m^3)	–	110.0	157.0	186.0
Abs. coeff. (kg/m^{2}s$^{0.5}$)	≈ 0	0.018	0.184	1.979

Predicted height of rise as well as predicted moisture profiles in the matrix show to be in good agreement with the measurements, showing the validity of the numerical model. In the next section we will use the developed model as a tool for investigating the effects of different fracture and matrix flow parameters on moisture imbibition in fractured porous media.

5.3 *Discussion*

It is worthwhile to examine the influence of the fracture aperture and of the matrix properties on the imbibition process in more detail. Therefore, we simulate free uptake in four fractured materials with distinct different material properties. The first material is a Belgian crinoidal limestone, further shortly indicate as blue stone. Free water uptake in this material is so slow, that the simulation on the blue stone can be regarded as a reference case when unsaturated moisture flow in the matrix is ignored. The second material is common concrete, while the third material corresponds to the ceramic brick used in the experimental set-up. The last material is a French limestone: Lavoux. This material is characterised by a very high water absorption coefficient. As can be seen in Table 1, giving an overview of the most import imbibition properties of the four materials, the magnitude of the water absorption coefficient of Lavoux is 10 times higher than the one of ceramic brick, while the capillary moisture content is of the same magnitude.

Figure 10 shows the evolution of the height of the waterfront in the fracture during a capillary uptake experiment in an initially dry sample. The sample measures $0.08 \times 0.08\,m^2$, with a fracture in the middle of the sample. The simulations are performed for the four materials and for three different fracture apertures: 0.02 mm, 0.1 mm and 0.2 mm. The corresponding maximum heights of rise are respectively 0.739 m, 0.148 m and 0.074 m. The figure clearly shows how the progress of the waterfront in the fracture is determined by the material properties of the matrix. Moisture flow from fracture to matrix results in a retardation of the waterfront in the fracture itself. Due to their low capacity, this feature is far more pronounced for small fractures. This is clearly visible

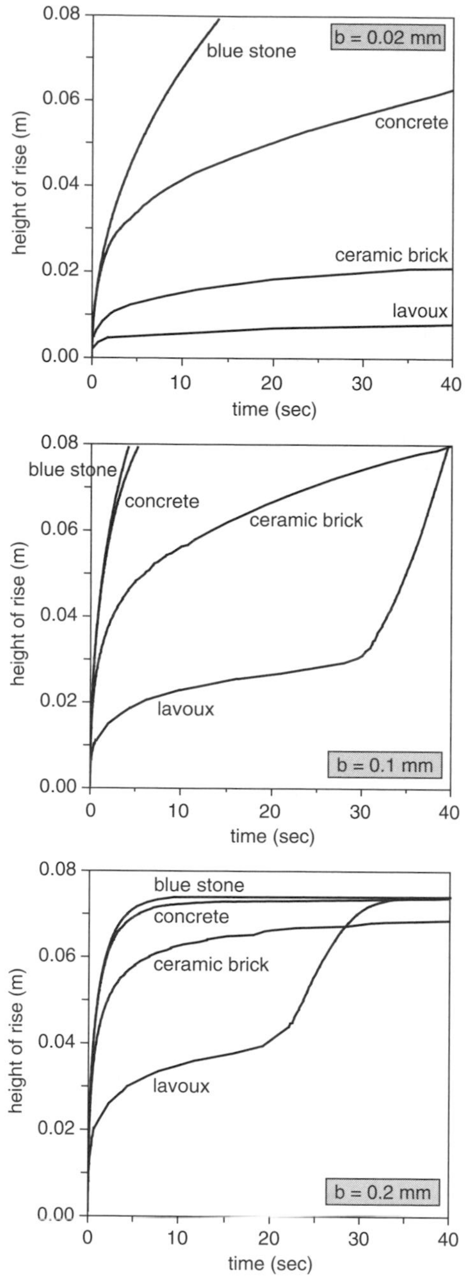

Figure 10. Height of the waterfront in the fracture during a capillary uptake experiment in four different fractured materials. Top: fracture aperture of 0.02 mm; middle: 0.1 mm and bottom: 0.2 mm.

when we compare in Figure 10 blue stone with concrete. Where, in the case of an aperture of 0.02 mm an important retardation in the waterfront in the concrete fracture can be observed, there is hardly any difference between evolution of the waterfront in the fracture of blue stone and concrete for the bigger apertures. In other words, the bigger the fracture aperture, the higher the absorption coefficient of the surrounding matrix has to be to influence the rise of the waterfront in the fracture. In the case of Lavoux, the absorption coefficient of the matrix is so high that even for a fracture with an aperture of 0.2 mm the rise of the waterfront in the fracture quickly stagnates. The water filled part of the fracture then mainly acts as an extra water source for the water uptake in the matrix. Only when the material is capillary saturated (for the present case – see Figure 10 – after approximately 31 seconds for a fracture aperture of 0.1 mm and 22 seconds for a fracture aperture of 0.2 mm) the flow from fracture to matrix drops to zero and the waterfront in the fracture can rise again. As also noticed in the experiment with the artificial fracture, this means that if the absorption coefficient of the matrix is high compared to the inflow capacity of the fracture, the moisture profiles in the matrix will run ahead of the waterfront in the fracture.

6 CONCLUSIONS

Microfocus X-ray computer tomography is used to measure 3D crack aperture data in a cracked concrete specimen. Using thresholding, segmentation and vectorisation techniques, the data are converted in a 3D network of tubes. Liquid filling processes and the permeability can be simulated using a static network methodology. The simulation results highlight preferential wetting and by-passing phenomena likely to happen in fractured concrete. These phenomena are not captured in nowadays simulation programs for coupled processes in cracked concrete.

A numerical model to simulate unsaturated moisture transport in fractured porous media has been presented. The model combines a discrete fracture model for moisture flow in a variable aperture fracture with a finite element model for unsaturated flow in the porous matrix. Movable nodes were introduced on the boundary of the finite element mesh to follow the moisture front in the fracture. Simulations of water uptake in fractured brick samples showed to be in good agreement with experiments monitored with X-ray radiography. The X-ray technique proved to be very informative since the evolution with time of the moisture profiles in the matrix as well as the waterfront in the fracture could be followed.

A parameter analysis showed that the height of rise of the waterfront in the fracture is strongly related to the water absorption coefficient of the matrix and the inflow capacity of the fracture. If the water absorption coefficient of the matrix is high in relation to the fracture aperture, a strong retardation and sometimes a stagnation of the waterfront in the fracture is observed and moisture profiles in the matrix run ahead of the waterfront in the fracture.

ACKNOWLEDGEMENT

Funding for this work was provided by KUL OT 00/17 and KUL IDO/99/004. The measurements with the X-ray equipment were performed at the department of Metallurgy and Materials Engineering of the KU-Leuven.

REFERENCES

Bear, J. & Bachmat, Y. 1975. *Introduction to modelling of transport phenomena in porous media.* Dordrecht: Kluwer Academic Publishers.

Carmeliet, J. Descamps, F. & Houvenaghel, G. 1999. Multiscale network for simulating liquid water and water vapour transfer properties of porous building materials. *Transport in porous media* 35: 67–88.

Delerue, J-F. Perrier, E. Yu, ZY. & Velde, B. 1999[a] *Physics and Chemistry of the Earth*, 24: 639.

Delerue, J-F. Perrier, Timmerman, A. & Rieu, M. 1999[b] in Modelling of transport process in soils, J. Feyen and K. Wiyo, editors, Wageningen Press, The Netherlands.

Durner, W. 1994. Hydraulic conductivity estimations for soils with heterogeneous pore structure. *Water Resources Research* 30: 211–223.

Gerke, HH. & van Genuchten, MTh. 1993. A dual-porosity model for simulating the preferential movement of water and solutes in structured porous media. *Water Resources Research* 29(2): 305–319.

Hvistendahl Karlsen, K. Lie, KA. Risebro, NH. & Froyen, J. 1998. A front-tracking approach to a two-phase fluid flow model with capillary forces. *In Situ* 22(1): 59–89.

McLaren, RG. Forsyth, PA. Sudicky, EA. VanderKwaak, JE. Schwartz, FW. & Kessler, JH. 2000. Flow and transport in fractured tuff at Yucca Mountain; numerical experiments on fast preferential flow mechanisms. *J Contam Hydrol* 43: 211–238.

Roels, S. Vandersteen K. & Carmeliet, J. 2003. Measuring and simulating moisture uptake in a fractured porous medium. Advances in Water Resources (in press).

Therrien, R. & Sudicky, EA. 1996. Three-dimensional analysis of variably-saturated flow and solute transport in discretely-fractured porous media. *J Contam Hydrol* 23: 1–44.

Van Besien, T. Roels, S. & Carmeliet, J. 2002. X-ray attenuation technique for the analysis of moisture flow in porous building materials (in prep.).

Vandersteen, K. Busselen, B. Van Den Abeele, K. & Carmeliet, J. 2002a. Quantitative characterization of fracture apertures using microfocus computed tomography. Special Publication of the Geological Society of London (in press).

Vandersteen, K. Carmeliet, J. & Feyen, J. 2002b. A network modeling approach to derive unsaturated hydraulic properties of a rough-walled fracture. Transport in Porous Media (in press).

Whitaker, S. 1977. Simultaneous heat, mass and momentum transfer in porous media: a theory of drying porous media. *Advances in heat transfer* 119–203.

Computational Modelling of Concrete Structures, Bićanić et al. (eds)
© 2003 Swets & Zeitlinger, Lisse, ISBN 90 5809 536 3

Drying shrinkage, creep and cracking of concrete: from coupled material modelling to multifield structural analyses

Stefan Grasberger & Günther Meschke
Institute for Structural Mechanics, Ruhr University Bochum, Germany

ABSTRACT: The paper describes a 3D coupled thermo-hygro-mechanical model that is formulated within the framework of thermomechanics of partially saturated porous media in the sense of the BIOT-COUSSY-theory. According to the coupled state equations arising from this theory, moisture transport and the constitutive pre- and postcracking properties are coupled by means of macroscopic coefficients. They are determined on the one hand by relating microscopic and macroscopic quantities and on the other hand by exploiting the symmetry relations that are connected to the existence of a macroscopic potential, yielding a model that is based on a minimum of material parameters. In addition to shrinkage-induced deformations also creep is considered in this extended version of the model. As a representative example, a numerical simulation concerning the long-term degradation of an inner tunnel lining is described in the paper.

1 INTRODUCTION

In nano-porous cementitious materials such as concrete, saturation-dependent internal stresses, acting on the nano- and micro-scale, develop as a consequence of molecular adsorption and capillary condensation. When subjected to drying, they may lead to severe cracking in concrete structures when the material strength is exhausted (Colina and Acker 2000; Sadouki and Wittmann 2001). While in engineering practice drying shrinkage is accounted for by means of shrinkage strains ε_s depending on an empirically determined shrinkage coefficient (CEB-FIP 1990), recent progress in computational durability mechanics (see, e.g. (Ulm et al. 1999; deBorst et al. 2001)), together with appropriate numerical methods open the perspective of a more fundamental approach to obtain not only a better insight into the degradation mechanisms resulting from the interaction between mechanical, thermal and hygral loading but also to provide more reliable estimates for the (residual) life-time of concrete structures.

In the paper, a 3D coupled thermo-hygro-mechanical elastoplastic damage model for finite element analyses of concrete structures formulated within the framework of the BIOT-COUSSY-theory (Biot 1941; Coussy 1995) is presented. In accordance with the hygral processes acting on the various levels of the nano-porous skeleton (nano, micro and capillary pores), the effect of shrinkage is taken into consideration by means of a macroscopic capillary pressure which represents, on a macroscopic level, hygrally induced stresses of various sources (Meschke and Grasberger 2002). In addition to deformations and cracking resulting from drying shrinkage, the effect of cracks on the moisture transport, the moisture-dependence of the strength and stiffness of concrete and deformations resulting from long-term creep are considered in this model. Since long-term creep is associated with dislocation-like processes in the nano-pores of the cement gel which are prestressed by hygrally induced stresses (disjoining pressure) (Bažant et al. 1997), creep depends on the moisture content. This coupling between moisture transport and creep deformations is also considered in the model.

2 THERMO-HYGRO-MECHANICAL MATERIAL MODELLING OF CONCRETE

The coupled thermo-hygro-mechanical material behaviour of concrete is described within the context of thermodynamics of deformable porous media based on the BIOT-COUSSY-theory (Biot 1941; Coussy 1995). In the model, concrete is assumed to consist of the matrix material (subscript s) – a mixture of cement paste and the aggregates – and the pores, which are partially saturated by liquid water (subscript l) and an ideal mixture of water vapour and dry air (subscript g). In order to account for the complex interaction processes on the nano- and micro-level between physically bonded and capillary water, respectively, and the matrix material, a functional relation between the macroscopic capillary pressure p_c – defined as the difference of the averaged pressures of the gaseous phase and the liquid phase – and the liquid saturation S_l is employed. Provided that there is thermodynamic equilibrium between the mixture of water vapour and dry air and the external atmosphere, it is often assumed that the gaseous phase is at constant atmospheric pressure, taken as zero (Bear and Bachmat 1991). Therefore, for the sake of simplicity, the capillary pressure is expressed as $p_c = -p_l$ in what follows.

2.1 *State equations*

Coupled phenomena on the microlevel of cementitious materials are described in a macroscopic frame-work using state variables. For the present 3D model, the function of free energy Ψ depends on three external variables (ε, m_l, T) and six internal variables (ε^p, ε^f, ϕ_l^p, ψ, α_R, α_{DP}). Following (Coussy 1995; Ulm and Coussy 1996) an additive split into two parts yields

$$\begin{aligned}\Psi &= \mathcal{W}(\boldsymbol{\varepsilon}-\boldsymbol{\varepsilon}^p-\boldsymbol{\varepsilon}^f, m_l-\rho_l\phi_l^p, \psi, T)\\ &+ \mathcal{U}(\alpha_R, \alpha_{DP}),\end{aligned} \tag{1}$$

whereby

- ε denotes the linearised strain tensor,
- ε^p is the tensor of plastic strains,
- ε^f are the flow strains corresponding to long-term creep effects,
- m_l denotes the liquid mass content variation,
- ρ_l is the mass density of the liquid phase,
- ϕ_l^p stands for the non-recoverable portion of the porosity occupied by the liquid phase,
- $\psi = 1 - d$ is the integrity with d denoting the isotropic damage parameter $0 \leq d \leq 1$,
- T denotes the absolute temperature and
- α_R and α_{DP} characterise the inelastic pre- and post-failure behaviour of concrete in tension (subscript R) and compression (subscript DP).

From the entropy inequality, the state equations are obtained as

$$\begin{aligned}&\boldsymbol{\sigma} = \frac{\partial \mathcal{W}}{\partial(\boldsymbol{\varepsilon}-\boldsymbol{\varepsilon}^p-\boldsymbol{\varepsilon}^f)}; \quad p_l = \frac{\partial \mathcal{W}}{\partial(\frac{m_l}{\rho_l}-\phi_l^p)};\\ &S = -\frac{\partial \mathcal{W}}{\partial T}; \quad q_R = -\frac{\partial \mathcal{U}}{\partial \alpha_R}; \quad q_{DP} = -\frac{\partial \mathcal{U}}{\partial \alpha_{DP}},\end{aligned} \tag{2}$$

where $\boldsymbol{\sigma}$ is the total stress tensor, p_l is the liquid pressure and S is the entropy. q_R and q_{DP} are the thermodynamic forces conjugate to α_R and α_{DP}, which determine the damage-dependent size of the damage (f_R) and loading (f_{DP}) surface in the stress space.

Inserting (2) into the CLAUSIUS-DUHEM inequality gives the intrinsic dissipation as

$$\begin{aligned}\mathcal{D}_{\text{int}} &= \boldsymbol{\sigma}:\dot{\boldsymbol{\varepsilon}}^p + \boldsymbol{\sigma}:\dot{\boldsymbol{\varepsilon}}^f + \frac{1}{2}\boldsymbol{\sigma}:(\dot{\psi^{-1}})\,\boldsymbol{\mathcal{D}}_u:\boldsymbol{\sigma}\\ &+ p_l\dot{\phi}_l^p + q_R\dot{\alpha}_R + q_{DP}\dot{\alpha}_{DP} \geq 0,\end{aligned} \tag{3}$$

where $\boldsymbol{\mathcal{D}}_u = \boldsymbol{C}_u^{-1}$ denotes the undrained elastic compliance tensor. Differentiation of $(2)_1 - (2)_3$ yields

$$\begin{aligned}\mathrm{d}\boldsymbol{\sigma} &= \boldsymbol{C}_u^{\text{ed}}:(\mathrm{d}\boldsymbol{\varepsilon}-\mathrm{d}\boldsymbol{\varepsilon}^p-\mathrm{d}\boldsymbol{\varepsilon}^f)\\ &- \psi M\boldsymbol{B}\left(\frac{\mathrm{d}m_l}{\rho_l}-\mathrm{d}\phi_l^p\right)\\ &+ \boldsymbol{\Lambda}_u\,\mathrm{d}\psi - \boldsymbol{A}_u\,\mathrm{d}T,\end{aligned} \tag{4}$$

$$\begin{aligned}\mathrm{d}p_l &= \psi M\left(\frac{\mathrm{d}m_l}{\rho_l}-\mathrm{d}\phi_l^p\right)\\ &- \psi M\boldsymbol{B}:(\mathrm{d}\boldsymbol{\varepsilon}-\mathrm{d}\boldsymbol{\varepsilon}^p-\mathrm{d}\boldsymbol{\varepsilon}^f)\\ &+ \Xi\,\mathrm{d}\psi + L\,\mathrm{d}T,\end{aligned} \tag{5}$$

$$\begin{aligned}\mathrm{d}S &= \frac{C_u}{T_0}\,\mathrm{d}T + \boldsymbol{A}_u:(\mathrm{d}\boldsymbol{\varepsilon}-\mathrm{d}\boldsymbol{\varepsilon}^p-\mathrm{d}\boldsymbol{\varepsilon}^f)\\ &+ s_l\,(\mathrm{d}m_l-\rho_l\,\mathrm{d}\phi_l^p)\\ &- L\left(\frac{\mathrm{d}m_l}{\rho_l}-\mathrm{d}\phi_l^p\right)+\Pi\,\mathrm{d}\psi.\end{aligned} \tag{6}$$

The symbols introduced in (4)–(6) represent the mixed partial derivatives of the free energy and can be interpreted as follows:

- $\boldsymbol{C}_u^{\text{ed}} = \psi\boldsymbol{C}_u$ denotes the undrained fourth-order stiffness tensor,
- the term $\psi M\boldsymbol{B}$ represents the hygro-mechanical couplings with M as the isotropic BIOT modulus and $\boldsymbol{B} = b\|$ as the second-order tensor of tangential BIOT coefficients b,
- $\boldsymbol{\Lambda}_u$ is the undrained second-order tensor describing the coupling mechanisms between damage evolution and the total stress increment,
- $\boldsymbol{A}_u = \boldsymbol{C}_u^{\text{ed}}:\|\alpha_{t,u}$ denotes the undrained second-order thermo-mechanical coupling tensor with $\alpha_{t,u}$ as the undrained thermic dilatation coefficient,
- Ξ is a coupling coefficient connected with the change of the liquid pressure due to damage evolution,
- $L = 3\psi M\alpha_{t,u}$ characterises the thermo-hygral coupling mechanisms,
- C_u denotes the undrained volume heat capacity and T_0 the reference temperature,
- s_l is the internal entropy of the liquid phase and
- $T_0\Pi$ represents the latent heat due to damage evolution.

Inserting (5) into (4) yields an alternative drained formulation for the differential stress tensor as

$$\begin{aligned} d\boldsymbol{\sigma} &= \mathbb{C}^{\text{ed}} : (d\boldsymbol{\varepsilon} - d\boldsymbol{\varepsilon}^p - d\boldsymbol{\varepsilon}^f) \\ &\quad - \boldsymbol{B}\, dp_l + \boldsymbol{\Lambda}\, d\psi - \boldsymbol{A}\, dT, \end{aligned} \tag{7}$$

with the drained stiffness tensor

$$\mathbb{C}^{\text{ed}} = \psi \left[\mathbb{C}_u - Mb^2 (\mathbf{1} \otimes \mathbf{1})\right] = \psi \mathbb{C}, \tag{8}$$

the drained thermo-mechanical coupling tensor

$$\boldsymbol{A} = \boldsymbol{A}_u - 3\alpha_{t,u} M \boldsymbol{B} = \mathbb{C}^{\text{ed}} : \mathbf{1}\alpha_t, \tag{9}$$

and the drained tensor

$$\boldsymbol{\Lambda} = \boldsymbol{\Lambda}_u + \Xi \boldsymbol{B}, \tag{10}$$

respectively.

2.2 *Identification of coupling coefficients*

In order to develop a model that is based on a minimum of material parameters, the coupling coefficients introduced in Subsection 2.1 are identified in what follows. On the one hand, the poroelastic hygro-mechanical coefficients b and M are determined by relating stress and strain quantities defined on the meso-level to respective homogenised quantities on the macro-level (Subsection 2.2.1). On the other hand, the coefficients related to damage phenomena $\boldsymbol{\Lambda}$ and Ξ are identified by exploiting the symmetry relations that are connected to the existence of a macroscopic potential (Subsection 2.2.2).

2.2.1 *Poroelastic coefficients*

The starting point of the derivation is the following expression of the volumetric macroscopic stress $\sigma = \text{tr}\boldsymbol{\sigma}/3$ in terms of the solid matrix stress σ_s and of the liquid pressure p_l

$$\sigma = (1-\phi)\sigma_s - \phi_l p_l, \tag{11}$$

whereby ϕ_l denotes the part of the porosity ϕ that is filled with liquid water. Differentiation of (11) yields

$$d\sigma = (1-\phi)\, d\sigma_s - \sigma_s\, d\phi - \phi_l\, dp_l - p_l\, d\phi_l. \tag{12}$$

Following (Coussy 1995), the variation of the porosity $d\phi$ can be related to the macroscopic volumetric strain $d\varepsilon = \text{tr}\, d\boldsymbol{\varepsilon}$ and to the incremental volumetric strain of the solid matrix $d\varepsilon_s = d\varepsilon_s^e + d\varepsilon_s^p$ as

$$d\phi = (1-\phi)(d\varepsilon - d\varepsilon_s). \tag{13}$$

Inserting (13) together with the constitutive equation for the matrix

$$d\sigma_s = K_s\, d\varepsilon_s^e + dK_s \varepsilon_s^e \tag{14}$$

into (12) yields the following formulation for the incremental volumetric macroscopic stress

$$\begin{aligned} d\sigma &= (1-\phi)K_s\, d\varepsilon + (1-\phi)\varepsilon_s^e\, dK_s - K_s\, d\phi \\ &\quad - (1-\phi)K_s\, d\varepsilon_s^p - S_l\phi\, dp_l - \phi p_l\, dS_l, \end{aligned} \tag{15}$$

with S_l denoting the liquid saturation defined as $S_l = \phi_l/\psi$. Moreover, the mass conservation

$$m_l = (1+\varepsilon)\rho_l\phi_l - (\rho_{l,0}\phi_{l,0}) \tag{16}$$

together with the incremental constitutive equation of the liquid phase,

$$dp_l = K_l \frac{d\rho_l}{\rho_l}, \tag{17}$$

with K_l as the bulk modulus of water, is used to express the variation of the porosity $d\phi$ in terms of macroscopic quantities. Differentiation of (16) together with (17) yields

$$\frac{dm_l}{\rho_l} = \frac{\phi S_l}{K_l}\, dp_l + S_l\phi\, d\varepsilon + \phi\, dS_l + S_l\, d\phi, \tag{18}$$

whereby the approximation $\varepsilon \ll 1$ is used. Inserting the macroscopic state equation for the liquid pressure (5) (neglecting the thermal part) into (18) results in

$$\begin{aligned} d\phi &= \left[\frac{b}{S_l} - \phi\right] d\varepsilon^e \\ &\quad + \left[\frac{1}{\psi M S_l} - \frac{\phi}{K_l} - \frac{\phi}{S_l}\frac{\partial S_l}{\partial p_l}\right] dp_l \\ &\quad - \phi\, d\varepsilon^p + \frac{d\phi_l^p}{S_l} - \frac{\Xi}{\psi M S_l}\, d\psi, \end{aligned} \tag{19}$$

using the assumption $S_l = S_l(p_l)$. Finally, from inserting (19) into (15), the following expression for the differential volumetric macroscopic stress is obtained:

$$\begin{aligned} d\sigma &= \left[K_s\left(1 - \frac{b}{S_l}\right)\right] d\varepsilon^e \\ &\quad + \left[\frac{\phi K_s}{K_l} - \phi S_l - \frac{K_s}{\psi M S_l} + \left(\frac{\phi K_s}{S_l} - \phi p_l\right)\frac{\partial S_l}{\partial p_l}\right] dp_l \\ &\quad + (1-\phi)\varepsilon_s^e\, dK_s + \frac{\Xi K_s}{\psi M S_l}\, d\psi \\ &\quad + K_s\, d\varepsilon^p - (1-\phi)K_s\, d\varepsilon_s^p - \frac{K_s}{S_l}\, d\phi_l^p. \end{aligned} \tag{20}$$

Comparing the first term on the right hand side of (20) with the volumetric part of the first term of state equation (7) results in an expression for the tangential BIOT coefficient

$$b = S_l \left[1 - \psi \frac{K}{K_s}\right], \tag{21}$$

which includes the expression $b = S_l$ suggested by (Coussy 1995) for the special case of poroelastic materials with incompressible matrix behaviour.

From a comparison of the respective second terms an expression for the BIOT modulus $\overline{M} = \psi M$ is obtained as

$$\overline{M} = \left[\phi\left(1 - \frac{S_l p_l}{K_s}\right)\frac{\partial S_l}{\partial p_l} + \frac{\phi S_l}{K_l} + \frac{S_l(b - \phi S_l)}{K_s}\right]^{-1} \tag{22}$$

see (Schrefler and Zhan 1993; Lewis and Schrefler 1998) for a similar formulation. For cementitious materials, expression (22) can be replaced by

$$\overline{M} \approx \left[\phi \frac{\partial S_l}{\partial p_l}\right]^{-1}. \tag{23}$$

In the special case of a fully saturated material ($S_l = 1$), (22) yields the classical relation (Coussy 1995; Lewis and Schrefler 1998)

$$\overline{M}_{S_l=1} = \left[\frac{\phi}{K_l} + \frac{(b - \phi)}{K_s}\right]^{-1}. \tag{24}$$

The third and fourth term of (20) are associated with damage mechanisms, represented in a macroscopic way by the third term of state equation (7). Therefore the following relation should hold

$$\frac{1}{3}(\boldsymbol{\Lambda} : \mathbf{1})\, \mathrm{d}\psi = (1 - \phi)\varepsilon_s^e \mathrm{d}K_s + \frac{\Xi K_s}{\psi M S_l}\, \mathrm{d}\psi. \tag{25}$$

Finally, the last three terms on the right hand side of (20) have no counterpart on the macroscopic level. Hence, they are set to zero

$$K_s\, \mathrm{d}\varepsilon^p - (1 - \phi)K_s\, \mathrm{d}\varepsilon_s^p - \frac{K_s}{S_l}\, \mathrm{d}\phi_l^p = 0, \tag{26}$$

yielding the well-known relation

$$\mathrm{d}\varepsilon^p = (1 - \phi)\, \mathrm{d}\varepsilon_s^p + \mathrm{d}\phi^p. \tag{27}$$

2.2.2 *Damage couplings*

The derivation of the damage-related coupling coefficient $\boldsymbol{\Lambda}$ is based on the MAXWELL symmetries that are connected with the existence of a macroscopic potential. The drained state equation (7) can be interpreted as the differentiation of the function $\boldsymbol{\sigma} = \boldsymbol{\sigma}(\varepsilon^e, p_l, \psi, T)$ (Coussy 1995). Hence, the following relations can be derived

$$\begin{aligned}
\frac{\partial \boldsymbol{C}^{\mathrm{ed}}}{\partial p_l} &= -\frac{\partial \boldsymbol{B}}{\partial \boldsymbol{\varepsilon}^e}; & \frac{\partial \boldsymbol{C}^{\mathrm{ed}}}{\partial \psi} &= \frac{\partial \boldsymbol{\Lambda}}{\partial \boldsymbol{\varepsilon}^e};\\
\frac{\partial \boldsymbol{B}}{\partial \psi} &= -\frac{\partial \boldsymbol{\Lambda}}{\partial p_l}; & \frac{\partial \boldsymbol{C}^{\mathrm{ed}}}{\partial T} &= -\frac{\partial \boldsymbol{A}}{\partial \boldsymbol{\varepsilon}^e};\\
\frac{\partial \boldsymbol{B}}{\partial T} &= \frac{\partial \boldsymbol{A}}{\partial p_l}; & \frac{\partial \boldsymbol{\Lambda}}{\partial T} &= -\frac{\partial \boldsymbol{A}}{\partial \psi}.
\end{aligned} \tag{28}$$

Using the MAXWELL symmetries $(28)_2$, $(28)_3$ and $(28)_6$ together with expression (21) for the BIOT coefficient, expression (8) for the drained stiffness tensor and expression (9) for the drained thermo-mechanical coupling tensor, the drained tensor $\boldsymbol{\Lambda}$ can be expressed as (Meschke and Grasberger 2002)

$$\begin{aligned}
\boldsymbol{\Lambda} = {} & \boldsymbol{C} : (\boldsymbol{\varepsilon} - \boldsymbol{\varepsilon}^p - \boldsymbol{\varepsilon}^f)\\
& + \frac{K}{K_s}\int_{p_l} S_l\, \mathrm{d}p_l \mathbf{1} - \boldsymbol{C} : \mathbf{1}\alpha_t T.
\end{aligned} \tag{29}$$

The identification of the coupling coefficient Ξ is accomplished by inserting (29) into relation (25) and afterwards splitting the result into a hygral and a non-hygral portion, yielding

$$\Xi = \frac{\overline{M} S_l K}{K_s^2}\int_{p_l} S_l\, \mathrm{d}p_l \approx 0. \tag{30}$$

This simplified assumption seems to be a reasonable approximation for cementitious materials.

2.3 *Effective stresses*

The concept of effective stress (Fillunger 1936; von Terzaghi 1936) is a generally accepted approach in soil mechanics for the determination of stresses in the skeleton of fully saturated soils. In addition to the original proposal of (von Terzaghi 1936), several alternative suggestions for the definition of effective stresses exist, taking the compressibility of the matrix material or the porosity into account (see e.g. (Biot 1941; Biot and Willis 1957; Nur and Byerlee 1971; Bishop 1973)).

Based on the relevance of the concept of effective stress for the analysis of fully saturated soils, this concept has also been adapted for the description of partially saturated soils. Early formulations introduced the capillary pressure in the (elastic) effective stress definition (Bishop 1959). However, difficulties to obtain satisfactory agreements with experimental results have motivated the use of two independent stress fields for the constitutive modelling of unsaturated soils (see e.g. (Bishop and Blight 1963; Alonso et al. 1990)). In these approaches, two separate yield surfaces, one being related to total stresses and one related to capillary pressure variations, are used for the description of the behaviour of partially saturated porous media in the inelastic range.

As far as the numerical modelling of partially saturated cement-based materials is concerned, the assumption of (elastic) effective stresses seems not to be well suited for the description of shrinkage-induced cracks using stress-based crack-models. However, the concept of plastic effective stress first introduced at a macroscopic level by (Coussy 1989) for saturated porous media (see (Coussy 1995) for details), allows to overcome these difficulties in the framework of poroplasticity – porodamage models. The proposed form of the plastic effective stress is the same as the classical BIOT-type, however, a plastic effective stress coefficient is used. A similar form has been derived from micromechanical considerations by (Lydzba and Shao 2000). This concept has been recently extended to partially saturated materials (Burlion et al. 2000; Grasberger and Meschke 2001), and is also adopted in the present formulation.

2.3.1 *Elastic effective stresses*

Starting point for the derivation of the elastic effective stress tensor $\boldsymbol{\sigma}'^e$ is the integrated form of (7). Inserting expression (8) for the drained stiffness tensor $\boldsymbol{C}^{\mathrm{ed}}$, expression (9) for the drained thermo-mechanical coupling tensor $\boldsymbol{A}$, expression (21) for the tangential BIOT coefficient b, expression (29) for the drained coupling tensor $\boldsymbol{\Lambda}$ and the relation $p_c = -p_l$, the drained formulation of the increment of the total stress tensor is obtained as

$$\begin{aligned} \mathrm{d}\boldsymbol{\sigma} \;=\; & \psi \boldsymbol{C} : (\mathrm{d}\boldsymbol{\varepsilon} - \mathrm{d}\boldsymbol{\varepsilon}^p - \mathrm{d}\boldsymbol{\varepsilon}^f) \\ & + \boldsymbol{C} : (\boldsymbol{\varepsilon} - \boldsymbol{\varepsilon}^p - \boldsymbol{\varepsilon}^f)\,\mathrm{d}\psi \\ & + S_l \left[1 - \psi \frac{K}{K_s}\right] \mathbf{1}\,\mathrm{d}p_c \\ & - \frac{K}{K_s} \int_{p_c} S_l(p_c)\,\mathrm{d}p_c \mathbf{1}\,\mathrm{d}\psi \\ & - \boldsymbol{C} : \mathbf{1}\alpha_t T\,\mathrm{d}\psi - \psi \boldsymbol{C} : \mathbf{1}\alpha_t\,\mathrm{d}T. \end{aligned} \tag{31}$$

Integration of (31) results in

$$\begin{aligned} \boldsymbol{\sigma} \;=\; & \psi \boldsymbol{C} : (\boldsymbol{\varepsilon} - \boldsymbol{\varepsilon}^p - \boldsymbol{\varepsilon}^f) \\ & + \left[1 - \psi \frac{K}{K_s}\right] \int_{p_c} S_l(p_c)\,\mathrm{d}p_c \mathbf{1} - \boldsymbol{A}T, \end{aligned} \tag{32}$$

yielding the following definition of the elastic effective stress tensor

$$\boldsymbol{\sigma}'^e = \psi \boldsymbol{C} : (\boldsymbol{\varepsilon} - \boldsymbol{\varepsilon}^p - \boldsymbol{\varepsilon}^f) - \boldsymbol{A}T, \tag{33}$$

with

$$\boldsymbol{\sigma} = \boldsymbol{\sigma}'^e + \left[1 - \psi \frac{K}{K_s}\right] \int_{p_c} S_l(p_c)\,\mathrm{d}p_c \mathbf{1}. \tag{34}$$

2.3.2 *Plastic effective stresses*

The plastic effective stress tensor $\boldsymbol{\sigma}'^p = \boldsymbol{\sigma}'$, defined as

$$\boldsymbol{\sigma}' = \boldsymbol{\sigma} - b^p p_c \mathbf{1}, \tag{35}$$

characterises the thermodynamic force associated with the plastic strain rate (Coussy 1995). In contrast to the elastic effective stress tensor, $\boldsymbol{\sigma}'$ represents the macroscopic counterpart to matrix-related micro-stresses with the coefficient b^p as the plastic counter-part of the BIOT coefficient b. By relating stress quantities on the meso-scale to respective macroscopic quantities, a possible identification of b^p as a function of the integrity ψ, the porosity ϕ and the liquid saturation S_l can be accomplished as

$$b^p = \psi \phi S_l(p_c), \tag{36}$$

see (Grasberger and Meschke 2001) for details.

2.4 *Multisurface damage-plasticity theory*

According to the concept of multisurface damage-plasticity theory, mechanisms characterised by the degradation of stiffness and inelastic deformations are controlled by four threshold functions defining a region of admissible stress states in the space of plastic effective stresses $\boldsymbol{\sigma}'$

$$\mathbb{E} = \{(\boldsymbol{\sigma}', q_k) |\; f_k(\boldsymbol{\sigma}', q_k(\alpha_k)) \leq 0, \;\; k = 1, ..., 4\}. \tag{37}$$

In (37), the index $k = 1, 2, 3$ stands for an active cracking mechanism associated with the damage function $f_{\mathrm{R},k}(\boldsymbol{\sigma}', q_{\mathrm{R}})$ and $k = 4$ represents an active hardening/ softening mechanism in compression associated with the loading function $f_{\mathrm{DP}}(\boldsymbol{\sigma}', q_{\mathrm{DP}})$.

Cracking of concrete is accounted for by means of the RANKINE criterion, employing three failure surfaces perpendicular to the axes of principal stresses

$$f_{\mathrm{R,A}}(\boldsymbol{\sigma}', q_{\mathrm{R}}) = \sigma'_A - q_{\mathrm{R}}(\alpha_{\mathrm{R}}) \leq 0, \quad A = 1, 2, 3. \tag{38}$$

In (38), the subscript A refers to one of the three principal directions and $q_{\mathrm{R}}(\alpha_{\mathrm{R}}) = -\,\partial \mathcal{U} / \partial \alpha_{\mathrm{R}}$ denotes the softening parameter.

The ductile behaviour of concrete subjected to compressive loading is described by a hardening/ softening DRUCKER-PRAGER plasticity model

$$f_{\mathrm{DP}}(\boldsymbol{\sigma}', q_{\mathrm{DP}}) = \sqrt{J_2} - \kappa_{\mathrm{DP}} I_1 - \frac{q_{\mathrm{DP}}(\alpha_{\mathrm{DP}})}{\beta_{\mathrm{DP}}} \leq 0, \tag{39}$$

with $q_{\mathrm{DP}}(\alpha_{\mathrm{DP}}) = -\partial\mathcal{U}/\partial\alpha_{\mathrm{DP}}$ as the hardening/softening parameter. The determination of the model parameters κ_{DP} and β_{DP} is based on the ratio of the biaxial and the uniaxial compressive strength of concrete f_{cb}/f_{cu} as (Meschke et al. 1998)

$$\kappa_{\mathrm{DP}} = \frac{1}{\sqrt{3}}\left(\frac{f_{cb}/f_{cu}-1}{2f_{cb}/f_{cu}-1}\right), \tag{40}$$

$$\beta_{\mathrm{DP}} = \sqrt{3}\left(\frac{2f_{cb}/f_{cu}-1}{f_{cb}/f_{cu}}\right), \tag{41}$$

whereby f_{cb}/f_{cu} is approximately equal to 1.16. The fracture energy concept is employed to ensure mesh-objective results in the post-peak regime. Details of the material model are found in (Meschke et al. 1998). For an efficient implementation of the multisurface model based on an algorithmic formulation in the principal stress space reference is made to (Meschke 1996).

2.5 *Evolution equations for dissipative processes*

The evolution equations of the tensor of plastic strains $\dot{\varepsilon}^p$, of the reciprocal value of the integrity (ψ^{-1}); of the plastic porosity occupied by the liquid phase $\dot{\phi}_l^p$ and of the internal variables $\dot{\alpha}_R$ and $\dot{\alpha}_{\mathrm{DP}}$ are obtained from the postulate of stationarity of the dissipation functional (Govindjee, Kay, and Simo 1995) as

$$\dot{\boldsymbol{\varepsilon}}^p = (1-\beta)\sum_{k=1}^{4}\dot{\gamma}_k\frac{\partial f_k}{\partial\boldsymbol{\sigma}'}, \tag{42}$$

$$(\psi^{-1})^{\cdot} = \beta\sum_{k=1}^{4}\dot{\gamma}_k\frac{\dfrac{\partial f_k}{\partial\boldsymbol{\sigma}'}:\boldsymbol{C}_u:\dfrac{\partial f_k}{\partial\boldsymbol{\sigma}'}}{\dfrac{\partial f_k}{\partial\boldsymbol{\sigma}'}:\boldsymbol{\sigma}'}, \tag{43}$$

$$\dot{\phi}_l^p = \sum_{k=1}^{4}\dot{\gamma}_k\frac{\partial f_k}{\partial\boldsymbol{\sigma}'}:\mathbf{1}b^p, \tag{44}$$

$$\dot{\alpha}_{\mathrm{R}} = \sum_{A=1}^{3}\dot{\gamma}_{\mathrm{R,A}}\frac{\partial f_{\mathrm{R,A}}}{\partial q_{\mathrm{R}}}, \tag{45}$$

$$\dot{\alpha}_{\mathrm{DP}} = \dot{\gamma}_{\mathrm{DP}}\frac{\partial f_{\mathrm{DP}}}{\partial q_{\mathrm{DP}}}, \tag{46}$$

together with the loading/unloading condition

$$f_k(\boldsymbol{\sigma}', q_k) \le 0;\quad \dot{\gamma}_k \ge 0;\quad \dot{\gamma}_k\, f_k(\boldsymbol{\sigma}', q_k) = 0. \tag{47}$$

The parameter $0 \le \beta \le 1$ contained in (42) and (43) allows a simple partitioning of effects associated with inelastic deformations due to the crack-induced misalignment of the asperities of the crack surfaces, resulting in an increase of inelastic strains ε^p, and deterioration of the microstructure, resulting in a decrease of the integrity ψ. An elastoplastic model ($(\psi^{-1})^{\cdot} = 0, \dot{\varepsilon}^p \neq \mathbf{0}$), and a damage model ($(\psi^{-1})^{\cdot} \neq 0, \dot{\varepsilon}^p = \mathbf{0}$) are recovered as special cases by setting $\beta = 0$ and $\beta = 1$, respectively.

2.6 *Long-term creep*

Consideration of long-term or flow creep effects is accomplished in the framework of the microprestress-solidification theory (Bažant et al. 1997). The evolution law of the flow strains is based on a linear relation between the rate $\dot{\varepsilon}^f$ and the stress tensor $\boldsymbol{\sigma}$ as

$$\dot{\varepsilon}^f = \frac{1}{\eta_f(S_f)}\boldsymbol{\mathcal{G}}^{\mathrm{ed}}:\boldsymbol{\sigma}, \tag{48}$$

with the fourth-order tensor $\boldsymbol{\mathcal{G}}^{\mathrm{ed}} = E(\boldsymbol{C}^{\mathrm{ed}})^{-1}$ and Young's modulus E. The viscosity η_f is a decreasing function of the microprestress S_f and can be written as (Bažant et al. 1997)

$$\frac{1}{\eta_f(S_f)} = cpS_f^{p-1}, \tag{49}$$

where c and $p \ge 1$ are positive constants. According to (Bažant et al. 1997), the microprestress relaxation is connected to changes of the disjoining pressure. Consequently, variations of the internal pore humidity h due to drying, which entail a changing disjoining pressure, lead to a change of the microprestress S_f. This mechanism partially explains the PICKETT effect (Pickett 1942), also called drying creep. An investigation of creep-related phenomena in the context of the proposed thermo-hygro-mechanical model will be shown in a follow-up publication.

3 MOISTURE AND HEAT TRANSPORT

Starting with a simplified nonlinear diffusion approach, in which the different moisture transport mechanisms in liquid and in vapour form are represented by means of a single macroscopic moisture-dependent diffusivity (Bažant and Najjar 1972), the relation between the moisture flux $\mathbf{q}_l$ and the spatial gradient of the capillary pressure ∇p_c is given by

$$\boldsymbol{q}_l = \frac{\boldsymbol{k}}{\mu_l}\cdot\nabla p_c. \tag{50}$$

In (50), $\boldsymbol{k}$ denotes the intrinsic liquid permeability tensor and μ_l is the viscosity of water. According to the hypothesis of dissipation decoupling (Coussy et al. 1998), possible couplings between heat and moisture transport are disregarded in the present formulation.

In order to account for the dependence of the moisture transport properties on the nonlinear material behaviour of concrete, $\boldsymbol{k}$ is additively decomposed into two portions as

$$\boldsymbol{k} = k_r(S_l)\left[k_t(T)\,k_\phi(\phi)\,\boldsymbol{k}_0 + \boldsymbol{k}_d(\alpha_{\mathrm{R}})\right], \qquad (51)$$

one related to the moisture flow through the partially saturated pore space and one related to the flow within a crack, respectively (Snow 1969). This approach is consistent with the smeared crack concept. In (51), $\boldsymbol{k}_0$ denotes the initial isothermal permeability tensor, k_r is the relative permeability, k_t accounts for the dependence of the isothermal moisture transport properties on the temperature and k_ϕ describes the relationship between the permeability and the porosity. Furthermore, $\boldsymbol{k}_d$ is the permeability tensor relating plane POISEUILLE flow through discrete fracture zones to the degree of damage in the continuum model, see (Meschke and Grasberger 2002) for details.

Using again the hypothesis of dissipation decoupling, the relation between the heat flux $\boldsymbol{q}_t$ and the gradient of the temperature ∇T can be described by a linear heat conduction law reading

$$\boldsymbol{q}_t = -D_t \nabla T, \qquad (52)$$

whereby $D_t(T, S_l, \phi)$ denotes the effective thermal conductivity.

4 NUMERICAL APPLICATION

Large tunnels often exhibit crown cracks in the inner linings caused by restrained stresses due to cooling and drying of the surface. In order to demonstrate the capability of the proposed 3D model to reproduce the coupled thermo-hygro-mechanical behaviour of concrete structures, a numerical simulation of the long-term degradation of an inner tunnel lining is presented in this section. The effect of creep is disregarded in this analysis.

Figure 1 contains the geometric dimensions of a cross section of the analysed tunnel. One half of the tunnel lining is discretised by means of 1832 finite volume elements. The assumed material parameters are summarised in Table 1 and the following relationship between the capillary pressure and the liquid saturation is used (van Genuchten 1980; Baroghel-Bouny et al. 1999):

$$p_c(S_l) = 18.62\left[S_l^{-2.27} - 1\right]^{1-1/2.27} \quad [\mathrm{N/mm^2}]. \qquad (53)$$

Moreover, the initial temperature is specified as $T_0 = 17.0°\mathrm{C}$, the initial pore humidity as $h_0 = 0.93$ and the initial porosity as $\phi_0 = 0.25$.

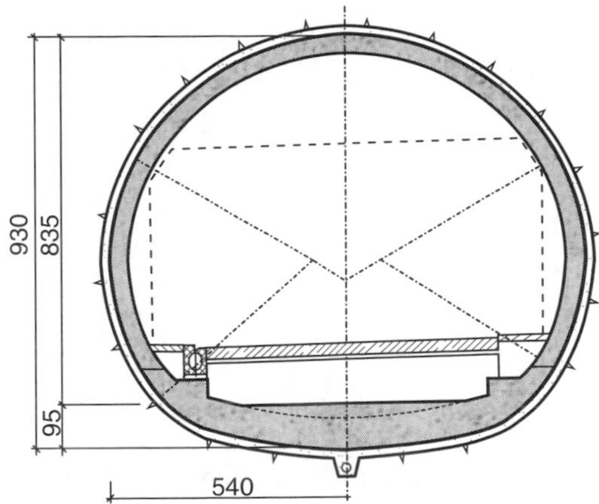

Figure 1. Numerical simulation of a tunnel lining: Geometry of the investigated tunnel (dimensions in [cm]).

Table 1. Numerical simulation of a tunnel lining: Concrete material parameters.

Young's modulus	E	$= 37000\,\mathrm{N/mm^2}$
Poisson's ratio	ν	$= 0.2$
Uniaxial tensile strength	f_{tu}	$= 4.1\,\mathrm{N/mm^2}$
Uniaxial compressive strength	f_{cu}	$= 50.0\,\mathrm{N/mm^2}$
Fracture energy tension	G_f	$= 0.1789\,\mathrm{N/mm}$
Fracture energy compression	G_c	$= 8.947\,\mathrm{N/mm}$
Initial liquid permeability	k_0	$= 2.0 \times 10^{-21}\,\mathrm{m^2}$
Dry thermal conductivity	$D_{t,0}$	$= 1.85\,\mathrm{W/(mK)}$
Thermic dilatation coefficient	α_t	$= 1.1 \times 10^{-5}\,\mathrm{K^{-1}}$

After application of the self-weight of the construction ($g = 25\,\mathrm{kN/m^3}$), the inner tunnel lining is exposed to cyclic changes of the hygral and thermal environmental conditions based on realistic meteorological data according to (Rocznik 1995). In accordance with the characteristics of the New Austrian Tunnelling Method (NATM), no forces from the surrounding rock formation are assumed within the investigated time period of five years.

4.1 *Damage evolution*

Figure 2 illustrates the calculated evolution of the crown cracks represented by the scalar damage parameter d after 1, 2 and 5 years, respectively. During the first year some cracks start to open at the top of the inner lining due to the arising temperature and moisture gradients. As the drying proceeds, these cracks are further opening, whereby the predicted crack spacing is about 20 ~ 25 cm. According to the stress distribution owing to the self-weight of the construction, the lower part of the tunnel lining remains undamaged.

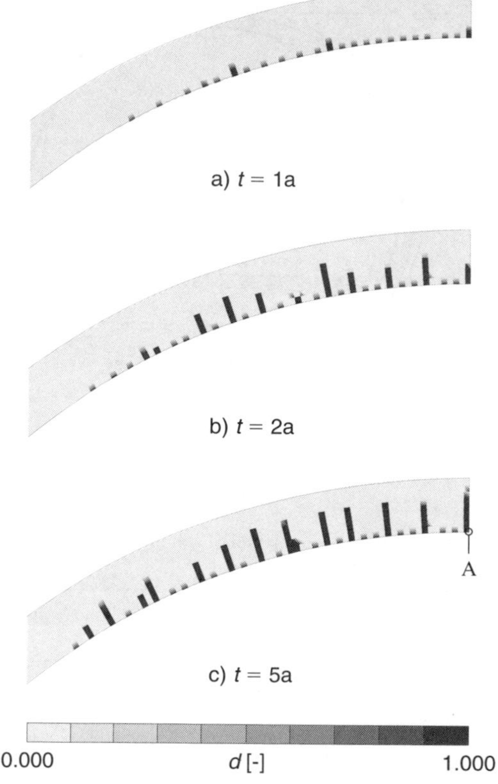

Figure 2. Numerical simulation of a tunnel lining: Calculated distribution of the scalar damage measure d at the crown.

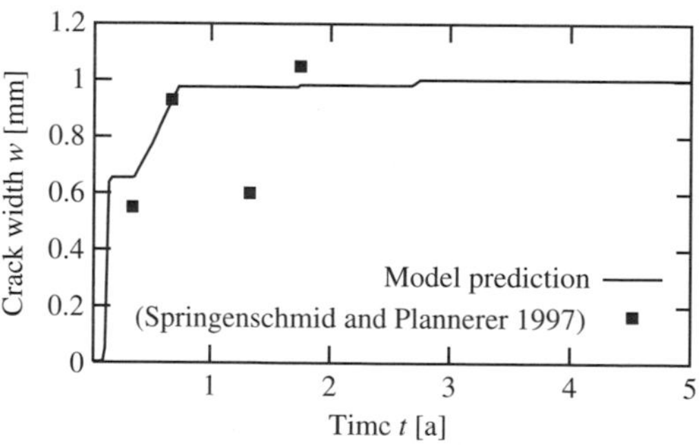

Figure 3. Numerical simulation of a tunnel lining: Comparison of the calculated evolution of the crack width w at Point A at the crown and measurements by (Springenschmid and Plannerer 1997).

Figure 3 contains a comparison of the predicted evolution of the crack width at the top of the inner lining (Point A in Figure 2) with reported measurements by (Springenschmid and Plannerer 1997). A good quantitative correlation concerning the maximum crack width is obtained. However, the seasonal crack closure is not represented by the model.

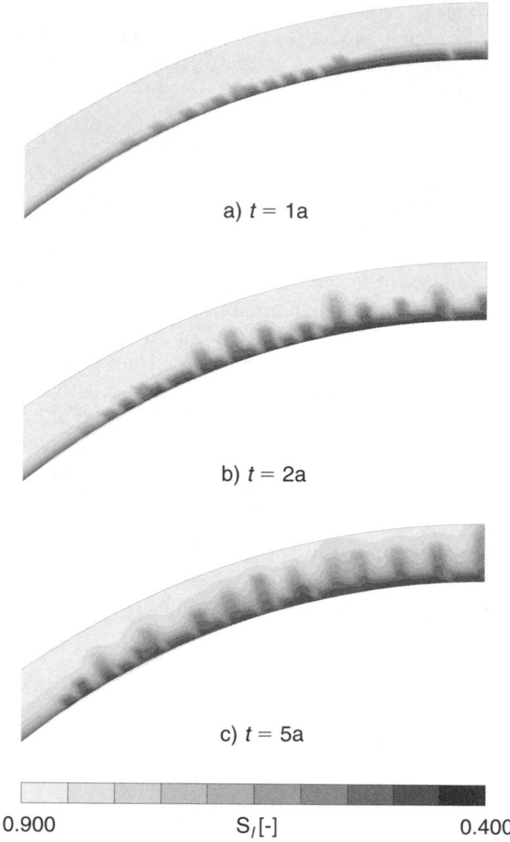

Figure 4. Numerical simulation of a tunnel lining: Calculated distribution of the liquid saturation S_l at the crown.

4.2 *Moisture distribution*

Figure 4 illustrates the predicted evolution of the drying process by means of the liquid saturation S_l. According to the strong interdependencies between moisture transport and the nonlinear material behaviour of concrete, the drying process at the top of the inner tunnel lining is clearly dominated by the influence of the growing crown cracks.

5 CONCLUSIONS

In the paper, a 3D coupled thermo-hygro-mechanical finite element model for numerical analyses of concrete structures, subjected to external loading and to changing hygral and thermal conditions, was proposed. The model was formulated within the framework of thermomechanics of partially saturated porous media in the sense of the BIOT-COUSSY-theory. It was shown, that relatively few material parameters are sufficient to fully describe the assumed coupling mechanisms on a macroscopic level. A strength criterion defined in the space of plastic effective stresses – in connection with a new formulation for the plastic effective stress coefficient – allows not only for the prediction of shrinkage-induced cracks but also for the moisture-dependence of the strength of cementitious materials. As a representative application, the numerical simulation of an inner tunnel lining exposed to realistic seasonal changes of the humidity and the temperature was described in the paper. The numerical results were compared to reported in-situ measurements on existing crown cracks in tunnel linings. It was shown, that the proposed finite element model is not only capable of reproducing shrinkage-induced cracks but also to predict the time of opening of the cracks, the location of the cracks and the width of the cracks in accordance with in situ observations. Furthermore, it was shown, that consideration of the influence of cracks on the moisture transport properties is crucial for realistic prognoses of shrinkage cracks in concrete structures.

ACKNOWLEDGMENTS

Financial support was provided by the German Science Foundation (DFG) through the graduate program *Computational Structural Dynamics* and in the framework of project A9 of the collaborative research center SFB 398. This support is gratefully acknowledged.

REFERENCES

Alonso, E., A. Gens, and A. Josa (1990). A constitutive model for partially saturated soils. *Géotechnique* 40, 405–430.

Baroghel-Bouny, V., M. Mainguy, T. Lassabatère, and O. Coussy (1999). Characterization and identification of equilibrium and transfer moisture properties for ordinary and high-performance cementitious materials. *Cement and Concrete Research* 29, 1225–1238.

Bažant, Z., A. Hauggaard, S. Baweja, and F.-J. Ulm (1997). Microprestress-solidification theory for concrete creep. I: Aging and drying effects. *Journal of Engineering Mechanics (ASCE)* 123, 1188–1194.

Bažant, Z. and L. Najjar (1972). Nonlinear water diffusion in nonsaturated concrete. *Materials and Structures* 5, 3–20.

Bear, J. and Y. Bachmat (1991). *Introduction to Modeling of Transport Phenomena in Porous Media*. Dordrecht: Kluwer Academic Publisher.

Biot, M. (1941). General theory of three-dimensional consolidation. *Journal of Applied Physics* 12, 155–165.

Biot, M. and D. Willis (1957). The elastic coefficients of the theory of consolidation. *Journal of Applied Mechanics* 24, 594–602.

Bishop, A. (1959). The principle of effective stress. *Teknisk Ukeblad* 39, 859–863.

Bishop, A. (1973). The influence of an undrained change in stress on the pore pressure in porous media of low compressibility. *Géotechnique* 23, 435–442.

Bishop, A. and G. Blight (1963). Some aspects of effective stress in saturated and partly saturated soils. *Géotechnique* 13, 177–197.

Burlion, N., F. Bourgeois, and J. Shao (2000). Coupling damage – drying shrinkage: experimental study and modelling. In V. Baroghel-Bouny and P.-C. Aïtcin (Eds.), *Proceedings of the International RILEM Workshop on Shrinkage of Concrete*, Number 17 in RILEM Proceedings.

CEB-FIP (1990). *Model Code 1990 (Design Code)*. CEB.

Colina, H. and P. Acker (2000). Drying cracks: Kinematics and scale laws. *Materials and Structures* 33, 101–107.

Coussy, O. (1989). A general theory of thermoporoelastoplasticity for saturated porous materials. *Transport in Porous Media* 4, 281–293.

Coussy, O. (1995). *Mechanics of porous continua*. Chichester: John Wiley & Sons.

Coussy, O., R. Eymard, and T. Lassabatère (1998). Constitutive modeling of unsaturated drying deformable materials. *Journal of Engineering Mechanics (ASCE)* 124, 658–667.

deBorst, R., J. Mazars, G. Pijaudier-Cabot, and J. van Mier (2001). Chapter: Durability, time dependence and coupled problems. In R. de-Borst, J. Mazars, G. Pijaudier-Cabot, and J. van Mier (Eds.), *Fracture Mechanics of Concrete Structures*. Balkema.

Fillunger, P. (1936). *Erdbaumechanik?* Wien.

Govindjee, S., G. Kay, and J. Simo (1995). Anisotropic modelling and numerical simulation of brittle damage in concrete. *International Journal for Numerical Methods in Engineering* 38, 3611–3634.

Grasberger, S. and G. Meschke (2001). Numerical simulation of coupled thermo-hygro-mechanical processes within concrete. In Z. B. F.-J. Ulm and F. Wittmann (Eds.), *Creep, Shrinkage and Durability Mechanics of Concrete and Other Quasi-Brittle Materials*, Oxford, UK, pp. 165–170. Elsevier.

Lewis, R. and B. Schrefler (1998). *The Finite Element Method in the Static and Dynamic Deformation and Consolidation of Porous Media*. Chichester: John Wiley & Sons.

Lydzba, D. and J. Shao (2000). Study of poroelasticity material coefficients as response of microstructure. *Mechanics of Cohesive-Frictional Materials* 5, 149–171.

Meschke, G. (1996). Consideration of aging of shotcrete in the context of a 3D viscoplastic material model. *International Journal for Numerical Methods in Engineering* 39, 3123–3143.

Meschke, G. and S. Grasberger (2002). Numerical modeling of coupled hygro-mechanical degradation of cementitious materials. *Journal of Engineering Mechanics (ASCE)*. In press.

Meschke, G., R. Lackner, and H. Mang (1998). An anisotropic elastoplastic-damage model for plain concrete. *International Journal for Numerical Methods in Engineering* 42, 703–727.

Nur, A. and J. Byerlee (1971). An exact effective stress law for elastic deformation of rock with fluids. *Journal of Geophysical Research* 76, 6414–6419.

Pickett, G. (1942). The effect of change in moisture-content of the creep of concrete under a sustained load. *Journal of the American Concrete Institute* 13, 333–355.

Rocznik, K. (1995). *Wetter und Klima in Deutschland: Ein meteorologisches Jahreszeitenbuch mit aktuellen Wetterthemen*. Stuttgart: Hirzel.

Sadouki, H. and F. Wittmann (2001). Numerical investigations on damage in cementitious composites under combined drying shrinkage and mechanical load. In R. de Borst, J. Mazars, G. Pijaudier-Cabot, and J. van Mier (Eds.), *Fracture Mechanics of Concrete Structures*, Volume 1, pp. 95–98. Balkema.

Schrefler, B. and X. Zhan (1993). A fully coupled model for water flow and airflow in deformable porous media. *Water Resources Research* 29, 155–167.

Snow, D. (1969). Anisotropic permeability of fractured media. *Water Resources Research* 5, 1273–1289.

Springenschmid, R. and M. Plannerer (1997). Firstrisse in der Innenschale großer Tunnel – Ursachen und Wege zur Vermeidung. *Betonund Stahlbetonbau* 92, 68–72, 109–112.

Ulm, F.-J. and O. Coussy (1996). Strength growth as chemo-plastic hardening in early age concrete. *Journal of Engineering Mechanics (ASCE)* 122, 1123–1132.

Ulm, F.-J., P. Rossi, I. Schaller, and D. Chauvel (1999). Durability scaling of cracking in HPC structures subject to hygromechanical gradients. *Journal of Structural Engineering (ASCE)* 125, 693–702.

van Genuchten, M. (1980). A closed-form equation for predicting the hydraulic conductivity of unsaturated soils. *Soil Science Society of America* 44, 892–898.

von Terzaghi, K. (1936). The shearing resistance of saturated soils and the angle between the planes of shear. In A. C. et al. (Ed.), *International Conference on Soil Mechanics*, Volume 1, pp. 54–56. Harvard University.

Computational Modelling of Concrete Structures, Bićanić et al. (eds)
© 2003 Swets & Zeitlinger, Lisse, ISBN 90 5809 536 3

Micromechanics of shotcrete elasticity

Christian Hellmich & Herbert Mang
Vienna University of Technology (TU Wien), Institute for Strength of Materials, Vienna, Austria

ABSTRACT: Understanding of the chemomechanical behavior of shotcrete (sprayed concrete) is a prerequisite for better design or failure risk assessment of civil engineering structures where this material is involved. Macroscopic material laws for shotcrete are preferably based on hydration degree-dependent material functions. For a given shotcrete mixture (i.e., for fixed water-cement ratio (w/c) and aggregate-cement ratio (a/c)), these functions are determined directly from macroscopic experiments. In a purely macroscopic context, consideration of changes in the w/c-ratio (as often encountered in large tunnel projects), and/or in the a/c-ratio would require additional experiments.

As an alternative, such hydration degree-dependent material functions can be predicted by considering quantitative information at the microlevel of the cementitious material, in the framework of continuum micromechanics. In this contribution, we focus on the dependence of the elastic stiffness on the hydration degree. We employ a two step homogenization procedure: within a representative volume element of 50 μm to 100 μm, cement grains (clinker), hydrates, water, and air build up a polycrystal. At a larger scale (5 to 10 cm), the aggregates are represented as spherical inclusions in a matrix built up by the aforementioned polycrystal. w/c and a/c are the input parameters for the model. On the basis of seven microelastic properties which are the same for different shotcretes, the chosen approach allows for prediction of hydration degree-stiffness relationships for different shotcrete mixtures. The high precision of these predictions is shown by comparing them to experimental values obtained from shotcretes with corresponding w/c and a/c values.

Ranging of w/c between 0.4 and 0.6, which may be encountered at the tunnel site, result in changes of shotcrete stiffness differing by a factor of up to two.

1 INTRODUCTION – INDUSTRIAL CONTEXT

In the New Austrian Tunneling Method, a shotcrete shell is used as primary support for the freshly excavated stretch of the tunnel. The challenge lies in mastering the interaction between the viscous rock which, in general, is moving inwards, and the hydrating shotcrete shell. This interaction is studied on site by means of elaborated monitoring devices (e.g., for the displacements of the tunnel shell). They allow for a qualitative prognosis of the rock stiffness in front of the tunnel face (Steindorfer and Schubert 1997) as well as for estimation of the forces in the tunnel shell (Rokahr and Zachow 1997). At a more elaborate stage, this estimation can be achieved by means of a hybrid method (Hellmich et al. 1999) where (i) displacement vector fields are approximated from measured displacement vectors at discrete points of the tunnel shell and (ii) these fields are prescribed as boundary values for a 3D Finite Element structural model of the tunnel shell.

The structural computation requires an elaborate material model for shotcrete, accounting for mechanical properties which change because of hydration. Models based on time-dependent stiffness and strength evolutions (Rokahr and Lux 1987; Schubert 1988; Swoboda et al. 1994; Meschke 1996) must be calibrated separately for different tunnel shells. In other words, these models are structure-dependent, which obviously limits their range of application. A constitutive model based on hydration degree-dependent mechanical properties (Hellmich et al. 1999; Sercombe et al. 2000; Lechner et al. 2001) is valid for different tunnel structures built up by one well-defined shotcrete material, characterized by constant water-cement and aggregate-cement ratios. Remarkably, on the tunnel site, the water-cement ratio may range between 0.4 and 0.6; moreover, it is difficult to quantify w/c during the construction process. Whereas the macroscopic hydration kinetics, accounted for in the aforementioned model by the chemical affinity concept (Ulm and Coussy 1996), remains roughly unaffected by the w/c-ratio (Hellmich et al. 2001), the influence of this ratio on the elasticity, strength, and creep properties may be remarkable, as, in general, is the case for concrete (Byfors 1980; Bažant and Baweja 1996).

Herein, we focus on the influence of w/c on the elastic properties of shotcrete. More precisely, with different water-cement ratios and aggregate-cement ratios as model input, we will predict the elasticity of different shotcrete mixtures on the basis of seven microelastic properties which are the same for all concretes and shotcretes. As an analysis tool, we employ continuum micromechanics (Zaoui 1997b), which recently has been applied successfully to the investigation of different aspects of concrete material behavior, such as chemomechanical couplings related to the alkali silica reaction (Lemarchand 2001; Lemarchand and Dormieux 2001), the influence of leaching on the elastic and frictional behavior of concrete (Constantinides and Ulm 2002; Lemarchand et al. 2002), and prediction of the elastic properties of cement pastes, mortars, and concretes, based on a detailed chemical definition of the cement (Bernard et al. 2002). When monitoring the loading level of a shotcrete tunnel shell, details of the chemical composition of the shotcrete cement are presently not available to tunnel engineers on site. This motivates the development of a simpler, yet reliable micromechanical model for shotcrete elasticity. Having in mind certain similarities between concrete and hard biological tissues like bone, we here adapt and modify a micromechanical model for the ultrastructural stiffness of mineralized tissues (Hellmich and Ulm 2002): A two-step homogenization procedure comprises (i) a polycrystal composite, and (ii) a matrix-inclusion composite where the matrix properties result from homogenization step I.

2 MICROMECHANICAL REPRESENTATION

The first step in applying continuum micromechanics to a material is the so-called representation of the material. This step comprises the definition of the homogeneous constituents of the material (the material phases) as well as their assemblages (Zaoui 1997a). If the characterization of the material requires scale transitions between more than two levels, as is the case for the complex material shotcrete, different representative volume elements (RVEs) must be defined. Each RVE characterizes the material at a specific observation scale. The characteristic length of such an RVE, ℓ, must be both considerably larger than the dimensions of heterogeneities within the RVE, d, and significantly smaller than the dimensions of a structure built up by the material defined on the RVE, $\mathcal{L}$; in mathematical terms, this separation-of-scales requirement reads as: $d \ll \ell \ll \mathcal{L}$.

Herein, we introduce two RVEs (Fig.1): An RVE with a characteristic length scale ℓ_p of about 50 μm to 100 μm represents the cement paste (characteristic volume V_p, Fig. 1(a)). It consists of four material phases with d_p of 1 to 10 microns: (i) the cement grains (clinker), (ii) water, (iii) hydrates, and (iv) air. In a highy disordered arrangement, each phase is adjacent to all other phases, building up the isotropic cement paste. In terms of micromechanics, this is represented by spherical phase inclusions embedded in a (fictitious) matrix with the material properties of the paste itself (polycrystal estimate, see the chapter on homogenization). On the other hand, the material shotcrete, defined on an RVE V_s with ℓ_s of 5 cm to 10 cm, can be regarded as a (connected) matrix built up by the cement paste with inclusions of aggregate, as it is often proposed in the literature, e.g. (Bongers and Rutten 1998) (Fig. 1(b)). d_s lies typically in the 1 cm range. The spherical shape of the inclusions represents the approximate isotropy of shotcrete.

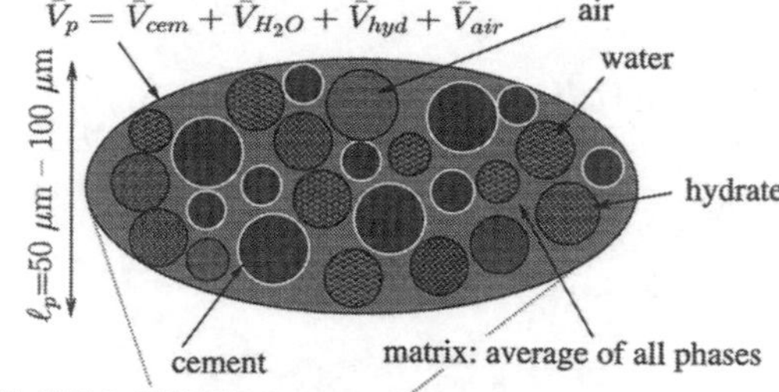

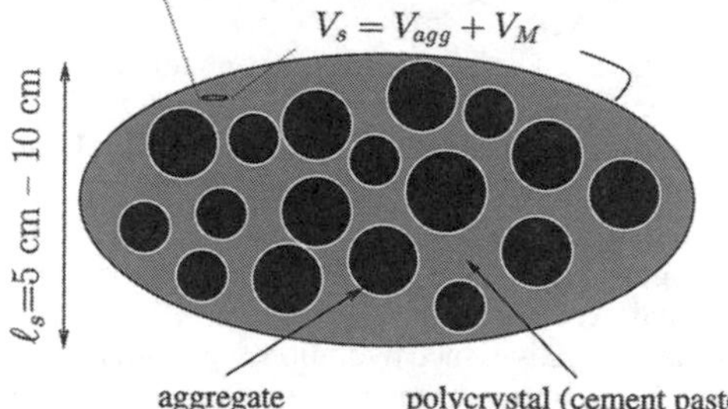

Figure 1. Micromechanical representation of shotcrete elasticity by a two-step homogenization procedure: (1) polycrystal built up of cement grains (clinker), water, hydrates, and air; (2) matrix-inclusion composite: matrix = polycrystal, inclusion = aggregate.

For engineering purposes, the volume fractions of the constituents of the cement paste can be determined as a function of the w/c-ratio and the degree of hydration, as reported by (Acker 2001), Fig. 2. There, the volumes of cement, hydrates, water, and air are normalized with respect to the volume of cement at the beginning of the hydration; i.e. $\bar{V}_{cem}(\xi = 0) = 1$ (Fig. 2).

Mathematically, these "dimensionless volumes" read as

$$\bar{V}_{cem}(\xi) \quad = \quad 1 - \xi \tag{1}$$

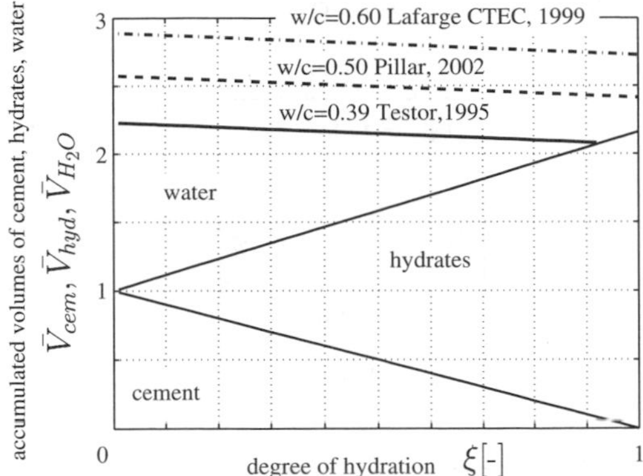

Figure 2. Volumes of cement paste components as a function of the degree of hydration, according to (Acker 2001); plotted for three different shotcretes.

$$\bar{V}_{H_2O}(\xi) \quad = \quad \frac{\rho_{cem}}{\rho_{H_2O}} \langle w/c - 0.42\xi \rangle \tag{2}$$

$$\bar{V}_{hyd}(\xi) \quad = \quad \frac{\rho_{cem}}{\rho_{hyd}} \xi \tag{3}$$

Here, $\langle . \rangle$ stands for the McAuley brackets, $\langle x \rangle = 1/2 \times (x+|x|) \cdot \rho_{cem} = 3.15\,\text{kg/dm}^3$, $\rho_{H_2O} = 1.0\,\text{kg/dm}^3$, and $\rho_{hyd} = 1.46\,\text{kg/dm}^3$ (Acker 2001) are the real mass densities of cement, water, and hydrates. Fig. 2 reflects the stoichiometry of the hydration reaction: For $w/c < 0.42$ ($\bar{V}_{H_2O}(\xi = 0) = (w/c)/(\rho_{H_2O}/\rho_{cem}) = 1.34$), the hydration cannot be completed ($\xi(t = \infty) < 1$). t denotes the time elapsed from the spraying of shotcrete. For $w/c > 0.42$, on the other hand, the hydration can be completed, and at complete hydration ($\xi(t = \infty) = 1$), free water is left in the shotcrete. It can also be seen from Fig. 2 that the volume of hydrates (reaction products), is approximately 10% smaller than the sum of the volumes of the reactants, water and cement (chemical shrinkage or Le Chatelier contraction, e.g. (Aitcin et al. 1997)). On the other hand, the total paste volume, which is $\bar{V}_p(\xi = 0) = \bar{V}_{cem}(\xi = 0) + \bar{V}_{H_2O}(\xi = 0)$ at the beginning of the hydration, changes only about 0.01% during hydration (autogeneneous shrinkage, see e.g. (Acker and Ulm 2001)); hence, $\bar{V}_p(\xi) \approx const. = \bar{V}_p(\xi = 0)$ for the present modeling purposes. Volume reduction through the formation of hydrates, at constant volume of the cement paste, leads to the formation of air pores within the cement paste during the hydration reaction; with the volume

$$\begin{aligned} \bar{V}_{air}(\xi) &= \bar{V}_{cem}(\xi = 0) + \bar{V}_{H_2O}(\xi = 0) \\ &\quad - \bar{V}_{cem}(\xi) - \bar{V}_{H_2O}(\xi) - \bar{V}_{hyd}(\xi) \end{aligned} \tag{4}$$

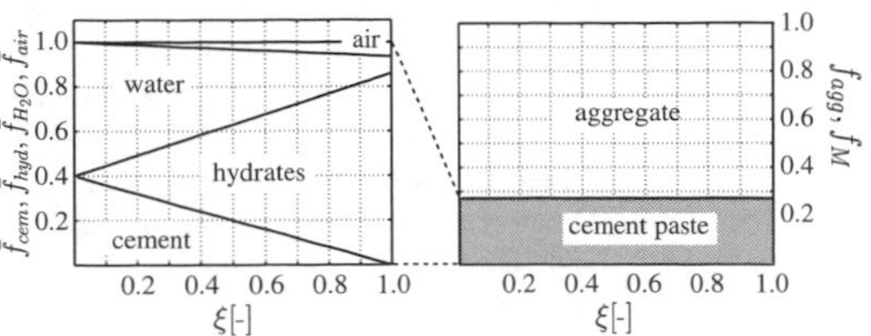

Figure 3. Volume fractions of shotcrete mixtures during hydration; plain shotcrete tested by (Pillar 2002): w/c = 0.50; a/c = 3.8.

The volume fractions of the phases in the cement paste (RVE V_p) are given by

$$\bar{f}_r(\xi) = \frac{\bar{V}_r(\xi)}{\bar{V}_p}; \quad \bar{f}_{cem} + \bar{f}_{H_2O} + \bar{f}_{hyd} + \bar{f}_{air} = 1, \tag{5}$$

with $r \in [cem, H_2O, hyd, air]$.

On the shotcrete level, RVE V_s is composed of the volume of the cement paste which is constant during hydration, and the volume of aggregates (Fig. 1(b)). Hence, the volume fraction of aggregates, f_{agg}, is also constant during hydration. f_{agg} can be determined from the water-cement ratio and the aggregate-cement ratio, w/c and a/c, as well as from the real mass densities of aggregate, water, and cement, ρ_{agg}, ρ_{H_2O}, and ρ_{cem}, through

$$\begin{aligned} f_{agg} &= \frac{V_{agg}}{V_{cem} + V_{H_2O} + V_{agg}} \\ &= \frac{(a/c)/\rho_{agg}}{1/\rho_{cem} + (w/c)/\rho_{H_2O} + (a/c)/\rho_{agg}} \end{aligned} \tag{6}$$

V_{agg}, V_{cem}, and V_{H_2O} are the volumes of aggregate, cement, and water in the RVE V_s at the beginning of the hydration. $\rho_{agg} = 2.5\,\text{kg/dm}^3$ is a typical value of (calcareous) shotcrete aggregate. The remaining volume of the RVE V_s is filled with the cement paste matrix (Fig.1(b)),

$$f_M = 1 - f_{agg} \tag{7}$$

The volume fractions entering the micromechanical model are illustrated in Fig. 3 as a function of the degree of hydration for a typical shotcrete mixture.

3 STIFFNESS PROPERTIES OF THE PHASES

Table 1 summarizes the elastic properties of the components of cement paste and shotcrete. Average cement stiffness values are taken from (Acker 2001). The stiffness of water is well-known, the stiffness of air is negligible with respect to the other stiffness values.

Table 1. Intrinsic isotropic stiffness values, valid for different types of shotcretes.

Phase	Bulk modulus k [GPa]	Shear modulus μ [GPa]
Cement (clinker)	$k_{cem} = 116.7$	$\mu_{cem} = 53.8$
Water	$k_{H_2O} = 2.3$	$\mu_{H_2O} = 0$
Hydrates	$k_{hyd} = 14.1$	$\mu_{hyd} = 8.9$
Air	$k_{air} = 0$	$\mu_{air} = 0$
Aggregate	$k_{agg} = 41.7$ [25.0; 58.3]*	$\mu_{agg} = 19.2$ [11.5; 26.9]*
Phase	**Young's mod. E [GPa]**	**Poisson's ratio $\nu[-]$**
Cement (clinker)	$E_{cem} = 140$	$\nu_{cem} = 0.30$
Hydrates	$E_{hyd} = 22$	$\nu_{hyd} = 0.24$
Aggregate	$E_{agg} = 50$ [30; 70]*	$\nu_{agg} = 0.30$

* From sensitivity analysis.

First experimental results for the elastic properties of hydrates were achieved recently by nanoindentation tests (Acker 2001; Constantinides and Ulm 2002). They showed two discernible types of calcium silicate hydrates with respect to elasticity. Calcium aluminates, which play an important role in low-sulfate shotcrete cements (Eichler 1999), were not yet tested. Since the precise composition of shotcrete clinker varies from one product to another, and is normally not known on the tunnel site, we here use as values for the hydrate stiffness mean values for the softer hydration products, which are predominant at w/c around 0.5 (Bernard et al. 2002). Respective values are taken from (Constantinides and Ulm 2002), see Table 1. Stiffness values for limestone aggregate can be found in the open literature (Wesche 1974; Mehlhorn 1996). Of all the phase stiffness values, the ones of limestone exhibit the largest scattering. To get insight into possible effects of the stiffness variation of limestone, we conduct a sensitivity analysis varying Young's modulus of limestone by more than a factor of two, from 30 GPa to 70 GPa.

4 LOCALIZATION AND HOMOGENIZATION I: CEMENT PASTE

The average local strains in the four phases, $\boldsymbol{\varepsilon}_{cem}$, $\boldsymbol{\varepsilon}_{H_2O}, \boldsymbol{\varepsilon}_{hyd}$, and $\boldsymbol{\varepsilon}_{air}$, are related to the homogeneous strains imposed at the boundary of $V_p, \boldsymbol{E}_p$, by the average 4th-order localization (or concentration) tensors $\boldsymbol{A}_{cem}, \boldsymbol{A}_{H_2O}, \boldsymbol{A}_{hyd}$, and $\boldsymbol{A}_{air}$, e.g. (Zaoui 1997b; Zaoui 2002),

$$\begin{aligned} \boldsymbol{\varepsilon}_{cem} &= \mathbf{A}_{cem} : \mathbf{E}_p, \quad \boldsymbol{\varepsilon}_{H_2O} = \mathbf{A}_{H_2O} : \mathbf{E}_p, \\ \boldsymbol{\varepsilon}_{hyd} &= \mathbf{A}_{hyd} : \mathbf{E}_p, \quad \boldsymbol{\varepsilon}_{air} = \mathbf{A}_{air} : \mathbf{E}_p \end{aligned} \tag{8}$$

An estimate of the homogenized stiffness tensor is given by the classical relation of continuum micromechanics,

$$\mathbf{C}_p^{est} = \langle \mathbf{c} : \mathbf{A}^{est} \rangle_{V_p} = \sum_r \bar{f}_r \mathbf{c}_r : \mathbf{A}_r^{est} \tag{9}$$

where $r \in [hyd, H_2O, air, cem]$ is the phase index, $\langle (.) \rangle_V = (1/V) \int_V (.)\, dV$ is the volume average, $\boldsymbol{c}_r$ denotes the stiffness tensor of phase r; and $\boldsymbol{A}_r^{est}$ is an estimate of the localization tensor of phase r. $\boldsymbol{A}_r^{est}$ incorporates the volume fractions, the elasticity, and the morphology of the phases in the model. $\boldsymbol{A}_r^{est}$ can be derived from Eshelby's problem (Eshelby 1957) of an inclusion perfectly bonded to a surrounding infinite matrix. In the present case of a polycrystal, the estimates of the localization tensors are given in an implicit form, which includes $\boldsymbol{C}_p^{est}$ (self-consistent scheme), e.g. (Zaoui 1997b):

$$\begin{aligned} \mathbf{A}_r^{est} = & \left[\mathbf{I} + \mathbf{S}_r^{Esh} : (\mathbf{C}_p^{est,-1} : \mathbf{c}_r - \mathbf{I})\right]^{-1} \\ & : \left\langle \left[\mathbf{I} + \mathbf{S}^{Esh} : (\mathbf{C}_p^{est,-1} : \mathbf{c} - \mathbf{I})\right]^{-1} \right\rangle_{V_p}^{-1} \end{aligned} \tag{10}$$

where $\boldsymbol{I}$ is the 4th-order unity tensor, $I_{ijkl} = 1/2 \times (\delta_{il}\delta_{jk} + \delta_{ik}\delta_{jl})$, and $\boldsymbol{S}_r^{Esh}$ is the Eshelby tensor of phase r. $\boldsymbol{S}_r^{Esh}$ depends on the shape of the inclusion characterizing phase r, and of the elasticity properties of the matrix.

Since both phases and the homogenized material are isotropic, the stiffness tensors $\boldsymbol{c}_{cem}, \boldsymbol{c}_{H_2O}, \boldsymbol{c}_{hyd}, \boldsymbol{c}_{air}$, and $\boldsymbol{C}_p^{est}$ read:

$$\begin{aligned} \mathbf{c}_{cem} &= 3k_{cem}\mathbf{K} + 2\mu_{cem}\mathbf{J}, \\ \mathbf{c}_{H_2O} &= 3k_{H_2O}\mathbf{K} + 2\mu_{H_2O}\mathbf{J}, \\ \mathbf{c}_{hyd} &= 3k_{hyd}\mathbf{K} + 2\mu_{hyd}\mathbf{J}, \\ \mathbf{c}_{air} &= 3k_{air}\mathbf{K} + 2\mu_{air}\mathbf{J}, \\ \mathbf{C}_p^{est} &= 3k^{est}\mathbf{K} + 2\mu^{est}\mathbf{J} \end{aligned} \tag{11}$$

where $k_{cem}, \mu_{cem}, k_{H_2O}, \mu_{H_2O}, k_{hyd}, \mu_{hyd}, k_{air}, \mu_{air}, k^{est}$, and μ^{est} are the bulk moduli and the shear moduli of the phases and the homogenized material, respectively; $K_{ijkl} = 1/3\delta_{ij}\delta_{kl}$ is the volumetric part of the 4th-order unity tensor $\mathbf{I}$, and $\mathbf{I}$, and $\mathbf{J} = \mathbf{I} - \mathbf{K}$ is the deviatoric part. The morphology of isotropic homogenized materials can be reasonably represented by phases filling up spherical inclusions. The respective Eshelbian tensor reads as, e.g. (Zaoui 1997b)

$$\mathbf{S}^{Esh}_{cem} = \mathbf{S}^{Esh}_{H_2O} = \mathbf{S}^{Esh}_{hyd} = \mathbf{S}^{Esh}_{air} = \mathbf{S}^{Esh}_{sph} = \alpha^{est}\mathbf{K} + \beta^{est}\mathbf{J} \tag{12}$$

with

$$\alpha^{est} = \frac{3k^{est}}{3k^{est} + 4\mu^{est}} \qquad \beta^{est} = \frac{6(k^{est} + 2\mu^{est})}{5(3k^{est} + 4\mu^{est})} \tag{13}$$

Substitution of (11) to (13) into (9) and (10) yields the homogenized moduli in the form

$$k^{est} = \left(\frac{\bar{f}_{cem} k_{cem}}{1 + \frac{\alpha^{est}(k_{cem} - k^{est})}{k^{est}}} + \frac{\bar{f}_{H_2O} k_{H_2O}}{1 + \frac{\alpha^{est}(k_{H_2O} - k^{est})}{k^{est}}} + \frac{\bar{f}_{hyd} k_{hyd}}{1 + \frac{\alpha^{est}(k_{hyd} - k^{est})}{k^{est}}} + \frac{\bar{f}_{air} k_{air}}{1 + \frac{\alpha^{est}(k_{air} - k^{est})}{k^{est}}} \right) \times \left(\frac{\bar{f}_{cem}}{1 + \frac{\alpha^{est}(k_{cem} - k^{est})}{k^{est}}} + \frac{\bar{f}_{H_2O}}{1 + \frac{\alpha^{est}(k_{H_2O} - k^{est})}{k^{est}}} + \frac{\bar{f}_{hyd}}{1 + \frac{\alpha^{est}(k_{hyd} - k^{est})}{k^{est}}} + \frac{\bar{f}_{air}}{1 + \frac{\alpha^{est}(k_{air} - k^{est})}{k^{est}}} \right)^{-1} \tag{14}$$

and

$$\mu^{est} = \left(\frac{\bar{f}_{cem} \mu_{cem}}{1 + \frac{\beta^{est}(\mu_{cem} - \mu^{est})}{\mu^{est}}} + \frac{\bar{f}_{H_2O} \mu_{H_2O}}{1 + \frac{\beta^{est}(\mu_{H_2O} - \mu^{est})}{\mu^{est}}} + \frac{\bar{f}_{hyd} \mu_{hyd}}{1 + \frac{\beta^{est}(\mu_{hyd} - \mu^{est})}{\mu^{est}}} + \frac{\bar{f}_{air} \mu_{air}}{1 + \frac{\beta^{est}(\mu_{air} - \mu^{est})}{\mu^{est}}} \right) \times \left(\frac{\bar{f}_{cem}}{1 + \frac{\beta^{est}(\mu_{cem} - \mu^{est})}{\mu^{est}}} + \frac{\bar{f}_{H_2O}}{1 + \frac{\beta^{est}(\mu_{H_2O} - \mu^{est})}{\mu^{est}}} + \frac{\bar{f}_{hyd}}{1 + \frac{\beta^{est}(\mu_{hyd} - \mu^{est})}{\mu^{est}}} + \frac{\bar{f}_{air}}{1 + \frac{\beta^{est}(\mu_{air} - \mu^{est})}{\mu^{est}}} \right)^{-1} \tag{15}$$

There is no explicit solution for $k^{est}(\bar{f}_{cem}, \bar{f}_{H_2O}, \bar{f}_{hyd}, \bar{f}_{air})$ and $\mu^{est}(\bar{f}_{cem}, \bar{f}_{H_2O}, \bar{f}_{hyd}, \bar{f}_{air})$. Therefore, these relationships are determined numerically. We here employ a two-membered evolution strategy (Schwefel 1977), representing a simple and powerful tool to solve non-linear minimization problems, see (Hellmich and Ulm 2002) for details.

5 LOCALIZATION AND HOMOGENIZATION II: SHOTCRETE

The average local strains in the two phases, i.e. in the matrix and in the inclusions of aggregates, are related to the homogeneous strains imposed at the boundary of the RVE V_s, $\mathbf{E}_s$, through

$$\boldsymbol{\varepsilon}_M = \mathbf{A}_M : \mathbf{E}_s, \qquad \boldsymbol{\varepsilon}_{agg} = \mathbf{A}_{agg} : \mathbf{E}_s, \tag{16}$$

where $\mathbf{A}_{agg}$ and $\mathbf{A}_M$ denote average 4th-order concentration (or localization) tensors. The morphology to be considered here is the one of a polycrystalline matrix phase, in which the aggregate inclusions are embedded. For this situation, estimates of the localization tensors are suitably provided by the Mori-Tanaka scheme, in an explicit form:

$$\mathbf{A}^{est}_{agg} = \left[\mathbf{I} + \mathbf{S}^{Esh}_{agg} : (\mathbf{c}_M^{-1} : \mathbf{c}_{agg} - \mathbf{I})\right]^{-1} : \left\langle \left[\mathbf{I} + \mathbf{S}^{Esh} : (\mathbf{c}_M^{-1} : \mathbf{c} - \mathbf{I})\right]^{-1} \right\rangle_{V_s}^{-1} \tag{17}$$

$$\mathbf{A}^{est}_M = \left\langle \left[\mathbf{I} + \mathbf{S}^{Esh} : (\mathbf{c}_M^{-1} : \mathbf{c} - \mathbf{I})\right]^{-1} \right\rangle_{V_s}^{-1} \tag{18}$$

This model ensures the connectivity of the matrix phase and accounts for the interaction of the aggregate inclusions.

The stiffness of the aggregates reads as

$$\mathbf{c}_{agg} = 3k_{agg}\mathbf{K} + 2\mu_{agg}\mathbf{J}, \tag{19}$$

whereas the stiffness of the matrix, $\mathbf{c}_M$, follows from the first homogenization step:

$$\mathbf{c}_M = 3k^{est}(\bar{f}_{cem}, \bar{f}_{H_2O}, \bar{f}_{hyd}, \bar{f}_{air})\mathbf{K} + 2\mu^{est}(\bar{f}_{cem}, \bar{f}_{H_2O}, \bar{f}_{hyd}, \bar{f}_{air})\mathbf{J} \tag{20}$$

The Eshelby tensor in Eqs. (17) and (18) reads as (Eshelby 1957):

$$\mathbf{S}^{Esh}_{agg} = \mathbf{S}^{Esh}_{sph} = \alpha_M\mathbf{K} + \beta_M\mathbf{J} \tag{21}$$

with

$$\alpha_M = \frac{3k_M}{3k_M + 4\mu_M} \qquad \beta_M = \frac{6(k_M + 2\mu_M)}{5(3k_M + 4\mu_M)} \tag{22}$$

Finally, the estimate for the homogenized shotcrete stiffness becomes

$$\mathbf{C}^{est}_s = \langle \mathbf{c} : \mathbf{A}^{est} \rangle_{V_s} = \sum_r f_r \mathbf{c}_r : \mathbf{A}^{est}_r \tag{23}$$

6 PREDICTIVE CAPABILITIES OF THE MODEL

k_{cem}, μ_{cem}, k_{H_2O}, μ_{H_2O}, k_{hyd}, μ_{hyd}, k_{air}, μ_{air}, k_{agg}, and μ_{agg} are intrinsic material constants. By intrinsic we mean that they are the same for different shotcrete mixtures, Table 1. These shotcrete mixtures differ only in composition, i.e. in the a/c and w/c ratios occurring in the equations for the volume fractions (5) and (6), involving the relations (1) to (4). w/c and a/c serve as input for the micromechanical model. The model, then, permits prediction of the stiffness properties of the shotcrete, i.e., the ensemble of elastic constants of an isotropic material, $\boldsymbol{C}_s^{est}$.

The predictive capabilities of the model are shown by (1) calculating hydration degree – stiffness relationships for a shotcrete mixture with given w/c and a/c; and (2) comparing them to corresponding relationships obtained from experiments. The evaluation of respective experiments will be described in detail in the next two paragraphs. Good agreement between experimental values and model predictions is shown in Fig. 4 for different shotcrete mixtures. The correlation coefficient between model estimates and experimental values amounts to a value as high as $r^2 = 98.0\%$. Notably, the variation of Young's modulus of the aggregate (limestone) within [30; 70] GPa only has a minor effect on this good agreement. A sensitivity analysis shows that for $E_{agg} = 30$ GPa and $E_{agg} = 70$ GPa, the correlation coefficient still amounts to $r^2 = 98\%$ and $r^2 = 97\%$, respectively.

Experimental determination of $E(\xi)$ requires testing of $\xi(t)$ and $E(t)$ at the same hydration states for one shotcrete mixture. Values of elastic stiffness E at different time instants after spraying t_i can be obtained from flexural resonant frequency tests on $4 \times 4 \times 16$ cm beams supported on two synthetic strips, through the standard relationship, e.g. (Mang and Hofstetter 2000),

$$E_s^{exp} = \frac{12\omega^2 \rho l^4}{\pi^4 b^2} \tag{24}$$

ω is the (measured) fundamental flexural resonant frequency, ρ is the shotcrete mass density, $l = 16$ cm is the length of the specimen, and $b = 4$ cm is the side length of the quadratic cross section of the specimen. Such tests on a shotcrete mixture with w/c = 0.48 and a/c = 5.3 were commissioned by the Institute for Strength of Materials to Lafarge Centre Europe Central, Mannersdorf, Austria, see the ordinates of data points in Fig. 4. These stiffness values were measured at specimens of t_i = 2, 4, 8, 12, 24, 168, 672 h age, hydrating under isothermal conditions. The corresponding values for the degree of hydration, $\xi(t_i)$, can be determined from temporal integration of a kinetics law of the Arrhenius type (Ulm and Coussy 1996),

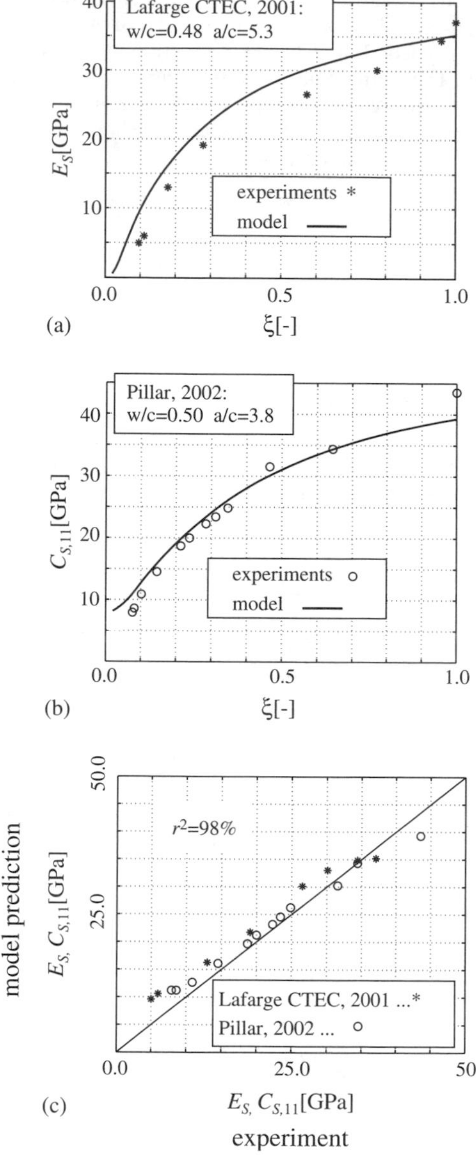

Figure 4. Comparison between model predictions and experiments: E (or C_{11})–ξ-relationships for (a) shotcrete tested at Lafarge CTEC Mannersdorf, Austria, and (b) shotcrete tested by (Pillar 2002) (c) direct comparison of stiffness values: model predictions versus experiments.

$$\xi(t=0) = 0; \quad \xi(t_i) = \int_{t=0}^{t_i} \tilde{A}(\xi) \exp\left(-\frac{E_a}{RT_0}\right) dt \tag{25}$$

with $T_0 = (273.15+20)$ K as the absolute (isothermal) temperature, and $E_a/R = 4000$ K (Byfors 1980) as the ratio of the activation energy and the universal gas constant. Application of Eq. (25) requires knowledge of the normalized chemical affinity $\tilde{A}(\xi)$. Following the procedure of (Ulm and Coussy 1996), $\tilde{A}(\xi)$ was determined from an adiabatic test on the same type of shotcrete. The values $\xi(t_i)$ are shown as the abscissae of the data points in Fig. 4(a).

Alternatively, stiffness values can be determined from ultrasonic pulse velocity tests where a longitudinal pulse of high frequency is transmitted through a specimen and its velocity is determined from the time of propagation. The first (axial) component of the stiffness tensor can be determined according to, e.g. (Fedorov 1968),

$$C_{s,11}^{exp} = \rho v_1^2 \tag{26}$$

with v_1 as the longitudinal pulse velocity. (Pillar 2002) investigated v_1 at $t_i = 3$, 3.5, 4.8, 6, 7.9, 9.6, 12, 24, 72, 168, 672 h age at isothermal conditions, see the ordinates in Fig. 4(b) for corresponding stiffness values. Under the same isothermal conditions, strength tests were conducted. They allow for the determination of the degree of hydration, through (Ulm and Coussy 1996; Hellmich et al. 1999),

$$\xi(t_i) = \xi_0 + (1 - \xi_0)\frac{f_c(t_i)}{f_c(t = \infty)} \tag{27}$$

with $\xi_0 \approx 5\%$ as the percolation threshold (Acker 1988; Ulm and Coussy 1996), and f_c as the uniaxial compressive strength of shotcrete.

7 GENERAL MODEL CHARACTERISTICS

Remarkably, the good agreement between model predictions and experiments (Fig. 4) holds for very different hydration stages, i.e. for very differently composed materials in terms of the volume fractions, see Fig. 3. Having thus gained confidence in the predictive qualities of the model, we investigate the effects of w/c varying from 0.4 to 0.6 on the ξ–E-relationships, Fig. 5(a). This range of w/c results in Young's moduli of shotcrete differing by a factor of up to two. A small w/c-ratio leads to large moduli even at the very beginning of hydration. Also at complete hydration, w/c has a remarkable influence on Young's modulus. Poisson's ratio of shotcrete (Fig. 5(b)), on the other hand, increases, for given ξ, with increasing w/c. This is because water exhibits the largest Poisson's ratio of all shotcrete components. For $\xi \rightarrow 1$, v_s tends towards one value, irrespective of w/c.

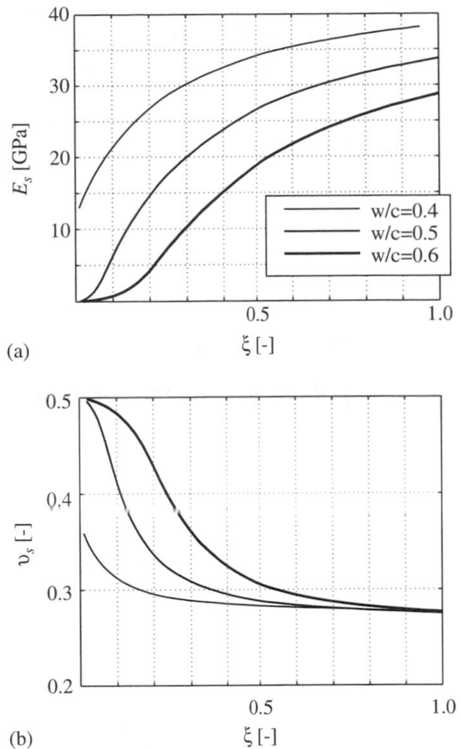

Figure 5. Model characteristics: influence of different w/c ratios on stiffness-hydration degree relationships (for a/c = 5 = const.): (a) Young's modulus, (b) Poisson's ratio.

8 SUMMARY AND CONCLUSIONS

Realistic relations between the degree of hydration and the dynamic stiffness of shotcrete can be predicted by a continuum micromechanics model based on seven (micro-)stiffness constants which are the same for different shotcrete mixtures. The water-cement ratio and the aggregate-cement ratio serve as input for the model. The model comprises a two-step homogenization procedure: Firstly, at the level of the cement paste, cement grains, hydrates, water, and air build up a polycrystal. Secondly, at the larger scale of shotcrete, the aggregates are represented by spherical inclusions in a matrix built up by the aforementioned polycrystal (cement paste).

The model reveals that stiffness values differ by a factor of up to two because of w/c ranging between 0.4 and 0.6. Such changes in w/c may be encountered at the tunnel site. This suggests a closer inspection of the current w/c-ratio on the tunnel site and incorporation of respective changes in the elastic stiffness of shotcrete into tunnel monitoring systems for stress quantification, such as the ones reported by

(Hellmich et al. 1999; Macht et al. 2003). This will provide the tunnel engineer with additional information, supporting his decisions for the further tunnel advance.

ACKNOWLEDGEMENTS

The first author gratefully acknowledges fruitful discussions and informative communications with Franz-Josef Ulm, Massachusetts Institute of Technology, Cambridge, USA, and Eric Lemarchand, Lille University of Sciences and Technology, Villeneuve d'Ascq, France. Furthermore, thanks are due to Nora Pillar, Universidade Federal de Santa Catarina, Florianopolis, Brazil, for the communication of experimental results.

REFERENCES

Acker, P. (1988). Comportement mécanique du béton: apports de l'approche physico-chimique [Mechanical behavior of concrete: a physico-chemical approach]. Technical Report Res. Rep. LCPC 152, Laboratoires des Ponts et Chaussées, Paris, France. In French.

Acker, P. (2001). Micromechanical analysis of creep and shrinkage mechanisms. In F.-J. Ulm, Z. Bažant, and F. Wittmann (Eds.), *Creep, Shrinkage, and Durability Mechanics of Concrete and other Quasi-brittle Materials – Proceedings of the Sixth International Conference CONCREEP-6@MIT,* pp. 15–26. Elsevier, Amsterdam.

Acker, P. and F.-J. Ulm (2001). Creep and shrinkage of concrete: physical origins and practical measurements. *Nuclear Engineering and Design* 203, 148–158.

Aitcin, P.-C., A. Neville, and P. Acker (1997). Integrated view of shrinkage deformation. *Concrete international – September 1997,* 35–41.

Bažant, Z. and S. Baweja (1996). Short form of creep and shrinkage prediction model B3 for structures of medium sensitivity. *Materials and Structures* (29), 587–593.

Bernard, O., F. Ulm, and E. Lemarchand (2002). A multiscale micromechanics – hydration model for the early-age elastic properties of cement-based materials. *Cement and Concrete Research.* Submitted for publication.

Bongers, J. and H. Rutten (1998). Concrete in multiaxial compression – a multilevel analysis. *Heron* 43(3).

Byfors, J. (1980). Plain concrete at early ages. Technical report, Swedish Cement and Concrete Research Institute, Stockholm, Sweden.

Constantinides, G. and F.-J. Ulm (2002). The effect of two types of C–S–H on the elasticity of cement-based materials: results from nanoindentation and micromechanical modeling. Personal communication.

Eichler, K. (1999). Bindemittel- und Verfahrens-technologie von modernem Spritzbeton [Binding agents and process technology of modern shotcrete]. In W. Kusterle (Ed.), *Spritzbeton-Technologie* 99, Innsbruck, Austria, pp. 195–202. Institut für Baustofflehre und Materialprüfung, University of Innsbruck, Innsbruck, Austria. In German.

Eshelby, J. (1957). The determination of the elastic field of an ellipsoidal inclusion, and related problems. *Proceedings of the Royal Society London, Series A* 241, 376–396.

Fedorov, F. (1968). *Theory of elastic waves in crystals.* Plenum Press, New York.

Hellmich, C., J. Macht, R. Lackner, H. Mang, and F.-J. Ulm (2001). Phase transitions in shotcrete: from material modelling to structural safety assessment. In S. Bernard (Ed.), *International Conference on Engineering Developments in Shotcrete,* Hobart, Tasmania, pp. 173–184. Balkema, Rotterdam.

Hellmich, C., H. Mang, and F.-J. Ulm (1999). Hybrid method for quantification of stress states in shotcrete tunnel shells: combination of 3D *in-situ* displacement measurements and thermochemoplastic material law. In W. Wunderlich (Ed.), *CD-ROM Proceedings of the 1st European Conference of Computational Mechanics,* Munich, Germany. Also published in *Computers & Structures,* 79: 2103–2115, 2001.

Hellmich, C. and F.-J. Ulm (2002). A micromechanical model for the ultrastructural stiffness of mineralized tissues. *Journal of Engineering Mechanics (ASCE)* 128(8), 898–908.

Hellmich, C., F.-J. Ulm, and H. A. Mang (1999). Multisurface chemoplasticity I: Material model for shotcrete. *Journal of Engineering Mechanics (ASCE) 125*(6), 692–701.

Lechner, M., C. Hellmich, and H. Mang (2001). Short-term creep of shotcrete – thermochemo-plastic material modelling and nonlinear analysis of a laboratory test and of a NATM excavation by the finite element method. In P. Vermeer, S. Diebels, W. Ehlers, H. Herrmann, S. Luding, and E. Ramm (Eds.), *Continuous and discontinuous modelling of cohesive-frictional materials, Lecture Notes in Physics,* Volume 568, pp. 47–62. Springer, Berlin.

Lemarchand, E. (2001). *Contribution de la micromécanique à l'étude des phénomènes de transport et de couplage poromécanique dans les milieux poreux: Application aux phénomènes de gonflement des géomateriaux [Contribution of micromechanics to the study of transport and poromechanical coupling phenomena: application to swelling phenomena in geomaterials].* Ph.D. thesis, L'École Nationale de Ponts et Chaussées, Marne-la-Vallée, France. In French.

Lemarchand, E. and L. Dormieux (2001). A micromechanical approach to the modeling of swelling due to alkali-silica reaction. In J. Tassoulas (Ed.), *Proceedings of the 14th Engineering Mechanics Conference, May 2000.*

Lemarchand, E., F.-J. Ulm, and L. Dormieux (2002). Effect of inclusions on friction coefficient of highly filled composite materials. *Journal of Engineering Mechanics (ASCE)* 128(8), 885–888.

Macht, J., R. Lackner, C. Hellmich, and H. Mang (2003). Quantification of stress states in shotcrete shells. In G. Beer (Ed.), *Numerical Simulation in Tunneling.* Springer, Berlin. To appear.

Mang, H. and G. Hofstetter (2000). *Festigkeitslehre [Strength of materials].* Springer, Vienna. In German.

Mehlhorn, G. (Ed.) (1996). *Der Ingenieurbau: Grundwissen – Werkstoffe, Elastizitätstheorie [Basic knowledge in Civil Engineering: Materials and Elasticity Theory],* Volume 4. Ernst & Sohn, Berlin, Germany. In German.

Meschke, G. (1996). Consideration of aging of shotcrete in the context of a 3D viscoplastic material model. *International Journal for Numerical Methods in Engineering* 39, 3123–3143.

Pillar, N. (2002). *Determination of early age properties of fibre reinforced shotcrete to predict the cracking behavior.* Ph.D. thesis, University of New South Wales, Sydney, Australia.

Rokahr, R. and K. Lux (1987). Einfluß des rheologischen Verhaltens des Spritzbetons auf den Ausbauwiderstand [Influence of the rheological behavior of shotcrete on the lining resistance]. *Felsbau* 5, 11–18. In German.

Rokahr, R. and R. Zachow (1997). Ein neues Verfahren zur täglichen Kontrolle der Auslastung einer Spritzbetonschale [A new method for the daily monitoring of the stress intensity of a sprayed concrete lining]. *Felsbau – Rock and Soil Engineering* 15(6), 430–434. In German.

Schubert, P. (1988). Beitrag zum rheologischen Verhalten von Spritzbeton [Contribution to the rheological behavior of shotcrete]. *Felsbau* 6, 150–153. In German.

Schwefel, H. (1977). *Numerische Optimierung von Computer-Modellen mittels der Evolutionsstrategie [Numerical optimization of computer models by means of the evolution strategy]* (First ed.). Basel, Switzerland: Birkhäuser. In German.

Sercombe, J., C. Hellmich, F.-J. Ulm, and H. A. Mang (2000). Modeling of early-age creep of shotcrete. I: model and model parameters. *Journal of Engineering Mechanics (ASCE) 126*(3), 284–291.

Steindorfer, A. and W. Schubert (1997). Application of new methods of monitoring data analysis for short term prediction in tunnelling. In J. Golser, W. Hinkel, and W. Schubert (Eds.), *Tunnels for People, Proceedings of the World Tunnel Congress 1997,* Volume 1, Vienna, Austria, pp. 65–69. Balkema, Rotterdam.

Suquet, P. (Ed.) (1997). *Continuum micromechanics.* Springer, Wien, New York.

Swoboda, G., A. Moussa, and N. Hafez (1994). Two and three dimensional modelling of layered shotcrete lining. In P. Lee, L. Tham, and Y. Cheung (Eds.), *Proceedings of the International Conference on Computational Methods in Structural and Geotechnical Engineering,* Hong Kong, China, pp. 1077–1084. China Translation & Printing Services, Hong Kong, China.

Testor, M. (1995). Trockenspritzbeton mit neuen Bindemitteln – Temperatureinfluss, Staube-influss und Rückprallreduktion [Dry-mix shotcrete with new cements – influence of temperature and of dust, reduction of rebound]. Master's thesis, University of Innsbruck, Innsbruck, Austria. In German.

Ulm, F.-J. and O. Coussy (1996). Strength growth as chemo-plastic hardening in early age concrete. *Journal of Engineering Mechanics (ASCE)* 122(12), 1123–1132.

Wesche, K. (1974). *Baustoffe für tragende Bauteile [Building materials for structural components]* (3 ed.). Wiesbaden, Germany: Bauverlag. In German.

Zaoui, A. (1997a). Matériaux hétérogéne et composites [Heterogeneous materials and composites]. Lecture Notes from École Polytechnique, Palaiseau, France. In French.

Zaoui, A. (1997b). *Structural morphology and constitutive behavior of microheterogeneous materials,* pp. 291–347. In Suquet (1997).

Zaoui, A. (2002). Continuum micromechanics: Survey. *Journal of Engineering Mechanics (ASCE)* 128(8), 808–816.

Computational Modelling of Concrete Structures, Bićanić et al. (eds)
© 2003 Swets & Zeitlinger, Lisse, ISBN 90 5809 536 3

Drying of porous media: numerical and experimental approach

D. Jankovic
Delft University of Technology, Faculty of Civil Engineering and Geo-Sciences, Microlab, Delft, the Netherlands

J.G.M. van Mier*
Swiss Federal Institute of Technology (ETH), Institute for Building Materials, Switzerland

ABSTRACT: Drying of porous media, such as cement paste, is observed through different types of experiments at meso- and micro-scales. Important aspect is the transport of water through the porous material. The results from Nuclear Magnetic Resonance (NMR) drying experiments have been numerically modeled by means of Lattice Gas Automaton (LGA) from Statistical Physics. It is shown that numerical modeling with LGA can be an approximate fit to the moisture flow, through the capillaries, in the cement paste samples. For the sake of comparisons, temporal and spatial scales are yet to be found.

1 INTRODUCTION

The behaviour of concrete at early age, and possible micro-crack phenomena, which take place before the application of any mechanical load, influence the durability of concrete structures. The cause of micro-cracks could be the moisture flow in the porous cement structure, which develops due to the differential volume changes and subsequent differential pressures. The physical-chemical processes, in the period of concrete hardening under various environmental conditions, that might cause drying (under high or very low temperatures), freezing and thawing and others, initiate moisture flow. In this research, we focus on drying, as a potential cause of moisture flow and drying shrinkage, which may result in micro-cracking.

Figures 1 and 2 show images of drying shrinkage micro-cracking in Environmental Scanning Electron Microscope (ESEM) experiments, on the top surface of $10 \times 10\,\text{mm}^2$ young cement paste samples. The experiment is performed on very young cement paste sample (4 days of age). In this experiment, a new method of cement paste sample preparation is used, and the preliminary results are described in (Jankovic et al. 2002). After casting 2-mm thick samples in a specially created mould, the samples were polished to the thickness of 1 mm, using a tool, particularly created for this purpose (Jankovic et al. 2002). The sample was wet cured at 20°C and 95% relative humidity (RH). Very thin samples of about 1 mm (1.2 mm in this example) were dried, in order to observe the

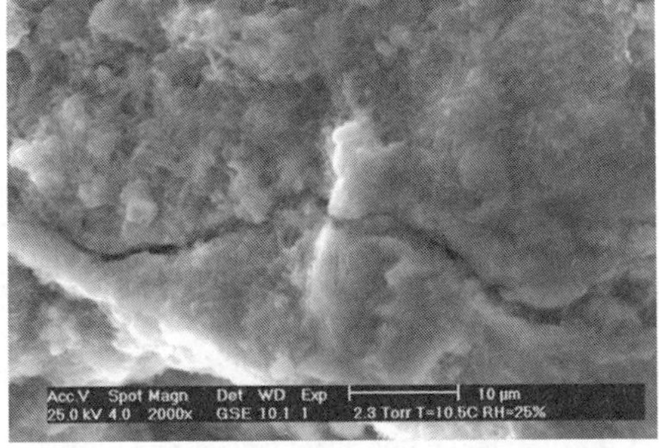

Figure 1. Image taken after drying to 10% and short re-wetting to 25% RH for the image improvement.

Figure 2. Image taken after 20 min of re-wetting to 92% RH. Closure of the crack is visible around the middle part of the crack.

* Formerly at Microlab, TU Delft.

drying mechanisms that are dependent only on the material properties (so-called real mechanisms) and not geometry (Wittmann 1982, Neubauer 1997).

Drying of the sample was performed in ESEM chamber, where the RH was lowered from 100% to 10%, maintaining a constant sample temperature, while slowly reducing the chamber pressure. After the drying period of a few hours, in the artificial ESEM climate, micro-cracks were observed at the end of the experiment at 10% RH (Figs 1, 2). At this point, it is difficult to say more about the period of time when cracking occurred. This is due to the fact that, in ESEM, one small spot of the sample surface is observed, which does not necessarily mean that cracks will occur there. Due to imperfections of the ESEM, such as stage rotation problems, we are still not able to make images at different RH at different spots or to maintain constant observations of the whole sample in order to find a potential crack. Very rapid crack development is observed in drying experiments with acoustic emission (Shiotani et al., in press).

After re-wetting from 10% to 25% and further to 92% RH, the micro-crack started to close. The closure was only partial. Swelling and the microstructure growth, which occurred after re-wetting could not close the crack completely. It is again a question of time if and when the crack could be healed. Regarding previous research experience from the literature, some residual cracks would always be present as an inelastic material deformation. If loaded, these residual cracks might cause later on a catastrophic fracture of the material. It is interesting to mention, that these types of cracks occurred only in the wet cured samples (cured in water at 20°C), while dry cured samples (sealed cured at 20°C and 95% RH), possibly gained sufficient strength, and dried out without cracking.

Due to these potential cement-microstructure deformations at very early age, it is logical to focus concrete research on very small scales (meso-, micro-level), with a potential inclusion of the nano-scale. The application of diverse suitable experimental as well as numerical methods, is a necessity. Usage of electron microscopy is already an advantageous way of microstructure observations. Unfortunately, it is still rather cumbersome to experimentally observe the early age concrete behaviour (deformations) at a very small scale (microscopes like ESEM and similar microscope types are still in the developing phase). To obtain the deformations indirectly, we can use digital image correlation techniques (Jankovic et al. 2002), in order to connect the deformation changes with the decrease of relative humidity.

Other methods, like Nuclear Magnetic Resonance (NMR), can be used as a helpful experimental tool in getting the total moisture content and its distribution, in a specific sample volume, at a certain time (Valckenborg et al. 2001). Additionally, NMR can be used to define the size of the pores with respect to the speed of drying.

Modeling of drying and drying shrinkage may be beneficial in defining different parameters, which might influence deformations, due to moisture flow and drying. Needless to say, that these models have to be based on the physics of fluid flow. One of these models, applied in this research is the Lattice Gas Automaton (LGA), a type of Cellular Automata, originating from Statistical Physics. In the existing analyses, LGA models moisture flow and does not include detailed cement microstructure. Different models of the cement microstructure, such as C–S–H models (Wittmann 1973), already exist and will not be discussed here. Goal of the present analysis is to model moisture flow and drying in order to determine similarities with the real moisture flow, as measured in drying experiments, using Nuclear Magnetic Resonance (NMR).

1.1 *NMR: experimental approach to moisture flow*

Gravimetric sampling is the usual (destructive) method for the analysis of moisture content, which is nowadays successfully, replaced by, among others, Nuclear Magnetic Resonance (NMR). NMR is used as a nondestructive method for measuring moisture transport in porous building materials. NMR is applied for first time in the 1960s (Fatt et al. 1967), to determine saturation. It enabled the insight to total moisture content of sample as well as distribution of water in porous materials. The NMR works on the principle of measuring of the absorbed energy, which is proportional to the present distinguishable nuclei of H and hence water. For the moisture transport in time, the size of the pores becomes important. The size classification of pores is into capillaries (size up to 10 μm) and voids (gel pores, size of 10 nm and less), Scheidegger 1960.

The complex network of pores, in different shapes and sizes, has important influence on the speed of drying. It easily influences shrinkage (and creep) due to the status of water presence. This is especially crucial for capillary pores, which present the water-filled space between C–S–H elements (Bažant et al. 1982).

Another important differentiation between the size of pores (capillaries and voids) can be made with respect to the importance of the pore walls. In voids, walls do not play any role, while in capillaries, walls are important for hydrodynamic phenomena in their interior (Scheidegger 1960).

In the tests, the NMR measurement technique was applied to mortars (Valckenborg et al. 2001) pure cement paste, and lately to cement paste combined with additional aggregates (glass pearls, $d = 0.5$ mm, in the amount of 35%), see Bisschop et al., in press.

In these NMR experiments (Bisschop et al., in press), drying of the cement paste cylindrical sample

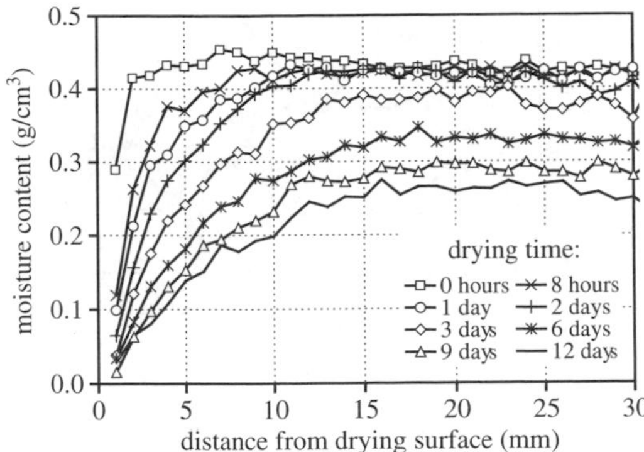

Figure 3. NMR drying of pure cement paste sample (Bisschop et al., in press).

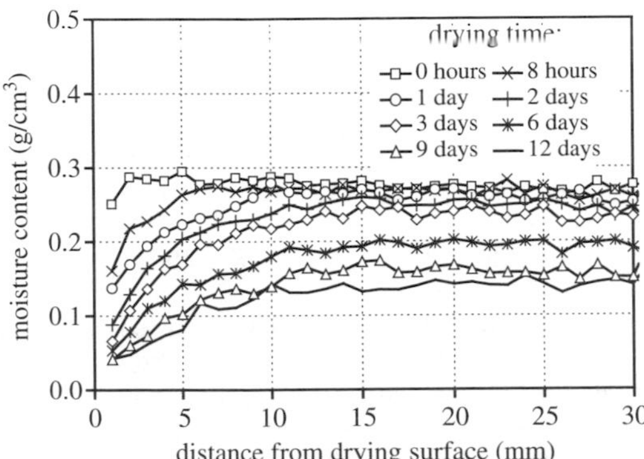

Figure 4. NMR drying of cement paste sample with solid particles (Bisschop et al., in press).

(cored from a standard prism), was applied from one side (the top of the sample) while all other sides were sealed (Figs 3, 4).

During drying experiments with mortar (Valckenborg et al. 2001), it was noticed that only capillary pores were emptied in 20 hours time. The duration of the experiment was limited to 3 days. The conclusion from the results was that water could not be taken out from the voids. In the experiments with cement paste (Bisschop et al., in press), drying lasted 12 days. No report was presented on the drying of pores and pore-scales in that experiment.

The NMR results from the experiments in Figures 3 and 4 were chosen for the comparison with the numerical modeling. In order to make comparisons with the NMR results, it would be necessary either to include parameters from the macroscopic flow (velocity, viscosity, Reynolds number and so on), or to observe the trend in moisture flow and drying behaviour. Since extracting of the parameters from the mentioned NMR experiments, which might help in the numerical analysis, is rather complex and doubtful, the trend of the two moisture flows is compared.

2 NUMERICAL APPROACH TO MOISTURE FLOW: LATTICE GAS AUTOMATON

2.1 *General notation for steady state*

An innovative, numerical way of approaching and describing the problem of fluid flow in complex porous media, dates from 1980s and 90s. The meth ods are called Lattice Gas Automata (LGA), FHP 2-D models (Frisch 1986, Frisch 1987, d'Humieres 1987). Derived from Statistical Physics, Lattice Gas Automata follow the definitions and basic rules of Cellular Automata. The population of fluid particles in triangular lattice propagates and collides according to certain rules, while conserving mass and linear momentum.

Lattice Gas Automaton (FHP model) is often described as a numerical solution of the Navier-Stokes equation. The similarity between LGA and Navier-Stokes equation, for the incompressible fluid flow, is obtained by the introduction of the Chapman-Enskog expansion (Wolfram 1986). This expansion gives the macroscopic behaviour of a fluid by averaging the microscopic (discretized) forms, for mass (density) and momentum (Eqs 1, 2) over the LGA area. The density and linear momentum equation (Eqs 1, 2) are as follows:

$$\rho(x) = fm \sum_i N_i(x) \tag{1}$$

$$\rho u(x) = fm \sum_i N_i(x) c_i \tag{2}$$

where $\rho(x)$ is a density per node; $\rho(x) = \rho$ for homogeneous case; u is mean velocity; ρu is momentum; f is the number of nodes per unit area ($f = 2/(\sqrt{3}/l^2)$, $l = 1$); m is a unit mass; c_i is the velocity of a single particle in any of 6 directions i ($c = l/\tau$) along the link length l in time τ expressed as $c_i = (cos(\pi/3)i, sin(\pi/3)i)$, $i = 0, \ldots, 5$. $N_i(x)$ is the average particle population of the cell expressed as the Fermi-Dirac distribution in general, as follows:

$$N_i(x) = \sum_i \frac{1}{1 + e^{(h+qc_i)}} \tag{3}$$

where h and q are LaGrange multipliers, nonlinear functions of ρ and u. N_i is the probability of particle (i) arriving at node x with velocity c_i. In the case of unit mass (m) and isotropic velocity distribution, averaging over the lattice area (f), the probability N_i will be $N_i = d$ as in Equation 4

$$N_i(x) = \frac{\rho}{7fm} = d \tag{4}$$

where ρ is a density (for the homogeneous case, $\rho = nd$; n is a number of particles per node); while m and f are mass and area, as already mentioned in Equations 1 and 2. The probability of a particle leaving the node x will be $N_i'(x)$. Propagation of fluid particles is defined as conservation of the mean population in the equality of probabilities N_i and N_i' as follows:

$$N_i(x + \tau c_i, t + \tau) = N_i'(x, t) \tag{5}$$

where τ is the time step (usually chosen as 1); in the steady state, the term t vanishes. The macroscopic equations, obtained through the averaging of the mentioned equations, have close similarity to Navier-Stokes equation for incompressible fluid ($\rho =$ *constant*), in Equation 6 and supplemental continuity equation, Equation 7, as follows:

$$\delta_t \mathbf{u} + (\mathbf{u}\nabla)\mathbf{u} = -\nabla P + \nu \nabla^2 \mathbf{u} \tag{6}$$

$$\nabla \mathbf{u} = 0 \tag{7}$$

where $\mathbf{u}$ is velocity vector; P is pressure; and ν is kinematic viscosity. More details about Chapman-Enskog expansions, as well as the lack of the Galilean invariance in LGA, in the simulation of Navier-Stokes equations, can be found in other references (Wolfram 1986).

2.2 *Lattice gas model (FHP)*

The so-called "center hexagonal FHP model", with 7 particles per cell, is used for our numerical experiment. Maximum one particle is at rest, with zero velocity, while the other six fluid particles can propagate in six directions, displaced by their velocity directions ($c_0 - c_5$). All fluid particles have unit speed, mass and equal initial density ($d_0 = d$), for all particles in each cell, regardless of any of 6 directions.

In order to simulate the porous medium, LGA can be populated, beside fluid particles, also with solid particles in different sizes and percentages. Solid particles are located at fixed positions on the vertices. Applying the conservation law of mass and momentum (Eqs 1, 2), followed by particle propagation and collision, the FHP model produces the result of a macroscopic flow of the real fluid. In the first step, fluid particles propagate, in order to collide. After the collision step, particles continue to propagate further, until the next collision. Collisions are specified as prescribed rules such as: collisions among fluid particles, collisions of fluid with the solid particle, and collisions of fluid particles with the boundaries.

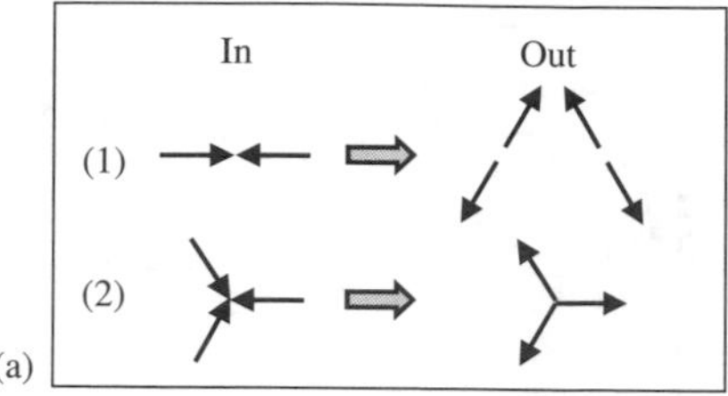

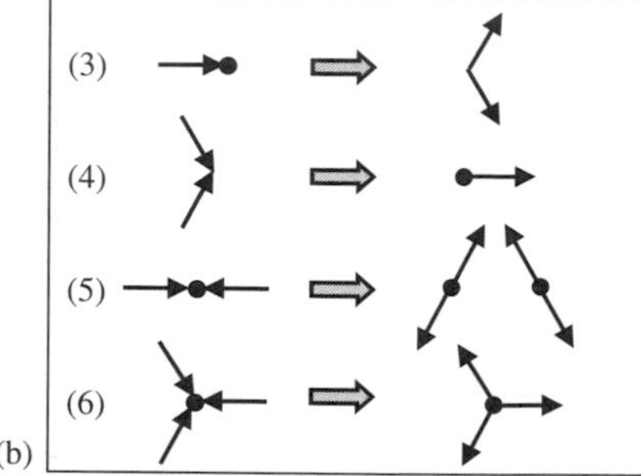

Figure 5. (a) Collisions without any particles at rest (FHP1 rule). Configuration of particles before and after the collision. (b) Double and triple collision with the particle at rest.

A collision of fluid particles, with any solid is used as either specular- or bounce-back reflection (Jankovic et al. 2001a, Lavallee 1991). The fluid particle collisions are given in the so-called look-up tables, where all possibilities of the particles configuration, at the nodes ($2^7 = 128$), and possible collisions, are presented. Some collision rules in the look-up tables are deterministic, while others have a probability of p ($p = 1/2$), depending on the possible rotations (see for example rule (1) and (5) in Figure 5). In the case of a probabilistic rule, the 8-bit is introduced as a switch, between two probabilities.

In our model, the LGA is used to simulate moisture flow during drying of a 2-D homogenous and heterogeneous sample. Drying can be treated simply as mass transfer. The flow occurs due to the difference in the density concentrations, which leads to a paradox called "uphill diffusion" (Pihlajavaara 1965), from lower to higher concentration. In the model, the difference, in the density concentration, is created by "inputting" lower density particles in the higher density environment (Jankovic et al. 2001a).

2.3 *Collision rules (look-up table)*

Standard collision rules include collisions in the 6-particle model, such as FHP1 rule (collisions of 2 and 3 particles) and 7-particle model (all possible 76 collisions from 2 to 5 fluid particles, including collisions with a particle at rest), d'Humieres, 1987.

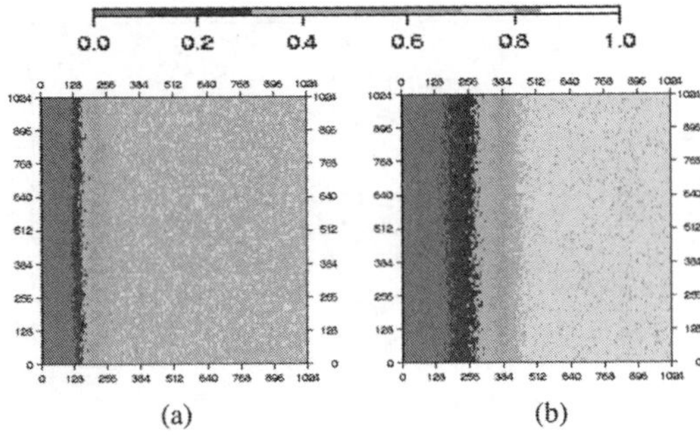

Figure 6. Moisture distribution in the homogeneous sample, (a) after 1000, and (b) after 4500 LGA steps.

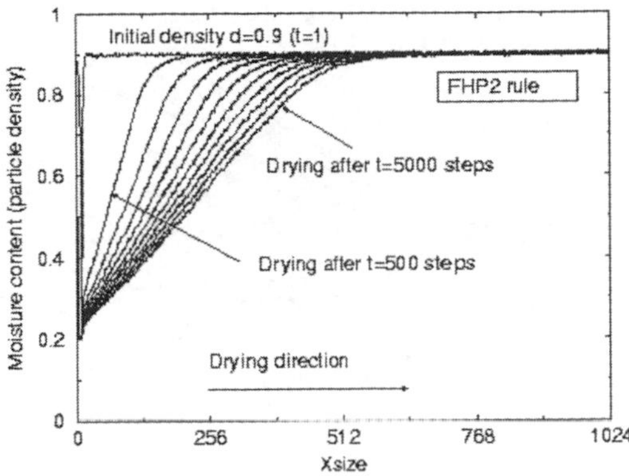

Figure 7. LGA simulation of drying of the homogeneous 1024×1024 sample, with FHP2 collision rule. Periodic boundary conditions,with wall on the left-hand side. Drying in maximum of 5000 steps.

In this research, the so-called FHP2 collision rule, with higher viscosity, is applied in order to keep a low Reynolds number to model porous media. Superposition of the rules in Figures (5a) and (5b) results in the rule FHP2. Usually, all FHP collision rules are defined in the "look-up" tables, listed as the input and output set of $n = 6$ velocities (expressed in Boolean notation as either 0 or 1), per each of 128 lines. The FHP2 rule includes 22 collisions: double, triple and collisions with the particle at rest, which amounts to 17.2% collisions (out of 128 velocity configurations).

2.4 *Boundary conditions*

Boundary conditions (BC) are essential for the LGA, but also significant for the modeling of drying. The boundary is defined in the vertical and horizontal direction, as periodic or as a wall (barrier) condition. In the presented examples, the wall is always located on the left-hand side. In the vertical direction, both periodic and top/bottom wall conditions were applied and compared.

Collision of fluid particles with solid particles is also treated as a boundary problem, where fluid particles, after being bounced from solid, reflect either as bounce-back ($r = 0$) or they have specular-response ($r = 1$). Both cases have been studied.

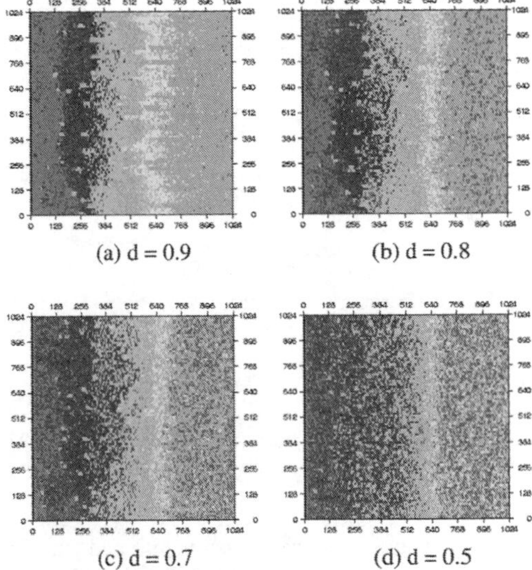

Figure 8. Moisture distribution in the heterogeneous sample, with different initial density d, after 1000 LGA steps.

3 LGA ILLUSTRATIONS

The size of the LGA mesh is 1024×1024 nodes in X and Y direction. In the triangular mesh, a non-orthonormal way of coordinate labeling is used, such that the x coordinates, in the odd rows, are shifted for $l/2$, in order to form a mesh under angle of 60°. Although the size of the NMR sample was $40 \times 20\,\text{mm}^2$ (ratio = 2:1), the LGA mesh is chosen to have equal number of nodes in horizontal and vertical direction. This provides an easier running of the averaging analysis, in the presence of the higher percentage of solid particles.

Two simulations of drying will be presented: drying of homogenous (Figs 6) and heterogeneous sample (Figs 8). The percentage of solid particles for the homogenous sample is 3%, occupying the size of only one LGA node. For the heterogeneous sample, the size of the solid particles is increased to 20×20 LGA nodes. The total percentage of solid population is 0.01%. In both cases, the solid particles are placed at random.

Different analyses have been performed, in order to find the best fit for the boundary conditions, that would give the closest numerical approximation to the experiment. In all presented cases in (Figs 6–10),

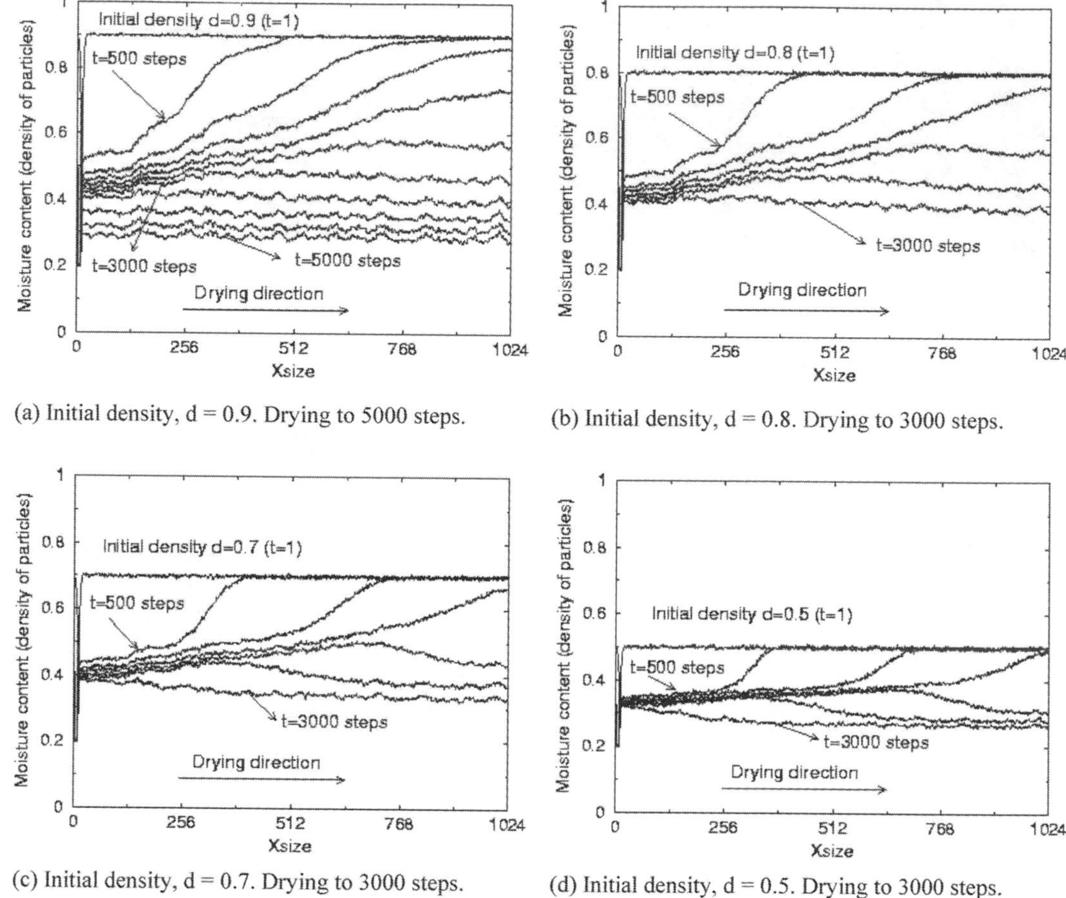

(a) Initial density, d = 0.9. Drying to 5000 steps.

(b) Initial density, d = 0.8. Drying to 3000 steps.

(c) Initial density, d = 0.7. Drying to 3000 steps.

(d) Initial density, d = 0.5. Drying to 3000 steps.

Figure 9. Moisture LGA profiles for different initial densities. Applied FHP2 collision rule. Periodic BC in vertical direction. Solid wall on the left-hand side. $r = 0$ (on the wall) and $r = 1$ (collision with solid particles).

boundary conditions were as mentioned in 2.4. Regarding the collision with the solid wall, the bounce-back boundary rule is applied in all cases. Two possibilities of fluid-solid collision (bounce-back or specular-reflection) were applied and compared. The LGA works on the principle of averaging over area and time in steps. In that sense, this is a truly statistical approach. The time averaging is made at every 500 steps (Figs 6–9) and 250 steps (Fig. 10). Regarding the averaging over the LGA area, we make an average in the vertical direction. By increasing the size of the sample, the noise in the results reduces (see Jankovic et al. 2001a).

3.1 *Drying of homogeneous LGA sample*

In Figure 6, the homogenous sample with 3% of small, solid particles, illustrates drying from the left-hand side. The initial density of particles is $d = 0.9$. Drying curves (Fig. 7) indicate regular smooth drying. The smoothness of the curves is due to using a larger LGA size. Due to the application of the FHP2 collision rule, drying goes very slowly. The analysis ran only for 5000 steps, since the tendency was to examine the shape of the drying curves. We want to compare the shape and trend in this "drying" analysis (Fig. 7) to the NMR drying experiment (Fig. 3).

3.2 *Drying of heterogeneous LGA sample*

In Figure 4, the NMR result of drying of a heterogeneous sample is given. Due to the presence of solid particles (35%), the content of cement paste was reduced to 65%. This means that, there will be less cement paste to dry, which will result in quicker drying. This logic was followed in the numerical analysis.

Drying in the Lattice Gas Automaton, with 0.01% of solid particles, is given in Figures 8. In order to

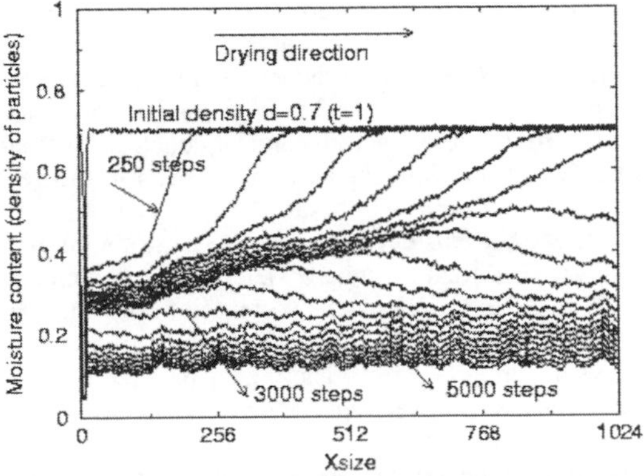

Figure 10. Initial density, $d=0.7$. Averaging over 250 steps. Maximum drying to 5000 steps. Applied FHP2 rule. Periodic BC in vertical direction, wall on the left-hand side. $r=0$ on the wall and $r=1$ on the solid particles.

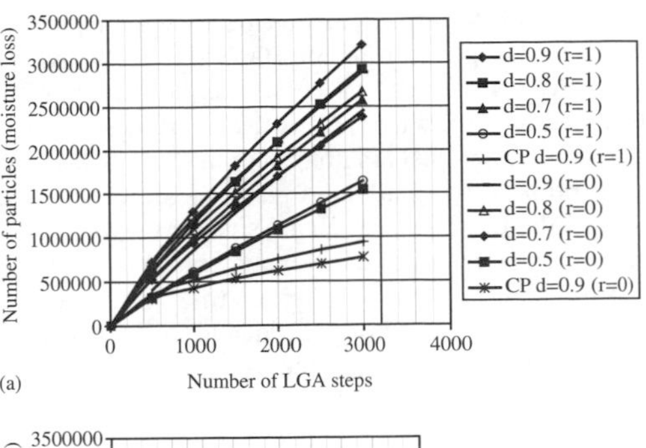

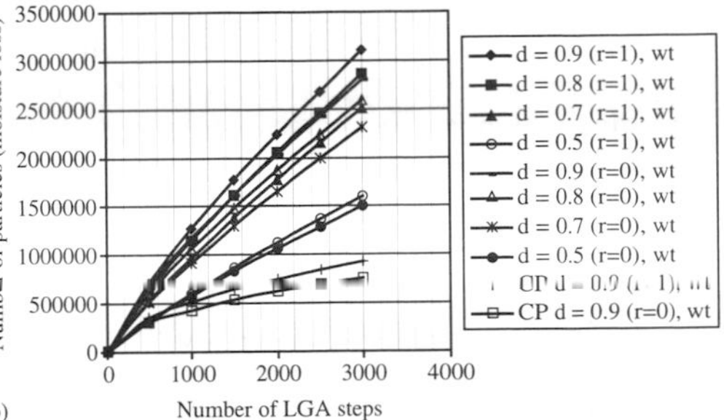

Figure 11. Moisture loss, for the samples with bigger solid particles and different initial densities ($d=0.9$ to $d=0.5$) with various types of boundary conditions ($r=1$, specular-reflection and $r=0$, bounce-back reflection) between solid and fluid particles. CP = cement paste with 3% of small (insignificant) solid particles, treated as a homogeneous sample. (a) with periodic BC in vertical direction, and (b) with the wall on top and bottom.

simulate quicker drying and less percentage of present moisture in the sample, the initial density was reduced from 0.9 to 0.5.

The difference in the results can be observed in the diagrams (Fig. 9), where the LGA results are averaged every 500 steps. The sample with $d=0.9$ was dried to 5000 steps (Fig. 9a) and to 3000 steps with the density from $d=0.8$ to 0.5 (Figs 9b, d).

The following can be observed: if the difference between the initial density (0.9, 0.8, 0.7 and 0.5) and the ultimate density (0.2) was smaller, than the drying was quicker, and smaller number of steps is necessary to dry the sample. In the case of $d=0.9$ (Fig. 9a), 5000 steps are necessary to get to $d=0.3$, while in the Figure 9d, only 3000 steps will give approximately the same drying condition. Some spurious behavior, in the drying curves, can be observed in the case of smaller initial densities (Figs 9c, d). The reason might be the lowering of the initial density, compared to the ultimate density, which was fixed to 0.2. On the other hand, drying to the lower fixed values ($d=0.05$ instead of $d=0.2$) in the smaller steps, would cause quicker drying (Fig. 10).

4 MOISTURE LOSS

Moisture content (loss) and moisture distribution, as presented in the drying curves in the LGA sample, depend on:

- different boundary conditions (collisions) of fluid particles with the solid wall: specular- or bounce-back reflection, or periodic boundary conditions
- boundary conditions with the solid obstacles
- initially prescribed density of the fluid particles
- content and size of solid particles (a simple representation of the cement/concrete microstructure)
- prescribed collision rules.

Some of the mentioned parameters have been compared in Figure 11. The same boundary condition (bounce-back, $r=0$) is kept on the solid wall, while the conditions with the solid obstacles run with either $r=0$ or $r=1$, specular-reflection.

The drying analysis shows that the number of lost particles decreases, as the speed of drying decreases. In the case of "pure cement paste" (*CP*), for the initial density ($d=0.9$), drying goes rather slowly and the moisture loss is also slow. Wall barrier on the top and bottom slowed the process of drying but not significantly (Fig. 11b).

In the experiments, the speed of drying would depend, among other parameters, on the w/c-ratio. In the LGA analysis, the w/c-ratio can be taken into account by applying different collision rules, and subsequently different number of collisions. Lower w/c-ratio means slower drying (see for example FHP2 rule, in Fig. 7), while higher w/c means quicker drying (application of FHP5 rule, in Jankovic et al. 2001a).

On the other hand, the moisture loss increases, when more solid particles are added (cases of different density d). This implies that as more/bigger obstacles are added, larger porous interface zones are created around the obstacles, which speeds up the moisture flow and drying, when and if the interfaces are connected. Cement paste is not as compact as in the case when no or very small solid particles are added. The use of solid particles, which are smaller in size, would create slower drying and lower percentage of cracking.

Implementing the collision rule of specular-reflection ($r = 1$), among fluid and solid particles, moisture loss slightly increases, compared to the case of $r = 0$ (bounce-back reflection), see Figure 11. The difference in moisture loss is due to the different surfaces of the obstacles. If obstacles are rough, than a bounce-back rule ($r = 0$) can be applied. If they are perfectly smooth, the specular-reflection rule ($r = 1$) is more applicable, implying that smooth surfaces do have much lower resistance than rough surfaces.

5 DISCUSSION AND CONCLUSIONS

In this research, we have showed that the use of a 2-D, isotropic FHP hexagonal Lattice Gas Automaton, with maximum of 6+1 fluid particle at rest (per node), can produce good comparisons with NMR experiments on drying of the cement paste samples. There are still many open questions to which we have to address.

Some remarks should be made regarding the type of LGA (6 and 7-particle LGA model). The maximum number of particles per node (6 or 7) influences the Reynolds number R_e (Rivet et al. 1986) which is significant in modeling of porous media. In order to model porous media correctly, R_e needs to be very low. In that case, the viscosity term should be higher, by introducing a lower number of collisions. As a result of this discussion, FHP2 rule, with a low number of collisions is applied.

In general, collision rules influence drying behaviour in the model. New collision rules may improve the moisture gradient, during the modeling of drying.

At this moment, no temporal or spatial scale is derived from the analyses. Regarding the temporal and spatial scale, it is important to say that the LGA model has no intrinsic temporal or spatial scale. By implying the law of similarity between the real and fluid in the model, where R_e number must be the same, it would be possible to find out about the temporal and spatial scale. It is important to keep in mind that, due to the lack of Galilean invariance in the microscopic derivations of Navier-Stokes equation, re-scaling had been applied (Frisch et al. 1987) such that new re-scaled R_e number is as follows:

$$R_e = \frac{u_0 l_0 g(\rho_0)}{v(\rho_0)} \tag{8}$$

The Reynolds number, re-scaled in this way, plays the same role as R_e number in real fluids. Re-scaling of the R_e number and subsequently of the space, time, velocity and pressure, makes smooth transformation of incompressible macrodynamical equation for Lattice Gas, into classical Navier-Stokes, for the real fluid.

Due to the existence of different sizes of solid particles (in different percentages), moisture gradients as well as drying mechanisms (compare Figs 7 and 9a) will be influenced. The size of the solid particles also plays an important role. The bigger the size of solid particles the quicker the drying, and the lower the moisture gradient. Due to the smaller size of particles and smaller circumferencial, interfacial transition zone (compared to bigger size of particles in Jankovic et al. 2001b), the moisture gradient around the particles is much less steep.

The number of averaged LGA steps has an effect on moisture gradient. If the number of averaged steps is 250 (Fig. 10), instead of 500 (Fig. 9c), increase in moisture gradient in the early phase is noticed. Taking into account the experiments on drying shrinkage cracking (Bisschop et al. 2001), this can be explained by the fact that drying in the early phase (surface drying) is more critical for the possible crack occurrence, than later stages of drying.

Use of periodic or wall boundary conditions, has an effect on the drying simulation, while the use of bounce-back or specular-reflection, has an effect on the speed of drying as well as on the moisture gradient (Fig. 11).

The distinction between capillaries and voids may be important in the numerical simulations, regarding the boundary conditions. Maybe the combination of both bounce-back and specular-reflection can be the right solution for this type of analysis, although it is still not clear how these matters are related.

Time seems to be the key factor in the experiments as well as in modeling. Some progress should be expected from the experiments in the ESEM. Deformations, due to drying at the small scale, can properly model standard drying experiments at the larger scale. Our experiments on drying are always limited in time. Basically, drying in cementitious materials is a never finished process, but the initial phase, where rapid crack propagation may occur (within hours) is considered very important.

ACKNOWLEDGEMENT

The present study is supported by the Dutch Technology Foundation (STW) and the Priority Program Materials Research (PPM) which is gratefully acknowledged.

REFERENCES

Bažant, Z.P. & Raftshol, W.J. 1982. Effect of Cracking in Drying and Shrinkage Specimens. *Cement and Concrete Research* 12: 209–226.

Bisschop, J., Pel, L. & Van Mier, J.G.M. 2001. Effect of Aggregate Size and Paste Volume on Drying Shrinkage Microcracking in Cement-based Composites. In F.-J. Ulm, Z.P. Bažant, F.H. Wittmann (eds), *Creep, Shrinkage and Durability Mechanics of Concrete and other Quasi-Brittle Materials; Proc. of the 6th intern. Conf. ConCreep-6, August 2001, Cambridge, USA*: 75–80. Amsterdam: Elsevier.

Bisschop, J., Pel, L. & Van Mier, J.G.M. In press. Mechanisms of Drying Shrinkage Microcracking in Concrete. (The extended paper of contribution to ConCreep-6, Boston, USA), to appear in *Cement and Concrete Research* (2003).

d'Humieres, D. & Lallemand, P. 1987. Numerical Simulation of Hydrodynamics with Lattice Gas Automata in Two Dimensions. *Complex Systems* 1: 599–632.

Fatt, I. & Saraf, D.N. 1967. *Soc. Pet. Eng. AIME 7, 235.*

Frisch, U., Hasslacher, B. & Pomeau, Y. 1986. Lattice-Gas Automata for the Navier-Stokes Equation. *Physical Review Letters* 14(56): 1505–1508.

Frisch, U., d'Humieres, D., Hasslacher, B., Lallemand, P., Pomeau, Y. & Rivet, J.-P. 1987. Lattice Gas Hydrodynamics in Two and Three Dimensions. *Complex Systems* 1: 648–707.

Jankovic, D., Küntz, M. & Van Mier, J.G.M. 2001a. Numerical Analysis of Moisture Flow and Concrete Cracking by means of Lattice Type Models. In R. de Borst, J. Mazars, G. Pijaudier-Cabot & J.G.M. Van Mier (eds), *Fourth International Conference on Fracture Mechanics of Concrete and Concrete Structures; Proc. intern. conf., Cachan, France, May–June, 2001*. 1: 231–238. Rotterdam: A.A. Balkema Publishers.

Jankovic, D. & Van Mier, J.G.M. 2001b. Crack Development in Concrete due to Moisture Flow. *HERON* 46(3): 169–180.

Jankovic, D. & Van Mier, J.G.M. 2002. Preliminary Investigation of Drying Shrinkage Cement Paste Specimens. In R. Pyrz, J. Schjodt-Thomsen, J.C. Rauche, T. Thomsen and L.R. Jensen (eds), *International Conference on New Challenges in Mesomechanics; Proc. intern. conf., Aalborg University, Denmark, August 2002*. 1: 265–271.

Lavallee, P., Boon, J.P. & Noullez, A. 1991. Boundaries in Lattice-Gas Flows. *Physica D* 47: 233–240.

Neubauer, C.M. 1997. *On the Chemistry, Microstructure, and Deformation Properties of Cement Pastes: Towards a New Strategy for Controlling Drying Shrinkage*. Ph.D. Thesis. Northwestern University, Chicago, IL.

Pihlajavaara, S.E. 1965. *On the Main Features and Methods of Investigation of Drying and Related Phenomena in Concrete*. Ph.D. Thesis. University of Helsinki, Finland, Julkaisu 100 Publication, Helsinki.

Rivet, J.P. & Frisch, U. 1986. Lattice Gas Automaton in the Boltzmann Approximation. *Compt. Rend. Acad. Sci. Paris*: 302(2): 267–272.

Scheidegger, A.E. 1960. *The Physics of Flow through Porous Media*. 2nd Ed. University of Toronto Press. Toronto.

Shiotani, T., Bisschop, J. & Van Mier, J.G.M. In press. Temporal and Spatial Development of Drying Shrinkage in Cement-based Materials. *Engineering Fracture Mechanics*.

Valckenborg, R.M.E., Pel, L., Hazrati, K., Kopinga, K. & Marchand, J. 2001. Pore Water Distribution in Mortar during Drying as determined by NMR. *Materials and Structures. Dec. 2001*, 34: 599–604.

Wittmann, F.H. 1973. Interaction of Hardened Cement Paste and Water. *Journal of the American Ceramic Society* 56(8).

Wittmann, F.H. 1982. Creep and Shrinkage Mechanisms. In Z.P. Baant and F.H.Wittmann (eds), *Creep and Shrinkage in Concrete Structures*: 129–161. New York: Wiley.

Wolfram, S. 1986. Cellular Automaton Fluids 1: Basic Theory. *Journal of Statistical Physics* 45: 471–526.

© 2003 Swets & Zeitlinger, Lisse, ISBN 90 5809 536 3

Thermodynamics of open systems with application to chemomechanical problems

Ellen Kuhl & Paul Steinmann
Chair for Applied Mechanics, University of Kaiserslautern, Kaiserslautern, Germany

ABSTRACT: A theoretical framework for chemomechanical problems within the theory of open system thermodynamics is presented. The governing equations, the balance of mass and momentum of the reactant, are discretized with the help of the finite element method. While the classical spatial motion approach is applied to analyze the evolution of the density and the movement of physical particles relative to the ambient space, the material motion problem allows for the evaluation of the movement of physical particles relative to the ambient material. The basic features of the spatial and the material motion framework in the context of chemomechanical coupling will be discussed with the help of selected computational examples.

1 INTRODUCTION

The basic concern of the present work is the derivation of a general theoretical and computational framework for the thermodynamics of open systems typically encountered in chemomechanical applications. Thereby, the chemical reactant is considered as an open system which is allowed to constantly gain or lose mass. In the case of chemically reacting concrete, dissolution of portlandite ($Ca(OH)_2$) or ettringite and progressive decalcification of the calcium silicate hydrates (CSH) represent typical examples of changes in mass of the cementitious skeleton, see e.g. (Coussy 1995), (Ulm and Coussy 1995), (Kuhl, Bangert, and Meschke 2000). Consequently, the classical conservation law of mass has to be recast into a balance equation balancing the rate of change of the mass of the reactant with an in- or rather outflux of matter, e.g. through moisture transport in the context of calcium leaching (Ulm, Torrenti, and Adenot 1999), (Torrenti, Mainguy, Adenot, and Tognazzi 1998), (Carmeliet 1998). In the most general case of open system thermodynamics, additional volume source or sink terms might be accounted for as well. The enhanced mass balance typically has a direct impact on the balance of momentum, kinetic energy, energy and entropy, see e.g. (Epstein and Maugin 2000), (Kuhl and Steinmann 2002a). Considering these higher order balance equations in open system thermodynamics, a strict distinction can be made between their volume specific and their mass specific format which is of no particular relevance in classical thermodynamics of closed systems.

The theory of open systems is closely related to the theory of mixtures (Truesdell and Toupin 1960), (Bowen 1976), particularly to the theory of porous media (Coussy 1995), (Nield and Bejan 1999), (de Boer 2000). If chemically-induced changes in mass are allowed to take place, we typically deal with the theory of reactive porous media, see e.g. (Coussy 1995), (Ulm, Coussy, and Hellmich 1998), (Kuhl and Meschke 2002). Thereby, every individual component of the porous medium can be interpreted locally as an open system undergoing a thermomechanical exchange with the outside world represented by the remaining constituents of the mixture. The loss or gain of mass of one constituent is thus compensated by the others, while the overall mass of the mixture itself remains constant. The theory of mixtures is based on the fundamental assumption that the individual constituents are superposed locally at each point in space. However, if the individual components are spatially separated and the overall behavior is primarily determined by one constituent alone, it proves reasonable to confine attention to only a part of the overall matter present by making use of the theory of reactive open systems.

Changes in the density field furnish a typical example of the local rearrangement of material inhomogeneities which can be characterized most elegantly in the material setting, see e.g. (Maugin 1993), (Gurtin 2000), (Steinmann 2002a), (Steinmann 2002b) or for the particular case of open system thermodynamics (Epstein and Maugin 2000), (Kuhl and Steinmann 2002c). From a computational point of view, the material motion problem naturally lends itself to the material force method, which can be applied to visualize the tendency of inhomogeneities in the density field to move relative to the ambient material (Kuhl and Steinmann 2002b).

Within the present contribution, the spatial motion problem of reactive open system thermodynamics will be discretized within a two-field finite element formulation with the reactant's density and the deformation as primary unknowns. Once the discrete solution of the spatial motion problem has been determined, the material motion problem can be evaluated in a straightforward way. The numerical results of both, the spatial and the material motion problem, will be discussed for a one-dimensional academic model problem as well as for a single edge notched specimen under tensile loading.

This contribution is organized as follows. After introducing the basic kinematics, the balance equations and the constitutive equations as the key ingredients of open system thermodynamics in chapters 2, 3 and 4, we will derive the corresponding weak form in chapter 5 followed by its temporal and spatial discretization and the corresponding linearization in chapters 6, 7 and 8. Chapter 9 briefly summarizes the ideas of the material force method, that will be further elaborated in the examples given in chapter 10.

2 KINEMATICS

To set the stage, we briefly summarize the underlying kinematics of geometrically nonlinear continuum mechanics. Let $\mathcal{B}_0$ and $\mathcal{B}_t$ denote the reference and current configuration occupied by the body of interest at time t_0 and $t \in \mathbb{R}$, respectively. The kinematic description is basically characterized through the deformation map φ mapping the material placement $\boldsymbol{X}$ of a physical particle in the material configuration $\mathcal{B}_0$ to its spatial placement x in the spatial configuration $\mathcal{B}_t$.

$$\boldsymbol{x} = \boldsymbol{\varphi}(\boldsymbol{X}, t): \quad \mathcal{B}_0 \times \mathbb{R} \rightarrow \mathcal{B}_t \tag{1}$$

The corresponding deformation gradient $\boldsymbol{F}$ characterizes the linear tangent map from the material tangent space $T\mathcal{B}_0$ to the spatial tangent space $T\mathcal{B}_t$.

$$\boldsymbol{F} = \nabla_X \boldsymbol{\varphi}(\boldsymbol{X}, t): \quad T\mathcal{B}_0 \rightarrow T\mathcal{B}_t \tag{2}$$

Its determinant defines the related Jacobian J as $J = \det \boldsymbol{F} > 0$. Moreover, we introduce the left Cauchy–Green tensor $\boldsymbol{b} = F \cdot F^t$ as a characteristic spatial strain measure. In what follows, $\mathrm{D}_t = \partial_t\{\bullet\}|_X$ will denote the material time derivative of a quantity $\{\bullet\}$ at fixed material placement $\boldsymbol{X}$. Accordingly, the spatial velocity $\boldsymbol{v} = \mathrm{D}_t\boldsymbol{\varphi}(\boldsymbol{X}, t)$ can be understood as the material time derivative of the deformation map $\boldsymbol{\varphi}$. For the classical spatial motion problem, we shall apply a formulation which is entirely related to the material frame of reference $\mathcal{B}_0$. Thus, $\nabla_X\{\bullet\}$ and $\mathrm{Div}\,\{\bullet\}$ denote the gradient and the divergence of any field $\{\bullet\}$ with respect to the material placement $\boldsymbol{X}$. All the relevant kinematic quantities are illustrated in figure 1.

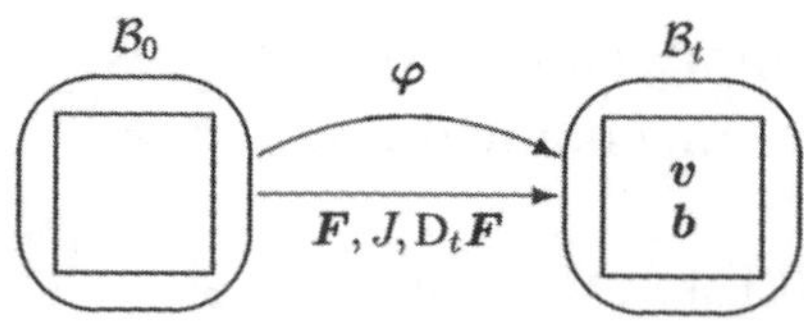

Figure 1. Spatial motion problem: Kinematics.

3 BALANCE EQUATIONS

The present chapter briefly summarizes the balance equations for open system thermodynamics which constitute the basis for the finite element analysis to be derived later on. Unlike classical closed systems typically found in traditional mechanical applications, open systems exhibit a permanent exchange of mass, momentum, energy and entropy with their environment. As a natural consequence, the classical balance of mass has to be enhanced by an additional mass flux and a mass source. Nevertheless, the impact of the open system on the higher balance equations is twofold. On the one hand, an explicit exchange with the outside world can take place which manifests itself in the corresponding Neumann boundary conditions and the non-classical volume source terms. On the other hand, the newly added or in-flowing mass typically carries a certain amount of momentum, energy and entropy which has to be accounted for as well. The latter, however, is only visible in the volume specific format of the balance equations. Herein, we shall strictly stick to the mass specific format which is free from all the explicit mass dependent influences. Apart from the non-standard Neumann boundary conditions and the non-standard source term, it takes the familiar form known from classical closed system thermodynamics. The balance of mass

$$\mathrm{D}_t \rho_0 = \mathrm{Div}\,\boldsymbol{R} + \mathcal{R}_0 \tag{3}$$

furnishes the fundamental relation in open system thermodynamics. It balances the rate of change of the reference density ρ_0 with a possible influx of mass $\boldsymbol{R}$ and a mass source $\mathcal{R}_0$. The mass specific balance of momentum

$$\rho_0\, \mathrm{D}_t\, \boldsymbol{p} = \mathrm{Div}\, \bar{\boldsymbol{\Pi}}^t + \bar{\boldsymbol{b}}_0 \tag{4}$$

balances the rate of change of the mass specific momentum density $\boldsymbol{p} = v$ with the reduced momentum flux $\bar{\boldsymbol{\Pi}}^t$ and the reduced momentum source $\bar{\boldsymbol{b}}_0$. Moreover, we can introduce the mass specific balance of internal energy which can be derived as the difference of the total energy balance and the balance of kinetic energy, a weighted version of the above stated balance of momentum. As a consequence, the rate of change of the mass specific internal energy I is required to be in equilibrium with the reduced non-mechanical energy flux $\bar{\boldsymbol{Q}}$, the reduced non-mechanical energy source $\bar{\mathcal{Q}}_0$ and the internal power $\mathsf{p}_0^{int} = \bar{\boldsymbol{\Pi}}^t : \mathrm{D}_t\boldsymbol{F}$.

$$\rho_0\,\mathrm{D}_t I = -\mathrm{Div}\,\bar{\boldsymbol{Q}} + \bar{\mathcal{Q}}_0 + \mathsf{p}_0^{int} \tag{5}$$

Finally, the balance of entropy can be introduced as the rate of change of the mass specific entropy S being in equilibrium with the reduced entropy flux $\mathrm{Div}\,\bar{\boldsymbol{H}}$, the reduced entropy source $\bar{\mathcal{H}}_0$ and the internal entropy production h_0^{int} which is required to be non-negative throughout the entire thermodynamical process as $\mathsf{h}_0^{int} \ge 0$.

$$\rho_0\,\mathrm{D}_t S = -\mathrm{Div}\,\bar{\boldsymbol{H}} + \bar{\mathcal{H}}_0 + \mathsf{h}_0^{int} \tag{6}$$

By making use of the following relations between the reduced entropy flux $\bar{\boldsymbol{H}}$ and source $\bar{\mathcal{H}}_0$ and the reduced non-mechanical energy flux $\bar{\boldsymbol{Q}}$ and source $\bar{\mathcal{Q}}_0$ as $\bar{\boldsymbol{H}} = \bar{\boldsymbol{Q}}/\theta + \boldsymbol{S}$ and $\bar{\mathcal{H}}_0 = \bar{\mathcal{Q}}_0/\theta + \mathcal{S}_0$ we can derive the classical spatial motion version of the Clausius–Duhem inequality.

$$\begin{aligned}\bar{\mathsf{d}}_0 \;=\; & \bar{\boldsymbol{\Pi}}^t : \mathrm{D}_t\boldsymbol{F} - \rho_0\,\mathrm{D}_t\Psi - \rho_0\,S\,\mathrm{D}_t\theta \\ & + [\mathrm{Div}\,\boldsymbol{S} - \mathcal{S}_0]\,\theta - \bar{\boldsymbol{Q}}\cdot\nabla_X \ln\theta \ge 0\end{aligned} \tag{7}$$

Herein, we have introduced the free energy Ψ through the classical Legendre–Fenchel transform as $\Psi = I - \theta S$. Moreover, we have included the non-standard extra terms $\boldsymbol{S}$ and $\mathcal{S}_0$ taking into account the explicit exchange of entropy with the outside world.

Remark 3.1
Within the framework of chemomechanics, in particular in the context of leaching, the balance of mass (3) typically expresses the evolution of the moisture content in terms of the moisture flux.

4 CONSTITUTIVE EQUATIONS

Finally, the set of governing equations has to be closed by introducing appropriate constitutive assumptions for the mass flux $\boldsymbol{R}$, the mass source $\mathcal{R}_0$, the reduced momentum flux $\bar{\boldsymbol{\Pi}}^t$ and the reduced momentum source $\bar{\boldsymbol{b}}_0$. The particular choice of constitutive equations basically defines the coupling between the density and the deformation problem. Paralleling the moisture transport law for isotropic materials, the mass flux $\boldsymbol{R}$ is related to the gradient of the density $\nabla_X\rho_0$ weighted by a mass conduction coefficient R_0.

$$\boldsymbol{R} = \boldsymbol{R}(\rho_0, \boldsymbol{\varphi}) = R_0\,\nabla_X\rho_0 \tag{8}$$

In the context of chemomechanical leaching, the mass flux $\boldsymbol{R}$ is typically replaced by the moisture flux, the density gradient $\nabla_X\rho_0$ is replaced by the gradient of the capillary pressure and R_0 is interpreted as the moisture permeability including water vapour and liquid water transfer. A possible mass source $\mathcal{R}_0$ can be assumed to depend on the reference density ρ_0 as well as on the deformation $\boldsymbol{\varphi}$, e.g. through the following constitutive assumption,

$$\mathcal{R}_0 = \mathcal{R}_0(\rho_0, \boldsymbol{\varphi}) = \left[\frac{\rho_0}{\rho_0^*}\right]^{-m} \Psi_0 - \Psi_0^* \tag{9}$$

whereby Ψ_0 denotes the free energy $\Psi_0 = \rho_0\Psi$ as a function of the deformation $\boldsymbol{\varphi}$ while ρ_0^*, Ψ_0^* and m represent additional material parameters. In the context of an open-pored cellular matrix (Gibson and Ashby 1997), the free energy Ψ_0 is typically characterized through the elastic free energy, e.g. of Neo–Hooke type $\Psi_0^{\mathrm{nh}} = [\lambda\,\ln^2 J + \mu[\boldsymbol{b} : \boldsymbol{1} - 3 - 2\ln J]]/2$, weighted by the relative density $[\rho_0/\rho_0^*]^n$ such that $\Psi_0 = [\rho_0/\rho_0^*]^n\,\Psi_0^{\mathrm{nh}}$. Herein, λ and μ are the classical Lamé constants. Moreover, the exponent n typically varies between $1 \le n \le 3.5$ according to the actual porosity of the open-pored ground substance. This particular choice of the free energy defines the reduces momentum flux as $\bar{\boldsymbol{\Pi}}^t = \rho_0\mathrm{D}_F\Psi$.

$$\bar{\boldsymbol{\Pi}}^t = \bar{\boldsymbol{\Pi}}^t(\rho_0, \boldsymbol{\varphi}) = \left[\frac{\rho_0}{\rho_0^*}\right]^n \bar{\boldsymbol{\Pi}}^{\mathrm{nh}_t} \tag{10}$$

The reduced momentum flux $\bar{\boldsymbol{\Pi}}^t$ can thus be interpreted as the classical Neo Hookean first Piola Kirchhoff stress tensor $\bar{\boldsymbol{\Pi}}^{\mathrm{nh}_t} = [\mu\boldsymbol{F} + [\lambda\ln J - \mu]\boldsymbol{F}^{-t}]$ weighted by the actual relative density $[\rho_0/\rho_0^*]^n$. For the sake of simplicity, the reduced momentum source $\bar{\boldsymbol{b}}_0$ is assumed to vanish identically.

$$\bar{\boldsymbol{b}}_0 = \bar{\boldsymbol{b}}_0(\rho_0, \boldsymbol{\varphi}) = \boldsymbol{0} \tag{11}$$

5 WEAK FORM OF THE COUPLED PROBLEM

While for classical chemomechanical applications, e.g. the leaching of portlandite, the capillary pressure and the deformation furnish the primary unknowns, we will introduce the reference density ρ_0 and the spatial deformation $\boldsymbol{\varphi}$ as primary unknowns within the mechanics of open systems. They are governed by the scalar-valued balance of mass (3) and by the vector-valued mass-specific balance of momentum (4) which can be cast into the residual statements

$$\begin{aligned} \mathrm{r}^{\rho}\,(\rho_0,\boldsymbol{\varphi}) &= 0 && \text{in } \mathcal{B}_0 \\ \mathbf{r}^{\varphi}\,(\rho_0,\boldsymbol{\varphi}) &= \mathbf{0} && \text{in } \mathcal{B}_0 \end{aligned} \tag{12}$$

with the residuals r^{ρ} and $\mathbf{r}^{\varphi}$ defined in the following form.

$$\begin{aligned} \mathrm{r}^{\rho} &= \quad \mathrm{D}_t\,\rho_0 - \mathrm{Div}\,\boldsymbol{R} \;-\mathcal{R}_0 \\ \mathbf{r}^{\varphi} &= \rho_0\,\mathrm{D}_t\boldsymbol{p} \;-\mathrm{Div}\,\bar{\boldsymbol{\Pi}}^t - \bar{\boldsymbol{b}}_0 \end{aligned} \tag{13}$$

Herein, the boundary $\partial\mathcal{B}_0$ of the material domain can be decomposed into disjoint parts $\partial\mathcal{B}_0^{\rho}$ and $\partial\mathcal{B}_0^{r}$ for the density problem and equivalently into $\partial\mathcal{B}_0^{\varphi}$ and $\partial\mathcal{B}_0^{t}$ for the deformation problem. While Dirichlet boundary conditions are prescribed on $\partial\mathcal{B}_0^{\rho}$ and $\partial\mathcal{B}_0^{\varphi}$,

$$\begin{aligned} \rho_0 - \rho_0^{presc} &= 0 && \text{on } \partial\mathcal{B}_0^{\rho} \\ \boldsymbol{\varphi} - \boldsymbol{\varphi}^{presc} &= \mathbf{0} && \text{on } \partial\mathcal{B}_0^{\varphi} \end{aligned} \tag{14}$$

Neumann boundary conditions can be given for the mass flux and the tractions on $\partial\mathcal{B}_0^{r}$ and $\partial\mathcal{B}_0^{t}$

$$\begin{aligned} \boldsymbol{R}\;\cdot\boldsymbol{N} - [\,r^{closed} + \bar{r}^{open}\,] &= 0 && \text{on } \partial\mathcal{B}_0^{r} \\ \bar{\boldsymbol{\Pi}}^t\cdot\boldsymbol{N} - [\,\boldsymbol{t}^{closed} + \bar{\boldsymbol{t}}^{open}\,] &= \mathbf{0} && \text{on } \partial\mathcal{B}_0^{t} \end{aligned} \tag{15}$$

with $\boldsymbol{N}$ denoting the outward normal to $\partial\mathcal{B}_0$. As a prerequisite for the finite element discretization, the coupled set of equations has to be reformulated in weak form. To this end, the residual statements of the balance of mass and momentum (12) and the corresponding Neumann boundary conditions (15) are tested by the scalar- and vector-valued test function $\delta\rho$ and $\delta\boldsymbol{\varphi}$, respectively.

$$\begin{aligned} \mathrm{g}^{\rho}\;(\delta\rho;\rho_0,\boldsymbol{\varphi}) &= 0 && \forall\;\delta\rho\;\text{ in } H_1^0(\mathcal{B}_0) \\ \mathrm{g}^{\varphi}\,(\delta\boldsymbol{\varphi};\rho_0,\boldsymbol{\varphi}) &= 0 && \forall\;\delta\boldsymbol{\varphi}\;\text{ in } H_1^0(\mathcal{B}_0) \end{aligned} \tag{16}$$

Thereby, the weak forms g^{ρ} and g^{φ} expand into the following expressions.

$$\begin{aligned} \mathrm{g}^{\rho} = &\int_{\mathcal{B}_0} \delta\rho \quad \mathrm{D}_t\rho_0 \quad \mathrm{d}V + \int_{\mathcal{B}_0} \nabla_X\delta\rho\;\cdot\boldsymbol{R}\,\mathrm{d}V \\ &- \int_{\partial\mathcal{B}_0^r} \delta\rho\;\;[r^{closed} + \bar{r}^{open}]\,\mathrm{d}A - \int_{\mathcal{B}_0} \delta\rho\;\;\mathcal{R}_0\,\mathrm{d}V \\ \mathrm{g}^{\varphi} = &\int_{\mathcal{B}_0} \delta\boldsymbol{\varphi}\cdot \quad \rho_0\mathrm{D}_t\boldsymbol{p} \quad \mathrm{d}V + \int_{\mathcal{B}_0} \nabla_X\delta\boldsymbol{\varphi} : \bar{\boldsymbol{\Pi}}^t\,\mathrm{d}V \\ &- \int_{\partial\mathcal{B}_0^t} \delta\boldsymbol{\varphi}\cdot[\boldsymbol{t}^{closed} + \bar{\boldsymbol{t}}^{open}]\,\mathrm{d}A - \int_{\mathcal{B}_0} \delta\boldsymbol{\varphi}\cdot\;\;\bar{\boldsymbol{b}}_0\,\mathrm{d}V \end{aligned} \tag{17}$$

6 TEMPORAL DISCRETIZATION

For the temporal discretization of the governing equations (16), we partition the time interval of interest $\mathcal{T}$ into n_{step} subintervals $[t_n, t_{n+1}]$ as

$$\mathcal{T} = \bigcup_{n=0}^{n_{step}-1} [t_n, t_{n+1}] \tag{18}$$

and focus on a typical time slab $[t_n, t_{n+1}]$ for which $\Delta t\;:\;= t_{n+1} - t_n > 0$ denotes the actual time increment. Assume, that the primary unknowns ρ_{0_n} and $\boldsymbol{\varphi}_n$ and all derivable quantities are known at the beginning of the actual subinterval t_n. In the spirit of implicit time marching schemes, we now reformulate the set of governing equations in terms of the unknowns $\rho_{0_{n+1}}$ and $\boldsymbol{\varphi}_{n+1}$.

$$\begin{aligned} \mathrm{g}^{\rho}_{n+1}\;(\delta\rho;\rho_{0\,n+1},\boldsymbol{\varphi}_{n+1}) &= 0 \quad \forall\,\delta\rho\;\text{in } H_1^0(\mathcal{B}_0) \\ \mathrm{g}^{\varphi}_{n+1}\,(\delta\boldsymbol{\varphi};\rho_{0\,n+1},\boldsymbol{\varphi}_{n+1}) &= 0 \quad \forall\,\delta\boldsymbol{\varphi}\;\text{in } H_1^0(\mathcal{B}_0) \end{aligned} \tag{19}$$

Without loss of generality, we shall apply the classical Euler backward time integration scheme in the sequel. In combination with the following approximations of the first order material time derivatives $\mathrm{D}_t\rho_0$ and $\mathrm{D}_t\boldsymbol{p}$ as

$$\begin{aligned} \mathrm{D}_t\rho_0 &= \frac{1}{\Delta t}\,[\,\rho_{0\,n+1} - \rho_{0\,n}\,] \\ \mathrm{D}_t\boldsymbol{p} &= \frac{1}{\Delta t}\,[\,\boldsymbol{p}_{n+1} \;-\boldsymbol{p}_n\;\,] \end{aligned} \tag{20}$$

we obtain the following semi-discrete weak forms of the balance of mass and momentum.

$$\begin{aligned} \mathrm{g}^{\rho}_{n+1} = &\int_{\mathcal{B}_0} \delta\rho \quad \frac{\rho_{0\,n+1} - \rho_{0\,n}}{\Delta t} + \nabla_X\delta\rho\cdot\boldsymbol{R}_{n+1} \quad \mathrm{d}V \\ &- \int_{\partial\mathcal{B}_0^r} \delta\rho\;\;[r^{closed}_{n+1} + \bar{r}^{open}_{n+1}]\mathrm{d}A - \int_{\mathcal{B}_0} \delta\rho\mathcal{R}_{0n+1}\;\mathrm{d}V \\ \mathrm{g}^{\varphi}_{n+1} = &\int_{\mathcal{B}_0} \delta\boldsymbol{\varphi}\;\cdot\rho_0\,\frac{\boldsymbol{p}_{n+1} - \boldsymbol{p}_n}{\Delta t} + \nabla_X\delta\boldsymbol{\varphi} : \bar{\boldsymbol{\Pi}}^t_{n+1} \quad \mathrm{d}V \\ &- \int_{\partial\mathcal{B}_0^t} \delta\boldsymbol{\varphi}\cdot[\boldsymbol{t}^{closed}_{n+1} + \bar{\boldsymbol{t}}^{open}_{n+1}]\mathrm{d}A - \int_{\mathcal{B}_0} \delta\boldsymbol{\varphi}\cdot\bar{\boldsymbol{b}}_{0n+1}\mathrm{d}V \end{aligned} \tag{21}$$

7 SPATIAL DISCRETIZATION

The following spatial discretization is essentially characterized through a $\mathcal{C}^0$-continuous interpolation of the density field ρ_0 in combination with the standard $\mathcal{C}^0$-continuous interpolation of the deformation field $\boldsymbol{\varphi}$ (Kuhl and Steinmann 2002d). In the spirit of the finite element method, the domain of interest $\mathcal{B}_0$ is discretized into n_{el} elements $\mathcal{B}_0^e$.

$$\mathcal{B}_0 = \bigcup_{e=1}^{n_{el}} \mathcal{B}_0^e \tag{22}$$

The geometry $\boldsymbol{X}^h$ of each subset is interpolated elementwise in terms of the local basis functions N_X and the discrete node point positions $\boldsymbol{X}_j$ of all $j = 1, \ldots, n_{eX}$ element nodes as $\boldsymbol{X}^h|_{\mathcal{B}_0^e} = \sum_{j=1}^{n_{eX}} N_X^j \boldsymbol{X}_j$. According to the isoparametric concept, the trial function $\boldsymbol{\varphi}^h$ is interpolated on the element level with the same basis function N_φ as the element geometry. Following the Bubnov–Galerkin approach, similar basis functions are applied to interpolate the test function $\delta\boldsymbol{\varphi}^h$.

$$\begin{aligned}
\delta\rho^h|_{\mathcal{B}_0^e} &= \sum_{i=1}^{n_{e\rho}} N_\rho^i \;\; \delta\rho_i && \in \; H_1^0(\mathcal{B}_0) \\
\delta\boldsymbol{\varphi}^h|_{\mathcal{B}_0^e} &= \sum_{j=1}^{n_{e\varphi}} N_\varphi^j \;\delta\boldsymbol{\varphi}_j && \in \; H_1^0(\mathcal{B}_0) \\
{\rho_0}^h|_{\mathcal{B}_0^e} &= \sum_{k=1}^{n_{e\rho}} N_\rho^k \;\; \rho_k && \in \; H_1(\mathcal{B}_0) \\
\boldsymbol{\varphi}^h|_{\mathcal{B}_0^e} &= \sum_{l=1}^{n_{e\varphi}} N_\varphi^l \;\; \boldsymbol{\varphi}_l && \in \; H_1(\mathcal{B}_0)
\end{aligned} \tag{23}$$

While the element set of deformation nodes $j = 1, \ldots, n_{e\varphi}$ typically corresponds to the set of node point positions $j = 1, \ldots, n_{eX}$, the element set of density nodes $i = 1, \ldots, n_{ep}$ can generally be chosen independently. The discretization of the gradients of the test functions $\nabla_X \delta_{\rho^h}$ and $\nabla_X \delta\boldsymbol{\varphi}^h$ and the gradients of the trial functions $\nabla_X \rho^h$ and $\nabla_X \boldsymbol{\varphi}^h$ is straightforward and results in the following expressions.

$$\begin{aligned}
\nabla_X \; \delta\rho^h \; |_{\mathcal{B}_0^e} &= \sum_{i=1}^{n_{e\rho}} \; \delta\rho_i \qquad \nabla_X N_\rho^i \\
\nabla_X \; \delta\boldsymbol{\varphi}^h |_{\mathcal{B}_0^e} &= \sum_{j=1}^{n_{e\varphi}} \; \delta\boldsymbol{\varphi}_j \otimes \nabla_X N_\varphi^j \\
\nabla_X \; \rho_0^h \; |_{\mathcal{B}_0^e} &= \sum_{k=1}^{n_{e\rho}} \; \rho_k \qquad \nabla_X N_\rho^k \\
\nabla_X \; \boldsymbol{\varphi}^h \; |_{\mathcal{B}_0^e} &= \sum_{l=1}^{n_{e\varphi}} \; \boldsymbol{\varphi}_l \quad \otimes \nabla_X N_\varphi^l
\end{aligned} \tag{24}$$

With the above discretizations, the discrete algorithmic balance of mass and momentum can be rewritten as

$$\begin{aligned}
\mathrm{r}_I^\rho \, ({\rho_0}^h_{n+1}, \boldsymbol{\varphi}^h_{n+1}) &= 0 \qquad \forall \;\; I = 1, \ldots, n_{n\rho} \\
\mathbf{r}_J^\varphi \, ({\rho_0}^h_{n+1}, \boldsymbol{\varphi}^h_{n+1}) &= \mathbf{0} \qquad \forall \;\; J = 1, \ldots, n_{n\varphi}
\end{aligned} \tag{25}$$

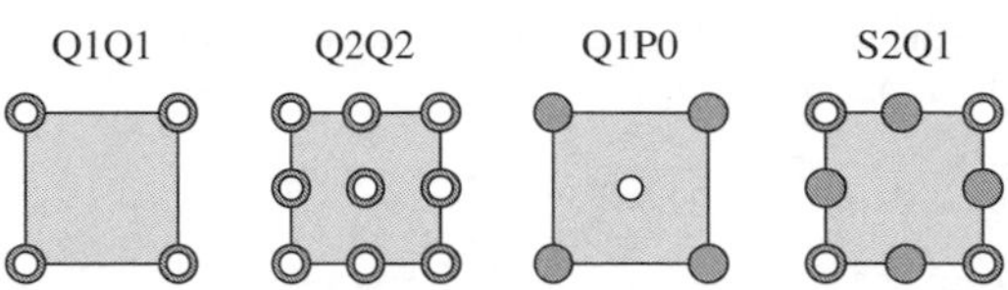

Figure 2. Alternative finite element discretizations.

whereby the discrete residua r_I^ρ and $\mathbf{r}_J^\varphi$ expand into the following forms.

$$\begin{aligned}
\mathrm{r}_I^\rho &= \mathop{\mathbf{A}}_{e=1}^{n_{el}} \int_{\mathcal{B}_0^e} N_\rho^i \quad \frac{\rho_{0n+1} - \rho_{0n}}{\Delta t} + \nabla_X N_\rho^i \cdot \boldsymbol{R}_{n+1} \quad \mathrm{d}V \\
&\quad - \int_{\partial\mathcal{B}_0^{er}} N_\rho^i [r_{n+1}^{closed} + \bar{r}_{n+1}^{open}] \mathrm{d}A - \int_{\mathcal{B}_0^e} N_\rho^i \mathcal{R}_{0n+1} \mathrm{d}V \\
\mathbf{r}_J^\varphi &= \mathop{\mathbf{A}}_{e=1}^{n_{el}} \int_{\mathcal{B}_0^e} N_\varphi^j \quad \rho_0 \frac{\boldsymbol{p}_{n+1} - \boldsymbol{p}_n}{\Delta t} + \nabla_X N_\varphi^j \cdot \bar{\boldsymbol{\Pi}}_{n+1} \quad \mathrm{d}V \\
&\quad - \int_{\partial\mathcal{B}_0^{te}} N_\varphi^j \, [\boldsymbol{t}_{n+1}^{closed} + \bar{\boldsymbol{t}}_{n+1}^{open}] \mathrm{d}A - \int_{\mathcal{B}_0^e} N_\varphi^j \bar{\boldsymbol{b}}_{0n+1} \, \mathrm{d}V
\end{aligned} \tag{26}$$

Herein, the operator $\mathbf{A}$ symbolizes the assembly of all element contributions at the element density nodes $i = 1, \ldots, n_{e\rho}$ and the element deformation nodes $j = 1, n_{e\varphi}$ to the overall residuals at the global density and deformation node points $I = 1, \ldots, n_{n\rho}$ and $J = 1, \ldots, n_{n\varphi}$. Figure 2 depicts a selection of possible finite element discretizations with the first index being related to the $n_{e\varphi}$ deformation nodes illustrated through the dark circles while the second index is related to the $n_{e\rho}$ density nodes indicated through the white circles.

8 LINEARIZATION

The discrete residual statements characterizing the mechanics of open systems (25) represent a highly nonlinear coupled system of equations which can be solved efficiently within the framework of a monolithic incremental iterative Newton–Raphson solution strategy. To this end, we perform a consistent linearization of the governing equations at time t_{n+1}

$$\begin{aligned}
\mathrm{r}_{I\;n+1}^{\rho\;k+1} &= \mathrm{r}_{I\;n+1}^{\rho\;k} + \mathrm{d}\,\mathrm{r}_I^\rho \doteq 0 \;\; \forall \, I = 1, \ldots, n_{np} \\
\mathbf{r}_{J\;n+1}^{\varphi\;k+1} &= \mathbf{r}_{J\;n+1}^{\varphi\;k} + \mathrm{d}\,\mathbf{r}_J^\varphi \doteq \mathbf{0} \;\; \forall \, J = 1, \ldots, n_{np}
\end{aligned} \tag{27}$$

whereby the iterative residua dr_I^ρ and $\mathbf{dr}_J^\varphi$ take the following format.

$$\begin{aligned}
\mathrm{dr}_I^\rho &= \sum_{K=1}^{n_{n\rho}} \mathrm{K}_{IK}^{\rho\rho} \; \mathrm{d}\rho_K + \sum_{L=1}^{n_{n\varphi}} \mathbf{K}_{IL}^{\rho\varphi} \cdot \mathrm{d}\boldsymbol{\varphi}_L \\
\mathbf{dr}_J^\varphi &= \sum_{K=1}^{n_{n\rho}} \mathbf{K}_{JK}^{\varphi\rho} \; \mathrm{d}\rho_K + \sum_{L=1}^{n_{n\varphi}} \mathbf{K}_{JL}^{\varphi\varphi} \cdot \mathrm{d}\boldsymbol{\varphi}_L
\end{aligned} \tag{28}$$

In the above definitions, we have introduced the iteration matrices

$$\begin{aligned} \mathbf{K}^{\rho\rho}_{IK} &= \frac{\partial \mathbf{r}^{\rho}_{I}}{\partial \rho_K} & \mathbf{K}^{\rho\varphi}_{IL} &= \frac{\partial \mathbf{r}^{\rho}_{I}}{\partial \boldsymbol{\varphi}_L} \\ \mathbf{K}^{\varphi\rho}_{JK} &= \frac{\partial \mathbf{r}^{\varphi}_{J}}{\partial \rho_K} & \mathbf{K}^{\varphi\varphi}_{JL} &= \frac{\partial \mathbf{r}^{\varphi}_{J}}{\partial \boldsymbol{\varphi}_L} \end{aligned} \tag{29}$$

which take the following abstract representations.

$$\begin{aligned} \mathbf{K}^{\rho\rho}_{IK} &= \mathop{\mathsf{A}}_{e=1}^{n_{el}} \int_{\mathcal{B}_0^e} N^i_\rho \frac{1}{\Delta t} N^k_\rho \,\mathrm{d}V - \int_{\mathcal{B}_0^e} N^i_\rho \,\partial_{\rho_0}\mathcal{R}_0\, N^k_\rho \,\mathrm{d}V + \int_{\mathcal{B}_0^e} \nabla_X N^i_\rho \cdot R_0 \,\nabla_X N^k_\rho \,\mathrm{d}V \\ \mathbf{K}^{\rho\varphi}_{IL} &= \mathop{\mathsf{A}}_{e=1}^{n_{el}} - \int_{\mathcal{B}_0^e} N^i_\rho \,\partial_F \mathcal{R}_0 \cdot \nabla_X N^l_\varphi \,\mathrm{d}V \\ \mathbf{K}^{\varphi\rho}_{JK} &= \mathop{\mathsf{A}}_{e=1}^{n_{el}} \int_{\mathcal{B}_0^e} \nabla_X N^j_\varphi \cdot \partial_{\rho_0} \bar{\boldsymbol{\Pi}}^t \, N^k_\rho \,\mathrm{d}V \\ \mathbf{K}^{\varphi\varphi}_{JL} &= \mathop{\mathsf{A}}_{e=1}^{n_{el}} \int_{\mathcal{B}_0^e} N^j_\varphi \,\rho_0 \frac{1}{\Delta t^2} \boldsymbol{I} \, N^l_\varphi \,\mathrm{d}V + \int_{\mathcal{B}_0^e} \nabla_X N^j_\varphi \cdot \partial_F \bar{\boldsymbol{\Pi}}^t \cdot \nabla_X N^l_\varphi \,\mathrm{d}V \end{aligned} \tag{30}$$

Note the striking similarity of the resulting set of equations for open system thermodynamics compared to the discretized equations for hygromechanical unsaturated porous media, compare e.g. (Carmeliet 1998). The specification of the partial derivatives of the mass source $\mathcal{R}_0$ and the reduced momentum flux $\bar{\boldsymbol{\Pi}}^t$ with respect to the primary unknowns ρ_0 and $\boldsymbol{\varphi}$ depends on the particular choice of the constitutive equations. For the constitutive equations suggested in section 4, we obtain the following expressions,

$$\begin{aligned} \partial_{\rho_0}\mathcal{R}_0 &= [n-m]\left[\frac{\rho_0}{\rho_0^*}\right]^{-m} \frac{1}{\rho_0}\Psi_0 \\ \partial_F \mathcal{R}_0 &= \left[\frac{\rho_0}{\rho_0^*}\right]^{-m} \bar{\boldsymbol{\Pi}}^t \\ \partial_{\rho_0}\bar{\boldsymbol{\Pi}}^t &= n \frac{1}{\rho_0}\bar{\boldsymbol{\Pi}}^t \\ \partial_F \bar{\boldsymbol{\Pi}}^t &= \left[\frac{\rho_0}{\rho_0^*}\right]^{n} [\mu \boldsymbol{I}\overline{\otimes}\boldsymbol{I} + \lambda \boldsymbol{F}^{-t}\otimes\boldsymbol{F}^{-t} \\ &\quad -[\lambda \ln J - \mu]\,\boldsymbol{F}^{-t}\underline{\otimes}\boldsymbol{F}^{-1}] \end{aligned} \tag{31}$$

whereby the component representations of the non-standard dyadic products read $\{\bullet\overline{\otimes}\circ\}_{ijkl} = \{\bullet\}_{ik}\otimes\{\circ\}_{jl}$ and $\{\bullet\underline{\otimes}\circ\}_{ijkl} = \{\bullet\}_{il}\otimes\{\circ\}_{jk}$. Finally, the solution of the system of equations (27) renders the iterative update for the increments of the global unknowns ρ_I and φ_J.

$$\begin{aligned} \Delta\,\rho_I &= \Delta\,\rho_I + \mathrm{d}\,\rho_I & \forall\, I &= 1,\ldots,n_{n\rho} \\ \Delta\boldsymbol{\varphi}_J &= \Delta\boldsymbol{\varphi}_J + \mathrm{d}\boldsymbol{\varphi}_J & \forall\, J &= 1,\ldots,n_{n\varphi} \end{aligned} \tag{32}$$

Remark 7.1
Within the chemomechanical context, the evolution of the density ρ_0 and the evolution of the deformation $\boldsymbol{\varphi}$ induce time scales which typically differ by orders of magnitude. To avoid the related numerical difficulties caused by the resulting ill-conditioned system matrices, the balance of momentum is usually evaluated in a quasi-static sense, i.e. the mechanical forces are interpreted as an average loading on the structure. Consequently, the first terms in the discrete momentum residual $N^j_\varphi\rho_0[\boldsymbol{p}_{0n+1} - \boldsymbol{p}_{0n}]\Delta t$ and in the tangent operator $N^j_\varphi\rho_0/\Delta t^2 \boldsymbol{I} N^l_\varphi$ in equations $(26)_2$ and $(30)_4$ vanish identically.

9 SPATIAL AND MATERIAL FORCES

The classical spatial motion problem defined above is based on the idea of following physical particles from a fixed material position $\boldsymbol{X}$ through the ambient space. In contrast to this, within the material motion problem, physical particles are followed through the ambient material at fixed spatial position $\boldsymbol{x}$. The material motion problem is thus typically applied to study the local rearrangement of material inhomogeneities or heterogeneities.

Equation $(26)_2$ defines the discrete residuum of the spatial momentum balance in terms of the spatial inertia forces, the spatial internal forces, the spatial surface forces and the spatial volume forces. At chemo-mechanical equilibrium, the discrete spatial forces $\mathbf{f}^{\varphi\ h}_{sur J}$ acting on the global node points can thus be calculated according to the following equation.

$$\mathbf{f}^{\varphi\ h}_{sur J} = \mathop{\mathsf{A}}_{e=1}^{n_{el}} \int_{\mathcal{B}_0^e} N^j_\varphi \,\rho_0 \frac{\boldsymbol{p}_{n+1} - \boldsymbol{p}_n}{\Delta t} \,\mathrm{d}V + \int_{\mathcal{B}_0^e} \nabla_X N^j_\varphi \cdot \bar{\boldsymbol{\Pi}}_{n+1} \,\mathrm{d}V - \int_{\mathcal{B}_0^e} N^j_\varphi \,\bar{\boldsymbol{b}}_{0\,n+1} \,\mathrm{d}V \tag{33}$$

Recall, that these discrete spatial node point forces are energetically conjugate to spatial variations of the node point positions. Unlike the spatial surface forces, the discrete material surface forces are energetically conjugate to material variations of the node point positions. They are defined through the material momentum balance which basically follows from a complete projection of the spatial momentum balance onto the material manifold. Generally speaking, the material motion problem can be introduced in complete duality to the classical spatial motion problem. However, as a fundamental difference to the spatial motion problem, the Neumann boundary conditions of the material motion problem cannot be considered as given input data. Correspondingly, the discrete material surface forces $\mathsf{F}^{\Phi\ h}_{sur J}$

$$
\begin{aligned}
\mathsf{F}^{\Phi\ h}_{sur J} = \mathop{\mathbf{A}}_{e=1}^{n_{el}} \int_{\mathcal{B}_t^e} & N_\Phi^j\, \rho_t \frac{\boldsymbol{P}_{n+1} - \boldsymbol{P}_n}{\Delta t} \, \mathrm{d}v \\
+ \int_{\mathcal{B}_t^e} & \nabla_x N_\Phi^j \cdot \bar{\boldsymbol{\pi}}_{n+1} \, \mathrm{d}v \\
- \int_{\mathcal{B}_t^e} & N_\Phi^j \, \bar{\boldsymbol{B}}_{t n+1} \, \mathrm{d}v
\end{aligned}
\tag{34}
$$

acting on the global node points can only be computed in a post processing calculation once the spatial motion problem has been solved. Their definition in terms of the material inertia forces, the material internal forces and the material volume forces parallels the definition of the discrete surface forces of the spatial motion problem of equation (33). For the quasi-static case, the discrete mass specific momentum density $\boldsymbol{P}_{n+1}$, the discrete reduced momentum flux $\bar{\boldsymbol{\pi}}_{n+1}$, and the corresponding momentum source $\bar{\boldsymbol{B}}_{tn+1}$ that are essentially needed to compute the discrete material node point forces are related to their spatial motion counterparts through the following transformation formulae (Kuhl and Steinmann 2002b).

$$
\begin{aligned}
\boldsymbol{P}_{n+1} &= - \boldsymbol{F}^t_{n+1} \cdot \boldsymbol{p}_{n+1} \\
\bar{\boldsymbol{\pi}}_{n+1} &= - j\, \boldsymbol{F}^t_{n+1} \cdot \bar{\boldsymbol{\Pi}}_{n+1} \cdot \boldsymbol{F}^t_{n+1} \\
&\quad + \rho_{t n+1}\, \Psi_{n+1}\, \boldsymbol{F}^t_{n+1} \\
\bar{\boldsymbol{B}}_{t n+1} &= - j\, \boldsymbol{F}^t_{n+1} \cdot \bar{\boldsymbol{b}}_{0 n+1} \\
&\quad - \rho_{t n+1}\, \partial_{\rho_0} \Psi_{n+1} \nabla_X \rho_{0 n+1} \\
&\quad - \partial_{\boldsymbol{\Phi}} [\rho_{t n+1}\, \Psi_{n+1}]
\end{aligned}
\tag{35}
$$

10 EXAMPLES

The features of the derived open system approach will be elaborated by means of a one-dimensional model problem, for which the additional benefit of the material force method will be illustrated. Finally, we present the numerical analysis of a single edge notched specimen under tensile loading.

10.1 *One-dimensional model problem*

First, we shall consider the academic example of a one-dimensional bar of unit size with Lamé constants of $\lambda = 0$ and $\mu = 0.5$. The specimen has a reference density of $\rho_0^* = 1$ and a reference free energy of $\Psi_0^* = 2$, while the two exponents are chosen to $n = 2$ and $m = 3$. In order to trigger an imperfection, the reference free energy Ψ_0^* has been reduced to $\Psi_0^* = 1$ in an area of 20% of the total specimen length in the middle of the bar. The bar is discretized with 100 Q1Q1 elements according to figure 2. We apply a constant load of $f = 1$ and simulate 50 time steps of $\Delta t = 0.1$ each, until the density has converged to its final equilibrium distribution. To study the influence of moisture transport, we gradually increase the mass conduction coefficient R_0.

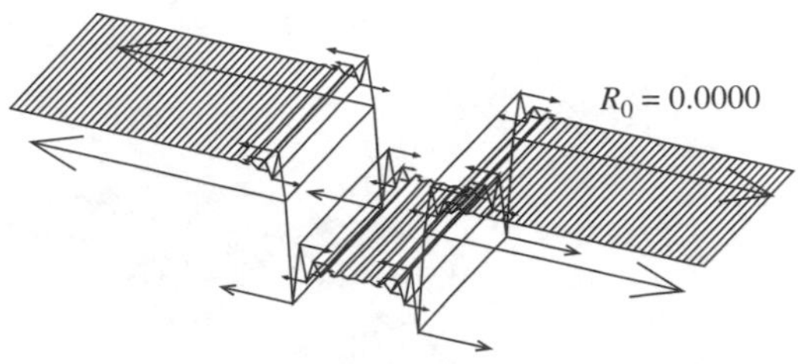

Figure 3. Density ρ_0 and material volume forces F^{Φ}_{vol}.

The density distribution resulting from $R_0 = 0$ for which diffusive effects are completely suppressed is depicted in figure 3. The discontinuity in the reference free energy Ψ_0^* obviously results in a non-homogeneous density distribution. Typically, the applied C^0-continuous finite element formulation fails to capture the analytical solution of a hat-type density distribution and shows characteristic spurious oscillations close to the discontinuity. To illustrate the tendency of the material to form a homogeneous distribution of matter, we have included the volume contribution to the material forces

$$
\mathsf{F}^{\Phi\ h}_{vol J} = \mathop{\mathbf{A}}_{e=1}^{n_{el}} \int_{\mathcal{B}_t^e} N_\Phi^j \, \bar{\boldsymbol{B}}_{t n+1} \, \mathrm{d}v
$$

in figure 3. The material volume forces indicate the tendency to smooth sharp density contours aiming at an equal concentration throughout the entire bar. This observation coincides with the numerical findings of (Müller and Maugin 2002), who observed that material forces always try to render the total body more homogeneous.

Next, we shall therefore allow for mass diffusion, e.g. through moisture transport, whereby the intensity of the mass flow is primarily governed by the mass conduction coefficient R_0. Figure 4 illustrates the resulting density distributions for gradually increasing values of R_0. The increase of the mass conduction coefficient clearly smoothes the density profile and widens the area that is affected by the initial imperfection. The smoother the density distribution, the smaller the related density gradients. The resulting material forces thus decrease with an increasing mass flux. For values of R_0 larger than a critical value, a homogeneous density distribution can be found, compare figure 4, bottom.

In summary, the volume contribution to the material forces F^{Φ}_{vol} can be interpreted as driving force for the movement of material inhomogeneities in the interior

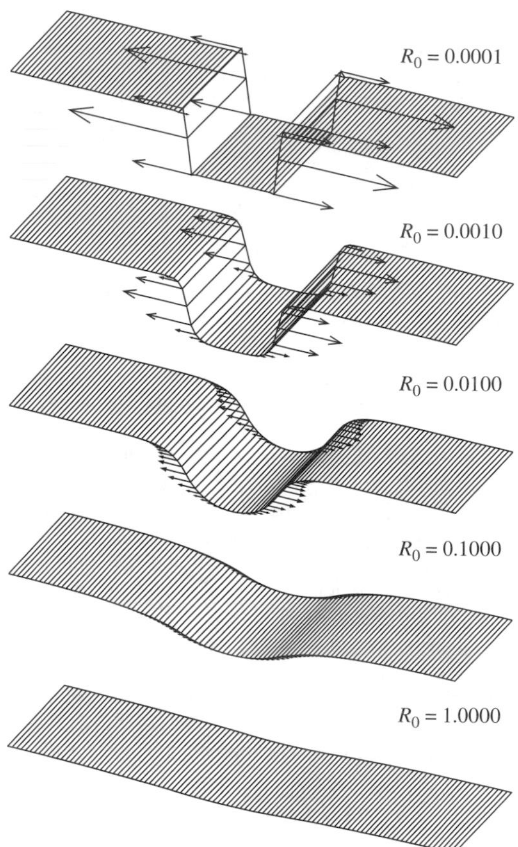

Figure 4. Density ρ_0 and material volume forces $\mathbf{F}^{\Phi}_{vol}$.

of the considered domain. They indicate the tendency of the considered matter to flow through the material with the aim of equilibrating initially unequal density distributions.

10.2 *Single edge notched specimen*

The second example is originally borrowed from linear elastic fracture mechanics. It represents the qualitative study of the density evolution in a single edge notched specimen subjected to tensile loading. The specimen is of rectangular shape with a length of $l = 225$ and a width of $w = 75$ units. The corresponding finite element discretization consisting of 3565 node points and 1140 finite elements is depicted in figure 5, left. The density and the deformation are interpolated with equal order by applying quadratic shape functions throughout, compare the Q2Q2 element illustrated in figure 2. A close-up of the heavily defined mesh at the crack tip is given in figure 6, left. The Lamé constants are chosen to $\lambda = 138.89$ and $\mu = 208.33$. The two exponents characterizing the density evolution are taken as $n = 2$ and $m = 3$, the reference density is chosen to $\rho_0^* = 1.0$, the mass flux is neglected as $R_0 = 0$ and the value of the reference free energy has been chosen to $\Psi_0^* = 0.50$. The specimen is loaded by a ramp load over 10 time steps of $\Delta t = 0.05$ each, by an incrementally increasing spatial force of $\Delta \mathbf{f}^{\varphi}_{sur} = 250$ applied on the upper and lower boundary. Once the final load level of $\mathbf{f}^{\varphi}_{sur} = 2500$ is reached, the load is held constant for another 50 time steps of $\Delta t = 0.05$ while the density distribution converges towards the chemomechanical equilibrium state. The resulting density contours depicted in figure 5 indicate enormous changes in density close to the crack tip while the material basically remains unaltered near the stress free crack faces.

Finally, we will turn to the solution of the material motion problem and analyze the material node point forces. In particular, we are interested in the evolution of the material forces at the crack tip which increase during the loading phase but then relax towards a remarkably lower value at chemomechanical equilibrium. The spatial evolution of the discrete material surface forces $\mathbf{F}^{\Phi}_{sur}$ and the corresponding discrete volume forces $\mathbf{F}^{\Phi}_{vol}$ is depicted in figure 6. Note, that for the sake of visualization, the nodal volume forces have been magnified by a factor 50 in comparison to the corresponding surface terms.

Recall, that in the context of linear elastic fracture mechanics, the material surface force on the crack tip $\mathbf{F}^{\Phi}_{sur}$ essentially corresponds to the vector-valued J-integral as proposed by (Rice 1968), however, pointing in the opposite direction. Discrete material surface forces thus point in the direction of a potential energy increase upon replacement of the material node point positions. They thus indicate, that changes in density opposite to this direction would reduce the potential energy.

In contrast to the material surface forces, the discrete material volume forces $\mathbf{F}^{\Phi}_{vol}$ are essentially orthogonal to the isolines of the density contour plot. They typically indicate the tendency to equilibrate density concentrations. Both, the surface and volume forces increase upon loading but show a relaxation at constant applied load. As time evolves, energy is dissipated locally at the crack tip. In the context of fracture mechanics this effect which manifests itself in the temporal decrease of the discrete material node point forces is typically referred to as stress shielding. The example of the notched specimen has shown the relevance of material forces in chemomechanical applications. Material forces can be utilized to locate areas of potential material apposition or deposition. They thus provide a better understanding of chemomechanically induced changes in the spatial density distribution.

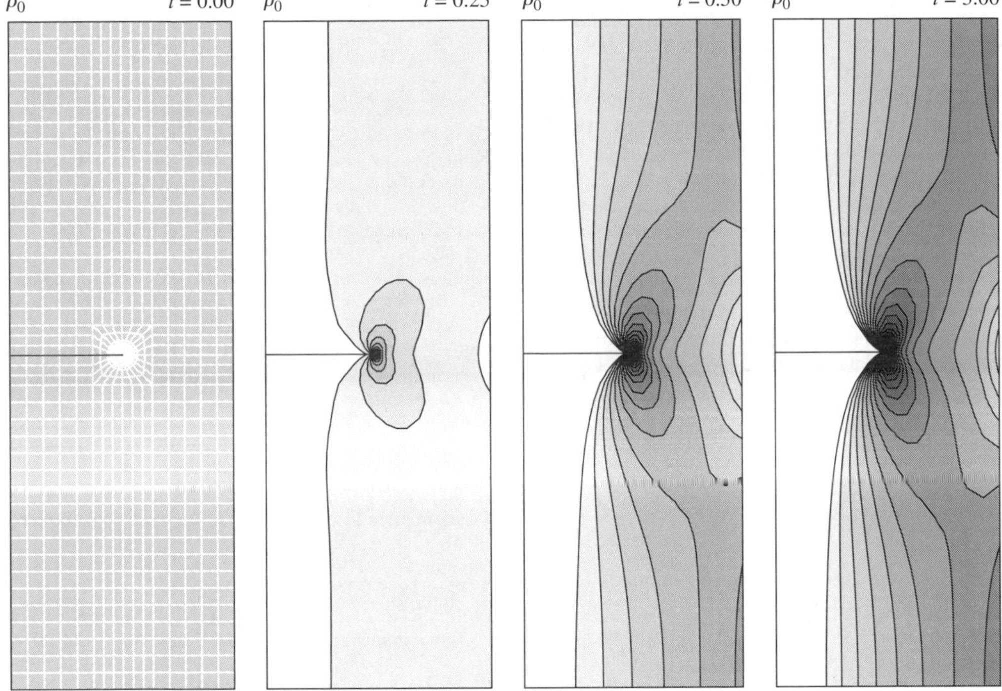

Figure 5. Notched specimen – changes in density ρ_0.

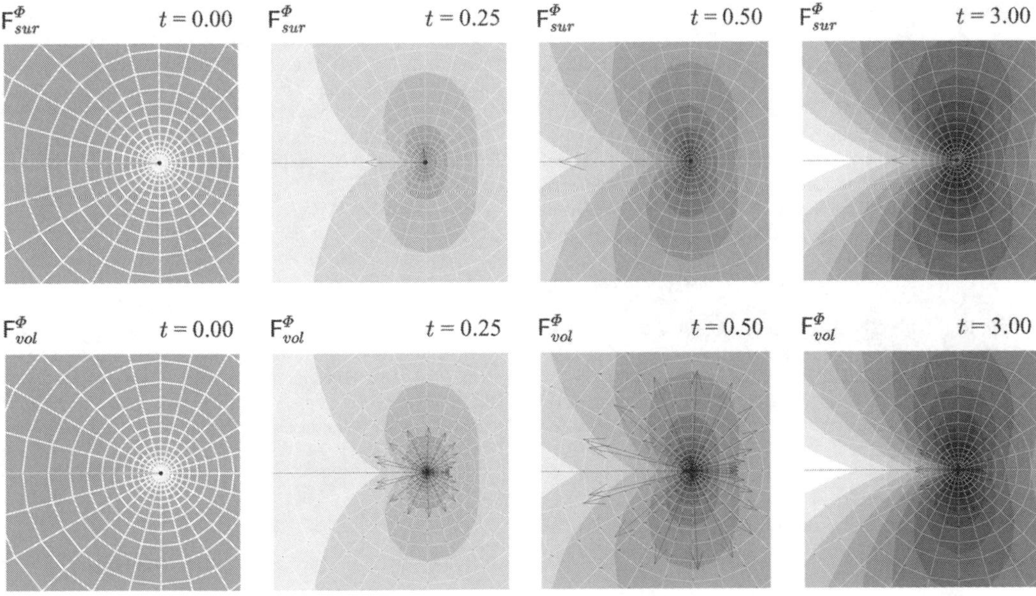

Figure 6. Notched specimen – evolution of material surface forces $\mathbf{F}^{\Phi}_{sur}$ and material volume forces $\mathbf{F}^{\Phi}_{vol}$.

11 CONCLUSIONS

A general theoretical and computational framework for the thermodynamics of open systems has been presented which is particularly suited for bio- or chemo-mechanical applications. The classical spatial motion formulation of open system thermodynamics typically defines the evolution of the density and the deformation field, whereby the latter basically characterizes the movement of physical particles relative to the ambient space. The material motion problem, however, characterizes the movement of physical particles relative to the ambient material, and can thus be applied to study the local rearrangement of material inhomogeneities. Numerical results for both, the spatial and the material motion problem, have been derived and discussed for a one-dimensional model problem as well as for a single edge notched specimen under tensile loading.

ACKNOWLEDGEMENTS

Support from the German Science Foundation DFG through the habilitation grant KU-1313 is gratefully acknowledged. Moreover, we would like to thank Lamberto Giovanni Chuisa for fruitful discussions which have contributed greatly to the understanding of the non-mechanical energy sources $\bar{Q}$ and $\bar{Q}_0$ in the energy balance of equation (5).

REFERENCES

Bowen, R. M. (1976). Theory of mixtures. In A. C. Eringen (Ed.), *Continuum Physics – Volume III – Mixtures and EM Field Theories,* pp. 1–127. Academic Press, New York – San Francisco – London.

Carmeliet, J. (1998). A poroviscoelastic damage model for hygromechanical damage processes in unsaturated porous media. In R. de Borst, N. Bićanić, H. Mang, and G. Meschke (Eds.), *Proc. EURO-C 1998*, pp. 559–566. Balkema, Rotterdam.

Coussy, O. (1995). *Mechanics of Porous Continua.* John Wiley & Sons.

de Boer, R. (2000). *Theory of Porous Media – Highlights in the Historical Development and Current State.* Springer Verlag, Berlin – Heidelberg – New York.

Epstein, M. and G. A. Maugin (2000). Thermomechanics of volumetric growth in uniform bodies. *Int. J. Plasticity 16,* 951–978.

Gibson, L. J. and M. F. Ashby (1997). *Cellular Solids* (2nd ed.). Cambridge University Press.

Gurtin, M. E. (2000). *Configurational Forces as Basic Concepts of Continuum Physics.* Springer Verlag, New York – Berlin – Heidelberg.

Kuhl, D., F. Bangert, and G. Meschke (2000). An extension of damage theory to coupled chemo-mechanical processes. In *Proc. ECCO-MAS 2000.* CIMNE, Barcelona, Spain.

Kuhl, D. and G. Meschke (2002). Higher order transport models for the simulation of dissociation–diffusion processes in reactive porous media. In H. A. Mang, F. G. Rammerstorfer, and J. Eberhardsteiner (Eds.), *Proc. WCCM* V. Vienna, Austria.

Kuhl, E. and P. Steinmann (2002a). Mass- and volume specific views on thermodynamics for open systems. *Internal Report, LTM, University of Kaiserslautern J02–03.*

Kuhl, E. and P. Steinmann (2002b). Material forces in open system mechanics. *Internal Report, LTM, University of Kaiserslautern J02–07.*

Kuhl, E. and P. Steinmann (2002c). On spatial and material settings of thermo-hyperelastodynamics for open systems. *Acta Mechanica, in press.*

Kuhl, E. and P. Steinmann (2002d). Theory and numerics of geometrically nonlinear open system mechanics. *Int. J. Num. Meth. Eng., accepted.*

Maugin, G. A. (1993). *Material Inhomogenities in Elasticity.* Chapman & Hall, London.

Müller, R. and G. A. Maugin (2002). On material forces and finite element discretizations. *Comp. Mech. 29,* 52–60.

Nield, D. A. and A. Bejan (1999). *Convection in Porous Media.* Springer Verlag, Berlin – Heidelberg – New York.

Rice, J. R. (1968). A path independent integral and the approximate analysis of strain concentration at notches and cracks. *J. Appl. Mech. 35,* 379–386.

Steinmann, P. (2002a). On spatial and material settings of hyperelastodynamics. *Acta Mechanica 156,* 193–218.

Steinmann, P. (2002b). On spatial and material settings of thermo-hyperelastodynamics. *J. Elasticity 66,* 109–157.

Torrenti, J.-M., M. Mainguy, F. Adenot, and C. Tognazzi (1998). Modelling of leaching in concrete. In R. de Borst, N. Bićanić, H. Mang, and G. Meschke (Eds.), *Proc. EURO-C 1998,* pp. 531–538. Balkema, Rotterdam.

Truesdell, C. and R. Toupin (1960). The classical field theories. In S. Flügge (Ed.), *Handbuch der Physik,* Volume III/1. Springer Verlag, Berlin.

Ulm, F.-J. and O. Coussy (1995). Modelling of thermomechanical coupling of concrete at early ages. *J. Eng. Mech. 121,* 785–794.

Ulm, F.-J., O. Coussy, and C. Hellmich (1998). Chemoplasticity: A review of evidence. In R. de Borst, N. Bićanić, H. Mang, and G. Meschke (Eds.), *Proc. EURO-C 1998,* pp. 421–439. Balkema, Rotterdam.

Ulm, F.-J., J.-M. Torrenti, and F. Adenot (1999). Chemoporoplasticity of calcium leaching in concrete. *J. Eng. Mech. 125,* 1200–1211.

Computational Modelling of Concrete Structures, Bićanić et al. (eds)
© 2003 Swets & Zeitlinger, Lisse, ISBN 90 5809 536 3

Computational modeling of transport mechanisms in reactive porous media – application to calcium leaching of concrete

Detlef Kuhl & Günther Meschke
Institute for Structural Mechanics, Ruhr University Bochum, Bochum, Germany

ABSTRACT: The paper is concerned with a dissociation-diffusion model for calcium leaching of cementitious materials formulated within the framework of the theory of reactive porous media. Main attention is paid to the critical investigation of different diffusion models for electrolyte solutions. This models can take the interaction of migrating electrically charged dissolution products by the electrostatic coupling, the relaxation effect and the electrophoretic phenomenon into account. It is demonstrated, that for reliable life time prognoses of concrete structures subjected to chemically aggressive media the consideration of ion-ion interactions in concentrated electrolyte solutions, resulting in a concentration dependent diffusion coefficient, is necessary. A representative numerical analysis of calcium leaching illustrates the properties and restrictions of the investigated ion diffusion models.

1 INTRODUCTION

Long term degradation of concrete structures is significantly accelerated by the attack of chemically aggressive species (e.g. (Coussy and Ulm 2001)). In particular, if concrete is permanently exposed to an aqueous environment, calcium is dissolved from calcium hydroxide and ettringite as well as from calcium silicate hydrates (csh). Calcium dissolution results in the increase of the porosity, the degradation of the stiffness and strength of concrete and, consequently, to a reduction of the life time of concrete structures (e.g. (Carde and François 1997)). In order to predict the life time of concrete structures subjected to chemical attack, several chemo-mechanical damage models have been developed (e.g. (Gérard 1996; Gérard 2000), (Gérard et al. 1998), (Ulm et al. 1999), (Le Bellégo et al. 2000), (Mainguy and Coussy 2000), (Mainguy et al. 2000) and (Kuhl et al. 2000; Kuhl et al. 2002)). These models are based on the assumption of a local chemical equilibrium between the chemically bounded calcium of the skeleton constituents and the calcium ions solved within the pore solution. According to this hypothesis, calcium leaching is controlled by the transport properties of the macroscopic material, namely the diffusivity of the pore solution and the porosity of the cementitious material.

Although the transport of the dissolved species is of crucial relevance for the diffusion controlled degradation mechanism, little attention is paid to a realistic description of the calcium ion diffusion through the pore fluid. Diffusive ionic transport is characterized by the mass balance of ions together with a concentration dependent diffusion coefficient. As suggested by e.g. (Chatterji 1999), a macroscopic equivalent of Kohlrausch's law (Kohlrausch 1900), describing the concentration dependent molar conductivity, is applied by (Kuhl et al. 2002) as a conceptual model to consider ion-ion interactions of the calcium ion transport in porous media. The comparison of transport models using a constant diffusion coefficient of calcium ions in water according to Fick's law (Fick 1855) and a concentration dependent diffusion coefficient demonstrates, that the chemical degradation is significantly underestimated by a constant diffusion coefficient. Consequently, the life time of concrete structures is overestimated.

In electrochemistry the ion-ion interaction phenomena in electrolyte solutions (e.g. the solution of concrete constituents in water) are well known. The theory of ionic clouds by (Debye and Hückel 1923a; Debye and Hückel 1923b; Debye 1927) had a significant impact on the research on electrolyte diffusion. Based on the Debye-Hückel theory, electrolyte diffusion models ((Hartley 1931), (Onsager and Fouss 1932)) and their simplifications for diluted solutions ((Onsager 1927), (Kohlrausch 1900)) can be developed. These models are able to capture ion-ion interactions on a well-founded theoretical basis.

The goal of the present paper is to apply the theory of electrolyte solutions to calcium leaching in order to improve the transport models used in computational models for life time prognoses of concrete structures subjected to external chemical attack. In Section 2 the phenomenological calcium leaching model for cementitious materials is briefly reviewed. Concentration dependent diffusion coefficients based on the theory of ionic clouds are presented in Section 3 and validated by means of experimental data in Section 4. In Section 5 a representative numerical analysis of calcium leaching using the investigated diffusion models is presented.

2 DISSOCIATION-DIFFUSION MODEL

Calcium leaching of cementitious materials is modeled within the framework of reactive porous media. A chemical damage model developed (Kuhl et al. 2002) which is based on the theory of porous media (e.g. (de Boer 2000) and (Ehlers 2002)) and on the phenomenological chemical equilibrium model by (Gérard 1996), will be combined with different electrolyte diffusion models investigated previously within the context of interacting calcium ion and hydroxide ion diffusion by (Kuhl 2002). In this section the phenomenological calcium leaching model by (Kuhl et al. 2002) is briefly reviewed.

2.1 *Skeleton dissolution by dissociation*

If the cementitious skeleton is in contact with a pore solution with a reduced ion concentration compared to the equilibrium state, portlandite (calcium hydroxide) $Ca(OH)_2$ ettringite $Ca_6Al_2O_6(SO_4)_3 \cdot 32H_2O$ and calcium silicate hydrates (csh) $n\mathrm{CaO} \cdot \mathrm{SiO_2} \cdot n\mathrm{H_2O}$ are dissolved at different states of the chemical degradation (see e.g. (Taylor 1997) and (Berner 1988)). During the dissolution cations Ca^{2+} and anions OH^-, $Al(OH)_4^-$, SO_4^{2-}, SiO_4^- are solved in the pore water. Since in all stages of the dissolution process calcium ions are dissolved within the pore fluid, the calcium ion concentration of the pore fluid is used as primary variable.

2.2 *Governing equations of calcium leaching*

Calcium leaching of cementitious materials is governed by the macroscopic mass balance of double charged calcium cations Ca^{2+} solved in the aqueous pore fluid.

$$\mathrm{div}\boldsymbol{q} + [\phi\, c]^{\cdot} + \dot{s} = 0 \quad \text{in} \quad \Omega \tag{1}$$

c is the molar concentration of calcium ions within the pore fluid and $\dot{s}$ is the molar calcium ion production by dissociation. $\phi = \phi_0 + \phi_c$ is the total porosity of the material in terms of the initial porosity ϕ_0 and the chemically induced porosity ϕ_c. The macroscopic calcium ion mass flux $\boldsymbol{q}$ is calculated based on the microscopic mass flux $\boldsymbol{D} \cdot \boldsymbol{\gamma}$ by multiplication with the volume fraction of the pores ϕ, where $\boldsymbol{\gamma}$ is the driving force of diffusion and $\boldsymbol{D}$ is the second order diffusion coefficient tensor.

$$\boldsymbol{q} = \phi\, \boldsymbol{D} \cdot \boldsymbol{\gamma}\,, \quad \boldsymbol{\gamma} = -\nabla c\,, \quad \boldsymbol{D} = D(c)\, \boldsymbol{1} \tag{2}$$

The determination of the concentration dependent diffusion coefficient $D(c)$ for electrolyte diffusion is discussed in detail in Section 3. Associated initial values, Neumann – and Dirichlet boundary conditions complete the non-linear initial boundary value problem of calcium leaching.

2.3 *Evolution of matrix dissolution*

The evolution of the chemically induced porosity ϕ_c is controlled by the internal variable κ_c characterizing the current chemical equilibrium state between the skeleton and the pore solution. Chemical damage propagates, if the chemical reaction criterion

$$\Phi_c = \kappa_c - c \leq 0 \tag{3}$$

is violated. The associated Kuhn-Tucker conditions

$$\Phi_c \leq 0, \quad \dot{\kappa}_c \leq 0, \quad \Phi_c \dot{\kappa}_c = 0 \tag{4}$$

distinguish between a non-linear diffusion problem ($\Phi_c \leq 0$) and a reaction-diffusion problem ($\dot{\kappa}_c < 0$). The chemically induced porosity is proportional to the dissolved amount of calcium $s_0 - s(\kappa_c)$ of the skeleton.

$$\phi_c(s(\kappa_c)) = \frac{M}{\rho}\, [s_0 - s(\kappa_c)] \tag{5}$$

The calcium concentration of the skeleton $s(\kappa_c)$ is defined as a function of the internal variable κ_c in the phenomenological chemical equilibrium model by (Gérard 1996) and (Delagrave et al. 1997).

$$\begin{aligned} s(\kappa_c) = s_0 - [1-\alpha_c] s_h \left[1 - \frac{\kappa_c}{10\bar{c}} + \frac{\kappa_c^2}{400\bar{c}}\right] \\ - \frac{s_0 - s_h}{1 + \left[\dfrac{\kappa_c}{c_p}\right]^{n_c}} - \frac{\alpha_c s_h}{1 + \left[\dfrac{\kappa_c}{c_{csh}}\right]^{m_c}} \end{aligned} \tag{6}$$

α_c, n_c, m_c, c_p, c_{csh} and s_h are model parameters, c_0 and s_0 represent the equilibrium concentrations of the sound material and $\bar{c} = 1\ mol/m^3$ is the physical dimension of κ. Based on equation (6) the molar calcium ion production can be calculated:

$$\dot{s} = \frac{\partial s}{\partial \kappa_c} \frac{\partial \kappa_c}{\partial c}\, \dot{c} \tag{7}$$

3 ELECTROLYTE DIFFUSION

In this section, the concentration dependent diffusion coefficient $D(c)$ (equation (2)) of electrolyte solutions is developed within the theory of ionic clouds (Debye-Hückel theory (Debye and Hückel 1923a; Debye and Hückel 1923b; Debye 1927)). Out of the broad range of diffusion models available from physical chemistry (see e.g. (Robinson and Stokes 1955), (Horvath 1985), (Bockris and Reddy 1998) and (Barthel et al. 1998)) the electrolyte diffusion models by (Nernst 1888), (Hartley 1931) and (Onsager and Fouss 1932) are investigated. As proposed by (Kuhl 2002) the underlying activity coefficient is estimated by neglecting the activity of the solvent (water) and the differences between the molal and the molar activity coefficients.

3.1 *Dissociation of single electrolytes in water*

Substances of hardened cement paste (calcium hydroxide, calcium silicate hydrate, ettringite) in contact with the pore water are dissociated in electrically charged products, denoted as cations (positively charged) and anions (negatively charged). This type of chemical reactions is investigated by means of a generalized chemical reaction: One mole of a solute B (electrolyte, e.g. calcium hydroxide) is dissociated in contact with the solvent A (water) into ν_+ mole cations B_+ and ν_- mole anions B_-.

$$B \rightleftharpoons \nu_+ \, B_+^{z+} + \nu_- B_-^{-z-}, \quad A \rightleftharpoons A \tag{8}$$

ν_+ and ν_- are the stoichiometric coefficients. z_+ and z_- represent the valency of the ions. The dissolved ions and the solvent constitute the electrolyte solution, where the principle of electrical neutrality should be satisfied.

$$\nu_+ \, z_+ - \nu_- \, z_- = 0 \tag{9}$$

The amount of solved ions within the pore fluid is measured by the number of moles per volume of the solvent (molar concentration c).

3.2 *Thermodynamics of electrolyte diffusion*

The chemical potential or the partial Gibbs free energy of the solute (μ_B) is expressed in terms of the chemical potential at the standard state (μ_B^0) and the activity of the solute a_B.

$$\mu_B = \mu_B^0 + R\,T \, \ln a_B \tag{10}$$

Herein, R is the universal gas constant (for natural constants see Table 1) and T is the total temperature. The activity a_B is given as function of the mean molar activity coefficient $y_\pm$, the molar electrolyte concentration c and the stoichiometric coefficients (Bockris and Reddy 1998).

Table 1. Natural constants, see e.g (Atkins 1998).

	Value	Unit	Quantity
e	$1.60218 \cdot 10^{-19}$	C	elementary charge
F	$9.64853 \cdot 10^{4}$	C/mol	Faraday
k	$1.38066 \cdot 10^{-23}$	J/K	Bolzmann
N	$6.02214 \cdot 10^{23}$	$1/mol$	Avogadro
R	8.31451	$J/Kmol$	universal gas constant
ε_0	$8.85419 \cdot 10^{-12}$	C^2/Jm	vacuum electric permittivity

$$a_B = y_\pm^{\nu_+ + \nu_-} \; \nu_+^{\nu_+} \; \nu_-^{\nu_-} \; c^{\nu_+ + \nu_-} \tag{11}$$

The diffusional force (driving force of diffusion) $\boldsymbol{f}_B = -c \nabla \mu_B$ associated with an electrolyte solution is given by the gradient of the chemical potential of the solute multiplied by the molar concentration. Finally, the molar flux of the solute $\boldsymbol{q} = -C_B \boldsymbol{f}_B$ is proportional to the driving force of diffusion.

$$\boldsymbol{q} = -C_B \, R\,T \, [\nu_+ + \nu_-] \left[1 + c \, \frac{d \ln y_\pm}{dc} \right] \nabla c \tag{12}$$

Herein, the term on the left hand side of ∇c can be interpreted as a concentration dependent diffusion coefficient $D(c)$.

$$\boldsymbol{q} = -D(c) \, \nabla c = D(c) \, \mathbf{1} \cdot \boldsymbol{\gamma} \tag{13}$$

In the following, the concentration dependent diffusion coefficient $D(c)$ is investigated in a chronological sequence based on the limiting molar conductivities of the involved ions, the dissociation quantities, natural constants and properties of the solvent.

3.3 *Infinitely diluted solutions*

The limiting value of the diffusion coefficient for infinitely diluted electrolyte solutions is given by (Nernst 1888).

$$D = D^N = D_0 = \frac{R\,T}{F^2} \, \frac{z_+ + z_-}{z_+ \, z_-} \, \frac{\lambda_+^0 \, \lambda_-^0}{\lambda_+^0 + \lambda_-^0} \tag{14}$$

Herein, F is the Faraday constant, λ_+^0 is the limiting equivalent conductivity of the cations and, λ_-^0 is the limiting equivalent conductivity of the anions (see Table 2). Nernst's diffusion coefficient represents the electrostatic coupling of cation and anion migration enforced by the principle of electrical neutrality. It corresponds to the thermodynamics of diffusion for ideal solutions ($y_\pm = 1$).

Table 2. Constants associated with the investigated ions and the solvent at $T = 298\,K$, see e.g. (Atkins 1998).

	OH^-	Ca^{2+}	Cl^-	Na^+	SO^{2-}	H_2O	Unit	Quantity
z	1	2	1	1	2		–	valency
λ^0	$1.992 \cdot 10^{-2}$	$5.950 \cdot 10^{-3}$	$7.635 \cdot 10^{-3}$	$5.010 \cdot 10^{-3}$	$8.000 \cdot 10^{-3}$		Sm^2/mol	conductivity
$a_\pm$	$1.33 \cdot 10^{-10}$	$0.99 \cdot 10^{-10}$	$1.81 \cdot 10^{-10}$	$0.95 \cdot 10^{-10}$	$2.58 \cdot 10^{-10}$		m	ionic radius
η						$8.91 \cdot 10^{-4}$	kg/ms	viscosity
ε_r						$7.854 \cdot 10^1$	–	relative permittivity

3.4 *Relaxation effect in concentrated solutions*

If the mean free distance between migrating ions gets smaller at higher concentrations, repelling or attracting electrostatic forces interact with the Brownian motion of the ions. This interaction results in the relaxation (or activity) effect which reduces the macroscopic ionic drift. The relaxation effect is considered by the Nernst-Hartley diffusion model (Hartley 1931).

$$D = D^{NH} = D_0 \left[1 + c \frac{d \ln y_\pm}{dc} \right] \tag{15}$$

Hartley has modified the Nernst diffusion coefficient of infinitely diluted ionic solutions by the factor $cd\ln y_\pm/dc$ defined in terms of the mean molar activity coefficient $y_\pm$ and the molar concentration c. The Nernst-Hartley diffusion model is justified by the thermo-dynamics of diffusion, compare equation 12.

The approximation of the molar activity coefficient

$$\ln y_\pm = -\frac{z_+ \, z_- \, e^2}{8 \, \pi \, \epsilon \, k \, T} \frac{\kappa}{1 + \kappa \, a} \tag{16}$$

which is based on the Debye-Hüuckel theory (Debye and Hückel 1923a; Debye and Hückel 1923b; Debye 1927), has been demonstrated to be well suited within the concentration range of calcium leaching of calcium hydroxide (Kuhl 2002). In equation (16) e is the elementary charge, $\varepsilon = \varepsilon_0 \varepsilon_r$ is the permittivity of the solvent, k is the Bolzmann constant and a is the ion size parameter (or the mean ionic diameter) of the electrolyte. In equation (16) $1/\kappa$ is the Debye-Hückel length (diameter of the ionic cloud)

$$\kappa = \sqrt{\frac{2 \, N \, e^2}{\epsilon \, k \, T} \, I} \tag{17}$$

which can be calculated in terms of the Avogadro number $N = R/k$, the ionic strength $I = [\nu_+ z_+^2 + \nu_- z_-^2] c/2$ and previously introduced constants. From equation (17) follows, that κ is proportional to the square root of the molar concentration,

$$\kappa = A_2 \sqrt{c}, \quad A_2 = \sqrt{\frac{N \, e^2}{\epsilon \, k \, T} \, [\nu_+ z_+^2 + \nu_- z_-^2]} \tag{18}$$

where A_2 is introduced as a constant of proportionality. The relaxation term of equation (15) is obtained as the derivative of the mean molar activity coefficient (16) with respect to the mean molar concentration.

$$c \frac{d \ln y_\pm}{dc} = A_3 \frac{\kappa}{[1 + \kappa \, a]^2}, \quad A_3 = \frac{-z_+ z_- \, e^2}{16 \, \pi \, \epsilon \, k \, T} \tag{19}$$

3.5 *Electrophoretic effect in concentrated solutions*

If the mobility of individually migrating cations and anions is significantly different, an additional ion-ion interaction phenomenon influences the electrolyte diffusion. Caused by the viscosity of the solvent, moving ions interact with the pore water and the ions in their vicinity due to the flow field. This phenomenon, known as the electrophoretic effect, is represented by the extended Nernst-Hartley diffusion coefficient proposed by (Onsager and Fouss 1932).

$$D = D^{OF} = [D_0 + \triangle] \left[1 + c \frac{d \ln y_\pm}{dc} \right] \tag{20}$$

The empirical electrophoretic correction of the diffusion coefficient (Onsager and Fouss 1932)

$$\triangle = A_1 \frac{\kappa}{1 + \kappa \, a} \tag{21}$$

can be computed with the factor of proportionality

$$A_1 = -\frac{k \, T}{6 \, \pi \, \eta} \frac{\left[z_+ \, \lambda_-^0 - z_- \, \lambda_+^0 \right]^2}{z_+ \, z_- \, \left[\lambda_+^0 + \lambda_-^0 \right]^2} \tag{22}$$

using the viscosity of the solvent η.

3.6 *Five parameter electrolyte diffusion model*

Summarizing previous sections, the Onsager-Fouss diffusion model can be identified as a five parameter electrolyte diffusion model

$$D^{OF} = D_0 + \frac{A_1 \kappa}{1 + \kappa a} + \frac{D_0 A_3 \kappa}{[1 + \kappa a]^2} + \frac{A_1 A_3 \kappa^2}{[1 + \kappa a]^3} \tag{23}$$

using the parameters $D_0, A_1, A_2 A_3$ and a. This general diffusion model can be specialized to the Nernst-Hartley ($A_1 = 0$) and the Nernst ($A_1 = A_3 = 0$) diffusion models.

$$D^{NH} = D_0 + \frac{D_0 A_3 \kappa}{[1+\kappa a]^2}, \quad D^N = D_0 \tag{24}$$

For an overview of models within the family of diffusion models (23) see (Kuhl 2002).

3.7 *Simplifications*

As proposed by (Onsager 1927), in diluted electrolyte solutions the mean ionic diameter a can be neglected compared to the diameter of the ionic cloud $1/\kappa$. Onsager's approximation D_O^{OF} of the diffusion coefficient D^{OF} (23)

$$D_O^{OF} = D_0 + [A_1 + D_0 A_3] A_2 \sqrt{c} + A_1 A_2^2 A_3 \, c \tag{25}$$

includes the constant Nernst diffusion coefficient $D^N = D_0$, a term proportional to the square root of the molar concentration $[A_1 + D_0 A_3] A_2 \sqrt{c}$ and a term proportional to the molar concentration $A_1 A_2^2 A_3 c$. The term $A_1 A_2^2 A_3 c$ captures the interaction between relaxation and electrophoretic effects. If, furthermore, this interaction is neglected, an expression equivalent to the empirical conduction law by (Kohlrausch 1900) and (Kohlrausch and von Steinwehr 1902) is obtained.

$$D_K^{OF} = D_0 + D_{0c} \sqrt{c} \tag{26}$$

Based on the formulation of Kohlrausch's law within the context of the the Debye-Hückel theory the Kohlrausch constant D_{0c} can be identified as:

$$D_{0c} = A_1 \, A_2 + D_0 \, A_2 \, A_3 \tag{27}$$

Unfortunately, above approximations (25, 26) of the concentration dependent diffusion coefficient are not sufficiently accurate within the relevant concentration range of calcium leaching of cementitious materials (Kuhl 2002).

4 VALIDATION OF DIFFUSION MODELS

The presented models for electrolyte diffusion by Nernst D^N (14, 24), Nernst-Hartley D^{NH} (15, 24) and Onsager-Fouss D^{OF} (20, 23) are applied to calculate the concentration dependent diffusion coefficient $D(c)$ of several electrolytes. The model results are compared to experimental results collected by (Robinson and Stokes 1955). In particular, the diffusion coefficients for sodium chloride (NaCl), taken as an example of an univalent electrolyte (1:1) with a small electrophoretic coefficient $z_+\lambda_-^0 - z_-\lambda_+^0 = 2.625 \cdot 10^{-3} Sm^2/mol$, sodium sulfate ($Na_2SO_4$), taken as an example of a polyvalent electrolyte (2:1) with a small electrophoretic coefficient $z_-\lambda_+^0 - z_+\lambda_-^0 = 2.02 \cdot 10^{-3} Sm^2/mol$, calcium chloride ($CaCl_2$), taken as an example of a polyvalent electrolyte with a moderate electrophoretic coefficient $z_+\lambda_-^0 - z_-\lambda_+^0 = 9.32 \cdot 10^{-3} Sm^2/mol$, and, finally, the polyvalent electrolyte calcium hydroxide ($Ca(OH)_2$) with a large electrophoretic coefficient $z_+\lambda_-^0 - z_-\lambda_+^0 = 33.89 \cdot 10^{-3} Sm^2/mol$ are investigated (compare Table 2). Since experimental results are only available for the aqueous NaCl, Na_2SO_4 and $CaCl_2$ electrolyte solutions, the models are validated on the basis of these electrolytes. Furthermore, the diffusion models are applied to calculate the diffusion coefficient associated with the aqueous $Ca(OH)_2$ electrolyte solution in order to point out the characteristics of ionic transport during chemical degradation of cementitious materials. The constants $D_0, A_1, A_2 A_3$ and a of the five parameter electrolyte diffusion model (Section 3.6) are summarized in Table 3 for $T = 298\,K$. In Figure 1 the diffusion coefficients are plotted as function of the molar electrolyte concentration.

From Table 3 it is obvious, that the limiting diffusion coefficient of electrolyte solutions D_0 and the related diffusion coefficients of independently migrating single ionic species (see e.g. (Atkins 1998)) are significantly different. Compared to the diffusion coefficient of single calcium ions $D_{Ca^{2+}} = 0.7918 \cdot 10^{-9}\, m^2/s$, the diffusion coefficient of the aqueous calcium hydroxide electrolyte solution $D_0 = 1.829 \cdot 10^{-9}\, m^2/s$ is more than two times larger. Consequently, the use of $D_{Ca^{2+}}$ for the simulation of calcium leaching leads to improper results of the life time prognosis of concrete structures.

For aqueous sodium chloride, sodium sulfate and calcium chloride electrolyte solutions the Onsager-Fouss model D^{OF} the correlation between the model results and the experimental results is excellent. The results obtained from the Nernst-Hartley model D^{NH} are close to D^{OF} for sodium chloride and sodium sulfate. Due to the large electrophoretic coefficient, the Onsager-Fouss and Nernst-Hartley models yield considerably different results for calcium chloride and calcium hydroxide solutions. For all investigated electrolytes Nernst's constant diffusion coefficient is identical to the Onsager-Fouss and Nernst-Hartley diffusion models only for infinitely diluted solutions.

Finally, the differences between the Nernst, Nernst-Hartley and Onsager-Fouss models are investigated. As the electrophoretic and relaxation effects are neglected, the Nernst model defines a constant diffusion coefficient, whereas the remaining two models are characterized by a strong dependence of the diffusion coefficient on the concentration c. According to the

Table 3. Parameters associated with aqueous $CaCl_2$, $Ca(OH)_2$, NaCl and Na_2SO_4 electrolyte solutions at $T = 298\,K$ (for the ion size parameter a see (Bockris and Redd 1998) and (Robinson and Stokes 1955)).

	$CaCl_2 \rightleftharpoons Ca^{2+}{+}2Cl^-$	$Ca(OH)_2 \rightleftharpoons Ca^{2+}{+}2OH^-$	$NaCl \rightleftharpoons Na^+{+}Cl^-$	$Na_2SO_4 \rightleftharpoons 2Na^+{+}SO^{2-}$	eq.	unit
$z_+ : z_-$	2 : 1	2 : 1	1 : 1	1 : 2		
$\nu_+ : \nu_-$	1 : 2	1 : 2	1 : 1	2 : 1		
D_0	$1.33502145 \cdot 10^{-09}$	$1.82903379 \cdot 10^{-09}$	$1.61023236 \cdot 10^{-09}$	$1.22990707 \cdot 10^{-09}$	(14)	$\frac{m^2}{s}$
A_1	$-5.76509839 \cdot 10^{-19}$	$-2.10188565 \cdot 10^{-19}$	$-1.05571120 \cdot 10^{-20}$	$-2.95285642 \cdot 10^{-21}$	(21)	$\frac{m^3}{s}$
A_2	$1.80048775 \cdot 10^{+08}$	$1.80048775 \cdot 10^{+08}$	$1.03951208 \cdot 10^{+08}$	$1.80048775 \cdot 10^{+08}$	(18)	$\sqrt{\frac{m}{mol}}$
A_3	$-3.56975479 \cdot 10^{-10}$	$-3.56975479 \cdot 10^{-10}$	$-1.78487740 \cdot 10^{-10}$	$-3.56975479 \cdot 10^{-10}$	(19)	m
a	$4.73000000 \cdot 10^{-10}$	$4.25000000 \cdot 10^{-10}$	$3.97000000 \cdot 10^{-10}$	$5.46000000 \cdot 10^{-10}$		m

small electrophoretic coefficient of NaCl and Na_2SO_4 electrolyte solutions the electrophoretic ion-ion interaction is only a secondary effect and, consequently, negligible for the associated diffusion coefficients. In contrast, the electrophoretic effect has a significant influence on the diffusion of dissociated calcium chloride and calcium hydroxide. This is manifested in the difference of the results obtained from the D^{OF} and the D^{NH} model. It can be explained by the fact, that the electrophoretic effect is controlled by the difference of the limiting molar conductivities λ^0 of the involved cations and anions (compare equation (21) and Table 2). This effect is particularly pronounced for the calcium hydroxide. Due to the proton jump mechanism the hydroxide ions (OH^-) are able to move significantly faster than the calcium ions. As a consequence, the limiting molar conductivity of hydroxide ions is more than three times larger than the limiting molar conductivity of calcium ions. Hence, the electrophoretic effect is crucial for the realistic modeling of calcium hydroxide leaching of cementitious materials.

According to the previous investigations, the Onsager-Fouss model is suggested for the realistic modeling of charged mass transport through the pore space of cementitious materials. The diffusion coefficient of the calcium hydroxide solution is reduced by ion-ion interactions in concentrated solutions by approximately 30%. Consequently, higher order transport models are an essential basis for the reliable prognosis of long term degradation of cementitious materials subjected to chemically aggressive media.

5 SIMULATION OF CALCIUM LEACHING

The investigated diffusion models are further evaluated by means of the numerical analysis of calcium leaching of a one dimensional cementitious bar illustrated in Figure 2. The numerical analysis is based on the weak formulation of the calcium mass balance, the finite element discretization, the implicit time integration and the Newton-Raphson iteration (Kuhl 2002). The bar is discretized by linear finite elements (for a convergence study using h-and p-refinement strategies see (Kuhl et al. 2002)) and integrated within the time interval $t \in [0,500\,\Delta t]$with the implicit midpoint rule using the time step $\Delta t = 150\,d$. Data of the phenomenological chemical equilibrium model and the chemical damage model are used according to (Kuhl et al. 2002) as $\phi_0 = 0.2$, $c_0 = 20.7378\,mol/m^3$, $c_p = 19\,mol/m^3$, $c_{csh} = 1.5\,mol/m^3$, $s_0 = 15\,kmol/m^3$, $s_h = 9\,kmol/m^3$, $\alpha_c = 0.565$, $n_c = 85$, $m_c = 5$ and $M/\rho = 0.035\,m^3/kmol$. The initial conditions are given by $c = c_0$ for $X \in \Omega = [0, L]$ and the calcium concentration is prescribed at the Neumann boundary $X \in \Gamma_q = [0]$ as illustrated in Figure 2.

Figure 3 shows the normalized calcium concentration of the pore fluid c/c_0 , the normalized concentration of calcium chemically bounded in the skeleton s/s_0 and the normalized dissolution rate $\dot{s}_{\text{norm}}$ (the maximum of the dissolution rate is used as reference value) as functions of space and time. In Figure 3 the black color characterizes a value of one and white represents zero. The dissociation-diffusion mechanism is briefly explained by means of the results obtained by the standard transport model of independently migrating Ca^{2+} ions: The concentration c/c_0 in space and time allows a clear distinction between the chemical loading and unloading regimes. Reaction-diffusion is active within the first 400 time steps. The material degradation can be represented by the concentration s/s_0. Regions of the virgin and fully degenerated material are indicated by $s/s_0 = 1$ and $s/s_0 = 0$, respectively. Between these zones the cementitious material is dissolved by the chemically aggressive pore fluid. In the chemical unloading regime the skeleton calcium concentration remains constant, since calcium leaching is irreversible. The calcium dissolution rate $\dot{s}_{\text{norm}}$ illustrates the pronounced front of calcium hydroxide (portlandite) dissolution and of the calcium silicate hydrates (csh) dissolution front. In comparison to the csh dissolution front, the calcium hydroxide front

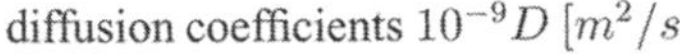

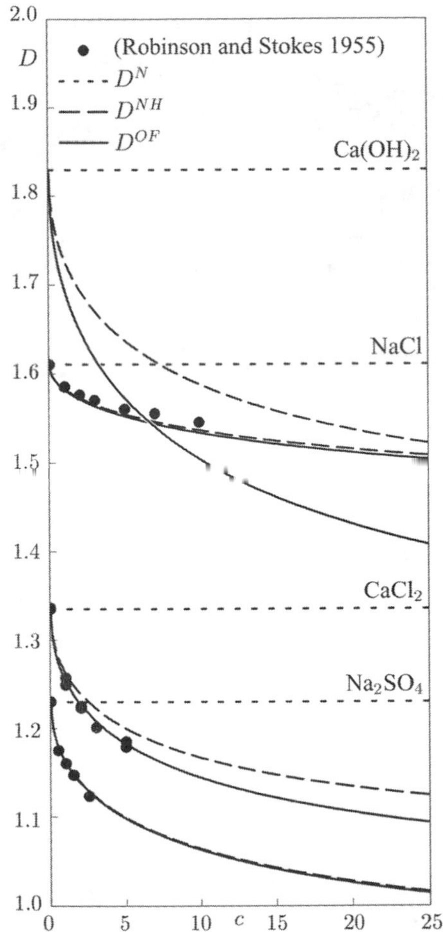

Figure 1. Diffusion coefficients $10^{-9}D[m^2/s]$ for calcium chloride $CaCl_2$, calcium hydroxide $Ca(OH)_2$, sodium chloride NaCl and sodium sulfate Na_2SO_4 aqueous electrolyte solutions at $T = 298\,K$ as function of the concentration $c[mol/m^3]$.

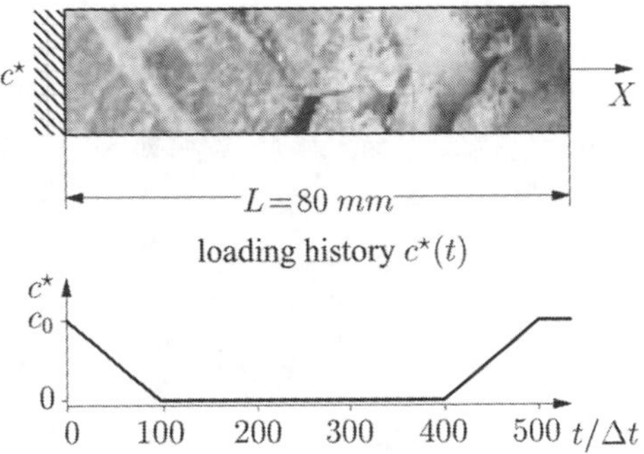

Figure 2. Calcium leaching of a cementitious bar – geometry and loading history.

shows a significantly higher dissolution rate. The calcium hydroxide front moves through the cementitious specimen, whereas the movement of the csh front is restricted to the boundary zone. Between both pronounced dissolution fronts a significantly smaller amount of ettringite is dissolved. Furthermore, three characteristics, corresponding to the end of the chemical loading (char l), the beginning of the chemical unloading (char u) and the end of the portlandite dissolution (char p), are observed. Whenever a characteristic meets a dissolution front, the velocity of the front, and consequently the time scale of the system is changed.

The results of the finite element simulations confirm the conclusions from the investigations of the diffusion models in Section 4. The standard diffusion model $D_{Ca^{2+}}$ yields a considerably slower chemical degradation compared to electrolyte diffusion models. This means, that the life time of concrete structures is overestimated by a factor two by the standard diffusion model. As expected, the differences between the electrolyte diffusion models are less pronounced. However, the differences between these models are best identified by means of the distances of the characteristics (char l) and (char p). From the three investigated electrolyte diffusion models, the slowest chemical degradation and, consequently, the longest life time prognosis is obtained by the Onsager-Fouss model. The differences between the Onsager-Fouss und Nernst-Hartley resulting from the electrophoretic effect are considerable (see also Figure 1). The application of Nernst's limiting diffusion coefficient leads to a considerable underestimation of the life time of concrete structures. In the context of the modeling of calcium leaching, characterized by an averaged level of concentration of $c \approx 10\,mol/m^3$ between the portlandite front and the boundary, the differences between the investigated models are significant. Consequently, it is recommended to calculate the concentration dependent diffusion coefficient by means of the more realistic Onsager-Fouss model. However, for a conservative design of concrete structures under chemical attack, the Nernst-Hartley or the Nernst electrolyte diffusion models may also be used. In contrast, the diffusion coefficient of independently migrating calcium ions leads to prognoses of a significantly longer life time of concrete structures. Consequently, unexpected chemically induced structural collapse may occur, if the chemo-mechanical structural analysis is performed by means of the standard diffusion model.

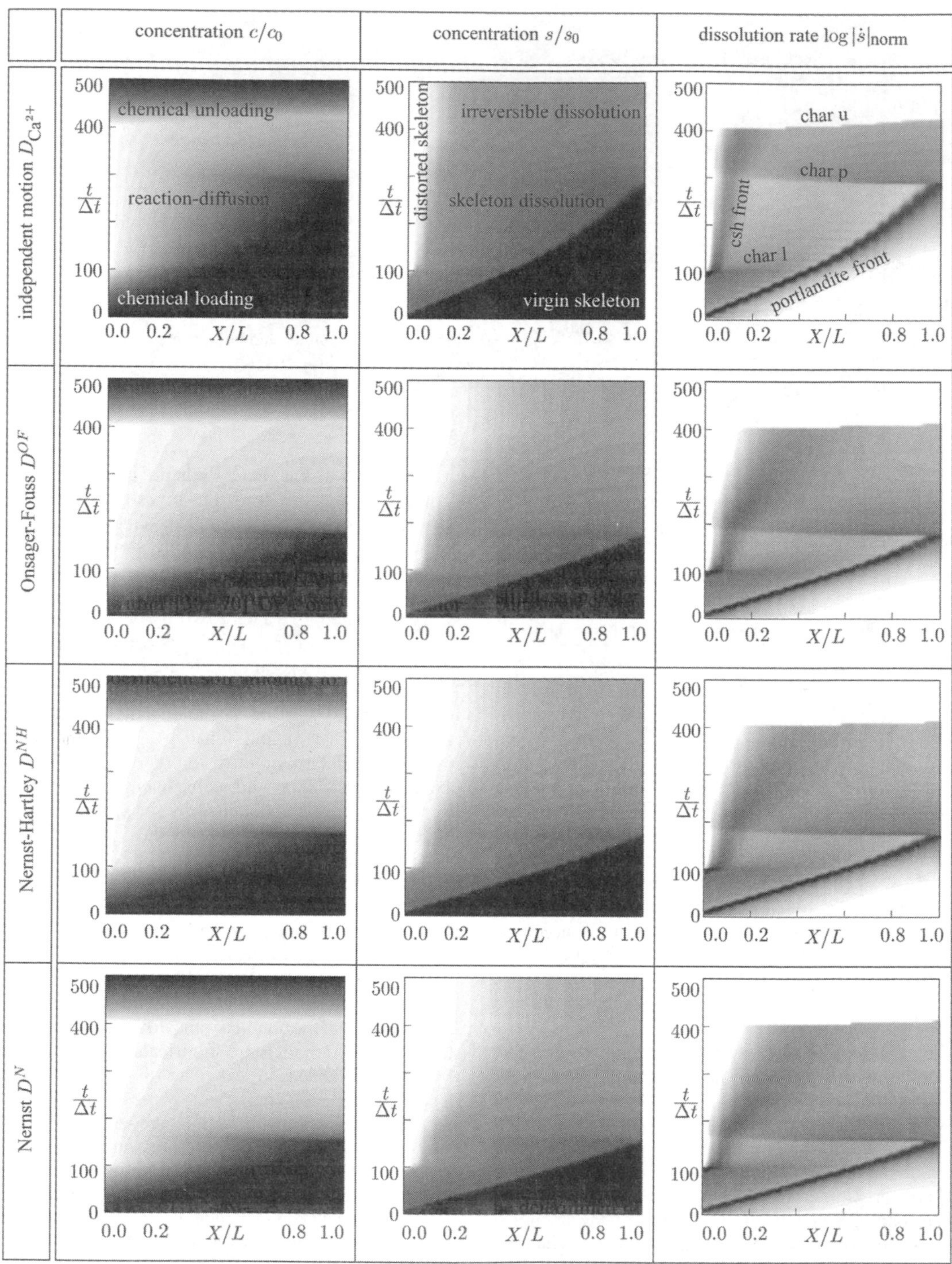

Figure 3. Calcium leaching of a cementitious bar – results.

6 CONCLUSIONS

A dissociation-diffusion model for calcium leaching of cementitious materials was presented. Main attention was devoted to the discussion and critical investigation of different electrolyte diffusion models based on the Debye-Hückel theory of ionic clouds. These models consider ion-ion interactions due to the electrostatic coupling of migration cations and anions (Nernst 1888), the relaxation effect (Hartley 1931) and the electrophoretic phenomenon (Onsager and Fouss 1932). Based on the comparison of the electrolyte diffusion coefficients obtained from these models with experimental results for several univalent and polyvalent electrolytes, the detailed study of ion-ion interaction phenomena and the results of the numerical analysis, the following recommendations are given:

- It is necessary to use at least Nernst's limiting diffusion coefficient as the basis for reliable simulations of calcium leaching.
- More accurate diffusion models are obtained, if ion-ion interactions are considered according to the Onsager-Fouss model.
- As a consequence of the proton jump mechanism associated with the transport of OH^- ions, the electrophoretic effect noticeably affects the diffusion of dissolved calcium hydroxide.

ACKNOWLEDGMENTS

Financial support was provided by the German National Science Foundation (DFG) in the framework of project A9 of the collaborative research center (SFB) 398. This support is gratefully acknowledged.

REFERENCES

Atkins, P. (1998). *Physical Chemistry*. Oxford: Oxford University Press.

Barthel, J., H. Krienke, and W. Kunz (1998). *Physical Chemistry of Electrolyte Solutions, Modern Aspects*. New York: Springer-Verlag.

Berner, U. (1988). Modelling the incongruent dissolution of hydrated cement minerals. *Radiochimica Acta 44-45*, 387–393.

Bockris, J. and A. Reddy (1998). *Modern Electrochemistry 1. Ionics*. New York: Plenum Press.

Carde, C. and R. François (1997). Effect on the leaching of calcium hydroxide from cement paste on mechanical properties. *Cement and Concrete Research 27*, 539–550.

Chatterji, S. (1999). Evidence of variable diffusivity of ions in saturated cementitious materials. *Cement and Concrete Research 29*, 595–598.

Coussy, O. and F.-J. Ulm (2001). Elements of durability mechanics of concrete structures. In F.-J. Ulm, Z. Bažant, and F. Wittmann (Eds.), *Creep, Shrinkage and Durability Mechanics of Concrete and other Quasi-Brittle Materials*, Amsterdam, pp. 3993–4009. Elsevier Science.

de Boer, R. (2000). *Theory of Porous Media. High-lights in the Historical Developement and Current State*. Berlin: Springer-Verlag.

Debye, P. (1927). Report on conductivity of strong electrolytes in dilute solutions. In C. Desche (Ed.), *The Theory of Strong Electroloytes. A General Discussion*, Volume 23, pp. 334–340. Transactions of the Faraday Society.

Debye, P. and E. Hückel (1923a). Zur Theorie der Elektrolyte. I. Gefrierpunktserniedrigung und verwandte Erscheinungen. *Physikalische Zeitschrift 24*(9), 185–206.

Debye, P. and E. Hückel (1923b). Zur Theorie der Elektrolyte. II. Grenzgesetz für die elektrische Leitfähigkeit. *Physikalische Zeitschrift 24*(15), 305–325.

Delagrave, A., B. Gérard, and J. Marchand (1997). Modelling the calcium leaching mechanisms in hydrated cement pastes. In K. Scrivener and J. Young (Eds.), *Mechanics of Chemical Degradation of Cement-Based Systems*, pp. 30–37. London: Chapman & Hall.

Ehlers, W. (2002). Foundations of multiphasic and porous materials. In W. Ehlers and J. Bluhm (Eds.), *Porous Media. Theory, Experiments and Numerical Applications*, pp. 3–86. Berlin: Springer.

Fick, A. (1855). Ueber Diffusion. *Annalen der Physik und Chemie 94*(1), 59–86.

Gérard, B. (1996). *Contribution des Couplages Mécanique-Chimie-Transfert dans la Tenue a long Terme des Ouvrages de Stockage de Déchets Radioactifs*. Ph. D. thesis, Laboratoire de Méchanique et Technologie. E.N.S. de Cachan.

Gérard, B. (2000). Hydrolysis of cement-based materials: A review. In M. Setzer (Ed.), *Pore Solution in Hardened Cement Paste*, pp. 271–304. Freiburg: Aedificatio Publishers.

Gérard, B., G. Pijaudier-Cabot, and C. La Borderie (1998). Coupled diffusion-damage modelling and the implications on failure due to strain localisation. *International Journal of Solids and Structures 35*, 4107–4120.

Hartley, G. (1931). Theory of the velocity of diffusion of strong electrolytes in dilute solution. *The London, Edinbourgh and Dublin Philosophical Magazine, and Journal of Science 12*(77), 473–488.

Horvath, A. (1985). *Handbook of Aqueous Electrolyte Solutions. Physical Properties, Estimation and Correlation Methods*. Chichester: Ellis Horwood Limited Publishers.

Kohlrausch, F. (1900). Über das elektrische Leitvermögen von Lösungen der Alkali-Jodate und eine Formel zur Berechnung von Leitvermögen. *Sitzungsberichte der königlich preussischen Akademie für Wissenschaften*, 1002–1008.

Kohlrausch, F. and H. von Steinwehr (1902). Weitere Untersuchungen über das Leitvermögen von Elektrolyten aus einwertigen Ionen in wässriger Lösung. *Sitzungsberichte der königlich preussischen Akademie für Wissenschaften*, 581–587.

Kuhl, D. (2002). Ion diffusion models based on the Debye-Hückel-Onsager theory applied to calcium leaching of cementitious materials. *Journal of Engineering Mechanics*. submitted.

Kuhl, D., F. Bangert, and G. Meschke (2000). An extension of damage theory to coupled chemomechanical processes. In E. Oñate (Ed.), *European Congress on Computational Methods in Applied Science and Engineering*, Barcelona.

Kuhl, D., F. Bangert, and G. Meschke (2002). Coupled chemo-mechanical deterioration of cementitious materials. Part I: Modeling. Part II: Numerical methods and simulations. *International Journal of Solids and Structures*. submitted.

Le Bellégo, C., B. Gérard, and G. Pijaudier-Cabot (2000). Chemo-mechanical effects in motar beams subjected to water hydrolysis. *Journal of Engineering Mechanics 126*, 266–272.

Mainguy, M. and O. Coussy (2000). Propagation fronts during calcium leaching and cloride penetration. *Journal of Engineering Mechanics 126*, 250–257.

Mainguy, M., C. Tognazzi, J.-M. Torrenti, and F. Adenot (2000). Modelling of leaching in pure cement paste and motar. *Cement and Concrete Research 30*, 83–90.

Nernst, W. (1888). Zur Kinetik der in Lösung befindlichen Körper. I. Theorie der Diffusion. *Zeitschrift für physikalische Chemie 2*, 613–637.

Onsager, L. (1927). Report on a revision of the conductivity theory. In C. Desche (Ed.), *The Theory of Strong Electroloytes. A General Discussion*, Volume 23, pp. 341–356. Transactions of the Faraday Society.

Onsager, L. and R. Fouss (1932). Irreversible processes in electrolytes. Diffusion, conductance, and viscous flow in arbitrary mixtures of strong electrolytes. *Journal of Physical Chemistry 35*, 2689–2778.

Robinson, R. and R. Stokes (1955). *Electrolyte Solutions. The Measurement an Interpretation of Conductance, Chemical Potential and Diffusion in Solutions of Simple Electrolytes*. London: London Butterworths.

Taylor, H. (1997). *Cement Chemistry*. London: Thomas Telford.

Ulm, F.-J., J.-M. Torrenti, and F. Adenot (1999). Chemoporoplasticity of calcium leaching in concrete. *Journal of Engineering Mechanics 125*, 1200–1211.

Computational Modelling of Concrete Structures, Bićanić et al. (eds)
© 2003 Swets & Zeitlinger, Lisse, ISBN 90 5809 536 3

A micromechanical analysis of the observed kinetics of ASR-induced swelling in concrete

Eric Lemarchand, Luc Dormieux* & Djimédo Kondo
LML-CNRS UMR 8107, USTL, France

ABSTRACT: Alkali-Silica Reaction induced macroscopic swelling is one of the leading research activity in Durability Mechanics of concrete structures. This phenomenon, resulting from a chemical reaction occuring in the vicinity of the aggregates, is characterized by the pressurization of a gel filling the porosity directly available. Based on experimental evidences, we consider that the ASR gel is trapped in randomly distributed cracks surrounding aggregates. Therefore, our purpose is to develop a micromechanical approach capable to explain the very first (elastic) part of the observed macroscopic swelling kinetics. For a set of cracks of identical radii, we show that the gel pressure rate is uniform in the saturated cracks when considering an isotropic macroscopic state of stress. Besides, the crucial effect of the kinetics of the filling process on the observed swelling kinetics is emphasized. At last, we study the consequences of an isotropic macroscopic stress on the initial distribution of the cracks. The resulting macroscopic swelling is discussed.

1 INTRODUCTION

Alkali-Silica Reactions are well-known deleterious phenomena occuring in the pore space of concrete materials. Macroscopically, the observed swelling is attributed to the formation of a gel, as a result of the combination of alcalis (Na^+, Cl^-), hydroxyles (OH^-) and reactive silica (Si^{2+}) constituting a part of some aggregates, which progressively fills the available pore space. Two scenarii, namely the through-solution and the topochemical mechanisms, have been proposed to describe this internal chemical process. The chemical reactions leading to the gel formation are now quite well understood, but these two invoked mechanisms differ from one another as regard the interpretation of the filling process of the available porosity. In the framework of micromechanics, it is of paramount importance to clarify this question since each of these mechanisms is associated to a morphological information on the distribution of the local pressures within the solid matrix. Based on experimental evidences (Larive 1998), we will consider that the local mechanism is essentially topochemical in nature, *i.e.* the gel is trapped in cracks surrounding the aggregates. Thus, as soon as local saturation is achieved, the ASR gel is pressurized and may damage the surrounding solid matrix.

* LMSGC-CNRS UMR 113, ENPC, France

Our purpose in the present study is to develop a micromechanical reasoning accounting for the pressurization of cracks related to the continuous gel formation and leading to classical poroelastic equations. The question of the pressure-induced crack propagation will be addressed in a further study. More specifically, accounting for the filling process, poroelasticity equations will lead to the identification of an effective stress of Bishop type, *i.e.* controlled by a saturation degree. In particular, we will see how this saturation degree controls the non linear macroscopic strain, whatever the kinetic of the gel formation. At last, the effect of an isotropic macroscopic loading applied before ASR occur is analysed.

Attention has to be made on the fact that aggregates are completely disregarded in the present work. A more detailed analysis can be derived from (Lemarchand et al. 2002), where the aggregates are geometrically and mechanically incorporated. At the same scale as the cracks, in order to account for their interaction.

2 EXPERIMENTAL EVIDENCES

The present section aims at interpretating experimental tests performed on concrete samples subjected to Alkali-Silica Reactions (Larive 1998) in order to retain one of the two principal local mechanisms considered in the open litterature. By its chemical definition, the trough-solution mechanism considers

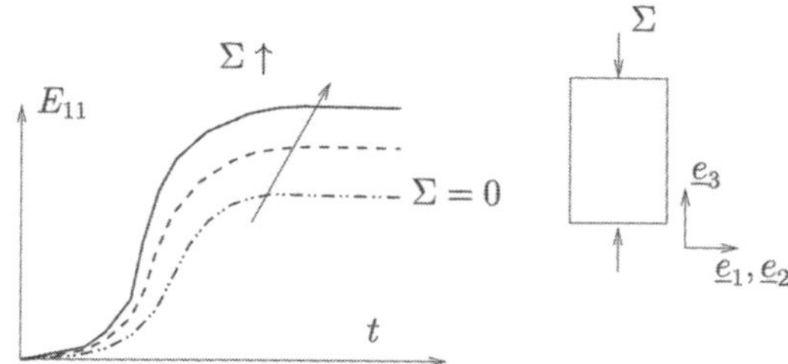

Figure 1. Macroscopic strains for different loadings.

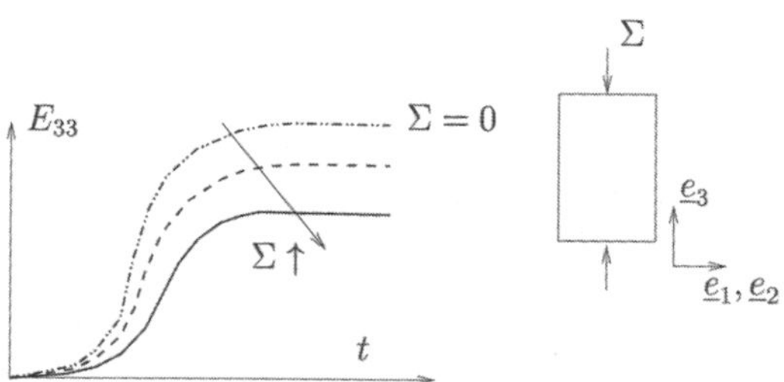

Figure 2. Macroscopic strains for different loadings.

that the ASR gel is formed everywhere in the whole connected porosity of the cement paste. By contrast, the topochemical mechanism suggests that the gel is trapped in existing cracks located at the aggregate/cement paste interface. While it seems difficult to discriminate one of these mechanisms with respect to the other when dealing with macroscopic free swelling conditions (Lemarchand 2001), application of a macroscopic stress can prove, by comparing with experimental results, that the through-solution mechanism is not suficient to explain the macroscopically observed response of an ASR-subjected concrete sample.

Indeed, Larive (Larive 1998) showed that application of an oriented compressive stress leads to a decrease (resp. increase) of the macroscopic swelling in the direction parallel (resp. perpendicular) to the stress application. This is physically attributed to the progressive desactivation of the gel formation sites.

In other words, the corresponding pores (cracks) being progressively closed, the reactants are not able to combine in order to create the gel. In the range of the stresses applied in these tests, it is clearly impossible to expect a shape transformation of the spherical pores of the connected porosity. By contrast, due to their very oriented (flat) geometry, the cracks can easily be closed by such a loading. Therefore, the local mechanism seems to be essentially topochemical.

Hence, following a topochemical mechanism, which states a gel formation in cracks located in the vicinity of the aggregates (Interfacial Transition Zone), we consider a set of cracks, randomly

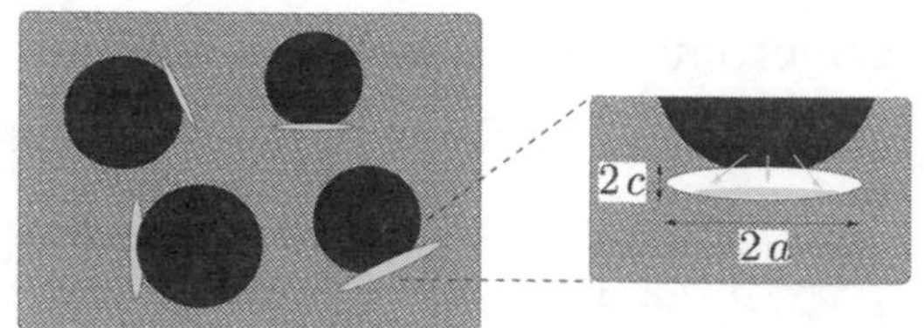

Figure 3. Formation and filling processes of ASR gel.

distributed, but always "in contact"[1] with a solid grain. Cracks are modelled as oblate spheroids. Therefore, the initial cracking state of a concrete sample is constituted by different i-families of cracks, each of them being defined by an initial radius $a_{i,0}$ an initial aspect ratio $X_0^i = c_{i,0}, /a_{i,0}(2a_{i,0} =$ initial crack diameter, $2c_{i,0} =$ initial crack opening, initial crack volume $V_{i,0}^c = \frac{4\pi}{3} X_0^i a_{i,0}^3)$ and an unit normal $n_{i,0}$.

In these conditions, the gel fills the cracks, through the solid/crack interface, along a circular area (see Fig. 3), proportional to the square of the crack radius. Accordingly, the rate of gel volume formation $\dot{\beta}$ in a crack of the i-family reads:

$$\dot{\beta}_i(t) = \kappa(t)\, a_i^2 \tag{1}$$

where $\kappa(t)[m \cdots^{-1}]$ is a chemical coefficient, directly related to the chemical kinetic controlling the gel formation. *A priori*, κ may depend on local mechanical conditions on the crack, but this possibility will be disregarded in the sequel.

Furthermore, we restrict our analysis to non propagating cracks, *i.e.* each crack keeps its initial length: $a_i(t) = a_i(t = 0) = a_{i,0}$.

3 CHEMOMECHANICAL COUPLINGS

We aim hereafter at expressing the mechanical consequences of the local chemistry (gel formation). These chemomechanical couplings develop as soon as a crack is saturated by the ASR gel. According to the ellipsoidal shape of the pores and the kinetics of the gel formation given in (1), we are able to determine the time required to fill a pore. Indeed, the characteristic time t_f^i of the filling process, which corresponds to the time required to saturate the volume of a crack of the i-family, defined by the couple (a_i, X_0^i), depends on the initial aspect ratio of a crack:

$$\frac{4\pi}{3} a_i^3 X_0^i = \int_0^{t_f^i} \dot{\beta}_i d\tau \;\rightarrow\; t_f^i = K^{-1}\left(\frac{4\pi}{3} a_i X_0^i\right) \tag{2}$$

[1] Still, the aggregates are not incorporated into our modelling. They explain here the physical origin of the cracks.

where $K(t) = \int_0^t \kappa(\tau)d\tau$. In particular, if κ is assumed constant, the characteristic time t_f^i depends linearly on the initial aspect ratio X_0^i:

$$t_f^i = \alpha\, a_i\, X_0^i \tag{3}$$

where $\alpha = 4\pi/3\kappa$ is a constant scalar.

As long as $t < t_f^i$, the unpressurized gel is filling the j cracks families satisfying $x_o^j \geq X_0^i$. On the other hand, as t_f^i is reached, the cracks of the i-family are saturated by the gel. The volume rate of one of these cracks can be expressed from the knowledge of the rate of the strain in the considered crack:

$$\dot{V}_c^i = \frac{4\pi}{3} a_i^3 X^i\, \boldsymbol{\delta} : \dot{\boldsymbol{\varepsilon}}_i \tag{4}$$

Since the crack is saturated by the gel, its volume rate can equally be expressed as:

$$\dot{V}_c^i = \dot{\beta}_i \tag{5}$$

Therefore, combining (4) and (5), we finally get:

$$\dot{\beta}_i = \frac{4\pi}{3} a_i^3\, X^i\, \boldsymbol{\delta} : \dot{\boldsymbol{\varepsilon}}_i \tag{6}$$

(6) clearly relates the rate of gel formation and the local deformation of an i-family crack. The latter can be estimated with the help of classical micromechanics approaches (Lemarchand 2001). In the line of micromechanical arguments developed by Eshelby (Eshelby 1957), $\dot{\varepsilon}_i$ can be estimated as the local strain in an ellipsoidal inclusion-like crack of the i-family, of elastic moduli $\mathbb{C}^c = 0$ embedded in an elastic medium of moduli $\mathbb{C}^s$, which may be caused by a prestress $\dot{\sigma}_i^p$, possibly acting in the crack and by a macroscopic strain rate $\dot{\boldsymbol{E}}$ applied at infinity. Provided that the crack volume fraction is small enough (*i.e.* what is referred to as the *dilute scheme*), application of Eshelby's solution to this inhomogeneity problem yields:

$$\dot{\boldsymbol{\varepsilon}}_i = \langle\mathbb{A}\rangle_i : \left(\dot{\boldsymbol{E}} - \mathbb{P}_i : \dot{\boldsymbol{\sigma}}_i^p\right)\ ;\ \langle\mathbb{A}\rangle_i = [\mathbb{I} - \mathbb{S}_i]^{-1} \tag{7}$$

where $\mathbb{P}_i = \mathbb{S}_i : \mathbb{S}^s$, $\mathbb{S}_i$ is the Eshelby tensor which depends on the compliance moduli of the solid matrix $\mathbb{S}^s = \mathbb{C}^{s^{-1}}$, the orientation and the aspect ratio of the i-family inclusions.

Restricting our purpose to very flat cracks, i.e. satisfying $X^i \ll 1$, we assume that the variation of the unit outward normal can be neglected ($\eta_i \approx \eta_{i,0}$), so that the prestress acting in the cracks of the i-family is nothing but the rate of the gel pressure $\dot{\sigma}_i^p = -\dot{p}_i\boldsymbol{\delta}$, so that:

$$\dot{\boldsymbol{\varepsilon}}_i = [\mathbb{I} - \mathbb{S}_i]^{-1} : \left(\dot{\boldsymbol{E}} + \dot{p}_i\, \mathbb{S}_i : \mathbb{S}^s : \boldsymbol{\delta}\right) \tag{8}$$

Then, taking advantage of the relation; $[\mathbb{I} - \mathbb{S}_i]^{-1} : \mathbb{S}_i = [\mathbb{I} - \mathbb{S}_i]^{-1} - \mathbb{I}$, (8) writes:

$$\dot{\boldsymbol{\varepsilon}}_i = [\mathbb{I} - \mathbb{S}_i]^{-1} : \dot{\boldsymbol{E}} + \dot{p}_i \left([\mathbb{I} - \mathbb{S}_i]^{-1} - \mathbb{I}\right) : \mathbb{S}^s : \boldsymbol{\delta} \tag{9}$$

Combining (6) and (9), we finally derive the relation linking the rate of gel formation and the rate of the loading parameters, that is, the gel pressure p_i acting in the cracks of the i-family and the macroscopic strain $\boldsymbol{E}$:

$$\dot{\beta}_i = \frac{4\pi a_i^3}{3}\, \boldsymbol{\delta} : \left[\mathbb{T}_i : \dot{\boldsymbol{E}} + \dot{p}_i \left(\mathbb{T}_i - X^i\mathbb{I}\right) : \mathbb{S}^s : \boldsymbol{\delta}\right] \tag{10}$$

where $\mathbb{T}_i = X^i[\mathbb{I} - \mathbb{S}^i$.

In particular, for very flat cracks, it has been shown (for instance (Deudé et al. 2002)) that the fourth order tensor $\mathbb{T}_i$ tends toward a finite constant value $\mathbb{T}_i^0$:

$$\lim_{X^i \to 0} \mathbb{T}_i = \lim_{X^i \to 0} X^i [\mathbb{I} - \mathbb{S}_i]^{-1} = \mathbb{T}_i^0(\nu^s) \tag{11}$$

where ν^s is the Poisson ratio of the solid matrix.

Let us introduce the classical deviatoric/spherical decomposition of the second order strain rate tensor:

$$\dot{\boldsymbol{E}} = \dot{\boldsymbol{E}}_d + \frac{1}{3}\dot{E}_v \boldsymbol{\delta} \tag{12}$$

where $\dot{\boldsymbol{E}}_d$ stands for its deviatoric part, and $\dot{E}_v = tr\dot{E}$ is the first strain invariant.

Therefore, provided that the cracks are flat enough, (10) can be expressed as:

$$\dot{\beta}_i = \frac{4\pi a_i^3}{9} T^0(\nu^s) \left[\dot{E}_v + \frac{\dot{p}_i}{k_s}\right] + \frac{4\pi\, a_i^3}{3}\, \boldsymbol{\delta} : \mathbb{T}_i^0 : \dot{\boldsymbol{E}}_d \tag{13}$$

where $T^0(V_s) = \boldsymbol{\delta} : \mathbb{T}_i^0 : \boldsymbol{\delta} = \frac{4(1-V_s^2)}{\pi(1-2V_s}$ whatever the crack distribution in the solid matrix.

Provided that the applied macroscopic stress is purely isotropic, the macroscopic strain is spherical ($\dot{\boldsymbol{E}}_d = 0$) for an isotropic distribution of cracks, so that (13) degenerates into:

$$\dot{\beta}_i = \frac{4\pi\, a_i^3}{9} T^0(\nu^s) \left[\dot{E}_v + \frac{\dot{p}_i}{k_s}\right] \tag{14}$$

It is interesting to note that (14) combined with (1) leads to:

$$\kappa = \frac{4\pi\, a_i}{9} T^0(\nu^s) \left[\dot{E}_v + \frac{\dot{p}_i}{k_s}\right] \tag{15}$$

Therefore, for a given kinetics of the gel formation, (15) allows us to state that, in a purely istropic problem, the rate of the gel pressure $\dot{p}_i$ is identical in each gel-producing crack family having the same radius

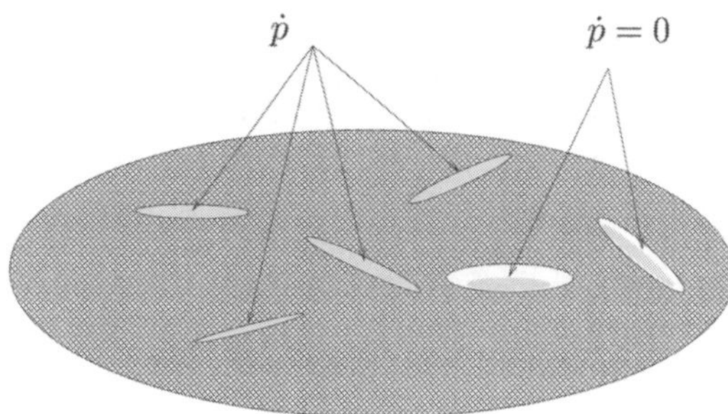

Figure 4. Uniform crack radii.

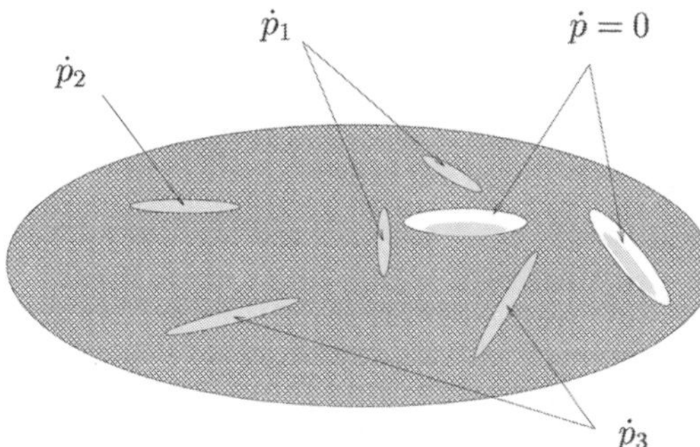

Figure 5. Different crack radii.

($\forall_i, a_i = a$), even for different aspect ratios (see Fig. 4). However, families having different crack radii are subjected to different rates of gel pressures. Besides, even for a set of identical cracks, the spherical part of the macroscopic tensor $\mathbb{T}_i^0 : \dot{\boldsymbol{E}}_d$ in (13) controls the possible deviation to this result. More precisely, out of a purely isotropic situation (cracks distribution, macroscopic applied stress), *i.e.* for a non-hydrostatic macroscopic strain, the rate of the gel pressure within the cracks differs from one family to another throughout concrete materials (see Fig. 5).

4 CRACKS STATISTICAL REPRESENTATION

In a macroscopic point of view, a statistic description of the cracks is due. In the present study, we restrict our purpose to an isotropic distribution of the cracks. Let $N(a, X_0)dad\, X_0$ be the number of mesocracks per macroscopic volume having a radius in between $[sa, a + da]$ and an aspect ratio in between $[X_0, X_0 + dX_0]$. The initial macroscopic crack density reads:

$$\mathcal{N}_0 = \int\int N(a, X_0)da\, dX_0 \tag{16}$$

Assuming an uncoupled description with respect to the crack radius and the initial aspect ratio, the initial crack density may be rewritten under the following form:

$$N(a, X_0) = n(a)\,\nu(X_0) \tag{17}$$

For the sake of simplicity, we consider from now on an uniform crack radius distribution, *i.e.* $\forall_i, a_i = a$. Therefore combining (16) and (17), we finally obtain:

$$\mathcal{N}_0 = \int \nu(X_0)\, dX_0 \tag{18}$$

where $\nu(X_0)dX_0$ represents the number of cracks defined by an initial aspect ratio $X_0 \in [X_0, X_0 + dX_0]$. Let $\delta(X_0) = \nu)(X_0)a^3$ be the initial crack density parameter associated to the aspect ratio X_0. The volume fraction of mesocracks (volume of mesocracks over initial volume) having an initial aspect ratio lying in between X_0 and $X_0 + dX_0$ is given by:

$$f_c dX_0 = \frac{4\pi}{3}\delta(X_0)\, X dX_0 \tag{19}$$

where X denotes the current value of the aspect ratio (as opposed to X_0 which denotes the initial one). According to (18), we also note that:

$$d_0 = \mathcal{N}_0 a^3 = \int \delta(X_0)dX_0 \tag{20}$$

where d_0 is the crack density parameter introduced by Budiansky and O'Connell (Budiansky and O'Connell 1976).

5 CHEMOPOROELASTICITY

The present section is devoted to the macroscopic poroelastic behavior of ASR-subjected concrete materials by means of a micromechanical approach, which has already proven its efficiency when dealing with classical Biot poroelasticity (Dormieux et al. 2002). Concrete is here defined as a two phase heterogeneous material, composed of a solid elastic matrix (s) and cracks (c) possibly under pressure. Denoting by $\dot{\boldsymbol{\sigma}}_\alpha$ and $\dot{\boldsymbol{\varepsilon}}_\alpha$, the rates of the local stress and strain fields in phase $\alpha(\alpha = s, c)$ the heterogeneous local behavior may be expressed as:

$$\dot{\boldsymbol{\sigma}}_\alpha = \mathbb{C}^\alpha : \dot{\boldsymbol{\varepsilon}}_\alpha + \dot{\boldsymbol{\sigma}}^p_\alpha \quad ; \quad \mathbb{C}^\alpha = \begin{cases} \mathbb{C}^s & \text{in}\,\Omega_s \\ 0 & \text{in}\,\Omega_c \end{cases} \tag{21}$$

and $\dot{\boldsymbol{\sigma}}^p_\alpha$ denotes the rate of a prestress tensor possibly acting in the α-phase, depending on the filling state of each crack:

$$\dot{\boldsymbol{\sigma}}^p_\alpha = \begin{cases} 0 & \text{in} \quad \Omega_s \\ -\dot{p}_i \dfrac{\langle t - t_f^i \rangle_+}{\|t - t_f^i\|}\boldsymbol{\delta} & \text{in} \quad \Omega_{c,i} \end{cases} \tag{22}$$

where $\langle a \rangle_+$ denotes the positive part of a.

According to Levin's theorem, equation (21) at the microlevel can equally be expressed at the macrolevel. Indeed, the rate of the macroscopic stress writes:

$$\dot{\Sigma} = \mathbb{C}^{\text{hom}} : \dot{\boldsymbol{E}} + \dot{\Sigma}^p \tag{23}$$

where $\dot{\Sigma}^p = \langle \dot{\sigma}^p : \mathbb{A} \rangle_\Omega$ and $\mathbb{C}^{\text{hom}} = (\langle \mathbb{C} : \mathbb{A} \rangle_\Omega$ ($\langle \cdot \rangle_\Omega$ being the average operator over domain Ω) denote respectively the rate of the macroscopic prestress tensor and the macroscopic tensor of elastic tangent moduli.

1. $\mathbb{C}^{\text{shom}}$ corresponds to the one of the solid matrix, affected by the cracks. Observing that the average strain concentration tensor in the non interacting cracks of the i-family is $\mathbb{A}_i = [\mathbb{I} - \mathbb{S}_i]^{-1}$, one obtains:

$$\mathbb{C}^{\text{hom}} = \mathbb{C}^s : \left(\mathbb{I} - \sum_i f_c^i \left[\mathbb{I} - \mathbb{S}_i \right]^{-1} \right) \tag{24}$$

Recalling the random distribution of the (flat) cracks, and using (19), the integral form of (24) reads:

$$\mathbb{C}^{\text{hom}} = \mathbb{C}^s : \left(\mathbb{I} - \frac{1}{3} \int \delta(X_0)\, dX_0 \int_0^{2\pi} d\phi \int_0^{\pi} \mathbb{T}^0(\theta, \phi) \sin\theta \, d\theta \right) \tag{25}$$

in which (11) has been used. As in (Deudé et al. 2002), we introduce the fourth order tensor $\mathbb{Q}$, defined as:

$$\mathbb{Q} = \frac{1}{3} \int_0^{2\pi} d\phi \int_0^{\pi} \mathbb{T}^0(\theta, \phi) \sin\theta d\theta \tag{26}$$

For an isotropic cracks distribution (Deudé et al. 2002), $\mathbb{Q}$ is an isotropic tensor:

$$\mathbb{Q} = Q_1(\nu^s)\, \mathbb{J} + Q_2(\nu^s)\, \mathbb{K} \tag{27}$$

where $\mathbb{J} = 1/3\boldsymbol{\delta} \otimes \boldsymbol{\delta}$ and $\mathbb{K} = \mathbb{I} - \mathbb{J}$, with:

$$\begin{cases} Q_1(\nu^s) = \dfrac{16}{9} \dfrac{(1 - \nu^{s^2})}{(1 - 2\nu^s)} = \dfrac{4\pi}{9} T^0(\nu^s) \\[2ex] Q_2(\nu^s) = \dfrac{16}{45} \dfrac{(1 - \nu^s)(5 - \nu^s)}{(2 - \nu^s)} \end{cases} \tag{28}$$

Introducing (20) and (26) in (25) yields:

$$\mathbb{C}^{\text{hom}} = \mathbb{C}^s : (\mathbb{I} - d_0\, \mathbb{Q}) \tag{29}$$

Taking advantage of the classical micromechanical definition of the anisotropic Biot tensor (Dormieux et al. 2002):

$$\boldsymbol{B} = \boldsymbol{\delta} : \left(\mathbb{I} - \mathbb{C}^{s^{-1}} : \mathbb{C}^{hom} \right) \tag{30}$$

it is readily seen that (29) leads to:

$$\boldsymbol{B} = d_0\, \boldsymbol{\delta} : \mathbb{Q} \tag{31}$$

Thus, for isotropic cracks distribution, (31) writes:

$$\boldsymbol{B} = b\, \boldsymbol{\delta}\ ;\ b = Q_1(\nu^s) d_0 = \frac{4\pi}{9} T^0(\nu^s) d_0 \tag{32}$$

2. Using (22) together with the fact that the rate of the gel pressure is uniform in the cracks when considering uniform crack radii, the rate of the macroscopic prestress $\dot{\Sigma}^p$ reads:

$$\dot{\Sigma}^p = -\dot{p}\, \boldsymbol{\delta} : \langle \mathbb{A} \rangle_{pr} \tag{33}$$

in which $\langle \mathbb{A} \rangle_{pr}$ stands for the average of the strain concentration tensor in the pressurized cracks. The density $\mathcal{N}_f$ of the cracks effectively filled with a pressurized gel is a function of time. Following the definition of the characteristic time of the filling process given in (3), it is defined as:

$$\mathcal{N}_f(t) = \int_0^{\frac{3K(t)}{4\pi a}} \nu(X_0)\, dX_0 \tag{34}$$

When κ is a constant, this expression simplifies into:

$$\mathcal{N}_f(t) = \int_0^{\frac{t}{\alpha a}} \nu(X_0)\, dX_0 \tag{35}$$

$\langle \mathbb{A} \rangle_{pr}$ can be determined by integration over all gel-filled cracks, that is, for $X_0 \leq \frac{3K(t)}{4\pi a}$:

$$\begin{aligned} \langle \mathbb{A} \rangle_{pr} &= \frac{1}{4\pi} \int_0^{\frac{3K(t)}{4\pi a}} [\mathbb{I} - \mathbb{S}]^{-1} f_c\, dX_0 \sin\theta d\theta d\phi \\ &= \mathbb{Q} \int_0^{\frac{3K(t)}{4\pi a}} \delta(X_0) dX_0 = a^3 \mathcal{N}_f(t)\, \mathbb{Q} \end{aligned} \tag{36}$$

Hence, the final expression of $\dot{\Sigma}^p$ is:

$$\dot{\Sigma}^p = -\mathcal{N}_f(t)\, a^3\, \dot{p}\, \boldsymbol{\delta} : \mathbb{Q} \tag{37}$$

Using the definition of Biot's coefficient (31), (37) writes:

$$\dot{\Sigma}^p = -S_r(t)\, \dot{p}\, \boldsymbol{B} \tag{38}$$

In (38), $S_r(t) = \frac{\mathcal{N}_f(t)}{\mathcal{N}_0}$ has the physical meaning of a saturation degree accounting for the filling state of the crack porosity. As long as the gel does not saturate any family of cracks, we have $\mathcal{N}_f = 0$, *i.e.* $S_r = 0$. By contrast, as soon as the gel saturates the whole crack porosity, $\mathcal{N}_f = \mathcal{N}_0$, *i.e.* $S_r = 1$.

Evidently, for an isotropic distribution of cracks, (38) reads:

$$\dot{\boldsymbol{\Sigma}}^p = -S_r(t)\,\dot{p}\,b\,\boldsymbol{\delta} \tag{39}$$

As a matter of fact, combining (23) and (38), the macroscopic poroelastic constitutive law writes:

$$\dot{\boldsymbol{\Sigma}} = \mathbb{C}^{\text{hom}} : \dot{\boldsymbol{E}} - \dot{p}\,S_r\,\boldsymbol{B} \tag{40}$$

which gives for randomly distributed cracks:

$$\dot{\boldsymbol{\Sigma}} = \mathbb{C}^s : (\mathbb{I} - d_0\,\mathbb{Q}) : \dot{\boldsymbol{E}} - \dot{p}\,S_r\,b\,\boldsymbol{\delta} \tag{41}$$

5.1 *Free swelling case*

We are interested here in the free swelling behavior (no macroscopic stress applied) of concrete samples containing a random distribution of cracks having the same radius. The question of the geometrical evolution of the cracks induced by application of a macroscopic isotropic stress will be addressed in the following paragraph.

Following these assumptions, the macroscopic strain rate is isotropic ($\dot{\boldsymbol{E}} = \frac{1}{3}\dot{E}_v\boldsymbol{\delta}$) so that the macroscopic state equation (41) reduces to:

$$\frac{\dot{p}}{k^s} = \frac{(1-b)}{b\,S_r}\,\dot{E}_v \tag{42}$$

Besides, introducing (32) into (15), one obtains:

$$\mathcal{N}_0\dot{\beta} = b(\dot{E}_v + \frac{\dot{p}}{k^s}) \tag{43}$$

Then, combining (42) and (43), it is possible to derive the time evolution of the macroscopic volume strain rate with respect to the rate of the gel formation:

$$\dot{E}_v(t) = \dot{\beta}(t)\,\mathcal{N}_0\,\frac{S_r(t)}{1 - b\,(1 - S_r(t))} \tag{44}$$

Figure 6 presents the variation of the ratio $\dot{E}_v(t)/\dot{\beta}(t)\mathcal{N}_0$ as a function of the saturation degree. This result deserves some comments:

1. $\dot{\beta}(t)\mathcal{N}_0$ can be interpreted as the macroscopic volume strain rate, once the filling process is full-filled.
2. if the solid matrix is incompressible ($b = 1$), the kinetic of swelling is controlled by the one of the gel formation, *i.e.* $\dot{E}_v(t) = \dot{\beta}(t)\mathcal{N}_0$.
3. for a compressible matrix, the non-linear macroscopic strain rate depends on the time-dependent saturation degree $S_r(t) < 1$. (44) actually represents the link between the kinetics of gel production (characterized by $\dot{\beta}$) and that of swelling (characterized by $\dot{E}_v$). As long as the filling process is going on, the kinetics of swelling is not a direct mechanical counterpart of the chemical process. The kinetics of gel production must be corrected by a factor taking the filling process into account.

Figure 7 shows the evolution of the macroscopic strain with respect to time. We observe that the curvature of the first part of the classical *s*-curve of ASR-subjected concrete materials can be associated to the progressive filling of the pores (cracks).

5.2 *Isotropic state of stress*

Let us consider the case of an isotropic macroscopic compressive (resp. tensile) loading defined by $\Sigma = \Sigma\delta(\Sigma < 0$ (resp. $\Sigma > 0$)) Application of such loading may have a non negligible effect on the local chemical process. Since the question is still open, we will simplify our purpose by considering that this macroscopic stress is applied on a concrete sample before ASR occurs. Evidently, Eqn. (42) and (44) are still valid, provided that the crack density $\mathcal{N}_0$, associated to an unloaded sample, be replaced by a stress-dependent initial crack density $\mathcal{N}_0(\Sigma)$. Hence, we are looking for the description of the morphological transformation of the initial state of the cracks, *i.e.* the distribution of the initial aspect ratios.

This question is easily addressed through a micromechanical reasonning. Indeed, for non-propagating

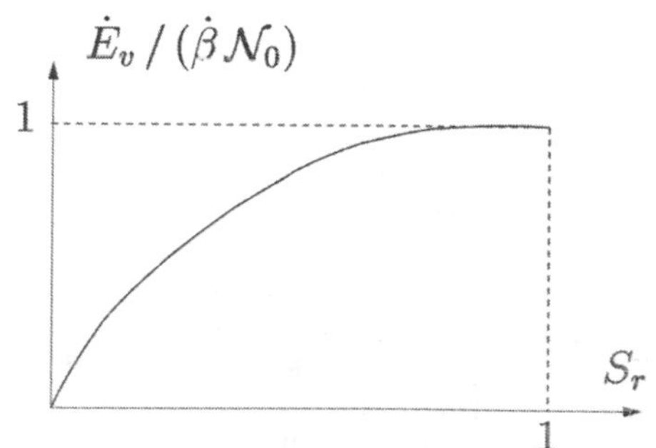

Figure 6. Mechanical impact of the filling process.

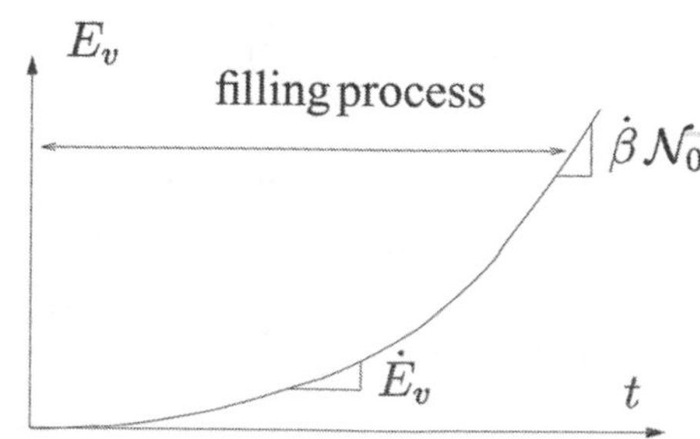

Figure 7. Macroscopicstrain evolution.

cracks, (4) allows us to express the evolution of the aspect ratio of a crack of the i-family as:

$$\dot{X}^i = X^i\,\boldsymbol{\delta} : \dot{\boldsymbol{\varepsilon}}_i \tag{45}$$

Before ASR oocurs, the cracks are empty, so that $\forall i, \dot{p}_i = 0$. Therefore from (8), the local rate of strain in a crack of the i-family reads (dilute scheme):

$$\dot{\boldsymbol{\varepsilon}}_i = [\mathbb{I} - \mathbb{S}_i]^{-1} : \dot{\boldsymbol{E}} \tag{46}$$

In the dilute scheme approximation, a crack being embedded in an infinite medium having the elastic moduli of the solid matrix, the macroscopic stress and strain rates are related by $\dot{\boldsymbol{E}} = \mathbb{S}^s : \dot{\boldsymbol{\Sigma}} (\mathbb{S}^s = \mathbb{C}^{s^{-1}})$, so that (46) finally writes:

$$\dot{\boldsymbol{\varepsilon}}_i = [\mathbb{I} - \mathbb{S}_i]^{-1} : \mathbb{S}^s : \dot{\boldsymbol{\Sigma}} \quad ; \quad \dot{\boldsymbol{\Sigma}} = \dot{\Sigma}\,\boldsymbol{\delta} \tag{47}$$

The aspect ratio evolution law with respect to the macroscopic applied stress is finally given by combining (15) and (47).

$$\dot{X}^i = \frac{T^0(\nu^s)}{3\,k^s}\dot{\Sigma} \tag{48}$$

Integration of (48) leads to the crack aspect ratio change:

$$X^i(\Sigma) = \langle X_0^i + \frac{T^0(\nu^s)}{3\,k^s}\Sigma\rangle_+ \tag{49}$$

(49) clearly expresses how application of an isotropic compressive (resp. tensile) stress is able to reduce (resp. increase) the initial aspect ratio of a crack[2]. Taking advantage of the definition of $T^0(V^s)$, (49) writes:

$$X^i(\Sigma) = \langle X_0^i + \delta\,\Sigma\rangle_+ \quad ; \quad \delta = \frac{4\,(1-\nu^{s^2})}{\pi\,E^s} \tag{50}$$

where E^s is the solid matrix Young's Modulus.

Hence, application of a macroscopic stress (before ASR occurs) leads to a new definition of the initial cracking state of the material. A quantitative formulation of this modified initial state is given by (50). This result is of primary importance since the initial aspect ratio distribution determines the duration of the filling process, according to (2) and (3). (50) indicates that the aspect ratio distribution is modified by a translation of intensity $\delta\Sigma$, illustrated in figure 8 for a compressive ($\Sigma < 0$) loading. More precisely, the new initial crack density distribution is defined by:

$$\nu(X_0, \Sigma) = \nu(X_0 - \delta\,\Sigma, 0) \tag{51}$$

where $\nu(X_0, 0) \equiv \nu(X_0)$ denotes as previously the aspect ratio distribution when no macroscopic stress is applied.

[2] for tensile stresses, in addition to the non-propagating assumption, the validity of this result requires that the aspect ratio remains very small after the stress application.

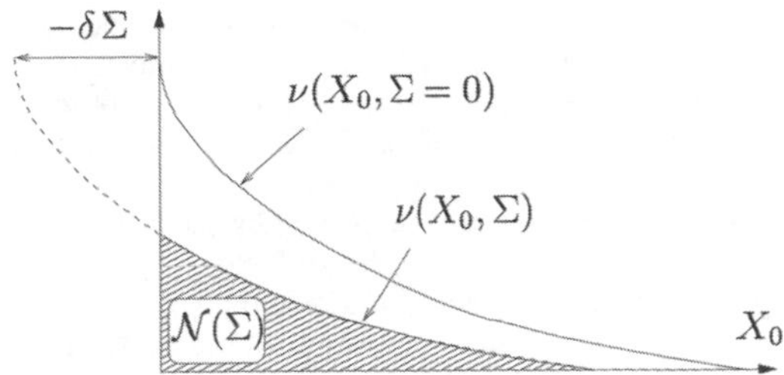

Figure 8. Crack density distribution evolution ($\Sigma < 0$).

Therefore, after application of a macroscopic stress, the cracked pore space likely to be filled by the ASR gel is modified such that the characteristic time of the filling process will be decreased (resp. increased) for a compressive (resp. tensile[2]) stress. More specifically, for an isotropic compressive stress, the crack density $\mathcal{N}_0$ in (44) should be replaced by the new initial crack density $\mathcal{N}(\Sigma)$ defined as ($\Sigma < 0$):

$$\mathcal{N}(\Sigma) = \mathcal{N}_0 - \int_0^{-\delta\Sigma} \nu(X_0)\,dX_0 \tag{52}$$

Typically, at the end of the filling process, the macroscopic volume strain rate reads $\dot{E}_v = \dot{\beta}\mathcal{N}(\Sigma) < \dot{\beta}\mathcal{N}_0$, which means that the elastic swelling strain is a decreasing function of the applied confining pressure.

Besides, in the calculation of the saturation degree $S_r(t)$ which also appears in (44), the density of filled cracks $\mathcal{N}_f$ defined in (34) or (35) should now be determined by integration over the distribution $\nu(X_0, \Sigma)$ defined in (51). For instance, when κ is a constant, one obtains:

$$\mathcal{N}_f(\Sigma, t) = \int_0^{\frac{t}{\alpha a}} \nu(X_0 - \delta\,\Sigma)\,dX_0 \tag{53}$$

while the saturation degree takes the form:

$$S_r(\Sigma, t) = \frac{\mathcal{N}_f(\Sigma, t)}{\mathcal{N}(\Sigma)} \tag{54}$$

6 CONCLUSIONS

The question of the mechanical response (swelling) of concrete materials undergoing an internal pressurization has been adressed in the present paper in the context of Alkali-Silica Reactions. ASR gel being assumed located in unconnected cracks located in the vicinity of some (reactive) aggregates, we developed a micromechanical reasoning allowing us to clarify the elastic part of the observed swelling kinetics. A detailed analysis of the cracks propagation process has been developed by the authors and will be the subject of a further paper. In the context of poroelasticity, we emphasized herein the role of the filling process on the macroscopic swelling of concrete materials. In particular, we find out that the non linear macroscopic elastic swelling is more controlled by the kinetics of the filling process than the one of the gel formation.

Moreover, we showed, in a quantitative way, how application of a macroscopic isotropic stress may change the crack distribution within concrete. In term of macroscopic swelling, this leads to a decrease of the elastic volume strain. Due to the fact that it is not possible to derive an analytical solution, application of an uniaxial stress has not been addressed herein. Indeed, the deviatoric part of the macroscopic strain rate will introduce a non-uniform distribution of gel pressures in the cracks. In a formal way, this situation will lead to a generalisation of Biot's poroelastic theory.

REFERENCES

Budiansky, B. and R. O'Connell (1976). Elastic moduli of a cracked solid. *Int. J. Solids Structures* 12, 81–97.

Deudé, V., L. Dormieux, D. Kondo, and S. Maghous (2002). Micromechanical approach to non-linear poroelasticity application to cracked rocks. *J. Eng. Mech.* 128(8), 848–855.

Dormieux, L., A. Molinari, and D. Kondo (2002). Micromechanical approach to the behavior of poroelastic materials. *J. Mech. Phys. Solids* 50, 2203–2231.

Eshelby, J. (1957). The determination of the elastic field of an ellipsoidal inclusion, and related problems. *Proc. Roy. Soc. London A241*, 235–251.

Larive, C. (1998). *Apports combinés de l'expérimentation et de la modélisation à la comprehension de l'alcali-reaction et de ses effets mécaniques*. Ph.D. thesis, Laboratoires des Ponts et Chaussées, Paris, France.

Lemarchand, E. (2001). *Contribution de la micromécanique à l'étude du couplage poromécanique et du transport dans les milieux poreux : Application au gonflement des géomatériaux*. Ph. D. thesis, LMSGC, Ecole nationale des ponts et chaussées, Champs sur Marne, France.

Lemarchand, E., L. Dormieux, and F.-J. Ulm (2002, March). Elements of Micromechanics of ASR-induced Swelling in Concrete Structures. *J. of Conc. Sc. Engng.* 4, 12–22.

Computational Modelling of Concrete Structures, Bićanić et al. (eds)
© 2003 Swets & Zeitlinger, Lisse, ISBN 90 5809 536 3

Modelling the splitting effects of corrosion in reinforced concrete

K. Lundgren
Department of Structural Engineering, Chalmers University of Technology, Gothenburg, Sweden

ABSTRACT: The corrosive products that forms when reinforcement in concrete corrodes occupy a larger volume than the steel it was formed of, thus leading to splitting stresses acting on the concrete. The mechanical behaviour of rust was studied by a combination of analyses and test results found in the literature. This led to the assumption that rust behaves like a granular material, i.e. its stiffness increases with the stress level. This mechanical behaviour, and the volume increase of the corrosive products compared with the virgin steel, were modelled in a corrosion layer. Several corrosion cracking tests carried out by various researchers were analysed using the corrosion model in finite element analyses. The concrete was modelled with non-linear fracture mechanics, using a smeared rotating crack model. The corrosion level that caused cracking of the cover in the tests and in the analyses was compared, showing a reasonable good agreement. The model was used to study the effect of uniform or localised corrosion. It was shown that for localised corrosion, less average corrosion penetration was needed to crack the cover than for uniform corrosion. Furthermore, the crack pattern differed. From these analyses, it was concluded that axisymmetric analyses appear to be a sufficient level of modelling when studying uniform corrosion. If localised corrosion is to be studied, three-dimensional models need to be used.

1 INTRODUCTION

The corrosive products that form when reinforcement in concrete corrodes occupy a larger volume than the steel it was formed of. This leads to splitting stresses acting on the concrete, which might cause splitting or spalling of the cover. This affects the structure in several ways: On the compressed side of a structure, spalling of the cover will lead to a decrease of the inner lever arm, which in turn decreases the bending moment capacity. Furthermore, the interaction between the reinforcement and the concrete is influenced. Due to the effect on the bond mechanism, the deformations increase, and if the corrosion takes place in certain parts of the structure, as at supports and at splices, the load-bearing capacity will be influenced. The effect of the volume increase is not so easy to take into account when calculating the load-bearing capacity of corroded reinforced concrete structures. A model of the corrosion together with a model of the bond mechanism was developed by the author in earlier work, see Lundgren (2002). Here, further work has been done with the corrosion model.

2 MODELLING STRATEGY

The modelling method used is specially suited for detailed three-dimensional finite element analyses, where both the concrete and the reinforcement are modelled with solid elements. Special interface elements are used at the surface between the reinforcement bars and the concrete to describe a relation between the traction **t** and the relative displacement **u** in the interface. The physical interpretations of the variables are as shown in Figure 1.

The interface elements describe the behaviour of the rust, and the interaction between the reinforcement and the concrete. To describe the interaction between the reinforcement and the concrete, the bond model presented in Lundgren & Gylltoft (2000) is used. The behaviour of the rust is described by a corrosion model, see Lundgren (2002). The corrosion model is here

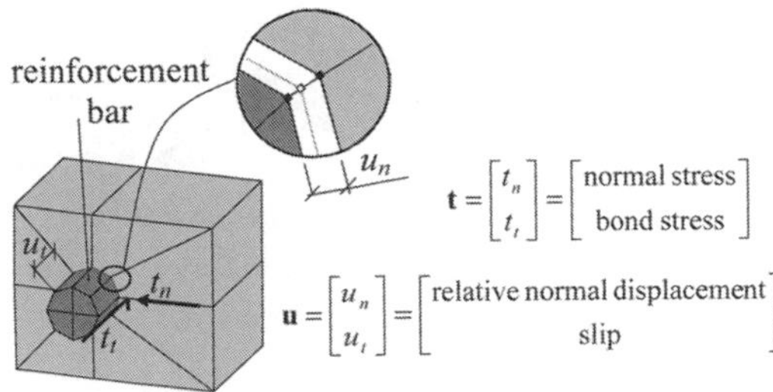

Figure 1. Physical interpretation of the variables t_n, t_t, u_n and u_t.

further developed. The corrosion model and the bond model can be viewed as two separate layers around a reinforcement bar. However, to reduce the number of nodes required to model a structure, they are integrated in one interface element. Due to equilibrium between the two layers, the traction **t** is the same in the bond and in the corrosion layer. The deformations are related as:

$$u_n = u_{ncor} + u_{nbond} \quad (1)$$

$$u_t = u_{tbond}, u_{tcor} = 0 \quad (2)$$

where the index *cor* means for the corrosion layer, and the index *bond* means for the bond layer.

3 CORROSION MODEL

The volume increase of the corrosion products compared to the virgin steel was modelled in a corrosion layer, as described in Lundgren (2002). The model presented here is the same, but the calibration of it was changed slightly. The volume of the rust relative to the uncorroded steel, and the corrosion penetration as a function of the time is given as input. The corrosion is then modelled by taking time steps. The physical interpretation of the variables in the corrosion model is shown in Figure 2. The free increase of the radius; i.e. how much the radius would increase if the normal stresses were zero is calculated from the condition that the rust (corresponding to the dark grey area in Figure 2) is v times the volume of the steel that has corroded (corresponding to the medium grey area in Figure 2), which results in:

$$a = -r + \sqrt{r^2 + (v-1)\cdot(2rx - x^2)}\,. \quad (3)$$

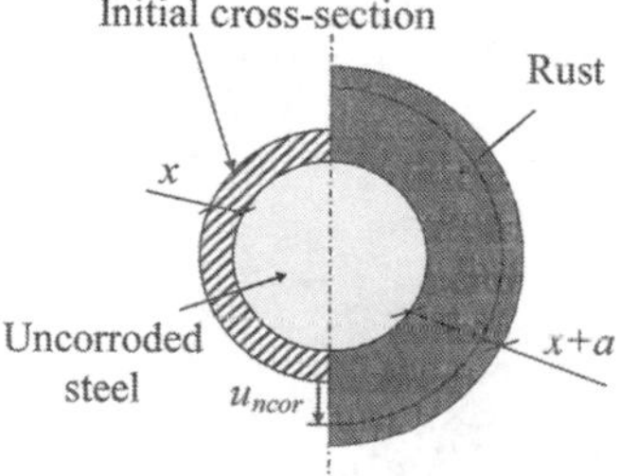

Figure 2. Physical interpretation of the variables in the corrosion model.

Thus, the parameter v is the volume of the rust relative to the uncorroded steel. This is an important parameter in the model. The value depends on which corrosion product forms. Values from 2.2 to 6.4 are given in the literature; see Table 1. Most often, a value of 2.0 has been used in analyses of corrosion splitting; see for example Coronelli (1998) and Molina et al. (1993). The latter value was also chosen here. It is then common to simply set the free increase of the radius equal to the corrosion penetration, i.e. $a = x$. This is very close to equation (3) for small corrosion penetrations, see Figure 3, where the two expressions are compared. However, due to the surrounding structure, the rust cannot expand freely. The real increase of the radius is u_{ncor}, corresponding to a strain in the rust:

$$\varepsilon_{cor} = \frac{u_{ncor} - a}{x + a} \quad (4)$$

From this strain in the rust, the normal stresses in the layer are determined.

3.1 *Evaluation of the mechanical behaviour of rust*

To investigate the mechanical behaviour of rust, corrosion cracking tests found in the literature were studied. A combination of experimental results found in the literature, and finite element analyses were used,

Table 1. The volume of the rust relative to the uncorroded steel according to, for example, Tuutti (1982).

Corrosion product	Volume increase
Fe_3O_4	2.2
$Fe(OH)_2$	3.8
$Fe(OH)_3$	4.2
$Fe(OH)_3, 3H_2O$	6.4

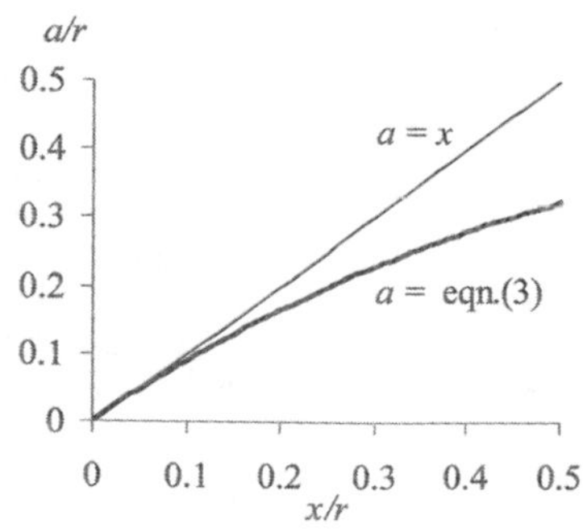

Figure 3. The free increase of the radius versus the corrosion penetration.

as outlined in Figure 4. The corrosion penetration that caused cracking of the concrete cover has been measured by several researchers, for various types of specimens. These specimens were analysed here using finite element modelling and non-linear fracture mechanics for the concrete. Only the concrete of the specimens was modelled, with a hole where the reinforcement was situated. A normal deformation was applied at the hole, resulting in a normal stress versus deformation curve typical for the specimen. From each analysis, the normal stress and deformation of the hole that caused cracking of the cover were used. This in combination with the experimentally measured corrosion penetration that caused cracking results in one point in a stress versus strain diagram.

To evaluate the strain in the rust, the deformation of the corrosion layer when the cover was cracked, $u_{ncor,cr}$, needs to be determined. It was calculated from the deformation of the hole that caused cracking of the cover, δ_{cr}, taking into account that the reinforcement bar and the bond layer are also slightly compressed due to the normal stress. The deformation of the reinforcement bar due to a normal pressure $-t_n$ acting on it can be calculated as

$$\delta_{reinf.} = \frac{t_n}{E_s} \cdot r \cdot \left(1 - \nu_s - 2\nu_s^{\,2}\right) \tag{5}$$

where E_s and ν_s are the Young's modulus and the Poison's ratio for steel. They were assumed to be 200 GPa and 0.3, respectively. The deformation of the bond layer can be calculated as

$$u_{nbond} = \frac{t_n}{D_{11}(u_{nbond} \leq 0)} \tag{6}$$

where $D_{11}(u_{nbond} \leq 0)$ is the stiffness D_{11} in the bond layer; u_{nbond} will be smaller than zero since the bond layer will be compressed without any slip applied.

The deformation $u_{ncor,cr}$ can then be calculated as

$$u_{ncor,cr} = \delta_{cr} - \delta_{reinf,cr} - u_{nbond,cr}\,. \tag{7}$$

These effects were assumed to be negligible in earlier work, Lundgren (2002); however, they did have some influence on the results, leading to slightly smaller evaluated strains in the rust. In Figure 5a, the modified results are shown as points in a stress versus strain diagram. Still, it appears reasonable to assume that the rust behaves like a granular material; i.e. its stiffness increases with the stress level. The mechanical behaviour of the rust at loading could be described with the equation:

$$t_n = K_{cor} \cdot \varepsilon_{cor}^{\;p} \tag{8}$$

Experimental results

The corrosion penetration at cracking measured in various corrosion cracking tests found in literature

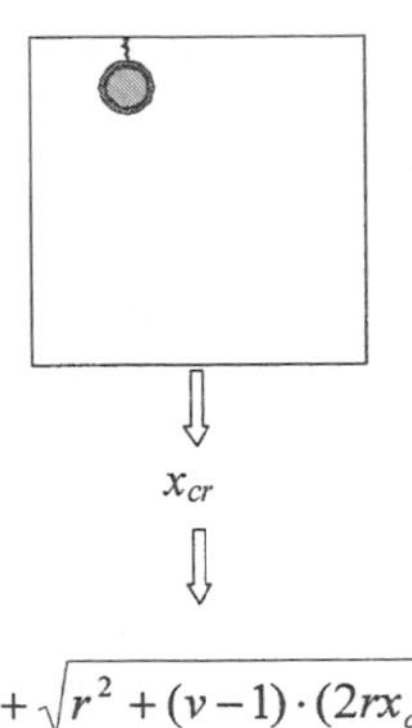

Theoretical results

Analyses of the various corrosion cracking tests give the expansion and the normal splitting stress at cracking.

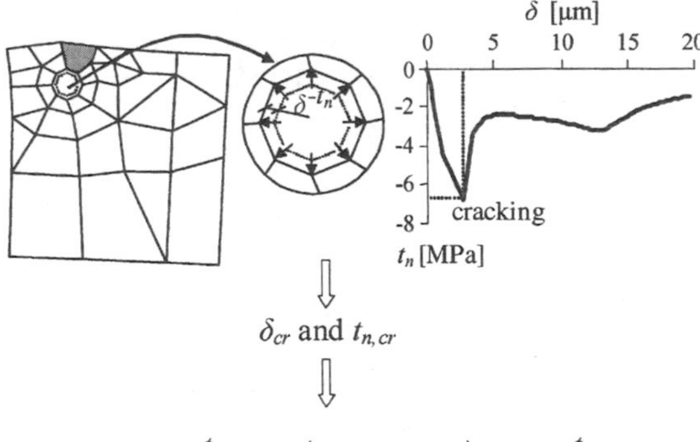

δ_{cr} and $t_{n,cr}$

$$u_{ncor,cr} = \delta_{cr} - \underbrace{\frac{t_{n,cr}}{E_s} \cdot r \cdot \left(1 - \nu_s - 2\nu_s^{\,2}\right)}_{\text{reinforcement}} - \underbrace{\frac{t_{n,cr}}{D_{11}\left(u_{nbond} \leq 0\right)}}_{\text{bond layer}}$$

$$\varepsilon_{cor,cr} = \frac{u_{ncor,cr} - a_{cr}}{x_{cr} + a_{cr}}$$

Figure 4. Schematic view of how one point in the stress versus strain diagram for the rust is obtained from a combination of experimental and theoretical results.

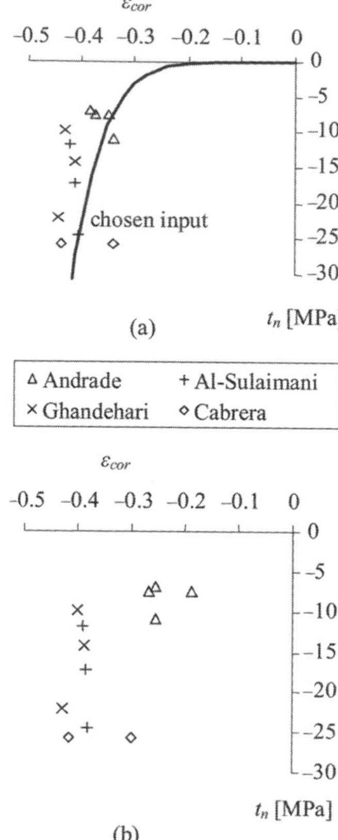

Figure 5. Normal stress versus strain in the rust evaluated from a combination of experimental results and analyses; (a) Assuming that the rust applies a pressure on the surrounding structure already from the beginning, together with the chosen input; and (b) Assuming that the rust fills the pores close to the reinforcement bar before applying stresses on the surrounding structure.

The value of K_{cor} was chosen to $K_{cor} = 14.0$ GPa, and the parameter p was set to $p = 7.0$.

3.2 *Alternative evaluation of the mechanical behaviour of rust*

Several researchers, for example Noghabai (1998) and Petre-Lazar (2000), have assumed that the corrosive products are infinitely stiff. Petre-Lazar (2000) points out the possibility for the rust to fill up the pores close to the reinforcement bars before starting to apply stresses in the structure. He therefore divides the mechanical action of the corrosion process in two phases:

1. In the first phase, the porosity of a mortar-rebar interface is filled with corrosion products without applying any stresses on the surrounding concrete.
2. When the porosity of the mortar-rebar interface is filled, normal stresses are applied to the surrounding concrete.

This means that a certain increase of the radius occurs before any normal stresses are applied to the surrounding concrete. Petre-Lazar (2000) gives a method for how to estimate this distance (here denoted f):

$$f = t_{in} \cdot \Phi \tag{9}$$

where t_{in} is the thickness of the mortar-rebar interface. Petre-Lazar (2000) used a value of 40 μm, based on observations with scanning electron microscope. Φ is the porosity of the mortar, calculated as half of the porosity of the cement paste:

$$\Phi = \left(1 - \frac{1 + 1.31\alpha}{1 + 3.2vct}\right) \Big/ 2 \tag{10}$$

where α is the level of hydratation and vct is the water–cement ratio.

The effect of this distance f can be taken into account in the evaluation of the mechanical properties of the rust by changing equation (7) to

$$u_{ncor,cr} = \delta_{cr} - \delta_{reinf,cr} - u_{nbond,cr} + f \tag{11}$$

The strain in the rust at cracking for different test specimens is recalculated in this way; thus including the effect of filling up the pores close to the reinforcement bars before starting to apply stresses in the structure. The estimated values of the water cement ratio and hydratation level are given in Table 2, where results from both calculation methods are summarised. The results from the analyses including the effect of the pores are also shown in a stress versus strain diagram in Figure 5b. If the corrosive products were infinitely stiff, all points in this diagram would be close to the zero strain line. However, this is not the case; instead large strains are evaluated. Comparing the results calculated with the two different assumptions (i.e. Figure 5a and 5b), there appears to be more scatter in the results where the effect of filling up the pores is included. Therefore, the modelling method where the stresses of the corrosive products are applied already from the beginning, i.e. without the effect of the pores, was chosen.

4 COMPARISON WITH CORROSION CRACKING TESTS

Tests of various kinds with corroding reinforcement were analysed with finite element models, using the

Table 2. Evaluation of normal stress and strain in the rust.

Reference	Test no.	f_{cc} MPa	r mm	x_{cr} µm	$t_{n,cr}$ MPa	δ_{cr} µm	vct –	α –	f µm	$\delta_{reinf,cr}$ µm	$D_{11}(u_{nbond} < 0)$ N/m³	$u_{nbond,cr}$ µm	$u_{ncor,cr}$ [1] µm	$u_{ncor,cr}$ [2] µm	$\varepsilon_{cor,cr}$ [1] –	$\varepsilon_{cor,cr}$ [2] –
Ghandehari	CsRs	40	5	159	−22.2	10.5	0.47	0.75	4.5	−0.27	$5.1 \cdot 10^{12}$	−4.3	15.1	19.6	−0.44	−0.43
	ClRl	40	10	92	−14.3	12.5	0.47	0.75	4.5	−0.37	$5.1 \cdot 10^{12}$	−2.8	15.6	20.1	−0.41	−0.39
	CsRl	40	10	72	−9.80	7.5	0.47	0.75	4.5	−0.25	$5.1 \cdot 10^{12}$	−1.9	9.66	14.2	−0.43	−0.40
Cabrera	opc	42	6	68	−25.8	16.0	0.55	0.80	5.5	−0.40	$5.2 \cdot 10^{12}$	−4.9	21.4	26.8	−0.34	−0.30
	pfa	42	6	192	−25.8	16.0	0.72	0.80	7.8	−0.40	$5.2 \cdot 10^{12}$	−4.9	21.4	29.1	−0.44	−0.42
		30	5	115	−24.5	15.0	0.55	0.80	5.5	−0.32	$4.7 \cdot 10^{12}$	−5.3	20.6	26.1	−0.40	−0.38
Al-Sulaimani		30	7	102	−17.3	13.3	0.55	0.80	5.5	−0.31	$4.7 \cdot 10^{12}$	−3.7	17.3	22.8	−0.41	−0.38
		30	10	93	−11.8	11.5	0.55	0.80	5.5	−0.31	$4.7 \cdot 10^{12}$	−2.5	14.3	19.8	−0.42	−0.39
Andrade	I	43	8	17.9	−6.83	2.75	0.50	0.80	4.6	−0.14	$5.2 \cdot 10^{12}$	−1.3	4.20	8.78	−0.38	−0.25
	II	43	8	14.4	−7.50	2.80	0.50	0.80	4.6	−0.16	$5.2 \cdot 10^{12}$	−1.4	4.39	8.97	−0.35	−0.19
	III	43	8	21.3	−10.8	4.50	0.50	0.90	3.6	−0.23	$5.2 \cdot 10^{12}$	−2.1	6.80	10.4	−0.34	−0.25
	IV	43	8	17.4	−7.50	2.80	0.50	0.90	3.6	−0.16	$5.2 \cdot 10^{12}$	−1.4	4.39	8.02	−0.37	−0.27

[1] Without the effect of the porosity, eqn. (7).
[2] With the effect of the porosity, eqn. (11).

described corrosion layer and bond model in special interface elements between the reinforcement and the concrete. In all analyses, the concrete was modelled with a constitutive model based on non-linear fracture mechanics. In analyses using the smeared crack concept, a rotating crack model based on total strain was used; see TNO (1998). For the axisymmetric models, four radial cracks were assumed.

From the various measured compressive strengths, an equivalent compressive cylinder strength, f_{cc}, was evaluated. Other necessary material data for the concrete were estimated according to the expressions in CEB (1993) from f_{cc}. For the tests where information about storage and age at testing was available, shrinkage was included in the models in an approximate way, by applying uniform shrinkage calculated according to CEB (1993). Creep was not included, since it was not possible to combine with the chosen material model. The constitutive behaviour of the reinforcement steel was modelled by the Von Mises yield criterion with associated flow and isotropic hardening. The elastic modulus of the reinforcement was assumed to be 200 GPa.

Corrosion cracking tests have been carried out by several researchers. Analyses of several such corrosion cracking tests were carried out, to compare the corrosion level that caused cracking of the cover in the tests, and in the analyses. Here follows a brief presentation of all the analysed tests.

Andrade et al. (1993) carried out corrosion cracking tests with specimens of size 15 cm · 15 cm · 38 cm. Here, slices of them were analysed, assuming fixed boundaries in the third direction, corresponding to a plane strain assumption. In the tests where the reinforcement bar was placed symmetric, discrete crack elements were used, while for the test where the reinforcement bar was placed in a corner, the smeared crack model was used. In the tests, the covers were cracked for corrosion penetrations from 14.4 to 21.3 μm, while in the analyses it ranged from 15.0 to 27.5 μm.

Al-Sulaimani et al. (1990) and Cabrera & Ghoddoussi (1992) have carried out pull-out tests on corroded reinforcement bars concentrically placed in concrete blocks. Here, the first part of their experiments was studied, i.e. when the reinforcement corroded until the specimens were cracked. The test specimens were approximated with axisymmetric finite element models. Ghandehari et al. (2000) used test specimens with corroded reinforcement bars in concrete cylinders. Three of their specimens were cracked during the accelerated constant current corrosion: with cylinder diameters of 100 and rebar diameters of 9.5 and 20 mm, and cylinder diameters of 150 mm and rebar diameters of 20 mm. Axisymmetric finite element models were used to model the specimens.

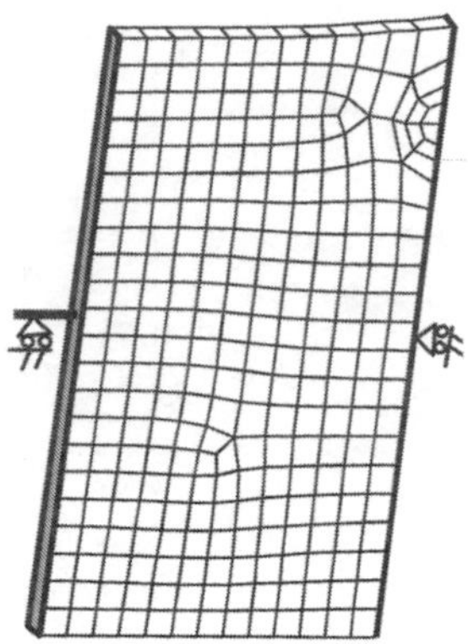

Figure 6. Deformed finite element model used for analyses of the test specimen of Liu & Weyers (1998) with cover 25 mm.

Liu & Weyers (1998) used slabs with chloride admixed in the concrete. Three of them cracked, all of them stored outside. The slabs had the dimensions $1180 \cdot 1180 \cdot 216\,mm^3$, and contained five rebars with diameter 16 mm with covers of 25, 51 and 76 mm. The corrosion was measured both by corrosion rate measurements and by the weight loss method when cracking had occurred. Here the corrosion at cracking measured by the weight loss method was used. Slices of them were analysed, using also the symmetry lines so that only half the reinforcement bar and halfway the distance to the next reinforcement bar was modelled, see Figure 6.

Rasheeduzzafar et al. (1992) have carried out accelerated corrosion tests and measured the time to cracking for specimens with cover to bar ratios ranging from 0.63 to 4.53. The corrosion penetration at cracking was measured by the weight loss method, and each value is an average of three tests. Here, slices of the tests were analysed, using the symmetry line so that only half the reinforcement bar was modelled. Hence, the models looked similar to the one shown in Figure 6, but without the support shown on the left side in the figure.

Almusallam et al. (1996) carried out cantilever bond tests. Their specimens were $152 \cdot 254 \cdot 279\,mm^3$, and had a 12 mm diameter bar with 102 mm embedment length. They were subjected to accelerated corrosion to varying levels; thereafter bond tests were conducted, and the corrosion penetration was measured by the weight loss method. Here is only the corrosion part of their tests analysed. The finite element mesh used for the analyses is shown in Figure 7, together with cracks that appeared in the analyses due to corrosion. At a corrosion penetration of about 160 μm in the analysis, a crack appeared from the reinforcement bar to the upper edge in Figure 7. For a corrosion penetration of 210 μm, the right crack reached the edge for the

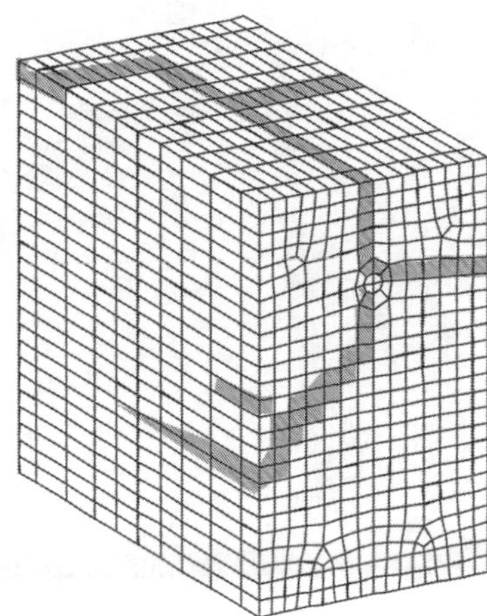

Figure 7. Finite element mesh used for the analyses of the cantilever bond tests of Almusallam et al. (1996). Grey marked regions indicate cracks; results from an analysis with an applied corrosion penetration of 500 μm.

corrosion only, and for corrosion penetrations larger than 375 μm, the inclined crack reached the left edge in the figure. This corresponds rather well with the findings in the tests, where the first crack (to the upper edge) was noted at a corrosion penetration of 140 μm.

Specimens with transverse reinforcement tested by Coronelli (1998) were analysed. He carried out beam tests following the recommendations by RILEM, but with reduced amount of transverse reinforcement. The specimens were subjected to accelerated corrosion to levels ranging from zero to 300 μm. Thereafter, the beams were loaded in four point bending, and the mid span deformation and the slip of the reinforcement were measured. The corrosion stage of one of his test series was analysed here, the fourth series where the cover was reduced to 35 mm. The transverse reinforcement was modelled with embedded reinforcement, i.e. complete interaction between the concrete and the reinforcement was assumed, while the main reinforcement was modelled with solid elements, surrounded by interface elements describing the special models for corrosion and bond. In the analyses, the beams were cracked due to the corrosion at a corrosion penetration about 30 μm. This can be compared with the tested beams, where the bottom cover cracked at a corrosion penetration around 20 μm.

The results from all of the analyses of corrosion cracking tests are summarised in Figure 8, where the corrosion penetration at cracking obtained in the analyses divided by the corrosion penetration at cracking measured in the experiments is plotted versus the cover to bar ratio. Since the tests of Andrade et al. (1993), Ghandehari et al. (2000), Al-Sulaimani et al. (1990), and Cabrera & Ghoddoussi (1992) were used

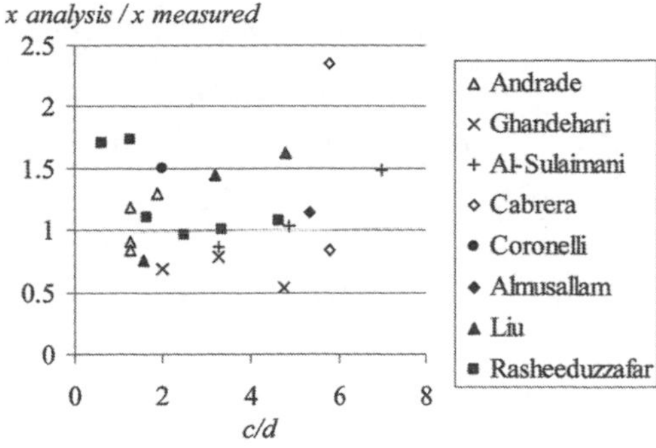

Figure 8. Comparison between corrosion penetration at cracking obtained in the analyses and in experiments versus the cover to bar ratio.

to give the structural behaviour of the rust in section 3, a good correlation between measured and analysed values are expected for these. However, also the analyses of the tests done by Liu & Weyers (1998), Rasheeduzzafar et al. (1992), Almusallam et al. (1996), and Coronelli (1998) show a good agreement with the measured value of the corrosion penetration at cracking. The main parameters of the corrosion model influencing these results are the ones describing the structural behaviour of the rust; i.e. the stiffness K_{cor} and the parameter p. Since these analyses show a reasonable agreement with test results, especially when considering the large scatter that is obtained in such tests, these parameters appear to be well adjusted.

5 CORROSION CRACKING

5.1 *The effect of uniform or localised corrosion on cracking*

Important information for designers is how much corrosion that is required to crack the concrete cover. This, and the effect of uniform or localised corrosion on cracking, was investigated. Two three-dimensional models with a reinforcement bar with diameter 16 mm and a cover of 25 mm respectively. 76 mm in one direction and large cover in the other directions was analysed, see Figure 9. The result was then compared with axisymmetric analyses, with the same reinforcement bar diameter and covers, assuming four cracks. In both the three-dimensional and the axisymmetric analyses, the concrete and steel were modelled as described in section 4. A normal strength concrete was chosen, with a compressive strength of 40 MPa and the maximum aggregate size 16 mm. Neither shrinkage, nor creep was included in these analyses. The effect of non-uniform corrosion was investigated in the three-dimensional models by applying corrosion only in parts of the interface between the reinforcement and

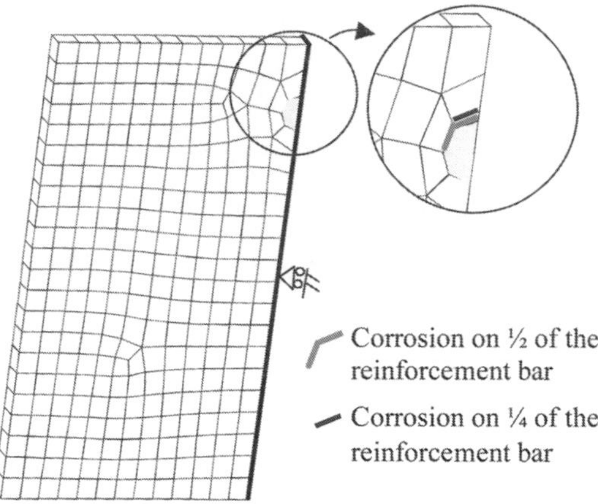

Figure 9. Finite element mesh used to study the effect of localized corrosion.

Table 3. Average corrosion penetration at cracking in various analyses.

	Average corrosion penetration at cracking			
Cover mm	Axisym. μm	3D[1] μm	3D[2] μm	3D[3] μm
25	25	22	20	12
76	110	110	100	60

[1] Uniform corrosion.
[2] Corrosion on ½ of the reinforcement bar.
[3] Corrosion on ¼ of the reinforcement bar.

the concrete. Two different cases were studied; corrosion on half of the reinforcement bar, and corrosion on one fourth of the reinforcement bar. In both cases, it was the part of the reinforcement bar closest to the edge that was assumed to corrode. The reason for this choice is that whether the corrosion is initiated due to chlorides or carbonatisation, it is the part of the reinforcement bar closest to the edge that will start to corrode first. Petre-Lazar (2000) showed that there were more corrosive products at a crack, than at the rest of the surface around a reinforcement bar.

The results are summarised in Table 3. As can be seen, the average corrosion penetration that causes cracking of the cover is approximately the same in the axisymmetric analyses, as in the three-dimensional analyses, when uniform corrosion is assumed. Also when half of the reinforcement bar was assumed to corrode, similar results were obtained. However, when all the corrosion was assumed to take place on only one fourth of the reinforcement bar, less average corrosion penetration was needed to crack the cover. Furthermore, the crack pattern differed between the different analyses. For all the studied cases, the first

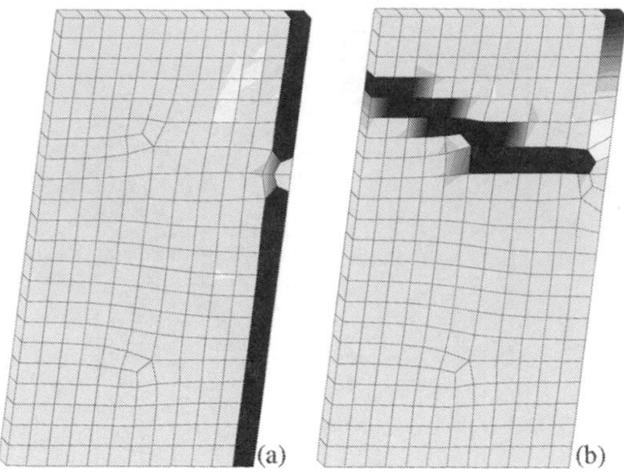

Figure 10. Crack pattern for (a) uniform corrosion and (b) localised corrosion combined with a large cover. Dark areas indicate cracks.

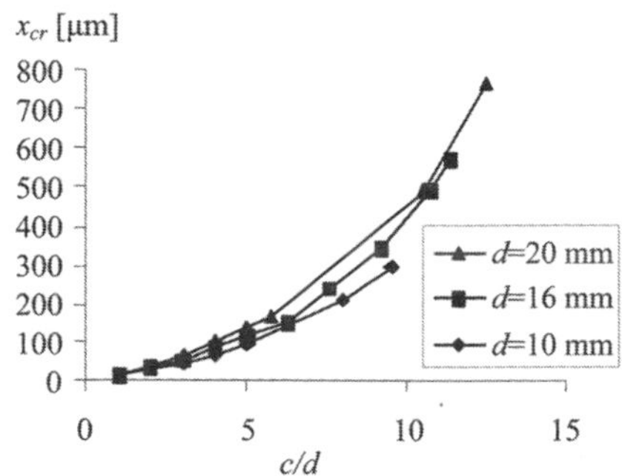

Figure 11. Corrosion penetration that causes cracking of the cover; uniform corrosion and a concrete with a compressive strength of 40 MPa and the maximum aggregate size 16 mm.

crack appeared from the reinforcement bar to the upper edge. For uniform corrosion, this crack then continued to grow. For the large cover and localised corrosion, inclined cracking appeared at a larger corrosion penetration, see Figure 10. The first crack was then closed, and in reality, this would most likely have caused spalling of the cover.

From these analyses, it was concluded that axisymmetric analyses appear to be a sufficient level of modelling when studying uniform corrosion. If localised corrosion is to be studied, three-dimensional models need to be used.

5.2 *Uniform corrosion*

To study the effect of uniform corrosion, a parameter study using axisymmetric analyses was done. Again, neither shrinkage, nor creep was included in these analyses. The corrosion penetration that causes cracking of the cover for various cover to bar diameter ratio is shown in Figure 11. These results are for a concrete

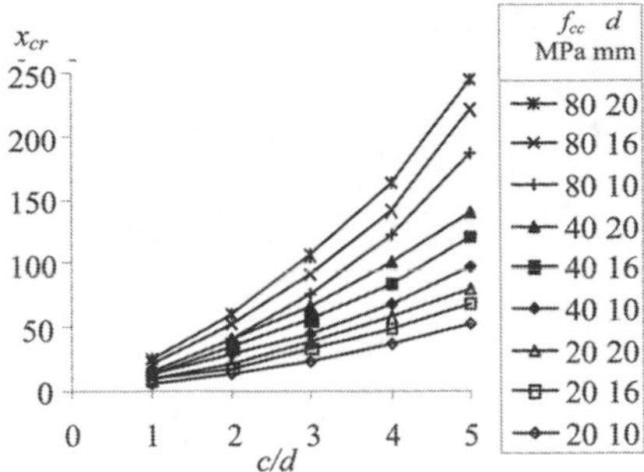

Figure 12. Corrosion penetration that causes cracking of the cover for various concrete qualities. Uniform corrosion, maximum aggregate size is 16 mm.

with a compressive strength of 40 MPa and the maximum aggregate size 16 mm. As can be seen, the results differ slightly for different bar diameters. The points shown in the diagram is the last ones where cracking was obtained; for larger cover to bar diameter ratios, the cover was not cracked regardless of how much corrosion that was applied. However, these points are of limited practical use, since they correspond to very large cover to bar diameter ratios, a lot larger than is used in practice.

To show the effect of the quality of the concrete, three different types were used. They all had the maximum aggregate size 16 mm, but the compressive strength was varied: 20, 40, and 80 MPa. The results for the cover to bar ratios that are of practical interest are shown in Figure 12. As can be seen, the quality of the concrete has a strong influence on how much corrosion the cover can withstand before it is cracked. Note, however, that it is not actually the compressive strength that has this influence; the main controlling parameter is the fracture energy of the concrete. That has, however, been calculated from the compressive strength and the maximum aggregate size according to the expressions in CEB (1993). For the three concrete types used here, the fracture energy was 48.7, 79.2, and 128.6 N/m.

6 CONCLUSIONS

The mechanical behaviour of rust was studied by a combination of analyses and test results found in the literature. This led to the assumption that rust behaves like a granular material, i.e. its stiffness increases with the stress level. This mechanical behaviour, and the volume increase of the corrosive products compared with the virgin steel, were modelled in a corrosion layer. By taking the mechanical behaviour of the rust itself into account, the rust gets a certain strain, and thereby the full volume increase will not be imposed on the concrete. Comparisons between analyses of corrosion cracking tests and test results found in literature indicate reasonable agreement.

The model was used to study the effect of uniform or localised corrosion. It was shown that for localised corrosion, less average corrosion penetration was needed to crack the cover than for uniform corrosion. Furthermore, the crack pattern differed. From these analyses, it was concluded that axisymmetric analyses appear to be a sufficient level of modelling when studying uniform corrosion. If localised corrosion is to be studied, three-dimensional models need to be used.

7 NOTATIONS

a	Free increase of the radius due to corrosion
a_{cr}	Free increase of the radius when the cover cracks
c	Cover
d	Diameter of reinforcement bar
D_{11}	The stiffness D_{11} in the bond layer
E_s	Young's modulus for steel
f	Possible increase of the radius before any normal stresses are applied to the surrounding concrete
f_{cc}	Concrete compressive strength
K_{cor}	Parameter describing the mechanical behaviour of the rust
p	Parameter describing the mechanical behaviour of the rust
r	Diameter of reinforcement bar
t_{in}	The thickness of the mortar – rebar interface
t_n	Normal stress
t_{ncr}	Normal stress that cracks the cover
t_t	Bond stress
u_n	Relative normal deformation
u_{ncor}	Normal deformation in the corrosion layer; i.e. increase of the radius due to corrosion
$u_{ncor,cr}$	The deformation of the corrosion layer when the cover cracks
u_{nbond}	Normal deformation in the bond layer
$u_{nbond,cr}$	Normal deformation in the bond layer when the cover cracks
u_t	Slip
v	Volume rust/volume steel
vct	Water–cement ratio
x	Corrosion penetration
x_{cr}	Corrosion penetration when the cover cracks
α	Level of hydratation
δ_{cr}	The deformation of the hole that causes cracking of the cover
δ_{reinf}	The deformation of the reinforcement bar due to a normal pressure
$\delta_{reinf,cr}$	The deformation of the reinforcement bar when the cover cracks

ε_{cor}	Strain in the rust
$\varepsilon_{cor,cr}$	Strain in the rust when the cover cracks
ν_s	Poison's ratio for steel
Φ	The porosity of the mortar

REFERENCES

Almusallam A.A., Al-Gahtani A.S. & Aziz A.R. 1996. Effect of reinforcement corrosion on bond strength. *Construction and Building Materials* 10(2): 123–129.

Al-Sulaimani G.J., Kaleemullah M., Basunbul I.A. & Rasheeduzzafar 1990. Influence of corrosion and cracking on bond behavior and strength of reinforced concrete members. *ACI Structural Journal* 87(2): 220–231.

Andrade C., Alonso C. & Molina F.J. 1993. Cover cracking as a function of bar corrosion. 1. Experimental test. *Materials and Structures* 26(162): 453–464.

Cabrera J.G. & Ghoddoussi P. 1992. The effect of reinforcement corrosion on the strength of the steel/concrete "bond". In (ed.), *Bond in Concrete, Proceedings of an International Conference*, Riga. CEB.

CEB 1993. *CEB-FIP Model Code 1990*. Lausanne, Switzerland: Bulletin d'Information 213/214.

Coronelli D. 1998. *Bar corrosion in steel-concrete bond: material and structural effects in R/C*. Ph.D. Thesis. Milano: Structural Engineering Department, Politecnico di Milano.

Ghandehari M., Zulli M. & Shah S.P. 2000. Influence of corrosion on bond degradation in reinforced concrete. In (ed.), *Proceedings EM2000, Fourteenth Engineering Mechanics Conference*, Austin, Texas. ASCE.

Liu Y.P. & Weyers R.E. 1998. Modeling the time-to-corrosion cracking in chloride contaminated reinforced concrete structures. *ACI Materials Journal* 95(6): 675–681.

Lundgren K. 2002. Modelling the effect of corrosion on bond in reinforced concrete. *Magazine of Concrete Research* 54(3): 165–173.

Lundgren K. & Gylltoft K. 2000. A model for the bond between concrete and reinforcement. *Magazine of Concrete Research* 52(1): 53–63.

Molina F. J., Alonso C. & Andrade C. 1993. Cover cracking as a function of rebar corrosion. 2. Numerical-model. *Materials and Structures* 26(163): 532–548.

Noghabai K. 1998. *Effect of Tension Softening on the Performance of Concrete Structures*. Ph.D. Thesis. Luleå, Sweden: Division of Structural Engineering, Luleå University of Technology.

Petre-Lazar I. 2000. *Evaluation du comportement en service des ouvrages en beton arme soumis a la corrosion des aciers (Aging assessment of concrete structures submitted to steel corrosion, in French)*. Ph.D. Thesis. Quebec: Department of Civil Engineering, Laval University.

Rasheeduzzafar, Al-Saadoun S.S. & Al-Gahtani A.S. 1992. Corrosion cracking in relation to bar diameter, cover, and concrete quality. *Journal of Materials in Civil Engineering* 4(4): 327–342.

TNO 1998. *DIANA Finite Element Analysis, User's Manual release 7*. Hague: TNO Building and Construction Research.

Tuutti K. 1982. *Corrosion of steel in concrete*. Swedish Cement and Concrete Research Institute.

Computational Modelling of Concrete Structures, Bićanić et al. (eds)
© 2003 Swets & Zeitlinger, Lisse, ISBN 90 5809 536 3

Creep deformation of FRP-plated R/C tensile elements using solidification theory

M. Savoia, B. Ferracuti & C. Mazzotti
DISTART – Structural Engineering, University of Bologna, Bologna, Italy

ABSTRACT: A numerical study on R/C tensile elements reinforced by external FRP plates and subject to long-term loading is presented. Relative slips between concrete, steel bars and FRP plates are considered, governed by interface shear stress – slip laws. Solidification theory and exponential algorithm are used to obtain incremental constitutive equations for concrete as well as for steel-concrete and FRP-concrete interface laws. Cohesive normal stresses across transverse cracks in concrete are considered. The incremental differential system is transformed into a non linear algebraic system by a finite difference discretization with respect to axial coordinate. Two different problems are studied: a) for short term loadings, the increase of tension stiffening effect due to plating; b) for long term loadings, the influence of plating on creep deformation.

1 INTRODUCTION

A great attention has been recently devoted to the behavior under service loadings of R/C structures retrofitted by FRP composites (plates or sheets) (see for instance Sato et al. 1999, Lee et al. 1999, Tripi et al. 2000). Correct prediction of deformability and crack opening requires both short term and long term behaviors be considered. Creep behavior and evolution of delayed deformation with time are very important aspects, since they may have significant consequences on the structural behavior. In the case of a retrofit intervention, for example, design errors and unreliable interventions with time can be done by ignoring delayed deformations (both of the structure to be retrofitted and of strengthening material). Creep effects of concrete and of bonding interfaces between concrete, steel and composites must be taken into account. Nevertheless, experimental or numerical studies devoted to time-dependent behavior of FRP reinforced R/C elements are very few at the moment (Plevris & Triantafillou 1994, Savoia et al. 2002). Theoretical and numerical studies require the complex phenomenon of load transfer between steel-concrete and plate-concrete interfaces close to specimen ends or transverse concrete cracks be correctly modeled.

With reference to tensile elements, the authors recently showed that FRP-retrofitting may significantly reduce crack width in concrete and may contribute to preserve concrete tension stiffening at axial load levels significantly higher than for unplated members.

In the present study, the role of creep strains on the serviceability behavior of R/C tensile elements retrofitted by FRP-plates is studied. Concrete member, steel bars and FRP plates are considered as different elements with bonding interfaces defined by appropriate bond-slip laws. For long-term behavior of concrete, a constitutive creep model based on Bazant' solidification theory (Bazant & Prasannan 1993) and recently improved by the authors (Savoia et al. 2002) is adopted. Solidification theory is used also to extend to long-term case both plate-concrete interface law and non linear steel-concrete bond-slip law. In the first case, creep compliance of both adhesive and concrete cover are considered. The governing equations are then written in incremental form to account for creep deformations without performing time convolution.

The incremental (with respect to time) governing equations are written in the form of a non linear system of differential equations (with respect to the element axial coordinate). A finite difference scheme is then used to reduce these equations to a non linear incremental algebraic system. Time integration is then performed making use of exponential algorithm (Bazant 1971), allowing for the use of variable time steps, small at the beginning of the loading and large after months or years, as is required to study delayed deformations. A predictor-corrector Newton-Raphson procedure is also used due to non linearity of equations. The formation of concrete tensile cracks when stress reaches tensile strength is prescribed by imposing appropriate internal boundary conditions, where concrete stress is written as a function of crack width amplitude according to fictitious crack model and continuity of steel and FRP plates is prescribed. Tests with force control, displacement control or crack width control may be simulated. The convergence properties of the proposed integration scheme are also studied.

Numerical simulations are performed to estimate the effectiveness of FRP strengthening under service loadings. For short term loadings, FRP-plating is shown to significantly reduce crack width, so preserving concrete tension stiffening effect (with respect to unplated case) in the cracked stage. For long term loadings, retrofit with FRP plates reduces creep deformation and, consequently, axial elongation of R/C tensile member with respect to unreinforced case.

2 GOVERNING EQUATIONS OF PROPOSED MODEL

The study concerns R/C tensile members subject to pure tension and strengthened by means of composite FRP plates glued along their external faces.

A displacement-based model was recently proposed by the authors (Ferretti & Savoia 2002), where transverse cross-section of concrete element and steel bars were assumed to remain plane after deformation, whereas Bernoulli beam theory was adopted for FRP plates, allowing to describe transverse deflection and corresponding transverse slip between concrete and plates. Subsequent analyses showed that, due to the very small thickness of FRP plates with respect to concrete, plate flexural thickness is negligible and has no significant effect, at least for loadings far from those causing peeling failure. Hence, in the present study a simplified kinematic model is assumed, where concrete, steel bars and FRP plates are subject to axial displacements only, denoted by u_c, u_s and u_p, respectively (see Fig. 1).

The governing equations are written in incremental form, in order to allow delayed deformations due to creep be taken into account. The general time increment may correspond to an external action increment (prescribed axial load or axial displacement) or to a constant action (classical creep or relaxation problems).

Hence, incremental compatibility, equilibrium and constitutive equations can be written in the form:

$$\frac{d\dot{u}_c}{dx} = \frac{1}{E_c(t)\,A_c}\dot{N}_c\,, \qquad \frac{d\dot{N}_c}{dx} = c_s\,\dot{\tau}_s + 2b_p\,\dot{\tau}_p \tag{1}$$

$$\frac{d\dot{u}_s}{dx} = \frac{1}{E_s\,A_s}\dot{N}_s\,, \qquad \frac{d\dot{N}_s}{dx} = -c_s\,\dot{\tau}_s \tag{2}$$

$$\frac{d\dot{u}_p}{dx} = \frac{1}{E_p\,A_p}\dot{N}_p\,, \qquad \frac{d\dot{N}_p}{dx} = -b_p\,\dot{\tau}_p \tag{3}$$

where dot denotes derivative with respect to time, subscripts c, s and p denote quantities related to concrete, steel bars and composite plates, respectively, N, A and E stand for axial force, area and Young modulus, τ_p and τ_s are concrete/plates and concrete/steel interfacial shear stresses, b_p is the plate width, $c_s = n_b\pi\,\Phi(n_b, \Phi =$ number and diameter of steel bars).

Incremental FRP-concrete and steel-concrete interface relations are also introduced, giving shear stresses as a function of relative slips between different materials:

$$\dot{\tau}_s = -f_s(\dot{s}_s, t) = -f_s(\dot{u}_s - \dot{u}_c, t) \tag{4a}$$

$$\dot{\tau}_p = -f_p(\dot{s}_p, t) = -f_p(\dot{u}_p - \dot{u}_c, t) \tag{4b}$$

In the present study, non linear steel-concrete interface law is adopted, whereas linear plate-concrete relation is retained. Both laws are extended to cover creep deformation, as described in the following section.

Finally, viscoelastic behavior of concrete in tension is considered, whereas non viscous linear elastic moduli for steel and FRP plates are adopted.

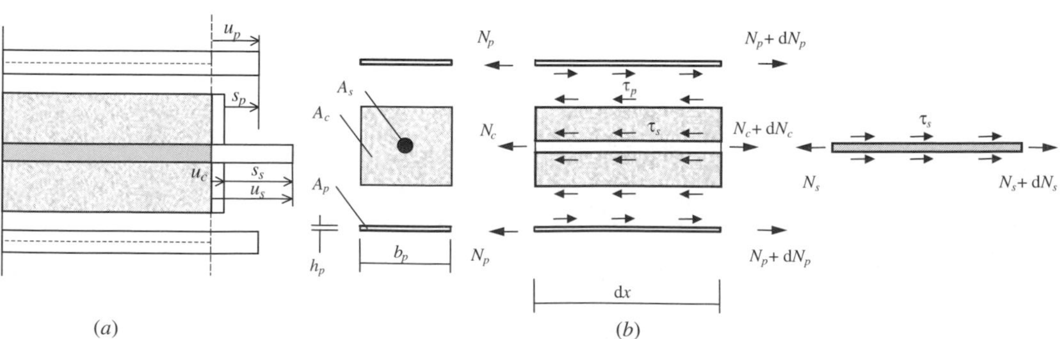

Figure 1. Notation adopted for (a) displacements and (b) axial forces and interfacial shear stresses.

3 VISCOELASTIC LAW FOR CONCRETE ACCORDING TO SOLIDIFICATION THEORY

Linearly viscoelastic time dependent behavior of concrete is considered. Incremental constitutive equation is written making use of solidification theory and exponential integration algorithm is used to obtain the finite time step equation.

3.1 *The creep law*

In the form originally proposed in Bazant & Prasannan (1993), solidification theory is based on the basic assumption that age-dependent material is considered as a varying composite whose components, a solidified (i.e. load carrying) material with volume fraction $v(t)$ and the complementary non solidified part with volume fraction $1 - v(t)$ are characterized by age-independent properties. Ageing of concrete with time is then governed by the adopted solidification law $v(t)$. The deposited layers of solidified constituent are assumed to be subject to the same total strain increment $d\varepsilon_c$. Creep properties of solidified layers are defined through the micro-relaxation function $\Psi(t-t')$. Hence, the relation at time t between the (micro)stress $s[v(t)]$ (corresponding to a layer whose solidification started at $t' = \tau$) and total strain ε_c is given by:

$$s(\tau,t) = \int_{\tau}^{t} \Psi(t-t')\, d\varepsilon_c(t') \; . \tag{5}$$

By superposition, using eqn (5) and global equilibrium condition for concrete, the relation between the macroscopic (applied) stress σ_c and the macroscopic strain ε_c can be written as:

$$\sigma_c(t) = \int_{\tau}^{t} \Psi(t-t')\, v(t')\, d\varepsilon_c(t') \, . \tag{6}$$

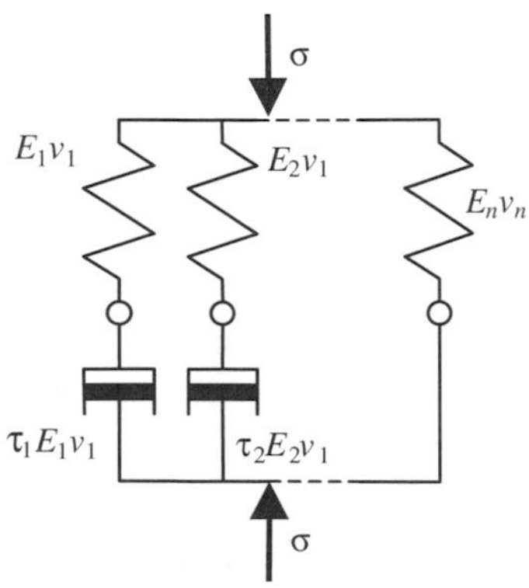

Figure 2. Maxwell chain composed of N units equivalent to the proposed creep model obtained from solidification theory.

In order to obtain an incremental relation from eqn (6), a Dirichlet series expansion for the microscopic relaxation function is introduced:

$$\Psi(t-t') = \sum_{\mu=1}^{n} E_\mu\, e^{-(t-t')/\tau_\mu} \, , \tag{7}$$

where n is the number of terms adopted in the expansion (analogous to the number of units in a Maxwell chain, see Figure 2), whereas $E_\mu, \tau_\mu (\mu = 1, \ldots, n)$ are the elastic moduli and the characteristic relaxation times of the units. Substituting eqn (7) into (6), straightforward algebra gives:

$$\sigma_c(t) = \sum_{\mu=1}^{n} \sigma_\mu(t) \, , \tag{8}$$

where the stress $\sigma_\mu(t)$, corresponding to the μth unit, is the solution of the differential equation:

$$\dot{\sigma}_\mu(t) + \frac{\sigma_\mu(t)}{\tau_\mu} = E_\mu\, v_\mu(t)\, \dot{\varepsilon}_c(t) \quad (\mu=1,\ldots, n) \, . \tag{9}$$

In the present model, two solidification functions only, $v_1(t), v_n(t)$, are adopted. Function $v_1(t)$ governs ageing of first $n - 1$ units, whereas $v_n(t)$ modulates ageing of the last unit, which is a degenerated purely elastic unit (i.e., $\tau_n \to \infty$), in order to ensure an asymptotic value of creep strain for $t \to \infty$, as prescribed by CEB Model Code (1990). The adopted solidification functions are defined as:

$$v_1(t) = \sqrt{e^{s_1(1-\sqrt{28/t})}} \, , \qquad v_n(t) = \sqrt{e^{s_n(1-\sqrt{28/t})}} \, , \tag{10}$$

where time is expressed in days and s_1, s_n are free parameters. These expressions are analogous to the time evolution law of concrete Young modulus given by CEB Model Code (1990).

For the numerical applications, 9 units are adopted, relaxation times are selected according to $\tau_\mu = 0.003{\cdot}10^\mu (\mu = 1, \ldots, n-1)$, whereas parameters E_μ defining elastic moduli and s_1, s_N for solidification functions are obtained by best fitting procedure from CEB Model Code (1990) relaxation curves for concrete in compression (see Savoia et al. 2002). These parameters are used here to predict creep in tension. Of course, this extension should be verified, even if it is normally adopted in the literature.

3.2 *The numerical formulation according to exponential algorithm*

Incremental formulations are very useful in creep problems, since they do not require storage of the whole stress history (as required in the solution of convolution integrals), but the last value only of state variables must be stored before the subsequent (finite) time step. Nevertheless, classical constant step integration methods cannot be adopted to obtain results over the entire time interval of practical interest since, using small time steps (as it is necessary at the beginning of loading) requires a great computational effort to predict the behavior after several months or years.

In the present study, the exponential algorithm is adopted, originally proposed in Zienkiewicz & Watson (1966), Bazant (1971), Bazant & Wu (1974) for linear viscoelasticity and later extended to non linear problems (Carol & Murcia 1989, Mazzotti & Savoia 2001, 2003). It is based on the exact integration of differential equation (9) over the time step $\Delta t_r = t_r - t_{r-1} (r = 1, \ldots, N)$, where $t_0, t_1, \ldots, t_N$ are the time instants where solution must be evaluated. By multiplying both sides of eqn (9) by the integration function:

$$f_\mu(t) = e^{\int_{t_{r-1}}^{t} \frac{dt'}{\tau_\mu}} \tag{11}$$

and performing integration over the time step Δt_r yield:

$$\sigma_\mu(t_r) = e^{-f_\mu(t_r)}\left[\sigma_{\mu,r-1} + \int_{t_{r-1}}^{t_r} E_\mu v_\mu(t') e^{-f_\mu(t')} d\varepsilon(t')\right] \tag{12}$$

where $\sigma_{\mu,r-1}$ is the value of stress of μth Maxwell unit at the previous (t_{r-1}) time instant. If Δt_r is sufficiently small or the variation of the ageing function $v_\mu(t)$ over the time step is small enough, then the r.h.s. of eqn (12) can be integrated analytically, so obtaining the stress at the t_r time instant:

$$\sigma_{\mu,r} = \sigma_{\mu,r-1}\, e^{-\frac{\Delta t_r}{\tau_\mu}} + E_\mu\, v_{\mu,r-1/2}\, \lambda_{\mu,r}\, \Delta\varepsilon_r \tag{13}$$

where:

$$\lambda_{\mu,r} = (1 - e^{-\frac{\Delta t_r}{\tau_\mu}})\frac{\tau_\mu}{\Delta t_r}, \qquad v_{\mu,r-1/2} = \tfrac{1}{2}\left(v_{\mu,r-1} + v_{\mu,r}\right) \tag{14}$$

are quantities depending on time instant t_r and rth time step Δt_r.

Substituting eqn (13) into (8) gives the following incremental pseudo-elastic stress-strain relation at time t_r:

$$\Delta\varepsilon_r = \frac{1}{E_r''}\Delta\sigma_r + \Delta\varepsilon_r'', \tag{15}$$

where:

$$E_r'' = \sum_{\mu=1}^{N} \lambda_{\mu,r} E_\mu v_{\mu,r-1/2}, \qquad \Delta\varepsilon_r'' = \frac{1}{E_r''}\sum_{\mu=1}^{N}\left(1 - e^{-\frac{\Delta t_r}{\tau_\mu}}\right)\sigma_{\mu,r-1}. \tag{16}$$

In the numerical procedure, time steps are chosen equally spaced in the log-scale, so that few time steps are required to cover the whole period of interest (years). Convergence properties of exponential algorithm for non linear problems are discussed in Mazzotti & Savoia (2001a), with reference to concrete damage in compression.

Eqn (15) can be easily used to compute the strain increment during the time step Δt_r, since pseudo-elastic modulus E_r'' and pseudo-deformation $\Delta\varepsilon_r''$ depend on quantities known after the previous time step $(r - 1)$ or function of time.

3.3 *Cohesive equation for transverse cracks in concrete*

The behavior of concrete in tension is assumed to be linearly viscoelastic up to the tensile strength f_{ct}. When stress reaches tensile strength, the formation of a transverse crack is imposed, which is modeled by means of fictitious crack model, postulating tensile stress in concrete σ_{ct} be a function of the fictitious crack opening w. Cohesion across the crack may be particular significant in the presence of plating due to the reduced crack width.

In order to take into account the partial closing of existing cracks when formation of the last crack occurs (in a displacement-controlled or in a crack width-controlled test), the cyclic $\sigma_{ct} - w$ relation proposed by Hordijk (1992) is adopted.

4 STEEL-CONCRETE INTERFACE LAW

It is well recognized that creep deformations can affect also the behavior of steel-concrete interface; the main effect of creep deformation is the increasing with time of slip and of anchorage length close to specimens ends or concrete cracks. The experimental results presented by Sato et al. (1987) show the time-dependent changes of bond stresses, steel strains and concrete strains along the longitudinal axis of R/C members in tension.

In the present study, solidification theory is used to define a time-dependent bond-slip steel-concrete law. Five units are adopted, where the last one is a degenerated unit, so that $\tau_5 \to \infty$ and $\lambda_{5,r} \to 1$. If, as a first stage, a linear shear stress-slip law is adopted, the integration of governing relation according to exponential algorithm gives the following incremental law:

$$\Delta\tau_{s,r} = K_s''(t)\left[\Delta s_{s,r} - \Delta s_{s,r}''\right], \tag{17}$$

where $s_s = u_s - u_c$ is the steel-concrete slip (see Figure 1), τ_s is the corresponding interfacial shear stress, and:

$$K_s''(t) = K_0 \cdot \left(\sum_{\mu=1}^{4} \lambda_{\mu,r}\, k_\mu\, \nu_{\mu,r-1/2} + k_5\, \nu_{5,r-1/2} \right), \tag{18}$$

where K_0 is the instantaneous stiffness for $t_0 = 28$ days time at loading and $k_1, \ldots k_5$ are non dimensional coefficients.

Eqns (17), (18) can be obtained following a procedure analogous to that described in the previous Section. Two solidification functions v_1, v_5 are adopted, selected in the same form as reported in eqns (10). Moreover, stiffness constants $K_1 \ldots K_5$ are defined following the same procedure adopted for concrete Maxwell units and using, for best fitting, the same relaxation curves. The residual term $\Delta s_{s,r}''$ in incremental eqn (17) is defined, in analogy to eqn (16), as:

$$\Delta s_r'' = \frac{1}{K_{s,r}''} \sum_{\mu=1}^{4} \left(1 - e^{-\Delta t_r / \tau_\mu} \right) \tau_{s\mu,r-1}. \tag{19}$$

The creep law (17) is then extended to cover non linear bond-slip behavior, as typically assumed for instantaneous loadings. Non linearity is introduced as a modification of elastic stiffness K_0, which is substituted by a tangent stiffness K_{tg} depending on slip level s_s. Starting from classical non linear bond slip relation proposed by CEB Model Code (1990):

$$\tau_s = \left(\frac{\tau_{s,\max}}{s_{s,\max}^{\alpha}} \right) s_s^{\alpha} \tag{20}$$

where, typically, $\tau_{s,\max} = 1.5 \cdot f_{ck}^{1/2}$, $s_{s,max} = 0.6$ mm and $\alpha = 0.4$, eqn (20) is rewritten in incremental form as:

$$\Delta\tau_s = K_{tg}(s_s)\, \Delta s_s, \tag{21}$$

where:

$$K_{tg}(s_s) = \left(\frac{\tau_{s,\max}}{s_{s,\max}^{\alpha}} \right) \alpha s_s^{\alpha-1}. \tag{22}$$

Setting $f_{NL}(s_s) = K_{tg}(s_s)/K_0$, incremental bond-slip eqn (17) is extended to non linear range as:

$$\Delta\tau_{s,r} = f_{NL}(s_{s,r})\, K_s''(t)\left[\Delta s_{s,r} - \Delta s_{s,r}''\right]. \tag{23}$$

Assumption adopted to obtain eqn (23) is analogous to that currently used for non linear creep modeling of concrete, i.e. non linearity is taken into account as a non linear function multiplying linear creep strain (Carol & Bazant 1993). This assumption is considered valid for medium stress levels, whereas for higher stresses the non linear creep mechanism is more complex (Mazzotti & Savoia 2001b, 2003). Analogous assumption is introduced by CEB-FIP (1997), even if a different creep function is introduced.

5 CREEP LAW FOR FRP PLATE-CONCRETE INTERFACE

The plate-concrete interface creep law must take shear deformation of both adhesive and external concrete cover into account. Numerical and experimental tests showed that a 30 ÷ 50 mm depth cover participates to interface shear compliance (Brosen & Van Gemert 1997, Savoia et al. 2003). Only taking that contribution into account, the actual decay length of end effects (about 80 ÷ 100 mm) can be explained, whereas 10 mm length is obtained by considering adhesive compliance only.

In the present study, solidification theory is used to define shear compliances of both adhesive and concrete, leading to incremental equations analogous to those obtained in previous Sections (see eqn (17)):

$$\Delta s_{A,r} = \frac{1}{K_A''(t)} \Delta\tau_{p,r} + \Delta s_{A,r}'', \qquad \Delta s_{C,r} = \frac{1}{K_C''(t)} \Delta\tau_{p,r} + \Delta s_{C,r}'', \tag{24}$$

where subscripts A and C stand for terms related to adhesive and concrete, respectively, whereas τ_p is the plate-concrete shear stress. Total plate-concrete slip increment $\Delta s_p = \Delta s_A + \Delta s_C$ is then obtained in the form:

$$\Delta s_{p,r} = \left(\frac{1}{K_A''(t)} + \frac{1}{K_C''(t)} \right) \Delta\tau_{p,r} + \Delta s_{p,r}'', \tag{25}$$

where $\Delta s_{p,r}'' = \Delta s_{A,r}'' + \Delta s_{C,r}''$.

According to eqn (25), different creep formulations for adhesive and concrete can be adopted. In the present work, non ageing two-term creep formulation is introduced for adhesive whereas, for concrete compliance contribution, an expression similar to that adopted for concrete in tension is used.

6 THE SOLUTION OF NON LINEAR INCREMENTAL EQUATIONS

6.1 *The numerical algorithm*

Substituting eqns (4) in eqns (1)–(3) and adopting incremental constitutive eqns (15), (23), (25) leads to a system of incremental equations for the rth time increment, which is rewritten in the following compact form:

$$\frac{\mathrm{d}\Delta\mathbf{y}_r(x,t)}{\mathrm{d}x} = \mathbf{A}_{tg}(\mathbf{y}_r, x, t)\ \ \Delta\mathbf{y}_r(x,t) + \Delta\boldsymbol{\alpha}_r''(x,t)$$

$$0 \le x \le L, \tag{26}$$

where $\Delta\mathbf{y}_r$ is the vector collecting the increments at the rth step of the vector $\mathbf{y}$ collecting the unknown function:

$$\mathbf{y}^T = \{N_c, N_s, N_p, u_c, u_s, u_p\} \tag{27}$$

and $\Delta\alpha_r''(x,t)$ is the vector collecting terms due to creep deformation (see eqns (15), (23), (25)). The vector is known when obtaining the solution for the rth time increment, because it is only related to the solution at the previous $(r-1)$th step.

Moreover, the boundary conditions to be prescribed at the two ends $x = 0, L$ of the tensile member can be written as:

$$\mathbf{B}_a\Delta\mathbf{y}_r(0) + \mathbf{B}_b\Delta\mathbf{y}_r(L) = \Delta\boldsymbol{\alpha}_r \tag{28}$$

Vector $\Delta\boldsymbol{\alpha}_r$ takes different forms if axial force increment ΔN or, alternatively, axial elongation ΔL is prescribed, in order to simulate creep or relaxation tests, respectively.

Incremental boundary value problem (26), (28) is solved here via finite difference method, by writing spatial derivatives (along x), i.e. the l.h.s. of eqn (26), by finite difference approximation (see Ferretti & Savoia 2002, for details). The interval $[0, L]$ is divided using J nodes at uniform distance h, i.e., $0 = x_1 < x_2 < \cdots < x_J = L$. For the general j-th interior mesh point $(1 < j < J)$, the derivative in Eqn (26) is replaced by the central finite difference approximation $(\Delta\mathbf{y}_{j+1} - \Delta\mathbf{y}_j)/h$ centered at $x_{j+1/2}$. Finite difference equations, together with boundary conditions (28), lead to the non-linear system (see Figure 3):

$$\begin{bmatrix} \mathbf{S}_1 & \mathbf{R}_1 & & & & \\ & \cdots & \cdots & & & \\ & & \mathbf{S}_j & \mathbf{R}_j & & \\ & & & \cdots & \cdots & \\ & & & & \mathbf{S}_J & \mathbf{R}_J \\ \mathbf{B}_a & & & & & \mathbf{B}_b \end{bmatrix} \begin{Bmatrix} \Delta\mathbf{y}_{1,r} \\ \Delta\mathbf{y}_{2,r} \\ \cdots \\ \cdots \\ \Delta\mathbf{y}_{J-1,r} \\ \Delta\mathbf{y}_{J,r} \end{Bmatrix} = \begin{Bmatrix} \mathbf{0} \\ \mathbf{0} \\ \cdots \\ \cdots \\ \mathbf{0} \\ \Delta\alpha_r \end{Bmatrix} + \begin{Bmatrix} \Delta\alpha''_{1,r} \\ \Delta\alpha''_{2,r} \\ \cdots \\ \cdots \\ \Delta\alpha''_{J-1,r} \\ \Delta\alpha''_{J,r} \end{Bmatrix} \tag{29}$$

where:

$$\mathbf{S}_j = -\frac{1}{h}\mathbf{I} - \frac{1}{2}\mathbf{A}_{tg}(\mathbf{y}_{j,r}, x_j, t),$$

$$\mathbf{R}_j = -\frac{1}{h}\mathbf{I} + \frac{1}{2}\mathbf{A}_{tg}(\mathbf{y}_{j+1,r}, x_{j+1}, t)\quad j = 1, 2, \ldots, J \tag{30}$$

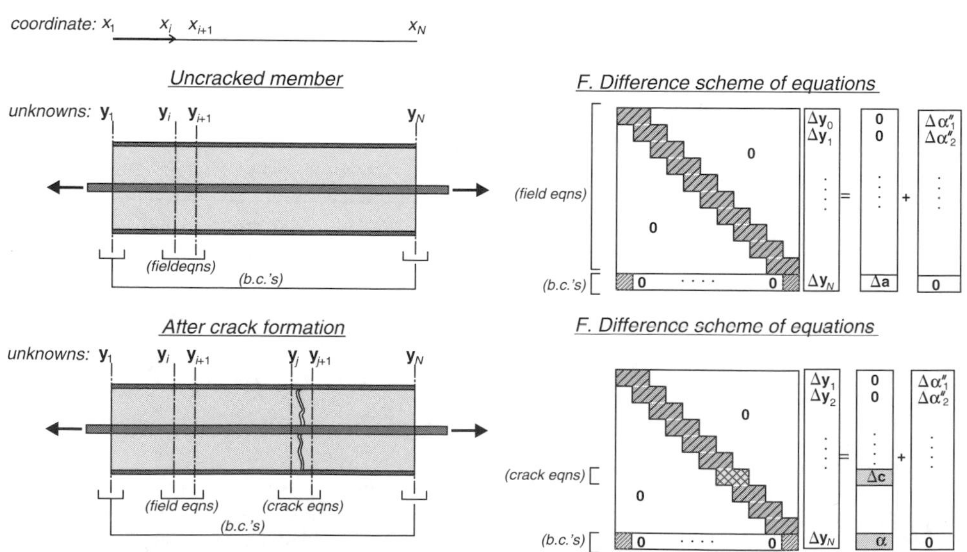

Figure 3. Finite difference scheme for the solution at the rth time increment: field equations, boundary conditions and crack equation imposing the presence of a transverse crack in concrete.

and **I** is the identity matrix. The second term at r.h.s. of Eqn (29) contains the prescribed (at rth step) values related to creep deformations, as obtained from solidification theory.

When tensile stress in concrete reaches the strength f_{ct} in a transverse section, say that corresponding to jth node, formation of a crack in concrete is imposed, with tensile stress depending on crack width amplitude increment $\Delta w = \Delta u_{c,j+1} - \Delta u_{c,j}$ according to cohesive crack law (see section 3.3), while continuity is preserved for steel and FRP-plate axial displacements. This condition is introduced by replacing the jth row of system in eqn (29) by an internal boundary condition, imposing the modified equilibrium and compatibility conditions. This equation can be written in the form (see Figure 3 and Ferretti & Savoia 2002 for details):

$$\mathbf{B}_c \Delta \mathbf{y}_{r,j} + \mathbf{B}_d \Delta \mathbf{y}_{r,j+1} = \Delta \mathbf{c}_j (w, \Delta w) \qquad (31)$$

where $\Delta \mathbf{c}_j(w, \Delta w)$ is a incremental vector whose non zero terms depend on cohesive tensile stress in concrete obtained from crack width through fictitious crack equation.

6.2 *Solution strategies*

To solve the incremental non linear system (29), for a given increment of applied action (axial force N or elongation ΔL), the solution is obtained iteratively by using modified Newton-Raphson method. According to eqn (29), the simplest forward incremental formulation has been adopted. To obtain an accurate solution, small step must be used, in order to avoid error accumulation. The step size has been calibrated, with reference to instantaneous loading, by comparison with a solution obtained by a non incremental Newton-Raphson solution procedure, i.e. by adopting total values of unknown variables and a secant value for the non linear matrix. With reference to the first example reported in Section 7, in Figure 4 comparison of force-axial displacement laws obtained by adopting secant matrix and incremental formulation with two different load steps is reported.

In order to adopt larger increments, for the creep problem where the adoption of an incremental formulation is mandatory, improved solution procedures are under study. Nevertheless, it is worth remembering that closed form integration over the time step performed by the exponential algorithm provides, in the case of weakly non linear problems, for a computational effective solution.

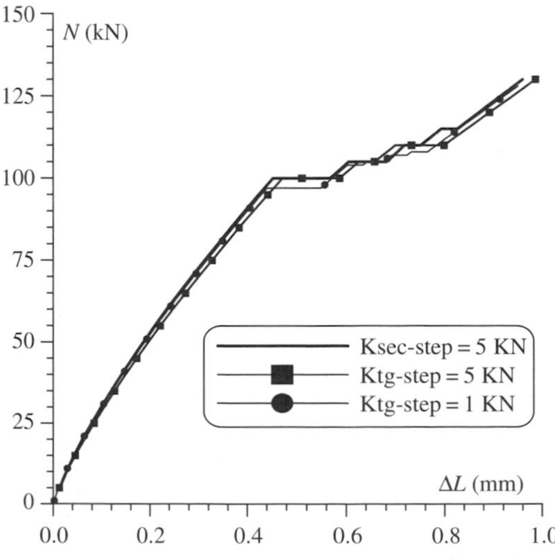

Figure 4. Convergence of the proposed incremental scheme.

7 TENSION STIFFENING IN FRP PLATED R/C ELEMENTS: SHORT TERM LOADING

Short term loading of a R/C tensile element strengthened by a couple of FRP plates is studied first. Geometrical and mechanical properties are reported in Figure 5.

In the first case, axial force is applied at the two ends at the steel bars. Axial force versus axial elongation curves are reported in Figure 6(a). Plated and unplated

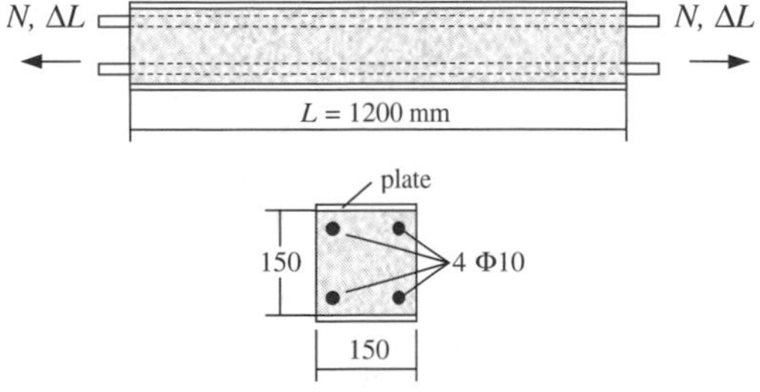

$L \times b \times h$ (mm)	$1200 \times 150 \times 150$
E_c (N/mm^2)	33640
f_c (N/mm^2)	36.4
f_{ct} (N/mm^2)	3.64
A_s (mm^2)	314
E_s (N/mm^2)	200000
$b_p \times h_p$ (mm)	$150 \times$ (1.016 or 0.33 or 0)
E_p (N/mm^2)	230000
h_g (mm)	1.5
E_g (N/mm^2)	1212.9

Figure 5. Geometry and mechanical properties of the FRP-plated R/C tensile element considered in numerical examples.

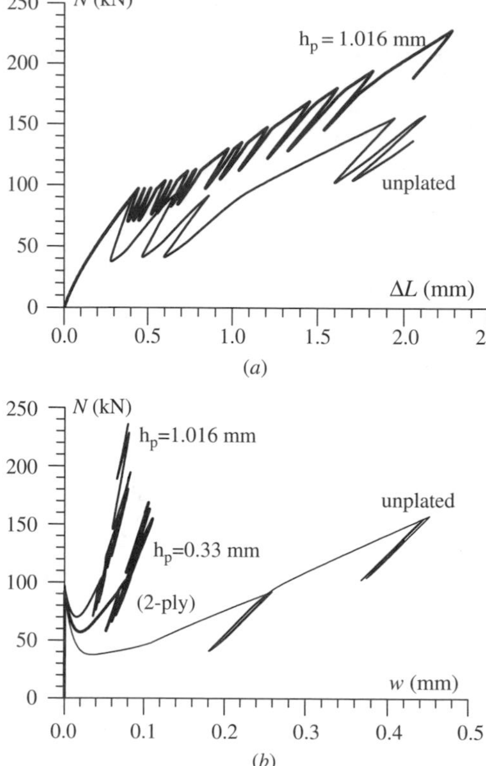

Figure 6. Short-term loading of a FRP plated R/C tensile member for different values of plate thickness h_p: (*a*) axial elongation and (*b*) crack width vs. axial force.

case are compared. After cracking, the test has been simulated by controlling width amplitude of the last transverse crack of concrete. The numerical algorithm is very stable: by controlling the crack amplitude, which is a monotonically increasing parameter, it is possible to follow the well known snap-back path due to elastic unloading in regions close to concrete cracks. The solution is not affected by numerical strain localization phenomena.

The Figure shows, for the unplated case, the classical formation of the primary cracking (for about $N = 83$ kN), corresponding to a significant stiffness loss. On the contrary, in the plated case, concrete cracking is much more gradual. For a given value of axial elongation (say 1.5 mm), a much higher number of cracks can be noted (3 crack for the unplated case and 11 for the strengthened element).

Figure 6(b) shows the evolution of width of the first crack during the simulated test. Unloading and subsequent reloading cycles are due to formation of other transverse cracks. It is worth noting that plating strongly reduces crack width which does not increase significantly when increasing axial load.

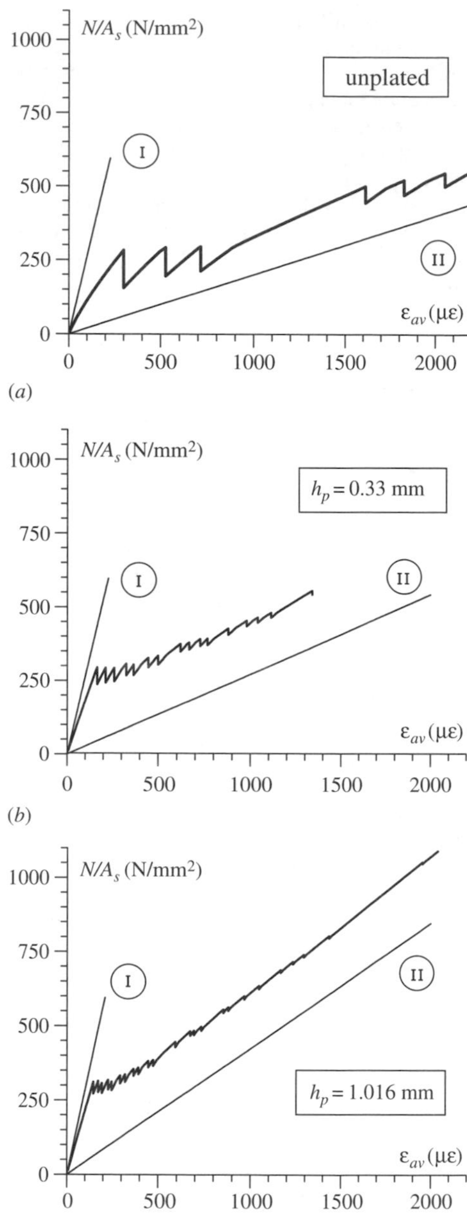

Figure 7. Short term loading: effect of plating on tension stiffening in tensile members.

In the second case, the test is simulated by controlling axial displacement of both steel bars and FRP plates, so that the two end sections are in State II condition. Figures 7(a–c) show axial force – elongation curves in three different cases: $h_p = 1.016$ mm, 0.33 mm and the unplated case. For each case, State I condition and State II condition lines are reported.

The figures confirm the significant increase of tension stiffening due to the presence of FRP plates: cracks are much more "smeared" along the member (many cracks with smaller width), so preserving a significant stiffness in the cracked range.

8 LONG TERM CREEP DEFORMATION

A creep test on the same tensile element depicted in Figure 5 has been numerically simulated. Axial load is prescribed at the two ends of steel bars.

In the first phase, a loading ratio equal to 3.47 kN/min has been prescribed, until the attainment of maximum load. Then, the load is mentained constant for about 30 years. Two different loading levels are considered: a) a maximum loading level equal to 60 kN, corresponding to 2.30 MPa tensile stress in concrete (no transverse cracking); b) a maximum load equal to 95 kN, with formation of the first transverse crack in concrete.

Figures 8 show the axial force – elongation curves for the plated members with $h_p = 0.33$ mm (Fig. 8b), $h_p = 1.016$ mm (Fig. 8c), and for the unplated member (Fig. 8a), compared with the corresponding short term curves. Test are simulated with force control condition.

A significant reduction of creep deformation for FRP-reinforced elements can be found.

Finally, for the $h_p = 1.016$ mm case, the profiles along the member of most significant stress and displacement variables are reported in Figures 9a–f. Dashed line stands for quantities at the end of loading (before the first crack formation), solid line after crack formation, and thick solid line after the period of sustained loading. Figure 9c shows that, due to crack formation, stress in plate increases about seven time with respect to (uncracked) state I solution, whereas strain in steel bars increses about three times. Hence, plates give a high contribution to axial stiffness in the cracked range. Moreover, due to significant concrete creep strain, axial stress in both steel bars and plates increases significantly with time. Finally, Figures 9e–f show that steel-concrete slip s_s and transmission length of end effects increase with time, whereas maximum values of corresponding shear stresses decrease significantly.

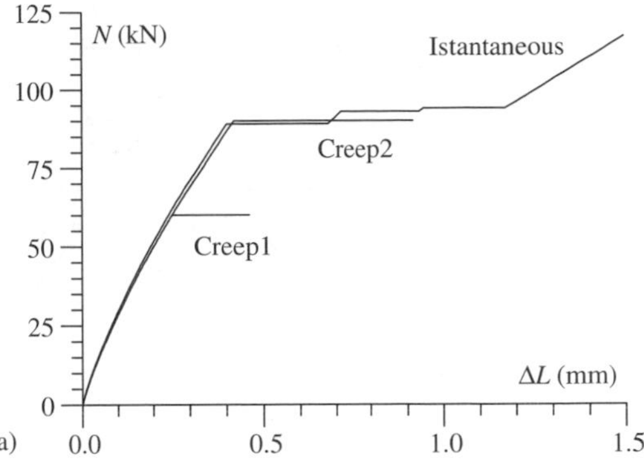

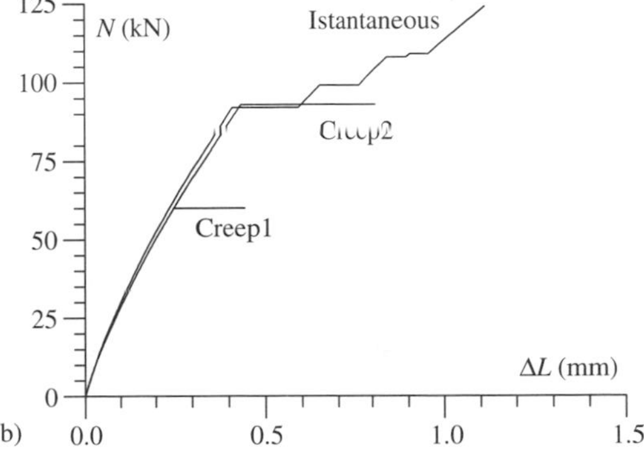

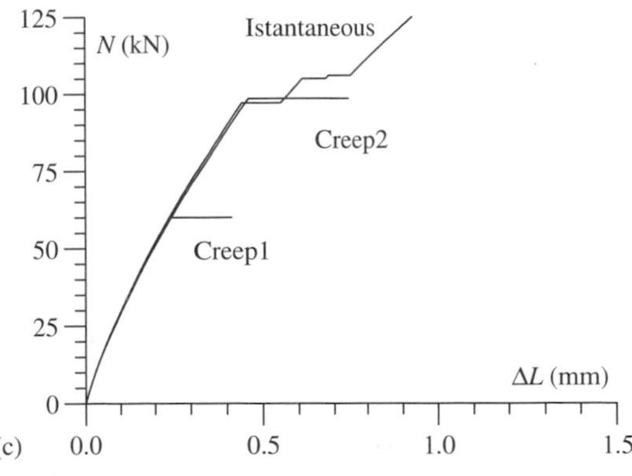

Figure 8. Long term force-elongation evolution with time, compared with short term results. (*a*) unplated, (*b*) $h_p = 0.33$ mm; (*c*) $h_p = 1.016$ mm.

9 CONCLUSIONS

An incremental model for the study of tensile RC members strengthened by means of FRP plates is presented. Concrete cracking and creep deformations of concrete, steel/concrete interface and FRP/concrete interface are considered. Their constitutive laws are defined making use of solidification theory. The numerical solution is obtained through a FDM discretization and a Newton-Raphson solution procedure is adopted. Test under force control, displacement control or crack amplitude control can be simulated. Several numerical examples are presented, where tension stiffening improvement due to the presence of plating is studied, for both short term and long term loadings. FRP-plating is effective for both preserving stiffness in the cracking range and strongly reducing crack width opening.

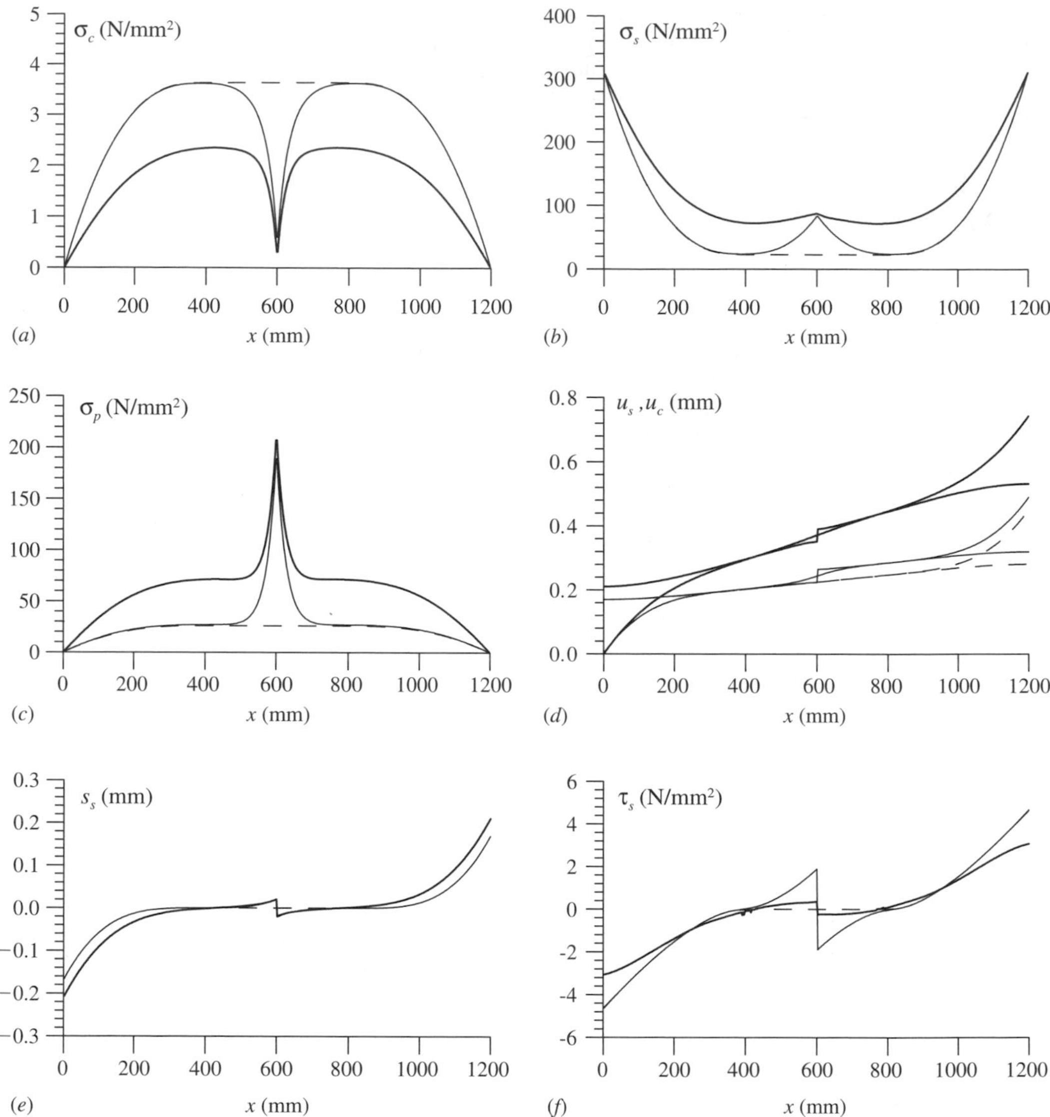

Figure 9. Long term creep effect on FRP-reinforced tensile member ($h_p = 1.016$ mm): (*a*) concrete stress; (*b*) steel stress; (*c*) stress in FRP plates; (*d*) Steel and concrete displacements; (*e*) steel-concrete slip; (*f*) steel-concrete bond shear stress (– – –: before concrete crack; ———: after crack; ———: long time values: 30 years).

ACKNOWLEDGMENT

The financial supports of MIUR (Ministry of Education, University and Research) adn C.N.R. (National Council of Research), 40% Grant on "*Theoretical and Experimental Analysis of Composite-Concrete Bonding for RC Members Reinforced by Composite Plates*" and PAAS Grant on "*Structural Effects of Rheological Properties of Composites*" are gratefully acknowledged.

REFERENCES

Bazant, Z.P. 1971. Numerically stable algorithm with increasing time steps for integral-type ageing creep. *Proc. 1st Int. Conf. Struct. Mech. in Reactor Tech.*, vol. 3, Paper H2/3.

Bazant, Z.P. & Prasannan, S. 1993. Solidification theory for concrete creep. I: Formulation; II: Verification and application. *J. Eng. Mech. ASCE* 115: 1691–1703, 1704–1725.

Bazant, Z.P. & Wu, S.T. 1974. Rate-type creep law of aging concrete based on Maxwell chain. *Materials and Structures* 7: 45–60.

Brosens, K. & Van Gemert, D. 1998. Plate end shear design for external CFRP laminates. In H. Mihashi & R. Keitetsu (eds), *FRAMCOS-3 Proceedings, Gifu, Japan:* 1793–1804.

Carol, I. & Murcia, J. 1989. A model for the non linear time-dependent behaviour of concrete in compression based on a Maxwell chain with exponential algorithm. *Materials and Structures* 22: 176–184.

CEB-FIP. 1992. *Model Code 1990, Bulletin d'Information n. 230,* London: T. Telford.

CEB-FIP. 1997. *CEB Serviceability models. Behaviour and modelling in serviceability limit states including repeated and sustained loads, Bulletin d'Information n. 235.* Lausanne, CEB.

Ferretti, D. & Savoia, M. 2002. Cracking evolution in R/C tensile members strengthened by FRP-plates. *Eng. Fract. Mech.* In press.

Hordijk, D.A. 1992. *Local Approach to Fatigue of Concrete.* PhD Thesis, Delft.

Lee, Y.J., Boothby, T.E., Bakis, C.E. & Nanni, A. 1999. Slip modulus of FRP sheets bonded to concrete. *J. Compos. for Construction ASCE* 3: 161–167.

Mazzotti, C. & Savoia, M. 2001a. An incremental non linear creep damage model for concrete in compression. In Z. Waszczyszyn (ed.), *ECCM-2001 Conference Proceedings, Cracow, 26–29 June 2001*: 1–16.

Mazzotti, C. & Savoia, M. 2001b. Experimental study of non linear creep of concrete at high stress level. In F.J. Ulm, (ed.), *CONCREEP-6,* Cambridge, August 2001. Elsevier: Amsterdam. 259–264.

Mazzotti, C & Savoia, M. 2002. Nonlinear Creep, Poisson's Ratio, and Creep-Damage Interaction of Concrete in Compression, *ACI Materials J.* 99(5): 450–457.

Mazzotti, C. & Savoia, M. 2003. Non linear creep damage model for concrete under uniaxial compression. *J. Eng. Mech.* In publication.

Plevris, N. & Triantafillou, T.C. 1994. Time-dependent behavior of RC members strengthened with FRP laminates. *J. Struct. Eng. ASCE* 120: 1016–1042.

Sato, R., Ujike, I., Akita, K. & Kasai, N. 1987. Properties of bond under sustained loading at an early age. *Transactions of the Japan Concrete Institute* 9: 257–264.

Sato, Y., Shouji, K., Ueda, T. & Kakuta, Y. 1999. Uniaxial tensile behavior of reinforced concrete elements strengthened by carbon fiber sheets, in Dolan, Rizkalla & Nanni (eds.), *FRPRCS-4 International Symposium, ACI SP-188*: 697–710.

Savoia, M., Ferretti, D. & Mazzotti, C. 2002. Creep behavior of RC tensile elements retrofitted by FRP plates. In Saadatmanesh & Ehsani (eds.), *ICCI '02 Conference Proceedings, San Francisco, June 10–12 2002*: 1–12.

Savoia, M., Ferracuti, B. & Mazzotti, C. 2003. Non linear bond-slip law for FRP-concrete interface. In K.H. Tan (ed.), *FRPRCS-6 Conference Proceedings, Singapore, 8–10 July 2003*: 1–10.

Tripi, J.M., Bakis, C.E., Boothby, T.E. & Nanni, A. 2000. De-formation in concrete with external CFRP sheet reinforcement. *J. Compos. for Construction ASCE* 4: 85–94.

Zienkiewicz, O.C. & Watson, M. 1966. Some creep effects in stress analysis with particular references to concrete pressure vessels. *Nucl. Eng. Design* 4: 406–412.

Computational Modelling of Concrete Structures, Bićanić et al. (eds)
© 2003 Swets & Zeitlinger, Lisse, ISBN 90 5809 536 3

Modelling drying and loading effect in structural concrete repair

Laurent Molez & Yves Berthaud
Laboratoire de Mécanique et Technologie, ENS Cachan/CNRS/Université Paris, France

Benoît Bissonnette & Denis Beaupré
Centre de Recherche sur les Infrastructures en Béton, Université Laval, Canada

ABSTRACT: Drying shrinkage may be a significant cause of deterioration of thin concrete repairs. Shrinkage induced stresses can be partially relieved by tensile creep. Hydration effects and materials properties may have a significant influence on concrete repair behaviour. In this paper, a numerical tool that takes into account these different phenomena is presented. The aim of this numerical model is to obtain an analysis tool to complete experimental tests. Experimental results obtained on small lab specimens permit to calibrate numerical coefficients. Numerical results on larger specimens show a good agreement with experimental data.

1 INTRODUCTION

Second half of twentieth century is marked by an intense activity of construction of concrete infrastructures. These structures have a limited lifespan, and several of them require repairs today. Thus the field of concrete repairs knows, since ten years, a significant rise in the sector of construction. For owners, achieving durable repairs is of primary importance because the costs involved are generally very high. Data presented at table 1 give an idea of the extent of work in the field of repairs. Experience shows, however, that design of durable concrete repairs can be as complex as the design of new structures because each damaged structure imposes its own set of conditions (Emmons and Vaysburd 1994).

Table 1. Value of construction work in some countries.

	Billion of			
	France[1] €	UK[2] £	Canada[3] $ CAD	USA[4] $ US
Building				
new	47,3			551
repairs	44			294
Nonbuilding				
new	17			103
repairs	9,5			67
All				
repairs		14,9	16	

[1] Ministère de l'Équipement (2001).
[2] Cusson (2001).
[3] Emberson and Mays (1996).
[4] US department of commerce (2000).

In this study, concrete repair is defined as a thin concrete layer ($\simeq$50 to 150 mm) cast on an existing structure to replace the deteriorated part close to the surface. These two concrete layers have different mechanical and physical characteristics. The old one is quasi-stable with time. The new one change with time (hydration, drying, shrinkage, creep …). In the figure 1, Saucier et al. (1997) resume the effect of hygral, thermal and mechanical change. Shrinkage is identified to be the most severe parameter (Saucier et al. 1991; Bissonnette 1997; Laurence 2001). In the other hand, tensile creep can partly relax the induced

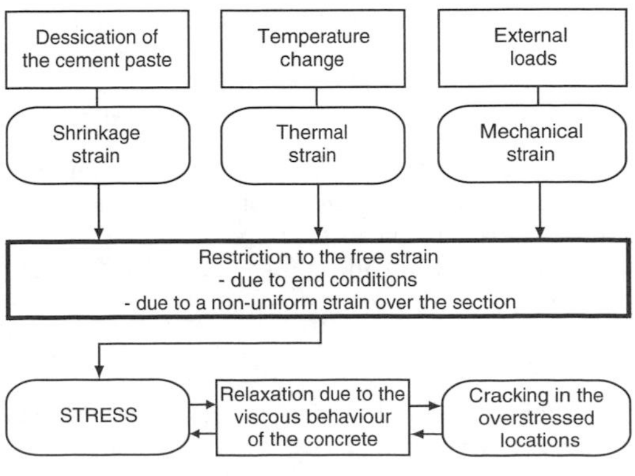

Figure 1. Schematical illustration of the stress build-up in repairs.

strains (Bissonnette 1997). The cracking of repair materials can cause the most serious deterioration processes leading to repair failures, since the aggressive agents can then penetrate through these cracks.

We first present the numerical model developed in this research study. Hydration process is not modelled, but its effects are taken into account in the evolution of physical and mechanical characteristics. Endogenous shrinkage is considered. The drying is modelled by a non-linear diffusion equation, and drying shrinkage is assumed to be proportional to relative humidity changes. Mechanical behaviour is introduced with respect of basic and drying creeps. A nonlinear continuum damage model is applied to concrete rupture. In second time, numerical results are compared to experimental ones from previous study (Molez et al. 2001; Molez 2003).

2 NUMERICAL MODEL

2.1 *Hydration effects*

Hydration mechanisms are not modelled, but their effects have to be taken into account. Degree of hydration ξ can be evaluated from the experimental evolution of compressive strength $f_c(t)$ (Ulm 1999):

$$\xi(t) = \xi_0 + (1 - \xi_0)\frac{f_c(t)}{f_c(\infty)} \tag{1}$$

where ξ_0 is the percolation threshold as defined by Acker (1987) and $f_c(\infty)$ is the compressive strength at an infinite time.

Thereafter, the evolution of strength will be expressed according to the following equation:

$$R(t) = R(\infty)\frac{t^{m_1}}{m_2 + t^{m_1}} \tag{2}$$

where $R(\infty)$ is the strength at an infinite time, and m_1, m_2 parameters determined from experimental results (figure 2).

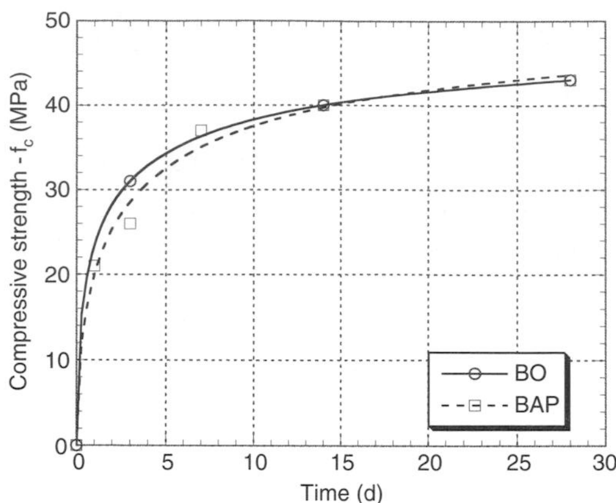

Figure 2. Experimental (dots) and calculated (lines) evolution of compressive strength for an ordinary concrete (BO) and a self-levelling concrete (BAP).

2.2 *Thermal effects*

We assumed that ambient temperature is constant. The thickness of new concrete layer is small then temperature growth due to chemical reactions is low. Consequently, in all this study, thermal effects are neglected.

2.3 *Drying model*

Mass conservation equation, expressed according to the mass water content w, is:

$$\frac{\partial w}{\partial t} = -\operatorname{div}(J_v + J_l) + \frac{\partial w_{ad}}{\partial t} \tag{3}$$

where J_v is the flux of water in vapour form and J_l the flux of water in liquid form. The derivative term of w_{ad} represent the variation of water mass content due to self-desiccation.

Using Fick law and ideal gas law, vapour flux can expressed according to the relative humidity h_r. Similarly, using Darcy law and Kevin law, the liquid flux can be written in terms of h_r. In additions, experimental results (Baroghel-Bouny 1994; Therrien 2001) show that desorption isotherms are linear (*i.e.* the slope of the curve $w = w(h_r)$ is constant) for $0.4 < h_r < 0.95$. Then we obtain the classical diffusion equation:

$$\frac{\partial h_r}{\partial t} = \operatorname{div}(D(h_r)\operatorname{grad} h_r) + \frac{\partial h_{ad}}{\partial t} \tag{4}$$

Thereafter, we will neglect the variation of relative humidity $\partial h_{ad}/\partial t$ due to hydration reactions. However associated shrinkage (part of endogenous shrinkage) is introduced further.

The macroscopic diffusion coefficient depend on h_r. When h_r is important, humidity transfers takes place in water of capillary spaces. In opposition, when relative humidity is low, transfer is of vapour form and surface diffusion occurs in areas of hindered adsorption. According to theses observations, we choose the following expression of $D(h_r)$ (Bažant 1988):

$$D(h_r) = D_0(t)\left(a + \frac{1-a}{1+\left(\frac{1-h_r}{1-h_c}\right)^n}\right) \tag{5}$$

where a, n and h_c are adjustable parameters specific to material. h_c would be the humidity threshold where transfer change from first phenomenon to the second. Those parameters can be identified with experimental data for weight loss (figure 3).

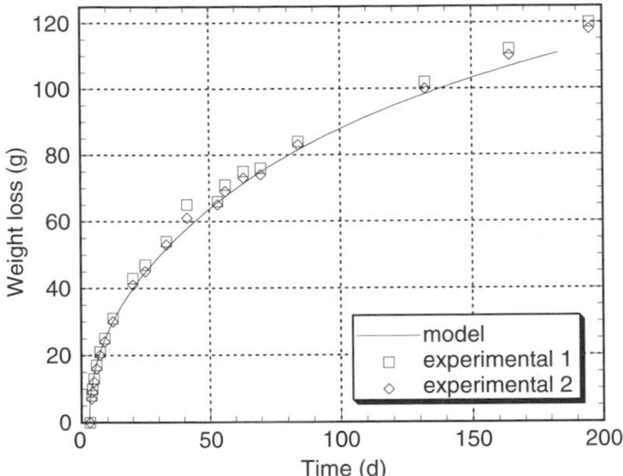

Figure 3. Comparison of experimental and calculated weight loss.

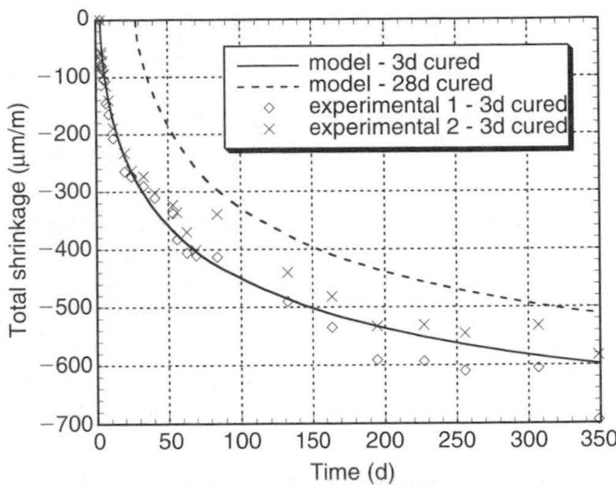

Figure 4. Evolution of total shrinkage. Comparison of experiments and numerical simulation.

Structure of the porous network evolves during hydration. Some results (Therrien 2001) show that diffusion coefficient is dependent on maturity of concrete. However, it is difficult to obtain an experimental relation $D(\xi)$. So, we fit the evolution of D to the inverse of the evolution of strength.

$$D_0(t) = D_0(t_\infty)\frac{m_2 + t^{m_1}}{t^{m_1}} \tag{6}$$

Boundary conditions of convective type are imposed at the surface Γ in contact with air at relative humidity h_r^a. Convective flux is then written:

$$J_\Gamma = \beta(h_r^\Gamma - h_r^a) \tag{7}$$

2.4 *Shrinkage*

Shrinkage of concrete is due to chemical reactions and drying of porous media. Thus, an endogenous shrinkage and a drying shrinkage can be identified.

In numerical simulations, experimental results of endogenous shrinkage are directly used. At each step, a homogeneous deformation is imposed in repair concrete layer.

Drying shrinkage is obtained from variation of relative humidity through a coefficient of hydrous dilatation:

$$\underline{\underline{\dot{\varepsilon}_{ds}}} = \alpha_h(t)\dot{h}_r\underline{\underline{1}} \tag{8}$$

This relation does not reflect micro-mechanisms of shrinkage (capillary depression, Gibbs-Bangham theory and disjoining pressure theory), but a theoretical study (Lassabatere 1994) has shown that it gives a not so bad approximation.

According to concrete structuration, $\alpha_h(t)$ should depend on the maturity of material. Some experimental results (Therrien 2001) show this dependence. However, it is difficult to obtain an experimental evolution law. Thus, α_h is considered constant and determined for each concrete (figure 4).

2.5 *Creep*

Creep of concrete can be separated in two parts: basic creep and drying creep. The origins of compressive or tensile creep are discussed in Bažant (1988), Bissonnette (1997), or Molez (2003).

Basic creep

Basic creep can be written according to viscoelasticity theory. A compliance function J is defined by:

$$\varepsilon(t) = J(t_0, t)\sigma_0 \tag{9}$$

where t_0 is the age at loading σ_0, and $t - t_0$ the load duration.

For a stresses history, creep strain can be written:

$$\underline{\underline{\varepsilon}}(t) = \underline{\underline{\varepsilon}}(t_0) + \int_{t_0}^{t} \underline{\underline{\underline{\underline{B}}}} J(\tau, t_0)\underline{\underline{\dot{\sigma}}}(\tau)\mathrm{d}\tau \tag{10}$$

where B is the fourth order tensor of Poisson's terms.

Adopting a Dirichlet series expression for compliance function (Bažant 1988):

$$J(t_0, t) = \frac{1}{C_0(t_0)} + \sum_{\mu=1}^{n} \frac{1}{C_\mu(t_0)}\left(1 - \exp(-\frac{t - t_0}{\tau_\mu})\right) \tag{11}$$

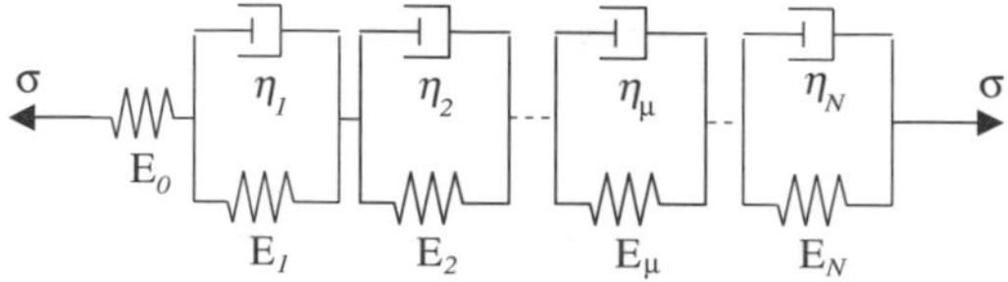

Figure 5. Kelvin chain equivalent to the proposed model.

makes it possible to free from storage of stresses history, and strain can be calculated from the preceding step. Previous expression can be explained in terms of Kelvin chain (figure 5) with ageing modulus $E_\mu = C_\mu - \tau_\mu \dot{C}_\mu$ and constant dashpot viscosity $\eta_\mu = \tau_\mu C_\mu$.

Strain evolution is calculated step by step. For step $[t_n, t_{n+1}]$, one can write:

$$\underline{\underline{\Delta\varepsilon}} = \underline{\underline{\varepsilon}}(t_{n+1}) - \underline{\underline{\varepsilon}}(t_n)$$

If C_μ and $\dot{\sigma} = \Delta\sigma/\delta t$ are constant during this step, integral can be easily calculated (see Molez (2003) for detailed calculus):

$$\underline{\underline{\Delta\varepsilon}} = \left[\frac{1}{C_0(t_n + \Delta t/2)} + \sum_{\mu=1}^{n} \frac{1 - \left(1 - \exp(-\frac{\Delta t}{\tau_\mu})\right) \tau_\mu/\Delta t}{C_\mu(t_n + \Delta t/2)}\right] \underline{\underline{B}}\,\underline{\underline{\Delta\sigma}} + \sum_{\mu=1}^{n} \left(1 - \exp(-\frac{\Delta t}{\tau_\mu})\right) \underline{\underline{\Delta\varepsilon^*_\mu}}(t_n) \tag{12}$$

with the recursive expression:

$$\underline{\underline{\Delta\varepsilon^*_\mu}}(t_n) = \frac{1 - \exp(-\frac{\Delta t}{\tau_\mu}) \tau_\mu/\Delta t}{C_\mu(t_n - \Delta t/2)} \underline{\underline{B}}\,\underline{\underline{\Delta\sigma}} + \underline{\underline{\Delta\varepsilon^*_\mu}}(t_{n-1}) \exp(-\frac{\Delta t}{\tau_\mu}) \tag{13}$$

Thus the increment of deformation is calculable according to the increment of load and the history of loading defined in previous step. The increment of basic creep is obtained by subtracting the increment of instantaneous strain from previous equation. We obtain:

$$\underline{\underline{\Delta\varepsilon_{bc}}} = \left[\sum_{\mu=1}^{n} \frac{1 - \left(1 - \exp(-\frac{\Delta t}{\tau_\mu})\right) \tau_\mu/\Delta t}{C_\mu(t_n + \Delta t/2)}\right] \underline{\underline{B}}\,\underline{\underline{\Delta\sigma}} + \underline{\underline{\Delta\varepsilon^{hist}}} \tag{14}$$

Characteristic relaxation times τ_μ are fixed a priori according to a logarithmic series (Bažant 1988; Granger 1997):

$$\tau_\mu = \tau_1\, 10^{\mu-1} \tag{15}$$

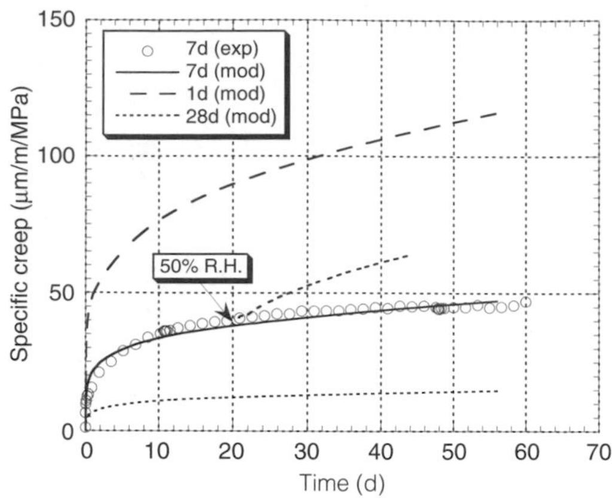

Figure 6. Experimental (dots) and calculated (lines) basic creep of an ordinary concrete.

and C_μ parameters are obtained by the method of least squares. An example of results is given in figure 6.

Drying creep

Creep experimental tests show that an extra creep can be measured when test specimen is allowed to dry. This phenomenon is called drying creep. Different interpretations are proposed. Bažant and Chern (1985) suggest a mechanism of stress-induced shrinkage. On the contrary, drying creep can be interpreted as a drying-induced creep (Bažant 1988). These different mechanisms can be written mathematically as a single equation:

$$\underline{\underline{\dot{\varepsilon}_{dc}}} = \kappa_{fs} \underline{\underline{B}}\,\underline{\underline{\sigma}} |\dot{h}_r| \tag{16}$$

Coefficient κ_{fs} is determined from basic and drying creep test. We can notice that the sign of $(\dot{\varepsilon}_{dc})_{ij}$ depends only on the sign of $(\sigma)_{ij}$. So in uniaxial traction, we notice an increase of drying creep whether it is in moistening or drying conditions.

Then, total creep can be obtained by adding basic and drying creep (figure 7).

2.6 *Mechanical damage*

Total strains ε_{total} are supposed to be the sum of instantaneous strains ε_i, shrinkage strains ε_s and creep strains ε_c:

$$\varepsilon_{total} = \varepsilon_i + \varepsilon_s + \varepsilon_c \tag{17}$$

The stress-strain relation of damaged material is written:

$$\underline{\underline{\sigma}} = \underline{\underline{\underline{\underline{C^d}}}} (\underline{\underline{\varepsilon_{total}}} - \underline{\underline{\varepsilon_s}} - \underline{\underline{\varepsilon_c}}) = \underline{\underline{\underline{\underline{C^d}}}}\, \underline{\underline{\varepsilon_i}} \tag{18}$$

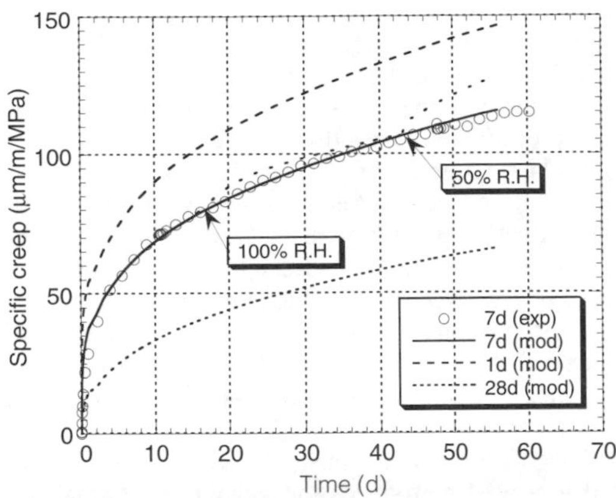

Figure 7. Experimental (dots) and calculated (lines) total creep (basic + drying creep) of an ordinary concrete.

In the present analysis, we use an isotropic scalar damage model (Mazars 1984). In this model, the mechanical effect of progressive micro-cracking due to external loads is described by a single internal variable which degrades the Young's modulus of the material. The constitutive relations are:

$$\sigma_{ij} = (1-d)\Lambda_{ijkl}(\varepsilon_i)_{kl} \tag{19}$$

where σ_{ij} and $(\varepsilon_i)_{kl}$ are the components of the stress and strain tensors respectively ($i,j,k,l \in [1,3]$), Λ_{ijkl} are the initial stiffness moduli, and d is the damage variable. The material is initially isotropic, with E and ν the initial Young's modulus and Poisson's ratio respectively. Damage is a function of the positive strains which means that it is mainly due to micro cracks opening in tension mode. In order to avoid ill-posedness due to strain softening, the mechanical model has to be enriched with an internal length (Pijaudier-Cabot and Bažant 1987).

3 NUMERICAL EXAMPLES

In this section, we present some results of numerical simulations of concrete repairs. Test configuration is the same one as that used in previous experimental study (Molez et al. 2001; Molez 2003).

Figure 8 shows the geometry of the investigated concrete repair. A 2D analysis of one half of the beam, using four-noded plane stress elements, was performed (figure 9).

Examples of calculated humidity distribution, corresponding deformation and induced damage are shown in figures 12, 13 and 14.

Numerical results can be compared to experimental ones. During experimental tests, crack width and deflection has been measured for different configurations: drying only, drying and loading at 33% of ultimate load, and flexural testing. Those results are compared in figures 10 and 11.

These numerical results show a good agreement with experimental ones.

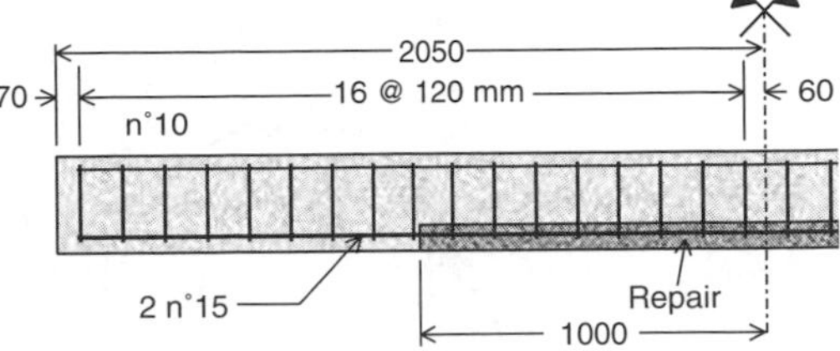

Figure 8. Test specimens.

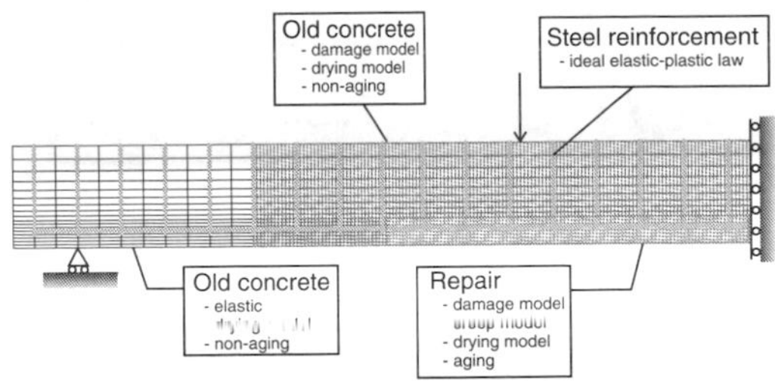

Figure 9. FE mesh.

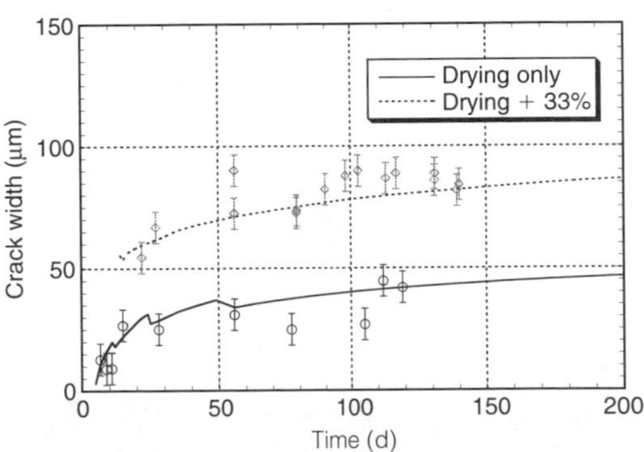

Figure 10. Comparison of experimental (dots) and calculated (lines) average crack width for a repaired beam submitted to drying and loading.

4 CONCLUSIONS

To be able to make a complex analysis (to complete experimental study) of the behaviour of structural concrete repairs, the main phenomena acting in those structures have been integrated in a finite elements code.

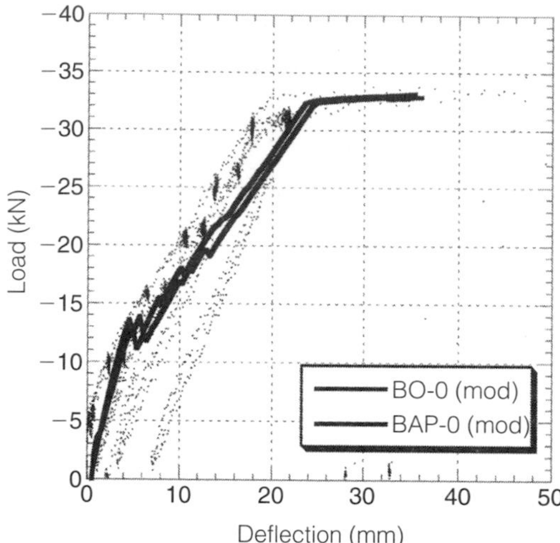

Figure 11. Flexural behaviour: experimental (dots) and calculated (lines) results.

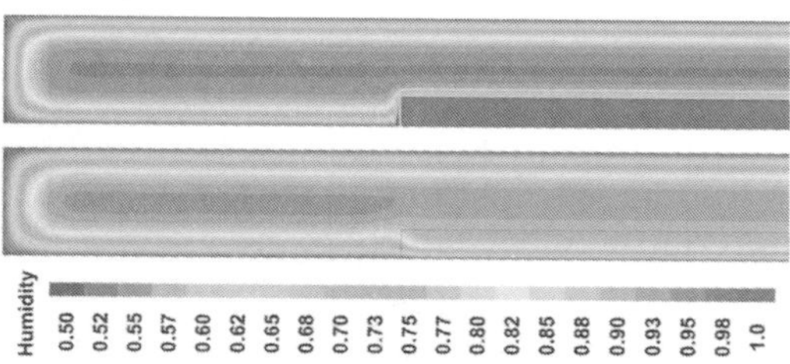

Figure 12. Distribution of humidity: initial state and after 497 days of drying.

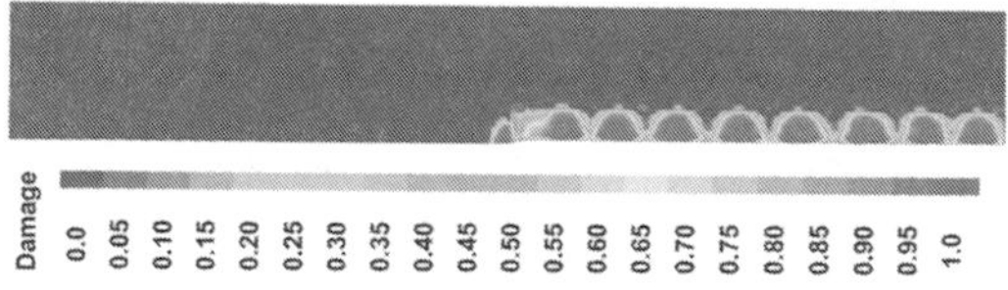

Figure 13. Induced damage distribution: final state.

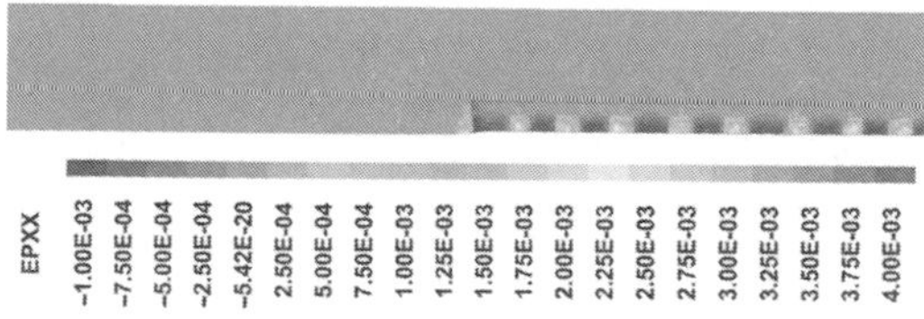

Figure 14. Induced deformation distribution: final state.

So, we can take into account the evolution of mechanical properties during hydration process and induced volume change (endogenous shrinkage). The drying of the material is modelled by a non linear law. Furthermore, the diffusion law is coupled with the evolution of hydration process by a diffusion coefficient which depends on the maturity of the material. Drying shrinkage is connected with the variation of humidity by a proportionality law that is experimentally verified for relative humidity included between 0.5 and 0.95. Viscoelastic behaviour is modelled by a Kelvin chain. The effects of hydration are taken into account with variable coefficients of the Kelvin chain. The influence of drying on creep is introduced by means of an extra term proportional to variation of relative humidity and imposed stress. A damage model is used to reproduce the mechanical effect of progressive micro-cracking of concrete.

The identification phase shows that the numerical tool well reproduces the behaviour of small lab specimens, and also bigger structures.

REFERENCES

Acker, P. (1987). *The mechanical behavior of concrete: contributions of the physico-chemical approach.* Thèse de doctorat, École nationale des ponts et chaussées.

Baroghel-Bouny, V. (1994). *Caractérisation des pâtes de ciment et des bétons – Méthodes, analyse, interprétations.* Thèse de doctorat, École nationale des ponts et chausées.

Bažant, Z. P. (Ed.) (1988). *Mathematical modeling of creep and shrinkage of concrete.*

Bažant, Z. P. and J. C. Chern (1985). Concrete creep at variable humidity: constitutive law and mechanism. *Materials and structures 18*(103).

Bissonnette, B. (1997). *Le fluage en traction: un aspect important de la problématique des réparations minces en béton.* Ph. D. thesis, Université Laval, Québec, Canada.

Cusson, D. (2001). La réparation durable des structures de béton passe par l'éducation et le transfert de la technologie. *Innovation en construction 6*(3). IRC, CNRC-NRC.

Emberson, N. K. and G. C. Mays (1996). Significance of property mismatch in the patch repair of structural concrete. Part 1,2,3. *Magazine of concrete research 48*(174).

Emmons, P. H. and A. M. Vaysburd (1994). Factor affecting the durability of concrete repair: the contractor's viewpoint. *Construction and building materials 8*(1).

Granger, L. (1997). *Calcul des ouvrages de construction*, Chapter 14: Ouvrages en béton soumis aux effets du fluage, pp. 614–654. Hermès. sous la direction de M. Prat.

Lassabatere, T. (1994). *Couplage hydromécanique en milieu poreux non saturé avec changement de phase: application au retrait de dessiccation.* Thèse de doctorat, Ecole nationale des ponts et chaussée, Paris, France.

Laurence, O. (2001). *La fissuration due au retrait restreint dans les réparations minces en béton: apports combinés de l'expérimentation et de la modélisation.* Thèse de doctorat, Université Laval, Canada, École nationale des ponts et chaussées, France.

Mazars, J. (1984). *Application de la mécanique de l'endommagement au comportement non linéaire et à la rupture de béton de structure.* Thèse de doctorat, Université Paris VI, France.

Ministère de l'Équipement (2001). Grands agrégats économiques de la construction. France.

Molez, L. (2003). Thèse de doctorat, École normale supérieure de Cachan, France et Université Laval (Québec), Canada.

Molez, L., Y. Berthaud, B. Bissonnette, and D. Beaupré (2001, 28 may–1 june). Structural concrete repair behavior: coupled effect of drying and loading. In de Borst, Mazars, Pijaudier-Cabot, and van Mier (Eds.), *Fracture mechanics of concrete structures*, pp. 263–270. FraMCoS: A. A. Balkema Publishers.

Pijaudier-Cabot, G. and Z. P. Bažant (1987). Nonlocal damage theory. *J. of Engrg. Mech. 113*.

Saucier, F., F. Claireaux, D. Cusson, and M. Pigeon (1997). The challenge of numerical modeling of strains and stresses in concrete repairs. *Cement and concrete research 27*(8).

Saucier, F., M. Pigeon, and M. Cameron (1991). Air-void stability, part v: temperature, general analysisand performance index. *ACI materials journal 88*(1).

Therrien, J. (2001). Étude de l'évolution des propriétés de transfert hygrométrique d'un mortier au jeune *â*ge.

Ulm, F.-J. (1999). Couplages thermochémomécanique dans les bétons :un premier bilan. Mémoire d'habilitation à diriger des recherches, Université de Marne la Vallée. Rapport provisoire.

US department of commerce (2000). 1997 economic census, construction. Economics and statistics administration, US Census Bureau.

Computational Modelling of Concrete Structures, Bićanić et al. (eds)
© 2003 Swets & Zeitlinger, Lisse, ISBN 90 5809 536 3

A non-saturated porous medium approach for the modelling of concrete behaviour submitted to high temperatures

A. Al Najim, F. Meftah & A. Mebarki
Laboratoire de Mécanique – LaM, Université de Marne La Vallée, France

ABSTRACT: A computational analysis of hygro-thermal and mechanical behaviour of concrete structure at high temperature is presented. The evaluation of thermal, hygral and mechanical performance of this material, including damage effect, needs the knowledge of the heat and mass transfer processes. These are simulated within the framework of a coupled model where non-linearities due to high temperature and pressure are accounted for. In the model the concrete was treated as partially saturated porous material. The model shows that it is a useful tool for analysis of the transient behaviour when the structure subjected to a sever conditions. An application of this model to a High Performance concrete structure will be shown.

1 INTRODUCTION

In nuclear engineering applications, in safety evaluation in tall buildings and in tunnels, in particular the recent major fires in the European tunnels (Channel, Mont-Blanc, Great Belt Link, Tauern), assessment of concrete performance will be of great interest (Bazant & Thonguthai, 1978, 1979). In these accidents, tunnels presented extensive damage to the concrete elements in the affected sections. Part of the concrete lining was almost completely removed by spalling. In such severe conditions, complex analysis is required to deal with the coupled heat and mass transfer that can occur, involving both liquid and vapour, and related mechanical effects (Bazant & Kaplan, 1996).

Loading and hydro-thermic state can induce a strong micro-structural changes that alter the mechanical behaviour of concrete and will lead to an additional significant degradation of material, loss of stiffness and strength of the material. This degradation is mainly due to different coexisting coupled processes at the material and structural level.

At the material level, as chemical process, due to the dehydration of cement paste in the micro-level (Lin, 1996); as thermal process, that results from the difference in thermal dilatation coefficient between the aggregates and the cement paste, in particular in concrete with siliceous aggregates (Ulm, 1999); and as hygral process, which is due to in one hand, the release and evaporation of a significant amounts of water, and when the concrete shows a low permeability to the liquid and gas that fill the pore space, therefore the movements of these fluid phases due to the temperature distribution can induce a pressure build-up (Anderberg, 1997), and in another hand, the shrinkage stresses resulting from the release of water.

At the structural level, as mechanical process, which is due to, in one hand to the mechanical loads, and in another hand to the elevated temperatures which induces a temperature gradient in the structural element, and when there is a restriction to the thermal dilatation due to the kinematics constraints, therefore a self-equilibrated stresses will be rise (Bazant, 1997, Nechnech et al. 2001).

All, or a part of, these phenomena's, when they combined together may cause internal stresses above the maximum allowable tensile stress of the structure. This will be the cause of cracks propagation and a possible explosive spalling of the structure.

Damaging processes in concrete ascribed to both combined mechanical loads and the thermo-hygral effect. The behaviour of concrete is modelled by a softening plasticity coupled to damage in which the effective stress concept has been introduced (Nechnech et al. 2001). Two isotropic damage variables, which describe stiffness degradation, are introduced in the relationship between the stress and the effective stress. One scalar damage variables is related to the degradation processes due to temperature, while the second one will be for the mechanical damage variable. In this model, the evolution of the mechanical damage variable is related to the physically observed crack opening represented by the plastic strain. This makes it convenient to calibrate the damage evolution from experimental results naturally based on the stress, without introducing an additional damage surface in the strain space. Furthermore, the stiffness recovery due to the closing of previously opened cracks when loading ranges between the tensile and the compressive states is considered (Nechnech et al. 2001). A multi-surfaces criterion is used to describe the induced anisotropy. In the model, a new phenomenological term has been introduced in the definition of the internal average pore pressure, which will be exerted on the solid skeleton, to account for the progressive dissipation of the meniscus formation due to the effects of cracks opening.

The theoretical approach used here is based on the general balance equations of (mass, linear momentum and energy) of continuum mechanics and results in a fully coupled numerical model to simulate transient phenomena involving heat-water and dry air transfer in deforming porous materials. An example about a one-dimensional drying problem is solved to demonstrate the potentialities of the model in describing hygro-thermal behaviour of concrete.

2 MATHEMATICAL MODEL

2.1 *Introduction*

When higher than the standard temperatures are considered, several non-linear phenomena should be taken into account in the modelling. Such phenomena involve not only heat conduction and vapour diffusion, but also heat convection, latent heat transfer due to water phase change (evaporation or condensation) and the dehydration processes inside the pores. Moreover liquid water flow caused by pressure gradients or capillary suction, gradients of adsorbed water content and the air diffusion may be taken also. For this reasons moisture transfer in porous media cannot be usually interpreted with a purely diffusive model. Indeed a formulation based on diffusion-like equations implies the use of transport coefficients that lump together different transport mechanisms, masking their dependence on the different controlling variables (Gawin et al. 1996).

2.2 *General assumptions*

A good design of a concrete structure should take all these phenomena into account. For this purpose the model that will be used, performed in considering the concrete as partially saturated porous material (Gawin & Schrefler 1996). This multiphase system consisting of a solid phase (the porous skeleton), where the voids of skeleton are filled partly with liquid water and the residual part of voids filled with a gas mixture (Gawin et al. 1996).

It is assumed that the liquid phase is made of a single constituent (pure water), which exist in various form (free or capillary water, adsorbed and chemically bound water), neglecting the presence of soluble air, and that the gaseous phase can be considered as an ideal mixture of an incondensable ideal gas (dry air) and a condensable one (water vapour). Moreover the local thermodynamic equilibrium hypothesis is assumed to hold on the grounds that the times scale of the modelled phenomena (that are quite slow) is substantially larger than the relaxation time required to reach the equilibrium locally. In addition, in the following, only small displacements are considered, so that second order effects (i.e. the need to account for the deformation of the reference frame) can be neglected.

2.3 *Governing equations*

The general approach to modelling non-equilibrium processes, including heat and mass transfer in a porous medium, is to start from an appropriate set of balance equations and to supplement them with an appropriate set of constitutive and state equations as well as by some thermodynamic relationships (Lewis & Schrefler, 1998).

2.3.1 *Balance equations*

Such balance equations can be obtained by using the procedure of space averaging of the balance equations, written for the individual constituents of the medium. The theoretical framework is based on the works of Whitaker (1977), Bear (1988), Bear & Bachmat (1986, 1990), Hassanizadeh & Gray (1979, 1980) and Lewis & Schrefler (1998).

The final mathematical model consists of four balance equations: mass balance equation of water species (both liquid water and vapour, taking phase change and dehydration processes into account); mass balance equation of dry air; an enthalpy balance equation of the whole multiphase medium, which considers the latent heat of phase change and dehydration processes. The deformability of the skeleton is to be taken into account though a linear momentum balance equation of the multiphase medium (Gawin et al. 1999).

Water species (liquid + vapour) mass balance equation:

$$\begin{aligned} &S^w \phi \frac{\partial \rho^w}{\partial t} + \phi \rho^w \frac{\partial S^w}{\partial t} + S^w \rho^w \frac{\partial \phi}{\partial t} + S^w \phi \rho^w div \mathbf{v}^s \\ &+ div \left(S^w \phi \rho^w \mathbf{v}_s^w \right) = -S^g \phi \frac{\partial \rho^{gw}}{\partial t} - \phi \rho^{gw} \frac{\partial S^g}{\partial t} - S^g \rho^{gw} \frac{\partial \phi}{\partial t} \\ &- S^g \phi \rho^{gw} div \mathbf{v}^s - div \left[S^g \phi \rho^{gw} \mathbf{v}_s^g \right] - div \left[S^g \phi \rho^{gw} \mathbf{v}_s^{gw} \right] = 0 \end{aligned} \tag{1}$$

Dry air mass balance equation:

$$\begin{aligned} &S^g \phi \frac{\partial \rho^{ga}}{\partial t} + \phi \rho^{ga} \frac{\partial S^g}{\partial t} + S^g \rho^{ga} \frac{\partial \phi}{\partial t} + S^g \phi \rho^{ga} div \mathbf{v}^s \\ &\quad + \mathrm{div} \left[S^g \phi \rho^{ga} v_s^g \right] + \mathrm{div} \left[S^g \phi \rho^{ga} v_g^{ga} \right] = 0 \end{aligned} \tag{2}$$

Energy balance equation (enthalpy balance equation):

$$\left(\rho C_p\right)_{eff}\frac{\partial T}{\partial t}-div\left[\lambda_{eff}\,grad\,(T-T_0)\right]$$
$$+\left[S^w\phi\rho^w C_p^w\mathbf{v}_s^w+S^g\phi\rho^g C_p^g\mathbf{v}_s^g\right]\cdot grad\,(T-T_0) \quad (3)$$
$$=-\dot{m}_{phase}\Delta H_{phase}+\dot{m}_{dehydr}\Delta H_{dehydr}$$

$$\rho=(1-\phi)\rho^s+\phi S^w\rho^w+\phi S^g\rho^g \quad (4)$$

Linear momentum equation (equilibrium equation):

$$div\,\sigma+\rho g=0 \quad (5)$$

where ρ^s, ρ^w, ρ^{gw}, ρ^{ga}, $\rho =$ densities of solid phase, liquid water, water vapour, dry air and the averaged density of the mixture; S^w, $S^g =$ degrees of saturation in volume of water and gas; $\mathbf{v}^s$, $\mathbf{v}_s^w$, $\mathbf{v}_s^g =$ macroscopic velocities of solid phase, liquid phase and gaseous phase; $\mathbf{v}_g^{gw}$, $\mathbf{v}_g^{ga} =$ relative velocities of the diffusing species of water vapour and dry air; $\dot{m}_{phase}$, $\dot{m}_{dehydr} =$ phases change due to (evaporation or condensation) and dehydration process; C_p^W, C_p^g, $C_p =$ specific heats of liquid water, gas mixture and the effective specific heat of porous material at constant pressure (taking into account all the phase); $\lambda_{eff} =$ effective thermal conductivity of porous material (taking into account all the phase); ΔH_{phase}, $\Delta H_{dehydr} =$ specific enthalpies of vaporisation and of dehydration; $\sigma =$ total stress tensor; $\phi =$ porosity and $t =$ the time. It should be noted that the specific body force g (usually corresponding to gravity acceleration) will be assumed negligible in the following but in the same time we will keep its term in the following equations.

The summation of the mass balances of liquid water and water vapour, allowing to eliminate the mass source term $\dot{m}_{phase}$ related to phase change in the equation of water species, but this term will be present in the energy balance equation then we could use either liquid water or water vapour mass balance equation to define this term of phase change.

2.3.2 *State equations*

For all the gaseous constituents, dry air and water vapour, the equation of state of prefect gases can be applicable:

$$p^{ga}=\frac{\rho^{ga}RT}{M_a} \quad (6)$$

$$p^{gw}=\frac{\rho^{gw}RT}{M_w} \quad (7)$$

and from Dalton's law:

$$p^g=p^{gw}+p^{ga} \quad (8)$$

then,

$$M_g=\frac{p^{gw}}{p^g}M_w+\frac{p^{ga}}{p^g}M_a \quad (9)$$

$$\rho^g=\frac{M_g}{M_w}\rho^{gw}+\frac{M_g}{M_a}\rho^{ga} \quad (10)$$

Let us now assume that there is no dissipation occurs during the phase change, this turns to assume that the liquid water (both capillary and adsorbed) and its water vapour remain in permanent local thermodynamic. Hence, by considering that the specific (mass) Gibbs functions (or thermodynamic potentials) of the two phases are equal at each time and using the conditions of equilibrium of vapour with bulk liquid on a flat surface ($p^{gw}=p^{gws}$, $p^w=p^{gws}+p^{ga}$) considered at the same temperature, one could get the following relationship:

$$p^l=p^{gws}(T)+p^{ga}+\frac{\rho^w RT}{M_w}\ln\left(\frac{p^{gw}}{p^{gws}(T)}\right) \quad (11)$$

where $p^{gws} =$ water vapour saturation pressure, which depend only upon temperature, obtained from empirical correlation with a tabulated results which links the water vapour saturation pressure with temperature:

$$p^{gws}(T)=\exp\left(6.4075+\frac{16.82669T}{228.73733+T}\right) \quad (12)$$

It is assumed that phase changes (water evaporation/condensation) are rapid with respect to the characteristic time typical of this type of problem. Therefore, they can be considered in local equilibrium, giving rise to a set of equilibrium restrictions that must be satisfied at all times. This assumption has added advantage that phase changes do not appear explicitly and the number of equations is thereby reduced (Gens & Olivella 2001).

If an open-dry porous medium is exposed to moist air, the weight of the solid will be increased because the moisture is adsorbed on the inner surfaces of the pores, starting with the finest one. In the cases of interest here, the water is usually present as a condensed liquid that, because of the surface tension, is separated from its vapour by a concave meniscus (capillary water) which could be defined as the pressure difference between the gas phase and the liquid phase:

$$p^c=p^g-p^w \quad (13)$$

Damage evolution

For damage evolution we will use an isotropic damage model. The damage variable, associated with thermal

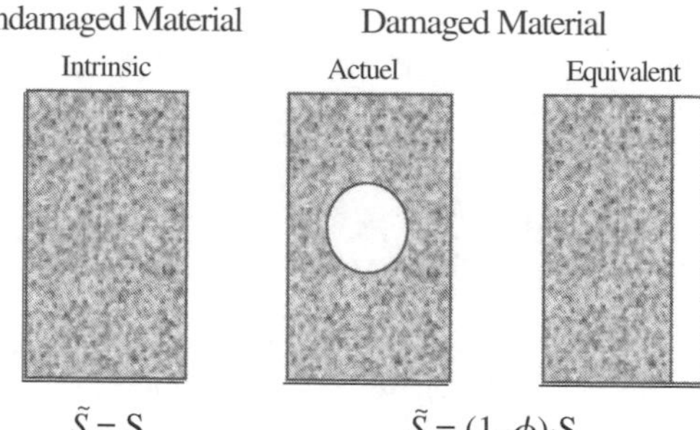

Figure 1. Schematic representation of damage on the effective resisting surface area.

or/and mechanical degradation processes of concrete, is interpreted as the surface density of materials defects (Kachanov, 1986, Ju, 1989), and will be defined as the ratio of the area occupied by created micro-cracks, to the total area (see Figure 1).

This definition requires that the damage variable is a non-decreasing parameter, since the rate of reduction of the effective resisting section area $\tilde{S}$ will continuously increase until failure occurs.

Coupling of Thermo-mechanical damage

In this work of the field problem THCM, the damage would be partitioned into a purely thermal ϕ_T and purely mechanical ϕ_M one. Thermal damage due to temperature only is induced by physico-chemical changes in the solid skeleton. These changes could be, the dehydration process in the matrix; the evolution of porosity due to the matrix thermal expansion; the shrinkage and the all other phenomena which may cause an evolution of porosity. Mechanical damage which produce the micro-cracks could be due to all the stresses that applied on the solid matrix, such as the external loads and also the internal pore pressure build-up. All these causes may lead to a decrease in the mechanical properties of material, namely the stiffness and strength. Then we could define the overall substantial link between the thermal and mechanical field equations by the definition of the apparent volume fraction of the skeleton $1 - \phi$ as:

$$1-\phi = (1-\phi_0)(1-\phi_T)(1-\phi_M) \tag{14}$$

where ϕ_0 represent the initial porosity (initial damage).

The state equation for the solid phase can be obtained by volume averaging of the partial stress tensors for all the phases through the concept of effective stress. We could have a relation between the applied stress on the skeleton $\boldsymbol{\sigma}$, effective stress in the solid matrix $\tilde{\boldsymbol{\sigma}}$ and the average pore pressure applied on the solid matrix p^s.

$$\boldsymbol{\sigma} = (1-\phi)\tilde{\boldsymbol{\sigma}} - \phi p^s \mathbf{I} \tag{15}$$

It should be noted that $\tilde{\boldsymbol{\sigma}}$ is different from Bishop's effective stress, modified by Biot & Willis (1957), usually used in THCM problems.

We could use also the following relationship:

$$S^w + S^g = 1 \tag{16}$$

Normally the definition of the average pore pressure applied on the solid matrix is:

$$p^s = S^w p^c + p^g \tag{17}$$

But when the creation of cracks will be appeared simultaneously with the heating process, the influence of micro-cracks which will be initiated in contact with the pores, where the capillary meniscus occur, will be of great interest on the formation (occurrence) of this capillary meniscus in these fine pores. For this reason we will try to take this effect of cracks in the definition of the solid pressure p^s (averaged pore pressure), by introducing a phenomenological term, which try to take into account this effect on the formation and the action of the capillary meniscus:

$$\tilde{p}^s = \left[(1-\phi_T)(1-\phi_M)\right]^{\alpha} S^w p^c + p^g \tag{18}$$

It should be noted that the action of this treatment is different with that rolled by the evolution of the total porosity ϕ which is already tacked into account in the mass balance equations (1), (2) and in (15) which will affect the evolution of the quantities S^π and p^π of the fluid phase in question inside the pore. Therefore, this evolution did not affect the existence of the meniscus phenomena, which depend strongly on the geometry of the pore. That means this phenomena will be continued in spite of the serious changes in the structural geometry of the ex fines pores, when they will be in contact with the micro-cracks, and that is not very realistic.

2.3.3 *Constitutive equations*

The solution of balance equations needs relationships to connect the fluxes of the conserved quantities (such as mass, momentum and energy) with the gradients of state variables. Model closure then requires constitutive equations describing the specific behaviour of the considered material, i.e. relations expressing the intrinsic permeability K, relative permeabilities k_{rw} and k_{rg}, the effective diffusivity D_{eff}, the effective thermal conductivity λ_{eff}, the saturation S as well as stress-strain relationship as a function of the chosen state variables $\mathbf{u}$, p^w, p^{ga} and T.

2.3.3.1 Constitutive equations for description of hygro-thermal state of concrete

In a previous part, in the energy balance equation the heat flux has been already expressed by using Fourier's law. The generalised Darcy's law (already include the Newton's law of viscosity) for multiphase porous medium gives respectively, average velocities of liquid water and gas phase:

$$\hat{\mathbf{v}}_s^w = S^w \phi \mathbf{v}_s^w = -\frac{Kk_{rw}}{\mu_w}\left[gradp^w - \rho^w g \right] \tag{19}$$

$$\hat{\mathbf{v}}_s^g = S^g \phi \mathbf{v}_s^g = -\frac{Kk_{rg}}{\mu_g}\left[gradp^g - \rho^g g \right] \tag{20}$$

where $\hat{\mathbf{v}}_s^w$, $\hat{\mathbf{v}}_s^g$ = volume-averaged velocities of the liquid water and gas phase; p^w, p^g = pressures of liquid water and gas phase; μ_w, μ_g = dynamic viscosity of liquid water and gas phase; k_{rw}, k_{rg} = relative permeability of liquid water and gas phase and K = intrinsic permeability of the medium.

For the binary gas mixture transport in porous medium (dry air and water vapour), the Fick's law gives the following relative average velocities of the diffusing species:

$$\begin{aligned} \hat{\mathbf{v}}_g^{ga} = S^g \phi \mathbf{v}_g^{ga} &= -\frac{\rho^g}{\rho^{gw}} \frac{M_w M_a}{(M_g)^2} D_{eff}\, grad\left(\frac{p^{ga}}{p^g}\right) \\ &= -\frac{\rho^g}{\rho^{ga}} \frac{M_w M_a}{(M_g)^2} D_{eff}\, grad\left(\frac{p^{gw}}{p^g}\right) = S^g \phi \mathbf{v}_g^{gw} = -\hat{\mathbf{v}}_g^{gw} \end{aligned} \tag{21}$$

where $\hat{\mathbf{v}}_g^{ga}$, $\hat{\mathbf{v}}_g^{gw}$ = non-advective (diffusive) fluxes of dry air and water vapour phase; M_a, M_w, M_g = molar masses of dry air, water vapour and their mixture; D_{eff} = effective diffusivity of vapour in air.

When the data of the sorption isotherms would be available, and through equations (11) and (13) the relationship between S, p^w and p^{gw} is easily obtained. In this way the saturation liquid water S can be expressed as a function (experimentally determined) of capillary pressure p^c and temperature T:

$$S = S(p^c, T) \tag{22}$$

The gas relative permeability for concrete, similarly as for most capillary porous materials, may be described by the formula of the form (Couture et al. 1996):

$$k_{rg} = 1 - \left(\frac{S}{S_{cr}}\right)^{A_g} \quad \text{for } S < S_{cr} \tag{23}$$

where S_{cr} = critical saturation value, above which there is no gas flow in the medium; A_g = constant.

A sharp increase of the capillary water flux is observed, when the relative humidity RH reaches the value higher than about 75% (Bazant & Najjar, 1972). Such kind of behaviour can be described by the following relationship (Gawin et al. 1999):

$$k_{rw} = \left[1 + \left(\frac{1-RH}{0.25}\right)^{B_w}\right]^{-1} \cdot S^{A_w} \tag{24}$$

where A_w, B_w = constants. It should be noted that the term in square brackets has a form originally proposed in (Bazant & Najjar, 1972).

2.3.3.2 Constitutive equations for thermo-hygro-mechanical behaviour of concrete

A phenomenological approach is usually applied for description of changes of concrete physical properties during complex hygro- thermal and mechanical phenomena at high temperature. It means that all these changes are expressed as a function of temperature, moisture content and gas pressure.

There is a need for mechanistic (structural) mathematical models which assume certain physical models of the phenomena analysed. Significant increase in the concrete diffusivity characteristic at high temperature is mainly generated by arising micro-cracks and by changes of material inner structure, as well as by crack-opening due to high gas pressure values. As a result, it depends not only upon temperature, moisture content and gas pressure, as assumed for a phenomenological approach, but also upon a degree of cracks development, which may be described by use of damage parameter ϕ_M.

However one can expect that a joint effect of temperature, gas pressure and material damaging (crack development) on the intrinsic material permeability, K, may be described by (Gawin et al. 1998):

$$K = K_0 \cdot 10^{A_T \cdot (T - T_0)} \cdot \left(\frac{p^g}{p_0^g}\right)^{A_p} \cdot 10^{A_{\phi M} \cdot \phi_M} \tag{25}$$

The effective coefficient of vapour inside the pores of partially saturated concrete, may be described by (Gawin et al. 1999):

$$D_{eff} = \phi (1-S)^{A_v} f_s D_{v0} \left(\frac{T}{T_0}\right)^{B_v} \frac{p_0}{p^g} \tag{26}$$

The effective thermal conductivity of partially concrete, may be described by (Gawin et al. 1999):

$$\lambda_{eff} = \lambda_d (T) \cdot \left(1 + \frac{4\phi\rho^w}{(1-S)\rho^s}\right) \tag{27}$$

$$\lambda_d = \lambda_{d0} \cdot \left[1 + A_\lambda (T - T_0)\right] \tag{28}$$

where A_T, A_p, A_{ϕ_M}, A_v, B_v, A_λ = material constants; D_{v0} = diffusion coefficient of vapour species in the air at a reference temperature T_0 and pressure p_0; f_s = structure factor coefficient; λ_{d0} = thermal conductivity of dry material.

2.3.3.3 Constitutive equations for mechanical behaviour of concrete

The constitutive stress-strain relationship for the solid skeleton has the form:

$$\begin{aligned}\sigma &= (1-\phi_0)(1-\phi_T)(1-\phi_M)\mathbf{E}^s : \varepsilon^e \\ &= (1-\phi_T)(1-\phi_M)\mathbf{E}_0 : \varepsilon^e \\ &= \mathbf{E} : (\varepsilon - \varepsilon^p - \varepsilon^T)\end{aligned} \tag{29}$$

where $\mathbf{E}^s$, $\mathbf{E}_0$, $\mathbf{E}$ = fourth order stiffness tensors for each, solid skeleton, porous media (macroscopically initial measured), current damaged porous media respectively; $\boldsymbol{\varepsilon}^e$ = elastic strain; $\boldsymbol{\varepsilon}^p$ = irreversible (plastic) strain; $\boldsymbol{\varepsilon}^T$ = thermal expansion strain.

Thermal damage variable

In a similar way to mechanical damage, the thermal damage of heated concrete is described in terms of experimentally determined relations $E(T)$, which link the temperature T to the current elastic modulus E, then combining this relation with the phenomenological relation $E(T) = (1-\phi_T)E_0$ (Ulm, 1999, Nechnech et al. 2001), one can determine the thermal damage function as:

$$\phi_T = 1 - \frac{E(T)}{E_0} \tag{30}$$

It should be noted that the function $E(T)$ is obtained by the uniaxial compressive and tensile tests on specimens that have been heated at different temperatures and then cooled progressively in order to avoid additional degradation due to quick cooling.

Mechanical damage variable

The evolution of this damage is considered to be of an exponential form (see equation (31)) of the cumulated plastic strain κ_x governing the cracking process and then the global shape of the softening curve.

$$1-\phi_M^x = \exp(-c_x \kappa_x) \tag{31}$$

$$\dot{\kappa} = \sqrt{(2/3)\dot{\varepsilon}^p : \dot{\varepsilon}^p} \tag{32}$$

where x = t, c traction and compression respectively; c_x = a material parameter identified from experimental stress-strain curve (Nechnech et al. 2001).

In order to describe properly the deference in the behaviour of concrete in compression and in tension, the damage variable is separated into a compressive one and tension one. The crack closure effect is taken into account by multiplying the damage variable in tension with a scalar weight factor p. This scalar reflects the elastic stiffness recovery during the unloading process from tension state to compression state:

$$\begin{aligned}&1-\phi_M = (1-\phi_M^c)\cdot(1-p\cdot\phi_M^t) \\ &p = \frac{\sum_{i=1}^{i=3}\langle\tilde{\sigma}_i\rangle}{\sum_{i=1}^{i=3}|\tilde{\sigma}_i|} \quad with\ \langle\tilde{\sigma}_i\rangle = \frac{(\tilde{\sigma}_i + |\tilde{\sigma}_i|)}{2}\end{aligned} \tag{33}$$

Coupling between plasticity and damage

Once micro-cracks are initiated, local stresses are redistributed over the effective area of the undamaged material micro-bonds (see Figure 2). Thus, effective stresses in undamaged material points are higher than nominal stresses. Accordingly, it appears reasonable to state that the plastic flow occur in the undamaged material micro-bonds by means of effective quantities Ju (1989). Therefore, the characterisation of plastic response should be formulated in the effective stress space in terms of effective quantities. The homogenised stress tensor $\boldsymbol{\sigma}$ is replaced by the effective stress tensor $\tilde{\boldsymbol{\sigma}}$ in the stress space yield criterion. However, when plasticity theory is used to describe concrete behaviour in a phenomenological way, nominal stresses are considered to drive the cracking process in all the material, and therefore plastic strain would represent the cracks opening. In this approach, softening plasticity is considered to govern cracks occurrence and evolution. Therefore, damage reflects the accompanying stiffness degradation, which justifies again the choice of the evolution

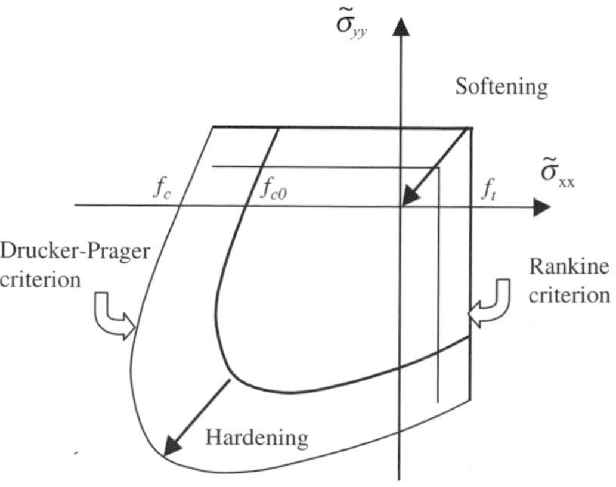

Figure 2. Drucker-Prager and Rankine criterion in the principal stress space (2D).

law adopted for the damage variable. Hence, the yield surface F is given as following:

$$F(\sigma,\kappa,T)=f(\sigma)-\tau(\kappa,T)\leq 0 \tag{34}$$

rewriting this equation in the space of the effective stresses such as:

$$F(\sigma,\kappa)=(1-\phi_T)(1-\phi_M)f(\tilde{\sigma})-\tau(\kappa,T)\leq 0 \tag{35}$$

or by introducing the effective strength of the material:

$$\tilde{\tau}(\kappa,T)=\frac{\tau(\kappa,T)}{(1-\phi_T)(1-\phi_M)} \tag{36}$$

$$F(\tilde{\sigma},\kappa,T)=f(\tilde{\sigma})-\tilde{\tau}(\kappa,T)\leq 0 \tag{37}$$

where f = function of the stress tensor invariant; τ, $\tilde{\tau}$ = the nominal and the effective strength of the material (as a function of the hardening/softening parameter κ and temperature T), and we could obtained τ from a uniaxial tension or compression test.

In order to reproduce a suitable behaviour in compression and in tension, a non-smooth multi-surface criterion is used to describe the non-symmetrical material behaviour in compression and tension. The yield surfaces employed are: Drucker-Prager criterion in compression and Rankine criterion in tension (see Figure 2).

Drucker-Prager criterion is written as:

$$F_c(\tilde{\sigma},\kappa_c,T)=J_2(\tilde{s})+\beta_1 I_1(\tilde{\sigma})-\beta_2\tilde{\tau}_c(\kappa_c,T)\leq 0 \tag{38}$$

where $J_2(\tilde{s})$ = second invariant of the deviatoric stress tensor, $I_1(\tilde{\sigma})$ = first invariant of the stress tensor, and $\beta_i(i=1,2)$ are two material parameters.

Rankine criterion is written as:

$$F_t(\tilde{\sigma},\kappa_t,T)=\tilde{\sigma}_1-\tilde{\tau}_t(\kappa_t,T)\leq 0 \tag{39}$$

where $\tilde{\sigma}_I$ = maximal principal stress in tension.

The nominal strengths are defined by:

$$\tau_x=f_0^x\left[(I+a_x)\exp(-b_x\kappa_x)-a_x\exp(-2b_x\kappa_x)\right] \tag{40}$$

where $a_x(T)$, $b_x(T)$ = are material parameters identified from unaxial test at different temperature; f_0^x = initial tensile (x = t) or compressive (x = c) strength at room temperature T_0 (Nechnech 2000).

Furthermore, concrete is a frictional material which makes it necessary to adopt a non-associated plastic flow to control the dilatancy, since the damage here is considered to be isotropic. A plastic potential is then introduced:

$$G_c=J_2(\tilde{s})+\beta_g I_1(\tilde{\sigma})-\beta_2\tilde{\tau}_c(\kappa_c,T) \tag{41}$$

$$G_t=F_t \tag{42}$$

where β_g = parameter chosen to give proper dilatancy.

The plastic strain rate tensor is then obtained by:

$$\dot{\varepsilon}^p=\dot{\lambda}_c\frac{\partial F_c}{\partial\tilde{\sigma}}+\dot{\lambda}_t\frac{\partial F_t}{\partial\tilde{\sigma}} \tag{43}$$

where λ_x is the plastic multiplier associated with the plastic potential function G_x in tension or in compression.

3 NUMERICAL EXAMPLES

A one-dimensional example is solved to demonstrate the potentialities of the model in describing hygro-thermal behaviour of concrete at high temperature. The example deals with a concrete wall of 30 cm thickness subject to transient heating from one side (Gawin et al. 1999). On the external surface a mixed boundaries conditions (BCs) for energy exchange with the outer atmosphere are considered. These BCs involving both convective and radiative exchange with a heat exchange coefficient $\alpha_c = 18$ W/m^2/K, and $e\sigma_0 = 5.1\times 10^{-8}$ W/m^2/K^4 respectively. An ambient air temperature increases according to the standard ISO fire curve is considered. For the mass exchange convective BCs, with $p^{gw} = 1000$ Pa and mass exchange coefficient $\beta_c = 0.018$ m/s are assumed. The analysed material is a High Performance Concrete, and it is characterised by the following parameters (at initial state):

1. dry state apparent density $\rho_0 = 2590$ kg/m^3,
2. initial porosity $\phi_0 = 0.1$,
3. initial thermal conductivity of dry material $\lambda_d = 1.67$ W/m/K,
4. initial intrinsic permeability $K_0 = 10^{-19}$ m^2,
5. Young's modulus $E = 14$ GPa and Poisson's ratio $\nu = 0.25$.

Simulation was made during the first 10 min of transient behaviour, before temperature at the surface reaches the critical point of water, i.e. 647.3 K (374.15°C), using a one-dimensional mesh of 30 elements and a time step of 1 second.

3.1 *Results and discussion*

The results of computations at time station t = 1, 2, 4, 6, 8, and 10 min are presented over a thickness of 8 cm only. In the remaining part of the wall the gradient of temperature is negligible.

Thermal gradient, that means a temperature difference between the central part of the wall and the surface, increased with the surface temperature rising

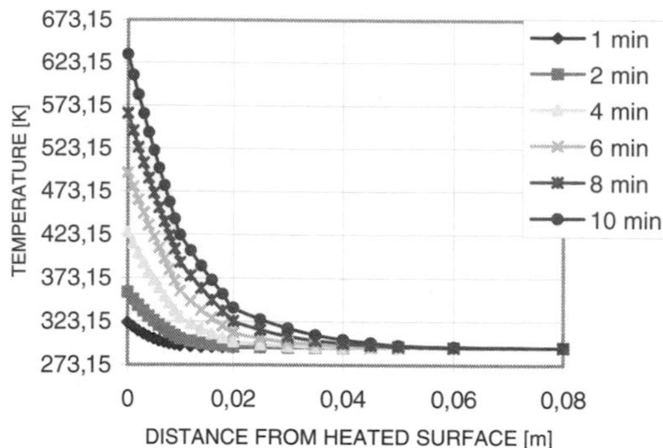

Figure 3. Evolution of the temperature.

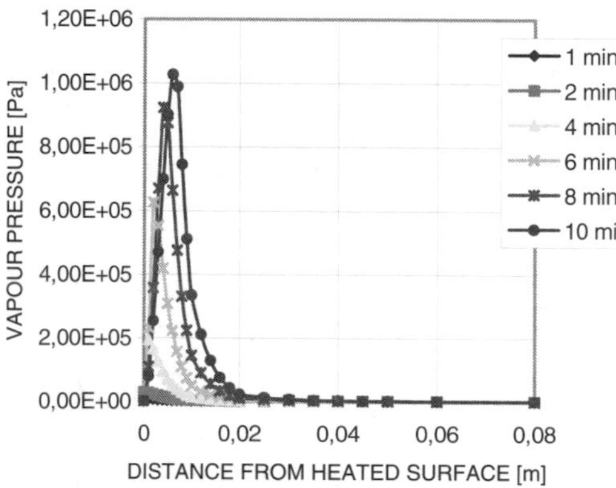

Figure 4. Evolution of the water vapour pressure.

which goes up to a maximum value of about 632 K (see Figure 3).

Two extreme processes can be considered for the transport of moisture through the heated wall. The first process corresponds to the diffusion of an increased water vapour pressure towards the outer atmosphere. In fact water vapour pressure continuously forms in a zone close to the heated surface. It results mainly from the rapid evaporation of water inside the wall. The causes of this evaporation are: in one hand, where the water temperature approaches and passes through the boiling point and in another hand, due to a thermodynamic imbalance between the external humidity and that prevailing within the wall. This will entail a diffusion process of the water vapour and of the dry air through the wall in order to maintain the equilibrium between liquid and vapour. The maximum pressure in this zone increases gradually up to 1.1 MPa after 10 min and the peak moves slowly toward the internal part of the wall (see Figure 4).

The second process corresponds to the Darcean convective transport of the liquid water. This one is caused by the gradient of the liquid underpressure, generated mainly by the intense evaporation occurring at the boundary of the wall (see Figure 5).

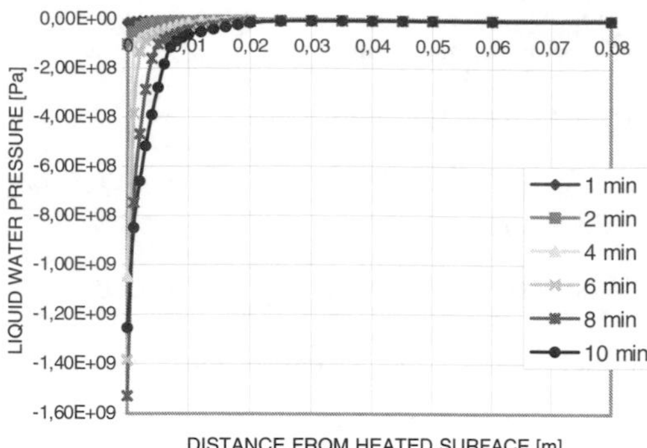

Figure 5. Evolution of the liquid water pressure.

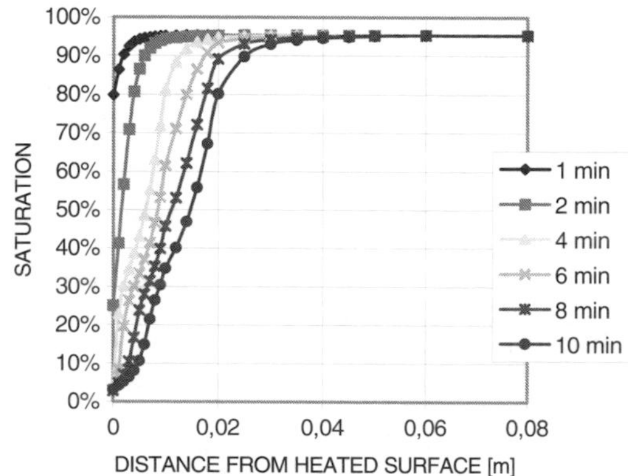

Figure 6. Evolution of the saturation.

It is also observed that the desaturation of concrete is restricted in a zone close to the heated surface. It is caused mainly by the two processes of moisture transport (mentioned above) which lead to the extraction of liquid water from the heated concrete wall (see Figure 6).

From the Figure 7 we could see that the behaviour of dry air may be influenced by different phenomena. The two major ones are the desaturation and the peak of the water vapour pressure effects. We could see also from the figure that the general behaviour of the curves take the same general behaviour as that due to the desaturation (see Figure 6). In fact the extraction of water from a zone will lead to an expansion in the volume of gas in that zone. This expanded volume will act to come in more molecules of dry air to fill it. Also, it should be noted that the permeability of the material is a very important factor which affect the flux. Therefore, a very low permeability will lead to come in less quantity of a dry air molecules in the

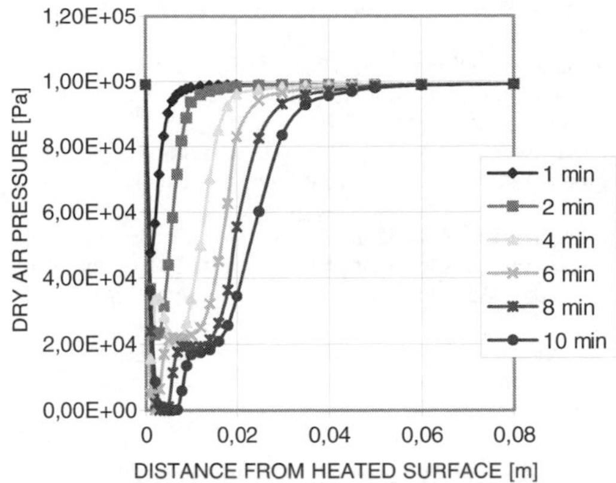

Figure 7. Evolution of the dry air pressure.

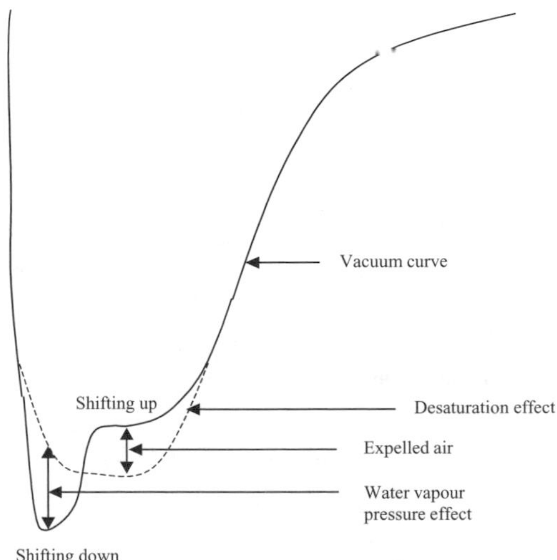

Figure 8. Schematic representation to the effect of water vapour pressure on the dry air pressure.

expanded volume and will be submitted to a vacuum as a result.

We could distinguish from Figure 7 the second important role, is that the peak of water vapour pressure could affect also the behaviour of the dry air. In time stations t = 1, 2 min, the increased water vapour pressure has not a clearly effect on the behaviour of the dry air because its peak was not very important (see Figure 4). But in the rest of time, when the peaks start to take high values, we could observe another behaviour of the dry air. This behaviour is represented by the appearance of two plateaux. These plateaux are performed by shifting down the vacuum curve of the dry air, in the zone where the peaks of water vapour pressure are increased, and shifting up this curve in the adjacent zone where the dry air will be expelled

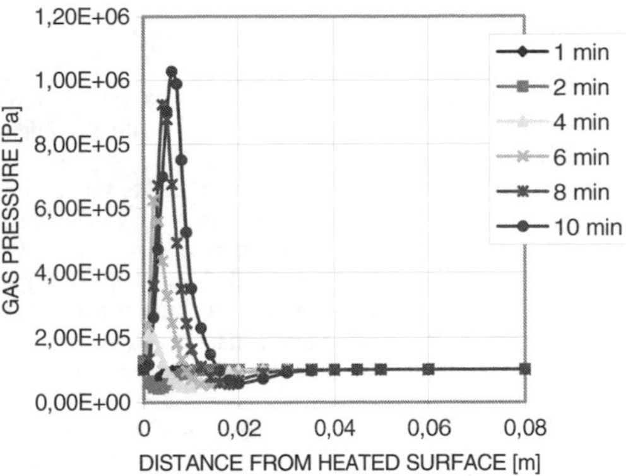

Figure 9. Evolution of the gas pressure.

to (see Figure 8). Finally we observed also that the vacuum zone was be acted to accelerate a little bit the desaturation process.

A general idea to the overall behaviour of the gas pressure could be seen in Figure 9.

4 CONCLUSIONS

The ensuing coupled model consists in solving the linear momentum balance equation for the whole multiphase medium, mass balance equations for aqueous and non aqueous fluid phases and the energy balance equation. The model is valid below the critical point of water, i.e. 647.3 K (374.15°C), because above this temperature capillary water cannot exist as a liquid, thus also capillary pressure has no more physical meaning. The results presented here don't pretend to be of any definitive character. The main objectives of this application is to analyse the capability of the proposed model to provide a realistic prediction of the behaviour of concrete structure at elevated temperature. An extension to these results concerning the mechanical and the damage effects will be introduced recently.

REFERENCES

Anderberg, Y. 1997. Spalling phenomena of HPC and OC. *Proc., In Workshop on Fire Performance of High-Strength Concrete, NIST Spec. Publ.* 919, L. T. Phan, N. J. Carino, D. Duthinh, and E. Garboczi, (eds), National Institute of Standards and Technology, Gaithersburg, Md., 69–73.

Bazant, Z. P. & Najjar, L. J. 1972. Nonlinear water diffusion in nonsaturated concrete. *Matériaux Constructions, Paris*, 5(25): 3–20.

Bazant, Z. P. & Thonguthai, W. 1978. Pore pressure and drying of concrete at high temperature. *In J. Engng. Mech. Div. ASCE*. 104: 1059–1079.

Bazant, Z. P. & Thonguthai, W. 1979. Pore pressure in heated concrete walls: theoretical prediction. *In Mag. of Concr. Res*. 31(107): 67–76.

Bazant, Z. P. & Kaplan, M. F. 1996. Concrete at High Temperatures: Material Properties and Mathematical Models. Harlow: Longman.

Bazant, Z. P. 1997. Analysis of pore pressure: thermal stresses and fracture in rapidly heated concrete. *Proc. In Workshop on Fire Performance of High-Strength Concrete, NIST Spec. Publ.* 919, L. T. Phan, N. J. Carino, D. Duthinh, and E. Garboczi, (eds), National Institute of Standards and Technology, Gaithersburg, Md., 155–164.

Bear, J. 1988. Dynamics of fluids in porous media New York: Dover.

Bear, J. & Bachmat, Y. 1986. Macroscopic modelling of transport phenomena in porous media, 2: applications to mass momentum and energy transfer. *In Transp. in Porous Media* 1: 241–269.

Bear, J. & Bachmat, Y. 1990. Introduction to modelling of transport phenomena in porous media. Dordrecht: Kluwer.

Biot, M. A. & Willis, P. G. 1957. The elastic coefficients of the theory of consolidation. *In J. Appl. Mech*. 24: 594–601.

Couture, F., Jomaa, W., & Puiggali, J.-R. 1996. Relative permeability relations: a key factor for a drying model. *In Transp. Porous Media* 23: 303–335.

Ulm, F.-J., Coussy, O., & Bazant, Z. P. 1999. The Chunnel fire I: Chemoplastic Softening in Rapidly Heated Concrete.

Gawin, D., Baggio, P., & Schrefler, B. A. 1996. Modelling of heat and moisture transfer in deformable porous building materials. *Arch. Civil Engng*. 42(3): 325–349.

Gawin, D. & Schrefler, B. A. 1996. Thermo-hydro-mechanical analysis of partially saturated porous materials. *In Engineering Computations* 13(7): 113–143.

Gawin, D., Majorana, C. E., & Schrefler, B. A. 1999. Numerical analysis of hygro-thermic behaviour and damage of concrete at high temperature. *In Mech. Cohes.-Frict. Mater.* 4: 37–74.

Gens, A. & Olivella, S. 2001. THM phenomena in saturated and unsaturated porous media. *Environmental Geomechanics* 5: 693–717.

Hassanizadeh, M. & Gray, W. G. 1979. General conservation equations for multiphase systems: 1 Averaging technique. *In Adv. Water Res*. 2: 131–144.

Hassanizadeh, M. & Gray, W. G. 1979. General conservation equations for multiphase systems: 2 Mass, momenta, energy and entropy equations. *In Adv. Water Res*. 2: 191–203.

Hassanizadeh, M. & Gray, W. G. 1980. General conservation equations for multiphase systems: 3 Constitutive Theory for Porous Media. *In Adv. Water Res*. 3: 25–40.

Ju, J. W. 1989. On Energy-Based Coupled Elastoplastic Damage Theories: Constitutive Modelling and Computational Aspects. *In J. of Solids and Structures* 25(7): 803–833.

Kachanov, L. M. 1986. Introduction to continum damage mechanics. Dordrecht, The Netherlands. Martinus: Nijhoff Publishers.

Lewis, R. W. & Schrefler, B. A. 1998. The Finite Element Method in the Static and Dynamic Deformation and Consolidation of Porous Media. Chichester: Wiley & Sons.

Lin, W.-M., Lin, T. D., & Powers-Couche, L. J. 1996. Microstructure of fire-damaged concrete. *ACI Mat. J.* 93(3): 199–205.

Nechnech, W., Meftah, F., & Reynouard, J. M. 2001. An elasto-plastic damage model for plain concrete subjected to high temperature. *In Engineering structures* 24: 597–611.

Whitaker, S. 1977. Simultaneous heat mass and momentum transfer in porous media: a theory of drying. *In Advances in heat transfer* 13. New York: Academic Press.

Computational Modelling of Concrete Structures, Bićanić et al. (eds)
© 2003 Swets & Zeitlinger, Lisse, ISBN 90 5809 536 3

Numerical models for coupling creep and fracture of concrete structures

M. Omar, G. Pijaudier-Cabot & A. Loukili
R & DO, Laboratoire de Génie Civil de Nantes Saint Nazaire, Ecole Centrale de Nantes, Nantes cedex, France

ABSTRACT: Numerical formulations to analyse basic creep and stress relaxation of concrete are presented in this paper, with a view to the computer implementation of the models. The numerical models are calibrated so that they fit experimental data. Then, coupling with time independent non local damage is discussed. The paper underlines the different approaches that can be envisioned in order to incorporate damage into the time dependent creep functions: the effective stress approach, and a second one where material damage acts on the time independent part of the creep functional. Some computer simulations have been carried out in order to judge the performance of the numerical models. Results show that using the effective stress approach, the kinetics of creep in bending beams is strongly correlated with the internal length when damage occurs. Also, the global behaviour of bending beams subjected to relaxation prior to failure is strongly influenced by the internal length variation and the loading rate effect. Results confirm the influence of the damage behaviour on creep or relaxation of concrete measured in the field of non-linear visco-elasticity associated to concrete fracture.

1 INTRODUCTION

In common practice, it is usually assumed that linear visco-elasticity takes place for low load levels and that the instantaneous mechanical behaviour is elastic. On the other hand, for high load levels, cracks grow and interact with visco-elasticity. This interaction between creep and crack propagation in concrete structures have drawn the attention of many authors, among them Rüsch (1960), who has carried out the first experimental results related to this subject. Thus, the residual capacity of a structure subjected to creep solicitation prior to failure decreases when increasing the creep loading duration.

Gettu & Bazant (1992) have also studied the loading rate effects and the scale effects on fracture growth. They found that the material becomes more brittle with decreasing rate of loading, the peak loads decrease, and the size of the fracture process zone for a crack propagation in mode I decreases too.

Numerical modeling for coupling the mechanical behaviour and the visco-elastic behaviour of the material assumed as a continuum medium, and its non-linear instantaneous behaviour have been tackled by many authors using a plasticity model coupled to a linear creep model (De Borst et al. 1993). This instantaneous behaviour can also be described using a smeared crack model (Rots et al. 1988) or a damage model (Omar 2000).

In this paper, two methods capable to predict non-linear creep at a moderately high stress level using two creep-damage models were compared. Creep is modeled on the basis of two distinct rheological models (generalized Kelvin chain model and generalized Maxwell chain model (Bazant 1974)). The Kelvin chain model is coupled in series to a damage model, where material damage acts on the time independent part of the creep functional. In the Maxwell chain model, damage is incorporated into the time dependent creep functions using the effective stress approach.

In order to simplify this comparison, we use a one-dimensional model implemented in a finite element code where the tested structures consist in prismatic specimens subjected to flexion.

These two numerical models are described and their implementation in the finite element code is discussed. Some numerical simulations dealing with creep or relaxation applied loading on three points bending concrete beams are performed.

2 NON LOCAL DAMAGE

The mechanical effect of progressive micro-cracking due to external loads is described by a single internal variable which degrades the Young's modulus of the material (Mazars 1984). The constitutive relations are:

$$\sigma_{ij}=(1-d)C_{ijkl}\varepsilon_{kl} \tag{1}$$

where σ_{ij} and ε_{ij} are the components of the stress and strain tensors respectively ($i,j,k,l \in [1, 3]$), C_{ijkl} are the initial stiffness moduli, and d is the damage variable. The material is initially isotropic, with E and ν the initial Young's modulus and Poisson's ratio respectively. For the purpose of defining damage growth, the equivalent strain is introduced first:

$$\tilde{\varepsilon}=\sqrt{\sum_{i=1}^{3}\left(\langle\varepsilon_i\rangle_+\right)^2} \tag{2}$$

where $\langle\varepsilon_i\rangle_+ = 0$ is the "Macauley bracket" and ε_i are the principal strains. In order to avoid ill-posedness due to strain softening, the mechanical model is enriched with an internal length (Pijaudier-Cabot & Bazant 1987, Mazars & Pijaudier-Cabot 1996). The non-local variable $\bar{\varepsilon}$ is defined as:

$$\bar{\varepsilon}(x)=\frac{1}{V_r(x)}\int_{\Omega}\psi(x-s)\tilde{\varepsilon}(s)ds \quad \text{with } V_r(x)=\int_{\Omega}\psi(x-s)ds \tag{3}$$

where Ω is the volume of the structure, $V_r(x)$ is the representative volume at point x, and $\Psi(x-s)$ is the weight function:

$$\psi(x-s)=\exp\left(-\frac{4\|x-s\|^2}{l_c^2}\right) \tag{4}$$

l_c is the internal length of the non local continuum. $\bar{\varepsilon}$ is the variable that controls the growth of damage according to the following conditions:

$$F(\bar{\varepsilon})=\bar{\varepsilon}-\kappa \tag{5a}$$

$$\text{if } F(\bar{\varepsilon})=0 \text{ and } \dot{\bar{\varepsilon}}>0 \text{ then } \begin{cases}\dot{d}=h(\kappa)\\ \dot{\kappa}=\dot{\bar{\varepsilon}}\end{cases} \text{ with the condition } \dot{d}\geq 0$$

$$\text{else} \begin{cases}\dot{d}=0\\ \dot{\kappa}=0\end{cases} \tag{5b}$$

κ is the softening parameter and takes the largest value $\bar{\varepsilon}$ of ever reached during the previous loading history at a given time and at the considered point in the medium. Initially $\kappa=\kappa_0$, where κ_0 is the threshold of damage. The damage evolution law h(κ) in (eq. 5b), is defined in an integrated form. In particular, the damage growth in tension is:

$$d=1-\frac{\kappa_0(1-A_t)}{\kappa}-\frac{A_t}{\exp(B_t(\kappa-\kappa_0))} \tag{6}$$

where constants A_t, B_t are model parameters which must be obtained from a fit of experimental data (Le Bellégo et al. 2002).

3 BASIC CREEP COUPLED TO DAMAGE

The creep phenomenon is modeled using the visco-elasticity theory. The calculation of stress (strain) generated by a variable strain (stress) is greatly facilitated by the principle of superposition (Bazant 1988). This principle is usually assumed to apply to concrete within the service stress range. We consider here basic creep only, which is the deformation of concrete without any hygral exchange with the surrounding medium. Constitutive law of linear visco-elasticity theory gives a tool for a correct description of concrete behaviour under variable loading. Material considered may be described with one of the rheological models proposed in literature. A characteristic function (integral equation) describes the relation between the stress and the strain and takes a form of a compliance function in the case of the Kelvin chain model (Fig.1) or a relaxation function in the case of the Maxwell chain model (Fig. 2) (Bazant 1988).

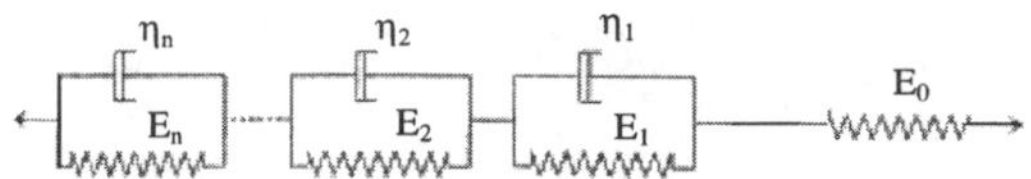

Figure 1. Generalized Kelvin model.

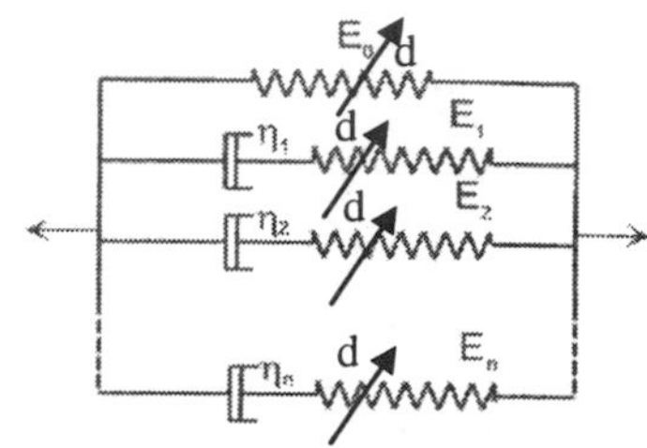

Figure 2. Generalized Maxwell model.

3.1 *Kelvin chain model*

The material response at the time t may be written as:

$$\varepsilon(t)=J(t_0,t)\cdot\sigma(t_0)+\int_{t_0}^{t}J(\tau,t)\dot{\sigma}(\tau)d\tau \tag{7}$$

Numerical computations of long term behaviour in structure are greatly accelerated by a substitution of the characteristic function with appropriate Dirichlet series. This representation is widely used in practice and in some recommendations even required (Rilem 1995). The creep function or compliance of a visco-elastic material can be expressed in the form of Dirichlet series as follows:

$$J(\tau,t)=\frac{1}{E_0(t_0)}+\sum_{\mu=1}^{n}\frac{1}{E_\mu}\left[1-\exp\left(-\frac{t-\tau}{\tau_\mu}\right)\right] \tag{8}$$

When the compliance function (eq. 8) is introduced into the superposition integral (eq. 7) the total strain may be written:

$$\varepsilon(t)=\int_{\tau=to}^{\tau=t}\frac{\dot{\sigma}(\tau)}{E_0(t)}d\tau+\int_{\tau=to}^{\tau=t}\sum_{k=1}^{\infty}\frac{1}{E_K}\left[1-\exp(-\frac{t-\tau}{\tau_K})\right]\dot{\sigma}(\tau)d\tau \quad (9)$$

then it can be subdivided as:

$$\varepsilon(t)=\varepsilon^{e}(t)+\varepsilon^{cr}(t) \quad (10)$$

in which ε^{e} is the instantaneous strain, ε^{cr} is the basic creep strain. By identification and differentiation, the creep strain can be written:

$$\varepsilon^{cr}(t)=\int_{\tau=to}^{\tau=t}\sum_{k=1}^{n}\frac{1}{E_k}\left[1-\exp(-\frac{t-\tau}{\tau_k})\right]\dot{\sigma}(\tau)d\tau \quad (11)$$

Differentiating equation 11 the creep strain increment is obtained:

$$\Delta\varepsilon^{cr}=\int_{\tau=to}^{\tau=t}\sum_{k=1}^{n}\frac{1}{E_k\cdot\tau_k}\left[\exp(-\frac{t-\tau}{\tau_k})\right]\dot{\sigma}(\tau)d\tau\cdot\Delta t \quad (12)$$

In a Kelvin chain model (Fig. 1), for $n \geq 8$, results obtained conform well with observed material behaviour. Numerical creep analysis can then be performed by subdividing the total time interval of interest into time steps Δt and discrete times t_r:

$t_r=t_0+r\Delta t$ et $\tau=t_0+q\Delta t$ avec $r\in[1,r_{max}]$ et $q\in[1,r-1]$.

$$\Delta\varepsilon^{cr}(t+\Delta t)=\sum_{q=1}^{r-1}\sum_{k=1}^{8}\frac{1}{E_k\cdot\tau_k}\left[\exp(-\frac{r\Delta t-q\Delta t}{\tau_k})\right]\Delta\sigma_q\cdot\Delta t \quad (13)$$

Then, the creep strain and the instantaneous strain at time $(t+\Delta t)$ can be written as:

$$\begin{cases}\varepsilon^{cr}(t+\Delta t)=\varepsilon^{cr}(t)+\Delta\varepsilon^{cr}(t+\Delta t)\\ \varepsilon^{e}(t+\Delta t)=\varepsilon(t+\Delta t)-\varepsilon^{cr}(t+\Delta t)\end{cases} \quad (14)$$

The elastic stiffness is written as a function of the instantaneous strain:

$$E(d)=\frac{E\cdot(1-d)\cdot\varepsilon^{e}}{\varepsilon} \quad (15)$$

In equation 15 we consider that damage acts on the time independent part of the creep functional. Coupling between damage and creep is then accomplished.

3.2 *Maxwell chain model*

The total concrete stress is decomposed into a sum of local stress in all branches of the model. For the time t this gives:

$$\sigma^{t}(t)=\sum_{\mu=0}^{n}\sigma^{t}_{\mu}(t) \quad (16)$$

$\sigma^{t}(t)$ is the effective stress applied to the Maxwell chain, it can also be written with the help of the relaxation function as:

$$\sigma^{t}(t)=R(t_0,t).\varepsilon(t_0)+\int_{t_0}^{t}R(\tau,t)\cdot\dot{\varepsilon}(\tau)\cdot d\tau \quad (17)$$

where t_0 is the age of loading.

In order to derive an incremental constitutive law, it is assumed that the relaxation function is decomposed into Dirichlet series:

$$R(\tau,t)=E_0(\tau)+\sum_{\mu=1}^{n}E_{\mu}(\tau).\exp(-\tau_{\mu}(t-\tau)) \quad (18)$$

Note that the formulation here is purely one-dimensional. Parameters of the rheological model for the process remain in constant proportion for each branch of the model:

$$\frac{E_{\mu}(t)}{\eta_{\mu}(t)}=const. \quad (19)$$

τ_{μ} and E_{μ} are respectively the relaxation time and the modulus of elasticity of each branch μ in the generalized Maxwell model. In general $E_{\mu}(t_0)$ can be calculated using the least squares method. As for τ_{μ}, however, they cannot be calculated from measured creep data but must be suitably chosen in advance.

In order to incorporate damage, which is a degradation of the elastic stiffness, into the time dependent creep functions, we use the effective stress approach. It is a classical approach, in which we consider a time independent damage. Mechanical damage is controlled by the total strain. As a first step, the computational implementation follows the classical one employed for a visco-elastic model. For next time steps, we can write an incremental formulation of the effective stress:

$$\Delta\sigma^{t}=\sum_{\mu=1}^{n}\sigma_{hist\mu}(t)+\Delta\varepsilon\left[E_0(t_0)+\frac{1}{\Delta t}\sum_{\mu=1}^{n}\tilde{E}_{\mu}\right] \quad (20)$$

where:

$$\tilde{E}_{\mu}=\tau_{\mu}\cdot E_{\mu}(t_0)\cdot\left(1-e^{-\Delta t/\tau_{\mu}}\right) \quad (21)$$

$$\sigma_{hist\mu}(t)=\sigma^{t}_{\mu}(t)\cdot\left(e^{-\Delta t/\tau_{\mu}}-1\right) \quad (22)$$

$\sigma^{t}_{\mu}(t)$ is the effective stress in the branch μ calculated at the preceding time step. Then we calculate the effective stress σ^{t} at the current time step $(t+\Delta t)$: $\sigma^{t}(t+\Delta t)=\sigma^{t}(t)+\Delta\sigma^{t}$. Damage can also be updated since it is an explicit function of the total strain. The stress σ is calculated classically as follows:

$$\sigma=\sigma^{t}(1-d) \quad (23)$$

and the creep strain is accounted for at each time step as:

$$\varepsilon^{cr}(t)=\varepsilon(t)-\varepsilon^{e}(t)=\varepsilon(t)-\frac{\sigma(t)}{E(1-d)} \tag{24}$$

The number of elements in the Dirichlet series n influences significantly accuracy of calculation. In the case of concrete, for $n \geq 4$, material behavior is well represented.

4 VISCO-ELASTIC PARAMETERS CALIBRATION

Practical use of the models requires an appropriate choice of the models parameters. The incremental formulation is based on the substitution of the creep or relaxation function with Dirichlet series. These Dirichlet series are completely defined when the number of terms constituting series n as well as set of values for $E_{\mu}(t)$ and τ_{μ} are known. The aim of this paragraph is to present a procedure of parameters identification which is used to derive the values of a rheological model of material from the characteristic material function. In both the two models (Kelvin model and Maxwell model) we assume a non ageing material.

To find the characteristic values of the rheological model used, two steps are necessary:

1. Choice of the material function for constitutive law (creep function $J(t_0,t)$ in the case of Kelvin model, relaxation function $R(t_0,t)$ in the case of Maxwell model).
2. Identification of discrete parameters n, E_{μ} and τ_{μ} describing material function.

Concerning the Kelvin chain model the parameters of the compliance function could be identified directly on the basis of creep functions existing in literature, using the Least Square Method as a procedure of identification.

As for the Maxwell chain model, the relaxation function may be either found in literature in an existing formulation or, for most cases, identified from creep functions using the integral Fredholm equation (eq. 25) which is the general form of relation between creep function and relaxation function.

$$J(t,t).R(t_0,t)-\int_{t_0}^{t}\frac{\partial J(\tau,t)}{\partial \tau}R(t_0,t)d\tau=1 \tag{25}$$

Such a formulation possesses the difficulty of a numerical integration over the whole loading history. A simplification consisting of limiting consideration to a given finite period is used. The whole time history $[t_0,t]$ is decomposed into finite times $[t_i, t_{i+1}]$.

Table 1. Visco-elastic parameters of Kelvin chain model for the concrete B11 at $t_0 = 28$ days.

E_{μ} [MPa]	τ_{μ} [days]
$E_0 = 33{,}170$	
$E_1 = 786{,}100$	$\tau_1 = 0.002$
$E_2 = 816{,}760$	$\tau_2 = 0.02$
$E_3 = 346{,}900$	$\tau_3 = 0.2$
$E_4 = 183{,}070$	$\tau_4 = 2$
$E_5 = 102{,}280$	$\tau_5 = 20$
$E_6 = 52{,}950$	$\tau_6 = 200$
$E_7 = 36{,}100$	$\tau_7 = 2000$
$E_8 = 17{,}520$	$\tau_8 = 20{,}000$

After discretisation and integration, we obtain the expression of the relaxation function in its final form:

$$R(t_0,t_k)=\frac{2+R(t_0,t_0)\left(J(t_1,t_k)-J(t_0,t_k)\right)+\sum R(t_0,t_i)\left(J(t_{i+1},t_k)-J(t_{i-1},t_k)\right)}{J(t_k,t_k)-J(t_{k-1},t_k)} \tag{26}$$

The process of identification of parameters n, E_{μ} and τ_{μ} is carried out in a calibration procedure. Dirichlet series resulting from these parameters should approach satisfactorily the relaxation function $R(t_0,t)$ known at this step by its formula or by discrete values obtained as a result of the procedure of inversion.

At this stage, it is necessary to provide the number n of the Dirichlet series. It is concluded that, using such an approach the concrete behaviour is well represented for $n \geq 4$ in the case of Maxwell chain model ($n \geq 8$ for Kelvin chain model). The next step consist of identifying the $(n+1)$ values of parameters $E_{\mu}(t_0)$ and τ_{μ}. The relaxation time τ_{μ} have been chosen according to a geometrical progression (Bazant 1974) $\tau_{\mu} = 10^{\mu-1}$. After this assumption the $(n+1)$ parameters $E_{\mu}(t_0)$ are obtained using the least square method (De Faget De Casteljau 1990). In this method the linear system given in equation 27 should be solved:

$$[A]\{E_{\mu}\}=\{B\} \tag{27}$$

where $\{E_{\mu}\}$ is the desired set of variables; [A] is a matrix written with relation to time and τ_{μ}; {B} is a vector depending on values of the relaxation function over the time period under analysis $[t_0,t]$.

Table 1 gives the values of the parameters E_{μ} and τ_{μ} of Kelvin chain model corresponding to concrete B11, which is the material considered in this study. These parameters are given in (Granger 1995) where the age of loading t_0 is considered 28 days.

Table 2 shows the values of parameters of Dirichlet series for the Maxwell chain model corresponding to the same concrete B11. These parameters are calibrated on the base of relaxation function inversed from the creep function of Granger (Granger 1995) for the same material.

Table 2. Visco-elastic parameters of Maxwell chain model for the concrete B11 at $t_0 = 90$ days.

E_μ [MPa]	τ_μ [days]
$E_0 = 15{,}550$	
$E_1 = 7921$	$\tau_1 = 1$
$E_2 = 4246$	$\tau_2 = 10$
$E_3 = 5574$	$\tau_3 = 100$
$E_4 = 5061$	$\tau_4 = 1000$

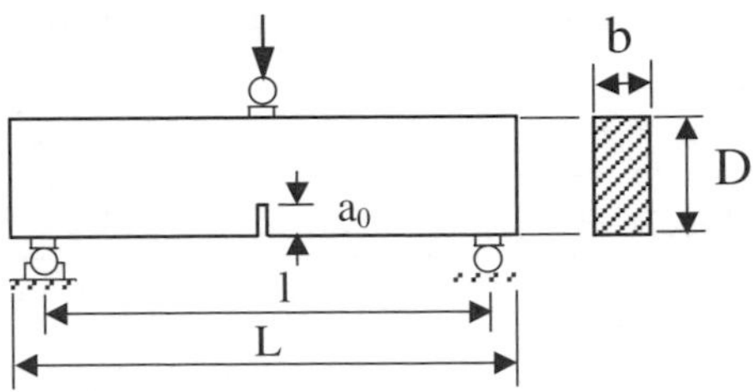

Figure 3. Three points bend test on notched specimen.

5 IDENTIFICATION OF MECHANICAL PARAMETERS

The calibration of the damage model parameters should rely on sufficiently robust methods in order to obtain an effective utilization. In this study, attention is focused on the model response in tension only. That is why the parameters, which should be involved in the identification procedure, are the constants A_t and B_t, the initial damage threshold κ_0, the internal length of the material l_c, the Young's modulus E and the Poisson ratio ν.

The identification of the internal length l_c which appears in the damage model constitutive equations, as well as other classical model parameters, should be achieved on the base of experimental tests. Saouridis (1988) demonstrated the possibility of identification of the mechanical parameters in tension using three points bending tests.

This study presents a procedure based on three points bend experiments carried out on geometrically similar notched beams of various dimensions. The purpose of the procedure is to get a good description of the stiffness, the maximum loading, the energy of fracture and the scale effects. The process of calibration has to rely on inverse analysis, where in a first step a manual calibration is performed, then an automatic calibration will lead to find an optimal set of material parameters.

5.1 *Experimental tests*

Three points bend tests on geometrically similar notched beams having various dimensions have been carried out in order to identify the parameters of the damage model. These tests are also called size effect tests. The prismatic specimens used in this study have the following dimensions: D1:$100 \times 100 \times 350\,mm^3$, D2:$100 \times 200 \times 700\,mm^3$, D3:$100 \times 400 \times 1400\,mm^3$, where the span to height ratio is $l/D = 3$ and the notch length $a_0 = 0.15D$. Figure 3 shows a typical 3 points bend test on notched specimen.

The material used in all experiments is the concrete B11 which is a synthetic concrete similar to that one used for the construction of the nuclear power station Civaux 1 in France. The formulation of this concrete is given in (Granger 1995). Its mechanical properties were determined on uniaxial compressive tests and splitting tests on cylindrical specimens having 160 mm in diameter and 320 mm in height. The dynamic modulus of elasticity E_{dyn} and the Poisson ratio ν are determined using the Grindosonic apparatus. The findings indicate that $E_{dyn} = 39{,}000$ MPa and $\nu = 0.28$.

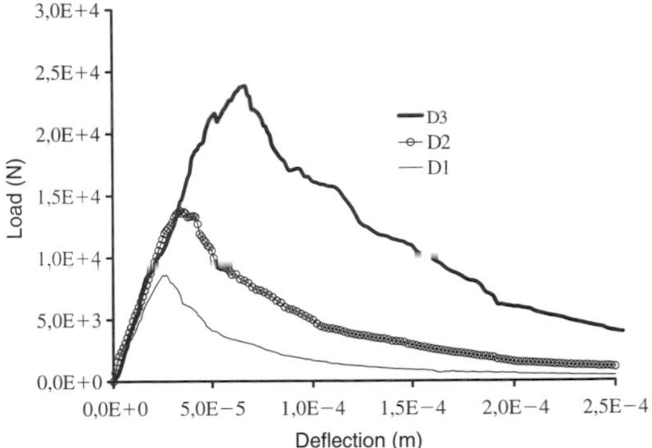

Figure 4. Three points bend testing of notched beams – experimental results for the small (D1), medium (D2) and large (D3) sizes.

Figure 4 shows the experimental load-deflection responses for the three different dimensions. Since 3 specimens of each size were tested for this purpose, the curves reported here represent the average of the experimental responses obtained on the 3 specimens of each size.

5.2 *Manual calibration*

First we intend to determine the Young's modulus and the Poisson ratio of the material. Concerning the Poisson ratio, we assume $\nu = 0.28$ as indicated above. The Young's modulus could be obtained from the fit of the experimental data. A calculation based on a linear elastic model was used in order to identify the Young's modulus. It appears that the value $E = 39{,}000$ MPa is satisfactory, it describes well the linear part of the experimental load-deflection curves.

Table 3. Mechanical parameters-results of manual calibration.

E [MPa]	ν	κ_0	A_t	B_t	l_c [m]
39,000	0.28	5×10^{-5}	0.85	30,000	0.075

Table 4. Optimal set of damage model parameters.

E [MPa]	ν	κ_0	A_t	B_t	l_c [m]
39,000	0.28	5×10^{-5}	0.81175	25748	0.075

The next step consists of a non-linear calculation taking into consideration the damage evolution in the tested specimen. As an initial guess, the internal length l_c is assumed equal to $3d_a$, where d_a is the maximum aggregate size of the concrete B11 (here $d_a = 25$ mm) (Bazant & Pijaudier-Cabot 1989). Then keeping l_c, E, and ν constants, the initial damage threshold and the two parameters A_t and B_t are varied only so that the numerical responses corresponding to the three specimen sizes are as consistent as much possible, with experimental results. Results of this manual calibration are reported in table 3.

These parameters listed above were then used as an entry within an optimization procedure. The model parameters are optimized so that numerical results, obtained on the three different sizes, are as close as possible to experimental results. For this purpose, a Levenberg-Marquardt algorithm has been implemented (Le Bellégo et al. 2002). The optimal set of mechanical parameters is given in table 4.

This final set of parameters could also be more improved by adjusting the internal length while the other parameters being kept constant, and then the more appropriate internal length could be chosen by analyzing the sensitivity of the model parameters on the size effect law.

In this contribution we won't deal with the final step and we keep for the following numerical simulations the set of parameters given in table 4.

Figure 5 shows the experimental load-deflection responses of the three different sizes with their computational fits obtained using the set of parameters in table 4.

6 RESULTS AND DISCUSSIONS

In this paragraph, a comparison between the two damage-creep approaches is performed. Some other simulations underlining the internal length effect and the influence of loading rate on the behaviour of the structure are presented. Moreover, the influence of the relaxation duration has been the focus of several calculations. The mechanical parameters used in the 2 models are those listed in table 4. Concerning the visco-elastic parameters: table 1 indicates parameters used in the case of Kelvin chain model, and table 2 gives those proposed for the Maxwell chain model. All calculations were carried out on the medium prismatic beams, which corresponds to the dimension used in coming experiments.

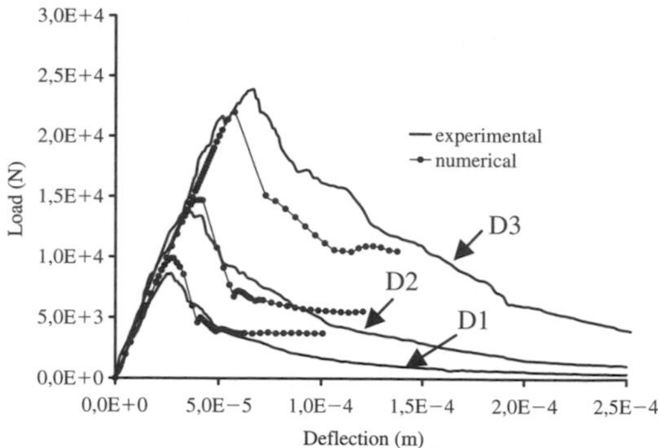

Figure 5. Experimental and computational results for the three different sizes (small: D1, medium: D2, large: D3).

In the following simulations we consider beams subjected to a constant displacement applied at mid span of the beam. Two kinds of loading are proposed. In the first one, relaxation starts in the pre-peak range, at about 85% of the peak load. In the second one, relaxation starts in the post-peak range for the same percentage of peak load. These simulations have been performed using the two damage-creep models proposed in paragraph 3. One can plot the relaxation curves $P(t)/P_0$, where P_0 is the load corresponding to relaxation start. Figure 6 shows results obtained using the Maxwell chain-damage model. A review of literature indicates that these results are entirely consistent with those obtained by Bazant & Jirasek (1992). The finding reveals that the relaxation curve starting in post-peak range lies below the curve starting in the pre-peak range. The relaxation curves obtained using the Kelvin chain-damage model are reported in Figure 7. A close comparison between results of the two numerical models reveals a contradiction between them. Actually, with the Kelvin chain-damage model the relaxation curve starting in the post-peak range lies over the curve starting in the pre-peak range.

For the Kelvin chain model, the evolution of the basic creep strain is calculated as a function of the applied stress (eq.13). In the case of Maxwell chain model this creep strain is calculated as a difference between the total strain and the elastic strain (eq. 24). The second method dealing with the effective stress approach is more realistic.

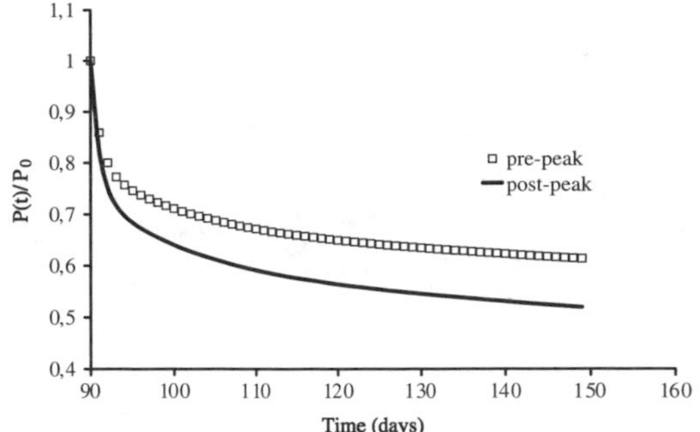

Figure 6. Relaxation curves – Maxwell chain model.

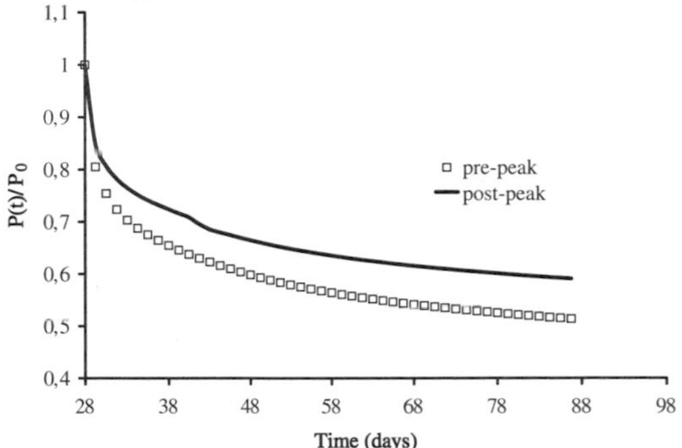

Figure 7. Relaxation curves – Kelvin chain model.

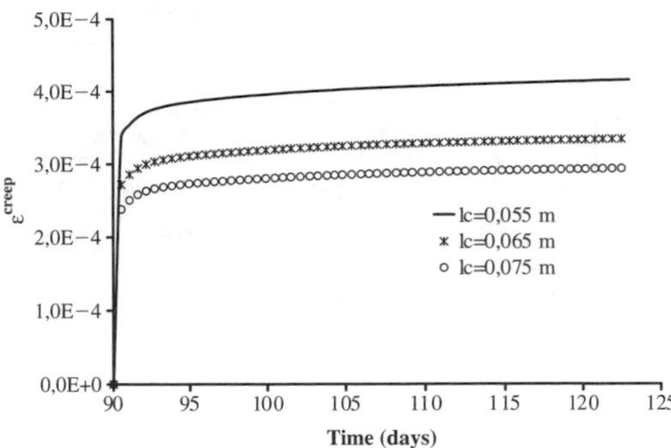

Figure 8. Creep strain evolution – internal length effect.

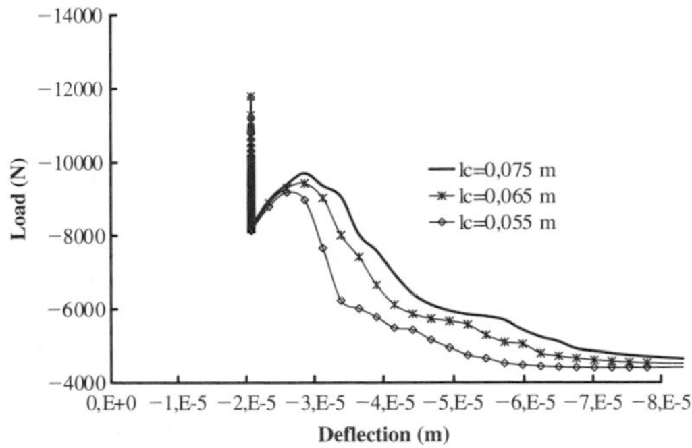

Figure 9. Load-Deflection responses – influence of the internal length on residual capacity.

The following simulations were performed using the Maxwell chain-damage model alone.

In order to evaluate the influence of the material internal length on the response of the structure, several calculations have been conducted on three points bending beams using three different values for l_c ($l_c = 0.055$ m, $l_c = 0.065$ m, $l_c = 0.075$ m). Three kinds of loading conditions are presented:

In the first case we compute the response of a beam subjected to a constant displacement at mid span, in this simulation we intend to determine the internal length effect on the kinetic of basic creep strain. Figure 8 shows the creep strain evolution versus time. It is observed that the kinetics of time dependent effects is strongly affected by the internal length when damage occurs, the longer the internal length, the smaller the creep strains. In order to underline this influence the applied displacement was chosen in the post-peak range of loading.

The second case corresponds to undergoing experiments in which beams are subjected to a constant displacement at mid-span for a given period of time, and then loaded up to complete failure. This test is aimed at obtaining the influence of relaxation on the capacity of bending beams. Figure 9 displays the load-deflection curves corresponding to each value of l_c. The results are quite comparable to those observed in the case of time independent material damage. The residual capacity of the structure decreases with a shorter material internal length.

Thus, the internal length has a great influence on the model prediction, which underlines the necessity of an accurate determination of this model parameter, and a good knowledge of its variation due to creep.

In addition to the internal length, another parameter appears to be important, it is the loading rate. It has been observed from previous studies that the rate of loading exerts a strong influence on the fracture behaviour of concrete. It has been also established that strength decreases when decreasing rate of loading.

Rate sensitivity is considered in monotonic loading in Figure 10. In this kind of simulations the beam is subjected to an increasing displacement applied at mid span till failure. P_c is the peak load corresponding to the smallest displacement rate used in these simulations and P is the calculated load. This figure shows that the peak load increases about 20% when the rate is increased by 10 times its smallest value, and about

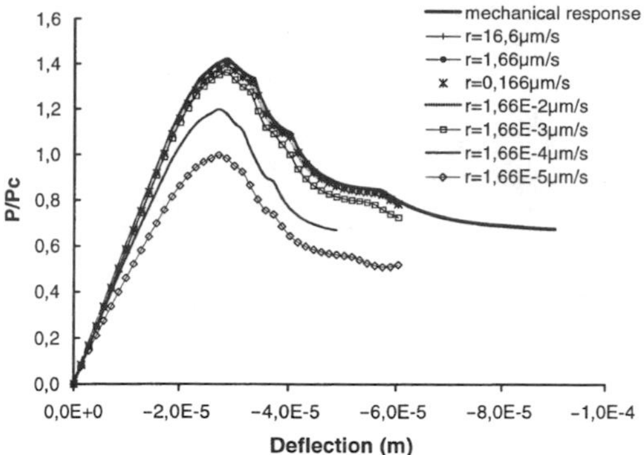

Figure 10. Loading rate effect on the Load-Deflection curves.

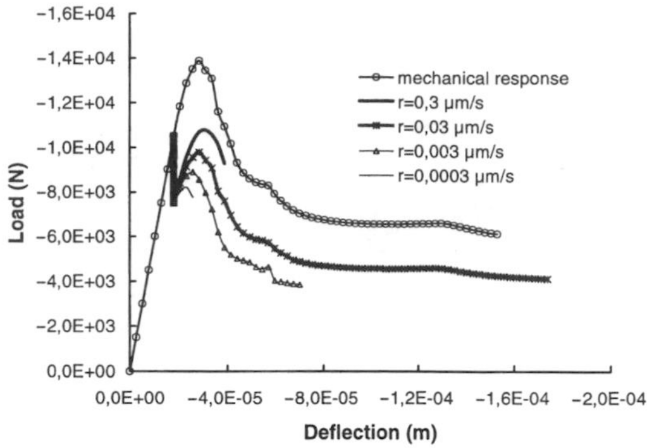

Figure 11. Load-Deflection responses – influence of loading rate on the residual capacity.

40% when it's increased by 10^5 this same smallest value.

We observe that the Maxwell chain-damage model reproduces qualitatively well observed results concluded. However, due to the choice of constants, there is a range of loading rates where rate sensitivity decreases and we see that there is a critical displacement rate over which the rate effect vanishes and the load-deflection curve approach the mechanical response of the structure where creep phenomenon doesn't interfere. In this case, this critical displacement rate is about $1{,}66 \times 10^{-2}\mu$m/s.

Another set of simulations investigating rate effects in concrete fracture was performed. The loading history is as follows: first a constant displacement is applied at mid span in the pre-peak range of loading for 10 days, then loading is increased up to failure. This simulation was repeated several times, increasing each time the rate of loading after the relaxation phase. Again, the peak loads increases when increasing the rate of loading. This peak load becomes smaller due to the relaxation (Fig. 11). This is clearly observed if we compare results to those obtained in the previous simulation (Fig. 10), where the peak load corresponding to a loading rate of 0,166 µm/s is about 14000 N. Considering the same conditions and the same constants in the numerical model, we can see that after relaxation the peak load corresponding to a loading rate of 0,3 µm/s is about 11000 N. The 10 days of relaxation have reduced the residual capacity of a three points bend beam.

Same as the rate of loading, the relaxation duration has a great effect on the residual capacity of the beams. In the set of numerical simulations shown in Figure 12, the relaxation started in the pre-peak range of loading at about 70% of the peak load of the mechanical response. The applied loading is a displacement that was held constant for three different periods of time (10 days, 90 days, and 120 days). After the load decreases from its initial value (70% of peak load) to some lower value corresponding to the relaxation end, the applied displacement is increased with a constant rate till failure.

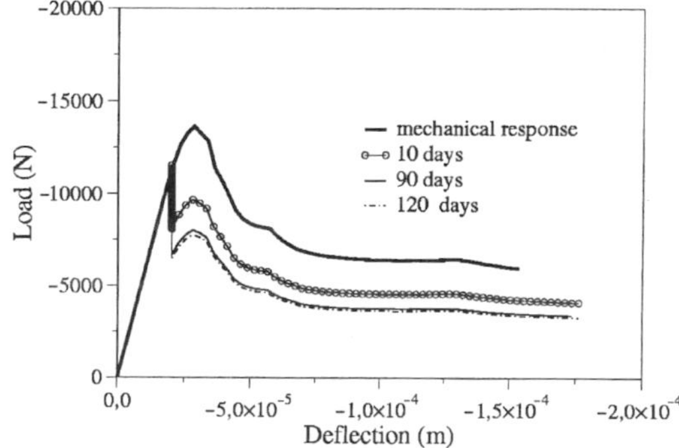

Figure 12. Load-Deflection responses – influence of relaxation duration on the residual capacity.

Figure 12 reveals that the residual capacity of a three points bending beam subjected to relaxation prior to failure decreases as the relaxation duration increases.

7 CONCLUSIONS

Two numerical models for coupling creep and damage in concrete structures have been proposed and implemented in a finite element code. The models parameters are calibrated so that they fit experimental data. The compliance function given in Granger (1995) has been chosen as a basis of identification of visco-elastic parameters. A set of size effect experiments has been carried out on notched specimens in order to calibrate the damage model parameters. A comparison between the predictions of the two damage-creep models was performed. The model based on the Maxwell chain dealing with the effective stress approach yields better results with significant savings in computing time compared to Kelvin chain model. The study was restricted to the use of concrete B11. Some interesting effects of the material internal length on the behaviour of concrete in flexion are underlined. Furthermore, this study has examined the loading rate influence on the capacity of bending beams and also the variation of this capacity due to the relaxation duration.

Results of these numerical simulations performed using the Maxwell chain-damage model allow some concluding remarks about the model:

1. The kinetics of creep strain is greatly influenced by the internal length of the material. The shorter the internal length, the greater the creep strain.
2. The influence of the internal length on the response of a three points bending beam subjected to relaxation prior to failure is comparable to that of a time independent material damage. The residual capacity decreases with l_c.
3. Findings concerning the rate dependent behaviour and the fracture of specimens at different loading rates, show that the model predictions agree with previous studies reported in literature.
4. The residual capacity, of a beam subjected to relaxation for a given period of time and then loaded up to failure, is clearly reduced due to relaxation. Also the duration of relaxation loading has an important effect on this capacity, it decreases when the duration is greater.

It is clear that much additional work should be required before a complete understanding of the interaction between fracture of concrete and its time dependent behaviour. Undergoing experiments will bring more information and will help explaining this interaction.

ACKNOWLEDGMENTS

Financial support from EDF-R&D and from the French ministry of research are gratefully acknowledged.

REFERENCES

Bazant, Z.P. 1974. Rate type creep law of aging concrete based on Maxwell chain. Materials and Structures 7 (37): 45–60.

Bazant, Z.P. 1988. Mathematical modeling of creep and shrinkage of concrete. John Wiley & Sons Ltd.

Bazant, Z.P. & Pijaudier-Cabot, G. 1989. Measurement of the characteristic length of non local continuum. J. Engineering and Mechanics, ASCE 115: 755–767.

Bazant, Z.P. & Jirasek, M. 1992. R-curve modeling of rate effect in static fracture and its interference with size effect, in Fracture Mechanics of Concrete Structures. Proc. Int. Conf. On Fracture Mechanics of Concrete Structures. Breckenridge, Colorado, June, ed. by Bazant Z.P., Elsevier Applied Science, London: 918–923.

De Borst, R., Van Den Boogaard, A.H. & Sluys, L.J. 1993. Computational issues in time-dependent deformation and fracture of concrete. Proc. The 5th Int. RILEM Symp. on Creep and shrinkage of concrete, Barcelona 1993: 309–325.

De Faget De Casteljau, P. 1990. Le lissage. Hermes, Paris: 142 p.

Gettu, R. & Bazant, Z.P. 1992. Rate effects and load relaxation: static fracture of concrete. ACI Materials J. 89 (5): 456–468.

Granger, L. 1995. Comportement différé du béton dans les enceintes de centrales nucléaires: Analyse et modélisation. Ph.D. Thesis ENPC, (in french).

Le Bellégo, C., Dubé, J.F., Pijaudier-Cabot, G. & Gérard, B. 2003. Calibration of non local model from size effect tests. Eur. J. of Mechanics A/Solids, in press.

Mazars, J. 1984. «Application de la mécanique de l'endommagement au comportement non linéaire et à la rupture du béton de structures». Thèse de doctorat d'état, Université Paris VI, (in french).

Mazars, J. & Pijaudier-Cabot, G. 1996. From damage to fracture mechanics and conversely: a combined approach. Int. J. Solids and Structures 33: 3327–3342.

Omar, M. 2000. Couplage déformations différées – fissuration du béton. DEA, Ecole Centrale de Nantes, (in French).

Pijaudier-Cabot, G. & Bazant, Z.P. 1987. Non local damage theory. J. Engineering and Mechanics, ASCE, 113: 1512–1533.

Rilem 1995. Technical committee 107. Creep and shrinkage prediction models: principles of their formulation, guidlines for characterizing concrete creep and shrinkage in structural design codes or recommendations. Materials and Structures 28 (175): 52–55.

Rots, J.G. 1988. Computational modelling of concrete fracture. Dissertation, Delft University of Technology, Delft, the Netherlands.

Rüsch, H. 1960. Research toword a general flexural theory for structural concrete. ACI J. 57 (1): 1–28.

Saouridis, C. 1988. Identification et numérisation objectives des comportements adoucissants: une approche multiéchelle de l'endommagement du béton. Ph.D. thesis Université de Paris 6, (in french).

Computational Modelling of Concrete Structures, Bićanić et al. (eds)
© 2003 Swets & Zeitlinger, Lisse, ISBN 90 5809 536 3

A transient thermal creep model for concrete

C.J. Pearce & N. Bićanić
Department of Civil Engineering, University of Glasgow, Glasgow, UK

C. Nielsen
Building Division, Danish Technological Institute, Denmark

ABSTRACT: A constitutive model for concrete subject to transient temperatures is proposed. A thermal-mechanical strain component is included that is irreversible and only occurs during first heating of concrete under load. The model also includes thermal damage and temperature dependent coefficient of free thermal strain. The proposed model is applied to a number of illustrative examples that demonstrate its predictive capability. The proposed model may be easily incorporated into existing thermal-mechanical analysis frameworks.

1 INTRODUCTION

When concrete is heated to several hundred degrees Celsius, its behaviour is altered significantly, becoming highly non-linear. This is an important issue to include when concrete is to be modelled and analysed under simultaneous loading and heating. The mechanical properties such as strength and modulus of elasticity are generally found to be decreasing with temperature, while high temperatures are found to significantly enhance the time dependent creep characteristics. Furthermore, a constitutive model also has to include the important experimental finding that alterations in strength, stiffness and creep depend on the combined load and temperature history.

The temperature dependent behaviour of concrete was summarised in the finishing document of RILEM technical committee 44-PHT (Schneider 1986 and Schneider 1988), including most of the test data published on concrete subject to high temperatures until the mid eighties. Furthermore, Khoury et al. (1985a) published a comprehensive review on the experimental evidence of transient creep of concrete and Bažant and Kaplan (1996) gave a thorough description of all aspects of concrete at high temperatures. At present RILEM Technical Committee 129-MHT continues the task of standardizing the procedures for high temperature tests on concrete.

Heating of concrete specimens beyond 100 °C under sealed conditions is practically impossible and thus, shrinkage of the cement paste is normally included in the test results (i.e. cannot be isolated). For an unloaded transient high temperature test, the specimen undergoes thermal expansion. Performing the same test under sustained loading the total strains, excluding the elastic response, differ significantly from the unloaded situation. This difference is normally referred to as the "transient creep" or "load induced thermal strain" and it is recognised to be much larger than the basic creep recorded under isothermal conditions. Furthermore, this strain component is found only for first heating and it is considered as a quasi-instantaneous response closely connected with the change in temperature and moisture transport (Schneider 1986, Khoury et al. 1985a, Thelandersson 1987).

Several investigations have dealt with the modelling of transient strains during combined heating and loading. Thelandersson (1982) and Schneider (1988) suggested uniaxial expressions for the transient creep, with the former applying the theory of plasticity and extending the model to 3D. Khennane and Baker (1992, 1993) also adopted the theory of plasticity and based their numerical implementation on the work of de Borst and Peeters (1989). Thelandersson (1987) proposed a model, decomposing the transient creep into viscous creep and thermo-mechanical strain.

Bažant and Chern (1987) extended their original creep model to include the effect of temperature alterations, suggesting that both heating and cooling yield transient creep. Furthermore, the model by Bažant and Chern couples numerical calculations of temperature and moisture conditions, making it rather complex for practical design applications.

Recently Thienel and Rostasy (1996) proposed a general constitutive model of the transient strains, based on their own experimental biaxial results.

The coupling of thermo-mechanical formulations with mass transport has been undertaken by Bazant and Kaplan (1996), Gawin et al. (1999) and Tenchev et al. (2001) amongst others. Such formulations either consider moisture as a single phase or the liquid and gas phases separately. Such an approach enables the build up of gas pressures, due to rapid heating or as a result of sealed conditions, to be captured and therefore enables the effective stress to be realistically predicted. Furthermore, including the effect of mechanical processes, such as cracking, on the heat and mass transport allows phenomena such as spalling to be realistically modelled. A fully coupled hygral-thermal-mechanical approach obviously comes at a price, as the number of variables increases and the coupling becomes more involved.

Whichever modeling approach is adopted, the effect of temperature on the mechanical properties, either through a direct dependence on temperature or indirectly via the effect of dehydration of the concrete, should be included in order to capture the complex response of concrete under combined thermal and mechanical loading.

Here a purely thermal extension to an isothermal model is suggested which captures the observed thermal effects but keeps the model simple enough to be readily included in current analysis frameworks. It is the intention of future work to include the proposed model into an existing hygral-thermal-mechanical formulation; this paper, however, is primarily concerned with the effect of temperature on the constitutive law and in particular the phenomenon of transient thermal creep.

2 STRAIN DECOMPOSITION

The total strain rate is decomposed as follows:

$$\dot{\boldsymbol{\varepsilon}} = \dot{\boldsymbol{\varepsilon}}^{m}(T,\boldsymbol{\sigma}) + \dot{\boldsymbol{\varepsilon}}^{ft}(T) + \dot{\boldsymbol{\varepsilon}}^{c}(T,\boldsymbol{\sigma}) + \dot{\boldsymbol{\varepsilon}}^{tm}(T,\boldsymbol{\sigma}) \tag{1}$$

where the strain rate components are the mechanical strain rate ($\dot{\boldsymbol{\varepsilon}}^{m}$) which can contain an elastic and an inelastic part, the free thermal strain rate ($\dot{\boldsymbol{\varepsilon}}^{ft}$), the creep strain rate and the thermo-mechanical strain rate. The last two terms can be combined to form the load induced thermal strain (abbreviated *lits* by Khoury et al. 1985b). The thermo-mechanical strain component is recognised to appear only during the first heating and not during subsequent cooling or heating cycles (Khoury et al. 1985a). Thus, the thermo-mechanical strain is an irrecoverable strain component that is crucial to the response of a concrete member subject to high temperatures since it may lead to severe tensile stresses during cooling (Nielsen et al. 2001).

One could also add a shrinkage strain component to eq. (1). However, since all experimental high temperature data are reported from unsealed test conditions, the shrinkage component is contained in the thermal strain. Furthermore, shrinkage is assumed to be independent of loading.

In the following sections each of the components of eq. (1) are described and models for the strain rates are presented. For the purely mechanical component, the work of Govindjee et al. (1995) and in particular the work of Meschke et al. (1998) is investigated. The proposed isothermal model is then extended to include temperature dependency of the elastic and inelastic parts based in part on the work of Stabler (2000). The thermal expansion and the load induced thermal strains are accounted for by adopting the models suggested by Nielsen et al. (2002).

3 MECHANICAL STRAIN

The non-linear behaviour of concrete may be modelled using either the theory of plasticity or damage mechanics. Concrete experiences the loss of both strength and stiffness under loading and exhibits irreversible strains as well as hardening/softening phenomena. Initially, an isothermal model will be presented and this will then be extended to include thermal effects.

3.1 *Isothermal model*

Govindjee et al. (1995) presented an anisotropic damage model for concrete based on similar concepts to the theory of plasticity. The principle of maximum dissipation was adopted to define the evolution of damage, and a damage function defined in the stress space is utilised to define the limit of linear elastic behaviour and therefore the onset of damage. Meschke et al. (1998) utilised a broadly similar approach to Govindjee et al., except their model includes both plasticity and damage. Once again, the principle of maximum dissipation is adopted but now both plastic and damage processes evolve and therefore the degradation of the elastic moduli, the development of irreversible strains and hardening/softening effects are all included within one framework.

Ignoring, for the moment, the effect of temperature on the mechanical behaviour, the strain rates may be decomposed into elastic and plastic parts as

$$\dot{\boldsymbol{\varepsilon}} = \dot{\boldsymbol{\varepsilon}}^{e} + \dot{\boldsymbol{\varepsilon}}^{p} \tag{2}$$

and the Helmholtz free energy function is written as

$$\Psi(\boldsymbol{\varepsilon}, \boldsymbol{\varepsilon}^{p}, \boldsymbol{E}, \alpha) = \frac{1}{2}(\boldsymbol{\varepsilon} - \boldsymbol{\varepsilon}^{p}) : \boldsymbol{E} : (\boldsymbol{\varepsilon} - \boldsymbol{\varepsilon}^{p}) + S(\alpha) \tag{3}$$

where $\boldsymbol{\varepsilon}$, $\boldsymbol{\varepsilon}^{p}$, $\boldsymbol{E}$ and α are the state variables. In particular, $\boldsymbol{\varepsilon}$ is the strain tensor, $\boldsymbol{\varepsilon}^{p}$ is the plastic strain tensor and $\boldsymbol{E}$ is a 4th order tensor containing the elastic moduli. The last term of eq. (3) is the contribution from the hardening/softening process to the free energy and thus α is the vector of internal variables controlling hardening/softening. The constitutive relationships are given by:

$$\begin{aligned}\boldsymbol{\sigma} &= \frac{\partial\Psi}{\partial\boldsymbol{\varepsilon}} = \boldsymbol{E}:(\boldsymbol{\varepsilon}-\boldsymbol{\varepsilon}^p) = \boldsymbol{E}:\boldsymbol{\varepsilon}^e \\ \sigma^p &= -\frac{\partial\Psi}{\partial\boldsymbol{\varepsilon}^p} = \boldsymbol{E}:\boldsymbol{\varepsilon}^e = \boldsymbol{\sigma} \\ \boldsymbol{Y} &= -\frac{\partial\Psi}{\partial\boldsymbol{E}} = -\frac{1}{2}\boldsymbol{\varepsilon}^e\otimes\boldsymbol{\varepsilon}^e \\ \boldsymbol{\beta} &= -\frac{\partial\Psi}{\partial\boldsymbol{\alpha}} = -\frac{\partial S}{\partial\alpha}\end{aligned} \tag{4}$$

where $\boldsymbol{\sigma}$ is the Cauchy stress tensor, $\boldsymbol{\sigma}^p$ is the plastic stress tensor (which is equal to the Cauchy stress tensor), $\boldsymbol{Y}$ is the so-called damage energy release rate and $\boldsymbol{\beta}$ is the thermodynamic force conjugate to $\boldsymbol{\alpha}$ (the exact meaning of which will become apparent). The purely mechanical part of the dissipation rate, $\mathcal{D}_{\text{mech}}$, can be written as

$$\begin{aligned}\mathcal{D}_{\text{mech}} &= \boldsymbol{\sigma}:\dot{\boldsymbol{\varepsilon}} - \dot{\psi} = \boldsymbol{\sigma}:\dot{\boldsymbol{\varepsilon}} \\ &\quad - \left(\frac{\partial\psi}{\partial\boldsymbol{\varepsilon}}\dot{\boldsymbol{\varepsilon}} + \frac{\partial\psi}{\partial\boldsymbol{\varepsilon}^p}\dot{\boldsymbol{\varepsilon}}^p + \frac{\partial\psi}{\partial\boldsymbol{E}}\dot{\boldsymbol{E}} + \frac{\partial\psi}{\partial\boldsymbol{\alpha}}\dot{\boldsymbol{\alpha}}\right) \geq 0\end{aligned} \tag{5}$$

Utilising eq. (4), eq. (5) can be rewritten as

$$\begin{aligned}\mathcal{D}_{\text{mech}} &= \boldsymbol{\sigma}:\dot{\boldsymbol{\varepsilon}}^p + \boldsymbol{Y}:\dot{\boldsymbol{E}} + \beta\cdot\dot{\alpha} \\ &= \boldsymbol{\sigma}:\dot{\boldsymbol{\varepsilon}}^p - \boldsymbol{\varepsilon}^e:\dot{\boldsymbol{E}}:\boldsymbol{\varepsilon}^e + \beta\cdot\dot{\alpha} \geq 0\end{aligned} \tag{6}$$

Denoting the fourth order compliance tensor as $\boldsymbol{C}$, recognising that $(\boldsymbol{C}:\boldsymbol{E}) = \boldsymbol{0}$ and recalling that $\boldsymbol{\sigma} = \boldsymbol{E}:\boldsymbol{\varepsilon}^e$, eq. (6) may be rewritten as

$$\mathcal{D}_{\text{mech}} = \boldsymbol{\sigma}:\dot{\boldsymbol{\varepsilon}}^p + \boldsymbol{\sigma}:\dot{C}:\boldsymbol{\sigma} + \beta\cdot\dot{\alpha} \geq 0 \tag{7}$$

The onset of plasticity and damage is defined by a set of n damage/yield surfaces such that admissible stress states are only those that satisfy the following inequality:

$$f_k(\boldsymbol{\sigma},\beta_k) \leq 0 \quad (k = 1,\ldots,n) \tag{8}$$

where β_k is a stress like variable (conjugate to α_k) that defines the current position of the damage/yield surfaces.

The Helmholtz free energy equation (eq. (3)) led to the constitutive relations in eq. (4); however, these equations by themselves are insufficient to fully characterise the non-linear material behaviour. The evolution of the internal variables must also be derived and this achieved by utilising the principle of maximum entropy production, which for purely mechanical processes is equivalent to maximising dissipation. Therefore the problem is one of finding values of $\boldsymbol{\sigma}$ and β_k which maximises the expression in eq. (7) but which also satisfy the constraints of eq. (8). This can be expressed mathematically by recasting the constrained optimisation problem to one of an unconstrained problem by constructing the associated Lagrangian:

$$\begin{aligned}\mathcal{L}(\boldsymbol{\sigma},\boldsymbol{\beta}) &= -\mathcal{D}_{\text{mech}} + \sum_{k=1}^{n}\lambda_k\phi_k \\ &= -\boldsymbol{\sigma}:\dot{\boldsymbol{\sigma}}^p - \frac{1}{2}\boldsymbol{\sigma}:\dot{\boldsymbol{C}}:\boldsymbol{\sigma} - \beta:\dot{\alpha} + \sum_{k=1}^{n}\lambda_k f_k\end{aligned} \tag{9}$$

where λ_k are the Lagrange multipliers. The optimal solution is obtained by finding the stationary value of $\mathcal{L}$ with respect to $\boldsymbol{\sigma}$ and β_k:

$$\begin{aligned}\frac{\partial\mathcal{L}}{\partial\boldsymbol{\sigma}} &= 0 \Rightarrow \dot{\boldsymbol{\varepsilon}}^p + \dot{\boldsymbol{C}}:\boldsymbol{\sigma} = \sum_{i=1}^{n}\lambda_k\frac{\partial f_k}{\partial\boldsymbol{\sigma}} \\ \frac{\partial\mathcal{L}}{\partial\beta_k} &= 0 \Rightarrow \dot{\alpha}_k = \sum_{k=1}^{n}\lambda_k\frac{\partial f_k}{\partial\beta_k}\end{aligned} \tag{10}$$

The loading/unloading conditions are defined as:

$$f_k \leq 0, \quad \lambda_k \geq 0 \quad \text{and} \quad \sum_{k=1}^{n}\lambda_k f_k = 0 \tag{11}$$

Eq. (10) represents the evolution laws for $\boldsymbol{\varepsilon}^p$, $\boldsymbol{C}$ & $\boldsymbol{\alpha}$. If the discussion is briefly restricted to elastoplasticity, such that there is no damage and therefore no change in $\boldsymbol{E}$ or $\boldsymbol{C}$ (i.e. $\dot{\boldsymbol{C}} = 0$), eq. (10)[1] reduces to the classical evolution law for associated plasticity:

$$\dot{\boldsymbol{\varepsilon}}^p = \sum_{k=1}^{n}\lambda_k\frac{\partial f_k}{\partial\boldsymbol{\sigma}} \tag{12}$$

On the other hand, if one considers that there is only damage evolution present and no plasticity, eq. (10)[1] becomes:

$$\dot{\boldsymbol{C}}:\boldsymbol{\sigma} = \sum_{k=1}^{n}\lambda_k\frac{\partial f_k}{\partial\boldsymbol{\sigma}} \tag{13}$$

where eq. (13) represents the evolution equation for pure damage and was the approach taken by Govindjee et al. (1995). Meschke et al. (1998), on the other hand, start from eq. (10)[1] and arbitrarily split the plastic and "damage" strain rates by introducing a scalar parameter, η, where $0 \leq \eta \leq 1$:

$$\begin{aligned} \dot{\boldsymbol{\varepsilon}}^p &= (1-\eta)\sum_{k=1}^{n}\lambda_k \frac{\partial f_k}{\partial \boldsymbol{\sigma}} \\ \dot{\boldsymbol{C}} : \boldsymbol{\sigma} &= \eta \sum_{k=1}^{n}\lambda_k \frac{\partial f_k}{\partial \boldsymbol{\sigma}} \end{aligned} \tag{14}$$

It is clear from eq. (14) that setting $\eta = 1$ represents the case where pure damage and no irreversible strains will occur; on the other hand, setting $\eta = 0$ will result in plastic strains only and no damage. Meschke et al. (1998) illustrated that a value of $\eta = 0.4$ showed reasonable agreement with experimental evidence. This simple but effective method of separating the plastic and "damage" strains is the one adopted here.

The rate of the compliance tensor is determined by pre-multiplying eq. (14)[2] by $\boldsymbol{\sigma}$, which leads to:

$$\begin{aligned} \boldsymbol{\sigma} : \dot{\boldsymbol{C}} : \boldsymbol{\sigma} &= \eta \sum_{i=1}^{n}\lambda_i \left(\boldsymbol{\sigma} : \frac{\partial f_i}{\partial \boldsymbol{\sigma}}\right) \\ &= \eta \sum_{i=1}^{n} \frac{\left(\boldsymbol{\sigma} : \frac{\partial f_i}{\partial \boldsymbol{\sigma}}\right)\left(\frac{\partial f_i}{\partial \boldsymbol{\sigma}} : \boldsymbol{\sigma}\right)}{\left(\frac{\partial f_i}{\partial \boldsymbol{\sigma}} : \boldsymbol{\sigma}\right)} \end{aligned} \tag{15}$$

Therefore,

$$\dot{\boldsymbol{C}} = \eta \sum_{i=1}^{n}\lambda_i \frac{\left(\frac{\partial f_i}{\partial \boldsymbol{\sigma}} \otimes \frac{\partial f_i}{\partial \boldsymbol{\sigma}}\right)}{\left(\frac{\partial f_i}{\partial \boldsymbol{\sigma}} : \boldsymbol{\sigma}\right)} \tag{16}$$

This equation illustrates that the change in the compliance tensor $\boldsymbol{C}$ is dependent on the damage/yield function gradient and therefore will result in an anisotropic update of the compliance tensor.

The damage/yield functions defining the space of admissible stress states have so far been left unspecified and will not be discussed in great detail here. In this work, the Rankine or maximum principal stress criteria will be adopted for the tensile stress regime and a Drucker-Prager criteria will be adopted for the compressive stress regime. The Rankine criteria will be discussed briefly here and is defined as:

$$\sigma_1 - f_t^T + \beta_1(\alpha_1) \leq 0 \tag{17}$$

where $\boldsymbol{\sigma}_1$ is the major principal stress and f_t^T is the tensile strength of concrete at temperature T. For the moment it is assumed that T is constant. The reduction of strength observed in concrete after a crack has been initiated will be modelled here using a fracture energy controlled softening law. The use of this material property coupled with a length scale to control the softening law is a very common approach to solve the problem of mesh dependency that is inherent in the analysis of materials which exhibit softening behaviour. This is also the technique adopted by Meschke et al. and investigated by Govindjee et al. A hyperbolic softening law of the form:

$$\beta_1 = f_t^T \left(1 - \frac{1}{(1+\alpha_1/\alpha_u)^2}\right) \tag{18}$$

is suggested by Meschke et al.

The energy dissipated per unit volume, in the crack zone, can be determined from the following integration:

$$\begin{aligned} \frac{G_f^T}{l_c} &= \int_0^\infty (f_t^T - \beta_1)d\alpha_1 \\ &= \int_0^\infty \frac{f_t^T}{(1+\alpha_1/\alpha_u)^2}\, d\alpha_1 = f_t^T \alpha_u \end{aligned} \tag{19}$$

where G_f^T is the fracture energy per unit area for a given temperature T and l_c is the characteristic length associated with the crack zone. Therefore

$$\alpha_u = \frac{G_f^T}{f_t^T l_c} \tag{20}$$

Full algorithmic details of this coupled plasticity/damage model can be found in Meschke et al. (1998).

3.2 *Thermo-mechanical model*

Experiments of both "hot" concrete and "cold" concrete (i.e. concrete cooled after exposure to high temperatures) shows that there is a general reduction of the elastic modulus (Schneider 1988, Bazant and Kaplan 1996). Here, the thermal damage that results in a loss of stiffness will be modelled via a simple isotropic thermal damage model and follows the work of Stabler (2000). A scalar thermal damage variable ω is introduced into the formulation so that the Helmholtz free energy takes the following form:

$$\Psi(\boldsymbol{\varepsilon}, \boldsymbol{\varepsilon}^p, \boldsymbol{E}, \omega, \boldsymbol{\alpha}) = \frac{1}{2}(1-\omega)\boldsymbol{\varepsilon}^e : \boldsymbol{E} : \boldsymbol{\varepsilon}^e + S(\boldsymbol{\alpha}) \tag{21}$$

and the thermodynamic properties become:

$$\begin{aligned} \boldsymbol{\sigma} &= \frac{\partial \Psi}{\partial \boldsymbol{\varepsilon}} = (1-\omega)\boldsymbol{E} : (\boldsymbol{\varepsilon} - \boldsymbol{\varepsilon}^p) = (1-\omega)\boldsymbol{E} : \boldsymbol{\varepsilon}^e \\ \boldsymbol{\sigma}^p &= -\frac{\partial \Psi}{\partial \boldsymbol{\varepsilon}^p} = (1-\omega)\boldsymbol{E} : \boldsymbol{\varepsilon}^e = \boldsymbol{\sigma} \\ \boldsymbol{Y} &= -\frac{\partial \Psi}{\partial \boldsymbol{E}} = -\frac{1}{2}(1-\omega)\boldsymbol{\varepsilon}^e \otimes \boldsymbol{\varepsilon}^e \\ \boldsymbol{Y}_\omega &= -\frac{\partial \Psi}{\partial \omega} = \frac{1}{2}\boldsymbol{\varepsilon}^e : \boldsymbol{E} : \boldsymbol{\varepsilon}^e \\ \beta &= -\frac{\partial \Psi}{\partial \alpha} = -\frac{\partial S}{\partial \alpha} \end{aligned} \tag{22}$$

It can be seen from eq. (22)[1] that the constitutive law includes the scalar damage parameter, ω, where $0 \leq \omega \leq 1$. Such a formulation allows for damage to occur as a result of both thermal and mechanical loading. It is important to note that in general $\boldsymbol{E}$ is not the original tensor of elastic moduli as it can be affected by purely mechanical damage and that the term $(1-\omega)\boldsymbol{E}$ represents the degraded elastic moduli due to both thermal and mechanical damage. Furthermore, it should be noted that, $Y_\omega \dot{\omega}$ is considered a part of the mechanical dissipation, as it is a result of the loss of mechanical stiffness despite the fact that it is a result of exposure to elevated temperatures. Thus the mechanical dissipation becomes:

$$\mathcal{D}_{\text{mech}} = \boldsymbol{\sigma} : \dot{\boldsymbol{\varepsilon}}^p + \boldsymbol{\sigma} : \dot{\boldsymbol{C}} : \boldsymbol{\sigma} + Y_\omega \dot{\omega} + \beta \cdot \dot{\alpha} \geq 0 \quad (23)$$

where Y_ω is termed the thermal damage energy release rate.

In order to capture the experimentally observed direct dependence of the thermal damage on temperature an empirical approach is adopted such that the thermal damage function (Stabler 2000) is given by:

$$g = T - T_0 + k_T(\omega) \quad (24)$$

where T is the absolute temperature, T_0 is the reference temperature and k_T is a hardening parameter which controls the extent of the thermal damage surface. Initially, k_T is set to zero which then grows as thermal damage occurs. The loading/unloading conditions are defined as:

$$g \leq 0, \quad \dot{\omega} \geq 0 \quad \text{and} \quad \dot{\omega} g = 0 \quad (25)$$

The thermal damage parameter ω is now defined in an integral sense as a function of the temperature and will reflect the variation of the elastic modulus with temperature. Here a simple function is defined:

$$\begin{aligned} \omega &= 0.2\,\theta - 0.01\,\theta^2 && \text{for } 0 \leq \theta \leq 10 \\ \omega &= 0.0 && \text{for } 0 \leq \theta \leq 10 \end{aligned} \quad (26)$$

where θ represents a dimensionless temperature defined by:

$$\theta = \frac{T - T_0}{100°\text{C}} \quad (27)$$

The thermally damaged elastic modulus according to eq. (26) is shown in Figure 1.

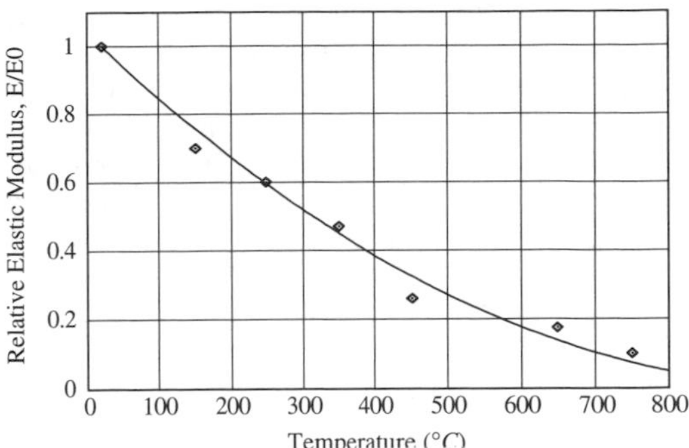

Figure 1. Relative elastic modulus vs. temperature. Current representations compared to experimental evidence (Schneider 1988).

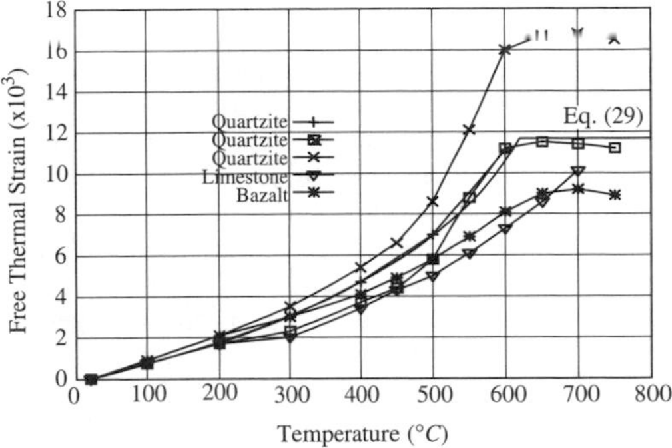

Figure 2. Experimental free thermal strain vs. temperature for various concretes (Schneider 1986) and proposed free thermal strain representation.

4 FREE THERMAL STRAINS

The free thermal strain of concrete is mainly influenced by the type and amount of aggregate. Furthermore, it is highly non-linear (Figure 2), depending on the thermal stability of the aggregate. It is assumed that $\boldsymbol{\varepsilon}^{ft}$ is a function of temperature only although it is recognised to depend on the heating rate as well. Hence, in case of high heating rates the chemical reactions and the moisture transport processes within the cement paste may not develop fully, affecting the recorded thermal strain.

The isotropic thermal strain is defined here as

$$\dot{\boldsymbol{\varepsilon}}^{ft}_{ij} = \alpha \dot{T} \delta_{ij} \quad (28)$$

where α is the coefficient of free thermal strain.

As suggested by Nielsen et al. (2001) the temperature dependent coefficient α for quartzite normal concrete is expressed by the following function

$$\begin{aligned} \alpha &= \frac{6 \times 10^{-5}}{7 - \theta} && \text{for } 0 \leq \theta \leq 6 \\ \alpha &= 0 && \text{for } \theta > 6 \end{aligned} \quad (29)$$

In Figure 2, this expression is compared with test data taken from Schneider (1986). Note that significant experimental variation in the thermal strain response exists, and therefore eq. (29) is modelling a correct trend rather than fitting specific experimental values. It is seen how the experimental thermal strains level off when the temperature exceeds 600°C. Bažant and Kaplan (1996) suggested that this behaviour is particularly associated with quartzite aggregate, showing a significant increase in the thermal strain between 500°C and 600°C (due to the conversion of quartz crystals) where after the strain-temperature curve flattens out.

5 LOAD INDUCED THERMAL STRAIN

The load induced thermal strain, corresponding to the last two terms of eq. (1), is determined experimentally by measuring total strains during first time heating on a concrete specimen under sustained loading. The free thermal strain, recorded on an unstressed specimen, and the initial elastic strain are subtracted from the total strain to give the load induced thermal strain (or transient creep) as a function of temperature.

In the rheological model proposed by Thelandersson (1987) the uniaxial load induced thermal strain rate included a dashpot and a thermo-mechanical element corresponding to temperature dependent creep ($\dot{\boldsymbol{\varepsilon}}^{c}$) and thermo-mechanical strain ($\dot{\boldsymbol{\varepsilon}}^{tm}$) respectively. It is recognised that these two components cannot be separated experimentally and thus the subdivision in two components is mainly theoretical and has little practical implications for short duration heating as stated by Schneider (1988). However, one reason for the subdivision is that the two components have different characteristics. For instance, the creep component takes place under both heating and cooling whereas thermo-mechanical strain is only experienced during first heating and not during subsequent cooling or reheating.

Thelandersson (1987) stipulated that the creep strain is minor compared with the thermo-mechanical strain for practical test conditions even though steady-state creep is found to be enhanced by high temperatures. However, if the temperature increase is sufficiently slow the two components may be of comparable magnitude. Nielson et al. (2002) illustrated, by means of a small analytical example, that even for a very slow heating rate of 0.05°C/min the creep strain corresponds to only 15–20% of the thermo-mechanical strain. This value drops significantly for increased heating rates. As practical experimental values for the heating rate are typically above 1°C/min (Schneider (1986) reported 2°C/min), it seems plausible to assume that the creep component of the load induced thermal strain is neglected when interpreting the test results.

In the constitutive model presented here the creep component is entirely neglected. However, in the case of long-term scenarios with loading and heating, the temperature dependent creep strain should be included in the calculations.

Schneider (1986) reported several test series of load induced thermal strain of concrete tested under sustained uniaxial compressive load and constant heating rate. The test series had been carried out at various laboratories throughout the 1970's, see Schneider (1986). During the 1980's extensive experimental research took place at Imperial College in London. The tests included creep under both transient conditions (Khoury et al. 1985b) and isothermal conditions (Khoury et al. 1986). Moreover, Khoury et al. used the term basic creep for the isothermal (steady-state) experiments and load induced thermal strain (*lits*) for transient temperature experiments. Khoury et al. (1985b) suggested that a *lits* master curve exists, largely independent of aggregate type, concrete strength, age and initial moisture content. Recently, Terro (1998) applied a fourth order polynomial to fit this master curve based on the Imperial College data.

The general formulation of the thermo-mechanical strain rate by Thelandersson (1987) reads

$$\dot{\boldsymbol{\varepsilon}}_{ij}^{tm} = (\gamma_1 \boldsymbol{\sigma}_{kk}\delta_{ij} + \gamma_2 \boldsymbol{\sigma}_{ij})\dot{T} \tag{30}$$

where the two coefficients γ_1 and γ_2 need to be determined experimentally. For the unixaxial case, Thelandersson assumed proportionality between $\boldsymbol{\varepsilon}^{tm}$ and $\boldsymbol{\varepsilon}^{ft}$, an assumption also adopted by others due to its simplicity (de Borst and Peeters 1989, Khennane and Baker 1993).

Thelandersson also expressed Eq. (30) as:

$$\dot{\boldsymbol{\varepsilon}}_{ij}^{tm} = (\gamma_v \boldsymbol{\sigma}_m \delta_{ij} + \gamma_2 \boldsymbol{\sigma}'_{ij})\dot{T} \tag{31}$$

where $\boldsymbol{\sigma}_m = \boldsymbol{\sigma}_{kk}/3$ is the hydrostatic stress, $\boldsymbol{\sigma}'_{ij} = \boldsymbol{\sigma}_{ij} - \boldsymbol{\sigma}_m\delta_{ij}$ is the deviatoric stress tensor and $\gamma_v = 3\gamma_1 + \gamma_2$. Thus, the first term in the bracket (eq. (31)) represents the volumetric part of the thermo-mechanical strain and the second term represents the deviatoric part. A still clearer physical understanding is achieved by applying the following substitutions, similar to those suggested by de Borst and Peeters (1989):

$$\gamma_1 = -\nu_c \frac{\beta}{f_c^{T_0}}, \quad \gamma_2 = (1 + \nu_c)\frac{\beta}{f_c^{T_0}} \tag{32}$$

where ν_c is the transient creep Poisson's ratio and β is the coefficient of uniaxial thermo-mechanical strain. Inserting eq. (32) into eq. (30) yields

$$\dot{\boldsymbol{\varepsilon}}_{ij}^{tm} = \frac{\beta}{f_c^{T_0}}((1 + \nu_c)\boldsymbol{\sigma}_{ij} - \nu_c \boldsymbol{\sigma}_{kk}\delta_{ij})\dot{T} \tag{33}$$

A different assumption is adopted here, i.e. the bi-parabolic thermo-mechanical strain model was presented by Nielsen et al. (2002) of the form:

$$\varepsilon^{tm} = \frac{\sigma}{f_c^{T_0}} y$$

$$y = \begin{cases} A\theta^2 + B\theta; & \text{for } 0 \leq \theta \leq \theta^* = 4.5 \\ C(\theta - \theta^*)^2 + A(2\theta - \theta^*) + B\theta; & \text{for } \theta > \theta^* \end{cases} \quad (34)$$

where θ^* is a dimensionless transition temperature between the two expressions (470°C). The reason for applying a combination of two parabolic expressions instead of following Terro (1998), who applied a 4th order polynomial expression, is to capture the rather abrupt change in behaviour detected experimentally detected around the transition temperature. Furthermore, the bi-parabolic model keeps the formulation simple with A and B governing the initial part and C the subsequent part of the $\varepsilon^{tm} - \theta$ relationship.

The simplicity of the model is also maintained by applying a direct proportionality with the stress level unlike Schneider's model where ε^{tm} is a non-linear function of stress. A direct proportionality with stress is plausible (Khoury et al. 1985b, Thienel 1993).

It is seen that the parabolic model captures the experimental behaviour in a satisfactory manner, Figure 3. The two sets of parameters (A, B, C) applied in Figure 3 describe an upper (Set 2) and lower (Set 1) curve. There is relatively good correspondence with the master curve by Khoury et al. (1985b) and Terro (1998). The two limiting parabolic curves in Figure 3 define a range where the thermo-mechanical strain is likely to be found for normal concretes with various aggregate types.

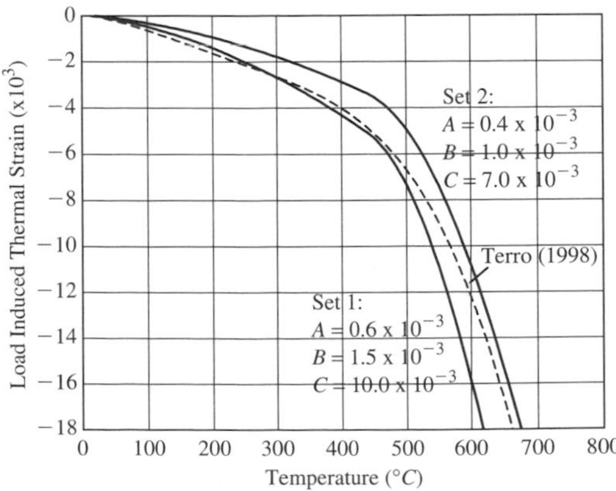

Figure 3. Load induced thermal strain. Parabolic model (Nielsen et al. 2002) compared to Terro (1998) and experimental data for quartzite concrete taken from Schneider (1986).

The coefficient of uniaxial thermo-mechanical strain, β, can be determined from eq. (34) as:

$$\dot{\varepsilon}^{tm} = \frac{\sigma}{f_c^{T_0}} \frac{dy}{d\theta} \frac{d\theta}{dT} \dot{T} = \frac{\sigma}{f_c^{T_0}} \beta \dot{T} \quad (35)$$

where

$$\beta = 0.01 \begin{cases} 2A\theta + B; & \text{for } 0 \leq \theta \leq \theta^* \\ 2C(\theta - \theta^*) + 2A\theta^* + B; & \text{for } \theta > \theta^* \end{cases} \quad (36)$$

It is noted that eq. (35) does not include a dependence on the stress variation ($\dot{\sigma}$). This reflects the fact that, under a constant temperature ($\dot{T} = 0$), a variation of stress would not result in a change in the thermo-mechanical strain.

All experimental tests for load induced thermal strains are performed on concrete specimens under sustained compressive loading and then subject to transient heating. The authors are currently unaware of any similar tests performed under sustained tensile loading and therefore it is not possible to determine whether β adopted for compressive stress conditions could be used for tensile stress conditions as well. Thus eq. (33) is rewritten as:

$$\dot{\varepsilon}_{ij}^{tm} = \frac{\beta^-}{f_c^{T_0}}((1 + \nu_c)\sigma_{ij}^- - \nu_c \sigma_{kk}^- \delta_{ij})\dot{T} + \frac{\beta^+}{f_t^{T_0}}((1 + \nu_c)\sigma_{ij}^+ - \nu_c \sigma_{kk}^+ \delta_{ij})\dot{T} \quad (37)$$

where β^- and β^+ are the coefficients of compressive and tensile uniaxial thermo-mechanical strain respectively. σ^- and σ^+ are the negative and positive projections of the stress tensor. At present, it seems reasonable to set β^+ to zero until more experimental evidence is made available on the load induced thermal strain due to tensile loading is available. Therefore eq. (37) becomes:

$$\dot{\varepsilon}_{ij}^{tm} = \frac{\beta^-}{f_c^{T_0}}((1 + \nu_c)\sigma_{ij}^- - \nu_c \sigma_{kk}^- \delta_{ij})\dot{T}\,\mathrm{sgn}(\dot{\omega}) \quad (38)$$

where $\mathrm{sgn}(\dot{\omega})$ has been introduced to recognise the irreversible nature of thermo-mechanical strains.

Eq. (38) illustrates that ν_c is analogous to Poisson's ratio. To the authors' knowledge the bulk of experiments on the load induced thermal strain do not include lateral deformation measurements so that not much information is available on ν_c. However, the experiments of Ehm (1986) showed that load induced thermal strains have a lateral component similar to that observed in conventional creep tests.

6 NUMERICAL RESULTS

First, transient test data reported by Thelandersson (1987) are reproduced using the proposed model. The tests consisted of concrete cylinders subject to various sustained stress levels and constant heating rate. The results are shown in Figure 4 and it is noted that the proposed model for free thermal strain fits the experimental data very well. As the stress level increases, the predictions by the proposed model fit the experimental results better, but in all cases the general trend observed by the experiments is captured very well.

Figure 5 illustrates, qualitatively, the effect of transient thermal strains on a simply supported beam under point load. Two separate scenarios are included: first the beam is heated (without load), loaded, cooled and then unloaded; second, the beam is loaded, heated, cooled and then unloaded. The first scenario illustrates the effect of thermal damage on the stiffness of the beam by comparing this response to the initial response of the second scenario. The second scenario is analysed with and without the inclusion of the thermal-mechanical strain term. The result of including transient thermal creep is to cause a significant increase in the deformation of the beam upon loading, a large proportion of which is not recovered upon cooling and unloading.

Figure 6 illustrates the response of a uniaxial concrete member subject to a constant rate of heating and fully restrained against expansion. Curves 1 and 2, in Figure 6, depict the influence of the limiting parabolic curves. Curve 3 uses the same material properties as Curve 2 but with a prestress applied. Curve 4 illustrates how the restraining stress would develop if the thermal-mechanical strain effect were ignored. Curve 5 is identical to Curve 2 except that the member is cooled down from 480°C. These are compared to experimental findings (Curves 6 & 7). The first vertical line represents the transition temperature on the bi-parabolic definition of the uniaxial coefficient of thermal-mechanical strain and the second vertical line represents the temperature at which the free thermal strains cease to increase.

It is noted that the applied model parameters are not fitted to match the test results. However, it is clear that the proposed model qualitatively captures

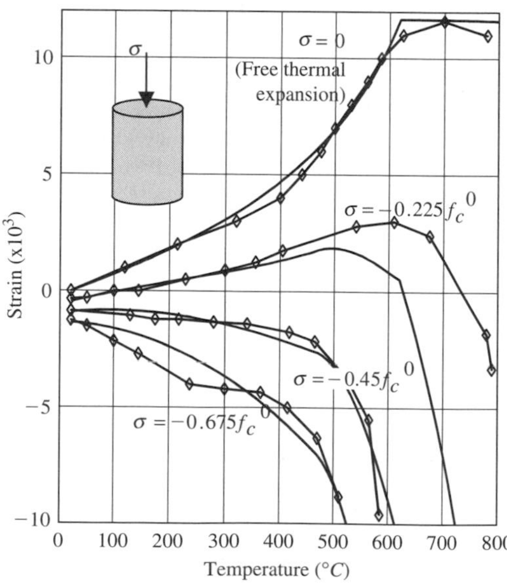

Figure 4. Total strains versus temperature: numerical predictions compared to experimental results from Thelandersson (1987).

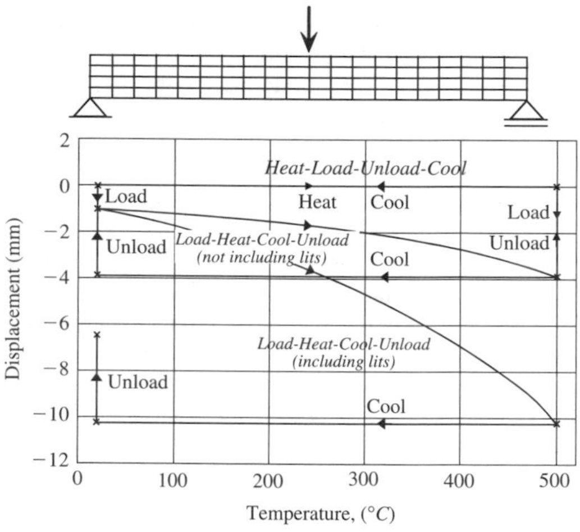

Figure 5. Qualitative analysis of simply supported beam subject to different load and heat scenarios: (a) heat-load-cool-unload; (b) load-heat-cool-unload.

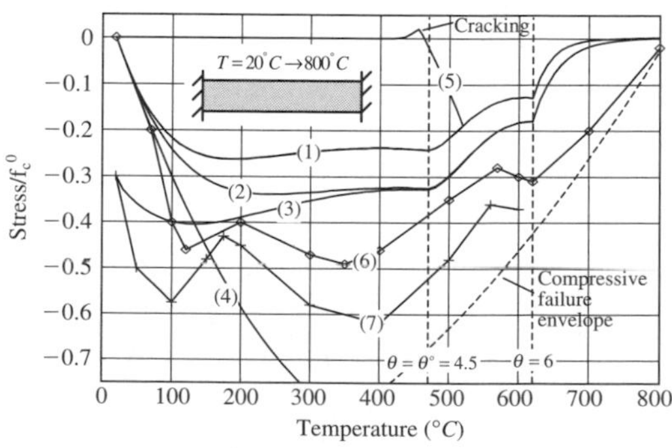

Figure 6. Load induced thermal strain. Numerical predictions compared to experimental results. (1) Set 1, Figure 3; (2) Set 2, Figure 3; (3) Set 2 plus initial prestress $\sigma = -0.3 f_c^0$; (4) No thermo-mechanical strains; (5) Set 2 plus cooling from 480°C; (6) & (7) Experimental results, Ehm (1986) and Schneider (1986). ($E_0/f_c^0 = 500$).

the experimental behaviour, even though the experimental stress peaks are more pronounced and shifted along the temperature axis compared with the proposed model. Especially for lower temperatures, the model does not predict the build-up of stresses quickly enough, indicating that α is too small relative to β.

It is clear to see from Figure 6 the effect that transient creep has on the response of the concrete member, whereby the stresses are significantly relaxed as compared to the response obtained when the thermal-mechanical strains were omitted. Moreoever, Curve 5 illustrates the effect that transient thermal creep can have on such restrained structural concrete members; on cooling the thermal-mechanical strains are not recovered and therefore there is significant potential for cracking to occur. The observed response for Curve 5 was obtained by localizing strain in one finite element; a Rankine yield criterion was adopted for the tensile response.

7 CONCLUSIONS

A simple constitutive model for concrete behaviour under transient temperatures has been proposed based upon a rheological model suggested by Thelandersson (1987). The total strain is decomposed into pure mechanical strain, free thermal strain and load induced thermal strain. The last term is often termed transient creep and includes temperature dependent conventional creep and thermal-mechanical strain; here, conventional creep is ignored.

The thermal-mechanical strain is irreversible and occurs only during first heating due to the combination of stress and temperature increase. During heating these strains relax the stresses arising from thermal gradients and incompatibilities between the aggregate and the cement paste. Significantly, however, the irreversibility of the thermal-mechanical strain leads to the build up of tensile restraining stresses during cooling.

The model introduces a parabolic expression to describe the uniaxial thermal-mechanical response, whereby β^- is a bi-linear function of temperature. Furthermore, a transient creep Poisson's ratio is introduced to describe the corresponding lateral thermal-mechanical response.

Loss of strength and stiffness due to mechanical processes is included via a coupled plasticity/damage formulation (Meschke et al. 1998) and thermal damage is included via the introduction of thermal damage parameter (Stabler 2000).

Finally, it is realized that more experimental data is needed both under uniaxial conditions, including recordings of axial as well as lateral strains, and especially under multiaxial conditions. Also the thermal-mechanical response in tension is still investigated experimentally.

ACKNOWLEDGEMENTS

The third author was employed as a Marie Curie Fellow funded by the European Commission under FP5, the Fifth Framework Programme (contract no. HPMF-CT-1999-0171).

REFERENCES

Bazant, Z.P. & Chern, J.C. 1987. Stress-induced thermal and shrinkage strains in concrete. *J. Engng. Mech.* 113(10): 1493–1511.

Bazant, Z.P. & Kaplan, M.F. 1996. *Concrete at high temperatures, Material properties and mathematical models.* Essex, England: Longman Group Ltd.

De Borst, R. & Peeters, P.P.J.M. 1989. Analysis of concrete structures under thermal loading. *Comp. Meth. Appl. Mech. Engng.* 77: 293–310.

Ehm, C. 1986. *Versuche zur festigkeit und verformung von beton unter zweiaxialer beanspruchung und hohen temperaturen.* PhD thesis, Heft 71, IBMB, Technical University of Braunschweig, Germany.

Gawin, D., Majorana, C.E. & Schrefler, B.A. 1999. Numerical analysis of hygro-thermal behaviour and damage of concrete at high temperature. *Mech. Cohes.-Frict. Mater.* 4: 37–74.

Govindjee, S., Kay, G.J. & Simo, J.C. 1995. Anisotropic modelling of and numerical simulation of brittle damage in concrete. *Int. J. Numer. Meth. Engng.* 38: 3611–3633.

Khennane, A. & Baker, G. 1992. Thermoplasticity model for concrete under transient temperature and biaxial stress. *Proceedings of Royal Society* 439: 59–80.

Khennane, A. & Baker, G. 1993. Uniaxial model for concrete under variable temperature and stress. *J. Engng. Mech.* 119(8): 1507–1525.

Khoury, G.A., Grainger, B.N. & Sullivan, P.J.E. 1985a. Transient thermal strain of concrete: literature review, conditions within specimens and behaviour of individual constituents. *Mag. of Concr. Res.* 37(132): 131–144.

Khoury, G.A., Grainger, B.N. & Sullivan, P.J.E. 1985b. Strain of concrete during first heating to 600°*C* under load. *Mag. of Concr. Res.* 37(133): 195–215.

Meschke, G., Lackner, R. & Mang, H.A. 1998. An anisotropic elasto-plastic damage model for plain concrete. *Int. J. Numer. Meth. Engng.* 42:703–727.

Nielsen, C.V., Pearce, C.J. & Bicanic, N. 2001. Theoretical model of high temperature effects on uniaxial concrete member under elastic restraint. *Mag. Concr. Res.* 54(4): 239–249.

Nielsen, C.V., Pearce, C.J. & Bicanic, N., 2002. Constitutive model of transient thermal strains for concrete at high temperatures. Submitted to *J. Engng. Mech.*

RILEM TC 129-MHT 1998. Part 7: Transient creep for service and accident conditions. *Mater. Struct, RILEM* 31: 290–295.

Schneider, U. 1986. *Properties of materials at high temperatures, concrete, 2nd Edition*, RILEM Technical Committee 44-PHT, Technical University of Kassel, Kassel, Germany.

Schneider, U. 1988. Concrete at high temperatures – a general review. *Fire Safety Journal* 13: 55–68.

Stabler, J. 2000. *Computational Modelling of thermo-mechanical damage and plasticity in concrete*. PhD thesis, The University of Queensland, Australia.

Tenchev, R., Li, L.Y. & Purkiss, J.A. 2001. Finite element analysis of coupled heat and moisture transfer in concrete subjected to fire. *Numerical Heat Transfer, Part A* 39: 685–710.

Terro, M.J. 1998. Numerical modeling of the behaviour of concrete structures in fire. *ACI Struct. J.* 95(2): 183–193.

Thelandersson, S. 1982. On the multiaxial behaviour of concrete exposed to high temperature. *Nucl. Engng. and Design* 75: 271–282.

Thelandersson, S. 1987. Modelling of combined thermal and mechanical action in concrete. *J. Engng. Mech.* 113(6): 893–906.

Thienel, K.-C. 1993. *Festigkeit und verformung von Beton bei hoher temperatur und biaxialer beanspruchung – versuche und modellbildung*. PhD thesis, Heft 104, IBMB, Technical University of Braunschweig, Germany.

Thienel, K.-C. & Rostasy, F.S. 1996. Transient creep of concrete under biaxial stress and high temperature. *Cem. Concr. Res.* 26(9): 1409–1422.

Computational Modelling of Concrete Structures, Bićanić et al. (eds)
© 2003 Swets & Zeitlinger, Lisse, ISBN 90 5809 536 3

Prestressed bridges under corrosion attack

Břetislav Teplý, David Lehký & Drahomír Novák
Faculty of Civil Engineering, Brno University of Technology, Czech Republic

Eva M. Eichinger & Johann Kollegger
Institute for Structural Concrete, University of Technology, Vienna, Austria

Radomír Pukl & Vladimír Červenka
Cervenka Consulting, Prague, Czech Republic

Mark G. Stewart
School of Engineering, University of Newcastle, Australia

ABSTRACT: The reliability of existing prestressed bridges damaged by corrosion of the prestressing tendons or wires is investigated. Results from material tests are used to determine the mechanical properties of the prestressing steels. A stochastic nonlinear assessment of precast prestressed concrete bridge girder is performed using the newly developed software package *SARA* (*S*tructural *A*nalysis and *R*eliability *A*ssessment). This package was specifically designed for the purpose of nonlinear reliability assessment of reinforced concrete structures. Results from computer simulations of the prestressed bridge girder with corroded reinforcement are presented. Load carrying capacity and reliability index of the deteriorated bridge structure are evaluated and compared to the undamaged state.

1 INTRODUCTION

In recent years a lot of progress has been made in accurately determining the actual load carrying capacity and reliability of existing bridge structure. In this context deterioration of concrete and consequently corrosion of mild or prestressing reinforcement are serious aspects influencing service life and reliability of reinforced concrete structures. Furthermore, special corrosion problems may arise in older bridge structures with quenched and tempered prestressing wires that are sensible to hydrogen induced stress corrosion cracking (SCC). In this case large uncertainties are related to the deterioration of the materials and the parameters describing the material behavior.

Thus, the determination of the load carrying capacity of a structure should be based on advanced solution methods taking into account the uncertainties of the relevant variables on the resistance side (material properties) as well as on the loading side is of utmost importance.

Computational models taking into account uncertainties give useful information about a structure and its behavior in various states of its life-time. This is especially important if the experimental investigation of an existing structure could be unsuitable or even impossible. Prestressed bridges damaged by hydrogen induced SCC represent such a case.

In this paper the reliability of an old precast prestressed bridge girder damaged by hydrogen induced SCC is investigated. The software package *SARA* for stochastic nonlinear assessment of reinforced concrete structures is used taking into account rupture of some of the prestressing steels due to hydrogen induced stress corrosion. Deterministic as well as stochastic computer simulation of the precast prestressed bridge girder was performed in the undamaged state and at different levels of corrosion attack in order to assess and compare reliability of the bridge girder for various degrees of corrosion of the prestressing wires.

2 CORROSION OF PRESTRESSING STEELS

To guarantee a certain reliability level of prestressed concrete structures over their service life the durability of the prestressing is crucial. The main reasons for damage to prestressing wires and strands are wrong treatment of the steels during transport and storage, cracking of concrete, manufacturing errors as well as corrosion mainly induced by chlorides from e.g. de-icing salts. Furthermore, for post-tensioned structures insufficient injection of the ducts may be a main durability problem.

Additionally, the durability of prestressed structures is influenced by the sensitivity of the prestressing steels used to hydrogen induced stress corrosion cracking (SCC). The brittle fracture mode should be considered in this case. The loss of bearing capacity is abrupt and hence more dangerous as compared to the other kinds of degradation.

Hydrogen induced stress corrosion unlike other types of steel corrosion can take place not only if the alkalinity of the concrete is reduced by chlorides or carbonation but also in the so-called "passive" environment without corrosive substances being present. Even without the presence of oxygen hydrogen is produced and absorbed by the sensitive steel.

In contrary to cold-drawn steels quenched and tempered wires of the types SIGMA oval and NEPTUN N40, whose chemical and mechanical properties are rather similar, are quite critical to hydrogen induced SCC. This is also due to the different structure of quenched and tempered steels compared to cold-drawn steels (Fig. 1).

From the SIGMA-steels produced before 1965 – the so-called "old type" – it is known that some batches tend to hydrogen induced SCC. In this context steels with oval cross-sections are more sensitive than steels with circular cross-sections. The main cause is the relatively small cross-sectional area compared to the surface area.

SIGMA steels which were produced after 1965 show a higher resistance to hydrogen induced SCC as their chemical composition is different from the steels produced before 1965. Higher contents of chromate and lower contents of silicon, carbon and manganese increased the resistance to hydrogen induced SCC. Nevertheless, as recent studies show (Bertram et al., 2002) also quenched and tempered SIGMA steels produced in the period of 1965 to 1978 might not be completely insensitive to this phenomenon.

Independent from other influencing parameters it was shown that very high strengths increase the sensitivity of all steel types to hydrogen induced SCC (Mietz, 1999).

Structural reliability considering deterioration and degradation effects were modeled and studied e.g. for mild steel corrosion by Stewart (1997), Vu et al. (2000) and Novák et al. (2000), for SCC on a cross-section level by Izquierdo et al. (2001). It has to be stated that to the author's best knowledge no reliable and well-calibrated model for hydrogen induced SCC analysis exist. Therefore, in the study presented in this paper some ad-hoc situations with failed wires are introduced and the degradation process in time is not directly taken into account. The influence of rupture of the prestressing wires due to hydrogen induced SCC on the structural behavior along the loading path and on the reliability is studied.

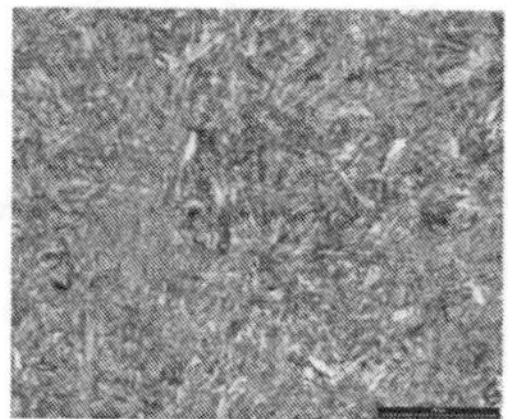
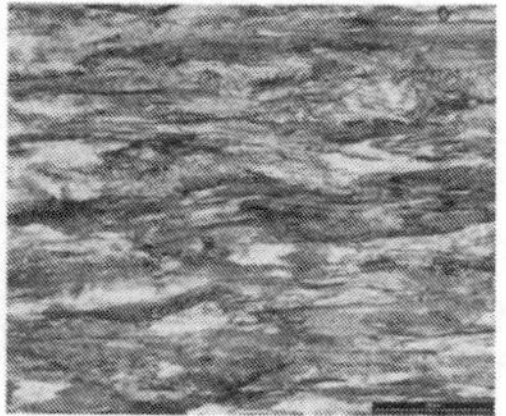

Figure 1. Structure of quenched/tempered steel (left) and cold-drawn steel (right), 1000 × enlarged.

3 PROPERTIES OF OLD QUENCHED AND TEMPERED PRESTRESSING STEELS

To guarantee accurate and realistic models of the variables on the resistance side information on the actual material properties of old prestressing steels damaged by corrosion is a very important basis for a numerical simulation. This is especially true when an existing bridge is assessed and the load carrying capacity or the reliability has to be determined.

To identify the properties of the post-tensioning wires the chemical composition of the wires was determined and tensile, fatigue, relaxation and corrosion tests were performed. The chemical properties of the post-tensioning wires were examined using emission spectrometry. The composition of the quenched and tempered steel wires shows quite high contents of silicon and manganese and matches the so-called "old" type of SIGMA steel produced before 1965.

Realistic estimates of the material parameters in comparable existing bridge structures – like the precast concrete beam presented in this publication – could be derived. The wires were taken from bridges built in the 1950s to 1970s, which had to be demolished because concerns about their durability and ultimate strength were raised. Only the most important results are summarized in the following while detailed results of this study are given in Eichinger et al. (2002).

3.1 *Tensile tests*

As suspected some minor corrosion does not effect the strength properties of the steels. Only if strong corrosion is present the ductility of the steel is influenced. As the tests showed strain and contraction values are reduced, while yield and tensile strength are not effected. The material parameters in Section 5.3 (Tab. 2) of this paper are based on these tensile test results.

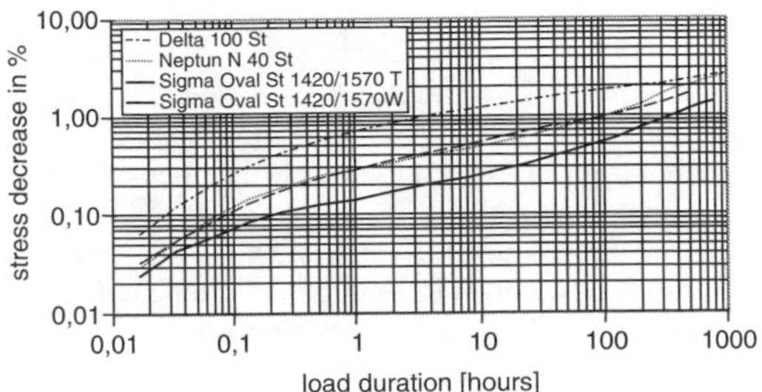

Figure 2. Results of relaxation tests.

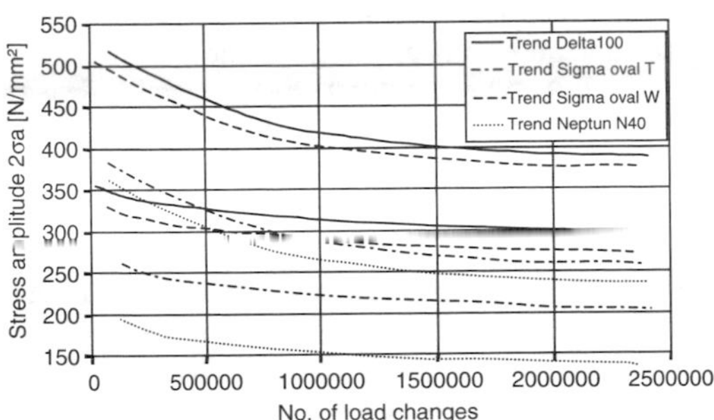

Figure 3. Results of fatigue tests.

3.2 *Relaxation tests*

The relaxation tests were performed according to ISO 15630–3 with a loading equivalent to 70% of the effective tensile strength and a maximum allowable decrease in stress of 2.5%. The results are shown in Figure 2.

All wires examined passed the test. This indicates that the relaxation properties of the tested steels were much higher at the time of production as the steels have been in use in the bridge structures for more than 30 years.

3.3 *Fatigue tests*

The influence of corrosion is clearly shown in the fatigue tests, as corrosion considerably decreases the fatigue strength. The results of the fatigue tests in form of trend lines for the different types of steels examined are presented in Figure 3.

In the fatigue test the cold-drawn wires DELTA 100 showed far better results than the quenched and tempered steels, but their fatigue strengths still are under the values allowable for new steels.

3.4 *Corrosion tests*

To show whether the steels were sensible to hydrogen embrittlement two different types of corrosion tests

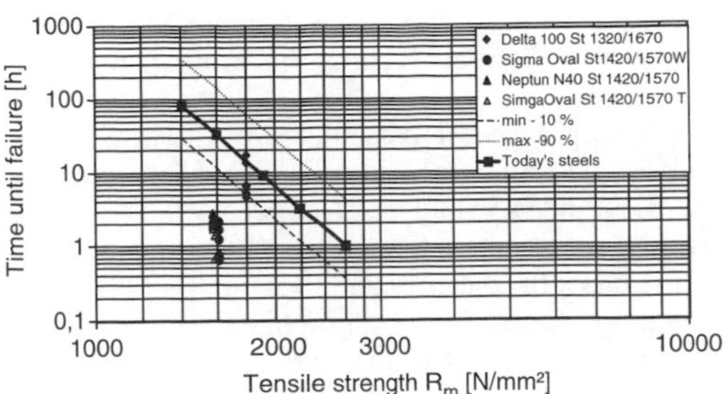

Figure 4. Results of corrosion tests according to ISO 15630-3.

were performed. In the FIP test a concentrated ammoniumthiocyanate (NH_4SCN) solution is used and the specimens are loaded with a constant load equivalent to 80% of the tensile strength. The testing temperature is 50°C. The time until failure of the specimen is recorded, after 500 hours the test is stopped. The FIP test was developed for post-tensioning strands and due to the highly concentrated solution usually leads to a rather quick failure of the steel tested. Furthermore, DIBt tests were performed. This tests lasts up to 2000 hours and a thinned NH_4SCN solution is employed.

The results of the FIP tests are shown in Figure 4 and compared to values which are common for steels produced today. The higher sensitivity of quenched and tempered wires in comparison to cold-drawn steel is clearly visible. Only the cold-drawn steel Delta 100 gives satisfying results comparable to those for new steels.

4 NONLINEAR STOCHASTIC SIMULATION

Recent tools for computer simulation enable to investigate numerical models of a structure comparable to experimental testing of the real structure. Concrete structures damaged due to corrosion of reinforcement can be efficiently investigated using appropriate computer simulation. Since the knowledge on the parameters influencing the corrosion process is rather poor, application of probabilistic methods is desirable. Furthermore, stochastic analysis enables to determine sensitivity of the structural response to input variables and to assess reliability of the structure.

The software package *SARA* (*S*tructural *A*nalysis and *R*eliability *A*ssessment) Studio (Bergmeister et al. 2002) developed for advanced reliability assessment of reinforced concrete structures was employed in the numerical simulation of prestressed bridge structures with corroded reinforcement and reliability assessment taking into account deterioration processes. *SARA* Studio is an integration of the nonlinear finite element software ATENA into the probabilistic package FREET (Pukl et al. 2002), containing supporting tools.

The program ATENA enables realistic simulation of reinforced concrete structures. It is based on advanced material models with damage concepts and a smeared crack approach. The fracture of concrete is modeled by the crack band theory with fracture-energy-related softening. A user-friendly graphical interface offers an excellent support to the user during all stages of nonlinear analysis. This analysis is usually performed for deterministic input parameters. The conceptually object-oriented structure of the program enables "randomization" of the nonlinear analysis using a Microsoft COM-interface for exchange of input data and results.

The multi-purpose probabilistic software package FREET ensures generation of stochastic input data for the nonlinear calculation, evaluation of the stochastic response and assessment of the reliability index of a structure. As the nonlinear analysis is computationally very intensive, efficient statistic methods are employed in order to achieve accurate results with a rather small number of realizations.

A special type of numerical probabilistic simulation called Latin Hypercube Sampling (LHS) makes it possible to decrease the number of necessary simulations significantly. The range of the cumulative distribution function of each random variable is divided into N intervals of equal probability $1/N$ (N is the number of simulations). The regularity of sampling over the range of cumulative distribution functions ensures efficient sampling of random variables.

Statistical correlation among input random variables can also be considered. A stochastic optimization technique called Simulated Annealing is utilized to adjust random samples in such a way that the resulting correlation matrix is as close as possible to the target (user-defined) correlation matrix. Note that the approach allows working also with a non-positive definite matrix on the input side, which can be the result of lack of knowledge of the user. This technique generates samples as close as possible to a positive definite matrix (mathematically and physically correct).

As already mentioned the software package *SARA* – Studio combines the programs FREET and ATENA in order to allow for a probabilistic fracture analysis of engineering structures. The basic aim of statistical nonlinear fracture analysis is to obtain the estimation of the structural response statistics (failure load, deflections, cracks, stresses, etc.). The procedure can be itemized as follows:

1) Uncertainties are modeled as random variables described by their probability density functions (PDF). The result of this step is the set of input parameters for ATENA computational model – random variables described by mean value, variance and other statistical parameters (generally by PDF).
2) Random input parameters are generated according to their PDF using LHS sampling. Statistical correlation among the parameters is imposed using simulated annealing.
3) Generated realizations of random parameters are used as input for ATENA computational model. The complex nonlinear solution is performed and results (structural response) are saved.
4) Previous two steps are repeated N-times (N is the number of simulations used). At the end of the whole simulation process the resulting set of structural responses is statistically evaluated. The results are: histogram, mean value, variance, coefficient of skewness, empirical cumulative probability density function of structural response.

5 ILLUSTRATIVE EXAMPLE

As an application example a precast prestressed bridge girder damaged by hydrogen induced SCC is simulated. The material parameters for the prestressing steel are based on the results of the tests described above. The influence of a rupture of the corroded prestressing steels on the ultimate load and reliability of the structure is investigated, several alternatives are compared.

Breakage of the prestressing steel damaged by corrosion and the renewal of bond between concrete and prestressing steel in a certain distance from the rupture point was modeled using the software package ATENA (Pukl et al. 2002). For this example probabilistic approach to the solution of the problem is important due to two main reasons:

(i) The majority of input parameters are uncertain or random and
(ii) the reliability of the structure and the influence of deterioration of materials on the reliability over the life-time of the structure has to be studied.

As an example a bridge element with a cross-section frequently used was selected – a precast prestressed girder with dimensions similar to the AASHTO Type IV, capable of spanning distances in the 21 to 30 m (70 to 100 ft) range (Taly, 1998).

5.1 *Geometry*

The cross-section of the girder is shown in Figure 5. The length of the girder is considered to be 23.4 m. The prestressing in the inner part of the girder is realized with the third-point harping of the steels (see FEM mesh in Fig. 6, detail Fig. 7). Because of the limited thickness of the web only the centrally located prestressing steels in three rows can be harpened, while the others have to remain straight.

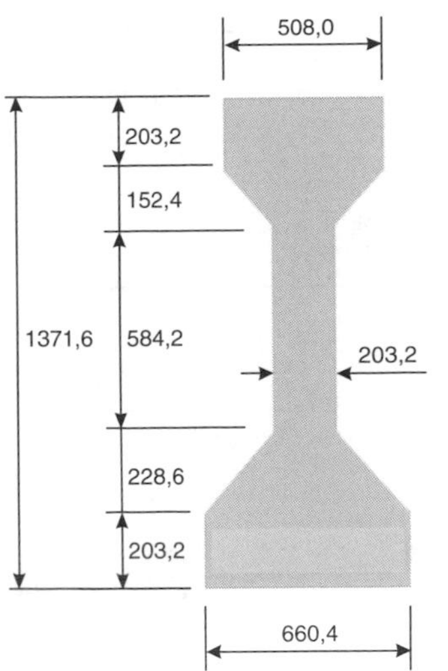

Figure 5. Cross-section of bridge girder (dimensions in mm).

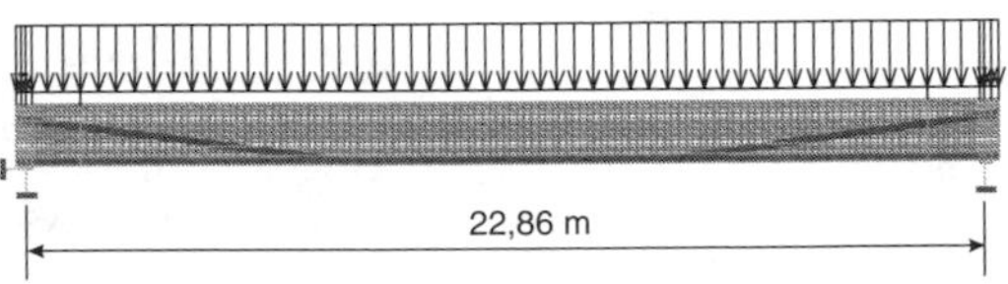

Figure 6. Beam model with longitudinal reinforcement, bearings and distributed load.

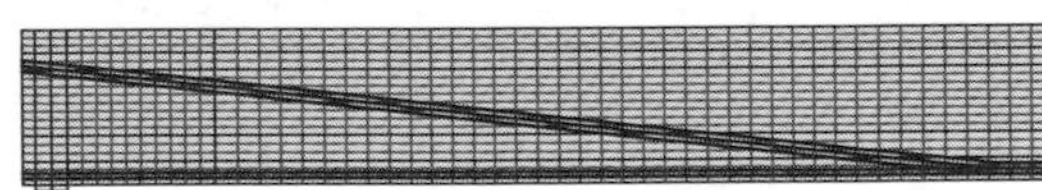

Figure 7. Detail of harping of the prestressing.

5.2 *Concrete*

Basic material properties of concrete are summarized in Table 1, which contains the relevant statistical parameters.

Nonlinear behavior of concrete in both tension and compression including softening was considered according to features offered by ATENA software. As many parameters concerning concrete are based mainly on the value of compressive strength and ATENA also offers some default values they will not be not listed all here.

5.3 *Prestressing steel*

Properties of the prestressing steel is described in Table 2 by their mean values and statistical parameters. Elasto-plastic stress-strain diagram for prestressing steel was considered. Note that losses of prestressing due to creep and shrinkage and other factors were not calculated within the computational model, they were just estimated based on relevant literature sources, as the purpose of this study is not necessarily a complex treatment of this problem.

5.4 *Bearings*

Boundary conditions have been carefully considered – see finite element mesh in Figure 6. Material of bearings was considered to be perfectly elastic with a modulus of elasticity of 100 GPa and a Poisson ratio v equal to 0.3. Modeling of supports in this way is near to the reality and prevents excessive cracking near the supports which might appear if simplified boundary conditions are considered and would reflect an unrealistic behavior of the structure.

5.5 *Load*

As the purpose of the study is to show the influence of degradation on ultimate load capacity of the

Table 1. Concrete properties.

	Symbol	Unit	Mean	Coeff. of Variation	Distribution
Poisson's ratio	v	–	0.20	0.05	LN
Tensile strength	f_t	MPa	2.32	0.18	Weibull
Compressive strength	f_c	MPa	25.50	0.10	LN
Specific fracture energy	G_F	N/m	57.90	0.20	Weibull
Uniaxial compressive strain	ε_c	–	0.0017	0.15	LN
Reduction of strength	c_{Red}	–	0.80	0.06	Rect.
Critical comp. displacement	w_d	m	0.0005	0.10	LN
Specific material weight	ρ	MN/m^3	0.023	0.10	N

Table 2. Properties of prestressing steel.

	Symbol	Unit	Mean	Coeff. of Variation	Distribution
Modulus of elasticity	E_s	GPa	195.0	0.03	LN
Yield stress	f_y	MPa	1582	0.07	LN
Prestressing force	F	MN	0.0392	0.04	N
Area of pretressing	A_s	mm^2	2561	0.001	N

Figure 8. Typical calculated crack pattern at ultimate load steels.

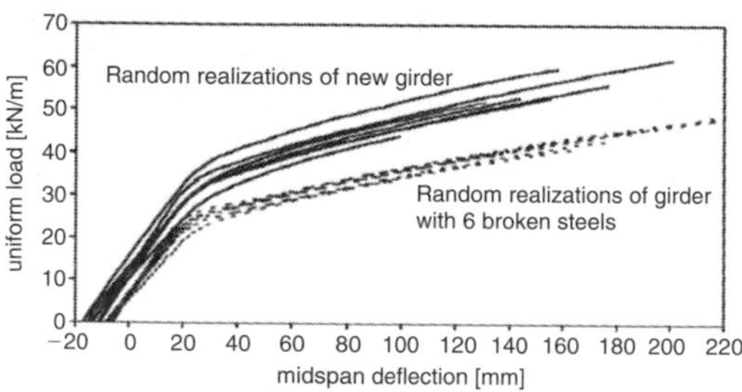

Figure 9. Load-deflection diagram.

girder, only one uniformly distributed loading was considered. This load was alternatively modeled as deterministic or random. After prestressing this loading was increased in increments until a collapse (peak load) was achieved.

5.6 *Finite element model*

Using the ATENA software package the FEM computational model was developed, the mesh and arrangement of prestressing are shown in Figure 6 and Figure 7.

The girder was solved as 2D plane stress problem with variable thickness along the height of girder. The model consists of about 3000 finite elements with over 4500 nodes. The numerical solution was performed using Newton-Raphson algorithm. The load was increased in small increments and in each increment the equilibrium of internal and external forces was iteratively achieved. To model the real technological process first the prestressing was applied in a few steps, see Figure 9.

Then continuous loading was increased until peak – in Figure 8 the fracture process zone can be seen by graphical representation of cracks. The loading procedure is documented mainly by the load-deflection curve in Figure 9, where deflection is monitored at midspan of the girder and the load is represented by the appropriate reaction force.

5.7 *Corrosion of prestressing steel*

Three levels of corrosion attack due to hydrogen induced SCC were assessed in the present study. For the sake of a transparent recognition/comparison of the degradation effects firstly an undamaged girder was analyzed, i.e. no degradation was considered.

It was estimated that in a bridge girder with quenched and tempered prestressing steels at certain age some prestressing steel could be broken due to hydrogen induced SCC. Therefore, rupture of steel caused by hydrogen induced SCC without bond renewal was modeled in two alternatives: with 2 or 6 broken prestressing steels (only the second case is reported here). The area of the prestressing steel was reduced appropriately over the whole length of the steel.

5.8 *Randomization*

For each of the cases described above 40 random realizations were generated using Latin Hypercube Sampling based on the statistical properties given in Tables 1 and 2. As an example the random distribution of the concrete compressive strength is shown in Figure 10.

Correlation between random variables representing concrete properties was introduced by a prescribed correlation matrix as shown in Table 3 (upper right triangle). Using Simulated Annealing the optimal correlation matrix for 40 random realizations was generated by FREET as shown in the lower left triangle of Table 3.

6 SELECTED RESULTS

The results from all realizations of all models were evaluated and statistic parameters of the response (ultimate load) were estimated and compared. The determined ultimate load represents the actual resistance of the analyzed structure. The reliability index

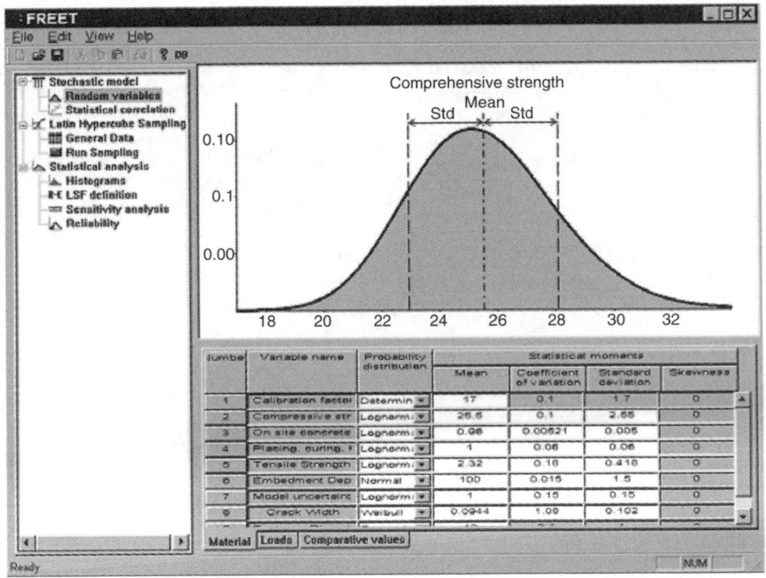

Figure 10. Random distribution of concrete compressive strength in FREET environment.

Table 3. Correlation matrix (upper right triangle = prescribed, lower left triangle = generated for 40 realizations).

Variable	E_c	f_t	f_c	G_f	ε_c
E_c	1.0	0.7	0.9	0.5	0.9
f_t	0.69	1.0	0.8	0.9	0.6
f_c	0.9	0.78	1.0	0.6	0.9
G_f	0.5	0.89	0.61	1.0	0.5
ε_c	0.9	0.61	0.89	0.49	1.0

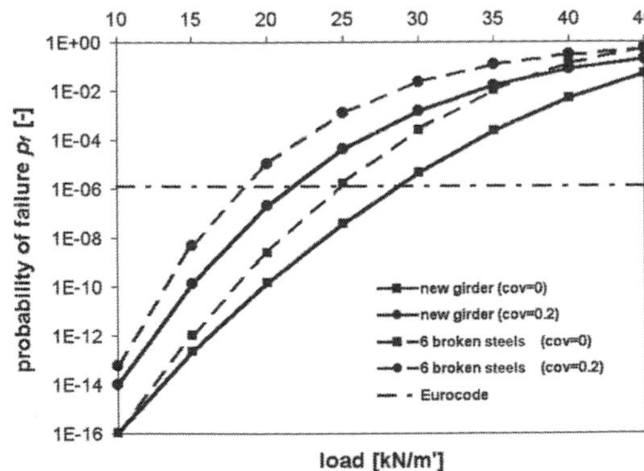

Figure 11. Probability of failure.

was assessed according to Pukl et al. (2003) under the assumption of normal distributions of resistance and load. The coefficients of variation of the load were assumed equal to 0.0 and 0.2.

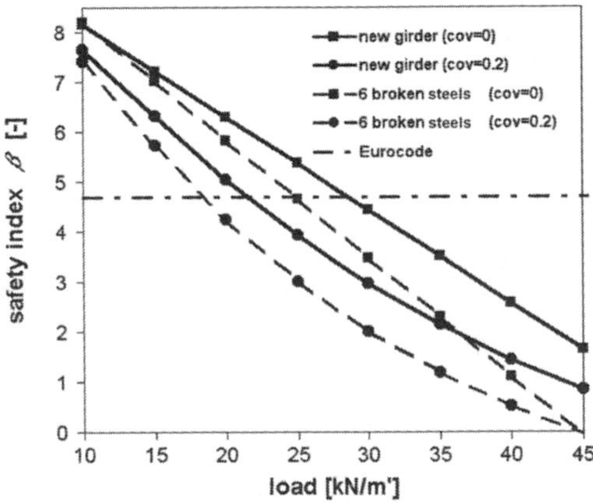

Figure 12. Reliability index.

Load-deflection diagrams (Fig. 9), failure probability (Fig. 11) and reliability index plots (Fig. 12) are presented comparing the undamaged state and the case with 6 broken prestressing steels.

7 CONCLUSIONS

Detailed information on status and behavior of a precast bridge girder with corroded prestressing reinforcement during service life could be efficiently obtained by combining experimental investigations with advanced numerical modeling. The reliability of the structure exposed to corrosion could be assessed using efficient probabilistic methods documenting recent possibilities of computational modeling of reinforced concrete structures. The range of acceptable load relevant to the EC reliability requirements may be considerable in the case of corrosion effect.

ACKNOWLEDGEMENT

The financial support by grant of the Grant Agency of Czech Republic (GAČR) No. 103/02/1030 is greatly appreciated.

REFERENCES

Bertram, D., Hartz, U., Isecke, B., Jungwirth, D., Litzner, H.-U., Manleitner, S., Nürnberger, U., Riedinger, H., Rustler, G. und Schießl, P. (2002). Gefährdung älterer Spannbeton-bauwerke durch Spannungsrißkorrosion an vergütetem Spannstahl in nachträglichem Verbund. *Beton- und Stahlbetonbau.* Vol. 97, No. 5.

Mietz, J. (1999). Wasserstoffinduzierte Spannungsrisskorrosion an vergüteten Spannstählen. *Bauingenieur*, Vol. 74, pp. 403–411.

Eichinger, E.M., Winter, H. und Kollegger, J. (2002). *Untersuchungen an Spannstählen der Typen NEPTUN N40, SIGMA oval und DELTA 100.* Report No. 2/2002, Institute for Structural Concrete, University of Technology, Vienna.

Taly, N. (1998). *Design of Modern Highway Bridges.* McGraw-Hill, New York.

Izquierdo, D., Andrade, C. and Tanner, P. (2001). Reliability analysis of corrosion in post-tensional tendons. *Proceedings Safety, Risk, Reliability – Trends in Engineering.* Malta, pp. 1001–1006.

Novák, D., Teplý, B., Lawanwisut, W. & Keršner, Z. (2000). The modelling of RC beam deterioration: a sensitivity study. In Melchers & Stewart (eds), *Applications of Statistics and Probability*, pp. 307–314, Balkema, Rotterdam.

Stewart, M.G. (1997). Time-dependent reliability of RC structures. *Journal of Structural Engineering,* ASCE, Vol. 123, No. 7, pp. 896–903.

Vu, K.A.T. & Stewart, M.G. (2000). Structural reliability of concrete bridges including improved chloride-induced corrosion models. *Structural Safety*, No. 22, pp. 313–333.

Pukl, R., Tepý, B., Novák, D. & Stewart M.G. (2002). Modeling of precast prestressed bridge girder under corrosion attack. *Beton TKS*, Vol. 3/2002, pp. 50–54 (in Czech).

Bergmeister, K., Strauss, A., Eichinger, E.M., Kollegger, J., Novak, D., Pukl, R. & Cervenka, V. (2002). Structural analysis and safety assessment of existing concrete structures. *Proceedings The first fib Congress* 2002. Osaka, pp. 47–54.

Pukl, R., Novák, D. & Bergmeister, K. (2003). Reliability assessment of concrete structures. *Proceedings Euro-C* 2003. St. Johann, Austria.

Computational Modelling of Concrete Structures, Bićanić et al. (eds)
© 2003 Swets & Zeitlinger, Lisse, ISBN 90 5809 536 3

A chemo-mechanical modelling framework of concrete: Coupling between calcium leaching and continuum damage mechanics

B. Nedjar, V.H. Nguyen
Laboratoire d'Analyse des Matériaux et Identification, Cité Descartes, Marne-la-vallée Cedex 2, France

J.M. Torrenti
IRSN/DIR/Pg, Fontenay-aux-Roses Cedex, France

ABSTRACT: We present in this work a phenomenological chemo-mechanical model to represent the degradation of concrete-like materials. The chemical behaviour is described by the nowadays well known simplified calcium leaching approach. And the mechanical damage behaviour is described by a recently proposed continuum damage modelling framework which involves the gradient of damage quantity. The problem of numerically integrating the equations at hand is addressed within the context of the finite element method. Special care is taken to resolve the highly nonlinear equation governing the calcium dissolution-diffusion part of the problem. The algorithmic design is based on a Newton-type iterative method where use is made of a recently proposed *relaxed* linearization procedure. Finally, numerical simulations are compared with experimental results on cement paste.

1 INTRODUCTION

Durability mechanics of concrete is not exclusively affected by damage induced by mechanical loads. The life-time of such a material may also depend on the environmental conditions. As a typical example, concrete is commonly employed in radioactive waste disposal, and then, concrete containment structures must ensure the load-bearing capacity over extend periods depending on the level of radioactivity. It is then necessary to take into account the coupling between chemical and mechanical effects when one wants to design a model to represent the behaviour of concrete for such applications.

We present in this work a chemo-mechanical model to represent the degradation behaviour of concrete materials. Use is made of phenomenological descriptions for both the mechanical and chemical damage processes. On one hand, the chemical behaviour is described by the nowadays well known simplified approach of calcium leaching, see (Gérard 1996; Carde et al. 1996; Torrenti et al. 1998; Ulm et al. 1999; Bangert et al. 2001) among many other references. And on the other hand, the mechanical damage is described by using a continuum damage mechanics framework which involves the gradient of damage quantity. In this latter, the formulation and the evolution equations relative to the damage part of the modelling are issued from a recent formulation of the principle of virtual power where the power of the internal forces is chosen to depend, besides on the classical term involving the strain rates, also on the damage rate and the gradient of the damage rate, see for example (Frémond and Nedjar 1993; Frémond and Nedjar 1996; Nedjar 2001).

It is shown how the thermodynamics of irreversible processes is crucial in the characterization of the dissipative phenomena and in setting the convenient forms for the constitutive relations.

The paper is organized as follows. In section 2, we outline the basic equations describing the coupled problem. The constitutive equations are given within the continuum thermodynamics framework and the different coupling effects are discussed. In section 3, we give a model example for the damage part of the modelling. As far as numerical formulation is concerned, a numerical design to implement the equations at hand is addressed within the context of the finite element method in section 4. Special care is taken when numerically integrating the strongly nonlinear calcium dissolution-diffusion equation of the problem. Finally, the results of a numerical simulation going from the calcium leaching to the mechanical loadings are compared with experimental results in section 5.

2 THE COUPLED PROBLEM. OUTLINES OF THE GOVERNING EQUATIONS

During the last few years, different models have been developed to describe the coupling between chemical and mechanical effects. In this work, use is made of a phenomenological approach. We first outline the basic equation decribing the calcium leaching process. Then the mechanical part is given which is restricted within the present work to an elastic-damage behaviour. And finally, we give the required constitutive relations with some discussions of how couplings can be taken into account.

2.1 *Calcium leaching part of the problem*

For cement-based materials, the kinetics of calcium leaching is determined from the mass conservation of calcium which describes the evolution of the amount of calcium in the liquid phase of the material. One has the following macroscopic balance of the calcium ion mass:

$$\mathbf{divq} + \dot{\overline{(\phi c)}} + \dot{s} = 0, \tag{1}$$

where $(\dot{})$ denotes the time derivative and div [.] is the divergence operator. And where, c denotes the calcium concentration in the liquid phase, s denotes the calcium concentration in the solid phase, ϕ is the porosity in the material, and $\mathbf{q}$ denotes the calcium mass flux vector. The rate $\dot{s}$ denotes the calcium mass production due to the dissolution of the solid phase.

According to (Gérard 1996), this simplified approach focuses on the evolution of one variable only, namely: the calcium concentration c. That is, all the quantities s, ϕ and $\mathbf{q}$ may depend on the (state) variable c.

Typically, the function $s(c)$ represents the equilibrium between the calcium in the solid phase and the calcium in the liquid phase c. As an example, for the case of an accelerated calcium leaching process using ammonium nitrate, a plot of this function proposed by (Tognazzi 1998) is given in Figure 1 where two dissolution regimes are clearly distinguished: the dissolution of the portlandite and the progressive decalcification of the C-S-H.

Experimental observations also show that the porosity ϕ increases with decreasing calcium concentration c. Figure 2 shows a plot of this dependence for the case of calcium leaching process also under ammonium nitrate solution. See (Tognazzi 1998; Torrenti et al. 1998) for example.

2.2 *Mechanical damage part of the problem*

The phenomenological response due the mechanical actions is characterized by progressive loss of stiffness with increasing loading leading to eventual

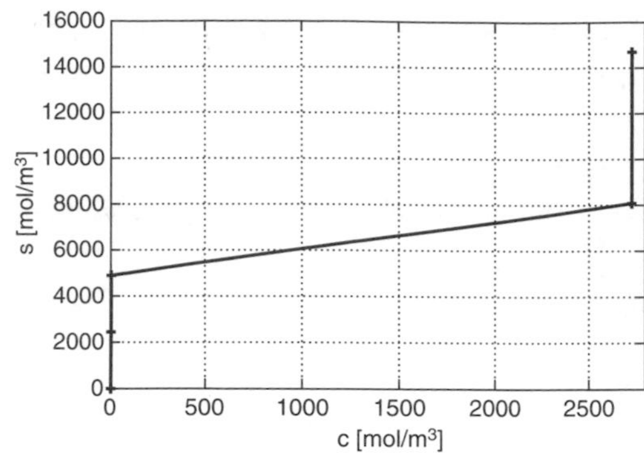

Figure 1. Chemical equilibrium $s(c)$ for accelerated ammonium nitrate calcium leaching.

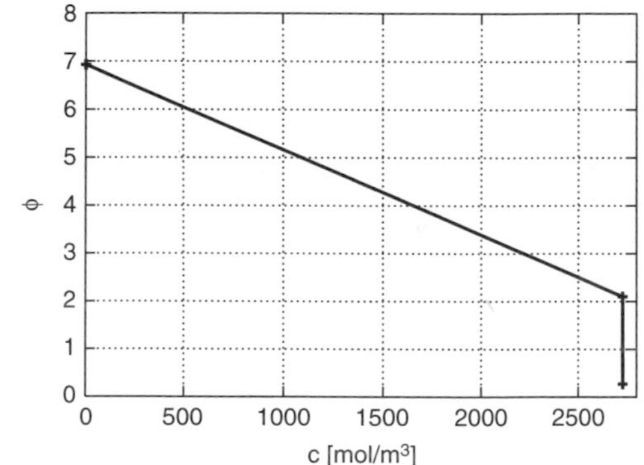

Figure 2. Porosity ϕ versus the calcium concentration c.

failure. This degradation inevitably leads to the so called strain softening. And from the numerical point of view, classical *local* models exhibit unacceptable mesh dependence in the presence of strain softening. Several attempts have been investigated to circumvent these nowadays well known difficulties. Among the most effective enhancements, higher order formulations have been proposed by many authors these last few years.

In this work, use is made of the gradient damage formulation developed in (Frémond and Nedjar 1993; Frémond and Nedjar 1996), see also (Nedjar 2001). Here we recall the main steps of this formulation. For more details, we refer to the above referenced papers. Let the scalar $\beta(\mathbf{x}, t)$ be the damage quantity with value 1 when the material is undamaged and value 0 when it is completely damaged. The basic idea of the theory we use is to modify the power of the internal forces $\mathcal{P}_{\text{int}}$ as follows. For a domain Ω with boundary $\partial\Omega$, we write:

$$\mathcal{P}_{\text{int}} = -\int_{\Omega} \boldsymbol{\sigma} : \dot{\boldsymbol{\epsilon}}(\mathbf{u})\, \mathrm{d}\Omega - \int_{\Omega} (B\dot{\beta} + \mathbf{H}.\mathbf{grad}\dot{\beta})\, \mathrm{d}\Omega, \tag{2}$$

where σ is the stress tensor and $\dot{\varepsilon}(\mathbf{u})$ is the strain rates tensor ($\mathbf{u}$ being the displacement vector), B is the internal work of damage and $\mathbf{H}$ is the flux vector of internal work of damage.

Taking the classical expression for the power of the external forces $\mathcal{P}_{ext}$, the principle of virtual power in quasi-statics gives two sets of balance equations:

$$\begin{cases} \mathrm{div}\boldsymbol{\sigma} + \mathbf{f} = \mathbf{0} \quad \text{in } \Omega, \qquad \boldsymbol{\sigma}\mathbf{n} = \mathbf{T} \quad \text{on } \partial\Omega, \\ \mathrm{div}\mathbf{H} - B = 0 \quad \text{in } \Omega, \qquad \mathbf{H}.\mathbf{n} = 0 \quad \text{on } \partial\Omega, \end{cases} \tag{3}$$

where $\mathbf{n}$ is the outer normal to $\partial\Omega$, $\mathbf{T}$ is the prescribed surfacic traction force vector and $\mathbf{f}$ is the volumetric body force. Equation $(3)_2$ is nonclassical. It is intended to describe the mechanical damage within the solid.

The coupled problem (1) and (3) must be completed with adequate constitutive equations. For this, use is made of the thermodynamics of irreversible processes.

2.3 *Constitutive equations*

With the choice made in (2), and confining our attention to isothermal processes, the Clausius-Duhem inequality reads:

$$\begin{aligned} \mathcal{D} = \boldsymbol{\sigma}:\dot{\boldsymbol{\epsilon}} \; + \; B\dot{\beta} \; + \; \mathbf{H}.\mathbf{grad}\dot{\beta} \\ + \; \mu\dot{c} \; - \; \mathbf{q}.\mathbf{grad}\mu \; - \dot{\psi} \geq 0, \end{aligned} \tag{4}$$

which must be satisfied for any admissible process. In (4), ψ denotes the free energy function which characterizes the state of the material. μ is the chemical potential of calcium associated to the calcium concentration c. Assuming for the sake of clarity that there is no dissipation due to plasticity, we assume the free energy function of the general form: $\psi \equiv \psi(\boldsymbol{\varepsilon}, c, \beta, \mathbf{grad}\beta)$. Substituting this latter into (4) and using the Coleman's method, see for example (Coleman and Gurtin 1967), one obtains that the following three inequalities must be satisfied:

$$\mathcal{D}_1 = (\boldsymbol{\sigma} - \frac{\partial\psi}{\partial\boldsymbol{\epsilon}}):\dot{\boldsymbol{\epsilon}} + (\mu - \frac{\partial\psi}{\partial c})\dot{c} \geq 0, \tag{5}$$

$$\mathcal{D}_2 = (\mathbf{H} - \frac{\partial\psi}{\partial(\mathbf{grad}\beta)}).\mathbf{grad}\dot{\beta} + (B - \frac{\partial\psi}{\partial\beta})\dot{\beta} \geq 0, \tag{6}$$

$$\mathcal{D}_3 = -\mathbf{q}.\mathbf{grad}\mu \geq 0. \tag{7}$$

In (6), $\mathcal{D}_2$ represents the dissipation due to damage. Inequalities (5) and (6) give:

$$\sigma = \frac{\partial\psi}{\partial\epsilon} \qquad \mu = \frac{\partial\psi}{\partial c} \qquad \mathbf{H} = \frac{\partial\psi}{\partial(\mathbf{grad}\beta)} \tag{8}$$

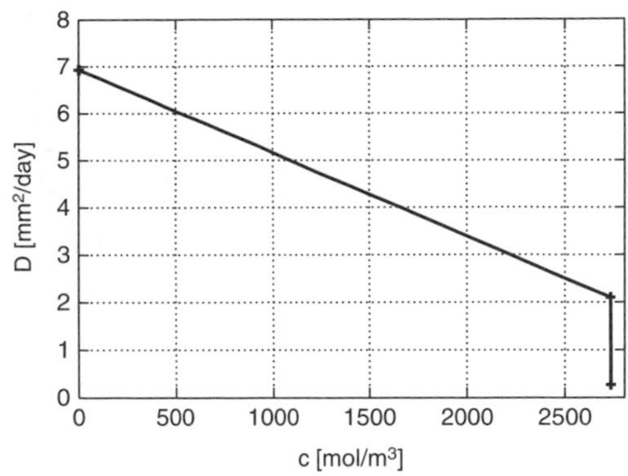

Figure 3. Apparent conductivity $\tilde{D}$ versus the calcium concentration c.

together with the reduced inequality

$$\mathcal{D}_2 = (B - \frac{\partial\psi}{\partial\beta})\dot{\beta} \geq 0. \tag{9}$$

Noting that this latter must be *strict* when damage occurs. The choice $B = \partial_\beta\psi$ is then unacceptable. We postulate then the existence of a damage potential of dissipation $\chi(\dot{\beta}; \ldots)$, *i.e.* a positive, convex and subdifferentiable function with $\chi = 0 \Leftrightarrow \dot{\beta} = 0$, such that

$$B - \frac{\partial\psi}{\partial\beta} = \frac{\partial\chi}{\partial\dot{\beta}} \quad \Longrightarrow \quad B = \frac{\partial\psi}{\partial\beta} + \frac{\partial\chi}{\partial\dot{\beta}} \tag{10}$$

The convexity of χ is sufficient to satistfy the inequality (9) (and hence (6)) for any admissible process. Note that in the definition of χ, other variables have the possibility to act as parameters.

In (7), $\mathcal{D}_3$ represents the dissipation due to diffusion of calcium. Following similar arguments as in (Gérard 1996), use of Fick's law together with the relation $(8)_2$, leads to the following constitutive relation for the calcium mass flux vector:

$$\mathbf{q} = -\tilde{D}(c)\mathbf{grad}c, \tag{11}$$

where $\tilde{D}(c)$ is the apparent calcium ion conductivity. With this latter, the calcium diffusion problem to be solved is obtained by susbstituting (11) into (1) to give:

$$\dot{\overline{(\phi(c)\; c)}} + \dot{s}(c) = \mathrm{div}\left[\tilde{D}(c)\mathbf{grad}c\right], \tag{12}$$

This equation is highly nonlinear. On one hand because of the relations $s(c)$ and $\phi(c)$ shown in Figures 1 and 2, respectively. And on the other hand, because the apparent calcium conductivity $\tilde{D}$ also depends nonlinearly on c. A plot of this dependence is shown in Figure 3 for the case of a calcium leaching using ammonium nitrate (Tognazzi 1998).

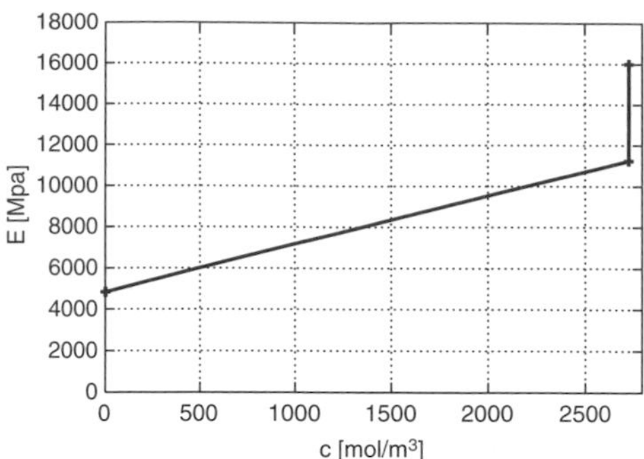

Figure 4. Young's modulus E as a function of c.

2.4 *Chemo-mechanical couplings*

The interactions between calcium leaching and mechanical damage have been investigated by many authors, see (Gérard 1996; Carde 1996; Torrenti et al. 1999; Ulm et al. 1999; Bangert et al. 2001; Le Bellégo et al. 2001) among others.

On one hand, calcium leaching leads to a decrease of mechanical properties. And in the other hand, mechanical damage increases the apparent conductivity of the calcium in the liquid phase (Torrenti et al. 1998). However, and for the sake of simplicity, this latter effect will not be considered for the moment in the present work. We consider only the case when calcium leaching affects the mechanical properties of the material. As an illustration, Figure 4 shows how the calcium concentration affects the Young's modulus $E(c)$ of a cement-based material (Torrenti et al. 1998).

3 AN ELASTIC-DAMAGE MODEL

To make matters as concrete as possible, we first give an elastic-damage model. For the mechanical part of the free energy function ψ, we choose the following form already studied in (Nedjar 2001; Nedjar 2002a):

$$\psi_{\text{mech}} = \frac{1}{2}\beta\boldsymbol{\epsilon}:\mathbf{C}(c):\boldsymbol{\epsilon} + \frac{1}{2}k(\mathbf{grad}\beta)^2, \tag{13}$$

where the first term constitutes the simplest model where damage is coupled to elasticity, $\mathbf{C}(c)$ is the classical rank four Hooke's elasticty tensor ($\lambda(c)$ and $\mu(c)$ being the Lamé's coefficients) where the Young's modulus $E(c)$ is given, for instance, by Figure 4. And in the second term, k is the factor of influence of damage gradient.

For the damage dissipation function χ in (10), we choose the following rate-independent damage form (also proposed in (Nedjar 2001; Nedjar 2002a)):

$$\begin{aligned}\chi(\dot{\beta}) =& -\dot{\beta}\left[W(c) + \frac{1}{2}\boldsymbol{\epsilon}:\mathbf{C}(c):\boldsymbol{\epsilon} - \beta^n S(\boldsymbol{\epsilon})\right] \\ &+ I_-(\dot{\beta}),\end{aligned} \tag{14}$$

where $W(c)$ is the initial damage threshold which here depends on the calcium concentration $c, n > 0$ is a parameter that controls the hardening/softening material behaviour. When $n > 2$, the material exhibits hardening while when $n < 2$ it exhibits softening. $S(\varepsilon)$ is a source of damage which depends on the mechanisms that govern the damage growth within the material, and $I_-(\dot{\beta})$ is the indicator function of the subset $[-\infty, 0]$ which forces $\dot{\beta}$ to be negative or zero valued ($I_-(x) = 0$ if $x \leq 0$, and $I_-(x) = +\infty$ if $x > 0$). This latter is introduced to take into account the fact that damage is an irreversible process.

The choice of source of damage function $S(\boldsymbol{\varepsilon})$ is not unique. It can depend on the total deformation or part of it following the material under consideration. Here $S(\boldsymbol{\varepsilon})$ depends on the positive strains as proposed in (Nedjar 2001).

Substitution of (13) and (14) into $(8)_1$, $(8)_3$ and (10) gives:

$$\begin{aligned}&\boldsymbol{\sigma} = \beta\mathbf{C}(c):\boldsymbol{\epsilon}, \qquad \mathbf{H} = k\mathbf{grad}\beta, \\ &B \in \beta^n S(\boldsymbol{\epsilon}) - W(c) + \partial I_-(\dot{\beta}).\end{aligned} \tag{15}$$

In $(15)_3$, $\partial I_-(\dot{\beta})$ is the generalized derivative of $I_-(\dot{\beta})$. It is a mathematical reaction which forces $\dot{\beta}$ to be negative or equal to zero, i.e. $\partial I_-(x) = \{0\}$ if $x < 0$ and $\partial I_-(0) = [0, +\infty]$.

Then, substitution of the last results into (3) gives the equations of the mechanical part of the problem:

$$\begin{cases} \text{div}(\beta\mathbf{C}(c):\boldsymbol{\epsilon}) + \mathbf{f} = \mathbf{0} & \text{in } \Omega, \\ \boldsymbol{\sigma}\mathbf{n} = \mathbf{T} & \text{on } \partial\Omega, \end{cases} \tag{16}$$

$$\begin{cases} -k\Delta\beta + \beta^n S(\boldsymbol{\epsilon}) + \partial I_-(\dot{\beta}) \ni W(c) & \text{in } \Omega, \\ k\mathbf{grad}\beta.\mathbf{n} = 0 & \text{on } \partial\Omega, \end{cases} \tag{17}$$

where Δ denotes the Laplacian operator.

Note that within the present work, the chemo-mechanical coupling is present only through the Young's modulus $E(c)$ and the damage threshold $W(c)$. For the application that follows, $E(c)$ is given by Figure 4, and $W(c) = (E(c)/E_0)W_0$, where E_0 and W_0 are the initial corresponding values when no leaching and no damage occur.

4 VARIATIONAL FORMULATIONS AND NUMERICAL DESIGN

The variational formulations of the local form of the governing equations (12), (16) and (17) play a central role in the numerical solution of the coupled boundary value problem. As the calcium leaching is a slow process, we choose to decouple the chemical and the mechanical parts of the problem in the actual numerical implementation. That is, the transient problem of calcium leaching is solved first to obtain the calcium concentration field. Then, this latter is injected into the coupled elastic-damage sub-problem to obtain the displacement and damage fields.

Of course, a fully coupled resolution of the field equations can easily be performed from the developments that follow. We first describe the numerical resolution of the calcium leaching problem. The elastic-damage sub-problem is treated next.

4.1 *The calcium leaching part of the problem*

The weak form of the calcium dissolution-diffusion equation (12) is given by:

$$\int_\Omega \delta c \frac{\partial \tilde{S}(c)}{\partial t}\, d\Omega + \int_\Omega \mathbf{grad}\delta c . \tilde{D}(c) \mathbf{grad} c \, d\Omega = 0, \tag{18}$$

which must hold for all admissible variations δc. For simplicity, Dirichlet boundary conditions are assumed on the entire boundary $\partial\Omega$. Here and in all what follows we adopt the notation

$$\tilde{S}(c) = s(c) + \phi(c)c, \tag{19}$$

which is a given function together with $\tilde{D}(c)$.

Many strategies have been developed to solve this problem. The numerical resolution is based on an iterative procedure of the discrete version of (18). Typically, this requires a linearization and, accordingly, one solves a sequence of successive linear problems. However, one can expect severe numerical difficulties and use of refined mesh is necessary because of the strong nonlinearities exhibited by $\tilde{S}(c)$ and $\tilde{D}(c)$ functions, see Figures 1–3. Another way is to treat this equation as a Stefan-like problem as in (Ulm et al. 2002), or resolve it via the finite volumes method as in (Mainguy et al. 2000).

In this work instead, and motivated by a recent algorithm proposed in (Nedjar 2002b) within the context of nonlinear heat problems involving phase change, we employ a *relaxed* linearization as described below.

4.1.1 *Relaxed linearization*

The procedure will concern the highly nonlinear functions $\tilde{S}(c)$ and $\tilde{D}(c)$. The key idea proceeds in three steps:

- In the first step, use is made of the reciprocal forms of the precedent two functions. That is, we introduce the functions:

$$c = \mathcal{S}(\tilde{S}) \quad \text{and} \quad c = \mathcal{D}(\tilde{D}), \tag{20}$$

  which can be easily deduced from $\tilde{S}(c)$ and $\tilde{D}(c)$ respectively.

- In the second step, the linearizations of the above functions $\mathcal{S}$ and $\mathcal{D}$ are given by:

$$\begin{aligned} c^{(i+1)} &= \mathcal{S}(\tilde{S}^{(i)}) + \mathcal{S}'(\tilde{S}^{(i)})\Delta\tilde{S}^{(i)}, \\ c^{(i+1)} &= \mathcal{D}(\tilde{D}^{(i)}) + \mathcal{D}'(\tilde{D}^{(i)})\Delta\tilde{D}^{(i)}, \end{aligned} \tag{21}$$

  where, $C^{(i+1)} = C^{(i)} + \Delta c^{(i)}$, $\Delta\tilde{S}^{(i)} = \tilde{S}^{(i+1)}_{-}\tilde{S}^{i}$, $\Delta\tilde{D}^{(i)} = D^{(i+1)}_{-}\tilde{D}^{(i)}$, and where S' and $\mathcal{D}'$ are the derivatives of $\mathcal{S}$ and $\mathcal{D}$ with respect to their argument. The superscripts (i) and $(i+1)$ refer to the iteration number.

  The identities (21) are equivalently written as:

$$\begin{aligned} \Delta\tilde{S}^{(i)} &= \frac{1}{\mathcal{S}'(\tilde{S}^{(i)})}\left[\Delta c^{(i)} + (c^{(i)} - \mathcal{S}(\tilde{S}^{(i)}))\right], \\ \Delta\tilde{D}^{(i)} &= \frac{1}{\mathcal{D}'(\tilde{D}^{(i)})}\left[\Delta c^{(i)} + (c^{(i)} - \mathcal{D}(\tilde{D}^{(i)}))\right]. \end{aligned} \tag{22}$$

- And in the third step, the latter incrementations (22) are relaxed by replacing the quantities $1/\mathcal{S}'(\tilde{S}^{(i)})$ and $1/\mathcal{D}'(\tilde{D}^{(i)})$ by constant quantities μ and γ in all the domain and during the whole iterative process as:

$$\begin{aligned} \Delta\tilde{S}^{(i)} &= \mu\left[\Delta c^{(i)} + (c^{(i)} - \mathcal{S}(\tilde{S}^{(i)}))\right], \\ \Delta\tilde{D}^{(i)} &= \gamma\left[\Delta c^{(i)} + (c^{(i)} - \mathcal{D}(\tilde{D}^{(i)}))\right]. \end{aligned} \tag{23}$$

The relaxation parameters μ and γ must satisfy the respective conditions:

$$\mu \leq \frac{1}{\max \mathcal{S}'(\tilde{S})} \qquad \gamma \leq \frac{1}{\max \mathcal{D}'(\tilde{D})} \tag{24}$$

where $\mathcal{S}'(\tilde{S})$ and $\mathcal{D}'(\tilde{D})$ are determined from the $S(\tilde{S})$ and $\mathcal{D}(\tilde{D})$ curves. In our numerical examples, we use $\mu = 1/\max \mathcal{S}'(\tilde{S})$ and $\gamma = 1/\max \mathcal{D}'(\tilde{D})$.

4.1.2 *Resolution algorithm*

To solve the problem (18) using the precedent algorithmic procedure, we first perform a finite difference scheme in time. In this paper, use is made of a backward-Euler scheme as follows. Consider a typical time sub-interval $[t_n, t_{n+1}]$, then, starting from a known conveged state $(c_n, \tilde{S}_n, \tilde{D}_n)$ at time $t = t_n$, we look for the new state $(c_{n+1}, \tilde{S}_{n+1}, \tilde{D}_{n+1})$ at time $t = t_{n+1}$ by solving:

$$\int_\Omega \delta c \frac{\tilde{S}_{n+1}}{\Delta t}\, d\Omega + \int_\Omega \mathbf{grad}\delta c . \tilde{D}_{n+1} \mathbf{grad} c_{n+1}\, d\Omega = \int_\Omega \delta c \frac{\tilde{S}_n}{\Delta t}\, d\Omega, \tag{25}$$

where $\Delta_1 = t_{n+1} - t_n$. A *classical* linearization of (25) gives:

$$\int_\Omega \delta c \frac{\Delta \tilde{S}^{(i)}_{n+1}}{\Delta t}\, d\Omega + \int_\Omega \mathbf{grad}\delta c.\mathbf{grad} c^{(i)}_{n+1} \Delta \tilde{D}^{(i)}_{n+1}\, d\Omega + \int_\Omega \mathbf{grad}\delta c.\tilde{D}^{(i)}_{n+1} \mathbf{grad}\Delta c^{(i)}_{n+1}\, d\Omega = R^{(i)}_{n+1} \tag{26}$$

where the residual $R^{(i)}_{n+1}$ at iteration (i) is given by

$$R^{(i)}_{n+1} = \int_\Omega \delta c \frac{\tilde{S}_n - \tilde{S}^{(i)}_{n+1}}{\Delta t}\, d\Omega - \int_\Omega \mathbf{grad}\delta c.\tilde{D}^{(i)}_{n+1} \mathbf{grad} c^{(i)}_{n+1}\, d\Omega. \tag{27}$$

In (26) and (27), subscripts refer to the time step and superscripts refer to the iteration within the time step.

The *relaxed* linearization of the problem (18) is then obtained by replacing the incrementations (23) into the first and second terms of the left hand side of (26):

$$\begin{aligned}\int_\Omega \delta c \frac{\mu \Delta c^{(i)}_{n+1}}{\Delta t}\, d\Omega &+ \int_\Omega \mathbf{grad}\delta c.\gamma \mathbf{grad} c^{(i)}_{n+1} \Delta c^{(i)}_{n+1} d\Omega \\ &+ \int_\Omega \mathbf{grad}\delta c.\tilde{D}^{(i)}_{n+1} \mathbf{grad}\Delta c^{(i)}_{n+1}\, d\Omega \\ &= R^{(i)}_{n+1} \\ &- \int_\Omega \frac{\mu}{\Delta t}(c^{(i)}_{n+1} - \mathcal{S}(\tilde{S}^{(i)}_{n+1}))\, d\Omega \\ - \int_\Omega \mathbf{grad}\delta c.\mathbf{grad} c^{(i)}_{n+1}\, &\gamma (c^{(i)}_{n+1} - \mathcal{D}(\tilde{D}^{(i)}_{n+1}))\, d\Omega.\end{aligned} \tag{28}$$

At the end each iteration, the updating procedure is accomplished as follows:

$$\begin{aligned} c^{(i+1)}_{n+1} &= c^{(i)}_{n+1} + \Delta c^{(i)}_{n+1}, \\ \tilde{S}^{(i+1)}_{n+1} &= \tilde{S}^{(i)}_{n+1} + \mu \left[c^{(i+1)}_{n+1} - \mathcal{S}(\tilde{S}^{(i)}_{n+1}) \right], \\ \tilde{D}^{(i+1)}_{n+1} &= \tilde{D}^{(i)}_{n+1} + \gamma \left[c^{(i+1)}_{n+1} - \mathcal{D}(\tilde{D}^{(i)}_{n+1}) \right]. \end{aligned} \tag{29}$$

From the finite element point of view, the update $(29)_1$ is accomplished at the nodal level while the updates $(29)_2$ and $(29)_3$ are performed at the integration points. For completness, Table 1 summarizes the algorithmic iterative scheme to solve the calcium leaching part of the problem.

Table 1. Iterative scheme to solve the calcium leaching part of the problem.

1. Initialization from a converged solution at $t = t_n$: $i = 0$, $c^{(0)}_{n+1} = c_n$, $\tilde{S}^{(0)}_{n+1} = \tilde{S}_n$, $\tilde{D}^{(0)}_{n+1} = \tilde{D}_n$
2. Integrate the element matrices and residuals from (28) and (27)
3. Assemble and solve for a new calcium concentration increment: $\Delta c^{(i)}_{n+1}$
4. Update the nodal values of calcium concentration by using the update $(29)_1$
5. Update locally at each integration point by using the formulae $(29)_2$ and $(29)_3$
6. Set $i \longleftarrow i+1$ and go to **2**

4.2 *The elastic-damage part of the problem*

Following standard arguments, the two sets of balance equations (16) and (17) are equivalent to the following two weak forms:

$$\int_\Omega \nabla^s(\delta \mathbf{u}) : \beta \mathbf{C}(c) : \nabla^s(\mathbf{u})\, d\Omega = G_{ext}(\delta \mathbf{u}), \tag{30}$$

$$\left\{ \begin{aligned} \int_\Omega \mathbf{grad}\delta\beta.k\mathbf{grad}\beta\, d\Omega &+ \int_\Omega \delta\beta\, S(\boldsymbol{\epsilon})\beta^n\, d\Omega \\ &= \int_\Omega \delta\beta\, W(c)\, d\Omega, \\ \text{with } \dot{\beta}(\mathbf{x},t) \le 0, \quad \forall \mathbf{x} &\in \Omega, \forall t, \end{aligned} \right. \tag{31}$$

which must hold for all admissible variations $\delta\mathbf{u}$ and $\delta\beta$, of displacement and damage, respectively. In (30), $G_{ext}(\delta\mathbf{u})$ is a short hand notation for the virtual work of the external loading, and ∇^s denotes the symmetric gradient operator. In (31) the imposition of the constraint $(31)_2$ reflects the presence of the mathematical reaction $\partial I_-(\dot{\beta})$ in $(17)_1$.

The strategy we use to solve numerically this problem is the one proposed in (Nedjar 2002a) that improves the one used in (Nedjar 2001; Frémond and Nedjar 1996). Here the constraint $(31)_2$ is treated with an augmented Lagrangian technique.

In order to circumvent the restriction imposed by $(31)_2$, we first introduce an additional variable r over Ω and which we define as the *damage history* variable:

$$r(\mathbf{x},t) = \min_{\tau \le t}\{\beta(\mathbf{x},\tau)\}. \tag{32}$$

Next, the constraint $(31)_2$ is replaced by the following equivalent one:

$$\beta(\mathbf{x},t) - r(\mathbf{x},t) \le 0, \quad \forall \mathbf{x} \in \Omega, \forall t. \tag{33}$$

Table 2. Nested augmented Lagrangian scheme for the elastic-damage part of the problem.

1. Initialization:
set $\eta^{(0)} = \langle \eta + \varepsilon(\beta - r) \rangle$ from the last step,
$k = 0$.
2. Solve the coupled equations (30) and (34) for $\mathbf{u}^{(k)}$ and $\beta^{(k)}$
3. Check for constraint satisfaction
IF $(\beta^{(k)}(\mathbf{x}) - r(\mathbf{x}) \leq TOL \; \forall \mathbf{x} \in \Omega)$ THEN
Convergence. EXIT
ELSE
Augment:
$\eta^{(k+1)} = \langle \eta^{(k)} + \varepsilon(\beta^{(k)} - r) \rangle$
$k \longleftarrow k + 1$
GOTO **2**
END IF

Then, the damage part of the sub-problem (31) is rewritten in the following single form:

$$\int_\Omega \mathbf{grad}\delta\beta.k\mathbf{grad}\beta \, d\Omega + \int_\Omega \delta\beta \, S(\boldsymbol{\epsilon})\beta^n \, d\Omega + \int_\Omega \delta\beta \, < \eta + \varepsilon(\beta - r) > \, d\Omega = \int_\Omega \delta\beta \, W(c) \, d\Omega, \quad (34)$$

where $<.>$ is the Macauley bracket defined as $<x> = \frac{1}{2}[x + |x|]$, ε is a *fixed* penalty parameter, and is a *penalty regularization* parameter. The idea of this method is to search the correct η in an iterative process. Equation (34) is solved with fixed current estimate $\eta^{(k)} \geq 0$ of η at iteration (k) and then, use is made of the following update formula

$$\eta^{(k+1)} = < \eta^{(k)} + \varepsilon(\beta - r) > . \quad (35)$$

This update procedure has been motivated by a similar strategy developed in (Simo and Laursen 1992) for the treatment of frictionless contact problems. Notice that it is also necessary to solve the coupled problem (30) and (34) in an iterative manner. In this work, use is made of the nested iterative scheme, as refered to in (Simo and Laursen 1992). That is, the coupled problem is solved completely before the update (35) is performed. For the sake of clarity, Table 2 summarizes the conceptual steps involved during the resolution. Recall that within this step of the algorithmic resolution, the field of the calcium concentration c is considered as a fixed given data.

In the scheme considered here, ξ is fixed during the procedure. In practice ξ is chosen to be as large as possible without including ill-conditioning. The satisfaction of the constraint (32) (or equivalently $(31)_2$) can be improved even if ξ is undersized through repeated application of the augmentation procedure which only changes $\eta^{(k)}$.

The advantage of the present resolution strategy is, on one hand, the possibility to use a monolithic (or simultaneous) scheme to solve the coupled equations in the inner-loop step (step 2 of Table 2) and, on the other hand, to preserve the structure of the actual algorithmic treatments of finite strains elastic and elastoplastic problems widely investigated in the literature, taking then advantage of their good convergence properties when Newton-Raphson solution schemes are adopted.

Within the inner-loop in step 2 of Table 2, the coupled nonlinear equations (30) and (34) are solved here by means of the Newton-Raphson method. Accordingly, this requires the linearization of these equations about a known state $\{\mathbf{u},\beta\} = \{\mathbf{u}^{(i)}, \beta^{(i)}\}$ at each *inner* iteration (i). For more details about this linearization together with its finite element discretization, we refer to (Nedjar 2002a).

4.3 *Discretization of the elastic-damage part*

From the nature of the governing equations (30) and (34), damage is treated as a continuous field. From the finite element viewpoint, damage is required at the nodal level. Moreover, the same shape functions are used for the displacement and the damage fields. As usual, consider a discretization of the domain $\Omega \subset \mathbb{R}^{ndim}$ ($n_{\text{dim}} = 2$ or 3 is the space dimension) into a collection of finite elements Ω_e. The interpolations of the displacement and damage fields over a typical element Ω_e take the form:

$$\mathbf{u}_e(\boldsymbol{\zeta}, t) = \sum_{A=1}^{n^e_{\text{node}}} N^A(\boldsymbol{\zeta})\mathbf{d}^e_A(t), \qquad \beta_e(\boldsymbol{\zeta}, t) = \sum_{A=1}^{n^e_{\text{node}}} N^A(\boldsymbol{\zeta})\beta^e_A(t), \quad (36)$$

where $\mathbf{d}^e_A(t) \in \mathbb{R}^{ndim}$ and $\beta^e_A(t) \in \mathbb{R}$ denote the nodal displacements and damage, respectively, associated with node $A.N^A(\boldsymbol{\zeta})$ are the shape functions.

Here the coupled elastic-damage sub-problem is solved by using a monolithic scheme. The element contributions to the tangent stiffness matrix associated with the element nodes are written as:

$$\mathbf{K}^{AB}_e = \begin{bmatrix} \mathbf{K}^{AB}_{e11} & \mathbf{K}^{AB}_{e12} \\ \mathbf{K}^{AB}_{e21} & K^{AB}_{e22} \end{bmatrix} \in \mathbb{R}^{(n_{\text{dim}}+1)\times(n_{\text{dim}}+1)}, \quad (37)$$

for $A, B = 1, \ldots n^e_{\text{node}}$. In this matrix, the first column (row) is associated with the n_{dim} components of the nodal displacements while the second column (row) is associated with the nodal damage:

$$
\begin{aligned}
\mathbf{K}_{e11}^{AB} &= \int_{\Omega_e} (\mathbb{B}[N^A])^T \mathbf{C}(c)\,(\mathbb{B}[N^B])\,\mathrm{d}\Omega_e,\\
\mathbf{K}_{e12}^{AB} &= \int_{\Omega_e} (\mathbb{B}[N^A])^T \mathbf{C}(c):\boldsymbol{\epsilon}\,N^B\,\mathrm{d}\Omega_e,\\
\mathbf{K}_{e21}^{AB} &= \int_{\Omega_e} N^A\,\beta^n \left[\frac{\partial S(\boldsymbol{\epsilon})}{\partial \boldsymbol{\epsilon}}\right]^T (\mathbb{B}[N^B])\,\mathrm{d}\Omega_e,\\
K_{e22}^{AB} &= \int_{\Omega_e} (\nabla[N^A])^T.k(\nabla[N^B])\,\mathrm{d}\Omega_e\\
&+ \int_{\Omega_e} N^A\{n\beta^{(n-1)}S(\boldsymbol{\epsilon})+\varepsilon\tilde{H}\}N^B\,\mathrm{d}\Omega_e,
\end{aligned}
\tag{38}
$$

where $\nabla[N^A]$ and $\mathbb{B}[N^A]$ are the usual discrete gradient and symmetric gradient operators, respectively. $\tilde{H}$ is an homographic approximation of the Heaviside function $H(\eta+\varepsilon(\beta-r))$.

The damage history variable $r(\mathbf{x},t)$ and the penalty regularization parameter $\eta(\mathbf{x},t)$ are stored in an element data base at the integration points level during the iterative process.

5 NUMERICAL VALIDATION

As an illustration of the modelling framework developed in this paper, we give in this section numerical simulations within the context of the finite element method. The model has been implemented in an enhanced version of the CESAR-LCPC finite element software, see (Humbert 1989). The first section concerns the simulation of a calcium leaching test under ammonium nitrate while the second section concerns mechanical tests on the chemically degraded samples of the first section. All the simulations are compared with experimental results obtained on cement paste.

5.1 *Calcium leaching under ammonium nitrate solution*

We consider a cylindrical specimen of radius $R=15\,\text{mm}$ and height $h=30\,\text{mm}$, and submitted to calcium leaching under ammonium nitrate solution. The finite element mesh discretization we use is given in Figure 5.

For the numerical simulation, an axisymmetric analysis is performed. The boundary conditions consist on zero mass fluxes on the top and bottom boundaries (the specimen is isolated there), and on imposed calcium concentration $c=0[mol/m^3]$ on the rest of the boundary (see Figure 5).

The characteristic material data used during the computation are the smoothed versions of the $s(c), \phi(c)$ and $\tilde{D}(c)$ curves shown in Figures 1–3

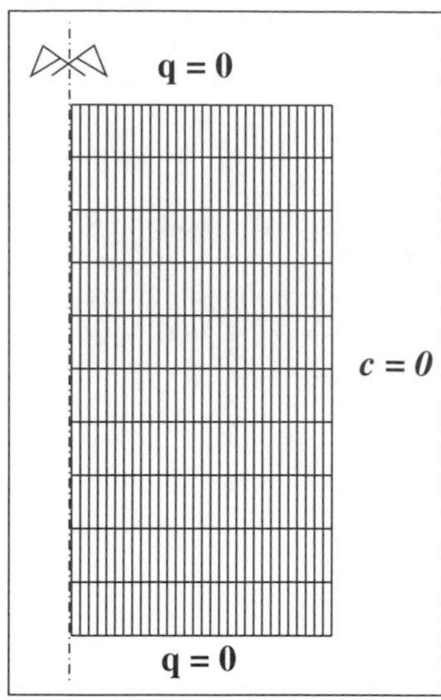

Figure 5. Finite element mesh and boundary conditions.

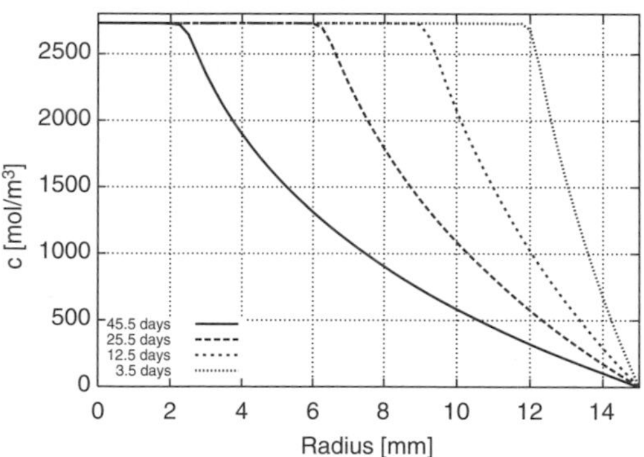

Figure 6. Calcium concentration profile at different times.

to circumvent evident numerical difficulties due to the δ-Dirac type behaviour. Note that for calcium leaching under almost demineralized water, this smoothing procedure is not necessary.

To initiate the computation, the initial condition for the calcium concentration in the liquid phase is given by $c_0(\mathbf{x}, t=0)=2730\,[mol/m^3]$ for all $\mathbf{x}\in\Omega$, *i.e.* which corresponds to the initial value $s_0(\mathbf{x}, t=0)=14700\,[mol/m^3]$ for the calcium concentration in the solid phase.

Figure 6 shows a sequence of computed calcium concentration profiles at different times of degradation. The corresponding profiles of calcium concentration in the solid phase $s(c)$ are given in Figure 7. In this latter, one can clearly identify the two dissolution fronts of the portlandite and the C-S-H phases.

In Figure 8, we plot the evolution of the computed degraded depth versus the square root of time

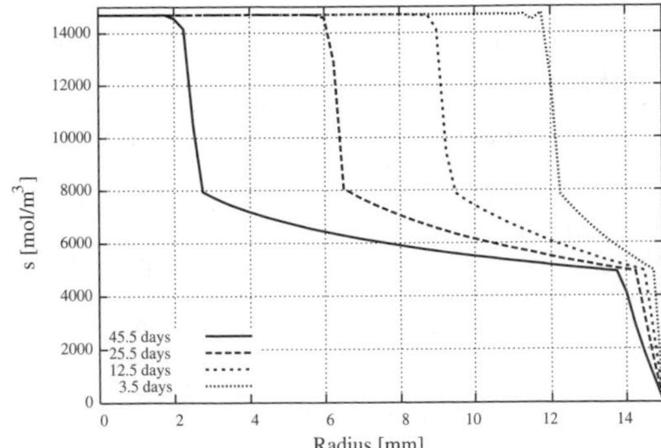

Figure 7. Corresponding calcium concentration in the solid phase.

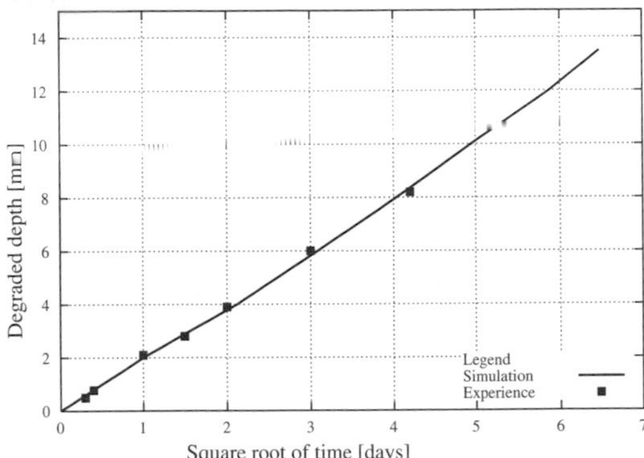

Figure 8. Degraded depth versus the square root of time. Comparison with experimental data.

(in days). As expected, an almost linear relation is obtained and, moreover, it is in good adequation with the corresponding experimental results from (Carde 1996).

5.2 *Mechanical damage. Compression at different levels of calcium leaching degradation*

The precedent cylindrical specimen is submitted to compressive tests at different levels of calcium leaching degradation depths obtained from the results of the last section. Use is made of the elastic-damage model of section 3. Again, note that the chemo-mechanical coupling is present only through the Young's modulus $E(c)$ and the first (mechanical) damage threshold $W(c)$.

For the elastic material data, the $E(c)$ function is the one given in Figure 4, with $E(c) = E_0 = 16000\,MPa$ when the material is unleached and undamaged, and the Poisson's ration is $\nu = .2$.

For the mechanical damage parameters, the first damage threshold follows the same evolution as the

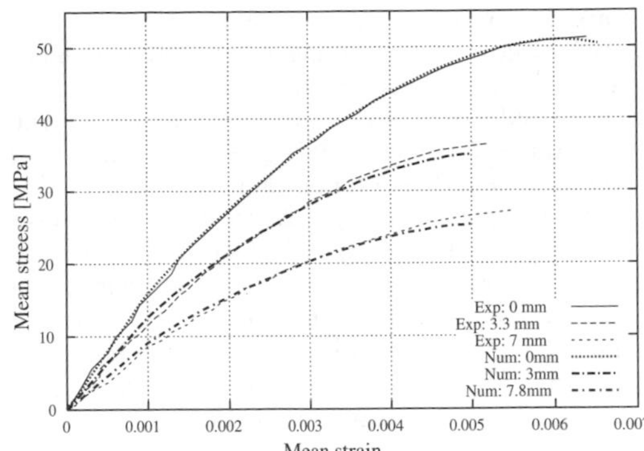

Figure 9. Mean stress versus mean strain curves at different degraded depths.

Young's modulus: $W(c) = (E(c)/E_0)W_0$, where here $W_0 = 5.310^{-4}\,MPa$.

The factor of influence of damage gradient is $k = .1\,MPa.mm^2$. The source of damage function $S(\varepsilon)$ (see (14) and (17)) depends here on the positive strains and is given by:

$$S(\boldsymbol{\epsilon}) = \frac{1}{2}\left[2\mu(c)\boldsymbol{\epsilon}^+ : \boldsymbol{\epsilon}^+\lambda(c) \ < \mathrm{tr}[\boldsymbol{\epsilon}] >^2\right], \quad (39)$$

where tr denotes the trace operator.

In this work, the exponent parameter has been adapted to vary with the damage variable: $n \equiv n(\beta)$. That is, to make it decrease when damage grows, *i.e.* when β decreases. For the sake of simplicity, we have made the following choice:

$$n(\beta) = n_0\beta \quad (40)$$

with $n_0 = 10$ in the following computations. Of course, other functions instead of (40) can be easily adapted. One can think for example to make n depend on the source of damage function: $n = n(S(\varepsilon))$. Finally, the penalty parameter ε we have used in all the following computations is $\varepsilon = 10^4$.

The result of the compression tests for different levels of calcium leaching degradation are shown in Figure 9. They are superposed with the experimental results of (Carde 1996), see also (Carde et al. 1996), for different degradation depths. Again, one can see the good correspondance between the numerical simulations and the experimental test results.

6 CONCLUSION

We have presented in this work a phenomenological modelling of chemo-mechanical degradation to describe the degradation of cement-based materials. Use has been made of the well known simplified approach of calcium leaching coupled with a continuum damage model involving the gradient of damage quantity. We have succesfully simulated the experimental results of (Carde 1996) as this was done with other proposed models in (Carde et al. 1996; Gérard 1996; Ulm et al. 1999) for example.

One of our contributions was to numerically solve the calcium leaching part of the problem with an algorithmic treatment based on a *relaxed* linearization procedure due to strong nonlinearity of the problem. Although its implementation was decoupled from the elastic-damage part of the problem, a fully coupled resolution procedure can be easily performed.

Of course, other chemo-mecanical couplings than the present $E(c)$ and $W(c)$ couplings can be adapted. For instance, apparent calcium conductivity must also depend on mechanical damage. All the possible coupling effects are actually under investigation together with the inclusion of the plasticity in the modelling.

On the experimental point of view, we are planning degradation tests on concrete and fully coupled degradation-creep tests.

REFERENCES

Bangert, F., D. Kuhl, and G. Meschke (2001). Finite element simulation of chemo-mechanical damage under cyclic loading conditions. In R. de Borst et al. (Ed.), *Fracture Mechanics of Concrete Structures*, pp. 145–152. Swets and Zeitlinger, Lisse.

Carde, C. (1996). *Caractérisation et modélisation de l'altération des propriétés mécaniques due à la lixiviation des maériaux cimentaires*. Ph. D. thesis, INSA Toulouse.

Carde, C., R. François, and J. M. Torrenti (1996). Leaching of both calcium hydroxyde and C-S-H from cement paste: modeling the mechanical behaviour. *Cement and Concrete Res.* **28**(*6*), 1257–1268.

Coleman, B. D. and M. E. Gurtin (1967). Thermodynamics with internal variables. *J. Chem. Phys.* **47**, 597–613.

Frémond, M. and B. Nedjar (1993). Endommagement et principe dea puissances virtuelles. *C. R. Acad. Sci. Paris serie II* **317**, 857–864.

Frémond, M. and B. Nedjar (1996). Damage, gradient of damage and principle of virtual power. *Int. J. Solids Struct.* **33** *(8)*, 1083–1103.

Gérard, B. (1996). *Contribution des couplages mécaniques-chimie-transfert dans la tenue à long terme des ouvrages de stockage de déchets radioactifs*. Ph. D. thesis, ENS Cachan.

Humbert, P. (1989). CESAR-LCPC, un code général de calcul par éléments finis. *Bull. Liais. Labo. des Ponts et Chaussées* **119**, 112–116.

Le Bellégo, C., B. Gérard, and G. Pijaudier-Cabot (2001). Mechanical analysis of concrete structures submitted to an aggressive water. In R. de Borst et al. (Ed.), *Fracture Mechanics of Concrete Structures*, pp. 239–246. Swets and Zeitlinger, Lisse.

Mainguy, M., C. Tognazzi, J. M. Torrenti, and F. Adenot (2000). Modelling of leaching in pure cement paste and mortar. *Cement and Concrete Research* **30**, 83–90.

Nedjar, B. (2001). Elastoplastic-damage modelling including the gradient of damage: formulation and computational aspects. *Int. J. Solids Struct.* **38**, 5421–5451.

Nedjar, B. (2002a). An augmented lagragian treatment of continuum damage problems involving the gradient of damage. In H. Mang, F. Rammerstrofer, and J. Eberhardsteiner (Eds.), *Proc. WCCM V, Fifth World Congress on Computational Mechanics*. Vienna University of Technology, Austria, ISBN 3-9501554-0-6.

Nedjar, B. (2002b). An enthalpy-based finite element method for nonlinear heat problems involving phase change. *Computers and Structures* **80**, 9–21.

Simo, J. C. and T. A. Laursen (1992). An augmented lagrangian treatment of contact problems involving friction. *Computers and Structures* **42***(1)*, 97–116.

Tognazzi, C. (1998). *Couplage fissuration-dégradation chimique dans des matériaux cimentaires : caractérisation et modélisation.* Ph. D. thesis, INSA Toulouse.

Torrenti, J. M., F. Adenot, C. Tognazzi, S. Danese, and S. Poyet (1998). Application du modéle de dégradation du béton au cas des milieux fissurés et au couplage avec la mécanique. In *Concrete: from material to structure proceedings of the international RILEM conf.* Arles.

Torrenti, J. M., O. Didry, J. P. Ollivier, and F. Plas (1999). *La dégradation des bétons – couplage fissuration-dégradation chimique*. Hermés.

Torrenti, J. M., M. Mainguy, F. Adenot, and C. Tognazzi (1998). Modelling of leaching in concrete. In R. de Borst, N. Bicanic, H. Mang, and G. Meschke (Eds.), *Proc. Euro-C 98, Computational Modelling of Concrete Struct.*, pp. 531–538. Balkema, Rotterdam, The Netherland.

Ulm, F. J., E. Lemarchand, and F. H. Heukamp (2002). Elements of chemomechanics of leaching of cement-based materials at different scales. *Engng. Frac. Mechanics in press.*

Ulm, F. J., J. M. Torrenti, and F. Adenot (1999). Chemoporoplasticity of calcium leaching in concrete. *Journal of Engineering Mechanics (ASCE)* **125***(10)*, 1200–1211.

Computational Modelling of Concrete Structures, Bićanić et al. (eds)
© 2003 Swets & Zeitlinger, Lisse, ISBN 90 5809 536 3

Thermal degradation in heterogeneous concrete materials

K. Willam, I. Rhee & Y. Xi
University of Colorado at Boulder, USA

ABSTRACT: In this paper we will examine the interaction of thermal and mechanical damage processes in heterogeneous materials such as concrete. After a brief introduction of the effect of coupling on the conditioning of algebraic and differential problems, we will address two topics: (a) Interaction of Thermal and Mechanical Damage. In the first part we will explore the formulation of thermal and mechanical damage at the continuum level of observations, when thermal softening is interacting with mechanical degradation. To this end we will contrast two thermoelastic formulations which will illustrate the effect of stress/strain on the effective thermal expansion. (b) Thermal Stress and Deformation Studies. In the second part we will consider thermal effects in heterogeneous composites with applications to concrete materials when thermal softening leads to massive degradation of their resistance.

1 INTRODUCTION

In this paper we examine thermal deterioration effects in heterogeneous two-phase materials such as concrete. Starting from continuum thermodynamics we consider different forms of mechanical and thermal coupling, and in particular the stress dependence of thermal expansion.

The thermal degradation formulation is applied to model based simulations of two-phase concrete composites. This application illustrates the effects of thermoelastic coupling when mismatch of the two-phase material introduces localized degradation at the interfaces between the two constituents. The numerical results will be compared with experimental observations on concrete specimens which are subjected to combined thermal and mechanical load histories in order to explore thermal and mechanical interaction effects in steady and transient response regimes.

2 MOTIVATION

In the aftermath of extreme events in the recent past the performance of cement-based materials has come to forefront in fire safety assessment of tunnels and high rise buildings. An accurate mathematical characterization of cement-based composites, which describes thermal softening and dimensional changes due to changes in the temperature environment has been the focus in the high temperature reactor industry for over thirty years. It is indeed unfortunate that several accidents in the recent past demonstrate the needs for a more comprehensive understanding of cement-based materials exposed to fire environments. A successful predictive model will have great implications for high temperature analysis and design of cement-based material systems with direct implication to fire safety assessment of concrete structures. The main deficiency of current constitutive models is the fact that they are calibrated from isothermal conditions in which the transitional effects of variable temperature, humidity, chemical and mechanical loading are neglected. Moreover, no established theory exists at present how to construct high temperature constitutive relations and cohesive interface properties that are based on micro- and mesostructural features of the adherent bulk materials.

In the back of these fundamental issues on multiscale analysis and design of high temperature materials are the fields of fire protection and fire resistance of cement-based materials. Thereby, the exploratory study is intended as a proof of concept for multiscale engineering to reduce ablation and spall effects in thermal shock problems when fire safety issues are considered.

The performance of cement-based materials under elevated temperatures are very complicated and difficult to characterize. With increasing temperature there is a decrease of compressive strength, density, thermal conductivity, thermal diffusivity in concrete [Thelandersson 1987; Schneider 1988; Thienel & Rostasy 1995; Phan 1996; Bazant & Kaplan 1996; Shin et al. 2002; Poon et al. 2001] which are caused by three main processes taking place at elevated temperatures:

(a) Phase transformation processes – Loss of free water occurs at about 100°C. Decomposition of calcium hydroxide takes place at about 450°C.
(b) Evolution processes in pore structure – The volume and surface of pores increase up to a temperature of approximately 500°C, and decrease with further increase of temperature.
(c) Thermo-hygro-mechanical coupled processes – The thermo-mechanical coupling is associated with the temperature gradient upon rapid heating causing severe thermal stress in concrete. The thermo-hygral coupling is associated with the multiphase transport of water in liquid phase, moisture in vapor phase, and air in gas phase.

3 COUPLING OF PARTITIONED SYSTEMS

Coupled diffusion equations provide an appropriate structure of the governing transport equations of heat and moisture, which need to be extended in order to include the effects of elastic and inelastic degradation. In fact, cracking due to high temperature effects will result in significant variation of the *effective* material properties, while stable crack extension will in turn affect the transport properties.

The coupling of thermal diffusion with the quasi-static equations of motion may be written as a system of first ODE after semi-discretization with finite elements in space:

$$\begin{bmatrix} \mathbf{C}_{TT} & \mathbf{C}_{TU} \\ \mathbf{C}_{UT} & \mathbf{C}_{UU} \end{bmatrix} \begin{bmatrix} \dot{\mathbf{T}} \\ \dot{\mathbf{U}} \end{bmatrix} + \begin{bmatrix} \mathbf{K}_{TT} & \mathbf{K}_{TU} \\ \mathbf{K}_{UT} & \mathbf{K}_{UU} \end{bmatrix} \begin{bmatrix} \mathbf{T} \\ \mathbf{U} \end{bmatrix} = \begin{bmatrix} \mathbf{f}_T \\ \mathbf{f}_E \end{bmatrix} \tag{1}$$

Here the coupled system of first order differential equations has been expressed in terms of the discrete thermal and mechanical nodal variables [**T**, **U**] that interconnect the response of finite elements in 3-D space and time. For example, the four-node simplex tetrahedron has four degrees of freedom per node, three displacement degrees of freedom, $\mathbf{U} = [u(\mathbf{x}_i, t), v(\mathbf{x}_i, t), w(\mathbf{x}_i, t)]$, one scalar temperature, $T = T(\mathbf{x}_i, t)$. The forcing term $\mathbf{f}(t)$ on the right hand side is comprised of conjugate components which describe the mechanical and environmental load histories. The viscous and the elastic material properties **C**, **K** and the conjugate capacity and diffusion properties are highly nonlinear at the macro-scale, but they become increasingly simple (constant) at the micro-structural scales. At least, this is the main conjecture behind multi-scale material engineering that the geometric complexities of the micro-structural features compensate for the elementary format of material behavior.

3.1 *Algebraic coupling*

For the sake of argument, let us first consider the static case of a thermal and an elastic partition, when the differential relations reduce to the algebraic form,

$$\begin{bmatrix} \mathbf{K}_{TT} & \mathbf{K}_{TU} \\ \mathbf{K}_{UT} & \mathbf{K}_{UU} \end{bmatrix} \begin{bmatrix} \mathbf{T} \\ \mathbf{U} \end{bmatrix} = \begin{bmatrix} \mathbf{f}_T \\ \mathbf{f}_U \end{bmatrix} \tag{2}$$

In this case, the Schur theorem states that the determinant of the monolithic system is the product of the determinants of the diagonal partition and its Schur complement,

$$\det \mathbf{K} = \det \mathbf{K}_{UU} \det[\mathbf{K}_{TT} - \mathbf{K}_{TU}\mathbf{K}_{UU}^{-1}\mathbf{K}_{UT}] \tag{3}$$

Assuming $\det \mathbf{K}_{UU} > 0$, the coupling partitions $\mathbf{K}_{UT}$, $\mathbf{K}_{TU}$ lead to a reduction of the product of the determinants of the diagonal partitions which is an upper bound,

$$\det \mathbf{K} \le \det \mathbf{K}_{UU} \det \mathbf{K}_{TT} \tag{4}$$

In other terms, mechanical degradation and thermal softening of each property is compounded by the coupling of the two fields. This fact is highlighted by the inverse of the partitioned matrix which clearly demonstrates the negative effect of coupling on the positiveness of the inverse, $\mathbf{K}^{-1}$,

$$\begin{bmatrix} [\mathbf{K}_{TT} - \mathbf{P}_{TT}]^{-1} & \mathbf{Q}_{TU}[\mathbf{P}_{UU} - \mathbf{K}_{UU}]^{-1} \\ [\mathbf{P}_{UU} - \mathbf{K}_{UU}]^{-1}\mathbf{Q}_{UT} & [\mathbf{K}_{UU} - \mathbf{P}_{UU}]^{-1} \end{bmatrix} \tag{5}$$

where,

$$\begin{aligned} \mathbf{P}_{TT} &= \mathbf{K}_{TU}\mathbf{K}_{UU}^{-1}\mathbf{K}_{UT}, & \mathbf{P}_{UU} &= \mathbf{K}_{UT}\mathbf{K}_{TT}^{-1}\mathbf{K}_{TU} \\ \mathbf{Q}_{TU} &= \mathbf{K}_{TT}^{-1}\mathbf{K}_{TU}, & \mathbf{Q}_{UT} &= \mathbf{K}_{UT}\mathbf{K}_{TT}^{-1} \end{aligned} \tag{6}$$

Consequently, it is erroneous to think that thermo-mechanical coupling will improve the shear banding tendencies of softening plasticity and vice versa. In other terms, the conditioning of the monolithic system depends to a great extent on the strength of the coupling partitions, $\mathbf{K}_{UT}, \mathbf{K}_{TU}$. In fact, the spectral conditioning of the partitioned system,

$$cond(\mathbf{K}) \mapsto \frac{\lambda_{max}(\mathbf{K})}{\lambda_{min}(\mathbf{K})} \tag{7}$$

deteriorates proportionally to the magnitude of the coupling partitions, $||\mathbf{K}_{UT}|| \times ||\mathbf{K}_{TU}||$. Another observation of the deleterious coupling effect is the convergence indicator of an iterative solver using successive substitution. In line with the conditioning argument, the strength of the off-diagonal partitions determines the rate of convergence of staggered iterative solvers,

$$\rho(\mathbf{K}) \mapsto \frac{||\mathbf{K}_{UT}|| \times ||\mathbf{K}_{TU}||}{||\mathbf{K}_{UU}|| \times ||\mathbf{K}_{TT}||} \le 1 \tag{8}$$

3.2 *Differential coupling*

Direct integration of the coupled transient thermomechanical relations advances the solution in the time interval $\Delta t = t_{n+1} - t_n$. Using the single step alpha-method for the normalized time interval $\alpha \in [0, 1], \exists\, t_{n+\alpha} = t_n + \alpha \Delta t$ and $\frac{d\alpha}{dt} = \frac{1}{\Delta t}$, we may evaluate the nodal displacements and nodal temperatures in terms of the generalized nodal coordinates $\mathbf{d} = [\mathbf{T}, \mathbf{U}]^t$ their rates at any stage within the time interval,

$$d_{n+\alpha} = [1 - \alpha] d_n + \alpha d_{n+1}$$
$$\dot{d}_{n+\alpha} = \frac{1}{\Delta t}[d_{n+1} - d_n]$$

Collocation of the coupled first order equations at the time $t = t_{n+\alpha}$,

$$\mathbf{C}\,\dot{\mathbf{d}}_{n+\alpha} + \mathbf{K}\,\mathbf{d}_{n+\alpha} = \mathbf{f}_{n+\alpha}$$

leads to the incremental algebraic relation and the recursion algorithm to advance the coupled system in the time step Δt:

$$\mathbf{B}_{n+1}\,\mathbf{d}_{n+1} = \mathbf{B}_n\,\mathbf{d}_n + \Delta t \mathbf{f}_{n+\alpha}$$

where

$$\mathbf{B}_{n+1} = \mathbf{C} + \alpha \Delta t \mathbf{K}$$
$$\mathbf{B}_n = \mathbf{C} - [1 - \alpha] \Delta t \mathbf{K}$$

According to the "LAX Theorem" consistency and stability are necessary and sufficient conditions for convergence.

- $\alpha = 0$ the Euler Forward marching strategy is explicit if $\mathbf{C} = diag$.
- $\alpha = 0$ the Euler Backward marching strategy is fully implicit exhibiting high numerical damping.
- $\alpha = \frac{1}{2}$ the midstep/trapezoidal rule is known under the name of Crank-Nicolson.
- $\alpha = \frac{2}{3}$ this algorithm is known under the name of Galerkin or finite element in time method since it is obtained by weighting the heat balance equation in the time interval, i.e. $\mathbf{d}(\mathbf{x}, t) = \mathbf{N}(\alpha)\mathbf{N}_I(\mathbf{x})\mathbf{d}_I(t)$.

Spectral analysis of the partitioned two-field system leads to eigenvalues $\lambda^{\min}..\lambda_i\ldots\lambda^{\max}$ which typically span 4–6 orders of magnitude in spatial finite elements applied to thermal diffusion and elasticity problems. Thereby the fundamental thermal modes dominate the long term temperature response, while the high frequency behavior is mobilized in thermal shock problems.

For the stability analysis of the alpha-method we resort to the "Modal Amplification" argument of a SDOF or rather a system of two coupled thermoelastic TDOF. For stability analysis it suffices to examine the bounded response of the homogeneous PDE which is governed by the exponential decay function. For a SDOF the argument reduces to,

$$d_{n+1} = e^{-\lambda \Delta t}\, d_n$$

where $e^{-\lambda\Delta t} \Rightarrow A(\lambda \Delta t) = \frac{1-[1-\alpha]\lambda\Delta t}{1+\alpha\lambda\Delta t}$. For stability the homogeneous response needs to be bounded, i.e. for $\lambda > 0 \rightarrow |T_{n+1}| \leq |T_n|$.

A-stability infers that the amplification function satisfies $|A| \leq 1$. Thus values of $\alpha < 0.5$ restrict $\Delta t < \Delta t_{stabil} = \frac{2}{[1-2\alpha]\lambda}$, while $\alpha \geq 0.5$ do not restrict the time step $\Delta t_{stabil} \rightarrow \infty$. L-stability infers that $|A| \rightarrow 0$ for $\lambda \Delta t \rightarrow \infty$. Note: Only the backward algorithms $\alpha = 1$ provide L-stability and strong damping of the high frequency contributions.

Stability analysis of coupled MDOF systems generalizes the argument above requiring that the "Amplification Operator" remains bounded,

$$\mathbf{A}(\lambda\Delta t) = \mathbf{B}_{n+1}^{-1}\,\mathbf{B}_n$$

Thus $\alpha \geq 0.5$ values provide a sufficient condition for bounded solutions of MDOF. In case of time steps which satisfy the conditional stability argument $\Delta t_{stabil} = \frac{2}{[1-2\alpha]\lambda_{max}}$, where the maximum frequency drives the CFL condition of the first order system.

For a detailed study of accuracy and stability of higher order α-methods we refer to Argyris, Vaz & Willam [1977, 1981], and to Felippa, Park & Farhat [2001] for a discussion of stability of partitioned systems. We conclude that the stability arguments of the coupled partitioned system are the same as those of the unpartitioned system as long as the solution of the two field problem is advanced simultaneously within a monolithic approach. In fact, it is interesting that the staggered solution method does not change the stability argument above of the $\alpha \geq 0.5$-methods as long as the computational flow uses the governing first order differential relations to evaluate the rates at the time steps.

In summary, the algebraic solution of the governing differential equations does not impose additional stability restrictions when coupled processes are considered. However, we need to keep in mind that the stability arguments are based on linear and positive definite assumptions of each component. Therefore the earlier arguments of degradation of the algebraic properties must be taken into account in the evaluation of "Modal Amplification", whereby the conditioning and positiveness of the coupled system decreases strongly with the strength of the coupling partitions, $||\mathbf{K}_{UE}|| \times ||\mathbf{K}_{UE}||$.

3.3 *Thermomechanical coupling*

Although the thermodynamic setting is well established, there are a number of issues related to the

Table 1. First & second derivatives of the free energy potential in terms of the thermal and mechanical parameters.

ψ_0	ϵ	θ	d
$\partial\psi_0$	$[1-d]\frac{\partial\psi_0}{\partial\epsilon}:\dot{\epsilon}$	$[1-d]\frac{\partial\psi_0}{\partial\theta}\dot{\theta}$	$-\psi_0\dot{d}$
$\partial^2\psi_0$	$\frac{1}{2}[1-d]\dot{\epsilon}:\frac{\partial^2\psi_0}{\partial\epsilon\otimes\partial\epsilon}:\dot{\epsilon}$	$\frac{1}{2}[1-d]\dot{\theta}\frac{\partial^2\psi_0}{\partial\theta^2}\dot{\theta}$	$\dot{d}\frac{\partial y}{\partial d}\dot{d}$
Cplg	$[1-d]\dot{\theta}\frac{\partial^2\psi_0}{\partial\theta\partial\epsilon}:\dot{\epsilon}$	$-\dot{d}\frac{\partial\psi_0}{\partial\theta}\dot{\theta}$	$-\dot{d}\frac{\partial\psi_0}{\partial\epsilon}:\epsilon$

coupling effects of thermal softening in the bulk material. Aside from the thermoelastic Gough-Joule effect of cooling under expansion in the first law of energy balance, there are important interactions in the mechanical description when transitional effects are considered (non-isothermal, non-isentropic). For illustration the classical free Helmholtz energy format of thermoelasticity is based on the additive expansion,

$$\psi(\epsilon, d, \theta) = [1-d_\epsilon]\psi_\epsilon(\epsilon) + [1-d_\theta]\psi_\theta(\theta)$$

which decouples the evolution of mechanical $[1-d_\epsilon]\psi_\epsilon$ and thermal damage $[1-d_\theta]\psi_\theta$. This is entirely different when compared to the interaction effects of the product expansion when we consider the representation of isotropic tensor functions,

$$\psi(\epsilon, d, \theta) = [1-d]\psi_0(tr\epsilon, tr\epsilon^2, tr\epsilon^3, \theta)$$

where the scalar damage measure is now a function of mechanical strain as well as temperature, $d = d(\epsilon, \theta)$. In this case the Helmholtz free energy expands into an irreducible set of strain invariants, $(tr\epsilon)$, $(tr\epsilon^2)$, $(tr\epsilon^3)$, the thermomechanical scalar damage variable, $[1-d]$, and the absolute temperature, θ. First and second derivatives of the free energy potential define the thermal and mechanical constitutive relations, tabulated below:

The second derivatives of the truncated Taylor series illustrate the interaction of temperature, damage and elastic moduli in the tangential stiffness tensor,

$$\mathbb{E}^{tan} = \frac{1}{2}[1-d]\frac{\partial^2\psi_0}{\partial\epsilon\otimes\partial\epsilon}$$

The thermal stress tensor,

$$\boldsymbol{\beta}^{tan} = [1-d]\frac{\partial^2\psi_0}{\partial\theta\partial\epsilon}$$

and the heat capacity,

$$c_v^{tan} = \frac{1}{2}[1-d]\frac{\partial^2\psi_0}{\partial\theta^2}.$$

The thermomechanical damage variable $[1-d(\epsilon, \theta)]$ reflects the interacting effects of mechanical and thermal degradation as opposed to the uncoupled format $[1-d_\epsilon]$ and $[1-d_\theta]$ of the additive energy expansion above. In fact, the thermoelastic integrity of the material is represented well by the product form,

$$[1-d] = [1-d_\epsilon][1-d_\theta] \tag{9}$$

Moreover, it is the strain dependence of the thermal stress or rather the dependence of the thermal expansion on stress, $\alpha = \alpha(\theta, \sigma)$ which leads to the dramatic reduction of axial stress due to spalling when confined concrete specimens are subjected to high temperature sweeps $20 \leq T \leq 600°C$. Figure 1 reproduced from Anderberg & Thelandersson [1976, 1987] illustrates the dramatic effect of axial restraint on the thermal stress when the temperature in the concrete sample is increased to very high levels when spalling sometimes leads to explosive failures. Figure 2 shows the effect of axial loading on the thermal expansion of the cylindrical test article indicating that the free thermal expansion reverts to contraction under the presence of axial compression.

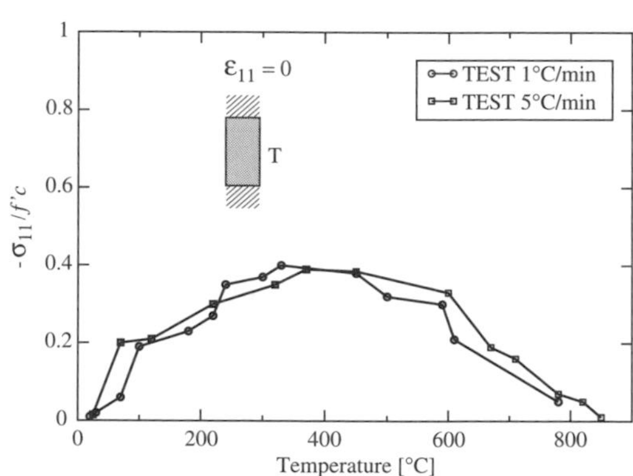

Figure 1. Axial Strains in Concrete Specimens σ_a due Increasing Temperature [Anderberg & Thelandersson, 1976].

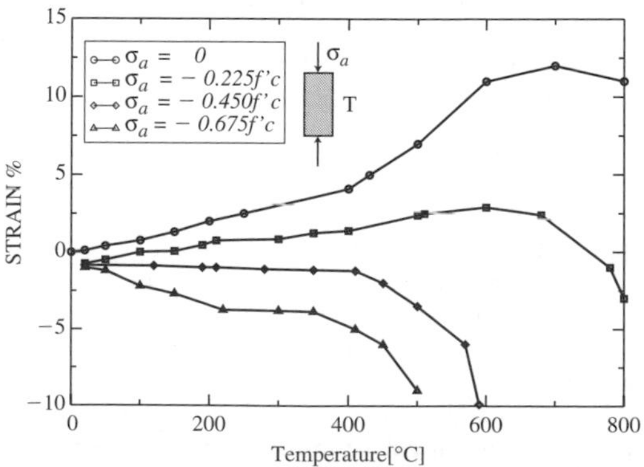

Figure 2. Thermal Stresses in Concrete Specimens σ_a due Increasing Temperature [Anderberg & Thelandersson, 1976].

In sum, the dependence of thermal expansion on the stress level is an important effect which needs to be addressed especially when transient high temperature excursions are considered.

4 CONTINUUM THERMODYNAMICS

In multiphysics there is a need to start from first principles in order to establish the governing equations of each field and their interrelationships. In the case of thermomechanical problems it is natural to start from the principles of continuum thermodynamics which guide the development of coupling between the mechanical and thermal fields, and which provide the constitutive format of the underlying state equations including interface degradation.

4.1 *First law of thermodynamics*

The conservation law of energy states that the sum of mechanical and thermal energies remains constant. In other terms, the rate of mechanical and thermal energy contributions must be a total differential such that the integral through any load cycle remains zero,

$$\oint [dT + dW^{int} - dW^{ext} - dQ\,]dt = 0$$

Here the different terms denote the kinetic energy is $T = \frac{1}{2}\int \rho \mathbf{v}\cdot \mathbf{v} d\Omega$. the internal energy per unit mass $W^{int} = \int \rho u\, d\Omega$ and the power of external body forces and surface tractions $dW^{ext} = \int \rho \mathbf{b}\cdot \mathbf{v} d\Omega + \int \mathbf{t}\cdot \mathbf{v}\, d\Gamma$. The rate of heat flow into the body, $dQ = \int \rho r\, d\Omega - \int \mathbf{n}\cdot \mathbf{q}\, d\Gamma$, includes the internal heat production, r, per unit mass and the heat flux, $\mathbf{q}$, into surface of the body with the outward normal $\mathbf{n}$. The principle of energy balance asserts there exists a state variable, u such that

$$\frac{d}{dt}\int \rho u\, d\Omega = dQ + \int \boldsymbol{\sigma} : d\boldsymbol{\epsilon}, d\Omega$$

where $dW^{ext} - dT = \int \boldsymbol{\sigma} : d\boldsymbol{\epsilon}, d\Omega$

For the sake of argument we assume small displacement gradients, thus there is no need to distinguish between the deformed and undeformed configuration, between material and time derivatives, and between spatial and material descriptions. In the case of small deformations, using the divergence theorem and invoking conservation of mass, linear and angular momenta,

- Conservation of Mass: $\dot{\rho} + \rho\, div(\dot{\mathbf{u}}) = 0$
- Conservation of Linear Momentum:

$$div\boldsymbol{\sigma} + \rho \mathbf{b} = \rho \ddot{\mathbf{u}}$$

- Conservation of Angular Momentum: $\boldsymbol{\sigma} = \boldsymbol{\sigma}^t$

The local conservation of energy reduces to,

$$\rho\dot{u} = \boldsymbol{\sigma} : \dot{\boldsymbol{\epsilon}} - div\mathbf{q} + \rho r \qquad (10)$$

4.2 *Second law of thermodynamics*

While the first law of thermodynamics is a statement of interchangeablity of energies which holds for an equilibrium state, the second law of thermodynamics defines the direction of non-equilibrium processes. E.g. heat flows from the warm side to the cold, mechanical dissipation is converted into heat. The thermodynamic state variable is the entropy $S = \int \rho s\, d\Omega$, the rate of which is always greater than the rate of heating divided by the absolute temperature and the inflow of heat through the boundary Γ. Thus the global for of the Clausius-Duhem inequality states, that

$$\frac{d}{dt}\int \rho s\, d\Omega \geq \int \frac{\rho r}{\theta}\, d\Omega - \int \frac{\mathbf{n}\cdot\mathbf{q}}{\theta}\, d\Gamma$$

Using the divergence theorem the local form of the second law of thermodynamics results in the Clausius-Duhem inequality which states that the internal entropy production rate per unit mass must be positive or zero in the case of full reversibility.

$$\rho\dot{s} - \frac{\rho r}{\theta} + \frac{div\mathbf{q}}{\theta} - \frac{\mathbf{q}\cdot\nabla_x\theta}{\theta^2} \geq 0$$

4.3 *Local Clausius-Duhem inequality*

Substituting conservation of energy and multiplying the Clausius-Duhem inequality by the absolute temperature θ, the second law reduces to the standard form of the internal energy $u = u(\boldsymbol{\epsilon}, s)$,

$$\rho\theta\dot{s} - \rho\dot{u} + \boldsymbol{\sigma} : \dot{\boldsymbol{\epsilon}} - \frac{\mathbf{q}\cdot\nabla_x\theta}{\theta} \geq 0$$

Another form of the entropy production inequality may be formulated in the form of the local dissipation inequality,

$$\mathcal{D}^{int} = \boldsymbol{\sigma} : \dot{\boldsymbol{\epsilon}} - \rho\dot{u} + \rho\theta\dot{s} \geq 0$$

when we disregard the contribution of the conduction inequality $\frac{\mathbf{q}\cdot\nabla_x\theta}{\theta} < 0$. Note the dissipation inequality must be zero in the case of reversible equilibrium processes.

4.4 *Helmholtz free energy: $\psi = \psi(\boldsymbol{\epsilon}, \theta)$*

The question is which variables are needed to described thermoelasticity and elastic damage in concrete materials. For the sake of argument let us begin with thermoelasticity which is well-understood, at least that is what we tend to believe.

In order to eliminate the entropy as an independent thermodynamic variable we replace the internal energy u by the Helmholtz free energy,

$$\psi(\boldsymbol{\epsilon},\theta) = u - \theta s \quad \text{with} \quad \dot{\psi} = \dot{u} - \theta\dot{s} - s\dot{\theta}$$

In this case the Clausius-Duhem inequality reduces to the following form,

$$\boldsymbol{\sigma} : \dot{\boldsymbol{\epsilon}} - \rho\dot{\psi} - \rho s\dot{\theta} - \frac{\mathbf{q}\cdot\nabla_x\theta}{\theta} \geq 0$$

The total differential of the Helmholtz free energy leads to:

$$\dot{\psi} = \frac{\partial\psi}{\partial\boldsymbol{\epsilon}} : \dot{\boldsymbol{\epsilon}} + \frac{\partial\psi}{\partial\theta}\dot{\theta}$$

Substituting into the Clausius-Duhem inequality,

$$[\boldsymbol{\sigma} - \rho\frac{\partial\psi}{\partial\boldsymbol{\epsilon}}] : \dot{\boldsymbol{\epsilon}} - \rho[s+\frac{\partial\psi}{\partial\theta}]\dot{\theta} - \frac{\mathbf{q}\cdot\nabla\theta}{\theta} \geq 0$$

Since the strain rate and the temperature rate can be controlled independently, the so-called Coleman relations yield the constitutive relations as,

$$\boldsymbol{\sigma} = \rho\frac{\partial\psi}{\partial\boldsymbol{\epsilon}}$$

and $s = -\frac{\partial\psi}{\partial\theta}$ where we have assumed that the thermal process must be positive-semidefinite, $\frac{\mathbf{q}\cdot\nabla_x\theta}{\theta} < 0$. This heat condition inequality is satisfied by the material law of Fourier,

$$\mathbf{q} = -\mathbf{k}\cdot\nabla_x\theta \quad \rightarrow \frac{\nabla_x\theta\cdot\mathbf{k}\cdot\nabla_x\theta}{\theta} \geq 0$$

as long as the thermal conductivity $\mathbf{k}$ remains positive definite. Thereby, the internal dissipation inequality takes the form of the local entropy production inequality

$$\mathcal{D}^{int} = \boldsymbol{\sigma} : \dot{\boldsymbol{\epsilon}} - \rho\dot{\psi} - \rho s\dot{\theta} \geq 0$$

4.5 *Isotropic energy expansion*

Let us start from the traditional decomposition of the Helmholtz free energy expansion into the sum of a mechanical and thermal part. Omitting the mass density from the energy potential below and replacing the absolute temperature by the relative temperature $\Delta T = T - T_o$, we have

$$\psi = \psi(\boldsymbol{\epsilon}, \Delta T)$$

According to the representation theorem of scalar-valued tensor functions we can expand the potential in terms of the irreducible set of three strain invariants, $(tr\boldsymbol{\epsilon}), (tr\boldsymbol{\epsilon}^2), (tr\boldsymbol{\epsilon}^3)$, and the temperature, ΔT. For the sake of argument we adopt a cubic polynomial expansion to describe small excursions from an equilibrium reference state,

$$\psi(\boldsymbol{\epsilon}, \Delta T) = b_0 + b_1(tr\boldsymbol{\epsilon}) + b_2(tr\boldsymbol{\epsilon})^2 + b_3(tr\boldsymbol{\epsilon}^2) + b_4(tr\boldsymbol{\epsilon}^2)(tr\boldsymbol{\epsilon}) + b_5(tr\boldsymbol{\epsilon}^3)$$

where the polynomial coefficients depend on temperature, $b_i = b_i(\Delta T)$. This leads to the nonlinear thermoelastic stress-strain relationship:

$$\sigma = \frac{\partial\psi}{\partial\boldsymbol{\epsilon}} = [[b_1 + 2b_2(tr\boldsymbol{\epsilon}) + b_4(tr\boldsymbol{\epsilon}^2)]\mathbf{1} + [2b_3 + 2b_4(tr\boldsymbol{\epsilon})]\boldsymbol{\epsilon} + 3b_5\boldsymbol{\epsilon}^2]$$

where the five elastic coefficients $b_i = b_i(\Delta T)$ are a function of temperature. For the sake of argument we restrict the discussion to linear elastic behavior when $b_4 = b_5 = 0$. Using the classical notation for linear isotropic thermoelasticity, $b_1 = -3K\alpha\,[\Delta T], 2b_2 = \Lambda$, $b_3 = G$,

$$\boldsymbol{\sigma} = \Lambda\,(tr\boldsymbol{\epsilon})\,\mathbf{1} + 2G\,\boldsymbol{\epsilon} - 3K\alpha\Delta T\mathbf{1}$$

here K denotes the elastic bulk modulus, α the traditional coefficient of thermal expansion, while Λ, G, denote the Lamé constants of linear elasticity. A convenient way for the thermoelastic stress-strain relation is to decompose the strain into volumetric and deviatoric components, $\boldsymbol{\epsilon} = \mathbf{e} + \frac{1}{3}[tr\boldsymbol{\epsilon}]\mathbf{1}$ which yields the form,

$$\boldsymbol{\sigma} = K[\,tr(\boldsymbol{\epsilon}) - \,tr(\boldsymbol{\alpha})\Delta T]\mathbf{1} + 2G\mathbf{e}$$

where $\mathbf{1}$ denotes the unit second order tensor. The elastic bulk and shear moduli K, G are in general a function of temperature, $G = \frac{E}{2[1+\nu]}; K = \frac{E}{3[1-2\nu]}$.

Total differentiation of the stress expression leads to rate format of hyperelasticity,

$$\dot{\boldsymbol{\sigma}} = \frac{\partial^2\psi}{\partial\boldsymbol{\epsilon}\otimes\partial\boldsymbol{\epsilon}} : \dot{\boldsymbol{\epsilon}} + \frac{\partial^2\psi}{\partial\boldsymbol{\epsilon}\partial\Delta T}\Delta\dot{T}$$

where the gradients of the thermoelastic constitutive relationship entail,

$$\frac{\partial\boldsymbol{\sigma}}{\partial\boldsymbol{\epsilon}} = [\Lambda\,\mathbf{1}\otimes\mathbf{1} + 2G\,\mathbb{1}]$$

$$\frac{\partial\boldsymbol{\sigma}}{\partial T} = \frac{\partial}{\partial T}([\Lambda\,(tr\boldsymbol{\epsilon})\,\mathbf{1} + 2G\boldsymbol{\epsilon}] + \frac{\partial}{\partial T}(3K\alpha\Delta T)\mathbf{1}$$

The $\mathbb{1}$ symbol denotes the symmetric fourth order unit tensor. The thermoelastic stress rate is driven by the rate of deformation and the rate of temperature,

$$\dot{\boldsymbol{\sigma}} = \mathbb{E}^{iso} : \dot{\boldsymbol{\epsilon}} - \boldsymbol{\beta}^{iso}\dot{T}$$

The tangential elastic stiffness tensor reads for isothermal conditions,

$$\mathbb{E}^{iso} = \frac{\partial^2 \psi}{\partial \boldsymbol{\epsilon} \otimes \partial \boldsymbol{\epsilon}} = [\Lambda \mathbf{1} \otimes \mathbf{1} + 2G \mathbb{1}]$$

while the stress-temperature tensor is comprised of two contributions, the traditional one for constant strain, $d\epsilon = 0$, and two terms which depend on the state of strain and the temperature dependence of the elastic stiffness properties

$$\boldsymbol{\beta}^{iso} = -\frac{\partial^2 \psi}{\partial \boldsymbol{\epsilon} \partial \Delta T} = \beta_1 \mathbf{1} - \beta_2 (tr \boldsymbol{\epsilon}) \mathbf{1} - \beta_3 \boldsymbol{\epsilon}$$

The product form assumes that the mass density remains constant, while the three thermal stress coefficients $\beta_1 = \beta_1(\Delta T), \beta_2 = \beta_2(\Delta T), \beta_3 = \beta_3(\Delta T)$ exhibit temperature effects as

$$\begin{aligned} K &= K_0 + \tfrac{\partial K}{\partial \Delta T}[T - T_0] \\ G &= G_0 + \tfrac{\partial G}{\partial \Delta T}[T - T_0] \\ \Lambda &= \Lambda_0 + \tfrac{\partial \Lambda}{\partial \Delta T}[T - T_0] \\ \alpha &= \alpha_0 + \tfrac{\partial \alpha}{\partial \Delta T}[T - T_0] \end{aligned}$$

Consequently, differentiation with respect to temperature leads to the following contributions,

$$\beta_1 = 3[K\alpha + \tfrac{\partial K}{\partial T}\alpha[T - T_0] + K\tfrac{\partial \alpha}{\partial T}[T - T_0]]$$

Note that $\frac{\partial \sigma^\epsilon}{\partial T}$ introduces additional contributions to the thermal stress due to the temperature sensitivity of the elastic stiffness properties

$$\begin{aligned} \beta_2 &= \tfrac{\partial \Lambda}{\partial T} < 0 \text{ for } \tfrac{\partial E}{\partial T} < 0 \text{ and } \nu = const. \\ \beta_3 &= 2\tfrac{\partial G}{\partial T} < 0 \text{ for } \tfrac{\partial E}{\partial T} < 0 \text{ and } \nu = const. \end{aligned}$$

Observe that the last two contributions vanish in the case of elastic properties which are temperature independent. In this case only the first term in β_1 remains and reduces to the traditional thermal stress term $-3\rho K\alpha \mathbf{1}$. We also note that the thermal stress tensor β^{iso} depends not only on the temperature, but also on the level of strain. Thereby β_3 introduces strain induced anisotropy into the thermal stress tensor in spite of the assumed isotropy we started from. Thinking of concrete materials, let us examine two cases:

(a) Thermal stress rate at zero strain rate:

Consider a kinematically constrained situation where $\dot{\epsilon} = 0, \boldsymbol{\epsilon} = 0$, then the thermal stress increment only involves β_1

$$\dot{\boldsymbol{\sigma}} = -\boldsymbol{\beta}\dot{T} = -3[K\alpha + \tfrac{\partial K}{\partial T}\alpha[T - T_0] + K\tfrac{\partial \alpha}{\partial T}[T - T_0]]\mathbf{1}\dot{T}$$

This shows that reduction of the bulk modulus with increasing temperature diminishes the effect on thermal compression, while an increase of thermal expansion with temperature increases thermal compression.

(b) Thermal stress rate at zero strain rate:

Considering the effect of axial contraction on the thermal stress response due to increasing temperature, let us isolate the effect of increasing temperatures and the dependence on the elastic stiffness properties,

$$\dot{\boldsymbol{\sigma}} = -\boldsymbol{\beta}\dot{T} = [\tfrac{\partial \Lambda}{\partial T}(tr\boldsymbol{\epsilon})\mathbf{1} + 2\tfrac{\partial G}{\partial T}\boldsymbol{\epsilon}]\dot{T}$$

The combined effect decreases the thermal compression because of the decreasing elastic properties and the negative state of strain. This is in line with the argument of transitional thermal creep and stress dependent thermal expansion advocated by Thelandersson [1987] and by Bazant and Chern [1987] who explained the large reduction of thermal compression under increasing temperature when compared to free thermal expansion. It is intriguing that their thermoelastic concrete model led to the same form of the thermal stress tensor derived from the second derivatives of the Helmholtz free energy expansion, except they started from the representation of tensor functions without adhering to the thermodynamic foundation.

5 VOLUMETRIC-DEVIATORIC DAMAGE

Let us start from the specific free energy per unit volume in terms of elastic strain, temperature and two damage variables for volumetric and deviatoric degradation. Assuming linear behavior in the effective properties [Carol, Rizzi & Willam 2002], the energy contributions expand into quadratics,

$$\Psi = \tfrac{1}{2}\boldsymbol{\epsilon}_{el} : \mathbb{E} : \boldsymbol{\epsilon}_{el} - \tfrac{1}{2}c\tfrac{(\Delta T)^2}{T_o}$$

where $\boldsymbol{\epsilon}_{el} = \boldsymbol{\epsilon} - \alpha \Delta T$

Volumetric and deviatoric decomposition of thermoelastic damage:

$$\Psi_{el} = \tfrac{1}{2}[1 - d_{vol}]K^o[tr(\epsilon - \boldsymbol{\alpha}\Delta T)]^2 + [1 - d_{dev}]G^o[tr(\mathbf{e}^2)]$$

(a) Elastic Stress:

$$\boldsymbol{\sigma} = \frac{\partial \Psi}{\partial \boldsymbol{\epsilon}_{el}} = [1 - d_{vol}]K^o[tr(\boldsymbol{\epsilon} - \boldsymbol{\alpha}\Delta T)]\mathbf{1} + [1 - d_{dev}]2G^o\mathbf{e}$$

(b) Internal Entropy:

$$s = -\frac{\partial \Psi}{\partial \Delta T} = c\frac{\Delta T}{T_o} + [1 - d_{vol}]K^o[tr(\boldsymbol{\alpha})]\Delta T[tr(\boldsymbol{\alpha})]$$

(c) Conjugate Thermodynamic forces:

$$\mathcal{Y}_{vol} = -\frac{\partial \Psi}{\partial d_{vol}} = \tfrac{1}{2}K^o[tr(\boldsymbol{\epsilon} - \boldsymbol{\alpha}\Delta T)]^2, \quad \mathcal{Y}_{dev} = -\frac{\partial \Psi}{\partial d_{dev}} = G^o[tr(\mathbf{e}^2)]$$

Modified Dissipation Inequality:

$$\dot{\mathcal{D}} = \mathcal{Y}_{vol}\ \dot{d}_{vol} + \mathcal{Y}_{dev}\ \dot{d}_{dev} \geq 0$$

This is satisfied when $\dot{d}_{vol} > 0$, and $\dot{d}_{dev} > 0$ since thermodynamic forces are strictly positive because of their quadratic expressions.

Let us assume volumetric degradation due to thermal dependence of the bulk modulus and the thermal expansion, $K = K(T), \alpha = \alpha(T)$, [Nenech, Meftah & Reynouard, 2002], where the bulk and thermal damage variables are defined as

$$[1 - d_{vol}] = [1 - d_K][1 - d_\alpha]$$
$$d_K = 1 - \frac{K(T)}{K^o}$$
$$d_\alpha = 1 - \frac{\alpha(T)}{\alpha^o}$$

In contrast, we assume that the deviatoric degradation of the shear modulus is primarily due to mechanical damage neglecting temperature effects,

$$G = [1 - d_{dev}]G^o, \text{ where } d_{dev} = 1 - \frac{G(\mathbf{e})}{G^o}$$

Two independent damage functions define the onset of thermal and elastic degradation,

$$F_{vol} = f(\mathcal{Y}_{vol}) - r(d_{vol}) \leq 0$$
$$F_{dev} = f(\mathcal{Y}_{dev}) - r(d_{dev}) \leq 0$$

Two Energy Demand Functions:

$$f(\mathcal{Y}_{vol}) \to \mathcal{Y}_{vol} = \Psi^o_{vol}$$
$$f(\mathcal{Y}_{dev}) \to \mathcal{Y}_{dev} = \Psi^o_{dev}$$

Two Energy Resistance Functions:

$$r_{vol} = \tfrac{1}{2}[1 - d_{vol}]K^o[tr(\boldsymbol{\epsilon} - \boldsymbol{\alpha}\Delta T)]^2$$
$$r_{dev} = r[1 - d_{dev}]^{p_{dev}-1}$$

where deviatoric damage initiates at

$$r^o_{dev} = \tfrac{1}{2}\tau^o_s\gamma^o_s, \text{ and } p_{dev} = \frac{r^o_{dev}}{g^{II}_f/h_{dev}} < 1$$

Two Independent Loading Conditions:

$$F_{vol} = 0 \to \mathcal{Y}_{vol} = r(d_{vol}) \text{ and } \dot{d}_{vol} > 0$$
$$F_{dev} = 0 \to \mathcal{Y}_{dev} = r(d_{dev}) \text{ and } \dot{d}_{dev} > 0$$

Two Independent Consistency Conditions:

$$\dot{F}_{vol} = 0, \qquad \frac{\partial F_{vol}}{\partial \mathcal{Y}_{vol}}\dot{\mathcal{Y}}_{vol} + \frac{\partial F_{vol}}{\partial d_{vol}}\dot{d}_{vol} = 0$$
$$\dot{F}_{dev} = 0, \qquad \frac{\partial F_{dev}}{\partial \mathcal{Y}_{dev}}\dot{\mathcal{Y}}_{dev} + \frac{\partial F_{dev}}{\partial d_{dev}}\dot{d}_{dev} = 0$$

where the rate of the volumetric thermodynamic forces involves,

$$\dot{\mathcal{Y}}_{vol} = K^o[tr(\boldsymbol{\epsilon} - \boldsymbol{\alpha}\Delta T)][tr(\dot{\boldsymbol{\epsilon}} - \boldsymbol{\alpha}\Delta\dot{T})],$$

Assuming linear dependence of the bulk modulus and the thermal expansion on temperature,

$$K(T) = K^o[1 - h_K\Delta T] \text{ and } d_K = h_K\Delta T$$
$$\alpha(T) = \alpha^o[1 - h_\alpha\Delta T] \text{ and } d_\alpha = h_\alpha\Delta T$$

The evolution equation for volumetric damage accounts for thermally induced damage:

$$\dot{d}_{vol} = \dot{d}_K + \dot{d}_\alpha - \dot{d}_K d_\alpha - d_K\dot{d}_\alpha$$
$$\dot{d}_{vol} = h_{vol}\Delta\dot{T}$$

where the temperature dependence of the bulk modulus and the thermal expansion reduce for linear thermal dependence,

$$\dot{d}_K = -\frac{1}{K_o}\frac{\partial K}{\partial T}\Delta\dot{T} = h_K\Delta\dot{T}$$
$$\dot{d}_\alpha = -\frac{1}{\alpha_o}\frac{\partial\alpha}{\partial T}\Delta\dot{T} = h_\alpha\Delta\dot{T}$$
$$h_{vol} = h_K + h_\alpha - h_K[h_\alpha\Delta T] - h_\alpha[h_K\Delta T]$$

The rate of the deviatoric thermodynamic force and damage involve,

$$\dot{\mathcal{Y}}_{dev} = 2G^o\mathbf{e} : \dot{\mathbf{e}}, \text{ and } \mathcal{H}_{dev} = \frac{\partial r_{dev}}{\partial d_{dev}} = -\frac{\partial F_{dev}}{\partial d_{dev}}$$

The evolution equation for deviatoric damage variable accounts here only for mechanical damage:

$$\dot{d}_{dev} = \frac{2G^o}{\mathcal{H}_{dev}}\mathbf{e} : \dot{\mathbf{e}}$$

Starting from the stress-strain-damage relation:

$$\boldsymbol{\sigma} = [1 - d_{vol}]K^o[tr(\boldsymbol{\epsilon} - \boldsymbol{\alpha}\Delta T)]\mathbf{1} + [1 - d_{dev}]2G^o\mathbf{e}$$

we obtain the differential form as follows,

$$\dot{\boldsymbol{\sigma}} = [1 - d_{vol}]K^o[tr(\dot{\boldsymbol{\epsilon}} - \boldsymbol{\alpha}\Delta\dot{T})]\mathbf{1} - \dot{d}_{vol}K^o[tr(\boldsymbol{\epsilon} - \boldsymbol{\alpha}\Delta T)]\mathbf{1} + [1 - d_{dev}]2G^o\dot{\mathbf{e}} - \dot{d}_{dev}2G^o\mathbf{e}$$

Substituting the rate of thermoelastic damage yields the stress rate in terms of strain and temperature rates,

$$\dot{\boldsymbol{\sigma}} = [1 - d_{vol}]K^o[\mathbf{1}\otimes\mathbf{1}] : [\dot{\boldsymbol{\epsilon}} - \boldsymbol{\alpha}\Delta\dot{T}] + \left[[1 - d_{dev}]2G^o\mathbf{1} - \frac{4G_o^2}{\mathcal{H}_{dev}}\mathbf{e}\otimes\mathbf{e}\right] : \dot{\mathbf{e}} - h_{vol}K^o[tr(\boldsymbol{\epsilon} - \boldsymbol{\alpha}\Delta T)]\mathbf{1}\Delta\dot{T}$$

which yields the tangential stiffness relations:

$$\dot{\boldsymbol{\sigma}} = \mathbb{K}^{tan}_{ed} : [\dot{\boldsymbol{\epsilon}} - \boldsymbol{\alpha}\Delta\dot{T}] + \mathbb{G}^{tan}_{ed} : \dot{\mathbf{e}} - \beta_{vol}\mathbf{1}\Delta\dot{T}$$

where the tangential bulk modulus tensor is defined as,

$$\mathbb{K}_{ed}^{tan} = [1 - d_{vol}] K^o \mathbf{1} \otimes \mathbf{1}$$

The contribution from the temperature dependent properties is,

$$\beta_{vol} = h_{vol} K^o [tr(\boldsymbol{\epsilon} - \boldsymbol{\alpha} \Delta T)]$$

The volumetric stress-strain rate relation involves a strain driven and a temperature driven component,

$$\dot{\boldsymbol{\sigma}}_{vol} = [\dot{\sigma}_{vol}^{\epsilon} - \dot{\sigma}_{vol}^{T}] \mathbf{1}$$

where

$$\dot{\sigma}_{vol}^{\epsilon} = K^o [1 - d_{vol}] [tr \dot{\boldsymbol{\epsilon}}]$$

and

$$\dot{\sigma}_{vol}^{T} = [[1 - d_{vol}][tr \boldsymbol{\alpha}] - h_{vol} [tr \boldsymbol{\alpha}] \Delta T] \, [\Delta \dot{T}]$$

Note that the volumetric stress rate is driven by volumetric strain rate and the temperature rate, whereby the third term on the rhs reflects the temperature dependence of the bulk modulus and the coefficient of thermal expansion inherent in h_{vol}.

The deviatoric stress-strain rate relations read as follows,

$$\dot{\mathbf{s}} = \mathbb{G}_{ed}^{tan} : \dot{\mathbf{e}}$$

where

$$\mathbb{G}_{ed}^{tan} = \left[[1 - d_{dev}] 2 G^o \mathbb{1} - \frac{4 G_o^2}{\mathcal{H}_{dev}} \mathbf{e} \otimes \mathbf{e} \right]$$

Neglecting the temperature dependence of the hardening/softening modulus defines the rate of change of deviatoric energy with regard to shear damage:

$$\mathcal{H}_{dev} = \frac{\partial r_{dev}}{\partial d_{dev}} = -r_{dev}^o [p_{dev} - 1][1 - d_{dev}]^{p_{dev} - 2}$$

Alternatively we could formulate directly the shear damage in terms of,

$$h_{dev} = \frac{2G^o}{\mathcal{H}_{dev}}$$

which reduces the tangent shear tensor to,

$$\mathbb{G}_{ed}^{tan} = [1 - d_{dev}] 2 G^o \mathbb{1} - 2 G_o h_{dev} \mathbf{e} \otimes \mathbf{e}$$

6 MODEL PROBLEM

For illustration we consider the two-phase composite shown in Figure 3. The RVE is subjected to a transient temperature sweep considering mechanical and thermal degradation. The objective of the exercise is to explore the effect of temperature dependence of the mechanical and thermal properties and to compare the numerical results with the experimental observations of Anderberg & Thelandersson [1976]. This permits us to investigate the effect of strain onto the thermal stress tensor and to assess the performance of the volumetric-deviatoric damage model.

6.1 *Temperature sweep*

The results of the transient temperature analysis of the heterogeneous particle composite is shown in Figure 4 which depicts the temperature distribution in the RVE at the time when the surrounding air temperature reaches 800 C. The results are based on the following heat transfer properties of the aggregate and the cement matrix.

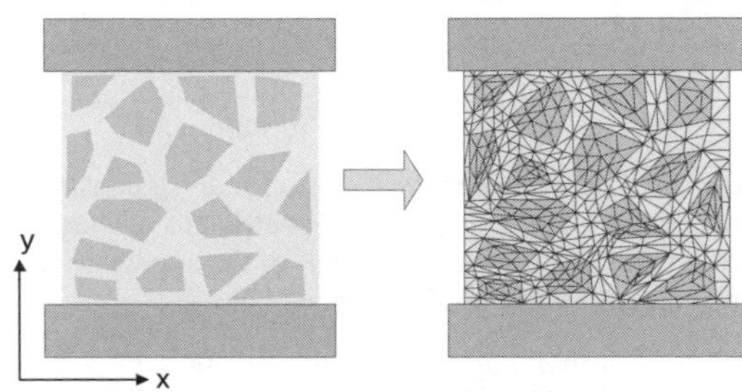

Figure 3. Two-phase particle RVE subjected to vertical confinement during external temperature sweep.

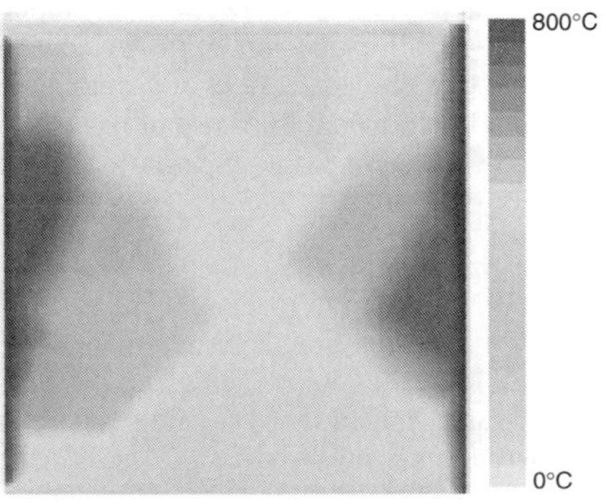

Figure 4. Temperature distribution in two-phase RVE at 800 C (transient analysis).

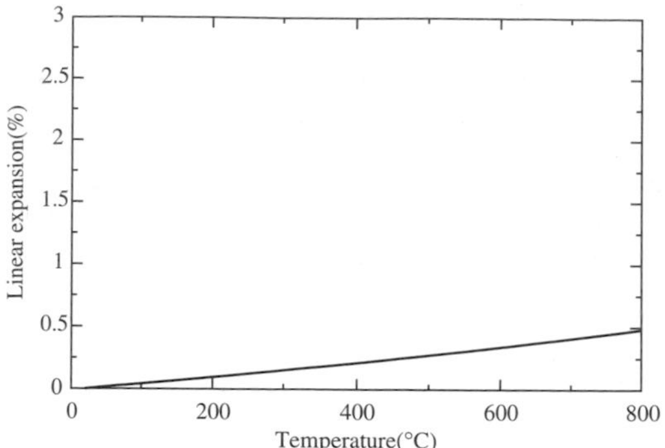

Figure 5. Coefficient of thermal aggregate expansion [%].

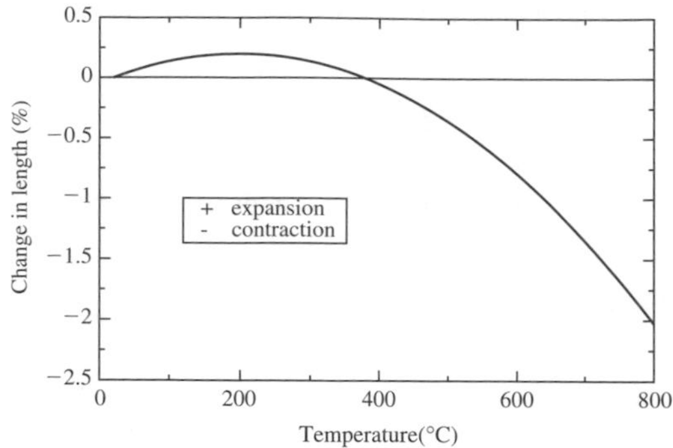

Figure 6. Coefficient of thermal matrix expansion [%].

6.1.1 *Aggregates*

Thermal conductivity, $k_a = 2.4 \times 10^{-3}\,\mathrm{kW}/(mm \cdot \mathrm{C})$ Convection coefficient, $h_a = 0.0\,\mathrm{kW}/(mm^2 \cdot \mathrm{C})$ Specific heat, $c_a = 1170$ (kW · h)/(kg · C) Mass density, $\rho_a = 1.92 \times 10^{-6}\,\mathrm{kg}/mm^3$.

6.1.2 *Cement matrix*

Thermal conductivity, $k_m = 10^{-4}\,\mathrm{kW}/(\mathrm{mm} \cdot \mathrm{C})$ Convection coefficient, $h_m = 2 \times 10^{-5}\,\mathrm{kW}/(mm^2 \cdot \mathrm{C})$ Specific heat, $c_m = 1170$ (kW · h)/(kg · C) Mass density, $\rho_m = 1.92 \times 10^{-6}\,\mathrm{kg}/mm^3$.

6.2 *Thermomechanical damage study*

The thermomechanical degradation analysis considers the temperature effects on the coefficient of thermal expansion and the elastic moduli of the aggregate and the cement matrix as summarized below.

6.2.1 *Thermal coefficient of expansion: α*

Matrix: $\alpha_m^0 = 15.5 \times 10^{-6}/\mathrm{C}$
$\alpha_m = 0.00555557\,\alpha_m^0$ (200-T)
Aggregates: $\alpha_a^0 = 9.0 \times 10^{-6}/\mathrm{C}$
$\alpha_a = \alpha_a(e^{0.05T/100} - 0.01)$

Figure 5 for the aggregate illustrates the effect increasing expansion with increasing temperature shown. The thermal expansion of the cement matrix was modified to include the effect of drying shrinkage which results in an overall decrease of the coefficient of expansion with increasing temperature as indicated in Figure 6.

6.2.2 *Elastic modulus: E(T)*

$\mathrm{E} = E_0(0.03921 + e^{-0.002T})$

Figure 7 illustrates the degradation of the elastic stiffness properties with increasing temperature. Thereby it was assumed that Poisson's ratio exhibits little temperature sensitivity.

6.2.3 *Material properties*

$E_m = 25000$ Mpa, $E_a = 75000$ Mpa, $\nu_m = \nu_a = 0.2$
$\tau_o = 7.0$ Mpa, $g_f^{II} = 2.20$ N/mm

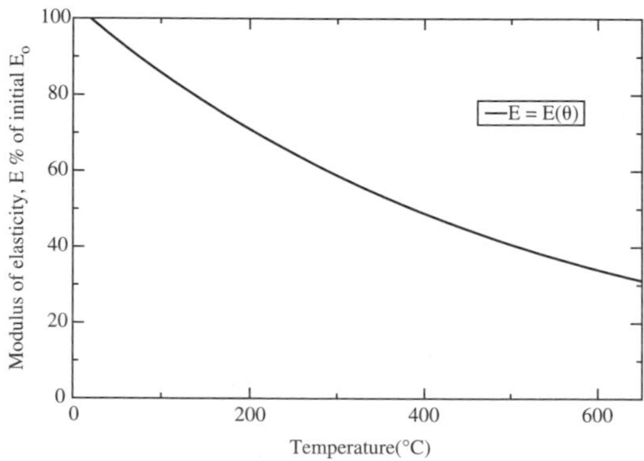

Figure 7. Reduction of elastic modulus with rising temperature, $E = E(T)$.

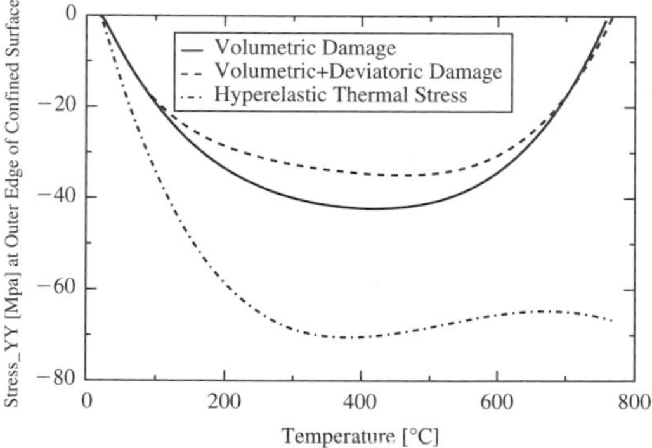

Figure 8. Variation of σ_{yy} at outer edge of confined surface.

Using these mechanical properties Figure 8 illustrates the axial thermal stress σ_{yy} near the top left corner of the restrained RVE specimen. The figure compares the predictions of the volumetric-deviatoric damage model as compared to the hyperelastic formulation which includes strain induced effects in the thermal expansion. Qualitatively, the trend of the

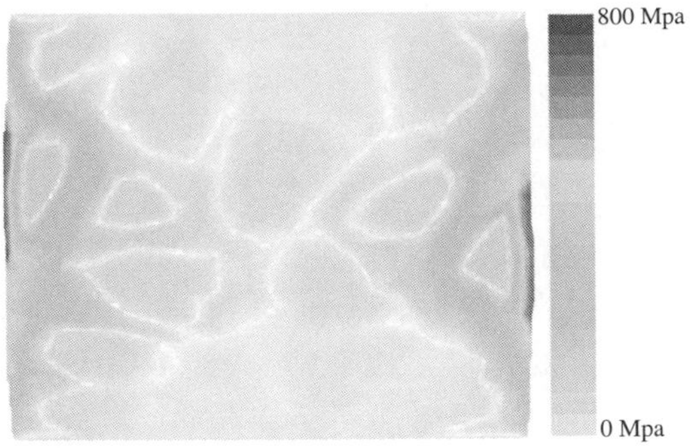

Figure 9. σ_{yy} Contours in RVE at temperature $t = 800\,C$.

volumetric-deviatoric damage model reproduces the stress reversal in Figure 1 at high temperatures, while the hyperelastic model does not show this reversal. Figure 9 shows the overall distribution of axial stress within the heterogeneous RVE.

7 CONCLUSIONS

The paper addressed subtle issues of thermomechanical coupling when the notion of a linear coefficient of thermal expansion is forfeited. The volumetric-deviatoric damage model which incorporates thermal as well as mechanical degradation in a rational manner appears to be capable in reproducing thermal stress effects under extreme thermal sweeps.

REFERENCES

Anderberg, Y. and J. Thelanderssson (1976). Stress and deformation characteristics of concrete at high temperatures. Technical report, Lund Institute of Technology, Lund, Sweden.

Argyris, J.H., V.L. and K. Willam. Higher order methods for transient diffusion analysis. *Computer Methods in Applied Mechanics and Engineering 12.*

Argyris, J.H., V.L. and K. Willam. Integrated finite element analysis of coupled thermoviscoplastic problems. *Journal of Thermal Stresses 4.*

Bazant, Z. and J.-C. Chern (1987). Stress-induced thermal shrinkage strain in concrete. *Journal of Engineering Mechanics, ASCE 113*(10), 1493–1511.

Bazant, Z. and M. Kaplan (1996). *Concrete at High Temperatures, Material Properties and Mathematical Models.* Burnt Mill, England: Longman Group Limited.

Carol, I., E. Rizzi, and K. Willam (1994). A unified theory of elastic degradation and damage based on a loading surface. *International Journal of Solids and Structures 31*(20), 2835–2865.

Carol, I., E. Rizzi, and K. Willam (2002). An “extended” volumetric/deviatoric formulation of anisotropic damage based on a pseudo-log rate.

Felippa, C., P. K. and C. Farhat (2001). Partitioned analaysis of coupled mechanical systems.

Phan, L. (1996). Fire performance of high strength concrete: A report of the state-of-the-art. Technical report, Building and Fire Research Laboratory, National Institute of Standards and Technology, Maryland.

Poon, C.-S., A. S.-A. M. and Y.-L. Wong. Comparison of the strength and durability performance of normal and high strength pozzolanic concretes at elevated temperatures. *Cement and Concrete Research 31.*

Schneider, U. (1988). Concrete at high temperature - a general review. *Fire Safety Journal 13*(1), 55–68.

Shin, K-Y., K. S.-B. K. J.-H. C. M. and P.-S. Jung. Thermo-physical properties and transient heat transfer of concrete at elevated temperature. *Nuclear Engineering and Design 212.*

Thelandersson, S. (1987). Modeling of combined thermal and mechanical action in concrete. *ASCE Journal of Engineering Mechanics 113*(6), 893–906.

Thienel, K.-C. and F. Rostasy. Strength of concrete subjected to high temperature and biaxial stress: Experiments and modeling. *Materials and Structures (RILEM) 28.*

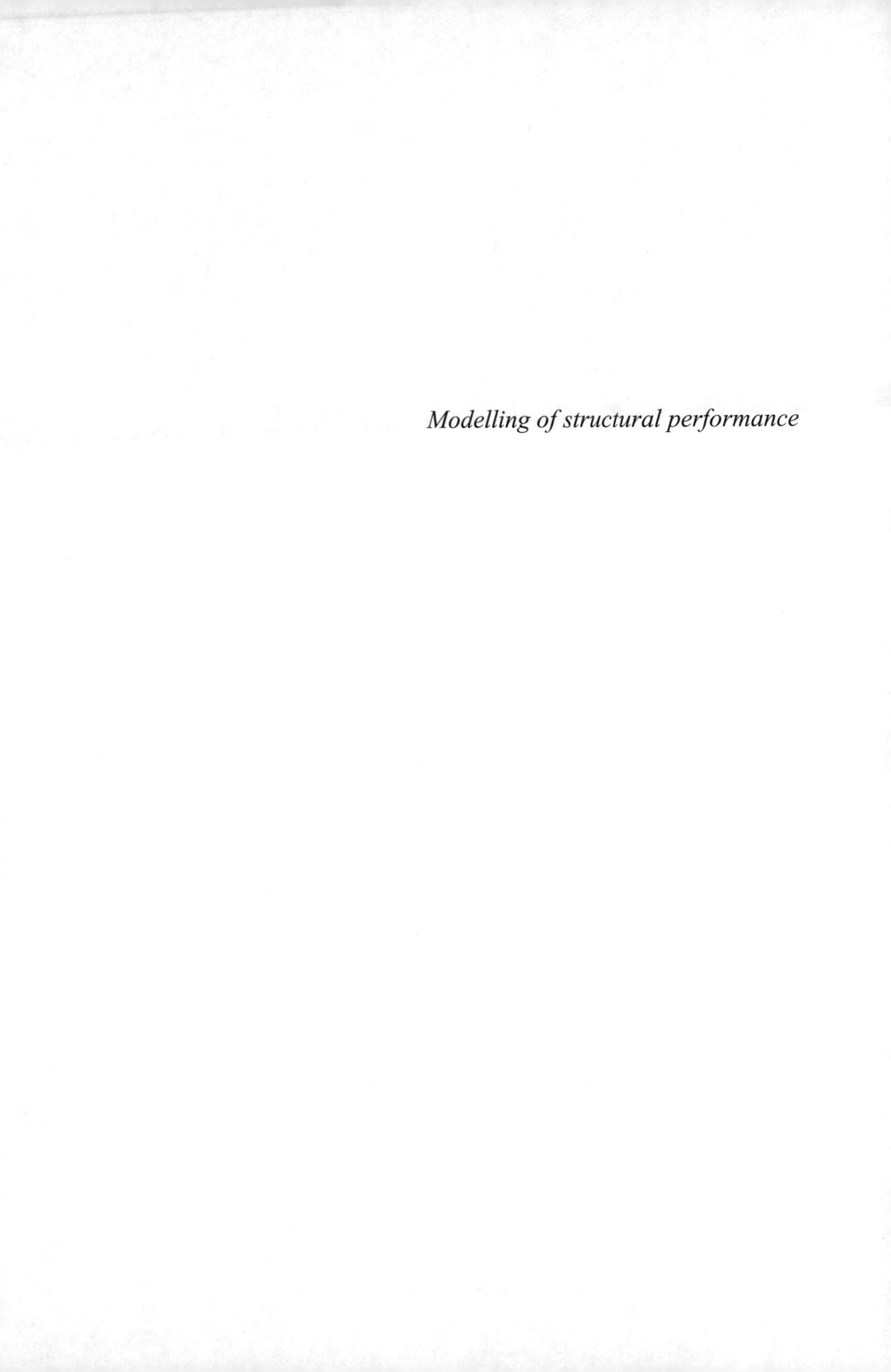

Modelling of structural performance

Computational Modelling of Concrete Structures, Bićanić et al. (eds)
© 2003 Swets & Zeitlinger, Lisse, ISBN 90 5809 536 3

Keynote paper: Stochastic models for deformation and failure of quasibrittle structures: recent advances and new directions

Zdeněk P. Bažant
Northwestern University, Evanston, Illinois, USA

Drahomír Novák
Brno University of Technology, Brno, Czech Republic

ABSTRACT: The paper presents an apercu of the current problems of probabilistic failure analysis of concrete structures and proposes a new approach to finite element estimation of loads of very small failure probability. The conference presentation begins by brief comments on the problem of interplay of the nonlocal characteristic length, associated with the deterministic models for cohesive fracture or softening damage, with the autocorrelation length of the random field of local material strength, a problem on which important results were achieved in Delft by de Borst, Carmeliet & Gutiérez. The rest of the paper is focused on the problem of loads of extremely low failure probability (such as 10^{-7}), which should form the basis of rational design. For quasibrittle structures, this problem can be based on the nonlocal generalization of Weibull statistical theory. A salient aspect of the theory is the statistical-deterministic size effect. In the typical case of failures at crack initiation, the distribution tail can be checked by scaling the structure to a very large size. In the limit, the randomly located fracture process zone becomes infinitely small compared to the structure size (i.e., a point), which means that failure at one point causes the whole structure to fail. In that limit case, the tail of distribution of the failure load cannot be anything but Weibull (which follows from Fisher and Tippett's condition of form stability of the extreme value distributions). The existing stochastic finite element methods fail this fundamental requirement, which means that their far-off probability structure cannot be realistic. A new stochastic finite element formulation which guarantees the probability tail of structural strength and the large-size asymptotic size effect to be of the Weibull type is presented. A numerical example demonstrates a good representation of the statistical data of Koide et al. Furthermore, a simple formula for the mean size effect, recently derived by asymptotic matching, is discussed. Finally, a recent more fundamental derivation of the formulas for the mean size effect and for the standard deviation and entire probability distribution of the failure loads, which has been based on the nonlocal generalization of Weibull theory, is reviewed.

1 INTRODUCTION

Probabilistic mechanics is of paramout importance for progress in concrete design because the errors due to the uncertainty of the currently used empirical safety factors and reliability indices are much larger than the errors of finite element analysis. Stochastic finite element analysis has been intensely studied in the 1980s and significant progress has been achieved. However, the probabilistic structure of the existing formulations suffices only for the computation of the first and second moment statistics, that is, the means and the standard deviations. This is adequate for the design against excessive deflections and flexibility, for which probabilities between 10^{-1} and 10^{-2} are generally acceptable, but not for the very small tail probabilities required for safety against failure.

The probabilistic structure of the existing stochastic finite element method is at present not yet adequate for yielding the failure loads of extremely small probabilities such as 10^{-7}, on which the design of civil engineering structures must be based. The problem is not the numerical computation of such loads from the given statistical distributions of material properties, but the very formulation of these statistical distributions themselves. These properties must be incorporated in a way that would yield realistic far-off tails of the distributions as dictated by the known established results of the extreme value statistics (Fisher & Tippett 1928, Tippett 1925, Peirce 1925, Fréchet

1927, von Mises 1936, Gnedenko 1943, Gnedenko & Kolmogorov 1954, Epstein 1948, Freudenthal 1956, 1968, 1981, Saibel 1969, Gumbel 1954, 1958).

Concrete as well as many other materials behave in a quasibrittle manner. Their failure zone, called the fracture process zone (FPZ) is not negligibly small, as in brittle materials, but it does not behave in a plastic manner, as in brittle ductile materials. Instead, it undergoes distributed cracking, or strain-softening. When the structure is of a similar size as the FPZ, it behaves in an almost plastic manner, with no size effect, no localized failure, and no weakest-link-type randomness. When the structure is far larger than the FPZ, it fails in an almost perfectly brittle manner, and it also follows the weakest-link statistical model unless the location of fracture front is restricted by a notch or previous stable growth of a large crack. The effective size of the fracture process zone, or the characteristic length of the material, is a key parameter whose ratio to the structure size governs the brittleness or non-brittleness of response and, evidently, the size effect. Since the field of random strength must be autocorrelated, the characteristic length interacts with the autocorrelation length of the random field.

After recalling the basic facts about the cohesive, crack band and nonlocal models of quasibrittle failure (Hillerborg et al. 1976, 1983, Petersson 1961, Bažant 1976, 1984, Bažant et al. 1984, 1985, 2002, Pijaudier-Cabot & Bažant 1987, Peerlings et al. 1996, Bažant & Planas 1998), Weibull (1939) theory of perfectly brittle failure, and size effect theories (Bažant 1987, 1999, 2001, Bažant & Chen 1987), the conference presentation will briefly review statistical generalization of nonlocal finite element models for softening damage and especially the recent studies in Delft (see Appendix I) clarifying the interplay of the characteristic length for nonlocal analysis and the autocorrrelation length of the random local strength field. The nonlocal generalization of Weibull statistical theory of brittle failure will then be discussed and the limitations of the statistical size effect identified. The rationale behind the amalgamated energetic-statistical size effect law will be examined and the underlying asymptotic matching argument discussed. A new derivation of this size effect from the probabilistic nonlocal theory will be outlined. Finally, a generalization of nonlocal finite element analysis to extreme value statistics will be proposed and demonstrated by numerical examples.

2 WEIBULL THEORY OF RANDOM STRENGTH AND ITS LIMITATIONS

The Weibull statistical theory of random strength and size effect is based on the weakest link model for the failure of a chain consisting of links whose strengths are statistically independent random variables (Fig. 1b). In many two- or three-dimensional structures, the failure occurs as soon as one small element of material (representative volume V_r) fails. From the probabilistic viewpoint, such structures are analogous to the weakest link model for a chain. In the continuum limit of infinitely many infinitely small links, the weakest link model leads to the following cumulative distribution of failure probability:

$$P_f(\sigma_N) = 1 - e^{-\int_V c[\boldsymbol{\sigma}(\boldsymbol{x}, \sigma_N)]\mathrm{d}V(\boldsymbol{x})} \tag{1}$$

$$c(\boldsymbol{\sigma}) = \sum_{I=1}^{3} \frac{P_1[\sigma_I(\boldsymbol{x})]}{V_r} \tag{2}$$

where $\sigma_I(\boldsymbol{x})$ = principal stress just before failure at point of coordinate vector $\boldsymbol{x}(I = 1, 2, 3)$, V = volume of structure, and $c(\boldsymbol{\sigma})$ = function giving the spatial concentration of failure probability of the material ($=V_r^{-1}$ × failure probability of material representative volume V_r) (Freudenthal 1968, Bažant & Planas 1998); $c(\boldsymbol{\sigma})$ = concentration function (spatial density of failure probability); and $P_1(\sigma_I)$ = failure probability (cumulative) of the smallest possible test specimen, of volume V_r, subjected to stress σ_I. Eq. (1) gives the failure probability of the structure, provided that the structure (with the loading system) is of such a geometry that the failure (the maximum load) occurs as soon as a macroscopic crack initiates (this is called positive geometry – a geometry for which the energy release function increases with the crack length). Eq. (1) can be derived by noting that the survival probability, $1 - P_f$, of a chain of N links is the joint probability that all the links survive; this implies that $1 - P_f = (1 - P_1)^N$. Of practical interest are only very small failure probabilities P_1 and P_f, and so we may write $\ln(1 - P_f) = N \ln(1 - P_1) \approx -NP_1$ or $1 - P_f = e^{-NP_1}$.

As proven mathematically by Fisher & Tippett (1928) and also justified by Freudenthal on physical grounds (based on an analysis of material flaws), the low-probability tail of $P_1(\sigma)$ must be a power law:

$$P_1(\sigma) = \left\langle \frac{\sigma - \sigma_u}{s_0} \right\rangle^m \tag{3}$$

(Weibull 1939) where m, s_0, σ_u = material constants (m = Weibull modulus, usually between 5 and 60). The threshold σ_u cannot be negative and is typically taken as 0 (because it is next to impossible to identify σ_u from the available test data unambiguously since different positive σ_u can give almost equally good fits of data). Substitution of 3 into Eq. (1) yields what came to be known as the Weibull distribution. This distribution (for $\sigma_u = 0$) to simple expressions for the mean of σ_N as a function of m and the coefficient of variation ω of σ_N;

$$\overline{\sigma}_N = s_0 \, \Gamma\left(1 + \frac{1}{m}\right)\left(\frac{V_r}{V}\right)^{1/m} \propto D^{-n_d/m} \tag{4}$$

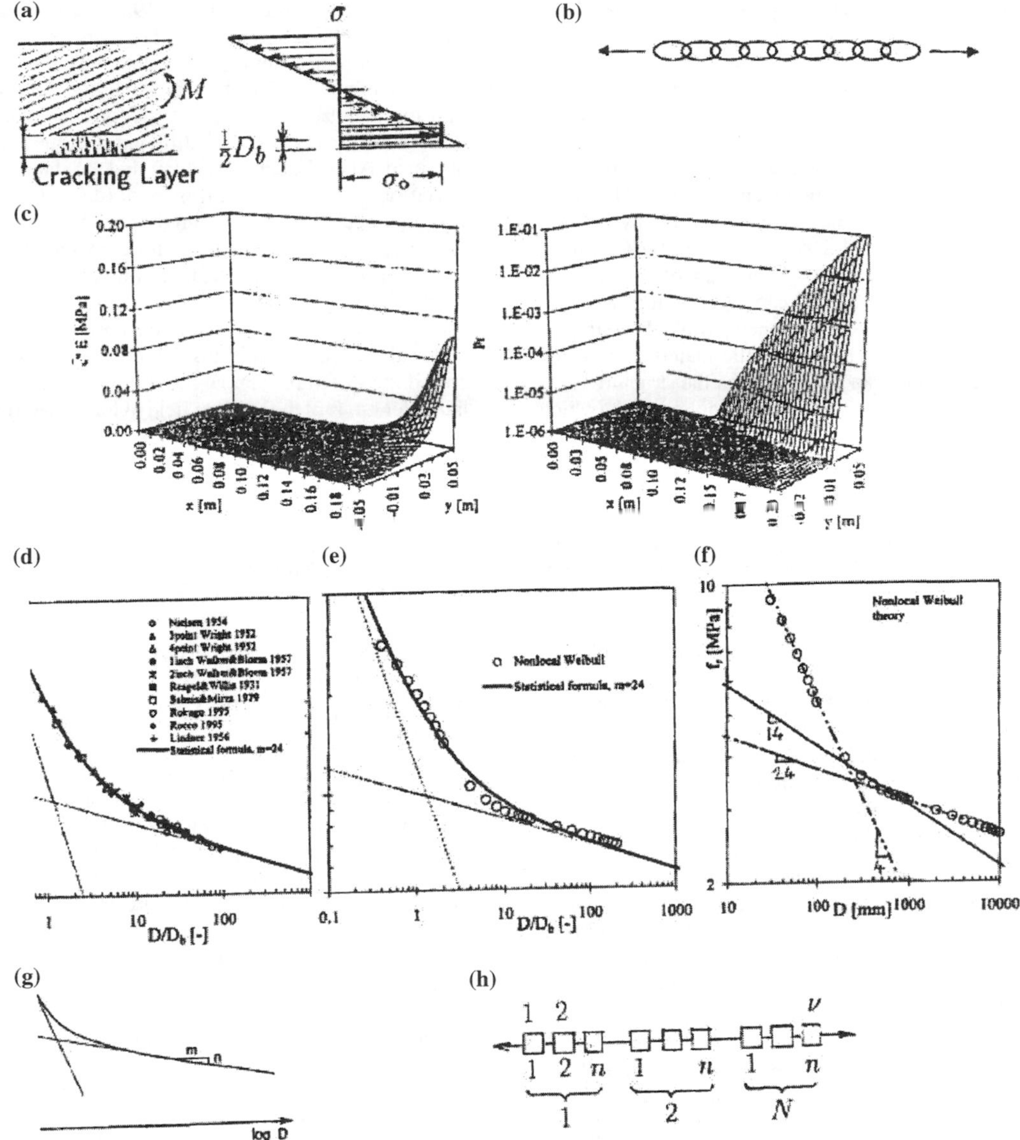

Figure 1. (a) Approximate effect of stress redistritution due to cracking in boundary layer. (b) Chain structure. (c) Field of inelastic strain in beam flexure (linear scale) and corresponding density field of contributions to failure probability (log-scale). (d,e) Relative flexural strength versus relative size for 10 test series from the literature, and results of nonlocal Weibull calculations, fit by energetic-statistical formula for crack initiation. (f) Differences in apparent Weibull moduli m corresponding to nonlocal Weibull calculations in different size ranges. (g) Energetic-statistical size effect for failures at crack initiation. (h) Chain subdivided into segments.

$$\omega = \left(\frac{\Gamma(1+2m^{-1})}{\Gamma^2(1+m^{-1})} - 1 \right)^{1/2} \quad (5)$$

where Γ is the gamma function, and $n_d = 1, 2$ or 3 for uni-, two- or three-dimensional similarity. Eq. (4) represents a power-law size effect on the mean nominal strength $\bar{\sigma}_N$ (if $\sigma_u = 0$). Since there is no size effect on ω, the expression for ω in (4) is normally used to identify m from tests.

Eq. (4) for $\bar{\sigma}_N$ suggests adopting the value $\sigma_W = \sigma_N (V/V_0)^{1/m}$ for a uniformly stressed specimen as a size-independent stress measure. Considering from this viewpoint a large crack-tip plastic zone in metals, Beremin (1983) proposed the idea of the so-called Weibull stress:

$$\sigma_W = \left(\sum_k \sigma_j{}^m \, V_j / V_r \right)^{1/m} \tag{6}$$

where $V_j (j = 1, 2, \ldots N_W)$ are the elements of the plastic zone having maximum principal stress σ_{Ij}. Ruggieri & Dodds (1996) replaced the sum in (6) by an integral (see also Lei et al. 1998). However, Eq. (6) can be applied only if the crack at the moment of failure is still microscopic, that is, small compared to the structural dimensions. Therefore, the concept of Weibull stress is not useful the case for quasibrittle materials in which the process zone can be as large as the structure.

When the structure geometry is not positive (which for example occurs in beams with tensile reinforcement, or when an adjacent compressed zone stabilizes a crack, as in gravity dams), a large crack must develop before the failure can occur. This precludes Weibull-type statistical analysis. Although rigorous probabilistic modeling seems prohibitively difficult, the case of negative geometry is not very important because the size effect is predominantly energetic (deterministic). So, when the size effect is mainly statistical, the violations of statistical independence of material elements at different locations have a negligible effect, and when it is not, the question of statistical independence of these elements becomes irrelevant.

In the case of quasibrittle structures, applications of the classical Weibull theory face a number of fundamental objections:

1) The fact that the size effect on $\bar{\sigma}_N$ is a power law means that the functional equation (1) is satisfied, and this implies the absence of any characteristic length. But this cannot be true if the material does contain sizable inhomogeneities, as does concrete.
2) The energy release due to stress redistributions caused by a macroscopic FPZ or a stable crack growth before P_{max} gives rise to a deterministic size effect, which is ignored. Thus the Weibull theory can be valid only if the structure fails as soon as a microscopic crack becomes macroscopic.
3) Every structure is mathematically equivalent to a uniaxially stressed chain or bar of a variable cross section, which means that the structural geometry and failure mechanism are ignored.
4) The size effect differences between the cases of two- and three-dimensional similarities ($n_d = 2$ or 3) are often much smaller than predicted by Weibull theory (because, for example, a crack in a beam causes failure only if it spreads across the full width of the beam).
5) Many tests of quasibrittle structures show a much stronger size effect than predicted by Weibull theory (e.g., diagonal shear failure of reinforced concrete beams; Walraven & Lehwalter 1994, Walraven 1995, Iguro et al. 1985, Shioya & Akiyama 1994, and many flexure tests of plain beams cited in Bažant & Novák 2000a,b).
6) When Weibull exponent m is identified by fitting the standard deviation of σ_N for specimens of very different sizes, very different m values are obtained. Also, the size effect data and the standard deviation data give very different m (e.g., $m = 12$ was obtained with small concrete specimens while the large-size asymptotic behavior corresponds to $m = 24$ (Bažant and Novák 2000); Fig. 1f (m varies from 4.2 to 24.2).
7) The classical theory neglects spatial correlations of material failure probabilities (which is admissible only if the structure is far larger than the autocorrelation length l_a of the random field of local material strength).

3 OVERVIEW OF NONLOCAL GENERALIZATION OF WEIBULL THEORY

One can discern three approaches to generalizing the Weibull theory in various ways, and to various degrees, the capture the effect of a large FPZ and quasibrittleness.

1) One classical approach is represented by various phenomenological models for load sharing (or parallel coupling of links) (Daniels 1945, Grigoriu 1990). Although these generalizations can simulate some effects of a large FPZ, they are not generally applicable and cannot capture the effect of structure geometry. Calibrating the model for one structure geometry, one cannot predict the behavior for another geometry.
2) Another classical approach attempts to overcome the problem of LEFM crack-tip singularity, which causes the classical Weibull integral to diverge for $m > 4$ (this includes all realistic m values); Beremin (1983), Ritchie, Becker, Lei et al. (1998), Lin, Evans, McClintock, Phoenix (1978), etc. For example, one can exclude from the domain of Weibull integral a finite circular zone surrounding the crack tip, in order to make the integral convergent, or one can consider plastic blunting of the stress profile ahead of the crack tip, or one can average the failure probability spatially. These approaches have been shown useful for tough metals with a moderately large yielding zone at the crack tip, but are doubtful when the effective FPZ length is of the same order of magnitude as the structure size (which is typical for reinforced concrete) and are not completely general (e.g., they cannot be applicable for crack initiation from a smooth surface).

3) The most recent approach is the nonlocal Weibull theory (Bažant & Xi 1991, Bažant & Novák 2000a,b). This is a general theory which has as its limit cases *both* the classical Weibull theory and the deterministic nonlocal continuum damage mechanics developed for finite element analysis of quasibrittle materials. The energetic size effect is implied by this theory as the asymptotic case for not too large structure sizes.

The nonlocal concept was proposed for elasticity in the 1960s (Kröner 1961, Eringen 1965, Kunin, Edelen) and later extended by Eringen et al. to hardening plasticity. In the 1980s, it was adapted to strain-softening continuum damage mechanics and strain-softening plasticity (Bažant 1984, Bažant et al. 1984, Pijaudier-Cabot & Bažant 1987), with three motivations:

1) to serve as a computational "trick" (localization limiter) eliminating spurious mesh sensitivity and incorrect convergence of finite element simulations of damage;
2) to reflect the physical causes of nonlocality, which are: (a) material heterogeneity, (b) energy release due to microcrack formation, and (c) microcrack interactions; and
3) to simulate the experimentally observed size effects that are stronger than those explicable by Weibull theory.

Because of material heterogeneity, the macroscopic continuum stress at a point material must depend mainly on the average deformation of a representative volume of the material surrounding that point, rather than on the local stress or strain at that point.

In the deterministic nonlocal theory for strain softening damage or plasticity, the spatial averaging must be applied only to the inelastic part ϵ'' of the total strain ϵ (or some of its parameters), rather than to the total strain itself (Pijaudier-Cabot & Bažant 1987). Accordinly, the cumulative failure probability $P_1(\sigma)$, considered in the classical Weibull theory as a function of the local stress tensor σ at continuum point x, must be replaced by a function of a nonlocal variable (Bažant & Xi, 1991, Bažant & Novák 2000a,b). The nonlocal stress is not acceptable because it decreases with increasing average strain. A suitable nonlocal variable is the nonlocal strain or, more precisely, the nonlocal inelastic part of strain. The material failure probability is thus defined in the nonlocal Weibull theory as

$$P_1 = \langle \bar{\sigma}/s_0 \rangle^m, \quad \bar{\sigma}(\boldsymbol{x}) = \boldsymbol{E} : [\epsilon(\boldsymbol{x}) - \bar{\epsilon}''(\boldsymbol{x})] \tag{7}$$

$$\bar{\epsilon}''(\boldsymbol{x}) = \int_V \alpha(\boldsymbol{s} - \boldsymbol{x}) \epsilon''(\boldsymbol{s}) \mathrm{d}V(\boldsymbol{s}) \,/\, \bar{\alpha}(\boldsymbol{x}) \tag{8}$$

in which $\boldsymbol{E}$ = initial elastic moduli tensor; $\alpha(\mathbf{s} - \mathbf{x})$ = a bell-shaped nonlocal weight function whose effective spread is characterized by characteristic (material) length l_0; and $\bar{\alpha}(\boldsymbol{x})$ = normalizing factor of $\alpha(\boldsymbol{s} - \boldsymbol{x})$.

The nonlocality makes the Weibull integral over a body with crack tip singularity convergent for any value of Weibull modulus m, and it also introduces into the Weibull theory spatial statistical correlation. Numerical calculations of bodies with large cracks or notches showed that the randomness of material strength is almost irrelevant for the size effect on the mean σ_N, except theoretically for structures extrapolated to sizes less than the inhomogeneity size in the material (Bažant & Xi 1991). Therefore, the energetic mean size effect law for the case of large cracks or large notches remains unaffected by material randomness. Intuitively, the reason is that a significant contribution to Weibull integral comes only from the FPZ, the size of which remains constant if the structure size is increased. The same reason applies to the boundary layer of cracking (Fig. 1a), and is documented by the inelastic strain field in (Fig. 1c left, linear scale) and the field of the density of contribution to the Weibull integral (right, log-scale) obtained by Bažant & Novák (2000a) in nonlocal beam flexure analysis.

A special case in which the statistical size effect is important is the failure at crack initiation in a very large structure, much larger than the inhomogeneity size. This is the case of bending of very thick plain concrete beams or plates, for example the flexural failure of an arch dam about 10 m thick (Bažant & Novák 2000a,b; Fig. 1d,e).

The asymptotic limits of the mean nonlocal Weibull size effect are, for $D \to 0$, the deterministic energetic size effect and, for $D \to \infty$, the mean classical Weibull size effect. Their asymptotic matching approximation leads to the following approximate formula for the mean size effect (Bažant 2001, Fig. 1 g,d,e):

$$\sigma_N = \sigma_0 \left[\left(\frac{D_b}{\eta D_b + D} \right)^{rn/m} + \frac{r D_b}{\eta D_b + D} \right]^{1/r} \tag{9}$$

where $rn/m < 1$; η, r = empirical constants. The special case for $\eta = 0$ was shown to fit the bulk of the existing test data on the modulus of rupture and closely agree with numerical predictions of the nonlocal Weibull theory over the size range 1:1000 (Bažant & Novák 2000b). Aside from the two aforementioned asymptotic limits, the formula also satisfies, as a third asymptotic condition, the requirement that the deterministic size effect on the modulus of rupture must be recovered for $m \to \infty$.

As for the coefficient of variation ω_N of the nominal strength σ_N, it can be proven analytically in general that, according to the nonlocal generalization of Weibull theory, it is independent of size D. In other words, ω_N exhibits no statistical size effect, is defined by the same expression as in the classical Weibull theory (Eq. 4), and is fully determined by the value of Weibull modulus m. Numerically this fact was demonstrated in Bažant & Novák (2000a).

4 GENERALIZATION OF STOCHASTIC FINITE ELEMENT METHOD (SFEM) FOR VERY LOW PROBABILITY FAILURES

In comparison to SFEM, the nonlocal generalization of Weibull statistical theory (Bažant & Xi, 1991, Bažant & Novák 2000a,b, Bažant 2002a,b) has two limitations:

1) It does not yield the statistics of stiffness, deflections and stresses during the loading process; and
2) the failure probability is not related to the probability that the first eigenvalue λ_1 of the tangential stiffness matrix K_t of the structure, K_t, becomes nonpositive. Properly, matrix K_t should loose positive definiteness at the onset of failure.

SFEM has become a powerful tool for calculating the statistics of deflections and stresses of arbitrary structures (e.g., Schuëller 1997a,b, Kleiber & Hien 1992, Ghanem & Spanos 1991, Liu et al. 1987, Deodatis & Shinozuka 1991, Shinozuka & Deodatis 1988, Takada 1990). A critical appraisal, however, suggests that SFEM, as it now exists, is not an adequate tool for determining failure loads of very small probability, on which structural design must be based. In this regard, the nonlocal generalization of Weibull statistical theory has three important advantages:

1) In the limit of infinite size, the nonlocal Weibull theory reduces to the classical (local) Weibull theory, which is not true for the SFEMs in their contemporary form.
2) The nonlocal characteristic length ℓ, in contrast to autocorrelation length in SFEMs L_a, has a clear physical meaning and can be easily evaluated from the size effect tests using simple LEFM-based formulas; it ensues as the transitional size D_0 obtained as the intersection of the asymptotes of an optimally matched size effect law, times a shape factor known from LEFM.
3) The nonlocal Weibull theory is simpler, since an autocorrelated random field is not needed (a certain kind of spatial correlation is implied by the characteristic length of the nonlocal averaging operator).

The first point is a fundamental one. It is related to the far-out tail of the probability distribution of the tangential stiffness, which governs the failure loads of very small probability. The following physical argument is pertinent in this regard (Bažant 2001):

Imagining the structure to be scaled up to infinity size ($D \to \infty$), the FPZ becomes infinitely small compared to the structure size D (i.e., a point in the dimensionless coordinates $\boldsymbol{\xi} = \boldsymbol{x}/D$). In that case, failure (of a structure of positive geometry) must occur right at fracture initiation. Therefore, the classical (local) Weibull theory must apply, and the failure load of very low probability then depends *only* on the far-off tail of the local strength distribution at a point of the structure. Thus, extrapolation to very large sizes is a way to identify the far-off tail of the local strength distribution.

On the other hand, the existing SFEMs have not been shown to converge to Weibull theory and to reproduce the Weibull size effect as $D \to \infty$, neither analytically nor computationally. This important requirement is not ensured by the existing SFEMs.

The requirement that the Weibull size effect must be approached for infinite size has some implications for the structure of the tail of the probability distribution of the stiffness coefficients, deflections and stresses. Normally, the material stiffness characteristics are assumed in SFEM to have the Gaussian or log-normal distributions. Since the probability distribution of the structural tangential stiffness matrix is essentially a weighted sum of the elemental distributions (i.e., the distributions of the stiffness characteristics of a small representative volume of the material), the distribution of the structural stiffness coefficients may be expected to be Gaussian, with the exception (1) of the far-out tail of probability distribution, and (2) of the states of damage localization in one or several finite elements which may (though need not) occur just before reaching the peak load and dominates the structural stiffness.

During the loading process, the maximum load (which represents a failure state under the conditions of load control), is reached when the tangential stiffness matrix of the structure, $\boldsymbol{K}_t$, ceases being positive definite, i.e., when the first eigenvalue λ_1 of $\boldsymbol{K}_t$ ceases being positive (e.g. Bažant & Cedolin 1991, ch. 4, 10 and 13). Therefore,

$$\text{Failure probability}(u) = \text{Prob}\,(\,\lambda_1(u) \leq 0\,) \qquad (10)$$

As indicated here, the failure probability and the first eigenvalue are regarded as functions of the load-point displacement u (since, in order to achieve computational stability, it is u, rather than P, which needs to be controlled during loading).

Now it is important to realize that, in the limit of infinite size, the distribution of extreme values is not arbitrary, not something that can be left to empirical observations. Rather, it must be the Weibull distribution, exactly, and nothing else. Consequently, it is necessary to satisfy, in the limit of infinite size, the following condition:

$$\text{Tail of}[\,\text{Prob}\,(\,\lambda_1(u) \leq 0\,)] = F_W[P(u)] \qquad (11)$$

Here $F_W(P)$ is the cumulative Weibull distribution function (Weibull 1939). This distribution has a power-law tail and a threshold, which can normally be taken as zero (excluding negative values). F_W is here properly considered as an implicit function of the controlled displacement u because the tangential stiffness in the direction of loading changes its sign at maximum load.

Imposing condition (11) can be justified by ruling out all the other possibilities, which are as follows. In a population of N statistically independent random variables X_i $(i = 1, 2, \ldots N)$ with arbitrary but identical statistical distributions $\text{Prob}(X_i \leq x) = P_1(x)$, henceforth called the elemental distribution ($x = \sigma/s_0$ = scaled stress, X_i = scaled random strength), the distribution of $Y_N = \min_{i=1}^{N} X_i$ for very large N has the general expression:

$$P_N(y) = 1 - e^{-NP_1(y)} \qquad (12)$$

where $P_N(y) = \text{Prob}(\min_{i=1}^{N} X_i \leq y)$; $P_N(y) = P_f$ = failure probability of structure, provided that the failure of one element causes the whole structure to fail. As proven by Fisher and Tippett (1928), there exist three and only three asymptotic forms (or limiting forms for $N \to \infty$) of the extreme value distribution $P_N(y)$:

1) Fisher-Tippett-Gumbel distribution:

$$P_N(y) = 1 - e^{-e^{y}} \qquad (13)$$

2) Fréchet distribution:

$$P_N(y) = 1 - e^{|y|^{-m}} \qquad (14)$$

3) Weibull distribution:

$$P_N(y) = 1 - e^{-y^{m}} \qquad (15)$$

(Case 1 is usually called the Gumbel distribution, but Fisher and Tippett derived it much earlier and Gumbel gave them credit for it.) Case 3 is obtained if the elemental distribution $P_1(y)$ has a power-law tail with a finite threshold (the simplest case is the rectangular probability density function, for which $m = 1$). Case 1 is obtained if $P_1(y)$ has an infinite exponentially decaying tail, and case 2 if $P_1(y)$ has an infinite tail with an inverse power law (such as $|\sigma|^{-m}$) (see also Bouchaud & Potters 2000).

Fisher & Tippett (1928) based their proof on three arguments: (1) The key idea is that the extreme of a sample of $\nu = Nn$ independent identical random variables x (the strengths of the individual links of a chain) can be regarded as the extreme of the set of N extremes of the subsets of n variables, e.g., the strengths of n links of a chain (Fig. 2g). (2) As both $n \to \infty$ and $N \to \infty$, the distributions of the extremes of samples of sizes n and Nn must have a similar form if an asymptotic form exists. This implies that these distribution must be related by a linear transformation in which only the mean and the standard deviation can change; i.e., $\sigma' = a_N\sigma + b_N$ where a_N and b_N are functions of N ($N \sim$ structure size). Although an asymptotic distribution of the extremes, as a limit for $N \to \infty$, does not exist, an asymptotic *form* (or shape) of the extreme value distribution should exist, i.e., the asymptotic distribution form should be stable with regard to increasing N. Thus the argument of a joint probability of survival of all N segments of the chain yields for the asymptotic form of the cumulative distribution of the survival probability $F(\sigma) = 1 - P_f = 1 - P_N$ of a very long chain the recursive functional equation:

$$F^N(\sigma) = F(a_N\sigma + b_N) \qquad (16)$$

which is called the stability postulate of extreme value distribution (a_N, b_N are coefficients depending on N). Fisher and Tippett proved that this functional equation for unknown function F has three and only three types of solution, and that they are given by (13)–(15). By substituting these forms into functional equation (16), one can check that indeed this equation is satisfied. The substitutions further give the dependence of a_N and b_N on N, which in turn characterizes the dependence of the mean and the standard deviation of each asymptotic distribution on N ($N \sim$ structure size).

The infinite negative tails of P_N of the Fréchet distribution and the Fisher-Tippett-Gumbel distribution are not acceptable for describing the strength. Therefore, these two distributions are ruled out. So, in the case of strength, there is no other acceptable tail distribution but Weibull distribution (which is what lends support to the condition (11)).

In SFEMs, there are of course techniques, such as the importance sampling, for calculating the failure loads of extremely small probability. Unfortunately, though, the fact that the probability structure of the existing SFEMs do not yield Weibull power-law tails, nor lead to the Weibull power-law size effect when the structure size is scaled up to infinity, means that the calculations of loads of a very small failure probability, such as 10^{-7} (and probably not even 10^{-3}), are unrealistic.

In design codes, the safety factors relate the mean failure load prediction (roughly the same as the median, or failure probability 0.5) to the failure load with a desired extremely low probability, typically about 10^{-7}. This is illustrated by the upper arc in Fig. 2k, spanning about 6.5 orders of magnitude. Since the existing experimental validations of SFEMs have been confined mainly to the standard deviation, the current SFEMs, with their exponential tails, might be realistic for calculating only loads of failure probability no less than about 10^{-2}. So, an empirical safety factor spanning 5 orders of magnitude is needed to relate this load to the failure load of the desired probability such as 10^{-7} (see the lower arc in the figure).

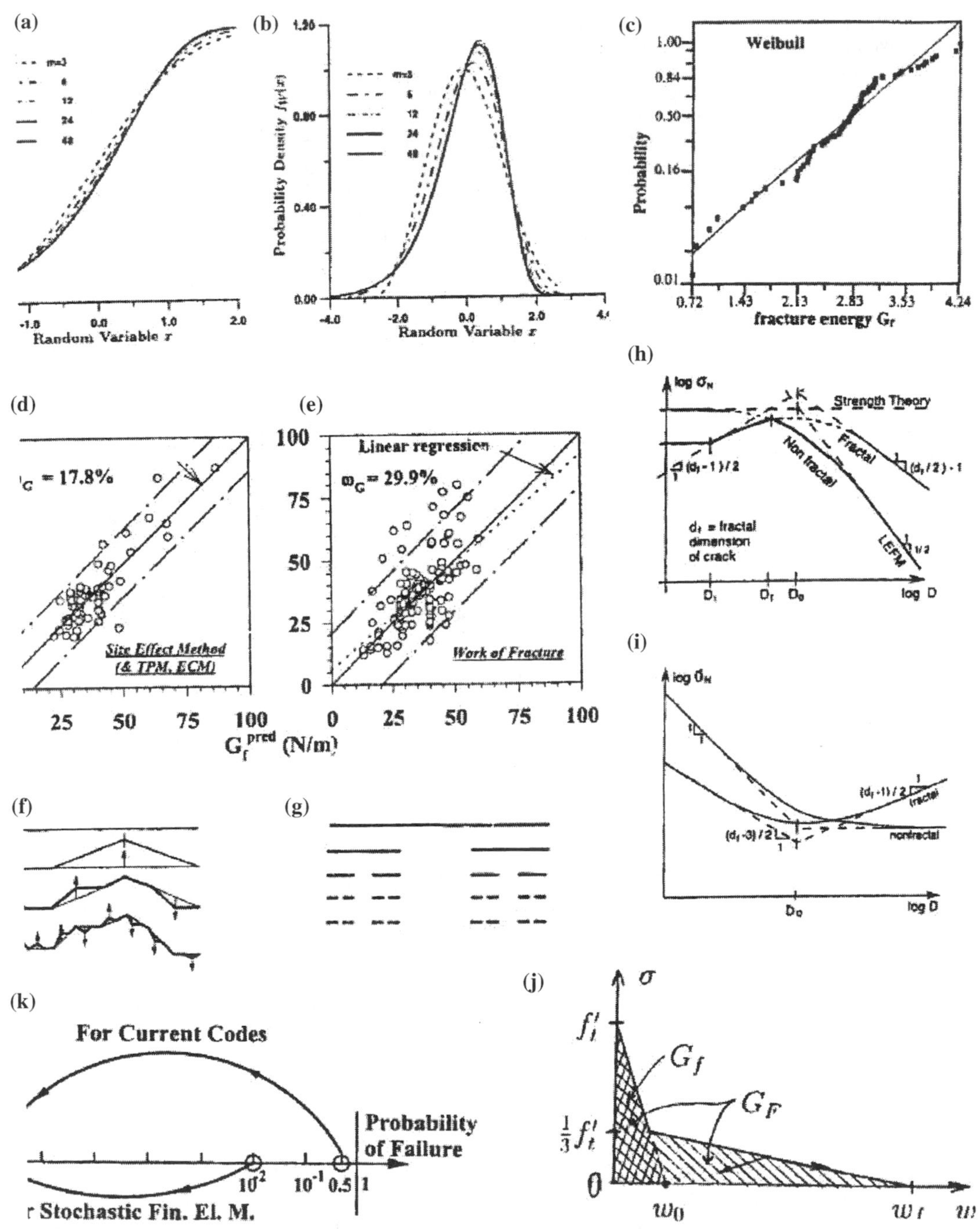

Figure 2. (a,b) Cumulative Weibull distribution and its density, for various Weibull moduli m. (c) 77 size effect test data on fracture energy from the literature, plotted on Weibull probability paper. (d,e) Plots of measured versus predicted values of size effect fracture energy (77 test series) and Hillerborg fracture energy (161 test series). (f,g) Invasive affine fractality of crack surface and lacunar fractality of microcracks. (h,i) Unreasonable size effects for large cracks and crack initiation ensuing from the hypothesis of fractal size effect. (j) Softening stress-crack separation curve of cohesive crack model of concrete. (k) Safety factors relating failure probability to calculations.

Comparing the lengths of the two arcs, one gets the sobering impression that, in terms of safety against failure, not too much is gained by the use of SFEMs if a proper tail probability structure is not enforced.

5 PRACTICAL FINITE ELEMENT APPROACH

In a primitive approach, the structure would have to be subdivided a very large number ν of finite elements having the fixed size of the characteristic volume of material. For a very large structure, ν would be a very large number, easily one billion. Therefore, such a primitive approach is impossible.

To achieve a proper tail probability with a manageable number of finite elements, the following idea, which is advanced in more detail in Novák, Bažant & Vořechovský (2003), will now be briefly outlined. The idea is to simulate by finite elements directly Fisher & Tippett's (1928) fundamental stability postulate of extreme value distributions. This can be done as follows.

Similar to the crack band model (Bažant 1982, Bažant & Oh 1983), the finite elements used in computations have a fixed size in the relative coordinate $\boldsymbol{\xi} = \boldsymbol{x}/D$. The structure is subdivided into a fixed number N of finite elements (called the macroelements) the size of which increases in proportion to structure size D. Each macroelement is imagined to consist of n microelements of a fixed actual size and a strength represented by an independent stochastic variable. The subdivision number is defined as $n = V_{el}/l^{n_d}$ where V_{el} is the macroelement volume, l^{n_d} is the characteristic volume of the material, l = characteristic length and n_d = number of spatial dimensions in which the structure is scaled up ($n_d = 1,2,3$ for one-, two- and three-dimensional similarity). Then the strength for each of these N subsets of n stochastic strength variables is simulated statistically, according to Weibull distribution, and for each subset the extreme (minimum strength) is selected to be the representative statistical property (the strength) of the finite element (macroelement). These N extremes of the subsets of n variables are then used in numerical finite element analysis of the whole structure.

This procedure ensures that the extreme value statistics is correctly approached, with one crucial advantage – the number n of finite elements (macroelements) remains reasonable from the computational point of view. Although N increases with the structure size, the determination of the extreme from the subdivision of each macroelement does not add to the computational burden since it is carried out analytically, outside the finite element analysis.

As already pointed out, one basic hypothesis of the classical Weibull theory of structural strength is the statistical independence of the strengths of the individual characteristic volumes l^{n_d}, where l is the characteristic length. The strength of each of these volumes can be described by Weibull distribution with Weibull modulus m and scale parameter s_0 (the threshold being taken as zero, as usual). Each of the aforementioned macroelements, whose characteristic size is L_0 and characteristic volume $L_0^{n_d}$, may be imagined to be discretized into n characteristic volumes l^{n_d}. This consideration provides the statistical properties of the macroelement. Since we are interested only in very small tail probabilities, we may simply substitute in these equations the tail approximation of the elemental (generic) Weibull distribution with a certain modulus and scale parameter. The tail approximation is the power function σ^m (times a constant), and its substitution leads for the strength of the macroelement again to Weibull distribution but with a different modulus and scale parameter, and thus with a different mean and variance, which are expressed according to Eq. (4).

6 NUMERICAL EXAMPLE OF FOUR-POINT-BEND FLEXURAL TESTS

By this time, abundant experimental evidence on the statistical size effect on plain concrete beams has been accumulated. Koide et al. (1998, 2000) recently reported tests of 279 plain concrete beams under four-point bending, aimed at determining the influence of the beam length L on the flexural strength of beams. These excellent statistical data permit a comparison of the cumulative probability distribution function of the maximum bending moment M_{max} at failure (Bažant & Novák 2000b, Novák et al. 2001). Beams of three different bending spans, 200, 400 and 600 mm (series C of Koide et al.) are shown in Fig. 3, along with the cracks obtained by deterministic finite element calculations (carried out with the code ATENA, Červenka 2002). The cross-sections of all the beams were kept constant (0.1 m × 0.1 m) and the experiments showed how M_{max} decreases with an increasing span (Fig. 4). To describe the size effect of the span, Koide et al. proposed a modification of the Weibull theory.

Koide et al. reported the compression strength of the concrete, but unfortunately not the direct tensile strength and the fracture energy of concrete. The experimental results plotted in Fig. 4 give the mean value for each size. The double logarithmic plot of M_{max} versus the span is approximately a straight line of slope $D^{-n_d/m}$, where n_d is the spatial dimension and m is the Weibull modulus. Since the depth and width of the beam are not increased, the problem is properly analyzed as one-dimensional, and then the overall slope of the experimental data in the figure is matched best with the value $m = 8$, which is unusually low for

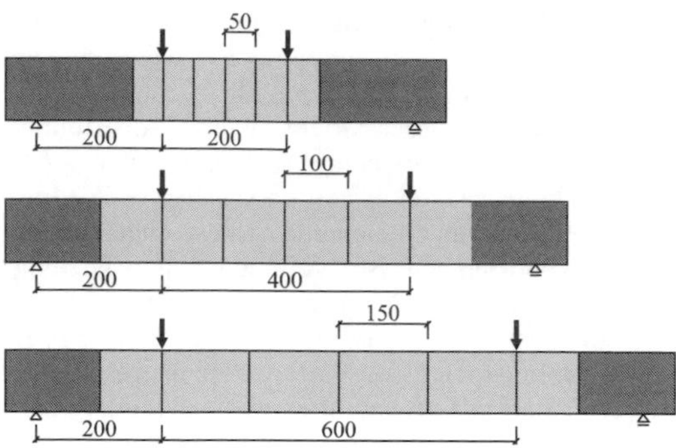

Figure 3. Koide's beams of bending span 200, 400 and 600 mm, series C.

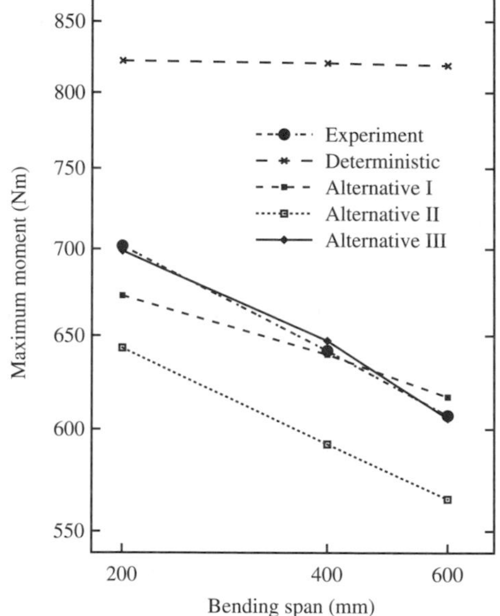

Figure 4. Comparison of means of Koide's data and ATENA Deterministic and Statistical simulations.

concrete and implies an unusually high coefficient of variation of the scatter of flexural strength.

Deterministic simulation with nonlinear fracture mechanics software (made with ATENA) indicate that no appreciable size effect is present. This is no surprise. According to fracture mechanics, the deterministic size effect on flexural strength of beams, whether unnotched or notched, is almost nil if the beam depth is not varied because the energy release function is almost independent of the beam span. This is useful in view of our focus on the statistical size effect. It allows a purely statistical analysis of the test data in Fig. 4, reflecting the fact that, the longer the beam, the higher is the probability of encountering in it a material element of a given low strength.

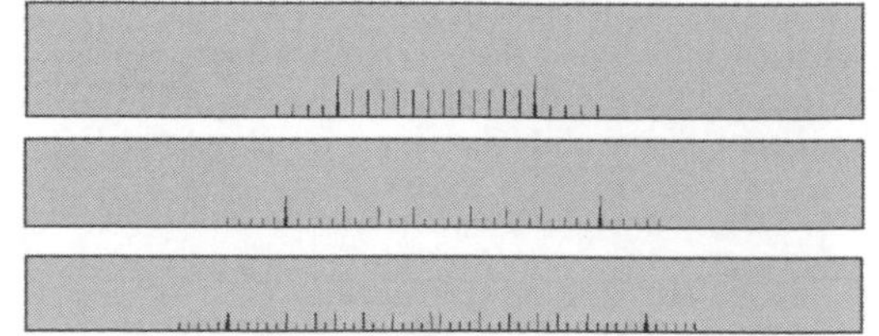

Figure 5. Deterministic cracks for sizes 20, 40 and 60.

The force increments applied on the beams were prescribed in numerical simulation in order to avoid a nonsymmetrical bending moment distribution when material randomness causes the crack pattern (Fig. 5) to become nonsymmetric. Because of load control, the load-deflection curves, including the peak and postpeak response, were calculated using the arc length method.

The probabilistic version of nonlinear fracture mechanics software ATENA (Červenka & Pukl 2002) was utilized to simulate the tests of Koide et al. by finite elements in the sense of extreme value statistics. This was made possible by integrating ATENA with the probabilistic software FREET (Novák et al. 2002, 2003). In this simulation, the finite element mesh is defined by using only 6 stochastic macroelements of strip-like form, placed in the central region of the test beams in which fracture initiates randomly; see Fig. 6. The strips suffice for simulating the Weibull size effect. We imagine n elements per macroelement of width L_0, while the finite element meshes for all the sizes are identical (except for scaling by a horizontal stretch).

The characteristic length is considered to be 50 mm, which is approximately three-times the maximum aggregate size. The Weibull modulus is taken as $m = 8$, and the scale parameter is $s_0 = 1.0$ MPa. The statistical parameters of the strength of the macroelements, imagined to consist of $N = L_0/l$ material elements each. For the three sizes (spans) considered here, $M = 50$, 100, 150 mm and $N = 1, 2, 3$.

For the stochastic finite element simulations, a stochastic computational model with $N = 6$ random tensile strength variables is defined for each beam size (span). These 6 variables are characterized by 16 random simulations based on the method of Latin hypercube sampling, using simulations by the FREET and ATENA softwares (Novák et al. 2003, Vořechovský & Novák 2003, Pukl et al. 2003). The statistical characteristics of the ultimate load can then be evaluated. The mean values of nominal strength obtained from a statistical set of the maximum load are determined first. The random cracking pattern at failure is shown in Fig. 6, as obtained for four realizations of three progressively improved alternatives of solution. To illustrate the randomness of failure, the corresponding

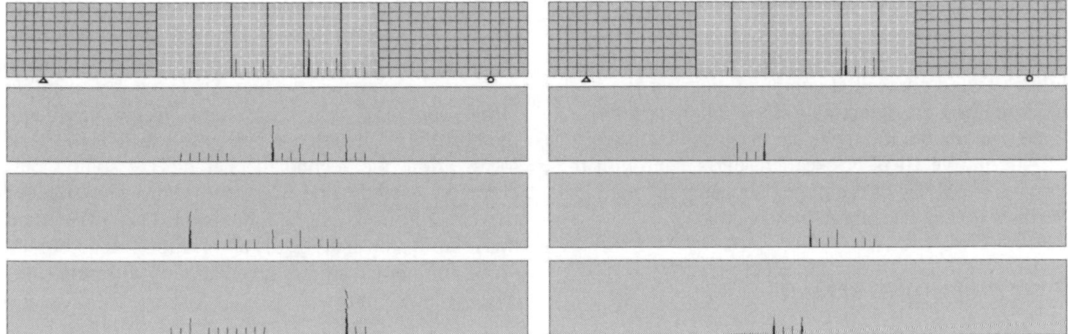

Figure 6. Macroelements and examples of random crack initiation for the first size; left: random tensile strength only, right: random and correlated tensile strength and fracture energy.

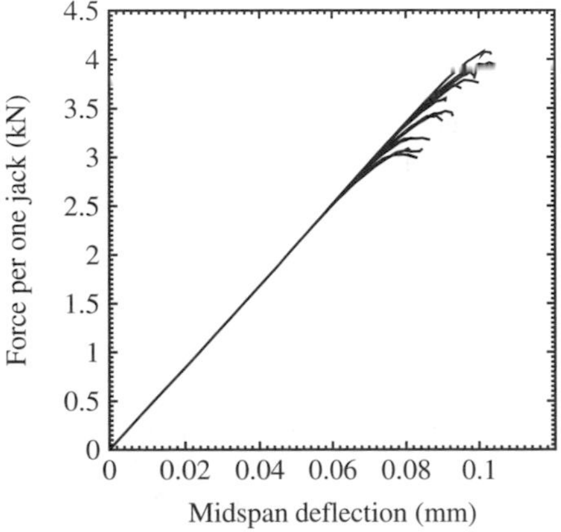

Figure 7. Random load deflection curves.

random load-deflection curves are shown in Fig. 7. The three alternatives, for which the results are presented in Fig. 4, are the following:

Alternative I: A pure Weibull-type statistical approach is the first alternative studied. Only the random scatter of tensile strength is considered, the generic mean value of tensile strength being fixed as 3.7 MPa. The resulting size effect curve obtained by probabilistic simulation is found to have a smaller slope than the experimental data trend, in spite of the fact that an unusually low Weibull modulus ($m = 8$) is used. To explain it, note that the Weibull theory strictly applies only when the failure occurs right at the crack initiation, before any (macroscopically) significant stress redistribution with energy release takes place. This is not the case for concrete, a coarse material relative to the beam depth used. So, a nonnegligible fracture process zone must form before a macroscopic crack can form and propagate, dissipating the required fracture energy G_f per unit crack surface. Therefore, the beam, analyzed by nonlinear fracture mechanics (the crack band model, approximating the cohesive crack model) does not fail when the first element fails (as required by the weakest link model imitating the failure of a chain). Rather, it fails only after a group of elements fails, and several groups of failing elements can develop before the beam fails; see Fig. 6. The finite element simulations are able to capture this behavior thanks to the cohesive nature of softening in a crack or crack band, reflecting the energy release requirement of fracture mechanics.

Alternative II: To overcome the aforementioned problem and match the size effect data, we must take into account the randomness of fracture energy G_f. But we cannot ignore the statistical correlation of G_f to tensile strength. For lack of available data, we simply assume a very strong correlation, characterized by correlation coefficient 0.9. Such a correlation tends to cause the (macroscopic) crack propagation to begin earlier than in Alternative I. The result is shown in Fig. 4 as Alternative II. The resulting slope of the simulated size effect curve is now close to the slope of experimental data. However, the entire curve is shifted down, i.e., all the beams are weaker than they ought to be. The strong correlation between the tensile strength and fracture energy is seen to cause the macroelements with a lower tensile strength to be more brittle. Therefore, the failure must localize into these macroelements.

Alternative III: In seeking a remedy, we must realize that Koide et al. measured neither the tensile strength nor the fracture energy, and that our foregoing estimates may have been too low. So, our only option is a heuristic approach. While keeping Alternative II, we are free to shift the size effect curve up by increasing the generic mean value of tensile strength and the fracture energy value. We increase them to 4 MPa and 100 N/m, respectively, and this adjustment is found to furnish satisfactory results; see Fig. 4. Although the size effect of Alternative III in the double logarithmic plot is not as straight as the trend of data, the differences from the data are negligible. These small differences may have been easily caused, for example, by an insufficient size of the data set or by weak stability of load-controlled computations near the peak load, making the detection of the peak inaccurate.

In conclusion, it may be emphasized that the result of Alternative III is in excellent agreement with the previous analysis of the same data using the nonlocal Weibull theory (Bažant & Novák 2000b).

7 CONCLUDING REMARK

The factors of safety for designing against the risk of failure of concrete structures have doubtless much larger errors than the errors stemming from the inadequacies of the finite element analysis. Therefore, it would make little sense to strive for improvements of the finite element analysis with the underlying constitutive and fracture laws without at the same time addressing the limitations of the statistical theory. The present practice of finite element analysis, and even its stochastic generalization, cannot realistically cope with the statistical risk of failure because it does not provide the correct probability structure of the far-off tail of the failure probability distribution as a function of the applied load. The present analysis attempts to identify the problem and propose a way, doubtless not the only way, of achieving the correct extreme value distribution of failure probability for structures that fail in a quasibrittle manner.

APPENDIX I INTERPLAY OF DETERMINISTIC AND STOCHASTIC LENGTH SCALES

Progress in the understanding of the uncertainties concrete failure and fracture behavior with reliability aspects has been achieved in many works; e.g. Shinozuka (1972), Mihashi & Izumi (1977), Mazars (1982), Bažant & Xi (1991), Breysse (1990), Breysse & Renaudin (1996), Carmeliet (1994), Carmeliet & Hens (1994), de Borst & Carmeliet (1996), Gutiérez (1999), Gutiérez & de Borst (1999, 2000, 2001, 2002) and others. A finite element reliability method for gradient-enhanced damage models has been formulated by Gutiérez & de Borst (1999). They considered the damage threshold as an autocorrelated random field and applied their model to quasibrittle damage in tensile double-edge-notched specimens and pull-out of steel anchors from concrete simulating damage threshold as a random field. The analysis furnished the most likely localization patterns corresponding to a chosen failure criterion and the influence of deterministic and statistical length scales on the statistical properties of structural response, size effect and damage accumulation.

The characteristic length of a nonlocal continuum governs the deterministic scaling and the autocorrelation length of random field in stochastic finite elements governs the statistical scaling. De Borst & Carmeliet (1996) showed that both the characteristic length and the correlation length are needed – the first to avoid localization, the second to characterize spatial randomness. A salient question is the relationship of these two characteristic length. Franziskonis (1998) studied this relationship analytically and Gutiérez & de Borst's (2002) numerical studies revealed very different roles of these two lengths.

APPENDIX II STATISTICAL PROPERTIES OF NONLOCAL GENERALIZATION OF WEIBULL THEORY FOR SIZE EFFECT IN QUASIBRITTLE STRUCTURES

The statistical properties of the nonlocal generalization of Weibull theory (Eq. 7) can be derived analytically. This derivation, which is more fundamental than the direct asymptotic matching approach (Bažant & Novák 2000c), Bažant (2002) and is presented in full in Bažant (2002c), will now be briefly outlined.

Considering the nonlocal averaging domains in a nonlocal model of a structure to be analogous to the links of a chain, one may calculate the the failure probability P_f of a structure is given by $-\ln(1-P_f)=\int_V\langle\hat{\sigma}\ (\boldsymbol{x})/s_0\rangle^m\ \mathrm{d}V(\boldsymbol{x})/V_r$ (Bažant & Xi 1991) where $V=$volume of structure, $V_0=$representative volume of material, $m=$Weibull modulus, $s_0=$scaling parameter, $\sigma(\boldsymbol{x})=$maximum principal stress at point of coordinate vector $\boldsymbol{x}$, $\hat{\sigma}=$nonlocal stress, $V_r=$representative volume of the material for which the Weibull parameters m and s_0 have been experimentally identified; and $\langle ..\rangle$ denotes the positive part of the argument. It is convenient to introduce dimensionless coordinates and variables by setting $\xi=D\xi$, $V_0=l^{n_d}$, $V=l^{n_d}v$, $\mathrm{d}V(\mathbf{x})=l^{n_d}\mathrm{d}v(\boldsymbol{\xi})$, $\sigma(\mathbf{x})=\sigma_N S(\boldsymbol{\xi})$ and $l=V_r^{1/n_d}$ where $Df=$size (characteristic dimension) of the structure; $n_d=$number of spatial dimensions in which the structure is scaled ($n_d=1$, 2 or 3 for one-, two- or three-dimensional scaling); $l=$characteristic length of material; $\boldsymbol{\xi}=\boldsymbol{x}/D=$dimensionless coordinate vector; and $\sigma_N=P/bD=$nominal strength of structure ($P=$maximum load, $b=$width of structure). We consider geometrically similar structures of different sizes D, for which the corresponding points have the same dimensionless coordinate $\boldsymbol{\xi}$; then (Bažant 2002c)

$$P_f=1-\mathrm{e}^{-(\sigma_N/s_D)^m},\quad s_D{}^m=\frac{s_0{}^m}{H^m}\left(\frac{l}{D}\right)^{n_d} \tag{17}$$

$$H^m=\int_v\langle\hat{S}(\boldsymbol{\xi})\rangle^m\mathrm{d}v(\boldsymbol{\xi}) \tag{18}$$

The nonlocal stress $\hat{S}(\xi)$ used above does not permit determining the size effect analytically. Therefore, we restrict attention to large enough structures such that the nonlocal averaging domain, roughly of the same size as the fracture process zone (the zone of distributed cracking or localized damage), is small enough compared to D, though not necessarily negligible.

It can be shown in general for various definitions of the nonlocal stress and verified by the example of a three-point bend beam that the first two terms of the asymptotic expansion of the dimensionless nonlocal integral H as a power series in l/D may be written as $H = H_0(1 - c_m l/D)$, the error being of the order of $(l/D)^2$; here H_0 and c_m are dimensionless constants for geometrically similar structures. Since the term $(1 - c_m l/D)$ becomes negative for small enough l/D (which would render (17) physically meaningless), it is suitable to introduce the following equivalent approximation (which maintains the same order of asymptotic accuracy, the error being of the second order in l/D):

$$H^m = H_0{}^m(1 - c_m l/D)^m \approx H_0{}^m(1 + mrc_m l/D)^{-1/r} \tag{19}$$

(Bažant 2002c) where r is an arbitrary positive empirical constant. Eq. (17) then yields:

$$\sigma_N = [-\ln(1 - P_f)]^{1/m} s_D \quad (D \gg l) \tag{20}$$

$$s_D = \frac{s_0}{H_0}\left(\frac{l}{D}\right)^{n_d/m}\left(1 + mrc_m\frac{l}{D}\right)^{1/r} \tag{21}$$

(Bažant 2002c). This expression represents the large-size size effect law of nominal strength of structure with any specified failure probability P_f. For $P_f = 0.5$ it represents the large-size size effect law for the median nominal strength.

Eq. (21) gives for any D a real value of σ_N which, for any fixed P_f, decreases monotonically with D through the entire size range $D \in (-\infty, \infty)$. However, the limiting nominal strength for $D \to 0$ is infinite. From a purely empirical viewpoint, this might not be considered as objectionable because unreasonably large σ_N might result only for a hypothetical structure size D much smaller than the aggregate size. However, we prefer the small-size asymptotic properties to agree with the theoretical small-size asymptotic properties of the cohesive (or fictitious) crack model or the crack band model, or the nonlocal damage model, which imply that the value of $\hat{\sigma}_N$ for $D \to 0$ should be finite and should be approached linearly in D (Bažant 2001, 2002); this may be achieved by replacing $1/D$ with $1/(\eta l + D)$, which has no effect on the large size asymptotic properties (η = empirical coefficient of the order of 1). With this replacement, Eq. (21) becomes:

$$s_D = \frac{s_0}{H_0}\left(\frac{l}{\eta l + D}\right)^{\frac{n_d}{m}}\left(1 + mrc_m\frac{l}{\eta l + D}\right)^{\frac{1}{r}} \tag{22}$$

(Bažant 2002c) where $D \gg l$. It may be noted that the mean size effect law for mean σ_N implied by this result is not identical to (9). However, the difference is barely distinguishable in data fitting.

Similar to the classical Weibull theory, the mean and standard deviation of σ_N may be calculated as:

$$\overline{\sigma}_N = \int_0^\infty \sigma_N \mathrm{d}P_f(\sigma_N) = s_D \Gamma(1 + 1/m) \tag{23}$$

$$\delta_N{}^2 = \int_0^\infty \sigma_N{}^2 \mathrm{d}P_f(\sigma_N) - \overline{\sigma}_N{}^2$$

$$= s_D{}^2\Gamma(1 + 2/m) - \overline{\sigma}_N{}^2 \tag{24}$$

Accordingly, the coefficient of variation of σ_N (for $D \gg l$)

$$\omega_N = \frac{\delta_N}{\overline{\sigma}_N} = \sqrt{\frac{\Gamma(1 + 2/m)}{\Gamma^2(1 + 1/m)} - 1} \tag{25}$$

is asymptotically independent of structure size D, and is given by the same expression as in Weibull theory. But this is true only for large enough D/l.

While Eq. (23) with (22) gives a realistic size effect formula for the mean strength throughout the full size range $D \in (0, \infty)$, the coefficient of variation in (25) and the entire probability distribution P_f given by (17) or (20) with (22) are certainly invalid for small enough D/l. The reason is that the failure of small structures with a cohesive crack or crack band approaches, for $D \to 0$, the case of elastic body with a plastic crack for which the failure probability has an entirely different structure. Since the failure of such a small structure must be non-propagating (i.e., simultaneous along the entire failure surface), the failure probability should obey Daniels' "fiber-bundle" (parallel coupling) model rather than extreme value statistics. So, for $D \to 0$, the size effect on the mean nominal strength must asymptotically vanish. The failure probability distribution according to Daniels' model is (in agreement with the central limit theorem of probability) the gaussian distribution, except for the far-out tails (this may explain why the old studies based on measuring ω_N gave for concrete $m \approx 12$ instead of the correct value $m \approx 24$ based on size effect measurements). Denoting $\sigma_N \propto F_g(P_f)$ = inverse of cumulative gaussian (normal) distribution for the small size limit, one may conjecture roughly the following general size effect law giving the nominal strength of any specified failure probability P_f for the full size range (Bažant 2002c):

$$\sigma_N = s_D \left\{ \frac{l^u}{l^u + D^u} F_g(P_f) + \frac{D^u}{l^u + D^u} \left[-\ln(1 - P_f) \right]^{1/m} \right\} \quad (26)$$

where u is a certain suitable positive exponent. According to Daniels' (1945) model (for infinitely many fibers in the bundle), the gaussian cumulative distribution to which $F_g(P_f)$ is inverse should be such that its mean σ_N is independent of D but its coefficient of variation is proportional to $1/\sqrt{D}$.

ACKNOWLEDGMENT

Partial financial support under Grant ONR-N00014-02-I-0622 from Office of Naval Research (monitored by Dr. Yapa D.S. Rajapakse) is gratefully acknowledged. The second author gratefully acknowledge the support from the Grant Agency of the Czech Republic under the grant No. 103/03/1350. Authors thanks to M. Vořechovský for numerical simulation of four-point-bend flexural tests.

REFERENCES

Bažant, Z.P. 1976. Instability, ductility, and size effect in strain-softening concrete. *J. Engrg. Mech. Div., Am. Soc. Civil Engrs.* 102 (EM2), 331–344; disc. 103, 357–358, 775–777, 104, 501–502.

Bažant, Z.P. 1984. Size effect in blunt fracture: concrete, rock, metal. *J. of Engrg. Mech. ASCE* 110, 518–535.

Bažant, Z.P. 1986. Mechanics of distributed cracking. *Appl. Mech. Reviews ASME* 39, 675–705.

Bažant, Z.P. 1987. Fracture energy of heterogeneous material and similitude. Preprints, SEM-RILEM *Int. Conf. on Fracture of Concrete and Rock* (held in Houston, Texas, June 1987), ed. by S.P. Shah and S.E. Swartz, publ. by SEM (Soc. for Exper. Mech.) 390–402.

Bažant, Z.P., Editor 1992. *Fracture Mechanics of Concrete Structures*, Proc., First Intern. Conf. (FraMCoS-1), held in Breckenridge, Colorado, June 1–5, Elsevier, London (1040 pp.).

Bažant, Z.P. 1993. Scaling laws in mechanics of failure. *J. of Engrg. Mech. ASCE* 119 (9), 1828–1844.

Bažant, Z.P. 1997. Scaling of quasibrittle fracture: Asymptotic analysis. *Int. J. of Fracture* 83 (1), 19–40.

Bažant, Z.P. 1998. Size effect in tensile and compression fracture of concrete structures: computational modeling and design. *Fracture Mechanics of Concrete Structures* (3rd Int. Conf., FraMCoS-3, held in Gifu, Japan), H. Mihashi and K. Rokugo, eds., Aedificatio Publishers, Freiburg, Germany, 1905–1922.

Bažant, Z.P. 1999. Size effect on structural strength: a review. *Archives of Applied Mechanics* (Ingenieur-Archiv, Springer Verlag) 69, 703–725 (75th Anniversary Issue).

Bažant, Z.P. 2001a. Size effects in quasibrittle fracture: Apercu of recent results. *Fracture Mechanics of Concrete Structures* (Proc., FraMCoS-4 Int. Conf., Paris), R. de Borst et al., eds., A.A. Balkema Publishers, Lisse, Netherlands, 651–658.

Bažant, Z.P. 2001b. Probabilistic modeling of quasibrittle fracture and size effect. (principal plenary lecture, Proc., 8th Int. Conf. on Structural Safety and Reliability (ICOSSAR), held at Newport Beach, Cal., 2001), R.B. Corotis, ed., Swets & Zeitlinger (Balkema), 1–23.

Bažant, Z.P. 2002c. Statistical properties of nonlocal generalization of Weibull theory for size effect in quasibrittle structures. Preliminary Report to ONR No. 02-12-C699s, Dept. of Civil Engrg., Northwestern University, Evanston, Dec. (also submitted to JEM-ASCE, and in condensed form to Proc., ICASP, Berkeley, July 2003).

Bažant, Z.P. & Chen, E.-P. 1997. Scaling of structural failure. *Applied Mechanics Reviews ASME* 50 (10), 593–627.

Bažant, Z.P. & Frangopol, D.M. 2002. Size effect hidden in excessive dead load factor. *J. of Structural Engrg. ASCE* 128 (1), 80–86.

Bažant, Z.P. & Kazemi, M.T. 1990. Size effect in fracture of ceramics and its use to determine fracture energy and effective process zone length. *J. of American Ceramic Society* 73 (7), 1841–1853.

Bažant, Z.P. & Novák, D. 2000a,b. Probabilistic nonlocal theory for quasibrittle fracture initiation and size effect. I. Theory, and II. Application. *J. of Engrg. Mech. ASCE* 126 (2), 166–174 and 175–185.

Bažant, Z.P. & Novák, D. 2000c. Energetic-statistical size effect in quasibrittle failure at crack initiation. *ACI Materials Journal* 97 (3), 381–392.

Bažant, Z.P. & Novák, D. 2001a. Proposal for standard test of modulus of rupture of concrete with its size dependence. *ACI Materials Journal* 98 (1), 79–87.

Bažant, Z.P. & Novák, D. 2001b. Nonlocal Weibull theory and size effect in failures at fracture initiation. *Fracture Mechanics of Concrete Structures* (Proc., FraMCoS-4 Int. Conf., Paris), R. de Borst et al., eds., Swets & Zeitlinger (A.A. Balkema Publishers), Lisse, Netherlands, 659–664.

Bažant, Z.P. & Planas, J. 1998. *Fracture and Size Effect in Concrete and Other Quasibrittle Materials*. CRC Press, Boca Raton, Florida.

Beremin, F.M. 1983. A local criterion for cleavage fracture of a nuclear pressure vessel steel. *Metallurgy Transactions A* 14, 2277–2287.

Breysse, D. 1990. A probabilistic formulation of the damage evaluation law. *Struct. Safety* 8, 311–325.

Breysse, D. & Renaudin, P. 1996. On the influence of local disorder on size effect. *Size-Scale Effects in the Failure Mechanisms of Materials and Structures* (Proc., IUTAM Symp. held in 1994 at Politecnico di Torino), ed. A. Carpinteri, E & FN Spon, London, 187–199.

Carmeliet, J. 1994. On stochastic descriptions for damage evolution in quasi-brittle materials. *DIANA Computational Mechanics*, G.M.A. Kusters and M.A.N. Hendriks (eds.)

Carmeliet, J. & Hens, H. 1994. Probabilistic nonlocal damage model for continua with random field properties. *ASCE J. of Engrg. Mech.* 120(10), 2013–2027.

Červenka, V. & Pukl, R. 2002. ATENA – Computer program for nonlinear finite element analysis of reinforced concrete structures, Červenka Consulting, Prague.

de Borst, R. & Carmeliet, J. 1996. Stochastic approaches for damage evolution in local and non-local continua. In *Size-Scale Effects in the Failure Mechanism of Materials and Structures*, (ed. A. Carpinteri), E & FN Spon, London, 242–257.

Deodatis, G. & Shinozuka, M. 1991. Weighted integral method. II. Response variability and reliability. *J. of Engrg. Mech. ASCE* 117 (8), 1865–1877.

Engelund, S. & Rackwitz, R. 1992. On predictive distribution functions for the three asymptotic extreme value distributions. *Structural Safety* 11, 255–258.

Epstein, B. 1948. Statistical aspects of fracture problems. *J. of Applied Physics* 19, 140–147.

Fisher, R.A. & Tippett, L.H.C. 1928. Limiting forms of the frequency distribution of the largest and smallest member of a sample. *Proc., Cambridge Philosophical Society* 24, 180–190.

Frantzisconis, G.N. 1998. Stochastic modelling of heterogeneous materials – A process for the analysis and evaluation of alternative formulations. *Mechanics of Materials* 27, 165–175.

Fréchet, M. 1927. Sur la loi de probabilité de l' écart maximum. *Ann. Soc. Polon. Math.* 6, 93.

Freudenthal, A.M. 1956. Physical and statistical aspects of fatigue. In *Advance in Applied Mechanics*, Vol. 4, Academic Press, 117–157.

Freudenthal, A.M. 1968. Statistical approach to brittle fracture. Chapter 6 in *Fracture*, Vol. 2, ed. by H. Liebowitz, Academic Press, 591–619.

Freudenthal, A.M. 1981. *Selected Papers*, *Am. Soc. of Civil Engrs.*, New York.

Freudenthal, A.M. & Gumbel, E.J. 1953. On the statistical interpretation of fatigue tests. *Proc. of the Royal Society (London)* A, Vol. 216, 309–332.

Freudenthal, A.M. & Gumbel, E.J. 1956. Physical and statistical aspects of fatigue. In *Advances in Applied Mechanics*, Vol. 4, Academic Press, 117–157.

Ghanem, R.G. & Spanos, P.D. 1991. *Stochastic finite elements: a spectral approach*. Springer-Verlag, New York.

Gnedenko, B.V. 1943. Sur la distribution limite du terme maximum d'une série aléatoire. *Ann. Math.* 44, p. 423.

Gnedenko, B.V. & Kolmogorov, A.N. 1954. *Limit distributions of sums of random variables*. Addison Wesley, Cambridge, Massachussetts.

Gumbel, E.J. 1954. *Statistical theory of extreme values and some practical applications*, Nat. Bureau of Standards, Washington, D.C., Appl. Math. Series No. 33.

Gumbel, E.J. 1958. *Statistics of Extremes*. Columbia University Press, New York.

Gutiérez, M.A. 1999. *Objective simulation of failure in heterogeneous softening solids*. Dissertation, Delft University of Technology.

Gutiérez, M.A. & de Borst, R. 1999. Deterministic and stochastic analysis of size effects and damage evaluation in quasi-brittle materials. *Archive of Applied Mechanics* 69, 655–676.

Gutiérez, M.A. & de Borst, R. 2000. Stochastic aspects of localized failure: material and boundary imperfections. *Int. J. of Solids and Structures* 37, 7145–7159.

Gutiérez, M.A. & de Borst, R. 2001. Deterministic and probabilistic material length scales and their role is size-effect phenomena. *Report*, Koiter Institute, Delft University of Technology, Delft, Netherlands; submitted for publication.

Gutiérez, M.A. & de Borst, R. 2002. Deterministic and probabilistic material length-scales and their role in size-effect phenomena. *ICOSSAR01, CD-ROM Proc. of Int. Conf. on Structural Safety and Reliability*, Newport Beach, California, Balkema.

Hillerborg, A., Modéer, M. & Petersson, P.E. 1976. Analysis of crack formation and crack growth in concrete by means of fracture mechanics and finite elements. *Cem. Concr. Res.* 6, 773–782.

Hinch 1991. *Perturbation Methods*. Cambridge University Press, Cambridge, UK.

Iguro, M., Shiyoa, T., Nojiri, Y. & Akiyama, H. 1985. Experimental studies on shear strength of large reinforced concrete beams under uniformly distributed load, *Concrete Library International, Japan Soc. of Civil Engrs.* No. 5, 137–154 (translation of 1984 article in Proc. JSCE).

Kleiber, M. & Hien, T.D. 1992. The stochastic finite element method. J. Wiley, New York.

Koide, H., Akita, H. & Tomon, M. 1998. Size effect on flexural resistance due to bending span of concrete beams. Fracture Mechanics of Concrete Structures (Proc., 3rd Int. Conf., FraMCoS-3, held in Gifu, Japan), H. Mihashi and K. Rokugo, eds., Aedificatio Publishers, Freiburg, Germany, 2121–2130.

Koide, H., Akita, H. & Tomon, M. 2000. Probability model of flexural resistance on different lengths of concrete beams. Application of Statistics and Probability (Proc., 8th Int. Conf., ICASP-8, held in Sydney, Australia 1999), R.E. Melchers and M.G. Stewart, eds., Balkema, Rotterdam, Vol. 2, 1053–1057.

Lei, Y., O'Dowd, N.P., Busso, E.P. & Webster, G.A. 1998. Weibull stress solutions for 2-D cracks in elastic and elastic-plastic materials. *Int. J. of Fracture* 89, 245–268.

Liu, W.K., Mani, A. & Belytschko, T. 1987. Finite element methods in probabilistic mechanics. *Probabilistic Engineering Mechanics* 2 (4), 201–213.

von Mises, R. 1936. La distribution de la plus grande de *n* valeurs. *Rev. Math. Union Interbalcanique* 1, p. 1.

Novák, D. et al. 2002. FREET – Feasible reliability engineering efficient tool. Report to Červenka Consulting, Prague, Brno University of Technology, Faculty of Civil Engineering, Institute of Structural Mechanics.

Novák, D., Rusina, R. & Vořechovský 2003. Small-sample statistical analysis – software FREET. *Proc., ICASP 9*, San Francisco, USA, in press.

Novák, D., Vořechovský, M., Pukl, R. & Červenka, V. 2001. Statistical nonlinear analysis – size effect of concrete beams. *Fracture Mechanics of Concrete and Concrete Structures* (Proc., 4th Int. Conference, FraMCoS-4), Cachan, France, Balkema, Lisse, Netherlands.

Peerlings, R.H.J., de Borst, R., Brekelmans, W. & Vre, J. 1996. Gradient enhanced damage for quasibrittle materials. *Int. J. of Num. Meth. in Engrg.* 39, 3391–3403.

Peirce, F.T. 1926. Tensile tests of cotton yarns – V. The weakest link. *J. Textile Inst.* 17, 355–368.

Pukl R., Strauss A., Novák, D., Červenka V. & Bergmeister K. 2003. Probabilistic-based assessment of concrete

structures using non-linear fracture mechanics. *Proc., ICASP 9*, San Francisco, USA, in press.
Ruggieri, C. & Dodds, R.H. 1996. Transferability model for brittle fracture including constraint and ductile tearing effects – a probabilistic approach. *Int. J. of Fracture* 79, 309–340.
Saibel, E. 1969. Size effect in structural safety. *Proc., Int. Conf. on Structure, Solid Mechanics and Engineering Design in Civil Engineering Materials*, Southampton, Part I, 125–130.
Schuëller, G.I. 1997a,b. Structural reliability – recent advances. (Freudenthal lecture), *Proc., ICOSSAR-97, Kyoto, Japan*, Balkema, Rotterdam, Vol. I, 3–35.
Schuëller, G.I. 1997. A state-of-art report on computational stochastic mechanics. *Probabilistic Engineering Mechanics* 12 (4), 199–321.
Shinozuka, M. & Deodatis, G. 1988. Response variability of stochastic finite elements. *J. of Engrg. Mech. ASCE* 118 (3), 499–519.
Shioya, Y. & Akiyama, H. 1994. Application to design of size effect in reinforced concrete structures. In *Size Effect in Concrete Structures* (Proc. of Intern. Workshop in Sendai, 1993), ed. by H. Mihashi, H. Okamura and Z.P. Bažant, E & FN Spon, London, 409–416.
Takada, T. 1990. Weighted integral method in multi-dimensional stochastic finite element analysis. *Prob. Engrg. Mech.* 5 (4), 158–166.
Tippett, L.H.C. 1925. On the extreme individuals and the range of samples. *Biometrika* 17, 364.
Vořechovský & Novák, D. 2003. Small-sample statistical analysis – software FREET. *Proc., ICASP 9*, San Francisco, USA, in press.
Walraven, J. 1995. Size effects: their nature and their recognition in building codes. *Studi e Ricerche* (Politecnico di Milano) 16, 113–134.
Walraven, J. & Lehwalter 1994. Size effects in short beams loaded in shear. *ACI Structural Journal* 91 (5), 585–593.
Weibull, W. 1939. The phenomenon of rupture in solids. Proc., Royal Swedish Institute of Engineering Research (Ingenioersvetenskaps Akad. Handl.) 153, Stockholm, 1–55.

Computational Modelling of Concrete Structures, Bićanić et al. (eds)
© 2003 Swets & Zeitlinger, Lisse, ISBN 90 5809 536 3

Keynote paper: Regularized saw-tooth softening

J.G. Rots
Delft University of Technology, Delft, The Netherlands

S. Invernizzi
Politecnico di Torino, Torino, Italy; visiting research fellow, Delft University of Technology

ABSTRACT: After a brief discussion on crack models, it is demonstrated that cracking is often accompanied by snaps and jumps in the load-displacement response which complicate the analysis. This paper provides a solution by simplifying nonlinear crack models into sequentially linear saw-tooth models, either saw-tooth tension-softening for unreinforced material or saw-tooth tension-stiffening for reinforced material. A linear analysis is performed, the most critical element is traced, the stiffness and strength of that element are reduced according to the saw-tooth curve, and the process is repeated. This approach circumvents the use of incremental-iterative procedures and negative stiffness moduli and is inherently stable. The main part of the paper is devoted to a regularization procedure that provides mesh-size objectivity of the saw-tooth model. The procedure is similar to the one commonly used in the smeared crack framework but, in addition, both the initial tensile strength and the ultimate strain are rescaled. In this way, the dissipated fracture energy is invariant with respect not only to the mesh size, but also to the number of saw-teeth adopted to discretize the softening branch. Furthermore, the algorithm for non-proportional loading has been fully implemented and results from a practical case are shown.

1 INTRODUCTION

Structures have the ability to redistribute stresses upon cracking or crushing. A crack may initiate, propagate and snap to a free surface, whereupon stresses are redistributed and a new crack or crush may start. For large-scale structures, like unreinforced masonry facades, the unbalance between elastic energy stored in the structure versus fracture energy consumed in crack localization is large, which makes this redistribution process brittle, involving so-called snap-backs in the quasi-static load-displacement response. This behavior is also typical of reinforced structures. In fact a local snap occurs whenever a crack crosses a reinforcing bar, giving a downward path under displacement or arc-length control followed by a new ascending path after redistribution. Tracing snap-back and snap-through response is a delicate matter in incremental-iterative nonlinear analysis (Crisfield 1984). The negative tangent stiffness gives rise to convergence problems, independent of the constitutive crack model adopted. Sometimes jump-over techniques are tried, but these cannot guarantee equilibrium and stability.

In this paper a sequentially linear saw-tooth model is elaborated to overcome these problems. The model, introduced by Rots (2001), replaces the downward stress–strain curve by a saw-tooth curve, either saw-tooth tension-softening for unreinforced material or saw-tooth tension-stiffening for reinforced material. A linear analysis is performed, the most critical element is traced, the stiffness and strength of that element are reduced according to the saw-tooth curve, and the process is repeated. This procedure is both simple and stable as the secant stiffness of the saw-teeth is always positive. Although a route was outlined to achieve mesh-size objectivity, this issue was not yet fully understood at that time. In the present paper, this issue will be fully solved by a fracture energy based procedure that adjusts both the ultimate strain of the saw-tooth diagram as well as the peak of the saw-tooth diagram to the mesh size. Furthermore, the algorithm to use the sequentially linear model in combination with non-proportional loading will be implemented and tested. Before coming to that, first some general comments on crack models are given.

2 COMMENTS ON SOME CRACK MODELS

For quasi-brittle materials like concrete and masonry a variety of crack models exists. In the Euro-C conference tradition, debates take place with regard to the pro's and contra's of the various approaches, often from a theoretical point of view, e.g. De Borst et al.

(1998). In the present section, a few models are discussed from a practical user's point of view, for applications to real structures. No completeness is claimed. Only models experienced by the author(s) are discussed. Important classes like higher-order, non-local and embedded discontinuity models, are left out.

2.1 *Decomposed-strain based smeared crack models*

Smeared crack models start from the notion of a continuum with strain, representing either the width of an individual localized crack, or an average crack strain in case of distributed cracking in reinforced concrete. It has been proven to be effective to decompose the strain into a part that belongs to the crack and a part that belongs to the material at either side of the cracks, see e.g. de Borst & Nauta (1986), Rots (1985). The strain-decomposition allows elastic-softening behavior to be described in an elegant way, both for mode I and mode II including secant unloading/reloading. Another advantage is that the strain of the material between the cracks can be sub-decomposed into e.g. an elastic part, a creep part, a plasticity part and/or a thermal part (de Borst 1987) which allows for the combination of cracking with other nonlinear phenomena. Furthermore, the crack strain can be decomposed into the local crack strains of a number of cracks at different orientations. This makes it possible to handle non-orthogonal multi-directional cracking. Though the model is thus conceptually attractive, the difficulty lies in the implementation and parameter choices. We mention two practical disadvantages. First, the algorithmic aspects are complicated. The model needs internal iterations to handle state changes like unloading, closing and re-opening, to handle nonlinear softening diagrams and to handle the combination with plasticity for compression strut action. When multiple cracks occur this internal iteration procedure may fail and even substepping cannot always guarantee success. Secondly, the user's choices for shear retention functions and inter-crack threshold angles are not obvious, which may e.g. result in a too stiff response, stress-locking or uncontrollable principal stress rotation (Rots 1988).

2.2 *Total-strain based models*

The difficulties mentioned above gave rise to the development of simplified total-strain based models, e.g. Feenstra et al. (1998). The material is described via stress-total strain relations in either fixed or rotating axes. For the fixed version, shear retention parameters are required, whereas the rotating version uses an implicit shear term to provide coaxiality between the rotating principal stress and strain. Advantages are as follows. First, these models are attractive to practicing engineers who think in terms of stress–strain relations and can plug these in directly. There is no need to input abstract yield functions or sophisticated crack laws. Secondly, the concept of a stress–strain relation cannot only be used for softening or stiffening in tension, but also for nonlinear behavior and crushing in compression. This provides a natural combination of cracking and crushing whereby unconfined 2D cases, confined 3D cases and various coupling effects can be accounted for. Thirdly, from an implementation and algorithmic point of view, the model is attractive. It is purely explicit and does not require internal iterations. After updating a number of internal state variables, the stress is computed directly from the strain. Local convergence problems or closing/re-opening difficulties do not exist. The disadvantages are as follows. Due to the fact that stress–strain relations are used, either in a fixed orthogonal system or in the rotating orthogonal principal stress system, non-orthogonal cracks cannot be modelled. Furthermore, combinations with creep, shrinkage and thermal strains are not straightforward due to the description in terms of lumped total strain.

2.3 *Plasticity based crack models*

A unified approach for tension and compression can also be achieved by taking a plasticity based concept. Rankine-Von Mises and Rankine-Drucker Prager crack-crush models including softening were developed by Feenstra & de Borst (1996). The general theory of elasto-plasticity provides a concise format for the implementation and algorithmic aspects, while often quadratic convergence is achieved. Another advantage is that these models can be extended to orthotropic materials like masonry, via e.g. Rankine-Hill (Lourenco et al. 1997). Also, the additive strain decomposition of elasto-plasticity allows for an extension to include creep, shrinkage and thermals strains and rate effects (Van Zijl et al. 2001). A practical disadvantage of plasticity based models, is the inherent elastic unloading/reloading. A secant approach back to the origin, as included in the decomposed as well as total-strain based smeared crack models, is a better approximation of the real unloading behavior for quasi-brittle materials. Figure 1 gives an example in tension.

Especially when the strain is (far) beyond the ultimate strain of the softening branch, the differences between elastic and secant unloading are pronounced. Positions far beyond the ultimate strain of the softening diagram occur frequently when real, large-scale structures are considered. Then, the element dimensions and consequently the crack band width h are large, giving a small value of ε_u in Fig. 1 according to the crack band model of Bazant and Oh (1983). The consequences of elastic versus secant unloading emerge for cyclic loadings (Feenstra & Rots 2001), but also for non-proportional monotonic loadings. The

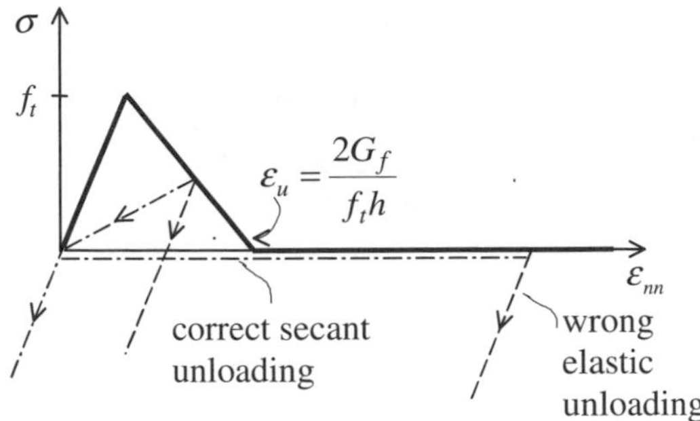

Figure 1. Secant and elastic unloading options.

authors experienced this when analyzing buildings that cracked initially under dead load and were subsequently subjected to a different load, e.g. wind load or imposed settlements. Upon adding this second load, the initial cracks unloaded and closed which resulted in a too stiff behavior when Rankine plasticity was used. The plasticity models can be extended with damage options to improve this behavior.

2.4 *Other continuum approaches*

Damage models (e.g. Mazars and Pijaudier-Cabot 1989) describe the degradation process by some scalar damage variable. The unloading is naturally of secant type. Isotropic versions of damage models are attractive in case of localized tensile fracture, as all stress components shrink to zero, preventing stress-locking. For reinforced concrete, where the compression strut action parallel to the crack is crucial, isotropic damage models are less practical. Here, extensions to anisotropic damage are required. Such models come closer to smeared crack models that are inherently anisotropic as only the material in the direction normal to a crack degrades. Another class is the micro-plane models (Bazant and Pratt 1988). Here, normal and shear stresses across a fixed set of microplanes are monitored. These models are also capable of unifying tension and compression. The multiple planes suggest a similarity with the non-orthogonal smeared crack model. However, in the non-orthogonal smeared crack model a set of planes is not predefined, but the cracks/planes emerge during the process, in the direction normal to the principal stress when this principal stress violates a tension cut-off. A model geared towards use for heavily reinforced structures is the four-way fixed crack model by Maekawa and co-workers, see e.g. Pimanmas & Maekawa (2001). Here, two sets of orthogonal cracks can be present, with a non-orthogonal angle between them. Good results were reported for non-proportional loading and path-dependence in case of pre-cracked reinforced structures.

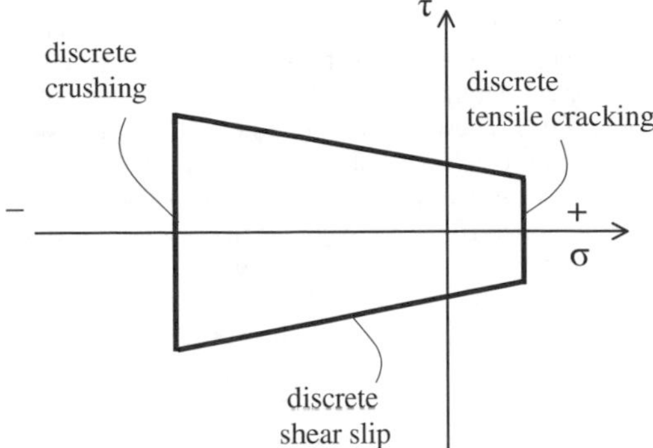

Figure 2. Three domains for interface behaviour.

There is room for future research to make the decomposed multi-directional model of section 2.1 more robust, by allowing only a limited number of cracks under fixed angles from the first arisen crack. Taking 45 degrees, a 4-plane crack model for 2D situations results, which is sufficient for many engineering RC situations. It allows for non-orthogonal cracking, but does not loose the conceptual attractiveness of the decomposed formulation. The implementation becomes compact and internal iterations and algorithmic aspects are under better control as the maximum number of cracks is constrained and "hesitating" cracks under small angles do not occur. For 3D cases and shells, the 45 degree pre-assumption results in a 9-plane crack model. Here, positive elements from smeared cracking, micro-plane models and Maekawa's model are combined.

2.5 *Discontinuum interface models*

Predefined interface elements can be used as potential planes for discrete cracks, discrete crush zones, shear slip planes or bond shear zones, e.g. Carol et al. (2001), Rots (1994) for masonry. This approach does not spread the softening over crack bands and therefore avoids problems with local instabilities and spurious modes. The elements define a relation between tractions and relative displacements across the interface, having a normal and shear component. In analogy to the total-strain based models of section 2.2, total relative-displacement based relationships can be used, for the tension, compression and shear domains, see Fig. 2. Such model is explicit and robust, it does not require internal iterations. The stresses are computed directly from the total relative displacements, taking internal states like loading/unloading and possible coupling effects between the three domains into account. Adequate solutions were found for pier-wall connections and complete prefab buildings (e.g. Frissen & Rots, 1999).

Alternatively, decomposed interface models can be used. In analogy to the plasticity models of section 2.3, plasticity based interface models were formulated by e.g. Stankowski et al. (1993) and Lourenco & Rots (1997). These models are able to elegantly combine tensile cracking and compression crushing with non-associated Coulomb friction for the shear domain. Advantages and disadvantages are along the same lines as for the plasticity based continuum approaches of section 2.3. Often, an initial dummy penalty stiffness is given to the interface elements in order to keep them initially inactive before the cut-off is reached. With the decomposed formulations, this high dummy stiffness continues playing a role in the nonlinear regime because of the additive displacement decomposition, which may unnecessarily complicate the analysis. In fact, the use of interface elements is already a "decomposition" as such, with the elastic bulk behavior at either side of the discontinuity being decoupled from this discontinuity. When the interface elements represent real joints that have a (non-dummy) stiffness with a physical meaning, like mortar joints in masonry, the issue is less relevant.

3 SNAPS AND JUMPS

Independent of the constitutive model selected, softening analyses are complicated because of snap-type response. Users in FE analysis of concrete and masonry structures know that convergence problems at global level may occur due to softening and negative tangent stiffness, independent of the local constitutive model adopted. This was already identified by Crisfield (1984) at the first Euro-C conference. A crack initiates, propagates and may either snap to a free surface or be arrested by a rebar, whereupon stresses are redistributed and a new crack or crush may start. This is accompanied by drops in the quasi-static load-displacement response, see also e.g. Carpinteri (1986). After reaching a valley, a new ascending path may be found up to a new local peak. The user cannot afford to miss these peaks as it might be the global peak for structural failure. The jumpy response is difficult to trace, although arc-length and indirect control schemes provide possibilities (e.g. de Borst 1987b).

This is not an incidental issue. We identify three examples. Figure 3 shows a reinforced tension bar. Bond-slip was modeled using an explicit tau-delta relation and smeared cracking was adopted. Both the experimental and numerical result shows four local peaks associated with the consecutive development of four primary cracks. Here, a subtle perturbation of strength in combination with delicate arc-length control provided a solution. Considering that such tension bar is only a minor component of a real structure, tracing the full response of structures would become unwieldy.

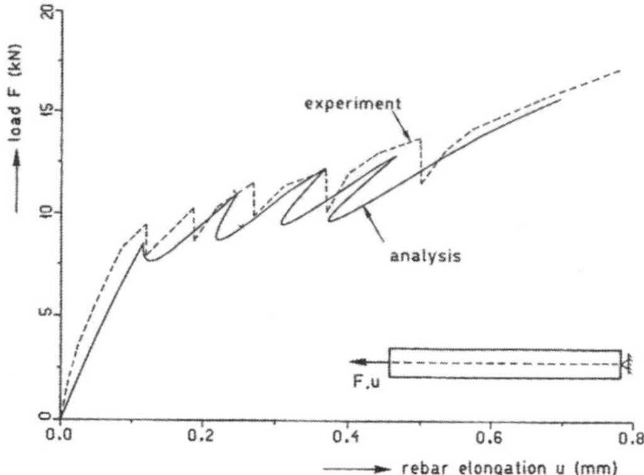

Figure 3. Response of reinforced tension bar (Rots 1988).

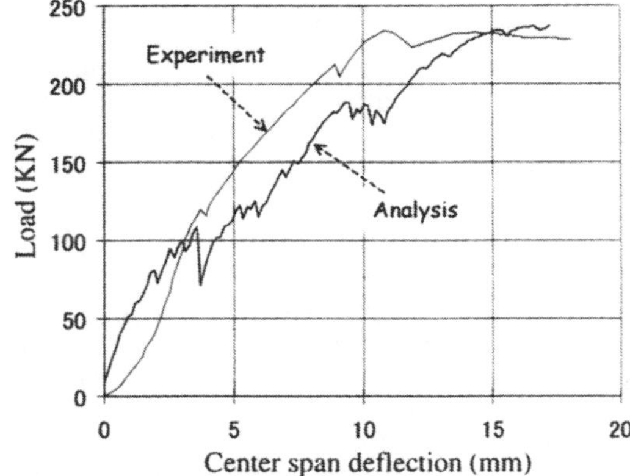

Figure 4. Response of pre-cracked RC structure (Pimanmas & Maekawa, 2001).

Figure 4 reprints a result by Pimanmas & Maekawa (2001) for a pre-cracked reinforced structure. Indeed, many local peaks are displayed, both in the experiment and computation. This behavior is typical of many reinforced structures. The peaks cannot be simply jumped over in load or displacement control.

Figure 5 shows a result for crack propagation in a large-scale unreinforced masonry facade subjected to settlement, see e.g. Boonpichetvong & Rots (2003). A global peak is visible and subsequently some sharp peaks in the valley, associated with the crack that discontinuously snapped from window to window. Only for discrete cracking, arc-length control provided a solution whereas all smeared trials failed and an unconverged jump-over was found. The difficulty here lies in the large scale (the masonry facade is 180 times larger than e.g. the SEN notched beam benchmark), which gives a large amount of stored elastic energy versus moderate consumption of fracture energy upon crack propagation, resulting in a brittle response and sharp snap-backs.

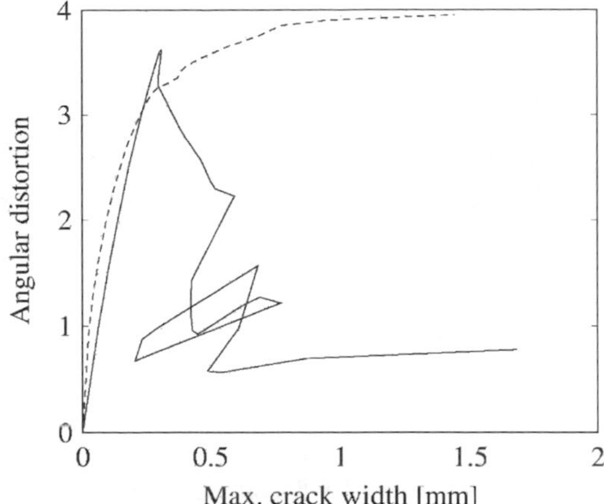

Figure 5. Response of large-scale masonry facade. Smeared (dashed) and discrete (drawn).

4 SEQUENTIALLY LINEAR APPROACH WITH SAW-TOOTH SOFTENING

4.1 *Global procedure*

The above snap-type responses inspired the idea to capture them in an entirely different fashion, via secant approximations with restarts from the origin. The softening diagram is approximated by a saw-tooth curve and linear analyses are carried out sequentially (Rots 2001). The global procedure is as follows. The structure is discretized using standard elastic continuum elements. Young's modulus, Poisson's ratio and a tensile strength are assigned to the elements. Subsequently, the following steps are sequentially carried out:

- Add the external load as a unit load.
- Perform a linear elastic analysis.
- Extract the critical element from the results. The critical element is the element for which the principal tensile stress is most close to its current strength. This principal tensile stress criterion is widely accepted in mode-I fracture mechanics of quasi-brittle materials.
- Calculate the critical global load as the unit load times the current strength divided by stress of the critical element.
- Extract also a corresponding global displacement measure, so that later an overall load-displacement curve can be constructed.
- Reduce the stiffness and strength, i.e. Young's modulus E and tensile strength f_t of the critical element, according to a saw-tooth tensile softening stress–strain curve as described in the next section.
- Repeat the previous steps for the new configuration, i.e. re-run a linear analysis for the structure in which E and f_t of the previous critical element are reduced.
- ... Repeat again, etc. ...
- Construct the overall load-displacement curve by connecting all load-displacement sets consecutively found in the above steps. This is similar to procedures for fracture analysis on lattices (Schlangen & Van Mier 1992), where little beam elements are removed rather than continuum elements reduced.
- Plot deformed meshes. These plots reveal the fracture localization because the series of critical weakened elements will display the largest strains, representing crack width.

4.2 *Saw-tooth softening model via stepwise reduction of Young's modulus*

The outcome of the above scheme heavily depends on the way in which the stiffness and strength of the critical elements are progressively reduced. This constitutes the essence of the model. A very rough method would be to reduce E to zero immediately after the first, initial strength is reached. This elastic perfectly brittle approach, however, is likely to be mesh dependent as it will not yield the correct energy consumption upon mesh refinement (Bazant and Cedolin 1979). In this study, the consecutive strength and stiffness reduction is based upon the concept of tensile strain softening, which is fairly accepted in the field of fracture mechanics of concrete (Bazant and Oh, 1983).

The tensile softening stress–strain curve is defined by Young's modulus E, the tensile strength f_t, the shape of the diagram, e.g. a linear or exponential diagram, and the area under the diagram. The area under the diagram represents the fracture energy G_f divided by the crack band width h, which is a discretisation parameter associated with the size, orientation and integration scheme of the finite element. Although there is some size-dependence, the fracture energy can be considered to be a material property. This softening model usually governs nonlinear constitutive behavior in an incremental-iterative strategy. Please note that here we adopt the curve only as a "mother" or envelope curve that determines the consecutive strength reduction in sequentially linear analysis. In the present study, attention is confined to a linear softening diagram, but extension to any other shape of the diagram is possible. For a linear softening diagram, the ultimate strain ε_u of the diagram reads:

$$\varepsilon_u = \frac{2G_f}{f_t h}, \tag{1}$$

In a sequentially linear strategy, the softening diagram can be imitated by consecutively reducing Young's modulus as well as the strength. Young's modulus can e.g. be reduced according to:

$$E_i = \frac{E_{i-1}}{a}, \quad \text{for } i = 1 \text{ to } n \tag{2}$$

with i denoting the current stage in the saw-tooth diagram, i-1 denoting the previous stage in the saw-tooth diagram and a being a constant. When a is taken as 2, Young's modulus of a critical element is reduced by a factor 2 compared to the previous state. n denotes the amount of reductions that is applied in total for an element. When an element has been critical n times, it is removed completely in the next step. This complete removal can be done explicitly so that "a hole in the mesh" occurs for full cracks, or it can be approximated by maintaining the element but giving it a very low residual Young's modulus for reasons of computational convenience (e.g. 10^{-6} times the initial Young's modulus). The reduced strength f_{ti} corresponding to the reduced Young's modulus E_i is taken in accordance with the envelope softening stress–strain curve:

$$f_{ti} = \varepsilon_u E_i \frac{D}{E_i + D}. \quad (3)$$

with

$$E_i = \frac{E}{a^i}, \quad (4)$$

and

$$D = \frac{f_t}{\varepsilon_u - \frac{f_t}{E}}, \quad (5)$$

being the tangent to the tensile stress–strain softening curve. Note that this is the softening curve in terms of stress versus *total* strain, i.e. the sum of elastic strain and crack strain of an imagined cracked continuum. The diagram includes the initial rising branch, which is steep compared to the downward slope in case of small-scale elements (small crack band width) and/or high fracture energy. As an example, Fig. 6 shows the envelope softening curve and the corresponding saw-tooth curve for an initial Young's modulus E of 38000 N/mm^2, initial tensile strength f_t of 3 N/mm^2, fracture energy G_f of 0.06 N/mm, crack band width h of 5 mm, factor a equal to 2 and number of reductions n equal to 10.

Strictly speaking, the sequentially linear approach corresponds to a set of unconnected lines of different slope, starting from the origin up to the current strength. For reasons of presentation, the lines have been connected into one discontinuous curve, by letting the stress drop vertically from the peaks to a residual level at which the new rising branch of reduced slope passes.

The model is simple. It always provides a solution: the secant saw-tooth stiffness is always positive, so that ill-conditioning or divergence does not appear in sequentially linear analysis. A physical explanation to the model is that fracture is a gradual separation process whereby the net cross section that connects material, and thus the stiffness, is gradually reduced. An advantage of the model is that the regular notions of fracture mechanics, like the principal tensile stress criterion, the envelope strength and fracture energy are maintained which helps in reaching realistic energy consumption and toughness as observed in experiments.

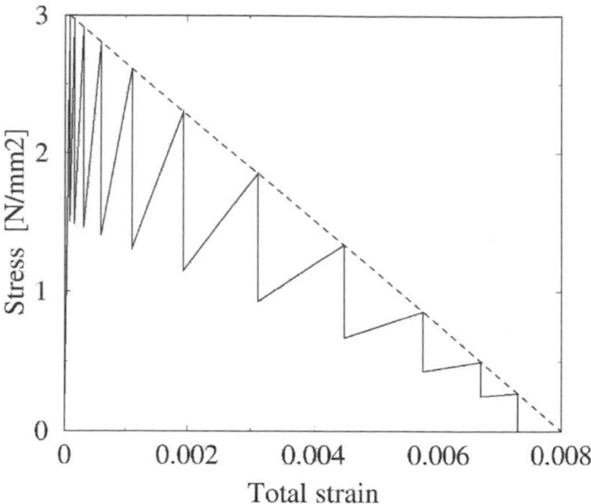

Figure 6. Envelope softening stress–strain diagram (dashed) and saw-tooth approximation (drawn).

The model has been elaborated here for continuum crack methods, using softening stress–strain curves. Sequentially linear saw-tooth approaches can also be employed for discrete crack methods with stress-relative displacement curves in interface elements or embedded representations. Unloading of softening zones is possible and inherently of the secant type, because of the reduction of elasticity.

4.3 *Alternative definition of saw-tooth diagram*

In the previous section, the starting point for defining the saw-tooth diagram was a consecutive reduction of Young's modulus via the factor a while the corresponding reduced strength was subsequently determined from the envelope curve according to (3). In first trials of the model, the starting point was taken alternatively by reducing the strength from f_t to zero in n steps, while the corresponding reduced Young's modulus was subsequently computed from the envelope curve.

The decreasing line of the envelope curve is in fact split into n equidistant portions. When n is assumed to be 10 and when the element is critical for the first time, the strength is reduced to 90% of the original strength. When it is critical for the second time, the strength is reduced to 80% of the original strength, etc. The resulting saw-tooth diagram for this alternative approach is depicted in Fig. 7, using the same parameters as subject to Fig. 6. Other alternative ways of stepwise reduction might be used as well.

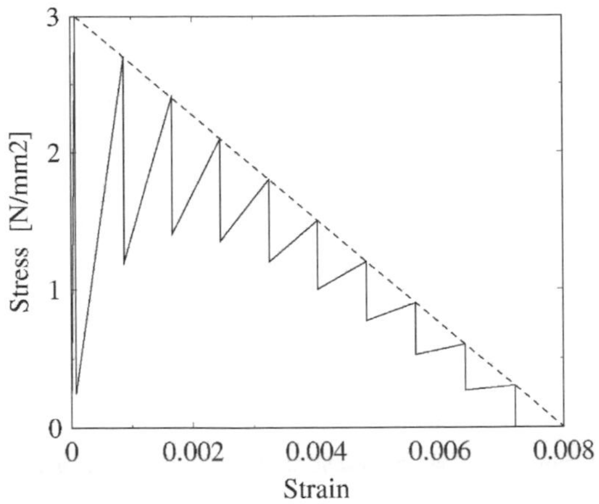

Figure 7. Envelope softening stress–strain diagram (dashed) and alternative saw-tooth approximation (drawn).

5 ANALYSIS OF NOTCHED BEAM

5.1 *Geometry and meshes*

A symmetric notched beam of total length 500 mm, span 450 mm, height 100 mm, thickness 50 mm and notch depth 10 mm was selected for analysis. The distance between the loading points in the symmetric four-point loading scheme is 150 mm. On purpose a regular straight mesh was adopted, so that the solution is not affected by any disturbing effect due to zig-zag crack band paths. Five different meshes were used, referred to as very coarse, coarse, medium, fine and very fine respectively. These meshes have a symmetric center crack band of 20 mm, 10 mm, 5 mm, 2.5 mm and 1.25 mm width respectively. The element height for the central elements was taken the same as the element width, so that the amount of center elements over the depth of the beam amounts to 5, 10, 20, 40 and 80 respectively. The ultimate strain ε_u of the envelope input curve was adapted to h according to Eq. (1). Four-node linear elements were used.

These were integrated using a two by two Gaussian scheme, except for the elements in the center band which were integrated using a single center-point integration.

This means that, if an integration point of an element in this center band reaches a local peak in the saw-tooth curve, Young's modulus of the entire element can be reduced. The meshes are depicted in Fig. 8. The material parameters are the same as mentioned in chapter 4 for defining Figs 6 and 7. Poisson's ratio was taken as 0.2. The model of Fig. 6 was adopted. Ten teeth were assumed. The factor a in Eq. (2) was taken as 2 with the factor n equal to 10, i.e. the stiffness of an element can be reduced 10 times according to Eq. (2), assuming the element becomes 10 times critical in a global sense.

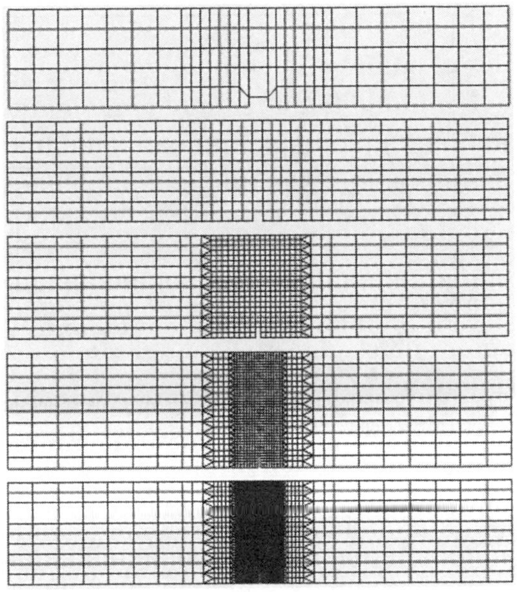

Figure 8. Meshes considered in the analysis, respectively referred in the following as very coarse, coarse, medium, fine and very fine.

Beyond that, the element is removed. Alternatively, a five teeth discretization ($a = 4$ in Eq. (2), and n equal to 5) and a twenty teeth discretization ($a = \sqrt{2}$ in Eq. (2), and n equal to 20) will be considered in parameter studies.

5.2 *Results – Unregularized – Ten teeth*

Figures 9–13 show the results for the five meshes in terms of the total load (sum of the two loads) versus displacement at the loading points. Figure 14 summarizes all results together in one plot. The curves are constructed by connecting the critical loads and the corresponding displacements for all linear analyses that have been executed sequentially. As a comparison, the reference curve (see Rots 1993) from a nonlinear softening analysis with the same parameters is included in the graphs. This reference curve appeared to be almost identical for all meshes, except for the coarse mesh where a minor deviation near peak occurred.

For all meshes, the load-displacement curve from sequentially linear analysis is irregular. This is because the process of elements becoming critical is discontinuous. If two elements have almost the same ratio of stress versus current strength, still the most critical one is selected first, but in the next step the other one will fail very soon, maybe at a lower load and lower displacement than in the previous step. This behavior is known also from sequentially linear analysis on lattices. In the present study, the irregular saw-tooth

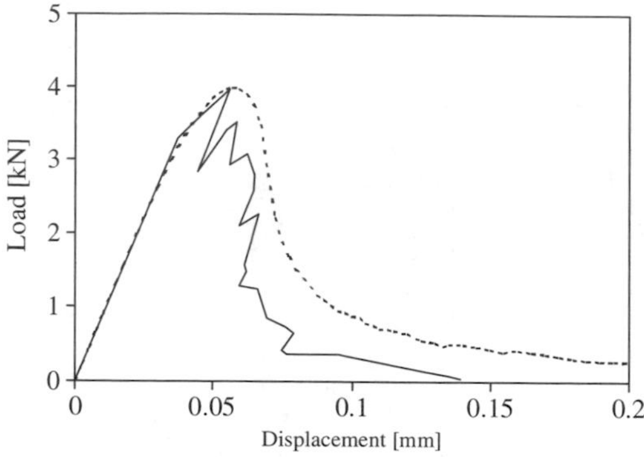

Figure 9. Load-displacement diagram, very coarse mesh, unregularized, ten teeth.

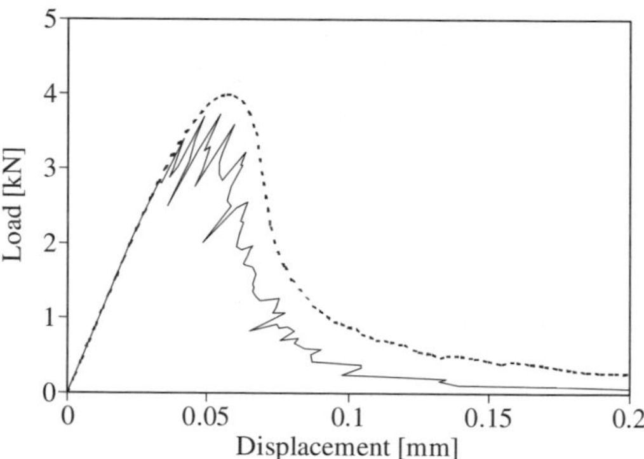

Figure 10. Load-displacement diagram, coarse mesh, unregularized, ten teeth.

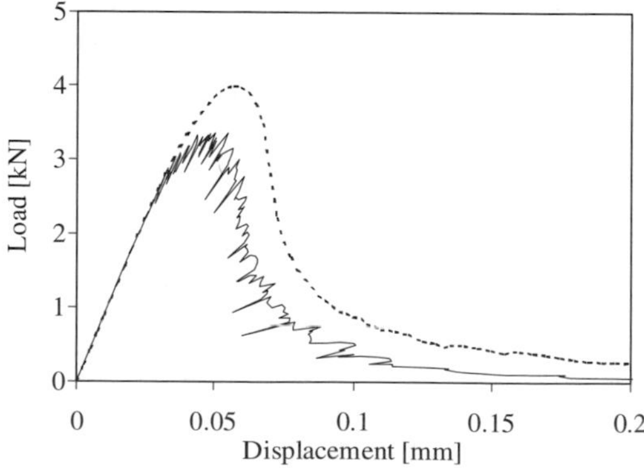

Figure 11. Load-displacement diagram, medium mesh, unregularized, ten teeth.

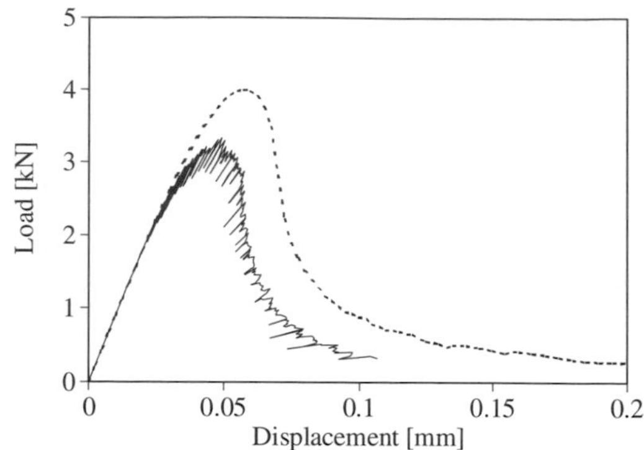

Figure 12. Load-displacement diagram, fine mesh, unregularized, ten teeth.

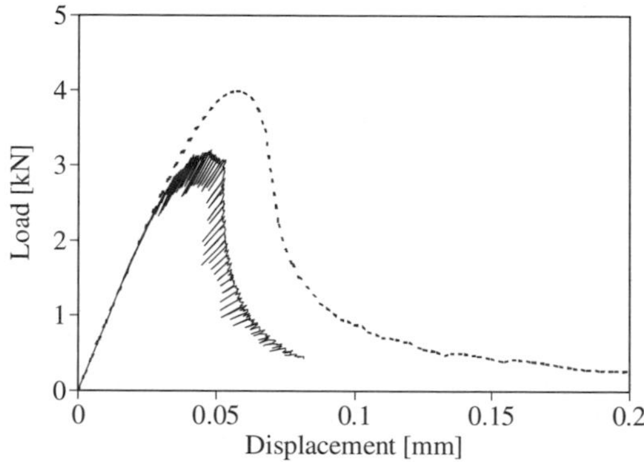

Figure 13. Load-displacement diagram, very fine mesh, unregularized, ten teeth.

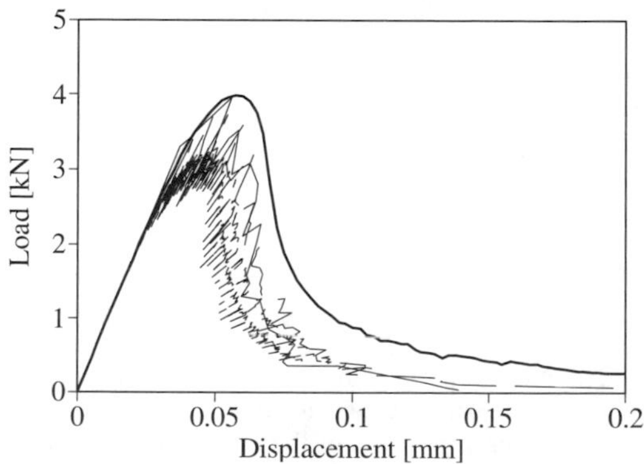

Figure 14. Load-displacement diagrams, all meshes together, unregularized, ten teeth.

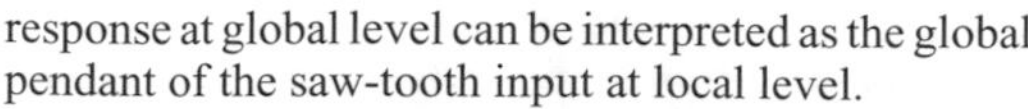

response at global level can be interpreted as the global pendant of the saw-tooth input at local level.

With increasing mesh fineness, the curves become smoother. Surprisingly, the envelope of the curve for the coarse mesh appears to resemble the reference curve most closely, both in terms of the maximum load and post-peak toughness. Please note that the model is especially derived with a view towards large-scale

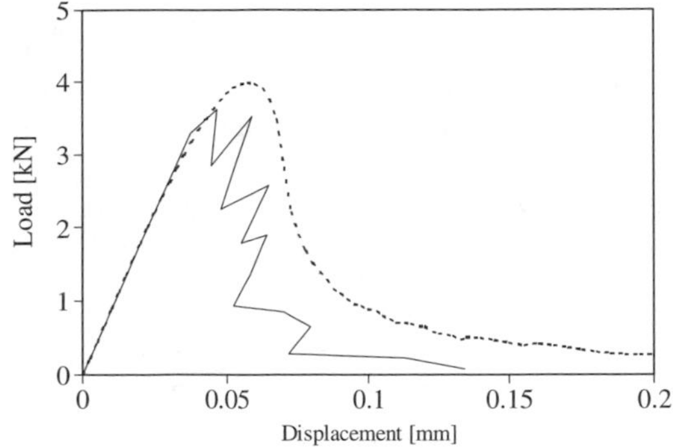

Figure 15. Load-displacement diagram, very coarse mesh, unregularized, five teeth.

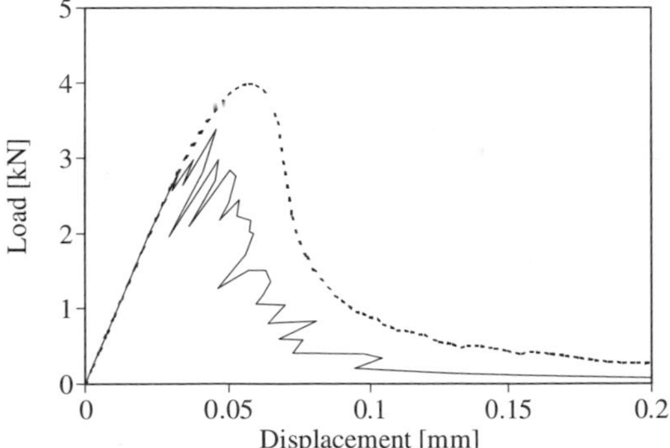

Figure 16. Load-displacement diagram, coarse mesh, unregularized, five teeth.

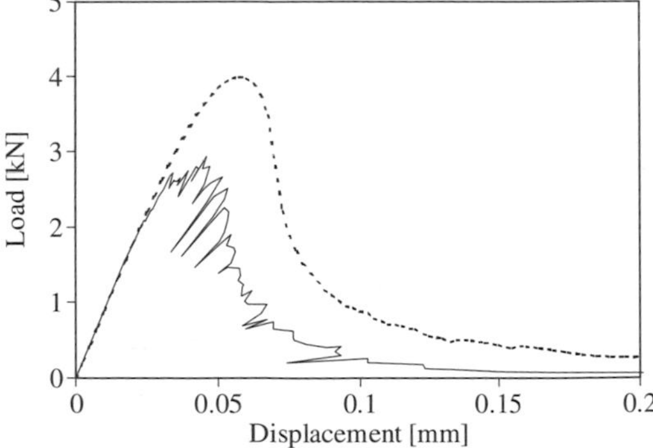

Figure 17. Load-displacement diagram, medium mesh, unregularized, five teeth.

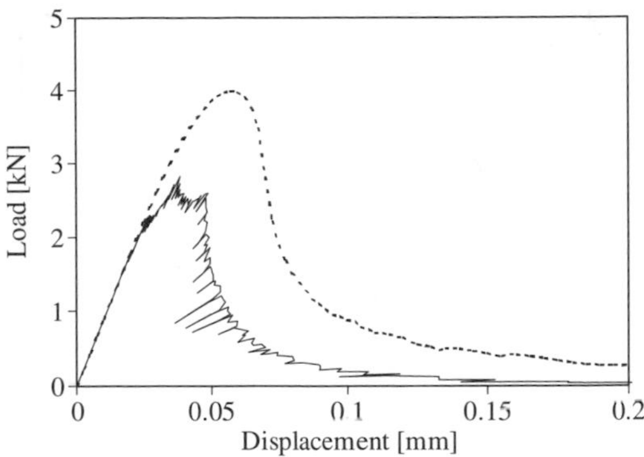

Figure 18. Load-displacement diagram, fine mesh, unregularized, five teeth.

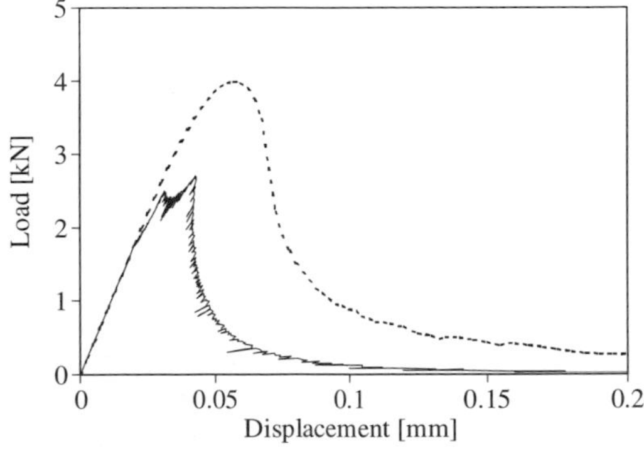

Figure 19. Load-displacement diagram, very fine mesh, unregularized, five teeth.

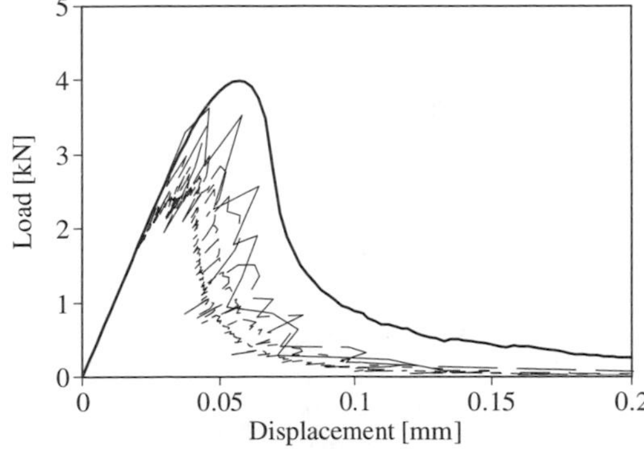

Figure 20. Load-displacement diagrams, all meshes together, unregularized, five teeth.

brittle fracture analysis, i.e. large element sizes, which is thus promising. With increasing mesh fineness, however, the peak in the load-displacement behavior becomes lower and lower compared to the correct peak from the nonlinear reference analysis that is to be mimicked. This demonstrates that sequentially linear models are in principle not objective with respect to mesh refinement. This is because increasing mesh

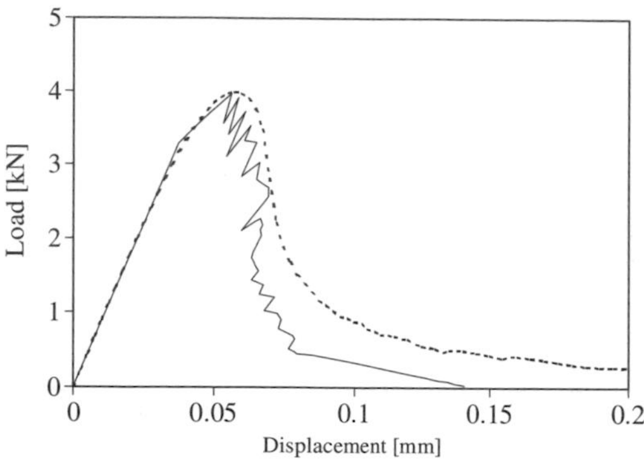

Figure 21. Load-displacement diagram, very coarse mesh, unregularized, twenty teeth.

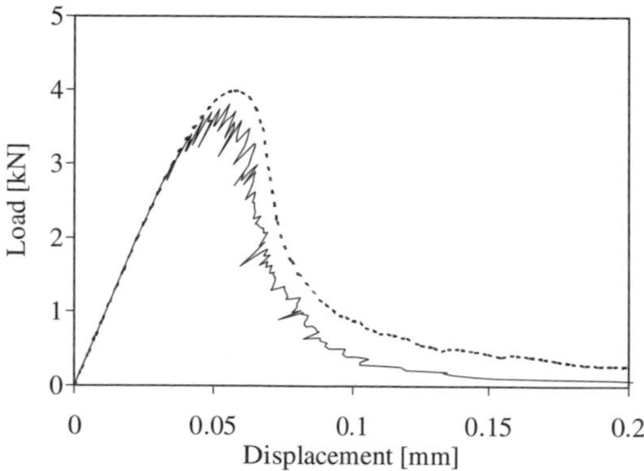

Figure 22. Load-displacement diagram, coarse mesh, unregularized, twenty teeth.

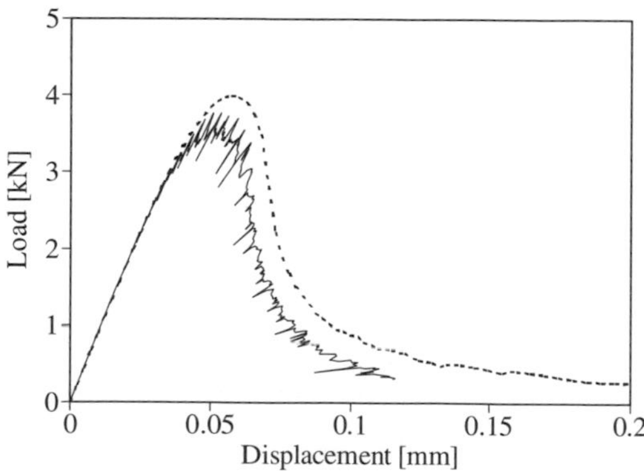

Figure 23. Load-displacement diagram, medium mesh, unregularized, twenty teeth.

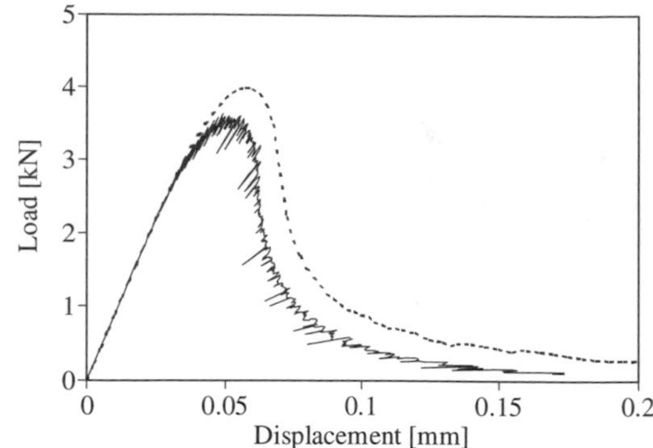

Figure 24. Load-displacement diagram, fine mesh, unregularized, twenty teeth.

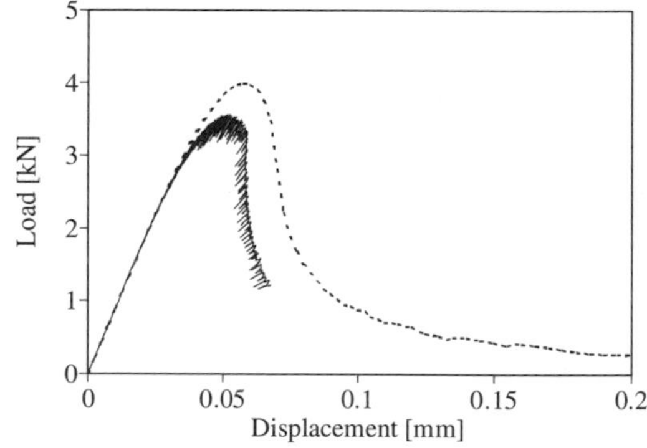

Figure 25. Load-displacement diagram, very fine mesh, unregularized, twenty teeth.

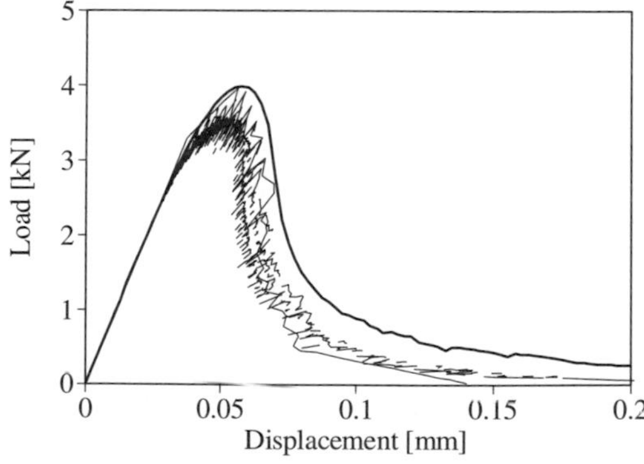

Figure 26. Load-displacement diagrams, all meshes together, unregularized, twenty teeth.

fineness leads to sharper stress peaks at the crack band tip, so that the strength at the crack band tip is reached earlier than for a coarse mesh. In fact, a kind of "zip-fastener" effect is introduced. With a sudden full removal of elements, i.e. a diagram with only one saw-tooth instead of the present ten saw-teeth, this effect would be even more pronounced. That was described already by Bazant & Cedolin (1979).

5.3 *Results – Unregularized – Five or twenty teeth*

The previous section showed that the ten-teeth result was not objective with respect to mesh refinement. It is interesting to study the effect of the number of teeth on this observation. To this end, the analyses were repeated using a five-teeth and twenty-teeth approximation, corresponding to stepwise reductions of Young's modulus with a factor a equal to $a=4$ and $a=\sqrt{2}$ respectively. For the five-teeth approximation, Figs 15–19 show the individual results for the five meshes and Fig. 20 collects them into one graph. For the twenty-teeth approximation, Figs 21–25 show the individual results for the five meshes and Fig. 26 collects them into one graph.

The curves show a regular trend. For the five teeth approximation, the load-displacement graph for a given mesh is located below the load-displacement graph of the ten-teeth approximation for the same mesh. The deviation from the reference curve becomes larger. For the twenty-teeth approximation, the graph for a given mesh is located above the graph of the ten-teeth approximation for the same mesh. For the very coarse mesh, the envelope curve now approximates the reference curve quite well (Fig. 21), but for finer meshes still a significant underestimation occurs (e.g. Fig. 25). The mesh-sensitivity still exists. For the five-teeth model, the spread in the collected graphs is significant (Fig. 20) and larger than for the ten-teeth model (Fig. 14). For the twenty-teeth model, the spread in the collected graphs still exists (Fig. 26), though it is less than for the ten-teeth model (Fig. 14). In conclusion, it can be said that the results of the unregularized model are mesh-sensitive. An increase in the number of teeth provides only a partial, insufficient remedy. Upon further teeth refinement, the mesh inobjectivity would become less, but for current hardware standards the amount of linear analyses and computing time required would become excessive.

It is noted that the simulation of the very end of the softening tail does not improve so much, even with the twenty teeth approximation.

Better results with respect to this, could instead be achieved using the alternative definition of the saw-tooth approximation of Fig. 7.

6 MESH REGULARIZATION

The above results call for a mesh regularization procedure. The concept of smeared crack basically assumes that the localized crack is distributed over a continuum finite element, provided that the crack opening δ is equal to the element strain ε times the so called crack band width h (for lower-order elements often equal to the element size). In order to achieve mesh-size objectivity, the ultimate strain ε_u in smeared crack models is usually adjusted to h according to Eq. (1) for linear softening, Bazant & Oh (1993). In the previous section, it appeared that such adjustment is not sufficient to guarantee mesh-size objectivity for the case of the sequentially linear model. In fact, due to the saw-tooth approximation of the softening curve, the dissipated energy is always less than the theoretical one, i.e. the one referring to the smooth "mother" softening curve. Moreover, the underestimation of the dissipated energy depends not only on the number of teeth, but also on the mesh size, since the ultimate strain depends on the crack band width. When finer meshes are considered, i.e. for a small value of h, the slope of the linear softening branch decreases and the area underestimation becomes more important. Therefore, the load-displacement diagrams obtained with the sequentially linear analysis give a worst approximation of the nonlinear reference curve for finer meshes and for a coarser saw-tooth approximation, see Fig. 27. A route to achieve mesh-objectivity was indicated (Rots 2001), by adjusting not only the ultimate strain of the diagram to h but also the peak of the diagram. However, the details of the approach were not yet understood at that time.

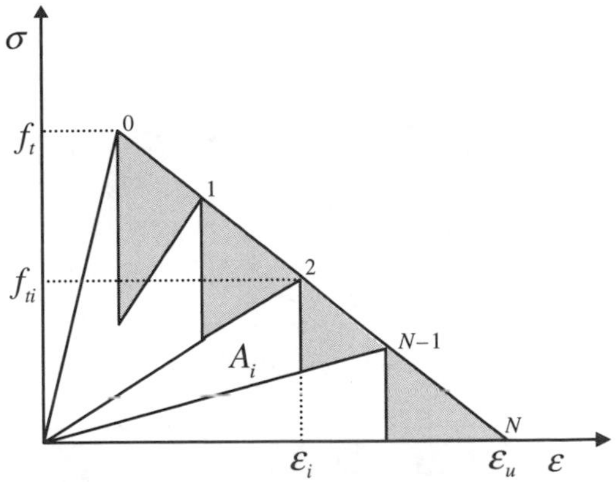

Figure 27. Saw-tooth softening approximation scheme, the underestimated area is shown in gray.

In order to provide a correct regularization procedure and achieve mesh independence, it is first of all necessary to provide a useful expression for the actual area beneath the saw-tooth curve. Referring to the scheme in Fig. 27, the formula is the following:

$$A=\sum_{i=0}^{N-1} A_i=\sum_{i=0}^{N-1}\frac{1}{2}\varepsilon_i f_{ti} b_i\,, \qquad (6)$$

where the index i refers to the triangular decomposition of the whole area.

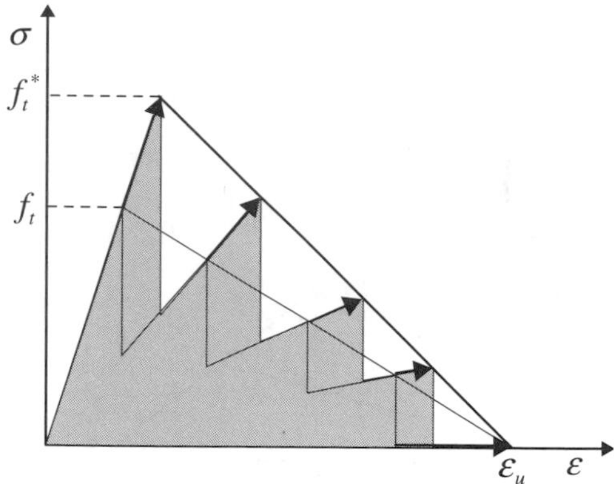

Figure 28. Regularization scheme with the tensile strength update.

The parameter b_i varies depending on the saw-tooth approximation method. In the case of stepwise Young's modulus reduction (Fig. 6), it is the following:

$$b_i = \begin{cases} \left(1 - \frac{1}{a}\right) & 0 \le i < N-1 \\ 1 & i = N-1 \end{cases}, \tag{7}$$

while for the alternative formulation (Fig. 7) it reads:

$$b_i = \left(1 - \frac{E_{i+1}}{E_i}\right), \tag{8}$$

with E_i being Young's modulus at step i.

The basic idea, thus, is to update the tensile strength, or the ultimate strain, or even both, in order to keep the dissipated energy invariant. In other words, the area A^*, under the updated constitutive law, becomes invariant and equal to:

$$A^* = \frac{G_f}{h}. \tag{9}$$

Eq. 9 shows clearly that not only the number of teeth, but also the mesh size (i.e. the crack band width h) comes into play.

Three main options can be pursued. The first choice could be to update only the tensile strength, as shown in Fig. 28.

In this case the updated strength f_t^* is equal to:

$$f_t^* = k \cdot f_t, \tag{10}$$

where k can be determined numerically in such a way that the new area satisfies Eq. 9.

This approach is not the most convenient since, although the results are mesh independent, it over-estimates the peak load. Results obtained with the ten-teeth approximation are shown in Fig. 29.

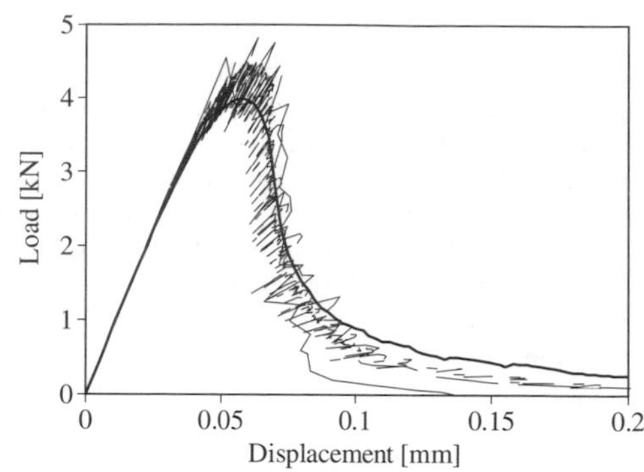

Figure 29. Load displacement diagrams, tensile strength regularization, five different meshes, ten teeth. All the curves are coherent but overestimate the reference peak load (continuous line).

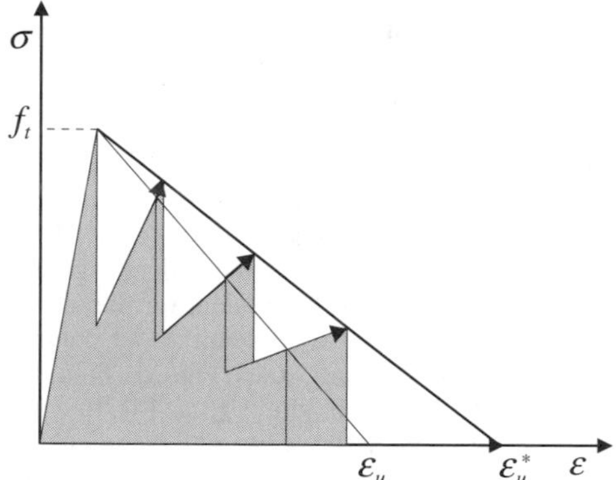

Figure 30. Regularization scheme with the ultimate strain update.

The second option is to update the ultimate strain, following the scheme depicted in Fig. 30. In this case the updated ultimate strain ε_t^* is equal to:

$$\varepsilon_t^* = k \cdot \varepsilon_t, \tag{11}$$

where k can be again determined numerically in such a way that the new area satisfies Eq. 9.

This provides again the mesh independence, but is still not the optimal choice, since the peak load is now underestimated. Results obtained with the ten teeth approximation are shown in Fig. 31.

The third option is to increase both the tensile strength as well as the ultimate strain. Equations 10 and 11 must now hold simultaneously, and k is again determined from Eq. 9. This regularization scheme is shown in Fig. 32, where it is worth noting that for linear softening it is equivalent to prescribe a regularization with constant softening modulus D.

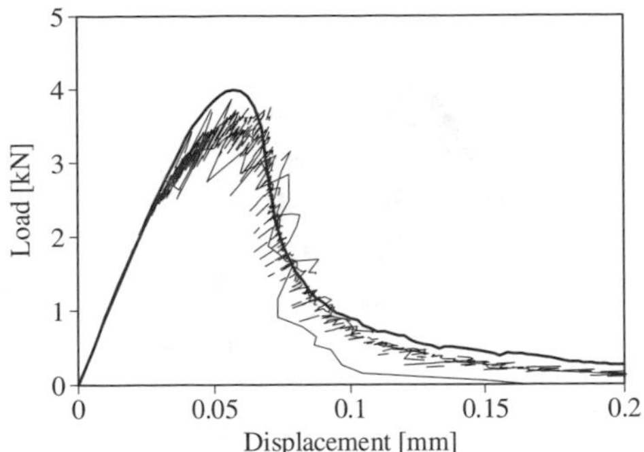

Figure 31. Load displacement diagrams, ultimate strain regularization, five different meshes, ten teeth. All the curves are coherent but underestimate the reference peak load (continuous line).

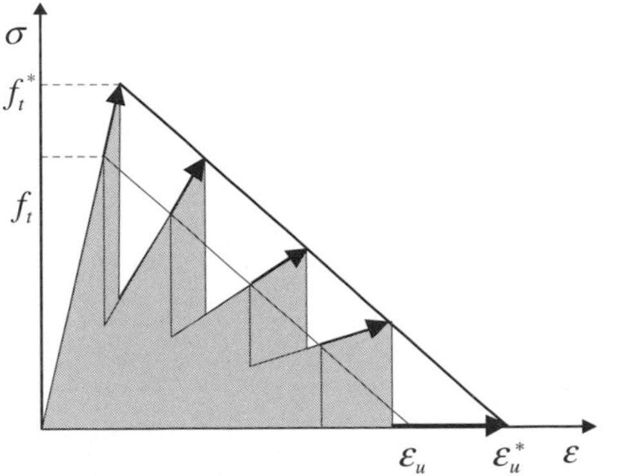

Figure 32. Regularization scheme with both the ultimate strain and the tensile strength update, keeping constant the softening modulus.

The third regularization approach is particularly convenient also because, after some analytical manipulation of Eq. 9, a closed form expression for the parameter k can be obtained. In the case of stepwise Young's modulus reduction, k is the following:

$$k = \sqrt{\frac{\dfrac{G_f}{h}}{\displaystyle\sum_{i=0}^{N-1} \frac{1}{2} \frac{f_{ti}^2}{E_i} b_i}}, \tag{12}$$

while, for the alternative formulation (Fig. 7), the expression is:

$$k = \sqrt{\frac{\dfrac{G_f}{h}}{\displaystyle\sum_{i=0}^{N-1} \frac{1}{2} \varepsilon_i f_{ti} \left(1 - \frac{E_{i+1}}{E_i}\right)}}. \tag{13}$$

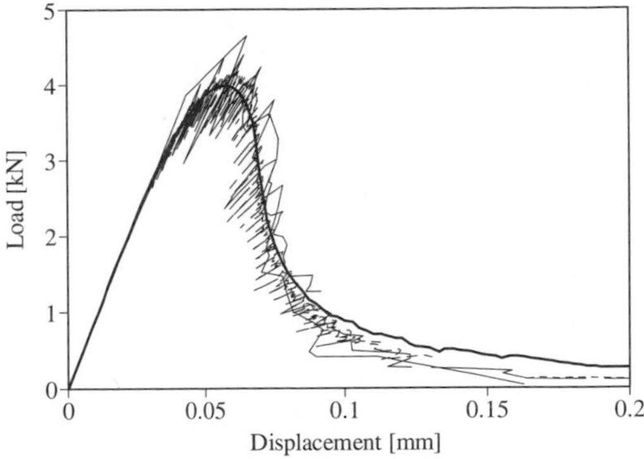

Figure 33. Load displacement diagrams, combined ultimate strain and tensile strength regularization, five different meshes, ten teeth. All the curves are coherent and closely resemble the reference peak load (continuous line).

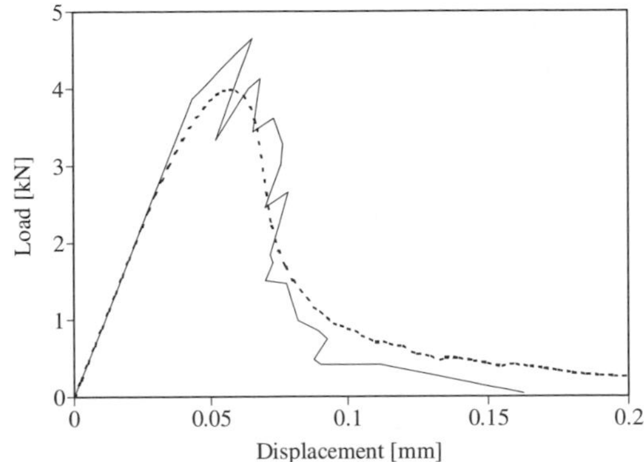

Figure 34. Load-displacement diagram, combined regularization, very coarse mesh, ten teeth.

The third available regularization provides the best results with respect to the nonlinear reference analysis. Figure 33 shows the results, again for stepwise Young's modulus reduction and ten teeth. Again, the mesh independence is evident. Now, both the peak load and the shape of the nonlinear reference curve are closely resembled.

For the third, combined ultimate strain and tensile strength regularization, the curves of Fig. 33 for the five different meshes are also depicted individually, in Figures 34–38, for the ten-teeth approximation.

Although all curves are close to the reference result, it is clear that upon increasing mesh fineness the curve becomes smoother and even more close to the reference result.

Finally, the third, combined tensile strength and ultimate strain regularization has been tested also for the five-teeth and twenty-teeth approximation. Figures 39–44 show the results for five teeth and Figures 45–50 for twenty teeth.

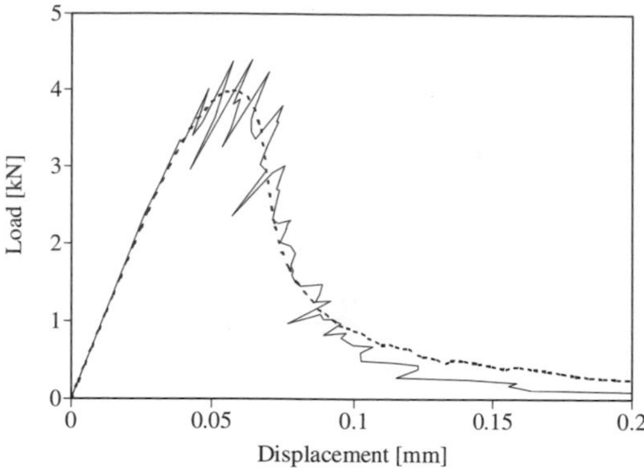

Figure 35. Load-displacement diagram, combined regularization, coarse mesh, ten teeth.

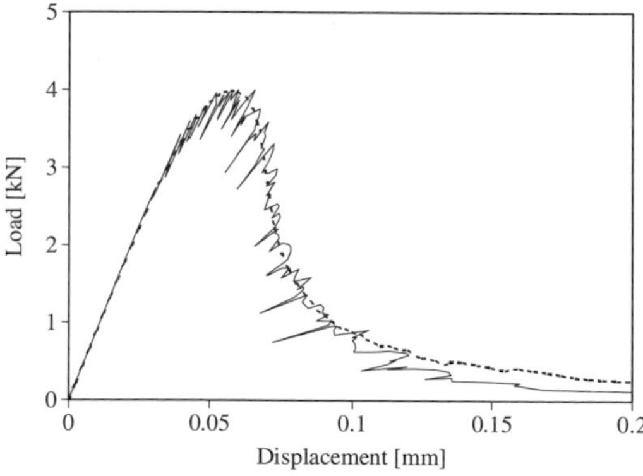

Figure 36. Load-displacement diagram, combined regularization, medium mesh, ten teeth.

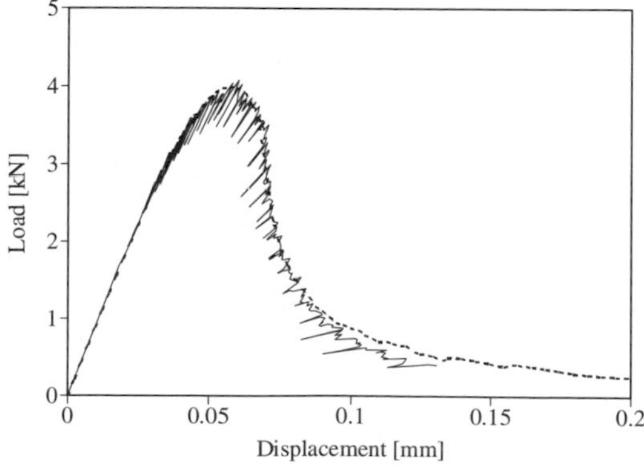

Figure 37. Load-displacement diagram, combined regularization, fine mesh, ten teeth.

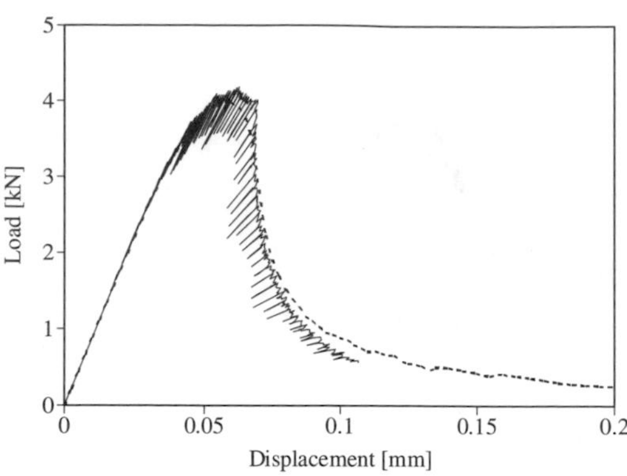

Figure 38. Load-displacement diagram, combined regularization, very fine mesh, ten teeth.

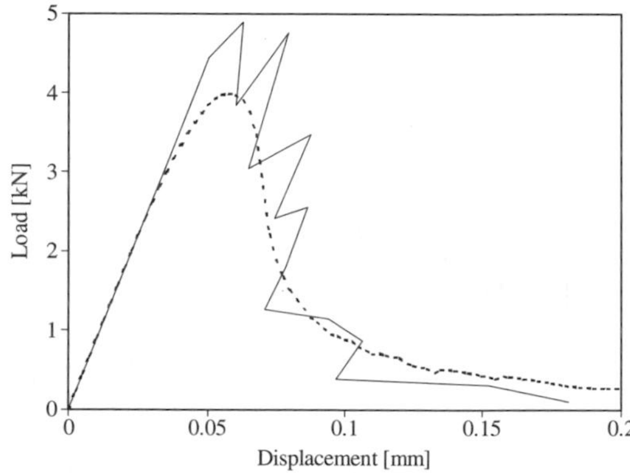

Figure 39. Load-displacement diagram, combined regularization, very coarse mesh, five teeth.

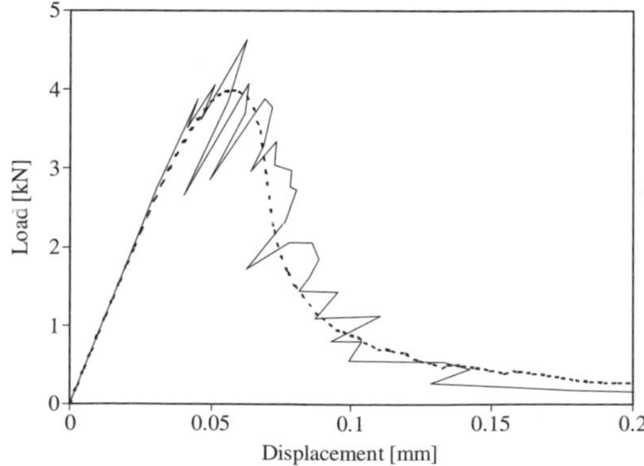

Figure 40. Load-displacement diagram, combined regularization, coarse mesh, five teeth.

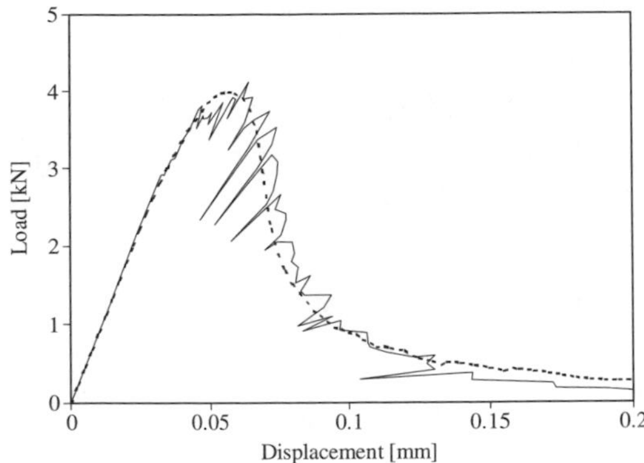

Figure 41. Load-displacement diagram, combined regularization, medium mesh, five teeth.

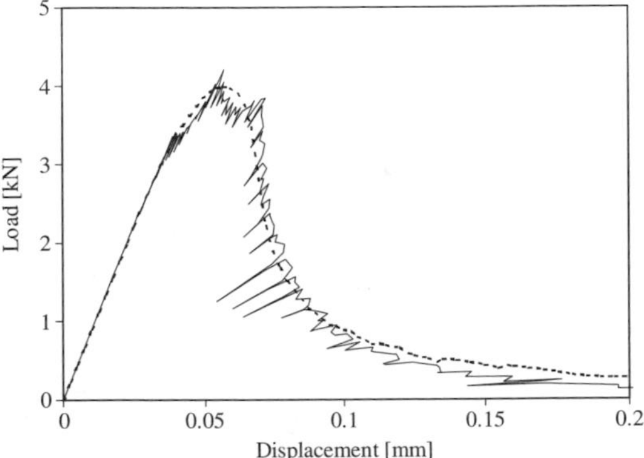

Figure 42. Load-displacement diagram, combined regularization, fine mesh, five teeth.

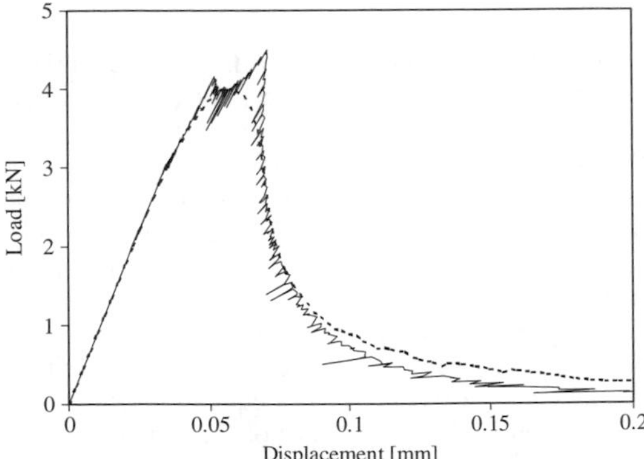

Figure 43. Load-displacement diagram, combined regularization, very fine mesh, five teeth.

All results are mesh size-objective. It is apparent that the five-teeth model performs still relatively accurate, though it shows some irregularity near peak. The twenty-teeth model shows the most accurate approximation, even a little better than the ten-teeth model.

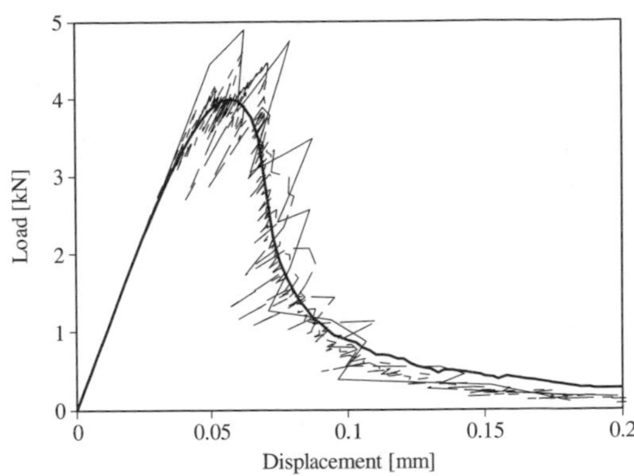

Figure 44. Load-displacement diagrams, combined regularization, all meshes together, five teeth.

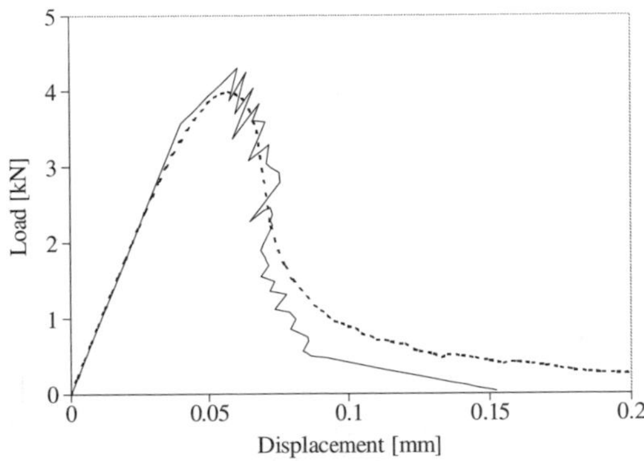

Figure 45. Load-displacement diagram, combined regularization, very coarse mesh, twenty teeth.

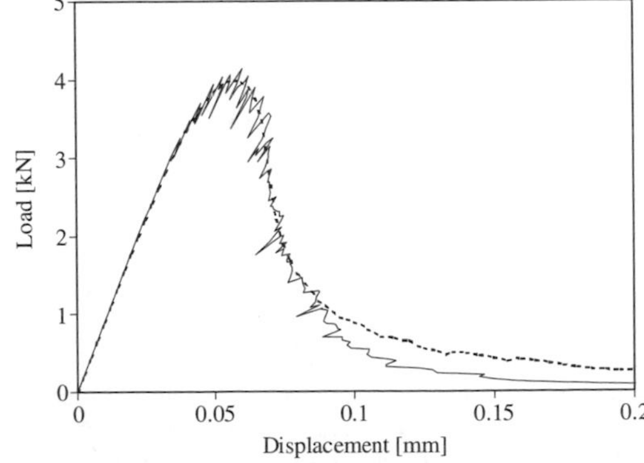

Figure 46. Load-displacement diagram, combined regularization, coarse mesh, twenty teeth.

It is concluded that the third, combined tensile strength and ultimate strain regularization technique

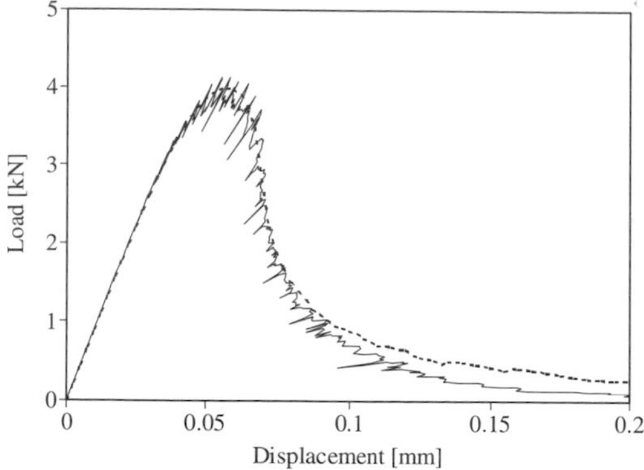

Figure 47. Load-displacement diagram, combined regularization, medium mesh, twenty teeth.

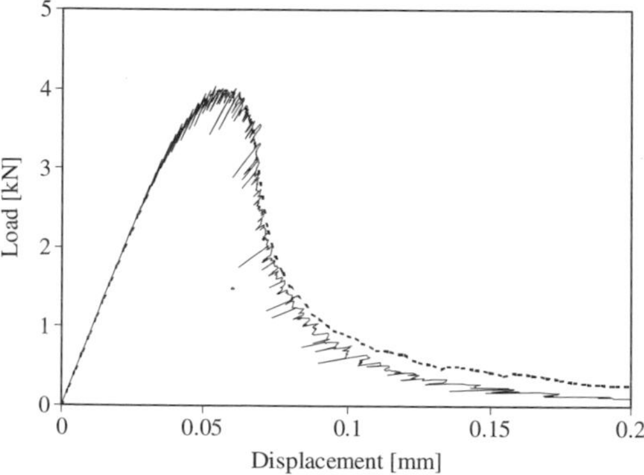

Figure 48. Load-displacement diagram, combined regularization, fine mesh, twenty teeth.

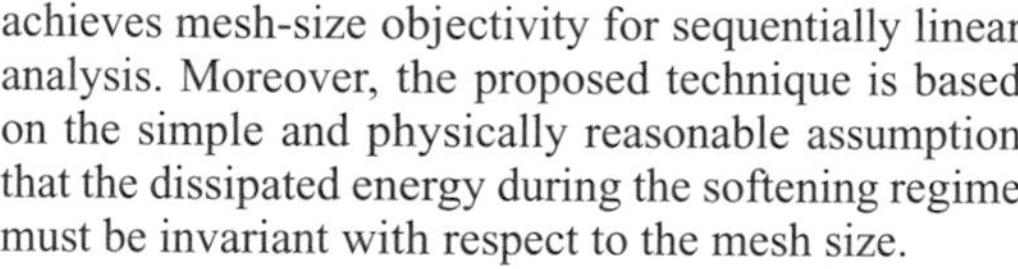

achieves mesh-size objectivity for sequentially linear analysis. Moreover, the proposed technique is based on the simple and physically reasonable assumption that the dissipated energy during the softening regime must be invariant with respect to the mesh size.

7 NON PROPORTIONAL LOADING OF MASONRY FAÇADE

This section considers a masonry façade subjected to settlements. For the geometry and parameters, the reader is referred to Boonpichetvong & Rots (2003) elsewhere in this Volume. Note that the dimensions of the facade and the crack band width are approximately 150 times larger than those for the small-scale notched beam. A ten-teeth model with $a = 2$, $n = 10$ is used. A difficulty with this problem is that it involves a non-proportional loading scheme. First, dead weight is added and subsequently the settlement trough. It

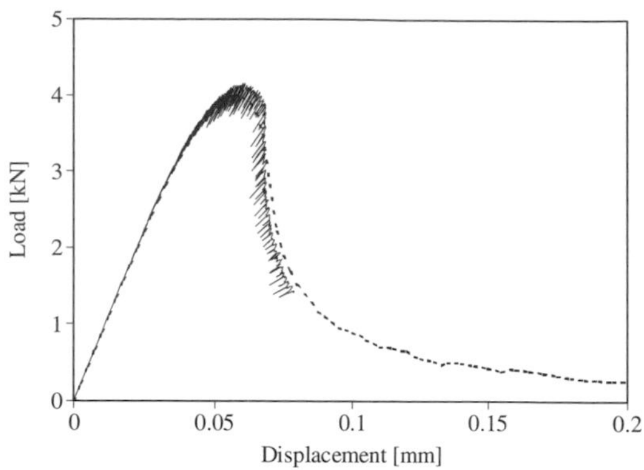

Figure 49. Load-displacement diagram, combined regularization, very fine mesh, twenty teeth.

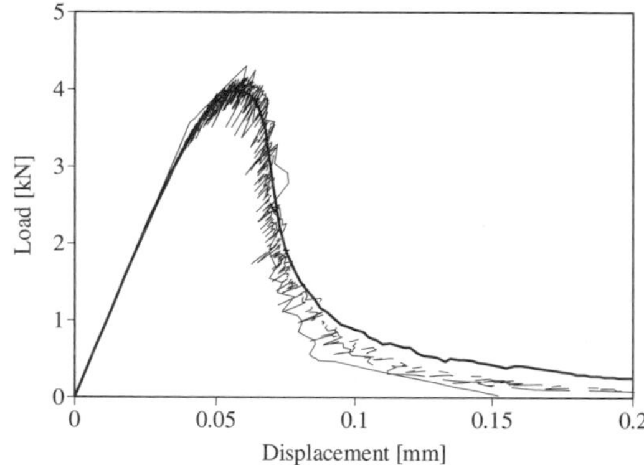

Figure 50. Load-displacement diagrams, combined regularization, all meshes together, twenty teeth.

appears that non-proportional loading for sequentially linear analysis on lattices or other structures has not yet received much attention in literature. A way to solve this is to run sequentially linear analyses for the various load sets in an incremental fashion. First, the first load set is analyzed sequentially linear and the results are frozen. From this frozen situation, the sequentially linear analysis for the second load set is undertaken whereby initial stresses, strains and displacements are accounted for. This process requires the new critical element for the second (or subsequent) load set to be determined again on the basis of the principal stress criterion.

The principal stresses can be derived on the basis of:

$$\begin{cases} \sigma_{xx} = \sigma_{xx,0} + \lambda \Delta \sigma_{xx} \\ \sigma_{yy} = \sigma_{yy,0} + \lambda \Delta \sigma_{yy} \, , \\ \sigma_{xy} = \sigma_{xy,0} + \lambda \Delta \sigma_{xy} \end{cases} \tag{14}$$

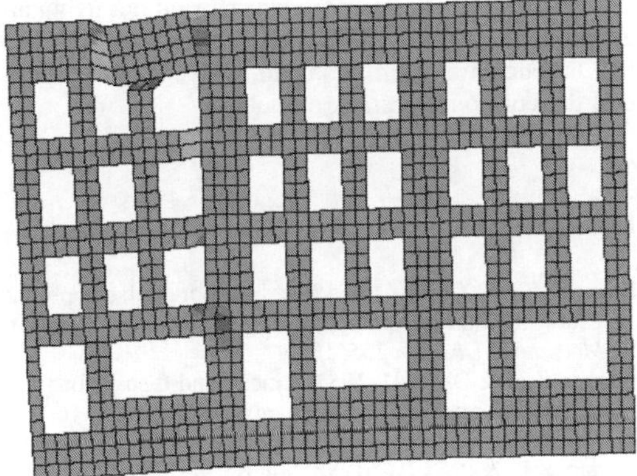

Figure 51. Deformed mesh in the hogging region of the settlement trough. The colors emphasize the crack path across the façade.

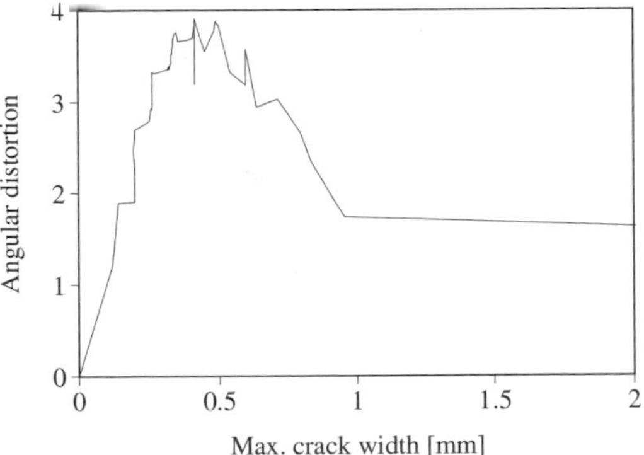

Figure 52. Load-displacement curve for the masonry façade subjected to non-proportional load.

for a plane stress situation, with $_0$ referring to the preceding load set(s) and Δ referring to the increment of the current load set. The set of Eq. 14 is of quadratic type and some algebraic manipulation provides a couple of cumbersome closed form solutions for the sequentially linear multiplier, where only the positive one can be considered. It is worth noting that when $\Delta\sigma_{xy} = \sqrt{\Delta\sigma_{xx} \cdot \Delta\sigma_{yy}}$ special expressions of the solution must be considered to avoid singularity.

This approach has been implemented in the sequentially linear procedure, and a practical case is shown regarding the study of the typical Amsterdam masonry façade.

The deformed mesh in Fig. 51 shows the obtained fracture pattern when the façade is in the hogging region of the settlement trough. Figure 52 shows the zoomed diagram describing the evolution of the angular distortion as a function of the maximum crack width, with the resulting snap behavior that accompanies the sudden crack propagation across the façade. Although further analyses are still under study, the results provide a stable alternative to the jumpy response of Fig. 5. Other results for an equivalent proportional loading can be found in Rots (2001).

8 DISCUSSION AND FUTURE DIRECTIONS

Above, a sequentially linear continuum model for concrete fracture has been proposed. The model approximates an envelope softening stress–strain curve by a saw-tooth diagram. In each linear analysis, a critical element is traced by comparing element stress with current element strength, i.e. with the current peak in the saw-tooth diagram. Next the stiffness and strength of the critical element are reduced according to the subsequent tooth of the diagram, and the process is repeated. The charm of this model is that words like iteration, ill-conditioning and divergence do not appear in the vocabulary. The stiffness of the linear analyses is always positive definite as the saw-teeth at local level are always rising. A notched beam analysis shows the ability of the method to reproduce the behavior of a nonlinear reference analysis on the same parameters. A large scale façade analysis demonstrated the ability to capture snap-type response.

A mesh regularization procedure has been developed. The best regularization is achieved by adjusting both the tensile strength and the ultimate strain of the saw-tooth diagram to the mesh size, while keeping the dissipated energy invariant. When only the tensile strength or only the ultimate strain is adjusted, the results are mesh-objective too, but the approximation of the nonlinear reference result is less accurate. Furthermore, the objectivity and convergence upon increasing the number of saw-teeth was demonstrated.

Current and future activities to extend the model include the following. We mention some points. First, objectivity with regard to mesh orientation should be studied, in addition to mesh size. The strategy followed for nonlinear analyses by Oliver (1989) and Jirasek & Zimmermann (2001) to accurately compute the crack band width and to use a non-local (smoothed) principal stress criterion at the "crack tip" can be adopted in the same way for sequentially linear analysis. Second, extension of the model to anisotropic degradation is possible. The present formulation is isotropic because Young's modulus is reduced for all directions. Extension of the saw-tooth model to fixed planes of orthotropy is possible, including memory of unloading/reloading and full crack closure and re-opening. Third, implementation of the sequentially linear model for interface elements is underway. A cut-off criterion for tension, shear and compression, as sketched in Fig. 2, followed by saw-tooth softening for the domains is implemented. Also for the

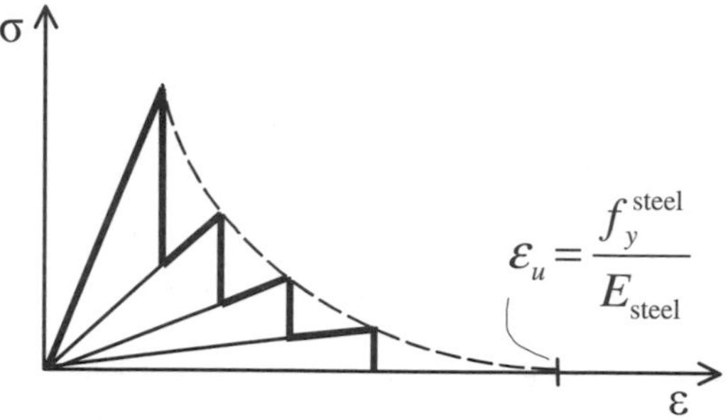

Figure 53. Saw-tooth tension-stiffening diagram for reinforced concrete.

continuum version, extension with saw-tooth softening in compression is possible. Fourth, and important for engineering applications on RC structures, is the extension towards saw-tooth tension-stiffening rather than tension-softening. This approach fits the nature of tension-stiffening as the tension-stiffening curve is already a saw-tooth serrated type of curve, depending on details of reinforcing lay-out and reinforcing percentages. The response of the reinforced tension bar in Fig. 3 then becomes the global pendant of a local saw-tooth tension-stiffening input diagram as sketched in Fig. 53.

The ultimate strain in Fig. 53 is determined by the yield strength of the reinforcing steel divided by Young's modulus of the steel. When the steel yields, the concrete in between the cracks cannot contribute any more to the tensile carrying capacity. The approach fits global practice in RC engineering, where non-linear behavior is sometimes captured via repeating linear analyses using reduced Young's moduli at areas of anticipated cracking.

Finally, it is mentioned that the required CPU-time for sequentially linear continuum analysis can be large, depending on the number of elements and saw-teeth adopted and the number of cracks or crushes emerging. On the other hand, the approach always "converges", as the secant stiffness for the saw-teeth at local level is always positive, so is the secant stiffness for the total system. Complex solution techniques like arc-length procedures, to pass the drops and snaps as for the RC structure in Fig. 4, are no longer required and the user is protected from spending time and effort in that. The required man-hour time to steer the analyses is low. In addition, more intelligent solvers can be used that re-use the decomposition process from a previous linear analysis, as only a single stiffness value in the overall matrix is altered upon the subsequent linear analysis.

ACKNOWLEDGEMENT

Financial support from Delft Cluster, COB and the Netherlands Technology Foundation STW is acknowledged. The research was carried out using an adapted version of DIANA. The authors are indebted to Dr. Feenstra from Stanford University for general dis-cussions on crack models.

REFERENCES

Bazant, Z.P. & Cedolin, L. 1979. Blunt crack band propagation in finite element analysis. ASCE *J. Engineering Mechanics Division* 105(2): 297–315.

Bazant, Z.P. & Oh, B.H. 1983. Crack band theory for fracture of concrete. RILEM *Materials and Structures* 16(93): 155–177.

Bazant, Z.P., & Pratt, P.C. 1988. Microplane model for brittle plastic material. ASCE *J. of Engng. Mechanics* 114: 1672–1702.

Boonpichetvong, M. & Rots, J.G. 2003. Settlement damage of masonry buildings in soft-ground tunneling. This Volume.

Borst, R. de & Nauta, P. 1986. Non-orthogonal cracks in a smeared finite element model. *Engrg. Comp.* 2: 35–46.

Borst, R. de 1987. Smeared cracking, plasticity, creep and thermal loading – A unified approach. *Comp. Meth. Appl. Mech. Engng.* 62: 89–110.

Borst, R. de. 1987b. Computation of post-bifurcation and post-failure behavior of strain-softening solids. *Computers & Structures* 25(2): 211–224.

Borst, R. de, Bicanic, N., Mang, H. & Meschke, G. (eds.) 1998. *Computational modelling of concrete structures; Proc. EURO-C 1998.* Rotterdam: Balkema.

Carol, I., Lopez, C.M. & Aguado, A. 2001. Micromechanical analysis of quasi-brittle materials using fracture-based interface elements. *Int. J. Num. Meth. Engng.* 52(1–2): 193–215.

Carpinteri, A. 1986. Mechanical damage and crack growth in concrete: plastic collapse to brittle fracture, Martinus Nijhoff Publishers, Dordrecht.

Crisfield, M.A. 1984. Difficulties with current numerical models for reinforced-concrete and some tentative solutions. *Proc. Int. Conf. Computer Aided Analysis and Design of Concrete Structures*, eds. F. Damjanic, N. Bicanic et al., Part I, 331–358.

Feenstra, P.H. & de Borst, R. 1996. A composite plasticity model for concrete. *Int. J. Solids Structures* 33: 707–730.

Feenstra, P.H., Rots, J.G., Arnesen, A., Teigen, J.G. & Hoiseth, K.V. 1998. A 3D constitutive model for concrete based on a co-rotational concept. In R. de Borst et al. (eds.), *Computational modelling of concrete structures; Proc. EURO-C 1998:* 13–22. Rotterdam: Balkema.

Feenstra, P.H. & Rots, J.G. 2001. Comparison of concrete models for cyclic loading. ASCE Committee Report *Modelling of Inelastic Behavior of RC Structures under Seismic Loads*, pp. 38–55, ASCE, USA.

Frissen, C.N. & Rots, J.G. 1999. Numerical simulation of stability of masonry structures. Collection of working reports CUR Committee A33 Masonry Mechanics. Internal Report, TNO Building and Construction Research, Delft.

Jirasek, M. & Zimmermann, T. 2001. Embedded crack model: II. *Int. J. for Numerical Methods in Engineering* 50(6): 1269–1290.

Lourenco P. 1996. Computational strategies for masonry structures. PhD thesis, Delft University of Technology, Delft, The Netherlands.

Lourenco, P., de Borst, R. & Rots, J.G. 1997. A plane stress softening plasticity model for orthotropic materials. *Int. J. for Num. Methods in Engng.* 40: 4033–4057.

Lourenco, P. & Rots, J.G. 1997. A multi-surface interface model for the analysis of masonry structures. *ASCE J. Engng. Mech.* 123(7): 660–668.

Mazars, J. & Pijaudier-Cabot, G. 1989. Continuum damage theory – application to concrete. *ASCE J. of Eng. Mechanics* 115: 345–365.

Oliver, J. 1989. A consistent characteristic length for smeared cracking models. *Int. J. for Numerical Methods in Engineering* 28: 461–474.

Pimanmas, A. & Maekawa, K. 2001. Multi-directional fixed crack approach for highly anisotropic shear behavior in pre-cracked RC members. *J. Materials. Conc. Struct., Pavements,* JSCE, No. 669/V-50, 293–307, Japan Society of Civil Engineers.

Rots, J.G., Nauta, P., Kusters, G.M.A. & Blaauwendraad, J. 1985. Smeared crack approach and fracture localization in concrete. *HERON* 30(1). 1–48, The Netherlands.

Rots, J.G. 1988. Computational modeling of concrete fracture. PhD thesis, Delft University of Technology, Delft, The Netherlands.

Rots, J.G. 1993. The smeared crack model for localized mode-I tensile fracture. In F.H. Wittmann (ed.), *Numerical models for fracture mechanics:* 101–113. Rotterdam: Balkema.

Rots, J.G. 1994 (Ed.). Structural masonry – An experimental/numerical basis for practical design rules. CUR report 171, CUR, Gouda (in Dutch). English version published by Balkema, Rotterdam in 1997.

Rots, J.G. 2000. Settlement damage predictions for masonry. In L.G.W. Verhoef and F.H. Wittmann (eds.), Maintenance and restrengthening of materials and structures – Brick and brickwork, Proc. Int. Workshop on Urban heritage and building maintenance: 47–62. Freiburg: Aedificatio.

Rots, J.G. 2001. Sequentially linear continuum model for concrete fracture. Fracture mechanics of concrete structures (eds. R. de Borst, J. Mazars, G. Pijaudier-Cabot and J.G.M. van Mier), Vol. 2, A.A. Balkema Publishers, 2001, pp. 831–839.

Schlangen, E. & Van Mier, J.G.M. 1992. Experimental and numerical analysis of micro-mechanisms of fracture of cement-based composites. Cement & Concrete Composites 14: 105–118.

Stankowski, T., Runesson, K. & Sture, S. 1993. Fracture and slip of interfaces in cementitious composites. I. Characteristics, and II: Implementation. *ASCE J. Struct. Engng.* 119: 292–327.

Zijl. G.P.A.G. van, de Borst, R. & Rots, J.G. 2001. A numerical model for the time-dependent cracking of cementitious materials. *Int. J. for Num. Methods in Engng.* 52(7): 637–654.

Computational Modelling of Concrete Structures, Bićanić et al. (eds)
© 2003 Swets & Zeitlinger, Lisse, ISBN 90 5809 536 3

Keynote paper: Optimization of jet-grouted support in NATM tunneling

Ch. Pichler, R. Lackner & H.A. Mang
Institute for Strength of Materials, Vienna University of Technology, Karlsplatz, Vienna, Austria

L. Martak
Magistrat der Stadt Wien (MA 29), Wilhelminenstraße, Vienna, Austria

ABSTRACT: According to its designation, horizontal jet grouting (HJG) is performed horizontally at the tunnel face in order to form a support structure of improved soil. HJG is employed to reduce the surface settlements in urban tunneling when buildings and other infrastructure sensitive to settlements are encountered. The yet unanswered question arising from engineering practice of HJG refers to the optimal design of the jet-grouted support structure and the underlying construction scheme. In this paper, numerical studies in the context of plane-strain finite element (FE) analyses are presented, comparing different layouts of the support structure as well as different construction schemes. Moreover, the influence of the soil properties is investigated. On the basis of the numerical results, the optimal layout of the jet-grouted support is identified for various soil stratifications including a homogeneous soil, a relatively soft layer of sand over a stiff layer of gravel, and the reverse situation. Sophisticated thermochemomechanical material models accounting for the strength and stiffness growth as well as chemical shrinkage and creep of cemented soil and shotcrete have been employed in the numerical investigations leading to the results reported in this work.

1 INTRODUCTION

The New Austrian Tunneling Method (NATM) has proved to be a very economic and flexible mode of construction. When driving tunnels according to the NATM, after the excavation of a cross section of a tunnel, shotcrete is applied onto the tunnel walls, constituting a thin and flexible shell. The NATM is characterized by a strong interaction of the hardening/creeping shotcrete shell and the viscous soil, exerting pressure on the lining. The creep properties of shotcrete are the source of the deformations required for the activation of the load-carrying capacity of the surrounding soil formation. In urban areas, however, these deformations must be minimized in order to avoid damage of buildings and infrastructure. In order to achieve this goal, the load-carrying capacity of soil surrounding the tunnel may be improved by means of grouting techniques or soil freezing. Recently, horizontal jet grouting (HJG) is used for this purpose. As implied by the designation, HJG is performed horizontally at the tunnel face. It is characterized by a support structure of the form of a sector of a conical frustum made of cemented soil in front of the tunnel face. Applications of HJG are reported in (Guatteri et al. 1994) (Fitl 1997) (Liebsch and Haberland 1998) (Gioda and Locatelli 1999) (Wittke et al. 2000).

The objective of this paper is to answer questions raised by engineers dealing with practical applications of HJG:

1. How does the layout of the jet-grouted support structure, i.e., the number and the position of jet-grouted columns, influence surface settlements?
2. Is it possible to reduce the time lag between the end of jet-grouting and the continuation of the excavation in order to decrease the construction time for the tunnel?
3. How do settlements of a tunnel supported by both a jet-grouted support structure and a shotcrete lining compare to the ones of a tunnel which is only supported by a shotcrete lining?
4. Does the optimal layout of the support structure (optimal with respect to surface settlements and economic effort) depend on the soil stratification?

The answer of these questions is attempted on the basis of plane-strain finite element (FE) analyses. The restriction to plane-strain analyses is enabled by the dominant load transfer in the circumferential direction of the stiff cone-shaped support structure made of jet-grouted columns. In this paper, a typical cross section characterized by low overburden is considered: the vertical distance between the top of the tunnel and the ground surface is chosen less than one tunnel

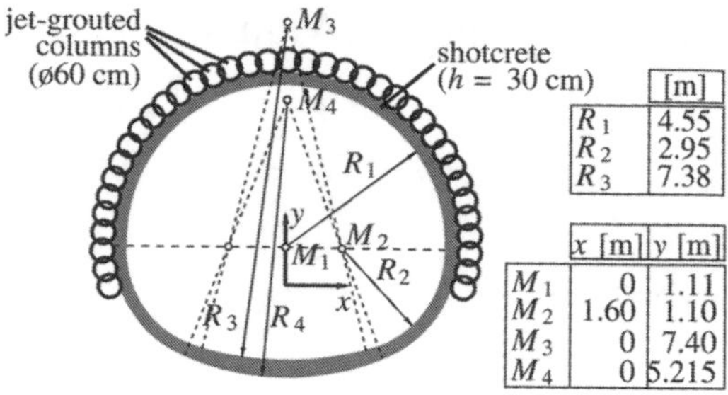

Figure 1. Cross section with support structure consisting of 37 jet-grouted columns.

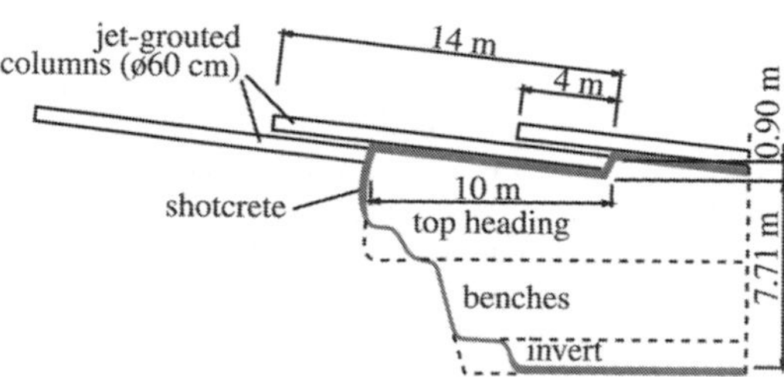

Figure 2. Longitudinal section.

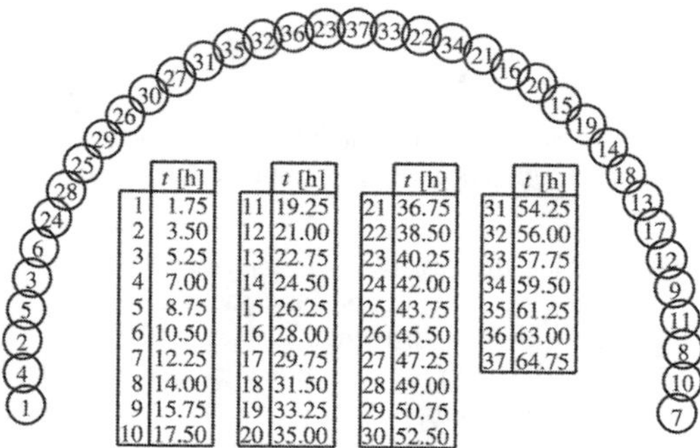

Figure 3. Jet-grouting scheme for the construction of a support structure consisting of 37 jet-grouted columns (t [h] indicates the time instant of the end of construction of the respective column, measured from the begin of construction of column 1).

diameter. The maximum number of jet-grouted columns of the support structure considered in this work is 37 (see Figure 1). Each column has a length of 14 m (see Figure 2). The overlapping between the different support structures in the longitudinal direction is 4 m. From the experience of previous applications of HJG (Liebsch and Haberland 1998), the diameter of each column was chosen as 60 cm.

During construction of the support structure, a certain jet-grouting scheme must be followed in order to avoid adjacent "fresh" columns. Figure 3 shows this

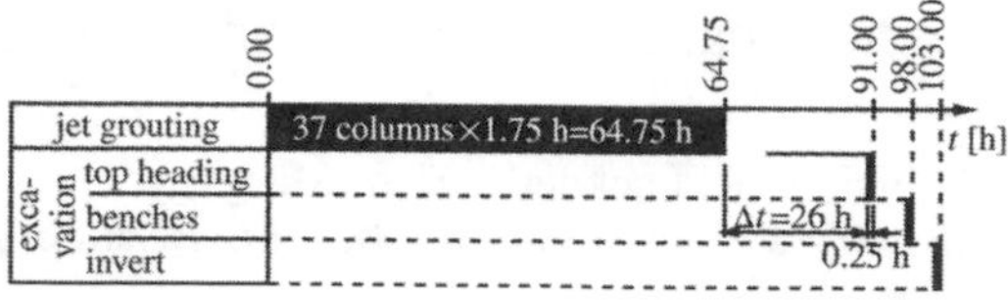

Figure 4. Time table for grouting of 37 columns and excavation at the cross section considered in the numerical analyses.

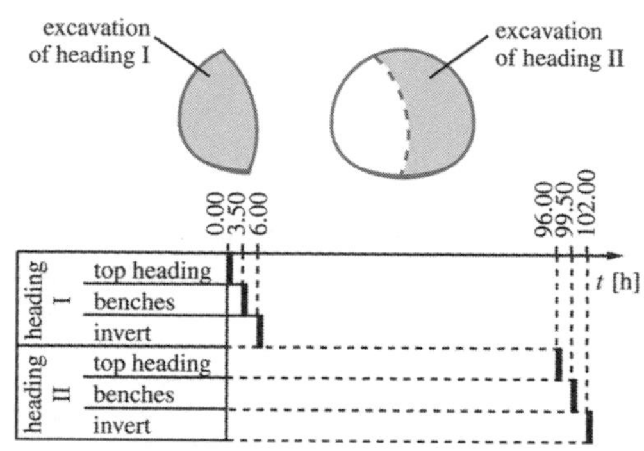

Figure 5. Time table for construction of NATM tunnel without jet-grouted support (staggered excavation scheme).

scheme assuming a construction time of 1.75 hours for each of the 37 columns.

After jet grouting, the tunnel is excavated according to Figure 2. The time span between the end of jet grouting and the begin of the excavation is denoted as Δt (see Figure 4). Large time spans result in a delay of construction work. Continuation of the excavation right after jet grouting, on the other hand, results in loading of the young jet-grouted soil mass. Its low strength and stiffness at the beginning of the hydration process could even cause larger settlements than the ones that would be obtained without jet grouting. In the numerical studies, the layout of the jet-grouted support structure (Figure 3) and, hence, the time associated with jet grouting will be modified.

The reduction of settlements by different layouts of the jet-grouted supports is compared with a tunnel excavation supported by a shotcrete lining only. In this case, however, the full-face excavation is replaced by a staggered excavation scheme. The respective time table employed in the numerical analysis is contained in Figure 5.

2 MATERIAL MODELS

An essential prerequisite for reliable results are realistic material models for the ground and the different cement-based materials, i.e., shotcrete and cemented soil.

The mechanical behavior of the soil is described by means of a multi-surface plasticity model consisting of the Drucker-Prager model for the mechanical behavior under compressive loading and a tension-cut-off for description of the material response under tensile loading. The model is extended to viscoplasticity by means of the viscoplastic material law by (Duvaut and Lions 1972).

In order to account for chemomechanical coupling observed at early age of cement-based materials, a chemomechanical approach is chosen for the description of the involved cement-based materials (Ulm and Coussy 1995) (Ulm and Coussy 1996). Within this material model, dissipative phenomena at the micro-level of the material are accounted for by means of (internal) state variables and energetically conjugate thermodynamic forces related to these variables via state equations. The rates of the internal state variables are related to the corresponding thermodynamic forces by means of evolution equations.

As is typical for cement-based materials, four dissipative phenomena govern the material behavior:

1. *Hydration*, resulting in chemical shrinkage strains, aging elasticity, and strength growth. The extent of the chemical reaction, i.e., of the hydration process, is described by the degree of hydration ξ. This variable is obtained by relating the mass of reaction products per unit volume, m, to the mass of reaction products at the end of hydration, m_∞. Hence, $\xi = m = m_\infty$, and $0 \leq \xi \leq 1$.
2. *Microcracking* of hydrates which are the result of the hydration process yields plastic strains ε^p. The state of microstructural changes resulting from microcracking is described by hardening variables $\mathcal{X}$.
3. Stress-induced *dislocation-like processes* within the hydrates result in flow (or long-term) creep strains ε^f (Wittmann 1982). The state of respective microstructural changes is described by the viscous flow γ (Ulm 1998).
4. Stress-induced *microdiffusion of water* in the capillary pores between the hydrates result in viscous (or short-term) creep strains ε^v (Ruetz 1966) (Wittmann 1982) (Ulm 1998).

So-called intrinsic material functions, i.e., functions which neither depend on field nor boundary conditions serve as input for the chemomechanical material model. Figure 6 shows intrinsic material functions describing uniaxial compressive strength f_c and Young's modulus E for cemented soil as functions of ξ. These function were determined in an experimental program initiated at Vienna University of Technology (Brandstätter et al. 2002). A detailed description of the employed chemomechanical material model can be found in (Hellmich et al. 1999) (Sercombe et al. 2000) (Lackner et al. 2002).

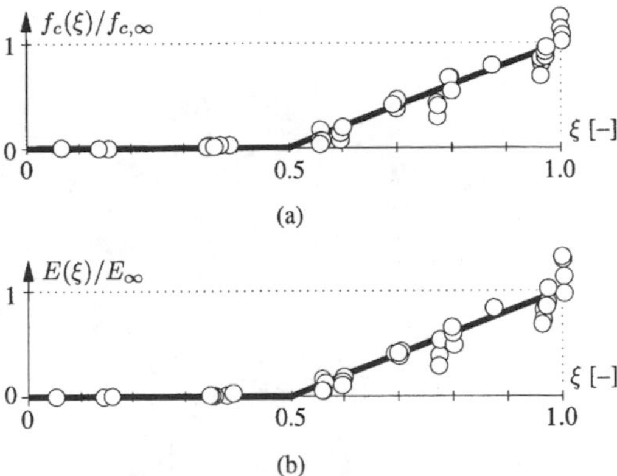

Figure 6. Intrinsic material function for (a) compressive strength $f_c(\xi)$ and (b) Young's modulus $E(\xi)$ of cemented soil.

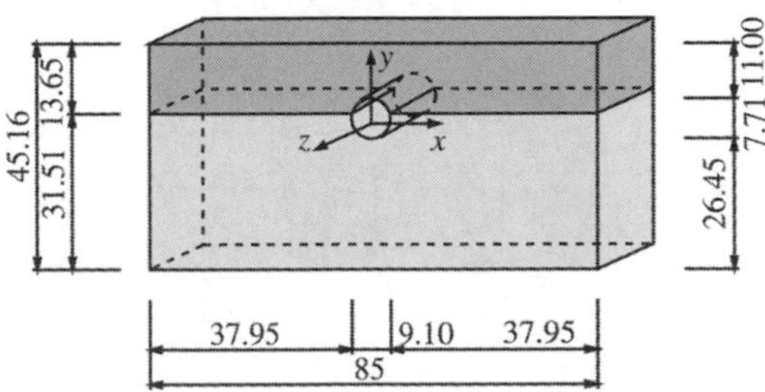

Figure 7. Geometric dimensions (in [m]) of the employed structural model.

3 STRUCTURAL MODEL AND FE DISCRETIZATION

For the simulation of (a) the excavation process, (b) the construction of the jet-grouted support structure, and (c) the installation of the shotcrete lining the plane-strain model depicted in Figure 7 was employed. In order to study the influence of the soil stratification on the performance of the jet-grouted support, two layers of soil were considered. The structural model was discretized by means of four-node finite elements. Two FE meshes, one for the tunnel supported by the jet-grouted support structure and full-face excavation (4188 finite elements, see Figure 8(a)), and one for the tunnel with staggered excavation (4344 finite elements, see Figure 8(b)), were designed.

The jet-grouting process is simulated by changing the material properties in the finite elements representing the jet-grouted column from *in situ* soil to young jet-grouted soil mass. According to Figure 6, this change results in negligibly small strength and stiffness of the cemented soil. Since the original

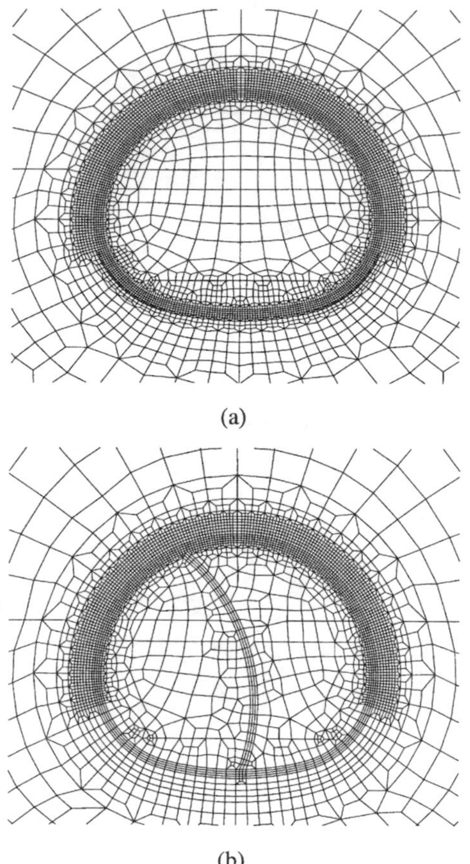

(a)

(b)

Figure 8. FE meshes for (a) full-face excavation (in case of HJG) and (b) staggered excavation.

microstructure of the soil is destroyed during jet grouting, disregard of strength and stiffness of the young jet-grouted soil mass is appropriate.

4 NUMERICAL RESULTS

In order to gain insight into the influence of HJG on surface settlements, the two layers of soil in Figure 7 are first assumed to consist of the same material (homogeneous condition), with Young's modulus $E = 20\,\mathrm{N/mm^2}$, Poisson's ratio $\nu = 0.35$, an angle of internal friction $\varphi = 27.5°$ and a cohesion $c = 0.015\,\mathrm{N/mm^2}$. Two analyses were performed:

- In the first analysis, a staggered excavation, which is only supported by a shotcrete lining, was investigated.
- In the second analysis, a tunnel supported by a support structure consisting of 37 jet-grouted columns was analyzed.

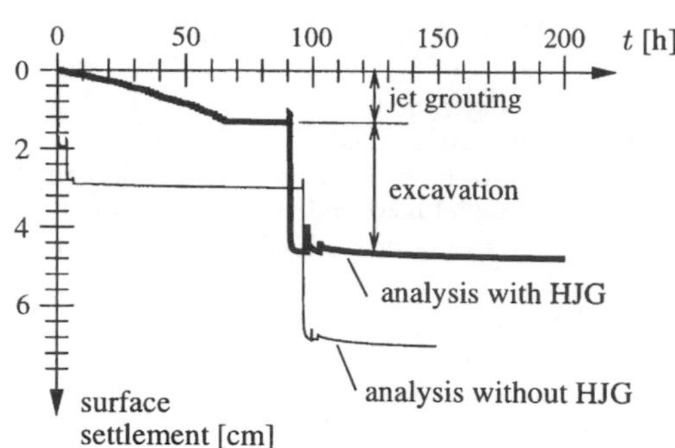

Figure 9. Surface settlement in homogeneous soil.

Figure 9 shows the evolutions of the surface settlement at $x = 0$ obtained from both analyses. The jumps in the evolution of the settlements refer to different excavation steps, namely, excavation of the top heading, the benches, and the invert. Modeling of jet grouting by changing the material properties in finite elements representing jet-grouted columns from soil to young jet-grouted soil mass results in a settlement of approximately 13 mm prior to the excavation process. Settlements resulting from the excavation process of the tunnel amount to 7.1 cm for the staggered excavation, 3.5 cm for the tunnel supported by the jet-grouted support structure. Hence, based on the numerical results, application of HJG leads to a reduction of the settlements caused by the excavation of the cross section of approximately 50%. Together with the settlements resulting from jet grouting, however, the surface settlement amounts to approximately 70% of the surface settlement obtained without jet grouting.

Whereas jet grouting resulted in a reduction of surface settlements of 30%, the time required for the installation of the jet-grouted support structure as well as the time lag between the end of jet grouting and the begin of excavation, Δt, increase the construction time of the tunnel. From an economic point of view, acceleration of the ground-improvement work is highly desirable. Basically, HJG provides two possibilities to achieve this goal:

- On the one hand, the time span between the end of jet grouting and the begin of the excavation, Δt, may be reduced.
- On the other hand, the number of jet-grouted columns may be reduced.

In the following subsections, both modes of acceleration of the tunnel construction will be investigated.

4.1 *Variation of* Δt

First, the time span between the end of jet grouting and the begin of the excavation Δt is reduced from 26 hours (see Figure 4) to 8 hours. The obtained histories

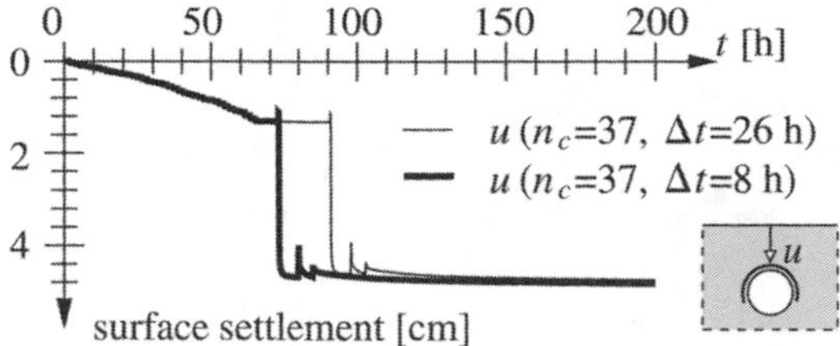

Figure 10. Surface settlement in homogeneous soil: variation of Δt.

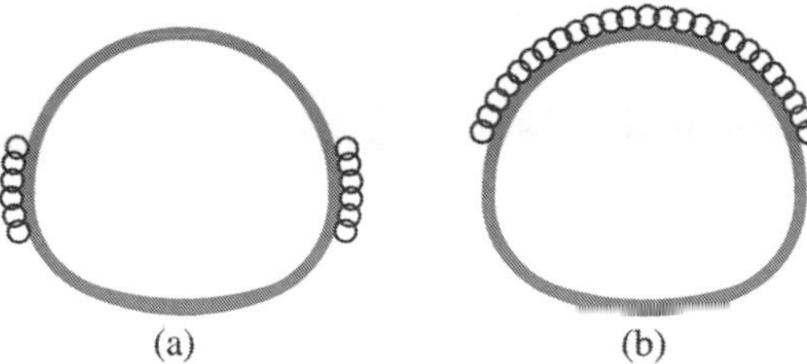

Figure 11. Jet-grouted support structure with (a) $n_c = 12$ and (b) $n_c = 25$.

of the surface settlement are shown in Figure 10. The reduction of Δt is found to have only a marginal influence on the total settlements. The insensitivity of the surface settlements on Δt can be attributed to the relatively long time span required for construction of the jet-grouted support structure. Most jet-grouted columns are considerably older than Δt when the excavation sequence is initiated.

4.2 *Variation of the layout of the jet-grouted support structure*

The second possibility to accelerate the construction progress is to reduce the number of jet-grouted columns. In addition to the layout of the support structure with 37 jet-grouted columns, $n_c = 37$ (Figure 1), two additional layouts are considered in the following study (see Figure 11). The number of jet-grouted columns chosen for the additional layouts are $n_c = 12$, and 25, respectively, resulting in a construction time of $12 \times 1.75\,\text{h} = 21\,\text{h}$ and $25 \times 1.75\,\text{h} = 43.75\,\text{h}$, respectively. The settlements prior to the excavation process caused by the weakening of soil in consequence of jet grouting show a strong dependence on the length of the horizontal projection of the jet-grouted support structure. For the support structure with $n_c = 12$, characterized by a horizontal projection that is only slightly larger than the diameter of a jet-grouted column, a settlement of 3 mm was obtained (Figure 12). For the support structure with $n_c = 25$, characterized by a horizontal projection that is roughly equal to the horizontal dimension of the tunnel cross-section, the analysis resulted in a settlement of 12 mm (Figure 12). The reduction of the number of jet-grouted columns resulted in total surface settlements exceeding the settlement obtained in case of a tunnel excavation without HJG.

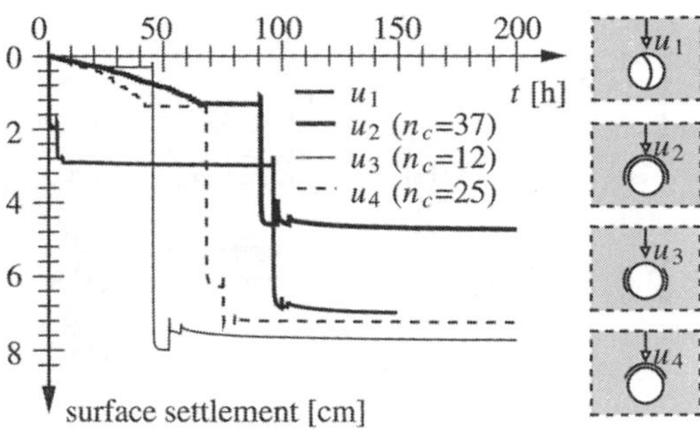

Figure 12. Surface settlement in homogeneous soil: variation of the layout of the jet-grouted support structure ($\Delta t = 26$ h).

Table 1. Material properties for soil.

	Gravel	Silty sand
Young's modulus E [N/mm^2]	80	20
Poisson's ratio ν [−]	0.32	0.35
Cohesion c [N/mm^2]	0	0.015
Angle of internal friction φ	32.5°	27.5°
Relaxation time τ [h]	0.015	0.015

4.3 *Variation of soil properties*

In the numerical studies presented so far, homogeneous soil conditions were considered. In general, such conditions do not occur. In this subsection, the optimal layout of the jet-grouted support structure with respect to the surface settlement and the economic effort is determined for different soil stratifications. Two types of soil are considered (see Figure 7). The respective soil parameters are listed in Table 1. In addition to the silty sand employed in the previous analyses, a layer of gravel will be considered in the following investigation.

4.3.1 *Stiff layer over a relatively soft layer*

The geological situation dealt with subsequently is characterized by a stiff layer of gravel over a relatively soft layer of sand. Figure 13 shows the evolution of the surface settlements. The settlement obtained for the staggered excavation scheme and the support structure with $n_c = 12$ amounts to approximately 3.3 cm. This settlement is only slightly larger than the settlement obtained for the support structure with $n_c = 37$.

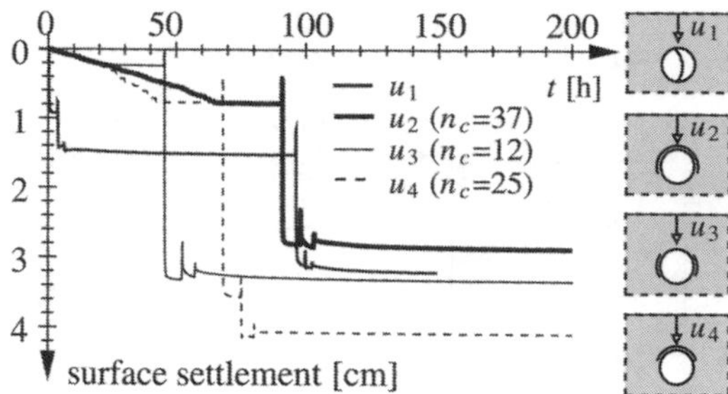

Figure 13. Surface settlement in layered soil: stiff layer of gravel over a relatively soft layer of sand ($\Delta t = 26$ h).

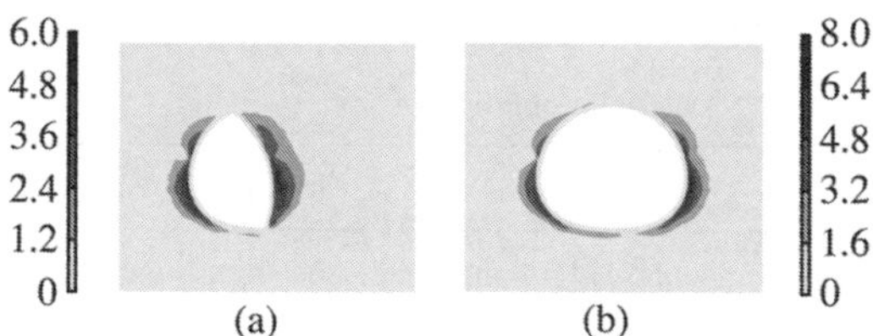

Figure 14. Volumetric plastic strain [%] for the staggered excavation scheme after excavation of (a) heading I and (b) heading II.

From an economic point of view, the staggered excavation scheme appears to be the best solution for the chosen soil stratification.

Figure 14 shows the zones of plastic loading in the soil by means of the volumetric plastic strain for the staggered excavation scheme. For the (compressive) primary stress state $|\sigma_y| > |\sigma_x| = |\sigma_z|$, with $\sigma_x = \sigma_z = K_0\sigma_y = (1 - \sin\varphi)\,\sigma_y$, where K_0 denotes the lateral earth pressure at rest (Jaky 1944), the excavation of the cross section of the tunnel leads to large stresses in the side walls during the respective excavation step. These are the regions where the first plastic deformations of the soil occur. The employed associated plasticity model (Drucker-Prager) evokes plastic dilation, compressive stresses and, hence, an increase of the load-bearing capacity of the soil in terms of the stress deviator. This leads to a limitation of the plastic zone around the tunnel as observed in Figure 14. For the analyses characterized by consideration of HJG (see Figure 15), the zones of plastic loading of the soil are concentrated around the footings of the jet-grouted support structure. This structure leads to an increase of the confinement of the soil. This has the same effect as plastic dilation, namely, an increase of the load-bearing capacity of the soil.

4.3.2 *Soft layer over a relatively stiff layer*

The geological situation considered in the previous study is altered such that a soft sand layer is located over a relatively stiff layer of gravel. Figure 16 shows the evolution of the surface settlement obtained from the numerical analyses. According to the obtained results, all three layouts for the jet-grouted support structure lead to smaller settlements, as compared to the staggered excavation scheme, resulting in a total settlement of 5.6 cm. The final settlement obtained with the support structure with $n_c = 12$ is only slightly larger than the settlement obtained with the support structure characterized by $n_c = 37$. The jet-grouted support structure with $n_c = 25$, on the other hand, yields an increase of settlements by approximately 1 cm. This indicates the high efficiency of the jet-grouted columns located at the benches of the tunnel, constituting areas of significant plastic material response (see, e.g. Figure 14).

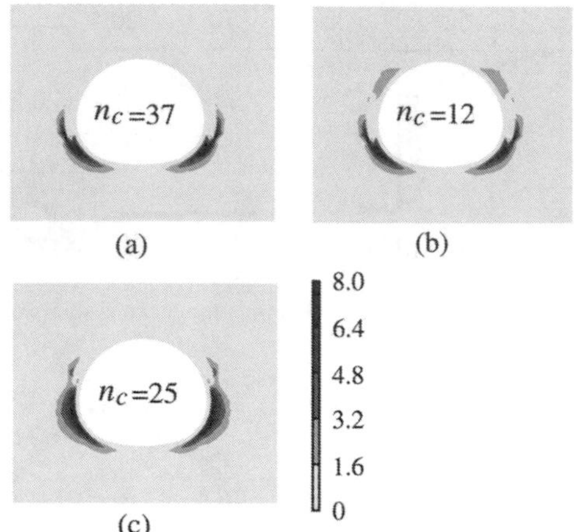

Figure 15. Volumetric plastic strain [%] for the excavation with HJG: (a) 37, (b) 12, and (c) 25 jet-grouted columns.

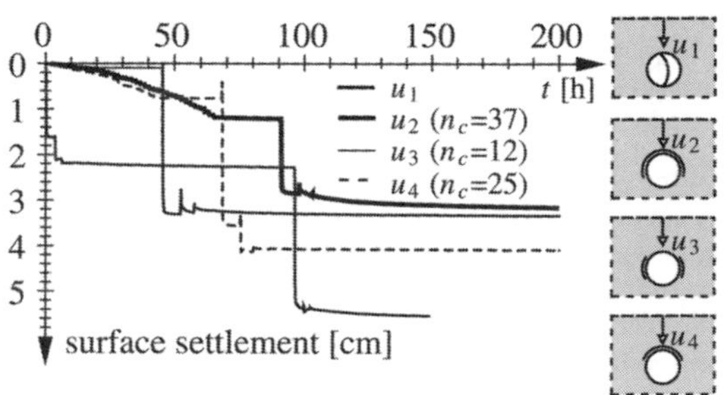

Figure 16. Surface settlement in layered soil: soft layer of sand over a relatively stiff layer of gravel ($\Delta t = 26$ h).

Also for the staggered excavation scheme the zones of significant plastic material response are located at the benches (Figure 17). For the jet-grouted support structure, these zones are situated at the footings (Figure 18). Since the soft soil layer is now located above the the stiff gravel layer, the plastic zones extend

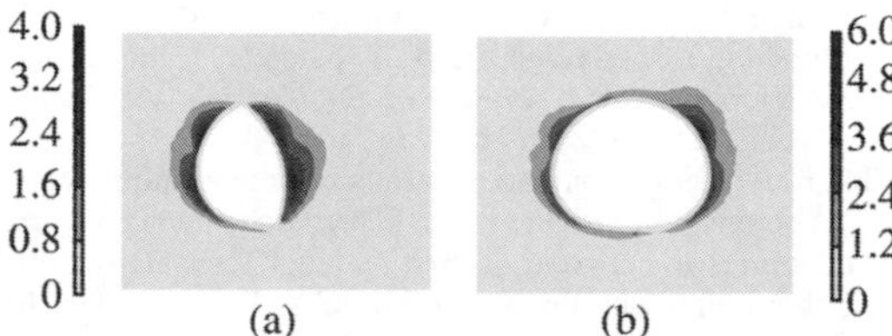

Figure 17. Volumetric plastic strain [%] for the staggered excavation scheme after excavation of (a) heading I and (b) heading II.

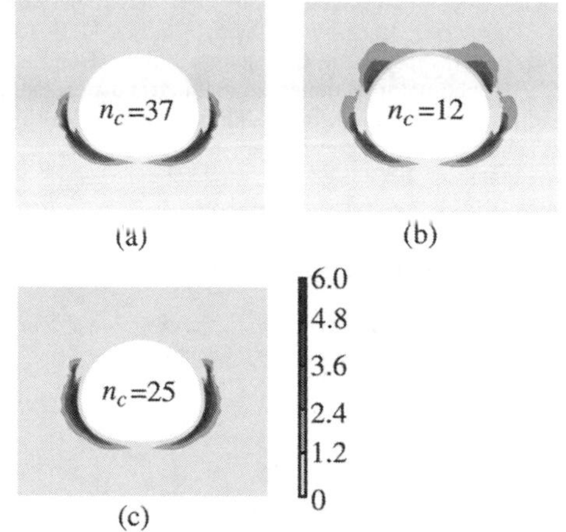

Figure 18. Volumetric plastic strain [%] for the excavation with HJG: (a) 37, (b) 12, and (c) 25 jet-grouted columns.

further upward for the case of no increase of the confinement by the jet-grouted support structure ($n_c = 12$, see Figure 18(b)).

The narrow bands of plastic material response (Figures 18(a) and (b)) at the benches can be attributed to the increase of the horizontal convergence in consequence of the excavation of the benches, resulting in a reduction of the confinement of the soil.

5 CONCLUSIONS

On the basis of sophisticated material models for cement-based materials and soil, several questions arising from practical application of HJG were addressed in this paper. Based on the numerical results, the following conclusions can be drawn:

1. Settlements in consequence of weakening of the soil structure during jet grouting strongly depend on the length of the horizontal projection of the jet-grouted support structure.
2. The time lag between the end of jet grouting and the begin of the excavation, Δt, was found to have only a marginal influence on the settlement. This can be attributed to the fact that most jet-grouted columns are considerably older than Δt at the beginning of the excavation.
3. In homogeneous soil conditions, HJG with 37 jet-grouted columns resulted in a reduction of the settlement by approximately 30%, as compared to the settlement following from a tunnel excavation with a staggered excavation scheme and no jet-grouted support. The reduction of the number of jet-grouted columns from 37 to 12 and 25, respectively, resulted in total surface settlements exceeding the settlement obtained in case of a tunnel excavation without HJG.
4. For a layer of stiff gravel over a relatively soft sand layer, the settlement obtained for the staggered excavation scheme is only slightly larger than the settlement that occurs in case of a support structure with 37 jet-grouted columns. From an economic point of view, the staggered excavation scheme appears to be best solution for this type of soil stratification. For the simulation of a soft sand layer over a relatively stiff layer of gravel, on the other hand, the settlement obtained for a support structure with 37 jet-grouted columns amounts to approximately 60% of the settlement for the staggered excavation scheme. Almost the same result was obtained for 12 jet-grouted columns.

The results presented in this paper were obtained from numerical analyses assuming isothermal conditions, with $T = 40°C$ in the jet-grouted columns and the shotcrete lining. For the consideration of non-isothermal conditions employing thermochemo-mechanical material models for jet-grouted soil mass and shotcrete, the reader is referred to (Pichler et al. 2003). Missing material parameters for jet-grouted soil mass and shotcrete can also be found in the aforementioned reference.

REFERENCES

Brandstätter, C., R. Lackner, and H. Mang (2002). Bestimmung von Materialeigenschaften zementgebundener Werkstoffe mittels Temperaturmessungen: Anwendung im Rahmen des Düsenstrahlverfahrens [Evaluation of properties of cementitious materials by means of temperature measurements: Application to jet grouting]. *Bauingenieur 77*(2), 51–57. In German.

Duvaut, G. and J. Lions (1972). *Les Inequations en Mechanique et en Physique [The inequalities in mechanics and physics]*. Paris: Dunod. In French.

Fitl, D. (1997). Der Tunnel Frankfurter Kreuz [The Frankfurter Kreuz tunnel]. *Felsbau – Rock and Soil Engineering 15*(5), 347–350. In German.

Gioda, G. and L. Locatelli (1999). Back analysis of the measurements performed during the excavation of a shallow tunnel in sand. *International Journal for Numerical and Analytical Methods in Geomechanics 23*, 1407–1425.

Guatteri, G., P. Mosiici, A. Koshima, and V. DoroAltan (1994). Application of jet grouting to tunnel portals and top headings in NATM tunnelling: Brazilian experience. In A. Bell (Ed.), *Grouting in the ground*, London, England, pp. 455–471. Thomas Telford, London.

Hellmich, C., F.-J. Ulm, and H.A. Mang (1999). Multisurface chemoplasticity I: Material model for shotcrete. *Journal of Engineering Mechanics (ASCE) 125*(6), 692–701.

Jaky, J. (1944). A nyugalmi nyomas tenyezoje [The coefficient of earth pressure at rest]. *Magyar Mernok es Epitesz-Egylet Kozlonye*. In Hungarian.

Lackner, R., C. Hellmich, and H. Mang (2002). Constitutive modeling of cementitious materials in the framework of chemoplasticity. *International Journal for Numerical Methods in Engineering 53*(10), 2357–2388.

Liebsch, H. and C. Haberland (1998). Soft ground tunnelling in urban areas by NATM. *Felsbau – Rock and Soil Engineering 16*(2), 78–90.

Pichler, C., R. Lackner, Y. Spira, and H. Mang (2003). Thermochemomechanical assessment of ground improvement by means of jet grouting in tunneling. *Journal of Engineering Mechanics (ASCE)*. Accepted for publication.

Ruetz, W. (1966). Das Kriechen des Zementsteins im Beton und seine Beeinflussung durch gleichzeitiges Schwinden [Creep of cement in concrete and the influence of simultaneous shrinkage on this type of creep]. *Deutscher Ausschuß für Stahlbeton, Heft 183*. In German.

Sercombe, J., C. Hellmich, F.-J. Ulm, and H. A. Mang (2000). Modeling of early-age creep of shotcrete. I: model and model parameters. *Journal of Engineering Mechanics (ASCE) 126*(3), 284–291.

Ulm, F.-J. (1998). Couplages thermochémomécaniques dans les bétons : un premier bilan. [Thermochemomechanical couplings in concretes: a first review.] Technical report, Laboratoires des Ponts et Chaussées, Paris, France. In French.

Ulm, F.-J. and O. Coussy (1995). Modeling of thermochemomechanical couplings of concrete at early ages. *Journal of Engineering Mechanics (ASCE) 121*(7), 785–794.

Ulm, F.-J. and O. Coussy (1996). Strength growth as chemo-plastic hardening in early age concrete. *Journal of Engineering Mechanics (ASCE) 122*(12), 1123–1132.

Wittke, W., B. Pierau, and C. Erichsen (2000). Der Einsatz von Hochdruckinjektionen zur Baugrundverbesserung und für den Tunnelbau im Lockergestein [Application of jet grouting in ground improvement and tunneling in granular material]. In G. Riedmüller, W. Schubert, and S. Semprich (Eds.), *Düsenstrahlverfahren: Möglichkeiten und Grenzen der Anwendung*, Christian Veder Kolloquium, Graz, Austria, pp. 155–183. Gruppe Geotechnik Graz, Graz. In German.

Wittmann, F. (1982). *Creep and shrinkage mechanisms*, Chapter 6, pp. 129–161. Wiley, Chichester.

Computational Modelling of Concrete Structures, Bićanić et al. (eds)
© 2003 Swets & Zeitlinger, Lisse, ISBN 90 5809 536 3

A note on the design of fibre-reinforced shotcrete linings for underground support

G. Barla, F. Barpi, C. Bertolino, B. Chiaia
Department of Structural Engineering and Geotechnics, Politecnico di Torino, Torino, Italy

ABSTRACT: The ability of sprayed concrete to hold the rock mass around an opening, avoiding progressive loosening and destabilization, is widely recognized. On the other hand, design of shotcrete supports is still more of an art than a science, with the majority of designs based on experience, on empirical recommendations or on monitoring and adapting during construction. In order to improve both the mechanical and the durability characteristics of sprayed linings, a recent trend is to replace the ordinary mesh reinforcement with steel fibres. It is well-known, in fact, that the usage of high-strength fibres (e.g. steel fibres) in the concrete mixture permits to maintain a high compressive strength while considerably improving the toughness and thus the structural ductility. In this paper, the evaluation of the strength and ductility characteristics of fibre-reinforced shotcrete linings is performed by adopting suitable theoretical and numerical models. Analyses show that failure of a shotcrete support is generally a two-step process, initially generated by loss of adhesion to the rock, followed by flexural failure. Flexural strength and toughness play a critical role in the post-peak region once adhesion is lost. After analysing the stress state in traditional underground configurations, the bridged crack model is used in order to define the correct softening curve of the material, based on the fibre content. By means of a numerical code based on the cohesive crack model, the post-peak behavior of the structure is investigated. It is shown that the fibre reinforcement greatly enhances the flexural and punching shear strength, as well as the energy absorption capacity.

1 INTRODUCTION

We define as shotcrete, or spritzbeton, a particular concrete mixture carried under pressure into a closed pipe system, projected against the application surface where it realizes a supporting shell.

Fibre reinforced shotcrete, as a particular case, in addition to the classical ingredients – cement, water, aggregates and additives, contains a reinforcing phase made of fibres which provides the material with enhanced mechanical (e.g. ductility and fracture energy) and durability (related to the reduced cracking extent due to plastic shrinkage) properties.

One of the most important field of application for the shotcrete is the field of undergroung engineering. Excavation of tunnels and galleries, in fact, provokes the alteration of the stress field in the rock mass, with consequent stress redistribution and strain development. Without appropriate confinement, large deformation may converge towards equilibrium or induce unacceptable damage which requires a reinforcing structure.

In the following, attention will be focused onto fibre-reinforced shotcrete, possibly accompanied by a radial bolting system, as a composite sustaining structure. The strength and ductility aspects of this material, together with its energy absorption capabilities, will be investigated through analytical and numerical methods.

The theoretical analysis of the shear and flexural behavior of shotcrete shells will provide closed-form analytical solutions which are helpful in determining the safety factor of composite underground linings. The numerical analysis will permit to translate the tools of nonlinear fracture mechanics (e.g., of the cohesive crack model) into design rules, under the hypothesis of a monocoque support. The geotechnical aspects related to the excavation of a circular tunnel will be investigated through the FLAC® commercial code, whereas, by means of the C.CRA.P. algorithm (developed at Politecnico di Torino), the post-peak response of the reinforcing shell will be modelled, by considering the entire stress-strain relation of fibre-reinforced shotcrete.

2 COLLAPSE MECHANISMS OF SHOTCRETE LININGS

The mechanical characteristics of rock masses, which affect the stress state of the supporting shell, depend both on the mechanical properties of the rock and on the distribution of discontinuities. Within a fractured rock mass, loads may transfer to the shell by means of unstable rock wedges comprised between adjacent discontinuities.

The mechanical characteristics of rock masses, which affect the stress state of the supporting shell, depend both on the mechanical properties of the rock and on the distribution of discontinuities. Within a fractured rock mass, loads may transfer to the shell by means of unstable rock wedges comprised between adjacent discontinuities.

The principal collapse mechanisms which can derive are listed below:

1. loss of adhesion (Fig. 1.a);
2. bending collapse (Fig.1.b);
3. direct shear collapse (Fig. 1.c);
4. punching shear collapse (Fig. 1.d).

As can be noticed in Fig. 1, two additional collapse mechanisms can be individuated, related to the overall loading system:

5. compression collapse (Fig. 1.e);
6. tension collapse (Fig. 1.f).

These collapse mechanisms refer to an equivalent continuum description of the rock-support system. Therefore, they are global damage mechanisms, which cannot be excluded in the presence of fractured rock masses although less specifical than modes 1–4.

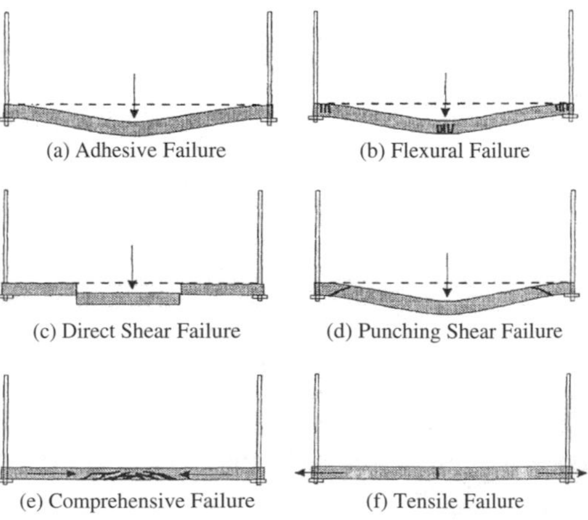

Figure 1. Principal collapse mechanisms for a composite shotcrete support in rock (Barrett & McCreath, 1995).

2.1 *Bending behavior of a shotcrete shell: limit analysis and structural response*

In the presence of a bolting system, the in situ behavior of a fibre-reinforced shotcrete lining under the action of a unstable rock mass, can be modelled as in Fig. 2. The first peak, at a very small value of deformation, corresponds to the loss of adhesion under a substantial elastic behavior. In the subsequent stage, under displacements of the order of some millimeters, the bending behvior prevails with the progressive formation of fractures in the shell. Ultimate strength is reached in correspondence of a prevailing membrane behavior.

Elastic bending model by Stille (1992)

The elastic analysis of the shotcrete lining under bending moment (second stage of Fig. 2) has been carried out by Stille (1992) who suggestes the following formula:

$$M_f = \left(\frac{t^2}{6}\right) \cdot V_f \cdot \sigma_t \cdot \frac{1}{k}, \quad (1)$$

where M_f is the limit elastic bending moment of the fibre-reinforced shell, t is the shell thickness, V_f the volume percentage of fibres, σ_t is the material tensile strength and k is an empirical coefficient taking into account the effective contribution given by the fibres. This coefficient can be calculated according to the following equation:

$$k = 150 \cdot V_f + 1.05. \quad (2)$$

Elastic bending model by Di Tommaso (1990)

Di Tommaso (1990) applies a purely elastic model to the evaluation of the bending capacity of shotcrete linings. The ultimate bending moment for

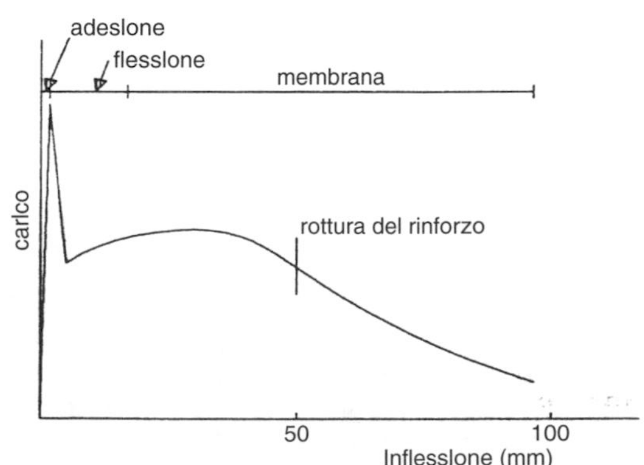

Figure 2. Bending behavior of a shotcrete shell element (Stille, 1992).

a fibre-reinforced shell is calculated according to the classical beam formula:

$$M_u = f_{fd} \cdot \frac{lt^2}{6}, \tag{3}$$

where f_{fd} is the design flexural strength taking into account fibre contribution, whereas l and t are, respectively, the width and the thickness of a shell strip subjected to bending.

Limit state plate analysis by Barrett e McCreath (1995)

The Authors apply limit state design to the calculus of the shotcrete shell which is considered as a bidimensional structure. The model is illustrated in Fig. 3. The shotcrete panel is modelled as a plate and the radial bolts are considered as vertical constraints. Therefore, the shotcrete shell is divided into a set of rectangular panels, with known dimensions, loaded by the weight of rock wedges. In Fig. 4 the moment distribution under such loading is shown: percentages are referred to the moment M_0 equal to:

$$M_0 = \frac{1}{8} \cdot q \cdot (s - c)^2, \tag{4}$$

where q is the distributed load, s is the spacing of the radial bolts and c is the width of the bolt anchorage plate.

The load-carrying capacity of the layer, C_{flex}, with reference to the bending effects, is given by the following expression which is a function of the flexural strength, σ_{flex}, of the bolt spacing, s, and of the thickness t:

$$C_{flex} = \sigma_{flex} \cdot \frac{t^2}{6} \cdot \frac{s}{2}. \tag{5}$$

In addition, the Authors provide, for comparison, an equivalent expression communicated by Holmgren (1993), which updates equation (5) in the presence of reinforcing fibres:

$$C_{flex} = 0.9 \cdot \frac{I_{10/5} + I_{30/10}}{200} \cdot \sigma_{flex} \cdot \frac{t^2}{6} \cdot \frac{s}{2}. \tag{6}$$

The two parameters $I_{10/5}$ e $I_{30/10}$ represent the two factors of flexural toughness, determined according to ASTM C1018-92.

Cohesive-hinge analysis by Marti (1999)

The analysis permits to evaluate the ultimate bending moment of fibre-reinforced composites, in the framework of limit analysis coupled with cohesive crack approach.

Marti (1999) provides two different analysis: the general one explicitly considers the extent of the compressed zone (modelled by a stress-block) within a plastic hinge in a bent element of height h. The second one neglects the extent of the compressed area.

Figure 5 represents the scheme referring to the complete analysis: the plastic hinge, within the element of

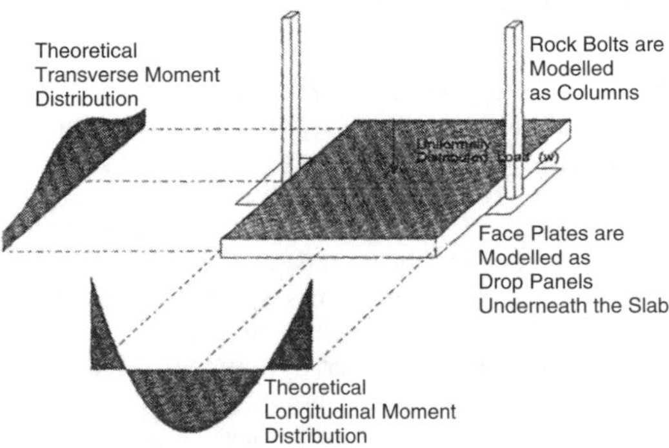

Figure 3. Plate model used for the analysis of bending collapse of shotcrete (Barrett e McCreath, 1995).

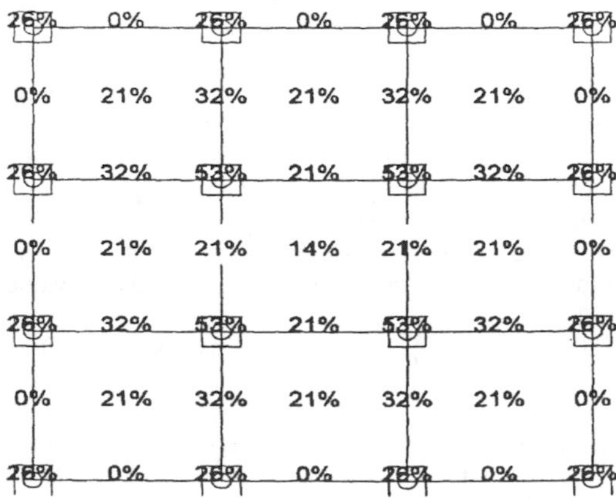

Figure 4. Principal bending moment values in the shell (percentages referred to M_0 (Barrett e McCreath, 1995).

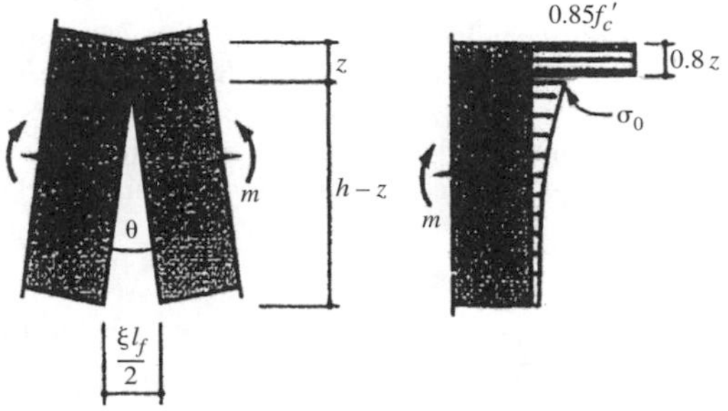

Figure 5. General analysis of a cohesive hinge (Marti, 1999).

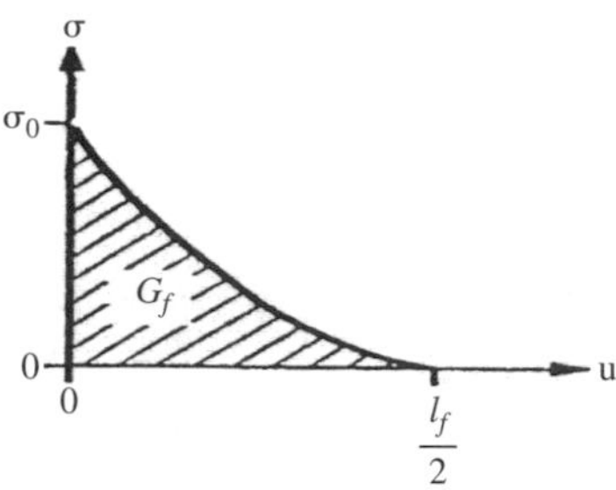

Figure 6. Softening curve for tension (Marti, 1999).

height h, is characterized by a compressed zone of height z and by a finite rotation θ (l_f = length of the fibres).

It is assumed that constant compressive stresses equal to $0.85f_c'$ (f_c': cylinder compressive strength of the shotcrete) act upon the extension $0.8z$ (stress block) of the cross section, and that a parabolic distribution of cohesive tensile stresses acts upon the remaining area of the section (Fictitious Crack Model, see Fig. 6).

When the parameter ξ (crack opening parameter) attains values comprised between 0 and 1, the height z of the compressed zone and the ultimate bending moment m can be calculated according to (see the reference scheme in Fig. 5):

$$z = \frac{h}{1 + \frac{2.04 \cdot f_c'}{\sigma_0 \cdot (3 - 3\xi + \xi^2)}}, \tag{7}$$

$$m = 0.68 \cdot f_c' \cdot z \cdot \left[0.6z + (h - z) \cdot \frac{6 - 8\xi + 3\xi^2}{12 - 12\xi + 4\xi^2}\right]. \tag{8}$$

The value σ_0 (Fig. 6) to be used into equation (7) can be evaluated from the following formula:

$$\sigma_0 = \frac{V_f \cdot l_f \cdot \tau_b}{2 \cdot d_f}, \tag{9}$$

where V_f is the volume percentage of fibres, τ_b represents the limit bond stress between fibres and concrete ($\tau_b = 2 \cdot f_{ct} = 0.6 \cdot f_c^{2/3}$, where f_{ct} is the concrete tensile strength), and d_f is the diameter of the fibres.

When $\xi > 1$, instead, the following formulas should be used:

$$z = \frac{h}{1 + \frac{2.04 \cdot f_c' \cdot \xi}{\sigma_0}}, \tag{10}$$

$$m = 0.68 \cdot f_c' \cdot z \cdot \left[0.6z + \frac{h - z}{4\xi}\right] \tag{11}$$

Notice that, the value $\xi = 1$ implies that all the fibres across the cracked section (cohesive hinge) have undergone pull-out from the surrounding concrete.

The simplified analysis neglects the extent of the compressed zone. Therefore, equations (8) and (11) become:

$$0 \le \xi \le 1: \quad m = \left[\frac{6 - 8\xi + 3\xi^2}{12}\right] \cdot \sigma_0 \cdot h^2, \tag{12}$$

$$\xi > 1: \quad m = \frac{\sigma_0 \cdot h^2}{12 \cdot \xi^2}. \tag{13}$$

In conclusion, if we consider a generic cross section of the fibre-reinforced shotcrete shell as the element of height h, we need to decide the best approach (considering or neglecting the compressive zone) also with regard to the efficiency of the adhesion to the rock. Thereby, on the basis of the value of z, the ultimate bending moment can be calculated with the appropriate formula. As a rule, the design moments shall be sufficiently smaller than this limit state.

Comparison between the different analyses

Table 1 reports the formulas, relative to the ultimate bending moments, obtained according to the previous approaches, by using the reference values described in the same Table, (i.e., the thickness t of the layer is equal to 20 cm, and the volumetric percentage of steel fibres is equal to 1% in all cases). Different symbols are used for the same mechanical quantity (e.g., the bending moment), in order to maintain the original choice of the different Authors.

Comparing the various theories, we observe that the approaches by Di Tommaso (1990) and Marti (1999) provide similar results. Instead, the ultimate bending moment obtained according to the theory by Barrett & McCreath (1995) is considerably higher. This is due to adopting the bidimensional plate theory (instead of beam theory) and to the beneficial presence of the radial bolts which has not been taken into account by the other Authors.

Moreover, there is a clear difference between the load-carrying capacity calculated according to Stille's theory (1992) and the other theories. The k^{-1} parameter, used by Stille to take explicitly into account the bridging action of the fibres, affects directly the value of the elastic bending moment.

Finally, it is important to stress that, since the effect of fibres is essentially confined to the post-peak softening branch (where they provide the shotcrete with an unexpected reserve of ductility), adopting a linear elastic approach does not permit to consider the beneficial role of the reinforcement, unless a fracture mechanics analysis is carried considering the enhanced fracture energy of the composite.

Table 1. Comparison of different theories for bending.

Stille: elastic approach	$M_f = \left(\frac{t^2}{6}\right) \cdot V_f \cdot \sigma_t \cdot \frac{1}{k}$	$t = 20\,\text{cm}$ $V_f = 1\%$ $\sigma_t = 4.22\,\text{MPa}$ $M_f = 10.8\,\text{KN·m}$
Di Tommaso: elastic approach	$M_u = f_{fd} \cdot \frac{lt^2}{6}$	$t = 20\,\text{cm}$ $V_f = 1\%$ $f_{fd} = 3.24\,\text{MPa}$ $M_u = 21.6\,\text{KN·m}$
Barrett, McCreath: limit state approach (plate theory)	$C_{flex} = \sigma_{flex} \cdot \frac{t^2}{6} \cdot \frac{s}{2}$	$t = 20\,\text{cm}$ $V_f = 1\%$ $\sigma_{flex} = 7.11\,\text{Mpa}$ $s = 1.2\,\text{m}$ $C_{flex} = 28.4\,\text{KN·m}$
Marti: limit state approach (beam theory)	Presence of the compressed zone ($\xi = 0.5$): $m = 0.68 \cdot f_c' \cdot z \cdot \left[0.6z + (h - z) \cdot \frac{6 - 8\xi + 3\xi^2}{12 - 12\xi + 4\xi^2}\right]$	$t = h = 20\,\text{cm}$ $V_f = 1\%$ $f_c' = 50.4\,\text{MPa}$ $z = 0.007\,\text{m}$ $\text{m} = 19\,\text{KN·m}$
	Absence of the compressed zone ($\xi = 0.5$): $m = \left[\frac{6 - 8\xi + 3\xi^2}{12}\right] \cdot \sigma_0 \cdot h^2$	$t = h = 20\,\text{cm}$ $V_f = 1\%$ $\sigma_0 = 2.11\,\text{MPa}$ $m = 19.3\,\text{KN·m}$

3 NUMERICAL APPROACHES TO EVALUATE THE LOAD-CARRYING CAPACITY OF SHOTCRETE LININGS

Two different numerical codes have been used to evaluate the load-carrying capacity of a fibre-reinforced circular shotcrete support. The geotechnical aspects related to the excavation operations and the soil-structure interaction have been analysed by means of the finite-difference code FLAC®. Afterwards, the post-peak behavior of the fibre-reinforced critical cross sections has been analysed with the finite-element code C.CRA.P (Cohesive CRAck Program), developed at Politecnico di Torino by one of the Authors (Barpi, 1996). This code is able to follow the entire load history of a fibre-reinforced shotcrete section, from the linear elastic ascending branch to the softening stage, since it is based on the fictitious crack model.

3.1 *The FLAC® numerical code (vers. 3.40)*

The commercial code FLAC® (developed by the ITASCA Consulting Group™), is commonly used in geotechnical engineering, especially in the presence of excavation operations. It is based on an explicit finite-difference algorithm and permits simulation of two-dimensional plane stress, plane strain or axi-symmetrical conditions. In our simulation, we adopted the plane strain module, which is close to the situation of long tunnels excavation (constant cross section and loads acting in the plane). A deep circular hole was considered, inspired by an existing engineering situation encountered in the Frasnadello/Antea gallery in San Pellegrino Terme (BG-Italy).

The coupling coefficient k_0 was put equal to 1, 0.3, 0. The cases when $k_0 = 0.3$ and $k_0 = 0$, which are typical of shallow tunnels or of stratified and fractured rock patterns with considerable stress release around the hole, were chosen in order to analyse the role of tension stresses (as well as their extent) arising in the shotcrete support.

The Mohr-Coulomb plastic criterion was adopted in the analyses. Due to the symmetry, the simulation was carried out only for a quarter of the domain (Fig. 7). The fibre-reinforced shotcrete support was simulated by assigning the pertinent mechanical properties to a shell of elements at the boundary of the hole. By applying localized rock gravity loads, tension stresses arise in the upper part of the support, although the plastic zone around the hole is always of limited size.

3.2 *The finite element code C.CRA.P. (Cohesive CRAck Program)*

The finite element program C.CRA.P., developed at Politecnico di Torino by one of the Authors (Barpi, 1996) was implemented to simulate the entire behavior

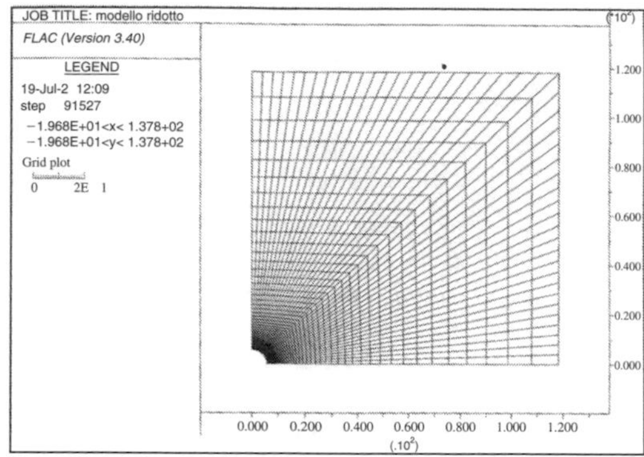

Figure 7. Undeformed mesh for the FLAC simulations of rock-support interaction.

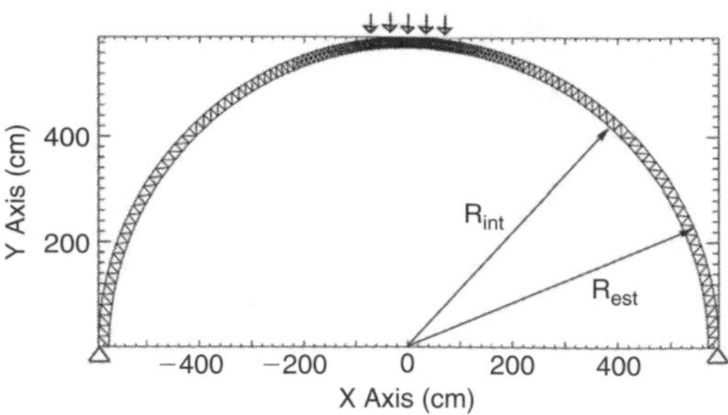

Figure 8. The model for crack propagation used in C.CRA.P. under the action of a localized gravity load.

of fibre-reinforced shotcrete linings in the framework of the cohesive crack model.

Since the loading conditions are those typical of Mode I fracture propagation, the crack path can be determined a priori by symmetry reasoning.

Moreover, in the cohesive crack model, stress singularities are absent, since the maximum principal stress at the apex of the fictitious crack is equal to σ_u (ultimate tensile strength of concrete), and remains constant during crack propagation.

This permits to define a subset of the finite element mesh, named 'rosette', which is constructed upon a moving reference system, with the origin coincident with the fictitious crack tip, which follows crack propagation as a gauge area where the stresses are highest. The rosette is made by an odd number of regions in order to have, in any loading situation, at least one element symmetrical to the x-axis of the moving system.

The crack propagation condition is that the maximum principal stress in the centroid of such element attains the ultimate tensile strength of the material.

Due to the symmetry conditions, numerical analyses have been carried out on an archetype of tunnel support, by modelling only one half of the circular lining. The resulting circular arch was generated, by means of GiD 6.1.2, possessing the internal radius $R_{int} = 5.7$ m and the external radius $R_{est} = 5.9$ m, i.e. a 20 cm thick shotcrete support was modelled. As shown in Fig. 8, the boundaries of the arch were constrained symmetrically.

As a feedback from the FLAC® simulations of the rock-support interaction, an initial notch has been generated into the C.CRA.P. model, in correspondence of the zones where the highest tensile stresses developed. The subsequent crack propagation starts from there, permitting to follow the post-peak behavior of such archetype of support.

Numerical analyses yield the history of the applied load as a function of the vertical displacement δ_y (measured at the top of the support) and as a function of the crack mouth opening displacement (*CMOD*). The experimental material data have been selected, in the literature, among the most important researches on fibre-reinforced concrete. They are mentioned in Table 2. In this Table, the most important results found during a wide research carried by Bekaert on fibre-reinforced shotcrete are reported. In particular, we refer to a linear cohesive law (e.g. linear dependence of the bridging stress σ on the crack opening displacement w).

In our simulations we loaded five nodes, symmetrically with respect to the top of the support. This condition is corresponding to a localised punching load due, for instance, to the self-weight of a unstable rock wedge.

The following diagrams (Figs. 9, 10) show the applied load as a function of the vertical displacement δ_y at the top, and of the *CMOD*. Notice that, due to the symmetry conditions, the horizontal displacement at the top node is zero.

In Fig. 11, an enlargement of the crack propagation zone is depicted. Notice the special finite elements belonging to the rosette and the crack propagated to a certain extent through the thickness of the shotcrete layer.

The maximum load that the support is able to sustain is one of the numerical outputs. It is plotted in Fig. 12. For instance, considering a volumetric percentage of fibres equal to 1.5%, the maximum punching load that the support can resist results equal to almost 12 tons.

The first points of the curves shown in Figs. 9 and 10 correspond, instead, to the load of first crack formation P_u, which is a function of concrete ultimate tensile stress (σ_u), of the cross-section area of the support (b, h), and of the radius of the hole (R) according to the following formula (which is valid under the hypothesis of infinite axial stiffness):

$$P_u = \frac{\sigma_u \cdot b \cdot h^2 \cdot 6 \cdot 0.182}{R}. \tag{14}$$

The shape of the curve beyond the first crack load depends deeply on the bridging action of the fibres.

Table 2. Experimental material data adopted in the C.CRA.P. simulations (data by Bekaert Robocon, 1996).

Steel fibre content (Dramix® ZC 30/.50)	[%]([kg/m^3])	0(0)	1(78.5)	1.5(117.5)
Elastic Modulus E	[MPa]	30400	33100	34800
Poisson coefficient ν	[-]	0.25	0.25	0.25
Extension of the area	[m]	1	1	1
Ultimate tensile strength σ_u	[MPa]	3.63	4.22	5.55
Fracture energy G_f	[N/m]	150	5200	6300
Critical crack opening w_c	[m]	0.0000826	0.002464	0.00227
Specific weight γ	[kg/m^3]	2265	2279	2355

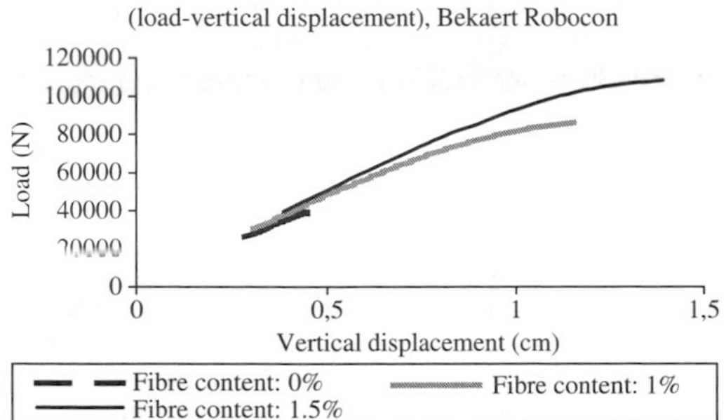

Figure 9. Numerical response (applied load vs. vertical displacement).

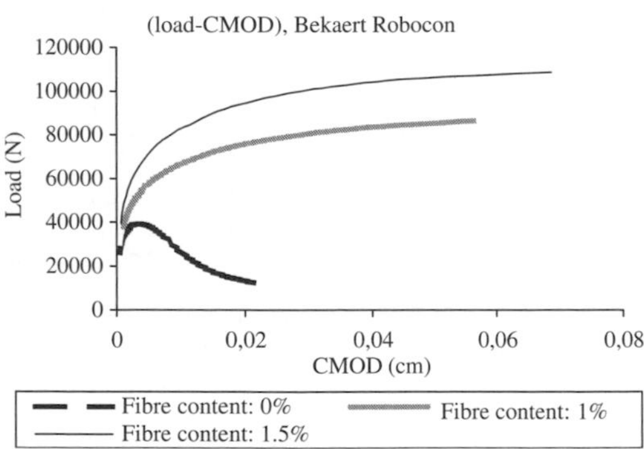

Figure 10. Numerical response (applied load vs. CMOD).

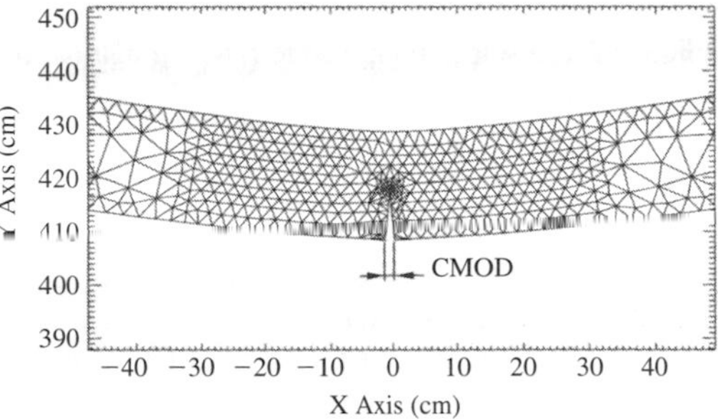

Figure 11. Enlargement of the crack propagation zone.

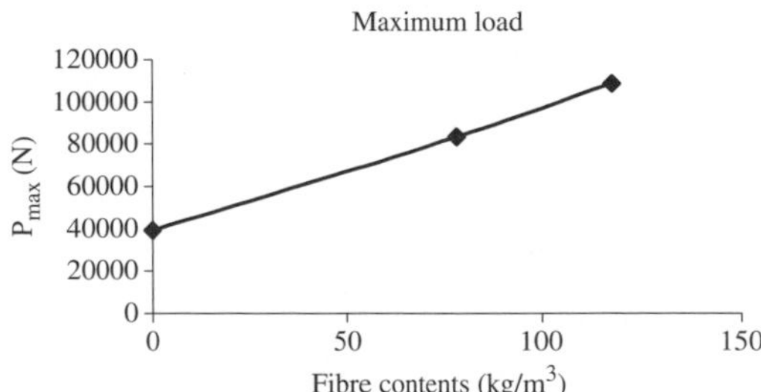

Figure 12. Increase of the maximum load as a function of the fibre content (numerical simulations by C.CRA.P.).

Considering, as an example, the situation depicted in Fig. 10, we can notice that the action of the fibres not only raises the load-carrying capacity of the support, but, more importantly, improves the fracture energy of the shotcrete and favours the transition from a brittle behavior towards a ductile behavior of the structure. Notice, in fact, that the post-peak response is very close to a horizontal (pseudo-plastic) branch. This means that, once the maximum load P_{max} is attained, further deformation of the shotcrete support is possible under constant, or slightly decreasing load. Of course, this result is possible because numerical analyses are carried by C.CRA.P. under crack length control.

The analysis of various numerical simulations permits to put into evidence the optimal volumetric percentage of steel fibres with respect to strength and ductility. With reference to Fig. 10, for example, a big difference (of the shape and of the absolute values) is present between the curves corresponding to the absence of fibres and the 1% content. On the contrary, the curves corresponding to 1% and 1.5% are remarkably similar. This implies that excessive quantities of fibres may result in a very expensive mix design, without a corresponding increase of performances.

Finally, once the maximum load bearing capacity of a fibre-reinforced shotcrete shell has been fixed (see, e.g., Fig. 12), we can calculate the maximum

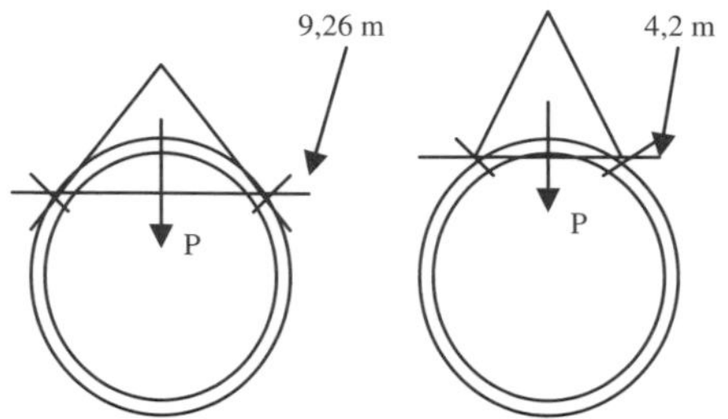

Figure 13. Qualitative representation of the potentially unstable rock wedges.

height of a unstable rock wedge pushing against the top of the hole (see Fig. 13). Consider the worst situation for the support, i.e., the presence of a pyramidal block whose size is comparable to the characteristic size of the hole (e.g., diameter of the hole = 11.8 m). For a volumetric percentage of fibres equal to 1.5% (117.5 kg/m^3; $P_{max} = 12$ tons), the maximum height of the wedge sustainable by the shotcrete support alone is equal to 1.5 m. In the presence of lower fibre contents (e.g., $V_f = 30\,\text{kg/m}^3 = 0.38\%$; $P_{max} = 6$ tons), which are more common for undergroung mixtures, the maximum height of the rock wedge becomes equal to 0.8 m.

Considering a less dramatic situation, e.g. a punching rock wedge with a smaller size, comparable for instance to the gallery radius (equal to 5.9 m), the following values of the maximum height are provided by the analysis:

- $V_f = 117.5\,\text{kg/m}^3 = 1.5\%$; $P_{max} = 12\,\text{t} \Rightarrow h_{max} = 3.4\,\text{m}$
- $V_f = 30\,\text{kg/m}^3 = 0.38\%$; $P_{max} = 6\,\text{t} \Rightarrow h_{max} = 1.8\,\text{m}$.

4 CONCLUSIONS

The role of concrete supports in tunnel engineering is not only that of sustaining loads but also that of stabilizing the rock mass around the hole by controlling its deformations. The load-bearing action, in fact, is very often accomplished by the rock itself. The shotcrete lining, behaving like a membrane, permits redistribution of stresses around the hole and contributes to avoid localized stress releases and separation of unstable rock blocks.

Thereby, the requirement of ductility for rock supports seems to be a very important issue in underground engineering. In this context, the principal characteristics of a fibre-reinforced shotcrete lining is the high fracture energy accompanied by the capacity to develop large deformations (provided the thickness of the layer is not too high). This permits to maintain remarkable load-carrying capabilities well beyond the maximum load, as shown by the numerical simulations reported in this paper.

Our results indicate that the fibre-reinforced shotcrete lining, in the presence of sufficiently good geomechanical conditions of the rock, provides excellent mechanical performances and is therefore perfectly capable to play the role of the final gallery support. The maximum sustainable loads have been shown to be high. This result, accompanied by the excellent post-peak performances, implies that high safety factors can be reached also with respect to unpredictable rock instabilities and localized punching loads. At the same time, adding a fraction of short fibres to the mixture provides an efficient method to reduce shrinkage cracking, improving the durability of the structure

Adequate structural theories should be adopted to take into account the ductility of the material at the design stage, otherwise the beneficial effects of fibre-reinforcement are missed. As a conclusion of this paper, we suggest that a step forward would be represented by implementing into the commercial codes commonly used for underground engineering (e.g., FLAC®) the softening constitutive law. In this way, a complete analysis of the rock-support interaction will be possible for the entire loading history.

REFERENCES

Barpi F. (1996), Modelli numerici per lo studio dei fenomeni fessurativi nelle dighe, Tesi di Dottorato, Dipartimento di Ingegneria Strutturale, Politecnico di Torino.

Barrett S.V.L., McCreath D.R. (1995), Shotcrete support design in blocky ground: towards a deterministic approach, Tunnelling and Underground Space Technology, V. 10, N. 1, pp. 79–89.

Bertolino C. (2002), Il calcestruzzo spruzzato fibrorinforzato nelle costruzioni in sotterraneo, Master Thesis, Dipartimento di Ingegneria Strutturale e Geotecnica, Politecnico di Torino.

Carpinteri A. (1999), Nonlinear crack models for nonmetallic materials, Kluwer Academic Publishers.

Di Tommaso A. (1990), Le opere preliminari di sostegno in spritzbeton fibroarmato, Officine Maccaferri.

Holmgren J. (1993), Tunnel linings of steel fibre reinforced shotcrete, Proceedings of the 5th International Congress on Rock Mechanics, Melbourne.

Karihaloo B. (1995), Fracture mechanics and structural concrete, Longman Scientific & Technical, London.

Marti P., Pfyl T., Sigrist V., Ulaga T. (1999), Harmonized test procedures for steel fiber-reinforced concrete, ACI Materials Journal, V. 96, N. 6, pp. 676–685.

Stille H. (1992), Rock support in theory and practice, Proceedings of the International Symposium on Rock Support, Sudbury, Canada, 16–19 June.

Vandewalle M. (1992), Tunnelling the world, Bekaert SA, Zwevegem, Holland.

Computational Modelling of Concrete Structures, Bićanić et al. (eds)
© 2003 Swets & Zeitlinger, Lisse, ISBN 90 5809 536 3

Nonlinear local analysis of reinforced concrete plates

B. Belletti, P. Bernardi, R. Cerioni & I. Iori
Department of Civil and Environmental Engineering and Architecture, University of Parma, Italy

ABSTRACT: A nonlinear method for the analysis of reinforced concrete plate elements subjected to general combined loads is proposed. The behaviour of the composite material "reinforced concrete" is described by means of a nonlinear model, which is based on equivalent uniaxial stress-strain curves in the uncracked stage, and on a recently proposed "physical approach" in the cracked stage (so-called PARC which means Physical Approach for Reinforced Concrete). This approach includes a set of realistic constitutive and interaction laws for the materials (steel and concrete). For a given plate element a relationship between generalized forces and global deformations of the cross-section is developed in terms of stiffness matrix. The solution of the nonlinear problem is carried out numerically through an incremental-iterative procedure based on a thickness subdivision into layers. The reliability and the effectiveness of the proposed model is shown performing comparisons with well-documented experimental tests reported in scientific literature.

1 INTRODUCTION

In these last years, as a consequence of a more and more frequent building of daring reinforced concrete structures made of high-performance materials, refined numerical procedures for design have been developed. These theoretical models are able to evaluate the real behaviour of a structure, not only as a whole, but also at a local level, providing useful advice for the design, both in terms of safety at ultimate states and in terms of good functionality at serviceability. In particular, in the case of complex structures (such as folded plates and shell elements adopted in bridge structures, large span roof elements and off-shore constructions), stress analysis up to failure must take into account several phenomena such as mechanical and geometrical nonlinearity, crack pattern development, stress concentration. Generally, for these structures the simplified design methods are not able to provide enough information to satisfy all the necessary checks.

Nonlinear finite element analysis is becoming one of the more common tools for studying critical details in the design of reinforced concrete structures. A careful assessment of the actual structural behaviour, at serviceability and ultimate states, will depend on the capability of the finite element program of modelling the phenomena occurring after cracking, such as tension stiffening, concrete compression softening, aggregate interlock, dowel-action, residual cohesion between crack surfaces, etc.

In this paper a constitutive relationship for the analysis of nonlinear behaviour up to failure of reinforced concrete plate elements subjected to membrane and flexural loads is presented. For a portion of plate, the relationship between membrane and flexural generalized forces and global strains is developed in terms of stiffness matrix. This is the first step of a research, which aims to implement the proposed stiffness matrix into nonlinear finite element programs able to predict the behaviour of structures which can be modelled as the assembly of reinforced concrete plates.

The plate thickness is subdivided into any number of reinforced concrete layers, each of them behaves as a membrane element described by a constitutive model for reinforced concrete subjected to plane stress recently proposed by the Authors (Belletti et al. 2001). This constitutive relationship is based on a nonlinear approach, which involves for uncracked reinforced concrete equivalent uniaxial stress-strain curves, while for cracked reinforced concrete a physical approach recently proposed by Authors and called PARC, based on realistic constitutive laws for materials and on steel-concrete interaction. Compatibility and equilibrium conditions are effectively modelled through a local approach which adopts as fundamental quantities governing the problem after cracking the crack opening and sliding and the axial strain of the concrete strut generated between cracks. The general hypotheses adopted in the formulation allow the proposed model to analyse plates reinforced with rectangular or skew bar meshes, or general rebar arrangement. For the design at the service and ultimate states, as load increases, displacements, stresses and strains in concrete and steel, crack patterns and crack opening widths can be predicted and compared with maximum values requested by codes (Belletti et al. 2002, Belletti 2002).

The validity of the proposed method has been verified by comparing the results of numerical analyses with those attained from significant experimental test programs available in literature (Cardenas & Sozen 1968) and with theoretical results obtained by some simplified methods frequently used in the design procedures.

2 STIFFNESS MATRIX OF A R/C PLATE ELEMENT

A concrete plate element, (see Figure 1), having unitary dimensions in x–y plane and thickness t, and reinforced by a generic order of steel bars, is examined. The element is subdivided into n number of reinforced concrete layers through thickness t. In this approach each concrete layer is assumed to be subjected to plane stress and the actual stress distribution along the cross section is modelled by a piecewise constant approximation. In addition, it has been hypothesised that the transverse shear stresses do not affect the inplane biaxial behaviour of concrete. The geometrical characterisation of the cross-section and of the reinforcement, as well as the mechanical properties are supposed to be given and the internal forces:

$$\{Q\} = \{\{Q_1\}\ \{Q_2\}\}^T = \{\{n_x\ n_y\ n_{xy}\}\{m_x\ m_y\ m_{xy}\}\}^T,$$

must be referred to the middle plane of cross-section, (Figure 1). The kinematic model, (Figure 1), is based on a linear distribution of the normal strains ε_x, ε_y and of the shear strain γ_{xy}, while the strains ε_z, γ_{xz}, γ_{yz} have been neglected. As a result, the strain field of the i-th layer can be expressed by the generalised deformation parameters $\{d\}$ of the section, that is:

$$\{\varepsilon_{xy}^{(i)}\} = \begin{Bmatrix} \varepsilon_x^{(i)} \\ \varepsilon_y^{(i)} \\ \gamma_{xy}^{(i)} \end{Bmatrix} = \begin{Bmatrix} \varepsilon_{x0} \\ \varepsilon_{y0} \\ \gamma_{xy0} \end{Bmatrix} + z^{(i)} \begin{Bmatrix} \chi_x \\ \chi_y \\ \chi_{xy} \end{Bmatrix} =$$

$$= \left[[I]\ z^{(i)}[I]\right] \begin{Bmatrix} \{d_1\} \\ \{d_2\} \end{Bmatrix} = [G]\{d\}$$

where:

$$\{d\} = \{\{d_1\}\ \{d_2\}\}^T = \{\{\varepsilon_{x0}\ \varepsilon_{y0}\ \gamma_{xy0}\}\ \{\chi_x\ \chi_y\ \chi_{xy}\}\}^T$$

are the deformation parameters;

$$\{d_2\} = \{\chi_x\ \chi_y\ \chi_{xy}\}^T = \left\{-\frac{\partial^2 w}{\partial x^2}\ \ -\frac{\partial^2 w}{\partial y^2}\ \ -2\frac{\partial^2 w}{\partial x \partial x}\right\}^T$$

are the curvatures, and $[I]$ is the identity matrix.

The equivalence conditions in the middle plane of cross-section between the internal forces $\{Q\}$ and the stresses generated by $\{d\}$ are imposed. Then, taking into account the nonlinear local behavioural laws

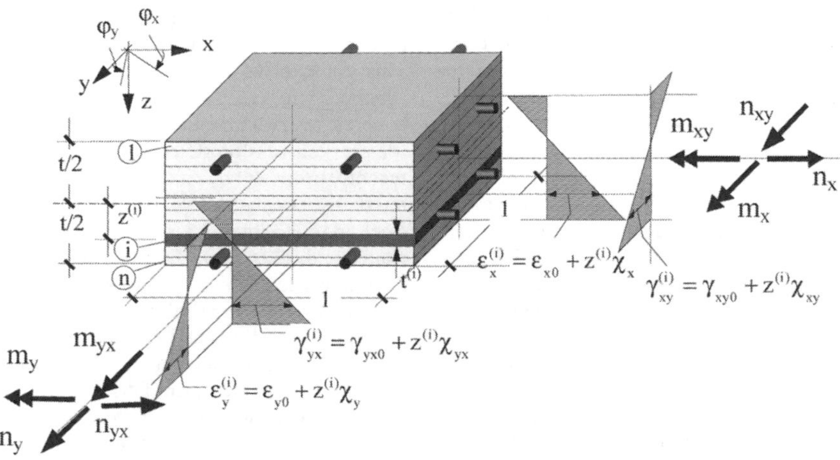

Figure 1. Outline of the examined plate element: subdivision in layers and assumed kinematic model (strains ε_x, ε_y, γ_{xy} and acting internal forces n_x, n_y, n_{xy}, m_x, m_y, m_{xy}).

both for uncracked and cracked reinforced concrete, we have:

$$\{Q\}=\begin{Bmatrix}\{Q_1\}\\ \{Q_2\}\end{Bmatrix}=\begin{Bmatrix}\begin{Bmatrix}n_x\\ n_y\\ n_{xy}\end{Bmatrix}\\ \begin{Bmatrix}m_x\\ m_y\\ m_{xy}\end{Bmatrix}\end{Bmatrix}=\sum_{i=1}^{n}t^{(i)}\begin{Bmatrix}\begin{Bmatrix}\sigma_x^{(i)}\\ \sigma_y^{(i)}\\ \tau_{xy}^{(i)}\end{Bmatrix}\\ z^{(i)}\begin{Bmatrix}\sigma_x^{(i)}\\ \sigma_y^{(i)}\\ \tau_{xy}^{(i)}\end{Bmatrix}\end{Bmatrix}=$$

$$=\sum_{i=1}^{n}t^{(i)}\begin{bmatrix}\left[D_{xy}^{(i)}\right] & z^{(i)}\left[D_{xy}^{(i)}\right]\\ z^{(i)}\left[D_{xy}^{(i)}\right] & z^{(i)2}\left[D_{xy}^{(i)}\right]\end{bmatrix}\begin{Bmatrix}\{d_1\}\\ \{d_2\}\end{Bmatrix}=$$

$$=\sum_{i=1}^{n}t^{(i)}\begin{bmatrix}k_{11} & k_{12}\\ k_{21} & k_{22}\end{bmatrix}\begin{Bmatrix}\{d_1\}\\ \{d_2\}\end{Bmatrix}=[k]\{d\},$$

where [k] is the stiffness matrix of the plate element and:

$$\left\{\sigma_{xy}^{(i)}\right\}=\begin{Bmatrix}\sigma_x^{(i)}\\ \sigma_y^{(i)}\\ \tau_{xy}^{(i)}\end{Bmatrix}=\left[D_{xy}^{(i)}\right]\begin{Bmatrix}\varepsilon_x^{(i)}\\ \varepsilon_y^{(i)}\\ \gamma_{xy}^{(i)}\end{Bmatrix}.$$

$$\left[D_{x,y}^{(i)}\right]=\left[T_\varepsilon^{(i)}\right]^T\left[D_{1,2}^{(i)}\right]\left[T_\varepsilon^{(i)}\right],$$

being $[D_{1,2}^{(i)}]$ evaluated, in accordance with symmetry orthotropic axes 1–2, by two different behavioral models (uncracked, Figure 2 and cracked reinforced concrete, Figure 3; in Figures 2, 3 apex$^{(i)}$ which indicates stiffness matrix of i-th layer is omitted).

Uncracked reinforced concrete

Concrete is assumed to be subjected to a biaxial state of stress, which is described by means of two equivalent uniaxial states following the approach proposed

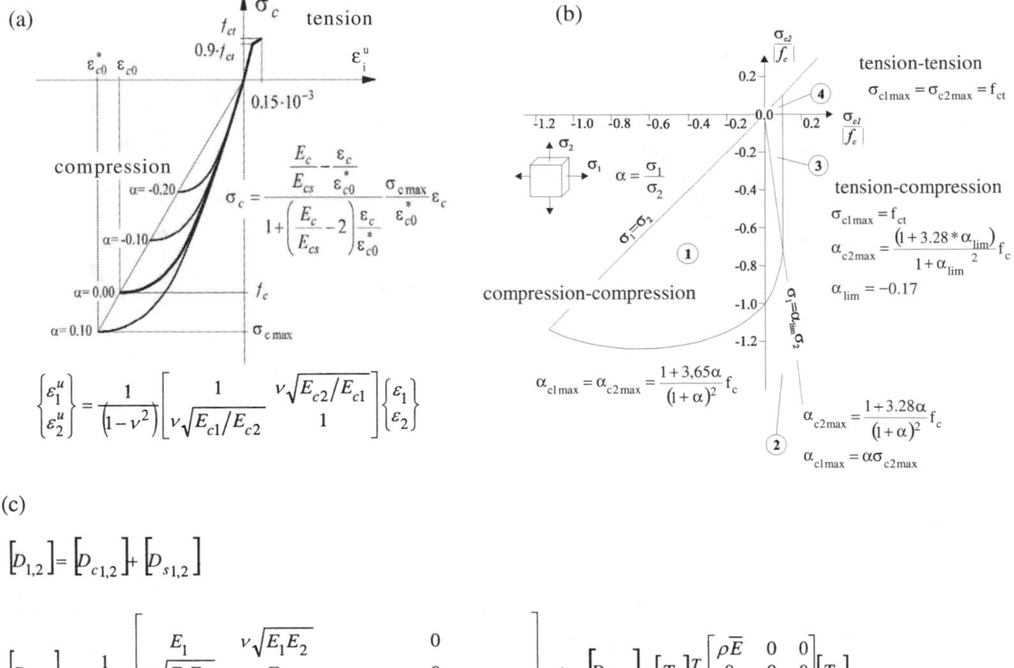

(c)

$$\left[D_{1,2}\right]=\left[D_{c1,2}\right]+\left[D_{s1,2}\right]$$

$$\left[D_{c1,2}\right]=\frac{1}{1-\nu^2}\begin{bmatrix}E_1 & \nu\sqrt{E_1E_2} & 0\\ \nu\sqrt{E_1E_2} & E_2 & 0\\ 0 & 0 & \frac{1}{4}\left(E_1+E_2-2\nu\sqrt{E_1E_2}\right)\end{bmatrix}\quad;\quad\left[D_{s1,2}\right]=\left[T_s\right]^T\begin{bmatrix}\rho\bar{E} & 0 & 0\\ 0 & 0 & 0\\ 0 & 0 & 0\end{bmatrix}\left[T_s\right]$$

Figure 2. Stress-strain relationships for uncracked concrete: (a) equivalent uniaxial curves; (b) strength domain; and (c) stiffness matrix (omitting apex$^{(i)}$ which indicates stiffness matrix of i-th layer).

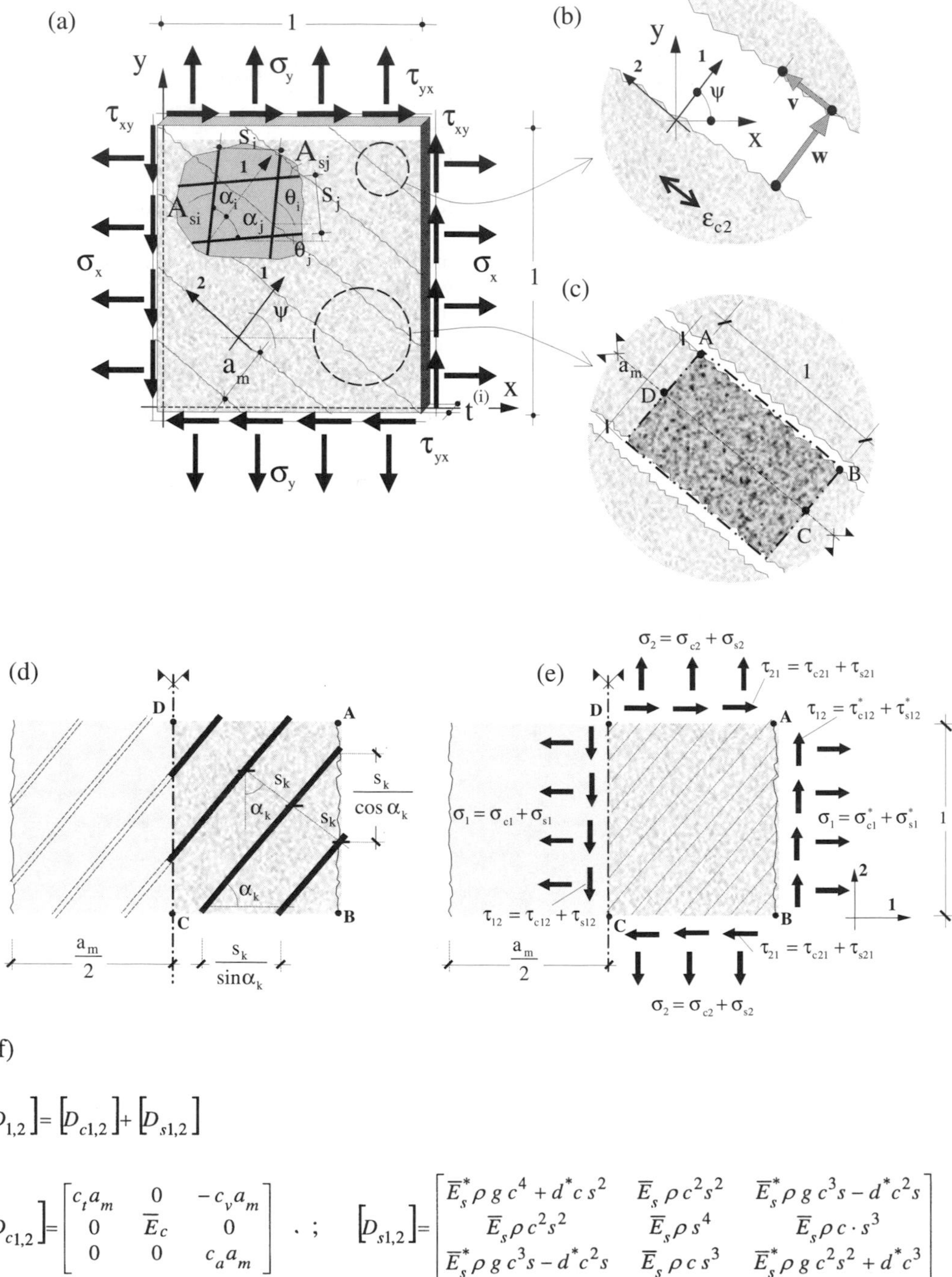

Figure 3. Cracked concrete: (a) reinforced-concrete membrane element subjected to plane stresses; (b) the quantities governing the local mechanical and kinematic behaviour; (c) basic geometrical element; (d) spacing of reinforcing bars and smearing of the steel bars; (e) stress field in the basic geometrical element; and (f) stiffness matrix of the cracked reinforced-concrete membrane element (omitting apex$^{(i)}$ which indicates stiffness matrix of i-th layer).

by Darwin & Pecknold (1977). If 1–2 denote the principal strain directions, the ultimate strength values, σ_{c1max} and σ_{c2max}, of the "equivalent" uniaxial curves, are determined through the analytical biaxial strength envelope suggested by Kupfer et al. (1969), (Figure 2b). The equivalent uniaxial curves in compression are based on the Sargin stress-strain law, while concrete in tension is modelled as a bilinear elastic brittle material (Figure 2a). Once the uniaxial stress-strain curve for the current strain has been determined, the secant moduli E_1 and E_2 are evaluated and the concrete stiffness matrix $[D_{c1,2}]$ takes the form reported in Figure 2c. Stiffness matrix $[D_{1,2}]$ of uncracked layer is obtained by adding steel contribution $[D_{s1,2}]$ to concrete matrix $[D_{c1,2}]$, Figure 2c.

Cracked reinforced concrete

When the principal tensile stress becomes greater than concrete tensile strength, the crack forms and the material stiffness matrix have to be formulated accordingly (Figure 3, Belletti et al. 2001). The material's behaviour is still considered orthotropic, with the orthotropic axes coincident with the directions at right angles (1) and along (2) to the crack surface (Figure 3b). Imposing the equilibrium and compatibility conditions, combined with the constitutive laws and the interaction relationships for the mechanically characteristic unit (called "basic element") of the composite material (Figure 3c), the stiffness matrix is determined (Figure 3f). The strain field is expressed as a function of three fundamental variables (Figure 3b): the crack width w, the crack slip v and the concrete axial strain ε_{c2} in the direction parallel to the crack (neglecting the other two concrete strains ε_{c1} and γ_{c12}). In the coordinate system of the orthotropic axes, the strain vector is given by:

$$\{\varepsilon_{1,2}\} = \{\varepsilon_1 \ \varepsilon_2 \ \gamma_{12}\}^T = \left\{\frac{w}{a_m} \ \varepsilon_{c2} \ \frac{v}{a_m}\right\}^T,$$

a_m being the crack spacing, which is assumed constant and related to bar spacing (Leonhardt & Schelling 1974, Iori & Dei Poli 1985) or to maximum aggregate size (plain concrete).

The stress field in the co-ordinate system 1–2 results from the superposition of the concrete stresses and of the equivalent steel stresses, these latter developed by the smeared reinforcement (Figure 3d):

$$\{\sigma_{1,2}\} = \{\sigma_1 \ \sigma_2 \ \tau_{12}\}^T,$$

where σ_1, σ_2 are the normal stresses referred to principal axes 1–2, respectively, while τ_{12} represents the shear stress along the crack (Figure 3e). The stiffness matrix for the cracked reinforced concrete $[D_{1,2}]$ (Figure 3f), takes into account all the fundamental resistant mechanisms concerning cracked reinforced concrete strength, such as tension stiffening (coefficient g), dowel action (coefficient d), bridging effect (coefficient c_t), confinement and aggregate interlock (coefficients c_v and c_a). The constitutive relationship adopted for steel close and far from a crack (with the secant moduli $\bar{E}_s$ and E_S, respectively) is the well-known bilateral elastic-hardening curve, while for concrete the softened curve has been implemented.

In case of free deformations of the middle plane of the plate in the x–y plane, the internal forces $\{Q_1\}$ are equal to zero, and then from the first sub-set of equations we can write:

$$\{d_1\} = -[k_{11}]^{-1}[k_{12}]\{d_2\}$$

which substituted into the second sub-set of the fundamental equations leads to:

$$\{Q_2\} = \left([k_{22}] - [k_{21}][k_{11}]^{-1}[k_{12}]\right)\{d_2\} = [k_f]\{d_2\}$$

where $[k_f]$ is the stiffness matrix relative to flexural and torsional behaviour of the plate element.

3 IMPLEMENTATION OF THE PROPOSED MODEL AND SOLUTION PROCEDURE

The proposed model has been implemented into a computer program capable of evaluating the behavior (in terms of stress and strain fields, crack width, etc., for each layer) up to failure of a plate element subjected to a general combination of internal forces $\{Q_2\}$. The numerical solution of the nonlinear problem is obtained through an incremental-iterative procedure, whose flow-chart is shown in Figure 4. Starting from the stiffness matrix of the plate element, corresponding to the strain field obtained from the previous load step (assuming linear elasticity only in the first load step), the procedure computes the vector of curvatures $\{d_2\}$

Then, after having found the middle-plane membrane deformations $\{d_1\}$, the global and local strains $\{\varepsilon_{xy}^{(i)}\}$ and $\{\varepsilon_{1,2}^{(i)}\}$ are determined, and subsequently a check on the possible violation of material's ultimate values is carried out. If maximum strains are not reached, the stress field is calculated through the constitutive laws, and then the equilibrated internal generalized forces $\{Q_2\}_{eq}$ are evaluated. These latter are compared with the value of the acting generalized internal forces $\{Q_2\}$: if their difference $\{Q_2\} - \{Q_2\}_{eq}$ is smaller than a prefixed tolerance (here indicated with e), the check is positive, and the convergence is considered reached. Otherwise, starting from the updated stiffness matrix, computed again on the basis of the field strain relative to the previously step (current field

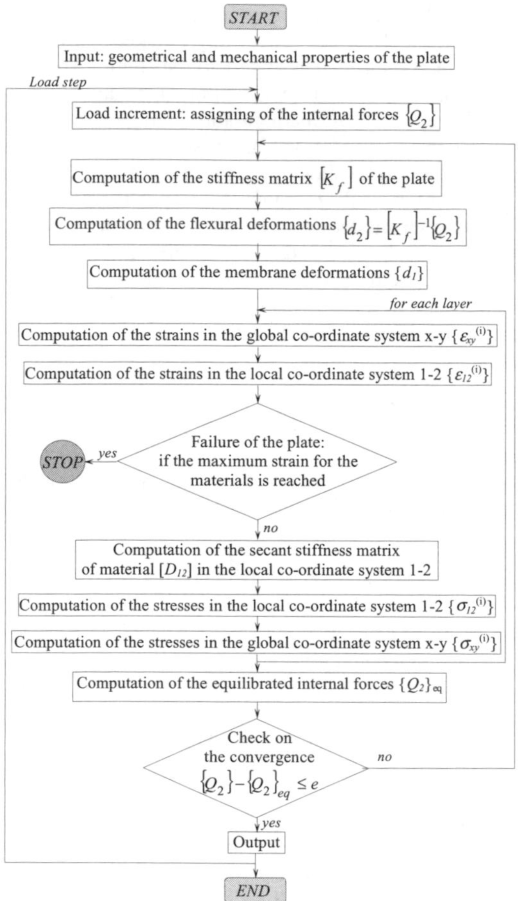

Figure 4. Flow chart of numerical procedure.

strain), the above-mentioned operations are repeated until convergence is reached.

4 COMPARISONS BETWEEN EXPERIMENTAL AND NUMERICAL RESULTS

In order to judge the capability of the proposed non linear procedure, some experimental results of slabs, subjected to uniform bending moment and tested by Cardenas & Sozen (1968), are compared with those numerically attained. These slabs, circular as well as rectangular, characterized by different reinforcement ratios and different orientations of reinforcement with respect to principal moment axes, have represented a test for several theoretical models proposed by other authors (Hu & Schnobrich 1991, Hand et al. 1973).

Only the rectangular slabs are taken into account in this paper; among them the B7, B10, B11, B12 slabs, subjected to uniaxial moment, the B15, B16, B17 slabs, subjected to pure torsion, and the B27, B28, B33 slabs, subjected to combined bending and torsion, are considered.

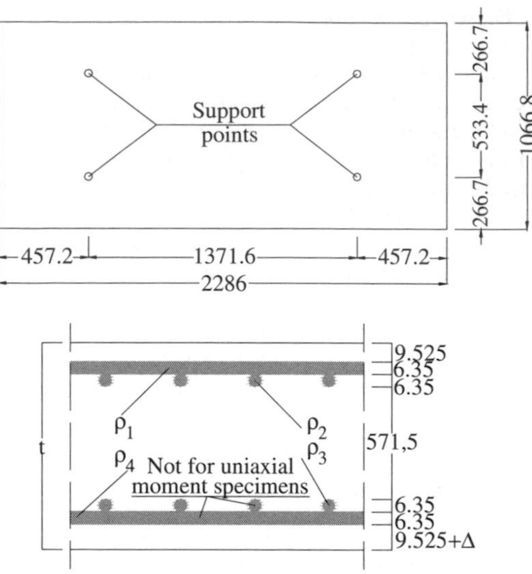

Figure 5. Geometrical features of slabs (Cardenas & Sozen 1968).

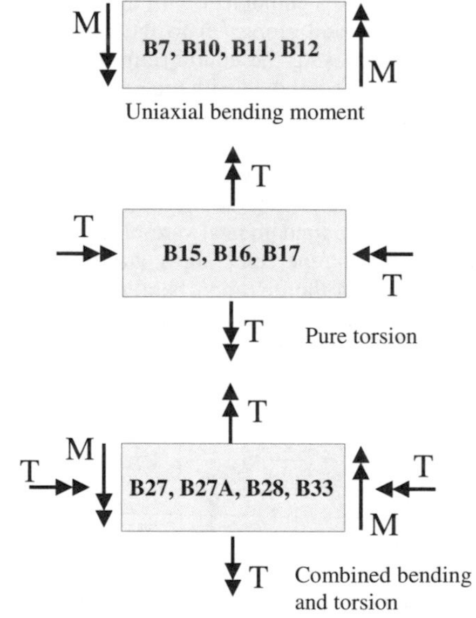

Figure 6. Loading of slabs (Cardenas & Sozen 1968).

Geometrical features of slabs are reported in Figure 5, while applied loads are shown in Figure 6. Table 1 resumes material properties and loading of specimens.

Figures 7, 8 show comparisons between experimental and numerical results in terms of principal unit moment versus principal curvature.

Table 1. Geometrical and mechanical properties of slabs (Cardenas & Sozen 1968).

	Thickness		Concrete	Reinforcement					Loads
Slab	t (mm)	Δ (mm)	fc (MPa)	fy (MPa)	ρ_1 and ρ_4	ρ_2 and ρ_3	ϕ_1 and ϕ_4 (deg)	ϕ_2 and ϕ_3 (deg)	Ratio of T/M
B7	105.1	3.5	35.5	344.8	0.00790	0.00862	135.0	45.0	0
B10	105.1	3.5	33.9	344.8	0.00810	0.00810	0.0	90.0	0
B11	104.6	3.0	33.1	344.8	0.00794	0.00433	157.5	67.5	0
B12	104.6	3.0	35.6	328.3	0.00794	0.00433	67.5	157.5	0
B15	103.8	2.2	36.2	330.3	0.00800	0.00873	135.0	45.0	∞
B16	102.6	1.0	32.6	333.1	0.00810	0.00884	90.0	0.0	∞
B17	102.3	0.7	38.1	350.3	0.00812	0.00886	157.5	67.5	∞
B27	103.1	1.5	36.9	311.7	0.00806	0.00879	135.0	45.0	0.45
B28	103.6	2.0	38.7	328.3	0.00802	0.00875	157.5	67.5	0.45
B33	103.3	1.7	34.0	316.5	0.00804	0.00219	0.0	90.0	0.45

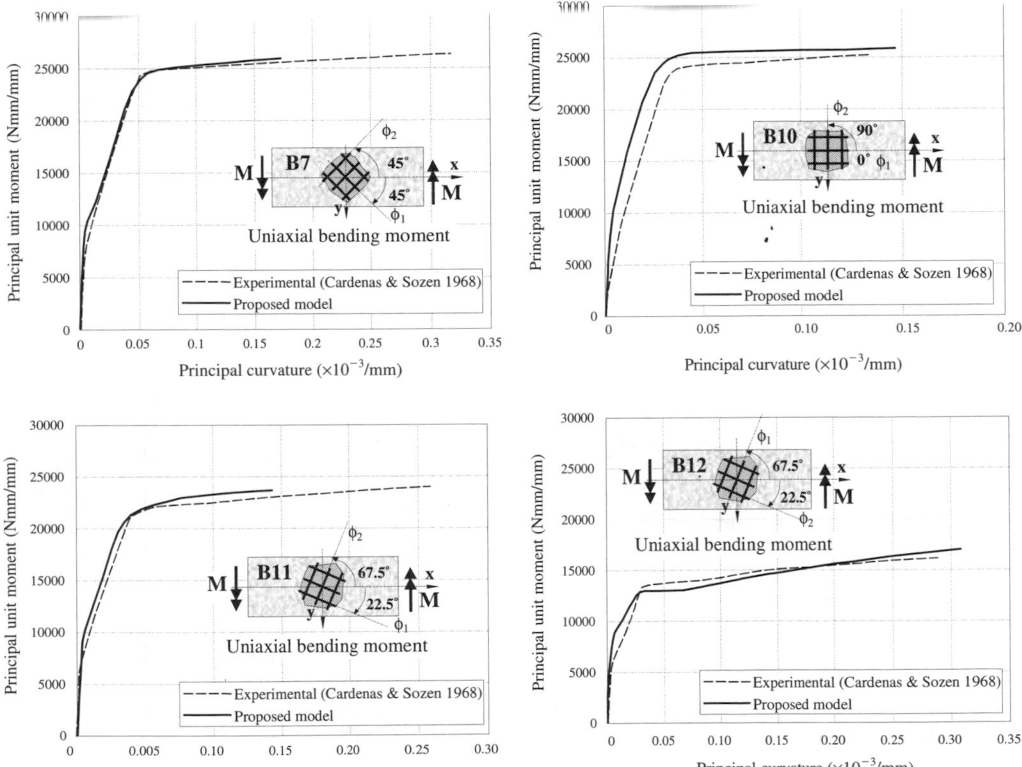

Figure 7. Comparison between experimental (Cardenas & Sozen 1968) and numerical curves of slabs subjected to uniaxial moment: principal moment versus principal curvatures.

The simple case of uniaxial bending moment (Figure 7) makes it possible to understand how reinforcement ratios and steel arrangement affect ultimate slab capacity.

Ultimate moment of B7 specimen, reinforced with almost symmetric reinforcement ratios ρ_1 and ρ_2 inclined of ± 45 degrees with respect to principal directions, is lightly greater than that of B10 characterized

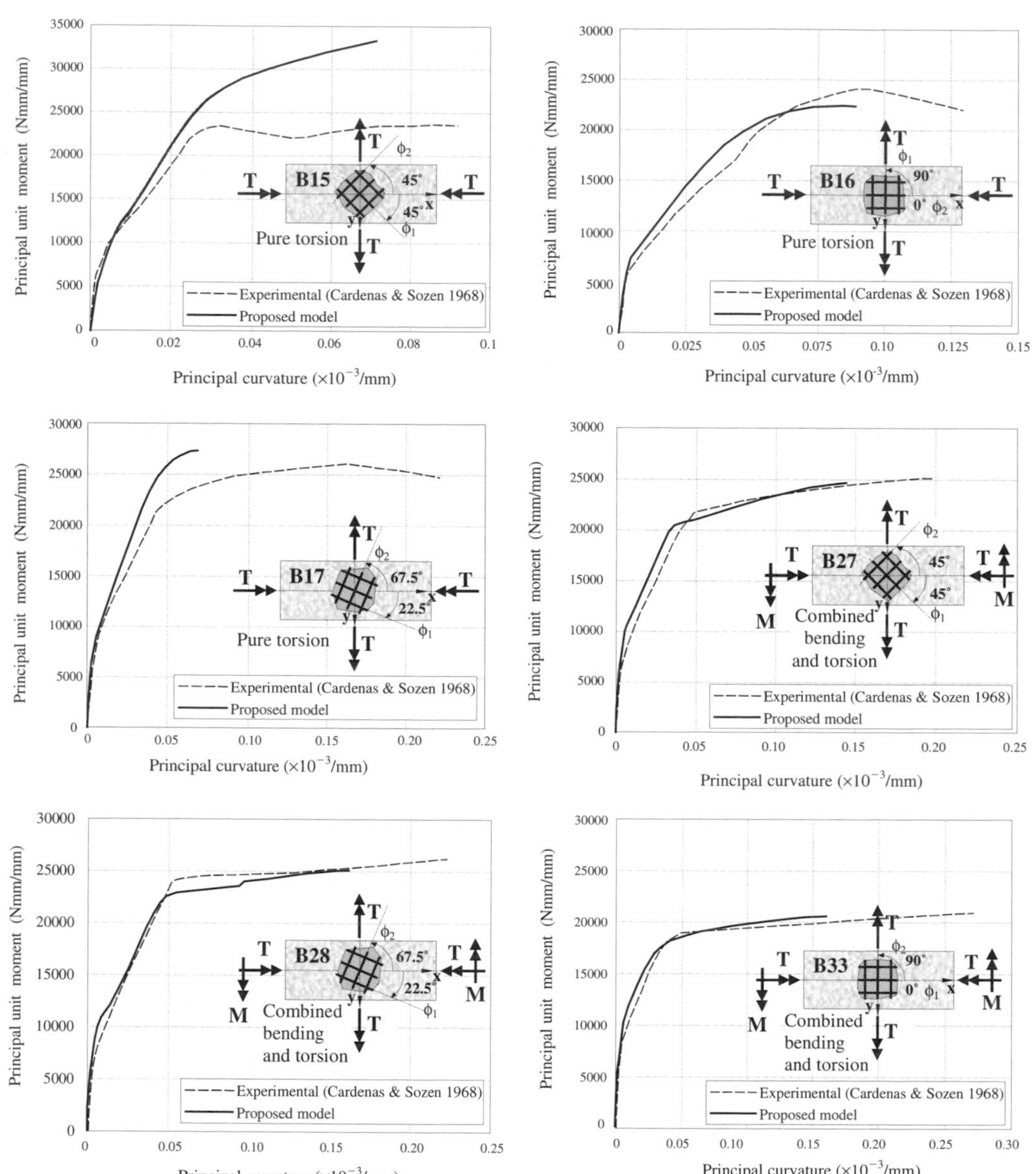

Figure 8. Comparisons between experimental (Cardenas & Sozen 1968) and numerical curves of slabs subjected to pure torsion and combined bending and torsion moment: principal moment versus principal curvatures.

by similar reinforcement ratios but different bar orientation, ρ_1 reinforcement being placed along principal direction.

Owing to reinforcement ratios ρ_2 equal to an half of B7 and B10 reinforcement ratios, ultimate moments of B11 and B12 slabs are lower than those of B7 and B10. Furthermore B11 ultimate bending moment is greater than that of B12 because of different angle between maximum principal direction of load and bar orientation ϕ_1, equal to 22.5 and 67.5 degrees, respectively.

5 DESIGN REMARKS

After showing the good agreement between theoretical predictions and experimental observations, numerical results are compared with those obtained by some of the simplified methods used in case of rectangular bar mesh, such as steel arrangement of slabs tested by Cardenas and Sozen.

The Leitz (1923) and Baumann (1972) approaches are adopted for these comparisons. Eurocode prescriptions make use of Leitz or Wood (1968) methods, which are simple equilibrium-based approaches which, starting from different hypotheses, reach the same issues. Cracks are at right angles to bars and reinforcement is designed by Leitz and Wood methods (Figure 9a) in order to guarantee a bending capacity of reinforcement greater than or equal to the normal moment acting on the least favoured plane. Comparisons between experimental results and those obtained from Leitz's method could be very interesting above all to evaluate the accuracy of current design procedures and the increase in bending capacity provided by the concrete and its interaction with steel. Baumann's approach takes into account compatibility of bar strains but neglects strain in concrete and crack slip unlike PARC model; cracks form at an angle which depends on bar directions (Figure 9b), reinforcement ratios and principal moment ratios. Unlike some theoretical model available in scientific literature, Leitz and Baumann methods consider only the two cases of fully-elastic or fully-yielded steel, so in order to provide useful information for the design in terms of yielding moment definition, only the first case has been taken into account, until yielding is reached by one of the steel layers which constitute slab reinforcement. In order to follow the same hypothesis adopted by these approaches, bar resistant moment is evaluated assuming the internal lever arm $d = t - c_m$ equal both for $\rho_1 - \rho_4$ and $\rho_2 - \rho_3$ reinforcements, being c_m the medium value of top and bottom covers. Table 2 summarises the results of this comparison in terms of coefficient R.

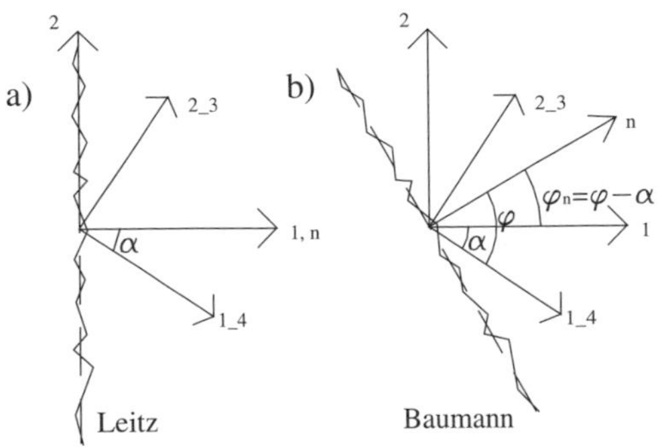

Figure 9. Leitz a) and Baumann b) crack patterns.

Adopting Leitz's method, the coefficient R will be chosen between R_{1-4} and R_{2-3}, evaluated as follows:

$$\frac{1}{R_{1-4}} = \frac{m_{1-4}}{m_1} = \cos^2\alpha + k\sin^2\alpha + \frac{(1-k)}{2}\sin(2\alpha)$$

$$\frac{1}{R_{2-3}} = \frac{m_{2-3}}{m_1} = \sin^2\alpha + k\cos^2\alpha - \frac{(1-k)}{2}\sin(2\alpha)$$

α being the angle between 1–4 and 1 directions (see Figure 9a), m_{1-4} and m_{2-3} bar resistant moments at yielding, $k = m_2/m_1$, m_1 and m_2 principal moments.

Through Baumann's method R_{1-4} and R_{2-3} coefficients assume the following expression:

$$\frac{1}{R_{1-4}} = \frac{(1+\tan\varphi\cdot\tan\alpha) + k(1-\tan\varphi/\tan\alpha)\tan^2\alpha}{1+\tan^2\alpha}$$

$$\frac{1}{R_{2-3}} = \frac{(1+\tan\varphi\cdot\tan\alpha) - k(1-\tan\varphi/\tan\alpha)}{1+\tan^2\alpha}\cdot\frac{\tan\alpha}{\tan\varphi}$$

ϕ being the root of the following equation nearest to α value:

$$\tan^4\varphi + \frac{\tan^{-1}\alpha + k\tan\alpha}{1-k}\tan^3\varphi - \\ - \frac{\tan\alpha + k\tan^{-1}\alpha}{\lambda(1-k)}\tan\varphi - \frac{1}{\lambda} = 0$$

Table 2. Comparison in terms of coefficient R between experimental and theoretical results.

	$\frac{\bar{m}_{1-4}}{\rho_{1-4}\cdot t\cdot f_y\cdot 0.9\cdot d}$	$\frac{\bar{m}_{2-3}}{\rho_{2-3}\cdot t\cdot f_y\cdot 0.9\cdot d}$	α	$k = \frac{m_2}{m_1}$	R Leitz	R Baumann	R PARC	R Experimental
B7	23839	26012	45	0	1	1.014	1.06	1.04
B10	24442	24442	0	0	1	1	1.10	1.0
B11	23713	12931	22.5	0	0.828	0.85	0.92	0.84
B12	22575	12311	67.5	0	0.828	0.935	1.10	1.01
B15	22511	24282	0	−1	1	1	1.25	1.03
B16	22281	24317	45	−1	1	1.02	1.00	1.08
B17	23401	25534	22.5	−1	0.707	0.868	1.13	1.04
B27	20910	22804	46.5	−0.148	1.03	1.03	1.01	1.06
B28	22082	24092	43.5	−0.148	0.97	0.99	1.06	1.08
B33	21264	5792	21	−0.148	2.60	1.5	3.46	3.03

with $\lambda = A_{2-3}/A_{1-4}$ and A_{1-4}, A_{2-3} are the areas steel bars.

The minimum value of R coefficient should come from Leitz's method, which doesn't take into account the increase in stiffness provided by aggregate interlock, dowel action, concrete, etc. For this reason the experimental R coefficients are greater than Leitz's previsions for all analysed slabs, while Baumann and PARC models provide R coefficients more similar to the experimental values.

6 CONCLUSIONS

A theoretical model which assumes a subdivision of thickness into any number of reinforced concrete layers has been adopted for the analysis of reinforced concrete plates subjected to membrane loads and bending and torsion moments. Each layer behaves as a membrane element which can be analysed by PARC model, a recent constitutive model for reinforced concrete subjected to plane stress proposed by the Authors.

In order to check the capability of the numerical procedure of predicting the behaviour of reinforced concrete slabs, numerical results are compared with well-documented experimental tests reported in scientific literature. The good fitting proves that the proposed model could be a useful tool for the design of this structural typology. In order to give some design remarks, the numerical yielding moment evaluated by the proposed model has been compared with the yielding capacity computed by other simplified approaches. The next step of this research will be the implementation of the stiffness matrix proposed, which ties generalized stresses to generalized strains, into finite element programs. The aim is to provide a useful tool for the design of whole structures which can be modelled by the assembly of reinforced concrete plates.

REFERENCES

Baumann, T. 1972. Zur Frage der Netzbewehrung von Flächentragwerken. *Der Bauingenieur*, 47(10).

Belletti, B. 2002. Analisi non lineare ad elementi finiti di tegoli di copertura in conglomerato armato precompresso. Proceedings of *XIV CTE Congress*, Mantova 7–9 November 2002, Italy.

Belletti, B., Cerioni, R. & Iori, I. 2001. A Physical Approach for Reinforced Concrete (PARC) Membrane Elements. In *Journal of Structural Engineering*, ASCE, Vol.127(12), 1412–1426.

Belletti, B., Cerioni, R. & Iori, I. 2002. Theoretical and Experimental Analyses of Precast Prestressed Concrete Roof Elements for Large Span. *Proceeding of First fib Congress 2002*, 13–19 October 2002, Japan.

Cardenas, A. & Sozen, M.A. 1968. Strength and behaviour of isotropically and nonisotropically reinforced concrete slabs subjected to combinations of flexural and torsional moments. *Civil Engineers Studies*, SRS 336, University of Illinois, Urbana, IL.

Darwin, D. & Pecknold, D.A. 1977. Non-linear biaxial stress-strain law for concrete. *Journal of the Engineering Mechanics Division*, Proceedings of the American Society of Civil Engineers, Vol.103, no.EM2, pp.229–241.

Gambarova, P.G. & Dei Poli, S. 1978. Influenza delle caratteristiche di armatura sulla fessurazione di piastre di c.a. con armatura a maglia quadrangolare. *Studi e Ricerche*, Vol. 14–77.

Hand, F.R., Pecknold, D. A. & Schonobrich, W.C. 1973. A layered finite element nonlinear analysis of reinforced concrete plates and shells. *Journal of Structural Division*, Proceedings of the American Society of Civil Engineers, Vol. 99, no. ST7, 1491–1505.

Hu, T.H. & Schnobrich, W.C. 1991. Nonlinear finite element analysis of reinforced concrete plates and shells under monotonic loading. *Computers and Structures*, Vol. 38(5/6), 637–651.

Iori, I. & Dei Poli, S. 1985. Sulla torsione di elementi strutturali in cemento armato. *L'Industria Italiana del Cemento*, No.2, 121–129.

Kupfer, H., Hilsdorf, H.K. & Rusch, H. 1969. Behavior of Concrete Under Biaxial Stresses. *Proceedings of ACI Journal*, Vol. 66(8), 656–666.

Leitz, H. 1923. Eisenbewehrte Platten bei allgemeinem Biegungszustand. *Die Bautechnik*, Helf 16.

Leonhardt, F. & Schelling, G. 1974. Torsionsversuche an Stahlbetonbalken. *D.A.f. St.*, Heft 239.

Wood, R.H. 1968. The reinforcement of slabs in accordance with a predetermined field of moment. Concrete, no. 2.

Computational Modelling of Concrete Structures, Bićanić et al. (eds)
© 2003 Swets & Zeitlinger, Lisse, ISBN 90 5809 536 3

On the behaviour of R/C beams without shear reinforcement

B. Belletti, P. Bernardi, R. Cerioni & I. Iori
Department of Civil and Environmental Engineering and Architecture, University of Parma, Italy

ABSTRACT: The behaviour of beams without shear reinforcement has been analysed through a non-linear finite element procedure, in which a constitutive model recently proposed by the Authors, (PARC), has been implemented into a FE program. The stiffness matrix has been evaluated assuming non-linear orthotropic relationships for plain and reinforced concrete, subjected to plane stresses. The main contributions to shear strength after cracking, such as dowel action, aggregate interlock, bridging effect, compression of concrete struts, have been taken into account. The plane stress analysis has been conducted adopting a mesh of quadratic membrane elements, representing the middle plane of the section. The proposed formulation has been verified by performing comparisons with well-documented experimental tests reported in scientific literature. The non-linear analysis has been able to accurately represent the response of the structures to the applied loads, and to investigate the influence of each stiffening contribution on the shear resistance evaluated at a given cracked vertical section.

1 INTRODUCTION

The behaviour of beams containing little or no shear reinforcement under bending and shear has been widely discussed (Kani, 1964, Reineck, 1991, Kim & White 1991, Zararis & Papadakis 2001), and a review of the design models and analytical methods for these particular elements has been recently carried out by the ACI-ASCE Committee 445 (1998). The interest in this problem rests on the commonly accepted conviction that the comprehension of the shear failure mechanism in elements without shear reinforcement could improve the knowledge of concrete contribution on shear strength of beams with web reinforcement. Furthermore, it has to be remembered that some important types of structural elements, such as, for example, hollow core slab panels, are cast without web reinforcement and in this case shear capacity is completely left to the concrete.

The research efforts have been focused in the past mostly on the analysis of shear collapse, but, because of its brittle nature, experimental observations have not yet provided enough information about the fundamental shear failure mechanism, despite the large number of laboratory investigations. Also the current design procedures proposed by international codes for the shear design of beams without web reinforcement, show that this problem still remains not satisfactorily solved. In fact, they are based on semi-empirical formulations, usually expressed as functions of the main parameters influencing the failure of these elements, such as the concrete tensile strength, the longitudinal reinforcement ratio, the effective depth of the member, the shear span-to-depth ratio, the axial force or amount of prestressing, the maximum aggregate size, etc. However, considerable differences exist between predictions resulting from the various equations and, moreover, they often supply an assessment of the shear capacity which is quite far from the corresponding value experimentally observed.

In this work, a non-linear finite element procedure for the analysis of beams without web reinforcement is presented, where the non-linear mechanical behaviour has been modelled through the PARC stiffness matrix, recently proposed by Belletti et al. (2001). This matrix has been found for cracked reinforced concrete elements subjected to in-plane forces, and is based on a smeared, fixed-angle crack approach, which means that the crack orientation remains fixed in the direction of first cracking. This modelling allows the computation of the shear stresses along the crack plane and the shear slip that consequently occurs, which is a fundamental feature for a correct valuation of shear strength, above all in beams without shear reinforcement, where most of the elements are constituted by plain concrete. Important mechanisms, such as aggregate interlock, confinement and dowel action, can in this way, be adequately taken into account. On the other hand, rotating-angle models have always revealed some deficiencies in modelling shear-critical beams containing very little or no transverse reinforcement (Vecchio, 2000), and it has become inevitable to pass to a fixed-angle crack formulation (Vecchio et al., 2001, Wang & Hsu, 2001).

The PARC matrix has been implemented into a finite element program (ABAQUS), following a procedure that provided accurate results also for the study of other structural elements (Belletti et al. 2002). The NLFEA was able to supply the stress and strain fields, as well as the pattern and width of cracks, useful for the design at the service and ultimate states. The reliability and capability of the proposed method have been verified by comparing NLFEA results with those of experimental tests carried out on beams without web reinforcement by Leonhardt & Walther, 1962.

In the considered set of beams the influence of beam sizes on shear strength was studied, this being one of the most discussed topical subjects in technical literature (Kani, 1967, Walraven & Lewalther, 1994). The comparisons have regarded load-deformation response, calculation of strength and failure mode, investigated through the examination of crack patterns corresponding to failure load. The proposed NLFEA has been able to describe with effectiveness the real behaviour of the analysed beams in all their parts, and to provide also further insights about the influence of each shear contribution (dowel action, aggregate interlock, bridging effect, concrete in tension and compression) to shear strength evaluated in a given cracked section. At the end, comparisons have been carried out by computing the ultimate shear force of the same beams through the semi-empirical equation proposed by EC2.

2 PARC CONSTITUTIVE MATRIX

A constitutive model recently proposed by the Authors and called PARC (Physical approach for reinforced concrete), is adopted as material constitutive matrix for plain and reinforced concrete subjected to plane stresses. This stiffness matrix is evaluated by assuming a non-linear orthotropic behaviour both in the uncracked and in the cracked stages, evaluating the stress field in each given point as the sum of the stresses in the concrete and those in the steel. More detailed, for uncracked concrete the 1, 2 orthotropic axes are assumed to be coincident with principal direction of stress. The actual biaxial state of stress is duplicated by two equivalent uniaxial curves (Darwin & Pecknold, 1977), Figure 1a, modelled with maximum strength values derived from an analytical biaxial strength envelope based on that suggested by Kupfer et al. (1969), Figure 1b. The orthotropic constitutive matrix in the local 1, 2 coordinate system is thus determined and added, for those elements constituted also by the smeared reinforcement, to the steel stiffness matrix (Fig. 1c), expressed by means of the transformation matrix $[Ts]$, function of the angle between the direction of the longitudinal steel bars and the 1 axis.

When maximum principal stress reaches concrete tensile strength, crack pattern develops, with fixed orientation and constant spacing a_m evaluated by an "a priori" method based on transmission length for concrete + steel layers, and on maximum aggregate size (a_{max}) in the case of plain concrete.

In the cracked reinforced concrete the orthotropic axes 1, 2, are assumed coincident with the stress principal directions of the first cracking, thus turning out, respectively, parallel and perpendicular to fixed cracks, Figure 1d. The variables governing the problem become the opening w and the sliding v of crack surfaces, as well as the normal strain ε_{c2} of the concrete along the crack direction, Figure 1e. All the main phenomena occurring after cracking are directly or indirectly expressed in terms of these variables; particularly, steel and concrete constitutive laws, softening of cracked concrete in compression, bridging effect (c_t), tension stiffening (g), aggregate interlock (c_a, c_v), and dowel action (d). The stiffness matrix in the 1, 2 co-ordinate system is then determined again from the superposition of the matrix containing all the concrete contributions and the steel matrix (for reinforced elements), or it will be given only from the concrete matrix for all the others. The respective expressions are reported in Figure 1g, and the explanation of the single coefficients is widely described in Belletti et al., 2001. From the PARC matrix the stress vector $\{\sigma_{12}\}$ in the local system is thus determined, and, operating with the transformation matrix $[T_\varepsilon]$, function of the angle ψ between the 1 axes and the x axes parallel to element side (Fig. 1d), the stresses in the x, y co-ordinate system are provided:

$$\left\{\sigma_{xy}\right\} = \left\{\sigma_x \; \sigma_y \; \tau_{xy}\right\}^t = [T_\varepsilon]^t \left\{\sigma_1 \; \sigma_2 \; \tau_{12}\right\}^t .$$

3 FINITE ELEMENT IMPLEMENTATION

The described PARC constitutive matrix, has been inserted in the general-purpose ABAQUS program. ABAQUS makes it possible through the definition of a user's sub-routine, including a list of instruction needed to calculate in each integration point the nine terms of material constitutive matrix $[D_{x,y}]$ referred to the local system of the element, which is then necessary to integrate the element's stiffness matrix $[k_e]$ as follows:

$$[k_e] = \int_V [B(x,y)]^t [D_{x,y}][B(x,y)] dV,$$

where $[B(x,y)]$ is geometric matrix of elements. The subsequent operations of numeric integration, assembly and solution of the non-linear system of equations

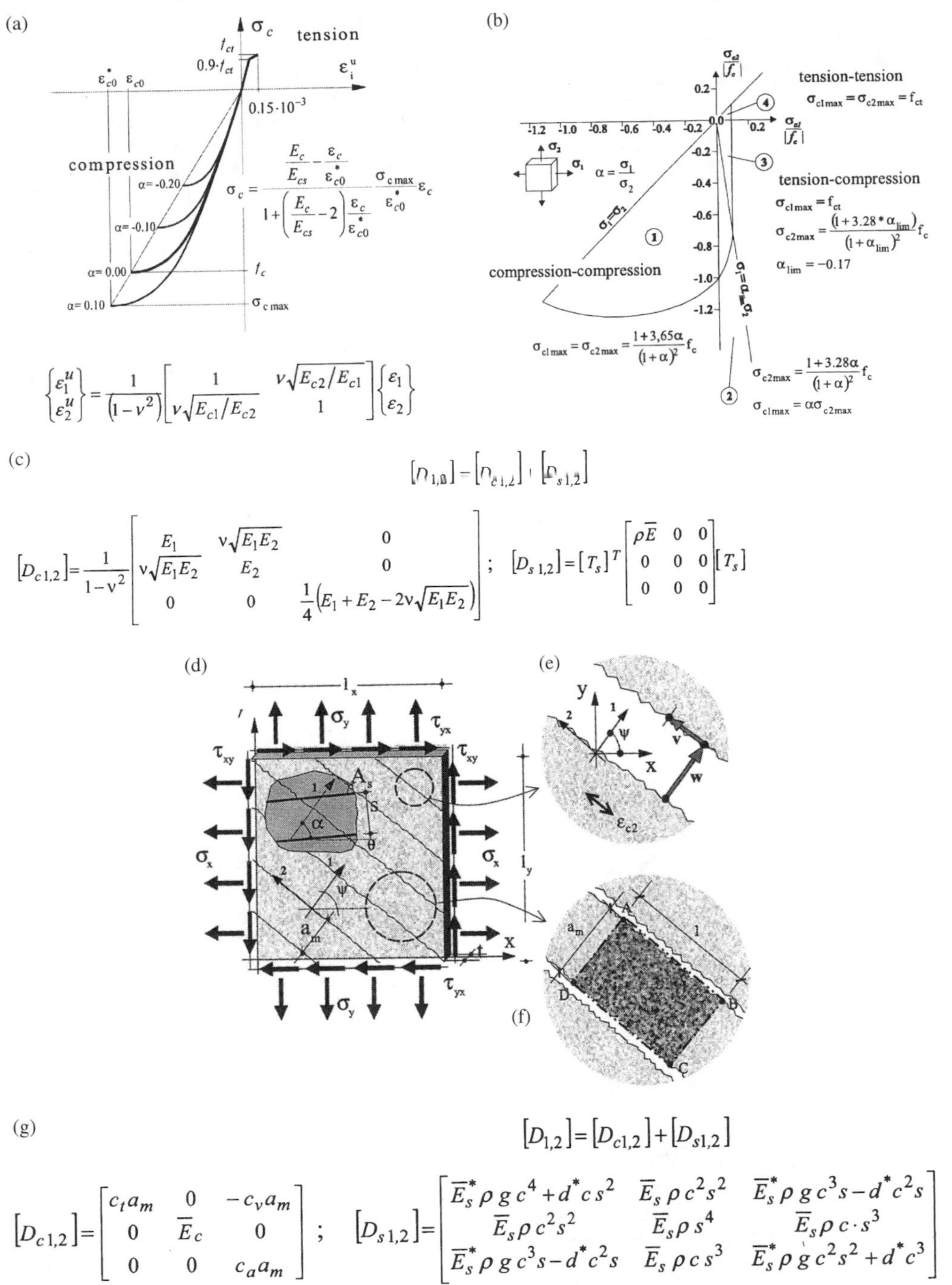

Figure 1. Basic equations ruling PARC constitutive matrix for uncracked concrete (equivalent uniaxial curves (a); analytical biaxial strength envelope (b); stiffness matrix for concrete and steel in local co-ordinate system (c)) and for cracked concrete (reinforced-concrete membrane element subjected to plane stresses (d); fundamental variables governing the problem (e); basic geometrical element (f), and stiffness matrix (g)).

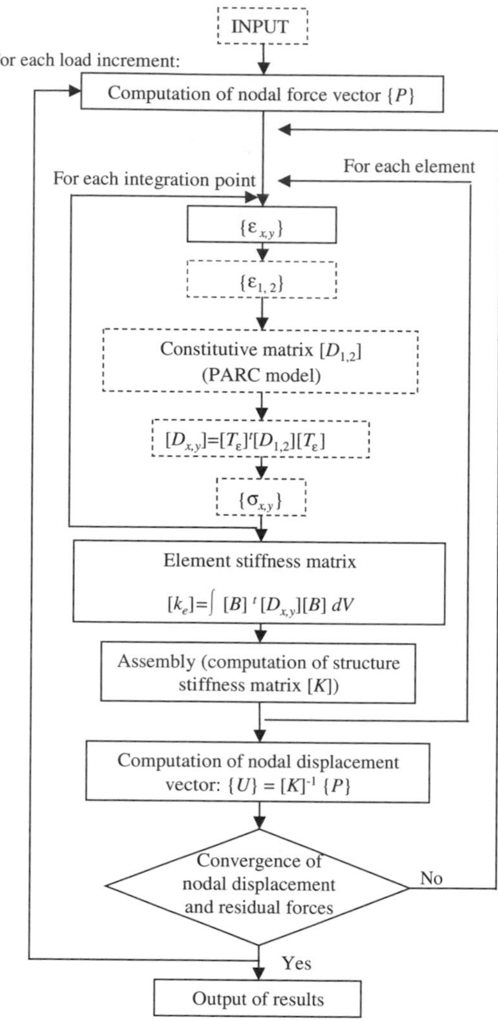

Figure 2. Flow-chart of non-linear finite element procedure.

are executed by ABAQUS. The flow-chart of the procedure is synthesized in Figure 2, where the operations introduced by the user are indicated within broken lines.

The finite element chosen for all the analysis is the quadratic isoparametric membrane element, characterized by 8 nodes and four Gauss points, the minimum required in a reduced integration using Gauss quadrature law.

Reduced integration has been preferred to full, because of the lower computational cost and for the reason that a low-order rule generally tends to soften an element, thus countering the overly stiff behaviour associated with the FEM approach. The choice of this element allows one to perform a plane stress analysis.

The advantages of implementing the constitutive matrix in this FE program regard not only the greater computational capacity, but also the wide possibilities in the display of results.

In fact, ABAQUS gives the user the possibility of verifying not only the stress and strain fields resulting from the static analysis, but also other variables, still chosen and defined by the user, which can help to complete the general understanding of structural behaviour, such as crack opening widths, or stresses and strains in concrete and steel. It is also possible to separate each single contribution constituting the stiffness matrix and follow its development during the whole load history.

Through these potentialities, the user is able to solve very complex problems without excessive requirement of computer time, obtaining the quantities needed for the design.

4 COMPARISON BETWEEN NUMERICAL AND EXPERIMENTAL RESULTS: BEAMS "C" AND "D" (LEONHARDT & WALTHER, 1962).

The NLFEA has been applied to the study of a set of beams without shear reinforcement tested by Leonhardt and Walther in 1962. These beams belong to series originally named "D" and "C", in which the influence of size-effect on shear resistance was examined.

The "D" series was constituted by four beams, characterized by a geometrical similarity as D1 : D2 : D3 : D4 = 1 : 2 : 3 : 4. Each beam was reinforced by two steel bars, having different diameters, varying with the same proportionality law. Also the "C" series was originally constituted by four beams (however, in the current paper only three of them will be analysed), whose geometrical properties were determined with the same law: C1 : C2 : C3 : C4 = 1 : 2 : 3 : 4, except for the bases, standing in a ratio of 1 : 1.5 : 2 : 2.5.

The diameters of the steel bars were equal to 16 mm for all the beams, but their number has been made to vary so as to guarantee a constant geometrical percentage (equal to 1.33%). Their ratio was 1 : 3 : 6 : 9. The geometrical features of all the beams and their steel arrangement are reported in Table 1, together with the mechanical properties of concrete and steel.

All the beams have been tested on two points load, placed symmetrically with respect to midspan (Fig. 3).

The analyses have been performed through the Finite Element approach, requiring the mechanical behaviour to be ruled by the PARC matrix.

The adopted meshes of membrane elements are represented in Figure 3, where it can be observed that only half of each beam has been modelled, taking advantages of the structural symmetry.

Table 1. Geometrical and mechanical properties of the analysed beams (Leonhardt e Walther, 1962).

Beam	L (cm)	a (cm)	d (cm)	b (cm)	A_s	Concrete f_c (MPa)	f_{ct} (MPa)	a_{max} (mm)	Reinforcement f_y (MPa)	ρ_s(%)
D1	52	21	7	5	$2\phi 6$	37.2	3.75	15	468.9	1.71
D2	104	42	14	10	$2\phi 12$	37.2	3.75	15	443.4	1.66
D3	104	63	21	15	$2\phi 18$	38.5	3.8	15	429.1	1.62
D4	208	84	28	20	$2\phi 24$	35.2	3.9	15	456.6	1.67
C1	100	45	15	10	$1\phi 16$	39.0	4.0	30	433.0	1.33
C2	200	90	30	15	$3\phi 16$	39.0	4.0	30	433.0	1.33
C3	300	135	45	20	$6\phi 16$	39.0	4.0	30	433.0	1.33

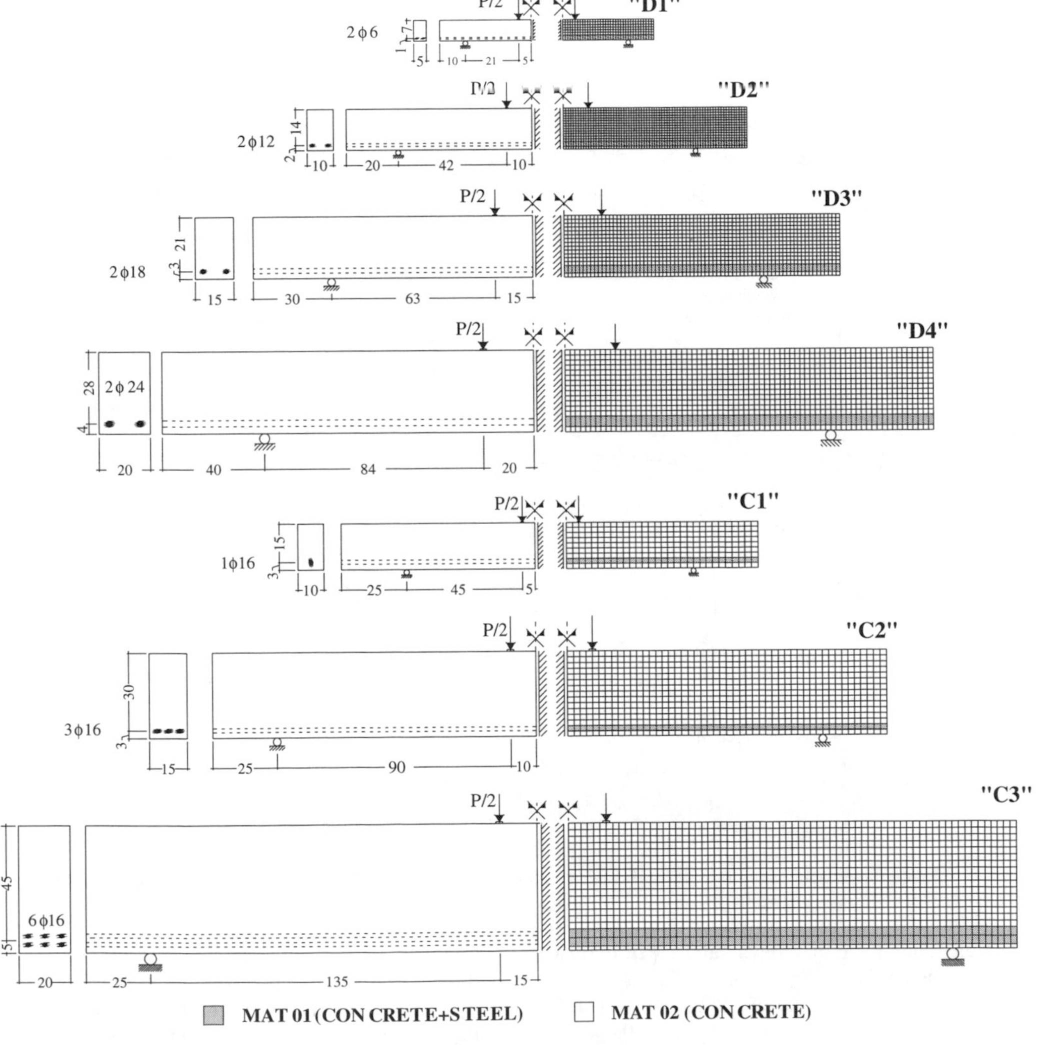

Figure 3. Set of "D" and "C" beams (Leonhardt e Walther, 1962): geometrical features (all dimensions in cm), steel arrangement and loading conditions beside the corresponding mesh adopted in the NLFEA analysis, with the subdivision in two materials (white for elements constituted by plain concrete, grey for elements constituted by concrete and smeared steel).

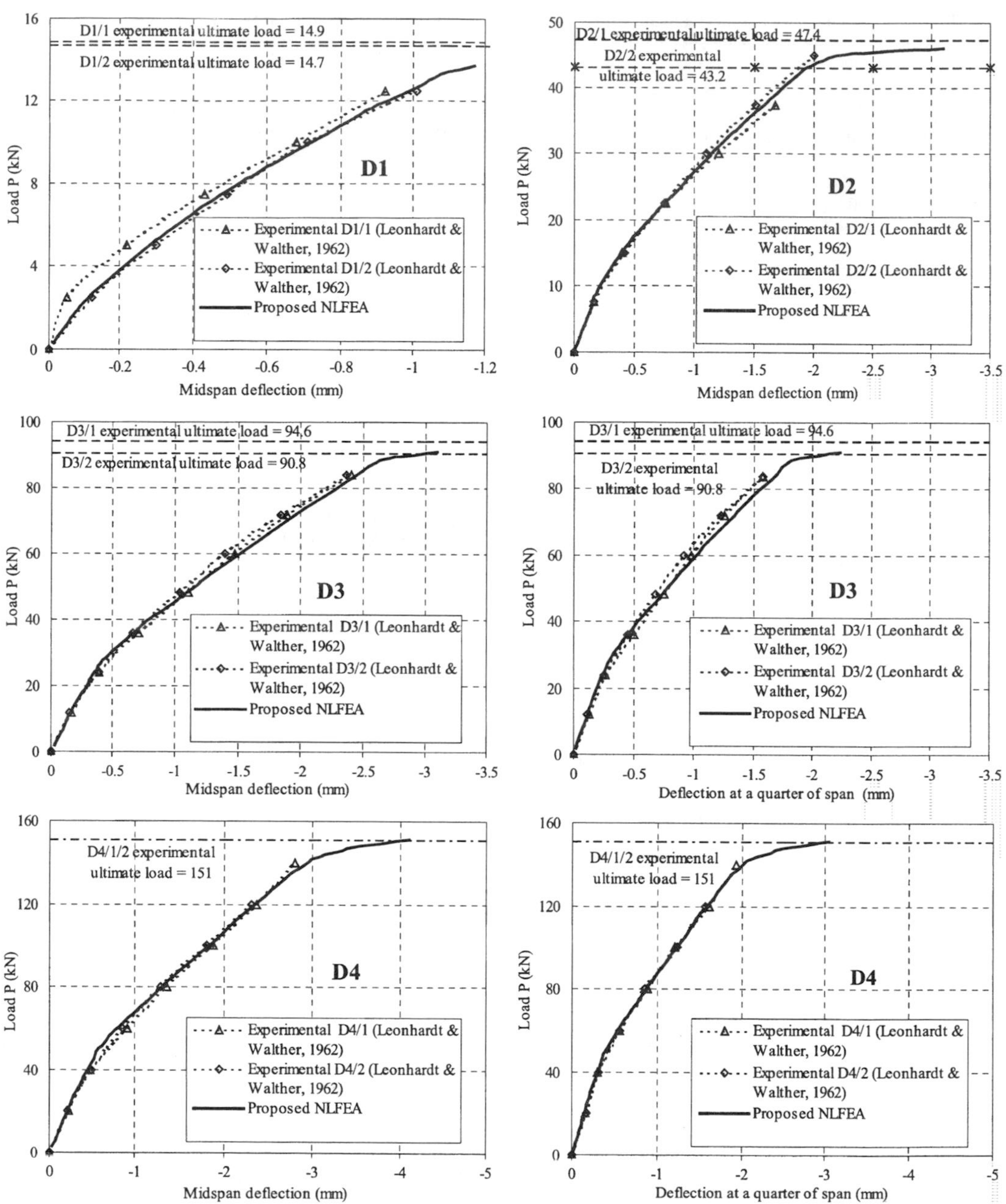

Figure 4. Comparisons between numerical and experimental results in terms of applied load against deflection at midspan or at a quarter of span for "D" beams.

The elements have then been subdivided into two groups, denoted by different colors in Figure 3: "MAT01" represents all the elements constituted by plain concrete, while "MAT02" includes those elements characterized by the contemporaneous presence of the concrete and the smeared steel reinforcement.

In the following, comparisons between the numerical responses, as obtained from the NLFEA analysis, and the experimental results are performed in terms of applied load – midspan deflection (or deflection at a quarter of span) for each beam of the "D" (Fig. 4) and "C" (Fig. 5) series, showing good correlations both

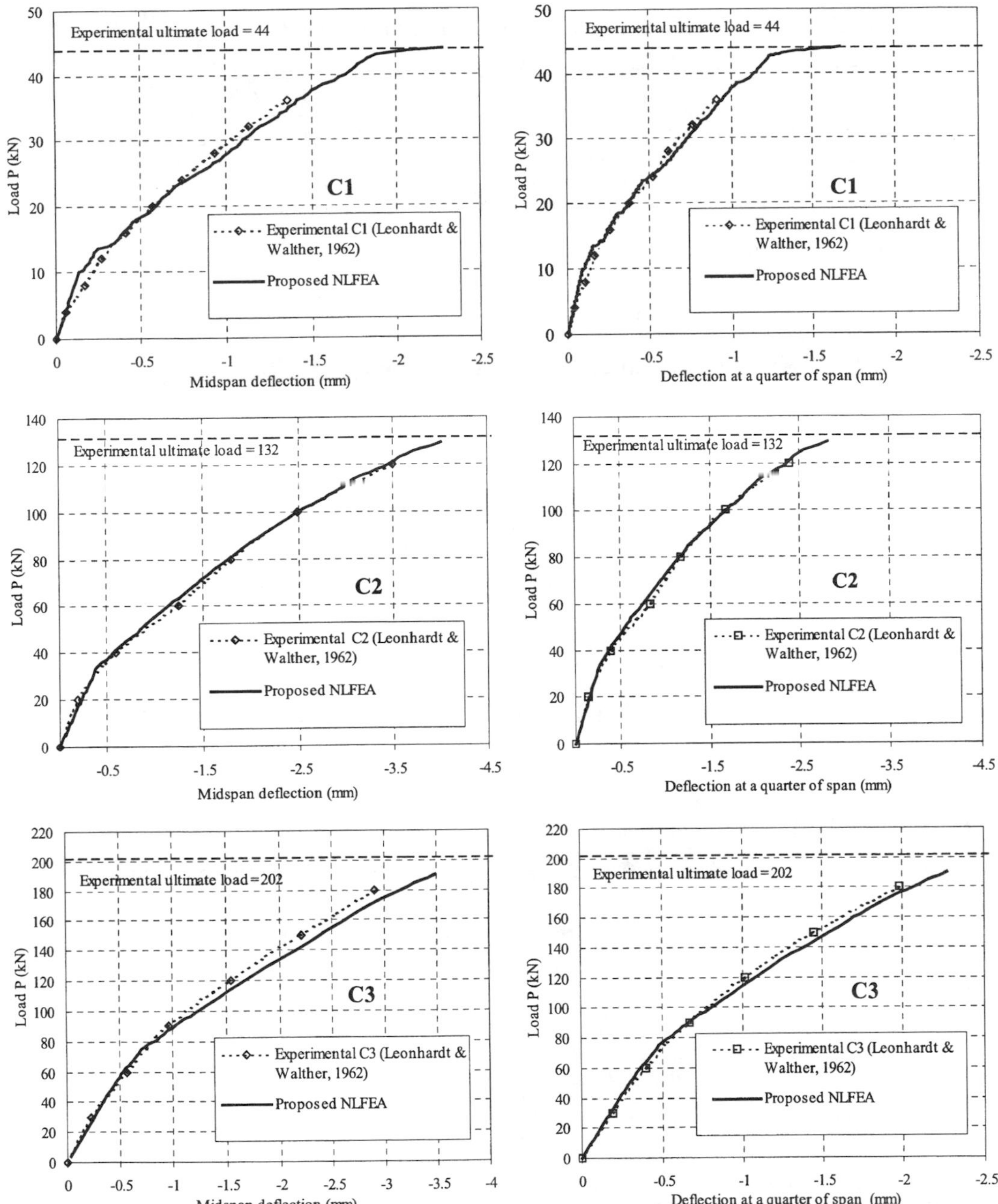

Figure 5. Comparisons between numerical and experimental results in terms of applied load against deflection at midspan or at a quarter of span for "C" beams.

at serviceability (cracking load), and at ultimate limit state (failure load). In Table 2, the numerical value of the ultimate load has been compared not only with the corresponding experimental one, but also with the ultimate resistant shear force obtained from the formula of Eurocode 2 (1992), calculated considering the characteristic values ($\gamma_c = \gamma_s = 1$). As can be observed, the ratio between the NLFEA ultimate load and the

Table 2. Comparisons between experimental and theoretical (NLFEA and EC2) ultimate load at failure.

Beams	$P_{u,exp}$ (kN)	$P_{u,theo}$ (kN)		$P_{u,theo}/P_{u,exp}$ (kN)	
		NLFEA	Eurocode 2 ($\gamma_c = 1$)	NLFEA	Eurocode 2 ($\gamma_c = 1$)
D1	14.8	14.00	16.5	0.95	1.11
D2	45.3	45.12	63.0	0.99	1.39
D3	89.8	91.2	138.2	1.01	1.53
D4	151.0	151.0	219.8	1	1.45
C1	44	44	65.1	1	1.48
C2	132.0	129.3	175.1	0.98	1.32
C3	202.0	189.9	309.9	0.94	1.53

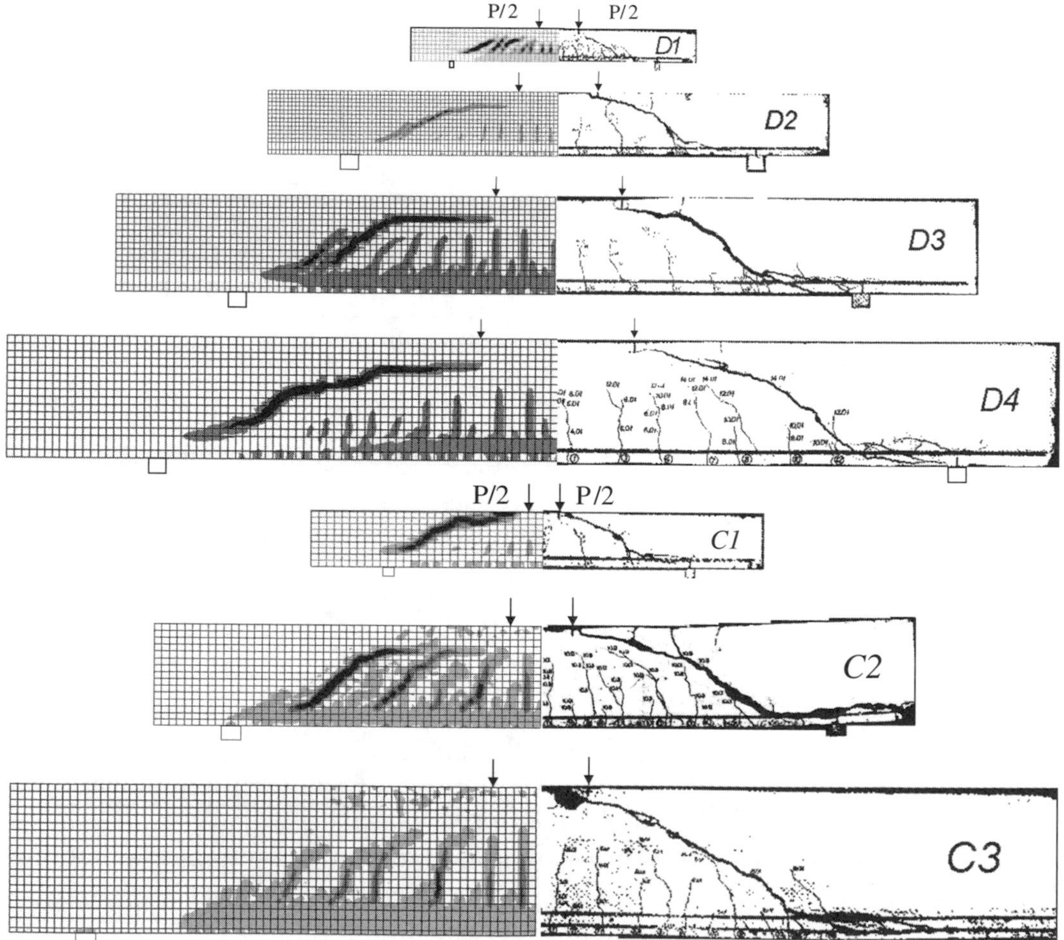

Figure 6. Comparisons between the crack pattern obtained from experimentation (left side) and from numerical analysis (right side) for the beams of Series "D" and "C".

experimental value is very close to the unity, also varying the beam's size, while Eurocode 2 semi-empirical formula provides values quite far from experimental observations.

As already underlined, the implementation of the PARC matrix in the FE program ABAQUS has made it possible to display also other important parameters, which can help to investigate better the local behaviour of the structures. Among them, particularly significant is the monitoring of the crack width, which allows one to reproduce the development of the crack pattern for each load step.

In Figure 6 the crack distributions experimentally observed by Leonhardt and Walther are compared with those numerically attained in correspondence of the failure of the beam (last load increment of FE analysis). As can be observed from Figure 6, the model has shown a good capability of reproducing the diagonal tension crack experimentally observed for these beams, responsible for the brittle shear failure.

In the same way, the contribution to shear strength provided by all the mechanisms occurring before and after cracking for plain or reinforced concrete can be evaluated and displayed in the output file. These contributions can be expressed in terms of shear stresses τ_{xy} in correspondence of every integration point; so, for a single section, when equilibrium is reached the amount of all resistant mechanisms on the section is equal to acting shear force V.

In Figure 8, equilibrium has been verified in correspondence of A-A, B-B and C-C sections of D4 beam and, through a histogram, internal stresses have been expressed in terms of percentage of external force V. These relative percentages depend on section position with respect to crack pattern, for example owing to lack of cracked integration points, shear strength of A-A section is principally due to uncracked plain or reinforced concrete. In the case of cracked integration

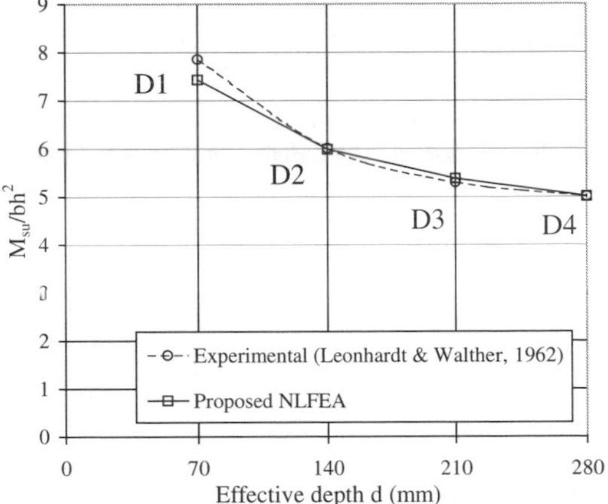

Figure 7. Comparisons between the experimental value of the dimensionless ratio M_{su}/bh^2 and that obtained from NLFEA for beams of different sizes belonging to set "D".

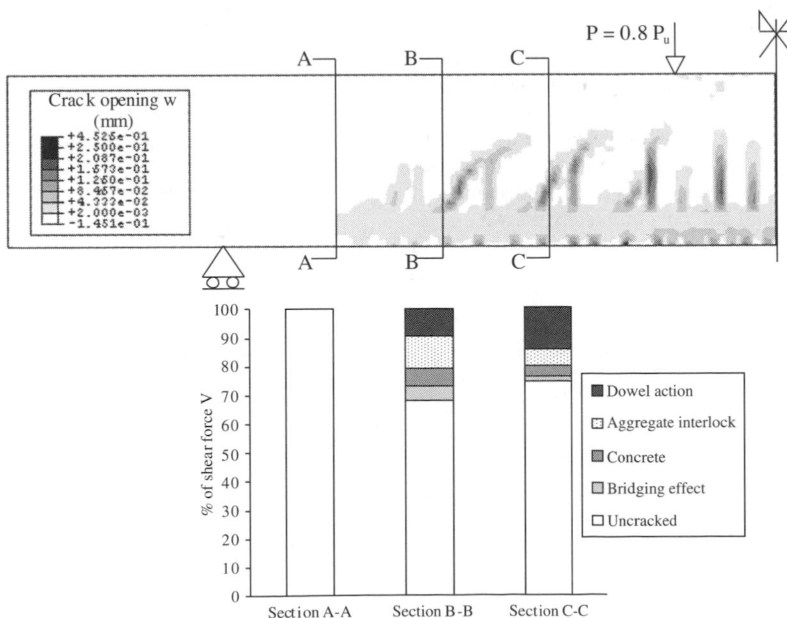

Figure 8. D4 beam: percentages of each shear stiffening contribution with respect to the total shear force V in three different vertical sections, characterized by different crack openings, at a given load stage.

points, dowel action, aggregate interlock and bridging effect between cracks appear the main phenomena which affect the shear strength.

It should be noted that the magnitude of some of these contributions, which govern the interface behaviour, is strictly linked to crack opening values; for this reason, comparing B-B and C-C percentages, they are greater along B-B section where crack opening values are lower than C-C section. This remark partly explains why the beams of C and D series having large dimensions are characterized by a ratio M_{su}/bh^2 lower than small ones (Fig. 7). Indeed, when beam dimensions increase, crack opening values also increase, causing a lost in effectiveness of all these contributions which are inversely proportional to crack opening, such as aggregate interlock and bridging effect.

5 CONCLUSIONS

In this work a non-linear analysis of beams without shear reinforcement has been carried out, implementing the recent proposed PARC constitutive matrix in a FE program. The model, based on a smeared, fixed-angle approach, has been able to simulate accurately the behaviour of a set of beams tested by Leonhardt & Walther (1962), both in terms of load-deflection response and in terms of cracking and ultimate load. The potentialities of the procedure have been applied to evaluate other features useful to understand the local behaviour of the examined structures. Thus, the crack pattern corresponding to each load increment was displayed, as well as the influence of each single stiffening contribution to support the shear force in a given cracked section. The procedure is proposed as an effective tool not only for the design, but also for an improved understanding of those mechanisms which lead to shear failure, a problem discussed for a long time and still not completely known.

REFERENCES

ASCE-ACI Committee 445.1998. Recent Approaches to Shear Design of Structural Concrete. *Journal of Structural Engineering*, ASCE, Vol. 124 (12), 1375–1417.

Belletti, B., Cerioni, R. & Iori I. 2001. A Physical Approach for Reinforced Concrete (PARC) Membrane Elements. In *Journal of Structural Engineering*, ASCE, Vol.127 (12), 1412–1426.

Belletti, B., Bernardi, P., Cerioni, R. & Iori, I. 2002. Some applications of "PARC" constitutive matrix for reinforced concrete structures. *Proceedings of ABAQUS Users' Conference*, Newport.

Darwin, D. & Pecknold, D.A. 1977. Non-linear biaxial stress-strain law for concrete. *Journal of the Engineering Mechanics Division*, Proceedings of the American Society of Civil Engineers, Vol. 103, (EM2), 229–241.

Eurocode 2. 1992. Design of concrete structures, part 1: General rules and rules for building. Thomas Telford, London.

Kani, G.N.J. 1964. The Riddle of Shear Failure and its Solution. *Journal of the American Concrete Institute, Proceedings*, Vol. 61 (4), 441–467.

Kani, G.N.J. 1967. How Safe are Our Large Reinforced Concrete Beams? *ACI Structural Journal*, Vol. 64 (3), 128–141.

Kim, W. & White, R.N. 1991. Initiation of Shear Cracking in Reinforced Concrete Beams with No Web Reinforcement. *ACI Structural Journal*, Vol. 88 (3), 301–308.

Kupfer, H., Hilsdorf, H.K. & Rusch, H. 1969. Behavior of Concrete Under Biaxial Stresses. *Proceedings of ACI Journal*, Vol. 66(8), 656–666.

Leonhardt, F. & Walther, R. 1962. Schubversuche an einfeldrigen Stahlbetonbalken mit und ohne Schubbewehrung. *Deutscher Ausschuss für Stahlbeton*, Heft 151, W. Ernst u. Sohn, Berlin.

Reineck, K.H. 1991. Ultimate shear force of structural concrete members without transverse reinforcement derived from a mechanical model. *ACI Structural Journal*, Vol. 88 (5), 592–602.

Vecchio, F.J. 2000. Analysis of Shear-Critical Reinforced Concrete Beams. *ACI Structural Journal*, Vol. 97 (1), 102–110.

Vecchio, F.J., Lai, D. & Ng, J. 2001. Disturbed Stress Field Model for Reinforced concrete: Validation. *Journal of Structural Engineering*, ASCE, Vol. 127 (4), 350–358.

Walraven, J.C. & Lewalther, N. 1994. Size Effects in Short Beams Loaded in Shear. *ACI Structural Journal*, Vol. 91 (5), 585–593.

Wang, T. & Hsu, T.C. 2001. Non-linear finite element analysis of concrete structures using new constitutive models. *Computers & Structures*, Vol. 79, 2781–2791.

Zararis, P.D. & Papadakis, G.C. 2001. Diagonal Shear Failure and Size Effect in RC Beams without Web Reinforcement. *Journal of Structural Engineering*, ASCE, Vol. 127 (7), 733–742.

Computational Modelling of Concrete Structures, Bićanić et al. (eds)
© 2003 Swets & Zeitlinger, Lisse, ISBN 90 5809 536 3

Settlement damage modelling of historical buildings

M. Boonpichetvong & J.G. Rots
Structural Mechanics Department, Faculty of Architecture, Delft University of Technology, Delft, the Netherlands

ABSTRACT: The application of fracture mechanics to predict the settlement damage in historical structures is presented in this paper from which useful and practical design information can be derived. Various continuum crack models are tested in a large-scale façade problem. Some observations in numerical results indicate the need for reliable numerical techniques to handle snaps and jumps in load-displacement response. In addition, it is investigated whether the inclusion of long-term creep improves the cracking response.

1 INTRODUCTION

Ground movement induced by tunnelling activity may considerably affect the serviceability and safety of neighbouring structures. Analytical and empirical approaches are available for evaluating the cracking damage of surrounding structures subjected to settlement (for example, the limiting tensile strain method by Burland and Wroth 1975). Potts and Addenbrooke (1996) employed elastic beam elements to represent an overlying structure on a soil continuum. However, due to many provided assumptions, the direct application of these simplified methods in very sensitive cases such as fragile historical masonry buildings raise the need for more detailed analysis.

Large-scale fracture analyses (Rots 2000, Boonpichetvong and Rots 2002) were proven to be a promising alternative for assessing the possible settlement damage in historical masonry buildings. Fracture mechanic allows for predicting crack initiation, crack propagation and snap-through of cracks and complete separation in masonry façades subjected to ground movement while calibration with instrumentation data in situ is carried out. So far, one of the most difficult tasks in simulating the response of historical structures is the idealisation of the initial condition of structures before the activation of external mechanical loadings. The influence of ambient and creep effects may significantly alter the response of historical structures.

In this paper, we present the difficulties encountered in large-scale fracture through our benchmark problem. Recent attempts in simulating the effect of the initial long-term creep on the cracking response of structures subjected to ground movements is described. Finally, brief recommendations are given for possible enhancements of the suitable crack model subjected to non-proportional loadings. All the examples in this study have been calculated with the DIANA finite element program.

2 HISTORICAL MASONRY FAÇADE

The layout of the chosen historical masonry façade for the benchmark study is shown in Figure 1. This is a block of three house units. The total height of the house units is 15.5 m, starting from the foundation level at 1.5 m below green field, a ground floor level and three levels above that. The length of a house unit is 6.8 m. The three house units together have a length of 20.4 m, so that the length over height ratio is 1.3. A uniform thickness of one brick (220 mm) is adopted here for the whole façade. The opening pattern shows two large

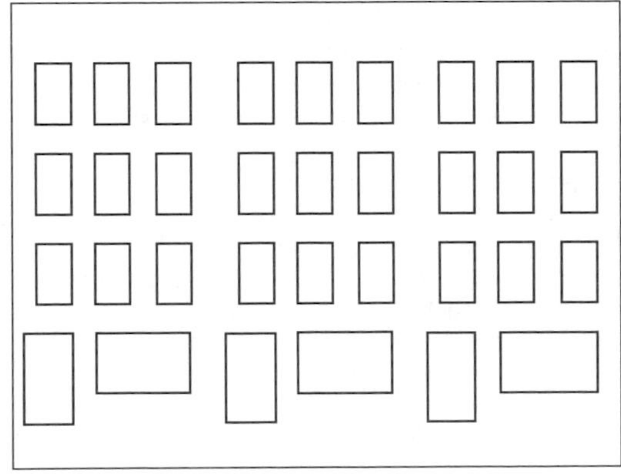

Figure 1. Three-house unit historical masonry façade.

openings at ground floor and a regular pattern of three window openings at the three floors above. Above the window openings at ground floor, lintels in the form of steel beams are present to distribute the vertical load to either side of the opening. The load-bearing and house-separating walls perpendicular to the façade have not been included in the model. Also their connections to the façade and other 3D effects have been omitted in the first instance. Although it is well understood that 3D effects can modify the damage pattern, the context of this study will be limited to 2D response.

3 LARGE-SCALE FRACTURE ANALYSES OF FAÇADE SUBJECTED TO SETTLEMENT

3.1 *Overview of modelling approach*

The currently available design practice for evaluating settlement effects on the surrounding structures due to soft-ground tunnelling can be classified into two types: uncoupled analysis and coupled analysis. In the first class, the greenfield settlement determined from empirical equations is directly imposed on the building model, while the second class allows for the full interaction between the above structure and the underlying soil. To account for the situation of historical masonry buildings founded on fragile piles, Rots (2000) introduced the semi-coupled scheme in which so-called bedding interface elements were employed to represent a simplified "smeared" model for fragile pile foundation. Boonpichetvong and Rots (2002) extended the investigation for coupled analysis. Figure 2 indicates the possible cracking damage in the historical masonry façade on shallow foundation by such preliminary coupled analyses for hogging type situation.

With the current emphasis on understanding the building response for extreme situations i.e. hogging and sagging type situations, it is decided to take the simpler model as shown in Figure 3 in which the soil and foundation system underneath the masonry building are replaced by the equivalent elastic soil-foundation medium. Although this model does not represent all features of real soil-foundation system, it serves as a good academic model to study building response subjected to ground movement. This simplified model facilitates a future study on the influence of initial pre-existing cracks in the historical building. Such influence of the present building condition including existing cracks and long-term creep on the subsequent short-term response of the historical structures during subsurface building activities is important and should not be neglected.

3.2 *Outline of the analyses*

Eight-node quadratic plane stress elements are employed to represent the masonry continuum and

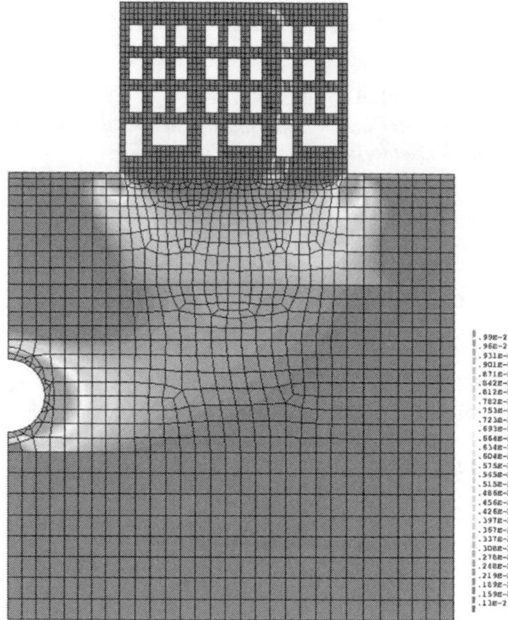

Figure 2. Example of possible settlement damage in the selected historical building.

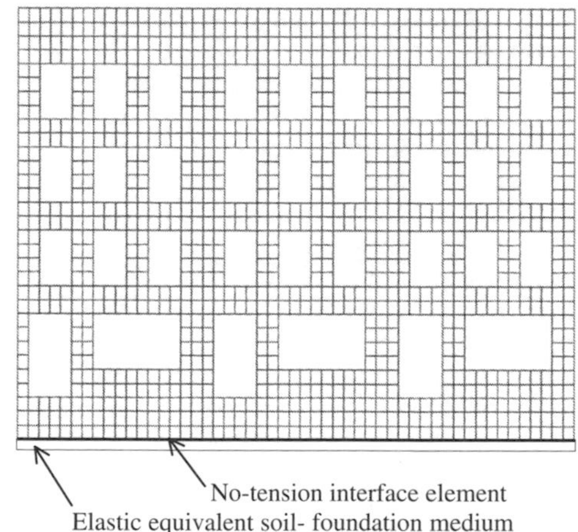

Figure 3. Simplified building model.

soil-foundation medium. The full, three by three integration scheme is adopted to reduce the danger of so-called spurious kinematics modes when softening occurs. No-tension interface elements with high stiffness in compression are inserted between the building model and the equivalent soil-foundation medium. This is reasonable, as we would ignore tensile resistance of soil boundary underneath buildings. The elastic-softening properties of the masonry are assumed as follows. Young's modulus (E) and Poisson's ratio for the masonry are adopted as 6000 N/mm^2 and 0.2 respectively. Fracture properties of the masonry are taken with a tensile strength (f_t) = 0.3 N/mm^2, fracture energy (G_f) = 0.05 N/mm with a linear softening diagram. Mass density of masonry is 2400 kg/m^3. As for the soil-foundation medium with thickness of 500 mm, the elastic behaviour is assumed with Young's modulus = 75 N/mm^2 and Poisson's ratio = 0.35.

The loading scheme starts with the activation of self-weight and live load of 5 kN/m at each floor. Subsequently, the settlement trough is applied incrementally. Vertical and horizontal ground displacements are applied to the bottom of the equivalent soil foundation medium by employing the expected "greenfield" settlement based on the analysis of case records (e.g. Mair et al. 1996), the immediate transverse settlement trough (S_v) and horizontal ground movement (S_h) at the horizontal distance x from the tunnel center line is well described by an inverted Gaussian distribution curve as:

$$S_v = \frac{0.31VD^2}{i}\exp\left[\frac{-x^2}{2i^2}\right]$$
$$S_h = \frac{x}{z_o}S_v \qquad (1)$$

where i is the horizontal distance from the tunnel centerline to the point of inflection on the settlement trough ($i = Kz_o$. K is the trough width parameter for tunnels and is taken as 0.5 in this study. Z_o is the depth to the tunnel axis, which is 20 m herein. Tunnel diameter (D) is assumed to be 6.5 m. Since the horizontal and vertical ground movements are increased proportionally with the ground loss (V), numerical results can be directly referred to the observation on site if available.

3.3 *Standard façade response*

Firstly, we employ the decomposed-strain fixed smeared crack model (Rots 1988) in which shear retention factor β is taken as 0.01 to reduce the so-called false stress locking. The first case performed is the situation in which this benchmark façade is located at point of inflection (x = 10 m. in equation 1) and only subjected to vertical ground displacement. The structured mesh of element size 400 by 500 mm is used to represent masonry façade (Figure 3). Because the quadratic elements have a linear strain distribution and cracks will localise at one side of an element thus the crack bandwidth is taken by an average of 225 mm.

Figure 4 illustrates the relationship between the angular distortion (Boscardin and Cording 1989) and the maximum crack width. Due to the non-proportional loading pattern, several unloading cracks

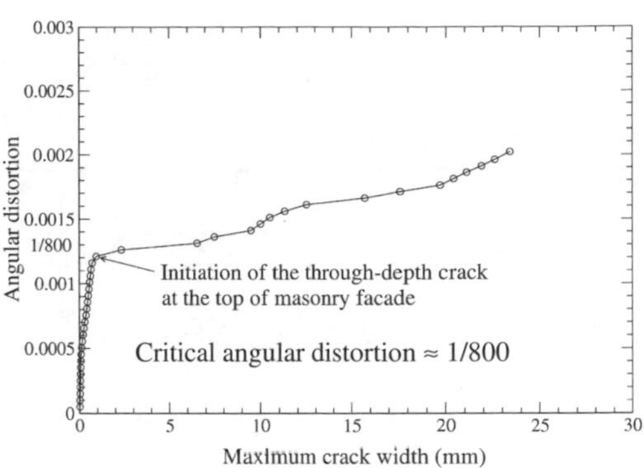

Figure 4. Standard façade response in the simplified analysis.

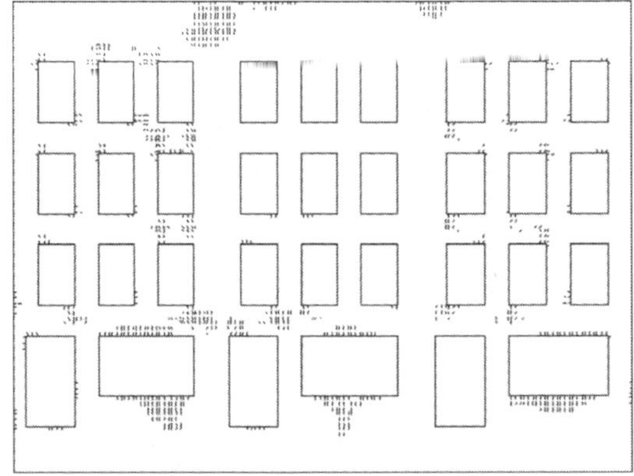

Figure 5. Predicted crack pattern at the critical angular distortion = 1/800.

are found at the first increment of prescribed settlement. These cracks occur at the window opening at the ground-floor level during the initialised state (by self-weight and live load). At each added settlement increment, new cracks propagate from the bottom part of the façade up towards the top. Along the window corners, several active cracks arise due to the high stress concentration. The response of this standard façade shows only relatively small crack widths until the angular distortion reaches the critical value, which is around 1/800. This critical angular distortion corresponds to 3.0% ground loss. The crack pattern at the critical angular distortion is shown in Figure 5. A large crack band exists at the top of the façade due to the pattern of principal tensile stress at the top prior to the onset of fully open cracks (Figure 6).

It is discovered that at the critical angular distortion, the fully open crack emerges at the top of façade i.e. the crack of which the ultimate strain is beyond the end of the descending branch, the softening is completed and no more tensile stress can be transferred. Beyond

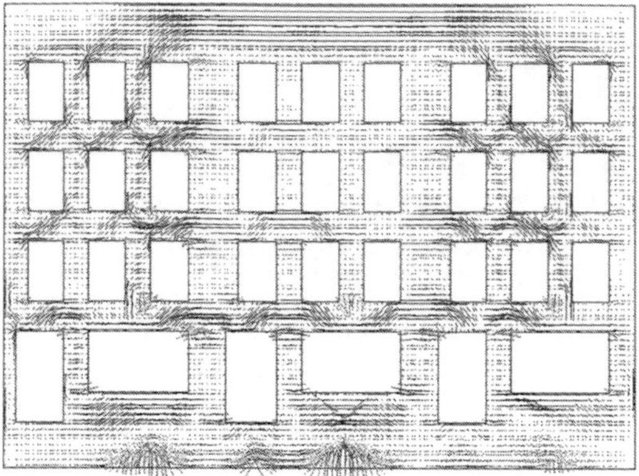

Figure 6. Principal tensile stress before the emergence of the fully open cracks.

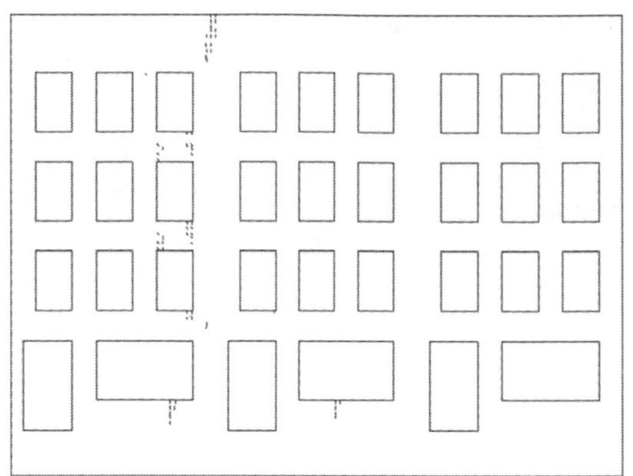

Figure 8. Fully open cracks just after the critical angular distortion.

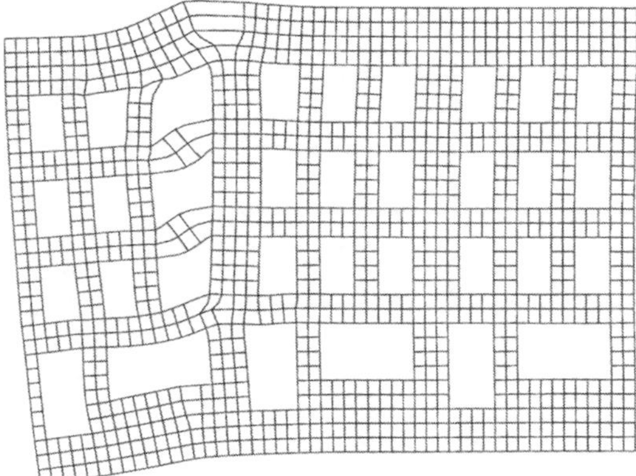

Figure 7. Localisation shown at the top of façade just after the critical angular distortion leading to unstable response.

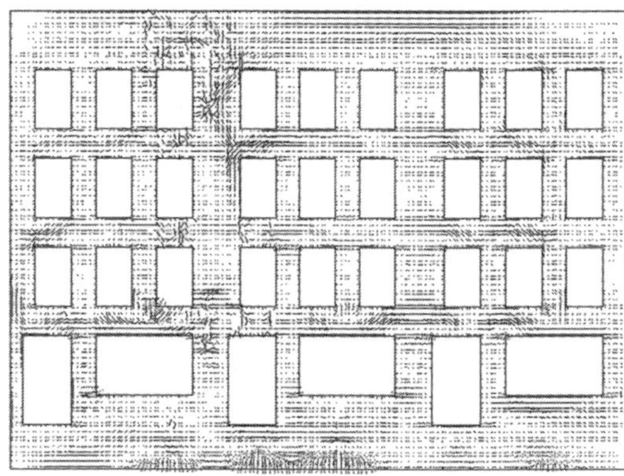

Figure 9. Stress locking just after the critical angular distortion.

this threshold, there is a sharp localisation i.e. localised vertical crack as shown in Figures 7 and 8.

After this critical stage, the angular distortion is almost constant in the second branch of the façade response. Along this branch, there is a jump of the crack width in which the convergence performance becomes very poor. The difficulty to achieve a good convergence is partly due to the significant residual stress (stress locking shown in Figure 9) in the fully open crack zone and mainly due to the significant unbalance between high elastic energy stored in the façade and low fracture energy of masonry. The snap-back response is subsequently revealed in section 3.6 by a discontinuum crack model.

3.4 *Performance of continuum crack models in large-scale fracture*

Various smeared crack models are then additionally tested in the same problem outlined in section 3.3 to reveal the performance of each model in this large-scale fracture. These comprise the total strain fixed smeared crack, the total strain rotating smeared crack and the Rankine plasticity crack model. The detailed formulations of these crack models are described respectively in Feenstra et al. (1998) and Feenstra (1993). The overall response of the façade simulated by each crack model are plotted in Figure 10.

All these tests are performed with the same structured mesh of quadratic elements. Similar problems are encountered during testing all these crack models i.e. convergence is hardly achieved in the second branch of the façade response. A zone of false stress locking at the top of façade is still present in case of using the total strain fixed crack model. Stress locking is less in case of the total strain rotating crack model (see Figure 11). This feature matches with a relatively better convergence compared to the other three crack models.

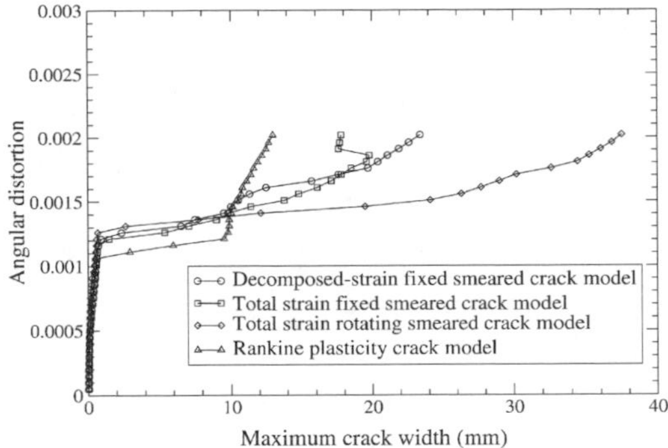

Figure 10. Façade response by various continuum crack models.

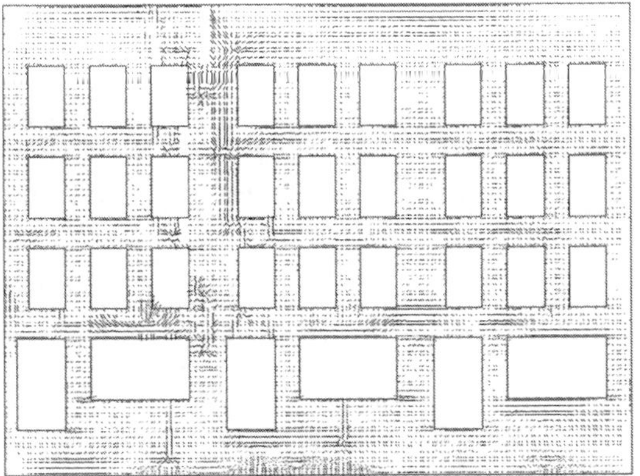

Figure 11. Reduction of stress locking zone by total strain rotating smeared crack model.

For the analysis using the Rankine plasticity-based crack model, a tangent stiffness scheme was adopted and several undesirable negatives pivots were detected pointing the question of the true equilibrium path. Unless branch switching techniques or indirect arc-length control techniques (de Borst 1986) are employed, the stability and uniqueness condition cannot be guaranteed. With the other smeared crack analyses, the positive secant modulus rather than the true negative softening modulus was adopted to set up the structural stiffness matrix, which to some extent avoided these problems. Another aspect of plasticity-based models is the inherent elastic unloading/reloading especially for the situation encountered in non-proportional loadings. A secant approach back to the origin, as included in the decomposed as well as total-strain based smeared crack models, is a better approximation of the real unloading behaviour for quasi-brittle materials. Figure 12 gives an example in tension. The fully elastic unloading inherent to plasticity models may lead to incorrect results especially

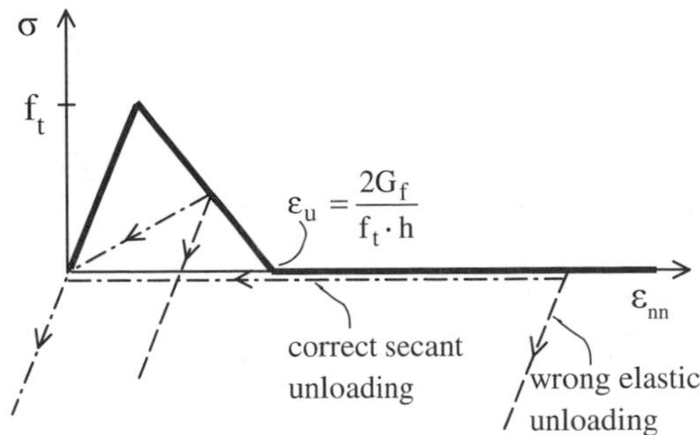

Figure 12. Secant and elastic unloading options.

when the material is brittle and/or the structure i.e. h is large.

3.5 *Mesh refinement test*

Based on the evaluation of continuum crack models in section 3.4, the total strain-based crack model is selected for the mesh refinement test. By maintaining the density of the parental finite element mesh, mesh refinement is done by splitting each eight-node element (CQ16M) into four four-node quadrilateral elements (Q8MEM) and further dividing each four-node element into two three-node triangular (T6MEM) elements. A two by two integration scheme is applied for the four-node quadrilateral elements and a one-point integration scheme for the three-node triangular elements. The element size of both elements becomes 200 by 250 mm. Since the vertical localised crack is expected to govern the final failure mode, a crack bandwidth of 200 mm is adopted for both meshes.

From our observation, the convergence performance is improved by using the triangular mesh whereas those of four-node and eight-node quadrilateral elements are comparable. Again, in all tests, the response in the second branch faces convergence difficulties. It was proved that problems regarding mesh dependency could not be fully solved by fracture-energy based approaches (e.g. Sluys 1992) in the smeared crack models. Nevertheless, for the present case, this approximation is considered to be reliable enough for practical interpretation. As shown in Figure 13, the limit points of façade response predicted by using different refined meshes are in the same magnitude and the threshold of approximately 0.0013 can be interpreted as the value that triggers notable damage for design guidelines.

3.6 *Discrete crack analysis*

To circumvent the inherent nature of stress locking for smeared crack analyses, the problem in section 3.3 is re-analysed by placing interface elements at the

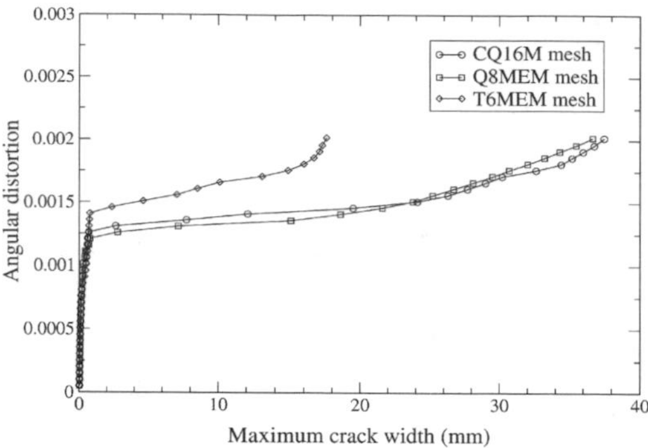

Figure 13. Façade response in mesh refine test.

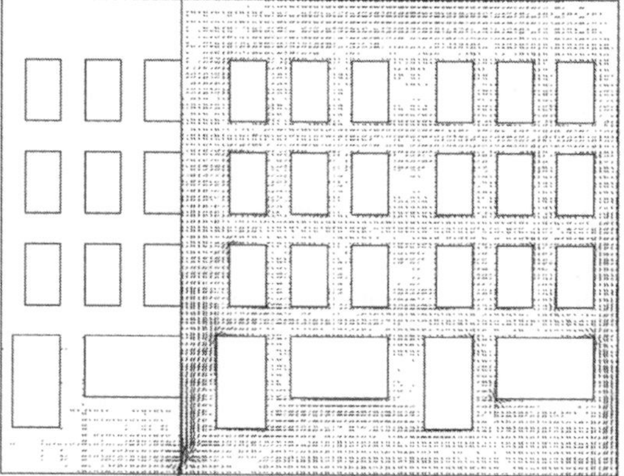

Figure 14. Residual principal tensile stress in discrete crack analysis.

vertical line of the final crack predicted by smeared crack analyses. By adopting such discrete crack corrector analysis together with the arc-length control technique, we could overcome the stress locking (Figure 14) and the associated convergence problem. Snap-back response (Figures 15 and 16) could be predicted by this approach, which was not successful for the smeared crack concept.

The strong and sharp snap-back response is indeed the brittle-failure characteristics for large-scale masonry structures. This is because of the amount of elastic energy stored in the wall versus fracture energy consumed during crack propagation. However, even with the discrete crack model, much effort is required and the approach is still not attractive. The authors believe that the proper technique to handle the highly brittle behaviour for large-scale fracture should be developed. Application of a new technique like the sequentially linear continuum concept (Rots 2002) is appealing. At this stage, while such technique is still under development, the use of the total strain rotating smeared crack model is recommended to determine

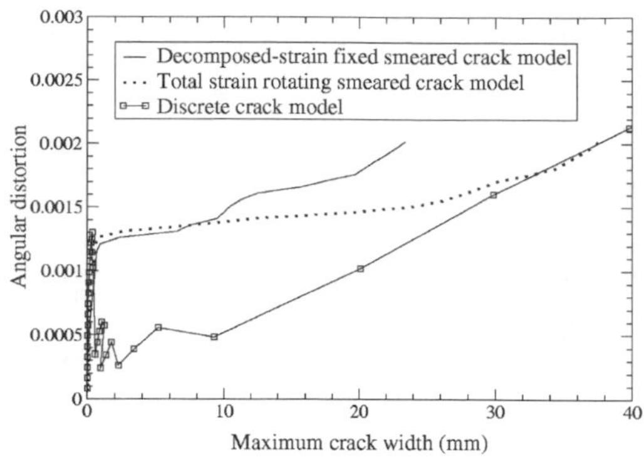

Figure 15. Computed angular distortion versus maximum crack width for the discrete crack analysis, compared with those by smeared crack analysis.

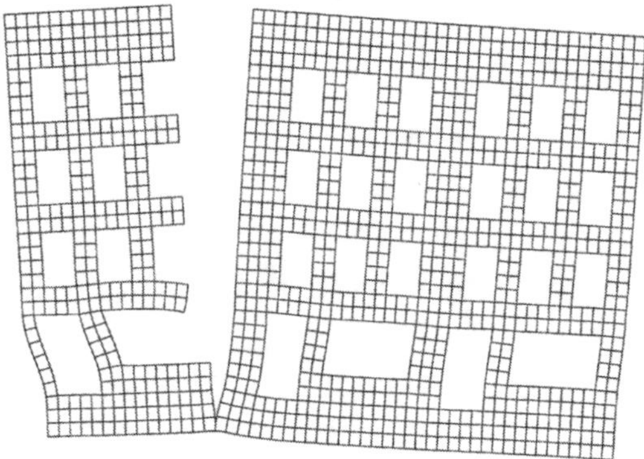

Figure 16. Incremental deformation during snap back response in discrete crack analysis.

the serviceability and safety of structures subjected to ground movement.

3.7 *Hogging and sagging response*

The response of the façade in section 3.3 will be highly altered depending on the location from the tunnel axis. The two critical situations are the hogging type situation (convex curvature) and sagging type situation (concave curvature). The overall response of the chosen historical façade is summarized in Figures 17 and 18.

Wall A and B are for the hogging type situation ($x = 10$ m. in equation 1), with and without the effects of horizontal ground movement transferred to the building respectively. On the other hand, Wall C and D are for the sagging type situation ($x = 0$ m. in equation 1), with and without the effects of horizontal ground movement transferred to the building respectively. It is noted that horizontal ground strain

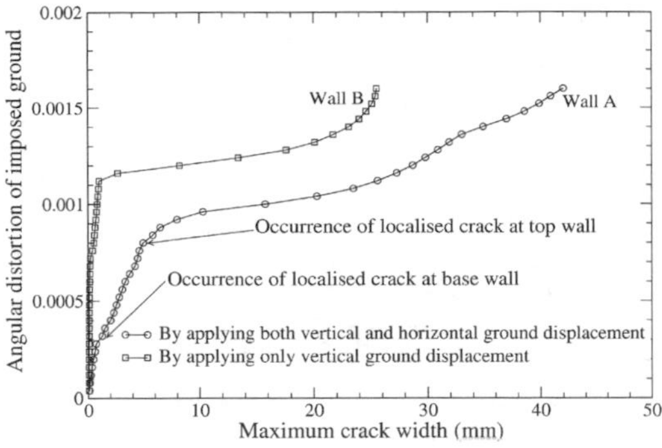

Figure 17. Façade response in the hogging type situation.

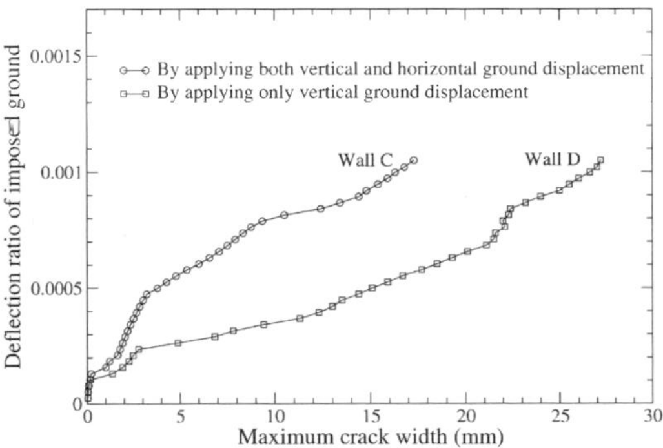

Figure 18. Façade response in the sagging type situation.

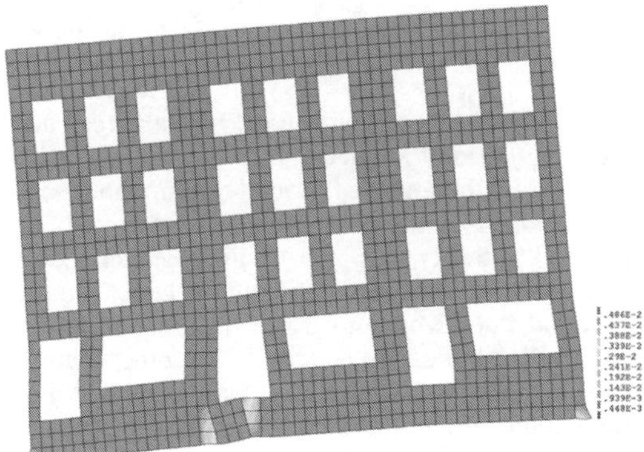

Figure 19. Deformed pattern of Wall A at 0.8% ground loss (imposed ground angular distortion = 0.0003).

considerably affects the façade response, though in a different way for hogging and sagging.

For the hogging type situation, the horizontal tensile ground strain causes an increase of the cracking damage in the structure. Localised cracking of Wall A is initiated at the bottom part of façade by the horizontal tensile ground strain at 0.8% ground loss (Figure 19)

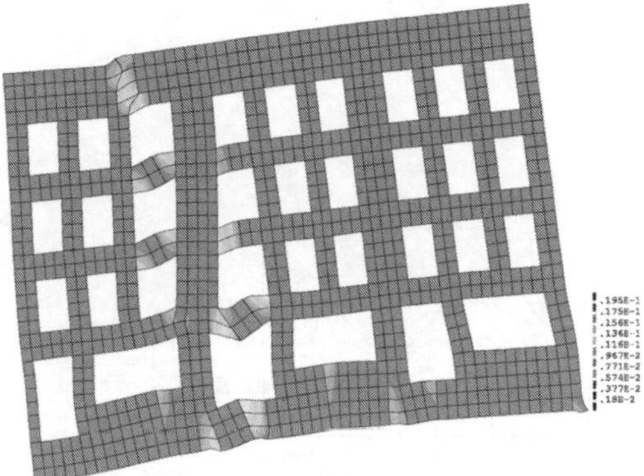

Figure 20. Deformed pattern of Wall A at 2% ground loss (imposed ground angular distortion = 0.00075).

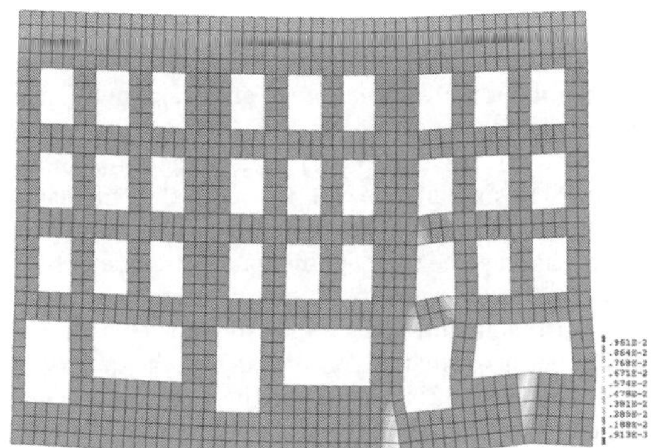

Figure 21. Deformed pattern of Wall D at 0.8% ground loss (imposed ground deflection ratio = 0.0002).

and reaches the top part of façade causing high distortion at 2% ground loss (Figure 20). When the influence of horizontal ground strain is neglected (Wall B), the first localised crack is detected instead at the top part at 3% ground loss (Figure 7).

For the sagging type situation, the horizontal compressive ground strain tends to restrain the cracking mechanism of the façade subjected to concave curvature at the bottom part. Neglecting horizontal compressive ground strain yields the extreme result in which cracking is found to propagate from the bottom part up towards the top as shown for the response of Wall D in Figure 21.

4 CREEP EFFECT ON SETTLEMENT DAMAGE OF MASONRY FAÇADE

To determine the long-term effect such as the initial creep of masonry on the subsequent settlement damage, we adopt the decomposed-strain based crack

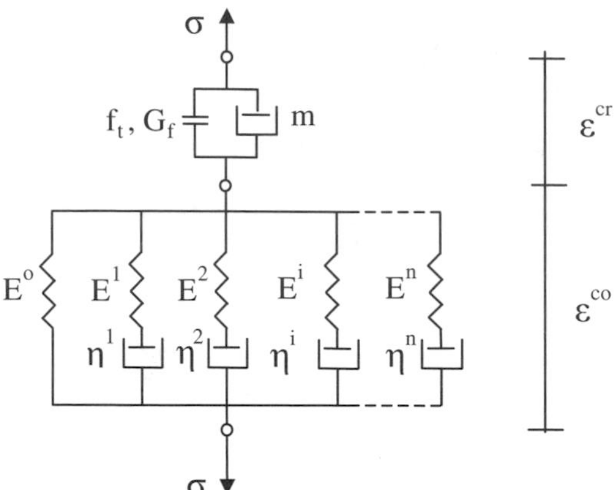

Figure 22. Adopted linked series of Maxwell chain and crack model.

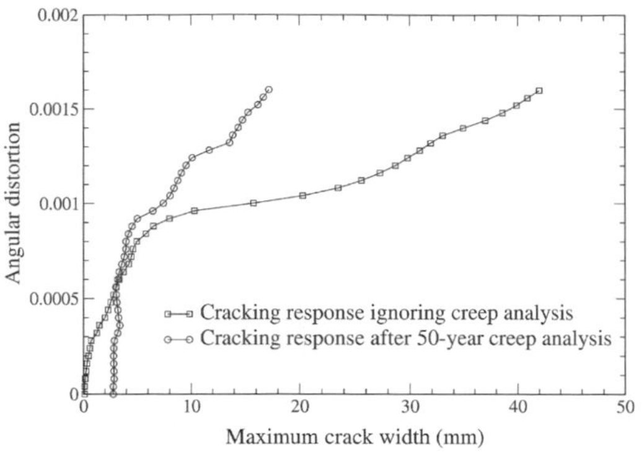

Figure 23. Predicted effect of creep on deformability in Wall A.

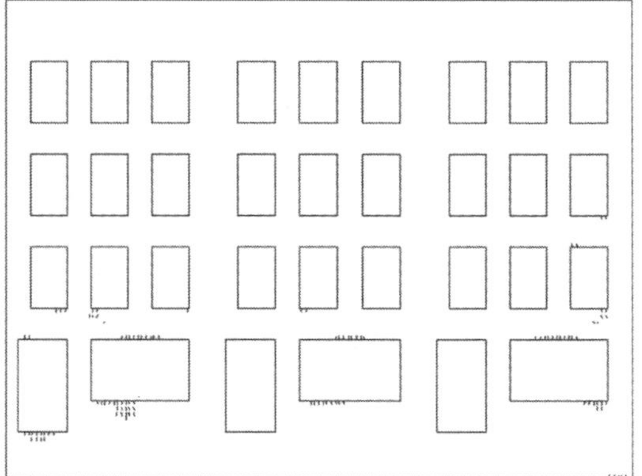

Figure 24. Initial crack pattern of Wall A at t = 0.

model in combination with visco-elasticity to allow for the interaction between creep and cracking. Visco-elastic behaviour based on the CEB-FIP90 model (Muller and Hilsdorf 1993) is assumed for intact masonry to simulate the creep process. The inclusion of rate dependency in the crack model may be important in case that the rate of immediate settlement is high according to poor quality of the tunnelling process or the insufficient grouting around tail void. This topic is still under investigation. Assuming that the quality of tunnelling process is reliable and the rate of immediate settlement process is relatively slow such that the crack rate dependency can be neglected, the adopted series model can be reduced to a combination of Maxwell chain (de Borst and van den Berg 1986) and rate-independent ($m = 0$) decomposed-strain fixed smeared crack model. Based on the observation in section 3.4, this crack model is expected to perform better than plasticity-based crack models in handling unloading, closure and (re)opening of the pre-existing cracks initially induced by long-term live loads and dead load of the building. This is important as upon applying the settlement trough, the initial cracks by the long-term loadings may be re-activated or they may be arrested while new cracks form under an angle with the initial cracks. It is noted that the total-strain smeared crack model cannot yet be combined with visco-elasticity, which explains the choice for the decomposed smeared crack model in this section.

As an example, we will consider the creep effect on settlement damage of Wall A in section 3.7. It is found that under sustained loading of the initial live load and dead load, the long-term 50-year creep analysis indicates the increased deformational capacity of Wall A as shown in Figure 23. The façade response after completion of the relaxation process is governed by the Maxwell element without dashpot resistance which results into the combined mechanism of the elastic-cracking model with $E = E^o$. With creep activated, the façade then becomes much more flexible, it can follow the imposed ground deformations more easily and less settlement damage is observed. The predicted initial crack pattern with creep simulation shows that cracks occurred around the large openings at ground floor which are formed during stress and strain redistribution by the relaxation process in the masonry (Figures 24 and 25). It was found that these initial cracks are not dominant and hardly affect the final failure mode of Wall A.

After the imposition of ground movement on Wall A, the final fully open crack pattern with creep activated indeed confirms the enhanced flexibility by relaxation of masonry (Figure 26) and cracking seems to be more distributed. For the analysis without creep (Figure 27), cracking is more concentrated leading to the higher degree of settlement damage. While the results by omitting creep analysis from this example may be conceived as conservative solution, the authors emphasise that this may not be generally valid. In certain circumstances, the

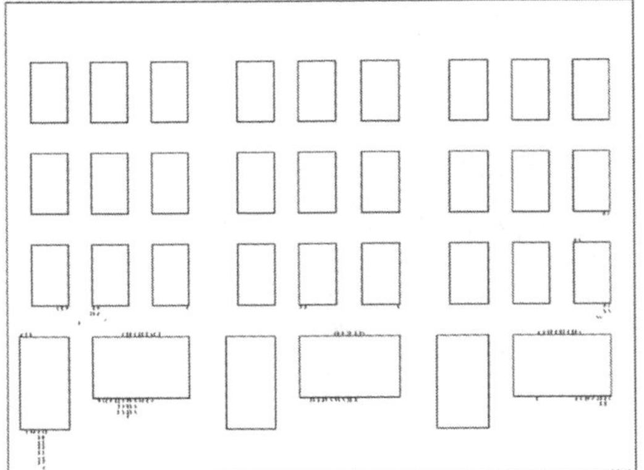

Figure 25. Initial crack pattern of Wall A at t = 50 years.

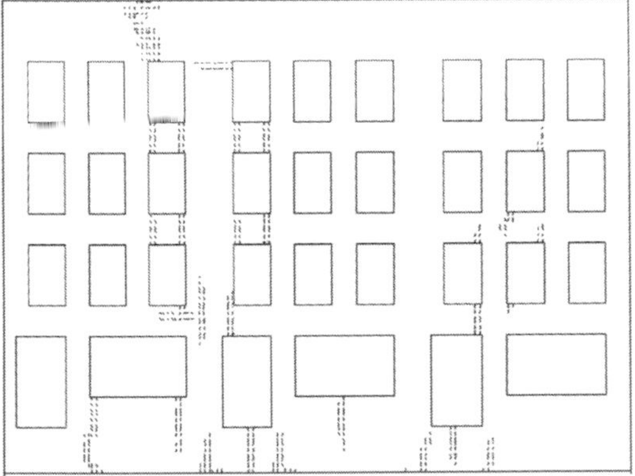

Figure 26. Final fully open cracks in Wall A with creep process.

induced initial cracks by non-proportional loadings may be important and trigger different failure modes depending on loading path and history.

5 FUTURE WORK

The effect of the initial pre-existing cracks and creep may significantly change the building response upon the action of ground movement. Existing cracks may unload while new cracks form under a non-orthogonal angle from the existing cracks. The influence of such mutual crack interaction can be best described by multi-plane approaches such as a multi-directional fixed crack approach (e.g. Rots 1988) or the microplane model (Bazant and Prat 1988). These classes of crack models are able to record and transfer the state variables over fixed planes and capture crack interaction.

The multiple planes in the microplane model suggest a similarity with the non-orthogonal smeared crack

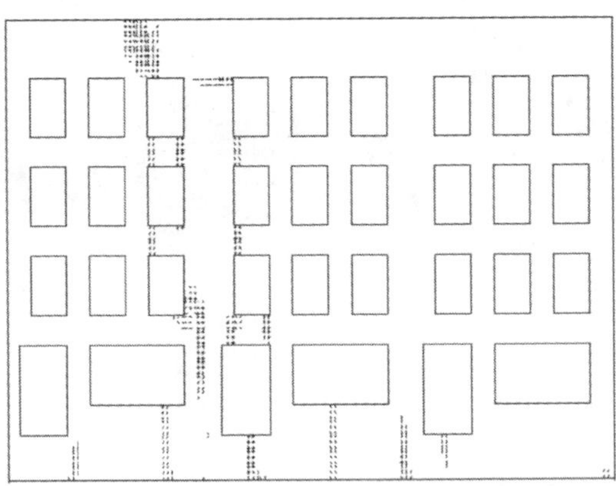

Figure 27. Final fully open cracks in Wall A without creep process.

model. However, in the non-orthogonal smeared crack model a set of planes is not predefined, but the cracks/planes emerge during the process, in the direction normal to the principal stress when this principal stress violates a tension cut-off. Recent progress has been reported in adopting the four-way crack scheme to derive the effect of initial anisotropy of pre-cracks on the residual performance of RC structures (Pimanmas and Maekawa 2001). Here, two sets of orthogonal cracks can be present, with a non-orthogonal angle between them. Good results were reported for non-proportional loading and path-dependency in case of pre-cracked reinforced structures, also including shear and compression non-linearity.

Further studies are required to compare the pro's and contra's of the various multi-plane formulations for applications to large-scale buildings under non-proportional load. In the authors' opinion, there is room for future research to make the decomposed multi-directional model more robust, by allowing only a limited number of cracks under fixed angles from the first arisen crack. Taking 45 degrees, a 4-plane crack model for 2D situations results, which is sufficient for many engineering also masonry situations. It allows for non-orthogonal cracking, but does not loose the conceptual attractiveness of the decomposed formulation providing for coupling with creep, thermal and other effects. The implementation becomes compact and internal iterations and algorithmic aspects are under better control as the maximum number of cracks is constrained and "hesitating" cracks under small angles do not occur. For 3D cases and shells, the 45 degree pre-assumption results in a 9-plane crack model. In this way, positive elements from smeared cracking, micro-plane models and Maekawa's model are combined. In addition, a 45° fixed crack scheme is appealing as it fits with the standard unit damage

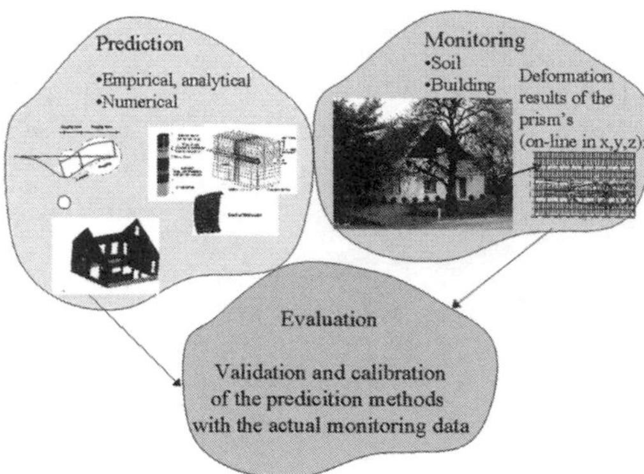

Figure 28. Full scale cycle of the Sophia railway test site (Netzel 2002).

patterns governed by the preferential damage orientation on the mesostructural characteristics of historical masonry (Massart 2001).

To validate the enhanced approach, verification will be performed by example of four masonry houses at the COB site of the Sophia Railway tunnel nearby Rotterdam. This is one of the available monitoring programs for Dutch soft-ground tunnelling which is still under process (Figure 28, for more details see Netzel 2002). The full-scale monitoring data of ground and buildings response during and after TBM passage will allow for the verification of numerical and analytical predictions.

6 CONCLUSIONS

The response of historical masonry structures subjected to ground movement is simulated computationally by using a fracture mechanics approach. Smeared and discrete crack models have been tested. The numerical results indicate the subtle snap and jump behaviour in the building response. With the use of arc-length control scheme in combination with a discrete crack model, the snap-back can be simulated. For continuum crack models, a sequentially linear concept is suggested to solve the brittle response.

Subsequently, the effect of long-term creep on the deformational capacity of the selected structure has been demonstrated. It is revealed that the relaxation process in the structure may be beneficial to follow the ground movements.

ACKNOWLEDGMENTS

The authors gratefully acknowledge the financial support by Delft Cluster Theme 1 'Soil and Structures' and by the Netherlands Center for Underground Structures CUR/COB Committee F220.

REFERENCES

Bazant, Z.P. and Prat, P.C. 1988. Microplane model for brittle-plastic material I.Theory & II.Verification, *J. Eng. Mechs*, ASCE, Vol.114, 1672–1702.

Boonpichetvong, M. & Rots, J.G. 2002. Settlement damage of masonry buildings in soft-ground tunnelling, In Max A.N. Hendriks and J.G. Rots (eds), *Finite Elements in Civil Engineering Applications*, Proceedings of the Third DIANA World Conference, 285–293. Lisse: Balkema.

Boscardin, M.D. & Cording, E.J. 1989. Building response to excavation induced settlement. *ASCE Journal of Geotechnical Engineering*, 115(1): 1–21.

Burland J.B. & Wroth C.P. 1975. Settlement of buildings and associated damage. *Proc. Conf. Settlement of Structures*, Cambridge, 611–654.

de Borst, R. & van den Berg 1986. Analysis of creep and cracking in concrete members, In Z.P. Bazant (ed), *Creep and Shrinkage of Concrete: Mathematical Modelling*, Proceedings of the Fourth RILEM International Symposium, 527–538.

de Borst, R. 1986. *Non-linear analysis of frictional materials*, Ph.D. Dissertation, Delft University of Technology, Delft, the Netherlands.

Feenstra, P.H. 1993. *Computational aspects of biaxial stress in plain and reinforced concrete*, Ph.D. Dissertation, Delft University of Technology, Delft, the Netherlands.

Feenstra, P.H., Rots, J.G., Arnesen, A., Teigen, J.G., & Høiseth, K.V. 1998. A 3D constitutive model for concrete based on a co-rotational concept, In R. de Borst et al. (eds), *Proc. Int. Conf. Computational Modeling of Concrete Structures*: 13–22. Rotterdam: Balkema.

Mair, R.J., Taylor, R.N., & Burland, J.B. 1996. Prediction of ground movements and assessment of risk of building damage due to bored tunnelling. In R.J. Mair and R.N. Taylor (eds), *Geotechnical aspects of underground construction in soft ground*: 713–718. Rotterdam: Balkema.

Massart, Th., Bouillard, Ph., Geers, M.G.D., & Peerlings, R.H.J. 2001. A 2D anisotropic damage model for masonry walls, In Z. Waszczyszyn and J. Pamin (eds), *Proc. 2nd European Conference on Computational Mechanics*, Cracow, Poland.

Muller, H.S. & Hilsdorf HK. 1993. Constitutive relations for structural concrete, Selected justification notes, *CEB Bulletin d'information* 213, CEB 1993.

Netzel, H. 2002. Full scale test on the building response due to TBM-tunnelling in Dutch soft soil, *Proc. 2nd Int. Conference on Soil-Structure Interaction in Urban Civil Engineering*, 271–275, Zurich, Switzerland.

Pimanmas, A. & Maekawa, K. 2001. Multi-directional fixed crack approach for highly anisotropic shear behavior in pre-cracked RC members. *J. Materials, Conc. Struct., Pavements*, JSCE, No. 669/V-50, 293–307, Japan Society of Civil Engineers.

Potts, D.M. & Addenbrooke, T.I. 1996. The influence of an existing surface structure on the ground movements due to tunnelling, In R.J. Mair and R.N. Taylor (eds),

Geotechnical aspects of underground construction in soft ground: 573–578. Rotterdam: Balkema.
Rots, J.G. 1988. *Computational modelling of concrete fracture*, Ph.D. Dissertation, Delft University of Technology, the Netherlands.
Rots, J.G. 2000. Settlement damage predictions for masonry, In L.G.W. Verhoef and F.H. Wittmann (eds), *Maintenance and restrengthening of materials and structures – Brick and brickwork*, Proc. Int Workshop on Urban heritage and building maintenance, 47–62.
Rots, J.G. 2002. Comparative study of crack models, In Max A.N. Hendriks and J.G. Rots (eds), In Max A.N. Hendriks and J.G. Rots (eds), *Finite Elements in Civil Engineering Applications*, Proceedings of the Third DIANA World Conference, 17–28. Lisse: Balkema.
Sluys, L.J. 1992. *Wave propagation, localisation and dispersion in softening solids*, Ph.D. Dissertation, Delft University of Technology, Delft, the Netherlands.

Computational Modelling of Concrete Structures, Bićanić et al. (eds)
© 2003 Swets & Zeitlinger, Lisse, ISBN 90 5809 536 3

A direct stiffness technique for the analysis of concrete-steel composite beams

G. Ranzi, M.A. Bradford & B. Uy
School of Civil & Environmental Engineering, The University of New South Wales, Sydney, Australia

ABSTRACT: This paper presents a method for the finite element analysis of concrete-steel composite beams using the direct stiffness method. Conventional analyses of these beams based on routine prescribed interpolation functions causes a curvature-locking phenomenon in the presence of partial shear interaction, and in many cases can lead to erroneous results. The technique developed herein does not cause undesirable curvature locking, and provides a framework for the inclusion of material nonlinearities for a comprehensive and robust technique of analysis.

1 INTRODUCTION

Composite concrete-steel beams are used frequently in order to utilise best the strength and stiffness attributes of the two materials. The structural mechanics of these beams is influenced not only by the properties of the concrete and steel components, but also by the shear connection at their interface. When the finite load-slip relationship at this interface (that exists with headed stud connectors) is incorporated into the analysis of composite beams, the solutions become quite complex, and indeed the more traditional techniques in matrix stiffness analysis may produce convergence difficulties and/or erroneous results.

The seminal and widely-cited solution published in 1952 by Newmark, Siess and Viest demonstrated that the midspan deflection of a simply supported beam under uniformly distributed loading was far more complex than the familiar $5wL^4/384EI$ because of the influence of the elastic shear connection that allows for slip between the concrete and steel. The finite element (FE) treatment of the problem is also difficult, owing to a curvature-locking phenomenon that can occur using general displacement-type approaches, such as those of Arizumi et al. (1981) and Daniels & Crisinel (1993). Dall'Asta & Zona (2000) have quantified the locking phenomenon that occurs with the use of interpolating polynomials of various orders that satisfy various degrees of continuity in their displacements and their derivatives to develop stiffness matrices. For example, in an 8-DOF element with the appropriate lengthwise cubic transverse deformations and linear axial deformations, the vector of nodal freedoms is

$$\langle w_{10}, w_{20}, v_0, \boldsymbol{\theta}_0, w_{1L}, w_{2L}, v_L, \boldsymbol{\theta}_L \rangle^{\mathrm{T}}$$

where w and v are the axial and transverse deflections respectively, $\theta = dv/dz$, the subscripts 1 and 2 denote the concrete and steel respectively and 0 and L denote $z = 0, L$. If h is the distance between the centroids of the concrete and steel, then from routine structural mechanics the slip is

$$s = w_2 - w_1 + h\theta \tag{1}$$

and when substituted into the interpolation polynomials for w_1, w_2 and θ produces

$$s = s_0 + (s_L - s_0 - s_{0L})\xi + s_{0L}\xi^2 \tag{2}$$

where

$$s_{0,L} = w_2^{0,L} - w_1^{0,L} + h\theta^{0,L};\ s_{0L} = 3h\left(\theta_0 + \theta_L + \frac{2(v_0 - v_L)}{L}\right)$$
$$\xi = z/L \tag{3}$$

Full interaction is achieved when $s \to 0\ \forall z \in [0,L]$ and so $s_0 = s_L = s_{0L} = 0$. The transverse deformation using a cubic interpolation polynomial is

$$v = v_0 + L\theta_0\xi - (3v_0 + 2L\theta_0 - 3v_L + L\theta_L)\xi^2 + (2v_0 + L\theta_0 - 2v_L + L\theta_L)\xi^3 \tag{4}$$

and substituting the second of Equations 3 into this gives

$$v = v_0 + L\theta_0\xi - \frac{L}{2}(\theta_0 + \theta_L)\xi^2 \tag{5}$$

so that v has been reduced in order from $0(z^3)$ to $0(z^2)$ and hence θ is $0(z)$ and so the curvature is constant ($0(z^0)$). This locking of the curvature arises from the combination of inconsistent field equations, with Dall'Asta and Zona (2001) demonstrating the oscillatory trends that may occur.

The latter study also explored using a 10-DOF element, with mid-length nodes for w_1 and w_2, or a 16-DOF element with one internal node for v, θ and three internal nodes for w_1, w_2, and showed that difficulties are still encountered with these higher order elements. This paper treats the FE problem from a different direction, and in deference to prescribing interpolation polynomials it starts with an internal solution for the slip in a similar fashion to Newmark's technique to develop an element using the direct stiffness method. The method is described first, and the use of this efficacious element is demonstrated for statically determinate and indeterminate beams.

2 GOVERNING EQUATIONS

The composite beam is shown in Figure 1, which is assumed to be subjected to a known loading regime. The concrete and steel are assumed to be elastic with moduli E_c and E_s respectively, with the elastic shear connection constitutive relationship between the interface shear flow q and slip s being $q = ks$, where k is the connector modulus. It is assumed that the curvature κ is equal in the two materials and that the Euler-Bernoulli theory of bending is valid. Recourse will be made to an invariant reference position at the top of the concrete slab.

The axial forces in each material are

$$N_{c,s} = \int_{c,s} E_{c,s}\varepsilon_{c,s}\,\mathrm{d}A_{c,s} \tag{6}$$

where the strains in each material are

$$\varepsilon_{c,s} = (y - y_{c,s})\kappa \tag{7}$$

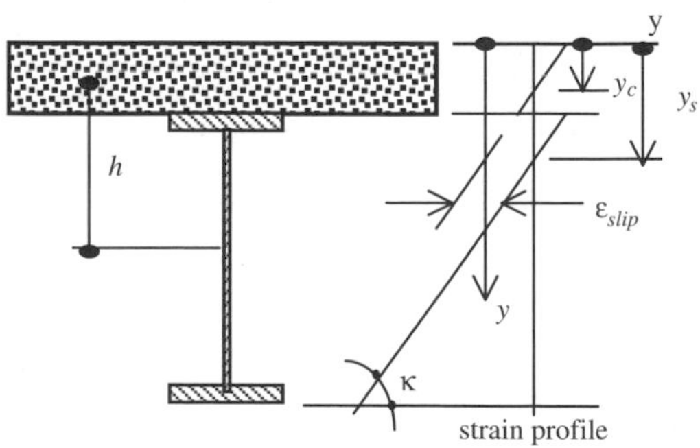

Figure 1. Composite cross-section.

in which y is the distance from the top and $y_{c,s}$ are defined in Figure 1.

Substituting Equation 7 into Equation 6 produces

$$N_{c,s} = (B_{c,s} - y_{c,s}A_{c,s})E_{c,s}\kappa \tag{8}$$

where $B_{c,s}$ is the first moment of area of the concrete or steel below the reference level. By using the condition of pure bending of the cross-section, $N_c + N_s = 0$ which produces the recurrence relationship

$$y_c = \frac{\overline{BE} - y_s A_s E_s}{A_c E_c}; \qquad y_s = \frac{\overline{BE} - y_c A_c E_c}{A_s E_s} \tag{9}$$

where $\overline{BE} = B_c E_c + B_s E_s$. The slip strain between the concrete and steel shown in Figure 1 is

$$\varepsilon_{slip} = \frac{\mathrm{d}s}{\mathrm{d}z} = \varepsilon_c - \varepsilon_s = (y_c - y_s)\kappa \tag{10}$$

so that

$$y_c = \frac{\overline{BE}}{\overline{AE}} - \frac{A_c E_c}{\overline{AE}\kappa}\varepsilon_{slip}; \; y_s = \frac{\overline{BE}}{\overline{AE}} - \frac{A_c E_c}{\overline{AE}\kappa}\varepsilon_{slip} + \frac{\varepsilon_{slip}}{\kappa} \tag{11}$$

where $\overline{AE} = A_c E_c + A_s E_s$.

The internal bending moment within the cross-section is given by

$$M_{\text{int}} = \int_A \sigma y\,\mathrm{d}A = \int_{A_c} E_c\varepsilon_c y\,\mathrm{d}A_c + \int_{A_s} E_s\varepsilon_s y\,\mathrm{d}A_s \tag{12}$$

which on using Equations 7 gives

$$M_{\text{int}} = \overline{EI}\kappa - (B_c E_c y_c + B_s E_s y_s)\kappa \tag{13}$$

where $\overline{EI} = E_c I_c + E_s I_s$ and $I_{c,s}$ are the second moments of area of the concrete and steel about the reference level. Substituting Equations 11 into Equation 13 produces

$$M_{\text{int}} = \left(\overline{EI} - \frac{\overline{BE}^2}{\overline{AE}}\right)\kappa + \frac{\overline{BE}}{\overline{AE}}\varepsilon_{slip}A_c E_c - B_c E_c \varepsilon_{slip} \tag{14}$$

Figure 2 shows a free body diagram of the slab of length δz at the interface. Horizontal equilibrium requires that

$$\left(N_c + \frac{\mathrm{d}N_c}{\mathrm{d}z}\delta z + q\delta z\right) - N_c = 0 \tag{15}$$

and so

$$\frac{\mathrm{d}N_c}{\mathrm{d}z} + q = 0 \tag{16}$$

which yields

$$\frac{\mathrm{d}N_c}{\mathrm{d}z} = -ks(z) \tag{17}$$

Substituting the first of Equations 9 into the first of Equations 11 for the slab then gives

$$N_c = B_c E_c \kappa - A_c E_c \frac{\overline{BE}}{\overline{\overline{AE}}}\kappa - \frac{\overline{\overline{\overline{AE}}}}{\overline{\overline{AE}}}\varepsilon_{slip} \tag{18}$$

where $\overline{\overline{AE}} = A_cE_cA_sE_s$, and so

$$\frac{\mathrm{d}N_1}{\mathrm{d}z} = \left(B_c E_c - \frac{A_c E_c \overline{BE}}{\overline{\overline{AE}}}\right)\frac{\mathrm{d}\kappa}{\mathrm{d}z} - \frac{\overline{\overline{\overline{AE}}}}{\overline{\overline{AE}}}\frac{\mathrm{d}^2 s}{\mathrm{d}z^2} \tag{20}$$

and using the equilibrium condition in Figure 2 gives

$$\left(B_c E_c - \frac{A_c E_c \overline{BE}}{\overline{\overline{AE}}}\right)\frac{\mathrm{d}\kappa}{\mathrm{d}z} - \frac{\overline{\overline{\overline{AE}}}}{\overline{\overline{AE}}}\frac{\mathrm{d}^2 s}{\mathrm{d}z^2} = -ks \tag{21}$$

The internal moment is equal to the applied moment $M(z)$, so from Equation 13 after rearrangement

$$\kappa = \gamma M + \alpha\varepsilon_{slip} \tag{22}$$

and so

$$\frac{\mathrm{d}\kappa}{\mathrm{d}z} = \gamma\frac{\mathrm{d}M}{\mathrm{d}z} + \alpha\frac{\mathrm{d}^2 s}{\mathrm{d}z^2} \tag{23}$$

where

$$\alpha = \frac{B_c E_c \overline{AE} - A_c E_c \overline{BE}}{\overline{AE}\,\overline{EI} - \overline{BE}^2},\quad \gamma = \frac{\overline{AE}}{\overline{AE}\,\overline{EI} - \overline{BE}^2} \tag{24}$$

Substituting Equation 23 into Equation 21 produces the generic differential equation for the partial interaction given by

$$\overline{\alpha}\left(\frac{\mathrm{d}^2 s}{\mathrm{d}z^2}\right) - ks = \alpha\left(\frac{\mathrm{d}M}{\mathrm{d}z}\right) \tag{25}$$

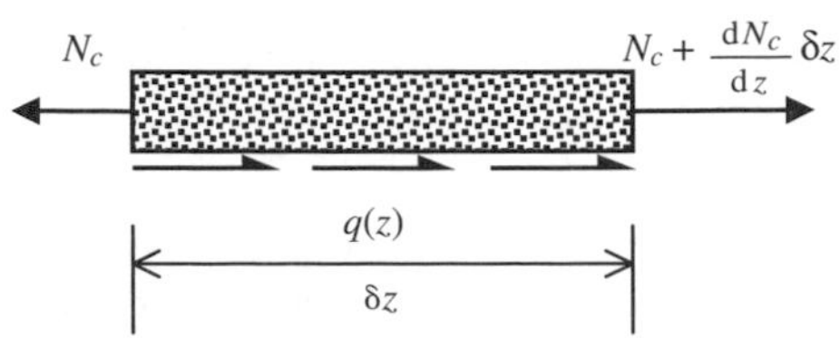

Figure 2. Free body diagram of slab at interface.

where

$$\overline{\alpha} = -\frac{(B_c E_c)^2 A_s E_s + (B_s E_s)^2 A_c E_c - \overline{\overline{AE}}\,\overline{EI}}{\overline{AE}\,\overline{EI} - \overline{BE}^2} \tag{26}$$

The formal solution of Equation 25 produces

$$s = C_1 e^{\nu z} + C_2 e^{-\nu z} - \frac{\alpha}{k}\left(\frac{\mathrm{d}M}{\mathrm{d}z}\right) \tag{27}$$

$$\varepsilon_{slip} = \nu C_1 e^{\nu z} - \nu C_2 e^{-\nu z} - \frac{\alpha}{k}\left(\frac{\mathrm{d}^2 M}{\mathrm{d}z^2}\right) \tag{28}$$

in which

$$\nu^2 = \frac{k}{\overline{\alpha}} \tag{29}$$

and C_1 and C_2 are constants of integration. When Equation 28 is substituted into Equation 22 and successively integrated with respect to z, then

$$\kappa = \gamma M + \alpha\nu C_1 e^{\nu z} - \alpha\nu C_2 e^{-\nu z} - \frac{\alpha^2}{k}\frac{\mathrm{d}^2 M}{\mathrm{d}z^2} \tag{30}$$

$$\theta = \gamma\int M\,\mathrm{d}z + \alpha C_1 e^{\nu z} + \alpha C_2 e^{-\nu z} - \frac{\alpha^2}{k}\int\frac{\mathrm{d}^2 M}{\mathrm{d}z^2}\mathrm{d}z + \hat{C}_1 \tag{31}$$

$$v = \gamma\int\int M\,\mathrm{d}z + \frac{\alpha C_1 e^{\nu z}}{\nu} - \frac{\alpha C_2 e^{-\nu z}}{\nu} - \frac{\alpha^2}{k}\int\int\frac{\mathrm{d}^2 M}{\mathrm{d}z^2}\mathrm{d}z + \hat{C}_1 z + \hat{C}_2 \tag{32}$$

where $\hat{C}_1$ and $\hat{C}_2$ are constants of integration. The axial force in the concrete is given after appropriate substitution into Equation 16 as

$$N_c = \gamma_1\kappa + \gamma_2\varepsilon_{slip} \tag{33}$$

in which

$$\gamma_1 = B_c E_c - A_c E_c \frac{\overline{BE}}{\overline{\overline{AE}}} \tag{34}$$

and

$$\gamma_2 = -\frac{\overline{\overline{\overline{AE}}}}{\overline{\overline{AE}}} \tag{35}$$

3 DIRECT STIFFNESS FE APPROACH

Firstly, the vectors of the nodal freedoms and actions at the ends of the element are defined respectively as

$$\vec{Q} = \langle V_0, M_0, N_0, V_L, M_L, N_L\rangle^{\mathrm{T}} \tag{36}$$

$$\vec{q} = \langle v_0, \theta_0, s_0, v_L, \theta_L, s_L\rangle^{\mathrm{T}} \tag{37}$$

where the subscripts denote the ends $z = 0$ and L.

Starting initially from a bending moment field specified by

$$M = R_0 z - M_0 \quad (38)$$

the direct stiffness procedure requires this field to be used in Equations 30 to 32 for an initially unloaded element whose freedoms are all restrained fully except for one, for which a unit value is assigned. Following this procedure gives

$$\kappa = \gamma(R_0 z - M_0) + \alpha \nu C_1 e^{\nu z} - \alpha \nu C_1 e^{-\nu z} \quad (39)$$

$$\theta = \gamma\left(\frac{R_0 z^2}{2} - M_0 z\right) + \nu C_1 e^{\nu z} + \nu C_1 e^{-\nu z} + \hat{C}_1 \quad (40)$$

$$v = \gamma\left(\frac{R_0 z^3}{6} - \frac{M_0 z^2}{2}\right) + \frac{\alpha C_1 e^{\nu z}}{\nu} + \frac{\alpha C_1 e^{-\nu z}}{\nu} + \hat{C}_1 z + \hat{C}_2 \quad (41)$$

with the slip becoming

$$s = C_1 e^{\nu z} + C_2 e^{-\nu z} - \frac{\alpha}{k} R_0 \quad (42)$$

Imposing a unit vertical displacement at $z = 0$ produces a kinematic field defined by the vector

$$\vec{q} = \langle 1,0,0,0,0,0 \rangle^{\mathrm{T}} \quad (43)$$

and when this is substituted into Equations 40 to 42 the constants of integration C_1 and C_2 may be determined (using $s_0 = s_L = 0$), $\hat{C}_1$ and $\hat{C}_2$ may be found (using $v_0 = 1$, $v_L = 0$) and R_0 and M_0 may be found (using $\theta_0 = \theta_L = 0$). Similarly, when

$$\vec{q} = \langle \delta_{1n}, \delta_{2n}, \ldots, \delta_{6n} \rangle^{\mathrm{T}} \quad (44)$$

is imposed for $n = 2, \ldots, 6$ (where δ_{ij} is the Kronecker delta) all of the constants of integration may be obtained at each step, and hence the corresponding entries in the load vector. The statement of the stiffness relationships then becomes

$$\vec{Q} + \begin{Bmatrix} -wL/2 \\ -wL^2/12 \\ -wL\omega \\ wL/2 \\ wL^2/12 \\ wL\omega \end{Bmatrix} = \begin{bmatrix} k_{11} & k_{12} & k_{13} & k_{14} & k_{15} & k_{66} \\ & k_{22} & k_{23} & k_{24} & k_{25} & k_{26} \\ & & k_{33} & k_{34} & k_{35} & k_{36} \\ & & & k_{44} & k_{45} & k_{46} \\ & \text{sym} & & & k_{55} & k_{56} \\ & & & & & k_{66} \end{bmatrix} \vec{q} \quad (45)$$

where the entries k_{ij} are given in Ranzi et al. (2003) and where

$$\omega = \frac{\nu\alpha(e^{-\nu L}+1)(\gamma_1\alpha+\gamma_2)/2 - \alpha(e^{-\nu L}-1)(Lk/12 - \gamma_1\alpha/L - \gamma_2/L)}{k(e^{-\nu L}-1)} \quad (46)$$

with w being the uniformly distributed load.

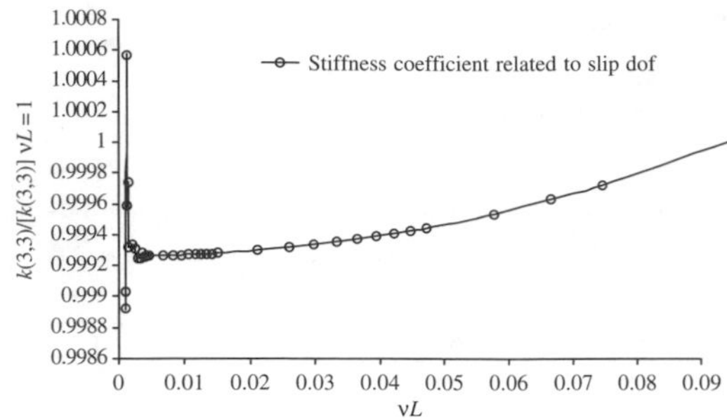

Figure 3. Behaviour of stiffness coefficient for the exponential form of the stiffness matrix.

Owing to the dependence of k_{ij} on terms involving exponential functions, some numerical instabilities were encountered in the direct stiffness technique, as shown in Figure 3 where the slip freedom k_{33}, normalised with respect to that with $\nu L = 1$, is plotted as a function of the dimensionless stiffness parameter νL. It is evident that the method can become unstable numerically when $\nu L < 0.05$. So that the technique remains robust, the slip in Equation 42 was expanded as a Taylor series truncated at the eighth term, viz.

$$s = C_1 \sum_{j=1}^{7} \frac{(\nu z)^j}{j!} + C_2 \sum_{j=1}^{7} (-1)^j \frac{(\nu z)^j}{j!} - \frac{\alpha}{k} R_0 \quad (47)$$

with the stiffness entries (corresponding to those in Equation 46) being given by Ranzi et al. (2003). When $\nu L > 0.05$, the exponential forms for k_{ij} were adopted, while when $\nu L \leq 0.05$, the polynomial ones based on Equation 47 were used. In all applications of the method, it was found that the solutions were the same with $\nu L = 0.05$, regardless of its actual value for the element. It is worth noting that the dimensionless stiffness parameter νL derived in this paper is identical to the parameter χ of Girhammar & Pan (1993), but which is expressed in terms of centroidal values rather than the reference level values used here.

4 APPLICATION TO A SIMPLY SUPPORTED BEAM

A simply supported beam of length 12 m was investigated using the present technique, using one element. To expedite calculation of the section properties, the slab was idealised as an elastic rectangle of width 600 mm, depth 300 mm and of modulus 20 kN/mm^2, while the steel was idealised as an elastic rectangle of width 60 mm, depth 300 mm and of modulus 200 kN/mm^2. The beam was cast shored (Oehlers & Bradford 1995) and supported a uniformly distributed load of 35 kN/m.

The effect of the degree of interaction has been investigated by varying the stiffness parameter νL,

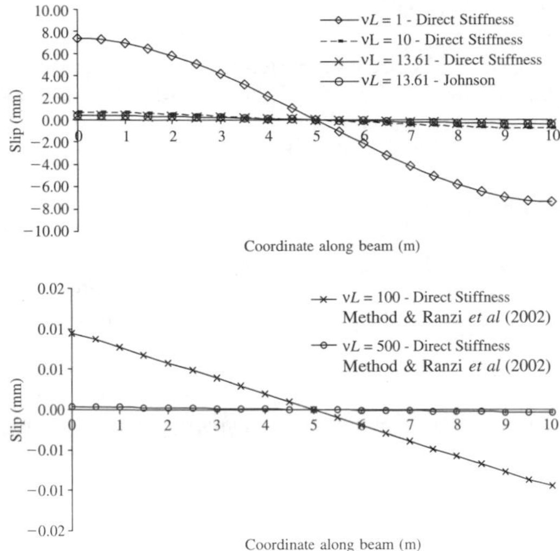

Figure 4. Slip along a simply supported beam.

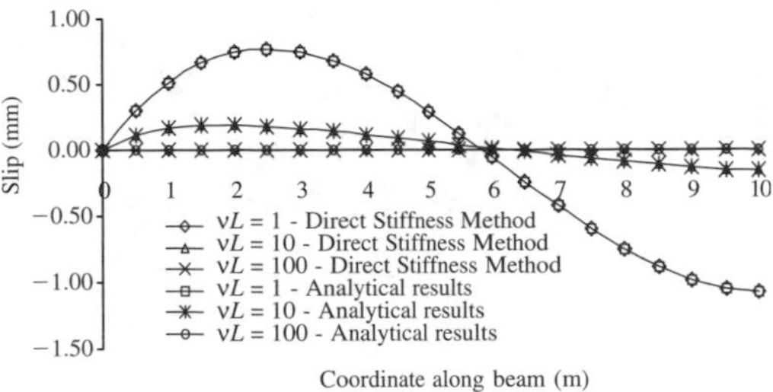

Figure 5. Slip along a propped cantilever.

with particular emphasis on the ability of the element to overcome the locking phenomenon that has been demonstrated by Dall'Asta & Zona (2001) to be significant when $\nu L > 10$. Figure 4 shows the variation of the slip along the beam for a variety of stiffnesses, and in particular for a value of νL = 13.61 that has been worked as a practical example in Johnson (1994), using a connector stiffness and spacing of 150,000 kN/m and 180 mm, for which a value of $k = 150{,}000/0.180 = 833\times10^6$ kN/m^2 has been adopted. Johnson's example is based directly on the closed form Newmark solution, and the results of the present method are identical with those in Johnson's text. Moreover, the efficacy of the technique has been demonstrated for very high values of the stiffness parameter ($\nu L = 100$ and 500) that would cause locking in Equation 5, but which does not present a complication with the direct stiffness approach.

5 APPLICATION TO PROPPED CANTILEVER

A propped cantilever with the same geometric properties as considered in the previous example has been studied when subjected to a uniformly distributed load of 10 kN/m. The numerical results shown in Figure 5 are identical to those obtained by the closed form technique of Ranzi et al. (2002) that is based on the formal solution of the governing differential equations.

6 APPLICATION TO A CONTINUOUS BEAM

In a continuous concrete-steel composite beam, the portion of the member in the vicinity of the internal support experiences negative bending that may cause the concrete to crack, and a simplifying assumption may be to take its elastic modulus as zero. The location of the point of inflexion that defines the negative bending zone is unknown *a priori*, and requires iteration. Using the direct stiffness method, the modulus E_c is taken as constant throughout, the inflexion point then determined using a simple algorithm, and the value of E_c is taken as zero in the negative region in the numerical model to determine a new inflexion point. The procedure is then repeated until the point of inflexion is converged upon within a predetermined tolerance.

The analysis of a three span composite beam (each span being of 10 m) that is continuous over two internal supports has been undertaken using the direct stiffness approach. The uniformly distributed load is 10 kN/m and the cross-section is the same as that of the previous examples, and the connection modulus was taken as 12,000 kN/m^2. The results shown in Figure 6 demonstrate that the iterative technique to identify the inflexion points converges sensibly after only four iterations.

7 CONCLUSIONS

This paper has presented an alternative method to the usual stiffness-based methods to derive an element for the FE analysis of composite concrete-steel beams with partial interaction. The technique is a direct stiffness approach, which starts from the formal solution of the governing differential equations, and develops the stiffness relationship for the element by imposing successive unit deformations upon the freedoms. It is superior to the more familiar technique of starting with prescribed polynomial interpolation functions, in that it is devoid of the curvature-locking phenomenon that occurs with the use of elements based on interpolation functions.

The numerical method has been validated against closed form solutions for a simply supported beam and for a propped cantilever, and was shown to give reliable results over the full range of elastic shear

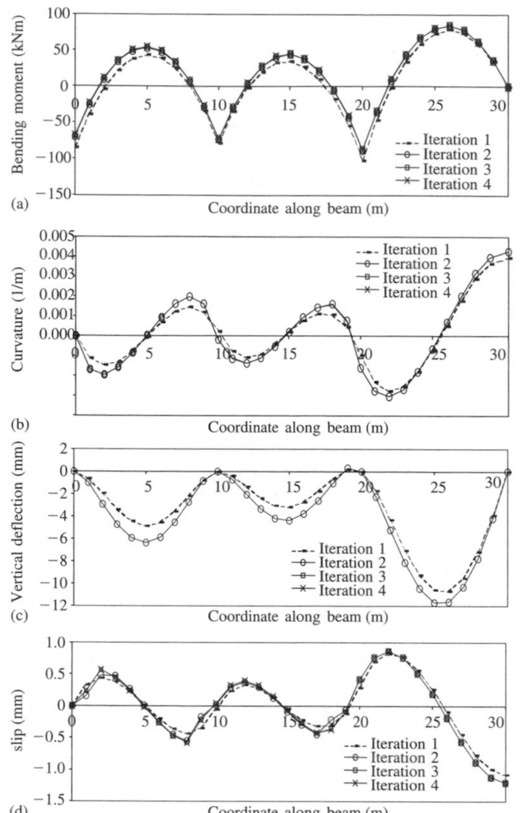

Figure 6. Three-span continuous beam with concrete cracking. (a) Bending moment diagram, (b) Curvature diagram, (c) Vertical deflection, (d) Interface slip.

connection stiffnesses. It was also used to study a continuous composite beam, and a simple technique to allow for concrete cracking was addressed and demonstrated. The technique is also amenable, with little modification, for incorporating material nonlinearities such as shrinkage, creep and yielding.

ACKNOWLEDGEMENT

This work has been supported in part by a Professorial Fellowship awarded to the second author by the Australian Research Council.

REFERENCES

Arizumi, Y., Hamada, T. & Kajita, T. 1981. Elastic-plastic analysis of composite beams with incomplete interaction by the finite element method. *Comps. & Structs.* 14(5–6): 453–462.

Dall'Asta, A. & Zona, A. 2000. Nonlinear analysis of composite beams by a displacement approach. *5th Int. Conf. on Computational Structures Technology*, Leuven, Belgium, 337–348.

DallAsta, A. & Zona, A. 2001. Finite elements for the analysis of composite beams with interlayer slip. *XVIII Italian Workshop on Steel* (CTA), Venice, Italy.

Daniels, B.J. & Crisinel, M. 1993. Composite slab behavior and strength analysis. Part I: Calculation procedure, J. Struct Eng., ASCE. 119(1):16–35.

Girhammar, U.A. & Pan, D. 1993. Dynamic analysis of composite members with interlayer slip. *Int. J. Solids & Structs.* 30(6):797–823.

Johnson, R.P. 1994. *Composite Structures of Steel and Concrete.* Oxford:Blackwell Scientific Publishers.

Newmark, N.M., Siest, C.P. & Viest, I.M. 1952. Tests and analysis of composite beams with incomplete interaction. *Proc. Society of Experimental Stress Analysis.* 9(1):75–92.

Oehlers, D.J. & Bradford, M.A. 1995. *Steel and Concrete Composite Structural Members: Fundamental Behaviour.* Oxford:Pergamon.

Ranzi, G., Bradford, M.A. & Uy, B. 2002. A general method of analysis of composite beams with partial interaction. Submitted for publication.

Ranzi, G., Bradford, M.A. & Uy, B. 2003. A direct stiffness analysis of a composite beam with partial interaction. *UNICIV Report*, The University of New South Wales, Sydney, Australia.

Computational Modelling of Concrete Structures, Bićanić et al. (eds)
© 2003 Swets & Zeitlinger, Lisse, ISBN 90 5809 536 3

Size effect of sandstone and concrete

Stefan L. Burtscher & Johann Kollegger
Institute for Structural Concrete, Vienna University of Technology

ABSTRACT: The compressive strength is often regarded as a material property. However, from experimental investigations it is known that for many types of fracture the strength is dependent on the size of the structure. Some investigations were done in the past on size effect in tension. Very few experiments were performed in compression, and aside from that most investigations show a very limited size range. In this paper investigations on sandstone and concrete are presented on larger size ranges. The size range of the sandstone series was 1:32 (here only 1:8 can be presented) and for the concrete series the size range was 1:16.

The experimental results show a size effect on the nominal strength for the sandstone and the concrete tests. The Size Effect Law by Bažant and the Multifractal Scaling Law by Carpinteri were applied to the test data. The tends of the laws are compared to the trends of the data.

1 INTRODUCTION

Experimental results on materials like concrete or rock showed that the specimen size has an influence on nominal strength and ductility. Therefore the experimentally determined strength is not a material property, but is dependent on the size of the structure. In the case of concrete, cylinders with a diameter of 15 *cm* and a length of 30 *cm* are often used. With the strength determined on these small specimens structures that are many scales larger are designed. The strength of the structure could only be determined by testing of the structure itself. Usually the size of structures exceeds the capability of testing machines. In the laboratory structures are usually scaled versions of real structures. The experimental results of small scale specimens can give information or insight into the structural behavior, but the influence of the size on the structural strength and ductility cannot be captured.

In the last three decades size effect has been studied intensively by numerous researchers. Most of the experiments were performed on various specimen geometries on concrete under tensile loading. The experiments were used to verify the size effect models. The most important models are the Size Effect Law by Bažant (Bažant 1984) and the Multifractal Scaling Law by Carpinteri (Carpinteri 1994) or (Carpinteri and Ferro 1994). One main disadvantage of these laws are that they have to be fitted for each material, geometry and each type of test.

1.1 *The size effect law by Bažant*

The Size Effect Law (SEL) was presented by (Bažant 1984). This theory is based on stress redistributions. In this theory small sizes are bound to plasticity limit, and large sizes follow the Linear Elastic Fracture Mechanics. The SEL was derived for geometrically similar structures with notches as

$$\sigma_N = \frac{B f_t}{\sqrt{1 + \frac{D}{D_0}}}, \tag{1}$$

where f_t is the tensile strength, D is a characteristic dimension of the structure, B is constant and D_0 is the transition size from plasticity limit to Linear Fracture Mechanics, which is also constant. The constants B and D_0 have to be determined by fitting of experimental data.

Compression failure is often fitted to the SEL, see equation 1. It was often argued that the failure mode of compressed concrete specimens can be considered as resulting from local tensile mechanisms, or from a combination of tensile and shear mechanisms depending on specimen geometry. In a paper by (Bažant and Xiang 1997) a dependence of the strength on size was derived as

$$\sigma_N = k_c D^{-2/5} . \tag{2}$$

In the analysis of (Bažant and Xiang 1997) it was assumed that the crack width is proportional to the material length and therefore constant and that the spacing of axial micro cracks is proportional to $D^{-1/5}$ (Bažant et al. 1993), which leads to a slope of 2/5 for the LEFM asymptote in the logarithmic plot. For the fitting of parameters both laws will be used for comparison in section 5 of this paper.

1.2 *The Multifractal Scaling Law by Carpinteri*

The Multifractal Scaling Law (MFSL) proposed by Carpinteri (Carpinteri 1994), (Carpinteri and Ferro 1994) writes as

$$\sigma_N = f_c\sqrt{1+\frac{l_{ch}}{d}} = \sqrt{A+\frac{B}{D}}\,, \tag{3}$$

where σ_N is the nominal compressive stress, f_c is the compressive strength of an infinitely large specimen and l_{ch} represents a internal material length and f_c is the strength for infinitely large specimens. Alternatively the second expression can be used for data fitting, where A and B are the constants that have to be fitted from experimental data.

In this model the mechanical properties are strictly connected with the considered scale of observation. For larger scales, the influence of disorder progressively vanishes.

1.3 *Aim of the study*

The purpose of this experimental investigation was the further investigation of the size influence on the strength of granular materials under compressive loading. Such phenomena can be observed on large size ranges more clearly than on small size ranges. The smallest size is limited by the representative volume. The largest size is limited by the maximum load and dimension of the testing machine. The Column Tester at the Vienna University of Technology allows to increase the size range because of its high testing load and large dimensions.

1.4 *Testing facilities*

The experiments were performed with the Column Tester and the D2 machine in the joint Laboratory of the Institute for Structural Concrete and the Institute for Steel Structures at the Vienna University of Technology. Additional tests were performed in the Inova machine at the Department of Structural Mechanics at CTU Prague.

The experimental investigation on the size effect is a difficult task, because one has to deal with different sizes of geometrically equal specimens, where the same boundary conditions have to be met for all specimens and specimen sizes. All geometric dimensions of the specimens have to be scaled by the same factor. The larger the factor from the smallest to the largest specimen, the more pronounced is the effect of size on the observed quantity. When multiplying all the specimen dimensions by a scale factor the length changes linearly with the size factor, the cross section changes by the order of two with the size factor and the volume by the order of three. This means practically that the testing frame has to be adapted to the length of each specimen type and that the maximum load changes by the order of two with the size factor from one size to the other (disregarding the size effect). Due to strongly varying maximum loads different testing frames had to be used.

At the Institute of Structural Concrete at the University of Technology, Vienna a massive frame structure (Schwarz 1987) is available. In order to carry out the size effect tests the frame was equipped with 4 servo hydraulic actuators, a computer supported data acquisition system, equipment for implementation of the actuators and setup for column testing, see figure 1. It is now possible to perform experiments with closed loop control on specimens with a length from 40 *cm*

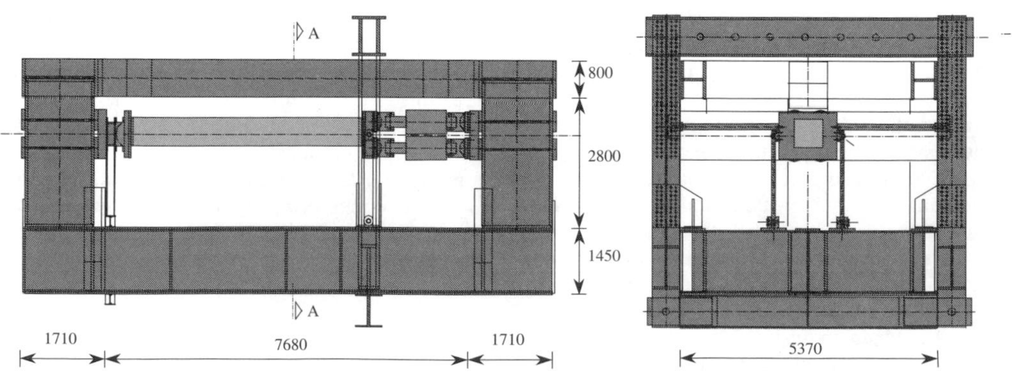

Figure 1. The Column Tester with equipment for column length of 5.0 m.

to 5 m, a cross section up to 120 × 100 cm and with a maximum load of 17000 kN. The same experimental set-up was used for spun concrete columns (Burtscher et al. 2002). This testing frame allows to test real size structures. Hence it is possible to compare the structural behavior of real size structures with laboratory size structures.

1.5 *Size effect experiments*

Most of the investigations were performed under tensile loading on compact test specimens and direct tension on dog bone specimens. Few experimental investigations were performed on the size effect of concrete under compressive loading. Experiments on unreinforced concrete found in literature were performed by Sener et al. (Sener et al. 1999), Hollingworth (Hollingworth 1998) and on reinforced concrete columns by Němeček (Němeček 2000) and Bažant et al. (Bažant and Kwon 1994). Additionally compressive tests on notched PEEK (poly-ether-ether-keton) specimens were performed by (Bažant et al. 1999). The size range in the experiments was not exceeding 1:4. In this paper a size effect series on sandstone with a size range of 1:32 (but here only a size range of 1:8 can be presented) and a size effect series on concrete with a size range of 1:16 are presented (Burtscher 2002) in figure 2. To indicate the huge geometric dimensions and the large size range of the sandstone and concrete series performed, the specimens of the test series are compared to the specimens from test in literature, see figure 3. The dimensions and the reinforcement of the specimens were all scaled by a factor for every size, thus geometrical similarity was fulfilled.

2 SOURCES FOR SIZE EFFECT ON STRUCTURAL STRENGTH

According to (Bažant and Planas 1998) there are six different sources for size effect in concrete:

Boundary layer effect is due to the fact that the concrete surface layer adjacent to the formwork has a smaller relative content of large aggregates, in comparison to the interior of the member. This layer has different properties, most important is the smaller stiffness. The thickness of this layer is dependent on the maximum aggregate and is independent of the structure size. The size effect is due to the fact that in a smaller member the boundary occupies a large portion of the cross section, while in a large member it occupies a small part of the cross section.

A second type of boundary layer effect arises because of the different elastic properties between the boundary layer and the interior. Normal stresses parallel to the boundary cause transverse stresses in the interior, while at the surface these stresses are zero.

The third type is due to the Poisson effect, that causes the surface layer to be nearly in plane stress, while the interior is nearly in plane strain.

Diffusion phenomena such as heat conduction or pore water transfer. The diffusion half times are proportional to the square of the size of the structure. The diffusion process changes the material properties, produces inelastic strains and cracking, e.g. drying may lead to cracking in the surface layer. Due to different drying times the extent and density of cracking could be different in small and large members.

(a) Specimen of concrete series on sizes S (diameter 5 cm) to XXL (diameter 80 cm)

(b) Specimen of test series on sizes XS to XXL smallest specimen in hand)

Figure 2. Overview of specimen of the concrete series and the test series on sizes XS to XXL on sandstone.

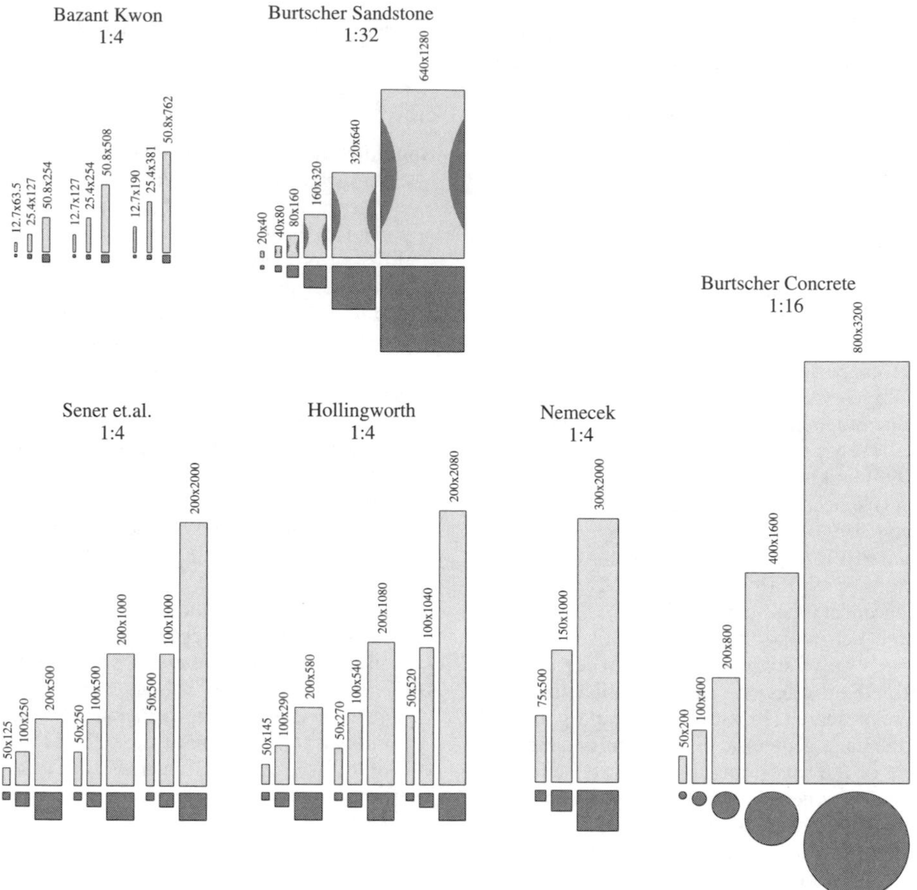

Figure 3. Comparison of different specimens for size of effect tests in compression [mm].

Hydration heat or phenomena associated with chemical reactions. This effect is related to the previous one due to the half times of dissipation of the hydration heat. Thicker members heat to higher temperatures and may cause cracking, induce drying, accelerate chemical reactions and alter the material properties.

The statistical size effect is the oldest known source of size effect and is based on the statistical distribution of the strength in the material. This size effect is originated by Weibull (1939) and is based on the model of the chain, whose maximum load is determined by the lowest strength in the chain. Therefore the longer the chain the lower the maximum load of the chain.

Fracture mechanics size effect is due to the release of stored energy into the fracture front. It is stated in literature that this effect is the most important source of size effect. Laws describing this source are given for tension and compression in section 1.1.

Fractal nature of crack surface. The size effect due to fractality of the fracture surface is still controversially discussed. A law describing this source is given in section 1.2.

Additionally due to preparation of specimens an additional source especially for small concreting heights (below 20 to 30 cm) is the:

Top layer strength. When fresh concrete is poured into a formwork the layer below the top surface has lower strength after hydration. After casting the water ascends from the concrete to the top of the surface. Thus more water is in the top layer, which is a source for lower strength.

This source can affect the strength of concrete columns that were produced in horizontal position.

Table 1. Sandstone specimens used for size effect tests.

Specimen	Number	Dimensions [mm]	Eccent. [mm]	Weight kg	Testing frame
XXL	3	640 × 640 × 1280	32	1150	Column Tester
XL	3	320 × 320 × 640	16	145	Column Tester
L	6	160 × 160 × 320	8	18	D2
M	6	80 × 80 × 160	4	2.3	D2

3 SANDSTONE EXPERIMENTS

3.1 *Advantages of sandstone*

In order to determine the fracture mechanics size effect of a granular material it is important that the material is the same in the whole volume of each specimen. In the case of concrete this is difficult to achieve because of the boundary layer effect, the microcracks due to hydration effects and due to diffusion phenomena, see section 2. These phenomena have different influence for different sizes and therefore may change the material response, which is not wanted for such kind of tests. These effects are introduced into the concrete during the moulding process. To avoid such unwanted effects, tests on sandstone were performed. This material has several advantages.

- The material shows no or a vanishing boundary layer effect (of first type in section 2). No hydration or diffusion effects of different kind took place in the material.
- Preliminary tests showed that the scatter of strength values was very low for different specimen.
- The average grain size is 0.5 mm (average pore size is 0.6 mm), which allows to test very small specimen sizes, without going beyond the conditions for a representative volume.
- The stress-strain response is nearly linear up to peak load.

The big disadvantage of sandstone is that the material parameters can scatter strongly, when inclusions or other inhomogeneities are present. There were inhomogeneities in the sandstone. But due to a proper sawing scheme all visible inhomogeneities were outside the notched region and did therefore not affect the strength. In the notched region the sandstone showed a very homogeneous structure.

3.2 *Specimen dimensions and geometry*

In the *test series on sandstone* specimens of 6 different sizes from XS to XXL with a total size range of 1:32 were tested. Currently only the sizes from M to XXL are available, thus the size range is only 1:8. All specimens for the *test series on sand-stone* were sawed in the quarry at St. Margarethen im Burgenland from one block with the dimensions 5.0 × 1.3 × 1.3 *m*. The dimensions of the specimens varied from 80 × 80 × 160 *mm* (M-size specimens) to 640 × 640 × 1280 *mm* (XXL-size specimens), see table 1. All specimens had the same geometric dimensions just scaled by a factor (similarity in three dimensions), see table 1. All specimen sizes from XS to XXL are in figure 2 (smallest specimen in hand) and one can see the impressive size range. From the XXL and the XL size 3 specimens were tested and 6 from the L and M sizes.

When a prismatic specimen is tested under compression the failure mode often is the spalling of edges near the loading surface. The specimen is influenced in that region by the boundary conditions. If failure occurs in that region the damage that precedes the failure is influenced by the interface and thus may affect the maximum load or the post peak behavior. Additionally and even more important is that it cannot be distinguished if the failure is due to improper boundary conditions (e.g. uneven loading surface) or by exceeding the material strength. Thus the specimen should be notched, that the location where fracture starts is predefined and is not located on the specimen ends. This makes assessing and in certain cases the control of the test easier. For tests on specimens loaded in tension or for three point bending tests notches are often used. In compression this is more problematic because the fracture band propagates inclined to the loading direction (Bažant et al. 1999). The inclination of the notch has to be equal to the propagating crack band. In this investigation smooth notches were used to circumvent this problem. In a smooth notch the conditions are also closer to a real structure, because there is no sharp notch tip where the stresses become theoretically infinite. The notch was investigated in a *preliminary test series* and is shown in figure 4.

3.3 *Loading arrangement*

All the specimens were loaded with a small eccentricity. The eccentricity was also scaled with the specimen size and was e = D/20 from the centroid of the notched cross section. This was done to circumvent failure by axial splitting, which is a failure mode that does not show a size effect. Additionally, the exact centric loading of a column is difficult to establish experimentally and due to the heterogeneity of the material the stress

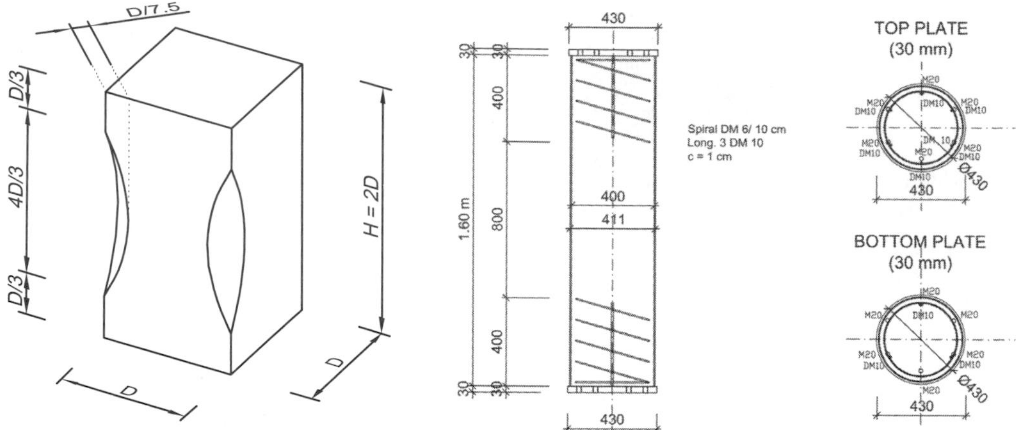

Figure 4. Geometry of the specimens with notch of sandstone series (left). Geometry and reinforcement of specimen XL of concrete series (right).

Figure 5. Installing of specimen 80b in the Column Tester (left). Specimen 80b prior to testing (right).

states in the cross section are never uniform. Another advantage is that due to the higher stress concentration the surface where the failure starts was predefined and could be observed during experiment.

In all experimental set-ups the loading platens were placed on longitudinal hinges on both ends of the specimen. The position of the force was defined by the axis of the hinge. Due to the rotating capability in one direction no moment could be transferred or generated during the experiment, i.e. rotation of the loading platen was possible without moment build up. To introduce the eccentricity the specimen was shifted in direction normal to the hinge axis by the amount of the desired eccentricity. The load did not change the position and did not rotate during experiment.

3.4 *Interface between loading platen and specimen*

In a *preliminary test series* various systems were investigated to provide a proper load transfer from loading platen to the specimen. For the size effect tests a layer of teflon and double sided sandpaper was used. The thickness of the teflon was 0.75 *mm*. The teflon layer was used to prevent stress peaks on the sand-stone loading face, due to an eventually bumpy surface of the sandstone. In the locations of the bumps stress peaks could arise. Teflon yields at very low stresses and therefore prevents stress concentrations on the loading surfaces of the specimen. The friction between steel and teflon was very low, which could lead to a translation of the specimen on the interface during experiment. This was intensified by eccentric loading of the specimen, especially when the post peak region was entered. Therefore the double sided sand paper was positioned between the loading plates and the teflon layer. The thickness of the teflon layer was not changed with specimen size, because the unevenness of the surface was assumed to be the same for all specimens.

Table 2. Geometric dimensions, eccentricity, weight testing machine of the concrete size effect series.

Specimen	Number	Diameter [*cm*]	Length [*cm*]	Eccent. [*cm*]	Weight [*kg*]	Testing Machine
XXL	3	80	320	2.	4020	Column Tester
XL	3	40	160	1.	500	Column Tester
L	3	20	80	0.5	65	Column Tester
L	2	20	80	0.5	65	Inova Prague
M	3	10	40	0.25	8	Inova Prague
M	3	10	40	0.25	8	D2
S	6	5	20	0.125	1	D2

4 CONCRETE EXPERIMENTS

In the concrete series 5 different sizes from S to XXL with a total size range of 1:16 were tested. All dimensions were scaled for every size by the same factor, thus geometrical similarity was provided. The aim of the series was to find the influence of size on strength, namely the size effect.

4.1 *Specimen dimensions*

The columns tested had a circular cross section. The diameter of the cross sections varied from 5 *cm* to 80 *cm*, see table 2 for the test program. The specimen length was always four times the diameter. All dimensions were scaled by a scaling factor. Therefore the specimens of the different sizes were all geometrically equal. In order to prevent failure and to ensure a proper load transfer from the loading platen to the specimen, the specimens were equipped with steel plates on both ends and spiral reinforcement, see figure 4. The spiral reinforcement was placed over the height of the diameter on both ends of the specimen and was scaled for the different sizes.

It is favorable to test specimens over a large size range to determine the influence of strength on size. The size of the smallest specimen is limited, because it has to be larger than the representative volume. Thus the smallest dimension has to be a multiple of the maximum aggregate size (8 *mm*). The ratio of the smallest dimension (50 *mm*) and the aggregate size was 6.25. It is stated in literature that this ratio is high enough to describe a representative volume. The maximum size of the largest specimen is limited by the capability of the testing equipment. The largest specimen tested in this series was the XXL size with a maximum load of 14380 *kN*. The size range from size S to XXL was 1:16. If the size range of the series would have been increased from 1:16 to 1:32 the maximum load of the XXXL size (not regarding the size effect) would have been approximately 58000 *kN*, which is beyond testing capabilities at Vienna University of Technology. The XXL size specimen of this series increased the effort for producing the formwork, the handling and the testing enormously. The weight of one XXL specimen was 4280 *kg* including steel plates, see table 2. For the smallest size the number of specimens was increased, because small dimensions and the small eccentricities lead to larger scatter of the experimental values. The specimens of sizes XXL, XL and L were tested in the Column Tester and in the Inova machine. The specimen of size M were tested in the Inova machine and the D2 machine.

4.2 *Specimen preparation*

To minimize sources of size effect that affect the material for the different sizes, namely diffusion phenomena, hydration heat and top layer strength already mentioned in section 2 special demands on moulding and formwork were necessary. All specimens were cast using one batch of concrete. 7 m^3 were necessary to produce the specimens. The formwork was made of cardboard tube with 400 μm PVC foil inside, see specimen XXL in figure 2. The foil was advantageous for several reasons:

- The formwork did not absorb water at the beginning, which would have led to drying of the concrete surface at very early ages.
- No moisture gradient could develop and therefore no diffusion phenomena could take place.
- The water that was available for the concrete to hydrate was for every specimen the water that was added during mixing.
- The cardboard tube was a good isolator. Therefore no temperature gradient could develop, that could influence the hydration and produce regions with different microcracks. However the temperature due to hydration was higher in the XXL specimens than in the S size specimens.

The formwork was removed between one hour and 8 hours before testing. The specimens were sealed during curing and no diffusion phenomena could take place before testing.

The specimens were produced in a standing position to minimize influence of the lower strength of the concrete below the surface, which is especially pronounced when specimens are produced horizontally. In the XXL specimen the region with lower strength was inside the reinforced region. For the sizes L, M and S the weak part would have been in the unreinforced region and the strength of the specimen would have been reduced due to the preparation process. This was prevented by adding an additional cardboard tube with a length of 20 *cm* at the top. This cardboard tube and the concrete was removed approximately 30 minutes after casting of the specimens.

The concrete was brought with a concrete transporter and was filled into the formwork with a concrete pump. For compaction vibrating units with diameter 4 *cm* were used for the XXL specimens and diameter 2 *cm* for XL, the L sizes. The size M and S were compacted using a shaking table. In the following week the concrete was watered and covered with foil to prevent drying at young age. With the specimens, concrete cylinders with a diameter of 10 *cm* and a height of 20 *cm* were produced. The formwork of the concrete cylinders was the same as for the big specimens.

4.3 *Loading arrangement*

All specimens were tested under eccentric loading. The eccentricity was scaled with the specimen size and was always $e = D/40$, see table 2. For the tests on concrete the same testing equipment was used as for the test series performed on sandstone.

4.4 *Interface between loading platen and specimen*

The loaded faces of the specimens had to obey special accuracy and evenness. With the concrete face itself this would not have been possible, therefore steel plates were used on both ends to provide the specimens with an even surface. Additionally reinforcement was inserted in a height of the specimen diameter to increase the strength in that region and compensate inaccuracies due to lack of right angles of the steel plates and disturbances from the testing machine.

5 RESULTS

5.1 *Results from sandstone experiments*

One of the main issues of the size effect theory is the influence of strength on the size of the specimen. The strength was determined on the compressed side as maximum stress according to equation

$$\sigma_{notch} = \frac{F}{A_{notch}} \pm \frac{F \cdot e_{notch} \cdot y_{notch}}{I_{notch}} . \qquad (4)$$

In figure 6 the strength is plotted versus specimen size. The results are displayed with linear (left) and logarithmic scaling (right) of the axis. In the plots a decrease of strength with increasing specimen size could be determined. Thus a size effect was present in the test series. The mean strength decreases from the M (8 cm) to the XXL (64cm) sizes from 23.1 Mpa to 19.4 Mpa, which corresponds to a strength decrease of 16% for a size range of 1:8.

After testing the fracture patterns of the specimens were determined. All specimens of this series showed a fracture of a transversally propagating band and were therefore used for comparison of the results. Except specimen 64b, which showed fracture only at the end of the specimen. 64b had no cracks in the notched region. It was concluded that an inhomogeneity with higher strength was in the notched region. Thus the specimen did not fail in the weaker cross section, but towards the ends. The specimen 64b was excluded for further comparison of results.

In figure 6 the Size Effect Laws by Bažant (SEL, SEL-compression) and the Multifractal Scaling Law by Carpinteri (MFSL) were fitted to the data. For simplicity and easier comparison the SEL for compression was used in a similar format as the SEL. The laws and the parameters for the data fit is given in table 3. The normalized strength versus size is given in linear (a) and double logarithmic scaling (b). In figure 6 the laws are also drawn for sizes that were larger than the tested ones. The SEL and the SEL-compression laws are on the same line inside the tested range and show only small differences at the maximum size plotted. The trend from the SEL and SEL-compression follows better the data than the trend from the MFSL.

5.2 *Results from concrete experiments*

The concrete tests were performed on specimens with circular cross section and a height to diameter ratio of 4:1. 5 different sizes were tested which were all geometrically equal. The size range was 1:16. The strengths of the specimens were calculated on the compressed side. All tests were performed between the 48th and the 75th day after moulding. The increase of concrete strength in this time range was determined on accompanying tests on concrete cylinders. It was assumed that the strength of the cylinders increased by the same ratio as the strength of the specimens. The strength were all calculated for the 52nd day after moulding and plotted in figure 7. The SEL and the MFSL were fitted to the data and plotted in figure 7. The parameters for the data fit is given in table 3. The results are similar to the results of the sandstone series. The trend of the large sizes shows clearly a decreasing strength with size. The difference between SEL and SEL-compression is little for small and large sizes. In the range of experimental data both laws are on a line.

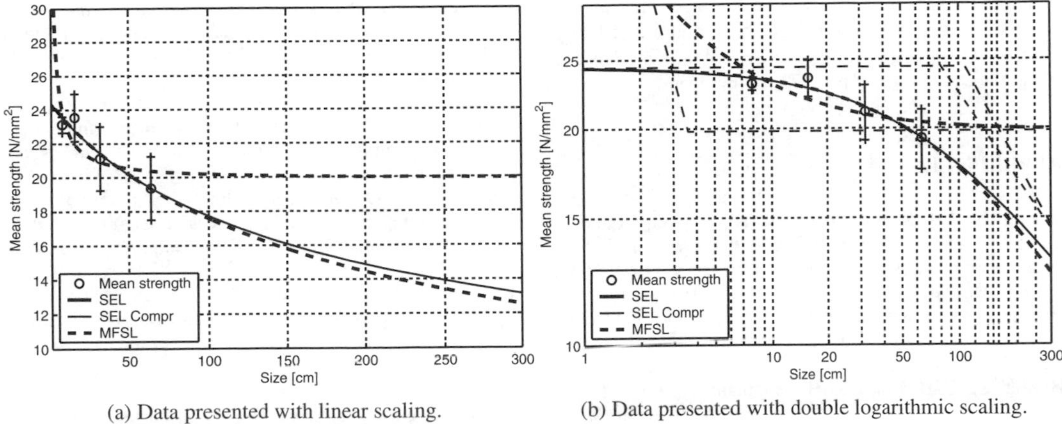

(a) Data presented with linear scaling.

(b) Data presented with double logarithmic scaling.

Figure 6. The mean strength of the specimens of the sandstone series and data fit with MFSL, SEL and SEL-compression.

Table 3 Table of parameters of different size effect laws for sandstone tests and concrete tests. D_0, l_{ch} in cm, Bf_t, f_c in N/mm^2.

SEL	SEL compression	MFSL
$\sigma_N = \frac{Bf_t}{\left(1+\frac{D}{D_0}\right)^{\frac{1}{2}}}$	$\sigma_N = \frac{Bf_t}{\left(1+\frac{D}{D_0}\right)^{\frac{2}{5}}}$	$\sigma_N = \sqrt{A + \frac{B}{D}}$
	Parameters of sandstone series	
$Bf_t = 24.4$	$Bf_t = 24.5$	A = 393.1, $f_c = 19.8$
$D_0 = 107.9$	$D_0 = 80.3$	B = 1424.6, $l_{ch} = 3.62$
r = 0.87	r = 0.89	r = 4.2
	Parameters of concrete series	
$Bf_t = 40.7$	$Bf_t = 40.8$	A = 1166.5, $f_c = 34.15$
$D_0 = 140.1$	$D_0 = 104.8$	B = 2861.4, $l_{ch} = 2.45$
r = 10.1	r = 10.1	r = 14.7

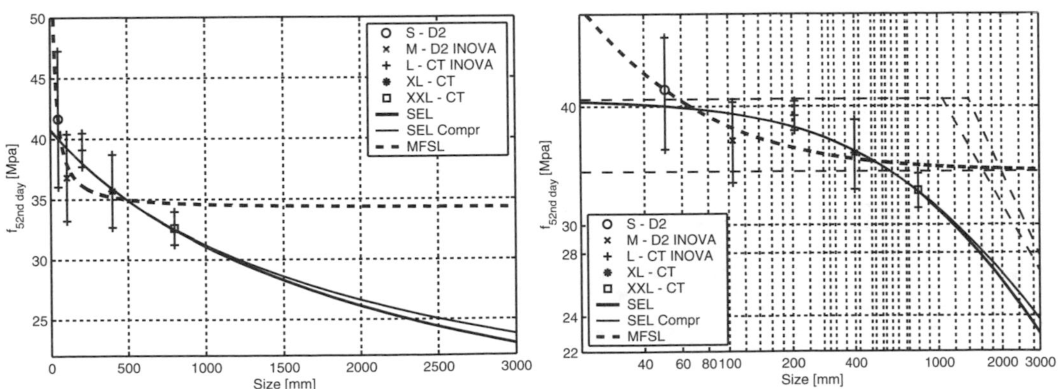

Figure 7. The mean strength of the specimens of the concrete series at 52nd day and data fit with MFSL, SEL and SEL-compression.

When extrapolating the laws to larger sizes one can see clearly in the double logarithmic plot, that the SEL and the SEL-compression follow the trend indicated by the data. The MFSL does not follow this trend.

REFERENCES

Bažant, Z., J.-J. H. Kim, I. M. Daniel, E. Becq-Giraudon, and G. Zi (1999). Size effect on compression strength of fiber composites failing by kink band propagation. *International Journal of Fracture 95*, 103–141.

Bažant, Z. and Y. Kwon (1994). Failue of slender and stocky reinforced concrete columns: tests of size effect. *Materials and Structures, RILEM 27*.

Bažant, Z., F.-B. Lin, and H. Lippmann (1993). Fracture energy release and size effect in borhole breakout. *Int. J. Numerical and Analytical Methods in Geomechanics 17*, 1–14.

Bažant, Z. and J. Planas (1998). *Fracture and size effect in concrete and other quasibrittle materials*. CRC Press LLC.

Bažant, Z. and Y. Xiang (1997). Size effect in compression fracture: Splitting crack band propagation. *J. Eng. Mech.-ASCE 123*(2), 162–172.

Bažant, Z. P. (1984). Size effect in blunt fracture: Concrete, rock, metal. *Journal of Engineering Mechanics 110*(4), 518–535.

Burtscher, S. L. (2002). *Size Effect of Concrete and Sandstone in Compression*. Ph. D. thesis, Institute for Structural Concrete, Vienna University of Technology, Heft 3. ISBN 3-902089-02-4.

Burtscher, S. L., J. Kollegger, G. Rinnhofer, and A. Barnas (2002). Schleuderbetonstützen mit hoher Tragfähigkeit für Bauvorhaben in Österreich. *OIB aktuell 3. Jahrgang*(3), 9–13.

Carpinteri, A. (1994). Scaling laws and renormalization groups for strength and thoughness of disordered materials. *International Journal of Solids and Structures 31*(3), 291–302.

Carpinteri, A. and G. Ferro (1994). Size effects on tensile fracture properties: A unified explanation based on disorder and fractality of concrete microstructure. *Materials and Structures 27*, 563–571.

Hollingworth, S. (1998). *Structural and mechanical properties of high strength concrete*. Ph. D. thesis, Disvision of Civil Engineering, Cardiff, University of Wales.

Němeček , J. (2000). *Modeling of compressive softening of concrete*. Ph. D. thesis, CTU Prag, in: CTU Reports. ISBN 80-01-02298-6.

Schwarz, L. (1987). Neue Möglichkeiten der Experimentellen Tragwerksuntersuchung an der Abteilung für Angewandte Modellstatik im Stahlbau der TU-Wien. *Stahlbau Rundschau 68*, 2–8.

Sener, S., B. Barr, and H. Abusiaf (1999). Size-effect tests in unreinforced concrete columns. *Magazine of Concrete Research 51*(1), 3–11.

Computational Modelling of Concrete Structures, Bićanić et al. (eds)
© 2003 Swets & Zeitlinger, Lisse, ISBN 90 5809 536 3

Simulation of headed anchor failure

Gianluca Cusatis, Giovanni Di Luzio, & Marco Rota
Dept. of Structural Engrg., Technical University (Politecnico) of Milan, Milan, Italy

ABSTRACT: This work concerns the numerical simulation of the failure behavior of headed studs embedded in concrete and subjected to tensile loading. The attention is directed to the analysis of a single anchor placed in a specimen of undamaged concrete without reinforcement and far from the specimen edges. The simulations are carried out by means of two different constitutive models: 1) a smeared crack model with nonlocal regularization (nonlocal microplane model M4) and 2) a discrete model simulating the concrete mesostructure (shear-confinement lattice model). The ability of both models to predict the anchor ultimate load is checked through the comparison of the numerical results with some available experimental data and various design formulas taken from the literature (ACI-method, VAC-method, CC-method and Theoretical method). Both the nonlocal microplane model and the shear-confinement lattice model are able to predict the ultimate load of the headed anchor under tension and to reproduce the well-known cone-shaped failure of the embedding concrete.

1 INTRODUCTION

In the practice of concrete structural design, tensile loads are often transferred into concrete by using metallic anchoring systems. Nowadays various types of anchors such as cast-in situ anchors (headed studs) and post-installed anchors (expansion anchors, undercut anchors and adhesive anchors) are available. They mainly differ on the load transferring mechanism, which can be governed by friction or by interlocking effect (Eligehausen et al. 2001). Single anchors subjected to tensile loading and located far from edges can experience basically three different types of failure: 1) a ductile failure due to the yielding of the fastening device, 2) pull-out failure occurring by sliding of the fastener out from concrete and 3) brittle failure of concrete due to the extraction of a concrete cone. A complete review of the mechanisms of anchor failure appear in Eligehausen et al. (2001).

Pull-out failure is usually prevented by installing fasteners with a sufficient embedment depth. Moreover, the concrete capacity related to the cone extraction must be predicted accurately, in order to insure a ductile failure, achieved by setting the fastener yielding capacity lower than the concrete capacity (ductile design) or to guarantee a sufficiently low probability of failure (brittle design). In this work the ultimate load for cone extraction is predicted by means of two triaxial constitutive models: the nonlocal microplane model (Di Luzio 2001) and the shear-confinement lattice model (Cusatis 2001; Cusatis et al. 2002) reviewed in the next section. The numerical results are compared with some available experimental data and with the predictions of the more widely used design formulas such as ACI-method (ACI Committee 349 1997), VAC-method (Klingner and Mendonca 1982), CC-method (Fuchs et al. 1995) and Theoretical method (Eligehausen and Ozbolt 1992). All these methods are based on the assumption that a constant tensile stress is transmitted by the failure cone surface radiating from the bearing edge of the anchors. The angle between this conical surface and the concrete surface is assumed equal to 45 deg in the ACI-method and it varies from 28 deg (small embedment depth) to 45 deg (large embedment depth) in the VAC-method. The CC-method and the Theoretical method assume a 35 deg failure cone and take also into account the size effect on the ultimate nominal stress, the former according to linear elastic fracture mechanics (LEFM), the latter according to and non-linear fracture mechanics (Bažant's Size Effect Law).

2 A BRIEF REVIEW OF THE CONSTITUTIVE MODELS ADOPTED FOR CONCRETE

In order to simulate correctly the concrete capacity related to cone extraction, the constitutive model must reproduce all aspects of concrete behavior including hardening response under triaxial compression and softening fracture under tension. The two models adopted in this work, the nonlocal microplane model and the shear-confinement lattice model, have shown to have such capability.

Nonlocal Microplane Model M4. The microplane model for concrete has been developed at Northwestern University by Bažant and coworkers in the last twenty years. The last version M4 is able to reproduce well all the triaxial behavior of concrete for both compressive and tensile dominant stress state (Bažant et al. 2000; Caner and Bažant 2000).

The microplane constitutive law is defined by a relation between stresses and strains acting on a plane called the microplane, having an arbitrary orientation characterized by its unit normal $\vec{n}$. The basic hypothesis, which ensures stability of the softening behavior (Bažant 1984), is the kinematic constraint, consisting in the fact that the strain vector on the microplane is considered to be the projection of the strain tensor.

The static equivalence (or equilibrium) of stresses between the macro and micro levels is achieved through the principle of virtual work (Bažant 1984) written for the surface of a unit hemisphere. The macroscopic stress tensor is computed by imposing that the virtual work of macro-stresses (continuum stresses) within a unit sphere must be equal to the virtual work of micro-stresses (microplane stress components) regarded as the traction on the surface of the sphere. The integral physically represents a homogenization of different contributions coming from planes of various orientations within the material. The constitutive law is then formulated at level of each microplane by splitting the normal component in its volumetric and deviatoric component $\sigma_N = \sigma_V + \sigma_D$, $\varepsilon_N = \varepsilon_V + \varepsilon_D$. The microplane relations are incremental elastic $d\sigma_V = E_V d\varepsilon_V$, $d\sigma_D = E_D d\varepsilon_D$, $d\sigma_M = E_T d\varepsilon_M$, $d\sigma_L = E_T d\varepsilon_L$ and the non linear behavior is obtained by imposing that the stresses do not exceed given stress-strain boundaries which may be regarded as strain dependent yield limits. More details appear in Bažant et al. (1996) and Bažant et al. (2000).

Since the microplane model is a continuum constitutive law, in the sense that relates the stress tensor to the strain tensor, it may suffer of pathological mesh sensitivity during fracture simulation. The local formulation gives fracture propagation behavior depending on the FE mesh size and orientation. It can also lead to problems like localization instabilities, incorrect convergence to zero energy dissipation, directional bias of the mesh and incorrect simulation of the size effect.

Usually in FE element analysis these types of problems are avoided by making the constitutive model to be dependent upon the finite element characteristic size. This approach is acceptable when the fracture propagates in pure mode I or, more in general, when the crack path is known in advance. Unfortunately this not the case of the anchor failure in which the crack pattern is quite complex and the fracture deviates during propagation.

A general and physically justified approach used to regularize softening continuum models is the *nonlocal generalization.* The nonlocal formulation of a constitutive model, in general, consists in replacing a certain local variable $f(\mathbf{x})$, that characterizes the material damage or softening by its nonlocal counterpart $\overline{f}(\mathbf{x})$, the nonlocal operator being defined as:

$$\overline{f}(\mathbf{x}) = \int_V \widehat{\alpha}(\mathbf{x}, \boldsymbol{\xi}) f(\boldsymbol{\xi}) dV(\boldsymbol{\xi}) \tag{1}$$

where V = volume of the structure, $\mathbf{x}, \boldsymbol{\xi}$ = coordinates vectors, and $\hat{\alpha}(\mathbf{x}, \boldsymbol{\xi})$ is the normalized nonlocal weight function:

$$\widehat{\alpha}(\mathbf{x}, \xi) = \frac{\alpha(\mathbf{x} - \boldsymbol{\xi})}{\int_V \alpha(\mathbf{x} - \boldsymbol{\xi}) dV(\boldsymbol{\xi})} \tag{2}$$

The weight function $\alpha(\mathbf{x} - \boldsymbol{\xi})$ can be, for instance, taken as a bell-shaped function defined as

$$\alpha(\mathbf{x}, \boldsymbol{\xi}) = \begin{cases} \left(1 - \frac{|\mathbf{x}-\boldsymbol{\xi}|^2}{R^2}\right)^2 & \text{if } 0 \le |\mathbf{x} - \boldsymbol{\xi}| \le R, \\ 0 & \text{if } R \le |\mathbf{x} - \boldsymbol{\xi}|. \end{cases} \tag{3}$$

where R is a parameter proportional to the material characteristic length l_{ch}:$R = \rho_0 \cdot l_{ch}$. The coefficient ρ_0 is determined so that the volume under function α be equal to the volume of the uniform distribution ($\rho_0 = 0.9375$ for 1D, $\rho_0 = 0.9086$ for 2D, $\rho_0 = \sqrt[3]{35}/4 = 0.8178$ for 3D).

The standard nonlocal formulation can be refined (for reasons that can be found in Planas and Guinea (1996)) by introducing the so-called "over-nonlocal formulation", in which:

$$\hat{f}(\mathbf{x}) = m\overline{f}(\mathbf{x}) + (1 - m) f(\mathbf{x}) \tag{4}$$

where $\hat{f}(\mathbf{x})$ is the over-nonlocal variable, $\overline{f}(\mathbf{x})$ is the nonlocal variable and m is an empirical coefficient $m > 1$.

The microplane model M4 has been generalized in the nonlocal sense in Di Luzio (2001) by assuming the microplane stress-strain boundaries as functions of the over-nonlocal microplane strains coming from the projection of the over-nonlocal macro-strain tensor. The complete formulation appears in Di Luzio (2001).

Shear-Confinement Lattice Model. This is a lattice-type model which is based on the simulation of the concrete mesostructure by a system of interacting aggregate particles. Starting from the basic concrete properties and the granulometric distribution of the aggregates it is possible, by using a random procedure, to set the position of each aggregate piece through the given concrete specimen. Only aggregate sizes

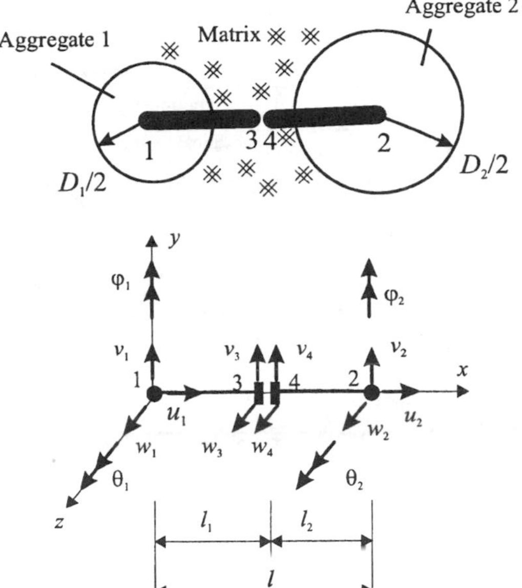

Figure 1. Lattice strut connecting two adjacent aggregates.

larger than a certain limit are considered in order to have reasonable computational time during numerical simulations. Each aggregate interacts with the surrounding aggregates through the embedding mortar, simulated by inter-aggregates links. These links are defined according to a Delaunay triangulation of the center points of the aggregates.

The contact layer of mortar between two adjacent aggregate particles can transmit both normal and shear stresses which are assumed to be functions of normal and shear strains which, in turn, are defined as $\dot{\varepsilon}_N = (\dot{u}_2 - \dot{u}_1)/l, \dot{\varepsilon}_M = (\dot{v}_2 - \dot{v}_1 - l_2\dot{\vartheta}_2 - l_1\dot{\vartheta}_1)/l$ and $\dot{\varepsilon}_L = (\dot{w}_2 - \dot{w}_1 + l_2\dot{\varphi}_2 + l_1\dot{\varphi}_1)/l$. Here u_1, u_2 are the displacements of the two aggregate in the direction of the line connecting their centers, v_1, v_2, w_1 and w_2 are the transversal displacements and $\vartheta_1, \vartheta_2, \varphi_1$ and φ_2 are the rotations of the aggregates (Fig. 1). All the strain components are assumed uniformly smeared over the length l of the connection.

The stress-strain relations are given as

$$\sigma_N = \sigma\varepsilon_N/\varepsilon \quad \sigma_M = \sigma\alpha\varepsilon_M/\varepsilon \quad \sigma_L = \sigma\alpha\varepsilon_L/\varepsilon \tag{5}$$

in which σ (*effective stress*) is assumed to be a function of $\varepsilon = (\varepsilon_N^2 + \alpha\varepsilon_T^2)^{1/2}$ (*effective strain*), $\omega = \arctan[\varepsilon_N/(\alpha\varepsilon_T)]$ (*coupling strain*) and $\varepsilon_T = (\varepsilon_M^2 + \varepsilon_L^2)^{1/2}$ (*total shear strain*). The symbol α is a constant material parameter which represents the ratio between shear and normal stiffness of the connecting link.

The initial linear elastic behavior is captured by using the incremental relation $\dot{\sigma} = E\dot{\varepsilon}$. The non linear behavior is simulated by using the concept of stress-strain boundary, already introduced for the microplane model formulation, which leads to the inequality $0 \leq \sigma \leq \sigma_b$ where

$$\sigma_b = \sigma_0(\omega)\exp\left\{\frac{K(\omega)}{\sigma_0(\omega)}\langle\varepsilon - \varepsilon_0(\omega)\rangle\right\} \tag{6}$$

In Eq. (6) $\sigma_0(\omega)$ is the effective strength, $\varepsilon_0(\omega) = \sigma_0(\omega)/E$ is the elastic strain limit and $K(\omega)$ governs the stress evolution in the non linear range. $K(\omega)$ is negative (softening behavior) for tension, shear-tension and shear with low compression. On the contrary $K(\omega)$ is positive (hardening behavior) for compression and shear with high compression. For tensile dominant stress states the post-peak behavior at the meso-level is also assumed to be sensitive to the confining stress transversal to the connecting strut between particles. The whole formulation and the validation of the shear-confinement lattice model can be found in Cusatis (2001) and Cusatis et al. (2002).

3 NUMERICAL SIMULATION OF HEADED STUD FAILURE

In this section the non-local microplane model and the shear-confinement lattice model are used to simulate the experimental ultimate loads measured for the extraction of a cast-in situ anchor with stud diameter equal to 12 mm, embedment length equal to $h_{ef} = 65$ mm and diameter of the head equal to 24 mm. The basic material properties of the embedding concrete are the following: cement content $c = 276\ \mathrm{kg/m^3}$, water-cement ratio $w/c = 0.71$, aggregate-cement ratio $a/c = 7.0$. The assessment of the parameters of the constitutive models is done by fitting the used mechanical properties of the concrete: compressive strength $f_c' = 30.8$ MPa, tensile strength $f_t' = 2.86$ MPa, Young modulus $E_c = 35$ GPa and fracture energy $G_F = 64.5$ N/m (estimated). The usual Poisson ratio of concrete ($v_c = 0.18$) has been assumed. The mechanical behavior of the steel stud has been modeled as linear elastic with Young modulus equal to $E_s = 210$ MPa and Poisson ratio equal to $v_s = 0.3$. This assumption is justified by the fact that the experimental ultimate load $N_u = 44.1$ kN leads to a stress in the steel rod ($N_u/(0.25\pi\phi_s^2) = 389.9$ MPa) lower than the yielding strength.

For the numerical simulations carried out by the microplane model an axialsymmetric mesh of CST finite elements is used (Fig. 2a). The diameter of the concrete block is assumed to be equal to 23 cm. The mesh is constrained in the direction of the applied load with sliding supports located at a distance of $2d_{ef} = 13$ cm from the axis of the stud. This distance is usually considered large enough in order to prevent

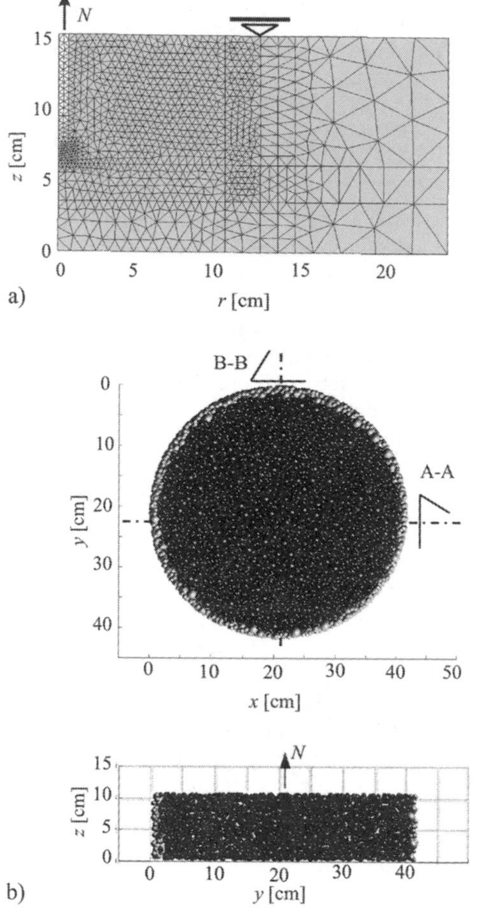

Figure 2. Discretization of the headed stud system: a) axialsymmetric mesh for microplane model simulations and b) tridimensional aggregate distribution for lattice model simulations.

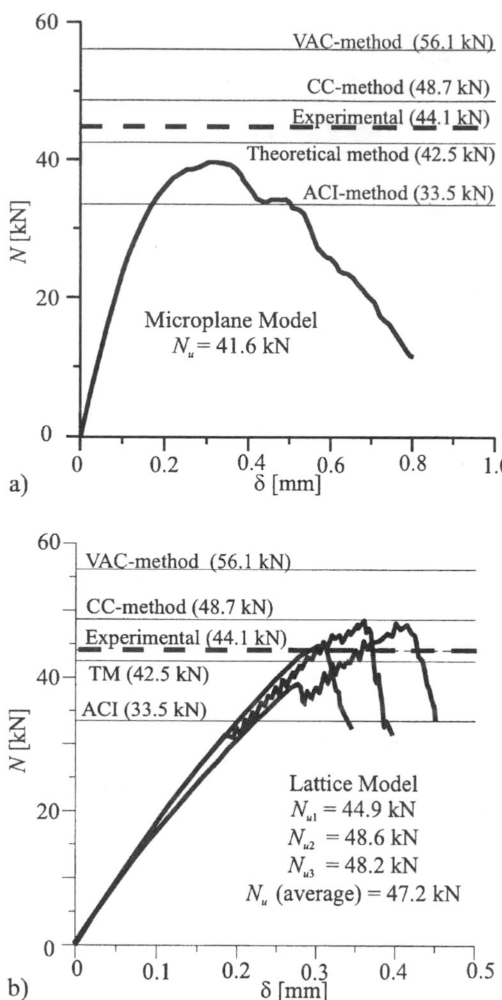

Figure 3. Numerical results: a) microplane model response and b) lattice model response.

the cone extraction to be influenced by the boundary conditions (Ožbolt et al. 1999). The same specimen geometry is adopted for the simulations through the lattice model but in this case the finite element mesh is fully tridimensional. Fig. 2b shows a typical tridimensional aggregate system from which the mesh is constructed by Delaunay triangulation.

In both cases the response of the anchor is computed by imposing the load under displacement control in order to get the post-peak behavior.

In Fig. 3a the response of the non-local microplane model is plotted along with the experimental ultimate load and the predictions of the design formulas. The numerical prediction (41.6 kN) is 5.7% lower than the experimental data (44.1 kN). Among the design formulas, the one which gives an ultimate load closest to the prediction of the microplane model is the theoretical method. Fig. 3b shows the numerical results obtained by the lattice model. The load-displacement curves obtained from three different random generations of the aggregate distributions are compared with the experimental results and the ultimate loads given by the design formulas. The mean of the computed ultimate loads is equal to 47.2 kN which is about 7% higher than the experiment. The ultimate load given by the CC-method is in close agreement with the one predicted by the lattice model.

The predictions of the numerical models actually comprise the experimental result, and their scatter is much lower than that of the design formulas, one of which (VAC-method) seems particularly unsafe.

The ultimate load obtained through the microplane model differs from that given by the lattice model of

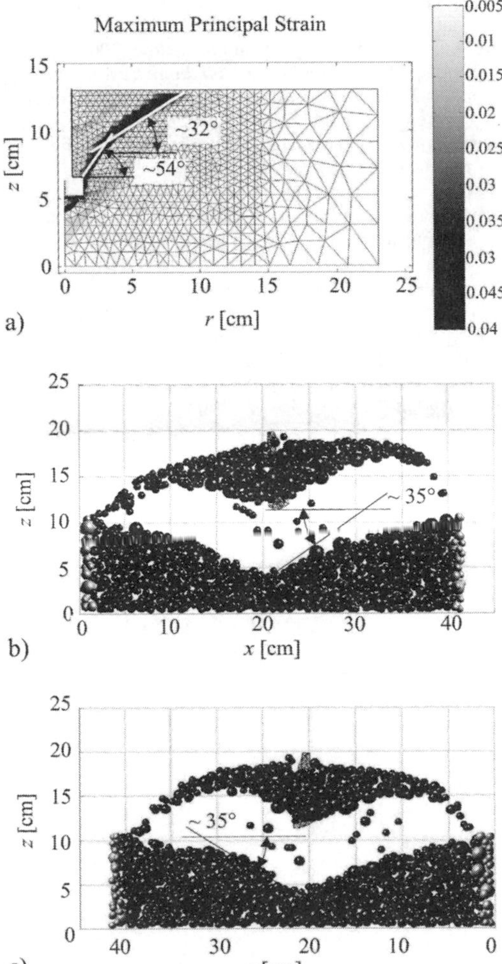

Figure 4. Cone shaped failure obtained by a) microplane model and b) lattice model.

about 12%. This difference can be understood by analyzing the size and shape of the extracted concrete cone at failure predicted by the two numerical models.

In Fig. 4a the contours of the maximum principal strain computed by the microplane model at failure are shown. One can see that the strain localizes in a narrow band starting from the bearing edge of the fastener and propagating towards the upper surface of the concrete specimen. This localization band represents the fracture propagation which leads to the concrete cone extraction. Thus we can infer that the microplane model predicts a fracture which starts to propagate with an angle of about 54 deg. After a while fracture deviates and the angle becomes of about 32 deg (Fig. 4a). The average of these angles is 43 deg, which gives a behavior between the 45-deg method (ACI-method) and the 35-deg method (CC-method).

The failure mode predicted by the lattice model is represented in Figs. 4b, c in which the amplified deformed configuration of the aggregate particles at failure is plotted for the sections A-A and B-B, respectively (see Fig. 2b). Also in this case the numerical model is able to reproduce the localization of damage into a discrete fracture. From the pictures in Figs. 4b, c it is difficult to infer a clear value for the inclination angle of the fracture surface. This is due to its irregularity which, in turn, depends on the randomness of concrete mesostructure. The fracture inclination tends to decrease as the propagation proceeds, an approximate average of the inclination can be estimated as about 35 deg (Figs. 4b, c). This fact explains why the predicted ultimate loads are close to the prediction of the 35-deg method (CC-method). Moreover, what clearly comes out from the comparison between Fig. 4a and Figs. 4b, c is that shear confinement lattice model leads to a failure cone bigger that the one obtained by the microplane model and consequently to a higher ultimate load.

4 CONCLUSIONS

The analysis of the failure of an anchoring system has been carried out by using the nonlocal microplane model and the shear-confinement lattice model. The prediction of the models are in satisfactory agreement with the experiments. They predict the ultimate experimental load with an accuracy of less than 10% even if the model calibration has been based upon very few data (Young modulus, compressive strength, tensile strength). The qualitative shape of the concrete cone extracted at failure is also predicted by both models. Among the commonly used design formulas the CC-method and the Theoretical method (both 35-deg methods) seem to give the best predictions.

ACKNOWLEDGMENT

The work of the first and third authors of this study was partially supported under Grant COFIN 2000 (*Mechanical Deterioration and Durability of Normal and High-Performance Concrete Structures*) from the Italian Government to Technical University (Politecnico) of Milan. The work of the first and second authors was further partially supported under the European Commission ANCHR Project at Politecnico di Milano.

REFERENCES

ACI Committee 349 (1997). *ACI manual of concrete practice. Part 4: bridges, substructure, sanitary and other special structures. Structural properties* (ACI International Edition ed.), Volume 4 of *Code requirements for nuclear safety related concrete structures.* Detroit, U.S.

Bažant, Z. P. (1984, Jan.). Microplane model for strain controlled inelastic behavior. In C. Desai and R. Gallagher (Eds.), *Mechanics of Engineering Materials,* Arizona, Tucson, pp. 45–59. J. Wiley, London.

Bažant, Z. P., F. C. Caner, I. Carol, M. D. Adley, and S. A. Akers (2000). Microplane model M4 for concrete. I: Formulation with work-conjugate deviatoric stress. *J. Engrg. Mech., ASCE 126(9),* 944–953.

Bažant, Z. P., Y. Xiang, and P. C. Prat (1996). Microplane model for concrete. I: Stress-strain boundaries and finite strain. *J. Engrg. Mech., ASCE 122(3),* 245–254.

Caner, F. C. and Z. P. Bažant (2000). Microplane model M4 for concrete. II: Algorithm and calibration. *J. Engrg. Mech., ASCE 126(9),* 954–961.

Cusatis, G. (2001, December). *Tridimensional random particle model for concrete.* Civil Engineering PhD Thesis, DIS – Politecnico di Milano, Milan, Italy.

Cusatis, G., Z. Bazant, and L. Cedolin (2002, August). Shear-Confinement Lattice Model for Concrete Damage in Tension and Compression. Technical Report 02-8/C699c, North-western University, Evanston, U.S.

Di Luzio, G. (2001, December). *A new nonlocal microplane model for fracture and damage in concrete.* Civil Engineering PhD Thesis, DIS – Politecnico di Milano, Milan, Italy.

Eligehausen, R., I. Hofacker, and S. Lettow (2001). *Fastening technique: current status and future trends,* Volume 1, pp. 11–27. University of Stuttgart, Stuttgart, GE: RILEM.

Eligehausen, R. and J. Ozbolt (1992). *Influence of crack width on the concrete cone failure load* (Z. P. Bažant ed.), pp. 876–881. Elsevier Applied Science.

Fuchs, W., R. Eligehausen, and E. Breen (1995, January–February). Concrete Capacity Design (CCD) approach for fastening to concrete. *ACI Structural Journal* (92 S-9), 73–94.

Klingner, R. E. and J. A. Mendonca (1982, July–Aug.). Tensile Capacity of Short Anchor Bolts and Welded Studs: A Literature Review. *ACI Journal (4)*, 270–279.

Ožbolt, J., R. Eligehausen, and H. Reinhardt (1999). Size effect on the concrete cone pull-out load. *International Journal of Fracture (95)*, 391–404.

Planas, J., E. M. and G. Guinea (1996). Basic issue of non-local models: uniaxial modeling. Technical Report 96-jp03, Departamento de Ciencia de Materiales, ETS de Ingenieros de Caminos, Universidad Politecnica de Madrid, Ciudad Universitaria sn., 28040 Madrid, Spain.

Computational Modelling of Concrete Structures, Bićanić et al. (eds)
© 2003 Swets & Zeitlinger, Lisse, ISBN 90 5809 536 3

Structural modeling of reinforced concrete slabs subjected to falling rock impacts

Ph. Berthet-Rambaud
Cemagref, Grenoble & CETE Lyon, France

Y. Timsah
Beirut Arab University, Beirut, Lebanon

L. Daudeville, J. Mazars
Laboratoire Sols Solides Structures, Grenoble, France

ABSTRACT: In rocky mountains region where the falling rocks constitute a major hazard to civil structures, the protective structural systems are usually composed of over dimensioned reinforced concrete elements. This is mainly due to the lack of knowledge of the dynamic effects caused by these rock falls. A thick backfilling layer that prevents the direct impact of falling rocks by constituting a damping medium commonly covers the roof slab of protection structures. This allows the design of the slab with static dead loads (backfilling and rocks weights). Recent experiments were performed in France on a new type of protection system characterized by a roof slab without damping medium (no backfilling) and simply supported on vertical elements by a set of "fuse" steel supports. The roof slab resists directly the falling rocks impacts, which cause limited local damage to the impact zone, in case of field impact, or the yielding of steel supports for boundary impact cases. The main advantage of this new protection system is to allow continuous uses of the structure through the repair of the damaged zones in the roof slab or the replacement of the "fuse" steel support after each impact. The aim of the present study is to predict the structural response of the new proposed system by a rigorous three-dimensional modeling of the roof slab and its supporting elements. The analysis introduces the impact load in a way similar to that of the performed experiment, and a stress–strain concrete relationship that allows a realistic representation of the concrete behavior under dynamic loads and its corresponding damages. The comparison of the experimental measurements with those obtained from the present analysis proves the accuracy of the built model in predicting the real behavior of the protection structure.

1 INTRODUCTION

The continuous expansion of urban zones in mountain regions increases the need of protection systems that protect the civil structures and infrastructures from natural hazards such as snow avalanches, falling rocks, and soil displacements. The dimensioning and designing of reinforced concrete protection systems is based, in the current codes of practice, on approximate method that consider the straining forces as static forces. The use of approximate static analysis methods leads either to over-dimensioned protection systems (massive concrete elements in usual gallery protection systems against rock fallings), or to under dimensioned systems.

The optimal dimensioning of protection systems (strength + limited cost) should be based on the following conditions:

- Considering the dynamic feature of the loading.
- Using behavior properties for the structural materials that allow an accurate description of the stress–strain relationship variation under the applied dynamic loads.
- Performing numerical structural analysis that includes realistic problem data (three dimensional geometry, generation of dynamic effects due to impacts, non-linear behavior …).

The present study focuses on the gallery rock falling protection system, conventionally composed of reinforced concrete sub-structural elements (walls, columns, and foundations) and a roof slab covered by a thick backfilling layer (Figure 1). The roof slab is rigidly connected to sub-structural elements, and the

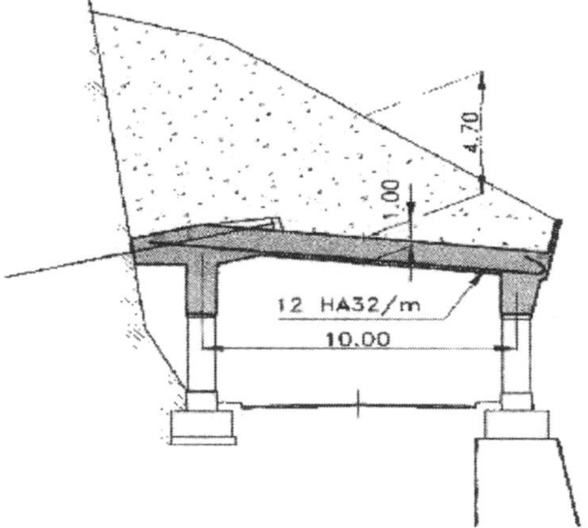

Figure 1. Conventional protective road gallery system.

backfilling layer constitutes a damping medium, and allows therefore, the design of the system with only static dead loads (own weight, backfilling and rocks loads).

The structure is not designed any more to resist the impact of blocks but especially to support the backfilling layer. This solution has the main disadvantage of producing over dimensioned reinforced concrete elements. The foundations, which must be dimensioned consequently, cause often some site construction problems as for the case of a raised slope, a weak soil ….

Considering that the request for this type of equipment will be increasing, an investigation was carried out to improve the design and limit costs. The basic idea was to eliminate the backfilling layer and to use a semi-probabilistic approach with the notion of "acceptable damage" to the structure.

For the purpose of finding an optimal solution, a new system was proposed in France by the consulting company "Tonello IC", which consists of a roof slab pin supported (no continuity) on the sub-structural elements. The roof slab is subjected to the direct impact of falling rocks and slab reactions are transmitted to the

Figure 2. New system, gallery "les Essariaux" and a fuse support.

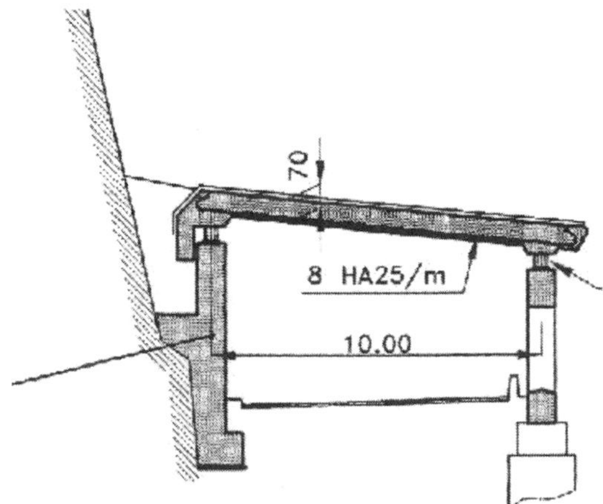

Figure 3. Schematic view of the new system.

sub-structures throughout ductile steel supports that act as dissipating energy fuses and protect the sub-structural elements (Figures 2 and 3). The slab is then designed in order to resist directly to a falling rock impact that causes a local damage limited to the shock zone in case of field impact.

The first example of this protection system was built in 1999 at "les Essariaux" between Albertville and Chamonix in French Alps.

2 EXPERIMENTS

During June and July 2001, the Tonello IC company has performed experiments in collaboration with the University of Chambery on a one third reduced scale system model of the roof slab of the proposed new system in order to evaluate the response and the performance of this system.

The experiment consisted of a concrete slab ($12 \times 4.4 \times 0.28$ m) set on two lines of 11 steel fuses (12.6 cm

Figure 4. Construction of the experimental slab.

high and spaced at 1.14 m) and impacted by a 450 kg reinforced concrete cubic block (58 cm side) falling from different heights to various locations of the slab.

The slab dimensioning and reinforcement was performed using a simplified method based on the principle of momentum and energy conservations. Assuming that the governing flexural deformed shape is the first mode of vibration for the impacted slab, the slab mass is substituted by an equivalent mass M*:

$$M^* = ?y_i^2 dm/y_0 \quad (1)$$

Where y_0 is the amplitude of the flexural deformation of the slab, y_i is the displacement of the partial mass dm of the slab.

Assuming also that after the impact there were neither a bounce of the projectile nor a penetration into the slab, the principle of momentum and energy conservations leads to the following relation:

$$E\frac{m}{M^*+m} = P_u\,(0.5f_e+f_p) \quad (2)$$

Where m represents the projectile mass. The right hand side represents the strain energy; P_u is the peak impact, f_e and f_p represent the elastic and plastic strains respectively.

The concrete block was released to fall freely from 15 and 30 m high, and to impact the slab, almost on one of its faces. The impact velocity varied from 17.2 to 24.2 m/s and the kinetic energy from 67.7 to 135 kJ (Figures 4 and 5).

Three impacts were carried out: the first and the second from 15 and 30 m high in the inner part of the slab and the third from 30 m on the edge of the slab (the support line).

The slab response to the different impacts was obtained through general displacement measurements and with strain gauges.

These relatively large-scale experiments provide interesting and complete data that allow better understanding of the structural response to impact loads.

Figure 5. Rock fall test.

They also allow the evaluation of the numerical analysis performance of the modeled structure.

3 NUMERICAL ANALYSIS

A realistic prediction of the structural response through numerical analysis requires rigorous three-dimensional finite elements modeling of the different structural components. In the present study, the finite elements code Abaqus was used. The explicit module of this code allows highly non-linear transient dynamic analysis of phenomena like impacts.

Abaqus offers also the possibility of managing several interactive entities (the slab and the block in the present case). The analysis can, therefore, introduce the impact in a way similar to that of the experiment, managing only the impact characteristic.

3.1 *Finite element modeling*

The slab and the block were modeled with volumetric finite elements (Figure 6) with different degree of mesh refinement (essentially at the impact zones of the slab). The reinforcement was represented by bar elements or with the Abaqus "rebar" option that uses a superposition of stiffness matrix method.

The steel supports were modeled using non-linear springs in a first (preliminary) analysis, then with volumetric elements in a second (refined) analysis (Figure 7).

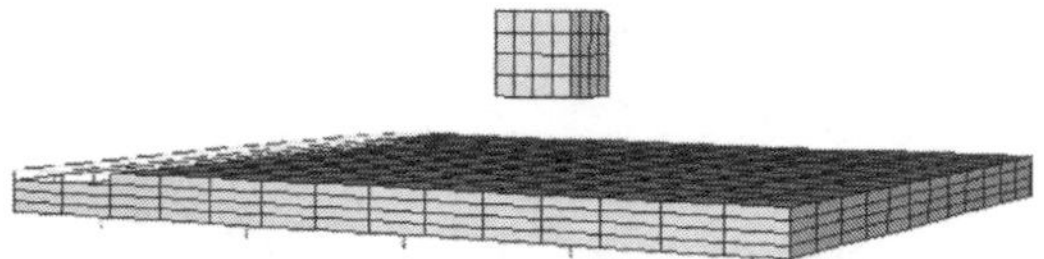

Figure 6. Three dimensional modeling of the slab and the block.

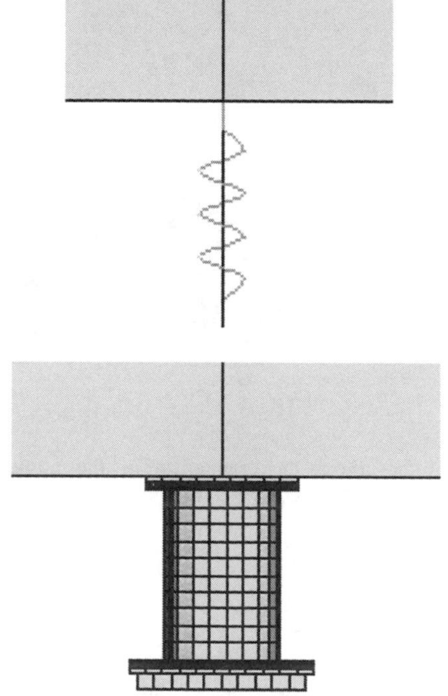

Figure 7. Non-linear spring and 3D model of a steel support.

The constitutive behavior of the non-linear spring was identified through the knowledge of both the cross-section area of the steel support and its stress–strain relationship.

3.2 *Constitutive behavior modeling of concrete*

For an accurate simulation of the structural response, it is necessary to use a realistic representation of the materials behavior under dynamic loads. For the concrete, the behavior properties must include some phenomena that are related to the damage under dynamic loads such as decrease in material stiffness due to cracking, stiffness recovery related to closure of cracks, and inelastic strains concomitant to damage.

The stress–strain relationship was represented in the numerical analysis by the PRM (Pontiroli-Rouquand-Mazars) (Pontiroli 1995) damage model that uses one scalar damage variable D, which is the damage

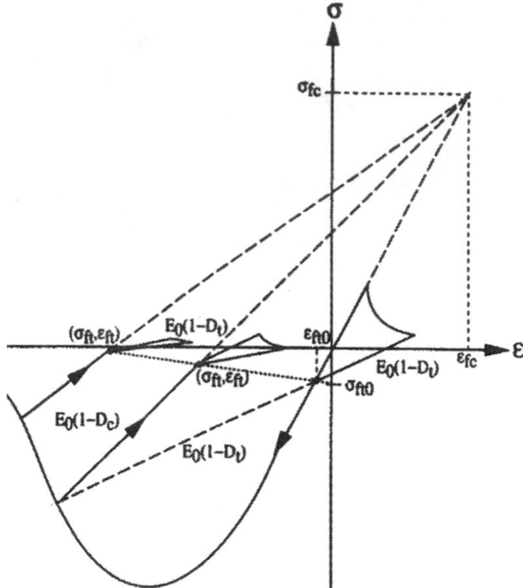

Figure 8. PRM concrete behavior.

indicator. The one-dimensional expression of stress–strain relationship is the following:

$$(\sigma-\sigma_{fl}) = E_0(1\text{-}D)(\varepsilon - \varepsilon_{fl}) \tag{3}$$

Where E_0 is the Young modulus, σ_{fl} is the crack closure stress in tension and ε_{fl} is the irreversible strain corresponding to σ_{fl}.

A similar expression is used with tensors to describe the three-dimensional states.

D includes damages due to compression and tension; its value varies from 0 (for uncracked material) to 1 (for macro-cracked material). The variation of D is governed by the equivalent strain $\tilde{\varepsilon}$ (Mazars 1984 and 1986):

$$\tilde{\varepsilon} = \sqrt{\sum_{i=1,3} \langle \varepsilon_i \rangle_+^2} \tag{4}$$

Where $\langle \cdot \rangle_+$ denotes the positive part and ε_i are the principal strains.

A typical stress–strain response produced by this model for a uni-axial alternate loading in multi traction-compression steps is given in Figure 8. The three dimensional version of this model (used in the numerical analysis) was implemented in Abaqus-Explicit using an external Fortran subroutine.

The stress–strain relationship for the reinforcing bars is considered as simply elastic plastic, with hardening or without hardening.

The material parameters of the model were identified from classical data for concrete ($f_c = 30$ MPa,

Table 1. Experimental and numerical vertical displacements.

Test and impact point	Measurements (mm)	Model results (mm)
15 m high – inner part	14.5	15.2
30 m high – inner part	23.0	23.6
30 m high – support line	21.5	21.2

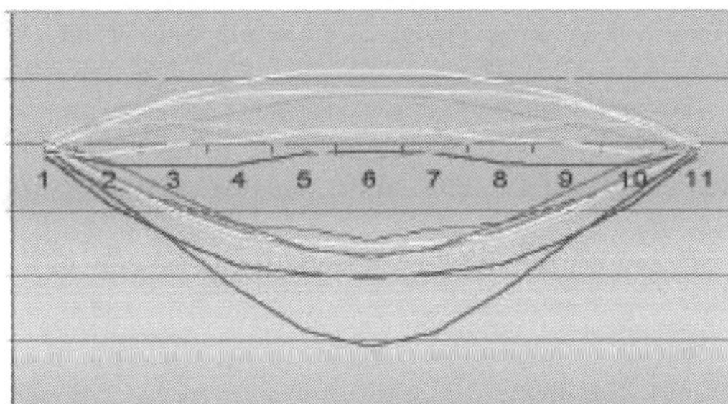

Figure 9. Transversal profile of vertical displacements in the slab versus time.

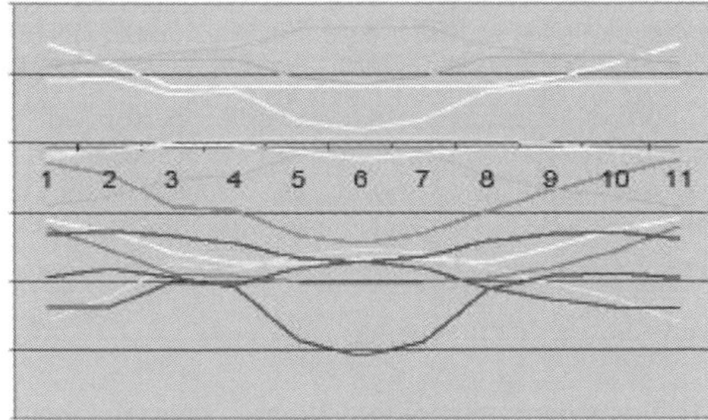

Figure 10. Longitudinal profile of vertical displacements in the slab versus time.

$E_0 = 30$ GPa) and rebars ($f_s = 500$ MPa) where f_c and f_s are the ultimate strengths of concrete and steel.

The used regularization technique, in order to avoid mesh dependency, is the Hillerborg's one (Hillerborg 1976).

4 ANALYSIS RESULTS

A wide range of results was obtained from the different loadings cases (impacts locations and falling heights), the modeling of steel supports (non-linear springs versus 3D complete model), and the reinforcement modeling (rebar versus bar elements).

The comparison of the analysis results with the experiment measurements showed global agreement, which indicates a good performance of the model in predicting the real behavior of the structure.

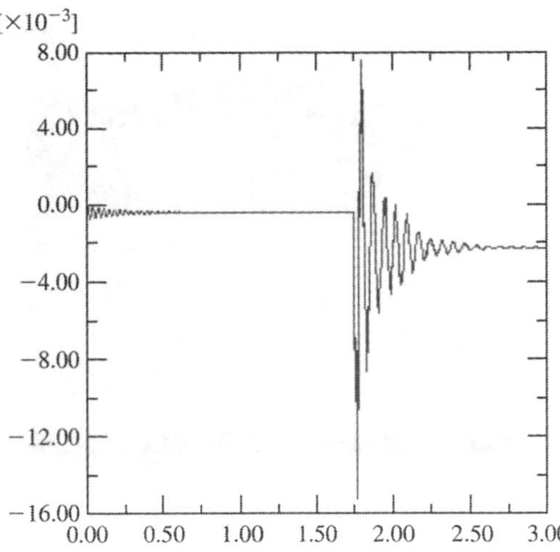

Figure 11. Variation with time of the vertical displacement of the slab at the impact point.

As an illustration of this performance Table 1 gives the values of maximum vertical displacements of the slab at the different impact points.

Figures 9 and 10 show the transversal and longitudinal evolution with time of the slab deflections, whereas the vertical displacement of the slab impact point is shown in Figure 11.

The displacements obtained from the third test (impact on the support line) were equivalent for the two models of supports. The 3D modeling gave a deformed shape similar to that obtained in reality (Figure 12).

The advantage of this type of analysis, in addition to traditional analysis output (strain, stress, internal forces, reactions …), is to allow obtaining additional type of results throughout the implemented subroutine. In particular evaluating the damage states by mapping the values of D in the slab. Figure 13 shows the damage localization, i.e. the values of damage on the sub-face of the slab due to different impacts.

The repartition of damages within the slab thickness can also be obtained. Whereas further investigations has to be performed into the relationship between D and the real state damage, initial comparison of experimental and numerical results for the described tests showed already an interesting agreement between damage repartition obtained from the analysis and that observed with the experimentation as shown in Figure 14 (spalling, radial cracking …).

5 CONCLUSION

The evaluation of the performance of the new proposed system for reinforced concrete galleries was the main

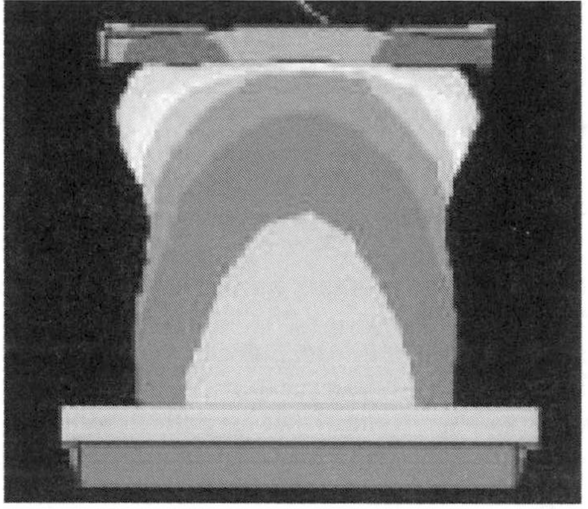

Figure 12. Deformed shape of the experimental and modeled steel support.

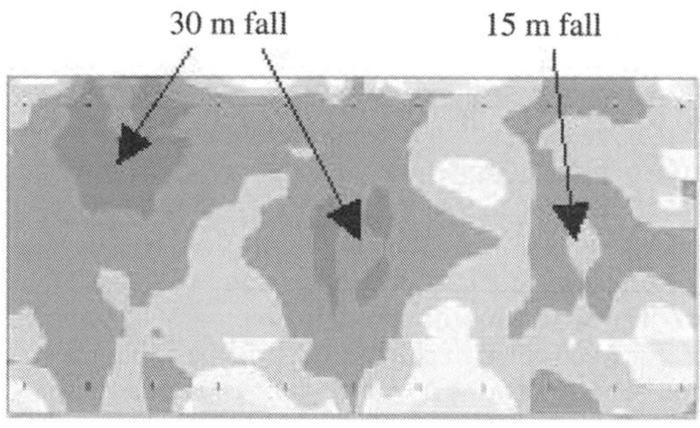

Figure 13. Mapping of damage distribution in the slab.

aim of this study. This was achieved by a numerical analysis including: a 3D model of the structural elements (block, slab, and the steel supports), a damage mechanics based model for concrete, and a realistic representation on the straining actions (different impacts of the projectile).

The numerical analysis allowed also the evaluation of the influence of different parameters as the impact velocity of the block, the block mass, the bearing capacity of the fuses, the material characteristics ….

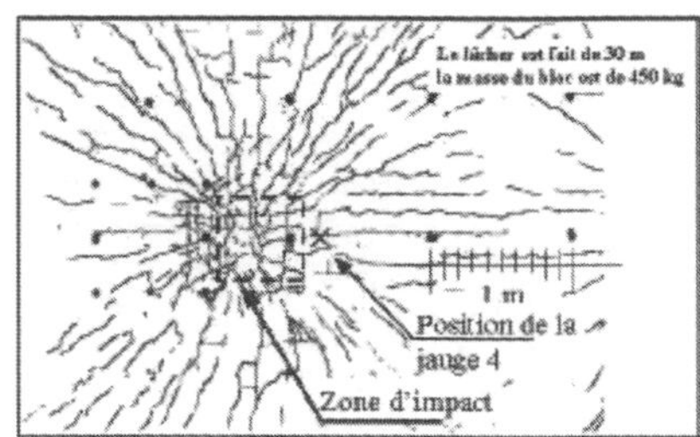

Figure 14. Damage pattern on the sub-face of the slab.

The analysis results proved to be in agreement with the experimental measurements, as well as representing the damage states under the different cases of loadings.

Based on the obtained results it can be concluded that:

- The new proposed system well fits its uses: optimal dimensioning, ability to resist blocks impacts that cause reparable damages, this would be done without affecting the serviceability state of the structure.
- The numerical modeling represents a powerful mean in predicting the behavior of this type of structures, and can therefore be used for their design.

Further research must however be performed to include additional cases such as the response of a repaired slab to new impacts, the structure response when subjected to several consecutive impacts.

ACKNOWLEDGEMENT

The financial support of the PIR program from French Ministry of Research is greatly appreciated.

The authors thank their colleagues of ESIGEC Chambery for this fruitful collaboration.

REFERENCES

Mazars, J. 1984. Application de la mécanique de l'endommagement au comportement non linéaire et à la rupture du béton. Thèse de doctorat d'état, Univ. Paris VI, Paris, France.

Mazars, J. 1986. A description of micro- and macro-scale damage of concrete structures. *Engineering Fracture Mechanics* 25 (5/6): 729–737.

Pontiroli, C. 1995. Comportement au souffle des structures en béton armé, analyse expérimentale et modélisation. Thèse de l'ENS Cachan, Cachan, France.

Hillerborg, A., Moder, M. & Petersson, P.E. 1976. Analysis of crack formation and growth in concrete by means of fracture mechanics and finite elements. *Cement and Concrete Research* 6: 773–782.

Computational Modelling of Concrete Structures, Bićanić et al. (eds)
© 2003 Swets & Zeitlinger, Lisse, ISBN 90 5809 536 3

Predicting tunneling-induced settlement damage for a concrete frame structure with a masonry façade

K.S. Douglas,* J.G. Rots, H.D. Netzel
Delft University of Technology, Delft, the Netherlands

S.L. Billington
Stanford University, Stanford, California, USA

ABSTRACT: This paper presents a case study of a concrete frame building with a masonry façade for which settlement damage due to Tunnel Boring Machine (TBM) tunneling was predicted. Two frame and two continuum models were analyzed to study the effect of soil settlement on building deflections, moments, and the formation of cracks in the masonry façade. Of particular interest is the comparison of predicted damage for each model.

1 INTRODUCTION

Tunneling activities, such as those involved with the construction of subways and other underground infrastructures, induce ground deformations above the bored tunnel. These deformations may lead to aesthetic and serviceability damage in the surrounding structures. This is especially true of masonry buildings, where differential settlements can lead to cracking in the walls or façades.

Techniques exist to protect buildings from settlement damage. This typically involves either reducing the ground settlements through methods such as compensation grouting, or improving the response of the structure by using tie rods, structural propping or jacking, or underpinning. However, these techniques are both expensive and disruptive. Thus, as the construction of subways and other underground activity increases in urban historical centers, research into improving methods for predicting tunneling-induced settlement damage in masonry buildings is becoming particularly important (e.g. Burd et al. 2000, Netzel et al. 2000, Rots 2000).

This paper discusses a case study to predict settlement damage due to TBM-tunneling in a concrete frame building with a masonry façade. An example of such a building is provided in Figure 1. To provide a thorough damage assessment, four models were studied. These included two frame models from a transverse and longitudinal section of the building, and two continuum models of the masonry façade along the longitudinal frame. In one of the continuum models, the frame and façade were modeled together to study their interaction during settlement, while in the other continuum model the façade was considered to act independently of the frame. The analyses focused on the deflections of the building, changes in load and moment distribution during settlement, and the formation of cracks in the façade.

Figure 1. Example of a typical concrete frame building with a masonry façade.

2 CASE STUDY MODEL

The structure considered in this work is a 36-m high concrete frame building with a masonry façade that

* Stanford University, Stanford, California, USA

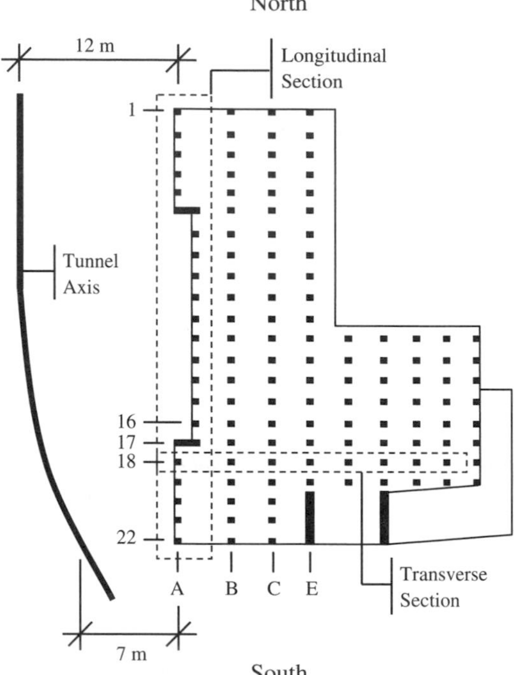

Figure 2. Map of the building with respect to the bored tunnel. Labels refer to column lines for each section.

consists of seven floors and a basement. A schematic plan of the structure is provided in Figure 2. Two sections of the building were analyzed for settlement damage – a longitudinal section parallel to the tunnel, and a section transverse to the tunnel. These sections are highlighted in Figure 2.

The building has a regular grid of columns that support the concrete floor slabs and beams. Column spacing is 5 to 6 m in the longitudinal direction and 7 to 8 m in the transverse direction. The columns lead to footings, which are further supported by groups of timber piles. Pile groups also support the concrete basement wall.

3 ANALYTICAL MODELS

Two types of analyses were performed to predict settlement damage on the building. The first was an analysis on the concrete frame for a transverse and longitudinal section of the building. The second analysis was a cracking analysis on the longitudinal masonry façade of the building. Each analysis was performed using the DIANA finite element program. It should be noted that in each model, only the basement through the fourth floor was modeled. This decision was based on the fact that the majority of the masonry façade only extends up through the fourth floor. Furthermore, the models are purely two-dimensional; three-dimensional effects such as the additional stiffness from out-of-plane beams and walls were omitted.

Table 1. Material properties for masonry and concrete.

Material	Concrete	Masonry
Young's modulus (N/mm^2)	31,000	5000
Poisson's Ratio	0.2	0.2
Density (kg/m^3)	2500	2400
Tensile Strength (N/mm^2)	N/A[1]	0.25
Fracture Energy, G_f (N/mm)	N/A[1]	0.03

[1]Concrete was modeled as linear elastic.

3.1 *Material properties*

Material properties for the masonry and concrete are summarized above in Table 1. For the analysis of the façade, the masonry was assumed to remain linear elastic in compression, but to soften linearly in tension after reaching its tensile strength. The concrete in each model was idealized as linear elastic to help simplify the analysis; cracking and reinforcement were not taken into account.

Pile groups under the footings and basement wall were represented by a single pile with a stiffness and cross-sectional area equivalent to the pile group. The representative pile was modeled as shown in Figure 3. Two truss elements, 6 m long, formed the representative pile. In addition, 2-noded spring elements were located between the truss elements and at the base of the pile to represent the skin friction from the soil along the pile shaft and the soil stiffness at the pile toe, respectively. The skin friction spring is a parallel spring, while the pile toe spring is in series with the lower truss element.

The pile springs account for the interaction between the piles and the soil. For a single pile of the building, the stiffness of these spring elements was derived from data in Figure 4 relating the force on a pile to its deformation into the soil. Other parameters for a single timber pile included an average diameter of 180 mm, Young's Modulus of 12,000 N/mm^2, and a density of 700 kg/m^3. The equivalent stiffness of the representative pile was determined by multiplying the Young's modulus and spring stiffness by the number of piles in the group, which ranged from 26 to 40 piles.

3.2 *Frame analysis*

Two frame sections, one longitudinal and one transverse, were analyzed for settlement damage. The

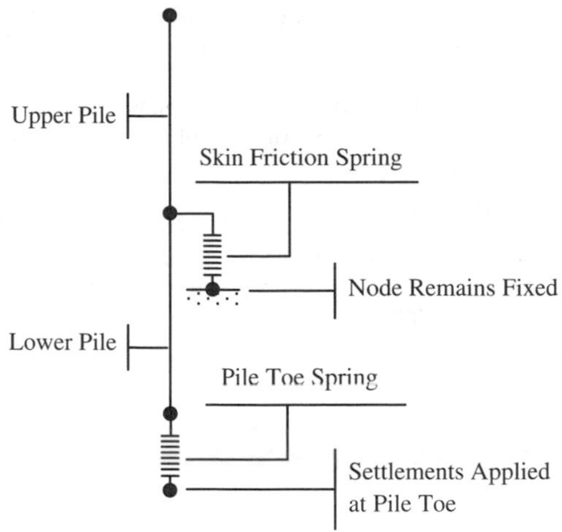

Figure 3. Model of the building piles.

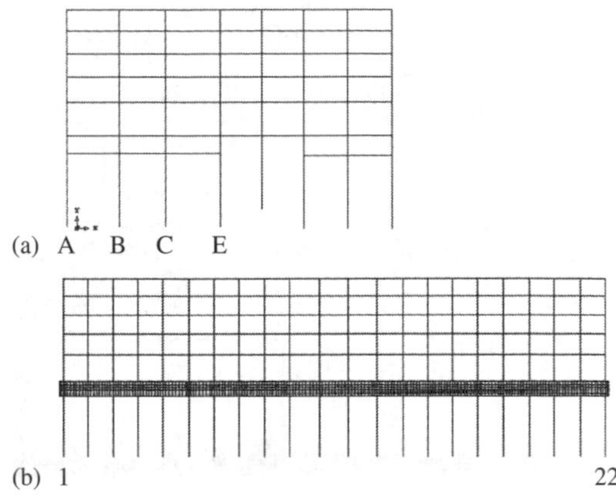

Figure 5. Frame models for the transverse section (a) and the longitudinal section (b). Labels refer to column lines.

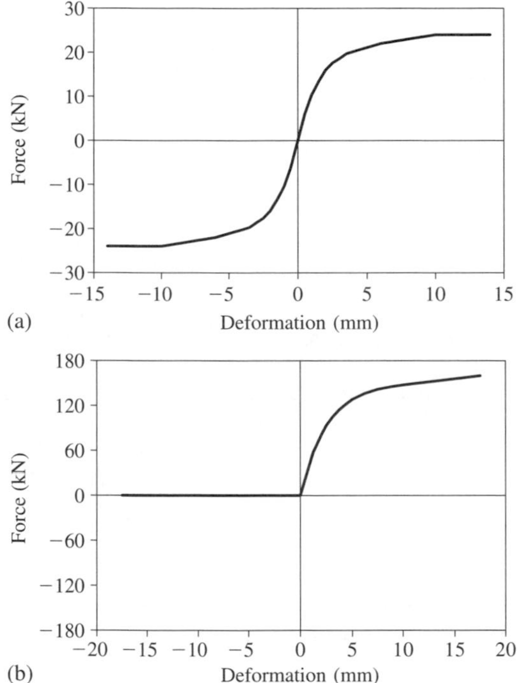

Figure 4. Relationship between the force on a pile to its displacement into the soil. Resistance to displacement is provided along the pile shaft (a) and at the pile toe (b).

location of each section is shown in Figure 2, and the frame models are presented in Figure 5. For the transverse frame, the two-way floor slabs of the structure were supported by concrete girders which spanned between the columns. In general, the floor slabs were 125 mm thick and the girders had a rectangular cross section of 425 by 350 mm. Thus the beams in each of the frame analyses were modeled as T-beams with an effective flange width. Columns were typically square or rectangular with a cross-sectional area between 0.6 and 0.9 m^2.

For the longitudinal frame, concrete girders were not present between the columns at the edge of the building. Thus, the beams in the longitudinal frame model represent only the floor slabs along the edge of the building. These floor slabs were similarly given an effective width. The columns of the concrete frame were square or rectangular, but their cross sections varied greatly, ranging from 0.2 to 5.1 m^2. These columns framed into the basement wall, which was 550 mm thick and contained massive columns averaging 2 m wide by 1.6 m thick. A foundation beam, 500 mm deep by 1065 mm wide, ran along the underside of the basement wall.

Beams, columns, and floor slabs were modeled with 3-noded Mindlin-Reissner beam elements, and the basement wall and basement columns in the longitudinal frame were modeled with 8-noded plane stress elements. A 2×2 integration scheme was used for the plane stress elements.

In the longitudinal frame, beam elements were connected to the top and bottom of the basement wall to represent the ground level floor slab and foundation beam, respectively. Additionally, beam elements were used to model the footings beneath the basement wall columns. These beam elements were overlaid node-for-node onto the plane stress elements of the basement wall. The ground floor columns and piles in turn shared the same nodes as these beam elements and the basement wall elements. It is important to emphasize that the pile spring elements were the only nonlinear elements in the frame analyses – the concrete frame and basement wall were modeled as linear elastic.

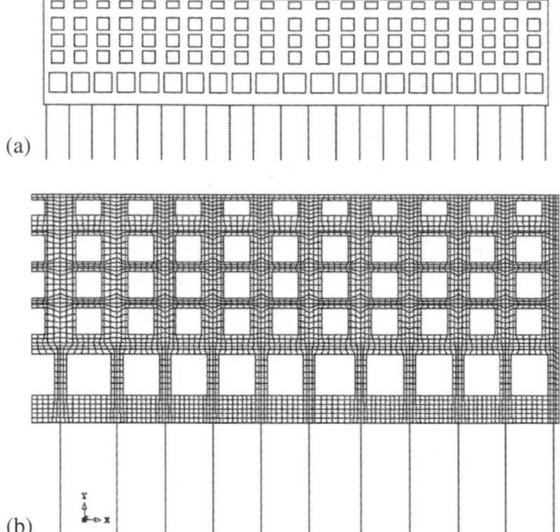

Figure 6. Model of the masonry façade (a) and the finite element mesh used (b).

3.3 *Analysis of the masonry façade*

In addition to the frame analyses, a cracking analysis was performed on two continuum models of the masonry façade outside of the longitudinal frame. An example of these models is shown in Figure 6.

In the first model, the masonry façade was considered to act independently of the concrete frame (i.e. no anchoring to the frame). Thus the façade was supported solely by the basement wall and piles. These were modeled in the same manner as for the longitudinal frame discussed in the previous section.

In the second model, the interaction between the frame and façade was studied by anchoring the façade to the columns of the frame. Anchoring was accomplished by overlaying beam elements representing the columns onto the façade. The beam elements, which were also 3-noded Mindlin-Reissner elements, were rigidly connected node-for-node to the plane stress elements of the masonry. The concrete slabs framing into the columns were not modeled in this study.

For each analysis, the masonry was represented with 8-noded plane stress elements, with element sizes ranging from 430 by 250 mm to 820 by 500 mm. All elements had a thickness of 300 mm and used a 2×2 integration scheme.

3.4 *Crack model*

A total-strain rotating crack model developed by Feenstra et al. (1998) was used for the analysis of the masonry façade. This model offers the advantages of simplicity in its application, and has been shown to provide less stress locking and relatively better convergence for large-scale fracture than other plasticity-based and fixed smeared crack models (Boonpichetvong & Rots 2002). A full Newton-Raphson iteration scheme and an average crack bandwidth of 450 mm were employed for the analysis.

3.5 *Settlements*

The settlement trough in the transverse direction of the building is based on an inverted Gaussian curve of the form (Netzel 2001):

$$w_{trans}(x) = w_{\max} \exp\left(\frac{-x^2}{2i^2}\right) \quad (1)$$

where

$$i = K(z_o - z)$$

$$w_{\max} = \frac{0.313 f V_l D^2}{i}$$

and x = distance from the tunnel axis; i = distance from the point of inflection of the settlement trough to the tunnel axis; K = trough width parameter, which depends on soil type; z_o = depth of the tunnel axis from the ground surface; z = depth at which the settlement trough is being evaluated; f = curve factor for the increase of volume loss along curves (dependent on the curve radius of the horizontal alignment); V_l = volume loss; and D = diameter of the tunnel. The longitudinal settlement trough is in turn derived from the transverse trough.

For this study, the following values were used: $K = 0.8$ (for the subsurface settlements at the pile toe level), $z_o = 24$ m, $z = 13$ m, $f = 1.55$ at the South end of the building and 1.0 at the North end, $V_l = 0.0081$, and $D = 7.05$ m.

It is important to point out that the soil-structure interaction is modeled with the simplified approach of the pile spring elements. Advanced modeling with a fully coupled soil-structure model can influence the actual settlement profile of the soil and the differential displacement along the base of the structure (Burd et al. 2000, Liu et al. 2000). Furthermore, horizontal ground movements were not taken into account in this study.

3.6 *Loading*

In each analysis, the dead load was applied first to simulate the initial state of the structure. For the frame analyses, the dead load included the weight of the structure, a permanent dead load of 1 kN/m^2, and a mobile load of 4 kN/m^2. The weight of the structure was composed not only of the weight of the modeled elements, but also point loads from out-of-plane beams and point loads along the top of the model from the absent upper floors. Furthermore, the permanent and mobile loads were assumed to act on all floors simultaneously to simplify the analysis.

For the analysis of the façade with no anchoring to the frame (referred to as the independent façade), the only load applied to the façade was the weight of the masonry. However, point loads were applied along the basement wall at the column lines to account for the weight of the concrete frame. For the anchored façade (i.e. anchored to the concrete frame), the columns were loaded with their original loads from the frame analysis. In addition, point loads were applied to the columns to account for the loads from the floor slabs.

In all cases, displacements during application of the dead load were suppressed to better determine the response of the structure to settlement.

After the dead load had been applied, settlements were imposed on the structure. Settlements were only applied to the bottom node of the pile toe springs. The shape of the settlement trough in each analysis was assumed to remain constant during the analysis while the magnitude of settlement was incremented up to the value given by Equation 1. However, as pointed out by Rots (2000), this situation differs from reality in that the shape, position, and magnitude of the trough are changing continuously as boring proceeds.

4 COMPUTATIONAL RESULTS

4.1 *Transverse frame*

4.1.1 *Displacements*

The displaced shape of the transverse frame after settlement is shown in Figure 7. Settlement causes the frame to experience shearing action in two directions: horizontally as the right half of the frame is pulled to the left, and vertically as the left half of the frame settles.

Displacements for selected floors are compared to the settlement trough in Figure 8. The building itself is subjected entirely to hogging deformation (i.e. the settlement trough is concave downward). It is interesting to note that in this case, the displacements at all three levels of the building are essentially identical and that they more or less follow the settlement trough. Thus the frame itself is quite flexible and susceptible to settlement. The slightly lower displacement at column line A (see Figure 5) could be due to the resistance provided by the structure or to a change in the load distribution through these columns and piles.

4.1.2 *Load distribution*

Changes in the structure's load distribution can be seen by examining the forces in the piles. These are provided in Figure 9. Note that the upper and lower pile references refer to the sections of the pile above and below the node connecting the skin friction spring, respectively (see Figure 3).

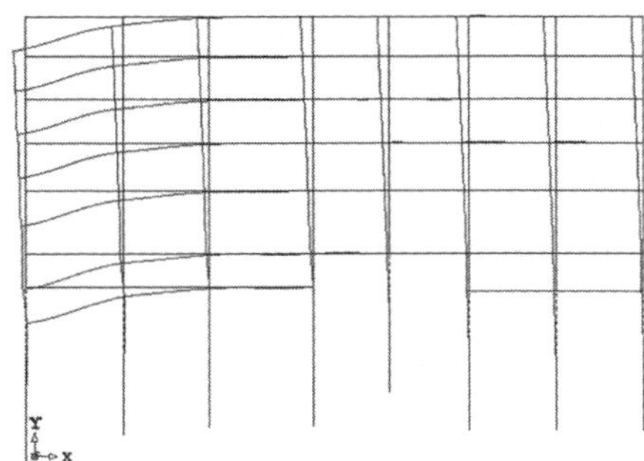

Figure 7. Displaced shape of the transverse frame after settlement.

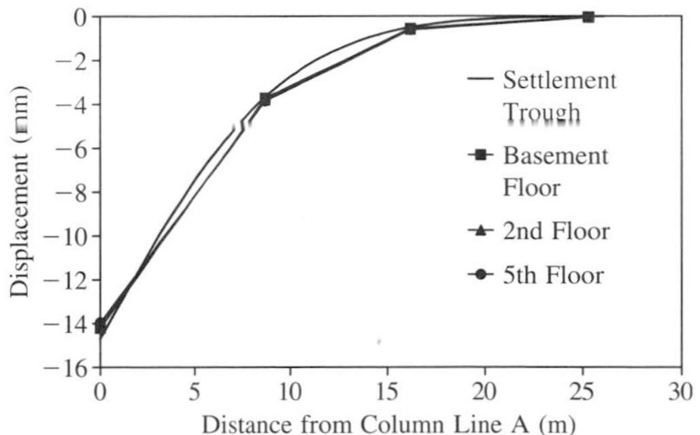

Figure 8. Floor displacements for the transverse frame.

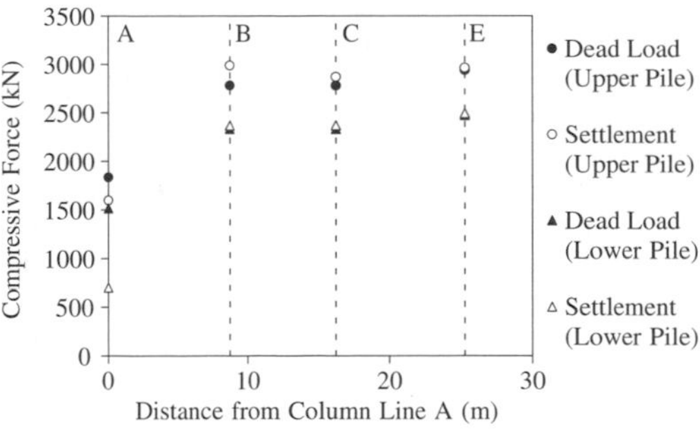

Figure 9. Pile forces in the transverse frame.

Settlement causes the load of the structure to shift slightly to the right, unloading the pile group at column line A and then distributing extra load to pile groups further down the settlement trough. It is also interesting to note from Figure 9 that the effect of settlement on the structure's load distribution more or less dissipates 25 m from column line A, which is approximately halfway into the structure.

4.1.3 *Beams*

Imposition of the settlement trough mostly affects the basement floor beams from column lines A to E, with

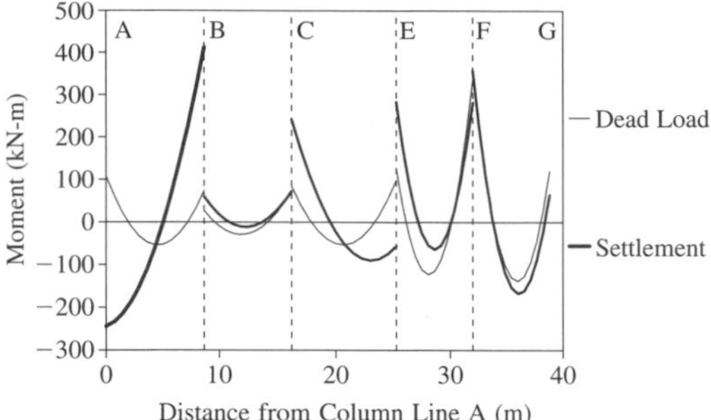

Figure 10. Moments for the basement floor beams of column lines A–G of the transverse frame.

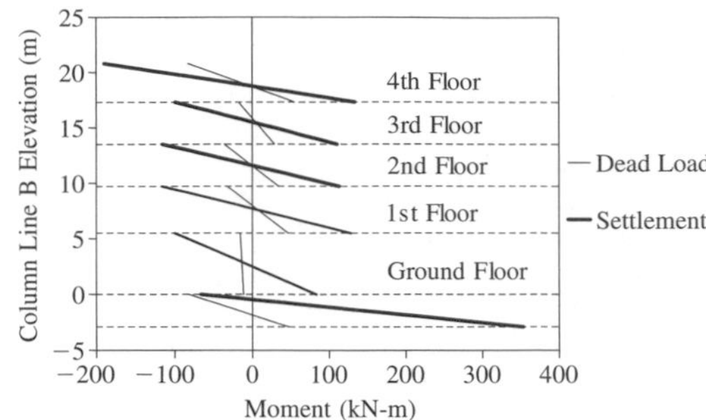

Figure 11. Moments for column line 16 in the transverse frame at selected floors.

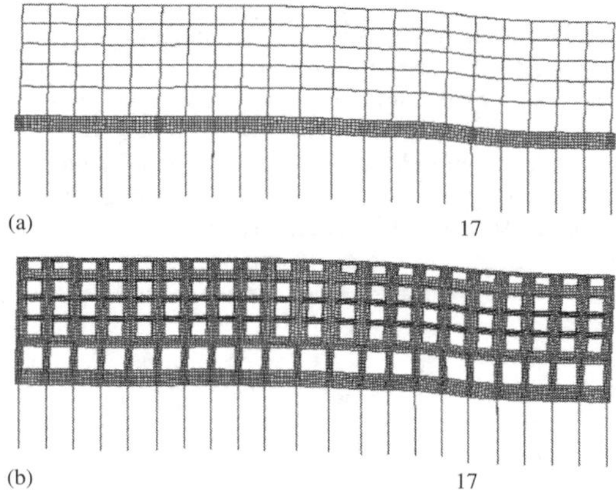

Figure 12. Displaced shape of the longitudinal frame (a) and façade models (b) due to settlement only.

the moments in the upper floors largely unaffected by settlement. The exception to this were the upper floor beams in the first bay of the building, where moments at column B increased by 50% to 420 kN-m. However, the critical floor was clearly the basement floor, for which the moments are detailed in Figure 10. The moments at column B in the first bay of the structure were more than 5 times greater after settlement.

The moment capacity of a typical beam section located in the ground floor of the building is 1190 kN-m for the mid-section moments and 530 kN-m at the ends. These values are conservative estimates since reinforcement in the floor slab (i.e. the flange of the beam) was ignored in determining the moment capacity and ground-level beams will have a smaller cross section than basement beams. However, given these numbers, the beams of the building at all floors have enough moment capacity to resist the affects of settlement with a sufficient margin of safety.

The distribution of shear force shows the same trends as the moment distribution, with the upper floors being mostly unaffected by settlement. For the basement floor however, shear forces more than doubled for the first and third bays, reaching values up to 142 kN. In spite of this increase in shear, the capacity of the basement floor beams (estimated at 300 kN) should be adequate to prevent shear failure with an acceptable safety margin.

4.1.4 *Columns*

The elements most affected by settlement in the transverse section were the columns. Column moments were typically two to three times greater at all floor levels after settlement for columns A through E. An example of the shifts in moment is provided in Figure 11 for column line B. Note that the moment for the basement column in Figure 11 was seven times greater after settlement. Due to the shift in load distribution previously discussed, this particular column supported an extra 120 kN after settlement, which is approximately a 5% increase in the normal force. Although this is a small increase in the normal force, the combined moment-axial force effect should be a point of attention with respect to stability.

4.2 *Longitudinal frame and façade*

In the following sections, the results for the longitudinal frame and the two façade models will be presented together to aid in the comparison of these models in settlement damage prediction.

4.2.1 *Displacements*

The displaced shape of each model is presented in Figure 12. In addition, the displacements along the top of the fourth floor are compared to the settlement trough in Figure 13. It is important to note that the irregular shape of the settlement trough is due to the stepped shape of the longitudinal section of the building (Figure 2) and the different curve factors applied to Equation 1. As shown in Figure 13, the effects of the jagged settlement trough are smoothed out as a result of the massive linear-elastic basement wall.

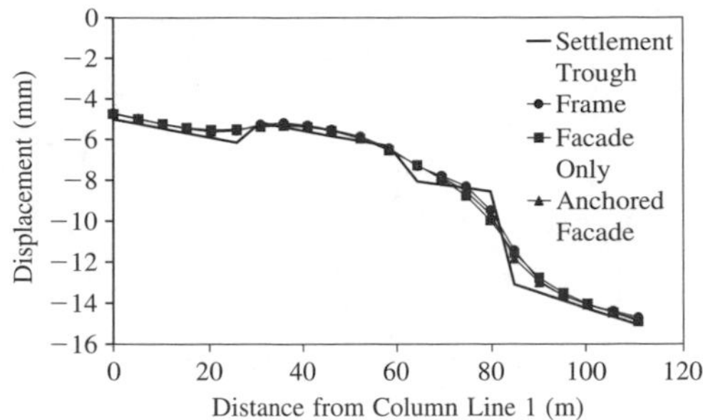

Figure 13. Fourth-floor displacements for the longitudinal section.

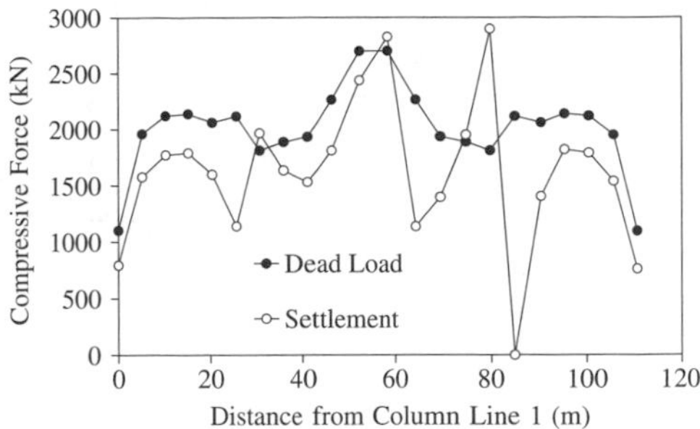

Figure 14. Forces in the lower half of the piles after settlement for the anchored façade.

The building experiences three shear regions at 25, 64 and 80 m, which correspond to the jumps in the settlement trough at these locations. With the exception of a 0.5 mm discrepancy at 80 m, the fifth floor displacements for each model are essentially identical and more or less follow the settlement trough. Thus, as with the transverse frame, the building acts in a flexible manner.

One important observation was that there were no significant discrepancies between the displacements of the fifth floor and those of the ground floor in any of the longitudinal models. In fact, the basement wall experienced the same deformations due to settlement in each model, and these deformations were transmitted directly to the top floor.

It is possible that cracking of the masonry in the anchored façade model (discussed in Section 4.2.3) reduced the stiffness of the façade enough to allow the structure to deform in the same manner as the frame. However, recall that these analyses account for a simplified soil-structure interaction. If a fully coupled soil-structure interaction were taken into account, the stiffer anchored façade might have reduced the differential displacements along the basement wall and throughout the building.

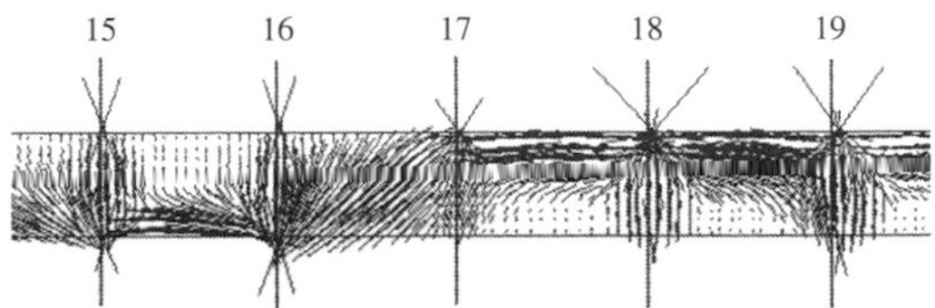

Figure 15. Principle compressive stresses in the basement wall after settlement for the longitudinal section.

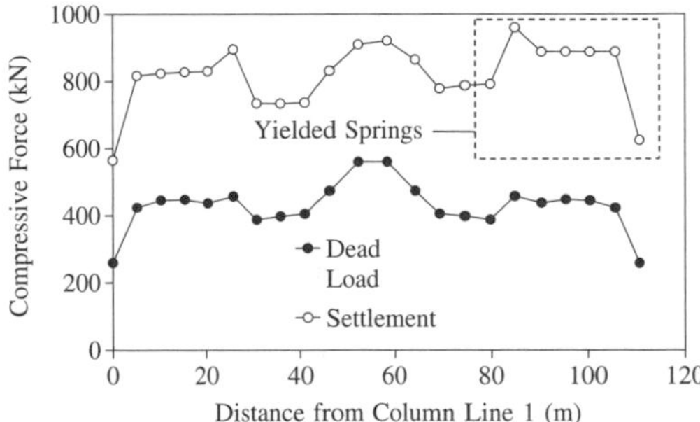

Figure 16. Forces in the skin friction springs for the anchored façade.

4.2.2 *Load distribution*

The load distribution behavior for the three longitudinal models was identical, and therefore the following discussion will focus only on the results for the anchored façade model.

The shearing action at the right side of the building caused the lower half of the pile group at column line 17 (85 m) to unload completely, as shown in Figure 14. Consequently, the loads that were previously carried by this pile group are redistributed through the basement wall to the pile group at column line 16 (80 m). This shift can be clearly seen in Figure 15 with the flow of principle compressive stresses in the basement wall. The principle compressive stresses in the region of the unloaded pile group decreased by more than half, from 0.8 to 0.3 N/mm^2. Similar shifts in load distribution can also be found in the other shear regions at 25 and 64 m.

Aside from the shear regions in the building, settlement generally causes the pile toe springs and the lower half of the piles to unload by approximately 400 kN. This unloading is more or less balanced by an increase in the skin friction spring forces, which are provided in Figure 16. In fact, the skin friction springs of the last seven pile groups at the right of the building (column lines 16 through 22) actually yielded during settlement. Consequently, the soil along the length of these pile groups would provide no resistance to further load or settlement.

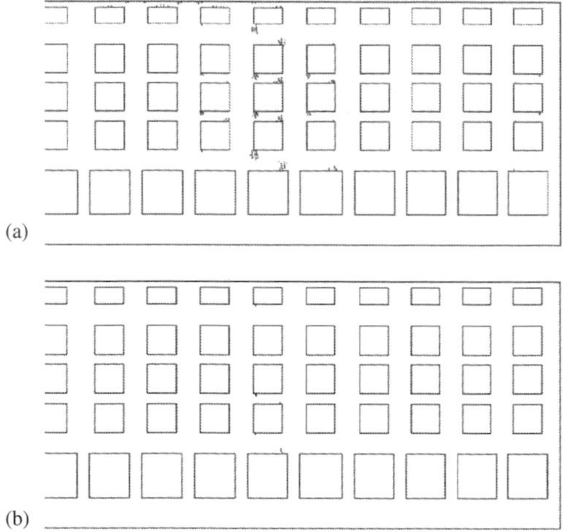

Figure 17. Crack pattern (a) and fully open cracks (b) for the independent façade.

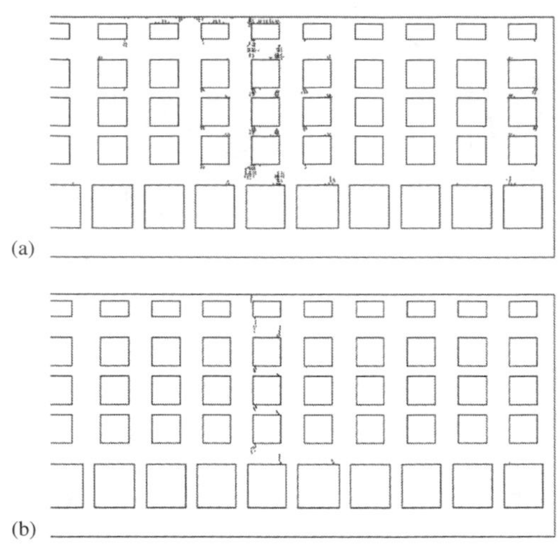

Figure 18. Crack pattern (a) and fully open cracks (b) for the anchored façade.

4.2.3 *Cracking*

In the following discussion, attention will be paid first to the cracking damage on the independent façade. These results will then be compared against the damage sustained by the façade when anchored to the frame.

For the independent façade analysis, cracks concentrated in the shear region between column lines 16 and 17. This corresponds to the drop in the settlement trough at 80 m in Figure 13. The predicted crack pattern for the right half of the building is illustrated in Figure 17. In addition, the fully open cracks are also presented. Fully open cracks are defined as those cracks for which softening has completed and no more tensile stress can be transferred.

Cracking actually begins during the first settlement increment in the middle of the building above the ground floor windows and at the ends of the building in the corners of the third floor windows. Soon after, cracks begin to form at the corner of the third floor window in the shear region. Cracking continues to intensify in this region and also along the top of the ground floor windows. Eventually, the cracks spread to the upper windows, leading to the final crack pattern shown. Minor cracking also occurred at the opposite end of the building where another jump occurred in the settlement trough (at 25 m in Figure 13). These cracks also formed mostly at the corners of the second and third floor windows.

When the façade was anchored to the frame, the interaction between them actually amplified the cracking damage. Cracks were observed at both ends of the building after applying the dead load. However, these dead load cracks are not realistic since the masonry façade would most likely have been constructed after the concrete frame was in place. As the dead load cracks intensified, others began to form in the shear region at the corners of the upper floor windows and at the top of the ground floor windows. Finally, cracking spread to the top of the fourth floor windows. The crack pattern for the anchored façade is presented in Figure 18, along with the fully open cracks. In addition to these cracks, a similar crack pattern was observed in the shear region at the other end, although to a lesser extent.

As these crack patterns show, the anchored façade sustained more damage as a result of settlement than when the façade was modeled independently. In fact, the fully open cracks almost extend through the depth of the building. The difference in cracking damage between the two models is also apparent in a plot of the maximum crack widths, which is shown in Figure 19. (Crack widths were calculated by multiplying the crack strain with the average crack bandwidth.) Significant softening of the anchored façade began at approximately 9 mm of settlement, whereas the independent façade remained relatively stiff throughout the settlement process. Note that the crack widths presented in Figure 19 represent the maximum crack width in the façade at any given level of settlement. The location of the maximum crack may in fact vary during the settlement process.

The increased damage of the anchored façade is likely a result of the restraint that the columns impose on the masonry. Since the basement wall experiences the same magnitude of deformation in both models, it will impose the same deformations onto the façades. The added restraint on the masonry in the anchored façade induces higher tensile stresses, which in turn leads to more cracking. Again, it is

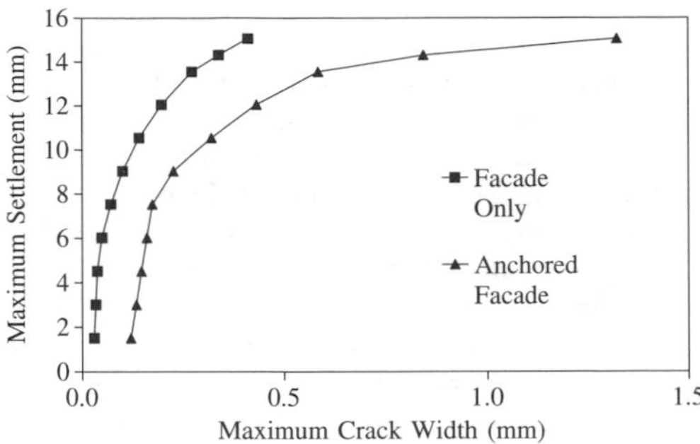

Figure 19. Maximum crack width in the façade as a function of the maximum settlement along the structure.

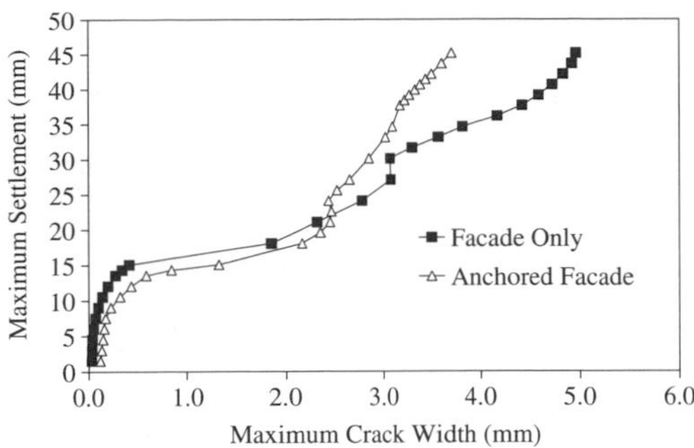

Figure 21. Maximum crack widths in the façade at magnified settlement levels.

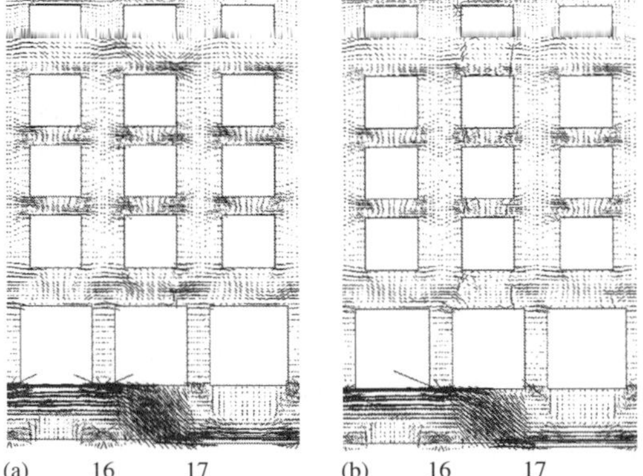

Figure 20. Principle tensile stresses for the independent façade (a) and the anchored façade (b) after settlement.

important to emphasize that these effects are a result of the simplified coupling between the soil settlement and the stiffness and weight of the structure (see Section 4.2.1).

Another significant difference between the two models was the increased stress locking (Rots 1988) that was observed for the anchored façade. This stress locking can be seen with a plot of the principle tensile stresses for the crack region in Figure 20. In addition, the diagonal stress flow in the basement wall and between the windows clearly indicates the shearing action on the building. Stress locking was observed to concentrate in areas of fully open cracks, which explains why the anchored façade experienced it to a greater extent. Furthermore, stress locking has been found to lead to poor convergence (Boonpichetvong & Rots 2002b, Rots 2000). For this study, in which a total strain rotating crack model was used, convergence was only achieved using an energy norm convergence criteria of 10^{-3}.

4.2.4 *Increasing the settlements*

The previous results are based on the prescribed settlements of Equation 1. However, it is also interesting to allow settlement to continue beyond this point and observe the effect on the predicted crack damage. To this end, settlement was allowed to continue to three times the prescribed values. A plot of the crack widths during this extended settlement phase for the two façade models is presented in Figure 21.

The first thing to note in Figure 21 is that both façades experience a plateau in the crack width response during settlement. This plateau has been associated with through depth cracking of masonry façades and increased stress locking, leading to poor convergence (Rots 2000). Rots circumvented this poor convergence by following the smeared crack analysis with a discrete crack analysis, in which an arc-length procedure was employed to trace the response. Although stress locking was found to increase with the development of fully open cracks in this study, convergence was achieved for both models during this plateau by using an energy norm criteria of 10^{-3}.

Another interesting observation is that both models experienced a jump in the maximum crack width with increased settlement. For the independent façade this jump occurred between 25 and 30 mm of settlement, and for the anchored façade it occurred between 20 and 25 mm of settlement. The jump coincided with the sudden formation of fully open cracks at the left side of the building, and the unloading of previously open cracks at the right side of the building. This sudden formation of new cracks and unloading of old cracks is illustrated for the independent façade in Figure 22 with a plot of the fully open cracks before and after the jump occurred. In essence, this behavior suggests that the maximum crack width in the façade occurred at different locations before and after the jump.

It is also interesting to note that the maximum crack widths in the anchored façade after the jump were less than those for the independent façade. This

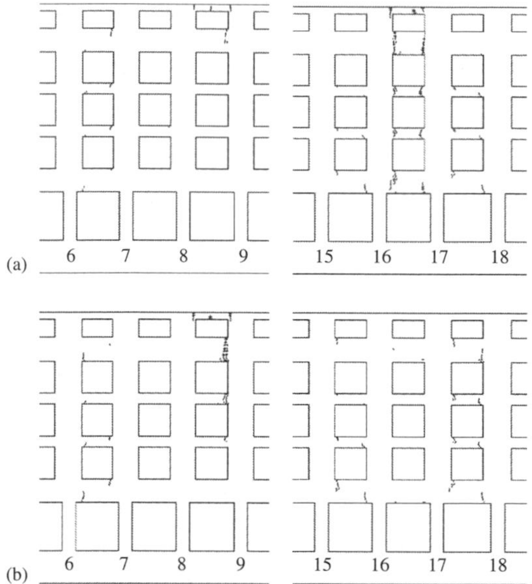

Figure 22. Fully open cracks for the independent façade before (a) and after (b) the jump in the maximum crack width.

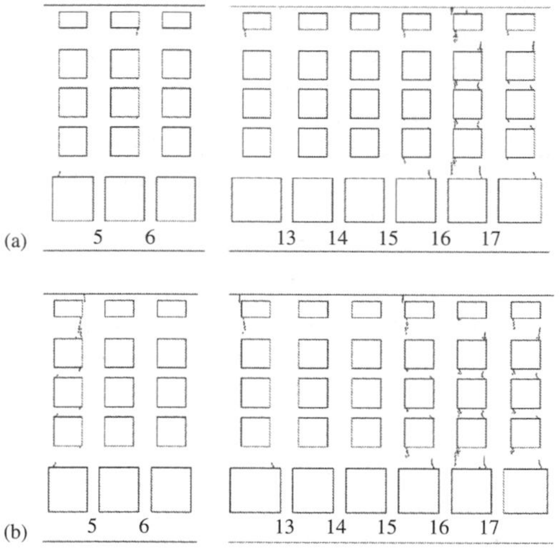

Figure 23. Fully open cracks for the anchored façade before (a) and after (b) the jump in the maximum crack width.

appears to be a result of the anchored façade's ability to redistribute stresses with the help of the linear elastic concrete frame. As shown in Figure 23, the formation of new, fully open cracks occurred not only at the left side of the building (column lines 4 to 6), but also in the middle of the building (column lines 12 to 13) and to a significant extent between column lines 15 and 18.

5 CONCLUSIONS

The analysis of the concrete frame building with a masonry façade in this work indicated that the structure was relatively flexible in both the longitudinal and transverse directions. Although the beams should maintain a sufficient margin of safety against shear and moment increases during settlement, the columns must be particularly considered with respect to stability issues resulting from the combined effect of increases in moment and axial force. Cracking of the façade concentrated on the right side of the building, which was subjected to significant shearing action. Crack widths reached 0.4 to 1.3 mm for the two façade models studied.

However, the results obtained are sensitive to the analysis assumptions. These assumptions included the use of simplified soil-structure interactions and a linear elastic model for concrete. The overall effect of these assumptions was that more cracking was observed for the stiffer anchored façade model. A conservative estimate of the building's true cracking response will likely be somewhere between the results of the two façade models.

ACKNOWLEDGEMENTS

The authors would like to acknowledge the project organization of the North-South metro line in Amsterdam for their support and contact regarding this study.

REFERENCES

Boonpichetvong, M. & Rots, J.G. 2002a. Settlement damage of masonry buildings in soft-ground tunneling. In Max A.N. Hendriks & Jan G. Rots (eds.), *Finite Elements in Civil Engineering Applications, Proc. of the 3rd DIANA World Conference 2002*. Tokyo: Balkema.

Boonpichetvong, M. & Rots, J.G. 2002b. Large-scale fracture analysis of masonry buildings subjected to ground movement. *Advanced Technologies for the Development of Civil Engineering, Proc. of the 8th National Convention on Civil Engineering*. 1:65–70.

Burd, H.J., Houlsby, G.T., Augarde, C.E., & Liu, G. 2000. Modelling tunnelling-induced settlement of masonry buildings. *Proc. Instn. Civ. Engrs Geotech. Engng.* 143(Jan): 17–29.

Feenstra, P.H., Rots, J.G., Arnesen, A., Teigen, J.G., & Hoiseth, K.V. 1998. A 3D constitutive model for concrete based on a co-rotational concept. In R. de Borst et al. (eds.), *Computational modelling of concrete structures; Proc. EURO-C 1998*. Rotterdam: Balkema.

Liu, G., Houlsby, G.T., & Augarde, C.E. 2000. 2-dimensional analysis of settlement damage to masonry buildings caused by tunneling. *The Structural Engineer* 79(1): 19–25.

Netzel, H. 2001. Empirisch, analytische predicties van grandverrorningen en gebouwrespons. COB-committee F220 Praktijkonderzoek Sophiaspoortunnel (in Dutch).

Netzel, H. & Kaalberg, F.J. 2000. Numerical damage risk assessment studies on adjacent buildings due to TBM-tunnelling in Amsterdam. *Proc. of GEOENG 2000*. Melbourne.

Rots, J.G. 1988. *Computational modeling of concrete fracture*, Ph.D. Dissertation, Delft University of Technology, Delft, the Netherlands.

Rots, J.G. 2000. Settlement damage predictions for masonry. In L.G.W. Verhoef and F.H. Wittmann (eds.), *Maintenance and restrengthening of materials and structures – brick and brickwork; Proc. Int. workshop on urban heritage and building maintenance*. Freiburg: Aedificatio.

Computational Modelling of Concrete Structures, Bićanić et al. (eds)
© 2003 Swets & Zeitlinger, Lisse, ISBN 90 5809 536 3

Bending behaviour simulations of fibre reinforced beams with rebars by using the bridged crack model

Alberto Carpinteri, Giuseppe Ferro & Giulio Ventura
Department of Structural and Geotechnical Engineering, Politecnico di Torino, Italy

ABSTRACT: The bridged crack model has been demonstrated to be an efficient numerical tool for investigating the behaviour of structural elements in bending. In the model, Linear Elastic Fracture Mechanics concepts are used to determine the equilibrium and the compatibility equations of a beam segment subjected to bending in presence of a mode I crack. Recently, the model has been extended to include the presence of closing stresses as a function of the crack opening in addition to steel reinforcement closing tractions. The ability of the model to analyse the mechanical behaviour of members having both steel and fibre reinforcements makes it a practical tool in structural design. In fact, it allows for studying the ratio of steel to fibre reinforcement percentages in order to optimise mechanical behaviour and costs. Some new techniques, recently proposed for reducing significantly the computing time in presence of cohesive stresses, are presented in the paper and an approximate criterion to account for concrete crushing is introduced as well.

1 INTRODUCTION

The rapid evolution of high-performance concretes and their importance in engineering applications requires the development of computational models suitable to describe their mechanical properties. In particular, the mechanical interaction of fibers added to the concrete matrix is to be accounted for, extending the analysis beyond the simple durability effect or a qualitative assertion of improved crack toughness of the matrix.

The bridged crack model has been originally proposed in (Carpinteri 1981a), (Carpinteri 1984) and in (Bosco and Carpinteri 1995) for RC beams, reformulated in (Carpinteri and Massabó 1996), (Carpinteri and Massabó 1997) for unreinforced concrete members with cohesive closing stresses and extended to the simultaneous presence of both steel and fibers reinforcements in (Carpinteri et al. 2002). The bridged crack model applies Fracture Mechanics concepts to the study of beams in bending. The ability of dealing at the same time with steel reinforcements and closing stresses in the matrix results in a very flexible computational model, capable of modelling a wide range of concrete materials. Moreover, while limit state analysis yields only the ultimate load, the bridged crack model reveals in addition scale effects, instability phenomena and brittle-ductile transition of the structural element.

In the paper the theoretical model is briefly recalled and the computational problem of an effective evaluation of the cohesive stresses is addressed. In fact, the interaction between the bar traction and the cohesive stresses requires a complex iteration structure accounting for the general, user-defined, cohesive law at the same time. This is reflected into a considerably larger computing time compared to the case when only steel bars or cohesive stresses are present. A very significant reduction in the computing time can be obtained analyzing the integration domains of the terms accounting for the cohesive contribution.

Eventually the problem of introducing a limit to the compressive stresses in the concrete is first introduced in this paper. The percentage of steel reinforcement inducing concrete crushing is determined. This is reformulated as an upper limit to the value of the brittleness number characterizing the bridged crack model. In the present approach this limit is established by limit state analysis and is restricted to absence of cohesive stresses. A general fracture mechanics model for concrete crushing including the cohesive stresses contribution is currently under development.

2 THE BRIDGED CRACK MODEL

The bridged crack model can be applied for evaluating the monotonic bending of a cracked reinforced

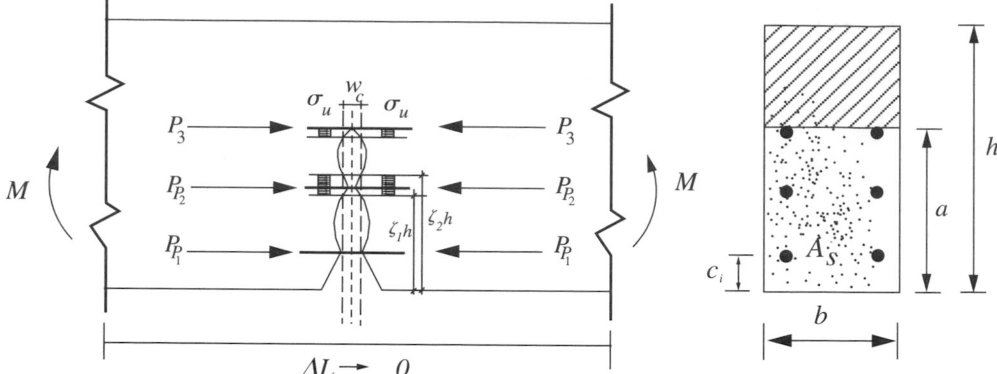

Figure 1. Model scheme of a cracked beam cross section.

concrete beam assuming as control parameter the crack depth at a fixed cross section. The model accounts for both the main reinforcement (steel bars) and a secondary reinforcement. It can be physically interpreted as the nonlinear tension behaviour of the concrete or the presence of reinforcement fibers.

Linear Elastic Fracture Mechanics is assumed for the matrix with a crack propagation condition ruled by the comparison of the stress intensity factor K_I to the matrix toughness K_{IC}. The stress intensity factor is computed from the solutions reported in the stress intensity factors handbooks (Okamura et al. 1975); (Tada et al. 1963).

The adopted model scheme is reported in Figure 1 along with the used symbols: the section width b and height h, the crack depth a, the positions of the steel bars c and the relevant tractions P_i. The geometric dimensions are converted into nondimensional quantities, after dividing by the height h. The nondimensional crack depth is denoted by $\xi = a/h$ and $\zeta = x/h$ represents the generic nondimensional abscissa from the bottom part of the cross section.

The distribution of the discrete forces P_i and the continuous ones $\sigma(w)$ applied to the crack surfaces represent the bridging mechanism of the steel bars and cohesive stresses respectively. The nondimensional position of the i-th steel reinforcement is denoted by $\zeta_i = c_i/h$, while function $\sigma(w)$ represents the constitutive relation of the cohesive and/or fibers traction, w being the crack opening at a generic point along the crack. Function $\sigma(w)$ is assumed to be zero for w greater than the critical value w_c. Its actual expression can be derived from experimental results or model codes.

The constitutive relation for the reinforcement bars is assumed as rigid-plastic with no upper limit to the maximum deformation. This model describes both the bar yielding and the relative slip between the matrix and the reinforcement.

The maxima for the bridging actions are defined by the ultimate traction $P_{Pi} = A_i\sigma_y$ for the bars and by the ultimate stress $\sigma_0 = \gamma\sigma_u$ for the fibers (or matrix), A_i being the i-th bar area, γ the volumetric percentage of fibers, σ_y and σ_u the minimum between yielding and sliding stress for the bars and fibers respectively.

With reference to Figure 1, let K_I be the stress intensity factor at the crack tip. By the superposition principle, it is given by the sum of the stress intensity factors K_{IM} due to the bending moment, K_{Ii} due to the m reinforcement bar traction P_i and $K_{I\sigma}$ due to the distributed closing stresses $\sigma(w)$ along the crack (Carpinteri et al. 2002), (Ferro 2002):

$$K_{IC} = K_{IM} - \sum_{i=1}^{m} K_{Ii} - K_{I\sigma}; \tag{1}$$

In (1) the stress intensity factors can be expressed in the form:

$$K_{IM} = \frac{M_F}{bh^{1.5}} Y_M(\xi) \quad ; \quad K_{Ii} = \frac{P_i}{bh^{0.5}} Y_P(\xi, z_i); \tag{2}$$

$$K_{I\sigma} = \sum_{i=1}^{n_c} \frac{1}{bh^{0.5}} \int_{\zeta_{1i}}^{\zeta_{2i}} \sigma_0\left(w\left(\zeta\right)\right) Y_P\left(\xi, \zeta\right) bh\, \mathrm{d}\zeta; \tag{3}$$

where the functions Y_M and Y_P, are reported in (Okamura et al. 1975); (Tada et al. 1963); (Carpinteri and Massabó 1996). In (3) n_c is the number of cohesive zones, where $\sigma_0(w) \neq 0$. These zones are defined over the intervals $[\zeta_{1_i}, \zeta_{2_i}]$, $i = 1 \ldots n_c$.

Let ρ be the bar reinforcement percentage and define the brittleness numbers $N_P^{(1)}, N_P^{(2)}$ and the critical crack opening for the cohesive stresses $\tilde{w}_c$ as:

$$N_P^{(1)} = \rho \frac{\sigma_y h^{0.5}}{K_{IC}} \quad ; \quad N_P^{(2)} = \gamma \frac{\sigma_u h^{0.5}}{K_{IC}}; \tag{4}$$

$$\tilde{w}_c = \frac{E w_c}{K_{IC} h^{0.5}}. \tag{5}$$

Substituting (2) and (3) in (1), the following nondimensional equilibrium equation is obtained:

$$\tilde{M}_F = \frac{1}{Y_M(\xi)}\left(1+N_P^{(1)}\sum_{i=1}^{m}\tilde{\rho}_i\tilde{P}_i Y_P(\xi,z_i)+ + N_P^{(2)}\sum_{i=1}^{n_c}\int_{\zeta_{1_i}}^{\zeta_{2_i}}\tilde{\sigma}_0(\tilde{w})\,Y_P(\xi,\zeta)\,d\zeta\right) \tag{6}$$

where:

$$\tilde{M}_F = \frac{M_F}{K_{IC}bh^{1.5}} \quad ; \quad \tilde{P}_i = \frac{P_i}{P_{P_i}}; \tag{7}$$

$$\tilde{\rho}_i = \frac{\rho_i}{\rho} \quad ; \quad \tilde{\sigma}_0 = \frac{\sigma_0(w(\zeta))}{\gamma\sigma_u}. \tag{8}$$

The equilibrium equation (6) gives the bending propagation moment as a function of the bar traction and of the closing stresses. These quantities depend on the crack opening profile by means of the constitutive equations.

The crack opening at a general nondimensional abscissa ζ can be determined by summing the three contributions of the bending moment, bars traction and closing stresses. The nondimensional opening, evaluated at the crack propagation bending moment $M = M_F$, has the following expression:

$$\tilde{w} = \tilde{w}_M - \tilde{w}_P - \tilde{w}_\sigma = 2\tilde{M}_F\int_\zeta^\xi Y_M(x)Y_P(x,\zeta)\,dx+ - 2N_P^{(1)}\sum_{i=1}^{m}\tilde{\rho}_i\tilde{P}_i\int_{\max(\zeta,\zeta_i)}^{\xi} Y_P(x,z_i)Y_P(x,\zeta)\,dx+ - 2N_P^{(2)}\sum_{i=1}^{n_c}\int_{h_1(\zeta)}^{\xi}\int_{\zeta_{1_i}}^{h_2(x)}\tilde{\sigma}_0 Y_P(h_2(x),y)\,dy Y_P(x,\zeta)\,dx, \tag{9}$$

where:

$$h_1(\zeta) = \max(\zeta,\zeta_{1_i}); \quad h_2(x) = \min(x,\zeta_{2_i}). \tag{10}$$

By introducing the rigid-plastic constitutive equation for the bars, the displacement evaluated at $\zeta = \zeta_i$, $i = 1\ldots m$ equals zero if $P_{P_i} - P_i < 0$, i.e. if $1 - P_i < 0$. Let H be the Heaviside step-function. The diagonal matrix $[H_P] = \mathrm{diag}\left(H(1-\tilde{P}_i)\right)$, $i = 1\ldots m$ allows for expressing the vector of the openings at the reinforcement bars as:

$$\{\tilde{w}\} = [H_P]\left(\{\tilde{\lambda}_M\}\tilde{M} - [\,\tilde{\lambda}\,]\{\tilde{P}\} - \{\tilde{w}_\sigma\}\right); \tag{11}$$

where the elements of the above vectors and matrices are (i, $j = 1\ldots m$):

$$\{\tilde{\lambda}_M\}_i = 2\int_{z_i}^{\xi} Y_M(x)Y_P(x,z_i)\mathrm{d}x \tag{12}$$

$$[\,\tilde{\lambda}\,]_{ij} = 2N_P^{(1)}\tilde{\rho}_j\int_{\max(z_i,z_j)}^{\xi} Y_P(x,z_i)Y_P(x,z_j)\,\mathrm{d}x \tag{13}$$

$$\{\tilde{w}_\sigma\}_i = \tilde{w}_\sigma(z_i). \tag{14}$$

Equation (11), with $\tilde{M}$ given by (6), is a non-linear integral equation in the unknowns $\tilde{w}, n_c$, $[\zeta_{1_i}, \zeta_{2_i}]$, $i = 1\ldots n_c$. Its solution for a fixed crack depth ξ allows for the determination of the opening function, the crack propagation bending moment $\tilde{M}_F$ through (6) and the relative rotation of the cross section, given in nondimensional form by:

$$\tilde{\phi} = \phi\frac{Eh^{0.5}}{K_{IC}} = \tilde{\phi}_M - \sum_{i=1}^{m}\tilde{\phi}_I - \tilde{\phi}_\sigma = = 2\tilde{M}_F\int_0^\xi Y_M^2(\zeta)\mathrm{d}\zeta+ - 2N_P^{(1)}\sum_{i=1}^{m}\rho_i P_i\int_{z_i}^{\xi} Y_P(\zeta,z_i)Y_M(\zeta)\mathrm{d}\zeta+ - 2N_P^{(2)}\sum_{i=1}^{n_c}\int_{\zeta_{1_i}}^{\xi}\int_{\zeta_{1_i}}^{h_2(x)}\tilde{\sigma}_0 Y_P(h_2(x),y)\,\mathrm{d}y Y_M(x)\,\mathrm{d}x. \tag{15}$$

3 EFFICIENT EVALUATION OF THE CLOSING STRESSES CONTRIBUTION

The opening contributions $\tilde{w}_\sigma$ and the rotation $\tilde{\phi}_\sigma$ due to the closing stresses arising from the traction resistance of the matrix and/or of the fibers, eqs. (9), (15), can be evaluated by solving two double integrals. By dropping the constant terms and assuming:

$$g_2 = \max(\zeta_1, \min(x,\zeta_2) - \delta_3) \tag{16}$$

$$g_1 = \min(\xi, \max(\zeta,\zeta_1) + \delta_3) \tag{17}$$

$$h_2(x) = \min(x,\zeta_2) \tag{18}$$

the two integrals present the following form:

$$I_\phi = \int_{\zeta_1}^{\xi}\int_{\zeta_1}^{g_2}\tilde{\sigma}_0(y)Y_P(h_2(x),y)\,\mathrm{d}y Y_M(x)\,\mathrm{d}x \tag{19}$$

$$I_w = \int_{g_1}^{\xi}\int_{\zeta_1}^{g_2}\tilde{\sigma}_0(y)Y_P(h_2(x),y)\,\mathrm{d}y Y_P(x,\zeta)\,\mathrm{d}x \tag{20}$$

δ_3 being the *cut-off* defined by (Carpinteri and Massabó 1996) to overcome the non-convergence of the integral. The integrals (19) and (20) may be evaluated by iterated integrals library routines, allowing for the computation of integrals in the general form:

$$\int_a^b\int_{g_1(x)}^{g_2(x)} f(x,y)\,\mathrm{d}y\,\mathrm{d}x, \tag{21}$$

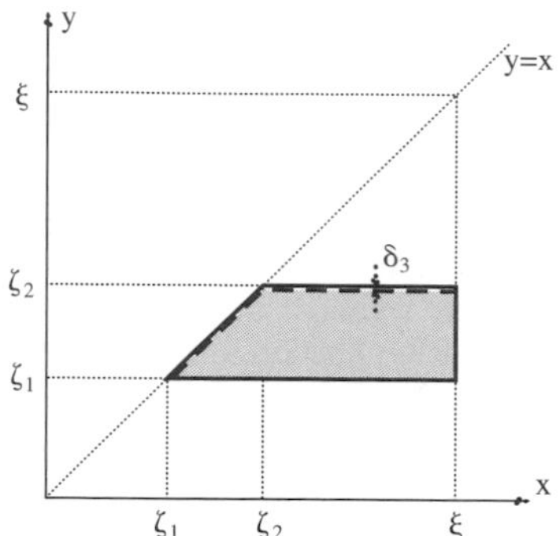

Figure 2. Rotation integration domain, Eq. (19).

$g_1(x)$ and $g_2(x)$ being the functions giving the lower and upper limits for the innermost integral. Adaptive quadrature for (21) is usually computationally expensive, being constructed through two nested adaptive quadratures for univariate functions. This solves the general problem, but generates very inefficient arrangements of the quadrature points when applied to standard domains.

By observing the integration limit functions in (19) and (20) it is however noted that the integration domains can be expressed as an assemblage of standard rectangular and triangular domains. The integration limit functions are represented in Figure 2 for I_ϕ and in Figure 3 for I_w. The integral I_w refers to the evaluation of the crack opening at an abscissa ζ for a stress distribution in the interval $[\zeta_1, \zeta_2]$. In the figures the dotted line represents the cutoff.

The quadrature can be much more efficient when carried out by standard quadrature rules on rectangular and triangular domains (Stroud 1971) or adaptive libraries (Cools and Haegemans) for standard regions. The speedup in the computation by comparing the IMSL iterated integral routine vs. the Cools package (Cools and Haegemans) with the present approach has been observed about a factor of 100.

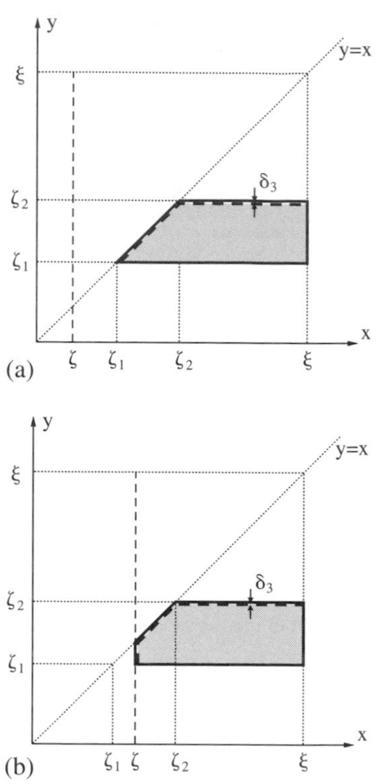

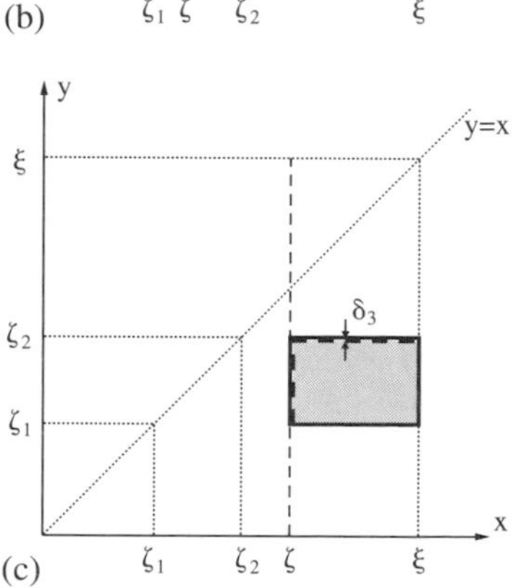

Figure 3. Crack opening integration domains in the cases: (a) $\zeta < \zeta_1$, (b) $\zeta_1 < \zeta < \zeta_2$, (c) $\zeta > \zeta_2$. Opening evaluated at ζ, Eq. (20).

4 CONCRETE CRUSHING AND MAXIMUM REINFORCEMENT

The bridged crack model described so far does not include any criterion for accounting for the strength of the concrete in compression. In the present approach this limit is introduced by limit state analysis (CEB-FIB model) and is restricted to absence of cohesive stresses. A general fracture mechanics model for concrete crushing including the cohesive stresses contribution is currently under development.

In the limit state analysis model, the crushing of concrete in bending is attained when the deformation reaches the critical value $\varepsilon_{cu} = 0.0035$. Concrete is simulated by a nonlinear law, which may be simplified assuming a rectangular stress block in compression, whose height is $0.8d$, d being the distance of the neutral axis from the upper edge of the rectangular section. The maximum compressive stress in concrete is given in this model by $0.85\sigma_{cu}$, where σ_{cu} is the compressive strength of concrete. This is illustrated in Figure 4, where all the quantities used in the present derivation are defined as well.

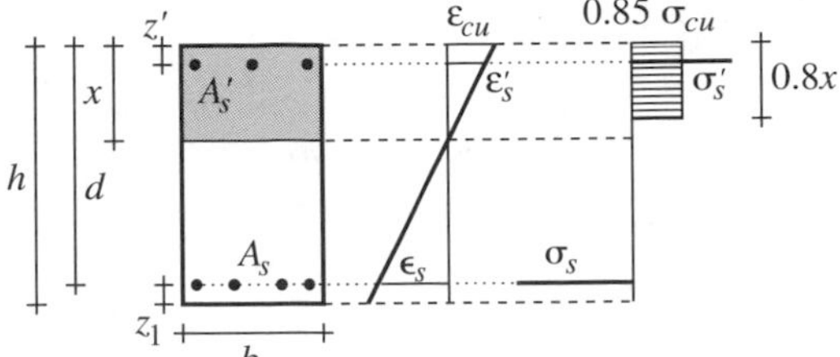

Figure 4. CEB-FIB model for a rectangular section in bending.

With reference to Figure 4, the following equations hold:

- linear deformation field

$$\varepsilon_s = \frac{d-x}{x}\varepsilon_{cu} \tag{22}$$

- equilibrium (rotation and translation)

$$M = b\,0.8\,x\,0.85\,\sigma_{cu}(d-0.4\,x) - \sigma_s\eta A_s(d-z') \tag{23}$$

$$\sigma_s A_s = b\,0.85\,\sigma_{cu}\,0.8\,x + \sigma_s\eta A_s \tag{24}$$

where $\eta = A's/A_s$. The steel bars are assumed with linear elastic behavior till the yielding stress σ_y when the strain ε_y is reached. Then, perfectly plastic behavior up to the rupture strain ε_{su} (usually $\varepsilon_{su} = 0.01$) is assumed.

Several observations can be made at this point, allowing for a solution of the above equations in closed form. First of all, as the strain in concrete is fixed to the crushing value, the problem has solution only for a tension reinforcement area greater than a minimum value A_{sc} to be determined. Moreover for any $A_s > A_{sc}$ a different value of the bending moment M at crushing is obtained. A qualitative plot of the bending moment as a function of the neutral axis position is sketched in Figure 5, and it is observed a monotonically increasing function of the position x. As a consequence, because of the equilibrium, the bending moment at concrete crushing is a monotonically increasing function of the tension reinforcement area A_s. When $x \to d, As \to \infty$ while the reinforcement stress decreases to zero.

The minimum bending moment producing the crushing collapse of concrete is therefore obtained when the maximum allowable strain ε_{yu} in the tension reinforcement is present. This condition is assumed here, being both a safe and optimal design condition at the same time.

It will be therefore assumed $A_s = A_{sc}$. The value of A_{sc} can be computed by the condition that the deformation and stress in the tension steel are the rupture ones, i.e. $\varepsilon_s = \varepsilon_{su}, \sigma_s = \sigma_y$. From Eq. (22) the neutral

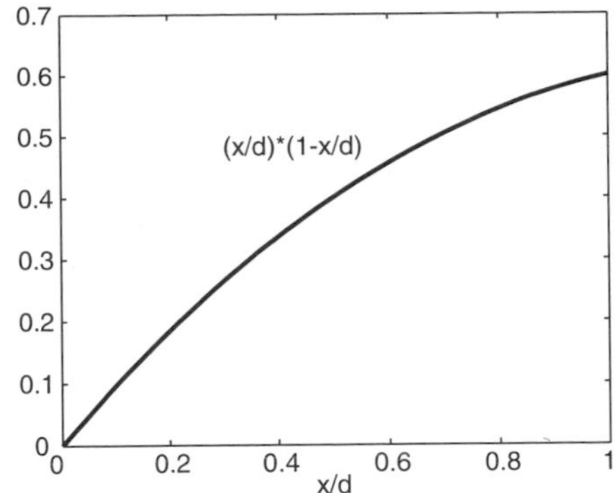

Figure 5. Trend of the bending moment curve as a function of the nondimensional neutral axis position.

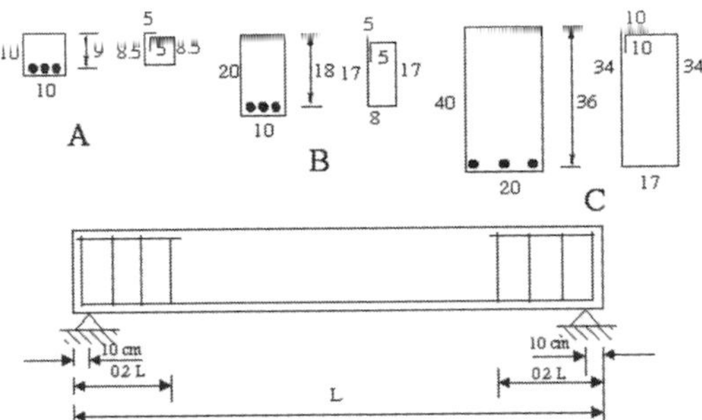

Figure 6. Geometry of the beams tested by Carpinteri et al. (1999).

axis position is determined as:

$$x = \frac{\varepsilon_{cu}}{\varepsilon_{su} - \varepsilon_{cu}}h, \tag{25}$$

and, upon substitution of the usual values for the maximum deformation,

$$x = 0.259\,h \tag{26}$$

A final consideration holds for the stresses in the compression reinforcement steel. It is observed that for usual beam geometries ($z' \ll h$) and steel type, the compression reinforcement yields significantly before concrete crushing, so that $\sigma_s' = \sigma_y$ can be assumed.

The minimum reinforcement area inducing concrete crushing failure is obtained after substitution of (26) into (24)

$$A_{sc} = \frac{0.176}{1-\eta}bd\frac{\sigma_{cu}}{\sigma_y}; \qquad \eta < 1. \tag{27}$$

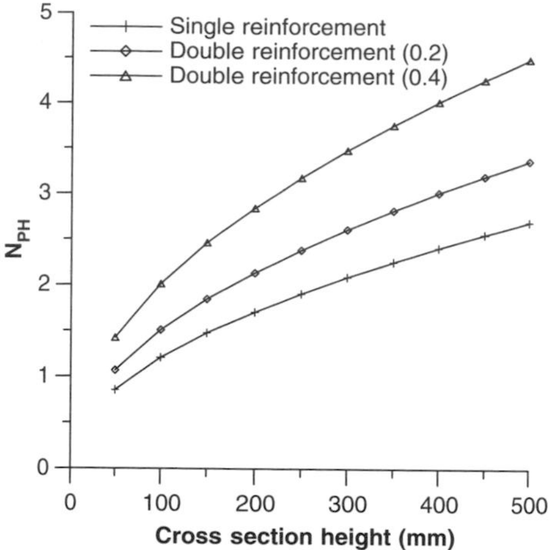

Figure 7. Maximum brittleness number for single and double reinforced beams.

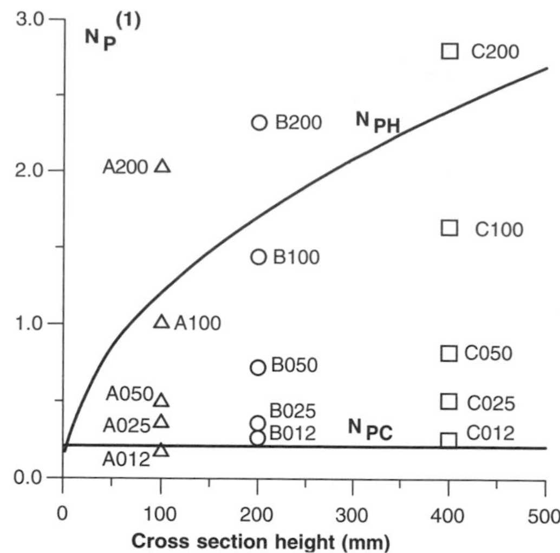

Figure 8. Brittleness numbers for the tested beams and curve of concrete crushing limit $N_{PH}^{(1)}$.

From this equation, observing that $h = d + z_1$ and dividing by bh both terms, the critical reinforcement area ratio ρ_{sc} is obtained, and upon substitution into the brittleness number definition it follows:

$$N_{PH}^{(1)} = 0.176\frac{1-\zeta_1}{1-\eta}\frac{\sigma_{cu}h^{0.5}}{K_{IC}}; \qquad \eta < 1. \tag{28}$$

In Figure 7 a plot of $N_{PH}^{(1)}$ is reported assuming the data $\sigma_{cu} = 48.2\,\mathrm{Nmm}^{-2}$, $K_{IC} = 63.4\,\mathrm{Nmm}^{-3/2}$, $\zeta_1 = 0.1$, $\eta = 0, 0.2, 0.4$, by varying the cross section height between 50 and 500 mm.

The value N_{PH} represents the upper limit of applicability for the bridged crack model both for beams with single or double reinforcement. Beams having $N_P^{(1)} > N_{PH}^{(1)}$ will exhibit failure for concrete crushing.

On the other hand the model allows for the determination of the minimum reinforcement so that the rupture is ductile. This has been determined in (Bosco and Carpinteri 1992), where the Authors demonstrated the following relation between the minimum brittleness number $N_{PC}^{(1)}$ and the compressive strength of concrete:

$$N_{PC}^{(1)} = 0.1 + 0.0023\sigma_{cu} \tag{29}$$

σ_{cu} being expressed as $\mathrm{N/mm}^2$. A beam having a brittleness number lower than the limit expressed by (29) exhibits a brittle failure because of insufficient reinforcement. In this case, when the crack develops at the lower edge and crosses the reinforcement, it is immediately yielded and strained to rupture, so that the peak load is higher than the yielding branch, Figure 11.

In this way a region of brittleness number values where a beam has ductile behavior is naturally defined:

$$N_{PC}^{(1)} < N_P^{(1)} < N_{PH}^{(1)} \tag{30}$$

This has been confirmed comparing (30) to the experimental results provided by Bosco and Carpinteri (Carpinteri et al. 1999). In (Carpinteri et al. 1999) the Authors examine three series of reinforced beams by varying the cross section and the reinforcement area. The tests were performed on 35 beams of classes A, B and C, with cross-sectional area equal to 100 × 100, 100 × 200 and 200 × 400 mm, respectively, Figure 6.

The examined reinforcement percentages are 0.12%, 0.25%, 0.50%, 1.00%, 2.00%. The beams are labelled with the letter of the series, the reinforcement percentage and, occasionally, with the slenderness ratio L/h, e.g. A025-6 means a series A beam with 0.25% reinforcement and $L/h = 6$. In Figure 8 the brittleness numbers computed for the experimental tests are plotted together with the curve of N_{PH}, Eq. (28), and the value of N_{PC}, Eq. (29). All the beams above the N_{PH} curve failed experimentally because of concrete crushing, while the beams with $N_P < N_{PC}$ failed for insufficient reinforcement, Figures 10 and 11. All the above concepts about minimum and maximum reinforcement can be immediately translated into minimum and maximum reinforcement area, the brittleness number $N_P^{(1)}$ being proportional to this quantity. Equating the definition of $N_P^{(1)}$ to Equations 28 and 29,

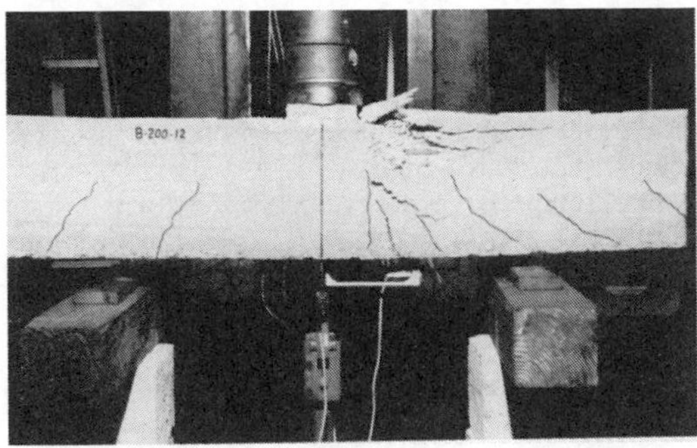

Figure 9. Brittle failure mode (crushing in compression) of the beam B200.

Figure 10. Brittle failure mode (insufficient reinforcement) of the beam B012.

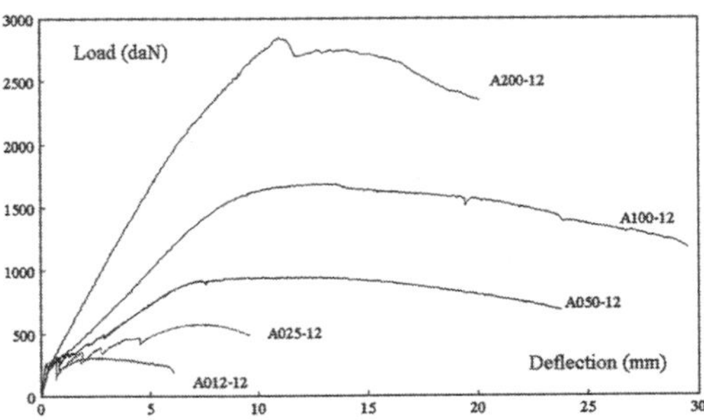

Figure 11. Load–deflection curves for series A beams. The curve A012 evidences brittle failure for insufficient reinforcement.

two curves are determined enclosing the reinforcement percentages for a given section height so that the mechanical behavior is ductile. An example is reported in Figure 13 for the data.

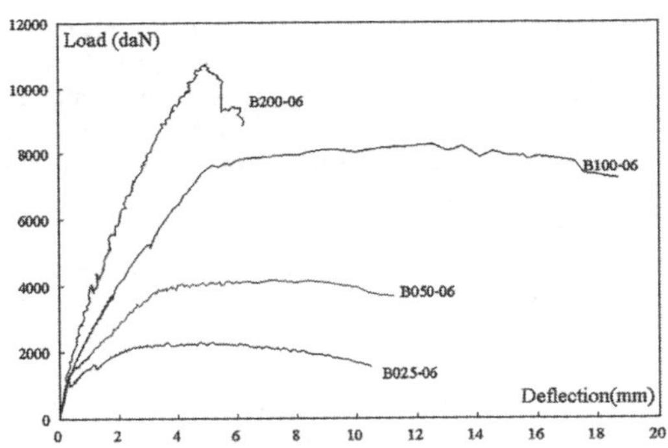

Figure 12. Ductile to brittle behavior in the load–deflection curves for series B beams.

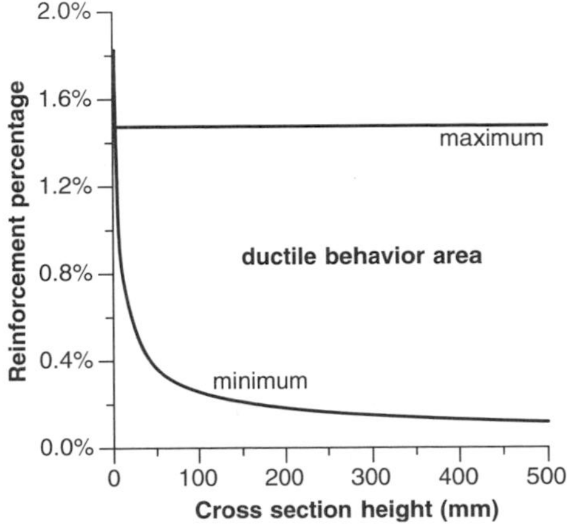

Figure 13. Reinforcement percentages for ductile cross sections.

5 SIMULATION OF EXPERIMENTAL RESULTS FOR FIBER REINFORCED BEAMS

The bridged crack model has been used to simulate some experimental results reported in (Swamy and Al-Ta'an 1981). Two different three point bending tests have been considered, related to beams having the same geometry and tension steel reinforcement. The two tests are denominated DR30 and DR32 in (Swamy and Al-Ta'an 1981) and differ in the fiber volume ratio. While the beam DR30 is made by plain concrete, the beam DR32 has a 1% fiber volume ratio. The beams have rectangular cross section (130 × 203 mm) and a total length of 2500 mm. The distance between the supports is 2250 mm. The tension reinforcement is constituted by 2 ϕ 12 of high-strength steel bars (yield strength = 617 N/mm^2). The steel percentage is $\rho = 0.99\%$ of the cross section area, as reported by the Authors, if the area is computed assuming as height the distance between the bars and the upper side of the cross section (176 mm, see Figure 1 in (Swamy and Al-Ta'an 1981)). According to the model presented in this paper, the steel percentage is referred to the total cross section area, so that the steel percentage to be considered is $\rho = 0.86\%$.

The model produces as a result the values of the nondimensional bending moment and rotation as functions of the crack depth. For comparison with the experimental results these values have been converted into displacement versus load diagrams. The conversion is realized by considering the scheme of Figure 14, where a three point bending test is depicted. The displacement at midspan of the beam is supposed as given by the elastic part plus a rigid part due to the concentrated rotation of the cracked section. From the definition of nondimensional bending moment and rotation, we can write:

$$M_F = \tilde{M} K_{\mathrm{IC}} b h^{1.5}, \tag{31}$$

$$\phi = \frac{\tilde{\phi} K_{IC}}{E h^{0.5}}. \tag{32}$$

Consequently, the vertical load and displacement at midspan are given by:

$$P = \frac{4 M_F}{L}, \tag{33}$$

$$\delta = \delta_{el} + \delta_{\phi} = M_F \frac{L^2}{48 E^{\star} I} + \frac{\phi L}{4}, \tag{34}$$

where the inertia is related to the total cross section, L is the span length, and $E^{\star} = E/2.2$, E being the conventional 28 days static modulus. This assumption is known in the literature, e.g. (Carpinteri 1981b), (Jenq and Shah 1986), and takes into account the nonlinear material behavior in the zone ahead the crack tip. For the beam DR30 the modulus $E^{\star}$ is then considered equal to 12.786 KN/mm^2.

Observing the experimental results reported in (Swamy and Al-Ta'an 1981), it has been noticed that the load vs. displacement diagram for the beam DR30 must be referred to a tension steel with a resistance higher than the reported one (617 N/mm^2). In fact, the maximum load corresponds to a bending moment at midspan whose value is higher than the effective height of the cross section times the traction at the steel reinforcement. For this reason a tension steel reinforcement strength equal to 725 N/mm^2 has been considered in the simulations.

Therefore, for the beam DR30, assuming a matrix toughness $K_{IC} = 35.76$ N/mm$^{3/2}$, the brittleness number $N_P^{(1)}$ is given by:

$$N_P^{(1)} = \frac{\rho \sigma_y \sqrt{h}}{K_{\mathrm{IC}}} = 2.484. \tag{35}$$

The experimental and computed load vs. displacement diagrams are reported in Figure 15. The numerical simulation has been carried out using the bridging option, as the fibers were not present in this case. The results of the model do not catch the progressive decrease of the tangent modulus due to concrete damage, although reproduce closely the qualitative behavior of the structural member as well as the limit load. The latter result is due to assumed steel strength.

The mechanical parameters adopted for the beam DR30, have been used for the simulation of the test

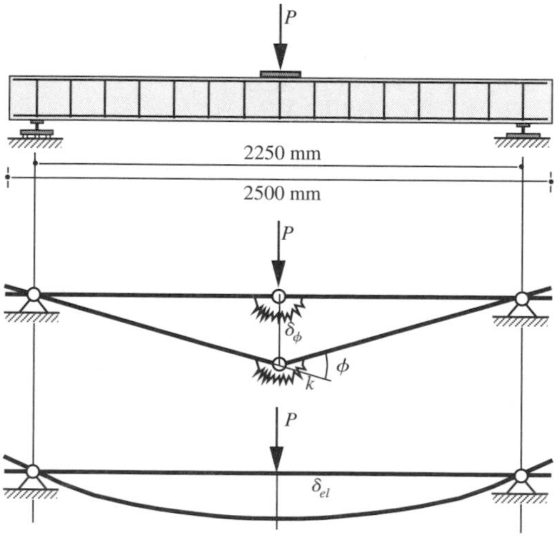

Figure 14. Details of test beams and plastic hinge formation.

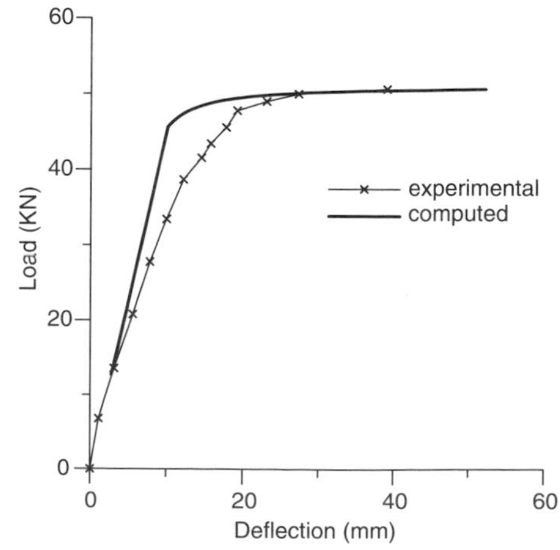

Figure 15. Experimental and computed load vs. deflection diagrams for the beam DR30.

DR32, where crimped steel fibers have been added to concrete. The steel fibers are characterized by a length equal to 50 mm, a diameter equal to 0.5 mm, a strength of 1050 N/mm^2 and are mixed with concrete with a percentage of 1%. A first reduction factor of $1/3$ is to be assumed for the fiber volume percentage, for taking into account the random spatial distribution of the fibers inside the matrix. This simple assumption is based on the observation that the fibers are fully effective if their direction is coaxial with the beam axis, and totally ineffective if they are directed along the other two orthogonal spatial directions. In addition, a further reduction must account for the fact that the fibers are not bisected by the crack, but intersected at a random point along their length (Li et al. 1991). For this reason the "effective density" of the fibers to be introduced in the model has been assumed equal to 0.2%, i.e. $1/5$ of the real one.

These assumptions provide a brittleness number $N_P^{(2)}$:

$$N_P^{(2)} = \frac{\gamma\sigma_u\sqrt{h}}{K_{\rm IC}} = 0.84. \qquad (36)$$

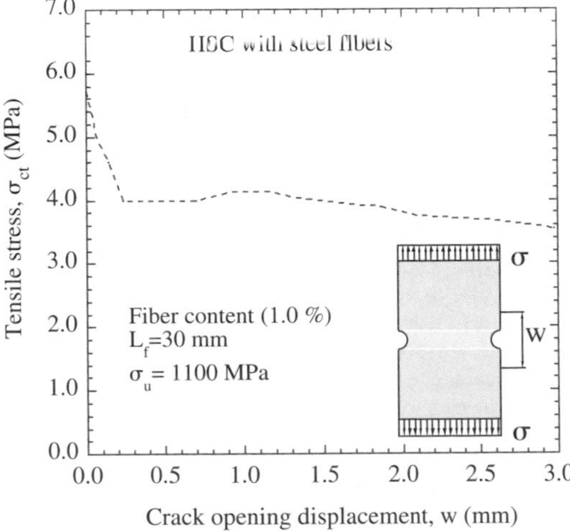

Figure 16. Response of high-strength concrete with fibers.

Finally, a rigid-perfectly plastic law is considered for the fibers. This assumption finds justification considering experimental results on similar steel fibers (Noghabai 2000), see Figure 16. As the experimental load vs. displacement diagrams have always positive slope with no slope discontinuities, the tension steel does not enter softening behavior and the fibers does not break along the crack depth, i.e. the crack opening is always less than the critical value w_c and the cohesive zone extends along the whole crack depth. In fact, when the critical opening w_c is achieved, the load vs. displacement diagram exhibits a softening branch, or at least a slope discontinuity. This is compatible with the fibers used in the experiment, as the crack opening at the maximum load is about 5 mm.

Consequently, the data used for running the numerical simulation of the beam DR32 are the two brittleness numbers (35) and (36), as the parameter $\tilde{w}_c$ is not influent, being $\tilde{w} < \tilde{w}_c$. Figure 5 reports the experimental load vs. displacement graphs compared to the computed ones, in both the cohesive or bridging option. The two curves are almost coincident, although in the two cases the nondimensional diagram bending moment versus crack depth markedly differs, Figure 5. As is expected for a given nondimensional moment, the crack depth is higher in the cohesive option. With reference to Figure 5, once the value $\tilde{M} = 1$ is fixed, a crack depth $\xi = 0.4376$ (bridging option) or $\xi = 0.5446$ (cohesive option) is found.

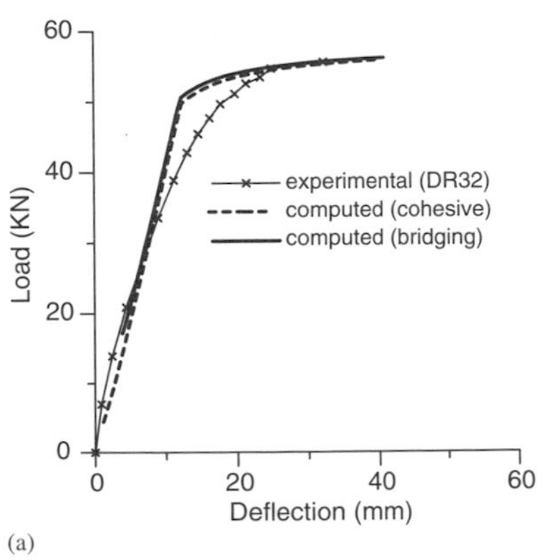

(a)

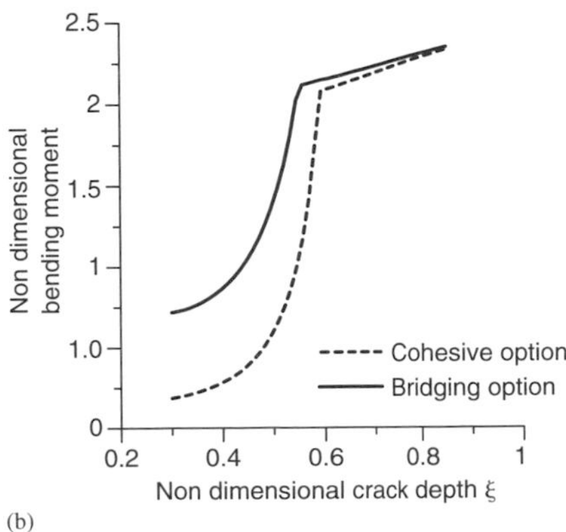

(b)

Figure 17. (a) Experimental and computed load vs. deflection diagrams for the beam DR32; (b) Nondimensional bending moment vs. crack depth diagrams for the beam DR32.

6 CONCLUSIONS

The bridged crack model has recently extended to the simultaneous presence of conventional steel bar reinforcement and fibers mixed into the cementitious matrix. The two reinforcements act at different scales and influence each other in the global structural response. Compared to classical limit state analysis, the introduction of Fracture Mechanics concepts in the modelling of reinforced concrete members allows for determining ductile-brittle transitions, scale effects and the contribution of fibers and, in general, nonlinear matrix tensile behavior. A contribution for improving the numerical implementation of the model has been presented in the paper, allowing for a much faster evaluation of the integrals accounting for the fiber/matrix tensile behavior. Moreover, crushing of the concrete is introduced in the model by determining an upper bound to the brittleness number. When coupled to the results presented for the minimum reinforcement (Carpinteri et al. 1999), this allows for defining a ductility domain. The ductility domain allows for designing the collapse mechanism for a beam in bending by simply evaluating its brittleness number. This extension to the model is here presented as limited to the case of conventional steel bar reinforcement, and is being extended to the general case, where fibers and/or tensile resistance of the matrix are present as well. Several tests demonstrate the ability of the model in reproducing experimental data.

REFERENCES

Bosco, C. and A. Carpinteri (1992). Softening and snap-through behavior of reinforced elements. *J. Eng. Mechanics (ASCE) 118*, 1564–1577.

Bosco, C. and A. Carpinteri (1995). Discontinuous constitutive response of brittle matrix fibrous composites. *Journal of the Mechanics and Physics of Solids 43*, 261–274.

Carpinteri, A. (1981a). A fracture mechanics model for reinforced concrete collapse. In *Proceedings of the I.A.B.S.E. Colloquium on Advanced Mechanics of Reinforced Concrete*, Delft, pp. 17–30.

Carpinteri, A. (1981b). Static and energetic fracture parameters for rocks and concrete. *Materials and Structures 14*, 151–162.

Carpinteri, A. (1984). Stability of fracturing process in rc beams. *Journal of Structural Engineering (ASCE) 110*, 544–558.

Carpinteri, A., G. Ferro, C. Bosco, and M. Elkatieb (1999). Scale effects and transitional failure phenomena of reinforced concrete beams in flexure. In A. Carpinteri (Ed.), *Minimum Reinforcement in Concrete Members,* Volume 24 of *ESIS Publications*, pp. 1–30. Elsevier Science Ltd.

Carpinteri, A., G. Ferro, and G. Ventura (2002). Size effects on flexural responce of reinforced concrete elements with a nonlinear matrix. *Engineering Fracture Mechanics*. In press.

Carpinteri, A. and R. Massabó (1996). Bridged versus cohesive crack in the flexural behavior of brittle-matrix composites. *International Journal of Fracture 81*, 125–145.

Carpinteri, A. and R. Massabó (1997). Continuous vs. discontinuous bridged-crack model for fiber-reinforced materials in flexure. *International Journal of Solids and Structures 34*, 2321–2338.

Cools, R. and A. Haegemans. Cubpack: a package for automatic cubature; framework description. Submitted for publication.

Ferro, G. (2002). Multilevel bridged crack model for high-performance concretes. *Theoretical and Applied Fracture Mechanics 38*, 177–190.

Jenq, Y. and S. Shah (1986). Crack propagation in fiber-reinforced concrete. *Journal of Engineering Mechanics (ASCE) 112*, 19–34.

Li, V., Y. Wang, and S. Backers (1991). A micromechanical model of tension-softening and bridging toughening of short random fiber reinforced brittle matrix composites. *Journal of the Mechanics and Physics of Solids 39*, 607–625.

Noghabai, K. (2000). Beams of fibrous concrete in shear and bending: experiment and model. *Journal of Structural Engineering (ASCE) 126*, 243–251.

Okamura, H., K. Watanabe, and T. Takano (1975). Deformation and strength of cracked member under bending moment and axial force. *Engineering Fracture Mechanics 7*, 531–539.

Stroud, A. (1971). *Approximate Calculation of Multiple Integrals*. Englewood Cliffs, N.J.: Prentice Hall.

Swamy, R. and S. Al-Ta'an (1981). Deformation and ultimate strength in flexure of reinforced beams made with steel fiber concrete. *ACI Journal 5*, 395–405.

Tada, H., P. Paris, and G. Irwin (1963). *The Stress Analysis of Cracks Handbook*. St. Louis, Missouri: Paris Productions Incorporated (and Del Research Corporation).

Computational Modelling of Concrete Structures, Bićanić et al. (eds)
© 2003 Swets & Zeitlinger, Lisse, ISBN 90 5809 536 3

Application of a constitutive model for concrete to the analysis of a precast segmental tunnel lining

B. Winkler
Hilti AG, New Business & Technology, Numerical Simulation, Schaan, Principality of Liechtenstein; formerly Research Associate, Institute for Structural Analysis and Strength of Materials, University of Innsbruck, Innsbruck, Austria

G. Hofstetter & H. Lehar
Institute for Structural Analysis and Strength of Materials, University of Innsbruck, Innsbruck, Austria

ABSTRACT. A constitutive model for concrete, based on the smeared crack approach and formulated within the framework of the theory of plasticity, was used to perform a nonlinear FE-analysis of a permanent tunnel lining made of hexagonal precast segments. During the construction work of the lining hairline cracks were detected on the inner surface of some precast segments, running parallel to the longitudinal axis of the tunnel lining. They were supposed to be mainly caused by the installation process of the lining. In order to gain more information about the origin of these cracks, a nonlinear numerical analysis of the installation process of the lining was performed. The results of the numerical simulation, showing under which conditions cracks are initiated, will be presented and discussed.

1 INTRODUCTION

The development of computer-aided analysis procedures for the numerical simulation of the nonlinear behaviour of plain and reinforced concrete structures by the finite element method (FEM) has been the focus of extensive research for more than thirty years. Within this context the employed constitutive model for plain concrete plays an important role. The latter contains the mathematical description of the material behaviour up to failure, an essential part of which is the model for cracking.

Surveys over the large number of smeared crack models, which have been developed in the last three decades, can be found e.g. in Isenberg (1993) and Hofstetter & Mang (1995). However, the earlier crack models were characterized by the deficiency of mesh-dependent results. Especially for plain concrete and slightly reinforced concrete refinement of the finite element mesh resulted in the decrease of the numerically predicted ultimate load without achieving convergence. The reason for this shortcoming is given by the fact that after attaining the tensile strength of concrete, in the post-peak region of the material behaviour a unique stress-strain relationship does not exist. In recent years improved crack models, ensuring objective results, have been proposed. Referring to Feenstra & de Borst (1993) a relatively simple approach, which is suited for ultimate load analyses of large concrete structures, is characterized by adjusting the constitutive relation for the post-peak range according to the size of the finite elements by means of the specific fracture energy.

The concrete model, proposed by Feenstra & de Borst within the framework of the theory of plasticity, was extended by coupling damage due to tensile stresses with damage due to compressive stresses for mixed tension-compression loading and by introducing an isotropic scalar damage model for unloading and reloading. The constitutive model was validated by means of well-known tests taken from the literature and by laboratory tests on L-shaped panels, which have been performed at the University of Innsbruck by Winkler et al. (2001). In recent years the L-shaped panel has become a popular benchmark test for the validation of the mathematical description of the post-cracking behaviour of concrete. The constitutive model for plain concrete was then employed to perform an analysis of a permanent tunnel lining made of hexagonal precast segments. Considering these subjects the paper includes (i) the validation of the employed material model for concrete and (ii) a nonlinear FE-analysis of a permanent tunnel lining made of hexagonal precast segments.

2 MATERIAL MODEL

The constitutive model for concrete subjected to biaxial stress states, proposed in Feenstra (1993), serves as the basis for the description of the material behaviour of plain concrete. It was extended by taking into account the coupling of damage due to tensile stresses with damage due to compressive stresses for biaxial tension-compression loading and considering unloading and reloading in the post-peak range of the material behaviour.

The model is summarized briefly in this section. Based on the theory of plasticity, it is characterized by a composite yield surface and suitable softening laws to describe tensile and compressive failure. Hence, cracking is represented in a smeared manner. According to Feenstra & de Borst (1996) the composite yield function consists of the Rankine (principal stress) criterion f_1 to limit the tensile stress and of a Drucker-Prager yield function f_2 to describe the compression regime (Fig. 1).

Assuming small strains, the strain rate vector can be subdivided into an elastic part $\dot{\boldsymbol{\varepsilon}}^e$ and a plastic part $\dot{\boldsymbol{\varepsilon}}^p$ by

$$\dot{\boldsymbol{\varepsilon}} = \dot{\boldsymbol{\varepsilon}}^e + \dot{\boldsymbol{\varepsilon}}^p . \tag{1}$$

The evolution of the plastic part of the strain rate vector is determined by Koiter's generalized flow rule for multi-surface plasticity:

$$\dot{\boldsymbol{\varepsilon}}^p = \sum_i \dot{\lambda}_i \frac{\partial g_i}{\partial \sigma} . \tag{2}$$

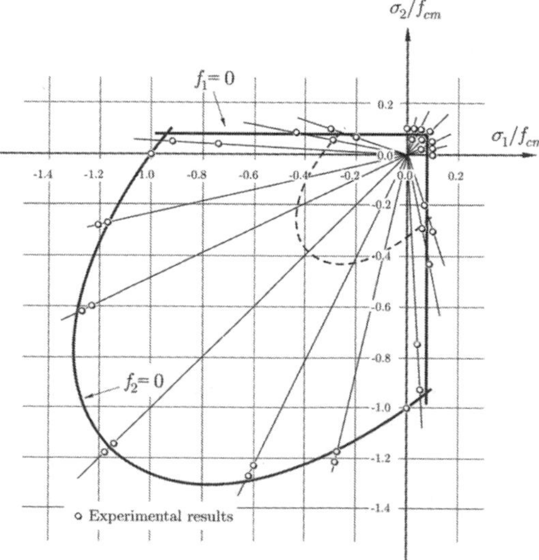

Figure 1. Comparison of the composite yield function at ultimate strength with the results of the experimental investigations of Kupfer et al. (1969).

Assuming associated plasticity, the plastic potential functions g_i are equal to the yield functions f_i.

The plastic multipliers λ_i and the yield functions f_i have to fulfill the Kuhn-Tucker conditions

$$f_i \leq 0, \quad \dot{\lambda}_i \geq 0, \quad \dot{\lambda}_i f_i = 0 . \tag{3}$$

For plane stress states with the stress vector $\sigma^{\mathrm{T}} = \{\sigma_{11}, \sigma_{22}, \sigma_{12}\}$ the two yield functions are given as:

$$\begin{aligned} f_1 &= \left(\tfrac{1}{2}\,\boldsymbol{\sigma}^{\mathrm{T}} \mathbf{P}_1 \boldsymbol{\sigma}\right)^{1/2} + \alpha_1 \boldsymbol{\pi}^{\mathrm{T}} \boldsymbol{\sigma} - \beta_1 \bar{\sigma}_1(\kappa_1) \\ f_2 &= \left(\tfrac{1}{2}\,\boldsymbol{\sigma}^{\mathrm{T}} \mathbf{P}_2 \boldsymbol{\sigma}\right)^{1/2} + \alpha_2 \boldsymbol{\pi}^{\mathrm{T}} \boldsymbol{\sigma} - \beta_2 \bar{\sigma}_2(\kappa_2) . \end{aligned} \tag{4}$$

A comparison with the experimental data of Kupfer et al. (1969) shows that the two yield functions match the ultimate strength data of concrete in the regions of biaxial tension and biaxial compression. In Figure 1 the stresses are normalized with respect to the uniaxial compressive strength f_{cm}. The factors α_1 and β_1 in (4) result from the Rankine (principal stress) criterion f_1 to

$$\alpha_1 = \tfrac{1}{2}, \qquad \beta_1 = 1 . \tag{5}$$

Using the ratio β_c between the biaxial compressive strength and the uniaxial compressive strength, the factors α_2 and β_2 for the Drucker-Prager yield function f_2 can be determined by

$$\alpha_2 = \frac{\beta_c - 1}{2\beta_c - 1}, \qquad \beta_2 = \frac{\beta_c}{2\beta_c - 1} . \tag{6}$$

According to Kupfer et al. (1969) $\beta_c \approx 1.16$. In (4) the projection matrices $\mathbf{P}_1, \mathbf{P}_2$ and the projection vector $\boldsymbol{\pi}$ are given by

$$\mathbf{P}_1 = \begin{bmatrix} \tfrac{1}{2} & -\tfrac{1}{2} & 0 \\ -\tfrac{1}{2} & \tfrac{1}{2} & 0 \\ 0 & 0 & 2 \end{bmatrix}, \quad \boldsymbol{\pi} = \begin{bmatrix} 1 \\ 1 \\ 0 \end{bmatrix}, \quad \mathbf{P}_2 = \begin{bmatrix} 2 & -1 & 0 \\ -1 & 2 & 0 \\ 0 & 0 & 6 \end{bmatrix}, \tag{7}$$

and the equivalent stresses $\bar{\sigma}_1(\kappa_1)$ and $\bar{\sigma}_2(\kappa_2)$ stand for the uniaxial tensile strength and the uniaxial compressive strength in terms of the internal variables κ_1 and κ_2. The latter can be considered as damage indicators.

Considering the experimental data of Kupfer et al. (1969) for combined tensile and compressive loading, fracturing of the material due to compressive loading in one direction results in a reduction of the maximum attainable tensile stress in lateral direction. This effect is taken into account by coupling damage caused by compressive loading with damage caused by tensile loading.

The coupling of the internal variables κ_1 and κ_2 allows to model the reduction of the tensile strength under mixed tension-compression loading. Beginning with the initial yield surface under compressive loading f_2, indicated by the broken curve in Figure 1, upon further compressive loading the material will be damaged such that the attainable tensile stress in lateral direction decreases from the uniaxial tensile strength to zero. Figure 2 shows the comparison of the computed ultimate strength curve considering the coupling effect of the damage parameters κ_1 and κ_2 with the experimental results of Kupfer et al. (1969).

Concerning the evolution of the irreversible material behaviour within the framework of the theory of plasticity, the internal variables κ_1 and κ_2 can be determined according to Feenstra & de Borst (1996) by assuming that the equivalent stresses are related to energy dissipation via a work-hardening hypothesis

$$\dot{W}_i^p = \boldsymbol{\sigma}^{\mathrm{T}} \dot{\boldsymbol{\varepsilon}}^p \equiv \bar{\sigma}_i(\kappa_i)\dot{\kappa}_i . \tag{8}$$

Using the scalars ζ_{ij} for representing the coupling of damage in the material at different stress states the work-hardening hypothesis can formally be expressed as

$$\dot{W}_i^p = \sum_j \zeta_{ij} \frac{\bar{\sigma}_i}{\bar{\sigma}_j} \, \boldsymbol{\sigma}^{\mathrm{T}} \left[\dot{\lambda}_j \frac{\partial f_j}{\partial \boldsymbol{\sigma}} \right] = \bar{\sigma}_i \, \dot{\kappa}_i . \tag{9}$$

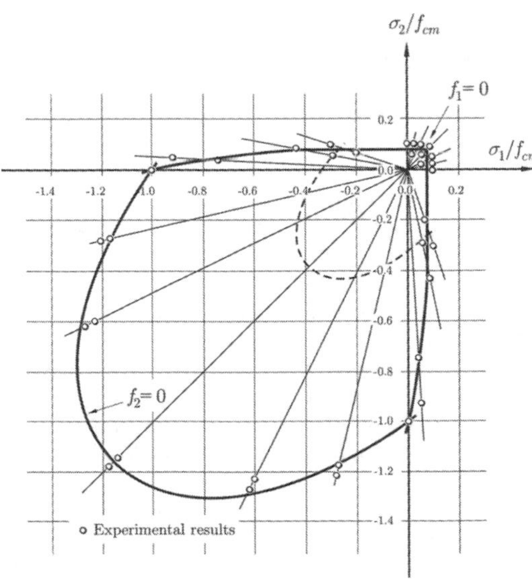

Figure 2. Computed ultimate strength curve considering the coupling effect of the damage parameters κ_1 and κ_2.

Considering the Euler's theorem of homogeneous functions the internal variables κ_i can be determined by

$$\dot{\kappa}_i = \sum_j \zeta_{ij} \, \beta_j \dot{\lambda}_j \tag{10}$$

with β_1 resulting from the Rankine (principal stress) criterion f_1 and β_2 from the Drucker-Prager yield function f_2. In this way, the evolution of the internal variables is determined by the plastic strain rate and can be considered as a measure of the progressive fracturing of the material.

The equivalent stress $\bar{\sigma}_1(\kappa_1)$ is described by means of an exponential softening law for plain concrete (Fig. 3) by

$$\bar{\sigma}_1(\kappa_1) = f_{ctm} \, \exp\left[-\frac{\kappa_1}{\kappa_{1u}} \right] \tag{11}$$

with f_{ctm} as the uniaxial tensile strength. The parameter κ_{1u} is given in terms of the specific fracture energy G_f for tensile failure and the equivalent length h as

$$\kappa_{1u} = \frac{G_f}{h \, f_{ctm}} . \tag{12}$$

In a direct tension test the equivalent length represents the length of the specimen and in a FE-analysis h is related to a representative length of the element.

The equivalent stress $\bar{\sigma}_2(\kappa_2)$ is described by means of a parabolic hardening law and an exponential softening law (Fig. 4) and is determined for $\kappa_2 < \kappa_{2e}$ by

$$\bar{\sigma}_2(\kappa_2) = \frac{f_{cm}}{3} \left[1 + 4 \frac{\kappa_2}{\kappa_{2e}} - 2 \frac{\kappa_2^2}{\kappa_{2e}^2} \right] \tag{13}$$

and for $\kappa_2 \geq \kappa_{2e}$ by

$$\bar{\sigma}_2(\kappa_2) = f_{cm} \, \exp\left[-\frac{(\kappa_2 - \kappa_{2e})^2}{(\kappa_{2u})^2} \right] \tag{14}$$

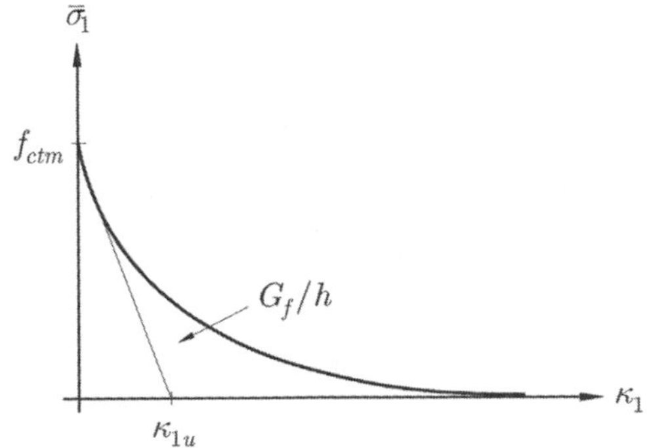

Figure 3. Uniaxial material behaviour in tension

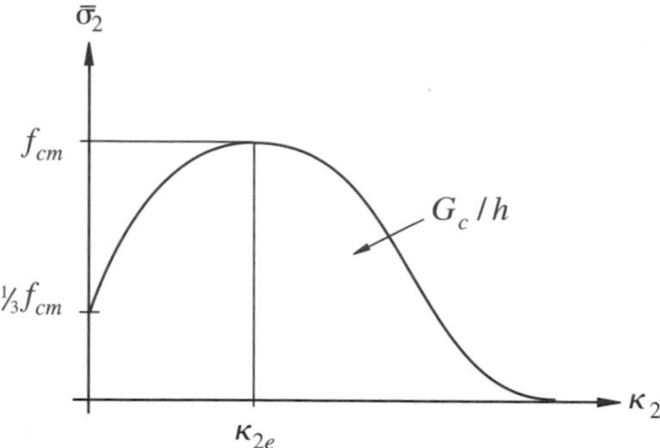

Figure 4. Uniaxial material behaviour in compression.

with f_{cm} as the cylindrical compressive strength. The parameters κ_{2e} and κ_{2u} are given in terms of the specific fracture energy G_c for compressive failure, the equivalent length h and the modulus of elasticity E_c as

$$\kappa_{2e} = 0.0022 - \frac{f_{cm}}{E_c}, \quad \kappa_{2u} = \frac{2}{\sqrt{\pi}} \frac{G_c}{h f_{cm}}. \tag{15}$$

The introduction of the specific fracture energies G_f and G_c for tensile and compressive failure of concrete, respectively, and of the equivalent length h yields objective results with respect to the employed mesh size.

In the context of a finite element analysis the equivalent length in (12) and (15) depends on the type, the size, the shape and the integration scheme of the finite element. However, in this study the equivalent length is simply determined as the square root of the area of the respective finite element.

During ultimate load analyses of plain and reinforced concrete structures material points can undergo cyclic response due to stress redistribution. Referring to Menrath (1999) these local stress relief states are described by an isotropic scalar damage model for unloading and reloading conditions. A complete crack closure in case of unloading in the post-peak region is avoided by introducing a permanent crack width, assumed as 20% of the crack opening at the beginning of the unloading process.

Considering reinforced concrete the post-cracking behaviour of reinforced concrete differs considerably from the respective behaviour of plain concrete. If cracking occurs tensile stresses are transferred between reinforcement and concrete by bond, resulting in the tension stiffening effect. Details for the extension of the described material model to reinforced concrete and the interaction between tension softening and tension stiffening can be found in Winkler (2001).

The described material model for concrete has been implemented into the finite element program ABAQUS within the framework of the return mapping algorithms. To this end a user-defined subroutine, defining the constitutive behaviour at the integration point level, was written in FORTRAN 77.

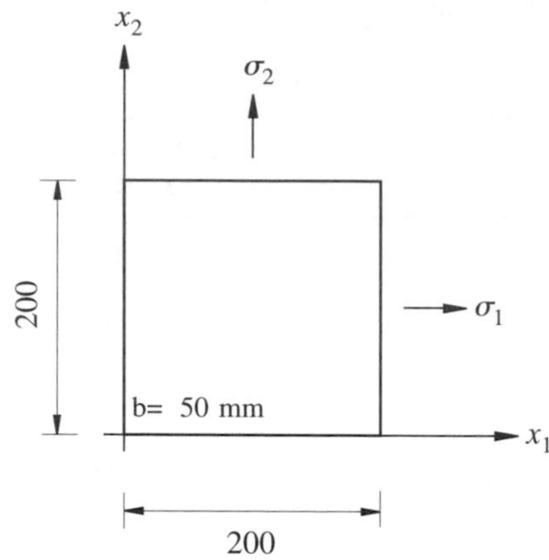

Figure 5. Geometry of the specimens used for the experimental investigations of Kupfer et al. (1969).

3 VALIDATION

The described computational model for concrete was validated by means of well-known tests taken from the literature. As an example for the validation the experimental investigations of Kupfer et al. (1969) on plate-type specimens with the dimensions of 200 × 200 × 50 mm will be discussed (Fig. 5). These specimens were tested under proportional monotonically increasing biaxial loadings with constant stress ratio σ_1/σ_2 for each test.

The results of the experimental investigations contain the respective stress-strain curves for concrete under biaxial compression, combined tension and compression and biaxial tension. In the following the combined tension and compression load cases will be simulated numerically.

Concerning the experimental results the uniaxial tensile strength was set to 2.84 N/mm^2 and the uniaxial compressive strength to 31.10 N/mm^2. The Poisson's ratio amounted to 0.195 and the tangent modulus of elasticity to 33000 N/mm^2. The fracture energies for tensile and compressive failure were assumed to 0.10 N/mm and to 10 N/mm, respectively.

The results of the numerical analyses are shown in Figure 6 in terms of stress-strain curves. Figure 6a shows the stress-strain curves related to the compressive stress. Figure 6b contains the stress-strain curves for the tensile stress. The comparison of the experimentally and numerically determined compressive

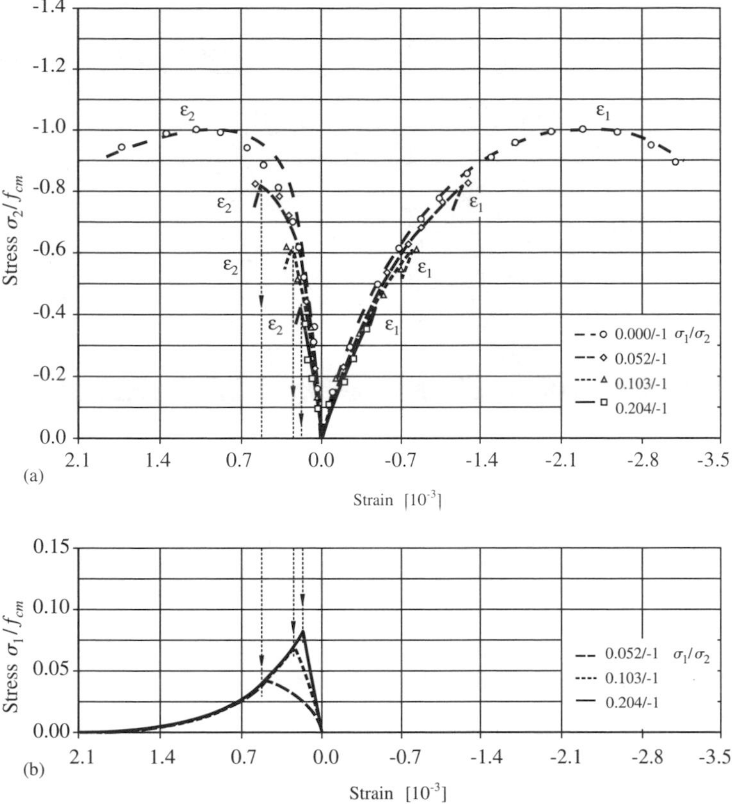

Figure 6. Computed stress-strain curves for concrete subjected to combined tension and compression in terms of (a) compressive stresses and (b) tensile stresses compared with the experimental results of Kupfer et al. (1969).

stress-strain curves demonstrates good correspondence between numerical and experimental results.

Beginning with the stress ratio $\sigma_1/\sigma_2 = 0.0/-1.0$ and ending with the ratio $\sigma_1/\sigma_2 = 0.204/-1.0$ the ultimate compressive stress decreases as the applied tensile stress in lateral direction is increased (Fig. 6a).

The influence of damage coupling can be seen in Figure 6b in terms of the maximum attainable tensile stress, which decreases with increasing compressive stress.

Supplementary to various well-known examples taken from the literature the employed material model was checked by laboratory tests on L-shaped panels performed at the University of Innsbruck by Winkler et al. (2001).

The experimental investigations consisted of a series of L-shaped panels, made of plain concrete and three series of reinforced L-shaped panels with different layout of the reinforcement, which were loaded until failure.

The comparison between experimental and computed results included the load at the initiation of cracking and the load-displacement curves in the pre- and post-peak region. Additionally, the experimentally determined crack patterns were compared with the computed crack propagation and damage behaviour of the material.

4 PRECAST SEGMENTAL TUNNEL LINING

The employed constitutive model for concrete was used to perform an analysis of a permanent tunnel lining made of hexagonal precast segments. Referring to Vigl (2000) this tunnel lining was part of an upgrading project for two hydropower plants in Slowenija, where two new pressure tunnels were required. Within the scope of this project the so-called double shield TBM technology, combined with a monoshell segmental lining built of hexagonal precast segments, was used. The lining is composed of four segments per ring of uniform hexagonal shape with a thickness of 200 mm, which were fabricated in two Slowenijan factories. Depending on the specific requirements the

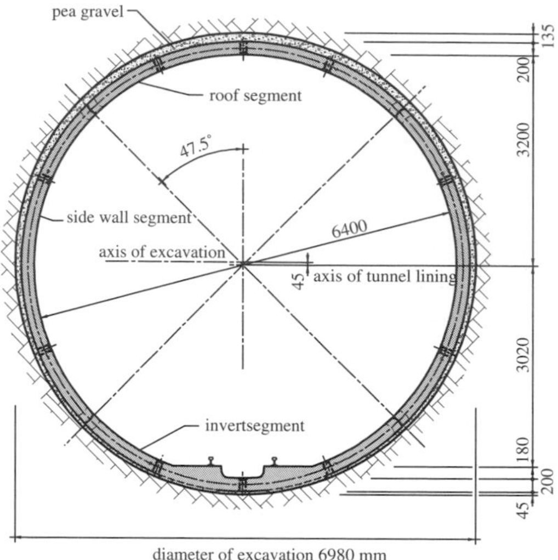

Figure 7. Cross section of the tunnel lining (Reference: Archives of the Vorarlberger Illwerke AG).

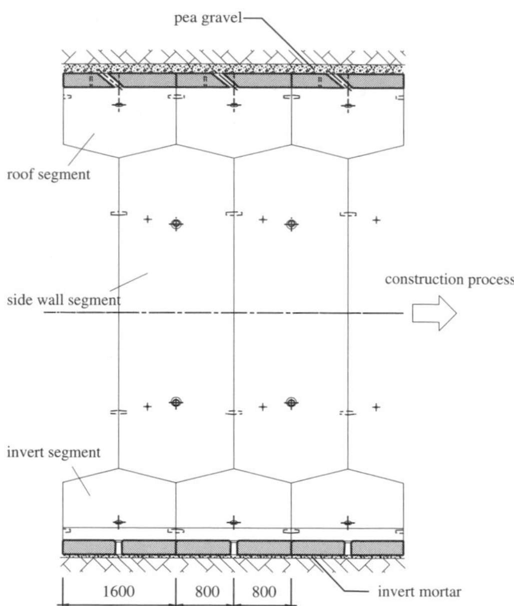

Figure 8. Section of the tunnel lining in longitudinal direction (Reference: Archives of the Vorarlberger Illwerke AG).

used segments differed in small details. The invert segment contains an invert culvert, invert pads and an invert grouting hole. The side wall segments and the roof segment are nearly identical, differing only in the arrangements of the required grouting holes (Fig. 7).

The segments are installed under the protection of the tail shield of the TBM in a circular form and kept in position by the help of the contact pressure of the TBM. To achieve stabilization of the circular lining ring the invert section of the lining is backfilled with mortar.

Afterwards the gap between the remaining part of the lining and the rock mass is backfilled with gravel (Figs. 7 and 8). The backfilled gravel provides a certain instantaneous bedding stiffness between the installed segments and the rock mass. Such a segmental lining system is designed to act as an arch load bearing system, which pre-dominantly carries normal compression forces. Hence, only a minimum quantity of reinforcement is needed for the loads, carried during production, handling, transport and installation. A detailed description of the construction work can be found in Vigl & Jäger (2000).

During the construction work of the permanent tunnel lining hairline cracks were detected on the inner surface of some precast segments, running parallel to the longitudinal axis of the tunnel lining. They were supposed to be mainly caused by the installation process of the lining. In order to gain more information about the origin of these cracks, a nonlinear numerical analysis of the installation process of the lining was performed.

Taking advantage of symmetry for the numerical investigation only half of the permanent tunnel lining was modelled using continuum elements for the precast segments and the backfilled gravel (Fig. 9). Concerning the width of the precast segments of 1600 mm and the gaps between the single segments in the longitudinal direction of the tunnel lining a plane stress state was assumed.

The diameter of the excavation amounts to 6.98 m, the inner diameter of the tunnel lining to 6.40 m. The axis of the excavation is located 45 mm above the axis of the tunnel lining.

The generated mesh consisted of 1158 finite elements, using 822 elements for the discretization of the precast segments and 336 finite elements for the discretization of the backfilled gravel. The employed elements were eight-node quadrilateral elements with reduced integration. The inner contour of the rock mass was modelled by a rigid surface. The mechanical connection between the precast segments and the backfilled gravel and between the backfilled gravel and the rock mass, respectively, was modelled by contact surfaces disregarding possible effects due to friction. Assuming a considerable stiffness and stabilization effect due to the gearing of the segments with respect to the direction parallel to the longitudinal axis of the tunnel lining, the joints between the single segments, running parallel to the longitudinal axis of the tunnel lining, were not modelled.

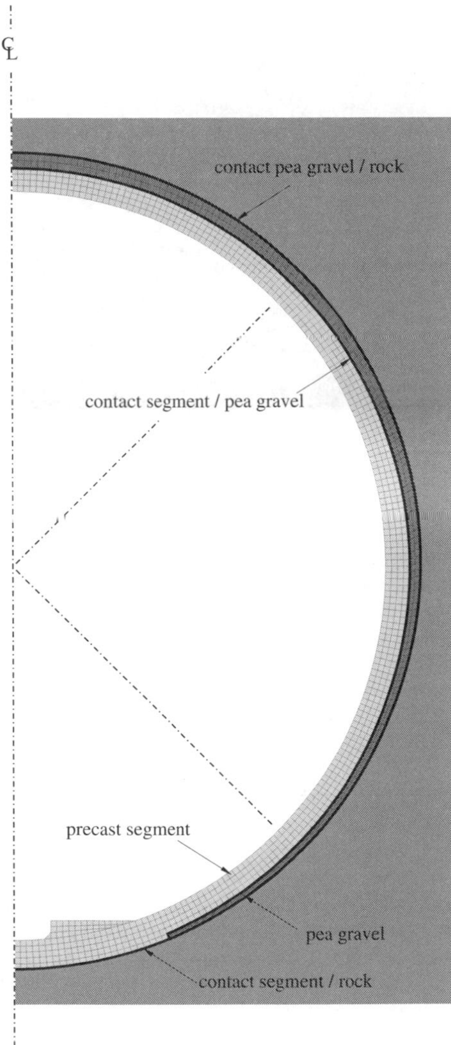

Figure 9. Discretization of the permanent tunnel lining.

The reinforcement of the segments was modelled by means of rebar-elements. These are one-dimensional elements, employing plasticity theory to describe the nonlinear material behaviour. However, since only a minimum quantity of reinforcement, required for the loads, carried during production, handling, transport and installation, is provided, failure is governed by the material behaviour of plain concrete.

For the precast segments concrete qualities B35 and B45 according to DIN (1988) were used. The uniaxial tensile strength was set to 3.00 N/mm^2. Referring to CEB-FIP (1991) the fracture energy for tensile failure was assumed to 0.095 N/mm.

Concerning the installation process there are two essential loadings, the dead weight of the segments, given as 25 kN/m^3 and the dead weight of the backfilled gravel, given as 19 kN/m^3.

Considering the installation procedure of the segments, in a first step the segments are kept in position by the help of the contact pressure of the TBM. This was modelled by fixing the continuum elements, representing the precast segments. During this fixation step the dead weight of the segments and the dead weight of the backfilled gravel were activated. In the following step the fixations of the segments were removed. Thus, the dead weight of the segments and the dead weight of the backfilled gravel were stressing the already stabilized tunnel lining. Neglecting geometric imperfections of the lining this load case resulted only in small bending moments acting in the invert segment and in the roof segment, respectively. Assuming a completely backfilled and compacted gravel layer and the regularly bedding of the segments the predicted tensile stresses are smaller than the uniaxial tensile strength. Hence, no cracks should appear.

However, the observed hairline cracks on the inner surface of some precast segments, running parallel to the longitudinal axis of the tunnel lining, indicate additional loads acting during the installation process, which cause the mentioned cracks.

Thus, a further numerical analysis was performed dealing with the situation before the gap between the lining and the rock mass will be backfilled with gravel. Concerning the installation of the segments they are kept in position by the help of the contact pressure of the TBM. Due to this pressure and the gearing of the segments with respect to the direction parallel to the longitudinal axis of the tunnel lining a considerable part of the dead weight of the segments will be transferred to the neighbouring ring of segments, which already has been stabilized by the backfilled gravel. The remaining part of the dead weight acts on the invert segment, which is not yet bedded at that moment. Considering this load case additional bending moments will arise in the invert segment.

For the numerical simulation of this load case the already described finite element model for the tunnel lining was used, just disregarding the backfilled gravel and the mortar bedding of the invert segment. Neglecting the load transfer to the neighbouring ring of segments due to the gearing effects with respect to the longitudinal direction the structure was loaded by the total dead weight of the segments, applying the latter in incremental steps.

In the numerical simulation the first cracks were predicted on the inner surface of the invert segment. These cracks were observed at 52% of the total dead weight of the segments and were located in the cantilevering part of the invert segment. They are recorded in Figure 10 in terms of the internal variable κ_1,

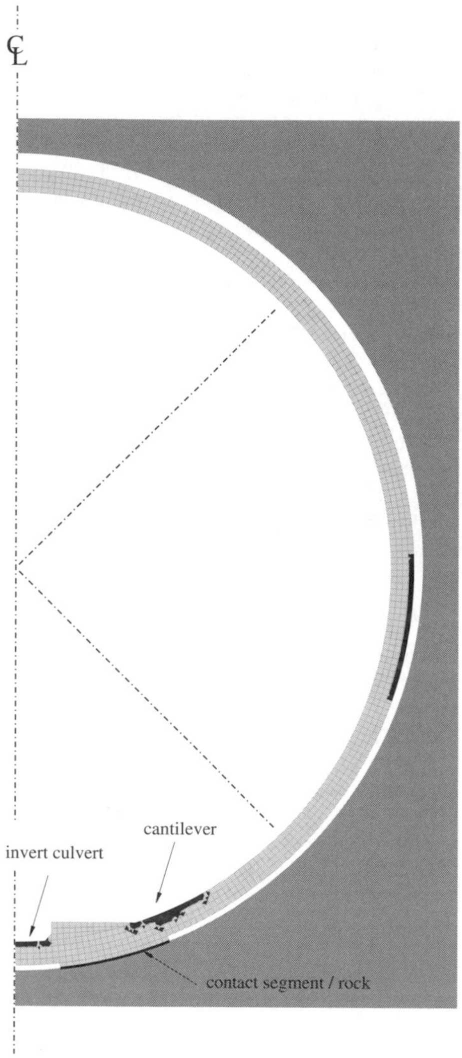

Figure 10. Computed tensile damage.

representing the damage indicator for tension. The material behaviour of those regions, which are shown in different intensities of grey shading in Figure 10 and 11, is characterized by tension softening, i.e., $\kappa_1 > 0$.

They represent the partly damaged regions (dark-grey colour) up to the heavily damaged zones (black colour) of the segments. At 55% of the total dead weight additional cracks were predicted. These cracks appeared on the inner surface of the invert segment in the region of the invert culvert. Since they were caused by stress concentrations at the corner of the invert culvert, they will be prevented in practise by rounding off the corner region. Finally, at 95% of the total dead weight of the segments cracks were predicted on the outer surface of the side wall segment.

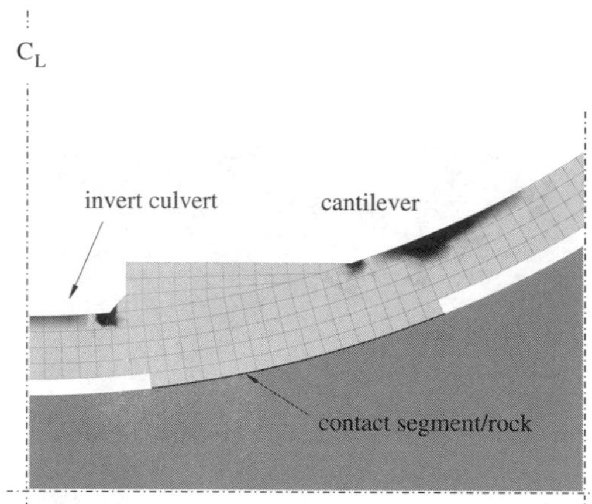

Figure 11. Computed tensile damage in the region of the invert culvert.

In Figure 10 the material damage after application of the total dead weight is displayed. The heavily damaged zones can be observed in the cantilevering part of the invert segment. This region is enlarged in Figure 11 by using a refined damage scale. In contrast to Figure 10, where all damaged regions are displayed roughly in black colour, Figure 11 shows the damage propagation in detail, using different intensities of grey shading.

Concerning the results of the FE-analysis the numerically determined crack propagation demonstrates good correspondence with the cracks observed at the construction site. The computed cracking loads are in agreement with the crude assumption, that approximately half of the dead weight of the segments will be transferred to the already stabilized neighbouring ring of segments, otherwise cracks will occur. This explains, why hairline cracks are detected only for some precast segments.

5 CONCLUSIONS

The aim of this contribution was to apply a constitutive model for concrete, which was validated by means of well-known tests taken from the literature and by tests on L-shaped panels, to perform a nonlinear FE-analysis of a permanent tunnel lining made of hexagonal precast segments.

The numerically determined damage behaviour demonstrates good correspondence with the hairline cracks, observed at the construction site.

REFERENCES

CEB-FIP (1991). *Model Code 1990, Bulletin d'information.* Lausanne, Comité Euro-International du Béton (CEB).

DIN 1045 (1988). Beton und Stahlbeton, Bemessung und Ausführung. *Deutsches Institut für Normung, Ausgabe Juli 1988.*

Feenstra, P.H. (1993). Computational aspects of biaxial stress in plain and reinforced concrete. PhD-thesis, Delft University of Technology, Netherlands.

Feenstra, P.H. & de Borst, R. (1993). Aspects of robust computational modelling for plain and reinforced concrete. *Heron* 38(12).

Feenstra, P.H. & de Borst, R. (1996). A composite plasticity model for concrete. *International Journal of Solids and Structures*, 33: 707–730.

Hofstetter, G. & Mang, H.A. (1995). *Computational mechanics of reinforced concrete structures.* Braunschweig/Wiesbaden: Vieweg & Sohn.

Isenberg, J. (1993). *Finite element analysis of reinforced concrete structures, Vol. II.* New York: American Society of Civil Engineers.

Kupfer, H., Hilsdorf, H.K. & Rüsch, H. (1969). Behavior of concrete under biaxial stresses. *ACI Journal* 66: 656–666.

Menrath, H. (1999). Numerische Simulation des nichtlinearen Tragverhaltens von Stahlverbundträgern. PhD-thesis, University of Stuttgart, Germany.

Vigl, A. (2000). Honeycomb segmental tunnel linings - simple, economical, successful. *Felsbau* 18(6): 24–31.

Vigl, A. & Jäger, M. (2000). Tunnel Plave-Doblar, Slowenien, 7 m Doppelschild-TBM mit einschaligem, hexagonalen Volltübbing-Auskleidungssystem. *In Österreichischer Betontag* 2000, 97–100, Wien.

Winkler, B. (2001). Traglastuntersuchungen von unbewehrten und bewehrten Betonstrukturen auf der Grundlage eines objektiven Werkstoffgesetzes für Beton. PhD-thesis, University of Innsbruck, Austria.

Winkler, B., Hofstetter, G. & Niederwanger, G. (2001). Experimental verification of a constitutive model for concrete cracking. *Proceedings of the Institution of Mechanical Engineers, Part L, Materials: Design and Applications* 215: 75–86.

Computational Modelling of Concrete Structures, Bićanić et al. (eds)
© 2003 Swets & Zeitlinger, Lisse, ISBN 90 5809 536 3

From elementary tests to industrial applications: validation of a non local isotropic damage model

L. Jason & G. Pijaudier-Cabot
R&DO, Ecole Centrale de Nantes, Nantes, France

S. Ghavamian
EDF, DRD, Clamart, France

A. Huerta
LaCan, UPC, Barcelona, Spain

ABSTRACT: In computational analyses, a hierarchy of steps has to be followed prior to industrial applications. The constitutive response, computations on structural elements and comparisons with experiments are among these steps. For these, benchmarks are required in order to proceed gradually from elementary to structural tests. It is these benchmarks which are presented in this contribution. The basis of the model is the isotropic damage model developed by Mazars. The simple scalar constitutive relation is coupled with a gradient non local approach to avoid spurious strain localization. First applications are elementary uniaxial tests. Quadratic convergence of the iterative algorithm is checked. To proceed further, a three point bending test applied to a RC beam and Nooru Mohamed's test are studied to highlight non local interest. Finally, a representative part of a containment building of a nuclear power plant is modelled.

1 INTRODUCTION

In computational analyses, the final goal is generally to simulate a global real industrial application. Indeed, there are situations where experimental investigation is so difficult that numerical study remains the most convenient solution to provide information. Nevertheless, in such cases, the validity of the constitutive law is a relevant question that cannot be eluded. Regarding computational costs, one has to justify the choice of the model (linear, nonlinear, enriched, …). That is why a clear methodology has to be defined before launching heavy studies.

The validation of the numerical implementation of a model generally follows different steps. First, the model is tested on simple applications where experimental results are available. Once it succeeds in simulating these elementary tests, it is then tried in situations where the difficulty becomes harder. The ultimate goal is to gradually tend to real "industrial" cases for which experimental investigation is not easily possible. At each step of the validation, benchmarks are required in order to proceed further and to make sure that the model will be able to provide solutions for real industrial applications.

In this contribution a set of benchmarks is presented. It aims at proposing a gradual validation of the gradient non local version of the isotropic damage model initially developed in Mazars (1984). It has to be noted that Mazar's model is only taken as an example here and that the tests could also be run with some other constitutive relations.

2 ISOTROPIC DAMAGE MODEL

2.1 *Local formulation*

The basis of the model is the damage model developed by Mazars. It describes the constitutive behavior of concrete by introducing a scalar variable D that quantifies the influence of micro-cracking. It results in a degradation of the elastic stiffness:

$$\sigma_{ij} = (1-D)C_{ijkl}\varepsilon_{kl} \quad (1)$$

where σ_{rs} and ε_{rs} are respectively the stress and strain components rs, C_{ijkl} is the fourth order elastic tensor and D is the damage variable.

The damage variable, ranging from zero for the virgin material to one for completely damaged one, differentiates the mechanical responses in traction and in compression by introducing two scalars D_t and D_c:

$$D = \alpha_t D_t + \alpha_c D_c \tag{2}$$

D_t and D_c are the tensile and compressive part of the damage. They are expressed by:

$$D_{t,c} = 1 - \frac{\kappa_0(1 - A_{t,c})}{\kappa} - \frac{A_{t,c}}{\exp[B_{t,c}(\kappa - \kappa_0)]} \tag{3}$$

where $A_{t,c}$ $B_{t,c}$ and κ_0 are the five parameters of the model and κ is the history variable.

The weights α_t and α_c are computed from the strain tensor ε. They are defined as functions of the principal values of the strains ε^t_{ij} and ε^c_{ij}, due to positive and negative stresses.

$$\alpha_{t,c} = \left(\sum_{i=1}^{3} \frac{<\varepsilon_i^{t,c}><\varepsilon_i>_+}{\varepsilon_{eq}^{\ 2}} \right)^{\beta} \tag{4}$$

In uniaxial tension, $\alpha_t = 1$ and $\alpha_c = 0$. In uniaxial compression $\alpha_c = 1, \alpha_t = 0$. The exponent β reduces the effect of damage on the response of the material under shear, compared to tension.

The evolution of damage is based on an equivalent strain ε_{eq} which characterizes the material extension during loading

$$\varepsilon_{eq} = \sqrt{\sum_{i=1}^{3} (<\varepsilon_i>_+)^2} \tag{5}$$

where $<\varepsilon_i>_+$ are the positive principal strains.

The loading surface f is classically defined by:

$$f(\varepsilon_{eq}, \kappa) = \varepsilon_{eq} - \kappa \tag{6}$$

where the history variable $\kappa = \mathrm{Max}_{/t}(\varepsilon_{eq}, \kappa_0)$.

Finally, the evolution of damage has the following form:

$$\text{if } f(\varepsilon_{eq}, \kappa) = 0 \text{ and } \dot{f}(\varepsilon_{eq}, \kappa) = 0 \text{ then } \begin{cases} D = \alpha_t D_t + \alpha_c D_c \\ \kappa = \varepsilon_{eq} \end{cases} \text{where } \dot{D} \geq 0 \tag{7}$$

$$\text{otherwise} \begin{cases} \dot{D} = 0 \\ \dot{\kappa} = 0 \end{cases}$$

2.2 *Non local gradient approach*

Let us consider for example a bar with a variable section under tension loading whose boundary conditions are given in figure 1. In the damage distribution (figure 2), D localizes only in the most loaded element. In the same way, if we consider the extreme case of an infinitely fine mesh, this geometrical localization yields a fracture without energy dissipation, phenomenon that is physically unacceptable. The model thus exhibits a mesh dependency related to its "local" formulation. This is a well known problem for softening model and is due to "strain localization" as mentioned in Askes (2000) and Peerlings et al. (1996) among others. After micro cracks processes have taken place in the first part of the inelastic behavior of concrete, a macroscopic mechanism develops in the form of a damage band.

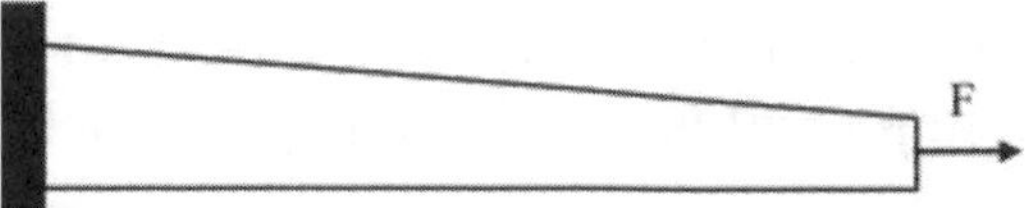

Figure 1. Bar under tensile loading. Geometry and boundary conditions.

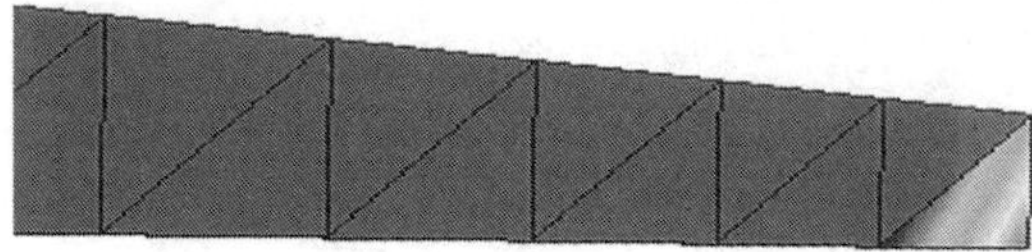

Figure 2. Damage distribution under tensile loading. Zoom on the extremity of the bar. Dark zones correspond to virgin material and clear ones to damage material.

The strains tend to localize in a specific area of finite dimension (macrocrack). The stress response of a material point cannot be described locally but has also to consider its neighborhood in order to take into account microcracks spatial interaction. This does not appear in the local formulation and a regularization technique has to be added.

Different techniques exist to limit strain localization (Pijaudier-Cabot & Bazant 1987, Peerlings et al. 1996). Nevertheless, in every case, the principal is to include a spatial information, in the form of a characteristic length.

Here a gradient approach is chosen. The method is based on the definition of a non local strain tensor expressed as the weighted average of ε over a volume V

$$\bar{\varepsilon}_{ij} = \frac{1}{V} \int_V g(\xi) \varepsilon_{ij}(x + \xi) dV \tag{8}$$

where g is the weight function.

The integral equation is then approximated with Taylor series and after some transformations, the characteristic law is given by:

$$\varepsilon_{ij} = \bar{\varepsilon}_{ij} - c\nabla^2 \bar{\varepsilon}_{ij} \tag{9}$$

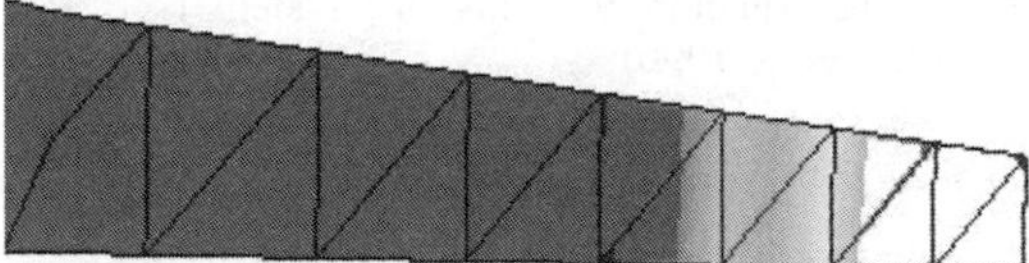

Figure 3. Non local response. Physical localization of the damage. Clear zones correspond to damage material.

where ∇^2 denotes the second order gradient operator. The constant c is proportional to a length square and characterizes non local interaction.

The numerical problem is thus governed by the equilibrium equation and by the definition of the non local strains

$$\begin{aligned} &\nabla.\sigma + f = 0 \\ &\bar{\varepsilon}_{ij} - c\nabla^2 \bar{\varepsilon}_{ij} = \varepsilon_{ij} \end{aligned} \tag{10}$$

where f represents an external load density.

Once non local strains have been computed, internal variables are obtained replacing the local equivalent strain of equation 5 by the non local quantity

$$\bar{\varepsilon}_{eq} = \sqrt{\sum_{i=1}^{3} (<\bar{\varepsilon}_i>_+)^2} \tag{11}$$

The main advantages of the present gradient approach are to provide a solution to mesh dependency problem and to fit to every local constitutive law (isotropic or anisotropic) because it provides a non local strain tensor from which any invariant can be derived in order to control plasticity or damage.

Figure 3 illustrates the non local response of the bar considered in figure 2. Damage is no more geometrically but physically localized.

3 ELEMENTARY TESTS

The first step of the validation is elementary tests. They are applications on a simple element and aim at evaluating how the model behaves under usual loading.

3.1 *Monotonic and alternated uniaxial loading*

Monotonic tension is the simplest way to check the validity of the constitutive law.

The numerical response is first compared with an experimental tensile test performed by Gopalaratnam et al. (1985). Figure 4 gives the axial stress-strain curve. The isotropic damage model reproduces the global tensile behavior of concrete. Experimental and numerical peak positions are about the same. That validates the choice for the damage threshold κ_0. Finally,

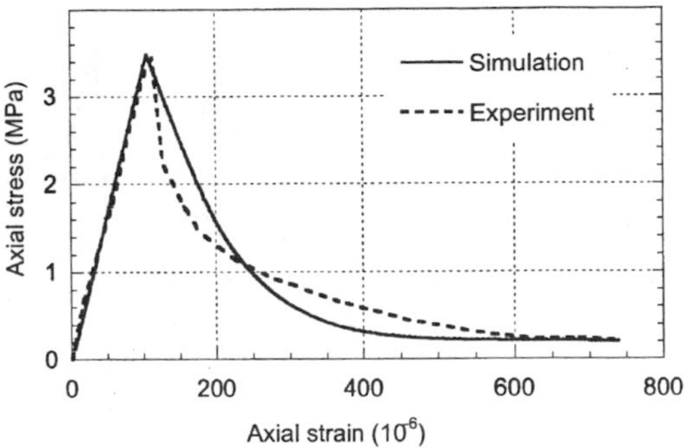

Figure 4. Experimental and numerical results for tensile loading. Comparison of the axial strain – stress response.

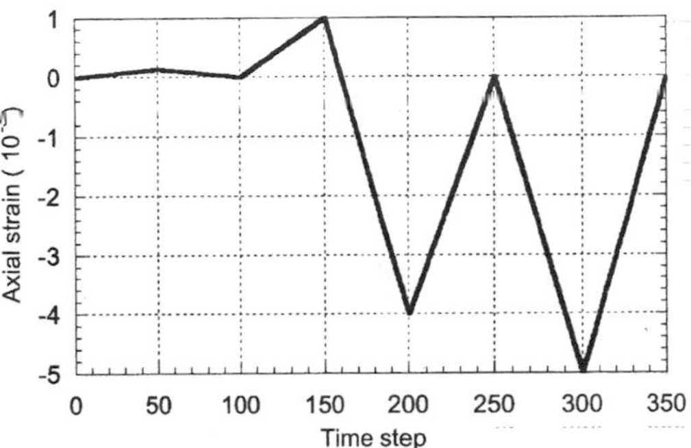

Figure 5. Loading path for the alternated tension – compression test.

the post-peak behavior is not exactly fitted. Nevertheless, one can consider that the model provides an interesting solution for monotonic uniaxial loading.

To go further in the validation, alternated tension compression application is tested. The test is still performed on a simple element. Figure 5 illustrates the loading path. The element is twice loaded – unloaded in tension and twice in compression. Figures 6 and 7 show the numerical response.

This result yields some remarks. First, for total unloading, zero stress corresponds to zero strain. Mazars' model as described in this contribution is a pure damage model. "Plasticity" is not taken into account that is why irreversible strains do not appear. The unloading is elastic with a slope equal to the damaged Young's modulus E_d.

$$E_d = (1 - D)E_0 \tag{12}$$

where E_0 is the virgin Young's modulus.

After the second loading in tension, the damage variable reaches its maximal value. As it cannot decrease, unloading in tension and compression is linear.

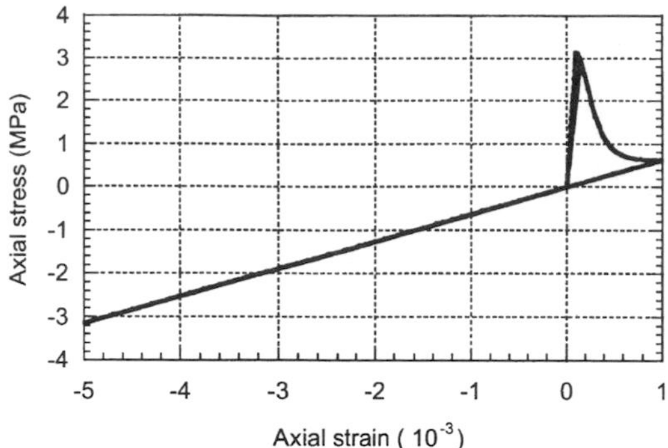

Figure 6. Stress strain curve for alternated tension compression loading.

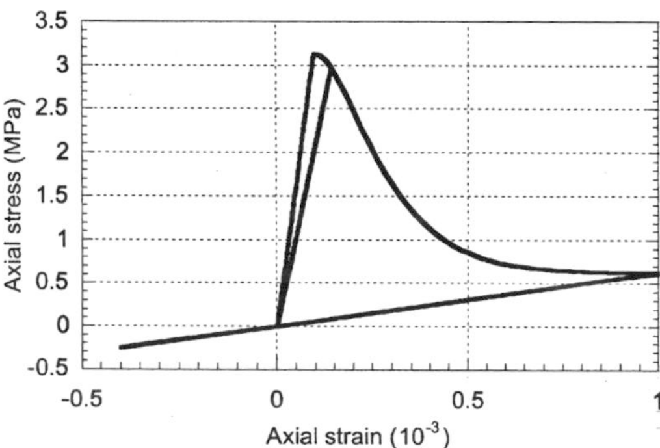

Figure 7. Stress strain curve for alternated loading. Tensile part.

Experimentally, crack closing plays a key role in alternated loading. During unloading, microcracks gradually close and the material stiffness increases while the damage keeps a constant value. The model does not include this process and that is why the slope recovery cannot be captured. As a conclusion, the isotropic constitutive law is not completely adapted to cyclic loading.

3.2 *Numerical quality: convergence*

The ability to reproduce an experimental result is typically the principal point used for evaluating the performance of a numerical model. However, this is not the only quality by which effectiveness of the numerical implementation is also an important step in the validation process. On structural tests with fine mesh and complex loading, the efficiency of the iterative algorithm directly influences computational costs.

The complete model was implemented in the finite element code "Code_Aster" at EDF using a full Newton-Raphson procedure. The discretization of the governing equation 9 yields the system (see also Peerlings et al. 1996):

$$\begin{aligned} f_{ai} &= f_{ae} \\ K_{ee}\overline{\varepsilon}^{n} - f_{e} &= 0 \end{aligned} \tag{13}$$

where superscript n denotes for nodal values and

$$\begin{aligned} f_{ai} &= \int B^{T}\sigma d\Omega \quad f_{ae} = \int N^{T}\tau d\Gamma + \int N^{T} f d\Omega \\ K_{ee} &= \int (c\overline{B}^{T}\overline{B} + \overline{N}^{T}\overline{N}) d\Omega \overline{\varepsilon}^{n} \quad f_{e} = \int \overline{N}^{T}\varepsilon d\Omega \end{aligned} \tag{14}$$

τ represents the boundary conditions on stresses. Two interpolation functions are defined, N for the displacement field u and one for the non local strains

$$\begin{aligned} u &= N u^{n} \\ \overline{\varepsilon} &= \overline{N}\,\overline{\varepsilon}^{n} \end{aligned} \tag{15}$$

B includes the partial derivatives of the interpolation function N with respect to the coordinates. In order to avoid stress oscillations, the non local strain tensor is discretised with interpolation functions that are linear while the interpolation of displacements is quadratic.

Equation 13 is the linearized to build the incremental iterative Newton-Raphson solution procedure. The system to be solved is finally written as:

$$\begin{bmatrix} K_{aa} & K_{ae} \\ K_{ea} & K_{ee} \end{bmatrix} \begin{pmatrix} \delta u_{k}^{n} \\ \delta \overline{\varepsilon}_{k}^{n} \end{pmatrix} = \begin{Bmatrix} f_{ae} - f_{ai,k} \\ f_{e,k} - K_{ee}\overline{\varepsilon}_{k}^{n} \end{Bmatrix} \tag{16}$$

with subscript k denotes the iteration number. The matrices K is defined with four elementary parts.

$$\begin{aligned} K_{aa} &= \int B^{T}\frac{\partial\sigma}{\partial\varepsilon} B\, d\Omega \quad K_{ae} = \int B^{T}\frac{\partial\sigma}{\partial\overline{\varepsilon}}\overline{N}\, d\Omega \\ K_{ea} &= -\int N^{T} B\, d\Omega \quad K_{ee} = \int (c\overline{B}^{T}\overline{B} + \overline{N}^{T}\overline{N})\, d\Omega \end{aligned} \tag{17}$$

δu_k is the corrective term in the displacement field and $\delta\varepsilon_k$ in the non local strain tensor:

$$\begin{aligned} u_{k+1} &= u_{k} + \delta u_{k} \\ \overline{\varepsilon}_{k+1} &= \overline{\varepsilon}_{k} + \delta\overline{\varepsilon}_{k} \end{aligned} \tag{18}$$

The derivatives of the stress with respect to the strain (K_{aa}) and the non local strain (K_{ae}) play a key role in the convergence of the iterative alogrithm.

In the modified Newton-Raphson scheme for example, the elastic law is used to compute the derivatives:

$$\sigma_{ij} = 2\mu\varepsilon_{ij} + \lambda\varepsilon_{kk}\delta_{ij} \tag{19}$$

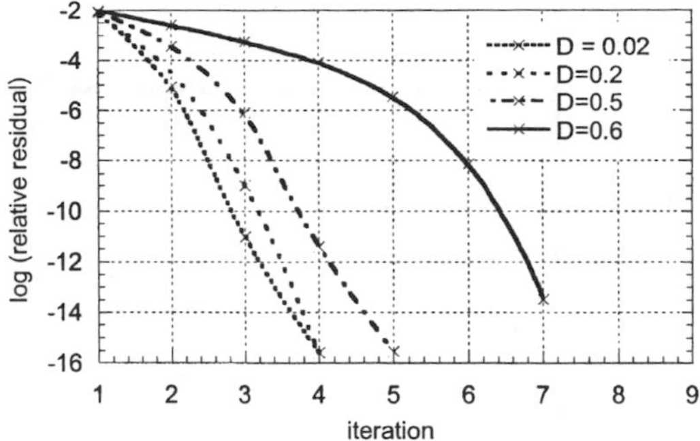

Figure 8. Evolution of the relative residuals for the Newton Raphson scheme. Uniaxial tensile test.

This method yields a linear convergence.

If an exact derivation of the constitutive relation is performed, the convergence is quadratic. Less iterations are required to converge but the derivation has to be done carefully regarding the extreme sensitivity of the algorithm on small errors. In the damage model proposed by Mazars, the so called consistent tangent operator is difficult to derive properly. The differentiation of the factors α_t and α_c (Equation 4) are particularly uneasy. That is why the variation of these coefficients is neglected in this contribution. Consequently, quadratic convergence is only expected for radial loading (α_t and α_c keep a constant value).

To check the calculation of this derivative, figure 8 shows the evolution of the relative residual during the iterations for several initial states of damage. The computation is an uniaxial monotic tensile test.

The parabola which is approximately followed by the residual illustrates the quadratic convergence of the iterative algorithm and validates the good implementation of the consistent tangent operator.

4 STRUCTURAL TESTS

Elementary tests as described in part 3 are not sufficient to totally validate the isotropic damage model. The main reason is that a simple element simulation does not highlight the non local interest of the formulation. The following step is thus structural tests. They are usual applications involving three dimensional problems. Two structural cases are extracted from benchmarks proposed by EDF (Ghavamian 1999).

4.1 *Three point bending test*

The first structural application is a 3D computation of a reinforced concrete beam. The geometry and the load system are presented in figure 9. Figure 10 depicts the steel distribution.

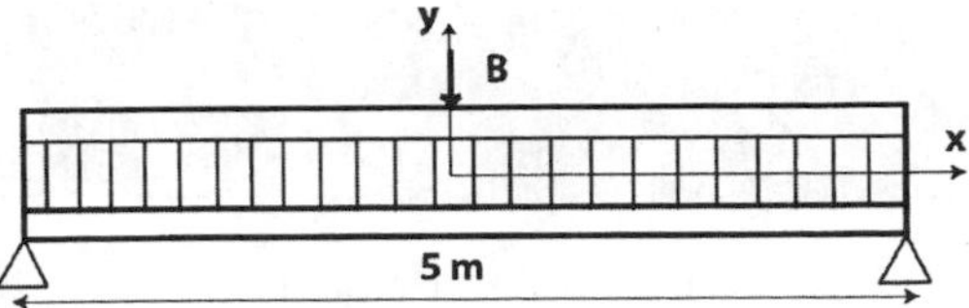

Figure 9. Three point bending test on a RC beam.

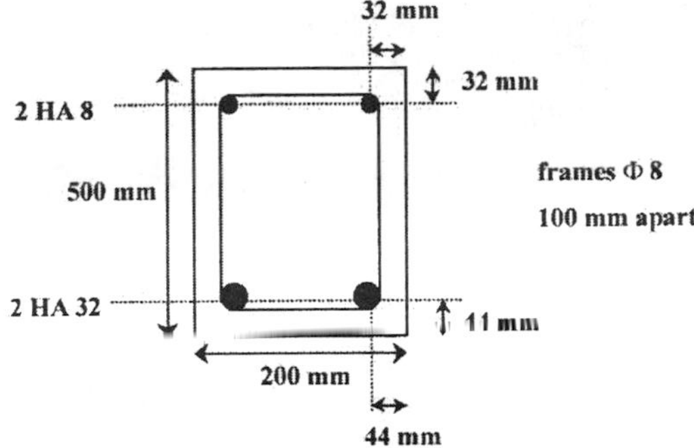

Figure 10. Steel distribution in the reinforced concrete beam.

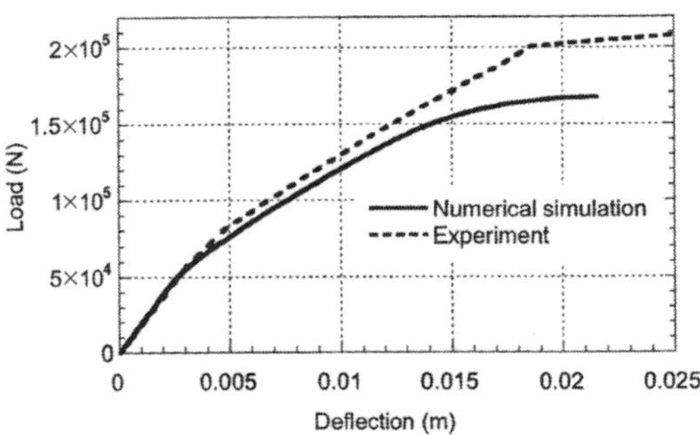

Figure 11. Load deflection curve. Three point bending test.

The aim of this test is to evaluate the three dimensional performance of the model. Figure 11 provides the numerical load deflection curve and a comparison with experiments. The results show qualitative and quantitative similarities: the elastic part is well reproduced and the inelastic behavior fits well. The role of the steel in the global response is also underlined by the non apparition of the softening branch.

For bending tests, the isotropic non local gradient model seems thus to be sufficient to simulate the structural response.

This test gives also the opportunity to highlight the interest of the non local formulation. Figures 12 and 13 show the damage distribution in the half beam for local and non local simulations. It is a longitudinal cross

Figure 12. Local damage distribution for three point bending test in the half beam. Clear zones correspond to damaged material. 20 longitudinal element mesh.

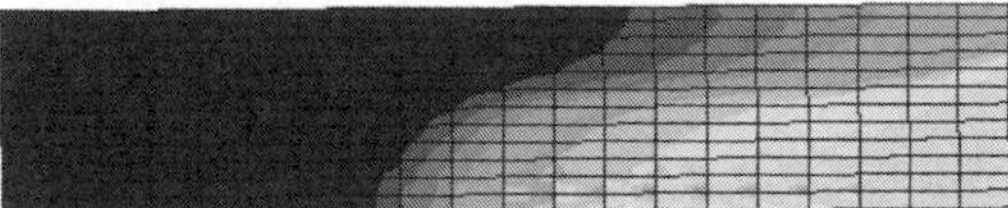

Figure 13. Non local damage distribution for three point bending test in the half beam. Clear zones correspond to damaged material. 20 longitudinal element mesh.

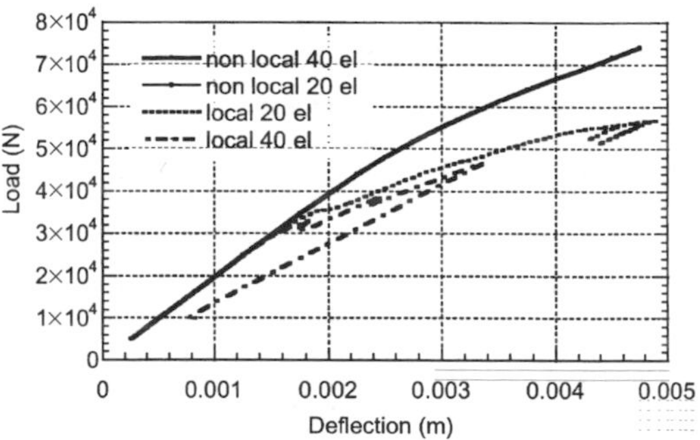

Figure 14. Local and non local computations for 20 and 40 element mesh. Load-deflection curve for three point bending test.

section at mid-span. In the local case, damage geometrically localizes, as already mentioned in part 1. The distribution is dependant on the smoothness and the orientation of the mesh. On the contrary, the non local approach gives a continuous zone of localization which is no more influenced by the mesh. Figure 14 illustrates the differences between non local and local approaches in term of load deflection curves regarding the number of elements in the longitudinal cross section of the half beam.

The simulations with 20 or 40 longitudinal element mesh (the transversal mesh is not modified) give exactly the same result when the non local gradient approach is used. On the contrary, if a local computation is considered, one can observe a mesh dependency and the apparition of characteristic numerical snap backs which correspond to unrealistic partial unloading of the beam.

In this case, including a gradient approach is thus necessary to avoid a spurious strain and damage localization.

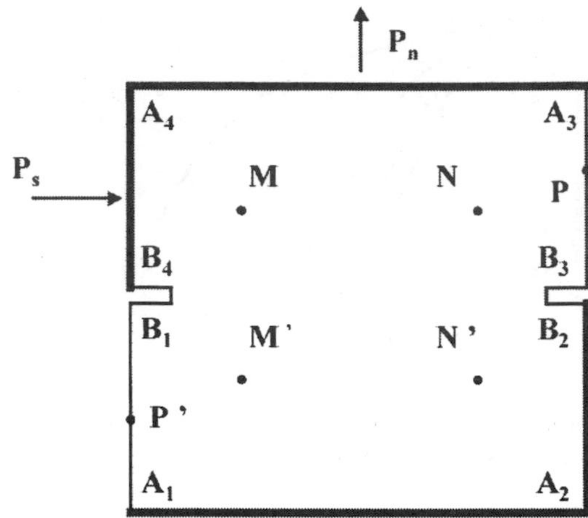

Figure 15. Nooru-Mohamed test. Geometry and boundary conditions.

4.2 *Nooru-Mohamed test*

The following structural test evaluates the ability of the model to respent mixed mode fracture of concrete (Nooru-Mohamed 1992). It is symmetrical double edge notch plate subjected to tensile (P_n) and shearing (P_s) strain, as illustrated in figure 15.

The boundary conditions are restrainers in x and y directions, on the lower part and on the lower half of the right-hand side. The upper face $A_3 - A_4$ is kept in the horizontal position and the left face $A_4 - B_4$ is kept in the vertical position. Plane stress conditions are considered (2D simulation).

An indirect control is imposed on the displacement. The shearing (δ_s) is measured at points P and P'. The elongation (δ_n) is obtained from the displacements of points M, M', N, N'.

$$\begin{aligned} \delta_s &= \delta_s^P - \delta_s^{P'} \\ \delta_n &= (\delta_n^M - \delta_n^{M'} + \delta_n^N - \delta_n^{N'})/2 \end{aligned} \tag{20}$$

For the loading path A, the specimen is first subjected to the controlled displacement of $A_3 - A_4$ until an average vertical displacement of 2.2 mm (δ_n) is reached. It is a test in tension to highlight model I fracture.

For the loading path B, the specimen is first subjected to shearing until the shearing force reaches 5 kN (P_s) while P_n keeps a constant zero value.

The shearing force is then maintained at a constant level and an increasing vertical displacement is applied.

Figure 16 presents the evolution of the damage distribution during tensile loading. Crack initiation from the notches (weak points of the structure) is well reproduced. The development of cracks is not symmetric

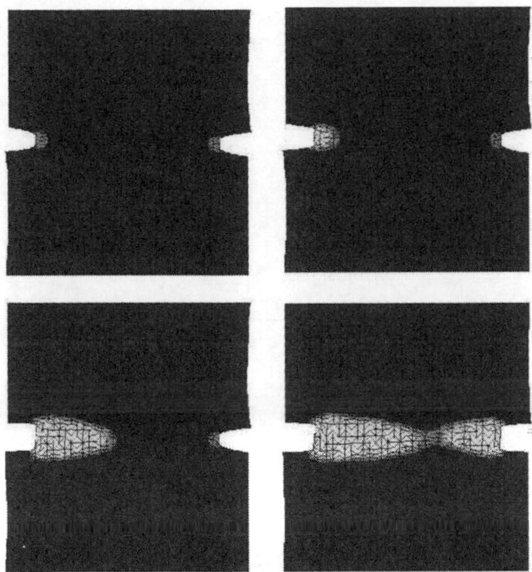

Figure 16. Evolution of the damage distribution in the structure. Mode I fracture.

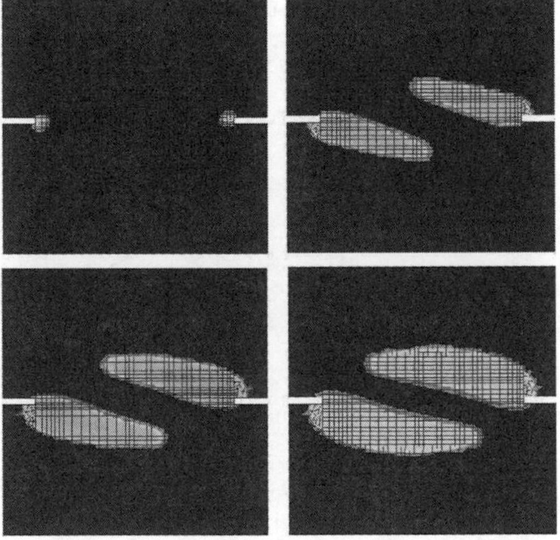

Figure 17. Evolution of damage distribution for loading path B. Clear zones correspond to damaged ones.

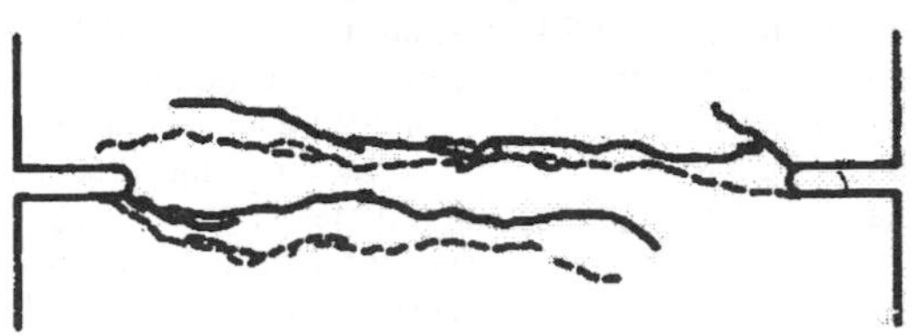

Figure 18. Experimental crack pattern for loading path B.

because the boundary conditions are not too. Cracks propagate following the horizontal axis of the structure and characterize mode I fracture.

Figure 17 illustrates the damage distribution for loading path B.

Compared to the experimental crack pattern in figure 18 (Nooru-Mohamed 1992), the numerical results seem to provide a good approximation of the experimental data.

5 FINAL APPLICATION

The numerical simulations using the non local gradient isotropic model have so far revealed some advantages and drawbacks of the formulation. The model reproduces well the behavior of concrete in simple tension, simple compression or three point bending test (parts 3.1 and 4.1). Moreover, the numerical implementation exhibits a quadratic convergence for radial loading with an analytical consistent tangent operator (part 3.2).

Nevertheless, for some other applications, like alternated loading (parts 3.1), the model is too simple to take into account special phenomena like crack closing or plasticity.

Taking into account these results, the next step is the simulation of a representative part of a containment building of a nuclear power plant.

5.1 *Presentation of the simulation*

Here is presented an application called PACE 1300 recently developed by Electricité de France. This test is a Representative Structural Volume of a containment building of a French 1300 MW nuclear power plant. Figure 19 illustrates the position of the extracted volume in the global structure. It has almost all the features of the reinforced containment building: concrete, vertical and horizontal longitudinal passive bars, transversal passive bars and prestressed tendons in two directions.

The geometry has been chosen to include a sufficient number of components to be representative of the general behaviour. It is large enough to avoid

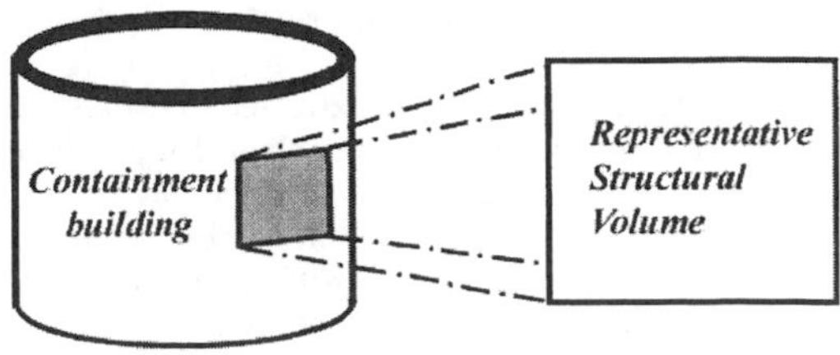

Figure 19. Representative Structural Volume of a containment building. Principle.

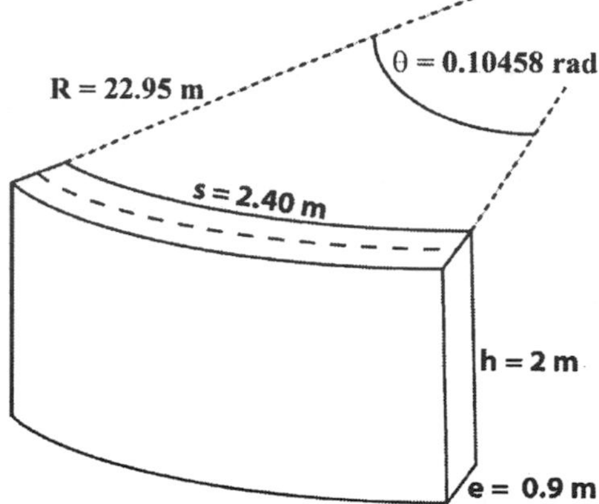

Figure 20. Structural Representative Volume. Geometry.

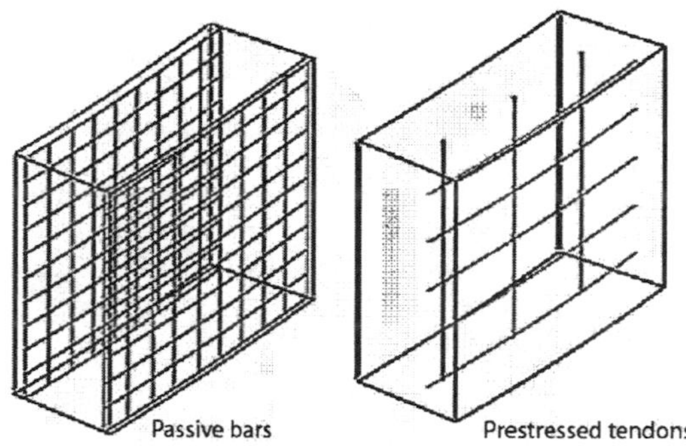

Figure 21. Passive steel bars and prestressed tendons in the Representative Structural Volume.

boundary problems and small enough to enable non linear studies to be performed with fine meshes. With these restrictions, the structural representative volume includes eleven internal and external horizontal passive bars, ten internal and external vertical passive bars, ten internal and external vertical passive bars, five prestressed horizontal tendons, three prestressed vertical tendons and twenty four frames uniformly distributed in the volume. The dimensions of the problem are given in figure 20 and the steel distribution in figure 21. Table 1 provides information about steel properties.

The material parameters are extracted from existing data about a French nuclear power plant. They include concrete, passive bars and tendons properties.

The behavior of the representative volume has to be as close as possible to the in situ situation. It supposes the following boundary conditions (figure 22): a prestress on SB and SH in z direction, on SG and SD in θ_1 and θ_2 directions and a cinematic relation on SG and SD which imposes no displacement in θ_1 and θ_2 directions. The loading is applied in the form of a radial pressure on the internal face SI (simulation of integrity tests). The self weight of the structure is also taken into account.

With these conditions, the mesh has been defined with 16500 hexa20 elements for concrete.

Figures 23 and 24 illustrate the orthoradial ($\sigma_{\theta\theta}$) stress distribution and the damage distribution respectively in the representative volume for an internal pressure of 0.8 MPa (twice the pressure that the structure has to withstand during integrity tests).

One can notice a decrease of the orthoradial stress around the tendons which clearly underlines their role in the global response of the structure. Concerning the damage distribution, the volume exhibits strongly

Table 1. Steel bar properties.

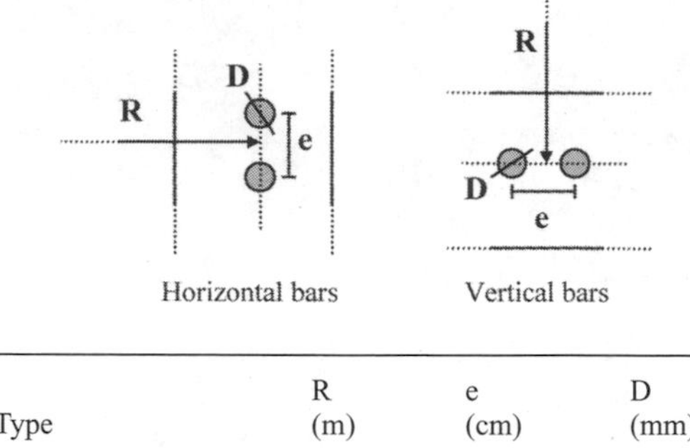

Type	R (m)	e (cm)	D (mm)
Horizontal internal passive bars	22.60	20	20
Horizontal external passive bars	23.35	20	20
Vertical internal passive bars	22.60	27.297	25
Vertical external passive bars	23.35	27.170	25
Horizontal tendons	23.15	40.5	40.5
Vertical tendons	22.95	80	40.5
Frames*	x	x	3.685

*Frames are uniformly distributed in the representative volume.

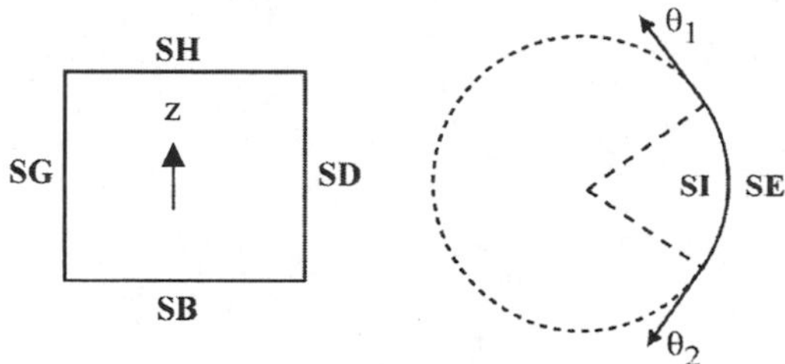

Figure 22. Boundary conditions. Definition.

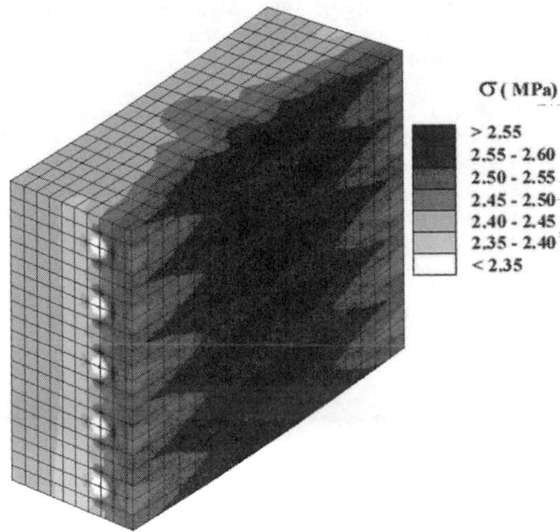

Figure 23. Stress distribution in the representative structural volume.

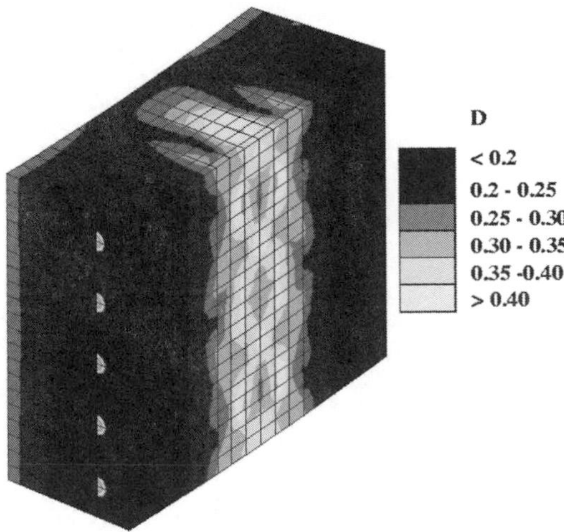

Figure 24. Damage Distribution in the representative structural volume.

damaged zones located in the middle of the external surface. A damage localisation near the vertical tendons due to the application of the prestress is obtained, probably due to the definition of the boundary conditions. As expected, the behaviour of the structure is symmetric.

This test shows that an isotropic damage model is able to provide results for complex "industrial" applications. The simulation highlights significant zones where damage and stress develop in a preferential way. It thus give interesting information for further studies and comparisons with other models.

6 CONCLUSION

In this contribution, a set of benchmarks was proposed to define a global methodology for the validation of constitutive laws. It has applied on an isotropic damage model implemented in the finite element code "Code_Aster" at EDF. This description was coupled with a non local gradient approach to avoid spurious strain localization, mesh dependency and physically unacceptable results.

Elementary tests represent the first step of the validation. They are basic applications on which the ability of the model to reproduce an experimental behavior is evaluated. For our chosen model, they underlined a good prediction in uniaxial tension or compression but also the difficulty to simulate alternated loading (crack closure effect and irreversible strains).

These simple applications give also the opportunity to show the efficiency of the iterative algorithm. In our case, the quadratic convergence of the Newton-Raphson procedure and the definition of the consistent tangent operator were checked.

Structural tests are the second step in the validation process and aim at evaluating more complex problems like mesh dependency or non local interest. They generally reproduce classical laboratory tests. In this contribution, a three pont bending test on a RC beam and a Nooru-Mohamed application showed that the isotropic model can be sufficient to reproduce a structural behavior. The importance of the non local formulation was also underlined.

Once the previous steps have been fulfilled, the final goal is to consider real "industrial" cases. One can then estimate how the model behaves in complex situations. For the isotropic model, a representative structural volume of a containment building was chosen. The simulation gave first results in the form of damage and stress distributions that could be used for further studies (evaluation of the structural integrity, permeability …).

ACKNOWLEDGMENTS

Partial financial support from EDF and from the EU through MAECENAS project (FIS5-2001-00100) is gratefully acknowledged. The authors would like to thank EDF for scientific support toward the developments in the FE code "Code_Aster".

REFERENCES

Askes, H. 2000. Advanced Spatial Discretisation strategies for localised failure. Mesh adaptivity and meshless methods. PhD Thesis, Delft University of Technology Faculty of Civil Engineering and Geosciences.

Ghavamian, S. 1999. Evaluation tests on models of non-linear behaviour of cracking concrete, using the dimensional modelling. Benchmark EDF/Division R&D, CR – MMN 99/232.

Gopalaratnam, V. S. & Shah, S. P. 1985. Softening Response of plain Concrete in Direct tension. *ACI Journal*, May–June 1985:310–323.

Mazars, J. 1984. Application de la mécanique de l'endommagemnt au comportement non linéaire et à la rupture du béton de structure. PhD Thesis, Université Paris VI, Paris, France.

Nooru-Mohamed, M. B. 1992. Mixed-mode fracture of concrete: an experimental approach. PhD Thesis, Delft University of Technology.

Peerlings, R. H. J. De Borst, R. Brekelmans, W. A. M. & De Vree, J. H. P. 1996. Gradient Enhanced damage for quasi brittle materials. *International Journal for numerical methods in engineering*, Vol. 39: 3391–3403.

Pijaudier-Cabot, G. & Bazant, Z. P. 1987. Nonlocal damage theory. *Journal of Engineering Mechanics*, Vol. 113 No 10:1512–1533.

Computational Modelling of Concrete Structures, Bićanić et al. (eds)
© 2003 Swets & Zeitlinger, Lisse, ISBN 90 5809 536 3

Modelling the behaviour of RC beams retrofitted with CARDIFRC®

B.L. Karihaloo and F.J. Alaee
Division of Civil Engineering, School of Engineering, Cardiff University, Queen's Buildings, Newport Road, Cardiff, UK

ABSTRACT: A new retrofitting technique based on a material (CARDIFRC®) compatible with concrete has been developed at Cardiff University. It overcomes some of the problems associated with the current techniques based on externally bonded steel plates and FRP (fibre-reinforced polymer) laminates which are due to the mismatch of their tensile strength and stiffness with that of the concrete structure being retrofitted.

CARDIFRC® is characterised by high tensile/flexural strength and high energy-absorption capacity (i.e. ductility). The special characteristics of CARDIFRC® make it particularly suitable for repair, remedial and upgrading activities (i.e. retrofitting) of existing concrete structures. It has been shown that damaged reinforced concrete beams can be successfully strengthened and rehabilitated in a variety of different retrofit configurations using pre-cast CARDIFRC® strips adhesively bonded to the prepared surfaces of the damaged beams. To predict the moment resistance of the beams retrofitted in this manner two analytical models are described in the present paper. The results of the analytical models are found to be in very good agreement with the test results.

1 INTRODUCTION

A new technique of retrofitting damaged and/or under-strength reinforced concrete beams has been recently developed by the authors. In this technique thin, pre-cast strips of high-performance fibre-reinforced cementitious composite, also developed at Cardiff and designated CARDIFF™, are adhesively bonded to the tension and, if necessary, other faces of the RC beam. This technique ensures that the ultimate failure of the retrofitted beams will occur in the gradual, flexural mode. It therefore overcomes some of the problems associated with the existing retrofitting techniques, based on externally bonded steel plates and FRP laminates, which arise primarily due to the mismatch of their tensile strength and stiffness with that of concrete.

The performance of current techniques of rehabilitation and strengthening (the collective term retrofit, which implies the addition of structural components after initial construction, captures both rehabilitation and strengthening) using externally bonded steel plates and fibre-reinforced plastic (FRP) laminates has been extensively investigated (Ahmed and Gemert 1999; El-Refaie et al. 1999; Fanning and Kelly 1999; Yagi et al. 1999). The technique of retrofitting using externally bonded steel plates has gained widespread popularity, being quick, causing minimal site disruption and producing only minimal change in section size. However, several problems have been encountered with this technique, including the occurrence of undesirable shear failures, difficulty in handling heavy steel plates, corrosion of the steel, and the need for butt joint systems as a result of limited workable lengths (Jones et al. 1988; Ziraba et al. 1994; Hussain et al. 1995).

FRP materials as thin laminates or fabrics would appear to offer an ideal alternative to steel plates. They generally have high strength to weight and stiffness to weight ratios and are chemically quite inert, offering significant potential for lightweight, cost effective and durable retrofit (Nanni 1995; Büyüköztürk 1998). Retrofitting using FRP is also vulnerable to undesirable brittle failures due to a large mismatch in the tensile strength and stiffness with that of concrete.

CARDIFRC® is a fibre reinforced cementitious composite which is characterised by high tensile/flexural strength and high energy-absorption capacity (i.e. ductility). The special characteristics of CARDIFRC® make it particularly suitable for repair, remedial and upgrading activities (i.e. retrofitting) of existing concrete structures. The key advantage of CARDIFRC® mixes for retrofitting is that unlike steel and FRP, their tensile strength, stiffness and coefficient of linear thermal expansion are comparable to that of the material of the parent member. Two typical optimum mix proportions of CARDIFRC® (Mix I and

Table 1. Mix proportions for optimised CARDIFRC® Mix I and Mix II (per m^3).

Constituents (kg)	Mix I	Mix II
Cement	855	744
Microsilica	214	178
Quartz sand		
9–300 μm	470	166
250–600 μm	470	–
212–1000 μm	–	335
1–2 mm	–	672
Water	188	149
Superplasticiser	28	55
Fibres		
6 mm	390	351
13 mm	78	117
Water/cement	0.22	0.20
Water/binder	0.18	0.16

Table 2. Typical material properties of CARDIFRC® Mix I and Mix II.

Material properties	Mix I	Mix II
Indirect tensile strength (MPa)	24	25
Fracture energy (J/m^2)	12,210	12,380
Compressive strength (MPa)	207	185

Mix II) are given in Table 1, with the typical material properties shown in Table 2.

Several studies have previously been undertaken at Cardiff into the feasibility of using CARDIFRC® for the rehabilitation and strengthening of damaged RC flexural members (Karihaloo et al., 2000, 2002; Alaee et al., 2001, 2002). To predict the moment resistance and the load-deflection behaviour of the beams retrofitted with CARDIFRC® two analytical models have been developed (Alaee and Karihaloo, 2002a, b). One model is based on the classical strength theory, but takes into account fully the tensile contributions (i.e. pre-peak and post-peak) from concrete and CARDIFRC®. The second analytical model is based on fracture mechanics concepts. This model mimics the initiation and growth of the flexural crack that eventually leads to the failure of the retrofitted beam.

Both computational models predict ultimate moment capacity and load-deformation behaviour that are in excellent agreement with test results. It is also shown that such an agreement would not have been possible without the inclusion of proper tension softening response of both concrete and CARDIFF®, thus demonstrating conclusively the importance of fracture mechanics concepts to the design of RC structures.

A rheological study was recently carried out in Cardiff to optimise HPFRCCs. The aim was to achieve good workable mixes with a very low water/binder ratio and a high volume fraction of steel fibre, in order that the resulting material, in its hardened state, will be very ductile with a relatively high tensile strength. As a result of many trial mixes and testing, the mixes shown in Table 1 are the optimised ones. Two different mixes (designated CARDIFRC®, Mix I and Mix II) of high-performance fibre reinforced concrete differing mainly by the maximum size of quartz sand used in the mix have been developed using novel mixing and fibre dispersion procedures. These procedures are described in the patent application GB 0109686.6.

Brass-coated steel fibres diameter 0.16 mm, 6 mm or 13 mm long are used to prevent corrosion. The optimised grading of quartz sands leads to a considerable reduction in the water demand without loss in workability. All materials used in Table 1 are available commercially.

A volume fraction of 6% short and long fibres is used, comprising 5% short fibres and 1% long fibres for Mix I, and 4.5% short fibres and 1.5% long fibres for Mix II. The specimens were hot-cured at 90°C for seven days. The strengths attained have been found to be the equivalent of standard 28-day water curing at 20°C. Table 2 shows the material properties of the optimised mixes. The Young Modulus of CARDIFRC® is around 50 GPa.

2 TEST BEAMS

Two types of beam differing only by the reinforcement were used for stages I and II of the experimental programme. The beams in stage I were reinforced in flexure only with a single 12 mm rebar, whereas in the beams tested in stage II, stirrups were also provided in the shear spans of the beams. All the beams were made from a standard concrete mix and were 1200 mm long, 100 mm wide and 150 mm deep.

2.1 *Stage I*

Of the thirty-two beams used in stage I, four were tested to failure as control beams to compare with the performance of those retrofitted with CARDIFRC® strips. The remaining twenty-eight beams were preloaded to approximately 75% of the above failure load to induce flexural cracking. In addition to parameters such as the material (Mix I or II) and thickness of retrofit strips (16 or 20 mm), four different configurations of retrofitting were investigated, as shown in Fig. 1a–d.

2.2 *Stage II*

Of the fourteen beams produced for stage II, three were tested without any repair as control beams. These beams were tested to failure under four-point bending over a span of 1100 mm. The spacing between the applied loads was 400 mm. The remaining eleven

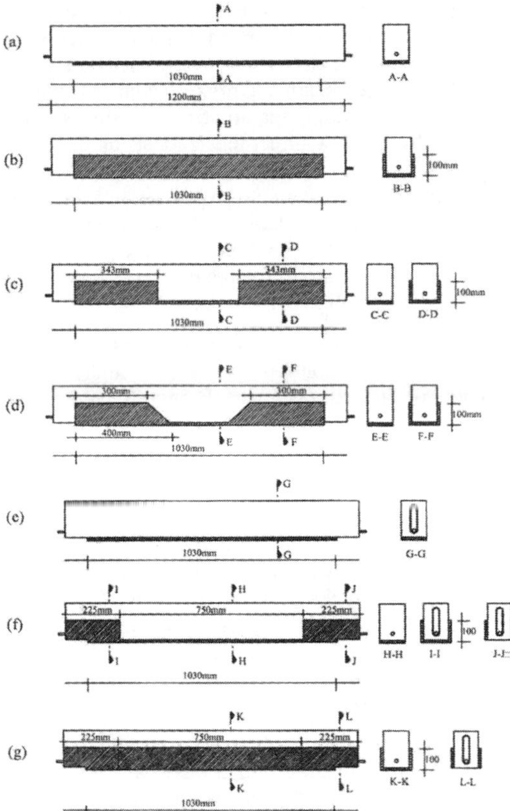

Figure 1. Configurations of retrofitting. Beams retrofitted with: (a) one strip on the tension face, (b) one strip on the tension face and two strips on the sides, (c) one strip on the tension face and four rectangular strips on the sides, (d) one strip on the tension face and four trapezoidal strips on the sides, (e) one strip on the tension face, (f) one strip on the tension face and four short strips on the sides covering the supports and the ends of the tension strip, and (g) one strip on the tension face and four short and two continuous strips on the sides, fully covering the supports and the tension strip sides.

beams were pre-loaded in the same manner as the control beams to approximately 75% of the failure load. To improve the flexural behaviour of the damaged beams three configurations of retrofitting strips were investigated in this stage, as shown in Fig. 1e–g.

2.3 *Casting of strips*

The retrofit materials, CARDIFRC® Mix I and Mix II were cast as flat strips in 1030 mm long and 100 mm wide steel moulds with a well-oiled base and raised border whose height could be adjusted to give 16 or 20 mm thick plates. The moulds were filled on a vibrating table at 50 Hz frequency and smoothed over with a float. To ensure a uniform thickness (within 1 mm) a glass panel was located on top of the raised border. The strips were left to cure in the moulds for 24 hours at 20°C before demoulding. The retrofit strips were then hot-cured at 90°C for a further 9 days (including one day for raising and one day for lowering the temperature). The short rectangular and trapezoidal side strips, were cut from the long cast strips to the required size using a diamond saw.

2.4 *Adhesive bonding*

To improve the bond between the retrofit strips and the damaged beams, all contacting surfaces were carefully cleaned and roughened. An angle grinder was used to create a grid of grooves approximately 3 mm deep at a spacing of 50 mm on the contacting surfaces of the damaged beams.

The retrofit strips were bonded to the prepared surfaces of the damaged concrete beams with a commercial thixotropic epoxy adhesive. The two parts of the adhesive were thoroughly mixed and applied to the tension side of the damaged beam with a serrated trowel to a uniform thickness of 3 mm. The strips were placed on the adhesive and evenly pressed. To ensure good adhesion, pressure must be applied to the strips during the hardening of the adhesive (24 hours) in accordance with the manufacturer's recommendation.

For the retrofitted beam with more than one strip, the beam was turned on its side to which the strip was bonded in the same manner as above. After another 24 hours, this procedure was repeated on the other side of the damaged beam. In practice, to ensure good adhesion between the strips and the damaged beam pressure can be applied using G-clamps.

3 FIRST ANALYTICAL MODEL

To predict the moment resistance and the load-deflection behaviour of the control and retrofitted beams two analytical models have been developed. In the first model the strain hardening as well as tension softening of both concrete and CARDIFRC® in tension have been taken into account. The stress-strain relationships of materials were assumed to be according to the test results or the proposed stress-deformation diagrams of the Model Code CEB-FIP (1993). Based on this code, steel is assumed to be perfectly elasto-plastic [Fig. 2(a)], whereas a parabolic relation is used for concrete in compression [Fig. 2(b)]. Table 3 shows the values of the relevant parameters assumed in the model. For concrete, the compressive strength was measured experimentally, and the remaining parameters were calculated from the relations proposed by CEB-FIP. The yield stress f_y and the modulus of elasticity E_s of steel were obtained from tension test on rebars.

Tensile failure of concrete and CARDIFRC® is always a discrete phenomenon. Therefore, to describe this behaviour a stress-strain and a stress-crack opening relation should be used for the uncracked and cracked sections, respectively. For normal concrete in tension the stress-deformation behaviour proposed by CEB-FIP was assumed [Fig. 2(c)], whereas the behaviour of CARDIFRC® in tension was modelled based on the theory of fracture mechanics and a few available test results [Fig. 2(d)]. The parameters assumed for modelling the behaviour of concrete and CARDIFRC® in tension can be found in Tables 3 and 4, respectively. For concrete, the direct tensile strength f_{ctm} was estimated from the splitting test results and the remaining parameters were again calculated from the relations proposed by CEB-FIP. For CARDIFRC® the tensile strength of the matrix f_{tp} was estimated from the splitting test results of the mix without fibres. However, the specific fracture energy G_F and the modulus of elasticity E were directly measured using the notched beam and prism specimens, respectively. The remaining parameters in Table 4 were obtained from a few direct tension tests on dog-bone shape specimens.

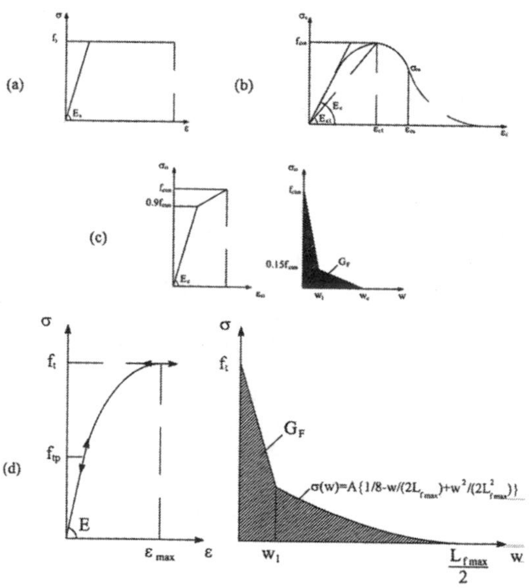

Figure 2. Stress-deformation diagrams assumed in the model for: (a) steel, (b) concrete in compression, (c) concrete in tension, and (d) CARDIFRC® in tension. The constant A depends on the aspect ratio and volume fraction of fibre and the fracture toughness of cementitious matrix.

The moment resistance of a section retrofitted by CARDIFRC® can be calculated based on the distribution of stresses caused by bending. To determine the strain distribution along the height of the section the following assumptions are made:

- Plane sections remain plane after bending. In other words, the distribution of strain through the full height of the beam is linear (Bernoulli hypothesis);
- The bond between the retrofit strips and the original beam is perfect and there is no sliding at the interface (deformation compatibility). This assumption was fully validated by tests.

The stress distribution in concrete and CARDIFRC® strips cannot be assessed directly from the value of strain after cracking, as the constitutive relations are expressed in terms of stress-crack opening rather than stress-strain. Using the following assumptions, the evaluation of the crack opening from the strain distribution becomes possible:

- The crack opening at the tension retrofit strip (w) is the product of the strain at this level (ε_f), and an effective length of retrofit strip (L_{eff}),
- The dominant flexural crack tip is located at the level of the neutral axis. The faces of this crack open in a linear manner [Fig. 3(a)].

Table 4. Parameters for modelling the tensile behaviour of CARDIFRC®.

	CARDIFRC® in tension					
Stage	f_t MPa	f_{tp} MPa	E GPa	G_F N/mm	w_1 mm	ε_{max} –
I&II	18	7.2	50	13	0.5	7.6×10^{-4}

Table 3. Parameters for modelling the behaviour of concrete and steel.

	Concrete										
	In compression					In tension				Steel	
Stage	f_{cm} MPa	E_c GPa	E_{c1} GPa	ε_{c1} –	ε_{cu} –	f_{ctm} MPa	G_F N/mm	w_1 mm	w_c mm	f_y MPa	E_s GPa
I	45	35.6	20.7	0.0022	0.00315	4.0	0.0725	0.017	0.128	544	205
II	47	36.1	21.4	0.0022	0.00330	3.5	0.0675	0.015	0.128	500	203

In fact, the strain over the effective length of retrofit strip (L_{eff}) is released in the form of a local crack. To determine L_{eff}, the length of strain-free part of the retrofit strip should be calculated. If the tensile stress carried by the cracked strips is ignored in comparison with the tensile stress transferred by the reinforcement, the shear stress at the interface is dependent on the shear stress applied by the reinforcement, as shown in Fig. 3(b). Assuming the shear stress at the level of reinforcement is distributed at 45°, a length of retrofit strip (L_{eff}) is stress-free and consequently strain-free. The deformation of this length of strip is localised in the crack opening. Therefore, to calculate the crack opening of the tension retrofit strip, the strain at this level (ε_f) can be multiplied by this effective length (L_{eff}) i.e. twice the distance between the reinforcement and the tension strip. It can be seen that by using this method the stress distribution in the repair material can also be worked out from the strain distribution. Due to the fact that the crack opening displacements (i.e. crack widths) of the test beams were too small for accurate measurement, the crack openings calculated from the above method could not be compared directly with measured values. However, the consequences of the above assumptions to the calculation of the moment resistance and the load deflection response of the beams will become clear when we compare the model and test results later in this paper.

To evaluate the moment resistance of the control beams and the beams retrofitted with different configurations of CARDIFRC® strip, which fail in flexure, a programme was written. Figure 4 illustrates the flowchart of this programme. First, a strain in the top concrete fibre and a neutral axis depth are assumed. Then, the linear strain distribution along the height of the beam is defined in terms of these assumed values.

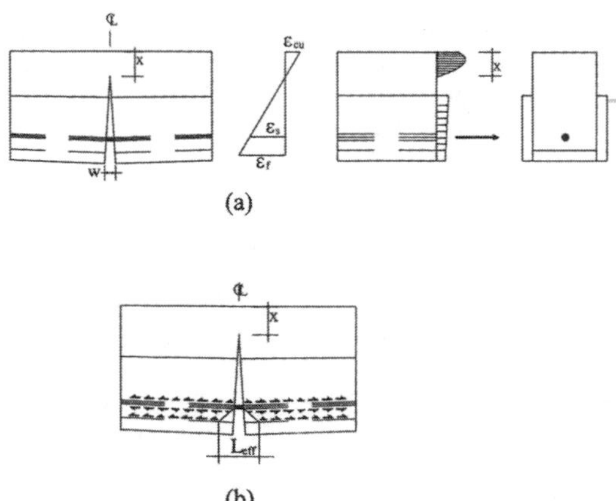

Figure 3. (a) Modelling of the flexural crack in the middle of the beam strengthened with three strips, (b) effective length of strip for calculation of crack opening.

The depth between the top compression fibre and the neutral axis is divided into 10 sections. The average strain over each section is calculated assuming piecewise linear fibre strain. The compressive stress can now be found using the concrete stress-strain relation. Multiplying this by the area of the section gives the compressive force. A similar calculation is made to determine the tensile forces in the concrete in tension, in the retrofit strips and in the reinforcing steel. As mentioned before, to determine the tensile stress of cracked concrete and retrofit strips the crack opening should also be calculated.

Having calculated all the forces the neutral axis is adjusted until the sum of the compressive forces equals the sum of the tensile forces. When this is achieved, the moment is determined by summing the compressive and tensile forces times their moment arms about a single point. In addition, the curvature of the beam can also be easily worked out using the strain in the top concrete fibre and the neutral axis depth. This process is repeated for different assumed strains in the top fibre of concrete. The maximum moment resistance of the section occurs when either the moment reduces for an increase in the top fibre strain, or the top fibre concrete strain exceeds the ultimate strain of concrete in compression (ε_{cu}).

Due to the fact that all the beams tested are statically determinate, their bending moment diagrams at any stage of loading are uniquely defined. This information can be combined with the moment-curvature diagram of sections to produce the curvature diagram of the beam at different load levels by dividing the beam span into a number of intervals. The moment-area method is then used to compute the slope and the deflection of any point of the beam.

3.1 *Model results*

Figure 5 compares the maximum moment resistance of the beams predicted by the analytical model with the three- and four-point bend test results of stage I, respectively. It should be emphasised that the present model is only applicable to beams which fail in flexure.

Figure 6(a) compares the typical load-deflection response of the beams retrofitted with one continuous and four rectangular 20 mm strips [Fig. 1(c)] with the model predictions. The same comparison is made in Fig. 6(b) for the beams retrofitted with three 16 mm continuous strips [Fig. 1(d)]. It can be clearly seen that the model predictions are in very good agreement with the test results, especially before the attainment of the maximum load.

Figure 7 compares the maximum load carrying capacity of the stage II beams (with shear reinforcement) with the model predictions. It can be seen that the model predictions are again in good agreement with the test results. Of course, the failure load of

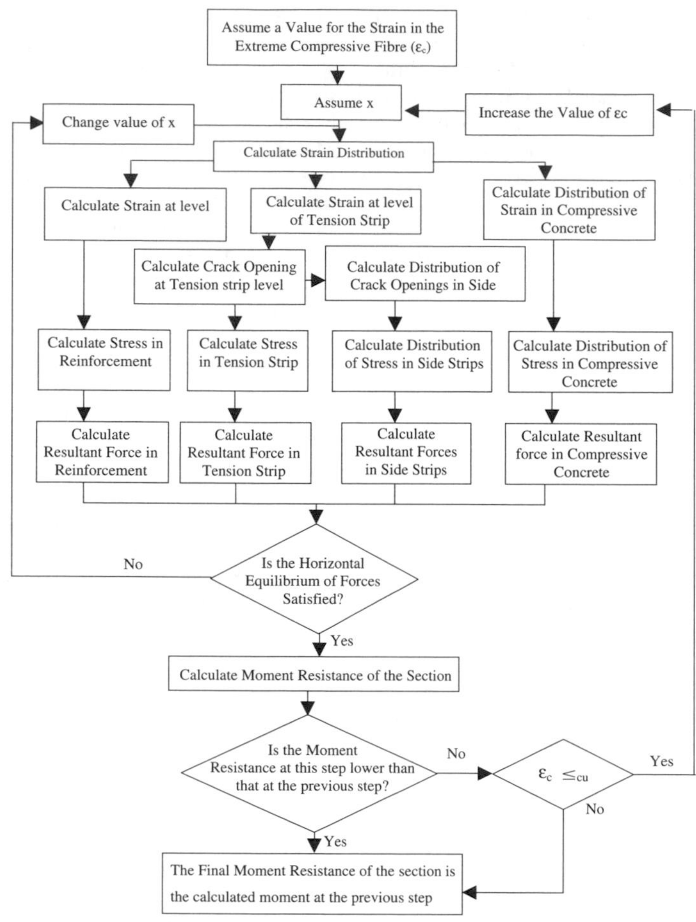

Figure 4. Flowchart of the programme for calculating the moment resistance of the beams.

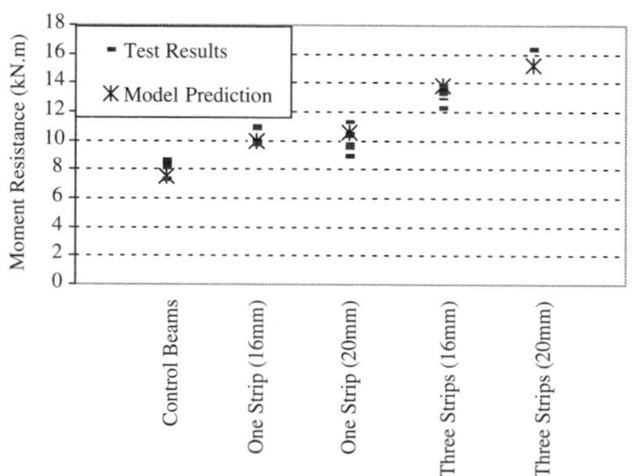

Figure 5. Comparison of the moment resistance of the stage I beams with the predictions of the first analytical model.

some beams retrofitted with 20 mm strips is lower than that predicted by the model. This is likely to be the result of the poor quality of some 20 mm strips used for retrofitting the beams.

4 SECOND ANALYTICAL MODEL

Fracture mechanics studies the response and failure of structures as a consequence of crack initiation and propagation. To model the response of a structure based on this theory, it is therefore necessary to have a good knowledge of the details of the formation and growth of the cracks in the structure.

Let us review the sequence of the crack growth in the beams retrofitted with different configurations of retrofit strip. Of course, to model the flexural failure of the beams we will only concentrate on the behaviour of the beams, which failed in flexure and ignore those

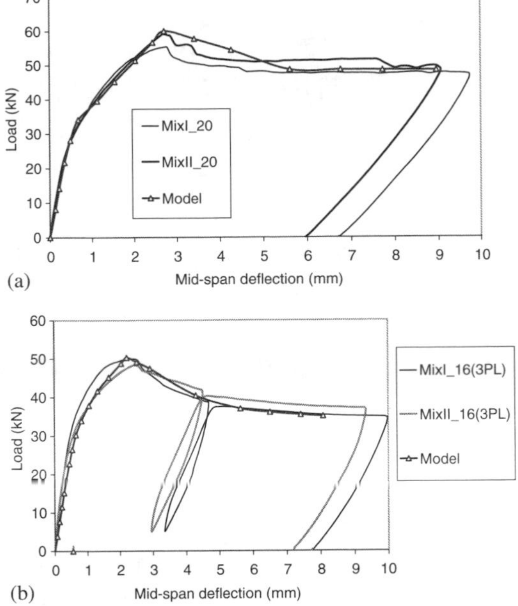

Figure 6. Comparison of the typical load-deflection response of the retrofitted beams with that predicted by the model (stage I): (a) beams retrofitted with one continuous and four rectangular 20 mm strips under four-point bending, and (b) beams retrofitted with three continuous 16 mm strips under three-point bending.

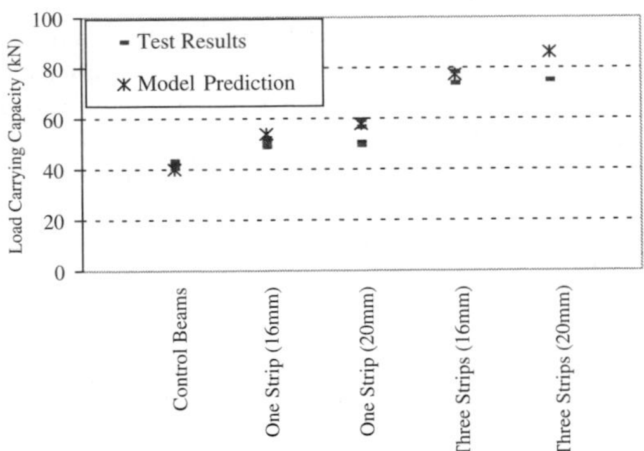

Figure 7. Comparison of the load carrying capacity of the stage II beams with that predicted by the first analytical model.

(very) few retrofitted beams that failed in shear or in a combination of shear and flexural modes.

As mentioned earlier, all the beams were pre-cracked prior to the retrofitting. During the process of pre-loading, a few flexural cracks appeared in the beams and opened up. These cracks closed (became invisible to the naked eye), when the load was removed from the beams [Fig. 8a].

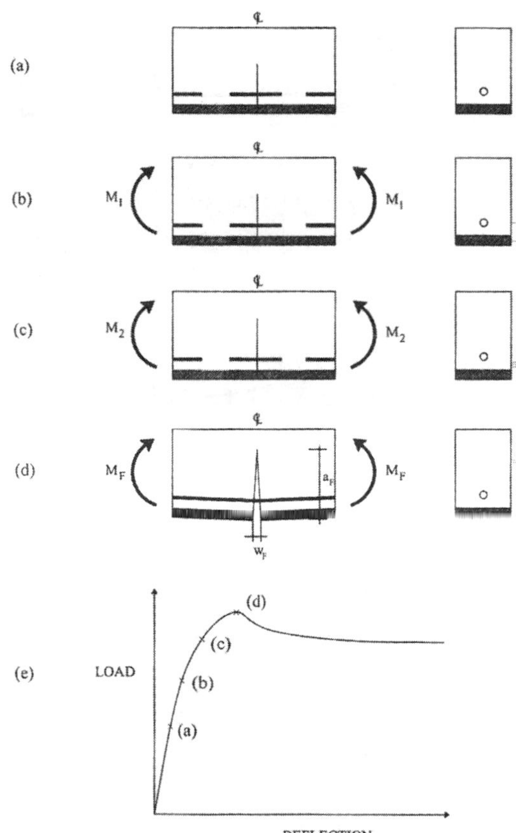

Figure 8. Sequence of crack growth in the middle part of the beams retrofitted with one strip: (a) existing crack in the pre-damaged beam, (b) reopening of the existing crack in the pre-damaged beam, (c) propagation and penetration of the dominant crack into the tension strip, (d) opening of the dominant crack, and (e) load levels corresponding to the observed crack growth sequence.

After the beams were retrofitted, they were loaded again, but this time to failure. As the applied load was increased, the existing hairline cracks, induced by pre-loading, became visible again [Fig. 8b]. One of these cracks in the pre-damaged beam opened and with a further increase in load this crack opened further and propagated into the tension strip [Fig. 8c]. In none of the test beams was there any evidence of this crack branching into the interface. The pre-existing crack and the new crack, formed in the tension strip, continued to open until the maximum load was reached [Fig. 8d]. At this stage the crack could be seen across the entire width of the tension strip. However, no attempt was made to measure the crack mouth opening displacement at the maximum load because it was too small for accurate measurement. The approximate

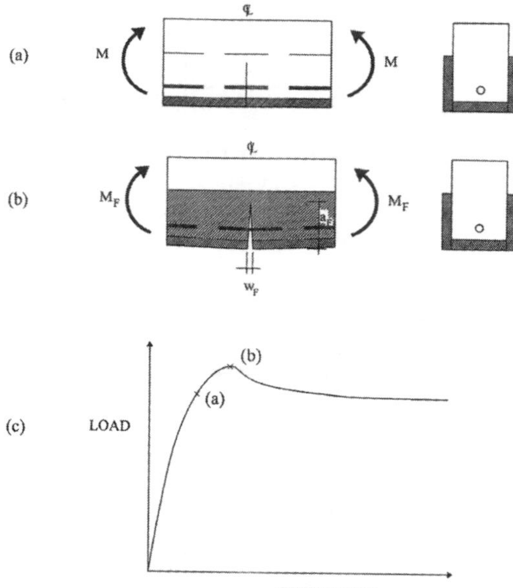

Figure 9. Cracking in the middle part of the beams retrofitted with three strips. (a) when the pre-existing crack penetrated into the tension strip. For clarity the side strip is not shown. Note that the stages (a) and (b) of Figure 8 are obscured by the side strips; (b) at maximum load, and (c) load levels corresponding to stages (a) and (b).

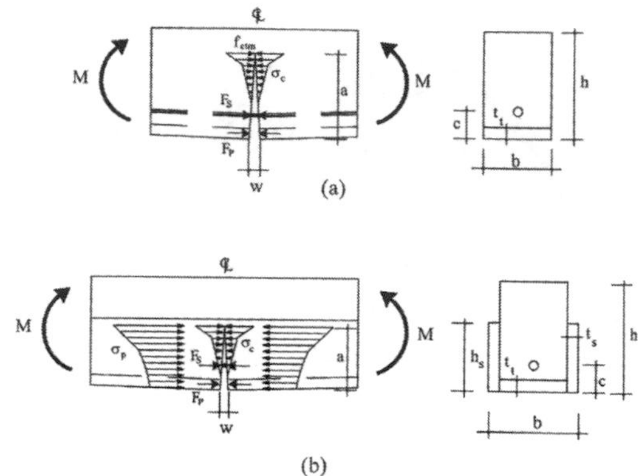

Figure 10. Free body diagram of the dominant flexural crack in retrofitted beams. Beams retrofitted with (a) one tension strip, and (b) one tension and two side strips.

load levels corresponding to the above sequence of crack growth are shown schematically in Fig. 8e.

In the beams retrofitted with a tension strip and side strips the load increased further until the crack in the tension strip extended along the side strips. However, the crack was still within the side strips when the maximum load was reached [Fig. 9b]. It extended beyond the side strips and into the concrete beam after the attainment of the maximum load. The approximate load levels are shown schematically in Fig. 9c.

To model the flexural failure of the retrofitted beams, let us consider the free body diagram of a part of the retrofitted beam containing the dominant flexural crack and study the effect of different loads on it [Fig. 10]. It can be seen that in addition to the moment due to the applied load, there are three loads which appear in the free body diagram. The first load is due to the bridging force across the crack faces from the reinforcing steel, the second is from the post-peak tension softening response of concrete, and the third is from the bridging stresses in the retrofit strips. It is evident that only the moment tends to open the crack and the remaining loads tend to close it.

To simplify the computations, it is assumed that the crack profile is always linear. Therefore, a crack can be specified by its depth a and the opening w at its mouth. To calculate the moment resistance of the retrofitted beams M_F, the condition of the dominant crack at the maximum load is crucial. In fact, M_F cannot be found, unless the crack depth (a_F) and the crack mouth opening at the maximum load (w_F) are computed.

The bridging force from steel F_S is the most significant closing force on the crack faces. There is no definite relationship between the geometrical properties of the crack (w and a) and the bridging force in steel. However, there are some empirical relations, which will be discussed later.

As shown in Fig. 10, the residual tensile stress in concrete σ_c, acts as a closure pressure on the dominant crack faces. This stress varies from f_{ctm} (the tensile strength of concrete) at the crack tip, to nothing where the crack opening equals to the critical crack opening of concrete w_c [see Fig. 2c]. If the crack mouth opening w and the crack depth a are given [Fig. 9], the distribution of this bridging stress can be easily worked out from the tension softening diagram of concrete [Fig. 2c].

The effect of the retrofit strips on the dominant crack is very similar to that of the concrete. CARDIFRC® with its tension softening behaviour produces a closure pressure on the crack faces, which again is a function of the crack opening [Fig. 2d].

Next we relate the applied moment M to the crack depth a and crack mouth opening w.

As shown in Fig. 10, closure pressures exerted by steel, concrete and retrofit strips counteract the opening action of the applied moment. As the stress at the crack tip is finite, the net stress intensity factor at the crack tip must vanish. In fact, this requires that the crack faces close smoothly near the tip. This would seem to violate the assumption of linear crack opening that we have made. However, we shall proceed in the spirit of Barenblatt cohesive crack model (Barenblatt 1959) and assume that the zone c near the

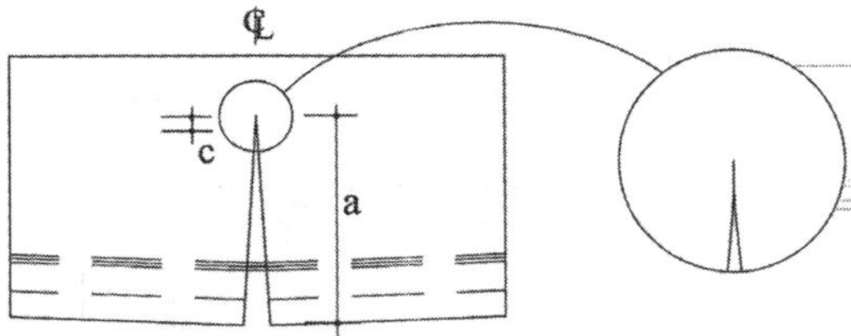

Figure 11. Smooth closure of crack faces.

tip of the crack where the faces close smoothly is very much smaller than the total length of crack $c << a$, as shown in Fig. 11. We shall consider later the implications of this assumption. The net K_I at the crack tip is obtained by superposing the stress intensity factors produced at the crack tip by the applied moment (K_{IM}), and the closure forces exerted by steel (K_{IS}), concrete (K_{Iconc}), tension retrofit strip ($K_{I(t-strip)}$), and side retrofit strips ($K_{I(s-strip)}$) (if they are used). The condition of finite stress at crack tip, i.e. $K_I = 0$ is therefore

$$K_{IM} - K_{IS} - K_{Iconc} - K_{I(t-strip)} - K_{I(s-strip)} = 0 \qquad (1)$$

The negative sign in front of the last four terms indicates that the stress intensity factor produced by the applied moment is reduced by the various closure forces. K_{IM} can be calculated from the formula given by Guinea et al. (1998) and the remaining stress intensity factors can be evaluated directly or with integration from the relation proposed by Tada et al. (1985). For ease of reference, the expressions for the stress intensity factors appearing in (1) are given in Appendix I.

In addition to the condition of smooth closure of crack faces at its tip, we must consider the compatibility of crack opening displacement of a retrofitted beam (Leung, 1998). The crack opening displacement can again be written as the vectorial sum of the contributions from the applied bending moment and the closure forces exerted by steel, concrete and retrofit strips. The compatibility condition of the crack opening need be satisfied only at the level of the steel reinforcement, because of the assumed known (i.e. linear) variation along the length of the crack:

$$(w_S)_M - (w_S)_S - (w_S)_{conc} - (w_S)_{t-strip} - (w_S)_{s-strip} = w_S \qquad (2)$$

where $(w_S)_i$'s are the crack opening displacement at the level of the steel bar produced by the applied bending moment and the closure forces exerted by steel, concrete, tension strip, and side strips, respectively. Note that the crack opening w_S at the level of the reinforcement is not known, but is to be determined.

Each term in the left-hand side of (2) can be expressed in terms of the corresponding compliance coefficients. For instance, the crack opening at the level of the steel bar produced by the applied moment is

$$(w_S)_M = \lambda_{SM} M \qquad (3)$$

where λ_{SM} (the compliance coefficient) is the crack opening at the level of steel when a unit bending moment is applied to the crack. The compliance coefficients can be computed from energy principles and Clapeyron's theorem (Bosco and Carpinteri, 1992). These are listed in Appendix II.

The relationship between the geometrical properties of a crack and the bridging force exerted by steel crossing this is quite complicated. Although a number of approximate, semi-theoretical and empirical formulae have been proposed to relate these parameters, none of them has been found to be consistent with the test results. Three methods for the calculation of the bridging force exerted by steel are presented below.

Bosco and Carpinteri (1992) assumed that the displacement discontinuity in a cracked cross section at the level of reinforcement is zero up to the moment of yielding or slippage of the reinforcement

$$F_S = A_S f_y \qquad (4)$$

where f_y is the yield stress of the reinforcing bar and A_S the area of steel. The steel force is therefore uncoupled from the crack opening displacement at the level of steel.

In another study, Kaar and Mattock (1963) expressed the tensile stress in steel as a function of the crack width w_S at the level of deformed bar reinforcement

$$f_S = \frac{F_S}{A_S} = \frac{11876.5 w_S}{A^{1/4}} \qquad (5)$$

where A is the area of concrete surrounding each bar ($A = A_e/n$ where n is the number of bars). In eqn (5), f_S, w_S and A are in MPa, mm and mm^2, respectively.

Lange-Kornbak and Karihaloo (1999) conducted an experimental programme and compared the test results with approximate nonlinear fracture mechanical prediction of the ultimate capacity of three-point bend, singly-reinforced concrete beams. They found that Kaar's formula (5) overestimates the crack opening by the factor of 3.5 to 4. Due to the relative similarity between the properties of the beam in their study and those in present experimental programme, Kaar's formula was improved by calibrating the results against the test data with the least variation. In the present sets of tests the beams retrofitted with one 16 mm tension strip and tested in four-point bending showed the least variation in the moment resistance (see, Fig. 14). This calibration suggested the following relationship between the tensile stress in steel and the crack width at the level of deformed bar

$$f_S = \frac{F_S}{A_S} = \frac{4.8 \times 11876.5 w_S}{A^{1/4}} \tag{6}$$

It will be noticed that the right hand side of eqn (6) has increased by a factor of 4.8 in comparison with the original Kaar's formula. As a result, the bridging force exerted by steel reinforcement in this case is higher than that based on Kaar's formula (5) and it can therefore be expected that this modification increases the moment resistance of the beams in comparison with that based on Kaar's formula.

As shown in the previous sections, from the fracture mechanics point of view, two equations should be simultaneously satisfied for a flexural crack; the equation of smooth closure of the crack faces (1) and the equation of the crack opening compatibility (2). The three unknown parameters in these two equations are the crack depth a, the crack mouth opening w and the applied moment M which were obtained as follows in an enumerative optimisation procedure.

- First, the equation of smooth closure of the crack faces (1) was solved for all the possible crack depths and crack mouth openings, and the corresponding moment M was calculated.
- Next, the equation of the crack opening compatibility (2) was solved for the same ranges of values of a and w.
- The results of the previous two stages were compared and the intersection of these two sets identified. This set satisfies both the smooth closure condition (1) and the crack opening compatibility equation (2).
- Among the set of values identified in the previous stage, the maximum moment was selected. In fact, this moment M_F is the moment resistance of the beam, and the corresponding crack length a and crack mouth opening w describe the condition of the dominant flexural crack at the maximum moment.

A flowchart of the above procedure is given in Fig. 12.

4.1 *Model results*

Figures 12 and 14 compare the moment resistance of the beams predicted by the model with the test results. It can be seen that the trend of the model results for both stage I and stage II beams is in agreement with the test results. However, as expected, the model based on Carpinteri's assumption overestimates the moment resistance of the beams and that based on Kaar's formula (5) underestimates it significantly. It is also seen that the model results based on the modified Kaar formula (6) are in very good agreement with the test

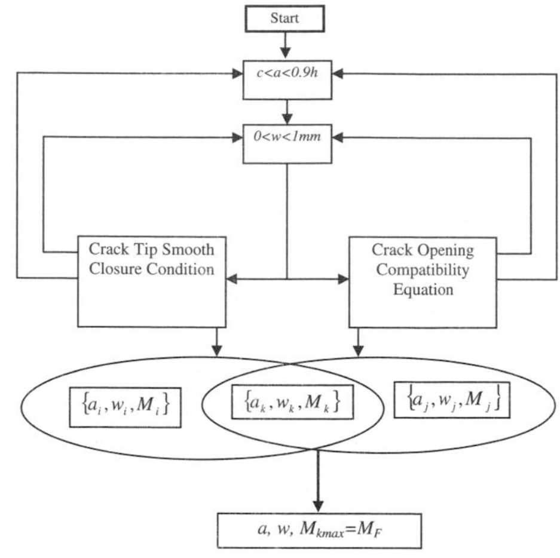

Figure 12. Flowchart of the programme for calculating the moment resistance of the beams.

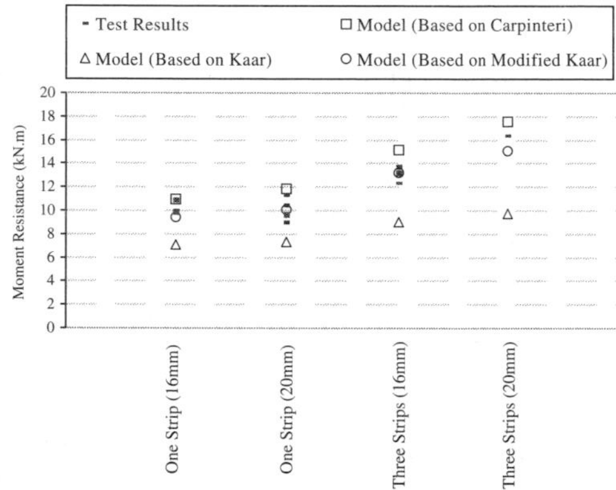

Figure 13. Comparison of the moment resistance of the stage I beams with the predictions of the second model.

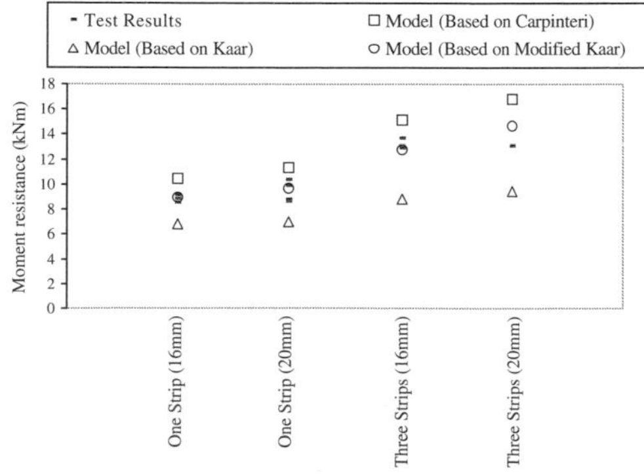

Figure 14. Comparison of the moment resistance of the stage II beams with the predictions of the second model.

results. It should be noted that only for the single beam retrofitted with three 20 mm strips is the prediction lower than the test result. This is not surprising when one realises that this beam failed in shear. As mentioned before, the model developed above is based on the behaviour of a flexural crack; it is not capable of predicting the load at which the beam fails in shear.

5 ASSUMPTION OF LINEAR CRACK OPENING WITHOUT SMOOTH CLOSURE OF CRACK TIP

The assumption of smooth closure of crack faces at its tip means that the stress at the crack tip is finite and therefore the net stress intensity factor at the crack tip must vanish. However, the stress is finite when there is no structural cracking and the material is in a linear elastic state. In concrete, there is generally a lot of microcracking in the cement paste matrix as a result of which the pre-peak response of concrete is non-linear. The small non-linear zone of microcracking encloses the large fracture process zone, as shown in Fig. 15. Thus, the stress at the crack tip need not be finite, or in other words can well exceed the tensile strength of the paste and lead to microcracking.

Therefore, eqn (1) has to be replaced with

$$K_{IM} - K_{IS} - K_{Iconc} - K_{I(t-strip)} - K_{I(s-strip)} = K_{Ic} \qquad (7)$$

where K_{Ic} is fracture toughness of the hardened cement paste (hcp). In fact, the stress intensity factor at the crack tip is large enough to reach the fracture toughness of the paste K_{Ic} and therefore create some micro-cracks in the cement paste. Note that the description of hcp by just one fracture parameter is justified because it behaves essentially as an ideal brittle material because of its homogeneous structure.

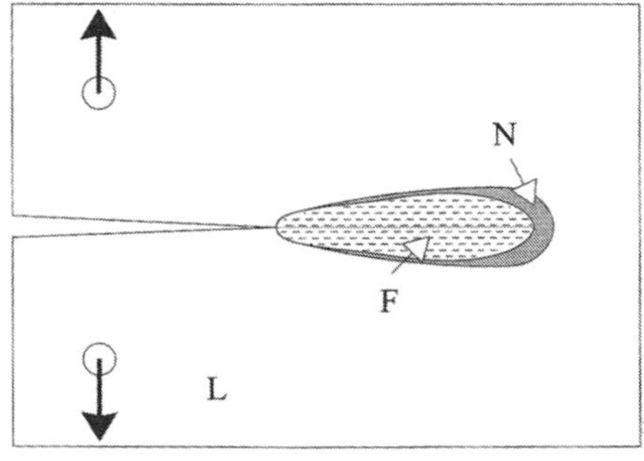

Figure 15. Distinguishing features of fracture in a quasi-brittle material. L refers to linear elastic region, N to nonlinear zone, and F to fracture process zone (After ACI Report 446.1R.91).

Assuming K_{Ic} to be 0.45 MPa$\sqrt{m}$ (Ohigishi et al., 1986), the crack mouth opening w, crack depth a and the moment resistance M_F of the retrofitted beams were again worked out. It was observed that the moment resistances predicted on the basis of the two assumptions (1) and (7) are very close to one another. The maximum difference between the moment predictions for the retrofitted beams was 1.7%. The closeness of the results is due to the smallness of the fracture toughness K_{Ic} of the hardened cement paste. In fact, as Fig. 10 shows the size of the region in which the concrete behaves nonlinearly is very small in comparison with the size of its fracture process zone. Therefore, the inclusion of this region in the calculations i.e. the assumption of non-zero stress intensity factor at the end of the fracture process zone, only marginally affects the model results.

6 CONCLUSIONS

The new technique using the CARDIFRC® strip bonding system is a promising method for improving the flexural and shear behaviour, as well as the serviceability of damaged concrete beams. It does not suffer from the drawbacks of the existing techniques, which are primarily a result of the mis-match in the properties between the concrete and the repair material.

The mechanical properties of CARDIFRC® mixes I and II are very similar, therefore there is no real difference in the behaviour of the beams retrofitted with either of these mixes.

The moment resistance and load-deflection response of the beams retrofitted using this technique can be predicted analytically by either of the two models presented here, providing that the strain hardening and tension softening response of concrete and CARDIFRC® are properly taken into account.

The technique described in this paper may be used when there is a need to improve the durability of existing concrete structures, as CARDIFRC® mixes are very durable because of their highly dense microstructure. Research is currently being undertaken to study the fatigue, shrinkage and creep properties of CARDIFRC® and the performance of concrete structures retrofitted with CARDIFRC® under dynamic, thermal and hygral loads.

APPENDIX 1

The stress intensity factor produced at the crack tip by the moment M is (Guinea et al., 1998)

$$K_{IM} = \frac{M}{h^{3/2}b} Y_M(\xi) \qquad (8)$$

where ξ is a/h and $Y_M(\xi)$ is

$$Y_M(\xi) = \frac{6\xi^{1/2}(1.99 + 0.83\xi - 0.31\xi^2 + 0.14\xi^3)}{(1-\xi)^{3/2}(1+3\xi)} \tag{9}$$

The stress intensity factor produced by the concentrated forces F_S is (Tada et al., 1985)

$$K_{IS} = \frac{2F_S}{b\sqrt{\pi a}} F(c/a,\xi) \tag{10}$$

where the function $F(c/a,\xi)$ is given by

$$F(c/a,\xi) = \frac{3.52(1-c/a)}{(1-\xi)^{3/2}} - \frac{4.35-5.28\,c/a}{(1-\xi)^{1/2}} + \left[\frac{1.30-0.30(c/a)^{3/2}}{(1-(c/a)^2)^{1/2}} + 0.83 - 1.76\,c/a\right] \times [1-(1-c/a)\xi] \tag{11}$$

To work out K_{Iconc}, the residual stress in concrete is discretized into infinitely many concentrated forces. Each force is applied over small crack segments of length dy at a distance y from the bottom of the crack (see, Fig. 16). The magnitude of each force is

$$dF_c = \sigma_c(w_y)b\,dy \tag{12}$$

where

$$w_y = w(1-y/a) \tag{13}$$

In the above equations, w_y is the crack opening at the level of the point load dF_c. Now, using (10) the stress intensity factor produced by this point load can be calculated

$$dK_{Iconc} = \frac{2dF_c}{b\sqrt{\pi a}} F(y/a,\xi). \tag{14}$$

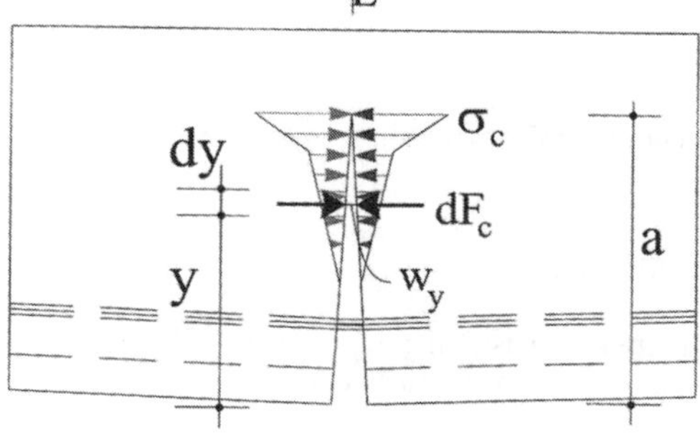

Figure 16. Towards the discretisation of the bridging stresses in concrete.

Finally, this relation is integrated over the length of the crack in concrete

$$K_{Iconc} = \int_{t_t}^{a} \frac{2\sigma_c(w_y)}{\sqrt{\pi a}} F(y/a,\xi)dy \tag{15}$$

To calculate $K_{I(t-strip)}$ the magnitude of the closure force F_P exerted by the tension retrofit strip is calculated using the corresponding crack opening w_P at the level of F_P

$$F_P = \sigma_P(w_P)bt_t \tag{16}$$

where

$$w_P = w(1-t_t/2a) \tag{17}$$

Substitution of (16) into (10) gives

$$K_{I(t-strip)} = \frac{2t_t\sigma_P(w_P)}{\sqrt{\pi a}} F(t_t/2a,\xi) \tag{18}$$

Finally, $K_{I(s-strip)}$ can be determined in the same manner as K_{Iconc}. Dividing each of the side strips into infinitesimal elements of height dy, the closure force on each element of the two side strips is

$$dF_{SP} = 2\sigma_P(w_y)t_s dy \tag{19}$$

Substitution of (19) into (10) gives the stress intensity factor produced by this point load at the crack tip

$$dK_{I(s-strip)} = \frac{4\sigma_P(w_y)t_s}{b\sqrt{\pi a}} F(y/a,\xi)dy \tag{20}$$

To calculate $K_{I(s-strip)}$, this relation is integrated as follows

$$K_{I(s-strip)} = \int_0^{\alpha} \frac{4\sigma_P(w_y)t_s}{b\sqrt{\pi a}} F(y/a,\xi)dy \tag{21}$$

where α is $\min\{a, h_S\}$ (i.e. the smaller of a or h_S).

APPENDIX II

The compliance coefficient due to a unit moment (λ_{SM}) can be rewritten as (Bosco and Carpinteri, 1992)

$$\lambda_{SM} = \frac{2}{hbE}\int_{c/h}^{\xi} Y_M(\xi\xi)Y_P(c/h,\xi)d\xi \tag{22}$$

where

$$Y_P(c/h,\xi) = F(c/a,\xi)\frac{2}{\sqrt{\pi\xi}} \tag{23}$$

In (22), E is the Young modulus of concrete, and $F(c/a, \xi)$ is given by (11).

Similarly, the compliance coefficient due to a unit steel force is given by (Bosco and Carpinteri, 1992)

$$\lambda_{SS} = \frac{2}{bE} \int_{c/h}^{\xi} Y_P^2(c/h,\xi)d\xi \quad (24)$$

The crack opening $(w_S)_S$ is then obtained as

$$(w_S)_S = \lambda_{SS} F_S \quad (25)$$

To work out the crack opening at the level of the steel bar produced by tension softening of concrete $(w_S)_{conc}$, the continuous closure stresses exerted by concrete are again discretized into infinitely many concentrated loads dF_c, each acting on an infinitesimal crack length dy (see, Fig. 15). The crack opening $(w_S)_{dF_c}$ produced by one of these concentrated loads is calculated by using the compliance method. The corresponding compliance coefficient λ_{S-dF_c} can again be computed from energy principles (Bosco and Carpinteri, 1992) considering the point load dF_c acting simultaneously with the steel force F_S

$$\lambda_{S-dF_c} = \frac{8}{E\pi b} \int_{\beta}^{a} \frac{F(c/a,\xi)F(y/a,\xi)}{a} d\xi \quad (26)$$

where β is $\max\{y, c\}$ (i.e. the larger of y and c).

Multiplying this compliance factor by dF_c from (12) and changing the variable of integration to ξ gives

$$(w_S)_{dF_c} = \frac{8\sigma_c(w_y)dy}{E\pi} \times \int_{\beta/h}^{\xi} \frac{F(c/a,\xi)F(y/a,\xi)}{\xi} d\xi \quad (27)$$

To calculate the crack opening at the level of reinforcement due to the tension softening of concrete, (27) is integrated over the length of the crack in concrete

$$(w_S)_{conc} = \int_{t}^{a} \left[\frac{8\sigma_c(w_y)}{E\pi} \int_{\beta/h}^{\xi} \frac{F(c/a,\xi)F(y/a,\xi)}{\xi} d\xi \right] dy \quad (28)$$

Using the same procedure as above, the compliance coefficient due to a unit force in the tension retrofit strip $\lambda_{S-(t-strip)}$ can be written as

$$\lambda_{S-(t-strip)} = \frac{8}{E\pi\pi} \int_{c}^{a} \frac{F(c/a,\xi)F(t_t/2a,\xi)}{a} d\xi \quad (29)$$

Multiplying this compliance factor by F_P from (16) and changing the variable of integration to ξ gives

$$(w_S)_{t-strip} = \frac{8\sigma_P(w_P)t_t}{E\pi} \times \int_{c/h}^{\xi} \frac{F(c/a,\xi)F(t_t/2a,\xi)}{\xi} d\xi \quad (30)$$

To calculate $(w_S)_{s-strip}$, exactly the same procedure as that adopted for concrete is followed. Equation (26) in this case becomes

$$\lambda_{S-dF_{SP}} = \frac{8}{E\pi b} \int_{\beta}^{a} \frac{F(c/a,\xi)F(y/a,\xi)}{a} d\xi \quad (31)$$

Multiplying this compliance coefficient by the magnitude of the point load dF_{SP} from (19) gives the crack opening produced by this point load

$$(w_S)_{dF_{SP}} = \frac{16\sigma_P(w_y)t_S dy}{E\pi\pi} \times \int_{\beta/h}^{\xi} \frac{F(c/a,\xi)F(y/a,\xi)}{\xi} d\xi \quad (32)$$

Again, to compute the crack opening at the level of steel due to the forces exerted by the side strips, (32) is integrated over the length of the cracked side strips

$$(w_S)_{s-strip} = \int_{0}^{a} \left[\frac{16\sigma_P(w_y)t_S}{E\pi b} \int_{\beta/h}^{\xi} \frac{F(c/a,\xi)F(y/a,\xi)}{\xi} d\xi \right] dy \quad (33)$$

ACKNOWLEDGEMENT

This work is supported by the UK EPSRC grant GR/R11339.

REFERENCES

ACI Committee 544 (1988). Measurement of Properties of Fibre Reinforced Concrete, *ACI Structural Journal*, 85(6), 583–593.

Ahmed, O. and Gemert, D. V. (1999). Behaviour of R.C. beams strengthened in bending by CFRP laminates. *Proceedings of Structural faults & repair – 99:8th International Conference*, London, CD-Rom.

Alaee, F. J., Benson, S. D. P. and Karihaloo, B. L. (2001). Strengthening of RC Beams with High-Performance Cementitious Composites, *Proceedings International Conference on Civil Engineering*, Technical Committee ICCE-2001 (eds), Interline Publishing, Bangalore, India, 1–13.

Alaee, F. J., Benson, S. D. P. and Karihaloo, B. L. (2002). High-Performance Cementitious Composites for Retrofitting, *International Journal of Materials and Product Technology*, 17(1/2), 17–31.

Alaee, F. J. and Karihaloo B. L. (2002a). Retrofitting of RC beams with CARDIFRC®, *ASCE Journal of Composites for Construction*, in press.

Alaee, F. J. and Karihaloo, B. L. (2002b). Fracture model for flexural failue of beams retrofitted with CARDIFRC®, *ASCE Journal of Enigneering mechanics* (submitted).

Barenblatt, G. I. (1959). On Equilibrium Cracks Forming During Brittle Fracture (in Russian) *Prikladnaya Matematika I Mekhanika (P M M)*, 23, 434–444.

Bosco, C. and Carpinteri, A. (1992). Fracture Mechanics Evaluation of Minimum Reinforcement in Concrete Structures. In *Application of Fracture Mechanics to Reinforced Concrete* (Proc Int Workshop, Turin, Italy 1990), Carpinteri, A. (ed.), Elsevier, London, 347–377.

Büyüköztürk, O. and Hearing, B. (1998). Failure behaviour of precracked concrete beams with FRP. *Journal of Composites for Construction*, ASCE, 2(3), 138–144.

CEB-FIP Model Code (1993). Lausanne, Switzerland.

El-Refaie, S. A., Ashour, A. F. and Garrity, S. W. (1999). Flexural capacity of R.C. beams strengthened with external plates. *Proceedings of Structural faults & repair – 99:8th International Conference*, London, CD-Rom.

Fanning, P. and Kelly, O. (1999). Shear strengthening of R.C. beams: An experimental study using CFRP plates. *Proceedings of Structural faults & repair – 99:8th International Conference*, London, CD-Rom.

Guinea, G. V., Pastor, J. Y., Planas, J. and Elices, M. (1998). Stress Intensity Factor, Compliance and CMOD for a general three-point-bend beam. *International Journal of Fracture*, 89, 103–116.

Hussain, M., Sharif, A., Basenbul, I. A., Baluch, M. H. and Al-Sulaimani, G. J. (1995). Flexural behaviour of precracked reinforced concrete beams strengthened externally by steel plates. *ACI Structural Journal*, 92(1), 14–22.

Jones, R., Swamy, R. N. and Charif, A. (1988). Plate Separation and anchorage of reinforced concrete beams strengthened by epoxy-bonded steel plates. *The Structural Engineer*, 66(5), 85–94.

Kaar, P. H. and Mattock, A. H. (1963). High Strength Bar as Concrete Reinforcement, Part 4. Control of Cracking. *Journal of the PCA Research and Development Laboratories*, 7(1), 15–38.

Karihaloo, B. L., Alaee, F. J. and Benson, S. D. P. (2002). A New Technique for Retrofitting Damaged Concrete Structures, *Proceedings of the Institution of Civil Engineers., Buildings & Structures*, 152(4), 309–318.

Karihaloo, B. L., Benson, S. D. P., Didiuk, P. M., Fraser, S. A., Hamill, N. and Jenkins, T. A. (2000). Retrofitting damaged RC beams with high-performance fibre-reinforced concrete. *Proceedings of Concrete Communication Conference*, British Cement Association, Birmingham, 153–164.

Lange-Kornbak, D. and Karihaloo, B. L. (1999). Fracture Mechanical Prediction of Transitional Failure and Strength of Singly-Reinforced Beams. In *Minimum Reinforcement in Concrete Members*, ESIS Publication 24, Carpinteri, A. (ed.), Elsevier, London, 31–42.

Leung, C. K. Y. (1998). Delamination Failure in Concrete Beams Retrofitted with a Bonded Plate. In *Fracture Mechanics of Concrete Structures Volume 3*, Mihashi, H., Rokugo, K. (eds), AEDIFICATIO Publishers, Freiburg, Germany, 1783–1792.

Nanni, A. (1995). Concrete repair with externally bonded FRP reinforcement: examples from Japan. *Concrete International*, 97, 22–26.

Ohgishi, S., Ono, H., Takatsu, M. and Tanahashi, I. (1986) Influence of Test Conditions on Fracture Toughness of Cement Paste and Mortar, In *Fracture Toughness and Fracture Energy of Concrete* (Proc. Int. Conf., Lausanne, Switzerland 1985) Wittmann, F. H. (ed.), Elsevier, Amsterdam, 281–290.

Tada, H., Paris, P. and Irwin, G. (1985). The Stress Analysis of Cracks Handbook. Del Research Corporation, St. Louis, Missouri, Part 2, 16–17.

Yagi, K., Tanka, K. and Otaguro, H. (1999). Durability of carbon fibre sheet for repair and retrofitting. *Proceedings of Structural faults & repair – 99:8th International Conference*, London, CD-Rom.

Ziraba, N., Baluch, M. H., Basunbul, I. A., Sharif, A., Azad, A. K. and Al-Sulaimani, G J. (1994). Guidelines toward the design of RC beams with external plates. *ACI Structural Journal*, 91(6), 639–646.

Computational Modelling of Concrete Structures, Bićanić et al. (eds)
© 2003 Swets & Zeitlinger, Lisse, ISBN 90 5809 536 3

Benchmarking of non-linear finite element analyses for reinforced concrete deep beams

I. M. May
Department of Civil and Offshore Engineering, Heriot-Watt University, Edinburgh, UK

C. T. Morley
Department of Engineering, University of Cambridge, Cambridge, UK

Y. Y. Chen
Department of Civil and Offshore Engineering, Heriot-Watt University, Edinburgh, UK

ABSTRACT: This paper describes a series of analyses carried out using the commercial finite element package, DIANA, to analyse three different types of reinforced concrete continuous two span deep beams, S1, S2 and S3. The spans were each of 1340 mm the depth of the beam 650 mm and the width 90 mm, the difference between the three types of beam was in the reinforcement patterns. Test results obtained over a number of years for nominally identical beams were available, 6 results for S1, 2 for S2 and 5 for S3. The tests were carried out at the University of Cambridge, and the non-linear finite element analysis by a separate team at Heriot-Watt University.

The data provided for the analysis was the geometric data for the beams, the reinforcement patterns and the yield strength of the reinforcement and the cube strength and tensile strength of the concrete. Also provided was the loading arrangement used in the tests. No further communications about the experiments was permitted. The beams were analysed and on completion results compared with the experimental data. No subsequent analyses have been carried out although after inspection of the experimental data various "improvements" that could be implemented in the analyses were noted. Whilst the analyses were carried out using DIANA, many of the conclusions apply to any non-linear reinforced concrete finite element analysis.

1 INTRODUCTION

The last 40 years has seen dramatic developments in non-linear finite element analysis for reinforced concrete structures. There has been the introduction of numerous material models for concrete and reinforcement and various techniques for modeling the interaction and effects of the composite action of the concrete and reinforcement. Many of these models have an array of material constants, etc., the values of which are difficult to ascertain. There are also a number of other modeling assumptions that have to be made before the analysis of a structure can be carried out. Most of the software developed using these models has been validated against experimental results often giving excellent agreement. However, this agreement has sometimes been considered to be partly dependent on the analyst's choice of mesh, material parameters, etc., since the analyses are frequently carried out well after the structures have been tested, with the guidance of the experimental data, the parameters of the model can be manipulated to fit into the experimental results. Criticisms against such nonobjective practice are well known.

Ideally a set of benchmarks against which to test new models should be available. However, unlike many structural problems, it is not possible to devise a set of theoretical solutions to use as benchmarks. Therefore analysts must resort to experimental data. In order to validate software using such data it would be preferable to be able to carry out a blind analysis of the tests. This would give more confidence in the use of software, particularly if some indication of the possible scatter of experimental data were available. The Authors would define a blind analysis one in which only the details of the structure, as tested, the material data and the type, but not the value, of loads are available.

For the analyses described the finite element package DIANA, TNO [1996], has been applied to study a series of two-span continuous deep beam tested at the University of Cambridge.

2 EXPERIMENTAL SETUP AND DETAILS

The experimental program consisted of a series of tests of groups of up to three deep beams, namely S1, S2 and S3 tested annually at the University of Cambridge. The tests were designed to study the relationships between load carrying capacity of the deep beam and the role of the stirrups. In each series nominally identical beams were tested, only the material properties varied, Table 1.

The dimensions of the three deep beams are the same, and are shown in Fig 1, with a depth of 650 mm and two continuous spans of 1340 mm each. All the beams have a thickness of 90 mm.

All specimens have the same main reinforcement layout, which is shown in Fig 2, with two 10 mm diameter bars centred 23 mm and two 12 mm diameter bars centred 35 mm from the top, four 8 mm diameter bars in the web, and two 12 mm diameter bars centred 25 mm from the bottom. Three additional 6 mm diameter vertical stirrups were centred on each support point to make a cage. It should be noted that:

In S1 no additional vertical stirrups were placed in the failure zone.

In S2 additional 6 mm diameter stirrups were placed at 150 mm spacing.

In S3 additional 8 mm diameter stirrups were placed at 130 mm spacing.

The main bending reinforcement was high yield steel; the web reinforcement mild steel. All three series of beams were cast using concrete with 10 mm maximum aggregate size, ordinary portland cement, and w/c ratio of 0.58, and were tested when 14 days old.

An Amsler testing machine was used to apply increments of the load at mid span of a simple-supported steel spreader beam, which was situated on top of the concrete deep beam so that the loading on the concrete beam remained symmetrical about the centre line throughout the experiments.

Dial gauges were placed at three positions to measure the vertical displacement at mid span and the horizontal displacement at the lower edge and the upper edge. The strains in the reinforcement were recorded

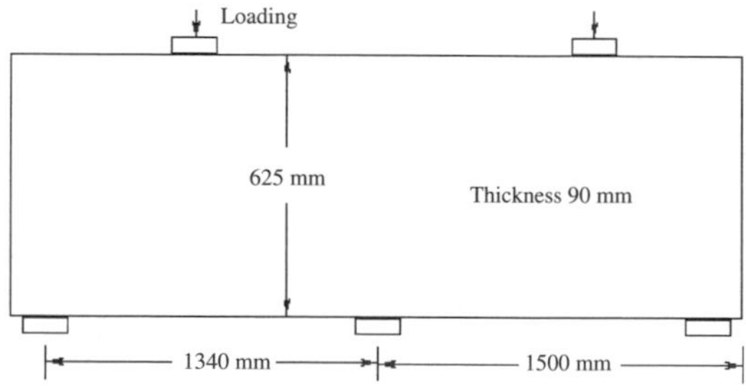

Figure 1. Details of test specimen.

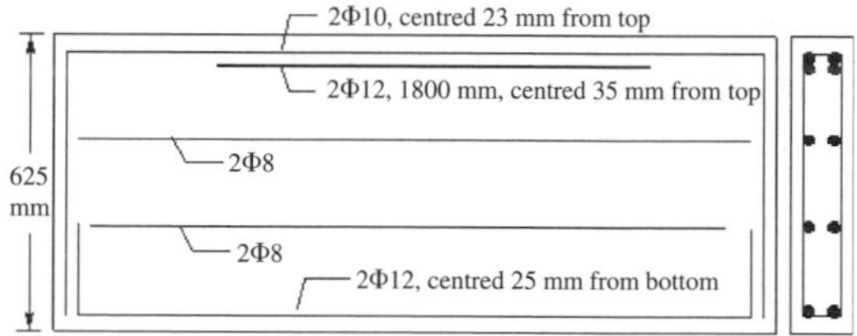

Figure 2. Details of main reinforcement.

Table 1. Details of tests and summary of experimental results.

		Concrete strength (N/mm^2)		Experimental results (kN)		
Year	Beam	Cube compress	Split tensile	Flex. cracks ①	Shear cracks ②	Failure ③
1995	S1	C40	N/A	N/A	350	450
1996	S1	39.4	3.10	235	345	497
1997	S1	36.0	2.75	300	250	370
1998	S1	42.1	3.60	240	296	420
1999	S1	50.4	3.90	250	300	>400
2000	S1	47.8	3.80	250	350	>400
1995	S2	26.0	2.40	200	N/A	470
1996	S2	36.9	2.70	250	300	450
1995	S3	33.7	2.70	200	295	640
1997	S3	42.2	2.95	300	N/A	>450
1998	S3	31.4	2.60	230	300	700
1999	S3	44.8	3.40	150	300	700
2000	S3	N/A	N/A	N/A	N/A	>600

①②③ represent the load at first appearance of flexural cracks, shear cracks and the failure load respectively.

electronically. Four strain gauges were placed on the top and bottom of the 12 mm diameter bars, 100 mm and 550 mm respectively from the central of the beam. Strain gauges were also placed on the web 8 mm diameter bar 400 mm from the central of the beam.

Load cells were placed under the outer support plate to measure the reaction forces. Details of the position of the measurements taken are given in Fig 3.

3 FINITE ELEMENT MODELING

Since the structure and the loading condition are symmetric, only half the beam was modeled. The FE mesh used in the analysis is shown in Figs 4 and 5. The concrete was modeled by 8-noded iso-parametric plane elements with reduced Gauss integration. The loading plate at the top of the deep beam and the supporting plates were also modeled using plane stress elements with reduced integration schemes. The nonlinear response of concrete was modeled using the Rankine/Von Mises and multi-directional fixed smeared crack models available in DIANA. The main reinforcements and all stirrups were modeled using embedded reinforcement elements with 3-point Gauss integration. These elements are considered to be uniaxial elements embedded into the iso-parametric parent element, and do not have any degrees of freedom of their own. The displacements of the reinforcing bar are compatible with the displacements of the parent element, hence perfect bond between reinforcement and concrete are implied. A perfectly plastic Von Mises yield criterion was used to model the post-yielding behaviour of reinforcement. It should be noted that special tying schemes were adopted at the supporting plate areas in order to reduce the local stress concentration and hence avoid numerical difficulties.

The material properties used in the analysis are shown in Table 2.

In the non-linear analysis, in order to avoid numerical problems and local stress concentration, the loading and support steel plates were modeled using a Von Mises perfect plasticity model with a slightly higher compressive strength compared to that of the surrounding concrete.

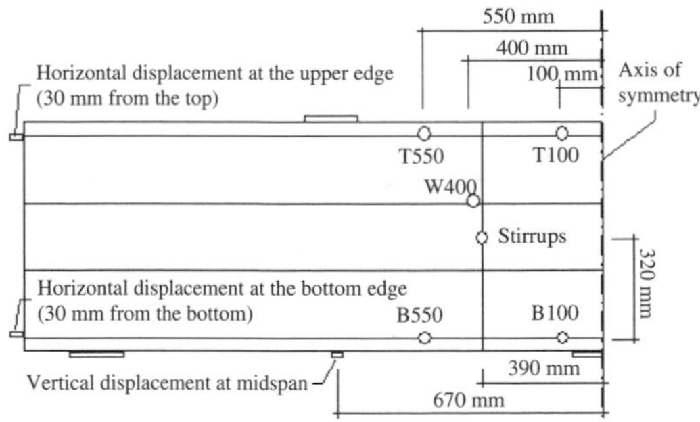

Figure 3. Location of instrumentation.

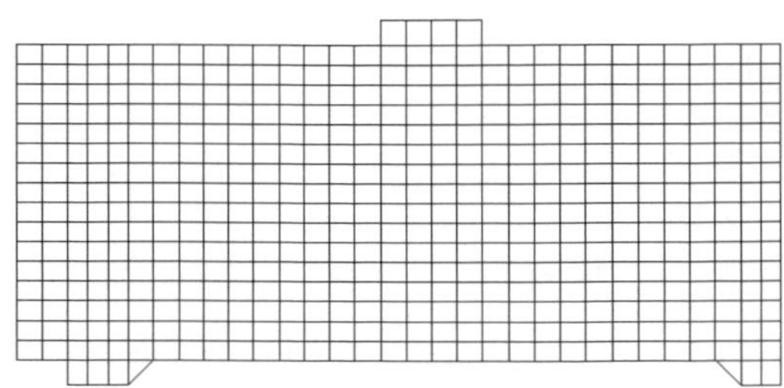

Figure 4. Finite element mesh – concrete.

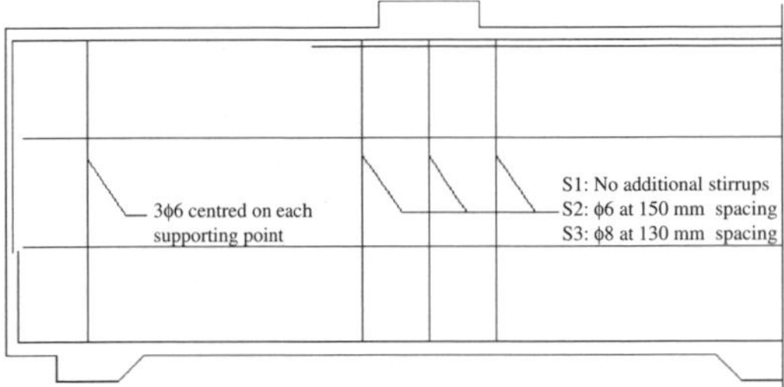

Figure 5. Finite element mesh – reinforcement.

Table 2. Material properties for deep beams.

	Concrete	
f_{cm}	28.0	N/mm^2
E_c	3.04×10^4	N/mm^2
γ	0.15	-
$f_{ct,m}$	2.50	N/mm^2
G_c①	5.00	Nmm/mm^2
G_f②	0.050 (S1) 0.065 (S2) 0.065 (S3)	Nmm/mm^2
Threshold angle α	60.0	Degree
Shear retention β	0.20	
	Reinforcement and Stirrups	
f_{sy}③	470 (12 mm dia) 530 (10 mm dia) 355 (8 mm dia) 355 (6 mm dia)	N/mm^2
E_s	1.95×10^5 (12 mm dia) 2.00×10^5 (10 mm dia) 1.95×10^5 (8 mm dia) 1.95×10^5 (6 mm dia)	N/mm^2

① Parabolic strain softening, ② Hordijk, [1991], tension softening ③ Perfect plasticity.

4 DISCUSSION OF ANALYTICAL RESULTS

4.1 *Structural behaviour and failure mechanisms*

4.1.1 *Load-displacement relationship and crack development*

The analytical load-displacement relationships at the loading point for the three series of deep beams are shown in Fig 6. It should be noted that in the following figures the lines showing cracking are normal to the cracks and give an indication of the crack width.

According to the analysis results, the structural behaviour of series S2 and S3 can be divided into four stages, however series S1 only has the first three stages.

Stage One (Figs 6 and 7).

After the initial elastic deformation stage, the flexural cracks develop at the bottom of the beam running towards the loading plate, and also at the top of the beam running towards the central supporting plate. The crack patterns and load-displacement curves for all three beams are very similar. It appears that, as expected, at this stage the vertical stirrups do not have much influence on the overall structural behaviour.

Stage Two (Figs 6 and 8).

This stage marked the sudden appearance of diagonal cracks at the middle of the deep beam running towards the loading plate and central supporting plate. These diagonal cracks caused a sudden drop in the applied load. In the case of series S1, the drop of the load is almost 100 kN, whereas in S2 & S3 the drop in the load is about 50 kN. It should be noted that the occurrence of the diagonal cracks in series S1 (with no stirrups) appeared much later when compared with the other two series. It seems that the additional vertical stirrups were unable to prevent the onset of the diagonal cracks, or may even trigger the occurrence earlier. Such a phenomenon might be explained by the relative large spacing of the web reinforcements, which in turn has little influence on the initial cracking criterion of concrete.

Stage Three (Figs 6 and 9).

The development of the flexural and diagonal cracks continued throughout this stage. In series S1, the diagonal cracks appeared to localize as the applied load continued to increase. It should be noted that in series S1 the diagonal cracks developed into a rather zigzag pattern. The development of flexural cracks also appeared to be very limited. In comparison, the diagonal cracks in series S2 and S3 appeared to be more distributed and widespread, due to the effect of the vertical stirrups. The flexural cracks continued to develop, expanded into nearby areas and began to bend towards the loading plate.

Stage Four (Figs 6 and 10)

This stage marked the gradual development of a further diagonal crack from the end support plate towards the loading plate in series S2 and S3.

The primary diagonal cracks continued growing and developing into multiple branches and becoming even more widespread until failure occured.

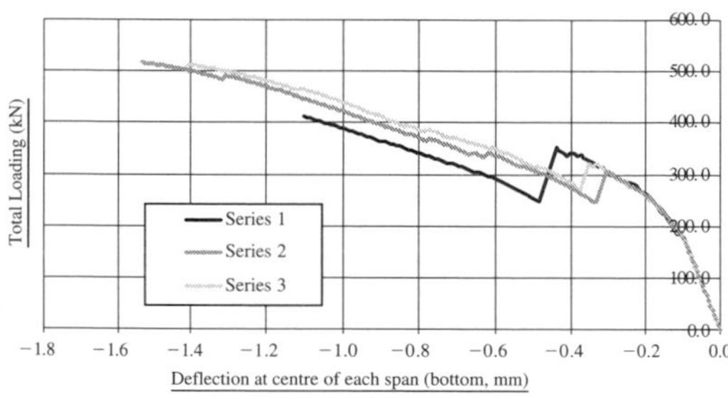

Figure 6. Load displacement curves.

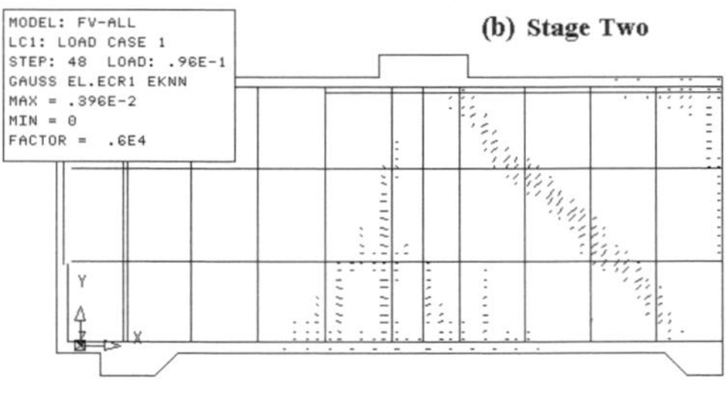

Figure 8. Crack patterns – Stage 2.

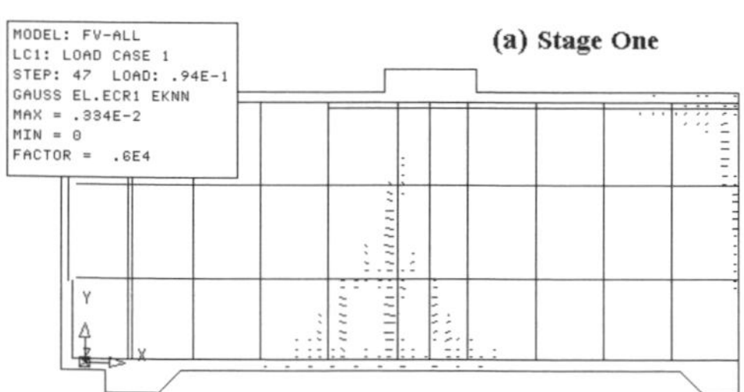

Figure 7. Crack patterns – Stage 1.

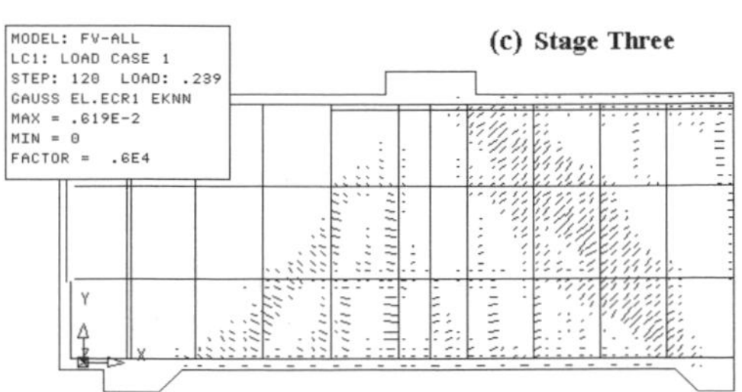

Figure 9. Crack patterns – Stage 3.

4.1.2 *Failure Mechanisms*

4.1.2.1 Beam S1

The tensile stresses in the two 8 mm diameter longitudinal reinforcement bars in the web of the deep beam reached the yield strength, and the reinforcement bars yielded at the position where they intercepted the primary diagonal cracks, Fig 11. The deformed shape of the series S1 beam at the last loading step is shown in Fig 12 which is somewhat different to that of beams S2 and S3, Figs 14 and 15.

It is concluded that the yielding of the two 8 mm diameter web reinforcement bars triggered further development of the primary diagonal cracks, which inevitably lead to the failure of beam S1.

4.1.2.2 Beam S2

The two 8 mm diameter longitudinal reinforcement bars in the web of the deep beam did not reach the tensile yield strength. However, the three 6 mm diameter vertical stirrups between the loading plate and the central supporting plate yielded at the position where they intercepted the primary diagonal cracks, Fig 13.

The deformed shape of series S2 beam at the last loading step is shown in Fig 14. Hence, it is suspected that any increase in deflection of the beam would result in local compression failure just above the central supporting plate.

4.1.2.3 Beam S3

Neither the two 8 mm diameter longitudinal reinforcement bars in the web of the beam nor the four 8 mm diameter stirrups between the loading plate and the central supporting plate reached the tensile strength.

The deformed shape of beam S3 at the last loading step is very similar to that of series S2, Fig 15. Hence, it is assumed that any increase in deflection of the beam would result in local compression failure above the

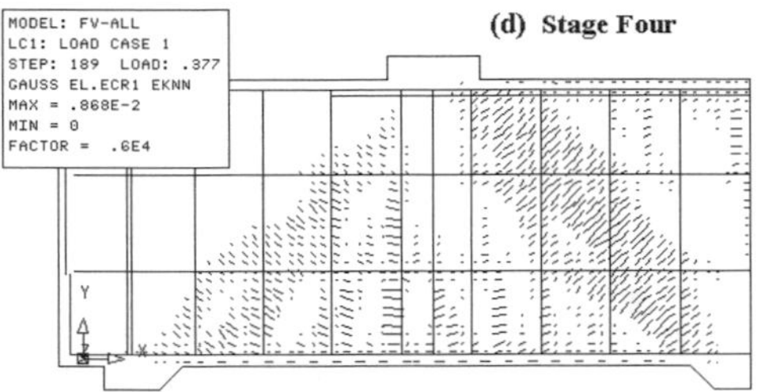

Figure 10. Crack patterns – Stage 4.

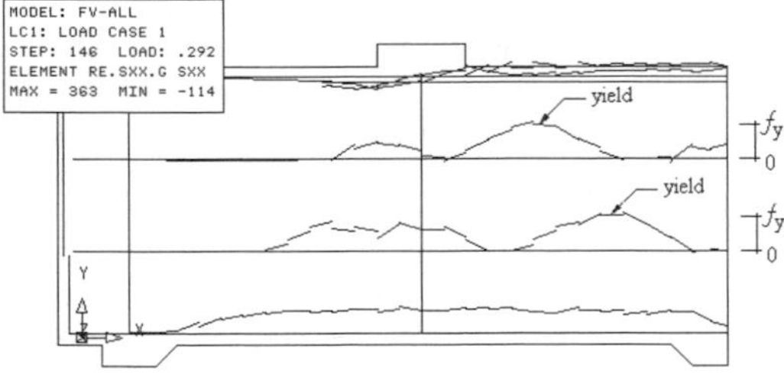

Figure 11. Longitudinal reinforcement stresses Beam S1.

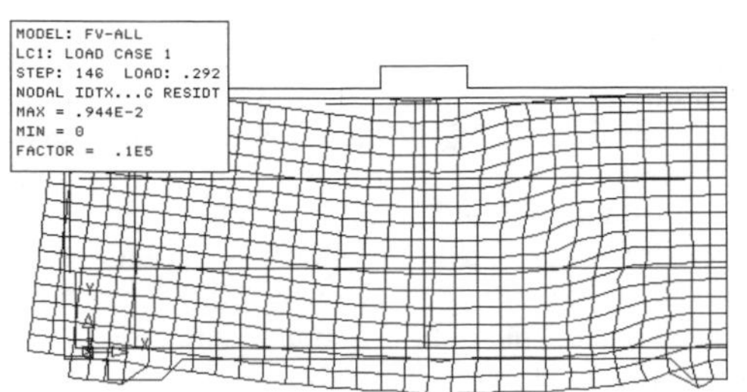

Figure 12. Deformed shape at last loading step Beam S1.

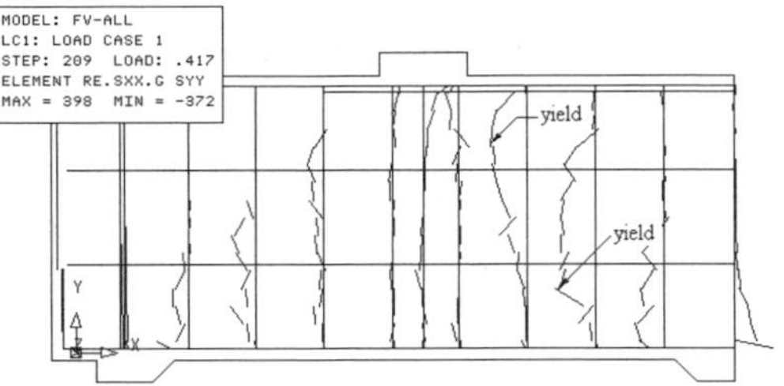

Figure 13. Stirrup reinforcement stress Beam S2.

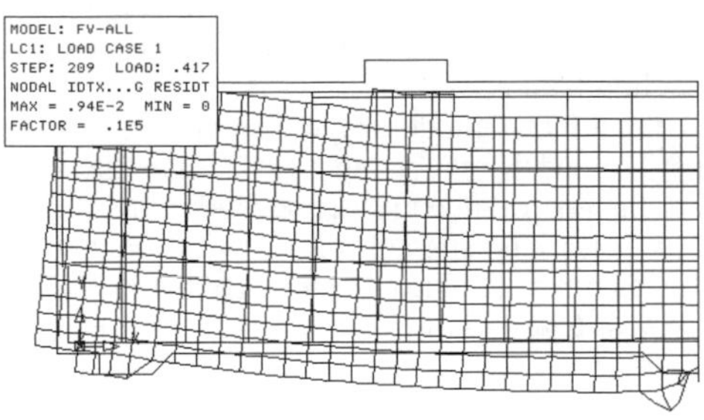

Figure 14. Deformed shape at last loading step Beam S2.

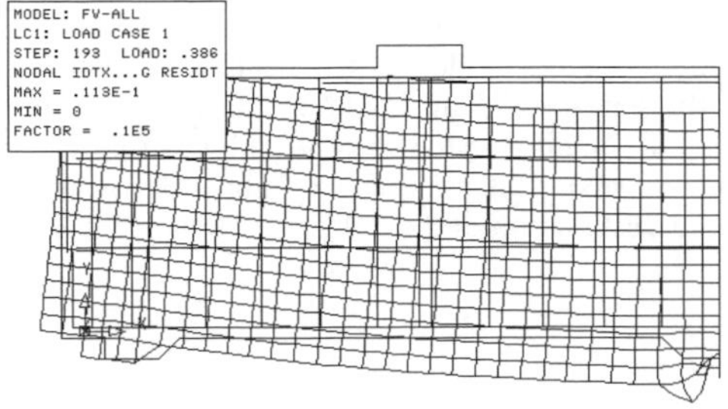

Figure 15. Deformed shape at last loading step Beam S2.

central supporting plate, which would lead to ultimate failure.

5 COMPARISON WITH EXPERIMENTAL RESULTS

5.1 *General comments*

One problem associated with making comparisons between FE simulation and experiment is the variability of the experimental results for nominally identical specimens, mainly due to the variability of the concrete properties. A summary of experimental results is given in Table 1, detailing the concrete tensile and compressive strengths from control specimens, the test load values at the occurrence of cracks and failure, indicates the problem.

It was reported that the failure of the deep beams usually occurred with a sudden widening of a pre-existing shear crack, accompanied by marked spalling of the concrete near the loading plates – S1 being more sudden than S2 and S3 – followed by a marked drop in the applied load. Comparing the analysis with the experimental results, the FE simulation performed fairly well, it did predict the sudden widening of the primary shear crack when the web reinforcement and stirrups yielded in the series S1 and S2 beams. However, the analytically predicted local compression failure near the central supporting plate in S2 and S3 was not reported during the experiment, which may due to the fact that the concrete at these support is in a tri-axial compression state, which was not modeled. Hence such local failure should be considered as a modeling failure.

As for the magnitude of the experimental failure load, it is noted that the strength of S2 was little if any greater than S1 despite the increased shear reinforcement, whereas the strength of S3 (with more shear steel) was typically much greater than that of S1 and S2 (typical 650 kN rather than 450 kN – an increase of about 45%). This contrasts markedly with the FE simulation, where S2 and S3 have much the same maximum load, about 25% greater than that of S1, although the predicted magnitude of failure load for S1 at about 410 kN seems quite reasonable. As mentioned above, if the local compression failure was considered to be due to inadequate modeling, which could be avoided, then the series S3 beam should be able to carry more load until yielding of the stirrups and web reinforcement occurred. This possibly explains the reason for the underestimation of the failure strength of S3.

5.2 *Beam S1*

In addition to the data in Table 1, a full set of experimental results for beam S1 in 1997 was provided by the

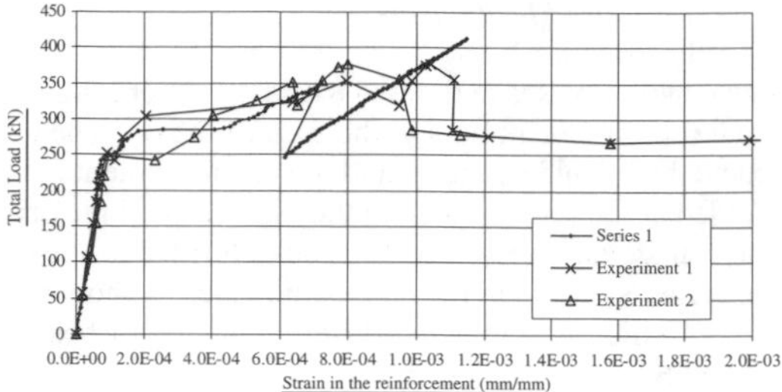

Figure 16. Load-strain relationship, position B550, Beam S1.

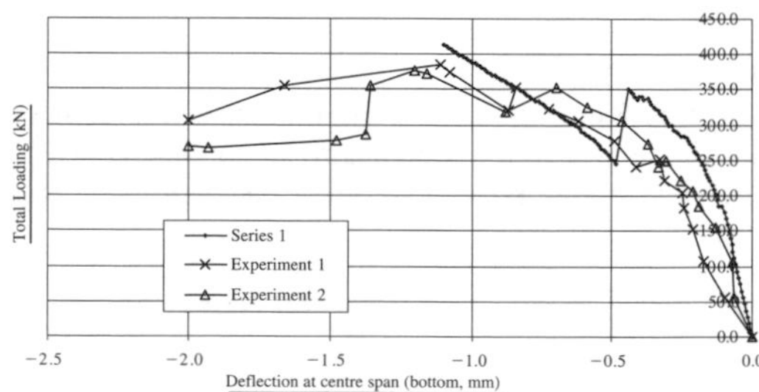

Figure 17. Load-displacement relationship, midspan. Beam S1.

University of Cambridge after the analyses had been carried out. This was considered by the University of Cambridge to be the best set of experimental data for various reasons.

It should be noted that with the S1 beam tested in 1997 the shear cracks were first observed at around 250 kN, with flexural cracking not occurring until the next load stage 300 kN, Table 1. Such observations were neither in line with the FE simulation, nor with the experimental observations in other years. However, for the same beam in the other span, the shear cracks were initiated around 350 kN, which is similar to the FE prediction, Fig16, where the legends "Experiment 1" and "Experiment 2" are for the two spans of beam S1 tested in 1997. Hence, it is possible that in the 1997 S1 beam there are some variations in the material in one of the spans of the beam, which triggered the shear crack prematurely.

To compare the FE prediction with the experimental results initially the overall structural behaviour of the beam is considered. Fig 17 shows the load – mid span displacement measured at the bottom edge. Figs 18 and 19 show the load – horizontal in-plane displacement relationship at the top and bottom edges respectively. It is noted that the predictions given by the FE simulation are fairly reasonable until the sudden initiation of the primary diagonal crack. After the occurrence of the diagonal crack, the load in the FE prediction continued to increase while in the experiment the beam was only

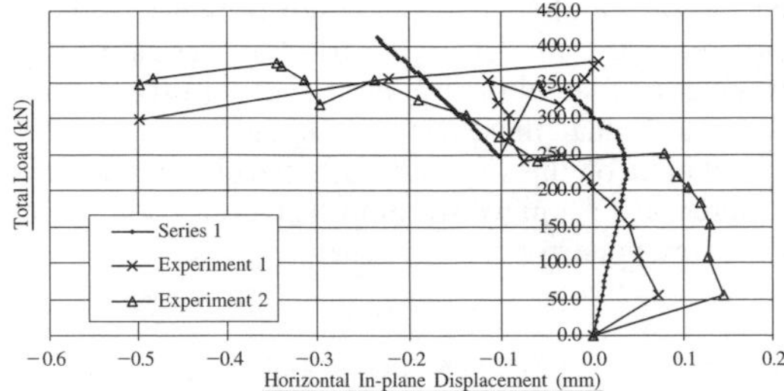

Figure 18. Load-horizontal displacement relationship, upper edge. Beam S1.

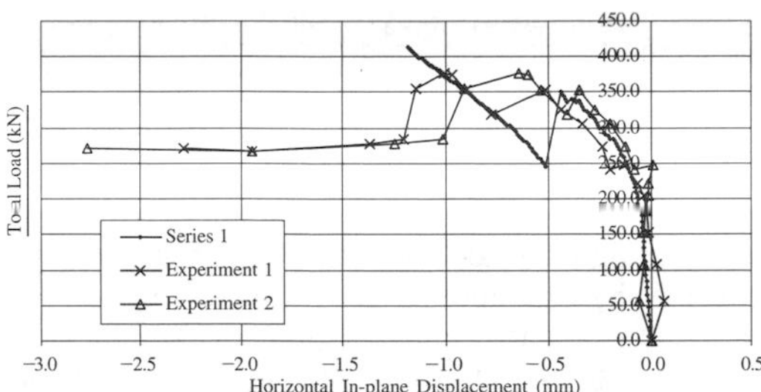

Figure 19. Load-horizontal displacement relationship, bottom edge. Beam S1.

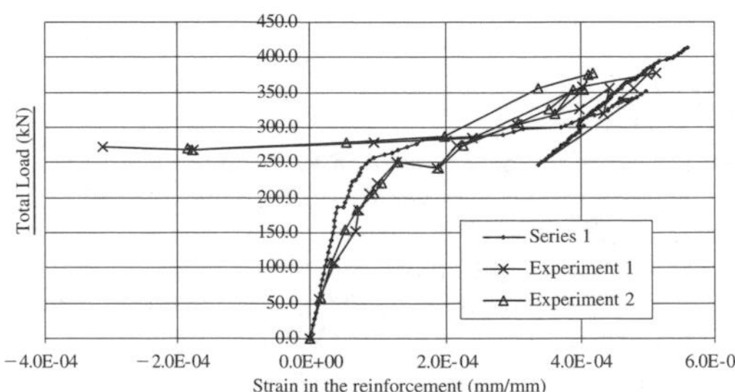

Figure 20. Load-strain relationship, position T100, Beam S1.

able to take a small increase in the load and then failed completely.

Secondly, the strains measured at positions T100 and B550 are considered. In this part of beam the reinforcement is under tension throughout the loading process. Fig 20 represents the load–longitudinal top steel strain relationship near the axis of symmetry, position T100; and Fig 21 is the load–longitudinal bottom steel strain relationship near the mid span, position B550. Again the FE prediction and the experimental results are similar up to the point at which the primary diagonal crack occurs.

Finally, the load–web reinforcement strain relationship at position W400 is considered, Fig 22. The prediction and the experiment all agree on the sudden

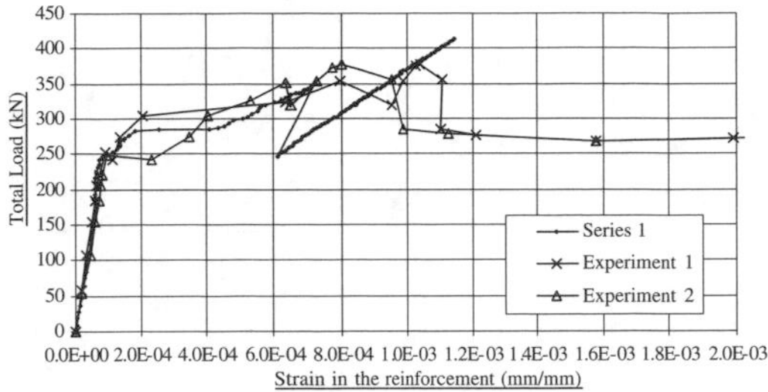

Figure 21. Load-strain relationship, position B550, Beam S1.

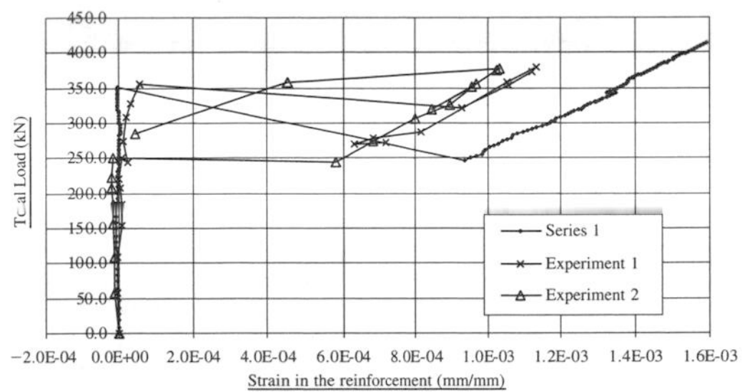

Figure 22. Load-strain relationship, position W400, Beam S1.

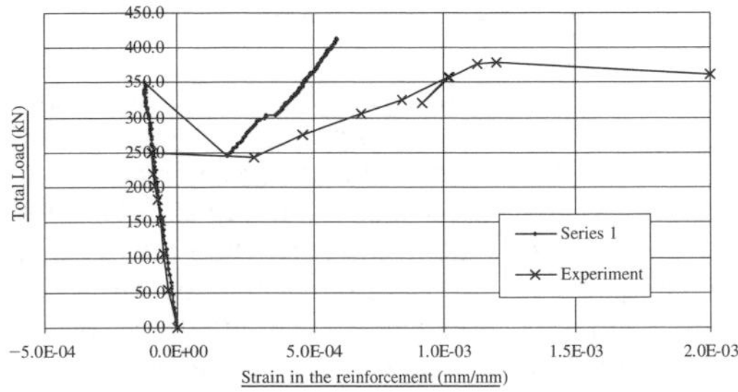

Figure 23. Load-strain relationship, position T550, Beam S1.

increase of the tensile strains in the web reinforcements when the primary diagonal crack suddenly appears. Fig 23 represents the load–longitudinal top steel strain relationship near the loading plate, position T550, and Fig 24 is the load–longitudinal bottom steel strain relationship near the central support, position B100. Again the analytical prediction and the experimental results agree on the major shift of the reinforcement strain from compression into tension when the primary diagonal crack occurs.

From the above comparisons, it is noted that the FE simulation has reasonably successfully modeled the structural behaviour of beam S1 before the yielding of the web reinforcements, which subsequently caused the sudden widening and the localization of

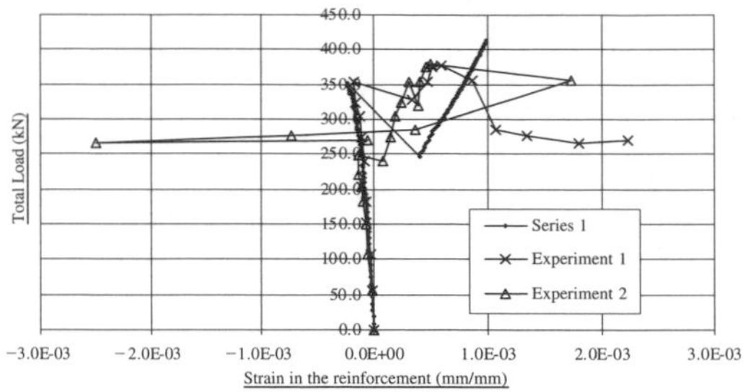

Figure 24. Load-strain relationship, position B100. Beam S1.

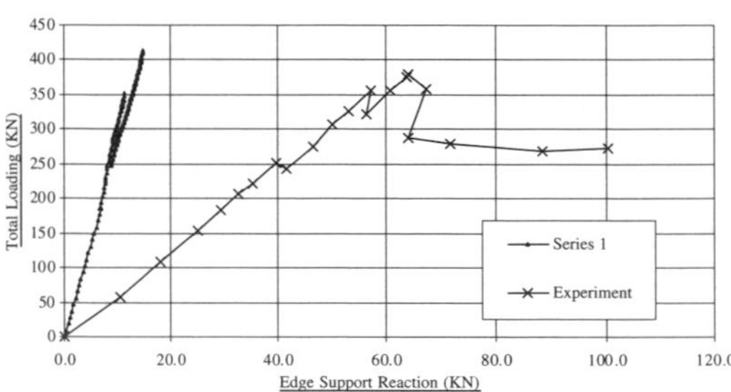

Figure 25. Load-outer support reaction. Beam S1.

the primary diagonal crack. After yielding of the web reinforcement, the FE simulation produced a stiffer response, which may due to the fact that a smeared crack model could not appropriately model the genuine localization of the diagonal crack.

Another interesting comparison, Fig 25 is the comparison of load – outer support reaction relationship, which shows significant differences between the FE results and the experiment. It was initially suspected by Morley [2001], who carried out an independent analysis, that the difference was mainly caused by the choice of the element type. (Surprisingly a 4-noded iso-parametric model gave a much closer estimation when compared with an 8-noded model.) It was later discovered that there was a bug in DIANA in the calculation of the reaction forces which caused the difference – a timely warning that even commercial codes are not error free!

6 SUMMARY AND CONCLUSIONS

In conclusion, it has been shown that the agreement between the experimental and numerical results before the localization of the primary diagonal crack is reasonable, unsurprisingly, since a smeared cracking model is quite capable of modeling a distributed fracture problem.

However, the failure mechanism of the beams can only be partly established from the FE simulation, mainly due to the inadequacy of the smeared model in representing localized fractures. Such problem could be solved by incorporating interface elements into the mesh to model the genuine fracture localization.

A strategy of a smeared-crack-prediction analysis followed by a discrete-crack-correction process was suggested by Rots [1988], which could lead to safe and correct solutions for these type of localized fracture problem. However, this approach was not tried for.

REFERENCES

Hordijk D. A., "Local Approach to Fatigue Of Concrete", PhD Thesis, Delft University Of Technology, 1991.

Morley C., Private communication, Department of Engineering, University of Cambridge, 2001.

Rots J. G., "Computational Modeling Of Concrete Fracture", PhD Thesis, Delft University Of Technology, 1988.

TNO Building and Construction Research, "DIANA User's Manual, Non-linear Analysis".

Computational Modelling of Concrete Structures, Bićanić et al. (eds)
© 2003 Swets & Zeitlinger, Lisse, ISBN 90 5809 536 3

RC structures with textile reinforcement and strengthening

F. Steinigen, B. Möller, W. Graf & A. Hoffmann
Institute of Structural Analysis, Dresden University of Technology, Germany

ABSTRACT: Since 1999 new types of technical textiles produced by the warp knit technique have been developed and tested within the scope of the Collaborative Research Centre 528 at the Dresden University of Technology. These textiles are used to reinforce or post-strengthen RC structures. In the case of the investigations presented here fibres comprised of alkali-resistant glass (AR glass) were used. In this paper new computational models for the numerical simulation of the load-bearing behaviour of textile-strengthened reinforced concrete structures are presented and the models are verified on the basis of experimental results.

1 INTRODUCTION

A reinforced concrete structure is subjected to a variety of external influences during its life cycle, leading to material and bond deterioration, ageing, corrosion, fatigue and structural damage. The result of these effects is to reduce the load-bearing capacity of the structure. A change in use of existing structures is frequently accompanied by increases in load. Structural strengthening may therefore be necessary during the service life of the structure.

A new method of strengthening reinforced concrete structures is the use of textile mats in a fine concrete matrix. The textile mats are comprised of several layers of rovings, which are aligned in arbitrary directions (Fig. 1a). The textile mats are embedded in a fine concrete matrix, whereby approx. 2 mm of fine concrete are reckoned for each textile layer (Fig. 1b). The orientation of the textile fibres (glass, polypropylene, aramide fibres) may be appropriately matched to the loading action so as to ensure multi-axial load-bearing behaviour.

Besides investigations concerning material and manufacturing problems, the reliable implementation of this new technology requires the development of algorithms and mechanical conceptual models for the purpose of numerical simulation and safety assessment. A mechanical model describing this new type of multi-layer system (reinforced concrete structure combined with textile strengthening) is described, e.g. in [Graf, Möller, Hoffmann & Steinigen 2001]. This paper deals in particular with the modelling of the load-bearing behaviour of textile-strengthened fine concrete and the treatment of cyclic loading processes.

2 DAMAGE PHENOMENA

2.1 *Old reinforced concrete structures*

Damage phenomena in the case of reinforced concrete under monotonic loading include, in particular, crack formation, reinforcement yield and concrete rupture due to exceedance of the concrete compressive strength. In the case of concrete, for example, a reduction in the elasticity moduli due to micro-crack formation occurs under cyclic loading. A numerical

(a)

(b)

Figure 1. A textile mat and textile strengthening [Curbach & Jesse 2002].

model which takes these damage phenomena into account is presented in Section 3.1.

2.2 *Load-bearing behaviour of textile reinforcement in a fine concrete matrix*

In order to manufacture textile mats so-called multi-filament yarns (rovings) are used. These are comprised of 400 (155 tex) to 2000 filaments (1200 tex).The thickness of a filament lies between 13.5 and 24.4 μm. The fineness of a roving is specified in tex (grams per kilometre length). The glass filaments, which are differently stretched, are treated with a sizing agent which influences the force transfer between the filaments. The maximum sustainable filament tensile force is dependent on the sizing agent as well as the strain velocity.

In tensile tests the rovings exhibit linear-elastic behaviour up to about 60% of the rupture load. Beyond this point a weak nonlinearity develops up to the rupture load. In yarn tensile tests under repeated loading and unloading (up to 60% of the rupture load) the rovings behave in a linear-elastic manner [Wulfhorst 2002]. The long-term behaviour of the rovings has also been investigated in creep and relaxation tests [Wulfhorst 2002]. In the creep test (15 hours under 75% of the rupture load) a strain increase of 0.27% was determined. In the relaxation test (20 hours under 1% strain) a stress reduction of 13% was registered, whereby the stress relaxation did not converge.

In the case of a roving embedded in a fine concrete matrix the matrix penetrates the outer filaments (boundary fibres). The latter are well-bonded to the matrix. The internal filaments (core fibres) only contribute to load transfer by virtue of cohesive and frictional forces between the boundary and core fibres. In the uncracked fine concrete the boundary fibres are subjected to a significantly larger load than the core fibres. In the fractured state the boundary fibres suffer extensive elongation, finally resulting in fibre failure. The tensile forces are sustained by the core fibres, which transfer the forces between the cracks via bond stresses into the boundary fibres and the matrix.

In contrast to steel reinforcement, glass fibre rovings represent a material with a highly inhomogeneous cross-section. On attainment of the maximum stress the material fails in a brittle manner. Owing to the small diameter of the rovings (<1 mm) the ratio between their surface area and their cross-sectional area is very much larger than in the case of steel reinforcement, thus ensuring an excellent bond between the matrix and the boundary fibres.

2.3 *Bond between the old structure and the strengthening layer*

The composite action between the old structure and the strengthening layer plays an important role in the load-bearing behaviour of textile-strengthened reinforced concrete structures. Three types of failure have been identified in the bond zone by way of tests [Curbach, Baumann & Beyer 2002]:

- failure in the old concrete
- failure in the old concrete/fine concrete interface
- failure in the fine concrete in the first reinforcement layer (delamination)

In the case of post-strengthening the textile-reinforced fine concrete is applied to a pre-damaged structure. The different age of the structural components leads to additional loading as a result of concrete creep and shrinkage.

3 NUMERICAL MODEL

3.1 *FE Model*

A reinforced concrete structure with textile post-strengthening is advantageously approximated as a layered computational model for the numerical simulation, see [Möller, Graf & Kluger 1997]. This gives rise to a multi-layered system comprised of a reinforced concrete old structure and a strengthening layer comprised of endless fibre textile in the fine concrete matrix. Compared with old RC structure the strengthening layers are very thin.

For the computational model comprised of layers of concrete, reinforcement steel and textile strengthening on the surface, special kinematics are introduced. In each strengthening layer a linear kinematic shape function is chosen, whose independent kinematic degrees of freedom are assigned to a reference plane in each case. This multi-reference plane model (MRM) with parallel reference planes is described, e.g. in [Möller, Beer, Graf & Hoffmann 2000].

In the MRM-based finite element development an extended variational formulation is adopted. In the case of physical nonlinearity the existence of a potential may only be assumed for a differential loading and/or constraint change. In the transition region between a basic state (BS) and a differential neighbouring state (NS) the stationary condition applies

$$\frac{1}{2}\delta(d^2\Pi_h) = \delta\sum_{RP=1}^{R}\left(\sum_{e=1}^{n}\left[\frac{1}{2}\sum_{m=0}^{s}\left\{\int_{V^{e,m}}(d\boldsymbol{\sigma}_{el}^{m})^T\, d\boldsymbol{\varepsilon}_{el}^{m}\, dV + 2\int_{V^{e,m}}(d\boldsymbol{\sigma}_{el}^{m})^T\, d\boldsymbol{\varepsilon}_{0}^{m}\, dV\right\} - \int_{O_v^{r,e}}(d\mathbf{p}^{r,e})^T\, d\bar{\mathbf{v}}^{r}dO - \int_{O_p^{r,e}}(d\mathbf{p}^{r,e})^T\, d\mathbf{v}^{r}dO + \int_{O_p^{r,e}}(d\bar{\mathbf{p}}^{r,e})^T\, d\mathbf{v}^{r}dO\right]\right) = 0 \quad (1)$$

with R – number of reference planes RP in the system; n – number of elements per reference plane; s – number of layers m per element; r,e – element boundaries; V – volume; $\bar{\mathbf{p}}^{r,e}$ external forces along the interface $O_p^{r,e}$ of the element; $\bar{\mathbf{v}}^r$ prescribed displacements of interfaces $O_v^{r,e}$, $\mathbf{p}^{r,e}$; – conjugated forces corresponding to $\bar{\mathbf{v}}^r$; $\boldsymbol{\sigma}^m$–strain-dependent stress component in the layer; $\boldsymbol{\varepsilon}_{el}^{m^{el}}$–strains in the layer and $\boldsymbol{\varepsilon}_0^m$–initial strains in the layer.

This 2-D form leads to folded-plate elements with six degrees of freedom per node (combined slab and plate action). In each reference plane a discretization is undertaken with layered hybrid finite elements with assumed stress distribution (stress shape function within the element, displacement shape function on the element boundary). In the numerical formulation the free values of the internal force shape function are pre-eliminated on the element level. Analogous to the displacement element formulations of the FEM, only the nodal displacements remain as unknowns. The internal forces within an element are directly determined from the internal force shape functions and not from the derivatives of the interpolated displacement field. The FE meshes are mutually congruent and the FE nodes lie above each other in the direction of the normals.

The composite action between the reinforced concrete layers and the post-reinforcement layers is modeled by means of an interface. The stiffness matrix of the interface is developed on the basis of the differential form of the minimum principle of potential energy

$$d\Pi_p = d\Pi_{p,i} + d\Pi_{p,a} \tag{2}$$

No external loads act on the interface, and the following holds

$$d\Pi_{p,a} = 0 \quad \text{and} \qquad d\Pi_p = d\Pi_{p,i} = \int_V (d\boldsymbol{\sigma})^T d\varepsilon \tag{3}$$

The differential deformation increases $d\varepsilon$ yield the differential relative displacement increases $d\Delta$.

$$d\Pi_p = \int_A (d\boldsymbol{\sigma})^T d\Delta dA \tag{4}$$

With the aid of a nonlinear bond matrix **V**, which describes the relationship between the shear and tensile stresses and the relative displacements at a material point, the tangential stiffness matrix for the interface $\mathbf{K}_T^I$ is developed, see [Graf, Möller, Hoffmann & Steinigen 2001].

For incremental loading stages the following holds for the global system with the tangential system stiffness matrix $\mathbf{K}_T$ at the beginning of the increment.

$$\mathbf{K}_T\Delta\mathbf{q} - \Delta\mathbf{R} - \Delta\mathbf{R}_K = \mathbf{0} \tag{5}$$

with $\Delta\mathbf{R}$ vector of the incremental components due to element load, boundary load, nodal load, initial deformations and prescribed displacements

$\Delta\mathbf{R}_K$ vector of the correction forces arising from the nonlinear material law, composite behaviour and crack formation

$\Delta\mathbf{q}$ vector of the incremental increases in nodal displacements

The tangential system stiffness matrix is constructed from

$$\mathbf{K}_T^h + \mathbf{K}_T^I = \mathbf{K}_T \tag{6}$$

with $\mathbf{K}_T^h$ tangential stiffness matrix derived from the hybrid energy functional, Eqn. (1)

$\mathbf{K}_T^I$ tangential stiffness matrix for the interface, Eqn. (2)–(4)

The selected FE formulation – hybrid finite elements with assumed stress distribution combined with an interface element (element displacement shape function) – leads to a mixed hybrid approach. The respective loading process is computed using a incremental iterative technique. A modified Newton-Raphson method or a path following algorithm is applied for this purpose.

Endochronic material laws are applied for general loading, unloading and repeated loading processes, taking into account the accumulated material damage during the load history [Möller, Graf & Kluger 1997].

The concept of smeared, fixed cracks in the concrete leads to an anisotropic continuum with stress restrictions (crack continuum). Depending on the principal stresses, up to two non-orthogonal cracks may form at a material point. The composite action between steel and concrete in the cracked concrete state is described by a modified stress–strain relationship for the steel or concrete (tension stiffening).

3.2 *Textile-reinforced fine concrete*

For the purpose of modelling the textile-reinforced fine concrete two model variants were considered:

(I) fine concrete and textile treated separately
(II) homogenised as a single continuum

In the separate modelling case (I) the endochronic material law according to Bažant is adopted for the fine concrete. The textile is modelled in each case as uniaxially smeared under the assumption of a nonlinear elastic-brittle material law. The roving is treated as being homogeneous, i.e. no distinction is made between the boundary and core fibres. The effect of boundary fibre rupture due to crack formation in the

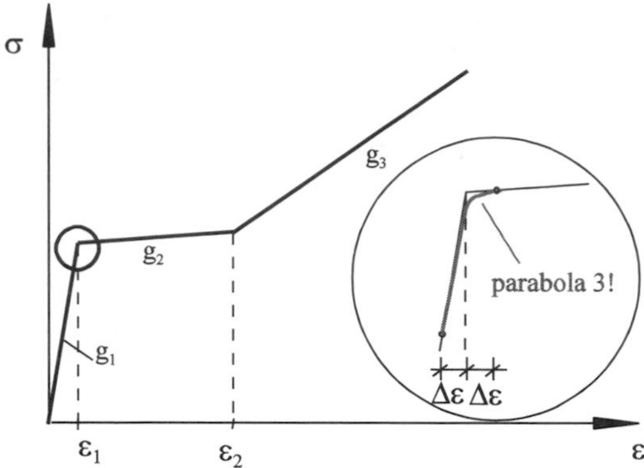

Figure 2. Stress–strain relationship model (II), schematic.

matrix is taken into account using a strain-dependent damage function $f_A(\varepsilon)$.

$$f_A(\varepsilon) = \begin{cases} 1 & \text{for} \quad \varepsilon \le \varepsilon_a \\ 1-\left(\dfrac{\varepsilon-\varepsilon_a}{\varepsilon_e-\varepsilon_a}\right)\cdot\dfrac{A_0-A_K}{A_0} & \text{for} \quad \varepsilon_a < \varepsilon \le \varepsilon_e \\ \dfrac{A_K}{A_0} & \text{for} \quad \varepsilon_e < \varepsilon \end{cases} \tag{7}$$

with A_0 cross-sectional area of the textile
A_K cross-sectional area of the core fibres
ε_a matrix strain at the commencement of crack formation
ε_e matrix strain at the end of crack formation
ε previously experienced maximum strain

The following holds for the effective textile cross-sectional area

$$A(\varepsilon) = f_A(\varepsilon)\cdot A_0 \tag{8}$$

In the treatment of the textile-reinforced fine concrete as a homogeneous material (II) special stress–strain relationships are adopted for different yarn finenesses and different fibre contents, as derived from uniaxial tensile tests [Curbach & Jesse 2002].

In the case of separate modelling of the fine concrete and the textile it is possible to take account of the biaxial load-bearing behaviour of the fine concrete and provide estimates of the fracture state and fracture angle. The attained stress level in the textile is known. Complete bonding is thereby assumed between the fine concrete and the textile (boundary fibres). A reduction in the effective fibre content is accounted for by Eqn. (8). A possible tension stiffening effect (as known in the case of reinforced concrete) may be taken into consideration as far as the concrete is concerned.

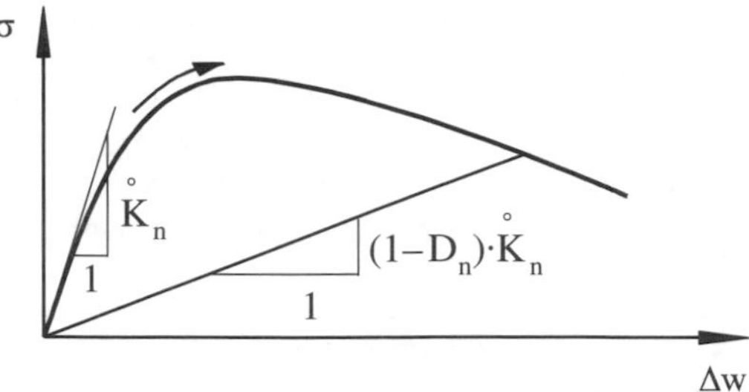

Figure 3. Stress vs. relative displacement over the interface.

The above-mentioned effects are also accounted for in the second modelling variant, as the stress–strain relationships are determined from tensile tests. A disadvantage of modelling variant II is a restriction to monotonic loading and the fact that the biaxial load-bearing behaviour of the fine concrete is (at present) neglected.

3.3 *Interface, monotonic loading*

Under cyclic loading, damage occurs in the reinforcement layer in the fine concrete matrix and the textile structure. This is accompanied by a degeneration of composite action between the old concrete and the textile reinforcement. This damage and additional plastic deformations may be described by a combination of plasticity and continuum damage theory, e.g. [Haufe 2001; Pölling 2000; Döbert, Mahnken & Stein 2000].

For verification purposes only the results of tests on delamination and shear failure under monotonic loading are available. For this reason the analysis was carried out under the following assumptions:

1. Only monotonic loading occurs.
2. Damage is adequately described by the leading diagonal elements of the composite matrix **V**, a degradation due to lateral strain is not accounted for.

Under these assumptions the damage process may be treated as a thermodynamic process in which plastic dissipation is neglected. Damage may thus be described alone using continuum damage theory as orthogonal elasto-damage with the damage variable D_s in the interface plane and D_n normal to the interface plane (Fig. 3).

For the presumed elasto-damage the HELMHOLTZ-Energy Ψ, the corresponding stresses σ_n normal to the interface plane and the stresses τ_s in the interface plane may be formulated as follows

$$\sigma_n = \frac{\partial\Psi}{\partial\Delta w} = (1-D_n)\cdot\mathring{K}_n\cdot\Delta w \tag{9}$$

$$\tau_s = \frac{\partial\Psi}{\partial\Delta u} = (1-D_s)\cdot\mathring{K}_s\cdot\Delta u \tag{10}$$

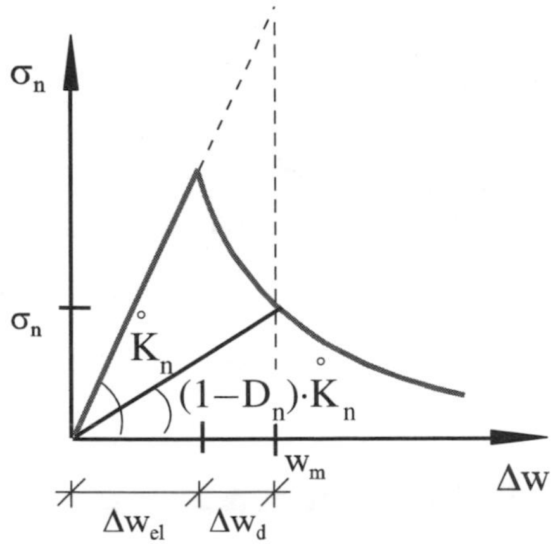

Figure 4. Damage evolution [Haufe 2001].

The loading and unloading functions are defined for the damage state

$$f_n(\Delta w, w_m) = \Delta w - w_m \leq 0 \tag{11}$$

$$f_s(\Delta u, u_m) = \Delta u - u_m \leq 0 \tag{12}$$

and the threshold values w_m and u_m for controlling the load history are introduced.

$$w_m = \max\{\Delta w\} \quad \text{and} \quad u_m = \max\{\Delta u\}$$

Evolution equations may be given for the damage variables (Fig. 4), e.g. for

$$D_n(\Delta w_d) = \begin{cases} 0 & \text{for } \Delta w \leq \Delta \\ 1 - \dfrac{\sigma_n}{(\Delta w_{el} + \Delta w_d) \cdot \mathring{K}_n} d_{n,\,test} & \text{for } \Delta w > \Delta \end{cases} \tag{13}$$

The constant scalar $d_{n,\,test}$ serves for the purpose of matching to uniaxial experimental results.

In the event that the limiting values Δw^* and Δu^* are very small, delamination and shear failure may be approximately modeled as quasi-brittle damage, neglecting post-damage behaviour. A distinction must then be made between three damage states for the interface:

(1) Neither shear failure nor delamination has occurred. The composite action along the interface remains intact.
(2) The stresses normal to the interface have exceeded a maximum. Delamination occurs and composite action is completely annihilated. The interface is incapable of transferring normal or shear stresses.
(3) The shear sustainability is exceeded. The adhesive action is nullified and tensile normal stresses may no longer be transferred. If compressive normal stresses exist along the interface, shear stresses may still be transferred by friction.

These three damage states are modeled as follows:

case (1): – $d_{s,\,test} = 0$, $d_{n,\,test} = 0$
– the allocation of elements to $\mathbf{V}_T(\boldsymbol{\sigma}, D_n, D_s)$ is dependent on the stress level attained along the interface

case (2): – $d_{s,\,test} = 1$ and $d_{s,\,test} = 1$; it thus follows that $\mathbf{V}_T(\boldsymbol{\sigma}, D_n, D_s) = 0$
– the shear stresses along the interface are set to zero independent of the relative displacements Δu and Δv
– the normal stresses are given by σ_{zz} (Δw)

case (3): – $d_{s,\,test} = 1$
– $d_{n,\,test} = 0$ in the case of compressive normal stresses; $d_{n,\,test} = 1$ in the case of tensile normal stresses
– the tensile normal stresses along the interface are set to zero independent of the relative displacement Δw; compressive normal stresses are transferable
– the shear stresses are given by $\tau_{zx}(\Delta u)$ (parabola f_2 in subsection shear failure) in the case of compressive normal stresses; in the case of tensile normal stresses the shear stresses are set to zero

The stress/relative displacement relationships are approximated in such a way that they may be fitted to the respective experimental curves.

Shear failure

A modified Coulomb criterion is applied as a rupture condition for the contact surface between the old and fine concrete

$$f_s = \tau_{zx}^2 + \tau_{zy}^2 - \tau_{max}^2(\sigma_{zz}, \Theta) \leq 0 \tag{14}$$

The maximum transferable shear stress τ_{max} is dependent on the stress σ_{zz} normal to the contact surface. In order to take account of a possible dependency between the maximum sustainable shear stress and the reinforcement direction of the textile reinforcement the angle Θ is additionally introduced.

Nonlinear elastic-brittle behaviour is assumed for the relationship between the composite stress and the relative displacement (Fig. 5).

If the contact surface is compacted ($\sigma_{zz} < 0$), larger shear stresses may be transferred for the same relative displacement. In order to take account of this effect an enlargement function $V_\tau(\sigma_{zz})$ is introduced (Fig. 6). This is normalized on $V_\tau(\sigma_{zz} = 0) = 1$.

The following then holds for the shear stress

$$\tau_{zx}(\Delta u, \sigma_{zz}) = V_\tau(\sigma_{zz}) \cdot \tau_{zx}(\Delta u, \sigma_{zz} = 0) \tag{15}$$

together with

$$V_\tau(\sigma_{zz}) = \begin{cases} V_\tau(\sigma_{zz}) & \text{für } \sigma_{zz} < 0 \\ 1 & \text{für } \sigma_{zz} > 0 \end{cases} \tag{16}$$

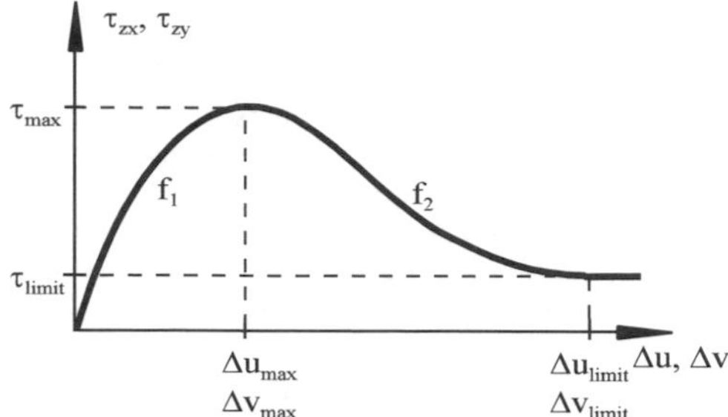

Figure 5. Bond stress vs. relative displacement for $\sigma_{zz} = 0$.

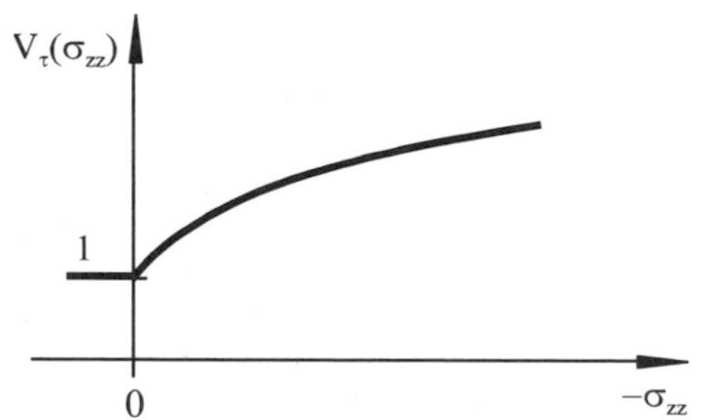

Figure 6. Enlargement function $V_\tau(\sigma_{zz})$.

Delamination
In order to account for delamination the following stress criterion is adopted as a rupture condition

$$f_D = \sigma_{zz} - \sigma_{zz,\,max} \leq 0 \tag{17}$$

The mathematical description of the stress/relative displacement relationship is similar to the description of the shear behaviour, with the additional restriction that the residual strength (σ_{limit}) is set to zero to account for the fact that tensile forces may no longer be transferred following separation (delamination) of the layers.

The tangential bond matrix $\mathbf{V}_T(\boldsymbol{\sigma}, \mathbf{d})$ is approximately filled as follows

$$\mathbf{V}_T(\boldsymbol{\sigma},\mathbf{d}) = (\mathbf{E}-\mathbf{d})^T \begin{bmatrix} \frac{\partial}{\partial u}\tau_{zx}(\Delta u,\sigma_{zz}) & 0 & 0 \\ 0 & \frac{\partial}{\partial v}\tau_{zy}(\Delta v,\sigma_{zz}) & 0 \\ 0 & 0 & \frac{d}{dw}\sigma_{zz}(\Delta w) \end{bmatrix} = \tag{18}$$

$$(\mathbf{E}-\mathbf{d})^T \begin{bmatrix} V_\tau(\sigma_{zz})\frac{\partial}{\partial u}\tau_{zx}(\Delta u,\sigma_{zz}=0) & 0 & 0 \\ 0 & V_\tau(\sigma_{zz})\frac{\partial}{\partial v}\tau_{zy}(\Delta v,\sigma_{zz}=0) & 0 \\ 0 & 0 & \frac{d}{dw}\sigma_{zz}(\Delta w) \end{bmatrix}$$

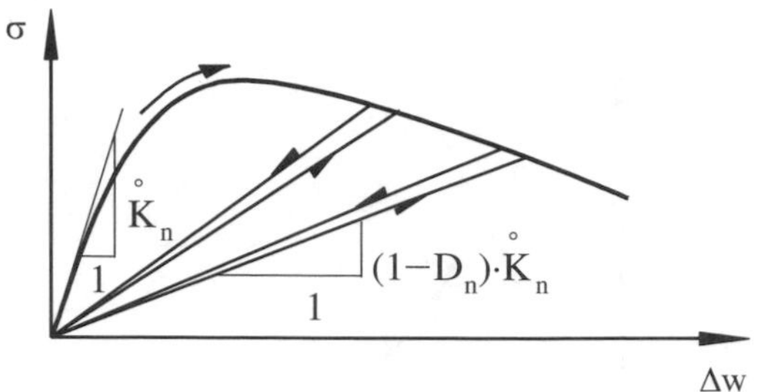

Figure 7. Stress vs. relative displacement over the interface, cyclic loading.

together with

$$\mathbf{E} = \text{diag}\,(1 \quad 1 \quad 1)$$
$$\mathbf{d} = \text{diag}\,(D_s \quad D_s \quad D_n)$$

Following holds for the damage parameters D_s and D_n

$$D_s = \begin{cases} 0 & f_S \leq 0 \wedge f_D \leq 0 \\ 1 & f_S > 0 \vee f_D > 0 \end{cases} \tag{19}$$

$$D_n = \begin{cases} 0 & \text{for } (\sigma_{zz} < 0) \vee (D_s = 0 \wedge f_D \leq 0) \\ 1 & \text{for } f_d > 0 \vee (D_s = 1 \wedge \sigma_{zz} \geq 0) \end{cases} \tag{20}$$

3.4 *Interface, cyclic loading*

The multi-surface model described in [Haufe 2001] is capable of simulating unidirectional damage per definition by the scalar damage model. According to [Haufe 2001] the application of the multi-surface model with isotropic damage under cyclic loading (threshold loading) leads to a physical unmotivated degradation of the lateral components of the material tensor. This effect is avoided along the interface in the present analysis due to the fact (as described in Section 3.2.) that the degradation resulting from lateral strain is neglected and the elements of the bond matrix **V** become uncoupled. An extension of the described method to anisotropic damage is a simple matter.

3.5 *Long-term behaviour*

Provided the long-term concrete compressive stresses do not exceed a value of approx. 40% of the compressive strength, linear creep may be assumed. In this case an approach in accordance with CEB-FIP Model Code 90 is adopted.

The assumption of linear behaviour permits the application of superposition for determining creep behaviour as a result of variable concrete stresses.

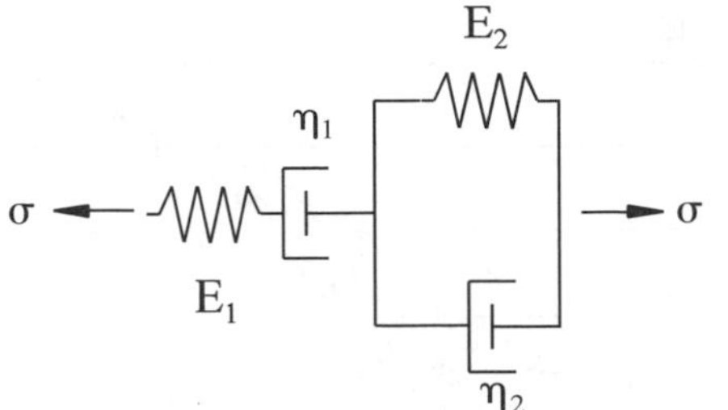

Figure 8. Burgers-element.

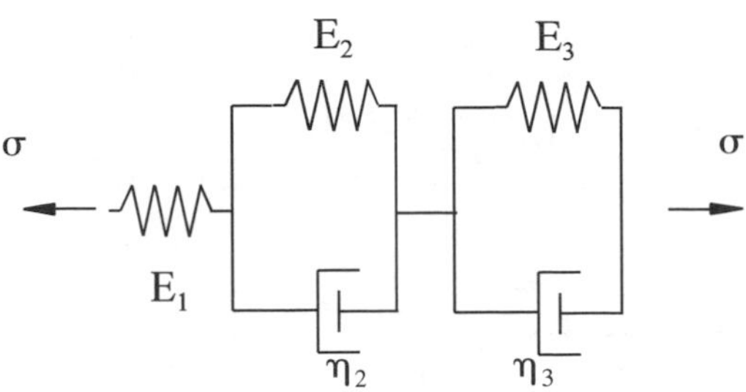

Figure 9. Rheological model for glass fibre rovings.

The overall strain due to creep-induced stresses (introduced at time t_0) follows from

$$\varepsilon_{c\sigma}(t, t_0) = \sigma_c(t_0)\cdot\Phi(t, t_0) + \int_{\tau=t_0}^{\tau=t} \Phi(t, \tau)\frac{\partial\sigma_c(\tau)}{\partial\tau}\,d\tau \qquad (21)$$

together with

$$\Phi(t, t_0) = \frac{1}{E_c(t_0)} + \frac{\varphi(t, t_0)}{E_c}$$
$$\Phi(t, \tau) = \frac{1}{E_c(\tau)} + \frac{\varphi(t, \tau)}{E_c} \qquad (22)$$

$\varphi(t,\ t_0)$ creep coefficient
E_c 28-day elasticity modulus, $E_c = 10^4 \cdot (f_{cm})^{1/3}$ [MPa]
$E_c(t_0)$ elasticity modulus at time t_0
$E_c(\tau)$ elasticity modulus at time τ

For the numerical evaluation of Eqn. (21) the time axis is subdivided into time intervals $\Delta t = t_i - t_{i-1}$, and σ_c (τ) is assumed to be constant during a time increment.

$$\varepsilon_{c\sigma}(t_k, t_0) = \sigma_c(t_0)\cdot\Phi(t_k, t_0) + \sum_{i=1}^{i=k}\Phi(t_k, t_{i-1})\cdot\Delta\sigma(t_{i-1}) \qquad (23)$$

As an alternative, rheological models may be applied. In this case Burgers model is implemented for the concrete and fine concrete (Fig. 8).

The corresponding differential equation is

$$\frac{E_2}{\eta_1}\sigma + \left(1 + \frac{E_2}{E_1} + \frac{\eta_2}{\eta_1}\right)\frac{\partial\sigma}{\partial t} + \frac{\eta_2}{E_1}\frac{\partial^2\sigma}{\partial t^2} = E_2\frac{\partial\varepsilon}{\partial t} + \eta_2\frac{\partial^2\varepsilon}{\partial t^2} \qquad (24)$$

The creep behaviour of glass fibre rovings may be modelled using a Hooke element and two Kelvin elements in sequence (Fig. 9).

The corresponding differential equation is

$$\left(\frac{E_3+E_2}{\eta_3} + \frac{E_2E_2}{\eta_3E_1}\right)\frac{\partial\sigma}{\partial t} + \left(1 + \frac{E_2}{E_1} + \frac{\eta_2}{\eta_3} + \frac{\eta_2E_3}{\eta_2E_1}\right)\frac{\partial^2\sigma}{\partial t^2}$$
$$+ \frac{\eta_2}{E_1}\frac{\partial^3\sigma}{\partial t^3} = \frac{E_2E_3}{\eta_3}\frac{\partial\varepsilon}{\partial t} + \left(\frac{\eta_2E_3}{\eta_3} + E_2\right)\frac{\partial^2\varepsilon}{\partial t^2} + \eta_2\frac{\partial^3\varepsilon}{\partial t^3} \qquad (25)$$

3.6 *Modification of the system*

In order to realistically determine the limit state of the serviceability and the load-bearing capacity of RC structures with textile post-strengthening it is necessary to take account of all relevant non-linear effects during the entire life cycle. As soon as non-linearities occur, however, the superposition of system reactions (stresses, internal forces, displacements, deformations), which occur at different times, is no longer possible.

For the realistic modelling of system behaviour it is thus necessary to correctly account for system modifications during the life cycle of the loading process.

Generally speaking, a system modification implies the modification of structures due to the addition, removal, or alteration of cross-sectional or structural components and their boundary conditions (including supports) as well as a modification of system geometry. System modification may occur during the production process as well as during the service life of a structure. Examples of this include the construction of structures in stages and the post-strengthening of already damaged old structures.

In order to correctly account for a system modification the loading process to which the structure is subjected must be traced. The system modification become effective at known points of time during the loading process; the system reactions may be determined at arbitrary times during the loading process.

In the case of post-strengthening, stress-free structural components are added to the deformed and

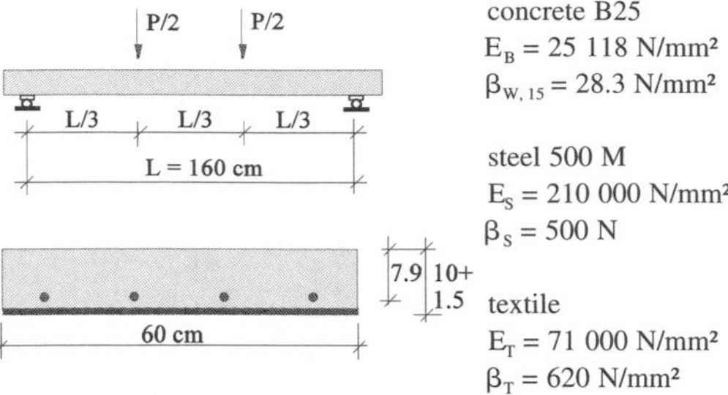

Figure 10. System and geometry.

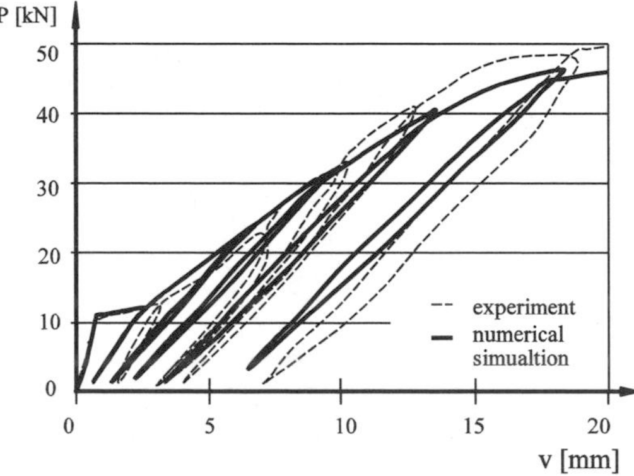

Figure 11. Load-displacement relationship, unstrengthened.

pre-loaded system. The latter are first subjected to loading following the application of subsequent external loads. The removal of loaded components from the structure results in a redistribution of stresses and internal forces.

4 EXAMPELS

4.1 *Single-span plate under cyclic loading*

In order to verify the developed models, plate experiments [Strizke, Curbach, Wagner & Bösche 2002] were simulated numerically; the geometry and loading on a typical plate are shown by way of example in Figure 10.

One of the plates, both a steel reinforcement mat R513, was strengthened by 3 layers of a biaxial structure (NWM-022-00) in a 10 mm fine concrete matrix. The load-displacement relationships obtained from experiments and the numerical simulation for the unstrengthened and textile-strengthened reinforced concrete plate are shown in Figures 11 and 12. Satisfactory agreement is obtained between the corresponding curves.

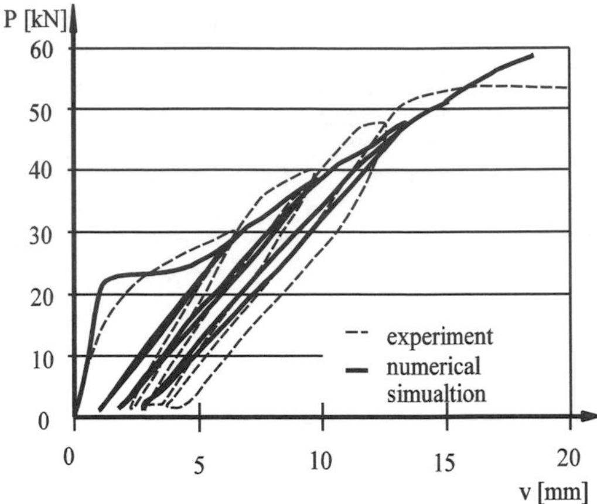

Figure 12. Load-displacement relationship, textil-strengthened.

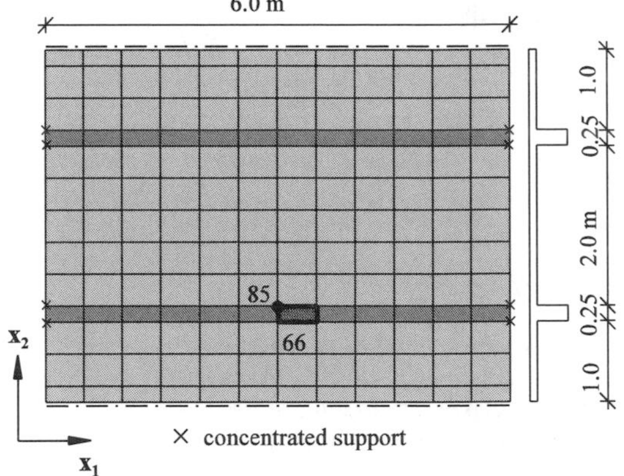

Figure 13. T-beam floor construction and FE mesh.

4.2 *T-beam floor construction*

A T-beam floor construction was designed for a live load of 5 kN/m^2 (Fig. 13). At the same time an additional load of 1.5 kN/m^2 was taken into consideration. Owing to a change of use it is necessary to dimension the floor for a live load of 10 kN/m^2.

For the numerical simulation of the load-bearing behaviour of the construction without additional strengthening 156 layered hybrid finite folded plate elements with assumed stress distribution were used with 9 layers in the floor plate and 15 layers in the supporting beam (Fig. 14).

After applying the dead weight and the additional load the live load was increased in increments up to system failure (Fig. 15).

The ultimate load of the unstrengthened reinforced concrete construction is p = 15.5 kN/m^2 (Fig. 16a). The fracture state in layer 5 (underside of the floor

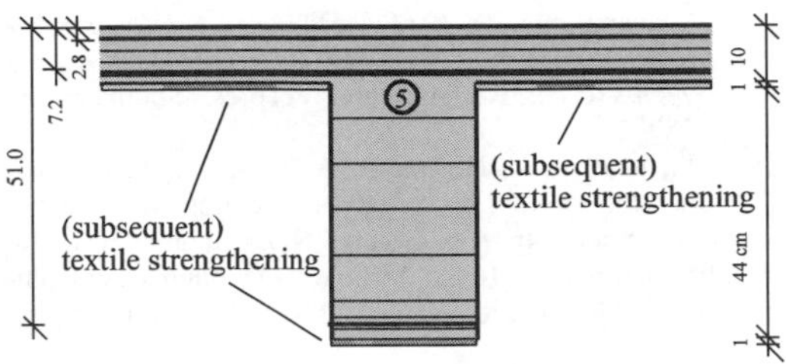

thickness of steel layers (mm):	0.221 (plate longitud. u/b)
	0.05 (plate cross u/b)
	3.08 (beam b)
textile layer (mm):	0.2

steel: S500 $E_S = 210\,000$ N/mm² $R_S = 500$ N/mm²
textile: 620 tex glass $E_G = 74\,000$ N/mm² $R_G = 1400$ N/mm²
concrete: B25

Figure 14. Layered model and material properties.

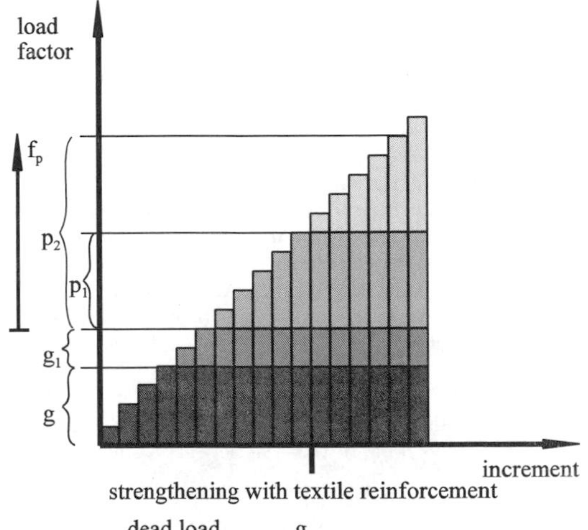

dead load g
additional load $g_1 = 1.5$ kN/m²
live load (old) $p_1 = 5.0$ kN/m²
live load (new) $p_2 = 10.0$ kN/m²

Figure 15. Loading process.

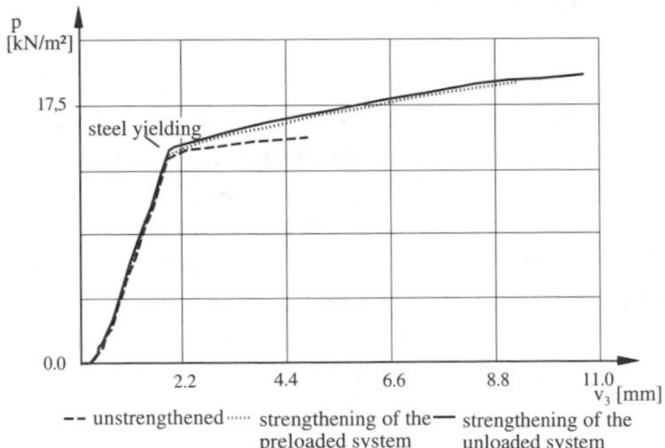

(a) Load-displacement relationship, node 85

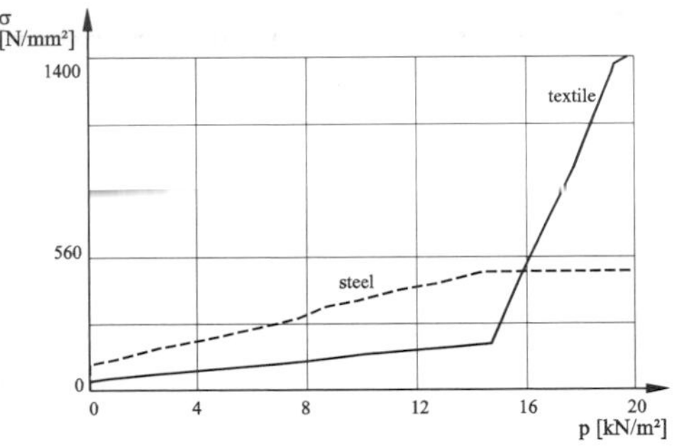

(b) Stresses in the steel and textile reinforcements element 66, (strengthening of the unloaded system)

Figure 16. Selected result of the numerical analysis.

Table 1. Ultimate loads of numerical simulation.

	Unstrengthened	Strengthening of the preloaded system	Strengthening of the unloaded system
Ultimate load p [kN/m²]	15.5	19.8	20.3

plate) for a load level p_2 of $p = 10\,\text{kN/m}^2$ is shown in Figure 13. The direction of all cracks is about 90° to the support direction of the supporting beams (in the x_2-direction). In the plate the stresses in the x_1-direction dominate (T-beam action) compared to the stresses in the x_2-direction (continuous beam action). Moreover, cracks do not form above the supporting beams in the x_1-direction. For this reason textile reinforcement is only applied to the underside of the construction (Fig. 14). It is differentiated between strengthening of the unloaded or preloaded construction. In the preloading case the system is subjected to dead weight, additional load and a live load p_1 of $p = 5\,\text{kN/m}^2$ before the textile reinforcement acts as strengthening. This is a modification of the system during the life cycle. In the unloaded case the modification of the system (strengthening with textile reinforcement) is effected at the beginning of the life cycle. The ultimate load in both cases increases in comparison with the unstrengthened construction (Fig. 16a and Table 1). The ultimate load attained with strengthening of the preloaded system is smaller owing to the fact that the T-beam floor is already damaged before textile reinforcement is applied.

5 CONCLUSIONS

The presented deterministic models for numerical simulation of the load-bearing behaviour of constructions with textile post-reinforcement are the basis for the application of the fuzzy-probabilistic safety concept, which has been developed by the project E3 of SFB 528. The fuzzy-probabilistic safety concept contains the probabilistic safety assessment as special case.

An extension of the deterministic model to dynamic structural behaviour is intended.

ACKNOWLEDGEMENTS

Authors gratefully acknowledge the support of the German Research Foundation (DFG) within the framework of the Collaborative Research Centre (SFB) 528.

REFERENCES

Curbach, M. & Jesse, F. 2002a. Experimentelle Entwicklung des einaxialen Stoffgesetzes von textilbewehrten Beton. In SFB *528, Arbeits- und Ergebnisbericht 1999–2002*, S. 165–206, TU Dresden.

Curbach, M., Baumann, L. & Beyer, R. 2002b. Flächige Übertragung der Schubspannungen vom Altbeton in den textilbewehrten Beton. In *SFB 528, Arbeits- und Ergebnisbericht 1999–2002*, S. 207–246, TU Dresden.

Döbert, C., Mahnken, R. & Stein, E. 2000. Numerical simulation of interface debonding with a combined damage/friction constitutive model. *Computational Mechanics* 25 pp. 456–467.

Graf, W., Möller, B., Hoffmann, A. & Steinigen, F. 2001. Numerical simulation and experimental measurments of RC structures with textile reinforcement. In Y. Villacampa Esteve, G.M. Carlomagno & C.A. Brebbia (eds), *Computational Methods and Experimental Measurements (CMEM X)*, Alicante, WITpress: Southampton, pp. 387–396.

Graf, W., Möller, B., Hoffmann, A. & Steinigen, F. 2002. Numerical simulation of RC structures with textile reinforcement. In A.H. Mang, F.G. Rammerstorfer & J. Eberhardsteiner (eds.), Fifth World Congress on Computational Mechanics. Vienna, 2002, 10 pp., http://wccm.tuwien.ac.at.

Haufe, A. 2001. *Dreidimensionale Simulation bewehrter Flächentragwerke aus Beton mit der Plastizitätstheorie*, Univ. Stuttgart, Diss.

Möller, B., Graf, W. & Kluger, J. 1997. Endochronic material modelling in nonlinear FE-analysis of folded plates. In P. Anagnostopoulos, G.M. Carlomagno & C.A. Brebbia (eds.), *Computational Methods and Experimental Measurements (CMEM VIII)*, Rhodos, Computational Mechanics Publ.: Southampton, pp. 97–106.

Möller, B., Beer, M., Graf, W. & Hoffmann, A. 2000. RC-folded plate structures with textile reinforcement. In Onate, E. (ed.), *ECCOMAS Barcelona*, CD-ROM.

Möller, B., Graf, W., Hoffmann, A. & Steinigen, F. 2002a. Numerische Simulation des Tragverhaltens textilverstärkter Bauwerke. In *SFB 528, Arbeits- und Ergebnisbericht 1999–2002*, S. 405–448, TU Dresden.

Möller, B., Graf, W., Hoffmann, A. & Steinigen, F. 2002b. Nachträgliche Verstärkung von Stahlbeton-Konstruktionen mit textiler Bewehrung. In D. Dinkler (Hrsg.), *Baustatik – Baupraxis, 8. Fachtagung*, TU Braunschweig, 2002, Bericht, S. 161–173.

Pölling, R. 2000. *Eine praxisnahe, schädigungsorientierte Materialbeschreibung von Stahlbeton für Strukturanalysen.*, Ruhr-Universität Bochum, Diss.

Strizke, J., Curbach, M., Wagner, A. & Bösche, A. 2002. Verstärkung von Platten und Plattenbalken mit textilbewehrtem Beton. In *SFB 528, Arbeits- und Ergebnisbericht 1999–2002*, S. 359–404, TU Dresden.

Wulfhorst, B. 2002. Garn- und Textilkennwerte (Stoffgesetze, Geometrie). In *SFB 532, Arbeits und Ergebnisbericht 1999–2002*, S. 163–194, RWTH Aachen.

Computational Modelling of Concrete Structures, Bićanić et al. (eds)
© 2003 Swets & Zeitlinger, Lisse, ISBN 90 5809 536 3

Optimisation of strut-and-tie model in deep beams with openings

J. Parsons & H. Guan
School of Engineering, Griffith University Gold Coast Campus, Queensland, Australia

ABSTRACT: Reinforced concrete deep beams find useful applications in tall buildings and foundations. Over the past two decades, as the need for deep beams continued to grow throughout the construction industry, numerous design models for deep beams have been suggested. However even the latest design codes still offer little insight into the design of deep beams in particular when complexities exist in the beams like openings. A method commonly suggested for the design of deep beams with openings is the strut-and-tie model which is primarily used to represent the actual load transfer mechanism in a structural concrete member under ultimate load. In this study, the development of the strut and tie model is transformed to the topology optimisation problem of continuum structures. The optimal strut-and-tie model is generated by gradually removing inefficient material from an over-designed area. During the optimisation process, both the stress and displacement constraints are satisfied and the performance of progressive topologies is evaluated. The influence on the strut-and-tie model in relation to different size, location and number of openings in deep beams is examined herein.

1 INTRODUCTION

In the engineering research field, fast, accurate and reliable methods of design are increasingly sort after. As new technology and ideas are generated, better design methods maybe established. In recent years structural optimisation has become an area emerging that has the possibility of modifying classical design.

Deep beam is a type of non-flexural member, which is generally defined as a member that has a span to depth ratio of less than 5. In various forms of construction, openings in the web area of deep beams are frequently provided for essential services and accessibility, for example door openings, windows, ventilation ducts and heating pipes (Ashour & Rishi 2000). Numerous investigations have been conducted into various attributes of deep beams (Kong 1990, Tan et al. 1995, 1997a, b, c, 1998, 2002a, Foster & Gilbert 1998). However only limited research has been done on deep beams with openings (Almeida & Pinto 1999, Maxwell & Breen 2000, Ashour & Rishi 2000, Tan et al. 2002b). The presence of openings induces areas of discontinuity into the deep beams, which only enhances the complexity of the nonlinear stress distribution along the depth of the beams. This in turn further complicates the design of deep beams and even the latest design codes still offer little insight into the design. A method commonly suggested for the design of deep beams with openings is the strut-and-tie model, which is primarily used to represent the actual load transfer mechanism in a structural concrete member under ultimate load. However many of the ways used in deriving the strut-and-tie model can be laborious and complex, because the compression strut from the loaded area generally separates and tracks around the opening before joining together again before the supports. This is especially true when predicting the correct strut-and-tie model for members with complex loading and geometric conditions (various location and size of openings). Hence it would be advantageous if a simple and effective method of generating the strut-and-tie model can be derived.

In this study, the development of the strut-and-tie model is transformed to the topology optimisation problem of continuum structures. The optimal strut-and-tie model is generated by gradually removing inefficient material from an over-designed area. During the optimisation process, both the stress and displacement constraints are satisfied and the performance of progressive topologies is evaluated. The influence on the strut-and-tie model in relation to different size, location and number of openings in deep beams is examined, through the investigation on four simply supported deep beams under single and two concentrated loads. The optimal strut-and-tie models achieved are also compared with published experimental crack patterns.

2 METHODOLODY

2.1 *Optimisation with stress and displacement constraints*

In this study the strut-and-tie model of deep beams with openings shall be found by means of the Evolutionary Structural Optimisation (ESO) method (Xie & Steven 1997), with both von Mises stress and displacement constraints.

The ESO method is based on the simple concept that by slowly removing inefficient material from an over-designed structure, the residual shape evolves in the direction of making the structure better. In the case of deep beams, when lowly stressed material is removed the residual shape should represent the strut-and-tie model.

In the finite element analysis of a deep beam, the von Mises stress of each element $\sigma_{\rm VM}^{\rm e}$ and the maximum von Mises stress of the structure $\sigma_{\rm VM,max}$ are evaluated. A deletion criterion can be obtained using the rejection ratio RR_i, a number expressed as a percentage, in conjunction with $\sigma_{\rm VM,max}$. This deletion criterion is the stress at which all elements with a lower stress are deemed insignificant. As such, an element is removed if its $\sigma_{\rm VM}^{\rm e}$ is less than the deletion criterion or:

$$\sigma_{\rm VM}^{\rm e} \leq RR_i \times \sigma_{\rm VM,\,max} \tag{1}$$

A small value of RR_i (0.1%) is used in this study to ensure that only a small number of elements are removed each time, where i indicates the iteration number.

Quite often in a structure the displacement has to be limited to a certain value. Through the induction of a displacement constraint, a particular maximum displacement value can be specified in the optimisation procedure. When this displacement value is reached the optimisation procedure will be terminated. The displacement control is performed by using a displacement sensitivity number which is express as:

$$\alpha_{d,i} = \sum_{k=1}^{L}\sum_{j=1}^{M} \lambda_{jk} \cdot |\mathbf{u}_{ij}^{T} \cdot \mathbf{K}_i \cdot \mathbf{u}_{ik}| \tag{2}$$

where $\alpha_{d,i} =$ displacement sensitivity number; $\mathbf{u}_{ik} =$ displacement vector of the ith element due to load case $\mathbf{P}_k$; $\mathbf{u}_{ij} =$ displacement vector of the ith element due to the virtual unit load $\mathbf{F}_j$; $\mathbf{K}_i =$ stiffness matrix of the ith element; $L =$ total number of load cases; $M =$ total number of displacement constraints; $\lambda_{jk} = |\mathrm{u}_{jk}|/\mathrm{u}_{jk}^{*} =$ weighting parameter indicating the contribution of the jth displacement constraint under the kth load case; $|\mathrm{u}_{jk}| =$ change in displacement; and $\mathrm{u}_{jk}^{*} =$ prescribed displacement value.

In the optimisation process, the sensitivity number $\alpha_{d,i}$ is calculated for each of the elements that satisfies the stress condition (Equation 1). To minimise the displacement change $|\mathrm{u}_{jk}|$, a number of elements with the lowest $\alpha_{d,i}$ are removed.

The finite element analysis followed by systematic removal of lowly stressed elements forms an optimisation cycle where RR_i remains constant. Such cycle or iteration is continued until no more elements are deleted. To proceed to the next iteration, RR_i is increased by adding the evolution ratio ER, which is also taken as 0.1%. The repeated cycle of optimisation process continues until a desired topology is obtained.

2.2 *Selection of optimal strut-and-tie model*

As the optimisation cycle progresses, the resulting topology improves with increase in iteration. To identify the final topology which can be translated to the optimal strut-and-tie model, a performance index PI_d in terms of the nodal displacement is used (Liang et al. 2000), or:

$$PI_d = \frac{u_o V_o}{u_i V_i} \tag{3}$$

where $u =$ nodal displacement at the point where the displacement limit is imposed; V = volume of the structure; and subscripts o and i = the original stage and the ith iteration, respectively.

During the optimisation process, PI_d measures the efficiencies of the progressive topologies. As the optimisation procedure continues, the number of iterations increases, and PI_d also increases, until a certain point where the efficiency or performance of the model deteriorates. The maximum value of PI_d corresponds to the most efficient topology, which would lead to the optimal strut-and-tie model.

3 NUMERICAL EXAMPLES

3.1 *Details of deep beams*

In this investigation three factors viz. the number, size and location of openings in deep beams are examined. This is done through the use of four examples, covering single/double, square/rectangle openings, located at mid-height and bottom of the beams. All beams are simply supported and are designated as DB1, DB2, DB3 and DB4. Detailed dimensions of the four beams are summarised in Table 1. The configurations of the beams are shown in Figure 1.

3.2 *Results*

During the optimisation process, the progressive topologies at various iterations are recorded. For the

Table 1. Dimensions of deep beams.

Beam number	DB1	DB2	DB3	DB4
Number of openings	1	2	1	2
Size of openings (mm × mm)	200 × 200	200 × 200	225 × 150	225 × 150
Location of openings from bottom (mm)	400	400	100	100
Number of point load(s)	1	1	2	2
Magnitude of load(s) (kN)	825	618	97.5×2	97.5×2
Young's modulus (×10^4MPa)	3.1451	3.1451	3.0966	3.0966
Poisson's ratio	0.2	0.2	0.15	0.15
Length of beam (mm)	1950	1950	1325	1325
Height of beam (mm)	1000	1000	750	750
Thickness of beam (mm)	100	100	100	100

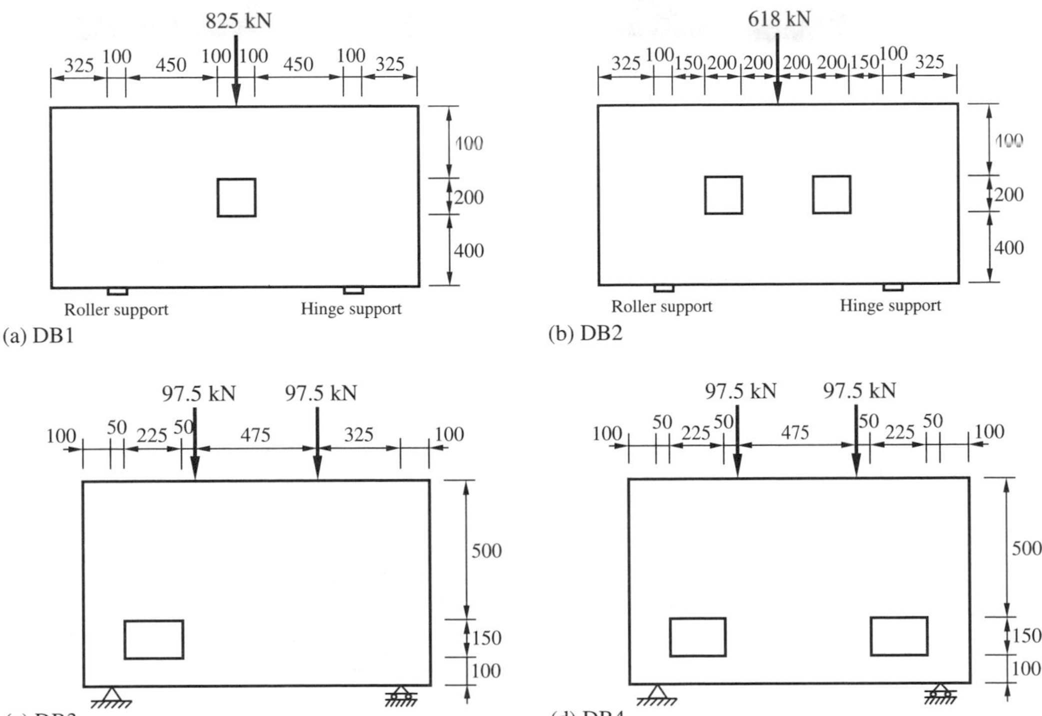

Figure 1. Configurations of deep beams.

four beams, topologies at selected iterations are presented in Figures 2, 4, 5 and 6 together with the final topology, which is determined based on the performance index curve. Also presented for all the beams are the optimal strut-and-tie models translated from the final topologies. Note that in the strut-and-tie models, the solid lines represent the tension ties while the dash lines denote the compression struts. The experimental crack patterns are also included if available. As a typical example, the optimisation history of DB1 including the performance index curve and the volume variation is illustrated in Figure 3.

Note that in the optimisation process, intermediate mesh refinement is often used to improve the strut-and-tie model at reasonable computational cost. This is done by refining either the entire mesh or only the area in the vicinity of openings, at certain iteration. This helps to rectify computational difficulties associated with singularity.

3.3 *Comparison and discussion*

Of the identical dimensions and under the mid-span point load, both DB1 and DB2 have square web opening(s) at mid-height of the beam. DB1 has a single

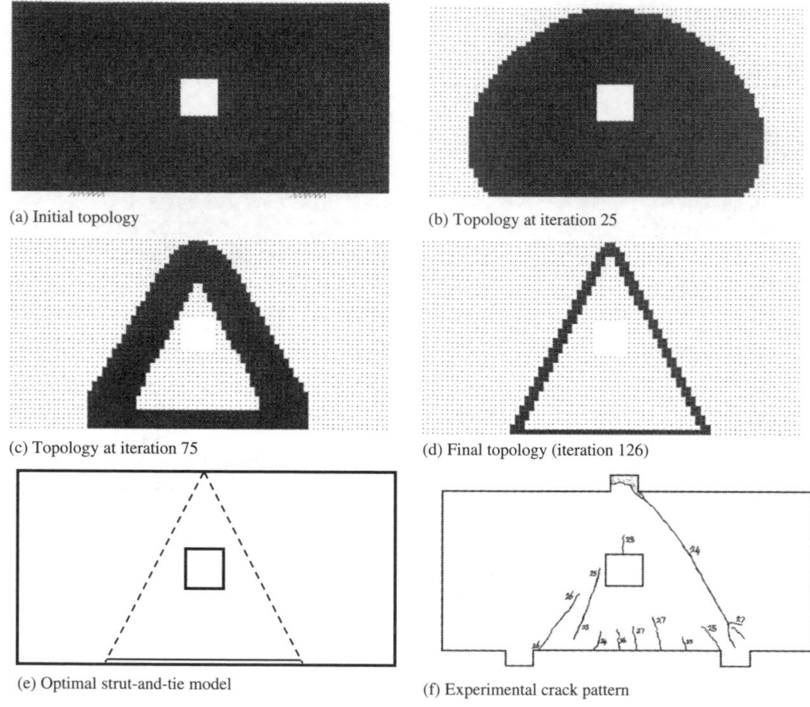

(a) Initial topology (b) Topology at iteration 25

(c) Topology at iteration 75 (d) Final topology (iteration 126)

(e) Optimal strut-and-tie model (f) Experimental crack pattern

Figure 2. Topologies and strut-and-tie model for DB1.

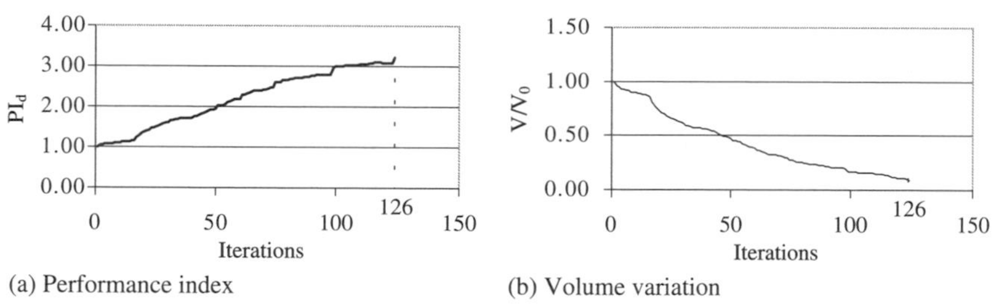

(a) Performance index (b) Volume variation

Figure 3. Optimisation history of DB1.

opening at mid-span whereas DB2 has two openings falling in the compression load transfer paths. When the opening is located away from the load path, the compression transfer takes the shortest path between the load and support points, as indicated in DB1. When the opening falls in the compression transfer path, the strut-and-tie model revolutionizes. As can been seen in DB2, the path are re-routed around the opening, thus introducing tension ties around the opening. As logic would tend to suggest that the compression struts pass the opening on the left- and right-hand sides, while the tension ties connect the compression zones above and below the openings.

Also of the same dimensions and under two point loads, both DB3 and DB4 have rectangle web opening(s) close to the bottom of the beam. If only one opening is present in the deep beam as the case of DB3, one side (without opening) of the strut-and-tie model becomes relatively simple, as the compression transfer path runs directly from the loaded area to the support. If two openings are present the strut-and-tie model becomes more complicated on both sides. However as expected, DB4 is symmetrical, hence the strut-and-tie model is consistent on both sides. This factor allows the designer to model only half the beam, saving on computational time.

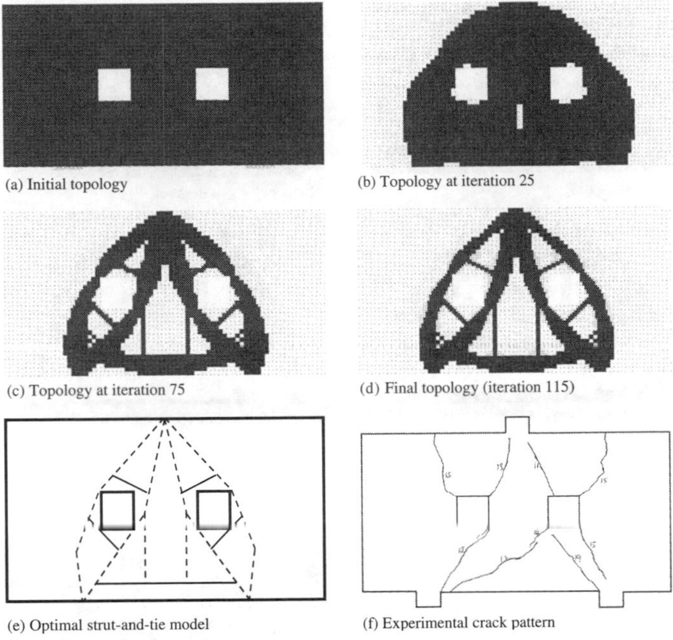

(a) Initial topology (b) Topology at iteration 25

(c) Topology at iteration 75 (d) Final topology (iteration 115)

(e) Optimal strut-and-tie model (f) Experimental crack pattern

Figure 4. Topologies and strut-and-tie model for DB2.

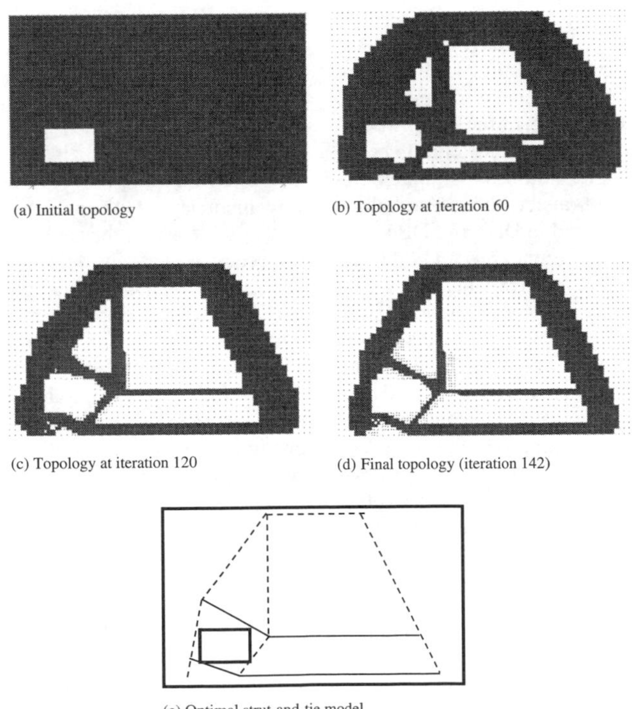

(a) Initial topology (b) Topology at iteration 60

(c) Topology at iteration 120 (d) Final topology (iteration 142)

(e) Optimal strut-and-tie model

Figure 5. Topologies and strut-and-tie model for DB3.

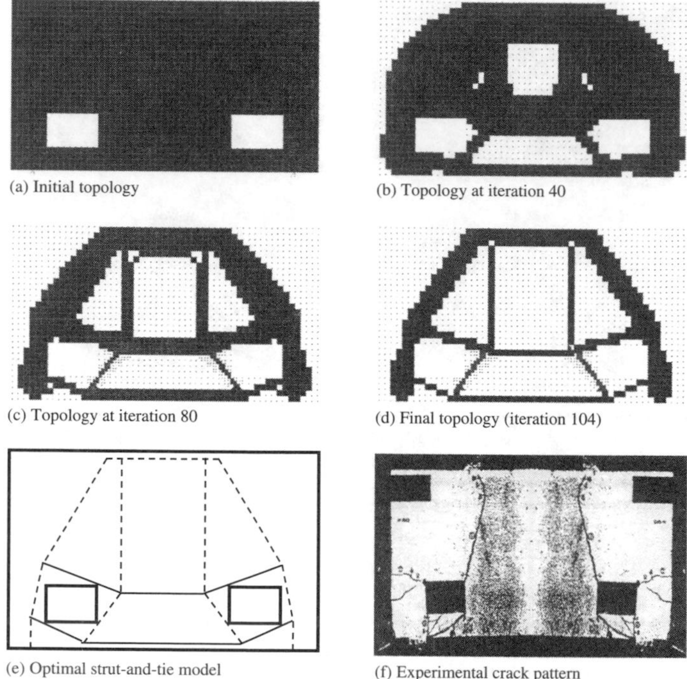
(a) Initial topology
(b) Topology at iteration 40
(c) Topology at iteration 80
(d) Final topology (iteration 104)
(e) Optimal strut-and-tie model
(f) Experimental crack pattern

Figure 6. Topologies and strut-and-tie model for DB4.

Comparing DB1 and DB3 as well as DB2 and DB4, the differences in the strut-and-tie model are demonstrated when the position of the openings is moved from the mid-height to the bottom of the beam. When the opening(s) is lower in the beam there tends to be an addition of two horizontal tension ties that unite both left and right compression transfer, above and below the openings. This is illustrated in DB3 and DB4.

It should be noted that in Figures 2, 4 and 6, when experimental crack patterns are available, it can be seen that there are similarities between experimentally derived results and computer generated results. From various experimental observations made by different researchers over the years, it is generally accepted that the first cracks appear at the top and bottom of the openings provided that the opening is in the compression transfer zone. As the load increases these cracks propagate from the top of the opening(s) towards the loaded area(s) and from the bottom of the opening(s) towards the closest support. In addition to this diagonal or shear cracking, flexural cracking can occur at the bottom of the beams. However in relation to the distance that these flexural cracks extend, and the crack width at failure, they are generally considered to be minor issues, and the more predominate cracks that occur are shear related. Therefore it is safe to assume that the top and bottom of the opening are under high-tension stresses because that is where the first major cracks appear. It should also be noted that the bottom of the beam also experiences reasonably high-tension forces as well. Therefore it is logical to place steel reinforcement in these sections to counteract the low-tension capacity of concrete. Therefore as the strut-and-tie model is used to determine reinforcement layout it would be expected that tension ties would form in these places. As can be seen in all the beams under consideration, this is generally what happens. Thus it can be concluded that the computer generated strut-and-tie model does accurately predict where the steel reinforcement is to be placed.

In addition to these counter tension reinforcements other reinforcement may be required. Sometimes in deep beams reinforcement is provided around the opening to counteract cracking at regions of stress concentrations which would occur at the corners of the openings.

4 CONCLUSIONS

Based on the four deep beams being investigated, the follow findings are observed:

- If the location of the opening is in the compression transfer zone, it is re-routed around the sides of the opening, thus causing the formation of additional tension ties at the top and bottom of the opening;

- The presence of more than one opening complicates the strut-and-tie model;
- The size of the opening affects the effort required to generate the optimal strut-and-tie model;
- When the opening(s) are lower in the beams there tends to be an addition of a single horizontal tension tie that unites both left and right compression transfer, above the opening(s);
- The optimal strut-and-tie models generated compared favorably with experimental crack patterns, thereby suggesting the accuracy of the optimal strut-and-tie models;
- Overall the Evolutionary Structural Optimization (ESO) technique is suitable for generating the strut-and-tie model in deep beams with openings, as it is accurate and reliable in a variety of different situations;

In summary, through this investigation, the influence of the number, size and location of openings on the optimal strut-and-tie model in deep beams is examined in some detail. The study has provided some insights into various parameters that affect the load transfer mechanisms of deep beams with openings. The ESO method has proven to be accurate and efficient in generating the strut-and-tie model for a variety of deep beams with openings.

REFERENCES

Almeida, A. & Pinto, N. 1999. High strength concrete deep beams with web openings. *ACI Special publications* 1(186):567–613.

Ashour, A. & Rishi, G. 2000. Tests of reinforced concrete continuous deep beams with web openings. *ACI Structural J* 97(3):418–426.

Foster, S. & Gilbert, R. 1998. Experimental studies on high strength concrete deep beams. *ACI Structural J* 95(4):382–390.

Liang, Q.Q., Xie, Y.M. & Steven, G.P. 2000. Topology optimisation of strut-and-tie model in reinforced concrete structures using an evolutionary procedure. *ACI Structural J* 97(2):322–330.

Maxwell, B. & Breen, J. 2000. Experimental evaluation of strut-and-tie model applied to deep beam with opening. *ACI Structural J* 97(1):142–148.

Kong, F.K. 1990. *Reinforced Concrete Deep Beams*. Glasgow: Blackie and Sons Ltd.

Tan, K.H., Kong, F.K., Teng, S. & Guan, L. 1995. High-strength concrete deep beams with effective span and shear span variations. *ACI Structural J* 92(S37):395–405.

Tan, K.H., Kong, F.K. & Weng, L.W. 1997a. High strength concrete deep beams subject to combined top- and bottom-loading. *The Structural Engineer* 75(11):191–197.

Tan, K.H., Kong, F.K., Teng, S. & Weng, L.W. 1997b. Effect of web reinforcement on high strength concrete deep beams. *ACI Structural J* 94(5):572–582.

Tan, K.H., Kong, F.K., Teng, S. & Lu, H.Y. 1997c. Main tension steel in high strength concrete deep and short beams. *ACI Structural J* 94(6):752–768.

Tan, K.H., Kong, F.K. & Weng, L.W. 1998. High-strength reinforced concrete deep beams and short beams: shear design equations in North America and UK practice. *ACI Structural J* 95(3):318–329.

Tan, K.H., Tang, C.Y. & Tong, K. 2002a. A direct method for deep beams with web reinforcement. *Magazine of Concrete Research*.

Tan, K.H., Tong, K. & Tang, C.Y. 2002b. Consistent strut-and-tie modelling of deep beams with web openings. *Magazine of Concrete Research*.

Xie, Y.M. & Steven, G.P. 1997. *Evolutionary Structural Optimisation*. London: Springer-Verlag.

Computational Modelling of Concrete Structures, Bićanić et al. (eds)
© 2003 Swets & Zeitlinger, Lisse, ISBN 90 5809 536 3

Nonlinear design procedures for reinforced concrete plates and shells

Clemens Preisinger, Rossitza Popov & Johann Kollegger
Institute for Structural Concrete, Vienna University of Technology, Austria

ABSTRACT: This paper deals with reinforcement design procedures based on nonlinear finite element calculations. Iterative design algorithms according to the principles formulated in EC2 and DIN 1045-1 and their implementation in the nonlinear finite element program SEGNID are described. For the EC2 this is the method of iteration of sectional forces. Corresponding to the nonlinear design concept of the DIN 1045-1 the ultimate limit load iteration procedure is used. After calculating reinforcement distributions for a quadratic plate under a distributed force, comparisons are drawn between the different design methods.

1 INTRODUCTION

Design codes for concrete structures like the Eurocode 2 (EC2) or the German DIN 1045-1 admit nonlinear analysis methods to be used – not only for sectional reinforcement design but also for calculating sectional forces (e.g. EC2) or ultimate limit load computations (e.g. DIN 1045-1). However these design codes only give statements as to whether a given structure is sufficiently reliable or not. No hints are given concerning the design procedure.

The advantage of nonlinear design methods is, that the realistic behaviour of a structure near collapse (e.g. the redistribution of internal forces) can be taken into consideration when making decisions about the amount and placement of the reinforcement.

Using nonlinear analysis methods for determining the amount of reinforcement needed for a certain RC structure leeds to the following problem: The distribution of reinforcement chosen as the input for a nonlinear calculation influences the results (sectional forces, steel strains) on the basis of which the reinforcement has to be calculated. That is why the reinforcement can only be determined iteratively.

2 PREVIOUS WORK

Strategies for the combination of nonlinear FE-calculations and reinforcement design have been published by Wittek et al. (2000). Corresponding to the different formats of ultimate limit state design in EC2 and DIN 1045-1 two procedures are suggested:

a) iteration of sectional forces
b) ultimate limit load iteration.

Method a) is a consequence of EC2, which demands the fulfillment of $R_d \geq S_d$ in each point of a RC structure. The approach chosen in EC2 makes use of the lower bound theorem of the theory of plasticity, which states, that if a static system is in equilibrium with the yield condition fulfilled in every point, the corresponding external forces constitute a lower limit of the ultimate limit load.

Figure 1 shows the different steps of the procedure: A linearly elastic calculation of internal forces is used to determine starting values a_s^0 for the distribution of reinforcement in the structure. The necessary amount of reinforcment can be calculated e.g. with the help of the equations formulated by Baumann (1972). What then follows is a nonlinear calculation of sectional forces, that results in a reinforcement distribution a_s^1 which serves as input a_s^0 for the next iteration and so on. The algorithm stops, when the change of the amount and distribution of reinforcement between two successive steps falls below a user defined level.

The external loads used for the calculations can be either the design loads S_d or, as proposed in Wittek et al. (2000) some enhanced load level $\gamma.S_d$. The

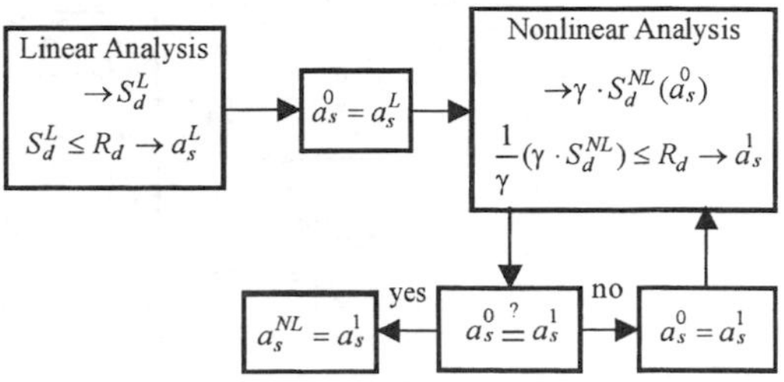

Figure 1. Iteration of internal forces – cross section design.

advantage of the latter variant being, that nonlinear effects – like the redistribution of internal forces due to the yielding of the reinforcement – develop to a greater extent as compared to the structure under design loads. This has a beneficial influence on the amount of reinforcement needed. On the side of the material mean values are used for the computation of sectional forces; design values for calculating the distribution of reinforcement.

In order to circumvent the duality of calculational and design material properties, the DIN 1045-1 takes a different approach to the problem of nonlinear reinforcement design as compared to the EC2: The safety of RC structures is not guaranteed on sectional level, but in a global manner. The condition for structural safety is primarily stated in terms of the global load factor for the design loads, that has to reach a value of at least $\gamma_R = 1.3$ at the ultimate limit state. For the material properties calculational values are chosen, that are given by the corresponding design values times 1.3. This approach is intended to give realistic results for the behaviour of the structure near collapse, and at the same time to satisfy the demands regarding safety. The method used in DIN 1045-1 leads to procedure b) the ultimate limit load iteration as described in Wittek et al. (2000) (see Figures 2a, b). Two variants, that differ in terms of the chosen initial reinforcement distribution, are at disposal: One possibility is to start the algorithm with reinforcement quantities calculated for each section from a linearly elastic analysis (Figure 2a). This means, that the reinforcement quantities have to be reduced in the course of the design process.

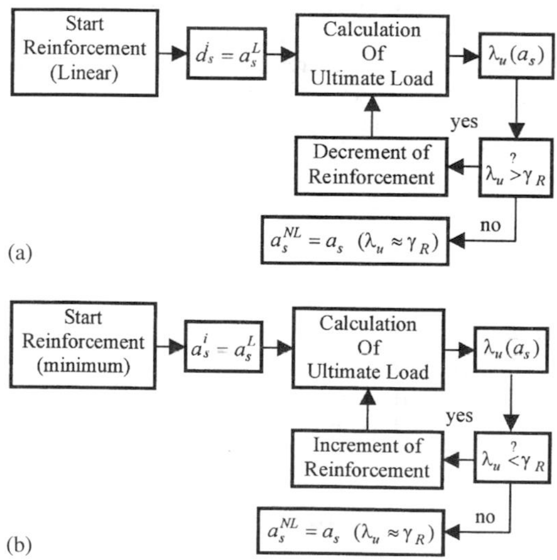

Figure 2. Iteration of ultimate limit load. (a) variant 1, (b) variant 2.

The other alternative is to start from the minimum reinforcement (Figure 2b), that can be determined from the height of the concrete cross section. In the course of calculation the external loads are increased until the ultimate limit state is reached. Now steel increments are given at the spots of the largest steel strains. Again a nonlinear limit state calculation is performed. The procedure stops when the load factor reached at collaps is sufficiently large.

In Kollegger (1991) an algorithm for the reinforcement design of plates and shells was presented that functions accordingly. It takes the sectional forces as well as the minimum reinforcement at an integration point as input. After each ultimate limit state calculation only the reinforcement layer with the largest steel strains gets incremented.

3 LAYERED SHELL ELEMENTS

For the calculations presented in this paper 9-noded isoparametric layered shell elements are used. Assuming a linear variation of the strains over the element height, the strain state $\varepsilon_{xx}(k), \varepsilon_{yy}(k), \gamma_{xy}(k)$ for layer k is given by:

$$\begin{aligned} \varepsilon_{xx}(k) &= \varepsilon_{xx} + \chi_{xx} \cdot z(k) \\ \varepsilon_{yy}(k) &= \varepsilon_{yy} + \chi_{yy} \cdot z(k) \\ \gamma_{xy}(k) &= \gamma_{xy} + \chi_{xy} \cdot z(k) \cdot 2 \end{aligned} \tag{1}$$

where $\varepsilon_{xx}, \varepsilon_{yy}, \gamma_{xy}, \chi_{xx}, \chi_{yy}$ and χ_{xy} are the strains respectively the curvatures in the middle plane. $z(k)$ designates the distance of layer k from the reference surface. With the strains given, assuming a plane stress state, the distributed internal forces of each layer can be calcultated via the material models for reinforcement and concrete. Sectional forces and moments are attained by integrating the layer contributions over the element thickness. The element stiffness matrix can be evaluated in a similar way by adding up the layer stiffness matrices.

3.1 *Concrete layers*

The physical nonlinearity of concrete under compression is dealt with by an elastic plastic material model with isotropic hardening. A Drucker-Prager yield condition and an associated flow rule are used. Since the material model is intended as the basis of a reinforcement design algorithm the concrete tensile strength as well as tension stiffening are neglected. The uniaxial comparative stress-strain curve incorporated into the model is a parabola-rectangle diagram.

For concrete cracking a smeared crack concept is adopted. By checking the principal stresses it is decided whether cracks open in one or two directions or not at all. Closing cracks are detected via the crack normal strain. According to the rotating crack concept, the crack directions continuously follow the orientation of the principal strains. It was shown by Baumann (1972) that this minimizes the internal strain energy of the concrete layers and thus delivers solutions which lie on the safe side. In order to speed up convergence of the equilibrium iteration, the so called "rotational stiffness matrix" derived by Akbar and Gupta has to be added in case of one crack. For concrete layers with one crack the stiffness matrix in the principal strain oriented coordinate system is simply

$$D_{c12} = \begin{bmatrix} 0 & 0 & 0 \\ 0 & E_{c2} & 0 \\ 0 & 0 & 0 \end{bmatrix} \tag{2}$$

with E_{c2} being the stiffness parallel to the crack. The rotational stiffness matrix in the xy-oriented coordinate system can be expressed as follows, where θ designates the cracks inclination towards the x-axis and σ_{c2} the compressive stress in crack direction:

$$D_{d12} = \frac{-\sigma_{c2}}{\sqrt{2(\varepsilon_{xx}-\varepsilon_{yy})^2+\gamma_{xy}^2}} \begin{bmatrix} \sin^2 2\theta & -\sin^2 2\theta & -\sin 2\theta \cdot \cos 2\theta \\ -\sin^2 2\theta & \sin^2 2\theta & \sin 2\theta \cdot \cos 2\theta \\ -\sin 2\theta \cdot \cos 2\theta & \sin 2\theta \cdot \cos 2\theta & \cos^2 2\theta \end{bmatrix} \tag{3}$$

No stiffness is added in case of concrete cracked twice.

3.2 *Reinforcement layers*

The parameters used to characterize the reinforcement are its distance from the middle surface, angle between reinforcement direction and global x-axis and layer thickness. Thus the reinforcment is conceived as being smeared over the longitudinal element dimensions. An elastic-plastic stress-strain relationship with linear hardening is assumed in compression and tension. The maximum steel strain is 25%, beyond which the steel stresses are set to zero.

4 DETERMINATION OF THE ULTIMATE LIMIT LOAD

For calculating the structural response to given external loads the usual iterative-incremental solution strategy is applied. In order to automatically determine the load factor corresponding to the ultimate limit state the external loads are increased step by step until no equilibrium can be found. Once no convergence is achieved the system is reset to the last equilibrium state, the load increment gets halfed and a new try starts. This process continues until the load increment falls below a user defined threshold.

5 DETAILS ON THE IMPLEMENTATION OF THE REINFORCEMENT DESIGN PROCEDURES

5.1 *Ultimate limit load iteration*

The concept for the ultimate limit load iteration described above has been integrated into the nonlinear finite element program SEGNID (Dinges et al. (1987)). It is based on the design algorithm suggested in Kollegger (1991). However in order to apply the procedure to structures as a whole the criterium for the stepwise incrementation of the reinforcement had to be extended: The following formula is used for all reinforcement layers of all elements for augmenting the reinforcement areas:

$$a_{s(1)} = a_{s(i-1)} \cdot (1 + f_{1(i)} \cdot f_{2(i)}) \tag{4}$$

where $a_{s(i)}$ designates the new amount of reinforcement, $a_{s(i-1)}$ that was computed in the last reinforcement iteration. Throughout an element the thickness of each layer is assumed to be constant. The factor $f_{1(i)}$ accounts for the difference between the load factor λ_{i-1} reached with the reinforcement $a_{s(i-1)}$ and the load factor λ_{Goal} for which structural reliability is sufficient:

$$f_{1(i)} = (\lambda_{Goal} - \lambda_{i-1})/\lambda_{i-1} \cdot C_1 \tag{5}$$

The userdefined constant C_1 controls the stepsize of the reinforcement increments. Since $f_{1(i)}$ goes towards zero as soon as λ_{i-1} reaches λ_{Goal}, which would result in very small incrementation steps near the design goal, a lower boundary of

$$|f_{1(i)}| \geq 20.0 \cdot \lambda_{TOL} \tag{6}$$

was chosen. Here λ_{TOL} is the user defined tolerance which is also used to determine whether or not the iteration goal is met. The procedure stops if the condition

$$\lambda_{TOL} \leq |(\lambda_{Goal} - \lambda_{i-1})/\lambda_{i-1}| \tag{7}$$

is fulfilled. The factor $f_{2(i)}$ given by

$$f_{2(i)} = (\varepsilon_{spl(i)}/\varepsilon_{spl,\max(i)})^{C_2} \tag{8}$$

is used to control the way in which reinforcement increments are distributed in those areas where the steel yields. $\varepsilon_{spl(i)}$ designates the plastic steel strain of a reinforcement layer at an integration point, $\varepsilon_{spl,max(i)}$ stands for the maximum plastic steel strain encountered in the structure. The factor C_2 has to be chosen by the user. Small positive values for C_2 (e.g. 0.001) result in an even distribution of reinforcement increments in the areas of yielding steel, whereas for high values (e.g. 2.0) increments concentrate in the regions of large plastic strains. Thus the parameter C_2 can be used to control the structures ductility: low values lead to a ductile collapse induced by failure of the reinforcement. With high values of C_2 the concrete gets utilized locally to a higher degree which results in a more brittle breakdown of the RC structure.

5.2 *Iteration of sectional forces*

With the initial reinforcement distribution given (which can be the minimum reinforcement or the reinforcement calculated from linearly elastic sectional forces) a first ultimate load calculation is performed – using mean material properties. The sectional forces $S_{d(i-1)}$ are then scaled to the load level demanded:

$$S_{d(i-1)scaled} = S_{d(i-1)} \cdot \lambda_{Goal}/\lambda_{i-1} \tag{9}$$

For external loads on design level λ_{Goal} is equal to 1.0. These sectional forces $S_{d(i-1)scaled}$ serve as input for the sectional reinforcement design algorithm used in Kollegger (1991). Ultimate limit load calculations and sectional reinforcement design steps are performed until the difference of reinforcement between two consecutive iteration steps falls below a user defined level. The procedure used here somewhat deviates from that suggested by Wittek et al. (2000): Each nonlinear calculation is carried through to the ultimate limit state, whereas in Wittek et al. (2000) a factor of 1.15 for the design loads is used. The latter variant has the advantage that one avoids the computational cost of limit state calculations. Yet the full extent of internal force redistribution can only be exploited by use of the first method.

6 NUMERICAL EXAMPLES

Figure 3 shows the layout of a quadratic plate for which reinforcement distributions were calculated according to the methods described above. Along the boundaries all degrees of freedom except for horizontal displacements are kept fixed. The design load is chosen as a distributed load of 60 kN/m^2. Due to symmetry only one quarter e.g. that framed by thick lines has to be inspected. Figure 3 displays in a sectional view the arrangement of the different reinforcement layers that run parallel to the plates edges.

For the chosen material parameters for concrete C25/30 and reinforcement steel BSt 500 see Table 1.

A bit of a problem is the value for the Youngs Modulus of concrete on design, mean and calculational level. For a parabola-rectangle-diagram the initial Youngs Modulus is given by

$$E_0 = 2 \cdot f_c/\varepsilon_c \tag{10}$$

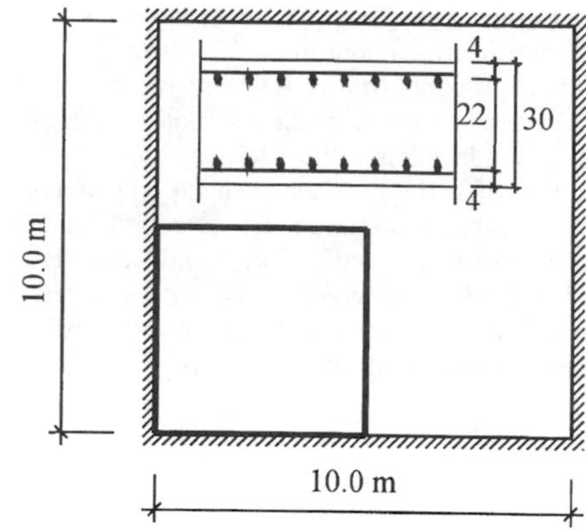

Figure 3. Quadratic plate – view and section.

Table 1. Quadratic plate – material properties.

Concrete N/mm^2	Reinforcement N/mm^2
$E_{0m} = 30500$	$E_s = 200000$
$E_{0d} = 14200$	$E_T = 35.96$
$E_{0R} = 13136$	$f_{ym} = 50.0$
$f_{cm} = -33.0$	$f_{yd} = 434.8$
$f_{cd} = -14.2$	$f_{yR} = 550.0$
$f_{cR} = -18.1$	

This evaluates with $f_{cd} = -14.2$ N/mm^2 and $\varepsilon_c = 2.0\%$ to 14200 N/mm^2. If the sigma-epsilon diagram of concrete is chosen to stay linearly elastic until the yield stress σ_y is reached then

$$E_0 = (2 \cdot f_c - \sigma_y)/\varepsilon_c \tag{11}$$

which evaluates for $f_{cR} = 18.1$ N/mm^2, $\sigma_y = 0.4 f_{cR}$ and $\varepsilon_c = 2.2\%$ to 13136 N/mm^2. This is less than half of the mean value of E_0 and thus results in excessive deflections. In case of structures where second order theory has to be applied the low value for E_0 used in the DIN 1045-1 may lead to an underestimation of the ultimate limit load and thus necessitates too large amounts of reinforcement.

The minimum reinforcement for a sectional height of 0.3 m and the material properties given in Table 1 is 3.52 cm^2/m in each reinforcement direction and layer. For the finite element discretisation of one quarter of the plate a mesh of 25 nine noded isoparametric layered shell elements was chosen. Each element measures 1 m × 1 m. This may seem a little bit course, predominantly for the regions near the boundaries where steep gradients of the internal moments occur. However the mesh chosen is sufficient to demonstrate the differences between the results of the design methods investigated.

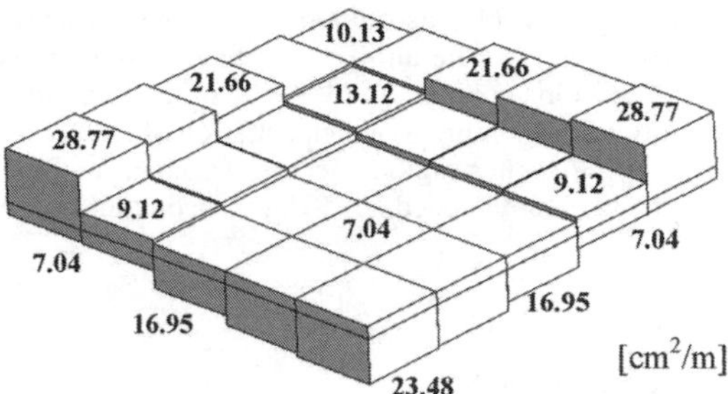

Figure 4. Reinforcement due to linear elastic calculation.

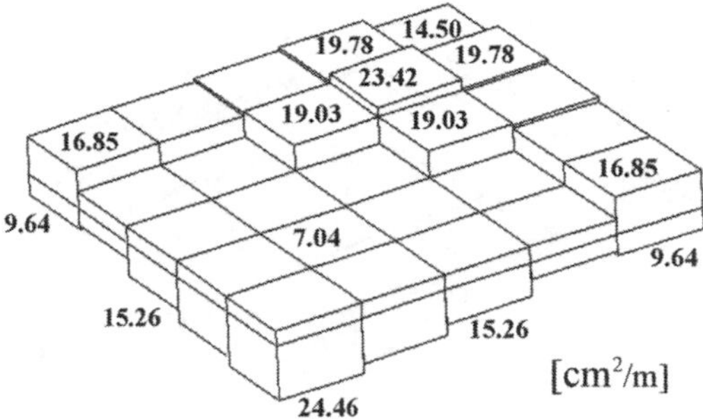

Figure 5. Reinforcement with iteration of sectional forces. Minimum reinforcement taken as initial reinforcement.

6.1 *Reinforcement based on linearly elastic sectional forces*

The result of a sectional reinforcement design for a quarter of the plate based on a linear elastic calculation according to EC2 or DIN 1045-1 is shown in Figure 4. The added reinforcement of the two directions for the upper and lower reinforcement layers are drawn as columns above and below the middle surface of the plate. The overall amount of reinforcement needed according to this design method is 497 kg for one quarter of the plate. A subsequent calculation with mean material values gives an ultimate limit load factor of 1.63 with regard to the design load of 60 kN/m^2.

6.2 *Reinforcement based on the iteration of sectional forces*

In a first try the minimum reinforcement served as the starting point of the iteration process. The result (see Figure 5) shows two substantial differences when compared with the linearly elastic design. The reinforcement is distributed more evenly along the plates boundaries. This is due to the redistribution of internal forces when the reinforcement starts to yield: Yielding starts in the midsides of the plate and progresses towards the corners. Because of the rising steel strains the concrete compression zone in the sections parallel to the boundary gets narrowed in until the concrete breaks. That is why near system collapse the moments at the supports have their maximum near the plates edges. The calculated reinforcement distribution reflects this fact.

A second point to be noticed is the higher amount of reinforcement needed in the edge regions of the plate. There the angle between reinforcement direction and principal moments has a value of approximately 45°. This and the fact that the principal moments along the plates diagonal have different signs makes the reinforcement rather inefficient. Due to the use of mean values for the calculation of sectional forces, and lower design values for sectional reinforcement design, high amounts of reinforcement are needed

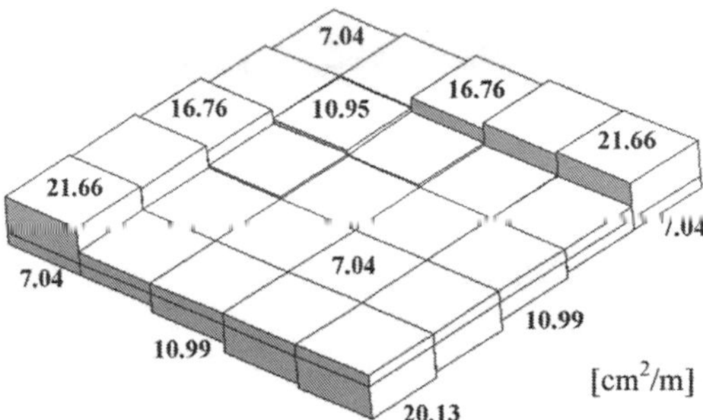

Figure 6. Reinforcement due to iteration of sectional forces. Result of linear elastic design taken as initial reinforcement.

there to ensure reliability on sectional level. The weight of reinforcement calculated amounts to 515 kg which is slightly more than for the linear elastic design variant. From this it can be seen, that the distribution of internal forces resulting from initial minimum reinforcement leads to poor results. With the method of iteration of sectional forces a once established load carrying mechanism serves as the basis for the following iterations until the design goal is met. In order to gain a more suitable distribution of internal forces the calculation was repeated using the reinforcement from the linearly elastic example of 6.1 as the starting point. Figure 6 shows the result attained, which now much more resembles that of the linear elastic calculation. The concentration of reinforcement near the corners has vanished. The total amount of reinforcement for one quarter of the plate is now 401 kg. This shows the influence of the initial reinforcement selected, yet the question remains how to determine this reinforcement distribution for optimum results.

6.3 *Reinforcement based on ultimate limit load iteration*

For the reinforcement design according to DIN 1045-1 the minimum reinforcement was chosen to initialize the iteration process. The parameter C_2 was set to 2.0.

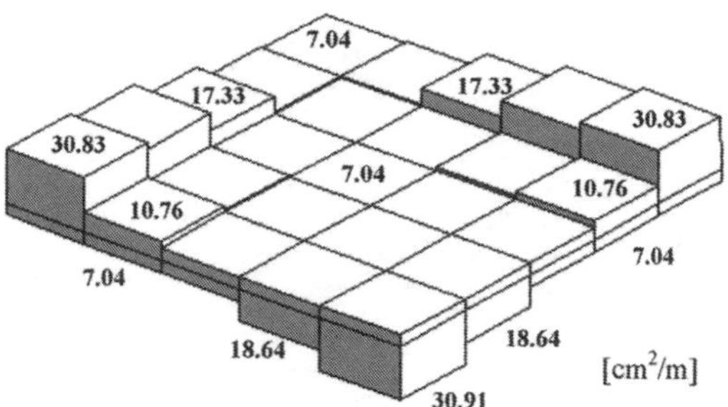

Figure 7. Reinforcement due to ultimate limit load iteration, $C_2 = 2.0$.

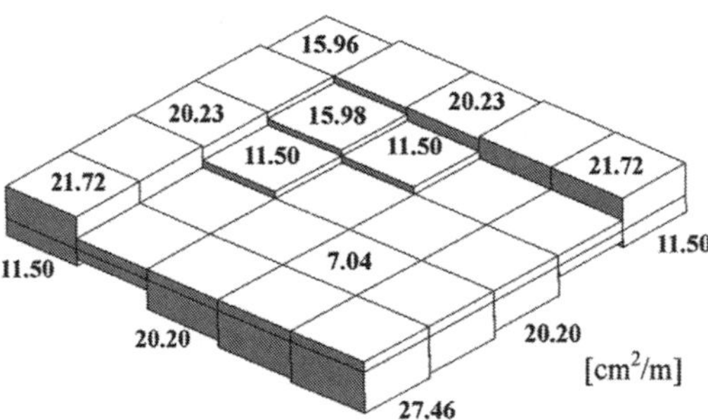

Figure 8. Reinforcement due to ultimate limit load iteration, $C_2 = 0.001$.

It can be seen from Figure 7 that the result resembles that of the linear elastic calculation. However the upper reinforcement is more concentrated at the midsides of the plate, where the plastic steel strains reach their maximum. The lower reinforcement in the center region of the plate is grouped in a relatively small area. The amount of reinforcement needed according to DIN 1045-1 is 413 kg.

Contrary to the example in 6.2 the initial reinforcement distribution chosen has no adverse effects on the results. The reason for this is, that the algorithm utilizes steel strains at the ultimate limit state as input. These show less dependence on the initial reinforcement than sectional forces.

Starting a new ultimate limit iteration with a C_2 value of 0.001 gives the reinforcement distribution shown in Figure 8. As mentioned above $C_2 = 0.001$ means, that the overall reinforcement increment in each step is evenly distributed among the steel layers which yield at the ultimate limit state. Analogous to the solution according to EC2 with initial minimum reinforcement (Figure 8) a concentration of reinforcement in the corner regions occurs. The total amount of reinforcement needed is 550 kg for a quarter of the plate.

7 CONCLUSIONS

From the above described examples the following conclusions can be drawn: The reinforcement distribution from an iteration of sectional forces according to the nonlinear dimensioning concept of EC2 depends on the starting reinforcement chosen. It is thus up to the user to select a suitable initial reinforcement distribution in the structure. In contrast to this, reinforcement design according to DIN 1045-1, using an ultimate limit state iteration, shows less dependency on the initial reinforcement. Regarding total amounts of steel needed, the calculations for EC2 and DIN 1045-1 result in 401 kg (EC2) and 413 kg (DIN 1045-1) for a quarter of the plate investigated. Which is a reduction of roughly 16% when compared with the results from a calculation based on the assumption of linear elastic material behaviour.

REFERENCES

Baumann, T. 1972. *Tragwirkung orthogonaler Bewehrungsnetze beliebiger Richtung in Flächentragwerken aus Stahlbeton,* Deutscher Ausschuß für Stahlbeton, Heft 217, Verlag von Wilhelm Ernst & Sohn.

Dinges, D., Schulz, J.-U., Kollegger, J. 1987. *SEGNID – FE-Programm zur Berechnung geometrisch und physikalisch nichtlinearer Strukturen*, Bericht Nr. 4, Massivbau, GH Kassel.

Kollegger, J. 1991. Beton- und Stahlbetonbau, *Algorithmus zur Bemessung von Flächentragwerk elementen unter Normalkraft- und Momenten beanspruchung*: 114–119, Verlag von Wilhelm Ernst & Sohn.

Wittek, U., Meiswinkel, R. & Lang, C. 2000. *Beitrag zu nichtlinearen Dimensionierungskonzepten von Flächentragwerken aus Stahlbeton*, Bericht 2, Universität Kaiserslautern.

Computational Modelling of Concrete Structures, Bićanić et al. (eds)
© 2003 Swets & Zeitlinger, Lisse, ISBN 90 5809 536 3

Fire resistance of SFRC thin plates

M. di Prisco, R. Felicetti, & M. Colombo
Politecnico di Milano

ABSTRACT: Fire resistance is the most important limitation to the thickness reduction of cross section profiles used in the production of precast roof elements made of Steel Fibre Reinforced Concrete. An experimental investigation was carried out on HPSFRC to assess the mechanical behaviour, when subjected to high temperature. The damage was first investigated providing uniaxial tensile tests after thermal cycles at increasing maximum temperatures. The mechanical behaviour was also identified by means of bending tests, after a thermal cycle, and also for hot specimens after a fast extraction from the oven. A damage evolution law depending on the maximum temperature and a strain invariant was identified. It was adopted to reproduce the fire behaviour of 3 SFRC bent plates subjected to three load levels and a standard fire exposition curve. The comparison shows a good fitting of the experimental results and a significant efficiency of steel fibre in fire resistance.

1 INTRODUCTION

The mechanical behaviour of steel fibre reinforced concrete when exposed to high temperature has not been carefully investigated up to now. It is not clear if steel fibre increases thermal conductivity, causing a reduction of concrete strength, as well as if residual tensile strength, guaranteed at 20°C by pull-out of steel fibre, is still controlled by pull-out at high-temperature and to what extent it is damaged. In fact, high temperature could cause the failure of fibre owing to the reduction of steel maximum strength. Finally, it is interesting to understand if high temperature causes an irreversible damage related only to the maximum temperature reached or dependent on the actual temperature. In the former case, the damage could be experimentally identified after a thermal cycle at assigned maximum temperature, by testing several specimens at room temperature.

The material here investigated is a siliceous high-strength concrete ($f_c = 75$ MPa) with a steel fibre content of 50 kg/m^3. The fibre is hooked, 30 mm long, with an aspect ratio of 45, and made of low-carbon steel ($f_{tk} = 1330$ Mpa). It was selected to produce a new prefabricated covering structure, where traditional diffused reinforcement (a steel mesh fabric) is substituted by steel fibre to support transversal bending, while prestressed reinforcement is still devoted to bear longitudinal bending (di Prisco & Felicetti, 1999; di Prisco, 2000; di Prisco et al., 2000). In the design of this structure, fire resistance is the most important limitation to the thickness reduction of the cross-section profiles.

In order to answer these questions, an experimental investigation was carried out. First, fixed-end notched specimens were tested in uniaxial tension after a thermal cycle at increasing maximum temperatures ($T_{max} = 20$, 200, 400, 600°C; Felicetti et al., 2000). Then, four-point bending tests were performed: the strain rate was increased in order to compare the results obtained on cooled specimens after a thermal cycle and those obtained from hot specimens, just extracted from the oven. This technique was recently proposed by Pimienta in compressive tests (Pimienta, 2001).

The mechanical behaviour detected was used to identify a damage evolution law which was able to describe the dependence of damage on the maximum temperature and a strain invariant according to crush-crack model (di Prisco & Mazars, 1996).

The damage evolution laws were adopted to model fire behaviour of 6 SFRC plates subjected to three load levels in a 4-point bend load scheme and exposed to a standard fire curve (Farina & Riccardi, 2001; di Prisco et al., 2002). The temperature inside the concrete plates was measured in ten positions placed in five layers along the plate thickness. The thermo-mechanical analysis was uncoupled in order to simplify the physics of the problem. After a thermal analysis, whose reliability was estimated on the basis of the experimental temperature profiles, a mechanical analysis was performed in order to predict the time needed to reach collapse, depending on the load level. The analysis was performed conserving the plane-sections beam model and assuming a structural characteristic length equal to the beam depth. It worth noting that only local constitutive laws are here introduced, although the authors, following a non-local approach, have already shown that the material characteristic length of high performance plain concrete, increases at high temperature (di Prisco et al., 1999).

2 EXPERIMENTAL TESTS

2.1 *Uniaxial tension test*

To identify the mechanical constitutive law in uniaxial tension after a thermal cycle, a set of fixed-end tests were carried out (Fig. 1). The specimens were sawn from an original plate (Fig. 2) and then introduced into an oven to be subjected to a thermal cycle. Five temperature thresholds were assumed as a reference: 20, 200, 400, 600, 800°C. The heating process was characterised by a rate of 30°C/h up to the reference temperature, a stabilisation phase at constant temperature (2 hours), a controlled cooling process to just 200°C (rate 12°C/h) and finally a natural cooling to room temperature.

The specimens were prismatic (Fig. 2), double notched and the tests were displacement controlled. Six LVDTs were adopted to measure crack opening, as shown in Figure 2. The test machine used was a closed-loop electro-mechanical press suitably equipped with four hand-adjustable ties to keep the press platens parallel.

The relative displacement rate imposed on the machine platens was 0.05 μm/s up to pull-out phase and then it was increased gradually 50 times.

Three TCC (Tension, Cold specimen, Cold environment) tests nominally identical were carried out. The results (Fig. 4; Farina & Riccardi, 2001; di Prisco et al., 2002) highlight a significant reduction in the peak strength at 400°C and an expected increase of the ductility for high temperature. In fact, pull-out residual strength is less affected by damage with respect to matrix tensile strength.

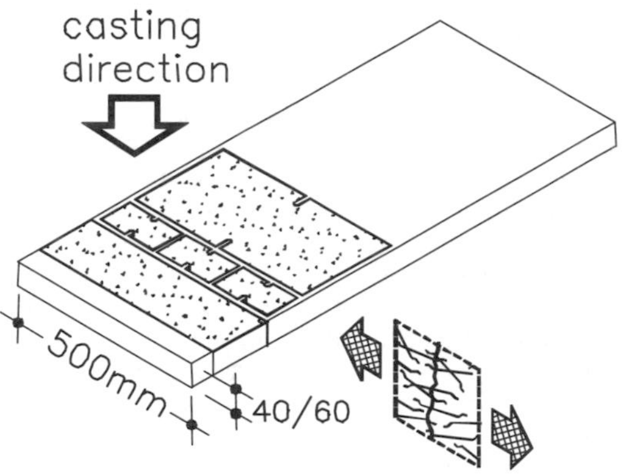

Figure 1. Specimen preparation.

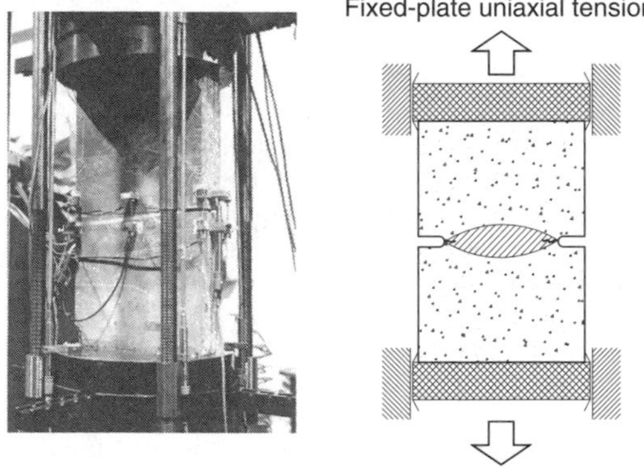

Figure 3. Fixed-plate uniaxial tension tests.

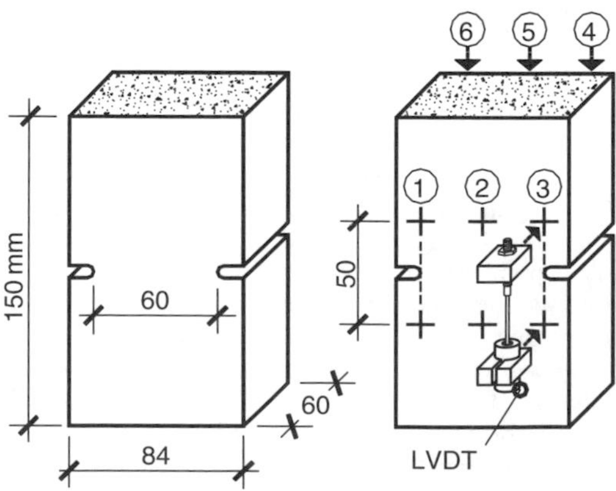

Figure 2. Specimen geometry and LVDT positions.

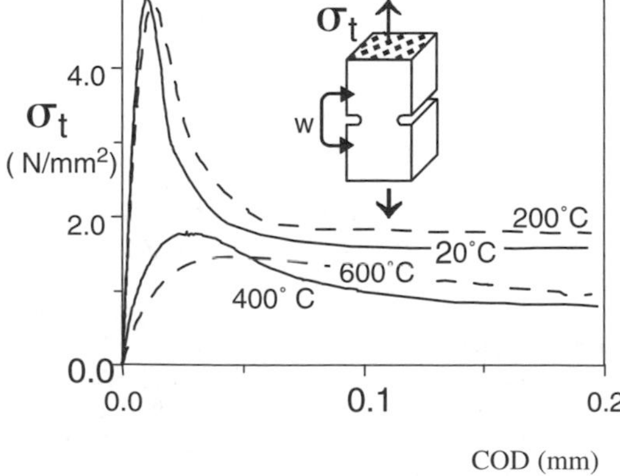

Figure 4. Uniaxial tension test: average curves at various reference temperatures.

The tests do not give any information on the influence of actual temperature on the residual strength, because the tests are performed at room temperature. This means that the mechanical characterisation is only significant if the damage introduced by high-temperature exposition is related to the maximum temperature reached, indicated as a reference.

2.2 *Bending tests*

In order to investigate the role of specimen temperature on the mechanical behaviour in tension, a comparative set of bending tests was planned and carried out in the prefabrication factory.

Four-point bending tests were adopted in the experimental programme. The choice was dictated with the aim of simplifying the experimental procedure as well as by the desire to identify the uniaxial tension by bending. Four materials were investigated assuming the hooked steel fibre type (low carbon: aspect ratio $l_f/d_f = 45$, $f_{tk} = 1300$ Mpa – high carbon: $l_f/d_f = 80$, $f_{tk} = 2300$ Mpa), both 30 mm long, the matrix strength ($f_c' = 40$ and 75 Mpa) and the fibre content as variables.

The material, regarded as a reference, was that already described in the introduction, obtained combining the high performance matrix with the most competitive fibre from an economical point of view (low-carbon). The other materials investigated were obtained by changing the fibre type (high-carbon) or the matrix strength ($f_c = 40$ Mpa); in the latter case a reduced fibre content was also considered (35 kg/m^3).

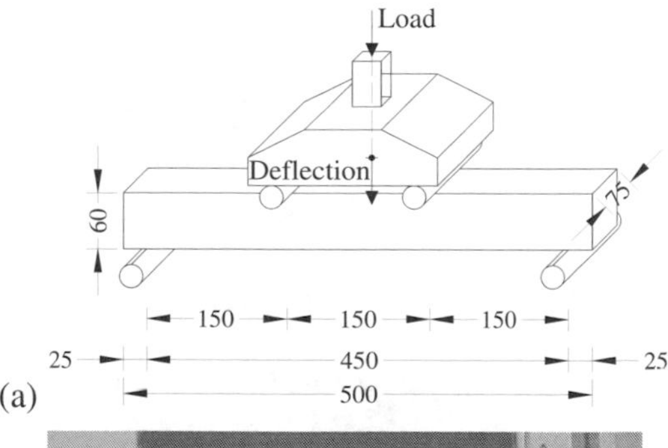

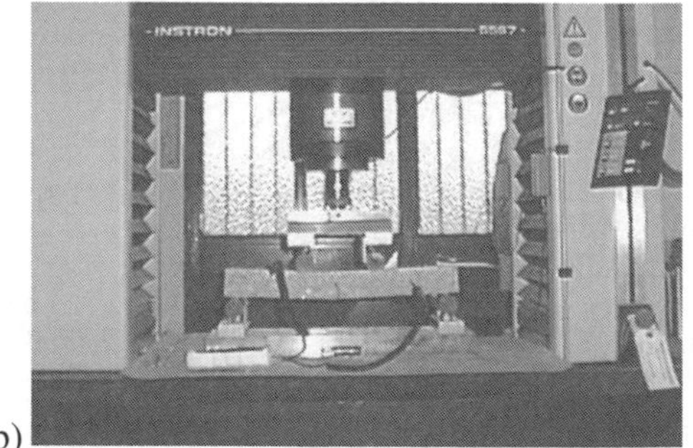

Figure 5. Four point bending tests: (a) geometry (b) load set-up.

The specimens were also prepared by sawing an original plate; their geometry is described in Figure 5a. An electro-mechanical closed-loop press was used: the tests were displacement controlled regarding the press stroke as feedback parameter (Fig. 5b). The specimens were not instrumented in order to accelerate and simplify the testing procedure. Crack opening displacements were estimated on the basis of stroke-displacement measures, knowing the crack position and assuming a rigid-body assumption. It is necessary to point out that only one crack propagates in the critical zone in the whole set of tests. In the cases investigated three tests nominally identical were tested.

The tests can be divided into three groups on the basis of the thermal treatment of each specimen group. Reference specimens tested at room temperature, without being exposed to any heating process. Specimens tested at room temperature after a thermal cycle (BCC: Bending, Cold specimens, Cold environment), characterised by various temperature thresholds (T = 200, 400, 600 and 800°C), as the previous uniaxial tension tests.

Finally, specimens tested at high temperature, just extracted from the oven after reaching the reference maximum temperature (BHC: Bending, Hot specimens, Cold environment).

The heating process had a 24-hour duration. After the reaching of the maximum temperature a 2 hour stabilisation phase was imposed. The BCC specimens were tested 48 hours after the last heating step (stabilisation at T_{max}): the specimens were left inside the oven the first day, and then conserved in laboratory conditions.

The bending tests were carried out at high displacement rate in order to reduce the heat-flow exchange with the environment for BHC specimens as much as possible. The displacement rate assumed was equal to 1 mm/min in the pre-peak stroke range, and 2 mm/min in the post-peak phase.

To estimate the displacement rate influence, some bending tests were performed with a reduced stroke speed. The selected speed was 0.1 mm/min up to a stroke of 1.5 mm, 0.25 mm/min in the following 1.5 mm step of stroke and 0.5 mm/min in the final step.

The difference, expressed in terms of average nominal strength in the range of crack opening displacement indicated, was calculated assuming the peak as the localisation point. It is shown in Table 1. Although the results seem affected by the displacement rate, it is difficult to highlight a clear trend.

The results are summarised by means of the average load versus stroke-displacement curves. The most significant result is the good accordance between BCC

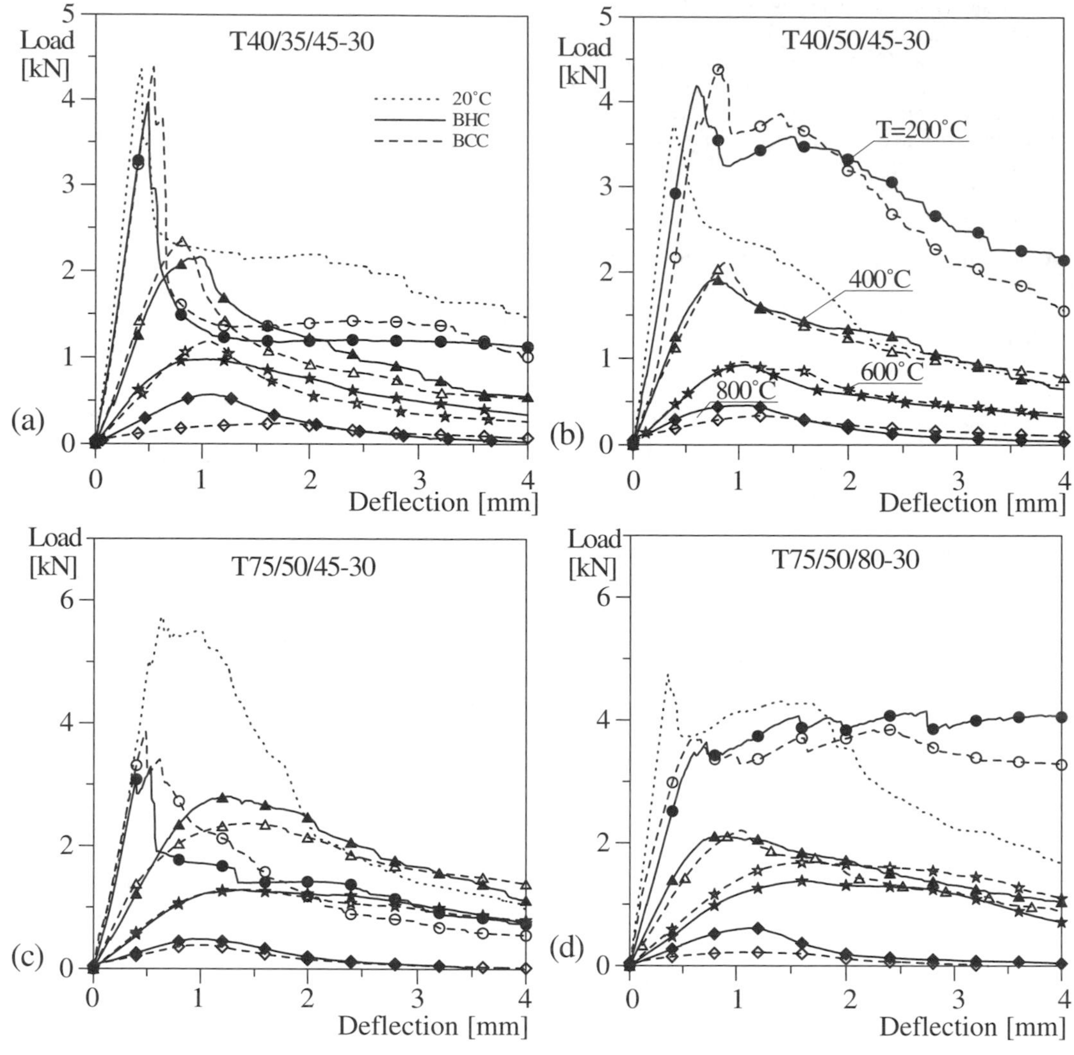

Figure 6. Bending test: average load vs. deflection curves at various references temperatures, with different testing modalities.

Table 1. Bending tests: displacement rate influence on post-cracking toughness related to standard-rate tests.

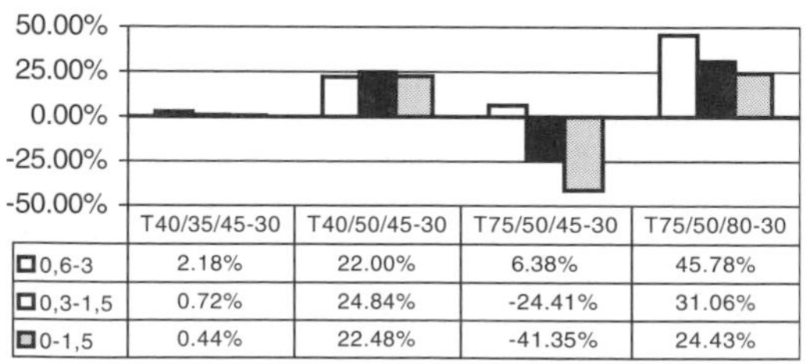

	T40/35/45-30	T40/50/45-30	T75/50/45-30	T75/50/80-30
0,6-3	2.18%	22.00%	6.38%	45.78%
0,3-1,5	0.72%	24.84%	-24.41%	31.06%
0-1,5	0.44%	22.48%	-41.35%	24.43%

and BHC tests. The two set of average curves show a weak mismatch, that gives a measure of the influence of the specimen temperature during testing.

Other important data characterising BHC tests is the temperature variation during testing. It is possible to estimate the temperature in the specimen by means of Eurocode 2 recommendations, taking into account the convection, radiation and diffusion phenomena by means of a finite-difference algorithm, as exemplified in Figure 7. The prevision fits the temperature changes, experimentally detected, quite well: in Figure 8 the theoretical fitting is proposed only for the initial temperature of 400°C (triangle symbols).

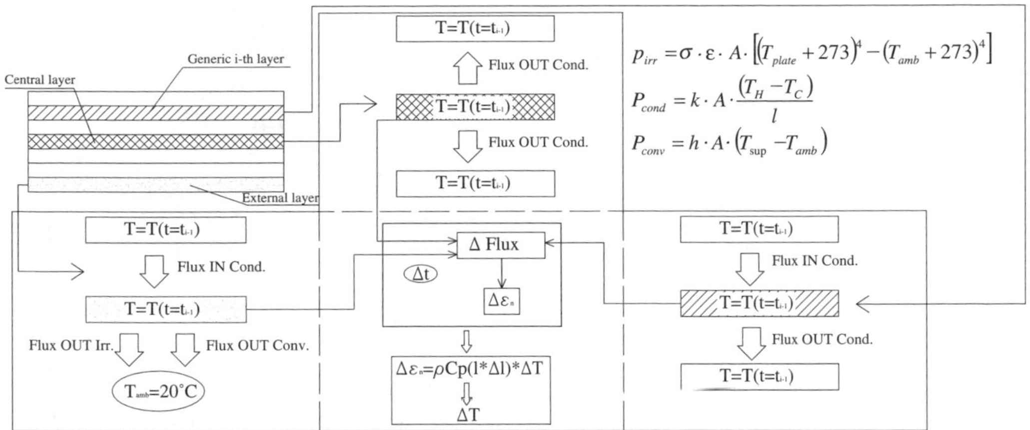

Figure 7. Flow chart of the thermal-identification algorithm.

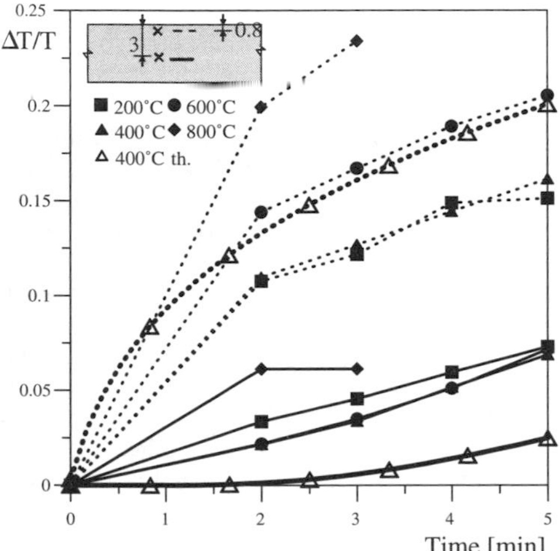

Figure 8. Experimental and theoretical temperature in the center of the specimen and close to the free surface during bending tests.

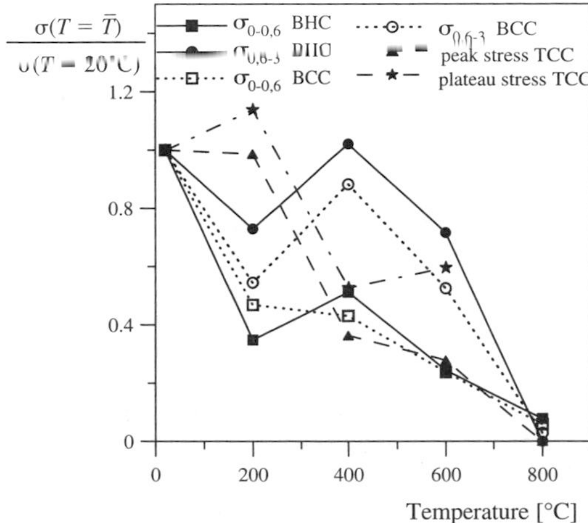

Figure 9. Experimental degradation of model parameters.

In this case the core temperature decreases by less than 5%, while the point close to the free surface decreases by about 20% in the testing time. The experimental measures related to the other reference temperatures, are proposed in the same figure.

A possible measure of the degradation can be computed dividing the strength corresponding to a temperature reference by the strength at room temperature, for the same crack opening displacement evaluated starting from the peak strength of each curve.

The degradation curves computed according to this procedure (Fig. 9), clearly show a larger strength reduction of the peak strength with respect to residual strength evaluated in the range 0.3–1.5 mm of crack opening displacement, always regarding the peak as the localisation point. It is interesting to note that as the degradation of the nominal strength is not monotonic as generally assumed: this effect seems more pronounced for bending than for uniaxial tension. Of course, the specimen deflection at the localisation point (Fig. 6) increases accordingly with the reference temperature up to beside 2 times the corresponding value at room temperature.

3 CONSTITUTIVE MODELING

Constitutive laws proposed in the literature to model SFRC mechanical behaviour, at room temperature, usually combine the common non-linear pre-peak behaviour of plain concrete, with a piecewise linear, or a hyperbolic branch to reproduce the residual

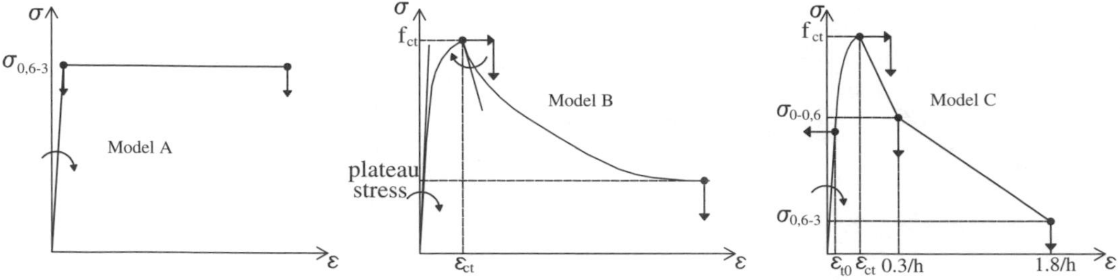

Figure 10. Local uniaxial-tension constitutive models.

strength induced by fibre pull-out mechanism in the post-cracking regime. It is evident that, for design purposes, the matrix behaviour and its related damage due to high temperature is less significant than residual strength, because it cannot be considered to compute the ultimate resistant bending moment and consequently the collapse time for an assigned fire exposure.

In this case the designer could be more interested to take into account a simplified constitutive model as an elasto-plastic one. It needs the identification of only one parameter to adjust the limit strength. In the present investigation this parameter is chosen on the basis of the nominal residual strength $\sigma_{0.6-3}$, averaged in the range 0.6–3 mm of crack opening displacement (Fig. 10). This parameter is computed as one third of the nominal stress. The constant 3 is related to the ratio between the elastic bending module of a rectangular bent section, $W_e = bh^2/6$, and the plastic bending module, $W_p = bh^2/2$, computed assuming the compressive force concentrated in the extrados fibre and a rigid-plastic tensile behaviour.

A smeard crack approach is adopted here; it needs a characteristic length in order to spread the strain on a finite material volume homogeneously, to conserve the toughness experimentally detected. For SFRC plates without any other steel reinforcement, two characteristic lengths can be introduced. A material characteristic length related to a fraction of the steel fibre length ($<l_f/5$), and a structural characteristic length related to the kinematic model are assumed. Usually, the larger value is introduced in the analysis. In fact, the first can be considered active when a F.E. analysis is carried out, with a finite element size significantly smaller than the reference volume. In this case the kinematic model, expressed by shape functions, does not constraint the possible deformation. Otherwise, when the assumed kinematic model prevents the real deformation inside of a structure piece (e.g. plane-section model), the structural characteristic length plays the key role and the other length is disregarded.

In the following, having adopted the plane-section kinematic model, only the structural characteristic length will be considered. Therefore, the smeared approach allow us to introduce a continuum description of localisation, expressing the tensile constitutive law in terms of tensile strain computed as crack opening displacement divided for the structural length. The length value here was always assumed equal to the specimen depth *h* of the bending tests.

With reference to residual bearing of fire damaged structures, it could be interesting to know not only the residual post-cracking strength at assigned crack opening, but also the peak strength in order to compute the cracking bending moment. In this case, starting from the experimental data presented, a privileged way could be represented by the identification of matrix behaviour directly from uniaxial tension test, because of the fine instrumentation there introduced. Two models are here proposed to trace the pristine response in uniaxial tension. The first (Farina & Riccardi, 2001; Fig. 10) adopts three parameters to describe the prepeak behaviour, and a re-elaboration of Hordjik's curve (Hordijk, 1991) to describe the softening behaviour, depending on other two parameters (Model B; Fig.10). The second model (Model C; Fig.10) introduces a linear elastic phase in the prepeak description and the choice involves the addition of one parameter, the elastic threshold ε_{t0}; on the contrary, softening is described by two linear branches, controlled by two parameters associated to two fixed crack opening thresholds. The last ones are selected as the mean values in two assigned COD ranges, in order to reproduce the crack situation of service limit states ($\sigma_{0-0.6}$; $0 < w < 0.6$ mm) and ultimate limit states ($\sigma_{0.6-3}$; $0.6 < w < 3$ mm). The former parameter was computed multiplying the nominal strength by 0.45.

Each parameter introduced is assumed variable with temperature and can be identified using the experimental tests carried out. A linear variation of the parameters was taken into account. In particular, adopting model C, it is possible to identify the elastic threshold ε_{t0} (Fig. 11), starting from uniaxial-tension test or bending of hot (BHC) or cold specimens (BCC).

Taking advantage of the relations introduced in the *crush-crack* model (di Prisco & Mazars, 1996), between irreversible tensile strains ε_{+}^{ir} and damage *D*,

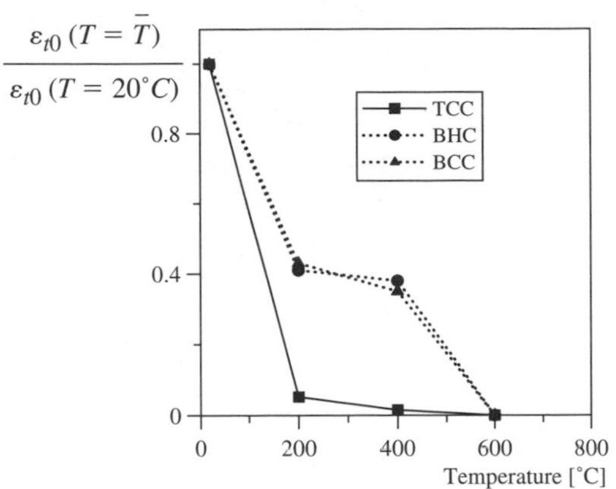

Figure 11. Damage threshold vs. temperature.

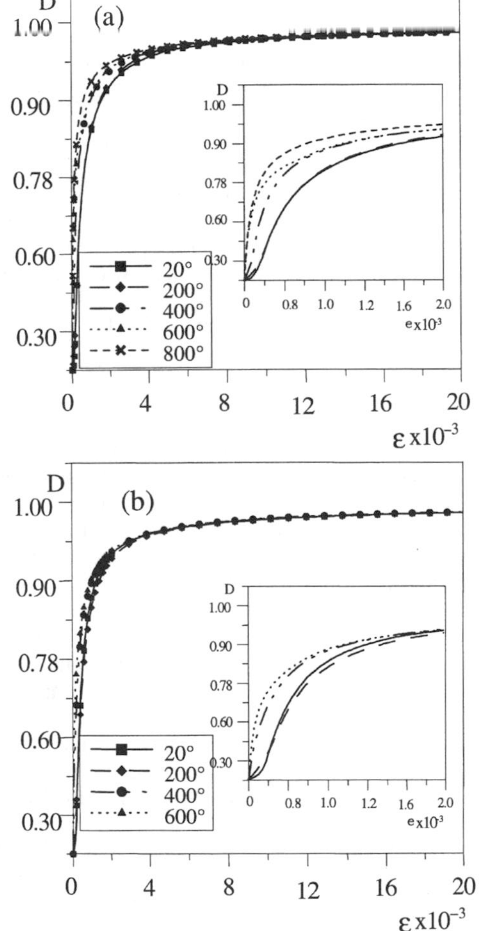

Figure 12. Damage evolution law according to model C (a) and model B (b).

assumed as a scalar internal variable, it is possible to consider damage as a function of a strain invariant and the maximum temperature T_{max}.

In the case of uniaxial tension, the strain invariant reduces to the total tensile strain ε and consequently the constitutive relations become:

$$\varepsilon = \varepsilon^{el} + \varepsilon_{+}^{ir} \tag{1}$$

$$D = D(\varepsilon, T_{max}) \tag{2}$$

$$\varepsilon_{+}^{ir} = \varepsilon_{t0}^{T=20°C} f(D) \tag{3}$$

$$f(D) = \frac{D}{\left(1 - D^{1.5}\right)^{1.6}} \tag{4}$$

The damage evolution laws can thus be identified for each T_{max} experimentally investigated, with reference to models C and B (Figs.12a,b).

The curve trend indicates a damage growth associated to the maximum temperature, but the whole curve set respects the physical constraint represented by a always positive damage increment.

4 FIRE TESTS ON THIN PLATES

As a check of the constitutive relations presented, a structural test is here proposed. The test consists in the bending of thin plates exposed to a standard fire according to the National recommendations (UNI 9502, 2001).

The oven was equipped to test three plates (180 × 60 × 6 cm; Fig. 13), made of the same HPS-FRC material (aforementioned T75/50/45-30), at the same time. The plates were instrumented with ten thermo-couples introduced as shown in Figure 14.

The plates were bent by means of concentrated loads, according to a 4-point bending scheme with a 570 mm long lever arm. Three load conditions were examined: 12% (only dead load), 24% and 40% of cracking bending moment estimated at room temperature (equal to 2.85 kNm for $f_{ctf} = 7.91$ Mpa).

Six SFRC plates were tested. The first three plates (test 1) were simply supported on the whole perimeter. They were tested with the aim of investigating only the thermal problem and to evaluate the reliability of the thermal parameters proposed for HPC by Codes, to compute temperature profiles when steel fibre is added to the mix. The other three plates (test 2) were supported only along the opposite short sides. The temperature vs. time curve, defining the fire exposure, is shown in Figure 15. A sudden drop of the curve revealed an explosive spalling of a small plate portion, which prevents the following of a correct temperature evolution versus testing time. Also the three chambers did not have exactly the same temperature, as revealed by a thermocouple introduced in each chamber. These facts complicated the result interpretation

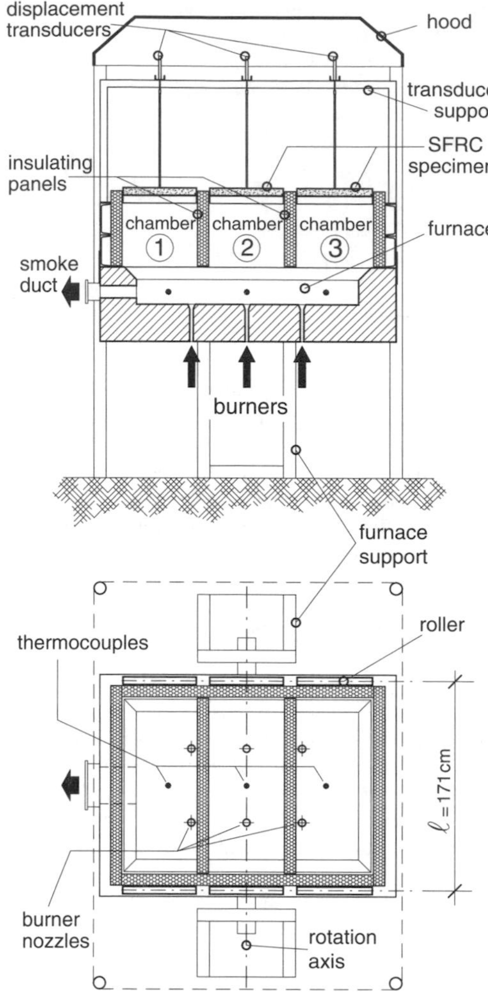

Figure 13. Oven set-up.

finalised to determine the time collapse and needed a modelling to estimate the real collapse time according to the accepted standard fire exposure curve.

Finally, three plain concrete plates made of the same concrete were also tested. Also in this case a spalling of a plate portion occurred. The collapse time detected were significantly reduced with respect to the previous SFRC plates. The values are resumed in Table 2.

4.1 *Temperature profiles*

By adopting the thermal parameters suggested in EC2 recommendations and by using the same algorithm described in Figure 7 with different boundary conditions, in order to take the temperature measured by the thermocouple introduced in the oven chamber into account, it is possible to fit the experimental

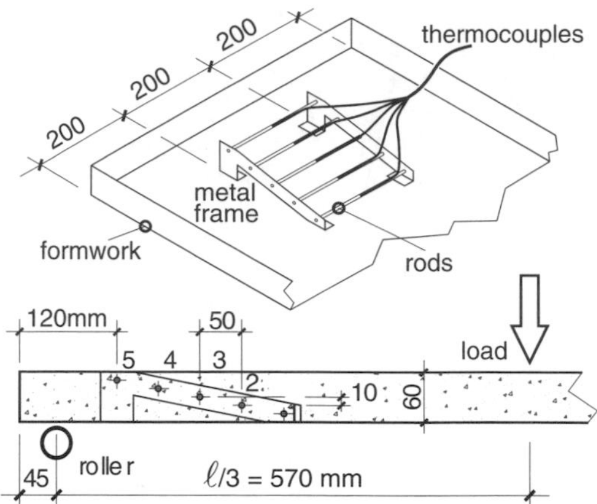

Figure 14. Thermocouples position in the plate section.

temperature profiles along the plate thickness (Fig. 16) quite well. The experimental scattering corresponding to six measures (two measures each plate for three plates) confirm the reliability of a thermal prevision when performed with HPC parameters. This means that steel fibre does not significantly affect the thermal properties of cementitious materials. The good fit suggests also a right way to compute the shifting of collapse times experimentally detected. In fact, it is possible to compute the according to the right fire exposure assumed and compare them with the experimental ones, in order to estimate the real shifting in the time axis of failure occurrence (Fig.17).

4.2 *Bearing capacity*

On the basis of an uncoupled solution of the thermomechanical problem, it is possible to calculate the resistant bending moment of the plate section when subjected to a standard fire exposure. After the computation of temperature profile, by using the finite-difference algorithm, damage associated to each plate layers for each single time is calculated.

A theoretical prevision was carried out following this procedure and adopting the three constitutive laws previously introduced. The whole set of parameters were identified starting from uniaxial tension tests to adjust the pre-peak constitutive behaviour and the tangent to the softening branch at the peak, just after localisation, and from BHC bent specimens to identify the residual behaviour related to fibre pull-out.

Only the analysis performed using model B took into account the real loading history; the other analyses assume to load the structure, characterised only by an assigned temperature profile, at each step.

The results highlight a significant contribution of the matrix strength up to twenty minutes (Fig. 18).

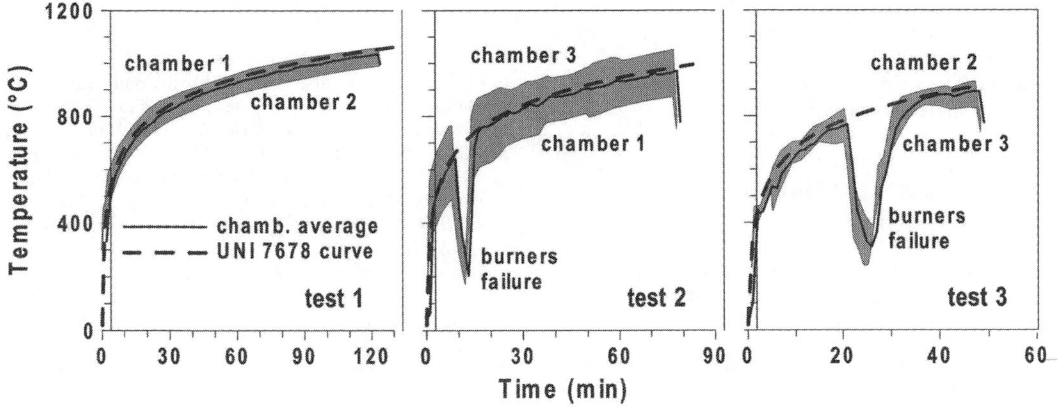

Figure 15. Fire curves: standard and experimental path.

Table 2. Collapse time for the considered load levels.

	Collapse time (min)		
Material	Max. load	Mean load	Min. load
PC	6	16	51
SFRC	36	50	65(°)

(°) Collapse due to partial spalling.

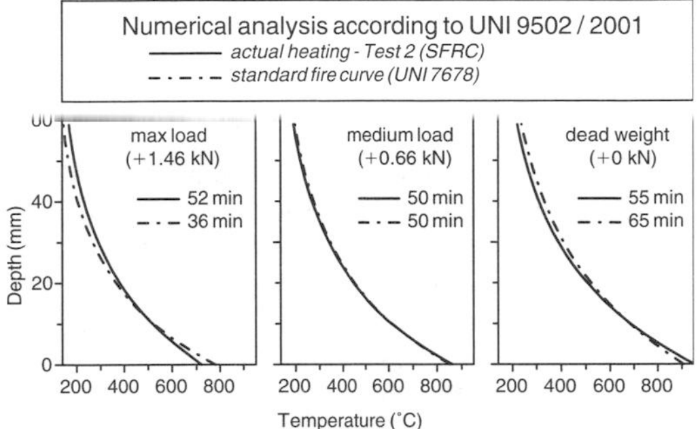

Figure 17. Temperature profiles used to shift collapse times.

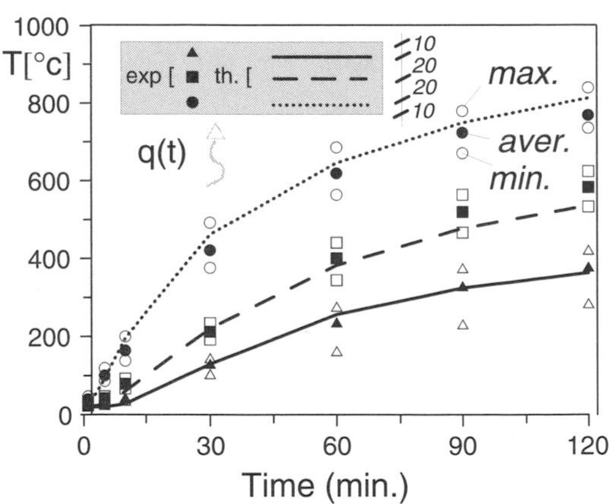

Figure 16. Temperature trends in plate layers.

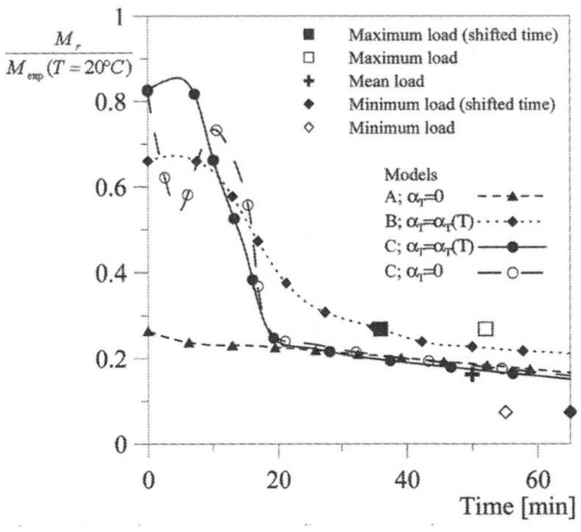

Figure 18. Resistent moment vs. fire exposure time.

The minimum design time-resistance is more than twenty minutes; therefore, for design purposes, the elasto-plastic approach (model A) gives reliable results. It is important to underline that this computation ignores completely thermal strains induced by the temperature growth. This means that the state of stress induced by the nonlinear temperature distribution along the thickness is quite negligible at that time.

The resistant bending moment computed according to models B and C give comparable results. It is interesting to observe in the first minutes of fire exposure how C model, without considering thermal strains, has characterised by a strange bulb in the bending moment decrease, related to the unexpected increase of nominal strength at about 400°C (Fig. 6c) for the investigated material. The worst fit of experimental data concerns the plate subjected only to dead loads.

5 CONCLUDING REMARKS

On the basis of an extended experimental investigation performed on mechanical behaviour of SFRC exposed to high temperature, the following conclusions can be highlighted:

- damage caused by high-temperature exposure seems strictly related to the maximum temperature reached in the material. Although the tests performed on cold and hot specimens showed some differences, the good correlation suggests to consider damage as a function of the maximum and not of the actual temperature;
- concrete matrix strength is more deteriorated than residual strength by high-temperature. As a consequence, the exposure to high temperature makes the mechanical behaviour significantly less brittle;
- three constitutive relationships are proposed to model uniaxial tensile behaviour according to a smeared crack approach. The identification of the parameters can be performed by means of bending and/or uniaxial tension tests. Experimental data confirm a not negligible difference between the two basic tests.
- the damage evolution law $D = D(\varepsilon, T_{max})$ cannot be reduced to the product of two separated functions $D = D(\varepsilon)D(T_{max})$ as assumed by other authors (Stabler, 2000);
- thermal properties seem scantly affected by steel fibre presence. A prevision of temperature profile based on HPC parameters suggested by Codes fits quite well experimental data;
- the modelling of bent plates exposed to fire performed by adopting the three constitutive models introduced in a plane-section approach gives an acceptable prevision of the time needed to reach collapse. For design values of time resistance, the elasto-plastic approach seems the most convenient even disregarding the anelastic strains.

ACKNOWLEDGEMENTS

The authors thank Magnetti Larco-Building for the technical support in the experimental investigation and Ms. C. Failla for the direction of the experimental investigation carried out in the factory.

REFERENCES

CEB-FIP bull.°228 1995. High Performance Concrete: 1–45.

di Prisco, M. & Mazars, J. 1996. Crush-Crack: a Non-local Damage Model for Concrete. *Mechanics of Cohesive-Frictional Materials and Structures* 1321–1347.

di Prisco, M. & Felicetti, R. 1999. HSC thin-web roof-elements: an experimental investigation on steel fibre benefits. *Proc. of the 5th Int. Symposium on Utilization of High Strength, High Performance Concrete:* 546–555. Sandefjord Norway.

di Prisco, M., Felicetti, R. & Gambarova, P.G. 1999. On the evaluation of the characteristic length in high strength concrete. In A. Azizinamini, D. Darwin & C. French (eds.), *High Strength Concrete*: 377–390. ASCE.

di Prisco, M. 2000. Design of SFRC precast roof elements. In M. di Prisco & G. Toniolo (eds.), *Structural Applications of steel fibre reinforced concrete*: 33–53. Milan: CTE Press.

di Prisco, M., Felicetti, R. & Iorio, F. 2000. FRHPC Precast roof elements: from constitutive to structural behaviour in bending. *Proc. of 5th RILEM Symp. on Fibre Reinforced Concrete BEFIB*: 233–243. Lyon, France.

di Prisco, M., Felicetti, R. & Iorio, F. 2001. The bending behaviour of HPC thin plates. In M. di Prisco & G. Plizzari (eds.), *Fracture Mechanics in HPC.* Brescia (in printing).

di Prisco, M., Failla, C. & Felicetti, R. 2002. On fire behaviour of steel fibre reinforced concrete, (in Italian), *Proc. 14° CTE Conf.* : 305–316. Mantova: CTE press.

Farina, L. & Riccardi, G. 2001. On fire behaviour of SFRC plates (in Italian): 1–446. Ms. Thesis, Politecnico di Milano.

Felicetti, R., Gambarova, P.G., Natali Sora, M.P., & Khoury, G.A. 2000. Mechanical behaviour of HPC and UHPC in direct tension at high temperature and after cooling. *Proc. 5th Symp. on Fibre-Reinf. Concr. BEFIB 2000*: 749–758. Lyon, France.

Hordijk, D. 1991. Local approach to fatigue of concrete. Ph.D.Thesis: 1–207. Delft University of Technology.

Pimienta, P. 2001. Évolution des caractéristiques des BHP soumis à des températures élevées: Résistances en compression et modules d'élasticité. *Cahiers du CSTB* n. 3353: 1–14.

prEN 1992-1-2 2000. Eurocode 2: Design of concrete structures. Part 1.2: General rules – Structural fire design.

Stabler, J.T. 2000. Computational Modelling of Thermo-Mechanical Damage and Plasticity in Concrete. PH.D. Thesis: 1–203. University of Queensland.

UNI 9502 2001. Analytical procedure to assess fire resistance of plain concrete, r/c and p/c, building elements.

Computational Modelling of Concrete Structures, Bićanić et al. (eds)
© 2003 Swets & Zeitlinger, Lisse, ISBN 90 5809 536 3

Reliability assessment of concrete structures

R. Pukl
Červenka Consulting, Prague, Czech Republic

D. Novák
Faculty of Civil Engineering, Brno University of Technology, Czech Republic

K. Bergmeister
Institute of Structural Engineering, University of Applied Sciences, Vienna, Austria

ABSTRACT: A probabilistic approach to the nonlinear analysis of concrete structures is presented. Finite element software ATENA was integrated with stochastic package FREET in order to randomize nonlinear simulation of engineering structures. The basic aim of the stochastic nonlinear analysis is to obtain an estimation of the structural response statistics (failure load, deflections, cracks, stresses, etc.). Reliability assessment of the analyzed structure is performed.

1 INTRODUCTION

Safety-based maintenance management of large engineering structures like bridges is recently subject of numerous research studies (Frangopol 2000, Enevoldsen 2001, Casas et al. 2002, Eichinger 2002). Objective assessment of safety and reliability of the structures should be based on realistic structural analysis. Therefore, advanced solution methods for structural analysis, namely nonlinear finite element simulation, should be employed in combination with probabilistic approach.

However, usual stochastic techniques like pure Monte Carlo method need large number of samples (thousands) in order to obtain acceptable good results, which is in strong contradiction to the demanding nonlinear finite element analysis used for each sample. Therefore, special procedures should be employed for the probabilistic simulation, which could satisfactorily estimate stochastic parameters for reliability assessment using only small number of samples (tens).

Furthermore, a system for reliability assessment should be supported with user-friendly interactive graphical environment in order to support the user during the whole demanding task – nonlinear stochastic analysis of concrete structures.

SARA software presented in this paper integrates the most advanced methods of nonlinear mechanics with appropriate reliability procedures into complex and versatile system, which is ready to use for practical applications. It can be employed in stochastic analysis of structures, sensitivity studies, optimization of structures, lifetime and degradation analysis, reliability assessment as well as identification of computer models with the real structures based on monitoring data (Bergmeister 2002).

2 NONLINEAR FINITE ELEMENT ANALYSIS

ATENA software is well established for realistic simulation of damage and failure of concrete and reinforced concrete structures in deterministic way (Červenka 1998, 2000, 2002). The constitutive relation in a material point (constitutive model) plays the most crucial role in the finite element analysis and decides how the structural model represents reality. Since concrete is a complex material with strongly non-linear response even under service load conditions, special constitutive models for the finite element analysis of concrete structures are employed (Červenka et al. 2001).

Tensile behavior of concrete is modeled (Fig. 1) by non-linear fracture mechanics combined with the crack band method (Hillerborg et al. 1976, Bažant & Oh 1983) and smeared crack concept. Main material parameters are tensile strength, fracture energy and shape of the stress-crack opening curve – the model proposed by Hordijk (1991) is employed. A real discrete crack is simulated by a band of localized strains (Fig. 2). The crack strain is related to the element size. Consequently, the softening law in terms of strains for the smeared model is calculated for each element individually, while the crack-opening law is preserved. This model is objective due to the energy formulation

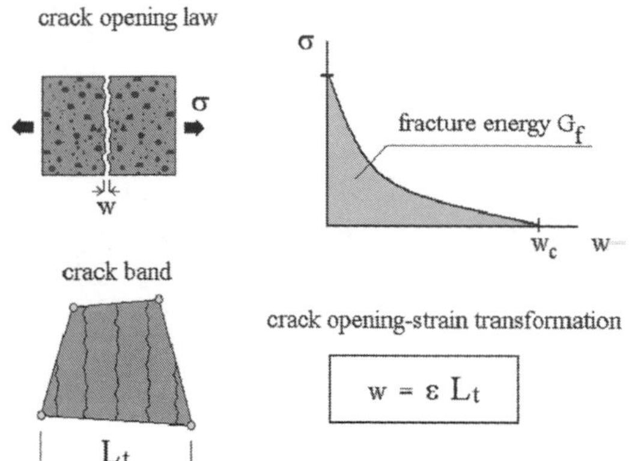

Figure 1. Smeared crack model for tensile behavior of concrete.

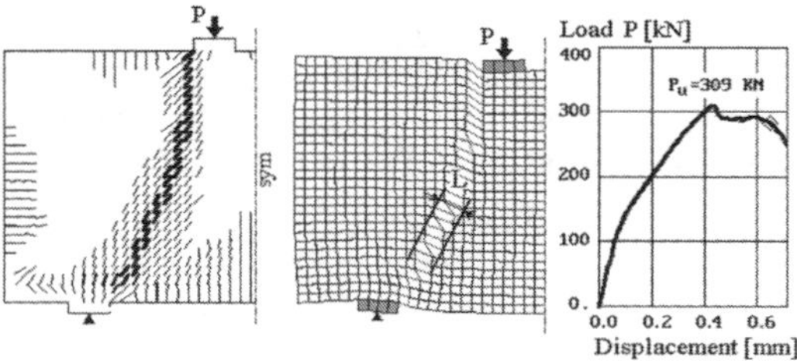

Figure 2. Example – crack band in a shear wall analysis.

and its dependency on the finite element mesh size is neglectable, which was confirmed by numerous studies (Červenka & Pukl 1995, Červenka 1998).

Two alternative crack models are recognized: fixed and rotated crack model (Rots & Blaauwendraad 1989). In the fixed crack model the crack direction is determined and fixed at the time of crack initiation. In the rotating crack model the crack direction is identical with a principal strain direction at every stage and rotate if the strain direction changes. The main difference in these crack models is the absence of shear stresses on the crack plane in the rotating crack model due coincidence of principal strain directions with the crack orientation, which makes the rotating crack model more simple. In the fixed crack model the shear resistance of the cracks is modeled by means of the variable shear retention factor, which reflects the aggregate interlock effect of cracked concrete.

Concrete in plane stress condition can be well described by a damage model. The model used in ATENA is based on the equivalent uniaxial law (Fig. 3), which covers the complete range of the plane stress behavior in tension and compression. The effect of biaxial stress state on the concrete strength is captured by the biaxial failure function due to Kupfer et al. (1969), Figure 4. For the tensile response (cracking) the crack band method described above is applied.

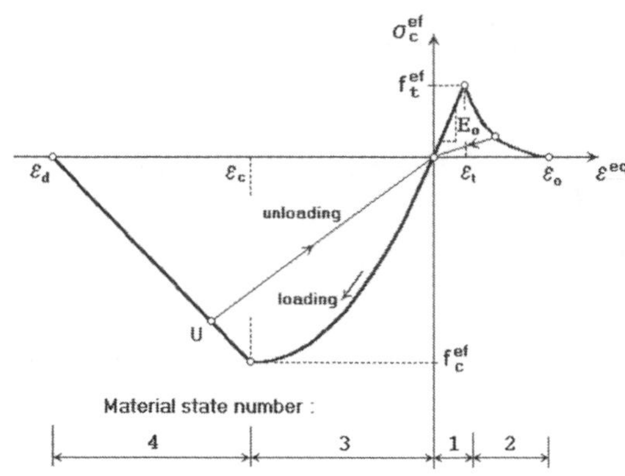

Figure 3. Equivalent uniaxial law.

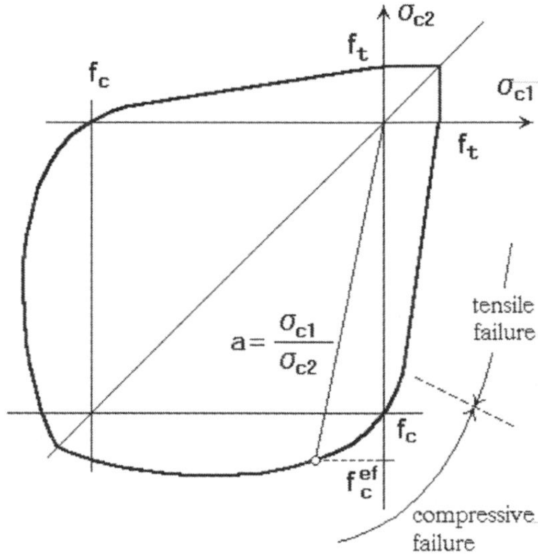

Figure 4. Biaxial failure function.

Similar method is applied for the compressive softening. Thus the complete softening behavior is based on an objective and mesh independent approach.

Concrete in three-dimensional stress state (Fig. 5) is covered by the theory of plasticity with a non-associated flow rule (Červenka & Červenka 1999). An important phenomenon in the three-dimensional behavior of concrete, which is reflected in the implemented material model according to Menétrey & Willam (1995), is a strong influence of the lateral stresses to the compressive strength, so called confinement. The strength is increasing with the hydrostatic compressive stress (Fig. 6).

The failure occurs when the deviatoric shear stress reaches the failure function. The plasticity theory describes also the plastic flow and volume change due to a distorsion and can model the volume increase of

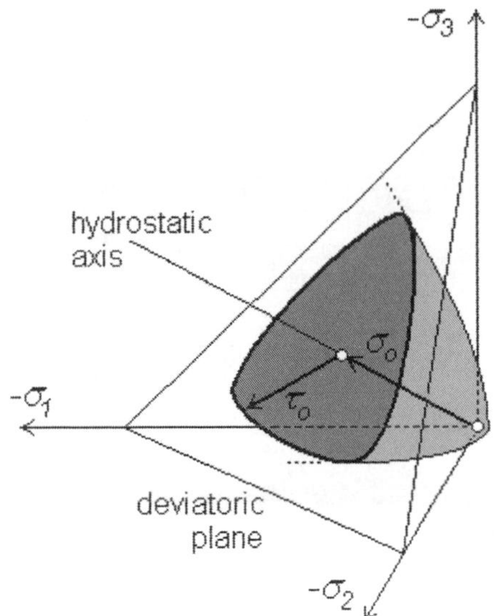

Figure 5. Concrete failure surface in 3D stress state.

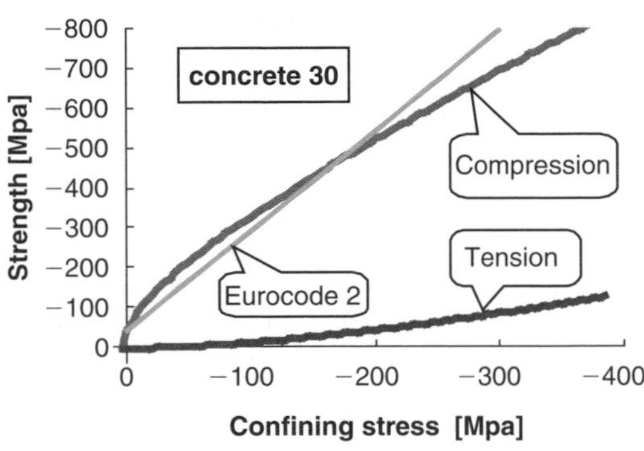

Figure 6. Strength increase under confinement effect.

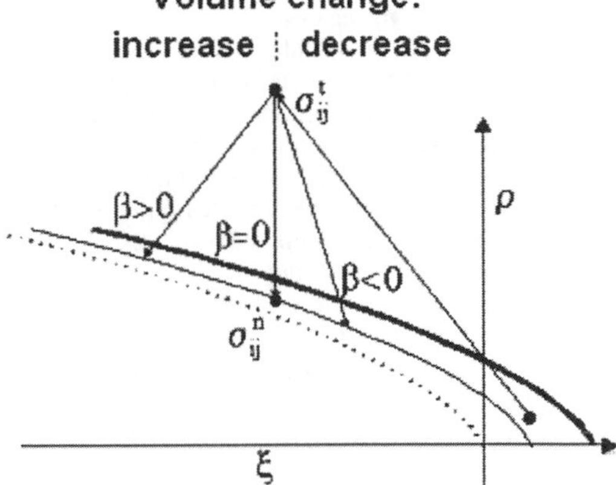

Figure 7. Return mapping algorithm.

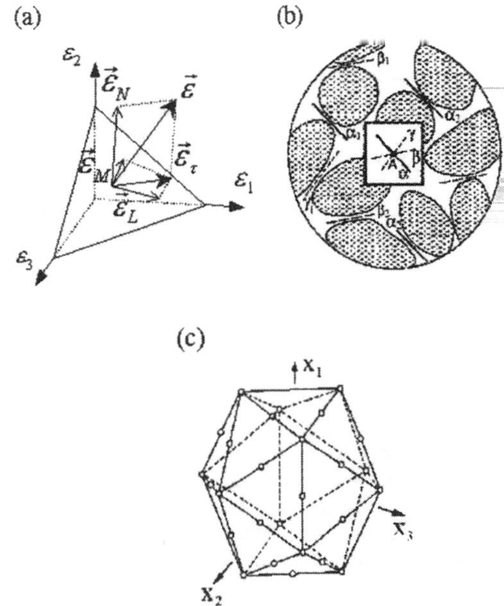

Figure 8. Fundamental concept of the microplane model.

concrete under plastic deformations. The predictor-corrector scheme with return mapping algorithm (Fig. 7) is used for this purpose.

The general three-dimensional behavior of concrete can be also well described by the microplane models (e.g. Pukl et al. 1998) based on a microstructural damage idea: constitutive laws are formulated on several planes with an arbitrary orientation – microplanes (Fig. 8). Stress–strain equations are defined independently for projection of the macroscopic strain tensor to the microplanes. The macroscopic material response (stress tensor in a material point) is consequently integrated over all microplanes by principle of virtual work. Microplane model by Bažant et al. (2000) is implemented in ATENA and can be used for solution of practical cases.

Many other material models are implemented in ATENA in order to support a successful simulation (reinforcement, steel, bond slip of reinforcing bars, interface, soil, etc.). Out of these models the bond between concrete and reinforcing bars is treated more in detail in another paper at this conference (Červenka et al. 2003).

An efficient solution of engineering problems based on the described material models is supported by user-friendly graphical environment (GUE). Native ATENA GUE is available for 2D and rotationally symmetrical problems. It supports the user during pre- and postprocessing and enables real-time graphical tracing and control during the analysis.

The preprocessing includes an automatic meshing procedure, which generates Q10, isoparametric quadrilateral and triangular elements. Reinforcement can be treated in form of reinforcing bars, prestressing cables or as smeared reinforcement given by reinforcement ratio and direction. The discrete reinforcement is fully independent on the finite element mesh.

The structure can be loaded with various actions: body forces, nodal or linear forces, supports, prescribed deformations, temperature, shrinkage, prestressing. These loading cases are combined into load steps, which are solved utilizing advanced solution methods: Newton-Raphson, modified Newton-Raphson or arc-length. Secant, tangential or elastic material stiffness can be employed in particular models. Line-search method with optional parameters accelerates the convergence of solution, which is controlled by energy-based and residua-based criteria.

The interactive solution control window enables graphical as well as numerical monitoring of the actual task, and supports user interventions during the analysis (user interrupt, restart). For the 3D interactive solution control and postprocessing a graphical console AtenaWin is available.

The graphical postprocessing can show cracks in concrete with their thickness, shear and residual normal stresses. User-defined crack filter is available for obtaining of realistic crack patterns. Other important values (strains, stresses, deflections, forces, reactions etc.) can be represented graphically as rendered areas, isoareas, and isolines, in form of vector or tensor arrow fields. All values can be also obtained in well-arranged numerical form.

In the necessary modifications of ATENA for implementation into SARA system it was a crucial point to keep all of the features described in the concise survey above available also for the repeated stochastic analysis. It was enabled due to versatile programming architecture and build up of ATENA system, which supersedes usual finite element packages.

ATENA works under Microsoft Windows operating system and its code is written in Microsoft Visual C++. It heavily uses MFC (Microsoft Foundation Class) and ATL (Active Templates Libraries) libraries, thereby ensuring high productivity in code development and high compatibility with other third-party PC-based software. The code has object-oriented architecture and is created in hierarchical manner (Fig. 9) and each software layer has its own DLL library (Dynamically Linked Libraries). Code and associated data are arranged in objects together in i.e. C++ classes. ATENA analytical core consists of several DLLs and a few control programs (Červenka et al. 2001).

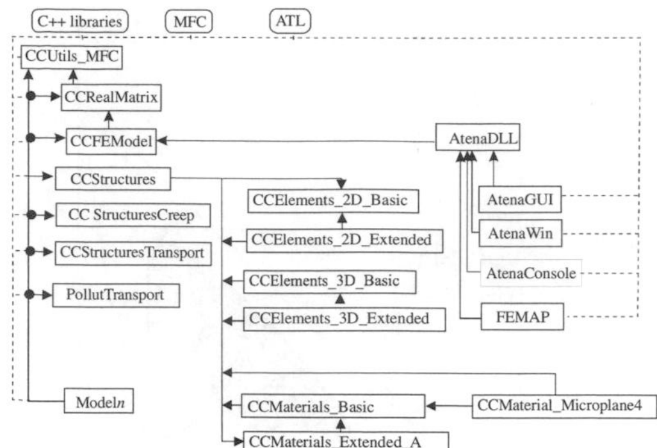

Figure 9. Layered structure of ATENA.

3 PROBABILISTIC APPROACH

A probabilistic software for statistical, sensitivity and reliability analysis of engineering problems FREET (Feasible Reliability Engineering Efficient Tool) is based on efficient reliability techniques. It is designed in the form suitable for relatively easy assessment of any user-defined computational problem written in C++ or FORTRAN programming languages, but it has been developed with a focus on the computationally intensive problems, which do not allow to perform thousands of samples (Novák et al. 2002, 2003).

A special type of numerical probabilistic simulation called Latin Hypercube Sampling (LHS) makes it possible to use only a small number of Monte Carlo simulations. This technique, originally proposed by McKay et al. (1979), appeared to be useful reliability technique until present days. LHS is a special type of Monte Carlo numerical simulation, which uses the stratification of the theoretical probability distribution function of input random variables.

The LHS is very efficient for the estimation of first two or three statistical moments of structural response. It requires a relatively small number of simulations – repetitive calculations of the structural response resulting from adopted computational model (tens or hundreds). The utilization of LHS strategy in reliability analysis can be rather extensive, it is not restricted just for the estimation of statistical parameters of structural response (Ayyub & Lai 1989, Novák et al. 1998).

There are generally two stages of LHS:

- Samples for each variable are strategically chosen to represent the variables distribution function;
- Samples are reordered to match required statistical correlation among variables.

The cumulative probability distribution functions (CPDF) for all random variables are divided into N equivalent intervals (N is a number of simulations);

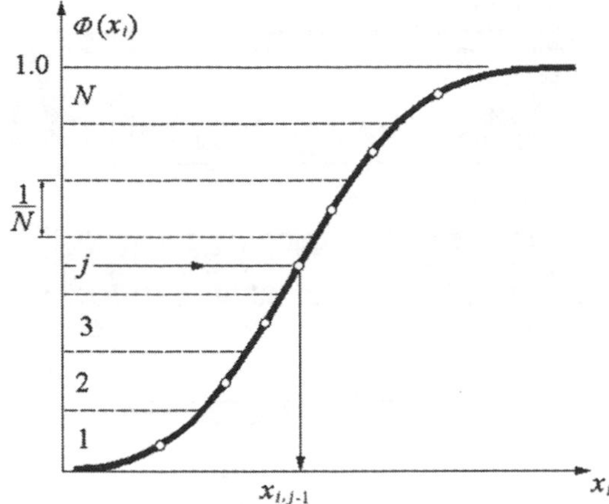

Figure 10. LHS – division of cumulative probability distribution function into layers.

centroids of intervals are then used in simulation process. This means that the range of the probability distribution function $\Phi(X_i)$ of each random variable X_k is divided into N intervals of equal probability $1/N$, see Figure 10. The representative parameters of variables are selected randomly based on random permutations of integers $1, 2, \ldots, j, \ldots, N$. Every interval of each variable must be used only once during the simulation.

Statistical correlation among input random variables can be considered. There are two problems concerning statistical correlation related to LHS: First, during sampling undesired correlation can be introduced between random variables especially in case of very small number of simulations, where the number of interval combination is rather limited.

Second task is to introduce prescribed statistical correlation between random variables defined by prescribed correlation matrix. Stochastic optimization technique called Simulated Annealing (Laarhoven & Aarts 1987, Otten & Ginneken 1989) is utilized in FREET in order to adjust random samples in such a way that resulted correlation matrix is as close as possible to the target (user-defined) correlation matrix. The imposition of the prescribed correlation matrix into sampling scheme can be understood as an optimization problem: The difference between prescribed and generated correlation matrices should be as small as possible. Two-steps evolution strategy (Mutation and Selection) of the stochastic optimization is employed. A suitable quality measure of the overall statistical properties as the norm of the difference of the matrices is introduced. This newly developed technique used in FREET is robust, efficient and very fast and has several advantages in comparison with former techniques (Vořechovský et al. 2002, Vořechovský & Novák 2002, 2003). Note that the approach allow to work also with nonpositive definite matrix on input, which can be the result of lack of knowledge of the user. This technique generates samples as close as possible to a positive definite matrix (mathematically and physically correct).

An important task in the structural reliability analysis is to determine the significance of random variables – how they influence a response function of a specific problem. There are many different approaches of sensitivity analysis; a summary of present methods is given in (Novák et. al. 1993). A sensitivity analysis can answer the question "what variables are the most important?". In this way the dominating and non-dominating random variables can be distinguished. The technique of sensitivity analysis based on nonparametric rank-order statistical correlation appeared to be suitable within the framework of Latin Hypercube Sampling. It uses Spearman correlation coefficient or Kendall's tau. This technique is distribution free and quite robust. Parallel coordinates representation (Wegman 1990) in FREET graphical user environment gives an insight into statistical structure of relationship between random input variables and response output variables.

Cornell´s reliability index could be calculated in FREET from the limit state (performance) function under assumption of normal probability distribution for both structural resistance and acting load. Reliability index is estimated from mean value and standard deviation of the limit state function. Histogram of safety margin as specified in limit state function definition can be visualized. Curve fitting of the most suitable mathematical model of probability distribution is also utilized for better prediction of theoretical failure probability. The results can be compared with the target reliability index, e.g. 4.7 for 1 year as specified by Eurocode (2001).

Program FREET is equipped with a user-friendly graphical user environment, which enables comfortable input of stochastic parameters as well as illustrative interpretation of results.

4 NONLINEAR STOCHASTIC SIMULATION

The programs FREET and ATENA are integrated in software package SARA (Structural Analysis and Reliability Assessment) in order to allow for a probabilistic nonlinear analysis of concrete structures (Bergmeister et al. 2002). An interactive graphical shell SARA Studio was developed in order to assure well-arranged data exchange and management as well as control of both mentioned programs and additional supporting tools. The whole process of the nonlinear stochastic simulation is controlled by the user due to commands and interfaces available in SARA Studio.

The material properties and other input parameters used in ATENA deterministic analysis are firstly defined. These values are exported to FREET, where they will be used as mean values for random distributions of the corresponding variables. Further stochastic parameters (variance, type of the probability density function) for selected variables are defined directly in FREET. The randomness of input variables reflects uncertainties and randomness of the input values regarding material properties, geometry of the structure, prestressing etc. Integrated database of stochastic parameters for various structural and material properties (concrete, reinforcing steel, prestressing, geometrical imperfections) is available in order to support the user in preparing of the stochastic input data. Correlation between random input.variables can be introduced in form of the correlation matrix. Number of samples is selected according to complexity of the problem to be solved and required quality of expected results. Already 8 samples could give a reasonable estimation of stochastic parameters of the structural response and reliability index prediction.

In the next step, sets of input parameters for the required number of samples are generated by FREET. SARA Studio prepares input data for the multiple analysis using ATENA. The single samples are consequently solved in ATENA under SARA Studio control. Selected results from the structural response from ATENA solution (ultimate load, deflection, maximum crack width etc.) are collected. Finally, obtained results are transferred to FREET and evaluated in form of histograms of structural response and sensitivity plots. Reliability index can be assessed.

The whole procedure can be itemized as follows:

1. Deterministic model of the structure is prepared and checked within ATENA.
2. Uncertainties and randomness of the input parameters are modeled as random variables described by their probability density functions (PDF). The result of this step is the set of input parameters for ATENA computational model – random variables described by mean value, variance and other statistical parameters (generally by PDF).
3. Random input parameters are generated according to their PDF using LHS sampling. Statistical correlation among the parameters is imposed using simulated annealing.
4. Generated samples of random parameters are used as inputs for ATENA computational model. The complex nonlinear solution is performed and selected results (structural response) are saved.
5. Previous two steps are repeated for all samples.
6. The resulting set of structural responses from the whole simulation process is statistically evaluated. The results are: histogram, mean value, variance, coefficient of skewness, empirical cumulative probability density function of structural response, sensitivity evaluation, reliability index assessment.

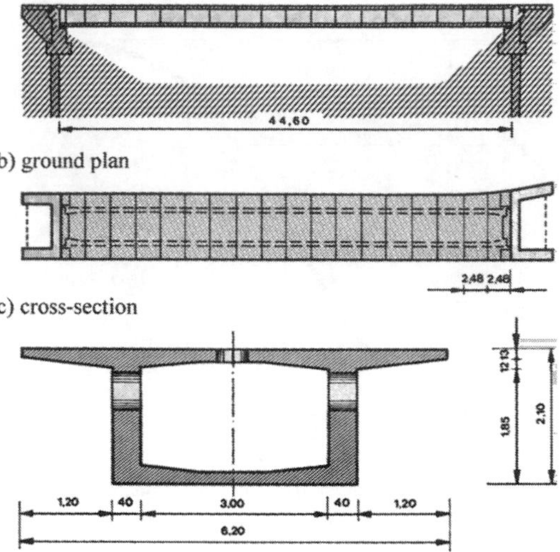

Figure 11. Box-girder bridge in Vienna.

The SARA system described above allows also degradation analysis of structures as demonstrated in another paper presented at this conference, which deals with reliability of prestressed bridge girder subjected to stress corrosion cracking (Teplý et al. 2003). In future it will be extended for optimization of structures and for identification of structural computer models with the real structures according to the measured data from monitoring. Next application area of SARA could be Performance Based Design.

5 APPLICATION EXAMPLES

The feasibility and outcomes of the stochastic fracture analysis are documented on two practical examples of statistical failure simulation and reliability assessment of existing bridge structures: segmental bridge in Vienna, Austria, with a length of 44.6 m, and cantilever beam bridge on the Brenner Motor-way in Italy with the total length of 167.5 m.

5.1 *Segmental bridge in Vienna*

The single span bridge in Vienna over Highway A23 is 25 years old. The length of the bridge is 44.60 m, the width 6.40 m and the height 2.10 m (Fig. 11).

Due to an ongoing construction project the bridge was recently demolished. During the demolition a range of non-destructive tests as well as finally a full-scale destructive load test were performed. The presented reliability assessment was a part of the

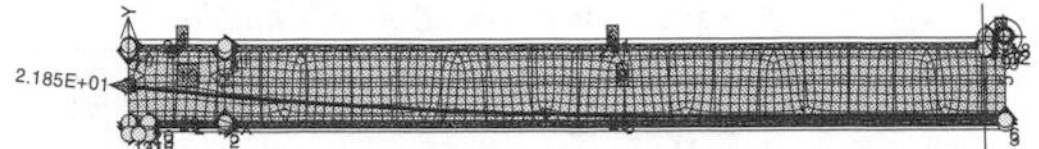

Figure 12. Finite element model of the box-girder bridge.

predictive numerical study (Pukl et al. 2002) for planning and optimizing the test setup. The fully post-tensioned box-girder bridge was made of 18 segments with lengths of 2.485 m each. The segmental joints were filled with epoxy resin. The segments were cast from concrete B500 and are reinforced with mild steel St 50. The post-tensioning tendons consist of 20 strands St 160/180. Symmetrical half of the box-girder was modeled for finite element analysis as shown in Figure 12.

Mean values of the material properties for concrete B500 were generated using ATENA defaults (based on recommendations by CEB, fib, RILEM etc.). The bridge was pre-stressed and then loaded with prescribed displacement in the middle of the span. The ultimate failure load and the descending branch were obtained. The bridge girder failed typically in the middle of the span. First the prestressing tendons yielded, tensile cracks developed, and consequently the concrete in the compression flange of the box girder crushed and split.

Two stochastic simulations – with 8 and 32 samples – were performed. Statistical parameters of input random variables (material properties – Table 1) were adjusted according to available literature sources (Mirza et al. 1979, Wittmann et al. 1994, Al-Harthy & Frangopol 1994, JCSS 2001) and SARA database.

Statistical correlation among random variables was considered according to the prescribed correlation matrix as shown in the upper triangle of Table 2. The correlation matrix generated by simulated annealing for 8 samples is shown in the lower triangle of Table 2. The correlation matrix generated for 32 samples contained almost the prescribed values.

Resulting load-displacement diagrams for 8 samples are shown in Figure 13, histogram of the ultimate loads for 32 samples is presented in Figure 14. Statistical characteristics of the ultimate load (i.e. resistance of the structure) were estimated (Table 3).

Considering different levels of the load (mean values) and two alternatives of variability (coefficient of variation 0.1 and 0.2), reliability index obtained in FREET is plotted in Figure 15. The horizontal line represents the reliability index (4.7 for 1 year) according to Eurocode (2001).

Table 1. Statistical properties of random variables.

Variable*	Units	Mean	CoV	Distribution
E_c	GPa	36.95	0.15	Lognormal
v	–	0.2	0.05	Lognormal
f_t	MPa	3.275	0.18	Weibull
f_c	MPa	42.5	0.10	Lognormal
G_f	N/m	81.43	0.20	Weibull
ε_c	–	0.0023	0.15	Lognormal
c_{Red}	–	0.8	0.06	Rectangular
w_d	m	0.0005	0.10	Lognormal
ρ	MN/m^3	0.023	0.10	Normal
E_s	GPa	200.0	0.03	Lognormal
f_y	MPa	1600.0	0.07	Lognormal
F_p	MN	21.85	0.04	Normal
A_s	m^2	0.0237	0.001	Normal

*Notation of random variables:
Concrete: E_c, Young's modulus of elasticity; v, Poisson's ratio; f_t, tensile strength; f_c, compressive strength; G_f, specific fracture energy; ε_c, uniaxial compressive strain at compressive strength; c_{Red}, reduction of compressive strength due to damage; w_d, critical compressive displacement; ρ, specific material weight.
Prestressing strands: E_s, Young's modulus of elasticity; f_y, yield stress; F_p, prestressing force; A_s, area of strands.

Table 2. Correlation matrix for concrete properties.

Variable*	E_c	f_t	f_c	G_f	ε_c
E_c	1	0.7	0.9	0.5	0.9
f_t	0.77	1	0.8	0.9	0.6
f_c	0.83	0.82	1	0.6	0.9
G_f	0.69	0.83	0.75	1	0.5
ε_c	0.84	0.77	0.83	0.69	1

*Notation of variables see Notes to Table 1.

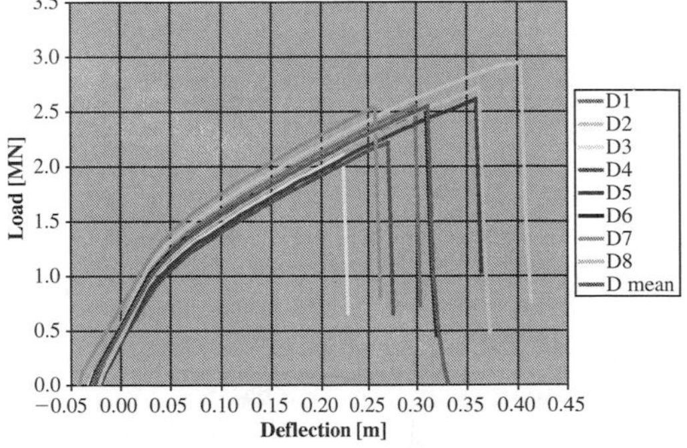

Figure 13. Load–deflection curves (8 samples).

5.2 *Cantilever beam motorway bridge*

The fully post-tensioned box-girder bridge built in 1969 is one of the objects of the Brenner Motorway in Italy. The Brenner Motorway consists of two separated lanes, each of them includes several objects of this type. The bridges are cast-in-place balanced cantilever beams with varying girder depth, Figure 16.

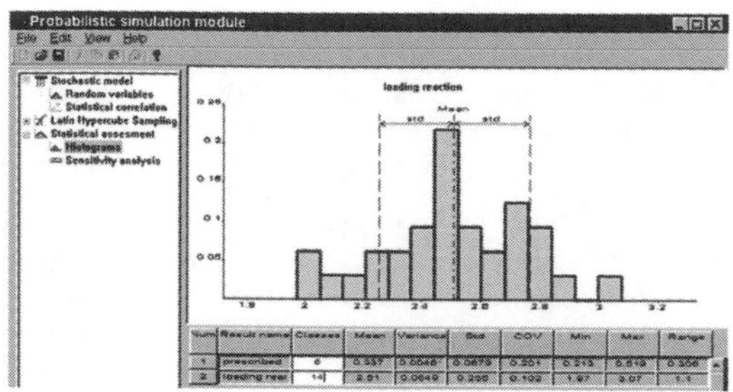

Figure 14. Histogram of ultimate loads (32 samples).

Table 3. Estimation of basic statistical parameters of the ultimate load.

Number of samples	Mean value (MN)	Variance $(MN)^2$	Standard deviation (MN)	Coefficient of variation
8	2.52	0.0707	0.266	0.105
32	2.51	0.0649	0.255	0.102

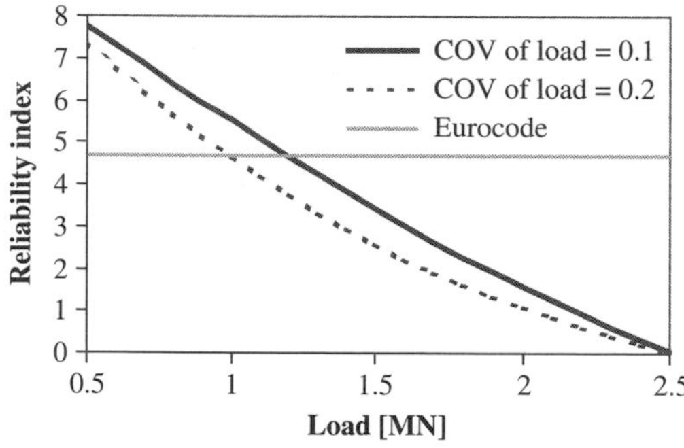

Figure 15. Reliability index plot.

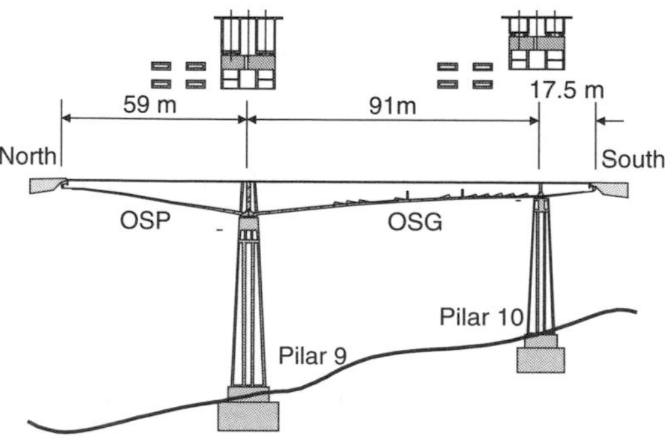

Figure 16. Bridge scheme.

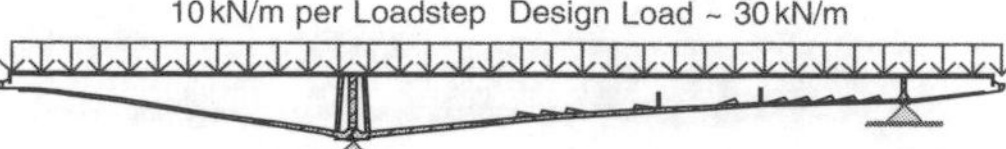

Figure 17. Geometrical shape for the finite element model.

Table 4. Basic random variables.

Variable*	Units	Mean	CoV	Distribution
E_c	GPa	37.0	0.15	Lognormal
f_t	MPa	3.26	0.18	Weibull
f_c	MPa	42.5	0.10	Lognormal
G_f	N/m	120.0	0.20	Weibull
ρ	MN/m^3	0.023	0.10	Normal
E_s	GPa	210.0	0.03	Lognormal
f_{ys}	MPa	500.0	0.05	Lognormal
f_{yp}	MPa	1350.0	0.20	Lognormal

*Notation of random variables:
Concrete: E_c, Young's modulus of elasticity; f_t, tensile strength; f_c, compressive strength; G_f, specific fracture energy; ρ, specific material weight;
Reinforcement: E_s, Young's modulus of elasticity (for both mild steel and prestressing tendons); f_{ys}, yield stress of mild steel; f_{yp}, yield stress of prestressing tendons.

The mid-span has a length of 91 m, the cantilever beams have a lengths of 59 m and 17.5 m. Total length of the bridge is 167.5 m. The lane slab has a width of 10.60 m and a thickness is of about 0.20 m. The lower girder slab has a width of 6.00 m and a thickness of about 0.20 m. The height of the box girder varies from 10.80 m over the middle support to 2.85 m in the mid-span. The bridge is cast from concrete B500 and is reinforced with mild steel BST 500. The post-tension tendons system consists of 211 strands of St 1350/1500.

The geometrical shape for the finite element model is shown in Figure 17. The bridge was prestressed and then loaded with a prescribed line load along the whole lane slab. The ultimate failure load and the descending branch were obtained in the analyses. The bridge girder failed typically next to the middle support. First the prestressed tendons yielded, tensile cracks developed in the upper flange of the box girder and finally shear failure occurred.

Stochastic simulations with 8 and 30 samples were performed, see also Pukl et al. (2003). Mean values of the concrete properties for nonlinear analysis were generated in ATENA from the cubic compressive strength. Statistical properties of the random variables (Table 4) originated mostly from the database available in the SARA system. The coefficient of variation for the yield strength of the prestressed cable was estimated rather large due to high uncertainties. Correlation between material parameters was introduced,

Table 5. Correlation of random variables.

Variable*	E_c	f_t	f_c	G_f
E_c	1	0.7	0.9	0.5
f_t	0.698	1	0.8	0.9
f_c	0.896	0.798	1	0.6
G_f	0.500	0.892	0.601	1

*Notation of random variables see Notes to Table 4.

Table 6. Estimation of statistical parameters of the ultimate line load.

Number of samples	Mean value (kN/m)	Variance (kNm^2)	Standard deviation (kN/m)	Coefficient of variation
8	234.3	388	19.69	0.05
30	235.0	324	18.00	0.08

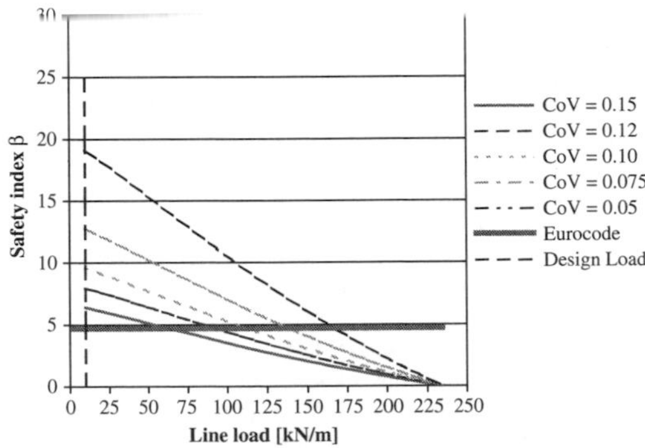

Figure 18. Reliability assessment of the ultimate line load.

the prescribed correlation matrix is shown in the upper triangle of Table 5. The lower triangle of Table 5 shows the correlation matrix generated by simulating annealing for 30 samples.

The stochastic simulation provided an estimation of the statistical characteristics of the ultimate load (resistance of the structure – Table 6) and of the displacements at midspan and at the end of each cantilever beam. Two different failure modes occurred in the analyses – with negative and positive vertical displacement at the end of the cantilever beams.

Reliability index for the ultimate load under consideration of different levels of the line load (mean values) is plotted in Figure 18, various values of coefficient of variation are compared. The horizontal line represents the target reliability index of 4.7 (for 1 year) as specified by Eurocode (2001).

Furthermore, reliability of the bridge related to the vertical displacements was investigated and compared with JCSS regulations (2001), see Figures 19 and 20.

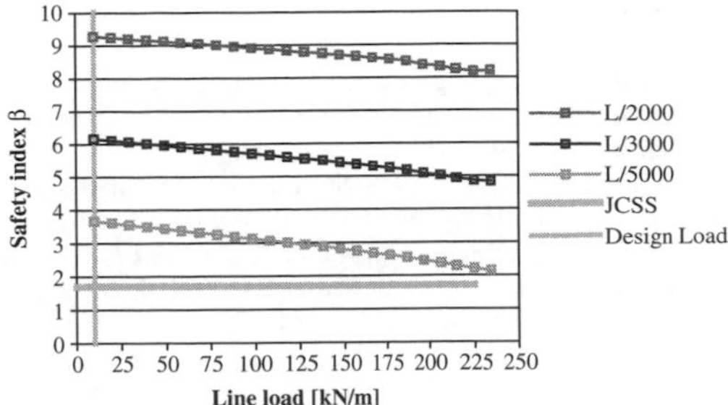

Figure 19. Reliability assessment of the displacement at mid-span.

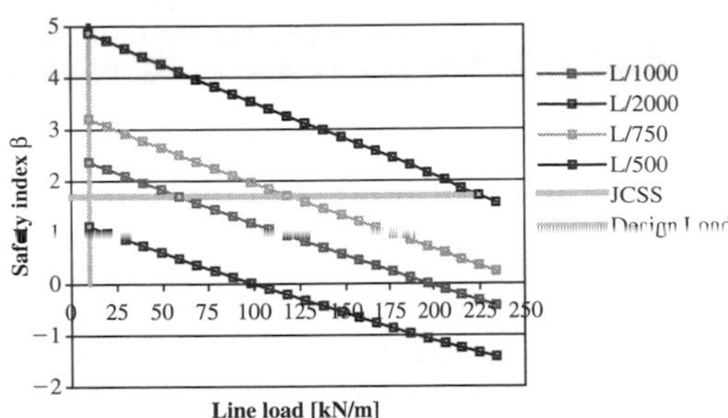

Figure 20. Reliability assessment of the displacement at the cantilever north.

It resulted that the end of the north cantilever is the critical point regarding the vertical displacements.

6 CONCLUSIONS

Methodology for probabilistic based assessment of concrete structures is introduced and documented on practical examples. The programs FREET and ATENA are combined in the software package SARA in order to support stochastic nonlinear analysis. This approach enables reliability assessment of the analyzed structure, which is going beyond the boundaries of design codes. It can lead to considerable cost saving as the reliability requirements can be targeted more precisely. Realistic behavior of concrete bridges is investigated from reliability point of view providing a new insight into bridge administration and decision-making process.

ACKNOWLEDGMENT

The financial support by grant of the Grant Agency of Czech Republic (GAČR) No. 103/02/1030 is greatly appreciated.

REFERENCES

Al-Harthy, A.S. & Frangopol, D.M. 1994. Reliability assessment of prestressed concrete beams. *Journal of Structural Engineering ASCE* 120 (1): 180–199.

Ayyub, B.M. & Lai, K.L. 1989. Structural Reliability Assessment Using Latin Hypercube Sampling. In: *The 5th International Conference on Structural Safety and Reliability (ICOSSAR'89)*, San Francisco, USA, Vol. I, Structural Safety and Reliability: 1177–1184.

Bažant, Z.P., Caner, F.C., Carol, I., Adley, M.D. & Akers, S.A. 2000. Microplane model M4 for concrete: I. Formulation with work-conjugate deviatoric stress. *Journal of Engrg. Mechanics ASCE* 126 (9): 944–953.

Bažant, Z.P. & Oh, B.H. 1983. Crack band theory for fracture of concrete. *Materials and Structures* 16: 155–177.

Bergmeister, K. 2002. Robust bridge design – maintenance and innovative solutions. In: *Betonářské dny 2002 (Czech noccrete days)*, ČBS, Pardubice, Czech Republic: 420–427.

Bergemeister, K., Strauss, A., Eichinger, E.M., Kolleger, J., Novák, D., Pukl, R. & Červenka, V. 2002. Structural analysis and safety assessment of existing concrete structures. In: *The first **fib** Congress*, Osaka, Japan, Session 11: 47–54.

Casas, J.R., Frangopol, D.M. & Nowak, A.S. 2002. *Bridge Maintenance, Safety and Management*. Proceedings of the First International Conference IABMAS'02, Barcelona, Spain: CIMNE.

Červenka, J. & Červenka, V. 1999. Three Dimensional Combined Fracture-Plastic Material Model for Concrete. In: *5th U.S. National Congress on Computational Mechanics*, Boulder, Colorado, USA.

Červenka, V. 1998. Simulation of shear failure modes of R.C. structures. In: *Computational modelling of concrete structures (Euro-C 1998)*, Badgastein, Austria: 833–838.

Červenka, V. 2000. Simulating a Response. *Concrete Engineering International* 4 (4): 45–49.

Červenka, V. 2002. Computer simulation of failure of concrete structures for practice. In: *The first **fib** Congress*, Osaka, Japan, Session 13: 289–304.

Červenka, V. & Pukl, R. 1995. Mesh Sensitivity Effects in Smeared Finite Element Analysis of Concrete Structures. In: F.H. Wittmann (ed.), *Second International Conference on Fracture Mechanics of Concrete Structures (FRAMCOS 2)*, Aedificatio, ETH Zürich, Switzerland: 1387–1396.

Červenka, V., Červenka, J. & Pukl, R. 2001. ATENA – an Advanced Tool for Engineering Analysis of Connections. In: *Connections between Steel and Concrete*, RILEM Publications, Ens, France: 658–667.

Červenka, V., Červenka, J. & Jendele, L. 2003. Finite element modelling of bond in reinforced concrete. In: *Computational modelling of concrete structures (Euro-C 2003)*, St. Johann in Pongau, Austria.

Eichinger, E.M. 2002. A probability-based approach to the safety evaluation of existing concrete bridges using stochastic finite elements. In: *4th International Ph.D. Symposium in Civil Engineering*, Munich, Germany.

Enevoldsen, I. 2001. Experience with probabilistic-based assessment of bridges. *Structural Engineering International* 11 (4): 251–260.

Enevoldsen, I., Jensen, F.M. & Bjerrum, J. 2002. Implementation of safety-based maintenance management for the side-walk at Storstroem bridge. In: *1st International Conference on Bridge Maintenance, Safety and Management (IAB-MAS'02)*, Barcelona, Spain.

Eurocode 2001. Basis of structural design. Brussels, Belgium: CEN.

Frangopol, D.M. 2000. Advances in life-cycle reliability-based technology for design and maintenance of structural systems. In: B.H.V. Topping (ed.) *Computational mechanics for the twenty-first century*. Edinburgh, UK, Saxe-Coburg Publications: 311–328.

Hillerborg, A., Modéer, M. & Peterson, P.E. 1976. Analysis of crack formation and crack growth in concrete by means of fracture mechanics and finite elements. *Cement Concrete Research* 6: 773–782.

Hordijk, D.A. 1991. *Local Approach to Fatigue of Concrete*. Ph.D. Thesis, Delft University of Technology, The Netherlands.

JCSS 2001. *Probabilistic Model Code*. Joint Committee on Structural Safety.

Kupfer, H., Hilsdorf, H.K. & Rüsch, H. 1969. Behavior of Concrete under Biaxial Stress, *Journal ACI*, 66 (8): 656–666.

Laarhoven, P.J & Aarts, E.H. 1987. *Simulated Annealing: Theory and Applications*. Netherlands: D. Reidel Publishing Company.

McKay, M.D., Conover, W.J. & Beckman, R.J. 1979. A Comparison of Three Methods for Selecting Values of Input Variables in the Analysis of Output from a Computer Code. *Technometrics* 21: 239–245.

Menétrey, P. & Willam, K.J. 1995. Triaxial failure criterion for concrete and its generalization. *ACI Structural Journal* 92 (3): 311–318.

Mirza, S.A., Hatzinikolas, M. & McGregor, J.G. 1979. Statistical descriptions of strength of concrete. *Journal of Structural Division ASCE* 105 (6): 1021–1037.

Novák, D., Teplý, B. & Shiraishi, N. 1993. Sensitivity analysis of structures: A Review. *Int. Conf. CIVIL COMP "93*, Edinburgh, Scotland: 201–207.

Novák, D., Teplý, B. & Keršner, Z. 1998. The Role of Latin Hypercube Sampling Method in Reliability Engineering. In: *Structural Safety and Reliability (ICOSSAR-97)*, Rotterdam, The Netherlands: Balkema.

Novák, D., Pukl, R., Vořechovský, M., Rusina, R. & Červenka, V. 2002. Structural reliability assessment of computationally intensive problems – non-linear FEM analysis. In: *AS-RANet Colloquium*, Glasgow, Scotland.

Novák, D., Rusina, R. & Vořechovský, M. 2003. Small-sample statistical analysis – software FREET. In: *9th International Conference on Applications of Statistics and Probability in Civil Engineering (ICASP9)*, Berkeley, California, USA, July 2003 (in print).

Otten, R.H.J.M. & Ginneken, L.P.P.P. 1989. *The Annealing Algorithm*. USA: Kluwer Academic Publishers.

Pukl, R., Ožbolt, J. & Eligehausen, R. 1998. Load-bearing Behavior of Bonded Anchors – Nonlinear Simulation. In: *Computational modelling of concrete structures (Euro-C 1998)*, Badgastein, Austria: 849–853.

Pukl, R., Novák, D. & Eichinger, E.M. 2002. Stochastic Nonlinear Fracture Analysis. In: *1st International Conference on Bridge Maintenance, Safety and Management (IAB-MAS'02)*, Barcelona, Spain.

Pukl, R., Strauss, A., Novák, D., Červenka, V. & Bergmeister, K. 2003. Probabilistic-based assessment of concrete structures using nonlinear fracture mechanics. In: *9th International Conference on Applications of Statistics and Probability in Civil Engineering (ICASP9)*, Berkeley, California, USA, July 2003 (in print).

Rots, J.G. & Blaauwendraad, J. 1989. Crack models for concrete: discrete or smeared? Fixed, multi-directional or rotating? *Heron* 34 (1), TU Delft, The Netherlands.

Teplý, B., Lehký, D., Novák, D., Eichinger, E.M., Kolleger, J., Pukl, R., Červenka, V. & Stewart, M.G. 2003. Prestressed bridges under corrosion attack. In: *Computational modelling of concrete structures (Euro-C 2003)*, St. Johann in Pongau, Austria.

Vořechovský, M. & Novák, D. 2002. Correlated Random Variables in Probabilistic Simulation. In: *4th International Ph.D. Symposium in Civil Engineering*, Munich, Germany.

Vořechovský, M. & Novák, D. 2003. Statistical correlation in stratified sampling. In: *9th International Conference on Applications of Statistics and Probability in Civil Engineering (ICASP9)*, Berkeley, California, USA, July 2003 (in print).

Vořechovský, M., Rusina, R. & Novák, D. 2002. A New Efficient Technique for Samples Correlation in Latin Hyper-cube Sampling. In: *VII International Scientific conference*, Košice, Slovak Republic.

Wegman, E.J. 1990. Hyperdimensional data analysis using parallel coordinates. *Journal of the American Statistical Association* 85 (411): 664–675.

Wittmann, F.H., Slowik, V. & Alvaredo, A.M. 1994. Probabilistic aspects of fracture energy of concrete. *Materials and Structures* (27): 499–504.

Computational Modelling of Concrete Structures, Bićanić et al. (eds)
© 2003 Swets & Zeitlinger, Lisse, ISBN 90 5809 536 3

Some advantages of 1D- instead of 2D- or even 3D-modelling for non-linear analysis of reinforced concrete frames

U. Pfeiffer & U. Quast
Technical University Hamburg-Harburg, Germany

ABSTRACT: Traditionally frames are designed following linear elastic analysis. However to calculate deflections, effects of imposed actions or second order effects the non linear behavior of reinforced concrete has strictly to be taken into account. Already a 1D-modelling of reinforced structures with a combination of the Transfer Matrix Method and the FE-method gives good and stable results. The background of this method, some advantages and problems as well as some examples of calculations will be shown.

1 INTRODUCTION

With the growing architectural demands, combined with the use of high strength concrete and new construction methods (e.g. joint-less types of constructions), an increasing quality of static calculations is required. In many cases, a usual linear calculation no longer meets these requirements delivering too large differences between the calculated and actual deformations and internal forces.

So far most of the work on non-linear concrete calculations has been done only in universities for research purposes. The effort especially for 2D- or 3D-modelling of structures has been often quite high, software packages are accordingly expensive and the elaborate calculations even with the use of expensive hardware takes long time.

Especially in the case of continuous beams and frames where the load is carried in only one direction, a 2D- or even a 3D-modelling of the structure only increases the effort and time for computations without an increase in accuracy in return. For these cases an appropriate non-linear 1D-modelling is accurate enough for practical purposes.

Some of the programs available in the market today provide modules to perform nonlinear calculations especially for one way spanned beam systems. Their calculation algorithms use mainly bar elements, are based on computations made with polynomial functions and in every iteration the stiffness of the cross section is calculated according to the current internal forces. The problem is that the accuracy of the results depend on the user's choice of bar discretization since with in each bar only continuous functions are used to describe the sometimes discontinuous functions of stiffness, curvature, etc.

2 COMBINATION OF NON-LINEAR SECTION ANALYSIS, TRANSFER MATRIX METHOD AND FINITE ELEMENT METHOD

For the dimensioning of the required reinforcement a cross section analysis is preferred and the action effects are therefore needed in terms of axial force and bending moments. The results of 1D-modelling with beam elements directly provide these values. The relationships between action effects and elongation and curvature as well as their gradients can be calculated by stress integration within the cross section applying an iterative method.

A very efficient tool for the non-linear calculation of reinforced concrete beams is the reduced transfer matrix method. With prior computed non-linear moment-curvature-relationship it is possible to use the correct bending stiffness and curvatures during the calculation of the beam. Besides it is not necessary to insert plastic hinges for considering beam regions with reinforcement being plastically strained.

A disadvantage of this method is that only a calculation of continuous beams or columns is possible and that the computation of more complex 2D-frames is too tedious for practical purposes. Nevertheless with the combination of non-linear cross-section analysis, the matrix transfer method and FEM-calculations it is possible to calculate the otherwise too complicated 2D-frames.

To do this every bar is considered rigidly supported and computed with the transfer matrix method. By this the stiffness matrix and the forces-vector can be established. The solution of the linear equation system will yield the deformation of the nodes.

The program Stab2D-NL was developed to provide a user-friendly combination of the methods described above, making possible a simple analysis of 2D-frames. Comparison between laboratory measurements and calculated results was performed to verify the satisfactory accuracy of the results provided be the algorithms.

In the following this combined method is explained in more detail.

3 TRANSFER MATRIX METHOD

3.1 *Basic procedure*

In the following a single span beam will be used to explain or review the transfer matrix method. The calculations begin with the division of the bar in a temporary number of segments of length dx. In the Figure 1 below a free body diagram of such a segment shows the internal forces acting on its left and right faces as well as the external loading along the length dx.

Simple equations describe the relationship between the internal forces left and right to the segment. When 2nd order effects are neglected the equations e.g. for shear and bending moment read as follows:

$$M_{right} = M_{left} - \left(\left(\Delta f \cdot \frac{dx}{3} + f\right) \cdot \frac{dx}{2} - Q_{left}\right) \cdot dx$$
$$Q_{right} = Q_{left} - \left(\Delta f \cdot \frac{dx}{2} + f\right) \cdot dx \qquad (1)$$

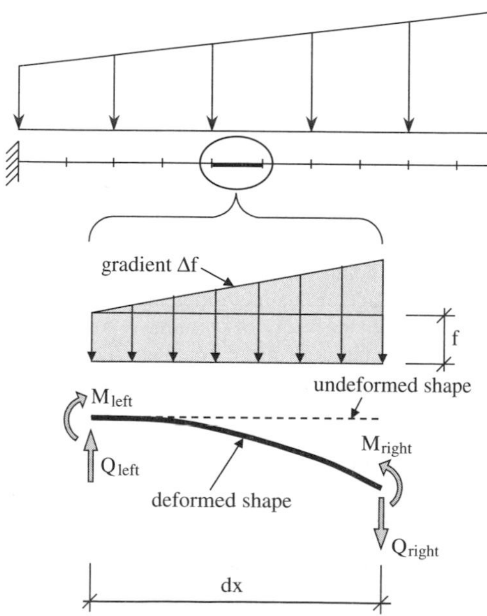

Figure 1. Beam with discretization in segments. Free body diagram of one segment.

Using traditional beam theory and the principle of virtual work similar expressions can be written for the displacement (w_{ver}, w_{hor}) and rotation (φ) magnitudes. However, when considering 2nd order effects (influence of axial forces) and the non-linear behaviour of reinforced concrete, additional terms need to be added to these equations. For example the expression for the vertical displacement Δw_{ver} becomes:

$$x_1 = \left(\left(\left(\Delta f \cdot \frac{dx}{5} + f\right) \cdot \frac{dx}{4} - Q_{left}\right) \cdot \frac{dx}{3} - M_{left}\right) \cdot \frac{dx^2}{2} \cdot \frac{1}{[EI_{II}]}$$
$$- \left(\left(\Delta f \cdot \frac{dx}{3} + f\right) \cdot \frac{dx}{2} - Q_{left}\right) \cdot dx \cdot \frac{1}{[GA]}$$
$$- (k_{\Delta T} + [k_0]) \cdot \frac{dx^2}{2} - \left[\frac{\partial k}{\partial N}\right] \cdot \Delta N \cdot \frac{dx^2}{6}$$
$$x_2 = 1 - \left(\left(\Delta n \cdot \frac{dx}{5} + n\right) \cdot \frac{dx}{4} + N_{left}\right) \cdot \frac{dx^2}{6} \cdot \frac{1}{[EI_{II}]}$$
$$+ \left(\left(\Delta n \cdot \frac{dx}{3} + n\right) \cdot \frac{dx}{2} + N_{left}\right) \cdot \frac{1}{[GA]}$$
$$\Delta w_{ver} = \frac{x_1}{x_2} \qquad (2)$$

Values inside square brackets are calculated individually for every segment according to the strain distribution and stiffness corresponding to the current internal forces N_x and M_y. The following values were also additionally introduced into the equations:

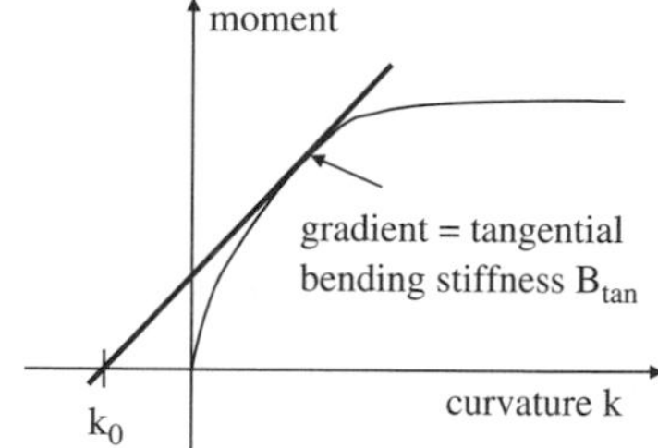

B_{tan} = tangential bending stiffness
k_0 = intersection on curvature-axis of tangent defining B_{tan}
S = shear stiffness (linear GA)
$\frac{\partial k}{\partial N}$ = change of curvature due to the change of normal force (non-linear behaviour of reinforced concrete)
$K_{\Delta T}$ = curvature, due to a temperature gradient

With these formulas it is possible to calculate the internal forces and deformations for the whole bar given the initial values at the left end. The initially unknown starting values for N_x, Q_z and M_y are calculated in an iterative manner.

For this purpose simultaneously to the normal calculation three additional calculations are performed in which each one of the internal forces is considered with an increment, i.e. a first calculation with $N_x + \Delta N/Q_z/M_y$; a second with $N_x/Q_z + \Delta Q_z/M_y$; and a third with $N_x/Q_z/M_y + \Delta M_y$.

In this way the elements of the Jacobian Matrix can be determined and the starting values calculated. Here the vector $(dw_x/dw_y/dw_z)$ represents the difference at the end of the bar between the deformation of the current iteration and the known boundary conditions.

In this way it is possible to improve the starting values at the left end of the bar which now yield: $N_x + dN_x; Q_z + dQ_z; M_y + dM_y$.

3.2 *Implementation of 3rd order effects*

The Transfer Matrix Method allows the simple implementation of 3rd order effects in the calculations. Initially each segment is calculated with an initial angle $\varphi = 0$ using traditional beam theory. For the axial deformation Δx a simple approach is taken and considered as sufficient. The most relevant calculation is the associated rotation of the internal forces and external loads, so that they keep an orthogonal position to the bar's axis also in the deformed shape.

Depending on the magnitude of the vertical deformation, the partition of the bar in segments can be modified so that when 3rd order effects become relevant smaller segments are used. In this way it is possible even in the case of large deformation to model the structure with only one bar element as shown in Figure 4.

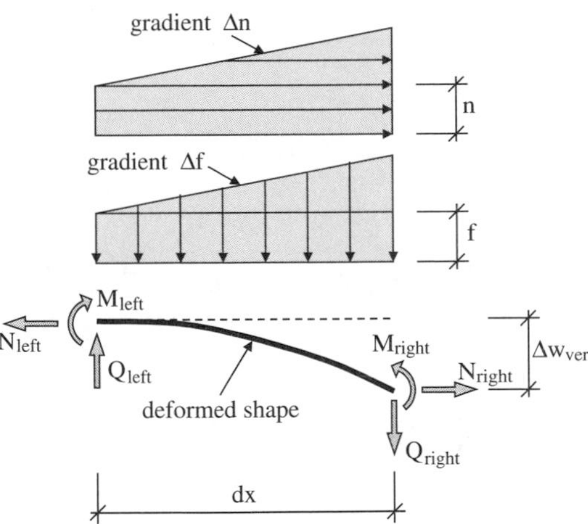

Figure 2. Free body diagram of a bar segment of length dx with additionally introduced values.

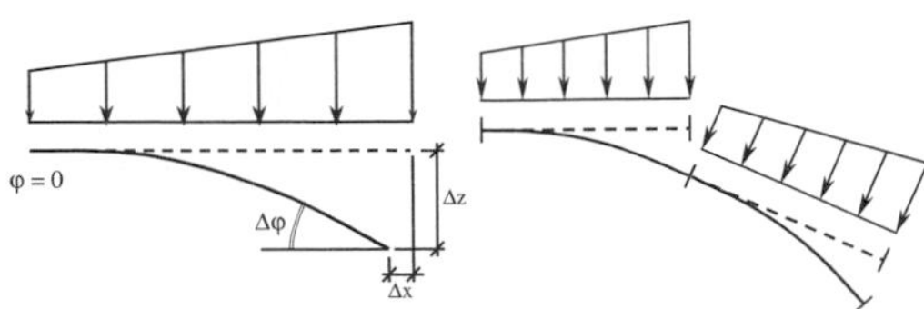

Figure 3. Calculation considering 3rd order effects.

3.3 *Calculating with tangential stiffness in the transfer matrix method*

For both the axial and bending stiffness, tangential values are used during the calculations. This course of action has the advantage of providing a good approximation of the moment-curvature relationship in a relatively large range. For a given bar discretization the tangent stiffness provides higher accuracy in the results then a calculation with secant stiffness.

The advantages of the tangential over the secant stiffness are specially noticeable in the plastic range, where the moment-curvature relationships approximate an almost horizontal curve, which also guarantees a fast convergence for this range.

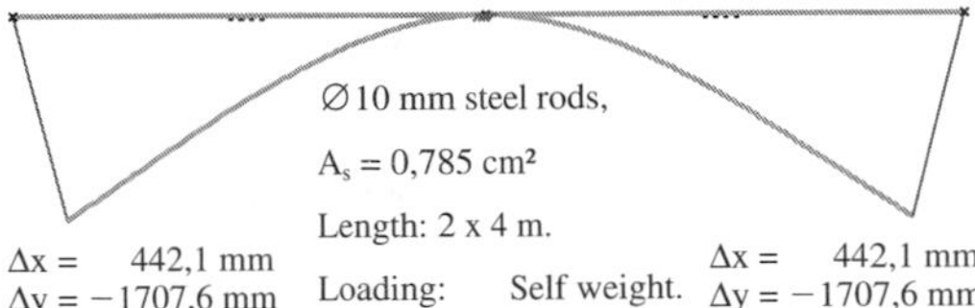

Figure 4. Deformation considering 3rd order effects.

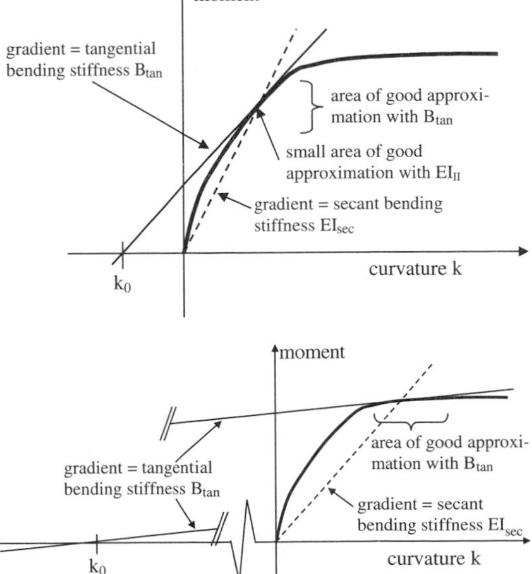

Figure 5. Comparison between tangent and secant stiffness.

3.4 *Calculation of the tangential stiffness*

The calculation is carried out using the algorithms of the program INCA2 under the premises of the Bernoulli-hypothesis and compatibility of stresses between reinforcement and the adjacent concrete (rigid bond). By integration of stresses for a chosen strain distribution the internal forces are calculated. The strain distribution is varied until the resultants match the desired internal forces. Following a small variation of the strain distribution of $\Delta\varepsilon_0$ and Δk_z allows the calculation of the axial and the bending stiffness as well as the corresponding values of k_0 und ε_0. For a later accurate calculation the determination of the values of $\partial k/\partial N$ and $\partial\varepsilon_0/\partial M_y$ is here also important.

Formulations providing a suitable modelling of the material behaviour are important for good quality results. Especially the values for young-modulus at the origin, the maximum stress and corresponding strain have to be properly chosen. For both reinforcement steel and concrete under compression the standards provide enough and sufficiently accurate curves. The tension stiffening effect of the cracked concrete is taken into account using the assumptions according to /Quast/ and /Espion/ which for practical usage return sufficient consistency between calculated values and experimentally measured deformations.

Moreover specific effects for concrete can be modelled by using a pre-strain for shrinkage as well as a simple modification of the stress-strain-diagram for the creep effect. Pre-stressed or additional supplemented cross-sections as shown in Figure 6 can be modelled too.

3.5 *Optimum bar discretization*

For every bar segment the stiffness B_{tan} und D_{tan} is individually calculated. These values vary from segment to segment when for example the concrete cross section cracks; when reinforcement starts yielding or by a change in the value of the axial force. Depending on the magnitude of these variations the bar is divided into a larger number of segments. Thus a faster calculation with fewer bar segments can be performed in areas of small stiffness variations while areas with large stiffness variations are calculated more accurately with a finer discretization as it can be seen in Figure 7.

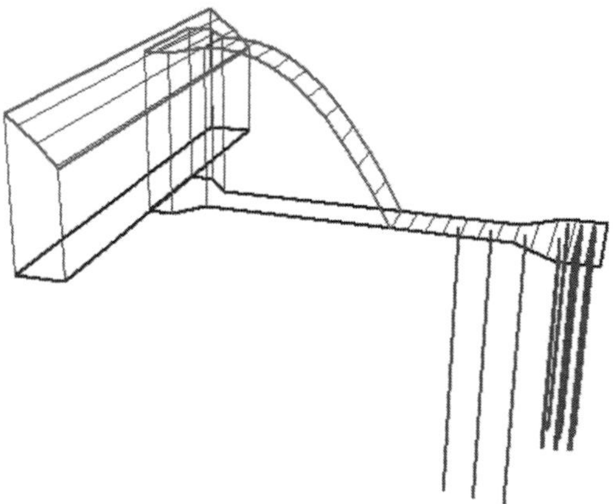

Figure 6. 3D-stress distribution of a reinforced concrete cross section with pre-stress and an additional supplemented slab.

3.6 *Advantages of the transfer matrix method*

As mentioned before it is pursued to keep an optimal bar discretization during the whole calculation. This program-supported automatic discretization minimises the calculation effort, increases accuracy, and makes the results no longer dependant of the user's choice of bar discretization.

The non-linear analysis of the cross-sections, necessary for the calculation of the bar elements, can be easily optimized in the transfer matrix method by using the strain distribution and the gradients (tangent stiffness) of the later iteration as the initial values for the new iteration with the current combination of internal forces.

In summary the following effects can be considered:

- each material behaviour, modelled with pre-defined usual stress-strain relationships or with

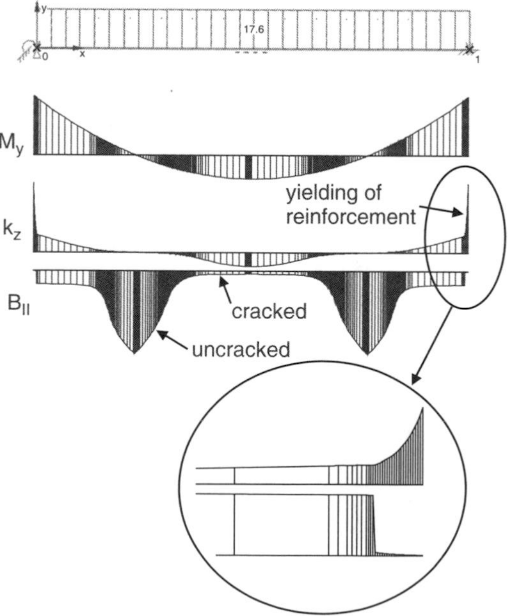

Figure 7. Discretization of the beam in segments depending on the changing of bending stiffness.

a spline-interpolation, inclusive time-dependent effects like shrinkage and creep
- non-linear relation between
  - moment and curvature
  - axial force and strain
  - moment and strain as well as for axial force and curvature (e.g. elongation of the centroid axis due to cracking of the cross-section)
  - 2nd and 3rd order effects
- deformation due to shear force with different assumptions (neglect, linear shear stiffness GA, reduction of GA according to the strain-distribution)
- continuous change of cross section
- not automatically considered effects are:
  - any exact 2D- or 3D-stress distribution
  - exact state of stresses in areas where concentrated loads are applied
  - corners of frames, supports
  - abrupt changes of cross section (curtailing of reinforcement, openings, etc.)
- non-linear distribution of axial stress and shear stress in the cross-section (e.g. load transfer between reinforcement and surrounding concrete, effective width of flanges)
- pliable bond between cross section elements (e.g. composite constructions with non rigid shear transfer, slip of reinforcement).

Even though all these effects are not considered automatically by the calculation method, it is possible in part to take them into account by a suitable modelling of the structural system and loading. For example, to consider the spread of a concentrated load as shown in Figure 8 a uniformly distributed load along a certain effective length can be modelled. The pressure at middle-supports can also be taken into account as a uniformly distributed load equivalent to the calculated reaction at the support. An automatically consideration of this pressure in the developed program Stab2D-NL is possible. By doing this the more realistic rounded moment-distribution over supports is taken into account.

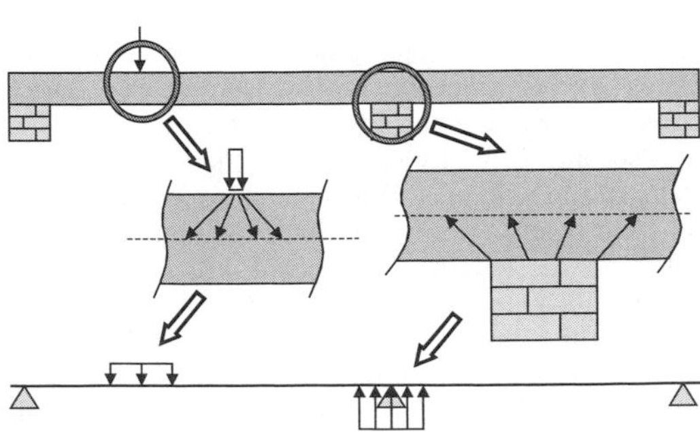

Figure 8. Modeling of a single force and a support.

Changes in the cross sections can be modelled with haunched bars. Other effects like the non uniform stress distribution in transversal direction (e.g. load transfer between reinforcement and surrounding concrete) can be very well neglected or like in the case of T beams these effects can be taken into account by a proper choice of the effective flange width.

4 FINITE ELEMENT METHOD

4.1 *Basic procedures*

The stiffness matrix is generally calculated with the principle of virtual work. The columns of the K-matrix can be interpreted as vectors of forces which give another possibility of calculating the matrix.

When a unit displacement is artificially added to the lower support as shown in the Figures 9–11, the multiplication $K \cdot u$ gives the ith column of the K-matrix. This vector consists of the forces needed to get the system equilibrium in this deformed position.

Vice versa with this consideration it is possible to get the K-matrix by putting a unit displacement to all DOF and calculate the necessary vector of

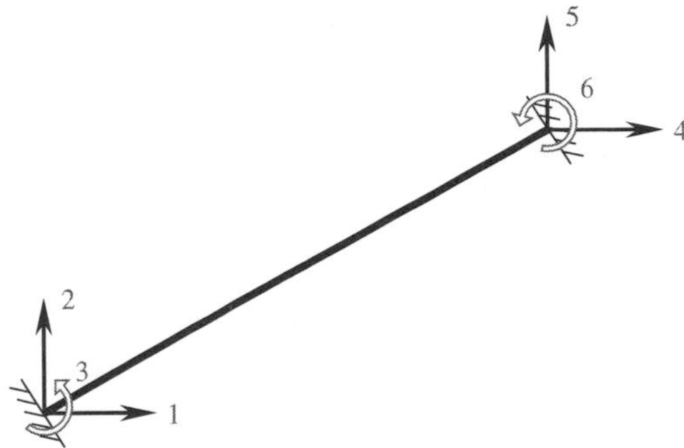

Figure 9. Undeformed system with degrees of freedom 1 to 6.

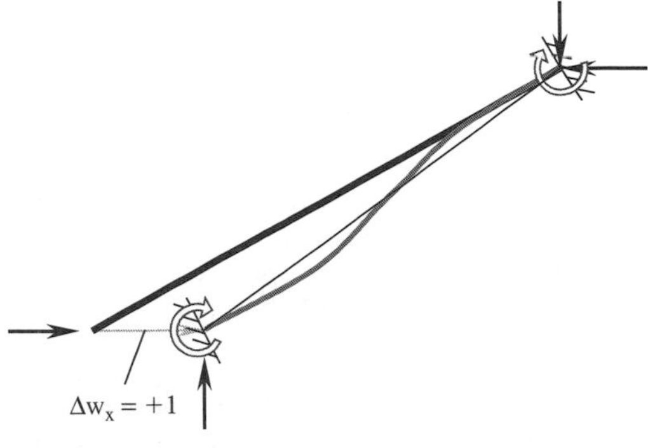

Figure 10. Deformed system for a displacement of degree of freedom number 1, $\Delta w_x = +1$.

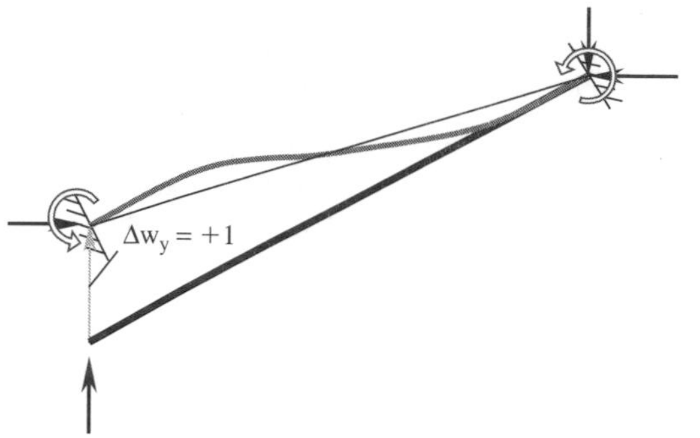

Figure 11. Deformed system for a displacement of degree of freedom number 2, $\Delta w_y = +1$.

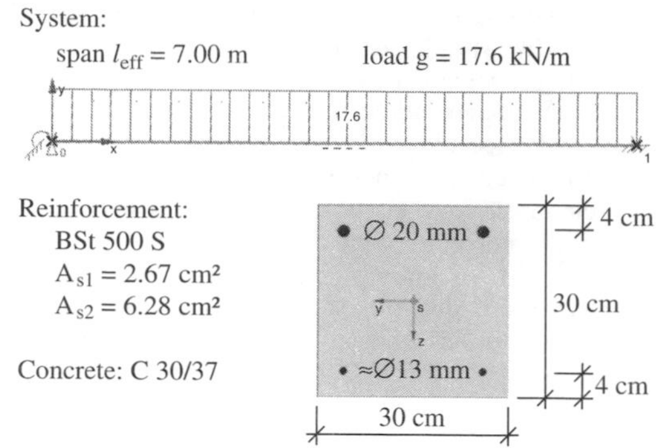

Figure 12. System, loads and cross-section.

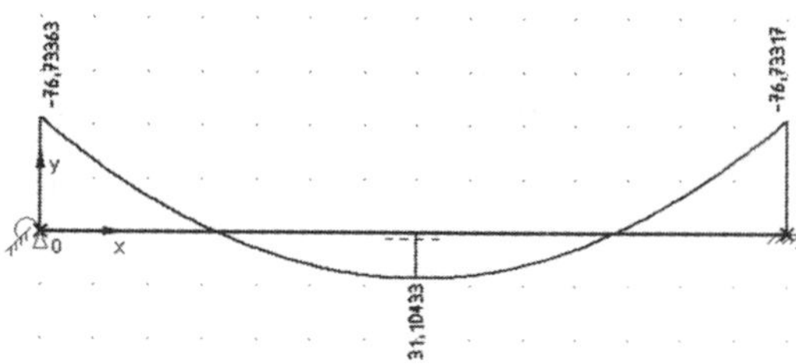

Figure 13. Results of a non-linear calculation.

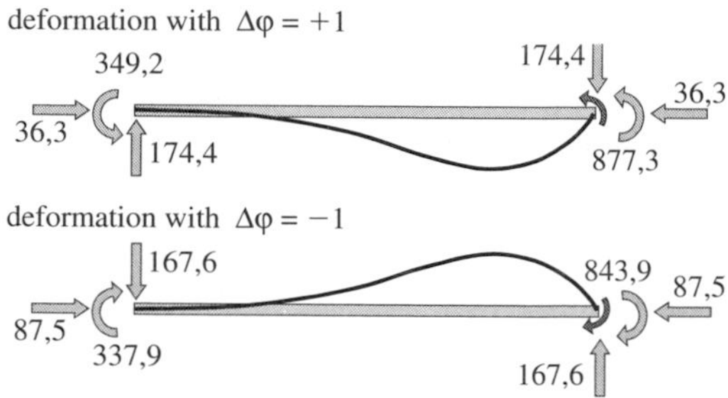

Figure 14. Anti-clockwise and clockwise unit rotation at DOF 6, resultant forces.

forces. Adding all the vectors together gives the stiffness-matrix.

The difficulty of this procedure is to calculate the vectors of forces in an easy and fast way with a high accuracy. A possible but time-consuming way is to put on a small real deformation at every DOF and compute the required forces. By doing this one calculation of the beam with rigid supports and six calculations of the beam with an unit displacement at each DOF is needed. Certainly this method gives a matrix of high accuracy but the effort is incredibly high.

A faster method involves the calculation of the transfer matrix method by itself. To compute the beam with rigid supports increments will be transferred parallel to the normal variables of state. By knowing this it is possible to calculate the changes of the starting values (e.g. $N_x/M_y/M_z$) to get a unit displacement at the end of the beam for each DOF. This method gives a less accurate but always good stiffness-matrix due to the linear approximation during the calculation of the increments.

4.2 *Problems of computing the K-matrix*

The difficulty is the fact that the K-matrix is not symmetric or an unambiguous computation is not possible. The reasons are not related to the implementation of algorithms, but it is the consideration of the effects like elongation of centroid axis or 3rd order theory. This is elaborated with the following example of a rigid supported single span beam.

By putting on a unit displacement at DOF 6 (Fig. 14) the beam starts cracking and then the system axis elongates. A compressive axial force is required to fulfill the boundary conditions due to the rigid supports at all other DOF.

By rotating DOF 6 anti-clockwise one gets a normal force in the beam of $N_x = -36.3$ kN. For a clockwise rotation one gets $N_x = -87.5$ kN. In both cases one can compute a normal force with the same sign. By this it is no longer possible to calculate a correct relation between angle of rotation and axial force. The other forces have a correct sign but the magnitude are partly different due to the non-linear material behavior.

A second difficulty comes with the required symmetry of the K-matrix. When the unloaded reinforced concrete beam is imposed a unit rotation at the end (DOF 6), the length of the beam will increase due to the cracks and the elongation of the centroid axis. To avoid the elongation (rigid supports as boundary conditions) there must be a compressive force. In contrast, the rotation will not appear when the beam with DOF 4 is imposed by a unit displacement. Therefore a moment is not required. However a bending moment is needed because of the requirement of the symmetric K-matrix. Equation 3 shows this, the important elements are marked with brackets.

These shown problems lead to an unsymmetric matrix which has to be equalized by computing the average values of upper and lower triangle-matrix.

$$K = \begin{pmatrix} \ddots & \circ & \circ & N_{4,1} & \circ & N_{6,1} \\ \circ & \ddots & \circ & 0 & \circ & Q_{6,2} \\ \circ & \circ & \ddots & 0 & \circ & M_{6,3} \\ \circ & \circ & \circ & N_{4,4} & \circ & [N_{6,4}] \\ \circ & \circ & \circ & 0 & \ddots & Q_{6,5} \\ \circ & \circ & \circ & [0] & \circ & M_{6,6} \end{pmatrix} \quad (3)$$

5 COMPARISON WITH EXPERIMENTS

The following two experiments from the literature will be explained and the measured deformations will be compared to the computed ones. For the calculations a PC with an AMD Athlon 1800 XP with 512 MB RAM was used.

5.1 *Example 1 – concrete slab*

The following results of experiments and calculations using 2D- or 3D-modelling were published in the "Journal of Engineering Mechanics" by M. Hörmann, H. Menrath & E. Ramm (May 2002).

Dimensions:
$\ell_{tot} = 2.29\,m \qquad \ell_{eff} = 2.04\,m$
$A_s = 2.13\,cm^2$
$b/h/d/d_1 = 48/10.2/8.1/2.1\,[cm]$

Concrete: $f_c = -33.16\,N/mm^2$

Reinforcement:
$f_y = 478\,N/mm^2 \qquad \varepsilon_y = 2.39\,mm/m$
$E = 200,000\,N/mm^2$
$E_{hardening} = 10,000\,N/mm^2$

The test-setup with concrete slabs and post-strengthened concrete slabs is shown in Figure 15.

The two tests (Fig. 16) with nominal identical parameters show differences in the deflection behaviour prior to yielding of steel and less pronounced after yielding. The differences before yielding can only be explained by different tension stiffening behaviour and after yielding by geometrical imperfections.

The authors of this article used different 2D- and 3D-meshes as well as different arrangements of reinforcement (concentrate on one element, spread over three elements). As shown in Figure 16 the 2D-modelling diverges significantly. Even an expensive 3D-modelling can not predict the yielding point and the post yielding behaviour (Fig. 17).

While the calculation of load deflection curve from "1D-beam-modelling" in Figure 18 using material properties as defined in the new German concrete standard 1045-1:2001-07 give a better result. Increasing the lever arm of the reinforcement by only 4 mm from 85 to 89 mm (concrete cover 4 mm less) gives in the

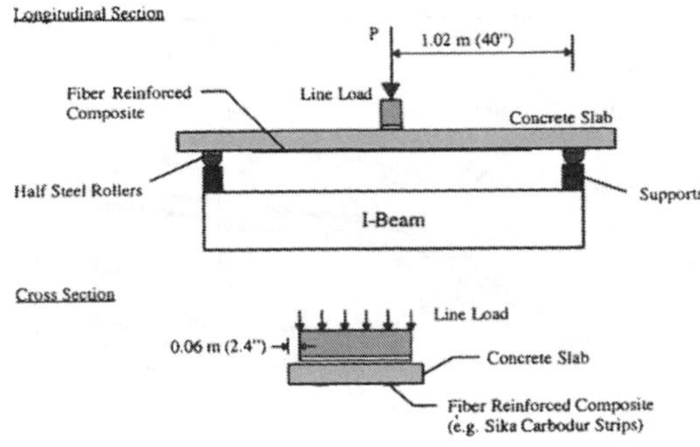

Figure 15. Test-setup.

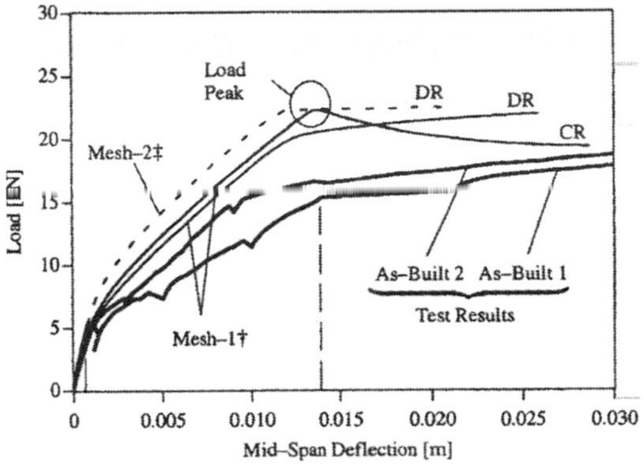

Figure 16. Mid-span deflection and 2D-modelling.

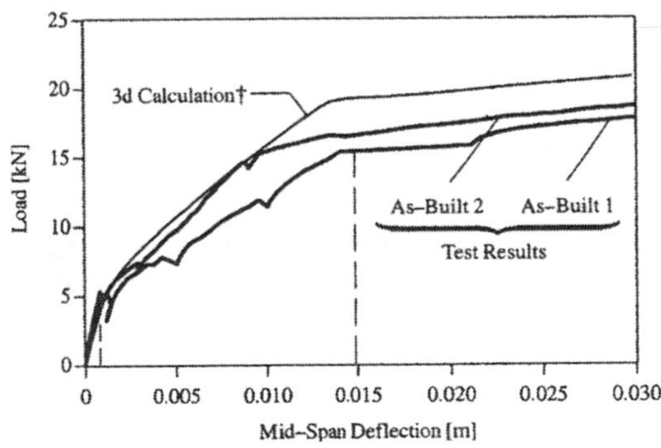

Figure 17. Mid-span deflection and 3D-modelling.

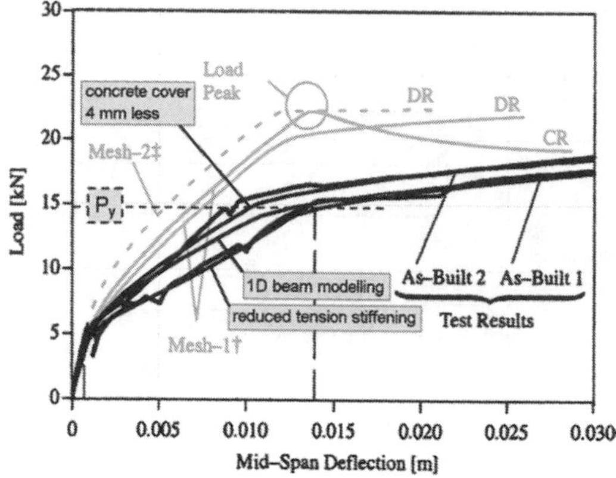

Figure 18. 1D-modelling with the program Stab2D-NL. Calculation-time: 6 seconds for 40 load-steps.

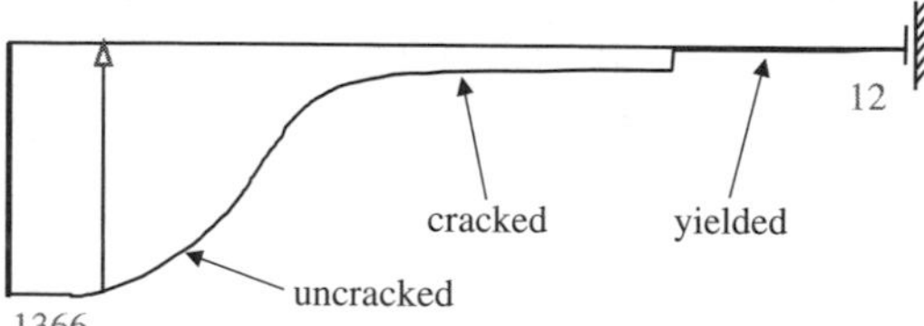

Figure 19. Tangential bending stiffness in kNm^2. Uncracked stiffness without reinforcement: EI = 1358, 3 kNm^2.

upper line and reducing the tension stiffening contribution produces the lower line. The differences in comparison with the FEM analysis are quite obvious. Instead of trying to match the test results it is more valuable to have an impression of the sensitiveness or scatter.

Moreover a 1D-modelling gives the values for curvature and bending stiffness to enable an impression of the area of uncracked, cracked and yielded parts of the beam as shown in Figure 19.

5.2 *Example 2 – beam with elongation of centroid axis and high shear force*

The following results of experiments were published by U. Kuhlmann & J. Ehmann, Institute for Construction and Design, University Stuttgart February 2001).

In this experiments reinforced, rectangular cross-sections with a constant normal force were tested. At the beginning the normal force was increased to N = 560 kN. After the complete unload to N = 0 kN a force of N = 340 kN was established and remained constant during the experiment (force controlled with

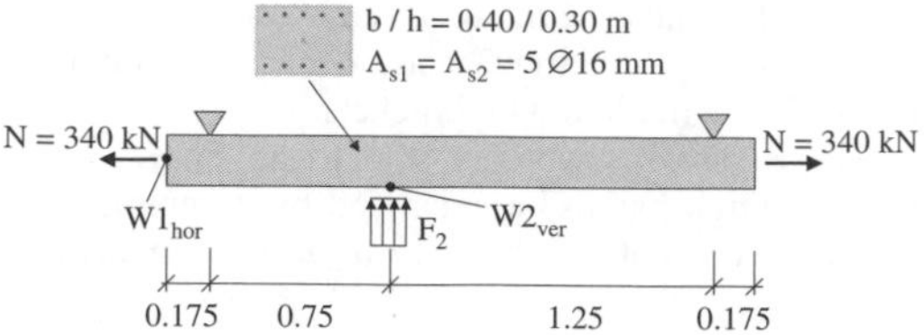

Figure 20. Test-setup. W1 and W2 record the displacements.

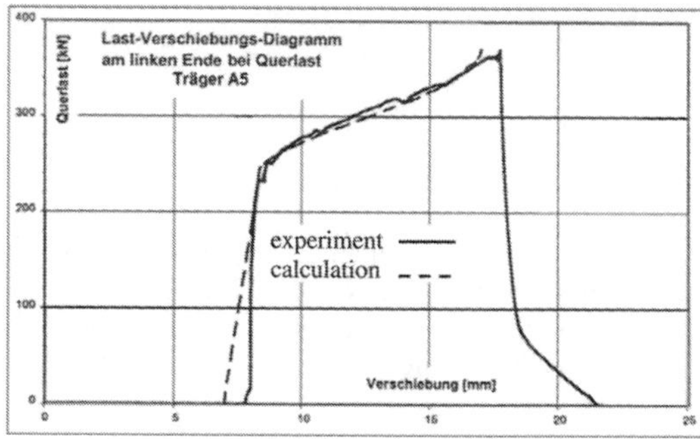

Figure 21. Load-deformation-graph at $W1_{hor}$ due to the elongation of system-axis.

recording the displacement $W1_{hor}$). The force F2 was increased until the maximum capacity of the hydraulic press was reached.

With the chosen arrangement of the measured deformations it is possible to compare the experimental values ($W1_{hor}$) with a calculation considering the elongation of the centroid axis due to bending moments.

The measured deformation at $W1_{hor}$ starts at around 8 mm due to the slip of the connecting screws. The calculated deformation (dashed) was then inserted in the graph for comparison (Fig. 21).

At the beginning part of the curve the beam is cracked already. The differences between computed and measured deformations probably are a fact of the temporarily normal force of N = 560 kN which produced a pre-damage of the structure with partly opened cracks. This effect is neglected in the calculation.

With a force $F_2 \approx 250$ kN the reinforcement starts to yield. The cracks in the cross-section are wide opened with a further elongation of the centroid axis.

The computed deformations at $W2_{ver}$ at the hydraulic press agree very well with the measured values.

With a force of $F_2 = 250$ kN a bending moment of M = 98.7 kNm is calculated. The ultimate bending moment for the given axial force of N = 340 kN is M_u = 115, 23 kNm which means only an increase of 17%. This would correspond to an increase of the

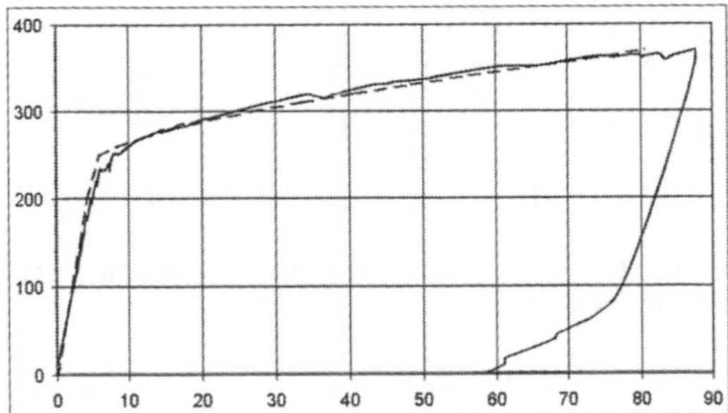

Figure 22. Load-deformation-graph at W_{ver}.

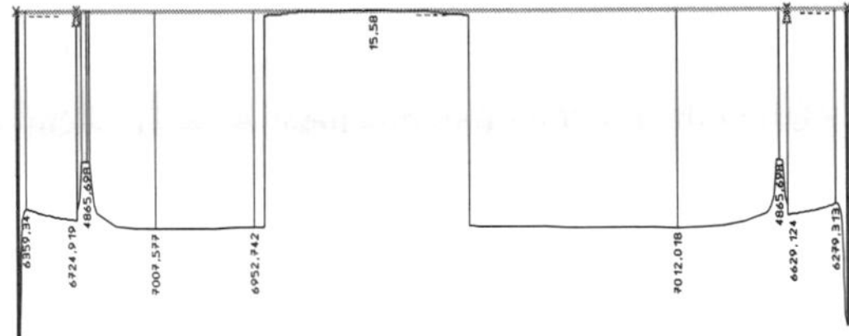

Figure 23. Distribution of bending stiffness at $F_2 = 350$ kN.

force to $F_2 = 292$ kN only. But due to the large deformation of the beam and the action of the axial force of N = 340 kN an additional bending moment with $\Delta M = 0.08\,m \cdot 340\,kN = 27.2$ kNm due to 2nd order effects appears. Only with this part an increase of the load until $F_2 = 370$ kN is possible. But this means that the cross-section with a load of $F_2 = 250$ kN and an according M = 98.7 kNm is very close at the ultimate limit state. Due to the yielding of reinforcement the bending stiffness is reduced to approximately 1/350 of the stiffness of the cracked zone.

With $F_2 = 320$ kN the calculated safety factor is $\gamma = 1,015$ only. This means, that for a large range of the force F_2 the calculation is done very close to the ultimate limit state. However, due to the formulation with tangential stiffness an acceptable calculation-time is reachable with 110 seconds for 40 load steps.

6 EXAMPLE FOR IMPROVEMENTS OF EXISTING STRUCTURES

With the shown method it is possible to compute larger structures faster, easier and better for practical purposes with a sufficient accuracy. New construction-types like joint-less bridges can be developed while existing structures can be calculated more exact to save money. Due to the redistribution of inner forces smaller cross-sections or less reinforcement is needed. A computation of existing structures for reuse and increase of the variable loads is possible too.

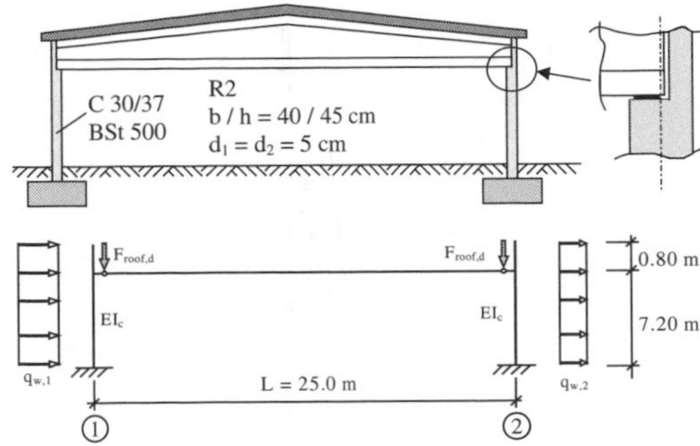

$q_{w1,d} = 3.6$ kN/m $\quad q_{w2,d} = 2.25$ kN/m
$F_{roof,d} = 462.5$ kN eccentricity each with e = 0.10 m
$F_{col,d} = 23.1$ kN (½ self-weight of the columns, at x = 7.20 m)

Figure 24. System and loads.

In the following a relevant example shows the potential of non-linear calculations in structural engineering.

6.1 *Non-linear calculation of connected columns*

Due to the large calculation efforts, non-linear calculations have mostly been used only for selected components. The example of a storage building consisting of prefabricated members can illustrate the potential of the non-linear methods when calculating the structure as a whole.

6.1.1 *System and loads*

The system consists of two columns and one simple beam which connects the columns. The two columns have the same cross-section with the dimensions b/h = 40/45 cm. The critical combination of load cases (wind as main load, unfavourable normal forces, snow reduced with $\psi_2 = 0,7$) shall be studied in more detail. Wind (luff and lee) and snow on the roof take effect as variable loads. The vertical loads due to the roof's self weight and snow are included in the force $F_{roof,d}$.

Moreover, additional inclination of the columns with $\gamma = 1/200$ for taking imperfections into account is considered with an equivalent horizontal load at x = 7.20 m. Since the system is symmetric, the deformation due to creep has no effect to the sideway deflection of the top of the column (point of load) and can therefore be neglected.

6.1.2 *Calculation and design*

There are three main possibilities for dimensioning the system:

1. Calculation of the inner forces (1st order theory) for the unconnected columns, calculation of the additional excentricity e_2 with the simplified model

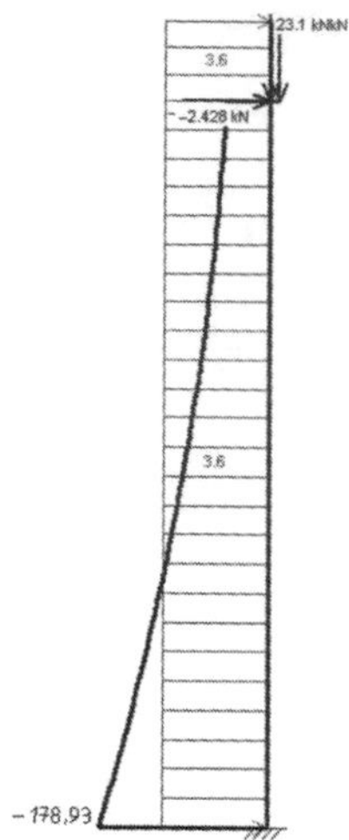

Figure 25. Load and resulting bending moment (th. 1st order).

column method, design of the cross-section, i.e. from Eurocode 2.

2. Calculation of the inner forces (1st order theory) for the complete system (connected columns), calculation of the additional excentricity e_2 as before, design.
3. Iterative choice of reinforcement, non-linear calculation of the inner forces of the complete system (2nd order theory, connected columns), check of safety.

6.1.3 *Option 1 – Unconnected columns*

For this alternative the left column (luff-side) is usually assumed to be most stressed. The load from wind and the excentricity of the roof turn in the same direction. Following results are obtained:

$$M_{1,Sd} = 3.6\tfrac{kN}{m}\cdot\frac{(8.0\,m)^2}{2} + 462.5\,kN\cdot 0.1\,m$$

$$+\frac{7.2\,m}{200}\cdot(462.5\,kN + 23.1\,kN) = 178.93\,kNm$$

$$e_2 = \frac{\ell_0^2}{2070\cdot d} = \frac{(2\cdot 7.2\,m)^2}{2070\cdot 0.40\,m} = 0.25\,m$$

$$M_{2,Sd} = 0.25\,m\cdot(462.5\,kN + 23.1\,kN) = 121.6\,kNm$$

$$M_{1,Sd} + M_{2,Sd} = 300.54\,kNm$$

$$N_{Sd} = -485.6\,kN \qquad \text{req}\,A_s = 27.86\,cm^2$$

6.1.4 *Option 2 – linear/non-linear method*

With a linear calculation of the complete system one finds that column 2 (lee-side) will get higher bending moments than the first one (even smaller than for option 1). The beam of the roof carries an axial force

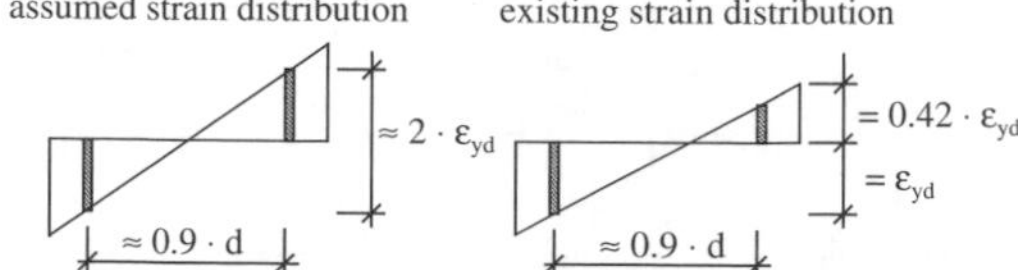

Figure 26. Strain-distribution in the cross-section for the yielding moment M_{yd}.

of $N = -11.9$ kN from the first to the second column. The resulting bending moment is:

$M_{1,Sd} = 128.9$ kNm
$M_{1,Sd} + M_{2,Sd} = 128.9$ kNm $+ 121.6$ kNm
$= 250.5$ kNm

Using the resulting reinforcement $A_s = 21.22\,cm^2$ for a non-linear calculation of the separated column 2, a deformation at the top of only $w = 3.26$ cm is obtained, considerably less than the assumed $e_2 = 25$ cm. Also, the maximum bending moment $M = 144.72$ kNm (for a curvature of $1/r = k_z = 3.349\cdot 10^{-3}\frac{1}{m}$) is much smaller than calculated.

The model column method used for standard cases gives a very safe result for the required reinforcement. The reason for this can be found in a bad approximation of the curvature of the cross-section for the yielding moment M_{yield} (Fig. 26) and the distribution of the curvature for the calculation of the deformation. From there one gets a much smaller deflection at the top of the column, and even at a reached moment of $M_y = 144.7$ kNm a stable equilibrium.

In summary, the calculation shown provides very safe results. Only by some additional computer-based non-linear calculations, a deeper view into the real behaviour of the system can be obtained and the results can be further improved.

6.1.5 *Option 3 – non-linear calculation of the complete system*

By means of a non-linear calculation of the complete system, the cross section of the columns can be reduced to 40 × 40 cm, and the reinforcement to req $A_s = 14.72\,cm^2$ (this also reduces the overall material consumption as well as transport costs and CO2-emission). The calculation effort is comparatively small, suitable programs provided. The load cases are realistically reproduced, and the system reserves are optimally utilized by redistribution of the forces. Figure 27 shows the results of the non-linear calculation. For clarification, only the results regarding the columns shown.

At column 2 the start of yielding of the reinforcement at the fixed support is clearly visible (Fig. 27, right). The curvature k_z and the strain ε_0 strongly increase, and the tangential bending stiffness is significantly reduced. Column 1, initially supported by

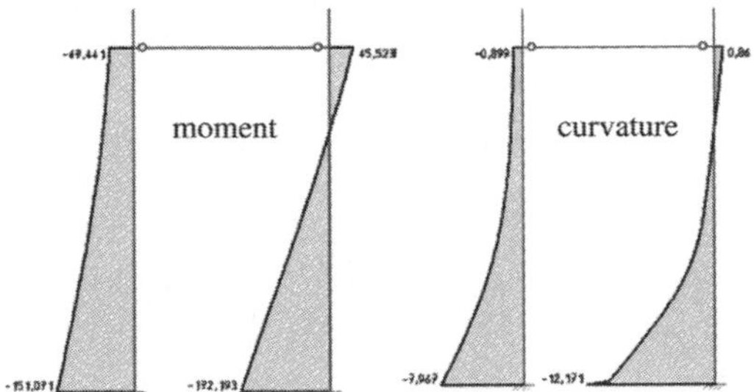

Figure 27. Bending moments and according curvature of columns 1 and 2.

the coupling force to column 2, will cause a reduction in this lateral supporting force as a result of the yielding of the reinforcement in column 2. A comparison of the moments shows that the differences between column 1 and 2 ($\Delta M = 172.19 - 151.07 = 21.12$ kNm) are smaller than for the previous calculation (option 2, $\Delta M = 128.91 - 93.26 = 35.65$ kNm). This shows the effect of redistribution of the inner forces from column 2 to column 1.

7 CONCLUSION

Currently with the developed Stab2D-NL program only 2D-frames can be calculated and only uniaxial bending is considered. Nevertheless with a slight extension of the algorithms also 3D-frames could be calculated. The computation of bending stiffness and the associated transfer matrix method for the calculation of bars is simple to implement into the existing algorithms. The difficulty lays in a good approximation of the rotation stiffness of the cracked cross section, which is only possible with an accurate modelling of the shear stress distribution. In the future an efficient method for calculating the torsion stiffness still must be developed.

However a lot of problems in the praxis can be modelled with a 2-dimensional system which can be analysed satisfactorily with the introduced algorithms in this work.

LITERATURE

Hörmann, M., Menrath, H. & Ramm, E. May 2002. Numerical Investigations of Fiber Reinforced Polymers Poststrengthened Concrete Slabs, Journal of Engineering Mechanics, Page 552–561.

Kuhlmann, U. & Ehmann, J. February 2001. Versuche zur Ermittlung der Querkrafttragfähigkeit von Verbundplatten unter Längszug ohne Schubbewehrung. Stuttgart, Germany.

Computational Modelling of Concrete Structures, Bićanić et al. (eds)
© 2003 Swets & Zeitlinger, Lisse, ISBN 90 5809 536 3

Reliability analysis of concrete structures

A. Strauss, K. Bergmeister, U. Santa
Institute of Structural Engineering, University of Vienna – BOKU, Vienna, Austria

ABSTRACT: The objective of this project is to create tools to enable professionals involved in the planing, design and analysis of structures, to use probabilistic based methods for the assessments of existing concrete structures. Input data of structural models are often very uncertain. Therefore, it is necessary to consider them to be random variables in a probabilistic risk assessment. Reability calculations using random variables or random fields and repetitive calculation of the problem are usually excessively time-consuming. Advanced simulation and approximation reliability techniques are needed in order to reduce the computational effort. Therefore, the project uses such an advanced simulation called the Latin Hypercube Sampling (LHS). The inclusion of this method in the existing nonlinear analysis program is required to estimate the statistical parameters for the structural response, for the sensitive analysis to calculate the failure load at a certain safety level and for Bayesian updating.

1 GENERAL

1.1 *Introduction*

In the following essay a possibility of the assessment of the reability of engineering structures is shown. The method is based on modern nonlinear FEM techniques, on intelligent sampling methods to include uncertainties and on information from monitoring and from stochastic models. Engineers and researcher are permanent escorted by uncertainties in their daily work. Therefore it is important to know about how to handle these uncertainties. The method presented in the following is a powerful tool for such a task.

1.2 *Sampling methods and techniques – LHS*

The multi-purpose probabilistic software for statistical, sensitivity and reliability analysis of engineering problem FREET, Novák (2002) is based on efficient reliability techniques. The probabilistic methods utilized in the program FREET were explained in more detail e.g. in Novák, Vorechovský, Červenka & Pukl (2001). They are summarized below. A special type of numerical probabilistic simulation called Latin Hypercube Sampling (LHS) makes it possible to use only a small number of simulations, Novák, Teplý & Keršner (1998). The range of the cumulative distribution function of each random variable is divided into N intervals of equal probability 1/N (N is the number of simulations). The regularity of sampling over the range of the cumulative distribution function ensures an efficient sampling of random variables. Statistical correlation among input random variables can be considered. A special stochastic optimization technique called simulated annealing is utilized to adjust random samples in such a way that the resulting correlation matrix is as close as possible to the target (user-defined) correlation matrix. Note that the approach allows working also with non-positive definite matrix on input, which can be the result of lack of knowledge of the user. This technique generates samples as close as possible to a positive definite matrix (mathematically and physically correct). The programs FREET and ATENA are integrated in order to allow for a probabilistic fracture analysis of engineering structures. The basic aim of statistical nonlinear fracture analysis is to obtain the estimation of the structural response statistics (failure load, deflections, cracks, stresses, etc.).

1.3 *Stochastic models*

The probability distribution functions for the basic input variables to these advanced simulation techniques can be obtained from stochastic models of materials (with and without degradation of strength), stochastical models for proof loading (permanent load, variable and traffic load) and from the comparison of recorded structural monitoring data. Therefore an additional part to this project is to create a database for the statistical description of the material properties and the load models. The database containing experimental and theoretical data from literature and provides the

Table 1. Stochastical model for concrete.

Concrete	Distribution type
Compressive strength	
Tensile strength	Normal
Modulus of elasticity	Log normal
Fracture energy	Weibull
Poison rate	
Parameters of Influence:	
Compactation, Curring, Hardening, Aggregate types, Spezimen sizes, Geometrical uncertainties, Degradation of concrete	

Table 2. Stochastical model for reinforcing steel.

Reinforcing steel	Distribution type
Yield strength	
Ultimate strength	Normal
Modulus of elasticity	Log normal
Strain R5	Weibull
Poisson rate	
Parameters of influence:	
Steeltype, Geometry of steel, Diameter of steelbars, Degradation, Corrosion	

Table 3. Stochastical model for structural steel.

Structural steel	Distribution type
Yield strength	
Ultimate strength	Normal
Modulus of elasticity	Log normal
Strain R5	Weibull
Poisson rate	
Parameters of Influence:	
Steeltype, Geometry of steel, Corrosion	

simulation methods with the necessary input. Table 1 to Table 6 show an overview of the stochastic models investigated with particular emphasis.

2 PROBABILISTIC ANALYSIS OF THE COLLE ISARCO BRIDGE

2.1 *Description of the modelling method*

In the following example dealing with the probabilistic approach, the non-linear analysis of a concrete structure is presented. The non-linear finite element software ATENA developed by Červenka Consulting in Prague is used for the simulation of damage and

Table 4. Stochastical model for prestressing steel.

Prestressing steel	Distribution type
Yield strength	Normal
Ultimate strength	Log normal
Modulus of elasticity	Weibull
Parameters of Influence:	
Steeltype, Geometry of steel, Diameter of steelbars, Degradation, Corrosion	

Table 5. Stochastical model for geometrical dimensions.

Geometrical dimensions	Distribution type
Concrete cover	
Effective depth	Normal
Dimension of columns	Gumbel
Slabs and beams	

Table 6. Stochastical model for model uncertainties.

Model uncertainties	Distribution type
	Normal
	Gumbel

failure in concrete and reinforced concrete structures. It is based on advanced material models including damage concept, smeared crack approach and crack band theory with fracture energy-related softening. The material properties as well as the load properties are accompanied more or less with uncertainties. These uncertainties cannot be avoided completely and their amount depend on the grade of information. Since these values are combined with the input parameters of a finite element analysis it is desirable to take these uncertainties also into account. These values considering uncertainties can be described by the mean value, the standard deviation and the distribution type. Thus, the programs FREET and ATENA were combined – Software SARA – in order to randomise the fracture analysis of engineering structures. The statistical nonlinear analysis provides an estimation of the structural statistic response (failure load, deflections, cracks, stresses, etc.), and enables the performance of a sensitivity and reliability analysis. The applicability of this advanced engineering method is demonstrated by the stochastic assessment of an existing pre-stressed bridge structure, see Figure 1 and Figure 2.

2.2 *Technical description*

The Prestressed Concrete Bridge to be assessed is located in Italy and was built in 1969. It is a full

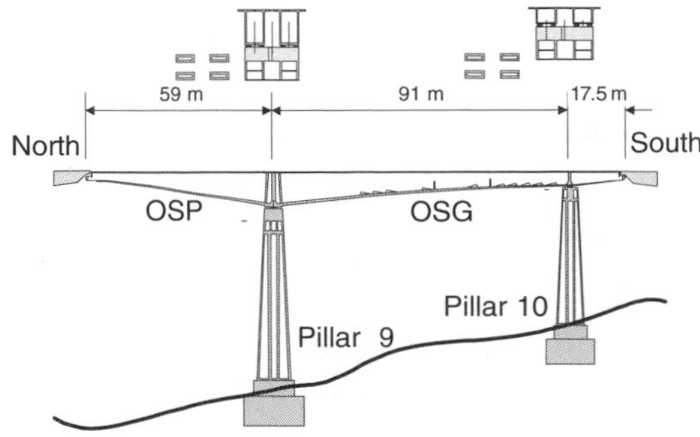

Figure 1. Longitudinal section of the cast – in Place Balanced Cantilever Beam – "Colle Isarco Bridge".

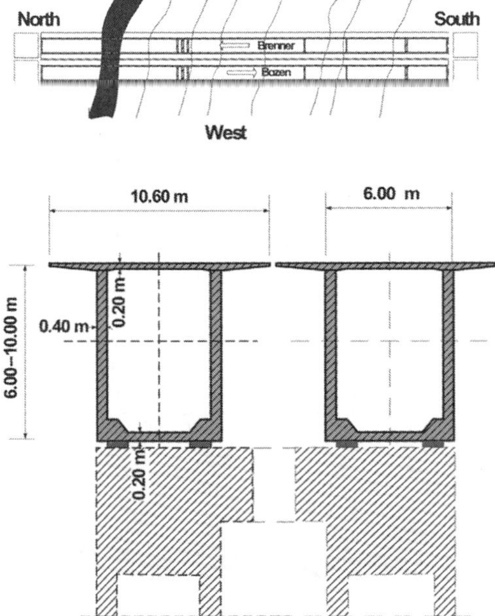

Figure 2. Ground plan and cross section of the cast – in place balanced cantilever beam – "Colle Isarco Bridge".

post-tensioned box girder bridge and is one of the objects of the Brenner Motorway. The Brenner Motorway consists of two separated lanes. Each of them includes several objects of the properties mentioned above. The bridges are "Cast – in Place Balanced Cantilever Beams" with a varying girder depth. The mid-span has a length of 91 m, the cantilever beam in the north has a length of 59 m and the cantilever beam in the south has a span length of 17.5 m. The total length of the bridge is 167.5 m. The slab of the lanes has a width of 10.60 m and a thickness of about 0.20 m. The lower girder slab has a width of 6.00 m

Table 7. Technical data of the bridge.

Random varible description	Unit	Value
Cantilever beam length – north	m	59.00
Midspan length	m	91.00
Cantilever beam length – south	m	17.50
Width of the lane plate	m	10.60
Width of the box girder	m	6.00
Height over the south support	m	4.50
Height over the middle support	m	10.80
Height at the north end	m	2.85
Whole construction length	m	167.50
Design line load	kN/m	10

and a thickness of about 0.20 m. The height of the box girder varies from 10.80 m, over the middle support, to 2.85 m, see Figure 1, Figure 2 and Table 7. At the bridge, which has been investigated, a permanent monitoring system is installed which collected data of the strain behaviour of the prestressed cable, data of the vertical displacements and data of the inclinometers. The aim of the monitoring system is similar to the performed stochastic analysis, to find out the realistic behaviour of the object during the lifetime. The bridge is cast of concrete B500 and is reinforced with mild steel BST 500. The post-tension tendons system consists of about 211 strands of St1350/1500. The bridge structure was analysed using the stochastic non-linear simulation package describe above.

2.3 *Material properties for the stochastic analysis*

The mean values of the material parameters for concrete B500 were generated using ATENA defaults (based on recommendations by CEB, fib, RILEM etc.). The statistical properties of the random variables used in this study were extracted from several sources Joint Committee on Structural Safety (2000), Mirza, Hatzinikolas & McGregor (1979), Wittmann, Slowik & Alvaredo (1994), Al-Harthy & Frangopol (1994). In a database, which is also available in the software SARA, useful statistical properties of materials were collected and evaluated. Some statistical properties of the random variables originated from this database, Strauss (2002). The parameters are summarized in Table 8. In reality the random variables are more or less correlated with each other. These phenomena are also considered in this analysis. Some information on correlation can be found in references Joint Committee on Structural Safety (2000) and Mirza, Hatzinikolas & McGregor (1979). You can find the prescribed correlation matrix in the upper triangle of Table 9. The lower triangle of Table 9 shows the correlation matrix generated by simulating annealing of 30 samples.

Table 8. Vector X of basic variables X_i.

Random variables description		Unit	Mean	Std	CoV	Distr
Young modulus of concrete	E_c	N/mm^2	37000	5540	0.15	LN
Tension strength of concrete	f_t	N/mm^2	3.26	0.586	0.18	W
Compressive strength of concrete	f_c	N/mm^2	42.5	4.25	0.1	LN
Fracture energy	G_f	N/m	120	24	0.2	W
Density of concrete	ρ	MN/mm^3	42.5	4.25	0.1	LN
Young modulus of reinforcement	E_s	N/mm^2	210000	63000	0.03	LN
Yield strength of reinforcement	f_y	N/mm^2	500	25	0.05	LN
Young modulus of prestressed steel	E_p	N/mm^2	210000	5540	0.03	LN
Yield strength of prestressed steel	f_p	N/mm^2	1350	270	0.20	LN

Table 9. Correlation of basic variables X_i.

	E_c	f_t	f_c	G_f	E_s	f_y	E_p	f_p
E_c	1	0.7	0.9	0.5	0	0	0	0
f_t	0.698	1	0.8	0.9	0	0	0	0
f_c	0.896	0.798	1	0.6	0	0	0	0
G_f	0.500	0.892	0.601	1	0	0	0	0
E_s	0	0	0	0	1	0	0	0
f_y	0	0	0	0	0	1	0	0
E_p	0	0	0	0	0	0	1	0
f_p	0	0	0	0	0	0	0	1

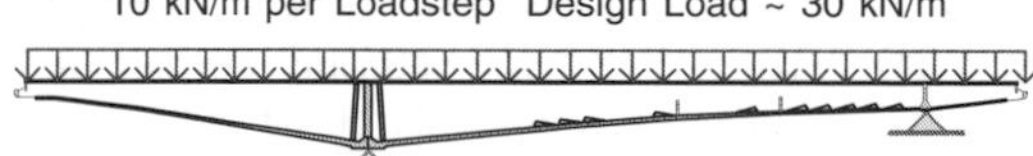

Figure 3. Geometrical structure for the finite element model.

2.4 *Non-linear finite element analysis and selected results*

The geometrical structure of the finite element model is shown in Figure 3. The bridge was prestressed and then loaded with a prescribed line load along the whole bridge. The ultimate failure load and the descending branch were obtained. The bridge girder failed typically next to the middle support. First the prestressed tendoms yielded, tensile cracks developed in the upper flange of the box girder and finally shear failure occurs.

Figure 4 shows a typical distribution of the principal strain shortly before failure on the deformed structure. The deformations are enlarged by a factor of 20.

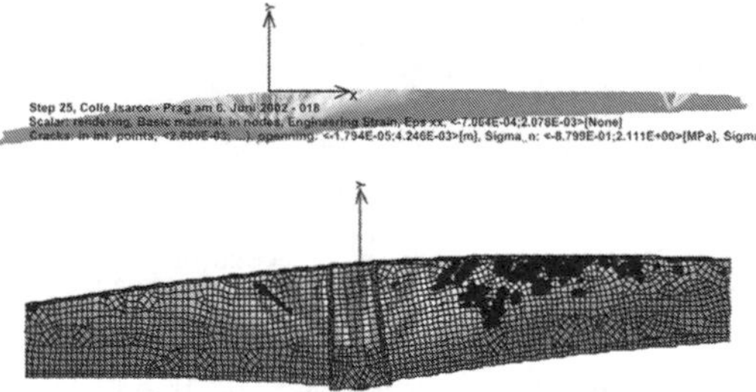

Figure 4. (a) Failure mode of the bridge, (b) Failure mode of the bridge – detail middlesupport.

For reliability analysis a proper analysis of the load S has to be done in order to work with the limit state function $g = R - S$. If g is normally distributed, the reliability index can be estimated from mean value and standard deviation regarding g as $\beta = \mu_g/\sigma_g$, see Figure 9. Stochastic simulations with 8 and 30 samples were performed and the structure response was

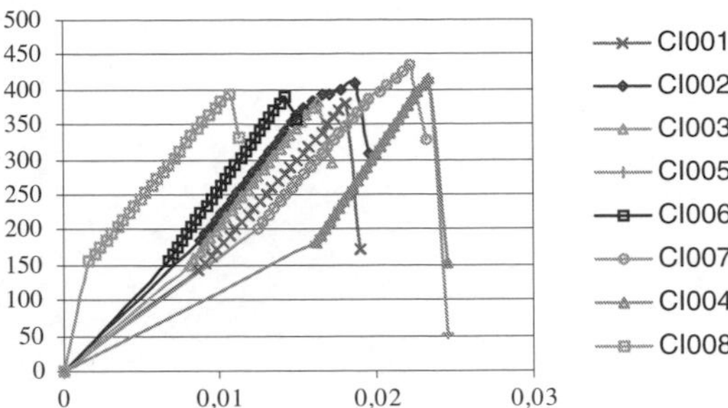

Figure 5. Midspan – vertical displacement; 25 Load steps; CoV = 0.586.

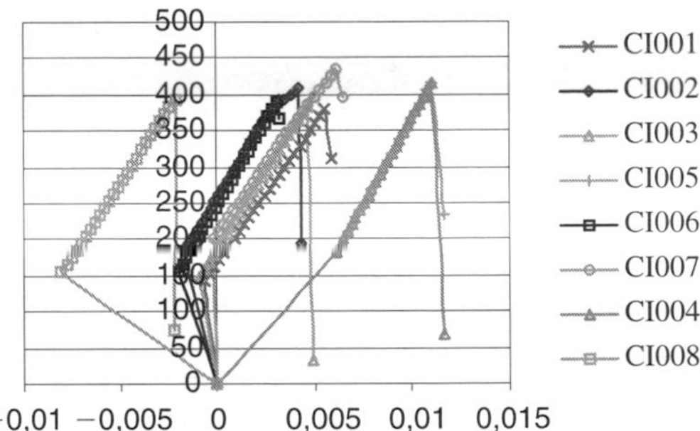

Figure 6. Vertical displacement at the cantilever 107.5 m; Load step 25; CoV = 0.792.

evaluated, see Figure 5 to Figure 7. The simulation provided an estimate of the statistical characteristics of the ultimate load and the displacements at midspan and at the end of each cantilever beam, see Figure 5 to Figure 7 and Table 10, i.e.

2.5 *Consideration of the displacement behavior*

As it is recognizable from Figure 5, the vector of displacement in midspan is gradually increasing up to the ultimate load of about 400 kN/m. The ultimate load also includes the body forces of the structure. The first part of the load – displacement curve originate from the body forces and the second part follows from the variable loads which are increased step by step during the calculation. Most of the simulations show a linear behavior of the load – deflection curve until reaching the ultimate load. In the area of the ultimate load a nonlinear behavior occurs. The mean value of the deflection derived from 8 simulations at the ultimate load is about 0.016 m. Figure 6 shows the behavior of the vertical deflection of the short cantilever beam. The characteristics of the displacement vectors is as similar as for the midspan. However a different behavior of the direction of the displacement can be seen.

Considering the displacements at the state of the body forces a grouping around the origin is visible but

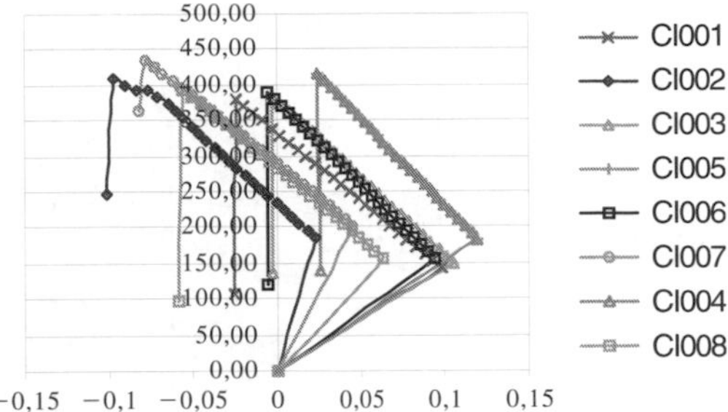

Figure 7. Vertical displacement; at cantilever 59 m: 25 Load steps; CoV = 0.770.

in addition greater displacements occurs also in the positive and negative region. This phenomenon also occurs at the longer cantilever beam and an amplification is recognizable with increasing load. This bearing gives the annoyance for the accurate analysis of this behavior with FREET. FREET also allows to observe the participation of the basic variables at the result sensitivity analysis. The use of this tool shows that there are two failure modes and that the responsible basic variable for these two modes is the yield strength of the prestressed steel.

Conclusion

The high scattering of the vertical displacements, especially at the end of the cantilever beams, results form the high estimated coefficient of variation for the yield strength of the prestressed cable due to the high uncertainties. From Figure 7 it is visible that two failure modes can occur due to this assumption – negative vertical displacement and positive vertical displacement.

2.6 *Consideration of the ultimate load*

The investigation of the ultimate load has the main task to verify the effectiveness of the sampling method LHS and the inclusion of the correlation tool. As you can see from Table 10, the calculation is performed with two different numbers of samples, 8 samples and 30 samples. It is noticeable that the higher number of simulations achieves a smaller increase of the accuracy. Due to this small progress of the results it can be said, that a small number of simulations, greater than the number of the basic variables and smaller than 100, provides rather exact results.

2.7 *Reability assessment ultimate load*

Since the results of the stochastic calculation provides information of the statistical characteristics of the ultimate load but not detailed description of the load model a parameter study – varying the coefficient of the load – was performed. The distribution function

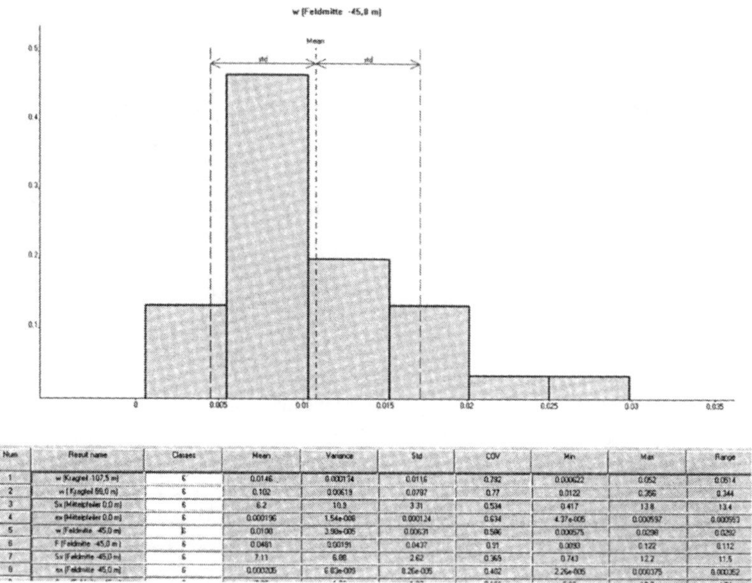

Num	Result name	Classes	Mean	Variance	Std	COV	Min	Max	Range
1	w (Kragteil 107,5 m)	6	0.0146	0.000134	0.0116	0.792	0.000622	0.052	0.0514
2	w (Kragteil 88,0 m)	6	0.102	0.00619	0.0787	0.77	0.0122	0.356	0.344
3	Sx (Mittelpfeiler 0,0 m)	6	6.2	10.9	3.31	0.534	0.417	13.8	13.4
4	ex (Mittelpfeiler 0,0 m)	6	0.000196	1.54e-008	0.000124	0.634	4.37e-005	0.000597	0.000553
5	w (Feldmitte -45,0 m)	6	0.0108	3.98e-005	0.00631	0.586	0.000575	0.0298	0.0292
6	F (Feldmitte -45,0 m)	6	0.0481	0.00191	0.0437	0.91	0.0093	0.122	0.112
7	Sx (Feldmitte -45,0 m)	6	7.11	6.88	2.62	0.365	0.743	12.2	11.5
8	ex (Feldmitte -45,0 m)	6	0.000205	6.83e-009	8.26e-005	0.402	2.26e-005	0.000375	0.000352

Figure 8. Statistical assessment, histogram for the displacement in midspan at load step 25.

Table 10. Statistical values of the ultimate load.

Number of samples	Mean value [kN/m]	Variance $[kN/m]^2$	Standard deviation [kN/m]	Coeff. of variation –
8	234.3	388	19.69	0.05
30	235	324	18.00	0.08

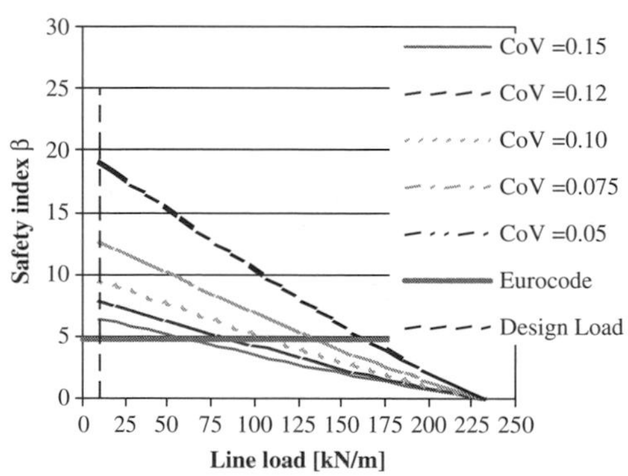

Figure 9. Reliability assessment of the load.

of the load follows with the mean value the values of the load steps. The determination of the reliability results from the comparison, "Cornell – Basler", of the ultimate load distribution with the load distribution. Figure 9 shows as expected, that the reliability index β decreases with increasing coefficient of variation.

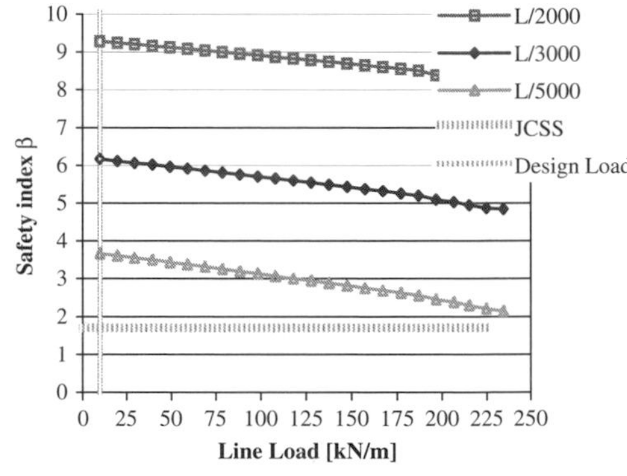

Figure 10. Reliability assessment of the vertical displacement at the midspan.

Despite of a high coefficient of variation, 0.15, the design load can be increased three times higher to reach the allowable safety limit recommended in Joint Committee on Structural Safety (2000).

2.8 *Reability assessment of the displacement in the midspan*

Beside the investigation of the ultimate load the vertical displacements are additional factors of interest. Therefore the formulation of the limit state functions for the serviceability were performed. The statistical characteristics of the displacement at each load step were compared with different limit states, see Figure 10. It can be seen that the vertical displacement in midspan isn't an authoritative value compared with the allowed values of Joint Committee on Structural Safety (2000).

2.9 *Reability assessment of the displacement at the cantilever beam with 59 m*

In the same way as in chapter 2.8 the investigation of the vertical displacement – service ability – was performed at the end of the long cantilever beam.

The reliability analysis showed that this response was an important factor – regarding the limit state – by the assessment of the service ability, see Figure11.

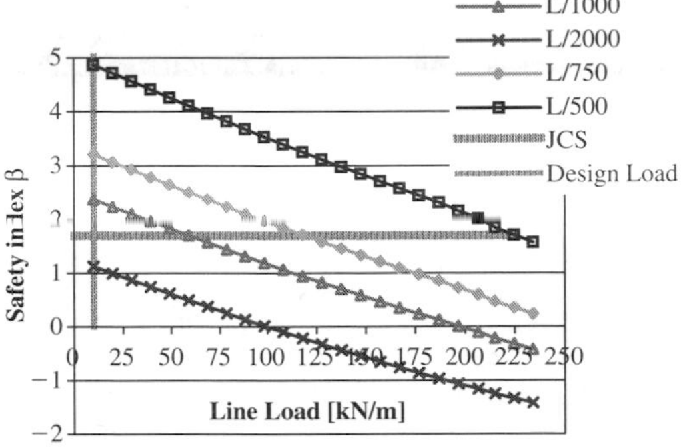

Figure 11. Reliability assessment of the vertical displacement at the end of the long cantilever beam – 59 m.

Conclusion

Under the assumption of a normal probability distribution for R and S, a simple possibility of utilization of obtained statistical characteristics of ultimate load and vertical displacement for reliability calculation can be drawn. The development of the reliability index for the ultimate load under the consideration of different levels of load S (mean values) is plotted in Figure 9. There are chosen different CoV of the load to demonstrate their influence. The horizontal line represents target reliability index as specified by Eurocode 1 (2001) (4.7 for 1 year). Additional an observation of the vertical displacements using the reliability concept was performed, see Figure 10 and Figure 11. It results that the end of the north cantilever beam is the authoritative element for the vertical displacements.

3 MONITORING

As already mentioned in chapter 2.2 a monitoring system is established at this investigated object. So it is looked ahead to compare the information extracted of the simulation with the monitoring data. This scheme allows to adjust in more detail the stochastic models – some of the parameters are assumed – and furthermore to evaluate the current probability of failure and the remaining lifespan. Therefore the problematic of the monitoring is sketched in the following paragraph.

The continued use of existing systems is, due to environmental, economical and socio-political assets,

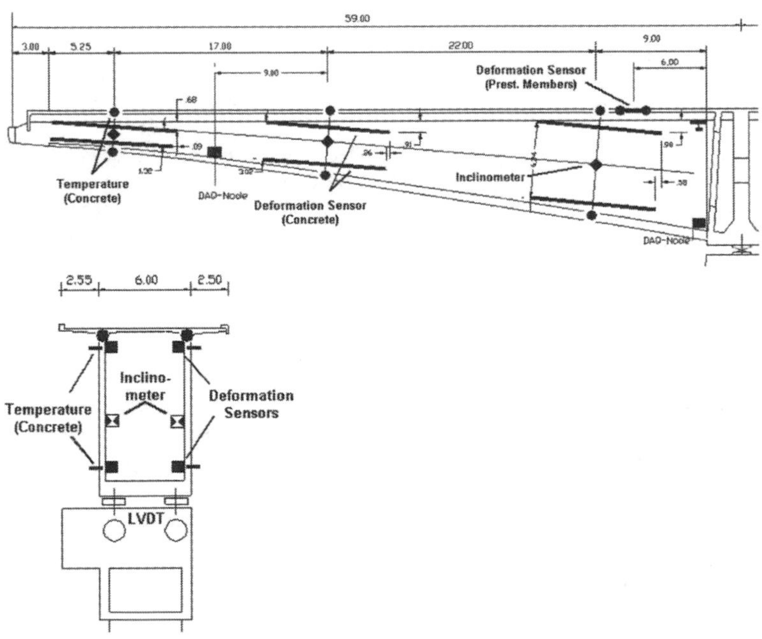

Figure 12. Instrumentation detail of one box girder.

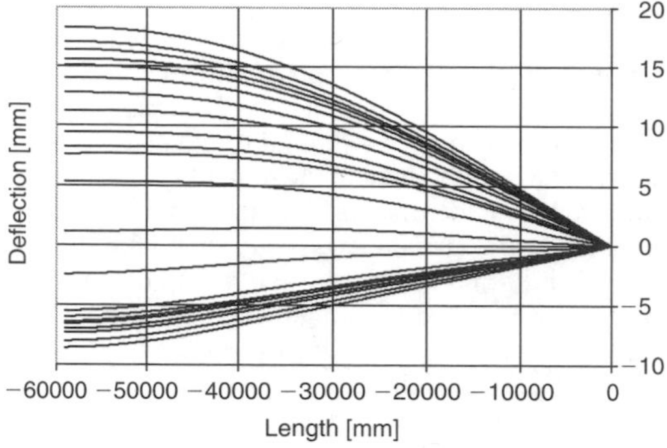

Figure 13. Evaluation of the deflection.

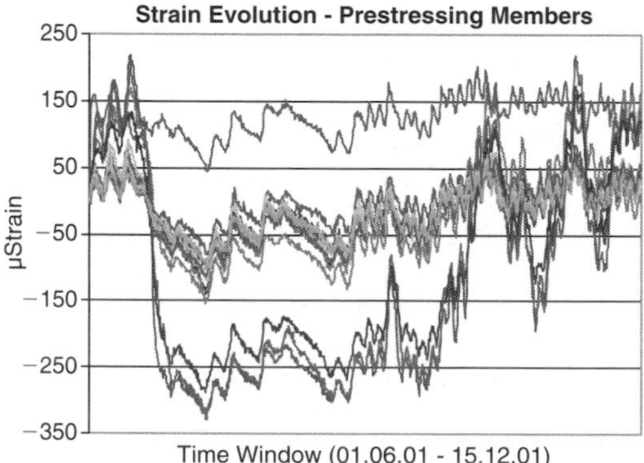

Figure 14. Evaluation of the strain.

of great significance growing larger every year. For the purpose of developing adequate life extension and replacement strategies, issues such as whole-life performance assessment rules, target safety levels and optimum maintenance strategies must be formulated and resolved from a lifetime reliability viewpoint and lifecycle cost perspective, Frangopol, (2000). In front of this background we find the necessity to identify the key parameters and procedures to verify and update the knowledge about the present condition of a structure with respect to a number of aspects. A visual inspection may yield a first qualitative, maybe purely intuitive impression. A better judgment would be based on the evaluation of quantitative information collected by adequate instrumentation and monitoring systems. This includes the observation of mechanical parameters such as loads, strains, displacements and deformations as well as durability aspects. On the Colle Isarco Viaduct (Brennero-Highway A22, Italy) a network of fibre optic SOFO-sensors Inaudi (1997) has been installed for strain and deformation measurements. 96 sensors with a base length of 10 m were installed on the concrete surface parallel to the neutral axis of the 8 box girders.

The displacement and curvature profile can be calculated from the local strain measurements of the sensor network through a deformation calculus algorithm as described in Vurpillot, Gaston, Benouaich, Clément & Inaudi (1998). Sensors with a base length of 0.5 m were installed to measure the strains of 16 selected prestressing members. Another 24 sensors with a base length of 8 and 10 m were installed on two piles of the structure. In order to separate the influence of thermal gradients in the structure and other agents (traffic, snow, wind), 60 thermocouples have been installed in the girders and on the piles. Further, 10 LVDTs have been used to instrument movements on the bearings. On 4 selected columns 12 electrochemical multiprobes have been installed for durability monitoring. These devices determine the concentration of free chlorides, corrosion current and the electrochemical potentials at different embedding levels, Zimmermann, Schiegg, Elsener & Böhni (1997).

4 CONCLUSION

Considering realistic material parameters as well as load parameters allows the design engineer to be much more precise in the analysis of the overall behaviour. The reliability index for existing structures depends on the material, on the load parameters, on the mechanical model and on the nonlinearity of the geometry and the material. This innovative concept "SARA" allows to analyse the safety of a structure including material, structural and stochastic nonlinearity and gives the possibility to extract the sensibility factors with the most important participation on the results. For the future a flexible safety margin of $\beta = 4.7$ to 2.5 for existing structures may be proposed, Ishida. The monitoring system combined with this simulation method allows under the use of proof loading procedures an accurate determination of the realistic behavior of the structure and allows statements about the remaining lifespan.

REFERENCES

Al-Harthy, A.S. & Frangopol, D.M. 1994. Reliability assessment of prestressed concrete beams, *Journal of Structural Engineering*, ASCE, Vol. 120, N. 1, pp. 180–199.

Eurocode 1 2001. Basis of structural design, *CEN, Brussels, Belgium.*

Frangopol, D.M. 2000. Bridge Health Monitoring and Life Prediction based on Reliability and Economy. Bauhaus-University Weimar, *International Workshop on the Present and Future in Health Monitoring*. Germany.

Inaudi, D. 1997. Fiber optic smart sensing, Optical Measurement Techniques and Applications: Editor Rastogi, P.K., Artech House, pp. 255–275.

Ishida, T. An integrated computational system of mass/energy generation, transport and mechanics of materials and structures. *Documentation report on the achievements in the Ph.D. thesis entitled.*

Joint Committee on Structural Safety 2000. *Probabilistic Model Code*, 12th draft.

Mirza, S.A., Hatzinikolas, M. & McGregor, J.G. 1979. Statistical descriptions of strength of concrete. *Journal of Structural Division*, ASCE, Vol. 105, N. 6, pp. 1021–1037.

Novák, D. 2002. FREET – Feasible Reliability Engineering Efficient Tool – program documentation. Brno University of Technology, faculty of Civil Engineering, Institute of Structural Mechanics/Červenka Consulting, Prague, Czech Republic.

Novák, D., Vorechovský, M., Červenka, V. & Pukl, R. 2001. Statistical Nonlinear Analysis – Size Effect of Concrete Beams. In R. de Borst (eds), *Fracture Mehanics of Concrete Structures (FraMCoS 4)*: 823–830. Rotterdam: Balkema.

Novák, D., Teplý, B. & Keršner, Z. 1998. The Role of Latin Hypercube Sampling Method in Reliability Engineering. *Structural Safety and Reliability (ICOSSAR – 97)*. Rotterdam: Balkema.

Strauss, A. 2002. Database for Engineering Stuctures. Dissertation at the Institut of Structural Engineering, Univ. Vienna, BOKU.

Vurpillot, S., Gaston, K., Benouaich, D., Clément, D., Inaudi, D. 1998. Vertical deflection of a pre-stressed concrete bridge obtained using deformation sensors and inclinometer measurements. *ACI Structural Journal*, Vol. 95, No. 5, pp. 518.

Wittmann, F.H., Slowik, V. & Alvaredo, A.M. 1994. Probabilitic aspects of fracture energy of concrete. *Materials and Structures*, N. 27, pp. 499–504.

Zimmermann, L., Schiegg, Y., Elsener, B., Böhni, H. 1997. Electrochemical techniques for monitoring the conditions of concrete bridge structures. *Repair of Concrete Structures, Proceedings of Int. Conference*.

Computational Modelling of Concrete Structures, Bićanić et al. (eds)
© 2003 Swets & Zeitlinger, Lisse, ISBN 90 5809 536 3

A displacement-based formulation for steel-concrete composite beams with shear lag

F.F. Sun
Department of Structural and Mechanical Engineering, University of Trento, Via Mesiano, Trento, Italy formerly at the Department of Building Engineering, Tongji University, Shanghai, China

O.S. Bursi
Department of Structural and Mechanical Engineering, University of Trento, Via Mesiano, Trento, Italy

ABSTRACT: A displacement-based formulation for composite beams with shear lag is proposed in this paper. In order to validate accuracy and robustness of the proposed formulation, some modelling aspects relevant to kinematic assumptions, shear warping function, shape functions, patch test and convergence rate are investigated in detail.

1 INTRODUCTION

In order to perform realistic modelling of steel-concrete composite beams, two kinematic aspects must be considered: the deformability of the shear connection and the non-uniform distribution of longitudinal displacements in the concrete slab, namely, the so-called shear lag effect. The deformability of the shear connection allows a slip at the slab-beam interface, which increases the global flexibility of the member and reduces the bending stresses in the slab as it happens with the shear lag.

Gjelsvik (1991) introduced an analog-beam method of analysis for beams with shear-lag effects and/or with shear deformations resulting from shear studs by replacing the real beam with an analog beam, where all the shear deformation is concentrated in a thin layer. In order to perform time-dependent analyses of the shear-lag effect in composite beams with flexible shear connection, Dezi et al. (2001) established a coupled system of integral-differential equations, which were numerically solved by the finite difference method and the step-by-step general method in time.

Up to now, in the literature, however, only beam elements for composite beams with flexible shear connection but without shear lag have been formulated. All these elements are based on the Newmark kinematic model (Newmark et al. 1951): the two parts of the composite beam are modelled as Euler-Bernoulli beam; the effects of the shear connection are accounted for by using an interface model with distributed bond, and the contact between the components is preserved.

Arizumi et al. (1981) proposed a 12 degrees of freedom (DOF) beam element that assures the continuity of the axial and transverse displacements and their first derivatives. Daniel & Crisinel (1993) presented a 10DOF beam element, assuring the least regularity for the solution (continuity for rotations and mean axial strain), and introduced two additional internal nodes only for axial displacements. In order to overcome the limitations of displacement-based formulations, Salari et al. (1998) adopted a force-based beam element and Ayoub & Filippou (2000) introduced a displacement-stress mixed element.

Shear lag, which exists in a variety of engineering structures, has long been of interest to researchers of structural engineering. In practical design the concept of effective width has been introduced, the basic idea of which is to replace the real wide flange on a beam with a narrower effective flange and then treats the replacement beam according to the Navier's assumption. This method is somewhat inaccurate and therefore, several methods have been proposed to account for shear lag, see Reissner (1941), Gjelsvik (1991), Dezi et al. (2001) and Sapountzakis & Katsikadelis (2001), among others.

In detail, Reissner (1941) assumed the shear lag effect as a parabolic longitudinal stress distribution in the flanges of box beams. Song & Scordelis (1990) developed a harmonic shear-lag analysis, using plane stress elasticity, for the stresses in flanges of isotropic I-, T-, and box cross sections. Bauchau (1985) derived eigenwarping functions for closed thin-walled sections made of curvilinear orthotropic material through the application of the Gram-Schmidt orthogonalization process. Koo & Cheung (1989) presented a mixed variational formulation for thin-walled beams with shear lag. Lopez-Anido & GangaRao (1996) proposed a warping solution for shear lag in straight thin-walled orthotropic composite beams by assembling macroelements through compatibility and equilibrium conditions at nodal lines. Sapountzakis & Katsikadelis (2001) proposed a non-linearly coupled plate-beam system, which was solved using an analog equation method (AEM).

In this paper, firstly a displacement-based beam element is obtained from the theorem of virtual displacement, by introducing a function of shear warping distribution of the concrete slab into the Newmark kinematic model. Secondly, in order to distinguish errors arising from kinematic assumptions from those due to finite element approximations, a closed-form solution of this problem is derived. Finally, several modelling problems are examined through numerical validation: choice of the shear warping function, locking phenomenon under high shear connection degrees, patch test and convergence rate.

2 STATEMENT OF THE PROBLEM

The prismatic composite beam, made of a steel beam and a concrete upper slab connected by shear connectors, is considered. A Cartesian coordinate system {O; X, Y, Z} is introduced, with the Z-axis parallel to the rectilinear axis of the undeformed composite beam. Under the assumption of a symmetric cross section, the coordinate plane YZ can be chosen to coincide with the symmetry plane of the beam as depicted in Figure 1.

Following Dezi et al. (2001), the assumption for the concrete slab in the Newmark model is modified to take into account the shear lag, while the remaining assumptions are kept: (1) the transverse deflections of the two beam components, i.e. the concrete slab and the steel beam, are the same; (2) the Euler-Bernoulli kinematic assumption is adopted for the steel beam; more specifically, plane sections in the steel beam remain plane after deformation. Therefore, the longitudinal displacement u_{zs} of the steel beam can be obtained as follows,

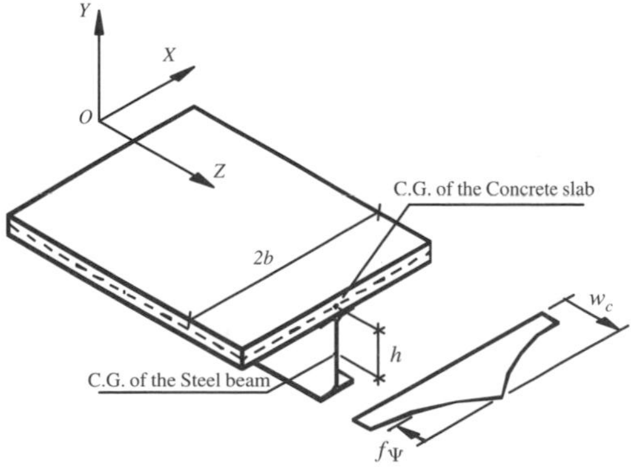

Figure 1. Assumption for the shear warping distribution.

$$u_{zs}(y,z) = w_s(z) - (y - y_s)v'(z) \tag{1}$$

where w_s and y_s are the longitudinal displacement and the vertical coordinate of the centroid of the steel beam, respectively; $v(z)$ is the deflection of the composite beam; the prime denotes derivative with respect to z.

In order to account for the shear lag effect, the variation of the longitudinal displacement of the concrete slab will be expressed as a product of a shape function $\psi(x)$ and an intensity function $f(z)$ of shear warping displacements. It is similar to the approach followed in the classical theory of thin-walled members (Gjelsvik 1981). In our context, ψ is a known function, which depends on the variable x only, so that the shape of shear warping displacements in cross sections is the same along the slab span and the warping displacements are constant along the slab depth. Hence, the longitudinal displacement u_{zc} of the concrete slab reads

$$u_{zc}(x,y,z) = w_c(z) - (y - y_c)v'(z) + f(z)\psi(x) \tag{2}$$

where w_c and y_c are the longitudinal displacement and the vertical coordinate of the centroid of the concrete slab, respectively.

According to the symmetry of loading and to the geometry of the composite beam, $\psi(x)$ is symmetric with respect to the Y axis and it is continuous along the slab width. Moreover, zero shear stresses at the slab edges require that

$$\psi_{,x}(\pm b) = 0 \tag{3a}$$

where b = half the slab width; and comma denotes derivative. In order to simplify the expression of the slab-beam interface slip, $\psi(x)$ is set equal to zero in the middle of the slab, i.e.

$$\psi(0) = 0 \tag{3b}$$

Hence, shear warping functions of polynomial form can be written as

$$\psi(x) = \sum_{n\geq 2} j_n [n\frac{|x|}{b} - \frac{|x|^n}{b^n}] \tag{4}$$

where j_n = predetermined constants.

The slab-beam interface slip ξ_z can be readily derived from Equations 1 and 2 (See Fig. 2) as follows,

$$\begin{aligned} \xi_z(z) &= u_{zs}(y_0, z) - u_{zc}(0, y_0, z) \\ &= w_s(z) - w_c(z) - v'(z)h \end{aligned} \tag{5}$$

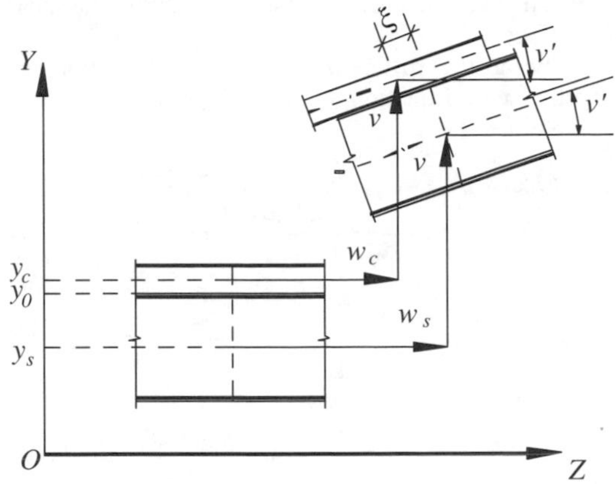

Figure 2. Kinematic relationship for the slip between the steel beam and the concrete slab.

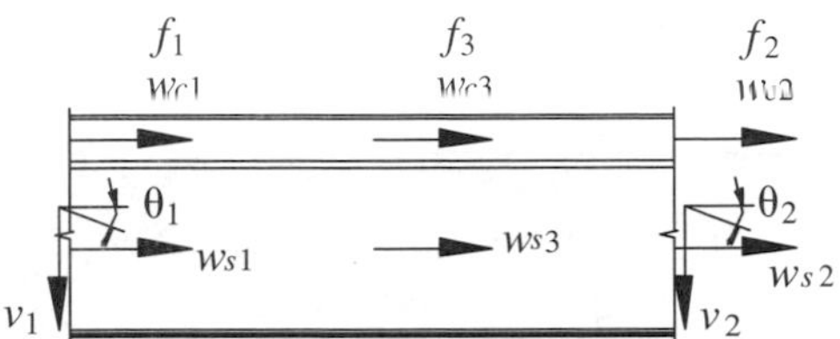

Figure 3. Nodal displacements.

where $y_0 =$ the vertical coordinate of the slab-beam interface; and $h = y_c - y_s$, is the distance between the centroids of the concrete slab and the steel beam.

Strain components can be evaluated as

$$\varepsilon_c(x,y,z) = w'_c - (y - y_c)v'' + f'\psi \tag{6a}$$
$$\gamma_c(x,y,z) = f\psi_{,x} \tag{6b}$$
$$\varepsilon_s(y,z) = w'_s - (y - y_s)v'' \tag{6c}$$

where ε_c and γ_c are the normal strain and shear strain of the concrete slab, respectively, and ε_s the normal strain of the steel beam.

Introducing linear constitutive relationships for concrete, steel and shear connection, the stresses of concrete and steel and the shear force of the shear connection can be obtained as follows

$$\sigma_c(x,y,z) = E_c\varepsilon_c = E_c[w'_c - (y - y_c)v'' + f'\psi] \tag{7a}$$
$$\tau_c(x,z) = G_c\gamma_c = G_c f\psi_{,x} \tag{7b}$$
$$\sigma_s(y,z) = E_s\varepsilon_s = E_s[w'_s - (y - y_s)v''] \tag{7c}$$
$$q_z(z) = p\xi_z = p[w_s - w_c - v'h] \tag{7d}$$

where E_c and G_c are the Young's modulus and shear modulus of concrete, respectively, E_s the Young's modulus of steel and p the stiffness of the shear connection.

The equilibrium conditions can be obtained from the theorem of virtual displacement in a variational form, equalizing the virtual work of internal stresses to that of external forces for each variation of the displacement functions. By assuming that the beam is subjected to quasi-static body and surface forces, b_{yi} and $s_{yi}(i = s, c$, denote forces applied to the steel part and the concrete part, respectively), the variational formulation can be expressed as

$$\begin{aligned}&\int_L [\int_{A_c} (\delta\varepsilon_c\sigma_c + \delta\gamma_c\tau_c)dA + \int_{A_s} \delta\varepsilon_s\sigma_s dA + \delta\xi_z q_z]dz \\ &= \int_L [\int_{A_c} \delta v b_{yc} dA + \int_{A_s} \delta v b_{ys} dA]dz + \int_L [\int_{\partial A_c} \delta v s_{yc} dx + \\ &+ \int_{\partial A_s} \delta v s_{ys} dx]dz + \left[\int_{A_c} \delta v s_{yc} dA + \int_{A_s} \delta v s_{ys} dA\right]_0^L .\end{aligned} \tag{8}$$

3 A BEAM ELEMENT FOR A COMPOSITE BEAM WITH SHEAR LAG

Following the displacement method, the displacement functions w_c, w_s and v, and the shear lag intensity function f are assumed as unknowns. Using the classical finite element approximation, a displacement-based beam element can be obtained from Equation 8. The displacement approach of the finite element method introduces a polynomial approximation of each displacement field at the interior of each element,

$$w_c = \phi_c(z)\mathbf{q}_c \tag{9a}$$
$$w_s = \phi_s(z)\mathbf{q}_s \tag{9b}$$
$$v = \phi_v(z)\mathbf{q}_v \tag{9c}$$
$$f = \phi_f(z)\mathbf{q}_f \tag{9d}$$

where $\phi_c(z), \phi_s(z), \phi_v(z), \phi_f(z) =$ vectors of shape functions; and $\mathbf{q}_c$, $\mathbf{q}_s$, $\mathbf{q}_v$, $\mathbf{q}_f =$ vectors of nodal displacement parameters.

Substituting Equations 5~7 and 9 into Equation 8 and integrating by parts, the following algebraic system is obtained,

$$\begin{bmatrix} \mathbf{k}_{cc} & \mathbf{k}_{cs} & \mathbf{k}_{cv} & \mathbf{k}_{cf} \\ \mathbf{k}_{sc} & \mathbf{k}_{ss} & \mathbf{k}_{sv} & \mathbf{0} \\ \mathbf{k}_{vc} & \mathbf{k}_{vs} & \mathbf{k}_{vv} & \mathbf{k}_{vf} \\ \mathbf{k}_{fc} & \mathbf{0} & \mathbf{k}_{fv} & \mathbf{k}_{ff} \end{bmatrix} \begin{bmatrix} \mathbf{q}_c \\ \mathbf{q}_s \\ \mathbf{q}_v \\ \mathbf{q}_f \end{bmatrix} = \begin{bmatrix} \mathbf{t}_c \\ \mathbf{t}_s \\ \mathbf{t}_v \\ \mathbf{t}_f \end{bmatrix}. \tag{10}$$

Different shape functions can be chosen to approximate the displacement field in the interior of each element. The simplest model providing the lowest regularity required involves a 10DOF beam element where the nodal parameters are

$$\mathbf{q}_c = [w_{c1}, w_{c2}]^T,\ \mathbf{q}_s = [w_{s1}, w_{s2}]^T,\ \mathbf{q}_v = [v_1, \theta_1, v_2, \theta_2]^T, \\ \mathbf{q}_f = [f_1, f_2]^T. \tag{11a}$$

According to Dall'Asta & Zona (2001), the contributions to the slip derived from axial and transverse approximations should be polynomials of the same order to avoid locking problems for cases of high shear connection degrees. Hence, two internal DOFs for w_c and w_s are needed for a 10DOF beam element. For the intensity function of the shear lag, two shape functions are considered: linear form and quadratic form, which lead to 12DOF and 13DOF beam elements, respectively. More specifically, for the 12DOF element the nodal parameters are

$$\mathbf{q}_c = [w_{c1}, w_{c3}, w_{c2}]^{\mathrm{T}},\ \mathbf{q}_s = [w_{s1}, w_{s3}, w_{s2}]^{\mathrm{T}},\ \mathbf{q}_v = [v_1, \theta_1, v_2, \theta_2]^{\mathrm{T}},\ \mathbf{q}_f = [f_1, f_2]^{\mathrm{T}}; \tag{11b}$$

while for the 13DOF element the nodal parameters read

$$\mathbf{q}_c = [w_{c1}, w_{c3}, w_{c2}]^{\mathrm{T}},\ \mathbf{q}_s = [w_{s1}, w_{s3}, w_{s2}]^{\mathrm{T}},\ \mathbf{q}_v = [v_1, \theta_1, v_2, \theta_2]^{\mathrm{T}},\ \mathbf{q}_f = [f_1, f_3, f_2]^{\mathrm{T}}. \tag{11c}$$

4 CLOSED-FORM SOLUTION FOR A COMPOSITE BEAM WITH SHEAR LAG

A closed-form solution can be derived for elastic prismatic uniform composite beams under uniform load g. This solution is helpful for identifying directly errors due to kinematic assumptions.

By substituting Equations 5–7 into Equation 8, integrating by parts and using fundamental variational calculus, the following differential equations are obtained

$$-E_c(A_c w''_c + f'' S_\psi) - p\xi_z = 0 \tag{12a}$$

$$-E_s A_s w''_s + p\xi_z = 0 \tag{12b}$$

$$(E_c I_c + E_s I_s) v^{\mathrm{IV}} + hp\xi'_z = g \tag{12c}$$

$$-E_c S_\psi w''_c - E_c I_\psi f'' + G_c I_{\mathrm{d}\psi} f = 0 \tag{12d}$$

where A_c, A_s, I_c and $I_s =$ the areas and the classical moments of inertia of the concrete slab and of the steel beam with respect to their own inertia principal axis, whereas

$$S_\psi = \int_{A_c} \psi \,\mathrm{d}A,\ I_\psi = \int_{A_c} \psi^2 \mathrm{d}A,\ I_{\mathrm{d}\psi} = \int_{A_c} \psi_{,x}^2 \mathrm{d}A$$

are the inertia of the concrete slab related to the cross-section warping.

Equations 12a and 12b represent the translation equilibrium conditions in the Z direction for slab and beam elements considered, respectively; Equation 12c is the equilibrium condition in the Y direction for the composite beam element; while Equation 12d imposes the equilibrium on the slab cross section between the shear stress τ_c and the normal stress σ_c, due to the shear lag effect.

Solving the differential equations Equation 12 together with Equation 4, a closed-form solution can be obtained:

$$w_c = (d_1\lambda_1^{-3}\lambda_3 e^{-\lambda_1 z} - d_2\lambda_1^{-3}\lambda_3 e^{\lambda_1 z} + d_3\lambda_2^{-3}\lambda_4 e^{-\lambda_2 z} - d_4\lambda_2^{-3}\lambda_4 e^{\lambda_2 z})\frac{1}{E_c S_\psi} + \frac{G_c I_{\mathrm{d}\psi}\rho_3}{6E_c S_\psi \rho_2} z^3 + d_5 \frac{G_c I_{\mathrm{d}\psi}}{2E_c S_\psi} z^2 + d_6 z + d_7 \tag{13a}$$

$$w_s = \frac{-1}{E_s A_s S_\psi}(d_1\lambda_1^{-3}\lambda_5 e^{-\lambda_1 z} - d_2\lambda_1^{-3}\lambda_5 e^{\lambda_1 z} + d_3\lambda_2^{-3}\lambda_6 e^{-\lambda_2 z} - d_4\lambda_2^{-3}\lambda_6 e^{\lambda_2 z}) - \frac{A_c G_c I_{\mathrm{d}\psi}\rho_3}{6E_s A_s S_\psi \rho_2} z^3 - d_5 \frac{A_c G_c I_{\mathrm{d}\psi}}{2E_s A_s S_\psi} z^2 + d_8 z + d_9 \tag{13b}$$

$$\begin{aligned} v = {} & \frac{h}{S_\psi (E_c I_c + E_s I_s)}(d_1\lambda_1^{-4}\lambda_5 e^{-\lambda_1 z} + d_2\lambda_1^{-4}\lambda_5 e^{\lambda_1 z} + d_3\lambda_2^{-4}\lambda_6 e^{-\lambda_2 z} + d_4\lambda_2^{-4}\lambda_6 e^{\lambda_2 z}) \\ & + \frac{g S_\psi \rho_2 - h A_c G_c I_{\mathrm{d}\psi}\rho_3}{24 S_\psi \rho_2 (E_c I_c + E_s I_s)} z^4 + d_5 \frac{G_c I_{\mathrm{d}\psi}(E_c A_c + E_s A_s)}{6h E_c E_s A_s S_\psi} z^3 \\ & + (d_6 - d_8 - \frac{A_c G_c I_{\mathrm{d}\psi}\rho_3}{p S_\psi \rho_2})\frac{z^2}{2h} + (-d_5 \frac{A_c G_c I_{\mathrm{d}\psi}}{p S_\psi} + d_7 - d_9)\frac{z}{h} + d_{10} \end{aligned} \tag{13c}$$

$$\xi_z = \frac{-1}{p S_\psi}(d_1\lambda_1^{-1}\lambda_5 e^{-\lambda_1 z} - d_2\lambda_1^{-1}\lambda_5 e^{\lambda_1 z} + d_3\lambda_2^{-1}\lambda_6 e^{-\lambda_2 z} - d_4\lambda_2^{-1}\lambda_6 e^{\lambda_2 z}) - \frac{A_c G_c I_{\mathrm{d}\psi}\rho_3}{p S_\psi \rho_2} z - d_5 \frac{A_c G_c I_{\mathrm{d}\psi}}{p S_\psi} \tag{13d}$$

$$f = \frac{\rho_3}{\rho_2} z - d_1 \frac{e^{-\lambda_1 z}}{\lambda_1} + d_2 \frac{e^{\lambda_1 z}}{\lambda_1} - d_3 \frac{e^{-\lambda_2 z}}{\lambda_2} + d_4 \frac{e^{\lambda_2 z}}{\lambda_2} + d_5 \tag{13e}$$

where $d_i (i = 1\text{–}10) =$ unknown coefficients which shall be determined by boundary conditions; $\lambda_i (i = 1\text{–}6)$, $\rho_2 =$ parameters relevant to geometrical and mechanical properties of the composite beam; $\rho_3 =$ a parameter of load effect relevant to the uniform load g. Lengthy expressions for these parameters can be found in Sun & Bursi (2002).

5 REMARKS ON MODELLING ASPECTS

5.1 *Errors from kinematic assumptions*

In view of a sound finite element (FE) modelling of a beam element, it is deemed necessary to check the errors inherent in the kinematic assumptions. Indeed, no discretization method can improve the precision of a model, once kinematic conditions and constitutive relationships have been established. Therefore, it is very important to explore this issue, before any implementation of the FE model.

The Newmark's kinematic model has been verified in the literature comparing numerical and experimental results. The assumption for shear lag in Equation 2 will be examined in the following section through comparison between results of the closed-form solution and those provided by a more realistic 3D FE model made with Abaqus (Hibbitt et al. 1998). One may observe from Equation 12 that x does not appear as an independent variable. Moreover, the shear warping function $\psi(x)$ does not appear explicitly, but only through S_ψ, I_ψ and $I_{d\psi}$, that define the geometric properties of the slab. Therefore $\psi(x)$ should be predefined independently.

The simplest shear warping function satisfying Equation 3 reads

$$\psi_1(x) = \frac{2|x|}{b} - \left(\frac{x}{b}\right)^2 \tag{14}$$

(See Equation 4). For the sake of simplicity Gjelsvik (1991) and Dezi et al. (2001) adopted this quadratic form, though they pointed out that a more complex function could be necessary. Proper selection of this function will be provided in the next section.

Moreover, Equation 2 implies that the transverse distributions of longitudinal displacements are of the same shape along the beam span. Together with the shape of $\psi(x)$ this requires the examination of due error types.

5.2 *Coupling of displacement fields*

The Euler-Bernoulli beam model is the simplest case of beam problems. When additional mechanical effects are introduced, difficulties both in the analytical and in the numerical solution may arise because, in general, coupling of different fields is introduced too. By means of closed-form solutions, we can examine the coupling phenomenon which accompanies the assumptions related both to the flexible shear connection and to the shear lag.

According to Equation 12d, letting the shear modulus of concrete G_c approach to infinity, the shear lag intensity f approaches to zero, and f'' vanishes at the limit point too. Hence, the beam changes to a composite beam without shear lag and the following differential equations are obtained from Equations 5 and 12:

$$-E_c A_c w''_c - p\xi_z = 0 \tag{15a}$$
$$-E_s A_s w''_s + p\xi_z = 0 \tag{15b}$$
$$(E_c I_c + E_s I_s) v^{IV} + hp\xi'_z = g \tag{15c}$$
$$\xi_z = w_s - w_c - v'h \tag{15d}$$

Similarly, let the stiffness of the shear connection p approach to infinity: then ξ_z approaches to zero accordingly (See Eq. 12). The shear force of the shear connection q_z, however, remains of non-zero limited value. Substituting $p\xi_z$ with q_z in Equation 12 and ξ_z with zero in Equation 5, differential equations for composite beams with shear lag and rigid shear connection are obtained as

$$-E_c(A_c w''_c + f'' S_\psi) - q_z = 0 \tag{16a}$$
$$-E_s A_s w''_s + q_z = 0 \tag{16b}$$
$$(E_c I_c + E_s I_s) v^{IV} + hq'_z = g \tag{16c}$$
$$-E_c S_\psi w''_c - E_c I_\psi f'' + G_c I_{d\psi} f = 0 \tag{16d}$$
$$w_s - w_c - v'h = 0\,. \tag{16e}$$

Furthermore, by eliminating the effect of the flexible shear connection in Equation 15, or the shear lag effect in Equation 16, differential equations for the Euler-Bernoulli beam are finally obtained

$$-E_c A_c w''_c - q_z = 0 \tag{17a}$$
$$-E_s A_s w''_s + q_z = 0 \tag{17b}$$
$$(E_c I_c + E_s I_s) v^{IV} + hq'_z = g \tag{17c}$$
$$w_s - w_c - v'h = 0\,. \tag{17d}$$

It is well-known that the closed-form solution of the Euler-Bernoulli beam under uniform load is of polynomial type. This solution is the simplest field that has to be reproduced by classical displacement-based finite elements, as standards beam elements employ polynomial-type shape functions. The fields in Equation 13 are sums of polynomials and exponential functions. The latter cannot be exactly reproduced with limited terms of polynomials. Setting ρ_3 as zero in Equation 13, it can be observed that the exponential functions exist in the homogeneous solution. This can be recognized as a result of the coupling of different fields in a one-dimensional beam formulation. The coupling implies that the finite element approximation for a set of coupled fields cannot be performed independently.

Closed-form homogeneous solutions of Equation 15 and 16 contain exponential functions too. It is very interesting to note that in both solutions the fields of w_c, w_s and v are of the same form

$$w_c = d_{c1}e^{-\alpha z} - d_{c2}e^{\alpha z} + d_{c3}z^2 + d_{c4}z + d_{c5} \tag{18a}$$
$$w_s = -d_{s1}e^{-\alpha z} + d_{s2}e^{\alpha z} + d_{s3}z^2 + d_{s4}z + d_{s5} \tag{18b}$$
$$v = d_{v1}e^{-\alpha z} + d_{v2}e^{\alpha z} + d_{v3}z^3 + d_{v3}z^2 + d_{v4}z + d_{v5} \tag{18c}$$

whereas the fields of f and ξ_z are equivalent in form

$$\frac{(-d_{f1}e^{-\alpha z} + d_{f2}e^{\alpha z})}{\alpha} + d_{f3} \tag{18d}$$

where $d_{ij}(i=c,s,v,f;\ \ j=1\text{–}5)=$ unknown coefficients. This means that coupling is identically introduced by both effects.

Coupling can be identified directly from the differential equations of such a one-dimensional problem. Though Equations 12, 15, 16 and 17 look similar to each other, only Equation 17 can be rewritten so that only one derivative appears for each field,

$$-E_cA_cw'''_c-q'_z=0 \tag{19a}$$

$$-E_sA_sw'''_s+q'_z=0 \tag{19b}$$

$$(E_cI_c+E_sI_s)v^{IV}+hq'_z=g \tag{19c}$$

$$w'''_s-w'''_c-v^{IV}h=0. \tag{19d}$$

Conversely, coupling can be easily seen in Equation 12d where different derivatives of the same field f, more specifically f'' and f, appear. Attention should be paid to coupling of different fields as a locking phenomenon may happen due to inappropriate finite element approximations.

5.3 *Locking problems*

Locking is a phenomenon of the finite element approximation and implies that the convergence rate is very low when meshes are refined. The cause of locking has been shown to be the inconsistent approximation of displacement fields. Thus the process of finite element approximation introduces some kinematic conditions which physically do not exist.

Dall'Asta & Zona (2001) pointed out a locking phenomenon occurring in the displacement-based formulation of composite beams without shear lag. In the case that the coupling between different displacement fields exists, such kind of locking problems are likely to occur if the interpolation of these fields is arbitrarily chosen.

The convergence rate is a measure of the performance of an element. However it cannot provide sufficient conditions for convergence. If the coupled displacement fields are inconsistent to each other, numerical difficulties in convergence may occur. A practical method, the patch test, is considered to be suitable to identify such problems in a finite element formulation (Zienkiewicz & Taylor 1989).

To ensure convergence, a finite element approximation must fulfill both consistency and stability conditions. The consistency requirements ensure that as the size h of the elements tends to zero, the approximation equations and their boundary conditions represent the exact differential equations and the boundary conditions (at least in a weak sense). The stability condition is simply translated as a requirement that the solution of the discrete equation system should be unique and avoid spurious mechanisms that may pollute the solution. For linear problems, it means simply that the stiffness matrix must be non-singular.

In common forms of patch tests, form C is necessary and sufficient for convergence (Zienkiewicz & Taylor 1989). In this form, a patch of elements is fixed only at the minimum number of nodal displacements necessary to obtain a physical solution, by eliminating the rigid body motions in the patch; and it is subjected only to prescribed natural boundary conditions. A solution is sought for the remaining nodal displacement values and compared to the assumed exact solution.

6 NUMERICAL VALIDATION

6.1 *Description of the worked example*

In this section, some results of numerical analysis on a composite beam with fixed ends are reported. The external restraints are such that the axial and vertical displacements, for beam and slab, as well as their rotations and the warping of the concrete slab, are prevented. Such a scheme may be representative of an internal span. The geometry and material properties shown in Figure 4 are taken after Dezi et al. (2001). In order to observe the effect of the shear connection on the performance of the proposed three elements, three shear connection degrees are considered herein, H, M and L which denote high, medium and low, respectively. The shear stiffness corresponding to each degree is shown in Figure 4.

Unlike the shear connection, the variation of the shear modulus G_c of concrete is rather small. Therefore, there is no need to consider the effects of this variation on the beam elements. Though very large values of the shear modulus of concrete may be used

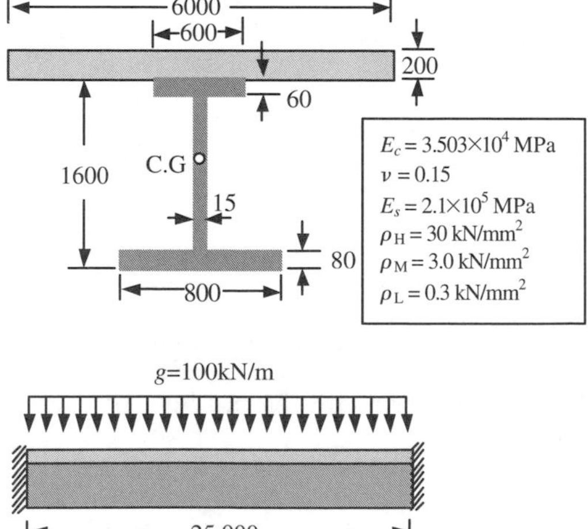

Figure 4. Data of the example (Dimensions in mm).

to obtain results of composite beams without shear lag, it is of little practical use after all.

6.2 *Validation of the kinematic assumptions*

Herein, the kinematic assumptions are validated through the comparison between the results of the closed-form solution and those of the finite element software Abaqus (Hibbitt et al. 1998).

The Abaqus code is able to reproduce a realistic distribution of normal stresses in the concrete slab by means of the 3D modelling of the composite beam. Because of symmetry, only one-fourth of the composite beam is modelled with Abaqus (See Fig. 5). The load is applied uniformly over the beam axis instead over the whole concrete slab in order to exclude the local loading effect on the stress distribution, which is not accounted for in the adopted kinematic model.

The steel beam is modelled with Element B33 (2-node cubic beam in space), while Element C3D20R (20-node quadratic brick element with reduced integration) is used for the concrete slab. The mesh size in the direction of the beam span is 125 mm while in the other two directions is 200 mm.

In order to simulate the kinematic relationship of the steel beam and the concrete slab, each node of beam elements (namely, the beam node) is not directly related to the corresponding node of brick elements (namely, the brick node). A virtual node is added sharing the position of the brick node. The virtual node is then connected both to the beam node with a rigid Beam connection, representing the kinematic relationship between the beam axis and the beam flange, and to the brick node, with a spring element Spring2 in the longitudinal direction and two kinematic constraint equations prohibiting any relative displacements in transverse and vertical directions, representing the shear connection (See Fig. 5).

To compute the ten unknown coefficients of the closed-form solution in Equation 12, Equation 14 is adopted and 10 boundary conditions at both ends are established,

$$w_c(0)=w_c(L)=0,\ \ w_s(0)=w_s(L)=0,\ \ v(0)=v(L)=0,$$
$$v'(0)=v'(L)=0,\ \ f(0)=f(L)=0 \tag{20}$$

where L = length of the beam span.

Figure 6 shows the good accuracy of the beam axis deflections by the closed-form solution compared with those provided by the Abaqus code and by Dezi et al. (2001). Figures 7 and 8 compare the longitudinal and transverse distribution of the normal stresses of the concrete slab with the shear connection degree M. Again the results are very close to those of Dezi et al. (2001) since both methods adopt the same kinematic assumptions, differing only in the solution method. In general, the stress results of the closed- form solution correlate well with those of the Abaqus code, implying that the adopted kinematic model is appropriate. Significant discrepancies occur merely at the corners of the concrete slab, where in reality stress concentrations exist, which are not taken into account in the adopted kinematic model. Discrepancies become smaller for composite beams with weaker shear connection. Moreover at the corners, normal stresses are

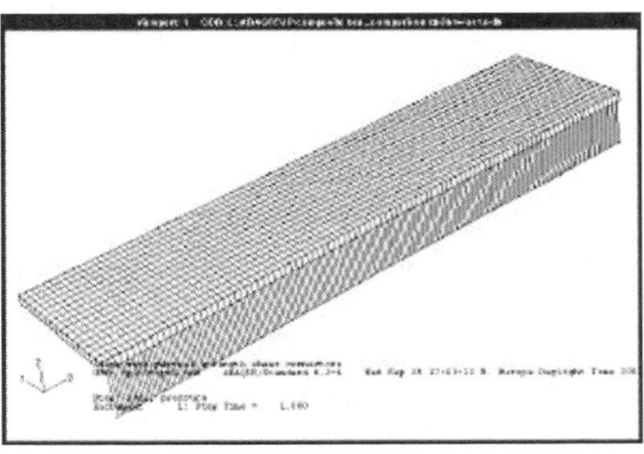

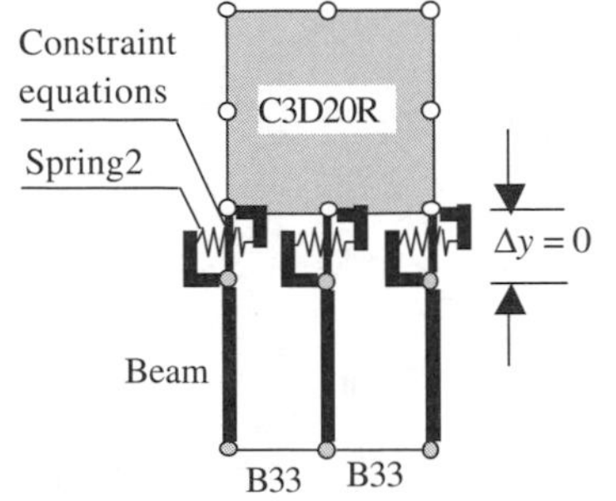

Figure 5. The finite element model of the composite beam with Abaqus.

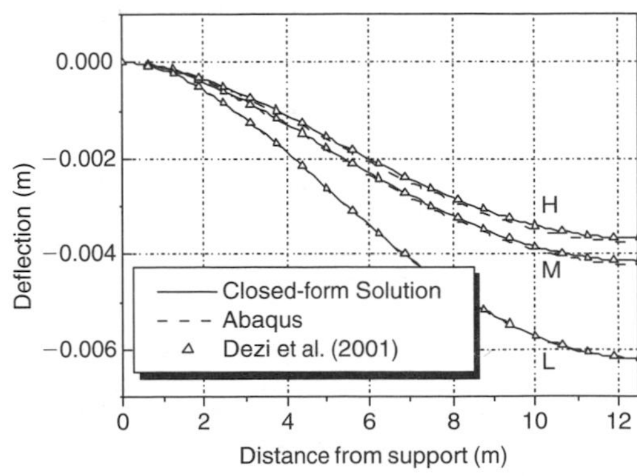

Figure 6. Deflections of the beam axis with different shear connection degrees.

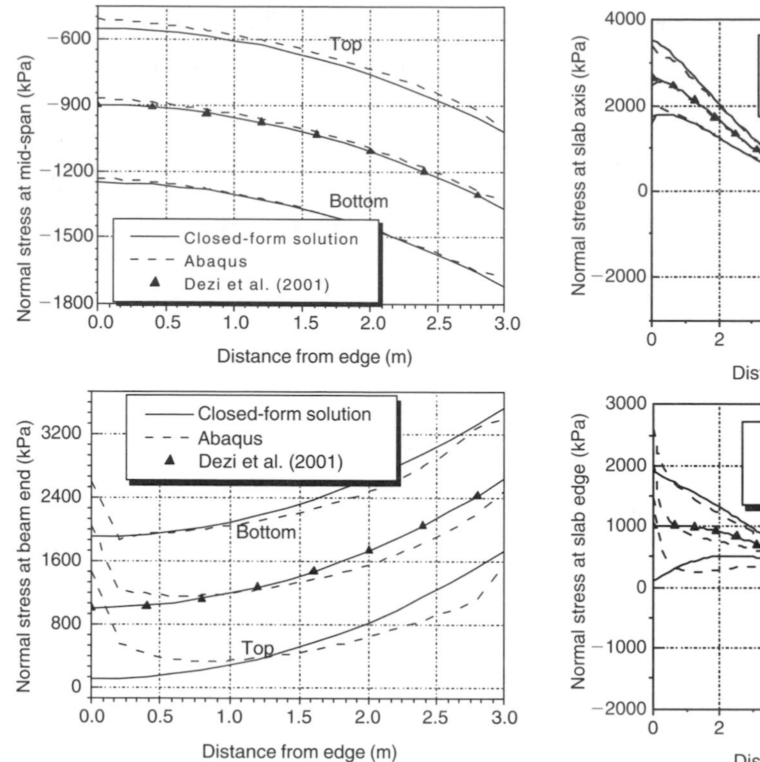

Figure 7. Transverse distribution of normal stresses at mid-span and ends with the medium shear connection degree.

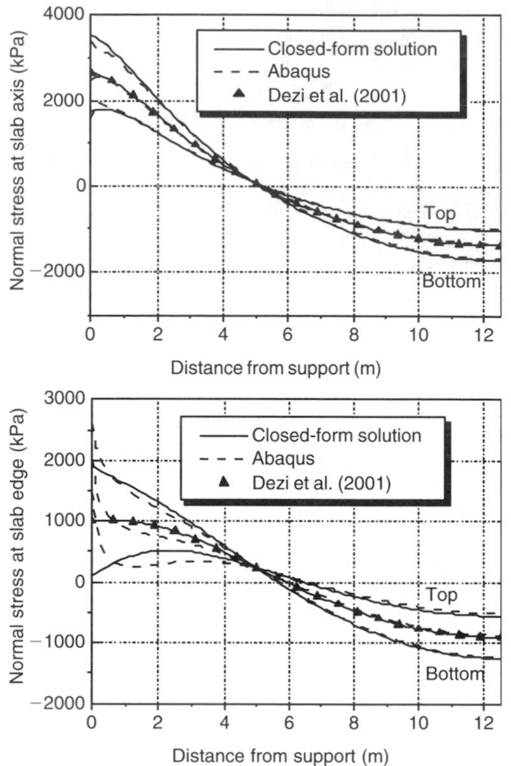

Figure 8. Longitudinal distribution of normal stresses at the ends and mid-span with the medium shear connection degree.

smaller than those at the slab axis (See Fig. 7) and therefore, such discrepancies can be neglected.

6.3 *Selection of the shear warping function*

Since the transverse distribution of the longitudinal stress of the concrete slab is predefined, the selection of the shear warping function may significantly affect the accuracy of the proposed formulation.

In general, the difference between the longitudinal stress at the slab edge and that at the slab axis at same height of a same section is non null. Thus $\psi(x)$ can be normalized without loss of generality by letting $\psi(0) = 1$. By deliberately selecting the values of $j_n(\mathrm{n} = 2, 3, 4 \ldots)$, a large variety of functions can be defined according to Equation 4 as illustrated in Figure 9.

However, better accuracy cannot be guaranteed by increasing the order of polynomial function for two reasons. The first one is that the shape of the shear warping may vary along the beam span as seen in Figure 7; therefore, a single shear warping function cannot ensure accuracy at all sections. By choosing $\psi_4(x)$ in Figure 9 as the shear warping function, transverse distributions at mid-span and at one end are shown in Figure 10. It can be seen that the distribution at the mid-span has been twisted compared to that depicted in Figure 7. The second one is that there is always a plateau at the slab edge in the transverse distribution because of the requirement of Equation 3a. Hence,

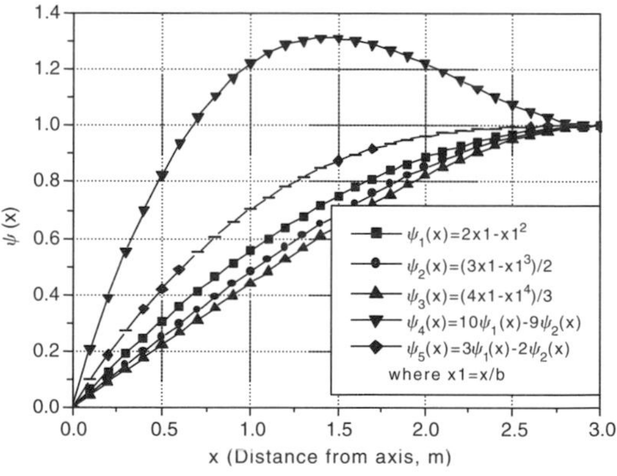

Figure 9. Shapes of possible shear warping functions.

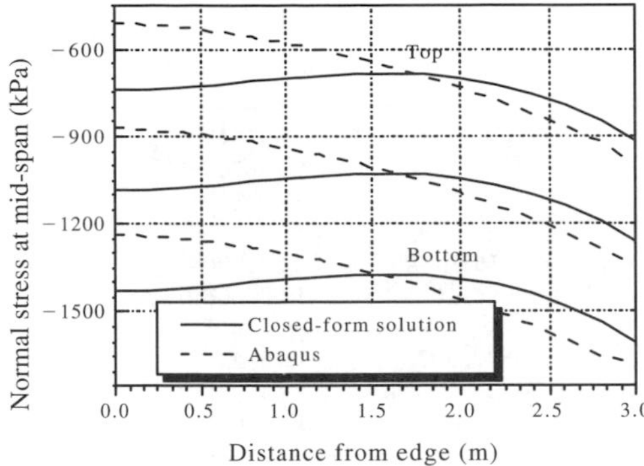

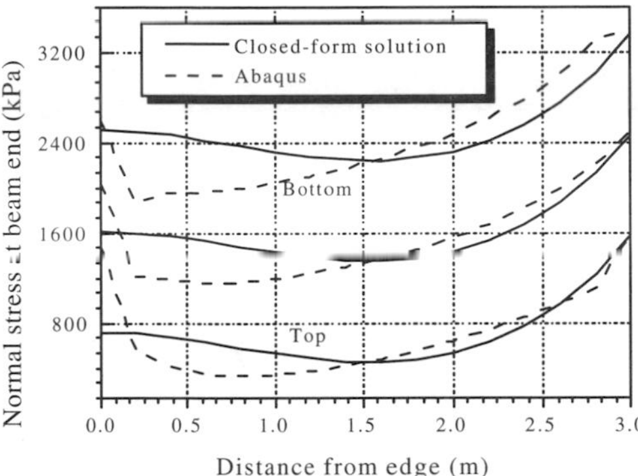

Figure 10. Transverse distribution of normal stresses at the ends and mid-span with $\psi_4(x)$ in Figure 9 and the medium shear connection degree.

the transverse distribution at end sections cannot be simulated by the proposed formulation with a selected accuracy.

Hence, it can be concluded that the quadratic form of the shear warping function in Equation 14 is acceptable, and no essential improvement can be expected by choosing higher order functions.

6.4 *Patch tests and convergence rate*

In order to perform the patch test, exact basic solutions, or so-called reference solutions of a problem should be derived in advance. Since it should be possible for a reference solution to be reproduced by classical finite elements, such a solution must be of polynomial form. Hence the closed-form solution in Equation 13, which contains exponential functions, cannot be used as a reference for patch tests. The reference solution, however, can be easily obtained by substituting assumed polynomials of all displacement fields into the homogeneous differential equations of the problem. Relationships of unknown coefficients are then obtained from identical conditions for the differential equations. Following this procedure, the reference solution of the proposed

Table 1. Coefficients of the reference solution for all cases of patch tests.

Case	d_1	d_2	d_3	d_4	d_5	d_6
1	$0.1/L$	0	0	0	0	0
2	0	0	$0.1/L$	0	0	0
3	0	0	0	0	$0.1/L^3$	0

formulation is obtained as follows

$$\begin{Bmatrix} w_c \\ w_s \\ v \\ f \end{Bmatrix} = d_1 \begin{Bmatrix} z \\ 0 \\ \dfrac{z^2}{2h} \\ 0 \end{Bmatrix} + d_2 \begin{Bmatrix} 1 \\ 0 \\ \dfrac{z}{h} \\ 0 \end{Bmatrix} + d_3 \begin{Bmatrix} 0 \\ z \\ -\dfrac{z^2}{2h} \\ 0 \end{Bmatrix} + d_4 \begin{Bmatrix} 0 \\ 1 \\ -\dfrac{z}{h} \\ 0 \end{Bmatrix}$$

$$+ d_5 \begin{Bmatrix} \dfrac{3E_sA_shz^2}{E_cA_c + E_sA_s} \\ -\dfrac{3E_cA_chz^2}{E_cA_c + E_sA_s} \\ z^3 - \dfrac{6E_cA_cE_sA_sz}{p(E_cA_c + E_sA_s)} \\ \dfrac{6E_cE_sA_sS_\psi h}{G_cI_{d\psi}(E_cA_c + E_sA_s)} \end{Bmatrix} + d_6 \begin{Bmatrix} 0 \\ 0 \\ 1 \\ 0 \end{Bmatrix} \tag{21}$$

where $d_i(i = 1\text{–}6)$ are coefficients independent to each other, which correspond to independent displacement fields: d_2, d_4 and d_6 correspond to rigid body motions; d_1 and d_3 correspond to constant strain modes; d_5 corresponds to a linear strain mode.

Three cases of constant and linear strain modes are considered for the patch test (See Tab. 1). The patch test is, in principle, required only for an infinitesimally small patch of elements. For the proposed formulation, the size of the patch, however, is not important and the test can be carried out on a patch of arbitrary dimensions. Therefore the geometry and the mechanical properties shown in Figure 4 are used in the patch test.

For brevity, only the results of Cases 1 and 3 of one-element patch test of form C are listed in Tables 2 and 3. It can be observed in Table 2 that the errors in the solutions of all the elements are only due to round-off. Hence, all the elements pass the test of the constant strain mode and this ensures convergence. The results of the 10DOF element which are gathered in Table 3 are associated with significant errors. As a result, the 10DOF element fails the patch test of the linear strain mode while the remaining elements pass it.

In order to examine the performance of the elements and to check possible locking phenomena, convergence rates of the elements are evaluated by means of energy errors. The energy error can be expressed as

$$e = \frac{|U - U_h|}{U} \tag{22}$$

Table 2. Results of one-element patch test – Case 1 of form C.

	Analytical	10DOF	12DOF	13DOF
w_{c1}	0.00000E+00	−9.65003E-18	1.37592E-16	7.89894E-17
w_{s1}	0.00000E+00	−9.07939E-18	1.36534E-16	7.77634E-17
v_1	0.00000E+00	*	*	*
θ_1	0.00000E+00	*	*	*
f_1	0.00000E+00	3.51736E-19	−8.22565E-19	2.82590E-18
w_{c2}	1.00000E-03	1.00000E-03	1.00000E-03	1.00000E-03
w_{s2}	0.00000E+00	*	*	*
v_2	1.16842E-02	1.16842E-02	1.16842E-02	1.16842E-02
θ_2	9.34737E-04	9.34737E-04	9.34737E-04	9.34737E-04
f_2	0.00000E+00	−3.51736E-19	8.22565E-19	1.73108E-18

where * denotes one of fixed nodal displacements.

Table 3. Results of one-element patch test – Case 3 of form C.

	Analytical	10DOF	12DOF	13DOF
w_{c1}	0.00000E+00	−7.21084E-07	3.85216E-18	9.98955E-18
w_{s1}	0.00000E+00	−6.38060E-20	3.78748E-18	9.89443E-18
v_1	0.00000E+00	*	*	*
θ_1	−2.03658E-07	*	*	*
f_1	−5.36391E-07	−5.36391E-07	−5.36391E-07	−5.36391E-07
w_{c2}	4.85861E-05	4.78650E-05	4.85861E-05	4.85861E-05
w_{s2}	−7.97922E-05	*	*	*
v_2	9.94908E-04	1.47805E-03	9.94908E-04	9.94908E-04
θ_2	1.19796E-04	1.19796E-04	1.19796E-04	1.19796E-04
f_2	−5.36391E-07	−5.36391E-07	−5.36391E-07	−5.36391E-07

where * denotes one of fixed nodal displacements.

where U, U_h = total strain energy of the composite beam according to the closed- form solution and to the beam element solution, respectively.

The energy errors of the beam elements with different shear connection degrees are plotted in a log-log plane in Figure 11. It can be observed that for each element, the weaker the shear connection the higher the convergence rate. Moreover, for each shear connection degree, the convergence rate of the 10DOF element is the lowest of the three elements, while that of the 13DOF element is the highest. The energy error of the 10DOF element is large especially for higher shear connection degrees and decreases with a low rate, exhibiting a locking phenomenon. The performance of the 13DOF element is favourable while the convergence rate of the 12DOF element degrades significantly when the shear connection stiffness increases, though the relevant errors shown in Figure 11 are still acceptable for practical use.

In displacement-based formulations, the results of stresses are less accurate than those of displacements. For the 13DOF element, at least 4 elements are needed to compute stresses of composite beam under uniform load (See Fig. 12), while in this case, only 2 elements are needed to compute displacements.

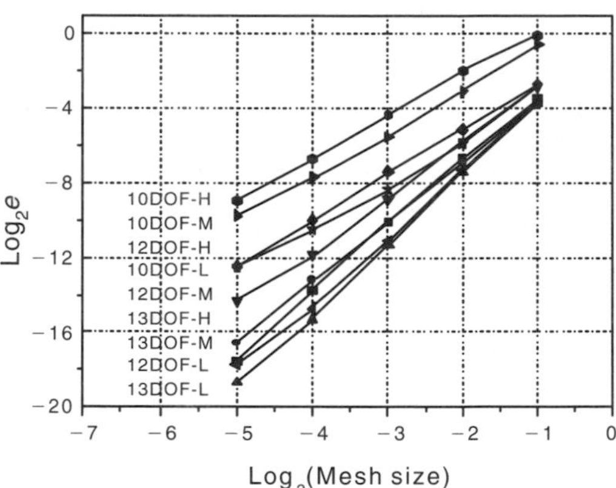

Figure 11. Energy errors of beam elements with different shear connection degrees.

It is worthy of noting that the distribution of stresses within one 13DOF element is a least-square fit of the distribution predicted by the closed-form solution (See Fig. 12); while this is not the case for the distributions both of the 10DOF and the 12DOF element. This indicates that stress oscillations exist both in the 10DOF

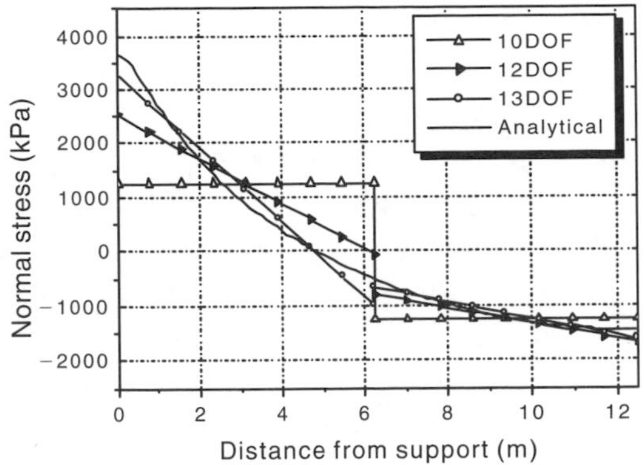

Figure 12. Longitudinal distribution of normal stresses at the centroid of the concrete slab of the composite beam with 4 elements.

and the 12DOF elements and that the displacement fields of these two elements are inconsistent. This has been recognized as a cause of locking (Prathap 1993).

It should be also noted that in the linear strain mode of the reference solution Equation 21, the fields of longitudinal displacements w_c and w_s are of quadratic form while the field of shear lag intensity f is constant. This is the reason why the 10DOF element fails and the 12DOF element passes the linear strain mode patch test. Moreover the stress distribution in Figure 12 indicates inconsistency of coupled displacement fields. Much attention should be paid to the results of patch tests in such cases and further research work on this topic is needed.

7 CONCLUSIONS

In this paper, a displacement-based formulation is proposed for composite beams with flexible shear connection and shear lag. The closed-form solution of this problem is derived, and the adopted kinematic model is verified by comparing the results of the closed-form solution to those provided by the Abaqus code. Furthermore, the shear warping function of quadratic form is shown to be accurate.

With regard to the coupling of displacement fields of the kinematic model, 3 beam elements, with different shape functions for longitudinal displacement and shear lag intensity of the concrete slab have been developed. By means of patch tests and of the convergence rate, the 10DOF element is found to exhibit locking problems for composite beams with a high shear connection degree; while the 13DOF element is validated and performs very well. The 12DOF element is not satisfactory, because it produces results with stress oscillations and degrades significantly for composite beams with very high shear connection degree, though it passes the linear strain mode patch test.

REFERENCES

Arizumi, Y., Hamada, S. & Kajita, T. 1981. Elastic-plastic analysis of composite beams with incomplete interaction by finite element method. *Computers & Structures* 14(5–6): 453–462.

Ayoub, A. & Filippou, F.C. 2000. Mixed formulation of nonlinear steel-concrete composite beam element. *Journal of Structural Engineering ASCE* 126(3): 371–381.

Bauchau, O.A. 1985. A beam theory for anisotropic materials. *J. Appl. Mech.* 52: 416–422.

Dall'Asta, A. & Zona, A. 2001. Locking in composite beams with weak shear connection, Internal Report, Universita degli Studi di Ancona, N.264.

Daniels, B.J. & Crisinel, M. 1993. Composite slab behaviour and strength analysis. Part I: calculation procedure. *Journal of Structural Engineering ASCE.* 119(1): 16–35.

Dezi, L., Gara, F. et al. 2001. Time-dependent analysis of shear-lag effect in composite beams. *Journal of Engineering Mechanics ASCE* 127(1): 71–79.

Gjelsvik, A. 1981. *The theory of thin walled bars*, New York, Wiley.

Gjelsvik, A. 1991. Analog-beam method for determing shear-lag effects. *Journal of Structural Engineering ASCE* 113(3): 557–574.

Hibbitt, Karlsson & Sorensen Inc. 1998. ABAQUS – User's manual, version 5.8, 1~2.

Koo, K.K. & Cheung, Y.K. 1989. Mixed variational formulation for thin-walled beams with shear lag. *Journal of Engineering Mechanics ASCE* 115(10): 2271–2286.

Lopez-Anido, R. & GangaRao, H.V.S. 1996. Warping solution for shear lag in thin-walled orthotropic composite beams. *Journal of Engineering Mechanics ASCE* 122(5): 449–457.

Newmark, N.M., Seiss, C.P. & Viest, IM. 1951. Test and analysis of composite beams with incomplete interaction. *Proc. Society for Experimental Stress Analysis*. New York, 9(1): 75–92.

Prathap, G. 1993. *The Finite Element Method in Structural Mechanics*. Dordrecht: Kluwer Academic Press.

Reissner, E. 1941. Least work solutions of shear lag problems. *J. Aeronautical Sci.* 8(7): 284–291.

Salari, M.R., Spacone, E. et al. 1998. Nonlinear analysis of composite beams with deformable shear connectors. *Journal of Structural Engineering ASCE* 124(10): 1148–1158.

Sapountzakis, E.J. & Katsikadelis, J.T. 2001. Analysis of prestressed concrete slab and beam structures. *Computational Mechanics* 27: 492–503.

Song, Q. & Scordelis, A.C. 1990. Formulas for shear-lag effect of T-, I- and box beams, *Journal of Structural Engineering ASCE* 116(5): 1306–1318.

Sun, F.F. & Bursi, O.S. 2002. Displacement-based formulation for steel-concrete composite beams with shear lag, Technical report 2002-11, Department of Mechanical and Structural Engineering, University of Trento, Italy.

Zienkiewicz, O.C. & Taylor, R.L. 1989. *The Finite Element Method*, vol. 1(4th edn). London: McGraw-Hill.

Computational Modelling of Concrete Structures, Bićanić et al. (eds)
© 2003 Swets & Zeitlinger, Lisse, ISBN 90 5809 536 3

Improvement of structural behaviour in concrete shells using optimisation techniques

A. Tomás & P. Martí
Structures and construction department, Technical University of Cartagena, Murcia, Spain

ABSTRACT: Although concrete shells may adopt any form, it could be interesting to know at what extent changes in their shape may avoid the appearance of bending moments or reduce them. The use of optimisation techniques may be effective in providing alternative geometric forms of shells that improve their mechanical behaviour, complying with the design conditions in an optimum way. In this paper, these techniques were used to find optimum geometrical designs having aesthetic shape similar to the form initially designed for the structure. As example, a shell based on Candela's blueprints was optimised under a state of predominant gravitational loads. The results confirm that significant improvements in the structural behaviour of the shell may be achieved with only slight changes in its form.

1 INTRODUCTION

Triangular distribution of stresses through a cross section is uneconomic since the maximum stress occurs on the extreme fibers. This is specially true for materials as concrete, whose resistance to tension is small compared to the resistance to compression. Therefore, the ultimate strength capacity of the cross section is bounded by the former resistance.

The structural behaviour of the shells, compared to that of other types of structures, is characterized by a higher mechanical efficiency. If appropriate designs are carried out, shells can support high loads and allow to cover important spans using less resistant material and/or thinner cross section. Moreover, their lightness and elegance are especially attractive from an aesthetic point of view leading some authors to use the term of "structural elegance" when referring to them (Ramm et al. 2000).

Although shells can adopt any form, the double-curvature shells are, without doubt, those presenting the best advantages (Candela 1985, Ortega & Arias 1998). Their particular behaviour is due to the arch-effect in two planes and, in contrast to the arch contained in only one plane, it allows to support different load configurations, mainly by means of membrane internal forces, with a very low risk of bending. Moreover, these surfaces have a practically inalterable form and are in equilibrium whatever the form and distribution of the loads. This implies that the shell structures designed to act as membranes are, by themselves, optimum structures. Unfortunately, as usually occurs in optimum systems, this high mechanical efficiency induces a structural behaviour which is extremely sensitive to small changes in certain response parameters. The classic example is the reduction in the buckling load of a shell when slight geometric imperfections appear. Owing to this ambivalent characteristic, the shell has been considered the "prima donna" of the structures (Ramm & Wall 2002).

The development of optimisation techniques was strongly boosted by the tremendous increase in computational and graphical capacities. These techniques represent an effective means to obtain alternative geometric forms of shells and improve their mechanical behavior, complying with the design conditions (stress constraints, construction conditions, etc.) in a optimum way (minimum weight, maximum stiffness, minimum stress level, etc.).

In this paper, optimisation techniques were applied to find optimum geometric designs close to a preconceived design, i.e. the resulting geometries should have an aesthetic shape similar to the form of the initially designed structure. It is known that slight changes in the form of this type of structures can introduce important improvements in their mechanical behaviour. To achieve this task, different objective functions can be used such as the strain energy, the weight of the structure or the tensile stress in both faces of the shell. In this study, the parameters that govern the geometry of the structure and the thickness of the shell were used as variables. The constraints referred to the minimum thickness of the shell, the tensile stresses in concrete, and several parameters of geometric control. Optimisation of a shell based on Candela's blueprints under predominant gravitational loads is given as example.

2 SHAPE OPTIMISATION OF CONCRETE SHELLS

Design process of the membrane state in a shell structure can be hindered by a series of factors such as the application of concentrated loads, the existence of free edge boundary conditions or the possible incompatibility that could appear between a given form of the shell and its thickness. Since these factors are strongly dependent on the form of the structure, it becomes essential to incorporate, in the design process, shape optimisation techniques that account for them. Therefore, the ultimate objective in shell design would be to find a form that, considering the specific properties of the material, satisfied the following design conditions (Ramm et al. 2000):

- Stresses and displacements are enclosed inside a pre-established interval,
- membrane stress state (or a close approximation of it) is needed,
- boundary conditions are fulfilled for all load configurations,
- buckling instability phenomena are avoided,
- sensitivity of the structural response to possible variations in the geometry is minimised, and
- the shell shape complies with aesthetic criteria.

Therefore, it could be stated that the aim of the structural optimum design is to obtain a design, that is a set of values for the *design variables*, which minimises an objective *function*, and which complies with a series of *constraints* that depend on the design variables.

The *design variables* of a structure can be: properties of the cross section of the elements (area, thickness, inertia moment, etc.); parameters of the geometry of the structure; topology of the structure (nodes and element connections), and properties of the material. The type of optimisation to be carried out will depend on the type of variables being considered. Generally, as the design of structures searches for minimum weight, the *objective function* is commonly related to the weight of the structure. Nevertheless, the weight is not always the determining factor, and the objective function could depend on other factors such as the construction cost, the reliability, the stiffness, etc. The *constraints* are the conditions the design must comply with.

In mathematical terms, the optimum design problem can be formulated as follows:

To find the vector $\mathbf{x}$ of the design variables which

$$\begin{array}{lll} \text{minimises:} & f(\mathbf{x}) & \\ \text{subject to:} & h_j(\mathbf{x}) = 0 & j = 1, 2, \ldots, m_i \\ & g_k(\mathbf{x}) \geq 0 & k = 1, 2, \ldots, m_d \\ & x_i^L \leq x_i \leq x_i^U & i = 1, 2, \ldots, n \end{array}$$

where $\mathbf{x}$ = n-dimensional vector of the design variables; $f(\mathbf{x})$ = objective function; $h_j(\mathbf{x}) = j$ equality design constraints; $g_k(\mathbf{x}) = k$ inequality design constraints; m_i = number of equality constraints; m_d = number of inequality constraints; n = number of variables; x_i^L = lower bound of the variable i; and x_i^U = upper bound of the variable i.

Once the problem formulated in this way, it can be solved, transparently for the user, by means of several different methods (Mathematical Programming, Genetic Algorithms, etc.).

With respect to shape optimisation, several authors have proposed suitable objective functions (Ramm & Mehlhorn 1991, Bletzinger & Ramm 1993, Ramm et al. 2000, Martí et al. 2001). As example, in order to find a state of membrane stresses, the highest principal tensile stress can be used as objective function. An alternative way is to substitute this condition for a constraint expressing that the appearance of tensions should be avoided in all points of the shell.

If improving the behaviour of the structure in case of instability phenomena is required, the buckling load can be used as objective function. As the response of shell structures is very sensitive to geometric imperfections, including the latter into the maximisation process of the buckling load is recommended. A first approach of the buckling load can be established by analyzing the initial stability through an eigenvalue analysis. However, in order to get a more realistic value, geometric nonlinear analysis is needed.

With the purpose of minimising displacements in the whole structure, a function called "Volumetric Displacement" (*VD*) may be defined according to the following expression (Robles & Ortega 2001)

$$VD = \sum_{i=1}^{n} d_i \times S_i \times Thick_i \tag{1}$$

where d_i = displacement vector modulus at each point i; S_i = area of influence at this point; and $Thick_i$ = average thickness of the structure at the mentioned area. With respect to the maximum displacement technique, the advantage of the *VD* function is to provide a wider view of displacements throughout the structure.

In order to reduce the bendings in the shell, it is very appropriate to minimise the strain energy, or equivalently, to maximise the stiffness of the structure

$$S_E = \frac{1}{2} \int \boldsymbol{\sigma} \boldsymbol{\varepsilon} \, \mathrm{d}V \tag{2}$$

where $\boldsymbol{\sigma}$ = stresses; and $\boldsymbol{\varepsilon}$ = strains, both calculated in all the points of the shell. Minimizing the strain energy leads to lower stresses and deformations in all points of the shell but, in an implicit way, allows to fulfill simultaneously the previous objectives. In other words, the behaviour of the structure is improved, together with

(i) a highest buckling load and (ii) a "relaxed" state of stresses close to the membrane state.

Considering that all the design conditions stated above can easily be satisfied by increasing the weight of the shell, it could be convenient in some cases to limit the weight to a maximum value, or even to introduce it in the optimisation process as an equality constraint.

The inequality constraints $g_k(\mathbf{x})$, commonly used to limit the tensile stresses and the displacements, can also be necessary to limit some parameters of shape, because the value of these parameters varies in the optimisation process. A distortion can occur in the geometry of the structure in such a way that its shape differs too much from the preconceived form, and from the desired aesthetic criteria.

3 EXAMPLE

In the following, the shape optimisation of a concrete shell structure is presented. It was designed for the entrance of the Universal Oceanographic Park of Valencia, Spain (Fig. 1). The roof shell is based on Felix Candela's blueprints. The geometry of the structure is formed by the intersection of three lobes whose mid-surfaces describe the shape of a hyperbolic paraboloid (hypar) (Solano 2001, Tomás et al. 2002). After analyzing the behaviour of Candela's model, several optimisation processes under a state of predominant gravitational loads have been carried out.

3.1 *CAD model*

To generate the CAD model, a global system of cylindrical coordinates with origin at the point of intersection of the three paraboloids was defined. The patches defining the mid-surface of the structure were derived from the coordinate system. The ANSYS program of finite elements was used to model the shell.

The following sequence of operations was programmed:

- Obtaining a set of points (*keypoints*) contained in the mid-surface,
- Linking the points by curved lines (*splines*), and
- Obtaining ruled surfaces (*Coons patches*) from the splines to generate the geometry of the structure.

On the whole, there were 1945 keypoints that, linked in sets of 6, create 432 splines which generated 216 Coons patches forming the model (Fig. 2).

In addition, the shell has been strengthened by two types of ribs with similar dimensions as used by Domingo et al. (1999). One of them spread from the support to the centre of the structure (main ribs). The other one surround a small central hole made in the shell to avoid problems when meshing, embracing a band of 200 mm wide (hole rib).

The implementation of the CAD model in ANSYS was carried out in such a way that the form of the structure can be modified by varying the values of some design parameters. The latter referred to (i) geometric control and (ii) thickness of the elements that form the structure.

3.2 *Analysis model*

Due to the symmetry of the structure, the different analyses were carried out on a sextant of the shell. Previously to **meshing** the surface of the CAD model, the thickness, material, element type and geometric characteristics of the elements were defined. A small circular hole of 100 mm diameter was made at the intersection of the lobes, in order to avoid meshing problems derived from the distortion of the elements generated in the area surrounding the center, which have very acute angles.

Figure 1. Shell under construction at the entrance of the Universal Oceanographic Park (Valencia, Spain).

Figure 2. CAD model of the structure.

The following steps were implemented in ANSYS: (i) generate the mesh of finite elements (ii) assign mechanical properties (iii) state the boundary conditions and (iv) apply loads on the structure.

The **material** of the structure was of concrete. The mid surface of the shell was provided with a reinforcement which was used to account for time-dependent effects of the concrete, since these effects can have a considerable influence when the thickness of the shell is small with respect to the other dimensions. Therefore, the contribution of the reinforcement was not considered in the analysis, excepted the effects of its density. The specific gravity.of the material was 25.00 kN/m^3, value commonly used for reinforced concrete. A characteristic cylinder compressive strength of 30 MPa was adopted, and the rest of mechanical properties were chosen following the recommendations of the Spanish structural concrete code (EHE): 20.00 MPa for the design compressive strength (f_{cd}), 1.35 MPa for the design tensile strength ($f_{ct,d}$), 0.20 for the Poisson's ratio and 28,576 MPa for the secant modulus referred to the concrete age of 28 days.

There are basically two different ways of formulating the **elements** used in concrete shells, those based on the degenerate solid approach and those based on a shell theory (Hofstetter & Mang 1995). We used an element belonging to the second type, the so-called *Shell93* in ANSYS program. The element had six degrees of freedom at each node: translations in the nodal x, y, and z directions and rotations about the nodal x, y, and z-axes. The deformation shapes were quadratic in both in-plane directions. The element had plasticity, stress stiffening, large deflection, and large strain capabilities. It may have variable thickness, which was assumed to vary smoothly over the area of the element, with the thickness input at the corner nodes. The thickness at the midside nodes was taken as the average of the corresponding corner nodes. The normal stress for this element varied linearly through the thickness, while the transverse shear stresses were assumed to be constant through it (ANSYS 2000).

Because of the symmetry of geometry and loads, the analysis of a lobe with an angle of 60° was carried out, applying symmetry boundary conditions to the nodes in the symmetry planes, and restricting the translations in the x, y, and z directions to the nodes in the foundation plane.

The applied loads were obtained from the Spanish standards (NBE AE-88), i.e. the weight of the structure and the distributed load (=1 kN/m^2). The action of the wind was not considered because of its slight contribution to the whole load, only a 3% of the gravitational loads. This percentage is a maximum value that was obtained adopting a simplified and safe hypothesis for introducing the wind into the analysis model (Solano 2001).

3.3 *Formulation of the optimum design problem of the hypar shell*

3.3.1 *Objective functions*

The objective functions were the following:

- Strain energy of the structure,
- weight of the hypar, and
- highest tensile stress at the nodes of the model.

3.3.2 *Design variables*

The following design variables whose initial values were proposed in the Candela's blueprints were used for the design of the entrance to the Oceanographic Park:

- K Constant in the equation of the mid-surface of the hypar. In the optimisation processes the initial value was 0,14 m^{-1}, being 0,13 m^{-1} and 0,17 m^{-1} the minimum and maximum values respectively.
- β Angle of inclination of the plane that defines the free edge of the hypar with respect to a horizontal line. Its initial value was 75°, allowing for a variation interval between 74° and 75°, since the design is very sensitive to this variable. With this interval, the structure height cannot be lower than 19 m.
- ω Angle between the two master axes of the hypar. Its initial value was 90° (equilateral hypar). The stated lower and upper limits were 84° and 91° respectively.
- e_1 Shell thickness. A minimum initial value of 60 mm for constructive conditions was chosen. In the optimisation processes, thickness was allowed to range from 60 to 80 mm depending on the case to investigate.
- e_2 Hole rib thickness. The initial value was 80 mm, stating as in the previous case, a minimum value of 60 mm.
- e_3 Main ribs thickness. The initial value was 350 mm, with a variation interval between 60 and 400 mm.

3.3.3 *Constraints*

The maximum stresses were restricted depending on the design strength of the material of the shell:

$$\sigma_t \leq f_{ct,d} \tag{3}$$

$$\sigma_c \leq 0,85 f_{cd} \tag{4}$$

where σ_t = tensile stress; and σ_c = compressive stress. Two shape parameters of the hypar, the height of the highest point of the free edge and the radius or distance from the z-axis to the support were also restricted. This was necessary because the values of these parameters tend to decrease during the optimisation process, distorting the geometry of the structure in such a way that its shape would depart significantly from the hyperbolic paraboloid. Furthermore, its aesthetic would not match the design criteria. The stated minimum values were 19 m for the height of the free edge and 11.5 m for the radius.

4 RESULTS

In a first stage, the analysis of the outputs of the model proposed by Candela for the Oceanographic Park allowed to obtain outstanding information, such as the value of stresses and displacements at the points of the structure, and its buckling load. In a second stage, several optimisation processes were executed with the purpose of improving the structural behaviour under different load combinations. The different optimisation processes of the initial model were classified in two groups, using the following criteria:

- Depending on the used objective function (strain energy, weight or tensile stress), and
- depending on the allowed minimum thickness (80 or 60 mm).

For each objective function, two optimisation processes were carried out depending on the allowed minimum thickness.

Moreover, a study of the buckling of the structure was realized for the initial model with three different thickness values and for the optimum designs, using two types of analysis: (i) initial stability or linear analysis and (ii) nonlinear analysis. The final values of the geometry variables in the different optimisation processes are shown in Table 1.

Finally, it was considered of interest to calculate the final values of two geometric parameters: (i) the height of the free edge of the hypar and (ii) the radius or distance in ground plan from the centre of the structure until one of its supports. The comparison of these parameters in the different processes could help in visualizing and evidencing the changes that have undergone in the geometry of the initial model.

It was observed that the value of the angle ω (angle between the master axes) decreased in all the optimisation runs, implying that the hypar was no longer equilateral. On the other hand, the height of the initial model decreased in all runs tending to the stated minimum value of 19 m. Regarding the thickness of the shell, it can be highlighted that its allowed minimum value was reached when the strain energy and the weight were optimised. However, when the maximum tensile stress was optimised, the thickness of the shell was near 160 mm, indicating the high cost of a form having a membrane behaviour when geometric constraints are used.

The final values of the objective functions are shown in Table 2, together with three additional parameters whose analysis and comparison could be useful: the shell thickness e_1, the maximum compressive stress $\sigma_{c,máx}$ and the maximum vertical displacement $U_{z,máx}$.

From the above results, it could be underlined that in all the optimisation processes the maximum compressive stresses were below 5 MPa and the maximum tensile stresses was lower than the design tensile strength of the concrete. In addition, the maximum vertical displacement of the structure was lower than 4 mm, which is in agreement with the results obtained by Ortega and Arias (1998) for this type of structures, where vertical displacements are below 10% of the shell thickness.

When tensile stress was the objective function, the value of the weight approximately doubled the values obtained in the other optimisation processes. However, no substantial decrease of the tensile stress was achieved, which confirms the high cost of a form with a membrane behaviour using geometric constraints restrictions mentioned previously.

Lastly, Table 3 shows the results of the studies of initial stability and nonlinear geometric analysis. These values correspond to the final designs in each process and for the initial model with thickness values of 60, 80 and 120 mm. The latter value was finally adopted in the construction of the hypar. The values indicate the buckling load factor expressed as multiple of weight of the shell.

Table 1. Optimisation processes. Final values of variables of geometry.

Process	t_1 (mm)	t_2 (mm)	t_3 (mm)	K (m^{-1})	ω (deg)	Radius (m)	Height (m)
Initial model	60.0	80.0	350.0	0.140	90.00	13.63	24.39
$SE(t_{min} = 60\,mm)$	61.2	60.0	362.1	0.158	85.42	11.87	19.04
$SE(t_{min} = 80\,mm)$	80.0	95.4	400.0	0.165	84.98	11.54	19.00
$W(t_{min} = 60\,mm)$	60.0	71.9	264.3	0.150	85.87	12.27	19.02
$W(t_{min} = 80\,mm)$	80.0	81.0	333.1	0.141	86.45	12.79	19.05
$\sigma_t(t_{min} = 60\,mm)$	159.4	80.9	384.6	0.139	86.53	12.91	19.01

where SE = strain energy; W = weight; σ_t = tensile stress; t_{min} = minimum thickness.

Table 2. Optimisation processes. Final values of objective functions, shell thickness (t_1), maximum compressive stress ($\sigma_{c,max}$) and maximum vertical displacement ($U_{z,max}$).

Process	Objective functions *SE* (N·m)	*W* (kN)	σ_t (MPa)	t_1 (mm)	$\sigma_{c,max}$ (MPa)	$U_{z,max}$ (mm)
Initial model	881.77	473.83	2.19	60.0	7.89	8.73
$SE(t_{min} = 60\,\text{mm})$	297.94	330.17	1.33	61.2	4.96	3.90
$SE(t_{min} = 80\,\text{mm})$	300.66	443.28	1.29	80.0	4.22	3.25
$W(t_{min} = 60\,\text{mm})$	318.29	309.54	1.26	60.0	4.89	3.94
$W(t_{min} = 80\,\text{mm})$	367.20	431.37	1.10	80.0	4.27	3.27
$\sigma_t(t_{min} = 60\,\text{mm})$	514.36	767.87	1.05	159.4	4.75	2.26

where SE = strain energy; W = weight; σ_t = tensile stress; t_{min} = minimum thickness.

Table 3. Buckling load factor (expressed as multiple of weight of the shell).

Process	Thickness (mm)	Buckling load factor Linear analysis	Nonlinear analysis
Initial model	60.0	8.65	7.86
Initial model	80.0	12.76	10.65
Initial model	120.0	22.29	17.93
$SE(t_{min} = 60\,\text{mm})$	61.2	17.38	15.35
$SE(t_{min} = 80\,\text{mm})$	80.0	25.61	22.22
$W(t_{min} = 60\,\text{mm})$	60.0	15.83	14.39
$W(t_{min} = 80\,\text{mm})$	80.0	22.42	19.90
$\sigma_t(t_{min} = 60\,\text{mm})$	159.4	53.43	44.66

where SE = strain energy; W = weight; σ_t = tensile stress; t_{min} = minimum thickness.

From the stability study we can highlight, on one hand, the high buckling load obtained, product of the huge stiffness of this type of structures. This was also confirmed by the obtained maximum vertical displacement of 8.7 mm in the initial model, and lower than 4 mm in the optimum designs. On the other hand, the buckling load was doubled in the optimum designs with respect to the initial models, highlighting that designing this type of concrete shell structures by using optimisation techniques provides an added value.

Finally, the initial model and a final design are compared in Figure 3. The latter was obtained using strain energy as objective function with a minimum thickness of 60 mm as constraint. Both geometries are intersected to provide a better perspective and shown the slight differences between them.

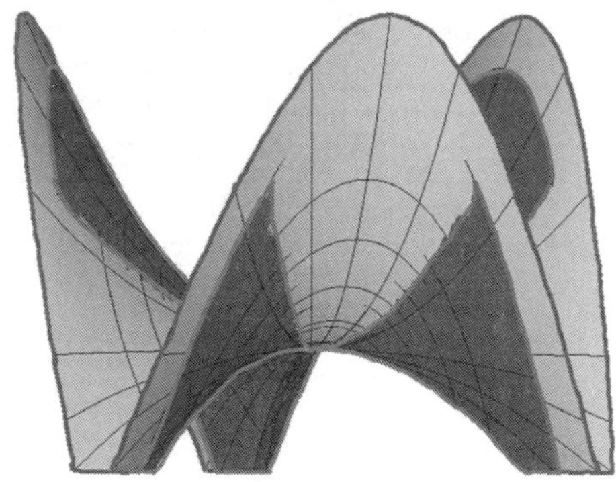

Figure 3. Intersection of initial model (in light shade) and a final design (in dark shade).

5 CONCLUSIONS

Traditionally, computers have been used, within the process of design of structures, to analyse the response of an user-defined structure and to check its safety for given applied loads. The use of optimisation techniques in the process of design of structures, widens the field of use of computers and allows the user to obtain optimum designs for stated design conditions.

In the present paper we have verified that with slight changes in the shape of a concrete shell, considerable improvements are obtained in its mechanical behaviour. In particular, we can underline three interesting aspects:

- If necessary, the membrane state of compressive stresses may be achieved.
- Deformations in the optimum shells decrease considerably to values lower than half of those in the initial design.
- The behaviour of the shell against instability improves significantly, since the buckling load is double the initial value.

REFERENCES

ANSYS, Inc. 2000. *Ansys elements reference*. Canonsburg: ANSYS, Inc.

Bletzinger, K.-U. & Ramm, E. 1993. Form finding of shells by structural optimization. *Engineering with computers* 9: 27–35

Candela, F. 1985. *In defense of the shapeliness and other writings*. Madrid: Xarait Libros. (In Spanish)

Domingo, A., Lázaro, C. & Serna, P. 1999. Design of a thin shell fibre reinforced concrete hypar roof. In R. Astudillo & A.J. Madrid (eds), *IASS; Proc. intern. cong.*, Madrid, 20–24 September 1999. Madrid: CEDEX.

Martí, P., Tomás, A. & Torrano, M.S. 2001. The application of optimization techniques in structural concrete teaching. *Structural concrete teaching; Proc. conf.*, Madrid, 18–19 October 2001. Madrid: ACHE. (In Spanish)

Ortega, N.F. & Arias, J.V. 1998. Comparison of mechanical efficiency between an hyperbolic paraboloid and an experimental model. In S. Idelsohn, E. Oñate & E. Dvorkin (eds), *Computational mechanics: new trends and applications*. Barcelona: CIMNE.

Ramm, E. & Mehlhorn, G. 1991. On shape finding methods and ultimate load analyses of reinforced concrete shells. *Engineering structures* 13: 178–198.

Ramm, E., Kemmler, R. & Schwarz, S. 2000. Form finding and optimization of shell structures. *Computation of shell & spatial structures; Proc. intern. coll.*, Chania-Crete, Greece, 4–7 June 2000.

Ramm, E. & Wall, W.A. 2002. Shells in advanced computational environment. In H.A. Mang, F.G. Rammestorfer & J. Eberhardsteiner (eds), *Computational mechanics; Proc. intern. cong.*, Vienna, 7–12 July 2002. Vienna: IACM.

Robles, S.I. & Ortega, N.F. 2001. Study of volumetric displacements of shells. *Journal of the IASS* 42(137): 139–147.

Solano, M.A. 2001. *Modelling, analysis and optimisation of a concrete shell structure*. Dissertation. Cartagena: UPCT. (In Spanish)

Tomás, A., Martí, P. & Solano, M.A. 2002. Shape optimisation of a concrete hypar shell. In J.M. Goicolea, C. Mota, M. Pastor & G. Bugeda (eds), *Numerical methods in engineering; Proc. cong.*, Madrid, 3–6 June 2002. Barcelona: SEMNI. (In Spanish)

Hofstetter, G. & Mang, H.A. 1995. Computational mechanics of reinforced concrete structures. Brunswick: Vieweg.

Computational Modelling of Concrete Structures, Bićanić et al. (eds)
© 2003 Swets & Zeitlinger, Lisse, ISBN 90 5809 536 3

Modeling of initially cracked reinforced concrete bridge decks

I.R. Pascu
Technical University of Construction, Bucharest, Romania

F. Toutlemonde
Laboratoire Central des Ponts et Chaussées (LCPC), Paris, France

ABSTRACT: This paper presents the finite element modeling of initially cracked reinforced concrete slabs, representative of composite bridge decks. Stresses in the reinforcement have to be precisely estimated to asses fatigue life of such existing structures, since it has been shown experimentally that lower transverse reinforcement stresses could be critical under fatigue. Numerical results are compared with data of the large-scale experiments performed at LCPC. Good agreement between non-linear finite element model behavior and experimental behavior is found. The computations confirmed that the main parameter influencing the stress distribution in the reinforcement turns out to be the presence of initial cracks due to concrete restrained shrinkage. A simplified elastic model with hinges is also proposed, which gives a safe estimation of reinforcement stresses with little computational cost, and with reduced inertia accounts for the correct structural flexibility.

1 RESEARCH SIGNIFICANCE

Composite bridges, made of a reinforced concrete deck connected to steel girders, are an economical solution in France for 50 to 100 m-span highway bridges. In 1997 in France, this solution represented 26% of the motorway and railway bridge deck surface built in the year. However, such decks have frequently been cracked, with transverse cracks going through the whole depth of the deck, due to concrete restrained shrinkage, Ducret & Lebet (1999). The effect of these cracks on the mechanical durability of the decks was identified as a major concern for bridge owners. It motivated specific recommendations for designers (Kretz et al. 1995) and an important research program carried out at the Structures Laboratory of LCPC (Ranc 1999).

Large-scale fatigue tests on six pre-cracked reinforced concrete slabs were performed with a high number of cycles representing 100 years of heavy motorway traffic, Toutlemonde & Ranc (2001a). In order to analyze obtained results in terms of deflections, stresses in concrete and reinforcement, and to finally proceed to fatigue verifications, a finite element modeling of the tested structure was developed using the finite element code CESAR-LCPC, developed at LCPC (Humbert et al. 2001).

2 EXPERIMENTAL DATA AND RESULTS

Specimens were designed as representative current zones of decks. Dimensions of the rectangular RC slabs were 5.00 m × 2.90 m × 0.18 m. The thickness (scaled about 2/3 as compared to current deck thickness) allows to keep a regular concrete mix-design with 20 mm as a maximum aggregate size. Two layers of reinforcement, one third on the upper side and two thirds as lower chord reinforcement, were provided in each direction. The longitudinal reinforcement ratio was 0.52%, close to the non-brittleness ratio in tension, which had currently been used in zero moment zones of bridges. The transverse ratio is 1.26% corresponding to classical transverse bending design.

The slabs were simply supported along the long sides with a 2.28 m clear span. The load was applied with a hydraulic actuator in the center of the slab, on a 0.09 m^2 steel plate (Fig. 1). The 10 to 86 kN load amplitude was designed, so that maximum classically computed stress variations in longitudinal and transverse lower reinforcement correspond to 90% of stress variations computed for a current bridge under the Eurocode fatigue load model, including the fact that 100 year traffic should be represented by 10 million cycles in the experiment.

Figure 1. Experimental setup for the fatigue bending test on the central strip of slabs, Toutlemonde & Ranc (2001b).

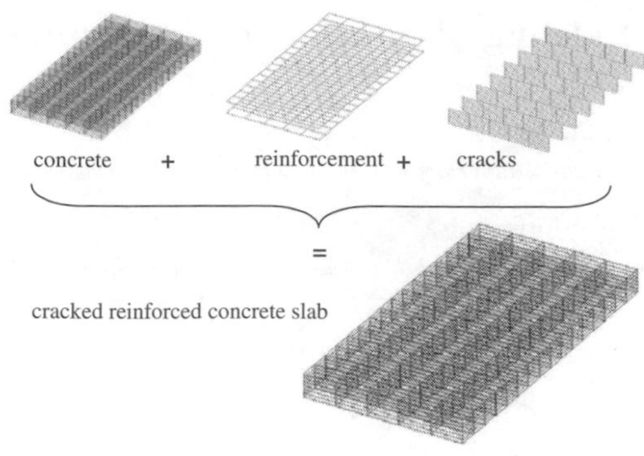

Figure 2. Finite element mesh of the RC slab.

Table 1. Mean material characteristics of the specimens.

	f_{cm} (28 d.) (MPa)	$f_{ctm,sp}$ (7 d.) (MPa)	E_c (28 d.) (GPa)	f_{yk} (MPa)
Mean	37.6	2.7	36.3	610
Stand. dev.	4.1	0.7	1.3	28

Material characteristics of tested RC slabs (identical C30/37 for all specimens) are given in Table 1.

In order to reproduce the effects of restrained shrinkage, specimens were first submitted to a general longitudinal tensile loading at the age of 7 days. This loading (removed afterwards) produced regularly spaced through-going cracks following the transverse reinforcement, having characteristic openings up to the desired 0.3 or 0.5 mm value. Such a pattern was considered as representative of the most critical zones of existing decks having up to 2 cracks per m with typical openings up to 0.5 mm.

Major mechanical consequences of this pre-cracking phase are the following:

- decrease of longitudinal stiffness, so that all tension stiffening effect vanishes, as soon as 0.1 to 0.2 mm crack openings are reached;
- increase of global flexibility (e.g. under the central load, deflection is multiplied by a factor of about 4 to 6 as compared to the theoretical value for the structure before cracking);
- modification of stress distribution for longitudinal and transverse reinforcing bars. This last point turns out to be the critical feature for fatigue verification of reinforced concrete bridge decks, and motivated the modeling presented hereafter.

In fact, strains measured during these tests pointed out that classically computed stresses, assuming homogeneity and linear elasticity of the slab, underestimate the stresses in lower transverse rebars by a 1.3 factor. Afterwards, following fatigue loading cycles on the central strip do not alter this moment distribution significantly. It has been shown that due to this distribution, stress variations in the lower transverse rebars become critical with respect to fatigue failure. In fact, for highly concentrated or high level loads, punching failure of concrete is obtained (Ranc 1999), as also observed by Youn & Chang (1998), and this mode is not so much affected by the initial transverse cracks. But, for the probable case of highly repeated loads at service level, fatigue failure of lower transverse rebars is obtained. This has been proved consistent with plain reinforcement characterization under fatigue, associated with the proper estimation of effectively applied stresses.

3 FINITE ELEMENT MODEL

The model developed is intended to give further insight in the understanding of the slab behavior after cracking. An explicit modeling of concrete, cracks and reinforcement is thus provided.

For reasons of symmetry, only a quarter of the slab is modeled (Fig. 2). The mesh contains 3497 nodes and 1082 elements. Concrete is represented by 480 20-nodes brick elements MTH20, disposed on six layers in the depth of the slab. An elasto-plastic model is used for concrete, with a Willam-Warnke 3 parameters criterion (Ulm 1996). Input parameters are given in Table 2. Self-weight is neglected, in order to compare directly computation results with experimental ones. Choices of the Young's modulus and of the tensile strength were not straightforward and are discussed later. Parameters describing the evolution of the yield criterion (a_0, b_0 and k) were chosen in order to give a plastic – softening behavior in tension. Namely for the level of loads applied experimentally (as under service loads) concrete in compression remained elastic, while cracking was observed on the side in tension

Table 2. Input parameters for the material models.

Parameter	Unit	Value
Concrete (solid elements)		
Young's Modulus	GPa	30
Poisson's ratio		0.2
Compressive strength	MPa	35
Tensile strength	MPa	0.5
Biaxial compressive strength	MPa	41.5
Initial yield surface		$a_0 = 1.0$
Final yield surface		$b_0 = 0.1$
Hardening coefficient		$k = 1000$
Reinforcement (associated bars)		
Young's Modulus	GPa	200
Poisson's ratio		0.3
Cracks (interface elements)		
Internal friction angle	°	80
Dilatancy angle	°	60
Stiffness	MPa	1

(especially longitudinally in the central strips, due to transverse bending).

The two grids of reinforcement are modeled with 362 3-nodes beam elements PT2. Isotropic elastic behavior was considered for the reinforcement, since it doesn't yield under service loads. Perfect bond is assumed with surrounding concrete.

Obtained "shrinkage" cracks in the experiment were regularly spaced, with major openings every 2 transverse re-bar position (i.e. approximately every 0.3 m). In the numerical model, these cracks are represented by 16 nodes contact elements FDH16 (Richer 1985). For the interface behavior, a Mohr-Coulomb model with non-associated flow rule is used. The joint elements are initially open. Since during the experimental tests almost no relative sliding was noticed, cohesion and friction angle of these elements were set to high values.

4 RESULTS

4.1 *Global behavior*

The first stage of the pre-cracked slabs bending tests, with a monotonous increase of the load F up to 86 kN, has been simulated using the TCNL modulus (solving of mechanical problems with material non linearity – plasticity – and contact elements) of CE-SAR-LCPC finite element software (Humbert et al. 2001). Load increments of 5 kN have been used which correspond to experimental data.

A first major objective of this computation consisted in describing correctly both the global load–deflection curve, and the deformed profiles of the slab longitudinally and transversally. Provided these global results

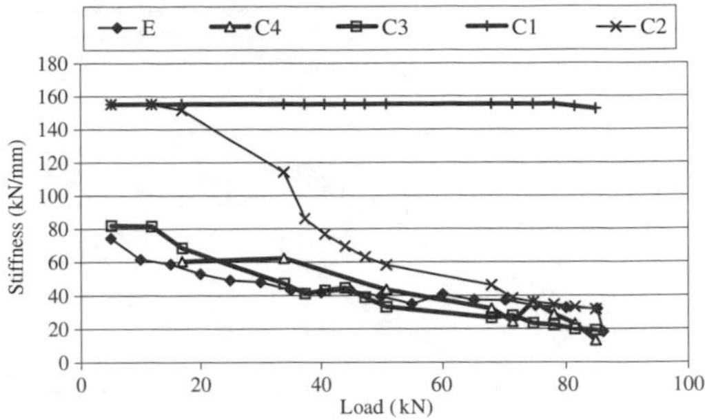

Figure 3. Flexural stiffness of the slab under central point loading. Computational results versus experimental data.

can be fitted, it is considered that the tangent elastic behavior of the structure is correctly captured.

Variation of concrete Young's modulus and tensile strength was applied with respect to parameters of Table 2, and the absence of contact elements was tested. Computations are designated as follows:

C1: no contact elements, $E = 30$ GPa, $f_{ct} = 3$ MPa
C2: no contact elements, $E = 30$ GPa, $f_{ct} = 0.5$ MPa
C3: contact elements, $E = 30$ GPa, $f_{ct} = 0.5$ MPa
C4: contact elements, $E = 20$ GPa, $f_{ct} = 3$ MPa
E: in superimposition of the computations, on following graphs, represents the experimental data.

For the six RC slabs, depending on the cracking pattern and crack openings, the experimental stiffness during the first central flexural loading range from 55 to 103 kN/mm. These values correspond to stage I, that is, before bending cracks appear longitudinally due to transverse bending moment. The mean value is 74.4 kN/mm.

The evolution of the global slab stiffness (applied load divided by the central deflection) is represented during this first loading (Fig. 3), for computations C1 to C4, and for the mean experimental values.

For computations C1 to C3, concrete Young's modulus was taken 20% lower than the mean experimental values on cylinders, in order to take into account the damage suffered by the concrete during the tensile preloading phase. However, the computations C1 and C2, without contact elements, give a too high value of the stiffness (about 155 kN/mm), even with this low value of Young's modulus (30 GPa). On the contrary, C3 computation with contact elements and $E_c = 30$ GPa gives a global initial stiffness of 82 kN/mm, which is in the range of the experimental values. The computation C4 with an even lower Young's modulus for concrete (20 GPa), which was intended to account for future damage in bending, gives a too low estimation of the initial stiffness (60 kN/mm).

Instead of the experimental value close to 3 MPa (Table 1) which was taken for computations C1 and

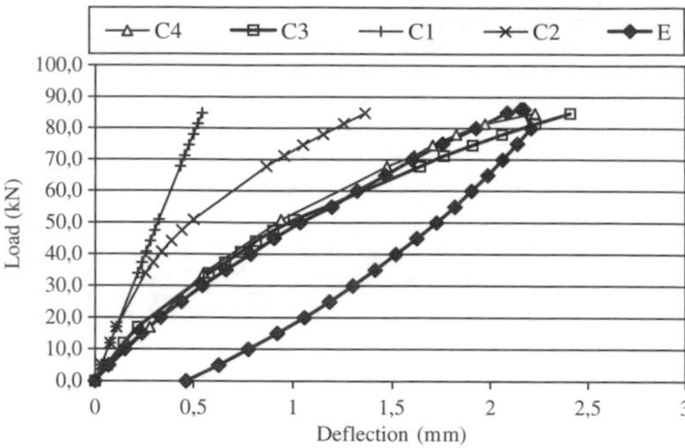

Figure 4. Computed vs. experimental load–deflection curves.

C4, the tensile strength was taken equal to a low value of 0.5 MPa for computations C2 and C3, because experimental results show a variation of the stiffness from the beginning. A greater value of f_{ct} maintains the stiffness constant at the initial value on a too great range of load. It is worth noting that the self-weight (which induces a tensile stress of about 0.6 MPa transversally at lower chord) and tensile stresses due to shrinkage are not directly taken into account.

The load–deflection curve (Fig. 4) gives a complementary validation of the parametric identification. Experimental results are clearly best approximated by variants with contact elements, C3 and C4. But the deeper examination of the stiffness values (Fig. 3) is better suited to show that for computation C4 the initial stiffness is too low and cracking begins too late. As a first conclusion, it turns out that the consequences of the initial pre-cracking phase shall be represented by explicit geometrical discontinuities, a local material stiffness not too much damaged, and significant decrease of concrete tensile strength.

It was recognized as possibly important to describe explicitly progressive damage within the structure, due to the repeated loading. Bond, concrete strength or stiffness degradation might a priori have to be accounted for. While fatigue failure of rebars is not reached, the results reported by (Ranc 1999) indicate that at the level of service loads – or not too much amplified service loads – reached in this program, these phenomena can be neglected. Adaptation of the structure takes place in a few cycles only by concrete cracking, which justifies the type of elasto-plastic modeling adopted. As a sign of this, the experimental deflection range is presented after 1 million cycles at 86 kN (after subtracting the residual deflection which is about 1.6 mm), in comparison to the first cycle (Fig. 5). It has varied very little. So there is no important damage for this load range, even after one million cycles. The residual deformation can be attributed to the cyclic creep. The stress range in the reinforcement also proved to vary very little. It is thus sufficient to

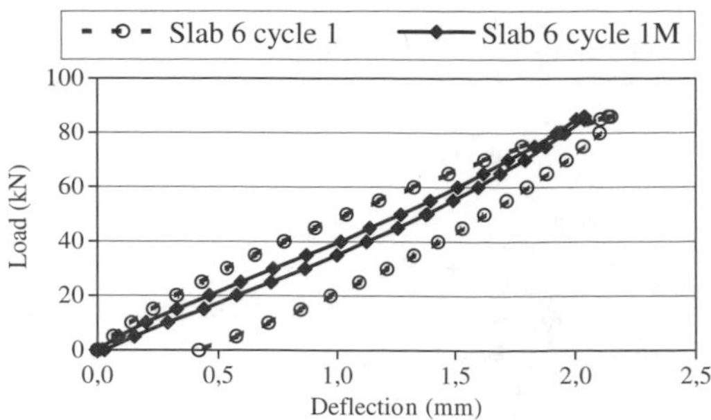

Figure 5. Experimental load–deflection curves for slab #6 at 1st cycle and after 1 million cycles at 86 kN, after (Ranc 1999).

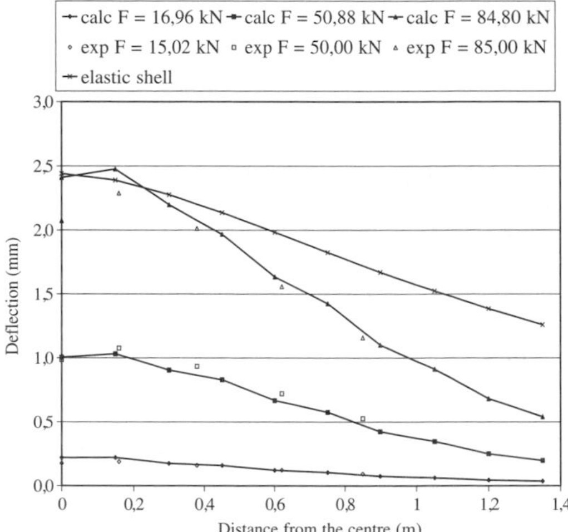

Figure 6. Axial longitudinal profile of the deformed slab: computations vs. experimental (slab #4 in (Ranc 1999)).

study the first loading cycle to capture the range of deflections.

The explicit accounting for transverse throughgoing cracks appears as necessary with respect to the global rigidity. The consequences in the longitudinal and transverse deflections along both directions of the slab are studied. First, the longitudinal deformed profile of the slab is considered for different load values (Fig. 6). The finite element computed profile (computation C3, i.e. with material data from Table 2) fits well the experimental values. The profile obtained by an elastic calculation with an isotropic model (scaled at the maximum value of the experimental deflection) is rather different. Although a lower global rigidity has been calibrated, the central band is relatively less loaded while the ends of the slab exhibit much higher deflections. The isotropic model clearly overestimates the spatial redistribution of the load and deformations,

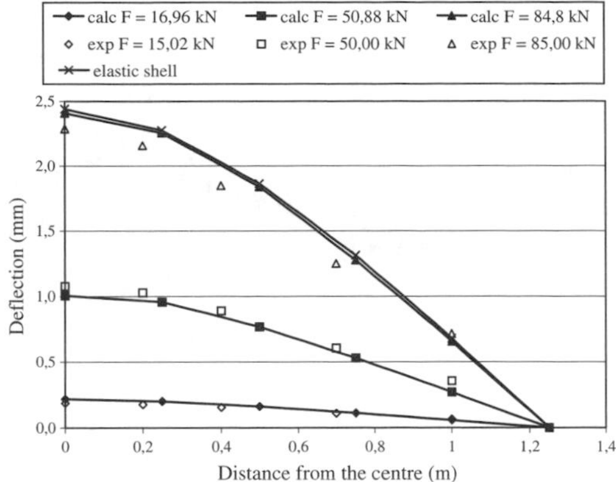

Figure 7. Axial transverse profile of the deformed slab: computations vs. experimental (slab #4 in (Ranc 1999)).

which only partially takes place due to the transverse cracks.

Similarly, the transversal deformation profile at the center of the slab is given in Figure 7. Experimental and finite element profiles fit well. Isotropic model deformations fit also well the experimental and the finite element data. However, the elastic deflection had to be multiplied by four to be scaled with the other curves in Figure 7.

4.2 *Stresses in the reinforcement*

After the global behavior of the slab is described, the major question concerning fatigue verifications consists in the correct prediction of stresses of the reinforcement. In fact, experimental observations indicate a significant discrepancy between the usual computations, based on the hypothesis of slab continuity and homogeneity, and the measured stresses. This is because the latter are influenced by both the initial transverse through-going cracks, and the progressively established longitudinal cracking pattern due to transverse bending.

The ratio of transverse to longitudinal stresses helps analyzing this redistribution due to cracking. In the stage II reinforced concrete calculations, stresses in the inferior reinforcement (derived from the bending moments given by an elastic isotropic analysis) turn out to be in the ratio $\sigma_{s,tr}/\sigma_{s,lo} = 0.55$, (Ranc 1999). The experimental measures gave rather the inverse ratio, with $\sigma_{s,tr} > \sigma_{s,lo}$. For example, in the first cycle at the maximum load F = 86 kN, $\sigma_{s,tr}/\sigma_{s,lo} \cong 1.20$ if the mean value of four gauges placed in the central zone is taken; and the ratio increases up to about 1.40 if only the gauges placed near the cracks are taken into account.

In the first cycle, an increase of the ratio with the load from 0.6 to 1.2 is noticed. On the contrary, after 1 million cycles the ratio is rather constant for the whole range of the load. The variation of the stress ratio in the first cycle should be linked to the development of the flexural cracks. After 1 million cycles at 86 kN, the crack pattern can be considered as stabilized, and there is no other redistribution. The stress ratio at F = 86 kN has increased for the first 1 million cycles, from 1.20 to 1.60. This increase is due to a slight decrease of the stress in the longitudinal reinforcement, while the stress in the transversal reinforcement remains almost constant. It may be assumed that the bond of longitudinal rebars, which has been significantly damaged during the pre-cracking phase, goes on degrading due to repeated cycles of longitudinal bending.

This analysis is illustrated in comparing experimental data with results of the computations C1 to C4. The comparison focuses on the stress ratio between transversal and longitudinal reinforcement (Fig. 8), stresses in transverse lower rebars (Fig. 9) and stresses in longitudinal lower rebars (Fig. 10). Experimental data come from slab #6, with 0.3-mm characteristic crack opening, which is considered as the limit of acceptable initial crack openings due to restrained

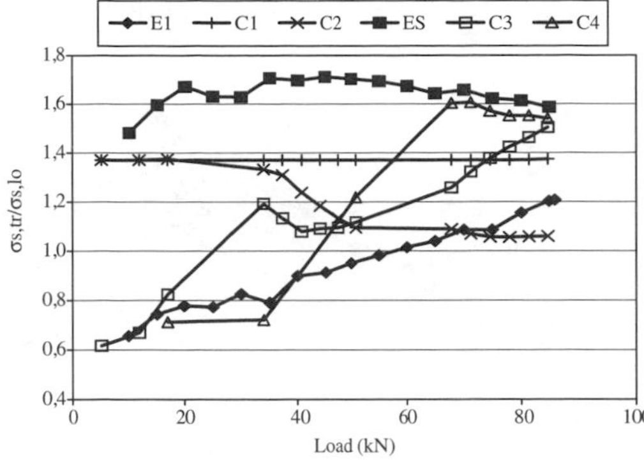

Figure 8. Stress ratio between transversal and longitudinal bottom reinforcement: computed versus experimental values.

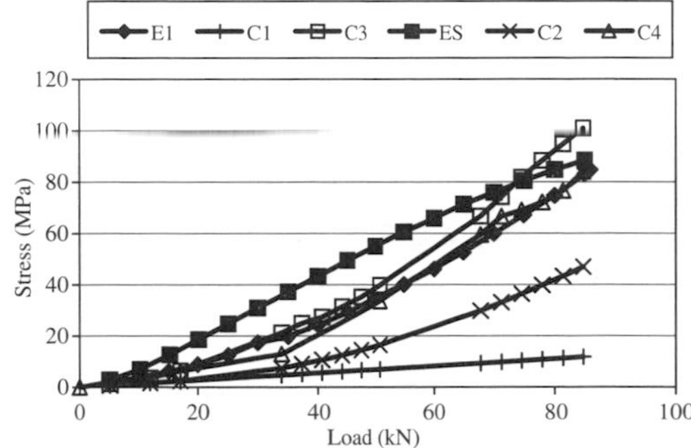

Figure 9. Stresses in the transversal bottom reinforcement: computed versus experimental values.

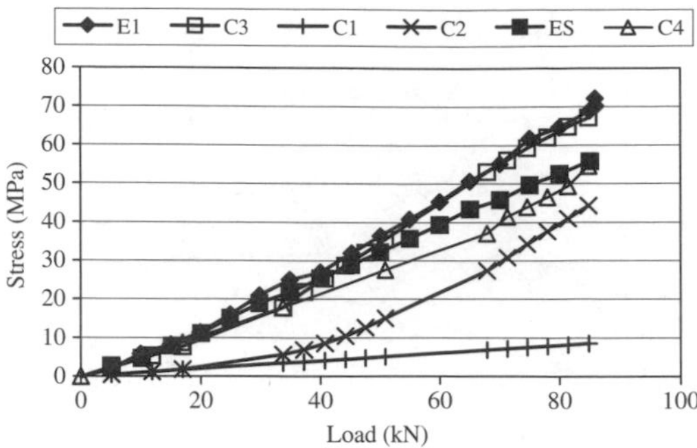

Figure 10. Stresses in the longitudinal bottom reinforcement: computed versus experimental values.

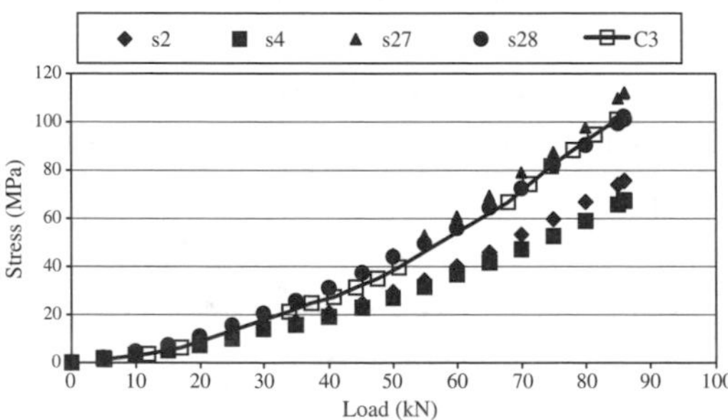

Figure 11. Stresses in transversal bottom reinforcement during the first loading cycle for 0-86 kN range, slab #6. C3 results versus stresses derived from strain gauges measurements.

shrinkage. Notation "E1" stands for the experimental results of slab #6 during the first loading cycle, "ES" stands for experimental "stabilized" values and corresponds to data of slab #6 after 1 million cycles, according to (Ranc 1999).

Models C1 and C2, without contact elements, give a bad estimation of the stress ratio and of the stress values. The predicted stress ratio has a value of about 1.4 before cracking, which corresponds approximately to the elastic bending moment ratio, and decreases down to about 1.10, which is opposite as compared to the experimental evolution of this ratio. Stress values are also significantly underestimated.

On the contrary, finite element models with contact elements C3 and C4 give a satisfactory estimation of the stress values in the reinforcement. In both cases, due to the initially opened transverse cracks, $\sigma_{s,tr}/\sigma_{s,lo}$ is predicted as low as about 0.6 before longitudinal cracking takes place, and the ratio increases up to about 1.55 which is closed to the experimental value after 1 Million cycle (stabilized crack pattern).

The computation with a higher Young's modulus gives slightly higher stress values, which appears as conservative. In fact, E1 and ES curves are based on

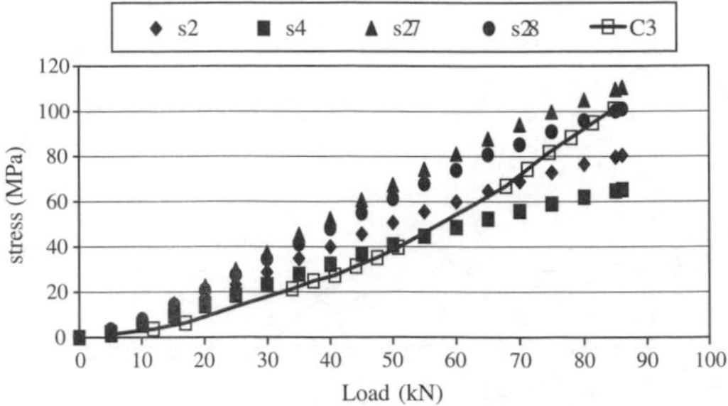

Figure 12. Stresses in transversal bottom reinforcement after 1 Million cycles for 0-86 kN range, slab #6. C3 results versus stresses derived from strain gauges measurements.

average indications of strain gauges located either near the reinforcement of opposite direction (s27 and s28 in Figs. 11–12), or at midpoint of the central mesh (s2 and s4 in Figs. 11–12).

Both at the first loading and after 1 million cycles, the signals of strain gauges close to the reinforcement (which helps initiating the position of cracks) are about 30% higher than those corresponding to the center of the mesh (located about 12 cm apart from the crack). This constant ratio is a sign of undamaged bond and constant anchoring length concerning transverse reinforcement, even after 1 Million cycles representative of service loads. Computation C3 (which corresponds to a first loading phase) fits quite well these high values, especially for the first loading.

Finally, the objective of computing the stresses of the reinforcement within the domain of service loads for a pre-cracked reinforced concrete slab has been reached successfully. The realistic position of transverse through-going cracks had to be reproduced with contact elements. The idealized rigid-plastic law of these elements could be a limiting factor for higher load levels, in the present context it is consistent with the high aggregate interlock and unobserved loss of shear transfer. It is noticeable that best agreement with experimental data is obtained for a value of Young's modulus only slightly lower than the experimental identification on cylinders, but for a low value of the tensile strength, which accounts for previous damage (first loading phase, shrinkage, effects of self-weight...), helps initiating flexural cracking, and may need to be further calibrated.

5 SIMPLIFIED ELASTIC MODEL

Finite element calculations with CESAR-LCPC code, including elasto-plastic model for concrete and contact elements for the initial cracks have thus shown the dominant influence of the shrinkage cracks on the stress distribution in the bottom reinforcement. If the

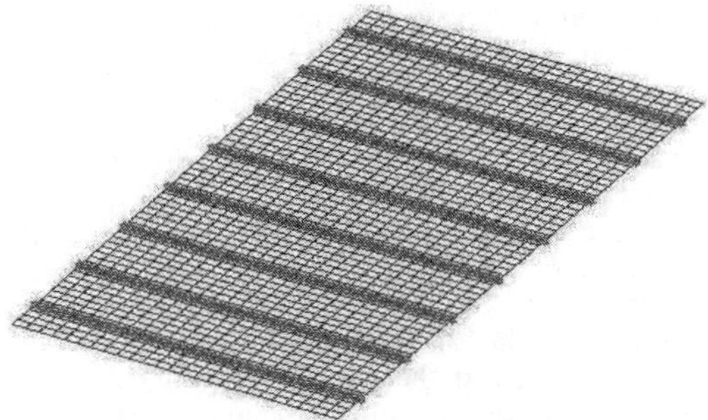

Figure 13. Mesh representing the slab with elastic shell elements and two nodes hinges.

reinforced concrete bridge decks are pre-cracked due to restrained shrinkage, stress calculations based on an elastic isotropic analysis, even with degraded material properties, give unsatisfactory results, especially underestimating the stresses in the transversal reinforcement. This underestimation may lead to unsafe fatigue verifications. However, non-linear finite elements calculations with contact elements are very costly, so a simpler method was searched to obtain a good estimation of the stresses in the reinforcement. In fact, the order of magnitude of the 1.3 factor indicated in Toutlemonde & Ranc (2001b), relating measured stresses in bottom transversal re-bars to their classical theoretical estimation, is to be justified. This "factor" is required, so that fatigue verifications could be based on a rational but simple method of moment and stress assessment in a cracked structure.

With this aim in view, different ways to combine a linear elastic calculation with the modeling of the shrinkage cracks were tested, as follows:

- The slab was modelled by a grid of orthogonal beams. The beams in the longitudinal direction were less stiff than the transversal beams.
- The slab was modelled with shell elements, but there were transversal "strips" of elements with low elastic properties at the same spacing as the shrinkage cracks.
- The slab was modelled with isotropic elastic shell elements, but at the place corresponding to the cracks, the shell elements were linked by two nodes hinge elements (zero stiffness). The mesh of this last variant is represented on Figure 13.

The bending moments obtained from the analysis were used to compute the stresses in the reinforcement. For this purpose, it was assumed that the reinforcement was the same that in the experiments and the usual rules for stage II reinforced concrete design were used.

Only the last variant gave realistic results for estimating the reinforcement stresses. These stresses are compared to experimental determinations of the stresses of the re-bars derived from the maximum strain measured for the different slabs. The simplified model as well as C3 and C4 calculations give a consistent ratio for both longitudinal and transverse directions, which implies a correct capturing of the anisotropy induced by initial cracks. The simplified model overestimates by 50 to 70% the mean experimental stresses and by about 20%, the maximum experimental stresses (Table 3). However, as for fatigue verifications maximum values should be used, and as the difference is always on the safe side, these results can be considered as satisfactory.

Table 3. Stresses of bottom reinforcement (MPa) for F = 86 kN.

	Transversal	Longitudinal
Exp. slab #1 after 1 M cycles	79	31.5
Exp. slab #2 after 1 M cycles	126	73
Exp. slab #3 after 1 M cycles	83	94
Exp. slab #4 after 1 M cycles	122	71
Exp. slab #5 after 1 M cycles	145	56
Exp. slab #6 after 1 M cycles	110	64
Mean experimental value	111	65
Minimum experimental value	79	31.5
Maximum experimental value	145	94
Isotropic elastic slab (Ranc 1999)	125	224
Ratio/exp. mean val.	1.13	3.45
Ratio/exp. max. val.	0.86	2.38
Elastic shells with hinges	168	113
Ratio/exp. mean val.	1.51	1.74
Ratio/exp. max. val.	1.16	1.20
Computation C3 (see section 4)	101	67
Ratio/exp. mean val.	0.91	1.03
Ratio/exp. max. val.	0.70	0.71
Computation C4 (see section 4)	84	55
Ratio/exp. mean val.	0.76	0.85
Ratio/exp. max. val.	0.58	0.59

Computation C3 with contact elements gives a more consistent estimation of both longitudinal and transverse stresses, however the prediction is correct on average but underestimates maximum recorded values by a factor 0.7. It shall be reminded that this "simplified" computation gives results within some minutes, while calculations with the model detailed in section 4 correspond to several tenth of hours on the same workstation.

It has also to be reminded that the "simplified" model actually tested does not predict correctly the global behavior of the structure. The computed central deflection is about 0.8 mm instead of 2.2 to 2.4 mm experimentally. In order to predict the correct structural flexibility, an equivalent reduced stiffness could be used for the shell elements. Romanian provisions give a factor of 0.3 for stage II cracked reinforced concrete as compared to non-cracked concrete Young's modulus and inertia. This factor would give a correct order of magnitude.

6 CONCLUSIONS

A non-linear finite element model has been presented, using the possibilities of CESAR-LCPC finite element software, which is able to predict the service behaviour of reinforced concrete slabs representative of cracked composite bridge decks, provided the initial cracks are modeled. The numerical results are in good agreement with the measured values, at the global scale (force–displacement curves) and at the local level (load versus reinforcement stress).

The quality of the results is related to the interface element. Due to the compression in the upper part, the element closes and relative vertical sliding displacements are very small, like in the experiment (10 to 20 μm). For the bottom part in tension, longitudinal rigidity only comes from the reinforcement, which is close to the model of hinges described in section 5.

The calculus confirmed that the effective stress in the transverse reinforcement is up to 30% greater than predicted by conventional design methods, which may lead to unsafe verifications regarding fatigue. The main influent factors are the initial cracks that induce a very strong anisotropy, while conventional design is based on isotropic behavior of the slab.

The quality of this numerical prediction is also related to the modeled transfer length of reinforcing bars. Here, bar elements are fixed to the nodes of concrete solid elements every two blocks (one solid element on both sides apart from the crack). This corresponds to about 0.15-m anchoring length in cracked concrete (longitudinal direction), and is consistent with crack spacing.

A "simplified" calculus was also performed using elastic shell elements for the concrete and 2 nodes spring elements to model initial cracks by hinges. Predictions of the deflection should account for a reduced "stage II" stiffness in order to be realistic. However, the bending moments obtained by the "simplified" model were used to compute the stresses in the reinforcement, by conventional reinforced concrete design methods. And the results constitute a proper (even if rather conservative) estimation of the maximum stresses in the reinforcement, with the aim of predicting the fatigue life of the slab, or computing the amount of transversal reinforcement required to avoid fatigue failure.

REFERENCES

Ducret J.-M. & Lebet J.-P. 1999. Effects of concrete hydration on composite bridges. In *Composite construction – conventional and innovative; Proc. IABSE symp. Innsbruck, sept. 1997.* Zürich: IABSE.

Humbert, P. et al. 2001. *CESAR-LCPC 3.3 User's Manual.* Paris: LCPC.

Kretz, T. et al. 1995. *Ponts mixtes, recommandations pour maîtriser la fissuration des dalles.* Bagneux: SETRA. Technical manual.

Ranc, G. 1999. *Résistance en fatigue des dalles de béton armé fissurées des ponts mixtes.* Paris: ENPC, Ph. D. Report.

Richer, S. 1985. *Résolution des problèmes de contact entre solides élastiques par la méthode des éléments finis.* Paris: ENPC, Ph. D. Report.

Seok-Goo Youn & Sung-Pil Chang. 1998. Behavior of composite bridge decks subjected to static and fatigue loading. *ACI Structural Journal* 95(3): 249–258.

Toutlemonde, F. & Ranc, G. 2001. Fatigue service life of transversally cracked reinforced concrete slabs representative of composite bridge decks. In N. Banthia et al. (eds), *Concrete under severe conditions; Proc. CONSEC'01*: 1157–1164. Vancouver: University of British Columbia.

Toutlemonde, F. & Ranc, G. 2001. Fatigue tests of cracked reinforced concrete slabs for estimating the service life of composite bridge decks. *Revue Française de Génie Civil* 5(4): 483–494.

Ulm, F.J. 1996. Un modèle d'endommagement plastique: application aux bétons de structure. *Etudes et Recherches des LPC* (OA19). Paris: LCPC.

Computational Modelling of Concrete Structures, Bićanić et al. (eds)
© 2003 Swets & Zeitlinger, Lisse, ISBN 90 5809 536 3

Design of the shotcrete tunnel lining of a metro station – safety considerations

H. Walter
IGT – Geotechnik und Tunnelbau, Consulting Engineers, Salzburg, Austria

ABSTRACT: In the course of the extension of a metro line in Vienna the station Taborstraße will be built. Two single track platform tunnels, a connection tunnel and an elevator tunnel will be driven through sediments following an NATM excavation scheme with part-excavations of the cross sections.

Time dependent three-dimensional finite element analyses using nonlinear material models have been performed, in order to predict the settlements due to the excavation of the tunnels and to design the shotcrete lining. Features of the viscoplastic shotcrete model are: a rapid development of strength and stiffness, cracking, and a pronounced creep at high stress levels.

At the tunnel intersections stress concentrations in the lining occur which would not be permitted according to the design rules of common standards. Variations of the material parameters of the shotcrete and the soil are used to show that safety margins comparable to those of the standards are present.

1 INTRODUCTION

In the course of the extension of the metro line U2 in Vienna the station Taborstraße will be built. The project is in the tender stage at the end of 2002 (Wiener Linien 2002), the line will go into operation in 2008. Client is the "Wiener Linien", the Vienna Transport Authority.

The station consists of two parallel single track platform tunnels with a cross section of 74 m^2 each. The platform tunnels are connected by a connection tunnel (cross section 59 m^2) from which an escalator tunnel (cross section 63 m^2) provides access to the street level. Two shafts, a large one at the western end of the station, the shaft "Taborstraße", and a small one at the eastern end between the platform tunnels, the shaft "Novaragasse", will be constructed as open cut prior to the platform tunnels. The tracks lie about 18 m below street level. The tunnels run partly below 5–7-storey-buildings, partly below streets. Figure 1 shows a plan of the station.

The tunnels will be driven through sediments of the Quarternary and Tertiary following an NATM excavation scheme. In order to minimise settlements the cross section of the platform tunnels will be divided into two main parts: A side drift, subdivided into top heading, bench and invert, see Figure 2, will be excavated first; then the remaining cross section will be excavated with the same subdivision.

Primary support will be a shotcrete lining with a thickness of 30 cm, and 25 cm at the temporary walls, respectively. The connection tunnel and the escalator tunnel will be subdivided into top heading, bench and invert, as well; the lining thickness is 30 cm, again. The linings will be reinforced with two layers of wire mesh.

Three-dimensional finite element analyses have been performed in order to predict the settlements due to the excavation of the tunnels and to design the shotcrete lining. Of special interest are the stress concentrations in the lining which develop at the intersections in the course of the excavation process.

The finite element model extends over the whole station area from the shaft "Taborstraße" with a length of 60 to 68 m in the direction of the line, a width of 80 m and a vertical span of 35 m. The chosen coordinate system has its origin at the height of "Wiener Null", the

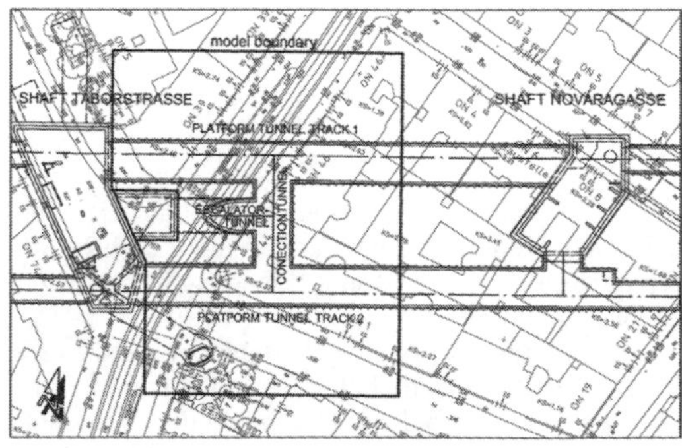

Figure 1. Plan view of the station and the model boundaries.

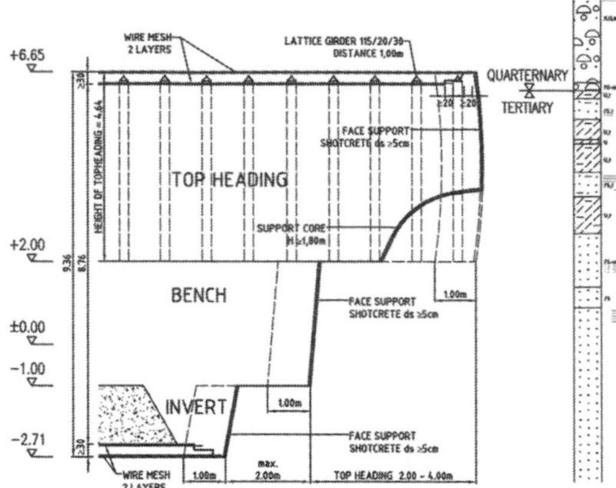

Figure 2. Excavation scheme: side drift of the platform tunnels.

local reference system, above the longitudinal axis of the connection tunnel in the middle between the two platform tunnels. Top of rail (TOR) is 15.0 m below the level of the basements or 12.5 m below "Wiener Null". The x-direction of the coordinate system is the direction of the line (eastwards positive), the z-direction is vertically upwards.

In the following the most important properties of the numerical model are described and some characteristic results are presented. A discussion of these results leads to the observation that a design of the shotcrete lining following the procedures in standards like the Eurocode (CEN 2001, CEN 2002) can hardly be performed using the internal forces resulting from the analysis.

Therefore, the last chapter deals with the implications of nonlinear material and structural behaviour on the interpretations of the design rules specified in the standards. With the help of additional analyses it will be shown that sufficient safety margins against failure of the structure are present if the shotcrete lining with the dimensions described above is used as support.

2 CONSTITUTIVE RELATIONS, MATERIAL PROPERTIES

2.1 *Soil*

The soil has been modelled by linear volume elements using the Mohr-Coulomb-material model. Three different soil layers were distinguished, with parameters according to Table 1.

The parameters given in the table are used in the analyses except for the following modifications: At a level of about 4 m below TOR the groundwater table has been assumed. (The groundwater table will be lowered to the level of the invert of the platform tunnels during construction of the station. The calculated settlements therefore do not contain the effects of the changes of the groundwater table.) Below this level not only the specific weight under water is used, but also the modulus of elasticity is increased artificially by a factor of 5 in order to simulate the stiffer soil behaviour during unloading below the tunnel.

The parameters given in the table were specified as characteristic values by experienced geotechnical engineers based on site investigations and former experience with constructions at similar geological conditions in Vienna. The parameters have been chosen as cautious estimates of the mean value in the sense of Eurocode 7 (CEN, 2001). The strength parameters do not explicitly contain any safety factors, neither do the stiffness parameters.

The soil above the level of the basements of the buildings has not been modelled. Instead, the dead load of the soil (70 kN/m^2) and the buildings (130 kN/m^2), respectively, have been specified as distributed loads at the level of the basements.

2.2 *Shotcrete*

The shotcrete lining consists of layered shell elements. A constitutive model developed by Meschke and Mang, and extended by additional creep and shrinkage terms, has been applied.

2.2.1 *Meschke-Mang-model*

Detailed descriptions of the Meschke-Mang-shotcrete model can be found in the literature (Meschke 1996; Meschke et al. 1996). It has already been applied for 3-D-tunnel analyses (Mang et al. 1994). The main features of the model are:

A strain-hardening Drucker-Prager loading surface with a time dependent hardening parameter to account for the compressive regime. Cracking of maturing shotcrete is accounted for in the framework of the smeared crack concept by means of three Rankine failure surfaces, perpendicular to the axes of principal stresses. Two independent hardening and softening mechanisms control the constitutive behaviour of shotcrete subjected to compressive and tensile stresses, respectively.

The increase of elastic stiffness during hydration of shotcrete as well as the time-dependent increase of compressive strength, tensile strength, and yield surface are all considered.

The extension of the inviscid elastoplastic model for aging shotcrete to viscoplasticity is based upon the model by Duvant and Lions.

A numerically efficient algorithmic formulation of multisurface viscoplasticity in principal axes results in a robust implementation for engineering applications.

Table 1. Material properties of the soil layers.

Soil layer		Quarternary	Tertiary (silt)	Tertiary (fine sand)
Description		Gravel, sandy	Silt, clay	Sand, silty
Lower boundary	m above TOR	~6	~3	–
Specific weight (wet)	kN/m^3	20	20	20
Specific weight (under water)	kN/m^3	11	11	11
Modulus of elasticity	MN/m^2	150	50	150
Poisson's ratio	1	0.28	0.35	0.30
Friction angle	°	35	25	27.5
Cohesion	kN/m^2	0	50	0

2.2.2 *Refined model*

The Meschke-Mang-model describes creep effects with one single parameter, the viscosity. At stress levels within the current yield surface, no creep or relaxation occurs. Therefore, the adaptability to experimental creep or relaxation data is limited.

In the course of a research project (Walter et al. 1996), experiments with shotcrete specimens were conducted at the Mining University Leoben, Austria. In the tests, shotcrete prisms were loaded at an age of only 6 hours. The specimens were subjected to a large variety of load paths, some with load control and some with displacement control.

In order to match the test results, it was necessary to extend the Meschke-Mang-model. A simple engineering approach was chosen:

A model already used for the description of the long-term shotcrete behaviour (Schubert, 1988) is used to define additional creep terms: It is based on the rate of flow-method (England & Illston, 1965).

For each of the principal stresses creep strain increments for the time interval $(t_i,\ t_{i+1})$ are calculated as

$$\Delta\varepsilon_{CR,u} = \sigma_{t_{i+1}} \cdot \Delta C(t) \cdot e^{k \cdot \sigma_{t_{i+1}}} + \Delta\varepsilon_d \tag{1a}$$

with

$$\Delta\varepsilon_d = \left(\sigma_{t_{i+1}} \cdot C_{d\infty} - \varepsilon_{d,t_i}\right) \cdot \left(1 - e^{\frac{-\Delta C(t)}{Q}}\right), \tag{1b}$$

$$\Delta C(t) = C(t_{i+1}) - C(t_i) \tag{1c}$$

and

$$C(t) = A \cdot t^{\frac{1}{3}}. \tag{1d}$$

In equations (1) denote

$\Delta\varepsilon_{CR,u}$ the increment of creep strains in the time interval from t_1 to t_{i+1}

$\sigma_{t_{i+1}}$ the principal stress at time t_{i+1}

and

$\Delta\varepsilon_u$ the increment of "delayed elastic strain", ε_d, i.e. of the reversible creep component, in the interval from t_i to t_{i+1}.

The other variables are model parameters:

A describes the amount of (irreversible) creep

$C_{d\infty}$ is the ultimate value for the reversible creep compliance

Q describes how fast the delayed elastic strains develop

k is used to introduce a nonlinearity for high stress levels.

With the help of a Poisson's ratio for creep, ν_{CR}, the creep strains in the principal directions are related to each other:

$$\varepsilon_{1,CR} = \varepsilon_{1,CR,u} - \nu_{CR}\left(\varepsilon_{2,CR,u} + \varepsilon_{3,CR,u}\right) \tag{2a}$$

$$\varepsilon_{2,CR} = \varepsilon_{2,CR,u} - \nu_{CR}\left(\varepsilon_{1,CR,u} + \varepsilon_{3,CR,u}\right) \tag{2b}$$

$$\varepsilon_{3,CR} = \varepsilon_{3,CR,u} - \nu_{CR}\left(\varepsilon_{1,CR,u} + \varepsilon_{2,CR,u}\right) \tag{2c}$$

In equations (2) the first index of the strain values denotes the index of the principal direction.

For the back transformation of the principal values of the creep components into global directions, the same transformation matrix as for the principal stresses is used.

The three-dimensional extension of the creep law is somewhat arbitrary. Due to the lack of experimental data for creep under multiaxial stress states this simple generalisation seems to be justified. As a first guess, the value of the creep Poisson's ratio has been set equal to the Poisson's ratio for shotcrete.

Table 2. Material parameters of shotcrete.

Parameter	Symbol	Unit	Value
Specific weight	γ	kN/m^3	25
Poisson's ratio	ν		0.20
Cylinder compressive strength at 28 days	$f_{cu}^{(28)}$	kN/m^2	17 000
Cylinder compressive strength at 1 day	$f_{cu}^{(1)}$	kN/m^2	4500
Yield stress at 28 days	$f_y^{(28)}$	kN/m^2	3000
Viscosity parameter	η	15 h	
Modulus of elasticity at 28 days	$E^{(28)}$	MN/m^2	25 000
Factor describing the amount of irreversible creep	A	$m^2/kN \cdot h^{-1/3}$	$1.0 \cdot 10^{-8}$
Ultimate value for the reversible creep compliance	$C_{d\infty}$	m^2/kN	$1.5 \cdot 10^{-7}$
Factor describing how fast the delayed elastic strains develop	Q	m^2/kN	$4.0 \cdot 10^{-8}$
Factor to introduce a nonlinearity for high stress levels	k		0

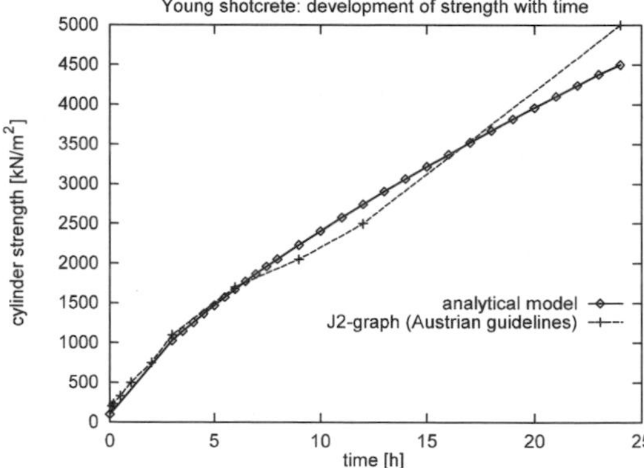

Figure 3. Development of the strength of young shotcrete with time.

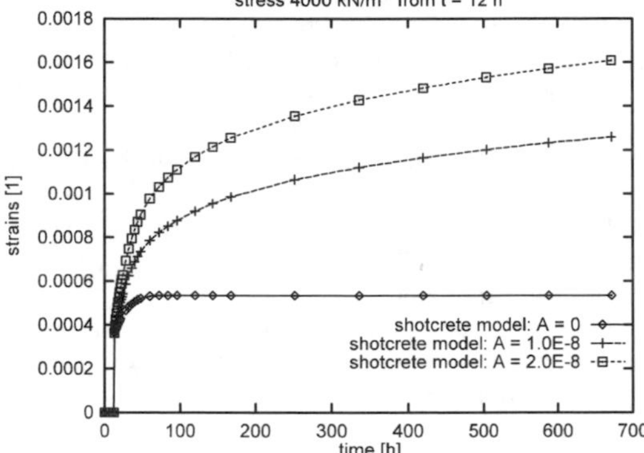

Figure 4. Development of strains at constant stress level.

Details of the implementation can be found in (Walter et al. 1996, Walter 1997). The refined model contains shrinkage terms according to the relation

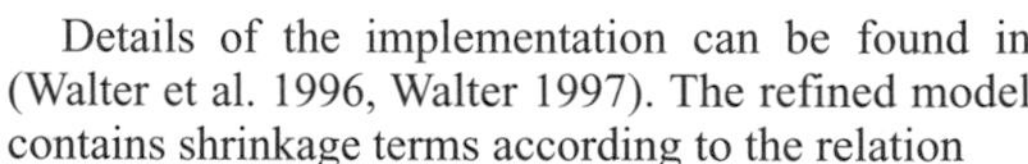

$$\varepsilon_{sh} = \varepsilon_{sh,\infty} \cdot \frac{t}{B+t} \tag{3}$$

with ε_{sh} being the volumetric shrinkage strains at time t, $\varepsilon_{sh,\infty}$ the ultimate value of the shrinkage strains and B a parameter describing the development of the shrinkage strains with time.

2.2.3 *Material parameters for shotcrete*

Table 2 shows the material parameters chosen for the shotcrete. They match the properties of the shotcrete type SpB 25(56)/J2 (Betonverein 1998) which has been specified as a minimum requirement in the tender documents, very well, see figure 3. Different amounts of creep have been specified in the course of a parameter study. The values given here yield a relatively small amount of creep strains and little stress reduction due to creep. Thus, a conservative stress level is achieved. Figure 4 shows the effects of varying the parameter A. (A = 0 results in using the original Meschke-Mang-model.)

3 EXCAVATION SEQUENCE, DISCRETIZATION IN TIME

The excavation of the tunnels is modelled by removal of the finite elements representing the excavated soil.

In order to keep the size of the model at a reasonable level two rounds, with a length of 1 m each, of the real excavation scheme have been combined to one fictitious round. In general, only one layer of finite elements has been used to simulate such a fictitious round. According to the construction schedule a progress of two rounds per day, i.e. one fictitious round with a length of 2 m in 24 hours, has to be expected. The size of the time steps in the analyses has been chosen as one time step per round, the size of one time step being 24 hours.

The excavation schemes have been simplified for the analyses: The excavation of one (fictitious) round of bench and invert has been combined to one step; the distance between the excavation of the top heading and bench + invert is 4 m or two (fictitious) rounds. The excavation of a round of the remaining cross section of the platform tunnels is simulated as one single step for top heading, bench and invert and removal of the temporary lining together. The minimum distance between excavation of the side drift and excavation of the remaining cross section is 4 months according to the construction schedule.

Both platform tunnels will be driven in the same direction in the analysis. The distance between the face of the top heading of the side drift and the face of the remaining cross section has been condensed to 36 m (18 days) for both platform tunnels. The distance between the drives of the two platform tunnels has been chosen as 16 m (8 days). Thus, the number of time steps required in the analyses could be kept small. Care has been taken that the distance between the drives is still large enough to prevent an artificial mutual influence.

The excavation of the connection tunnel and of the escalator tunnel is simulated in two phases, again: first the top heading is removed; two rounds later bench and invert are excavated together. The direction of the driving of the connection tunnel has been chosen in the direction from track 2 to track 1 because preliminary analyses have shown that this direction minimizes the settlements and inclinations of the buildings above the station. The excavation of the escalator tunnel starts at the connection tunnel. Before the excavation of the connection tunnel and after the end of the excavation process extra time steps with increased length have been added. They cover an additional period of one month each, with the goal to observe the amount of stress redistributions caused by creep of the shotcrete.

Shotcrete support is applied by stress-free activation of the shell elements simulating the newly added lining. It has been assumed that every newly excavated fictitious round is without shotcrete support and that the time of 24 hours is used just for excavation and mucking. The second round behind the face already contains shotcrete support. The shotcrete age at the end of this second round has been set to 18 hours, i.e. the hardening of the shotcrete starts 6 hours after the end of excavation and mucking. These assumptions result in a very conservative estimate of the deformations and the strength of the shotcrete support.

Perfect bond between soil and lining as well as between older and newer parts of the lining has been assumed.

Other means of support, like forepoling rods, lattice girders or temporary support of the face, have been neglected in the analyses.

4 RESULTS USING CHARACTERISTIC PARAMETERS

With the parameters of Tables 1 and 2 the station has been analysed. At most of the time steps convergence was achieved after 3 or 4 iteration steps using a tolerance of 1% for the residual forces.

4.1 *Settlements*

Figure 5 shows the calculated settlements at the level of the basements of the buildings. The basement walls are indicated on the plots.

Figure 5a shows one characteristic step, analysis step 29, during the excavation of the two platform tunnels: The top heading of the side drift of track 1 has proceeded 54 m into the model, the remaining cross section 18 m. The top heading of the side drift of track 2 has advanced 34 m into the model, and the first layer of the remaining cross section has been excavated.

Figure 5b shows an instant at the end of the excavation of the connection tunnel, Figure 5c depicts the settlements at the end of the analysis.

Directly above the face of the top heading of the first side drift the settlements are about 6 mm, they increase to 13 to 16 mm after the excavation of the remaining cross section. The maximum inclinations are about 1:700. The settlements increase to 21 mm above track 1, and to 24 mm above track 2 during the excavation of the connection tunnel. The excavation of the escalator tunnel creates additional settlements westerly of the connection tunnel. The maximum settlement above track 2 increases to 27 mm.

Of special interest are the maximum inclinations within the plan area of the buildings. They reach a maximum of 1:500 which is the maximum value allowed by the authorities.

The long term effects of creep of the shotcrete are small: The maximum settlements increase by less than 1 mm during the periods without excavation before commencing the connection tunnel and at the end of the analysis.

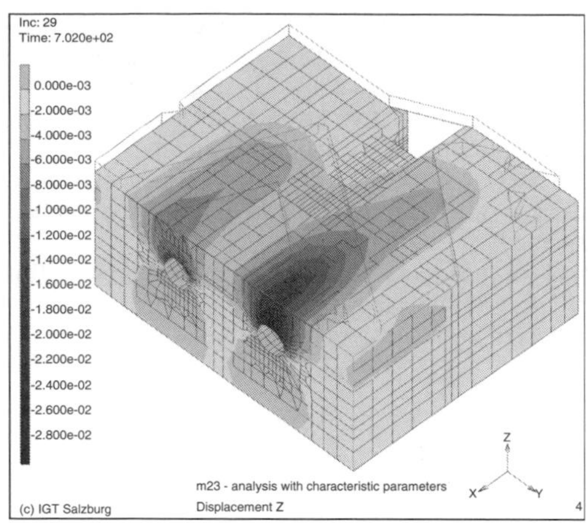

Figure 5a. Vertical displacements at analysis step 29.

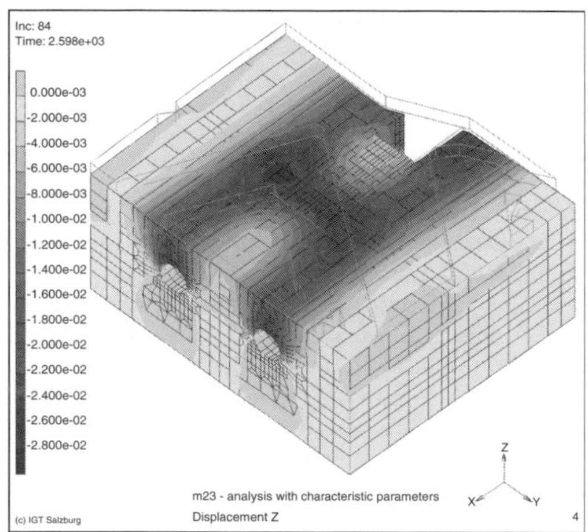

Figure 5b. Vertical displacements after excavation of the connection tunnel.

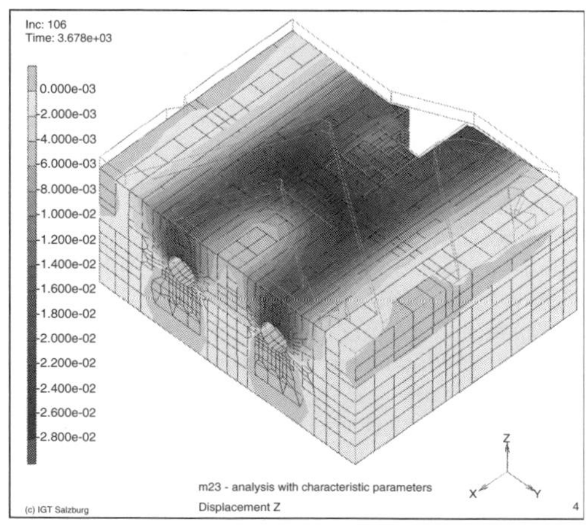

Figure 5c. Vertical displacements at the end of the analysis.

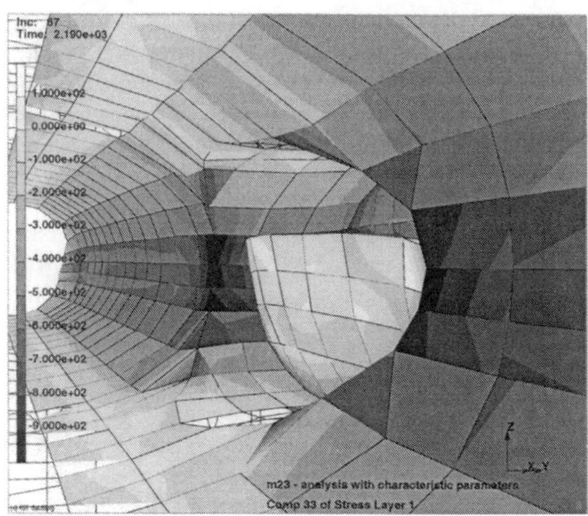

Figure 6. Vertical stresses in the soil at the beginning of the excavation of the connection tunnel. (Displacements magnified 50 times.)

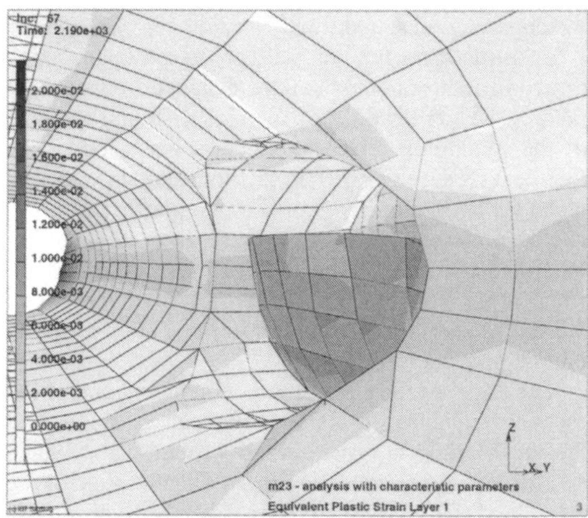

Figure 7. Equivalent plastic strains in the soil at the beginning of the excavation of the connection tunnel (analysis step 67). (Displacements magnified 50 times.)

4.2 *Stresses and plastic strains in the soil*

Figures 6 and 7 provide a perspective view of the intersection of the platform tunnel of track 2 with the connection tunnel the excavation of which has just begun. Figure 6 shows the vertical stresses in the soil. Stress concentrations at the corners of the intersections (increase of 300 to 400 kN/m^2) and stress relief at the free surfaces at the faces are clearly visible. The equivalent plastic strains of Figure 7 show slight plastifications throughout the side walls of the platform tunnel. They augment to about 1% at the unsupported faces whereas there is practically no increase at the corners of the intersection.

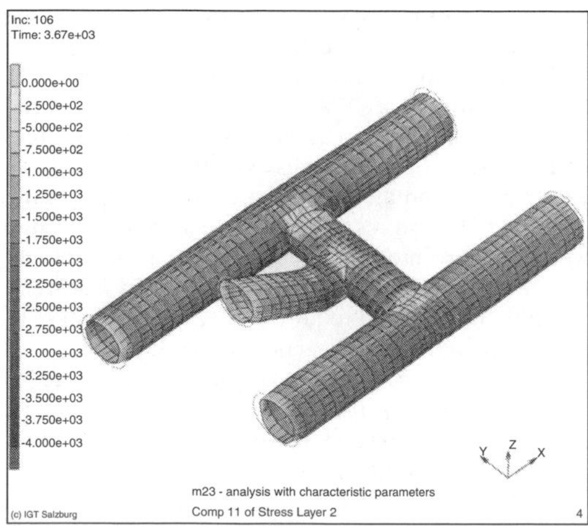

Figure 8. Normal stresses in the middle layer of the shotcrete lining in circumferential direction at the end of the analysis.

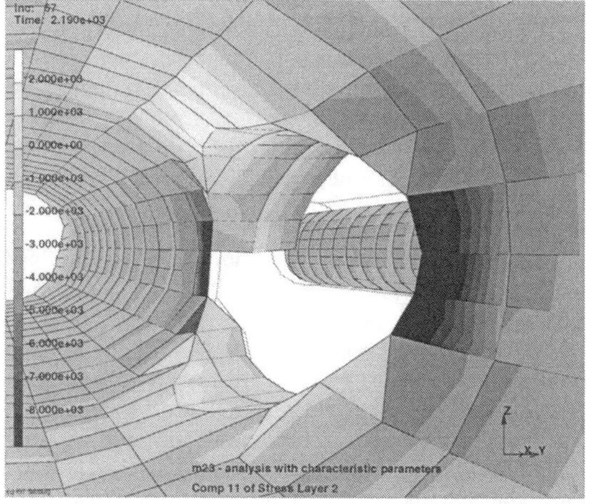

Figure 9a. Normal stresses in the middle layer of the shotcrete lining in circumferential direction at analysis step 67. (Displacements magnified 50 times.)

The load steps with the maximum plastic strains coincide with the load steps with the maximum displacements of the face. During excavation of the connection tunnel maximum displacements of 76 mm are calculated for the face of the bench and invert in analysis step 77; the maximum face displacements for the escalator tunnel are 56 mm, again at the bench.

4.3 *Stresses in the shotcrete lining*

The following figures contain the normal stresses in circumferential direction in the middle layer of the lining. The maximum compressive stresses occur at the side walls at the corners of the intersections. The largest stress increase due to the excavation occurs at the start of the driving of the connection tunnel in the lining of the platform tunnel of track 2. As can be observed in Figures 9a–c the application of the shotcrete lining in the connection tunnel together with creep effects diminishes the stresses. The stresses in the young shotcrete of the lining of the connection tunnel are considerably lower than those in the platform tunnel.

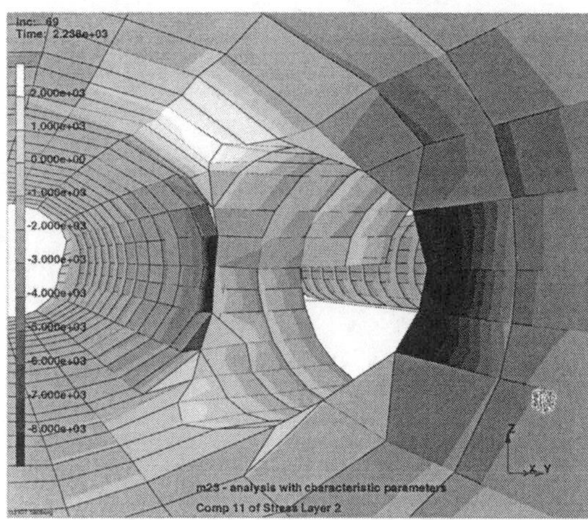

Figure 9b. Normal stresses in the middle layer of the shotcrete lining in circumferential direction at analysis step 69.

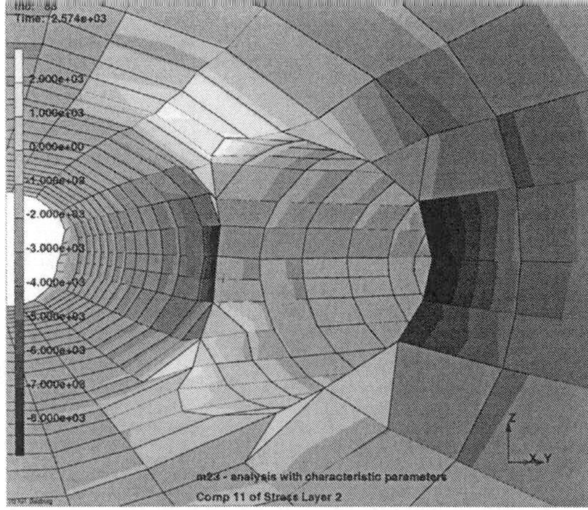

Figure 9c. Normal stresses in the middle layer of the shotcrete lining in circumferential direction at the analysis step 83.

The moments in the regions of maximum normal stresses in the middle layer are relatively small (Fig. 10). Obviously the stress level is so close to

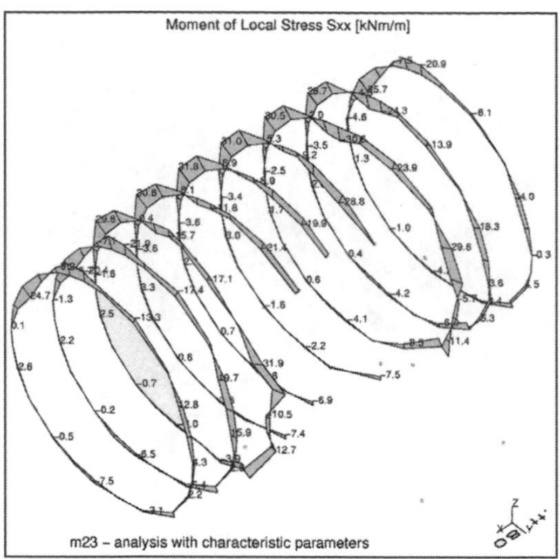

Figure 10. Bending moments in the lining resulting from stresses in circumferential direction at the end of the excavation of the connection tunnel.

the compressive strength there that the moments are reduced by creep effects and that stress redistribution within the cross section occurs.

The normal stresses in longitudinal direction increase in the vicinity of the intersections as well. However, the stress level is considerably lower than in circumferential direction. High tensile stresses at the roofs of the intersections match the corresponding tensile stresses in circumferential direction of the second part of the intersection.

Creep causes a stress reduction of about 10% during the observed time period in general. At the locations with stress levels close to the compressive strength the creep effects are more pronounced, as had to be expected.

5 DESIGN OF THE SHOTCRETE LINING, ADDITIONAL ANALYSES

The results of the previous chapter show that both, the soil and the shotcrete receive loads up to their load carrying capacity in the course of the excavation. It is clearly visible that the regions with high stress levels are locally confined. The plasticity-based mechanisms for dealing with stresses at yield built into the constitutive laws for soil and shotcrete enforce stress redistributions and prevent overloading of the soil or the shotcrete. Creep effects reduce the maximum stresses in the lining further. The good natured convergence behaviour at all analysis steps indicates that the "structure" – consisting of soil and shotcrete lining – is able to carry the applied loads and that the capacity for stress redistribution is sufficient.

According to the design rules of common standards, e.g. the Eurocodes (CEN 2001, CEN 2002), however, stress concentrations in the shotcrete lining with stress levels close to the compressive strength would not be permitted and the design be considered as unsafe. There are also cross sections where a standard design would require more reinforcement than the amount which can be placed on site. Additionally, Eurocode 7 (CEN 2001) would require a check on the safety of the soil in addition to the safety of the lining.

The question now arises whether the "structure" is safe in the sense of the standards, whether there is a sufficient safety margin against failure, i.e. whether the probability of failure is low enough.

In order to check the available safety margins a number of different approaches can be considered, among them:

- Analyses with characteristic parameters for shotcrete and soil, followed by a standard design of the reinforcement as described above.
- Analyses with increased loads, e.g. by increasing the specific weight and the building loads, and reducing the strength properties of the shotcrete.
- Decreasing the strength of the soil and reducing the strength properties of the shotcrete.
- Reducing only the strength of the shotcrete, but by an increased amount compared with the approach above.

The first approach is not applicable, at least not for all parts of the structure, for the reasons described above.

The second and the third approaches fit into the concept of applying partial safety factors to actions and resistances, respectively (CEN 2001, CEN 2002). The fourth approach resembles the concept of a global safety factor used in many older standards, and is similar to the first approach. In both approaches the safety of the soil is not investigated. The first approach has the advantage of yielding realistic displacements and deformations, whereas the fourth approach contains a thorough check on the load-redistribution capacity of the structure including also the soil.

In the opinion of the author the second approach is far off the physical reality. It has not been further investigated.

The third and fourth approaches are investigated by additional analyses. For the analyses the following safety factors have been chosen in accordance with (CEN 2001, CEN 2002) and applied to the strength parameters of the soil, cohesion and tangent of the angle of friction, and of the shotcrete, yield and ultimate stress in tension and compression.

A reinforcement has not been specified in the analyses in order to avoid an overestimation of the load

Table 3. Safety factors used in different approaches.

(Partial) safety factor	First approach	Second approach	Third approach	Fourth approach
Soil – cohesion	1.0	1.0	1.25	1.0
Soil – shearing resistance	1.0	1.0	1.25	1.0
Soil – dead weight	1.0	1.35	1.0	1.0
External loads	1.0	1.35	1.0	1.0
Analysis: shortcrete tensile strength	1.0	1.5	1.5	1.7
Analysis: shotcrete compressive strength	1.0	1.5	1.5	1.7
Internal forces and moments	1.35	1.0	1.0	1.0
Cross section design: Yield stress of reinforcement	1.15	1.0	1.0	1.0
Design: Compressive strength of the shotcrete	1.5	1.0	1.0	1.0

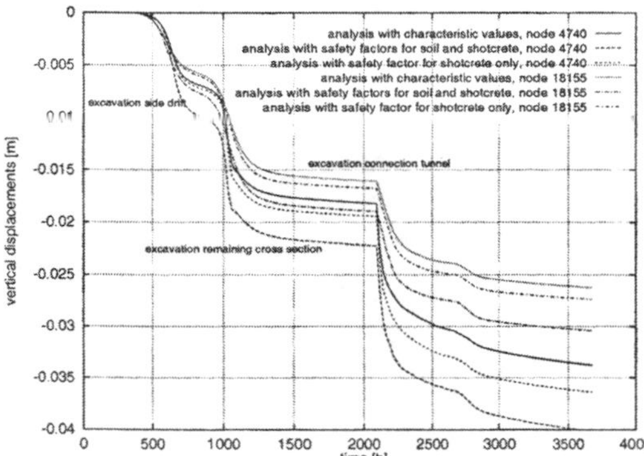

Figure 11. Vertical displacements at the roof of the tunnel (node 4740) and at basement level (node 18155) at the intersection of track 2 and the connection tunnel.

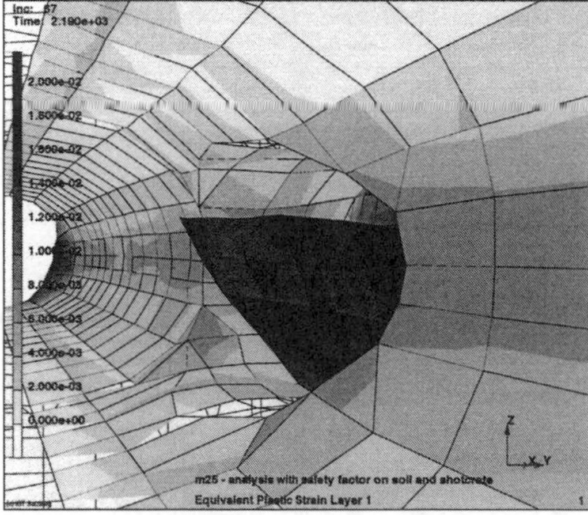

Figure 12b. Third approach: Normal stresses in circumferential direction in the middle layer of the lining at step 67.

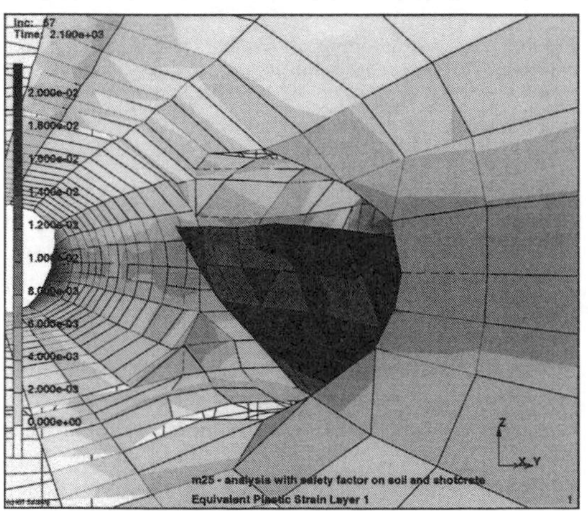

Figure 12a. Third approach: Equivalent plastic strains in the soil at step 67.

carrying capacity in tension. (The softening in the model is only a coarse approximation of the brittleness of the shotcrete in tension). Therefore, the safety factors for the resistance of the lining chosen for the analysis and those chosen for the design of the cross section are not directly comparable.

The differences between the approaches can be visualized by comparing the Figures 12 (for the third approach) and 13 (for the fourth approach) showing results for step 67, with the Figures 7 and 9a (for the first approach).

The displacements, especially the inward movement of the face of the bench, are higher than in the first approach. One reason is that higher plastic strains develop in the third approach because of the smaller yield surface; another reason is the higher stress level in the soil in the fourth approach because of the weaker lining. Both in terms of displacements and plastic

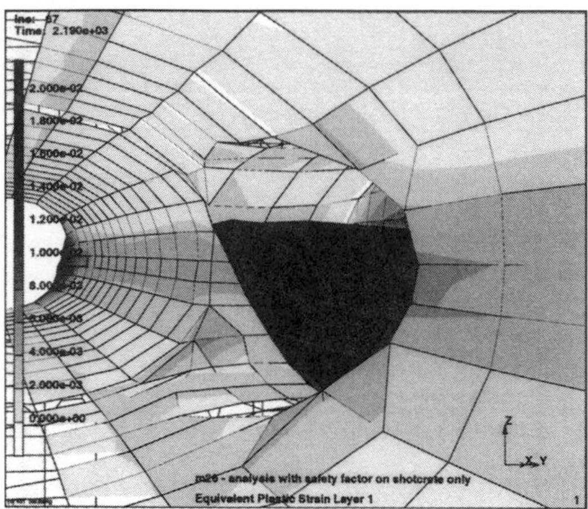

Figure 13a. Fourth approach: Equivalent plastic strains in the soil at step 67.

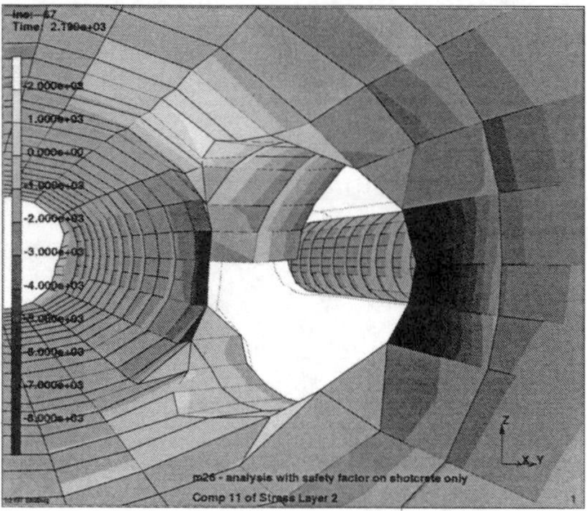

Figure 13b. Fourth approach: Normal stresses in circumferential direction in the middle layer of the lining at step 67.

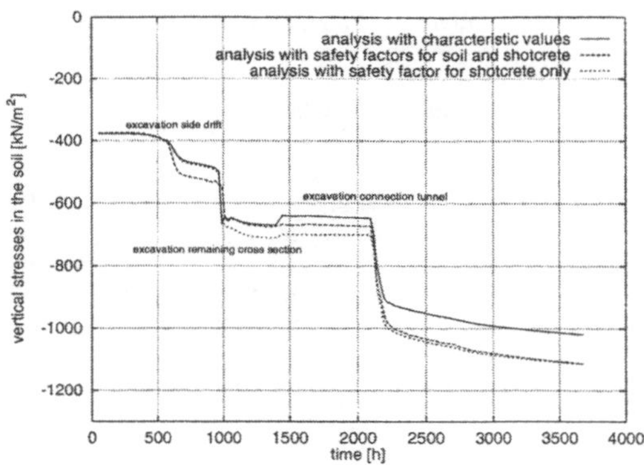

Figure 14. Vertical normal stresses at the bench in the corner with positive x-coordinate of the intersection of track 2 and the connection tunnel.

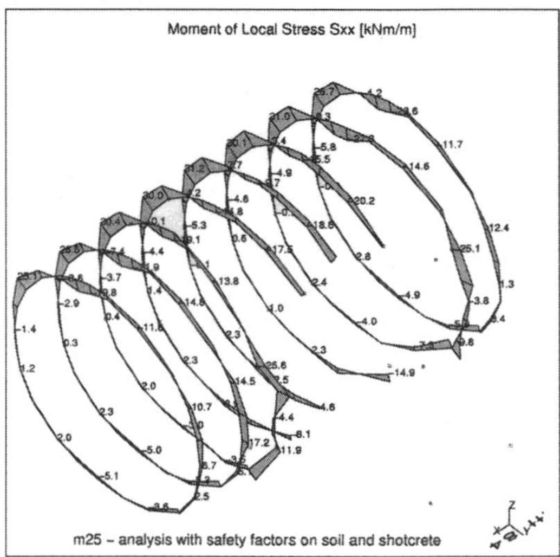

Figure 15a. Third approach (safety factor on soil and shotcrete): Moments in circumferential direction in the lining of track 2 in the vicinity of the intersection with the connection tunnel at the end of the excavation of the connection tunnel.

strains, the third approach yields more unfavourable results than the fourth approach.

The vertical stresses in the soil are very similar: the lowest stresses yields the first approach, there is almost no difference between the third and fourth approach (Fig. 14).

The smaller the compressive strength of the shotcrete, the lower are the maximum normal stresses and the wider is the area of increased stress levels in the shotcrete lining. Due to the local confinement of the stress maxima the differences are not very obvious on the plots. (More pronounced are the differences in the regular parts of the platform tunnels visible in the back of Figures 12b and 13b. The differences there originate from the different load carrying capacity of the soil and the resulting stress transfer to the lining.)

The moment diagrams in Figure 15 confirm the observations made at Figure 10. At cross sections with high normal stresses the moments are the smaller the smaller the compressive strength of the shotcrete because the whole cross section has reached its load carrying capacity. At cross sections with smaller normal forces the moments of the first and third approaches are very similar, only the fourth approach shows visibly smaller moments.

The number of iteration steps required for convergence is slightly increased in the third approach which

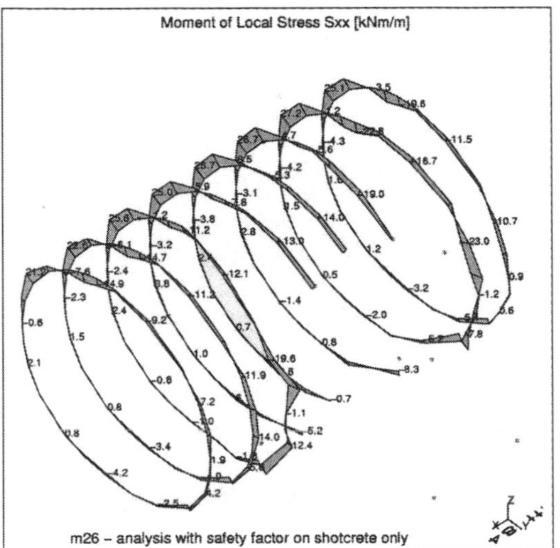

Figure 15b. Fourth approach (safety factor on shotcrete only). Moments in circumferential direction in the lining of track 2 in the vicinity of the intersection with the connection tunnel at the end of the excavation of the connection tunnel.

is obviously due to the larger amount of plastification at the free surfaces of the soil. None of the analyses indicates a divergent behaviour.

6 SUMMARY AND CONCLUSIONS

A first analysis of a metro station with characteristic material parameters showed small regions at the intersections of the tunnels, where the safety margins were not sufficient in terms of a standard design. Additional analyses with reduced strength parameters of soil and/or shotcrete confirmed that the regions with high stress levels are locally confined and do not endanger the convergence of the analyses. The shotcrete lining has been designed safely and safety margins comparable to those of the standards are present.

Some problems with the additional analyses still remain: Using safety margins on actions or resistances results in models which deviate from reality. There are many possibilities of applying safety margins, with results which are not directly comparable. Applying statistical methods (Thurner 2001) might be a remedy, but would currently be much too expensive for three-dimensional analyses.

REFERENCES

Betonverein 1998. *Richtlinie Spritzbeton, Anwendung und Prüfung.* Vienna: Österreichische Vereinigung für Betonbau und Bautechnik.

CEN 2001. *prEN 1997-1: Eurocode 7: Geotechnical design – Part 1: general rules*; final draft, October 2001, Brussels: European Committee for Standardization.

CEN 2002. *prEN 1992-1-1: Eurocode 2: design of concrete structures – Part 1: General rules and rules for buildings*; revised final draft, April 2002, Brussels: European Committee for Standardization.

England, G.L., & Illston, J.M. 1965. Methods of Computing Stress in Concrete from a History of Measured Strain. *Civil Engineering and Public Works Review* 60, pp. 513–517, 692–694, 846–847.

Meschke, G. 1996. Consideration of Ageing of Shotcrete in the Context of a 3-D Viscoplastic Material Model. *Int. Journal for Numerical Methods in Engineering* 39, 3123–3143.

Meschke, G., Kropik, C. & Mang, H.A. 1996. Numerical Analyses of Tunnel Linings by Means of a Viscoplastic Material Model for Shotcrete. *Int. Journal for Numerical Methods in Engineering* 39, 3145–3162.

Mang, H.A. (ed.) 1994. *Tunnelabzweigungen bei kriechaktiven Bodenverhältnissen*, Straß enforschung, Heft 426. Vienna: Bundesministerium für wirtschaftliche Angelegenheiten.

Schubert, P. 1988. Beitrag zum rheologischen Verhalten von Spritzbeton. *Felsbau* 6, 150–153.

Thurner, R. 2001. *Probabilistische Untersuchungen in der Geotechnik mittels deterministischer Finite Elemente-Methode.* Dissertation, Technical University Graz, Austria.

Walter, H. (ed.) 1996. *Praxisorientiertes Rechenmodell für Tunnel und Kavernen.* Final report to the Forschungsförderungsfonds der gewerblichen Wirtschaft. Vienna.

Walter, H. 1997. Application of a new shotcrete model in a 3 D-FE-analysis of a tunnel excavation. In S. Pietruszczak, G. N. Pande (eds.), *Numerical Models in Geomechanics, NUMOG 6; Proc. 6th intern. Symp. Montreal 2–4 July 1997.* Rotterdam: Balkema.

Wiener, L. 2002. *U-Bahn-Linie U2. Bauabschnitt U2/2. Ausschreibungsunterlagen.* Vienna.

Modelling of seismic and cyclic behaviour

Computational Modelling of Concrete Structures, Bićanić et al. (eds)
© 2003 Swets & Zeitlinger, Lisse, ISBN 90 5809 536 3

Keynote paper: Numerical modelling for earthquake engineering: the case of lightly RC structural walls

J. Mazars, A. Colombo & P. Kotronis
Laboratoire Sols, Solides, Structures INPG and RNVO network, Grenoble, France

N. Ile
INSA Lyon, France

G. Casaux & F. Ragueneau
LMT Cachan, France

M. Fischinger & D. Marusic
Institute of Structural and Earthquake Engineering, University of Ljubljana, Slovenia

ABSTRACT: Different type of analytical models exist to describe the non-linear behaviour of reinforced concrete structures. Based on the level of discretisation, they are often classified as refined or simplified ones. The reliability and the efficacy of the different classes of models in describing the global and the local behaviour of lightly RC structural walls subjected to seismic loadings are investigated in this paper. The results of an experimental campaign carried out as a part of the CAMUS 2000 research program are used as a reference.

1 INTRODUCTION

Reinforced concrete bearing walls with limited reinforcement ratios are commonly used in France and other European countries for building structures. Research up to date on this type of walls (e.g. CASSBA, CAMUS programs) has shown that such kind of constructions exhibited good behaviour under seismic loading, although their ductility might be limited due to the light reinforcement and the large wall regions having practically no reinforcement (Mazars 1998). Results of these research programs highlighted some unconventional mechanisms of earthquake resistance, such as the rigid block-type rotations of the walls (or part of) that take place at the interface between the foundation and the soil or at the level of cold construction joints. Consequences of this behaviour mechanism are the high frequency vertical vibrations excited in the walls due to the opening and closing of wide horizontal cracks, the conversion of part of the seismic energy into potential energy, etc.

On the other hand, in case of real structures such walls are likely to be loaded in the out of plane direction as well. Due to the out of plane loading, additional fluctuation of the axial force may arise and more damage is to be expected since cracks may not completely close at load reversal. It is clear that all these effects are hard to quantify within the framework of conventional elastic seismic analysis. Reliable numerical tools become more than necessary to assist engineers in the design phase.

Many analytical models have been proposed upto date for the non-linear analysis of reinforced concrete structures. They range from very refined and complex local ones to simplified global models. The level of refinement is usually correlated to the dimension of the problem. Refined analytical models are typically used in predicting the response of small structures or substructures, whereas simplified global models are useful for the dynamic response analysis of large structures.

In this paper the performance of both refined (a 3D finite element thin shell modellisation based on the use of shells, bars and solid elements) and simplified models (a fibre model and a beam model with multiple integration points) is evaluated by comparing the numerical with the experimental results of a five-storey lightly reinforced concrete wall submitted to dynamic loading. This experimental campaign was performed within the CAMUS 2000 research program presented hereafter.

1.1 *The CAMUS 2000 research program*

The 3-years combined experimental and numerical research program CAMUS 2000 was launched in 1998

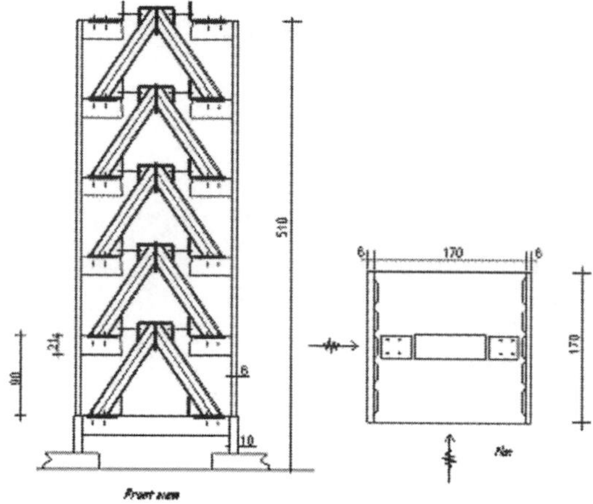

Figure 1. Layout of the specimen.

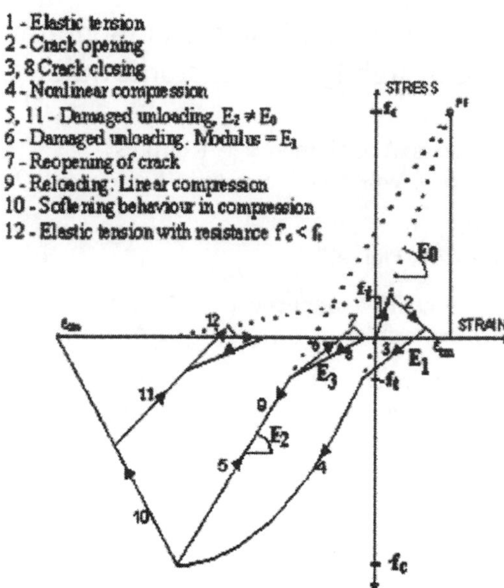

Figure 2. Uniaxial model: point initially in tension.

with the aim of evaluating the effects of torsion and the behaviour of lightly reinforced walls subjected to bi-directional motions.

As a part of the program two scaled (1/3) models representative of a five-storeys reinforced concrete building (Fig. 1) have been tested on the major Azalee shaking table of Commissariat à l'Energie Atomique (CEA) in the Saclay Nuclear Centre. The first structure (CAMUS 2000-1) has been subjected to a horizontal bi-directional excitation. A set of accelerograms was applied at increasing amplitude (effective acceleration equal to 0.15 g, 0.22 g, 0.25 g, 0.40 g, 0.55 g and 0.65 g). Structural stiffness was provided by two identical shear walls in one direction and a steel bracing system in the orthogonal direction. For the second test (CAMUS 2000-2), an in plane excitation was applied. A torsional response was caused by the asymmetry in the horizontal dimensions of the two walls.

2 3D THIN SHELL MODELLING

2.1 *Material modelling*

In order to achieve a good compromise between simplicity and accuracy a biaxial concrete model that provides acceptable representation of the cyclic inelastic behaviour of reinforced concrete under cyclic loading was used. This model (Merabet & Reynouard 1999) adopts the concept of a smeared crack approach with a possible double cracking only at 90°. It is based upon the plasticity theory for uncracked concrete with isotropic hardening and associated flow rule. Two distinct criteria describe the failure surface: Nadai in compression and bi-compression and Rankine in tension. Hardening is isotropic and an associated flow rule is used. When the ultimate surface is reached in tension, a crack is created perpendicularly to the principal direction of maximum tensile stress, and its orientation is considered as fixed subsequently. Each direction is then processed independently by a cyclic uniaxial law, and the stress tensor in the local co-ordinate system defined by the direction of the cracks is completed by the shear stress, elastically calculated with a reduced shear modulus μG, (with $0 < \mu < 1$, and μ being a function of the crack opening strain) to account for the effect of interface shear transfer:

$$\mu = 0.4 \quad \text{if} \quad \varepsilon_{cr} - \varepsilon_{res} - \varepsilon_{tm} = 2\,\varepsilon_{tm} \tag{1}$$

$$\mu = 0 \quad \text{if} \quad \varepsilon_{cr} - \varepsilon_{res} - \varepsilon_{tm} = 2\,\varepsilon_{tm} \tag{2}$$

$$\mu = 0 \text{ and } \sigma_{12} = 0 \quad \text{if} \quad \varepsilon_{cr} - \varepsilon_{res} - \varepsilon_{tm} = 4\,\varepsilon_{tm} \tag{3}$$

in which: ε_{cr} is the total strain, ε_{res} the residual strain after unloading in compression, ε_{tm} the crack opening strain and σ_{12} the shear stress.

The behaviour of a point initially under tension, which completely cracks prior to undergoing a reverse loading in compression, is illustrated in Figure 2. Similar laws describe the case of an initial compressed point or that of a point which has not totally cracked under a reverse loading. The model has been described in detail and verified elsewhere (Ile 2000), (Ile & Reynouard 2000), (Fleury 1996).

For steel, a cyclic model that can take into account the Bauschinger effect and buckling of reinforcing bars has been adopted. The monotonic branch is characterised by an initial linear branch followed by a plateau and hardening up to failure. The cyclic behaviour is described by the formulation proposed

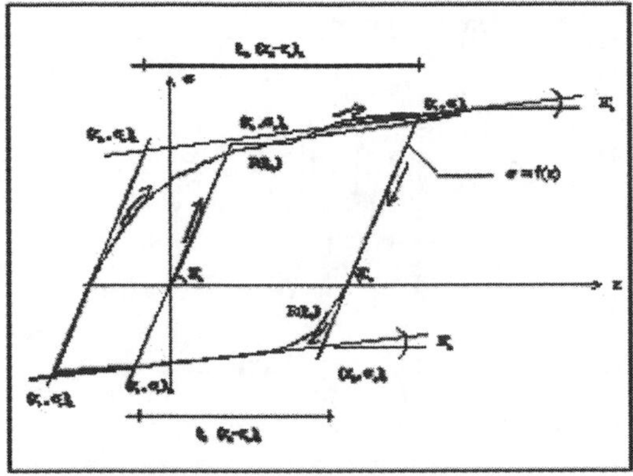

a) cyclic behaviour without buckling

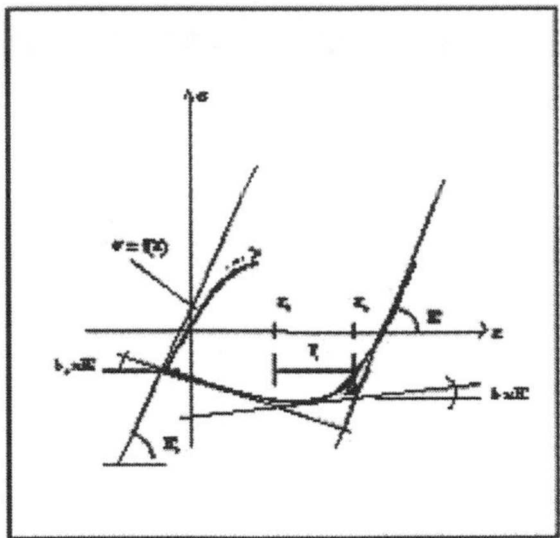

b) cyclic behaviour with buckling

Figure 3. Numerical model for steel under cyclic loading.

by Giuffré and Pinto and implemented by Menegoto & Pinto (1973). The steel model is presented in Figure 3.

The evaluation of steel parameters was quite simple, the measured elasticity modulus – 200000 MPa, yield stress – 664 MPa, failure stress – 733 MPa and strain at failure – 2.2%, being directly used. For concrete, the material properties were generally as provided by the CAMUS 2000-1 experimental data (Fouré 2001) and design codes: 23000 MPa for the initial elastic modulus, 34 MPa for the compressive strength and 2.6 MPa for the tensile strength. An initial value of 0.40 for the post-cracking parameter μ was assumed in the analysis.

2.2 *Model description*

The numerical analyses have been performed using the general-purpose finite element program CASTEM 2000 developed at CEA-Saclay (Millard 1993).

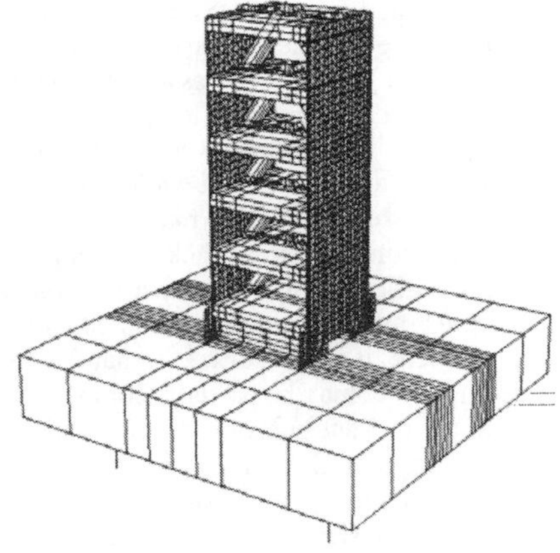

Figure 4. 3-D finite element mesh of the CAMUS 2001 specimen including the shaking table.

To predict the inelastic seismic response of the tested structures with sufficient accuracy, special care has been taken to create detailed models of the specimens, taking into account the necessary geometric characteristics, construction details and boundary conditions. An example of the 3-D finite element mesh used in the analyses is reported in Figure 4.

Due to the direction of the applied loading, in plane as well as out of plane behaviour of the walls needs to be analysed. Therefore, layered thin shell discrete Kirchoff triangles (DKT) are used to represent the two walls. The slabs, as well as the shaking table are modelled with solid eight-node brick elements, while four-node shell elements are used to model the steel I-shaped bracing system. A discrete modelling is adopted to represent the horizontal and vertical reinforcement through the use of two-node truss-bar elements. The structure is assumed fully restrained at all nodes along the base of the shear wall, but the shaking-table was considered as a rigid block fixed to 4 vertical restraining rods, the axial stiffness of each rod being estimated equal to 215 MN/m. Perfect bond was assumed to exist between concrete and reinforcement. The possibility of non-linear material behaviour was specified for all the wall concrete elements and reinforcing bar-elements, while the behaviour of the slabs, bracing system and shaking-table was considered as elastic.

Assuming a 2% critical damping factor (close to the measured value) for the first and second vibration mode, the damping parameters α and β were calculated and used subsequently to form the Rayleigh damping matrix $[C] = \alpha\ [M] + \beta\ [K]$, M and K being the mass and stiffness matrix. Despite the fact that a modal characterisation is theoretically correct only for linear elastic systems, the damping matrix [C] obtained in this way is assumed to remain constant throughout the loading cycle. Since with increasing damage, modal frequencies decrease due to cracking and reinforcement yielding, this assumption may lead to unrealistic over damping on lower modes. When several sequential input motions are considered in a seismic analysis, the best choice would probably be to reduce the Rayleigh damping matrix with increasing applied motion. The drawback of this approach relies on the difficulty of selecting a suitable value for the damping matrix. Notwithstanding the complexity of the problem, it seems however, that the viscous over damping effect may be partially compensated by the fact that the present non-linear model cannot take properly into account all sources of hysteretic damping: unilateral cracking, shear slip between the lips of the crack, bond–slip between steel and concrete, etc. In any case, since one of the aims of the present study was to test the capabilities of the existing cyclic model, all modelling decisions (including the choice of a constant damping matrix) were made before all calculations were executed and no tuning of the analysis was done in an attempt to obtain a better fit to the experimental results. To solve the non-linear equilibrium equations, a modified Newton–Raphson iteration solution scheme was used.

2.3 *Main results*

In order to take into account the effect of accumulation of damage a series of 4 non-linear time-history dynamic analyses were sequentially performed. The following PGA levels of the uncorrelated input signals applied during the test (in direction x and y) were considered in the numerical analyses: 0.22 g, 0.40 g, 0.55 g and 0.65 g. A large number of results, which are still under investigation, have been produced. For the sake of brevity, only the main results related to the last input motion (PGA = 0.65 g), which caused failure of the specimen are presented and discussed here.

The time history of the calculated roof displacement is compared in Figure 5 with the corresponding measured displacement. It can be observed that the maximum in-plane horizontal displacements are reasonably predicted (here given at the top of the left wall for both 0.55 g and 0.65 g), while the model generally underestimates the out of plane displacement.

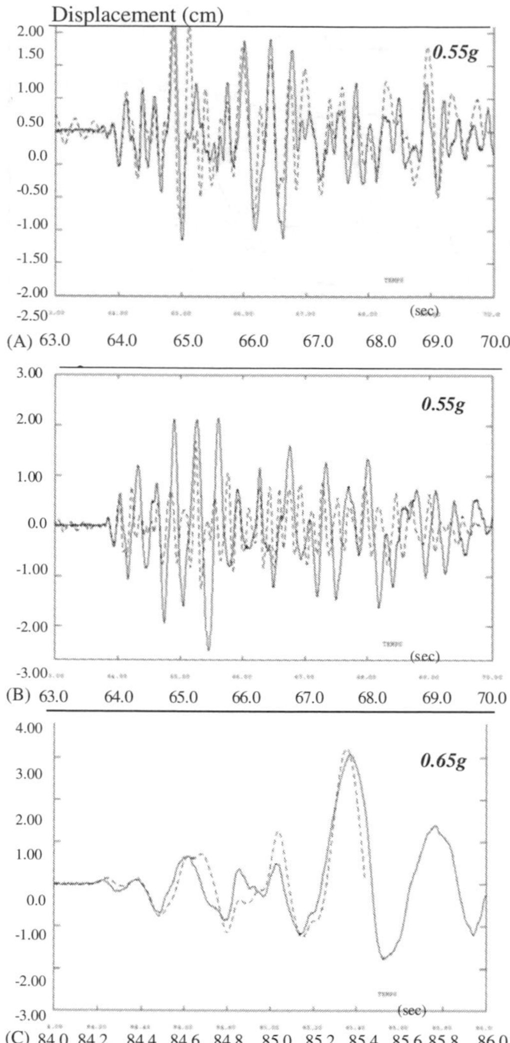

Figure 5. Comparison between calculated and measured horizontal top displacements: (A) in plane displacement (0.55 g) – left wall; (B) out-of-plane displacement (0.55 g). (C) in plane displacement (0.65 g) – left wall.

Actually, during this last test, the failure of the connection of the bracing system with the slab was also obtained, one branch of the bracing system being completely dislocated from the slab. This was probably due to a higher out of plane shear force than anticipated in design. Since the behaviour of this connection was considered as elastic in the analysis, the observed discrepancies between experimental results and analysis seem normal.

For the last test on CAMUS 2000-1, a large variation of the axial force has been observed, its amplitude, at the base of the first story, being comparable with the static axial force, which is about 165 KN for one wall. One part of the axial force variation, denoted here as "dynamic axial force", comes from the extension mode caused by the in plane bending of the walls: at maximum horizontal deflection, the neutral axis is at its maximum distance from the centre of the wall cross-section, the raising of masses is maximum and the dynamic variation of the axial force is a tensile force; the frequency of this vertical motion is two times that of the horizontal movement;

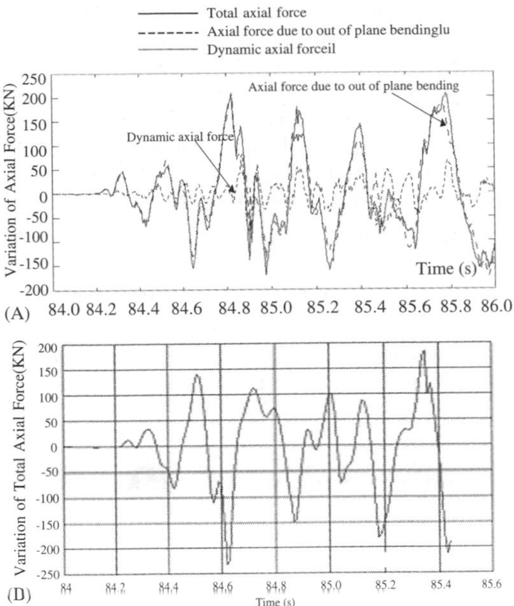

Figure 6. Variation of the axial force at the base of the first story (PGA = 0.65 g) for one wall: (A) experimental and (B) numerical results.

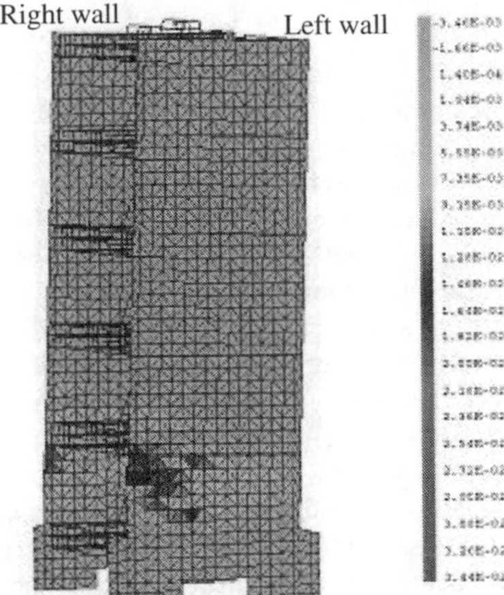

Vertical concrete strain contours at failure

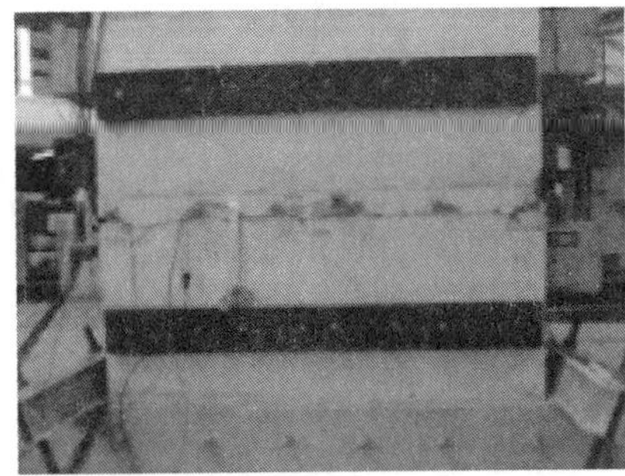

Figure 7. Comparison between test and analysis at failure: (A) numerical vertical concrete strain contour at failure, (B) crack pattern at the end of the test.

then, at cracks closure, when concrete recovers its stiffness, compression forces strongly increase and these shocks excite the vertical vibration mode of the system (shaking table + specimen), which is about 20 Hz. This variation was also observed during the previous CAMUS in plane seismic tests (Bisch & Coin 1998).

The other part of the axial force variation comes from the out of plane bending, which induces a complementary compressive axial force in one wall and a tensile axial force in the other. The total axial force variation in each wall is then obtained by summing the corresponding values of the dynamic axial force and of the axial force due to the out of plane bending. The different components of the experimental axial force variation in the left wall, as obtained from the last test (PGA = 0.65 g) are shown in Figure 6.

When comparing these results it can be noted that the total variation of the axial force is mainly due to the out of plane bending component which is several times larger than the dynamic axial force. This is due to the fact that the specimen was subjected perpendicular to the walls plane (y-direction) to a PGA level as high as that applied in the direction of the walls plane (x-direction) and to the fact that the frequency of the out of plane vibration mode was close to thatcorresponding to the in plane vibration mode (around 6 Hz).

For comparative purposes the variation of the total axial force in the left wall and at the base of the first story as deduced from analysis is presented in Figure 6B. As the analysis was stopped after rupture of one reinforcing bar, the analysis time scale is not as large as that corresponding to the previous figure. By comparing Figure 6A and B it can be noted that the maximum variation of the dynamic axial force is comparable with the measured values.

Local results as obtained from the dynamic analysis are seen in Figure 7. This figure depicts the damage distribution obtained just after one of the steel bars of the left wall has broken in tension.

It is to be noted that cracks are almost horizontally distributed in the two walls, but the crack opening is larger for the wall, which is in tension. The analysis also showed that damage is almost entirely concentrated just under the 2nd floor level, level where the main cracks develop. Actually, inspection after the test showed that the steel bars were broken below the second floor level and a large horizontal crack develops at the same level.

3 FIBRE MODEL

Non-linear dynamic analysis of civil engineering structures requires large scale calculations, implying delicate solving techniques. The necessity to perform parametrical studies led us to adopt special solutions in order to reduce the computational cost. The response of a structure submitted to severe loadings, depends on a strong interaction between "material" (local non-linearities), "structural"(geometry, mass distribution, joints) and "environment" (interaction of the structure with its support) effects. For concrete structures, the local material behaviours are the major sources of non-linearities. The wish to keep good prediction ability for the model guided us to use refined constitutive equations, taking into account the main physical phenomena (damage, inelasticity, crack-reclosure, . . .). At a structural level, the choice of a "simplified approach" has been made by applying simplifying assumptions compromising as little as possible the quality of results.

The choice of using a multifiber F.E. configuration combines the advantage of using beam type finite elements with the simplicity of uniaxial behaviour. Each finite element is a beam discredited into several fibres (see Figure 8). The F.E. code used is again CASTEM 2000.

A first series of calculations (modal analysis) have been performed in order to insure that the main phenomena of the problem have been well represented. Figure 9 gives a comparison of the multifiber approach with both the experimental and the fully 3D calculation results for the first three natural modes.

As it has been already discussed, the shaking table – due to its softness – must also be included to the mesh (using 3D elements or orthogonal beams for the two models respectively) to insure the good correlation of the calculations. The results obtained with the fiber model are as good as those obtained with the 3D FEM model, with a much less significant cost.

3.1 *Material modelling*

Material modelling for reinforced concrete structures submitted to dynamic loading has to take into account the decrease of stiffness as well as plasticity of the reinforcements till failure.

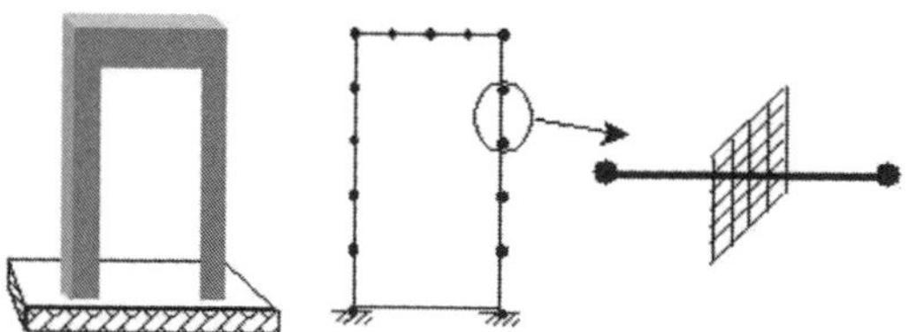

Figure 8. Multifiber discretisation principle.

In describing the non-linear behaviour of reinforcement rebars, a classical plasticity model taking into account a non-linear cinematic hardening (Armstrong et al., 1966) has been implemented. Response under uniaxial cyclic loading is presented in Figure 10.

Seismic loading, which includes cyclic aspects, produces micro cracking in concrete. Some major phenomena have to be taken into account:

- decrease in material stiffness as the micro cracks open,
- stiffness recovery as crack closure occurs, inelastic strains concomitant to damage.

To account for such a behaviour we adopt a continuum damage model (La Borderie 1991), which

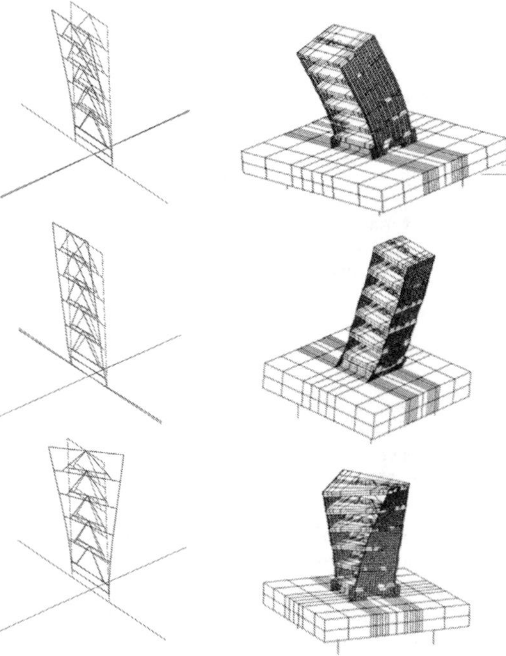

Figure 9. Natural modes: on the left fibre model on the right 3D model. From the top to the base: In plane: fibre: 6.00 hz, 3D: 6.14 Hz, exp: 6.00 Hz. Out of plane: fibre: 5.5 Hz, 3D: 6.07 Hz, exp.: 5.45 Hz. Torsion: fibre: 10.5 Hz, 3D: 10.73, exp.:

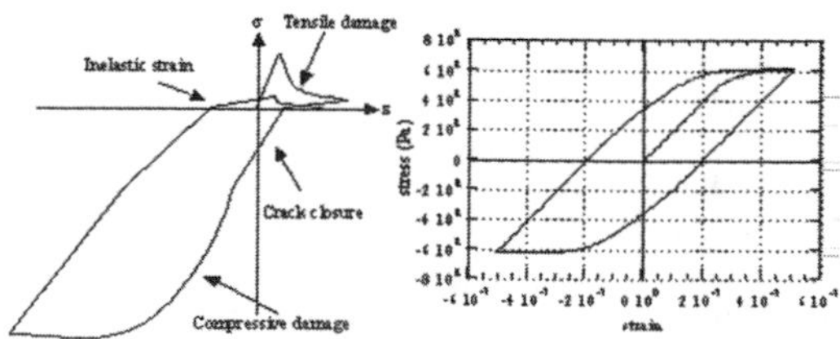

Figure 10. Uniaxial stress–strain relations for concrete and steel.

incorporates two scalar damage variables, one for damage due to tension D_1, the other for damage due to compression D_2 and which includes a recovery stiffness procedure and the description of isotropic inelastic strain. Constitutive equations are: total strain: $\varepsilon = \varepsilon^e + \varepsilon^{in}$

$$\varepsilon^e = \frac{\langle\sigma\rangle_+}{E(1-D_1)} + \frac{\langle\sigma\rangle_-}{E(1-D_1)} + \frac{\upsilon}{E}(\sigma - Tr(\sigma)\mathbf{1}) \tag{4}$$

$$\varepsilon^{in} = \frac{\beta_1 D_1}{E(1-D_1)}\frac{\partial f(\sigma)}{\partial\sigma} + \frac{\beta_2 D_2}{E(1-D_2)}\mathbf{1} \tag{5}$$

with ε^e elastic strains and ε^{in} inelastic strains. $\mathbf{1}$ denotes the unit tensor and $Tr(\sigma) = \sigma_{ij}$.

Damage criteria are expressed as: $f_i = Y_i - Z_i$ with Y_i, associates forces to damage and Z_i the hardening variable. The evolution laws for damage take the following form:

$$D_i = 1 - \frac{1}{1+[A_i(Y_i - Y_{0i})]^{B_i}} \tag{6}$$

where $f(\sigma)$ and σ_f are the crack closure function and the crack closure stress respectively. $<>+$ denotes the positive part of a tensor. E is the initial Young's modulus and ν the Poisson ratio. D_1 and D_2 are respectively the damage variables for traction and compression. β_1 and β_2 are material constants. Figure 10 gives the stress–strain response of that model for a uniaxial traction-compression-traction loading.

3.2 *Model description*

The finite element mesh has already been presented in Figure 9. The additional masses and the weight load of each floor are concentrated at each story. The stiffness of the springs below the shaking table is identified so as to feet the first eigenmodes measured on the virgin structure before the seismic loadings. Despite the lack of physical meaning, damping is generally introduced in the analysis through viscous forces generated by the means of a damping matrix. This is the classical Viscous Rayleigh damping matrix, derived from the general expression proposed by Caughey (1960). The two parameters allow calibrating the matrix by imposing the value of the damping ratio for two eigenmodes of the virgin structure. The Rayleigh damping coefficients have been adjusted to ensure a value of 1% on the first mode and 2% on the second mode. Great attention has been focused to keep these damping values as stable as possible throughout the analysis. This remark may become important for concrete structures where cracking induces loss in stiffness and by that way a shift of the fundamental frequency. Therefore the damping of the first eigen mode has been chosen so as to remain around the minimum constant range of the Rayleigh diagram.

3.3 *Main results*

Results are presented in terms of horizontal top displacements in the plane (x-direction) of the walls (Fig. 11) and global flexural moment in the y-direction (Fig. 12) for the signal corresponding to 0.55 g.

These results have been obtained without calibration with the experimental results. Work is in progress on the effects of damping and improvements of the modelling are carried out to account for torsion and 3D material behaviours using an enhanced beam formulation.

4 BEAM MODEL

The realistic dynamic description and modelling of material failure is one of the actual problems in structural mechanics. Analyses of failure processes require the use of complex FE discretisation and advanced constitutive models. For modelling of concrete it is necessary to capture several important phenomena such as development of anisotropy, damage and ductility (possibly softening). This implies the description of the concrete behaviour with refined constitutive models. Such complex analyses demand for large scale computing, which must be feasible from the

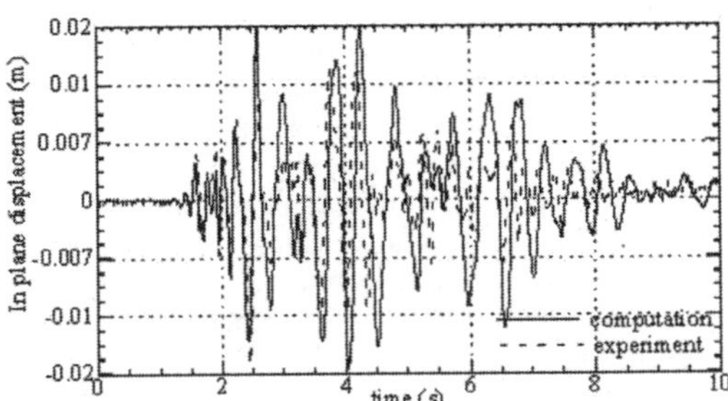

Figure 11. In plane top horizontal displacement: 0.55 g of maximum acceleration.

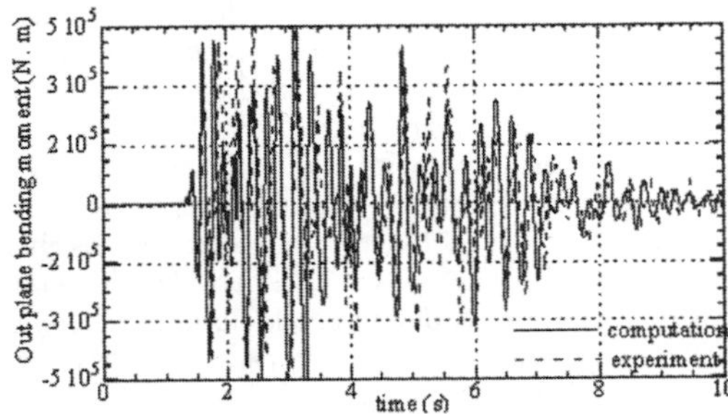

Figure 12. Out plane bending moment: 0.55 g of maximum acceleration.

view of both time and available resources. In this light the use of simplified modellisations coupled with explicit procedures became very popular for solving time dependent problems. Together with the limited number of numerical elements used in simplified approaches, the utilization of explicit procedures contributes to a general reduction of the time requested by the calculation, since they require no iterations and no tangent stiffness matrix. Even though they are in general conditionally stable, their stability is automatically assured by the small time increments required by the solution of dynamic problems.

To verify the efficiency of both simplified approaches and explicit methods in describing the seismic behaviour of reinforced concrete wall structures a three-dimensional finite element model of the building was realized by using only beam elements with multiple integration points. This model differs from the one presented in the previous paragraph in the fact that structural members are characterised by a homogeneous cross-section. The explicit version of the commercial computer code ABAQUS version 6.1 has been chosen for the analyses.

The main characteristics of the element chosen for the modelling of the spatial behaviour of the structure are listed below.

- The beams used are Timoshenko beams that allow for transverse shear deformation. The transverse shear deformation is treated as if the response were linear elastic, independent of the axial and bending responses, even for non-linear material behaviour.
- A lumped mass formulation is used.
- The cross-section of each beam is integrated numerically to obtain the force-moment/strain-curvature relations for the section, which allows complete generality in material response, since each point of a section is considered independently by the constitutive routines.

4.1 *Material modelling*

The elastic–plastic model implemented in ABAQUS has been used to describe the behaviour of steel members. The cyclic behaviour of concrete has been represented by using the PRM model (Mazars 1986, Pontiroli 1995, Rouquand & Mazars 2001). The analysis has been carried out with ABAQUS and the explicit version of the PRM model running in parallel.

The PRM model is based on constitutive laws able to account for the effects of stiffness deterioration, recovery of stiffness due to crack closure, effects of damage and strain rate. The main particularities of the model are given for a uniaxial loading as follows:

- Partition of strain and stress tensors:

$$\varepsilon = \varepsilon_d + \varepsilon_{ft} \quad \& \quad \sigma = \sigma_d + \sigma_{ft} \tag{7}$$

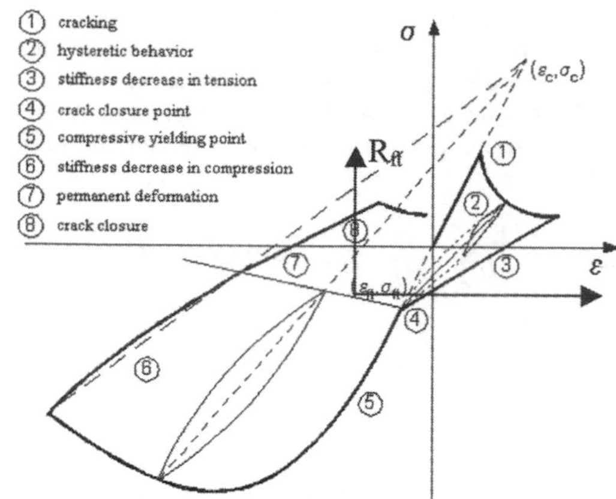

Figure 13. PRM model: stress-strain relationship.

- Combination of two modes of damage:

$$D = \alpha_t D_t + (1 - \alpha_t) D_c \tag{8}$$

α_t evolved in between 0 and 1 and the actual values depends on $(\varepsilon - \varepsilon_{ft})$

$$D_i = \text{fct}(\tilde{\varepsilon}, \varepsilon_{d0}, A_i, B_i) \quad \text{with } i = t, c \tag{9}$$

D_t damage in tension, D_c damage in compression. $\tilde{\varepsilon}$ is the equivalent deformation, (function of the local principal positive strain $\tilde{\varepsilon} = \sqrt{<\varepsilon_i>_+^2}$, Mazars 1986) which pilots the damage evolution after the initial threshold ε_{d0}, A_i, B_i are material parameters.

- Constitutive equations:

$$(\sigma - \sigma_{ft}) = \Lambda_0 \cdot (1-D) \cdot (\varepsilon - \varepsilon_{ft}) \tag{10}$$

In order to describe dissipation due to hysteretic loops (Fig. 13) a hysteretic stress term is added:

$$\sigma_{hyst} = (\beta_1 + \beta_2 D)\, E\, (1-D)\, (\varepsilon - \varepsilon_{ft})\, f(\varepsilon - \varepsilon_{ft}) \tag{11}$$

β_1 and β_2 are "Rayleigh" parameters and f is a function used to calibrate the evolution with the strain.

The problem becomes more simple if described in a new repair $R_{ft}(\sigma_{ft}, \varepsilon_{ft})$ as shown in Figure 13. The general 3D formulation of the model linking strain and stress tensors (in bold) is reported below:

$$\begin{aligned}(\boldsymbol{\sigma} - \boldsymbol{\sigma}_{ft}) &= \Lambda_0 (1-D) (\boldsymbol{\varepsilon} - \boldsymbol{\varepsilon}_{ft}) \\ &= \alpha_t (1-D_t) [\lambda_0 \,\text{trace}(\boldsymbol{\varepsilon} - \boldsymbol{\varepsilon}_{ft})\mathbf{1} + 2\mu_0 (\boldsymbol{\varepsilon} - \boldsymbol{\varepsilon}_{ft})] \\ &+ (1-\alpha_t)(1-D_c) [\lambda_0 \,\text{trace}(\boldsymbol{\varepsilon} - \boldsymbol{\varepsilon}_{ft})\mathbf{1} + 2\mu_0 (\boldsymbol{\varepsilon} - \boldsymbol{\varepsilon}_{ft})]\end{aligned} \tag{12}$$

where $\boldsymbol{\sigma}_{ft}$ and $\boldsymbol{\varepsilon}_{ft}$ are the crack closure stress and strain thresholds used to manage permanent effects; $\boldsymbol{\Lambda}_0$ is related to the initial mechanical characteristics.

One of the main advantages of this model is that it includes crack-closure effects, describes permanent strains, hysteretic loops and is expressed in an explicit form, compatible with the use of an explicit algorithm.

4.2 *Model description*

The same type of element has been used to represent the various structural components but with different sections.

- *Walls*. Rectangular section elements were chosen to model the concrete of the walls. 25 integration points characterize each section. The longitudinal reinforcement is accounted for by means of box section elements. Couples of bars at the same distance from the axis of the structural member were modelled by using the same numerical element – the area of the reinforcing bars has been transformed into an equivalent area of the box section. This choice allowed the representation of the walls to be done directly by the geometrical description of their axis. 16 integration points characterize each section. The *additional masses* placed along the walls have been modelled using concentrated masses.
- *Slab*. Rectangular and box section elements have been used to mesh the slabs. As for the walls, reinforcing bars have been accounted for through the use of box section. The *mass blocks* added to the floors to represent the dead load of the structure have been modelled by using fictive box sections: the density of the material has been chosen to represent the real weight of the blocks, whereas the stiffness of the elements has been limited, in order to avoid their influence on the dynamic behaviour of the structure.
- *Bracing system*. The bracing system has been modelled with I-section elements. The geometry of sections corresponds to the steel section adopted in reality. 13 integration points characterize each section.
- *Basement*. A network of elements has been used to simulate the stiffness of the basement in the three directions. As for the other structural elements, rectangular sections were adopted for concrete whereas the reinforcement has been modelled using box sections. The basement of the specimen is connected to the table through four circular section beams elements. A preload is applied to these components so as to reproduce the reality.
- *Shaking table*. A rigid body composed by rectangular section beams has been used to describe the behaviour of the shaking table. Circular beam elements have been chosen to simulate the deformability of the table. The *mass* of the table has been accounted for by defining an appropriate material density.

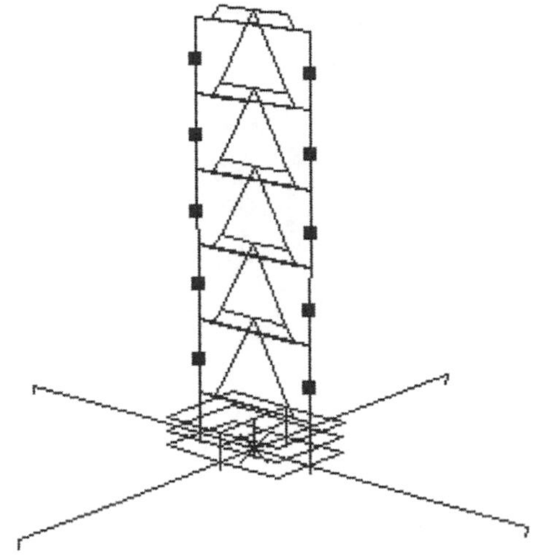

Figure 14. Simplified beam model.

An example of the mesh used in the analysis is reported in Figure 14.

4.3 *Main results*

A first verification of the model was carried out by performing a modal analysis. The first two frequencies obtained with the numerical model (5.5 and 6.02 Hz for the out-of-plane and the in-plane flexure) were similar to the measured ones (5.45 and 6.00 HZ respectively).

Seven seismic signals of increasing amplitude have been applied during the experimental campaign. Among them, only those corresponding to an intensity of 0.15 g, 0.4 g and 0.55 g have been considered in the numerical study. In spite of the simplified approach adopted in the analysis, numerical and experimental data were in good agreement both at global and at local level. Displacement time histories obtained for 0.55 g in the direction parallel to that of the walls are shown in Figure 15. The agreement at this stage is good; some differences have been detected from the experimental and the numerical displacements at 0.4 g. The reason for these dissimilarities can be found in the damage that might be suffered by the specimen during the application of the signals not considered in the analysis. The better agreement obtained for the signal of 0.55 g seems to verify this assumption. For the same intensity, the fair agreement between experimental and numerical axial load and moment time histories can also be observed in Figure 16 that represents an additional proof of the reliability of the model.

A complex state of stress inside the walls caused by the combination of bi-directional flexure and shear similar to that observed during the tests was also highlighted by the numerical analyses.

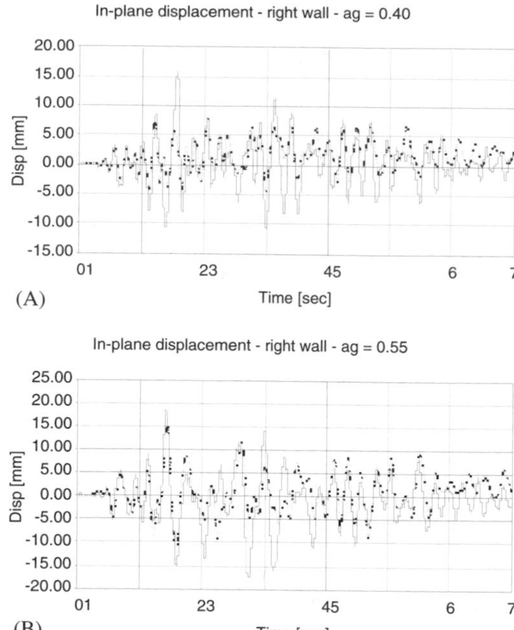

Figure 15. Comparison between calculated (dotted line) and measured (solid line) horizontal top in plane displacements: (A) signal of 0.4 g; (B) signal of 0.55 g.

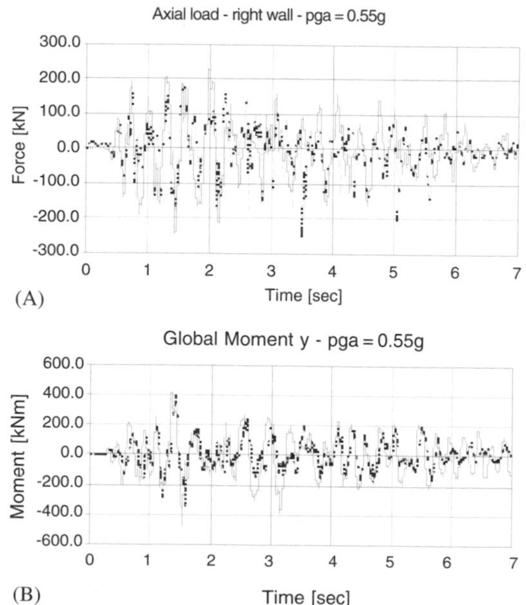

Figure 16. Comparison between calculated (dotted line) and measured (solid line) axial load in the right wall (A) and moment at the base of the walls (B) for a signal of 0.55 g.

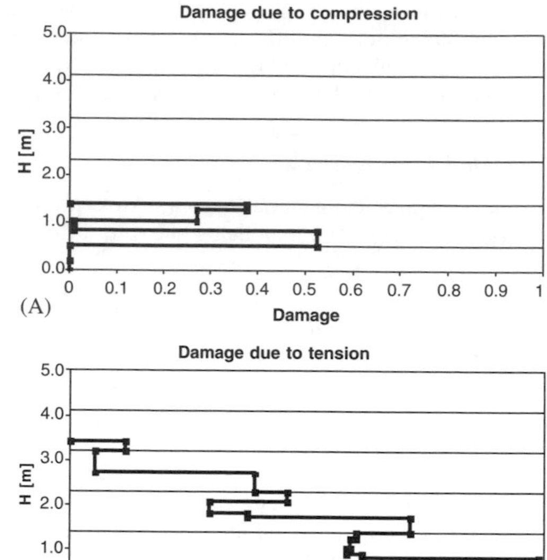

Figure 17. Damage evolution in one of the corners of the left wall: (A) contribution due to compression; (B) contribution caused by tension.

The damage theory implemented in the model used for concrete allowed the study of the variation of damage suffered in different points of the section. As an example, the damage suffered by the left wall calculated at one of the four corners of the section, for the 0.55 g excitation, is plotted in Figure 17. The effects of tension and compression are considered separately. From the graphs it can be observed that the effects of tension are maximum at the base of the wall, and decrease in the upper stories. On the contrary, compression causes damage only in the first story, ed in particular close to the basement and to the connection with the slab of the first story, where the maximum damage occurs, as it was observed during the tests.

These results have proved the ability of an explicit model and scheme of resolution to simulate the response of a mock-up submitted to a series of earthquake excitation. This was a first step to validate the modelling, however the main use will be, in the future, for dynamic problems at high velocity such as impacts.

Rock falls are common in mountain regions and protection systems have to be placed particularly on roads. Figure 18 gives an example of calculations performed with the PRM model used with an explicit scheme in order to analyse the efficiency of a reinforced concrete slab set up to protect a road and to play the role of a deflector for boulders. Experiments have been performed and the simulation is able to describe

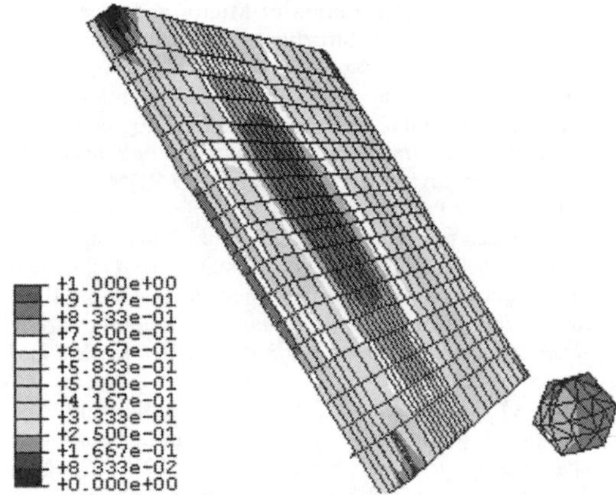

Figure 18. Inclined reinforced concrete slab submitted to rock fall – Simulation of damage induced (D_t into the slab and into the boulder with the PRM explicit model) by an impact of a boulder (450 kg from 30 m) in the center of the slab.

the main phenomena as tensile damage presented in Figure 18. This work is in progress.

5 CONCLUSIONS

The dynamic behaviour of a five-storeys shear wall building has been simulated in this study by using analytical models characterised by different levels of discretisation. The efficacy of a 3D finite element thin shell model, a beam and a fibre model have been verified by using the results of a shaking table test campaign performed within the CAMUS 2000 research programme.

As demonstrated by the results reported in the paper, not only the refined model but also the two simplified models were able to reproduce with good approximation the main features of the global response of the structure. This confirms that the level of discretisation and the type of numerical elements adopted in both simplified models are sufficient to describe the dynamic behaviour of wall structures. This is so also thanks to the stress–strain laws chosen to model the behaviour of steel and concrete. The accuracy of the results obtained for local indicators, e.g. elongation of rebars, damage index, etc, is reduced due to the hypothesis adopted in the modelling phase. The computation of damage in concrete can be cited as an example. The damage pattern evaluated with the 3D finite element thin shell model corresponds perfectly to that observed in the reality. On the contrary, by using the beam model only the general trend was represented.

The limited accuracy characterising the results obtained with simplified models of the same type of those considered in this paper is balanced by their reduced computational cost. This aspect assumes a fundamental importance when parametric studies or vulnerability analyses have to be performed (Negro & Colombo 1998).

The results of the experimental campaign indicate that the out-of-plane excitation can significantly increase the flexural demand of reinforced concrete walls with limited reinforcement, mainly because of the increase in the variation of the axial force, which is of the same order of magnitude than that due to permanent loads. Independently from the level of discretisation, the non-linear models considered in this study are able to realistically estimate the range of variation of the axial force. In addition to the ability in modelling the non-linear behaviour, the capacity to describe the variation of the axial load underlines the importance of a more close interaction between non-linear analysis and design.

Two ways of research are currently under investigation:

1. The ability of existing macro models with standard computer codes to model the global parameters of the response of a structure. The multiple-vertical-line-element model – MVLEM was chosen as the reference model. This study will be extended to the results of dynamic tests that will be performed at the shaking table facility of the LNEC. The specimen differs from the one considered in the study reported in this paper from the fact that the steel bracing system used to provide stiffness in the transversal direction is replaced by a more realistic reinforced concrete wall with openings.
2. The close agreement between the experimental and the analytical results obtained with the beam model verified the efficacy of explicit methods – the integration algorithms and more specifically the PRM model used to describe the behaviour of concrete – to simulate the response of structures subjected to dynamic loads. This validation allowed for the use of the same explicit model for dynamic problems of different nature. As an example, it is now used to study the effects of impact of blocks on reinforced concrete slabs.

REFERENCES

Armstrong, P.J & Frederick, C.O. 1966. A Mathematical Representation of the Multiaxial Bauschinger Effect. *G.E.G.B. Report RD/B/N 731.*

ASCE 1982. State-of-the-art report on finite element analysis of reinforced concrete. *ASCE special publication*, New York, N.Y.

Bisch, P. & Coin, A. 1998. The CAMUS research. *Proceedings of the 11th European Conference on Earthquake Engineering*, Paris, CD-ROM.

Bisch, P. & Coin, A. 2002. The "CAMUS 2000" research. *Proceedings of the 12th European Conference on Earthquake Engineering*, London, CD-ROM.

Caughey, T. 1960. Classical normal modes in damped linear systems. *Journal of Applied Mechanics*, 27, pp. 269–271.

Combescure , D., Queval J.C., Chaudat, T. & Sollogoub, P. 2002. Seismic Behaviour of Non Symmetric R/C Bearing Walls Specimen with Torsion. Experimental Results and Non Linear Numerical Modelling. *Proceedings of the 12th European Conference on Earthquake Engineering*, London, CD-ROM.

Fardis, M.N. 2000. Eurocode 8 – Present state, pre-normative and co-normative research needs (including design seismic action). *Proceedings of the Workshop: Mitigation of Seismic Risk Support to Recently Affected European Countries*, European Commission-JRC, Belgirate (VB), Italy.

Fleury, F. 1996. Prediction of the behaviour of reinforced concrete structures subjected to seismic loading: proposal for a global model for beam column joints integrating the behaviour of steel/concrete bond. *Ph.D. Thesis, Blaise Pascal – Clermont II University*, France.

Fouré, B. 2001. CAMUS 2000 Project – Control of execution of the specimens. *Report No. R112-9-348, CEBTP*, (in French).

Ile, N. 2000. Behaviour of reinforced concrete walls under seismic loading: contribution made by experiment and modelling to design. *Ph.D. Thesis, National Institute for Applied Sciences, Lyon*, France.

Ile, N. & Reynouard, J. M. 2000. Non-linear analysis of reinforced concrete shear wall under earthquake loading. *Journal of Earthquake Engineering*, Vol 4, No 2, 183–213.

La Borderie, Ch. 1991. Phénomènes unilatéraux dans un matériau endommageable : modélisation et application à l'analyse de structures en béton. *Ph. D. thesis*: Univ. Paris VI.

Mazars, J. 1986. A Description of Micro- and Macroscale Damage of Concrete Structures, *Engineering Fracture Mechanics*, V.25,No. 5/6.

Mazars, J. 1998. French advanced research on structural walls: An overview on recent seismic programs. *Proceedings of the 11th European Conference on Earthquake Engineering, Invited Lectures*, Paris, CD-ROM.

Menegoto, M. & Pinto, P. 1973. Method of analysis of cyclically loaded reinforced concrete plane frames including changes in geometry and non-elastic behaviour of elements under combined normal force and bending. *IABSE Symposium on resistance and ultimate deformability of structures acted on by well-defined repeated loads, Final report*, Lisbon.

Merabet, O. & Reynouard, J.M. 1999. Formulation d'un modèle elasto-plastique fissurable pour le béton sous chargement cyclique. *Contract Study EDF/DER, Final Report, No.1/943/002*, URGC-Structures, National Institute for Applied Sciences, Lyon, France.

Millard, A. 1993. CASTEM 2000, Manuel d'utilisation. *Report CEA-LAMBS No. 93/007*, Saclay, France.

Negro P. & Colombo, A. 1998. How Reliable are Global Computer Models? Correlation with Large-Scale Tests. *Earthquake Spectra*, Vol. 14, No. 3, pp. 441–467.

Rouquand, A. & Pontiroli, C. 1995. Some Considerations on Implicit Damage Models Including Crack Closure Effects and Anisotropic Behaviour. *Proceedings FRAMCOS-2*, Ed. F.H. Wittmann, AEDIFICATIO Publisher, Freiburg.

Rouquand, A. & Mazars, J. 2001. modèle incluant endommagement et dissipation hysteretique couplée, internal report CEG –DGA Gramat, France

Vulcano, A., Bertero, V.V. & Colotti, V. 1988. Analytical Modeling of R/C structural walls. *Proceedings of the 9th World Conference on Earthquake Engineering*, Vol.VI, Tokyo-Kyopto, Japan.

Computational Modelling of Concrete Structures, Bićanić et al. (eds)
© 2003 Swets & Zeitlinger, Lisse, ISBN 90 5809 536 3

Simulation of cyclically-loaded columns made with ductile cement-based composites

S.L. Billington
Stanford University, Stanford, CA, USA

J.K. Yoon
Earth Tech Corporation, Burnaby, BC, Canada

ABSTRACT: Simulations of two small-scale cantilever column experiments were performed as 2D plane stress analyses. The columns contained continuous, unbonded post-tensioning and were made of precast segments of concrete and a highly ductile fiber-reinforced cement-based composite referred to as Engineered Cementitious Composites (ECC). Nonlinear constitutive models for both the traditional concrete and the ECC were used. Embedded reinforcement elements modeled the mild reinforcement within each segment and truss elements were used to model the unbonded post-tensioning. The models were loaded under cyclic displacements up to drifts of ±4%. The modeling approach was able to simulate the experimental response with reasonable accuracy. The load vs. drift response and the amount of energy dissipation were captured well. The residual displacements and loss of tendon force over time due to column shortening were not as accurately predicted. Strengths and weaknesses of the simulation approach taken are discussed.

1 INTRODUCTION

A segmentally precast concrete bridge pier system that uses vertical unbonded post-tensioning and special precast segments of a highly ductile fiber-reinforced cement-based composite at high shear and flexural stress regions has been proposed (Fig. 1a). The cyclic response of 1/6-scale bridge columns of this system was determined experimentally (Fig. 1b) (Yoon & Billington, 2002). The distinguishing characteristics of the ductile composite, referred to as Engineered Cementitious Composites (ECC), and the unbonded post-tensioning together are that the bridge pier system can provide adequate energy dissipation during an earthquake and minimize post-earthquake damage and residual displacements.

ECC is composed of Portland cement, water, silica fume or fly ash, a low percentage by volume of polymeric fibers, and in most cases fine sand. ECC exhibits multiple, fine cracks upon loading in tension and can reach tensile strains of 3–6% while maintaining and/or increasing its load carrying capacity (summary in Li, 1998). This material also exhibits a unique unloading and reloading response when cycled between tension and compression (Fig. 2) (Kesner *et al.*, 2002), resulting in a higher amount of hysteretic energy dissipation, in particular in tension, than found in traditional concrete and fiber reinforced concrete (FRC).

A new constitutive model for two-dimensional (2D) nonlinear finite element simulations of structural components made with ECC was recently developed (Han *et al.*, 2002) and validated against small-scale cantilever beam experiments subjected to cyclic load (Fischer & Li, 2002a,b). The recent experiments on precast segmental concrete columns with unbonded post-tensioning subjected to cyclic lateral load (Yoon and Billington, 2002) offer a more complex system with which to test this newly developed constitutive model's predictive capabilities. The prediction of the experimental response of a selection of these more complex columns is presented here.

2 BACKGROUND

The experiments that were simulated in this research as well as background on precast, unbonded post-tensioned concrete bridge pier systems are discussed in detail in Yoon & Billington (2002) and Billington *et al.* (2001), respectively. Previous research on modeling unbonded post-tensioned concrete structural systems subjected to cyclic and/or seismic load has included simple single-degree-of-freedom models

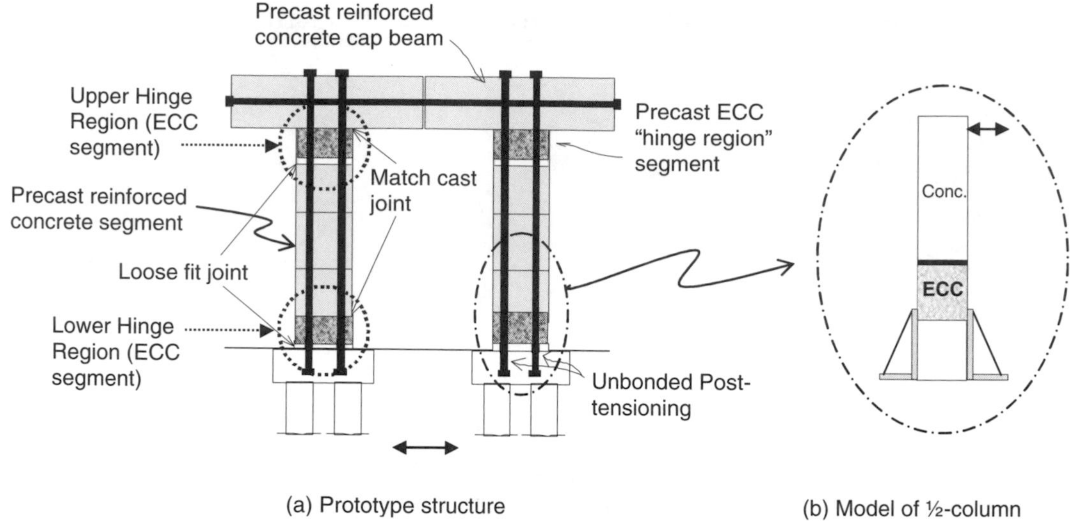

Figure 1. Prototype and model of ½ column.

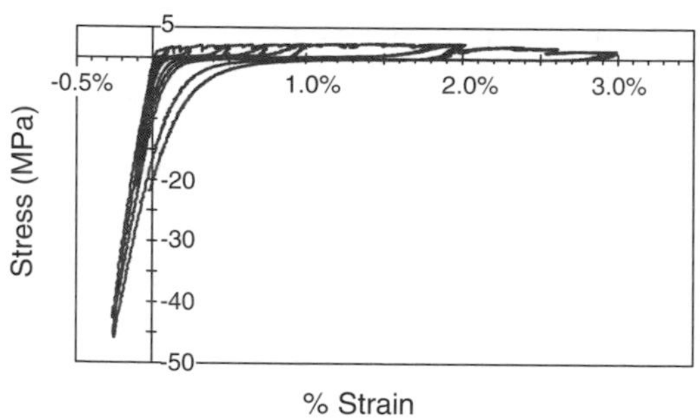

Figure 2. Reversed cyclic response of an ECC cylinder (experiment).

(e.g. Priestley & Tao, 1993) as well as frame models with phenomonological springs at connections (e.g. Cheok *et al.*, 1998) and with fiber elements (Kurama *et al.*, 1999, El-Sheikh *et al.*, 1999, 2000). Nonlinear finite element simulations on 2D models of unbonded post-tensioned concrete bridge piers was conducted by Kwan and Billington (2003a,b).

A new challenge in the simulation of the bridge columns in the research presented here is the simulation of the ECC material within the unbonded post-tensioned column. Kabele (2001) proposed a constitutive model for ECC, which has been used to simulate the cyclic response of for instance, shear beams and shear wall infill panels made of ECC. Kabele's model is based on the theoretical micromechanics of the material response. A more recently developed constitutive model for ECC adopted for this research is based on reversed cyclic loading experiments on ECC materials (Han *et al.*, 2002). With this model, it was found that the unique unloading and reloading behavior of ECC was particularly important to predict accurately the cyclic response of structural components where the ECC response dominated the component's hysteretic response. More specifically, where the main reinforcement of the component does not yield (e.g. when fiber-reinforced polymer reinforcement is used or where unbonded post-tensioning steel is used) the ECC response plays a more significant role in dictating the component response (Han *et al.*, 2002).

3 EXPERIMENTS

Seven specimens with a 200 mm × 200 mm cross-section were tested under quasi-static cyclic lateral loading to failure. Four specimens were 380 mm tall (from the point of fixity to the loading point) and three were 685 mm tall. In this paper, the results of two of the 685 mm tall specimens are presented. Full details on the experimental program are given in Yoon & Billington (2002).

Figure 3 shows a schematic diagram of the two test specimens with the labeling notation defined. Each specimen represents an upper or lower hinge region in a bridge pier (Fig. 1a) with the ECC segment located in the hinge region just above the fixity. Joints between segments were either match-cast and epoxy-filled or cast "loose-fit" and filled with an epoxy grout. The different joint types for the two specimens shown in Fig. 3a did not affect the experimental results (Yoon & Billington, 2002).

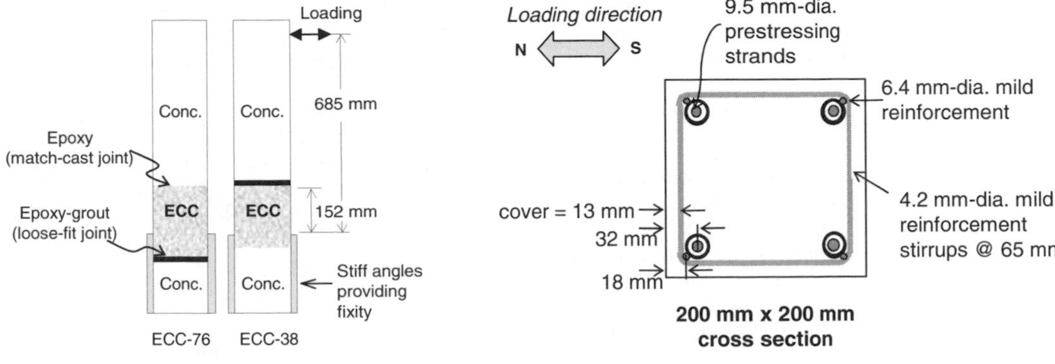

(a) Column Specimens

(b) Column segment cross-section (typ.)

Figure 3. Details of column specimens.

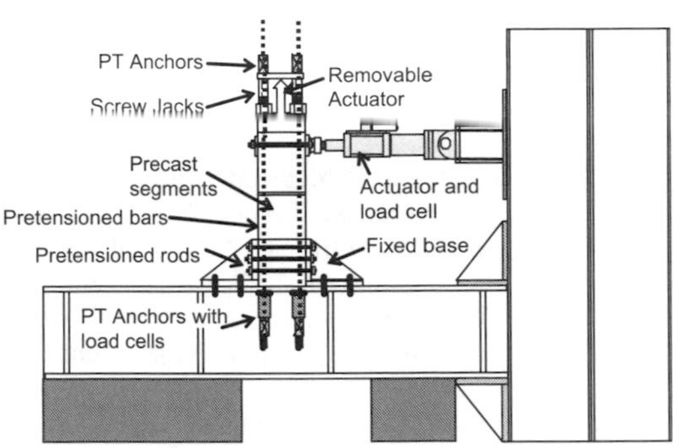

Figure 4. Test Set-up.

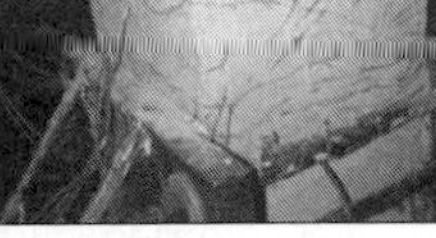

Figure 5. Cracking in ECC segments at 4% drift.

A schematic diagram of the test setup is shown in Fig. 4. Stiff bracing angles were used to provide base fixity. The fixed angles were 255 mm high and were joined together horizontally by six pretensioned rods, providing roughly 6.2 MPa of compression across the fixed segment. The pretensioning prevented gaps from forming between the specimen and the stiff base angles. The angles were bolted to the base testing beam with pretensioned bolts. The key variable between the two specimens was the depth of the ECC segment's embedment into the fixity (38 mm or 76 mm). The fixity was assumed to begin at the top of the stiff angles shown in Fig. 3. The 38 mm depth below the fixity did not provide enough depth for the mild steel within the segment to develop its yield strength. Therefore the steel also did not facilitate considerable spreading of cracking over the ECC segments. However the 76 mm depth below the fixity did provide adequate development length for the longitudinal mild reinforcement within the segment, and considerably more fine cracking was seen (Fig. 5).

The column capacity was the same for each specimen and the design was a roughly 1/6-th scale design of a circular bridge column design for the Santa Monica Viaduct in California, USA (Sritharan *et al.*, 1999). The typical cross-section is shown in Fig. 3b. Each column had four unbonded 9.5 mm-diameter seven-wire strands. In addition, all segments contained four 6.4 mm-diameter mild reinforcing bars (one in each corner). The unbonded post-tensioning was the only reinforcement to be continuous through the column (across the segments). Shear stirrups were designed according to Section 11.4 of ACI 318-99 (ACI 1999) resulting in 4.2 mm-diameter stirrups at 65 mm spacing. Seismic details such as additional transverse reinforcement for concrete confinement were not used in the design.

The column was designed to be partially prestressed with a prestressing force of 156 kN. An external axial load of 44.5 kN, representing dead load, was added to the prestressing force. The design moment capacity was estimated as 25.8 kN-m.

The lateral loading point of each column was connected to the actuator with a pin connection. The pin support at the column end was necessary to replicate an inflection point given that the specimens represented half of a bridge column that would experience reverse curvature.

The specimens were tested under cyclic, displacement-controlled loading up to failure or until the capacity of the actuator was reached.

4 SIMULATION APPROACH

The main objective of the research presented here was to assess the ability of the recently developed constitutive model for ECC (Han *et al.*, 2002) to predict accurately the local and global response of unbonded post-tensioned precast concrete columns. The assessment was made by comparing the simulated response to that of the recent experiments described in Section 3. Particular attention in this modeling was given to balancing accuracy and computational efficiency. For this reason, 2D nonlinear simulations were conducted. The modeling and analyses served as validation to use this simulation approach to design and predict the response of larger-scale experiments and full designs of bridge piers.

4.1 *Finite element models*

The finite element meshes for the two columns simulated are shown in Fig. 6. The models used eight-noded quadrilateral isoparametric plane stress elements for the ECC and concrete segments. The mild reinforcement was modeled with three-noded straight embedded reinforcement elements. With the embedded reinforcement, it is assumed that there is perfect bond between the plane stress elements and the embedded reinforcement elements. The stiff steel angles that provided a fixed base in the experiments were modeled as three-noded beam elements with stiffnesses equivalent to the actual steel angles. The bottom nodes of the base concrete segments were fixed using individual pin supports.

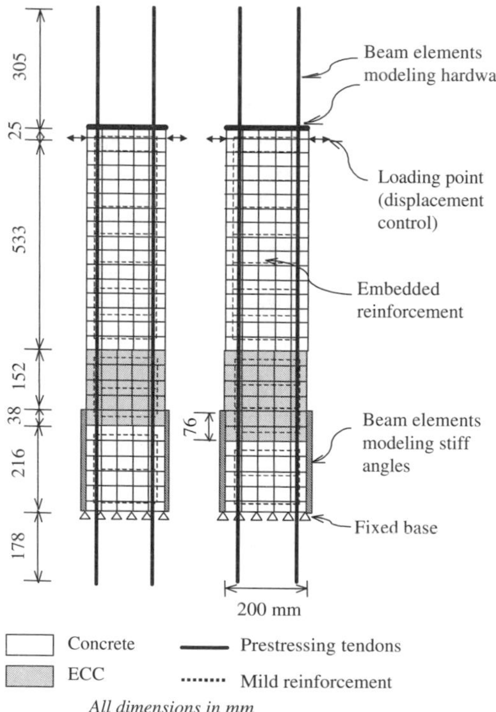

Figure 6. Finite element models.

Two-noded truss elements were used to model the unbonded post-tensioning tendons. The tendons extended 305 mm above and 180 mm below the column. The extended tendon lengths were the distances to the top and bottom sets of wedge anchors above and below the experimental specimen, respectively (Fig. 4). The 305 mm extension at the top was where adjustable screw jacks, split washers and metal plates were placed during the experiments; thus providing stiffness in this area. In preliminary simulations, plane stress elements with steel parameters were used to model the hardware. However, it was found that no significant movements occurred in that region and the plane stress elements were replaced by beam elements with equivalent stiffnesses. Vertical beam elements were used to connect the top and bottom of the tendons to the top and bottom surface of the column, respectively. Additional horizontal beam elements were placed across the top of the column to connect the vertical beams.

Two-noded truss elements were also used to model the horizontal prestressed base rods across the bottom segment.

4.2 *Constitutive models*

The top concrete segments were modeled as linearly elastic in both compression and tension because no cracking or crushing was anticipated throughout the simulation in this region. The bottom concrete segment between the stiff angles was modeled with a total-strain based rotating crack model (Feenstra *et al.*, 1998) with elastic-perfectly-plastic response in compression and linear tension-softening (fracture energy-based). Secant unloading was used. The model parameters and material properties for the concrete are given in Table 1.

The ECC segments were primarily modeled with the recently developed total-strain based rotating crack model for ECC (Han *et al.*, 2002). This model uses a multi-linear function in both compression and tension and has a polynomial unloading curve to reflect ECC's unique unloading with more energy dissipation than traditional concrete.

Figure 7 shows the stress-strain curves for ECC with unloading curves plotted up to ±4% drift using the material properties and parameters shown in

Table 1. The values for the parameters of a cyclic stress-strain curve of ECC were taken from the experimental results of cylinders subjected to reversed cyclic loading (tension-compression) (Kesner *et al.*, 2002).

The top row of ECC elements was modeled as linearly elastic to prevent localization due to a stress concentration between the concrete and ECC. The stress concentration at this region arises from an incompatible Poisson's effect between the two materials. A value for Poisson's ratio for ECC has not been researched. A value of 0.15 was assumed as Poisson's ratio for both the concrete and ECC. Given that the ECC has a modulus of elasticity roughly one half that of concrete, a stress concentration is expected and was observed at this interface in preliminary simulations. This stress concentration caused early localization at the interface, which was not expected nor seen in the experiments.

Table 1. Properties and parameters of concrete and ECC.

	Concrete	ECC
Young's modulus	27.6 GPa	13.8 GPa
Poisson's ratio	0.15	0.15
Tensile properties	$f_t = 3.5$ MPa $G = 72$ N-m/m^2	$\sigma_{T0} = 1.4$ MPa $\varepsilon_{T0} = 1.000$E-4 $\sigma_{T1} = 2.1$ MPa $\varepsilon_{T1} = 0.035$ $\varepsilon_{T2} = 0.1835$*
Compressive properties	$f_c = 34.5$ MPa	$\sigma_{C0} = 48.3$ MPa $\varepsilon_{C0} = 0.0035$ $\varepsilon_{C1} = 0.229$

* strain at zero stress.

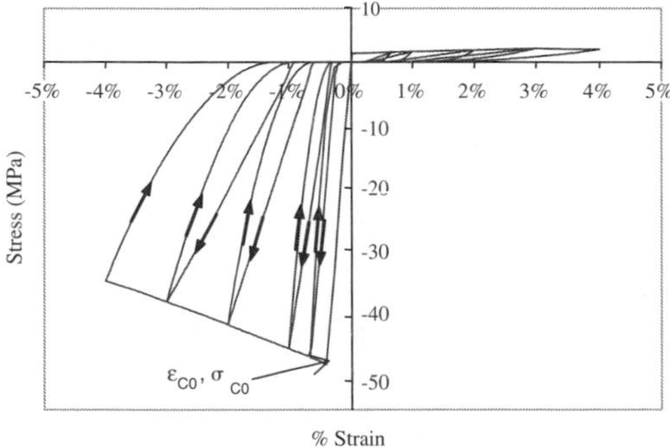

(a) Full stress-strain curve of ECC mode with cyclic loading

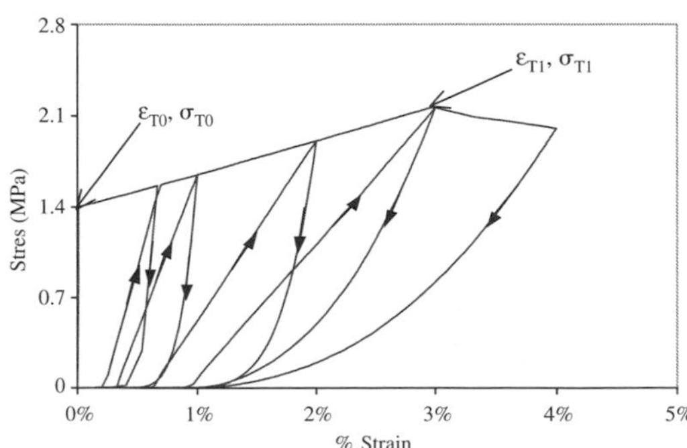

(b) Tensile stress-strain curve of ECC model with cyclic loading

Figure 7. ECC constitutive model.

The prestressing tendons and mild steel reinforcement was modeled as elastic-perfectly plastic without strain-hardening. The properties used for the reinforcement are shown in Table 2.

4.3 *Loading history*

Loading was applied through displacement steps in single cycles to drift levels of ±0.66%, ±1%, ±2%, ±3%, and ±4%. The regular Newton-Raphson iterative procedure was used. All of the analyzed models reached these drift level without numerical difficulties

5 SIMULATION RESULTS

5.1 *Strain contours*

Figures 8 and 9 show the principal tensile and compressive strain distributions from the simulated results of the two models at 4% drift. The cracking behavior of the column at this drift is shown in Fig. 5. The maximum tensile strain of the contour plot is the strain, ε_{t1}, at which the ECC is expected to localize and the maximum compressive strain of the contour plot is the peak compressive strain, ε_{co}. A localized region of cracking is predicted in both models at the 4% drift. In the experiment, localization was observed in ECC-38 at 4% drift and in ECC-76 at 6% drift. In both columns, there was distributed cracking throughout the ECC segment (Fig. 5) which was also observed in the simulations. Compressive damage was predicted in the simulation at 4% drift. Compressive damage was not visible in the testing of ECC-38 until after 4% drift. However compressive damage in the form of vertical (lateral) cracking in the compression region was observed at 4% drift in ECC-76 (Fig. 5).

Table 2. Properties of steel materials.

	Tendons	Longitudinal mild steel	Transvers mild steel
Young's modulus (GPa)	173	186	186
Poisson's ratio	0.3	0.3	0.3
Yield stress (MPa)	1760	500	400

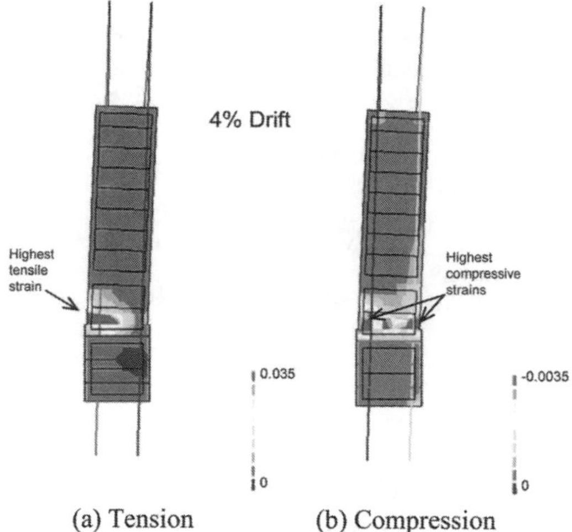

Figure 8. Principal strain contours for ECC-38.

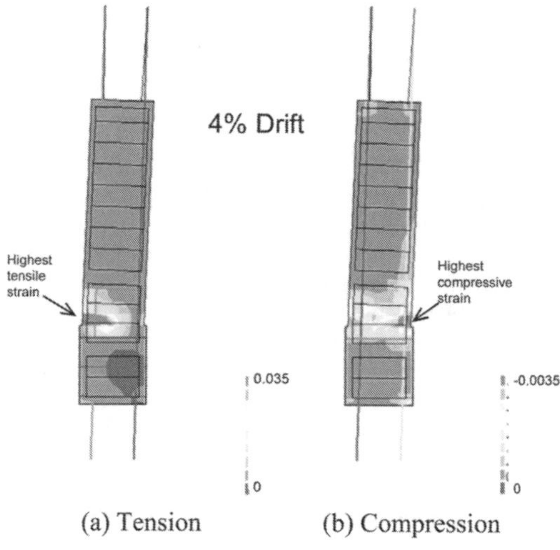

Figure 9. Principal strain contours for ECC-76.

5.2 *Load-drift response*

Figures 10 and 11 show the load-drift response up to 4% drift from both the simulations and the experiments for ECC-38 and ECC-76, respectively. The drift was calculated as the lateral displacement at the top of the column divided by the distance from that point down to the top of the stiff steel angles (the assumed point of fixity of the column). Each 1% drift represented roughly 7 mm of displacement at the loading point. Table 3 summarizes the initial stiffness, the peak load with corresponding drift and moment, and the residual displacement from the simulations and experiments.

There was no significant difference between the load vs. drift responses of the simulations of the two tall column models with the different embedded depth of the ECC segments below the fixity. The nonlinear response of both models resulted from cracking as observed with the concentrated tensile stresses close to the top of the stiff angle (Figs. 8–9). As expected with unbonded post-tensioning that has not yielded, the residual displacements are very low. In the simulations, the longitudinal mild reinforcement within the ECC segment yielded.

Compared with the experimental response, the peak loads obtained from the simulations were within ±8% of the experimental results. The most significant difference was with column ECC-38, with the 38 mm embedment. In this column, the simulation predicted higher loads (and therefore higher resisting moments) than measured in the experiments. This increase in strength is attributed to the model's "perfect bond" embedded reinforcement, which can carry significant load and eventually yield. In reality, the mild steel within these segments does not have sufficient development length to carry much load or to yield. Therefore, the mild steel in the experiment does not contribute significantly to the specimen's strength.

Looking specifically at the response of model ECC-38 (Fig. 10), the initial stiffness of the experiment and simulation was comparable. The final residual displacement from the experiment was four times larger than that of the simulation (Table 3 and Fig. 10). The residual displacements at this drift level are caused by

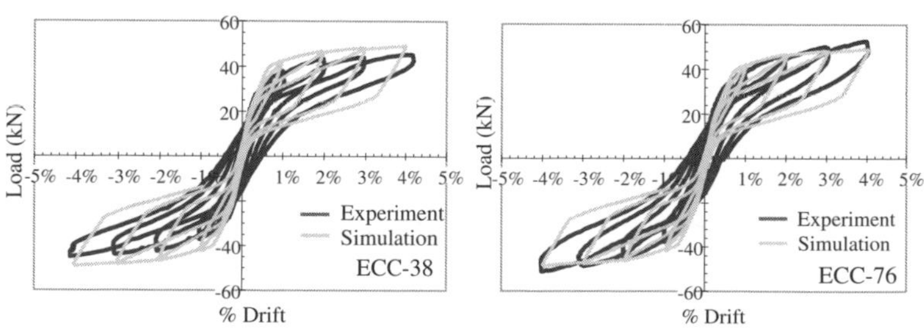

Figure 10. Load-drift response of Columns ECC-38 and ECC-76.

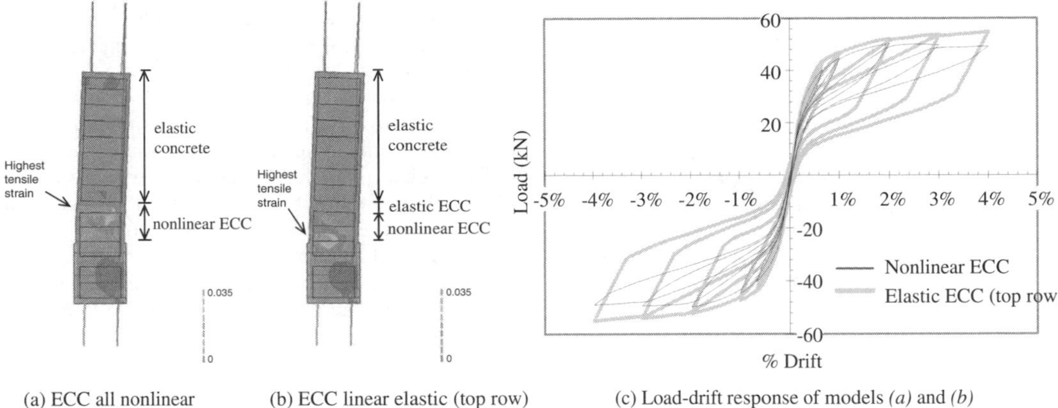

(a) ECC all nonlinear (b) ECC linear elastic (top row) (c) Load-drift response of models *(a)* and *(b)*

Figure 11. Effect of ECC-concrete interface stress concentration.

Table 3. Summary of numerical and experimental results.

	Results	Initial stiffness (MPa)	Peak load* (kN)	Corresponding moment (kN-m)	Residual displacement (%)
ECC-38	Exp	19.3	44.9	30.8	0.4
	Sim	17.9	49.0	33.6	0.1
ECC-76	Exp	18.6	51.6	35.4	0.4
	Sim	17.9	49.4	33.9	0.1

*All at 4% drift.

cracking damage of the ECC. The inability of the simulation to accurately predict the residual displacement is attributed to the ECC material model. As shown in Fig. 7, the ECC model itself does not capture any residual displacements, but rather goes through the origin with each cycle.

For ECC-76, the initial stiffness of the experiment and of the simulation was also similar (Table 3 and Fig. 10). In the experiment, ECC-76 showed no sign of losing load even after the 4% drift level was reached. However the simulation, although still increasing in load slightly, appears to have reached more of a plateau. The final residual displacement from the experiment was also four times larger than that of the simulation.

5.3 *Effect of ECC-concrete interface modeling*

As mentioned in Section 4.2, the top row of elements in the ECC segment was modeled as a linear elastic material to avoid the stress concentration between the ECC and concrete elements. These two materials were assumed to have the same Poisson's ratio but are known to have significantly different moduli of elasticity. Figure 11 compares the principal tensile strains and load-drift response of ECC-76 where all of the ECC is modeled with the nonlinear constitutive model (Fig. 11a,c) with a model where the top row of ECC elements is modeled as a linear-elastic material (Fig. 11b,c).

With all of the ECC modeled as nonlinear, the crack localization occurred at the top of the ECC segment, where the stress concentration due to lateral movement occurs. When the ECC-concrete interface is modeled with elastic ECC, the cracking strains are greatest at the base of the column, as expected. The results of the incorrect location of crack localization are a load-drift response with less energy dissipation and no internal mild steel yielding (Fig. 11c).

5.4 *Tendon loads*

A significant difference between the experiments and the simulation results was the tendon load change during loading. Figure 12 shows the change in North tendon force up to 4% drift in column ECC-38 during the experiments and simulations. The drift directions and the location of the North tendons are indicated on

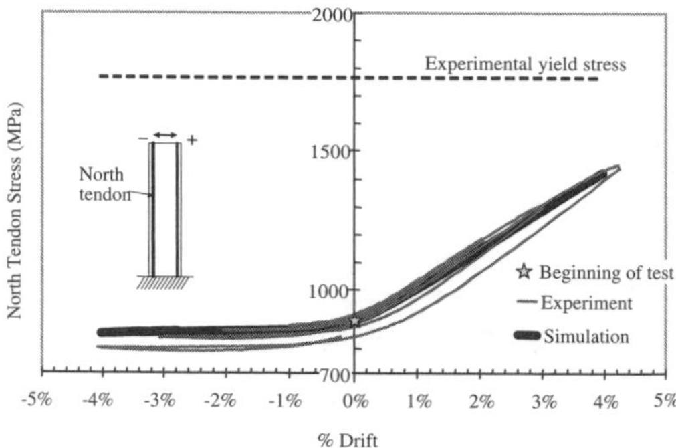

Figure 12. Tendon load changes.

the figure insert. The star on the y-axis represents the average stress value of the tendons at the beginning of the tests and the simulations.

The experiments showed that the load in each tendon decreased by roughly 10% after 4% drift rather than maintaining the initial prestress. The tendon loads in the finite element model consistently maintained the initial prestress. The decrease in experimental tendon load was due to the specimen shortening in the experiment under partially irrecoverable deformation. This effect was even greater at higher drifts when loads reached the compressive strength of the ECC (Yoon & Billington, 2002). The irrecoverable deformation was not captured in the finite element models because the strain in compression up to 4% drift remained elastic (no damage was modeled in the initial, elastic portion of the material model). Additionally, any shortening from cracks not closing completely was not modeled.

It should be noted that a limitation of the adopted tendon model in the simulations was that the 2-noded truss elements remain linear throughout the analysis. Any curvature of the tendons due to restraint from the ducts at the point of rigid body rotation at higher drifts could not be captured. Instead of tracing the tendon duct paths, which was the case for the experiments, the beam elements for tendons remain straight while connected to the top and the bottom nodes. The tendon members in the simulations therefore span a shorter distance than the tendons in the experiments. The straight tendons therefore will predict a later onset of yielding in the tendons than would be observed in the experiments. The simulations here were not taken to high enough drifts to cause yielding of the tendons.

6 CONCLUSIONS

The major findings from the presented research were:

- There was no significant difference between the responses of the simulated columns with different embedment depths.
- Using the adopted models, the peak load vs. drift response from the experiments and the simulations were similar for the column with the larger (76 mm) embedment.
- The peak load vs. drift response for the column with the smaller embedment (38 mm) exhibited higher strength in the simulation than in the experiment. The higher simulated strength is attributed to the modeling of the mild steel as having perfect bond and therefore contributing more strength in the simulation than possible in the experiment.
- The simulations were able to capture the distributed cracking and the onset of crack localization in the ECC segments.
- A stress concentration was evident at the interface between the ECC and concrete, which caused an inaccurate prediction of column response when both materials were modeled as nonlinear.
- The residual displacements and the loss of tendon force from the experiments were not captured in the simulations.

REFERENCES

ACI Committee 318 1999. *Building Code Requirements for Structural Concrete* (ACI 318-99) and *Commentary* (ACI 318R-99), Detroit, MI, American Concrete Institute.

Billington, S.L., Barnes, R.W. and Breen, J.E. 2001. Alternate Substructure Systems for Standard Highway Bridges. *ASCE J. Bridge Engineering* 6(2): 87–94.

Cheok, G.S., Stone, W.C. and Kunnath, S.K. 1998. Seismic Response of Precast Concrete Frames with Hybrid Connections. *ACI Structural Journal* 95(5): 527–539.

El-Sheikh, M., Sause, R., Pessiki, S. and Lu, L.-W. 1999. Seismic Behavior and Design of Unbonded Post-Tensioned Precast Concrete Frames. *PCI Journal* 44(3): 54–71.

El-Sheikh, M., Pessiki, S., Sause, R. and Lu, L.-W. 2000. Moment Rotation Behavior of Unbonded Post-Tensioned Precast Concrete Beam-Column Connections. *ACI Structural Journal* 97(1): 122–131.

Feenstra, P.H., Rots, J.G., Arnesen, A., Teigen, J.G. and Hoiseth, K.V. 1998. A 3D Constitutive Model for Concrete Based on Co-rotational Concept. In R. de Borst, N. Bićanić, H. Mang, G. Meschke (eds). *Computational Modelling of Concrete Structures, Proc. of the EURO-C Conference, Bad Gestein, March 1998*. Rotterdam: Balkema.

Fischer, G. and Li, V.C. 2002a. "Effect of Matrix Ductility on Deformation Behavior of Steel Reinforced ECC Flexural Members under Reversed Cyclic Loading Conditions," *ACI Structural Journal* 99(6): 781–790.

Fischer, G. and Li, V.C. 2002b. "Deformation Behavior of FRP Reinforced ECC Flexural Members under Reversed Cyclic Loading Conditions," *ACI Structural J.*, accepted, Feb.

Han, T.S., Feenstra, P.H., Billington, S.L. 2002, Simulation of Highly Ductile Fiber-Reinforced Cement-based Composite Components, *ACI Structural J.*, accepted, Nov.

Kabele, P. 2001. *Assessment of Structural Performance of Engineered Cementitious Composites by Computer Simulation*. CTU Report 4, Vol. 5, Prague.

Kesner, K.E., Billington, S.L. and Douglas, K.S. 2002. Cyclic Response of Highly Ductile Fiber-Reinforced Cement-based Composites. *ACI Materials Journal*, submitted for review, August.

Kurama, Y.C., Sause, R., Pessiki, S., and Lu, L.-W. 1999. Lateral Behavior and Seismic Design of Unbonded Post-Tensioned Precast Concrete Walls. *ACI Structural Journal* 96(4): 622–632.

Kwan, W.P. and Billington, S.L. 2003a. "Unbonded Post-tensioned Bridge Piers: Part I – Monotonic and Cyclic Analyses," *ASCE J. Bridge Engineering*, to appear, March.

Kwan, W.P. and Billington, S.L. 2003b. "Unbonded Post-tensioned Bridge Piers: Part II – Seismic Analyses," *ASCE J. Bridge Engineering*, to appear, March.

Li, V.C. 1998. Engineered Cementitious Composites – Tailored Composites Through Micromechanical Modeling, In N. Banthia, A. Bentur, and A. Mufti (eds), *Fiber Reinforced Concrete: Present and the Future*, Montreal: Canadian Soc. for Civ. Engrg.

Priestley, M.J.N. and Tao, J.R. 1993. Seismic Response of Precast Prestressed Concrete Frames with Partially Debonded Tendons. *PCI Journal* 38(1): 58–69.

Sritharan, S., Priestley, M.J.N. and Seible, F. 1999. Enhancing Seismic Performance of Bridge Cap Beam-to-Column Joints Using Prestressing. *PCI Journal* 44(4): 74–91.

Yoon, J.K. and Billington, S.L. 2002. *Experimental and Numerical Studies of Precast Unbonded Post-tensioned Bridge Columns with Engineered Cementitious Composites*, Research Report 02-03, Ithaca, NY: Cornell University.

Computational Modelling of Concrete Structures, Bićanić et al. (eds)
© 2003 Swets & Zeitlinger, Lisse, ISBN 90 5809 536 3

Non linear modelling of masonry infilled frames: Identification of the global model characteristics using local modelling

D. Combescure & P. Sollogoub
EMSI Laboratory, DEN/DM2S/SEMT, CEA Saclay, Gif-Sur-Yvette, France

F.Vita
Universita degli study di Roma La Sapienza, Facolta di Ingegneria, Via Eudossiana, Roma, Italy

P. Pégon
ELSA, Structural Mechanic Unit, Joint Research Centre, European Commission, Ispra (VA), Italy

ABSTRACT: The present paper aims at presenting an overview about the effects of the unreinforced masonry infills on the in-plane seismic behaviour of reinforced concrete frames and the global and local non linear models available for the seismic assessment of such type of structures. An extensive parametrical study has been performed with the refined non linear finite element models in order to identify the main properties of the equivalent diagonal struts-stiffness and strength – and to assess the shear forces induced by the masonry infill in the RC frame.

1 INTRODUCTION

In several European countries, the building structures are commonly made by reinforced-concrete frames infilled with unreinforced masonry panels. The infill panels, usually considered as non-structural elements, have a significant effect on the global seismic linear and nonlinear responses of R/C frame structure. On one side, they can increase the stiffness and the strength of the frame in a very significant way, for example, by changing the torsional behaviour of the structure or creating a soft storey mechanism. On the other side, they induce supplementary shear or normal forces in the surrounding frame which may conduct to a brittle failure of the reinforced concrete members. This is particularly true for the existing reinforced concrete structures poorly or not designed for a modern seismic action.

The present paper aims at presenting the non linear models and the local-to-global approach available for the seismic assessment of unreinforced masonry infilled RC frames.

Two levels of non linear modelling are used for the analysis of the infilled structures under in-plane loading. At the local level, each constituent has its own constitutive law and geometric finite element support. The main phenomena such as cracking and crushing of concrete and masonry could be reproduced by using the continuous damage and plasticity theories or semi-global models: 2D plasticity model for masonry, joint elements for the interface between masonry and RC frame, non linear fibre type model for the RC frames ... Once the constitutive laws validated on elementary tests, this level of modelling allows one to perform predictive calculations on structures with various geometries and material characteristics. The effect of the openings can also be estimated. However the cost of the computations does not allow extensive or dynamic studies and thus the global level – where equivalent diagonal struts and global beam elements with constitutive laws based on empirical rules reproduce respectively the behaviour of the masonry infills and the RC frames – represents the unique strategy for the analysis of complete civil engineering structures under extreme seismic loading. Such a modelling approach has already been validated on experimental results on one bay one storey infilled frames tested under static cyclic loading at LNEC in Lisbon and multistorey structures tested in JRC Ispra (Combescure, 1996 and 2000). The paper presents also the results of an extensive parametrical study performed with the refined non linear finite element models in order to identify the main properties of the equivalent diagonal struts-stiffness and strength – and to assess the shear forces induced by the masonry infill in the RC frame. The results of these parametrical studies have been summarized in simple analytical formulae and compared to several classical formulae available in the literature.

2 THE LOCAL TO GLOBAL MODELLING APPROACH

2.1 *Definition of the local and global modelling levels*

Two modelling approaches, global and local, are classically used to analyse the infilled frame structures under horizontal seismic loading. In the global approach, each masonry panel is often replaced by two trusses with an uniaxial behaviour law (Klingner et al., 1976). The complexity of the behaviour depends on the various phenomena taken into account by the model (pinching due to crack closure, crushing of masonry at the corners, decrease of stiffness due to cracking, etc...). The frame is modelled by beam and column elements with moment-curvature relationships or fibre type model. This approach allows to perform a large number of computations with dynamic or cyclic loading but the identification of the truss parameters is often based on empirical rules. In case of a modification in the panel characteristics, the limit of validity of the formulae may be reached.

In order to cope with this difficulty, it is proposed to use the refined material models not only to identify the parameters of the panel element but also to highlight the limitation of such a global model by studying, for example, the interaction between different infills in a multi-storey structure (Combescure 1996). In the local approach, each part (the frame and the infill panels) is discretized. Both materials – masonry and RC concrete – are considered as homogeneous media with an elastic or a non-linear, isotropic or anisotropic behaviour law. In this local modelling, the hypothesis made for the contact between the frame and the infill panels becomes important since the global stiffness is highly dependent on the presence of cracks at this interface. All the computations presented here have been performed with the Finite Element code CASTEM 2000 developed by CEA-France. Within such an environment, the user can easily compare the two levels of modelling and use them in a complementary way and it becomes easy to identify the parameters of the global models using refined modelling.

2.2 *Local modelling*

2.2.1 *R/C frame*

2D Timoshenko beam elements supporting a fibre type model have been used for the frame (Guedes, 1997). Each column has been discretized by 10 elements with 6 concrete fibre and 2 or 3 steel fibres, each fibre having 2 Gauss points. Simplified uniaxial laws have been considered for both concrete and steel: parabolic curve with a perfectly plastic plateau in compression and no strength in tension for concrete (similar to the concrete law of the French BAEL concrete code) and a plastic model with kinematic hardening for steel. Shear behaviour is assumed elastic. Note that such a modelling allows to know the shear forces and the axial forces in the frame and thus to quantify the effect of the presence of the masonry infill onto the surrounding frame.

2.2.2 *A masonry plasticity based model*

The experimental results and some previous studies have shown that the specimen failure is reached when masonry crushes (Pires, 1993), (Combescure, 1996). Furthermore, the ultimate strength depends on the number of cycles applied to the specimen. In order to cope with the former property, a plasticity-based model with two yield surfaces and softening behaviour in compression and traction has been developed. Details about this model and its numerical implementation are available in (Combescure, 1996 and 2000). As for the classical plasticity-based models, unilateral phenomena due to cracking is not considered. This fact has a minor importance also under cyclic loading since the main cracking is assumed localized at the interface between frame and infill and is modelled by a joint/interface element.

The Young modulus is identified with the results of diagonal test ($E = 4.G$ for $\nu = 0.2$ if G is calculated with the RILEM rules) whereas the compressive strength of the isotropic model of masonry is directly given by the compression tests on wallettes perpendicular to the holes. Note that the identification of the masonry parameters must be realized with tests performed on complete masonry wallettes since masonry has a very complex behaviour due to the difference of Poisson ratio between bricks and mortar.

2.2.3 *Contact modelling*

The modelling of the interface has a major influence on the failure pattern, the initial stiffness and the global strength. A plasticity-type joint model with a Coulomb yield surface is used (Snyman et al., 1991). While the sliding behaviour is governed by plasticity rules, the unilateral phenomenon is reproduced in tension: the joint opens without creating plastic strain. Associated or non associated plastic flow and dilatancy phenomenon can be considered. In our case, the dilatancy angle is assumed equal to zero. The considered tensile strength of the joint is equal to 10 percent of the masonry compressive strength. A classical value of 40 degrees taken from the results of the tests described by (Mehrabi et al., 1995) on mortar-brick interfaces is used for defining the Coulomb failure surface.

2.3 *Global modelling*

2.3.1 *Simplified modelling of the RC frames*

The mesh of the surrounding frame is reduced for the global computation: each column is discretized with one linear Bernoulli beam element with a reduced elastic stiffness (2/3rd of the elastic stiffness) placed between two Timoshenko elements supporting the non linear fibre model. A constant length equal to the column width has been considered for the plastic hinges.

2.3.2 *A global model for infill panels*

Since the work performed by (Klingner et al., 1976), the non-linear analysis of infilled frames have been usually performed by replacing each individual panel by two- or more-diagonal struts with a uniaxial compressive law. The model introduced in CASTEM 2000 is also supported by a truss element and the behaviour law (Figs. 1 and 2) is able to reproduce the classical models by choosing the appropriate parameters. The phenomena reproduced are the stiffness degradation due to cracking – mainly at the interface between the frame and the panel – the development of plastic strain and the softening due to crushing, the strength degradation under cyclic loading and the pinching associated with sliding. The strut has no tensile strength and the stress-strain curve under monotonic compressive loading is multilinear and may be identified by using the results given by refined modelling.

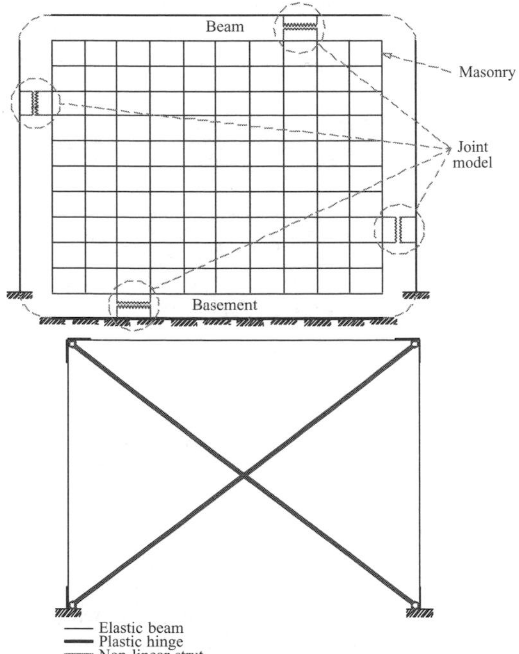

Figure 1. Local and global models of a one bay one storey infilled frame.

3 DESCRIPTION OF THE REFERENCE EXPERIMENTAL RESULTS

A series of 9 one-bay RC frames has been tested under cyclic loading in the framework of the SEISMIER research program supported by the European Commission. The geometric characteristics and the reinforcement details of the models, as well as the applied loads, tried to simulate the real conditions of a reinforced concrete frame located in the ground floor of an ordinary building (Fig. 3). The models had an height of 1.80 m and a length of 2.40 m. The columns and the beams cross sections have, respectively, 0.15 m × 0.15 m and 0.15 m × 0.20 m. The columns were reinforced with 8ϕ 10 longitudinal bars and ϕ8//0.04 hooks. The beams were reinforced with 6ϕ8 longitudinal bars and ϕ6//0.05 stirrups. The infill walls were built with 0.30 m × 0.20 m × 0.15 m horizontally hollow bricks, usual in Portugal, bedded using mortars with the proportions 1:4 in volume (cement: river sand). The materials used in the construction of the frames were a C20/25 concrete and a S400 steel. The models were built on reinforced concrete blocks with a 3.24 m × 0.74 m × 0.35 m volume. These concrete blocks were used to fasten the model to the shaking table.

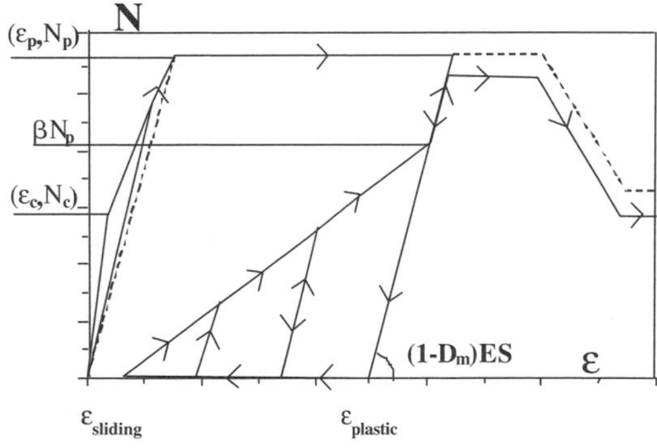

Figure 2. Global axial force-axial strain used for the global model of infill.

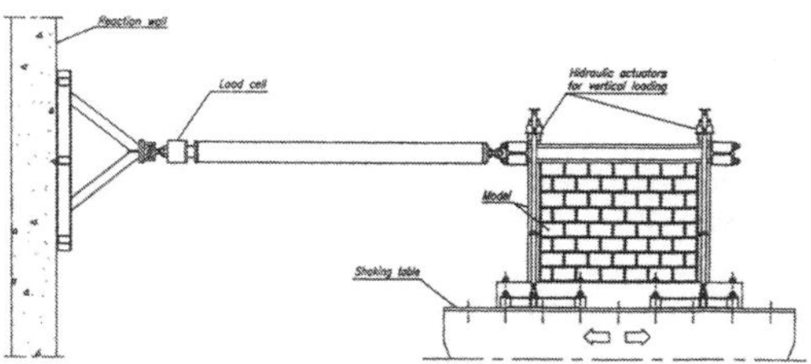

Figure 3. Experimental Set-up.

Table 1. Increase of strength and stiffness due to the masonry infill panel.

Specimen	Max. shear strength	Initial stiffness	Increase of stiffness due to infill
Bare frame			
I3	60 kN	4.8 kN/mm	/
I4	64 kN	4.99 kN/mm	/
Uniformly infilled frame			
I1	190 kN	206 kN/mm	×42.9
I7	250 kN	182.4 kN/mm	×38
I8	181 kN	155.7 kN/mm	×32.4
I9	193 kN	170 kN/mm	×35.4
Infilled frame with window			
I2	156 kN	50 kN/mm	×10.4
I5	123 kN	74.4 kN/mm	×15.5
I6	123 kN	75.1 mm	×15.6

The models were tested in the platform of a shaking table. The tests consisted basically in the application of a relative horizontal displacement history between the base and the top of the models. A vertical force of 100 kN was applied at the top of the columns. This force was kept approximately constant during the entire tests. Each stage of the tests consisted in the application of 2 complete sine waves of relative horizontal displacement between the base and the top of the models. The maximum amplitude of the imposed displacement increased from stage to stage of the tests (0.6 mm, 25 mm, 50 mm, 75 mm and 100 mm). The experimental set-up and the main test results are described in details in (LNEC, 1998).

Table 1 shows the maximum strength of the 9 specimens and the increase of initial stiffness due to the presence of the infill panel. Note the great influence of the infill panel for these frames. All the uniformly infilled frames had similar failure patterns: cracking occurs at the interface between the frame and the masonry panel, crushing of masonry in the corners of the wall and some cracking of masonry in the wall itself (Fig. 5). The force-displacement relationships are characterized by softening and important pinching after the beginning of masonry crushing (Fig. 4). Strength degradation under cyclic loading is also visible on the global force-displacement curves: after one cycle, the specimens do not find again the initial strength but a reduced strength.

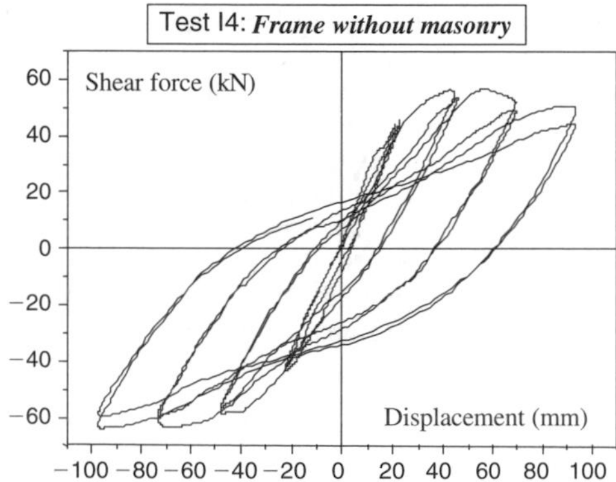

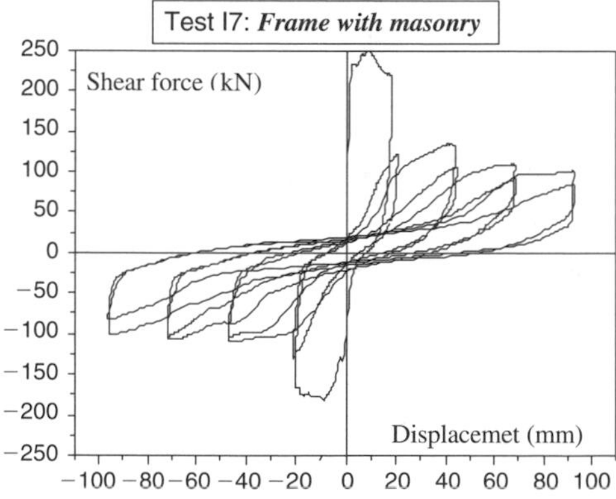

Figure 4. Shear force – top displacement curve (experimental results).

4 REFINED ANALYSIS OF ONE-BAY INFILLED FRAMES

4.1 *Study of the reference specimen*

In the present study, the Young modulus and the compressive strength of masonry has been taken equal to 4000 MPa and 2.2 MPa respectively which correspond to the characteristics determined using compression tests on masonry wallettes. The bricks were horizontally perforated with a void ratio of 60% and have an average compressive strength perpendicular

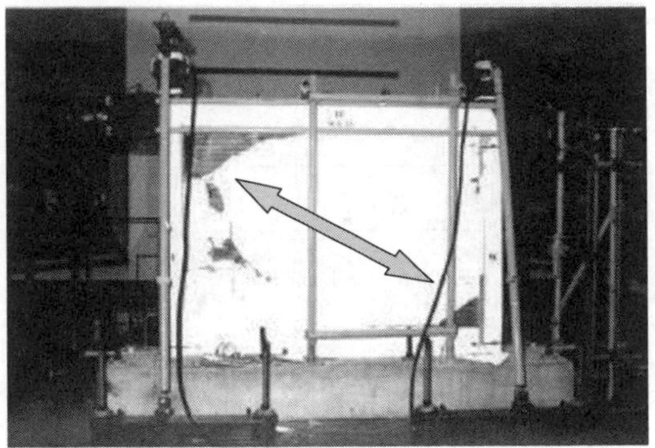

Figure 5. Damage state at the end of the tests showing the crushing of the diagonal strut.

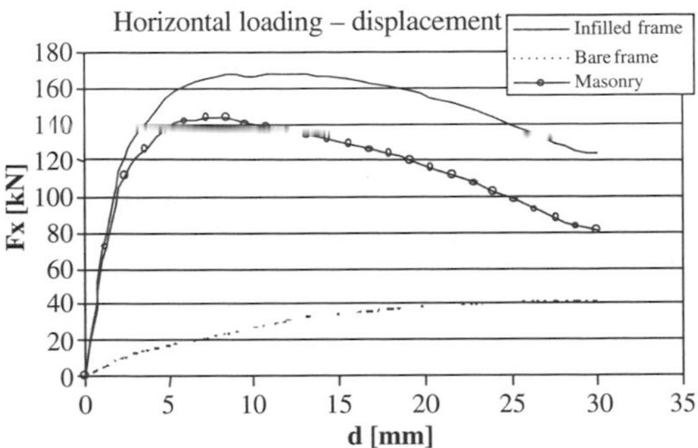

Figure 6. Force-displacement curve for the infilled frame.

to the holes equal to 4.8 MPa. Let remind the masonry panel has the same thickness than the RC columns (15 cm).

The analysis has been conducted under monotonic loading. The calculation gives the global force-displacement curve of the infilled frame (Fig. 6). The curve of the masonry infill is determined by doing the difference between the masonry infilled frame and the bare frame curves. The failure pattern observed during the tests is well captured by the refined modelling: an equivalent diagonal appears between two opposite corners and the maximum strength is reached when masonry begins to crush in the corners (Fig. 7). Failure is also characterized by the motion of the diagonal strut down to the base of the windward column.

During past earthquakes, failure of RC frames with limited damage in the infill has been observed for example in case of short columns. It is thus very interesting to know the interaction between the frame and the infill in the refined modelling. For this purpose, the distribution on the height of shear and axial forces and the evolution of their maximum in function of top displacement has been analysed. Fig. 7 shows also the distribution of shear force for the uniform infilled

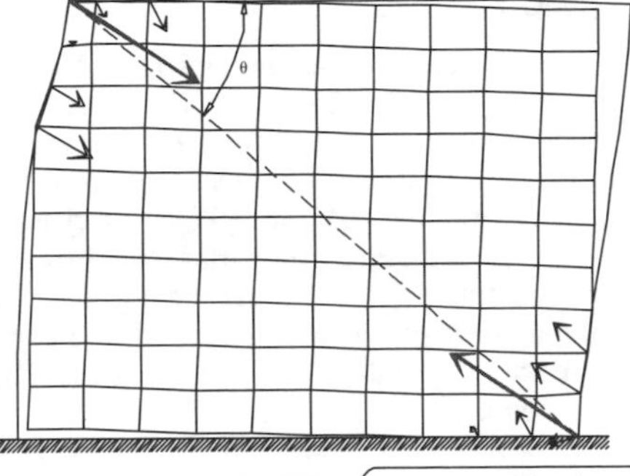

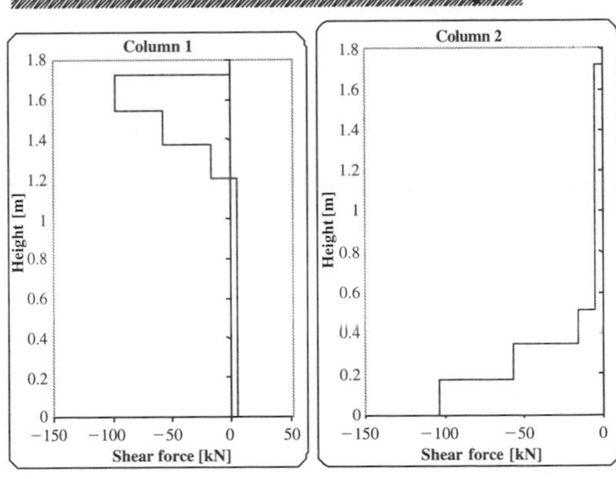

Figure 7. Resulting forces on the masonry and shear forces distribution in the columns corresponding to the maximum value of shear.

Table 2. Maximum values of shear force and axial force in the columns.

	Left column	Right column
Uniformly infilled frame		
Shear force	112 kN (4.98 MPa)	123 kN (5.47 MPa)
Axial force	39 kN	192 kN
Infilled frame with window		
Shear force	78.2 kN (3.47 MPa)	28.2 kN (1.25 MPa)
Axial force	41.3 kN	134.9 kN

frame. Note the increase of shear in the parts of the columns at the extremities of the diagonal strut. The maximum values of shear force and axial force are given in Table 2. In the present case, the computed values of shear stress are very high but the columns had sufficient stirrups to avoid brittle shear failure. For axial force, compression is taken as positive. Thus the maximum value is given for the right column which is in compression and the minimum value for the left column which is in tension. These values include the axial force corresponding to the vertical load which is 80.5 kN per column.

4.2 *Parametrical studies*

One of the main interests of the non linear model is to allow extensive parametrical studies in order to draw general conclusions and formulae for the identification of the mechanical characteristics of masonry infill with different mechanical and geometrical characteristics.

Due to the large number of parameters of each non linear constitutive law, the present analysis focus only on a couple of parameters which are the compressive strength f_c and the Young modulus E_m. The ratio C between the Young modulus E_m and strength f_c has been taken, in a first stage, equal to the value of the reference model (C = 1818) and, in a second stage, equal to 1000 which is a value widely used for masonry. The parameters are both proportional to the same coefficient n:

E_m = n.4000 MPa
f_c = n.2.2 MPa or n.4 MPa
n is equal to n = 1/3; 2/3; 1; 1.5; 2; 2.5 or 3.

A variation of the parameter n can also be representative of a variation of other parameters. For example, the variation of Young modulus E_m and of the compression strength f_c, maintaining constant the ratio between them is equivalent to the variation of the infill thickness t_w. Other parameters such as the relative stiffness between the masonry infill and the frame (E_m/E_c) depend also directly on n.

The parametrical analysis have two objectives:

- Predict the characteristics of the equivalent strut whose values can be compared to some reference formulae. The uncracked stiffness and the maximum strength are the 2 characteristics investigated in the present study.
- Assess the shear force induced by the masonry infill in the columns which can be expressed in percentage of the horizontal component of the diagonal compression in the strut (called additional shear force rate).

The force-displacement curve converted to an axial force-axial strain curve is used for the determination of the elastic stiffness, the cracked stiffness and the ultimate strength of the infill (Fig. 8). These characteristics are defined in accordance with the multilinear envelop curve of the global model. The elastic stiffness is the result of the difference of the results of the elastic calculations on the infilled and the bare frames. The ultimate strength of the global model is defined as 95% of the maximum strength given by the refined model and is reached for an axial strain for which the axial force given by the refined model is equal to 90% of the maximum strength. This point defines also the cracked stiffness. The elastic stiffness and the ultimate strength can be expressed in term of diagonal width if the infill thickness, the Young modulus and compressive strength of masonry are considered. The

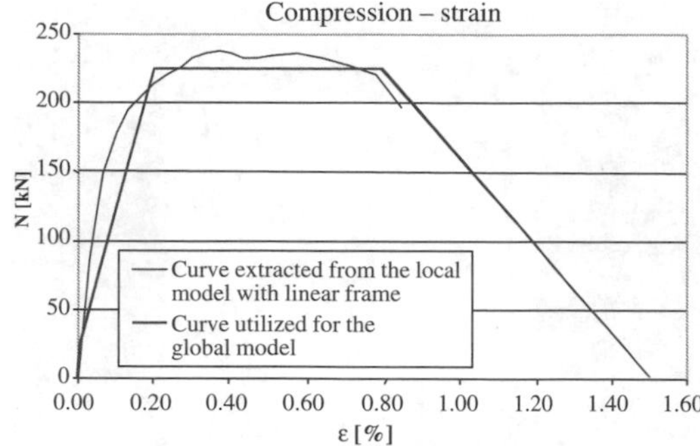

Figure 8. Computation of the axial force-axial strain of the global model with the results of the refined model.

diagonal width is given in percentage of the diagonal length D and is noted %D. One may notice the diagonal widths for cracked stiffness and ultimate strength are identical. Table 3 gives an example of determination of the strut characteristics in the case of a linear frame model.

The parametrical study has been conducted with 2 different models for the RC frame which are the non linear model of the frames tested in Lisbon and a linear elastic frame with a reduced Young modulus in order to take into account cracking (2/3 of the elastic Young modulus of the non linear model). Note that the stiffness and the strength of the masonry infill can strongly depend on the characteristics of the frame as illustrated in the following section.

4.2.1 *Uncracked stiffness*

Figure 9 shows the width of the uncracked diagonal strut versus the masonry Young modulus for the 2 models of frame. These numerical results show a very good agreement with the values given by the formulae of Smith (Smith, 1966) and Dawe and Seah (Dawe et al., 1989). These 2 analytical formulae underestimate only slightly the value of width found by the present parametrical study specially for the lower values of Young modulus E_m. In these formulae, the diagonal width w is function of the contact length between the masonry panel and the surrounding frame and so is directly function of the stiffness of the frame.

The classical formulae of Smith is:

$$w = \frac{\pi}{2\lambda_h}\cos\theta + \frac{\pi}{\lambda_l}\sin\theta \quad \text{(Smith, 1966)}$$

$$\text{with } \lambda_h = \sqrt[4]{\frac{E_m \cdot t_w \cdot \sin 2\theta}{4E_H \cdot I_H \cdot H}} \text{ and } \lambda_L = \sqrt[4]{\frac{E_m \cdot t_w \cdot \sin 2\theta}{4E_L \cdot I_L \cdot L}}$$

with E_w, t_w the Young modulus and the thickness of the masonry infill, E_H, I_H the Young modulus and the inertia of the beam, E_L, I_L the Young modulus and the inertia of the columns, H and L the height and length of

Table 3. Identification of the parameters of the global model using the model with linear frame.

Stage	Cracking	Crushing	Softening	Ultimate
Rel. Displ. d	0.09 mm	7.17 mm	28.2 mm	53.6 mm
Base shear force	18.9 kN	179.2 kN	179.2 kN	0 kN
Axial strain in the strut	0.003%	0.20%	0.79%	1.50%
Axial force in the strut	23.8 kN	225.8 kN	225.8 kN	0 kN
Strut width % D	52.6%	24.1%	24.1%	24.1%

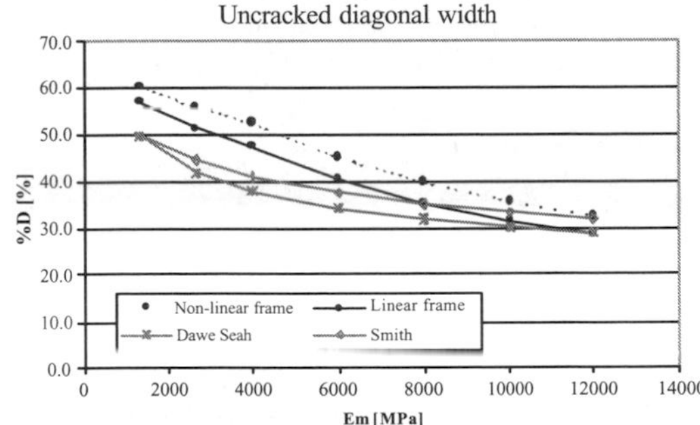

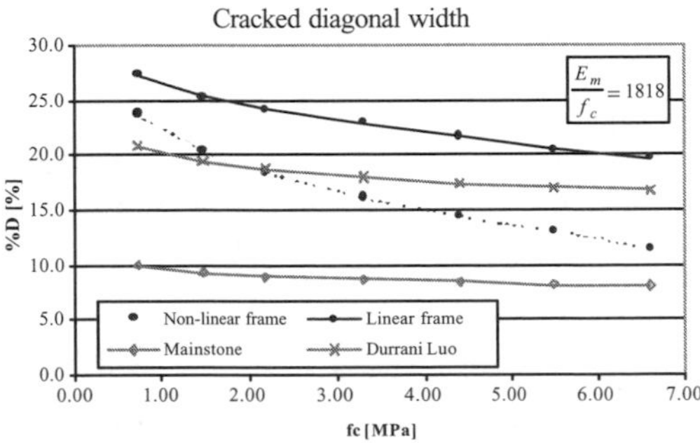

Figure 9. Widths of the uncracked and cracked sections versus the masonry Young modulus and compressive strength.

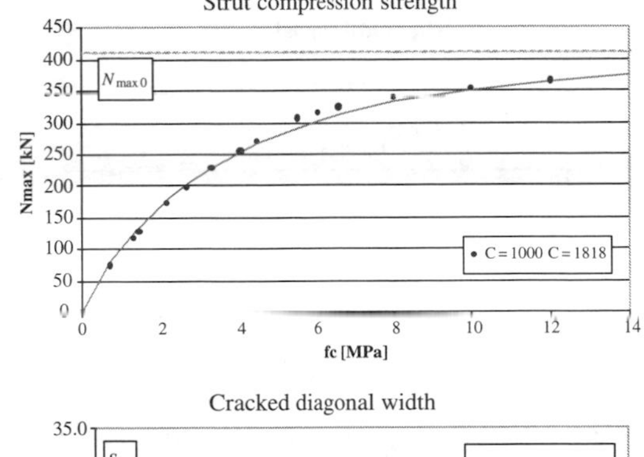

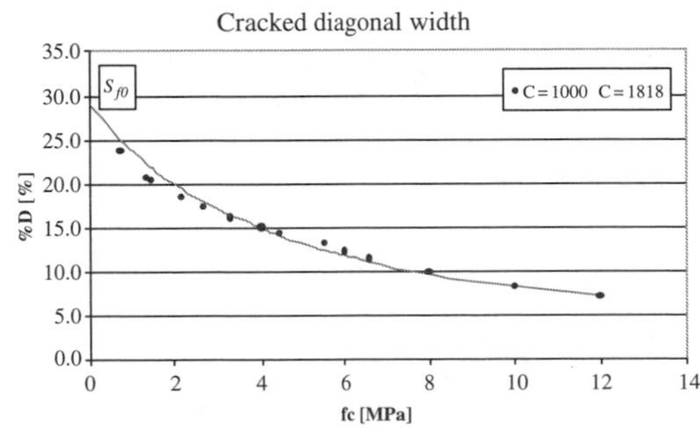

Figure 10. Strut compression strength and width of the cracked diagonal strut versus masonry compression strength (non linear frame).

the masonry infill and θ the inclination of the diagonal strut.

Dawe and Seah give a very similar expression:

$$w=\frac{\pi}{1.5\lambda_h}\cos\theta+\frac{\pi}{1.5\lambda_l}\sin\theta \quad \text{(Dawe et al., 1989)}$$

4.2.2 *Ultimate strength*

Figures 9, 10 and 11 show the width of the cracked diagonal strut versus the masonry compressive strength f_c or the Young modulus E_m for the 2 models of frame. At the opposite of the elastic stiffness, the diagonal width depends strongly on the model of frame (linear or non linear). The numerical results have been compared to the values given by the formulae of Mainstone (Mainstone, 1971) and Durrani and Luo (Durrani et al., 1994) which are respectively:

$$w=0.175(\lambda_H H)^{-0.4}\sqrt{H^2+L^2} \quad \text{(Mainstone, 1971)}$$

$$w=\gamma\cdot\sin\theta\sqrt{H^2+L^2} \quad \text{(Durrani et al., 1994)}$$

with $\gamma=0.32\cdot\sqrt{\sin 2\theta}\cdot\left(\frac{H^4\cdot E_m\cdot t_w}{m\cdot E_H\cdot I_H\cdot H}\right)^{-0.1}$ and

$$m=6\cdot\left(1+\frac{6E_L\cdot I_L\cdot H}{\pi\cdot E_H\cdot I_H\cdot L}\right)$$

These 2 formulae give very different estimation of width. The Durrani and Luo formulae which has been

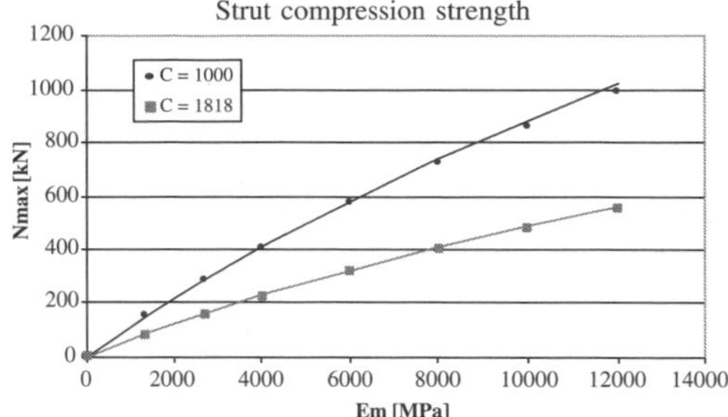

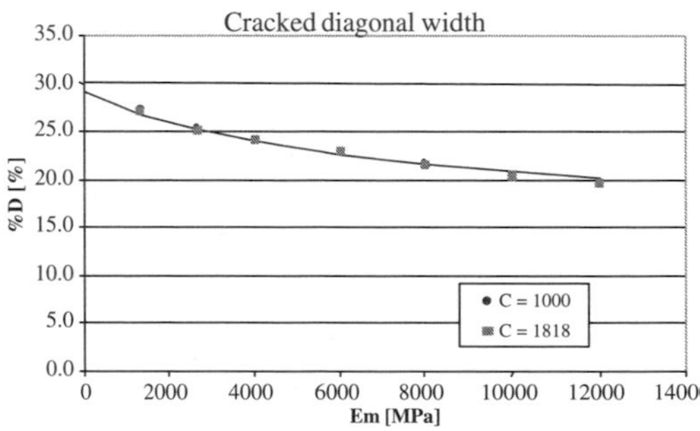

Figure 11. Strut compression strength and width of the cracked diagonal strut versus masonry Young modulus (linear frame).

calibrated on finite element models gives results similar to the results of the parametrical analysis but the Mainstone formulae seems to underestimate strongly the width and so the strength of the diagonal strut.

4.2.3 *Analytical formulae for the non linear and the linear frames*

The results of the parametrical studies have been reassumed in simple analytical formulae which can be used directly for the identification of the diagonal struts of the non linear global model.

The maximum axial force found with the non linear model for the RC frame has 2 properties:

a) for the high values of masonry strength, the axial strength tends to an horizontal asymptote: the maximum strength of the masonry infilled frame is limited by the strength of the frame and the yielding of the column in tension. The corresponding value of axial force is equal to:

$$N_{\max 0} = \frac{A_s \cdot f_y}{sin\theta}$$

with f_y the steel yield stress and A_s the total section of steel bars in the column section

b) for the low value of masonry strength, the width of the diagonal strut tends to 30% of its length.

The following formula fulfills these 2 conditions:

$$N_{\max}(f_c)=N_{\max 0}\cdot\left(1-(b\cdot f_c+1)^{-a}\right) \text{ with } b=\frac{0.3\cdot\sin(2\theta)\cdot D\cdot t_w}{a\cdot N_{\max 0}}$$

and $a = 2.22$

The value of a has been determined to minimize the difference between the analytical and the numerical results.

For a linear model for the frame, the infill strength is not anymore limited by the capacity of the surrounding frame. On Fig. 11, the section of the cracked diagonal strut $S_f(E_m)$ seems independent on the coefficient C. The strut compression strength N_{max} becomes:

$$N_{\max}(E_m)=\frac{E_m}{C}\cdot S_f(E_m)$$

In a first approximation, the diagonal width can be taken equal to a value between 20% and 30% of the diagonal length. For the present case, it has been chosen the following analytical formula:

$$N_{\max}(E_m)=\frac{d}{C}\cdot\left((b\cdot E_m+1)^a-1\right) \text{ with } b=\frac{0.3\cdot\sin(2\theta)\cdot D\cdot t_w}{a\cdot d}.$$

The error minimization gives $a = 0.7$ and $d = 355$ MN.

The application of the 2 analytical formulae to the infilled frames considered in the present study shows a very good agreement with the results of the parametrical analysis (Figs. 10 and 11).

4.2.4 *Shear forces induced in the RC frames*

The global model based on 2 diagonal struts is not able to catch the additional shear forces induced in the infill by the masonry infill since the struts are fixed directly to the nodes common to the beam and the columns. This is not the case for the refined computations which give the total shear force in the columns for both the infilled frame and the bare frame. So the additional shear force can be computed by doing the difference with the values found on the bare frame. The additional shear force has been compared to the horizontal projection of the axial force in the strut $N.\cos\theta$. In order to generalize the results of the present study, a dimensionless ratio named additional shear rate has been introduced:

$$\rho=\frac{T_{\text{infill}}}{\cos\theta\cdot N_{\max}}$$

So one may find the shear induced by the masonry infill in the column to be added to the shear directly given in the columns by the calculation performed with the global model:

$$T_{\text{total}} = T_{\text{column}} + T_{\text{infill}}$$

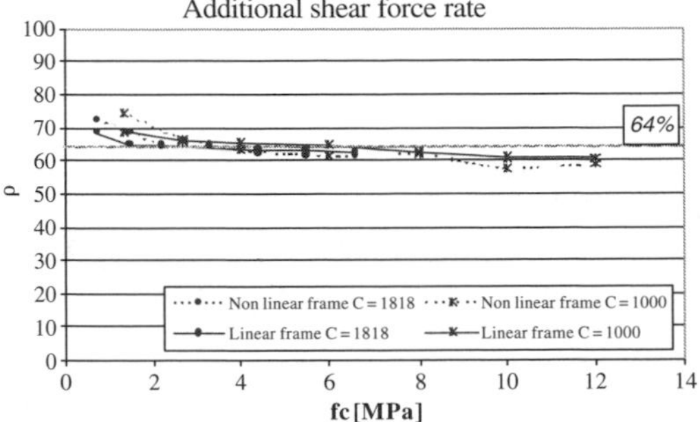

Figure 12. Additional shear force rate versus compressive strength f_c.

A remarkable result has been evidenced by the parametrical studies: for this particular geometry of infill panel, the additional shear rate does not depend on the masonry characteristics and is equal to 64% (Fig. 12). This rate is also almost independent of the modelling of the frame (linear or non linear).

5 CONCLUSIONS

The present paper has presented an application of a local-to-global approach to the modelling of masonry infilled frames under extreme horizontal loading. An extensive parametrical analysis has been performed on one-bay one-storey infilled frames with two levels of modelling: the non linear FEM model allows to understand the failure pattern under monotonic loading, to estimate the forces created by the infill panel in the surrounding frame and to identify the properties of the global model of infill while the global model is used for the analysis of complete structures under simplified static loading or dynamic loading. Simple rules are also given for the identification of the diagonal properties and the shear forces into the surrounding RC frame. The dispersion of the results given by the formulae available in the literature which can be used to characterize both the uncracked and the cracked diagonal struts must be highlighted. The values of diagonal width depend also strongly on the characteristics available for masonry (bricks, mortar or masonry). These results can be directly used for the assessment of existing structures which requires efficient and simple procedure of calibration of the model parameters.

REFERENCES

Combescure D. 1996. Modélisation du comportement sous chargement sismique des structures de bâtiment comportant des murs de remplissage en maçonnerie. *PhD Thesis prepared at JRC Ispra. Ecole Centrale Paris*. France.

Combescure D., Pegon P. 2000. Application of the local-to-global approach to the study of infilled frame structures under seismic loading. *Nuclear Engineering and Design 196* Vol.1, 17–40.

Dawe J.L., Seah C.K.1989. Analysis of concrete masonry infilled steel frames subjected to in-plane loads. *5th Canadian Masonry Symposium*. Vancouver. Canada.

Durrani A.J., Luo Y.H. 1994. Seismic retrofit of flat slab buildings with masonry infill. *NCEER Workshop on Seismic response of masonry infills*. San Francisco, USA.

Guedes J. 1997. Seismic behaviour of reinforced concrete bridges. Modelling, numerical analysis and experimental assessment. *PhD Thesis prepared at JRC Ispra, Universidade do Porto*,. Portugal.

Klingner R.E., Bertero V.V. 1976. Infilled frames in earthquake resistant construction. *Report 76-32. University of California*. Berkeley. USA.

LNEC. 1998. Study of the earthquake behaviour of masonry infilled reinforced concrete frames. Tests of model I1 to I9. *C3ES Research reports, Laboratorio Nacional de Engenharia Civil*, Lisbon, Portugal.

Mainstone R.J. 1971. On the stiffness and strength of infilled frames. *Proc. Inst. Civil Eng., Suppl. IV*, Paper 7360S, 57–90.

Mehrabi A.B., Shing P.B., Schuller M.P., Noland J.L. 1994. Performance of masonry infilled R/C frames under in-plane lateral loads. *Report CU/SR-94/6. University of Colorado*. Boulder. USA.

Pires F. 1993. Influencia das paredes de alvenaria no comportamento de estructuiras reticuladas de betao armado sujeitos a accoes horizontaes. *PhD Thesis. LNEC Report*. Lisbon. Portugal.

Smith B.S. 1966. Behaviour of square infilled frame. *Journal of the Structural Division*, Vol. 92, n. ST1, pp.381–403.

Computational Modelling of Concrete Structures, Bićanić et al. (eds)
© 2003 Swets & Zeitlinger, Lisse, ISBN 90 5809 536 3

Toward a high-quality FEM analysis of RC members subjected to reversed cyclic shear

H. Noguchi
Chiba University, Chiba City, Japan

ABSTRACT: The material models were improved especially in order to represent the cyclic behavior of RC structural members more precisely in two-dimensional FEM analytical program. The transition hysteretic models of tension-compression stresses of concrete, shear stiffness hysteretic models of cracked concrete, and realistic multi-curves hysteretic models of reinforcement were developed. These models could improve the precision of the stiffness and strength degradation, unloading and reloading stiffness, slip stiffness, residual strains and degradation characteristics under reversed cyclic loading in the hysteretic loops of materials. The revised FEM program was applied to RC panels subjected to cyclic shear and normal stresses and RC shear walls surrounded by columns and beams subjected to cyclic shear in order to verify the revised analytical models. The analytical results gave a correct estimation of unloading and reloading hysteretic shape of load–deflection curves and residual displacement as compared with the test results.

1 INTRODUCTION

In the earthquake resistant design for reinforced concrete (RC) buildings based on the inelastic displacement concept (AIJ, 1990), seismic input energy is absorbed by the ductile flexural yielding mechanisms of the whole moment resisting frames. In this design concept, it is important that specific necessary lateral loads, story drifts and structural member deformation can be kept within the design limitation during the earthquake in order to ensure the seismic performance. The correct analytical prediction of the energy absorbed by the structural members under the reversed cyclic loading at the earthquake is necessary to assure the design concept.

The FEM analysis has become a powerful tool to investigate the shear behavior of RC members, because the material properties can be directly reflected to the microscopic FEM analysis through the constitutive laws of materials. For grasping the influence of earthquake input, it is important to develop a method to examine analytically the behavior of RC members subjected to reversed cyclic loading. In the FEM analysis of RC members subjected to reversed cyclic loading, the target of the analytical model is to simulate the hysteretic behavior of the structural members correctly. The loading rate and loading hysteresis are not constant and simple but irregularly changed in the real earthquake loads. Therefore, it is required that the strain rate effect should be considered into the loading rate modeling and reversed cyclic loading model is necessary for the loading hysteresis modeling. But the strength degradation in the inelastic area under the cyclic loading that means the inelastic deformation capacity can not be grasped sufficiently in the previous analytical study of structural members, elastic-plastic response analysis of frames and design methods of deformation capacity. It is essential to clarify the inelastic characteristics of structural members and to establish the method to utilize the inelastic characteristics effectively to practice the structural performance design effectively and precisely.

In this study, more realistic analysis models such as the hysteresis model for the stress-strain relationship of concrete especially in tension-compression regions, the multi-curves model for reinforcement, and the histeresis model for shear characteristics of cracked concrete were incorporated into the two-dimensional non-linear FEM analysis programs developed previously by the author (Noguchi et al., 1999). Thus the characteristics of the behavior of RC under cyclic loading, such as the tangent stiffness for unloading and reloading, slip stiffness, residual strain and the deterioration under cyclic loading, could be simulated more precisely. Then the histeresis loop was reproduced and the effectiveness of those analysis were verified through the analysis of RC panels and RC shear walls surrounded by beams and columns which are subjected to combined stresses.

2 ANALYSIS MODELS

2.1 *Cyclic behavior rules of concrete and reinforcement*

In this analysis, according to the knowledge provided by the previous experiment about cyclic behavior of concrete and the models proposed by Naganuma & Ohkubo (2000), the cyclic hysteresis loops including unloading and reloading curves in tension and compression regions and the regions between tension and compression are defined as shown in Figures 1–4 using multi-curves. They can simulate real cyclic hysteresis behavior of concrete very well and also contribute to removing the difficulty of the convergence in solving the equations. And the cyclic hysteresis of reinforcement is defined as shown in Figure 5 by using the Menegotto-Pinto model proposed by Ciampi & Paolo (1982).

Order : O: E: C: D: P: C: X
P. E : Unloading Point
P. C : Common Point ($\sigma_C = 5/6\sigma_E$)
P. D : Stiffness Change Point ($\sigma_D = 1/2\sigma_E$)
P. P : Compression Side Residual Strain Point (Suggestion Type of Karsan & Jirsa (1969))
P. X : Envelope Curve Recovering Point
P. M : Maximum Strength Point

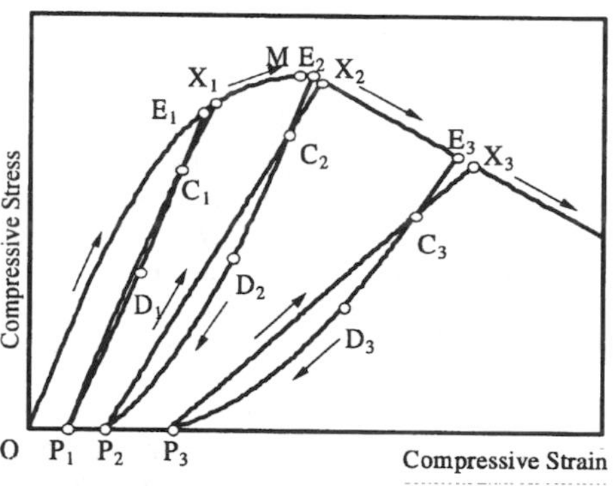

P. O-M : Compression Envelope Curve (Type of Saenz (1964))
P. E-D : The Straight Having the Slope According to a Compressive Strain(by Naganuma & Ohkubo(2000))
P. D-P : Second curve
P. P-X : Straight

Figure 1. Model of compression area history of concrete.

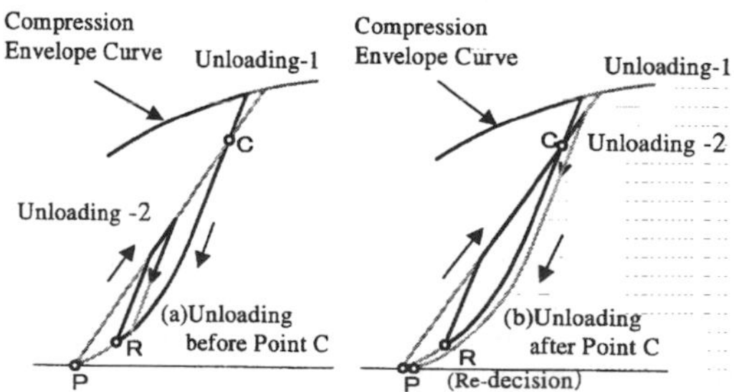

Figure 2. Unloading/reloading rule in compression area.

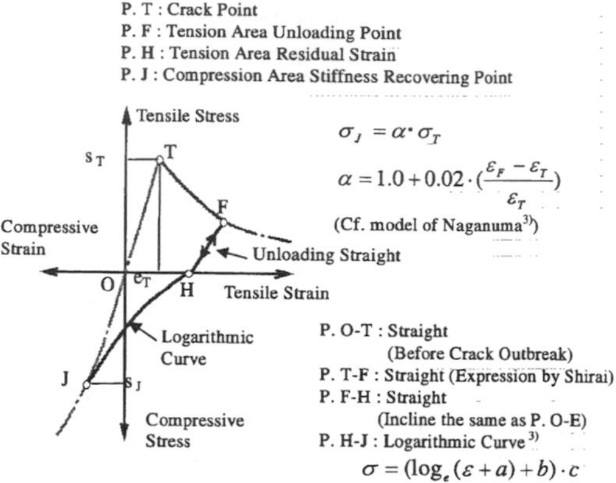

Figure 3. Model of tension area history of concrete.

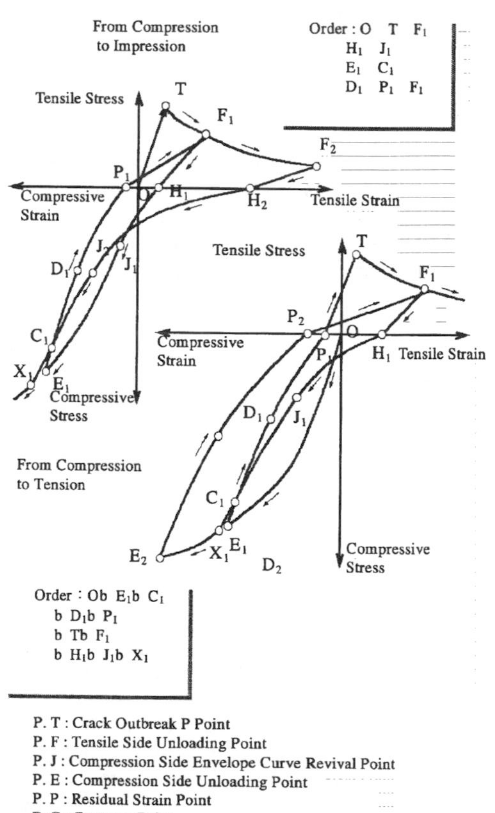

Figure 4. History between compression and tension of concrete.

The examples of proposed analytical hystresis loops are shown in Figure 6 as compared with the corresponding compressive stress–strain curves obtained in the previous tests by Karsan & Jirsa (1969) and Tanigawa & Kosaka (1978). As the reloading curve

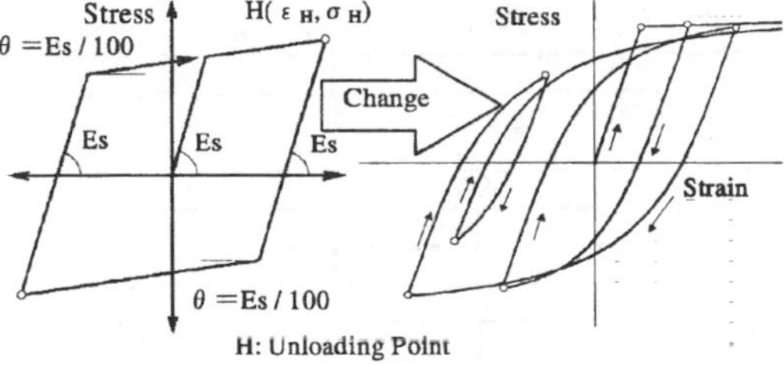

Figure 5. Behaving rule of reinforcing bar (left) traditional bi-linear (right) Menegotto-Pinto model.

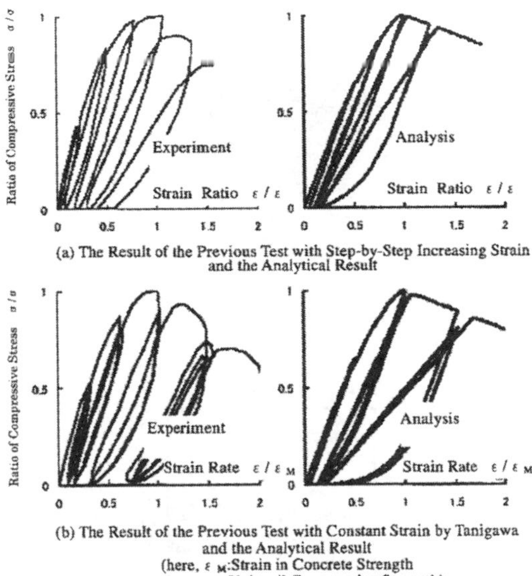

Figure 6. The tests and analyses under compressive cyclic loading.

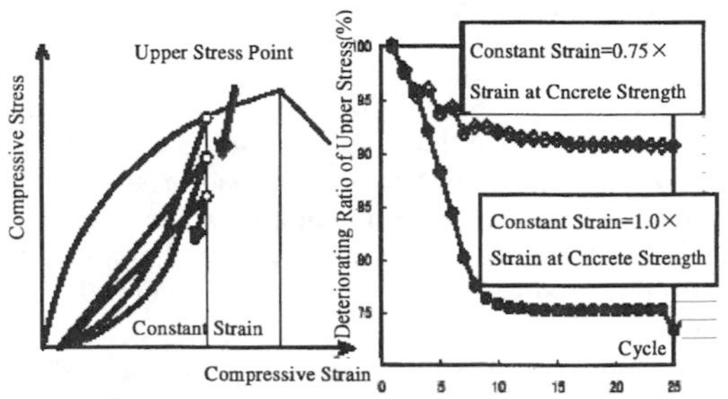

Figure 7. Degradative behavior of upper stress under cyclic load with constant strain in compression zone.

was represented by straight lines, the area of the analytical hysteresis loops was a little smaller hysteresis loop increased as the strain at the unloading point became high-leveled. From the comparison with Tanigawa's test result (1978) in Figure 6, the stress degradation from cyclic loading with constant strain can be represented. But the degradation level is smaller in the analysis results as compared with the test results.

The relationships between the reduction ratio of the upper limit stress and the number of cyclic loading with constant compressive strain in this model are shown in Figure 7. The reduction of the upper limit stress at each loading cycle becomes more remarkable as the number of loading cycle or the compressive strain increases. The reduction ratio tends to converge nearly at the tenth cycle independently of the magnitude of the upper limit strain.

2.2 *Damage model of RC elements*

As for concrete, 8-nodal isoparametric elements were used, and as for the failure criterion of compressive strength reduction factor of cracked concrete were modeled as follows.

The tension stiffening effect in the orthogonal direction of a crack depending on bond was represented by the Sato & Shirai's equation (1978) as shown in Figure 3. As for the shear stiffness along the crack surface, the Ihzuka & Noguchi's model (1992), which substituted the dowel stiffness of reinforcement for the equivalent stiffness of concrete, was superimposed on the concrete, the Kupfer & Gerstle's model (1973) was used. The directions of the principal axes were the same as the directions of principal stress before cracking of concrete. When the concrete element cracks, the principal axes were fixed and did not change again. These directions were calculated by the Mohr's stress circle at each loading step.

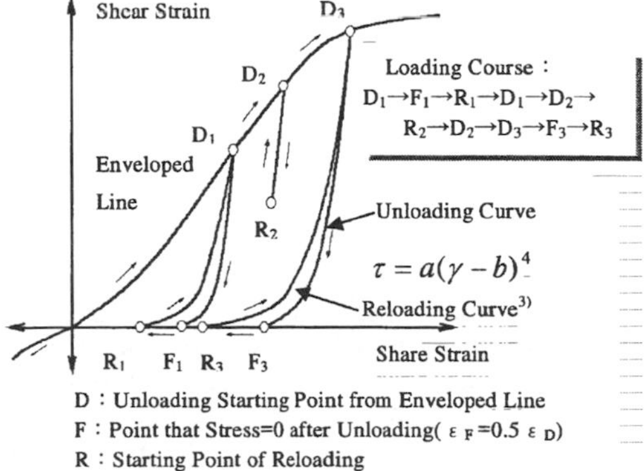

Figure 8. Modeling of shear behavior in the cracking direction.

Table 1. Specifications of RC Plate Test Specimen.

	Specimen Name		SE 8		SE 9		SE 10	
	Size of Specimen	(mm)	1625×1625×285					
	Measuring Range	(mm)	1524×1524					
	Loading Method		Positive and Negative Cyclic Shear Force				Cyclic Shear Force + Bi-axial Compressive Stress	
	a		0		0		- 1/3	
Concrete	Compressive Strength	(Mpa)	37.5		44.2		34	
	Compressive Strain	($\times10^{-3}$)	2.60		2.65		2.20	
	Cleavage Strength	(Mpa)	3.4		4.3		4.0	
Rein-forcing Bar	Method of Bar Arrangement		X	Y	X	Y	X	Y
	Rainforcing Ratio	(%)	2.94	0.98	2.94	2.94	2.94	0.98
	Bar Arrangement Interval	(mm)	72		72		72	
	Used Reinforcing Bar		20M	10M	20M	20M	20M	10M
	Yield Strength	(Mpa)	492	479	422	422	422	479

a : Ratio of Bi-axial Compressive Force to Loaded Shear Force

As for the nonlinear behavior characteristics of cracked concrete elements, tension stiffening, the shear stiffness along the crack surface and the stitutive law of the shear transfer proposed by the Okamura & Maekawa (1991) using the contact density function along the crack surface. The shear stiffness to the crack and the tensile strain normal to the crack. This is a model that the cross effects of G is calculated with equation $G = G_m + G_i$, here, G_m is the stiffness calculated with the Okamura & Maekawa's equation, and G_i is the dowel stiffness calculated with the Ihzuka's equation. Both of G_m and G_i is defined as a function of shear strain parallel compression/tension and shear is represented by the fluctuation of G. The relationships between shear stress and shear strain are shown in Figure 8. And as for the shape of hysteresis loop in unloading or reloading, the equation proposed by Naganuma & Ohkubo (2000) was used.

As for the compressive strength in the direction parallel to the crack, the Noguchi & Ihzuka's equation (1992) that considered the reduction of the compressive strength with parameters such as the crack width (tensile strain) and the uni-axial compressive strength was used.

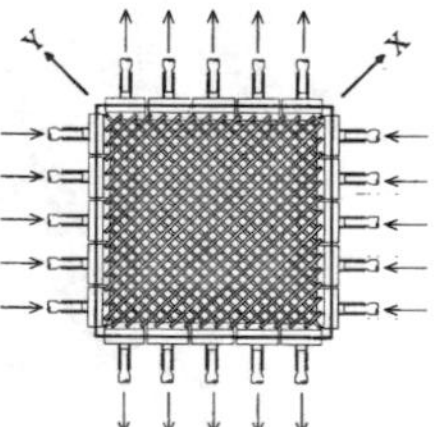

Figure 9. Outline of loading device in the panel test by Collins et al. (1987).

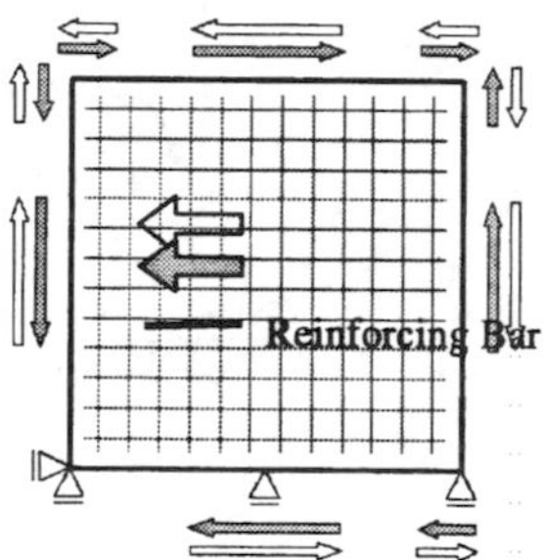

Figure 10. Boundary conditions of RC panels.

3 ANALYSIS OF RC PANELS

3.1 *Outline of the analysis*

The analytical test specimens were chosen from the panels specimens tested by Stevens, Uzumeri & Collins (1987), including SE8, SE9, which were subjected to reversed the cyclic shear loading (the rate of reinforcement is set as the parameter), SE10, which was subjected to shear loading and axial loading simultaneously. The specifications of the test specimens are shown in Table 1. The shape of the test specimens are shown in Figure 9, and the finite element mesh and the boundary condition were shown in Figure 10. As shown in Figure 9, because a pure shear stress field is assumed in the direction parallel to the reinforcement, it is supposed that the stress condition can be reproduced sufficiently with just one element. The two-dimensional non-linear FEM analysis program using the cyclic hysteretic models of materials, which have been introduced in Chapter 2. The reinforcement is represented as multi-layered elements. The loading divided to the nodes in the ratio of 1:4:1 along the

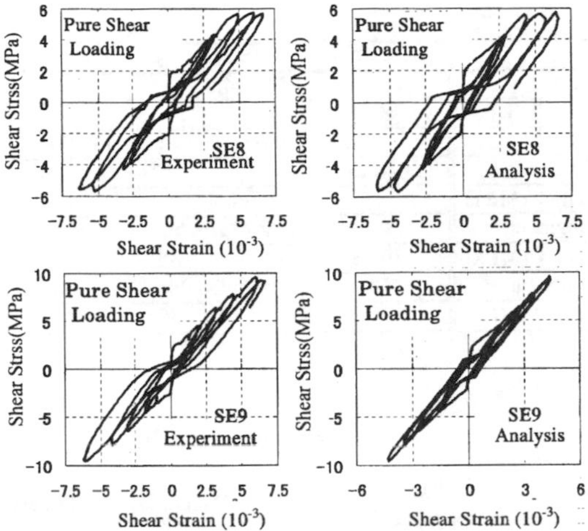

Figure 11. Shear stress–shear strain relation of SE8,9.

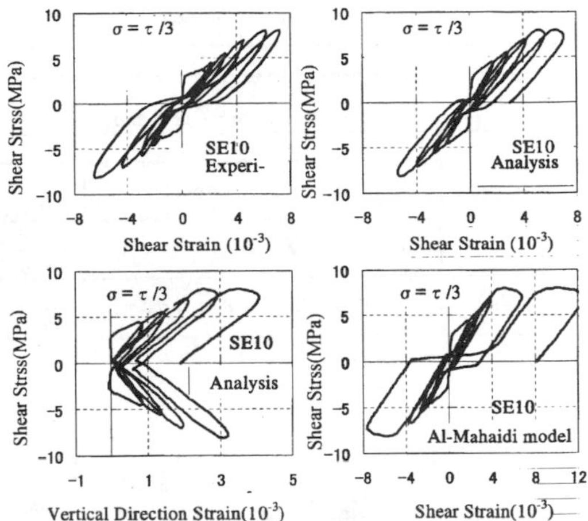

Figure 12. Shear stress–shear strain relation of SE10.

sides of the panels was applied in the direction of the arrow under loading control. According to the results of these analyses, equal fields of stress and strain were represented in each integral point.

3.2 *Analytical results*

The analytical results and the test results of specimens SE8 and SE9 are shown in Figure 11. The yielding of the bars causes the failure of specimen SE8, which was reinforced with different ratio in each direction, and the reversed cyclic loading was applied until the strain in reinforcement was much more than the yield strain. Good agreements were obtained in the yield strength, envelope curve, unloading and reloading curve, and the hysteresis in the slip region between the test results and the analytical results at each cycle. Especially at the 3rd and 4th cycle, where the shear stress was fixed as the stress amplitude in 6 MPa, the increase of the shear strain showed a great agreement with the test result. But about specimen SE9, which was reinforced with the same ratio in each direction, the shear strain is smaller than the test result. One of the reasons is considered as that the degradation of the concrete according to the loading cycle number wasn't considered in this program. The area of the hysterety is smaller than the test results, although the qualitative tendency can be represented.

In the experiment, at the cycle up to $\tau \cdot 9.55$, the concrete compressive failure of the concrete and the yielding of reinforcement occurred and the shear strain increased step by step. All of this made the loop shape of the history expanded. In the other hand, in the analysis, the strain didn't increase because the reinforcement didn't yield. To improve the hysteretic shape of the loop, the average stress-strain relationship of the embedded reinforcement should be checked up, and the suitable yielding stress of reinforcement should be used referencing the real behavior.

The shear stress–strain relationship and the relationship of shear stress and vertical strain of specimen SE10 are shown in Figure 12. The specimen SE10, which was reinforced with different ratio in each direction and had great slip at the cracks, is considered to be a good specimen for checking the transfer of the shear stresses. The analysis using the models introduced in Chapter 2 is compared with the analysis using the A1-Mahaidi model (1979) for the shear stiffness after the occurrence of cracks. The A1-Mahaidi model represented the reduce of the shear stiffness as a function of tensile strain normal to the cracks. But because this model doesn't support the unloading behavior, the stiffness after unloading was reduced too much and the strain is greatly lager as shown in lower right of Figure 12. In the other hand, the result of the proposed model (upper right) generally gave a good agreement with the test result.

4 ANALYSIS OF RC SHEAR WALLS

As shown in Chapter 3, it was recognized that the analytical result using the proposed constitutive model gave a good agreement with the test one of RC panels. In this chapter, RC shear walls were analyzed in order to discuss the validation of this FEM modeling for RC structural members.

4.1 *Outline of the test*

The analytical test specimens for analysis, including P2004, P2012 and P4012, is chosen from the RC shear walls subjected to reversed cyclic loading, which were tested by Aoyama, Kato & Katsumata in 1982.

Table 2. Details of shear wall specimens.

Specimen Name				P2004	P2012	P4012
Way of Making Test Body				Monolithic		
Concrete Compressive Strength			(MPa)	28.15	29.13	
Perimeter Column	Cloumn Size		(mm)	200 × 200		400×200
	Main Reinforcement	Bar arrangment		4-D13	12-D13	
		Yield Strength	(MPa)	360.9		
	Column Shear Rineforcement	Bar arrangment		D6-@75 Double		
		Yield Strength	(MPa)	391.3	355	
Wall	Vertical and Lateral Reinforcement	Bar arrangment		D6-@75 Double		
		Yield Strength	(MPa)	391.3	355	

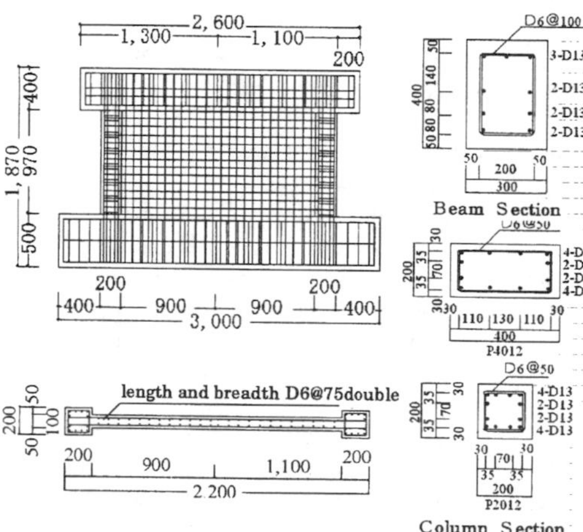

Figure 13. Details of shear wall specimen.

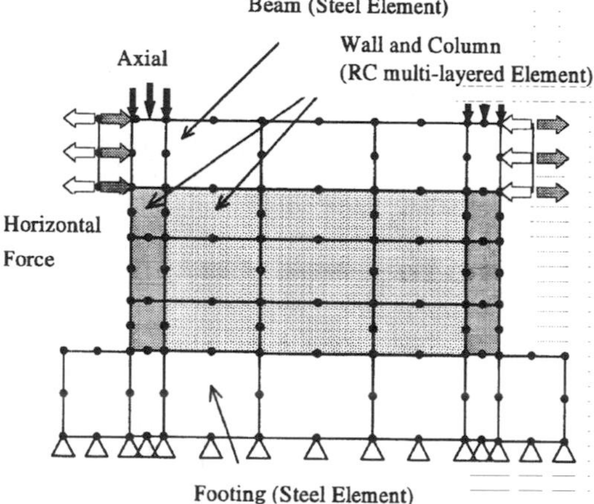

Figure 14. Finite element model and boundary condition (P2012).

The specifications of the test specimens are shown in Table 2, and the details of the specimens are shown in Figure 13. The parameters include the volume of reinforcement and the restraint to the walls offered by the frame around the walls, which was changed by the width of the column. Shear forces by push- and pull-over loadings of both upper beam-ends in the specimen were applied under an axial load of 118 kN in both side-columns respectively. Specimen P2004 with the remarkable shear slip failure at the top of the shear wall showed the deterioration of shear capacity, The concrete compressive failure along diagonal shear cracks at the middle of the shear wall in Specimen P2012 occurred after shear slip failure at the faces adjacent to bottom of side-columns in the test. On the other hand, Specimen P4012 reached to the failure by concrete compressive damages occurred at the middle of the shear wall.

4.2 *Outline of the analysis*

The division of the analytical elements and the boundary condition are shown in Figure 14. As the analyses for wall panels, reinforcements are represented as a part of multi-layered element, and it was considered that there is no slip occurred between concrete elements and reinforcement elements. As crack model, distributed crack model was used. In the same way as the test loading method, at the first stage, the distributed axial loading were applied to the columns, then, cyclic lateral loading were applied to the end of the beams under displacement control.

4.3 *Analytical results*

The relationship of shear loading and lateral displace is shown in Figure 15. Each of the specimen showed a good agreement in the stiffness under unloading or reloading and in the residual displacement, at the 2nd and 3rd cycle, under the reversed cyclic loading with the amplitude of displacement fixed, the reduce of the yield strength can be confirmed in the analysis even though a little larger reduction was obtained than the tests. But, as shown conspicuously at the 2nd and 3rd cycle, the parts of the hysteresis loop after the residual displacement point, extended inside the same parts of

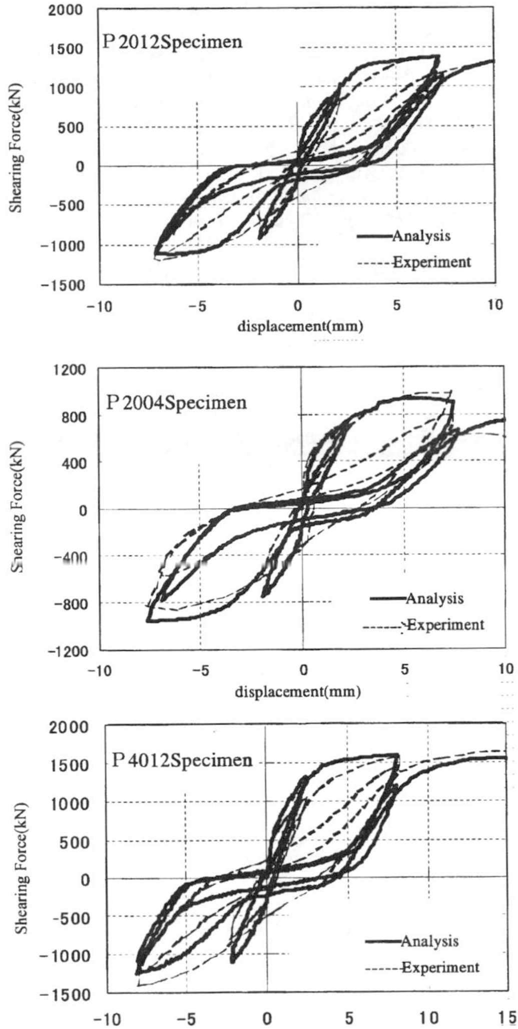

Figure 15. The relationship between shear force and horizontal displacement of RC shear walls.

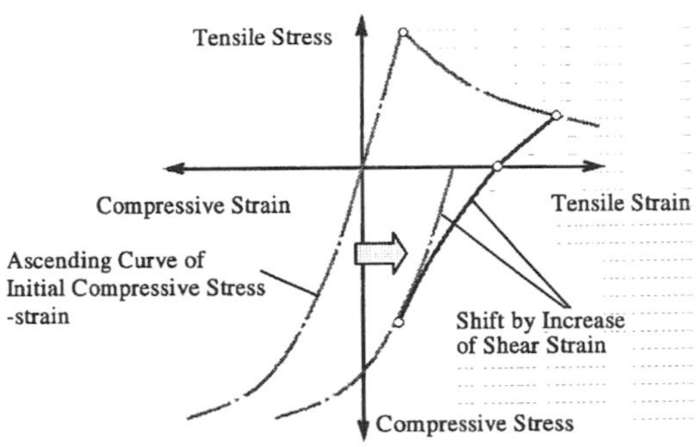

Figure 16. Effect of shear strain on the hysteresis curves.

the hysteresis loop of the experiments, and made the hysteresis loop like a contra "S" shape. It means that, in the analysis, the restoration of the stiffness on account of the aggregate interlock was under estimated than the tests. For this reason, the hysteretic models from the H point (residual displacement point in tension region) to J point (the restoring point in compression region), which is shown in Figure 3, should be reviewed. As shown in Figure 16, it is the model considering the strain shift effect to tension side, which is caused by the increase of the shear strain and the large shear strain.

The initial stiffnesses of each analytical specimen are 10% ~ 20% larger than the tests, it is because that the columns and the walls were fully fixed with the basement, so the slip at the foot of the column couldn't occur. If the slip can be represented using cracking elements, it is considered that the envelope curve will be closer to the test results.

5 CONCLUSIONS

The RC panels and shear walls were analyzed by two-dimensional FEM program using revised concrete and reinforcement hysteresis models considering the deterioration of the materials under cyclic loading. The analytical load-deformation, unloading and reloading hysteresis loop shape and residual displacement at each loading cycle gave a good agreement with the test results.

As the future subjects of the FEM analysis of cyclic loading, it is necessary to develop discrete crack elements which can represent cyclic loading hysteresis behavior including the shear slip along cracks and local crack opening. They will contribute to the development of the accurate FEM analysis of poorly reinforced concrete members with remarkable shear slip along cracks and crack opening.

ACKNOWLEDGEMENTS

The work reported in this paper was sponsored by Grant-in-Aid for Science Research (B) (2) (No. 14350293, Representative Researcher: Noguchi H., Chiba University) of Japan Society for the Promotion of Science. The author wishes to express his gratitude to Mr. Kashiwazaki T. of a research associate, Messrs. Kobayashi S. and Sakurai T. of former graduate students, and Mr. Yu Y. of a graduate student in Chiba University.

REFERENCES

Al-Mahaidi, R. S. H. 1979. Nonlinear Finite Element Analysis of Reinforced Concrete Deep Members. *Report*

No. 79-1. Department of Structural Engineering, Cornell University.

Aoyama, H., Kato, D. & Katsumata, H. 1982. An Experimental Study on Strength and Ductility of Infilled RC Shear Walls. *Summaries of Technical Papers of Annual Meeting* C-2: pp.1407–1410.: Architectural Institute of Japan.

Architectural Institute of Japan. 1990. *Design Guidelines for Earthquake Resistant Reinforced Concrete Buildings Based on Ultimate Strength Concept*: Architectural Institute of Japan.

Ciampi, V. & Paolo, E. 1982. Analytical Model for Concrete Anchorages of Reinforcing Bars under Generalized Excitations. *Report No.UCB/EERC-82/23*. Earthquake Engineering Research Center, University of California.

Ihzuka, T. & Noguchi, H. 1992. Nonlinear Finite Element Analysis of Reinforced Concrete Members with Normal to High Strength Materials. *Proceedings of the Japan Concrete Institute* Vol.14, No.2: pp. 9–14: Japan Concrete Institute.

Karsan, I. D. & Jirsa, J. O. 1969. Behavior of Concrete under Compressive Loadings. *Journal of the Structural Division* Vol.95, No.ST12: pp. 2543–2563.: American Society of Civil Engineers.

Kupfer, H. B. & Gerstle, K. H. 1973. Behavior of Concrete under Biaxial Stresses. *Journal of the Engineering Mechanics Division* Vol.99, No.EM4: pp.853–866: American Society of Civil Engineers.

Naganuma, K. & Ohkubo, M. 2000. An Analytical Model for RC Panels under Cyclic Stresses. *Journal of Structural and Construction Engineering* No.536: pp.135–142.: Architectural Institute of Japan.

Nozaki, Y. & Noguchi H. 1999. A Study on RC Models for FEM Analysis of RC Panels Subjected to Cyclic Loading. *Proceedings of the Japan Concrete Institute* Vol.21, No.3: pp.769–774.: Japan Concrete Institute.

Okamura, H. & Maekawa, K. 1991. *Nonlinear Analysis and Constitutive Models of Reinforced Concrete*: Gihodo Publication.

Saenz, L. P. 1964. Discussion of "Equation for the Stress-Strain Curves of Concrete" by Desayi and Krishnan. *Proceedings of American Concrete Institute* Vol.61, No.9: pp.1229–1235.: American Concrete Institute.

Sato, T. & Shirai, N. 1978. Elasto-Plastic Behavior of RC Shear Walls. *Summaries of Technical Papers of Annual Meeting* C-2: pp.1615–1618.: Architectural Institute of Japan.

Stevens, N. J., Uzumeri, S. M. & Collins, M. P. 1987. Analytical Modeling of Reinforced Concrete Subjected to Monotonic and Reversed Loadings: *Publication No.87-1*. Department of Civil Engineering, University of Toronto.

Computational Modelling of Concrete Structures, Bićanić et al. (eds)
© 2003 Swets & Zeitlinger, Lisse, ISBN 90 5809 536 3

PC members with bonded and un-bonded tendon under uniaxial cyclic load; experiment and analysis

M.R. Salamy & T. Higai
Yamanashi University, Yamanashi, Japan

T. Yoshioka
Oriental Construction Co., Tokyo, Japan

ABSTRACT: An experiment on the behavior of PC and RC members under cyclic uniaxial tension was conducted as a part of the activity of the PC Tank Sub-committee of The Japan Pre-stressed Concrete Contractors Association. The aim of the experiment was to study the effect of repeated loading that occurs during moderate earthquakes on the behavior and the residual cracks of pre-stressed tank structures and its simulation by finite element analysis. Subsequently, the above-mentioned experiment results were used to test the precision of an FEM analysis in predicting the cyclic behavior of the tested specimens. Although the overall analytical responses of the specimens are in good agreement with experiment, residual crack width was in some cases different than those observed in experiments. Analytical bond stress-slip relationship which has important effect in our analyses has been calculated based on CEB design code.

1 INTRODUCTION

Pre-stressed concrete structures are widely used in any kind of structures at the present with any combinations of different structural members. Pre-stressing force can be applied to columns, beams, slabs, foundations as well as membrane elements and thick shells. In this paper, PC members under cyclic load are studied in detail to evaluate reliability of analytical prediction of residual cracks occurrence due to the reversal loading.

This study consists of two parts; experiment and analysis of RC and PC members under direct cyclic tension. The experimental results of this study were mainly to investigate occurrence of the residual cracks after moderate earthquake in pre-stressed concrete water tanks which are also used here to verify and enhance the analytical prediction. An experiment on the behavior of pre-stressed and nonpre-stressed concrete members under cyclic uniaxial tension was conducted in 1998 as a part of the activity of the PC Tank Sub-committee of The Japan Pre-stressed Concrete Contractors Association (1998, 1999). The experiment studied mainly the effect of two parameters, reinforcement steel ratio and pre-stressing stress, by comparing the behavior of two series of specimens. The first series (designated A series) comprised four specimens that had the same reinforcement steel amount and varied the pre-stressing stress between 0 to 50 kgf/cm^2. The second series (B series) comprised three specimens with similar pre-stressing force as A series specimens, however with higher reinforcement ratio. An additional series (C series) was built to test the effect of the presence of a cast-in-place connection. This latter series is out of the scope of this paper and therefore it will not be referred to here.

The analytical prediction of the behavior of pre-stressed concrete members under cyclic uniaxial loading is still a challenging subject for design engineers, and as such, no method has been proved adequate for predicting the hysteresis behavior of these members. Subsequently, the above-mentioned experiment results were used to test the precision of an FEM analysis in predicting the cyclic behavior of the tested specimens (Salamy, 2002).

The analytical method was based on a specialized FEM package for analyzing concrete members (Cervenka, 2000). The analytical simulation of the pre-stressed grouted specimens did not match the experimental results, particularly the inclination of the descending and ascending branches of the load history. One reason, among others, was relevant to the constitutive model employed to simulate tensile behavior of concrete in unloading and reloading process. Although this model yields acceptable accuracy in analyzing PC or RC members under bending or bending–compression load condition, it manifested its lack of reliability in prediction of the behavior in direct tension. For this reason, rigorous constitutive model

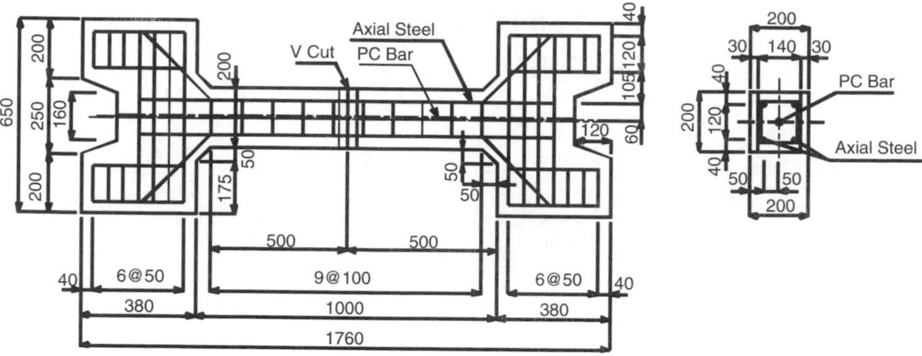

Figure 1. Specimen dimension and steel arrangement.

should be elaborated while the PC members are loaded in direct reversal tension.

Beside aforementioned experiment with grouted tendons, new experiment was initiated in order to investigate the bond effect on the behavior of the pre-stressed concrete members. Specimens that are identical to the specimens tested in the previous experiment, however without grouting the PC bars, were built and tested in a similar way.

The results of this study might be used in analysis and design of concrete members or PC structures that transmit loads primarily by direct tension rather than bending such as silos, tanks, shells, ties of arches, bridge trusses, braced frames and towers. Particular consideration should be given to structures such as tanks and water resources in the area with severe earthquake excitation and to keep the contents of those reservoirs safe and conserved, where residual cracks width play an important role.

As for analytical works, the two main approaches in FEM analysis to represent cracking in concrete structures are discrete crack approach and smeared crack approach wherein this study, the latter one is used. Based on analytical observation of this work as well as other works on RC and PC analysis, it seems this model yields adequate results for RC structures subjected to the cyclic loads but not well fitted to the results of the experiment particularly in descending and ascending branch inclination of load cycles for PC structures. In addition, the smeared crack model is capable to deal with fracture nature of concrete wherein cracks are modeled by changing the constitutive relations in the vicinity of the crack therefore it is treated as a continuum solid with a strain-softening nature.

2 EXPERIMENTAL INVESTIGATION

Experimental work of this study consists of two RC specimens along with five PC specimens with bonded and un-bonded tendons. In the 1998 experiment, loading jack reached to its limit (50 *tf*) before reaching to the failure load of specimens A-4 and B-4 (the experiment was interrupted after a few cycles in this case) therefore in new experiment, 100 *tf* jack was installed to overcome this. For this reason, specimen B-4 is not referred here due to the pre-matured result. Details of the specimens as well as loading apparatus are shown in Figures 1 and 2 and material properties of the specimens are given in Tables 1–3.

2.1 *Test results*

Test results of RC and PC specimens with or without grout are presented in this section. Experimental results of RC specimens are illustrated in Figure 3. As shown in the presented figures, RC members yielded less energy dissipation in each cycles than PC members. For comparison of the results of both grouted and non-grouted specimens (PC members), test results are summarized in Table 4 for both sets of experiments. The table shows concrete compressive strength at time of testing as well. It is noted that the higher compressive strength of the concrete (consequently the higher tensile strength) of the non-grouted specimens of the 2002 experiment (Ammar, 2002) caused higher load capacity comparing to the grouted specimens of the 1998 experiment.

In case of A-3, where the difference in compressive strength was not as high as the other specimens, the grouted specimen had higher loads than the non-grouted one. Observing the values of the loads and the difference in compressive strength of the concrete, it could be generally concluded that the grouted specimen had or should have had higher cracking, yielding and failure loads if the compressive strength were similar to that of the non-grouted specimens.

2.1.1 *Comparison of the behavior of two sets of the specimens under cyclic loading*

A comparison of the hysteresis behavior of all the specimens is given in Figures 4–7. Behaviors of each

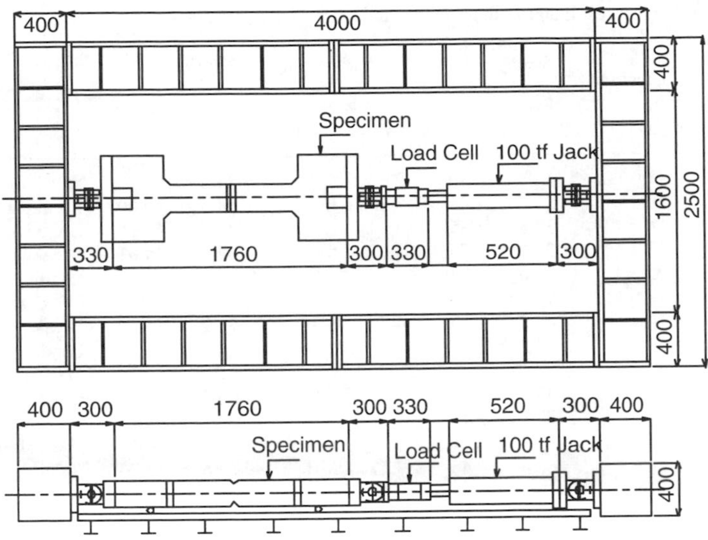

Figure 2. Uniaxial loading frame and setup

Table 1. Mechanical properties of concrete at different ages for both 1998 and 2002 experiments.

	At testing time			
Specimen	Age Day	f'_c MPa	f_t MPa	E_c GPa
A-2G	28	37.3	3.07	30.3
A-2NG	8	42.3	3.66	28.4
A-3G	63	41.3	3.89	30.7
A-3NG	12	44.6	3.37	29.4
A-4G	28	41.5	4.36	31.6
A-4NG	20	48.0	3.41	30.7
B-3G	49	37.2	4.43	30.0
B-3NG	15	45.8	3.85	29.2
B-4G	53	37.9	4.22	30.6
B-4NG	21	49.5	3.87	30.3

Table 2. Results of tension tests of reinforcement steel for both 1998 and 2002 experiments.

Specimen	Yielding strength MPa	Yielding strain μ	Modulus of elasticity GPa	Ultimate strength MPa
Grouted	365	1803	203	–
Grouted A-2	382	1872	204	–
Non-Grouted	362	2047	180	521

two identical specimens from the previous and the new experiments are compared and given in one figure for the overall loading history for the few first cycles for more clarification. In Figures 4–7 the following marks

Table 3. Details of the pre-stressing bars used in both 1998 and 2002 experiments.

Specimen	Bar diameter (mm)	Yielding strength (MPa)	Yielding strain (μ)	E (GPa)
A-2G	φ13	1077	5385	200
A-2NG	φ13			
A-3G	φ17	1019	5066	201
B-3G				
A-3NG	φ17			
B-3NG				
A-4G	φ23	1014	5066	200
B-4G				
A-4NG	φ23		7400	
B-4NG				

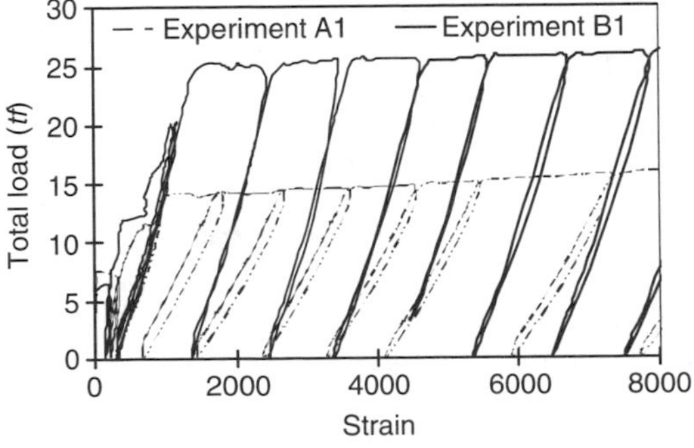

Figure 3. Experimental results of RC specimens A-1 and B-1.

Table 4. A comparison of test results for both grouted and non-grouted specimens.

Specimen	C.C.S.* f'_c MPa	Cracking P_{cr} (tf)	Steel yielding P_s (tf)	Steel yielding δ_s (mm)	PC bar yielding P_{pc} (tf)	PC bar yielding δ_{pc} (mm)	Failure P_F (tf)	Failure δ_F (mm)	Mode
A-2G	37.3	11.0	19.8	0.80	28.4	5.28	33.2	44.80	Breaking of PC bar
A-2NG	42.3	14.0	21.0	1.25	30.0	8.40	34.3	40.00	Breaking of PC bar
A-3G	41.3	18.5	32.3	1.55	35.3	3.13	45.2	44.95	Breaking of PC bar
A-3NG	44.6	17.7	26.9	0.75	39.5	9.60	44.0	44.61	Breaking of PC bar
A-4G	41.5	22.5	41.0	1.52	–	–	–	–	Loading jack reached capacity ($\delta = 3.8$ mm)
A-4NG	48.0	24.0	44.2	1.29	64.0	9.80	71.2	45.10	Breaking of PC bar
B-3G	37.2	18.0	44.2	1.95	46.5	2.80	–	–	Loading jack reached capacity ($\delta = 4.12$)
B-3NG	45.8	20.3	43.0	0.63	57.4	7.56	60.7	24.19	Breaking of PC bar
B-4G	37.9	24.0	51.8	2.10	–	–	–	–	Loading jack reached capacity ($\delta = 2.1$ mm)
B-4NG	49.5	28.6	62.4	0.45	77.7	8.10	79.6	17.70	Breaking of PC bar

*Concrete compressive strength at testing time.

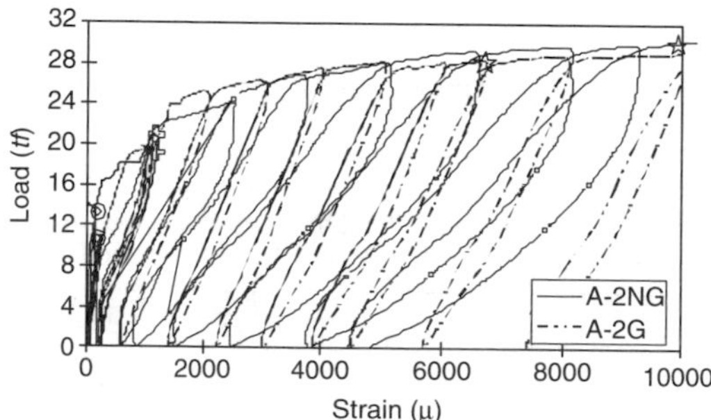

Figure 4. Experimental results of PC specimens A-2.

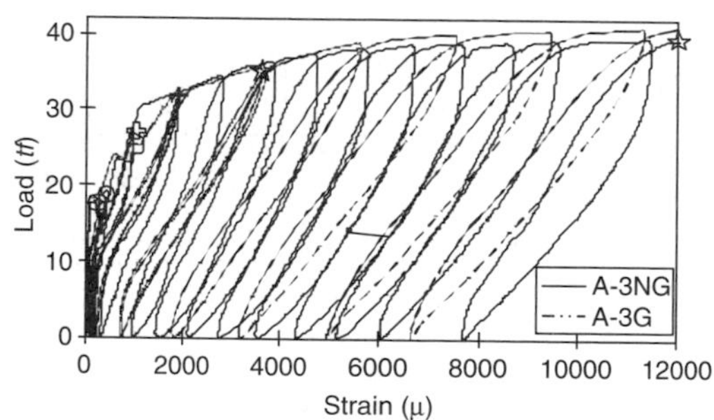

Figure 5. Experimental results of PC specimens A-3.

were used:

- ◎ To indicate Cracking load
- ✣ To indicate Steel yielding load
- ☆ To indicate PC bar yielding load

For A-2 specimens, changing the condition of the PC bars from grouted to non-grouted tendon affected

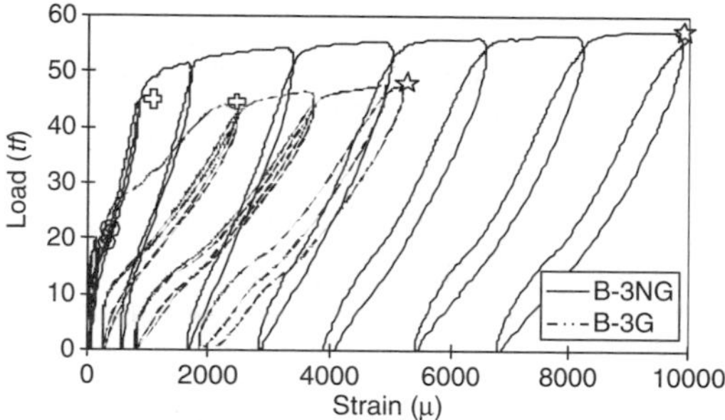

Figure 6. Experimental results of PC specimens B-3.

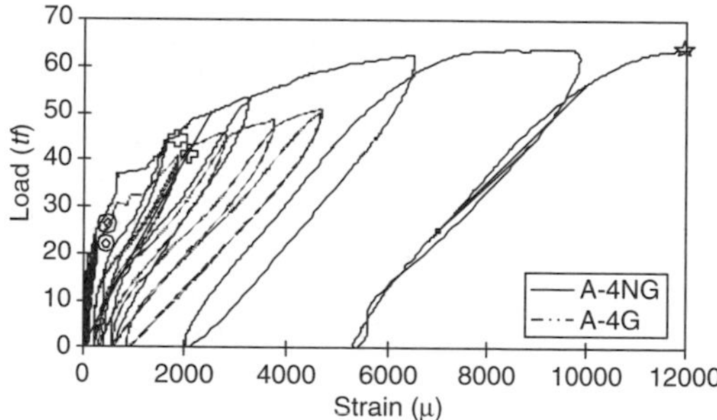

Figure 7. Experimental results of PC specimens A-4.

considerably the loading–unloading inclination of the load history, with much steeper inclination for the grouted specimen. However, this was not the case with A-3 and A-4 specimens where the inclinations were close for the grouted and non-grouted cases particularly at last cycles.

Another note-worthy observation is the great difference in the path of the descending and ascending branches of the load history for specimens A-2 and A-3. This phenomenon could be caused by the energy absorption of the materials in general; however, the difference (especially in A-2 specimen) is higher than to be attributed solely to energy absorption.

Specimens B-3 and B-4 showed in turn a noticeable difference between the grouted and non-grouted cases. Here, in contrast to the A series specimens, the non-grouted specimens had a steeper inclination of the load hysteresis than the grouted specimens. Energy absorption was less noticeable in comparison to the A series.

Based on experimental observation, crack patterns were somehow similar for non-grouted and grouted specimens especially after the first few cycles. However the non-grouted specimens showed more cracks in comparison with the grouted one. In addition, the usage of a higher capacity loading jack in the new experiment allowed reaching higher loads for the non-grouted specimens and thus the cracks were more severe in this case.

3 ANALYTICAL INVESTIGATION AND EVALUATION OF THE RESULTS WITH EXPERIMENT

Nonlinear finite element method is adopted here to analyze test specimens by means of a specialized FE code in concrete structures analysis (Cervenka, 2000). Finite element discretization as well as reinforcement arrangement are illustrated in Figures 8 and 9 respectively. To simulate the specimens as accurate as possible, connection between end plate and concrete body has been considered as interface element. The characteristic of this element is that there is no interaction in tension while full contact is assumed in compressive load. This assumption eliminates arbitrary cracks which occur in concrete of connecting part and neighboring regions due to the applied tension. In this section, analytical results of RC and PC specimens are presented and compared with experimental results. All specimens are subjected to cyclic load in tension zone without compressive external load. Details of the specimens along with loading apparatus have already been given in Figures 1 and 2. The fixed crack model with variable shear retention factor (exponential form) is adopted here. As we know, in the fixed crack model, once crack initiates in a finite element, the crack direction is calculated according to the principal stress direction. The crack direction is kept constant during further load increments and considered as the material axis of orthotropy. As a general case, principal stress directions need not to be coincide with axes of orthotropy and can rotate during loading

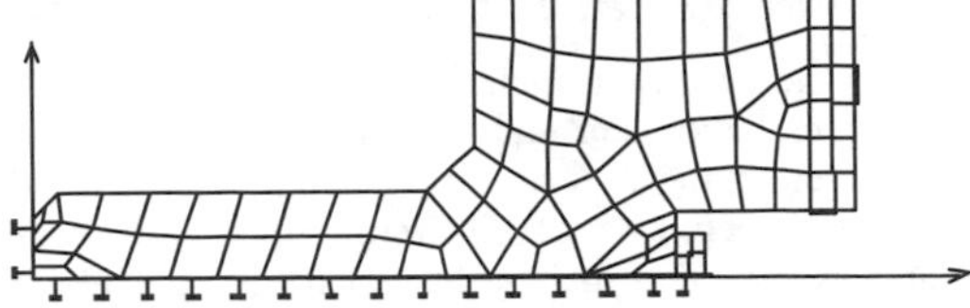

Figure 8. Finite element discretization.

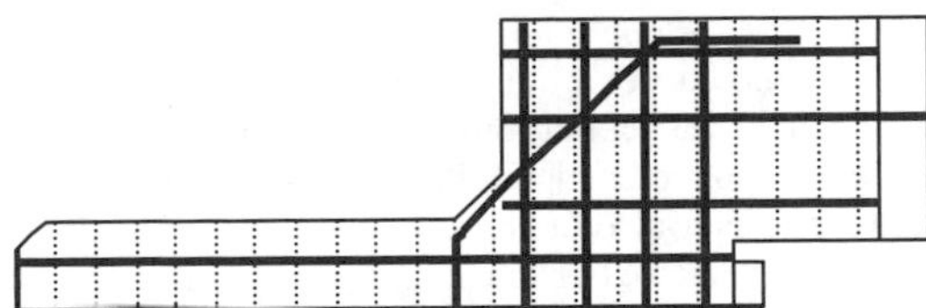

Figure 9. Reinforcement arrangement in FEM.

process. This assumption produces a shear stress in crack surface. In order to prevent the effects of this artificially existed shear stress in the analysis, a shear retention factor as a reduction coefficient is always applied in this model. This factor varies during analysis and is a function of crack width as well as smeared reinforcement ratio.

Alternatively, rotated crack model is presented where the direction of the principal stress coincides to the direction of the principal strain. Since crack direction rotates according to the principal stress direction, no shear stress is generated on the crack surface and just two principal components need to be defined. A considerable point in these two models is that in actuality, the angle of inclination of the concrete struts at failure lies between those assumed by the fixed crack model and the rotated crack model. Therefore, these two theories furnish the two boundaries for the true situation. However, only the fixed crack model is adopted in the present analyses.

In the analyses, perfect bond is assumed for smeared reinforcements and bar reinforcements are embedded within finite elements with sliding bond with aforementioned stress-slip relationship in both reinforcements and tendons. In all calculations, displacement control with Newton-Raphson solution procedure is adopted. Specimens are partly modeled by FEM due to symmetries in both global directions of X and Y (1/4 of specimens) and are subjected to both proportional load (external force) and non-proportional load (prestressing force). Tension stiffening factor by means of additional stress component is not used in the analyses. It is supposed that softening branch of tensile concrete after cracking is sufficient to explain tension-stiffening phenomenon if reinforcement ratio is taken into account. Responses to the applied cyclic load in terms of the average strain monitored in a distance of 80 cm of the middle part of the specimens.

3.1 *Concrete material*

Due to the nature of applied load in experiment, all specimens have been under tension so the behavior of concrete in the experiment procedure is dominated by tensile behavior of concrete. Since the maximum compression due to the pre-stressing is about $7\,MPa$, compressive stress never goes beyond elastic zone therefore only tension constitutive model is presented here although the full model of concrete in compression is applied in analysis.

All concrete members are assumed 2D elements with plane-stress condition and nonlinear cementitious material. Concrete in tension is modeled as a fracture type material after cracking while before cracking is linear elastic. After cracking however, a softening branch forms and it is assumed that the descending path follows an exponential function of crack width derived experimentally by Hordijk (1991) in Equation 1.

$$\frac{\sigma}{f_t} = \left\{1 + \left(c_1 \frac{w}{w_c}\right)^3\right\} \exp\left(-c_2 \frac{w}{w_c}\right) - \frac{w}{w_c}\left(1 + c_1^3\right) \exp\left(-c_2\right) \tag{1}$$

where w is the crack opening; w_c is the crack opening at the complete release of stress which is a function of fracture energy G_f defined by Equation 2; and σ is the normal stress in crack and f_t is the tensile strength of concrete in one dimension system or effective tensile strength in two dimension system. Values of the constants are, $c_1 = 3$, $c_2 = 6.93$.

$$w_c = 5.14\frac{G_f}{f_t} \tag{2}$$

This model is shown in Figure 10.

Concrete as a plain material is clearly different than concrete while cooperating with steel reinforcement. The difference manifests itself in descending or softening branch as a function of steel percentage (Okamura, 1991) as well as shear retention factor when the fixed crack model is adopted. Tensile response of the concrete based on the above-mentioned expression is illustrated in Figure 11. Since concrete tensile strength in experiment is obtained by means of Brazilian test which is not very accurate and always gives higher cracking stress than reality, tensile strength f_t was calculated based on design code equation (CEB, 1990) below.

$$f_t = 0.24 f_{cu}^{2/3} \tag{3}$$

where f_{cu} is the compressive strength of concrete cube sample at age of 28 days.

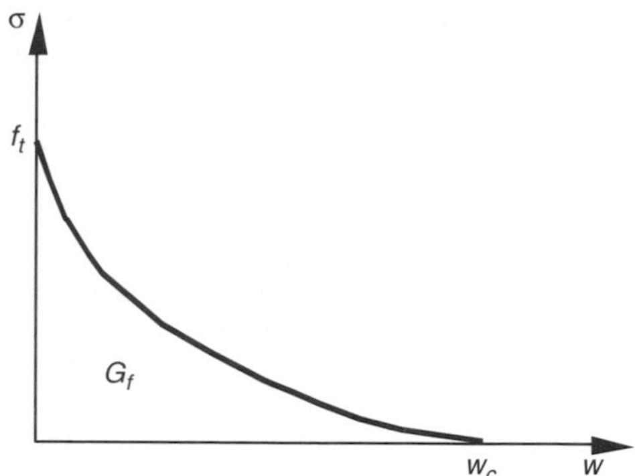

Figure 10. Concrete fracture model in tension.

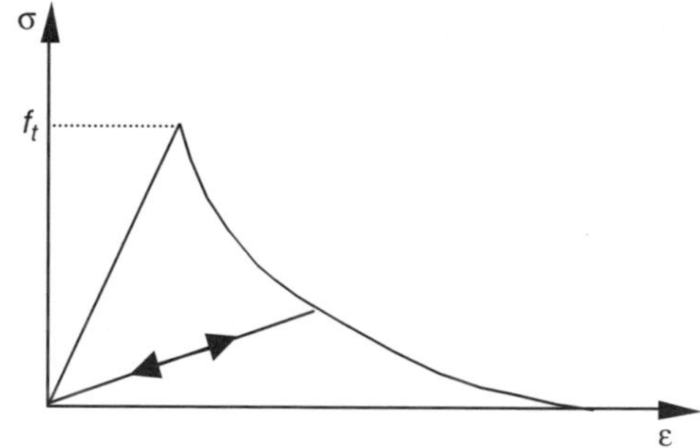

Figure 11. Concrete constitutive model in tension.

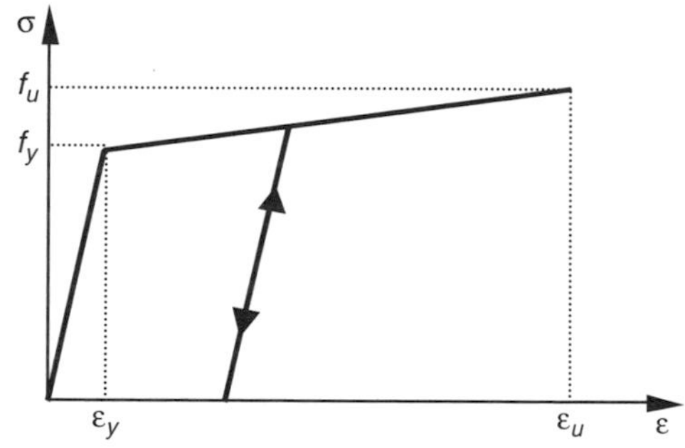

Figure 12. Steel constitutive model.

3.2 *Steel reinforcements and pre-stressing tendons*

Steel members including longitudinal reinforcements, stirrups and pre-stressing tendons have been characterized by an elasto-plastic behavior with bilinear stress–strain relationship (Fig.12). The model is adopted for cyclic tension and compression with Bauschinger effect. The stirrups are simulated as smeared steel while the longitudinal reinforcements and PC tendons are treated as discrete embedded reinforcement

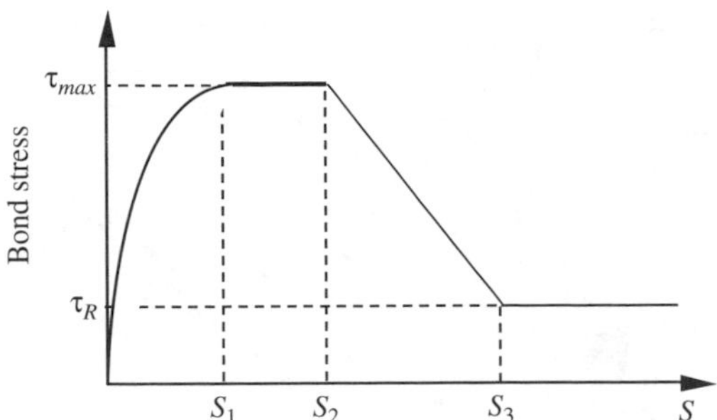

Figure 13. Bond stress-slip relationship (CEB).

within concrete element with entire bond properties. Tangential hardening stiffness of steel as well the shear stiffness of the reinforcement has set to zero for all kind of reinforcements.

3.3 *Bond constitutive model*

The behavior of reinforced concrete structures subjected to large alternate action is highly dependent on the interaction between steel and concrete. Beside the economical requirement of energy dissipation in such structures in seismic hazardous zone, analytical prediction will be unrealistic without considering bond sliding and deterioration between steel and reinforcement.

The first trial of this study was to analyze the members with perfect bond without sliding or energy dissipation characteristic. However, those analyses failed to predict the real behavior of structures while reversal load is applied. In this paper, analytical CEB bond stress-slip model for monotonic loading (CEB, 1990) has been adopted in analysis. This model can comply adequately response of RC elements under monotonic load condition as well as cyclic load condition while only tensile loading is applied. Assumed bond stress-slip relationship is illustrated in Figure 13 wherein unconfined concrete is assumed in bond condition.

Values of the parameters are $S_1 = S_2 = 0.6\,mm$ and $S_3 = 1.0\,mm$ due to the confinement condition. Shear stress of the bond model can be defined through Equations 4–6 below:

$$\tau = \tau_{\max}\left(\frac{S}{S_1}\right)^{\alpha} \tag{4}$$

where $\alpha = 0.4$ and

$$\tau_{\max} = 2.5\sqrt{f_{cu}} \tag{5}$$

$$\tau_f = 0.15\tau_{\max} \tag{6}$$

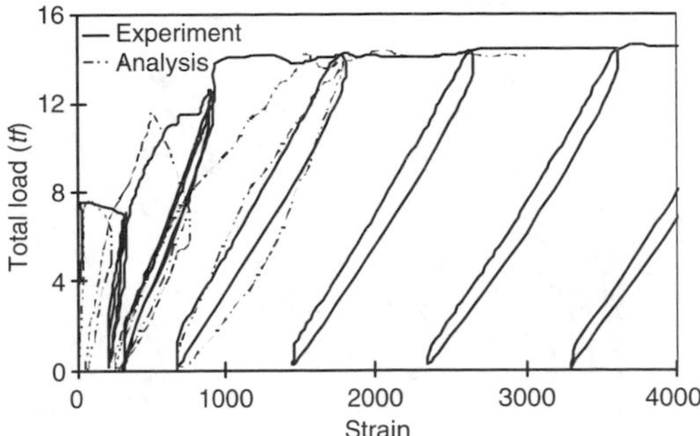

Figure 14. Response of RC specimen A-1.

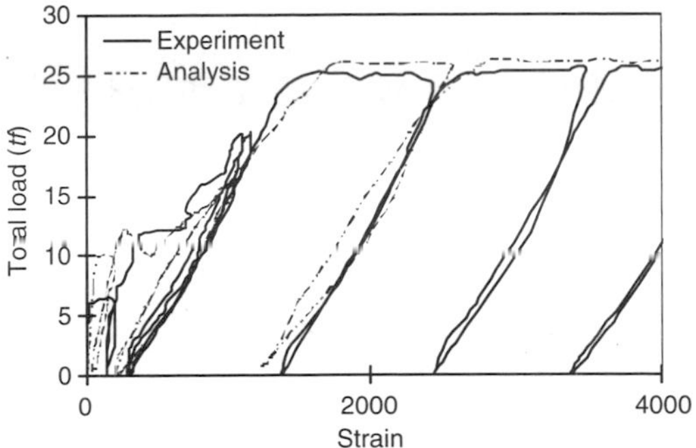

Figure 15. Response of RC specimen B-1.

3.4 *Analysis of the RC specimens A-1 and B-1*

These specimens are RC members under uniaxial cyclic load at the ending points. Responses subjected to the cyclic load in terms of the average strain ($\times 10^6$) are shown in Figures 14 and 15. Agreement is very good in most parts of the loading process including loading and unloading branches. In the first cycle of load as can be seen in the figure, unloading branch headed to the origin of loading point due to the applied constitutive equation for concrete in tension under reversing stress. However after second cycle where the concrete is already steadily cracked, due to the bond sliding, obtained residual cracks in analysis agree well with experiment. After third cycle where the behavior of structure is dominated by reinforcing steel, the response is much more adjusted with experimental observation. The results in this part show that the model can adequately predict the behavior of structures under cyclic load.

3.5 *Analysis of the PC specimens*

PC specimens A-2 and A-3 are picked up from two series for analysis here. Responses to the cyclic load in terms of the average strain are shown in Figures 16–19.

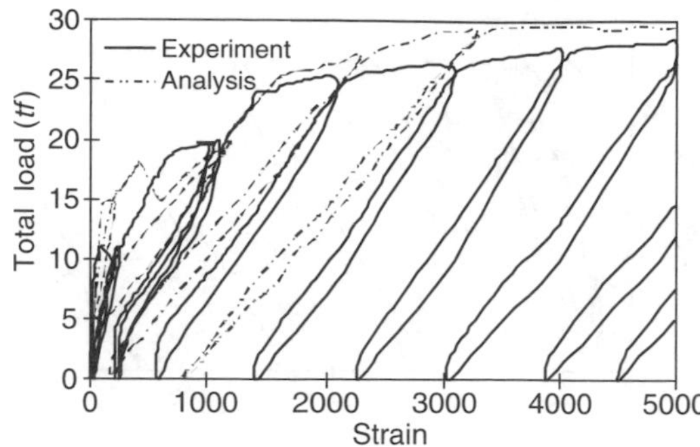

Figure 16. Response of grouted PC specimen A-2.

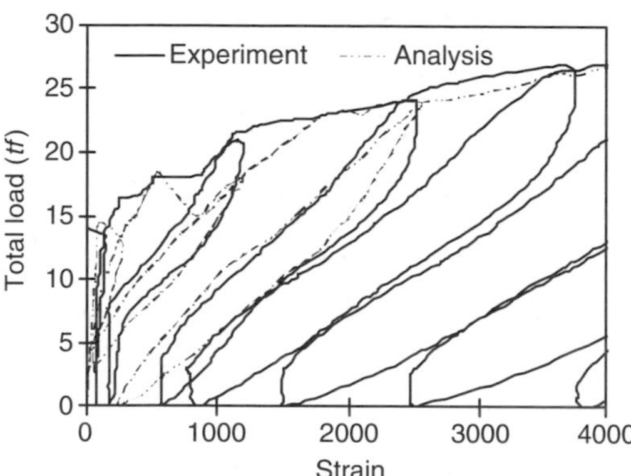

Figure 17. Response of non-grouted PC specimen A-2.

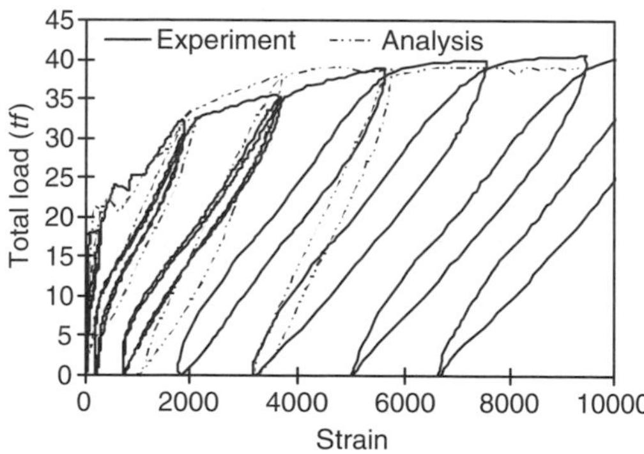

Figure 18. Response of grouted PC specimen A-3.

In contrary to the overall analytical responses of the specimens which are well fitted to the experiments, obtain crack widths are not in well agreement with experiment.

Looking closer at the results of the specimens with bonded tendon, A-2 shows unsafe results while crack widths in A-3 are conservative. On the other hand, all analytical results of PC specimens with un-bonded

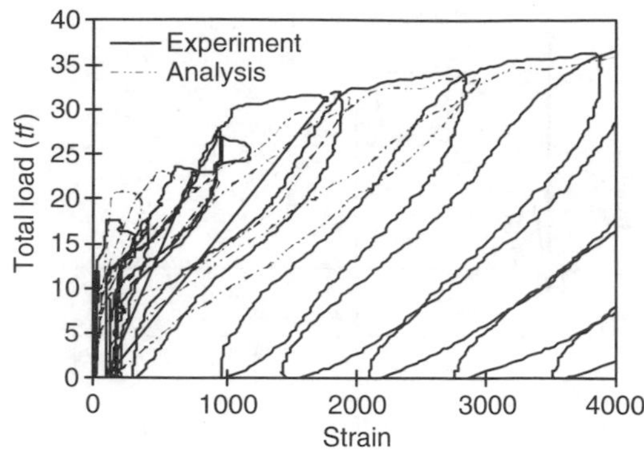

Figure 19. Response of non-grouted PC specimen A-3.

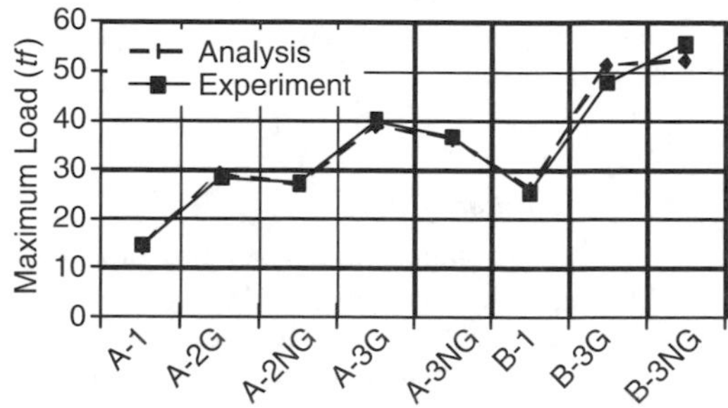

Figure 20. Load capacity of both sets of specimens.

tendon show softer behavior in unloading–reloading process with smaller residual cracks. However energy absorption obtained in analysis and experiment was almost similar for both sets of specimens.

Since the residual cracks are one of the main concerns of this study, it seems that simplified constitutive models cannot handle a reliable response of PC structures particularly in direct tension. Comparison between two sets of the specimens explains the importance of PC tendon's bond model in addition to the tensile constitutive model. As illustrated in the figures, overall load capacities as well as energy absorption of each cycle are well predicted by analysis. A comparison of the analytical and experimental results of all specimens in terms of overall load capacity is presented in Figure 20. It is worth noted here that unloading–reloading branches of the analyses are well agreed with experiment where the behavior of structure was dominated by steel response in all analyses.

3.6 *Unloading–reloading inclination in experiment and analysis*

Figures 21 and 22 show the inclination of the unloading–reloading path (represented as an average

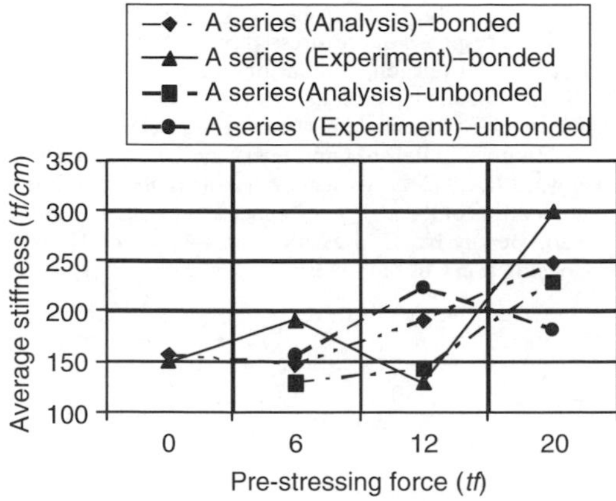

Figure 21. Last cycle average stiffness in terms of pre-stressing force A series specimens.

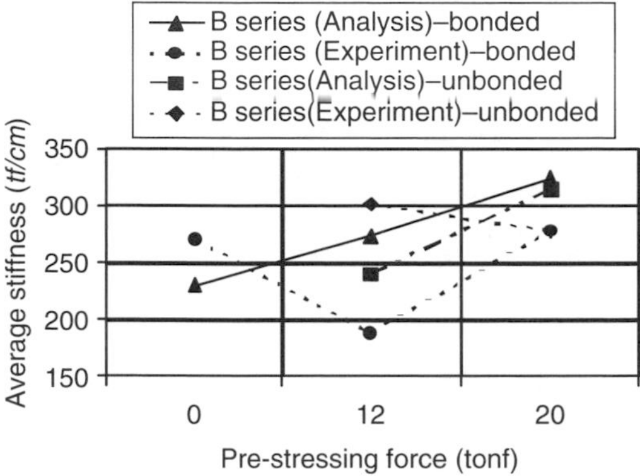

Figure 22. Last cycle average stiffness in terms of pre-stressing force B series specimens.

stiffness; *tf/cm*) related to the last circle of loading in analyses in terms of average strain ($\times 10^6$). As presented in these figures, the main differences between experiment and analysis are the change of unloading–reloading inclination in analyses from A-1 to A-4 and B-1 to B-4 which have a regular incremental form despite irregular changes of inclination in experimental response of the specimens. In these figures for instance, as far as pre-stressing force increases (for constant reinforcing ratio) inclination also steeping gradually. In contrary, experimental response varies uncoordinated to the pre-stressing force increments. This shows the complexity of the bond influence in PC elements.

4 CONCLUSION

The experimental results of two RC specimens A-1 and B-1 and five PC specimens A-2, A-3, A-4, B-3 and B-4 in two cases of bonded and un-bonded PC tendons along with their analytical results are presented in this paper. All specimens are subjected to cyclic load in only tension zone without compressive load. A specialized nonlinear finite element package for reinforced concrete structures has been used to analyze the specimens with bond characteristic for reinforcements. Applied constitutive equations for concrete are fixed due to the code limitation. Since all specimens are under direct cyclic tension, response of specimens was highly dependent to the constitutive model of material particularly concrete as well as reinforcement bond model. The latter one influenced the behavior of the structures due to the progressive deterioration of bond between concrete and reinforcement. Based on the experimental and analytical results of both RC and PC specimens, the following conclusions are drawn.

In case of RC specimens A-1 and B-1, the results were quite acceptable and had very good agreement with experiment including overall responses and residual cracks as well as energy dissipation of each cycles. It can be concluded that the employed constitutive equation and bond stress-slip relationship can adequately explain cyclic behavior of RC structures with acceptable accuracy for design.

Despite good prediction of load capacity in PC specimens, residual cracks due to the cyclic load are not well agreed with experiment. The results showed the complexity of bond influence in PC members. Accordingly a new versatile bond model should be for proposed for PC structures particularly in cyclic or dynamic load condition which comprise influences of pre-stressing, tendon/reinforcement ratio as well as energy dissipation due to bond sliding and deterioration. The need for such bond model is much more if the behavior of structure is dominated by tensile response of PC members and residual crack widths are of higher importance in practical design. The future works of this study is aimed in developing a new bond model for PC specimens under uniaxial tension. To increase precision of analysis in first cycles after concrete cracking, much realistic concrete tensile constitutive model is to be adopted.

ACKNOWLEDGMENTS

This study was supported by Japan Society for Promotion of Science (JSPS) fellowship and by Research Laboratory of Oriental Construction Company. This support is gratefully acknowledged. Any opinions expressed in this paper are those of the writers and do not reflect the views of sponsoring institutes. Special thanks go to Mr. Hasan Ammar and to Mr. Shinzo Eguchi for their help in experimental as well as analytical works.

REFERENCES

Ammar, H. 2002. Report on the new experiment of non-grouted pre-stressing bars, Research laboratories of Oriental Construction Co. Ltd., Japan.

CEB Comite EURO-International du Beton, 1990, CEB-FIP model code, Bulletin D'information No. 195.

Cervenka, V. 2000. ATENA program documentation, Cervenka Consulting, Prague.

Hordijk, D.A. 1991. Local approach to fatigue of concrete, Doctor Dissertation, Delft University of technology, The Netherlands.

Japan pre-stressed concrete contractors association, 1998. A study on PC members under axial tension, Report, PC Tanks Sub-committee, (Japanese).

Ohtani, S., Eguchi, S. & Yoshioka, T. 1999. Experimental study on response of pre-stressed concrete members subject to axial tension, Journal of structural engineering, JSCE, V. 45A, (Japanese).

Okamura, H. & Maekawa K. 1991. Nonlinear analysis and constitutive models of reinforced concrete, pp. 28–60.

Salamy, M.R. 2002. Analytical evaluation of the experiment observation of RC and PC specimens under uniaxial cyclic load, Report No. 1, Research laboratories of Oriental Construction Co. Ltd., Japan.

Computational Modelling of Concrete Structures, Bićanić et al. (eds)
© 2003 Swets & Zeitlinger, Lisse, ISBN 90 5809 536 3

Author index